Tropical Montane Cloud Forests

Science for Conservation and Management

This volume represents a uniquely comprehensive overview of our current knowledge on tropical montane cloud forests. It comprises 72 chapters divided into seven sections covering a wide spectrum of topics including cloud forest distribution, climate, soils, biodiversity, hydrological processes, hydrochemistry and water quality, climate change impacts, and cloud forest conservation, management, and restoration. The final chapter presents a major synthesis by some of the world's leading cloud forest researchers, which summarizes our current knowledge and considers the sustainability of these forests in an ever-changing world.

This book is a must-have for anyone interested in the study, conservation, management, or restoration of tropical montane cloud forests. It represents the state of the art with respect to our knowledge of cloud forest occurrence and status, as well as the biological and hydrological value of these unique forests. The presentation is academic but with a firm practical emphasis. It will serve as a core reference for academic researchers and students of environmental science and ecology, as well as practitioners (natural resources management, forest conservation) and decision makers at local, national, and international levels.

LEENDERT ADRIAAN (Sampurno) BRUIJNZEEL is a Professor of Land Use and Hydrology based at VU University, Amsterdam. He has 35 years of experience with forest hydrological research in the humid tropics, mostly in South-East Asia, the Caribbean, and Latin America. His main research interests include the water and nutrient dynamics of tropical forests, above all montane cloud forests; his other main research topics include the hydrological impacts of land-cover change (deforestation/reforestation) and erosion and sediment transport processes. Professor Bruijnzeel is the author of two other books and the co-editor of *Forests, Water and People in the Humid Tropics* (2005) also published by Cambridge University Press and UNESCO as part of the International Hydrology Series. In 2005 he received the prestigious Busk Medal from the Royal Geographical Society.

FREDERICK N. SCATENA is Professor and Chair of the Department of Earth and Environmental Science at the University of Pennsylvania. Since 1988 he has served as the Principal Co-PI of the National Science Foundation's Luquillo Long-Term Ecological Research Project and since 2009 as the Lead PI of the NSF Luquillo Critical Zone Observatory, both in Puerto Rico. Professor Scatena has worked in tropical forest hydrology for the past 32 years, on topics ranging from water supply to the influence of hurricanes on the biogeochemistry and water quality of Caribbean streams. He has been awarded the International Institute of Tropical Forestry Puerto Rican Ecology Award (2008), and various USDA Research Scientist Awards.

LAWRENCE S. HAMILTON is a Professor Emeritus (Cornell University) and former Senior Fellow at the East–West Center in Hawai'i. He has authored over 300 publications throughout his career and is known popularly as the "father of cloud forests" due to his successful campaign over many years to get these unique forests on the international conservation agenda. His awards include Forest Conservationist of the Year from the New York State Conservation Council (1969); the Environmental Achiever Award from Friends of UNEP (1987); the Packard International Parks Merit Award from the International Union for Conservation of Nature (IUCN) World Commission on Protected Areas (2003); and the prestigious King Albert Gold Medal for Mountain Conservation Leadership (2004). In 2008 he was named an Honorary Member of IUCN, and in the same year was profiled as one of the 20 global "Earth Movers" by IUCN.

INTERNATIONAL HYDROLOGY SERIES

The **International Hydrological Programme** (IHP) was established by the United Nations Educational, Scientific and Cultural Organization (UNESCO) in 1975 as the successor to the International Hydrological Decade. The long-term goal of the IHP is to advance our understanding of processes occurring in the water cycle and to integrate this knowledge into water resources management. The IHP is the only UN science and educational programme in the field of water resources, and one of its outputs has been a steady stream of technical and information documents aimed at water specialists and decision-makers.

The **International Hydrology Series** has been developed by the IHP in collaboration with Cambridge University Press as a major collection of research monographs, synthesis volumes, and graduate texts on the subject of water. Authoritative and international in scope, the various books within the series all contribute to the aims of the IHP in improving scientific and technical knowledge of freshwater processes, in providing research know-how and in stimulating the responsible management of water resources.

TITLES IN PRINT IN THIS SERIES

M. Bonell, M. M. Hufschmidt and J. S. Gladwell *Hydrology and Water Management in the Humid Tropics: Hydrological Research Issues and Strategies for Water Management*

Z. W. Kundzewicz *New Uncertainty Concepts in Hydrology and Water Resources*

R. A. Feddes *Space and Time Scale Variability and Interdependencies in Hydrological Processes*

J. Gibert, J. Mathieu, and F. Fournier *Groundwater/Surface Water Ecotones: Biological and Hydrological Interactions and Management Options*

G. Dagan, and S. Neuman *Subsurface Flow and Transport: A Stochastic Approach*

J. C. van Dam *Impacts of Climate Change and Climate Variability on Hydrological Regimes*

D. P. Loucks, and J. S. Gladwell *Sustainability Criteria for Water Resource Systems*

J. J. Bogardi, and Z. W. Kundzewicz *Risk, Reliability, Uncertainty and Robustness of Water Resource Systems*

G. Kaser, and H. Osmaston *Tropical Glaciers*

I. A. Shiklomanov, and J. C. Rodda *World Water Resources at the Beginning of the Twenty-First Century*

A. S. Issar *Climate Changes during the Holocene and their Impact on Hydrological Systems*

M. Bonell, and L. A. Bruijnzeel *Forests, Water and People in the Humid Tropics: Past, Present and Future Hydrological Research for Integrated Land and Water Management*

F. Ghassemi, and I. White *Inter-Basin Water Transfer: Case Studies from Australia, United States, Canada, China and India*

K. D. W. Nandalal, and J. J. Bogardi *Dynamic Programming Based Operation of Reservoirs: Applicability and Limits*

H. S. Wheater, S. Sorooshian, and K. D. Sharma *Hydrological Modelling in Arid and Semi-Arid Areas*

J. Delli Priscoli, and A. T. Wolf *Managing and Transforming Water Conflicts*

L. A. Bruijnzeel, F. N. Scatena, and L. S. Hamilton *Tropical Montane Cloud Forests*

Tropical Montane Cloud Forests

Science for Conservation and Management

L. A. Bruijnzeel

VU University, Amsterdam

F. N. Scatena

University of Pennsylvania, Philadelphia

L. S. Hamilton

Charlotte, Vermont

CAMBRIDGE
UNIVERSITY PRESS

CAMBRIDGE
UNIVERSITY PRESS

University Printing House, Cambridge CB2 8BS, United Kingdom

One Liberty Plaza, 20th Floor, New York, NY 10006, USA

477 Williamstown Road, Port Melbourne, VIC 3207, Australia

314-321, 3rd Floor, Plot 3, Splendor Forum, Jasola District Centre, New Delhi - 110025, India

79 Anson Road, #06-04/06, Singapore 079906

Cambridge University Press is part of the University of Cambridge.

It furthers the University's mission by disseminating knowledge in the pursuit of education, learning and research at the highest international levels of excellence.

www.cambridge.org
Information on this title: www.cambridge.org/9780521760355

First published 2010

A catalogue record for this publication is available from the British Library

Library of Congress Cataloging in Publication data
Tropical montane cloud forests : science for conservation and management / edited by L. A. Bruijnzeel, F. N. Scatena, L. S. Hamilton.
p. cm. – (International hydrology series)
ISBN 978-0-521-76035-5 (Hardback)
1. Cloud forest conservation. 2. Cloud forest ecology. 3. Cloud forests. 4. Mountain ecology–Tropics.
I. Bruijnzeel, Leendert Adriaan. II. Scatena, F. N. III. Hamilton, Lawrence S.
QH541.5.C63T765 2010
577.34–dc22

2010036467

ISBN 978-0-521-76035-5 Hardback

Contents

The color plates will be found between pages 100 and 101.

Contributors

Aiba S.
Faculty of Science,
Kagoshima University,
Kagoshima 890-0065,
Japan

Aide T. M.
Department of Biology,
P.O. Box 23360,
University of Puerto Rico,
San Juan,
Puerto Rico 00931,
USA

Ainuddin A. N.
Faculty of Forestry,
Universiti Putra Malaysia,
43400 UPM Serdang,
Malaysia

Alvarez-Aquino C.
Instituto de Genética Forestal,
Universidad Veracruzana,
Apartado Postal 51,
Xalapa,
Veracruz 91000,
Mexico

Ambrose K.
Formerly with Ecopar, Quito,
Ecuador; presently at CARE Canada,
9 Gurdwara Road,
Ottawa,
Ontario K2E 7X6,
Canada

Anhuf D.
Department of Physical Geography,
University of Passau,
Innstrasse 40,
94032 Passau,
Germany

Araoz E.
CONICET-Laboratorio de Investigaciones Ecológicas
de las Yungas,
Universidad Nacional de Tucumán,
Casilla de Correo 34 (4107),
Yerba Buena,
Tucumán,
Argentina

Arco M. del
Universidad de La Laguna,
La Laguna,
Tenerife,
Canary Islands,
Spain

Asbjornsen H.
Formerly with Iowa State University, Ames, USA;
presently at the Department of Natural Resourcer and the
Environment University of New Hampshire,
Rudmah Hall,
46 College Road,
Durham, NH 03824,
USA

Axmacher J. C.
Terrestrial Ecology Research Unit,
Department of Geography,
University College London,
Pearson Building,
Gower Street,
London WC1E 6BT,
UK

Aylward B.
Ecosystem Economics LLC,
P.O. Box 2062, Bend,
OR 97709,
USA

Bach K.
Faculty of Geography,
University of Marburg,
Deutschhausstrasse 10,
35032 Marburg,
Germany

Báez S.
Formerly with Ecopar, Quito,
Ecuador; presently at CONDESAN,
Digo de Briedà Clemonte Celi,
Quito,
Ecuador

Barradas V. L.
Instituto de Ecología,
Universidad Nacional Autónoma de México,
Apartado Postal 70-275,
Circuito Exterior,
Ciudad Universitaria,
04510 Mexico, D. F.,
Mexico

Bendix J.
Faculty of Geography,
Laboratory for Climatology and Remote Sensing,
University of Marburg,
Deutschhausstrasse 10,
35032 Marburg,
Germany

Benner J.
Department of Biological Sciences,
Stanford University,
371 Serra Mall,
Stanford, CA 94305,
USA

Boy J.
Geographical Institute,
Johannes Gutenberg University of Mainz,
Johann-Joachim-Becherweg 21,
55099 Mainz,
Germany

Brehm G.
Formerly with the University of Bayreuth, Bayreuth,
Germany; presently at the Institute for Special Zoology
and Evolutionary Biology of the Phyletic Museum,
Erbertstrasse 1, 07743 Jena,
Germany

Bruijnzeel L. A.
Faculty of Earth and Life Sciences,
VU University,
De Boelelaan 1085,
1081 HV Amsterdam,
the Netherlands

Bubb P.
United Nations Environmental Programme – World
Conservation Monitoring Centre,
219 Huntingdon Road,
Cambridge CB3 0DL,
UK

Burkard R.
Formerly with the Institute of Geography,
University of Bern,
Hallerstrasse 12,
CH-3012 Bern,
Switzerland

Calvo-Alvarado J. C.
Escuela de Ingeniera Forestal,
Instituto Tecnológico de Costa Rica,
Apartado 159-7950,
Cartago,
Costa Rica

Carilla J.
CONICET-Laboratorio de Investigaciones Ecológicas de las
Yungas,
Universidad Nacional de Tucumán,
Casilla de Correo 34 (4107),
Yerba Buena,
Tucumán,
Argentina

Catchpole D. J.
School of Geography and Environmental Studies,
University of Tasmania,
Private Bag 78, Hobart,
Tasmania 7001,
Australia

Cavelier J.
Formerly with the Gordon and Betty Moore Foundation,
Washington DC, USA; presently at The Global
Environmental Facility,
1818 H Street NW,
Washington, DC 20433,
USA

Cervantes-Pérez J.
Centro de Ciencias de la Tierra,
Universidad Veracruzana,
Francisco J. Moreno 207, Colonia Emiliano Zapata,
91090 Xalapa,
Veracruz,
Mexico

Chang S. C.
Institute of Natural Resources,
National Dong Hwa University,
974 Hualien,
Taiwan

Chen Y. T.
Institute of Natural Resources,
National Dong Hwa University,
974 Hualien,
Taiwan

Cueva K.
Ecopar,
Casilla 17-11-6706,
Quito,
Ecuador

Dalitz H.
Institute of Botany and Botanical Garden,
University of Hohenheim,
Garbenstrasse 30,
70599 Stuttgart,
Germany

Dallas S.
Monteverde Institute,
Monteverde,
Puntarenas,
Costa Rica

Das S.
Formerly with UNEP-WCRC,
Cambridge, UK; presently at Climate Change Division,
The Energy and Resources Institute,
Darbari Seth Block,
IHC Complex,
Lodhi Road,
New Delhi 110 003,
India

Davison G.
Formerly with WWF-Malaysia,
Petaling Jaya,
Malaysia; presently at National Parks Board Singapore,
1 Cluny Road,
Singapore 259569,
Singapore

DeLay J. K.
Department of Geography,
University of Hawai'i at Mānoa,
2424 Maile Way,
Honolulu,
HI 96822,
USA

Dohrenbusch A.
Silviculture and Forest Ecology of the Temperate Zones,
Burckardt Institute,
University of Göttingen,
Büsgenweg 1,
37077 Göttingen,
Germany

Downer C.
Andean Tapir Fund,
P.O. Box 456,
Minden, NV 89423,
USA

Eugster W.
ETH Zurich, Animal and Agroecosystem,
Institute of Plant Sciences,
Universitätsstrasse 2,
CH-8092 Zürich,
Switzerland

Fabian P.
Institute for Bioclimatology and Immission Research,
Technical University of Munich,
Am Hochanger 13,
85354 Freising,
Germany

Firestone M. K.
Ecosystem Science Division,
Department of Environmental Science,
Policy, and Management,
151 Hilgard Hall,
University of California,
Berkeley,
CA 94720,
USA

Fleischbein K.
Formerly with the Justus Liebig University of Giessen, Giessen, Germany; presently at the Leibniz Institute for Agricultural Engineering, Department 2,
Technology Assessment and Substance Cycles,
Max Eyth Allee 100,
14469 Potsdam,
Germany

Foster P.
Department of Earth Sciences,
University of Bristol,
Wills Memorial Building,
Bristol BS8 1RJ, UK

Frumau K. F. A.
Formerly with the VU University Amsterdam, the Netherlands; presently with the Air Quality and Climate Change Group,
Energy Research Centre of the Netherlands,
P.O. Box 1, 1755 ZG Petten,
the Netherlands

García-Santos G.
Formerly with the VU University Amsterdam, the Netherlands; presently at the Department of Geography,
University of Zurich,
Winterthurerstrasse 190,
CH-8057 Zürich,
Switzerland

Garnica-Sánchez Z.
Rainforest Alliance,
Smartwood Programme,
Oaxaca,
Mexico

Gehrig C.
Department of Systematic Botany,
Albrecht von Haller Institute of Plant Sciences,
Untere Karspüle 2,
University of Göttingen,
37073 Göttingen,
Germany

Gerold G.
Department of Landscape Ecology,
Institute of Geography,
University of Göttingen,
Goldschmidtstrasse 5,
37077 Göttingen,
Germany

Giambelluca T. W.
Department of Geography,
University of Hawai'i at Mānoa,
2424 Maile Way,
Honolulu,
HI 96822,
USA

Gil-Montero R.
CONICET,
Instituto de Estudios Geográficos,
Universidad Nacional de Tucumán,
Tucumán,
Argentina

Gingerich S. B.
Water Resources Division,
United States Geological Survey,
677 Ala Moana Boulevard,
Suite 415, Honolulu,
HI 96813,
USA

Goldstein G.
Department of Biology,
University of Miami,
1301 Memorial Drive,
Coral Gables,
FL 33124,
USA

Goller R.
Formerly with the University of Bayreuth, Germany; presently at Bayerisches Landesamt für Umwelt (Referat 104),
Hans Högn Strasse 12,
95030 Hof/Saale,
Germany

González J.
c/o Environmental Monitoring and Modelling Research Group,
King's College London, Strand,
London WC2R 2LS,
UK

Gradstein S. R.
Department of Systematic Botany,
Albrecht von Haller Institute of Plant Sciences,
Untere Karspüle 2,
University of Göttingen,
37073 Göttingen,
Germany

Granados-Ramirez R.
Instituto de Geografía,
Universidad Nacional Autónoma de México,
Circuito de la Investigación Científica,
Ciudad Universitaria,
04510 Mexico, D. F.,
Mexico

Grau H. R.
Laboratorio de Investigaciones Ecologicas de las Yungas,
Universidad Nacional de Tucumán,
Casilla de Correo 34 (4107),
Yerba Buena,
Tucumán,
Argentina

Guswa A. J.
Picker Engineering Program,
Smith College,
Northampton,
MA 01063,
USA

Häger A.
Formerly with the University of Göttingen, Germany;
presently at the School for Field Studies,
Centro de Estudios sobre Desarrollo Sostenible,
P.O. Box 150 4013,
Atenas,
Costa Rica

Hamilton L. S.
342 Bittersweet Lane,
Charlotte,
VT 05445,
USA

Hemp A.
Ecological Botanical Garden,
University of Bayreuth,
95440 Bayreuth,
Germany

Hemp C.
Department of Animal Ecology II,
University of Bayreuth,
95440 Bayreuth,
Germany

Herman D. J.
Ecosystem Science Division,
Department of Environmental Science, Policy, and Management,
151 Hilgard Hall,
University of California,
Berkeley,
CA 94720,
USA

Hertel D.
Plant Ecology,
Albrecht von Haller Institute of Plant Sciences,
University of Göttingen,
Untere Karspüle 2,
37073 Göttingen,
Germany

Hietz P.
Institute of Botany,
University of Natural Resources and Applied Life Sciences
(BOKU),
Gregor Mendel Strasse 33,
1180 Vienna,
Austria

Hofstede R.
Ecopar,
Casilla 17-11-6706,
Quito,
Ecuador

Hölscher D.
Tropical Silviculture and Forest Ecology,
Burckhardt Institute,
University of Göttingen,
Büsgenweg 1,
37077 Göttingen,
Germany

Holwerda F.
Formerly with the VU University Amsterdam, the Netherlands; presently at Department of Natural Resource Ecology and Management,
Iowa State University,
234 Science II, Ames,
IA 50010,
USA

Homeier J.
Plant Ecology,
Albrecht von Haller Institute of Plant Sciences,
University of Göttingen,
Untere Karspüle 2,
37073 Göttingen,
Germany

Hsia Y. J.
Institute of Natural Resources,
National Dong Hwa University,
974 Hualien,
Taiwan

Huth A.
Helmholtz Centre for Environmental Research, UFZ,
Department of Ecological Modelling,
Permoserstrasse 15,
04318 Leipzig,
Germany

Jarvis A.
International Centre for Tropical Agriculture (CIAT) and Biodiversity International,
AA 6713, Cali,
Colombia

Jones T. J.
Department of Biology,
University of Miami,
1301 Memorial Drive,
Coral Gables,
FL 33124,
USA

Kappelle M.
Formerly with The Nature Conservancy,
San José, Costa Rica;
presently with WWF-NL,
P.O.Box 7, 3700 AA Zeist,
the Netherlands

Katchpole S.
Department of Geology,
Smith College,
Northampton,
MA 01063,
USA

Kim E. M.
Department of Geology,
Smith College,
Northampton,
MA 01063,
USA

Kirkpatrick J. B.
School of Geography and Environmental Studies,
University of Tasmania,
Private Bag 78, Hobart,
Tasmania 7001,
Australia

Kitayama K.
Center for Ecological Research, Kyoto University,
509-3 Hirano 2-chome, Ohtsu,
Shiga 520-2113,
Japan

Knoblich K.
Institute of Applied Geosciences,
Justus Liebig University of Giessen,
Diezstrasse 15,
35390 Giessen,
Germany

Köhler L.
Plant Ecology,
Albrecht von Haller Institute of Plant Sciences,
University of Göttingen,
Untere Karspüle 2,
37073 Göttingen,
Germany

Krömer T.
Centro de Investigaciones Tropicales, Universidad Veracruzana,
Interior de la Exhacienda Lucas Martín,
Privada de Araucarias s/n,
Colonia 21 de Marzo, Xalapa,
C.P. 91019,
Veracruz,
México

Kroon L. J. M.
Meteorology and Air Quality Group,
Wageningen University and Research Centre,
P.O. Box 47,
6700 AA Wageningen,
the Netherlands

Kumaran S.
Formerly with WWF-Malaysia, Petaling Jaya,
Malaysia; presently at EnviroLogic Consulting,
18 Jalan 20/2,
Paramount Garden,
46300 Petaling Jaya,
Selangor,
Malaysia

Küppers M.
Institute of Botany and Botanical Garden,
University of Hohenheim,
Garbenstrasse 30,
70599 Stuttgart,
Germany

Kuraji K.
University Forest in Aichi,
University Forests,
Graduate School of Agricultural and Life Sciences,
The University of Tokyo,
Seto, Aichi,
Japan

Lawton R. O.
Department of Biological Sciences,
National Space Science Technology Center,
University of Alabama in Huntsville,
Huntsville,
AL 35899,
USA

Lee M. S.
WWF-Malaysia,
49 Jalan SS23/15,
47300 Petaling Jaya,
Selangor,
Malaysia

Letts M. G.
Department of Geography,
Water and Environmental Science Program,
University of Lethbridge, Lethbridge,
Alberta T1K 3M4,
Canada

Leuschner Ch.
Plant Ecology,
Albrecht von Haller Institute of Plant Sciences,
University of Göttingen,
Untere Karspüle 2,
37073 Göttingen,
Germany

Li H. M.
Xishuangbanna Tropical Botanical Garden,
Chinese Academy of Sciences, Mengla,
Yunnan Province 666303,
P. R. China

Liu W. J.
Xishuangbanna Tropical Botanical Garden,
Chinese Academy of Sciences,
88 Xuefu Road,
Kunming 650223,
P. R. China

Lizcano D.
Formerly with the University of Kent, Canterbury, UK; presently at Facultad de Ciencias Básicas,
Universidad de Pamplona,
Pamplona,
Norte de Santander,
Colombia

Lyaruu H. V. M.
Department of Botany,
University of Dar es Salaam,
Dar es Salaam,
Tanzania

Malmer A.
Department of Forest Ecology and Management,
Swedish University of Agricultural Sciences,
Umeå SE 90183,
Sweden

Marzol-Jaén M. V.
Department of Geography,
Universidad de La Laguna,
La Laguna, Tenerife,
Canary Islands,
Spain

Masse G.
Instituto Nacional de Estadísticas y Censos,
Buenos Aires,
Argentina

McJannet D. L.
CSIRO Land and Water,
120 Meiers Road,
Indooroopilly,
Queensland,
Australia

Membiela M. de
Departamento de Dendrocronología e Historia Ambiental,
Instituto Argentino de Nivología,
Glaciología y Ciencias Ambientales,
Mendoza,
Argentina

Méndez A. R.
Fondo de Financiamiento Forestal (FONAFIFO),
Apartado 594-2120 San José,
Costa Rica

Meng F. R.
Faculty of Forestry and Environmental Management,
University of New Brunswick, Fredericton,
New Brunswick E3B 6C2,
Canada

Meyer J.-Y.
Délégation à la Recherche,
Gouvernement de la Polynésie Française,
B.P. 20981,
Papeete, Tahiti,
French Polynesia

Molen M. K. van der
Formerly with the VU University Amsterdam,
Amsterdam, the Netherlands; presently at the Meteorology and Air Quality group, Wageningen University and Research Centre,
P.O. Box 47, 6700 AA Wageningen
the Netherlands

Motzer T.
Formerly with the University of Mannheim,
Mannheim, Germany; presently at PE International,
Bebelstrasse 68, 70193 Stuttgart,
Germany

Mulligan M.
Environmental Monitoring and Modelling Research Group,
Department of Geography,
King's College London, Strand,
London WC2R 2LS, UK

Munz N.
Department of Physical Geography,
University of Mannheim, L 9, 1-2,
68131 Mannheim,
Germany

Nadkarni N. M.
The Evergreen State College, Olympia,
WA 98505, USA

Nair U. S.
Earth Science System Center,
National Space Science Technology Center,
University of Alabama in Huntsville, Huntsville,
AL 35806,
USA

Newell S. E.
Department of Geology,
Smith College,
Northampton,
MA 01063,
USA

Nitta I.
Institute of Environmental Studies,
Graduate School of Frontier Sciences,
The University of Tokyo,
7-3-1 Hongo, Tokyo 113-0033,
Japan

Nullet M. A.
Department of Geography,
University of Hawai'i at Mānoa,
2424 Maile Way,
Honolulu, HI 96822,
USA

Obregon A.
Faculty of Geography,
Laboratory for Climatology and Remote Sensing,
University of Marburg,
Deutschhausstrasse 10,
35032 Marburg,
Germany

Oesker M.
Institute of Botany and Botanical Garden,
University of Hohenheim,
Garbenstrasse 30,
70599 Stuttgart,
Germany

Ohsawa M.
Institute of Environmental Studies,
Graduate School of Frontier Sciences,
The University of Tokyo,
7-3-1 Hongo,
Tokyo 113-0033,
Japan

Oord A. L.
c/o Faculty of Earth and Life Sciences,
VU University Amsterdam,
De Boelelaan 1085,
1081 HV Amsterdam,
the Netherlands

Ostertag R.
Department of Biology,
University of Hawai'i–Hilo,
200 W. Kawili Street, Hilo,
HI 96720,
USA

Owiunji I.
Formerly with the Wildlife Conservation Society,
Kampala, Uganda;
presently at the School of Environment,
University of Manchester,
Oxford Road, Manchester M13, PL, UK

Pedraza R.A.
Instituto de Genética Forestal,
Universidad Veracruzana,
Apartado Postal 551,
Xalapa, Veracruz 91000,
Mexico

Perumal B.
Formerly with Wetlands International,
Petaling Jaya, Malaysia; presently at the Global
Environment Centre,
78 Jalan SS2/72, 47300 Petaling Jaya,
Selangor,
Malaysia

Pielke Sr. R. A.
CIRES, University of Colorado,
Stadium 255-10, Boulder,
CO 80309,
USA

Plumptre A.
Albertine Rift Programme,
Wildlife Conservation Society,
P.O. Box 7487,
Kampala,
Uganda

Pounds J. A.
Monteverde Cloud Forest Reserve,
Monteverde,
Puntarenas,
Costa Rica

Puchet-Anyul C.
Instituto de Ecología,
Universidad Nacional Autónoma de México,
Apartado Postal 70-275,
Circuito Exterior,
Ciudad Universitaria,
04510 Mexico, D.F.,
Mexico

Pufall A.
Department of Geology,
Smith College,
Northampton,
MA 01063,
USA

Ramos-Palacios R.
Instituto de Ecología,
Universidad Nacional Autónoma de México,
Apartado Postal 70-275,
Circuito Exterior,
Ciudad Universitaria,
04510 Mexico, D.F.,
Mexico

Ray D. K.
Formerly with the University of Alabama in Huntsville, USA;
presently at the Department of Forestry and Natural Resources,
Purdue University,
West Lafayette, IN 47907,
USA

Reddell P.
Formerly with CSIRO Land and Water,
Tropical Forest Research Centre, Atherton,
Queensland,
Australia

Regmi A.
Formerly with the University of Alabama, Huntsville, USA;
presently at the Department of Forestry and Natural Resources,
Purdue University,
West Lafayette,
IN 47907,
USA

Rhodes A. L.
Department of Geology,
Smith College,
Northampton,
MA 01063,
USA

Rincón-Romero M. E.
Universidad del Valle,
Facultad de Ingenierias,
Escuela de Ingeniería Civil y Geomática,
Grupo de Investigación en Geomática Aplicada,
Calle 13 – Carrera 100,
Valle de Lili, Cali,
Colombia

Rollenbeck R.
Faculty of Geography,
Laboratory for Climatology and Remote Sensing,
University of Marburg,
Deutschhausstrasse 10,
35032 Marburg,
Germany

Roman L.
Formerly with the University of Pennsylvania,
Philadelphia, USA;
presently at ESPM,
University of California Berkeley,
McBride Laboratory,
140 Mulford Hall,
Berkeley,
CA 94720,
USA

Rubiano J.
Formerly with King's College London,
United Kingdom; presently at the Universidad Nacional de Colombia,
Carrera 32 Chapinero,
Vía Candelaria Palmira,
Valle del Cauca,
Colombia

Rüger N.
Formerly with UFZ Centre for Environmental Research,
Leipzig, Germany; presently at the Department of Forest Biometry and Systems Analysis,
Institute of Forest Growth and Computer Sciences,
Dresden University of Technology,
P.O. Box 117,
01735 Tharandt,
Germany

Ruiz-Jaen M. C.
Department of Biology,
University of Puerto Rico,
San Juan, Puerto Rico 00931 3360,
USA

Sánchez-Azofeifa A.
Earth and Atmospheric Sciences Department,
University of Alberta,
Edmonton,
Alberta T6G 2E3,
Canada

Sánchez Megía J.
National Meteorological Institute,
San Sebastian 77,
Santa Cruz de Tenerife,
Tenerife, Canary Islands,
Spain

Santiago L. S.
Department of Botany and Plant Sciences,
University of California,
2150 Batchelor Hall, Riverside,
CA 92521,
USA

Scatena F. N.
Department of Earth and Environmental Science, Hayden Hall,
University of Pennsylvania,
240 South 33rd Street,
Philadelphia,
PA 19104,
USA

Schawe M.
c/o Department of Landscape Ecology,
Institute of Geography,
University of Göttingen,
Goldschmidtstrasse 5,
37077 Göttingen,
Germany

Schmid S.
Formerly with the Institute of Geography,
University of Bern,
Hallerstrasse 12,
CH-3012 Bern,
Switzerland

Scholl M.
Water Resources Discipline,
United States Geological Survey,
12201 Sunrise Valley Drive, Reston,
VA 20192,
USA

Schrumpf M.
Formerly with the University of Bayreuth,
Germany; presently at the Max Planck Institute for
Biogeochemistry,
07745 Jena,
Germany

Shumiya T.
The Nature Conservation Society of Japan,
5-24 Sanbancho,
Tokyo 102-0075,
Japan

Siegwolf R.
Laboratory of Atmospheric Chemistry,
Stable Isotopes and Ecosystem Fluxes,
Paul Scherrer Institute,
CH-5232 Villigen,
Switzerland

Silver W. L.
Ecosystem Science Division,
Department of Environmental Science, Policy, and Management,
151 Hilgard Hall,
University of California,
Berkeley,
CA 94720,
USA

Suzuki M.
Department of Forest Sciences,
Graduate School of Agricultural and Life Sciences,
The University of Tokyo, Bunkyo-ku,
Tokyo,
Japan

Takizawa H.
College of Bioresource Sciences,
Nihon University, Fujisawa,
Kanagawa,
Japan

Takyu M.
Faculty of Regional Environmental Science,
Tokyo University of Agriculture,
Sakuragaoka 1-1-1, Setagaya-ku,
Tokyo 156-8502,
Japan

Tanaka N.
University Forest in Aichi,
Graduate School of Agricultural and Life Sciences,
The University of Tokyo,
11-44 Goizuka, Seto City,
Aichi 489-0031,
Japan

Tangtham N.
Department of Conservation,
Faculty of Forestry,
Kasetsart University,
Bangkok,
Thailand

Tantasirin C.
Department of Conservation,
Faculty of Forestry,
Kasetsart University,
Bangkok,
Thailand

Thompson A. W.
Ecosystem Science Division,
Department of Environmental Science, Policy, and
Management,
151 Hilgard Hall,
University of California, Berkeley,
CA 94707,
USA

Tobón C.
Formerly with the VU University Amsterdam,
the Netherlands; presently at the Departamento de Ciencias Forestales,
Universidad Nacional de Colombia,
Calle 59[a], no. 63-20,
Medellín,
Colombia

Tognetti S. S.
Environmental Science and Policy Consultant,
10211 Menlo Avenue,
Silver Spring,
MD 20912,
USA

Valarezo C.
Universidad Nacional de Loja,
Centro de Estudios de Postgrado,
Área de Desarrollo Rural,
Unidad Operativa de la Facultad de Ciencias Agricolas,
Loja,
Ecuador

Vázquez-Rodríguez P.
Instituto de Ecología,
Universidad Nacional Autónoma de México,
Apartado Postal 70-275,
Circuito Exterior,
Ciudad Universitaria,
04510 Mexico, D.F.,
Mexico

Villalba R.
Departamento de Dendrocronología e Historia Ambiental,
Instituto Argentino de Nivología,
Glaciología y Ciencias Ambientales,
Mendoza,
Argentina

Vitousek P.
Biogeochemistry Laboratory, Stanford University,
Stanford, CA 94305,
USA

Vugts H. F.
Faculty of Earth and Life Sciences,
VU University Amsterdam,
De Boelelaan 1085,
1081 HV Amsterdam,
the Netherlands

Wallace J. S.
CSIRO Land and Water,
Davies Laboratory,
Townsville, QLD 4811,
Australia

Wang C. M.
Southwest Forestry College,
Kunming 650224,
P.R. China

Wang C. P.
Taiwan Forestry Research Institute,
100 Taipei,
Taiwan

Wangda P.
Formerly with The University of Tokyo, Chiba,
Japan; presently at the Renewable Natural Resources Research Centre – Yusipang, Council for RNR Research of Bhutan,
Ministry of Agriculture,
P.O. Box 212,
Thimpu,
Bhutan

Wård Y.
Formerly with the Swedish University of Agricultural Sciences,
Umeå, Sweden; presently at Boliden Mineral AB,
Boliden Området, Kontorsvägen 1,
SE 993681 Boliden,
Sweden

Welch R. M.
Department of Atmospheric Science,
National Space Science Technology Center,
University of Alabama in Huntsville, Huntsville,
AL 35806,
USA

Wilcke W.
Formerly at University of Mainz,
Germany; presently at Institute of Geography,
University of Bern, Hallerstrasse 12,
CH-3012 Bern,
Switzerland

Wildpret W.
Universidad de La Laguna,
La Laguna, Tenerife,
Canary Islands,
Spain

Williams-Linera G.
Instituto de Ecología, A.C.,
Apartado Postal 63, Xalapa,
Veracruz 91000,
Mexico

Wolf J. H. D.
Faculty of Science,
Institute of Biodiversity and Ecosystem Dynamics (IBED),
Universiteit of Amsterdam, Kruislaan 318,
1098 SM Amsterdam,
the Netherlands

Wu J. T.
Research Center for Biodiversity,
Academia Sinica,
115 Taipei,
Taiwan

Wu M. J.
Institute of Natural Resources,
National Dong Hwa University,
974 Hualien,
Taiwan

Yeh C. F.
Institute of Natural Resources,
National Dong Hwa University,
974 Hualien,
Taiwan

Yerena E.
Departamento de Estudios Ambientales,
Universidad Simón Bolívar,
Caracas,
Venezuela

Zech W.
Institute of Soil Science and Soil Geography,
University of Bayreuth,
95440 Bayreuth,
Germany

Zhang Y. P.
Xishuangbanna Tropical Botanical Garden,
Chinese Academy of Sciences,
Mengla,
Yunnan Province 666303,
P. R. China

Foreword

While studying tropical ecology in the early 1980s I became fascinated by the intriguing lectures Tom van der Hammen and Antoine Cleef gave on the exuberance of cloud forests in the Tropics – it all sounded so magical, mythical, Shakespearean even; it seemed like one was part of the story of *Macbeth*, observing the three witches disappearing in the mist! I felt privileged when my teachers in Amsterdam subsequently sent me out to Colombia and Costa Rica to study the structure and composition of a special kind of tropical montane forests, the highland oak forests of the American Tropics.

I felt even more fortunate when colleagues at Costa Rica's Universidad Nacional took me out to get to know the cloud forest paradise of Monteverde. Just like Nalini Nadkarni and Nathaniel Wheelwright, who, back in the year 2000, edited their amazing book on the ecology and conservation of this lush cloud forest preserve, I was amazed by the extraordinary richness and complexity of the gnarled elfin woodlands at Monteverde. Only then did I begin to grasp the tremendous diversity of the different kinds of tropical mountain forests; whilst the Monteverde cloud forest has a relatively low stature, the Talamancan oak forests harbor trees reaching over 50 m tall. This sense of diversity and complexity became even stronger after visiting montane cloud forests in Puerto Rico, southern Mexico, Guatemala, Nicaragua, Venezuela, Colombia, Ecuador, Bolivia, northern Argentina, and Cuba – all these places belonging to only one continent. Now, try to imagine the added diversity and differences found among the cloud forests of Africa, South East Asia and Oceania! During my visit to Cuba it was a great honor for me to accompany the late Alwyn Gentry, one of the world's most knowledgeable botanists of tropical floras of all times. Alwyn died tragically in 1993 during an airplane crash, ironically when flying over montane cloud forest in Ecuador doing an assessment for conservation purposes.

In May of that same year, 1993, dozens of scientists from around the globe gathered in Puerto Rico to participate in the first-ever worldwide symposium on tropical montane cloud forests. Experts shared their research experiences and results which were published in 1995 as a 400+ page volume edited by Larry Hamilton, Jim Juvik, and Fred Scatena – two of whom are co-editors of the present volume as well. Just a few weeks later, in June 1993, over 50 scientists met at the New York Botanical Garden to attend the first Neotropical Montane Forest Biodiversity and Conservation Symposium. The proceedings of this meeting were also published in 1995 as a 700+ page book edited by Steve Churchill, Henrik Balslev, Enrique Forero, and Jim Luteyn and include much material on cloud forests. Only six years later I was happy to publish a 700-page volume in Spanish together with Alejandro Brown as part of the global International Mountain Year (2001), bringing together chapters on tropical cloud forests from all Spanish-speaking Latin American countries and the Caribbean.

In the new millennium, the time had come to present an updated state-of-knowledge review of tropical mountain cloud forests worldwide, meant to inform future research, biodiversity conservation, and sustainable development for nature's and humans' well-being. To make that happen, Jim Juvik, Sampurno Bruijnzeel, Fred Scatena, Larry Hamilton, and Philip Bubb teamed up to organize the Second International Symposium on Tropical Montane Cloud Forests, building on the 1993 Puerto Rico experience and taking into account the many subsequent scientific publications, regional meetings, and global policy efforts. This second worldwide symposium took place in Waimea, Hawai'i, July 27 – August 1 2004, and was attended by some hundred participants from 25 countries around the globe. The symposium's title was *Mountains in the Mist: Science for Conserving and Managing Tropical Montane Cloud Forests*. The present book is the prime concrete product stemming from this important gathering. It brings together a total of 72 chapters authored by over 170 scientists and reviewed by some 60 colleagues. The book includes chapters on such diverse topics as floristic and animal diversity, altitudinal zonation, hydrometeorology, nutrient cycling, ecophysiology and photosynthesis, climate change – exemplified by changes in cloud cover – conservation of endemic and threatened species, and restoration and management of cloud forest fragments.

As a scientist and conservationist, I am particularly happy to see that this new book not only summarizes current knowledge but also combines science and conservation practice in a single volume. The editors have done a wonderful job in providing a comprehensive and up-to-date look at tropical cloud forests by integrating purely biological/ecological case studies with research on the role of humans in these fragile ecosystems. Too often, scientific and conservation issues are published in separate volumes, illustrating the divergence between those who study cloud forest evolution and ecology, and those who try to conserve them. Clearly, however, we can only succeed in conserving these diverse and fragile ecosystems if we understand the underlying patterns and processes governing their structure and functioning, as well as cloud forest response to such global threats like forest conversion, climate change, fire, and invasive species. Now, thanks to the conviction and commitment of the editors, with the current volume in hand we can start to make a difference and develop science-based conservation strategies that take into account the ecological specifics of different kinds of cloud forests. The creation and consolidation of protected areas, the controlled development of ecotourism, as well as arrangements to ensure payment for environmental services from tropical montane cloud forests are just three promising strategies that should be implemented to help conserve cloud forests in the long run. These three conservation strategies, and several others, are discussed in this volume and should help set the stage for a global movement to protect, use, and even restore the world's remaining cloud forests. This will benefit both nature and humankind as people depend more and more on biodiversity and water resources for their continued survival.

I am convinced that the holistic approach of this book will stimulate greater worldwide awareness of all aspects of tropical cloud forests. For that reason, I would like to invite researchers, teachers, managers, practitioners, and students across the globe to carefully read this book and the messages it contains, and learn about the beauty, the complexity, the diversity, and the uniqueness of these forests; and learn about the extent to which we depend on tropical montane cloud forests for their water, plants, and animals, and the various regulatory ecosystem services they provide, such as erosion control, flood reduction, and pollination services. Hopefully, the book will be translated into the languages of the countries where most of the remaining tropical cloud forests are found: in Latin America, the Caribbean, Africa, Asia, and Oceania, thereby enabling wider access to this *opus magnum* for interested scientists, educators, extension workers, NGO personnel as well as forest and land managers.

Maarten Kappelle
(Lead Scientist for Latin America, The Nature Conservancy)

Preface

Just below an area of cloud forest on the Island of Hawai'i, an outstanding cadre of the principal cloud forest researchers and managers from around the world gathered from July 27 to August 1, 2004. Their purpose was to bring forward before peers the latest information on the occurrence, site conditions, ecological functioning (especially in terms of water, nutrient, and carbon dynamics), as well as threats to and management of these special ecosystems. As well as providing valuable interchange among cloud forest colleagues, and providing new partnerships in research, this event has subsequently resulted in this book. It presents much of the current state-of-the-art knowledge about cloud forests.

Only once previously had such an international convocation been effected. This was organized by the East–West Center and the International Institute of Tropical Forestry of the United States Forest Service, with support from the UNESCO International Hydrological Programme in 1993. It was held in San Juan, Puerto Rico and brought together some 44 scientists who were working in cloud forests on every continent, in 20 countries. The collected papers and a synthesis chapter were published in *Tropical Montane Cloud Forests* in 1995 by Springer-Verlag, edited by Lawrence Hamilton, James Juvik, and Fred Scatena. This event and this book stimulated several new university, governmental, and intergovernmental programs of research and education.

Some ten years later, the three conveners and editors, augmented by the energy of Sampurno Bruijnzeel, were moved into action at the insistence of Jim Juvik, to attempt the convening of another symposium, a "Puerto Rico Plus 10" event. Its purpose would be to capture previously omitted research and management, and to provide a vehicle for presentation of new work since 1993. This would also serve as a checkpoint to assess the progress that was called for in the 1995 IUCN publication, *A Campaign for Cloud Forests: Unique and Valuable Ecosystems at Risk*. This campaign called for the following actions: (i) worldwide inventory and mapping; (ii) raising awareness as to values, based on science; (iii) increased monitoring and benchmark establishment, especially in the face of global climate change; (iv) integrated and long-term research on ecosystem processes and elements; (v) applied research to answer management needs, plus management plans and sustainable land use practices; and (vi) more protection through their designation as formal Protected Areas. A significant milestone along the way to the Hawai'i Symposium was the publication for international awareness-raising of *Decision Time for Cloud Forests* in the IHP Humid Tropics Programme Series (no. 13). Moreover, the World Conservation Monitoring Centre (WCMC) in Cambridge, UK began the compilation of a directory of major cloud forest sites.

The four Symposium planners, augmented by Philip Bubb from WCMC, met in Vermont in October 2002 to lay the foundation for this event. They were not able to implement a summit gathering of cloud forests aficionados in 2003, but the following year the Second International Symposium became a reality on the island of Hawai'i, at the base of cloud forests. It bore the title: *Mountains in the Mist: Science for Conserving and Managing Tropical Montane Cloud Forests*. While three of the Symposium planners concentrated on program contents and participants, to Jim Juvik of the University of Hawai'i at Hilo must go the kudos for a superb job of all aspects of local organization, from field trips to Hawaiian *luaus*. One hundred and two participants benefited from the organizational skills of Jim and his support team as we met at an excellent venue, the Hawaiian Preparatory Academy at Waimea. It was a superb site for the exchange of information and ideas. The (mostly invited) participants came from 25 countries, but reported on work in many more. As well as four full days of formal papers being presented, two effective poster sessions brought in additional project participation. These contributions, supplemented by a few solicited additional chapters and minus ten chapters that did not pass muster upon peer review, made this book a reality. Acknowledgment of the fine work of the more than 60 peer reviewers – some of whom assessed and commented upon as many as five manuscripts – is given at the end of this Preface. The editors wish to thank Maarten Kappelle in particular for his substantial input to the Synthesis chapter in the arena of cloud forest management and conservation, and Mark Mulligan for adding his modeling skills to the chapter.

During the meeting two individuals were recognized for leadership by being presented with Distinguished Scientist Awards by the University of Hawai'i, Hilo: Robert Schemenauer and Lawrence Hamilton. Furthermore, a new DVD documentary called *Mountains in the Mist: Discovering Cloud Forest*, directed by Sampurno Bruijnzeel and produced by Halsundbeinbruch Film from Switzerland, and featuring cloud forest research at Monteverde, Costa Rica, was screened for the first time. Since then it has been viewed by around one million people all over the world while part of the proceeds go to the upkeep of the Monteverde Preserve. An all-day field excursion around the island of Hawai'i (led by Jim Juvik with the help of Tom Giambelluca), and early morning or late afternoon trips to a nearby cloud forest research site under the enthusiastic guidance of John DeLay kept participants linked to the ground.

The Symposium was sponsored by the UNESCO International Hydrology Programme, by IUCN's World Commission on Protected Areas (Mountain Biome) and by the Gordon and Betty Moore Foundation. Additional support enabling people from various parts of the world to participate in the Symposium was provided by the Department for International Development of the United Kingdom. (Forest Research Project R 7991), the United States National Science Foundation (project EAR-0309731), the Deutsche Forschungsgemeinschaft (project FOR 402), and the International Institute of Tropical Forestry of the United States Forest Service. Funding from the Gordon and Betty Moore Foundation and VU University Amsterdam has made financially possible the publication of the results of the Symposium by Cambridge University Press. To all of these, the editors express their deep gratitude. The junior editors also thank on behalf of all the authors, the massive effort of Sampurno Bruijnzeel, who, as lead editor, deserves the lion's share of the credit.

It is clearly displayed in the various chapters and syntheses in this book, that a great deal has been accomplished over the past decade, in implementing the six actions called for in the 1995 *Campaign for Cloud Forests.* The quality of the research and number of projects has increased greatly, and this volume represents an attempt to capture most of the relevant new material. It is pertinent to state that, prior to submission of the final book manuscript to the publishers, Sampurno has (single-handedly) updated all of the chapters to Spring 2010 with new relevant published research. Thus, although the meeting was held more than five years ago, this volume is truly up to date. More, however, remains to be done. It is hoped that this book will stimulate greater attention to, and more action for the amazing cloud forests of the world.

Lawrence S. Hamilton
Charlotte, Vermont, USA

Acknowledgements

List of chapter reviewers

S. Aiba
T. M. Aide[a]
R. Bain
J. Bendix[a]
H. J. Boehmer
J. Boy
G. Brehm
P. Bubb
J. C. Calvo-Alvarado
A. Cleef[a]
J. K. DeLay
C. Downer
W. Eugster[a]
P. Fabian
J. Fallas
K. Fleischbein
G. García-Santos
J. H. C. Gash
T. W. Giambelluca[a]
A. Hemp
D. Hertel[a]
P. Hietz[a]
D. Hoelscher[a]
R. Hofstede
F. Holwerda[a]
J. O. Juvik
R. O. Lawton[a]
Ch. Leuschner
M. Lubczynski
M. Kappelle
M. K. van der Molen[a]
T. Motzer
D. Mueller-Dombois
M. Mulligan[b]
N. M. Nadkarni[a]
U. S. Nair[a]
M. Oesker
B. Ostertag
I. Porras
R. Rollenbeck
N. Rüeger
L. Santiago[a]
F. Sarmiento[a]
J. Schellekens
R. S. Schemenauer[b]
M. A. Scholl
W. L. Silver[a]
C. J. Still
A. Taber
E. V. J. Tanner
C. Tobón[a]
S. S. Tognetti
P. M. Vitousek
M. J. A. Werger
B. Wickel
W. Wilcke[a]
J. H. D. Wolf[b]
K. Zaal
G. Zotz

[a] Two or more reviews.
[b] Five or more reviews.

Part I
General perspectives

L. A. Bruijnzeel, F. N. Scatena, and L. S. Hamilton

1 Setting the stage

F. N. Scatena
University of Pennsylvania, Philadelphia, Pennsylvania, USA

L. A. Bruijnzeel
VU University, Amsterdam, the Netherlands

P. Bubb and S. Das
UNEP – World Conservation Monitoring Centre, Cambridge, UK

BACKGROUND

The cloudy, wet, and generally difficult terrain of the world's Tropical Montane Cloud Forests (TMCF) has not only made them hydrologically and ecologically unique, but has historically given them some *de facto* protection compared to other tropical forests. In the late 1970s and early 1980s it became apparent that this *de facto* protection was diminishing and that TMCF in many parts of the world were rapidly becoming converted or fragmented and in need of protection (LaBastille and Pool, 1978; White, 1983; Stadtmüller, 1987). Indeed, by the early 1990s it was clear that TMCF were high on the list of the world's most threatened terrestrial ecosystems. Moreover, during the period 1981–1990, montane forests were being lost at a rate considerably greater than that estimated for lowland tropical forests (1.1% $year^{-1}$ vs. 0.8% $year^{-1}$, respectively; Doumenge *et al.*, 1995). It was also being recognized that the scientific information needed to manage and protect these unique but vulnerable ecosystems was generally lacking (Stadtmüller, 1987).

In response to this information need, an international Symposium on TMCF was held in San Juan, Puerto Rico between 31 May and 5 June 1993. The meeting resulted in a 27-chapter book on the world's TMCFs (Hamilton *et al.*, 1995), which included overview chapters on the hydrology and nutrient dynamics of TMCF (Bruijnzeel and Proctor, 1995) and the importance of TMCF for endemic and threatened birds (Long, 1995), as well as the first in-depth description of guidelines for managing and valuing an especially vulnerable type of TMCF – elfin cloud forest (Scatena, 1995) – and useful summary descriptions of the biogeography of TMCF in widely different settings, including a host of Pacific islands, Sri Lanka, NE Borneo, Mexico, SE Brazil, NW Argentina, and Perú.

Since the San Juan Symposium there have been many new local studies as well as several international initiatives aimed at increasing the understanding, appreciation, and protection of TMCF. The latter include the 1995 "Campaign for Cloud Forests" by the World Conservation Union (IUCN) (Hamilton, 1995) and the 1999 "Tropical Montane Cloud Forest Initiative" of UNEP–WCMC, WWF, IUCN, and UNESCO-IHP (Aldrich *et al.*, 2000) which has since evolved into a "Cloud Forest Agenda" designed to encourage new TMCF conservation actions (Bubb *et al.*, 2004). In addition, there have been various regional symposia as well as compilations of (mostly biological) information on montane forests (e.g. Churchill *et al.*, 1996; Nadkarni and Wheelwright, 2000; Kappelle and Brown, 2001; Kappelle *et al.*, 2001; Kappelle, 2004; Beck *et al.*, 2007; Gradstein *et al.*, 2008). TMCF hydrometeorological aspects have also been discussed at a series of Conferences on Fog and Fog Collection that have been held every three years since 1998 (Schemenauer and Bridgman, 1998; Schemenauer and Puxbaum, 2001; Rautenbach and Oliver, 2004; Biggs and Cereceda, 2007) whereas a recent overview of Andean studies has been provided by Tobón (2009). Whilst all of these activities have improved our knowledge of TMCF, most of the post-1993 efforts have had a Neotropical focus. Knowledge on the Asian and African TMCF has not only lagged behind considerably but has also remained limited to a few iconic mountains, like Mt. Kinabalu in Borneo (Kitayama, 1995; Kitayama *et al.*, 2000; Kitayama and Aiba, 2002; Kitayama and Nais, 2002) and Mt. Kilimanjaro in Tanzania (Røhr, 2003; Hemp, 2005, Hemp, this volume #12 and references therein).

Major threats to TMCF have been identified in several venues (Hamilton *et al.*, 1995; Aldrich *et al.*, 1997; Bruijnzeel and Hamilton, 2000; Kappelle and Brown, 2001; Bubb *et al.*, 2004; Beck *et al.*, 2007; Gradstein, 2008; cf. Mulligan, this volume). Conversion to agricultural and grazing lands, over-harvesting,

Tropical Montane Cloud Forests: Science for Conservation and Management, eds. L. A. Bruijnzeel, F. N. Scatena, and L. S. Hamilton. Published by Cambridge University Press.

alien invasions, roads, and various types of development are threats in all regions. Mining, fire, forest clearing for drug cultivation, and other activities can be locally important. The various symposia and venues cited above have also identified the following major information gaps in our understanding of TMCF:

- Inadequate information on the spatial distribution, biological richness, and ecological variation of TMCF at the continental, regional, mountain, and local scales.
- Inadequate information on the hydrometeorology and plant physiology of TMCF.
- Inadequate information on the nutrient and carbon dynamics of TMCF, especially in relationship to their productivity, resilience, and potential for restoration.
- Inadequate information on the hydrological and ecological consequences of converting TMCF to other forms of land use.
- Inadequate information on the influence of changes in regional land use and climate on the biodiversity and ecological functioning of TMCF.
- Inadequate information on the conservation status, restoration potential, and management strategies of different types of TMCF.

In view of the increased research efforts in TMCF and the need to summarize the increased understanding of their occurrence, value and functioning that has developed since the 1993 San Juan Symposium, the second International Symposium on Tropical Montane Cloud Forests was held in Waimea, Hawai'i in the summer of 2004. The purpose of this book is to report on advances made in various fields since 1993 as presented at the Waimea Symposium. The book is organized around the information gaps described above and includes a total of 72 chapters that range from inventories of biodiversity, through detailed investigations of TMCF hydrometeorological, physiological, and biogeochemical functioning, to studies of the impact of climate change, the potential for sustainable use, and various conservation strategies. The introductory section of the book consists of invited synthesis chapters on topics that are important to all types of TMCF: viz. their global extent and distribution, climatic variability and climate change, the eco-physiology of epiphytes, global and local variation in soil nutrient contents, nutrient cycling and nutrient limitation to forest productivity, and the state of TMCF restoration. An overview chapter of the hydrological functioning of TMCF was not included because such an overview had just been produced prior to the Waimea Symposium (Bruijnzeel, 2005) as part of a state-of-knowledge compilation of tropical forest hydrology (Bonell and Bruijnzeel, 2005). Therefore, the new information on TMCF hydrology, hydrometeorology, and plant physiology presented in the chapters of this book has been combined with earlier reviews (Bruijnzeel and Proctor, 1995; Bruijnzeel, 2005; Tobón, 2009) in the final Synthesis chapter.

The 11 chapters making up the second part of the book focus on regional floristic or animal diversity in TMCF from various parts of the world, including Africa and the Himalayas, two areas that remained under-discussed during the 1993 Symposium. This section also draws attention to the occurrence of Lowland Cloud Forests, a previously unstudied type of cloud forest (Gradstein *et al.*, this volume). Part II is designed to be a useful complement to the various overview publications on Neotropical TMCF diversity listed above.

The core of the book consists of Part III (hydrometeorology, 19 papers), Part IV (nutrient dynamics, eight papers), Part V (water use, photosynthesis, and the soil and water impacts of TMCF conversion to pasture, eight papers) and Part VI (effects of climate variability and climate change, seven papers). As this final topic was in its infancy during the 1993 Symposium (Lugo and Scatena, 1992; Benzing, 1998; Markham, 1998), this latter collection of papers provides a unique synthesis of the current state of play concerning the effects of climate variability and change on TMCF. Last, but certainly not least, Part VII (10 papers) focuses on the potential for, and approaches to the conservation, management, and restoration of cloud forests. This information was largely lacking in 1993 and includes such recent developments as payments for environmental services delivered by TMCF (Tognetti *et al.*, this volume), multi-stakeholder learning initiatives for sustainable forest use (Hofstede *et al.*, this volume), and community-based forest protection (Asbjornsen and Garnica-Sánchez, this volume). The book concludes with a Synthesis chapter that summarizes what we have learned since 1993 and identifies some of the more important remaining issues that need to be resolved to ensure future sustainable use and protection of TMCF.

GLOBAL DISTRIBUTION OF TMCF

Defining cloud forest

Although the biodiversity, ecological, and hydrological values of TMCF have been widely acknowledged, their definition and the delineation of their spatial distribution has remained both a persistent challenge and need (Stadtmüller, 1987; Campanella, 1995; Hamilton *et al.*, 1995; Ashton, 2003; Bach, 2004; Bubb *et al.*, 2004; Martin *et al.*, 2007; Mulligan, this volume). Historically, this problem has been confounded by a myriad of imprecise, overlapping and at times, contradictory definitions of TMCF (Stadtmüller, 1987). One of the major advances since the 1993 symposium has been the recognition and development of definitions for different TMCF types (Bruijnzeel and Hamilton, 2000; Bruijnzeel 2001). These forest types are described below and are based on forest structure, the degree of mossiness and leaf sclerophylly (Grubb, 1977; Frahm and Gradstein, 1991; Bach, 2004; Table 1.1), and observed contrasts in the fraction

Table 1.1 *Summary of key structural characteristics marking the chief tropical (montane) forest types distinguished in the present volume*

Forest formation[a]	LERF	LMRF/LMCF	UMRF	SACF
Canopy height	25–45 m	15–33 m	1.5–18 m	1.5–9 m
Emergent trees	Up to 67 m tall	Often absent, up to 37 m	Usually absent, up to 26 m	Usually absent, up to 15 m
Compound leaves	Abundant	Occasional	Rare	Absent
Principal leaf size class[b]	Mesophyllous	Meso-/notophyllous	Microphyllous	Nanophyllous
Leaf drip-tips	Abundant	Present	Rare or absent	Absent
Buttresses	Frequent and large	Uncommon, and small	Usually absent	Absent
Cauliflory	Frequent	Rare	Absent	Absent
Big woody climbers	Abundant	Usually absent	Absent	Absent
Bole climbers	Often abundant	Frequent to abundant	Very few	Absent
Vascular epiphytes	Frequent	Abundant	Frequent	Very rare
Non-vascular epiphytes (mosses, liverworts)	Occasional	Occasional/Abundant <10%/25–50%	Usually abundant 70–80%	Abundant >80%

[a] LERF, lowland evergreen rain forest; LMRF/LMCF, lower montane rain/cloud forest; UMCF, upper montane cloud forest; SACF, sub-alpine cloud forest.

[b] Leaf sizes according to the (1934) Raunkiaer classification system: mesophyllous, 4500–18 225 mm^2; notophyllous, 2025–4500 mm^2; microphyllous, 225–2025 mm^2; nanophyllous, <225 mm^2.

Source: Based on Whitmore (1998) and Frahm and Gradstein (1991).

of net precipitation reaching the forest floor (Bruijnzeel, 2005). Three general TMCF types are recognized (lower montane cloud forest, upper montane cloud forest, and sub-alpine cloud forest) within the widely adopted definition of cloud forests as "forests that are frequently covered in cloud or mist" (Stadtmüller, 1987; Hamilton *et al.*, 1995). These definitions also recognize the important influence of temperature and humidity levels on montane forest zonation.

With increasing elevation on wet tropical mountains, distinct changes occur in forest appearance and structure (Table 1.1). In general, the tall and often buttressed trees of the multi-storied *lowland rain forest* (main canopy height 25–45 m, with emergents up to 60 m), gradually give way to *lower montane forests.* With a mean canopy height of up to 35 m and emergent trees as high as 45 m, these lower montane forest can still be quite impressive. Yet, with two rather than three canopy layers, the structure of lower montane forest is simpler than that of lowland forest. The large buttresses and climbers that are so abundant in the lowland forest all but disappear while epiphytes (orchids, ferns, bromeliads) on branches and stems become more numerous with elevation (Whitmore, 1998). The change from lowland to lower montane forest is normally observed at the elevation where the average minimum temperature drops below 18 °C. At this threshold many lowland tree species are displaced by a floristically different assemblage of montane species (Kitayama, 1995). On large equatorial inland mountains this transition usually occurs at an altitude of 1200–1500 m.a.s.l. and coincides with the general range of incipient and intermittent cloud formation (Kitayama, 1995). As elevation increases, the trees become gradually smaller, moss cover on stems increases from 10% to 25–50%, and a *lower montane cloud forest* is observed. With further increases in elevation there is usually a very clear change from relatively tall (15–35 m) lower montane cloud forests to a distinctly shorter-statured (2–20 m) and much more mossy (70–80% bryophytic stem cover) *upper montane cloud forest* (Frahm and Gradstein, 1991). Although these two forest types are not separated by a known or distinct thermal threshold, there can be little doubt that the transition from lower to upper montane cloud forest coincides with the level where cloud condensation becomes most persistent (Grubb and Whitmore, 1966; Kitayama, 1995; cf. Schawe *et al.*, this volume). On large mountains in equatorial regions away from the ocean this belt of persistent clouds typically occurs at elevations of 2000–3000 m.a.s.l. (Grubb, 1977; Kitayama, 1995). On small oceanic island mountains and in some coastal mountains this change from lower to upper montane forests may occur at much lower altitudes (see below). Mosses also start to cover rocks and fallen trunks on the soil surface in the upper montane cloud forest zone. With increasing elevation and exposure to wind-driven fog and rain, tree stems become increasingly crooked and gnarled, and bamboos often replace palms as dominant undergrowth species (Kappelle, 1995). The eerie impression of this tangled mass, wet with fog and glistening in the morning sun, has given rise to names like "elfin" forest or "fairy" forest to the more stunted and dwarfed forms of these upper montane cloud forests (Stadtmüller, 1987).

A third major type of TMCF occurs at the elevation where the average maximum temperature falls below 10 °C. Here the upper montane forest gives way to still smaller-statured (1.5–9 m) and more species-poor *sub-alpine cloud forest* (or scrub) (Kitayama, 1995; Bach, 2004; Hemp, 2006). This forest type is characterized not only by its low stature and gnarled appearance but also by even smaller leaves, and a comparative absence of epiphytes. Mosses usually remain abundant and cloud incidence is still a paramount feature (Frahm and Gradstein, 1991; cf. Schawe *et al.*, this volume). On large equatorial mountains the transition to sub-alpine forest is generally observed at elevations between 2800 and 3200 m.a.s.l. (Kitayama, 1995; Hemp, this volume #12; Schawe *et al.*, this volume). As such, this type of forest is encountered only on the highest mountains, mostly in Latin America and Papua New Guinea, where it may extend to *c.* 3900 m.a.s.l. (Whitmore, 1998; cf. Hemp, this volume #12).

Climate and the distribution of TMCF

Although the different types of TMCF described in Table 1.1 are recognized in many regions, there is a wide variation in the elevations of TMCF. In general their distributions depend on the upper and lower bounds of clouds and the global, regional, and local factors that influence cloud formation. The transition from lower to upper montane cloud forest, and the thickness of the cloud forest belt itself, is primarily governed by the level of persistent cloud condensation (Grubb and Whitmore, 1966; Frahm and Gradstein, 1991; Kitayama, 1995; Hemp, this volume #12; Schawe *et al.*, this volume). The lower elevation of cloud formation, in turn, is determined by the moisture content and temperature of the atmosphere such that the more humid the uplifted air is, the sooner it will condense. With increasing distance from the ocean, air tends to be drier and therefore requires lower temperatures and higher elevations to reach its condensation point. Consequently, the associated cloud base and TMCF vegetation will occur at a higher elevation farther from the oceans. Likewise, at given moisture content, the condensation point is reached more rapidly for cool air than for warm air. Thus, at greater distance from the equator, the average temperature – and therefore the altitude at which condensation and TMCF occurs will be lower (Nullet and Juvik, 1994; Jarvis and Mulligan, this volume).

In addition to the elevation of the cloud *base*, the distribution and extension of the TMCF belt is also governed by the *upper level* of cloud formation which is influenced by global-scale atmospheric circulation. In general, in a large-scale atmospheric circulation pattern (the Hadley cell), heated air rises to great elevations in the equatorial zone, and flows poleward and eastward in the upper atmosphere as it cools. The cool dry air then descends in a broad belt in the outer tropics and sub-tropics from where it returns to the equator. This subsidence reaches its maximum expression at oceanic sub-tropical high-pressure centers and along the eastern margins of the oceanic basins. As the air descends and warms, it forms a temperature inversion that separates the moist layer of surface air (that is being cooled while rising) from the drier descending air above. This so-called "trade wind inversion" (TWI) has a tilted three-dimensional surface that generally rises toward the equator and from east to west across the oceans. Over the eastern Pacific Ocean, the TWI occurs at only a few hundred meters above sea level, e.g. off the coast of southern California. It rises to about 2000 m near Hawai'i and dissipates in the equatorial western Pacific (Nullet and Juvik, 1994). The consequences of the TWI for the occurrence of the upper boundary of montane cloud forest are profound and are another reason why the vegetation zonation on mountains situated away from the equator tends to become compressed (Stadtmüller, 1987). For instance, windward slopes in the Hawai'ian archipelago receive more than 6000 mm of rain per year below the inversion layer. However, at 1900–2000 m.a.s.l., montane cloud forest suddenly gives way to dry sub-alpine scrubs because the clouds are prevented from moving upward by the presence of the temperature inversion (Kitayama and Müller-Dombois, 1994a,b). One of the best-known examples of the TWI and its effect on vegetation zonation comes from the Canary Islands. Situated between 27° and 29° N, a daily "sea of clouds" develops between 750 m and 1500 m which sustains evergreen Canarian laurel forests in an otherwise rather arid environment (Ohsawa *et al.*, 1999; cf. Marzol *et al.*, this volume; García-Santos, 2007). However, the dry conditions prevailing at the level of the TWI also promote the occurrence of wildfires and thereby help to maintain the upper boundary of TMCF (Hemp, 2005; Martin *et al.*, 2007).

Superimposed on these global-scale moisture and temperature gradients are the more local processes that influence the temperature of the air column and thus the "starting point" for cooling. These include the influence of offshore sea surface temperatures, land–sea interactions involving the coastal plain, the size of a mountain and its orientation and exposure to the prevailing winds (Van Steenis, 1972; Stadtmüller, 1987; Jarvis and Mulligan, this volume). The interactions of these local and regional influences on the distributions of TMCF can be quite pronounced. For example, the sheer mass of large mountains exposed to intense radiation during cloudless periods is believed to raise the temperature of the overlying air enough to decrease the lapse rate and enable plants to extend their altitudinal range. This effect is commonly referred to as the "mass elevation" or "telescoping" effect and has been recognized for decades (Van Steenis, 1972; Whitmore, 1998; Figure 1.1). More recent research has indicated that low-stature, mossy, upper montane-looking forests that occur at relatively low elevations (<1000 m.a.s.l.) on some small coastal mountains can be ascribed to the high humidity of the oceanic air promoting cloud formation at very low elevations rather than to a steeper temperature lapse rate (Jarvis and Mulligan, this volume). Further support for this comes from the observation that the effect is most pronounced in areas with high rainfall and thus high atmospheric humidity (Van Steenis, 1972; Bruijnzeel *et al.*, 1993). Locally produced inversions can also influence the distributions of cloud

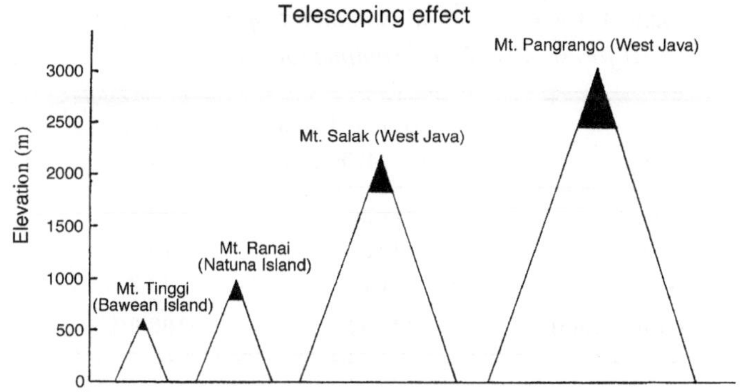

Figure 1.1. The telescoping effect of vegetation zonation on differently sized mountains (after Van Steenis, 1972).

forests, as has been described by Proctor *et al.* (2007) for Mt. Cameroon. Here, fires and active volcanism have created an inversion that limits the upper occurrence of (cloud) forest to a comparably low elevation of *c.* 2200 m despite the large size of this continental mountain (4095 m) and the presence of fertile soils.

Mapping cloud forests

As a result of the complex interactions and the major variations in global- and local-scale influence of climate, topography, and atmospheric circulation briefly described above, the mapping of actual and potential area of TMCF at the global scale is more than challenging (Bubb *et al.*, 2004). Ideally, a global TMCF map would explicitly account for multiple climatic and physiographic factors, including cloud frequency, wind and rainfall patterns, as well as aspect, latitude, altitude, the size of mountains, their distance from the sea, and local vegetation classification. Unfortunately, much of this information has not been available at the spatial scales needed, although this situation is changing rapidly with respect to topography, climate, and cloud occurrence (cf. Jarvis and Mulligan, this volume; Mulligan, this volume; Lawton *et al.*, this volume).

A map of the global distributions of TMCF cannot be compiled from existing national forest assessments because the term "cloud forest" is not commonly or consistently used. Moreover, at least 35 different names have been used to typify cloud forests in the past (Stadtmüller, 1987). In Africa, cloud forests have generally not been distinguished from the broader Afro-montane forest category (Hemp, this volume #12), whereas in Asia and the Pacific region the term "cloud forest" is rarely recognized outside of the "mossy forest" of the Philippines (Penafiel, 1995) and Malaysia (Kumaran *et al.*, this volume). In the Americas, the concept of cloud forest is relatively well known and approximate maps of the distributions of TMCF have been compiled for some Latin countries (see Kappelle and Brown, 2001). However, Mexico is the only country with a national vegetation classification scheme that explicitly includes a category corresponding to "cloud forest" (Vázquez-García, 1995).

Given the limitations in resources and data coverage, different approaches to mapping TMCF have been considered. Maps based on ground-measured *floristic or physiognomic characteristics* have been used to map TMCF at the spatial scales of a single mountain or landscape (Frahm and Gradstein, 1991; Bach, 2004; Hemp, this volume #12) and for mapping of ecotones of TMCF on either side of a ridge (e.g. Lawton and Dryer, 1980; Martin *et al.*, 2007). Other approaches have used elevation as a proxy for the climatic (temperature, rainfall, and fog incidence) and edaphic (soil water status, acidity) conditions that tend to be associated with cloud forest (e.g. Campanella, 1995; Bubb *et al.*, 2004). Whilst this approach is the least data-demanding and therefore particularly suitable for global-scale inventories, it also tends to lump different types of TMCF into a single category. Elevation-based predictions of TMCF occurrence are also complicated by the mass elevation effect (Figure 1.1) and need local or regional calibration. The most complex and data demanding techniques involve mapping according to a set of pre-defined *hydro-climatic characteristics*, such as cloud frequency or water balance (Mulligan, this volume; cf. Lawton *et al.*, this volume).

Areal estimates of cloud forest distribution based on these different techniques can differ substantially and a definitive global-scale map of actual or potential TMCF is still far from being developed. Nevertheless, our knowledge of TMCF distribution has improved significantly since the 1993 San Juan Symposium when the first global maps of TMCF were made by having symposium attendees place tags on maps to identify sites known to have TMCF. The map analysis provided below is a second iteration of work carried out by UNEP–WCMC as a product of the Mountain Cloud Forest Initiative referred to earlier (Aldrich *et al.*, 1997; Bubb *et al.*, 2004), and represents the first major attempt to calculate a global estimated cloud forest distribution. The overall approach followed was to first delineate the potential distribution of TMCF using pre-set altitudinal limits that are likely to include cloud forests of any kind between 30° N and 30° S (Table 1.2). These altitudinal limits were determined from a database of TMCF sites that was compiled at UNEP–WCMC from literature sources and expert opinion (Table 1.2). The database contains 650 records from 46 countries (Aldrich *et al.* 1997). This information was supplemented by cloud forest altitudinal ranges as reported for several Latin American countries in Kappelle and Brown (2001), and modified further on the basis of comments received from participants during the Waimea Symposium. The area of potential TMCF within these altitudinal bands is likely to include montane rain forests and drier forest types as well as TMCF. A Geographic Information System (GIS) was used to estimate the potential distribution of TMCF as follows:

- A GIS layer of the mountain areas within the altitudinal limits of TMCF for each major mountain range or region was produced from the GIS-base layer of the mountains of the world, using the definition of mountains proposed by

Table 1.2 *Regional groupings of tropical montane cloud forest and the lower and upper altitudinal limits used in the present analysis. See text for details*

Regional groupings	Lower and upper altitudinal limits of cloud forest used in the analysis (m.a.s.l.)
Mexico west of the isthmus of Tehuantepec	2200–3500
Mexico east of the isthmus of Tehuantepec and Central and South America	1800–3500
Caribbean islands and coastal Venezuelan mountains	above 660
Atlantic coast of Brazil	above 700
Galápagos islands and coastal Ecuadorian mountains	above 400
Hawai'i	1000–3000
African Atlantic coast countries	1000–3500
Ethiopia and Albertine Rift mountains, including DR Congo	2000–3500
Kenya and northern Tanzania	1500–3500
Southern Tanzania, Malawi, Mozambique and Réunion	1000–3500
Madagascar	1300–3500
Philippines	1200–3500
New Guinea	2000–3500
Continental South-East Asia, Borneo, Sumatra, Java and Sulawesi	1500–3500
South India, Sri Lanka, other insular Asia–Pacific islands, northern Australia	1000–3500

Kapos *et al.* (2000), i.e. (i) all land above 2500 m.a.s.l., (ii) land between 1500 and 2500 m.a.s.l. and having a slope $\geq 2°$, (iii) land between 1000 and 1500 m.a.s.l. and having a slope $\geq 5°$ or a local elevation range >300 m, and (iv) land between 300 and 1000 m.a.s.l. and having a local elevation range >300 m. This phase utilized the GTOPO30 Digital Elevation Model from the US Geological Survey's EROS Data Center, which has a horizontal grid spacing of 30 arc sec (*c.* 1 km; http://edc.usgs.gov/products/elevation/gtopo30/gtopo30.html).

- To convert the potential distribution to actual distribution, the mountain areas within the altitudinal limits of TMCF were combined with a map of tropical montane forest and tropical lowland evergreen forest taken from the UNEP–WCMC global forest inventory. This data-set is comprised of national and regional data from many sources that have been harmonized to display global forest cover in the early 1990s at a resolution of 1 km (Iremonger *et al.*, 1997). Areas of non-forest and non-humid forest were excluded from the TMCF analysis.
- Finally, these base layers were updated using MODIS satellite-based vegetation coverage for the year 2000 (VCF 2000) to represent the actual distribution of TMCF in 2000.

Table 1.3 *Area (km²) of estimated tropical montane cloud forest (TMCF) by continental region*

Region	Estimated area of TMCF (km²)	% of global TMCF
Americas	87 626	40.8
Africa	34 328	16.0
Asia	92 676	43.2
Global total	**214 630**	**100.0%**

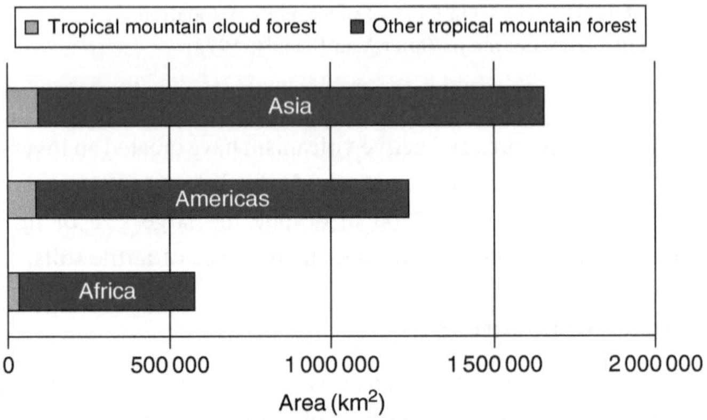

Figure 1.2. Estimated area (km²) of tropical montane cloud forest and tropical montane forest by continental region.

A significant source of error in this analysis is the variability of forest coverage and classifications used to compile the UNEP–WCMC global forest coverage (Iremonger *et al.*, 1997). This classification system divides the world's forest into 26 major types that reflect climatic zones as well as the principal types of trees. Pertinently, the coverage was derived from national sources and classifications that did not explicitly distinguish TMCF as a separate category, let alone made a distinction between lower or upper montane cloud forest.

A second source of error lies in using the GTOPO30 Digital Elevation Model in mountainous regions, which gives a single value of altitude for a 1-km grid cell. Thus, these altitudinal belts are based on kilometer-scale regional patterns and do not include the smaller-scale variations in aspect and exposure that are known to be important (Grubb and Tanner, 1976; Lawton and Dryer, 1980; Weaver, 1995; cf. Häger and Dohrenbusch, this volume).

Using the elevation- and publication-based approach described above, the estimated global area of TMCF (of all categories) is about 215 000 km² (Table 1.3 and Figure 1.2), which is approximately 0.14% of the Earth's land surface and 1.4% of the total area

Table 1.4 *Estimated area of tropical montane cloud forest as a percentage of all tropical forest and tropical mountain forest. Areas of tropical (mountain) forest based on Iremonger* et al. *(1997) and Kapos* et al. *(2000). The calculations of Kapos* et al. *(2000) of the areas of the world's mountain forest included altitudinal ranges from 300 m to above 4500 m.a.s.l.*

Region	All tropical forest (km^2)	TMCF as % of all tropical forest	Tropical mountain forest (km^2)	TMCF as % of all tropical mountain forest
Americas	7 762 359	1.1%	1 150 588	7.6%
Africa	4 167 546	0.8%	544 664	6.3%
Asia	3 443 330	2.7%	1 562 023	5.9%
Global total	**15 373 235**	**1.4%**	**3 257 275**	**6.6%**

of the world's tropical forests (Table 1.4). This estimate of the total area of TMCF is considerably less than the only other published figure of *c.* 500 000 km^2 of cloud forests in the humid tropics (Bockor, 1979). The estimate is also less than Kappelle and Brown (2001) their estimate of a potential extent of 750 000 km^2 for Latin America alone. The present estimate is also less than the recent estimates of hydro-climatically defined cloud forest using satellite measures of cloud frequency as derived by Mulligan (this volume) who also discusses several reasons for this discrepancy. Nevertheless, the present results are considered to represent a conservative estimate of the global extent of actual, not potential, TMCF and to accurately reflect their relative distribution (Figures 1.3–1.5).

Of all the TMCF mapped, 43% occur in Asia (including northern Australia and Oceania), 41% in the Americas (including the Hawai'ian archipelago), and 16% in Africa (Table 1.3). TMCF is also a relatively scarce habitat amongst all forest types in tropical mountain regions, occupying an estimated 7.6%, 6.3%, and 5.9% of the Tropical Mountain Forest biome in

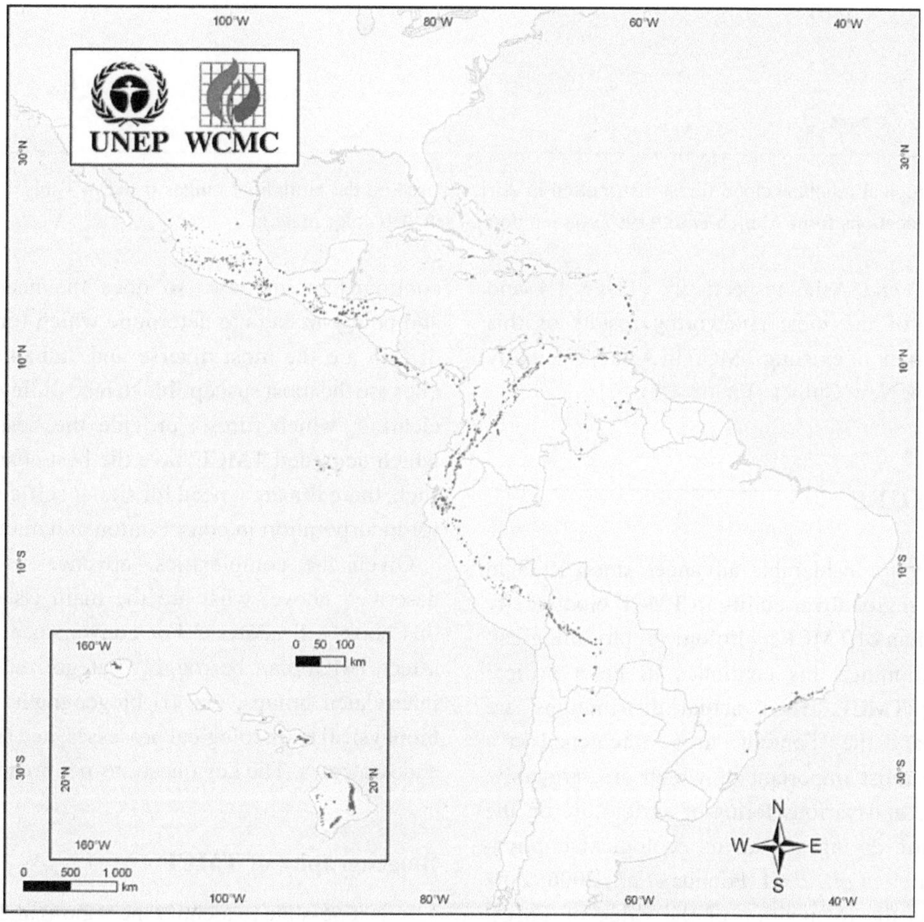

Figure 1.3. Estimated tropical montane cloud forest distribution in the Americas (including the Hawai'ian archipelago), based on the altitudinal ranges listed in Table 1.2 (areas in red) and known cloud forest site locations from Aldrich *et al.* (1997) (green dots). (See also color plate.)

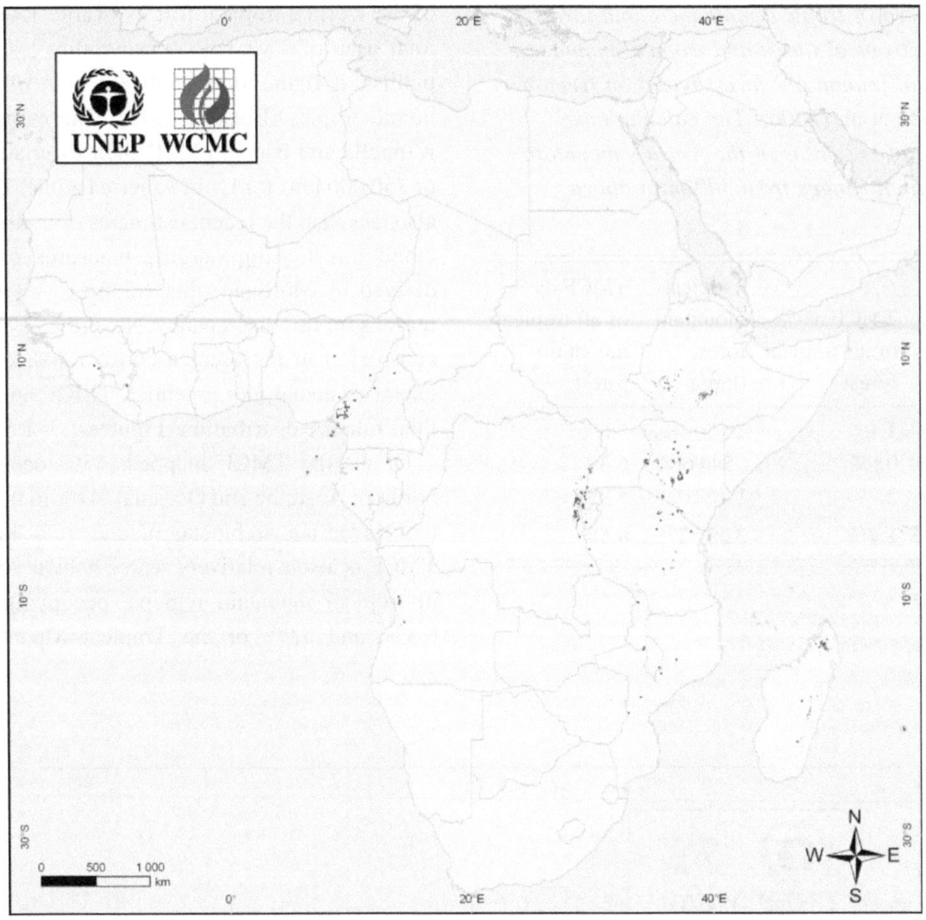

Figure 1.4. Estimated tropical montane cloud forest distribution in Africa, based on the altitudinal ranges listed in Table 1.2 (areas in red) and known cloud forest site locations from Aldrich *et al.* (1997) (green dots). (See also color plate.)

the Americas, Africa, and Asia, respectively (Table 1.4 and Figures1.3–1.5). One of the most noteworthy results of this analysis is the large extent of existing TMCF in Asia, principally in Indonesia and Papua New Guinea (Figure 1.5).

EMERGING ISSUES

Whilst there have been considerable advances since 1993 in our knowledge of the regional variability in TMCF biodiversity and in our understanding of TMCF hydrological, physiological, and ecological functioning, this continues to be a critical "decision time" for TMCF. Their actual distributions are still poorly defined and they continue to be threatened in a variety of ways, the most important of which are, arguably, conversion to pasture and various forms of agriculture on the one hand, and climatic drying and all its ecological implications on the other (Bubb *et al.*, 2004; Pounds *et al.*, 2006; Zotz and Bader, 2009). As the recognition of the value of TMCF as treasure houses of biodiversity, protectors against soil erosion, and providers of a steady supply of high-quality water continues to increase, so does the need for land managers and policy-makers to determine which forests under their jurisdiction are the most diverse and valuable biologically; which ones are the most susceptible to landsliding and soil erosion upon clearing, which forests provide the best water supplies, and which degraded TMCF have the best chances for recovery? As such, there is a great need for site-specific information on TMCF for incorporation in conservation and management plans.

Given the complexities, advances, and information gaps described above, what are the main issues and questions that this book will address? For convenience, the major issues that affect TMCF can be roughly categorized into three broad and interrelated groups, viz. (i) biogeography and biodiversity; (ii) biophysical and ecological processes; and (iii) management issues and strategies. The key questions per broad category include:

Biogeography of TMCF

- Is it possible to identify the regions and areas with the greatest diversity in TMCF flora and fauna? Do these relationships vary between "maritime" and "continental" settings?

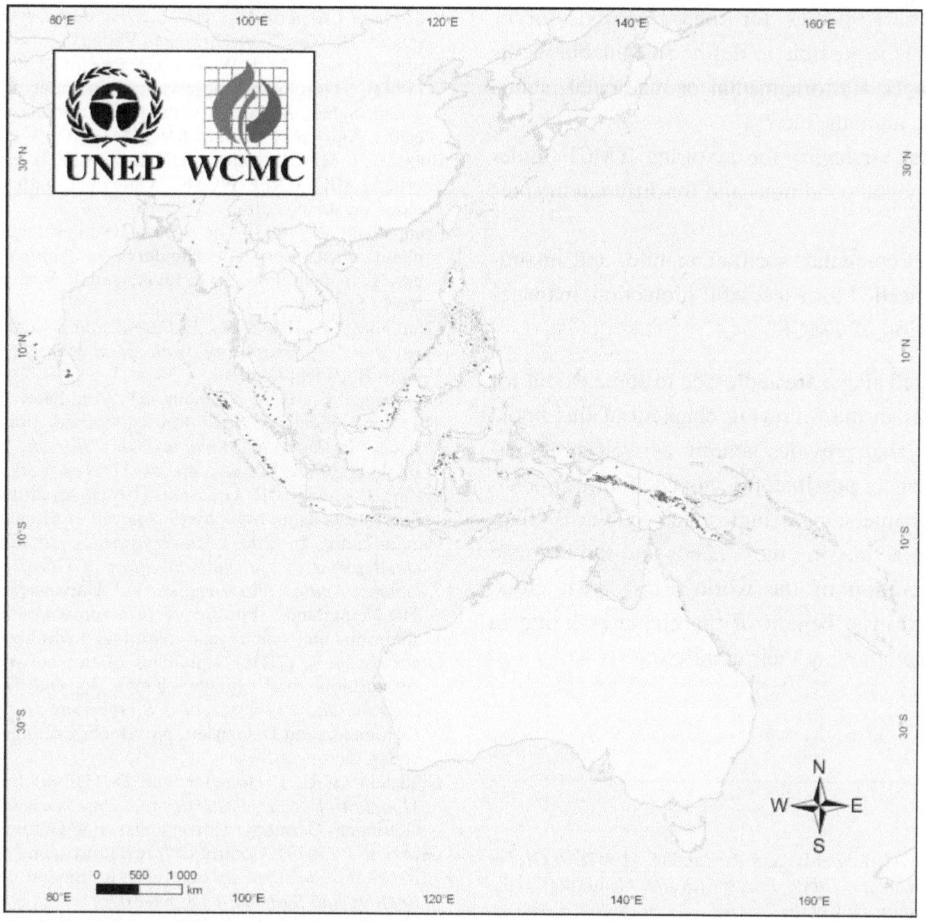

Figure 1.5. Estimated tropical montane cloud forest distribution in Asia (including northern Australia and Oceania), based on the altitudinal ranges listed in Table 1.2 (areas in red) and known cloud forest site locations from Aldrich *et al.* (1997) (green dots). (See also color plate.)

- Is it necessary on hydrological or ecological grounds to distinguish between lower and upper montane forests when mapping TMCF? And if so, what are the remotely sensed, modeled, and field data that can be used to distinguish between lower, upper, and sub-alpine TMCF?

Biophysical processes in TMCF

- What are the absolute and relative amounts of cloud water interception (*CWI*) and wind-driven rain (*WDR*) in different types of TMCF? Are there predictable regional and forest-type related patterns in *CWI* and *WDR*? How can these be measured and their spatial and temporal distributions modeled?
- What is the water use (evapotranspiration) and carbon uptake (photosynthesis) of different types of TMCF? Do similarly statured forests at different elevations cycle water, carbon, and nutrients at similar rates?
- How are annual and seasonal water yields and ecosystem services affected by converting different types of TMCF to pasture, annual crops, or coffee plantations?
- Are there important differences in soil nutrient levels, soil water status (e.g. degree of waterlogging), and aluminum or hydrogen toxicity between different types of TMCF? How do these differences relate to above-ground forest biomass and overall ecosystem productivity? How do soil resources change with land-cover change?
- How are different types of TMCF affected by climatic drying and a reduction in precipitation? What ecosystem component or function is most affected? Is the extent of climatic drying in TMCF mostly caused by local, regional, or global processes? Do these relationships vary between "maritime" and "continental" settings?

Management strategies for sustaining TMCF

- What is the present conservation status of TMCF? Where are these forests threatened the most and why?
- Which ecosystem components (e.g. ornamental plants, bryophytes, anoline species, large mammals) are the most vulnerable, and which are the most resilient to human activities or climatic drying?

- What are the best strategies for managing existing or degraded TMCF? Is it possible to define sustainable yields for specific products, e.g. ornamental or medicinal plants, mosses, fuelwood, animals, etc.?
- What are the best strategies for restoring TMCF under different environmental conditions and for different degrees of degradation?
- What are the environmental, socio-economic, and institutional conditions needed for successful protection, management, and restoration of TMCF?

All of the questions listed above are addressed to some extent for specific sites or regions in the following chapters of this book. The Synthesis chapter also provides generic as well as forest-type specific answers where possible. It is hoped that this book – through its overview chapters, case studies, and Synthesis chapter – will contribute to addressing these needs and the ultimate preservation and restoration of the world's last great cloud forests, both for the intrinsic benefit of the creatures living in them and for the people enjoying their manifold gifts.

REFERENCES

Aldrich, M., C. Billington, M. Edwards, and R. Laidlaw (1997). *A Global Directory of Tropical Montane Cloud Forests (draft).* Cambridge, UK: World Conservation Monitoring Centre.

Aldrich, M., P. Bubb, S. Hostettler, and H. Van de Wiel (2000). *Tropical Montane Cloud Forests: A Time for Action,* ArborVitae Supplement. Gland, Switzerland: WWF and IUCN.

Ashton, P. S., (2003). Floristic zonation of tree communities on wet tropical mountains revisited. *Perspectives in Plant Ecology, Evolution and Systematics* **6**: 87–104.

Bach, K. (2004). Vegetationskundliche Untersuchungen zur Höhenzonierung tropischer Bergregenwälder in den Anden Boliviens. PhD thesis, University of Göttingen, Göttingen, Germany.

Beck, E., J. Bendix, I. Kottke, F. Makeschin, and R. Mosandl (eds.) (2007). *Gradients in a Tropical Mountain Ecosystem of Ecuador.* Berlin: Springer-Verlag.

Benzing, D. H. (1998). Vulnerabilities of tropical forests to climate change: the significance of resident epiphytes. *Climatic Change* **39**: 519–540.

Biggs, A., and P. Cereceda (eds.) (2007). *Proceedings of the 4th International Conference on Fog and Fog Collection and Dew,* Pontificia Universidad Católica de Chile, La Serena (Chile), July 22–27, 2007.

Bockor, I. (1979). Analyse von Baumartenzusammensetzung und Bestandesstruckturen eines andinen Wolkenwaldes in Westvenezuela als Grundlage zur Wald-typengliederung. PhD thesis, University of Göttingen, Göttingen, Germany.

Bonell, M., and L. A. Bruijnzeel (eds.) (2005). *Forests, Water and People in the Humid Tropics.* Cambridge, UK: Cambridge University Press.

Bruijnzeel, L. A. (2001). Hydrology of tropical montane cloud forests: a reassessment. *Land Use and Water Resources Research* **1**: 1.1–1.18.

Bruijnzeel, L. A. (2005). Tropical montane cloud forests: a unique hydrological case. In *Forests, Water and People in the Humid Tropics,* ed. M. Bonell and L. A. Bruijnzeel, pp. 462–483. Cambridge, UK: Cambridge University Press.

Bruijnzeel, L. A., and L. S. Hamilton (2000). *Decision Time for Cloud Forests,* IHP Humid Tropics Programme Series No. 13. Paris: UNESCO, Amsterdam: IUCN-NL, Gland: WWF International (available in English and Spanish at: http://sea.unep-wcmc.org/forest/cloudforest/index.cfm).

Bruijnzeel, L. A., and J. Proctor (1995). Hydrology and biogeochemistry of tropical montane cloud forests: What do we really know? In *Tropical Montane Cloud Forests,* eds. L. S. Hamilton, J. O. Juvik, and F. N. Scatena, pp. 38–78. New York: Springer-Verlag.

Bruijnzeel, L. A., M. J. Waterloo, J. Proctor, A. T. Kuiters, and B. Kotterink (1993). Hydrological observations in montane rain forests on Gunung Silam, Sabah, Malaysia, with special reference to the '*Massenerhebung*' effect. *Journal of Ecology* **81**: 145–167.

Bubb, P., I. May, L. Miles, and J. Sayer (2004). *Cloud Forest Agenda.* Cambridge, UK: UNEP-WCMC. Online at: http://sea.unep-wcmc.org/forest/cloudforest/index.cfm.

Campanella, R. (1995). The role of GIS in evaluating contour-based limits of cloud forest reserves in Honduras. In *Tropical Montane Cloud Forests,* eds. L. S. Hamilton, J. O. Juvik, and F. N. Scatena, pp. 116–124. New York: Springer-Verlag.

Churchill, S. P., H. Balslev, E. Forero, and J. Luteyn (eds.) (1996). *Biodiversity and Conservation of Neotropical Montane Forests,* New York: New York Botanical Garden.

Doumenge, C., D. A. Gilmour, M. Ruiz-Perez, and J. Blockhus (1995). Tropical montane cloud forests: conservation status and management issues. In *Tropical Montane Cloud Forests,* eds. L. S. Hamilton, J. O. Juvik, and F. N. Scatena, pp. 24–37. New York: Springer-Verlag.

Frahm, J. P., and S. R. Gradstein (1991). An altitudinal zonation of tropical rain forests using bryophytes. *Journal of Biogeography* **18**: 669–676.

García-Santos, G. (2007). *An ecohydrological and soils study in a montane cloud forest in the National Park of Garajonay, La Gomera (Canary Islands, Spain).* PhD Thesis, VU University Ámsterdam, Ámsterdam, The Netherlands. [http://www.falw.vu.nl/nl/onderzoek/earth-sciences/geo-environmental-science-and-hydrology/hydrology-dissertations/index.asp].

Gradstein, S. R. (2008). Epiphytes of tropical montane forests: impact of deforestation and climate change. In *The Tropical Mountain Forest: Patterns and Processes in a Biodiversity Hotspot,* eds. S. R. Gradstein, J. Homeier, and D. Gansert, pp. 51–65. Göttingen, Germany: Universitätsverlag Göttingen.

Gradstein, S. R., J. Homeier, and D. Gansert (eds.) (2008). *The Tropical Mountain Forest: Patterns and Processes in a Biodiversity Hotspot.* Göttingen, Germany: Universitätsverlag Göttingen.

Grubb, P. J. (1977). Control of forest growth and distribution on wet tropical mountains: with special reference to mineral nutrition. *Annual Review of Ecology and Systematics* **8**: 83–107.

Grubb, P. J., and E. V. J. Tanner (1976). The montane forests and soils of Jamaica: a reassessment. *Journal of the Arnold Arboretum* **57**: 313–368.

Grubb, P. J., and T. C. Whitmore (1966). A comparison of montane and lowland rain forest in Ecuador. II. The climate and its effects on the distribution and physiognomy of the forests. *Journal of Ecology* **54**: 303–333.

Hamilton, L. S. (1995). *A Campaign for Cloud Forests.* Gland, Switzerland: IUCN.

Hamilton, L. S., J. O. Juvik, and F. N. Scatena (1995). The Puerto Rico tropical montane cloud forest symposium: introduction and workshop synthesis. In *Tropical Montane Cloud Forests,* eds. L. S. Hamilton, J. O. Juvik, and F. N. Scatena, pp. 1–16. New York: Springer-Verlag.

Hemp, A. (2005). Climate change driven forest fires marginalize the impact of ice cap wasting on Kilimanjaro. *Global Change Biology* **11**: 1013–1023.

Hemp, A. (2006). Continuum or zonation? Altitudinal diversity patterns in the forests on Mt. Kilimanjaro. *Plant Ecology* **184**: 27–42.

Iremonger, S., C. Ravilious, and T. Quinton (eds.) (1997). *A Global Overview of Forest Conservation. CD-ROM.* Cambridge, UK: World Conservation Monitoring Centre and Centre for International Forestry Research. Also available at www.unep-wcmc.org/forest/global_map.htm

Kapos, V., J. Rhind, M. Edwards, and M. F. Price (2000). Developing a map of the world's mountain forests. In *Forests in Sustainable Mountain Development: A State-of-Knowledge Report for 2000,* eds. M. F. Price and N. Butt, pp. 4–9. Wallingford, UK: CAB International. Also available at www.unep-wcmc.org/habitats/mountains/homepage. htm.

Kappelle, M. (1995). Ecology of mature and recovering Talamancan montane *Quercus* forests, Costa Rica. PhD thesis, University of Amsterdam, Amsterdam.

Kappelle, M. (2004). Tropical montane forests. In *Encyclopedia of Forest Sciences,* Vol. **4**, ed. J. Burley, pp. 1782–1783. Oxford, UK: Academic Press.

Kappelle M., and A. D. Brown (eds.) (2001). *Bosques Nublados del Neotropico.* Heredia, Costa Rica: INBIO.

Kappelle, M., A. D. Brown, M. Ataroff, and F. Sarmiento (2001). Results from the Andean cloud forest workshop held during the *4th Simposio*

Internacional de Desarrollo Sustentable en los Andes: La estrategia Andina para el siglo XXI, Universidad de los Andes, Merida, Venezuela, November 25–December 2, 2001.

Kitayama, K. (1995). Biophysical conditions of the montane cloud forests of Mount Kinabalu, Sabah, Malaysia. In *Tropical Montane Cloud Forests*, eds. L. S. Hamilton, J. O. Juvik, and F. N. Scatena, pp. 183–197. New York: Springer-Verlag.

Kitayama, K., and S. I. Aiba (2002). Ecosystem structure and productivity of tropical rain forests along altitudinal gradients with contrasting soil phosphorus pools on Mount Kinabalu, Borneo. *Journal of Ecology* **90**: 37–51.

Kitayama, K., and D. Müller-Dombois (1994a). An altitudinal transect analysis of the windward vegetation on Haleakala, a Hawaiian island mountain. I. Climate and soils. *Phytocoenologia* **24**: 111–133.

Kitayama, K., and D. Müller-Dombois (1994b). An altitudinal transect analysis of the windward vegetation on Haleakala, a Hawaiian island mountain. II. Vegetation zonation. *Phytocoenologia* **24**: 135–154.

Kitayama, K., and J. Nais (2002). Perspectives of the long-term ecological research on Mount Kinabalu. *Sabah Parks Nature Journal* **5**: 1–5.

Kitayama, K., N. Majalap-Lee, and S. Aiba (2000). Soil phosphorus fractionation and phosphorus-use efficiencies of tropical rainforests along altitudinal gradients of Mount Kinabalu, Borneo. *Oecologia* **123**: 342–349.

La Bastille, A., and D. J. Pool (1978). On the need for a sytem of cloud-forest parks in Middle America and the Caribbean. *Environmental Conservation* **5**: 183–190.

Lawton, R. and V. Dryer (1980). The vegetation of the Monteverde Cloud Forest Reserve. *Brenesia* **18**: 101–116.

Long, A. (1995). Restricted-range and threatened bird species in tropical montane cloud forests. In *Tropical Montane Cloud Forests*, eds. L. S. Hamilton, J. O. Juvik, and F. N. Scatena, pp. 79–106. New York: Springer-Verlag.

Lugo, A. E., and F. N. Scatena (1992). Epiphytes and climate change research in the Caribbean: a proposal. *Selbyana* **13**: 123–130.

Markham, A. (1998). Potential impacts of climate change on tropical forest ecosystems (guest editorial). *Climate Change* **39**: 141–143.

Martin, P. H., R. E. Sherman, and T. J. Fahey (2007). Tropical montane ecotones: climate gradients, natural disturbance, and vegetation zonation in the Cordillera Central, Dominican Republic. *Journal of Biogeography* **34**: 1792–1806.

Nadkarni, N. M., and N. T. Wheelwright (eds.) (2000). *Monteverde: Ecology and Conservation of a Tropical Cloud Forest*. New York: Oxford University Press.

Nullet, D. A., and J. O. Juvik (1994). Generalised mountain evaporation profiles for tropical and subtropical latitudes. *Singapore Journal of Tropical Geography* **15**: 17–24.

Ohsawa, M., W. Wildpret, and M. del Arco (1999). *Anaga Cloud Forest: A Comparative Study on Evergreen Broad-Leaved Forests and Trees of the Canary Islands and Japan*. Chiba, Japan: Laboratory of Ecology, Chiba University.

Penafiel, S. R. (1995). The biological and hydrological values of the mossy forests in the Central Cordiller Mountains, Philippines. In *Tropical Montane Cloud Forests*, eds. L. S. Hamilton, J. O. Juvik, and F. N. Scatena, pp. 266–273. New York: Springer-Verlag.

Pounds, A. J., M. Bustamante, L. A. Coloma, *et al.* (2006). Widespread amphibian extinctions from epidemic disease driven by global warming. *Nature* **439**: 161–167.

Proctor, J., I. D. Edwards, R. W. Payton and L. Nagy (2007). Zonation of forest vegetation and soils of Mount Cameroon, West Africa. *Plant Ecology* **192**: 251–269.

Rautenbach, H., and J. Oliver (eds.) (2004). *Proceedings of the 3rd International Conference on Fog and Fog Collection*. Ottawa, Canada: IDRC.

Røhr, P. C. (2003). A hydrological study concerning the southern slopes of Mt. Kilimanjaro, Tanzania. Ph.D. thesis, Norwegian University of Science and Technology, Trondheim, Norway.

Scatena, F. N. (1995). The management of Luquillo Cloud Forest ecosystems: irreversible decisions in a non-substitutable ecosystem. In *Tropical Montane Cloud Forests*, eds. L. S. Hamilton, J. O. Juvik, and F. N. Scatena, pp. 296–308. New York: Springer-Verlag.

Schemenauer, R. S., and H. A. Bridgman (eds.) (1998). *Proceedings of the 1st International Conference on Fog and Fog Collection*. Ottawa, Canada: IDRC.

Schemenauer, R. S., and H. A. Puxbaum (eds.) (2001). *Proceedings of the 2nd International Conference on Fog and Fog Collection*. Ottawa, Canada: IDRC.

Stadtmüller, T. (1987). *Cloud Forests in the Humid Tropics: A Bibliographic Review*. Tokyo: United Nations University, and Turrialba, Costa Rica: CATIE. Also available at www.unu.edu/unupress/unupbooks/80670e/80670E00.htm.

Tobón, C. (2009). *Los bosques andinos y el agua*, Serie Investigación y Sistematización 4. Quito, Ecuador: Programa Regional ECOBONA – INTERCOOPERATION – CONDESAN.

Van Steenis, C. G. G. J. (1972). *The Mountain Flora of Java*. Leiden, the Netherlands: E. J. Brill.

Vázquez-García, J. A. (1995). Cloud forest archipelagos: preservation of fragmented montane ecosystems in tropical America. In *Tropical Montane Cloud Forests*, eds. L. S. Hamilton, J. O. Juvik, and F. N. Scatena, pp. 314–332. New York: Springer-Verlag.

Weaver, P. L. (1995). The Colorado and dwarf forests of Puerto Rico's Luquillo Mountains. In *Tropical Forests: Management and Ecology*, eds. A. E. Lugo and C. Lowe, pp. 109–141. Berlin: Springer-Verlag.

White, F. (1983). *The Vegetation of Africa*. Paris: UNESCO–MAB.

Whitmore, T. C. (1998). *An Introduction to Tropical Rain Forests*, 2nd edn. Oxford, UK: Clarendon Press.

Zotz, G., and M. Y. Bader (2009). Epiphytic plants in a changing world: global change effects on vascular and non-vascular epiphytes. *Progress in Botany* **70**: 147–170.

2 Modeling the tropics-wide extent and distribution of cloud forest and cloud forest loss, with implications for conservation priority

M. Mulligan
King's College London, London, UK

ABSTRACT

This chapter uses a combination of remote sensing, computer modeling, and data assimilation to provide: (i) estimates of the global extent and distribution of "hydro-climatically" defined tropical montane cloud forests (TMCFs), and (ii) an initial assessment of the past and future impacts of climate change and land-use change upon them. The overall goal is to improve the understanding of cloud forests as an ecosystem and to assist in the geographic targeting of research, inventory, and conservation priorities. These hydro-climatically defined TMCFs might be better termed "significantly cloud-affected forests," since most – but not all – occur in areas of high elevation and high rainfall and show the structural characteristics typically associated with wetter TMCFs. The distribution of these forests was modeled on the basis of satellite-observed atmospheric cloud presence and/or modeled ground-level condensing conditions. Areas which experience these conditions >70% of the time gave the best fit with the UNEP–WCMC database of known cloud forest sites. Significantly cloud-affected forests have been estimated by this analysis as representing some 14.2% of all tropical forests and covering an area of 2.21 Mkm^2 between 23.5° N and 35° S. This figure is much greater than previous estimates for the area covered by TMCF based primarily on altitudinal criteria but is likely to be reduced if the exercise were to be repeated with higher spatial resolution data than are currently available. Many of the maps in this chapter are also available as interactive layers in the Google Earth viewer at www.ambiotek.com/cloudforests, from where the data can also be obtained as downloadable GIS files that are free for non-commercial use. Moreover, readers are encouraged to use the website to provide feedback on the representation of cloud forests in areas that they know. Readers can also add cloud forests that they know to the database from this website.

INTRODUCTION

Cloud forest as a conservation priority

Tropical montane cloud forests (TMCFs) are biologically and hydrologically unique and important ecosystems. Biologically they are home to a high diversity of plant and animal species including many that are endemic to cloud forests (Leo, 1995; Long, 1995; Salaman *et al.*, 2003). Numerous cloud forest species are threatened with extinction (Pounds *et al.*, 1999, 2006). Hydro-climatically, cloud forests tend to occur in headwater areas with high precipitation and low evaporation which consequently produce high volumes of streamflow. These forests are often found on steep mountain slopes, which are generally highly susceptible to erosion and mass movement in the absence of the protection provided by intact forest (Bruijnzeel, 2004; Sidle *et al.*, 2006). Because of the restrictive altitudinal (climatic) conditions that "produce" cloud forests, they are usually rather isolated and restricted in spatial extent and thus potentially sensitive to changes in climate and land use, as well as more difficult to protect compared with extensive lowland forest areas that are less penetrable on account of their higher area to perimeter ratios. Cloud forests also occur in some of the more agriculturally attractive and thus populated areas of the tropics, such as mountains and islands having climates and soils amenable to both agriculture and cattle raising (cf. Kappelle and Brown,

Tropical Montane Cloud Forests: Science for Conservation and Management, eds. L. A. Bruijnzeel, F. N. Scatena, and L. S. Hamilton. Published by Cambridge University Press.

2001). Montane zones, although topographically challenging, also tend to have denser road networks than some of the vast and climatically inhospitable remaining lowland rain forest blocks. Altogether this means that cloud forests present a rather challenging conservation target.

DEFINING CLOUD FORESTS

Cloud forests are, by definition, "forests affected by frequent and/or persistent ground-level cloud" (Grubb, 1977). The clouds significantly affect energy, light, and temperature regimes and import potentially large amounts of water as rainfall and horizontal precipitation (cloud water interception). Thus, the presence of ground-level cloud (i.e. fog) produces a very different environment to that in which other types of montane (and lowland) rain forest are found (Jarvis and Mulligan, this volume; cf. Frahm and Gradstein, 1991; Bruijnzeel and Veneklaas, 1998). Because there is a positive relationship between the presence of ground-level cloud and altitude, cloud forests tend to occur within the range of elevations between the locally defined lifting condensation level (LCL) for rising air and the high-altitude temperature minima at which tree-like vegetation is replaced by grassland and *paramo*. However, because sea-level temperature and saturated adiabatic lapse rates and their diurnality and seasonality are highly spatially variable across the tropics, the altitudinal bands at which ground-level cloud (and therefore TMCF) occurs are also likely to be highly variable (Jarvis and Mulligan, this volume; cf. Bruijnzeel and Veneklaas, 1998).

Whilst most TMCFs are found within the tropical belt (Bubb *et al.*, 2004), montane cloud forests are also known from the sub-tropics, e.g. the Canary Islands (Ohsawa *et al.*, this volume; García Santos, 2007, Marzol-Jaén *et al.*, this volume), the Himalayas (Wangda and Ohsawa, this volume), and the Pacific coasts of California, Chile, and Perú (Aravena *et al.*, 1989; Dawson, 1998). In addition, fog-affected forests do not always occur in truly montane environments, at least not in altitudinal terms, because they can be found below 300 m.a.s.l. (Jarvis and Mulligan, this volume; cf. Gradstein *et al.*, this volume) while montane environments are generally considered to be above 500 m.a.s.l. (Meybeck *et al.*, 2001).

Grubb's definition of cloud forest is seldom used to identify cloud forest in the field because the persistence and frequency of immersion in cloud is difficult to quantify without long-term observation (cf. Lawton *et al.*, this volume). In addition, cloud forests tend to differ in terms of their plant and animal communities and plant physiological and structural adaptations as well as hydro-climatic conditions (water inputs, evaporative losses, frequency of canopy wetting and soil waterlogging, soil nutrient status) resulting from differences in the frequency or persistence of the immersion in cloud (Stadtmüller, 1987; Frahm and Gradstein, 1991; Hamilton *et al.*, 1995; Bruijnzeel and Veneklaas, 1998; Bruijnzeel, 2001, 2005; cf. Roman *et al.*, this volume; Schawe *et al.*, this volume). The structural and physiological differences between, on the one hand, cloud forest and other montane forests not affected by fog, and between cloud forests subject to moderate and more persistent cloud, are increasingly well documented and various environmental controls have been suggested to explain these differences (Bruijnzeel *et al.*, 1993; Bruijnzeel and Veneklaas, 1998; Hafkenscheid, 2000; cf. Roman *et al.*, this volume; Benner *et al.*, this volume; Letts *et al.*, this volume).

There is not always a clear boundary separating cloud forest from neighboring forest formations and transitions may be gradual (e.g. Liebermann *et al.*, 1996; cf. Hemp, this volume #12). However, often there is a rapid increase in mossiness from forest not affected significantly by fog (classified as lower montane rain forest, LMRF *sensu* Grubb, 1977), via moderately cloud-affected forest tentatively classified as lower montane cloud forest (LMCF) by Bruijnzeel and Hamilton (2000), to strongly cloud-affected forest, classified as upper montane rain forest (UMRF) by Grubb (1977) and as upper montane cloud forest (UMCF) by Bruijnzeel and Hamilton (2000; cf. Frahm and Gradstein, 1991). Generally speaking, forest stature and stem diameter, leaf area and leaf size, all tend to decrease from LMRF via LMCF to UMCF whereas, apart from degree of mossiness, occurrence of gnarled or multi-stemmed trees, and abundance of certain indicator species (e.g. tree ferns) typically increase along the altitudinal sequence as well. However, because the specific characteristics of cloud forests result from the climatic implications of the frequency and persistency of ground-level cloud it is possible that they might also occur under conditions of lower cloud frequencies and persistence in areas subject to (very) high rainfall, particularly under wind-exposed conditions.

At the local scale, the hydrologically most significant cloud forests can be expected to be those that are highly cloud-affected and exhibiting the highest net precipitation to rainfall ratios and the lowest evaporation losses (Bruijnzeel, 2001, 2005). However, such forests are likely of much more limited spatial extent than less- or non-cloud-affected forests which, by virtue of their greater areal extent, will be of greater regional hydrological significance. A strict hydro-climatic definition of cloud forest might choose to include all forests with ground-level cloud frequency greater than zero since they are all "cloud-affected." Moreover, one could argue that if persistent ground-level cloud allows the development of a forest cover in areas that would otherwise be too dry, that this forest could be termed cloud forest even though it may not exhibit all of the characteristics of the high-rainfall TMCFs. However, given the cited contrast in hydrological and environmental–ecological characteristics of different types of cloud forest, it is arguably more useful to set a threshold of cloud frequency beyond which a forest may be considered to be "significantly" or "highly" cloud-affected. In this chapter, this threshold will be derived empirically on the basis of the cloud frequency and persistence at known and ecologically defined cloud forest sites.

"Forest" is generally defined in terms of a minimum areal extent and proportional cover with trees. FAO (2005) defined forest as "land spanning more than 0.5 ha, with trees higher than 5 m and a canopy cover of more than 10%." Bubb *et al.* (2004) used a minimum of 40% tree cover over 1-km^2 grid cells as a threshold between forest and non-forest. To provide estimates of cloud forest cover with a broader ecological interest in mind this chapter will consider (as per FAO, 2005) all areas with a tree cover >10% as forest. All areas with <10% cover are considered to contain hedgerows and isolated trees only. The FAO's tree height restriction to define forest is less pertinent in a cloud forest context since this would exclude some of the more stunted upper montane and elfin cloud forests.

Thus, whilst TMCFs are effectively generally defined as "significant aggregations of trees with certain characteristics and within certain defined altitudinal bands on tropical mountains or islands," this chapter defines them as: "aggregations of trees occupying more than 10% of the land area of interest exposed to an empirically defined frequency of ground-level cloud cover, irrespective of elevation and rainfall."

These matters of definition are important and, as will be shown later, have profound effects on derived estimates of the distribution and extent of TMCFs.

PREVIOUS ASSESSMENTS OF CLOUD FOREST DISTRIBUTION AND EXTENT

The most comprehensive assessment of the distribution of cloud forests throughout the tropics is that compiled under the auspices of UNEP–WCMC by Aldrich *et al.* (1997). This is a database comprising more than 560 point observations distributed throughout the tropics and representing areas that have been defined as cloud forests in the literature or by local experts (UNEP–WCMC, 2004). The UNEP–WCMC data-set is a valuable resource for better understanding the distribution and characteristics of cloud forests (cf. Jarvis and Mulligan, this volume) but it is limited because it: (i) represents only a sample of all cloud forests, (ii) represents cloud forests as a single point with no indication of spatial extent, and (iii) gives no indication of the intensity of cloud conditions, i.e. no separation is made between sites with LMCF and UMCF. There are also difficulties in terms of consistency of definition, interpretation, and terminology amongst different observers and of precision in geographic referencing of the sites. It is nevertheless the best assessment available to date.

Regional estimates of the distribution and extent of cloud forests have tended to be based on altitudinal limits. For example, all forests above 1800 m.a.s.l. in Honduras have been considered to be cloud forests (Campanella, 1995) vs. 1500 m in Malaysia (Kumaran *et al.*, this volume) and 1400 m in Costa Rica (Calvo-Alvarado *et al.*, this volume). UNEP–WCMC have estimated the global distribution and extent of TMCF by combining global forest cover assessments with global elevation data and regionally specific altitudinal limits for cloud forest, within a GIS context (Bubb *et al.*, 2004; Scatena *et al.*, this volume). The derived total cover of TMCF was estimated to be in the order of 215 000 km^2 (0.26% of the Earth's land surface, or 0.5% of the tropical land surface and 1.4% of the total area of all tropical forests). About 43% of these cloud forests are found in Asia, 41% in the Americas, and 16% in Africa (Scatena *et al.*, this volume). However, Jarvis and Mulligan (this volume) stress the very wide range of climatic and landscape situations (temperature, rainfall, altitude, distance to sea, mountain size) represented by the more than 560 observed UNEP–WCMC cloud forest sites. Because this climatic variability is not just controlled by altitude, altitude-based approaches to estimate cloud forest distribution will be able to indicate the major cloud forest areas but they are not likely to identify all cloud-affected forests and may thus represent an underestimate of the true cloud forest distribution and extent. Whilst Bubb *et al.* (2004) and Scatena *et al.* (this volume) provided regionally specific altitudinal limits to improve the estimates (cf. Table 1.2), such broad-brush approaches easily miss local variability in the height at which TMCFs are found, e.g. due to windward/leeward effects (Stadtmüller, 1987; Jarvis and Mulligan, this volume).

AIMS

This chapter marks an attempt to better define the extent of, and threats to, TMCFs on the basis of the best climatic, land cover, and cloud distribution data currently available. More specifically the chapter will provide the following:

- A tropics-wide map of the current distribution of TMCFs at a resolution of 1 km.
- An estimate of the distribution and extent of TMCF loss to date.
- A characterization of future climate change based on scenarios of at least two widely used general circulation models (GCMs) of the atmosphere, and
- An assessment of the conservation status of TMCFs and identification of priority areas for protection.

MODELING THE DISTRIBUTION OF TROPICAL MONTANE CLOUD FORESTS

Study area

This chapter focuses on TMCFs. Thus, the area of interest incorporates the tropics between latitudes 23.5° N and 23.5° S. However, because in the southern hemisphere a number of

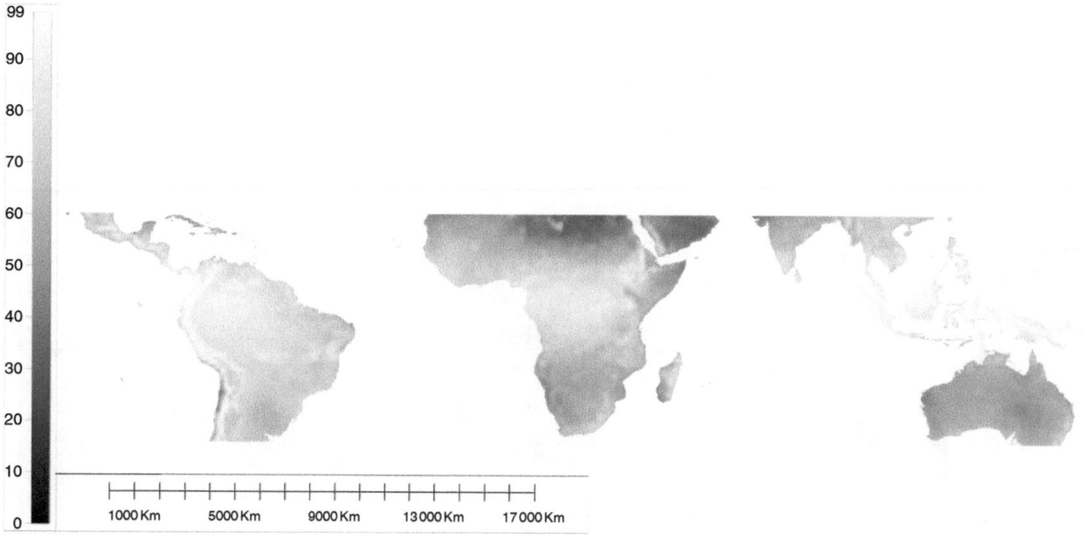

Figure 2.1. Global distribution of satellite-observed atmospheric (as opposed to ground-level) cloud frequency (%) in comparison with the distribution of UNEP–WCMC cloud forest sites (indicated as small black dots).

extensive cloud forests can be found southward of 23.5° S, the area considered has been extended to 35° S. All territorial areas discussed in the text refer to the areas of those countries within these bounds, not the entire areas of those countries.

Methods

Using a GIS-based model developed as an ESRI ARCGIS AML program, cloud forests were mapped on the basis of the following three factors:

(i) The frequency of atmospheric cloud cover as determined from the HIRS (High Resolution Infrared Radiation Sounder) sensor recording on the NOAA fleet of satellites (Jin *et al.*, 1996) since 1979 and assembled into a satellite cloud climatology by Wylie *et al.* (1994).

(ii) The frequency of ground-level condensing conditions (i.e. fog occurrence) generated by modelling the lifting condensation level (LCL, the altitude at which dew point is reached at the given atmospheric pressure and temperature) at a spatial resolution of 1 km on the basis of monthly climate data for temperature and diurnal temperature range (New *et al.*, 2000; Hijmans *et al.*, 2004), relative humidity (New *et al.*, 2000), and atmospheric pressure at sea level (Basnett and Parker, 1997). This information was then combined with the GTOPO30 digital elevation model (DEM) (USGS, 2005) for the derivation of areas where the calculated LCL occurs at ground (tree) level.

(iii) The presence of a forest vegetation cover as assessed independently from the GLC_2K (Global Land Cover 2000) database (Joint Research Centre, 2003), and Moderate Resolution Imaging Spectro-radiometer (MODIS) Vegetation Continuous Fields (VCF) imagery (Hansen *et al.*, 2003).

In the following sections these factors are discussed in further detail.

AD (I), THE FREQUENCY OF ATMOSPHERIC CLOUD

The HIRS data-set is a 22-year (1979–2001) satellite cloud climatology based on data from the NOAA satellites. It has been analyzed here to define long-term mean cloud frequency (number of cloudy observations as a percentage of the total number of observations) and persistence (represented as the coefficient of variation of monthly cloud frequency) across the tropics over an interpolated 1-km grid (the original data have a spatial resolution of 1 spherical degree approximately 111 km at the equator). This is then coupled with a tropics-wide assessment of the average climatic LCL level for each cell, taking account of seasonal (monthly) and diurnal variations in temperature and seasonal (monthly) variations in humidity and pressure at sea level, to produce a 1-km resolution assessment of ground- or vegetation-level cloud frequency. Figure 2.1 shows the global distribution of satellite-observed atmospheric (as opposed to ground-level) cloud frequency (%) in comparison with the distribution of UNEP–WCMC cloud forest sites (indicated as small black dots). Sometimes this atmospheric cloud will be in contact with the land surface and, even if it does not, it is usually associated with lower radiation inputs and higher rainfall. Thus, the cloud forests tend to be clustered in the areas of most frequently observed atmospheric cloud (the whitest areas, with >70% of observations having cloud). A further comparison indicates that all UNEP–WCMC

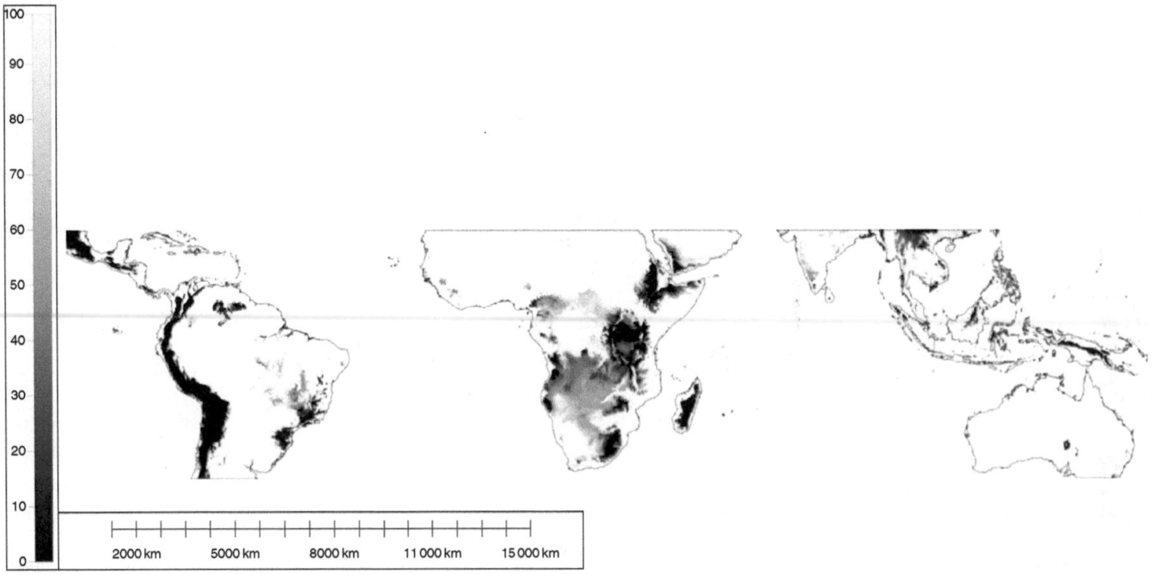

Figure 2.2. Modeled frequency of ground-level condensing conditions across the tropics (%).

observed cloud forest sites occur in areas of cloud frequency >50% but not all areas with cloud frequencies >50% are occupied by known cloud forests. Some 78% of the sites are associated with cloud frequencies >80% and 32% with cloud frequencies >95%. This range of cloud frequencies covers the full spectrum from LMCF through UMCF.

As indicated earlier, atmospheric cloud cover is likely to impact site environment because of the associated reductions in solar radiation and thus in temperature and evaporative losses, while rainfall tends to be increased with increased cloud cover. Together, this tends to produce increasingly wet conditions as cloud cover becomes more frequent and more persistent (cf. Schawe *et al.*, this volume). Arguably, these effects are even greater where the cloud is frequently at ground (vegetation) level, thereby providing additional inputs of horizontal precipitation (cloud water interception) and further reducing evaporation as a result of high atmospheric humidity and more prolonged periods of wet canopy conditions. Thus, a stronger association between cloud forest occurrence and the presence of ground-level fog rather than with atmospheric cloud is expected.

AD (II), THE FREQUENCY OF GROUND-LEVEL CONDENSING CONDITIONS

The propensity for ground-level condensation of atmospheric water vapor (i.e. occurrence of fog) was determined by calculating the LCL for each 1-km^2 grid cell throughout the tropics for four representative periods of the diurnal cycle (06:00, 12:00, 18:00, and 24:00 h) for each month using monthly data for mean sea-level pressure, temperature, and humidity. The climatic data are means over the last 30–40 years gridded at 1 km in the case of temperature and diurnal temperature range, and interpolated to 1 km (from coarser grids) in the case of humidity and mean sea-level pressure.

The process for calculating the frequency of ground-level condensing conditions from the gridded data-set involved the following five steps:

- Calculation of absolute humidity from ground-level temperature and relative humidity fields derived from the New *et al.* (2000) data-set.
- Calculation of dew point temperature using standard equations (cf. Foster, this volume).
- Calculation of the LCL, again using standard equations.
- Conversion of the LCL to meters elevation and assuming all altitudes in the GTOPO30 DEM above the local LCL to have the potential for ground-level condensation.
- Summing across the 48 time-steps representing the diurnal and seasonal cycles to obtain the frequency of condensing conditions as a climatologic average.

The results are shown in Figure 2.2 and indicate large areas with frequent ground-level condensing conditions at altitude, especially in South and Central America, but also in Africa (though with lower frequencies). Much less extensive areas of frequent ground-level condensing conditions were derived for South-East Asia.

AD (III), SENSITIVITY TO FOREST CLASSIFICATION THRESHOLD

Most forest distribution analyses take a binary classification approach to forest cover rather than a continuous (fractional) one. One of the difficulties of the binary classification approach

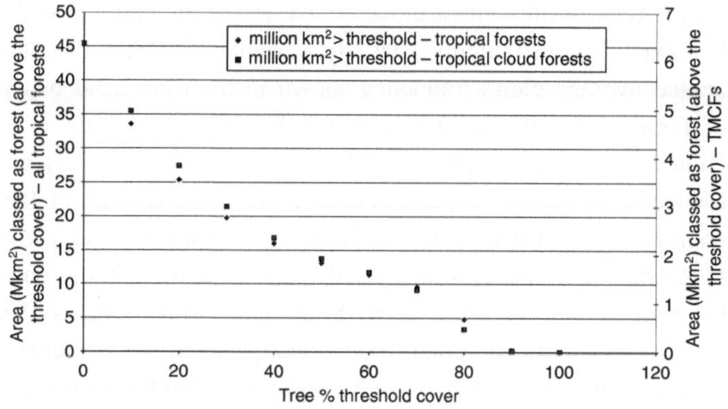

Figure 2.3. Areas classified as tropical montane (cloud) forest (in Mkm^2) for different threshold values of tree cover (%) using the binary approach.

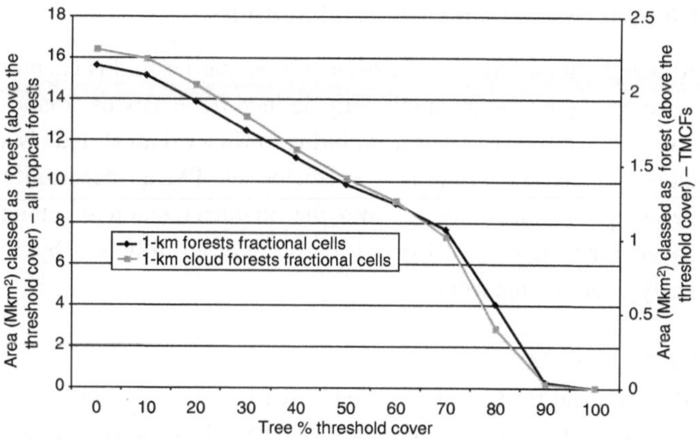

Figure 2.4. Areas classified as tropical montane (cloud) forest (in Mkm^2) for different threshold values of tree cover (%) fractional forest covers.

is choosing an appropriate threshold for classifying cells as forested or non-forested and then accounting for the consequences of that threshold in any areal analysis. Bubb *et al.* (2004) and Scatena *et al.* (this volume) used a threshold of 40% tree cover to separate forest and non-forest. In other words, any cell with a tree cover of less than 40% is classified as being entirely without forest and vice versa. The MODIS Vegetation Continuous Field (VCF) data (500-m resolution) have the advantage that one can either determine a threshold for the existence of forest within a cell or use the continuous fields to calculate forest area based on the actual fractional coverage per cell and thus avoid classification altogether.

THE BINARY CLASSIFICATION METHOD

Figure 2.3 summarizes the results of an analysis of the dependence of derived forest area on the threshold value used to define tree cover using MODIS_VCF data. The analysis was carried out twice, once for the full tropical window as used in this chapter (i.e. all forests between 23.5° N and 35° S) and again for only those areas classified as having cloud forest according to the analysis presented later in the chapter. There are large differences in derived total forest area for different threshold values, with sensitivity to threshold value being particularly pronounced in the 0–40% range (Figure 2.3). Furthermore, in all cases the area of forest cover derived with the binary classification method is very high: 0.3–45 Mkm^2 for all tropical forests and 0.02–6.34 Mkm^2 for modeled cloud forests, depending on the threshold value used. The results may also be dependent on cell size because the finer the cell size used in the analysis the more precisely the true fractional cover is represented (note that after testing the impact of this, for computational reasons a cell size of 1 km^2 was used although the original VCF data have a resolution of 500 m). At the 40% threshold (cf. Bubb *et al.*, 2004), the binary approach gave a total tropical forest area of 15.99 Mkm^2 of which 2.35 Mkm^2 (14.7%) can be classified as cloud forest. This sensitivity to the choice of forest-cover threshold indicates that significant potential overestimation may be introduced when using the binary threshold method to classify grid cells rather than summing their fractional forest covers as done in this chapter.

THE CONTINUOUS FIELDS METHOD

If one uses the same threshold approach to define forest but counts only those fractions of the forested cells occupied by trees (rather than the whole cells) as forest then there is a lower sensitivity to the magnitude of the threshold (especially in the range 0–60%) whereas significant changes in forest cover are only experienced for thresholds of 70–90% (Figure 2.4). This approach accounts much more precisely for the fragmented nature of most forests as a result of natural processes (tree-falls, landslides, river networks) and human activities (logging and clearing; cf. Aide *et al.*, this volume). The continuous fields method also produces measured forest areas that are some 30% lower in the case of all tropical forests (e.g. 11.16 Mkm^2 at the 40% threshold level) and a 32% lower value for cloud forest (1.6 Mkm^2 or 14.4%) than using the conventional binary approach (Table 2.1).

Table 2.1 summarizes the total tropical forest cover estimations derived with the various techniques. There are very large differences between the binary classification-based methods (using GLC_2K vegetation cover data) and the various MODIS_VCF threshold methods. The various threshold fractional methods give much lower estimates that are, in addition, much closer together. It may be argued that the fractional method with a 10% tree cover threshold for forest (as per FAO, 2005) is the most appropriate for the present analysis as it avoids hedgerows and isolated trees being counted as (cloud) forest in pixels without substantial forest cover. Also, forest cover is counted only as the actual area covered rather than the area of the entire cell, so that this method is also relatively insensitive to the cell size adopted in the analysis.

Cloud forests, cloud cover, and altitude

Although cloud forests have been defined earlier in this chapter as being exposed to an empirically defined frequency of ground-level cloud cover, it is worth exploring to what extent altitude is a more reliable (and simpler) predictor of TMCF occurrence before going much further into the characterization of cloud cover. First the case of Costa Rica is analyzed, followed by an analysis across the tropics.

COSTA RICA

Mulligan and Burke (2005a) produced a 1-km cloud climatology for Costa Rica based on 400 MODIS cloud mask images (MOD35 L2, http://modis-atmos.gsfc.nasa.gov/MOD35_L2/index.html) distributed diurnally and seasonally over four years. They combined this climatology with the climatically driven model of ground-level condensing conditions described above to find that the frequency of ground-level cloud can have a simple relationship with altitude on average at the national scale but that it is also highly spatially variable (Figure 2.5). The annual average cloud frequency shown in the right-hand panel of Figure 2.5 indicates a significant Atlantic cloud bank with cloud frequencies in excess of 80% for parts of the Atlantic lowlands and uplands, particularly in the northern part of the country. The left-hand panel of Figure 2.5 shows the equivalent HIRS data and indicates the spatially crude nature of those data by comparison with the MODIS product. This remains the weakest element of the pan-tropical analysis and work is under way to compile a global 1-km resolution cloud climatology equivalent to that produced for Costa Rica from MODIS imagery. Needless to say, this is a substantial analytical and computational undertaking.

As illustrated in Figure 2.6, there are areas in Costa Rica that are subject to fog from 300 m.a.s.l. upward, with a clear peak (with nearly 70% of all observations being cloudy) found between 1250 and 1350 m.a.s.l. above which ground-level cloud is consistently frequent (>60%). This supports the notion that cloud forest in Costa Rica is found at altitudes above *c.* 1400 m.a.s.l (cf. Calvo-Alvarado *et al.*, this volume; Lawton *et al.*, this volume). The relationship between ground-level cloud frequency and altitude is linear from the lowlands through to 1000 m, in line with the altitudinal control on ground-level condensing conditions. Above 1000 m there is a fairly persistent ground-level cloud bank. There is, however, a great deal of spatial variation in cloud frequency throughout the country within these altitudinal bands. This variability is greatest between 500 and 1000 m and decreases with altitude thereafter (Figure 2.6), reflecting differences in seasonal and diurnal variations in the LCL resulting from spatial variations in sea-level temperature, humidity, and pressure (cf. Foster, this volume; Lawton *et al.*, this volume). The 1400-m level marks the

Table 2.1 *Estimates of total tropical forest cover by different methods and percentage thresholds*

Calculation method	Km^2 of tropical forest
From GLC_2K (1 km) (classification)	37 675 473
From MODIS_VCF_40% (classification)	15 997 816
From MODIS_VCF_90% (classification)	285 132
From MODIS_VCF fractional 0% (continuous)	15 636 946
From MODIS_VCF fractional 10% (continuous)	15 153 384
From MODIS_VCF fractional 40% (continuous)	11 160 099
From MODIS_VCF fractional 90% (continuous)	266 444
FAO Forest Resources assessment "pan tropical"	15 710 000

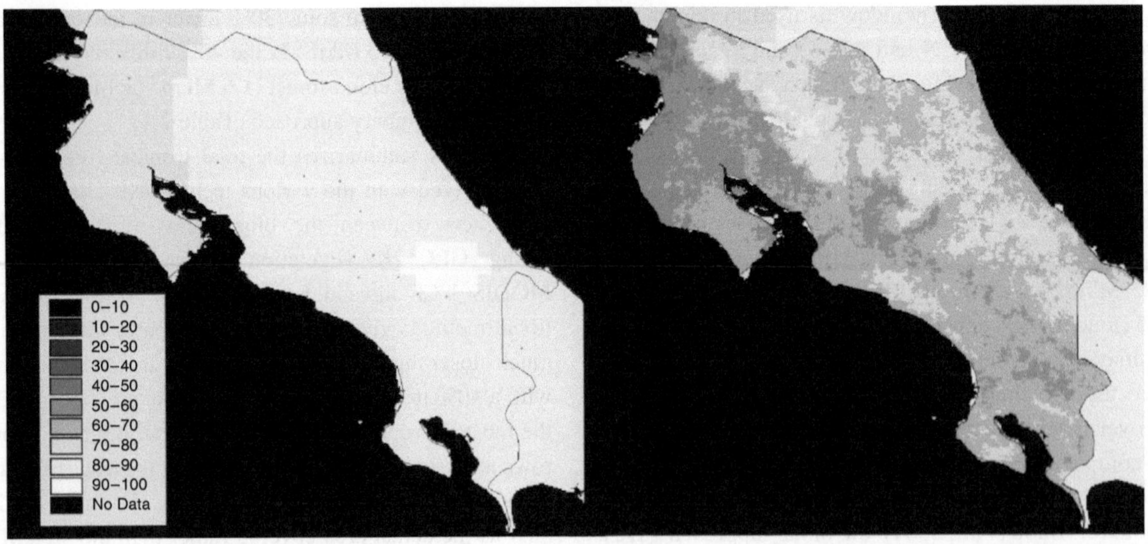

Figure 2.5. Comparison of cloud cover estimates for Costa Rica based on global HIRS (left-hand panel) and local MODIS MOD35 (right-hand panel) satellite imagery.

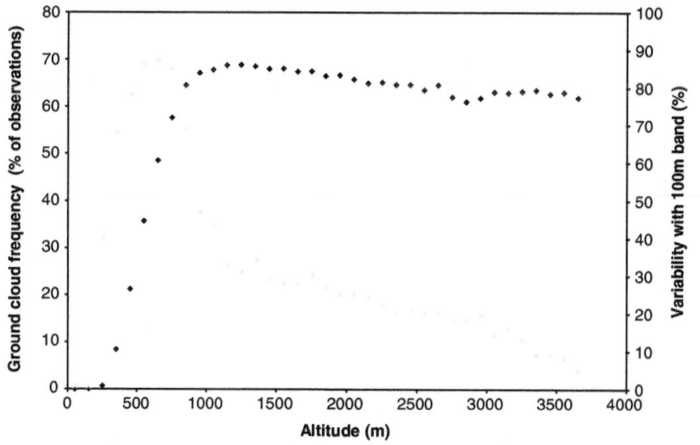

Figure 2.6. Ground-level cloud frequency versus altitude for Costa Rica. Means are for 100 m altitudinal bands for the entire country. Diamonds, cloud frequency; crosses, coefficient of variation of frequency.

lowest altitude at which cloud cover is both high on average and has low spatial variation around the country (Figure 2.6). Thus, in the case of Costa Rica, altitude and cloud forest presence are rather strongly related. However, Costa Rica occupies a narrow isthmus dominated by the NE trade winds bringing moisture-laden air that rises up the Atlantic slopes to produce cloud upon reaching the LCL. Conversely, climates over mountains in continental settings with more complex climate dynamics may show a different picture (cf. Jarvis and Mulligan, this volume).

TROPICS-WIDE ANALYSIS

The relationship between satellite (HIRS) observed atmospheric cloud frequency and the modeled frequency of ground-level cloud condensing conditions may be calculated across the tropics in the same way as for the Costa Rican analysis. By taking a large randomly located sample of points from the pan-tropical 1-km grid and considering these as being representative of tropical land areas, and by then extracting data from the same grid for those cells in which cloud forest is reported to be present according to the UNEP–WCMC database, it becomes possible to derive a better understanding of the potential relationship between cloud forest presence, cloud frequency, and altitude (as given in the GTOPO30 DEM). Figure 2.7a indicates that altitude is not a good surrogate for satellite-observed cloud frequency when viewed across the entire tropics. Although the minimum observed cloud frequency does increase linearly with altitude (areas close to sea level having cloud frequencies around 30% in the tropics), sites at a particular altitude can show a range of cloud frequencies from the observed minimum up to 100%, depending on other (climatic) factors. Nevertheless, at altitudes greater than 1400 m.a.s.l., cloud frequencies are generally greater than 65% (Figure 2.7a).

Modeled frequencies of ground-level condensation conditions (Figure 2.7b) show a clear increase with altitude from sea level to around 1400 m.a.s.l. but with a considerable range in the elevation at which a particular frequency of condensation conditions is achieved (as determined by local conditions). The level at which condensation conditions are persistent (as indicated by the presence of cloud forest points) varies from around 700 m upward depending on location, confirming that the ground-level fog frequency vs. altitude relationship varies considerably across the tropics. Figure 2.7c shows the satellite-observed cloud cover for 1-km cells in which cloud forest is known to be present according to the UNEP–WCMC database. These cells represent essentially a sub-set of the ones shown in Figure 2.7a and cover most of the range of altitudes found in the tropics. Very few known cloud forest locations have cloud frequencies less than 60% whereas a large number of tropical land locations have general cloud frequencies in the range 30–60%. Also, very few cloud forests occur at altitudes below 700 m. Once again, altitude per se is not a good discriminator of cloud forest occurrence throughout the tropics. Cloud forests are known to occur across all altitudes >700 m.a.s.l., but there will be many areas above that elevation that do not carry cloud forest.

MODEL ITERATIONS AND TESTING

The cloud forest distribution prediction model used here combines thresholds of atmospheric cloud frequency (satcloud), frequency of ground-level condensing conditions (fogfreq), and (actual fractional) forest cover to determine the extent and distribution of cloud forests. A number of model versions ("iterations") were applied using different thresholds for each of these variables and using different vegetation cover assessments (GLC_2K or MODIS_VCF). Each iteration produced a map of cloud forest distribution that was then verified visually against other global assessments and the author's knowledge of cloud forest distribution in regions of Latin America and South East Asia in particular. Validation against the UNEP–WCMC cloud forest locations was carried out for the significant iterations listed in Tables 2.2 and 2.3 for GLC_2K and MODIS_VCF data, respectively. For each iteration listed in Table 2.2, the algorithm used is given along with the percentage of UNEP–WCMC cloud forest sites predicted correctly as cloud forest by the model, plus the associated modeled extent of cloud forest (in Mkm^2 and as a proportion of all tropical forests based on GLC_2K which is closest to the FAO assessment of total tropical forest extent). Because no data-sets exist for the absence of cloud forests one has to rely solely on validation using presence records. The disadvantage of this approach is that the greater the area of cloud forest predicted by a model, the greater will be its potential "hit rate" of UNEP–WCMC cloud forest sites, and thus the better the validation will appear. Thus, the best model was considered to be that which covered the greatest proportion of UNEP–WCMC cloud forest sites while predicting the smallest global extent of cloud forests (as indicated by the performance index in Tables 2.2 and 2.3).

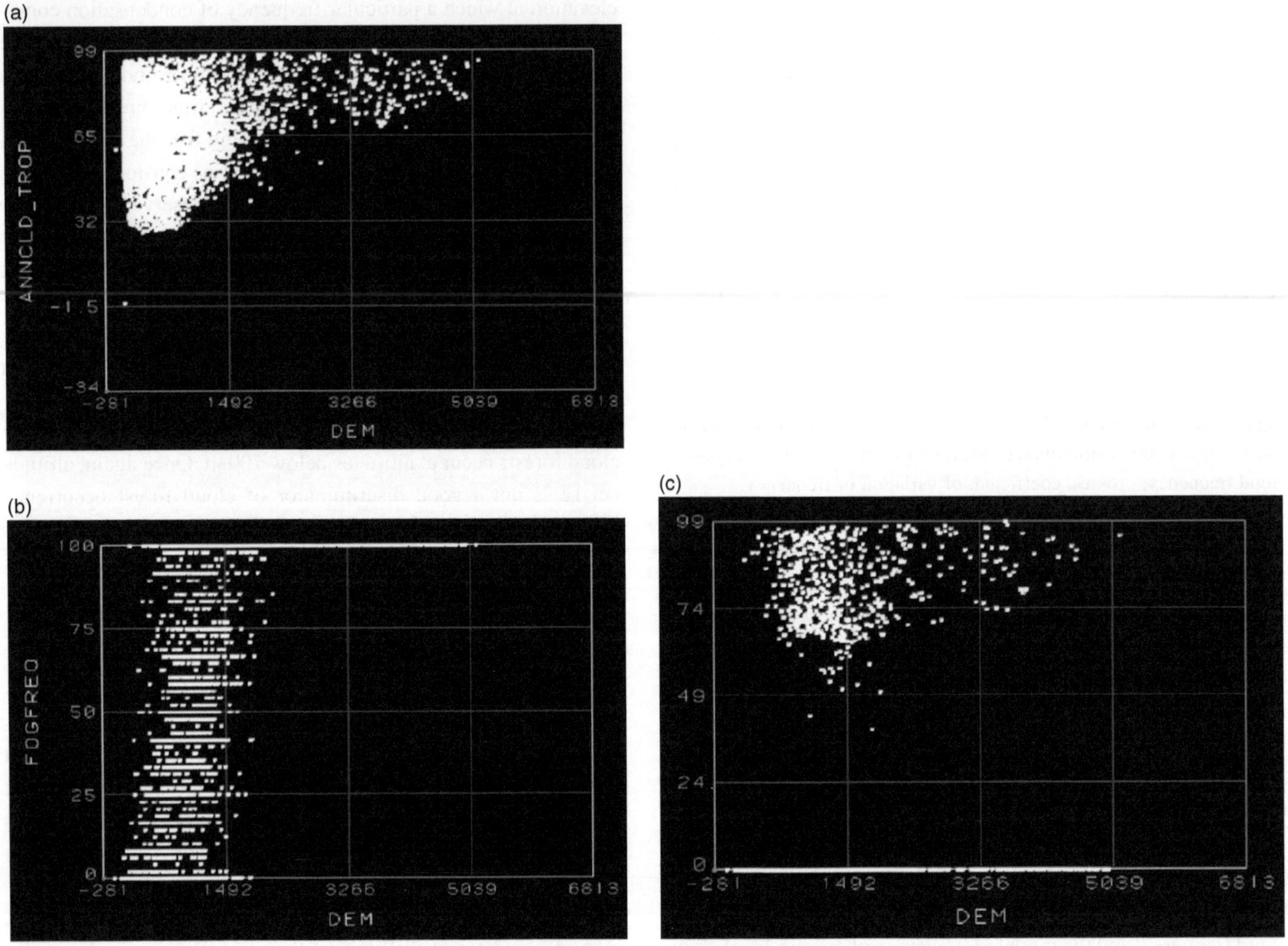

Figure 2.7. (a) Observed cloud frequency versus altitude for the tropics (top left); (b) modeled frequency of the LCL at ground level versus altitude for the tropics (bottom left); and (c) observed cloud frequency for tropical montane cloud forest cells (bottom right).

All of the data-sets used in this work have associated errors and uncertainties, both in their absolute values, in the smoothing or lack of representation of grid and sub-grid variation, and in the accuracy and precision of their geospatial referencing. It is important, therefore, to take these uncertainties into account in the modeling and validation process. This was achieved by assigning a margin of error to the validation results. If a model predicted cloud forest to be absent but a UNEP–WCMC record indicated that, in fact, one was present then a further validation took place. In this validation the distance from UNEP–WCMC cloud forest sites to the nearest modeled cloud forest cell was calculated. In addition, the number of UNEP–WCMC sites not predicted as cloud forest by the model but having modeled cloud forest within 5 km was calculated. It was assumed that model "failure" on those occasions is more likely to be a result of data errors than model errors. Excluding those records from the analysis produced a second assessment of model performance given in brackets in column 3 of Table 2.2.

MODELED GLOBAL EXTENT OF CLOUD FOREST

Estimates based on GLC_2K forest cover data

The global extents of cloud forest predicted by some of the model iterations (e.g. test_0 in Table 2.2) are clearly impossible and such models were rejected outright. For the others a performance index was calculated as the percentage of correctly predicted UNEP–WCMC sites (hit rate) per Mkm^2 of predicted cloud forest area. The greater the performance index, the better the model performance. The best performing model using GLC_2K seems to be test_8 (cloud forest occurs where the mean of atmospheric cloud and ground-level condensing frequency is greater than 70%). This model predicted 79% of the UNEP–WCMC sites correctly and cloud forest occurrence within 5 km of 93% of the UNEP–WCMC sites, whilst maintaining a tropics-wide area of potential cloud forest of 5.12 Mkm^2 (*c.* 13.6% of all

Table 2.2 *List of tests for defining cloud forests based on ground cloud frequency with validation results. Figures in brackets are the results for cloud forests within 5 km of predicted occurrence. satcloud, satellite observed atmospheric cloud frequency; fogfreq, percentage of time with ground-level condensing conditions*

Test name	Model (all models applied for GLC_2K only)	% of UNEP–WCMC correctly predicted	Global cloud forest area (million km^2)	Performance index	% of total forest in tropics (GLC_2K)
Test_0	satcloud>70	75% (87)	24.55	3.06 (3.55)	65.04
Test_1	satcloud>70 and altitude >500	67% (84)	8.64	0.08 (0.10)	22.96
Test_4	min(satcloud,fogfreq)>70	62% (77)	3.74	16.57 (20.57)	9.93
Test_4a	min(satcloud,fogfreq)>80	45% (53)	2.17	20.69 (24.37)	5.77
Test_4c	min(satcloud,fogfreq)>75	51% (62)	2.69	18.95 (23.03)	7.14
Test_5a	min(satcloud,fogfreq)>70 and windward exposed	27% (52.5)	1.74	15.48 (30.09)	4.63
Test_8	mean(satcloud,fogfreq)>70	79% (93)	5.12	15.42 (18.16)	13.59
Test_9	mean(satcloud,fogfreq)>80	60% (75)	3.76	15.93 (19.91)	10.00
Test_10	mean(satcloud,fogfreq)>90	44% (51)	1.97	22.29 (25.84)	5.24
Test_11	mean(satcloud,fogfreq)>95	25% (27)	1.04	23.92 (25.83)	2.77
Test_12	(satcloud*fogfreq)>70	56% (69)	3.06	18.30 (22.54)	8.12

Table 2.3 *Experiments with various characterizations using MODIS_VCF data*

Test name	% of UNEP–WCMC correctly predicted	Global cloud forest area (million km^2)	Performance index	% of total forest in tropics (GLC_2K)	% of total forest in tropics (MODIS_VCF40%)	% of total forest in tropics (MODISVCF_90%)	% of total forest in tropics (MODIS_fractional >10%)
Test8_vcf40	54% (76)	2.80	19.25	7.4	17.5	NA	NA
Test8_vcf40_500 m	NA	2.35	NA	6.3	14.7	NA	NA
Test8_vcf10_1 km_fractional	71% (81)	2.21	32.17	5.9	13.8	NA	14.2

GLC_2K-based tropical forests; Table 2.2). Test_4 also performed well, correctly predicting 62% of the UNEP–WCMC sites directly and cloud forest occurrence within 5 km for 77% of the UNEP–WCMC sites. This model gave a total cloud forest area of 3.74 Mkm^2 (*c.* 10% of total tropical forest area). Test_4 must be considered more physically meaningful in that it assumed cloud forest to occur only where both satellite-observed cloud and the propensity for ground-level condensation were greater than 70%, although this model is also highly sensitive to errors in satcloud or fogfreq. Test_5a represented a variant of test_4 with the added condition that cloud forest sites must also be exposed to the dominant wind direction (assessed as described in Mulligan and Burke, 2005a). Application of this model reduced the area of predicted cloud forest significantly whereas the hit rate of UNEP–WCMC sites dropped to an unacceptable level (27%), although some 52.5% of sites fell within a 5-km radius of modeled cloud forest using this algorithm (Table 2.2). Overall, the best model in terms of predictive success is test_8 and this is carried forward for use with the MODIS_VCF forest cover data below.

Table 2.4 *Estimates of total tropical montane cloud forest cover by different methods*

Calculation method	Km^2 of tropical montane cloud (affected) forest
From GLC_2K (1km) (classified)	n/a
From MODIS_VCF_40% (classified)	2 355 062
Bubb *et al.* (2004)	381 166
Scatena *et al.* (this volume)	215 000
From MODIS_VCF_90% (classified)	27 909
From MODIS_VCF fractional 0% (continuous)	2 276 296
From MODIS_VCF fractional 10% (continuous)	2 213 292
From MODIS_VCF fractional 40% (continuous)	1 601 670
From MODIS_VCF fractional 90% (continuous)	26 204

Estimates based on MODIS_VCF forest cover data

Table 2.3 shows the results of three experiments to model cloud forest distribution based on MODIS_VCF data. Test8_vcf40 is closest to the method used in Scatena *et al.* (this volume) applying a 40% tree cover threshold to define the presence of forest. This model produced some 2.8 Mkm^2 of cloud forest. This is only 7.4% of the total GLC_2K-based tropical forest area, but 17.5% of the total tropical forest area as obtained with the same method (MODIS_VCF 40% threshold). The model performed rather poorly under validation (54% overlap with UNEP–WCMC sites) although 76% of the validation sites were within 5 km of cloud forests identified by this model (Table 2.3). Test8_vcf40_500m shows the impact of using the maximum resolution MODIS data (500 m) instead of the 1-km data to be rather small. Thus, further computations continued to use the 1-km MODIS_VCF data to minimize computational constraints. The final test listed (test8_vcf10_1km_fractional) used a 10% threshold for forest cover as discussed earlier but it accumulated forest area not on the basis of whole cells but rather as the actually forested fractions of the cells. This test correctly predicted 71% of the UNEP–WCMC points as cloud forest and gave a total cloud forest area of 2.21 Mkm^2 or 14.2% of all tropical forests calculated using the same technique (MODIS fractional >10%), but as little as 5.9% of the (inflated) GLC_2K tropical forest area (Table 2.3). A comparison of the areas of cloud-affected forest obtained with the various parameterizations used in this study and by other studies is given in Table 2.4.

Summarizing, application of the continuous fields method coupled with a 10% threshold for tree cover to assess total TMCF extent gave a slightly lower value (2.21 Mkm^2 or 14.2% of the total tropical forest area estimated with the same method) than that using a threshold of 40% (2.80 Mkm^2; Table 2.3). These estimates are much higher than those given earlier by Bubb *et al.* (2004) (381 166 km^2 or 2.5% of all tropical forest) and by Scatena *et al.* (this volume) (215 000 km^2 or 1.4%) using a combination of altitudinal criteria and the binary classification approach with a 40% tree cover threshold. This is because the presently derived estimates consider all cloud-affected forests, not just those occurring within particular elevational bands.

Persistent errors

Some 71% of the known UNEP–WCMC cloud forest sites were predicted correctly using model version test8_vcf10_1km_fractional (Table 2.3). This figure was raised to 81% after inclusion of modeled forests occurring within 5 km of known UNEP–WCMC cloud forest sites (i.e. within the spatial error bounds of the various data-sets). Upon examination of the distribution of the 19% of the UNEP–WCMC cloud forest sites that were not predicted properly by the model (and more than 5 km away from a modeled cloud forest cell) it becomes clear that the mal-predicted sites are situated mostly on small-island chains, isolated islands, and various coastal locations (Figure 2.8). In addition, a few more continental sites were not predicted (e.g. on the eastern flanks of the Andes in Bolivia, some isolated areas in Mexico, and various sites in West and Central Africa; Figure 2.8).

The reasons for these errors may include the following:

- Errors in the UNEP–WCMC database, especially for cloud forests recorded in the early literature that may now be deforested (possibly the case for some of the West African sites).
- Positional uncertainty in one or more of the grids and, in combination with lack of data or too coarse data resolution, affecting coastal sites and small island sites in particular.
- The presence of apparent cloud forests for reasons other than frequent and persistent ground-level cloud (e.g. extremely high rainfall and wind exposure producing substantial amounts of wind-driven and horizontal precipitation). It is likely that high exposure in combination with high rainfall reduces the frequency of cloud necessary to produce a cloud forest. This may be true for some of the West African forests (parts of which fall in areas of very high rainfall, others of which may be affected by advected sea fog) and some coastal South-East Asian and South American forests (which may be exposed to rain- and cloud-bearing winds). The eastern Bolivian anomalies may be associated with extremely high rainfall cells (>4000 mm $year^{-1}$).

Sensitivity of model predictions to quality of topographic and cloud cover data

Before using the present cloud forest assessment in various analyses of the effects of changes in land use or climate, and

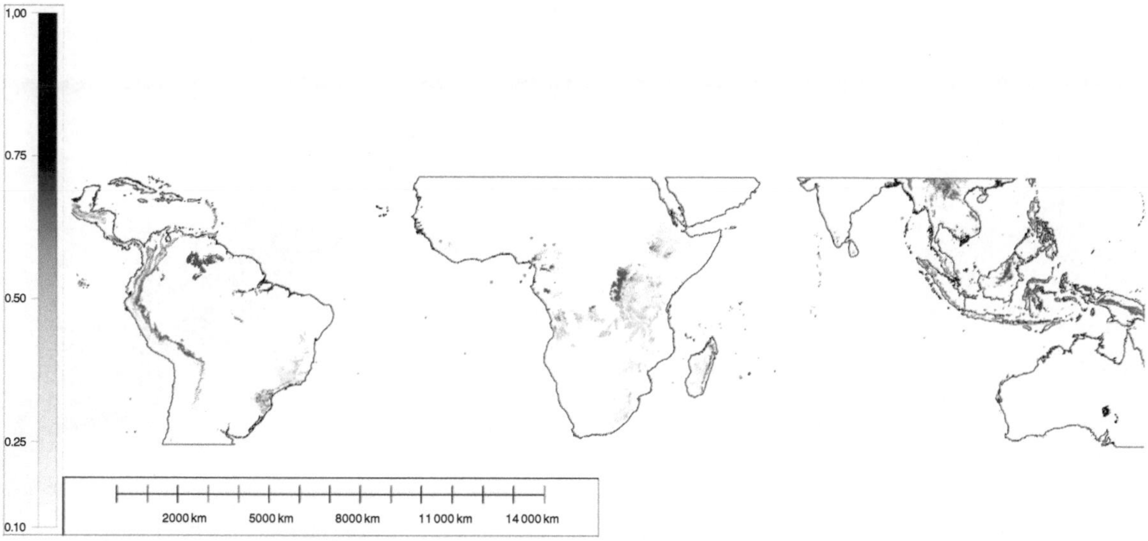

Figure 2.8. Persistently misclassified UNEP–WCMC-listed cloud forest sites (indicated by red crosses) that are also more than 5 km from the nearest modeled cloud forest. (See also color plate.)

drawing conclusions as to conservation priorities, it is important to test the sensitivity of the assessment to the limitations and uncertainty of the data-sets used for the calculation of atmospheric cloud frequency and ground-level condensing conditions. The HIRS data for atmospheric cloud frequency are much more spatially crude than the other data used in this analysis. Furthermore, the definition of ground-level cloud frequency is highly dependent upon land elevation relative to the LCL. At 1-km resolution the GTOPO30 DEM is also relatively crude considering the large elevation differences that may occur over short distances in steep mountainous terrain. Therefore, the same analysis as carried out for the global model was carried out for Costa Rica using finer resolution data-sets (notably the CGIAR-processed NASA Shuttle Radar Topography Mission (SRTM) DEM at 90-m resolution of Jarvis *et al.* (2004) and the MODIS cloud climatology of Mulligan and Burke (2005a) at 1-km original resolution but re-sampled to 90 m for compatibility with the DEM data). A comparison between these high spatial resolution runs and a standard resolution run using the HIRS and GTOPO30 data for Costa Rica allows one to better understand the reliability of the results and the extent to which estimates might be improved using better global data-sets.

Figure 2.9a shows the difference in cloud forest distributions predicted by model version test_8 with GLC_2K vegetation data and using topographic and cloud cover data-sets with resolutions of 1 km and 90 m, respectively. There is clearly very little difference between the two distributions because the HIRS data tend to overestimate cloud cover compared with MODIS imagery and therefore both reach the necessary threshold for atmospheric cloud frequency over much of high-altitude Costa Rica. In addition, the model seems to be more sensitive to the height of the LCL than to atmospheric cloud frequency in this region. Hence, the observed differences in predicted cloud forest distribution are largely due to small elevational differences between the 90-m SRTM and the 1-km GTOPO30 data-sets. The 90-m version of the model produced a cloud forest extent for Costa Rica that was about 82.7% of that estimated with the 1-km model.

Figure 2.9b shows the difference between the two cloud forest maps produced by test_4 with GLC_2K vegetation data for the two resolutions (cf. Table 2.2). Test_4 is much more dependent upon prescribed minimum atmospheric and ground-level cloud frequency values, hence the very different pattern of cloud cover from the high-resolution MODIS data compared with the coarser HIRS data produced some significant differences for the two resolutions. The 90-m model estimated only 39.3% of the cloud forest cover predicted by the 1-km model. This may explain the poorer validation performance of test_4 with the global data-sets (Table 2.2). It also means that it is less suitable for application until the higher-resolution cloud climatology is completed since model version test_4 is clearly more sensitive to the spatial resolution of data.

Figure 2.10 shows the modeled distribution of cloud forest for Costa Rica at the two resolutions for tests_4 and 8. The most accurate model in terms of representing the (well-) known cloud forests of Costa Rica (from the UNEP–WCMC coverage) was test_8 at 90-m resolution which successfully located 100% of the forests. Test_4 and test_8 at 1-km resolution proved equivalent since they both missed only two cloud forests on an isolated mountain in the far north of the country. The worst performing model (38% recovery) was the 90-m application of test_4. This version missed many of the Pacific cloud forests which are probably sustained by ground-level cloud spilling over from the

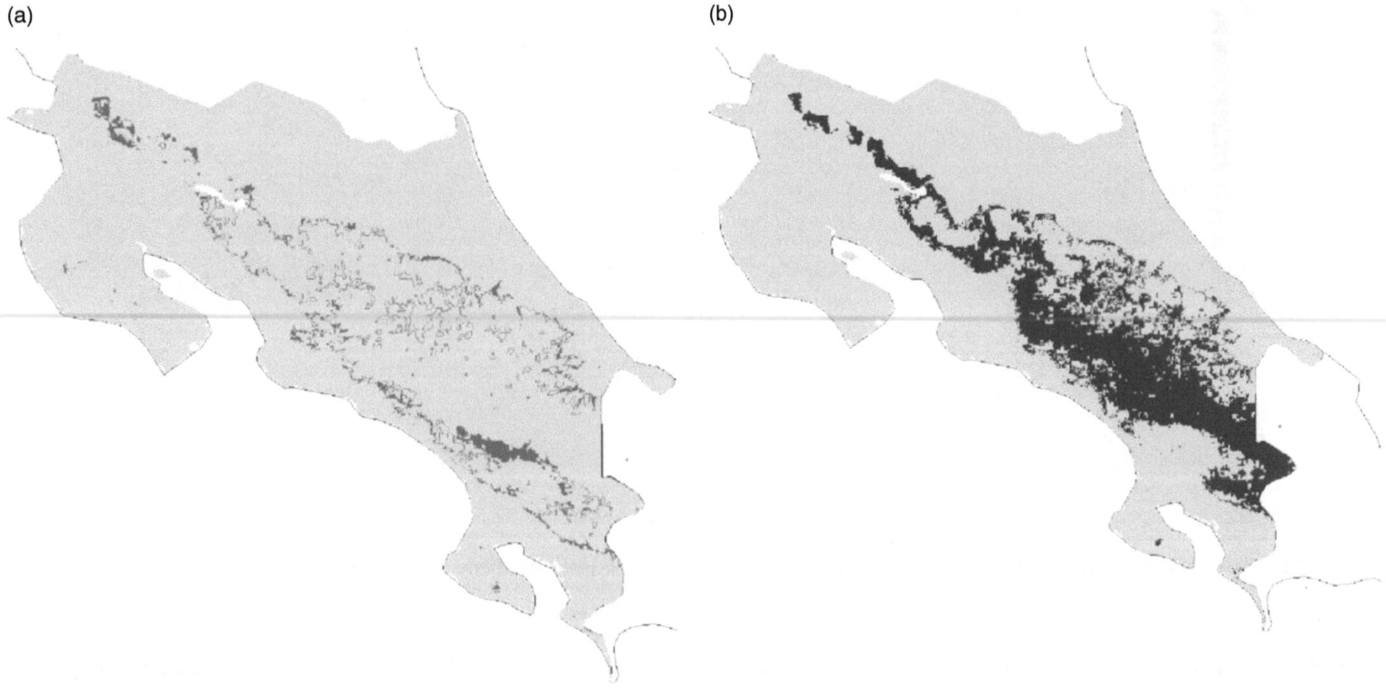

Figure 2.9. Differences in the area classified as cloud forest according to model simulations (a) test_8 and (b) test_4 using two different spatial resolutions (90 m and 1 km). The dark colors represent areas that were classified as cloud forest in the 90-m runs but not in the 1-km runs.

Atlantic slopes (cf. Lawton *et al.*, this volume) and this effect may not be measured well by the MODIS sensor. Therefore, test_4 may not be as appropriate for application with higher-resolution data-sets as test_8 whereas test_8 also has the advantage that it seems to perform well in locating cloud forests at the current resolution of available global-scale data. This is because test_4 is more sensitive to errors in both the DEM and the cloud climatology (since it takes the minimum of both), whereas test_8 takes a mean (and thus is less sensitive to errors in any one of these data-sets). Using test8_vcf10_1km_fractional, increased data resolution leads to a reduction of cloud-affected forest area in Costa Rica of around 17% (implying a correction factor of ×0.827). If this factor were applicable globally the predicted global cloud forest area would become 1.83 Mkm^2 (12% of tropical forests) (Mulligan and Burke, 2005a).

The best current model

In summary, the combination of satellite-measured cloud frequency (satcloud), modeled frequency of ground-level condensing conditions (fogfreq), and high-resolution continuous fields for tree cover (vcf10_1km_fractional) provides a good model to predict the distribution of TMCFs. The best results were obtained using a model which requires the mean of observed atmospheric cloud frequency and modeled frequency of ground-level condensation to be greater than 70% (in other words more than 70% of the time there should be satellite-visible cloud and/or ground-level condensing conditions). The best model predicted 71% of the UNEP–WCMC cloud forest sites correctly and cloud forest to be present within 5 km from 81% of the UNEP–WCMC sites (Table 2.3). The mal-predicted sites were all rather anomalous (isolated small islands, coastal sites, etc.) and thus there are clear reasons for their mal-prediction (Figure 2.8). The model also seems relatively insensitive to the resolution of cloud cover and topographic data. Therefore, this model will be used in the following sections as the most appropriate model for predicting cloud forest distributions and derived estimates of cloud forest loss, etc.

COMPARING DISTRIBUTIONS OF CLOUD-AFFECTED FORESTS

Figure 2.11 shows the distribution of cloud-affected forests according to the adopted model alongside the distribution of observed UNEP–WCMC cloud forest sites (indicated as red dots). It is noteworthy that there are very few areas in which the model predicts cloud forest to be present but no observed sites exist. Examples include a small lowland patch in central Brazil (perhaps due to the regular occurrence of radiation fog; see Gradstein *et al.*, this volume), some areas in southern Cameroon, Democratic Republic of Congo (mostly in the eastern mountainous part), Zimbabwe, South Africa, Lesotho, and the southern Democratic Republic of Congo. These are all areas in

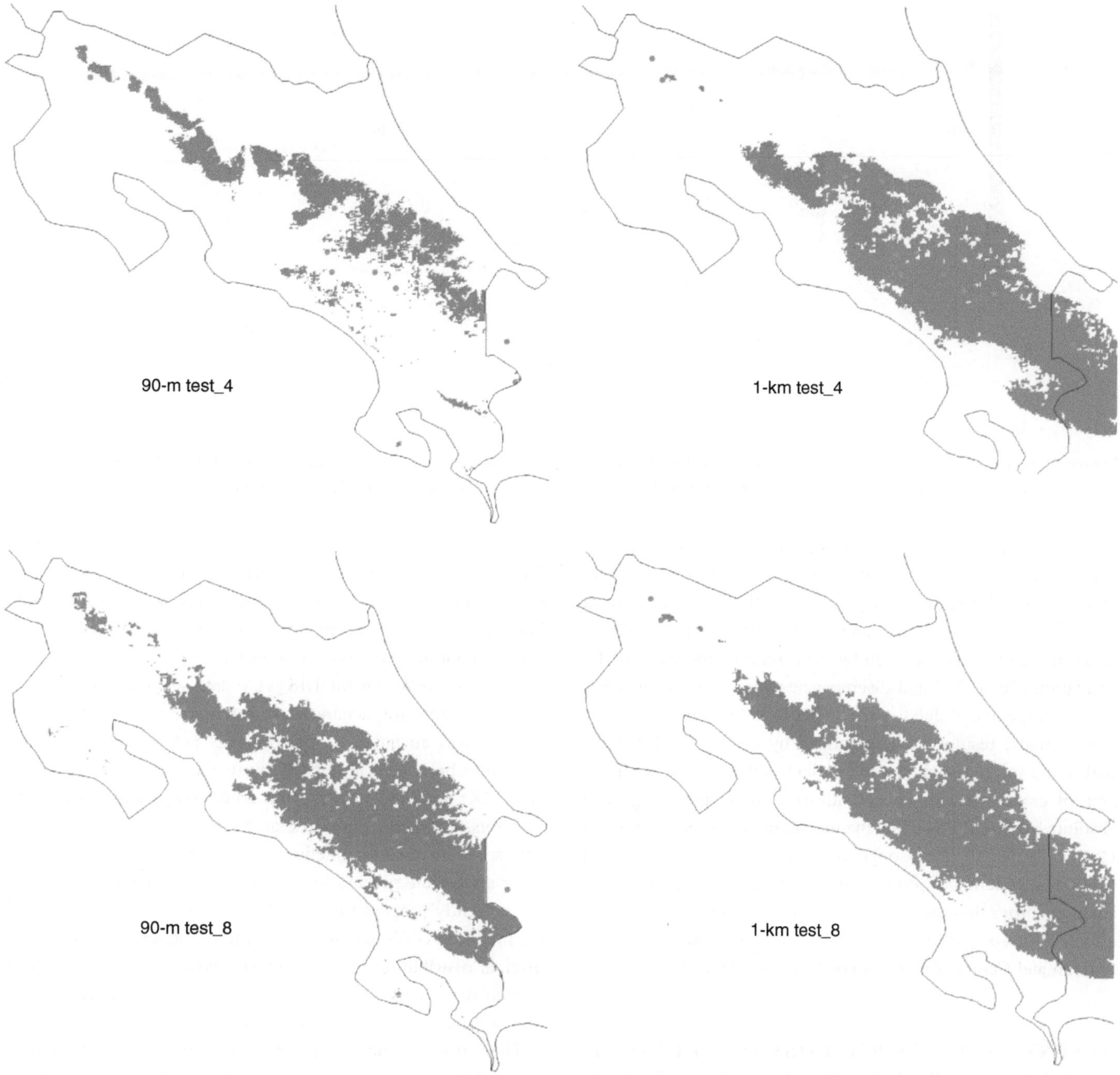

Figure 2.10. Modeled cloud forest distribution for Costa Rica. (See also color plate.)

which satellite-based cloud frequency is high enough for cloud forest to exist but where rainfall may be too low to produce the physical characteristics commonly associated with wetter cloud forests and thus for the forests to appear in the UNEP–WCMC database. In South-East Asia the model predicts cloud forests in Taiwan, northern Laos, and southern China bordering Laos, Vietnam, and Burma, although none or very few cloud forests are recorded in the UNEP–WCMC database as existing in those areas. These omissions probably reflect the need for further inputs to the UNEP–WCMC database for these areas. There are also likely other cloud forest areas that fall outside of the latitudinal boundaries used for this analysis.

Overall, the geographic distributions of cloud forests produced here (Figure 2.11) are similar to those of Bubb *et al.* (2004). The presently proposed distributions are more extensive but many areas have low fractional covers (i.e. they are highly fragmented). Considering intact forest only (with a cover of, say, >90%) would give a much more restricted spatial distribution. The presently modeled distributions are particularly more extensive than suggested by Bubb *et al.* (2004) and Scatena *et al.* (this

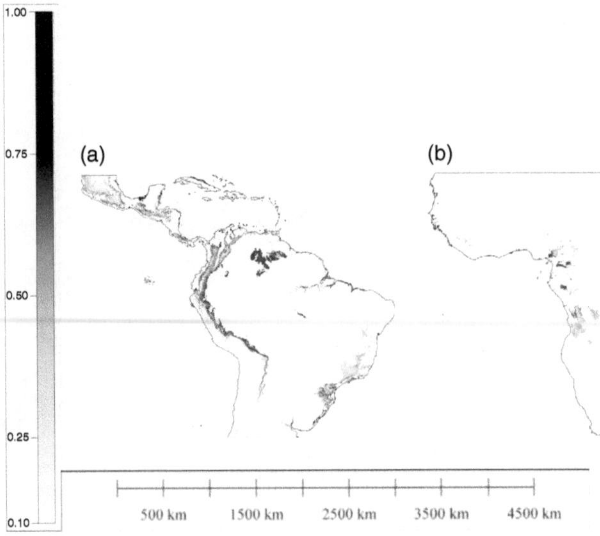

Figure 2.11. The distribution of tropical montane cloud forests (cloud affected tropical montane forests) in detail with UNEP–WCMC sites overlaid in red. The darker the shade the greater the fractional cover of forest within the 1-km pixel. (See also color plate.)

volume) in Africa and in mainland South-East Asia (southern China, Vietnam, Laos) but always in areas where there are some known cloud forest sites (cf. Liu *et al.*, this volume; Tanaka *et al.*, this volume; Wangda and Ohsawa, this volume). Confidence in these results is given by their good performance under validation (Table 2.3) and discrepancies with previous analyses are to a large extent due to differences in definition.

The model results are summarized by country in Table 2.5 with some surprises evident. Indonesia has the greatest predicted area of cloud-affected forest (both in terms of percentage and absolute extent), followed by the mountainous eastern part of the Democratic Republic of Congo, SE Brazil, Venezuela, Perú, and Colombia. In terms of proportion of national (tropical) territory, small countries dominate: Taiwan has the greatest area of cloud-affected forests (28%), followed by Laos, and various Central African and Central American countries (Table 2.5).

ASSESSING HISTORIC LOSS OF TROPICAL MONTANE CLOUD FOREST

Mulligan and Burke (2005b) conducted an assessment of the loss of cloud-affected forest using both GLC_2K and MODIS_VCF data. Here only the results based on the MODIS_VCF data are presented since these are considered to be better. Historic cloud forest loss was calculated as the difference in fractional forest cover between the Global Forest Watch (GFW) assessment of original forest cover at 8000 years B.P. (Bryant *et al.*, 1997) and the MODIS_VCF assessment of 2001. UNEP–WCMC developed the first detailed map of estimated forest cover "prior to the impact of modern man" (*c.* 8000 years ago) for GFW, using many global and regional biogeographic maps. The WCMC map is an indicator of original cover, depicting where forest might be expected to occur today in the absence of humans, on the basis of climate, topography, and other variables. Further details on the original forest cover mapping methodology can be found in Billington *et al.* (1996).

First, the GFW and MODIS_VCF data-sets were subjected to quality control for accurate geo-referencing. Next, the GFW data-set was amended to include areas not classified as forest originally but which are under forest cover today (according to GLC_2K) so as to improve the consistency between the two data-sets. All areas classified as forest in the modified GFW data-set were given a fractional cover of one (1), whereas all other areas were given a fractional cover of zero. By re-running the presently proposed model of cloud forest distribution using the modified GFW data for original forest cover, the potential or original distribution of cloud forests shown in Figure 2.12 was produced, with extensive coverage throughout the Andes, eastern Africa, Madagascar, and eastern Asia.

The top four countries in terms of potential (original) cloud-affected forest cover are (eastern) Brazil, Mexico, Indonesia, Perú and the (mountainous eastern) Democratic Republic of Congo, each with 0.3–0.4 Mkm^2 (Table 2.6).

By subtracting the MODIS_VCF-based current cloud forest cover from the areas with 100% original forest cover within the modified GFW map, estimates of fractional cloud forest loss per km^2 over the last 8000 years were derived. The assumption of 100% cover for all GFW forests is slightly more risky for montane forests which even at potential forest cover may have $<100\%$ cover at the 1-km scale because of the high spatial variability in complex mountain environments. Nevertheless, the presence of a few cloud forest areas which remain at 100% cover today (Figure 2.13) provides some evidence that this assumption can hold true.

Table 2.5 *Countries with more than 10% cloud forest coverage in rank order of percentage cover, and countries with highest cloud forest coverage in rank order of cloud forest extent (km²)*

Country name	Area of cloud forest (km²)	Cloud forests as a % of national (tropical) territory	Country name	Area of cloud forest (km²)	Cloud forests as a % of national (tropical) territory
Indonesia	251 881	11.5	Taiwan	5 472	27.6
Democratic Republic of Congo	236 262	8.6	Laos	77 304	27.3
Brazil	195 357	1.9	Burundi	6 916	21.6
Venezuela	165 853	15.5	Rwanda	6 292	21.4
Perú	158 960	10.4	Ecuador	63 323	21.3
Colombia	152 281	11.4	Papua New Guinea	107 862	20.1
Mexico	139 986	12.2	Costa Rica	11 389	19.0
Tanzania	111 645	10.1	Honduras	24 709	18.3
Papua New Guinea	107 862	20.1	Guatemala	21 756	16.4
Ethiopia	80 592	6.0	Venezuela	165 853	15.5
Angola	78 051	5.2	Uganda	42 899	15.1
Laos	77 304	27.3	China	57 518	14.1
Madagascar	70 622	9.7	Mexico	139 986	12.2
Bolivia	64 900	4.9	Indonesia	251 881	11.5
Ecuador	63 323	21.3	Colombia	152 281	11.4
Myanmar (Burma)	60 792	9.7	Perú	158 960	10.4
China	57 518	14.1	Malaysia	38 631	10.1
Cameroon	43 870	8.0	Tanzania	111 645	10.1
Uganda	42 899	15.1	Myanmar (Burma)	60 792	9.7
Zambia	39 670	4.4	Madagascar	70 622	9.7

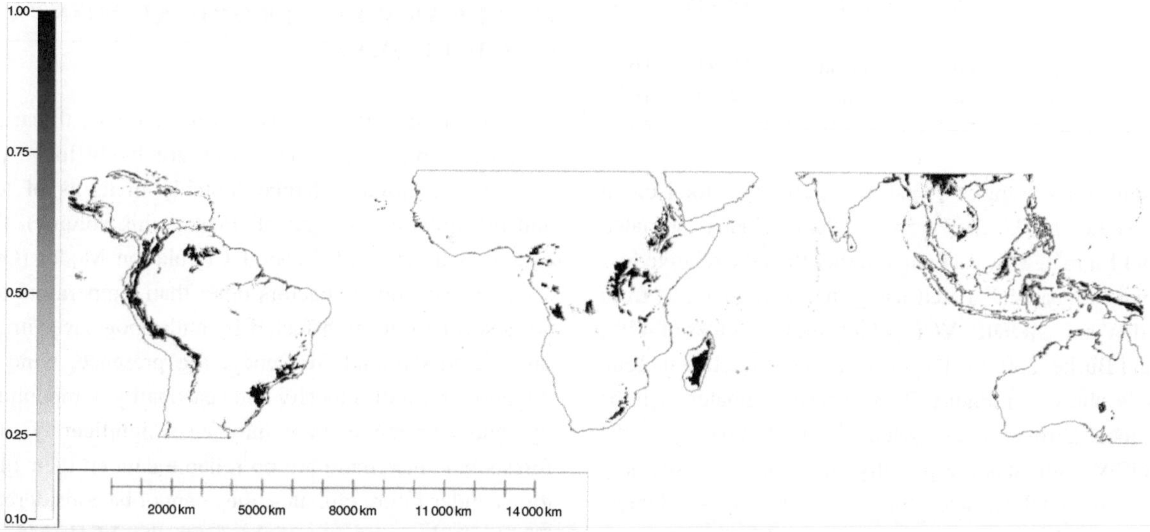

Figure 2.12. Climatically potential (original) distribution of cloud forests based on a combination of MODIS_VCF satellite imagery and modified Global Forest Watch (GFW) vegetation data.

Table 2.6 *Climatically potential ("original") cloud forest cover by country ranked first by total area and then by percentage of national (tropical) land territory*

Country (sorted by potential cloud forest)	Potential cloud forest (km²)	% of national territory originally cloud forest	Country (sorted by % of national territory)	Potential cloud forest) (km²)	% of national territory originally cloud forest
Brazil	435 426	4.3	Taiwan	8 442	42.5
Mexico	415 997	36.2	Madagascar	310 209	42.4
Indonesia	365 933	16.6	Burundi	13 027	40.8
Perú	348 898	22.8	Honduras	54 828	40.6
Democratic Republic of Congo	325 898	11.9	Laos	112 823	39.8
Madagascar	310 209	42.4	Uganda	111 054	39.2
Colombia	300 751	22.6	Ecuador	108 466	36.4
Ethiopia	282 098	21.1	Guatemala	48 150	36.4
Venezuela	219 445	20.4	Mexico	415 997	36.2
Papua New Guinea	162 227	30.2	Costa Rica	19 880	33.1
Bolivia	151 878	11.4	Rwanda	9 193	31.2
China	125 061	30.6	China	125 061	30.6
Laos	112 823	39.8	Papua New Guinea	162 227	30.2
Uganda	111 054	39.2	Peru	348 898	22.8
Ecuador	108 466	36.4	Colombia	300 751	22.6
Angola	100 129	6.7	Dominican Republic	12 492	21.1
Myanmar (Burma)	99 871	15.9	Ethiopia	282 098	21.1
South Africa	90 054	5.5	Venezuela	219 445	20.4
Tanzania	88 602	8.0	Indonesia	365 933	16.6
Cameroon	84 334	15.4	Panama	14 206	16.5

The present assessment of forest loss suggests a total cloud forest loss to date of 2.72 Mkm^2 (55% of the originally estimated cover of 4.93 Mkm^2). This is slightly higher than the relative loss for all tropical forests calculated using the same methods from modified GFW and MODIS_VCF (which showed a loss of 47%; Mulligan and Burke, 2005b). The distribution of fractional cloud forest loss is shown in Figure 2.13. These estimates will be subject to local errors in the potential forest cover (i.e. the modified GFW estimates), especially in areas with spatially variable and marginal climates for forest cover (e.g. Mexico, SW Andes). Additional error is associated with the assessment of the current forest area, but by and large the present results indicate the areas of known forest loss well.

Extensive loss of cloud-affected forest is inferred to have occurred throughout Latin America, with intensive losses in much of Mexico, the Colombian western and central Cordilleras, western Ecuador, Perú and Bolivia, and the Brazilian Atlantic forests. The most intact remaining Latin American cloud-affected forests are found on the eastern flanks of the Andes and southern Venezuela. Significant inferred loss of cloud-affected forest in Africa is concentrated in Madagascar, Ethiopia and Eritrea, South Africa, Uganda, Kenya, and Tanzania. In South-East Asia, the most extensive losses are inferred for Indonesia, southern China, Laos, and Vietnam.

Table 2.7 lists estimated cloud forest loss per country. According to this analysis, Mexico would have experienced the highest absolute loss of cloud-affected forest – closely followed by Brazil, Madagascar, Ethiopia, and Perú (all above 0.2 Mkm^2) – but estimates of the original forest cover for these countries may have been inflated because of their sometimes marginal climates. Expressed as a proportion of a country's original cloud forest cover, inferred losses have been greatest in some countries with dry lowland climates and limited original areas of wet (montane) cloud forest (e.g. Eritrea, Chile, Somalia) or in countries that have undergone extreme overall forest loss (e.g. Haiti, Madagascar).

In summary, more than half of the original cover of TMCFs has been converted to other land uses and some countries have very little of their original cloud forest resources remaining. The largest and most intact areas of cloud-affected forest remaining are to be found in southern Venezuela, eastern Congo and maritime South-East Asia (Borneo, Celebes, Irian Jaya and Papua New Guinea; Table 2.5 and Figure 2.11). Losses can be expected to continue in all cloud forest areas but particularly in those that have shown rapid loss to date (Table 2.7 and Figure 2.13).

POTENTIAL CLIMATE CHANGE IN AREAS WITH TROPICAL MONTANE CLOUD FOREST

Conversion to other land uses is not the sole threat to TMCFs. Because tropical montane forests are likely to be particularly sensitive to climate change (notably in terms of temperature and precipitation change; cf. Foster, this volume), this section examines a series of General Circulation Model (GCM) scenarios for the tropics. Factors other than temperature and rainfall are also likely to be affected by anthropogenic climate change. Such factors (cloud frequency, fog presence, wind speed and direction, rainfall intensity and seasonality, temperature diurnality, and extreme events) may also significantly affect cloud forests but since there are no reliable data on their likely trajectories under future climates, they cannot be considered here. The GCM results used are those from the ECHAM5 (Hamburg) model (DKRZ Model User Support Group, 1992) and the UK Hadley Centre Model (HADCM3; Cullen, 1993). Data were

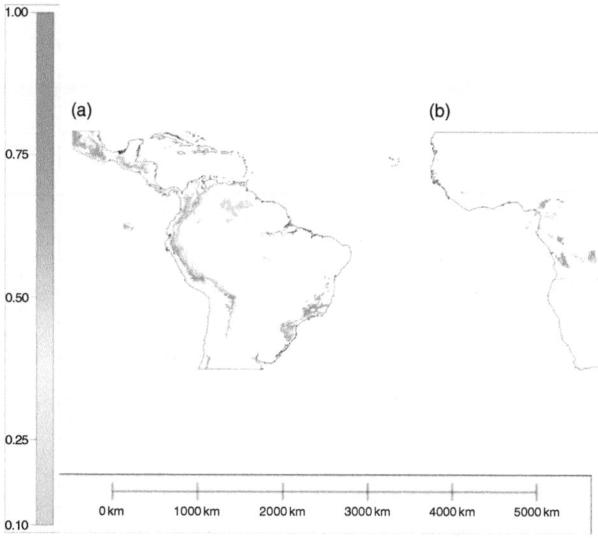

Figure 2.13. Fractional cloud-forest loss according to MODIS_VCF satellite product for (a) Latin America, (b) Africa, and (c) South-East Asia. The darker the shade of red, the greater the forest loss. Red, >60% forest loss; green, <25% forest loss. (See also color plate.)

Table 2.7 *Cloud forest loss by country sorted by area deforested and by percentage of original national cloud forest lost, according to MODIS_VCF satellite imagery*

Country	Deforested cloud forest (km^2)	Country	% of original national cloud forest deforested
Mexico	284 367	Eritrea	99.8
Brazil	264 942	Chile	94.1
Madagascar	239 918	Somalia	93.8
Ethiopia	230 020	Lesotho	89.0
Perú	200 500	South Africa	83.1
Colombia	157 181	Ethiopia	81.5
Democratic Republic of Congo	137 442	Haiti	78.8
Indonesia	114 353	Swaziland	77.3
Bolivia	90 387	Madagascar	77.3
Uganda	75 850	Tanzania	77.3
South Africa	74 846	Congo	76.2
Tanzania	68 455	Kenya	76.0
China	67 563	Malawi	75.7
Angola	64 210	Nigeria	74.6
Venezuela	60 345	Burundi	73.7
Kenya	56 069	Mozambique	72.1
Papua New Guinea	54 486	Zimbabwe	69.6
Ecuador	52 086	Rwanda	69.1
Cameroon	40 464	Zambia	68.6
Myanmar (Burma)	39 079	Mexico	68.4

provided by the IPCC Data Distribution Centre (http://ipcc-ddc.cru.uea.ac.uk/). The scenarios used were the SRES A2 CO_2 forecasts prepared for the IPCC's (Intergovernmental Panel for Climate Change) Third Assessment Report of 2001 which were the most recently produced scenarios in the public domain.

Data were available in the original model resolution (3.75 by 2.5 degrees for HADCM3 and 2.5 by 2.5 degrees for ECHAM5) as monthly values for the period of the simulation (100–150 years). All climate grid data were geo-referenced and converted to GIS layers in the geographic coordinate system and split into two groups according to time period (1950–2000) and (2000–2050). The mean values of the parameters of interest were calculated for each of these periods and the difference between the two means was calculated in order to provide a single map of projected change for temperature and another for rainfall. For the following analysis, the area covered by cloud forests was extracted and viewed in isolation from the global extent.

Figure 2.14 shows that, on average, cloud forest areas are expected to see temperature increases of 2–5 °C between 1950–2000 and 2000–2050. According to the HADCM3 model the most continental cloud forests (in the African interior) will be experiencing the greatest temperature change, whilst the small-island cloud forests are predicted to be subjected to the smallest changes. Results from the ECHAM5 model were similar in direction and geographic pattern but differed slightly in magnitude to those obtained with the HADCM3 model (Figure 2.15). For example, the ECHAM5 model predicted a consistent warming over cloud forest areas in the range of 1.3–4.2 °C by 2050, with the greatest warming again occurring in continental areas, in particular for some of the largest remaining blocks of intact forest as found in the eastern Democratic Republic of Congo and southern Venezuela (Figure 2.15).

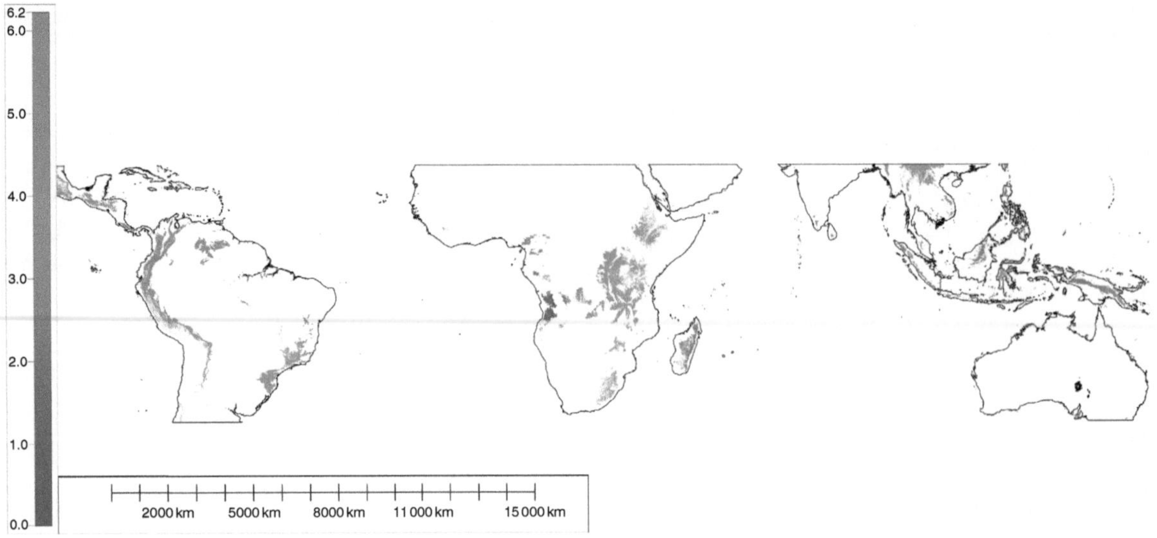

Figure 2.14. Temperature changes (1950–2000 vs. 2000–2050) in °C in the vicinity of cloud forests as predicted by the HADCM3 global circulation model. (See also color plate.)

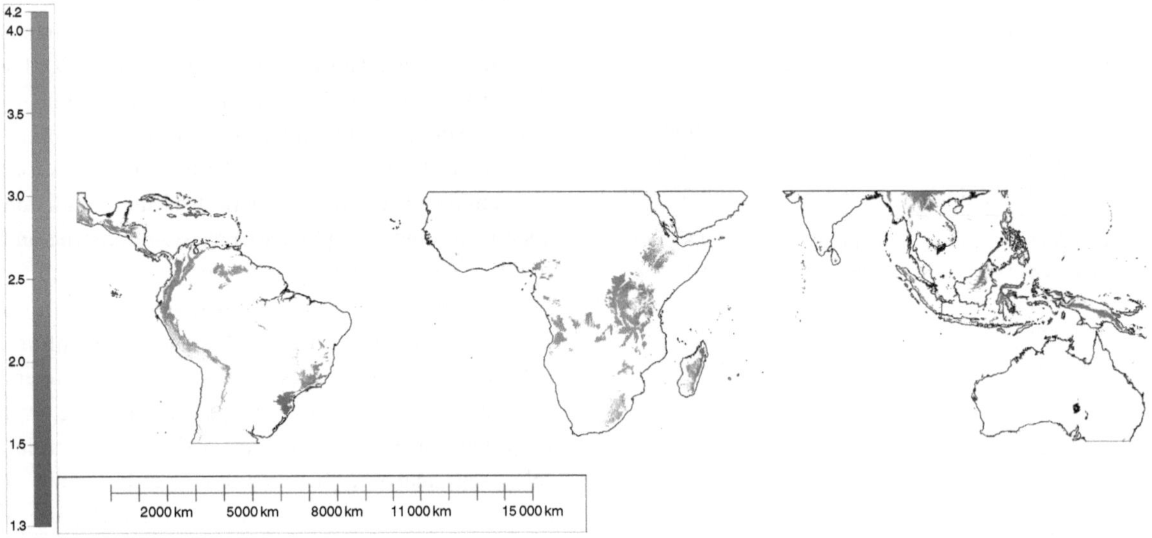

Figure 2.15. Temperature changes (1950–2000 vs. 2000–2050) in °C in the vicinity of cloud forests as predicted by the ECHAM5 global circulation model. (See also color plate.)

The GCM results for changes in rainfall are both more extreme and more uncertain than those for temperature (especially at the regional and sub-regional scale). Predicted changes in rainfall over cloud forest areas between 1950–2000 and 2000–2050 indicate a potential loss in rainfall (of 100 to 1000 mm year^{-1} over the period) over some areas (e.g. Colombia, Mexico, Central Africa) and a potential gain in others (e.g. Perú, Madagascar, and the South-East Asian islands (+100 to +1000 mm year^{-1} over the period) according to the HADCM3 model (Figure 2.16). These significant changes reflect changes in dominant sea-surface temperature and atmospheric circulation patterns and thus the precise location of spatially restricted cells of high precipitation.

In contrast to the predictions by the HADCM3 model, according to the ECHAM5 model almost all cloud forest areas will experience an increase in rainfall of a few hundred mm year^{-1}, with the exception of Madagascar, West Africa and a few of the South-East Asian islands (Figure 2.17). Thus, differences in predicted changes in rainfall between GCMs are large. Higher-resolution regional model runs would be desirable.

In summary, an examination of the outputs of the most recently available climate model scenarios indicates that the

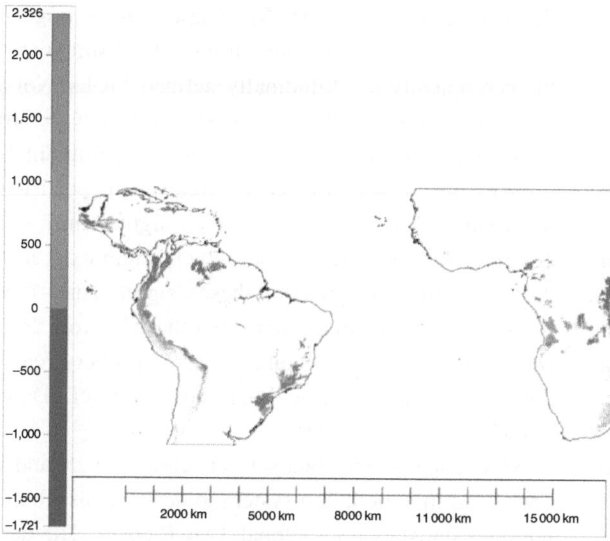

Figure 2.16. Changes in rainfall (1950–2000 vs. 2000–2050) in mm year^{-1} in the vicinity of cloud forests as predicted by the HADCM3 global circulation model. (See also color plate.)

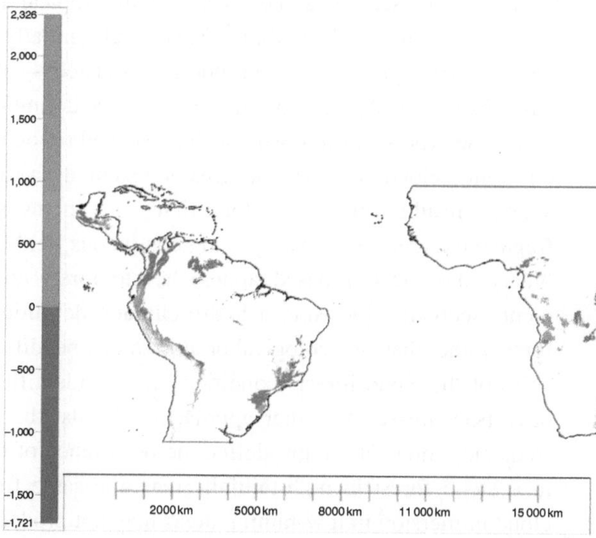

Figure 2.17. Changes in rainfall (1950–2000 vs. 2000–2050) in mm year^{-1} in the vicinity of cloud forests as predicted by the ECHAM5 global circulation model. (See also color plate.)

tropical land masses in general may be subject to increased temperatures of 1–4 °C and either significant rainfall increases or losses depending upon precise location. Cloud forest areas in particular will see lower magnitude warming (generally less than 3 °C). Rainfall projections over tropical mountains are still highly uncertain, with the HADCM3 model indicating a rainfall decline for the majority of sites whilst the ECHAM5 model mostly indicates a rainfall increase. Either way, the rainfall changes are likely to be very significant and potentially much more important (hydrologically and ecologically) than the projected temperature changes (Mulligan and Burke, 2005b; cf. Foster, this volume).

CONSERVATION STATUS OF TROPICAL MONTANE CLOUD FORESTS

It is of interest to examine cloud forest extent, distribution and loss through land-use change within the context of the protection (conservation) status of these areas by combining the data produced in this analysis with the world database of protected areas (WDPA Consortium, 2004). Some 5.8% of the total land area of the extended tropics is at least nominally protected under IUCN Categories I–VI (72.34 Mkm^2 of land area, 4.20 Mkm^2 of protected area) whereas 10.0% (or 1.57 Mkm^2) of the tropical *tree-covered* area (15.64 Mkm^2, based on MODIS_VCF_10%;

Table 2.1) is protected, compared with 14.7% (0.38 Mkm^2) of forest areas classified as cloud-affected forests in the present analysis. Thus, ~9% of the total tropical protected land area has cloud forest.

The proportion and the total area of cloud forest for each of the protected areas defined in the WDPA that are also within the tropics, have been ranked by the percentage of the protected area that is cloud forest, and by the area of cloud forest within a protected area. A number of protected areas contain mostly cloud forest, particularly in Venezuela (including the largest individual protected cloud forest area, Formaciones de Tepuyes), Costa Rica, the Philippines, Madagascar, Vietnam, Laos, Cameroon, and Indonesia. The largest individual protected cloud forests are found in Venezuela, Indonesia, and Colombia (see Mulligan and Burke, 2005b for details).

Table 2.8 gives a summary of cloud forest loss within protected areas and is presented in order to indicate particularly threatened cloud forest reserves. Forest loss may have occurred before or after the area became protected. Either way, the remaining forest is important since it represents protected forest that is a remnant of a much more extensive former cloud forest within the boundaries of the protected area (and thus potentially capable of regeneration). Protected areas with the greatest proportion of deforested cloud forest occur throughout Central America and the Caribbean, as well as in Thailand and Indonesia. Protected areas that have lost the greatest absolute areas of cloud forest are largely located in more continental settings, where one also finds some of the largest protected areas and the largest areas of cloud forest (Table 2.8). However, it is pertinent to note that the figures derived for individual protected areas may be sensitive to misclassification of forest in the GFW data-set or to cloud "contamination" of the MODIS_VCF estimates, especially in topographically very variable environments.

Finally, by examining, on a national basis, countries that have significant areas of cloud forest outside of protected areas it is possible to identify areas with further potential (and likely need) for forest protection. As shown in Table 2.9, the countries with the largest absolute areas of unprotected cloud forest (according to WDPA, 2004) are the Democratic Republic of Congo, Brazil, Indonesia, Perú, and Colombia. Similarly, those with the greatest proportion of their cloud forests unprotected (again according to WDPA, 2004) are Laos, Burundi, Papua New Guinea, Taiwan, Honduras, and Ecuador.

CONCLUSIONS

This study has adopted an empirically defined hydro-climatic characterization of cloud forest as forest under frequent or persistent exposure (>70% of the time) to ground-level cloud, independent of elevation and in a zone which extends beyond the tropics (down to 35° S). These forests might be termed "significantly cloud-affected forest" to distinguish them from the ecologically or altitudinally defined MCFs. Not all of these "significantly cloud-affected forests" will show the structural and ecological characteristics common to altitudinally defined TMCFs but all are affected by frequent or persistent exposure to ground-level cloud (fog), which significantly changes their environment compared to that which might exist in the absence of fog. Moreover, some of these "significantly cloud-affected forests" occur in areas that are otherwise too dry to support forests; they are thus "cloud forests" in a very real sense since they would not exist without the cloud (cf. Dawson, 1998; Hildebrandt *et al.*, 2007).

New tropics-wide data-sets on climate and land cover have been used in a modeling system to derive a distribution of – and national statistics for – global TMCF cover. The derived distribution has been tested against an observed data-set of over 560 cloud forest locations with a high level of success. Hydro-climatically defined cloud forests have been estimated by this study as representing some 14.2% of all tropical forests and covering an area of 2.21 Mkm^2. This is substantially more than derived from previous – but poorly validated – altitudinally based estimates. Apart from differences in the definition of cloud forest used (previous assessments have tended to focus primarily on wetter cloud forests), the greater extent derived from this analysis results from two factors: firstly, the inclusion of small, fragmented areas of forest by using continuous fields rather than binary classification-based approaches to forest cover assessment. Secondly, by using a hydro-climatic definition of cloud forest rather than an ecological or altitudinal one, different intensities of the cloud forest "condition" can be identified that may have been missed by other approaches. Whilst the 70% cloud frequency threshold might define the occurrence of upper montane cloud forest in high-rainfall areas, the same frequency of cloud immersion in low-rainfall areas may not produce similarly looking forest. Yet, such forests are cloud forests in a very real sense, both ecologically and hydro-climatically.

A series of tropical land-cover change assessments have shown the importance of using vegetation continuous fields (VCF) approaches for representing land cover rather than binary classification approaches. Using VCF-based vegetation data the extent of loss of TMCFs in the last 8000 years was calculated, both globally and by country. Around 55% of cloud forests were estimated to have been lost compared with 47% for all tropical forests. These losses have been widespread, with severe losses in a few areas. Very few large areas of intact cloud forest remain, notably in southern Venezuela, Borneo, Celebes, Papua New Guinea, and in the eastern Democratic Republic of Congo (DCR). About 14.7% (0.37 Mkm^2) of TMCFs as defined here are at least nominally protected (IUCN categories I–VI) compared with only 10.0% (1.57 Mkm^2) for tropical forests as a

Table 2.8 *Protected areas with the greatest proportion and greatest area of deforested cloud forest land*

Name	Country	% loss	Name	Country	Loss (km^2)
Cerro Tomabu	NIC	71.8	Mochima	VEN	37870
Mesa de Moropotente	NIC	69.0	Alto Orinoco-Casiquiare	VEN	2038
Cerro Tisey–Estanzuela	NIC	65.8	Jaua Sarisari±ama	VEN	722
Cerro El Arenal	NIC	64.5	Yushan	TWN	409
El Uyuca	HND	60.6	Murchison Falls	UGA	349
Cerro Quiabuc (Las Brisas)	NIC	58.6	San Rafael de Guasare	VEN	344
Cerro Azul de Copán	HND	58.3	Ta-Wu Mountain	TWN	283
Cerro Apante	NIC	57.9	Piedemonte Norte de la Cordillera Andina	VEN	266
El Pital	HND	53.3	Pico Bonito	HND	264
Volcán Yalí	NIC	52.9	Sierra de Aroa	VEN	257
Area Metropolitana de Caracas	VEN	51.6	La Paragua	VEN	250
Cordillera Dipilto y Jalapa	NIC	51.2	Sierra de Nirgua	VEN	231
Tepesomoto/Pataste	NIC	51.0	Celaque	HND	226
Yucul	NIC	50.2	Valle Nuevo	DOM	209
Cerro Kilamb	NIC	48.7	Formaciones de Tepuyes	VEN	206
Ramal de Datanli – Cerro El Diablo	NIC	48.4	Mudumalai	IND	199
Valle Nuevo	DOM	47.0	Henri Pittier	VEN	189
Salto del Río Yasica	NIC	45.3	Doi Inthanon	THA	165
La Tigra	HND	45.0	Cordillera Dipilto y Jalapa	NIC	164
Celaque	HND	43.9	Cusuco	HND	127
Macizo de Pe±as Blancas	NIC	43.6	Mae Yuam	THA	117
Mudumalai	IND	42.2	Pian Upe	UGA	104
Fila Cerro FrÝo–La Cumplida	NIC	39.1	Klong Wang Chao	THA	100
Taiwan Cycas	TWN	37.4	Cerro Azul de Copßn	HND	100
Ta-Wu Mountain	TWN	34.5	Area Metropolitana de Caracas	VEN	96
Yushan	TWN	30.6	Yob	ERI	93
Doi Inthanon	THA	25.9	Mae Wong	THA	91
Blue Mountain	IND	25.7	La Tigra	HND	82
Jaua Sarisari±ama	VEN	23.4	Doi Chiang Dao	THA	82
Pico Bonito	HND	23.2	Pico Pijol	HND	74
Atitlßn	GTM	21.0	Mesa de Moropotente	NIC	71
Cerro Guabule	NIC	20.5	Cerro KilambÚ	NIC	67
Sierra de Aroa	VEN	19.4	Macizo de Penas Blancas	NIC	67
Pico Pijol	HND	17.4	Sri Lanna	THA	64
Cusuco	HND	17.3	Waynad	IND	62
Henri Pittier	VEN	16.9	Namtok Mae Surin	THA	60
Phu Hin Rong Kla	THA	14.9	Phu Hin Rong Kla	THA	57
Sierra de Nirgua	VEN	14.5	Tepesomoto/Pataste	NIC	56
Doi Chiang Dao	THA	12.9	Lipan	UGA	50
Khawnglung	IND	12.7	Cerro Tisey–Estanzuela	NIC	48
Klong Wang Chao	THA	11.7	Doi Pha Chang	THA	45
Namtok Mae Surin	THA	11.7	Khlong Lan	THA	40
Waynad	IND	11.7	Simien Mountains	ETH	39
Silent Valley	IND	10.6	Atitlan	GTM	33
Sierra Kiragua	NIC	10.0	Taiwan Cycas	TWN	31
Cerro Cumaica–Cerro Alegre	NIC	10.0	Khao Laem	THA	28

Table 2.9 *Countries with significant areas of unprotected cloud forest ranked by proportion and area*

Country	% unprotected cloud forest	Country	Unprotected cloud forest (km^2)
Laos	23.3	Democratic Republic of Congo	216 291
Burundi	20.2	Brazil	186 655
Papua New Guinea	20.1	Indonesia	184 536
Taiwan	19.7	Perú	146 210
Honduras	17.4	Colombia	130 907
Ecuador	17.3	Mexico	130 806
Rwanda	17.1	Papua New Guinea	107 843
Guatemala	16.4	Tanzania	97 048
China	13.4	Angola	77 963
Uganda	13.3	Ethiopia	69 302
Mexico	11.4	Laos	66 076
Costa Rica	10.8	Madagascar	64 517
Colombia	9.8	Myanmar (Burma)	60 788
Myanmar (Burma)	9.7	China	54 797
Perú	9.5	Venezuela	54 438
Malaysia	9.3	Bolivia	53 561
Madagascar	8.8	Ecuador	51 608
Tanzania	8.8	Cameroon	40 248
Indonesia	8.4	Uganda	37 577
Lesotho	8.0	Zambia	36 762
Dominican Republic	7.9	Malaysia	35 672
Democratic Republic of Congo	7.9	Vietnam	27 793
Philippines	7.8	Philippines	26 769
Comoros	7.7	South Africa	26 537
Cameroon	7.4	Kenya	24 240
Vietnam	7.0	Honduras	23 439
Angola	5.2	Guatemala	21 713
Ethiopia	5.2	Argentina	14 399
Venezuela	5.1	Zimbabwe	13 784
Malawi	4.8	Thailand	12 143
Panama	4.7	Guyana	9 878
Zambia	4.1	Gabon	9 745
Bolivia	4.0	Mozambique	9 397
Guyana	4.0	India	7 260
Kenya	3.5595	Malawi	6 846
El Salvador	3.2410	Costa Rica	6 486
Gabon	3.2015	Burundi	6 445
Zimbabwe	2.8606	Congo	6 191
Sri Lanka	2.4328	Rwanda	5 036
São Tomé and Principe	2.2561	Dominican Republic	4 689
Nicaragua	2.2156	Panama	4 065
Haiti	1.9947	Taiwan	3 912
Thailand	1.9604	Nicaragua	3 393
Solomon Islands	1.8789	Lesotho	3 254
Brazil	1.8347	Sudan	3 204
South Africa	1.6290	Australia	2 146
Congo	1.5375	Sri Lanka	1 876
Swaziland	1.4851	Central African Republic	1 680
Vanuatu	1.4458	Chile	1 573
Mozambique	0.9769	Nigeria	1 209
Argentina	0.6249	Guinea	1 048
Guinea	0.3607	El Salvador	800
India	0.3566	Haiti	655
Jamaica	0.3399	Solomon Islands	562
Chile	0.3156	Swaziland	332
Equatorial Guinea	0.2722	Cuba	243
Central African Republic	0.2307	Vanuatu	203
Cuba	0.1812	Ivory Coast	180
Brunei	0.1750	Namibia	159
New Zealand	0.1679	Liberia	155
Liberia	0.1383	Comoros	146
New Caledonia	0.1359	Yemen	142
Nigeria	0.1122	Somalia	132
Sudan	0.1067	Cambodia	93
Ivory Coast	0.0476	Equatorial Guinea	85
Eritrea	0.0435	Eritrea	63

whole. Nine percent of protected lands in the tropics are cloud forests. There are some protected areas which are almost entirely cloud forest (located in Venezuela, Costa Rica, the Philippines, and Madagascar) and some protected cloud forests are very extensive (>6000 km^2), especially in Venezuela and Indonesia.

Conservation priorities should include: (i) protected areas that conserve the remnants of formerly potentially extensive cloud forest areas of which much of the original forest may have been lost; (ii) countries in which large areas of cloud-affected forest remain unprotected at the national level; and (iii) countries that have very small areas of (as yet unprotected) cloud-affected forest that could harbor important endemics.

Cloud forests are likely to be exposed to greenhouse-effect induced changes in temperature and precipitation over the next half century and beyond. An analysis of results obtained with two widely used global circulation models indicates that temperature increases will be greater for cloud forests in large continental interiors than in maritime environments. Rainfall change is much more spatially complex and uncertain, with some areas showing strong increases and others showing strong decreases. Either way, the strong climatic dependence of cloud forests and their location on steep altitudinal and thus climatic gradients means that those protected areas conserving cloud forests today may not have suitable climates for those same cloud forests in the future. The connection of protected areas with corridors along altitudinal and climatic gradients is thus an important priority for the long-term conservation of these ecosystems.

It is pertinent to keep in mind that the present assessment of the distribution and extent of cloud-affected forests in the tropics is modeled, not measured, although the model predictions were validated successfully against the best available and independent observational data-sets. The predicted extents and distributions may change with improved data-sets and methods. A project is underway to repeat this analysis with 90-m spatial resolution topographic data, 1-km resolution cloud frequency data and improved assessments of the impact of exposure to wind and rain. These data may change the detail of the cloud forest distribution, degree of deforestation, impacts of climate change, and conservation analysis presented here. Nevertheless, there is reason for some confidence that the present assessments are reasonable in pattern, if not detail, and that the main messages will not change upon further analysis. Work is also under way to develop assessments of cloud forest distribution based on other modelling and measurement techniques, such as the use of MODIS standard data products to estimate the thickness, top heights, and base heights of clouds and regional-scale atmospheric modeling approaches (cf. Nair *et al.*, 2008; Welch *et al.*, 2008; Lawton *et al.*, this volume; Nair *et al.*, this volume), or comparisons of measured and modeled fog water inputs and the distribution of characteristic fauna and flora. In the end, the distributions and extents of cloud forest will always depend on the precise definition of the ecosystem, which is different for the hydrologist, ecologist, and biogeographer.

Though the present analysis has been validated against the UNEP–WCMC database of more than 560 cloud forest sites it needs further validation for certain regions and countries (cf. Figure 2.11). Many of the maps in this chapter are available as interactive layers in the Google Earth viewer at http://www.ambiotek.com/cloudforests from where the data can also be obtained as downloadable GIS files. Readers are encouraged to use the cited website to provide positive or negative feedback on the representation of cloud forests in areas that they know. Readers can also add cloud forests that they know to the database.

ACKNOWLEDGEMENTS

This publication is in part an output from a research project funded by the United Kingdom Department for International Development (DfID) for the benefit of developing countries. The views expressed are not necessarily those of DFID ZF0216 (Forestry Research Programme). Thanks are due to Jon Palmer of the DfID Forest Research Programme for his support and to Philip Bubb and UNEP–WCMC for provision of the cloud forest database and other kind favors and to the developers and providers of all data-sets used in this work (acknowledged individually in the reference list). I also owe a great deal to all of the undergraduates, masters students, Ph.D. students, and other friends who worked alongside me in a variety of cloud forests as I have tried to make some sense of them over the years. Thanks particularly to Sampurno Bruijnzeel for his enthusiastic and energetic support of cloud forest research in general and of this work in particular.

REFERENCES

Aldrich, M., C. Billington, M. Edwards, and R. Laidlaw (1997). *A Global Directory of Tropical Montane Cloud Forests*. Cambridge, UK: UNEP-CMC.

Aravena, R., O. Suzuki, and A. Pollastri (1989). Coastal fog and its relation to ground-water in the IV region of northern Chile. *Chemical Geology* **79**: 83–91.

Basnett, T. A., and D. E. Parker (1997). Development of the Global Mean Sea Level Pressure Data Set GMSLP2, Hadley Centre for Climate Research Technical Note CRTN 79. Bracknell, UK: Hadley Centre for Climate Research.

Billington, C., V. Kapos, M. Edwards, S. Blyth, and S. Iremonger (1996). *Estimated Original Forest Cover Map: A First Attempt*. Cambridge, UK: UNEP-WCMC.

Bruijnzeel, L. A. (2001). Hydrology of tropical montane cloud forests: a reassessment. *Land Use and Water Resources Research* **1**: 1–18.

Bruijnzeel, L. A. (2004). Hydrological functions of tropical forests: not seeing the soil for the trees? *Agriculture Ecosystems and Environment* **104**: 185–228.

Bruijnzeel L. A. (2005). Tropical montane cloud forests: a unique hydrological case. In *Forests, Water and People in the Humid Tropics*, eds. M. Bonell and L. A. Bruijnzeel, pp. 462–483. Cambridge, UK: Cambridge University Press.

Bruijnzeel, L. A., and L. S. Hamilton (2000). *Decision Time for Cloud Forests*, IHP Humid Tropics Programme Series No. 13. Paris: UNESCO, Amsterdam: IUCN-NL, and Gland, Switzerland: IUCN.

Bruijnzeel, L. A., and E. J. Veneklaas (1998). Climatic conditions and tropical montane forest productivity: the fog has not lifted yet. *Ecology* **78**: 3–9.

Bruijnzeel, L. A., M. J. Waterloo, J. Proctor, A. T. Kuiters, and B. Kotterink (1993). Hydrological observations in montane rain forests on Gunung Silam, Sabah, Malaysia, with special reference to the '*Massenerhebung*' effect. *Journal of Ecology* **81**: 145–167.

Bryant, D., D. Nielsen, and L. Tangley (1997). *The Last Frontier Forests: Ecosystems and Economies on the Edge*. Washington, DC: World Resources Institute.

Bubb, P., I. May, L. Miles, and J. Sayer (2004). *Cloud Forest Agenda*. Cambridge, UK: UNEP–WCMC.

Campanella, R. (1995). The role of GIS in evaluating contour-based limits of cloud forest reserves in Honduras. In *Tropical Montane Cloud Forests*, eds. L. S. Hamilton, J. O. Juvik, and F. N. Scatena, pp. 116–124. New York: Springer-Verlag.

Cullen, M. J. P. (1993). The unified forecast/climate model. *Meteorological Magazine* **122**: 81–94.

Dawson, T. E. (1998). Fog in the California redwood forest: ecosystem inputs and use by plants. *Oecologia* **117**: 476–485.

DKRZ Model User Support Group (1992). *ECHAM3 Atmospheric General Circulation model.* Hamburg, Germany: DKRZ.

FAO (2000). *Global Forest Resources Assessment 2000: Part 1. Global Issues.* Rome: FAO. Also available at www.fao.org/documents/show_cdr.asp?url_file=/DOCREP/004/Y1997E/y1997e06.htm.

FAO (2005). *Terms and Definitions for the National Reporting Tables for FRA 2005.* Rome: FAO. Also available at www.fao.org/documents/show_cdr.asp?url_file=//docrep/007/ae156e/AE156E03.htm.

Frahm, J. P., and S. R. Gradstein (1991). An altitudinal zonation of tropical rain forests using bryophytes. *Journal of Biogeography* **18**: 669–678.

Grubb, P. J. (1977). Control of forest growth and distribution on wet tropical mountains: with special reference to mineral nutrition. *Annual Review of Ecology and Systematics* **8**: 83–107.

Hafkenscheid, R. L. L. J. (2000). Hydrology and biogeochemistry of tropical montane rain forests of contrasting stature in the Blue Mountains, Jamaica. Ph.D. Thesis, VU University Amsterdam, Amsterdam, The Netherlands. Also available at http://dare.ubvu.vu.nl/bitstream/1871/12734/1/tekst.pdf.

Hamilton, L. S., J. O. Juvik, and F. N. Scatena (1995). The Puerto Rico tropical cloud forest symposium: introduction and workshop synthesis. In *Tropical Montane Cloud Forests*, eds. L. S. Hamilton, J. O. Juvik, and F. N. Scatena, pp. 1–16. New York: Springer-Verlag.

Hansen, M., R. DeFries, J. R. Townshend, *et al.* (2003). *500 m MODIS Vegetation Continuous Fields.* College Park, MD: The Global Land Cover Facility.

Hijmans, R. J., S. E. Cameron, J. L. Parra, P. G. Jones, and A. Jarvis (2004). *The WorldClim Interpolated Global Terrestrial Climate Surfaces*, Version 1.3. Available at http://biogeo.berkeley.edu/

Hildebrandt, A., M. Al Aufi, M. Amerjeed, M. Shammas, and E. A. B. Eltahir (2007). Ecohydrology of a seasonal cloud forest in Dhofar. I. Field experiment. *Water Resources Research* **43**, W10411, doi:10.1029/2006WR005261.

Jarvis, A., J. Rubiano, A. Nelson, A. Farrow, and M. Mulligan (2004). *Practical Use of SRTM Data in the Tropics: Comparisons with Digital Elevation Models Generated from Cartographic Data*, Working Document No. 198. Cali, Colombia: International Centre for Tropical Agriculture.

Jin, Y., W. B. Rossow, and D. P. Wylie (1996). Comparison of the climatologies of high-level clouds from HIRS and ISCCP. *Journal of Climate* **9**: 2850–2879.

Joint Research Centre (2003). *Global Land Cover 2000.* Available at www.gem.jrc.it/glc2000.

Kappelle M., and A. D. Brown (eds.) (2001). *Bosques nublados del Neotropico.* Heredia, Costa Rica: INBIO.

Leo, M. (1995). The importance of tropical montane cloud forest for preserving vertebrate endemism in Peru: the Rio Abiseo National Park as a case study. In *Tropical Montane Cloud Forests*, eds. L. S. Hamilton, J. O. Juvik, and F. N. Scatena, pp. 198–211. New York: Springer-Verlag.

Letts, M. G., and M. Mulligan (2005). The impact of light quality and leaf wetness on photosynthesis in north-west Andean tropical montane cloud forest. *Journal of Tropical Ecology* **21**: 549–557.

Liebermann, D., M. Liebermann, R. Peralta, and G. S. Hartshorn (1996). Tropical forest structure and composition on a large-scale altitudinal gradient in Costa Rica. *Journal of Ecology* **84**: 137–152.

Long, A. (1995). Restricted-range and threatened bird species in tropical montane cloud forests. In *Tropical Montane Cloud Forests*, eds. L. S. Hamilton, J. O. Juvik, and F. N. Scatena, pp. 79–106. New York: Springer-Verlag.

Meybeck, M., P. Green, and C. Vörösmarty (2001). A new typology for mountains and other relief classes: an application to global continental water resources and population distribution. *Mountain Research and Development* **21**: 34–45.

Mulligan, M., and S. M. Burke (2005a). *DFID-FRP Project, ZF0216 Global Cloud Forests and Environmental Change in a Hydrological Context*, Final Report. London: Department for International Development. Also available at www.ambiotek.com/cloudforests.

Mulligan, M., and S. M. Burke (2005b). *DFID-FRP Project R7991, FIESTA: Fog Interception for the Enhancement of Streamflow in Tropical Areas*, Final Technical Report of AMBIOTEK/ King's College London Contributions. London: Department for International Development. Also available at www.ambiotek.com/fiesta.

Nair, U. S., S. Asefi, R. M. Welch, *et al.* (2008). Biogeography of tropical montane cloud forests. II. Mapping of orographic cloud immersion. *Journal of Applied Meteorology and Climatology* **47**: 2183–2197.

New, M., M. Hulme, and P. D. Jones (2000). *Global Monthly Climatology for the Twentieth Century: Data Set.* Oak Ridge, TN: Oak Ridge National Laboratory Distributed Active Archive Center. Also available at www.daac.ornl.gov.

Pounds, J. A., M. P. L. Fogden, and J. H. Campbell (1999). Biological response to climate change on a tropical mountain. *Nature* **389**: 611–614.

Pounds, J. A., M. Bustamante, L. A. Coloma, *et al.* (2006). Widespread amphibian extinctions from epidemic disease driven by global warming. *Nature* **439**: 161–167.

Salaman, P., T. M. Donegan, M. Mulligan, *et al.* (2003). A new species of wood-wren (Troglodytidae: Henicorhina) from the western Andes of Colombia. *Ornitología Colombiana* **1**: 4–21.

Sidle, R. C., A. D. Ziegler, J. N. Negishi, *et al.* (2006). Erosion processe sin steep terrain: truths, myths, and uncertainties related to forest management in Southeast Asia. *Forest Ecology and Management* **224**: 199–225.

Stadtmüller, T. (1987). *Cloud Forests in the Humid Tropics. A Bibliographic Review.* Tokyo: United Nations University, and Turrialba, Costa Rica: CATIE.

UNEP–WCMC (2004). *Database of Global Cloud Forest Point Data.* Cambridge, UK: UNEP–WCMC.

USGS (2005). *Global 30 Arc-Second Elevation Data Set (GTOPO30).* Available at http://edc.usgs.gov//products/elevation/gtopo30.html.

WDPA Consortium (2004). *World Database on Protected Areas 2004.* Cambridge, UK: World Conservation Union (IUCN) and UNEP-WCMC. (Source for this data-set was the Global Land Cover Facility.)

Welch, R. M., S. Asefi, J. Zeng, *et al.* (2008). Biogeography of tropical montane cloud forests. 1. Remote sensing of cloud base heights. *Journal of Applied Meteorology and Climatology* **47**: 960–975.

Wylie, D. P., W. P. Menzel, H. M. Woolf, and K. I. Strabala (1994). Four years of global cirrus cloud statistics using HIRS. *Journal of Climate* **7**: 1972–1986.

3 The climate of cloud forests

A. Jarvis
International Centre for Tropical Agriculture (CIAT), Cali, Colombia
M. Mulligan
King's College London, London, UK

ABSTRACT

This chapter analyzes the climatic conditions where cloud forests are reported. Spatial data-sets of climate, derived from the WorldClim database, were used to describe the climate in 477 cloud forest sites identified by UNEP–WCMC with 85% of the sites being found at altitudes between 400 and 2800 m.a.s.l., with an average altitude of slightly less than 1700 m. The range of altitudes at which cloud forests are found is impressive (220–5005 m). The climate of cloud forests is highly variable from site to site, with an average rainfall of *c*. 2000 mm year^{-1} and an average temperature of 17.7 °C. In addition, cloud forests are found in seasonal and aseasonal environments alike, both in terms of rainfall and temperature. There are some clear differences in the climates of cloud forests found in Africa, Latin America and the Caribbean, and Asia. Comparisons are made between the climate of cloud forest sites and of randomly generated sites covering forested areas throughout the montane tropics, with the aim of identifying the climatic variables most important in distinguishing cloud forests from other tropical forests. Cloud forests are found to be wetter (by 184 mm year^{-1} on average), cooler (by 4.2 °C on average), and less seasonally variable than other montane forests. The most statistically significant climatic differences between cloud forests and other montane forests in order of significance are: maximum temperature > mean temperature > rainfall > rainfall seasonality. Cloud forests are also almost completely confined to within 350 km from the nearest coast, and they are located closer to the coast than montane forests. Cloud forests further occupy more topographically exposed areas and differently sized mountains than montane forests. Thus, cloud forests tend to occur in fairly coastal climates and environments with lower maximum and mean temperatures and higher altitudes compared with other montane forests. Finally, the climatic representativity of 14 intensively studied sites (ISS) was analyzed; as a group, the sites provided a fair representation of the climates found in cloud forests, evenly covering the ranges in temperature and rainfall. The majority of cloud forest sites occur in regions with 2000–2600 mm of rainfall and annual mean temperatures of 14–18 °C, with five ISS clustering in this range (Mount Cameroon, Blue Mountains, San Francisco, Sierra de las Minas, East Maui). However, relatively dry cloud forest sites (< 1000 mm of rain year^{-1}) are under-represented, and some low-temperature sites (mean temperatures 10–13 °C) are also lacking.

INTRODUCTION

Tropical montane cloud forests (TMCF) are defined as tropical forests occurring in areas of frequent or persistent ground-level cloud (Grubb, 1977; Bruijnzeel and Proctor, 1995). They generally occur between 1200 and 2500 m.a.s.l. although they may be found below 500 m and even above 3500 m (La Bastille and Poole, 1978; Stadtmüller, 1987; cf. Hemp, this volume #12). In this chapter these forests are simply referred to as "cloud forests." Strictly speaking, the term montane does not always apply. Whilst the boundary between mountains and hills is usually set at 500 m, not all areas above 500 m elevation are mountains (e.g. some are plateaux), hence mountains proper are defined both by their altitude and by their topographic relief (Meybeck *et al.*, 2001).

Tropical Montane Cloud Forests: Science for Conservation and Management, eds. L. A. Bruijnzeel, F. N. Scatena, and L. S. Hamilton. Published by Cambridge University Press.

This chapter examines the variation in climatic conditions of areas in which cloud forests are known to occur. Since climate is likely to be a major control on both the hydrology and ecology of cloud forests, setting the climatic context for cloud forests is fundamental to understanding their structure and functioning, and thus the likely impacts of environmental (notably climatic) change upon them (cf. Foster, this volume). Though the prime interest here is not to predict cloud forest distribution by comparing climatic conditions at known cloud forest sites with those prevailing at sites with other montane forest types (cf. Mulligan, this volume), knowledge of climatic conditions will enable a better understanding of the overall conditions that form and maintain the cloud forest ecosystem. This includes examining the geographic context of sites in terms of such factors as distance to the nearest coast, mountain size, and topographic exposure, which are all known to affect temperature, humidity, rainfall, fog occurrence, wind speeds, etc. For example, TMCFs tend to be found at lower altitudes on small outlying mountains compared with large ones (the "mass-elevation" or "telescoping" effect), and this has been related to the fact that large mountain massifs, by their uptake of solar radiation and its slow release as long-wave radiation, may be better at warming the atmosphere above them than do smaller mountains (Schröter, 1926). It is also likely to be the result of a lowering of the cloud base because of the higher humidity levels prevailing in coastal and island areas (Van Steenis, 1972; Bruijnzeel *et al.*, 1993) where outlying small mountains often occur. Whichever is the case at a specific location, climate is likely to play an important role in determining forest type and stature, either directly or indirectly, e.g. through its effect on soil properties (notably degree of waterlogging and nitrogen content; cf. Roman *et al.*, this volume; Benner *et al.*, this volume).

Part one of this chapter uses climate data from a 1-km gridded climate database (WorldClim, available from http://www.worldclim.org; Hijmans *et al.*, 2005) for areas of known cloud forests represented in the World Conservation Monitoring Centre (WCMC) spatial database of protected areas (Aldrich *et al.*, 1997; UNEP–WCMC, 2004). The range of climatic and environmental conditions in these cloud forests are examined in a latitudinal and altitudinal context and compared with climates for areas covering the same range of altitudes and which support forest (as defined by the 1-km FRA2000 data-set; FAO, 2005), but which are not TMCF according to the UNEP–WCMC database. This is done in order to identify the conditions that characterize TMCF and to understand better the global variation of climatic and topographic contexts within which cloud forests exist.

Part two examines the climatic detail of a number of well-studied TMCF sites in Latin America, Africa, and South-East Asia (so-called intensively studied sites, ISS) in order to define the range of temperature and precipitation regimes in which these well-known cloud forests occur. These well-studied cloud forest areas are also placed within the wider climatic context for the known (but not necessarily well-studied) tropical cloud forests of part one. In this way, an indication is obtained of how well existing heavily studied sites represent the range of climatic characteristics in which TMCFs are found. Furthermore, climatic similarities and dissimilarities between well-studied sites are pointed out as a means of understanding differences in the interpretation of their hydrological and ecological structure and functioning (cf. Bruijnzeel and Proctor, 1995; Bruijnzeel, 2001; Roman *et al.*, this volume; Benner *et al.*, this volume).

METHODOLOGY

Objectives

Data-sets on cloud forest localities across the globe, gridded climatic data (WorldClim), and elevation data-sets are used in this chapter with the following objectives:

- To examine the relationship between the gridded climate data and ground-measured station data for stations within 5 km of UNEP–WCMC cloud forest sites in order to quantify how representative the WorldClim climates are for those measured on the ground, and their utility in the analysis to come.
- To look briefly into the latitudinal and longitudinal distribution of the cloud forest sites.
- To analyze the climate of areas containing known cloud forests and to compare the climate of these sites with all tropical forest sites.
- To compare the geographic setting of UNEP–WCMC cloud forest sites with all tropical montane sites in terms of distance from sea, size of mountain, and topographic exposure.
- To examine climatic differences (including cloud cover) with distance from sea, size of mountain, and topographic exposure to see to what extent these variables control climate, beyond the control exerted by altitude alone (as inherent in the WorldClim data because of their interpolation). Further, to examine whether the mass-elevation effect is apparent in the cloud forest distributions.
- To examine the distribution of ISS climates within the context of cloud forest climates as a whole.

Data-sets used

This analysis was carried out using the UNEP–WCMC database of global cloud forest point data (UNEP–WCMC, 2004), and global maps of monthly climatic variables from the WorldClim database (Hijmans *et al.*, 2005). The UNEP–WCMC database represents only well-known cloud forests, many of which are protected areas, and thus constitutes only a sample of areas potentially or actually under cloud forest (indeed, only nine of the 14 ISS used in the present analysis occur in the UNEP–WCMC database). Furthermore, the data are given as point locations for what may be large sites representing named places. For example, some data points are plotted as a mountain top when in reality the cloud forest is found in a band at lower altitudes on that mountain (P. Bubb, personal communication). A total of 526 cloud forest sites were identified in the UNEP–WCMC data-set at the time of writing of this chapter but because some sites occupied the same 1-km grid cell, or had dubious geographic coordinates, only 477 cells were considered to contain cloud forest for certain. Some additional uncertainty is introduced by the fact that not everyone agrees as to what is and what is not cloud forest, and names and definitions are legion (cf. Grubb, 1977; Stadtmüller, 1987; Frahm and Gradstein, 1991; Bruijnzeel and Hamilton, 2000; Hietz, this volume). Indeed, improving the availability and quality of data on cloud forest distribution is a high priority for this and much other related work (cf. Mulligan, this volume; Lawton *et al.*, this volume). In this chapter, TMCFs were simply taken as to occur in those areas defined as such by the UNEP–WCMC database.

In addition, the 1-km resolution gridded Global Forest Resources Assessment (FRA2000) of the UN Food and Agricultural Organization (FAO) (USGS, 2000) has been used alongside the SRTM (Shuttle Radar Topography Mission) 30-arc-second (1-km) digital elevation model (DEM) (USGS, 2004) to define all montane forests, "montane" being defined as all areas higher than 500 m after Meybeck *et al.* (2001), thereby incorporating all levels of topographic roughness (both true mountains and plateaux). Mountain size was also calculated using this same definition. The Digital Chart of the World (DCW) was used to define coasts (including the shores of the Great Lakes of Africa because these are large enough to produce coastal effects on the local climate; cf. Van der Molen *et al.*, 2006), for the calculation of distance from the coast.

The main climate database used here (WorldClim) was developed through the interpolation of data from climate stations covering the entire terrestrial surface, collated from various data sources including international climate databases, such as the Global Historical Climatology Network (GHCN), the FAO, the World Meteorological Organization (WMO), the International Centre for Tropical Agriculture (CIAT), R-HYdronet, as well as a number of country-level databases (Hijmans *et al.*, 2004). Only stations with more than 10 years of data were used to produce the database, with the majority of stations covering the period 1950–1990. This has resulted in some 46 000 stations for rainfall (*rf*), 26 000 for mean temperature (T_{mean}) and 18 000 for average daily maximum (T_{max}) and minimum (T_{min}) temperatures. Station data were interpolated using thin-plate smoothing splines (Hutchinson, 1995) with elevation (derived from the SRTM 30-arc-second elevation database) as a co-variable. The result is a continuous surface of monthly climate means (*rf*, T_{mean}, T_{max}, T_{min}) and mean diurnal range in temperature (T_{range}, calculated as the mean of monthly $T_{max} - T_{min}$), all with a grid resolution of 30 arc-seconds (approximately 1 km at the equator). Because the WorldClim database does not include data on cloud cover (one of the variables considered to be of considerable importance here; cf. Mulligan, this volume), these data were extracted from the smaller CIAT climate database (Jones, 1991) with 2273 stations in 111 countries throughout the world. Although wind-speed data are sometimes available in gridded climate products, wind-direction data for terrestrial areas are not. Thus, it was not possible to accurately characterize wind as a climatic variable in this chapter.

The expected error in Worldclim *rf* data is estimated to be <10 mm month^{-1} at the majority of places, while temperature errors are estimated at <0.3 °C in most areas (Hijmans *et al.*, 2005). Nevertheless, the production of the climate surfaces from relatively sparsely distributed station data fails to capture small-scale orographic processes, and so rainfall in some isolated, very wet areas is underestimated, whilst at some dry spots rainfall is overestimated. Given the significant topographic (and thus climatic) variability in cloud forest environments, slight errors in the geographic coordinates of the sites might also cause significant differences between the interpolated climate surface and the climate on the ground. Moreover, local variations in slope- and aspect-related radiation loads, wind funneling (affecting actually received amounts of rain and fog), cloud cover, and cloud base height are not incorporated in WorldClim. As such, application to individual areas or sites should be made with caution.

In all cases climatic, forest and topographic data were extracted only for tropical terrestrial areas (taken here as to lie between 23° N and 28° S – the latitudinal range of sites in the UNEP–WCMC TMCF database). However, the entire data-set for the tropics is very large (>2.3 million cells) and this creates some significant data-processing limitations. Hence, a randomly distributed sample of 53 519 cells representing montane forest sites (originally 60 000 but reduced somewhat to include only those cells for which data were available for all variables) was extracted for analysis and comparison with the UNEP–WCMC cloud forest data points (477 cells). Data processing was carried out using a combination of ARC-INFO, Arcview 3.2 and PCRASTER GIS systems.

Analyzing cloud forest climates

REPRESENTATIVITY OF THE WORLDCLIM DATA-SET

WorldClim station data are interpolated in a manner which does not retain values for the individual stations that form the basis of the interpolation (in order to produce smoothed rather than locally pitted and peaked surfaces). Before using this database one needs to ascertain how well the interpolated surfaces represent station data near cloud forest sites. Therefore, as a first step, a basic analysis was made to examine the proximity of cloud forest sites to the nearest climate station(s), and the representativity of the positions of the climate stations in the landscape in terms of topographic exposure. Next, to quantify the correlation between interpolated climatic and observed records, WorldClim climatic data for the 1-km cell above each ground station were compared with all available station data within a 5-km radius of the UNEP–WCMC cloud forest sites (150 stations for rainfall and 107 for temperature). Whilst it would have been better to compare WorldClim climates derived for UNEP–WCMC cloud forest points with independent station data, a literature review produced fewer than 10 comparable and long-term climate time-series. In addition, even where such data did exist, they were invariably measured at altitudes different to the average grid-cell altitude in the WorldClim database, and usually more indicative of micro-scale climates rather than the more macro-scale climates examined here. Thus, the current test is not an independent test of the quality of the WorldClim data-set (since the very station data that produced the surfaces were used), but rather a test of the extent to which the smoothed data-set still represents on-the-ground climate conditions near observed cloud forest sites.

GEOGRAPHIC DISTRIBUTION OF UNEP–WCMC CLOUD FOREST SITES

The latitudinal and longitudinal distributions of the 477 cloud forest sites in the UNEP–WCMC database were examined by producing frequency distributions of cloud forest occurrence in 1-degree classes of latitude and longitude. To avoid location bias introduced by the global distribution of land, the cloud forest site frequencies (grouped into 10-degree classes of longitude and 5-degree classes of latitude) were divided by the area of land within the tropics within each respective band of latitude and of longitude. Land area was calculated from the DCW combined with grids of the latitude and longitude classes. Further, to take account of any bias due to the latitudinal and longitudinal distribution of mountains (altitudes >500 m), the SRTM 30-arc-sec elevation database was used to calculate mountain areas and these were aggregated into the same latitudinal and longitudinal classes as the cloud forest frequencies. The frequencies were then converted to sites per unit mountain area.

CLIMATES OF UNEP–WCMC CLOUD FOREST SITES

For each identified cloud forest site the monthly climate data were extracted from the WorldClim climate surface, and general patterns in temperature and rainfall examined. A rainfall seasonality index was also calculated using a method developed by Markham (1970). The index ranges from 0 to 1, with 0 representing climates with no intra-annual variation, and 1 representing climates with all rain falling in one single month. For the analysis of cloud cover, the CIAT climate database was used and a simple analysis was made of average annual fractional cloud cover as a function of altitude, distance to the nearest coast, and mountain size to see whether the often cited "mass-elevation" effect was apparent in the data. To better understand the relationship between climate variables and distance from coast, the climate data were extracted for three natural break classes of mountain size (small: 72 000 km^2 on average, medium: 6.8 million km^2 on average, and large: 14 million km^2 on average) for comparison. Since initial analysis gave reason to suspect that the three mountain size classes corresponded to the three continents, the analysis was repeated for Latin America and the Caribbean only, so as to confirm whether any climate differences between mountain size classes were due to mountain size rather than the geographic configuration of each continent.

COMPARISON OF CLOUD FOREST CLIMATES WITH CLIMATES OF OTHER TROPICAL FORESTS

Climate data were extracted for the 477 raster cells with confirmed presence of cloud forest and compared with those extracted for all tropical forests (a random sample of 53 519 raster cells) and then for only tropical *montane* forests (also 53 519 raster cells). In each case, the data were examined for each variable by means of frequency distributions and tables. To assess the statistical significance of differences between distributions, a two-sample Kolmogorov–Smirnov (K–S) test was carried out on the distributions in S-Plus 2000 using a 95% confidence level for two independent samples. For a single sample of data, the K–S test was used to test whether or not the sample was consistent with a specified distribution function. For two samples of data, it was used to test whether or not these two samples might reasonably be assumed to come from the same distribution (note that the K–S test does not assume that the data population is normally distributed).

TOPOGRAPHIC AND GEOGRAPHIC SETTING OF CLOUD FORESTS

A number of climatically related topographic attributes were calculated using the SRTM 30-arc-sec elevation database. Distance from the coast was calculated using DCW data. Mountain size was approximated by calculating the area that falls inside the 500 m contour (in km^2). This gave large values for parts of large mountain chains and plateaux (primarily the Andes and the African highlands), and smaller values for isolated peaks.

Topographic exposure was calculated using the so-called toposcale procedure of Zimmermann (2004). Toposcale is an ARC-AML algorithm for multi-scale topographic exposure analysis. It applies circular moving windows with increasing radii to a DEM, and calculates the difference between the average elevation of the window and the elevation of the center cell of the window. The topographic exposure can be interpreted as a ridge or peak if the center cell in the moving window has a higher elevation than the average elevation of the cells in the surrounding window. Integration into a single multi-scale measure is achieved by starting with the standardized exposure values of the largest window, then adding standardized values from smaller windows where they exceed the values of the larger-scale map. In this analysis search radii of 3–15 cells were used (i.e. a maximum search radius of approximately 15 km). Although other techniques for the calculation of directional exposure are available (e.g TOPEX; Ruel *et al.*, 2002), these require wind-direction data which are generally not available at the global scale. Moreover, toposcale has practical advantages over such large grids because of its operation in ARC-INFO and its computational ease compared with TOPEX. The multi-scale approach of toposcale is also preferable in this case where the scale at which exposure is important is unknown. Next, the resulting distance to coast, mountain size, and topographic exposure data were extracted and compared for all tropical mountains and cloud forest sites. Data examination and testing for statistical significance of differences between the two distributions were carried out as described in the previous paragraph. Key climatic variables were examined as a function of topographic exposure. In each case a random sample of 53 519 cells in the montane tropics was extracted for analysis.

REPRESENTATIVITY OF INTENSIVELY STUDIED CLOUD FOREST SITES

The climates extracted for the 14 ISS were compared with the data for the 477 UNEP–WCMC cloud forest sites. To understand the interaction between climatic factors and to place the ISS within the wider context of all identified cloud forest sites, a summary table of ISS climates was produced, as well as a simple scatter-plot of T_{mean} versus rainfall, presenting the position of the ISS within the total range of TMCF sites. Although this probably captured the most important climatic factors, multivariate statistics were then applied so as to capture the variability across the remaining climatic factors. Principal components analysis (PCA) was applied to annual means of T_{mean}, T_{max}, T_{min}, and T_{range}, as well as to annual rainfall, and monthly seasonality of rainfall. The first two principal components were then plotted, again with the ISS highlighted amongst all UNEP–WCMC cloud forest sites. Finally, in order to assess the degree of climatic similarity between the 14 ISS, a cluster analysis was performed using Ward's methods and the variables cited above.

RESULTS

Representativity of the WorldClim data-set

It is encouraging to note that the average distance between cloud forest sites and the nearest climate station in the WorldClim database proved relatively small: on average 21 km for rainfall, vs. 38 km for T_{mean} and 53 km for T_{max} and T_{min}. The minimum distance was zero km while maximum distances were as high as 122 km for rainfall and 342 km for T_{mean} (367 km for T_{min} and T_{max}). Nevertheless, in a montane setting a station some 21 km from its nearest cloud forest can have a very different altitude, aspect, and exposure, and thus a very different climate. The climate stations were found across a range of topographic exposures, with 56% of the stations having neutral levels of topographic exposure, while 25% were located at topographically exposed sites and 19% at unexposed sites.

The comparison of gridded WorldClim data (WC) and station data (ST) for cells close to cloud forests showed good qualitative correspondence for the 130 stations with a rainfall of less than 3000 mm year^{-1} ($Rain_{WC} = 0.85Rain_{ST} + 230$, $r^2 = 0.75$), but for stations with more than 3000 mm of rain ($n = 18$) the relationship broke down ($Rain_{WC} = 0.22Rain_{ST} + 2430$, $r^2 = 0.26$). For all rainfall values the best relationship proved to be logarithmic ($Rain_{WC} = 1373.1\text{Ln}(Rain_{ST}) - 370$, $r^2 = 0.81$) indicating that locally high rainfalls are reduced in the interpolated data. For all 130 stations with annual rainfall below 3000 mm, the average residual indicated a remarkable conformity, with the gridded rainfall being only 10.5 mm less than the corresponding station rainfall as positive and negative residuals canceled each other out. However, individual stations did show large residuals, with the average absolute residual being 83.7 mm year^{-1}. For the 18 stations with rainfall in excess of 3 000 mm the average residual indicated that the gridded data were about 1204 mm year^{-1} lower than the equivalent station data. These 18 stations are distributed across altitudes ranging from 500 m to 2900 m, 69% of the stations being located at topographically exposed sites. Clearly, rainfall data are only sufficiently reliable for the present analysis in the case of stations receiving less than 3000 mm.

WorldClim annual mean temperatures also showed good qualitative correspondence with ground data ($T_{WC} = 0.9971{*}T_{ST}$, $r^2 = 0.87$; $n = 103$) and the average residual indicated T_{WC} to be only 0.05 °C higher than T_{ST}. The absolute residual showed individual grid cells to differ on average by around 0.05 °C compared with the stations underneath them, implying an interpolation error inferior to the expected measurement error of the temperature sensors themselves. As such, for temperature (which is a much simpler function of altitude than rainfall), the errors are low, both overall and at a station, thereby increasing confidence in interpolations for individual cloud forest sites. By and large, the WorldClim data constitute a good representation of the station data from which they were produced and where stations are within close proximity to cloud forests the WorldClim data will represent those forest climates well. However, where stations are further away or climatic gradients steep, the WorldClim data will be less representative.

Geographic distribution of UNEP–WCMC cloud forest sites

The 477 cloud forest sites of the UNEP–WCMC database are spread across 62 countries. Countries with more than 10 confirmed sites include: Indonesia, Mexico, Malaysia, Venezuela, Ecuador, Philippines, Papua New Guinea, Colombia, Honduras, Perú, Kenya, Sri Lanka, Costa Rica, and Panamá. The latitudinal distribution of the cloud forests shows a significant clustering close to the equator, with the majority of sites (83%) found in the northern hemisphere tropics. Arguably, this is to be expected because 60% of the tropical terrestrial surface and nearly 83% of tropical mountains (land above 500 m.a.s.l.) lie north of the equator. However, even when expressed per unit mountain area there are still proportionally more confirmed cloud forests north of the equator (by a factor of four). The longitudinal distribution of cloud forests is also highly unequal. Even when taking into account the land mass at each longitude, some longitudes have ten times the number of cloud forests sites as others, also after expressing the number of sites per unit area of mountains at each longitude. This inequality in latitudinal and longitudinal distributions may reflect disparities in sampling effort but is also likely to reflect climatic or topographic differences.

Climates of UNEP–WCMC cloud forest sites

Because of the high spatial variability of climate in tropical mountains, along with the smoothing induced by the climatic interpolation and inaccuracies in cloud forest classification and geographic position, there will be random errors in the distribution of climates across the UNEP–WCMC sites with confirmed cloud forest presence. Such errors are expected to be most pronounced at climatic extremes and for small sites located on steep topographic and climatic gradients. Nevertheless, one would expect the general patterns to be representative.

Figure 3.1 presents frequency histograms of altitude, annual rainfall, and rainfall seasonality, as well as of T_{mean}, T_{max}, and T_{min} at the 477 UNEP–WCMC cloud forest sites. Table 3.1 lists averages and maximum and minimum values of these variables for all the sites, whereas Table 3.2 gives the average climatic parameters separated by continent. In general, the histograms cover a wide range of climatic conditions. The average altitude of the cloud forest sites is slightly less than 1700 m.a.s.l., with 77% of all sites lying between 500 m and 2800 m. One site is registered as low as 22 m, but this is most likely an artifact of a small error in geographic coordinate over steep terrain. As many as 48 UNEP–WCMC cloud forest sites (10%) are registered below the 500 m altitude threshold used here to define montane forest. Some of these may, again, be due to erroneous coordinates, or represent island sites where cloud forests have been documented to occur at lower than usual altitudes. This may also occur due to considerable altitudinal variation within a 1-km grid cell. Within-cell variation reached as much as 5517 m in the Himalayas, and had a global upper quartile range of 550 m (Hijmans *et al.*, 2005), indicating that the cloud forest site might be in a higher elevation sector of the grid cell.

The average annual rainfall for all sites is *c.* 2000 mm (Table 3.1), with 94% of sites having between 800 mm and 3400 mm. The wettest site is located at Bukit Batu Bora in Malaysia, receiving 4500 mm annually, although high-rainfall sites are likely to be significantly underestimated in this analysis. T_{mean} for all sites is 17.7 °C, spanning a large range. The lowest value (allegedly *c.* 1 °C) concerns Puncak Jaya/Mount Carstensz in Irian Jaya (Indonesia) at 3800 m but this is very likely an example of location error in the cloud forest points data-set. The maximum value is 27.3 °C, at Mount Halcon (22 m) in the Philippines. Levels of rainfall seasonality also vary a great deal between sites, with the index showing a negative skew and ranging from just 0.01 (practically homogeneous rainfall distribution as found in Santa Cruz, Ecuador) to 0.76 (highly seasonal, with a long pronounced dry season, in the Simen Mountains in Ethiopia). The high coefficients of variation calculated for each climatic variable indicate the high degree of variability in climates between cloud forests sites.

Examining the climatic conditions for each continent (Table 3.2 and Figure 3.2), some clear differences become apparent. Cloud forest sites in Africa tend to be drier (average rainfall less than 1500 mm), with Asia tending to have significantly wetter cloud forests (*c.* 2150 mm of rain on average) and the Latin American and Caribbean (LAC) region experiencing a wide range of rainfall conditions (Figure 3.2). African cloud forests also have the highest rainfall seasonality and LAC ones the lowest. Average temperatures are fairly similar for each continent, but Asian cloud forests tend to have a lower diurnal temperature range (9.3 °C compared with 10.3 °C and 10.4 °C for Latin America and Africa, respectively) whereas African cloud forests also tend to be warmer. However, such differences are likely to be as much a function of differences in overall climate between the regions as differences in the climates favored by cloud forests, since the distribution of cloud forests is also limited by other, non-climatic factors.

Comparison of cloud forest climates with climates of other tropical montane forests

Comparing cloud forest climates with those for all tropical montane forests gives a clearer distinction of the specific climatic

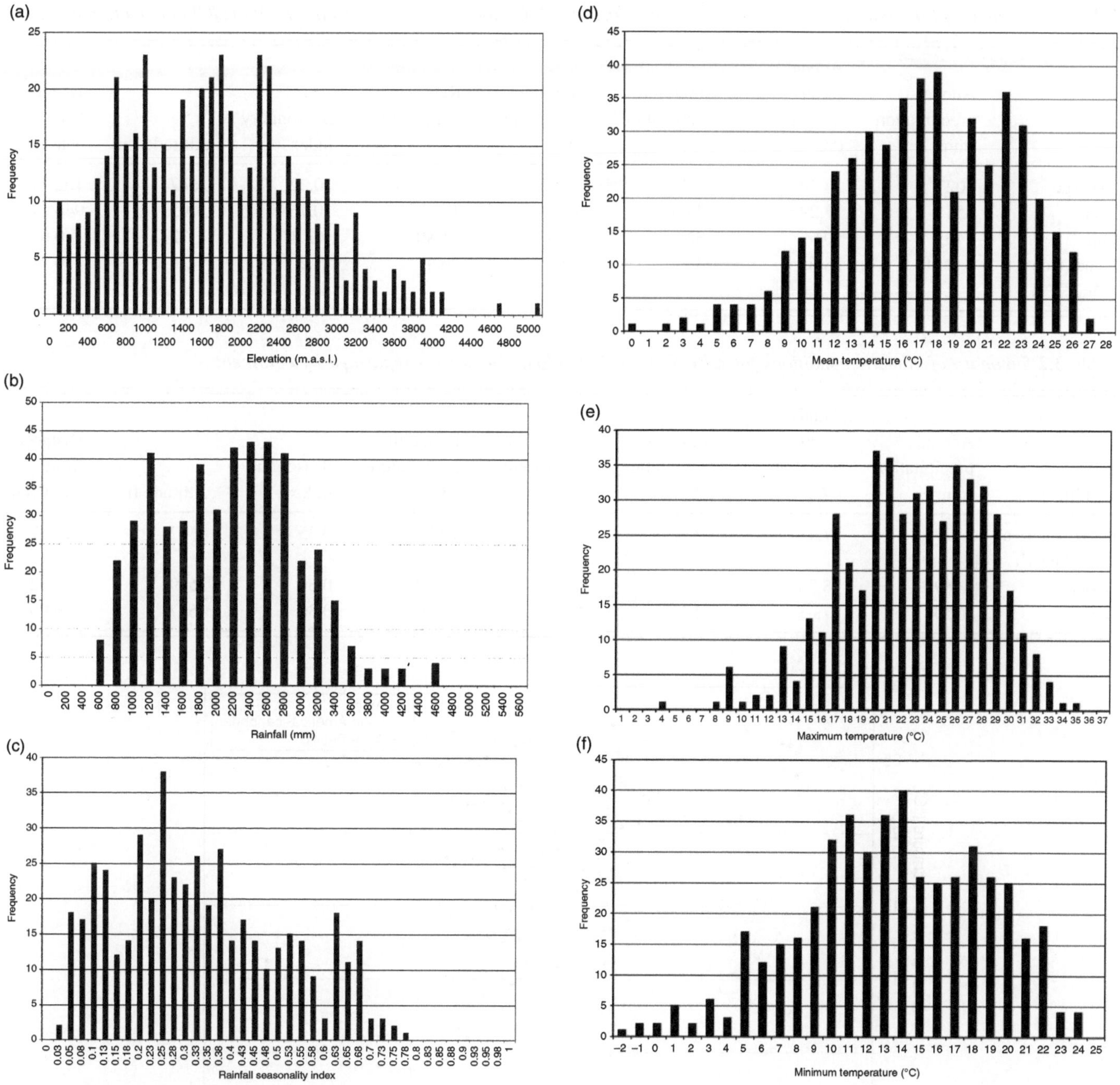

Figure 3.1. Frequency histograms for UNEP–WCMC cloud forest sites for (a) elevation, (b) annual rainfall, (c) rainfall seasonality index, (d) mean, (e) maximum, and (f) minimum temperatures.

characteristics of cloud forests compared with their nearest neighbors. The differences are presented as tabulated averages (Table 3.3), and the frequency distributions for cloud forests alongside those for tropical montane forests in general are shown in Figure 3.3. Annual rainfall for cloud forests is on average 184 mm higher than for other tropical montane forests (2027 mm vs. 1842 mm), with the minimum cloud forest rainfall being *c.* 320 mm higher than for other montane forests. Variations between sites are similar (CVs of *c.* 41% in both cases). Mean temperature is some 4.2 °C lower for cloud forests than for other montane forests. Site-to-site variability in all temperature variables is higher for cloud forests but rainfall seasonality is lower for cloud forests as is T_{range} (although it is more variable between sites).

Cloud forests are negatively skewed toward higher rainfall regimes (even though these are probably underestimated) compared with tropical montane forests in general. Rainfall seasonality for cloud forests is positively skewed (i.e. they are generally

Table 3.1 *Summary of annual climatic variables for all UNEP–WCMC cloud forest sites, using the WorldClim climate database*

	Annual precipitation (mm)	Annual mean temperature (°C)	Maximum temperature (°C)	Minimum temperature (°C)	Range in temperature (°C)	Seasonality index	Altitude (m)	Distance from coast (km)
Average	2027	17.70	22.41	12.95	9.47	0.32	1687	102.71
Max	4541	27.29	34.03	23.64	16.78	0.76	5005	946.71
Min	405	0.93	3.99	−2.07	1.80	0.01	22	0.05
Stand. Coeff. Var. (%)	41.0	28.6	23.2	40.3	26.4	57.2	56.1	120.7

Table 3.2 *Summary of climatic conditions for all UNEP–WCMC cloud forest sites separated by continent*

Continent	Annual precipitation (mm)	Annual mean temperature (°C)	Maximum temperature (°C)	Minimum temperature (°C)	Range in temperature (°C)	Seasonality index	Altitude (m)	Distance from coast (km)
Latin America + Hawai'i	1779	17.48	22.60	12.31	10.29	0.35	1747	121.57
Africa	1489	18.36	23.53	13.13	10.40	0.39	1619	135.22
Asia	2150	17.63	22.23	12.98	9.26	0.30	1687	56.90

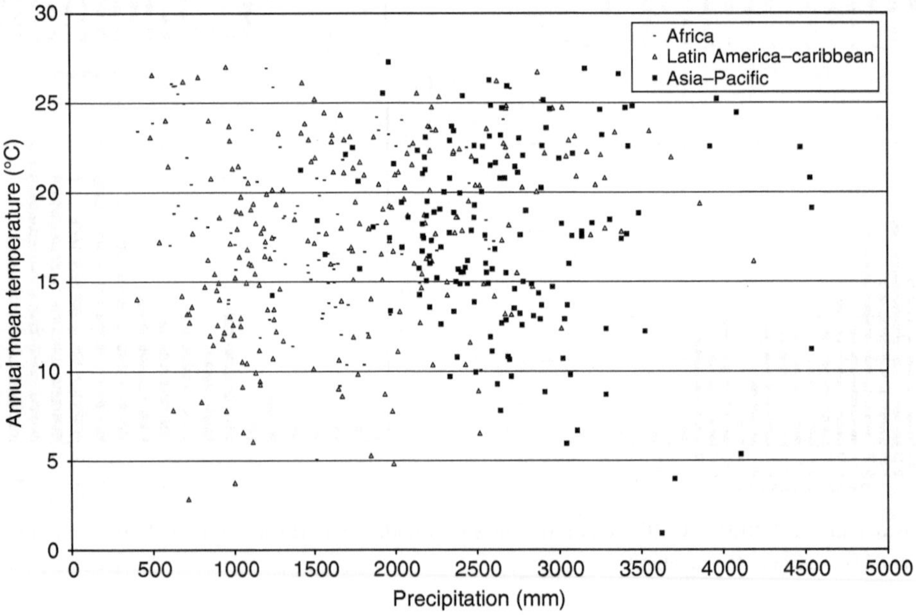

Figure 3.2. Annual climatic conditions at cloud forest sites separated by continent.

less seasonal) compared with tropical montane forests. T_{mean} is significantly negatively skewed for cloud forests compared with tropical montane forests in general (i.e. cloud forests occur in cooler environments) whereas the same patterns exist for T_{max} and T_{min}. Finally, T_{range} is positively skewed for cloud forests compared with all tropical montane forests (i.e. they occur in diurnally less variable environments).

The results of the Kolmogorov–Smirnov (K–S) tests (Table 3.4) indicate that the distributions for cloud forest and all tropical montane forests are all statistically dissimilar (p-values <0.05 in all cases), although some distributions are more different than others. The greatest differences occur in T_{max} (*KS* value 0.49), and T_{mean} (0.47) whereas the distributions are most similar for rainfall seasonality (0.19) and rainfall (0.25).

Table 3.3 *Climate data for cloud forest sites compared with all other tropical* montane *forests. Negative values indicate variable value is less for cloud forest compared with tropical montane forest in general and vice versa*

Cloud forest minus all tropical *montane* forest	T_{mean} (°C)	T_{max} (°C)	T_{min} (°C)	T_{range} (°C)	Annual rainfall	Rainfall seasonality index
Average	−4.2	−5.0	−3.4	−1.6	185	−0.06
Max	−0.7	−1.4	0.9	−2.4	−2316	−0.09
Min	2.6	1.0	4.2	−1.9	320	0.01
Coeff. Var.	15.0	11.6	20.4	4.6	−0.16	−0.58

Topographic and geographic setting of cloud forests

Describing and comparing the topographic and geographic settings of cloud forests may help to explain the climatic patterns observed, but also point to other potentially important factors that interact with these climatic factors to determine the distribution of cloud forests, such as distance to coast, or mountain size and site exposure. A correlation analysis between altitude, mountain size, topographic exposure, and distance to coasts showed low correlations between the variables, the highest correlation coefficient being 0.27 for altitude vs. exposure. The respective variables are therefore sufficiently independent and the climatic effects assigned to each variable in the following analyses are thus the result of that variable and not due to co-linearity with other variables.

ALTITUDE

The identified cloud forest sites had an average altitude of 1687 m.a.s.l. (Table 3.1) compared with 993 m for other montane tropical forests. The altitude of the cloud forests was highly variable between sites ($CV = 56\%$) and they occurred over a much wider range of altitudes compared with all tropical montane forests (Figure 3.4a). The K–S results for this variable indicated a significant difference between the two distributions ($KS = 0.43$; Table 3.4).

DISTANCE FROM COAST

The distance from the nearest coast at which known cloud forests are found was also highly variable (Figure 3.4b), averaging 103 km but ranging from just 50 m (for some island cloud forests) to nearly 950 km for Pico de Neblina in Brazil (Table 3.1). Asian sites are much closer to the coast (on average 57 km), with Latin American and African sites on average being situated over twice that distance from the coast (Table 3.2). This is likely to reflect the characteristics of available land in these regions as much as the tolerances of the cloud forest communities themselves. Cloud forests tend to be much closer to coasts than other tropical montane forests (Figure 3.4b, with average distances of 103 km and 445 km, respectively). Indeed, known cloud forests are almost completely confined to within 350 km from the nearest coast (Figure 3.4b) and occur relatively rarely in continental interiors (<5% of sites). The K–S analysis also shows a very significant difference between the cloud forest and general montane forest distributions ($KS = 0.48$; Table 3.4).

TOPOGRAPHIC EXPOSURE

A topographic exposure analysis indicated that the average exposure for all montane sites was −0.81 (slightly sheltered) vs. 96.3 (exposed) for cloud forest sites. The distributions are clearly different (Figure 3.4c) and a Kolmogrov–Smirnov test indicated that the difference between the two distributions is the most significant of all climate and topographic variables analyzed ($KS = 0.51$, Table 3.4). Since exposure is directionally dependent it would be useful to repeat the analysis for exposure relative to the (idealized) locally dominant wind directions throughout the tropics.

MOUNTAIN SIZE AND THE "MASS-ELEVATION" EFFECT

Average mountain size for the tropical montane forest group was determined at 8.24 million km^2, compared with 3.26 million km^2 for cloud forests. The associated frequency distributions (Figure 3.4d) are dominated by the broken distribution of mountain sizes which correspond to the Andes (the smallest cluster), the African Highlands, and the Himalayas. The K–S test indicated that size of mountain area is also a significant variable in distinguishing cloud forests sites from tropical montane forest sites ($KS = 0.44$, Table 3.4).

An analysis of the relationship between altitude and mountain size, and altitude and distance from coast yielded no significant correlation when all cloud forests sites were used, nor when the analysis was restricted to LAC cloud forest sites (r^2<0.1 in all cases). However, an analysis of the relationship between altitude and the climate variables for stations separated into classes of mountain size showed that the climate vs. altitude relationships differed for the three classes of mountain size distinguished here. Because the analysis might be confounded by the fact that the mountain size classes occupy different continents (which may have different altitude–climate relationships for other reasons than the mountain sizes themselves) the analysis was repeated just for the LAC region ($n = 30\,300$). Mountain size data for the latter conveniently separated into two equal-sized groups: (i) Andean (part of the

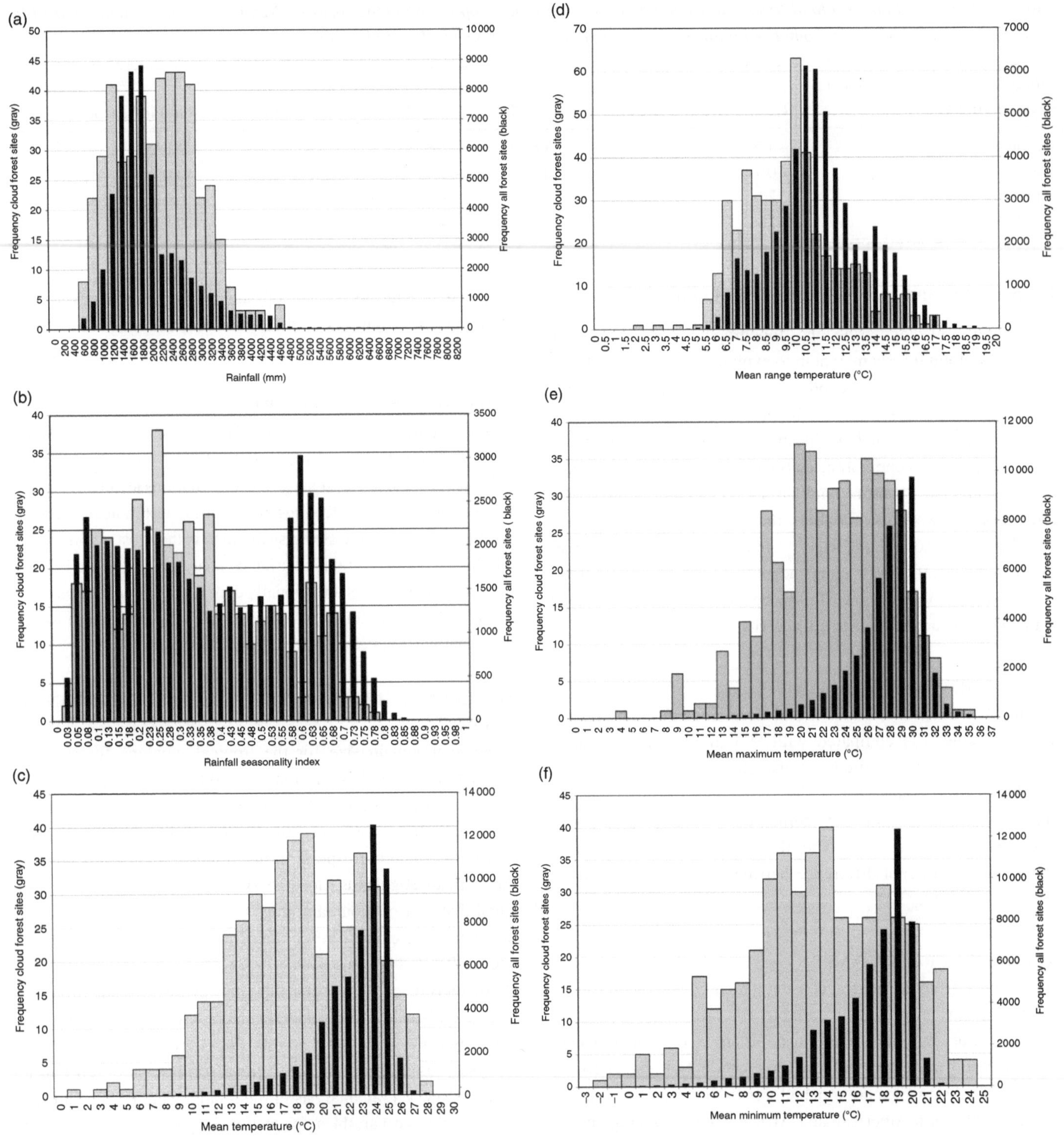

Figure 3.3. Frequency distributions of (a) rainfall, (b) rainfall seasonality index, (c) mean temperature, (d) diurnal temperature range, (e) maximum temperature, and (f) minimum temperature for UNEP–WCMC-listed cloud forest sites compared with tropical montane forest sites in general.

large mountain region that covers the entire Andes, average mountain area 14.6 million km^2) and (ii) non-Andean (smaller, isolated mountain ranges, islands etc., mean mountain area 1.72 million km^2). Altitude was, on average, lower for mountains in the "large" mountain category (899 m compared with 1081 m) and thus T_{mean} was also higher for this group (22.6 °C vs. 21.3 °C), although annual rainfall was lower (1517 mm for the large mountains and 2200 mm for the smaller mountains).

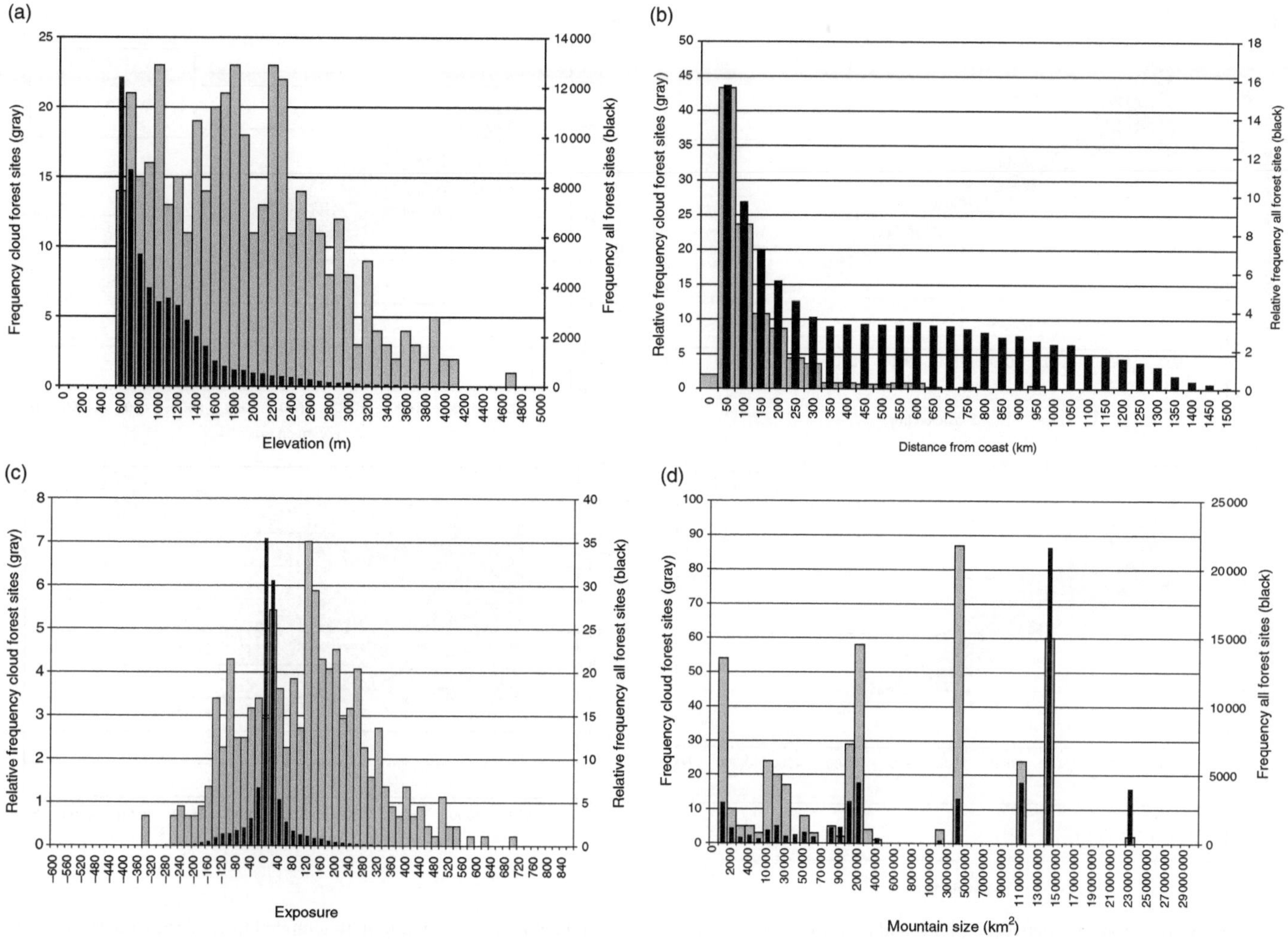

Figure 3.4. Frequency distributions of (a) elevation, (b) distance from coast, (c) topographic exposure, and (d) mountain size for UNEP–WCMC-listed cloud forest sites and tropical montane forest sites in general.

The altitude–climate relationships show some important differences between the two classes of mountain (Figures 3.5 and 3.6). The adiabatic lapse rate for T_{mean} is slightly less (but not statistically significant) for large mountains (0.51 °C 100 m^{-1}) compared with small mountains (0.53 °C 100 m^{-1}), although average T_{mean} at sea level is some 0.5 °C higher in the vicinity of large mountains. At a given altitude, temperature is generally higher for larger mountains and, especially, the difference in maximum or minimum temperature with mountain size increase with altitude (Figure 3.5). As a result, T_{range} also increases much more with altitude for larger mountains than for smaller ones. This conforms with the idea that small mountains have a steeper adiabatic lapse rate due to the "mass-elevation" effect, although the differences are small. Theoretically, this effect might be confounded if smaller mountains tend to occur nearer to coasts where climates tend to be more humid and cloud condensation (and thus cloud forest occurrence) tends to happen at lower elevations than further inland. However, a correlation analysis showed no

Table 3.4 *Results of the Kolmogorov–Smirnov statistical analysis of the distributions of climatic parameters at montane tropical forests* (n = *53 519) and UNEP–WCMC cloud forest sites* (n = *477)*

Variable	Kolmogorov–Smirnov
Rainfall	0.25
Mean temperature	0.47
Max temperature	0.49
Min temperature	0.38
Daily range temperature	0.34
Rainfall seasonality	0.19
Elevation	0.43
Mountain size	0.44
Distance from coast	0.48
Topographic exposure	0.51

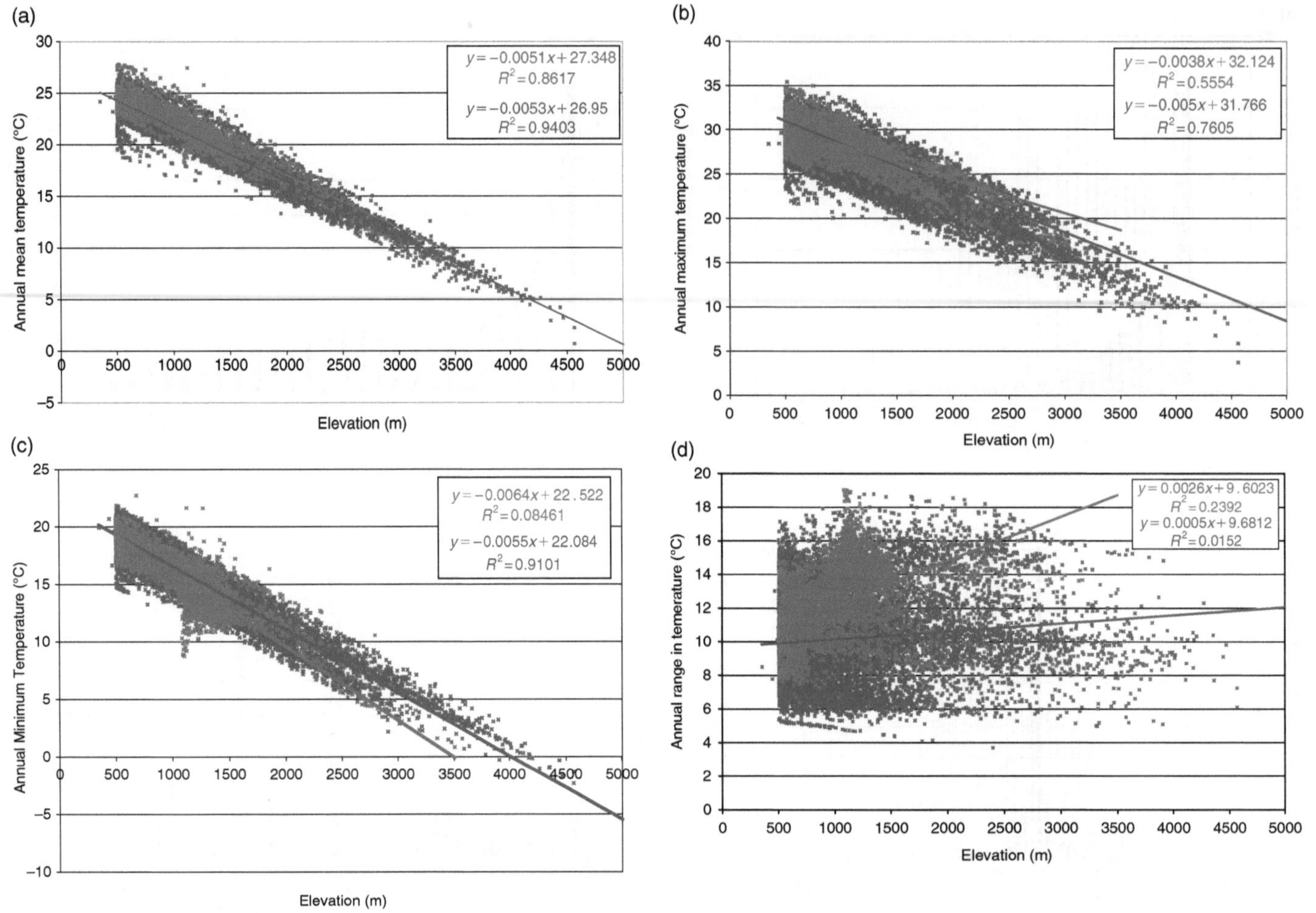

Figure 3.5. Relationships between altitude and (a) mean temperature, (b) maximum temperature, (c) minimum temperature, and (d) temperature range (all in °C) for large-sized (shown in green) and small-sized (shown in blue) mountains in Latin America. (See also color plate.)

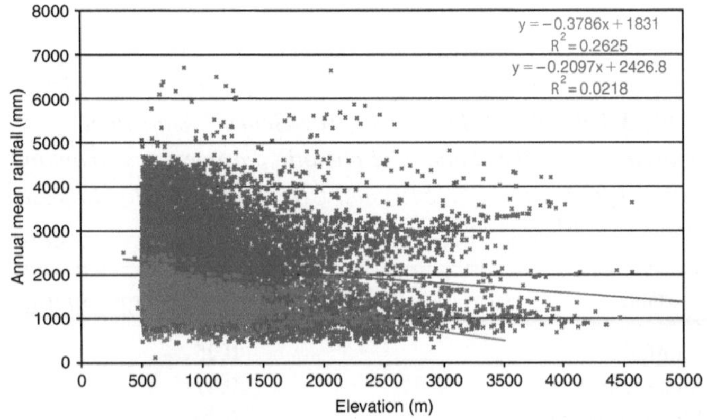

Figure 3.6. Relationships between altitude (m.a.s.l.) and mean annual rainfall (mm) for large-sized (shown in green) and small-sized (shown in blue) mountains in Latin America. (See also color plate.)

significant correlation between mountain size and distance to coast for the global tropical montane data-set ($n = 53\,519$, $r^2 = 0.19$) so this is unlikely. The difference in average lapse rates between large and small mountains suggests that clouds would form some 43 m lower on small mountains than on large mountains (assuming the same base temperature). The average difference of 0.5 °C in base temperatures between small and large mountains has more than double this effect, with clouds forming *c.* 100 m lower over small mountains. Together, the two effects cause cloud formation on average to be about 143 m lower over small mountains compared with larger mountains.

Variations in rainfall with altitude in the LAC cloud forest data were much more diverse than for temperatures (Figure 3.6) and no clear patterns emerged. Large mountains show a decrease in rainfall with altitude, particularly above 1500 m. Small mountain ranges, on the other hand, sometimes show no major change in rainfall with altitude and sometimes increased rainfall since at these smaller scales precipitable water is less likely to be limiting and orographic effects will dominate. Thus, the diversity of patterns reflects the diversity in orographic and meso-climatic settings found in mountainous Latin America and the Caribbean.

CLOUD COVER

Although annual average cloud cover fraction is highly variable at a given altitude (Figure 3.7a), the minimum observed cloud cover at an altitude increases with altitude so that one can find sites with low cloud cover at low altitudes, but much less so at high altitudes. Areas of high and low cloud cover are found at low altitudes but above 2000 m elevation the majority of sites have annual average cloud cover fractions greater than 0.5. There is also a general (but weak) decrease in average cloud cover with increasing distance from the coast. The minimum observed annual average cloud cover increases with distance from coast but the maximum decreases at the same time (Figure 3.7b).

Looking into the cloud-cover data on a frequency basis adds some clarity to these patterns. By aggregating altitude, distance from the sea, and mountain size into classes and then calculating the frequency of sites with annual average cloud cover greater than 50% (0.5 fraction) in each of these classes (Figure 3.8), it becomes clear that from 0 to 2000 m altitude the percentage of sites with cloud cover >50% decreases from 60% to around 40% but then increases dramatically to close to 100% from 2000 m to over 3000 m before falling off again (Figure 3.8a). In terms of distance to coast, 60% of the sites have >50% cloud cover at the coast, increasing to 80% at *c.* 100 km inland and then falling off to 30% at *c.* 500 km inland (Figure 3.8b). Sample sizes beyond 500 km inland were too small for any robust analysis.

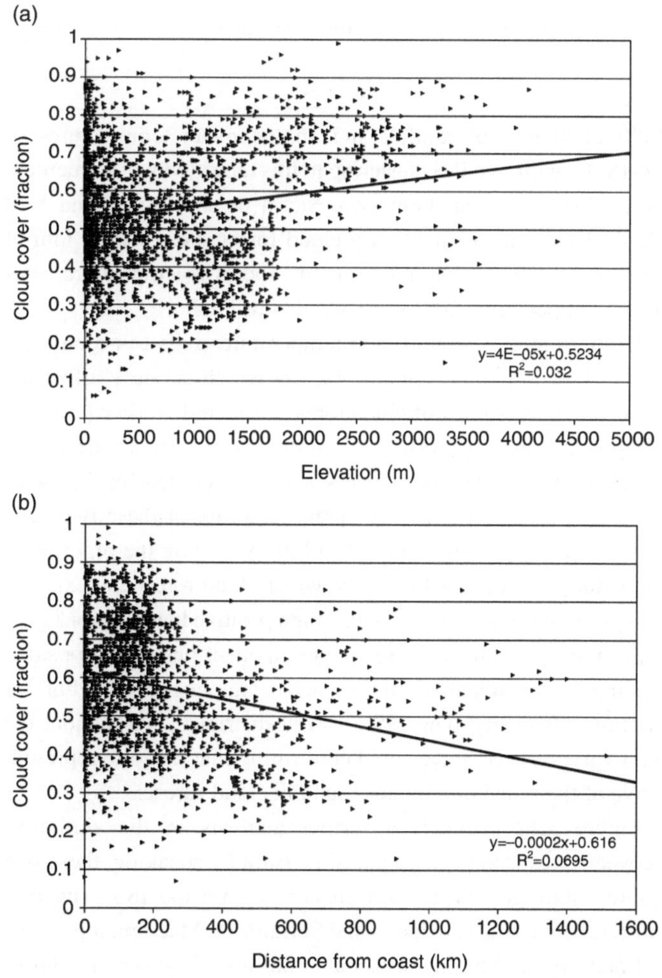

Figure 3.7. Change in cloud cover as a function of (a) altitude (m.a.s.l.) and (b) distance from nearest coast (km) according to the CIAT climatic database (n = 2273 stations).

REPRESENTATIVITY OF INTENSIVELY STUDIED CLOUD FOREST SITES

Intensively studied sites (ISS) with an established body of research were selected based on a review of cloud forest research, principally the studies listed in Hamilton *et al.* (1995), the key research sites selected by Bruijnzeel (2001), and the abstracts for the present volume. In all, 14 sites were selected (see Figure 3.9 for

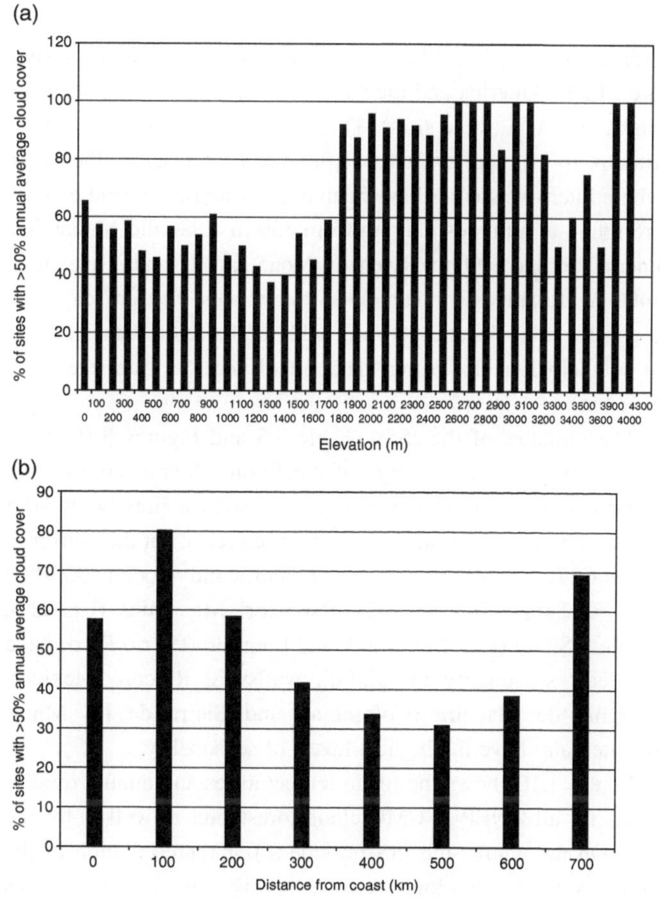

Figure 3.8. Frequency distribution of stations in the CIAT climatic database with high cloud cover vs. (a) elevation (m.a.s.l.) and (b) distance from coast (km).

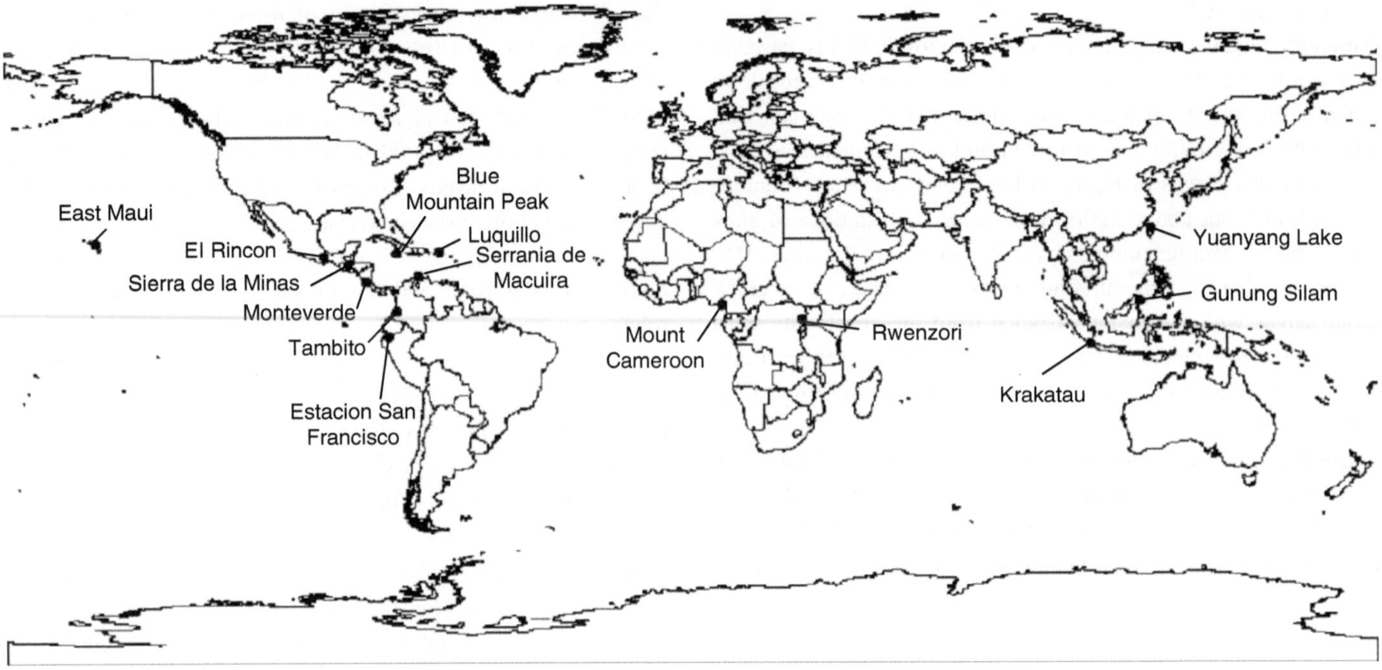

Figure 3.9. Distribution of intensively studied sites (ISS) with an established body of tropical montane cloud forest research.

locations), covering Maui in the Hawai'ian archipelago (one site), Latin America and the Caribbean (eight sites), Africa (two sites), and Asia (three sites). This list is not exhaustive, but does attempt to be representative of the many sites across the globe where intensive studies have been made. Intensively studied sites are found at altitudes ranging from 350 m (Luquillo Mountains, Puerto Rico) to 2700 m (Rwenzori Mountains, Uganda), averaging 1400 m. Relative to other cloud forest sites they are closer to coasts (average distance 41 km), with the Estacion San Francisco in southern Ecuador being the most distant ISS from the coast (141 km).

The climates of the ISSs (Table 3.5 and Figures 3.10–3.14) tend to cover a broad range of conditions. Monteverde (Costa Rica) and Krakatau (Indonesia) are the wettest sites (with 3105 mm and 3180 mm of rain year^{-1}, respectively), but they differ in terms of T_{mean}, Monteverde being significantly cooler (20.9 °C) than Krakatau (24.8 °C). The Rwenzori Mountains (Uganda), Gunung Silam (East Malaysia), and Luquillo (Puerto Rico) have the least seasonality in rainfall, whilst El Rincon (Mexico), Serrania de Macuira (Colombia), and Sierra de las Minas (Guatemala) have the highest levels of seasonality.

Figure 3.10 shows the mean temperatures and annual rainfall totals for all UNEP–WCMC cloud forest sites as well as for the ISS. On the whole, the ISS provide a fair representation of the climates found in cloud forests, evenly covering the ranges in temperature and rainfall (with the proviso that WorldClim rainfall estimates for some of the ISS may be underestimates). The majority of cloud forest sites occur in regions with 2000–2600 mm of rainfall and annual mean temperatures of 14–18 °C, with five ISS clustering in this range (Mount Cameroon, Blue Mountains, San Francisco, Sierra de las Minas, and East Maui). However, relatively dry cloud forest sites (<1000 mm of rain year^{-1}) are under-represented, and low-temperature sites (mean temperatures 10–13 °C) are also lacking.

Examining only rainfall and temperatures fails to account for many other important climatic factors which strongly affect the biology and hydrology of cloud forests. The multivariate principal components analysis performed on the seven variables described earlier provided two components which accounted for 72.5% of the total variance. Figure 3.11 presents a plot of these two variables, which provides greater detail in examining the representativity and similarity of ISS in the wider cloud forest context. The first principal component contains high positive loadings for T_{mean}, T_{max}, and T_{min}, whilst the second component has a high positive loading for mean annual rainfall, and a high negative loading for rainfall seasonality. The ISS are well distributed throughout the multivariate space, once again confirming that they are representative of the range of climates found in cloud forests.

Figure 3.12 provides a further analysis of the similarity between sites using cluster analysis. Broadly speaking, these two figures highlight three main clusters within the ISS. The first cluster, containing El Rincon and Serrania de Macuira, represents hot and dry, and highly seasonal conditions. The second cluster (Krakatau, Gunun Silam, Monteverde, East Maui, Luquillo, San Francisco, and Rwenzori) has average–high rainfall, relatively warm climates, and very low seasonality. The third and final

Table 3.5 *Summary of annual climatic variables for intensively studied sites, as derived from the WorldClim climate database*

Intensively studied site (ISS)	Lat	Long	Annual precipitation (mm)	Annual mean temperature (°C)	Maximum temperature (°C)	Minimum temperature (°C)	Range in temperature (°C)	Seasonality index	Altitude (m)	Topographic exposure index	Distance from coast (km)
Rwenzori Mountains, Uganda	0.250	29.950	1626	13.58	19.13	7.96	11.17	0.05	2703	−172.54	23.90
Mount Cameroon, Cameroon	4.117	9.167	2601	17.28	20.57	13.92	6.65	0.38	1449	MISSING	12.23
Gunung Silam, Malaysia	4.967	118.167	2350	23.69	27.33	20.01	7.32	0.10	611	MISSING	3.30
Blue Mountain Peak, Jamaica	18.042	−76.578	2611	14.19	17.05	11.28	5.77	0.22	2126	471.15	14.32
Munchique/ Tambito, Colombia	2.550	−76.950	2840	14.73	19.88	9.53	10.36	0.24	2108	157.31	80.41
Monteverde Cloud Forest Reserve, Costa Rica	10.283	−84.800	3107	20.89	25.73	16.01	9.73	0.36	1004	−89.08	28.24
Estacion San Francisco, Ecuador	−3.972	−79.079	2368	17.76	22.96	12.50	10.46	0.19	2048	−33.69	141.73
Serrania de Macuira, Colombia	12.171	−71.388	584	24.03	27.67	20.34	7.33	0.38	806	111.77	16.83
Sierra de las Minas Biosphere Reserve, Guatemala	15.150	−89.833	2321	14.26	18.73	9.73	8.99	0.42	2165	160.85	58.30
Luquillo, Puerto Rica	18.317	−65.750	2263	24.49	28.73	20.21	8.52	0.17	351	MISSING	6.48
East Maui, Hawai'i	20.800	−156.290	2581	17.76	21.54	13.94	7.60	0.23	1128	73.77	13.75
Yuanyang Lake, Taiwan	24.583	121.400	2774	12.73	15.31	10.13	5.18	0.28	1862	211.08	41.38
El Rincon, Sierra Norte, Oaxaca, Mexico	17.375	−96.333	1540	18.37	24.52	12.17	12.35	0.53	1309	−202.46	130.32
Krakatau, Indonesia	−6.146	105.438	3179	24.83	29.46	20.14	9.32	0.21	436	MISSING	0.70

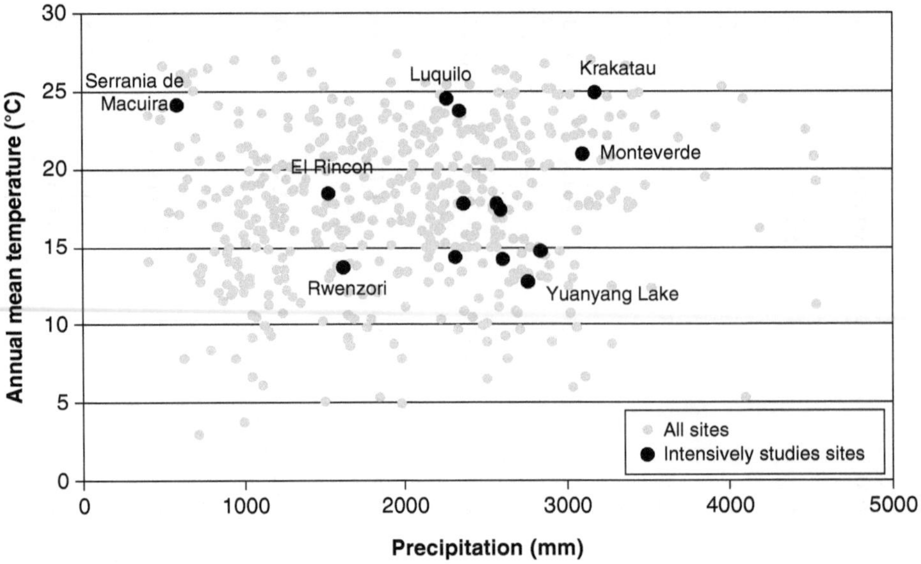

Figure 3.10. Annual rainfall (mm) vs. mean temperature (°C) at all UNEP–WCMC-listed cloud forest sites, with intensively studied sites highlighted.

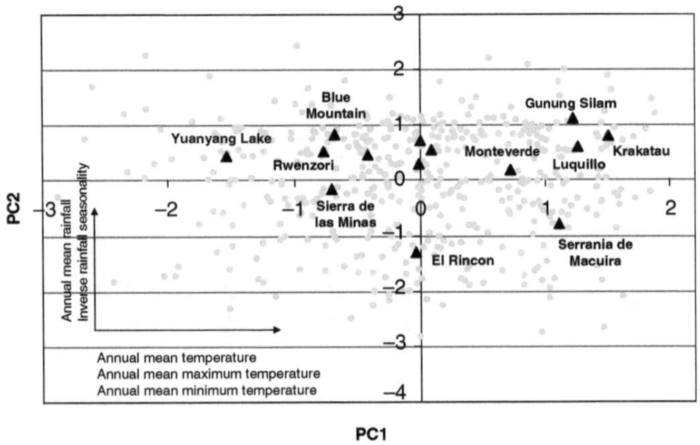

Figure 3.11. Principal component analysis plot of the first two components, placing intensively studied sites (black triangles) in the wider context of all UNEP–WCMC-listed cloud forest sites (gray circles). The labels in the bottom-left corner indicate the approximate significance of each component based on the variables that received the highest loadings.

cluster (Yuanyang Lake, Tambito, Blue Mountains, Sierra de las Minas, and Mount Cameroon) is characterized by lower temperatures, intermediate levels of seasonality, relatively high rainfall, and low diurnal ranges in temperature. The Hawai'ian site in East Maui is perhaps the most atypical cloud forest site (at least in terms of the variables examined here) whereas the Serrania de Macuira stands out as an extreme site with very low rainfall (600 mm) and high mean temperature (24.0 °C).

DISCUSSION AND CONCLUSIONS

Cloud forests in the UNEP–WCMC database show a preferential distribution toward northern hemisphere latitudes close to the equator and also cluster longitudinally. Both clusterings occur irrespective of any bias that might be introduced due to the uneven distribution of tropical mountains. The climatic conditions leading to the generation and maintenance of cloud forests must be prevalent in these areas.

It has been shown here through coupling of a series of global data-sets that 85% of the 526 known cloud forest sites are found at altitudes between 400 and 2800 m.a.s.l., with an average altitude of slightly less than 1700 m. The range of altitudes at which cloud forests are found is impressive (220–5005 m). The climate of cloud forests is highly variable from forest to forest, with an average rainfall of *c.* 2000 mm year^{-1} and an average temperature of 17.7 °C. In addition, cloud forests are found in seasonal and in aseasonal environments alike, both in terms of rainfall and temperature. There are some clear differences in the climates of cloud forests found in Africa, Latin America and the Caribbean, and Asia.

Compared with other tropical montane forests, cloud forests are wetter (by 184 mm year^{-1} on average) and cooler (by 4.2 °C on average) whereas individual cloud forests are also more climatically varied than tropical montane forests in general. At the same time, cloud forests are less seasonally variable in terms of rainfall than other montane forests. The most statistically significant climatic differences between cloud forests and other montane forests in order of significance are: maximum temperature > mean temperature > rainfall > rainfall seasonality.

Cloud forests are almost completely confined to within 350 km from the nearest coast, and are located closer to coasts than montane forests in general. Cloud forests also occupy more topographically exposed areas and differently sized mountains than montane forests in general. The order of significance for geographic differences between cloud forests and montane forests in general is: topographic exposure > distance from coast

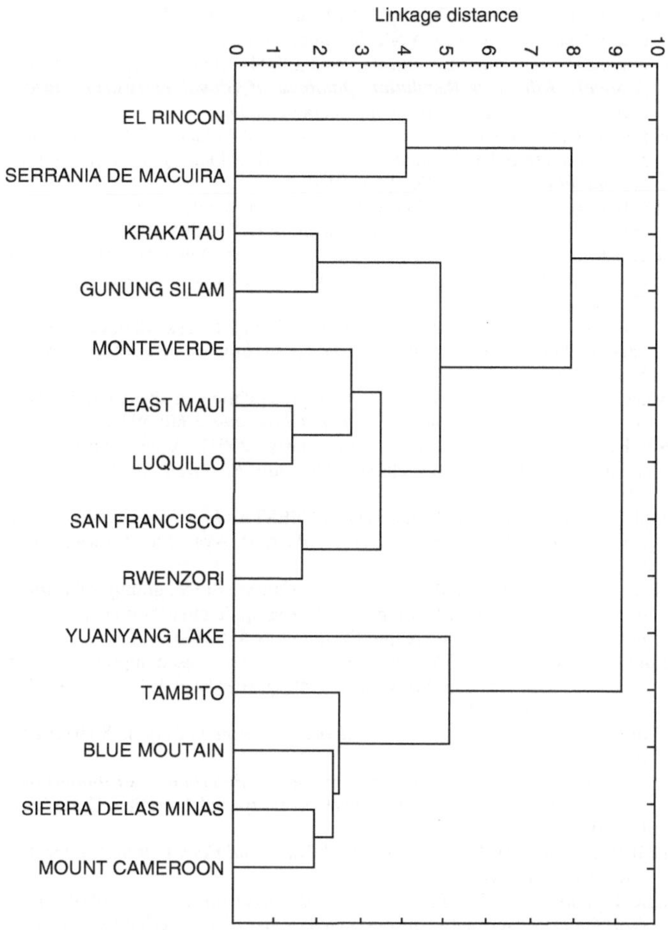

Figure 3.12. Multivariate cluster analysis (using Ward's method) of climatic conditions at intensively studied sites, using data from the World Clim database.

> mountain size > altitude, and these variables show little to no co-linearity. Overall, the order of statistical difference in geographic and climatic variables for cloud forests compared with montane forests in general as expressed by the Kolmogorov–Smirnov test is: topographic exposure (0.51) > distance from coast (0.49) > maximum temperature (0.49) > mean annual temperature (0.47) > mountain size (0.44) > altitude (0.43) >> rainfall (0.25) > rainfall seasonality (0.19).

Thus, cloud forests tend to occur in fairly coastal climates and environments with lower maximum and mean temperatures and higher altitudes compared with other montane forests. Where this is not the case, cloud forests occur in areas with higher rainfall coupled with lower rainfall seasonality than other montane forests. This is no paradigm shift compared to earlier qualitative descriptions (Stadtmüller, 1987; Hamilton *et al.*, 1995; Bruijnzeel and Veneklaas, 1998), but it is important that these trends are observed in the global data-sets used in the present analysis. The fact that topographic exposure is an important characteristic defining cloud forests is indicative that other climatic variables may be important (particularly wind speed), and further macro-scale analyses of cloud forest climates are required to better understand these relationships.

Analysis of the WorldClim data has shown that temperature at a given altitude is higher for larger mountains, which is in agreement with the "mass-elevation" effect, with smaller mountains having a steeper adiabatic lapse rate and thus a potentially lower cloud condensation level, leading in turn to a lowered occurrence of cloud forest. The presently used method of calculation of mountain size excluded mountains less than 500 m high (by definition) and this may have had implications for the observed lack of relationship between mountain size and distance from coast, as well as the frequency distribution of mountain sizes for tropical montane (cloud) forests. Large mountains also have a much greater diurnal range in temperature (i.e. lower minimum and higher maximum temperatures) at a given altitude, which probably results from the climatic effects of the mountain on local air masses. The effect is not small: the altitude at which a maximum temperature of 20 °C is reached is some 800 m lower for the smaller mountains in the LAC region than for the larger ones (see Figure 3.5).

High cloud cover (>60%) is frequent near sea level and above 2000 m.a.s.l. and within 100 km from the coast and this is likely to be one of the most significant climatic variables for cloud forests and the reason why altitude, temperature, and distance from coast are important to cloud forest distribution. Furthermore, the fact that cloud forests are shown here to occur preferentially at topographically exposed sites indicates that exposure to the climate is critical. Exposed sites where ground-level cloud is frequent are likely to receive additional water inputs from this source while evaporative losses are reduced (Bruijnzeel and Proctor, 1995; Bruijnzeel, 2001).

Whilst the data used here have been useful in demonstrating broad patterns, some important improvements are necessary to allow more regional analyses to be made and enhanced robustness of some of the results presented here, especially at the extremes of the climatic distributions. First and foremost, a database of cloud forest sites is needed that incorporates more of the actual cloud forests and includes more precise information on their extent as well as location and type (Mulligan, this volume; Bruijnzeel *et al.*, this volume). This could be achieved by coupling satellite-derived cloud-base altitudes with remotely sensed data on forest cover (cf. Nair *et al.*, 2008; Lawton *et al.*, this volume). Moreover, improvements in the incorporation of small-scale rainfall variability in the WorldClim database would be of use, as would be the availability of further climate parameters that may be of importance for cloud forest climates (notably wind speed and direction, fog inputs, wind-driven rain, and solar radiation). At the national to regional scale, incorporation of these variables has proved useful in identifying water budget "hot spots" in montane tropical areas, and in identifying priority areas for conservation (Mulligan and Burke, 2005).

ACKNOWLEDGEMENTS

The authors wish to thank Philip Bubb at UNEP–WCMC for providing access to the cloud forests site database, and for additional information as to the process used in the generation of this database. We also thank Robert Hijmans, the creator of the WorldClim data-set, for access to the raw station data. Further thanks go to Otoniel Madrid for providing some of the statistical analyses and to Sampurno Bruijnzeel whose editing greatly improved the chapter.

REFERENCES

Aldrich, M., C. Billington, M. Edwards, and R. Laidlaw (1997). *A Global Directory of Tropical Montane Cloud Forests*. Cambridge, UK: UNEP–WCMC.

Bruijnzeel, L. A. (2001). Hydrology of tropical montane cloud forests: a reassessment. *Land Use and Water Resources Research* **1**: 1.1–1.18.

Bruijnzeel, L. A., and L. S. Hamilton (2000). *Decision Time for Cloud Forests*, IHP Humid Tropics Programme Series No. 13. Paris: UNESCO, Amsterdam: IUCN-NL, and Gland, Switzerland: IUCN.

Bruijnzeel, L. A., and J. Proctor (1995). Hydrology and biochemistry of tropical montane cloud forests: what do we really know? In *Tropical Montane Cloud Forests*, eds. L. S. Hamilton, J. O. Juvik, and F. N. Scatena, pp. 38–78. New York: Springer-Verlag.

Bruijnzeel, L. A., and E. J. Veneklaas (1998). Climatic conditions and tropical montane forest productivity: the fog has not lifted yet. *Ecology* **29**: 3–9.

Bruijnzeel, L. A., M. J. Waterloo, J. Proctor, A. T. Kuiters, and B. Kotterink (1993). Hydrological observations in montane rain forests on Gunung Silam, Sabah, Malaysia, with special reference to the '*Massenerhebung*' effect. *Journal of Ecology* **81**: 145–167.

FAO (2005). Terms and definitions for the national reporting tables for FRA 2005. Rome: FAO. Also available at www.fao.org/documents/show_cdr.asp?url_file=//docrep/007/ae156e/AE156E03.htm.

Frahm, J. P., and S. R. Gradstein (1991). An altitudinal zonation of tropical rain forests using bryophytes. *Journal of Biogeography* **18**: 669–678.

Grubb, P. J. (1977). Control of forest growth and distribution on wet tropical mountains: with special reference to mineral nutrition. *Annual Review of Ecology and Systematics* **8**: 83–107.

Hamilton, L. S., J. O. Juvik, and F. N. Scatena (eds.) (1995). *Tropical Montane Cloud Forests*. New York: Springer-Verlag.

Hijmans, R., S. Cameron, and J. Parra (2004). *WorldClim*, Version 1.2: *A Square Kilometer Resolution Database of Global Terrestrial Surface Climate*. Available at http://biogeo.berkeley.edu.

Hijmans, R., S. Cameron, J. Parra, P. J. Jones, and A. Jarvis (2005). Very high resolution interpolated climate surfaces for global land areas. *International Journal of Climatology* **25**: 1965–1978.

Hutchinson, M. (1995). Interpolating mean rainfall using thin-plate smoothing splines. *International Journal of GIS* **9**: 305–403.

Jones, P. G. (1991). *The CIAT Climate Database*, Version 3.41, *Machine Readable Dataset*. Cali, Colombia: Centro Internacional de Agricultura Tropical (CIAT).

La Bastille, A., and D. Poole (1978). On the need for a system of cloud forest parks in Middle America and the Caribbean. *Environmental Conservation* **5**: 183–190.

Markham, C. G. (1970). Seasonality of precipitation in the United States. *Annals of the Association of American Geographers* **60**: 593–597.

Meybeck, M., P. Green, and C. Vörösmarty (2001). A new typology for mountains and other relief classes. *Mountain Research and Development* **21**: 34–45.

Mulligan, M., and S. M. Burke (2005). FIESTA: *Fog Interception for the Enhancement of Streamflow in Tropical Areas*. Available at www.ambiotek.com/fiesta/.

Nair, U. S., S. Asefi, R. M. Welch, *et al.* (2008). Biogeography of tropical montane cloud forests. II. Mapping of orographic cloud immersion. *Journal of Applied Meteorology and Climatology* **47**: 2183–2197.

Ruel, J., S. Mitchell, and M. Dornier (2002). A GIS-based approach to map wind exposure for windthrow hazard rating. *Northern Journal of Applied Forestry* **19**: 183–187.

Schröter, Ch. (1926). *Das Pflanzenleben der Alpen*. Zurich, Switzerland: Albert Raustein Publishers.

Stadtmüller, T. (1987). *Cloud Forests in the Humid Tropics: A Bibliographic Review*. Tokyo: The United Nations University, and Turrialba, Costa Rica: CATIE.

UNEP–WCMC (2004). *Database of Global Cloud Forest Data*. Cambridge, UK: UNEP–WCMC.

USGS (2000). *Global Forest Resources Assessment (FRA2000): 1-km Gridded Data*. Available at http://edcdaac.usgs.gov/glcc/fao/index.asp.

USGS (2004). *Shuttle Radar Topography Mission (SRTM) 30 Arc Second Gridded Global Elevation Data*. Available at http://srtm.usgs.gov/.

Van der Molen, M. K., A. J. Dolman, M. J. Waterloo, and L. A. Bruijnzeel (2006). Climate is affected more by maritime than by continental land use change: a multiple scale analysis. *Global and Planetary Change* **54**: 128–149.

Van Steenis, C. G. G. J. (1972). *The Mountain Flora of Java*. Leiden, the Netherlands: E.J. Brill.

Zimmermann, N. (2004). *The Toposcale and Topoclass AML*. Available at www.wsl.ch/staff/niklaus.zimmermann/programs/aml4_1.html.

4 Changes in mist immersion

P. Foster
University of Bristol, Bristol, UK

ABSTRACT

This chapter reviews recent research on changes in cloud formation at cloud forest and other sites. These changes are discussed in the context of the global hydrological cycle, global climate change, and tropical deforestation. After a simplified review of the basic theory on (changes in) cloud formation, a simple equation is derived governing the expected changes in the altitude of the lifting condensation level associated with given changes in humidity and temperature of the air. Using the ECHAM4 Global Circulation Model, predictions are made about changes in cloud base height for the next century. These prove to be consistent with an intensification of the hydrological cycle.

INTRODUCTION

How cloud characteristics will change in the coming centuries is of critical importance to cloud forests (Foster, 2001). Recent literature has concentrated mostly on the changes in the altitude of the cloud base, primarily because it is easily quantifiable in models and observation (cf. Lawton *et al.*, this volume). This chapter will continue that trend but notes that changes in the frequency of cloud formation and/or the water content of clouds (Hildebrandt and Eltahir, 2008) may be of equal or greater consequence than changes in the altitude of formation per se. Emphasis here will be on changes driven by increases in atmospheric CO_2 although it is recognized that aerosols, because of their key role in cloud formation and circulation patterns (Liepert *et al.*, 2004), should also be included in the prediction of future cloud formation regimes.

After discussing how clouds form and defining terms, a simple equation is derived for estimating the altitude of cloud formation and how this altitude may change in response to changes in surface land cover and/or temperature. Next, recent research results concerning observed or modeled changes in the hydrological cycle are reviewed in relation to global warming. The final section addresses research on shifts in cloudiness in cloud forest areas, including both observations and modeling, and looks in detail at the respective roles of deforestation and global warming.

THEORY: ESTIMATING THE LIFTING CONDENSATION LEVEL

One might expect condensation to occur when the vapor pressure of the air (e) equals the saturation vapor pressure (e_s), i.e. when relative humidity RH is 100%. However, for water vapor to condense into a droplet (rather than onto a plane) a much higher humidity is needed to make the transition energetically favorable. For an embryonic cloud water droplet of radius 0.01 μm to form, the theoretical critical RH is ~112% whereas RH in clouds rarely exceeds ~101% (Wallace and Hobbs, 1977). In practice, vapor condenses onto the surface of aerosols with radii much greater than a water molecule, thus decreasing RH_{crit} to realizable values. Increased aerosol concentrations, therefore, have a significant effect on cloud formation (Lohmann and Feichter, 2005).

Microphysics aside, low-altitude clouds tend to form when large-scale RH exceeds 80% (Slingo, 1987). Some general circulation models (GCMs) use this threshold to diagnose stratiform cloud presence which typically arises from large-scale lifting and diabatic cooling of air (e.g. the European Centre for Medium Range Weather Forecast model; Tiedtke, 1993). These models also follow the evolution of cumulus clouds, boundary layer clouds, and sometimes shallow cumulus clouds with various parameterizations to account for updraft, precipitation, mixing, and, in some models, aerosols (Lopez, 2007). Many of these cloud formation processes are therefore parameterized in GCMs, as their grid cells usually exceed 50 km (Trenberth *et al.*, 2007). Regional climate models, or meso-scale models, with grid

Tropical Montane Cloud Forests: Science for Conservation and Management, eds. L. A. Bruijnzeel, F. N. Scatena, and L. S. Hamilton. Published by Cambridge University Press.

cells less than 2–3 km, are able to explicitly model cumulus convection (Molinari and Dudek, 1992), although they still use parameterizations for other small-scale processes.

In this chapter, two cloud formation mechanisms relevant to cloud forests are considered: orographic uplift and convection. In each case, cooling via uplift reduces air temperatures, which in turn reduces e_s and increases RH, and thus triggers condensation. In the orographic case, winds encounter mountains and are forced upslope. The steadiness of the trade winds can then result in cloud formation at roughly the same altitude on a given mountain's slope every day. In the case of convective cloud formation, hot surfaces heat the air above them inducing uplift. In both cases, one can estimate the altitude at which condensation will occur by considering an unsaturated air parcel that is rising adiabatically through the atmosphere, i.e. without phase changes or energy exchanges to the surroundings. This rising parcel, with an initial temperature of T_0, will cool according to the dry adiabatic lapse rate $\Gamma_{dry} = -dT/dz = 9.8^\circ\ km^{-1}$ (Wallace and Hobbs, 1977, cf. the bold line labeled Γ_{dry} in Figure 4.1a). As the parcel cools, it will eventually reach a temperature at which its water vapor will condense to form liquid. This condensation altitude is estimated to be the level at which $RH = 100\%$ and is referred to as the lifting condensation level or LCL. Even though the caveats about energy considerations, discussed above, are ignored in the calculation of the LCL, observations show that the LCL predicts the observed altitudes of cumulus cloud formation (Craven *et al.*, 2002).

The LCL may thus be computed using thermodynamic relationships that define lapse rates (temperature change with altitude) and the dew point temperature of a parcel undergoing adiabatic ascent. Dew point temperature, T_{dew}, is the temperature to which a parcel must be cooled, isobarically, to reach saturation. Upon adiabatic uplift, T_{dew} decreases in response to atmospheric pressure drop as $\Gamma_{dew} = -dT_{dew}/dz$. Saturation will then be reached when $T = T_{dew}$, or graphically, when the dry adiabat intersects the Γ_{dew} line (i.e. the dashed line in Figure 4.1a). If the dew point lapse rate is assumed to be constant, then the LCL of an air parcel with temperature and dew point temperature at the surface of T_0 and $T_{dew,0}$ is (Bohren and Albrecht, 1998):

$$\mathrm{LCL} = (T_0 - T_{dew,0})/(\Gamma_{dry} - \Gamma_{dew}). \tag{4.1}$$

It should be noted that this expression relies on the assumption of a constant dew point lapse rate, which may not be accurate for all atmospheric thermodynamic profiles. This is unlikely to be significant, however, when analyzing temporal trends for given sites as done in this paper, but may be important when comparisons are made between sites.

If moisture-laden winds, forced upwards by a mountain, reach the altitude of the LCL, an orographic condensation will occur. However, in the case of convection, simply reaching the altitude of cloud formation will not ensure a cloud being formed because sustaining convective cloudiness also depends upon the stability of the atmosphere (Wallace and Hobbs, 1977). If the temperature profile of the atmosphere follows a curve like the thin line in Figure 4.1b (labeled Γ for the atmospheric or environmental temperature profile), it can be seen that a parcel that is perturbed upward from T_0 (and cools along the bold line, Γ_{dry}) will be cooler than the air around it at the LCL. Therefore, it will be denser and will sink back to where it started from. In other words, the atmosphere is stable, convection does not occur and clouds will not form. On the other hand, if the parcel receives a big enough upward push via turbulence or differential heating for the parcel to rise above the LCL, the parcel will then condense and subsequently follow the saturated adiabat, Γ_{sat}, the second bold line in Figure 4.1b. Most atmospheric soundings have $\Gamma_{sat} < \Gamma < \Gamma_{dry}$, such that at some altitude the atmospheric temperature profile will intersect the saturated adiabat. Above this intersection, the parcel will be warmer than its surroundings, and therefore less dense and positively buoyant. It will now move upward of its own accord, i.e. convection will be initiated (represented by the expanding bubbles in Figure 4.1b). The altitude at which the environmental lapse rate crosses over the saturated adiabat is referred to as the level of free convection, or the LFC. In order for convection to be sustained, air parcels must receive a large enough upward boost to reach this level.

Later, an estimate of Γ_{dew} will be needed, which can be from the Clausius–Clapeyron equation (Wallace and Hobbs, 1977) by substituting $T = T_{dew}$ and $e = e_s$ and rearranging:

$$1/T_{dew} = [1/T_{273} - R_v/L \ln e/e_{273}] \tag{4.2}$$

where R_v is the gas constant for water vapor, L the latent heat of vaporization, T_{273} a reference temperature of 273 K, and e_{273} the saturation vapor pressure at this temperature (or 6.11 mb). Taking the partial derivative of this equation with respect to pressure p, i.e. substituting $de/e = dp/p$ and applying hydrostatic equilibrium $1/p * dp/dz = -g/R_dT$ (where g is the gravitational constant and R_d is the gas constant for dry air and equal to 0.622 R_v), gives:

$$T_{dew} = -dT_{dew}/dz = T_{dew}{}^2/T^* g/(0.622^* L). \tag{4.3}$$

The impact on the LCL of changes in either the temperature or the moisture content of the surface air can be estimated by taking the derivative of Eq. (4.1):

$$\Delta\mathrm{LCL} = \Delta T_0/(\Gamma_{dry} - \Gamma_{dew}) - \Delta T_{dew,0}/(\Gamma_{dry} - \Gamma_{dew}) \tag{4.4a}$$

$$= \mathrm{LCL}(\Delta T_0 - \Delta T_{dew,0})/(T_0 - T_{dew,0}) \tag{4.4b}$$

where changes in Γ_{dew} have been ignored because they are small compared to the terms given above. In this formulation, T_0 and $T_{dew,0}$ are the base, or unperturbed temperatures, and ΔT_0 and $\Delta T_{dew,0}$ represent the climatic changes.

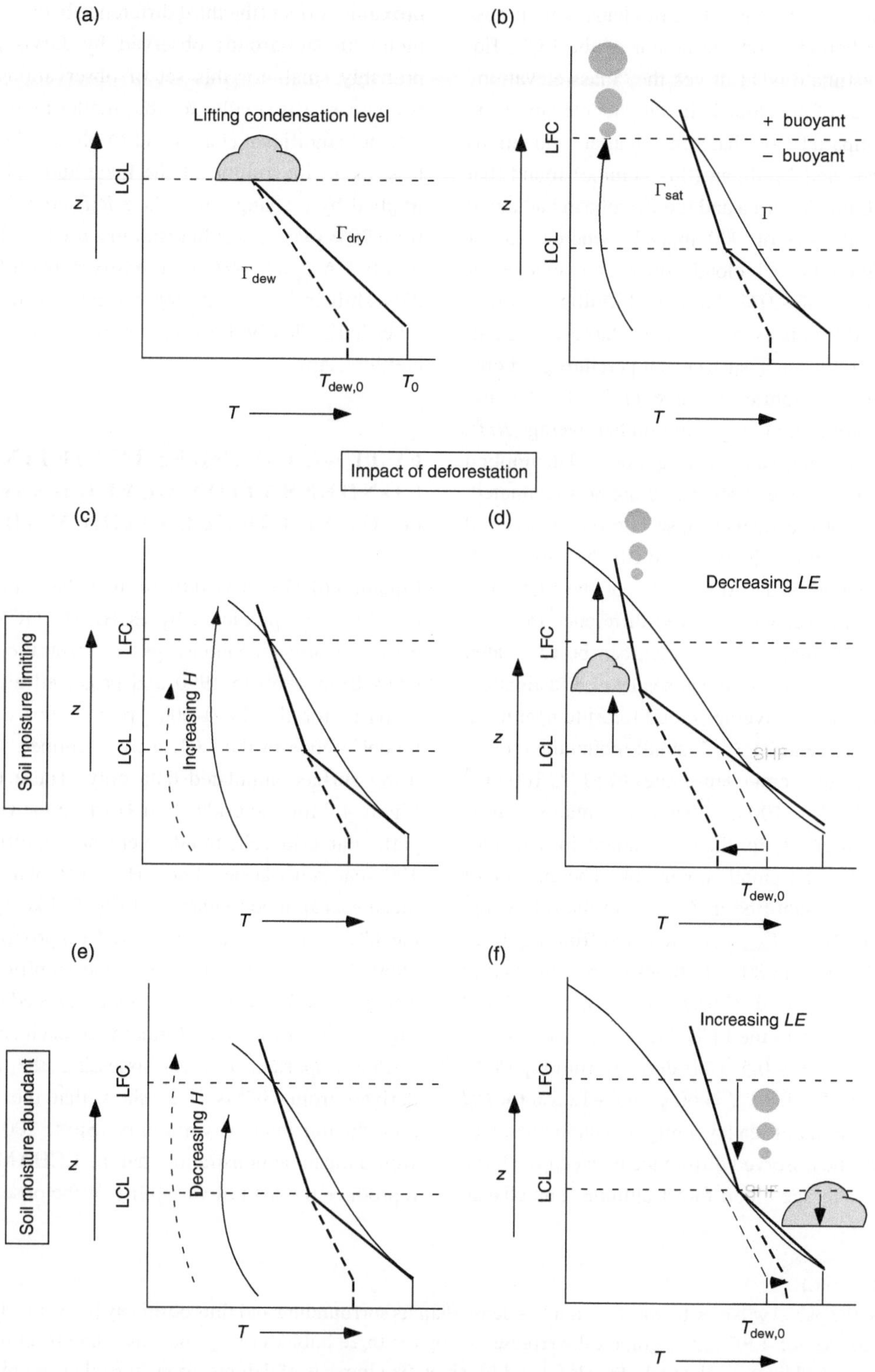

Figure 4.1. In panel (a) the change in temperature, Γ_{dry}, of an unsaturated parcel of gas being lifted through the atmosphere is shown as the bold line from its initial altitude and temperature, T_0, up to the lifting condensation level, LCL. The altitude of the LCL is determined by the intersection of Γ_{dry} with the line starting from $T_{dew,0}$ and ascending with Γ_{dew} to the point at which $T = T_{dew}$. After the rising parcel of gas reaches the LCL, it becomes saturated – represented by the gray cloud with a lower edge at the LCL. If the parcel of gas continues to rise, it will cool along the saturated adiabat, Γ_{sat} as shown in panel (b) – both adiabats are shown as bold lines in all panels. Also shown is the atmospheric lapse rate as the light solid line; this represents the actual temperature profile of the air at a theoretical location. In the tropics, environmental air is typically warmer than an adiabatically lifted parcel of gas in the vicinity of the LCL, meaning that the perturbed parcel will be denser than its surroundings and will settle back to its initial position, i.e. the atmosphere is stable or negatively buoyant. If the parcel receives a large enough perturbation that it rises to an altitude at

As a preliminary to assessing the effect of changes in atmospheric temperature or humidity on the height of the LCL, Eq. (4.4) can be used to estimate what drives the "mass-elevation" effect, i.e. the occurrence of the cloud belt at higher elevations on large mountains compared to that on smaller mountains (Richards, 1996). Jarvis and Mulligan (this volume) found that smaller mountains in Latin America and the Caribbean had cloud forest at an average altitude of 900 m.a.s.l., whereas larger mountains in the region hosted cloud forests on average at 1080 m.a.s.l., a difference of ~21%. Jarvis and Mulligan further noted three observed differences between the large and small mountain sub-groups: (i) steeper surface temperature gradients (by 4%), (ii) lower air temperatures at sea level (by 0.5 °C), and (iii) a greater coastal proximity (suggesting higher average *RH*), for the smaller mountains relative to the larger ones. The implied changes in LCL from these three observations are now estimated. To estimate the impact of the steeper lapse rate (i.e. Jarvis and Mulligan's first observation), it is assumed that the LCL responds linearly to change in Γ: ΔLCL ~4% of the LCL. The impact on the LCL of the second observed difference (a lower sea-level temperature on smaller mountains), can be calculated using Eq. 4.4. First, T_0, the temperature at sea level, is estimated. From the observed quantities – average cloud forest temperature, 22.6 °C (large mountains) and 21.3 °C (small mountains), average values of the LCL, and average lapse rates (0.51 °C 100 m^{-1} (large mountains) and 0.53 °C 100 m^{-1} (small mountains) (Jarvis and Mulligan, this volume) – T_0 for the large mountains becomes 28.1 °C vs. 26.1 °C for the small mountains. The impact of different temperatures is quantified in ΔT_0, so in the following, T_0 is taken to be 27 °C. Next, $T_{dew,0}$ is calculated from Eq. (4.2) for two values of *RH*, 60% and 80%, to bracket probable values of the local humidity: $T_{dew,0} = 19$ °C ($RH = 60\%$) and 23 °C ($RH = 80\%$). Finally, the change in the LCL due to the difference in sea-level temperature, $\Delta T_0 = +0.5$ °C, is derived from Eq. (4.4): ΔLCL = +6.2% of the LCL for $RH = 60\%$ and +12.5% for $RH = 80\%$. Thus, the warmer air at sea level might explain up to half (for the 80% *RH* case) the observed difference between the large and small mountains' average cloud forest altitude. The coastal proximity effect (the third difference between the large and small mountain sub-groups observed by Jarvis and Mulligan) was probably small for this set of observations: although distance from coast was smaller for the smaller mountains, the difference was not significant (Jarvis and Mulligan, this volume). Nonetheless, as an illustration of the magnitude of the change in LCL implied by a change in surface *RH*, note that a decrease in *RH* from 70% to 65% would result in a predicted increase in the LCL of 14%. As such, *RH* could easily explain half of the observed 21% shift in LCL while the warmer surface air can explain the other half. Clearly, however, more data are needed to substantiate this claim.

FUTURE CHANGES IN LIFTING CONDENSATION LEVEL BASED ON A GLOBAL CIRCULATION MODEL

Equation (4.1) was used to estimate the altitude of the cloud base (i.e. LCL) as predicted by ECHAM4 GCM-based simulations using historic greenhouse gas concentrations and sulfate emissions from 1860 to 1990 and projected along the IS92a IPCC scenario for the 1990–2049 period. For each grid cell in the model, trends in the LCL were examined as a function of year using surface simulated data only. These trends are shown in Figure 4.2 for one grid cell in Texas in the United States. For this particular grid cell, trends were not significant over the entire 190-year run (dashed line). However, when a 10-year running mean was applied to damp out ENSO-like cycles, and when only the CO_2-rich era was considered (approximated by the period 1960–2049), the trend did become significant at the 2 σ level. Those cells for which LCL trends exceeded 1 σ are plotted in Figure 4.3 for the central months of each season.

One important test for assessing the reliability of trends derived from GCMs is to show that the GCM is capable of reproducing internal variability (Stott, 2003). In this case, this would mean demonstrating that the ECHAM4-based LCL results reproduce the natural variability of the observed LCL. However,

Caption for Figure 4.1. (*cont.*)
which Γ and Γ_{sat} cross, the parcel will now be warmer, and less dense than its surrounding and thus positively buoyant – this altitude is referred to as the level of free convection LFC and is graphically represented by the three bubbles rising and expanding from the LFC. The LFC and LCL are shown as dashed lines in all panels. Panels (c) and (d) show two impacts of deforestation given that soil moisture is limited such that the sensible heat flux *H* (temperature) increases and the latent heat flux *LE* (moisture) decreases. Panel (c) shows that increasing *H* will increase upward convective flows, thus making it more likely for an updraft to reach the LFC and sustain cumulus cloud formation. In panel (d), less moisture is evaporated into the atmospheric surface layer, decreasing $T_{dew,0}$, represented as the horizontal arrow in the figure. This increases the altitudes of both the LCL and LFC, thus decreasing the likelihood of cumulus formation. The case where soil moisture is unlimited is shown in panels (e) and (f). In this case additional surface radiation will go into increasing *LE* and decreasing *H*. The latter will decrease the likelihood of uplift reaching the LFC, thus reducing the chances of cloud formation, while increasing *LE* will increase it by lowering the LCL and LFC. Thus, in both cases (limited and unlimited soil moisture), the chance for cloud formation can be either reduced or enhanced depending on which change dominates.

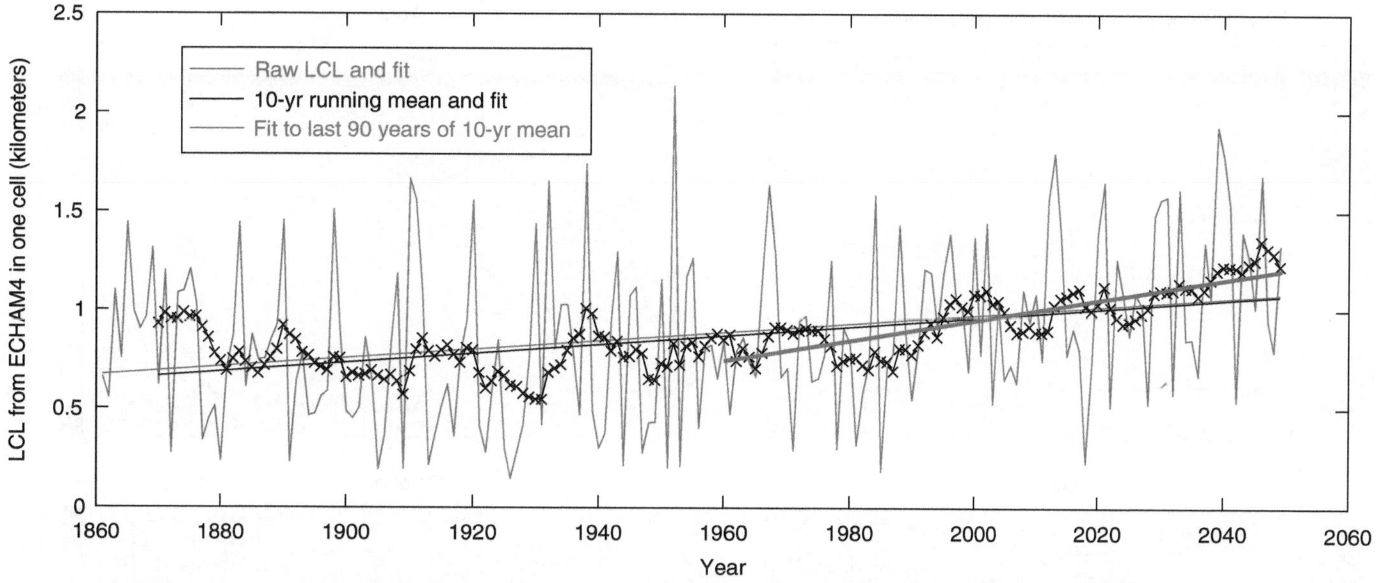

Figure 4.2. Changes over time in the lifting condensation level (LCL) in January indicated by the oscillating line as calculated by inserting ECHAM4's IS92a IPCC simulated monthly averaged output into Eq. (4.1) and shown for one grid cell in Texas between 1860 and 2049. The upper straight line represents a least-squares fit to the calculations. The jagged line (crosses) shows the 10-year running mean of the calculated LCL and a fit to it is shown as the upper straight line. The bold line shows a fit to the last 90 years (centered around 2005) of the10-year running mean as representative of the industrial era. Over these last 90 years the raw LCL changes by 0.19 km (±0.42), the 10-year running mean by 0.20 km (±0.20), and the fit to the last 90 years by 0.46 km (±0.16). The fit to the last 90 years of the 10-year running mean for the world as a whole is plotted in Figure 4.3 for January, April, July, and October.

such observations do not exist. Instead, trends predicted for a 10-year running mean over the years 1870–1959 were compared to those for 1960–2049, with the following results: (i) fewer cells had significant trends during the first period (45% vs. 61% of cells, averaged over all four central months), and (ii) those that did have significant trends had a smaller average magnitude of change (42 m vs. 55 m). This indicates that the trends since 1960 are more pronounced than natural variations and could be due to anthropogenic forcing.

Another caution regarding GCM results is that individual cells cannot be relied upon for accurate predictions. However, patterns can tell us something and one does see a suggestion that LCLs will rise in excess of 200 m over the Amazon, western Australia, and the western US for most seasons, whereas, equatorial Africa and parts of South-East Asia indicate a drop in the LCL (Figure 4.3). The presently derived amplitude of change is very similar to a previous GCM estimate using the GENESIS GCM and *RH* and warmth index contours (Still *et al.*, 1999).

The pattern of the presently predicted changes in LCL appears to coincide with patterns of the global hydrological cycle. This is especially apparent in July (Figure 4.3c), which shows latitudinal banding of the trend in LCL that follows the Hadley cell circulation, with the inter-tropical convergence zone (ITCZ) shifted north of the equator – as expected for July. The decrease in the LCL in the convergence and uplift zone (about 25° of latitude around the ITCZ) and the increase in the sub-tropical subsidence zone could be interpreted as an indication of a strengthening of the global circulation with regions of uplift showing more moisture and clouds (hence lower LCLs) and regions of subsidence becoming drier and less cloudy (increased LCLs) (Chen *et al.*, 2002). The more varied equatorial response in January, April, and October (Figure 4.3) may represent the fact that not only does the equatorial atmosphere circulate from north to south in Hadley cells, but also from east to west in Walker cells, such that there are regions of subsidence, with suppressed cloud formation, near the equator as well.

Figure 4.3, which shows trends for the period 1960–2049, may be considered as a proxy for the last few decades since it simply shows the reaction of the atmosphere to CO_2 forcings over a long period. Accepting this argument, the results portrayed in Figure 4.3 may be compared with recent observed changes in cloud base height. The pattern of Figure 4.3c is markedly different from that given in Figure 4 of Chernyck *et al.* (2001), which shows calculated trends in July cloud base height from 795 radiosonde time-series for 1964–1998 based on the derivatives of temperature and humidity profiles. They found, on average, a decrease in cloud base height of 44 m per decade, although it varied considerably with cloud amount. On the other hand, the results of Richardson *et al.* (2003), who looked at airport logs in the eastern US, are more similar to the present results than those of Chernyck *et al.*

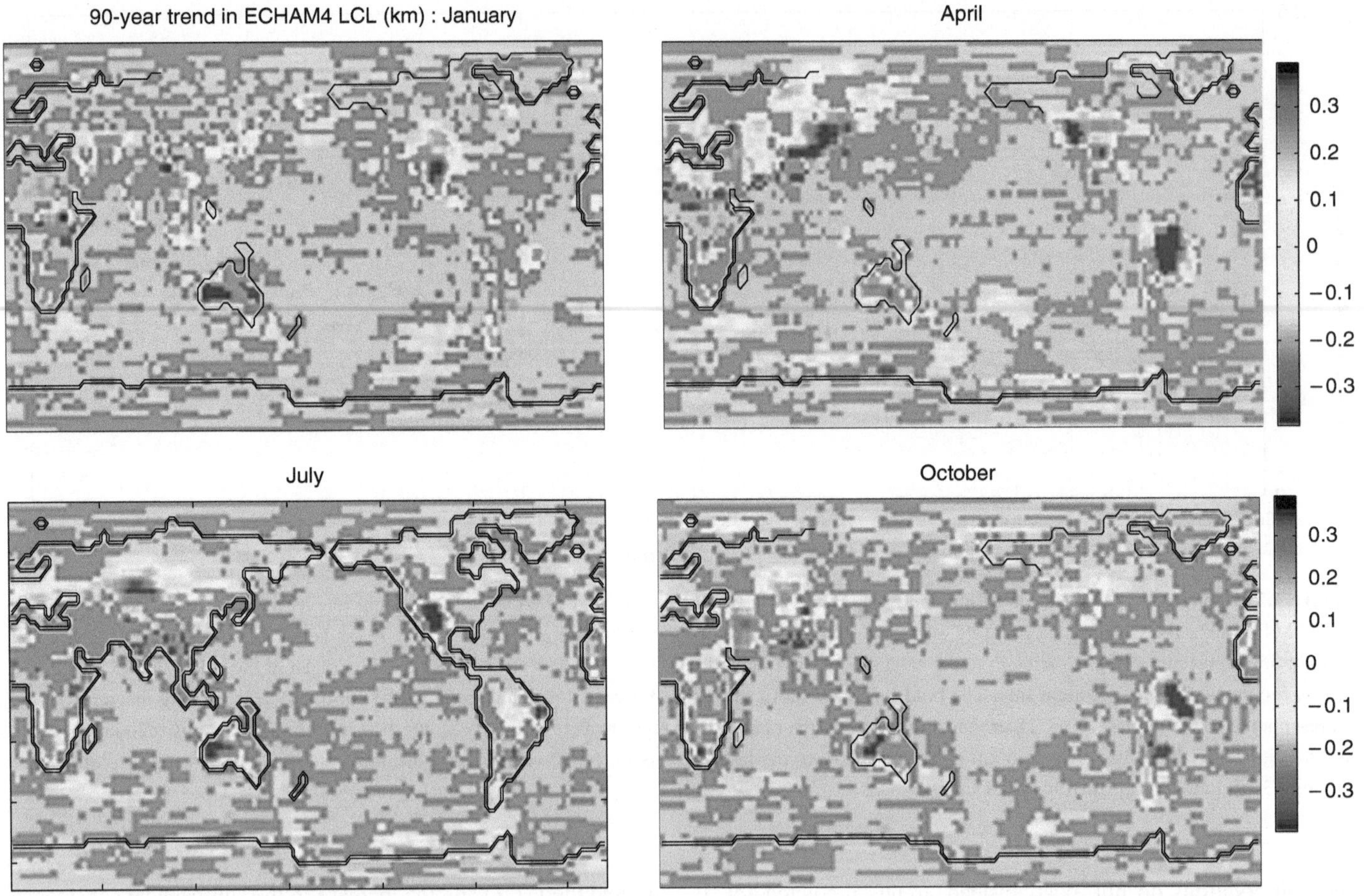

Figure 4.3. Ninety-year trends, centered on 2005, in the 10-year running mean of the lifting condensation level (LCL), as discussed under Figure 4.1. Cells with non-significant trends ($<\sigma$) are set to −999 m and appear in gray. In order to preserve strong color contrast, the scale in each panel is from −400 m to +400 m. In January, Central Africa has a trend more negative than −400 m and thus appears gray; the same is true in South-East Asia in April. Otherwise, all other gray cells had insignificant trends. July shows a strong similarity to global hydrological patterns. (See also color plate.)

(2001). Richardson *et al.* (2003) show six sites south of 37.5° N with decreasing cloud base heights and 18 sites northward of 37.5° N showing increasing heights. Both observational sets do show the latitudinal trends that are suggestive of changes in the Hadley cell. Two cloud forest sites where recent climate change has been implicated in changing the altitude of dry season cloud formation are Monteverde, Costa Rica (Nair *et al.*, 2008; Lawton *et al.*, this volume; Nair *et al.*, this volume) and Tenerife, Canary Islands (Sperling *et al.*, 2004; Marzol-Jaén *et al.*, this volume). Both areas are situated in regions of small to insignificant changes in currently predicted LCL (see Figure 4.3b for the dry season at Monteverde and Figure 4.3c for that in Tenerife) exemplifying how GCMs only capture broad trends and not site-specific responses.

One of the most striking features of Figure 4.3 is that the predicted trends in LCL are only large, and significant, over land cells. This is easily explained since both ΔT and ΔT_{dew} will be dampened over the oceans relative to the land, but even more importantly, air over the oceans is closer to saturation. This means that its *RH* probably will not change much in the event of climate change. It has long been argued that the short residence time of water in the atmosphere means that all air parcels have recently been close to saturation and thus more moisture will not increase *RH*. Constant *RH* in the event of changing climate has been found in seasonal observations at high latitude, after the eruption of Mount Pinatubo as well as in GCM models of global warming (Allen and Ingram, 2002 and references therein). Observations of vertically integrated water vapor content (both linear trends and inter-annual variability) can be explained using observed sea-surface temperature changes and assuming a constant *RH* (Trenberth *et al.*, 1995) – in concordance with analyses suggesting *RH* trends from 1976–2006 are very small (Dai, 2006).

If *RH* is to remain constant under climate change, the change in T_{dew} can be related to that in T, thus simplifying the equation for ΔLCL (i.e. Eq. 4.4). For $\Delta RH = 0$, $\Delta e/e = \Delta e_s/e_s$. According to the Clausius–Clapeyron equation one can write $\Delta e_s/e_s = L/R_v * \Delta T/T^2$. Taking the derivative of Eq. (4.2) yields

$\Delta T_{dew}/T_{dew}^2 = R_v/L \ \Delta e/e$, which in combination with the previous two equalities yields:

$$\Delta T_{dew} = (T_{dew}/T)^2 \Delta T \text{ at constant } RH. \quad (4.5)$$

Substituting this into Eq. (4.4) gives:

$$\Delta LCL = \Delta T(1 - (T_{dew}/T)^2)/(\Gamma_{dry} - \Gamma_{dew}), \text{ again at constant } RH. \quad (4.6)$$

When Eq. (4.6) was used to estimate the trend in LCL, as done for Figure 4.2, it was found that the LCL did not change much over the entire globe. This was similar to the results over the oceans in Figure 4.2, but dissimilar to the land results, and furthermore is consistent with Dai's (2006) conclusions that *RH* remained constant over the ocean but decreased slightly over land during the period 1976–2005. Thus, assuming a constant *RH* appears to be a poor assumption when estimating changes in the LCL over land.

CHANGES IN THE GLOBAL HYDROLOGICAL CYCLE

The equator annually receives more than two times the energy received at the poles. In the absence of atmospheric circulation, this would result in a very large difference in equatorial and polar temperatures ($\Delta T_{eq\text{–}poles}$). However, "fluid sunshine," in the form of the latent heat of evaporated water, flows out of the equatorial regions toward the poles, thus reducing $\Delta T_{eq\text{–}poles}$. In this simplistic model, the stronger the hydrological cycle, the smaller $\Delta T_{eq\text{–}poles}$ becomes. Large values of $\Delta T_{eq\text{–}poles}$ during the last glacial maximum are thus interpreted as being in concordance with a reduced hydrological cycle. During warmer climates, such as those of the Eocene and Cretaceous, $\Delta T_{eq\text{–}poles}$ appears to have been much reduced (Peirrehumbert, 2002).

It has generally been assumed that anthropogenic global warming will likewise lead to an enhancement of the global hydrological cycle, as higher air and sea-surface temperatures lead to greater evaporation, greater saturation vapor pressure in the atmosphere, and more latent heat transport toward the poles. This has been fairly well reproduced by suites of GCMs showing a nearly linear increase of 3.4% of global precipitation for each degree K of warming (Allen and Ingram, 2002). However, some GCM studies showed no change in the strength of the hydrological cycle (Allen and Ingram, 2002; Peirrehumbert, 2002 – for Eocene warming) or even hydrological damping upon warming (Yang *et al.*, 2003). Yang *et al.* (2003) argued that when temperature increases arise via CO_2-induced radiational cooling of the atmosphere, the hydrological cycle will be dampened because the energy balance in the atmosphere will require that latent heating of the atmosphere be reduced, and thus rainfall will drop. Similarly, complex interactions of aerosols with the hydrological cycle may potentially contribute to a dampening of the hydrological cycle in a warmed world (see also below; cf. Liepert *et al.*, 2004).

Similar to numerical model results, observational analysis studies are also ambiguous, with some indicating an accelerating hydrological cycle while others indicate a deceleration. Enhanced precipitation rates are observed over many land stations, but rainfall trends are both difficult to measure and different from one location to the next (Trenberth *et al.*, 2007). On a global scale, enhancement of rainfall might be expected over the uplift regions, including the ITCZ, and a decrease of rainfall might be expected in regions of subsidence, for instance over the great deserts in the world. Accompanying decreases in cloudiness will be enhanced outgoing global long-wave radiation (more clear skies) and decreased incoming short-wave radiation (fewer clouds to reflect). Wielicki *et al.* (2002) and Chen *et al.* (2002) claim to have detected all of these signals in various data-sets of the 1990s. It should be noted, however, that Trenberth *et al.* (2002) have urged caution with the interpretation of these data-sets due to significant problems with stability of satellites, recalibration issues and a strong ENSO event in the early 1990s.

Conversely, signals of a reduced hydrological cycle include observations that potential evaporation from the surface has decreased in the last few decades – although this may be due to decreases in surface radiation arising from increased cloudiness (Trenberth *et al.*, 2007). Liepert *et al.* (2004) and references therein show that GCM-based simulations which include the radiative and cloud microphysical effects of aerosols can reproduce global warming trends, with a reduced hydrological cycle.

Tropical temperature lapse rates (Γ), which are sometimes interpreted as indicating the intensity of the hydrological cycle, turn out to be equally challenging to interpret. Lapse rates from the surface to 700 hPa appear to have decreased for two decades prior to 1980 and to have increased for the two decades after. There are significant differences in interpretation of these observations as evidenced by Chen *et al.* (2002) arguing that the *increases* after 1980 indicate an intensification of the hydrological cycle, whereas Brown *et al.* (2000) interpreted the *decrease* prior to 1980 as being indicative of an enhanced water cycle. These seemingly contradictory statements can be reconciled by noting that the pre-1980 decrease in Γ arose from increasing temperatures at the 700-hPa level, which Brown *et al.* (2000) argued to come from more latent heat release, whereas the post-1980 increase in Γ derived from surface heating, resulting from enhanced subsidence and compressional heating – both driving forces could be indicative of hydrological enhancement. It is thus difficult to predict what a decreasing lapse rate would mean for the LCL. However, it does seem that decreasing Γ would decrease the chances of sustained convection by increasing the altitude of the LFC.

RECENT CHANGES IN THE LIFTING CONDENSATION LEVEL

Observed shifts

There are a few observations at cloud forest sites which indicate that local cloud banks have shifted. Pounds *et al.* (1999) have documented an increase in the number of mist-free days at Monteverde, Costa Rica since the 1970s. Since this occurred at the same time as a decrease in the diurnal temperature range (DTR), they concluded that the cloud bank is rising, thus reducing mist immersion and moisture input. Two groups have modeled this change in cloudiness and attribute it to two possibly complementary causes: global climate change and atmospheric warming and drying after deforestation (see below). In Puerto Rico, after Hurricane Hugo defoliated much of the island in September 1989, the cloud cap rose temporarily by several hundred meters to a position above the island's peak, until the leaves had grown back again (Van der Molen *et al.*, 2006). Finally, Sperling *et al.* (2004) observed an increased occurrence of low level clouds at Tenerife (cf. Marzol-Jaén *et al.*, this volume).

Deforestation and global climate change impacts

Initial modeling studies of the impact of Amazonian deforestation (conversion to pasture) on climate showed that deforestation would lead to decreased precipitation through a de-intensification of the hydrological cycle (Lean and Warrilow, 1989: Nobre *et al.*, 1991; Walker *et al.*, 1995). Observations however, did not confirm this hypothesis, and, indeed, increased precipitation and cumulus formation have been observed over deforested areas in Amazonia (Chu *et al.*, 1994; Gash and Nobre, 1997; Chen *et al.*, 2001). Models that consider inhomogeneous deforestation, i.e. cleared areas interspersed with intact forest, have concluded that the higher temperatures observed in pasture sites led to lower relative pressures, thus enhanced convergence ("forest wind") and uplift, resulting in enhanced cumulus cloud formation and rainfall (Roy and Avissar, 2002).

A recent study by Van der Molen *et al.* (2006) suggests that the same kind of differential heating is the controlling factor in driving cloud formation in eastern Puerto Rico. In this case, it is the difference between the ocean and the land surface that appears to be driving the circulation. The land temperatures rise more than the ocean temperatures, causing convergence, uplift, and cumulus formation over the land (Malkus, 1955). In Amazonian studies, conversion of tall lowland forest to pasture led to higher temperatures, increased sensible heat flux H, and thus enhanced uplift and cumulus formation (see also Figures 4.1c and d). However, in Puerto Rico the detailed field study of Van der Molen *et al.* (2006) observed that conversion of coastal forest to pasture resulted in *lower* temperatures and lower H as well as increased evaporation (latent heat flux LE; Figures 4.1e and f). This increase in LE arose from a combination of comparatively low water use by the coastal forest (possibly because of salt interference) and a high water table beneath the pasture which provided ample soil moisture for evaporation and thus a greater proportion of incoming solar radiation being absorbed by the surface. This led to an observed reduction in surface temperatures, a reduced sea breeze circulation, and less rainfall, in agreement with increasing water shortages experienced on the island concurrent with the deforestation (Van der Molen *et al.*, 2006). Thus, although there was more moisture being pumped into the air after forest conversion (Figure 4.1f), it could not compensate for the decrease in uplift (i.e. H; Figure 4.1e) and cloud formation decreased as a result. In regional climate simulations of Puerto Rico, the enhanced LE and lower temperatures of the grasslands resulted in a lowering of the cloud base height of about 100 m relative to the forest simulation (Figure 4.1f), although the authors noted that this comprised only about one cell in the vertical and that this result should be treated with caution (cf. Van der Molen *et al.*, this volume).

In contrast to the Amazonian case, Nair *et al.* (2003) found that forest conversion to pasture and a concurrent increase in sensible heat flux H led to *suppressed* cumulus formation, in climate simulations done over a flatland version of Costa Rica (and more recently over realistic topography in Ray *et al.*, 2006; cf. Nair *et al.*, this volume). This contrasting result arose because the accompanying decrease in LE resulted in a rise in the LCL and LFC via a decrease in $T_{dew,0}$ (Figure 4.1d). Similar cumulus reductions were seen over the pasture side of the Australian "bunny fence experiment" in summer when moisture was limiting. The result was reversed in winter when the bunny fence pasture had ample water and radiation reaching the surface enhanced evaporation. The increase in atmospheric water vapor led to enhanced cumulus formation in this case (Figure 4.1f; Ray *et al.*, 2003).

In summary, if the soil is dry, the extra radiation reaching the ground surface after deforestation will go into raising temperatures and H, and decreasing LE. If the increase in H is large enough to increase uplift above the LFC, cloud formation may be enhanced (Figure 4.1c). However, if the enhanced circulation is not sufficient to overcome the reduction in moisture arising from the decreasing LE and the subsequent lifting of the LCL and LFC, cumulus formation will be suppressed (Costa Rica – Nair *et al.*, 2003; Nair *et al.*, this volume; Western Australian bunny fence in summer – Ray *et al.*, 2003; Figure 4.1d). If soil moisture is not limiting, and the radiation reaching the surface following deforestation increases (and all else is held constant), then LE will increase and H will decrease. The increase in moisture (LE) but decrease in uplift (H) can lead to either suppressed cloud formation (Puerto Rico case, although this case is complicated by salt – Figure 4.1e) or enhanced cloud formation as the additional moisture will lower the LCL and the LFC (Australian bunny fence in winter – Ray *et al.*, 2003; Figure 4.1f).

Even at a given site, the role of deforestation on cloud formation is difficult to unravel. For instance, Ray *et al.* (2006) found that regional climate model simulations of lowland deforestation induced a rise in the cloud bank at Monteverde in Costa Rica at an altitude of about 1500 m.a.s.l. In contrast, using the same regional climate model, Pounds *et al.* (2006) found that lowland deforestation downwind of the Monteverde cloud forest had very little impact on cloud formation at the altitude of the cloud forest, even in the event of total deforestation. Pounds *et al.* (2006) further found that the deforestation impacts were much less than those induced by global climate change. Differences between the studies, such as the time of the analysis (daytime for Ray *et al.*, night-time for Pounds *et al.*) and defining cloudiness (clouds for Ray *et al.*, relative humidity for Pounds *et al.*) are undoubtedly part of the explanation for the differing conclusions.

CONCLUSIONS

The predicted pattern of change in lifting condensation levels (LCL) upon global warming as indicated by the ECHAM4 GCM gives hints of the patterns expected with an enhanced Hadley cell circulation, in agreement with observations indicating intensified uplift and subsidence (Chen *et al.*, 2002). Further research is needed to determine whether the present results are robust and includes looking at predictions of the LCL by other GCMs, as well as validating the LCL predictions of various GCMs against observations. Cloud base height data that do exist are sparse and do not compare in detail with the current predictions, although the observational studies do show latitudinal banding (Chernyhk *et al.*, 2001; Richardson *et al.*, 2003), as suggested in Figure 4.3c. Theoretical work is needed to link changes in the global hydrological cycle and temperature to shifts in the LCL and other cloudiness factors.

To further relate GCM results to specific sites, observations of *RH*, temperature, and cloud base height at various cloud forest sites around the tropics are needed (Nair *et al.*, 2008; Welch *et al.*, 2008; Lawton *et al.*, this volume; Rollenbeck *et al.*, this volume). To test the regional climate models that are being used to predict climate change implications for cloud forests (Still *et al.*, 1999; Karmalkar *et al.*, 2008), a detailed observational data-set documenting one mountain's cloudiness regime is sorely needed. Observational and modeling work is needed on water content and frequency of cloud formation changes (cf. Bendix *et al.*, 2004; Nair *et al.*, 2008; Welch *et al.*, 2008; Lawton *et al.*, this volume) as well as the impact of increased aerosol concentrations in the atmosphere (cf. Liepert *et al.*, 2004).

ACKNOWLEDGEMENTS

Many thanks are due to Udaysankar Nair and Sampurno Bruijnzeel for helpful comments.

REFERENCES

Allen, M.R., and W.J. Ingram (2002). Constraints on future changes in climate and the hydrological cycle. *Nature* **419**: 224–232.

Bendix, J., R. Rollenbeck, and W.E. Palacios (2004). Cloud detection in the tropics: a suitable tool for climate ecological studies in the high mountains of Ecuador. *International Journal of Remote Sensing* **25**: 4521–4540.

Bohren, C.F., and X. Albrecht (1998). *Atmospheric Thermodynamics*. Oxford, UK: Oxford University Press.

Brown, S.J., D.E. Parker, C.K. Folland, and I. Macadam (2000). Decadal variability in the lower-tropospheric lapse rate. *Geophysical Research Letters* **27**: 997–1000.

Chen, J., B.E. Carlson, and A.D. Del Genio (2002). Evidence for strengthening of the tropical general circulation in the 1990s. *Science* **295**: 838–841.

Chen, T. -C., J. -H. Yoon, K. J. St. Croix, and E. S. Takle (2001). Suppressing impacts of Amazonian deforestation by the global circulation change. *Bulletin of the American Meteorological Society* **82**: 2209–2216.

Chernykh, I. V., O. A. Alduchov, and R. E. Eskridge (2001). Trends in low and high cloud boundaries and errors in height determination of cloud boundaries. *Bulletin of the American Meteorological Society* **82**: 1941–1947.

Chu, P. -S., Z. -P. Yu, and S. Hastenrather (1994). Detecting climate change concurrent with deforestation in the Amazon Basin: which way has it gone? *Bulletin of the American Meteorological Society* **75**: 579–582.

Craven, J.P., R.E. Jewell, and H.E. Brooks (2002). Comparison between observed convective cloud-base heights and lifting condensation level for two different lifted parcels. *Weather and Forecasting* **17**: 885–890.

Dai, A. (2006). Recent climatology, variability and trends in global surface humidity, *Journal of Climate* **19**: 3589–3606.

Foster, P. (2001). The potential negative impacts of global climate change on tropical montane cloud forests. *Earth-Science Reviews* **55**: 73–106.

Gash, J.H. C., and C.A. Nobre (1997). Climatic effects of Amazonian deforestation: some results from ABRACOS. *Bulletin of the American Meteorological Society* **78**: 823–830.

Hildebrandt, A., and E. A. B. Eltahir (2008). Using a horizontal precipitation model to investigate the role of turbulent cloud deposition in survival of a seasonal cloud forest in Dhofar. *Journal of Geophysical Research* **113**, G04028, doi:10.1029/2008JG000727.

Karmalkar, A.V., R.S. Bradley, and H.F. Diaz (2008). Climate change scenario for Costa Rican montane forests. *Geophysical Research Letters* **25**, L11702, doi: 10.1029/2008GL033940.

Lean, J., and D.A. Warrilow (1989). Simulation of the regional climatic impact of Amazon deforestation. *Nature* **342**: 411–413.

Liepert, B.G., J. Feichter, U. Lohmann, and E. Roeckner (2004). Can aerosols spin down the water cycle in a warmer and moister world? *Geophysical Research Letters* **31**: L06207.

Lohmann, U., and J. Feichter (2005). Global indirect aerosol effects: a review. *Atmospheric Chemistry and Physics* **5**: 715–737.

Lopez, P. (2007). Cloud and precipitation parameterizations in modeling and variational data assimilation: a review. *Journal of Atmospheric Science* **64**: 3766–3784.

Malkus, J.S. (1955). The effects of a large island upon the trade-wind air stream. *Quarterly Journal of the Royal Meteorological Society* **81**: 538–550.

Molinari, J., and M. Dudek (1992). Parameterization of convective precipitation in mesoscale numerical models: a critical review. *Monthly Weather Review* **120**: 326–344.

Nair, U.S., R.O. Lawton, R.M. Welch, and R.A. Pielke Sr. (2003). Impact of land use on Costa Rican tropical montane cloud forests: sensitivity of cumulus cloud field characteristics to lowland deforestation. *Journal of Geophysical Research* **108** (D7): 4206–4218.

Nair, U. S., S. Asefi, R. M. Welch, *et al.* (2008). Biogeography of tropical montane cloud forests. II. Mapping of orographic cloud immersion. *Journal of Applied Meteorology and Climatology* **47**: 2183–2197.

Nobre, C. A., P. J. Sellers, and J. Shukla (1991). Amazonian deforestation and regional climate change. *Journal of Climate* **4**: 957–988.

Peirrehumbert, R. T. (2002). The hydrological cycle in deep time climate problems. *Nature* **419**: 191–198.

Peixoto, J. P., and A. H. Oort (1992). *Physics of Climate*. New York: American Institute of Physics.

Pounds J. A., M. P. L. Fogden, and J. H. Campbell (1999). Biological response to climate change on a tropical mountain. *Nature* **398**: 611–615.

Pounds, J. A., M. Bustamante, L. A. Coloma, *et al.* (2006). Widespread amphibian extinctions from epidemic disease driven by global warming. *Nature* **439**: 161–167.

Ray, D. K, U. S. Nair, R. M. Welch, *et al.* (2003). Effects of land use in Southwest Australia. I. Observations of cumulus cloudiness and energy fluxes. *Journal of Geophysical Research* **108**: 4414–4433.

Ray, D. K., U. S. Nair, R. O. Lawton, R. M. Welch, and R. A. Pielke Sr. (2006). Impact of land use on Costa Rican tropical montane cloud forests: sensitivity of orographic cloud formation to deforestation in the plains. *Journal of Geophysical Research* **111**: D02108, doi: 10.1029/2005JD006096.

Richards, P. W. (1996). *The Tropical Rain Forest*. Cambridge, UK: Cambridge University Press.

Richardson, A. D., E. G. Denny, T. G. Siccama, and X. Lee (2003). Evidence for a rising cloud ceiling in Eastern North America. *Journal of Climate* **16**: 2093–2098.

Roy, S. B., and R. Avissar (2002). Impact of land use/land cover change on regional hydrometeorology in Amazonia. *Journal of Geophysical Research* **107**: doi:10.1029/2000JD00026, 2002.

Slingo, J. M. (1987). The development and verification of a cloud prediction scheme for the ECMWF model. *Quarterly Journal of the Royal Meteorological Society* **113**: 899–927.

Sperling, F. N., R. Washington, and R. J. Whittaker (2004). Future climate change of the subtropical north Atlantic: implications for the cloud forests of Tenerife. *Climatic Change* **65**: 103–123.

Still, C. J., P. N. Foster, and S. H. Schneider (1999). Simulating the effects of climate change on tropical montane cloud forests. *Nature* **389**: 608–610.

Stott, P. A. (2003). Attribution of regional-scale temperature changes to anthropogenic and natural causes. *Geophysical Research Letters* **30**: 1728–1731.

Tiedtke, M. (1993). Representation of clouds in large-scale models. *Monthly Weather Review* **121**: 3040–3061.

Trenberth, K. E., B. A. Wielicki, A. D. Del Genio, *et al.* (2002). Changes in tropical clouds and radiation. *Science* **296**: 2095.

Trenberth, K. E., J. Fasullo, and L. Smith (2005). Trends and variability in column integrated atmospheric water vapor. *Climate Dynamics* **24**: 741–758.

Trenberth, K. E., P. D. Jones, P. Ambenje, *et al.* (2007). Observations: surface and atmospheric climate change. In *Climate Change 2007: The Physical Science Basis*. Contribution of Working Group I to the Fourth Assessment Report of the Intergovernmental Panel on Climate Change, eds. S. Solomon, D. Qin, M. Manning, *et al.*, pp. 235–336. Cambridge, UK: Cambridge University Press.

Van der Molen, M. K., A. J. Dolman, M. J. Waterloo, and L. A. Bruijnzeel (2006). Climate is affected more by maritime than by continental land use change: a multiple-scale analysis. *Global and Planetary Change* **54**: 128–149.

Walker, G. K., Y. C. Sud, and R. Atlas (1995). Impact of the ongoing Amazonian deforestation on local precipitation: a GCM simulation study. *Bulletin of the American Meteorological Society* **76**: 346–361.

Wallace, J. M., and P. V. Hobbs (1977). *Atmospheric Science: An Introductory Survey*. New York: Academic Press.

Wang, J., R. L. Bras, and E. A. B. Eltahir (2000). The impact of observed deforestation on the mesoscale distribution of rainfall and clouds in Amazonia. *Journal of Hydrometeorology* **1**: 267–286.

Welch, R. M., S. Asefi, J. Zeng, *et al.* (2008). Biogeography of tropical montane cloud forests. I. Remote sensing of cloud base heights. *Journal of Applied Meterology and Climatology* **47**: 960–975.

Wielicki, B. A., T. Wong, R. P. Allen, *et al.* (2002). Evidence for large decadal variability in the tropical mean radiative energy budget. *Science* **295**: 841–844.

Yang, F., A. Kumar, M. E. Schlesinger, and W. Wang (2003). Intensity of hydrological cycles in warmer climates. *Journal of Climate* **16**: 2419–2423.

5 Ecology and ecophysiology of epiphytes in tropical montane cloud forests

P. Hietz
University of Natural Resources and Applied Life Sciences (BOKU), Vienna, Austria

ABSTRACT

Because several reviews of epiphyte ecology and physiology are already available, this chapter focuses on what is particular about epiphyte ecology in cloud forests as opposed to lowland rain forests. The combination of high atmospheric humidity, frequent precipitation and low temperature results in much lower drought stress, which enables a denser colonization of exposed twigs with no "canopy soil" (humus), and the survival of small species, small individuals with little internal water storage, and species with no adaptations to drought. Because juveniles have the highest mortality with drought as the most important factor, water availability in most cases explains why cloud forest epiphytes do not occur in lowland forests. Species not occurring in cloud forests are limited by low temperatures (particularly in sub-tropical and upper montane areas) or excessive humidity (particularly bromeliads). The lower temperatures in cloud forests reduce night-time respiration, resulting in higher growth of non-vascular epiphytes. This, together with lower decomposition rates, leads to the formation of thick layers of canopy soil, providing a rooting substrate for vascular epiphytes. The very high living and dead epiphytic biomass plays an important role in the nutrient and hydrological cycles of tropical montane cloud forests.

INTRODUCTION

As even a casual visitor will note, epiphytes are very prominent in tropical montane cloud forests (TMCF). However, they can be abundant in other humid tropical forests too (cf. Gradstein *et al.*, this volume). This begs the question of whether and how epiphyte communities and their ecology in cloud forests differ from those in other forests. Vascular epiphytes in general have been the subject of several reviews and treatises (Lüttge, 1989; Benzing, 1990; Zotz and Andrade, 2001; Zotz and Hietz, 2001). Therefore, this chapter will focus on aspects in which epiphytes in cloud forests differ or could differ from those of other tropical forests. To some extent, conclusions about cloud forest epiphytes will be based on comparisons between lowland forests and montane rain forests in general.

Tropical montane cloud forests can be defined as evergreen mountain forests found in tropical areas that are frequently covered in fog or low cloud. Although occupying a distinct altitudinal zone, the zone of maximum cloud formation depends on local topography and climate and may range from less than 1000 m.a.s.l. to more than 3000 m.a.s.l. (Stadtmüller, 1987; Bruijnzeel and Veneklaas, 1998; cf. Jarvis and Mulligan, this volume). Whilst most TMCF may be readily recognizable as such, there are no clear definitions as to what constitutes "regular" or "frequent" cloud immersion, and the transition between "true" TMCF and other humid montane forests is not always clear (cf. Hemp, this volume #12; Mulligan, this volume). For instance, in Mexico the term *bosque mesófilo de montaña*, which was apparently derived from the French *forêt mésophile de montagne* as used by Lebrun in the 1930s (cited by Richards, 1952) for montane forests in the Congo, is often translated as cloud forest, although in many of these forests fog is rare during most of the year (Holwerda *et al.*, 2010; Ponette-González *et al.*, 2009; Lawton *et al.*, this volume) and their epiphyte communities are distinct from those of "proper" cloud forests. Between TMCFs, tree physiognomy may vary widely from tall forests with trees little smaller than those in lowland rain forests, to stands of gnarled trees a few meters high (so-called "elfin cloud forest", ECF), depending on local climatic and edaphic factors (Bruijnzeel and Veneklaas, 1998; Benner *et al.*, this volume). In the following, the term "montane rain forest" will be used whenever it is not clear whether the forest in question constitutes

Tropical Montane Cloud Forests: Science for Conservation and Management, eds. L. A. Bruijnzeel, F. N. Scatena, and L. S. Hamilton. Published by Cambridge University Press.

a "proper" cloud forest or not, or when discussing montane rain forests together with cloud forests. The term "cloud forest" as used here includes elfin forests, whereas the term "humid montane forests" includes forests that may not be considered proper rain or cloud forests, such as some of the Mexican *bosques mesófilos de montaña.*

Apart from their plant communities, TMCFs are best characterized by their hydrological behavior. The amount of net precipitation (crown drip) increases, and overall water use (evapotranspiration) decreases as one goes from lower montane rain forest (LMRF) below the cloud belt, via lower montane cloud forest (LMCF), to upper montane cloud forest (UMCF) and elfin cloud forest (Bruijnzeel, 2001; Bruijnzeel *et al.*, this volume). The associated changes in microclimatic conditions are likely to affect epiphytes more than any other plant life-form (Frahm and Gradstein, 1991) and Grubb *et al.* (1963) have suggested that epiphytes might be the most differentiating feature between different types of tropical montane forests. Cloud forests typically receive additional water from the interception of passing fog and low cloud, with the largest amounts associated with UMCF and ECF. In addition, estimated contributions by this "horizontal" or "occult" precipitation to total water input vary widely between sites (Bruijnzeel and Proctor, 1995; Bruijnzeel, 2001; Bruijnzeel *et al.*, this volume; cf. Häger and Dohrenbusch, this volume; Giambelluca *et al.*, this volume).

EPIPHYTE COMMUNITIES IN CLOUD FORESTS

Several authors have noted a "bulge" in the diversity or biomass of vascular (Gentry and Dodson, 1987a; Hietz and Hietz-Seifert, 1995; Ibisch, 1996; Hemp, 2001; Wolf and Flamenco-Sandoval, 2003; Krömer *et al.*, 2005; and Figure 5.1) and non-vascular (Frahm and Gradstein, 1991; Wolf, 1993c) epiphytes along altitudinal transects on tropical mountains. Along a transect from 30 to 2600 m.a.s.l. in Costa Rica diversity of vascular epiphytes was highest in a cloud forest at 1000 m, which to a large extent appears to result from the mid-elevation overlap of large-ranged species (Cardelús *et al.*, 2006). There was also a positive correlation between rainfall and epiphyte richness along the gradient. However, comparisons of species numbers between different forest types do require some caution as the size and density of epiphytes may differ and results tend to depend on the scale of the sampled area (Nieder *et al.*, 1997; cf. Häger and Dohrenbusch, this volume). For instance, in a cloud forest at El Triunfo in southern Chiapas, Mexico, a comparable plot size yielded two- to three-fold higher numbers of epiphytic and hemi-epiphytic species than in the lowland rain forest of Los Tuxtlas, Veracruz, although the species–area curve leveled off earlier (Figure 5.2). Conversely, in areas of several square

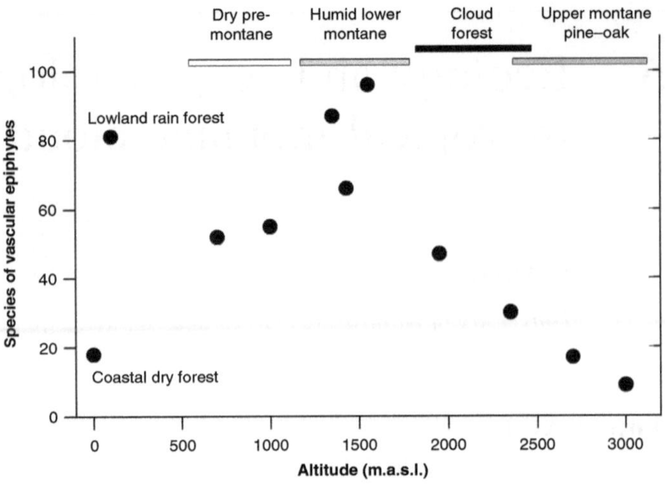

Figure 5.1. Diversity of vascular epiphytes in natural forests along an altitudinal gradient from coastal Veracruz along the Sierra Madre Oriental to the Cofre de Perote volcano, Veracruz, Eastern Mexico. Data for the lowland rain forest (Los Tuxtlas area, southern Veracruz) are from Hietz-Seifert *et al.* (1996). The humid montane forest with the highest diversity is that described by Flores-Palacios and García-Franco (2003) with some corrections for accidental epiphytes.

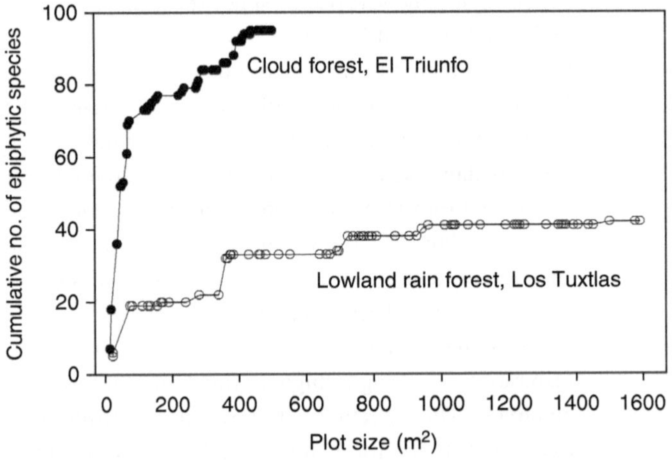

Figure 5.2. Species–area curve for epiphytes from a montane cloud forest (El Triunfo, Chiapas, Mexico) at ~2000 m.a.s.l. and for a lowland rain forest (Los Tuxtlas, Veracruz, Mexico) at 150 m.a.s.l. Both forests receive >2500 mm of rain year^{-1} (modified after Hietz and Wolf, 1996).

kilometers, 154 species were found in the lowland forest compared to 137 in the cloud forest, where collections had been made at a somewhat lower intensity (Ibarra-Manríquez and Sinaca, 1987; Long and Heath, 1991; plus several additions by the author). Considering the entire flora of Perú, the number of epiphytic species declines with altitude (Ibisch *et al.*, 1996), but because the available area also declines at higher elevations, absolute numbers do not tell us much about species density or diversity of sites at different altitudes. Sampling effort is an additional factor to reckon with in areas that have been poorly surveyed. Generally speaking, in cloud forests, individual trees tend to host a larger number of often small epiphyte individuals. As a consequence, more species are found in small plots and cloud forests have higher epiphyte species density, which is not the same as having higher diversity. What is largely lacking (and would be of substantial importance in a conservation context) is a compilation of information on the geographic distribution of cloud forest epiphytes, where endemism is supposed to be sparticularly high (Gentry, 1992).

Epiphyte communities in montane rain forests are always rich in ferns and orchids (Hietz and Hietz-Seifert, 1995; Ingram *et al.*, 1996; Nieder *et al.*, 2002; Krömer *et al.*, 2005). Filmy ferns (Hymenophyllaceae) are particularly common in cloud forests and also reach the canopy, whereas in other rain forests, if present at all, they are mostly restricted to the stem base and to humid ravines (Zotz and Büche, 2000; Hemp, 2001). Along elevational gradients, the diversity of epiphytic and hemi-epiphytic aroids is lower than in lowland forests, the contribution of ferns increases with altitude and, at least in Mexico, bromeliads peak in drier lower montane forests rather than in cloud forests (Hietz and Hietz-Seifert, 1995; Wolf and Flamenco-S, 2003; Krömer *et al.*, 2005; Cardelús *et al.*, 2006).

Along an altitudinal gradient in the Colombian Andes, there was a marked increase in non-vascular epiphyte communities around 2200 m.a.s.l., with the higher forests having a high frequency of fog as well as higher atmospheric humidity (Wolf, 1993b). The most distinguishing feature of the epiphytic vegetation in TMCF is the dense cover of non-vascular epiphytes on thick and many medium-sized branches. The most bryophyte-rich forests have also been described as "mossy" forests (Richards, 1952), which usually – but not necessarily – represents UMCF or ECF as some LMCF in high-rainfall areas can also be very mossy (cf. Nadkarni *et al.*, 2004; Köhler *et al.*, 2007). Bryophytes may be the most useful group to classify humid montane forests (including cloud forests) and, because many species have a wide distribution, they also permit pan-tropical comparisons (Frahm and Gradstein, 1991; Kürschner and Parolly, 1999).

Along an elevational gradient epiphytes will be affected most by differences in water availability – which is almost certainly highest in the cloud forest belt (Bruijnzeel 2001, 2005; Schawe *et al.*, this volume) – and by temperature, which declines linearly with altitude (Jarvis and Mulligan, this volume). At high elevations, low temperatures and occasional frosts will limit vascular epiphytes in particular, few of which venture into temperate zones with recurring frost (Zotz, 2005). However, altitudinal and moisture effects on epiphytes are complexly interrelated. Since non-vascular epiphytes have little control of transpiration and little internal water storage capacity, they should profit more from a continuous water supply than vascular epiphytes. Whilst maximum diversity of non-vascular epiphytes along an Andean transect in Colombia occurred between 2500 and 3000 m (i.e. within the cloud forest belt; Wolf, 1993c), maximum diversity of vascular epiphytes along the slope of the Mexican Sierra Madre Oriental is clearly below the zone of highest fog frequency (Figure 5.1). Where cloud condensation occurs at relatively low altitudes, such as in the Monteverde area of Costa Rica, both vascular and non-vascular epiphytes may find their optimum within the cloud forest belt (cf. Häger and Dohrenbusch, this volume; Köhler *et al.*, 2007). In Colombia, bryophyte diversity decreases within the cloud belt (Wolf, 1993c), while the highest epiphytic biomass is found in UMCF near the tree-line at 3700 m.a.s.l. (Hofstede *et al.*, 1993). Such observations indicate that optimum diversity and biomass do not necessarily coincide.

The abundance of non-vascular epiphytes in humid montane forests (or their lack in lowland forests) is only partly based on the availability of water (Zotz, 1999). Many lichens and bryophytes, being desiccation tolerant, are able to survive prolonged and recurring periods without water in the canopy in a largely inactive state. However, photosynthesis is only possible in a hydrated state, which does not last long for poikilohydric plants in climates with strong but short rains followed by intense sunshine, as is typical for lowlands. During humid nights the water status is improved and physiological activity resumed, but the high nocturnal temperatures in lowland forests lead to high respiratory carbon loss and a dangerously low total carbon balance (Zotz, 1999). By contrast, in many TMCFs, bryophytes are well watered during the day, ensuring positive photosynthesis, whereas carbon losses during the cool nights are low (cf. Letts *et al.*, this volume). The same explanation should also apply to poikilohydric filmy ferns which physiologically are more akin to mosses than to vascular plants (Hietz and Briones, 1998).

In many TMCFs, branches are covered with thick layers of bryophytes that decompose slowly, thereby contributing to the build-up of "canopy soils" (or canopy "humus"). Both bryophytes and canopy soil provide a rooting substrate storing potentially large amounts of water (Bohlman *et al.*, 1995; Köhler *et al.*, 2007; Köhler *et al.*, this volume) for use by many species of vascular epiphytes that would not be able to grow on bare branches. As such, canopy humus should facilitate the growth of vascular epiphytes in TMCF. In forests where only a fraction of the suitable bark surface is colonized, competition between epiphytes should not be intense, except perhaps for the scarce exceptionally suitable sites such as knotholes and branch forks where canopy humus tends to accumulate (Hietz, 1997; cf. Köhler *et al.*, 2007). Facilitation may be of importance, particularly for seedling establishment in the substrate held by epiphyte roots or even in bromeliad tanks (Ball *et al.*, 1991). On thick branches in TMCF, which are generally densely covered with epiphytes, facilitation should be the norm, but also competition for light and possibly other resources will be more intense.

EPIPHYTES AND WATER SUPPLY

Precipitation in TMCF is not always high (Jarvis and Mulligan, this volume). Water inputs in the form of intensive and short pulses, as common in lowland rain forests, is of relatively little use to epiphytes, which can access only intercepted water. However, in many TMCF a significant proportion of total precipitation may occur as prolonged drizzle that keeps the surfaces of epiphyte roots and leaves wet. Moreover, low temperatures and high atmospheric humidity translate into much lowered water vapor pressure deficits, which, together with high cloudiness, reduce water losses via evaporation (Bruijnzeel and Veneklaas, 1998; cf. Tobón *et al.*, this volume #26).

If water shortage resulting from the lack of soil contact is indeed the main factor limiting growth in most epiphytes (Zotz and Hietz, 2001), the abundance of epiphytes in TMCF is not surprising. The positive effect of canopy soil on the water supply of epiphytes was investigated in Monteverde, Costa Rica, where epiphytes from six systematic groups had consistently more negative $\delta^{13}C$ values when rooted in canopy soil compared to bare branches, indicating lower degrees of drought stress (Hietz *et al.*, 2002). Further support comes from the observations of Köhler *et al.* (2007) and Tobón *et al.* (this volume #26) that water loss through evaporation from canopy humus during rainless periods proceeds at slower rates than that from bryophytes.

If drought is indeed uncommon in TMCF canopies, one may expect fewer adaptations to drought in epiphytes from cloud forests compared to lowland or other, less wet montane forests. In Papua New Guinea, the incidence of crassulacean acid metabolism (CAM), a photosynthetic pathway typically found in drought-adapted succulents, declined with altitude from 25% of vascular epiphyte species in lowland forests to 0% in upper montane and sub-alpine forests at 2600–3600 m.a.s.l. (Earnshaw *et al.*, 1987), and along the Atlantic slope of eastern Mexico from 56% in pre-montane dry forests to 9% in a cloud forest at 1980 m (Hietz *et al.*, 1999). Perhaps surprisingly, a comparison of carbon isotope ratios between the same species of C3 epiphytes growing either in relatively dry pre-montane or in humid montane forests gave no significant difference (which would have indicated lower drought stress) in humid montane forests. However, the effect of decreasing temperatures with elevation on $\delta^{13}C$ values may have counteracted that of decreasing drought stress (Hietz *et al.*, 1999). Because the lack of water particularly limits small individuals with little internal or external water storage capacity relative to their transpiring surface (Zotz *et al.*, 2001), the presence of often very large numbers of species and individuals of small epiphytes such as pleurothalliid orchids (Bogh, 1992; Ingram *et al.*, 1996) in TMCF might also be a result of the better water supply conditions in TMCF.

In the Neotropics, bromeliads are among the most prominent epiphytes in many forests, but in TMCF their diversity and/or abundance often declines relative to that of orchids, ferns, and other groups. This family has two distinct adaptations to the irregular water supply in the canopy, which may be of little value or even a disadvantage in TMCF with their often abundant and regular water inputs. Bromeliad tanks can store external water over extended periods (Zotz *et al.*, 2003), but the necessary arrangement of the overlapping leaves causes a substantial proportion of the leaf area to become shaded which therefore does not receive sufficient light for positive photosynthesis, particularly in forests with low irradiation levels. Under such conditions, photosynthesis has to be maintained by other tissue. Water-absorbing trichomes, which cover the whole leaf surface in atmospheric bromeliads that do not form tanks, serve to rapidly absorb surface water over a large area. However, as long as the leaf surface is wet, the water film held by the trichomes greatly increases the resistance to gas diffusion, impeding photosynthetic CO_2 uptake (Benzing *et al.*, 1978). For instance, in central Veracruz (Mexico), the bromeliad *Tillandsia recurvata* is common in the dry and warm pre-montane forests and also in the dry highlands with occasional frost, but absent from the humid upper montane forests. If bromeliads are found in cloud forests at all, they tend to occupy the more exposed positions in the canopy where their leaf surface dries relatively fast. Because epiphytes in the more open canopy of secondary or disturbed forests are more exposed, they may be favored by disturbance. Indeed, in a montane rain forest region in Venezuela, bromeliads were the only group of epiphytes that was more diverse and more common in secondary vegetation compared to closed forest (Barthlott *et al.*, 2001). In Veracruz, Mexico, bromeliads also had higher densities in a disturbed forest than in a relatively intact forest (Hietz *et al.*, 2006), and higher biomass on isolated trees and on trees in a disturbed forest than in an intact forest plot (Flores-Palacios and García-Franco, 2003).

Many TMCF have a distinct dry season with higher sunshine and evaporation and reduced precipitation and humidity (Jarvis and Mulligan, this volume). Therefore, there can be little doubt that epiphytes in these TMCF will experience water stress (cf. Köhler *et al.*, this volume). Among ferns studied in a Mexican cloud forest, those found at more exposed positions and thinner branches were better adapted to avoid or tolerate drought (Hietz and Briones, 1998). In the same forest, after a long dry spell during the otherwise normal wet season of 1999, the water content of five out of eight epiphytic fern species had declined to <40% of saturation. Fronds of several ferns were wilted, but recovered after re-watering in the laboratory and also in the field. Wilting of vascular epiphytes has also been reported from montane rain forests in New Guinea (Brass, 1956, cited by Grubb and Whitmore, 1966).

Several authors (e.g. Grubb and Whitmore, 1966; Pócs, 1980) have suggested that water uptake from fog is important for cloud forest epiphytes, particularly non-vascular ones such as bryophytes (cf. Tóbon *et al.*, this volume #26). Liquid water is absorbed when deposited onto organs dedicated to water uptake such as roots or bromeliad trichomes and in a few cases similar structures have been described on the leaves of orchids (Raciborski, 1898) and ferns (Müller *et al.*, 1981). The same applies to many poikilohydric cryptogams, where cuticles, if present, are very inefficient at preventing water loss but will favor the uptake of liquid water through the entire leaf surface. Water may be absorbed as vapor directly from a saturated atmosphere or as liquid water if droplets are deposited on the leaf surface. Water uptake through the cuticle has been observed in a few non-epiphytes (Kerstiens, 1996), but the cuticles of many lowland epiphytes are particularly efficient barriers against water loss (Helbsing *et al.*, 2000) and even in several of the cloud forest ferns studied, uncontrolled (mainly cuticular) water loss is very low (Hietz and Briones, 1998). To measure the capacity of different cloud forest epiphytes to absorb liquid water through their leaves should be relatively simple, but this has not been attempted so far. There is some evidence that leaves take up liquid and gaseous water, with the latter probably playing a minor role (Breazeale *et al.*, 1950; Yates and Hutley, 1995). Recently, Burgess and Dawson (2004) showed that the leaves of *Sequoia sempervirens* (redwood) trees can absorb substantial quantities of fog water in California's coastal forests and the same may be true for vascular epiphytes in TMCF. Whereas non-vascular epiphytes can rapidly absorb deposited fog water, for vascular epiphytes the overriding effect of the high atmospheric humidity prevailing in cloud forest environments is mostly to reduce water loss.

Climate change is both expected (Still *et al.*, 1999; Lawton *et al.*, 2001; Foster, this volume) and to some extent observed, to result in an upward shift of the level of cloud formation on tropical mountains (cf. Pounds *et al.*, 1999; Ray *et al.*, 2006; Nair *et al.*, this volume). This may have a dramatic effect on vascular as well as non-vascular epiphytes, since most epiphytes in these forests are poorly adapted to drought (Zotz and Bader, 2009).

EPIPHYTE NUTRIENT SUPPLY

With most research efforts focusing on water limitations in epiphytes, there is relatively little evidence that nutrients limit epiphyte growth (Zotz and Hietz, 2001). Two experimental studies have compared the effects of water and nutrient supply on the growth of epiphytic bromeliads, both comparing plants of various sizes. A study in a Mexican humid montane forest suggested that after *Tillandsia guatemalensis* had developed water-storing tanks, nutrient supply became the primary limiting factor (Castro-Hernández *et al.*, 1999) although there were no comparisons with plants exposed to natural nutrient input levels. This contention is supported by the observation that in tank-forming *Tillandsia* drought stress decreased, as indicated by more negative $\delta^{13}C$ values. In addition, the C/N ratio increased with increasing plant size and tank volume, which also points to relatively greater limitation by water stress in small plants versus nitrogen limitation in large plants (Hietz and Wanek, 2003). However, a full factorial design testing the effects of light, nutrients, and water on *Vriesea sanguinolenta* from a lowland forest in Panama found that nutrients affected growth only in small and intermediate plants (Laube and Zotz, 2003).

Being without soil contact, epiphytes may be expected to be independent from the soil, but Gentry and Dodson (1987a) postulated epiphyte diversity and abundance in (mostly lowland) forests to be lower on nutrient-poor soils compared to more fertile soils. If true, this would imply that the nutrient status of the entire forest is of relevance for epiphytes, and one might find differences in epiphyte nutrient supply between cloud or montane forests and other forests. Such contrasts might become manifest via nutrients taken up by, and released from the host tree – or reflect differences in nutrient inputs from other (e.g. atmospheric) sources. The importance of host-tree nutrient supply on an epiphytic lichen was shown by a dramatic increase in the abundance of *Pseudocyphellaria crocata* in Hawaiian montane forests after phosphorus fertilization of the soil (Benner *et al.*, 2007). Generally speaking, nutrient concentrations in fog water are much higher than those in precipitation (Asbury *et al.*, 1994; Clark *et al.*, 1998; Hafkenscheid, 2000). If the water supply for epiphytes is more favorable in cloud forest than in other forest types, cloud forest epiphytes could be relatively more limited by nutrients, and even outgrow their nutrient supply, which would result in lower tissue nutrient concentrations. There are no studies comparing the growth of individual epiphyte species occurring in cloud and other forests, but a substantial data-set is available for concentrations of nitrogen, which is the one nutrient that is considered to be limiting photosynthesis and growth in montane ecosystems in general (Benner *et al.*, this volume; cf. Soethe *et al.*, 2008). Foliar (or stem in the case of cacti) nitrogen concentrations were about 50% higher in epiphytes from three humid montane forests (at least one of which classified as a cloud forest) than in three drier pre-montane forests in Mexico (Hietz *et al.*, 1999). Although this particular comparison may be problematic because the epiphyte communities were composed of different species, among the 10 species found in both pre-montane and humid montane forests, all had higher foliar nitrogen concentrations at the higher and wetter sites, on average by 57%.

What may be the cause of higher nitrogen concentrations in epiphytes from humid montane forests? Potential nitrogen sources for epiphytes include atmospheric wet and dry deposition, animal remains and droppings, host-tree leachates, nitrogen fixation, and nutrients released by the decomposition of host material. The importance of each is discussed briefly below.

Nitrogen supplied by animals is unlikely to be important (except for ant-hosting plants and perhaps tank bromeliads) as this would require animals to be present in higher numbers in TMCF than in other forests which is generally not the case (e.g. Kikuzawa *et al.*, 2002; Tanabe *et al.*, 2002). However, the fauna in bromeliad tanks and canopy soil helps to break down litter and thus will accelerate decomposition and nutrient release (Ngai and Srivastava, 2006).

Because foliar concentrations of nitrogen, as well of most other nutrients studied, generally decrease with altitude, and litter production decreases as well (Bruijnzeel and Proctor 1995; Tanner *et al.*, 1998; Benner *et al.*, this volume), the direct input of nitrogen from the host trees should, if anything, be lower in TMCF than in lowland rain forests. Concentrations of nitrogen and several other nutrients (sodium, magnesium, calcium, sulfur, potassium) in cloud water are about an order of magnitude higher than in rainwater (Asbury *et al.*, 1994; Clark *et al.*, 1998; Hafkenscheid, 2000; Chang *et al.*, 2002). Thus, where cloud water deposition contributes significantly to overall precipitation, this may represent an important nutrient source for cloud forest epiphytes. Comparison of the nutrient fluxes in incident precipitation and throughfall in several epiphyte-laden TMCF showed net retention in the canopy of, mostly, NH_4-N and/or NO_3-N but large net leaching losses from the canopy for base nutrients, especially potassium (Veneklaas, 1990; Asbury *et al.*, 1994; Clark *et al.*, 1998; Hafkenscheid, 2000).

Interestingly, several of the more intensively studied TMCF are situated in areas of recent volcanic activity, including Monteverde (Costa Rica), Colombia, and Hawai'i. A review of nutrient sources from outside the pedosphere (Clarkson *et al.*, 1986) does not mention volcanic sources at all, and very little consideration has been given to the potential supply of nutrients to epiphytes from volcanic emissions (except in Coxon and Nadkarni, 1995), even though volcanic soils are well known to be among the most fertile in the tropics. Grubb and Whitmore (1966) suggested volcanism to be a potential nutrient source for epiphytes in a montane forest in Ecuador and Veneklaas (1990) reported very high concentrations of calcium, magnesium, and sulfur in throughfall after the Nevado del Ruiz deposited volcanic dust onto the canopy of montane rain forests at 2550 and 3370 m in Colombia. Klinge (1962) analyzed the composition of minerals from canopy soils of several cloud forests in El Salvador. The bulk of the material consisted of more or less weathered pumice as in many terrestrial soils in the area, but also a number of very young volcanic minerals such as augite, hornblende, hypersthene, and young volcanic glass were found.

Nitrogen fixation from epiphylls on understory leaves were reported to contribute a substantial 2–5 kg ha^{-1} year^{-1} to overall nitrogen inputs to a Costa Rican pre-montane wet forest (Freiberg, 1998), with the fixation activity being particularly dependent on the water content of the epiphylls. In lowland rain forests, most non-vascular epiphytes in the upper canopy will be desiccated and thus inactive for most of the time, and the potential for nitrogen fixation by epiphylls in continuously wet cloud forests could be substantially higher. However, the density of epiphylls also depends on light and host leaf longevity, and epiphylls on leaves of canopy trees tend to be sparse, also in cloud forests. By contrast, the growth of lichens, including nitrogen-fixing species, on branches and exposed twigs can be impressive in montane rain forests and Forman (1975), using estimates of the biomass of nitrogen-fixing lichens in a Colombian cloud forest and published data on nitrogen fixation rates, calculated nitrogen inputs by epiphytic lichens as 1.5–8 kg ha^{-1} year^{-1}.

For terrestrially rooted plants, the bulk of nitrogen taken up is not derived from external sources but rather from nitrogen recycled from decomposing litter (Jordan, 1985). For epiphytes rooting on bark there is little potential for nitrogen recycling, but those in canopy soil appear to derive much of their nitrogen from the decomposition of epiphytic matter. Only a very small fraction of host-tree leaves remain on epiphyte-covered branches (Nadkarni and Matelson, 1991), thus most of the litter decomposing to canopy soil is derived from epiphytes. This should result in a large proportion of nitrogen being recycled within the canopy, which is also supported by isotopic signals (Hietz *et al.*, 2002). By contrast, non-vascular epiphytes, which absorb and lose nutrients through most of their surface and have a much higher turnover, retain nutrients only for a short time and will depend more on nutrients in precipitation passing through the canopy than on nutrients from canopy soil (Clark *et al.*, 1998; Hafkenscheid, 2000). The beneficial effect of canopy soil was even more significant for epiphyte nitrogen supply than for water relations in leeward cloud forest at Monteverde, with plants growing in canopy soil having 9–81% higher nitrogen concentrations than comparable plants on bare branches (Hietz *et al.*, 2002). Apart from potential differences in nutrient inputs between cloud and other forests (discussed above), the fact that many cloud forest epiphytes root in canopy soil, where the water supply and consequently the possibility for nutrient uptake are more continuous than on bare bark, will affect epiphyte nutrient supply.

Unless enriched by volcanic ash, canopy soils are usually largely of organic origin with a high carbon content and they have therefore even been termed epiphytic Histosols (Clark and Nadkarni, 2000). In contrast to most terrestrial Histosols, where nutrient concentrations can be extremely low (Roman *et al.*, this volume), canopy soils can have rather high nutrient contents and low C/N ratios. For example, for the previously cited LMCF at 1480 m at Monteverde, Clark and Nadkarni (2000) reported C/N ratios of 15.9 in canopy soil vs. 25.8 in the terrestrial A-horizon; Hietz *et al.* (2002) found values of 25.0 vs. 12.8, and Vance and Nadkarni (1990) of 17.3 vs. 15.6, respectively. Rates of nitrogen mineralization in the canopy soil were very low, but there was no evidence of nitrogen immobilization (Vance and Nadkarni, 1990).

Many of the cited data on nitrogen inputs may be considered fairly coarse estimates. The available data on nitrogen inputs and foliar concentrations suggest that availability of nitrogen is higher for LMCF epiphytes than for epiphytes in lowland rainforests, and also higher than in UMCF (cf. Benner *et al.*, this volume and discussion of Wolf (1993a) below). For individual epiphytes the contribution of specific sources such as nitrogen fixation can be assessed by analyzing nitrogen isotopes (Hietz and Wanek, 2001), but once different sources contribute to the overall nitrogen cycle, the isotopic signal soon becomes difficult to follow.

In contrast to nitrogen, little information is available for other nutrients in epiphytes and canopy soil. In the LMCF of Monteverde, Costa Rica, nutrient concentrations in dead canopy organic matter were 16‰ nitrogen, 0.7‰ phosphorus, 6‰ calcium, 3‰ potassium, and 1‰ magnesium (Nadkarni *et al.*, 2004). Under these conditions, epiphyte leaves had much lower nitrogen concentrations than leaves of terrestrially rooted trees (14 vs. 23‰), the concentration of phosphorus was about the same, and calcium, potassium, and magnesium were present in higher concentrations in epiphyte leaves. In a Colombian UMCF at 3700 m.a.s.l., concentrations of water-extractable nitrogen (mostly NH_4-N), potassium, and calcium were similar in canopy and terrestrial soils (cf. Rabatin *et al.*, 1993), but extractable phosphorus from canopy soil exceeded that from terrestrial soil more than 20-fold (Hofstede *et al.*, 1993). Nutrient concentrations in epiphytic bryophytes were generally highest near the foot of the trunks. Phosphorus concentrations in bryophytes were highest on the outer branches, where potassium concentrations were lowest; concentrations of nitrogen and calcium showed no trend with branch position. Along the same altitudinal transect in the northern Colombian Andes, concentrations of ash, nitrogen, phosphate, potassium, sulfate, and calcium were approximately two times higher in canopy soil material from LMRF at 1250–1980 m compared to UMCF at 2550–3670 m (Wolf, 1993a). With the exception of SO_4, concentrations of water-soluble nutrients were also higher in the LMRF canopy soils. No C/N ratios were given, but with a nitrogen concentration of 1.15% and 4.4% ash content, the C/N ratio should be around 40. This is substantially higher than the values reported for canopy soil in the LMCF at Monteverde and suggests that nitrogen in the cooler and wetter UMCF is indeed in shorter supply. A few recent studies further suggest that phosphorus may be more limiting to epiphyte growth than nitrogen. In several species a higher proportion of phosphorus than nitrogen is reabsorbed from senescent leaves (Zotz, 2005) and phosphorus rather than nitrogen appears to limit fruit production (Zotz and Richter, 2006). Fertilizing the terrestrial soil of a Hawaiian montane forest with phosphorus for 5.5 years resulted in a dramatic increase in both biomass and diversity of epiphytes, particularly in a forest on phosphorus-poor Oxisol, but fertilization with nitrogen and other nutrients had no effect (Benner and Vitousek, 2007; Benner *et al.*, 2007).

Summarizing, there are few data comparing foliar concentrations of nutrients other than nitrogen in epiphytes in different montane forest types, life-forms, or canopy positions. Such information should be relatively straightforward to obtain, however, and would help to shed light onto the sources and supply of nutrients whose cycles differ substantially from that of nitrogen (cf. Benner *et al.*, this volume).

IMPORTANCE OF EPIPHYTES IN THE CLOUD FOREST ECOSYSTEM

In montane rain forests, vascular epiphytes can account for 30% or more of the total forest flora and they unquestionably constitute a major component of the forest's floristic diversity (Gentry and Dodson, 1987b). Most epiphytes are zoophilous and support a large number of pollinators and, even if rates of herbivory are low, their contribution to the diversity of the canopy fauna is probably also substantial. Given the epiphytic biomass of several tons or even tens of tons per hectare (Coxon and Nadkarni, 1995; Köhler *et al.*, 2007) – which rivals and sometimes exceeds that of tree leaves – and the importance of fog interception in TMCF, several authors have stressed the importance of epiphytes to nutrient and water cycles in these forests (see among others the detailed review by Coxson and Nadkarni, 1995).

As to the potential effect of epiphytes on the canopy water balance, Rhoades (1995) stated that heavy loads of non-vascular epiphyte can greatly influence the hydrological "cycle" in forest canopies by intercepting large amounts of water. Pócs (1980) estimated that non-vascular epiphytes in a Tanzanian sub-montane forest intercepted 10%, and in an ECF 24% of total precipitation. Such claims would seem to be supported by the relatively low throughfall fractions (55–65% of incident rainfall) reported for a number of epiphyte-laden lower montane (cloud) forests (Cavelier *et al.*, 1997; Ataroff, 1998; Clark *et al.*, 1998; Fleischbein *et al.*, 2006; Fleischbein *et al.*, this volume). However, the former figures on the effect of epiphytes on water cycles were derived from calculations extrapolating estimates of epiphyte biomass and measurements of the absorptive capacity of collected epiphytes, and they contrast with several field studies that measured the actual hydrological properties of entire epiphyte-covered branches. Furthermore, in most cases, the interception of rainfall and fog water by epiphytes and canopy soil will be substantially lower than the theoretical maximum. It is pertinent to note that the live and dead epiphytic matter can only absorb the difference between current water content and saturated water content. When precipitation is frequent, field water content is continuously rather high, leaving little room to absorb additional moisture (Veneklaas *et al.*, 1990; Hölscher *et al.*, 2004; Köhler *et al.*, 2007; cf. Tobón *et al.*, this volume #26).

Veneklaas *et al.* (1990) found that an average of 81% of the rain falling on cut epiphyte-covered branches in a Colombian UMCF was intercepted initially. Part of this water drained off within the first few hours, i.e. the branches only delayed throughfall, whereas the remainder of the intercepted water was lost by evaporation. Overall interception by the forest canopy was 18.3% of incident rainfall (Veneklaas and Van Ek, 1990) but the authors made no attempt to estimate the contribution to total interception by epiphytes. An approximate number may be derived as follows: overall epiphytic biomass for the forest was very high but similar to that of the Monteverde cloud forest (44 and 33 t ha^{-1}, respectively; Hofstede *et al.*, 1993; Nadkarni *et al.*, 2004). The branch surface area within the inner tree crowns carrying thick mats of epiphytes at Monteverde has been estimated at 520 m^2 ha^{-1} (i.e. 5.2%; Nadkarni and Matelson, 1991). The projected area of all branches may be higher, but thin twigs will carry fewer epiphytes whereas some of the large branches will overlap when projected horizontally. Supposing that epiphyte-covered branches retain 50% of the rain falling on them (Veneklaas *et al.*, 1990), then the 5% surface area provided by epiphyte-covered branches would intercept 2.5% of the precipitation. Given the fact that overall throughfall in the Monteverde forest amounts to 65% of rainfall (not counting an estimated potential annual cloud water input of 886 mm; Clark *et al.*, 1998), it will be clear that most of the 35% interception loss reported for this forest must be due to other causes than its high epiphyte load (notably atmospheric factors such as low rainfall intensity, strong winds, advected energy; cf. Schellekens *et al.*, 2000).

The latter contention is supported by a recent hydrological study from an UMRF not affected much by fog at 2900 m.a.s.l. in the Cordillera Talamanca, Costa Rica (Hölscher *et al.*, 2004; Köhler *et al.*, this volume). The entire canopy of this old-growth forest intercepted *c.* 28% of bulk precipitation. Some epiphytic bryophytes (total epiphytic biomass was estimated at 3.4 t ha^{-1}) were shown to be able to absorb water up to 700% of their dry weight, but the net contribution of epiphytic materials to total interception loss was only 6% of total interception or less than 2% of bulk precipitation (Hölscher *et al.*, 2004). Chang *et al.* (2002) estimated fog interception by pendant epiphytic bryophytes alone as 36% of total atmospheric deposition in a subtropical coniferous forest in Taiwan. They measured the weight gain of air-dried mosses during short exposures to fog and extrapolated this to the stand level by multiplying times the estimated total bryophyte biomass. This procedure would only yield approximately correct values if all bryophytes are pendant and growing on exposed branches, but will result in a gross overestimation if the bryophytes grow in large mats on branches, and if fog interception in the interior of the forest is lower than for the exposed experimental plants. In addition, air-drying the samples prior to exposure will also produce higher absorption rates (Mulligan *et al.*, this volume #25; Tobón *et al.*, this volume #26).

CONCLUSIONS

Epiphyte communities in most cloud forests are luxuriant because of the nearly optimal water supply, but temperature becomes limiting for many vascular species at higher altitudes because of the risk of frost. Available evidence suggests that the supply of nitrogen, and possibly of other nutrients, for epiphytes is more favorable in lower montane cloud forests than in lowland forests, with the decomposition of canopy organic matter and nitrogen release being slowed down by lower temperatures and higher moisture content in upper montane cloud forest. Unquestionably, interception of rainfall and fog by epiphytes is vitally important for their water supply, but the amounts of water actually intercepted by epiphytes are unlikely to substantially affect net precipitation amounts arriving at the forest floor.

REFERENCES

Asbury C.E., W.H.McDowell, R. Trinidad-Pizarro, and R. Berrios (1994). Solute deposition from cloud water to the canopy of a Puerto Rican montane forest. *Atmospheric Environment* **28**: 1773–1780.

Ataroff, M. (1998). Importance of cloud water in Venezuelan Andean cloud forest water dynamics. In *Proceedings of the 1st International Conference on Fog and Fog Collection*, eds. R.S. Schemenauer and H.A. Bridgman, pp. 25–38. Ottowa, Canada: IDRC.

Ball, E., J. Hann, M. Kluge, *et al.* (1991). Ecophysiological comportment of the tropical CAM-tree *Clusia* in the field. I. Growth of *Clusia rosea* Jacq. on St John, US Virgin Islands, Lesser Antilles. *New Phytologist* **117**: 473–481.

Barthlott, W., V. Schmit-Neuerburg, J. Nieder, and S. Engwald (2001). Diversity and abundance of vascular epiphytes: a comparison of secondary vegetation and primary montane rain forest in the Venezuelan Andes. *Plant Ecology* **152**: 145–156.

Benner, J.W., and P.M. Vitousek (2007). Development of a diverse epiphyte community in response to phosphorus fertilization. *Ecology Letters* **10**: 628–636.

Benner, J.W., S. Conroy, C.K. Lunch, N. Toyoda, and P.M. Vitousek (2007). Phosphorus fertilization increases the abundance and nitrogenase activity of the cyanolichen *Pseudocyphellaria crocata* in Hawaiian montane forests. *Biotropica* **39**: 400–405.

Benzing, D.H. (1990). *Vascular Epiphytes: General Biology and Related Biota.* Cambridge, UK: Cambridge University Press.

Benzing, D.H., J. Seemann, and A. Renfrow (1978). The foliar epidermis in Tillandsioideae (Bromeliaceae) and its role in habitat selection. *American Journal of Botany* **65**: 359–365.

Bogh, A. (1992). Composition and distribution of the vascular epiphyte flora of an Ecuadorian montane rain forest. *Selbyana* **13**: 25–34.

Bohlman, S.A., T.J. Matelson, and N.M. Nadkarni (1995). Moisture and temperature patterns of canopy humus and forest floor soil of a montane cloud forest, Costa Rica. *Biotropica* **27**: 13–19.

Brass, L.J. (1956). Results of the Archbold Expeditions. No. 79. Summary of the Fourth Archbold Expedition to New Guinea (1953). *Bulletin of the American Museum of Natural History* **118**: 1–70.

Breazeale, E.L., W.T. McGeorge, and J.F. Breazeale (1950). Moisture absorption by plants from an atmosphere of high humidity. *Plant Physiology* **25**: 413–419.

Bruijnzeel, L.A. (2001). Hydrology of tropical montane cloud forests: a reassessment. *Land Use and Water Resources Research* **1**: 1–18.

Bruijnzeel, L.A. (2005). Tropical montane cloud forests: a unique hydrological case. In *Forests, Water and People in the Humid Tropics*, eds. M. Bonell and L.A. Bruijnzeel, pp. 462–483. Cambridge, UK: Cambridge University Press.

Bruijnzeel, L.A., and J. Proctor (1995). Hydrology and biogeochemistry of tropical montane cloud forests: what do we really know? In *Tropical Montane Cloud Forests*, eds. L.S. Hamilton, J.O. Juvik, and F.N. Scatena, pp. 38–78. New York:Springer-Verlag.

Bruijnzeel, L.A., and E.J. Veneklaas (1998). Climatic conditions and tropical montane forest productivity: the fog has not lifted yet. *Ecology* **79**: 3–9.

Burgess, S.S.O., and T.E. Dawson (2004). The contribution of fog to the water relations of *Sequoia sempervirens* (D. Don): foliar uptake and prevention of dehydration. *Plant, Cell and Environment* **27**: 1023–1034.

Cardelús, C.L., R.K. Colwell, and J.E. Watkins (2006). Vascular epiphyte distribution patterns: explaining the mid-elevation richness peak. *Journal of Ecology* **94**: 144–156.

Castro-Hernández, J.C., J.H.D. Wolf, J.G. Garcia-Franco, and M. Gonzalez-Espinosa (1999). The influence of humidity, nutrients and light on the establishment of the epiphytic bromeliad *Tillandsia guatemalensis* in the highlands of Chiapas, Mexico. *Revista de Biologia Tropical* **47**: 763–773.

Cavelier, J., M. Jaramillo, D. Solis, and D. de Leon (1997). Water balance and nutrient inputs in bulk precipitation in tropical montane cloud forest in Panama. *Journal of Hydrology* **193**: 83–96.

Chang, S.C., I.L. Lai, and J.T. Wu (2002). Estimation of fog deposition on epiphytic bryophytes in a subtropical montane forest ecosystem in northeastern Taiwan. *Atmospheric Research* **64**: 159–167.

Clark, K.L., and N.M. Nadkarni (2000). Epiphytic histosols. In *Monteverde: Ecology and Conservation of a Tropical Cloud Forest*, eds. N.M. Nadkarni and N.T. Wheelwright, pp. 34–35. New York: Oxford University Press.

Clark, K.L., N.M. Nadkarni, D. Schaeffer, and H.L. Gholz (1998). Atmospheric deposition and net retention of ions by the canopy in a tropical montane forest, Monteverde, Costa Rica. *Journal of Tropical Ecology* **14**: 27–45.

Clarkson, D.T., P.J.C. Kuiper, and U. Lüttge (1986). Mineral nutrition: sources of nutrients for land plants from outside the pedosphere. *Progress in Botany* **48**, 80–96.

Coxon, D.S., and N.M. Nadkarni (1995). Ecological roles of epiphytes in nutrient cycles of forest ecosystems. In *Forest Canopies*, eds. M. Lowman and N.M. Nadkarni, pp. 495–543. San Diego, CA: Academic Press.

Earnshaw, M.J., K. Winter, H. Ziegler, *et al.* (1987). Altitudinal changes in the incidence of crassulacean acid metabolism in vascular epiphytes and related life forms in Papua New Guinea. *Oecologia* **73**: 566–572.

Fleischbein, K., W. Wilcke, R. Goller, *et al.* (2006). Rainfall interception in a lower montane forest in Ecuador: effects of canopy properties. *Hydrological Processes* **19**: 1355–1371.

Flores-Palacios, A., and J.G. García-Franco (2003). Effect of isolation on the structure and nutrient budget of oak epiphyte communities. *Plant Ecology* **173**: 259–269.

Forman, R.T.T. (1975). Canopy lichens with blue–green algae: a nitrogen source in a Colombian rain forest. *Ecology* **56**: 1176–1184.

Frahm, J.P., and S.R. Gradstein (1991). An altitudinal zonation of tropical rain forests using bryophytes. *Journal of Biogeography* **18**: 669–676.

Freiberg, E. (1998). Microclimatic parameters influencing nitrogen fixation in the phyllosphere in a Costa Rican premontane rain forest. *Oecologia* **117**: 9–18.

Gentry, A.H. (1992). Tropical forest biodiversity: distributional patterns and their conservational significance. *Oikos* **63**: 19–28.

Gentry, A.H., and C.H. Dodson (1987a). Diversity and biogeography of neotropical vascular epiphytes. *Annals of the Missouri Botanical Garden* **74**: 205–233.

Gentry, A.H., and C.H. Dodson (1987b). Contribution of nontrees to species richness of a tropical rain forest. *Biotropica* **19**: 149–156.

Grubb, P.J., and T.C. Whitmore (1966). A comparison of montane and lowland rain forest in Ecuador. II. The climate and its effects on the distribution and physiognomy of the forests. *Journal of Ecology* **54**: 303–333.

Grubb, P.J., J.R. Lloyd, T.D. Pennington, and T.C. Whitmore (1963). A comparison of montane and lowland rain forest in Ecuador. I. The forest structure, physiognomy, and floristics. *Journal of Ecology* **51**: 567–601.

Hafkenscheid, R.L.L.J. (2000). Hydrology and biogeochemistry of tropical montane rain forests of contrasting stature in the Blue Mountains, Jamaica. Ph.D. thesis. VU University Amsterdam, Amsterdam, The Netherlands. Also available at http://dare.ubvu.vu.nl/bitstream/1871/12734/1/tekst.pdf.

Helbsing, S., M. Riederer, and G. Zotz (2000). Cuticles of vascular epiphytes: efficient barriers for water loss after stomatal closure? *Annals of Botany* **86**: 765–769.

Hemp, A. (2001). Ecology of the pteridophytes on the southern slopes of Mt. Kilimanjaro. II. Habitat selection. *Plant Biology* **3**: 493–523.

Hietz, P. (1997). Population dynamics of epiphytes in a Mexican humid montane forest. *Journal of Ecology* **85**: 767–775.

Hietz, P., and O. Briones (1998). Correlation between water relations and within-canopy distribution of epiphytic ferns in a Mexican cloud forest. *Oecologia* **114**: 305–316.

Hietz, P., and O. Briones (2001). Photosynthesis, chlorophyll fluorescence and within-canopy distribution of epiphytic ferns in a Mexican cloud forest. *Plant Biology* **3**: 279–287.

Hietz, P., and U. Hietz-Seifert (1995). Composition and ecology of vascular epiphyte communities along an altitudinal gradient in central Veracruz, Mexico. *Journal of Vegetation Science* **6**: 487–498.

Hietz, P., and W. Wanek (2001). Use of stable isotopes in canopy research. In *Tropical Ecosystems: Structure, Diversity and Human welfare*, eds. K.N. Ganeshaia, U. Uma Shaanker, and K.S. Bawa, pp. 412–415. New Delhi: Oxford–IBH.

Hietz, P., and W. Wanek (2003). Size-dependent variation of carbon and nitrogen isotope abundances in epiphytic bromeliads. *Plant Biology* **5**: 137–142.

Hietz, P., and J. H. D. Wolf (1996). Vascular epiphytes. In *How to Sample the Epiphytic Diversity of Tropical Rain Forests*, eds. S.R. Gradstein, P. Hietz, R. Lücking, *et al.*, pp. 60–63. New York: Ecotropica.

Hietz, P., W. Wanek, and M. Popp (1999). Stable isotopic composition of carbon and nitrogen and nitrogen content in vascular epiphytes along an altitudinal transect. *Plant, Cell and Environment* **22**: 1435–1443.

Hietz, P., W. Wanek, R. Wania, and N.M. Nadkarni (2002). Nitrogen-15 natural abundance in a montane cloud forest canopy as an indicator of nitrogen cycling and epiphyte nutrition. *Oecologia* **131**: 350–355.

Hietz P., G. Buchberger, and M. Winkler (2006). Effect of forest disturbance on abundance and distribution of epiphytic bromeliads and orchids. *Ecotropica* **12**: 103–112.

Hietz-Seifert, U., P. Hietz, and S. Guevara (1996). Epiphyte vegetation and diversity on remnant trees after forest clearance in southern Veracruz, Mexico. *Biological Conservation* **75**: 103–111.

Hofstede, R.G.M., J.H.D. Wolf, and D.H. Benzing (1993). Epiphytic biomass and nutrient status of a Colombian upper montane rain forest. *Selbyana* **14**: 37–45.

Hölscher, D., L. Köhler, A.I.J.M. Van Dijk, and L.A. Bruijnzeel (2004). The importance of epiphytes to total rainfall interception by a tropical montane rain forest in Costa Rica. *Journal of Hydrology* **292**: 308–322.

Holwerda, F., L.A. Bruijnzeel, L.E. Muñoz, M. Equihua, and H. Asbjornsen (2010). Rainfall and cloud water interception in mature and secondary lower montane cloud forests of central Veracruz, Mexico. *Journal of Hydrology* **384**: 84–96.

Ibarra-Manríquez, G., and C.S. Sinaca (1987). *Listados florísticos de México VII*. Los Tuxtlas, Veracruz: Estación de Biología Tropical, and Mexico City: Instituto de Biología, UNAM.

Ibisch, P.L. (1996). Neotropische Epiphytendiversität – das Beispiel Bolivien. *Archiv naturwissenschaflicher Dissertationen* **1**: 1–357.

Ibisch, P.L., A. Boegner, J. Nieder, and W. Barthlott (1996). How diverse are neotropical epiphytes? An analysis based on the 'Catalogue of the flowering plants and gymnosperms of Peru'. *Ecotropica* **2**: 13–28.

Ingram, S.W., K. Ferrell-Ingram, and N.M. Nadkarni (1996). Floristic composition of vascular epiphytes in a neotropical cloud forest, Monteverde, Costa Rica. *Selbyana* **17**: 88–103.

Jordan, J.F. (1985). *Nutrient Cycling in Tropical Forest Ecosystems*. Chichester, UK: John Wiley.

Kerstiens, G. (1996). Cuticular water permeability and its physiological significance. *Journal of Experimental Botany* **305**: 1813–1832.

Kikuzawa, K., S. Suzuki, K. Umeki, and K. Kitayama (202). Herbivorous impacts on tropical mountain forests implied by fecal pellet production. *Sabah Parks Nature Journal* **5**: 131–142.

Klinge, H. (1962). Über Epiphytenhumus aus El Salvador (Zentralamerika). I. Die Standorte und ihre gesamtökologische Situation. *Pedobiologia* **2**: 1–8.

Köhler, L., C. Tobón, K.F.A. Frumau, and L.A. Bruijnzeel (2007). Biomass and water storage of epiphytes in old-growth and secondary montane rain forest stands in Costa Rica. *Plant Ecology* **193**: 171–184.
Krömer, T., M. Kessler, S.R. Gradstein, and A. Acebey (2005). Diversity patterns of vascular epiphytes along an elevational gradient in the Andes. *Journal of Biogeography* **32**: 1799–1809.
Kürschner, H., and G. Parolly (1999). Pantropical epiphytic rain forest bryophyte communities: coeno-syntaxonomy and floristic-historical implications. *Phytocoenologia* **29**: 1–52.
Laube, S., and G. Zotz (2003). Which abiotic factors limit vegetative growth in a vascular epiphyte? *Functional Ecology* **17**: 598–604.
Lawton, R.O., U.S. Nair, R.A. Pielke Sr., and R.M. Welch (2001). Climatic impact of tropical lowland deforestation on nearby montane cloud forests. *Science* **294**: 584–587.
Long, A., and M. Heath (1991). Flora of the El Triunfo biosphere reserve, Chiapas, Mexico: a preliminary floristic inventory and the plant communities of Polygon I. *Annales del Instituto Biologico UNAM* **62**: 133–172.
Lüttge, U. (1989). *Vascular Plants as Epiphytes: Evolution and Ecophysiology*. Heidelberg, Germany: Springer-Verlag.
Müller, L., G. Starnecker, and S. Winkler (1981). Zur Ökologie epiphytischer Farne in Südbrasilien. I. Saugschuppen. *Flora* **171**: 55–63.
Nadkarni, N.M., and T.J. Matelson (1991) Fine litter dynamics within the tree canopy of a tropical cloud forest. *Ecology* **72**: 2071–2082.
Nadkarni, N.M., D. Schaefer, T.J. Matelson, and R. Solano (2004). Biomass and nutrient pools of canopy and terrestrial components in a primary and a secondary montane cloud forest, Costa Rica. *Forest Ecology and Management* **198**: 223–236.
Ngai, J.T., and D.S. Srivastava (2006). Predators accelerate nutrient cycling in a bromeliad ecosystem. *Science* **314**: 963.
Nieder, J., P.L. Ibisch, and W. Barthlott (1997). Biodiversidad de epífitas: una cuestión de escala. *Revista del Jardín Botánico Nacional* **17–18**: 59–62.
Nieder, J., S. Engwald, and W. Barthlott (2002). Patterns of neotropical epiphyte diversity. *Selbyana* **20**: 66–75.
Pócs, T. (1980). The epiphytic biomass and its effect on the water balance of two rain forest types in the Uluguru Mountains (Tanzania, East Africa). *Acta Botanica Academiae Scientiarum Hungaricae* **26**: 143–167.
Ponette-González, A.G., K.C. Weathers, and L.M. Curran (2009). Water inputs across a tropical montane landscape in Veracruz, Mexico: synergistic effects of land cover, rain and fog seasonality, and interannual precipitation variability. *Global Change Biology*, doi: 10.1111/j.1365-2486.2009.01985.x
Pounds, J.A., M.P.L. Fogden, and J.H. Campbell (1999). Biological response to climate change on a tropical mountain. *Nature* **398**: 611–615.
Rabatin, S.C., B.R. Stinner, and M.G. Paoletti (1993). Vesicular–arbuscular mycorrhizal fungi, particularly *Glomus tenue*, in Venezuelan bromeliad epiphytes. *Mycorrhiza* **4**: 17–20.
Raciborski, M. (1898). Biologische Mitteilungen aus Java. *Flora* **85**: 325.
Ray, D.K., U.S. Nair, R.O. Lawton, R.M. Welch, and R.A. Pielke Sr. (2006). Impact of land use on Costa Rican tropical montane cloud forests: sensitivity of orographic cloud formation to deforestation in the plains. *Journal of Geophysical Research* **111**: D02108.
Rhoades, F.M. (1995). Nonvascular epiphytes in forest canopies: worldwide distribution, abundance, and ecological roles. In *Forest Canopies*, eds. M. Lowman and N.M. Nadkarni, pp. 353–408. San Diego, CA: Academic Press.
Richards, P.W. (1952). *The Tropical Rain Forest*. Cambridge, UK: Cambridge University Press.
Schellekens, J., L.A. Bruijnzeel, F.N. Scatena, N.J. Bink, and F. Holwerda (2000). Evaporation from a tropical rain forest, Luquillo Experimental Forest, eastern Puerto Rico. *Water Resources Research* **36**: 2183–2196.
Soethe, N., W. Wilcke, J. Homeier, J. Lehmann, and C. Engels (2008). Plant growth along the altitudinal gradient: role of plant nutritional status, fine root activity, and soil properties. In *Gradients in a Tropical Mountain Ecosystem of Ecuador*, eds. E. Beck, J.Bendix, I. Kottke, F. Makeschin, and R. Mosandl, pp. 259–266. Berlin: Springer-Verlag.
Stadtmüller, T. (1987). *Cloud Forests in the Humid Tropics: A Bibliographic Review*. Tokyo: United Nations University, and Turrialba, Costa Rica: CATIE.
Still, C.J., P.N. Foster, and S.H. Schneider (1999). Simulating the effects of climate change on tropical montane cloud forests. *Nature* **389**: 608–610.
Tanabe, S., M.J. Toda, M.B. Lakim, and M.B. Mohamed (2002). Abundance, biomass, and composition of insect communities in various forests on Mount Kinabalu. *Sabah Paks Nature Journal* **5**: 219–237.
Tanner, E.V.J., P.M. Vitousek, and E. Cuevas (1998). Experimental investigation of nutrient limitation of forest growth on wet tropical mountains. *Ecology* **79**: 10–22.
Vance, E.D., and N.M. Nadkarni (1990). Microbial biomass and activity in canopy organic matter and the forest floor of a tropical cloud forest. *Soil Biology and Biochemistry* **22**: 677–684.
Veneklaas, E.J. (1990). Nutrient fluxes in bulk precipitation and throughfall in two montane tropical rain forests, Colombia. *Journal of Ecology* **78**: 974–992.
Veneklaas, E.J., and R. van Ek (1990). Rainfall interception in two tropical montane rain forests, Colombia. *Hydrological Processes* **4**: 311–326.
Veneklaas, E.J., R.J. Zagt, A. van Leerdam, *et al.* (1990). Hydrological properties of the epiphyte mass of a montane tropical rain forest, Colombia. *Vegetatio* **89**: 183–192.
Wolf, J.H.D. (1993a). Epiphyte communities of tropical montane rain forests in the northern Andes. II. Upper montane communities. *Phytocoenologia* **22**: 53–103.
Wolf, J.H.D. (1993b). Ecology of epiphytes and epiphyte communities in montane rain forests, Colombia. Ph.D. dissertation, Amsterdam: University of Amsterdam.
Wolf, J.H.D. (1993c). Diversity patterns and biomass of epiphytic bryophytes and lichens along an altitudinal gradient in the northern Andes. *Annals of the Missouri Botanical Garden* **80**: 928–960.
Wolf, J.H.D., and A. Flamenco-Sandoval (2003). Patterns in species richness and distribution of vascular epiphytes in Chiapas, Mexico. *Journal of Biogeography* **30**: 1689–1707.
Yates, D.J., and L.B. Hutley (1995). Foliar uptake of water by wet leaves of *Sloanea woollsii*, an Australian subtropical rainforest tree. *Australian Journal of Botany* **43**: 157–167.
Zotz, G. (1999). Altitudinal changes in diversity and abundance of non-vascular epiphytes in the tropics: an ecophysiological explanation. *Selbyana* **20**: 256–260.
Zotz, G. (2004). The resorption of phosphorus is greater than that of nitrogen in senescing leaves of vascular epiphytes from lowland Panama. *Journal of Tropical Ecology* **20**: 693–696.
Zotz, G. (2005). Vascular epiphyte in the temperate zones: a review. *Plant Ecology* **176**: 173–183.
Zotz, G., and J.L. Andrade (2001). *Ecología de epífitas y hemiepífitas*. In *Ecología de bosques lluviosos Neotropicales*, eds. M. Guariguata and G. Kattan, pp. 271–296. San Josó, Costa Rica: IICA.
Zotz, G., and M.Y. Bader (2009). Epiphytic plants in a changing world: global change effects on vascular and non-vascular epiphytes. *Progress in Botany* **70**: 147–170.
Zotz, G., and M. Büche (2000). The epiphytic filmy ferns of a tropical lowland forest: species occurrence and habitat preferences. *Ecotropica* **6**: 203–206.
Zotz, G., and P. Hietz (2001). The physiological ecology of vascular epiphytes: current knowledge, open questions. *Journal of Experimental Botany* **52**: 2067–2078.
Zotz, G., and A. Richter (2006). Changes in carbohydrate and nutrient contents throughout a reproductive cycle indicate that phosphorus is a limiting nutrient in the epiphytic bromeliad, *Werauhia sanguinolenta*. *Annals of Botany* **97**: 745–754.
Zotz, G., P. Hietz, and G. Schmidt (2001). Small plants, large plants: the importance of plant size for the physiological ecology of vascular epiphytes. *Journal of Experimental Botany* **52**: 2051–2056.
Zotz, G., P. Reichling, and F. Valladares (2003). A simulation study on the importance of size-related changes in leaf morphology and physiology for carbon gain in an epiphytic bromeliad. *Annals of Botany* **90**: 1–7.

6 Global and local variations in tropical montane cloud forest soils

L. Roman and F. N. Scatena
University of Pennsylvania, Philadelphia, Pennsylvania, USA

L. A. Bruijnzeel
VU University, Amsterdam, the Netherlands

ABSTRACT

Although soil resources are widely considered as a major factor that reduces the productivity, stature, and diversity of tropical montane cloud forests (TMCF), systematic comparisons of soil resources within and between TMCF are lacking. This study combines published reports on TMCF soils with new data on the soils and forest structure of the Luquillo Mountains in Puerto Rico to assess the current state of knowledge regarding global and local-scale variation in TMCF soils. At the global scale, soils from 33 TMCF sites and over 150 pedons are reviewed. Compared to soils in humid lowland tropical forests, TMCF soils are relatively acidic, have higher organic matter content, and are relatively high in total nitrogen and extractable phosphorus. Across all sites, significant correlations also exist between mean annual precipitation and soil pH and base saturation, but not between any soil chemical factor and canopy height, site elevation, or air temperature. Although comparisons between TMCF are limited by inconsistent sampling protocols, analysis of available data does indicates that lower montane cloud forests (LMCF) have taller canopies, higher soil pH, lower soil nitrogen, and higher C/N ratios than upper montane cloud forests (UMCF). Within an UMCF in NE Puerto Rico, the abundance of soil nitrogen, carbon, and potassium accounted for 25% to 54% of the variation in canopy height. However, as much as 68% of the variation in stand height could be accounted for when site exposure, slope gradient, and the percent coverage of surface roots were also included in the analysis. The importance of these local site factors is apparently related to interactions between exposure, slope steepness, and stand age on the accumulation of soil organic matter. This review indicates that soils investigations in TMCF need to explicitly include the influence of local site conditions and use consistent sampling and analytical methodologies. Nevertheless, the soils of upper and lower montane cloud forests apparently have distinct characteristics.

INTRODUCTION

Whilst there have been considerable advances in our understanding of the hydrology and ecology of tropical montane cloud forests (TMCF), less emphasis has been placed on understanding their soil resources. This study addresses TMCF soils at the global and local scale through a comparative analysis of published reports on TMCF soils, and new information on local-scale relationships between soils and forest structure in the Luquillo Mountains of Puerto Rico.

The unique features of TMCF are widely recognized and have puzzled natural scientists for decades. Compared to lowland and montane forests not frequently enveloped in fog or clouds, TMCF are typically shorter in stature, have smaller, thicker, and more leathery leaves, and gnarled stems and branches that tend to be covered with bryophytes and epiphytes (Stadtmüller, 1987; Whitmore, 1990). More recent assessments have divided TMCF into lower montane cloud forests (LMCF), upper montane cloud forests (UMCF), and sub-alpine cloud forests (SACF) (Bruijnzeel, 2005). This convention is followed here and considers that LMCF have 25–50% epiphytic cover on their stems vs. 70–80% in UMCF (cf. Frahm and Gradstein, 1991; see Bruijnzeel (2005) for detailed discussion).

Most researchers agree that multiple climatic and site factors influence the structural and functional characteristics of TMCF.

Tropical Montane Cloud Forests: Science for Conservation and Management, eds. L. A. Bruijnzeel, F. N. Scatena, and L. S. Hamilton. Published by Cambridge University Press.

It is pertinent to note that – with very few exceptions (e.g. Hafkenscheid, 2000) – most earlier reviews seeking an explanation for the distinct characteristics of TMCF (Grubb, 1977; Flenley, 1993; Bruijnzeel and Proctor, 1995; Bruijnzeel and Veneklaas, 1998) made no explicit distinction between conditions prevailing in LMCF and UMCF. As a result, a host of sometimes contradictory mechanisms have been proposed, including the following:

- Periodic water shortage, especially in the case of shallow and stony soils.
- Saturated soils and impeded root transpiration.
- Reduced photosynthesis associated with low radiation inputs and air temperatures.
- Limited nutrient uptake due to (a) climatically controlled reduction in transpiration rates, (b) extreme soil acidity and/or "free" aluminum levels, (c) low soil fertility, and/or (d) reduced decomposition and mineralization of litter.
- Exposure to strong winds or high UV-B radiation levels.
- Presence of phenolic compounds in leaves and soil moisture that interfere with metabolic processes.

Although six out of these ten possible mechanisms are directly related to surface soil resources, these early reviews could not identify an overriding soil chemical factor that uniquely characterizes *all* TMCF (Whitmore, 1990; Bruijnzeel and Proctor 1995). As such, there is a need to examine and quantify key differences (if any) in the soil characteristics of LMCF and UMCF.

DEFINING TMCF SOILS

Defining the spatial dimensions of TMCF ecosystems has proved to be a great challenge (Hamilton *et al.*, 1995; Mulligan, this volume; Scatena *et al.*, this volume). Defining TMCF soils may be an even greater challenge due to the abundance of detritus, roots, and soil micro-organisms that occur above the surface of the forest floor (Edwards, 1982; Cavelier and Peñuela, 1990; Nadkarni and Longino, 1990; Paoletti *et al.*, 1991). This review focuses on surface soils in all TMCF and recognizes differences between UMCF and LMCF where possible (see methods section below). It also recognizes, but does not address, the importance of aerial soils and micro-organisms in these ecosystems (cf. Benner *et al.*, this volume; Hietz, this volume).

Regardless of their precise definition, a review of the literature indicates that several generalizations are commonly made regarding TMCF soils. Namely, that these soils are: (i) waterlogged and anaerobic, (ii) high in soil organic matter (SOM), and (iii) highly acidic. Many also consider these soils to be limiting to above-ground forest productivity because they are either (iv) nutrient poor, or (v) inherently toxic (cf. Benner *et al.*, this volume). These generalizations are discussed briefly below.

Waterlogging and anaerobic conditions

Waterlogging and associated anaerobic conditions have often been cited as principal characteristics of TMCF soils (Herrmann, 1971; Hetsch and Hoheisel, 1976; Kapos and Tanner, 1985; Bruijnzeel *et al.*, 1993; Schawe *et al.*, this volume). In general, soil oxygen decreases with increasing SOM and precipitation, and TMCF soils can be anoxic for extended periods of time (Silver *et al.*, 1999; Silver *et al.*, this volume). Many studies have also found that increasing precipitation and waterlogging are related to changes in soil respiration, redox potential, nutrient content, and productivity in many tropical forests (Cavelier and Peñuela, 1990; Bohlman *et al.*, 1995; Austin and Vitousek, 1998; Raich, 1998; Silver *et al.*, 1999; Santiago *et al.*, 2000; Schuur and Matson, 2001; Austin, 2002; cf. Schawe *et al.*, this volume). Nevertheless, waterlogged and anaerobic soils can support tall and highly productive wetland forests (Lugo *et al.*, 1990). In addition, both observations and soil profile characteristics indicate that the soils from some TMCF sites rarely or never experience extensive waterlogging, even under high-rainfall conditions (Hafkenscheid *et al.*, 2002; Tobón *et al.*, this volume #52). These observations also suggest that waterlogged and anaerobic soils are not the sole cause of the lower above-ground productivity and stature of TMCF. However, the presence of these features may be useful in distinguishing between UMCF and LMCF.

Acidity and SOM

The presence of high SOM, humus, or peat has traditionally been used as a key defining characteristic of TMCF soils (Hamilton *et al.*, 1995). This abundance of SOM is generally explained by slow rates of litter decomposition (Vitousek and Sanford, 1986; Tanner *et al.*, 1998; Hafkenscheid, 2000). In turn, soil acidity and the reduced rates of nutrient cycling that are associated with slow decomposition rates have been suggested as direct or indirect mechanisms causing the perceived low productivity of TMCF (Grubb, 1977; Edwards, 1982; Vitousek, 1984; Hafkenscheid, 2000). Nevertheless, a wide variety of tall lowland forests with high above-ground productivity occurs on organically rich and highly acidic soils (Lugo *et al.*, 1990), suggesting that reduced SOM turnover and high soil acidity are not the sole cause of the relatively low stature and above-ground productivity of TMCF ecosystems (Bruijnzeel and Veneklaas, 1998). In addition, Röderstein *et al.* (2005) and Hertel and Leuschner (this volume) show that – whilst above-ground productivity can indeed decrease strongly with increasing elevation and decreasing temperatures – below-ground productivity becomes increasingly important, thereby dampening the effect of altitude on overall ecosystem productivity (Moser *et al.*, 2007, 2008). Again, most early comparisons of TMCF and their soils did not distinguish between UMCF and LMCF.

Nutrient limitation and toxicity

Both nitrogen and phosphorus limitation have been invoked as restrictors on above-ground productivity in TMCF (Vitousek,

1984; Vitousek and Sanford, 1986; Tanner *et al.*, 1990, 1992; Bruijnzeel *et al.*, 1993; Hafkenscheid, 2000; cf. Benner *et al.*, this volume) as well as in some lowland forests (Sollins, 1998; Tanner *et al.*, 1998). Some TMCF soils have also been considered toxic due to high levels of free aluminum and polyphenols (reviewed in Bruijnzeel and Proctor, 1995; Bruijnzeel and Veneklaas, 1998). Unfortunately, few studies have reported the abundance of free aluminum or polyphenols in TMCF soils (e.g. Grimm and Fassbender, 1981; Coxson *et al.*, 1992; Bruijnzeel *et al.*, 1993; Hafkenscheid, 2000). Comparative work in Jamaica indicated that the concentration of polyphenols is strongly correlated with dissolved oxygen concentration (DOC) levels in soil water; absolute concentrations of polyphenols were much larger in stunted UMCF compared to adjacent taller LMCF, and their influence on forest productivity may therefore be limited to UMCF (Hafkenscheid, 2000). This may partly explain why at least some LMCF soils are used in agriculture (Hamilton *et al.*, 1995). However, polyphenols have varied effects on ecosystem processes, not all of which are detrimental (Hättenschwiler and Vitousek, 2000). For example, Northup *et al.* (1995) demonstrated how phenolic compounds produced by the roots in stunted pine forests helped prevent leaching of scarce nitrogen. Others have argued that the ecological cost of producing these secondary compounds in a nutrient-poor environment may result in reduced above-ground stature (Hafkenscheid, 2000; cf. Hertel and Leuschner, this volume).

Explanations based solely on nutrient limitations or toxicity are also contradicted by the observation that *total* soil nitrogen levels can be higher in UMCF than in adjacent LMCF or lowland forests (Hafkenscheid, 2000; Cox *et al.*, 2002). Nevertheless, plant uptake of *available* nitrogen in UMCF may be influenced by phenols and the temporal variability of NO_3-N in some UMCF soils may be related to antecedent rainfall and stagnant water held in micro-pores (Hafkenscheid, 2000). Studies of nutrient turnover in fine litter in a Costa Rican LMCF also suggested that neither nitrogen nor phosphorus were limiting (Nadkarni and Matelson, 1992). Finally, recent studies in LMCF in southern Ecuador have demonstrated that soil nutrients are very heterogeneously distributed at the local scale, whereas observed concentrations showed no indications for possible limitation of growth due to nitrogen or phosphorus deficiencies (Wilcke *et al.*, 2001, 2002; cf. Soethe *et al.*, 2006). However, forests in the same area showed a marked shift in below-ground carbon allocation with increasing elevation (Röderstein *et al.*, 2005; Leuschner *et al.*, 2007; Moser *et al.*, 2008; Hertel and Leuschner, this volume). In addition, foliar concentrations of nitrogen, phosphorus and potassium in UMCF and ECF at 2400 and 3000 m.a.s.l. were much lower than in the LMCF, suggesting that at these higher elevations the uptake of nutrients was reduced more strongly than above-ground biomass. Furthermore, tree basal area increment rates were significantly correlated with nutrient concentrations in the organic layer whereas rooting patterns at higher elevations suggested a reduced capacity for spatial exploitation of soil nutrients (Soethe *et al.*, 2008a, b).

METHODS

Review of global data

Soil data for a given site were included in this global analysis of TMCF soils if the site met at least two of the following conditions: (i) presence of tropical or sub-tropical forest, with elevation and precipitation being within the characteristic range for TMCF (Jarvis and Mulligan, this volume); (ii) the publication in question mentioned frequent fog occurrence; and (iii) the location was either included in the list of known TMCF sites in Hamilton *et al.* (1995) or known by the authors. In all, soil data from 33 sites and over 150 pedons were compared, thereby updating the previous review of TMCF soils by Bruijnzeel and Proctor (1995) (Table 6.1).

Because the global data-set included a variety of forest types and sampling strategies, individual sites were further grouped and compared by forest type and sampling depth. Three forest types were distinguished initially – LMCF, UMCF, and SACF – following the definitions provided by Bruijnzeel (2005). Classifications were based on published site descriptions and our personal knowledge of the sites. However, because the number of sub-alpine sites was limited, the analysis was restricted to comparisons between all TMCF, and between UMCF and LMCF. Soil chemical data were also divided into two groups, viz. one in which the surface organic horizon was included, and the other in which only the soil below was sampled.

Luquillo field study

Comparisons of TMCF soil data are limited by a lack of standard sampling and analytical methods (Bruijnzeel and Proctor, 1995). To evaluate directly the extent to which soil and site characteristics could account for local variation in forest structure, a complementary study was conducted in which regional climate, bedrock geology, and forest type were all held constant. All sampling was done within the UMCF or elfin cloud forest zone of the Luquillo Experimental Forest of NE Puerto Rico. The study area is underlain by erosion-resistant hornfels formed by contact metamorphism of granodiorites and volcanoclastic sediments and typically receives between 4500 and 5000 mm of rainfall annually. The elfin cloud forests are restricted to the upper elevation ridge tops and steep side slopes between 750 and 1050 m.a.s.l. and have gnarled stems that vary in height between 2.3 and 11.5 m. On ridge tops these soils can have organic surface horizons up to 1 m thick. However, these organic surface layers typically thin away from the ridge tops such that soils on steep side slopes can have rocky

Table 6.1 *Locations and environmental characteristics of tropical montane cloud forest soil studies*

Location	Elevation (m)	Forest type[a]	Average annual temperature (°C)	Average annual precipitation (mm $year^{-1}$)	Canopy height (m)	Soil type	Reference
Bolivia, Cordillera Oriental, Yungas	1850–3050	LMCF UMCF, SACF	10–17	5300–2300	10–20	Humic Dystrudept, Typic Placorthod	Schawe *et al.* (2001)
Colombia, Cordillera Central	2550	LMCF UMCF	12.2	2115	25	Andic Humitropept	Veneklaas (1990)
Colombia, Cordillera Central	1453–2115	LMCF UMCF	7.7–12.2	1453–2115	25	Andic Humitropept, Andaquept	Thouret & Faivre (1989)
Colombia, Cauca, Tambito	1400–2250	LMCF	12–18	5300	8–15	Andept	Letts *et al.* (this volume)
Colombia, Serrania de Macuira	650			855	7–9	Inceptisol, granite	Cavelier (1988)
Costa Rica, Volcan Barva	1500	LMCF	17	4015	25–30	Humitropept	Grieve *et al.* (1990)
Costa Rica, Volcan Barva	2000	LMCF	14	4015	20–25	Tropofibrist	Heaney & Proctor (1990)
Costa Rica, Volcan Barva	2000	LMCF	14	4015	20–25	Tropofibrist, Vitrandept	Lieberman *et al.* (1996)
Costa Rica, Volcan Barva	2600	LMCF	11	4015	20–23	Vitrandept	Marrs *et al.* (1988)
Costa Rica, Monteverde	1550	LMCF	17	2000	12–25	Typic Dystrandept	Vance & Nadkarni (1990)
Ecuador, Loja and Zamora	1900–2450	LMRF LMCF UMCF	15–16.2	2100–3840	7.19–20	Humic Eutrudepts, Typic Dystropepts, Oxyaquic Eutrudepts, Humic Dystrudepts	Wilcke *et al.* (2001) Wilcke *et al.* (2003) Wilcke *et al.* (this volume)
Hawai'i, Mauna Loa	1130–1660	UMCF	13.1–15.5	2620–4320	5.3–11.6	Lithic Tropofolist, Histosol	Raich *et al.* (1997)
Hawai'i, Mauna Loa	1220–1675	UMCF	12–15.5	2700–4450			Vitousek *et al.* (1988)
Indonesia, Rakata Island	680–730	UMCF	~23	3000	5–15	Andeptic Troporthent	Newsome (1986) Shinagawa *et al.* (1986) Schlesinger *et al.* (1998)
Indonesia, Seram, Gunung Kobipoto	1000	LMRF	~22	2100		Typic Humitropept	Payton (1993)
Indonesia, Seram, Gunung Kobipoto Gunung Binaiya	1470–2820	UMCF LMRF	~9–19	2100	10–27	Lithic Tropofolist, Humic Kanhapludult, Typic Hapludalf, Histic Humaquept	Edwards *et al.* (1993)
Indonesia, Kalimantan Bukit Raya	1700–2100	LMCF UMCF	~14–16	4000		Aquept	Van Reuler (1987)
Jamaica, Blue Mountains	1550	UMCF LMCF		2900	5–18	Dystric Cambisol, Folic Histosol	Tanner (1977) Hafkenscheid (2000)

Malaysia, Sarawak, Gunung Mulu	1310	LMCF	~22	4750		Histosol	Tie *et al.* (1979)
Malaysia, Sarawak Gunung Mulu	1310–1860	LMCF UMCF	19–22	4750		Gleysol, Histosol	Martin (1977)
Malaysia, Sabah, Kinabalu	1800	LMCF	16–17	2300	25	Humic, podzol	Kitayama (1992)
Malaysia, Sabah, Kinabalu	2000–3400	LMCF UMCF SACF	9–17	2300	6–30	Histosol, Humaquic Dystropept, Humic podzol	Frahm (1990)
Malaysia, Sabah, Kinabalu	2170–3080	UMCF SACF	10.4–15.9	2376–2887	10–20	Histosol, Humus podzol, Dystropept	Kitayama *et al.* (1998) Kitayama & Aiba (2002)
Malaysia, Sabah Gunung Silam	790	LMCF	~24	2500	10–15	Dystropept, Humitropept	Proctor *et al.* (1988) Bruijnzeel *et al.* (1993)
Mexico, Oaxaca, Sierra Juarez	1500–2380	LMCF UMCF	11–15	2550		Folic Cambisol, Folic Stagnic Podzol, Folic Stagnosol	Alvarez *et al.* (2008)
Mexico, Oaxaca, El Rincon	1850	LMCF	20–22	1720	18	Typic Dystrudept	Bautista-Cruz & Castillo (2005)
Mexico, Veracruz, Xalapa	1250–1840	LMCF	14–18	1517–2200		Cambisol	Williams-Linera (2002)
Puerto Rico, Luquillo	750–1050	UMCF	18–20	4500–5000	3–5		Silver *et al.* (1999) Olander *et al.* (1998) Cox *et al.* (2002)
Tanzania, Mt. Kilamanjaro	2265	LMCF UMCF	7–14	900–2700	10>30	Lithic Endoaquand, Pachic Haplustand, Acrudoxic Fulvudand, Acrudoxic Melanud, and Histic Placaquand	Schrumpf (2004) Hemp (this volume #12)
Venezuela, Parque Nacional Henri Pittier	1140–1580	LMCF UMCF	16–19	1650–1975	13–35	Tropohumult, Typic Dystropept, Typic Tropohumult	Zinck (1986)
Venezuela, Santa Ana, Copey, Zumbaor	730–3100		9–23	1630–4460	7–9	Inceptisol	Cavelier (1988)
Venezuela, Isla Margarita Cerro Copey	500–700	LMCF	>22	1000	15–22		Sugden (1986)
Venezuelan Andes La Mucuy, Monte Zerpa La Carbonera San Eusebio	2420–2670	UMCF	11–14	1500	24–30	Aquic Humitropept	Grimm *et al.* (1981) Schwarzkopf (2003) Steinhardt (1979)

[a] LMCF, UMCF, SACF are lower montane, upper montane, and sub-alpine cloud forests, respectively.

Table 6.2 *Summary of site, tree stem, and soil characteristics for transects in the Luquillo elfin cloud forest, Puerto Rico*

Measurement		Average	Min	Max
Site characteristics	Elevation (m)	873	657	963
(per transect, except slope per sub-plot)	Slope (degrees)	24	9	40
	Aspect (degrees)	135	75	190
	Exposure (1–5 scale)	3	1	5
Tree stem characteristics	Height (m)	6.4	2.3	11.5
(per sub-plot)	DBH (cm)	13.4	2.6	40.8
	Epiphytes (1–5 scale)[a]	3.3	1.7	4.7
	Bromeliad count (# per tree)	2.5	0	16.7
	Aerial roots (1–5 scale)[a]	1.1	1	2
Soil characteristics	Depth (cm)	24.9	1.5	30.0
(per sub-plot)	% Rocks	11	0	60
	% Soil/litter	44	10	80
	% Roots	46	10	80
	pH (log scale)	4.59[b]	3.80	5.34
	Bulk density (g cc^{-1})	1.07	0.60	1.52
	SOM (%)	28.84	17.25	38.85
	% C	13.30	3.87	26.80
	% N	0.59	0.18	0.98
	Ca (meq 100 g^{-1})	0.99	0.25	6.79
	K (meq 100 g^{-1})	1.89	1.07	3.42
	Mg (meq 100 g^{-1})	0.78	<0.01	1.64

[a] Epiphyte and aerial root scale explained in text.
[b] For pH, median is given in place of average.

surfaces without any visible surface organic layer. Weaver *et al.* (1986), Scatena (1995), and Silver *et al.* (1999) provide further details of the study area and forest.

The sampling design involved measuring and comparing soils and tree structure in sub-plots within transects that spanned the range of elfin forest heights, exposures, and slopes. Ten transects were measured on two peaks (El Yunque and Pico del Este) that are covered by elfin cloud forests, underlain by similar hornfels, and 5 km apart (Table 6.2). Transects were located in undisturbed, mature forest, and were chosen to represent a range of canopy heights within the same overall forest type. At each transect, 3–5 sub-plots of 1 m^2 area were established at 5-m intervals along the contour. The underlying assumption of this design was that disturbance history, tree species effects, and bedrock geology remained relatively constant across sites. Published climatic, ecological, and geological information for the Luquillo Mountains support these assumptions (Sieders 1971; Wang *et al.*, 2002). A statistical comparison of the soil cation contents on the two mountain peaks indicated that extractable magnesium was the only variable to show a significant difference, thereby confirming the similarity in soil forming processes operating at the two peaks.

For each sub-plot, the height and diameter at 1.3 m (diameter at breast height, DBH) were measured on the three largest stems within 1 m from the center. Abundance of epiphytes and surface (e.g. aerial) roots growing from the stem were estimated on each of the three stems using a 1–5 percentage scale (1 = 0–20%, 2 = 20–40%, etc.). Bromeliads larger than fist size on each stem were also counted and recorded.

Three 30-cm deep, 2.54-cm diameter soil cores were collected in each 1-m^2 sub-plot to estimate bulk density, after which they were combined for chemical analysis. Thus, the analyses reported here reflect the combined organic and mineral soil layers down to 30 cm. The percentage of rocks, soil/litter, and roots at the ground surface were quantified by randomly sampling 10 times in each sub-plot. Soils were dried at 65 °C and ground. Soil pH was measured in a mixture of 8 g of dry, ground unsieved soil and 15 ml of distilled water. Soils were then passed through a 2-mm sieve, and analyzed at the University of Pennsylvania using previously reported methods (Johnson *et al.*, 1991). Briefly, SOM was determined as percent dry soil mass lost upon ashing in a 500 °C furnace. Soil carbon and nitrogen were measured in a Carlo-Erba elemental analyzer, and exchangeable cations were extracted in NH_4Cl and measured in a Perkin Elmer ICP-AES.

Statistical analysis of the Luquillo data was carried out on a sub-plot basis using a type of multivariate, multi-step, analysis described by Myers (1990). Analysis of variance (ANOVA) and

Tukey tests were used to determine differences between slope gradient (≤20°, 20–30°, and ≥30°) and stem height. Principal component analysis was used to reduce the tree structural measurements of average stem height, maximum stem height, and average DBH into one overall dimension called "tree size." Principal component analysis was also used to reduce these same characteristics, plus epiphyte coverage, bromeliad count, and aerial root coverage into one overall dimension called "tree characteristics." Forward stepwise regression (0.250 probability to enter, 0.100 probability to leave) was then used to assess the degree to which soil and site characteristics contributed towards tree size and tree characteristics. A similar forward stepwise regression was also carried out for average stem height, and for maximum stem height. These four responses (tree size, tree characteristics, average stem height, and maximum stem height) were first analyzed using all possible soil and site characteristics, and subsequently using soil characteristics only.

RESULTS

Trends in global TMCF soils data

Soil orders reported for the collated TMCF data included Inceptisols, Histosols, Ultisols, and Mollisols (Table 6.1). However, only seven of the 33 studies explicitly reported a standard classification for their soil and most studies did not distinguish between organic surface and mineral soil horizons and merely indicated the depth of sampling (typically ranging between 10 and 35 cm; Table 6.1). Although subsequent analysis allowed many of the soils to be accurately classified (Bruijnzeel and Proctor, 1995), the lack of standardized methods does limit quantitative comparisons between sites.

Published levels of TMCF soil pH varied in the range 2.2–7.5, with an overall mean of 4.16 (Figure 6.1a). Soil carbon ranged from 1.1% to 49.6%, with a mean value of 16.2% (Figure 6.1b). This suggests that most studies combined organic and mineral horizons in their analysis. Soil nitrogen (total N) ranged from 0.2% to 3.6% (Figure 6.1b). In contrast, the range of soil nitrogen levels reported for nine different lowland sites in Costa Rica, Malaysia, and Puerto Rico was only 0.2–0.4% (see Silver (1994) for references). TMCF sites also had a higher mean soil nitrogen concentration than the lowland sites: 0.99% vs. 0.3%. Extractable soil phosphorus ranged from 0.2 to over 100 μg g^{-1} in TMCF (Figure 6.1c), compared to 4.0–26.0 μg g^{-1} in lowland forests (Silver, 1994). The mean extractable phosphorus concentration in TMCF soils was twice that of lowland forest soils (27.4 vs. 10.3 μg g^{-1}), suggesting that phosphorus is not limiting ecosystem productivity in TMCF. Extractable calcium in TMCF soils ranged from <0.1 to 22.1 meq 100 g^{-1} (Figure 6.1d), compared to 0–12.4 meq 100 g^{-1} in lowland forest soils (Silver, 1994). However, the TMCF data-set had a lower mean calcium concentration: 2.6 vs. 4.9 meq 100 g^{-1}.

Separating and comparing the TMCF soils into geologically "young" and "old" groups or "volcanic" and "non-volcanic" groups did not reveal statistical differences in total carbon, nutrient content, or pH. However, the comparisons were severely limited by small sample sizes and differing methodologies and cannot be considered definitive. Nevertheless, several pedons suggested that limestone bedrock can influence soil calcium while ultramafic bedrock may influence soil magnesium. Some altitudinal studies also demonstrated the influence bedrock can have on the genesis of TMCF soils (e.g. Alvarez Arteaga *et al.*, 2008).

When comparing all TMCF sites, higher-elevation sites were cooler and there was an expected negative relationship between elevation and temperature. Nevertheless, statistically strong relationships between mean annual rainfall, canopy height, and most soil properties were elusive. Canopy height did have weak but significant negative correlations ($p < 0.01$) with elevation and temperature for both UMCF and LMCF when the sub-sets of mineral soils were compared. Systematic differences were also apparent between the mineral soils of LMCF and UMCF (Figure 6.2). In general, canopy height and pH were higher for LMCF, whereas total nitrogen, SOM, and C/N ratios were higher in UMCF. Furthermore, UMCF tend to have more subsurface carbon, possibly lower nitrogen, and higher N/P and C/P ratios. As is typical of most soils, the global data-set of TMCF soils and the sub-sets of mineral soils from UMCF and LMCF showed strong and significant ($p < 0.0001$) positive correlations between carbon and total nitrogen concentrations, and between extractable phosphorus and total nitrogen. Significant ($p < 0.01$) but only moderate to weak correlations were found between other soil characteristics, regardless whether they were combined or grouped into LMCF or UMCF types. It should also be noted that these correlations are presented for comparative purposes as no causal or mechanistic relationships can be inferred from the correlations alone.

Luquillo field study

The 45 sub-plots in the Luquillo Experimental Forest, Puerto Rico contained a range of site, tree, and soil characteristics (Table 6.2). As expected, surface soils were acidic (median pH 4.59) and rich in SOM (average 28.8%) and carbon (average 13.3%). Compared to the global data-set (Figure 6.1), the standard deviations of the Luquillo elfin cloud forest soils were lower for pH (by 0.33 log units), carbon (by 3.95%), total nitrogen (by 0.16%), exchangeable Ca (by 1.15 meq 100 g^{-1}), and exchangeable Mg (by 0.35 meq 100 g^{-1}). The mean values for all these parameters at Luquillo were within one standard deviation of the respective global means, with the exception of exchangeable potassium which was higher in the Luquillo Mountains.

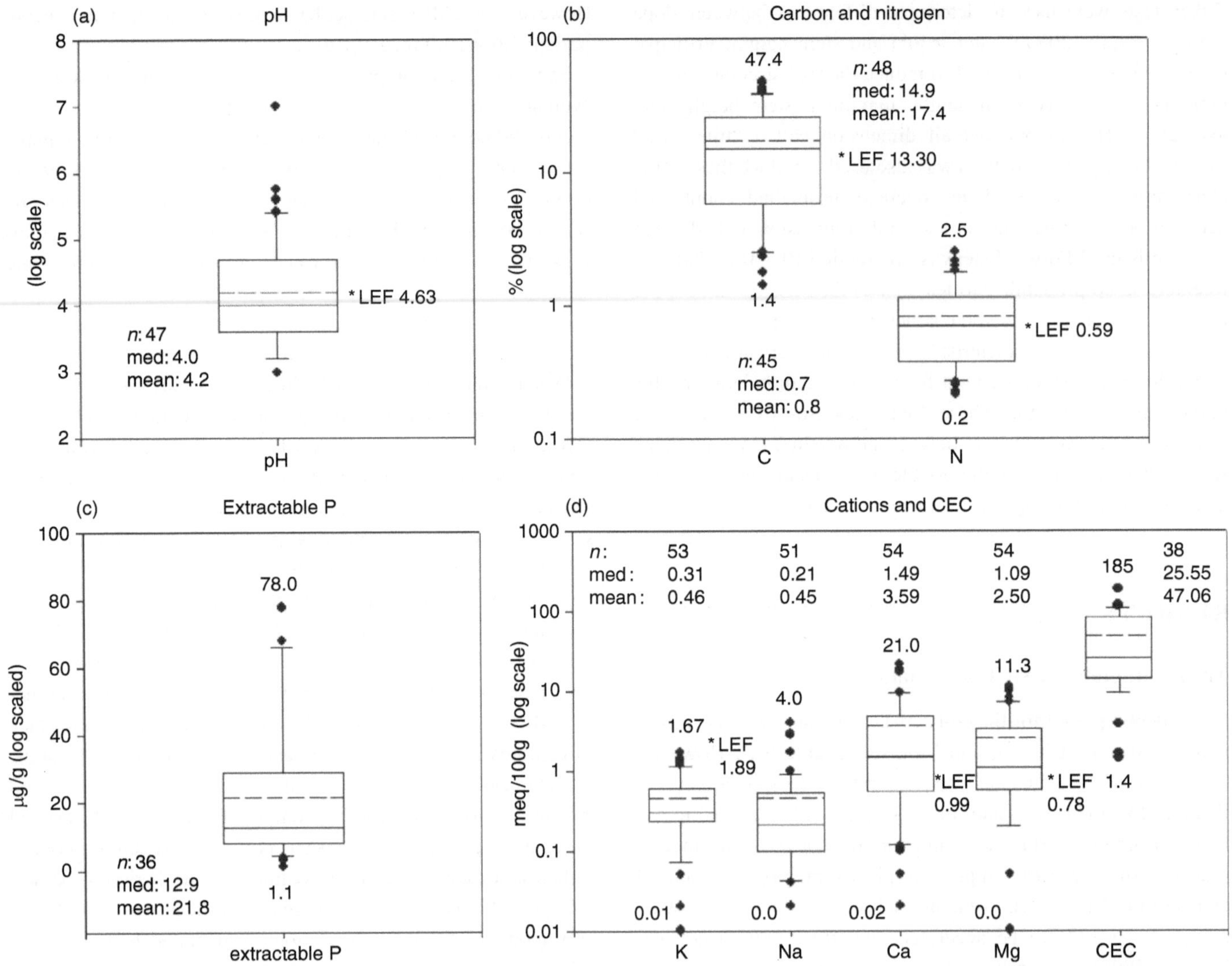

Figure 6.1. Chemical characteristics of the TMCF soils listed in Table 6.1 plus average values for the upper montane cloud forest soils of the Luquillo Experimental Forest (LEF) in Puerto Rico: (a) pH($_{H_2O}$), (b) carbon and total nitrogen (%), (c) extractable phosphorus (μg g^{-1}), and (d) exchangeable cations and cation exchange capacity (meq 100 g^{-1}). Box plots show median, 25th and 75th percentiles, outliers, and means (dashed lines). Values for the Luquillo soils are marked with an asterisk.

SOIL CORRELATIONS WITHIN LUQUILLO

As was observed for the global data-set, Luquillo cloud forest soils had significant ($p < 0.0001$) and strong correlations between concentrations of total carbon and nitrogen ($r^2 = 0.89$), and between SOM and carbon ($r^2 = 0.73$) or nitrogen ($r^2 = 0.88$). Again, significant ($p < 0.02$) but weak correlations were also found between exchangeable magnesium and potassium ($r^2 = 0.25$), between potassium and calcium ($r^2 = 0.15$), and between magnesium and calcium ($r^2 = 0.13$). In contrast to the global data-set, significant ($p < 0.01$) but moderate to weak correlations were found at Luquillo between exchangeable potassium and soil carbon ($r^2 = 0.15$), total nitrogen ($r^2 = 0.30$), and SOM ($r^2 = 0.44$). Soil pH correlated with carbon ($r^2 = 0.16$) and exchangeable calcium ($r^2 = 0.21$). The weaker correlations found using the global data-set may well reflect the wider range in soil types covered, as well as inconsistencies in sampling and laboratory analysis between studies.

RELATIONSHIPS BETWEEN TREE HEIGHT, SLOPE GRADIENT, AND SOM

In general, the shortest trees occurred on the steepest slopes (Figure 6.3a), and in areas with lower SOM ($r^2 = 0.15, p = 0.0090$) and soil carbon ($r^2 = 0.14$, $p = 0.0111$). The relationship between slope gradient and tree height became more pronounced when slope steepness was divided into three classes: ≤20°, 20–30°, and ≥30° (Figure 6.3b). Using these classes in an ANOVA, slope class was

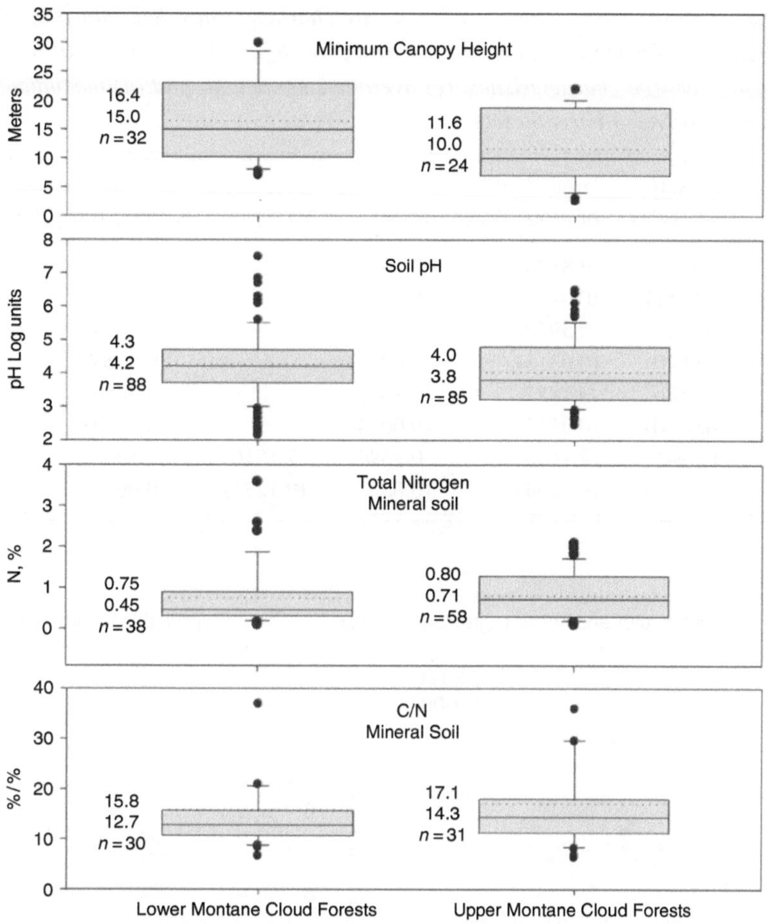

Figure 6.2. Forest canopy height (minimum reported) and soil characteristics by forest type for sites listed in Table 6.1 Comparisons for total nitrogen and C/N are only between mineral soils of comparable depths that are below the organic surface horizons.

significant in explaining average stem height ($p = 0.0284$). Tukey's test also indicated that trees on slopes $\leq 20°$ were significantly taller than those on slopes $\geq 30°$. Similar relationships existed between slope class and maximum tree height. Slope steepness did not appear as a major influencing factor in the multivariate analysis (Figure 6.4). This is apparently because of its cross correlation with other factors like soil depth, SOM content, and aspect.

STEPWISE REGRESSION ANALYSIS

Stepwise regression analysis (Table 6.3) indicated that between 56% and 68% of the variation in stem height within the Luquillo elfin forests could be accounted for by combinations of site exposure, surface root cover, soil carbon, exchangeable potassium and magnesium, and to a lesser extent, slope and total soil nitrogen. Average stem height, maximum stem height, and average DBH contributed approximately equally in the composite principal component variable called tree size (Figure 6.4). The abundance of epiphytes, bromeliads, and aerial roots contributed less to the variable tree characteristics and presumably rather reflect the age and disturbance history of the stand. Considering soil factors alone, only carbon, total nitrogen, and exchangeable potassium contributed and together explained 25% of the variation in tree size. These same variables explained 27% of the variation in tree characteristics, 38% of average stem height, and 54% of maximum stem height.

Stepwise regression analysis was also carried out using soil nutrient contents (e.g. kg ha^{-1}) as opposed to concentrations (e.g. mg g^{-1}). Because of the consistency in bulk density between sites, the resulting regressions and comparisons were similar regardless whether concentrations or contents were used. However, the resulting r^2-values tended to be slightly higher when the analysis was based on contents. Stepwise regressions were also conducted with and without the inclusion of ground cover as a variable, because ground-covering roots may be considered part of tree structure. In these cases, the r^2-values of the regressions were slightly greater when ground-covering roots were included but the general relationships and conclusions remained the same.

Table 6.3 *Results of forward stepwise regressions for all soil and site characteristics determined in 10 transects in the elfin cloud forest zone of the Luquillo Mountains, Puerto Rico, carried out with four different responses: (a) a composite variable for tree size, (b) a composite variable for tree characteristics, (c) average stem height, and (d) maximum stem height. Cell values are coefficients (top) and* p *values (between parentheses)*

Response	Slope (degrees)	Exposure (1–5 scale)	Ground cover of roots (%)	C (%)	N (%)	K (meq 100 g^{-1})	Mg (meq 100 g^{-1})	r^2
(a) Tree size	.	−0.7085	2.8824	0.1531	.	1.1702	1.3542	0.56
		(0.0051)	(0.0168)	(0.0059)		(0.0081)	(0.0559)	
(b) Tree	.	−0.7342	2.5964	0.1557	.	1.2074	1.4135	0.56
characteristics		(0.0040)	(0.0313)	(0.0055)		(0.0068)	(0.0480)	
(c) Avg. stem	−0.0297	−0.6586	3.1852	0.1681	.	1.0453	1.7055	0.63
height (m)	(0.3842)	(0.0310)	(0.0114)	(0.0034)		(0.0265)	(0.0182)	
(d) Max. stem	.	−0.5941	2.7053	0.3594	−7.1510	2.5093	.	0.68
height (m)		(0.0113)	(0.0164)	(0.0337)	(0.1271)	(0.0000)		

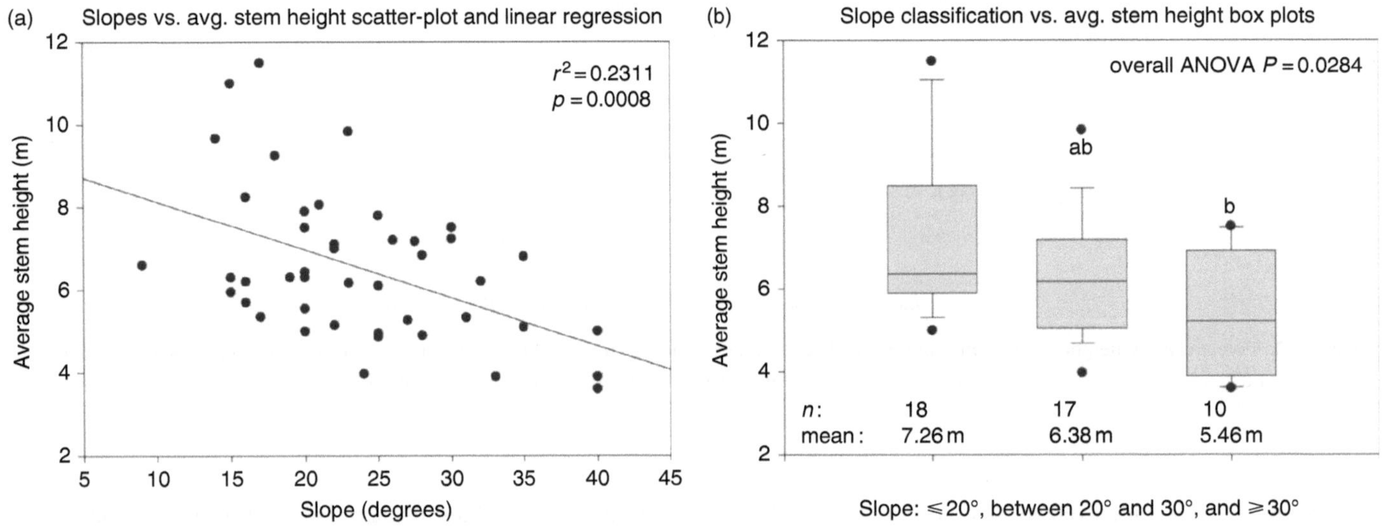

Figure 6.3. Relationships between slope gradient and average stem height in the Luquillo elfin cloud forests, Puerto Rico. (a) Box plots showing stem height distributions for each of three slope gradient classes; Tukey test levels not sharing the same letter are significantly different; (b) scatter-plot and linear regression between slope gradient and average stem height.

DISCUSSION AND CONCLUSIONS

How valid are the common generalizations about TMCF soils made earlier? The available data summarized above confirm that TMCF soils are generally more acidic and have more organic matter than the soils of lowland tropical forests. They may also show more inter-site variation. On average, they also have more total nitrogen and higher levels of extractable phosphorus and they are relatively rich in exchangeable cations. The available data also indicate that not all TMCF grow on peaty soils or soils that are inherently toxic or limiting to forest productivity. Comparisons across all sites also confirmed the results of site-specific studies on UMCF in Luquillo (this study) and LMCF in Ecuador (Wilcke *et al.*, 2002) that the organic layer in TMCF sites is highly variable but better developed in UMCF. Furthermore, the global data also suggest that if there are nutrient limitations to these forests, they are related to the plants' abilities to access the soil resource, rather than to the absolute size of the nutrient pools. However, the paucity of sound comparable soil data severely limits global-scale conclusions and supports the conclusions of earlier reviews that greater consistency in sampling, analysis and reporting is needed (Bruijnzeel and Proctor, 1995; Sollins, 1998). Including standardized soil classifications, making implicit distinctions between organic and mineral horizons, and adding information on geological substrate and local site conditions should be a standard element of future studies.

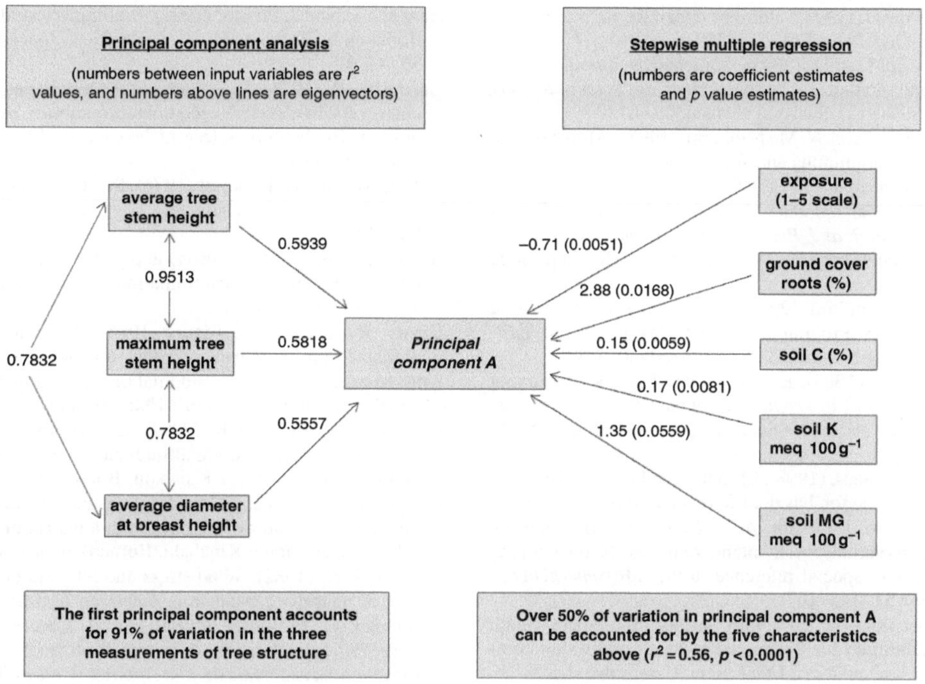

Figure 6.4. Principal component analysis and stepwise regression illustrating the links between tree size in elfin cloud forest in the Luquillo Mountains of Puerto Rico to environmental conditions. See also Table 6.3.

Nevertheless, the available data do indicate that there are distinct differences between UMCF and LMCF, both within and between sites. In general, UMCF have a higher frequency of waterlogging, better developed organic horizons, lower pH, and higher SOM, total nitrogen, total phosporus, and base cations than LMCF soils.

Within the UMCF of the Luquillo Mountains, amounts of soil nitrogen, carbon, and exchangeable potassium were able to account for 25–54% of the variation in stand height. As much as 68% of the variation in tree height could be accounted for when exposure, slope, and ground-cover roots were also included. The importance of these local site factors in explaining forest structure appears to be related to interactions between exposure, slope steepness, and the gradual accumulation of SOM and soil nutrients as stands age. Chronosequence studies of areas cleared during construction also indicate that UMCF in the Luquillo Mountains will not re-establish until the soils have regained sufficient SOM and exchangeable nutrients (Olander *et al.*, 1998). The latter study and the present correlation analysis of the Luquillo soil data-set both indicate that once TMCF are established, their soils accumulate SOM and become increasingly acidic and nutrient-rich over time. Both the correlation analysis, and the observations that steeper areas had more fallen and gnarled trees, suggest that the accumulation of SOM and the development of tree height is limited on steeper and more exposed slopes. Wind- and exposure-centered explanations of elfin cloud forest structure (Lawton, 1982) have received less attention in recent years. However, the importance of local exposure and site conditions demonstrated for the Luquillo Mountains suggests that these deserve greater consideration, at least in non-equatorial settings where wind speeds are often greater (cf. Jarvis and Mulligan, this volume).

ACKNOWLEDGEMENTS

LR and FNS wish to thank the University of Pennsylvania Nassau Undergraduate Research Fund and the Department of Earth and Environmental Science for travel funding. Field accommodations in Puerto Rico were provided by the U.S. Department of Agriculture Forest Service International Institute of Tropical Tropical Forestry. We are also thankful to P. S. Petraitis, A. H. Johnson, and W. Wilcke for their comments, to I. C. Baillie for help with soil classifications, to M. Dranoff for laboratory assistance, and to C. M. Maguire, A. S. Pike, and A. J. Moyer for field assistance.

REFERENCES

Alvarez Arteaga G., Garcia Calderon, N. E., Krasilnikov P. V., *et al.* (2008). Soil altitudinal sequence on base-poor parent material in a montane cloud forest in Sierra Juarez, Southern Mexico. *Geoderma* **144**: 593–612.

Austin, A. T. (2002). Differential effects of precipitation on production and decomposition along a rainfall gradient in Hawai'i. *Ecology* **83**: 328–338.

Austin, A. T., and P. M. Vitousek (1998). Nutrient dynamics on a precipitation gradient in Hawai'i. *Oecologia* **113**: 519–529.

Bautista-Cruz, A., and R. F. del Castillo (2005). Soil changes during secondary succession in a tropical montane cloud forest area. *Soil Science Society of America Journal* **69**: 906–914, doi:10.2136/sssaj2004.0130

Bohlman, S. A., T. J. Matelson, and N. M. Nadkarni (1995). Moisture and temperature patterns of canopy humus and forest floor soil of a montane cloud forest, Costa Rica. *Biotropica* **27**: 13–19.

Bruijnzeel, L. A. (2005). Tropical montane cloud forests: a unique hydrological case. In *Forests, Water and People in the Humid Tropics*, eds. M. Bonell and L. A. Bruijnzeel, pp. 462–483. Cambridge, UK: Cambridge University Press.

Bruijnzeel, L. A. and L. S. Hamilton (2000). *Decision Time for Cloud Forests*, IHP Humid Tropics Programme Series No. 13. Paris: IHP-UNESCO, Amsterdam: IUCN-NL, and Gland, Switzerland: WWF.

Bruijnzeel, L. A., and J. Proctor (1995). Hydrology and biogeochemistry of TMCF: what do we really know? In *Tropical Montane Cloud Forests*, eds. L. S. Hamilton, J. O. Juvik, and F. N. Scatena, pp. 38–45. New York: Springer-Verlag.

Bruijnzeel, L. A., and E. J. Veneklaas. (1998). Climatic conditions and tropical montane forest productivity: the fog has not lifted yet. *Ecology* **79**: 3–9.

Bruijnzeel, L. A., M. J. Waterloo, J. Proctor, A. T. Kuiters, and B. Kotterink (1993). Hydrological observations in montane rain forests on Gunung Silam, Sabah, Malaysia, with special reference to the '*Massenerhebung*' effect. *Journal of Ecology* **81**: 145–167.

Cavelier, J. (1988). The ecology of elfin rain forests in northern South America. Dissertation submitted for the annual research fellowship competition, Trinity College, University of Cambridge, Cambridge, UK.

Cavelier, J., and M. C. Peñuela (1990). Soil respiration in the cloud forest and dry deciduous forest of Serrania de Macuira, Colombia. *Biotropica* **22**: 346–352.

Cox, S. B., M. R. Willig, and F. N. Scatena (2002). Variation in nutrient characteristics of surface soils from the Luquillo Experimental Forest of Puerto Rico: a multivariate perspective. *Plant and Soil* **247**: 189–198.

Coxson, D. S., D. D. McIntyre, and H. J. Vogel (1992). Pulse release of sugars and polyols from canopy bryophytes in tropical montane rain forest (Guadeloupe, French West Indies). *Biotropica* **24**: 121–133.

Edwards, I. D., A. A. Macdonald, and J. Proctor (eds.) (1993). *Natural History of Seram, Maluku, Indonesia*. Andover, UK: Intercept.

Edwards, P. J. (1982). Studies of mineral cycling in a montane rain forest in New Guinea. V. Rates of cycling in throughfall and litter fall. *Journal of Ecology* **70**: 807–827.

Flenley, J. R. (1993). Cloud forest, the Massenerhebung effect, and ultraviolet insolation. In *Tropical Montane Cloud Forests*, eds. L. S. Hamilton, J. O. Juvik, and F. N. Scatena, pp. 150–155. New York: Springer-Verlag.

Frahm, J. P. (1990). The ecology of epiphytic bryophytes on Mt. Kinabalu, Sabah (Malaysia). *Nova Hedwigia* **51**: 121–132.

Frahm, J. P., and S. R. Gradstein (1991). An altitudinal zonation of tropical rain forests using bryophytes. *Journal of Biogeography* **18**: 669–676.

Grieve, I. C., J. Proctor, and S. A. Cousins (1990). Soil variation with altitude on Volcán Barva, Costa Rica. *Catena* **17**: 525–534.

Grimm, U., and H. W. Fassbender (1981). Ciclos bioquímicos en un ecosistema forestal de los Andes Occidentales de Venezuela. II. Producción y descomposición de los residuos vegetales. *Turrialba* **31**: 39–48.

Grubb, P. J. (1977). Control of forest growth and distribution on wet tropical mountains: with special reference to mineral nutrition. *Annual Review of Ecology and Systematics* **8**: 83–107.

Hafkenscheid, R. L. L. J. (2000). Hydrology and biogeochemistry of tropical montane rain forests of contrasting stature in the Blue Mountains, Jamaica. Ph.D. thesis, VU University Amsterdam, Amsterdam, The Netherlands. Also available at http://dare.ubvu.vu.nl/bitstream/1871/12734/1/ tekst.pdf.

Hafkenscheid, R. L. L. J., L. A. Bruijnzeel, R. A. de Jeu, and N. J. Bink (2002). Water budgets of two upper montane rain forests of contrasting stature in the Blue Mountains, Jamaica. In *Hydrology and Water Management in the Humid Tropics*, ed. J. S. Gladwell, pp. 399–424. Paris: IHP-UNESCO, and Panamá City: CATHALAC.

Hamilton, L. S., J. O. Juvik, and F. N. Scatena (1995). The Puerto Rico tropical cloud forest symposium: Introduction and workshop synthesis. In *Tropical Montane Cloud Forests*, eds. L. S. Hamilton, J. O. Juvik, and F. N. Scatena, pp. 1–23. New York: Springer-Verlag.

Hättenschwiler, S., and P. M. Vitousek (2000). The role of polyphenols in terrestrial ecosystem nutrient cycling. *Trends in Ecology and Evolution* **15**: 238–243.

Heaney, A., and J. Proctor (1990). Preliminary studies on forest structure and floristics on Volcan Barva, Costa Rica. *Journal of Tropical Ecology* **6**: 307–320.

Herrmann, R. (1971). Die zeitlichen Anderung der Wasserbindung im Boden unter verschiedenen Vegetationsformationen der Höhenstufen eines tropischen Hochgebirges (Sierra Nevada de Sta. Marta, Kolumbien). *Erdkunde* **25**: 90–102.

Hetsch, W., and H. Hoheisel (1976). Standorts- und Vegetationsgliederung in einem tropischen Nebelwald. *Allgemeine Forst- und Jagdzeitung* **147**: 200–207.

Johnson, C. E., A. H. Johnson, and T. G. Huntington (1991). Sample pool requirements for the determinations of change in soil nutrient pools. *Soil Science* **150**: 637–644.

Kapos, V., and E. V. J. Tanner (1985). Water relations of Jamaican upper montane rain forest trees. *Ecology* **66**: 241–250.

Kitayama, K. (1992). An altitudinal transect study of the vegetation on Mount Kinabalu, Borneo. *Vegetatio* **102**: 149–171.

Kitayama, K., and S. Aiba (2002). Ecosystem structure and productivity of tropical rain forests along altitudinal gradients with contrasting soil phosphorus pools on Mount Kinabalu, Borneo. *Journal of Ecology* **90**: 37–51.

Kitayama, K., S. Aiba, N. Majalap-Lee, and M. Ohsawa (1998). Soil nitrogen mineralization rates of rainforests in a matrix of elevations and geological substrates on Mount Kinabalu, Borneo. *Ecological Research* **13**: 301–312.

Lawton, R. O. (1982). Wind stress and elfin stature in a montane rain forest tree: an adaptive explanation. *American Journal of Botany* **69**: 1224–1230.

Leuschner, C., G. Moser, C. Bertsch, M. Röderstein, and D. Hertel (2007). Large altitudinal increase in tree root/shoot ratio in tropical mountain forests in Ecuador. *Basic and Applied Ecology* **8**: 219–230.

Liebermann, D., M. Liebermann, R. Peralta, and G. S. Hartshorn (1996). Tropical forest structure and composition on a large-scale altitudinal gradient in Costa Rica. *Journal of Ecology* **84**: 137–152.

Lugo, A. E., M. Brinson, and S. Brown (eds). (1990). *Forested Wetlands.* Amsterdam: Elsevier.

Marrs, R., J. Proctor, A. Heaney, and M. Mountford (1988). Changes in soil nitrogen mineralization and nitrification along an altitudinal transect in tropical rain forest in Costa Rica, *Journal of Ecology* **76**: 466–482.

Martin, P. J. (1977). *The Altitudinal Zonation of Forests along the West Ridge of Gunong Mulu.* Kuching, Malaysia: Sarawak Forest Department.

Moser, G., D. Hertel, and C h. Leuschner (2007). Altitudinal change in LAI and stand leaf biomass in tropical montane forests: a transect study in Ecuador and a pan-tropical meta-analysis. *Ecosystems* **10**: 924–935.

Moser, G., M. Röderstein, N. Soethe, D. Hertel, and C h. Leuschner (2008). Altitudinal changes in stand structure and biomass allocation of tropical mountain forest in relation to microclimate and soil chemistry. In *Gradients in a Tropical Mountain Ecosystem of Ecuador*, eds. E. Beck, J. Bendix, I. Kottke, F. Makeschin, and R. Mosandl, pp. 229–242. Berlin: Springer-Verlag.

Myers, R. H. (1990). *Classical and Modern Regression with Applications*, 2nd edn. Pacific Grove, CA: Duxbury Press.

Nadkarni, N. M., and J. T. Longino (1990). Invertebrates in canopy and ground organic matter in a Neotropical montane forest, Costa Rica. *Biotropica* **22**: 286–289.

Nadkarni, N. M., and T. J. Matelson (1992). Biomass and nutrient dynamics of fine litter of terrestrially rooted material in a Neotropical montane forest, Costa Rica. *Biotropica* **24**: 113–120.

Newsome, D. (1986). Aspects of soil surface characteristics on Rakata and Anak Krakatau. In *Krakatoa Centenary Expedition 1983 Final Report*, eds. M. Bush, P. Jones, and K. Richards, pp. 134–156. Hull, UK: Department of Geography, University of Hull.

Northup, R., R. A. Dahlgren, and Z. Yu (1995). Intraspecific variation of conifer phenolic concentration on a marine terrace soil acidity gradient: a new interpretation. *Plant and Soil* **171**: 255–262.

Olander, L. P., F. N. Scatena, and W. L. Silver (1998). Impacts of disturbance initiated by road contruction in a subtropical cloud forest in the Luquillo Experimental Forest, Puerto Rico. *Forest Ecology and Management* **109**: 33–49.

Paoletti, M. G., R. A. J. Taylor, B. R. Stinner, D. H. Stinner, and D. H. Benzing (1991). Diversity of soil fauna in the canopy and forest floor of a Venezuelan cloud forest. *Journal of Tropical Ecology* **7**: 373–383.

Payton, R. O. (1993). Soils of the Manusela National Park. In *Natural History of Seram, Maluku, Indonesia*, eds. I. D. Edwards, A. MacDonald, and J. Proctor, pp. 19–61. Andover, UK: Intercept Publishers.

Proctor, J., Y. F. Lee, A. M. Langley, W. R. C. Munro, and T. Nelson (1988). Ecological studies on Gunung Silam, a small ultrabasic mountain in Sabah, Malaysia. I. Environment, forest structure, and floristics. *Journal of Ecology* **76**: 320–340.

Raich, J. W. (1998). Aboveground productivity and soil respiration in three Hawaiian rain forests. *Forest Ecology and Management* **107**: 309–318.

Raich, J. W., A. E. Russel, and P. M. Vitousek (1997). Primary productivity and ecosystem development along an elevational gradient on Mauna Loa, Hawai'i. *Ecology* **78**: 707–721.

Röderstein, M., D. Hertel, and C. Leuschner (2005). Above- and below-ground litter production in three tropical mountain forests (South Ecuador). *Journal of Tropical Ecology* **21**: 483–492.

Santiago, L. S., G. Goldstein, F. C. Meinzer, J. H. Fownes, and D. Mueller-Dombois (2000). Transpiration and forest structure in relation to soil waterlogging in a Hawaiian montane cloud forest. *Tree Physiology* **20**: 673–681.

Scatena, F. N. (1995). The management of Luquillo elfin cloud forest ecosystems: irreversible decisions in a nonsubstitutable ecosystem. In *Tropical Montane Cloud Forests*, eds. L. S. Hamilton, J. O. Juvik, and F. N. Scatena, pp. 296–308. New York: Springer-Verlag.

Schlesinger, W. H., L. A. Bruijnzeel, M. B. Bush, *et al.* (1998). The biogeochemistry of phosphorus after the first century of soil development on Rakata island, Krakatau, Indonesia. *Biogeochemistry* **40**: 37–55.

Schrumpf, M. (2004). Biogeochemical investigations in old-growth and disturbed forest sites at Mount Kilimanjaro. Ph.D. thesis, University of Bayreuth, Bayreuth, Germany.

Schrumpf, M., G. Guggenberger, C. Valarezo, and W. Zech (2001). Tropical montane rain forest soils: development and nutrient status along an altitudinal gradient in the south Ecuadorian Andes. *Die Erde* **132**: 43–59.

Schuur, E. A. G., and P. A. Matson (2001). Net primary productivity and nutrient cycling across a mesic to wet precipitation gradient in Hawaiian montane forest. *Oecologia* **128**: 431–442.

Schwarzkopf, T. (2003). Biophysical characterization of Cloud Forest Vegetation in the Venezuelan Andes. Ph.D. thesis, Cornell University, Ithaca, New York.

Shinagawa, A., N. Miyauchi, T. Higashi, M. R. Djuwansah, and A. Sule (1986). The soils on the Krakatau Islands. *Memoirs of the Faculty of Agriculture, Kagoshima University* **22**: 101–130.

Sieders, V. M. (1971). *Geologic Map of the El Yunque Quadrangle, Puerto Rico*, U.S. Geological Survey Miscellaneous Geological Investigations Map I-658. Washington, DC: U.S. Department of the Interior.

Silver, W. L. (1994). Is nutrient availability related to plant nutrient use in humid tropical forests? *Oecologia* **98**: 336–343.

Silver, W. L., A. E. Lugo, and M. Keller (1999). Soil oxygen availability and biogeochemistry along rainfall and topographic gradients in upland wet tropical forest soils. *Biogeochemistry* **44**: 301–328.

Soethe, N., J. Lehmann, and C. Engels (2006). The vertical pattern of rooting and nutrient uptake at different altitudes of a south Ecuadorian montane forest. *Plant and Soil* **286**: 287–299.

Soethe, N., J. Lehmann, and C. Engels (2008a). Nutrient availability at different altitudes in a tropical montane forest in Ecuador. *Journal of Tropical Ecology* **24**: 397–406.

Soethe, N., W. Wilcke, J. Homeier, J. Lehmann, and C. Engels (2008b). Plant growth along the altitudinal gradient: role of plant nutritional status, fine root activity, and soil properties. In *Gradients in a Tropical Mountain Ecosystem of Ecuador*, eds. E. Beck, J. Bendix, I. Kottke, F. Makeschin, and R. Mosandl, pp. 259–266. Berlin: Springer-Verlag.

Sollins, P. (1998). Factors influencing species composition in tropical lowland rain forests: does soil matter? *Ecology* **79**: 23–30.

Stadtmüller, T. (1987). *Cloud Forests in the Humid Tropics: A Bibliographic Review*. Tokyo: United Nations University, and Turrialba, Costa Rica: CATIE.

Steinhardt, V. (1979). Untersuchungen über deu Wasser- und Nährstoffhaushalt eines andinen Wolkenwaldes in Venezuela. Göttinger Bodenkundliche Berichte **56**: 1–185.

Sugden, A. M. (1986). The montane vegetation and flora of Margarita Island, Venezuela. *Journal of the Arnold Arboretum* **67**: 187–232.

Tanner, E. V. J. (1977). Four montane rain forests of Jamaica: a quantitative characterization of the floristics, the soils, ad the foliar mineral levels, and a discussion of the interrelations. *Journal of Ecology* **65**: 883–918.

Tanner, E. V. J., V. Kapos, S. Freskos, J. R. Healey, and A. M. Theobald (1990). Nitrogen and phosphorus fertilization of Jamaican montane forest trees. *Journal of Tropical Ecology* **6**: 231–238.

Tanner, E. V. J., V. Kapos, and W. Franco (1992). Nitrogen and phosphorous fertilization effects on Venezuelan montane forest trunk growth and litterfall. *Ecology* **73**: 78–86.

Tanner, E. V. J., P. M. Vitousek, and E. Cuevas (1998). Experimental investigation of nutrient limitation of forest growth on wet tropical mountains. *Ecology* **79**: 10–22.

Tie, Y. L., I. C. Baille, C. M. S. Phang, and C. P. Lim (1979). *Soils of Gunung Mulu National Park*. Kuching, Sarawak, Malaysia: Department of Agriculture.

Vance, E. D., and Nadkarni, N. M. (1990). Microbial biomass and activity in canopy organic matter and the forest floor of a tropical cloud forest. *Soil Biology and Biochemistry* **22**: 677–684.

Van Reuler, H. (1987). Soil studies in the Bukit Raya nature reserve. In *Report of the 1982–1983 Bukit Raya Expedition*, ed. H. P. Nooteboom, pp. 7–23. Leiden, the Netherlands: Rijksherbarium.

Veneklaas, E. J. (1990). Rainfall interception and above-ground nutrient fluxes in Colombian montane tropical forest. Ph.D. thesis, University of Utrecht, Utrecht, the Netherlands.

Vitousek, P. M. (1984). Litterfall, nutrient cycling, and nutrient limitation in tropical forests. *Ecology* **65**: 285–298.

Vitousek, P. M., and R. L. Sanford (1986). Nutrient cycling in moist tropical forest. *Annual Review of Ecology and Systematics* **17**: 137–167.

Vitousek, P. M., P. A. Matson, and R. A. Turner (1988). Elevation and age gradients in Hawaiian montane rainforest: foliar and soil nutrients. *Oecologia* **77**: 565–570.

Wang, H., C. A. S. Hall, F. N. Scatena, N. Fetcher, and W. Wu (2002). Modeling the spatial and temporal variability in climate and primary productivity across the Luquillo Mountains, Puerto Rico. *Forest Ecology and Management* **61**: 1018–1026.

Waterloo, M. J. (1989). A hydrological study of the mass elevation effect on Gunung Silam, a small coastal ultrabasic mountain in Sabah, East Malaysia. M.Sc. thesis, VU University Amsterdam, Amsterdam, the Netherlands.

Weaver, P. L., E. Medina, D. Pool, *et al.* (1986). Ecological observations in the dwarf cloud forest of the Luquillo Mountains in Puerto Rico. *Biotropica* **18**: 79–85.

Whitmore, T. C. (1990). *An Introduction to Tropical Rain Forests*. Oxford, UK: Clarendon Press.

Wilcke, W., S. Yasin, C. Valerezo, and W. Zech (2001). Change in water quality during the passage through a tropical montane rain forest in Ecuador. *Biogeochemistry* **55**: 45–72.

Wilcke, W., S. Yasin, U. Abramowski, C. Valerezo, and W. Zech (2002). Nutrient storage and turnover in organic layers under tropical montane rain forest in Ecuador. *European Journal of Soil Science* **53**: 15–27.

Wilcke, W., H. Balladares, R. Stoyan, *et al.* (2003). Soil properties on a chronosequence of landslides in montane rain forest, Ecuador. *Catena* **53**: 79–95.

Williams-Linera, G. (2002). Tree species richness, complementarity, disturbance and fragmentation in a Mexican tropical montane cloud forest. *Biodiversity and Conservation* **11**: 1825–1843.

Williams-Linera, G., R. H. Manson, and E. I. Vera (2002). La fragmentación del bosque mesófilo de montaña y patrones de uso del suelo en la region oeste de Xalapa, Veracruz, México. *Madera y Bosques* **8**: 69–85.

Zinck, A. (1986). Los Suelos: caracteristicas y fragilidad de los suelos en ambiente de Selva Nublada – el ejemplo de Rancho Grande. In *La selva nublada de Rancho Grande Parque Nacional "Henri Pittier,"* ed. O. Huber, pp. 31–66. Caracas, Venezuela: Fondo Editorial Acta Cientifíca Venezolana.

7 Nutrient cycling and nutrient limitation in tropical montane cloud forests

J. Benner and P. M. Vitousek
Stanford University, Stanford, California, USA

R. Ostertag
University of Hawai'i at Hilo, Hawai'i, USA

ABSTRACT

In this chapter, the role of nutrient supply and cycling with respect to the characteristically low productivity of tropical montane cloud forests is investigated. Studies of nutrient stocks, turnover rates, and foliar nutrients all suggest that nitrogen supply to vegetation is lower in montane tropical forests than in lowland forests, whereas forest fertilization studies indicate that nitrogen and often phosphorus consistently limit above-ground productivity. Slow rates of nitrogen cycling, rather than low nitrogen inputs, appear to be responsible for the depressed nitrogen supply, and the high soil water content of many cloud-immersed montane forests is likely to be an important ultimate cause of the decreased rates of nitrogen cycling. Hydrological losses of biologically unavailable forms of nitrogen (such as dissolved organic nitrogen) may sustain nitrogen limitation over longer timescales.

INTRODUCTION

Regardless of location, tropical montane cloud forests (TMCF) worldwide share the same basic differences from lowland tropical forests and montane forests not affected by regular fog: lower productivity and diversity, lower canopy heights, thicker leaves with lower nutrient concentrations (especially of nitrogen), and higher soil organic matter and water content. In this chapter, the importance of nutrient availability in controlling this suite of traits is revisited, summarizing recent information on nutrient distribution, availability, and limitation in TMCF. To the extent that low nutrient availability contributes to the TMCF "syndrome," this chapter discusses whether it is an independent factor or a consequence of other factors that ultimately control the productivity of TMCF. It is suggested that, whilst low nutrient availability is an important proximate control of low productivity, leaf area, and forest stature, rarely is it an ultimate control of the TMCF syndrome. Those controls must rather be sought in the major environmental factors that differentiate TMCF from lowland forests and montane forests below the cloud belt, viz. lower temperatures, reduced incoming radiation and atmospheric evaporative demands, as well as frequent wetting of the vegetation by rain and fog, and, at exposed locations, frequent occurrence of strong winds (cf. Bruijnzeel, 2005; Jarvis and Mulligan, this volume). Finally, the chapter explores how these environmental factors individually and interactively influence the processes that regulate nutrient cycling and limitation in TMCF.

PATTERNS OF NUTRIENT CYCLING AND LIMITATION IN TROPICAL FORESTS

Soil carbon and nutrient stocks

Soil carbon content is higher in TMCF than in lowland forests and montane forests below the cloud belt, and increases along elevational gradients as above-ground biomass decreases (Marrs *et al.*, 1988; Grieve *et al.*, 1990; Jobbagy and Jackson, 2000; Raich *et al.*, 2006; cf. Roman *et al.*, this volume). As a result, an increasing fraction of ecosystem carbon stocks at higher altitudes is located below-ground (Kitayama and Aiba, 2002; Raich *et al.*, 2006), also because relative amounts of fine root biomass increase with elevation (Röderstein *et al.*, 2005; Moser *et al.*, 2008;

Tropical Montane Cloud Forests: Science for Conservation and Management, eds. L. A. Bruijnzeel, F. N. Scatena, and L. S. Hamilton. Published by Cambridge University Press.

Hertel and Leuschner, this volume). Total nitrogen content of the soil increases in parallel with that of total carbon, but soil stocks and available pools of other elements show no clear trend with altitude when all TMCF are grouped together (Bruijnzeel and Proctor, 1995). However, when a distinction is made between lower montane rain forest (LMRF, not affected significantly by fog) and lower montane cloud forest (LMCF, subject to moderate fog influence) on the one hand, and upper montane cloud forest (UMCF) and elfin cloud forest (ECF) (both strongly affected by fog) on the other, certain differences in soil nutrient stocks between the two groups become apparent (Roman *et al.*, this volume).

Nutrients in foliage and litterfall

Foliar and litter element concentrations can be used to indicate element availability in forests. Tanner *et al.* (1998) reviewed trends in foliar and litter concentrations of nutrients in tropical montane forests and concluded that for transects on a single mountain "foliar nitrogen decreases, and phosphorus and potassium usually decrease, but calcium and magnesium show no consistent trend" with increasing elevation. More recent studies generally support these trends (summarized for foliar nitrogen and phosphorus in Figure 7.1; see also Leuschner *et al.*, 2007; Moser *et al.*, 2007; Soethe *et al.*, 2008a,b). Although leaves in high-altitude tropical forests tend to be sclerophyllous, it is not clear whether sclerophylly represents an adaptation to low nutrients (e.g. Loveless, 1961) or to other selective pressures such as strong winds (Jaffe, 1980) or even periodically high UV-B levels (Flenley, 1995).

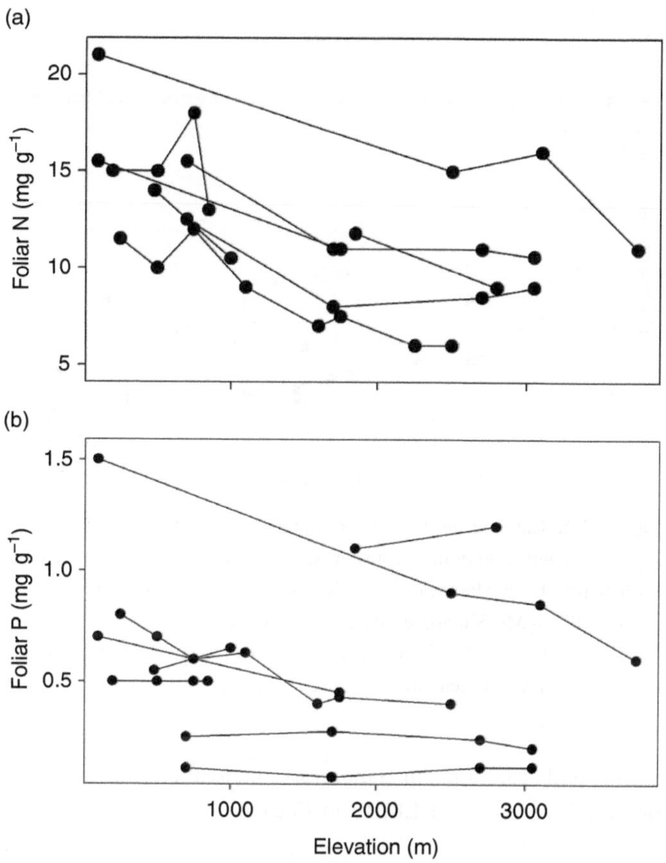

Figure 7.1. Patterns in foliar nitrogen (a) and phosphorus (b) concentrations on tropical mountains. All of the connected points represent sites sampled on a single mountain. Information from Tanner *et al.* (1998); Kitayama and Aiba (2002) – Mt. Kinabalu, Malaysia; Vera *et al.* (1999) – Colombia. "Elfin" forests (here defined as low-stature forests with a mean canopy height of less than 10 m) occur at the highest ultrabasic site on Mt. Kinabalu and the highest site of the New Guinea transect. The entire Hawaiian transect has a canopy height <10 m due to the young age of the substrate. There are no data on canopy height from the Malaysian transect.

Decomposition and soil nitrogen transformations

Slow mineralization of organic matter has been proposed as a primary mechanism limiting nutrient supply to vegetation in montane tropical forests (Grubb, 1971, 1977). Lower decomposition is indeed observed with increasing elevation on tropical mountains, for both litter decomposition (Bruijnzeel *et al.*, 1993; Vitousek *et al.*, 1994; Pendry and Proctor, 1996; Kitayama and Aiba, 2002) and soil organic matter turnover (Townsend *et al.*, 1995; cf. Leuschner *et al.*, 2007). At a given elevation, litter decomposition and nitrogen mineralization rates tend to be lower in forests of lesser stature and soil fertility (Hafkenscheid, 2000).

Most studies that have examined net nitrogen mineralization and net nitrification along elevational transects show that rates of nitrogen release decrease with elevation (Figure 7.2). To our knowledge, gross rates of nitrogen mineralization and nitrification have been measured across an elevational gradient only in Puerto Rico (Silver *et al.*, this volume); in the latter study, gross rates did not decline with increasing elevation.[1] Extractable ammonium concentrations generally are greater than extractable nitrate in TMCF soils (Marrs *et al.*, 1988; Vitousek *et al.*, 1988; Bruijnzeel *et al.*, 1993; Hafkenscheid, 2000). Studies of the mineralization of other elements are few. Phosphorus transformations in tropical forests are notoriously difficult to measure under realistic conditions due to the strength of abiotic phosphorus adsorption (Olander and Vitousek, 2004). Wilcke *et al.* (2002) found net immobilization of manganese and zinc, and slow release of

[1] Gross nitrogen mineralization refers to the total amount of nitrogen released in a given amount of time (regardless of whether it is subsequently taken up by microbes, etc.), whereas net nitrogen mineralization is the net accumulation of nitrogen in the soil over a given amount of time (accounting for uptake).

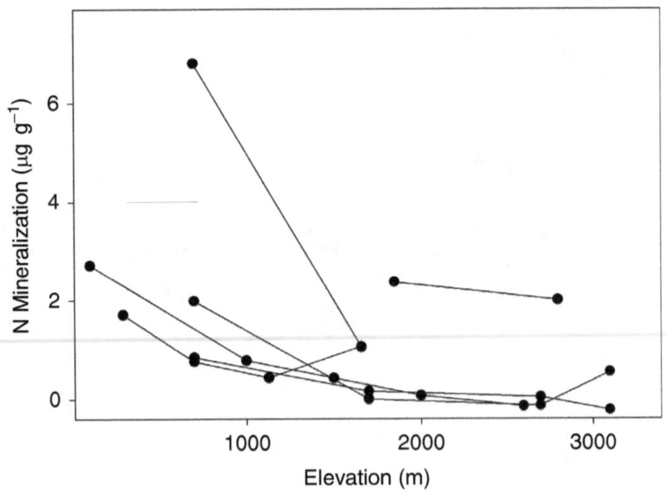

Figure 7.2. Rates of net nitrogen mineralization on tropical mountains. All of the connected points represent sites sampled on a single mountain. Information from Marrs *et al.* (1988) – Costa Rica; Kitayama and Aiba (2002) – Mt. Kinabalu, Malaysia; Raich *et al.* (1997) – Hawai'i; Vera *et al.* (1999) – Colombia. The highest Kinabalu ultrabasic site and the highest New Guinea site support elfin forest.

sulfur and calcium, in laboratory incubations of the soil organic layer in a tall LMRF in Ecuador.

ARE TROPICAL MONTANE FORESTS NUTRIENT-LIMITED?

A number of studies have determined nutrient limitation to montane tropical forests using fertilization experiments; however, not all of these studies were carried out in the cloud belt. The results of montane forest fertilization studies as summarized by Tanner *et al.* (1998) plus several more recent studies are listed in Table 7.1. The geographic distribution of fertilization studies is strongly biased towards the Hawaiian Islands, and it is quite possible that a similar suite of experiments in more diverse continental ecosystems would yield different results (although the same processes should be important). Experiments in Hawai'i across a chronosequence of soils developed from the same volcanic parent material showed a switch from nitrogen to phosphorus limitation from young to old soils (Vitousek *et al.*, 1993; Herbert and Fownes, 1995), while fertilization with nitrogen and phosphorus separately in Jamaican montane cloud forests resulted in trunk diameter growth increases in four tree species (Tanner *et al.*, 1990). In a Venezuelan cloud forest at 2500 m.a.s.l., fertilization with both nitrogen and phosphorus together increased trunk diameter growth, although nitrogen or phosphorus alone had no effect (Tanner *et al.*, 1992). In a study in a coastal elfin cloud forest in Colombia at 865 m.a.s.l., Cavelier *et al.* (2000) showed that trunk diameter growth was increased by nitrogen and phosphorus together, but not by either nitrogen or phosphorus alone; the authors speculated that the forest was primarily nitrogen-limited, but that an increase in the supply of nitrogen induced phosphorus limitation. In an upper montane cloud forest at 2425 m on Mt. Kinabalu (Sabah, Malaysia), fertilization with nitrogen and both nitrogen and phosphorus increased leaf emergence and litterfall, suggesting that leaf lifespan may decrease in response to a pulse of nutrients (Nomura and Kikuzawa, 2003). A similar reduction in leaf lifespan was found in response to nitrogen fertilization in a Hawaiian montane forest growing on young volcanic soils, on both individual leaf and whole-canopy levels (Cordell *et al.*, 2001; Harrington *et al.*, 2001). This decrease in leaf lifespan may relate to increases in specific leaf area (*SLA*) after fertilization (e.g. Harrington *et al.*, 2001), as leaf lifespan was highly correlated with *SLA* among evergreen and deciduous species in a Mexican cloud forest (Williams-Linera, 1999), as it is globally (Reich *et al.*, 1997).

Root productivity in response to fertilization has also been examined in tropical montane forests (Table 7.2). A study in Jamaica used root ingrowth cores amended with nitrogen, phosphorus, and both nitrogen and phosphorus to test whether root allocation in low-productivity Mor Ridge forest and higher productivity Mull Ridge forest responded to these nutrients (Stewart, 2000). Root growth in both forests increased in the +nitrogen cores, and in the Mor Ridge forest root growth also increased in the +phosphorus and +nitrogen and phosphorus cores, suggesting limitation by both nitrogen and phosphorus in the more stunted forest and by nitrogen alone in the taller forest. In the fertilization study on Mt. Kinabalu, fine root growth increased in response to nitrogen fertilization (Nomura and Kikuzawa, 2003). Root growth did not change in response to added nitrogen in a nitrogen-limited Hawaiian forest at 1176 m.a.s.l. (Ostertag, 2001), but the production of phosphatase enzymes increased significantly (Olander and Vitousek, 2000; Treseder and Vitousek, 2001), suggesting increased investment in obtaining phosphorus. Neither liming nor fertilization (nitrogen, +phosphorus, potassium, and magnesium) had a significant effect on fine root growth in either lower montane rain forest or elfin cloud forest in southern Ecuador (Soethe *et al.*, 2006). Studies from unfertilized tropical forests show that proportional allocation to fine root biomass and production are substantially greater in more stunted than in taller montane forests, which also suggests increased investment in below-ground resource capture at higher elevations or under more adverse soil conditions (Hafkenscheid, 2000; Röderstein *et al.*, 2005; Leuschner *et al.*, 2007; Soethe *et al.*, 2008a,b; Hertel and Leuschner, this volume; cf. Schawe *et al.*, this volume). In addition, rooting patterns with depth in UMCF and ECF in southern Ecuador suggested a reduced capacity for spatial exploitation of soil nutrients at higher elevations (Soethe *et al.*, 2008a,b).

Table 7.1 *Above-ground responses to expl erimental fertilization in tropical montane cloud forests*

Location	Fertilizer/method[a]	Substrate	Soil age (yr)	Trunk diameter increment response[b]	Foliar/litterfall response[b]	Reference
Hawai'i Puu Makaala	Complete	Lava	1500–3000	Significant increase (167%)	—	Gerrish *et al.* (1988)
Hawai'i Saddle Road	Complete	Lava	3400	Increase (25% – NS)	—	Gerrish *et al.* (1988)
Hawai'i Thurston lava tube	+N, +P, +NP	Ash	200	Significant increase due to N	Litterfall increased due to N	Vitousek *et al.* (1993)
Hawai'i Mauna Loa	+N, +P, +NP	'a'a lava	140	Significant increase due to N (357% in +N)	—	Raich *et al.* (1996)
Hawai'i Mauna Loa	+N, +P, +NP	pahoehoe lava	140	Significant increase due to N, P, and significant NP interaction	—	Raich *et al.* (1996)
Hawai'i Mauna Kea	+N, +P, +NP	Ash	20 000	Significant increase due to N	—	Vitousek and Farrington (1997)
Hawai'i Olaa	+N	Ash	2000	No significant increase	Litterfall increased (24% – NS)	Vitousek *et al.* (1987); Vitousek *et al.* (1994)
Hawai'i Kauai	+N, +P, +NP	—	4.1×10^6	Significant increase due to P	—	Herbert and Fownes (1995)
Venezuela	+N, +P, +NP	—	—	Significant increase (24%) in +NP treatment only	Litterfall increased in all 3 treatments; significant for +NP and +N only	Tanner *et al.* (1992)
Jamaica	+N, +P	—	—	Significant increase in +N (77%) and +P (32%) treatments	—	Tanner *et al.* (1990)
Colombia	+N, +P, +NP for 3 tree species	—	—	Significant increase for 2 species in +NP treatment only, no response for 1 species	—	Cavelier *et al.* (2000)
Kinabalu – LMF (1708 m.a.s.l.)	+N, +NP	—	—	—	Litterfall not affected; leaf emergence decreased in +N	Nomura and Kikuzawa (2003)
Kinabalu – UMF (2425 m.a.s.l.)	+N, +NP	—	—	—	Litterfall and leaf emergence increased in +N, +NP	Nomura and Kikuzawa (2003)
Kinabalu – sub-alpine forest (3080 m.a.s.l.)	+N, +NP	—	—	—	Litterfall increased in +N; leaf emergence not affected	Nomura and Kikuzawa (2003)
Puerto Rico – elfin forest	"Complete fertilizer" (after hurricane)	—	—	No response	Litterfall increased	Walker *et al.* (1996)

[a] "Complete fertilizer" includes N, P, K, B, Cu, Fe, Mn, Zn, Mg.

[b] NS = not significant. Significant increase "due to N" or "due to P" means overall effect of N or P was calculated.

Table 7.2 *Below-ground responses to experimental fertilization in tropical montane (cloud) forests*

Location	Fertilizer/method	Below-ground response	Reference
Jamaica – Mor Ridge (1550 m.a.s.l.)	Root ingrowth cores: +N, +P, +NP	Root growth increase in +N, +P, +NP cores	Stewart (2000)
Jamaica – Mull Ridge (1550 m.a.s.l.)	Root ingrowth cores: +N, +P, +NP	Root growth increase in +N core	Stewart (2000)
Hawai'i – 300 yrs (1176 m.a.s.l.)	Root ingrowth cores: +N, +P, +CaK	Root growth increase in +N core	Raich *et al.* (1994)
Hawai'i – 300 yrs (1176 m.a.s.l.)	+N, +P	No increase in root growth; phosphatase enzyme production increased in +N	Olander and Vitousek (2000); Ostertag (2001)
Hawai'i – 4.1 million yrs (1134 m.a.s.l.)	+N, +P	No increase in root growth; root turnover increased in +P	Ostertag (2001)
Kinabalu – LMF (1708 m.a.s.l.)	+N, +NP	Root growth suppressed in +N treatment	Nomura and Kikuzawa (2003)
Kinabalu – UMF (2425 m.a.s.l.)	+N, +NP	Root growth increase in +N treatment	Nomura and Kikuzawa (2003)
Kinabalu – sub-alpine forest (3080 m.a.s.l.)	+N, +NP	Root growth suppressed in +N, +NP treatments	Nomura and Kikuzawa (2003)

In sum, evidence from Hawai'i suggests that montane forests on younger soils are nitrogen-limited, while those on older soils are phosphorus-limited. Evidence from more species-rich forests on soils of undefined age indicates that both nitrogen and phosphorus can limit production on tropical mountains. The independent response of montane forests to calcium, potassium, or other elements has not been investigated outside of the Hawaiian Islands, where no growth responses were observed to nutrients other than nitrogen or phosphorus. Tanner *et al.* (1998) speculated that while montane tropical forests tend toward nitrogen limitation due to slow rates of decomposition and nitrogen turnover, lowland forests on highly weathered soils tend toward phosphorus limitation. Altitudinal patterns in soil nitrogen transformations and in foliar and litter nutrients are consistent with this suggestion (Vitousek, 1984). In addition, recent evidence from volcanic soils of varying ages along an elevation gradient in Costa Rica showed that total soil phosphorus and labile phosphorus increased with elevation (Titiz, 2005). However, too few fertilization studies have been carried out in lowland tropical forests to evaluate this pattern experimentally.

NUTRIENT LIMITATION: PROXIMATE OR ULTIMATE?

There is abundant evidence that nutrient supply limits the productivity of tropical montane forests – but does nutrient limitation represent an ultimate control over the stature and productivity of cloud forests, or is it a consequence of other factors (light, temperature, wind, atmospheric demand, or soil water status) that control ecological processes in these forests? Causes of differences in nutrient availability between tropical lowland forests and various kinds of montane forests including TMCF can be considered to fall into two broadly defined classes – processes that cause lower inputs of nutrients (or increase nutrient outputs) in montane forests, and those that cause slower cycling of nutrients. These classes interact, as will be discussed below, but the separation is useful because where inputs are so low (or outputs so large) that nutrients cannot accumulate within a system, nutrient limitation could be an ultimate control over the lower productivity of tropical montane forests relative to lowland forests.

The major pathways of nutrient inputs to forest ecosystems are chemical weathering of the underlying rock and soil, atmospheric inputs via precipitation and fog interception, dry deposition, and (for nitrogen) biological fixation. Weathering supplies most element inputs (other than nitrogen) to many soils – and rates of weathering depend upon many factors, including mineralogy, substrate age, temperature, rainfall surplus, and the generation of biological acidity. All else being equal (in mineralogy and substrate age), lower temperatures in montane environments should slow weathering in comparison to lowland forests, while high rainfall, soil water contents, and leaching rates should enhance weathering. All else rarely is equal, however. Mountains are dynamic features, maintained by rates of uplift that balance relatively high rates of soil loss by erosion and landsliding – and so soils in tropical montane regions generally are younger than those in lowland rain forests on geologically old and tectonically stable substrates (cf. Sidle *et al.*, 2006). These more youthful soils contain an abundance of weatherable minerals that have become depleted in most lowland soils – for example, Kitayama *et al.* (2000) reported more weatherable phosphorus in higher elevation soils on Kinabalu on soils derived from sedimentary rock, and stream-water fluxes of base cations are substantially greater from montane than lowland tropical forests (Bruijnzeel, 1991; McDowell and Asbury, 1994; Wilcke *et al.*, 2001; Liu *et al.*, 2003). Consequently, a difference in weathering is unlikely to cause systematically lower inputs of nutrients to montane as opposed to lowland forest soils, also because montane soils tend to be shallower and any nutrients released at the weathering front more readily accessible to roots than in the often much deeper lowland tropical soils (Baillie, 1989; Nepstad *et al.*, 1994).

Atmospheric deposition can be a substantial source of many nutrients – especially in coastal or insular regions where marine aerosols are abundant, and (in relative terms) on deeply leached soils where weatherable minerals have been exhausted (Chadwick *et al.*, 1999; Vitousek, 2004). Montane forests differ systematically from lowland forests in having a larger fraction of their inputs of water and nutrients in the form of fog water, which is substantially enriched in nutrients compared to precipitation (Clark *et al.*, 1998b; Hafkenscheid, 2000; Weathers *et al.*, 2000; Carrillo *et al.*, 2002; Chang *et al.*, this volume; Liu *et al.*, this volume). Consequently, montane forests generally should have higher rates of nutrient input via atmospheric deposition than lowland forests for a given rainfall.

In contrast, the potential for biological nitrogen fixation could be lower in montane than in lowland tropical forests. The distribution of woody legumes in particular is centered in the lowland tropics (Crews, 1999), a pattern that may contribute to the relatively nitrogen-rich nutrient status of lowland tropical forests (Martinelli *et al.*, 1999). However, woody plants with nitrogen-fixing mycorrhizal symbionts (e.g. *Morella*) are present in many montane tropical forests, and non-symbiotic fixation associated with epiphytic bryophytes, epiphylls, and lichens can provide a steady if not spectacular source of fixed nitrogen to cloud forest canopies (Forman, 1975; Freiberg, 1998; Matzek and Vitousek, 2003).

Overall, it can be concluded that differences in nutrient inputs probably contribute little to the lower nutrient availability and widespread nutrient limitation in tropical montane forests. Accordingly, differences in the rates at which nutrients cycle (or are lost) must be responsible for the low nutrient status of montane forests – and the causes of their nutrient deficiency must be sought not in the overall supply of elements, but in the processes that regulate rates of nutrient cycling within these forests. Rates of nutrient cycling are indeed slower in montane than in lowland forests – the combination of lower productivity and a greater accumulation of organic matter and organically bound nutrients in montane forest soils demonstrates that unequivocally, at least for biophilic nutrients like nitrogen (Roman *et al.*, this volume).

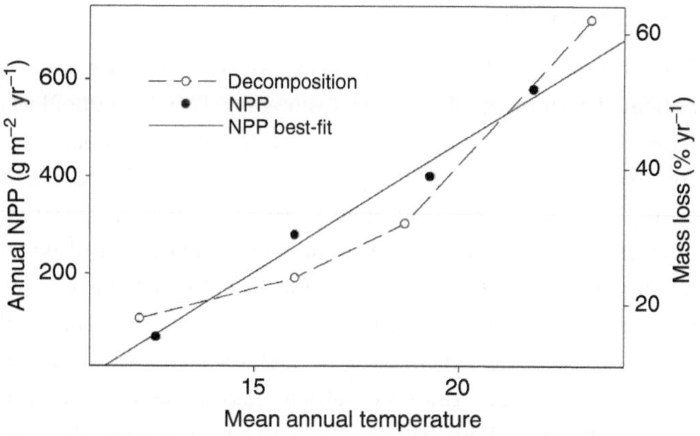

Figure 7.3. Above-ground net primary production (solid line and symbols) and decomposition (dashed line and open circles) of sites on a single lava flow on Mauna Loa Volcano, Hawai'i. The sites are arrayed along an elevational gradient from 290 to 1660 m.a.s.l.; temperature increases with decreasing elevation (revised from Raich *et al.*, 1997).

CAUSES AND CONSEQUENCES OF SLOWER NUTRIENT CYCLING IN TROPICAL MONTANE FORESTS

Reduced temperatures

All else being equal, the universal pattern of decreasing temperature at higher elevation should lead to decreasing rates of both primary production and decomposition in tropical montane relative to lowland forests (Jenny *et al.*, 1949; Grubb, 1977). If the effects of temperature on production and decomposition were identical, then changes in nutrient demand by plants would parallel changes in nutrient supply, and there would be no systematic change in nutrient limitation. However, production tends to increase linearly with increasing temperature, while decomposition increases exponentially (Figure 7.3) (Jenny *et al.*, 1949; Raich *et al.*, 1997) – with the net result that lower temperatures affect decomposition (and thus nutrient supply) to a greater extent than they do production (and thus nutrient demand). This pattern raises the possibility that slower nutrient cycling could lead to nutrient limitation. Such limitation should be transient, however; slower decomposition/mineralization should allow the accumulation of larger pools of soil organic matter and nutrients, as indeed observed in tropical montane environments, to the point where even slow rates of turnover should provide organisms with an adequate supply of biophilic nutrients. Transient nutrient limitation can be important during periods when nutrients are accumulating in biomass and soils following disturbance (Olander *et al.*, 1998) – and such transient effects could last substantially longer in montane forests due to their lower temperatures.

Excess soil water

Recent studies reinforce the importance of soil water content as a critical determinant of nutrient cycling in TMCF, particularly UMCF and ECF. Within the cloud belt on tropical mountains, soil water content rises sharply from LMCF to UMCF (see examples in Bruijnzeel and Proctor, 1995; Bruijnzeel, 2005; cf. Schawe *et al.*, this volume). The most important effect of water on nutrient availability in these wet forests is likely to be reduced litter decomposition and mineralization due to anaerobic soil conditions, perhaps accentuated by metal toxicity (e.g. manganese) and enhanced rates of nutrient losses via leaching (cf. McDowell and Asbury, 1994). To the extent that mineralization depends on decomposition, reduced decomposition with anaerobiosis should also slow rates of mineralization. Silver *et al.* (1999) identified substantial changes in soil O_2 and redox potential along an elevation and precipitation gradient in Puerto Rico; soil oxygen levels decreased to very low levels (soil O_2 concentrations <3%) for sustained periods of time in very wet ridge top ECF. Soil methane concentrations were inversely related to soil O_2, with very high methane levels in the cloud forest soils (up to 2.4×10^5 ppmv), suggesting strongly reducing soil conditions. The decrease in soil oxygen and increase in methane with increasing rainfall and elevation in Puerto Rico were accompanied by increases in soil total nitrogen, organic matter, pH, and extractable phosphorus. Inefficient metabolism, incomplete decomposition under anaerobic conditions, and microbial community composition could account for high soil organic matter contents, providing a positive feedback to soil water content by increasing soil water retention capacity.

Schuur and colleagues (Miller *et al.*, 2001; Schuur, 2001; Schuur and Matson, 2001; Schuur *et al.*, 2001) evaluated the effects of increasing precipitation and soil water content along a rainfall gradient from 2200 to 5000 mm year^{-1} at a single elevation in montane forests on Maui, Hawai'i. Not surprisingly, redox potentials in the soil were much lower and soil organic matter contents much greater at the wetter sites (Figure 7.4). Leaves of the dominant tree *Metrosideros polymorpha* decomposed 6.4 times more slowly at the wet end of the gradient than at the dry end of the gradient when they were both collected and decomposed in their site of origin. Some of this decrease was due to the lower quality of foliage at the wettest sites (higher lignin and lower nitrogen); when leaves from all sites along the gradient were decomposed at a common site, the wet-site leaves decomposed almost three times more slowly than the dry-site leaves. However, when leaves from a common site were decomposed across the gradient, the rate of decomposition was still 2.5 times slower at the wettest site. Clearly, both low litter quality and wet conditions reduce decomposition at high rainfall (Schuur, 2001). Schuur concluded that anaerobic soils were the primary cause of decreased decomposition at high annual precipitation sites, followed by litter quality.

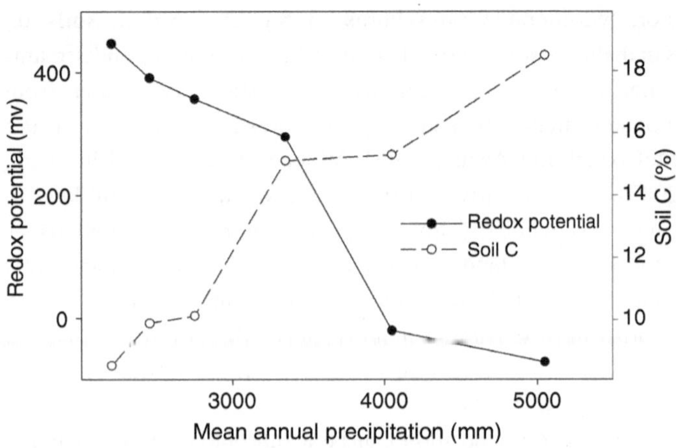

Figure 7.4. Mean soil redox potential (solid line and circles) and soil carbon concentrations (dashed line and open circles) in sites sampled along a rainfall gradient on Maui, Hawai'i (revised from Schuur *et al.*, 2001).

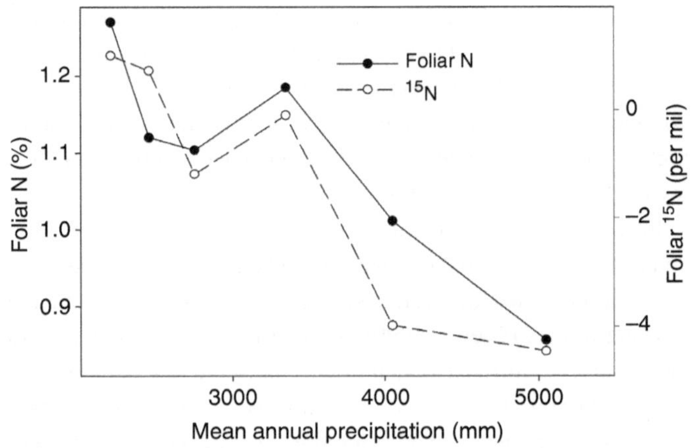

Figure 7.5. Foliar nitrogen concentrations (mean of 11 species) (solid line and circles) and foliar ^{15}N for *Metrosideros polymorpha* (dashed line and open circles) in sites along a rainfall gradient on Maui, Hawai'i (revised from Schuur *et al.*, 2001).

On the same gradient, foliar nitrogen concentrations decreased from dry to wet sites, and foliar ^{15}N became progressively more depleted (Figure 7.5) – leading Schuur *et al.* (2001) to conclude that nitrogen becomes progressively more limiting under wetter site conditions. Nitrogen limitation could result from slower litter decomposition/nitrogen mineralization – which could in turn drive a positive feedback to sclerophylly and a further reduction in decomposition. In addition, low redox conditions could desorb phosphorus from mineral surfaces, for example, by reductive dissolution of iron(III) minerals (Silver *et al.*, 1999; Miller *et al.*, 2001), thereby potentially increasing phosphorus availability and so increasing the likelihood of nitrogen limitation to primary production.

In addition to affecting nutrient availability by reducing decomposition, anaerobic soil conditions could also reduce nutrient uptake by inhibiting root growth (Weaver *et al.*, 1973). *Metrosideros polymorpha* populations on adjacent waterlogged and well-drained sites on the island of Maui (Hawai'i) were found to have a higher proportion of aerial roots on waterlogged soils (Santiago *et al.*, 2000), whereas Kitayama *et al.* (1997) found a correlation between rooting depth and soil redox conditions in Hawaiian montane forests across a soil age gradient. Häger and Dohrenbusch (this volume) reported a much higher incidence of tree-falls during storms on waterlogged cloud forest terrain in Costa Rica.

It is suggested here that the effects of persistently high soil water contents on nutrient availability are more sustained than those resulting from lower temperatures in UMCF environments. Furthermore, litter decomposition in anaerobic soils differs in kind, not just in rate, from decomposition in better-drained soils; it is both inefficient and incomplete, with the result that recalcitrant organic matter (and associated nutrients) can accumulate for longer periods in waterlogged than in cold soils. This accumulating nutrient sink in soils breaks the cycle of essential nutrients, and can induce long-term nutrient limitation (Vitousek, 2004).

Greater inputs of water via precipitation and fog interception coupled with reduced evaporative losses in UMCF and ECF also imply greater hydrological outputs (Bruijnzeel, 2005), and so a greater potential for leaching losses of nutrients. To the extent that particular nutrients limit the growth of organisms and occur in forms that those organisms can use, nutrients should be retained efficiently within ecosystems. However, dissolved organic forms of some nutrients (nitrogen in particular) include a large molecular weight fraction that is recalcitrant to decomposition and difficult for organisms to take up (Hedin *et al.*, 1995; Perakis and Hedin, 2001); this dissolved organic nitrogen (DON) represents the main form of hydrological nitrogen losses from many ecosystems (Perakis and Hedin, 2002). Where they are large enough, losses of DON can prevent the accumulation of soil nitrogen and thereby make nitrogen limitation a permanent rather than a transient phenomenon. The combination of large pools of recalcitrant organic nitrogen, inefficient decomposition, and large hydrological fluxes may create just these conditions in many tropical montane forests, although measurements of leachate losses of total nitrogen or DON from TMCF are still extremely rare (e.g. Hafkenscheid, 2000).

Reduced solar radiation

Although solar radiation intensity tends to increase with altitude, cloudiness and rainfall also increase, particularly on the windward sides of mountains (Cavelier *et al.*, 1996), so that tropical montane forests generally receive lower levels of solar radiation and fewer hours of bright sunshine each day (Richards, 1996; Jarvis and Mulligan, this volume; Schawe *et al.*, this volume). The low transpiration rates that have been measured in some types of TMCF may be caused by low atmospheric saturation deficits and radiant energy (Jane and Green, 1985; Holwerda, 2005; McJannet *et al.*, this volume), poorly developed root systems in waterlogged soils (Jane and Green, 1985; cf. Schawe *et al.*, this volume), and low total leaf area index (Santiago *et al.*, 2000; Moser *et al.*, 2007; Santiago *et al.*, this volume). Lower uptake of nutrients by poorly developed root systems may partially explain lower foliar nutrient levels in TMCF trees; this lower nutrient uptake could limit total leaf area and thus total canopy photosynthetic capacity. Nevertheless, the evidence that leaf-level photosynthetic capacity in montane trees is lower than in lowland trees is not conclusive. Aylett (1985) measured maximum photosynthetic rates of Jamaican montane trees (5–7 μmol CO_2 m^{-2} s^{-1}) and found them to be comparable to maximum rates in lowland forests; these photosynthetic rates were also similar to those found in three montane forest species from Puerto Rico (Fetcher *et al.*, 1996). Studies of non-structural carbohydrate pools in trees from both lowland and very high elevation (4810 m) tropical forests suggest that photosynthetic rates at these elevations are high enough to not limit growth, although specific evidence from TMCF is lacking (Körner, 2003; Hoch and Körner, 2005). The reduced levels of incoming photosynthetically active radiation (*PAR*) received by some TMCF might influence rates of nutrient cycling and induce limitation if low photosynthetic rates due to low incoming *PAR* make it economically necessary to have long-lived, sclerophyllous leaves (Bruijnzeel and Veneklaas, 1998; cf. Letts *et al.*, this volume).

Strong winds

Wind undoubtedly sculpts many TMCF canopies (cf. Häger and Dohrenbusch, this volume), but extreme winds also may have important consequences for nutrient cycling. Persistent wind may represent another selection pressure that favors sclerophylly (cf. Jaffe, 1980), but nevertheless major windstorms can alter nutrient dynamics in several other ways. For example, in Puerto Rican lower montane forest, total litterfall resulting from Hurricane Hugo was 1.9 times the mean annual litterfall during non-hurricane periods, and this litter had higher nitrogen, phosphorus, and potassium values than senesced litter (Lodge *et al.*, 1991). Nine years later, when Hurricane Georges hit the forest, total storm litterfall was 61–93% of annual litterfall inputs across a range of montane forest types (Ostertag *et al.*, 2003). In Hawai'i, the effects of Hurricane Iniki were intermediate between those of Hugo and Georges in Puerto Rico (Herbert *et al.*, 1999). These storm events represent substantial nutrient pulses (Lodge *et al.*, 1994), even if they are short-lived (Ostertag *et al.*, 2003). In addition, leaf litterfall after hurricanes in TMCF is lower for several years (Weaver, 1999). Fertilization of hurricane-damaged cloud forest therefore did not lead to increased diameter growth because the added nutrients went to leaf biomass and subsequently litterfall (Walker *et al.*, 1996). Besides defoliation, wind damage also increases the input of coarse woody debris to the forest floor, which may immobilize nutrients and depress net primary productivity on decadal timescales (Zimmerman *et al.*, 1995; Herbert *et al.*, 1999).

Epiphytes are a conspicuous feature in TMCF and constitute a substantial portion of the above-ground biomass in many montane forests (Nadkarni, 1984; Köhler *et al.*, 2007). Epiphytes can be stripped from branches during high winds (Nadkarni, 2000). Some of these epiphytes can re-root and survive on the forest floor, increasing carbon, nitrogen and phosphorus levels in the substrate beneath them (Pett-Ridge and Silver, 2002), which might provide important high-nutrient microsites. Epiphytes can also increase canopy capture and retention of atmospherically deposited nutrients. For example, epiphytic bryophytes in an LMCF at Monteverde, Costa Rica were shown to accumulate about 11–19% of the annual net nitrogen uptake by foliage, much of which comes in as inorganic nitrogen from atmospheric sources (Clark *et al.*, 1998a,b).

Biotic feedbacks

Several selection pressures may drive the development of sclerophyllous leaves in TMCF plants, including low nutrient availability (Loveless, 1961), occasional intense solar radiation (including UV-B; Flenley, 1995), and high wind (Jaffe, 1980). Whether sclerophyllous leaves are a result of low nutrient availability or a consequence of other selective factors, it is clear that traits associated with sclerophylly influence nutrient availability. Decomposition rates of leaf litter may be slow due to low litter nutrient concentrations as well as high lignin and phenolic contents (Bruijnzeel *et al.*, 1993; Schuur, 2001). In turn, slow decomposition and low-quality litter could cause microbes to become limited by nutrient supply – thereby enhancing the microbial immobilization of nutrients, and contributing to a positive feedback system that further lowers nutrient availability (Hobbie, 1992; Hobbie and Vitousek, 2000; Scowcroft *et al.*, 2000; Vitousek, 2004). In addition, the thick, waxy cuticle typical of sclerophyllous leaves may decrease nutrient leaching even if having such leaves does not directly improve nutrient use efficiency (Grubb, 1977; Turner, 1994).

CONCLUSIONS

The supply of nitrogen in particular limits the productivity of many tropical montane cloud forests, particularly those on younger substrates and under wetter conditions; phosphorus is often limiting as well, most likely on older soils and those derived from phosphorus-poor parent material. Nutrient limitation may contribute to the sclerophylly and low leaf area of such forests. Persistently high soil water contents and associated low oxygen levels contribute substantially to the development and maintenance of low nutrient availability in elfin and upper montane cloud forests by slowing decomposition and nutrient mineralization, by impeding root growth, and perhaps by enhancing losses of dissolved organic nutrients. A careful analysis of interactions between the disturbance regime of cloud forests, including wind and landslides (cf. Guariguata, 1990), and their nutrient dynamics could prove rewarding.

REFERENCES

Aylett, G. P. (1985). Irradiance interception, leaf conductance and photosynthesis in Jamaican upper montane rain forest trees. *Photosynthetica* **19**: 323–337.

Baillie, I. C. (1989). Soil characteristics and mineral nutrition of tropical wooded systems. In *Mineral Nutrients in Tropical Forest and Savanna Ecosystems*, ed. J. Proctor, pp. 15–26. Oxford, UK: Blackwell Scientific.

Bruijnzeel, L. A. (1991). Nutrient input–output budgets of tropical forest ecosystems: a review. *Journal of Tropical Ecology* **7**: 1–24.

Bruijnzeel, L. A. (2005). Tropical montane cloud forests: a unique hydrological case. In *Forests, Water and People in the Humid Tropics*, eds. M. Bonell and L. A. Bruijnzeel, pp. 462–483. Cambridge, UK: Cambridge University Press.

Bruijnzeel, L. A., and J. Proctor (1995). Hydrology and biogeochemistry of tropical montane cloud forests: what do we really know? In *Tropical Montane Cloud Forests*, eds. L. S. Hamilton, J. O. Juvik, and F. N. Scatena, pp. 38–78. New York: Springer-Verlag.

Bruijnzeel, L., and E. Veneklaas (1998). Climatic conditions and tropical montane forest productivity: the fog has not lifted yet. *Ecology* **79**: 3–9.

Bruijnzeel, L. A., M. J. Waterloo, J. Proctor, A. T. Kuiters, and B. Kotterink (1993). Hydrological observations in montane rain-forests on Gunung-Silam, Sabah, Malaysia, with special reference to the '*Massenerhebung*' effect. *Journal of Ecology* **81**: 145–167.

Carrillo, J. H., M. G. Hastings, D. M. Sigman, and B. J. Huebert (2002). Atmospheric deposition of inorganic and organic nitrogen and base cations in Hawaii. *Global Biogeochemical Cycles* **16**: 24: 1–16.

Cavelier, J., E. Tanner, and J. Santamaria (2000). Effect of water, temperature and fertilizers on soil nitrogen net transformations and tree growth in an elfin cloud forest of Colombia. *Journal of Tropical Ecology* **16**: 83–99.

Cavelier, J., D. Solis, and M. A. Jaramillo (1996). Fog interception in montane forests across the central cordillera of Panama. *Journal of Tropical Ecology* **12**: 357–369.

Chadwick, O. A., L. A. Derry, P. M. Vitousek, B. J. Huebert, and L. O. Hedin (1999). Changing sources of nutrients during four million years of ecosystem development. *Nature* **397**: 491–497.

Clark, K. L., N. M. Nadkarni, and H. L. Gholz (1998a). Growth, net production, litter decomposition, and net nitrogen accumulation by epiphytic bryophytes in a tropical montane forest. *Biotropica* **30**: 12–23.

Clark, K. L., N. M. Nadkarni, D. Schaefer, and H. L. Gholz (1998b). Atmospheric deposition and net retention of ions by the canopy in a tropical montane forest, Monteverde, Costa Rica. *Journal of Tropical Ecology* **14**: 27–45.

Cordell, S., G. Goldstein, F. Meinzer, and P. Vitousek (2001). Regulation of leaf life-span and nutrient-use efficiency of *Metrosideros polymorpha* trees at two extremes of a long chronosequence in Hawaii. *Oecologia* **127**: 198–206.

Crews, T. E. (1999). The presence of nitrogen fixing legumes in terrestrial communities: evolutionary vs. ecological considerations. *Biogeochemistry* **46**: 233–246.

Fetcher, N., B. L. Haines, R. A. Cordero, *et al.* (1996). Responses of tropical plants to nutrients and light on a landslide in Puerto Rico. *Journal of Ecology* **84**: 331–341.

Flenley, J. R. (1995). Cloud forest, the *Massenerhebung* effect, and ultraviolet insolation. In *Tropical Montane Cloud Forests*, eds. L. S. Hamilton, J. O. Juvik, and F. N. Scatena, pp. 150–155. New York: Springer-Verlag.

Forman, R. T. T. (1975). Canopy lichens with blue–green algae: a nitrogen source in a Colombian rain forest. *Ecology* **56**: 1176–1184.

Freiberg, E. (1998). Microclimatic parameters influencing nitrogen fixation in the phyllosphere in a Costa Rican premontane rain forest. *Oecologia* **117**: 9–18.

Gerrish, G., D. Mueller-Dombois, and K. W. Bridges (1988). Nutrient limitation and *Metrosideros* dieback in Hawai'i. *Ecology* **69**: 723–727.

Grieve, I. C., J. Proctor, and S. A. Cousins (1990). Soil variation with altitude on Volcan Barva, Costa Rica. *Catena* **17**: 525–534.

Grubb, P. J. (1971). Interpretation of the *Massenerhebung* effect on tropical mountains. *Nature* **229**: 44.

Grubb, P. J. (1977). Control of forest growth and distribution on wet tropical mountains: with special reference to mineral nutrition. *Annual Review of Ecology and Systematics* **8**: 83–107.

Guariguata, M. R. (1990). Landslide disturbance and forest regeneration in the upper Luquillo mountains of Puerto Rico. *Journal of Ecology* **78**: 814–832.

Hafkenscheid, R. L. L. J. (2000). Hydrology and biogeochemistry of tropical montane rain forests of contrasting stature in the Blue Mountains, Jamaica. Ph.D. thesis, VU University, Amsterdam, the Netherlands. Also available at http://dare.ubvu.vu.nl/bitstream/1871/12734/1/tekst.pdf.

Harrington, R. A., J. H. Fownes, and P. M. Vitousek (2001). Production and resource use efficiencies in N- and P-limited tropical forests: a comparison of responses to long-term fertilization. *Ecosystems* **4**: 646–657.

Hedin, L. O., J. Armesto, and A. Johnson (1995). Patterns of nutrient loss from unpolluted, old growth temperate forests: evaluation of biogeochemical theory. *Ecology* **76**: 493–509.

Herbert, D. A., and J. H. Fownes (1995). Phosphorus limitation of forest leaf area and net primary production on a highly weathered soil. *Biogeochemistry* **29**: 223–235.

Herbert, D. A., J. A. Fownes, and P. M. Vitousek (1999). Hurricane damage to a Hawaiian forest: nutrient supply rate affects resistance and resilience. *Ecology* **80**: 908–920.

Hobbie, S. E. (1992). Effects of plant species on nutrient cycling. *Trends in Ecology and Evolution* **7**: 336–339.

Hobbie, S. E., and P. Vitousek (2000). Nutrient limitation of decomposition in Hawaiian forests. *Ecology* **81**: 1867–1877.

Hoch, G., and C. Körner (2005). Growth, demography and carbon relations of *Polylepis* trees at the world's highest treeline. *Functional Ecology* **19**: 941–951.

Holwerda, F. (2005). Water and energy budgets of rain forests along an elevational gradient under maritime tropical conditions. Ph.D. thesis, VU University Amsterdam, Amsterdam, the Netherlands.

Jaffe, M. J. (1980). Morphogenetic responses of of plants to mechanical stimuli or stress. *BioScience* **30**: 239–243.

Jane, G. T., and T. G. A. Green (1985). Patterns of stomatal conductance in six evergreen tree species from a New Zealand cloud forest. *Botanical Gazette* **146**: 413–420.

Jenny, H., S. P. Gessel, and F. T. Bingham (1949). Comparative study of decomposition rates in temperate and tropical regions. *Soil Science* **68**: 419–432.

Jobbagy, E., and R. Jackson (2000). The vertical distribution of soil organic carbon and its relation to climate and vegetation. *Ecological Applications* **10**: 423–436.

Kitayama, K., and S. I. Aiba. (2002). Ecosystem structure and productivity of tropical rain forests along altitudinal gradients with contrasting soil phosphorus pools on Mount Kinabalu, Borneo. *Journal of Ecology* **90**: 37–51.

Kitayama, K., N. Majalap-Lee, and S. Aiba (2000). Soil phosphorus fractionation and phosphorus-use efficiencies of tropical rainforests along altitudinal gradients of Mount Kinabalu, Borneo. *Oecologia* **123**: 342–349.

Kitayama, K., E. Schuur, D. Drake, and D. Mueller-Dombois (1997). Fate of a wet montane forest during soil ageing in Hawaii. *Journal of Ecology* **85**: 669–679.

Köhler, L., C. Tobón, K. F. A. Frumau, and L. A. Bruijnzeel (2007). Biomass and water storage of epiphytes in old-growth and secondary montane rain forest stands in Costa Rica. *Plant Ecology* **193**: 171–184.

Körner, Ch. (2003). Carbon limitation in trees. *Journal of Ecology* **91**: 4–17.

Leuschner, Ch., G. Moser, C. Bertsch, M. Röderstein, and D. Hertel (2007). Large altitudinal increase in tree root/shoot ratio in tropical mountain forests of Ecuador. *Basic and Applied Ecology* **8**: 219–230.

Liu, W. Y., J. E. D. Fox, and Z. F. Xu (2003). Nutrient budget of a montane evergreen broad-leaved forest at Ailao Mountain National Nature Reserve, Yunnan, southwest China. *Hydrological Processes* **17**: 1119–1134.

Lodge, D. J., F. N. Scatena, C. E. Asbury, and M. J. Sánchez (1991). Fine litterfall and related nutrient inputs resulting from Hurricane Hugo in subtropical wet and lower montane rain forests of Puerto Rico. *Biotropica* **23**: 336–342.

Lodge, D. J., W. H. McDowell, and C. P. McSwiney (1994). The importance of nutrient pulses in tropical forests. *Trends in Ecology and Evolution* **9**: 384–387.

Loveless, A. R. (1961). A nutritional interpretation of sclerophylly based on differences in the chemical composition of sclerophyllous and mesophytic leaves. *Annals of Botany* **25**: 168–184.

Marrs, R., J. Proctor, A. Heaney, and M. Mountford (1988). Changes in soil nitrogen mineralization and nitrification along an altitudinal transect in tropical rain forest in Costa Rica. *Journal of Ecology* **76**: 466–482.

Martinelli, L. A., M. C. Piccolo, A. R. Townsend, *et al.* (1999). Nitrogen stable isotopic composition of leaves and soil: tropical versus temperate forests. *Biogeochemistry* **46**: 45–65.

Matelson, T. J., N. M. Nadkarni, and R. Solano (1995). Tree damage and annual mortality in a montane forest in Monteverde, Costa Rica. *Biotropica* **27**: 441–447.

Matzek, V., and P. Vitousek (2003). Nitrogen fixation in bryophytes, lichens, and decaying wood along a soil-age gradient in Hawaiian montane rain forest. *Biotropica* **35**: 12–19.

McDowell, W., and C. Asbury (1994). Export of carbon, nitrogen, and major ions from three tropical montane watersheds. *Limnology and Oceanography* **39**: 111–125.

McGroddy, M. and W. L. Silver (2000). Variations in belowground carbon storage and soil CO_2 flux rates along a wet tropical climate gradient. *Biotropica* **32**: 614–624.

Miller, A. J., E. A. G. Schuur, and O. A. Chadwick (2001). Redox control of phosphorus pools in montane forest soils in Hawaii. *Geoderma* **102**: 219–237.

Moser, G., D. Hertel, and Ch. Leuschner (2007). Altitudinal change in LAI and stand leaf biomass in tropical montane forests: a transect study in Ecuador and a pan-tropical meta-analysis. *Ecosystems* **10**: 924–935.

Moser, G., M. Röderstein, N. Soethe, D. Hertel, and Ch. Leuschner (2008). Altitudinal changes in stand structure and biomass allocation of tropical mountain forest in relation to microclimate and soil chemistry. In *Gradients in a Tropical Mountain Ecosystem of Ecuador*, eds. E. Beck, J. Bendix, I. Kottke, F. Makeschin, and R. Mosandl, pp. 229–242. Berlin: Springer-Verlag.

Nadkarni, N. (1984). Epiphyte biomass and nutrient capital of a neotropical elfin forest. *Biotropica* **16**: 249–246.

Nadkarni, N. (2000). Colonization of stripped branch surfaces by epiphytes in a lower montane cloud forest, Monteverde, Costa Rica. *Biotropica* **32**: 358–363.

Nepstad, D.C., C.R. De Carvalho, E.A. Davidson, *et al.* (1994). The role of deep roots in the hydrological and carbon cycles of Amazonian forests and pastures. *Nature* **372**: 666–669.

Nomura, N., and K. Kikuzawa (2003). Productive phenology of tropical montane forests: fertilization experiments along a moisture gradient. *Ecological Research* **18**: 573–586.

Northup, R., Z. Yu, R. A. Dahlgren, and K. A. Vogt (1995). Polyphenol control of nitrogen release from pine litter. *Nature* **377**: 227–229.

Olander, L. P., and P. M. Vitousek (2000). Regulation of soil phosphatase and chitinase activity by N and P availability. *Biogeochemistry* **49**: 175–190.

Olander, L. P., and P. M. Vitousek (2004). Biological and geochemical sinks for phosphorus in soil from a wet tropical forest. *Ecosystems* **7**: 404–419.

Olander, L. P., F. N. Scatena, and W. L. Silver (1998). Impacts of disturbance initiated by road construction in a subtropical cloud forest in the Luquillo Experimental Forest, Puerto Rico. *Forest Ecology and Management* **109**: 33–49.

Ostertag, R. (2001). The effects of nitrogen and phosphorus availability on fine root dynamics in Hawaiian montane forests. *Ecology* **82**: 485–499.

Ostertag, R., F. N. Scatena, and W. L. Silver (2003). Forest floor decomposition following hurricane litter inputs in several Puerto Rican forests. *Ecosystems* **6**: 261–273.

Pendry, C. A., and J. Proctor (1996). The causes of altitudinal zonation of rain forests on Bukit Belalong, Brunei. *Journal of Ecology* **84**: 407–418.

Perakis, S., and L. Hedin (2001). Fluxes and fates of nitrogen in soil of an unpolluted old-growth temperate forest, southern Chile. *Ecology* **82**: 2245–2260.

Perakis, S., and L. Hedin (2002). Nitrogen loss from unpolluted South American forests mainly via dissolved organic compounds. *Nature* **415**: 416–419.

Pett-Ridge, J. and W. L. Silver (2002). Survival, growth and ecosystem dynamics of displaced bromeliads in a montane tropical forest. *Biotropica* **34**: 211–224.

Raich, J. W., R. H. Riley, and P. M. Vitousek (1994). Use of root ingrowth cores to assess nutrient limitations in forest ecosystems. *Canadian Journal of Forest Research* **24**: 2135–2138.

Raich, J. W., A. R. Russell, T. E. Crews, H. Farrington, and P. M. Vitousek (1996). Both nitrogen and phosphorus limit plant production on young Hawaiian lava flows. *Biogeochemistry* **32**: 1–14.

Raich, J. W., A. E. Russell, and P. M. Vitousek (1997). Primary productivity and ecosystem development along an elevational gradient on Mauna Loa, Hawai'i. *Ecology* **78**: 707–721.

Raich, J. W., A. E. Russell, K. Kitayama, W. J. Parton, and P. M. Vitousek (2006). Temperature influences carbon accumulation in moist tropical forests. *Ecology* **87**: 76–87.

Reich, P. B., M. B. Walters, and D. S. Ellsworth (1997). From tropics to tundra: Global convergence in plant functioning. *Proceedings of the National Academy of Sciences USA* **94**: 13 730–13 734.

Richards, P. W. (1996). *The Tropical Rain Forest*, 2nd edn. Cambridge, UK: Cambridge University Press.

Röderstein, M., D. Hertel, and Ch. Leuschner (2005). Above- and below-ground litter production in three tropical montane forests in southern Ecuador. *Journal of Tropical Ecology* **21**: 483–492.

Santiago, L. S., G. Goldstein, F. C. Meinzer, J. H. Fownes, and D. Mueller-Dombois (2000). Transpiration and forest structure in relation to soil water-logging in a Hawaiian montane cloud forest. *Tree Physiology* **20**: 673–681.

Schuur, E. (2001). The effect of water on decomposition dynamics in mesic to wet Hawaiian montane forests. *Ecosystems* **4**: 259–273.

Schuur, E., and P. Matson (2001). Net primary productivity and nutrient cycling across a mesic to wet precipitation gradient in Hawaiian montane forest. *Oecologia* **128**: 431–442.

Schuur, E. A. G., O. A. Chadwick, and P. A. Matson (2001). Carbon cycling and soil carbon storage in mesic to wet Hawaiian montane forests. *Ecology* **82**: 3182–3196.

Scowcroft, P. G., D. R. Turner, and P. M. Vitousek (2000). Decomposition of *Metrosideros polymorpha* leaf litter along elevational gradients in Hawaii. *Global Change Biology* **6**: 73–85.

Sidle, R. C., Ziegler, A. D., Negishi, J. N., *et al.* (2006). Erosion processes in steep terrain: truths, myths, and uncertainties related to forest management in Southeast Asia. *Forest Ecology and Management* **224**: 199–225.

Silver, W. L., A. E. Lugo, and M. Keller (1999). Soil oxygen availability and biogeochemistry along rainfall and topographic gradients in upland wet tropical forest soils. *Biogeochemistry* **44**: 301–328.

Soethe, N., J. Lehmann, and C. Engels (2006). The vertical pattern of rooting and nutrient uptake at different altitudes of a southern Ecuadorian montane forest. *Plant Soil* **286**: 287–299.

Soethe, N., J. Lehmann, and C. Engels (2008a). Nutrient availability at different altitudes in a tropical montane forest in Ecuador. *Journal of Tropical Ecology* **24**: 397–406.

Soethe, N., W. Wilcke, J. Homeier, J. Lehmann, and C. Engels (2008b). Plant growth along the altitudinal gradient: role of plant nutritional status, fine root activity, and soil properties. In *Gradients in a Tropical Mountain Ecosystem of Ecuador*, eds. E. Beck, J. Bendix, I. Kottke, F. Makeschin, and R. Mosandl, pp. 259–266. Berlin: Springer-Verlag.

Stewart, C. G. (2000). A test of nutrient limitation in two tropical montane forests using root ingrowth cores. *Biotropica* **32**: 369–373.

Tanner, E., V. Kapos, S. Freskos, J. Healey, and A. Theobald (1990). Nitrogen and phosphorus fertilization of Jamaican montane forest trees. *Journal of Tropical Ecology* **6**: 231–238.

Tanner, E. V. J., V. Kapos, and W. Franco (1992). Nitrogen and phosphorus fertilization effects on Venezuelan montane forest trunk growth and litter-fall. *Ecology* **73**: 78–86.

Tanner, E., P. Vitousek, and E. Cuevas (1998). Experimental investigation of nutrient limitation of forest growth on wet tropical mountains. *Ecology* **79**: 10–22.

Titiz, B. (2005). Signatures in the soil: soil charcoal and phosphorus distribution patterns along an elevational gradient in a Costa Rican tropical rainforest. Ph.D. thesis, University of Denver, Denver, CO, USA.

Townsend, A. R., P. M. Vitousek, and S. E. Trumbore (1995). Soil organic matter dynamics along gradients in temperature and land use on the island of Hawaii. *Ecology* **76**: 721–733.

Treseder, K., and P. Vitousek (2001). Effects of soil nutrient availability on investment in acquisition of N and P in Hawaiian rain forests. *Ecology* **82**: 946–954.

Turner, I. M. (1994). Sclerophylly: primarily protective? *Functional Ecology* **8**: 669–675.

Vera, M., J. Cavelier, and J. Santamaria (1999). Tree leaf nitrogen and phosphorus reabsorption in a montane forest of the central Andes, Colombia. *Revista de Biologia Tropical* **47**: 33–43.

Vitousek, P. M. (1984). Litterfall, nutrient cycling, and nutrient limitation in tropical forests. *Ecology* **65**: 285–298.

Vitousek, P. M. (2004). *Nutrient Cycling and Limitation: Hawai'i as a Model System.* Princeton, NJ: Princeton University Press.

Vitousek, P., and H. Farrington (1997). Nutrient limitation and soil development: experimental test of a biogeochemical theory. *Biogeochemistry* **37**: 63–75.

Vitousek, P. M., L. R. Walker, L. D. Whiteaker, D. Mueller-Dombois, and P. A. Matson (1987). Biological invasion by *Myrica faya* alters ecosystem development in Hawaii. *Science* **238**: 802–804.

Vitousek, P., P. Matson, and D. Turner (1988). Elevational and age gradients in Hawaiian montane rainforest: foliar and soil nutrients. *Oecologia* **77**: 565–570.

Vitousek, P. M., L. R. Walker, L. D. Whiteaker, and P. A. Matson (1993). Nutrient limitation to plant growth during primary succession in Hawaii Volcanoes National Park. *Biogeochemistry* **23**: 197–215.

Vitousek, P. M., D. R. Turner, W. J. Parton, and R. L. Sanford (1994). Litter decomposition on the Mauna Loa environmental matrix, Hawaii: patterns, mechanisms, and models. *Ecology* **75**: 418–429.

Walker, L. R., J. K. Zimmerman, D. J. Lodge, and S. Guzmán-Grajales (1996). An altitudinal comparison of growth and species composition in hurricane-damaged forests in Puerto Rico. *Journal of Ecology* **84**: 877–889.

Weathers, K., G. Lovett, G. Likens, and N. Caraco (2000). Cloudwater inputs of nitrogen to forest ecosystems in southern Chile: forms, fluxes, and sources. *Ecosystems* **3**: 590–595.

Weaver, P. L. (1999). Impacts of Hurricane Hugo on the dwarf cloud forest of Puerto Rico's Luquillo Mountains. *Caribbean Journal of Science* **35**: 101–111.

Weaver, P. L., M. Byer, and D. Bruck (1973). Transpiration rates in the Luquillo Mountains of Puerto Rico. *Biotropica* **5**: 123–133.

Wilcke, W., S. Yasin, C. Valarezo, and W. Zech (2001). Change in water quality during the passage through a tropical montane rain forest in Ecuador. *Biogeochemistry* **55**: 45–72.

Wilcke, W., S. Yasin, U. Abramowski, C. Valarezo, and W. Zech (2002). Nutrient storage and turnover in organic layers under tropical montane rain forest in Ecuador. *European Journal of Soil Science* **53**: 15–27.

Williams-Linera, G. (1999). Leaf dynamics in a tropical cloud forest: phenology, herbivory, and life span. *Selbyana* **20**: 98–105.

Zimmerman, J. K., W. M. Pulliam, D. J. Lodge, *et al.* (1995). Nitrogen immobilization by decomposing woody debris and the recovery of tropical wet forest from hurricane damage. *Oikos* **72**: 314–322.

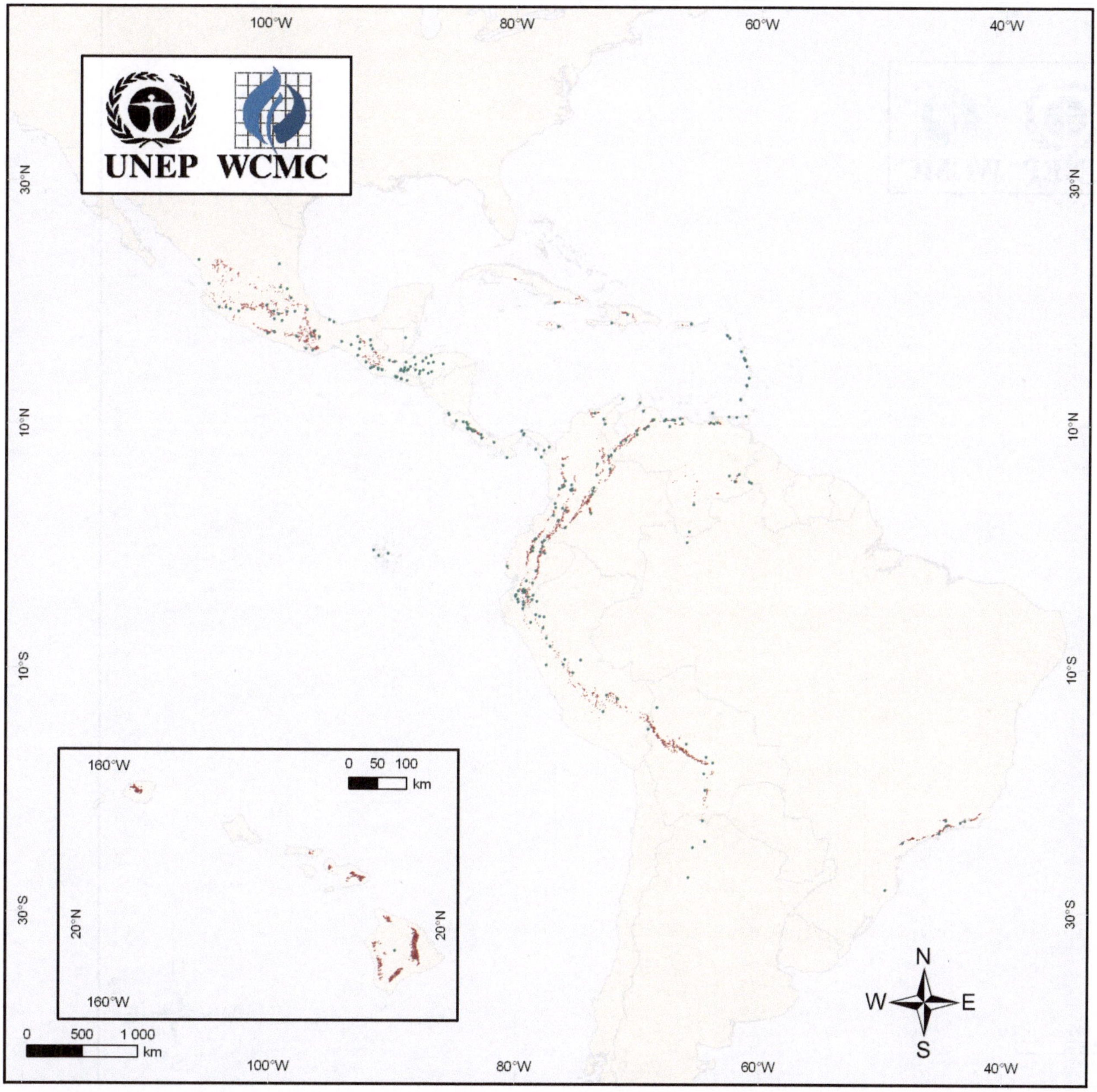

Figure 1.3. Estimated tropical montane cloud forest distribution in the Americas (including the Hawai'ian archipelago), based on the altitudinal ranges listed in Table 1.2 (areas in red) and known cloud forest site locations from Aldrich *et al.* (1997) (green dots).

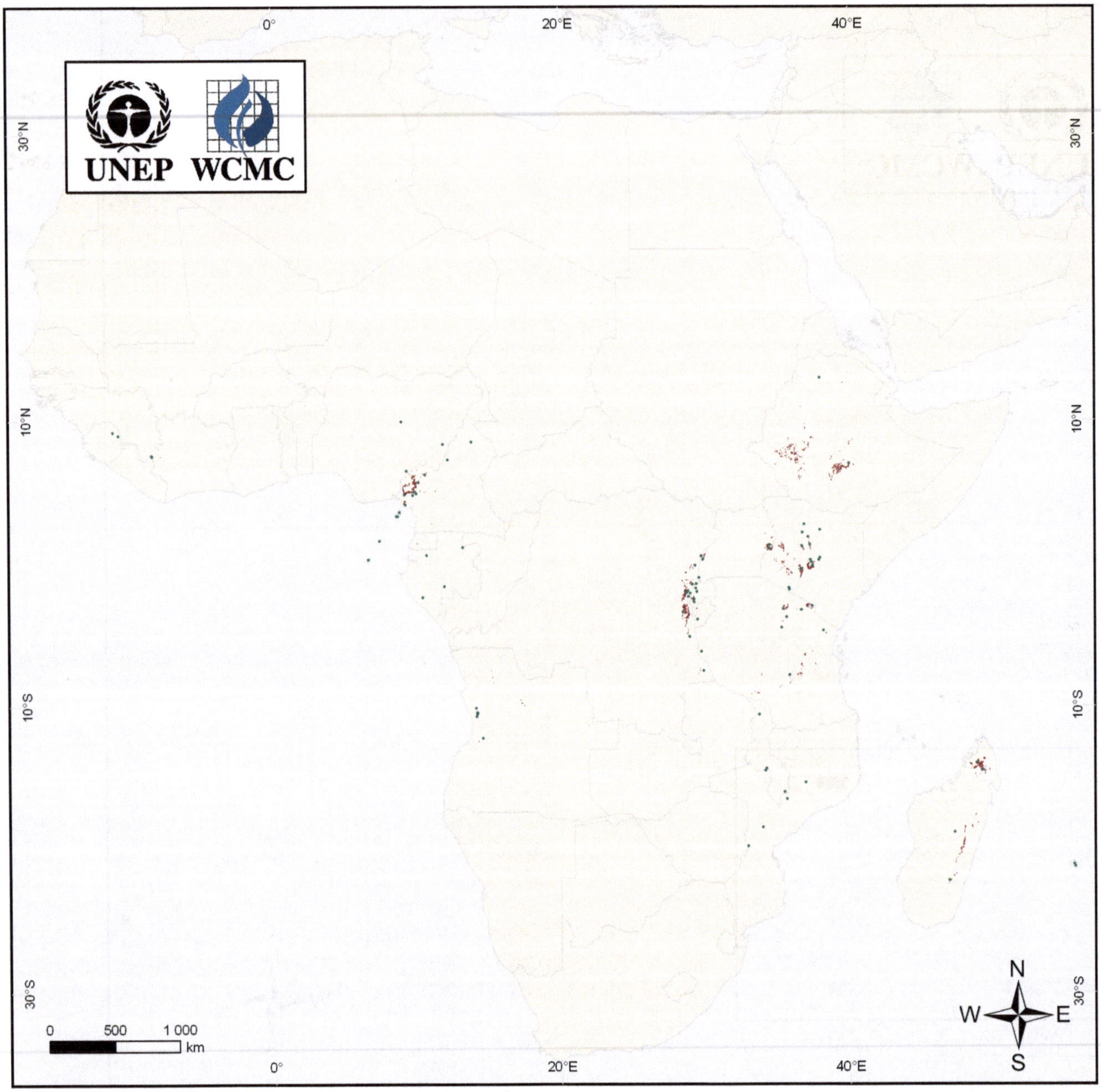

Figure 1.4. Estimated tropical montane cloud forest distribution in Africa, based on the altitudinal ranges listed in Table 1.2 (areas in red) and known cloud forest site locations from Aldrich *et al.* (1997) (green dots).

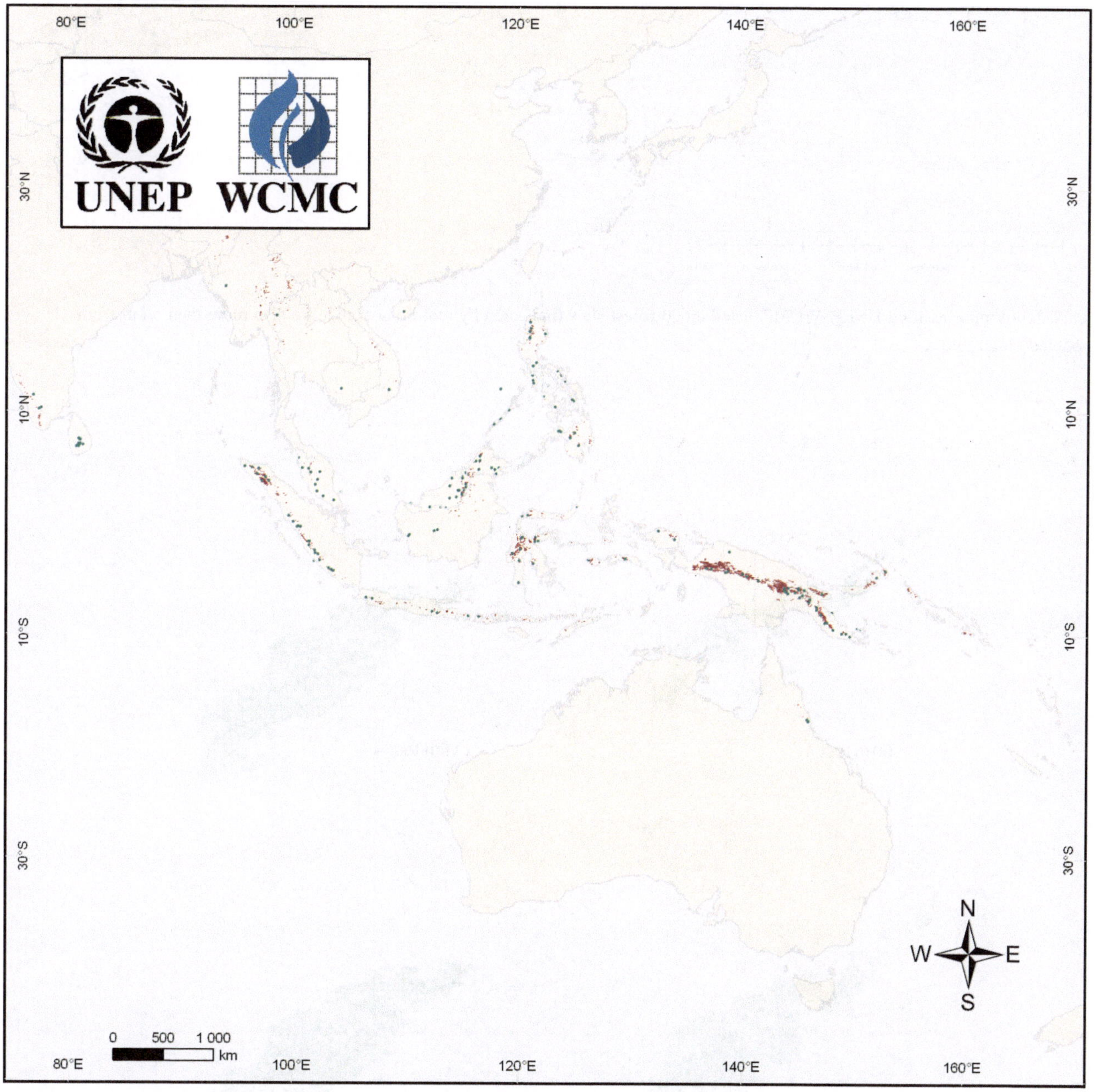

Figure 1.5. Estimated tropical montane cloud forest distribution in Asia (including northern Australia and Oceania), based on the altitudinal ranges listed in Table 1.2 (areas in red) and known cloud forest site locations from Aldrich *et al.* (1997) (green dots).

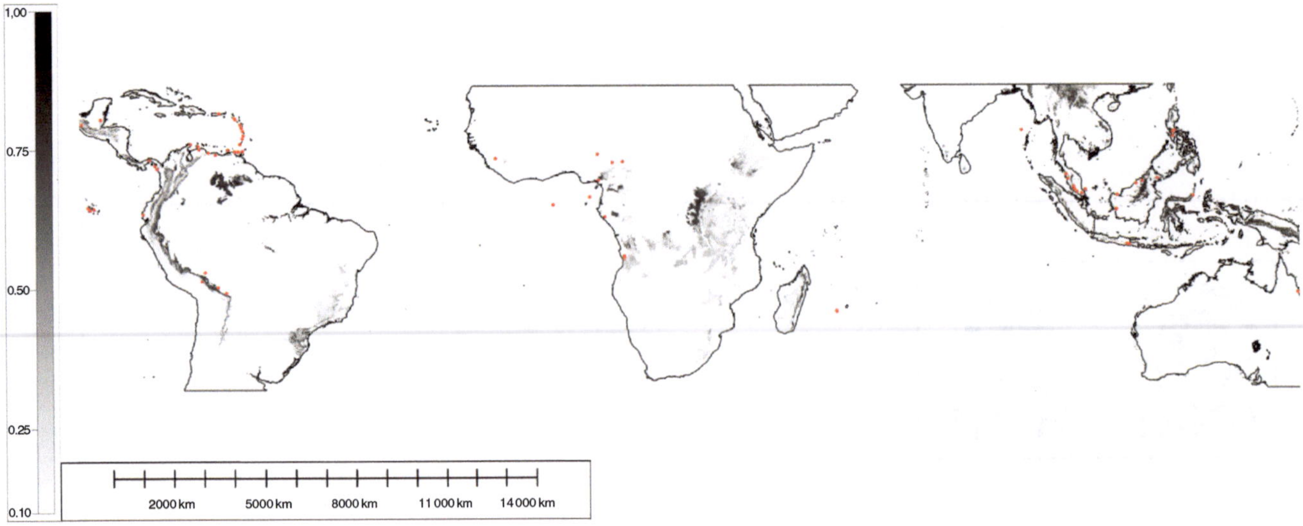

Figure 2.8. Persistently misclassified UNEP–WCMC-listed cloud forest sites (indicated by red crosses) that are also more than 5 km from the nearest modeled cloud forest.

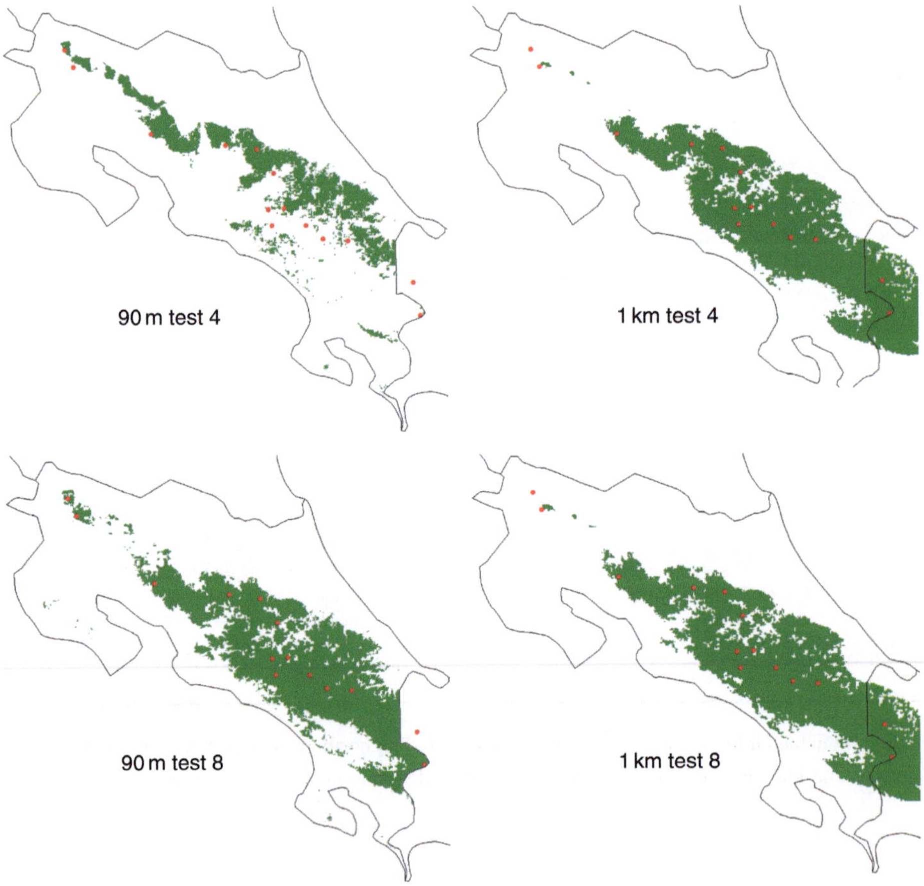

Figure 2.10. Modeled cloud forest distribution for Costa Rica.

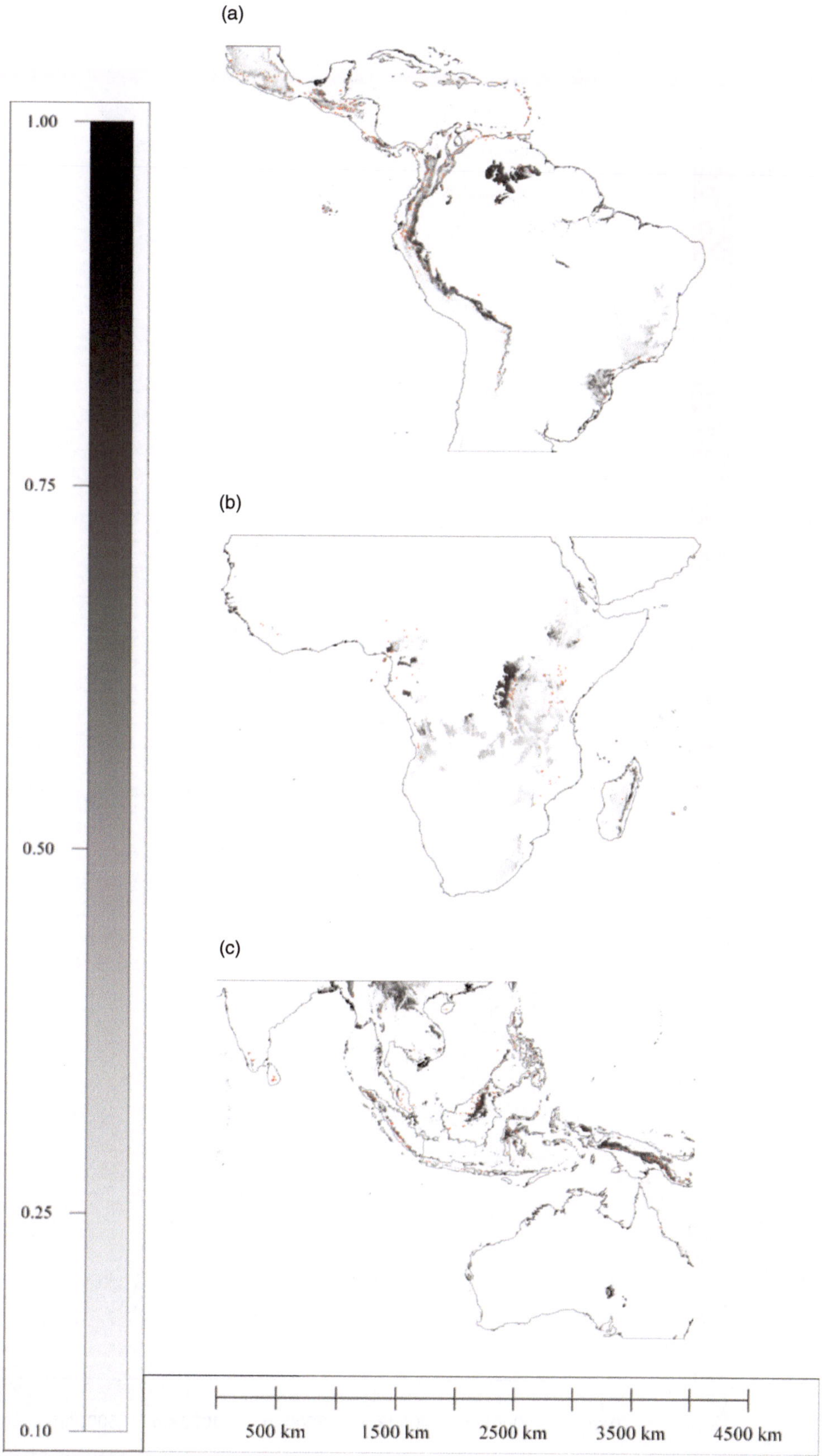

Figure 2.11. The distribution of tropical montane cloud forests (cloud affected tropical montane forests) in detail with UNEP–WCMC sites overlaid in red. The darker the shade the greater the fractional cover of forest within the 1-km pixel.

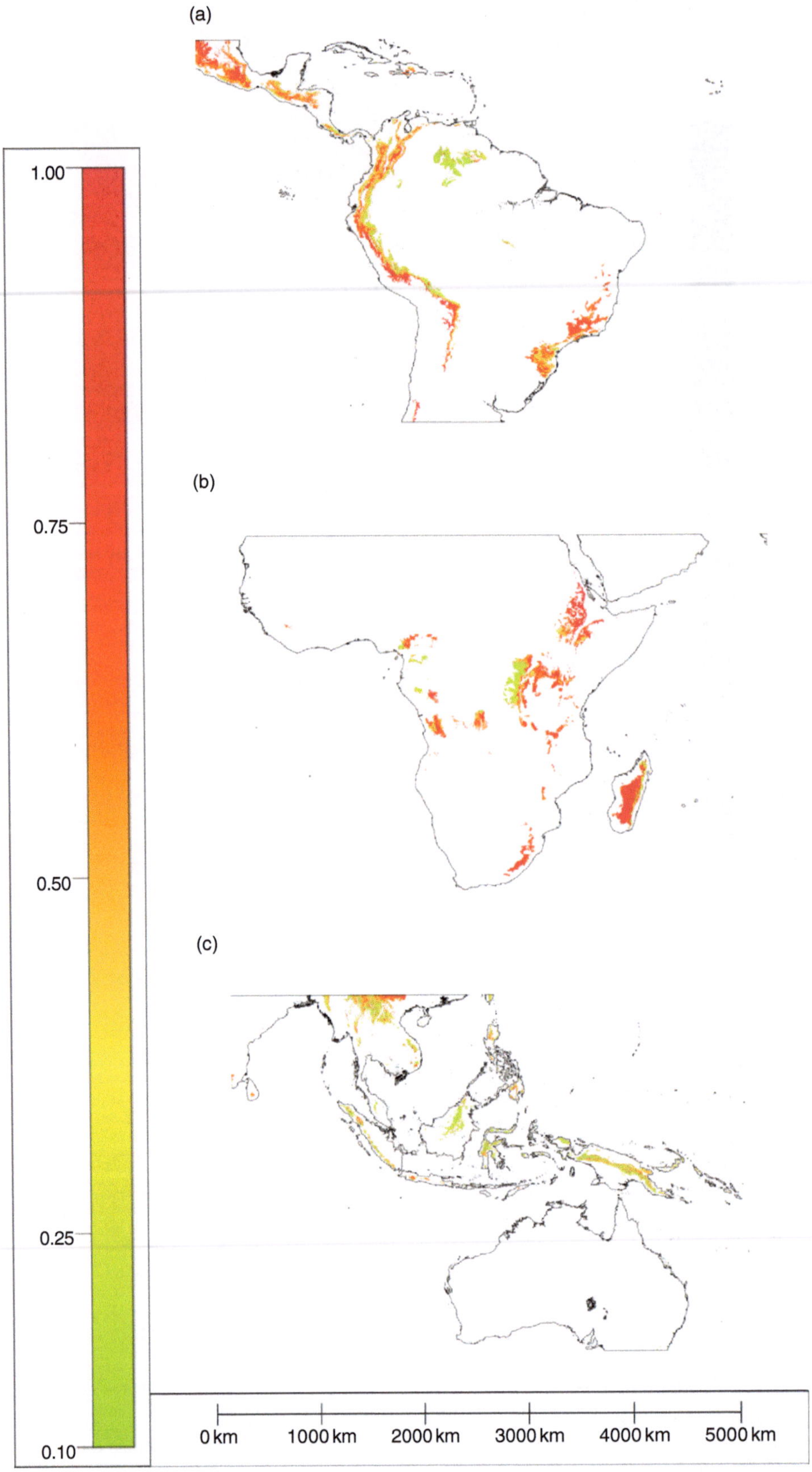

Figure 2.13. Fractional cloud-forest loss according to MODIS_VCF satellite product for (a) Latin America, (b) Africa, and (c) South-East Asia. The darker the shade of red, the greater the forest loss. Red, >60% forest loss; green, <25% forest loss.

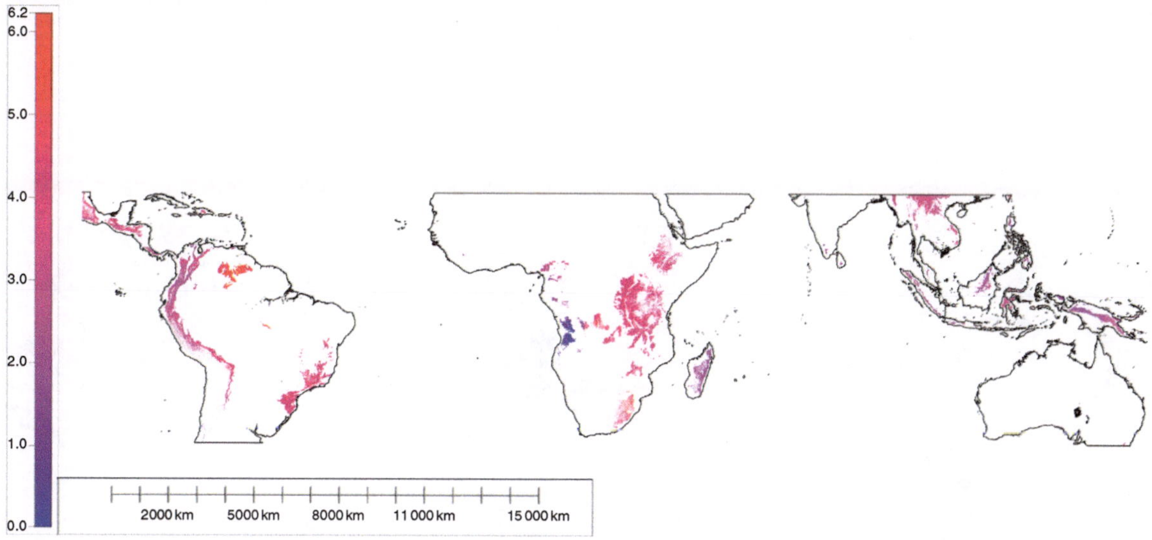

Figure 2.14. Temperature changes (1950–2000 vs. 2000–2050) in °C in the vicinity of cloud forests as predicted by the HADCM3 global circulation model.

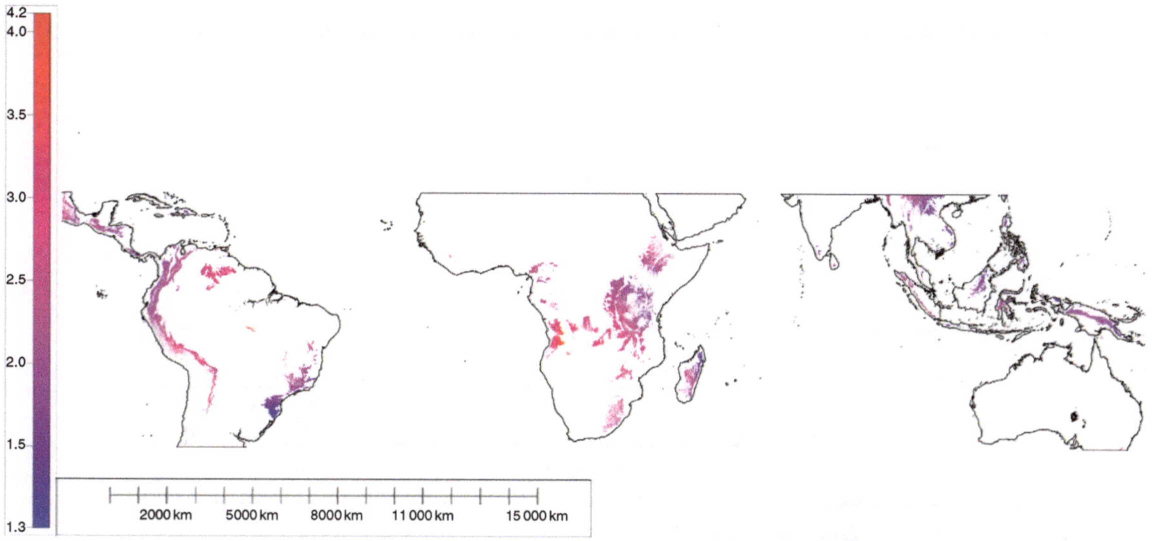

Figure 2.15. Temperature changes (1950–2000 vs. 2000–2050) in °C in the vicinity of cloud forests as predicted by the ECHAM5 global circulation model.

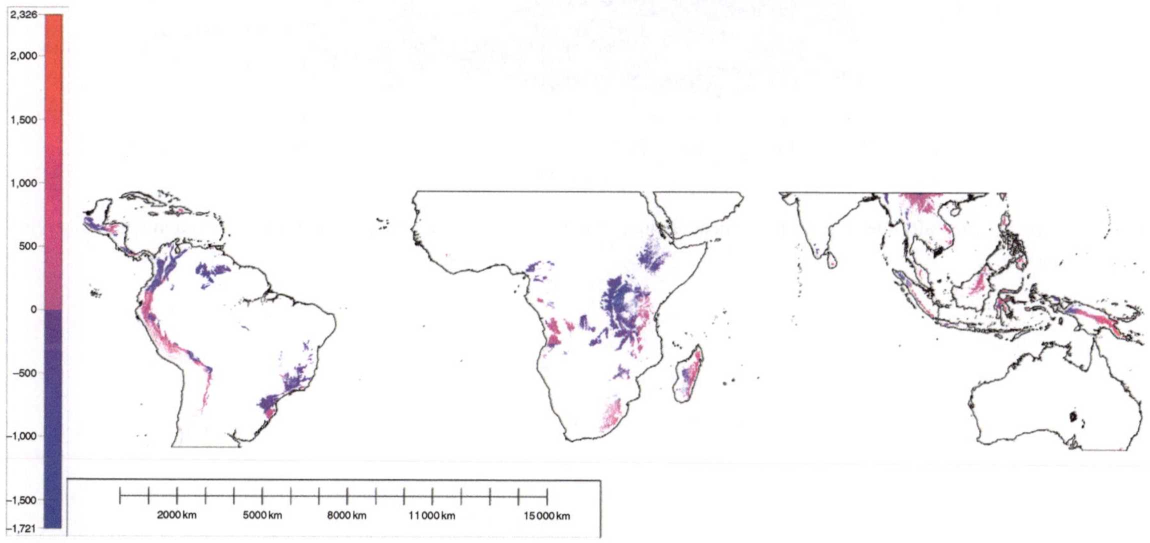

Figure 2.16. Changes in rainfall (1950–2000 vs. 2000–2050) in mm year^{-1} in the vicinity of cloud forests as predicted by the HADCM3 global circulation model.

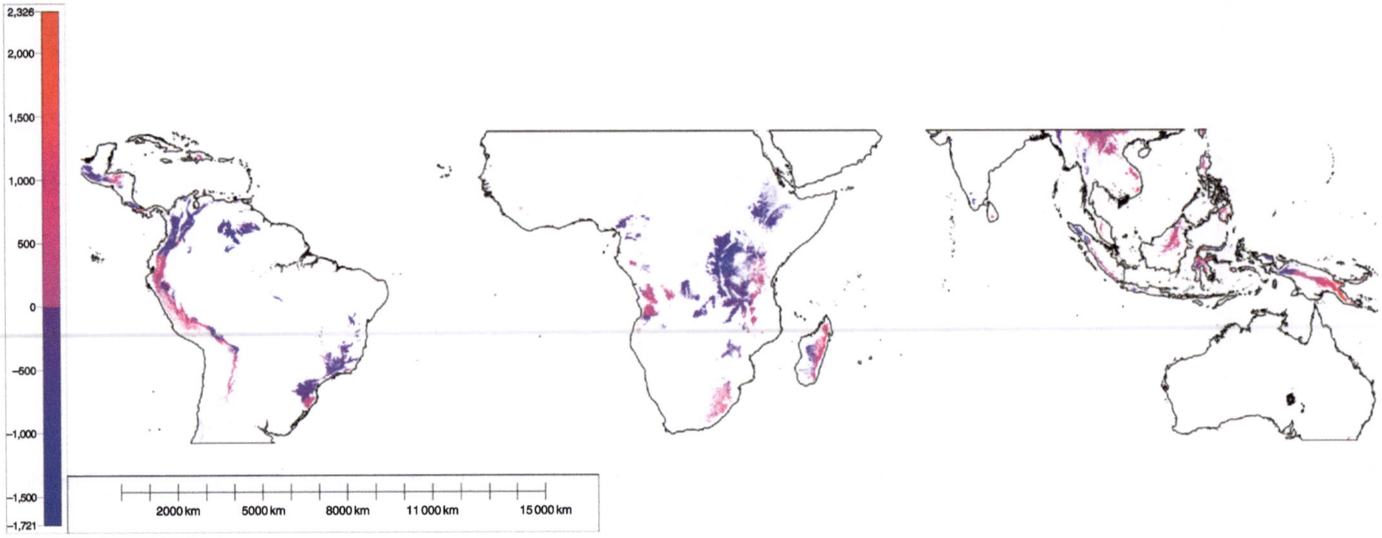

Figure 2.17. Changes in rainfall (1950–2000 vs. 2000–2050) in mm year^{-1} in the vicinity of cloud forests as predicted by the ECHAM5 global circulation model.

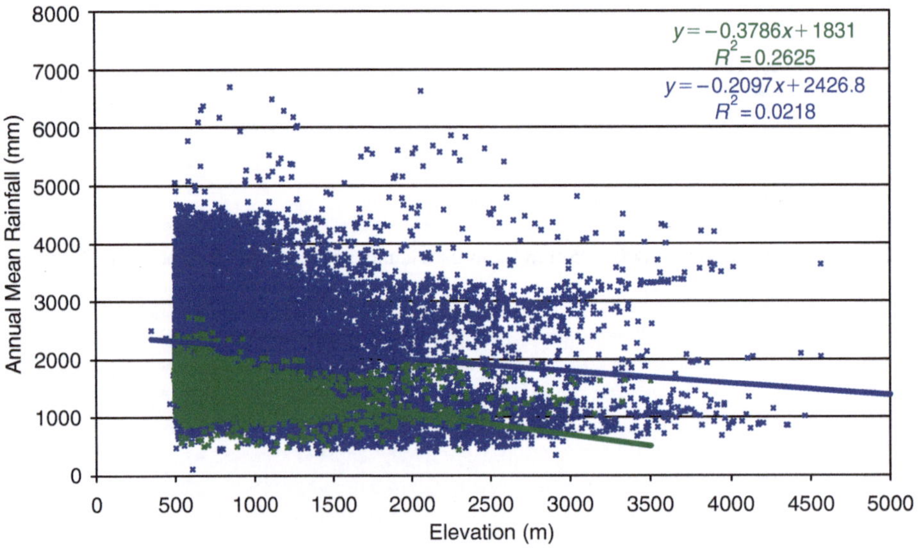

Figure 3.6. Relationships between altitude (m.a.s.l.) and mean annual rainfall (mm) for large-sized (shown in green) and small-sized (shown in blue) mountains in Latin America.

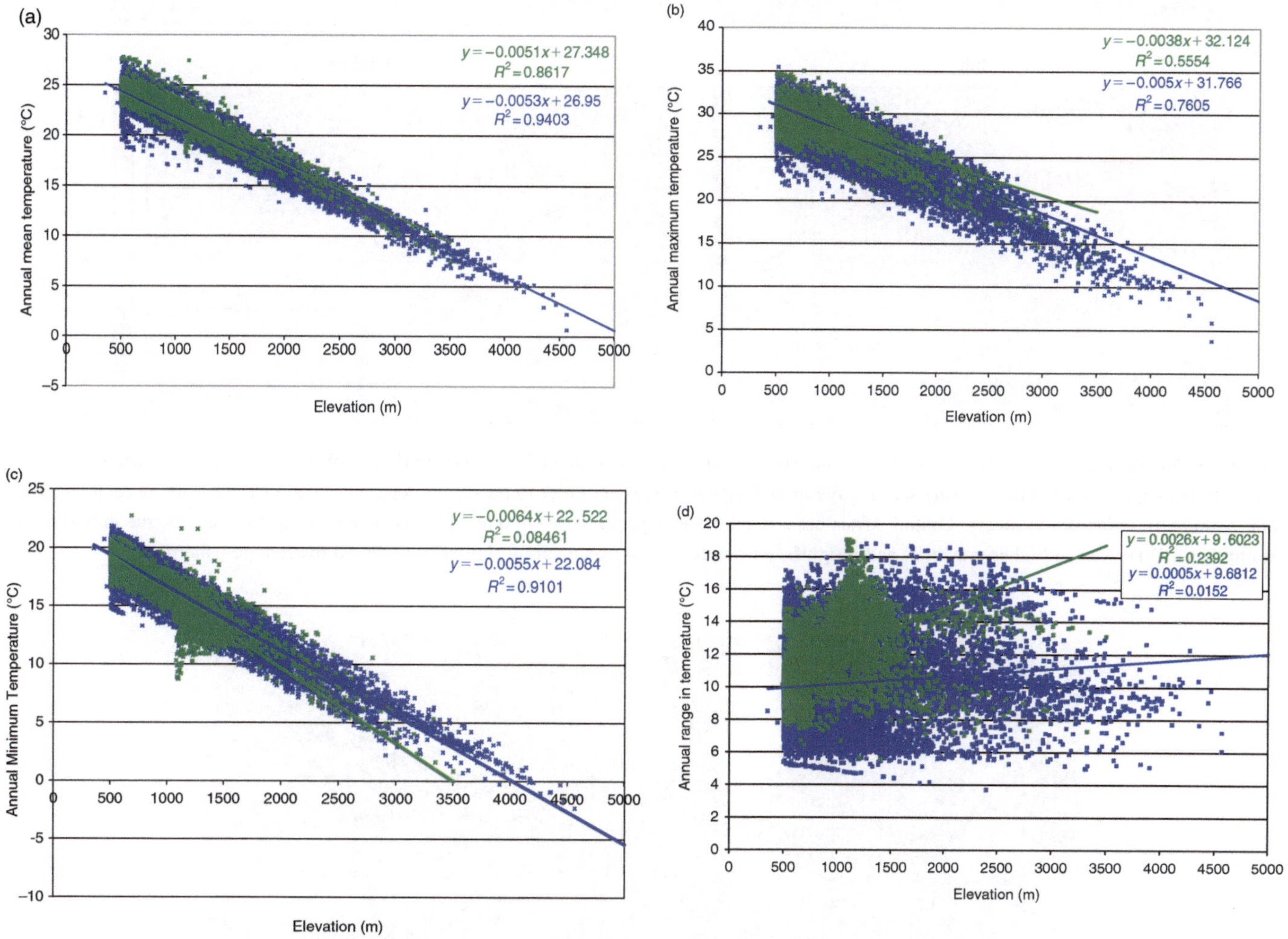

Figure 3.5. Relationships between altitude and (a) mean temperature, (b) maximum temperature, (c) minimum temperature, and (d) temperature range (all in °C) for large-sized (shown in green) and small-sized (shown in blue) mountains in Latin America.

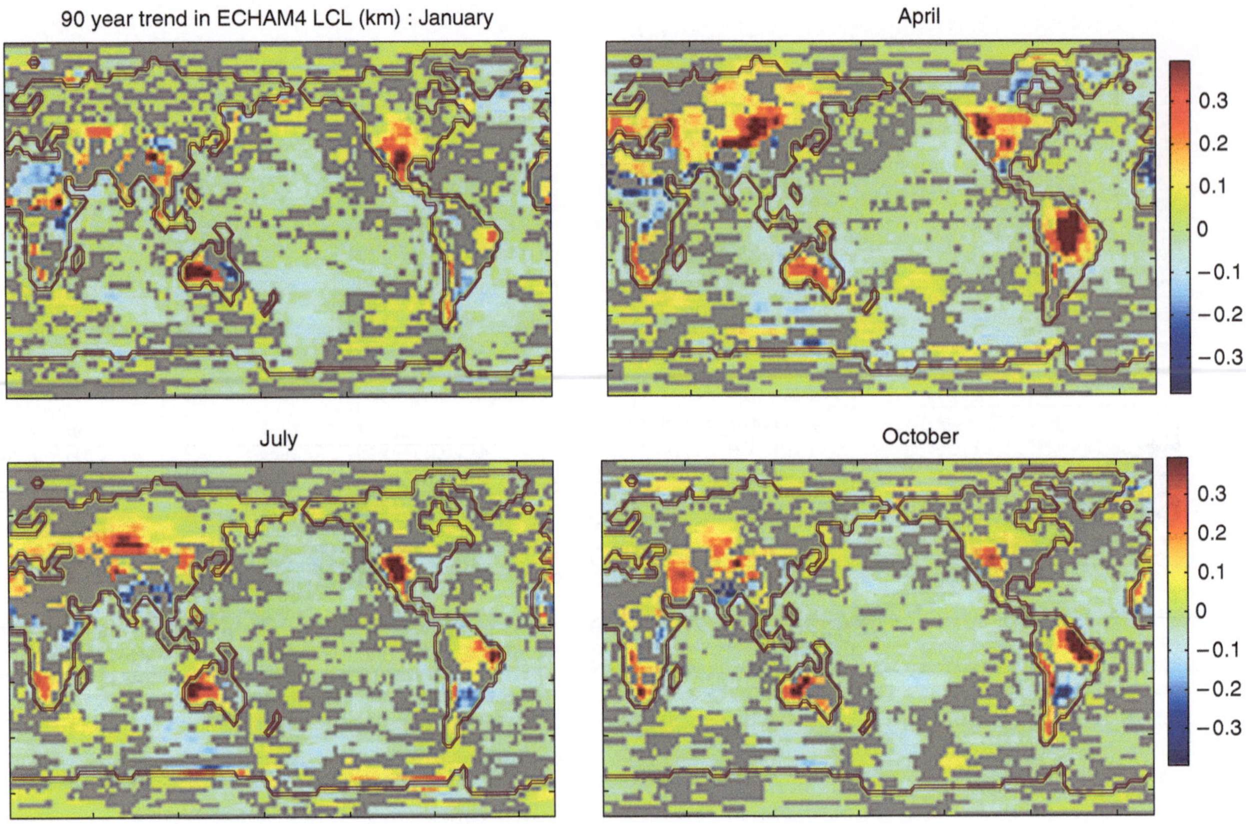

Figure 4.3. Ninety-year trends, centered on 2005, in the 10-year running mean of the lifting condensation level (LCL), as discussed under Figure 4.1. Cells with non-significant trends ($<\sigma$) are set to −999 m and appear in gray. In order to preserve strong color contrast, the scale in each panel is from −400 m to +400 m. In January, Central Africa has a trend more negative than −400 m and thus appears gray; the same is true in South-East Asia in April. Otherwise, all other gray cells had insignificant trends. July shows a strong similarity to global hydrologic patterns.

Figure 15.3. Moss hanging from the branches of a *Myrsine semiserrata* tree at 2540 m.a.s.l. indicating the start of cloud forest.

Figure 12.2. (a) Lower montane ravine forest at 1600 m.a.s.l. on the southern slope of Mt. Kilimanjaro. The tall trees with the bright bark are *Ekebergia capensis*. Location: Weru-Weru river gorge. (b) *Ocotea* forest, rich in vascular epiphytes and ferns, of the middle montane zone at 2400 m.a.s.l. on the southern slope. (c) Upper montane cloud forest at 2700 m.a.s.l. (all photographs by A. Hemp).

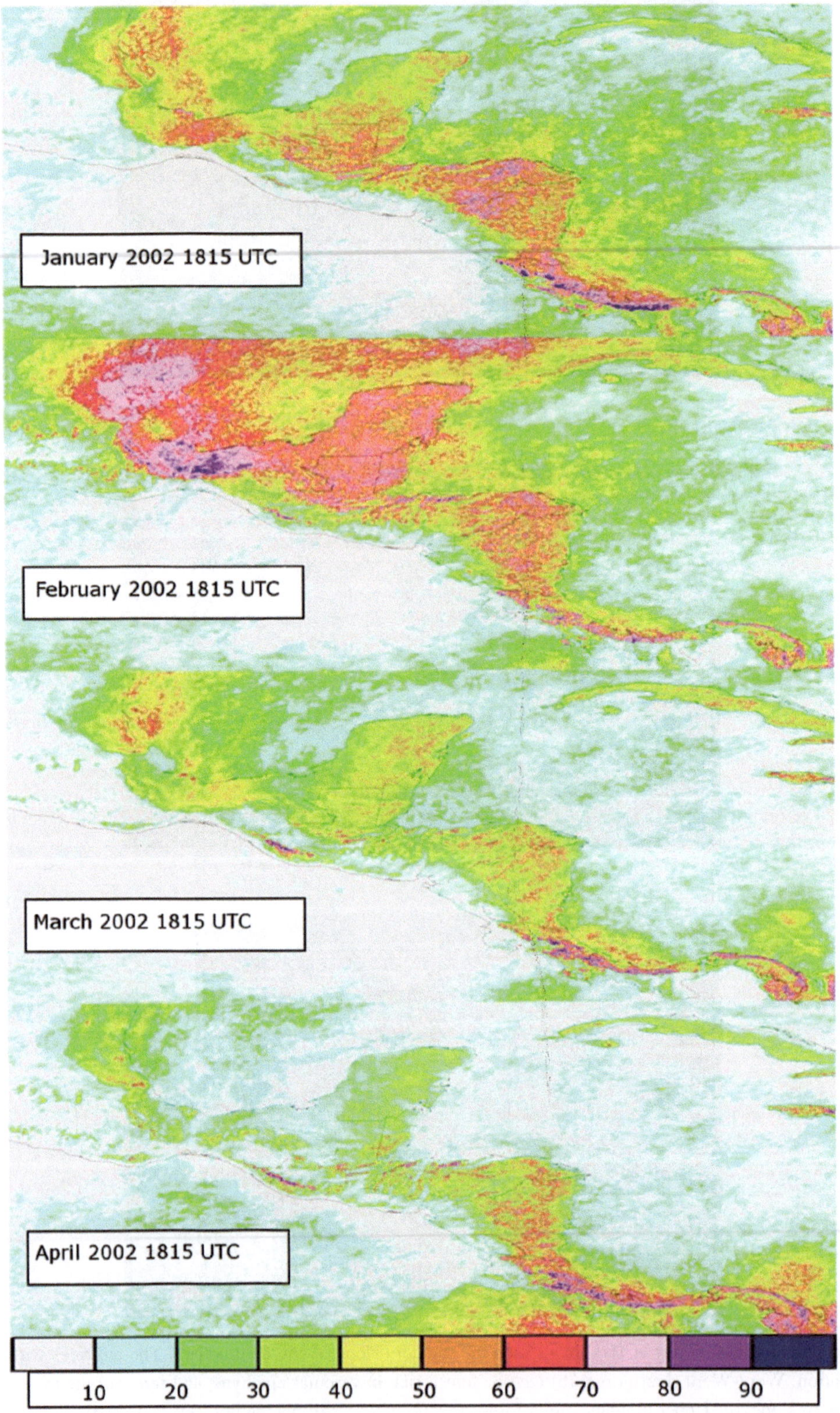

Figure 22.5. Proportion of the time covered by cloud at 18.15 UTC (~12.15 h local time) during the dry season months (January–April) of 2002 for the Central American region. Note that regions of trade-wind orography stand out as local maxima.

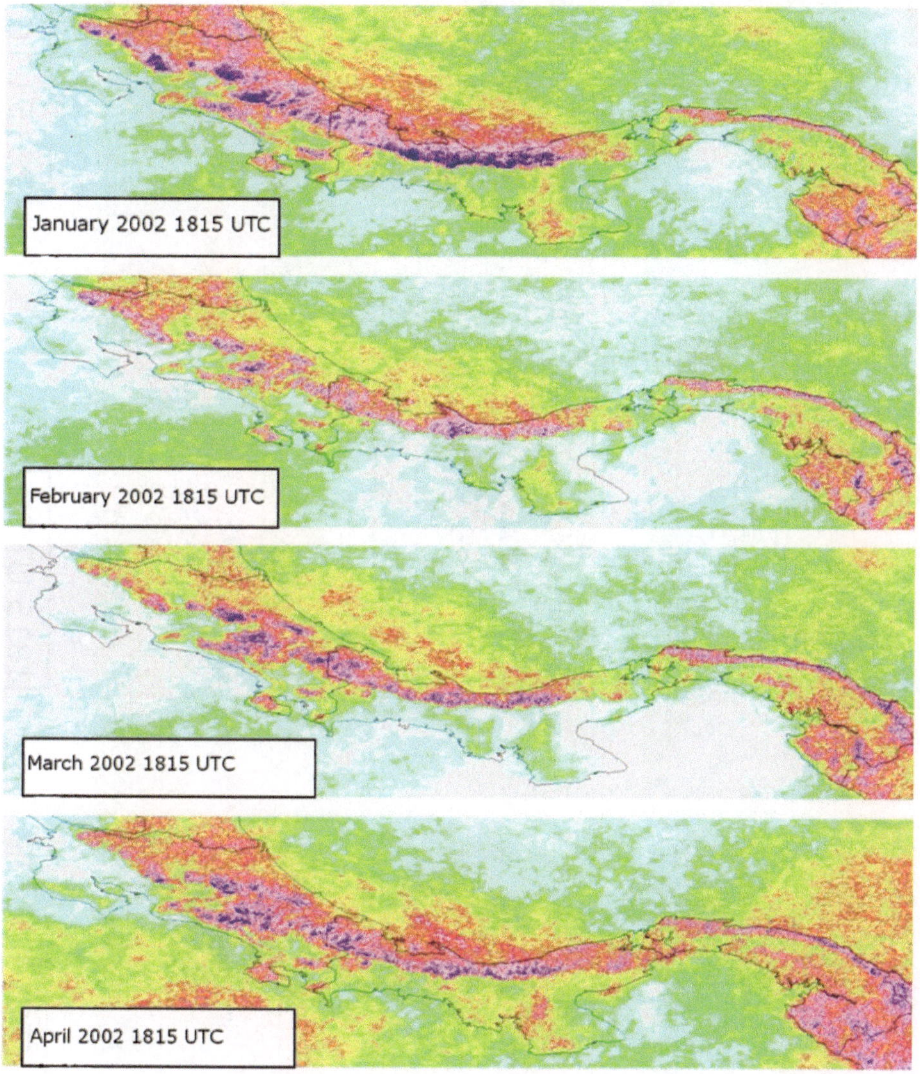

Figure 22.6. Proportion of the time covered by cloud at 18.15 UTC (~12.15 h local time) during the dry season months (January–April) of 2002 for Costa Rica and Panamá.

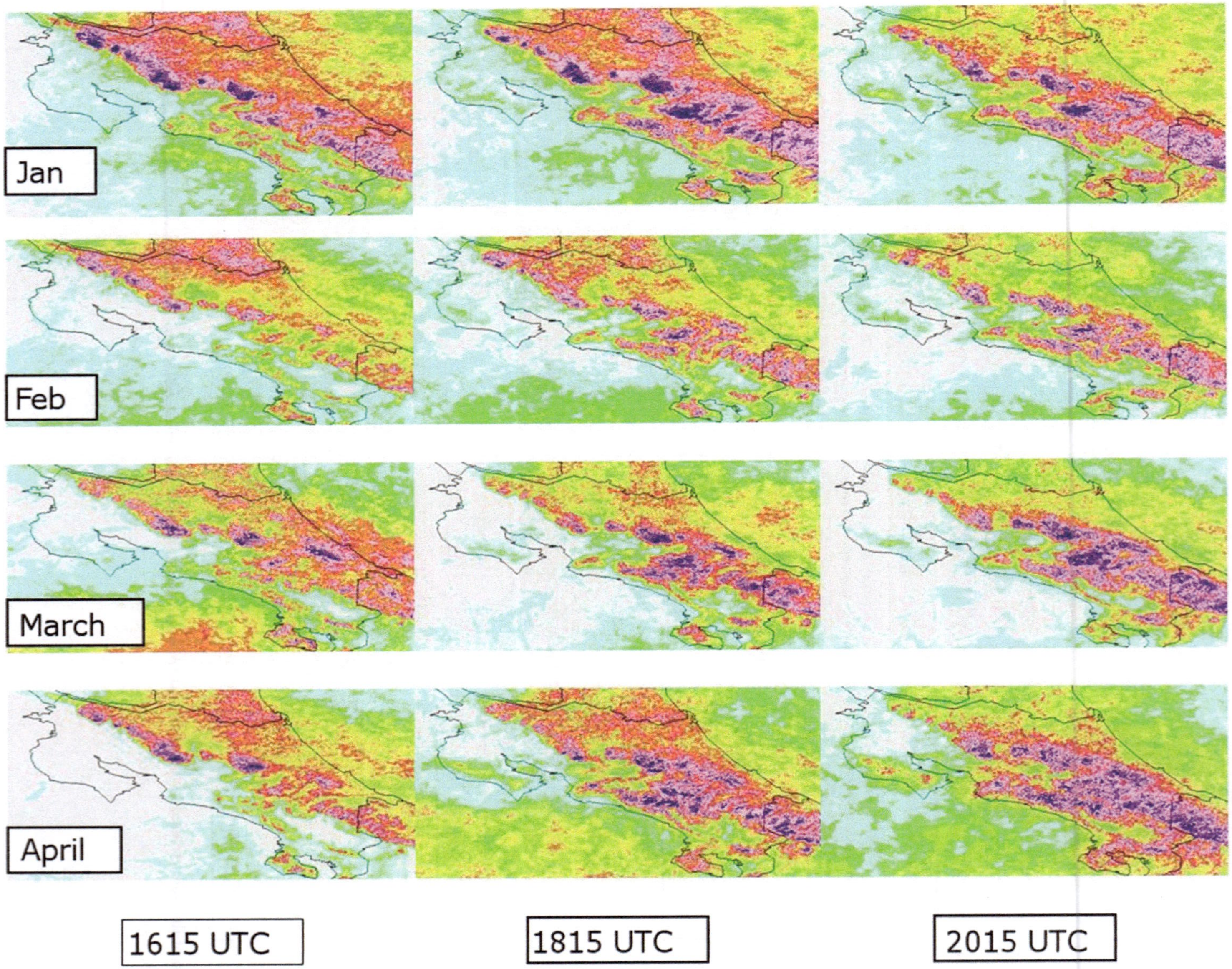

Figure 22.7. Proportions of the time covered by cloud at 16.15, 18.15, and 20.15 UTC for March 2003 in the Costa Rican region. Note the contrasting diurnal patterns of cloud development on the slopes of the various Costa Rican cordilleras.

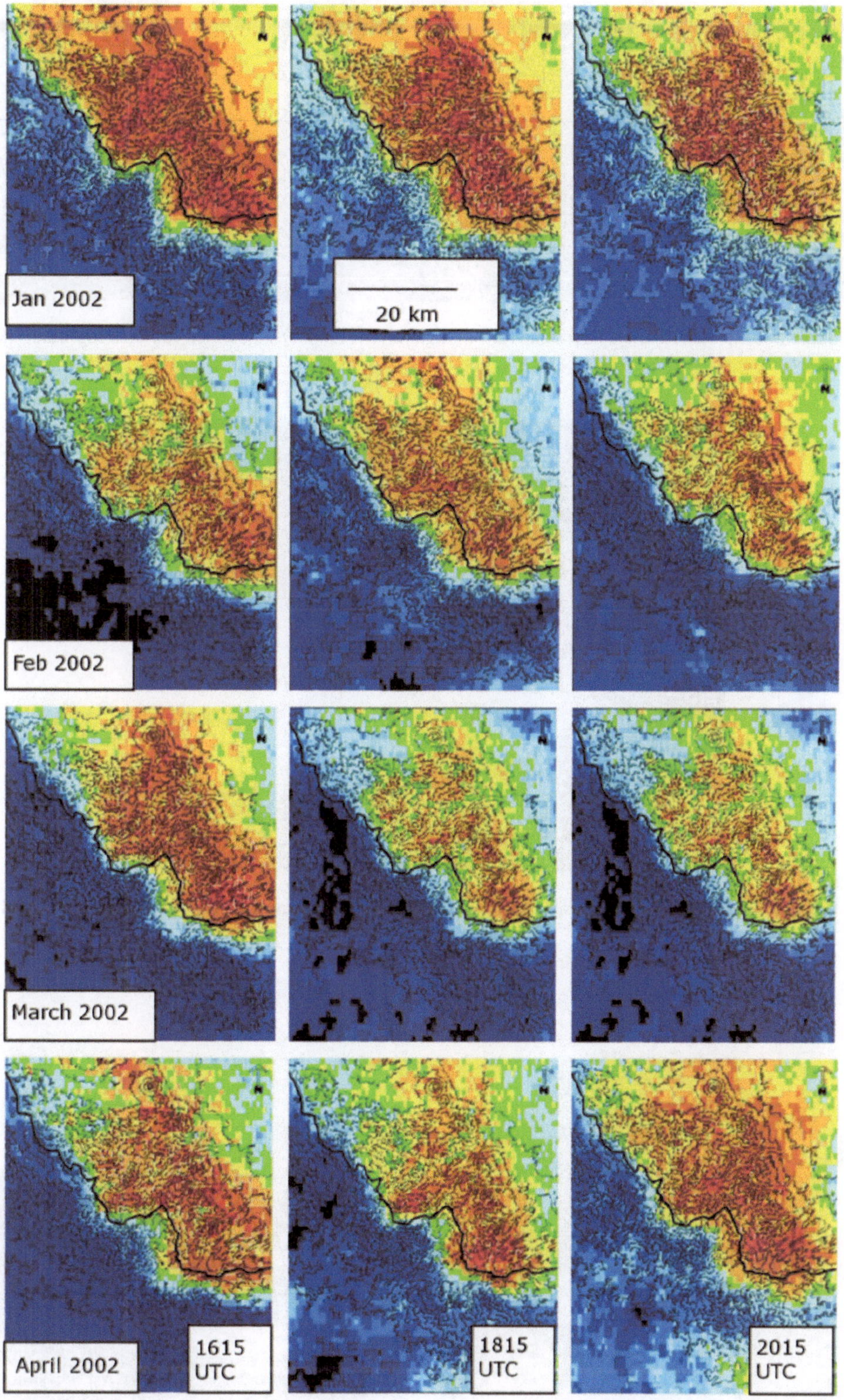

Figure 22.8. Proportion of time covered by cloud at 16.15, 18.15, and 20.15 UTC for the dry-season months (January–April) of 2002 for the central Cordillera de Tilarán. 100-m contour intervals and the course of the Continental Divide are shown.

Figure 37.3. Where the Garoé tree once stood, there now stands a 7-m high *til* (*Ocotea foetens*), which captures fog water like the former Garoé tree. See Figure 37.1 for location.

Figure 37.4. The "sea of clouds" covering the northern side of Tenerife. The Teide volcano can be seen in the background (3717 m.a.s.l.). Photo by Mª V. Marzol.

(a) (b)

Figure 37.6. Present-day fog harvesting experiments near the summits of El Hierro, above 1000 m.a.s.l. (a) An isolated 5-m high *Erica arborea* tree, located at the top of the Jinama cliffs, drips fog water onto the cemented ground beneath down to a water tank. (b) Eight *Erica* bushes capturing fog at El Cres. Photos by Mª V. Marzol.

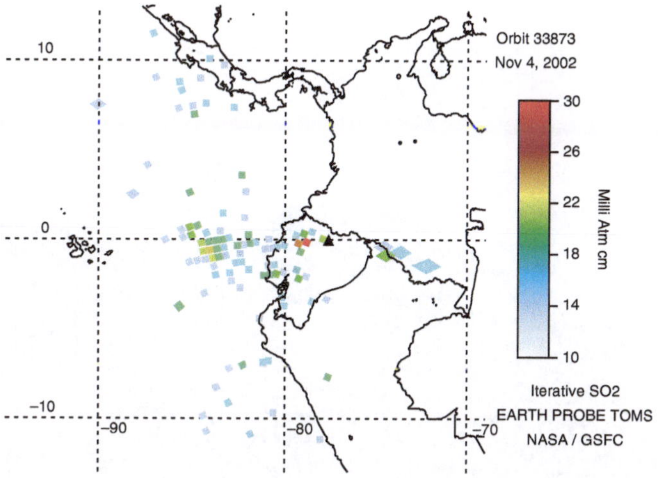

Figure 39.8. Sulfate emissions by the El Reventador volcano (black triangle) as observed by the TOMS satellite (diamonds, size proportional to atmospheric concentration). The gray lines show 10-day-backward trajectories, indicating the typical north-easterly trade winds that transported the emissions to the study area.

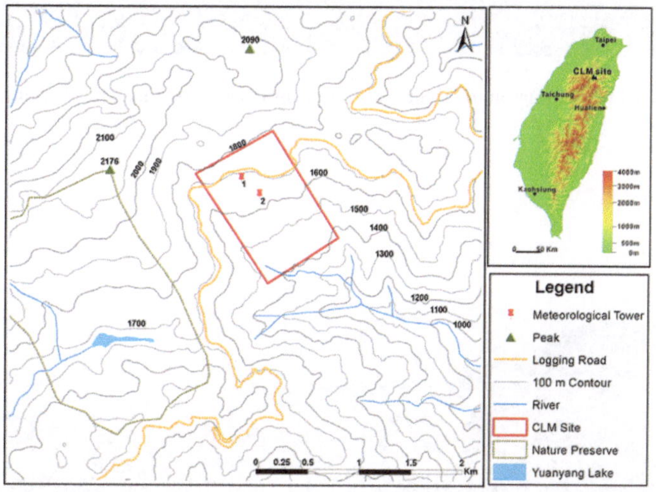

Figure 40.1. Topographic map of Taiwan showing the main cities and detailed contours of the Chi-Lan Mountain (CLM) study site.

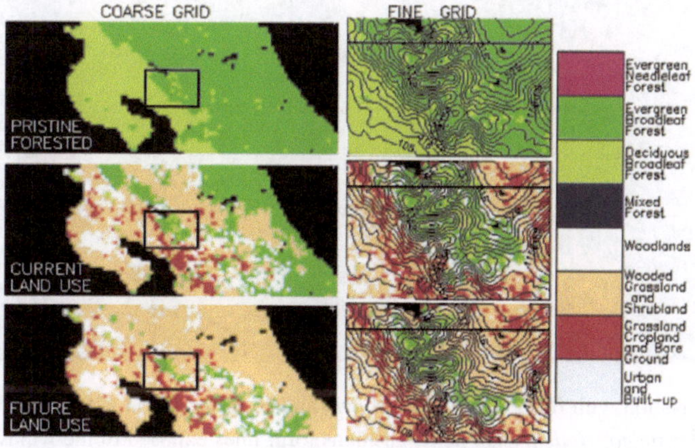

Figure 55.2. Outer coarse grid (left-hand panel) and inner fine grid (right-hand panel) for the three model simulations: (top) completely forested, (center) current conditions, and (bottom) all lowland and pre-montane regions deforested. Forested scenario had evergreen broadleaf forests on the wetter Caribbean, and deciduous broadleaf forests on the drier Pacific side. In the future deforestation scenario, deforestation is assumed to progress up to 1000 m.a.s.l. on the Caribbean side and 1400 m.a.s.l. on the Pacific side. The Pacific side currently is mostly deforested.

Figure 58.1. *Erica trimera* and *E. arborea* (flowering on the right-hand side) shrubs at 3200 m.a.s.l. substituting sub-alpine forest on Mt. Kilimanjaro (photograph by A. Hemp).

Figure 58.2. *Helichrysum* cushion vegetation with *H. newii*, *H. citrispinum*, and *H. forskahlii* at 4100 m.a.s.l. on Mt. Kilimanjaro. The grasses are *Pentaschistis borussica* and *Festuca abyssinica*. *Helichrysum newii* cushions form lines exposed to the windward direction. This growth form, which cannot be explained by solifluction, apparently enables the plants to capture fog moving upslope (photograph by A. Hemp).

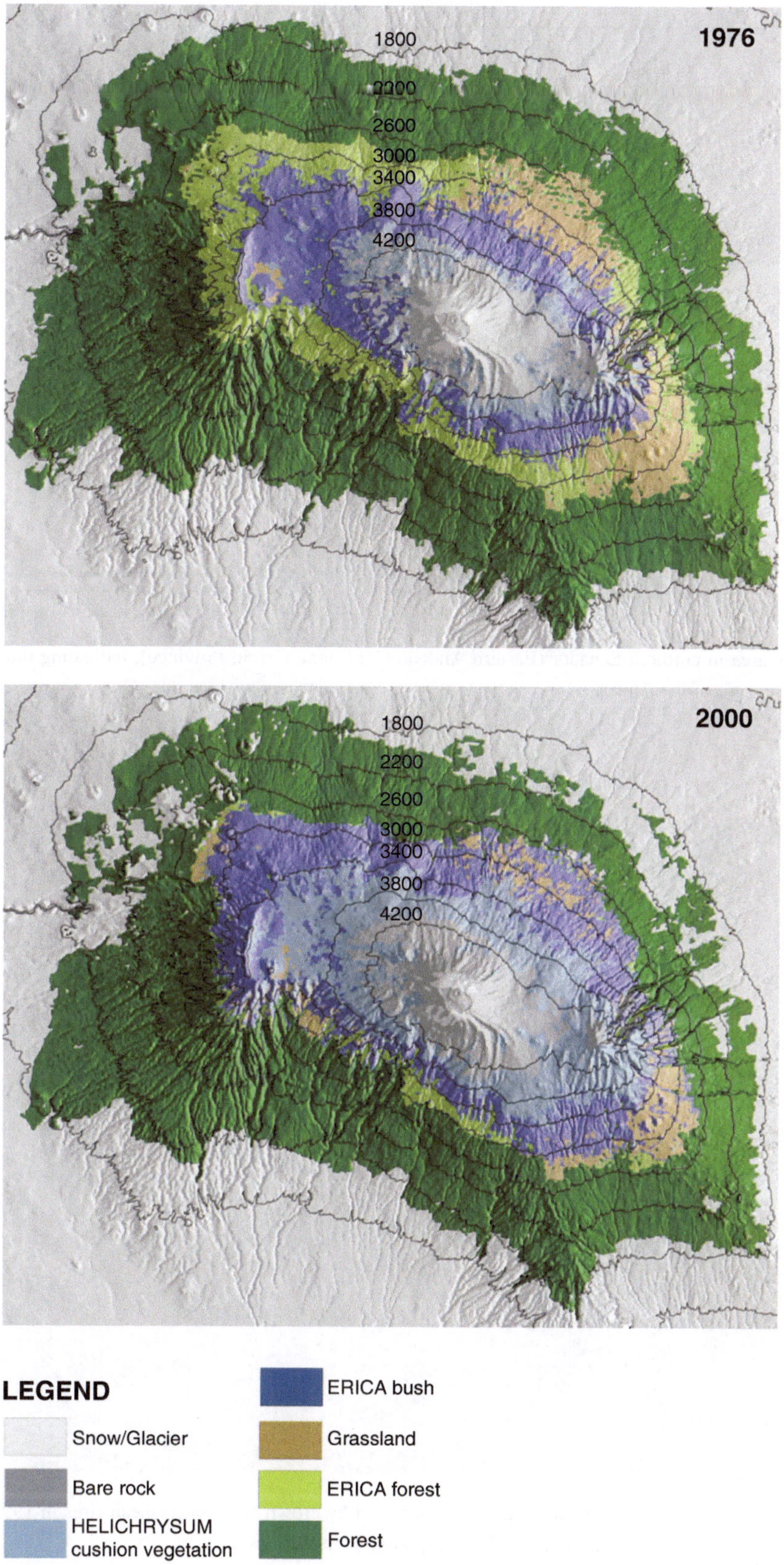

Figure 58.3. Changes in vegetation cover in the montane and alpine zone on Mt. Kilimanjaro. The maps result from supervised classification of Landsat MSS images taken on 24 January 1976 and Landsat ETM images taken on 29 January and 21 February 2000 (source: USGS/UNEP-GRID-Sioux Falls) using the software IDRISI 3.2. The entire sub-alpine ericaceous belt was shifted downslope and became replaced by *Helichrysum* cushion vegetation. Closed *Erica excelsa* forests were reduced to 17% of their land cover in 1976.

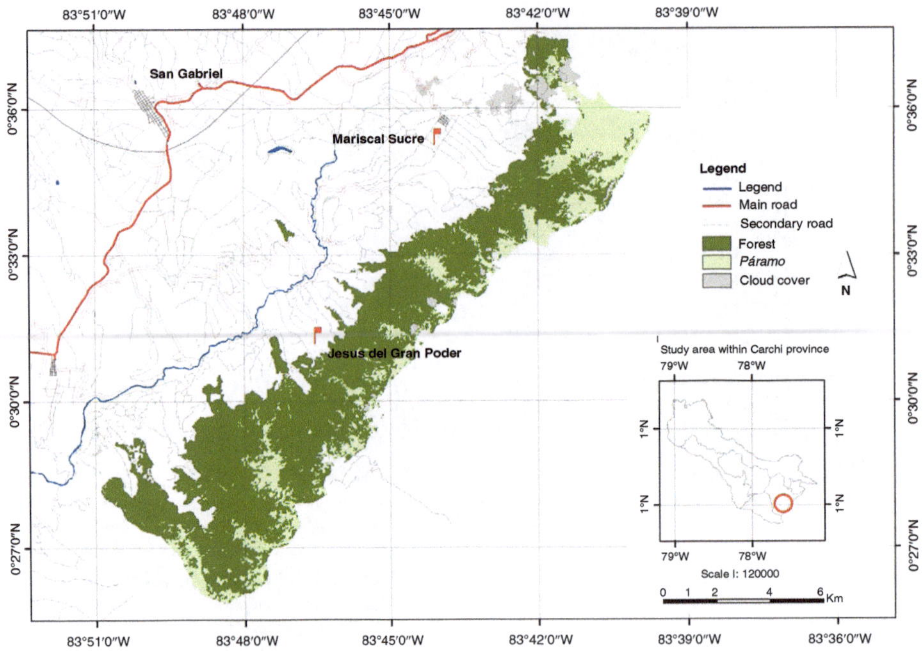

Figure 67.1. Map of the study area in northern Ecuador (Eastern Andean Cordillera, Carchi Province), indicating the present cover of *ceja andina* and *páramo* on the inter-Andean side of the Andes.

Figure 69.2. Large expanse of elfin cloud forest on karst topography affected by crown and subterranean fires in 1998 (photograph taken 2 years after the fires) in the Chimalapas region of Oaxaca, Mexico (photograph by H. Asbjornsen).

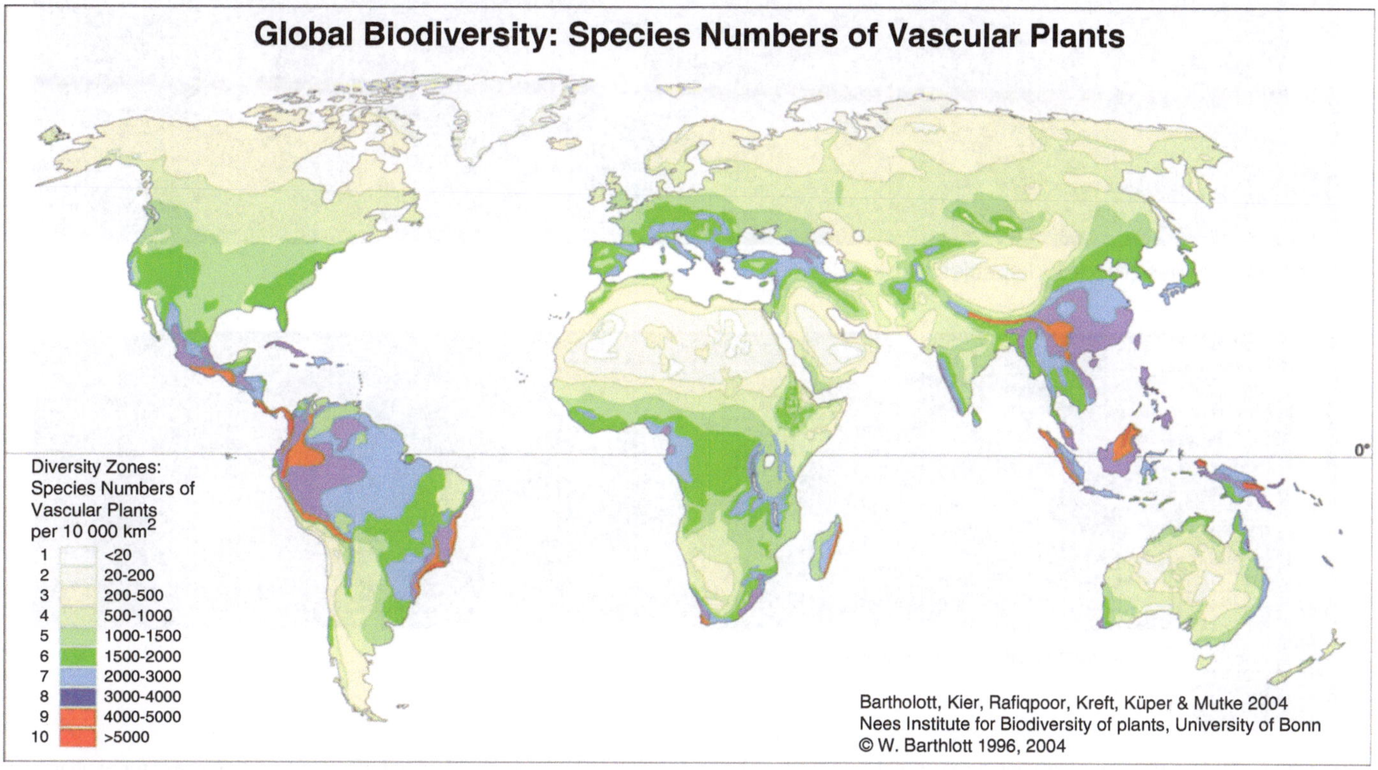

Figure 72.4. World map of species richness of vascular plants (Mutke and Barthlott, 2005).

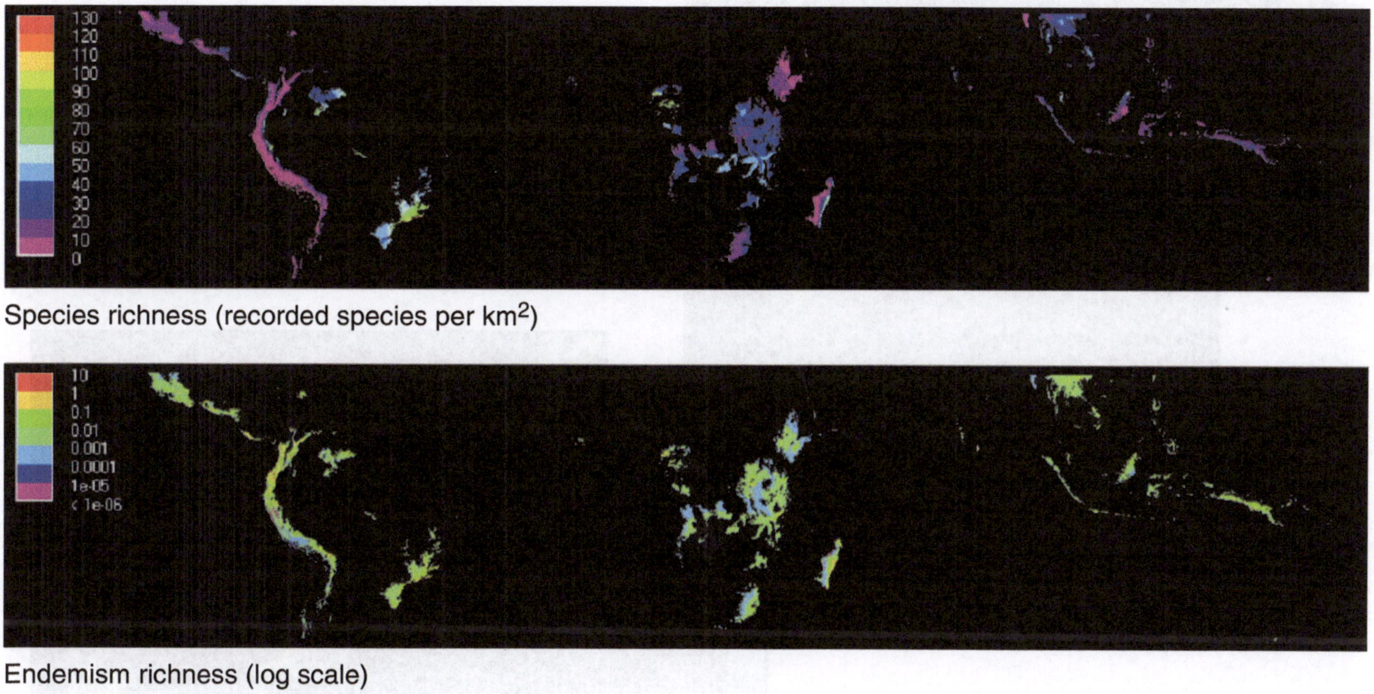

Figure 72.5. The distribution of species richness and endemism richness for amphibians in areas with cloud-affected forest. See text for explanation.

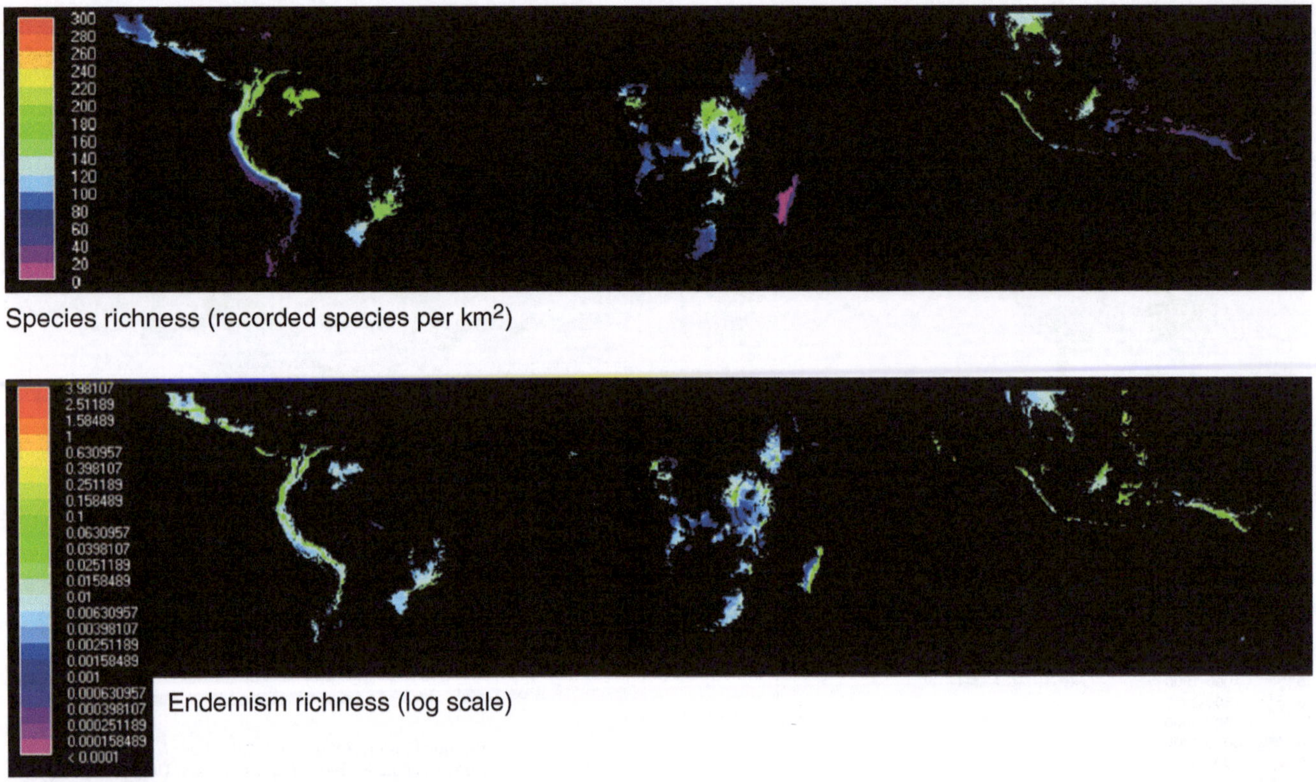

Figure 72.6. The distribution of species richness and endemism richness for mammals in areas with cloud-affected forest. See text for explanation.

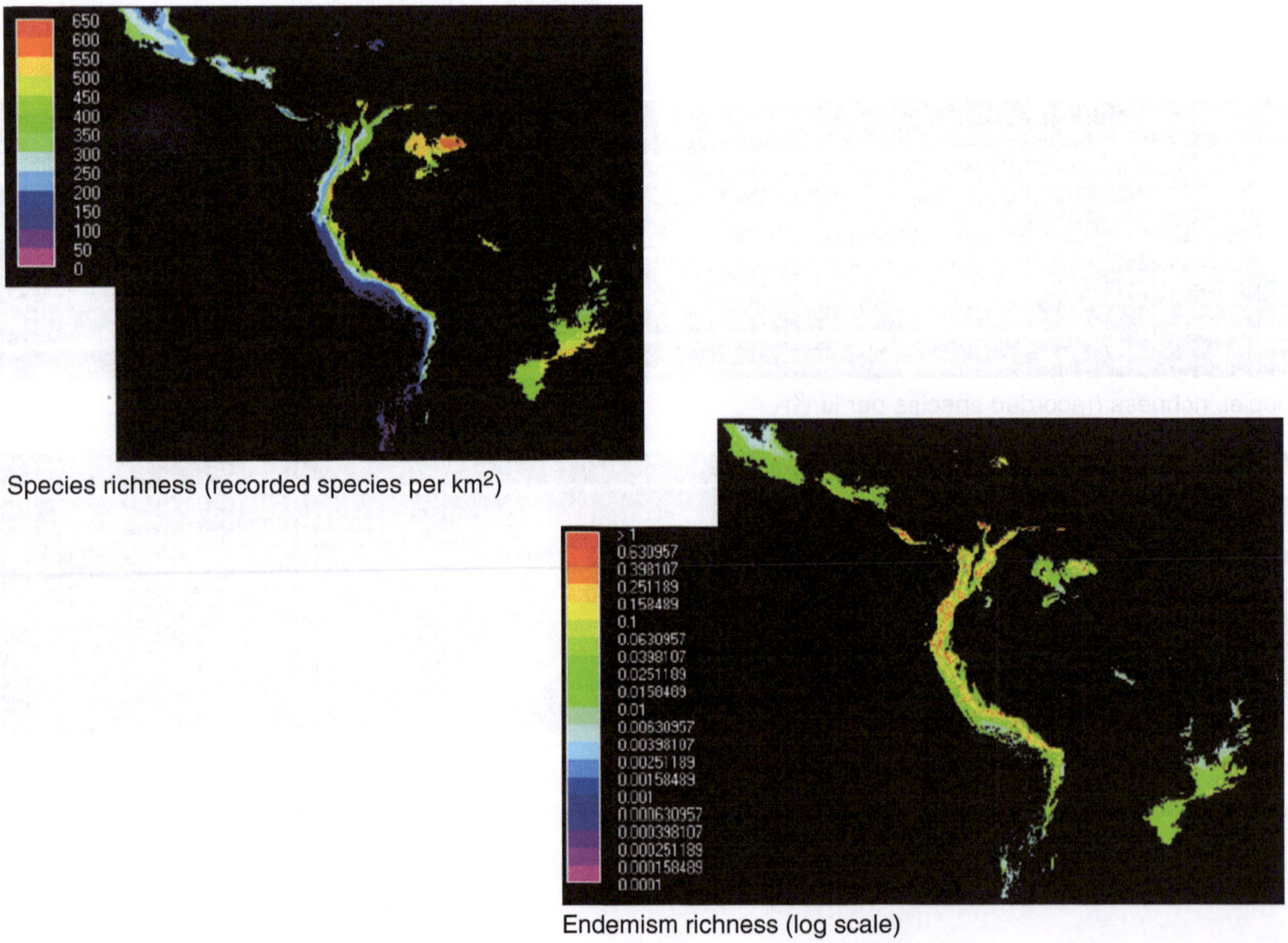

Figure 72.7. The distribution of species richness and endemism richness for birds in Latin American cloud-affected forests. See text for explanation.

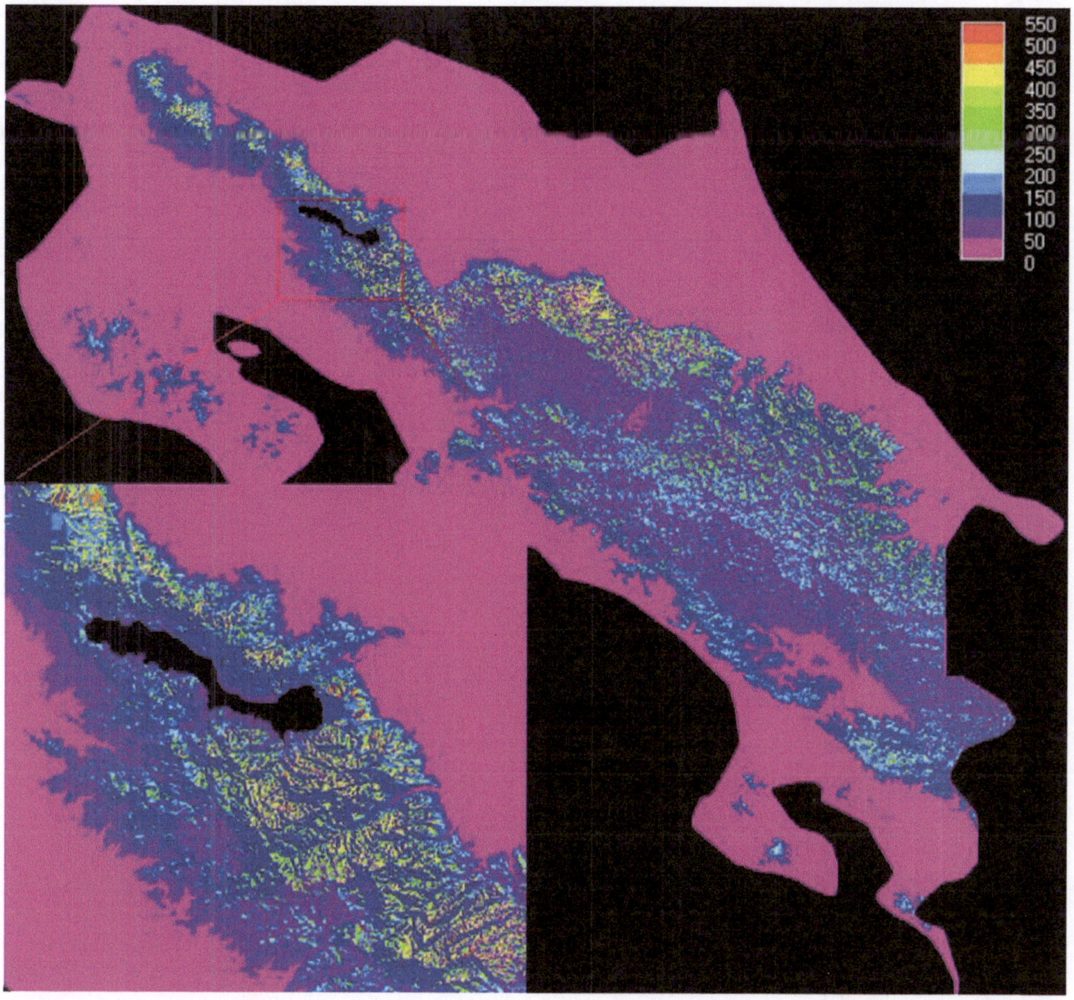

Figure 72.11. Modeled annual fog interception input (mm) at 90-m resolution across Costa Rica (after Mulligan and Burke, 2005b; reproduced with permission).

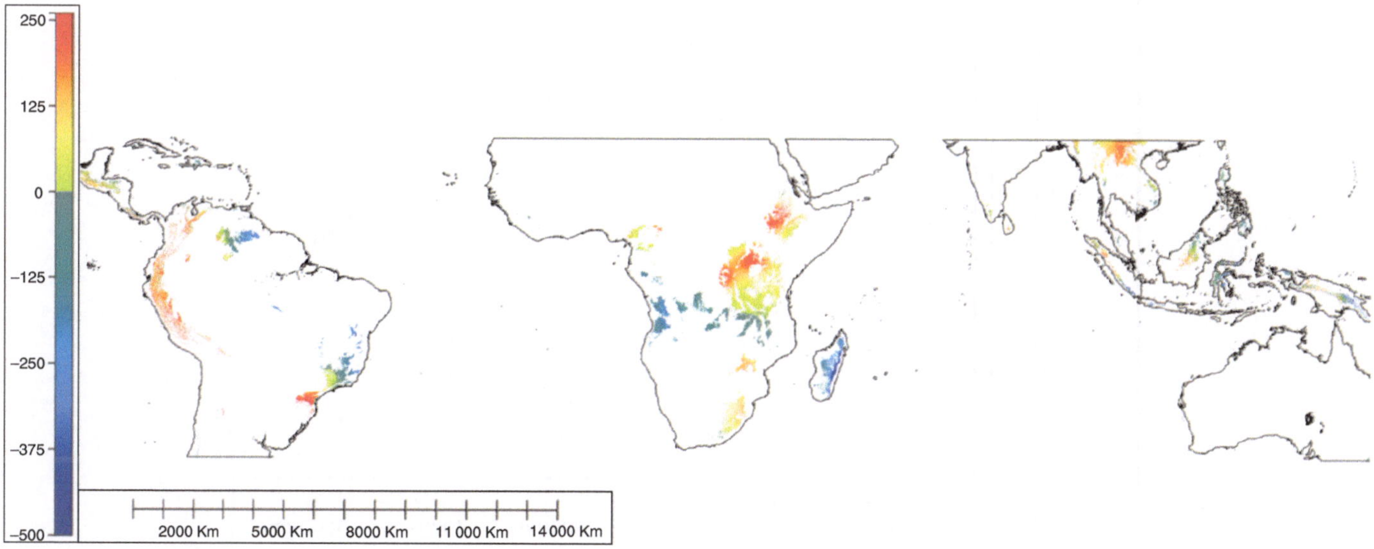

Figure 72.15. Impact of projected climate change (2000–2050) on the water balance (in mm year^{-1}) of cloud-affected forest areas across the tropics.

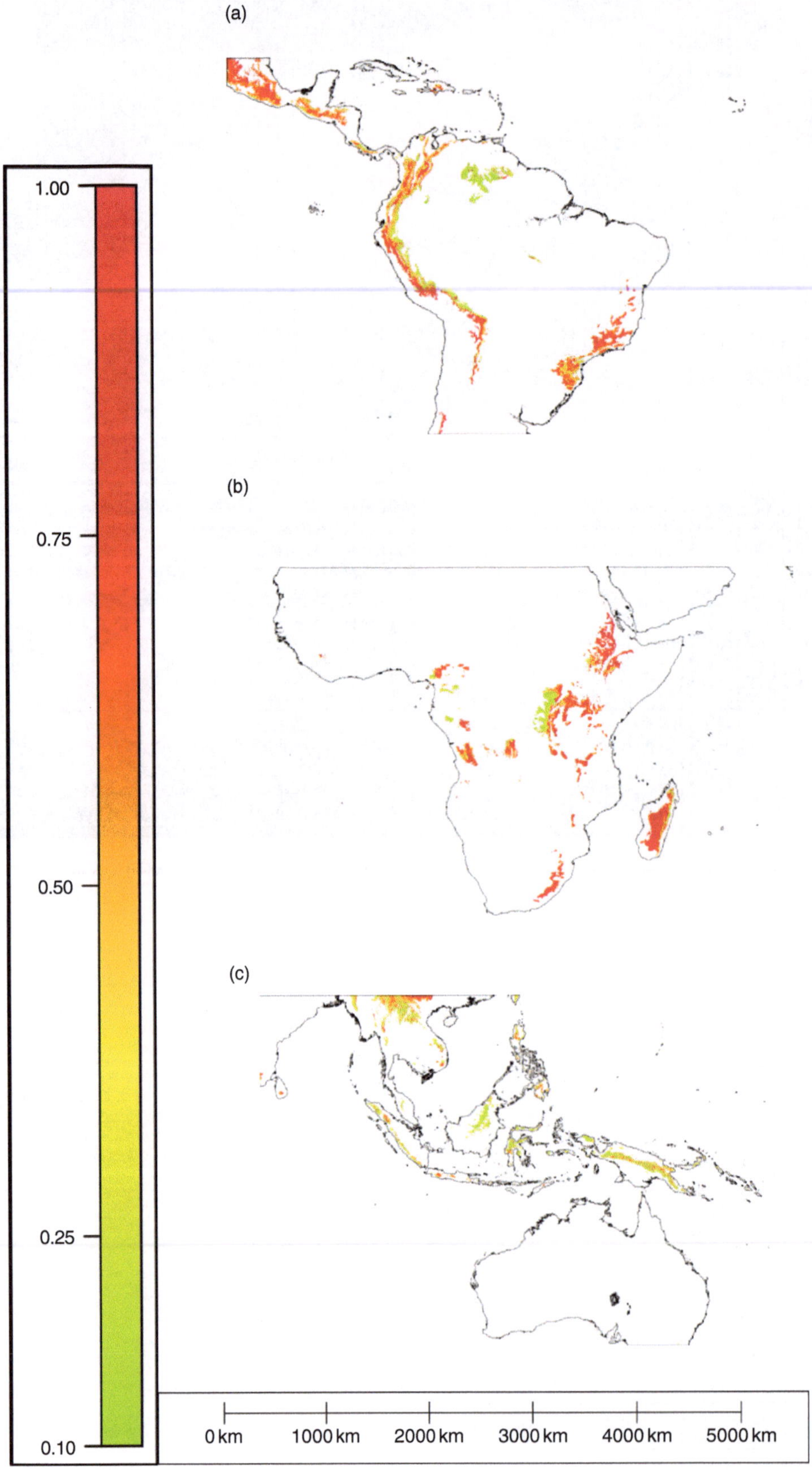

Figure 72.16. Fractional cloud-forest loss according to MODIS_VCF satellite product for (a) Latin America, (b) Africa, and (c) South-East Asia. The darker the shade of red, the greater the forest loss. Red: > 60% forest loss, green < 25% forest loss. (Reproduced from Figure 2.13).

8 What is the state of tropical montane cloud forest restoration?

T. M. Aide and M. C. Ruiz-Jaen

University of Puerto Rico, San Juan, Puerto Rico, USA

H. R. Grau

Universidad Nacional de Tucuman, Tucuman, Argentina

ABSTRACT

The conversion of tropical montane cloud forest (TMCF) to pastures and agricultural lands has been an important activity in this life zone for many years. Although forest clearing and grazing continues, in some areas, changing political, economic, and social drivers have led to the abandonment of marginal areas. These dynamics provide an excellent opportunity to study the rates of secondary succession and test different restoration strategies. The two major questions addressed in this review are: "What factors control rates of TMCF recovery once pastures or agricultural lands are abandoned?", and "What restoration strategies can be used to overcome barriers to regeneration and accelerate forest recovery?" To answer these questions a literature review was carried out. Because few restoration projects have been conducted in TMCF as such, the conclusions are mainly based on studies in tropical montane forests at large. Competition with invasive grasses and ferns and poor seed dispersal appear to be the most important factors limiting natural forest recovery. To overcome these barriers, one of the most cost-effective ways to accelerate recovery is to promote the establishment of shrubs, which help to shade out invasive grasses and ferns and create more appropriate conditions for seedling growth. Although this strategy can reduce competition, planting will also be required to recover a species composition similar to intact forest because most forest species are rarely dispersed far from forest stands.

INTRODUCTION

Tropical montane ecosystems have been transformed by human use for hundreds of years. The most common causes of forest transformation have been agriculture and grazing, but timber harvesting has also contributed to forest degradation. Although clearing for illegal crops – e.g. poppy, *Papaver somniferum* (Cavelier and Etter, 1995) or coca, *Erythroxylum coca* (Bedoya, 1997) – continues to threaten many areas of intact montane forest in Latin America, in many other regions, agricultural and grazing lands are being abandoned, often because of a loss of agricultural productivity associated with accelerated erosion. Soil erosion can be even worse where fire is part of the management practice (Cavelier *et al.*, 1998, 1999). For example, in the Sierra Nevada de Santa Marta in Colombia, a long history of land use, including fire, has resulted in the loss of >50 cm of soil, which greatly limits agriculture and forest recovery (Aide and Cavelier, 1994).

In recent years, rural to urban migration has been another major factor influencing the abandonment of montane lands (Aide and Grau, 2004). In some regions, armed conflicts have stimulated rural–urban migration. For example, in Colombia, approximately 2 million people have been displaced by civil unrest, and the majority of these people have migrated out of rural mountain regions (Chernick, 2000). Economic and social changes have also stimulated rural–urban migration in areas where small-scale agriculture on steep marginal lands cannot compete with large-scale industrial agriculture and imported products (Preston, 1996). In addition, many rural inhabitants, especially young people, are attracted to cities because of greater opportunities for education, social services, jobs, and an urban lifestyle (Harden, 1996; cf. Grau *et al.*, this volume).

Tropical Montane Cloud Forests: Science for Conservation and Management, eds. L. A. Bruijnzeel, F. N. Scatena, and L. S. Hamilton. Published by Cambridge University Press.

Where rural–urban migration continues, the associated reduction in human pressure on the ecosystems could permit forest recovery (Aide *et al.*, 2000; Guariguata and Ostertag, 2001; Aide and Grau, 2004). However, even when land is abandoned or grazing intensity decreases, some forests may not recover because the degraded system is "resilient." Resilience of non-forest vegetation is due to positive feedbacks that reinforce the maintenance of the degraded state of the system (Hobbs and Norton, 1996; Hartig and Beck, 2003; Suding *et al.*, 2004). For example, in many tropical areas, the grass *Melinis minutiflora* invades after fires and its resinous leaves are highly flammable, which promotes more fires and greatly inhibits forest regeneration (D'Antonio and Vitousek, 1992). In the South Ecuadorian Andes, the bracken fern *Pteridium arachnoideum* dominates degraded pastures and presents a major barrier to tree establishment (Hartig and Beck, 2003; Günter *et al.*, 2009).

This review addresses the following questions:

- Once pastures or agricultural lands are abandoned, what factors control rates of tropical montane (cloud) forest recovery?
- When sites do not recover (i.e. they are in the resilient degraded state), what restoration strategies can be used to overcome barriers and accelerate forest recovery?

To answer these questions, a literature review was conducted covering articles in *Biological Abstracts* (1997–2003), *Web of Science* (1990–2004), *Science Citation Index* (1997–2002), as well as The Scholarly Journal Archive (JSTOR), Springer, and Organization for Tropical Studies databases. Initially, the following key words were used: tropical, montane cloud forest, restoration, reforestation, rehabilitation, and recovery. Given that only a few articles were found with the explicit objective of active restoration in TMCF (as defined in Hamilton *et al.*, 1995), the search was expanded using the following keywords: montane forest, secondary succession, plantations, and afforestation. In addition, references within articles identified in these databases were used. Articles were evaluated to determine the major factors controlling the rate of forest recovery and what management activities were effective in accelerating forest recovery.

FACTORS CONTROLLING RATES OF FOREST RECOVERY AND MANAGEMENT SOLUTIONS

Studies of tropical montane (TMF) and tropical montane cloud forest (TMCF) have identified soils, fire, seed dispersal, colonization of pioneer species, and microhabitat conditions as the major factors controlling forest recovery. Plantations were also considered as a factor affecting forest recovery, but in other studies plantations have been considered mainly as a management tool (e.g. the first step in a reforestation project). The various observations are summarized below by first discussing how each factor affects rates of forest recovery and then presenting restoration and management responses that could ameliorate the negative effects of each factor and accelerate forest recovery.

Soils

PROBLEMS

Although saturated anoxic soils (Silver *et al.*, 1999), low nutrient availability (Tanner *et al.*, 1998; Stewart, 2000; Günter *et al.*, 2009; cf. Benner *et al.*, this volume; Roman *et al.*, this volume), compaction (Pedraza and Williams-Linera, 2003; Zimmermann and Elsenbeer, 2008), or extensive erosion (Aide and Cavelier, 1994; Slocum *et al.*, 2000) all can limit forest recovery, in the present survey, few studies have identified soils as a major limiting factor. In the context of TMCF, the prime soil characteristic reported as limiting forest regeneration was soil compaction associated with cattle grazing (Pedraza and Williams-Linera, 2003; cf. Zimmermann and Elsenbeer, 2008) or road construction (Olander *et al.*, 1998). Compacted soils have been associated with low sapling survivorship and growth (Pedraza and Williams-Linera, 2003). In contrast, disturbed areas along temporary roads or pipelines with low soil compaction had rapid woody vegetation recovery (Malizia *et al.*, 2004). An additional problem of compacted soils is that they are often colonized by grasses, which can increase susceptibility to fire, inhibit the colonization of woody species, and make the system resistant to forest recovery (Rosales *et al.*, 1997; Olander *et al.*, 1998; cf. Hartig and Beck, 2003; Zimmermann and Elsenbeer, 2008).

MANAGEMENT SOLUTION

If soils are a limiting factor at a regional scale, this can be a major challenge for restoration projects (Bradshaw, 1997). Although areas can be fertilized, and soil material added, or plowed in the case of compaction, these are expensive activities if large areas are to be recovered (Bradshaw, 1997). At any scale, an immediate management goal should be to reduce soil loss (Calle, 2003). For example, a common practice is to plant or leave strips of vegetation along the contours of a slope to reduce soil erosion (Calle, 2003; Pedraza and Williams-Linera, 2003). In relatively small areas, such as shallow landslips, fertilization with nitrogen and potassium has facilitated the establishment of tree species (Dalling and Tanner, 1995). At present, the most widely used technique for improving degraded soils has been to establish plantations because they increase soil nutrients and soil organic matter, and protect the soil from erosion (Lugo, 1992; Vanacker *et al.*, 2007). For gullied areas, a number of mechanical, but costly, techniques have been developed, but these are beyond the scope of this review (see e.g. Blaisdell, 1981; Hudson, 1995).

Figure 8.1. Remnant patches of montane forest in a matrix of degraded grasslands maintained by frequent fires in the Sierra Nevada de Santa Marta, Colombia.

Fire

PROBLEMS

In general, high rainfall and moisture in TMCF limit the number and extent of fires. Fires can occur in TMCF during extended dry periods (Asbjornsen *et al.*, 2005; Hemp 2005; Asbjornsen and Garnica-Sánchez, this volume; Hemp, this volume #60), but they appear to be more important in other high-elevation habitats (e.g. *paramo* and dry shrub/pine forest) where there is less precipitation (Hemp and Beck, 2001; Horn *et al.*, 2001). Although this may change in the future because of climatic drying (Hemp, 2005; Asbjornsen and Wickel, 2009; Foster, this volume), in most TMCF settings, fire is associated with agricultural and grazing activities (May, 1997; Scowcroft and Jeffery, 1999; Kessler, 2000). As in other montane systems, if fires occur frequently they can greatly reduce forest regeneration (Aide and Cavelier, 1994; Cavelier *et al.*, 1998; Ramirez-Marcial *et al.*, 2001; Duncan and Chapman, 2003a; Grau *et al.*, this volume) (Figure 8.1). In addition to the direct effect of killing most woody species, fires can indirectly affect forest recovery by increasing soil erosion and selecting for grasses and ferns. These effects will negatively affect forest regeneration and could lead to a resilient degraded state (Hartig and Beck, 2003; Suding *et al.*, 2004; Griscom *et al.*, 2008).

MANAGEMENT SOLUTION

Fortunately, fires are still rare in undisturbed TMCF (cf. Wård *et al.*, this volume), and given that agricultural activities are the major source of fires, if a region is being abandoned, then the frequency of fires should diminish. Where fires are a threat, they can be controlled with firebreaks (Aide and Cavelier, 1995), fire lookout and response teams, and education (Asbjornsen and Garnica-Sánchez, this volume). In some montane systems fire may be used as a restoration tool. For example, in Thailand, surface fires enhance seedling recruitment of deciduous dipterocarp forest and pines by reducing grasses and woody lianas (Werner and Santisuk, 1993). Similarly, in NW Argentina, fires of intermediate frequencies can favor the expansion of *Alnus acuminata* into montane grasslands by reducing competition between tree seedlings and grasses (Grau and Veblen, 2000).

Seed dispersal

PROBLEMS

Many studies have investigated the importance of seed dispersal as a limiting factor for natural regeneration (Aide and Cavelier, 1994; Holl *et al.*, 2000; Oosterhoorn and Kappelle, 2000; Zimmerman *et al.*, 2000; Cubiñá and Aide, 2001; Günter *et al.*, 2006). In virtually all studies, there is a rapid decrease in the diversity of species and total number of seeds with distance from the forest edge; seeds of most species are not detected more than 20 m from the forest edge. Small wind-dispersed species (often shrubs belonging to the Bignoniaceae and Asteraceae families) are an exception as their seeds can disperse further into the abandoned lands. Contrary to this strong pattern of dispersal limitation, recruitment patterns rarely show a clear procession from the forest edge toward the center of the abandoned lands, suggesting that seeding from adjacent forest is not always the limiting factor. Aide *et al.* (2000) suggested that some species colonize pastures when these are still actively used. If plants are capable of sprouting after being cut or grazed, they may survive for extended periods, and once the land is abandoned they can immediately dominate the area. Regardless of the dispersal patterns or time of colonization, the species that initially establish in these abandoned agricultural lands make up a very limited sub-set of montane forest species, suggesting that seed dispersal is a limiting factor for species composition (Oosterhoorn and Kappelle, 2000).

MANAGEMENT SOLUTION

Different techniques have been used to increase seed dispersal into abandoned lands. In many systems, remnant trees or shrubs have played an important role in attracting dispersers and accelerating regeneration (Slocum and Horvitz, 2000). Artificial perches also can attract birds and increase the seed rain below the perch (Aide and Cavelier, 1994), but they rarely increase seedling density (Holl *et al.*, 2000; Shiels and Walker, 2003).

A problem with artificial perches is that they do not produce enough shade to reduce the grass cover, which can limit the ability of seedlings to establish. Windbreaks (Harvey, 2000), fruit trees (Sarmiento, 1997a), and pioneer shrubs (Posada *et al.*, 2000) also attract bird dispersers and produce more shade than artificial perches, thus making them good candidates for increasing seed dispersal and seedling establishment. For example, in NW Argentinean cloud forests, exotic spiny shrubs (e.g. *Crataegus oxyacantha*) have facilitated the establishment of native species by attracting bird dispersers and protecting the new recruits from browsing by domestic animals (Malizia and Greslebin, 2000). However, if the goal of the project is to attain a species composition similar to intact forest, seed collection, nursery activities, and planting will all have to be major activities of a restoration project (Lamb and Gilmour, 2004).

Figure 8.2. A three-year-old abandoned pasture dominated by *Pennisetum clandestinum* in the Cordillera Central, Colombia.

Colonization of pioneer species

PROBLEMS

Although seed dispersal can be a limiting factor in many sites, grasses in abandoned pastures or early colonizers of abandoned agricultural lands (e.g. ferns) can inhibit forest recovery. In some cases, these species can even arrest succession. For example, introduced African grasses (e.g. *Melinis minutiflora, Pennisetum clandestinum, Setaria sphacelata*) often dominate montane pastures, and once the pastures are abandoned these grasses grow vigorously, impeding the establishment of woody species (e.g. Colombia – Posada *et al.*, 2000; Ecuador – Sarmiento, 1997a,b; Zahawi and Augspurger, 1999; cf. Günter *et al.*, 2009; Costa Rica – Holl, 1999; Hawai'i – D'Antonio and Vitousek, 1992) (Figure 8.2). When the previous land use has been logging or agricultural, ferns are often the first species to colonize the site. In the Dominican Republic, native ferns that colonize landslides have greatly expanded their local distribution by colonizing abandoned agricultural lands (Garcia *et al.*, 1994). These ferns stabilize soils, reduce erosion, and increase soil organic matter, but they can also inhibit the establishment of tree species (Slocum *et al.*, 2000). For example, *Dicranopteris pectinata* has colonized and dominated abandoned agricultural fields for >20 years because the fern produces a thick root mat on the soil surface, which acts as a physical barrier and inhibits the establishment of other species (Figure 8.3; Slocum *et al.*, 2000). A similar case has been documented for the bracken fern *Pteridium arachnoideum* in the South Ecuadorian Andes by Hartig and Beck (2003). At other sites, exotic trees and shrubs (e.g. *Ligustrum* spp. and *Citrus* spp.) have colonized abandoned agricultural lands and areas with a history of intense grazing (e.g. Grau and Aragon, 2000), and these species can also dominate a site for many generations (Schulenberg and Awbrey, 1997; Tecco and Rouges, 2000; Lichstein *et al.*, 2004).

Figure 8.3. *Dicranopteris pectinata* fernlands in the Cordillera Central, Dominican Republic.

Figure 8.4. A three-year-old abandoned pasture adjacent to the site depicted in Figure 8.2 which had low-intensity grazing and was dominated by Asteraceae shrubs.

MANAGEMENT SOLUTION

A diversity of approaches will be needed to eliminate or reduce the densities of these grasses, ferns, and exotic trees and shrubs. This constitutes a major challenge given that they often cover extensive areas. In Colombia, a fast-growing tree, *Montanoa quadrangularis* (Asteraceae), has been used to expand forest cover into pastures by shading grasses and enhancing seedling recruitment of forest species (Calle, 2003). Posada *et al.* (2000) demonstrated that low-density grazing reduced the cover of *Melinis minutiflora* and *Pennisetum clandestinum* in Colombia, and permitted the colonization of wind-dispersed shrub species that helped to shade out the grasses and create a microhabitat appropriate for montane forest species (Figure 8.4). In the case of the fern *Dicranopteris pectinata* in the Dominican Republic, fire (May, 2000) and manual removal (including the root mat; Slocum *et al.*, 2004) were techniques that allowed many native species to colonize. Surprisingly, the fern was slow in recovering, and the shade produced by shrubs was already sufficient to limit any future dominance, demonstrating that the fern-dominated resilient state was easily shifted to a recovering state. Eliminating invasive trees and shrubs will be much more difficult and costly, however, and must be accompanied by extensive planting of native species and long-term maintenance to avoid the recolonization of the exotic species.

Microhabitat conditions

PROBLEMS

At a more local spatial scale, microhabitat conditions can influence colonization, growth, and survivorship and, in turn, the composition of a recovering forest. Even if seeds disperse into a site, to colonize, seeds must avoid predators. Few studies have identified seed predation as a major limiting factor in montane forest, but ant (Pedraza and Williams-Linera, 2003), cattle (Alvarez-Aquino *et al.*, 2004; Griscom *et al.*, 2008), and rabbit (Holl and Lulow, 1997) herbivory can affect plant growth and survivorship in early stages of forest recovery. Other studies have shown that poor mycorrhizae colonization of roots has limited plant growth (Berish and Ewel, 1988; Rosales *et al.*, 1997; Holl *et al.*, 2000). Mycorrhizal limitation is most common in soils that are highly degraded (e.g. by mining; Miller and Jastrow, 1992). Other factors that can reduce growth and survivorship include: freezing at high elevation (Scowcroft *et al.*, 2000; cf. Hemp, this volume #12), low nutrient availability (Bruijnzeel and Veneklaas, 1998; Günter *et al.*, 2009) and photo-inhibition (Loik and Holl, 1999).

MANAGEMENT SOLUTION

The easiest way to avoid seed predators is to plant seedlings. Seedlings will still be vulnerable to herbivores, but few studies have identified herbivores as a limiting factor in abandoned agricultural lands (see Holl and Lulow, 1997). Furthermore, if seedlings are being produced for a restoration project they can be inoculated with mycorrhizae to promote better growth and survivorship. Other studies have shown that nurse plants can modify the microhabitat conditions and promote the establishment of more sensitive species (Rhoades *et al.*, 1998; Posada *et al.*, 2000; Scowcroft *et al.*, 2000; Calle, 2003; Weber *et al.*, 2008). For example, in Colombia seedling density and richness were higher below remnant trees in pastures in contrast to open areas in pastures (Calle, 2003), whereas Posada *et al.* (2000) found that low-density grazing promoted the establishment of wind-dispersed shrubs, which shade out the grass. In Ecuador, small nitrogen-fixing trees in pastures ameliorated microclimate conditions and increased soil nitrogen levels, which, in turn, favored the establishment and growth of montane species (Rhoades *et al.* 1998; cf. Weber *et al.*, 2008; Günter *et al.*, 2009). On Hawai'i, efforts to restore *Metrosideros polymorpha* and other native species in exotic grasslands have been unsuccessful because of high seedling mortality due to freezing. At this site, *Acacia koa* survived better than other species, and once established, it helped to reduce frost damage to seedlings planted below it (Scowcroft *et al.*, 2000).

PLANTATIONS

In the literature, plantations have been treated as both a factor limiting forest recovery and as a technique for accelerating forest recovery. Plantations can help ameliorate all the barriers discussed above. Plantations can improve soil conditions by protecting soils from erosion (Wiersum, 1984; Vanacker *et al.*, 2007), and improving their nutrient content by fixing nitrogen and adding organic matter (Lugo, 1992). In montane ecosystems, a landscape that has increased forest cover will be less vulnerable to fires than areas dominated by grasslands and agricultural areas. In addition, plantations can attract dispersers, which will increase plant diversity in the understory community (Parrotta, 1992). Furthermore, plantations will eliminate most grasses and ferns by producing shade, and thus improve the microclimate conditions for many forest species.

Although plantations can improve conditions for forest regeneration, the long-term impact on forest recovery depends on the species used, the management techniques, and the goals of the project. In Uganda, forest regeneration in pine plantations was faster than in grasslands, because recurrent fires set back forest recovery in the grasslands (Zanne and Chapman, 2001). However, when these grasslands were protected from fire, forest regeneration was similar to that within the plantation (Duncan and Chapman, 2003a,b). In Colombia, although plantations (e.g. *Pinus*, *Cupressus*, *Eucalyptus*, and *Alnus*) helped improve soil characteristics and native tree species established in the understorey, species diversity in secondary forest of similar age was much higher compared with plantations (Cavelier and Tobler, 1998; Cavelier and Santos, 1999; Murcia, 1997; cf. Hölscher *et al.*, this volume). These studies suggest that, in the absence of fire, natural regeneration is a better strategy for recovering biodiversity than establishing plantations.

If the motive of a project is to restore natural vegetation, are single-species plantations a useful restoration tool? The answer is positive if soils are highly degraded and plantation species can protect and improve soil conditions. Plantations can also be incorporated in a long-term restoration plan, where there is a need for short-term economic benefits (timber, carbon credits), but after harvesting, native species in the understory should be allowed to grow. Plantations may also be useful for shading out grasses or ferns and attracting dispersers of native species, but if the final goal is a diverse montane forest, plantation trees should not be established at high densities. In many other cases, plantations do not seem necessary or appropriate. Secondary forest of similar age will support a (much) greater diversity of native species in comparison with plantations, and at a much lower cost (cf. Hölscher *et al.*, this volume).

SYNTHESIS

Will tropical montane clould forest recover without intervention? In many cases it appears that TMCF will recover, and studies have estimated the recovery time to be between 65 and 200 years (Weaver, 2000; Kappelle, 2001; Luna *et al.*, 2001; Silver *et al.*, 2001). The recovery time will clearly depend on the kind, duration, and intensity of previous land use, but it will also depend on what is the measure of recovery. Although plant richness may recover in 65 years, it may take centuries to recover the species composition and fauna similar to intact forest. As shown by Köhler *et al.* (this volume), this is particularly true for the epiphyte layer in TMCF. Ruiz-Jaén and Aide (2005) have shown that the majority of restoration projects only measure a few structural variables and plant richness, and few studies incorporate measures of the fauna or ecosystems processes. As the field of restoration ecology matures, it is necessary to incorporate more variables (e.g. fauna – Murcia *et al.*, 2001; Medina *et al.*, 2002; Dunn, 2004; ecosystem process – Ruiz-Jaén and Aide, 2005) as measures of recovery.

In other cases, aggressive pioneer species, fire, degraded soils, or lack of seeds will produce feedbacks that can maintain the system in a degraded state with no or very slow recovery. How can the recovery process be accelerated and the shift from a resilient degraded state to a state of recovery be facilitated? The results of this review suggest that, in the majority of cases, competition with grass and ferns is the major factor limiting the initial stages of forest recovery in tropical montane systems, and seed dispersal of mature forest species limits later stages of recovery. To overcome the barrier of aggressive herbaceous species, low-density grazing or manual removal is often sufficient to initiate secondary succession. If native shrubs do not establish rapidly, fast-growing trees, possibly plantation species, can be used to shade out herbaceous species and attract dispersers. These strategies should produce a species-diverse secondary forest. If the goal of a project is to restore the species composition of a mature montane forest, it is recommended to establish a nursery of mature montane forest species and to plant these into the site once the herbaceous cover has been reduced.

HOW CAN RESTORATION ECOLOGY CONTRIBUTE TO THE FUTURE OF TMCF?

In many regions, old threats to TMCF, such as clearing for agriculture or grazing, or intensive logging continue (e.g. Mosandl *et al.*, 2008; cf. Mulligan, this volume). Although economic development in the lowlands may reduce land use intensity in many montane areas, development is also accompanied by expanding infrastructure, which can impact TMCF forest (e.g. in the form of wind turbines, communication towers, pipelines and roads; Young, 1994; Malizia *et al.*, 2004). A potentially much greater concern is the impact of global climate change on TMCF (Pounds *et al.*, 1999; Foster, 2001; Lawton *et al.*, 2001; Pounds and Puschendorf, 2004; Foster, this volume; Nair *et al.*, this volume). These impacts should be mitigated, but our collective restoration experience in TMCF is extremely limited. A recent review of tropical restoration showed that the majority of projects are located in the lowland forest zone, with only 13% conducted in TMCF (Meli, 2003).

On the other hand, if the trend of rural–urban migration continues in Latin America, large areas of marginal pastures and agricultural lands will be abandoned, and much of this area will be in TMCF. Puerto Rico is a dramatic example of rural–urban migration, agriculture abandonment, and forest recovery. Forest cover increased from <10% in the late 1940s to >40% at present, following socio-economic changes associated with economic globalization (Rudel *et al.*, 2000; Grau *et al.*, 2003). Although Puerto Rico is an extreme example, a similar process is occurring in other areas of Latin America (Zweifler *et al.*, 1994; Preston *et al.*, 1997; Southworth and Tucker, 2001; Rudel, 2002). How should these abandoned lands in the TMCF zone be managed? Most of these areas will be left to natural regeneration, if only due to financial limitations. In many cases, this will be an appropriate "management" strategy, particularly given the low cost. In other areas where aggressive herbaceous species have arrested or greatly slowed succession (i.e. resilient degraded sites), or areas of special conservation importance, active restoration could help to accelerate the recovery process.

As restoration projects expand forest cover, positive feedbacks should reinforce regeneration processes. For example, as forest cover and connectivity increase, fire frequency will decrease, and seed dispersal and animal populations should increase. Understanding the dynamics of these feedbacks, and developing dynamic models should be part of future restoration projects and be incorporated into the international research agenda, such as the Forest Landscape Restoration project (Maginnis and Jackson, 2005; www.unep-wcmc.org/forest/restoration/globalpartnership/index.htm). To meet these present and future restoration/management challenges, it is important to get beyond talking about restoration or applying results from other lifezones, and begin a coordinated effort to understand how TMCF can be restored. Payment for environmental services is a possible approach to stimulate future restoration efforts (e.g. Costa Rica – Rodriguez Zuñiga, 2003; cf. Tognetti *et al.*, this volume; Calvo-Alvarado *et al.*, this volume).

ACKNOWLEDGEMENTS

During the preparation of the manuscript the authors received support from the NASA-IRA program and the Dean of Graduate Studies and Investigation (DEGI) of the University of Puerto Rico. The International Institute of Tropical Forestry of the U.S. Forest Service assisted with the literature search and travel expenses. The comments of Becky Ostertag, Zoraida Calle, Sampurno Bruijnzeel, and Larry Hamilton improved the manuscript.

REFERENCES

Aide, T.M., and J. Cavelier (1994). Barriers to lowland tropical forest restoration in the Sierra Nevada de Santa Marta, Colombia. *Restoration Ecology* **2**: 219–229.

Aide, T.M., and H.R. Grau (2004). Globalization, migration, and Latin American ecosystems. *Science* **305**: 1915–1916.

Aide, T.M., J.K. Zimmerman, J. Pascarella, J. Marcano-Vega, and L. Rivera (2000). Forest regeneration in a chronosequence of tropical abandoned pastures: implications for restoration ecology. *Restoration Ecology* **8**: 328–338.

Alvarez-Aquino, C., G. Williams-Linera, and A.C. Newton (2004). Experimental native tree seedling establishment for the restoration of a Mexican cloud forest. *Restoration Ecology* **12**: 412–418.

Asbjornsen, H., and A.J. Wickel (2009). Changing fire regimes in tropical montane cloud forests: a global synthesis. In *Fire and Tropical Ecosystems*, ed. M.A. Cochrane, pp. 607–622. New York: Springer-Verlag.

Asbjornsen, H., C. Gallardo-Hernández, N. Velázquez-Rosas, and R. García-Soriano (2005). Deep ground fires cause massive above- and below-ground biomass losses in tropical montane cloud forests in Oaxaca, Mexico. *Journal of Tropical Ecology* **21**: 427–434.

Bedoya, E. (1997). Migración, falta de empleo y el impacto de la coca en los sistemas agrícolas tropicales y de montaña: un estudio de caso en el Perú. In *Desarrollo sostenible de ecosistemas de montaña: Manejo de areas frágiles en los Andes*, eds. M. Liberman and C. Baied, pp. 83–91. La Paz, Bolivia: The United Nations University.

Berish, C.W., and J.J. Ewel (1988). Root development in simple and complex tropical successional ecosystems. *Plant and Soil* **106**: 73–84.

Blaisdell, F.W. (1981). Engineering structures for erosion control. In *Tropical Agricultural Hydrology*, eds. R. Lal and E.W. Russell, pp. 325–355. New York: John Wiley.

Bradshaw, A.D. (1997). The importance of soil ecology in the restoration science. In *Restoration Ecology and Sustainable Development*, eds. K.M. Urbanska, N.R. Webb, and P.J. Edwards, pp. 33–64. Cambridge, UK: Cambridge University Press.

Bruijnzeel, L.A., and E.J. Veneklaas (1998). Climatic conditions and tropical montane forest productivity: the fog has not lifted yet. *Ecology* **79**: 3–9.

Calle, Z. (2003). *Restauración de suelos y vegetación nativa: ideas para una ganadería Andina sostenible*. Cali, Colombia: Centro para la Investigación en Sistemas Sostenibles de Producción Agropecuaria (CIPAV).

Cavelier, J. (1995). Reforestation with the native tree *Alnus acuminata*: effects on phytodiversity and species richness in the upper montane rain forest area of Colombia. In *Tropical Montane Cloud Forests*, eds. L.S. Hamilton, J.O. Juvik, and F.N Scatena, pp. 125–137. New York: Springer-Verlag.

Cavelier, J. and A. Etter (1995). Deforestation of montane forest in Colombia as result of illegal plantations of opium (*Papaver somniferum*). In *Biodiversity and Conservation of Neotropical Montane Forests*, eds. P. Churchill, H. Baslev, E. Forero, and J.L Luteyn, pp. 541–549. New York: New York Botanical Garden.

Cavelier, J. and C. Santos (1999). Effect of abandoned exotic and native species plantations on the natural regeneration of a montane forest in Colombia. *Revista de Biologia Tropical* **47**: 775–784.

Cavelier, J. and A. Tobler (1998). The effect of abandoned plantations of *Pinus patula* and *Cupressus lusitanica* on soils and regeneration of a tropical montane rain forest in Colombia. *Biodiversity and Conservation* **7**: 335–347.

Cavelier, J., T.M. Aide, C. Santos, A.M. Eusse, and J.M. Dupuy (1998). The savannization of moist forest in the Sierra Nevada de Santa Marta, Colombia. *Journal of Biogeography* **25**: 901–912.

Cavelier, J., T.M. Aide, J.M. Dupuy, A.M. Eusse, and C. Santos (1999). Long-term effects of deforestation on soil properties and vegetation in a tropical lowland forest in Colombia. *Ecotropicos* **12**: 57–68.

Chernick, M. (2000). Elusive peace: struggling against the logic of violence. *North American Congress on Latin America Report on the Americas* **34**: 34–37.

Cubiñá, A., and T. M. Aide (2001). The effect of distance from forest edge on seed rain and soil seed bank in a tropical pasture. *Biotropica* **33**: 260–267.

Dalling, J. W. and E. V. J. Tanner (1995). An experimental study of regeneration on landslides in montane forests in Jamaica. *Journal of Ecology* **83**: 55–64.

D'Antonio, C. M. and P. M. Vitousek (1992). Biological invasions by exotic grasses, the grass/fire cycle, and global change. *Annual Review of Ecology and Systematics* **23**: 63–87.

Duncan, R. S. and C. A. Chapman (2003a). Tree–shrub interactions during early forest succession in Uganda. *Restoration Ecology* **11**: 198–207.

Duncan, R. S. and C. A. Chapman (2003b). Consequences of plantation harvest during tropical forest restoration in Uganda. *Forest Ecology and Management* **173**: 235–250.

Dunn, R. R. (2004). Managing the tropical landscape: a comparison of the effects of logging and forest conversion to agriculture on ants, birds, and lepidoptera. *Forest Ecology and Management* **191**: 215–224.

Foster, P. (2001). The potential negative impacts of global climate change on tropical montane cloud forests. *Earth-Science Reviews* **55**: 73–106.

Garcia, R., M. Mejía, and T. Zanoni (1994). Composición floristica y principales asociaciones vegetales en la Reserva Cientifica Ebano Verde, Cordillera Central, República Dominicana. *Moscosoa* **8**: 86–130.

Grau, H. R. and M. R. Aragon (2000). Arboles invasores de la sierra de San Javier, Tucuman, Argentina. In *Ecologia de arboles exóticos en las Yungas Argentinas*, eds. H. R. Grau and M. R. Aragon, pp. 5–20. Tucuman, Argentina: Laboratorio de Investigaciones Ecologicas de las Yungas.

Grau, H. R. and T. T. Veblen (2000). Rainfall, fire, and vegetation dynamics in subtropical montane ecosystems in northwestern Argentina. *Journal of Biogeography* **27**: 1107–1121.

Grau, H. R., T. M. Aide, J. K. Zimmerman, *et al.* (2003). The ecological consequences of socioeconomic and land-use changes in post-agriculture Puerto Rico. *BioScience* **53**: 1159–1168.

Griscom, H. P., B. W. Griscom, and M. S. Ashton (2008). Forest regeneration from pasture in the dry tropics of Panama: effects of cattle, exotic grasses, and forest riparia. *Restoration Ecology* **17**: 117–126.

Guariguata, M. R. and R. Ostertag (2001). Neotropical secondary forest succession: changes in structural and functional characteristics. *Forest Ecology and Management* **148**: 185–206.

Günter, S., M. Weber, R. Erreis, and N. Aguirre (2006). Influence of distance to forest edges on natural regeneration of abandoned pastures: a case study in the tropical montane rain forest of Southern Ecuador. *European Journal of Forest Research*, doi: 10.1007/s10342–006–0156–0.

Günter, S., P. Gonzalez, G. Alvarez, *et al.* (2009). Determinants of successful regeneration of abandoned pastures in the Andes: soil conditions and vegetation cover. *Forest Ecology and Management* **258**: 81–91.

Hamilton, L. S., J. O. Juvik, and F. N. Scatena. (1995). The Puerto Rico tropical cloud forest symposium: Introduction and workshop synthesis. In *Tropical Montane Cloud Forests*, eds. L. S. Hamilton, J. O. Juvik, and F. N. Scatena, pp. 1–23. New York: Springer-Verlag.

Harden, C. (1996). Interrelationships between land abandonment and land degradation: a case from the Ecuadorian Andes. *Mountain Research and Development* **16**: 274–280.

Hartig, K. and E. Beck (2003). The Bracken fern (*Pteridium arachnoideum* Kaulf.) dilemma in the Andes of Southern Ecuador. *Ecotropica* **9**: 3–13.

Harvey, C. (2000). Windbreaks enhance seed dispersal into agricultural landscapes in Monteverde, Costa Rica. *Ecological Applications* **10**: 155–173.

Hemp, A. (2005). Climate change-driven forest fires marginalize the impact of ice cap wasting on Kilimanjaro. *Global Change Biology* **11**: 1013–1023.

Hemp, A. and E. Beck (2001). *Erica excelsa* as a fire-tolerating component of Mt. Kilimanjaro's forests. *Phytocoenologia* **31**: 449–475.

Hobbs, R. J. and D. A. Norton (1996). Towards a conceptual framework for restoration ecology. *Restoration Ecology* **4**: 93–110.

Holl, K. D. (1999). Factors limiting tropical moist forest regeneration in agricultural land: soil, microclimate, vegetation, and seed rain. *Biotropica* **31**: 229–242.

Holl, K. D., and M. E. Lulow (1997). Effects of species, habitat, and distance from edge on post-dispersal seed predation in a tropical rainforest. *Biotropica* **29**: 459–468.

Holl, K. D., M. E. Loik, E. H. V. Lin, and I. A. Samuels (2000). Tropical montane forest restoration in Costa Rica: obstacles and opportunities. *Restoration Ecology* **8**: 339–349.

Horn, S. P., L. M. Kennedy, and K. H. Orvis. (2001). Vegetation recovery following a high elevation fire in the Dominican Republic. *Biotropica* **33**: 701–708.

Hudson, N. W. (1995). *Soil Conservation*. London: Batsford.

Kappelle, M. (2001). Costa Rica. In *Bosques nublados del neotropico*, eds. M. Kappelle and A. D. Brown, pp. 301–370. Santo Domingo de Heredia, Costa Rica: Instituto Nacional de Biodiversidad.

Kessler, M. (2000). Observations on a human-induced fire event at humid timberline in the Bolivian Andes. *Ecotropica* **6**: 89–93.

Lamb, D. and D. Gilmour (2004). *Rehabilitation and Restoration of Degraded Forests*. Gland, Switzerland: World Conservation Union, IUCN.

Lawton, R. O., U. S. Nair, R. A. Pielke Sr., and R. M. Welch (2001). Climatic impact of tropical lowland deforestation on nearby montane cloud forests. *Science* **294**: 584–587.

Lichstein, J., H. R. Grau, and M. R. Aragon (2004). Recruitment limitation in secondary forests dominated by an exotic tree. *Journal of Vegetation Science* **15**: 721–728.

Loik, M. E., and K. D. Holl (1999). Photosynthetic responses to light for rainforest seedlings planted in abandoned pasture, Costa Rica. *Restoration Ecology* **7**: 382–391.

Lugo, A. E. (1992). Comparison of tropical tree plantations with secondary forest of similar age. *Ecological Monographs* **62**: 1–41.

Luna, I., A. Velazquez, and E. Velazquez (2001). Mexico. In *Bosques nublados del neotropico*, eds. M. Kappelle and A. D. Brown, pp. 183–229. Santo Domingo de Heredia, Costa Rica: Instituto Nacional de Biodiversidad.

Maginnis, S., and W. Jackson (2005). Balancing restoration and development. *ITTO Tropical Forest Update* **15**: 4–6.

Malizia, A., N. Chacoff, H. R. Grau, and A. D. Brown (2004). Vegetation recovery on a gas-pipeline track along an altitudinal gradient in the Argentinean Yungas forest. *Ecologia Austral* **14**: 165–178.

Malizia, L. R. and A. Greslebin (2000). Reclutamiento de especies arboreas bajo arbustos exoticos en la sierra de San Javier, Tucuman, Argentina. In *Ecologia de arboles exotícos en las Yungas Argentinas*, eds. H. R. Grau and M. R. Aragon, pp. 47–58. Tucuman, Argentina: Laboratorio de Investigaciones Ecologicas de las Yungas.

May, T. (1997). Fases tempranas de la sucesión en un bosque nublado de *Magnolia pallescens* después de un incendio (Loma de Casabalito, Reserva Científica Ebano Verde, Cordillera Central, República Dominicana). *Moscosoa* **9**: 117–144.

May, T. (2000). Respuesta de la vegetación en un "calimetal"de *Dicranopteris pectinata* después de un fuego, en la parte oriental de la Coordillera Central, República Dominicana. *Moscosoa* **11**: 113–132.

Medina, C. A., F. Escobar, and G. H. Kattan (2002). Diversity and habitat use of dung beetles in a restored Andean landscape. *Biotropica* **34**: 181–187.

Meli, P. (2003). Restauración ecológica de bosques tropicales: viente años de investigación académica. *Interciencia* **29**: 581–589.

Miller, R. M. and J. D. Jastrow (1992). The application of VA mycorrhizae to ecosystem restoration and reclamation. In *Mycorrhizal Functioning: An Integrative Plant Fungal Process*, ed. M. Allen, pp. 438–467. New York: Chapman and Hall.

Mosandl, R., S. Günter, B. Stimm, and M. Weber (2008). Ecuador suffers the highest deforestation rate in South America. In *Gradients in a Tropical Mountain Ecosystem of Ecuador*, eds. E. Beck, J. Bendix, I. Kottke, F. Makeschin, and R. Mosandl, pp. 37–40. Berlin: Springer-Verlag.

Murcia, C. (1997). Evaluation of Andean alder as a catalyst for the recovery of tropical cloud forests in Colombia. *Forest Ecology and Management* **99**: 163–170.

Murcia, C., G. Kattan, and A. Galindo (2001). Recovery of Bess beetles key to long-term restoration of Andean forest (Colombia). *Ecological Restoration* **19**: 254–255.

Olander, L. P., F. N. Scatena, and W. L. Silver (1998). Impacts of disturbance initiated by road construction in a subtropical cloud forest in the Luquillo Experimental Forest, Puerto Rico. *Forest Ecology and Management* **109**: 33–49.

Oosterhoorn, M., and M. Kappelle (2000). Vegetation structure and composition along an interior–edge–exterior gradient in a Costa Rican montane cloud forest. *Forest Ecology and Management* **126**: 291–307.

Parrotta, J. A. (1992). The role of plantation forests in rehabilitating degraded tropical ecosystems. *Agriculture, Ecosystems, and Environment* **41**: 115–133.

Pedraza, R. A. and G. Williams-Linera (2003). Evaluation of native tree species for the rehabilitation of deforested areas in a Mexican cloud forest. *New Forests* **26**: 83–99.

Posada, J. M., T. M. Aide, and J. Cavelier (2000). Cattle and weedy shrubs as restoration tools of tropical montane rainforest. *Restoration Ecology* **8**: 370–379.

Pounds, J. A., and R. Puschendorf. (2004). Clouded futures. *Nature* **427**: 107–109.

Pounds, J. A., M. P. L Fodgen, and J. H. Campbell (1999). Biological response to climate change on a tropical mountain. *Nature* **398**: 611–615.

Preston, D. (1996). People on the move: migrations past and present. In *Latin America Development: Geographical Perspectives*, ed. D. Preston, pp. 165–187. Harlow, UK: Addison Wesley.

Preston D., M. Macklin, and J. Warburton (1997). Fewer people, less erosion: the 20th century in southern Bolivia. *Geographical Journal* **163**: 198–205.

Ramirez-Marcial, N., M. Gonzalez-Espinosa, and G. Williams-Linera (2001). Anthropogenic disturbance and tree diversity in montane rain forest in Chiapas, Mexico. *Forest Ecology and Management* **154**: 311–326.

Rhoades, C. C., G. E. Eckert, and D. C. Coleman (1998). Effect of pastures trees on soil nitrogen and organic matter: implications for tropical montane forest restoration. *Restoration Ecology* **6**: 262–270.

Rodriguez Zuñiga, J. M. (2003). Paying for forest environmental services: the Costa Rican experience. *Unasylva* **212**: 31–33.

Rosales, J., G. Cuenca, N. Ramírez, and Z. de Andrade (1997). Native colonizing species and degraded land restoration in La Gran Sabana, Venezuela. *Restoration Ecology* **5**: 147–155.

Rudel, T. K. (2002). Paths of destruction and regeneration: globalization and forests in the tropics. *Rural Sociology* **67**: 622–636.

Rudel, T. K., M. Perez-Lugo, and H. Zichal (2000). When fields revert to forest: development and spontaneous reforestation in post-war Puerto Rico. *Professional Geographer* **52**: 386–397.

Ruiz-Jaén, M. C., and T. M. Aide (2005). Restoration success: how is it being measured? *Restoration Ecology* **13**: 569–577.

Sarmiento, F. O. (1997a). Arrested succession in pastures hinders regeneration of Tropandean forests and shreds mountain landscapes. *Environmental Conservation* **24**: 14–23.

Sarmiento, F. O. (1997b). Landscape regeneration by seeds and succcesional pathways to restore fragile tropandean slopelands. *Mountain Reseacrh and Development* **17**: 239–252.

Schulenberg, T. S., and K. Awbrey (1997). *A Rapid Assessment of the Humid Forest of South-Central Chuquisaca, Bolivia*, RAP Working Papers No. 8. Washington, DC: Conservation International.

Scowcroft, P. G. and J. Jeffrey (1999). Potential significance of frost, topographic relief, and *Acacia koa* stands to restoration of mesic Hawaiian forests on abandoned rangeland. *Forest Ecology and Management* **114**: 447–458.

Scowcroft, P. G., F. C. Meinzer, G. Goldstein, P. J. Melcher, and J. Jeffrey (2000). Moderating night radiative cooling reduces frost damage to *Metrosideros polymorpha* seedlings used for forest restoration in Hawaii. *Restoration Ecology* **8**: 161–169.

Shields, A. B., and R. F. Walker (2003). Bird perches increase forest seeds on Puerto Rican landslides. *Restoration Ecology* **11**: 457–465.

Silver, W. L., A. E. Lugo, and M. Keller (1999). Soil oxygen availability and biogeochemical cycling along elevation and topographic gradients in Puerto Rico. *Biogeochemistry* **44**: 301–328.

Silver, W. L., E. Marin-Spiotta, and A. E. Lugo. (2001). El Caribe. In *Bosques nublados del neotropico*, eds. M. Kappelle and A. D. Brown, pp. 155–181. Santo Domingo de Heredia, Costa Rica: Instituto Nacional de Biodiversidad.

Slocum, M. G., and C. C. Horvitz (2000). Seed arrival under different genera of trees in a neotropical pasture. *Plant Ecology* **149**: 51–62.

Slocum, M., T. M. Aide, J. K. Zimmerman, and L. Navarro (2000). La vegetación leñosa en helechales y bosques de ribera en la Reserva Cientifica Ebano Verde, Republica Dominicana. *Moscosoa* **11**: 38–56.

Slocum, M., T. M. Aide, J. K. Zimmerman, and L. Navarro (2004). Natural regeneration of subtropical montane forest after clearing fern thickets in the Dominican Republic. *Journal of Tropical Ecology* **20**: 483–486.

Southworth, J., and C. Tucker (2001). The influence of accessibility, local institutions, and socioeconomic factors on forest cover change in the mountains of western Honduras. *Mountain Research and Development* **21**: 276–283.

Stewart, C. G. (2000). A test of nutrient limitation in two tropical montane forest using root ingrowth cores. *Biotropica* **32**: 369–373.

Suding, K. N., L. G. Gross, and G. R. Houseman (2004). Alternative states and positive feedbacks in restoration ecology. *Trends in Ecology and Evolution* **19**: 46–53.

Tanner, E. V. J., P. M. Vitousek, and E. Cuevas (1998). Experimental investigation of nutrient limitation of forest growth on wet tropical mountains. *Ecology* **79**: 10–22.

Tecco, P. A., and M. Rouges (2000). El naranjo agrio (*Citrus aurantium*), exotica invasora de bosques maduros. In *Ecologia de arboles exotícos en las Yungas Argentinas*, eds. H. R. Grau and M. R. Aragon, pp. 37–45. Tucuman, Argentina: Laboratorio de Investigaciones Ecologicas de las Yungas.

Vanacker, V., F. von Blanckenburg, G. Govers, *et al.* (2007). Restoring dense vegetation can slow mountain erosion to near benchmark levels. *Geology* **35**: 303–306.

Weaver, P. L. (2000). Elfin woodland recovery 30 years after a plane wreck in Puerto Rico's Luquillo Mountains. *Caribbean Journal of Science* **36**: 1–9.

Weber, M., S. Günter, N. Aguirre, B. Stimm, and R. Mosandl (2008). Reforestation of abandoned pastures: silvicultural means to accelerate forest recovery and biodiversity. In *Gradients in a Tropical Mountain Ecosystem of Ecuador*, eds. E. Beck, J. Bendix, I. Kottke, F. Makeschin, and R. Mosandl, pp. 447–458. Berlin: Springer-Verlag.

Werner, W. L., and T. Santisuk (1993). Conservation and restoration of montane forest communities in Thailand. In *Restoration of Tropical Forest Ecosystems*, eds. H. Lieth and M. Lohmann, pp. 193–202. Dordrecht, the Netherlands: Kluwer.

Wiersum, K. F. (1984). Surface erosion under various tropical agroforestry systems. In *Proceedings of the Symposium on Effects of Forest Land Use on Erosion and Slope Stability*, eds. C. L. O'Loughlin and A. J. Pearce, pp. 231–239. Vienna: IUFRO, and Honolulu, HI: East–West Center.

Young, K. (1994). Roads and the environmental degradation of tropical montane forests. *Conservation Biology* **8**: 972–976.

Zahawi, R. A., and C. K. Augspurger (1999). Early plant succession in abandoned pastures in Ecuador. *Biotropica* **31**: 540–552.

Zanne, A. and C. A. Chapman (2001). Expediting reforestation in tropical grasslands: distance and isolation from seed sources in plantations. *Ecological Applications* **11**: 1610–1621.

Zimmerman, J. K., J. Pascarella, and T. M. Aide (2000). Barriers to forest regeneration in an abandoned pasture in Puerto Rico. *Restoration Ecology* **8**: 350–360.

Zimmermann, B., and H. Elsenberger (2008). Spatial and temporal variability of soil saturated hydraulic conductivity in gradients of disturbance. *Journal of Hydrology* **361**: 78–95.

Zweifler, M. O., M. A. Gold, and R. N. Thomas (1994). Land-use evolution in hill regions of the Dominican Republic. *Professional Geographer* **46**: 39–53.

Part II
Regional floristic and animal diversity

L. A. Bruijnzeel and L. S. Hamilton

9 Tropical montane cloud forests in Malaysia: current state of knowledge

S. Kumaran
WWF-Malaysia, Petaling Jaya, Selangor, Malaysia

B. Perumal
Wetlands International, Petaling Jaya, Selangor, Malaysia

G. Davison
WWF-Malaysia, Petaling Jaya, Selangor, Malaysia

A. N. Ainuddin
Universiti Putra Malaysia, Serdang, Malaysia

M. S. Lee
WWF-Malaysia, Petaling Jaya, Selangor, Malaysia

L. A. Bruijnzeel
VU University, Amsterdam, the Netherlands

ABSTRACT

Taking 1500 m elevation as the approximate lower limit of upper montane cloud forest occurrence in Malaysia, 0.72% of the total land area of the country may be considered to be potentially covered by tropical montane cloud forests (TMCF). Delineation of TMCF is not easy due to the interplay of climatic and topographic features (including the "mass-elevation" effect). Some studies have shown that the lifting condensation level (cloud base) occurs around 1200 m.a.s.l. in Peninsular Malaysia, at *c.* 1800–2000 m on large mountains like Mt. Kinabalu (Borneo), and as low as *c.* 700 m on small, outlying coastal mountains (e.g. Gunung Silam, Sabah). Published information on the biodiversity of TMCF in Malaysia is scattered. Important tree families in montane forests include Ericaceae, Myrtaceae, Coniferae, Fagaceae, and Lauraceae, which are often blanketed with epiphytes, ferns, bryophytes, and liverworts. About 3000–4000 vascular plant species are found in the montane forests of Peninsular Malaysia and Borneo. There is a distinct montane community amongst most classes of vertebrates and some invertebrates, but the fauna of TCMF is mostly a small sub-set of the overall montane fauna. About 18 mammal species in Borneo and 10 in Peninsular Malaysia are strictly montane but none are truly restricted to cloud forests. Bird diversity is well studied but limited information is available on the number of species that are cloud forest specialists. Patchy records are available for butterflies, moths, mollusks, and herpetofauna. Studies on the capture of cloud water in Malaysian cloud forests are rare. Cloud forest soils are reported to be acidic with frequent occurrence of peat accumulation but they have been studied at only a handful of locations and systematic mapping remains to be done. Although accessibility to TMCF is limited, on-site and off-site threats are on the increase. Fragmentation of cloud forest habitats by road construction leads to other developments associated with tourism, temperate agriculture, as well as township and telecommunication facilities.

INTRODUCTION

Malaysia has a total land area of about 33 015 780 ha (Table 9.1), of which *c.* 60% is under natural forest cover, mostly in hilly and mountainous areas above 900 m.a.s.l. (Economic Planning Unit,

Tropical Montane Cloud Forests: Science for Conservation and Management, eds. L. A. Bruijnzeel, F. N. Scatena, and L. S. Hamilton. Published by Cambridge University Press.

Table 9.1 *Extent of land areas in Malaysia by altitude class. Figures should be regarded as indicative*

Altitude class (m.a.s.l.)	Peninsular Malaysia (km^2)	Sarawak (km^2)	Sabah (km^2)	Total (km^2)	%
0–900	12 5954.34	11 4047.94	6 7994.47	**30 7996.75**	**93.29**
900–1200	4321.50	7065.27	3611.96	**1 4998.73**	**4.54**
1200–1500	1416.52	1946.55	1418.63	**4781.70**	**1.45**
1500–2100	546.29	857.74	864.62	**2268.65**	**0.69**
>2100	Trace	Trace	112.01	**112.01**	**0.03**
Total (km^2)	**PM 132 239**	**Sar 123 917**	**Sab 74 002**	**Total 330 158**	**100**
% land that is montane	**4.75**	**7.96**	**8.12**		
% of all Malaysian montane land	**40.05**	**37.53**	**22.42**	**100**	

Modified from: Economic Planning Unit (1993).

1993). Of the 19.5 Mha of forest, 14.4 Mha (74%) are reserved as "Permanent Forest Estate" (PFE) which includes 3.81 Mha (19% of all forest) of "Protection Forest" (Thang and Chappell, 2005). The environmental factors that lead to the establishment of cloud forest in Malaysia are only partially understood. Following Symington (1943), montane forests in Malaysia occur above 900 m.a.s.l. The term "cloud forest" is rarely used in the Malaysian context, and there is no clear and accepted definition of tropical montane cloud forest (TMCF) at the national or state level. There is also no comprehensive assessment of the state of information on TMCF in Malaysia although the chapter on montane habitats in McKinnon *et al.* (1996) provides a good summary of available information for mountains in Borneo (East Malaysia and Kalimantan).

This chapter seeks to give an overview of the state of knowledge with respect to cloud forest occurrence in Malaysia, its biodiversity, hydrometeorology, productivity, nutrient cycling, and soils, as well as the various threats faced by these unique ecosystems.

CLOUD FOREST OCCURRENCE AND EXTENT IN MALAYSIA

Although the extent of cloud forests in Malaysia has not been delineated and mapped systematically, there are numerous early accounts of their occurrence and characteristics (e.g. Ridley, 1914, 1915; Burgess, 1969; Whitmore and Burnham, 1969; Soepadmo, 1971; Whitmore, 1972; Martin, 1977; Stone, 1981). Burgess (1969) investigated ecological factors in hill and mountain forests and observed the often sharp boundary of the cloud forest zone, its depression on exposed ridges, and the widely varying altitude of the lower limit of cloud forests (often referred to locally as "montane Ericaceous forest" or "upper montane rain forest"; cf. Whitmore, 1984) on differently sized mountains, e.g. 1515 m.a.s.l. in the Main Range, 1212 m.a.s.l. on Mt. Tapis, and only 909 m.a.s.l. on Mt. Blumut. Similar observations were made by Martin (1977) on Gunung Mulu (2371 m.a.s.l.) in Sarawak where the elevation of mossy forest occurrence varied with site exposure and, to a lesser extent, geologic substrate (Tie *et al.*, 1979). Such contrasts partly reflect the complication that on small isolated mountains forest formations tend to be compressed while the associated lifting condensation level (LCL) occurs at a lower altitude than on larger mountains (Whitmore and Burnham, 1969). This phenomenon is usually called the "mass-elevation" or "telescoping" effect and refers to the higher air temperatures afforded by the greater mass of large mountains, leading in turn to higher LCL levels (cloud formation). Because the LCL associated with small coastal or island mountains is typically lower (600–700 m.a.s.l.) than further inland (>900 m.a.s.l.), some investigators also proposed a "coastal proximity" effect to capture the influence of the allegedly higher atmospheric moisture levels in coastal areas on the LCL (Van Steenis, 1972; Bruijnzeel *et al.*, 1993; cf. Jarvis and Mulligan, this volume).

To add to the complication of delineating cloud forest occurrence and extent in Malaysia, Ashton (1995) provided a detailed description of the altitudinal forest profile in Sabah and Sarawak, suggesting that there is no clearly distinguishable altitudinal zonation, as the changes in floristic composition of the forests with altitude were gradual and continuous. However, Martin (1977), Proctor *et al.* (1988a,b), and Kitayama (1992, 1995) all identified clearly distinguishable types of montane forests on Mts. Mulu (2371 m.a.s.l., Sarawak), Silam (890 m.a.s.l., Sabah), and Kinabalu (4095 m.a.s.l., Sabah), respectively, based on forest structure and, to a lesser extent, floristic composition. Whilst the latter authors only distinguished between lower montane and upper montane rain forests (LMRF and UMRF, respectively, of which UMRF is usually equated to TMCF; Whitmore, 1984), they all described a facies of tall, yet mossy forest occurring above the local LCL which might also be called lower montane cloud forest (LMCF). This distinction between tall LMCF and shorter-statured upper montane cloud forest (UMCF) proposed by Bruijnzeel and Hamilton (2000) is supported by observations on changes in the occurrence and frequency of bryophytes (Frahm and Gradstein, 1991) as well as in hydrological functioning (Bruijnzeel 2001, 2005) with elevation at numerous sites, including Malaysia (Frahm *et al.*, 1990; Bruijnzeel *et al.*, 1993; Kumaran, 2008). Furthermore, the two types of cloud forest also show contrasts in productivity (Proctor *et al.*, 1988b; Aiba *et al.*, this volume), nitrogen mineralization rates (Kitayama *et al.*, 1998), and soil formation (Tie *et al.*, 1979; Kitayama, 1992; Bruijnzeel *et al.*, 1993; cf. Roman *et al.*, this volume).

Taking an altitude of 900 m.a.s.l. as being approximately the lowermost level of the cloud base in Malaysia (cf. Symington, 1943), about 6.0% of the land would potentially support cloud forest (Table 9.1). Of the land area above 900 m, 8.7% lies within national parks and wildlife reserves (Davison, 1996). In Peninsular Malaysia, forests above 1000 m have been accorded protection forest status by the 14th National Forestry Council meeting on 28 December 1998, for regulating water resources, for biodiversity protection and to ensure soil stability. However, there are indications that the LCL level generally occurs at *c.* 1200 m.a.s.l. in both Peninsular Malaysia (Burgess, 1969) and Sarawak (Mulu area; Tie *et al.*, 1979) vs. *c.* 2000 m on (much larger) Mt. Kinabalu (Kitayama, 1995). This, and the topographic information given in Table 9.1 suggests that using the 900-m contour to delineate montane forests, and the 1200-m contour to delineate potential cloud forest occurrence, would produce major differences. Similarly, taking 1200 m (Burgess, 1969; Tie *et al.*, 1979) or 1500 m (Bubb *et al.*, 2004; cf. Table 1.2 in Scatena *et al.*, this volume) as approximating the lower limit of the potential distribution of TMCF in Malaysia, their area is estimated at 716 236 ha (2.17% of the total land area) and 238 066 ha (0.72%), respectively (Table 9.1). Although this requires further validation, the 1200-m contour is likely to mark the start of LMCF and the associated areal estimate cited above would therefore include both LMCF and UMCF, whereas in the case of the 1500-m contour this would be mostly UMCF. Whilst the above percentages may seem insignificant, expressed as a fraction of the total area under Protection Forest (3.81 Mha; Thang and Chappell, 2005) they become much more impressive at 18.8% and 6.2%, respectively.

These estimates should be compared with actually occurring cloud forest. The UNEP–WCMC database (Aldrich *et al.*, 1997; Bubb *et al.*, 2004) contains over 35 identified cloud forest sites for Malaysia but does not allow the estimation of an areal total. Mulligan (this volume) even estimated as much as 38 631 km^2 of cloud-affected forest for Malaysia based on modeled LCL levels, satellite-derived cloud frequency and forest cover, and topographic information. Given that this estimate is about 120% of the land area above the 900 m contour (Table 9.1), Mulligan's estimate may have to be regarded as an overestimate but at least it illustrates the frequent development of low cloud over these areas. In Sabah, sizeable areas of MCF are known only from Mts. Kinabalu (4095 m.a.s.l.), Trus Madi (2597 m.a.s.l.), and Tambuyukon (2597 m.a.s.l.). Applying the 1200-m limit (Tie *et al.*, 1979), one would expect cloud forests to occur in Sarawak on (and around) Mts. Mulu and Murud (2438 m.a.s.l.) as well as on Mt. Lawit (1767 m.a.s.l.) and (possibly) Mt. Dulit (1349 m.a.s.l.). Areas above 1200 and 1500 m in Sarawak make up 2.26% and 0.69% of the State's total area, respectively (Table 9.1).

CLOUD FOREST BIODIVERSITY

Malaysia's mountainous regions fall within a single biogeographic unit categorized by flora and fauna very distinct from other habitats. The mountain flora contains species with origins from both Asian and Australasian sources (Van Steenis, 1964; Cockburn, 1978) whereas the fauna is considered predominantly Asian. At least for birds and moths, the montane species are mostly of Himalayan origin (Holloway, 1970). Hence, it is postulated that at least some of the montane flora is of ancient Gondwanaland origin, while the fauna is more obviously and recently derived from mainland Asia.

Flora

Accounts of floristic composition of cloud forests are patchy and no detailed account is available for Malaysia as a whole. However, extensive data are available, amongst others, for Mts. Mengkuang Lebah ("Genting Highlands"), Tahan, and Benom in Peninsular Malaysia (Ridley, 1914, 1915; Soepadmo, 1971; Whitmore, 1972), for Mt. Mulu in Sarawak (Martin, 1977), and for Mts. Silam (Proctor *et al.*, 1988a) and, especially, Kinabalu (e.g. Smith, 1970; Van Steenis, 1972; Kitayama, 1992; Aiba and Kitayama, 1999). Trees in LMCF on the larger mountains may reach up to 25–30 m (Kitayama, 1992) but those on smaller outlying mountains (e.g. Gunung Silam) rather resemble UMCF (10–15 m) although they classify as LMCF on the basis of their floristic composition and dominant leaf size (Proctor *et al.*, 1988a). Canopy height of UMCF is typically 10–20 m (e.g. on Kinabalu) but on exposed summits and sharp ridges, the vegetation becomes gnarled and stunted, often being less than 10 m tall (Martin, 1977). On Kinabalu, similarly stunted sub-alpine cloud forest (SACF) occurs above *c.* 2800 m.a.s.l. (Kitayama, 1992). Important tree families include the Lauraceae and Fagaceae as being more frequent in LMCF and the Myrtaceae, Coniferae, and Ericaceae as the dominant ones in UMCF (Perumal and Lo, 1998). Epiphytes are characteristic, and branches, tree trunks and the forest floor are covered with ferns, mosses, and liverworts, with the highest incidence in the UMCF zone (Frahm *et al.*, 1990). Where the vegetation has been disturbed and light is abundant, tree ferns are common. Several species of rhododendron are present as shrubs and small trees in the mountains. Argent *et al.* (1988) described rhodendron occurrence and frequency on a series of mountains in Sabah,

Table 9.2 *Endemism amongst mammals and birds of the three largest Sundaic land masses*

	Borneo	Peninsular Malaysia	Sumatra mainland
Endemic mammals	39	8	–
Montane	11	2	–
Lowland	28	6	–
Endemic birds	41	3	18
Montane	23	3	13
Lowland	18	0	5

showing distinct altitudinal zonation for this genus to occur on Kinabalu.

The insectivorous pitcher plants are one of many attractive and conspicuous groups of plants occurring in the mossy forests. Almost half (32) of the world's pitcher plants (*Nepenthes* spp.) are found in Borneo, with the greatest proportion of Bornean species being montane, and occurring in cloud forests. Records show that 75% of the Bornean species are endemic and of the 10 species of pitcher plants occurring in Peninsular Malaysia, five are considered endemic, while four species (*N. macfarlanei, N. ramispina, N. gracillima*, and *N. sanguinea*) colonize mountain habitats between 1200 m and 2000 m on wet slopes, ridges, and summits of mossy forests (Clarke, 2002). The pitcher plants of Mt. Kinabalu in Borneo include the world's largest (*Nepenthes rajah*), found only there and on neighboring Mt. Tambuyukon (Phillipps and Lamb, 1996). It is estimated that in Peninsular Malaysia alone, more than 3000 species of all seed plants are confined to montane forests, suggesting that about 30% of plant species in Peninsular Malaysia inhabit the montane forests (Perumal and Lo, 1998). Mt. Kinabalu, one of the richest biologic diversity centers in South-East Asia, boasts an estimated 4000 species of vascular plants; at least half of the known Bornean species of figs, orchids, rhododendrons, mosses, and ferns are represented there (Smith, 1970).

Fauna

VERTEBRATES

Key factors influencing the distribution of montane forest vertebrates are reasonably well understood (Medway and Wells, 1976; Medway, 1977; Cranbrook, 1988). Their distribution is linked to the area of available habitat, which is correlated with the maximum height of the mountains. Fewer species occur on distant, isolated peaks than in the Main Range (in Peninsular Malaysia) or the central spine of Borneo (the island biogeography species syndrome). Of about 240 species of mammals in Borneo, 18 are montane (Payne *et al.*, 1985). Most are rodents, including squirrels and flying squirrels (six species) and rats (four species). All 18 species occur on Mt. Kinabalu (4095 m), nine on Trus Madi (2597 m), and eleven on Mt. Murud (2438 m) and the Kelabit Highlands (slightly lower at 2422 m but with more extensive montane forest), and so on down to Mt. Penrissen (1326 m) in western Sarawak, which has two or three montane species (Medway, 1977). Although primates such as macaques and gibbons occur in montane forest, and occasionally venture above 1500 m.a.s.l., (Davis and Payne in MacKinnon *et al.*, 1996), they are scarce and typically not resident in UMCF. Rodents, tree shrews, and squirrels form the bulk of the community. Perhaps only 10 mammals in Peninsular Malaysia are strictly montane specialists (Medway, 1983) and perhaps only two of these are endemic (Table 9.2), with none truly restricted to cloud forests. Amongst mammals of the region there are more endemics in the lowlands than in the mountains, but amongst birds it is the reverse. In both these groups, however, the older and more spectacular endemics (e.g. monotypic genera such as the partridge *Haematortyx*, or in Sumatra the forest rabbit *Nesolagus*) tend to be montane.

Distribution of birds is well documented, and species totals are known for many mountains in Peninsular Malaysia (Medway and Wells, 1976). The montane birds of Borneo are less well understood. The Mt. Kinabalu avifauna is the most diverse, as it is for montane mammals, and the only one with a sub-alpine element (for example, the island thrush (*Turdus poliocephalus*) which also extends to Sumatra). About 23 of Borneo's 41 endemic birds are montane, most of them headquartered in the higher mountains of the north-east (especially Kinabalu), and decreasing in numbers toward the south and west, where only a few survive in refugia such as Mt. Pueh (Smythies, 1999). In addition, there is a faunal discontinuity around the Sabah–Sarawak border. There are several lowland/montane pairs of taxa that have been considered sub-species, but now are (or should probably be) considered full species, e.g. forktails (*Enicurus* spp.), leafbirds (*Chloropsis* spp.), and partridges (*Rhizothera* spp.) (Davison, 1999; Moyle *et al.*, 2005). Many mammals and birds typical of lowland forest extend up into the montane zone, so the total fauna includes more than just the montane species totals listed in Table 9.3. A few lowland species, and some of the specialists in tall montane forest, extend above 1200 m into upper montane and elfin cloud forest facies. However, the number of species strictly dependent on elfin cloud forest is very small. Some that do occur only there are at the lower limit of their altitudinal tolerance, for example *Niltava sumatrana*, *Minla strigula*, and *Prinia atrogularis*, which all have wider distributions in montane continental Asia (Medway and Wells, 1976). These, and perhaps *Pyrrhula nipalensis*, would be susceptible to local extinctions if any rise in global temperature raised the lowermost limits of vegetation zones even slightly (cf. Foster, this volume).

Table 9.3 *Species richness amongst birds of lowland and montane forest in the Sundaic region. Totals for montane forest omit those predominantly lowland specialists that extend upward into montane forest as well*

	Borneo	Peninsular Malaysia	Sumatra	Java
Lowland forest residents	244	245	238	125
Montane forest residents	64	75	85	61
Migrants wintering mainly in lowland forest	27	50	31	20
Migrants wintering mainly in montane forest	3	6	4	3

Modified after Wells (1985).

In Peninsular Malaysia there are three endemic birds, all montane (*Arborophila campbelli*, *Polyplectron inopinatum*, and *Myiophoneus robinsoni*) (Jeyarajasingam and Pearson, 1999). At least the first two venture up into cloud forest. At mid-montane level (around 900 m.a.s.l.) there are several bamboo specialists, which are not strictly part of the cloud forest fauna but may venture there.

Higher mountains generally show more forest zones, with larger areas of each vegetation type (including cloud forest), and therefore greater potential for species packing. Thus, Mt. Kinabalu has at least three specialists in the elfin forest facies, both in stunted forest over ultramafic rocks as well as in the sub-alpine zone (*Turdus poliocephalus*, *Chlorocharis emiliae*, and *Bradypterus accentor*), though they also occur in tall forest down to 1500 or 1600 m.a.s.l. But the lower peaks in Peninsular Malaysia have none (Medway and Wells, 1976; Smythies, 1999). Many features are still unknown. Amongst these is the degree of genetic differentiation between isolated montane populations, and in Borneo the genetic distinctiveness (hence age) of montane–lowland pairs of taxa. Few plant–animal interactions have been studied in detail. Mixed flocking behavior has been reasonably well studied in the lowlands, but not in the mountains. Seasonality of montane food sources, both fruits and insects, is still poorly known. Many other topics for research could be listed.

INVERTEBRATES

Butterflies and moths are extremely diverse in South-East Asia. There are just over 1000 species of butterflies in Peninsular Malaysia, fewer in Borneo, but not many of these are exclusively montane (Corbet and Pendlebury, 1992). Strong decreases in the number of butterfly species have been observed from LMF to UMCF on Mts. Kinabalu (Holloway, 1978) and Mulu (Holloway, 1984). Amongst moths, species totals are far greater, perhaps 10 times greater for the larger and medium-sized moths alone. Species richness appears to be highest at mid-mountain levels, especially amongst the huge array of micro-moths, and population densities of any one species can be very low (Barlow, 1982). Wind drift may be an important factor influencing populations and their discovery (cf. Brehm, this volume).

Mollusks are poorly studied, but a small montane community is known to exist, including large forms such as members of the Ariophantidae, and *Bertia brookei* of highland Borneo, as well as minute forms of Diplommatinidae that, in the lowlands, are typically confined to and abundant on limestone but in the mountains are highly dispersed at low density (Tweedie, 1961; Davison, 1991). Peculiarities amongst montane invertebrates have been very poorly investigated. It is known that in some mountains of the region (Peninsular Malaysia, Sumatra, Java) giant earthworms occur (Cranbrook, 1995). They have seldom been seen, presumably emerging in cloud forest only when weather conditions are suitable, or perhaps mainly at night. At least some of these earthworms (e.g. *Metaphire musica*) are colored deep blue (Horst, 1883; Cranbrook, 1995), and allegedly they can produce mechanical sound (Vorderman, 1882). At the top of their food chain are giant leeches (known from mountains of Sumatra and Borneo) that feed by sucking the body contents from these or smaller worms (G. Davison, personal observation). The altitudinal distribution of the main groups of soil macrofauna (ants, termites, etc.) has been studied on Mt. Mulu by Collins *et al.* (1984). Strong decreases were observed in total numbers of soil invertebrates with elevation, with earthworms and some beetles replacing termites as the chief detritivores in LMRF at 1130 m.a.s.l. just below the cloud condensation level. In the peaty soils of the UMCF above 1300 m.a.s.l., beetles became dominant in association with earthworms, fly larvae, and cockroaches. In the most exposed, stunted vegetation zones, several groups fell away entirely (Collins *et al.*, 1984).

Studies by Inger and Stuebing (1992) of the montane herpetofauna revealed that at least 30 species of frogs are specific to montane habitats on Kinabalu. On Mt. Kajang (1038 m.a.s.l.), on Tioman Island off the coast of Peninsular Malaysia, *Ansonia tiomanica*, a frog species is found in and around cloud forests.

HYDROMETEOROLOGY

Cloud water interception (*CWI*) studies in Malaysian cloud forests are rare. Bruijnzeel *et al.* (1993) reported that the catch of a simple Grünow-type fog collector (sheltered against direct rainfall) was about 9% of ordinary rainfall at the summit of Mt. Silam, an isolated coastal mountain in eastern Sabah, during six weeks in the dry season. About 10% of the collector's catch actually produced canopy drip, suggesting direct contributions by *CWI* to soil water reserves to be modest at best, at least in the dry season when mosses are relatively dry and capable of absorbing large amounts of water. However, the daily cloud cap greatly reduced incoming radiation levels and evaporation losses, leading to more or less permanently saturated soils even during extended periods of drought (Bruijnzeel *et al.*, 1993). Nakashizuka *et al.* (1991) compared climatic conditions and altitudinal zonation of forest communities in Selangor, Peninsular Malaysia, and suggested that the lowered global radiation values recorded on low-peaked mountains from 800 m upwards were a factor determining the occurrence of cloud forests at lower altitudes (see also Kitayama *et al.*, 1999).

Recent *CWI* measurements made at two elevations on Mt. Brinchang (2031 m.a.s.l., Peninsular Malaysia) using a Juvik-type louvered aluminum cylindrical fog screen over 16 months suggested minimal contributions by low cloud (1.5% of incident rainfall) to a lower montane forest at 1600 m.a.s.l. vs. 8.6% in stunted summit UMCF at 2030 m.a.s.l. Whilst crown drip (throughfall) percentages in the two forests were comparable at 62.1% and 63.8%, amounts of stemflow differed vastly (2.2% and 30.6%; Kumaran, 2008).

SOILS, NUTRIENT DYNAMICS AND FOREST PRODUCTIVITY

The soils of the montane areas are predominantly (but not always) derived from highly weathered parent materials and are distinctly acidic (pH 3.0–5.5), particularly where surface peat develops (Burnham, 1974; Tie *et al.*, 1979). In the UMCF zone, peat is virtually continuous, even on very steep slopes up to 25°. Peat accumulations may reach depths up to 85 cm on gentle slopes and ridges (Tie *et al.*, 1979). Soils tend to become shallower at higher elevations, reflecting both the less intensive weathering associated with lower temperatures and soil rejuvenation by various forms of mass wasting (Burnham, 1974; Baillie 1989). C/N ratios decrease with elevation due to (strong) reductions in nitrogen-mineralization rates under the near-saturated conditions prevailing in, especially, the UMCF soils (Kitayama, 1992; Kitayama *et al.*, 1998; cf. Benner *et al.* this volume; Roman *et al.*, this volume). Some of the more dominant soils series that have been mapped in the montane areas in Malaysia (Paramananthan, 2000) include the Ulu Kali, Brinchang, Tanah Rata, Gunung Padang, Pa Sia, Baiayo, Kaintano, Mulu (Tie *et al.*, 1979), Umor, and Bareo. Whilst the (less wet and peaty) soils of the LMRF may classify as Dystropepts (Soil Survey Staff, 1998) or Cambisols (FAO, 1990), these are replaced by Histic Humaquepts or Humaquods in the wetter LMCF zone, and (mostly) by Histosols in the UMCF zone. Depending on the permeability of the geologic substrate, slope gradient, and rainfall the profiles may show distinct mottling (indicative of temporary saturation) in the LMCF zone. The almost permanent saturation of the sub-soils in the UMCF zone is often reflected by the grayish colors. The Soil Survey Staff (1998) possibly groups the UMCF soils as typic Haplosaprists or lithic Tropofolists while the FAO–UNESCO most likely classifies them as terric Histosols (FAO 1990). Further information on cloud forest soils is provided in Roman *et al.* (this volume).

Changes in nutrient cycling and productivity with elevation in Malaysian cloud forests have been studied on Gunung Silam (Proctor *et al.*, 1988b; Bruijnzeel *et al.*, 1993) and, especially, on Mt. Kinabalu (Kitayama *et al.*, 1998, 2000; Kitayama and Iwamoto, 2001, Kitayama and Aiba, 2002; Takyu *et al.*, 2003). Mt. Silam is a low (890 m.a.s.l.), coastal mountain underlain by ultramafic rocks and showing a strong compression of vegetation zonation (Proctor *et al.*, 1988a), whereas Mt. Kinabalu is much higher (4095 m.a.s.l.), underlain by sedimentary rocks (mostly graywacke) and, above 3000 m.a.s.l., granite, with patches of ultramafic rock in between (Kitayama, 1992). Mossy LMCF occurs at elevations as low as *c.* 700 m.a.s.l. on Mt. Silam vs. around 1800 m.a.s.l. on Kinabalu. As such, temperatures on Silam are much higher than on Kinabalu. On both mountains, nitrogen availability and productivity (both in terms of amounts of litterfall and trunk increment rates) decreased with elevation, with the lowest rates generally observed in the wettest cloud forest zone. Nevertheless, Bruijnzeel *et al.* (1993) showed that amounts of nutrients arriving at the forest floor of the stunted cloud forest on Mt. Silam in a readily soluble form greatly exceeded amounts taken up by the vegetation. They hypothesized that this might reflect the higher concentrations of possibly toxic polyphenols in the litter percolate of the cloud forest (cf. Benner *et al.*, this volume). On Kinabalu, the vegetation on sedimentary rock was capable of maintaining (estimated) stand-level net carbon assimilation rates at higher elevations (lower temperatures) by increasing foliar concentrations of nitrogen and phosphorus. In the case of forest on ultramafic rocks this was prevented by the phosphorus deficiency of the substrate (Kityama and Aiba, 2002; cf. Benner *et al.*, this volume). Observations on root biomass and root dynamics are needed on both mountains to obtain a more complete picture (cf. Kityama and Aiba, 2002; Hertel and Leuschner, this volume). Comparable observations in Sarawak are limited to the preliminary study of Proctor *et al.* (1983) measuring litterfall at various elevations on Gunung Mulu.

THREATS

Cloud forests in Malaysia are threatened by a range of economic activities that occur on-site as well as in adjacent lowlands. These include: construction of roads and other infrastructure, timber and mineral extraction, telecommunication facilities, tourist development (including golf courses), hydropower production, illegal plant collection, excessive and inappropriate tourism, and the encroachment of temperate agriculture. Some examples are township development in the Genting Highlands, highland agriculture (temperate vegetables and small fruits) and road construction in and around Cameron Highlands (Chan, 2006) and Mt. Kinabalu (Kitayama, 1995).

Such developments have caused major local changes even if they occupy small areas. For example, for the creation of tourism infrastructure at Genting Highlands, almost all local UMCF was removed. In the Cameron Highlands, although all forest in Reserves is theoretically protected by the Forestry Department, newspaper reports have repeatedly alleged the issuing of land titles by other government agencies. The Cameron Highlands Wildlife Sanctuary was degazetted in 1962, but this did not become widely known for more than 30 years (Davison, 1996). Because upper montane habitats tend to be aligned along the tops of mountain ranges and ridges, road construction along ridges can cause significant ecosystem damage even if removing only a small percentage of forest (cf. Scatena, 1995). Side-tipping of surplus earth from road cuttings along slopes causes damage to downslope vegetation, and there are many reports of landslides and mud slides during and long after road construction in the mountains (e.g. Jabatan Kerja Raya, 1995).

CONCLUSION

Although relatively small in area in absolute terms, cloud forests in the mountains of Malaysia play a disproportionately important role as habitats for flora and fauna (including many endemics), and in the provision of ecosystem services, not least of all a dependable supply of high-quality water. However, threats to cloud forests are on the increase, suggesting a need for concerted efforts in further documenting, studying, and protecting this priority ecosystem.

REFERENCES

Aiba, S., and K. Kitayama (1999). Structure, composition and species diversity in an altitude-substrate matrix of rain forest tree communities on Mount Kinabalu, Borneo. *Plant Ecology* **140**: 139–157.

Aldrich, M., C. Billington, M. Edwards, and R. Laidlaw (1997). *A Global Directory of Tropical Montane Cloud Forests.* Cambridge, UK: UNEP–WCMC.

Argent, G., A. Lamb, A. Phillipps, and S. Collenette (1988). *Rhododendrons of Sabah*, Sabah Parks Publication No. 8. Kota Kinabalu, Malaysia: Sabah Parks Trustees.

Ashton, P. S. (1995). Biogeography and ecology. In *Tree Flora of Sabah and Sarawak*, Vol. 1, eds. E., Soepadmo and K. M. Wong, pp. XLIII–LI. Kepong, Malaysia: FRIM, Sabah Forestry Department, and Sarawak Forestry Department.

Baillie, I. C. (1989). Soil characteristics and mineral nutrition of tropical wooded ecosystems. In *Mineral Nutrients in Tropical Forest and Savanna Ecosystems*, ed. J. Proctor, pp. 15–26. Oxford, UK: Blackwell Scientifics.

Barlow, H. S. (1982). *An Introduction to the Moths of South-East Asia.* Kuala Lumpur, Malaysia: Malayan Nature Society.

Bruijnzeel, L. A. (2001). Hydrology of tropical montane cloud forests: a reassessment. *Land Use and Water Resources Research* **1**: 1–18.

Bruijnzeel, L. A. (2005). Tropical montane cloud forests: a unique ecosystem. In *Forests, Water and People in the Humid Tropics*, eds. M. Bonell and L. A. Bruijnzeel, pp. 463–482. Cambridge, UK: Cambridge University Press.

Bruijnzeel L. A., and L. S. Hamilton (2000). *Decision Time for Cloud Forests*, IHP Humid Tropics Program Series No. 13. Paris: UNESCO, Amsterdam: IUCN-NL, and Gland, Switzerland: WWF.

Bruijnzeel, L. A., M. J. Waterloo, J. Proctor, A. T. Kuiters, and B. Kotterink (1993). Hydrological observations in montane rain forests on Gunung Silam, Sabah, Malaysia, with special reference to the '*Massenerhebung*' effect. *Journal of Ecology* **81**: 145–167.

Bubb, P., I. May, L. Miles, and J. Sayer (2004). *Cloud Forest Agenda.* Cambridge, UK: UNEP-WCMC.

Burgess, P. F. (1969). Ecological factors in hill and mountain forests of the States of Malaya. *Malayan Nature Journal* **22**: 119–128.

Burnham, C. P. (1974). Altitudinal changes in soils on granite in Malaysia. *International Congress on Soil Science* **10**: 290–296.

Chan, N. W. (ed.) (2006). *Cameron Highlands: Issues and Challenges in Sustainable Development.* Penang, Malaysia: School of Humanities, Universiti Sains Malaysia.

Clarke, C. (2002). *A Guide to the Pitcher Plants of Peninsular Malaysia.* Kota Kinabalu, Malaysia: Natural History Publications (Borneo).

Cockburn, P. F. (1978). The flora. In *Kinabalu, Summit of Borneo*, eds. D. M. Luping, Chin Wen, and E. R. Dindsley, pp. 179–190. Kota Kinabalu, Malaysia: The Sabah Society.

Collins, N. M., J. M. Anderson, and H. W. Vallack (1984). Studies on the soil invertebrates of lowland and montane rain forests in the Gunung Mulu National Park. *Sarawak Museum Journal* **30**(51): 19–33.

Corbet, A. S., and H. M. Pendlebury (1992). *The Butterflies of the Malay Peninsula*, 4th edn revised by J. N. Eliot. Kuala Lumpur, Malaysia: Malayan Nature Society.

Cranbrook, Lord (1988). *Key Environments: Malaysia.* New York: Pergamon Press.

Cranbrook, Lord (1995). A giant earthworm and other notes on the wildlife of Gunung Lawit, Terengganu. *Malaysian Naturalist* **49**: 22–25.

Davison, G. W. H. (1991). Terrestrial molluscs of Peninsular Malaysia. In *The State of Nature Conservation in Malaysia*, ed. R. Kiew, pp. 101–104. Kuala Lumpur, Malaysia: Malayan Nature Society.

Davison, G. W. H. (1996). *Land Use Planning for the Highlands: Protected Area in the Malaysian Mountains*, Report produced under Project MYS 333/95. Petaling Jaya, Malaysia: World Wide Fund for Nature Malaysia.

Davison, G. W. H. (1999). Notes on the taxonomy of some Bornean birds. *Sarawak Museum Journal* **75**: 289–299.

Economic Planning Unit (1993). *Malaysian National Conservation Strategy: Towards Sustainable Development*, Vol. 3, *Critical areas.* Kuala Lumpur, Malaysia: Economic Planning Unit.

FAO (1990). *Soil Map of the World*, revised legend, reprinted with corrections. World Soil Resources Report No. 60. Rome: FAO.

Frahm, J. P., and S. R. Gradstein, (1991). An attitudinal zonation of tropical rain forests using tryophytes. *Journal of Biogeography* **18**: 669–676.

Frahm, J. P., W. Frey, H. Kurschner, and M. Menzel (1990). *Mosses and Liverworts of Mt. Kinabalu*, Sabah Parks Publication No. 12. Kota Kinabalu, Malaysia: Sabah Parks Trustees.

Holloway, J. D. (1970). Biogeographical analysis of a transect sample of the moth fauna of Mount Kinabalu, Sabah, using numerical methods. *Biology Journal of the Linnean Society* **2**: 259–286.

Holloway, J. D. (1978). Butterflies and moths. In *Kinabalu, Summit of Borneo*, eds. D. M. Luping, Chin Wen, and E. R. Dindsley, pp. 255–265. Kota Kinabalu, Malaysia: Sabah Society.

Holloway, J. D. (1984). Notes on the butterflies of Gunung Mulu National Park. *Sarawak Museum Journal* **30**(51): 89–131.

Horst, R. (1883). New species of the genus *Megascolex* Templeton (*Perichaeta* Schmarda) in the collections of the Leyden Museum. *Notes Leyden Museum* **5**: 182–196.

Inger, R. F., and R. B. Stuebing (1992). The montane amphibian fauna of northwestern Borneo. *Malayan Nature Journal* **446**: 41–51.

Jabatan Kerja Raya. (1995). *Technical Report on the Investigation into the Debris Flow on the Slip Road to Genting Highlands.* Kuala Lumpur, Malaysia: Technical Investigation Committee, Ministry of Works.

Kitayama, K. (1992). An altitudinal transect study of the vegetation on Mount Kinabalu, Borneo. *Vegetatio* **102**: 149–171.

Kitayama, K. (1995). Biophysical conditions of the montane cloud forests of Mount Kinabalu, Sabah, Malaysia. In *Tropical Montane Cloud Forests*, eds. L. S. Hamilton, J. O. Juvik, and F. N. Scatena, pp. 183–197. New York: Springer-Verlag.

Kitayama, K., and S. I. Aiba (2002). Ecosystem structure and productivity of tropical rain forests along altitudinal gradients with contrasting soil phosphorus pools on Mount Kinabalu, Borneo. *Journal of Ecology* **90**: 37–51.

Kitayama, K., and K. Iwamoto (2001). Patterns of natural ^{15}N abundance in the leaf-to-soil continuum of tropical rain forests differing in N availability on Mount Kinabalu, Borneo. *Plant and Soil* **229**: 203–212.

Kitayama, K., S. Aiba, N. Majalap-Lee, and M. Ohsawa (1998). Soil nitrogen mineralization rates of rainforests in a matrix of elevations and geological substrates on Mount Kinabalu, Borneo. *Ecological Research* **13**: 301–312.

Kitayama, K., M. Lakim, and M. Zaini Wahab (1999). Climate profile of Mount Kinabalu during late 1995 – early 1998, with special reference to the 1998 drought. *Sabah Parks Nature Journal* **2**: 85–100.

Kitayama, K., N. Majalap-Lee, and S. Aiba (2000). Soil phosphorus fractionation and phosphorus-use efficiencies of tropical rainforests along altitudinal gradients of Mount Kinabalu, Borneo. *Oecologia* **123**: 342–349.

Kumaran, S. (2008). Hydrometeorology of tropical montane rain forests of Gunung Brinchang, Pahang Darul Makmur, Malaysia. Ph.D. thesis, Universiti Putra Malaysia, Serdang, Malaysia.

Jeyarajasingam, A., and A. Pearson (1999). *A Field Guide to the Birds of West Malaysia and Singapore.* Oxford, UK: Oxford University Press.

MacKinnon, K., G. Hatta, H. Halim, and A. Mangalik (eds.) (1996). *The Ecology of Indonesia Series*, Vol. 3, *The Ecology of Kalimantan.* Singapore: Periplus Editions.

Martin, P. J. (1977). *The Altitudinal Zonation of Forests along the West Ridge of Gunong Mulu.* Kuching, Malaysia: Sarawak Forest Department.

Medway, Lord (1977). *Mammals of Borneo: Field Keys and Annotated Checklist.* Kuala Lumpur, Malaysia: Malaysian Branch of the Royal Asiatic Society.

Medway, Lord (1983). *The Wild Mammals of Malaya (Peninsular Malaysia) and Singapore*, 3rd edn. Kuala Lumpur, Malaysia: Oxford University Press.

Medway, Lord, and D. R. Wells (1976). *The Birds of the Malay Peninsula*, Vol. 5, *Conclusion, and Survey of Every Species.* Kuala Lumpur, Malaysia: University of Malaya Press.

Moyle, R. G. M. Schilthuizen, M. A. Rahman, and F. H. Sheldon (2005). Molecular phylogenetic analysis of the white-crowned forktail *Enicurus leschenaulti* in Borneo. *Journal of Avian Biology* **36**: 96–101.

Nakashizuka, T., Z. Yusop, and A. R. Nik (1991). Altitudinal zonation of forest communities in Selangor, Peninsular Malaysia. *Journal of Tropical Forest Science* **4**: 233–244.

Paramananthan, S. (2000). *Soils of Malaysia: Their Characteristics and Identification*, Vol. 1, incorporating 100 standard description sheets. Kuala Lumpur, Malaysia: Academy of Sciences Malaysia and Param Agricultural Soil Surveys.

Payne, J., C. M. Francis, and K. Phillipps (1985). *Field Guide to the Mammals of Borneo.* Kota Kinabalu, Malaysia: Sabah Society, and Kuala Lumpur: WWF Malaysia.

Perumal, B., and F. S. J. Lo (1998). *Recorded Mountain Trees of Peninsular Malaysia: A Checklist*, Report produced under Project MYS 276/93 and MYS 339/96. Petaling Jaya, Malaysia: WWF Malaysia.

Phillipps, A., and A. Lamb (1996). *Pitcher-Plants of Borneo.* Kota Kinabalu, Malaysia: Natural History Publications (Borneo) in assocation with Royal Botanic Gardens, Kew and Malaysian Nature Society.

Proctor, J., J. M. Anderson, and H. W. Vallack (1983). Comparative studies on forests, soils, and litterfall at four altitudes on Gunung Mulu, Sarawak. *The Malaysian Forester* **46**: 60–73.

Proctor, J., Y. F. Lee, A. M. Langley, W. R. C. Munro, and T. Nelson (1988a). Ecological studies on Gunung Silam, a small ultrabasic mountain in Sabah, Malaysia. I. Environment, forest structure, and floristics. *Journal of Ecology* **76**: 320–340.

Proctor, J., C. Phillips, G. K. Duff, A. Heaney, and F. M. Robertson (1988b). Ecological studies on Gunung Silam, a small ultrabasic mountain in Sabah, Malaysia. II. Some forest processes. *Journal of Ecology* **77**: 317–331.

Ridley, H. N. (1914). On a collection of plants from Gunong Mengkuang Lebah, Selangor. *Journal of Federated Malay States Museums* **5**: 28–50.

Ridley, H. N. (1915). The botany of Gunong Tahan, Pahang. *Journal of Federated Malay States Museums* **6**: 127–202.

Scatena F. N. (1995). The management of Luquillo Cloud Forest ecosystems: irreversible decisions in a non-substitutable ecosystem. In *Tropical Montane Cloud Forests*, eds. L. S. Hamilton, J. O. Juvik, and F. N. Scatena, pp. 296–308. New York: Springer-Verlag.

Smith, J. M. B. (1970). Herbaceous plant communities in the summit zone of Mount Kinabalu. *Malayan Nature Journal* **24**: 16–29.

Smythies, B. E. (1999). *The Birds of Borneo*, 3rd edn, ed. G. W. H. Davison. Kota Kinabalu, Malaysia: Natural History Publications (Borneo).

Soepadmo, E. (1971). Plants and vegetations along the path from Tahan to Gunung Tahan. *Malayan Nature Journal* **24**: 118–124.

Soil Survey Staff (1998). *Keys to Soil Taxonomy*, 8th edn. Washington, DC: U.S. Department of Agriculture.

Stone, B. J. (1981). The summit flora of Gunung Ulu Kali (Pahang, Malaysia). *Federation Museum Journal* **26**(1), New Series, 1–157.

Symington, C. F. (1943). Foresters' manual of dipterocarps. *Malayan Forest Records* **16**. (Reprinted with plates and a historical introduction, University of Malaya Press, Kuala Lumpur, 1974.)

Takyu, M., S. I. Aiba, and K. Kitayama (2003). Changes in biomass, productivity and decomposition along topographical gradients under different geological conditions in tropical lower montane forests on Mount Kinabalu, Borneo. *Oecologia* **134**: 397–404.

Thang, H. C., and N. A. Chappell (2005). Minimising the hydrological impact of forest harvesting in Malaysia's rainforests. In *Forests, Water and People in the Humid Tropics*, eds. M. Bonell and L. A. Bruijnzeel, pp. 852–865, Cambridge, UK: Cambridge University Press.

Tie, Y. L., I. C. Baille, C. M. S. Phang, and C. P. Lim (1979). *Soils of Gunung Mulu National Park.* Kuching, Sarawak, Malaysia: Department of Agriculture, Sarawak.

Tweedie, M. W. F. (1961). On certain Mollusca of the Malayan limestone hills. *Bulletin of Raffles Museum* **26**: 49–65.

Van Steenis, C. G. G. J. (1964). Plant geography of the mountain flora of Mount Kinabalu. *Proceedings of the Royal Society of London Series B* **161**: 7–38.

Van Steenis, C. G. G. J. (1972). *The Mountain Flora of Java.* Leiden, the Netherlands: E. J. Brill.

Vorderman, A. G. (1882). Bijdrage tot de kennis van den Sondarie-worm. *Natuurkundig Tijdschrift voor Nederlandsch-Indië* **41**: 111–117.

Wells, D. R. (1985). The forest avifauna of western Malesia and its conservation. In *Conservation of Tropical Forest Birds*, eds. A. W. Diamond and T. E. Lovejoy, pp. 213–232. Cambridge, UK: ICBP.

Whitmore, T. C. (1972). The Gunong Benom Expedition. II. An outline description of the forest zones on northeast Gunong Benom. *Bulletin of the British Museum Natural History D* **23**: 11–15.

Whitmore, T. C. (1984). *Tropical Rain Forests of the Far East*, 2nd edn. Oxford, UK: Clarendon Press.

Whitmore, T. C., and C. P. Burnham (1969). The altitudinal sequence of forests and soils on granite near Kuala Lumpur. *Malayan Nature Journal* **22**: 99–118.

10 Montane cloud forests on remote islands of Oceania: the example of French Polynesia (South Pacific Ocean)

J.-Y. Meyer

Délégation à la Recherche, Gouvernement de la Polynésie française, Papeete, Tahiti, French Polynesia

ABSTRACT

Small, isolated patches of tropical montane cloud forest (TMCF) are found in many remote islands of the Pacific region (Oceania). French Polynesia comprises 37 high volcanic islands and islets, all located at more than 5000 km from the nearest continents. TMCFs are found on 12 of them (Huahine, Moorea, Raiatea, Tahaa, and Tahiti in the Society Islands, Fatu Hiva, Hiva Oa, Nuku Hiva, Tahuata, Ua Huka, and Ua Pou in the Marquesas Islands, and Rapa in the Austral Islands), with a total area of up to 8000 ha. Their current individual extent ranges from less than 20 ha (Huahine, Rapa, Tahaa) to *c.* 1000 ha (Hiva Oa, Nuku Hiva) and more than 5000 ha (Tahiti), the other islands having less than 100–200 ha each. TMCFs are located between 300–400 m.a.s.l. and up to 1600–1800 m.a.s.l. (Tahiti), but are more often found above 800–900 m.a.s.l. They are generally located on the upper slopes of valleys, on high-elevation plateaux, and in gullies and ridges below the summits, with annual rainfall ranging between 3000 and 8500 mm. TMCFs are floristically the most diverse of all plant communities in French Polynesia, with the highest endemism and number of biological types. Between 60% (Moorea, Tahiti, and Rapa) and more than 70% (Raiatea, Hiva Oa, Ua Pou, and Ua Huka) of the endemic vascular plant species are found in the TMCFs, and between 25% (Moorea and Rapa) and 50% (Hiva Oa, Nuku Hiva, and Ua Pou) of these endemics are restricted to these habitats. The TMCFs of the different archipelagos of French Polynesia share many common genera of flowering plants with other Polynesian island groups (e.g. Cook Islands, Samoa, Hawaiian Islands). Current threats include road construction, hydro-electricity development, grazing or trampling by feral ungulates (pigs, goats), and invasion by alien plant species, especially the tree *Miconia calvescens* in Tahiti and Moorea. Habitat conservation and invasive pest management is urgently needed to save the TMCFs in French Polynesia which still remain unprotected.

INTRODUCTION

Tropical montane cloud forests (TMCFs) in the Pacific region are normally found in small and isolated patches on rugged upland ridges and peaks of high volcanic oceanic islands (Merlin and Juvik, 1995; Mueller-Dombois and Fosberg, 1998). These "islands within islands" of restricted TMCFs occur over a diverse altitudinal range in response to a combination of atmospheric and topographic variables. In general, the cloud forest belt occurs at higher elevations on larger islands and reaches it lowest altitudinal expression on islands near the equator where the dual effects of precipitous mountains in close proximity to the sea, and the extremely humid tropical air (producing a very low lifting condensation level) combine to create TMCFs at elevations as low as 450–600 m.a.s.l. (e.g. Kosrae in the Federated States of Micronesia, Rarotonga in the Cook Islands, and some islands in Samoa and Fiji; Merlin and Juvik, 1995; Raynor, 1995; Watling and Gillison, 1995; Whistler, 1995). The numerous islands of French Polynesia, located in the South Pacific Ocean at more than 5000 km from the nearest continents (Figure 10.1) and scattered over an ocean surface the size of Europe present striking examples of remote small islands with TMCF. This chapter describes the current location and extent, the ecological and botanic characteristics, and the main past and current threats to the TMCFs in French Polynesia. It is based on published literature, personal communications of taxonomists, plant ecologists, and field botanists having worked in these islands, but also and mainly on the author's extensive field surveys and observations conducted in the last 10 years in all the high volcanic islands of French Polynesia.

Tropical Montane Cloud Forests: Science for Conservation and Management, eds. L. A. Bruijnzeel, F. N. Scatena, and L. S. Hamilton. Published by Cambridge University Press.

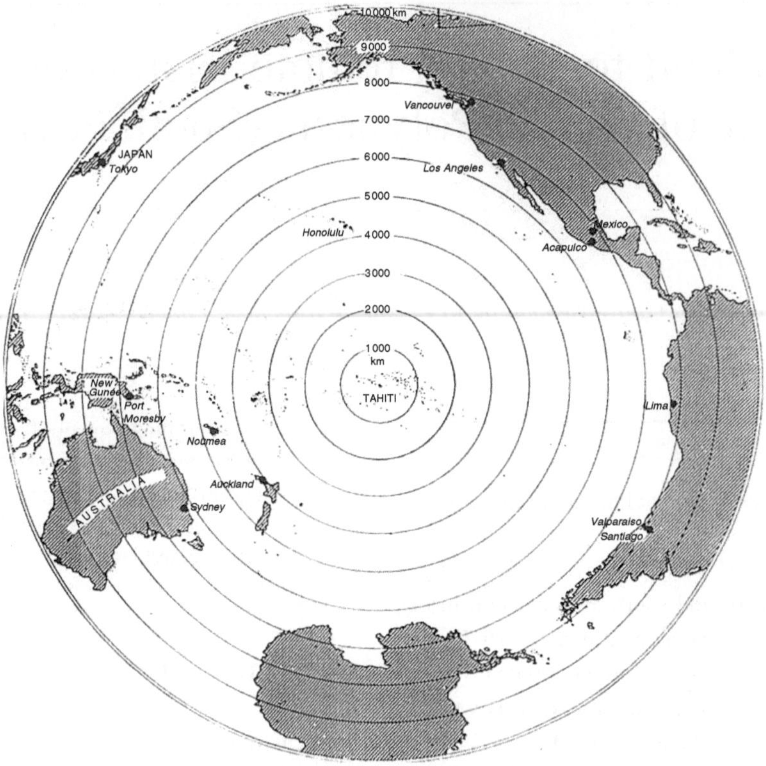

Figure 10.1. Location of the remote islands of French Polynesia in the South Pacific Ocean (centered on the main island of Tahiti).

LOCATION AND EXTENT OF TMCF IN FRENCH POLYNESIA

French Polynesia, a French overseas country located in the South Pacific Ocean between 7° and 35° S and 134° and 154° N, comprises a total of 121 tropical oceanic islands, including 84 atolls, raised atolls, and coral islets, and 37 high volcanic islands and rocky islets. These high islands often present a rugged topography and represent *c.* 80% of the terrestrial area of French Polynesia (2796 km^2 of a total area of 3521 km^2). They are found in four different archipelagos, namely the Austral, the Gambier, the Marquesas, and the Society Islands. TMCFs are found in 12 of these 37 high volcanic islands (Moorea, Huahine, Raiatea, Tahaa, and Tahiti in the Society Islands, Fatu Hiva, Hiva Oa, Nuku Hiva, Tahuata, Ua Huka, and Ua Pou in the Marquesas Islands, and Rapa in the Austral Islands). All of these 12 islands have an area of more than 40 km^2 and a summit reaching or extending above 600 m elevation (Figure 10.2).

The TMCFs represent an approximate total area of less than 8000 ha, i.e. less than 3% of the total high island terrestrial surface of French Polynesia (Table 10.1). The largest area of TMCF (>5000 ha) is found on Tahiti, the largest and highest island in the Society Islands (1045 km^2, highest summit at 2241 m.a.s.l.). Relatively large areas of TMCF (*c.* 1000 ha) are found on Nuku Hiva (340 km^2, 1227 m maximum elevation) and Hivao Oa (314 km^2, 1276 m elevation), the largest islands in the Marquesas. Smaller areas (<100–200 ha) are found in the high-elevation zones of Moorea and Raiatea in the Society Islands, and around the summits of Fatu Hiva, Tahuata, Ua Huka, and Ua Pou in the Marquesas Islands. Tiny patches of TMCF (<20 ha each) are found on the summits of the islands of Tahaa (590 m.a.s.l.) and Huahine (669 m.a.s.l.) in the Society Islands, and on the small island of Rapa (40 km^2, 650 m.a.s.l.) in the Austral Islands. Small areas of montane wet or rain forest exists on some of the highest summits of Bora Bora (located between 600 and 660 m.a.s.l.) in the Society Islands, and on Tubuai, Rurutu, and Raivavae (between 380 and 440 m.a.s.l.) in the Austral Islands but these are not considered true TMCFs (see descriptions below).

ECOLOGICAL DESCRIPTION OF TMCF IN FRENCH POLYNESIA

TMCFs in French Polynesia are characterized by high precipitation (>3000–4000 mm $year^{-1}$, up to 8500 mm $year^{-1}$ in the center of Tahiti; Pasturel, 1993), constant high humidity and cool conditions (e.g. 12 °C and 50% humidity on Mt. Aorai at 2000 m elevation in Tahiti, 19 °C and 75% on Toovii plateau at 900 m.a.s.l. in Nuku Hiva, and 22 °C and 85% on Temehani plateau at 600 m.a.s.l. in Raiatea; Papy, 1951), and a diurnal cloud cap. They have been described and classified for Tahiti by previous

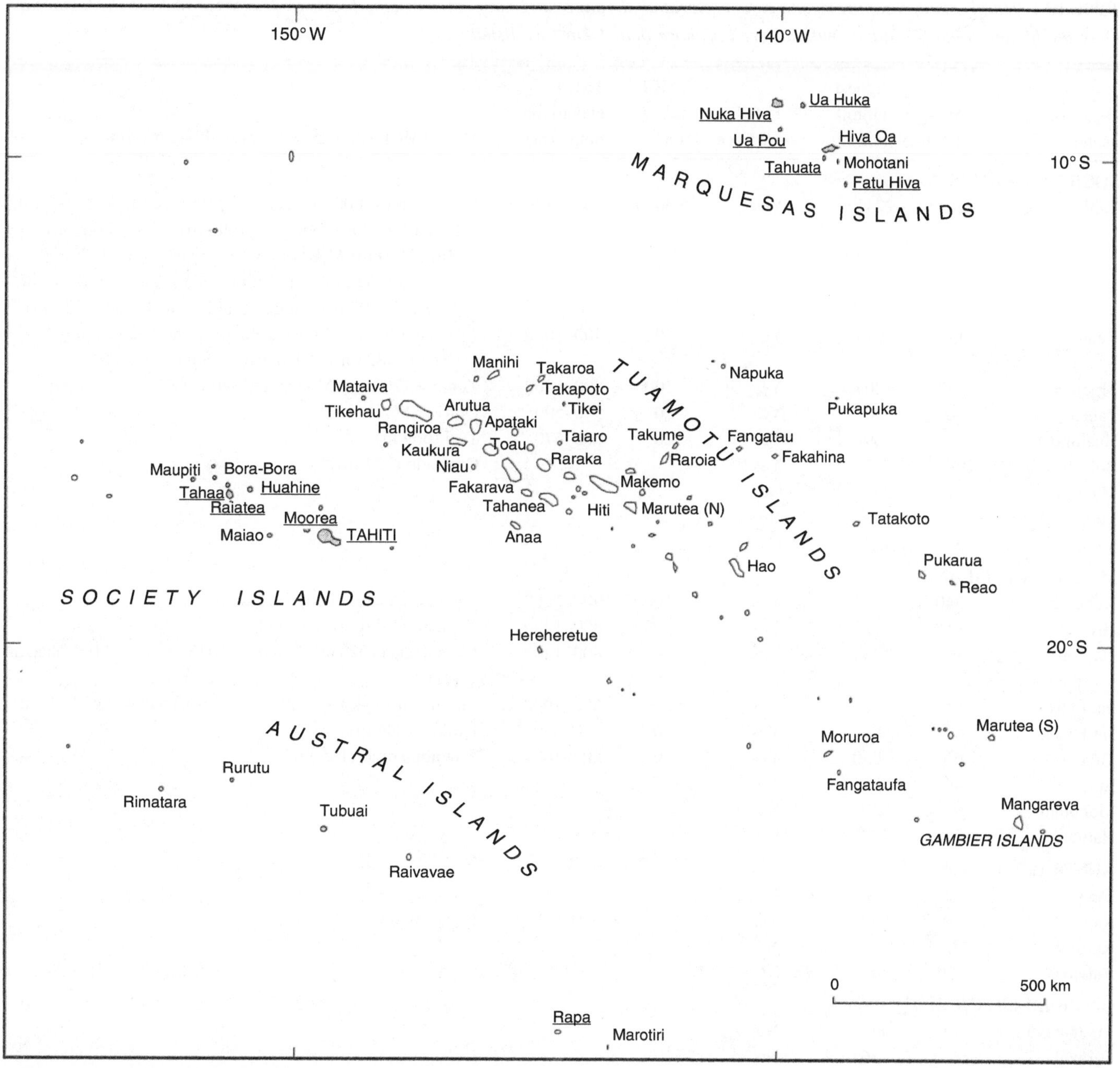

Figure 10.2. The different islands of French Polynesia (South Pacific Ocean), and location of the 12 islands with tropical montane cloud forest (underlined).

authors as (i) "valley head and swale formations" (Setchell, 1922), (ii) "*étage des hautes vallées et des secteurs très humides*" ("upper valley and very humid areas"; Papy, 1948), (iii) "*étage des forêts complexes des stations très humides et peu ensoleillées ou forêt hygrophyte*" ("complex forests of very humid places with little sunshine, or hygrophytic forest"; Papy, 1954), (iv) "*étage de la forêt pluvieuse*" ("rain forest"; Pétard, 1960), (v) "humid high-elevation forests of upper valleys and wet upper slopes and ridges" (Whittier, 1976), (vi) "*forêt de nuages*" (cloud forest; Florence, 1986) or (vii) "mossy forest or cloud forests" (Fosberg, 1992; Mueller-Dombois and Fosberg, 1998), "*groupement hygrophile de haute altitude*" ("high-elevation hygrophilous formation"; Florence, 1987), (viii) "*forêts ombrophiles d'altitude*" or "*groupements ombrophiles de haute altitude*" ("high-elevation ombrophilous forest"; Florence, 1993), (ix) "*végétation humide de moyenne et haute montagnes*" ("wet vegetation of middle and high mountains"; Florence, 2003), and in the Marquesas, as (x) "high altitude, cold climate vegetation" (Brown, 1931), (xi) "rain forest of the cloud zone" (Adamson, 1936), (xii) "upper-elevation wet forest and

Table 10.1 *Island location, approximate extent, elevational range, and main sites of cloud forest occurrence in French Polynesia. Only the higher volcanic islands with an area of more than 1 km² are listed*

High volcanic islands	Area (km²)	Highest summit (m.a.s.l.)	TMCF presence	TMCF area (ha)	TMCF elevational range (m)	Main TMCF site names (with highest elevation)
SOCIETY ISLANDS						
Tahiti	1045	2241	Yes	>5000	300–1800	Viriviriterai (1000 m), Terepo (1100 m), Marau (1493 m), Iviroa (1638 m), Ivirairai (1696 m), Aorai (2066 m), Pito Hiti (2110 m) Mauru (1361 m), Aramaoro (1530 m), Teamaa (1532 m), Urufau (1493 m), Tetufera (1799 m), Teatara (1197 m), Mairenui (1306 m) Ronui (1332 m)
Raiatea	171	1017	Yes	<200	400–1000	Temehani Rahi (770 m), Temehani Ute Ute (780 m), Oropiro (824 m), Toomaru-Tefatua (1017 m)
Moorea	142	1207	Yes	<100	800–1200	Tohiea (1207 m), Mouaputa (820 m)
Tahaa	90	590	Yes	<20	550–590	Ohiri (590 m)
Huahine	75	669	Yes	<20	600–670	Turi (669 m)
Bora Bora	29	727	Yes?[a]			Otemanu (727 m)?[a]
Maupiti	11	380	No			
Maiao	8	159	No			
Mehetia	2	435	No			
MARQUESAS ISLANDS						
Nuku Hiva	340	1224	Yes	<1000	900–1200	Tekao (1224 m), Toovii, Ooumu
Hiva Oa	315	1276	Yes	<1000	800–1200	Temetiu (1276 m), Feani (1015 m), Ootua (889 m)
Ua Pou	105	1203	Yes	<200	800–1200	Matahenua (1028 m), Pouakei (1035 m), Oave-Teavahaakiti (1203 m)
Fatu Hiva	85	1125	Yes	<200	650–1000	Mounanui (954 m), Tekou (1114 m),Touaouoho (1125 m)
Ua Huka	83	884	Yes	<50	750–880	Hitikau (884 m)
Tahuata	61	1050	Yes	<100	800–1000	Haaoiputeomo (1050 m)
Eiao	40	577	No			
Mohotani	9	520	No			
Hatutaa	7	420	No			
AUSTRAL ISLANDS						
Tubuai	45	422	No			
Rapa	40	650	Yes	<20	550–650	Perau (650 m)
Rurutu	32	389	No			
Raivavae	20	437	No			
GAMBIER ISLANDS						
Mangareva	14	441	No			
Taravai	4	256	No			
Akamaru	2	246	No			
FRENCH POLYNESIA				<8000	300–1800	

[a] The highest summit of Mt. Otemanu (727 m) remains unexplored because of the impossibility of access to the summit, even by helicopter. The other summits of Bora Bora (Mt. Pahia at 660 m, Mt. Hue at 620 m) have montane wet forests but not TMCF (J.-Y. Meyer, personal observation).

Source: Data based on personal field observations by the author (1995–2005) and on personal communications by other botanists.

high-elevation cloud forest" (Florence and Lorence, 1997), and (as noted earlier for Tahiti) as "*formations ombrophiles*" ("ombrophilous formations"; Florence, 1993). These forest are generally low-statured (between 5 and 10 m tall) and have a one-layer canopy with a multi-storied understory, and abundant epiphytic ferns, orchids, mosses, lichens, and terrestrial ferns.

TMCFs are usually found above 800–900 m elevation on the large islands (Moorea, Nuku Hiva, Hiva Oa, and the leeward side of Tahiti Nui; Florence, 1986) and up to 1600–1800 m elevation on Tahiti Nui (Florence, 1986), but they can be found at lower elevations on smaller islands (e.g. 450 m.a.s.l. in Raiatea; Papy, 1951, 1954; at 550 m.a.s.l. in Rapa; J.-Y. Meyer, personal observation); at 600 m.a.s.l. in Huahine; J.-F. Butaud, personal communication; at 650 m.a.s.l. in Fatu Hiva; J.-Y Meyer, personal observation; and at 750 m.a.s.l. on Ua Huka; J.-Y. Meyer, personal observation). Low-elevation cloud forests also occur on those sides of the largest island of Tahiti that are exposed to the dominant south-easterly trade winds (300–400 m.a.s.l. on the eastern and south-western sides of Tahiti Nui and on the peninsula of Tahiti Iti; Papy, 1954; Florence, 1983).

TMCFs are generally located on the upper slopes of valleys, on high-elevation plateaux, on precipitous volcanic necks (e.g. on the Marquesas Islands, Rapa), in the upper parts of ravines, and on ridges located below windswept summits that are covered by low, wet shrubland and sub-alpine communities. The latter have been referred to in various ways, including: (i) "*étage des hauts sommets*" (Papy, 1948, 1951–54) or "high-summits vegetation" (Whittier,1976), (ii) "*maquis des hauts sommets*" ("high summits scrub"; Florence, 1986), (iii) "*maquis sommitaux*" ("summit scrubs"; Florence, 1987, 1993), (iv) "mossy scrub" (Fosberg, 1992), and (v) "summit wet shrubland" (Florence and Lorence, 1997).

In Tahiti, the most important sites for TMCF are found near Mts. Marau (1493 m), Iviroa (1638 m), Ivirairai (1696 m), Aorai (2066 m), and Pito Hiti (2110 m) on the leeward side, and on Mts. Mauru (1361 m, see Meyer and Florence (1999) for a full description), Urufau (1493 m), Aramaoro (1530 m), Teamaa (1532 m), and Tetufera (1799 m) on the windward side. They are further found near Mts. Teatara (1197m), Mairenui (1306 m), and Ronui (1332 m) on the peninsula of Tahiti Iti (Table 10.1).

BOTANIC DESCRIPTION OF TMCF IN FRENCH POLYNESIA

TMCF is the vegetation type in French Polynesia that shows the highest number of biologic types. All TMCFs are characterized by an abundance of epiphytic ferns (especially Hymenophyllaceae or filmy ferns, e.g. *Hymenophyllum* and *Trichomanes,* and Grammitidaceae, e.g. *Grammitis* and *Calymnodon*), orchids (especially in the genera *Bulbophyllum, Calanthe, Corybas, Dendrobium, Habenaria, Liparis,* and *Phreatia*), mosses (e.g. the large golden tufts of *Spiridens* on tree fern trunks in Tahiti; Whittier, 1976), lichens, and terrestrial ferns (especially Aspleniaceae, Blechnaceae, Lycopodiaceae, and Polypodiaceae). Furthermore, small trees, shrubs, and herbs (mainly Rubiaceae, Gesneriaceae, and Urticaceae) grow in the understory.

Two broad types of TMCF can be recognized on the island of Tahiti, viz. (i) *Weinmannia parviflora* var. *parviflora – Alstonia costata* forest (Florence, 1983, 1986, 1993) with 7–10-m tall canopy, and (ii) *Ilex anomala – Streblus anthropophagorum* forest (Florence, 1993), with a canopy of 5–8 m and found in wetter habitats in gullies and on steep slopes. They are also characterized by the presence of the tree ferns *Cyathea* spp. and the native liana *Freycinetia impavida* (Papy, 1954), as well as by the endemic trees and shrubs *Ascarina polystachya, Astronidium* spp., *Coprosma taitensis, Glochidion* spp., *Metrosideros collina, Myrsine* spp., *Pipturus* spp., *Reynoldsia verrucosa,* the epiphytic orchids *Dendrobium* spp., *Phreatia tahitensis, Bulbophyllum tahitense, Liparis clypeolum,* and the terrestrial ferns *Marattia salicina, Diplazium harpeodes,* and *Blechnum* spp. Among the rare endemic flowering plants, the ground orchid *Calanthe tahitensis,* the small trees *Fitchia tahitensis, Fuchsia cyrtandroides, Melicope* spp., and *Sclerotheca* spp., the shrubs *Psychotria* spp., *Cyrtandra* spp., and *Scaevola tahitensis* and the sub-shrubs *Bidens* spp. and *Ophiorrhiza* spp. are found.

The summit shrublands are dominated by a *Metrosideros collina – Weinmannia parviflora* dwarf forest (2–3 m high), with *Vaccinium cereum, Styphelia* spp., *Astelia nadeaudii,* the epiphytic ferns *Elaphoglossum samoense* and *Selliguea feeoides,* and the ground fern *Paesia tahitensis.*

Some original sub-types of TMCF can be found on high-elevation plateaux on the island of Tahiti (at Viriviriterai between 800 and 1000 ma.s.l., at Terepo between 900 and 1100 m.a.s.l.) with a swamp forest dominated by *Metrosideros – Alstonia – Crossostylis biflora* and *Pandanus papenooense* and the rare endemic orchid *Corybas minutes.* Another example occurs on the island of Raiatea (at Temehani Rahi, Temehani Ute Ute between 400 and 780 m.a.s.l.; Papy, 1954; Meyer, 1996) in the form of a complex of scrub forest, scrub, and dwarf scrub and bogs (Fosberg, 1992; Mueller-Dombois and Fosberg, 1998) dominated by the island endemic *Weinmannia ovalifolia, Metrosideros collina* var. *temehaniensis, Cyclophyllum barbatum* var. *temehaniense, Pandanus temehaniensis, Psychotria* spp., *Bidens* spp. as well as the rare island endemics *Apetahia raiateensis, Geniostoma clavatum* (Meyer, 1996), and the recently discovered *Fitchia cuneata* (J.-Y. Meyer, personal observation 2005).

In the Marquesas, TMCFs are mainly found on the central mountains such as the Toovii–Tekao–Ooumu ridge on Nuku Hiva (see Meyer and Florence (1999) for a full description), the Temetiu–Feani ridge on Hiva Oa, the Oave–Teavahaakiti ridge on Ua Pou above 800 m elevation, and the Tekou–Touaouho ridge on Fatu Hiva above 650 m.a.s.l. They form a low, dense canopy, 3–5 m tall with *Crossostylis biflora, Weinmannia parviflora* var. *marquesana, Cheirodendron bastardianum, Ilex anomala, Metrosideros collina, Streblus anthropophagorum,* and *Freycinetia impavida.* Understory endemic trees and shrubs include *Alstonia marquesensis, Ascarina marquisensis, Claoxylon oomuense, Coprosma* spp., *Pipturus henryanus, Psychotria* spp., *Trimenia* spp., *Reynoldsia marchionicum, Geniostoma* spp.,

Table 10.2 *List of 55 genera of vascular flowering plants recorded at more than one cloud forest location in six Polynesian archipelagoes (Rapa in the Austral Islands, Rarotonga in the Cook Islands, the Hawaiian Islands, the Marquesas Islands, the Samoa Islands, and the Society Islands). The genera are listed according to their representation within any two to six of the selected archipelagoes*

Two archipelagos	Three archipelagos	Four archipelagos	Five archipelagos	Six archipelagos
Astronidium	*Alstonia*	*Ascarina*	*Astelia*	*Alyxia*
Cheirodendron	*Apetahia*	*Bidens*	*Cyrtandra*	*Coprosma*
Crossostylis	*Bulbophyllum*	*Claoxylon*	*Geniostoma*	*Freycinetia*
Cypholophus	*Calanthe*	*Dendrobium*	*Glochidion*	*Melicope*
Elatostema	*Carex*	*Dianella*	*Liparis*	*Metrosideros*
Oparanthus	*Eurya*	*Habenaria*	*Myrsine*	*Vaccinium*
Peristylus	*Fagraea*	*Meryta*	*Peperomia*	
Phreatia	*Fitchia*	*Omalanthus*	*Scaevola*	
Sclerotheca	*Gahnia*	*Pipturus*	*Weinmannia*	
	Hedyotis	*Pittosporum*		
	Ilex	*Psychotria*		
	Ixora	*Reynoldsia*		
	Korthalsella	*Streblus*		
	Machaerina	*Styphelia*		
	Macropiper			
	Pandanus			
	Santalum			

Source: Data for Samoa are based on Whistler (1995, 2002) and Cribbs and Whistler (1996), for Rarotonga on Wilder (1931) and MacCormack and Künzle (1995), and for Hawai'i on Wagner *et al.* (1990).

Melicope spp., and *Scaevola* spp. as well as the endemic herbs *Bidens* spp. and *Phyllanthus pacificus*. Associated with these rare endemic species are *Achyranthes marchionica*, *Apetahia* spp., *Hedyotis* spp., *Oparanthus* spp., and the ground orchids *Calanthe marquisensis* and *Habenaria marquesensis*. Tree ferns (*Cyathea* spp.) and the ground ferns *Polystichum marquesensis* and *Paesia rugulosa* are commonly found in the moist forest understory. In disturbed areas, the native climber *Freycinetia impavida* often forms dense thickets, especially on steep slopes, whereas the staghorn fern (*Dicranopteris linearis*) is dominant in more open areas. The wet summits of the highest peaks are covered by a shrubland dominated by *Metrosideros collina*, *Vaccinium cereum*, and *Styphelia tameiameiae* and the epiphytic ferns *Elaphoglossum* spp. and *Selliguea feeoides*.

On Rapa, TMCF is restricted to Mt. Perau between 550 m and 650 m elevation (J.-Y. Meyer, personal observation 2002). The native tree *Metrosideros collina,* the island endemic trees *Weinmannia rapensis* and *Corokia collenettei*, and the tree fern *Cyathea stokesii* are the dominant arborescent species. The island endemic trees and shrubs found in this small TMCF habitat include *Apetahia margaretae, Claoxylon collenettei, Coprosma* spp., *Eurya rapensis, Hedyotis rapensis, Meryta* spp., *Myrsine* spp., *Oparanthus coriaceus, Styphelia rapae*, and *Vaccinium rapae,* with the island endemic herbs *Astelia rapensis*, *Carex stokesii*, the very rare native epiphytic orchid *Liparis clypeolum*, the rare island endemic shrubs *Geniostoma rapense* and *Melicope margaretae*, and sub-shrubs *Haroldiella rapensis* and *Pilea occulta* found in wetter gullies. Small populations of *Pacifigeron rapensis*, one of the seven endemic genera found on Rapa, occur on ridges and cliffs. Many small endemic epiphytic ferns are found on mossy logs including *Elaphoglossum* spp., *Loxogramme parksii, Grammitis maireaui*, and *Calymnodon* spp.

FLORISTIC AFFINITIES BETWEEN ARCHIPELAGOS

Despite these high levels of endemism, especially at the species level, Pacific Island TMCFs share many common elements. Initial analysis of TMCF taxa at six island (or archipelago) locations in remote Oceania (namely Rarotonga in the Cook Islands, the Hawaiian Islands, Marquesas Islands, Samoa Islands, Society Islands, and Rapa in the Austral Islands) indicated 55 genera of vascular flowering plants recorded from more than one TMCF location. Six genera (10%) were found in all oceanic locations, and 46 genera (84%) in three or more of them (Table 10.2). This indicates a high taxonomic similarity between cloud forests throughout the Polynesian region.

Table 10.3 *Number of endemic plant taxa found in cloud forests of selected islands of French Polynesia*

ISLAND (archipelago)	Number of endemic genera found in TMCF/Total number of endemic genera in the island (name of the endemic genera)	Number of endemic plant species	Estimated number of endemic species found in TMCF (% of total)	Estimated number of endemic species restricted in TMCF (% of total)
Tahiti (Society)	2/3 (*Fitchia, Sclerotheca*)	224[a]	142[c] (63%)	85[c] (38%)
Moorea (Society)	2/2 (*Fitchia, Sclerotheca*)	76[a]	43[c] (56%)	19[c] (25%)
Raiatea (Society)	2/2 (*Apetahia, Fitchia*)	118[a]	86[c] (73%)	47[c] (40%)
Nuku Hiva (Marquesas)	3/5 (*Apetahia, Oparanthus, Plakothira*)	84[b]	54[c] (64%)	43[c] (51%)
Hiva Oa (Marquesas)	2/2 (*Apetahia, Oparanthus*)	82[b]	59[c] (72%)	42[c] (51%)
Ua Pou (Marquesas)	1/1 (*Apetahia*)	49[b]	35[c] (71%)	25[c] (51%)
Ua Huka (Marquesas)	1/1 (*Apetahia* extinct?)	42[b]	31[c] (71%)	15[c] (36%)
Rapa (Austral)	5/7 (*Apetahia, Fitchia, Haroldiella, Oparanthus, Pacifigeron*)	89[a]	51[c] (57%)	24[c] (27%)

[a] Data based on the "Nadeaud botanical data–base" (Florence 1993; J. Florence, personal communications).
[b] Data based on the "Vascular Flora of the Marquesas Islands Project" (http://ravenel.si.edu/botany/pacificislandbiodiversity/marquesasflora/).
[c] Unpublished data based on field observations by the author (1995–2005).

The completeness of this data-set could be improved by adding ferns and fern allies, such as mosses and bryophytes. Oceanic TMCFs are indeed likely to show high generic similarity for these taxa, because of their suitability for long-distance dispersal. Thus, these TMCFs contain numerous floristic elements with temperate biogeographic affinities (e.g. *Astelia, Coprosma, Vaccinium*), distinguishing them from the low-elevation rain forest flora that is dominated by tropical Malesian and South-East Asian species. Some endemic taxa confined to TMCFs in French Polynesia have peculiar floristic affinities: for example, the Tahitian taxa *Oreobolus* and *Fuchsia* are related to Hawaiian and South American taxa, respectively; the Marquesan species *Trimenia* and *Cheirodendron* with Fiji and Hawai'i, respectively; whereas an affinity with New Zealand is seen with *Hebe* and *Corokia* from Rapa (Meyer, 2004).

TMCF AS PLANT DIVERSITY HOT SPOTS AND REFUGIA FOR ENDEMIC ANIMALS

TMCFs are floristically the most diverse of all plant communities in French Polynesia, with the highest rate of endemism and the highest number of biologic types. Six of the 10 French Polynesian or south-eastern Polynesian endemic genera are restricted to TMCF (*Apetahia* in the Society and the Marquesas Islands, *Fitchia* in the Society Islands and Rapa, *Oparanthus* in the Marquesas Islands and Rapa, *Haroldiella* and *Pacifigeron* in Rapa, and *Sclerotheca* in the Society Islands). It is estimated that between 60% (e.g. in Moorea and Tahiti, Society Islands, and Rapa in the Austral Islands) and over 70% (e.g. in Raiatea, Society Islands, and Hiva Oa, Ua Pou, and Ua Huka, Marquesas Islands) of the endemic vascular plant species are found in the TMCFs, and that between 25% (Moorea, Rapa) and more than 50% (Hiva Oa, Nuku Hiva, Ua Pou) of these endemics are confined to these habitats (Table 10.3).

TMCFs are also the home of several endangered endemic animals such as tree snails (Partulidae) of the genera *Partula* spp. in Tahiti and *Samoana* spp. in the Marquesas (J.-Y. Meyer, personal observations), nesting sea birds (*Pterodroma* spp.) or land birds (*Ducula galeata* on Nuku Hiva, *Pomarea whitneyi* on Fatu Hiva) (J.-Y. Meyer, personal observations), aquatic insects (e.g. endemic damselflies belonging to the genus *Bedfordia*; R. Englund, personal communication) and terrestrial arthropods (e.g. spiders; R. Gillespie, personal communication; weevils belonging to the genus *Rhyncogonus*; E. Claridge, personal communication). In the Marquesas, a large number of endemic insects appear to be dependent on two most common host-plants found in the TMCF, *Metrosideros collina* and *Weinmannia* spp. (Mumford and Adamson, 1934). Other host-plants of importance are *Crossostylis biflora, Ilex anomala, Vaccinium cereum*, and the endemic Marquesan species of *Cyrtandra* and *Apetahia* (Adamson, 1939), all of them restricted to TMCF.

PAST AND CURRENT THREATS TO TMCF IN FRENCH POLYNESIA

Cloud forests in French Polynesia are located on high mountains in rugged terrain, thus being relatively undisturbed and inaccessible for humans. Natural disturbances such as cyclones and landslides are uncommon in French Polynesia and do not pose any major threat. Prior to European arrival, human disturbances were restricted to only a few activities (cultivation, fire?) as montane forests were considered *tapu* (i.e. forbidden) and the residence of gods (Bachinon, 1990). In deep gullies and valley heads, there are remnant groves of mountain plantain (*Musa troglodytarum*, syn. *M. fehi*), called *fe'i* in the Society Islands and *huetu* in the Marquesas. Plantain was (and still is) one of the main staple foods in French Polynesia along with breadfruit, taro, and sweet potato. According to MacDaniels (1947), "the best plants are found at altitudes of about 600 to 800 meters, near the heads of the valleys [. . .] and will grow at heights of 1200 meters." Pétard (1960) wrote that in Tahiti "*les fei abondent surtout entre 800 et 1200 m, ils atteignent des dimensions énormes*" whereas Oliver (1974) stated that "there are many varieties, that once grew plentifully in large thickets high up the mountain slopes at 600 to 900 meters." The increasing scarcity of mountain plantains at lower elevations may be due to the devastation wrought by feral cattle and pigs, introduced plants (MacDaniels, 1947), and the ravages of the introduced banana-borer *Cosmopolites sordidus* (Mumford and Adamson, 1934).

Current threats to montane forests in the islands include road construction (Mt. Marau, Toovii ridge on Nuku Hiva, Mt. Ootua on Hiva Oa), hydro-electricity development (Mt. Mauru), grazing or trampling by feral ungulates (pigs on Hiva Oa, goats on Rapa), and invasion by alien plant species (mainly the trees *Miconia calvescens, Psidium cattleianum,* and *Spathodea campanulata,* and the grass *Melinis minutiflora* on Tahiti and Moorea, *Syzygium cumini* on Hiva Oa, *Syzygium jambos* on Nuku Hiva, *Chrysobalanus icaco* on Raiatea and Fatu Hiva, and the shrub *Rubus rosifolius* in the Society Islands, Hiva Oa and Rapa; Meyer, 2004). The small invasive tree *Miconia calvescens* forms dense monospecific forest stands up to 1300 m elevation and poses a direct threat to the endemic species of TMCFs in Tahiti (Meyer and Florence, 1996) and the other high volcanic islands of French Polynesia.

Habitat conservation and invasive pest management (exclusion of feral ungulates, control of invasive plants) is urgently needed to protect the last of the TMCFs in French Polynesia. The Mt. Marau site (900–1400 m.a.s.l.) on Tahiti was first proposed as a natural park in 1976, and the Toovii–Tekao ridge was proposed to be protected as a natural preserve about 25 years ago, but these proposals have never been implemented. Both sites are currently threatened by new road construction activities. The only protected area that encompasses a small patch of cloud forest in French Polynesia remains the Vaikivi park and natural reserve on the small island of Ua Huka in the Marquesas. It was declared a protected area in 1995 as it is an important watershed area and offers refuge for rare endemic plant and animal species (Meyer, 1996).

CONCLUSIONS AND PERSPECTIVE

TMCFs represent less than 8000 ha or less than 3% of the terrestrial surface of the 37 high volcanic islands of French Polynesia. They are found in 12 of these islands and occur at different elevations, mostly between 300 and 1800 m.a.s.l. The majority of remaining TMCFs is concentrated on the large islands of Tahiti (Society Islands), and on Nuku Hiva and Hiva Oa (Marquesas Islands). TMCFs in French Polynesia contain a larger percentage of endemic species than other vegetation types (between 50% and 70% for most islands), and six of the 11 French Polynesian or south-eastern Polynesian endemic genera are restricted to these areas. They also act as refugia for numerous endemic birds and invertebrates. For a long time, they have remained virtually pristine. In the words of Mueller-Dombois and Fosberg (1998): "The cloud forest of most of the high mountains of Tahiti is probably as nearly an undisturbed example of Polynesian vegetation as still exists." However, because of their limited extent, TMCFs in French Polynesia are highly vulnerable to disturbance (notably by exotic invader plants) and extinction of endemic species. Research on the botanic (inventories, description of new species) and ecological (phyto-sociology, plant succession, resilience to invasion) aspects of TMCFs in French Polynesia has been limited and thus these are still poorly understood. TMCFs in Polynesian archipelagos and in the Pacific region exhibit high generic similarity but there is substantial scope for comparative studies (including molecular phylogenies) in these unusual forests.

During the last 10–15 years, field surveys and ecological assessments have been conducted in the TMCFs of all the high volcanic islands in French Polynesia, and proposals have been put forward for the protection of some of the most critical areas in the Society Islands (e.g. Mt. Marau, Mt. Mauru, Faufiru plateaux, Temehani plateaux, Viriviriterai and Terepo plateaux), the Marquesas Islands (Mt. Ootua, Temetiu–Feani ridge, Toovii–Tekao ridge), and Rapa (Mt. Perau) in the Austral Islands. It is hoped that this chapter will contribute to a better knowledge and conservation of the TMCFs in French Polynesia which represent botanic and ecological "jewels" in the remote tropical islands of Oceania.

ACKNOWLEDGEMENTS

I thank the following taxonomists, botanists, and field botanists for their personal communications on the TMCF flora in French Polynesia, and for their company during often memorable field trips and surveys: Dr. Jacques Florence (IRD, National Museum of Natural History, Paris), Jean-François Butaud (Service du Développement Rural, Tahiti), Dr. Tim Motley (New York Botanical Garden), Dr. Dave Lorence, Steve Perlman, and Ken Wood (National Tropical Botanical Garden, Hawai'i), Ravahere Taputuarai (Délégation à la Recherche, Tahiti), and Fred Jacq (Raiatea). *Mauruuru roa* and *Koutai* to my local guides, colleagues, and friends Henry Jaÿ, Maxime Chan, Elie Poroi, Christian Malinovsky, and Kiki Bordet (Te Rau Atiati, Tahiti), Emile Brotherson, Jean-Pierre Malet, and Romy Tavaearii (SDR, Raiatea), Teiki Richmond and late Rogatien Peterano (SDR, Hiva Oa), Benjamin Pukoki (SDR, Rapa), and Cerdan Faraire (Rapa). Many thanks to Matiu Prebble (Australian National University, Canberra) for revising the English of an early version of this chapter and for his comments, and to Mark Merlin, John Delay, and Mahuri Waite (University of Hawai'i at Manoa) for fruitful discussions and for sharing data on TMCF in Oceania.

REFERENCES

Adamson, A.M. (1936). *Marquesan Insects Environment*, Bernice P. Bishop Museum Bulletin No. 139. Honolulu, HI: Pacific Entomological Survey.

Adamson, A.M. (1939). *Review of the Fauna of the Marquesas Islands and Discussion of its Origin*, Bernice P. Bishop Museum Bulletin No. 159. Honolulu, HI: Pacific Entomological Survey.

Bachinon, P. (1990). *Tahiti entre Mythe et Réalités: Essai d'histoire géographique*. Paris: Editions du C.H.T.S.

Brown, F.B.H. (1931). *Flora of Southeastern Polynesia*, Vol. 1, *Monocotyledons*, Bernice P. Bishop Museum Bulletin No. 84. Honolulu, HI: Pacific Entomological Survey.

Cribbs, P., and Whistler, W.A. (1996). *Orchids of Samoa*. Kew, UK: Royal Botanic Gardens.

Florence, J. (1983). De Cuzent à nos jours: esquisse du paysage botanique actuel. In *Archipel de Tahiti: Recherches sur les Productions Végétales*, ed. G. Cuzent, pp. (167–172). Papeete, French Polynesia: Editions Haere Po No Tahiti.

Florence, J. (1986). Flore et végétation. In *Encyclopédie de la Polynésie française*, Vol. 2, *Flore et Faune Terrestres*, pp. 25–40. Papeete, French Polynesia: Christian Gleizal/Multipresse.

Florence, J. (1987). Endémisme et évolution de la flore de la Polynésie française. *Bulletin de la Société Zoologique de la France* **112**: 369–380.

Florence, J. (1993). La végétation de quelques îles de la Polynésie française. In *Atlas de la Polynésie française*, ed. J.F. Dupon, Planches 54–55. Paris: ORSTOM.

Florence, J. (2003). La Polynésie française: environnement phytogéographique, l'homme et ses plantes. In *Archéologie en Océanie Insulaire: Peuplement, Société et Paysages*, ed. C. Orliac, pp. 130–149. Paris: Editions Artcom.

Florence, J., and D.H. Lorence (1997). Introduction to the flora and vegetation of the Marquesas Islands. *Allertonia* **7**: 226–237.

Fosberg, F.R. (1992).Vegetation of the Society Islands. *Pacific Science* **46**: 232–250.

MacCormack, G. and J. Künzle (1995). *Rarotonga's Mountain Tracks and Plants. A Field Guide*. Rarotonga, Cook Islands: Cook Islands Natural Heritage Project.

MacDaniels, L.H. (1947). *A Study of the Fe'i Banana and its Distribution with Reference to Polynesian Migrations*. Honolulu, HI: Bernice P. Bishop Museum.

Merlin, M.D., and J.O. Juvik (1995). Montane cloud forest in the tropical Pacific: some aspects of their floristics, biogeography, ecology, and conservation. In *Tropical Montane Cloud Forests*, eds. L.S. Hamilton, J.O. Juvik, and F.N. Scatena, pp. 234–253. New York: Springer-Verlag.

Meyer, J.-Y. (1996). *Espèces et Espaces Menacés de la Société et des Marquises*, Contribution à la Biodiversité de Polynésie française No.1–5. Tahiti, French Polynesia: Délégation à l'Environnement/Délégation à la Recherche.

Meyer, J.-Y. (2004). Threat of invasive alien plants on native flora and forest vegetation of Eastern Polynesia. *Pacific Science* **58**: 357–375.

Meyer, J.-Y., and J. Florence (1996). Tahiti's native flora endangered by the invasion of *Miconia calvescens* DC. (Melastomataceae). *Journal of Biogeography* **23**: 775–781.

Meyer, J.-Y., and J. Florence (1999). Mont Mauru (Tahiti, Society Islands) and Toovii ridges (Nuku Hiva, Marquesas Islands), two natural areas of ecological interest in French Polynesia, proposed as PABITRA sites. In *19th Pacific Science Congress*, University of New South Wales, July 4–9, 1999, Sydney. Available at www.botany.hawaii.edu/ pabitra/Sites/sitedescriptions/french.htm.

Mueller-Dombois, D., and F.R. Fosberg (1998). *Vegetation of the Tropical Pacific Islands*. New York: Springer-Verlag.

Mumford, E.P., and A.M. Adamson (1934). Entomological researches in the Marquesas Islands. *Mémoires de la Societé Biogéographique* **4**: 219–234.

Oliver, D.L. (1974). *Ancient Tahitian Society*, Vol. **1**, *Ethnography*. Canberra, ACT: Australian National University Press.

Papy, H.R. (1948). Aperçu sommaire des étages de végétation à Tahiti. *Bulletin de la Société d'Histoire Naturelle de Toulouse* **83**: 217–222.

Papy, H.R. (1951). Sur un problème écologique posé par la végétation des hauts-sommets de Tahiti et du plateau Temehani à Raiatea (Océanie française). *Bulletin de la Société pour les Etudes Océaniennes* **84**(8): 183–191.

Papy, H.R. (1954). Tahiti et les Îles voisines, Part II, *La Végétation des Îles de la Société et de Makatea*. Toulouse, France: Laboratoire Forestier.

Pasturel, J. (1993). *La climatologie des îles. In Atlas de la Polynésie française*, Planche 43. ed. J.-F. Dupon, Paris: ORSTOM.

Pétard, P.H. (1960). Quelques plantes utiles de la Polynésie française: Ptéridophytes et Monocotylédones. Ph.D. thesis, University of Marseilles, Marseilles, France.

Raynor, B. (1995). Montane cloud forests in Micronesia: status and future management. In *Tropical Montane Cloud Forests*, eds. L.S. Hamilton, J.O. Juvik, and F.N. Scatena, pp. 274–283. New York: Springer-Verlag.

Wagner, W.L., D.R. Herbst, and S.H. Sohmer (1990). *Manual of the Flowering Plants of Hawai'i*. Honolulu, HI: University of Hawaii Press, Bishop Museum Press.

Watling, D. and A.N. Gillison (1995). Endangered species in low elevation cloud forest on Gau island, Fiji. In *Tropical Montane Cloud Forests*, eds. L.S. Hamilton, J.O. Juvik, and F.N. Scatena, pp. (333–342). New York: Springer-Verlag.

Whistler, W.A. (1995). The cloud forest of Samoa. In *Tropical Montane Cloud Forests*, eds. L.S. Hamilton, J.O. Juvik, and F.N. Scatena, pp. 353–362. New York: Springer-Verlag.

Whistler, W.A. (2002). *The Samoan Rainforest: A Guide to the Vegetation of the Samoan Archipelago*. Apia, Western Samoa: Isle Botanica.

Whittier, H.O. (1976). *Mosses of the Society Islands*. Gainesville, FL: University Press of Florida.

Wilder, G. (1931). *Flora of Rarotonga*. Honolulu, HI: Bernice P. Bishop Museum.

11 Tropical lowland cloud forest: a neglected forest type

S. R. Gradstein
University of Göttingen, Göttingen, Germany

A. Obregon
University of Marburg, Marburg, Germany

C. Gehrig
University of Göttingen, Göttingen, Germany

J. Bendix
University of Marburg, Marburg, Germany

ABSTRACT

Lowland cloud forests are well known from temperate regions but in the Tropics their occurrence has been overlooked. "Tropical lowland cloud forest" (TLCF) occurs in valleys in hilly coastal areas with high rainfall, leading to frequent occurrence of fog and dew, most likely due to nocturnal outgoing long-wave radiation related to drainage of cold air. TLCF resembles lowland rain forest but differs by its abundance of epiphytes, especially mosses and liverworts.

INTRODUCTION

Tropical montane cloud forests (TMCFs) occur within the cloud belts of montane and sub-montane regions at elevations that may differ between regions, but generally lie between 1200 and 3000 m.a.s.l. However, at island and coastal locations the level of cloud condensation tends to be much lower, with the lowest documented elevations being 300–500 m.a.s.l. (Hamilton *et al.*, 1995; Meyer, this volume). Characteristic of well-developed TMCF are the frequent incidence of fog and low clouds (Bruijnzeel, 2001; Jarvis and Mulligan, this volume) and the abundance and high species richness of epiphytes (mosses, liverworts, lichens, ferns, flowering plants; Frahm and Gradstein, 1991; Wolf, 1993). Indeed, it has been suggested that TMCF may have more species of epiphytes than any other forest type (Richards, 1996).

Inventories of epiphyte diversity in a number of moist lowland and montane forests in tropical America (Figure 11.1) have been carried out by the author and his students since the late 1970s (e.g. Van Reenen and Gradstein, 1983; Cornelissen and Ter Steege, 1989 ; Cornelissen and Gradstein, 1990; Montfoort and Ek, 1990; Wolf, 1993; Gradstein, 1995; Gradstein *et al.*, 2001; Holz *et al.*, 2002; Acebey *et al.*, 2003; Gradstein, 2003; Nöske *et al.*, 2003; Krömer and Gradstein, 2004; Holz and Gradstein, 2005; Gradstein, 2006; Nöske *et al.*, 2007; Krömer and Gradstein, this volume; cf. Schawe *et al.*, this volume). Special attention was paid in these studies to the non-vascular epiphytes (bryophytes, lichens), which had been neglected previously because of difficulties with species identification and the occurrence of many species in the higher and less accessible parts of the forest canopy. Because of their poikilohydric nature, epiphytic bryophytes and lichens are excellent ecological indicators, especially of atmospheric humidity and quality (Bates and Farmer, 1992; Gignac, 2001). Their abundance is strongly influenced by local climatic conditions, and the degree of "mossiness" is a useful parameter to characterize tropical rain forest types (Frahm and Gradstein, 1991).

These inventories revealed large differences between forest types in terms of species richness of epiphytic bryophytes. In most cases, species richness was higher in montane forests than in lowland forests, but in some lowland forests the number of epiphytic bryophyte species exceeded that of montane forest (Figure 11.2). For example, moist lowland forests at Saül (French Guiana) and Mabura Hill (Guyana) had up to 100–150% more bryophytic species than did moist sub-montane forest at about

Tropical Montane Cloud Forests: Science for Conservation and Management, eds. L. A. Bruijnzeel, F. N. Scatena, and L. S. Hamilton. Published by Cambridge University Press.

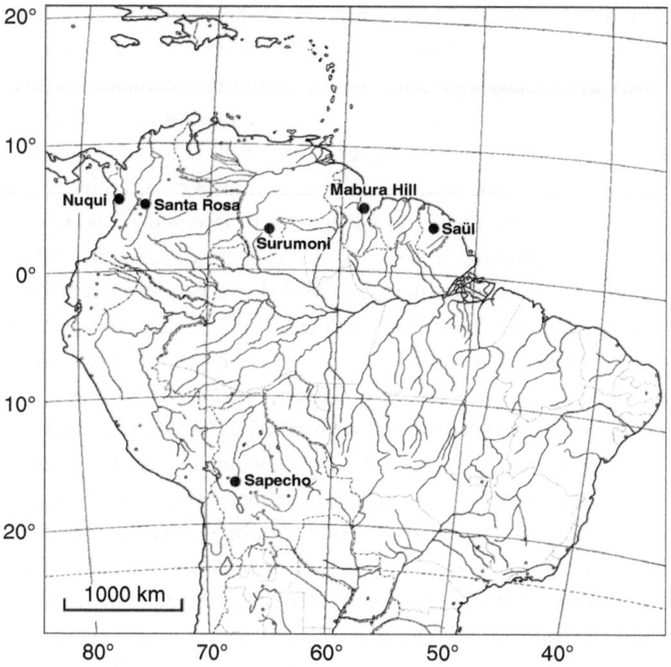

Figure 11.1. Map of tropical America showing location of study sites.

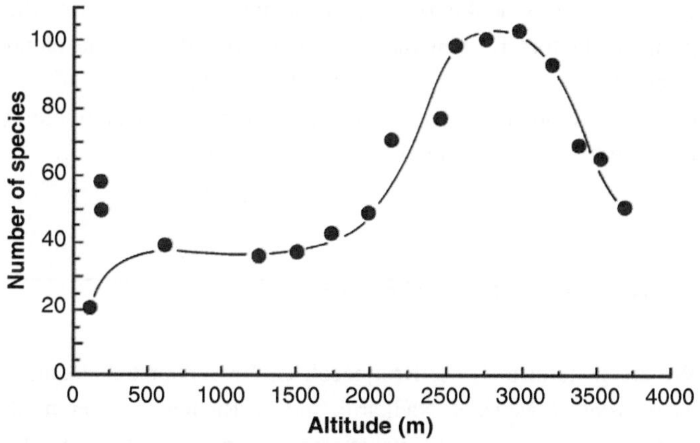

Figure 11.2. Species richness of epiphytic liverworts in moist evergreen forests of tropical America. Four trees were sampled at each location (Gradstein *et al.*, 1996, 2003). Data below 500 m.a.s.l. are from lowland forests at Surumoni, Venezuela (lowest number of species), Mabura Hill, Guyana (intermediate number of species), and Saül, French Guiana (highest number); data for 500–1000 m.a.s.l. elevation are from sub-montane forest at Sapecho, Bolivia; and above 1000 m.a.s.l. from montane forests at Santa Rosa, Colombia (based on Cornelissen and Gradstein, 1990; Montfoort and Ek, 1990; Wolf, 1993; Gradstein, 1995, 2006; Acebey *et al.*, 2002; S.R. Gradstein, unpublished data).

600 m.a.s.l. near Sapecho (Bolivia) or lower montane forest at about 1500 m near Santa Rosa (Colombia), and up to three times more species than found in moist Amazonian lowland forest at Surumoni (Venezuela) (Figure 11.2). At Surumoni, four sampled trees had only 20 species of liverworts and fewer than 10 species of moss, whereas at Mabura Hill and Saül the same number of sampling trees had at least 50 species of liverwort and about 30 of moss (Cornelissen and Ter Steege, 1989; Cornelissen and Gradstein, 1990; Montfoort and Ek, 1990; Gradstein, 1995; S.R. Gradstein, unpublished data; tree sampling methods followed Gradstein *et al.*, 1996, 2003). The greatest species richness was recorded at Saül, where four trees yielded about 100 species of bryophytes: 60 of liverworts and about 40 of mosses (Montfoort and Ek, 1990; Gradstein, 1995, 2006). Up to 50 bryophyte species occurred on a single tree. These figures are the highest ever recorded in tropical lowland rain forest and are similar to those recorded in lower montane cloud forest (LMCF) at about 2000 m.a.s.l. in Colombia (Figure 11.2; Wolf 1993; cf. Catchpole and Kirkpatrick (this volume) for equally diverse records of vascular epiphytes on a single tree in an LMCF at 2400 m.a.s.l. in eastern Perú).

The abundance of epiphytes and the degree of mossiness also varied considerably. Both in the lowland forests of Mabura Hill and Saül and in the montane forests in Colombia, the lower branches of the tree crowns were covered by dense mats of bryophytes, and vascular epiphytes were numerous. In the Amazonian lowland forest of Surumoni, however, epiphytes in the forest canopy were scarce and dense mats of bryophytes were virtually lacking. The marked differences in epiphyte diversity in the three investigated lowland forests were surprising, because elevation (100–200 m.a.s.l.) and annual precipitation (*c.* 2700 mm) at the respective sites were similar (Gradstein, 1995; Anhuf and Winkler, 1999). A possible explanation of the difference in epiphyte abundance and diversity relates to the high incidence of fog at the Mabura Hill and Saül sites, occurring typically during night and early morning (Montfoort and Ek, 1990; Gradstein, 2003, 2006).

To investigate fog occurrence in the valleys of French Guyana, meteorologic measurements were conducted at two sites: (i) the research station Saut Pararé (Nouragues Natural Reserve), where an automatic weather station including a visibility sensor was installed directly above the canopy (at 45 m above the ground) on a pylon of the "Canopy Operating Permanent Access System" (COPAS, 75 m.a.s.l.) in the valley of the Arataye river; and (ii) Saül, where climate stations were installed in two trees in the valley (245 m.a.s.l.) and on a ridge (330 m.a.s.l.). Preliminary results show that fog is a common phenomenon at night and in the early morning (Figure 11.3). During the first 90 days of operational use of the COPAS climate station, fog (visibility <1000 m) was registered on 88 days and visibility below 100 m (i.e. very dense fog) on 50 days. Fog occurrence showed a clear diurnal cycle: there is a significantly increased frequency after sunset (19:00 LST) and a steady increase throughout the night, with a maximum around sunrise (*c.* 90% of the days at 06:00 LST). Three hours after sunrise (09:00 LST), the fog is almost completely cleared by solar heating and during the rest of

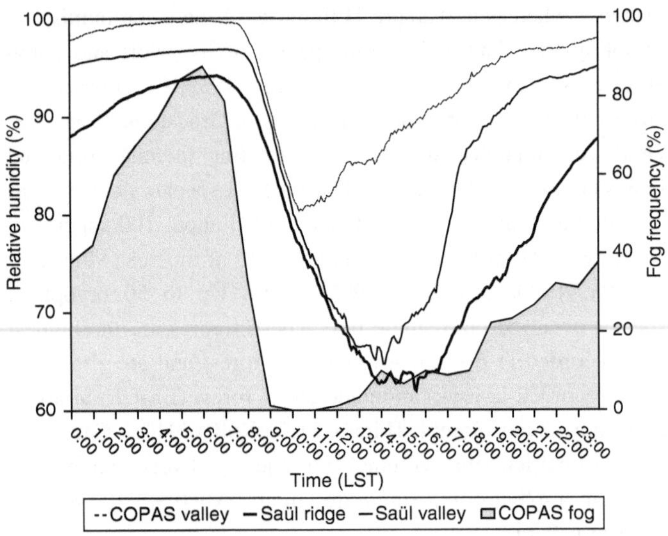

Figure 11.3. Diurnal course of fog (visibility <1 km) and average relative humidity at the COPAS valley site (canopy level) as well as average humidity for a valley and a ridge site at Saül, French Guiana.

the daytime, fog frequency is low (generally <10%). Recordings of relative humidity at COPAS are in close agreement with the visibility measurements and show high values close to saturation during hours of fog. An analysis of the mean diurnal cycle of relative humidity at Saül revealed markedly different patterns for the valley and the ridge (Figure 11.3). Whilst the valley site with its high abundance and diversity of epiphytic bryophytes was characterized by clearly enhanced relative humidity (especially at night), the ridge-top station generally showed much lower values. As such, it is to be expected that fog at Saül is mostly confined to the valleys and probably a rare phenomenon in more elevated parts of the terrain.

Fog development in these lowland valley forests is most probably due to long-wave radiation loss and cooling beyond dew point at night during times of negligible wind rather than to uplifting of air masses along mountain slopes (cf. Bruijnzeel *et al.*, 2005; Liu *et al.*, this volume). Fog formation may be enhanced by the following factors: undulating terrain with many small hills and creeks, saturation of air during the night and early morning because of low turbulence (very low wind speeds) or occurrence of heavy rain during the night, and waterlogging of valley-bottom soils (cf. Brouwer, 1996; Liu *et al.*, this volume). Because of the frequent occurrence of morning fog, the hills near Saül are called "*Monts la Fumée*" ("smoky mountains"). In contrast, fog is never observed at Surumoni where the local topography is relatively flat (Anhuf and Winkler, 1999).

These mossy tropical lowland forests with their high abundance of epiphytes and frequent fog may be called "tropical lowland cloud forest" (TLCF). Lowland cloud forests are well known from temperate regions, e.g. the coasts of California and northern Chile (Aravena *et al.*, 1989; Dawson, 1998), but in the humid tropics their occurrence has been largely overlooked. TLCFs are known from hilly coastal areas with high rainfall, where the already high atmospheric humidity is enhanced further due to the proximity to the sea. With increasing distance to the ocean, air tends to become drier (Bruijnzeel, 2001; cf. Jarvis and Mulligan, this volume). In addition to the two locations in the Guianas described here, TLCF has been observed in the Chocó along the Pacific coast of Colombia (Gradstein, 1995; Figure 11.1) and in northern Sulawesi (L. A. Bruijnzeel, personal communication). The TLCF in the Chocó is very mossy, with canopy branches covered by large mats of liverwort species such as *Bazzania hookeri, Herbertus divergens, Plagiochila* spp., and *Trichocolea tomentosa* (Gradstein, 1995). Elsewhere in tropical America these species exclusively occur in montane cloud forests. Occurrence of montane taxa in the Chocó is not restricted to liverworts, but has also been noted for higher (vascular) plants, including *Hedyosmum, Ilex, Podocarpus, Meliosma,* and *Brunellia* (Gentry, 1986). Occurrence of montane species of ferns in lowlands has been reported from very humid Amazonian lowland rain forest sites in eastern Ecuador and adjacent Perú (Kessler, 2001). It is suggested here that the high atmospheric humidity and precipitation, coupled with frequent incidence of fog, and the somewhat lower air temperatures along the Pacific coast relative to the Amazonian side of the Andes, constitute the principal factors responsible for the growth of montane cloud forest plant taxa at sea level in the Chocó. Further research is required to document other areas of tropical lowland cloud forest, both at inland and near-coastal locations.

ACKNOWLEDGEMENTS

We are grateful to L. A. Bruijnzeel, G. Gravenhorst, J. Jácomo, M. Kessler, S. Mori, N. Nadkarni, and M. Richter for references, helpful suggestions, and linguistic improvement of the text. The current project is funded by the German Research Foundation (DFG grants BE 1780/13–1 and GR 1588/12–1). We thank Philippe Gaucher (CNRS Guyane) for logistic help in French Guiana and maintenance of the COPAS Climate Station, and Dr. Rütger Rollenbeck and Sebastian Achilles for support during field work.

REFERENCES

Acebey, C., S.R. Gradstein, and T. Krömer (2003). Species richness and habitat diversification of bryophytes in submontane rain forest and fallows of Bolivia. *Journal of Tropical Ecology* **19**: 9–18.

Anhuf, D., and H. Winkler (1999). Geographical and ecological settings of the Surumoni–Crane-Project (Upper Orinoco, Estado Amazonas, Venezuela). *Anzeigen der Österreichischen Akademie der Wissenschaften, Mathematisch-Naturwissenschaftliche Klasse, Abteilung 1* **135**: 3–23.

Aravena, R., O. Suzuki, and A. Pollastri (1989). Coastal fog and its relation to ground-water in the IV region of northern Chile. *Chemical Geology* **79**: 83–91.

Bates, J.W., and A.M. Farmer (1992). *Bryophytes and Lichens in a Changing Environment*. Oxford, UK: Clarendon Press.

Brouwer, L.C. (1996). *Nutrient Cycling in Pristine and Logged Tropical Rain Forest, Guyana*. Georgetown, Guyana: Tropenbos.

Bruijnzeel, L.A. (2001). Hydrology of tropical montane cloud forests: a reassessment. *Land Use and Water Resources Research* **1**: 1–18.

Bruijnzeel, L.A., W. Eugster, and R. Burkard (2005). Fog as an input to the hydrological cycle. In *Encyclopaedia of Hydrological Sciences*, eds. M.G. Anderson and J.J. McDonnell, pp. 559–582. Chichester, UK: John Wiley.

Cornelissen, J.H., and S.R. Gradstein (1990). On the occurrence of bryophytes and macrolichens in different lowland rain forest types in Guyana. *Tropical Bryology* **3**: 29–35.

Cornelissen, J.H., and H. Ter Steege (1989). Distribution and ecology of epiphytic bryophytes and lichens in dry evergreen forest of Guyana. *Journal of Tropical Ecology* **5**: 131–150.

Dawson, T.E. (1998). Fog in the California redwood forest: ecosystem inputs and use by plants. *Oecologia* **117**: 476–485.

Frahm, J.-P., and S.R. Gradstein (1991). An altitudinal zonation of the tropical rain forest using bryophytes. *Journal of Biogeography* **18**: 669–678.

Gentry, A.W. (1986). Species richness and floristic composition of Chocó region plant communities. *Caldasia* **15**: 71–91.

Gignac, L.D. (2001). Bryophytes as indicators of climate change. *The Bryologist* **104**: 410–420.

Gradstein, S.R. (1995). Diversity of Hepaticae and Anthocerotae in montane forests of the tropical Andes. In *Biodiversity and Conservation of Neotropical Montane Forests*, eds. S.P. Churchill, H. Balslev, E. Forero, and J.L. Luteyn, pp. 321–334. New York: New York Botanical Garden.

Gradstein, S.R. (2003). Biodiversitätsforschung im tropischen Regenwald. In *Biodiversitätsforschung: Die Entschlüsselung der Artenvielfalt in Raum und Zeit*, eds. S.R. Gradstein, R. Willmann, and G. Zizka, pp. 95–110. Frankfurt, Germany: Senckenberg.

Gradstein, S.R. (2006). The lowland cloud forest of French Guiana: a liverwort hotspot. *Cryptogamie, Bryologie* **27**: 141–152.

Gradstein, S.R., P. Hietz, R. Lücking, *et al.* (1996). How to sample the epiphytic diversity of tropical rain forests. *Ecotropica* **2**: 59–72.

Gradstein, S.R., D. Griffin III, M.L. Morales, and N.M. Nadkarni (2001). Diversity and habitat differentiation of mosses and liverworts in the cloud forest of Monteverde, Costa Rica. *Caldasia* **23**: 203–212.

Gradstein, S.R., N.M. Nadkarni, T. Krömer, I. Holz, and N. Nöske (2003). A protocol for rapid and representative sampling of epiphyte diversity of tropical rain forests. *Selbyana* **24**: 87–93.

Hamilton, L.S., J.O. Juvik, and F.N. Scatena (eds.) (1995). *Tropical Montane Cloud Forests*. New York: Springer-Verlag.

Holz, I., and S.R. Gradstein (2005). Cryptogamic epiphytes in primary and recovering upper montane oak forests of Costa Rica: species richness, community composition and ecology. *Plant Ecology* **178**: 89–109.

Holz, I., S.R. Gradstein, J. Heinrichs, and M. Kappelle (2002). Bryophyte diversity, microhabitat differentiation and distribution of life forms in Costa Rican upper montane *Quercus* forest. *The Bryologist* **105**: 334–348.

Kessler, M. (2001). Pteridophyte species richness in Andean forests in Bolivia. *Biodiversity and Conservation* **10**: 1473–1495.

Krömer, T., and S.R. Gradstein (2004). Species richness of vascular epiphytes in two primary forests and fallows in the Bolivian Andes. *Selbyana* **24**: 190–195.

Montfoort, D., and R. Ek (1990). *Vertical Distribution and Ecology of Epiphytic Bryophytes and Lichens in a Lowland Rain Forest in French Guiana*. Utrecht, the Netherlands: Institute of Systematic Botany.

Nöske, N.M., S.R. Gradstein, H. Kürschner, G. Parolly, and S. Torrachi (2003). Cryptogams of the Reserva San Francisco (Province Zamora-Chinchipe, Southern Ecuador). I. Bryophytes. *Cryptogamie, Bryologie* **24**: 15–32.

Nöske, N.M., N. Hilt, F. Werner, *et al.* (2007). Disturbance effects on diversity in montane forest of Ecuador: sessile epiphytes versus mobile moths. *Basic and Applied Ecology* **9**: 4–12.

Richards, P.W. (1996). *The Tropical Rain Forest*, 2nd edn. Cambridge, UK: Cambridge University Press.

Van Reenen, G.B. A., and S.R. Gradstein (1983). Studies on Colombian cryptogams. XX. A transect analysis of the bryophyte vegetation along an altitudinal gradient on the Sierra Nevada de Santa Marta, Colombia. *Acta Botanica Neerlandica* **32**: 163–175.

Wolf, J. (1993). Diversity patterns and biomass of epiphytic bryophytes and lichens along an altitudinal gradient in the northern Andes. *Annals of the Missouri Botanical Garden* **80**: 928–960.

12 Altitudinal zonation and diversity patterns in the forests of Mount Kilimanjaro, Tanzania

A. Hemp
University of Bayreuth, Bayreuth, Germany

ABSTRACT

Based on an analysis of 600 vegetation plots using the method of Braun-Blanquet (1964) the altitudinal distribution of trees, shrubs, epiphytes, lianas, and herbs along transects spanning an elevation difference of 2400 m was studied in the montane (cloud) forest belt on the slopes of Mt. Kilimanjaro, Tanzania. Using the unidimensionally constraint clustering technique, significant discontinuities were observed to occur in parallel in the respective strata. Over 1200 species representing about half of the vascular flora on Mt. Kilimanjaro were found in the studied plots, making the forest belt the most important habitat with respect to plant diversity on the mountain. Species numbers for the different forest strata (except for epiphytes) peaked at two altitudes. The gap between the two diversity peaks can be explained by the long-lasting influence of humans on Kilimanjaro's landscape. Beside anthropogenic influence, precipitation (especially in the case of epiphytes), minimum temperature (in particular the occurrence of frost) and the variety of habitats were identified as key factors underlying the observed diversity patterns.

INTRODUCTION

Changes in floristic composition on high tropical mountains are well documented, notably for trees and selected taxa. Altitude (or factors related to altitude, such as temperature and precipitation) is by far the biggest determinant of diversity (e.g. Hamilton *et al.*, 1989; Gentry, 1995; Vázquez and Givnish, 1998). However, it is still a matter of debate whether altitudinal changes are continuous (e.g. Lieberman *et al.*, 1996; Vázquez and Givnish, 1998) or rather more discontinuous (e.g. Kitayama, 1992; Friis and Lawesson, 1993; Hemp, 2002; cf. Bach, 2004). Altitudinal trends in species richness are also debated. Several studies found a decreasing trend in species richness with altitude (e.g. Hamilton and Perrott, 1981: Kitayama, 1992; Vázquez and Givnish, 1998) whereas others detected a mid-altitude peak in species richness (e.g. Liebermann *et al.*, 1996; Beaman *et al.*, 2001; Hemp, 2001a), including the diversity of bryophytes and lichens (Wolf, 1993).

This chapter analyzes results from 600 vegetation plots on Mt. Kilimanjaro, one of the highest solitary mountains of the world, and an outstanding object for the study of altitudinal effects on species composition and diversity.

STUDY AREA

Mt. Kilimanjaro, a relic of an ancient volcano, rises from the savanna plains at 700 m elevation to a snow-clad summit at 5895 m.a.s.l. It is located *c.* 300 km south of the equator in Tanzania on the border with Kenya. Its climate is characterized by a bimodal rainfall pattern, with the "long rains" from March to May forming the main rainy season, and the "short rains" occurring around December (Paap, 1934; Coutts, 1969). Natural forests cover an area of about 1000 km^2 on Mt. Kilimanjaro (equivalent to *c.* 25% of the mountain's cross-sectional area), mostly between 1800 and 3200 m.a.s.l.

METHODS

Floristic and forest structural data have been collected since 1996 along 34 transects spanning an altitudinal range of 2400 m (Figure 12.1) and sub-divided in altitudinal intervals of 100 m. A total of 600 relevés (plots with a size of 1000 m^2 each)

Tropical Montane Cloud Forests: Science for Conservation and Management, eds. L. A. Bruijnzeel, F. N. Scatena, and L. S. Hamilton. Published by Cambridge University Press.

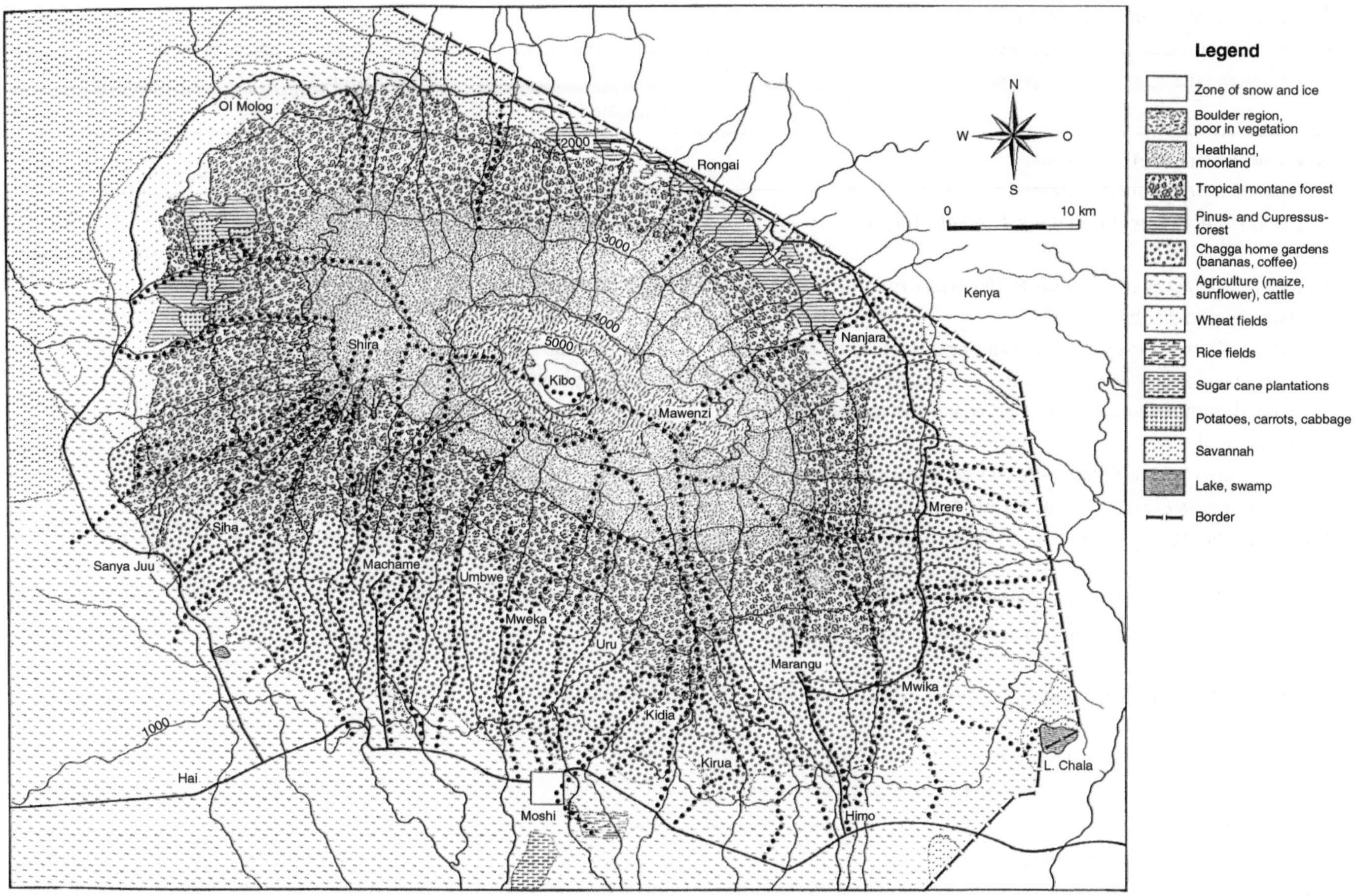

Figure 12.1. Land use and vegetation on Mt. Kilimanjaro. Dotted lines represent study transects.

were established in the various forests using the method of Braun-Blanquet (1964). Special attention was given to homogeneity. To include all important forest types present in a given altitudinal interval, the number of relevés per 100-m interval varied in accordance with the diversity of habitats present.

In each relevé a herb layer (<1 m tall), a shrub layer (1–10 m tall), and a tree layer (>10 m tall) were differentiated. Climbers (stranglers, scramblers, tendrillar plants, hook and root climbers and twiners; cf. Lind and Morrison, 1974) reaching into the shrub and tree layers were defined as lianas. Epiphytes (accidental, casual, hemi- and holo-epiphytes; cf. Kress, 1986) were treated as a separate layer. Above an elevation of about 3200 m the height of the canopy normally did not exceed 10 m. Such stands, also those occurring at lower elevations were not included in the "tree layer."

For the detection of different vegetation zones the "chronological clustering" approach (Legendre *et al.*, 1985) was used. Floristic similarities were calculated by Jaccard's coefficient of community, which uses presence/absence data, and by the Chord distance, which uses abundance data. Further details are given by Hemp (2005a).

RESULTS AND DISCUSSION

Forest zonation

Several statistically significant discontinuities in the floristic composition of the forests were detected, which occurred in parallel in all distinguished strata (Hemp, 2005a). Based on these discontinuities which are indicative of a transition from one zone to another, Table 12.1 summarizes the resulting altitudinal zonation of Mt. Kilimanjaro's forests.

Five forest zones could be distinguished on Mt. Kilimanjaro (Table 12.1), viz. (i) a sub-montane, (ii) a lower, and (iii) a middle montane forest zone; (iv) an upper montane forest zone, and (v) a sub-alpine forest zone. The latter three forest types correspond with the lower, upper, and sub-alpine rain forests, respectively, proposed by Grubb (1977). On the basis of presumed fog water inputs and observed forest structure (e.g. epiphyte abundance) the forests of the middle and upper montane zones and those of the sub-alpine zone may be defined as "cloud forests" (cf. Bruijnzeel and Hamilton, 2000). The present nomenclature for these zones differs slightly from that used in previous articles (Hemp, 2002, 2005b) in which the sub-alpine zone started at 2800 m.a.s.l.

Table 12.1 *Forest zones on Mt. Kilimanjaro*

	Altitudinal zone	Altitude (m.a.s.l.)	Forest zone (Grubb 1977)
Vegetation of the southern slope			
Remnants of *Erica trimera* forests	sub-alpine	3700–4000	subalpine
Pure *Erica excelsa* forest replacing *Hagenia* forest	upper montane	2800–3200 (up to 3500)	upper montane
Hagenia–Podocarpus forest with *Podocarpus latifolius, Hagenia abyssinica*, and *Prunus africana*	upper montane	2800–3200	
Podocarpus–Ocotea forest with prevailing *Podocarpus latifolius*	upper montane	2500–2800	lower montane
Ocotea–Podocarpus forest, consisting of *Ocotea usambarensis* associated with *Podocarpus latifolius* and the tree fern *Cyathea manniana*	middle montane	2200–2500	
Ocotea–Agauria forest, *Ocotea–Syzygium* forest consisting of *Ocotea usambarensis* associated with *Agauria salicifolia, Syzygium guineense, Macaranga kilimandscharica*, and *Polyscias fulva*	lower montane	1800–2200	
Remnants of gorge and riverine forest in deep valleys with *Ekebergia capensis, Strombosia scheffleri*, and *Leptonychia usambarensis*	lower montane	1500–1800	
Riverine forests with *Mitragyna rubrostipulata* and *Newtonia buchananii*	sub-montane	1100–1500	
Savanna forest with *Acacia, Terminalia, Grewia, Combretum*; riverine forests with *Ficus vallis-choudae* and *Lecaniodiscus fraxinifolius*	colline	700–1100	
Vegetation of the northern slope			
Juniperus–Podocarpus forest with *Juniperus procera, Podocarpus latifolius*, and *Hagenia abyssinica, Erica excelsa* forest	upper montane	2500–3100	upper montane
Cassipourea forest with *Cassipourea malosana, Teclea simplicifolia, Fagaropsis angolensis*, and *Olea capensis*	lower and middle montane	2000–2500	lower montane
Croton–Calodendrum forest with *Olea europaea* ssp. *africana, Croton megalocarpus, Calodendrum capense*, and *Diospyros abyssinica*	submontane	1600–2000	
Savanna forest with *Acacia, Terminalia, Grewia, Combretum*	colline	1400–1600	

However, because the forests between 2800 and 3200 m.a.s.l. consist of montane tree species (e.g. *Hagenia abyssinica, Rapanea melanophloeos* or *Erica excelsa*), it was considered more appropriate to include them in the upper montane zone even though these forests harbor many (sub-)alpine species in the understory. In addition, with the recent finding of remnant *Erica trimera* forest with 10-m high trees at 4000 m.a.s.l. it is now possible to better locate and define the sub-alpine zone. In contrast to the closely related *Erica excelsa, E. trimera* is a true sub-alpine element, which does not occur on Mt. Kilimanjaro inside the montane forest belt (Table 12.1).

Within the largely cultivated sub-montane and lower montane zones between 1000 and 1800 m.a.s.l. on the southern slopes of Mt. Kilimanjaro, forest is restricted to inaccessible, deep valleys and gorges (Figure 12.2a). These ravine forests strongly resemble the highly diverse wet (sub-)montane forest of the Pare and Usambara Mountains within the Eastern Arc (Hemp, 2006a) but differ completely from the montane forests of higher altitudes. Typical tree species include *Ekebergia capensis* (Meliaceae), *Heinsenia diervilleoides* and *Hallea rubrostipulata* (both Rubiaceae), *Newtonia buchananii* (Mimosaceae), *Leptonychia usambarensis* (Sterculiaceae), *Strombosia scheffleri* (Loganiaceae), *Dasylepis integra* (Flacourtiaceae), *Garcinia tanzaniensis* (Guttiferae; recently described by Verdcourt, 2006), and *Polyscias stuhlmannii* (Araliaceae) (Table 12.1). Many of the species in these ravine forests were previously not known from Mt. Kilimanjaro and were believed to be endemic to the Eastern Arc forests. Examples include *Polyscias stuhlmannii, Syzygium micklethwaitii* (= *S. sclerophyllum*; Myrtaceae), *Psychotria crassipetala* (Rubiaceae), and *Leptonychia usambarensis*.

In the lower and middle montane forests on the wet southern slopes, the dominant tree species is camphor (*Ocotea*

Figure 12.2. (a) Lower montane ravine forest at 1600 m.a.s.l. on the southern slope of Mt. Kilimanjaro. The tall trees with the bright bark are *Ekebergia capensis*. Location: Weru-Weru river gorge. (b) *Ocotea* forest, rich in vascular epiphytes and ferns, of the middle montane zone at 2400 m.a.s.l. on the southern slope. (c) Upper montane cloud forest at 2700 m.a.s.l. (all photographs by A. Hemp). (See also color plate.)

usambarensis, Lauraceae). Camphor forests (including potential camphor stands, where *Ocotea* trees have been depleted; Hemp, 2006a) cover an area of about 330 km^2. In the lower montane zone (1800–2200 m.a.s.l.) *O. usambarensis* is associated with *Agarista salicifolia* (Ericaceae), *Macaranga capensis* var. *kilimandscharica* (Euphorbiaceae), and *Polyscias fulva*. This zone is heavily influenced by illegal logging as well as the collecting of firewood and fodder (Lambrechts *et al.*, 2002; Hemp, 2006a).

The middle montane zone (2200–2500 m.a.s.l.) is the main habitat of camphor and pure stands still exist here (Figure 12.2b). Precipitation reaches its maximum in this zone (Figure 12.3), as also indicated by the wealth of epiphytes and ferns, in particular filmy ferns and tree ferns (Figure 12.2b; Hemp, 2001a, b; cf. Frahm and Gradstein, 1991). In gorges and along streams *Cornus volkensii* (Cornaceae) is an important constituent of the tree layer in this zone.

In the upper montane zone the gymnosperm *Podocarpus latifolius* starts to prevail, replacing *Ocotea* with increasing elevation between 2500 and 2800 m.a.s.l. Presumably due to a high frequency of fog, the stems are covered with bryophytes, often in the form of "moss balls" (Figure 12.2c). Between 2800 and 3200 m, *P. latifolius*, East African rosewood (*Hagenia abyssinica*, Rosaceae), and *Prunus africana* (Rosaceae) form the tree canopy. Mono-dominant stands of *Erica excelsa* (Ericaceae) are also pronounced in this zone, replacing *Podocarpus* and *Hagenia* forest after fire (Hemp and Beck, 2001; A. Hemp, this volume #58) and forming the current upper closed forest line at 3200 m. However, small remnant and burnt forests of this type indicate that the upper montane zone reached up to 3500–3600 m.a.s.l. only as recently as 1997 (Hemp, 2005b). Remnants of sub-alpine *Erica trimera* forest mark the former and potential upper closed forest line at 4000 m, representing the highest elevation forests in Africa known today.

The forests of the northern and western slopes of Mt. Kilimanjaro are completely different in species composition and structure due to the prevailing lower precipitation (Hemp, 2006a). On the western slopes below 1600 m and on the northern slopes below 2000 m the relatively dry sub-montane forest is dominated by wild olive (*Olea europaea* ssp. *cuspidata*, Oleaceae), *Croton megalocarpus* (Euphorbiaceae), *Calodendrum capense* (Rutaceae), and *Diospyros abyssinica* (Ebenaceae) (Table 12.1). The lower and middle montane forests (1600–2500 m) on the eastern, northern, and western slopes extend over about 280 km^2 and are characterized by the trees pillar wood (*Cassipourea malosana*, Rhizophoraceae), *Vepris simplicifolia* (Rutaceae), *Fagaropsis angolensis* (Rutaceae), and *Olea capensis*. Between 2500 and 3100 m cedar (*Juniperus procera*, Juniperaceae), *Podocarpus latifolius*, and (at higher altitudes) *Hagenia abyssinica* would be expected to be co-dominant tree species over an area of 120 km^2; however, only 40 km^{2}of cedar stands are left due to legal logging by sawmills up to the 1980s.

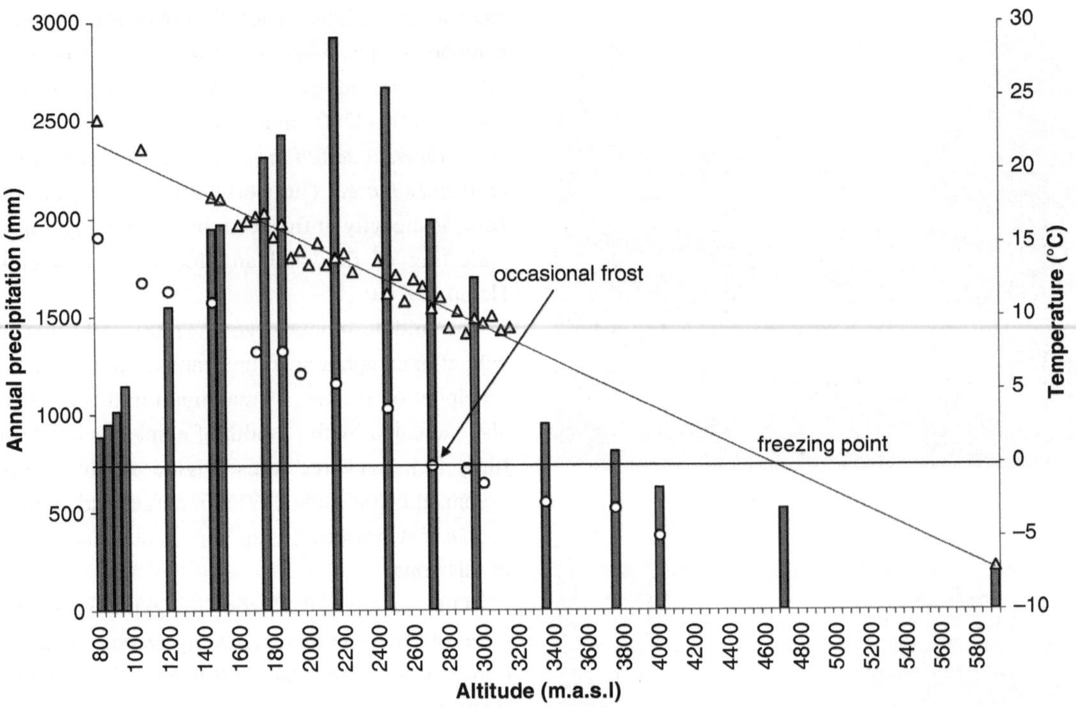

Figure 12.3. Mean annual precipitation (circles) and mean annual temperature (triangles) combined from three transects on the southern slope of Mt. Kilimanjaro. Mean annual temperature obtained from point measurements at 30 cm depth in the soil, except for the lowest and highest temperature values which were taken from Walter and Breckle (1984) and Thompson *et al.* (2002), respectively; mean annual minimum temperatures calculated from absolute minimum temperatures for the years 2001–2003 and the three transects; precipitation data for elevations above 1800 m.a.s.l. are from 1999 onward, and for elevations below 1800 m.a.s.l. from 1997 onward in addition to data from climate stations operated by the Tanzanian Meteorological Agency over various time periods.

A more detailed description of these forest types has been given by Hemp (2001a, 2006a).

Figure 12.4 illustrates the altitudinal zones and their main vegetation types as viewed from the western side of the mountain and showing the contrast between the dry northern and wet southern slopes. The vegetation of the northern slope exhibits the same major discontinuities at around 2500 m and 2800 m as on the southern slope (Hemp, 2005a). However, the zonation on the northern and southern slopes differs at lower elevations. Grubb and Whitmore (1966) suggested that low cloud reaching down to the ground (i.e. fog) was the most important factor determining the distribution of tropical montane rain forests in Ecuador. On Mt. Kilimanjaro, the combination of precipitation, fog incidence, and (minimum) temperatures appears decisive in this respect (Hemp, 2005a). Moisture availability seems to be the controlling factor at lower to intermediate altitudes, complemented by minimum temperature at higher elevations (Figure 12.3). This is illustrated by the difference in altitude at which the transition between dry *Calodendrum* and wetter *Cassipourea* forests on the south-western (1500 m.a.s.l.) and northern slopes (1900 m.a.s.l.) occurs (Figure 12.4). Precipitation on the lower slopes differs greatly between the two expositions (Hemp, 2006a) but becomes more or less equal at higher altitudes, where the same discontinuities were observed at similar elevations for both northerly and southerly expositions (Figure 12.4 and Table 12.1). The lifting level of cloud condensation occurs around 2100 m.a.s.l. on the wet southern slope vs. at *c.* 2500 m.a.s.l. on the drier northern slope (Figure 12.4). The extension of altitudinal zones on drier leeward slopes compared to wetter windward slopes of tropical mountains is a well-known feature and relates to differences in atmospheric humidity and temperature (cf. Grubb, 1971; Richards, 1996; Foster, this volume).

Annual water surplus (rainfall minus evaporation) appears to increase steeply at around 2100 m.a.s.l. due to the occurrence of more frequent low cloud above this elevation and marks the transition from lower to middle montane forest (or lower montane cloud forest *sensu* Bruijnzeel and Hamilton, 2000). Water surplus decreases again above 2400 m.a.s.l. due to the steep gradient in rainfall and is accompanied by a drop in mean minimum temperature (Figure 12.3), probably because of reduced nocturnal cloud cover. Together, the two climatic parameters are likely to cause the pronounced discontinuity at 2500 m.a.s.l. (Hemp, 2005a) marking the arrival of *Podocarpus* in significant numbers and representing the important boundary between lower and upper montane rain forests *sensu* Grubb

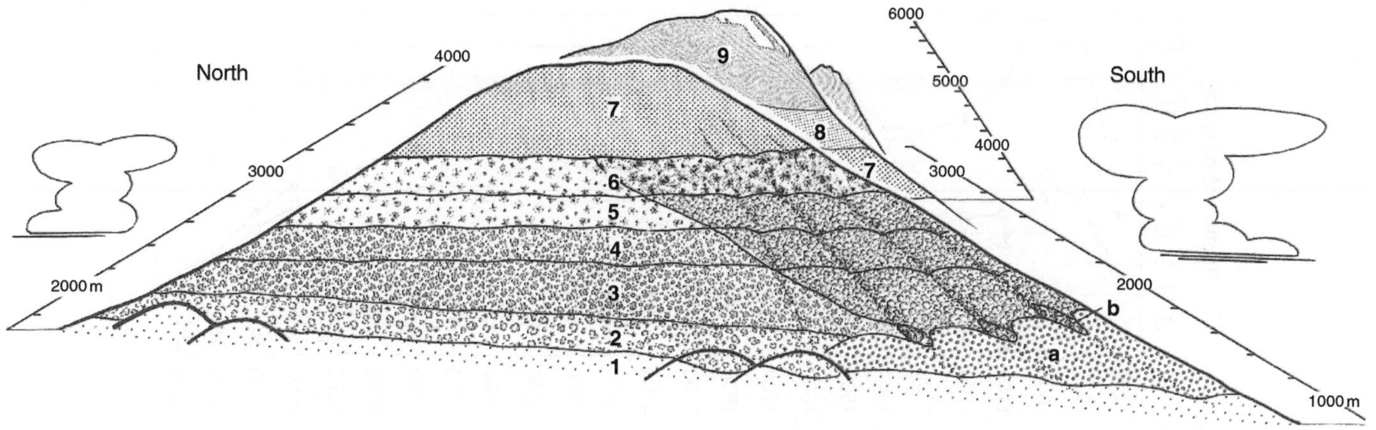

Figure 12.4. Schematic north–south profile showing the western slope of Mt. Kilimanjaro (Shira, Kibo, Mawenzi) with the following main altitudinal zones and vegetation types. 1, colline (savanna) zone; 2, sub-montane zone with *Croton–Calodendrum* forest; a, coffee–banana plantations in the sub-montane and lower montane zones on the southern slope; b, lower montane ravine forest on the southern slope; 3, lower montane zone with *Cassipourea* forests on the northern slope and *Agarista–Syzygium–Ocotea* forests on the southern slope; 4, middle montane zone with *Cassipourea* forests on the northern slope and *Ocotea* forests on the southern slope; 5, upper montane (cloud forest) zone with *Juniperus* forests on the northern slope and *Podocarpus–Ocotea* forests on the southern slope; 6, upper montane (cloud forest) zone with *Juniperus* forests on the northern slope and *Podocarpus* forests on the southern slope; 7, sub-alpine heathland (*Erica* bush); 8, lower alpine zone with *Helichrysum* cushion vegetation; 9, upper alpine and nival zone, mainly bare of vegetation. Note the different cloud condensation levels on the northern and southern slopes.

(1977) or between lower and upper montane cloud forests *sensu* Bruijnzeel and Hamilton (2000). The very moist altitudinal zone between 2100 and 2400 m is floristically characterized by the dominant occurrence of filmy and epiphytic ferns (Hemp, 2001a, b, 2006b) and other vascular epiphytes (Figures 12.2b and 12.6 below). Within the upper montane (cloud) forest zone the *Podocarpus–Ocotea* forests merge into the *Hagenia–Podocarpus* and *Erica excelsa* forests at 2800 m.a.s.l. From this altitude upward, nocturnal frost occurs during the cold season from June to August (Figure 12.3), which is probably one of the major factors causing the floristic discontinuity at this elevation. This climatic boundary is also important for the distribution of tropical insects such as the katydid *Aerotegmina kilimandjarica*, a canopy dweller (C. Hemp, 2001; C. Hemp, this volume) that does not cross this boundary.

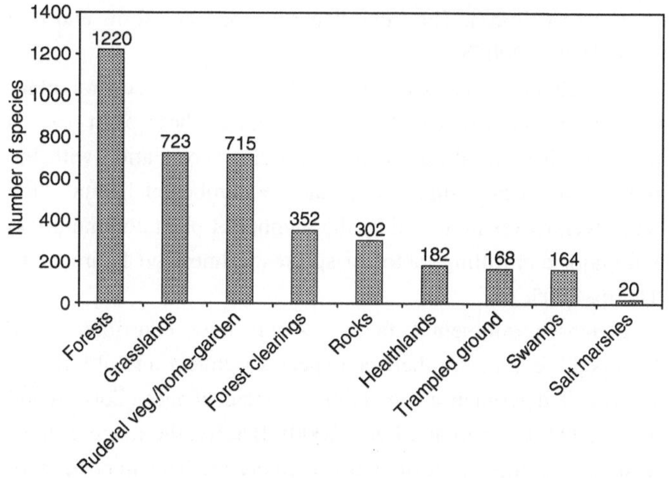

Figure 12.5. Species richness of the main vegetation formations (according to Hemp, 2001a) of Mt. Kilimanjaro, based on an evaluation of over 1400 vegetation plots.

Diversity

A total of 1223 plant species were recorded in the forest plots on Mt. Kilimanjaro. Seventeen percent of the forest flora (206 species) consisted of trees, 40% (485 species) occurred in the shrub layer, 15% (183 species) were lianas, 14% (172 species) were epiphytes and 70% (858 species) were found in the herb layer. It is pertinent to note that because species occurred in many cases in different layers, they were counted more than once; therefore the sum of the percentages exceeds 100%. These 1223 forest species represent about half of the whole vascular flora on Mt. Kilimanjaro (A. Hemp, unpublished data). Therefore, the forest belt is the most important habitat with respect to biodiversity on the mountain, followed by grasslands and vegetation of waste places, roadsides, and fallow arable land ("ruderal vegetation" in Figure 12.5).

Figure 12.6 shows the numbers of forest species for the different strata over an altitudinal range of 2400 m across the mountain, combining northern and southern slopes. The different strata, except epiphytes, display similar diversity curves with two peaks at 1000 and 1800 m. Any parallel trends for all strata, e.g. an increase with precipitation or a steady decrease from lowest to highest altitudes, as found in the Neotropics by Gentry (1982, 1988, 1995), were not observed on Mt. Kilimanjaro. The maximum species richness at 1000 m.a.s.l., for example, is not in

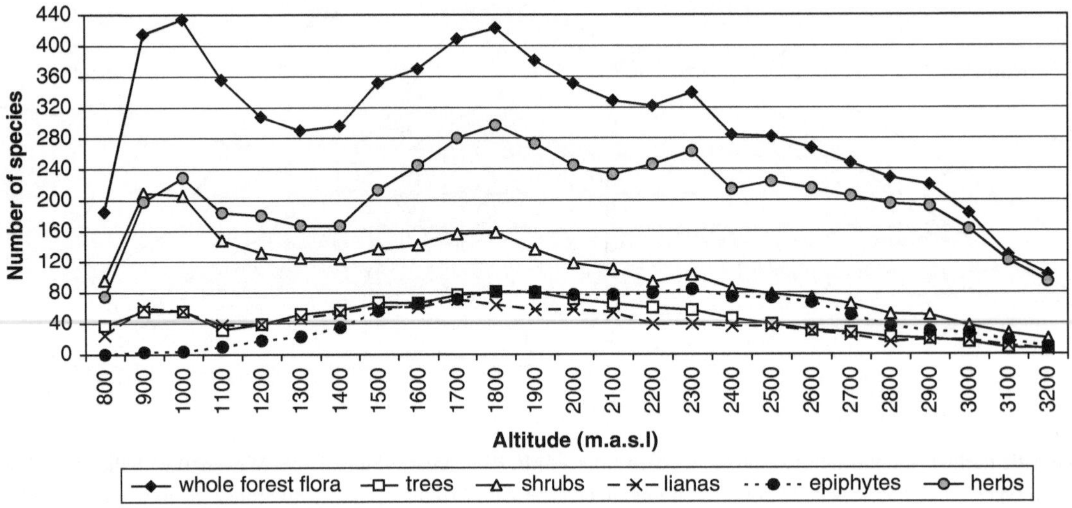

Figure 12.6. Altitudinal changes in the number of species of trees, shrubs, lianas, epiphytes, and herbs in the forests of Mt. Kilimanjaro, based on an evaluation of 600 vegetation plots.

agreement with the maximum of precipitation (at 2400 m.a.s.l.; Figure12.3). Only epiphyte abundance (mainly pteridophytes) showed a clear relationship with precipitation, which is not unexpected (Gentry and Dodson, 1987; Wolf, 1993; cf. Gradstein *et al.*, this volume). The same was true for filmy ferns and tree ferns (Hemp, 2006b).

Through numerical evaluation of the relevés, several plant associations were revealed. The numbers of these plant associations at different altitudes showed a clear correlation with the number of species, suggesting that the number of habitats, the beta-diversity (as indicated by the number of plant communities) is a major controlling factor of species richness on Kilimanjaro (Hemp, 2005c).

A non-environmental factor, which may contribute to a "humped" relationship between species richness and altitude, is the so-called geometric constraint on species ranges (Colwell and Hurtt, 1994; Colwell and Lees, 2000). Briefly, the range of a set of species along an altitudinal (or other) gradient may be geometrically constrained; for example, by sea level as the lowermost boundary and the top of a mountain as the uppermost boundary. However, because altitudinal diversity trends within the different forest strata (trees, lianas, shrubs, herbs versus epiphytes) are not congruent, it is unlikely that this effect is of major importance for the observed altitudinal species distribution. The gap in the species diversity between the two peaks at 1000 and 1800 m.a.s.l. is most likely an artifact related to the long-lasting human influence in the area. The *c.* 1000 km^2 of lower montane and sub-montane forests that have been converted to home-gardens for coffee and banana cultivation now constitutes the most densely populated zone on the mountain (Hemp, 2006c). However, new observations in small relicts of ravine forest within the cultivated belt suggest that the original forests may have been even richer than the forests below and above the coffee and banana belt, possibly possessing a comparable diversity to that found in the Eastern Arc Mountains (Hemp, 2006a). Therefore, the zone characterized by this obviously anthropogenically caused reduction in species richness could in fact represent the altitude with the main forest diversity on Mt. Kilimanjaro; it would also go some way toward explaining the low degree of endemism at this elevation. The shape of the species number vs. elevation curve suggests a maximum well above 440 species at an altitude of 1500–1600 m.a.s.l., raising overall forest diversity substantially.

The contradictory results of many transect studies with respect to the continuity or abruptness of changes in floristic composition and diversity appear to be, in many cases, caused by differences in sampling methods and intensities. For the detection of altitudinal vegetation zones it is helpful to select the sampling plots with regard to homogeneity of the vegetation (within a chosen minimum sampling area), and with regard to visible boundaries or transitions (centralized replicate sampling; Daubenmire, 1966; Mueller-Dombois and Ellenberg, 1974) as done in the phytosociological method of Braun-Blanquet (1964). Furthermore, sampling intensity should be commensurate with the variety of habitats encountered (see Hemp (2005a) for fuller discussion).

ACKNOWLEDGEMENTS

Field work was carried out with financial support from the Deutsche Forschungsgemeinschaft (grant no. He 2719/9–1) and with permission of the Tanzanian Commission for Science and Technology. For their support I owe gratitude to the Chief Park Wardens of Kilimanjaro National Park, Mr. Moirana and Mr. Mafuru, and to my research counterpart Mr. Jacob Mushi.

I also thank Luke Quentin and Simon Mathenge (both at Nairobi) and Dr. Bernard Verdcourt (Kew) for help with the identification of "difficult" species, and Professors Antoine Cleef (University of Amsterdam) and Sampurno Bruijnzeel (VU University Amsterdam) for valuable comments on the draft manuscript.

REFERENCES

Bach, K. (2004). Vegetationskundliche Untersuchungen zur Höhenzonierung tropischer Bergregenwälder in den Anden Boliviens. Ph.D. thesis, University of Göttingen, Germany.

Beaman, J. H., C. Anderson, and R. S. Beaman (2001). *The Plants of Mount Kinabalu,* Vol. 4, *Dicotyledon Families Acanthaceae to Lythraceae*. Kota Kinabalu, Malaysia: Natural History Publications (Borneo), in association with: Royal Botanic Gardens, Kew.

Braun-Blanquet, J. (1964). *Pflanzensoziologie*. Vienna: Springer-Verlag.

Bruijnzeel, L. A., and L. S. Hamilton (2000). *Decision Time for Cloud Forests, IHP Humid Tropics Programme Series No. 13*. Paris: IHP-UNESCO, Amsterdam: IUCN-NL, and Gland, Switzerland: WWF International.

Colwell, R. K., and G. C. Hurtt (1994). Nonbiological gradients in species richness and a spurious Rapoport effect. *American Naturalist* **144**: 570–595.

Colwell, R. K., and D. C. Lees (2000). The mid-domain effect: geometric constraints on the geography of species richness. *Trends in Ecology and Evolution* **15**: 70–76.

Coutts, H. H. (1969). Rainfall of the Kilimanjaro area. *Weather* **24**: 66–69.

Daubenmire, R. (1966). Vegetation: identification of typal communities. *Science* **151**: 291–298.

Frahm, J. P., and S. R. Gradstein (1991). An altitudinal zonation of tropical rain forests using bryophytes. *Journal of Biogeography* **18**: 669–676.

Friis, I., and J. E. Lawesson (1993). Altitudinal zonation in the forest tree flora of north-east tropical Africa. *Opera Botanica* **121**: 125–127.

Gentry, A. H. (1982). Patterns of Neotropical plant species diversity. *Evolutionary Biology* **15**: 1–84.

Gentry, A. H. (1988). Changes in plant community diversity and floristic composition on environmental and geographical gradients. *Annals of the Missouri Botanic Garden* **75**: 1–34.

Gentry, A. H. (1995). Patterns of diversity and floristic composition in Neotropical montane forests. In *Biodiversity and Conservation of Neotropical Montane Forests, Proceedings of the Neotropical Montane Forest Biodiversity and Conservation Symposium*, eds. S. P. Churchill, H. Balslev, E. Forero, and J. L. Lurteyn, pp. 103–26. New York: New York Botanical Garden.

Gentry, A. H., and C. H. Dodson (1987). Diversity and biogeography of neotropical vascular epiphytes. *Annals of the Missouri Botanic Garden* **74**: 205–233.

Grubb, P. J. (1971). Interpretation of the "*Massenerhebung*" effect on tropical mountains. *Nature* **229**: 44–45.

Grubb, P. J. (1977). Control of forest growth and distribution on wet tropical mountains: with special reference to mineral nutrition. *Annual Review of Ecology and Systematics* **8**: 83–107.

Grubb, P. J., and T. C. Whitmore (1966). A comparison of montane and lowland rain forest in Ecuador. II. The climate and its effects on the distribution and physiognomy of the forests. *Journal of Ecology* **54**: 303–333.

Hamilton, A. C., and R. A. Perrott (1981). A study of altitudinal zonation in the montane forest belt of Mt. Elgon, Kenya/Uganda. *Vegetatio* **45**: 107–125.

Hamilton, A. C., C. K. Ruffo, I. V. Mwasha, C. Mmari, and J.C. Lovett (1989). A survey of forest types on the East Usambaras using the variable–area tree plot method. In *Forest Conservation in the East Usambara Mountains Tanzania*, eds. A. C. Hamilton and R. Bensted-Smith, pp. 213–225. Gland, Switzerland: The IUCN Tropical Forest Programme.

Hemp, A. (2001a). Ecology of the pteridophytes on the southern slopes of Mt. Kilimanjaro. II. Habitat selection. *Plant Biology* **3**: 493–523.

Hemp, A. (2001b). Life form and strategies of forest ferns on Mt. Kilimanjaro. In *Life Forms and Dynamics in Tropical Forests*, eds. G. Gottsberger and S. Liede, pp. 95–130. Berlin: Dissertationes Botanicae.

Hemp, A. (2002). Ecology of the pteridophytes on the southern slopes of Mt. Kilimanjaro. I. Altitudinal distribution. *Plant Ecology* **159**: 211–239.

Hemp, A. (2005a). Continuum or zonation? Altitudinal gradients in the forests on Mt. Kilimanjaro. *Plant Ecology* **184**: 27–42.

Hemp, A. (2005b). Climate change-driven forest fires marginalize the impact of ice cap wasting on Kilimanjaro. *Global Change Biology* **11**: 1013–1023.

Hemp, A. (2005c). The impact of fire on diversity, structure and composition of Mt. Kilimanjaro's vegetation. In *Land Use Changes and Mountain Biodiversity*, eds. E. Spehn, M. Liberman, and C. Körner, pp. 51–68. Boca Raton, FL: CRC Press.

Hemp, A. (2006a). Vegetation of Kilimanjaro: hidden endemics and missing bamboo. *African Journal of Ecology* **44**: 305–328.

Hemp, A. (2006b). Ecology and altitudinal zonation of pteridophytes on Mt. Kilimanjaro. In *African Plants: Biodiversity, Ecology, Biogeography and Taxonomy*, eds. S. A. Ghazanfar and H. Beentje, pp. 309–337. Kew, UK: Royal Botanic Gardens.

Hemp, A. (2006c). The banana forests of Kilimanjaro: biodiversity and conservation of the agroforestry system of the Chagga Home Gardens. *Biodiversity and Conservation* **15**: 1193–1217.

Hemp, A., and E. Beck (2001). *Erica excelsa* as a component of Mt. Kilimanjaro's forests. *Phytocoenologia* **31**: 449–475.

Hemp, C. (2001). *Aerotegmina*, a new genus of African Listroscelidinae (Orthoptera: Tettigoniidae, Listroscelidinae, Hexacentrini). *Journal of Orthoptera Research* **10**: 125–132.

Kitayama, K. (1992). An altitudinal transect study of the vegetation on Mt. Kinabalu, Borneo. *Vegetatio* **102**: 149–171.

Kress, W. (1986). The systematic distribution of vascular epiphytes: an update. *Selbyana* **9**: 2–22.

Lambrechts, C., B. Woodley, A. Hemp, C. Hemp, and P. Nnyiti (2002). *Aerial Survey of the Threats to Mt. Kilimanjaro Forests*. Dar es Salaam, Tanzania: UNDP.

Legendre, P., S. Dallot, and L. Legendre (1985). Succession of species within a community: chronological clustering, with applications to marine and freshwater zooplankton. *American Naturalist* **125**: 257–288.

Liebermann, D., M. Liebermann, R. Peralta, and G. S. Hartshorn (1996). Tropical forest structure and composition on a large-scale altitudinal gradient in Costa Rica. *Journal of Ecology* **84**: 137–152.

Lind, E. M., and M. E. S. Morrison (1974). *East African Vegetation*. London: Longman.

Mueller-Dombois, D., and H. Ellenberg (1974). *Aims and Methods of Vegetation Ecology*. New York: John Wiley.

Paap, W. (1934). Die Niederschlagsverhältnisse des Schutzgebietes Deutsch-Ostafrika. *Aus dem Archiv der Deutschen Seewarte* **53**(3): 1–20.

Richards, P. W. (1996). *The Tropical Rain Forest*, 2nd edn. Cambridge, UK: Cambridge University Press.

Thompson, L. G., E. Mosley-Thompson, M. E., Davis, *et al.* (2002). Kilimanjaro ice-core records: evidence of Holocene climate change in tropical Africa. *Science* **298**: 589–593.

Vázquez, G. J. A., and T. J. Givnish (1998). Altitudinal gradients in tropical forest composition, structure, and diversity in the Sierra de Manantlán. *Journal of Ecology* **86**: 999–1020.

Verdcourt, B. (2006). A new species of *Garcinia* L. (Guttiferae) from Tanzania. *Kew Bulletin* **61**: 613–614.

Walter, H. and S.-W. Breckle (1984). *Spezielle Ökologie der Tropischen und Subtropischen Zonen*, Stuttgart, Germany: Fischer-Verlag.

Wolf, J. H. D. (1993). Ecology of epiphytes and epiphyte communities in montane rain forests, Colombia. Ph.D. thesis, University of Amsterdam, Amsterdam, the Netherlands.

13 The outstandingly speciose epiphytic flora of a single strangler fig (*Ficus crassiuscula*) in a Peruvian montane cloud forest

D. J. Catchpole and J. B. Kirkpatrick
University of Tasmania, Hobart, Tasmania, Australia

ABSTRACT

An unprecedented total of 190 holo-epiphytic and five hemi-epiphytic vascular plant species were collected from the canopy and the trunk of a single strangler fig (*Ficus crassiuscula*) in a Peruvian montane cloud forest. A large majority of the vascular species were rare in occurrence, suggesting a high proportion of patchily distributed species within the cloud forest. One hundred and fourteen of the vascular epiphyte species were orchids, 37 species were ferns, and 17 species were bromeliads. Seventy-three of the orchid species belonged to the Pleurothallidinae. Perú is one of the global strongholds of remaining unmodified tropical montane cloud forest. However, cloud forest studies on any topic emanating from Perú are relatively rare. The outstanding epiphyte species richness of a single tree highlights both the importance of cloud forests for global biodiversity, and the urgency for more research and conservation initiatives within the cloud forests of Perú.

INTRODUCTION

Peruvian montane cloud forests are one of the global strongholds of remaining unmodified tropical montane cloud forest (Mulligan, this volume). However, they have received comparatively little scientific attention until now (e.g. Frahm, 1987: Leo, 1995; Weng *et al.* 2004; Gomez-Peralta *et al.*, 2008) and are under increasing threat of deforestation (Young and León, 1995; Mulligan, this volume). The extinction rate of plant species through cloud forest removal in Perú is unknown because most cloud forests have not been botanically prospected (Honorio and Reynel, 2004). Epiphyte flora makes up a large component of plant diversity in Andean forests (Gentry and Dodson, 1987a; Kelly *et al.*, 1994; Bussmann, 2001; Nieder *et al.*, 2001, cf. Gradstein *et al.*, this volume; Krömer and Gradstein, this volume). Current understanding of Peruvian epiphyte diversity is limited due to the incomplete Peruvian catalogue, especially in the Orchidaceae (Van der Werff and Consiglio, 2004).

This chapter documents the epiphyte flora of a single *Ficus crassiuscula* emergent in a Peruvian lower montane cloud forest, and compares indices of diversity in this tree with those documented from previous work elsewhere.

STUDY AREA

The present study was conducted in a lower montane cloud forest at 2400 m.a.s.l. in the Yanachaga–Chemillén National Park, located on the Amazonian Flank of the central Peruvian Andes (10° 32′ S, 75° 21′ W). The average daily temperature is 13.5 °C, the average annual rainfall is 2750 mm with a distinct dry season, during which the frequency of canopy cloud immersion is reduced to 15% (D. Catchpole, unpublished data). The forest is characterized by a 15–25 m broken canopy on very steep slopes.

METHODS

A smooth-barked strangling *Ficus crassiuscula* (Moraceae) tree in a valley bottom was completely sampled for vascular epiphytes. The emergent tree had a height of 32 m, a crown diameter of 23 m, a diameter at breast height of 1.1 m, and supported seven principal ramifications in the vicinity of the apex of the

Tropical Montane Cloud Forests: Science for Conservation and Management, eds. L. A. Bruijnzeel, F. N. Scatena, and L. S. Hamilton. Published by Cambridge University Press.

14 m high trunk. The crown architecture did not conform to a standard architectural model, but resembled Rauh's model (Hallé *et al.*, 1978), with middle and lower branches showing a horizontal or drooping form.

Epiphyte sampling was conducted *in situ* from July 2003 until March 2004 using rope access techniques (Perry, 1978; Dial, 1994; cf. Köhler *et al.*, this volume). All individuals of each species of vascular epiphytes found on the tree were counted. Rhizomatous or clumping patches of one species were counted as one individual. Live plants were collected and placed in a greenhouse constructed at the study site in order to collect fertile material for identification. Herbarium vouchers were deposited in the Selva Central Herbarium (HOXA) of the Jardín Botánico de Missouri – Perú. In addition to species counts, Fisher's Alpha index (α) (Fisher *et al.*, 1943), the Shannon–Weaver Index (H') (Shannon and Weaver, 1949), and Simpson's index (D) (Simpson, 1949) were calculated as a measure of species richness.

RESULTS

One hundred and ninety holo-epiphytic vascular plant species and five hemi-epiphytic vascular plant species comprising 9022 individuals were recorded from the tree. The epiphyte species came from 60 genera of 18 families (Table 13.1). Eighty-five species came from just five genera (*Elaphoglossum* – 11 spp., *Lepanthes* – 10 spp., *Maxillaria* – 14 spp., *Pleurothallis* – 33 spp., *Stelis* – 17 spp.). Orchidaceae dominated the numbers of genera, species and individuals, with the Pleurothallidinae subtribe (Orchidaceae) accounting for 73 species and 25.5% of all individuals.

Fisher's α-value was 35.11, Shannon–Weaver's H' 3.45, and Simpson's D-value 0.93. Two species, a tank bromeliad (*Tillandsia* sp.) and a rhizomatous orchid (*Maxillaria notylioglossa*), represented *c.* 30% of all individuals. Fifty-nine species were represented by only one individual and 20 species were represented by only two individuals.

DISCUSSION

The species richness found in the *F. crassiuscula* host tree is unprecedented from a single tree and 50% richer than the highest value previously reported elsewhere (Table 13.2). This could be partly a result of the relatively large crown of the emergent tree and thorough sampling (inspecting every individual and translocating all unidentifiable individuals to a greenhouse), although few data exist for crown size in published epiphyte species richness accounts.

The species richness presented here is derived from a much smaller sampling area than an entire forest or 1-ha plot. With the only exception being an extensively collected Ecuadorian forest (153 spp. – Bussmann, 2001), the presently sampled single tree held more species of epiphytic orchids than did thoroughly inventoried forests (109 spp. – Gentry and Dodson, 1987b; 92 spp. – Ingram *et al.*, 1996; 96 spp. – Engwald *et al.*, 2000). This dominance of the Orchidaceae is a uniform trend in mesic tropical epiphyte communities and the familial composition of the present epiphyte community is similar to those found in other species-rich Neotropical montane areas (Bøgh, 1992; Ingram *et al.*, 1996; Engwald *et al.*, 2000; Bussmann, 2001; Wolf and Flamenco, 2003).

A high species richness within the Pleurothallidinae is a common phenomenon in Andean epiphyte floras (Bøgh, 1992; Bussmann, 2001). The diversity of the pleurothallid genera and species found in the presently sampled single tree contrasts with the relatively few species listed in Peruvian collections (Brako and Zarucchi, 1993). The epicenters of pleurothallid diversity in Bolivia suggested by Vásquez and Ibisch (2000) may need to be revised upon further collection. Indeed, the species diversity of many pleurothallid genera may be evenly spread across the Andes, rather than concentrated in epicenters.

A prominent feature of the community structure was dominance by a few species and the high contribution to species richness by species of low abundance. This suggests that many individuals have very patchy distributions. The 9000+ individuals from a single tree found in the present study constitute a much greater number than other published figures known to the authors (1001 – Barthlott *et al.*, 2001; 505 – Pupulin *et al.*, 1995).

At 35.11, the α-index derived in the present study is higher than values obtained for a montane plot of 175 m^2 in Ecuador (28.77; Bøgh, 1992); a montane plot of 0.1-ha plot in Venezuela (23.12; Engwald *et al.*, 2000); a montane single tree in Venezuela (15.86; Barthlott *et al.*, 2001); a lower montane single tree (only orchids) in Costa Rica (9.86; Pupulin *et al.*, 1995); and a 1.5-ha lowland forest plot in Venezuela (12.86; Nieder *et al.*, 2000).

Comparisons of single-tree collections with plot studies require some consideration of the effect of collection area. In plots, a higher percentage of species would be expected to come from the understorey compared to single-tree studies. In single trees, only one trunk is sampled in the understory relative to the large upper canopy area. Therefore, a higher diversity of species should be found in plot studies, which include whole trees. In view of this, the α-value of the *F. crassiuscula* host tree appears to be very high indeed.

The *F. crassiuscula* host tree also displayed a very high D-value (0.93), reflecting the dominance of total abundance by a few species (five spp. contributed 47.7% of individuals), although the two dominant species (efficient colonizers) were

Table 13.1 *Summary of the vascular epiphyte families found in the studied* Ficus crassiuscula *host tree*

Family	Genera	Species	Individuals	Relative abundance (%)
Araceae Juss.	2	3	23	0.25
Araliaceae Juss.	1	1	2	0.02
Aspleniaceae Newman	1	2	11	0.12
Asteraceae Martinov	1	1	1	0.01
Bromeliaceae Juss.	4	17	2275	25.22
Clusiaceae Lindl.	1	2	74	0.82
Crassulaceae J. St. Hil.	1	1	1	0.01
Elaphoglossaceae Pic. Serm.	1	12	807	8.94
Ericaceae Juss.	7	9	430	4.77
Grammitidaceae Ching	5	10	322	3.57
Hymenophyllaceae Link	1	2	24	0.27
Lentibulariaceae Rich.	1	1	11	0.12
Orchidaceae Juss.	24	114	4806	53.27
Oxalidaceae R. Br.	1	1	3	0.03
Piperaceae C. Agardh	2	6	114	1.26
Polypodiaceae Bercht. & J. Presl	4	8	43	0.48
Rubiaceae Juss.	1	1	25	0.28
Vittariaceae (C. Presl) Ching	2	2	7	0.08
Unidentified	–	2	43	0.48
Total 18	**60**	**195**	**9022**	

Table 13.2 *Vascular epiphyte species richness recorded for single trees throughout the Neotropics*

Species richness	Altitude (m.a.s.l.)	Location	Source
195	2400	Perú	Present study
126	1000	Costa Rica	Schuettpelz and Trapnell (2006)
109	2000	Ecuador	Nowicki (1998)
107	160	Mexico	Valdivia (1977)
83	1300	Bolivia	Krömer *et al.* (2005)
75	2100	Bolivia	Krömer *et al.* (2005)
74	45	French Guiana	Freiberg (1999)
66	2200–2600	Venezuela	Barthlott *et al.* (2001)
65	200	French Guiana	Freiberg (1996)
59	650	Bolivia	Krömer *et al.* (2005)
54	1800	Ecuador	Freiberg and Freiberg (2000)
51	1780–2000	Ecuador	Rudolph *et al.* (1998)
39	1350	Costa Rica	Pupulin *et al.* (1995)
24	1500	Costa Rica	Ingram *et al.* (1996)
21	465	Jamaica	Kelly *et al.* (1994)
21	80	Brazil	Kersten and Silva (2001)
12	100	Venezuela	Nieder *et al.* (2000)

represented by a large proportion of unfertile juvenile individuals. The only values of Simpson's D for other epiphyte communities available are those calculated with biomass by Hietz and Hietz-Seifert (1995) for a range of forest types. These are much lower ($D = 0.1$–0.54). The high H'-value of the *F. crassiuscula* host tree (3.45) reflects the high diversity amongst species of low abundance. This H' value was higher than that found in montane forest by Barthlott *et al.* (2001) (3.15). Again, the range of H' using biomass abundance as reported by Hietz and Hietz-Seifert (1995) was lower ($H' = 1.11$–2.7).

CONCLUSION

The alpha diversity for epiphytes of the presently sampled *F. crassiuscula* host tree is extraordinarily high and highlights the importance of the forest canopy for overall forest biodiversity. It also highlights the importance of a detailed epiphytic inventory with measures to obtain fertile material for estimating a realistic plant diversity for epiphyte-rich cloud forests. The diversity of the Orchidaceae is particularly high, although not surprising in the context of the total number of epiphyte species and individuals. The investigated tree lies in the Peruvian Andes, an area not previously known for its high epiphyte diversity due to the lack of herbarium vouchers. The high epiphyte diversity at this site validates the recommendation by Van der Werff and Consiglio (2004) to increase the area of protected Peruvian montane forests.

ACKNOWLEDGEMENTS

This research was supported financially by the Andrew J. Mellon Foundation, the Australian Ecological Society, and the Walter Guinness Charitable Trust. Some equipment was generously donated by Macpac. Logistic support was received from the Missouri Botanical Garden – Perú, Benjamin Kroll (Pro Naturaleza), the Andean Amazon Research Station, and the Universidad Nacional Agraria La Molina. Comments by three anonymous reviewers greatly improved the manuscript. Plant identification was assisted by Carlos Arias (Bromeliaceae), Edwin Becerra (Orchidaceae), Jorge Lingan (Araceae), Franco Mellado (Pteridophyta), Abel Monteagudo (Moraceae), Edgardo Ortiz (Ericaceae), Luis Valenzuela (Orchidaceae), and Rodolfo Vásquez (Araliaceae, Crassulaceae).

REFERENCES

Barthlott, W., V. Schmit-Neuerberg, J. Nieder, and S. Engwald (2001). Diversity and abundance of vascular epiphytes: a comparison of secondary vegetation and primary montane rain forest in the Venezuelan Andes. *Plant Ecology* **152**: 145–156.

Bøgh, A. (1992). The composition and distribution of the vascular epiphyte flora of an Ecuadorian montane rain forest. *Selbyana* **13**: 25–34.

Brako, L., and J. L. Zarucchi (eds.) (1993). *Catalogue of the Flowering Plants and Gymnosperms of Peru – Catálogo de las angiospermas del Perú.* St. Louis, MO: Missouri Botanical Garden.

Bussmann, R. W. (2001). Epiphyte diversity in a tropical Andean forest: Reserva Biologica San Francisco, Zamora-Chinchipe, Ecuador. *Ecotropica* **7**: 43–59.

Dial, R. (1994). Description of arborist methods for forest canopy access and movement. *Selbyana* **15**: 24–37.

Engwald, S., V. Schmit-Neuerberg, and W. Barthlott (2000). Epiphytes in rain forests of Venezuela: diversity and dynamics of a biocenosis. In *Results of Worldwide Ecological Studies, Proceedings of the 1st Symposium by the A.F.W. Schimper-Foundation*, eds. S. W. Breckle, B. Schweizer, and U. Arndt, pp. 425–434. Hohenheim, Germany: Verlag Günter Heimbach.

Fisher, R. A., A. S. Corbet, and C. B. Williams (1943). The relations between the number of species and the number of individuals in a random sample of an animal population. *Journal of Animal Ecology* **12**: 42–58.

Frahm, J. P. (1987). Ökologische Studien über die epiphytische Moosvegetation in Regenwäldern NO Perus. *Beihefte Nova Hedwigia* **88**: 143–158.

Freiberg, M. (1996). Spatial distribution of vascular epiphytes on three emergent canopy trees in French Guiana. *Biotropica* **28**: 345–355.

Freiberg, M. (1999). The vascular epiphytes on a *Virola michelii* tree (Myristicaceae) in French Guiana. *Ecotropica* **5**: 75–81.

Freiberg, M., and E. Freiberg (2000). Epiphyte diversity and biomass in the canopy of lowland and montane forests in Ecuador. *Journal of Tropical Ecology* **16**: 673–688.

Gentry, A. H., and C. H. Dodson (1987a). Contribution of non-trees to species richness of a tropical rain forest. *Biotropica* **19**: 149–156.

Gentry, A. H., and C. H. Dodson (1987b). Diversity and biogeography of neotropical vascular epiphytes. *Annals of the Missouri Botanical Gardens* **74**: 205–233.

Gomez-Perlata, D., S. F. Oberbauer, M. E. McClain, and T. E. Philippi (2008). Rainfall and cloud-water interception in tropical montane forests in the Eastern Andes of Central Peru. *Forest Ecology and Management* **255**: 1315–1325.

Hallé, F., R. A. A. Oldeman, and P. B. Tomlinson (1978). *Tropical Trees and Forests: An Architectural Analysis.* Berlin: Springer-Verlag.

Hietz, P., and U. Hietz-Seifert (1995). Composition and ecology of vascular epiphyte communities along an altitudinal gradient in central Veracruz, Mexico. *Journal of Vegetation Science* **6**: 487–498.

Honorio, E., and C. Reynel (2004). *Vacíos en la colección de la flora de los Bosques Húmedos del Perú.* Lima, Peru: Herbario de la Facultad de Ciencias Forestales, Universidad Nacional Agraria La Molina.

Ibisch, P., A. Boegner, J. Nieder, and W. Barthlott (1996). How diverse are neotropical epiphytes? An analysis based on the *Catalogue of Flowering Plants and Gymnosperms of Peru. Ecotropica* **2**: 13–28.

Ingram, S. W., K. Ferrel-Ingram, and N. M. Nadkarni (1996). Floristic composition of vascular epiphytes in a neotropical cloud forest, Monteverde, Costa-Rica. *Selbyana* **17**: 88–103.

Kelly, D. L., E. V. J. Tanner, E. M. N. Lughadha, and V. Kapos (1994). Floristics and biogeography of a rain-forest in the Venezuelan Andes. *Journal of Biogeography* **21**: 421–440.

Kersten, R. A., and S. M. Silva (2001). Composição florística e estrutura do componente epifítico vascular em floresta da planície litorânea na Ilha do Mel, Paraná, Brasil. *Revista Brasileira de Botânica* **24**: 213–226.

Krömer, T., M. Kessler, S. R. Gradstein, and A. Acebey (2005). Diversity patterns of vascular epiphytes along an elevational gradient in the Andes. *Journal of Biogeography* **32**: 1799–1809.

Leo, M. (1995). The importante of tropical montane cloud forest for preserving vertebrate endemism in Perú: the Río Abiseo National Park as a case study. In *Tropical Montane Cloud Forests*, eds. L. S. Hamilton, F. N. Scatena, and J. O. Juvik, pp. 198–211. New York: Springer-Verlag.

Nieder, J., S. Engwald, M. Klawun, and W. Barthlott (2000). Spatial distribution of vascular epiphytes (including hemiepiphytes) in a lowland Amazonian rain forest (Surumoni Crane Plot) of Southern Venezuela. *Biotropica* **32**: 385–396.

Nieder, J., J. Prosperí, and G. Michaloud (2001). Epiphytes and their contribution to canopy diversity. *Plant Ecology* **153**: 51–63.

Nowicki, C. (1998). Diversität epiphytischer und terrestrischer Pflanzen eines ecuadorianischen Bergnebelwaldes im Vergleich. Unpublished diploma thesis, University of Bonn, Bonn, Germany.

Perry, D. R. (1978). A method of access into the crowns of emergent and canopy trees. *Biotropica* **10**: 155–157.

Pupulin, F., E. Bianchi, M. Germani, D. Pedruzzi, and M. Wagner (1995). Orchid diversity and distribution on a tree at Reserva Forestal de San Ramon, Costa Rica. *Brenesia* **43**: 47–54.

Rudolph, D., G. Rauer, J. Nieder, and W. Barthlott (1998). Distributional patterns of epiphytes in the canopy and phorophyte characteristics in a western Andean rain forest in Ecuador. *Selbyana* **19**: 27–33.

Schuettpelz, E., and D. W. Trapnell (2006). Exceptional epiphyte diversity on a single tree in Costa Rica. *Selbyana* **27**: 65–71.

Shannon, C. E., and W. Weaver (1949). *The Mathematical Theory of Communication.* Urbana, IL: University of Illinois Press.
Simpson, E. H. (1949). Measurement of diversity. *Nature* **163**: 688.
Valdivia, P. E. (1977). Estudio botánico y ecológico de la región del Río Uxpanapa, Veracruz. IV. Las epifitas. *Biotica* **2**: 55–81.
Van der Werff, H., and T. Consiglio (2004). Distribution and conservation significance of endemic species of flowering plants in Peru. *Biodiversity and Conservation* **13**: 1699–1713.
Vásquez, R. and P. L. Ibisch (2000). *Orquídeas de Bolivia: Diversidad y estado de conservación. Subtribu Pleurothallidinae.* Santa Cruz de la Sierra, Bolivia: Editorial F.A.N.
Weng, C. Y., M. B. Bush, and M. R. Silman (2004). An analysis of modern pollen rain on an elevational gradient in Southern Peru. *Journal of Tropical Ecology* **20**: 113–124.
Wolf, J. H. D., and A. Flamenco (2003). Patterns in species richness and distribution of vascular epiphytes in Chiapas, Mexico. *Journal of Biogeography* **30**: 1689–1707.
Young, K. R., and B. León (1995). Distribution and conservation of Peru's montane forests: interactions between the biota and human society. In *Tropical Montane Cloud Forests*, eds. L. S. Hamilton, F. N. Scatena, and J.O. Juvik, pp. 363–376. New York: Springer-Verlag.

14 Comparative structure, pattern, and tree traits of laurel cloud forests in Anaga, northern Tenerife (Canary Islands) and in lauro-fagaceous forests of central Japan

M. Ohsawa
The University of Tokyo, Tokyo, Japan

T. Shumiya
The Nature Conservation Society of Japan, Tokyo, Japan

I. Nitta
The University of Tokyo, Tokyo, Japan

W. Wildpret and M. del Arco
University of La Laguna, La Laguna, Tenerife, Canary Islands, Spain

ABSTRACT

The evergreen broad-leaved (laurel) forest of the Canary Islands consists of species that in nearby central and southern Europe are found only as fossil remains. However, most of the evergreen tree species making up this Tertiary relic forest, have vicarious species in (South-) East Asian sub-tropical to warm-temperate rain forests, for example in southern Japan. The present study compares the eco-morphology of trees, species distribution, and community structure of the laurel cloud forests in Anaga, north-eastern Tenerife (Canary Islands), with the sub-tropical to warm-temperate rain forests in Kiyosumi, central Japan. It is shown that the Anaga forests still keep less-specialized morphologic tree traits common to the tropics such as hypsophyllary buds and sylleptic branching, while the East Asian Kiyosumi forests exhibit special adaptations to strong seasonality in temperature (colder and longer winters), such as deciduousness, scaled buds, and proleptic branching.

INTRODUCTION

The sub-tropical/warm-temperate rain forest, which constitutes the latitudinal equivalent of tropical mountain evergreen broad-leaved rain forest at latitudes up to 30–40° in both the northern and southern hemispheres, is fragmented and limited in areal extent to the eastern parts of the major continents (e.g. East Asia, south-eastern USA, Atlantic rain forest in south America, and coastal rain forests in Australia). This pattern is thought to reflect the extinction of these forests in other parts of the continents due to seasonal shortage of soil water under dry sub-tropical conditions, and the location of the critical isotherm of 0 °C (mean temperature of the coldest month; Walter, 1985; Ohsawa, 1995). An equivalent forest type occurs on tropical high mountains at low latitudes, in the form of upper and lower montane cloud forests (Ohsawa, 1995). The fragmented, remote, remnants of sub-tropical/warm-temperate rain forest are floristically diverse but share some common wide-ranging families, such as the Lauraceae, Myrsinaceae, and Aquifoliaceae (Ovington, 1983; Gutierrez and Ohsawa, 2000). Moreover, their dominant trees have common, convergent, eco-morphological traits, particularly in terms of leaf size (micro- to notophyllous), bud structure (naked/hypsophyllary/scaled buds), branching type (sylleptic/proleptic), and many other features (Walter, 1985; Hallé *et al.*, 1978; Nitta and Ohsawa, 1998, 2001).

During the Tertiary these forest types had a more or less continuous extension throughout the northern hemisphere, reaching up to central Europe, because fossil remains of evergreen broad-leaved trees have been found in central Europe (Bramwell and Bramwell, 1974; Nooteboom, 1975). The Macaronesian Islands (i.e. The Canaries, Azores, and Madeira; Walter, 1985) and the Colchis region on the eastern side of the Black Sea

Tropical Montane Cloud Forests: Science for Conservation and Management, eds. L. A. Bruijnzeel, F. N. Scatena, and L. S. Hamilton. Published by Cambridge University Press.

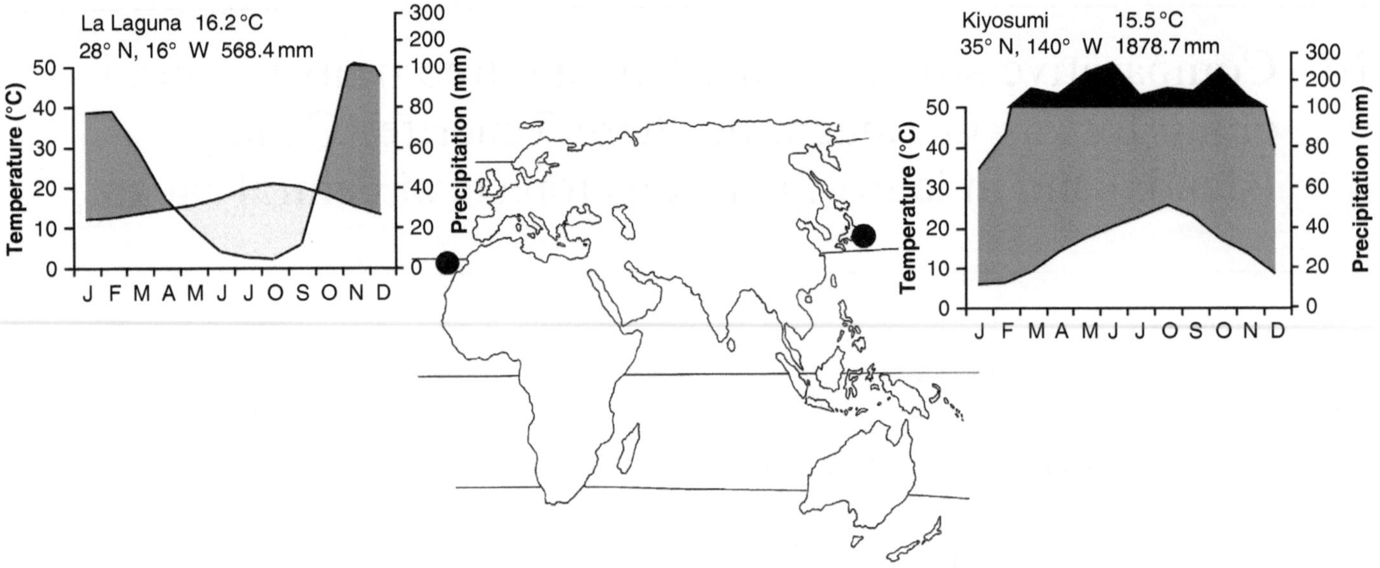

Figure 14.1. Location and climate diagrams for climatic stations around Anaga (Agua García, La Laguna), Tenerife (Canary Islands, Spain), and Kiyosumi, central Japan.

(Kikvidze and Ohsawa, 1999, 2001) support evergreen forests dominated by laurel-type leaf (*Laurocerasus*, *Ilex* spp., etc.) which are the only remnants of the former Tertiary evergreen broad-leaved forests of Europe. Since Schimper (1903) and Ruebel (1930), the East Asian evergreen broad-leaved forests have been considered the Asian equivalent of these ancient laurel forests (*laurisilvae*). In contrast to the Macaronesian laurel forests, which do not receive much rainfall but are continuously fed with moisture supplied by clouds and fog brought in by the trade winds (cf. García-Santos, 2007; Marzol-Jaén *et al.*, this volume), the sub-tropical/warm-temperate rain forests of East Asia are fed by abundant rainfall throughout the year; this condition is particularly favorable in the Japanese archipelago which – being surrounded by the Sea of Japan to the north-west and by the Pacific Ocean in the south-east – receives ample rainfall in both winter and summer (Satoo, 1983).

The Macaronesian and East Asian laurel forests have never been studied intensively in terms of their comparative ecology. The present chapter is a first attempt to compare these forest types in terms of their floristics, eco-morphology, and community structure. What follows below is mainly based on original data collected during intensive research in the Anaga cloud forest of northern Tenerife (Ohsawa *et al.*, 1999a) and in the Kiyosumi forest in central Japan (Ohsawa and Nitta, 2001).

STUDY AREAS AND METHODS

The Anaga cloud forest study site is located on a north-west facing slope along the crest (800–1200, m.a.s.l.) of the Anaga range, northern Tenerife (Canary Islands, Spain; *c.* 28° 28′ N, 16° 25′ W) whereas the Kiyosumi warm-temperate rain forest is situated in central Japan (300 m.a.s.l., 35° 9′ N, 140° 9′ E) (Figure 14.1). Mean annual temperature in Anaga at 985 m.a.s.l. is 13.6 °C and the mean temperature of the coldest month (April) is 9.9 °C (Eguchi *et al.*, 1999). Marzol-Jaén *et al.* (this volume) provide additional climatic data for a station at the summit of Anaga Rural Park (842 m.a.s.l., 28° 32′ N, 16° 14′ W; 1996–2005): average temperature 13.4 °C (16.5 °C in summer and 10.7 °C in winter), decreasing by *c.* 1.5 °C during times of fog. Average relative humidity is 83%, increasing on average to 99% during times of fog whereas wind speeds are high at this exposed site: on average 6.0 m s^{-1} (higher by 1.9 m s^{-1} during times of fog). Rainfall is only 470 mm $year^{-1}$ but this must be considered an underestimate in view of the prevailing strong winds (cf. García-Santos, 2007) whereas fog incidence as measured by Standard Fog Collector screens amounts to as much as 1214, mm $year^{-1}$. Fog is present for 41% of the time, being more frequent in summer (60%) than in winter (30%) and (on a diurnal scale) is more frequent at night (50%) than during the day (20%) (Marzol-Jaén *et al.*, this volume). Precipitation in the surroundings of Anaga is measured at Agua Garcia (800 m.a.s.l., 28° 28′ N, 16° 25′ W) and at La Laguna (547 m.a.s.l., 28° 29′ N, 16° 19′ W). Corresponding annual totals are 738 mm at Agua Garcia (Garcia Gallo and Wildpret de la Torre, 1990) and 568 mm at La Laguna (Ceballos *et al.*, 1976). Rainfall in Anaga is highly seasonal (Figure 14.1) but supplemented by fog in the cloud belt above ca. 700 m.a.s.l. (Marzol-Jaén *et al.*, this volume).

Average annual temperature at Kiyosumi is similar to that in Anaga at 14.2 °C while average rainfall is 2371, mm $year^{-1}$ and well distributed throughout the year (Tokyo University Forest,

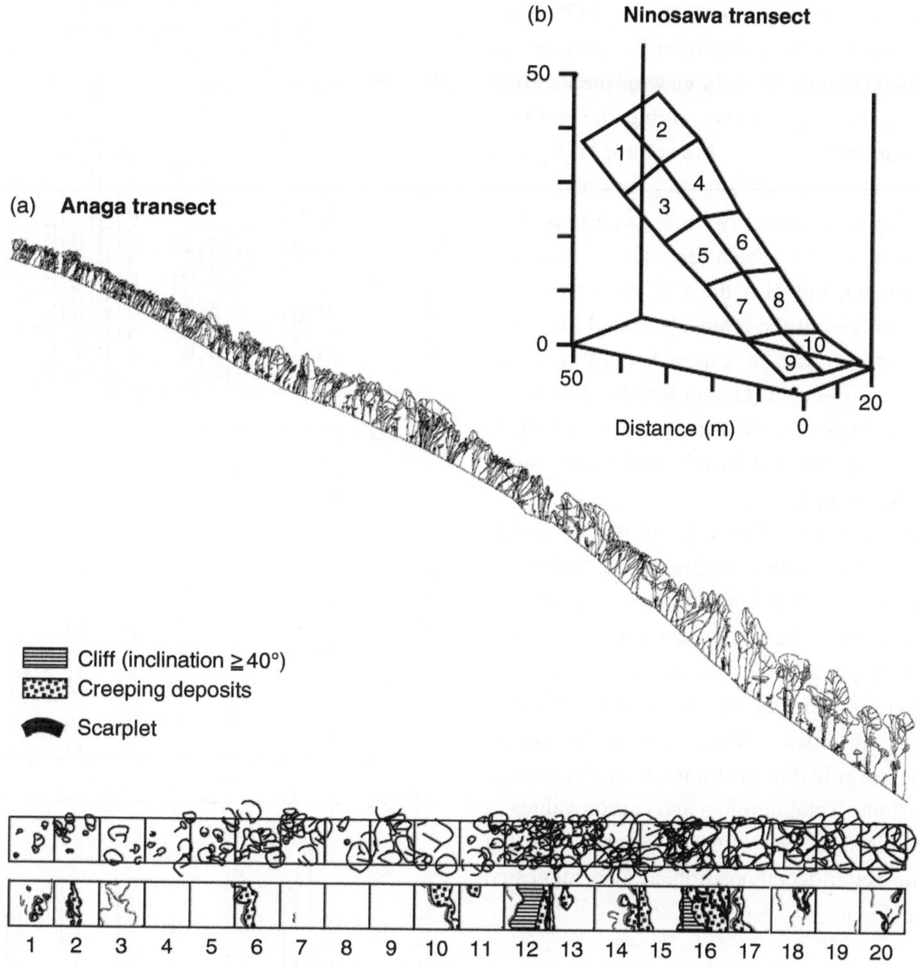

Figure 14.2. Forest transects at the Anaga and Kiyosumi sites. (a) Anaga forest profile diagram, crown projection map, and land-form map; (b) 3-D topographic map of the Kiyosumi transect. Numbers indicate vegetation sampling sub-quadrants (10 × 10 m each).

1992; Figure 14.1). Further comparative climatic information will be given later in this chapter.

To cover forest diversity along topo-edaphic gradients, a permanent transect was established from the ridge top to the lower slopes in both areas (Figure 14.2). The Anaga cloud forest transect was 10 m × 200 m, facing north (840–960 m.a.s.l.) and exposed to the trade winds, while the Kiyosumi rain forest transect at Ninosawa was 20 m × 50 m, facing south (350 m.a.s.l.). Slope gradients in Anaga ranged from 21 to 45 degrees vs. 2 (on the ridge top) to 39 degrees at Kiyosumi (Figure 14.2).

Each transect was divided into 10 m × 10 m sub-quadrats, and climatic conditions, topography, and soil conditions (including soil moisture and selected other soil features such as carbon and nitrogen contents) were determined (Tokyo University Forests, 1992; Ohsawa *et al.*, 1999b). Diameter at breast height was measured (DBH in cm), as well as height (H, in m) and other dimensions used in standard inventories (for trees > 1.3 m tall). Furthermore, heights of all saplings and seedlings were measured and their age estimated by counting the number of bud-scale scars.

RESULTS AND DISCUSSION

Comparison of climatic and topo-edaphic features at Anaga and Kiyosumi

The climate diagrams in Figure 14.1 clearly show the contrasting climatic conditions at Anaga and Kiyosumi. Both Anaga and Kiyosumi are coastal mountains under oceanic influence. However, whilst the mean annual temperatures are similar (13.4 °C for Anaga and 14.2 °C for Kiyosumi), the climate at Anaga is far milder (annual range of 6–8 °C) than at Kiyosumi (annual range of 18.5 °C) where the summer/winter monsoons form the dominant climatic factor. Mean annual precipitation at Kiyosumi (2371, mm) is not only three to five times that at Agua Garcia

(738 mm) and Anaga Rural Park summit station (470 mm; Marzol-Jaén *et al.*, this volume), but in addition the summers at Anaga are very dry as well (Figure 14.1). In view of the fact that the highest fog incidence at Anaga occurs during the summer (García-Santos, 2007; Marzol-Jaén *et al.*, this volume) it is clear that the fog is instrumental in maintaining an evergreen forest vegetation under such highly seasonal rainfall conditions. Not only is fog water captured by the vegetation (thereby contributing to soil water reserves), but also the fog and low cloud reduce the evaporative demand of the atmosphere and therefore evaporative losses. As shown for an exposed summit forest (*fayal–brezal*) on nearby La Gomera (García-Santos, 2007), the latter effect appears more important than the former. However, despite the extra inputs of fog water at Anaga, total water inputs are still far less than at Kiyosumi.

Further evidence for the moderating role of fog at Anaga comes from the soil water observations (Figure 14.3a). Soil water contents were generally above field capacity (pF < 2) during spring after the soil had been recharged by the winter rains but remained mostly between pF 2.5 and 3.0 throughout the dry season, suggesting no serious moisture stress (cf. García-Santos, 2007). As expected, pF values were slightly lower (i.e. conditions were wetter) in the lower half of the transect which received water draining from upslope. Total nitrogen and carbon values in the top 30 cm of soil at Anaga were slightly lower toward the lower slopes, suggesting better decomposition, as also suggested by the lower C/N ratio (Figure 14.3b–d).

Floristics and community structure

The tree floras at Anaga and Kiyosumi show several contrasting features. In the Anaga cloud forest, there are 17 tree species, all evergreens, while in the Kiyosumi rain forest, there are 42 species, including two coniferous, 22 (52%) evergreen broad-leaved, and 18 (43%) deciduous broad-leaved tree species (Table 14.1). The dominant family at Anaga is the Lauraceae (with three species), followed by the Ericaceae and Aquifoliaceae (each with two species), while at Kiyosumi Lauraceae and Fagaceae are co-dominant (both four species), followed by the Rosaceae, Ericaceae, Cornaceae, and Theaceae (each with three species).

Most of the common taxa (at the genus/family level) at Kiyosumi have an evergreen habit, but some families such as the Rosaceae, Leguminosae, and Caprifoliaceae have only deciduous species, whereas the Ericaceae and Oleaceae include both evergreen and deciduous species.

The topo-edaphic niches of the genus/family-level taxa indicate some contrasting community-ecological traits (Figure 14.4). For example, three dominant taxa at Anaga (Ericaceae – *Erica platycodon*, Myricaceae – *Myrica faya*, and Lauraceae – *Laurus azorica*) exhibit clear habitat segregation between ridges, upper slopes, and lower slopes, respectively, while *Ocotea foetens* and

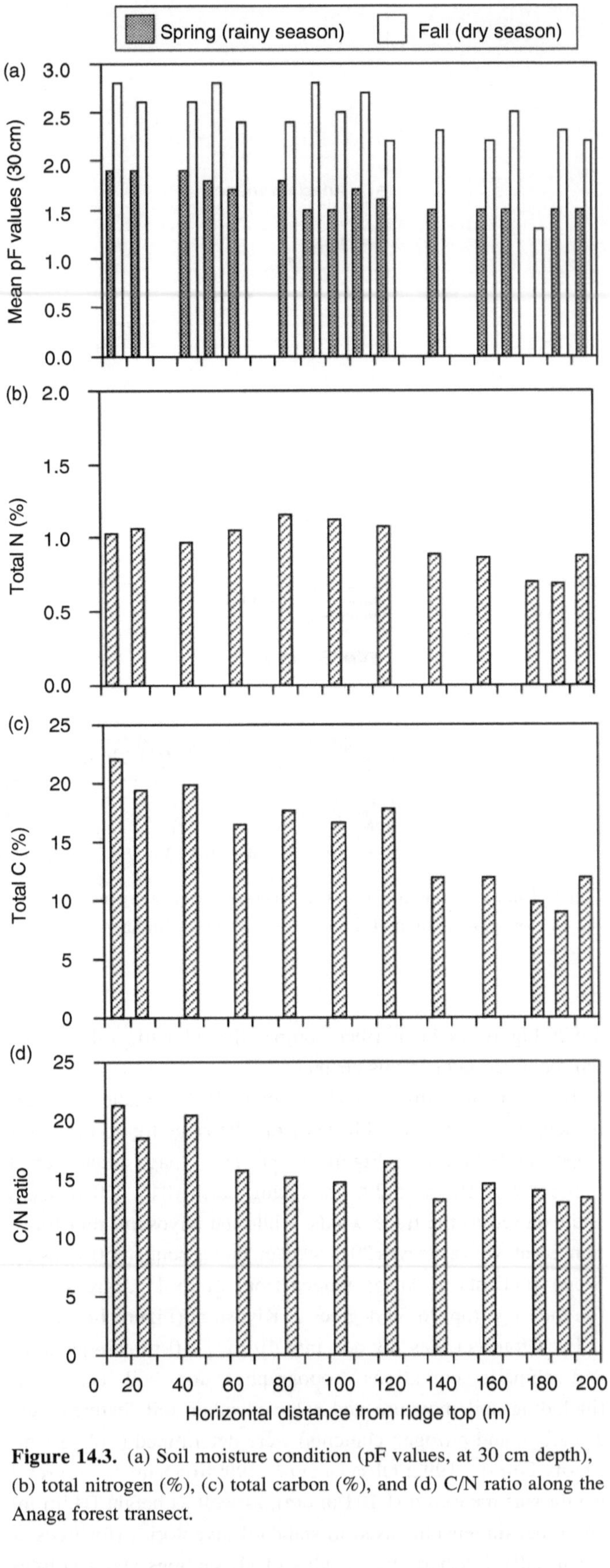

Figure 14.3. (a) Soil moisture condition (pF values, at 30 cm depth), (b) total nitrogen (%), (c) total carbon (%), and (d) C/N ratio along the Anaga forest transect.

Table 14.1 *Floristic composition of the Anaga and Kiyosumi transects. Species abundance is indicated by RBA (relative basal area, %) of all trees with height >1.3 m. Species are arranged according to taxa in common at both sites, and as one-sided taxa per site, respectively*

Anaga						Kiyosumi					
Species	Life form[a]	Bud type[b]	Family[c]	RBA(%)	D[d]	Species	Life form[a]	Bud type[b]	Family[c]	RBA(%)	D[d]
Common family or genus											
Laurus azorica	ET	HB	LAU	24.7	*	*Actinodaphne lancifolia*	ET	SB	LAU	0.0	
Ocotea foetens	ET	HB	LAU	2.0		*Cinnamomum japonicum*	ET	SB	LAU	0.2	
Persea indica	ET	HB	LAU	0.9		*Machilus thunbergii*	ET	SB	LAU	0.1	
						Neolitsea sericea	ET	SB	LAU	0.1	
Myrica faya	ET	HB	MYR	14.3	*	*Myrica rubra*	ET	HB	MYR	1.8	
Ilex canariensis	ET	HB	AQU	3.0		*Ilex rotunda*	ET	HB	AQU	1.0	
Ilex platyphylla	ET	HB	AQU	0.9		*Ilex integra*	ET	HB	AQU	0.6	
Prunus lusitanica	ET	SB	ROS	8.0		*Prunus jamasakura*	DT		ROS	0.8	
						Photinia villosa var. *laevis*	DT		ROS	0.1	
						Sorbus alnifolia	DT		ROS	0.1	
Erica platycodon	ES	HB	ERI	40.1	*	*Rhododendron obtusum* var. *kaempheri*	DS		ERI	0.0	
Erica arborea	ES	HB	ERI	0.7		*Rhododendron dilatatum*	DS		ERI	0.0	
						Vaccinium bracteatum	ES	HB	ERI	0.2	
Picconia excelsa	ES	HB	OLE	0.0		*Osmanthus heterophyllus*	ES	HB	OLE	0.1	
						Fraxinus lanuginosa form. Serrata	DT		OLE	0.2	
Teline canariensis	ES	NB	LEG	0.3		*Albizia julibrissin*	DT		LEG	1.4	
Ardisia bahamensis	ET	HB	MRS	3.3		*Myrsine seguinii*	ET	HB	MRS	+	
Viburnum rigidum	ES	HB	CAP	0.6		*Viburnum dilatatum*	DT		CAP		
Visnea mocanera	ET	HB	THE	+		*Eurya japonica*	ES	HB	THE	5.2	
Arbutus canariensis	ET	HB	ERI	+		*Pieris japonica*	ES	HB	ERI	0.1	
One-sided family or genus											
Rhamnus glandulosa	ET	NB	RHA	1.2							
Hypericum granulosum	ES	HB	HYP	0.0							
						Quercus acuta	ET	SB	FAG	31.9	*
						Quercus salicina	ET	SB	FAG	27.2	*
						Quercus glauca	ET	SB	FAG	5.4	
						Castanopsis sieboldii	ET	SB	FAG	7.0	
						Symplocos prunifolia	ET	HB	SYM	3.7	
						Illicium anisatum	ET	HB	ILL	1.0	
						Camellia japonica	ET	SB	THE	1.2	
						Cleyera japonica	ET	HB	THE	4.1	

Table 14.1 (*cont.*)

Anaga						Kiyosumi					
Species	Life form[a]	Bud type[b]	Family[c]	RBA(%)	D[d]	Species	Life form[a]	Bud type[b]	Family[c]	RBA(%)	D[d]
						Aucuba japonica	ES	HB	COR	0.0	
						Pittosporum tobira	ES	HB	PIT	0.2	
						Mallotus japonicus	DT		EUP	0.6	
						Zanthoxylum ailanthoides	DT		RUT	0.8	
						Styrax japonica	DT		STY	0.1	
						Celtis sinensis	DT		ULM	1.8	
						Cornus controversa	DT		COR	0.5	
						Cornus kousa	DT		COR	0.1	
						Diospyros japonica	DT		EBE	0.8	
						Acer mono	DT		ACE	0.3	
						Deutzia scabra	DS		HYD	0.0	
						Callicarpa japonica	DS		VER	0.0	
						Abies firma	C		PIN	0.8	
						Torreya mucifera	C		TAX	0.0	
Total				100.0						100.0	

[a] ET, evergreen tree; ES, evergreen shrub; DT, deciduous tree; DS, deciduous shrub; C, conifer.
[b] SB, scaled bud; HB, hypsophyllary bud; NB, naked bud.
[c] LAU, Lauraceae; MYR, Myricaceae; AQU, Aquifoliaceae; ROS, Rosaceae; ERI, Ericaceae; OLE, Oleaceae; LEG, Legnminosae; MRS, Myrsinaceae; CAP, Caprifoliaceae; THE, Theaceae; RHA, Rhamnaceae; HYP, Hypericaceae; FAG, Fagaceae; SYM, Symplocaceae; ILL, Illiciaceae; COR, Cornaceae; PIT, Pittosporaceae; EUP, Euphorbiaceae; RUT, Rutaceae; STY, Styracaceae; ULM, Ulmaceae; EBE, Ebenaceae; ACE, Aceraceae; HYD, Hydrangeaceae; VER, Verbenaceae; PIN, Pinaceae; TAX, Taxaceae.
[d] Dominant species in each transect, as indicated by *.

Persea indica (both Lauraceae) are found only in the lowermost quadrants which are the more humid and fertile parts of the slope (Figures 14.3 and 14.4).

At Kiyosumi, the vegetation gradient is strongly compressed but the ridge habitat is also characterized by Ericaceae spp., followed by Myricaceae – *Myrica rubra*, and, on the lower slopes, Lauraceae spp. However, all of these vicarious taxa are rather suppressed by the exclusive dominance of evergreen Fagaceae (*Quercus* spp. and *Castanopsis sieboldii*) throughout the transect (Figure 14.4).

Bud and leaf traits of major evergreen species at Anaga and Kiyosumi

Of the variety of eco-morphologic traits of trees, bud structural types (Nitta and Ohsawa, 1998), branching types (Nitta and Ohsawa, 2001), and leaf characteristics, such as leaf size and specific leaf weight (*SLW*; this also denote thickness of leaves as the predictor of drought tolerance; Ohsawa and Nitta, 1997, 2002; Tang and Ohsawa, 1999) were investigated. Table 14.2 inidicates bud-type spectra of evergreen broad-leaved trees for Anaga and Kiyosumi. Bud-type statistics clearly indicate a difference between the two sites, i.e. at Anaga there are two tree species with naked buds (NB), in which the shoot tip has no special protective organ during the unfavorable (winter) season, but this type is not found at Kiyosumi. Hypsophyllary buds (HB), in which the shoot tip is protected by a few immature foliar organs (so-called hypsophylls) when shoot elongation has ceased, were most abundant at Anaga (80%) vs. 59% at Kiyosumi. The most specialized scaled bud type (SB), where the bud is protected by scaly, specialized foliar organs (modified kataphylls; Romberger *et al*., 1993), was found for only one species (6.7%) at Anaga but was relatively abundant (nine species, 40.9%) at Kiyosumi.

Since these three types of buds represent a trend toward increased protection or morphologic adaptation to longer growth retardation of the shoot during the unfavorable cold season, and toward maximum growth efficiency during spring (Nitta and Ohsawa, 1998), the trend suggests that trees at Anaga are less specialized in their

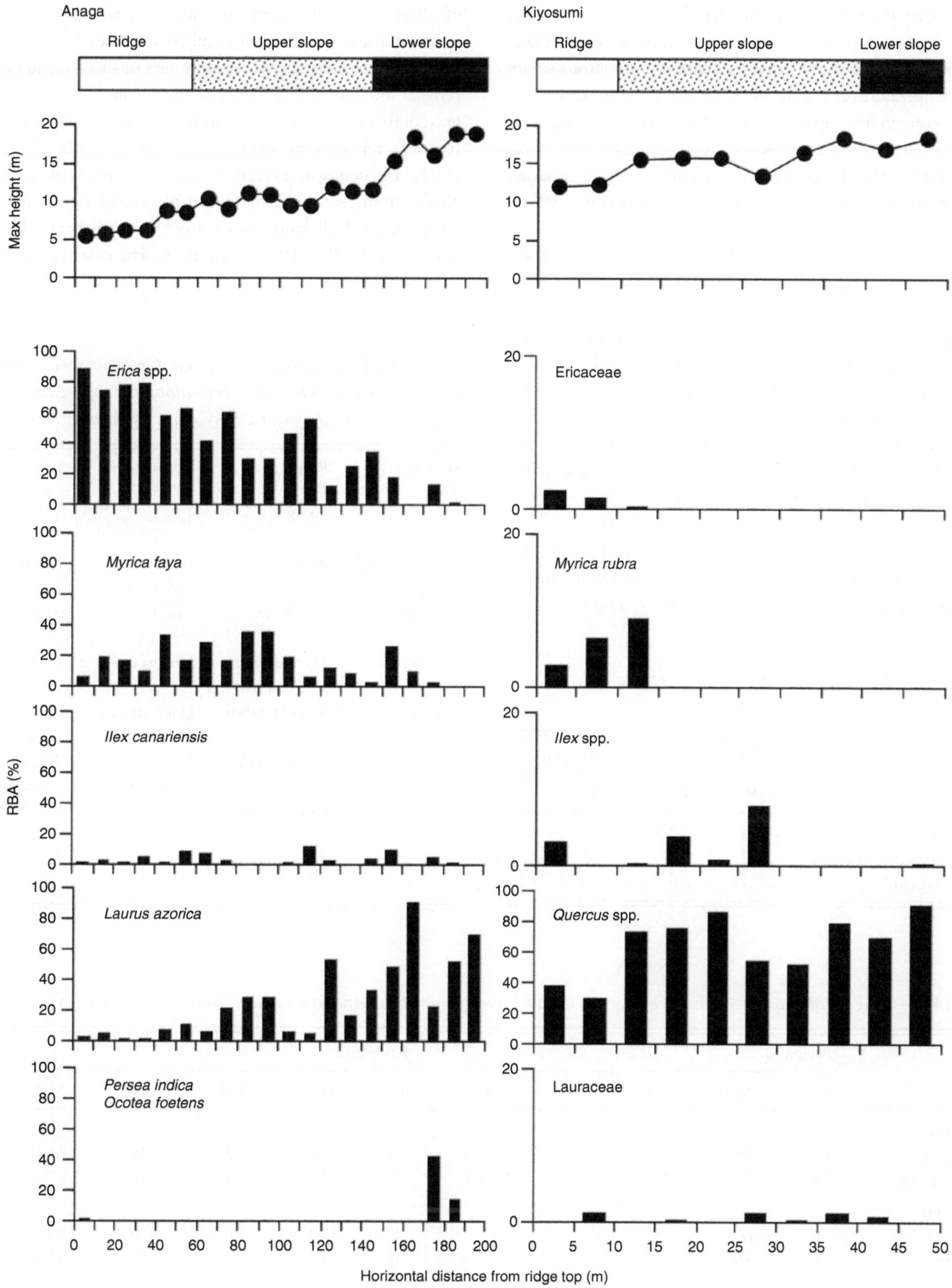

Figure 14.4. Topographic distribution of common taxa (at the genus/family level) along the Anaga and Kiyosumi forest transects. RBA is relative basal area of trees greater than 1.3 m.

shoot protection (in line with the milder prevailing temperatures as one would expect in the absence of low temperatures in Anaga), while the trees at Kiyosumi are more adapted to the strong seasonality in temperature (cf. Figure 14.1). Arguably, the branching types of the evergreen trees appear to support this trend: at Anaga, four out of 17 tree species had sylleptic branching (23.5%), in which axillary buds allowed continuing development with no episode of rest or inhibition (cf. Hallé *et al.*, 1978; Romberger *et al.*, 1993) vs. four out of 22 species at Kiyosumi (18%).

On the whole, leaf size and *SLW* for evergreen broad-leaved trees did not show significant differences between the two forests (Table 14.3). The similarity shown in both common taxa at the family/genus level and for all tree species indicates that they belong to the same forest formation with respect to leaf type. On the other hand, between habitat types (ridge, upper and lower slopes), there is a trend of decreasing *SLW* and increasing leaf size toward the lower, more humid slopes both at Anaga and Kiyosumi (Table 14.4), probably because of phenotypic adaptive plasticity of the leaf morphology (Bonan, 2002) reflecting better growth conditions (soil moisture and nutrients) on the lower slopes.

Although evergreen broad-leaved trees at the two sites exhibited similar leaf traits, the number of deciduous trees at Kiyosumi (19 species) is also nearly the same (Table 14.1). Some of these deciduous trees are successional species surviving among the evergreen canopy, others are trees favoring riparian zone or scree habitats. The occurrence of these deciduous trees reflects the transitional nature of the Japanese evergreen broad-leaved forests around their northern limit where this type of forest is replaced by deciduous forest at *c.* 40° N (Ohsawa, 1993). By contrast, evergreen fagaceous trees are completely absent from Anaga. Remote oceanic island floras like that at Anaga often lack barochorous trees (seed dispersal by weight only; Van der Pijl, 1982) such as the Fagaceae (Bramwell and Bramwell, 1974).

Table 14.2 *Bud-type spectra for all evergreen broad-leaved trees at the Anaga and Kisyosumi forest transects*

	Anaga		Kiyosumi	
Bud type	*N*	(%)	*N*	(%)
Naked bud (NB)	2	13.3	0	0.0
Hypsophyllany bud (HB)	12	80.0	13	59.1
Scaled bud (SB)	1	6.7	9	40.9

Table 14.4 *Leaf characteristics and dominant species amongst all evergreen broad-leaved trees along the topo-edaphic gradient at the Anaga and Kisyosumi forest transects*

SLW (g m^{-2})	**Ridge**	**Upper slope**	**Lower slope**
Anaga	113.0	130.3	83.7
Kiyosumi	93.8	129.7	125.1
Leaf area (cm^2)	**Ridge**	**Upper slope**	**Lower slope**
Anaga	10.7	18.3	44.2
Kiyosumi	9.5	18.9	24.6
Anaga	*Myrica faya*	*Laurus azorica*	*Ocotea foetens*
	Ilex canariensis		*Persea indica*
Kiyosumi	*Myrica rubra*	*Quercus acuta*	*Actinodaphne lancifolia*
	Ilex rotunda	*Quercus salicina*	*Cinnamomum japonicum*
	Ilex integra	*Quercus glauca*	*Machilus thunbergii*
	Pieris japonica		*Neolitsea sericea*

Table 14.3 *Average leaf characteristics of evergreen broad-leaved trees at the Anaga and Kisyosumi forest transects*

	Anaga					Kiyosumi					
	n	Mean	SD	Max	Min	*n*	Mean	SD	Max	Min	*t*-test
Common											
SLW (g m^{-2})	10	108.4	28.3	167.8	82.8	10	112.4	23.7	138.6	68.8	ns
Leaf area (cm^2)	13	21.1	16.7	53.0	0.0	10	19.9	12.3	43.7	7.3	ns
One-sided											
SLW (g m^{-2})	1	60.1				8	120.7	29.2	170.7	82.3	
Leaf area (cm^2)	1	16.9				8	17.8	3.8	23.6	12.5	
Total											
SLW (g m^{-2})	11	104.0	30.6	167.8	60.1	18	116.1	25.8	170.7	68.8	ns
Leaf area (cm^2)	14	20.8	16.1	53.0	0.0	18	19.0	9.3	43.7	7.3	ns

CONCLUSION

Leaf characteristics and topo-edaphic response of evergreen trees were similar in the Anaga cloud forest (Tenerife) and at Kiyosumi (Central Japan). At Kiyosumi, the mixture of evergreen and deciduous trees and an abundance of trees with scaled buds and proleptic branching habit represented a clear adaptation toward a high degree of seasonality in temperature or longer duration of dormant periods (Nitta and Ohsawa, 1998). The Anaga forest retains some of the less specialized tropical forest features, being composed exclusively of evergreen broad-leaved trees, having abundant naked and hypsophyllary buds, and only one exceptional species with scaled buds. Although the Anaga forest experiences prolonged rainless spells during summer, it receives abundant horizontal precipitation in the cloud zone which prevents excessive moisture stress.

The present study has clearly shown that bud characteristics are a useful tree trait to derive seasonal growth characteristics in comparative studies of forest formations.

ACKNOWLEDGEMENTS

We thank Professors Marinus Werger (University of Utrecht) and Sampurno Bruijnzeel (VU University) for their valuable comments and suggestions to improve the manuscript.

REFERENCES

Bonan, G. (2002). *Ecological Climatology: Concepts and Applications.* Cambridge, UK: Cambridge University Press.

Bramwell, D., and Z. Bramwell (1974). *Wild Flowers of the Canary Islands.* London: Stanley Thornes.

Ceballos, L., F. de Cordoba, and F. O. Medina (1976). *Estudio sobre la vegetacion y flora forestal de las Canarias Occidentales.* Santa Cruz de Tenerife: Cabildo Insular.

Eguchi, T., W. Wildpret, M. del Arco, and A. Reyes-Betancort (1999). Synoptic analysis of thermal and moisture conditions in Anaga, Tenerife, the Canary Islands. In *Anaga Cloud Forest*, eds. M. Ohsawa, W. Wildpret, and M. del Arco, pp. 23–38. Chiba, Japan: Chiba University.

Garcia Gallo, A., and W. Wildpret de la Torre (1990). Estudio floristico y fitosociologico del bosque de Madre del Agua en Agua Garcia (Tenerife). In *Homenaje al Profesor Dr. Telesforo Bravo*, Vol. **1**, pp. 307–347. La Laguna, Tenerife: Universidad de la Laguna.

García-Santos, G. (2007). *An ecohydrological and soils study in a montane cloud forest in the National Park of Garajonay, La Gomera (Canary Islands, Spain).* Ph.D. thesis, VU University Amsterdam, Amsterdam, the Netherlands. [http://www.falw.vu.nl/nl/onderzoek/earth-sciences/geo-environmental-science-and-hydrology/hydrology-dissertations/index.asp].

Gutierrez, E., and M. Ohsawa (2000). *Temperate Rainforests.* Farmington Hills, MI: The Gale Group.

Hallé, F., R. A. A. Oldeman, and P. B. Tomlinson (1978). *Tropical Trees and Forests: An Architectural Analysis.* Berlin: Springer-Verlag.

Kikvidze, Z., and M. Ohsawa (1999). Adjara: east Mediterranean refuge of Tertiary vegetation. In *Anaga Cloud Forest*, eds. M. Ohsawa, W. Wildpret, and M. del Arco, pp. 297–308. Chiba, Japan: Chiba University.

Kikvidze, Z., and M. Ohsawa (2001). Richness of Colchic vegetation: comparison between refugia of south-western and East Asia. *BMC Ecology* **1**: 1–10.

Nitta, I., and M. Ohsawa (1998). Bud structure and shoot architecture of canopy and understory evergreen broad-leaved trees at their northern limit in East Asia. *Annals of Botany* **81**: 115–129.

Nitta, I., and M. Ohsawa (1999). Shoot phenology, leaf life-span, and branch architecture of evergreen broad-leaved trees in Anaga cloud Forests. In *Anaga Cloud Forest*, eds. M. Ohsawa, W. Wildpret, and M. del Arco, pp. 247–262. Chiba, Japan: Chiba University.

Nitta, I., and M. Ohsawa (2001). Geographical transition of sylleptic/proleptic branching in three Cinnamomum species with different bud types. *Annals of Botany* **87**: 35–45.

Nooteboom, H. P. (1975). *Revision of the Symplocaceae of the Old World, New Caledonia Excepted.* Leiden, the Netherlands: Leiden University Press.

Ohsawa, M. (1993). Latitudinal pattern of mountain vegetation zonation in southern and eastern Asia. *Journal of Vegetation Science* **4**: 13–18.

Ohsawa, M. (1995). The montane cloud forest and its gradational changes in South-East Asia. In *Tropical Montane Cloud Forests*, ed. L. S. Hamilton, J. O. Juvik, and F. N. Scatena, pp. 254–265. New York: Spinger-Verlag.

Ohsawa, M., and I. Nitta (1997). Patterning of subtropical/warm-temperate evergreen broad-leaved forests in east Asian mountains with special reference to shoot phenology. *Tropics* **6**: 317–334.

Ohsawa, M., and I. Nitta (2001). Plant life and seasonality of evergreen broad-leaved trees. *Natural History of Chiba Prefecture* **5**: 565–579 (in Japanese).

Ohsawa, M., and I. Nitta (2002). Forest zonation and morphological tree-traits along latitudinal and altitudinal environmental gradients in humid monsoon Asia. *Global Environmental Research* **6**: 41–52.

Ohsawa, M., T. Shumiya, I. Nitta, *et al.* (1999a). Structure and differentiation of cloud forests along topographical gradients in Anaga Mountains, Tenerife, the Canary Islands. In *Anaga Cloud Forest*, eds. M. Ohsawa, W. Wildpret, and M. del Arco, pp. 67–118. Chiba, Japan: Chiba University.

Ohsawa, M., W. Wildpret, and M. del Arco (eds.) (1999b). Anaga cloud forest: a comparative study on evergreen broad-leaved forests and trees of the Canary Islands and Japan. In *Anaga Cloud Forest*, eds. M. Ohsawa, W. Wildpret, and M. del Arco, pp. 3–7. Chiba, Japan: Chiba University.

Ovington, J. D. (ed.) (1983). *Temperate Broad-Leaved Evergreen Forests*, Ecosystems of the World No. 10. Amsterdam: Elsevier.

Pijl, L. van der (1982). *Principles of Dispersal in Higher Plants*, 3rd revised and expanded edn. Berlin: Springer-Verlag.

Romberger, J. A., Hejnowicz, and J. F. Hill (1993). *Plant Structure: Function and Development.* Berlin: Springer-Verlag.

Ruebel, E. (1930). *Pflanzengesellschaften der Erde.* Berlin: Hans Huber.

Satoo, T. (1983). Temperate broad-leaved evergreen forests of Japan. In *Temperate Broad-Leaved Evergreen Forests*, Ecosystems of the World No. 10, ed. J. D. Ovington, pp. 169–189. Amsterdam: Elsevier.

Schimper, A. F. W. (1903). *Plant Geography upon a Physiological Basis.* Oxford, UK: Clarendon Press.

Shumiya, T., T. Ohtsuka, and M. Ohsawa (1999). Micro-landform and soil conditions along topographical transect in Anaga cloud forest. In *Anaga Cloud Forest*, eds. M. Ohsawa, W. Wildpret, and M. del Arco, pp. 55–66. Chiba, Japan: Chiba University.

Tang, C. Q., and M. Ohsawa (1999). Altitudinal distribution of evergreen broad-leaved trees and their leaf-size pattern on a humid subtropical mountain, Mt. Emei, Sichuan, China. *Plant Ecology* **145**: 211–233.

Tokyo University Forest (1992). Annual report of meteorological observations in the University of Tokyo (1985–1989). *Miscellaneous Information, Tokyo University Forests* 29: 141–204.

Walter, H. (1985). *Vegetation of the Earth and Ecological Systems of the Geo-biosphere*, 3rd edn. Berlin: Springer-Verlag.

15 Temperature and humidity as determinants of the transition from dry pine forest to humid cloud forests in the Bhutan Himalaya

P. Wangda and M. Ohsawa

The University of Tokyo, Tokyo, Japan

ABSTRACT

Air temperature and humidity, as well as soil moisture conditions were decisive in determining forest composition and structural changes along an altitudinal transect from a dry valley bottom at 1250 m.a.s.l. to a ridge top at 3550 m.a.s.l. in the Bhutan Himalaya. Mean annual temperature and soil moisture content were inversely related. Mean annual temperature decreased from 18.2 °C in the valley to 4.3 °C at the ridge top, with an average lapse rate of 0.62 °C $100\,m^{-1}$. Conversely, relative humidity, soil moisture content, and mean annual precipitation all increased with elevation, from 71.0%, 14.7%, and 584 mm in the valley to 93.5%, 73%, and 1576 mm at 3550 m. Based on quantitative vegetation data, climatic conditions, and observations on the presence of mosses and epiphytic vascular plants, the cloud-affected zone was defined as lying above *c.* 2500 m.a.s.l. and coincided with the transition from moist broad-leaved forest to moist/wet broad-leaved forest. Two floristic/physiognomic transitions were identified along the altitudinal gradient between broad-leaved and coniferous forests; the lower one at 2000 m a.s.l. between dry *Pinus roxburghii* forest and moist evergreen broad-leaved *Quercus lanata* forest (intervening with deciduous broad-leaved *Q. griffithii* in between). A critical soil moisture content of *c.* 20% marked the limit of downward extension of broad-leaved trees into the shade-intolerant but drought-tolerant pine zone. The second, upper transition at *c.* 2800 m.a.s.l. coincided with the transition from moist/wet evergreen broad-leaved forests dominated by *Q. semecarpifolia*, *Q. glauca*, and *Q. oxyodon*, to cold and humid *Tsuga dumosa* forest where lowest monthly mean temperatures fell below *c.* 0 °C. Forest biomass as indicated by maximum height (33 m), maximum diameter at breast height (DBH, 132.5 cm) and total basal area (BA, 101.7 $m^2\ ha^{-1}$) of cloud-affected forest at 3370 m.a.s.l. was significantly higher than that of dry pine forest (with height, DBH, and BA of 14.6 m, 35.5 cm, and 15.2 $m^2\ ha^{-1}$, respectively) at 1520 m.a.s.l., or even that of the mixed broad-leaved forests at intermediate elevations.

INTRODUCTION

The Himalaya stretches over 2500–3000 km from Mt. Nanga Parbat in Pakistan to Namcha Barwa in Nyingchi Prefecture of the Tibet Autonomous Region in China, with widths varying between 80 and 300 km between the southern foothills and the northern greater Himalaya (Barry, 1981; Gansser, 1983). Cloud forests are found mostly above 2000 m.a.s.l. and in the outermost ranges in the more humid eastern Himalaya, including the kingdoms of Nepal and Bhutan (Dobremez, 1976; Ohsawa, 1995). The great geographic and topographic diversity, combined with correspondingly diverse climatic conditions, contributes to Bhutan's outstanding range of biodiversity and ecosystem complexity (National Biodiversity Action Plan, 2002).

Climatically drier valleys within the Himalaya–Hindu Kush are found between Kabul in the west and Chang Jiang in the east and tend to run perpendicularly to the prevailing mountain orientation, cutting through the mountains roughly from north to south, including in Bhutan (Schweinfurth, 1992; cf. Figure 15.1a). These valleys differ from the *Dun* valleys of the Lesser Himalaya

Tropical Montane Cloud Forests: Science for Conservation and Management, eds. L. A. Bruijnzeel, F. N. Scatena, and L. S. Hamilton. Published by Cambridge University Press.

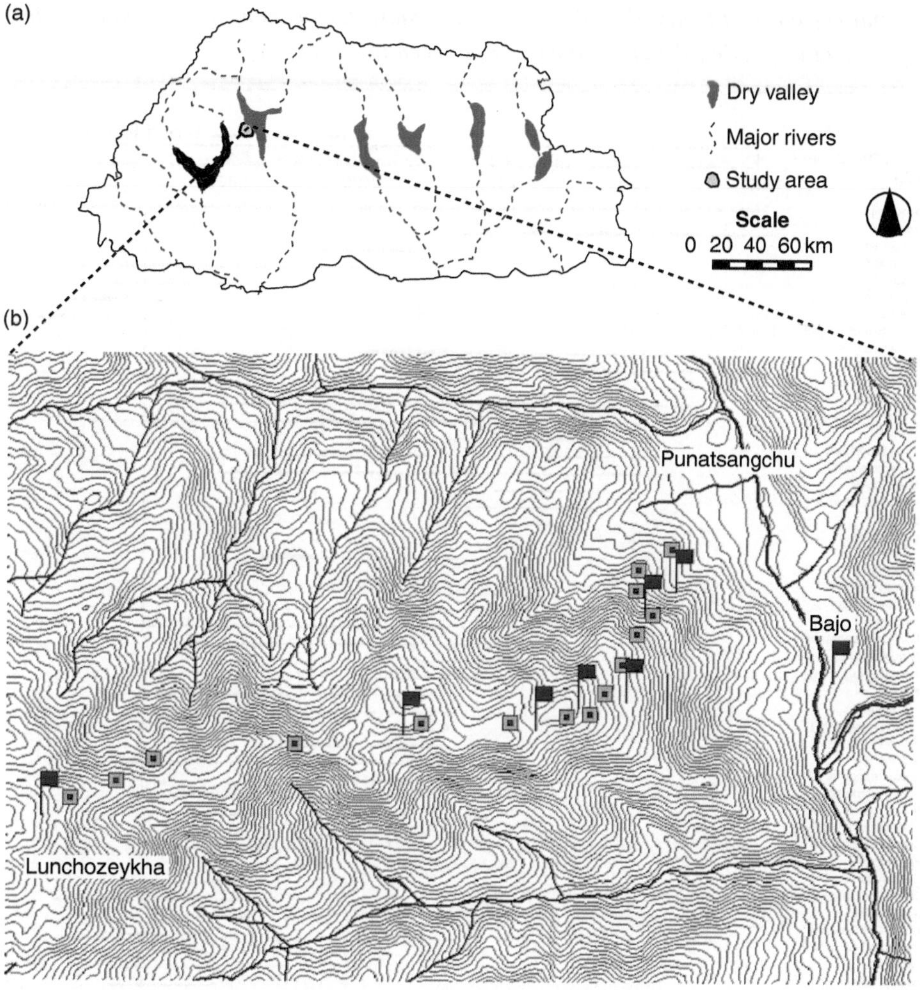

Figure 15.1. (a) Map of Bhutan showing dry valleys (shaded) along the major rivers; (b) location of the vegetation plots (squares with black center) and climatic stations (black flags). Contour interval = 40 m.

in India and Nepal that tend to run more parallel to the mountains; their drier condition is due to a rain-shadow effect (Higuchi *et al.*, 1982).

In the humid Bhutan Himalaya, dry valleys are associated with strong, hot valley winds and high temperatures that result in local dryness (Eguchi, 1987, 1997). The dry vegetation in the valley bottoms gradually gives way along the side slopes of the river valley to moister forest types and finally to cloud forest, with the elevations of the respective transitions depending upon overall humidity conditions in the valley (Ohsawa, 1991). Most of the population residing in and along the dry valley bottoms is dependent on the various forests for timber, fuelwood, and litter collection, and for cattle grazing as in other parts of the Himalaya (e.g. India – Tejwani, 1985; Nepal – Shrestha, 1988). Hence, the present study was conducted to clarify the influence of environmental factors (notably air temperature and humidity, as well as soil moisture content) on the location of the respective forest transitions, and to identify the cloud-affected forest zone along the altitudinal gradient.

MATERIALS AND METHODS

Study area

The study area was located in and around one of the typical dry valleys along the Punatsang-chu (Punatsang river) in West Central Bhutan (Figure 15.1a). A transect was established facing east and running between Bajo (1250 m.a.s.l., 27° 30′ N, 89° 52′ E) in the valley bottom, and Lunchozeykha (3550 m.a.s.l., 27° 28′ N, 89° 45′ E) on the ridge top (Figure 15.1b).

Based on the vegetation descriptions of Ohsawa (1991) and Wangda and Ohsawa (2006a) the vegetation consists of dry *Pinus roxburghii* forest at the valley bottom at *c.* 1500 m.a.s.l., grading into dry broad-leaved forest with evergreen *Quercus lanata* and deciduous *Q. griffithii* on the lower slopes between 2000 and 2500 m.a.s.l. In turn, this forest gives way to humid evergreen broad-leaved forest with *Q. semecarpifolia, Q. oxyodon*, and *Q. glauca* between 2640 and 2820 m.a.s.l. Finally, conifers associated with wet conditions (*Abies densa,*

Table 15.1 *Species distribution ranges with elevation along the study transect. Dominant species indicated by thick lines. CMT, mean temperature of the coldest months marking the upper limit of evergreen broad-leaved species*

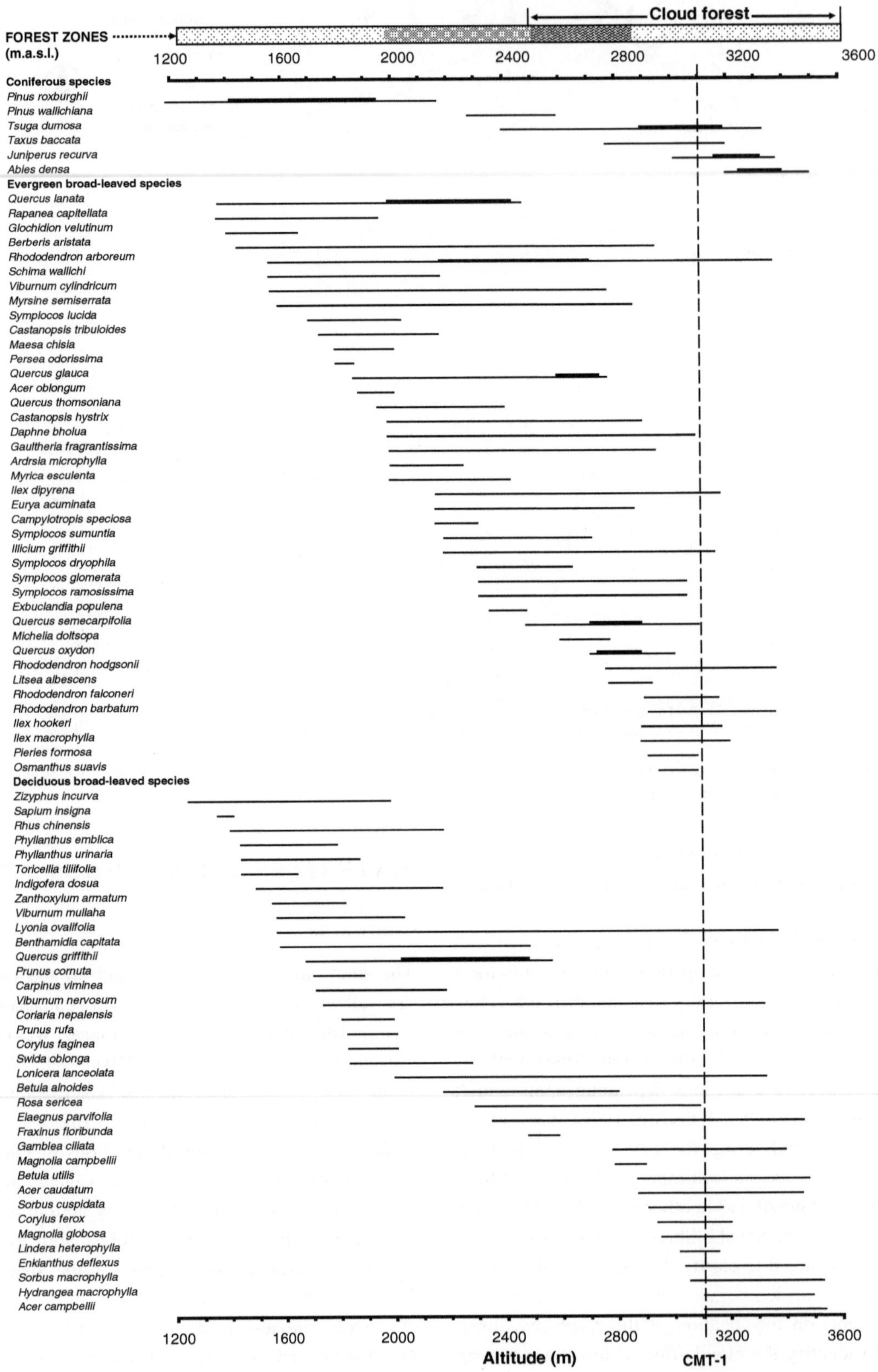

Tsuga dumosa, and *Juniperus recurva*) dominate the ridge tops. In the chir pine (700–2000 m.s.l.) and blue pine forest (2000–3000 m.a.s.l.) zone, the soils are moderately leached and have mostly shallow topsoils and moderate organic matter contents. Above the transition from temperate broad-leaved or blue pine forest to mixed conifer forest (2500–3300 m.a.s.l.) the soils are deep, bright reddish yellow to almost orange, silty and clay loams with high soil water contents. The soils at the tree-line (3800–4200 m.a.s.l.) are deep, dark, and friable, with common earthworms while the sub-soils are yellowish brown (Baillie *et al.*, 2004). Although the forests are to be conserved for watershed protection purposes (WWMP, 2007), the dry pine forest is affected by traditional human use, such as timber harvesting, and the collection of litter and firewood. The mid-altitude mixed broad-leaved forests are used for firewood and fodder collection, and the upper humid conifer forests for cattle grazing by yaks in winter and local cattle in summer.

Climatic data collection

Air temperature and relative humidity were monitored between 1999 and 2004 at elevations of 1250, 1450, 1500, 1960, 2210, 2320, 2650, 2820, 3185, and 3500 m.a.s.l. using H8 Pro Series sensors connected with Hobo Onset data-loggers (see Figure 15.1b for locations). Data were recorded at 1-hour intervals and downloaded every 6 months. Precipitation at the respective altitudes was monitored using tipping-bucket rainfall recorders with a resolution of 0.5 mm per tip (Ota Keiki Seisakusho Co., Tokyo). Secondary data from climate stations throughout Bhutan (Ministry of Agriculture, 1994–2001) were used for comparison. Soil moisture contents were measured manually down to 12 or 20 cm depth during each field visit using Hydro-sense (CD 620 + CS 620) Campbell Scientific soil water probes.

The climatic data were used to construct climate diagrams (Walter and Lieth, 1961–1967; Lieth *et al.*, 1999) to characterize the seasonality, duration of the dry period, and other climatic indices for the study area, such as the coldest month's mean temperature (CMT), temperature lapse rates with elevation, etc.

Vegetation survey and data analysis

Fifteen forest sampling plots were established along the transect between 1250 and 3550 m.a.s.l. (Figure 15.1b). Each plot was subdivided into sub-plots of 10 × 10 m and all species attaining a height above 1.3 m were identified and measured for height and diameter. Nomenclature of plants followed the *Flora of Bhutan* (Grierson and Long, 1983–2000), *Flowers of the Himalaya* (Polunin and Stainton, 1984), and *Flowers of the Himalaya: A Supplement* (Stainton, 1988).

Basal area totals per tree species were calculated from measurements of diameter at breast height (DBH, in cm) and corresponding numbers of individuals and expressed as the relative basal area (RBA) of each species. The dominant species in each plot was determined by dominance analysis (Ohsawa, 1984; Kikvidze and Ohsawa, 2002). Floristic composition was analyzed using a pivotal table in Microsoft Excel. The RBA data were subjected further to multivariate analysis using PC-ORD Multivariate Analysis of Ecological Data Version 4 and a cluster analysis was performed using the distance measure of Sorensen (Bray–Curtis method) (McCune and Mefford, 1999).

RESULTS AND DISCUSSION

Altitudinal vegetation change

The forest types and structural attributes changed along the altitudinal gradients from warm dry pine forest at the valley bottom (1250 m.a.s.l.) through warm moist evergreen broad-leaved forest (2200 m.a.s.l.) to humid cool conifer forest (3550 m.a.s.l.) (Wangda and Ohsawa, 2006a). The cloud-affected forest zone occurred in the transition zone (c. 2500 m. a.s.l.) between wet–moist evergreen broad-leaved forest and upper cool humid conifer forest. The floristic composition of the broad-leaved forest changed around 2500 m.a.s.l., from a dominance of evergreen *Quercus lanata* and deciduous *Q. griffithii* in drier habitats to evergreen *Q. glauca, Q. oxyodon*, and *Q. semecarpifolia* in moist to wet habitats (Table 15.1 below). From the forest structural perspective, maximum tree height, DBH, and total BA increased with elevation (Wangda and Ohsawa, 2006a, b) and were directly correlated with the various environmental changes along the altitudinal gradient as discussed in the following.

Altitudinal gradients in temperature, humidity, and rainfall

Annual mean temperature decreased with elevation at a mean lapse rate of 0.62 °C 100 m^{-1} from 18.2 °C at Bajo (1250 m.a.s.l.) to 4.3 °C at Lunchozeykha (3550 m.a.s.l.; Figure 15.2a). This is slightly higher than the average temperature lapse rate of 0.54 °C 100 m^{-1} derived for the humid slopes of the Bhutan Himalaya as a whole (as calculated from data listed in Ministry of Agriculture, 2003). This enhanced lapse rate probably reflects dry air moving up the dry valley slopes (Wangda and Ohsawa, 2006a; cf. Flohn, 1970). This contention is supported by the seasonal pattern of lapse rates, with higher values in the dry winter time compared to the wet summer months (Figure 15.2b). Monthly lapse rates correlated inversely with monthly mean relative humidity ($r = -0.80$, $p < 0.05$) indicating high temperature but low humidity as a result of high evapotranspiration (Figure 15.2b). The lowest monthly mean temperature of −1 °C (normally used as a thermal indicator for the upper limit of

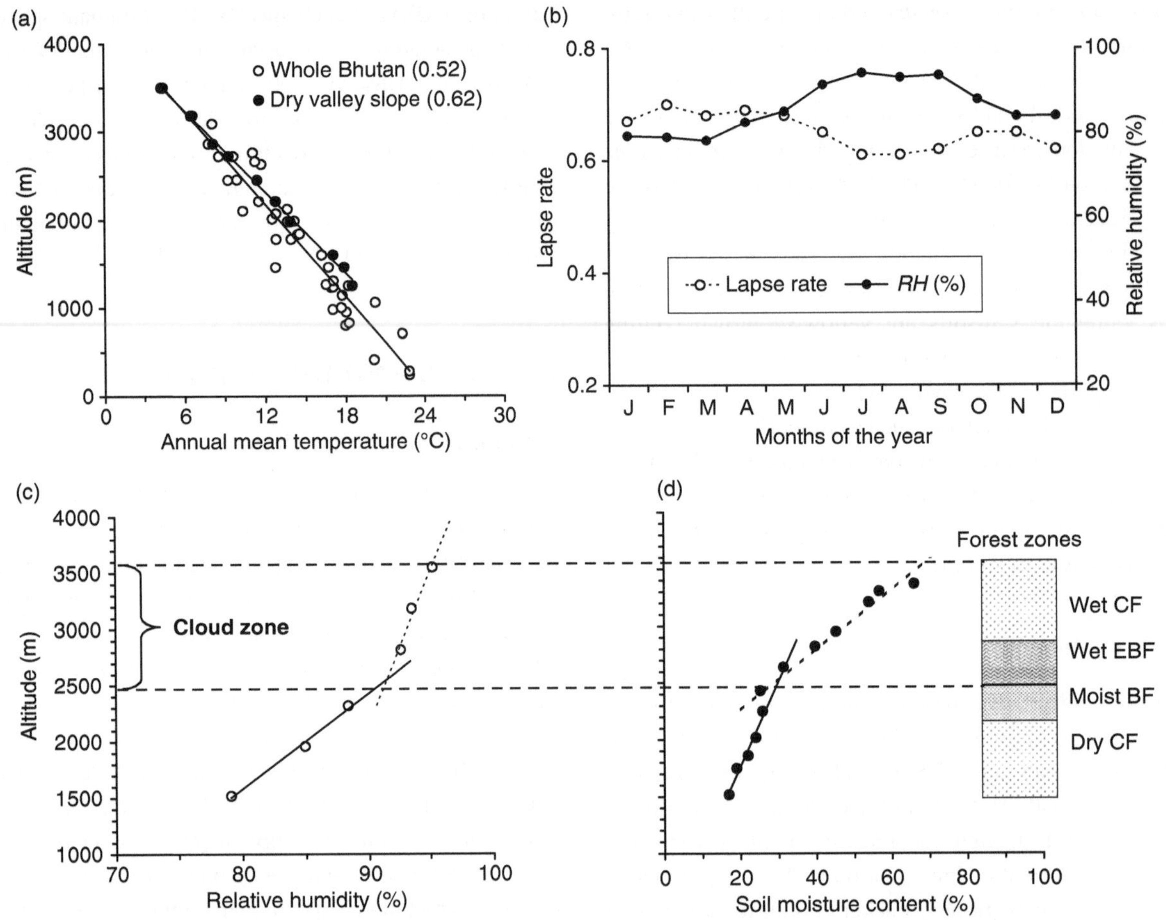

Figure 15.2. (a) Comparison of temperature lapse rates for humid slopes of the Bhutan Himalaya (1000–3500 m.a.s.l.) and for the study transect (1250–3550 m.a.s.l.); (b) seasonal variations in temperature lapse rates and overall average relative humidity along the study transect (after Wangda and Ohsawa, 2006a,b); (c) relative humidity and (d) soil moisture content vs. elevation. Forest zones are indicated on the right with cloud-affected zone above 2500–2700 m.a.s.l., BF, broad-leaved forest; EBF, evergreen broad-leaved forest; CF, conifer forest.

evergreen broad-leaved trees; Ohsawa, 1990), occurred at *c*. 3000 m.a.s.l. (cf. Table 15.1 below).

Both relative humidity of the air and soil moisture content exhibited marked changes with elevation (Figure 15.2c–d), from average values of 71.0% and 14.7%, respectively, at the valley bottom (1250 m.a.s.l.) to 93.5% and 73% at the ridge top (3550 m.a.s.l.). Similarly, mean annual precipitation totals increased from 572 mm at 1250 m.a.s.l. via 882 mm at 1450 m.a.s.l. to 1576 mm on the ridge (3185 m.a.s.l.). Thus, the slopes around the dry valley exhibited temperature, humidity, and rainfall gradients that can be expected to affect the distribution of the vegetation. The location of the cloud-affected zone was initially identified by the presence of biotic indicators such as mosses hanging from branches, mosses on tree trunks, and the presence of epiphytic vascular plants on the trees at elevations between *c*. 2500 m.a.s.l. and the ridge tops at 3370 m (Figure 15.3). This was confirmed by the abrupt changes in relative humidity (>90%)

Figure 15.3. Moss hanging from the branches of a *Myrsine semiserrata* tree at 2540 m.a.s.l. indicating the start of cloud forest. (See also color plate.)

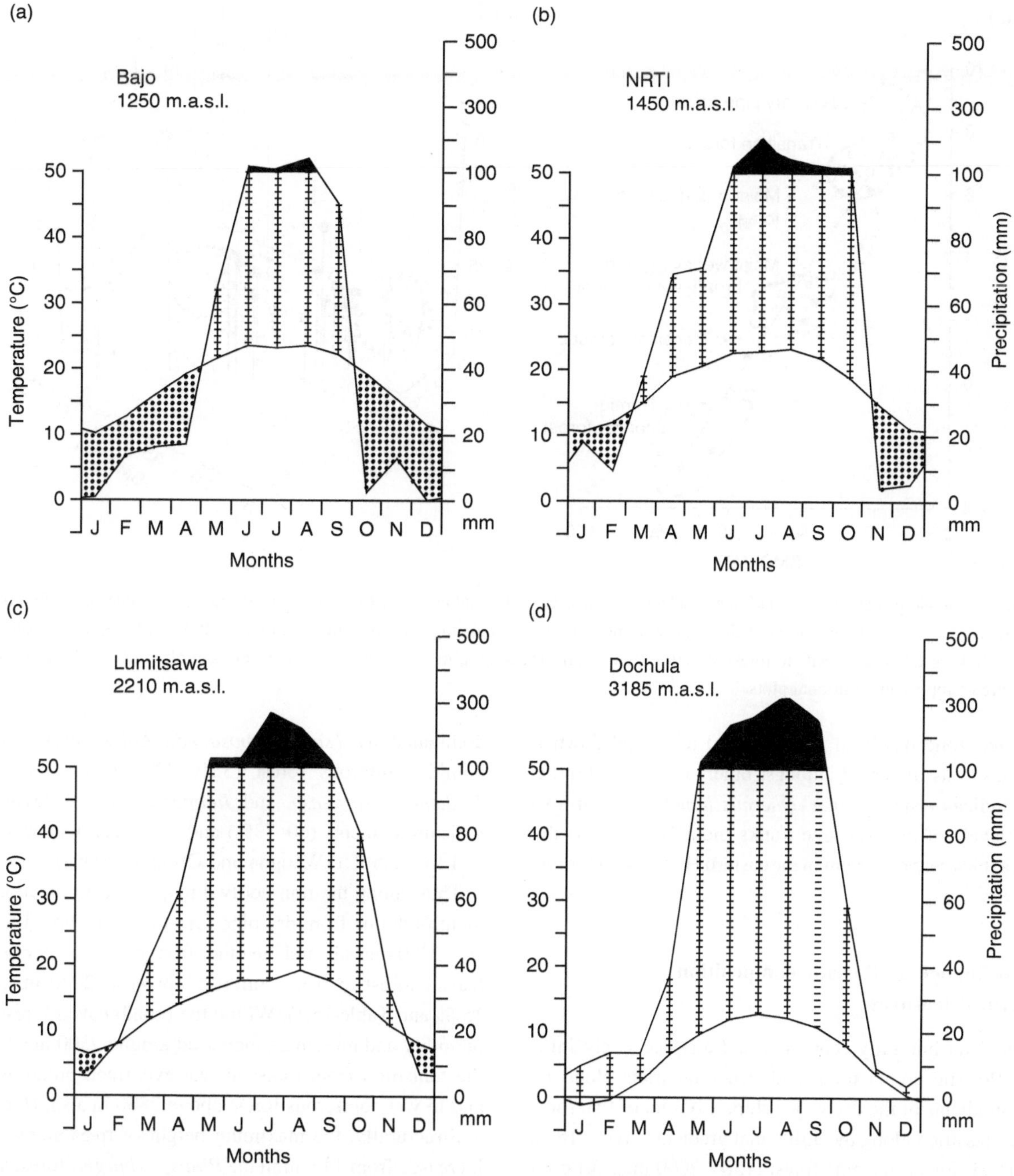

Figure 15.4. Climate diagrams for selected stations along the altitudinal transect indicating a pattern of decreasing length of dry spells with elevation. NRTI, Natural Resources Training Institute (located within the pine belt).

(Figure 15.2c) and soil moisture content (>30%) (Figure 15.2d) observed around 2500 m.

Differences in seasonal precipitation and temperature distribution patterns with elevation were reflected in differences in the duration and intensity of dry spells at different altitudes as shown in the respective climate diagrams (Figure 15.4). A prolonged dry spell was experienced from mid-October to mid-April at the valley bottom (1250 m) (Figure 15.4a), shortening to *c.* 4 months at 1450 m (Figure 15.4b), and to *c.* 2 months at 2210 m (Figure 15.4c). Conversely, in the cloud-affected zone, the increased precipitation and relative humidity levels combined with decreased temperatures maintained a wet environment throughout the year (Figure 15.4d).

Based on relative humidity and soil moisture content, as well as floristic composition and occurrence of mosses and epiphytes, the altitude of cloud-affected forest above the dry valleys of the study area was defined as lying above 2500 m.a.s.l. (Figure 15.2c–d). On wetter slopes elsewhere in the Bhutan Himalaya, the

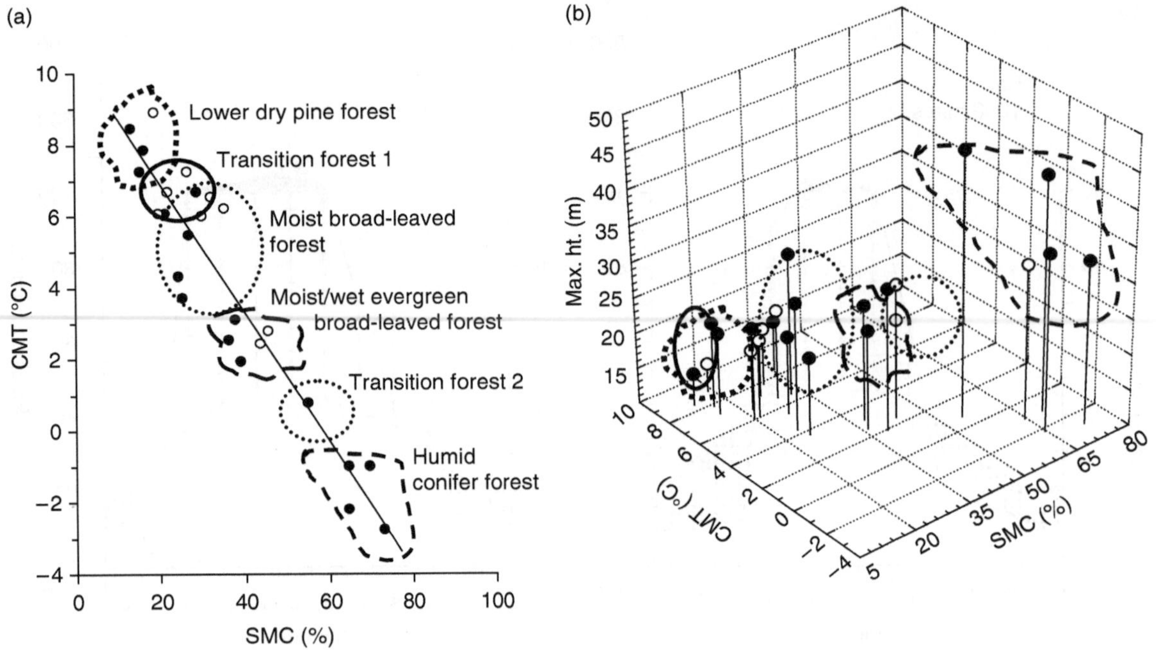

Figure 15.5. Relationships between vegetation and selected environmental variables along the study transect. (a) Forest types identified by cluster and dominance analysis vs. mean temperature of the coldest months (CMT) and average soil moisture content (SMC); (b) three-dimensional representation linking CMT and SMC to forest structure (maximum tree height, m). Closed circles represent sampling plots within the study area while open circles represent additional plots.

cloud-affected zone has been demonstrated to extend down to *c.* 2000 m.a.s.l., as indicated amongst others by the occurrence of certain indicator species (e.g. *Quercus lamellosa*) (Ohsawa, 1991) and layers of moss on tree trunks (notably *Quercus*) that may reach a thickness of *c.* 5 cm in the eastern Himalaya (Ohsawa, 1995).

Climatic influence on floristic composition and structural features

Temperature and humidity (both soil and air) decisively influenced the floristic composition and structural traits along the altitudinal gradients of the dry valley slopes. A total of five forest types were classified along the altitudinal gradient (Figure 15.5a, Table 15.1): (i) lower dry pine forest (1500–2000 m.a.s.l.) dominated by *Pinus roxburghii* and experiencing low soil moisture contents (15–20%) but high mean winter temperatures (6.7–8.4 °C); (ii) moist broad-leaved oak forest at mid-slope positions (2000–2540 m.a.s.l.) dominated by evergreen *Quercus lanata* and deciduous *Q. griffithii* associated with moderate soil moisture contents (21–35%) and mean winter temperatures of 3.1–6.1 °C, the upper part of this zone becoming gradually more affected by cloud; (iii) moist/wet evergreen broad-leaved forest (2650–2820 m.a.s.l.) dominated by *Q. semecarpifolia, Q. glauca,* and *Q. oxyodon* associated with relatively high soil moisture contents (37–40%) and low mean winter temperatures (1.0–2.5 °C); (iv) cool humid coniferous forest (2950–3210 m.a.s.l.) dominated by *Tsuga dumosa* and *Abies densa*; and (v) cold humid coniferous forest (3210–3370 m.a.s.l.) dominated by *T. dumosa, A. densa,* and *Juniperus recurva* having high soil moisture contents (60–73%) and low mean winter temperatures (−1 to −2.5 °C) (Wangda and Ohsawa, 2006a).

Thus, along the transect two major forest transitions could be identified: one from dry pine forest to moist broad-leaved forest at *c.* 2000 m.a.s.l., and the other from moist/wet evergreen broad-leaved forest to wet coniferous forest at 2820 m.a.s.l. (Figure 15.5a and Table 15.1). Within the broad-leaved forest zone, both humidity and mossiness increased around 2500 m.a.s.l., marking the transition from moist to wet evergreen broad-leaved forest and to wet coniferous forest above 2800 m.a.s.l. (Figure 15.5a).

Structurally, the maximum height of trees along the transect increased from 14.6 m in the *Pinus roxburghii* forest at 1520 m.a.s.l. to 33.0 m in *Abies densa* forest at 3370 m.a.s.l. (Figure 15.5b). Similarly, maximum DBH and total BA increased from 35.5 cm and 15.2 $m^2 ha^{-1}$ at 1520 m, to 133.0 cm and 101.7 $m^2 ha^{-1}$ at 3370 m. A direct relationship was observed between these forest structural traits and soil moisture content along the transect (Figure 15.5a, b; cf. Wangda and Ohsawa, 2006a).

A soil moisture threshold of *c.* 20% appeared to control the downslope extension of broad-leaved species into the pine zone below 2000 m.a.s.l. (Figure 15.2d). Further, the average temperature of the coldest month of −1 °C that is often used as an indicator of the upper limit of evergreen broad-leaved species in humid East Asian mountains (Ohsawa, 1990) coincided with the elevation (*c*. 3000 m.a.s.l.) at which evergreen broad-leaved species in the understorey of the coniferous forest reached their upper limit (Table 15.1).

CONCLUSIONS

The relative intactness of Bhutanese forests allowed the identification of changes in species composition and forest structure along the prevailing steep altitudinal and climatic gradients. Forest zonation in the study area is controlled principally by two environmental components: a mean winter temperature below −1 °C controlling the upslope extension of evergreen broad-leaved species above 3000 m.a.s.l., and low soil moisture content (*c*. 20%) controlling the downslope extension of broad-leaved species into the dry pine zone below 2000 m.a.s.l. In addition, the occurrence of fog and low cloud above 2500 m.a.s.l. marks a transition from moist to wet evergreen oak forest. In more humid areas elsewhere in the Bhutan Himalaya the lower boundary of cloud-affected forest is typically found around 2000 m.a.s.l. The higher cloud condensation level in the study area is thought to reflect upslope movement of warm dry air from the valleys. The local cloud belt extends to an elevation of 2800 m.a.s.l. and therefore includes both broad-leaved and coniferous forests.

ACKNOWLEDGEMENTS

We thank Professors Lawrence S. Hamilton and Sampurno Bruijnzeel as well as Dr. Zaal Kikvidze for comments and suggestions on this manuscript.

REFERENCES

Baillie, I. C., D. Tshering, Tshering Dorji, H. B. Tamang, and T. Dorji (2004). Regolith and soils in Bhutan, Eastern Himalayas. *European Journal of Soil Science* **55**: 9–27.

Barry, R. G. (1981). *Mountain Weather and Climate*. London: Routledge.

Dobremez, J. F. (1976). *Le Népal: Ecologie et biogéographie*. Paris: Editions du Centre National de la Recherche Scientifique.

Eguchi T. (1987). Synoptic analysis of temperature, precipitation and moisture conditions in the Bhutan Himalaya. In *Life Zone Ecology of the Bhutan Himalaya*, Vol. 1, ed. M. Ohsawa, pp. 209–247. Chiba, Japan: Chiba University.

Eguchi T. (1997). *Regional and Temporal Variations in Precipitation in the Eastern Part of the Himalayas*. Kochi, Japan: Faculty of Humanities and Economics, Kochi University.

Flohn, H. (1970). Beiträge zue Meteorologie der Himalayas. *Khumbu Himal* **7**: 25–45.

Gansser, A. (1983). *Geology of the Bhutan Himalaya*. Basel, Switzerland: Birkhauser Verlag.

Grierson A. J. C., and D. G. Long (1983–2000). *Flora of Bhutan*, Vol. 1, Parts 1, 2, 3, Vol. 2, Parts 1, 2, 3, and Vol. 3, Parts 1 and 2. Edinburgh, UK: Royal Botanic Gardens.

Higuchi, K., Y. Ageta, T. Yasunari, and J. Inoue (1982). Characteristics of precipitation during the monsoon season in high-mountain areas of the Nepal Himalaya. *International Association of Hydrological Sciences Publication* **138**: 21–30.

Kikvidze Z., and M. Ohsawa (2002). Measuring the number of co-dominants in ecological communities. *Ecological Research* **17**: 519–525.

Kimmins J. P. (1978). *Forest Ecology*. New York: Collier Macmillan.

Lieth, H., J. Berlekamp, S. Fuest, and S. Riediger (1990). *Climate Diagram World Atlas*. Leiden, the Netherlands: Backhuys.

McCune, B., and M. J. Mefford (1999). *PC-ORD: Multivariate Analysis of Ecological Data*, Version 4. Gleneden Beach, OR: MjM Software Design.

Ministry of Agriculture (1994–2001). *Meteorological Data of Bhutan for the Past 8 years (1994–2001)*. Thimphu Bhutan: Meteorological Section, Council for RNR Research of Bhutan, Ministry of Agriculture, Royal Government of Bhutan.

Ministry of Agriculture (2003). *Meteorological Data*.

Ohsawa, M. (1984). Differentiation of vegetation zones and species strategies in the subalpine region of Mt. Fuji. *Vegetatio* **57**: 15–52.

Ohsawa, M. (1990). An interpretation of latitudinal patterns of forest limits in south and East Asian mountains. *Journal of Ecology* **78**: 326–339.

Ohsawa, M. (1991). Montane evergreen broad-leaved forests of Bhutan Himalaya. In *Life Zone Ecology of the Bhutan Himalaya*, Vol. 2, ed. M. Ohsawa, pp. 89–156. Chiba, Japan: Chiba University.

Ohsawa, M. (1995). The montane cloud forest and its gradational changes in South-East Asia. In *Tropical Montane Cloud Forests*, eds. L. S. Hamilton, J. O. Juvik, and F. N. Scatena, pp. 254–265. New York: Springer-Verlag.

Polunin, O., and A. Stainton (1984). *Flowers of the Himalaya*. New Delhi: Oxford University Press.

Schweinfurth, U. (1992). Mapping mountains: vegetation in the Himalaya. *GeoJournal* **27**: 73–83.

Shrestha, T. B. (1988). *Development Ecology of the Arun River Basin in Nepal*. Kathmandu, Nepal: International Center for Integrated Mountain Development.

Stainton, A. (1988). *Flowers of the Himalaya: A Supplement*. New Delhi: Oxford University Press.

Royal Government of Bhutan (2002). *National Biodiversity Action Plan*. Thimphu, Bhutan: Royal Government of Bhutan, Ministry of Agriculture.

Royal Government of Bhutan (2003). *Statistical Yearbook 2002: Population*. Thimphu, Bhutan: Central Statistical Office, Planning Commission.

Tejwani, K. G. (1985). *Watershed Management in the Indian Himalayas*. Kathmandu, Nepal: International Center for Integrated Mountain Development.

Walter, H., and H. Lieth (1961–1967). *Climate Diagram World Atlas*. Jena, Germany: Fischer-Verlag.

Wangda, P., and M. Ohsawa (2006a). Gradational forest change along the climatically dry valley slopes of Bhutan in the midst of humid eastern Himalaya. *Plant Ecology* **186**: 109–178.

Wangda, P., and M. Ohsawa (2006b). Structure and regeneration dynamics of dominant tree species along altitudinal gradient in dry valley slopes of the Bhutan Himalaya. *Forest Ecology and Management* **230**: 136–150.

WWMP (2007). *Wang Watershed Management Project Final Report*. Thimphu, Bhutan: Royal Government of Bhutan, Ministry of Agriculture.

16 The importance of cloud forest sites in the conservation of endemic and threatened species of the Albertine Rift

I. Owiunji and A. J. Plumptre
Wildlife Conservation Society, Kampala, Uganda

ABSTRACT

The Albertine Rift extends from the northern tip of Lake Albert to the southern tip of Lake Tanganyika and encompasses the natural habitats within about 100 km of the border of the Democratic Republic of Congo on both sides. This region is the richest area for vertebrates in Africa, with more birds and mammals than any other region. Over 52% of Africa's birds have been recorded here and 39% of Africa's mammals. One hundred and twenty-five vertebrates are endemic to this region, including species such as the mountain gorilla and golden monkey. Information on total species richness and number of endemic and threatened species was compiled to prioritize sites within the Rift area for conservation. Six of the 40 sites in the Rift area that had species data are cloud forest sites: Rwenzori Massif, Bwindi Impenetrable National Park, Virunga Volcanoes, Nyungwe National Park, Kahuzi Biega National Park, and Itombwe Massif. The 40 sites were ranked both in terms of their total species richness and number of endemic and threatened species. All of the cloud forest sites ranked high in terms of endemic and threatened species. If only the cloud forests were to be conserved, some 94% of endemic mammals, 95% of endemic birds, 89% of endemic reptiles, and 69% of endemic amphibians would be protected. Each of these cloud forests is highly threatened by the high human population densities around them. The Itombwe Massif assumes the highest conservation priority at present as it is as yet unprotected.

INTRODUCTION

The Albertine Rift Region (ARR) stretches from the northern tip of Lake Albert to the southern tip of Lake Tanganyika (Figure 16.1) and encompasses the natural habitats within about 100 km of the eastern border of the Democratic Republic of Congo (DRC). The region is the most species-rich in mainland Africa for vertebrates and contains more endemic and threatened vertebrate species than any other site in Africa (N. Burgess, personal communication). Altitudes range from about 600 to 5100, m.a.s.l. and include a large variety of habitat types and ecosystems. Ecosystem types range from afro-alpine montane and submontane forests, savanna grasslands and woodlands, via hot springs, lava flows and the successional vegetation associated with it, to wetlands, and freshwater lakes. Montane forests, afro-alpine zones, and lakes are particularly important from a conservation perspective because these contain many of the endemic and threatened species of the region. Six cloud forest areas are found within the ARR, including the Rwenzori Massif, Bwindi Impenetrable National Park, Virunga Volcanoes (Mgahinga Gorilla, Volcanoes, and Mikeno sectors), Nyungwe National Park, Kahuzi Biega National Park, and the Itombwe Massif (see Figure 16.1 for locations). These range in altitude from 1700, to 5100, m.a.s.l. but all sites attain an altitude of at least 2900, m.a.s.l.

Human population densities in this region are among the highest in Africa, with densities reaching 400–600 people km^{-2} (Plumptre *et al.*, 2004). Consequently, there are great pressures on the natural habitats. Despite its biologic richness, financial resources to conserve sites within the region are largely lacking. Therefore, there is a distinct need to prioritize sites for conservation.

The Wildlife Conservation Society (WCS) has been working with many partners to compile existing data on species distributions within the ARR in order to identify priority areas for conservation and further research (Plumptre *et al.*, 2003). The available biologic information, both in published form or unpublished reports, has been compiled for 40 sites including the six cloud forest areas cited above and used to assess their relative importance for conservation. The sites include both protected

Tropical Montane Cloud Forests: Science for Conservation and Management, eds. L. A. Bruijnzeel, F. N. Scatena, and L. S. Hamilton. Published by Cambridge University Press.

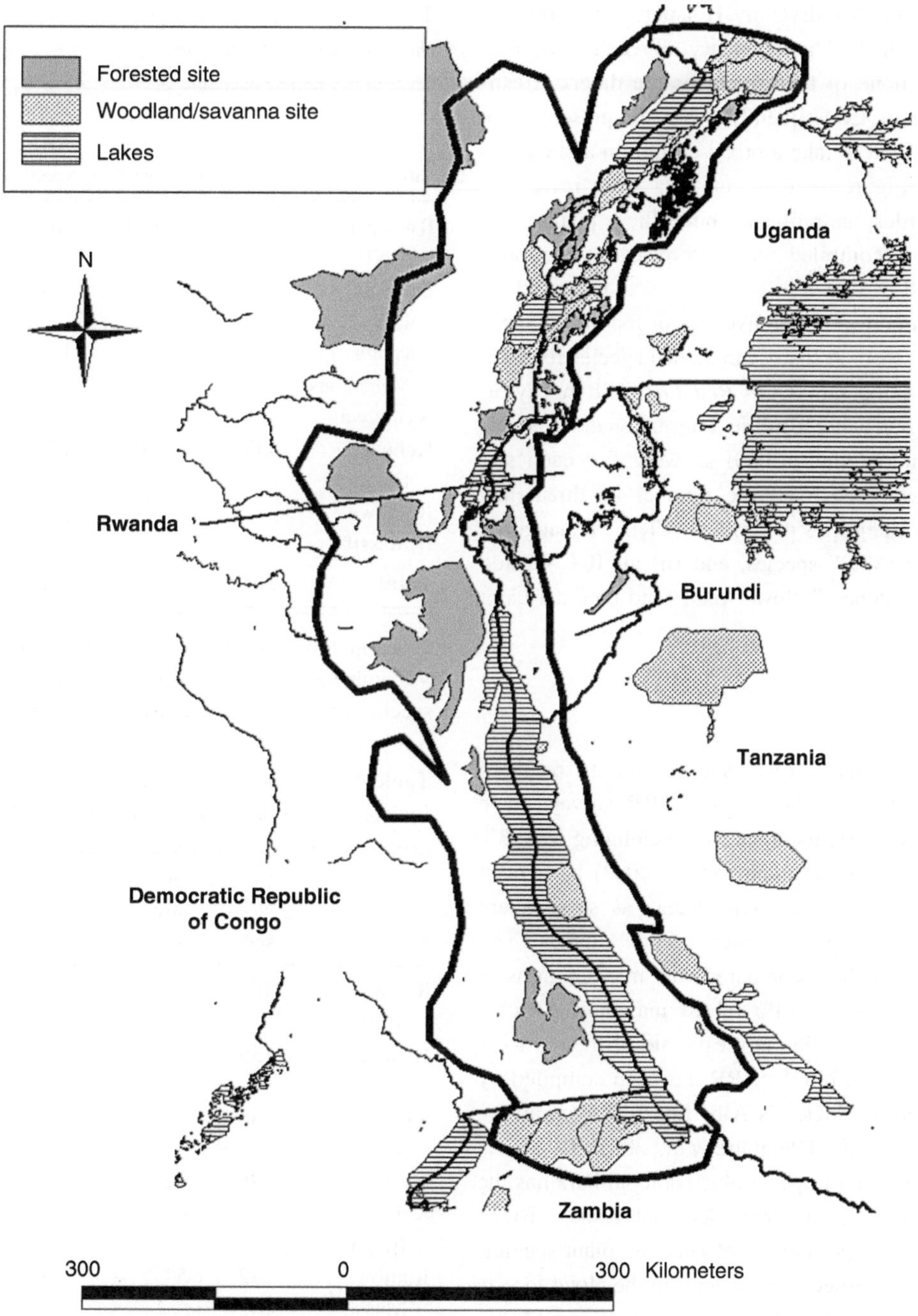

Figure 16.1. Map of the boundary of the Albertine Rift. Dark gray areas are forested protected areas or areas that may deserve protection, light gray areas are woodland/savanna protected areas.

areas and sites that are relatively undisturbed but not gazetted (cf. Figure 16.1). Two of the cloud forest sites – Kahuzi Biega National Park and the Itombwe Massif – have not been well surveyed to date, but it is believed that the patterns shown below reflect the reality of the situation in general, even though there could be changes in the numbers of certain species after future research. This chapter assesses the importance of cloud forests within the ARR to the conservation of endemic and threatened species in the area.

THE IMPORTANCE OF THE ALBERTINE RIFT FOR BIODIVERSITY CONSERVATION

The ARR has been identified as being an important region under several global priority-setting exercises. It has been identified as an "Endemic Bird Area" by BirdLife International (ICBP, 1992; Stattersfield *et al.*, 1998) and as an "Ecoregion" by the World Wide Fund for Nature (Olson and Dinerstein, 1998), and it will soon become part of the "Biodiversity Hotspots" identified by Conservation International. The boundary of each definition varies somewhat and none of them includes the diverse freshwater lakes in the region. Consequently, when the present work was started it was decided to take a broad definition and choose the area depicted in Figure 16.1. Seven taxa were reviewed: mammals, birds, reptiles, amphibians, butterflies, plants, and fish. Species lists were compiled for the terrestrial vertebrates and plants while a list of endemic butterflies was produced by T. Davenport. The literature on the diversity of fish in the large lakes was reviewed as well. A list of threatened species for each taxon was compiled using the IUCN *Red List* (Hilton-Taylor, 2000) for most taxa but using Birdlife International's analysis for birds (Birdlife International, 2000) as well. For each site two classifications were used: (i) the number of threatened species using IUCN's categories of "critical (CR)," "endangered (EN)," or "vulnerable (VU)" species, and (ii) all IUCN-listed species including "threatened," "lower-risk," and data-deficient species.

Plants

Higher plants have been relatively well surveyed in the forests of Uganda and Rwanda but elsewhere in the ARR surveys have been patchy. Currently 5793, plant species (belonging to 1537, genera and 233 families; see Plumptre *et al.*, 2003) have been recorded within the Rift but this will change as surveys are discovering new species regularly even within Uganda. The number of plant species is high compared with many regions of similar size and accounts for 14% of all mainland Africa's estimated plant species. A preliminary estimate of the numbers of endemic plant species within the ARR has been compiled by the Wildlife Conservation Society's Albertine Rift Programme and this now totals 551 species (Plumptre *et al.*, 2003). Considering the cloud forests, Bwindi Impenetrable National Park has the highest number of plant species recorded, but Kahuzi Biega National Park has the highest number of endemic plant species. About 1% of all 3010, recorded plant species in the cloud forests of the region is considered threatened to some degree (Table 16.1). By comparison, A. Hemp (this volume #12) reported 1223, vascular plants from the montane forest belt (>1800, m.a.s.l.) on Mt. Kilimanjaro in Tanzania, representing about half of all plant species found on the mountain as a whole.

Table 16.1 *Status of tree species in six cloud forest areas within the Albertine Rift Region*

Site	Species numbers	Tree species	ARE species	Threatened CR, EN, VU	IUCN-listed species
Rwenzori MNP	696	199	55	5	7
Bwindi INP	1405	393	74	18	22
Virunga Volcanoes	878	81	124	4	5
Nyungwe	1105	230	137	7	9
Kahuzi Biega	1171	218	145	9	12
Itombwe Massif	–	–	–	–	–
Total	**3010**	**554**	**283**	**35**	**45**

CR, critically endangered species; EN, endangered species; VU, vulnerable species; ARE, Albertine Rift Endemics; IUCN-listed species, threatened, lower risk, or data-deficient categories.

Table 16.2 *Status of mammal species for six cloud forest areas located within in the Albertine Rift*

Site	No. of species	No. of large mammals	ARE species	Threatened CR, EN, VU	IUCN-listed species
Rwenzori MNP	102	26	18	10	20
Bwindi INP	135	43	20	7	18
Virunga Volcanoes	86	34	18	6	16
Nyungwe	86	42	14	3	16
Kahuzi Biega	136	67	15	14	39
Itombwe Massif	72	54	4	10	24
Total	**260**	**90**	**36**	**23**	**58**

CR, critically endangered species; EN, endangered species; VU, vulnerable species; ARE, Albertine Rift Endemics; IUCN-listed species, threatened, lower risk, or data-deficient categories.

Mammals

The six cloud forests contain 260 (64.7%) species of mammals out of a total of 402 recorded in the ARR (representing, in turn, about 39% of all mammals of mainland Africa), with all of the 36 endemic species and 11 near-endemic species represented in the cloud forests. About 9% of all mammals are considered threatened to some degree (Table 16.2). The near-endemic species range into the west in the Ituri Forest, north of the Congo River, but for the most part are confined to the Albertine Rift. Most of the endemic mammals are shrews and rodents. Of the larger endemic mammals, the best known is the eastern gorilla with its two famous sub-species, the mountain gorilla

Table 16.3 *Status of bird species for six cloud forest areas located within in the Albertine Rift*

Site	No. of species	ARE species	Threatened CR, EN, VU	IUCN-listed species
Rwenzori MNP	270	21	4	6
Bwindi INP	381	24	6	9
Virunga Volcanoes	258	20	4	7
Nyungwe	280	26	7	11
Kahuzi Biega	335	29	14	19
Itombwe Massif	583	30	14	30
Total	**709**	**38**	**18**	**36**

CR, critically endangered species; EN, endangered species; VU, vulnerable species; ARE, Albertine Rift Endemics; IUCN-listed species, threatened, lower risk, or data-deficient categories.

Table 16.4 *Status of reptile species for six cloud forest areas located within the Albertine Rift*

Site	No. of species	ARE species	Threatened CR, EN, VU	IUCN-listed species
Rwenzori MNP	76	9	0	0
Bwindi INP	34	6	0	0
Virunga Volcanoes	43	7	0	0
Nyungwe	43	8	0	1
Kahuzi Biega	69	7	0	1
Itombwe Massif	35	5	0	0
Total	**117**	**13**	**0**	**2**

CR, critically endangered species; EN, endangered species; VU, vulnerable species; ARE, Albertine Rift Endemics; IUCN-listed species, threatened, lower risk, or data-deficient categories.

(*Gorilla beringei beringei*), and the eastern lowland or Grauer's gorilla (*Gorilla beringei graueri*). The Rwenzori duiker (*Cephalophus rubidus*) and the golden monkey (*Cercopithecus kandti*) are two other medium-sized endemic mammals. Among the cloud forests of the ARR, Bwindi Impenetrable National Park in Uganda has the highest number of endemic mammal species (Table 16.2) and is home to 300 of the estimated total mountain gorilla population of *c.* 700 individuals. Small mammals have been poorly surveyed and further effort could yield more species. At present there are no endemic families of small mammals that occur in the ARR, but there are three endemic genera of small mammals – *Rwenzorisorex, Delanymys*, and *Osbornictis*. These all occur in the cloud forests (Plumptre *et al.*, 2003).

Birds

At least 1061, bird species have been recorded in the ARR (about 52.3% of all recorded birds of mainland Africa) out of which 709 species (66.8% of all ARR birds) occur in the cloud forests. Forty-one species of endemic birds occur of which 38 species (92.7%) are found in the cloud forests (Table 16.3). Although birds constitute the most thoroughly surveyed group of animals in the ARR, new species continue to be added as migrant species and new range extensions are recorded. The Rwenzori tauraco is probably the most stunning of the endemic bird species, with its mantle of iridescent green, orange-yellow cheeks, iridescent blue back and tail, and bright red primary feathers. Found in 10 forest islands of the Albertine Rift, it is a good flagship for the montane forests. The African green broadbill is another species that is very rare, confined to only three sites within the Rift, and an endemic genus as well. The Itombwe Massif, one of the six major cloud forest areas and an unprotected area located west of the northern end of Lake Tanganyika, contains more endemic species than any other site in the ARR. The Itombwe Massif is the top priority for conservation in the ARR at present because it is highly threatened (it does not have any form of protection yet and human activities that threaten biodiversity cannot be controlled; cf. Tables 16.1 and 16.5 below). Three bird genera are endemic to the ARR, i.e. *Pseudocalyptomena, Graueria*, and *Hemitesia*, but no families are endemic.

Reptiles and amphibians

The ARR is home to a total of 175 recorded reptiles (about 14% of Africa's reptiles) and 119 amphibians (about 19% of Africa's amphibians). One hundred and seventeen reptiles (66.8% of all ARR reptiles) and 93 amphibians (78.2% of ARR amphibians) have been recorded in the ARR's cloud forests alone. Sixteen reptile species and 34 amphibians are endemic to the region, while 13 reptile species and 21 amphibians are endemic to the cloud forests. Both groups have been very sparsely surveyed and large areas within the ARR have no records. Among the cloud forest sites Rwenzori National Park has the highest number of reptile endemics at present, while Virunga Volcanoes has the highest number of amphibian endemics (Tables 16.4 and 16.5, respectively). The Rwenzori three-horned chamaeleon (*Chamaeleo johnstoni*) is found in many of the montane forests of the ARR, but the strange-horned chamaeleon (*Bradypodion*

Table 16.5 *Status of amphibian species for six cloud forest areas located within in the Albertine Rift*

Site	No. of species	ARE species	Threatened CR, EN, VU	IUCN-listed species
Rwenzori Massif	62	11	8	11
Bwindi INP	33	11	9	9
Virunga Volcanoes	47	16	12	14
Nyungwe	31	15	10	12
Kahuzi Biega	27	8	5	5
Itombwe Massif	23	16	13	15
Total	**93**	**28**	**16**	**31**

CR, critically endangered species; EN, endangered species; VU, vulnerable species; ARE, Albertine Rift Endemics; IUCN-listed species, threatened, lower risk, or data-deficient categories.

xenorhinus) is solely confined to the Rwenzori Mountains where it is very rare, having been over-collected for the wildlife trade. Three amphibian genera are endemic to the ARR, namely: *Laurentophryne*, *Chrysobatrachus*, and *Callixalus*. No endemic genera of reptiles have been found here and no endemic families of either reptiles or amphibians occur here.

Butterflies

The total number of butterfly species found in the ARR is unknown, as this information cannot be compiled until many more areas have been surveyed, particularly in the eastern part of the DRC. However, records for Uganda and Rwanda indicate that up to 1300 butterflies might occur in the ARR, about 35% of Africa's total. It is known further that 117 endemic species from 49 genera exist in the ARR (Plumptre *et al.*, 2003), of which 85% are forest-dependent. There are no known endemic families, but there is one genus restricted to the ARR, namely *Kumothales*.

THREATENED SPECIES

Among vertebrates, 35 mammal species are listed as threatened (critically endangered, endangered, or vulnerable) on the IUCN *Red List* (Hilton-Taylor, 2000) of which 23 species are found in the cloud forests (Table 16.2). Twenty-five species of birds are listed as threatened in the ARR of which 18 species occur in the cloud forests (Table 16.3). Only two threatened reptiles are currently listed for the ARR, both of which occur in the cloud forests (Table 16.4). Currently, revisions of IUCN categories are being made for reptiles and it is likely that many more species will become listed (S. Stuart, personal communication). The Global Amphibian Assessment, currently in progress, indicates that 16 amphibians in the Rift are threatened, of which 14 are Albertine Rift endemics and the remaining two are near-endemic species (Table 16.5). Two amphibian species, one endemic and one near-endemic, have only been recorded outside protected areas surveyed to date (S. Stuart, personal communication). Cloud forest amphibians are subject to massive declines worldwide, partly due to factors related to climatic change and partly due to habitat loss (Pounds *et al.*, 1999, 2006; Stuart *et al.*, 2004). Data on the African situation are particularly scarce (Stuart *et al.*, 2004).

THREATS TO SITE CONSERVATION WITHIN THE ALBERTINE RIFT

The ARR faces many challenges because the human population density is so high. Threats to conservation are mostly human-caused and are particularly intense in DRC at present because of the ongoing civil war and the breakdown of law and order. Both the Itombwe Massif and Kahuzi Biega National Park are undergoing severe destruction as a result of people settling in these areas to cultivate crops or mine for gold, diamonds, and columbo-tantalite (a mineral used in semiconductors and computer chips). As well as habitat loss, hunting of large mammals for meat is probably the second most severe threat, and large-mammal populations have declined drastically in the region since the late 1970s as a result of wars in Uganda, Rwanda, Burundi, and DRC. For example, in Murchison Falls (Uganda) and surroundings, elephant numbers have plummeted from 14 000 to merely 250, and in Queen Elizabeth (Uganda) from 2500 to 150, although numbers are slowly increasing again (Lamprey *et al.*, 2003).

An analysis of forest change over the past 15 years using satellite imagery shows that over 1500 km^2 has been lost to agricultural production in the forested areas of the ARR (Plumptre *et al.*, 2003). This represents 0.5% of the area of the ARR and 2.2% of the forested area of the region. Much of this loss has occurred outside protected areas. In spite of the fact that there is little official financial or political support for protection, many forests have survived simply because they have some protected status. Mulligan (this volume) estimated potential and actual cloud forest extent in Africa using a combination of modeled cloud lifting condensation levels, topographic information, and MODIS-VCF-based information on forest vegetation presence. His estimates of cloud forest loss are much higher than the presently cited figure, pointing to the need for better ground-based information.

Despite the various civil wars in the Great Lakes region, national parks staff have remained extremely dedicated to their jobs, working without pay for many months, or even years in the case of the DRC. Virunga National Park has lost over 100 rangers as a result of the armed conflict in the last six years and many were not paid for two to three years during this time. Recently the UN Foundation and UNESCO created an emergency fund to pay the salaries of these rangers in the World Heritage sites in DRC, which has helped, but places such as Virunga and Kahuzi Biega National Parks are still struggling to obtain funding despite being the two most species-rich parks for terrestrial vertebrates in Africa.

PRIORITIZING SITES FOR CONSERVATION

The available information on species richness, endemism, and numbers of threatened species (Tables 16.1–16.5) was used to rank sites for conservation priority. One assumption made is that by using these taxa it is possible to "capture" the total biodiversity of the sites. Tests of how well one taxon correlated with another for all sites yielded highly significant correlations for all pairings when considering endemic and threatened species. For total species richness, only birds were a poor predictor of reptiles and amphibians, but all other pairings were significant. Consequently, it is felt that these taxa are reasonable predictors of the diversity of each other and it can therefore be assumed that together they predict well the biodiversity of groups which have been very poorly surveyed. Each taxon was weighted equally rather than using total species numbers because plants would have dominated the results. Mean rankings were taken across the five taxa where data were available. Sites were ranked in terms of their species richness and number of endemic and threatened species. For both criteria they were grouped into high, medium, and low rankings. The results are summarized in Table 16.6. It should be noted that these rankings were made relative to each other. As such, a site that ranks low for both species richness and for endemic and threatened species is still richer than many important conservation areas in Europe or North America.

Eight sites ranked highly on both accounts (Table 16.6): Virunga, Semliki, Kibale, Bwindi Impenetrable, Nyungwe and Kahuzi Biega National Parks and the Itombwe Massif, and Lake Tanganyika. Other sites that are important for endemic and threatened species include the Rwenzori Mountains and Kibira National Parks, Kasyoha-Kitomi Forest Reserve, and Lakes Edward and George. The Itombwe Massif is a particularly important site for conservation as it is not gazetted under any form of protection at present. Virunga National Park is also a site of great importance because it is very species-rich and contains a high percentage of the endemic and threatened species (Tables 16.1–16.5). Thus, these two sites should be priorities for future conservation activities.

Table 16.6 *Relative rankings of sites within the Albertine Rift Region in terms of their "species richness" and number of "endemic and threatened" species*

	Endemic and threatened species		
Species richness	High	Medium	Low
High	Virunga NP	Murchison Falls NP	
	Itombwe Massif	Budongo FR	
	Kahuzi Biega NP	Kalinzu-Maramagambo	
	Semliki NP	Queen Elizabeth NP	
	Kibale NP	W. Lake Edward	
	Bwindi Impenetrable NP		
	Nyungwe FR		
	Lake Tanganyika		
Medium	Rwenzori Mts. NP	Lendu Plateau	Karuma WR
	Kasyoha-Kitomi FR	Semliki WR	Kyambura WR
	Kibira NP	Mahale Mts NP	Lake Albert
	Lakes Edward and George	Marungu Massif	
		Sumbu NP	
		Mweru-Wantipa NP	
Low	Echuya FR	Bugoma FR	Kagombe FR
	Mt. Kabobo	Mafuga FR	Kitechura FR
		Lake Kivu	Matiri FR
			Itwara FR
			Idjwi Island
			Bururi FR
			Lac Ruzizi NP
			Gombe Stream NP
			Mbizi FR

NP, National Park; FR, Forest Reserve; WR, Wildlife Reserve. Sites indicated in bold contain cloud forest.

CONSERVATION EFFORTS IN THE ALBERTINE RIFT

Several efforts have been initiated to ensure the conservation of the cloud forest sites within the ARR. These range from developing an overall framework and regional plans for long-term conservation, working regionally across international borders, to working with local communities to jointly conserve the forests.

Development of a strategic framework for conservation

In May 2003 a Strategic Framework for the conservation of the Albertine Rift was developed as the culmination of a two-year planning process supported by the John D. and Catherine T. MacArthur Foundation (ARCOS, 2003). A long-term vision for the strategic framework was developed: "By 2030 the ecological, economic and social values of biodiversity in the Albertine Rift are recognized and acknowledged locally, nationally, regionally and globally," with the following objectives:

- Appropriate policies and legislation for sustainable development and conservation of the Albertine Rift in place and enforced.
- Capacity of all implementing institutions and stakeholders increased.
- Ecological, economic, and social values of Albertine Rift biodiversity better understood.
- Attitudes and behavior of people changed through improved understanding of the value and importance of the Albertine Rift.
- Contribution of Albertine Rift biodiversity to human welfare enhanced locally, nationally, regionally, and globally.
- An effectively managed and representative protected area network in place.
- Regional network, cross-border collaboration, and partnership mechanisms enhanced.
- Long-term sustainable financing mechanisms established.

The ARR was divided into six planning units and each unit will develop regional plans that will define one goal, a set of objectives and specific activities to be implemented by all the different conservation partners. In this way it is planned that all stakeholders will work together to achieve the same goal and objectives.

Transboundary collaboration

For 12 years the International Gorilla Conservation Programme (IGCP) has been supporting transboundary collaboration and management in two of the cloud forests: Virunga Volcanoes and Bwindi Impenetrable–Sarambwe Parks. This successful program has helped reduce the negative impacts of the civil war between Uganda, Rwanda, and DRC over the past seven years and has led to increased coordination between managers of protected areas. The Wildlife Conservation Society (WCS) has recently started a similar program of transboundary work in the Greater Virunga Landscape (Plumptre *et al.*, 2003) between several parks in Uganda, north of Bwindi Impenetrable National Park, and the Virunga National Park in DRC. Together with IGCP's activities this will ensure coordinated management of all protected areas that straddle the Uganda–DRC border. WCS is also starting a process of transboundary coordination between Nyungwe National Park in Rwanda and Kibira National Park in Burundi. These transboundary efforts therefore are working in four of the cloud forests: Rwenzori Mountains, Virunga Volcanoes, Bwindi Impenetrable–Sarambwe and Nyungwe National Parks, and hence capture all the cloud forests which straddle international boundaries.

Involvement of local communities in park management

Involving communities in the management of protected areas has been piloted in Uganda for about 10–15 years. One park where several programs have been tried is the Bwindi Impenetrable National Park, one of the areas with cloud forest. Bwindi NP has pioneered a variety of means to involve the local communities in the management of the park; for example joint planning in formulation of the General Management Plan (Mapesa and Makombo, 2003), the establishment of a Mgahinga and Bwindi Impenetrable Forest Conservation Trust (MBIFCT) that supports community projects as well as supporting park management, and support to improving rural livelihoods. Crop-raiding in Bwindi Impenetrable National Park is now addressed by the park and community through the efforts of Human/Gorilla Conflict Resolution Programme (HUGO) committees. Community members are also allowed to extract natural resources from the park on a sustainable basis and are allowed to practice bee-keeping in designated areas known as multiple-user zones. Finally, the Uganda Wildlife Authority (UWA) has a revenue-sharing policy in which 20% of park entrance fees is paid to the local government surrounding the park. The money has mainly been used to develop social infrastructure such as schools, dispensaries, and roads. Priorities are decided upon by the local communities living around the parks (Mapesa and Makombo, 2003).

Relations between park authorities and local communities have greatly improved as a result of these various projects, and the incidence of illegal activities has declined. However, a survey examining the impact of Integrated Conservation and Development Plans (ICDP) around Bwindi and Maghinga showed that improved law enforcement, rather than the presence of ICDPs, is what people cite as the main reason illegal activities have declined (ITFC, in prep.).

CONCLUSION

The cloud forests of the Albertine Rift are some of the most important forests in the region for conservation of endemic and threatened species of vertebrates and plants. The Itombwe Massif in the DRC is the most threatened of these at present because of the lack of any protected status and because it is being steadily converted to agriculture. The Kahuzi Biega National Park in the DRC is similarly affected, with rebel groups and miners establishing themselves in the park, and killing large mammals for bushmeat. In Uganda and Rwanda the cloud forests are better protected, although there are still limited resources for their management and conservation. Involving communities and conservation across international borders are two techniques that are being used to reduce the incidence of illegal activities and to improve park–community relations.

REFERENCES

Albertine Rift Conservation Society (ARCOS) (2003). Albertine Rift strategic planning process: strategic framework workshop report. *Unpublished report.*

Birdlife International (2000). *Threatened Birds of the World.* Barcelona, Spain: Lynx Ediciones, and Cambridge, UK: Birdlife International.

Hilton-Taylor, C. (compiler). (2000). *IUCN Red List of Threatened Species.* Gland, Switzerland: IUCN.

Institute of Tropical Forest Conservation (ITFC) (in prep.) *Bwindi and Mgahinga National Parks in Uganda: Has 15 years of ICD Programming Succeeded in Increasing Support for Conservation among Local Communities?* Report of an assessment study of selected Integrated Conservation and Development strategies implemented in Bwindi Impenetrable National Park (BINP) and Mgahinga Gorilla National Park. BINP/Mbarara University, Ruhija.

International Council for Bird Preservation (ICBP) (1992). *Putting Biodiversity on the Map: Global Priorities for Conservation.* Cambridge, UK: ICBP.

Lamprey, R.H., E. Buhanga and J. Omoding (2003). *A Study of Wildlife Distributions, Wildlife Management Systems, and Options for Wildlife-Based Livelihoods in Uganda.* Kampala, Uganda: IFPRI/USAI.

Mapesa, M., and J. Makombo (2003) Bwindi Impenetrable National Park: a demonstration of benefits beyond boundaries. *Protected Areas in Uganda, 5th World Parks Congress Proceedings*, 2003, Durban, South Africa.

Olson, D.M., and E. Dinerstein (1998). The global 200: a representation approach to conserving the Earth's most biologically valuable ecoregions. *Conservation Biology* **12**: 502–515.

Plumptre, A.J., M. Behangana, T.R.B. Davenport, *et al.* (2003). *The Biodiversity of the Albertine Rift*, Albertine Rift Technical Report. No.3. Available at www.albertinerift.org\publications.

Plumptre, A.J., A. Kayitare, H., M. Gray, *et al.* (2004). *The Socio-Economic Status of People Living near Protected Areas in the Central Albertine Rift, Albertine Rift Technical Report* No.4. Available at www.albertinerift.org \publications.

Pounds, J.A., M.P.L. Fogden, and J.H. Campbell (1999). Biological response to climate change on tropical mountain. *Nature* **389**: 611–614.

Pounds, J.A., M. Bustamante, L.A. Coloma, *et al.* (2006). Widespread amphibian extinctions from epidemic disease driven by global warming. *Nature* **439**: 161–167.

Stattersfield, A.J., M.J. Crosby, A.J. Long, and D.C. Wege (1998). *Endemic Bird Areas of the World: Priorities for Biodiversity Conservation.* Cambridge, UK: Birdlife International.

Stuart, S.N., J.S. Chanson, N.A. Cox, *et al.* (2004). Status and trends of amphibian declines worldwide. *Science* **306**: 1783–1786.

17 The mountain tapir (*Tapirus pinchaque*) and Andean bear (*Tremarctos ornatus*): Two charismatic, large mammals in South American tropical montane cloud forests

J. Cavelier
The Global Environmental Facility, Washington DC, USA

D. Lizcano
University of Kent, Canterbury, Kent, UK

E. Yerena
Universidad Simón Bolívar, Caracas, Venezuela

C. Downer
Andean Tapir Fund, Minden, Nevada, USA

ABSTRACT

Mountain tapir and Andean bear are the two largest mammals in the tropical Andes, both charismatic, survivors of old lineages of mammal evolution, and threatened with extinction. Their current distribution is montane cloud forest and *páramos* (high-altitude grasslands) in South America, where they feed on fruits and vegetative parts, having an important role as seed dispersers. The distribution and population size of both species was analyzed based on expert opinion and a rule-and-base model within a GIS context. A gap analysis was performed to determine the animals their conservation status within National Parks. Areas of past and present distribution for the mountain tapir were 205 000 km^2 and 31 400 km^2, respectively. Based on existing suitable habitat, the population of mountain tapirs in their current range was estimated at 5700 individuals at the most. The overall population of Andean bears ranges between expert estimates of 5000–10 000 individuals. Several conservation recommendations emerge from the present analysis and review for the two species.

INTRODUCTION

The tropical cloud forests of the Andean region support some of the greatest plant and animal species diversity on Earth (Rahbek and Graves 2001; Barthlott *et al.*, 2005; Brehm *et al.*, 2005; Krömer *et al.*, 2005), with a remarkable number of endemic and rare species (Duellman, 1988; Kessler, 2002; Kattan and Franco, 2004). At the same time, the tropical Andes is one of the most endangered biodiversity "hot spots" (Myers *et al.*, 2000; Brooks *et al.*, 2002; cf. Mosandl *et al.*, 2008) due primarily to the conversion of forests to agriculture and cattle ranching, a process that continues today (Kattan *et al.*, 2004). Among the threatened species in these extreme habitats are the two largest mountain mammals in the Andes, the mountain tapir (*Tapirus pinchaque*) and Andean bear (*Tremarctos ornatus*). These two herbivores occupy some of the most remote remaining forest patches where deforestation has not yet reached. In spite of this isolation, the populations of these two animals are increasingly fragmented and pressured by hunting for their parts, mainly hooves and paws for medicinal and aphrodisiac purposes. If measures are not taken soon, the populations of these two animals may become too small and patchy to make the species viable. This chapter aims to review the available information on the two largest and charismatic mammal species of the tropical montane cloud

Tropical Montane Cloud Forests: Science for Conservation and Management, eds. L. A. Bruijnzeel, F. N. Scatena, and L. S. Hamilton. Published by Cambridge University Press.

forests in South America, and to analyze their current distribution, population size, and conservation constraints.

NATURAL HISTORY

Evolution

MOUNTAIN TAPIR

Currently there are four species of tapirs, one in South-East Asia (*Tapirus indicus*) and three in the Neotropics (*Tapirus terrestris, T. bairdii*, and *T. pinchaque*). In the past, the family Tapiridae had a prolific adaptive radiation with significantly more species than today (Prothero and Schoch, 1989; Holbrook, 1999). Together with other perissodactyls ("odd-toed" or "odd-hoofed" mammals like rhinoceros), tapirs declined in terms of their diversity and abundance during the late Cenozoic, in large degree due to climatic changes and the emergence of artiodactyl (even-hoofed like wild boar and deer) ruminants (Schoch, 1989). The extinction of tapirs and other large animals in North America also coincided with the migration of the first humans to this continent (Frison, 1998). Phylogenetic studies of the existing tapirs suggest that the Central American species (*T. bairdii*) got separated from the American lineage about 19–20 million years ago, and that the two South American species (*T. terrestris* and *T. pinchaque*) appeared later in the evolutionary history of this group (Ashley *et al.*, 1996). The appearance of these two species coincided with the formation of the Isthmus of Panama (Schoch, 1989) and the uplift of the Andes during the late Pliocene (Van der Hammen, 1982). Currently, mountain tapirs are found only in the Andes of Colombia (Acosta *et al.*, 1996; Lizcano and Cavelier, 2000a; Lizcano *et al.*, 2002), Ecuador (Downer, 1996, 1997), and northern Perú (Lizcano and Sissa, 2003).

ANDEAN BEAR

The Andean or spectacled bear, *Tremarctos ornatus*, is an intermediate-sized bear, showing sexual dimorphism in size, with males ($\leq$200 kg) being considerably larger than females ($\leq$65 kg; Peyton, 1999). Pelage is usually black with variations of dark red-brown phenotypes (Peyton, 1999). The English name, spectacled bear, and the Spanish name, *oso frontino*, come from the white to yellowish fur that often encircle the eyes. These markings are unique to each individual and can extend to forehead, muzzle, neck, and chest (Roth, 1964). Some other common names along its distribution range are *ukuku, ukumari*, and *oso real*. The Andean bear are probably the most arboreal of all living bears and are well adapted to climb trees and steep slopes (Peyton, 1999). Its large and massive jaw muscles plus its characteristic fourth premolar enable these bears to grind and crunch tough and fibrous foods (Davis, 1955; Thenius, 1976). Such features support the idea that Andean bear are possibly the most herbivorous of all living bear species (Peyton, 1999).

The chromosome number ($2n = 52$) of *T. ornatus* is unique among the rest of the living bear species which have a nearly identical karyotype and 74 chromosomes, except for the giant panda with its 42 chromosomes (Nash and O'Brien, 1987). The Andean bear and the panda are the basal taxa of the ursid radiation (Talbolt and Shields, 1996), making them the living representatives of the oldest lineages of bear evolution. Tremarctinae bears may have emerged some 3 million years later than the Ailuropodinae (the genus of the giant pandas), which appeared some 15–25 million years ago (O'Brien *et al.*, 1985). A possible affinity between the Andean bear and the panda bear is also suggested by the fact that both species are herbivorous foragers and neither possesses "typical" bear behavioral patterns (i.e. neither hibernates nor roars; O'Brien *et al.*, 1985). The arrival of Tremarctinae bears to South America may be dated back to the Early Pleistocene, about 1.8 million years ago (Soibelzon, 2005).

Habitat use and activity

MOUNTAIN TAPIR

The main habitats of the mountain tapir are Andean tropical upper montane cloud forest and *páramo* from north-eastern Colombia to northern Perú, at altitudes between *c.* 1700 and 4300 m.a.s.l. (Downer, 1996, 1997). In Colombia, tapirs are not present on the Western Cordillera or in the Sierra Nevada de Santa Marta, probably because of geographic barriers (notably deep dry valleys separating these mountain ranges from the Central and Eastern Cordilleras). Tapirs use *páramos* and open man-made grasslands associated with nearby forests. Their foot prints have also been observed in snow-capped mountains above 4500 m.a.s.l. Mountain tapirs tend to be more active in mature than in secondary forest. Nocturnal activity is higher during periods of full moon, compared with periods of new and quarter moons (Lizcano and Cavelier, 2000b). In Ecuador's Sangay National Park, mountain tapirs displayed a bimodal activity pattern, being most active during the early hours of the morning and the early hours of the evening (Downer, 1996).

ANDEAN BEAR

The Andean bear inhabits dry thorn forests, seasonal forest (evergreen to deciduous), humid to wet forests, including different kinds of cloud forests, *páramo*, and *puna*, the dry grasslands above the tree line in Peru and Bolivia (Yerena and Torres, 1994; Peyton, 1999). Most of the remaining cloud forest habitat for the Andean bear occurs on the eastern slopes of the eastern Andes (Peyton, 1999), facing the Orinoco, Amazon, and Paraná basins. It is recognized that the best habitats for Andean bear include humid to very humid montane or cloud forest (Peyton, 1999), which typically occupy a range between 1000 and 3000 m.a.s.l. Both mature and secondary forests provide food and shelter for Andean bear. The habitat use of the Andean bear has been described as seasonal, switching between forest and *páramo* when food

Table 17.1 *Selected families and genera of plant food items for Andean bear in montane forests*

Fruits and/or seeds	Vegetative parts (within parentheses part of the plant that is eaten)
Anacardiaceae: *Schinus*, Annonaceae: *Annona*, Arecaceae: *Catoblastus, Ceroxylon, Aiphanes*, Cecropiaceae: *Cecropia*, Clusiaceae: *Clusia*, Cyclanthaceae: *Ciclanthus*, Ericaceae: *Vaccinium, Gaultheria, Disterigma, Pernettya, Cavendishia*, Fabaceae: *Inga, Amburana*, Fagaceae: *Quercus*, Lauraceae: *Beilschmiedia, Ocotea, Aniba, Phoebe, Nectandra, Persea*, Moraceae: *Ficus*, Myrtaceae: *Psidium, Eugenia*, Proteaceae: *Panopsis*, Rosaceae: *Prunus, Rubus*, Sapotaceae: *Chrysophyllum, Lucuma*, Solanaceae: *Physalis*	Arecaceae (stem meristematic tissue): *Catoblastus, Dictyocaryum, Bactris, Ceroxylon, Geonoma, Iriartea*, Araceae (stem): *Anthurium*, Bromeliaceae: *Tillandsia, Guzmania, Pitcairnia, Vriesia, Aechmea*, Cyatheaceae (rhizome): not specified, Cyclanthaceae (stems): *Carludovica, Asplundia*, Dicksoniaceae: *Dicksonia*, Haloragaceae (stems): not specified, Orchidaceae (pseudobulbs): *Epidendrum*, Poaceae (shoots): *Chusquea, Guadua, Neurolepis*

Peyton (1980); Suárez (1985); Poveda (1986); Mondolfi (1989); Goldstein (1990); Rodriguez (1991); Rumiz *et al.* (1999); Figueroa and Stucchi (2002); Castellanos (2004).

availability in the forest is low (Peyton, 1980; Paisley, 2001; Paisley and Garshelis, 2006). Andean bears prefer high-elevation cloud forest and are not present in dry montane and Andean foothill forests (Rios-Uzeda *et al.*, 2006). Their activity patterns are mainly diurnal (Paisley and Garshelis, 2006). Andean bears make resting beds on the forest floor or on tree branches (Peyton, 1999; Castellanos, 2004). They are known to climb vines and trees in order to reach food (Peyton, 1999) and employ a destructive foraging technique that results in the creation of platforms of bent branches that have been described as tree nests (Peyton, 1987).

Diet and seed-dispersal role

MOUNTAIN TAPIR

Mountain tapirs eat a wide variety of plants (Downer, 2001). Although they have the potential to browse throughout the understory, they seem to concentrate on "eating grounds," which are often tree-fall gaps of different successional stages, shrubby vegetation between *páramo* and upper montane cloud forest, and open *páramos*, especially during the dry season. In Colombian montane forests, tapirs consume plant species like *Chusquea fendlerii* (Poaceae) and *Cavendishia nitida* (Ericaceae), which are both very abundant. Along tapir trails on ridges, the animals consume species like ferns of the genera *Blechnum* (Polypodiaceae), *Miconia* (Melastomataceae), and *Oreopanax* (Araliaceae), while on trails along creeks, the animals eat *Gunnera manicota* (Gunneraceae), *Plantago australis* (Plantiginaceae), and *Myrrhindendron pennellii* ("arracachuela"; Apiaceae). In the ecotone between *páramo* and upper montane cloud forest, tapirs consume *Lachemilla* sp. (Rosaceae). Observations of the feces of the mountain tapir suggest that although it may have some preferences (e.g. *Chusquea* sp. and *Myrrhindendron pennelli*; N. Monsalve, personal communication), tapirs browse a large number of other plant species as well (Lizcano, 1996; Lizcano and Cavelier, 2004).

Laboratory experiments and field observations in Ecuador have demonstrated that the seeds of 86 of the 205 vascular plants eaten by the mountain tapir germinated from the tapir's feces (Downer, 1999). In Sangay National Park the aster *Gynoxis* spp. (particularly *G. buxifolia*), was the most abundant food source (Downer, 2001), various ferns ranked second, horsetails (*Equisetum* spp.) third, and *Brachyotum* spp. (Melastomataceae) fourth (Downer, 1996). *Lupinus* spp. (Fabaceae) was the most sought-after food group (highest preference ratio) relative to its abundance in habitat (Downer, 1996, 1999, 2001). Evidence exists that the mountain tapir plays a vital role as a seed disperser in the northern Andes where it has mutualistically co-evolved with that plant community for *c.* 3 million years (Downer, 1999, 2001).

ANDEAN BEAR

This species is an opportunistic feeder, and may be considered somewhere between a frugivore/herbivore and a frugivore/omnivore (*sensu* Robinson and Redford, 1986), depending on resource availability (Dierenfeld, 1988). They are known to feed on an ample variety of plant species, depending on the habitat type where they are foraging and on the biogeographic region where they are located (Rivadeneira-Canedo, 2008). Plant items can be divided into two categories, depending on the part of the plant that is consumed: fruits which are believed to be the most important source of energy, and vegetative parts which are important when fruits are scarce and as a diet supplement (Table 17.1).

The Andean bear has been reported to feed occasionally on insects, rodents, birds, and livestock (Peyton, 1980; Goldstein, 1990). Its role as livestock predator or scavenger is still debated (Castellanos, 2003). Cattle parts are occasionally found within bear scats, and dead livestock remains are usually consumed by bears (Goldstein, 1991; Castellanos, 2003). The Andean bear may be the least aggressive of all bear species (Peyton, 1999), but *Tremarctos ornatus* is opportunistic and may be both a scavenger and an occasional predator.

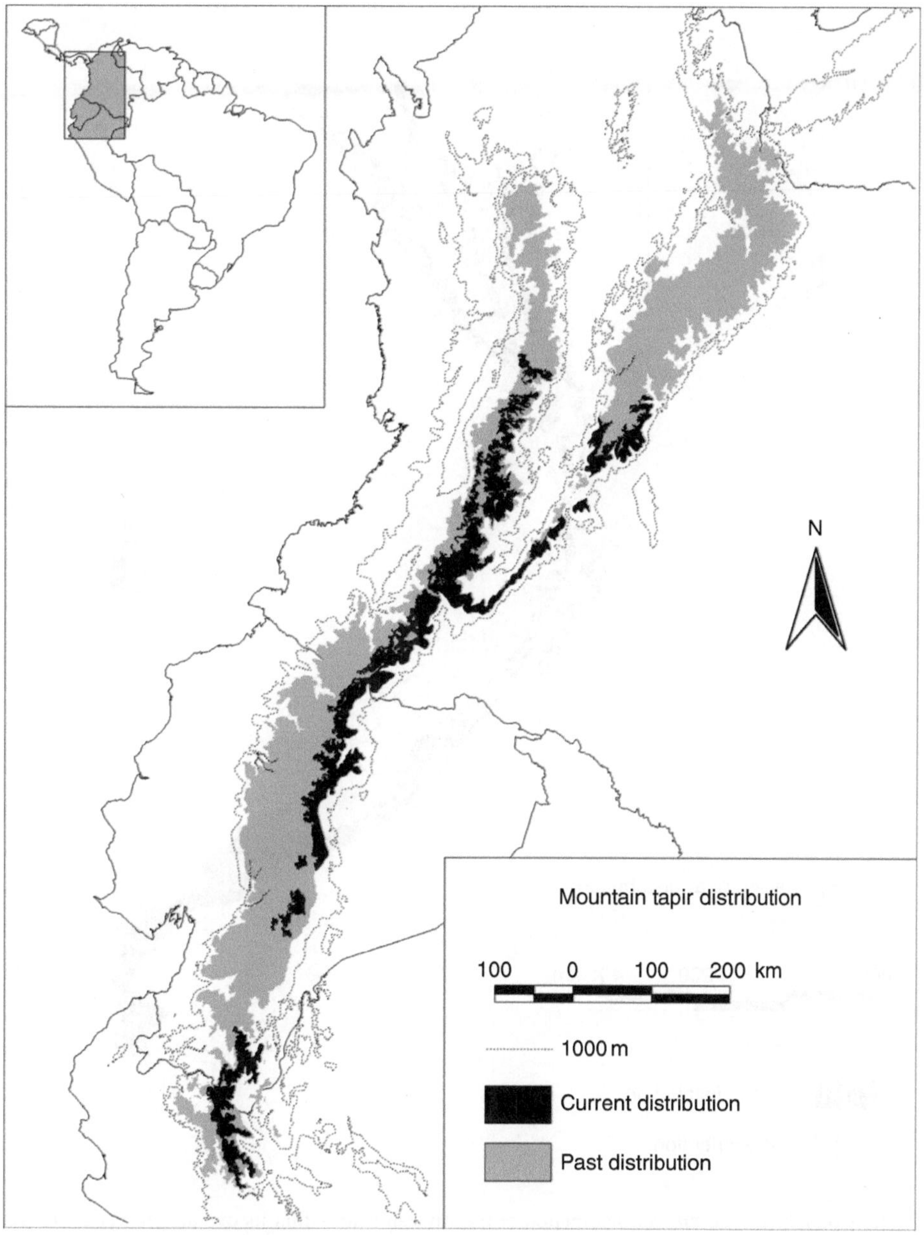

Figure 17.1. Map of past (pre-Columbian) and present distribution of the mountain tapir within its overall geographic range.

DISTRIBUTION AND POPULATION SIZE —

Distribution

MOUNTAIN TAPIR

The past (i.e. pre-Columbian) distribution of the mountain tapir in Colombia (Figure 17.1) was estimated by plotting those geographic locations (i.e. streams, peaks, ridges, farms, townships, counties) above 2000 m.a.s.l. with the name "danta" or "tapir" onto a map of the past distribution of montane cloud and rain forests. In doing so it was assumed that locations were named after the tapir because the species occurred in these localities at one or more points in time. Since these locations are situated between 2000 and 4000 m.a.s.l. where the only tapir is *T. pinchaque*, it is unlikely that the sites were named after a different species. The names of the geographic places were obtained by manually reviewing 144 topographic maps of the Andes of Colombia (scales 1 : 50 000–1 : 100 000). Additional locations were obtained from the *Diccionario Geografico de Colombia* (IGAC, 1996), records and reports in the scientific literature (Hershkovitz, 1954; Schauenberg, 1969; Velasco, 1991; Downer, 1996; Lizcano and Cavelier, 2000a; Lizcano *et al.*, 2002), and expert opinion (J. Hernandez, personal communication). Interviews with the heads of the natural resources offices of the various Andean

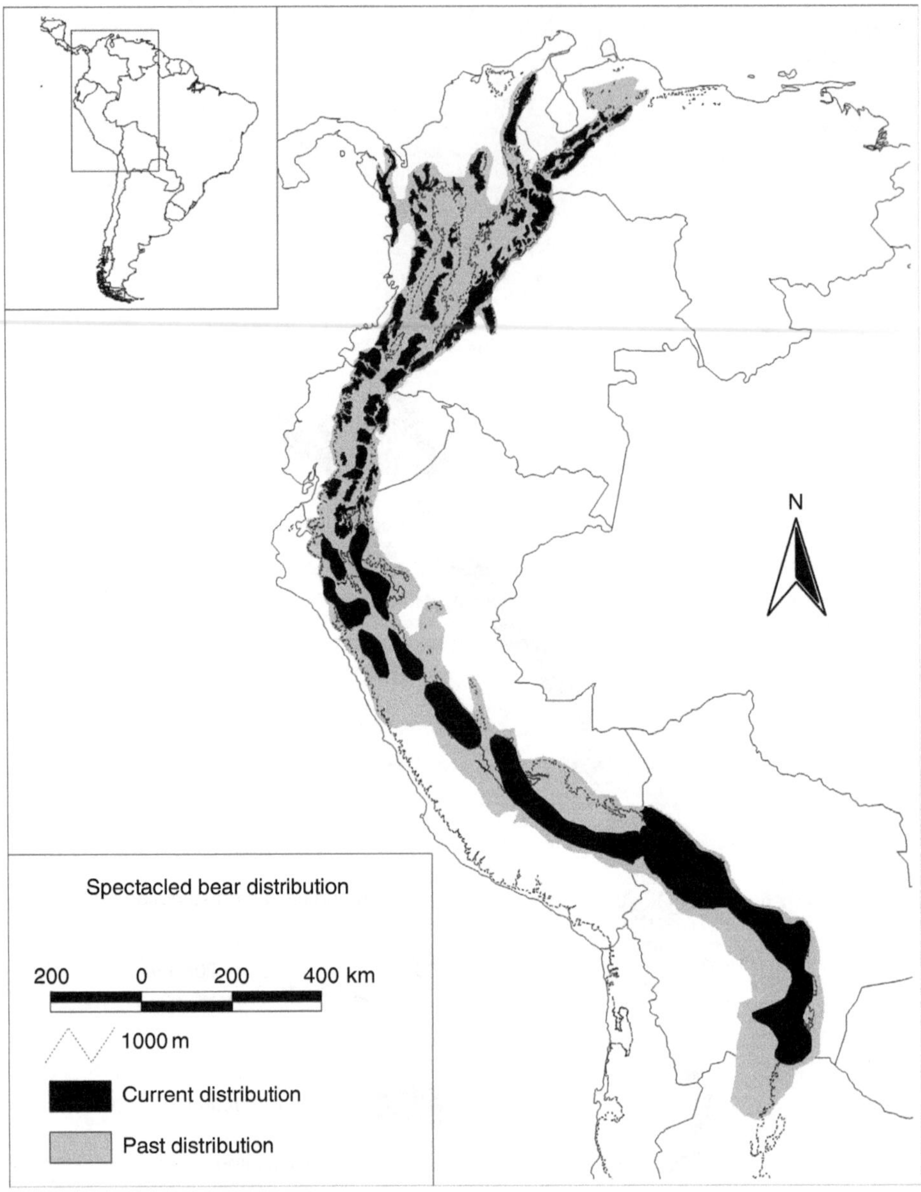

Figure 17.2. Map of past (pre-Columbian) and present distribution of the Andean bear within its overall geographic range.

municipalities as well as in-field verification also provided information on past and present distribution of the species. Mountain tapirs were considered absent from areas that contained less than 100 000 ha of montane rain forests in the pre-Columbian past and for which there were no geographic locations, records, or reports of their presence. The map of past distribution of montane rain forests of Colombia (1 : 1 500 000) was taken from Cavelier and Etter (1995). From this, only the montane rain/cloud forests and associated *páramos* between 2000 and 4000 m.a.s.l. were used as the conservatively estimated typical altitudinal distribution of *T. pinchaque* in Colombia (Downer, 1996; Lizcano *et al.*, 2002). For Ecuador and Perú, expert opinion was used, along with published information (Downer, 1996, 1997; Lizcano and Sissa, 2003), and a GIS-based analysis where tapir habitat was defined as forested areas in excess of 1000 ha between 2000 and 4000 m elevation. Currently there are mountain tapirs in the eastern and central Andes of Colombia, the eastern and western Andes (northern portion) of Ecuador, and the northern Andes of Perú, and in recent decades they have been reported as far south as the Huancabamba Depression, which may well have been the evolutionary southern limit of the species (Figure 17.1).

ANDEAN BEAR

The Andean bear's geographic range encompasses all of the Andes, from Venezuela to the Bolivia–Argentina border (Peyton, 1999). It is the only bear, together with the Asiatic sun bear (*Helarctos malayanus*), to reach the southern hemisphere (Figure 17.2). The Andean bear occurs throughout this portion of the

Andes, except for most large intra-montane valleys and high mountain plateaux. Such valleys and plateaux are usually densely populated by humans (Peyton, 1999; Suárez, 1999; cf. Hofstede *et al.*, this volume). In Venezuela the bear's range encompasses all of the Cordillera de Mérida and Sierra de Perijá (Mondolfi, 1989; Yerena, 1999). In Colombia, it includes nearby Andean mountain ranges such as the Sierra de Macarena and the Serranía del Darien (possibly even within Panamá), but excludes the Sierra Nevada de Santa Marta (Orejuela and Jorgenson, 1999; Rodriguez, 2001). In Perú, bear distribution does not reach some southern portions of the coastal and central Andean ranges (Peyton, 1999). In Bolivia, its distribution is restricted to the eastern slope of the Andean range (Rumiz and Salazar, 1999). The bear's altitudinal range goes from 250 m.a.s.l. up to the permanent snow line, at about 4750 m.a.s.l. (Peyton, 1999). Current distribution is more restricted than it was prior to European colonization. Drastic habitat alteration and hunting during the last 500 years or so have shaped the current distribution of the Andean bear. It is particularly interesting to note that no bear representation has been found in ancient Inca and Chimu art forms (Peyton, 1981). However, it was found to be an integral part of the Apolobamban culture in Bolivia, supporting the theory that the bear mythic theme in South America had pre-Columbian origins (Paisley, 2001).

Densities and home range

MOUNTAIN TAPIR

There are only two studies that have used radio-tracking on mountain tapirs. In Ecuador, the average home-range size of three adult tapirs (two animals were tracked for three years and one for two) was 8.8 km^2 of projectional flat area in the mountainous Sangay National Park (Downer, 1996). In Colombia, a pilot study using GPS collars in the central Andes determined the home range of an adult mountain tapir male, as calculated by the minimum convex polygon (MCP) method as 3.5 km^2, or 2.5 km^2 using the fixed kernel method (Lizcano and Cavelier, unpublished data). Differences in home range are even bigger when comparing the mountain tapir with Baird's tapir in Costa Rica, which has an average annual home range of 10.7 km^2 (Foerster, 2001), based on an extensive study of 26 animals over five years.

Limited evidence indicates that individual mountain tapirs shift their home ranges, especially when perturbed by livestock and hunters (Downer, 2002). Significant seasonal shifts within an individual's home range have been documented to occur in Sangay National Park (Downer, 1997). This is demonstrated when habitat use is correlated with the core area of a seasonal home range (Lizcano and Cavelier, 2004). Many local people living within mountain tapir distributions testify that mountain tapirs migrate to higher elevations during drier seasons and to lower elevations during wetter seasons. They also claim that lunar phase affects altitudinal movements. A correlation between lunar phase and tapir activity was found in both the central Andes of Colombia (Lizcano and Cavelier, 2000b) and in Sangay NP (C. Downer, unpublished data).

ANDEAN BEAR

No rigorous and empiric estimates of Andean bear density have been published. A theoretical estimation based on a body weight–density model for frugivorous–omnivorous mammals (Robinson and Redford, 1986) yielded an estimate of 1.26 individuals km^{-2} (Yerena, 1994). Interviews with experienced poachers in Venezuela about the number of bears living in two different and isolated forest patches yielded a figure of 0.04 bears km^{-2} (Yerena, 1994). Peyton (1999) indicates that the highest relative bear densities are to be found in wetter cloud forest between Colombia and Perú.

In Bolivia, a study showed home ranges of 7.4 and 6.6 km^2 for two young adult bears (Paisley, 2001) with one of them showing a larger (11 km^2) home range some months later (Rechberger *et al.*, 2001). Home range studies with wild-born bears are currently being undertaken in Ecuador. Preliminary home range calculations showed average home ranges of 66 km^2 for males ($n = 2$) and 14 km^2 for females ($n = 4$). Females have well-defined, stable, and overlapping home ranges, while males use movement corridors over much larger areas that are harder to define because of the difficulties in tracking them in mountain terrain (Castellanos, 2003, 2004). Such estimates are roughly approximate to theoretical estimation, based on a body weight–home range model (Harestad and Bunnel, 1979), which predicted a mean home range for Andean bear at 48.6 km^2 (Yerena, 1994).

Population size

MOUNTAIN TAPIR

To obtain estimates of past and present population size for the mountain tapir, the density estimates were multiplied by the area in which the species is believed to occur. Estimated density ranged from 1 individual per 551 ha (Lizcano and Cavelier, 2000a) to 1 individual per 587 ha (Downer, 1996). Areas of past and present distribution for the mountain tapir as obtained from Figure 17.1, were 205 000 km^2 and 31 400 km^2, respectively. Based on existing suitable habitat, the estimated population of mountain tapirs in all their current range would be between 5000 and 5700 individuals. However, the possibility exists that many of these intact habitat areas are without tapirs due to extirpation by the intense hunting pressure that is common in the Andes. Furthermore, a considerable number of conservationists and ecologists familiar with the northern Andes and its pervasive and uncontrolled hunting and livestock invasion would put the surviving number considerably lower (Downer, 1997, 2002). If only half of the remaining habitat is actually occupied by the

Table 17.2 *Estimation of the number of mountain tapirs within major portions of the Protected Area system within the geographic distribution range of the species*

Category	Name	Country	Area (km^2)	Low number of tapirs	High number of tapirs
National Park	Los Nevados	Colombia	98.503	17	18
National Park	Las Hermosas	Colombia	301.045	51	55
National Park	Nevado Del Huila	Colombia	1084.443	185	197
National Park	Sumapaz	Colombia	714.590	122	130
National Park	Cueva de los Guacharos	Colombia	39.816	7	7
National Park	Alto Fragua – Indi Wasi	Colombia	64.463	11	12
National Park	Cordillera de Los Picachos	Colombia	264.495	45	48
National Park	Purace	Colombia	618.197	105	112
National Park	Podocarpus	Ecuador	1069.749	182	194
Ecological Reserve	Cayambe-Coca	Ecuador	1438.606	245	261
National Park	Sumaco Napo Galeras	Ecuador	525.865	90	95
National Park	Antisana	Ecuador	617.042	105	112
National Park	Llanganatis	Ecuador	824.778	141	150
National Park	Sangay	Ecuador	1294.180	220	235
National Sanctuary	Tabaconas-Namballe	Peru	238.907	41	43
Total			9194.679	1567	1669

species, there may only be between 2675 and 2850 mountain tapirs left today.

ANDEAN BEAR

Larger Andean bear populations are believed to be concentrated in the cloud forests of the eastern slopes of the Eastern Cordillera, southward from the Ecuador–Colombia border (Peyton, 1999). According to Andean bear specialists, and based on its present distribution, vegetation maps, and relative abundance estimates, some 18 000 individuals may still exist in total (Peyton, 1999). Recent estimates based on genetic analysis suggest the following population numbers (Ruiz-García *et al.*, 2005): Venezuela, 910–1130; Colombia, 3600–6900; Ecuador 780–2780. Thus, the overall population might be between *c.* 5000 and *c.* 10 000. The Andean bear Action Plan of the World Conservation Union (IUCN) considers the minimum area for a conservation unit to maintain a viable Andean bear population without fairly intensive management as being 1900 km^2 of appropriate habitat (Peyton, 1999).

Gap analysis

MOUNTAIN TAPIR

A map of the mountain tapir's present distribution was overlaid with the map of Protected Areas above 2000 m.a.s.l. in the Colombian, Ecuadorian, and Peruvian Andes. Estimates of the population size of the mountain tapir inside protected areas were calculated by multiplying the areas of montane forest within the parks times the estimated densities (Table 17.2). Many Andean National Parks and Reserves also contain other vegetation types unsuitable for mountain tapirs, primarily lowland rain forest whereas a more precise estimate should also factor in the proportion of protected areas present as treeless *páramo*. However, the presently used method will serve to identify the status of the mountain tapir's most critical survival habitat: montane (cloud) forest.

According to Table 17.2, less than 10 000 km^2 (29.3%) within the distributional range of the mountain tapir (estimated at 31 400 km^2) falls within the Protected Area system, with the smallest area (239 km^2) in Perú. However, the Municipal Province of Ayabaca, Piura State, Perú, has recently declared an Ecological Reserve in the Cordillera de las Lagunillas (or Cordillera Piurana, as it is known) whereas the adjoining Huancabamba Province also wants to have its territory declared off-limits to mining activities and is preparing a proposal for Nature Reserve status.

ANDEAN BEAR

Some 50 habitat fragments have been recognized along the Andean bear geographic range, totalling around 260 000 km^2 (Peyton, 1999). The four largest fragments of continuous bear habitat occur on the eastern slopes of the Oriental Andes, within Perú and Bolivia; these two countries comprise over two-thirds of overall bear range (Peyton, 1999).

A recent estimation of Andean bear habitat availability from northern Perú up to Venezuela and Colombia, based on satellite images, states that there are 110 potential habitat fragments ranging in size from nearly 36 000 down to 113 km^2. All of

Table 17.3 *Existing or proposed "corridor" projects affecting Andean bear and its habitat, according to IUCN survey (iucn.sur.org) and authors' data. Names mostly reflect already existing protected areas*

Location	Project
Argentina/ Bolivia	Baritú–Tariquía (Regional Corridor for the Yungas)
Bolivia	Tariquía–Amboró, Amboró–Madidi
Bolivia/Perú	Amboró–Vilcabamba
Perú	Bahuaja Sonene–Alto Purús
Perú/Ecuador	Cóndor–Kutukú
Ecuador	Podocarpus–Yacuambí, Podocarpus–Sabanilla, Sangay–Llanganates, Cotacachi Cayapas–Awacachi, Cotacachi Cayapas–Cayapas Mataje, Chocó–Andino de Maquipucuna
Ecuador/ Colombia	Manabí–Chocó
Colombia	Páramo Bordoncillo–Laguna La Cocha, Guácharos–Puracé, Guácharos–Serranía Churumbelos, Corredor Bremen–Barbas (Quindío), Tatamá–Paraguas, Cordillera Central, Guantiva–Guanentá (Bosques Altoandinos de Roble)
Venezuela	Batallón–Sierra Nevada (Pueblos del Sur), Sierra Nevada–Guaramacal (Ramal de Calderas), Guache–Terepaima (Sierra de Portuguesa)

these 110 fragments total 208 086 km^2, 25% of which are under some Protected Area status (Rodriguez *et al.*, 2003). This number of potential fragments is twice that reported by Peyton (1999), but the two estimations are not entirely comparable because of differences in scale, data sources, and methodology. It is foreseeable that a more detailed survey would reveal many more fragments throughout the bear's distributional range. A similar study for the rest of the bear's distribution (i.e. the southern parts of the Andes) is not currently available, but undoubtedly the largest habitat fragments are to be found within the eastern Andean slopes of Perú and Bolivia.

THREATS AND RECOMMENDATIONS FOR CONSERVATION

MOUNTAIN TAPIR

Today, all four tapir species surviving in the world are threatened by hunting and habitat loss (IUCN, 2006). Probably the most threatened species is the mountain tapir. It is listed as "endangered" (IUCN, 2006) and is included as an Appendix I species ("most restricted") by CITES. Currently, mountain tapir populations are in decline because they are hunted for use both as food and for traditional medicine (Brooks *et al.*, 1997). An action plan for the rescue of this species has been published (Downer, 1997) and is currently being updated but awaits funding. This action plan outlines urgently needed conservation education programs combined with the promotion of more nature-harmonious lifestyles. Of particular concern is the increase in human adaptation of livestock culture and its impact on cloud forest and *páramo* habitats (Downer, 2002; cf. Baez *et al.*, this volume). This may jeopardize the quality and quantity of water supply for many species, including humans (cf. Buytaert, 2004; Buytaert *et al.*, 2006).

ANDEAR BEAR

Andean bear conservation is threatened by the accelerated pace of habitat conversion to agriculture (including illegal crops) and hunting (including illegal trading of bear parts) (Peyton, 1999). Habitat loss and fragmentation are the most important long-term threats (cf. Mulligan, this volume). However in some areas poaching may be an even larger threat to bear populations than deforestation (Yerena, 1999). Many bears are killed annually by professional poachers and by peasants to get food for their families, to use body parts for traditional medicine, to protect domestic animals, or to minimize raiding into crop fields (Peyton, 1999; Yerena, 1999).

The IUCN Bear Action Plan states several recommendations for Andean bear (Peyton, 1999), including:

- To strengthen institutions involved with the species and its habitat, at all social levels including government, communities, and private initiatives, with special emphasis on coordination, cooperation, and capacity building.
- To do more research and monitoring, on existing and potential bear habitat, relative population densities, relationship between diet and reproduction in the wild, role as seed dispersers for valuable trees, trends in deforestation, hunting mortality, economic loss from crop and livestock, and public attitudes towards bears.
- To create stewardship mechanisms for bear conservation so that communities co-existing with these animals and their habitat develop a real sense of ownership toward the natural

resources and the land, within a framework of sustainable development.

- To educate the public about the benefits of preserving both watersheds and bears, since watershed protection is essential for humans as well as for bears, especially in a cloud forest setting.
- To concentrate and expand management and conservation to the largest conservation units, and to the areas between them.

Some additional goals based on an eco-regional approach (Rodriguez *et al.*, 2003) would include reducing the rate of habitat loss and to increase connectivity among populations and habitat (Table 17.3). Such goals cannot be achieved without targeting the dramatic social, economic, and political conflicts afflicting rural communities throughout the Andes, and require policies and actions based on sustainable development criteria, land use planning, adequate legal framework, and mechanisms for real and effective participation of people.

REFERENCES

Acosta, H., J. Cavelier, and S. Londoño (1996). Aportes al conocimiento de la biología de la danta de montaña, *Tapirus pinchaque*, en los Andes Centrales de Colombia. *Biotropica* **28**: 258–266.

Ashley, M. V., J. E. Norman, and L. Stross (1996). Phylogenetic analysis of the perissodactylan family Tapiridae using mitochondrial cytochrome *c* oxidase (COII) sequences. *Journal of Mammalian Evolution* **3**: 315–326.

Barthlott, W., J. Mutke, M. D. Rafiqpoor, G. Kier, and H. Kreft (2005). Global centres of vascular plant diversity. *Nova Acta Leopoldina, Neue Folge* **92**, **342**: 61–83.

Brehm, G., L. M. Pitkin, N. Hilt, and K. Fiedler (2005). Montane Andean rain forests are a global diversity hotspot of geometrid moths. *Journal of Biogeography* **32**:1621–1627.

Brooks, K. M., R. E. Bodmer, and S. Matola (1997). *Tapirs, Status Survey and Conservation Action Plan.* Gland Switzerland: IUCN/SSC Tapir Specialist Group.

Brooks, T. M., R. A. Mittermeier, C. G. Mittermeier, *et al.* (2002). Habitat loss and extinction in the hotspots of biodiversity. *Conservation Biology* **16**: 909–923.

Buytaert, W. (2004). The properties of the soils of the South Ecuadorian páramo and the impact of land use changes on their hydrology. Ph.D. thesis, Catholic University of Leuven, Leuven, Belgium.

Buytaert, W., R. Célleri, B. de Bièvre, *et al.* (2006). Human impact on the hydroloogy of the Andean páramos. *Earth Science Reviews* **79**: 53–72.

Castellanos, A. (2003). Ecology of re-introduced Andean bears in the Maquipucuna Biological Reserve, Ecuador: conservation implications. *Re-introduction News* **23**: 32–34.

Castellanos, A. (2004). Andean bear research project in the Intag region, Ecuador. International Bear *News* **13**: 25–26.

Cavelier, J., and A. Etter (1995). Deforestation of montane forest in Colombia as result of illegal plantations of opium (*Papaver somniferum*). In *Biodiversity and Conservation of Neotropical Montane Forest*, eds. S. P. Churchill, H. Balslev, E. Forero, and J. L. Luteyn, pp. 125–137. New York: New York Botanical Garden.

Davis, D. D. (1955). Masticatory apparatus in the spectacled bear (*Tremarctos ornatus*). *Fieldiana: Zoology* **37**: 25–46.

Dierenfeld, E. (1988). Nutritional consideration in feeding the captive spectacled bear. In *Proceedings of the 1st International Symposium on the Spectacled Bear*, ed. M. Rosenthal, pp. 114–130. Chicago, IL: Chicago Park District Press.

Downer, C. C. (1996). The mountain tapir, endangered "flagship" species of the high Andes. *Oryx* **30**: 45–58.

Downer, C. C. (1997). Status and action plan of the mountain tapir (*Tapirus pinchaque*). In *Tapirs, Status Survey and Conservation Action Plan*, eds. D. M. Brooks, R. E. Bodmer, and S. Matola, pp. 10–22. Gland, Switzerland, and Cambridge, UK: IUCN/SSC Tapir Specialist Group.

Downer, C. C. (1999). Un caso de mutualismo en los Andes: observaciones sobre dieta-habitat del tapir de montaña. In *Manejo y conservacion de fauna silvestre en America Latina,* Parte VII, eds. T. G. Fang, O. Montenegro, and R. Bodmer, pp. 415–427. La Paz, Bolivia: Wildlife Conservation Society and Instituto de Ecologia.

Downer, C. C. (2001). Observations on the diet and habitat of the mountain tapir (*Tapirus pinchaque*). *Journal of Zoology*, London **254**: 279–291.

Downer, C. C. (2002). Ambito hogareño y utilización de hábitat del Tapir Andino e ingreso de ganado en el Parque Nacional Sangay, Ecuador. *Proceedings of the Conference on Biodiversity: Andean and Amazonian*, (Sept. (2001)). Munich, Germany and Cuzco, Perú: INKA.

Duellman, W. E. (1988). Patterns of species diversity in the anuran amphibians in the American tropics. *Annals of the Missouri Botanical Garden* **75**: 79–104.

Figueroa, J., and M. Stucchi (2002). *Situación actual del oso Andino en el Santuario Histórico de Machu Picchu (Cuzco, Perú)*. Lima, Perú: Informe a PROFANPE/GTZ.

Foerster, C. R. (2001). Results of a five-year telemetry study of Baird's tapir in Costa Rica. In *Proceedings of the 1st International Tapir Symposium*, Nov. (2001), ed. P. Medici, 9–10. San José, Costa Rica.

Foerster, C. R., and C. Vaughan (2002). Home range, habitat use, and activity of Baird's tapir in Costa Rica. *Biotropica* **34**: 423–437.

Frison, G. C. (1998). Paleoindian large mammal hunters on the plains of North America. *Proceedings of the National Academy of Sciences USA* **95**: 14 576–14 583.

Goldstein, I. (1990). Distribución y hábitos alimentarios del oso frontino, *Tremarctos ornatus*, en Venezuela. M.Sc. thesis, Universidad Simón Bolívar, Caracas, Venezuela.

Goldstein, I. (1991). Spectacled bear predation and feeding behavior on livestock in Venezuela. *Studies on Neotropical Fauna and Environment* **26**: 231–235.

Harestad, A. S., and F. L. Bunnel (1979). Home range and body weight: a re-evaluation. *Ecology* **60**: 389–402.

Hershkovitz, P. (1954). Mammals of northern Colombia, preliminary report No. 7: Tapirs (genus *Tapirus*), with a systematic review of American species. *Proceedings of the United States National Museum* **103** (3329): 465–496.

Holbrook, L. T. (1999). The phylogeny and classification of tapiromorph perissodactyls (Mammalia). *Cladistics* **15**: 331–350.

IGAC, Instituto Geografico Agustin Codazzi (1996). Diccionario geográfico de Colombia, CD-ROM edn. Bogotá, Colombia: IGAC.

IUCN (2006). Red List of Threatened Species. Gland, Switzerland: IUCN/SSC.

Kattan, G. H., and P. Franco (2004). Bird diversity along elevational gradients in the Andes of Colombia: area and mass effects. *Global Ecology and Biogeography* **13**: 451–458.

Kattan, G. H, O. L. Hernández, I. Goldstein, *et al.* (2004). Range fragmentation in the spectacled bear *Tremarctos ornatus* in the northern Andes. *Oryx* **38**: 155–163.

Kessler, M. (2002). The elevational gradient of Andean plant endemism: varying influences of taxon-specific traits and topography at different taxonomic levels. *Journal of Biogeography* **29**: 1159–1165.

Krömer, T., M. Kessler, S. R. Gradstein, and A. Acebey (2005). Diversity patterns of vascular epiphytes along an elevational gradient in the Andes. *Journal of Biogeography* **32**: 1799–1809.

Lizcano, D. J. (1996). Densidad poblacional y disponibilidad de hábitat de la danta de montaña (Tapirus pinchaque) en el Parque Regional Natural Ucumari y el Parque Nacional Natural los Nevados. B.Sc. thesis. Universidad de los Andes, Bogotá, Colombia.

Lizcano, D. J., and J. Cavelier (2000a). Densidad poblacional y disponibilidad de hábitat de la danta de montaña (*Tapirus pinchaque*) en los Andes Centrales de Colombia. *Biotropica* **31**: 165–173.

Lizcano, D. J., and J. Cavelier (2000b). Daily and seasonal activity of the mountain tapir (*Tapirus pinchaque*) in the Central Andes of Colombia. *Journal of Zoology, London* **252**: 429–435.

Lizcano, D. J., and J. Cavelier. (2004). Características químicas de salados y hábitos alimenticios de la danta de montaña (*Tapirus pinchaque* Roulin, 1829) en los Andes Centrales de Colombia. *Mastozoologia Neotropical* **11**: 193–201.

Lizcano, D.J., and A. Sissa (2003). Notes on the distribution and conservation status of mountain tapir (*Tapirus pinchaque*) in North Peru. *Tapir Conservation* **12**: 21–24.

Lizcano, D.J., V. Pizarro, J. Cavelier, and J. Carmona. (2002). Geographic distribution and population size of the mountain tapir (*Tapirus pinchaque*) in Colombia. *Journal of Biogeography* **29**: 7–15.

Mondolfi, E. (1989). Notes on the distribution, habitat, food habits, status and conservation of the Spectacled Bear (*Tremarctos ornatus* Cuvier) in Venezuela. *Mammalia* **53**: 525–544.

Mosandl, R., S. Günter, B. Stimm, and M. Weber (2008). Ecuador suffers the highest deforestation rate in South America. In *Gradients in a Tropical Mountain Ecosystem of Ecuador*, eds. E. Beck, J. Bendix, I. Kottke, F. Makeschin, and R. Mosandl, pp. 37–40. Berlin: Springer-Verlag

Myers, N., R.A. Mittermeier, C.G. Mittermeier, G.A. B. da Fonseca, and J. Kent (2000). Biodiversity hotspots for conservation priorities. *Nature* **403**: 853–858.

Nash, W.G., and S.J. O'Brien (1987). A comparative chromosome banding analysis of the Ursidae and their relationship to other carnivores. *Cytogenetics Cell Genetics* **45**: 206–212.

O'Brien, S.J., W.G. Nash, D.E. Wildt, M.E. Bush, and R.E. Benveniste (1985). A molecular solution to the riddle of the giant panda's phylogeny. *Nature* **317**: 140–144.

Orejuela, J., and J. Jorgenson (1999). Status and management of the spectacled bear in Colombia. In *Bears, Status Survey and Conservation Action Plan*, eds. C. Servheen, S. Herrero, and B. Peyton, pp. 168–179. Gland, Switzerland: IUCN/SSC Bear and Polar Bear Specialist Groups.

Paisley, S. (2001). Andean bears and people in Apolobamba, Bolivia: culture, conflict and conservation. Ph.D. thesis. Durrell Institute of Conservation and Ecology, University of Kent, Canterbury, UK.

Paisley, S., and D.L. Garshelis (2006). Activity patterns and time budgets of Andean bears (*Tremarctos ornatus*) in the Apolobamba Range of Bolivia. *Journal of Zoology, London* **268**: 25–34.

Peyton, B. (1980). Ecology, distribution and food habits of spectacled bears, *Tremarctos ornatus* in Peru. *Journal of Mammalogy* **61**: 639–652.

Peyton, B. (1981). Spectacled bears in Peru. *Oryx* **16**: 48–56.

Peyton, B. (1987). Habitat components of the spectacled bear. *International Conference on Bear Research and Management* **7**: 127–133.

Peyton, B. (1999). Status and management of the spectacled bear in Perú. In *Bears: Status Survey and Conservation Action Plan*, eds. C. Servheen, S. Herrero, and B. Peyton, pp. 182–193. Gland, Switzerland: IUCN/SSC Bear and Polar Bear Specialist Groups.

Poveda, F. (1986). El Oso Andino, especie para conservar. *Carta Ganadera, Colombia* **23**: 37–43.

Prothero, D.R., and R.M. Schoch (1989). Classification of the Perissodactyla. In *The Evolution of Perissodactyls*, eds. R.D. Prothero and R.M. Schoch, pp. 530–537. New York: Oxford University Press.

Rahbek, C., and G. Graves (2001). Multiscale assessment of patterns of avian species richness. *Proceedings of the National Academy of Sciences USA* **98**: 4534–4539.

Rechberger, J., R. Wallace, and H. Ticona (2001). Un movimiento de larga distancia de un oso andino (*Tremarctos ornatus*) en el norte del Departamento de La Paz, *Bolivia*. Ecologia *en Bolivia* **36**: 73–74.

Rios-Uzeda, B., H. Gomez, and R.B. Wallace (2006). Habitat preferences of the Andean bear (*Tremarctos ornatus*) in the Bolivian Andes. *Journal of Zoology, London* **271**: 271–278.

Rivadeneira-Canedo, C. (2008). Study of the Andean bear (*Tremarctos ornatus*) as a legitimate seed disperser and elements of its diet in the region of Apolobamba–Bolivia. *Ecología en Bolivia* **43**: 29–39.

Robinson, J., and K. Redford. (1986) Body size, diet and population density of Neotropical forest mammals. *American Naturalist* **128**: 665–680.

Rodriguez, D. (1991). Evaluación del hábitat del oso de anteojos (*Tremarctos ornatus*) en el P.N.N. Las Orquídeas. B.Sc. thesis, Universidad Nacional, Bogotá, Colombia.

Rodriguez, D. (2001). Programa nacional para la conservación y recuperación del oso andino (Tremarctos ornatus) especie amenazada de los ecosistemas andinos colombianos. Bogotá, Colombia: Ministerio del Medio Ambiente, Subdirección de Ecosistemas.

Rodriguez, D., F. Cuesta, I. Goldstein, *et al.* (2003). *Estrategia ecorregional para la conservación del oso andino en Los Andes del Norte*. Bogotá, Colombia: WWF-Colombia, Fundación Wii, EcoCiencia, WCS.

Roth, H.H. (1964). Ein beitrag zur Kenntnis von *Tremarctos ornatus* (Cuvier). *Zoological Garten* **29**: 107–129.

Ruiz-García, M., P. Orozco-terWengel, A. Castellanos, and L. Arias (2005). Microsatellite analysis of the spectacled bear (*Tremarctos ornatus*) across its range distribution. *Genes and Genetic Systems* **80**: 57–69.

Rumiz, D., and J. Salazar (1999). Status and management of the spectacled bear in Bolivia. In *Bears: Status Survey and Conservation Action Plan*, eds. C. Servheen, S. Herrero, and B. Peyton, pp. 164–168. Gland, Switzerland: IUCN/SSC Bear and Polar Bear Specialist Groups.

Rumiz, D., C. Eulert, and R. Arispe (1999). Situación del oso andino en los Parques nacionales Amboro y Carrasco, Bolivia. In *Manejo y conservación de Fauna Silvestre en America Latina*, eds. T. Fang, O. Montenegro, and R.E. Bodmer, pp. 375–382. La Paz, Bolivia: Instituto de Ecología.

Schauenberg, P. (1969). Contribution à l'étude du Tapir pinchaque, *Tapirus pinchaque* Roulin 1829. *Revue Suisse de Zoologie* **76**: 211–256.

Schoch, R.M. (1989). A brief historical review of perissodactyl classification. In *The Evolution of Perissodactyls*, eds. D.R. Prothero and R.M. Schoch, pp. 13–25. New York: Oxford University Press.

Soibelzon, L.H., E.P. Tonni, and M. Bond (2005). The fossil record of South American short-faced bears (Ursidae, Tremarctinae). *Journal of South American Earth Sciences* **20**: 105–113.

Suárez, L. (1985). Hábitos alimenticios y distribución estacional del oso de anteojos (Tremarctos ornatus) en el páramo suroriental del Volcán Antisana, Ecuador. B.Sc. thesis, Pontificia Universidad Católica del Ecuador, Quito, Ecuador

Suárez, L. (1999). Status and management of the spectacled bear in Ecuador. In *Bears: Status Survey and Conservation Action Plan*, eds. C. Servheen, S. Herrero, and B. Peyton, pp. 179–182. Gland, Switzerland: IUCN/SSC Bear and Polar Bear Specialist Groups.

Talbot, S., and G. Shields (1996). A phylogeny of the bears (Ursidae) inferred from complete sequences of three mitochondrial genes. *Molecular Phylogenetics and Evolution* **5**: 567–575.

Thenius, E. (1976). Zur stammesgeschichtlichen Herfunft von *Tremarctos* (Ursidae, Mammalia). *Zeitschrift für Saugetierkunde* **41**: 109–114.

Van der Hammen, T. (1982). Paleoecology of Tropical South America. In *Biological Diversification in the Tropics*, ed. G.T. Prance, pp. 60–67. New York: Columbia University Press.

Velasco, E. (1991). Distribución y uso de la fauna paramuna en los departamentos del Valle y Cauca, Colombia. M.Sc. thesis, Universidad Nacional, Heredia, Costa Rica.

Yerena, E. (1994). Corredores ecológicos en los Andes Venezolanos. In *Parques Nacionales y Conservación Ambiental*, No. 4, eds. S., and T. Amend, pp. 80–87. Caracas, Venezuela: Editorial Torino.

Yerena, L. (1999). Status and management of the spectacled bear in Venezuela. In *Bears: Status Survey and Conservation Action Plan*, eds. C. Servheen, S. Herrero, and B. Peyton, pp. 193–198. Gland, Switzerland: IUCN/SSC Bear and Polar Bear Specialist Groups.

Yerena, E., and D. Torres (1994). Spectacled bear conservation and dispersal corridors in Venezuela. *International Conference of Bear Research and Management* **9**: 169–172.

18 Cloud forests in East Africa as evolutionary motors for speciation processes of flightless Saltatoria species

C. Hemp
University of Bayreuth, Bayreuth, Germany

ABSTRACT

High mountains in East Africa are characterized by high biodiversity and a high degree of endemism, especially in the insect group Saltatoria (grasshoppers, locusts, and katydids). A conspicuous feature of most high mountains in the region is that they harbor arrays of closely related species within various flightless genera of Saltatoria. This chapter compares inventories of Saltatoria in different mountain massifs of contrasting geologic age in East Africa at the species and genus level. Speciation processes and possible migration events are discussed in detail for *Aerotegmina*, a highly specialized canopy dweller in cloud forests. Climatic fluctuations and topographic features within the landscape are the driving forces of speciation processes in flightless Saltatoria on East African high mountains. Cloud forests appear particularly suited to examine speciation processes because during times of more extensive forest cover they are sufficiently interconnected but at the same time isolated enough to allow speciation of taxa with narrow habitat demands. During migration events young volcanoes offered more niches for species than did geologically older areas, but at the genus level geologically young areas exhibited greater similarity. At the species level geographic neighborship is of greater relevance in terms of species congruence.

INTRODUCTION

East Africa is characterized by a series of old basement mountain ranges with an age of more than 30 million years. The Eastern Arc Mountains in Tanzania in particular have been hailed as "hot spots" of biodiversity and endemism, due to (more or less) constantly favorable climatic conditions and the presence of a permanent forest cover over very long periods of time (Rodgers and Homewood, 1982; Newmark and Senzota, 2003). However, during intensive studies in recent years it has become clear that comparatively young volcanoes such as Mt. Kilimanjaro and Mt. Meru in Tanzania, and Mt. Kenya in Kenya, also show a high diversity of plant and animal species (A. Hemp 2001, 2002; Lambrechts *et al.*, 2002; A. Hemp, this volume #12) including a high degree of endemism of flightless Saltatoria (grasshoppers, locusts, and katydids). Such findings are in contrast with the prevailing scientific opinion that these volcanoes have comparatively low biodiversity and endemism due to their young geologic age (Lovett, 1988, 1989; cf. Axmacher *et al.*, 2004).

The aim of this chapter is to highlight the role of montane areas (including cloud forests) on both young and old geologic substrates in East Africa, as motors of speciation processes for flightless Saltatoria species using the katydid genus *Aerotegmina* as an example (Figure 18.1). *Aerotegmina* is a typical inhabitant of cloud forests. The genus was erected after the discovery of *A. kilimandjarica* Hemp on Mt. Kilimanjaro (C. Hemp, 2001). Additional members of this genus were found later in the South and North Pare Mountains, on Mt. Meru, the Manyara Escarpment, Mt. Kenya, and Mt. Sabuk (C. Hemp, unpublished data; see Figure 18.2 for locations). It is hypothesized that past climatic fluctuations caused a more extensive forest cover during more humid times which enabled montane taxa like *Aerotegmina* to spread, while during drier periods grasslands prevailed with their typical fauna, such as *Parasphena* or *Phlesirtes* species.

METHODS

Orthopterologic field work was carried out in the years 1996–2005 on several mountain massifs in East Africa (Figure 18.2). These included in northern Tanzania the mountain ranges of the Eastern Arc (East and West Usambara Mountains, North and South Pare Mountains) and the volcanoes Mt. Meru, the Monduli

Tropical Montane Cloud Forests: Science for Conservation and Management, eds. L. A. Bruijnzeel, F. N. Scatena, and L. S. Hamilton. Published by Cambridge University Press.

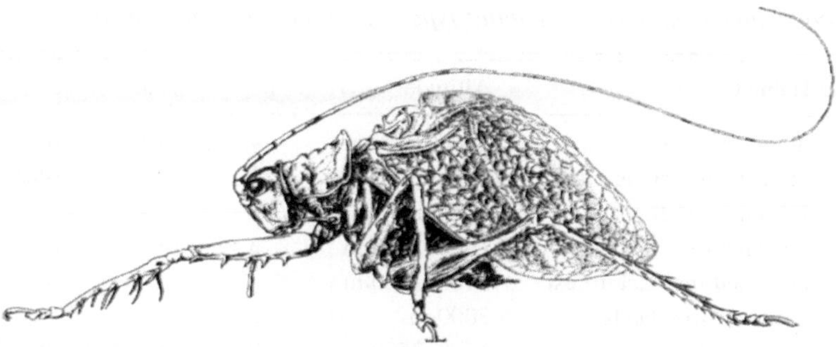

Figure 18.1. The katydid *Aerotegmina kilimandjarica* (Tettigoniidae, Listroscelidinae), a canopy dweller in cloud forests on Mt. Kilimanjaro.

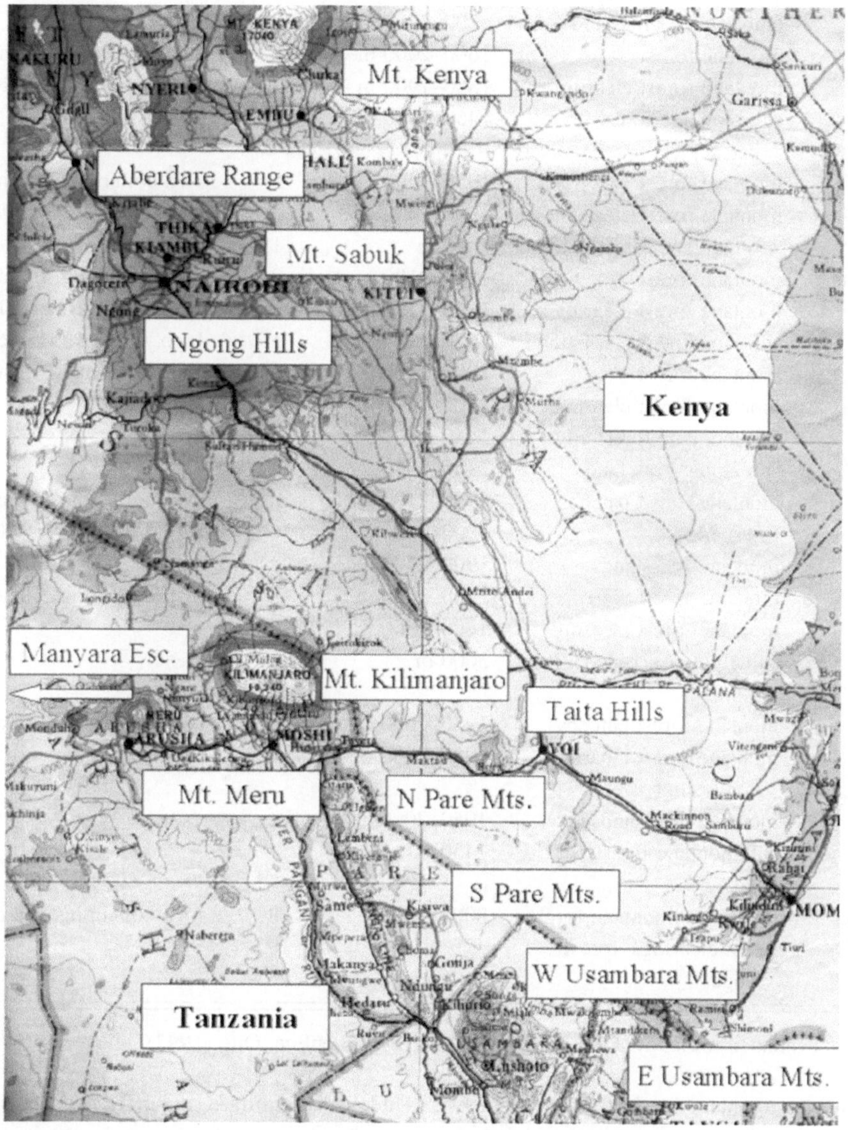

Figure 18.2. Map of East Africa, showing sampling locations.

Range, and Mt. Kilimanjaro. In Kenya the Chyulu Hills, Mt. Kenya (both of volcanic origin), the Ngong Hills, and the Aberdare Range were sampled. Most sampling plots were visited at least three times and during different seasons. Sampling plot size was chosen with respect to the minimum area of the vegetation, mostly 1000 m^2. Parallel plant sociologic relevés were available for each plot to guarantee that Saltatoria inventories were compared for corresponding plant communities. Beside forest communities also

Table 18.1 *Saltatoria collection sites: locality, habitat type, elevation, and source of data*

Locality	Habitat	Altitude	*n*	Data sources
Mt. Meru	montane forest	1700–1900 m	4	own collecting, literature
	montane forest clearing	1750 m	2	survey (Sjöstedt, 1909)
	sub-montane forest	1450 m	2	
Monduli Range	montane forest	1900–2000 m	3	own collecting
	disturbed montane forest	1900–2000 m	2	
	montane grasslands	2000 m	2	
Mt. Kilimanjaro	montane forest	1500–2500 m	10	own collecting, literature
	montane forest clearing	1800–2500 m	6	survey (Sjöstedt, 1909)
	sub-montane forest	1450–1500 m	2	
	(sub-)alpine grasslands	2700–4400 m	6	
North Pare Mts.	montane forest	1600–2000 m	4	own collecting, literature
	montane forest clearing	1850–1900 m	3	survey (e.g. Green, 1998)
	sub-montane forest patches	1450–1500 m	2	
South Pare Mts.	montane forest	1950–2400 m	6	own collecting, literature
	montane forest clearing	1950–2400 m	4	survey (e.g. Jago, 1981;
	sub-montane forest patch	1450 m	1	Green, 1998)
West Usambara Mts.	montane forest	1850–2100 m	8	own collecting, literature
	montane forest clearing	1850–2100 m	3	survey (e.g. Ragge, 1980;
	sub-montane forest patch	1450 m	1	Jago, 1981, 1994; OSF)
Mt. Kenya	montane forest	1800–2500 m	6	own collecting (Sjöstedt, 1912)
	montane forest clearing	1800–2500 m	6	
	sub-montane forest edge	1450 m	1	
	sub-alpine grasslands	3100–3300 m	4	
Ngong Hills	montane forest patches	1700–2100 m	8	own collecting, literature survey (e.g. Uvarov, 1941)
	forest edges	1900–2100 m	6	
	montane grasslands	1900–2100 m	3	
Aberdare Range	montane forest	1700–2500 m	5	own collecting, literature survey (OSF)
	montane forest clearing	1800–2500 m	5	
	sub-alpine grasslands	3000 m	1	
			4	
Chyulu Hills	montane forest patches	1950–2000 m	3	own collecting, literature survey (OSF)
	montane forest clearing, forest edge	1950–2000 m	6	
	montane grasslands	1950–2000 m	4	
	sub-montane forest	1450 m	1	
Outgroup				
East Usambara Mts.	colline-submontane forest	800–1000 m	8	own collecting, literature survey (OSF)
	forest clearings, forest edges		10	

n = number of plots evaluated for study; OSF: Orthoptera species file online, Otte (1997).

nearby open habitats (clearings, montane, and afro-alpine grasslands) were sampled at the respective altitudes. Species were collected by net sweeping and by shaking of trees. Tree-living katydids were recorded by their courtship song or, in the case of non-audible species, with the aid of a bat detector. Table 18.1 provides some basic data on the respective collection sites.

Cluster analysis was conducted with the SORT program (Ackermann and Durka, 2002) using the Sørensen and Jaccard indices of similarity. Genera and species selected from the data matrices were endemic species from single mountains, flightless species with restricted habitat requirements (mostly montane forest species bound to certain altitudes), and flightless species with restricted area of occurrence (such as a few geographically adjacent mountain ranges or massifs), and/or with narrow ecological demands). Species lists compiled for the respective mountainous areas were based on literature survey and own sampling effort (Table 18.1). All investigated genera, species, and subspecies are listed in Table 18.2.

Morphologically, the *Phlesirtes* species investigated in this chapter belong to three distinct lineages. Thus, in the cluster analysis at the genus level three different *Phlesirtes* lineages were erected instead of one genus *Phlesirtes*. Also *Gymnobothroides levipes* and *G. pullus* were not combined into one genus as each species consists of various sub-species typical for the investigated areas and reflecting historic relationships. Likewise, in the cluster analysis at the species level these sub-species were counted as species. The same was done for sub-species of *Heteracris brevipennis* and *Aerotegmina*. The colline zone stretches between 700 and 1100 m.a.s.l., followed by a submontane zone between 1100 and 1500 m.a.s.l. The montane zone is sub-divided into three sub-zones: lower montane (1500–2200 m), middle montane (2200–2500 m), and upper montane (2500–3200 m) (cf. A. Hemp, this volume #12). Above, the sub-alpine and alpine zones are located.

RESULTS AND DISCUSSION

Species-level analysis

Cluster analysis to investigate relationships at the species level between flightless Saltatoria from different mountain areas of contrasting geologic age gave nearly identical results regardless whether the Sørensen or Jaccard index of similarity was used. Figure 18.3 illustrates the cluster tree using the Jaccard index, indicating that relationships are strongest for adjacent areas while there is no relationship between geographically distant regions (cf. Figure 18.2).

Thus, Mt. Meru exhibits the closest faunistic similarity to the nearby Monduli range with which it is connected by a mountain ridge at least 1800 m high, thereby allowing an exchange of species within the montane zone (1500–3200 m) proper. Any differences found between the two areas may be primarily a result of insufficiencies in the Saltatoria inventory that further studies might reveal. Mt. Meru and Mt. Kilimanjaro are connected at a sub-montane level which explains the high degree of common flightless species. More than half of the flightless endemics (37 species) of the area are shared between the two volcanoes. The majority of sub-montane (1100–1500 m) and all colline (700–1100 m) elements are common to both massifs (and the Monduli range). Due to its huge isolated montane and afro-alpine zones, Mt. Kilimanjaro harbors quite a number of endemic species, including *Monticolaria kilimandjarica* Sjöstedt 1909, *Amytta kilimandjarica* Hemp 2001, *Aerotegmina kilimandjarica* Hemp 2001, *Phlesirtes kibonotense* Sjöstedt 1909, *P. kilimandjaricus* Sjöstedt 1909, and n.sp., and *Chromothericles kanga* Sjöstedt 1909, all in the montane zone (C. Hemp, 2005); and

Table 18.2 *Saltatoria species included in the present study*

Eumastacidae
Chromomastax jagoi Descamps 1974
Chromothericles decoloratus Descamps 1977
Chromothericles emaliensis Descamps 1977
Chromothericles kanga (Sjöstedt 1923)
Chromothericles sp. (2 species)
Chromothericles rubroornatus Descamps 1977
Chromothericles tvigga (Sjöstedt 1909)
Thamithericles tanzaniae Descamps 1977
Euschmidtia bidens Descamps 1964
Euschmidtia sp.
Euschmidtia uvarovi Descamps 1964
Lophothericles carinifrons (Karsch 1889)
Lophothericles crassipes Descamps 1977
Lophothericles kongoni (Sjöstedt 1909)
Lophothericles rubripes Descamps 1977
Lophothericles sp.
Lophothericles wardi Descamps 1977
Mastarammea karaseki (Ramme 1925)
Pieltainidia mira Ramme 1925
Plagiotriptus carli (C. Bolivar 1914)
Stenoschmidtia elegans Descamps 1967
Pyrgomorphidae
Loveridgacris impotens (Karsch 1888)
Parasphena keniensis Sjöstedt 1912
Parasphena meruensis Sjöstedt 1909
Parasphena teitensis Kevan 1949
Parasphena chyuluensis Kevan 1949
Parasphena kinangopa Uvarov 1938
Parasphena ngongensis Kevan 1949
Lentulidae
Altiusambilla modicicrus (Karsch 1896)
Rhainopomma montanum (Kevan 1950)
Rhainopomma pseudomontanum Hemp 2007
Rhainopomma uguenoensis Hemp 2007
Rhainopomma usambaricum (Ramme 1929)
Rhainopomma magnificum Jago 1981
Mecostiboides physalus Karsch 1896
Mecostibus leprosus Karsch 1896
Mecostibus sellatus Uvarov 1941
Usambilla emaliensis Jago 1981
Usambilla turgidicrus (Karsch 1896)
Usambilla olivacea Sjöstedt 1909
Acrididae
Coptacridinae
Parepistaurus deses deses Karsch 1896
Parepistaurus deses ssp.
Parepistaurus deses nairobii Green 1998
Parepistaurus deses vansomereni Kevan 1955
Parepistaurus intermedius Green 1998

Table 18.2 *(cont.)*

Parepistaurus jagoi Green 1998
Parepistaurus robertsoni Green 1998
Parepistaurus lindneri Kevan 1955
Parepistaurus sp. A Green 1998
Parepistaurus pygmaeus (Karny 1909)
Eyprepocnemidinae
Heteracris brevipennis brevipennis (I. Bolivar)
Heteracris brevipennis laticercus (Uvarov)
Heteracris brevipennis nyambeniensis (Grunshaw 1991)
Catantopinae
Aresceutica subnuda Karsch 1896
Aresceutica vansomereni Kevan 1956
Pseudophialosphera severini (Ramme 1929)
Hadrolecocatantops kilimandjaricus
Ixalidium bicoloripes Uvarov 1941
Ixalidium sp.
Ixalidium haematoscelis Gerstäcker 1869
Ixalidium sjöstedti Kevan 1950
Ixalidium transiens Ramme 1929
Kinangopa jeanelli Uvarov 1938
Pezocatantops kinangopi (Uvarov 1941)
Pezocatantops ngongi (Uvarov 1941)
Pezocatantops impotens (Johnston 1937)
Physocrobylus tessa Hochkirch 1996
Acridinae
Gymnobothroides levipes (Karsch 1896)
Gymnobothroides pullus pullus Karny 1915
Gymnobothroides keniensis Johnston 1937
Gymnobothroides pullus montanus Kevan 1950
Parodontomelus mazumbaiensis Jago 1983
Parodontomelus arachniformis Jago 1983
Afrophlaeoba usambaricus (Ramme 1929
Odontomelus chyulu Jago 1994
Odontomelus brachypterus (Gerstäcker 1869)
Odontomelus pareense Jago 1994
Odontomelus parabinervis Jago 1994
Odontomelus spodiopsis Jago 1994
Odontomelus pallidus Sjöstedt 1912
Odontomelus ixyonotus Jago 1994
Uganda kilimandjarica (Sjöstedt 1909)
Uganda darlingtonae Ritchie 1992
Tettigoniidae
Phaneropterinae
Dioncomena jagoi Ragge 1980
Dioncomena ornata Brunner von Wattenwyl 1878
Dioncomena tanneri Ragge 1980
Ectomoptera nepicauda Ragge 1980
Eulioptera monticola Ragge 1980
Eulioptera montana Ragge 1980
Eulioptera sp.
Euryastes jagoi Ragge 1980
Horatosphaga montivaga (Sjöstedt 1909)
Horatosphaga parensis Hemp 2002
Horatosphaga meruensis (Sjöstedt 1909)
Horatosphaga elgonis (Chopard 1938)
Horatosphaga sp.
Meruterrana elegans Sjöstedt 1912
Monticolaria kilimandjarica Sjöstedt 1909
Monticolaria meruensis Sjöstedt 1909
Peronura uguenoensis Hemp 2002
Meconematinae
Amytta olindo Hemp 2001
Amytta kilimandjarica Hemp 2001
Amytta sp. (4 species)
Amytta ukaimica Beier 1965
Afrophisis tanzanica Jin & Kevan 1991
Listroscelidinae
Aerotegmina kilimandjarica ssp. Hemp 2001
Aerotegmina shengenae Hemp 2006
Phlugidia usambarica Hemp 2002
Conocephalinae
Anthracites montium Sjöstedt 1909
Anthracites kilimandjaricus Sjöstedt 1909
Anthracites sp. (4 species)
Anthracites usambaricus Sjöstedt 1913
Phlesirtes kibonotense (Sjöstedt 1909)
Phlesirtes merumontanus (Sjöstedt 1909)
Phlesirtes kilimandjaricus (Sjöstedt 1909)
Phlesirtes meruense (Sjöstedt 1909)
Phlesirtes sp. (10 species)
Mecopodinae
Apteroscitus sp.
Gymnoscirtus unguiculatus (Karsch 1888)

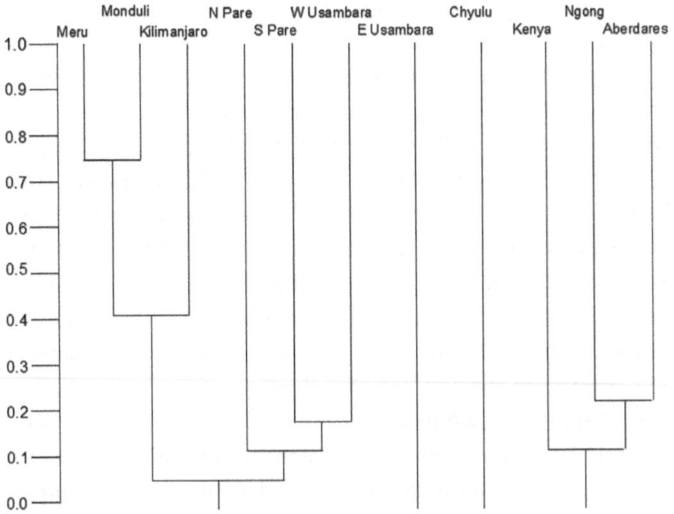

Figure 18.3. Cluster analysis (Jaccard's index of similarity) showing relationships of flightless Saltatoria at the species level. "Young" volcanic areas: Mt. Meru, Monduli Range, Mt. Kilimanjaro (Tanzania), Chyulu Hills, and Mt. Kenya (Kenya). Old basement mountains: North Pare Mountains, South Pare Mountains, West Usambara Mountains, and East Usambara Mountains.

Uganda kilimandjarica Sjöstedt 1909 and *Parasphena pulchripes* Gerstäcker 1869 in the upper montane and afro-alpine zone (>3700 m.a.s.l.). On Mt. Meru, endemics are also restricted to the montane zone (*Monticolaria meruensis* Sjöstedt 1909, *Chromothericles tvigga* Sjöstedt 1909, *Aerotegmina* n.ssp.).

The North Pare Mountains are situated adjacent to Mt. Kilimanjaro (Figure 18.2) but are faunistically very different from the Kilimanjaro/Meru area (Figure 18.3). Although "connected" by the Pangani River system which could serve as a vegetation bridge and thus function as a potential migration corridor, the area in between is very dry. Only a few colline and sub-montane elements are shared with the Mt. Kilimanjaro/ Mt. Meru area, such as *Odontomelus brachypterus* Gerstäcker 1869 and *Usambilla olivacea* Sjöstedt 1909. Furthermore, the topography drops below 800 m.a.s.l. Therefore, vegetation along the Pangani River may have been sparse even during more humid and warmer climatic periods in the past, thereby effectively inhibiting faunal exchange between the areas. Only during major climatic fluctuations with more humid and cooler weather conditions that would increase the area covered with montane forest could an exchange of stenoecic (species with narrow habitat demands) taxa have occurred, e.g. during the period between 11000 and 4000, years B.P. (Thompson *et al.*, 2002). Thus, the *Aerotegmina* species of the North Pare Mountains is genetically closest to *A. kilimandjarica* whereas the *Parepistaurus* species is a sub-species of *P. deses* Karsch 1896 of the Mt. Kilimanjaro/Mt. Meru area (Schultz, 2004). On the other hand, the North Pare Mountains harbor *Rhainopomma uguenoensis* Hemp 2007 (Lentulidae; Hemp *et al.*, 2007), an element considered typical for the Eastern Arc Mountains, while *Altiusambilla* is exclusively found in the montane zones of the Mt. Kilimanjaro/Mt. Meru area. Endemic to the North Pare Mountains according to present knowledge are the species *Peronura uguenoensis* Hemp 2002, *Amytta* n. sp., *Aerotegmina* n. ssp., *Chromothericles* n. sp., *Rhainopomma uguenoensis* Hemp 2007, *Parepistaurus deses* n. ssp, and two new *Phlesirtes* species.

Topographically the North Pare Mountains are separated by a dry savanna zone from the South Pare Mountains and connected only partly (as is the whole area from Mt. Kilimanjaro to the Indian Ocean) by the Pangani River system. Thus, common species are flightless Saltatoria typical for colline and submontane forests, such as *Usambilla olivacea* and *Rhainopomma magnificum* Jago 1981. Only one species of the montane zone is found on both mountain ranges: *Horatosphaga parense* Hemp 2002. Endemics of the South Pare Mountains include e.g. *Rhainopomma magnificum*, *Parepistaurus robertsoni* Green 1998, *Odontomelus parense* Jago 1995, *Amytta* n. sp., *Aerotegmina shengenae* Hemp 2006., and *Phlesirtes* n. sp.

The South Pare and West Usambara Mountains cluster as a block together although similarity is not very high (Figure 18.3). These two mountain ranges are also isolated from each other by dry savanna vegetation, sharing (via the Pangani River) only the species *Usambilla olivacea*. Probably due to the presence of wet forest communities on the north-eastern slopes of the South Pare Mountains (Gonja Forest) that connect the range with the West Usambara Mountains, also two species are found on both ranges: *Rhainopomma pseudomontanum* Hemp 2007 and *Parepistaurus intermedius* Green 1998. Endemics of the West Usambaras include *Chromothericles rubroornatus* Descamps 1977, *Mecostibus leprosus* Karsch 1896, *Parepistarus jagoi* Green 1998, *Parodontomelus mazumbaiensis* Jago 1983, *Afrophlaeoba longicornis* Jago 1983, *Dioncomena tanneri* Ragge 1980, *Euryastes jagoi* Ragge 1980, *Amytta ukaimica* Beier 1965, *Amytta* sp., *Gymnoscirtus* sp., *Anthracites usambaricus* Sjöstedt 1909, *Anthracites* sp., and *Phlesirtes* n. sp.

The East Usambara Mountains harbor a completely different species set partly due to the lack of a montane zone. Highest affinities within Saltatoria are found with coastal forms. The East Usambara Mountains are characterized by a high degree of endemism of Saltatoria species, especially within the families Tettigonoidea (*Dioncomena jagoi* Ragge 1980, *Ectomoptera nepicauda* Ragge 1980, *Amytta pellucida* Karsch 1888, *Phlugidia usambarica* Hemp 2002, *Anthracites usambaricus* Sjöstedt 1913) and Eumastacidae (*Chromomastax jagoi* Descamps 1973, *Mastrammea karaseki* Ramme 1925, *Pieltainidia mira* Ramme 1925, *Plagiotriptus carli* Bolivar 1914, *Stenoschmidtia elegans* Descamps 1967). However, the high degree of endemism might be due to insufficient sampling and additional collecting activity might reveal that species thought to be restricted to the East Usambaras might also occur in coastal forests. Although situated close to the West Usambara Mountains, only three flightless Saltatoria species are shared between these mountains: *Rhainopomma usambarica* Ramme 1929, *Parepistaurus pygmaeus* Karny 1909, and *Gymnobothroides pullus* Karny 1915. A small but deep furrow (Lwengeru Valley) of hot and dry savanna separates the two mountain ranges which must be regarded an effective barrier against exchange of species even during more humid climatic periods with increased forest cover.

The Chyulu Hills, located between Mts. Kilimanjaro and Meru in the south and the Kenyan Highlands to the north, do not share any species with either adjacent mountain area (Figure 18.3). Flightless species endemic to the Chyulu Hills include *Parasphena chyuluensis* Kevan 1949, *Odontomelus chyulu* Jago 1995, and *Phlesirtes* n. sp. However, some affinities exist with Mt. Kilimanjaro (Figure 18.3) as also stated by Kevan (1950).[1] Species restricted to the Chyulu Hills and the Mts. Kilimanjaro and Meru area are *Hadrolecocatantops kilimandjaricus* Ramme 1929) (also recorded from the Emali range) and *Eulioptera montana*

[1] When Kevan (1949) described *Parasphena chyuluensis* he remarked that it might also be a sub-species to *P. meruensis* Sjöstedt 1909. But since he did not study the holotype of *P. meruensis* he preferred giving it species status.

Ragge 1980. The latter two species have not been regarded in this analysis as they are fully winged (i.e. not flightless). The forest patches of the Chyulu Hills are still poorly investigated and might also harbor species analyzed in this chapter such as *Aerotegmina*, *Horatosphaga*, *Rhainopomma*/*Altiusambilla*.

A separate block is formed by the mountain ranges of the Kenyan highlands, with the highest affinities found between the Ngong Hills near Nairobi and the Aberdare range (Figure 18.3). Close to the Aberdare range, Mt. Kenya looms to afro-alpine heights (5199 m.a.s.l.). Species responsible for clustering Mt. Kenya, the Aberdares, and the Ngong Hills together are *Gymnobothroides keniensis* Johnston 1937, *Aresceutica vansomereni* Kevan 1956, and *Kinangopa jeanelli* Uvarov 1938. Common to both the Aberdares and the Ngong Hills are *Heteracris brevipennis brevipennis* Bolivar 1914 and *Odontomelus pallidus* Sjöstedt 1912. Like Mt. Kilimanjaro, Mt. Kenya is rich in endemic Saltatoria species, contradicting claims that young volcanic areas must be biologically poor due to their geologically young age. Thirteen flightless endemics were encountered on Mt. Kenya, e.g. *Parasphena keniensis* Sjöstedt 1912, *Pezocatantops impotens* (Johnston 1937), *Odontomelus spodiopsis* Jago 1994, *Meruterrana elegans* Sjöstedt 1912, and several as yet undescribed new species (including the genera *Aerotegmina*, *Amytta*, *Anthracites*, *Chromothericles*, *Horatosphaga*, and *Phlesirtes*). Endemic to the Aberdare Mountains are five species (*Parasphena kinangopa* Uvarov 1939, *Chromothericles decoloratus* Descamps 1977, *Pezocatantops kinangopi* Uvarov 1941, *Phlesirtes* n.sp, and *Uganda darlingtonae* Ritchie 1992) while Ngong Hills endemics include *Parasphena ngongensis* Kevan 1949, *Pezocatantops ngongi* Uvarov 1941, and *Phlesirtes* n.sp.

The Ngong Hills as well as the Chyulu Hills show a much reduced forest cover nowadays and thus stenoecic forest elements such as *Aerotegmina* may have gotten extinct. Forest patches in especially the western part of the Chyulu Hills were screened intensively for *Aerotegmina* but probably these are too small nowadays to guarantee survival for canopy dwellers requiring high levels of atmospheric humidity. Whilst the entire Chyulu Hills were still classified as "forest" on a topographic map dating from 1898 (Werther, 1898), forest patches in the investigated western parts are currently restricted to depressions along rivers where they are comparatively safe from recurrent fire. A similar loss of forest must be presumed for the Ngong Hills where today forest is restricted to the highest tops of the range and depressions along drainage lines.

Genus-level analysis

Evaluation of the data-set at the genus level (Figure 18.4) shows a different picture: first of all it is clear that affinities exist between all mountain ranges investigated. There are no isolated blocks as found at the species level (Figure 18.3). Thus, at the

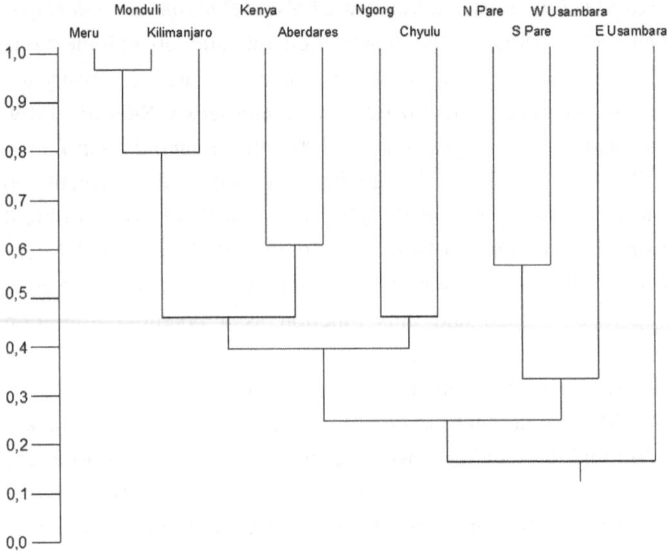

Figure 18.4. Cluster analysis (Jaccard's index of similarity) showing relationships of flightless Saltatoria at the genus level. Sites are as specified in Figure 18.3.

genus level geologic age seems to be more important for the development of relationships than geographic proximity. "Young" (few million years) areas like the volcanoes Mt. Meru, Mt. Kilimanjaro, Mt. Kenya, and the Chyulu Hills cluster more closely together, although geographic proximity is still of some importance (Mt. Kenya with the Aberdares, and the Chyulu Hills with Ngong Hills). On the other hand, the old basement areas (North and South Pare Mountains, West and East Usambara Mountains) form a separate block (Figure 18.4).

Due to the presence of topographic corridors in the Meru/Monduli/Kilimanjaro area these mountains cluster even more closely together at the genus level. Almost all genera are shared between Mt. Meru and the Monduli range and most with Mt. Kilimanjaro. Unexpectedly close affinities are seen between the high volcanoes Mts. Kilimanjaro/Meru and (more distant) Mt. Kenya. Also the Aberdare range reaches to afro-alpine heights and lying adjacent to Mt. Kenya it shares typical genera of these altitudes such as *Uganda* and *Parasphena*. Especially in the montane zone many genera are found on all the volcanoes, such as flightless *Horatosphaga*, *Phlesirtes*, *Chromothericles*, *Aerotegmina*, *Anthracites*, and *Amytta*, causing the clustering of these young mountains. Although lacking an afro-alpine zone and thus the associated typical genera such as *Uganda*, the Chyulu Hills are the next in terms of being related to the high volcanoes Mts. Meru, Kilimanjaro, and Kenya (Figure 18.4). Genera shared between these regions include *Parasphena*, *Phlesirtes*, and *Amytta*. The montane zone of the Ngong Hills also harbors many of the typical montane genera. The relationship of the Saltatoria fauna of the Chyulu and Ngong Hills might have turned out to be much closer to that of Mts. Meru, Kilimanjaro, and Kenya if the forest cover of the Hills would not have been reduced so much (e.g. canopy

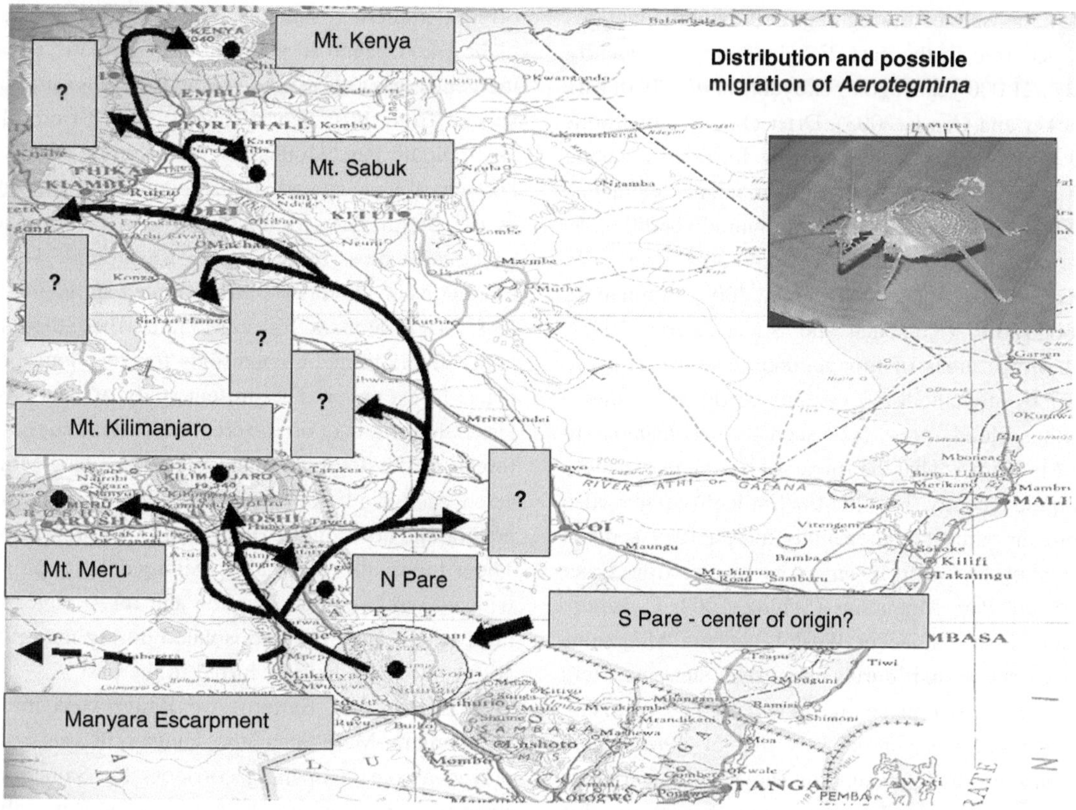

Figure 18.5. Map of East Africa showing distribution and possible migration corridors of *Aerotegmina*.

dwelling genera such as flightless *Horatosphaga*, *Aerotegmina*, *Anthracites*, or *Chromothericles* are not (yet) recorded for these mountain ranges).

The North and South Pare and West Usambara Mountains cluster as a separate block against the volcanoes (Figure 18.4), showing less relationship at the genus level despite being geographically close to the Mts. Meru–Kilimanjaro area (Figure 18.2). Common genera of the volcanoes, the Pare Mountains, and West Usambara Mountains are mostly sub-montane elements such as *Parepistaurus*, *Lophothericles*, *Ixalidium*, and *Odontomelus*. These taxa might have spread also during minor climatic fluctuations connecting sub-montane areas. The East Usambara Mountains plot separately from all other mountain ranges (Figure 18.4). These mountains lie close enough to the Indian Ocean to be under the favorable humid influence of the coastal climate and past climatic fluctuations probably did not have much effect on forest extension (Rodgers and Homewood, 1982) and thus on Saltatoria migration. Furthermore, the East Usambara Mountains are not high enough to have a stable montane zone with the typical associated taxa although the moist forests on the mountain ridges and summits may be regarded as cloud forests. Instead, these mountains are characterized by coastal lowland taxa such as *Stenoschmidtia*, *Pieltainidia*, *Mastarammea*, *Chromomastax*, *Phlugidia*, *Dioncomena*, *Afrophlaeoba*, or *Parodontomelus*. Genera shared with the more inland situated mountain complexes include *Parepistaurus*, *Rhainopomma*, *Ixalidium*, *Amytta*, and *Anthracites*.

The cloud forest genus *Aerotegmina*

Species of the genus *Aerotegmina* are typical cloud forest dwellers, being restricted to the canopy of trees in closed montane forest. It seems highly probable that this genus evolved in the montane forests of the South Pare Mountains where the morphologically most primitive species of the genus was found (Hemp, 2006). Also, genetic screening showed a more primitive status for the species from the South Pare Mountains compared with four investigated *Aerotegmina* from Mts. Kilimanjaro, Meru, and Kenya, as well as the North Pare Mountains (Schultz, 2004). To date, *Aerotegmina* has been recorded in seven isolated mountainous regions. The emerging distribution pattern (Figure 18.5) suggests migration events to have occurred during (wetter) times when the montane forest zones were interconnected, rather than migration in the form of accidental shifts (e.g. wind, birds).

The montane forests of the South Pare (and West Usambara) Mountains seem to have acted as a kind of "evolutionary motors" due to their special geographic position (C. Hemp and O. Schultz, unpublished data). These mountain ridges are located near enough to the coast to be under the influence of its stable climatic regime allowing continuous forest cover but they are also sufficiently far away to have become isolated during major climatic fluctuations known to have occurred in the past. For example, a major dry period occurred *c.* 21 000 years ago (Yokoyama *et al.*, 2000; Mix *et al.*, 2001; Barker and Gasse, 2003). Drier climatic conditions could have led to a forest retreat restricting forest to montane zones, thereby favoring speciation of montane elements. A major climatic fluctuation inducing more humid conditions is thought to have prevailed between 15 000 and 5000 years B.P. (de Menocal *et al.*, 2000; cf. Thompson *et al.*, 2002) culminating around 9000–6000 B.P. (Kutzbach and Street-Perrott, 1985). During that time huge areas of mountainous East Africa were probably covered by montane forest opening corridors for ancestors of stenoecious cloud forest taxa such as *Aerotegmina* to spread as far as Mt. Kenya. That the spread of *Aerotegmina* could have happened quite recently (i.e. after the last Ice Age) may also be inferred from the almost homogenous morphology and the close genetic relationship with *Aerotegmina* species from other areas than the South Pare Mountains (Schulz, 2004; C. Hemp, unpublished data). In the adjacent West Usambara Mountains *Aerotegmina* was not found during intensive surveys. Here *Euryastes jagoi* seems to replace *Aerotegmina.* The courtship song of *E. jagoi* resembles very closely that of *Aerotegmina* species while it also seems to occupy the same ecological niche (nocturnal activity, canopy dweller of montane forest) and this might explain why *Aerotegmina* could not establish here although the forest communities were found to be suitable for this taxon (A. Hemp, unpublished data). The genus *Euryastes* is monotypic and seems to have no close relatives among the other African Phaneropterinae (Ragge, 1980) thus underlining the hypothesis that the area around the West Usambara and South Pare Mountains might function as an evolutionary motor creating new taxa.

CONCLUSIONS

At the species level geographic neighborhood proved to be of major importance for the faunistic similarity of isolated mountains massifs in the study area. During even minor climatic fluctuations that produced areal changes in vegetation cover certain taxa of the sub-montane zones were able to spread more easily. Thus, genera such as *Odontomelus* or *Parepistaurus* occur in most mountainous areas whereas some of these genera contain species that occur on two or several adjacent situated mountain massifs. On the other hand, taxa restricted to montane habitats were able to spread only during major climatic changes affecting forest extension in the whole of East Africa, e.g. *Anthracites*, *Aerotegmina*, or *Amytta*. Whilst on the old basement mountain ranges of e.g. the Eastern Arc many niches were already occupied, geologically young areas such as Mt. Kilimanjaro, Mt. Meru, or Mt. Kenya with their enormous altitudinal gradients offered numerous habitats. Newly arriving taxa easily found niches, and began to speciate after climatic isolation. Thus, at the genus level the young volcanoes harbor proportionally more species from those genera that originated from old centers such as the South Pare Mountains. An example of how a cloud forest taxon may have spread is *Aerotegmina*, which probably evolved in the South Pare Mountains where the morphologically and genetically most basic species was found. The old basement mountains of the Eastern Arc complex stretching along the coast (e.g. East Usambara, Nguru Mountains) are thought to have been under stable climatic conditions for very long periods favoring dense forest cover. The present study of relationships of flightless Saltatoria taxa on different mountains suggests the montane forests of the South Pare and West Usambara Mountains to have acted as a kind of "evolutionary motor" due to their special geographic position (C. Hemp and O. Schultz, unpublished data). These mountain ridges are positioned near enough to the coast with its stable climatic regime and forest cover but sufficiently far away to have become isolated during major climatic fluctuations known to have occurred in the past.

Where the old Eastern Arc Range runs inland toward the Kilimanjaro–Meru area, the Pangani River system may have played a role in opening corridors for Saltatoria migration as riverine vegetation may have been used especially by taxa with broader habitat requirements such as *Rhainopomma* or *Parepistaurus.* These were thus able to spread even during minor climatic fluctuations (e.g. *Rhainopomma usambaricum* and *Parepistaurus pygmaeus* are shared between the East and West Usambara Mountains). Similarly, *Usambilla olivacea* (found from Mt. Kilimanjaro to the West Usambara Mountains) and *Rhainopomma magnificum* (from the Pare Mountains to the West Usambara Mountains) may have spread only recently along the Pangani River vegetation corridor as well.

ACKNOWLEDGEMENTS

Field work was carried out with financial support from the Deutsche Forschungsgemeinschaft and permission of the Tanzanian Commission for Science and Technology. For further support I owe gratitude to the Chief Park Warden of Kilimanjaro National Park, Mr. Moirana, and Mr. Mafuru and Mr. Jacob Mushi.

REFERENCES

Ackermann, W., and W. Durka (2002). *SORT 3.5*. Bayreuth, Germany: University of Bayreuth.

Axmacher, J.C., H. Tünte, M. Schrumpf, *et al.* (2004). Diverging diversity patterns of vascular plants and geometrid moths during forest regeneration on Mt. Kilimanjaro, Tanzania. *Journal of Biogeography* **31**: 895–904.

Barker, P., and F. Gasse (2003). New evidence for a reduced water balance in East Africa during the Last Glacial Maximum: implication for model-data comparison. *Quarternary Science Reviews* **22**: 823–837.

Bolivar, I. (1922). *Voyage de M. le Baron Maurice de Rothschild en Éthiopie et en Afrique Orientale Anglaise (1904–1905)*: Résultats scientifiques – Animaux articulés, Part 1, Orthopterès. Paris.

Chopard, L. (1954). Orthopteres Ensiferes. *Mémoires de l'IFAN* **40**: 25–100.

de Menocal, P., J. Ortiz, T. Guilderson, *et al.* (2000). Abrupt onset and termination of the African Humid Period: rapid climate responses to gradual insolation forcing. *Quaternary Science Reviews* **19**: 347–361.

Green, S.V. (1998). Revision of the African grasshopper genus *Parepistaurus* Karsch 1896 (Orthoptera Acrididae Coptacridinae). *Tropical Zoology* **11**: 259–332.

Hemp, A. (2001). Ecology of the pteridophytes on the southern slopes of Mt. Kilimanjaro. II. Habitat selection. *Plant Biology* **3**: 493–523.

Hemp, A. (2002). Ecology of the pteridophytes on the southern slopes of Mt. Kilimanjaro. I. Altitudinal distribution. *Plant Ecology* **159**: 211–239.

Hemp, C. (2001). *Aerotegmina*, a new genus of African Listroscelidinae (Orthoptera: Tettigoniidae, Listroscelidinae, Hexacentrini). *Journal of Orthoptera Research* **10**: 125–132.

Hemp, C. (2005). The Chagga home gardens: relict areas for endemic Saltatoria species (Insecta: Orthoptera) on Mt. Kilimanjaro. *Biodiversity and Conservation* **125**: 203–209.

Hemp, C. (2006). *Aerotegmina shengenae*, a new species of Listroscelidinae (Orthoptera: Tettigoniidae) from the Eastern Arc mountains of East Africa. *Journal of Orthoptera Research*: **15**: 99–103.

Hemp, C., and A. Hemp (2003). Saltatoria coenoses of high altitude grasslands on Mt. Kilimanjaro, Tanzania (Orthoptera: Saltatoria). *Ecotropica* **9**: 71–97.

Hemp, C., O. Schultz, A. Hemp, and W. Wägele (2007). New Lentulidae species from East Africa (Orthoptera: Saltatoria). *Journal of Orthoptera Research* **16**: 85–96.

Jago, N.D. (1981). A revision of the genus *Usambilla* Sjöstedt (Orthoptera, Acridoidea) and its allies. *Bulletin of the British Museum (Natural History) Entomology* **43**: 1–38.

Jago, N.D. (1994). *Odontomelus* I. Bolivar 1890 (Orthoptera Acridoidea Acrididae Acridinae): savanna–woodland grasshoppers with a major radiation of flightless species in Eastern Africa. *Tropical Zoology* **7**: 367–450.

Karny, H. (1912). Orthoptera Fam. Locustidae Subfam. Conocephalinae. *Genera Insectorum* **135–143**: 1–17.

Kevan, D.K.McE. (1949). New species of *Parasphena* Bolivar 1884 (Orthoptera, Acrididae, Pyrgomorphinae) from East Africa. *Journal of East Africa Natural History* **19**: 110–130.

Kevan, D.K.McE. (1950). Orthoptera from the hills of south-east Kenya. *Journal of the East Africa Natural History Society and National Museum* 1947–1948, **21**: 3–22.

Kutzbach, J.E., and F.A. Street-Perrott (1985). Milankovitch forcing of fluctuations in the level of tropical lakes from 18 to 0 kyr BP. *Nature* **317**: 130–134.

Lambrechts, C., B. Woodley, A. Hemp, C. Hemp, and P. Nnyiti (2002). *Aerial Survey of the Threats to Mt. Kilimanjaro Forests*. Dar es Salaam, Tanzania: UNDP.

Lovett, J.C. (1988). Endemism and affinities of the Tanzanian montane forest flora. *Monographs in Systematic Botany from the Missouri Botanical Garden* **25**: 591–598.

Lovett, J.C. (1989). Tanzania. In *Floristic Inventory of Tropical Countries*, eds. D.G. Campbell and H.D. Hammond, pp. 232–235. New York: New York Botanical Garden.

Mix, A.C., E. Bard, and R. Schneider (2001). Environmental processes of the Ice Age: land, oceans, glaciers (EPILOPG). *Quaternary Science Reviews* **20**: 627–657.

Newmark, W.D., and R.B.M. Senzota (2003). Power to detect trends in ecological indicators in the East Usambara Mountains, Tanzania. *African Journal of Ecology* **41**: 294–298.

Otte, D. (1997). *Orthoptera Species File*: *Tettigonoidea*. Philadelphia, PA: The Orthopterist´s Society and The Academy of Natural Sciences and Eades, D.C., Otte, D. and Naskrecki, P. *Orthoptera Species File Online*. Version 2.0/3.1.

Ragge, D.R. (1980). A review of the African Phaneropterinae with open tympana (Orthoptera: Tettigoniidae). *Bulletin of the British Museum (Natural History) Entomology* **40**: 1–192.

Rodgers, W.A., and K.M. Homewood (1982). Species richness and endemism in the Usambara mountain forests, Tanzania, *Biological Journal of the Linnean Society* **18**: 197–242.

Schultz, O. (2004). Molekulare Phylogenie und Radiation endemischer Orthopteren Ostafrikas. M.Sc. thesis, Ruhr-Universität, Bochum, Germany.

Sjöstedt, Y. (1909). 17. Orthoptera. Locustodea – Acridoidae. In *Wissenschaftliche Ergebnisse der Schwedischen Zoologischen Expedition nach dem Kilimanjaro, dem Meru und den umgebenden Massaisteppen Deutsch-Ostafrikas 1905–1906*, ed. Y. Sjöstedt, pp. 125–148 and 149–200.

Sjöstedt, Y. (1912). Über einige von Herrn Prof. E. Lönnberg in Britisch Ostafrika eingesammelte Orthopteren. *Arkiv för Zoologi* **7**(38): 1–26.

Thompson, L.G., E. Mosley-Thompson, M.E. Davis, *et al.* (2002). Kilimanjaro ice core records: evidence of Holocene climate change in tropical Africa. *Science* **18**: 589–593.

Uvarov, B.P. (1923). Notes on the Orthoptera in the British Museum. III. Some less known or new genera and species of the subfamilies Tettigoniinae and Decticinae. *Transactions of the Royal Entomological Society of London* **49**: 492–537.

Uvarov, B.P. (1941). New grasshoppers (Orthoptera, Acrididae) from Kenya. *Journal of the East African Natural History Society* **16**: 25–34.

Werther, W. (1898). *Karte des nordöstlichen Teiles von Deutsch-Ost-Afrika*. Berlin: Verlag Hermann Paetel.

Yokoyama, Y., K. Lambeck, P. De Deckker, P. Johnston, and L.K. Fifield (2000). Timing of the Last Glacial Maximum from observed sea-level minima. *Nature* **406**: 713–716.

19 Diversity of geometrid moths in two Neotropical rain forests

G. Brehm
University of Bayreuth, Bayreuth, Germany

ABSTRACT

The diversity of geometrid moths (Lepidoptera: Geometridae) was investigated in two tropical montane cloud forests in southern Ecuador and in central Costa Rica. The study covered an elevational range from *c.* 1000 to 2700 m.a.s.l. in both locations. Data were analyzed for eight sites, representing four elevations in each of the two study areas. A total of 770 species (4569 specimens) were sampled in Ecuador vs. 503 in Costa Rica (7303 specimens). Sampling was incomplete and it is expected that more species will be found in both areas. Moth diversity was extremely high in Ecuador and almost invariably higher than at comparable elevations in Costa Rica. Values of Fisher's α index ranged between 70 and 131 in Ecuador, and between 31 and 83 in Costa Rica. An analysis of rarefied species numbers led to very similar results. At a level of 390 rarefied specimens, the range was 131–179 species at the Ecuadorian sites, vs. 77–140 at the Costa Rican sites. Only 64 (5.3%) out of the total of 1209 species were common to both areas. The results of this study underline the fact that the tropical Andean mountains are one of the "hot spots" of local diversity of geometrid moths, and that the threatened Andean cloud forests should be given the highest priority in conservation policy. Similarly, countries like Costa Rica, which are already actively engaged in nature conservation, may find additional motivation in the present results to continue their efforts.

INTRODUCTION

Arthropods are the dominant group of animals in tropical rain forests in terms of species numbers. Their impact on these complex ecosystems must be enormous since they are involved in almost all ecological processes except primary production. However, knowledge of the biology and taxonomy of most arthropod taxa in tropical regions is very poor, and a large percentage of species remain to be taxonomically described. Moths play an important role both in rain forest food webs and as pollinators. However, even a basic knowledge of the diversity of most groups is lacking for many regions of the world. With *c.* 21 000 described species, the family Geometridae is one of the three most species-rich moth families in the world (Scoble, 1999), and more than 6000 species have been described from the Neotropics alone. Quantitative studies of the diversity of geometrid moths in South America were carried out in the southern Ecuadorian Andes by Brehm (2002), by Brehm *et al.* (2003a,b, 2005), and by Hilt *et al.* (2006, 2007). With 1266 species recorded, this was the highest number ever found regionally (Brehm *et al.*, 2005). Brehm *et al.* (2003b) further demonstrated that local diversity in southern Ecuador was extremely high, with up to 292 species recorded per site and values of Fisher's α index reaching up to 131. Diversity remained high along a broad elevational gradient ranging from 1040 to 2677 m.a.s.l. (Brehm *et al.*, 2003b).

A comparable study on the diversity of geometrid moths along an extended elevational transect in montane forests in central Costa Rica was initiated in 2003. In this chapter, an answer is sought to the following questions:

- Is the local diversity of geometrid moths in the Andes higher compared to Central America, as is the case for better-known groups of organisms such as birds (e.g. Blake and Loiselle, 2000; Ridgely and Greenfield, 2001) and plants (Jørgensen and León-Yánez, 1999; Hammel *et al.*, 2004)? Species checklists (e.g. Pitkin *et al.* (1996) for the geometrid sub-family Ennominae) suggest that moth diversity in Costa Rica could be considerably lower than in Ecuador. It was therefore expected to find lower local moth diversity in Costa Rica.

Tropical Montane Cloud Forests: Science for Conservation and Management, eds. L. A. Bruijnzeel, F. N. Scatena, and L. S. Hamilton. Published by Cambridge University Press.

- Does geometrid diversity remain as great at high elevations in Costa Rica as in Ecuador?
- How many species occur in both regions? Own observations in museum collections, distributional data on Neotropic geometrid genera (Pitkin, 2002), and the species checklist by Pitkin *et al.* (1996) led to the hypothesis that the overlap between regional faunas is relatively low. However, the extent of this broad-scale beta diversity is unknown so far.

MATERIAL AND METHODS

Study sites

Sampling was carried out along extensive elevational gradients in two study areas in Ecuador and Costa Rica (e.g. Brehm *et al.*, 2005, 2007; Hilt *et al.*, 2006, 2007). This chapter concentrates on montane forests and compares a selection of four paired sites from each area ($4 \times 2 \times 2 =$ a total of 16 sites). Elevation of the sites ranged from lower montane rain forest (at *c.* 1000 m.a.s.l.) to upper montane cloud forest (at *c.* 2750 m.a.s.l.), both in Ecuador and Costa Rica (Figure 19.1). The vegetation in both areas consists mostly of primary forest, but some selective logging had been carried out at lower elevations in Ecuador and at higher elevations in Costa Rica.

The study area in southern Ecuador is situated in the Eastern Cordillera of the Andes and forms part of the province of Zamora–Chinchipe, at the northern margin of the Podocarpus National Park. The study sites were located at elevations between 1040 m and 2677 m.a.s.l. between 3° 58′ S and 04° 07′S, and between 78° 58′ W and 79° 05′ W (Brehm and Fiedler, 2003). Rainfall in the area ranges between *c.* 2200 mm year^{-1} at lower elevations and 4700 mm year^{-1} at higher elevations (Bendix *et al.*, 2008). Results for alpha and beta diversity of 22 forest sites investigated in the same area were provided by Brehm and Fiedler (2003), and by Brehm *et al.* (2003a, b). Four paired sites (at elevations of 1040, 1380, 2005, and 2670 m.a.s.l.) were selected for this study (Figure 19.1).

The study area in Costa Rica is situated in the central Cordillera (Heredia province). The eight study sites were located within the Braulio Carrillo National Park (Barva transect) or close to its western boundaries (1070–2730 m.a.s.l.) between 10° 08′ N and 10° 16′ N, and between 84° 05′ W and 84° 07′ W at elevations of *c.* 1090, 1700, 2115, and 2730 m.a.s.l. (Figure 19.1). Rainfall in the lower montane rain forest zone on Volcán Barva below 2000 m.a.s.l. exceeds 4000 mm year^{-1} and decreases to *c.* 3300 mm year^{-1} at higher elevations. Lower montane (cloud) forest occurs at 2100 and upper montane cloud forest at 2600 m.a.s.l. (Heaney and Proctor, 1990). The role of environment and the "mid-domain effect" on moth species richness along the complete Barva transect was analyzed in detail by Brehm *et al.* (2007).

Sampling and identification of moths

Moths were sampled manually using portable light-traps consisting of two 15 W UV tubes (Sylvania blacklight-blue, F15W/T8/BLB and Philips TLD 15W 05) and a white gauze cylinder (height 1.60 m, diameter 0.70 m). Sampling methods were as described and discussed by Brehm and Axmacher (2006). Light-traps were run during the peak of moth activity at dusk between 18.30 and 21.30 h. The Ecuadorian sites were sampled two to

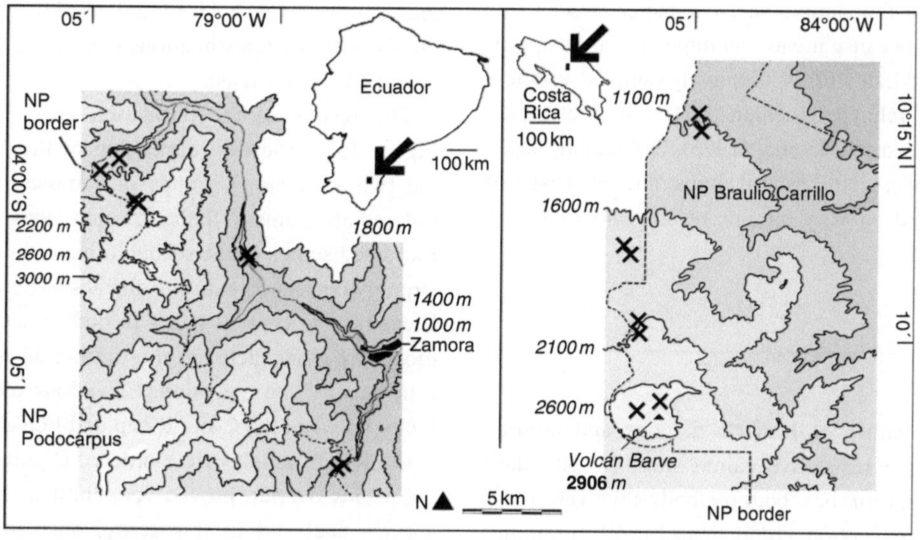

Figure 19.1. Study areas in Ecuador (left) and Costa Rica (right). In each area, the eight study sites are marked with crosses (×). Site positions were measured using the Global Positioning System. Other geographic information was taken from 1:50 000 topographic maps. Note that contour lines are spaced by 400 m in Ecuador and by 500 m in Costa Rica. NP, National Park.

four times in April, May, and December 1999 (Brehm, 2002; Brehm and Fiedler, 2003); those in Costa Rica were sampled twice, in April and May 2003. Further samples (two or three at each site) taken in 2004 were similar both in species composition and diversity. Combined samples from both years showed only slight deviations from the results presented here (Brehm *et al.*, 2007). In both areas, there is no strong seasonality, but the relatively short sampling periods may have had some influence on the results. Although the number of samples was low, a meaningful analysis appeared to be possible because each sample was large and contained at least 390 specimens.

Specimens were identified by the author as far as possible in the Natural History Museum in London (BMNH), the National Museum of Natural History in Washington DC (USNM), the American Museum of Natural History in New York (AMNH), and the Zoologische Staatssammlung München (ZSM). For both study areas, a high proportion (37–47%) of the species appear to be undescribed so far (Brehm *et al.*, 2005, 2007).

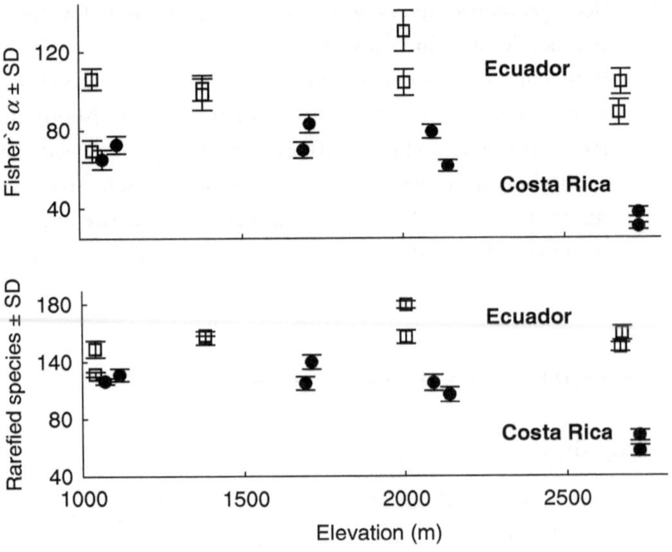

Figure 19.2. Local diversity of geometrid moths in two tropical montane forest areas in Ecuador and Costa Rica based on Fisher's α index and on rarefied species numbers (at the level of 390 specimens).

Statistical analysis

Fisher's α index was chosen as a measure for local diversity since it has been proven to be relatively insensitive to under-sampling (Magurran, 2004). The rarefaction method was used as an additional measure of local diversity because it is particularly useful in samples of different size (Hurlbert, 1971). Rarefied species numbers at the level of 390 specimens were calculated using a program developed by Kenney and Krebs (2000); 390 is the specimen number in the smallest sample.

Data on the moth ensembles of the sites were ordinated by non-metric multidimensional scaling (NMDS) using Statistica 5.5 software (StatSoft, Tulsa, OK, USA). The NESS index (Grassle and Smith, 1976) with the sample size parameter m set to its maximum was chosen because it was recommended as an appropriate abundance-based index for incompletely sampled, species-rich moth ensembles (Brehm and Fiedler, 2004). It was calculated using a program provided by Messner (1996). Ordinations based on the NESS index with m set to 1 as well as an ordination based on the Sørensen index led to very similar results and these are therefore not presented.

RESULTS

Figure 19.2 displays values of Fisher's α index and rarefied species numbers for the respective paired sites. Despite their different mathematic approaches, both methods gave very similar results. With one exception (1040 m.a.s.l.), all ensembles from Ecuador were more diverse than those from Costa Rica. Fisher's α index ranged between 70 and 131 for the Ecuadorian sites, and between 31 and 83 for the Costa Rican sites. Similarly, using a rarefaction level of 390 specimens, the range derived for the Ecuadorian sites was 131–179 species, vs. 77–140 species for the Costa Rican sites. Whilst geometrid moths in Ecuador remained highly diverse along the complete elevational gradient, diversity decreased markedly at the uppermost pair of sites in the upper montane cloud forest zone at 2730 m.a.s.l. in Costa Rica (Figure 19.2).

In Costa Rica 503 species of geometrid moths were found in the investigated gradient, whereas the total from Ecuador was 770, i.e. *c.* 50% more species, despite the fact that the number of individuals analyzed from Ecuador was considerably smaller than that from Costa Rica (4569 vs. 7303). In both cases, species numbers are underestimations (see Brehm *et al.*, 2005, 2007 for extended species lists).

The results of the ordination analysis are summarized in Figure 19.3. The first dimension of the NMDS clearly divides the two geographic groups of samples, whereas the sites are ordered according to their elevation along the second dimension. Extracted scores of the second dimension of the ordination were strongly correlated with site elevation (Ecuador: $r_S = 0.96$, Costa Rica: $r_S = 0.97$, $p < 0.001$ in both cases). This shows that the most important factor in the ordination is geographic position, followed by site elevation. Elevations did not correspond perfectly between the Costa Rican and Ecuadorian samples, i.e. the Costa Rican samples were ordered slightly "lower."

Species overlap proved very limited: from the total of 1208 species analyzed in this study, only 64 (5.3%) were shared between the areas. Although this number might be an underestimate due to the incomplete sampling, it seems likely that the order of magnitude of overlap will not be more than 10%.

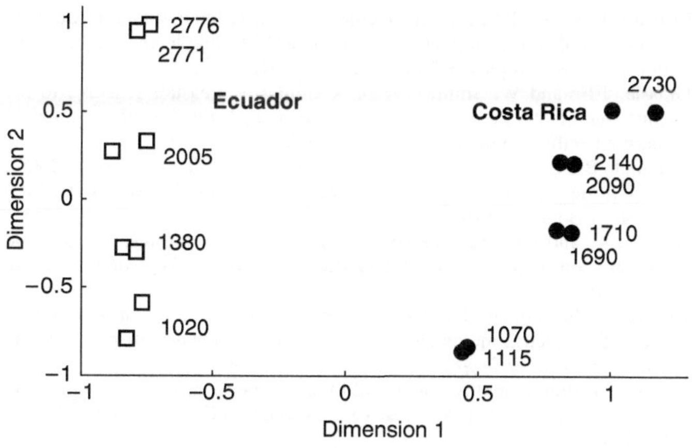

Figure 19.3. Non-metric two-dimensional scaling of eight moth samples from Ecuador and eight samples from Costa Rica based on the NESS dissimilarity index (m_{max} = 195). Information on elevations above mean sea level (in m) provided. Stress is low with an NESS value of 0.05, indicating low risk of artifactual patterns (Brehm and Fiedler, 2004).

DISCUSSION AND CONCLUSION

Diversity of geometrid moths is clearly higher in Ecuadorian than in Costa Rican montane forests. However, diversity in Costa Rica is still very high (Figure 19.2), and at all but the highest sites exceeds values for temperate or African sites (cf. Axmacher *et al.*, 2004; overview in Brehm *et al.*, 2005). However, moth diversity in montane Costa Rica is lower than at some lowland sites in South-East Asia (e.g. Barlow and Woiwod, 1989). Whilst there were no significant differences in species richness amongst lower montane (cloud) forests between elevations of 1100 and 2100 m.a.s.l. in Costa Rica (Figure 19.2), the highest site pair in the summit area of Volcán Barva showed an order of magnitude for Fisher's α index (31–38) that has also been recorded for some temperate forests (e.g. Thomas, 2002). It is nevertheless remarkable how rich the ensembles were even at this high elevation with its cold, wet, and windy climate (cf. Heaney and Proctor, 1990).

It might be argued that two samples per site in Costa Rica are not sufficient for a reliable analysis. However, at least 427 specimens were collected at each Costa Rican site, and combined results from four/six samples per elevation yielded similar, sometimes slightly higher diversity values (Brehm *et al.*, 2007) that do not affect any of the conclusions drawn here. The results thus indicate that the geometrid moth fauna in Central America is very diverse, but that the true diversity "hot spot" is clearly located in the Andes. Indeed, the Andean mountains are known as a "hyper hot spot" of diversity for many groups of organisms (e.g. Brummitt and Lughada, 2003; Brehm *et al.*, 2008). The present study supports this view for a very species-rich moth taxon. Various hypotheses potentially explain the high richness of the Andes compared with other regions (evolution and biogeography, climate history, biotic and abiotic factors, stochastic effects; see Brehm *et al.* (2008) for discussion).

Unfortunately, Andean and sub-Andean cloud forests have less than 4.5% and 6.4% of their original pre-Columbian extent currently protected (Armenteras *et al.*, 2003) and rates of montane forest conversion in the region are high (Bubb *et al.*, 2004; Mosandl *et al.*, 2008; Mulligan, this volume). The impact of deforestation on arthropods is largely unknown, but it must be assumed that specialized moth species will become extinct if their host-plants and habitats are destroyed (Brehm and Fiedler, 2005). Climate change might further accelerate extinction with lowland biotic attrition and range shift gaps, particularly for small-ranged species (Colwell *et al.*, 2008). The present study confirms that the Andean rain forests merit a high conservation priority because of their exceptional biodiversity and the rapid loss of the last remaining fragments. However, Central American forests should not be neglected since diversity there is still very high and species overlap with the Andes surprisingly small (cf. Figure 19.3). Furthermore, there is evidence that the proportion of endemic species increases with elevation (Brehm, 2002). Thus, the maintenance of an extended National Park system in Costa Rica's mountainous regions is not only important for watershed services (cf. Calvo-Alvarado *et al.*, this volume), but also for the protection of a unique component of the world's biodiversity. Further studies in both Central America and the Andes will improve our understanding of diversity patterns and help to better define priority areas for conservation on a regional scale (Myers *et al.*, 2000; cf. Leo, 1995; Catchpole and Kirkpatrick, this volume; Cavelier *et al.*, this volume; Mulligan, this volume).

ACKNOWLEDGEMENTS

I thank Sampurno Bruijnzeel, Jennifer Britton, Konrad Fiedler, and Jan Axmacher for helpful comments on the manuscript. Edgar Corrales, Adrienne Hogg, Jennifer Kay, Martin Knopf, Claudia Raabe, Rita Schneider, Vernon Sota, Dirk Süssenbach, and others helped with sampling, spreading and data-basing of the moths. Field work in Costa Rica was supported by Jürgen Kluge, the Arthropods of La Selva (ALAS) project (NSF grant DEB-0072702), the Instituto Nacional de Biodiversidad (INBio), and the Organization for Tropical Studies (OTS). Nature and Culture International (Loja) provided working facilities in Ecuador. Konrad Fiedler, Christoph Häuser, and Giovanni Onore provided administrative support. Colleagues at various institutions (see material and methods section) helped with the identification of the moths. The projects were funded by grants from the Deutsche Forschungsgemeinschaft (Br 2280/1-1, Fi 547/5-1, FOR 402/1-1 TP 15).

REFERENCES

Armenteras, D., F. Gast, and H. Villareal (2003). Andean forest fragmentation and the representativeness of protected natural areas in the eastern Andes, Colombia. *Biological Conservation* **113**: 245–256.

Axmacher, J., G. Holtmann, L. Scheuermann, *et al.* (2004). Diversity of geometrid moths (Lepidoptera: Geometridae) along an Afrotropical elevational rainforest transect. *Diversity and Distributions* **10**: 293–302.

Barlow, H.S., and I.P. Woiwod (1989). Moth diversity of a tropical forest in Peninsular Malaysia. *Journal of Tropical Ecology* **5**: 37–50.

Bendix, J., R. Rollenbeck, M. Richter, P. Fabian, and P. Emck (2008). Climate. In *Gradients in a Tropical Mountain Ecosystem of Ecuador*, eds. E. Beck, J. Bendix, I. Kottke, F. Makeschin, and R. Mosandl, pp. 63–73. Berlin: Springer-Verlag

Blake, J.G., and B.A. Loiselle (2000). Diversity of birds along an elevational gradient in the Cordillera Central, Costa Rica. *The Auk* **117**: 663–686.

Brehm, G. (2002). Diversity of geometrid moths in a montane rainforest in Ecuador. Ph.D. thesis, University of Bayreuth, Bayreuth, Germany. Available at http://opus.ub.uni-bayreuth.de/volltexte/2003/20

Brehm, G., and J.C. Axmacher (2006). A comparison of manual and automatic moth sampling methods (Lepidoptera: Arctiidae, Geometridae) in a rain forest in Costa Rica. *Environmental Entomology* **35**: 754–764.

Brehm, G., and K. Fiedler (2003). Faunal composition of geometrid moths changes with altitude in an Andean montane rain forest. *Journal of Biogeography* **30**: 431–440.

Brehm, G., and K. Fiedler (2004). Ordinating tropical moth ensembles from an elevational gradient: a comparison of common methods. *Journal of Tropical Ecology* **20**: 165–172.

Brehm, G., and K. Fiedler (2005). Diversity and community structure of geometrid moths of disturbed habitat in a montane area in the Ecuadorian Andes. *Journal of Research on the Lepidoptera* **38**: 1–14.

Brehm, G., J. Homeier, and K. Fiedler (2003a). Beta diversity of geometrid moths (Lepidoptera: Geometridae) in an Andean montane rainforest. *Diversity and Distributions* **9**: 351–366.

Brehm, G., D. Süssenbach, and K. Fiedler (2003b). Unique elevational diversity patterns of geometrid moths in an Andean montane rainforest. *Ecography* **26**: 456–466.

Brehm, G., L. Pitkin, N. Hilt, and K. Fiedler (2005). Tropical Andean rain forests are a global diversity hotspot of geometrid moths. *Journal of Biogeography* **32**: 1621–1627.

Brehm, G., R.K. Colwell, and J. Kluge (2007). The role of environment and mid-domain effect on moth species richness along a tropical elevational gradient. *Global Ecology and Biogeography* **16**: 205–219.

Brehm, G., J. Homeier, K. Fiedler, *et al.* (2008). Mountain rain forests in southern Ecuador as a hotspot of biodiversity: limited knowledge and diverging patterns. In *Gradients in a Tropical Mountain Ecosystem of Ecuador*, eds. E. Beck, J. Bendix, I. Kottke, F. Makeschin, and R. Mosandl, pp. 15–24. Berlin: Springer-Verlag

Brummitt, N. and E.N. Lughada (2003). Biodiversity: where's hot and where's not. *Conservation Biology* **17**: 1442–1448.

Bubb, P., I. May, L. Miles, and J. Sayer (2004). *Cloud Forest Agenda*. Cambridge, UK: UNEP–WCMC

Colwell, R.K., G. Brehm, C.L. Cardelús, A.C. Gilman, and J.T. Longino (2008). Global warming, elevational range shifts, and lowland biotic attrition in the wet tropics. *Science* **322**: 258–261.

Grassle, J.F., and W. Smith (1976). A similarity measure sensitive to the contribution of rare species and its use in investigation of variation in marine benthic communities. *Oecologia* **25**: 13–22.

Hammel, B.E., M.H. Grayum, C. Herrera, and N. Zamora (eds.) (2004). *Manual de plantas de Costa Rica*, Vol. 1, *Introducción*. St. Louis, MO: Missouri Botanical Garden.

Heaney, A., and J. Proctor (1990). Preliminary studies on forest structure and floristics on Volcan Barva, Costa Rica. *Journal of Tropical Ecology* **6**: 307–320.

Hilt, N., G. Brehm, and K. Fiedler (2006), Diversity and ensemble composition of geometrid moths along a successional gradient in the Ecuadorian Andes. *Journal of Tropical Ecology* **22**: 155–166.

Hilt, N., G. Brehm, and K. Fiedler (2007). Temporal dynamics of rich moth ensembles in the montane forest zone in southern Ecuador. *Biotropica* **39**: 94–104.

Hurlbert, S.H. (1971). The nonconcept of species diversity: a critique and alternative parameters. *Ecology* **52**: 577–586.

Jørgensen, P.M., and S. León-Yánez (eds.) (1999). *Catalogue of the Vascular Plants of Ecuador*. St. Louis, MO: Missouri Botanical Garden.

Kenney, A.J., and C.J. Krebs (2000). *Programs for Ecological Methodology*, Version 5.2. Setauket, NY: Exceter Software.

Leo, M. (1995). The importance of tropical montane cloud forests for preserving vertebrate endemism in Peru: the Río Abiseo National Park as a case study. In *Tropical Montane Cloud Forests*, eds. L. S. Hamilton, J. O. Juvik, and F. N. Scatena, pp. 198–211. New York: Springer-Verlag.

Magurran, A.E. (2004). *Measuring Biological Diversity*. Oxford, UK: Blackwell

Messner, S. (1996). Untersuchungen zur Biodiversität der Myrmecofauna (Formicidae) im Parc National de la Comoé (Elfenbeinküste). Diploma thesis, University of Würzburg, Würzburg, Germany.

Mosandl, R., S. Günter, B. Stimm, and M. Weber (2008). Ecuador suffers the highest deforestation rate in South America. In *Gradients in a Tropical Mountain Ecosystem of Ecuador*, eds. E. Beck, J. Bendix, I. Kottke, F. Makeschin, and R. Mosandl, pp. 37–40. Berlin: Springer-Verlag.

Myers, N., R.A. Mittermaier, C.G. Mittermaier, G A.B. da Fonseca, and J. Kent (2000). Biodiversity hotspots for conservation priorities. *Nature* **403**: 853–858.

Pitkin, L.M. (2002). Neotropical ennomine moths: a review of the genera (Lepidoptera, Geometridae). *Zoological Journal of the Linnean Society* **134**: 121–401.

Pitkin, L.M., R.A. Mora, and M.J. Scoble (1996). A checklist of the Ennominae (Geometridae) of Costa Rica: taxonomy for a national biodiversity inventory. *Gayana Zoology* **60**: 121–155.

Ridgely, R.S., and P.J. Greenfield (2001). *The Birds of Ecuador*. London: Croom Helm.

Scoble, M.J. (ed.) (1999). *Geometrid Moths of the World: A Catalogue (Lepidoptera, Geometridae)*. Collingwood, Australia: CSIRO Publishing.

Thomas, A.W. (2002). Moth Diversity in a Northeastern North American, Red Spruce Forest, Vol. 2, *The Effect of Silvicultural Practices on Geometrid Diversity (Lepidoptera, Geometridae)*. Fredericton, New Brunswick: Canadian Forest Service, Atlantic Forestry Service.

Part III
Hydrometeorology of tropical montane cloud forest

L. A. Bruijnzeel

20 Hydrometeorological patterns in relation to montane forest types along an elevational gradient in the Yungas of Bolivia

M. Schawe, G. Gerold, K. Bach, and S. R. Gradstein
University of Göttingen, Göttingen, Germany

ABSTRACT

Floristic composition, structure, and functioning of tropical montane forests depend on various abiotic and biotic factors, although the precise nature of the interactions is still a matter of debate. As part of an interdisciplinary project aiming to understand the relationship between vegetation and abiotic factors in the montane forest belt in the Yungas of Bolivia (Eastern Andes), hydrometerological observations were made along an elevational gradient. Earlier studies had revealed a decrease in biodiversity and forest stature with altitude. Three weather stations were in operation: at 1850 m.a.s.l. (lower montane forest, LMF), 2600 m.a.s.l. (upper montane cloud forest, UMCF), and 3050 m.a.s.l. (subalpine cloud forest, SACF) between October 2001 and October 2002. Precipitation increased strongly with elevation from 2310 mm year^{-1} at 1850 m.a.s.l. to 5150 mm year^{-1} at 3050 m.a.s.l. Compared to clear-sky conditions, reductions in short-wave radiation inputs by fog and clouds were estimated at 37% at 1850 m.a.s.l. vs. 58–62% at 2600 and 3050 m.a.s.l. However, intensities of photosynthetically active radiation (PAR) remained well above the light saturation point for local vegetation, and changes in PAR with elevation are therefore unlikely to control vegetation zonation. FAO reference evaporation rates decreased from 3.3 mm day^{-1} in the LMF zone to 1.4 and 1.3 mm day^{-1} in the UMCF and SACF zones, respectively. The present results suggest that the difference in forest stature between the LMF on the one hand, and the UMCF and SACF on the other, is dominated by a strong increase in precipitation excess, leading in turn to persistently saturated conditions, high acidity and leaching in the SACF.

INTRODUCTION

With increasing elevation, tropical montane rain forests often exhibit major changes in plant diversity (Wolf, 1993; Gentry, 1995; Kessler, 2000), stature (Körner *et al.*, 1983; Kitayama, 1992; Hemp, this volume #12; Häger and Dohrenbusch, this volume), and leaf size (Körner *et al.*, 1983). These changes are often so pronounced that they can be expressed in the form of distinctly different forest types with elevation (Grubb, 1974; Frahm and Gradstein, 1991; Kappelle *et al.*, 1995). The precise factors underlying these altitudinal changes are still a matter of debate. According to some, reductions in radiation due to fog and low cloud, coupled with decreases in temperature and increases in atmospheric humidity with elevation may limit photosynthetic activity, transpiration, and nutrient uptake (as reviewed by Grubb, 1977; Bruijnzeel and Veneklaas, 1998). Others have drawn attention to the fact that tree height and leaf area in some places showed an inverse relationship with the degree of soil water saturation (e.g. Hetsch and Hoheisel, 1976; Santiago *et al.*, 2000; cf. Häger and Dohrenbusch, this volume; Santiago *et al.*, this volume). Thus, unfavorable soil conditions caused by high precipitation excesses over evaporation may constitute an important factor determining forest stature on wet tropical mountains (cf. Benner *et al.*, this volume; Roman *et al.*, this volume). However, these and other climatic influences on forest stature have only rarely been documented, mostly due to the difficulty in establishing and operating meteorological and soil water measuring stations under the adverse topographic and logistic conditions characterizing so many cloud forest sites (Herrmann, 1971; Cavelier and Mejia, 1990; Bruijnzeel *et al.* 1993; Pendry and Proctor, 1996; Hafkenscheid, 2000; Holwerda, 2005; cf. Häger and Dohrenbusch, this volume). The objectives of the present chapter are to describe the changes in climate with elevation in

Tropical Montane Cloud Forests: Science for Conservation and Management, eds. L. A. Bruijnzeel, F. N. Scatena, and L. S. Hamilton. Published by Cambridge University Press. © Cambridge University Press 2010.

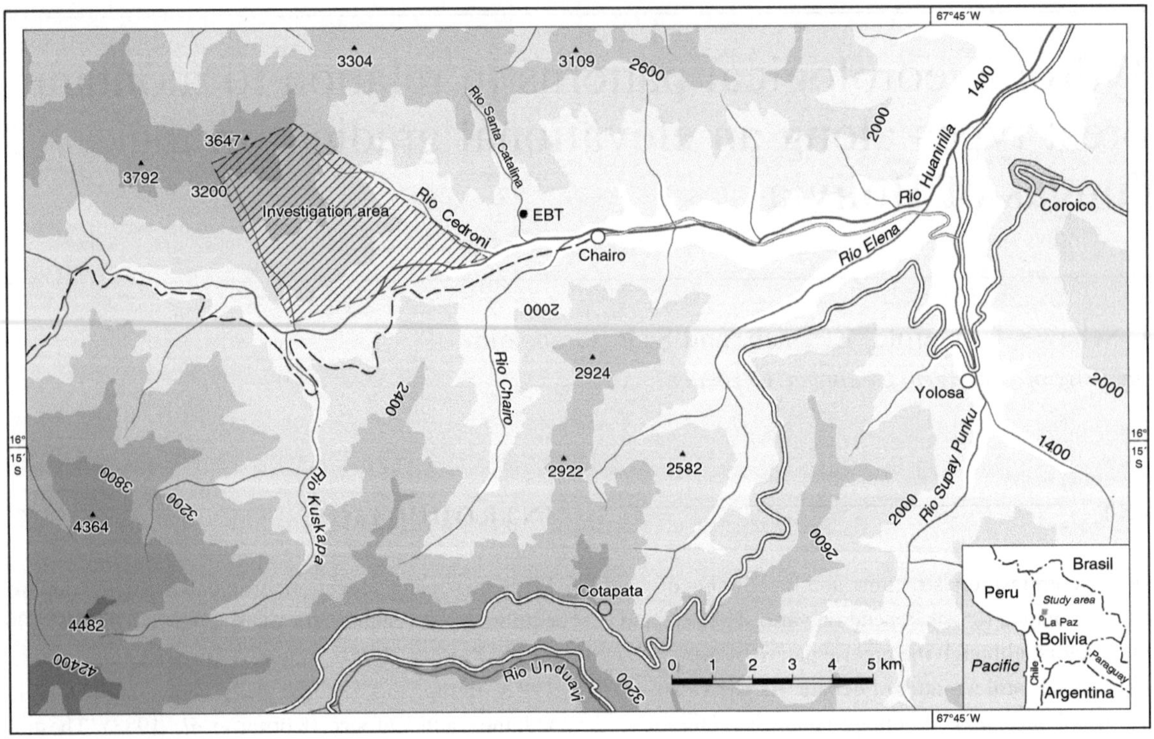

Figure 20.1. Location of the study area in the Cotapata National Park, Yungas, Bolivia.

the Yungas region of Bolivia and to evaluate their influence on forest stature along the elevational gradient.

STUDY SITE

The present study was carried out along a transect on a south-easterly orientated slope in the Bolivian National Park of Cotapata, in the north-eastern part of the Bolivian Andes and *c.* 80 km north of La Paz (16° 09′ S, 68° 55′ W). The transect extended from 1700 to 3400 m.a.s.l. and encompassed the entire vertical range of tropical montane forest types in the "Yungas of La Paz" region (Bach *et al.*, 2003; Figure 20.1). The longitudinal slope profile was mostly linear and had an average inclination of 25–30° although some areas exceeded 40°. The eastern Andes are part of the seasonal humid tropical zone (Köppen climatic type Aw) with a maximum of precipitation during the summer and short drier periods in winter (July, August). The substrate consisted of Ordovician meta-siltstones, slates and meta-sandstones (*Mapa geologico de Coroico*, Hoja 6045). Soils along the transect were characterized by high acidity, low cation exchange capacity, high aluminum saturation in the mineral horizons, and an organic layer ranging in thickness from 15 cm at 1850 m.a.s.l. to 30 cm at 3050 m.a.s.l. Degree of podzolization and water-logging increased with elevation. Dominant soil types were humic Dystrudepts (USDA classification) below 2150 m.a.s.l. and typic Placaquods within the cloud belt (above 2200 m) (Gerold *et al.*, 2003; Schawe *et al.*, 2007). The majority of the roots were found in the organic layer with a sharp decrease within the A-horizon. Aerial roots were common in the cloud belt (Schawe, 2005).

Based on vegetation structure and the floristic composition of species, three vegetation formations were recognized (Bach, 2004): lower montane forest (LMF), upper montane cloud forest (UMCF), and sub-alpine cloud forest (SACF). LMF extended up to 2150 m and had a dense, *c.* 20 m high canopy and emergent trees up to 35 m. *Myrsine coriacea* (Sw.) R.Br. ex Röm. and Schult. was the dominant tree, while the fern *Blechnum ensiforme* Liebm. and the aroid *Philodendron ornatum* Schott dominated the understory. Around 2150 m.a.s.l. there was a distinct transition to UMCF. This forest type was subject to persistent cloud incidence and was dominated by *Podocarpus oleifolius* D. Don ex Lamb., *Weinmannia crassiflora* Ruíz and Pavón, and *Clusia multiflora* Kunth. The canopy of the UMCF was closed and lower than that of the LMF, typically reaching up to 15 m. The abundance of epiphytes, mostly bryophytes (mosses, liverworts) in the UMCF was noteworthy; terrestrial (ground surface) bryophyte cover was estimated at about 45%. Above 2800 m.a.s.l. UMCF gave way to SACF, with the dominant tree species *Myrsine coriacea* and *Podocarpus rusbyi* J. Buchholz and N. E. Gray reaching a height of about 10 m and not forming a closed canopy. Moss cover on the soil surface increased to 75% and was dominated by *Sphagnum* species. Scleromorphic shrubs such as *Miconia* sp., *Gaultheria pernettyoides, Gynoxys* ssp., *Ilex* ssp., and *Escallonia myrtilloides* L.f. were abundant. Within the Melastomataceae, the proportion of scleromorphic species increased from 11% in the UMCF (at 2300 m.a.s.l.) to over 50% in the SACF (at 3050 m.a.s.l.). In all of the analyzed plant groups (ferns, melastoms, aroids, bromeliads, palms, cacti) α-diversity decreased significantly with elevation. The highest overall diversity was found at 1900 m.a.s.l. (Bach, 2004).

METHODS

Climatic measurements were carried out over a 1-year period (October 2001–October 2002) in forest clearings within each of the three vegetation formations, viz. at 1850 m.a.s.l. (LMF), 2600 m.a.s.l. (UMCF), and 3050 m.a.s.l. (SACF). The clearings were restricted to 2500 m^2 (LMF) and 900 m^2 (UMCF, SACF) because of National Park regulations and the relief situation. Rainfall (mm) was measured at 1 m above ground level with a tipping-bucket recording gage (ARG 100 EM) with 0.2-mm resolution per tip. Short-wave radiation (W m^{-2}) was measured with a pyranometer (8101, Schenk), net radiation (W m^{-2}) with a net radiometer (NR-lite, Kipp and Zonen), and photosynthetic photon flux density (PPFD, wavelength 400–700 nm, μmol m^{-2} s^{-1}) with a quantum sensor (DK-PHAR, Deka). All radiation sensors were placed at 2 m height and on support arms extending 1.5 m from the mast to avoid shading of the instruments. Air temperature (°C) and relative humidity (%) were also measured at 2 m height, with the sensors placed within a Gill-type radiation shield. Measured relative humidity values in excess of 100% were set to 100%. Wind speed (m s^{-1}) (A100 anemometer, Vector Instruments) was also recorded at 2 m height, i.e. well below the level of the surrounding canopy. It is recognized that the recorded wind speeds represent underestimates. Similarly, a cylindrical wire-harp type of fog sampler placed in the clearings did not capture any measurable quantities of water, whereas the surrounding vegetation also restricted the proper measurement of radiation at low sun angles (Schawe, 2005). Instruments were sampled at 10-min intervals except for wind speed which was sampled every minute. Data were processed by a Delta T data-logger system. Averages and their standard deviations were calculated over 30-min periods. Reference evaporation (potential evapotranspiration, *PET*) was calculated from short-wave radiation R_s, temperature, humidity, and wind speed data using the modified Penman equation advanced by Doorenbos and Pruitt (1988) to compare evaporative conditions between sites. The degree of reduction in R_s due to the presence of fog and clouds was estimated from a comparison with potential radiation values as calculated using a topography- and exposition-based radiation model (Böhner *et al.*, 1997). Direct radiation was calculated from the Sun's inclination vis-à-vis on-site slope gradient and exposition as taken from a digital elevation model (DEM) of the study area. Amounts of diffuse radiation were calculated from solar altitude and direct radiation, taking the value of air pressure into account.

Water content of the organic surface layer was measured daily for a period of three weeks during the dry season (24 September–12 October 2002) at 200 m elevation intervals between 1800 and 3000 m.a.s.l. At least three non-disturbed samples were taken at 10 cm depth per sampling site, weighed, and dried at 105 °C to constant weight and reweighed. Gravimetric water contents were expressed in grams of water per gram of oven-dried soil (g g^{-1}).

RESULTS

Radiation

The UMCF at 2600 m.a.s.l. received about 33% less short-wave radiation (R_s) than the LMF at 1850 m, whereas the SACF at 3050 m received *c.* 5% less than the UMCF (Table 20.1, Figure 20.2).

Table 20.1 *Radiation inputs and temperatures between October 2001 and October 2002 at different altitudes along the study transect. Mean daily short-wave radiation input and range* (R_s, R_s *range, MJ* m^{-2}, ± *SD), average and maximum daily photosynthetic photon flux density* (PPFD, *Max* PPFD, *μmol* m^{-2} s^{-1}), *average sum of daily photosynthetic photon flux density* (PPFD, *mol* m^{-2} day^{-1}), *average daily temperature* T *(°C), daily amplitude of* T *(°C), average minimum* T *and range, average maximum* T *and range*

	LMF (1850 m)	UMCF (2600 m)	SACF (3050 m)
R_s (MJ m^{-2})	15.5 ± 5.4	10.4 ± 4.3	9.9 ± 3.8
R_s range (MJ m^{-2})	4.6–29.8	3.3–28.1	3.1–24.8
PPFD (μmol m^{-2} s^{-1})	749	471	455
Max *PPFD* (μmol m^{-2} s^{-1})	1957	1475	1332
PPFD (mol m^{-2} day^{-1})	33.6	22.0	20.5
T (°C)	16.8 ± 1.7	12.7 ± 1.6	10.0 ± 1.5
Daily amplitude of *T* (°C)	9.8	6.6	5.9
Min *T* (°C)	13.3 (7.7–17.2)	10.2 (5.7–14.0)	7.4 (2.1–10.8)
Max *T* (°C)	22.8 (14.8–29.0)	16.8 (8.8–23.8)	13.3 (6.9–18.5)

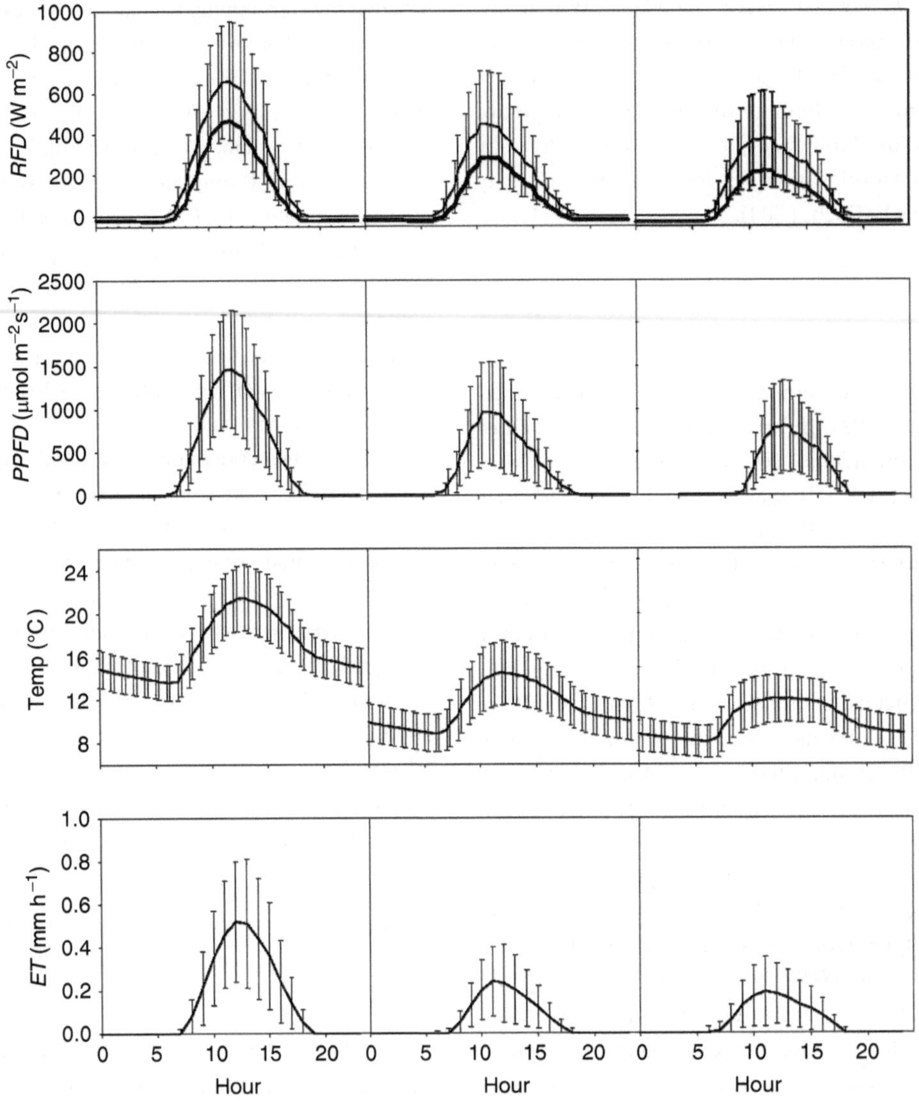

Figure 20.2. Average diurnal patterns of radiation flux densities (*RFD*), short-wave radiation (normal line), and net radiation (bold line) (all in W m^{-2}), photosynthetic photon flux density (*PPFD*, µmol m^{-2} s^{-1}), temperature (°C), and Penman evapotranspiration (*ET*, mm hr^{-1}) between 1 October 2001 and 1 October 2002 at three altitudes representing the studied vegetation formations. Vertical lines represent ± 1 standard deviation.

The average inferred reduction in R_s was 37% of the calculated potential value at the LMF site, 58% at the UMCF site, and 62% at the SACF (Figure 20.3). Thus, radiation reduction was most pronounced in the SACF where it exceeded 50% on 306 out of the 365 days of observations and reaching as much as 90% during a single 4-day period in the dry season.

Photosynthetic photon flux density (*PPFD*) also decreased with altitude (Table 20.1, Figure 20.2). Average daily *PPFD* decreased by 37% from the LMF to the UMCF zone, and by a further 7% in the SACF. In the LMF, *PPFD* exceeded 1000 µmol m^{-2} s^{-1} on average for 4.5 hours per day with a maximum (on average 1470 mol m^{-2} s^{-1}) occurring around noon; corresponding durations in the UMCF and SACF were 4 and 3.5 hours, with average maximum values (958 and 805 mol m^{-2} s^{-1}) occurring around 11:00 h and 11:30 h, respectively (Figure 20.2).

Temperature and humidity

Mean daily maximum, minimum and average temperatures at the three sites are given in Table 20.1. The overall decline in temperature between the LMF and SACF of 0.59 K 100 m^{-1} is close to the wet adiabatic lapse rate (about 0.56 K 100 m^{-1}; Lauer, 1975) but the recorded lapse rates differed between forest zones (0.54 K 100 m^{-1} between the LMF and UMCF and 0.62 K 100 m^{-1} within the cloud belt; Table 20.1). The diurnal range in temperature decreased with increasing altitude (Figure 20.2).

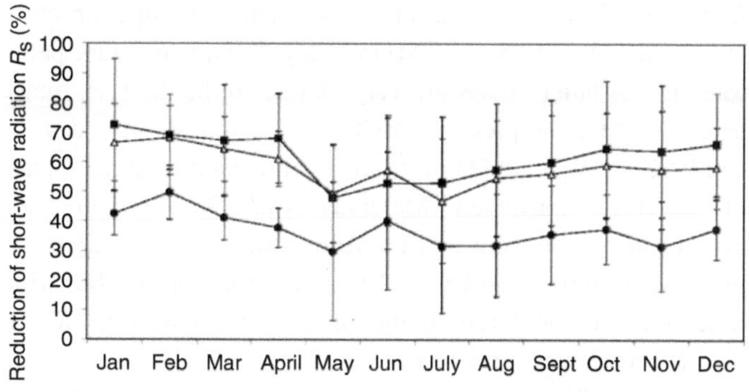

Figure 20.3. Average daily short-wave radiation (R_s, %) as a percentage of potential extraterrestrial radiation at altitudes of 1800 m (square line), 2600 (triangle line), and 3050 m (dot line) with standard deviation *s*.

Table 20.2 *Annual rainfall totals (mm), average hourly rainfall intensities (mm), annual total and average daily potential evapotranspiration* (PET, *mm) at different altitudes along the study transect between October 2001 and October 2002*

	LMF (1850 m)	UMCF (2600 m)	SACF (3050 m)
Rainfall (mm year^{-1})	2310	3970	5150
Rainfall intensities (mm hr^{-1})	0.3	0.5	0.6
Annual *PET* (mm year^{-1})	1190	462	403
Average daily *PET* (mm day^{-1})	3.3	1.3	1.1
PET range (mm day^{-1})	0.3–8.6	0.07–6.4	0.04–5.7

Average relative humidity increased slightly from 90% in the LMF to 96% in the UMCF and 98% in the SACF, whereas the average diurnal range decreased with altitude, from 30% at 1850 m.a.s.l. to 5% at 3050 m.a.s.l.

Precipitation

Annual rainfall (*P*) was high and increased considerably with altitude (Table 20.2). The UMCF site received 1.7 times more rainfall than the LMF site whereas *P* at the SACF site was 2.2 times that recorded at the LMF. Between the LMF and the UMCF, *P* increased by 208 mm 100 m^{-1} increase in elevation, and by 295 mm 100 m^{-1} between the UMCF and the SACF. The dry season lasted from April to October and the wet season from November to March (Figure 20.4). The number of days with <1 mm of rainfall decreased from 72% in the LMF to 60% in the SACF, but falls of 2–10 mm increased from 15% to 23% with increasing elevation. The length of continuously dry periods (mostly during August) decreased from 9 days at 1850 m.a.s.l. to 7 days at 2600 m.a.s.l. and 5 days at 3050 m.a.s.l.

Rainfall intensities were low throughout the transect but increased from 0.3 mm hr^{-1} in the LMF to 0.5 and 0.6 mm hr^{-1} in the UMCF and SACF, respectively (Table 20.2). Maximum daily rainfall occurrence in the LMF and UMCF was between 2:30 and 3:00 h, but in the SACF around 14:30 h (Figure 20.5) suggesting differences in the origin of the precipitation at the respective sites.

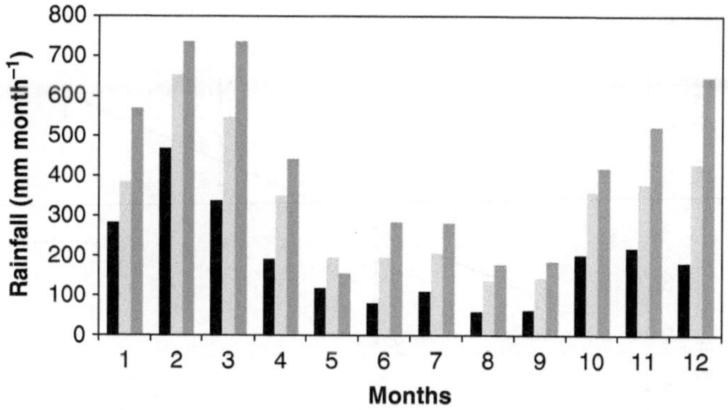

Figure 20.4. Monthly rainfall between 1 October 2001 and 1 October 2002 at 1800 m.a.s.l. (black bars), 2600 m.a.s.l. (light gray bars); and 3050 m.a.s.l. (dark gray bars) along the study transect.

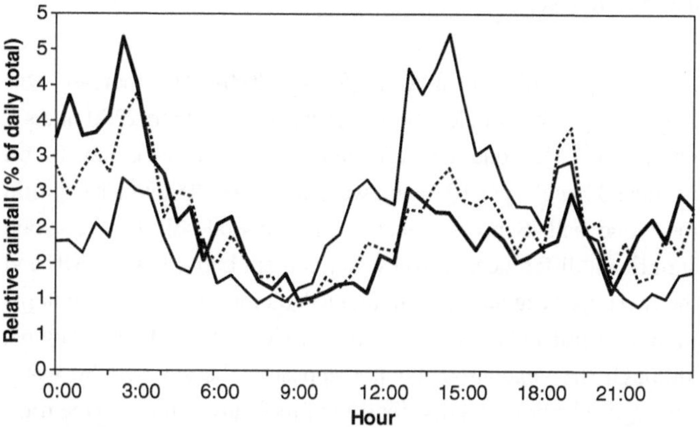

Figure 20.5. Average diurnal distribution of relative amounts of rainfall (% of daily total) between 1 October 2001 and 1 October 2002 at 1800 m.a.s.l. (normal line), 2600 m.a.s.l. (bold line), and 3050 m.a.s.l. (dotted line) along the study transect.

Reference evaporation

Daily reference evaporation (*PET*) values at the UMCF and SACF sites were very similar at 1.1–1.3 mm and almost three times lower than at the LMF site, although ranges in daily values were substantial at all three sites (Table 20.2). Average hourly *PET* was 0.14 mm in the LMF but only 0.05 mm in the UMCF and SACF (Figure 20.2). Rainfall exceeded *PET* 1.9 times in the LMF, vs. 8.6 times in the UMCF and 13.2 times in the SACF (Table 20.2).

Soil moisture

Gravimetric soil moisture content in the organic layer increased (although irregularly) with elevation from 2.5 g g^{-1} at 1800 m.a.s.l.

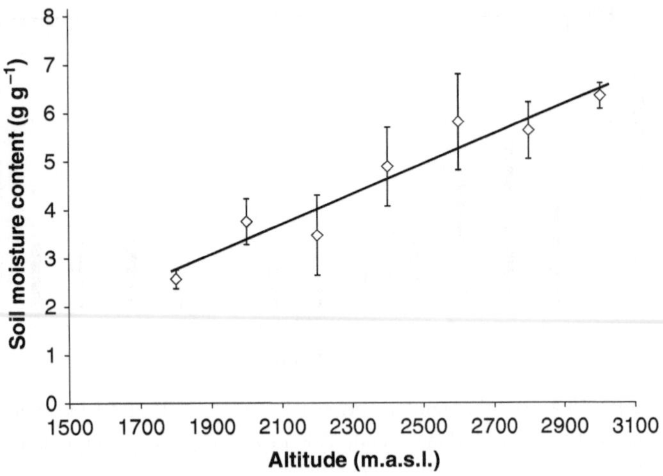

Figure 20.6. Moisture content of the organic layer vs. elevation along the study transect.

to 6.3 g g^{-1} at 3000 m.a.s.l. (Figure 20.6). The increase was 0.6 g g^{-1} per 100 m elevation interval from 1800 to 2200 m.a.s.l. within the LMF zone least affected by fog and low cloud, vs. on average 0.3 g g^{-1} per 100 m between 2200 and 3000 m.a.s.l. within the cloud belt. Sensors placed in the mineral soil also recorded significant differences in soil water content between the LMF and the two upper vegetation formations. Volumetric moisture contents in the mineral soil at 1850 m.a.s.l. were lowest in July and August, although the vegetation did not appear to be stressed by water shortage. The seasonal distribution of moisture content in the mineral soil in both the UMCF and the SACF was relatively constant. The soil in the SACF was saturated throughout the year, whereas in the UMCF the soil was saturated only during the rainy season (November until February) but remained at field capacity at other times.

DISCUSSION

Radiation

Daily inputs of short-wave radiation (R_s) in the study area decreased roughly by one-third from the LMF to the SACF due to the pronounced increase in fog and low cloud above 2200 m.a.s.l. Within the cloud belt, however, cloudiness increased only marginally with elevation (Table 20.1). The average R_s of 15.5 ± 5.4 MJ m^{-2} day^{-1} in the LMF at 1850 m.a.s.l. is comparable to the 13.2 MJ m^{-2} day^{-1} reported for an LMF at 1975, m.a.s.l. in southern Ecuador (4° S; Motzer *et al.*, this volume), and the 13.8 ± 4.6 MJ m^{-2} day^{-1} measured at 1850 m.a.s.l. in the Blue Mountains of Jamaica (18° N) under conditions of light cloud incidence (Hafkenscheid, 2000). Under the more cloudy conditions prevailing at 900–1050 m.a.s.l. in eastern Puerto Rico (18° N), Holwerda (2005) obtained average R_s values of 9.7 ± 4.0 MJ m^{-2} day^{-1} (stunted ridge-top elfin cloud forest) and 10.5 ± 4.8 MJ m^{-2} day^{-1} (upper montane palm forest). The latter values are very similar to the daily R_s totals determined for the present UMCF (10.4 ± 4.3 MJ m^{-2} day^{-1}) and SACF (9.9 ± 3.8 MJ m^{-2} day^{-1}) at much higher elevations (Table 20.1). Conversely, McJannet *et al.* (2007) reported an average insolation value of 12.6 MJ m^{-2} day^{-1} for an extremely wet UMCF in northern Queensland, Australia; for an LMF with limited cloud incidence in the same area an average R_s of 14.4 MJ m^{-2} day^{-1} was recorded.

Holwerda (2005) estimated R_s in the Puerto Rican cloud belt to be reduced by roughly 50% compared to potential solar radiation inputs, which is again similar to the presently estimated 58–62% and confirming the extreme cloudiness of the Bolivian UMCF and SACF sites.

Changes in photosynthetic photon flux density (*PPFD*) along the elevational gradient reflected the measured changes in R_s (Table 20.1). Average *PPFD* totals in Papua New Guinea varied from 38 mol m^{-2} day^{-1} at 1100 m.a.s.l. to 22 mol m^{-2} day^{-1} at 3480 m.a.s.l. (Körner *et al.*, 1983) and were similar to values observed in the present study area (33.6, 22.0 and 20.5 mol m^{-2} day^{-1} at 1850, 2600, and 3050 m.a.s.l., respectively). In a Jamaican UMCF, light saturation in the crowns of *Cyrilla racemiflora* and *Clethra occidentalis* trees occurred at *PPFD* values above 500 μmol m^{-2} s^{-1} (Aylett, 1985). According to Garcia-Nunez *et al.* (1995), light saturation levels required for photosynthetic activity in a Venezuelan MCF were 200–300 μmol m^{-2} s^{-1} for tree species adapted to gaps and 100–200 μmol m^{-2} s^{-1} for understory tree species. Taking the 500 μmol m^{-2} s^{-1} threshold proposed by Aylett (1985), *PPFD* values in the LMF exceeded this value for 64% of the time vs. 52% in the UMCF and SACF. Applying the lower limit of 200 μmol m^{-2} s^{-1} increased these percentages to 76% in the LMF, 72% in the UMCF, and 76% in the SACF. Even under fully overcast conditions the *PPFD* remained above 200 μmol m^{-2} s^{-1} for 8 h in the LMF, for 6.5 h in the UMCF, and for 7 h in the SACF.

According to Körner (1982), short periods of intense *PPFD* were more important for the growth of alpine sedges than low averaged *PPFD* levels occurring over longer periods. The presently found average daily maximum *PPFD* values (Table 20.1) were all well above the threshold values discussed above. Thus, the *PPFD* data suggest that despite the marked difference in radiation levels between the LMF and the two higher vegetation units, the reduction in *PPFD* is unlikely to be a prime cause for the reduction in forest structure with increasing elevation.

Evaporation

As observed in other tropical montane forest belts, reference evaporation *PET* decreased with altitude due to decreases in radiation, temperature, and vapor pressure deficit (Figure 20.2; Keig *et al.*,

1979; Kitayama, 1995; Bruijnzeel and Veneklaas, 1998; cf. McJannet *et al.*, this volume). Annual *PET* at the UMCF site was 61% lower than that for the LMF, and decreased an additional 13% in the SACF. It should be noted that *PET* is a measure of the water use of well-watered grass (Doorenbos and Pruitt, 1988) and therefore not directly comparable to actual evaporation (*ET*) from tall vegetation types like montane rain forests or cloud forests, where evaporation of intercepted rainfall and fog water come into play next to soil water uptake (transpiration). However, expressed as an annual total, the *PET* value for the LMF site (1190 mm) is within the range of *ET* values determined for a number of equatorial LMF sites with negligible fog incidence using water budget techniques (1155–1380 mm year^{-1}; Bruijnzeel and Proctor, 1995). Neither is it too dissimilar from the 1280 mm determined for an LMF at 1900–2200 m.a.s.l. in southern Ecuador (Fleischbein *et al.*, this volume) although the latter estimate included a high interception loss component of 39% of incident *P* that is not taken into account by the reference evaporation estimate.

Conversely, the annual *PET* values for the UMCF (461 mm year^{-1}) and SACF (403 mm year^{-1}) cannot be compared with water-budget-based *ET* values for similar forests elsewhere (310–390 mm year^{-1}) because only apparent *ET* values can be derived with the latter approach due to the confounding effect of unmeasured cloud water inputs (Bruijnzeel and Proctor, 1995). Based on direct measurements of water budget components Holwerda (2005) estimated annual *ET* for the cloud-affected upper montane palm and stunted ridge-top elfin cloud forests in Puerto Rico cited earlier at 674 mm and *c.* 480 mm, respectively. Transpiration rates for the two Puerto Rican forests were 1.33 ± 0.95 mm day^{-1} and 0.81 ± 0.97 mm day^{-1}, respectively (Holwerda 2005) and comparable to the daily *PET* values derived for the present cloud forest sites (Table 20.2). McJannet *et al.* (2007) derived an annual *PET* of 910 mm for an UMCF at 1560 m.a.s.l. in northern Queensland, a finding in line with the higher radiation total recorded for the site. However, despite this higher radiation input, transpiration (as measured with sapflow sensors) was low at 1.0 mm day^{-1} (McJannet *et al.*, this volume). Although these rates are low, even lower rates (<0.5 mm day^{-1}) of xylem water movement have been demonstrated in a Hawaiian cloud forest (Santiago *et al.*, 2000; cf. Santiago *et al.*, this volume). Whilst reductions in montane forest stature and productivity have been attributed by some to strongly reduced transpiration rates (Odum, 1970; Kitayama and Müller-Dombois, 1994), this idea was refuted on theoretical grounds by Grubb (1977) and Tanner and Beevers (1990). Moreover, recent evidence suggests that the decline in above-ground forest stature with elevation reflects increased investment in below-ground productivity (Leuschner *et al.*, 2007; Moser *et al.*, 2007; cf. Hertel and Leuschner, this volume), probably in response to gradually more adverse edaphic conditions (Benner *et al.*, this volume; Roman *et al.*, this volume; see also below).

Precipitation and soil moisture

The presently observed increase in *P* above 2000 m.a.s.l. is atypical for tropical regions. According to Lauscher (1976), *P* in tropical regions is often highest between 1000 m and 2000 m.a.s.l. Vis (1986) and Veneklaas and Van Ek (1990) reported *P* in the central Cordillera of the Colombian Andes to decrease from 3150 mm year^{-1} at 1700 m.a.s.l. to 1700 mm year^{-1} at 3050 m.a.s.l. On Mt. Kilimanjaro (Tanzania), maximum *P* occurs around 2300 m and drops markedly with elevation (Hemp, this volume #12), whereas on Mt. Kinabalu (East Malaysia) changes in *P* with elevation seemed less pronounced (Kitayama, 1992). However, the presently observed increase in *P* with altitude is matched by various other observations in the eastern Andes. For example, Weischet (1969) reported increased *P* between 2600 and 3300 m.a.s.l. in the eastern Cordillera of the Colombian Andes due to the occurrence of a secondary maximum of precipitation, caused in turn by heating of the high mountain plateau and forced convective precipitation on the upper slopes. Further, a marked increase in *P*, from *c.* 2000 mm at 1900 m.a.s.l. to *c.* 4000 mm at 2650 m.a.s.l. and 5–6000 mm at 3400 m.a.s.l., has been recorded in the Andes of south-eastern Ecuador (Bendix *et al.*, 2008).

The difference in timing of the diurnal maximum in *P* at different elevations in the study area (Figure 20.5) is attributed to a difference in origin of the precipitation. In the LMF and UMCF zones, maximum *P* occurred late at night and in the early morning, suggesting orographic influences. Conversely, the mid-afternoon maximum observed in the SACF indicates convective processes. It is proposed here that the heating of the Altiplano of Bolivia during the day causes large humid lowland air masses to rise and this eventually results in high precipitation on the higher slopes.

The pronounced changes in the climatic water balance (i.e. *P* minus PET) resulted in major changes in soil moisture content along the elevational gradient. In the LMF, waterlogging of soils was absent but in the UMCF the soils were saturated during the rainy season, whereas in the SACF soils were saturated throughout the year. Similar patterns of precipitation excess and soil wetness with elevation have been reported for northern Colombia (Herrmann, 1971), Mt. Kinabalu (Kitayama, 1992), and Puerto Rico (Silver *et al.*, 1999; cf. Holwerda, 2005).

Persistently wet conditions have been reported for some, but certainly not all, tropical cloud forest soils (Bruijnzeel and Proctor, 1995; Roman *et al.*, this volume), providing further illustration that not all cloud forests show equal hydrological behavior (Bruijnzeel, 2005). However, increases in the degree and duration of soil waterlogging are often paralleled by a decrease in montane forest stature (Herrmann, 1971; Hetsch and Hoheisel, 1976; Bruijnzeel *et al.*, 1993; Silver *et al.*, 1999; Santiago *et al.*, 2000; Häger and Dohrenbusch, this volume). Interestingly, such parallels have been observed both at high and low elevations, suggesting that precipitation excess overrides any temperature effects. Where waterlogging is particularly persistent – as in the current SACF – aerial roots tend to be particularly common (Schawe, 2005) as also observed in other very wet cloud forests (e.g. Gill, 1969; Lyford, 1969; Häger and Dohrenbusch, this volume).

Principal component analysis and multiple regressions between plant species diversity (α-diversity) and a "sesquioxide factor" for the A-horizon (included parameters: total content of aluminum, iron, and phosphorus, oxalate-soluble aluminum and iron, aluminum saturation, pH, and exchangeable protons) at 100-m altitudinal steps demonstrated a significant correlation (Schawe, 2005). The sesquioxide factor may be interpreted as an indirect indicator of the increased soil wetness and nutrient leaching potential with elevation. Based on the latter analysis, soil water status constitutes an important factor determining changes in plant species diversity along the elevational gradient (Schawe, 2005). The relationship between edaphic factors (including degree of waterlogging) and forest productivity is discussed in further detail by Leuschner *et al.* (2007), Moser *et al.* (2007), Soethe *et al.* (2008a,b), Benner *et al.* (this volume), Roman *et al.* (this volume) and Hertel and Leuschner (this volume).

CONCLUSION

Despite the marked differences in short-wave radiation, *PPFD* and *PET* between the lower montane forest and the upper montane and sub-alpine cloud forest formations, these hydrometeorological factors did not seem to have a direct influence on forest stature, composition or productivity. Rather, the effect was indirect in that the strong increase in precipitation excess with elevation caused major changes in the rate of litter mineralization (and therefore nitrogen availability), soil nutrient leaching, and degree of soil aeration (cf. Schawe, 2005; Schawe *et al.*, 2007). These are considered the principal factors responsible for the observed changes in vegetation structure and plant diversity along the elevational gradient.

ACKNOWLEDGEMENTS

The study was carried out in close collaboration with the Instituto de Ecología of the Universidad Mayor de San Andrés, La Paz and the National Park Administration (SERNAP) in Bolivia. Funding was received from the Deutsche Forschungsgemeinschaft (DFG) (Grants GE 431/12-3 to G. Gerold and GR 1588/4 to S.R. Gradstein). The manuscript benefited from comments by Sampurno Bruijnzeel and Fred Scatena.

REFERENCES

Aylett, G.P. (1985). Irradiance interception, leaf conductance and photosynthesis in Jamaican upper montane rain forest trees. *Photosynthetica* **19**: 323–337.

Bach, K. (2004). Vegetationskundliche Untersuchungen zur Höhenzonierung tropischer Bergregenwälder in den Anden Boliviens. Ph.D. thesis, University of Göttingen, Göttingen, Germany.

Bach, K., M. Schawe, S. Beck, *et al.* (2003). Vegetación, suelos y clima en los diferentes pisos altitudinales de un bosque montano de Yungas, Bolivia: primeros resultados. *Ecología en Bolivia* **38**: 3–14.

Bendix, J., R. Rollenbeck, M. Richter, P. Fabian, and P. Emck (2008). Climate. In *Gradients in a Tropical Mountain Ecosystem of Ecuador*, eds. E. Beck, J. Bendix, I. Kottke, F. Makeschin, and R. Mosandl, pp. 63–73. Berlin: Springer-Verlag.

Böhner, J., R. Köthe, and C. Trachinow (1997). Weiterentwicklung der automatischen Reliefanalyse auf der Basis von digitalen Geländemodellen. *Göttinger geographische Abhandlungen* **100**: 3–21.

Bruijnzeel, L.A. (2005). Tropical montane cloud forests: a unique hydrological case. In *Forests, Water and People in the Humid Tropics*, eds. M. Bonell and L.A. Bruijnzeel, pp. 462–483. Cambridge, UK: Cambridge University Press.

Bruijnzeel, L.A., and J. Proctor (1995). Hydrology and biogeochemistry of tropical montane cloud forests: what do we really know? In *Tropical Montane Cloud Forests*, eds. L.S. Hamilton, J.O. Juvik, and F.N. Scatena, pp. 38–78. New York: Springer-Verlag.

Bruijnzeel, L.A., and E.J. Veneklaas (1998). Climatic conditions and tropical montane forest productivity: the fog has not lifted yet. *Ecology* **79**: 3–9.

Bruijnzeel, L.A., M.L. Waterloo, J. Proctor, A.T. Kuiters, and B. Kotterink (1993). Hydrological observations in montane rain forests on Gunung Silam, Sabah, Malaysia, with special reference to the '*Massenerhebung*' effect. *Journal of Ecology* **81**: 145–167.

Cavelier, J., and C. Mejia (1990). Climatic factors and tree stature in the elfin cloud forest of Serrania de Macuira, Colombia. *Agricultural and Forest Meteorology* **53**: 105–123.

Doorenbos, J., and W.O. Pruitt (1988). *Yield Response to Water,* FAO Irrigation and Drainage Paper No. 33. Rome: FAO.

Frahm, J.P., and S.R. Gradstein (1991). An altitudinal zonation of tropical rain forests using bryophytes. *Journal of Biogeography* **18**: 669–678.

Garcia-Nunez, C., A. Azocar, and F. Rada (1995). Photosynthetic acclimation to light in juveniles of two cloud forest tree species. *Trees* **10**: 114–124.

Gentry, A.H. (1995). Patterns of diversity and floristic composition in Neotropical montane forests. In *Biodiversity and Conservation of Neotropical Montane Forests*, eds. S.P. Churchill, H. Balslev, E. Forero, and J.L. Luteyn, pp. 103–126. New York: New York Botanical Garden.

Gerold, G., M. Schawe, and L. Joachim (2003). Pedoökologische hypsometrische Varianz in ungestörten Bergregenwäldern der Anden (Yungas). *Geoökologie* **24**: 153–162.

Gill, A.M. (1969). The ecology of an elfin forest in Puerto Rico. VI. Aerial roots. *Journal of Arnold Arboretum* **50**: 197–209.

Gradstein, S.R., and B.O. van Zanten (1999). High-altitude dispersal of spores: an experimental approach. Abstract, XVI *International Botanical Congress.* St. Louis, MO: Missouri Botanical Garden.

Grubb, P.J. (1974). Factors controlling the distribution of forest-types on tropical mountains: new facts and a new perspective. In *Altitudinal Zonation in Malaysia, Transactions of the 3rd Aberdeen–Hull Symposium on Malaysian Ecology*, ed. J.R. Flenley, pp. 13–46. Hull, UK: Department of Geography, University of Hull.

Grubb, P.J. (1977). Control of forest growth and distribution on wet tropical mountains: with special reference to mineral nutrition. *Annual Review of Ecology and Systematics* **8**: 83–107.

Hafkenscheid, R.R.L.J. (2000). Hydrology and biogeochemistry of tropical montane rain forests of contrasting stature in the Blue Mountains, Jamaica. Ph.D. thesis, VU University Amsterdam, Amsterdam, the Netherlands. Available at http://dare.ubvu.vu.nl/bitstream/1871/12734/1/tekst.pdf.

Hetsch, W., and H. Hoheisel (1976). Standorts- und Vegetationsgliederung in einem tropischen Nebelwald. *Allgemeine Forst- und Jagdzeitschrift* **147**: 200–209.

Herrmann, R. (1971). Die zeitliche Änderung der Wasserbindung im Boden unter verschiedenen Vegetationsformationen der Höhenstufen eines tropischen Hochgebirges (Sierra Nevada de Sta Maria/Kolumbien). *Erdkunde* **25**: 90–112.

Holwerda, F. (2005). Water and energy budgets of rain forests along an elevation gradient under maritime tropical conditions. Ph.D. thesis, VU University Amsterdam, Amsterdam, the Netherlands. Available at www.falw.vu.nl/nl/onderzoek/earth-sciences/geo-environmental-science-and-hydrology/hydrology-dissertations/index.asp.

Kappelle, M., J.G. Uffelen, and A.M. Cleef (1995). Altitudinal zonation of montane *Quercus* forests along two transects in the Chirripó National Park, Costa Rica. *Vegetatio* **119**: 119–154.

Keig, G., P.M. Flemming, and J.R. McAlpine (1979). Evaporation in Papua New Guinea. *Journal of Tropical Geography* **48**: 19–30.

Kessler, M. (2000). Elevational gradients in species richness and endemism of selected plant groups in the central Bolivian Andes. *Plant Ecology* **149**: 181–193.

Kitayama, K. (1992). An altitudinal transect study of the vegetation on Mount Kinabalu, Borneo. *Vegetatio* **102**: 149–171.

Kitayama, K. (1995). Biophysical conditions of the montane cloud forests of Mount Kinabalu, Sabah, Malaysia. In *Tropical Montane Cloud Forests*, eds. L.S. Hamilton, J.O. Juvik, and F.M. Scatena, pp. 115–125. New York: Springer-Verlag.

Kitayama, K., and D. Müller- Dombois (1994). An altitudinal transect analysis of the windward vegetation on Haleakala, a Hawaiian island mountain. II. Vegetation zonation. *Phytocoenologia* **24**: 135–154.

Körner, C (1982). CO_2 exchange in the alpine sedge *Carex curvula* as influenced by canopy structure, light and temperature. *Oecologia* **53**: 98–104.

Körner, C., A. Allison, and H. Hilscher (1983). Altitudinal variation of leaf diffusive conductance and leaf anatomy in Heliophytes of montane New Guinea and their interrelation with microclimate. *Flora* **174**: 91–135.

Lauer, W. (1975). Klimatische Grundzüge der Höhenstufung tropischer Gebirge. *Tagungsbericht und Wissenschaftliche Abhandlungen* **40**: 76–90.

Lauscher, F. (1976). Weltweite Typen der Höhenabhängigkeit des Niederschlags. *Wetter Leben* **28**: 89–90.

Leuschner, Ch., G. Moser, C. Bertsch, M. Röderstain, and D. Hertel (2007). Large altitudinal increase in tree root/shoot ratio in tropical mountain forests of Ecuador. *Basic and Applied Ecology* **8**: 219–230.

Lyford, W.H. (1969). The Ecology of an elfin forest in Puerto Rico: soil, root, and earthworm relationships. *Journal of Arnold Arboretum* **50**: 210–224.

McJannet, M., J.S. Wallace, P. Fitch, M. Disher, and P. Reddell (2007). Water balance of tropical rainforest canopies in north Queensland. *Hydrological Processes* **21**: 3473–3784.

Moser, G., D. Hertel, and Ch. Leuschner (2007). Altitudinal change in LAI and stand leaf biomass in tropical montane forests: a transect study in Ecuador and a pan-tropical meta-analysis. *Ecosystems* **10**: 924–935.

Odum, H.T. (1970). Rain forest structure and mineral cycling homeostasis. In *A Tropical Rain Forest: A Study of Irradiation and Ecology at El Verde, Puerto Rico*, eds. H.T. Odum and R.F. Pidgeon, pp. H3–H52. Oak Ridge, TN: U.S. Atomic Energy Commission.

Pendry, C.A., and J. Proctor (1996). The causes of altitudinal zonation of rain forests on Bukit Belalong, Brunei. *Journal of Ecology* **84**: 407–418.

Santiago, L.S., G. Goldstein, F.C. Meinzer, J.H. Fownes, and D. Müller-Dombois (2000). Transpiration and forest structure in relation to soil waterlogging in a Hawaiian montane cloud forest. *Tree Physiology* **20**: 673–681.

Schawe, M. (2005). Hypsometerischer Klima- und Bodenwandel in Bergregenwaldökosystemen Boliviens. Ph.D. thesis, University of Göttingen, Göttingen, Germany.

Schawe, M., S. Glatzel, and G. Gerold (2007). Soil development along an altitudinal transect in a Bolivian tropical montane rainforest: podzolization vs. hydromorphy. *Catena* **69**: 83–90.

Silver, W., A.E. Lugo, and M. Keller (1999). Soil oxygen availability and biogeochemistry along rainfall and topographic gradients in upland wet tropical forest soils. *Biogeochemistry* **44**: 301–328.

Soethe, N., J. Lehmann, and C. Engels (2008a). Nutrient availability at different altitudes in a tropical montane forest in Ecuador. *Journal of Tropical Ecology* **24**: 397–406.

Soethe, N., W. Wilcke, J. Homeier, J. Lehmann, and C. Engels (2008b). Plant growth along the altitudinal gradient: role of plant nutritional status, fine root activity, and soil properties. In *Gradients in a Tropical Mountain Ecosystem of Ecuador*, eds. E. Beck, J. Bendix, I. Kottke, F. Makeschin, and R. Mosandl, pp. 259–266. Berlin: Springer-Verlag.

Tanner, W., and H. Beevers (1990). Does transpiration have an essential function in long-distance ion transport in plants? *Plant, Cell and Environment* **13**: 745–750.

Veneklaas, E.J., and R. van Ek (1990). Rainfall interception in two tropical montane rain forests, Colombia. *Hydrological Processes* **4**: 311–326.

Vis, M. (1986). Interception, drop size distributions and rainfall kinetic energy in four Colombian forest ecosystems. *Earth Surface Processes and Landforms* **11**: 591–570.

Weischet, W. (1969). Klimatologische Regeln zur Vertikalverteilung der Niederschläge in Tropengebirgen. *Die Erde* **100**: 287–306.

Wolf, J.H.D. (1993). Diversity patterns and biomass of epiphytic bryophytes and lichens along an altitudinal gradient in the northern Andes. *Annals of the Missouri Botanical Garden* **90**: 928–960.

21 Structure and dynamics of tropical montane cloud forests under contrasting biophysical conditions in north-western Costa Rica

A. Häger and A. Dohrenbusch
University of Göttingen, Göttingen, Germany

ABSTRACT

Forest structure and composition were investigated in seven 10 × 50 m plots along a 2.5-km transect crossing the Continental Divide between 1200 and 1500 m.a.s.l. on the windward and leeward slopes of the Monteverde Cloud Forest Reserve, Costa Rica. Seven climate stations (measuring rainfall, horizontal precipitation, throughfall, temperature, and soil water content) were installed along the transect. Epiphytic vegetation was compared between leeward forests at 1450 m.a.s.l. (near the ridge top) and at 1200 m.a.s.l. Fallen trees were recorded over one year in a 50-m wide strip along the entire transect. The highest basal areas (up to nearly 100 m^{-2} ha^{-1}) and the greatest canopy heights were found on the leeward slope between 1200 and 1450 m.a.s.l., and in the lowermost windward plot. The forest formations on the ridge and on the steep upper windward slope had basal areas of 30–40 m^{-2} ha^{-1} and tree heights of less than 15 m. Epiphytic biomass was very high at both investigated sites (32–39 t ha^{-1}). Close to the wet ridge top, epiphyte abundance and vascular species richness were greatest. The monitoring of forest dynamics resulted in a total of 5.2 recorded tree-falls ha^{-1} $year^{-1}$. Annual rainfall (March 2003–February 2004) ranged from 3690 mm on the leeward slope to 6390 mm on the windward side. Horizontal precipitation was estimated at 3560 mm at the ridge, compared to 330 mm at the lowest windward plot and 28 mm at 1200 m.a.s.l. on the leeward slope. Throughfall remained below incident rainfall on the lower leeward slope but was usually higher than rainfall around the ridge. Mean temperature at 1500 m.a.s.l. was 16 °C. The average daily wind speed at the ridge top was about 11 times higher than at the lower sites. Volumetric water content of the highly organic soils of the flat ridge top was around 80% for most of the year.

INTRODUCTION

The Tilarán mountain range forms the Continental Divide of the Central American isthmus in north-western Costa Rica and acts as a barrier against the north-easterly trade winds. Thus, the eastern (Atlantic) slopes receive more rain and moisture than the (Pacific) slopes facing west, which experience a more pronounced dry season with decreasing altitude, from January to mid-May. Gradients in precipitation, temperature and wind speed across the mountain range lead to a broad diversity of cloud forest habitats within a relatively small area (Lawton and Dryer, 1980). The objective of this research was to investigate interactions between biophysical conditions and forest structure in the Monteverde Cloud Forest Reserve (10° 17′ N, 84° 46′ W) between 1200 and 1500 m.a.s.l. on both sides of the Continental Divide (Figure 21.1). The climate and geology of the Monteverde area have been described by Clark *et al.* (2000), while a fuller account of the present results can be found in Häger (2006).

METHODS

Forest structure and composition

Stand structure (position, diameter at breast height (DBH), height, crown length, crown diameters in four directions, and stem shape of trees, plus incidence and dimensions of buttressed or stilt roots) as well as floristic composition were investigated in each of seven 10 × 50 m forest plots, considering all individuals with DBH > 5 cm. The plots were located along a 2.5-km

Tropical Montane Cloud Forests: Science for Conservation and Management, eds. L. A. Bruijnzeel, F. N. Scatena, and L. S. Hamilton. Published by Cambridge University Press.

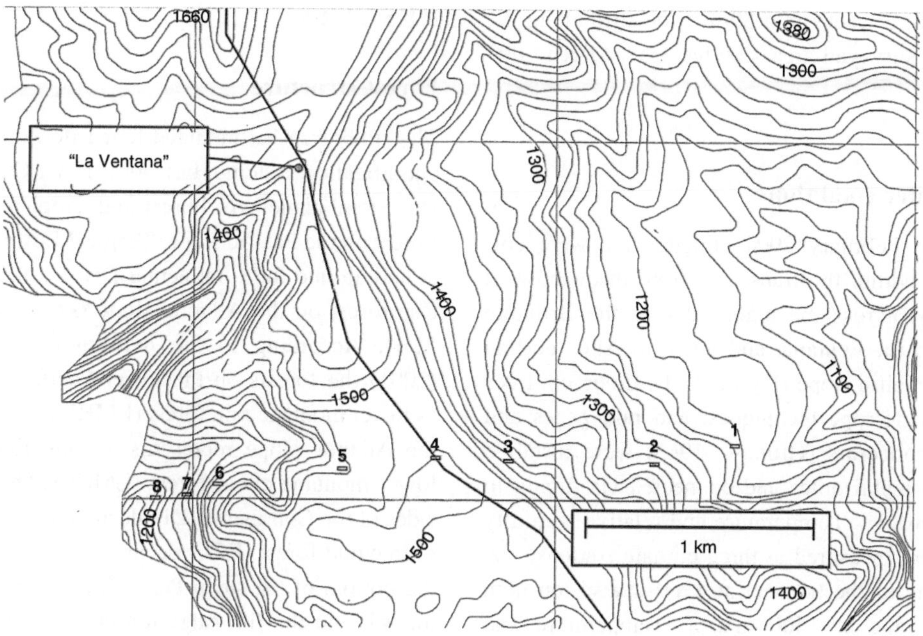

Figure 21.1. Map of the Monteverde area showing the locations of the eight forest plots. Solid line indicates the Continental Divide.

transect crossing the Continental Divide and covering both windward and leeward slopes. Plots were established at corresponding altitudes (1200 m, 1260–1320 m, and 1440–1450 m.a.s.l.) on the two slopes, with a central ridge-top site at 1500 m.a.s.l. (Figure 21.1). Tree species were identified by William Haber. Detailed measurements were taken of the stems and crowns of 15 trees from three different forest strata in five plots across the transect. The sample trees belonged to 12 different species from six families and included the six trees that were sampled for their epiphyte loading as described below. Approximate stand-level stem volumes were calculated using regression models, based on the tight relationship found between DBH (cm) and trunk volume (m^3) for the sample trees (for details see Häger, 2006):

$$\text{volume} = 3{*}10^{-5}\ \text{DBH}^{2,844};\ (r^2 = 0.98; n = 15)\ \text{for DBH from 5 to 80 cm}$$

$$\text{volume} = 9{*}10^{-4}\ \text{DBH}^2 + 0.0362\ \text{DBH} - 0.972;\ (r^2 = 0.95; n = 15)\ \text{for DBH} > 80\ \text{cm}$$

Profile diagrams of the forest plots were drawn to scale and digitized. Plant diversity for single plots was described by the Shannon index. To quantify the β-diversity of the transect area, the modification of Whittaker's index by Harrison *et al.* (1992) was used for paired comparisons. This measure ranges from 0 (no turnover in species between sites, identical set of species) to 100 (complete turnover, the compared sites have no species in common; Magurran, 2004).

Epiphytic biomass and distribution

Epiphytic vegetation was compared between two sites, viz. in a moist Windward Cloud Forest (*sensu* Lawton and Dryer, 1980) at 1450 m.a.s.l. on the Pacific slope but located just below the main ridge top and therefore effectively experiencing windward conditions, and in a drier forest at 1200 m.a.s.l. on the Pacific slope. One representative sample tree from the canopy, subcanopy, and understory layers was chosen in each plot (with DBH classes of >60 cm, 30–60 cm, and 15–30 cm, respectively). Sample trees were accessed either by choosing recently fallen individuals or using single-rope techniques. Epiphyte samples were taken from the stems, from the inner and outer parts of the crown (considering both vertical and horizontal branches), with a total of five to eight samples per tree. Epiphytic vegetation was removed over a length of 0.5 m around the stem or branch. In addition, from each tree one peripheral branch (diameter < 10 cm) was cut off to collect its complete epiphytic biomass. Samples were separated into bryophytes, ferns, herbs and bromeliads, woody plants, lianas or hemi-epiphytes, and crown humus. Vascular species were counted. Epiphytic biomass was dried at 70 °C to constant weight at the laboratory of the Monteverde Reserve. Digital photos with scaled ropes placed on the stems and in the crowns of the sampled trees allowed calculation of the bark surface for total epiphytic biomass projection per tree. Approximate epiphyte biomass (in t ha^{-1}) was estimated by multiplying single-tree values times the number of trees per diameter class in each plot, using data from the forest inventory.

Tree-falls

The number of fallen canopy and sub-canopy trees (with minimum heights of 10 m for the taller forest formations) were recorded at intervals of 2 weeks in a 50-m wide strip along the 2.5-km transect (total surveyed area 12.5 ha) from March 2003 until February 2004.

Climate and soil water conditions

From March 2003 to February 2004 climate data were collected at seven plots within the transect – from the windward Peñas Blancas valley, across the main ridge at the Brillante Gap, 1.3 km south to La Ventana, and down to Altos de San Luis on the leeward Pacific slope (Figure 21.1). Emergent trees were accessed by single-rope techniques, and rain gages and simple passive fog collectors (Grunow, 1963) were installed above the canopy on metal poles and connected via tubing to storage cans on the ground. Temperature and relative humidity above the canopy were measured at three climate stations, viz. at the lowermost plots of the transect at 1200 m.a.s.l. on both slopes and on the ridge at 1500 m.a.s.l. Temperature and relative humidity were compared during the dry season months March–April 2003, during the wet season months June–October 2003, and during the transitional months May 2003 and November 2003–January 2004 at the two lowest plots. Data were recorded by loggers with integrated sensors (Testo) every 30 min. In addition, minimum–maximum thermometers were installed inside the forest at all three climate stations. On the ridge top a resistance temperature sensor was used, the data of which were recorded every 10 min by a logger (Ecotec) from March through April 2003 and from June 2003 through February 2004 (the humidity sensor failed under the extremely moist conditions prevailing at this site). Wind speed above the canopy on the ridge was recorded with a cup anemometer during the same periods as for temperature whereas at the two lowest plots above-canopy wind speed was measured from March to June 2004.

Throughfall was measured at five fixed points within each 500-m^2 plot (at the center of 100-m^2 sub-plots) using rain gages with a diameter of 11 cm. All precipitation equipment was sampled at least every 2 weeks for 1 year. Wind damage affected several climate stations. For this reason complete annual data (March 2003–February 2004) for horizontal precipitation could only be obtained for plot 1 (Atlantic Slope, 1200 m.a.s.l.), plot 4 (ridge top, 1500 m.a.s.l.), plot 6 (Pacific slope, 1320 m.a.s.l.), and plot 8 (Pacific slope, 1200 m.a.s.l.). Complete annual throughfall data for the same period were obtained for plots 1, 4, and 8. Volumetric soil water contents at 5- and 30-cm depths were determined for 1 year at monthly intervals in two different sub-plots at each of the seven forest plots using the gravimetric method in the laboratory of the Reserve. Further experimental and instrumental details can be found in Häger (2006).

RESULTS

Forest structure

The study transect included four Life Zones from the Holdridge classification system (Hartshorn, 1983; CCAD, 2002), viz. Tropical Premontane Wet Forest and Tropical Lower Montane Wet Forest on the Pacific slope; Tropical Lower Montane Rain Forest at and around the Continental Divide; and Tropical Premontane Rainforest on the Atlantic slope (Peñas Blancas valley; Figure 21.1). According to the definitions proposed by Bruijnzeel (2001) the forests on the lower Pacific slope can be classified as lower montane rain forest (LMRF), whereas the lowest plot on the Atlantic slope represents a transition between LMRF and lower montane cloud forest (LMCF). The upper slopes on both sides of the Continental Divide are covered with LMCF, whereas elfin cloud forest (ECF) occurs on the windswept slope between the bottom of the steep Peñas Blancas Valley headwall at 1400 m.a.s.l. and the flat ridge top at 1500 m a.s.l. (cf. Lawton and Dryer, 1980).

Forest stature generally increased from the ridge to the more sheltered lower slopes, while stem density decreased (Table 21.1, Figure 21.2). The average canopy height of the ECF in the extremely wind-exposed and poorly drained ridge-top area at 1500 m.a.s.l. (plot 4) reached only about 10 m, with maximum tree heights of 15 m, vs. maxima exceeding 30 m below 1300 m.a.s.l. Stilt–rooted trees were very common on the partly swampy soils on the ridge. However, even lower basal area and tree heights were observed for the dense thicket on the steep headwall of the Peñas Blancas Valley, on the Atlantic side at about 1450 m.a.s.l. (plot 3). The impact of the constant trade winds is the most important dynamic factor shaping these forests (Lawton and Dryer, 1980). On the Pacific side, basal area and canopy height increased rapidly in sheltered sections behind the ridge top. The forest at 1450 m.a.s.l. (plot 5) was classified as a Windward Cloud Forest (*sensu* Lawton and Dryer, 1980) with a very high basal area and emergent trees up to 25 m. Compared with the stunted ECF of plot 3, stand volumes for the sheltered leeward forests of plot 5 (770 m^3 ha^{-1}) and plot 7 (850 m^3 ha^{-1}) and in the lower Peñas Blancas valley (plot 1; 680 m^3 ha^{-1}) were between four and almost six times larger (Table 21.1). Below 1300 m.a.s.l., maximum tree heights in excess of 35 m were measured on both slopes, with some emergent trees even reaching up to 50 m (e.g. *Cedrela tonduzii, Ulmus mexicana;* cf. Figure 21.2). Buttressed roots seemed to become more common with decreasing elevation and on the Pacific slope. Two-dimensional coverage by tree crowns, derived from crown diameter measurements, decreased from 3.1–3.4 m^2 m^{-2} at lower elevations to 2.4 m^2 m^{-2} on the ridge, with an absolute minimum of 1.6 m^2 m^{-2} on the upper windward slopes. Crown cover was generally related to canopy height (Table 21.1).

Table 21.1 *Basic site, forest structural, and rainfall data (March 2003–February 2004) for eight plots along a 2.5-km transect near Monteverde, north-western Costa Rica*

Plot	1	2	3	4	5	6	7	8
Altitude (m.a.s.l.)	1200	1320	1440	1500	1450	1320	1260	1200
Exposition	40°	60°	55°	80°/260°	260°	310°	275°	255°
Average inclination	11°	34°	51°	8°	15°	32°	36°	21°
Rainfall (mm $year^{-1}$)	6390	4400	3760	4310	4210	4590	–	3690
Horizontal precipitation (mm $year^{-1}$)	350	*	*	3560	*	240	–	28
Throughfall (mm $year^{-1}$)	4470	*	*	5100	*	*	–	2660
Stem density (ha^{-1})	1160	1560	3280	1940	1340	–	1280	1360
Basal area (m^2 ha^{-1})	56	54	33	40	99	–	70	67
Stand volume (m^3 ha^{-1})	680	450	150	240	770	–	850	690
Canopy (max) height (m)	20 (36)	14 (22)	6 (12)	10 (15)	18 (25)	–	20 (36)	20 (29)
Crown coverage (m^2 m^{-2})	3,4	2,8	2,5	2,5	3,6	–	3,5	3,5

* incomplete data. Forest inventory data from plot 7 at 1260 m a.s.l. on the Pacific slope were used in plot 6 as well after the latter was lost due to landslides in May 2004.

Floristic composition

Dominant families at lower elevations included the Sapotaceae, Fabaceae, Elaeocarpaceae, Moraceae, Sabiaceae, and Lauraceae, whereas the most common families of the elfin forest were the Araliaceae, Clusiaceae, Rubiaceae, and Myrsinaceae (*Ardisia* sp.). Fabaceae (e.g. *Inga* sp.), Moraceae (*Ficus* sp.), Lauraceae (*Ocotea* sp.), Meliaceae (*Guarea* sp.) and Melastomataceae were widespread along the transect. The most frequently encountered tree was *Elaeagia auriculata* (Rubiaceae) although it clearly lost importance below 1300 m.a.s.l. on the Pacific slope.

The diversity of woody plants (according to the Shannon index) ranged from 2.69 to 3.37 and the number of species from 21 close to the ridge, to 34 at the lowest leeward site for the given plot size (Table 21.2). Paired calculations of β-diversity between plots showed a pronounced species turnover ($\beta = 94$) going from the ridge (plots 4 and 5) to the lower Pacific slope (plots 7 and 8). The difference in species composition between the Atlantic slope (plots 1 and 2) and the ridge also gave high β-diversity values, but the change was more gradual ($\beta = 77$). The comparison of the lower Atlantic and Pacific slope plots resulted in $\beta = 79$.

Epiphytes

Calculated total epiphytic biomass ranged from an average of 39 t ha^{-1} at the higher Pacific slope (1450 m.a.s.l.) to 32 t ha^{-1} in the drier forest type 250 m lower down. These figures included epiphytic humus, ground-rooted lianas and hemi-epiphytes (Table 21.3). Epiphytic biomass excluding ground-rooted plants amounted to 32 and 26 t ha^{-1} at the respective sites. Differences between plots were found in terms of epiphyte distribution, composition, and diversity (Figure 21.3). Plant cover and vascular species richness were greater close to the mountain ridge, where horizontal precipitation input was up to 140 times higher (see below), and oscillations in relative humidity much smaller compared to the lowest leeward plot.

The average number of vascular epiphyte species per dm^2 of stem or branch surface at 1200 m.a.s.l. seemed low at 0.14 ± 0.12 (SD), compared with 0.29 ± 0.20 dm^{-2} close to the ridge at 1450 m. a.s.l. Whilst bromeliads, woody Ericaceae, and ferns were dominant at 1200 m, lianas and hemi-epiphytes were abundant at both sites. However, due to the decrease in occurrence of other vegetation forms, the latter constituted a higher proportion of the epiphytic vegetation on the sampled trees of the lower plot (Figure 21.3). About 80% of the estimated total epiphytic biomass for the drier forest was concentrated on the bigger inner branches of mature canopy trees. On the lower Pacific slope, vascular epiphytes decreased on the branches of sub-canopy and understory trees as well as on the stem surface of all individuals. At higher elevations, epiphyte cover, canopy humus, and epiphyte diversity were distributed more equally over the three tree layers (Figure 21.3).

In particular, bryophyte mass decreased considerably toward lower elevations on the Pacific slope. Whilst average bryophyte mass ranged between 305 and 391 g m^{-2} of bark surface on the three sampled trees at 1450 m.a.s.l., values declined to 18–110 g m^{-2} at 1200, m.a.s.l. Mosses accounted for 14–15% of total epiphytic biomass per m^2 of bark surface on canopy and sub-canopy trees at 1450 m.a.s.l., a figure that increased to 27% for the sampled understory tree. Corresponding figures for the lower, drier site were 3–4% and 43%, respectively.

Tree-falls

Forest dynamics are accelerated in the Tilarán mountains. From March 2003 until February 2004, 65 fallen canopy and sub-canopy trees were counted on the 12.5-ha sample strip (implying 5.2 fallen trees ha^{-1} $year^{-1}$). In January 2004 alone, 31 trees were downed

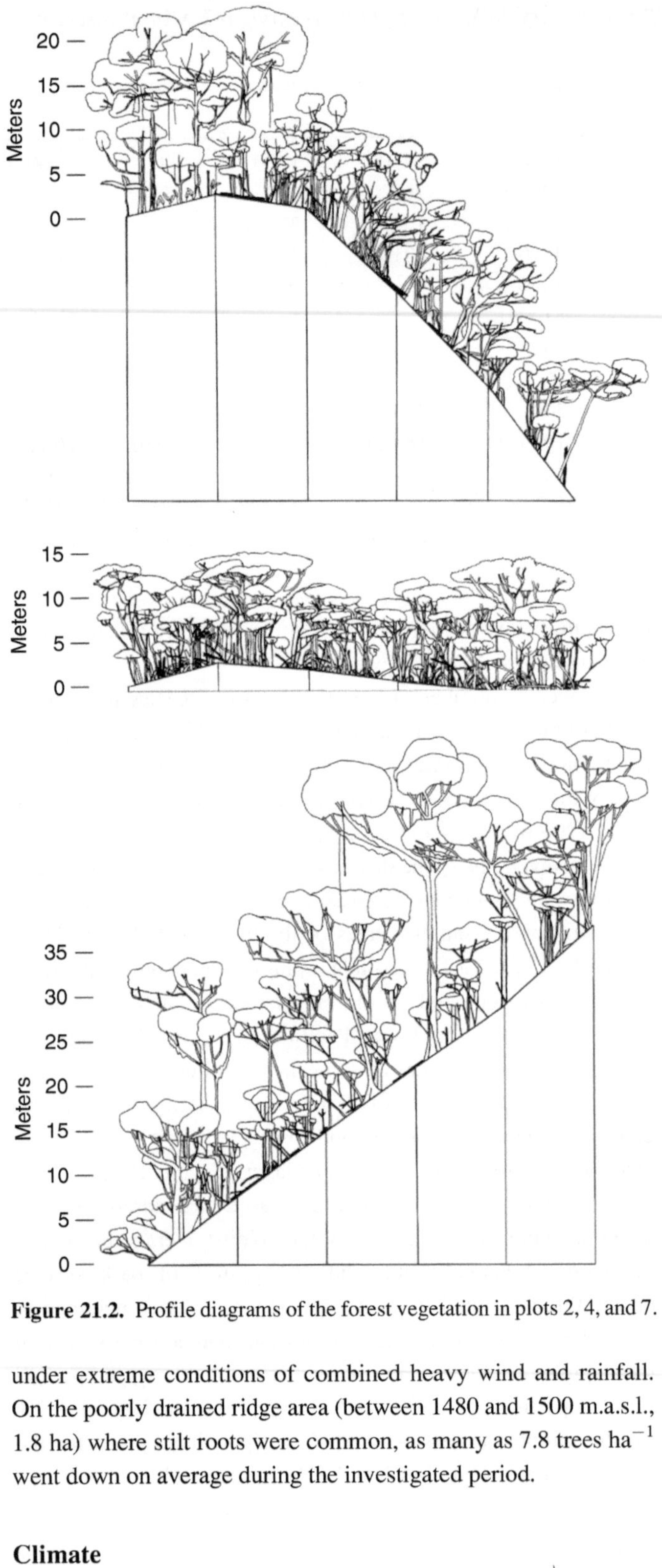

Figure 21.2. Profile diagrams of the forest vegetation in plots 2, 4, and 7.

under extreme conditions of combined heavy wind and rainfall. On the poorly drained ridge area (between 1480 and 1500 m.a.s.l., 1.8 ha) where stilt roots were common, as many as 7.8 trees ha^{-1} went down on average during the investigated period.

Climate

PRECIPITATION

The highest annual rainfall (*P*) (6390 mm) was recorded at the windward site at 1200 m.a.s.l. (Table 21.1). Conversely, the same altitude on the leeward side received only 3690 mm, which ranged

Table 21.2 *Diversity of woody plants in selected plots within the Monteverde transect. See Figure 21.1 for plot locations*

Plot	1	4	5	8
Number of families	21	20	14	20
Number of genera	23	26	18	30
Number of species	33	29	21	33
Shannon index	3.00	3.09	2.69	3.30

closer to the amount recorded at higher elevations (4310 mm at the ridge top). The dry season affected both sides with similar intensity. The driest month was April 2003, during which *P* ranged between 30 mm and 134 mm, with the highest amounts observed at the ridge position. Maximum monthly values of *P* occurred again at 1200 m.a.s.l. on the Atlantic side (880 mm) in December 2003. The present observations for the period until November 2003 are in agreement with those of other workers in the area (Pounds and Crump, 1994; Clark *et al.*, 2000). However, *P* remained extremely high during the subsequent dry season, with values in March 2004 exceeding 700 mm on the Atlantic slope and on the ridge top (about seven times the regional long-term average).

The highest annual input of horizontal precipitation (mist and wind-driven rain, *HP*) was derived for the ridge top (3560 mm; Table 21.1). At 1200 m.a.s.l. on the windward slope, *c.* 10% (350 mm) of the latter value was measured, vs. less than 1% at 1200 m.a.s.l. on the Pacific side (28 mm). *HP* was lowest during the less windy months of the rainy season (i.e. June, September and October; Figure 21.4). In the dry season of 2003 *HP* was important on the Atlantic slope and on the ridge whereas extremely high inputs were recorded in December (800 mm on the ridge top; Figure 21.4). This month is usually characterized by high horizontal inputs (Clark *et al.*, 2000).

As was the case with *HP*, throughfall (*TF*) varied with altitude and exposition (Figure 21.5). On the ridge, annual *TF* was 118% (5100 mm) of regular uncorrected rainfall. On the windward side, annual *TF* (4470 mm) was about 70% of *P* at the lowest climate station. Maximum relative plot-averaged *TF* occurred on the Atlantic slope at 1320 m.a.s.l., reaching up to 221% of *P*, in November 2003. However, individual *TF* collectors on the ridge top captured the highest absolute totals (up to 1700 mm in December). On the lower Pacific slope (1200–1300 m.a.s.l.), *TF* remained below *P* throughout the year (2660 mm or 72% of *P*), reflecting the small inputs of *HP* at this site. The spatial variation among *TF* collectors within a plot was smallest on the Pacific slope and most pronounced on the Atlantic slope and the ridge (Figure 21.5).

TEMPERATURE, RELATIVE HUMIDITY, AND WIND SPEED

Mean annual temperature was 16.2 °C at 1500 m.a.s.l. and declined at an average lapse rate of 0.6 °C 100 m^{-1} on the

Table 21.3 *Epiphytic biomass (dry weight) and diversity in a leeward cloud forest (1450 m.a.s.l.) and in a drier forest at 1200 m.a.s.l. on the Pacific slope at Monteverde*

Plot/ altitude	Epiphyte biomass	Sample tree	DBH	Height	Crown diameter	Tree surface	Epiphyte biomass	Humus	Number of species
	($t\ ha^{-1}$)	(No.)	(cm)	(m)	(m)	(m^2)	(kg)	(kg)	($n\ dm^{-2}$)
5		1	120	24.5	21	117	230 (217)	70	0.18
	39 (32)	2	60	16.3	9.8	49	104 (67)	22	0.35
1,450 m		3	28	11.2	4.2	9	13 (12)	2,5	0.36
8		1	90	32.2	15.5	94	313 (252)	81	0.14
	32 (26)	2	54	23.6	14.5	60	34 (26)	13	0.13
1,200 m		3	23	14.5	4.4	13	2 (1.5)	0,3	0.14

Numbers in brackets indicate values without lianas and hemi-epiphytes.

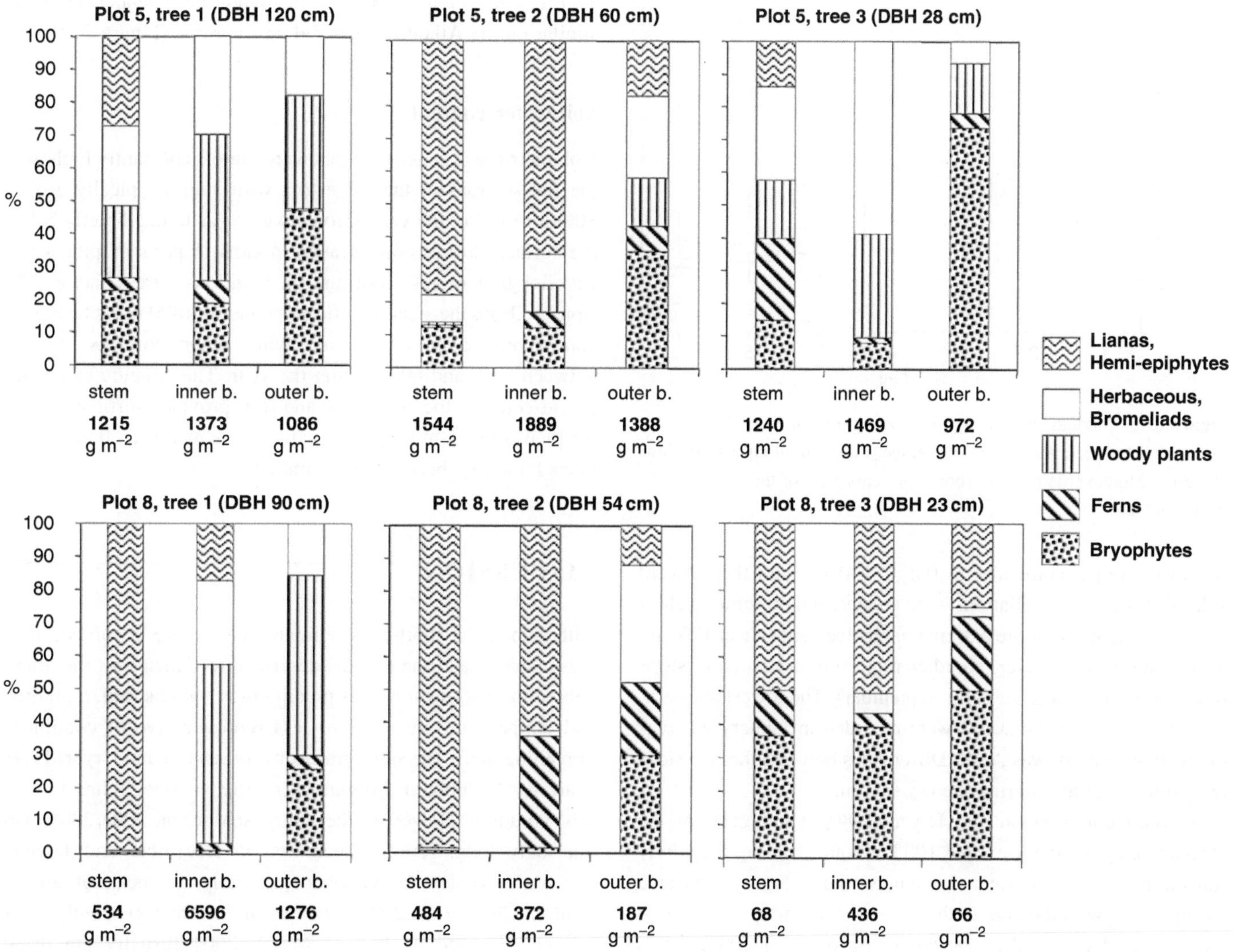

Figure 21.3. Distribution and composition of the epiphytic flora on sampled trees (stem, inner, and outer branches) in leeward forest at 1450 and 1200 m.a.s.l. Data normalized per m^2.

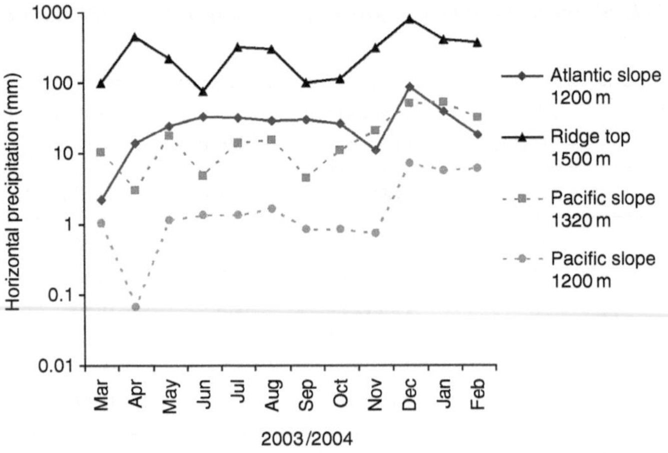

Figure 21.4. Distribution of horizontal precipitation along the Monteverde transect between March 2003 and February 2004.

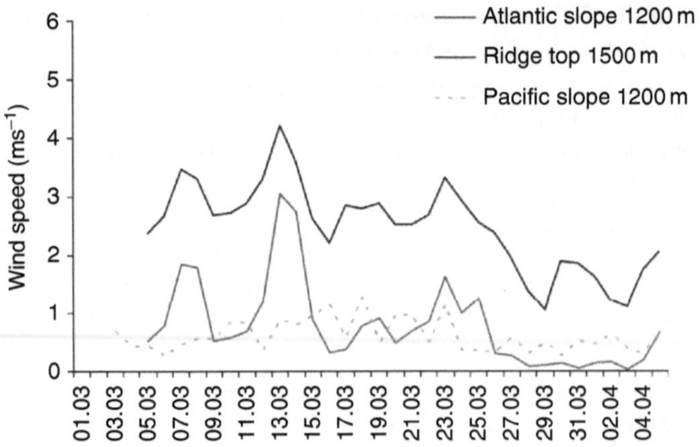

Figure 21.6. Average daily wind speeds above the canopy as measured at plots 1, 4, and 8 in March and April 2004.

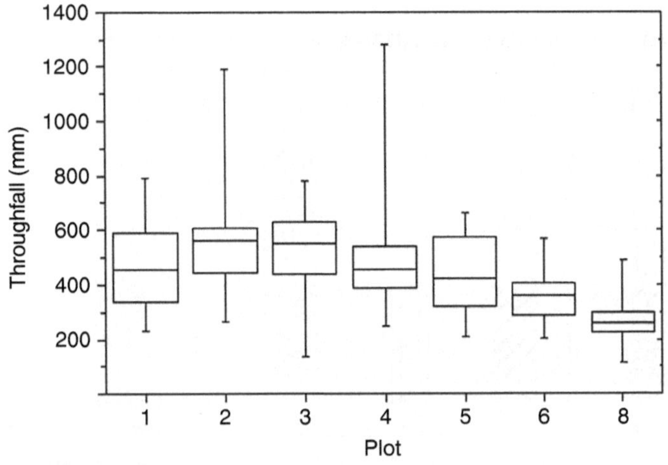

Figure 21.5. Median and 5th and 95th percentiles of monthly throughfall between August and December 2003 for all climate stations along the Monteverde transect (box represents 50% of the measurements).

windward slope, compared to 0.8 °C 100 m^{-1} on the leeward side. Temperature oscillations were greater, and minimum values of relative humidity more extreme on the leeward side at 1200 m. a.s.l., than on the cooler, cloudier ridge top and Atlantic slope throughout the year (except for September). The lowest temperatures (10 °C at 1500 m.a.s.l.) were recorded in January whereas the warmest month was April. Differences between the two sides tended to be greater during the dry season.

According to Lawton and Dryer (1980), wind gusts at the Brillante Gap regularly exceed 100 km hour^{-1} during the windy transitional season between December and February. Extreme storm events with damage to the cloud forest during the investigation period occurred at the beginning of April 2003 and in December 2003/January 2004, in combination with prolonged rainfall. Whilst only slight differences in wind speed were found between the lowest windward and leeward sites (Figure 21.6), daily values between March and June of 2004 were on average 11 times higher at 1500 m.a.s.l. Wind speed tendencies were similar on the Atlantic slope and on the ridge (Figure 21.6).

Soil water content

Volumetric soil water contents were almost constantly highest at the flat sections of the ridge top, with values typically around 80% for most of the year. Groundwater was found directly below the surface during rainy season months in parts of plot 4 and waterlogging might constitute an additional stress factor. The stony soils on the steep Pacific slope below 1400 m.a.s.l. showed consistently drier conditions, with water contents ranging between 16% and 60% during the year. The Atlantic slope was less affected by the dry season and soils provided sufficient water for plant growth at any time, with values measured below 1400 m.a.s.l. ranging between 49% and 81%.

DISCUSSION

Although only short-term climatic data were recorded, it is reasonable to assume that the described tendencies for the gradients of the various forms of precipitation, wind speed, humidity, and temperature are more or less representative of conditions prevailing along the study transect (cf. Lawton and Dryer, 1980; Clark *et al.*, 2000; cf. Frumau *et al.*, this volume; Schmid *et al.*, this volume). However, the rainy season of 2003/2004 was unusually long. Rainfall increased in December and January above the values for October (normally the peak of annual rainfall; Clark *et al.*, 2000), and extreme storm events with heavy rainfall occurred as late as March 2004 (usually the driest month). Rainfall amounts of up to seven times the long-term average for the dry season months, led to an accumulation of destructive events such as downed trees and landsliding on saturated slopes.

The forests in the study area show important differences in structure and composition that occurred in consonance with spatial changes in the physical environment (Table 21.1). When discussing variations in stand structure and composition, the relatively small plot size used in the present study must be considered, especially concerning the taller forests. Lamprecht (1986), for instance, indicated that a minimum area of *c.* 0.5 ha was necessary for the proper evaluation of species richness in a Venezuelan cloud forest. Although an unknown variation per forest type remains, there are clear trends in the presently found data, which allow one to distinguish different forest formations along the transect (Figure 21.2). The most outstanding ecological factors on the ridge top include cooler temperatures, frequent cloud cover and horizontal precipitation at vegetation level, seasonal waterlogging, and, above all, high wind speeds. The importance of frequent disturbance by strong winds at and around the ridge site has been discussed previously by Lawton (1982) and Lawton and Putz (1988). This is reflected by a lower canopy height and basal area compared to forests at lower and more sheltered elevations (Table 21.1), and by a distinctive species composition (cf. Table 21.2). Stand-level stem volume was found to be highest in the relatively flat section of the Peñas Blancas valley at 1200, m.a.s.l., and on the Pacific slope from 1200,–1450, m.a.s.l., where wind stress was much less (cf. Figure 21.6).

The distribution of forest formations may serve as an indicator of climate zones (cf. Hartshorn, 1983). In particular, the diversity and abundance of specialized epiphytic flora closely reflect prevailing atmospheric conditions (notably the frequence and duration of fog) and changes therein (Lugo and Scatena, 1992; Wolf, 1993; cf. Gradstein *et al.*, this volume). Epiphytic life forms can comprise up to 50% of the floristic richness in some Neotropical forests (Zotz and Andrade, 2002). Apart from atmospheric conditions, several factors determine the epiphyte cover of a tree at a given site and exposition, including its growth rate, bark characteristics, and crown architecture, and these make representative sampling for epiphyte biomass estimation difficult (cf. Köhler *et al.*, 2007; Köhler *et al.*, this volume). Nevertheless, clear differences in epiphyte biomass and diversity were demonstrated between the two investigated sites, such as a decrease in biomass and diversity on smaller trees, and the presence of fewer plants on the trunks in the drier forest at 1200 m.a.s.l. (Figure 21.3), where water deficiency may become a limiting factor for epiphytes (cf. Hietz, this volume).

Bryophytes in particular exhibited strong changes in biomass and distribution along the climatic gradient (Figure 21.3), with epiphyte decrease being paralleled by lower inputs of *HP* and increasing and more variable temperatures. According to Gradstein (2000), the growth of cloud forest bryophytes is much more luxuriant in montane wet forests than in forest formations below 1000 m.a.s.l. Nadkarni (1984) reported bryophytes to contribute 40% of the total epiphytic biomass in the elfin forest on the ridge near Brillante Gap, compared to only 6% in a leeward cloud forest between 1480 and 1530 m.a.s.l., 1 km to the west from the Continental Divide in the Monteverde Reserve (Nadkarni and Ingram, 1993). In both studies only canopy trees were sampled. Comparing the cited values to the results for the canopy and subcanopy trees from the present study, it can be concluded that the percentage of bryophytes decreases consistently with altitude and with (horizontal) distance from the Continental Divide in the Monteverde area. At 32 t ha^{-1}, the presently derived total epiphytic biomass at 1450 m.a.s.l. is very similar to the amount found by Nadkarni *et al.* (2004) for a leeward forest at 1480 m.a.s.l. (33.1 t ha^{-1}), compared to a value of 26 t ha^{-1} at 1200 m.a.s.l. on the leeward slope (Table 21.3). For a wet windward LMCF at 1490 m.a.s.l. in the adjacent headwater area of the Caño Negro, Köhler *et al.* (2007) obtained a total epiphytic biomass of 16.2 t ha^{-1}, much of which (>50%) consisted of bryophytes.

Information on epiphyte mass and diversity can be helpful in the prediction of the response of complex forest ecosystems to changes in the physical environment (Lugo and Scatena, 1992; cf. Tobón *et al.*, this volume #26). Climate changes that have begun to affect the cloud forest in Monteverde have already been discussed in relation to global warming and the El Niño effect (Pounds *et al.*, 1999) and to regional deforestation in the Caribbean lowlands (Lawton *et al.* 2001; Ray *et al.*, 2006), both of which tend to lead to less cloud immersion (cf. Foster, this volume; Nair *et al.*, this volume).

ACKNOWLEDGEMENTS

We thank the German Research Foundation (DFG) for financial support, the Tropical Science Center for the friendly cooperation, and Przemyslaw Walotek, William Haber, Julio Calvo, Dagoberto Arias, Conrado Tobón, and Arnoud Frumau for support.

REFERENCES

Bruijnzeel, L. A. (2001). Hydrology of tropical montane cloud forests: a reassessment. *Land Use and Water Resources Research* 1 1.1–1.18. Also available at http://www.luwrr.com.

Clark, K. L., R. O. Lawton, and P. R. Butler (2000). The physical environment. In *Monteverde: Ecology and Conservation of a Tropical Montane Cloud Forest*, eds. N. M. Nadkarni and N. T. Wheelwright, pp. 15–34. New York: Oxford University Press.

Comisión Centroamericana de Ambiente y Desarollo (2002). *Base de Datos SIG de Mesoamerica/CCAD–BM*. Available at www.ccad.ws/mapas/mapoteca.html.

Gradstein, S. R. (2000). Bryophytes. In *Monteverde: Ecology and Conservation of a Tropical Cloud Forest*, eds. N. M. Nadkarni and N. T. Wheelwright, pp. 39–69. New York: Oxford University Press.

Grunow, J. (1963). Weltweite Messungen des Nebelniederschlags nach der Hohenpeissenberger Methode. *International Association of Scientific Hydrology Publication* **65**: 324–342.

Häger, A. (2006). Einfluss von Klima und Topographie auf Struktur, Zusammensetzung und Dynamik eines Tropischen Wolkenwaldes in Monteverde, Costa Rica. Ph.D. thesis, University of Göttingen, Göttingen, Germany. Available at http://webdoc.sub.gwdg.de/diss/2006/haeger/haeger.pdf.

Harrison, S., S. J. Ross, and J. H. Lawton (1992). Beta diversity on geographic gradients in Britain. *Journal of Animal Ecology* **61**: 151–158.

Hartshorn, G. S. (1983). Plants. In *Costa Rican Natural History*, ed. D. H. Janzen, pp. 118–350. Chicago, IL: University of Chicago Press.

Köhler, L., C. Tobón, K. F. A. Frumau, and L. A. Bruijnzeel (2007). Biomass and water storage of epiphytes in old-growth and secondary montane rain forest stands in Costa Rica. *Plant Ecology* **193**: 171–184.

Lamprecht, H. (1986). *Waldbau in den Tropen: Die tropischen Waldökosysteme und ihre nachhaltige Nutzung*. Hamburg, Germany: Paul Parey.

Lawton, R. O. (1982). Wind stress and elfin stature in a montane rain forest tree: an adaptive explanation. *American Journal of Botany* **69**: 1224–1240.

Lawton, R. O., and V. Dryer (1980). The vegetation of the Monteverde Cloud Forest Preserve. *Brenesia* **18**: 101–116.

Lawton, R. O., and F. E. Putz (1988). Natural disturbance and gap-phase regeneration in a wind-exposed tropical cloud forest. *Ecology* **69**: 764–777.

Lawton, R. O., U. S. Nair, R. A. Pielke Sr., and R. M. Welch (2001). Climatic impact of tropical lowland deforestation on nearby montane cloud forests. *Science* **294**: 584–587.

Lugo, A. E., and F. N. Scatena (1992). Epiphytes and climate change in the Caribbean: a proposal. *Selbyana* **13**: 123–130.

Magurran, A. E. (2004). *Measuring Biological Diversity*. Oxford, UK: Blackwell.

Nadkarni, N. M. (1984). Epiphyte biomass and nutrient capital of a neotropical elfin forest. *Biotropica* **16**: 249–256.

Nadkarni, N. M., and S. W. Ingram (1993). Composition and distribution of epiphytic organic matter in a neotropical cloud forest, Costa Rica. *Biotropica* **25**: 370–383.

Nadkarni, N. M, R. O. Lawton, K. L. Clark, T. J. Matelson, and D. Schaefer (2000). Ecosystem ecology and forest dynamics. In *Monteverde: Ecology and Conservation of a Tropical Montane Cloud Forest*. eds. N. M. Nadkarni and N. T. Wheelwright, pp. 303–335. New York: Oxford University Press.

Nadkarni, N. M., D. Schaefer, T. J. Matelson, and R. Solano (2004). Biomass and nutrient pools of canopy and terrestrial components in a primary and a secondary montane cloud forest, Costa Rica. *Forest Ecology and Management* **198**: 223–236.

Pounds, J. A., and M. L. Crump (1994). Amphibian declines and climate disturbance: the case of the Golden Toad and the Harlequin Frog. *Conservation Biology* **8**: 72–85.

Pounds, J. A., M. P. L. Fogden, and J. H. Campbell (1999). Biological response to climate change on a tropical mountain. *Nature* **398**: 611–615.

Ray, D. K., U. S. Nair, R. O. Lawton, R. M. Welch, and R. A. Pielke Sr. (2006). Impact of land use on Costa Rican tropical montane cloud forests: sensitivity of orographic cloud formation to deforestation in the plains. *Journal of Geophysical Research* **111**: D02108, doi: 10.1029/2005JD006096.

Wolf, J. H. D. (1993). Diversity patterns and biomass of epiphytic bryophytes and lichens along an altitudinal gradient in the northern Andes. *Annals of the Missouri Botanical Garden* **80**: 928–960.

Zotz, G., and J. L. Andrade (2002). La ecología y la fisiología de las epífitas y las hemiepífitas. In *Ecología y conservación de bosques neotropicales*, eds. M. R. Guariguata and G. H. Kattan, pp. 271–296. Cartago, Costa Rica: Libro Universitario Regional.

22 Quantitative measures of immersion in cloud and the biogeography of cloud forests

R. O. Lawton and U. S. Nair

University of Alabama in Huntsville, Huntsville, Alabama, USA

D. Ray and A. Regmi

Purdue University, West Lafayette, Indiana, USA

J. A. Pounds

Monteverde Cloud Forest Reserve and Tropical Science Center, San José, Costa Rica

R. M. Welch

University of Alabama in Huntsville, Huntsville, Alabama, USA

ABSTRACT

Sites described as tropical montane cloud forests differ greatly, in part because observers tend to differ in their opinion as to what constitutes "frequent and prolonged" immersion in cloud. This definitional difficulty interferes with hydrological analyses, assessments of environmental impacts on ecosystems, and biogeographical analyses of cloud forest communities and species. Quantitative measurements of cloud immersion can be obtained on site, but the observations are necessarily spatially limited, although well-placed observers can examine 10–50 km of a mountain range under rainless conditions. Regional analyses, however, require observations at a broader scale. This chapter discusses remote sensing and modeling approaches that can provide quantitative measures of the spatio-temporal patterns of cloud cover and cloud immersion in tropical mountain ranges. These approaches integrate remote sensing tools of various spatial resolutions and frequencies of observation, digital elevation models, regional atmospheric models, and ground-based observations to provide measures of cloud cover, cloud base height, and the intersection of cloud and terrain. This combined approach was applied to the Monteverde region of northern Costa Rica to illustrate how the proportion of time the forest is immersed in cloud may vary spatially and temporally. The observed spatial variation was largely due to patterns of airflow over the mountains. The temporal variation reflected the diurnal rise and fall of the orographic cloud base, which was influenced in turn by synoptic weather conditions, the seasonal movement of the Intertropical Convergence Zone and the north-easterly trade winds. Knowledge of the proportion of the time that sites are immersed in clouds should facilitate ecological comparisons and biogeographical analyses, as well as land use planning and hydrological assessments in areas where intensive on-site work is not feasible.

INTRODUCTION

Tropical montane cloud forests (TMCF) may be threatened by changes in climate, particularly if the elevation of the cloud base is raised, be it due to global warming (Pounds *et al.*, 1999; Still *et al.*, 1999; Karmalkar *et al.*, 2008; Foster, this volume), or regional deforestation (Lawton *et al.*, 2001; Nair *et al.*, 2003; Ray *et al.*, 2006; Van der Molen *et al.*, 2006; Nair *et al.*, this volume). As with many montane habitats threatened by changes in climate, this threat has its roots in the insular or archipelago-like patterns of distribution of cloud forest (Vázquez-García, 1995; Bubb *et al.*, 2004; Mulligan, this volume). If the base of the cloud deck providing the immersion in cloud which

Tropical Montane Cloud Forests: Science for Conservation and Management, eds. L. A. Bruijnzeel, F. N. Scatena, and L. S. Hamilton. Published by Cambridge University Press.

characterizes TMCF is raised, then the area occupied by cloud forest must dwindle.

There are both physiological and island biogeographical problems faced by cloud forest organisms under "lifting cloud base" scenarios. At the physiological level it is important to know when the incidence and severity of stressing events begins to influence the performance of individuals and thus the demography of populations and composition of communities. As an alarming example, Pounds *et al.* (1999) have shown that the frequency of prolonged mist-free periods at the leeward margin of the Monteverde cloud forest of Costa Rica's Cordillera de Tilarán has increased in the past 15 years, that the upper margin of the altitudinal range of many bird species has increased on the leeward Pacific slopes, and that amphibian populations in the area have crashed (cf. Pounds *et al.*, 2006). It is equally important to know when island biogeographical effects come into play. The well-known observation that small islands hold fewer species than similar-sized areas on nearby mainland (MacArthur and Wilson, 1967) is likely to also hold for insular patches of TMCF, and the dwindling of their areas under a lifting cloud base scenario might be expected to increase rates of local extinction. Furthermore, if cloud forests decrease in size, the geography of most mountain ranges dictates they must fragment as well, which can only exacerbate problems of extinction and colonization.

However, it is difficult to assess the distribution of TMCF for two primary reasons. Firstly – although TMCF have been defined as forests subjected to "frequent and prolonged" immersion in cloud (Stadtmüller, 1987) – cloud forest sites differ in terms of their cloudiness. Ground-level cloudiness commonly varies continuously, and often dramatically, at a variety of spatial scales, reflecting local topographic influences (Malkus, 1955; Clark *et al.*, 2000), regional patterns of airflow (Van Steenis, 1972; cf. Mulligan, this volume; Jarvis and Mulligan, this volume), and synoptic-scale climatic patterns (Silver *et al.*, 2001). Secondly, people differ in what they consider "frequent and prolonged" immersion in cloud, and consequently in what they consider cloud forest. These problems are easy to notice in any regional or global survey of cloud forests, (e.g. Kappelle and Brown, 2001; Bubb *et al.*, 2004; Mulligan, this volume), and have contributed to the profusion of names given to cloud forests (Stadtmüller, 1987). TMCFs do differ in their degree of "mossiness" (Frahm and Gradstein, 1991) and relative amounts of net precipitation reaching the forest floor (Bruijnzeel, 2001; cf. Bruijnzeel *et al.*, this volume), presumably due to differences in exposure to, and immersion in, clouds. Thus, there is a clear need for quantitative measures of cloudiness that will allow comparison of conditions at different times at a given site, between sites within a mountain range, and between sites in different mountain ranges.

Cloud immersion data are also essential to scale up hydrological and ecological observations from local research sites to the large watersheds and regions that are the responsibility of governmental and private managers of natural resources. They are required to understand the local ranges of species and the biogeographical connections among mountain ranges, and they are needed to develop conservation strategies in tropical montane regions. Important measures of immersion in cloud include absolute or proportional time in cloud, and temporal patterns of immersion – diurnal, seasonal, and those dependent on weather, as well as cloud characteristics such as liquid water content, the distribution of droplet sizes, and the flux of cloud water reaching the vegetation. These can be measured *in situ* (e.g. Eugster *et al.*, 2006; Holwerda *et al.*, 2006; Schmid *et al.*, this volume) but this necessarily provides site-specific values, from which spatial extrapolation can be difficult in the absence of a statistically adequate spatial array of sampling stations (cf. Delay and Giambelluca, this volume). Establishing such an array would be expensive and operationally difficult. For this reason, it is important to develop remote sensing and modeling tools to assess cloud distribution and cloud characteristics, with the aim of providing quantitative regional maps of cloudiness (e.g. Li *et al.*, 2003; Bendix *et al.*, 2004; Mulligan, this volume). This chapter describes various approaches to the quantitative monitoring of both cloud base height and the spatio-temporal extent of cloud cover, with particular reference to the Monteverde region of northern Costa Rica and Central America, and with a focus on the dry season, when orographic clouds induced by the trade winds constitute an important input of water in many of the region's cloud forests (cf. Holder, 2003; Mulligan and Burke, 2005; Rhodes *et al.*, 2006) while at the same time reducing evaporative losses (cf. Schawe *et al.*, this volume).

DETERMINATION OF CLOUD BASE HEIGHTS

Maps of cloud cover are essential, but not sufficient to map the distribution of cloud forest. Cloud base elevations also need to be known. Cloud base elevation is directly observable from the ground, by ceilometers, radiosondes, or local observers. All have their limitations. Ceilometers are commercially available, but expensive. Many airports have them, and data are available from many of these – but (understandably) there are very few airports in areas with TMCF. Radiosondes are commonly used to estimate cloud conditions (Chernyk and Eskridge, 1996; Richardson *et al.*, 2002) but launching sondes under TMCF conditions can be difficult, and an ongoing launch program is quite expensive. Like data from ceilometers, radiosonde data are likely to be of greatest utility in assessing remotely sensed and modeled outputs of cloud conditions.

Of course, clouds can be seen directly on mountains, and from proper observation points local cloud base heights can be estimated by noting known topographic landmarks which are visible or obscured by cloud. A photographic monitoring program has been established to acquire documentary data on cloud

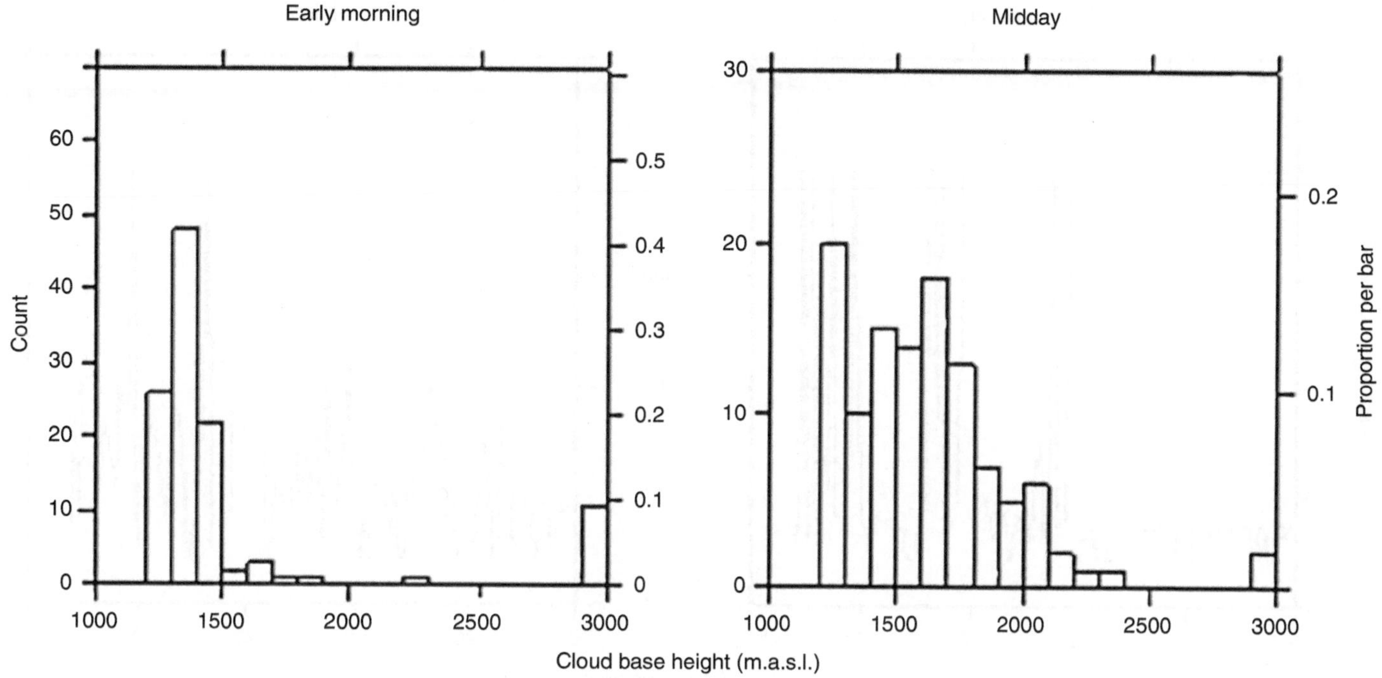

Figure 22.1. Frequency distributions of the cloud base height in the early morning (~06.00 h local time) and near noon along the Continental Divide at the Brillante Saddle in the central Cordillera de Tilarán for the dry season of 2003.

occurrence several times daily at two points along the Continental Divide in the Cordillera de Tilarán near Monteverde, northern Costa Rica. This chapter presents data for the Cerro Famosa site at which 15 landscape features are used to determine cloud base heights. These landscape features range in elevation between 1300 and 1800 m.a.s.l. and are situated at distances of 1–7 km from the photographic observation site at 1540 m.a.s.l. (1.6 km to the lee of the Continental Divide). Differences between the elevation of the cloud base and the elevation of the lowest landscape feature obscured by the cloud increase with distance between the cloud and the obscured feature and with the viewing angle from the horizontal. In the present example, the maximum distance between cloud and obscured feature is about 1 km for features at 1500 to 1600 m elevation and 3.5 km distance, but the largest errors occur with higher viewing angles when cloud bases are low. For example, if the knoll at 1300 m.a.s.l. and 1.2 km distance from the observation point is obscured by a cloud occurring at a distance of 0.6 km then the cloud base is at 1420 m, but if the cloud is 0.9 km distant the cloud base is 1360 m. The magnitude of such observation errors will depend upon the particulars of topography and the cloud bank at any given site. For rapid processing of observation photographs, a decision table was constructed that provides an estimated cloud base height for each feature that might be obscured. The lowest estimated base height is then considered the cloud base for that photograph. All direct observations, of course, are limited to a particular place and time, but turbulence in orographic airflow can produce substantial spatial and temporal variation in the cloud base (>100 m) at scales of 1 km and a few minutes. Therefore, an accuracy of 50–100 m was estimated for the cloud base determinations from the Cerro Famosa monitoring site.

In Costa Rica's Cordillera de Tilarán, photographic observations from the ground reveal that cloud base heights change diurnally and seasonally, and they also differ between sites within the mountain range. For example, in the broad saddle along the Continental Divide at the head of the Peñas Blancas valley (Brillante Gap), dry-season cloud base heights in the early morning were substantially lower, and less variable than those at midday (Figure 22.1). This presumably reflects increased transfer of sensible heat from the ground to the atmosphere as the day progresses, and the development of midday upslope convective winds on both windward and lee slopes. Cloud base heights in the early morning changed little as the dry season progressed, but periods with no clouds overhead below the trade-wind inversion (at ~2800–3000 m.a.s.l.) increased in frequency (Figure 22.2). Midday cloud base heights did increase, however, and became more variable as the dry season progressed (Figure 22.2). In the latter part of the dry season and in the transition to the rainy season (April–May), the interplay between the trade winds and the Intertropical Convergence Zone (ITCZ) involves an alternation of periods of trade-wind dominance and periods of calm days on which convective cells build increasingly well developed clouds (Clark *et al.*, 2000). On such days with well-developed convective cloud developing on upslope winds, forests above

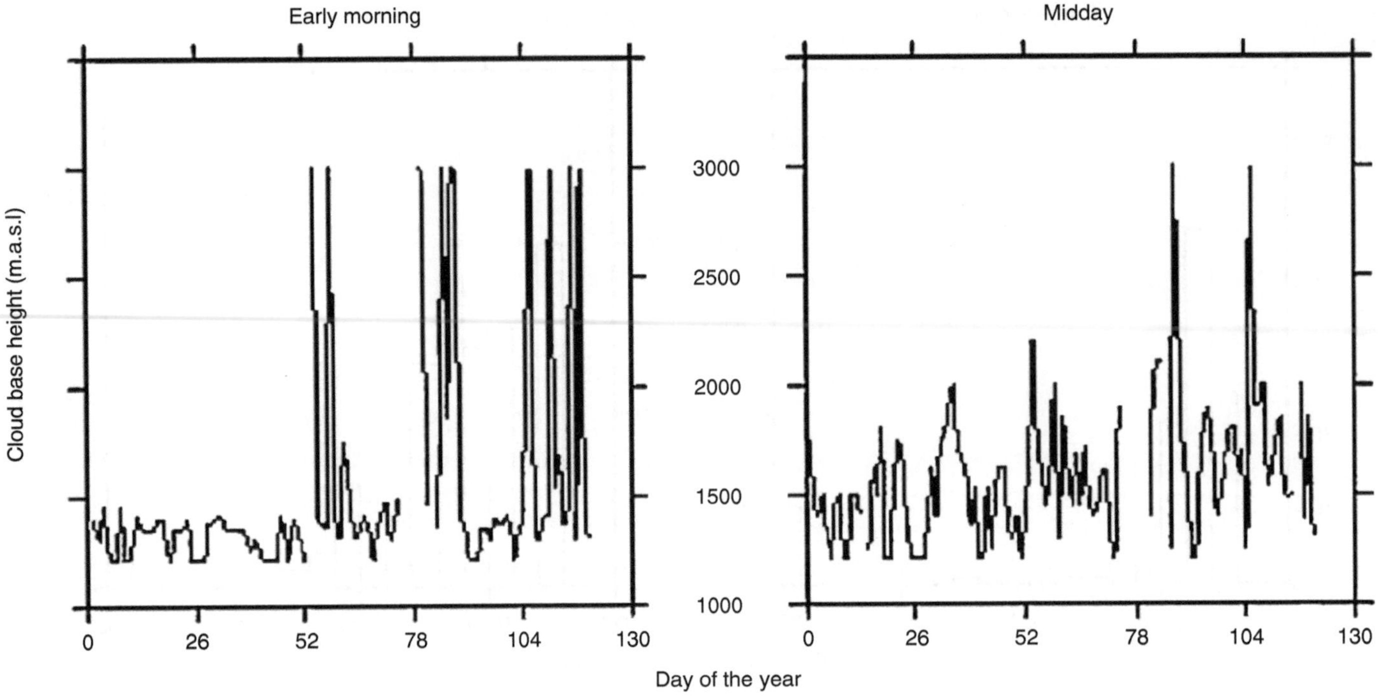

Figure 22.2. Daily cloud base height in the early morning (~06.00 h local time) and near noon along the Continental Divide at the Brillante Saddle in the central Cordillera de Tilarán for the dry season of 2003. Note the increasing frequency of clear early mornings in the latter half of the dry season, and the increase in cloud base height near noon as the dry season progresses.

1400 m.a.s.l. may be immersed in slowly moving convective cloud rather than in trade-wind-induced orographic cloud.

Cloud base heights can also be calculated from some satellite observations. Berendes *et al.* (1992) developed a semi-automated methodology to estimate cumulus base heights over flat surfaces using high spatial resolution Landsat imagery. The approach employs image processing techniques to match cloud edges with their corresponding shadow edges. Cloud base height then is estimated trigonometrically from the separation distance between the corresponding reference points. Accuracies in the order of 100 m are achieved using Landsat Multispectral Scanner (MSS) imagery, whereas using Landsat Thematic Mapper (TM) imagery a cloud base height accuracy of about 50 m can be achieved. This approach allows the calculation of the base heights of fair weather cumulus clouds in areas like the Llanuras de San Carlos upwind of the Monteverde cloud forests. However, this approach is limited to days with moderately abundant fair weather cumulus and by the availability of imagery with sufficient resolution to identify clearly distinctive features on small cumulus clouds. Nevertheless, the data from a limited array of such imagery were in accord with direct observations of cloud base heights against the windward edge of the Cordillera de Tilarán (not shown). The orographic cloud deck itself was observed to tilt up from the windward edge of the Cordillera to the Continental Divide: simultaneous direct and photographic observations showed that cloud bases at the Continental Divide were 100 to 500 m higher than where the orographic cloud deck first hit the Cordillera 20 km away. Observations on several days suggested the tilt to be most pronounced when the cloud deck was lowest.

THE USE OF SATELLITE IMAGERY TO OBSERVE CLOUD COVER

Imagery from a variety of satellite borne sensors is available to monitor changes in cloud cover. Sensors vary in the number of spectral bands available, in spatial resolution, and in the temporal availability of the images. The Geostationary Operational Environmental Satellite 8 (GOES-8) provides data continuously in five spectral channels, one in the visible (0.52–0.72 μm) and four in the infrared (IR) region (3.78–4.03 μm, 6.47–7.02 μm, 10.2–11.2 μm, and 11.5–12.5 μm). Spatial resolution at the center of the image (nadir) for channel 1 is ~1 km, vs. ~4 km for channels 2, 4, and 5, and 8 km for channel 3 (Anonymous, 2007). The visible channel 1 was used for the mapping of cumulus clouds in the present study. Landsat MSS imagery provides data at 57-m resolution in four spectral bands bands between 0.5 and 1.1 μm, while the Landsat TM provides data in seven channels in the visible, near-IR, and IR regions at a spatial resolution of 28.5 m (Rocchio, 2007). Landsat data, however, have a limited temporal availability; the orbits are such that a site is revisited only after 16 or 18 days (depending on the satellite carrying the instrument). In this chapter, both Landsat MSS and TM data are used to determine patterns of deforestation and cloudiness, and to geometrically calculate the base heights of cumulus clouds. Furthermore, the Moderate Resolution Imaging Spectrometer (MODIS) on the Terra and Aqua satellites provides data twice daily (~10.15 and 13.15 h local time) as the satellites pass overhead. MODIS imagery has 36 channels in the visible, near-IR, and IR regions, with spatial resolution varying from 250 m in the visible range to 1 km in the IR (Maccherone, 2007). The 250-m data were used to map the locations of the cumulus cloud fields and the orographic cloud deck at a finer spatial resolution than available from GOES. In addition, the 1-km MODIS data can be used to retrieve the effective radius of droplets in orographic cloud, by the method outlined by Han *et al.* (1994), and to estimate the occurrence of drizzling conditions using various assumptions concerning the distribution of drop sizes. The advantage of MODIS data over Landsat MSS or TM data, is their daily coverage and finer spectral discrimination; the disadvantage is insufficient spatial resolution to determine cloud base heights geometrically. Finally, IKONOS imagery was used, which provides a spatial resolution of 1 m in a panchromatic band, vs. 4 m in three visible bands and one IR band (GeoEye, 2007), to examine how vegetation varies locally in response to gradients in cloudiness. However, IKONOS data are expensive, and thus of limited utility in examining the spatio-temporal distribution of clouds over large areas.

The National Space Science and Technology Center (NSSTC), a NASA/University of Alabama in Huntsville facility, has dual GOES downlinks that archive full spatial resolution data. Daytime GOES-8 imagery over the Central American region is taken at 2-hour intervals. The frequency of total cloud cover and coverage by cumulus cloud fields was derived from the GOES-8 visible band via structural thresholding (Nair *et al.*, 1999). Firstly, the data were calibrated and normalized by the cosine of solar zenith angle to yield the albedo values after which the contrast of the images was enhanced by subtracting the background albedo, i.e. the smallest albedo value observed for each pixel from a time series of 10 consecutive days at the same time of day. This improves the detection of atmospheric boundary layer clouds over land. Next, a so-called "region-growing algorithm" (Levine and Shaheen, 1978; Chou *et al.*, 1994; Nair *et al.*, 1999) was used to identify cloudy pixels. This identifies statistically distinct clusters of pixels, and then assigns pixel membership to clusters. The mean gray levels of the clusters then were used to erect a threshold to distinguish cloudy pixels from the background to give total cloudiness. To examine the development of cumulus cloud fields, from which downwind orographic clouds often coalesce, a cloud number density was determined by segmentation (Kuo *et al.*, 1993), while a cumulus cloud field was identified by the presence of neighboring clouds (Nair *et al.*, 1999). Patterns of cloudiness were summed across a series of images to yield the proportion of time a given pixel was covered by cloud or by cumulus cloud fields.

MOUNTAINS AND THE PATTERN OF CLOUD COVER IN CENTRAL AMERICA

Daily GOES images show the strong dependence of Central American cloud cover patterns on terrain. The topography of the region is shown in Figure 22.3. Figure 22.4, a GOES image taken at midday (18.15 UTC) on 11 March 2002 and cropped to show Costa Rica and southern Nicaragua, shows a large area of congested cumulus clouds over the swampy lowlands near the mouth of the Rio San Juan, while cumulus cloud fields are extensive in south-eastern Nicaragua and northern Costa Rica south of Lake Nicaragua. The orographic cloud deck of the Cordillera de Tilarán appears about halfway between the eastern tip of Lake Nicaragua and the Gulf of Nicoya on Costa Rica's Pacific slope, whereas the orographic cloud decks on the large strato-volcanoes of the Cordillera Central are visible to the east. The Meseta Central, in the lee of the Cordillera Central, and north-western Costa Rica, in the lee of the Cordilleras de Tilarán and Guanacaste are conspicuously free of clouds. The high ridge of the Cordillera de Talamanca, in central eastern Costa Rica, extends above the trade-wind inversion, and is cloud free. The windward, Caribbean slopes of the Talamancas have few orographic clouds, but on the leeward side of the Talamancas and the coastal range toward the south-west, upslope winds, which are features of large-scale turbulence produced in the trade-wind flow by the large mountain range, do produce an extensive set of cumulus clouds (Figure 22.4).

In regional maps of dry-season cloud cover over Central America (Figure 22.5), both seasonal variation and terrain effects are apparent. In general, cloud cover declines from the early dry season (January and February) to the late dry season (March and April). In the late dry season, when the north-easterly trade winds most thoroughly dominate the regional climate, areas covered by cloud at midday on >50% of the days are restricted to mountain ranges forcing the establishment of orographic cloud decks. The core area of very cloudy conditions lies in southern Central America in an arc from north-western Costa Rica (Cordillera de Tilarán) to central Panamá (eastern Serranía de Tabasará), along the windward slope of a series of cordilleras in which only a few passes are lower than 1500 m.a.s.l. Most of this strip experiences cloud cover near noon on >60% of the days, while scattered sites are covered on >80% of the days. There are no sites in Nicaragua with this degree of cloud cover in the late dry season, and in Honduras only the Parque Nacional Agalta,

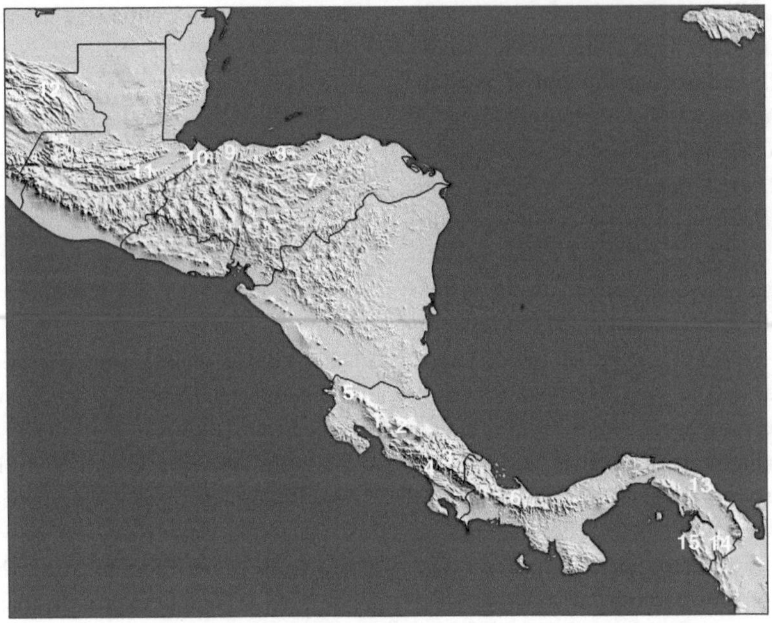

Figure 22.3. Topography of Central America, based on a digital elevation model with 1-km pixels. Numbers indicate mountains mentioned in the text: (1) Cordillera de Tilarán, (2) Cordillera Central, (3) Cordillera de Talamanca, (4) Fila Costeña, (5) Cordillera de Guanacaste, (6) Serranía de Tabasará, (7) Cerro Agalta, (8) Pico Bonito, (9) Cerro Texiguat, (10) Cerro Cusuco, (11) Sierra de las Minas, (12) Chiapan Highlands, (13) Serranía del Darien, (14) Cerro Pirré, and (15) Cerro Sapo.

Figure 22.4. GOES satellite image cropped to include Costa Rica and southern Nicaragua from 18.15 UTC (~12.15 h local time) on 11 March 2002. Note the extensive cumulus cloud fields (1) in southern Nicaragua and north-western Costa Rica, the orographic cloud banks on the Caribbean (windward) slopes of the Cordillera de Tilarán (2) and the Cordillera Central (3), and the clouds on the Pacific slope of the Fila Costeña (4) and the Cordillera de Talamanca. The main ridge of the Talamancas above ~2800 m.a.s.l. (5) rises above the trade-wind inversion and is free of cloud. The Meseta Central (6) and the Guanacaste lowlands (7) west of the Cordillera de Tilarán are conspicuously cloud-free.

P. N. Pico Bonito, the Refugio Vida Silvestre Rexiguat, and P. N. Cusuco, all with peaks above 2200 m.a.s.l., have extensive areas covered with midday cloud on more than 60% of late dry-season days. In Guatemala, the Sierra de las Minas and the ridges to the north have areas covered by cloud on more than 60% of late dry-season days, but the most extensive very cloudy areas occur along the string of large volcanoes parallel to the Pacific coast extending from Lake Atitlan into Mexico (Figure 22.5). During March and April few Mexican montane forests experience the degree of cloud cover seen in southern Central America. A few scattered sites on the northern margin of the Chiapan highlands are covered by cloud near noon on >60% of the days.

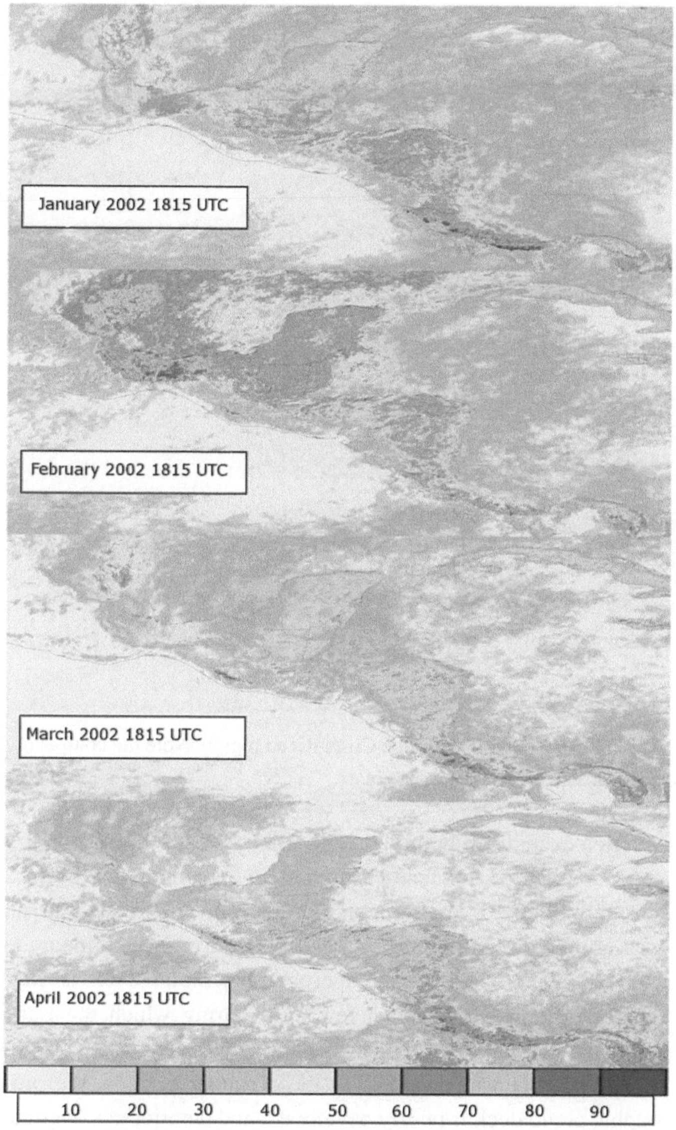

Figure 22.5. Proportion of the time covered by cloud at 18.15 UTC (~12.15 h local time) during the dry season months (January–April) of 2002 for the Central American region. Note that regions of trade-wind orography stand out as local maxima. (See also color plate.)

The Tuxtlas on the Veracruz coast have similar cloud cover, as does the Cofre de Perote area in Central Veracruz. The high ridges of the Sierra de Juarez of northern Oaxaca and the Sierra de Oaxaca along the Pacific coast are covered by cloud around midday on 40–60% of the days in the late dry season. Clearly, montane forests in northern Central America and Mexico are exposed to drier conditions in the late dry season than the montane forests in Costa Rica and Panamá.

The general distribution of cloud forests in Costa Rica and Panamá is well known (e.g. Myers, 1969; Kappelle, 1996; Nadkarni and Wheelwright, 2000), but a map of cloud cover frequency shows remarkable detail (Figure 22.6; see also Mulligan and Burke (2005) for similar patterns of modeled fog incidence). For example,

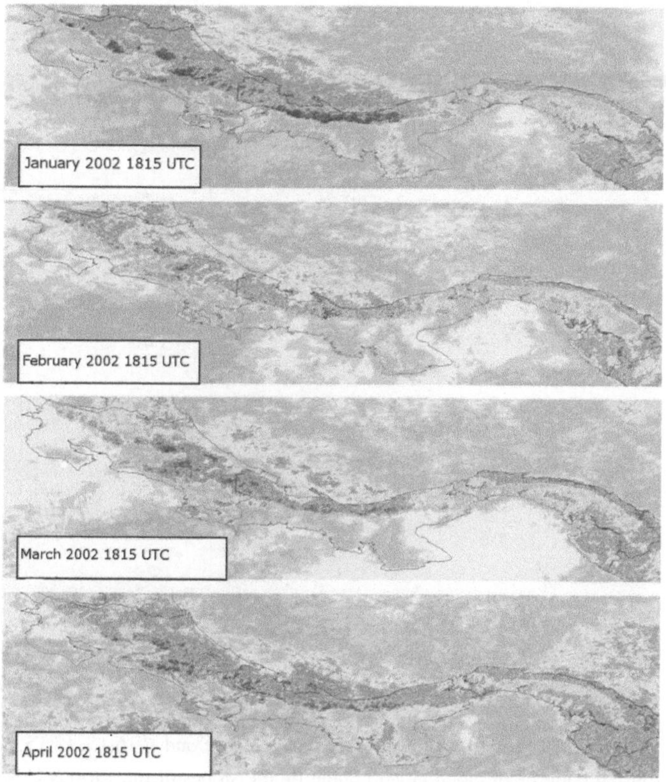

Figure 22.6. Proportion of the time covered by cloud at 18.15 UTC (~12.15 h local time) during the dry season months (January–April) of 2002 for Costa Rica and Panamá. (See also color plate.)

south of Lake Nicaragua, the isolated volcanoes of the Cordillera de Guanacaste (Volcan Cacao, V. Rincon del la Vieja, V. Miravalles, and V. Tenorio) in northern Costa Rica each support areas with midday cloud cover on >70 or 80% of late dry-season days. Also, within the long arc of cloudy mountains from the Cordillera de Tilarán to the Serrania de Tabasará of Panama, there are extremely cloudy patches separated by somewhat less cloudy regions. This variation must influence the abundance of epiphytic growth, particularly of bryophytes, but cloud impact on plant (and animal) distribution remains to be explored in detail (cf. Häger and Dohrenbusch, this volume; Köhler *et al.*, this volume). In eastern Panamá, the cloud forest sites noted by Myers (1969) have conspicuously high dry-season cloud cover. A long but narrow arc of cloudy conditions lies along the Serranía del Darien from Cerro Jefe in the west to Cerro Tacarcuna on the Colombian border. This serranía consists of a low ridge, but is situated directly along the Caribbean coast, where cloud bases should be low (Sugden, 1982a,b; Cavelier, 1988). Cloudy conditions are also marked on Cerro Pirre, Cerro Sapo, and along the Jaque–Imamadó divide in south-easternmost Panamá. The Serranía de Cañazas, to the lee of the Serranía del Darien, and the mountains near the southern end of the Azuero Peninsula are, however, clearly less cloudy at midday in the dry season (Figure 22.6).

Diurnal patterns of dry-season cloud development may differ between nearby mountain ranges, as illustrated by the mountains

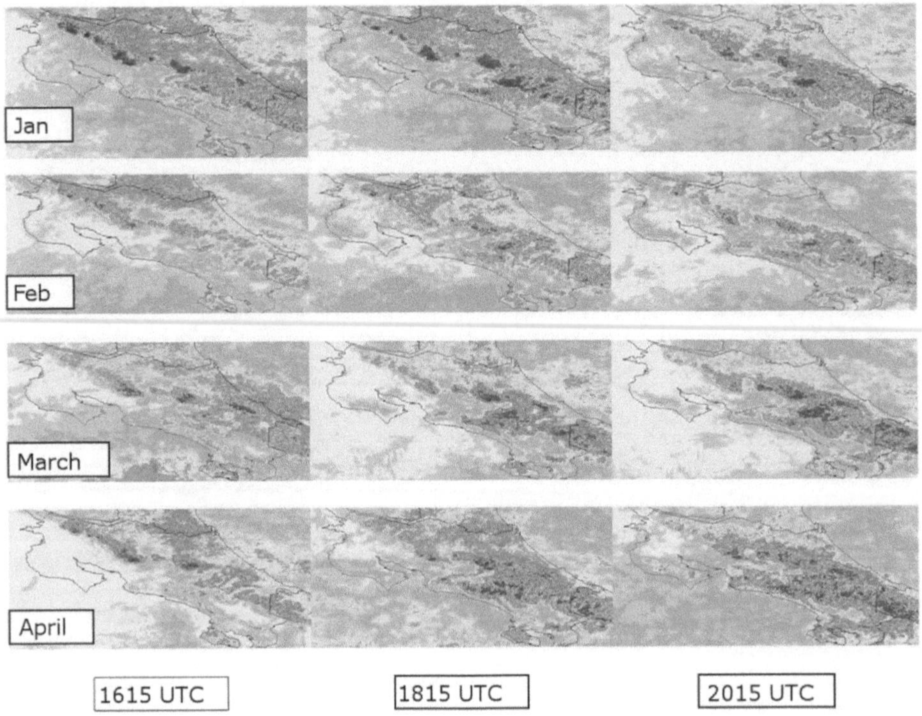

Figure 22.7. Proportions of the time covered by cloud at 16.15, 18.15, and 20.15 UTC for March 2003 in the Costa Rican region. Note the contrasting diurnal patterns of cloud development on the slopes of the various Costa Rican cordilleras. (See also color plate.)

of Costa Rica (Figure 22.7). Cloud cover increases in frequency over the course of the day, particularly in the late dry season, in the high Talamancas in central and southern Costa Rica, the main ridge of which lies between 2600 and 3800 m.a.s.l., with the higher peaks extending above the trade-wind inversion. In contrast, cloud cover changes much less during the day in the northern parts of the country in the lower (<1800 m.a.s.l.) and smaller (~50 km long) Cordillera de Tilarán, or on the volcanoes of the Cordillera de Guanacaste. In addition, this difference in diurnal development is more marked in the late dry season than in January, when frontal systems from North America more commonly influence Central American weather (Clark *et al.*, 2000). The Talamancas form a significant barrier to the trade winds, which are in part diverted through the lowland gap between northern Costa Rica and southern Nicaragua. The isolated volcanoes of the Cordillera de Guanacaste stand within this lowland gap and the Cordillera de Tilarán stands on its southern margin (cf. Figure 22.3). As such, cloud formation on these ranges is more strongly influenced by trade-wind orography, while cloud formation in the Talamancas depends more on local, convectively driven, upslope winds, with their typical diurnal course of development (Figure 22.7).

An examination of seasonal and diurnal trends in cloud cover frequency in the Cordillera de Tilarán shows that diurnal trends change as the dry season progresses (Figure 22.8). In January, cloud frequency declines little from mid-morning to mid-afternoon as during this time of year cold fronts from North America exert a strong influence (Clark *et al.*, 2000). In February and March, cloud frequency is high in mid-morning, but declines substantially by mid-afternoon. This decline is related to an oscillation between periods of a few days during which the trade winds blow strongly, and the orographic cloud deck is well developed, and periods of a few days during which the trade winds tend to stall over the course of the day, and local convection develops upslope winds. In these calmer periods, the orographic cloud deck is poorly developed, and deteriorates over the course of the day, although small cumulus clouds often develop on the updrafts provided by the upslope winds. In April, however, during similarly calm periods, the local convective cells become vigorous enough to develop substantial mid-afternoon cloud cover. This produces a diurnal pattern that has frequent mid-morning cloud cover produced by trade-wind orography, followed by a midday depression in cloud frequency due to stalling of the trade winds as local upslope winds develop, and a mid-afternoon increase in cloud frequency due to cumulus formation on the upslope winds (Figure 22.8). The dry season terminates in May, as the ITCZ moves northward and the upslope winds develop local thunderstorms.

Within the Cordillera de Tilarán, cloud frequency drops dramatically from the windward Atlantic side of the Continental Divide to the leeward Pacific side (cf. Figure 22.8). Since the regional airflow is dominated by the trade winds, this abrupt margin to cloudy conditions is produced by evaporation of the orographic clouds in the adiabatically heating air as the trade

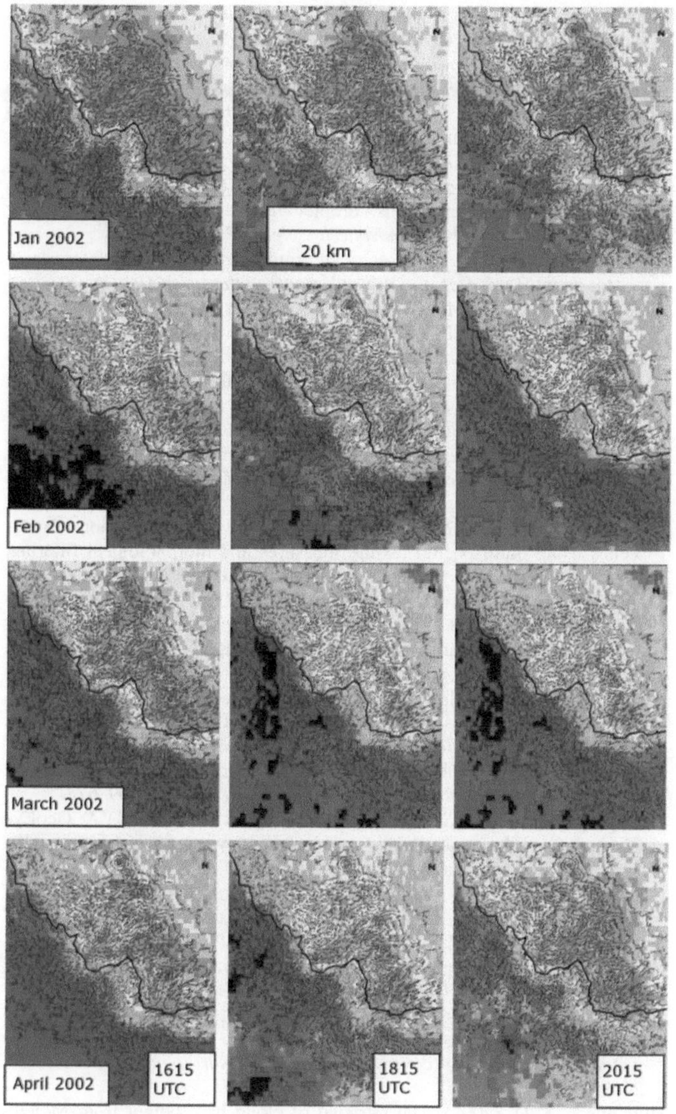

Figure 22.8. Proportion of time covered by cloud at 16.15, 18.15, and 20.15 UTC for the dry-season months (January–April) of 2002 for the central Cordillera de Tilarán. 100-m contour intervals and the course of the Continental Divide are shown. (See also color plate.)

winds descend along the leeward slope. Cloud frequency declines more gradually with decreasing elevation on the windward flank of the Tilarán range, and at the ends of the Range. This alone might make the transition from cloud forest to other forest types more gradual on the windward slope than on the leeward slope, but the issue is complicated by patterns of precipitation (Lawton and Dryer, 1980; Mulligan and Burke, 2005; cf. Häger and Dohrenbusch, this volume). Elevations below ~1300 m.a.s.l. on the windward slope are usually below the orographic cloud deck, but do not appear to be much (if any) drier than the forest immersed in cloud above. This is because light rains and drizzle from the orographic clouds above are common during the dry season (cf. Häger and Dohrenbusch, this volume) while insolation is dampened by the presence of the cloud deck as well (Lawton, 1990; cf. Schawe *et al.*, this volume).

These satellite observations of cloud cover frequency in the Cordillera de Tilarán can be readily combined with the cloud base height distributions measured *in situ*, if it is assumed that cloud base heights and cloud cover frequencies are independently distributed and the cloud deck is level. Multiplying the probability of base heights being below a given elevation times the frequency of cloud cover suggests that sites above 1600 m.a.s.l. on the windward slope are immersed in cloud about half of the time (42–56%) at midday during the middle of the dry season. This figure drops rapidly as elevations decline. Sites between 1500 and 1600 m.a.s.l. on the windward slope are immersed in cloud a little less than one-third of the time (24–32%), while sites between 1400 and 1500 m.a.s.l. are immersed 15–20% of the time. The observed tilt of the cloud base, however, means that the probability of immersion for a site near the windward margin is greater than for a site at the same elevation nearer the Continental Divide. Cloud base height is probably correlated with cloud cover, so that more accurate estimates of immersion will require the relationship between cloud base height and cloud cover to be clarified.

FUTURE RESEARCH

The most critical need is for information about cloud base height at a broad spatial scale and this requires more than ground-based observations at isolated points. There are two promising approaches to this problem: (i) the development of additional remote-sensing tools, and (ii) the refined use of regional atmospheric modeling. For example, MODIS data can deliver a cloud-top pressure measurement at 5-km resolution (Menzel *et al.*, 2002). With information on the atmospheric profile, cloud top height can be calculated. Then, if cloud thickness is known, the height of the cloud base can be calculated (Welch *et al.*, 2008). Three techniques for estimating cloud thickness may be considered: one which uses the relationship between cloud optical depth – available as a MODIS-derived product at 1-km resolution – and cloud thickness (Minnis *et al.*, 1997); another which uses effective cloud droplet size, also available from MODIS data, and assumed values of cloud liquid water content (Hutchinson, 2002); and a third which uses cloud optical depth and the assumption of moist adiabatic water content (Brenguier *et al.*, 2000).

Furthermore, as an alternative to direct observation, cloud bases may be estimated from the lifting condensation level (LCL) as derived from observationally constrained climate model outputs – reanalysis data, e.g. the model of the National

Center for Environmental Prediction and the National Center for Atmospheric Research (Kalnay *et al.*, 1996; Mulligan and Burke, 2005), or as simulated in regional atmospheric models (Nair *et al.*, 2008; see below). Reanalysis data are generated on a coarse scale (~2 degrees for the NCEP/NCAR model), but since many cloud forests are found in coastal or near-coastal mountain ranges (e.g. the cordilleras of Central America, the Caribbean, or the Indonesian-Melanesian region; Scatena *et al.*, this volume), reanalysis data might provide reliable and fairly accurate estimates of the LCL in the atmosphere moving onshore from upwind seas. Furthermore, there are several decades of reanalysis data available. However, the LCL is highly variable in space, influenced by a variety of factors (including land cover) that alter surface energy balances and thus the fluxes of heat and water from the surface to the lower atmosphere. As a result, the height of cloud bases changes quickly as air moves onshore, and moves over heterogeneous terrain.

To examine this kind of terrain- and land-cover effects on the development of clouds the Colorado State University Regional Atmospheric Modeling System (RAMS), a non-hydrostatic model useful for simulating atmospheric phenomena from the level of individual clouds to the meso-scale (Pielke *et al.*, 1992) may be used. RAMS uses finite difference methods for solving the various conservation equations governing the atmospheric flow. At the lower boundary, RAMS uses a multi-layer soil model (Tremback and Kessler, 1985) and a vegetation model (Avissar and Pielke, 1989) to represent the various land surface processes. RAMS has been used to simulate the impact of land cover change on atmospheric conditions, including cloud formation in Puerto Rico (Van der Molen *et al.*, 2006; Van der Molen *et al.*, this volume), and to simulate effects of sea surface temperature change on the LCL (Foster, this volume). RAMS has also been employed to simulate the impact of lowland forest conversion to pasture on cumulus cloud field characteristics upwind of adjacent mountains, and the base height of orographic clouds in northern Costa Rica (Lawton *et al.*, 2001; Nair *et al.*, 2003; Ray *et al.*, 2006; Nair *et al.*, 2008; cf. Nair *et al.*, this volume). Although model predictions are dependent on the parameterization of inputs, only such an approach allows changes in land uses or past and future climatic conditions to be evaluated in terms of their impact on cloud formation and cloud forest distribution and function.

REFERENCES

Anonymous (2007). *GOES Imager Tutorial*. Niwot, CO: Regional and Mesoscale Meteorology Branch, Cooperative Institute for Research in the Atmosphere, Colorado State University. Also available at http://rammb.cira.colostate.edu/training/goes_8/table_of_contents.asp.

Avissar, R., and R.A. Pielke (1989). A parameterization of heterogeneous land surfaces for atmospheric models and its impact on regional meteorology. *Monthly Weather Review* **117**: 2113–2136.

Bendix, J., R. Rollenbeck, and W.E. Palacios (2004). Cloud detection in the tropics: a suitable tool for climate ecological studies in the high mountains of Ecuador. *International Journal of Remote Sensing* **25**: 4521–4540.

Berendes, T., S.K. Sengupta, R.M. Welch, B.A. Wielicki, and M. Navar (1992). Cumulus cloud base height estimation from high spatial resolution Landsat data: a Hough transform approach. *IEEE Transactions in Geosciences and Remote Sensing* **30**: 430–443.

Brenguier, J.L., H. Pawlowska, L. Schuller, *et al.* (2000). Radiative properties of boundary layer clouds: droplet effective radius versus number concentration. *Journal of Atmospheric Science* **57**: 803–821.

Bruijnzeel, L.A. (2001). Hydrology of tropical montane cloud forests: a reassessment. *Land Use and Water Resources Research* **1**: 1–18.

Bubb, P., I. May, L. Miles, and J. Sayer (2004). *Cloud Forest Agenda.* Cambridge, UK: UNEP–WCMC. Also available at: http://sea.unep-wcmc.org/forest/cloudforest/index.cfm.

Cavelier, J. (1988). The ecology of elfin cloud forests in Northern South America. Dissertation submitted for the annual research fellowship competition, Trinity College, University of Cambridge, Cambridge, UK.

Chernykh, I.V., and R.E. Eskridge (1996). Determination of cloud amount and level from radiosonde soundings. *Journal of Applied Meteorology* **35**: 1362–1369.

Chou, J., R.C. Weger, J.M. Ligtenberg, *et al.* (1994). Segmentation of polar scenes using multispectral texture measures and morphological filtering. *International Journal of Remote Sensing* **15**: 1019–1036.

Clark, K.L., R.O. Lawton, and P.R. Butler (2000). The physical environment. In *Monteverde: Ecology and Conservation of a Tropical Cloud Forest*, eds. N. Nadkarni and N. Wheelwright, pp. 15–38. New York: Oxford University Press.

Eugster, W., R. Burkard, F. Holwerda, F.N. Scatena, and L.A. Bruijnzeel (2006). Characteristics of fog and fog-water fluxes in a Puerto Rican elfin cloud forest. *Agricultural and Forest Meteorology* **139**: 288–306.

Frahm, J.-P., and S.R. Gradstein (1991). An altitudinal zonation of the tropical rain forest using bryophytes. *Journal of Biogeography* **18**: 669–678.

GeoEye (2007). *IKONOS Product Guide*, Version 17. Dulles, VA: GeoEye Corporation. Also available at http;//geoeye.com/whitepapers_pdfs/GeoEye_IKONOS_Products_Guide_v17.pdf.

Han, Q., W.B. Rossow, and A.A. Lacis (1994). Near-global survey of effective droplet radii. I. Liquid water clouds using ISCCP data. *Journal of Climate* **7**: 465–497.

Holder, C.D. (2003). Fog precipitation in the Sierra de las Minas Biosphere Reserve, Guatemala. *Hydrological Processes* **17**: 2001–2010.

Holwerda, F., R. Burkard, W.E. Eugster, *et al.* (2006). Estimating fog deposition at a Puerto Rican elfin cloud forest site: comparison of the water budget and eddy covariance methods. *Hydrological Processes* **20**: 2669–2692.

Hutchinson, K.D. (2002). The retrieval of cloud base height from MODIS and three-dimensional cloud fields from NASA's EOS Aqua mission. *International Journal of Remote Sensing* **23**: 5249–5265.

Hutchinson, K.D., and A. Huang (2001). *Cloud Base Height Visible Infrared Imager/Radiometer Suite Algorithm Theoretical Basis Document*, v. 4. Lanham, MD: Raytheon Systems Co.

Kalnay, E., M. Kanamatsu, R. Kistler, *et al.* (1996). The NCEP/ NCAR 40-year reanalysis project. *Bulletin of the American Meteorological Society* **77**: 437–471.

Kappelle, M. (1996). *Los bosques de roble (Quercus) de la Cordillera de Talamanca, Costa Rica*. Heredia, Costa Rica: INBio.

Kappelle, M., and A.D. Brown (eds.) (2001). *Bosques nublados del neotropico.* Santo Domingo de Heredia, Costa Rica: INBio.

Karmalkar, A.V., R.S. Bradley, and H.F. Diaz (2008). Climate change scenario for Costa Rican montane forests. *Geophysical Research Letters* **25**, L11702, doi: 10.1029/2008GL033940.

Kuo, K.S., R.M. Welch, and R.C. Weger (1993). The three-dimensional structure of cumulus clouds over the ocean. I. Structural analysis. *Journal of Geophysical Research* **98**: 20 685–20 711.

Lawton, R.O. (1990). Canopy gaps and light penetration into a wind-exposed tropical lower montane rain forest. *Canadian Journal of Forest Research* **20**: 659–667.

Lawton, R.O., and V. Dryer (1980). The vegetation of the Monteverde Cloud Forest Preserve. *Brenesia* **18**: 101–116.

Lawton, R.O., U.S. Nair, R.A. Pielke Sr., and R.M. Welch (2001). Climatic impact of tropical lowland deforestation on nearby montane cloud forests. *Science* **294**: 584–587.

Levine, M.D., and S.I. Shaheen (1978). A modular computer vision system for picture segmentation and interpretation. *IEEE Transactions on Pattern Analysis and Machine Intelligence*, **3**: 540–556.

MacArthur, R.H., and E.O. Wilson (1967). *The Theory of Island Biogeography*. Princeton, NJ: Princeton University Press.

Maccharone, B. (2007). *MODIS Web. Data*. http://modis.gsfc.nasa.gov/data/

Malkus, J.S. (1955). The effects of a large island upon the trade-wind air stream. *Quarterly Journal of the Royal Meteorological Society* **81**: 538–550.

Menzel, W.P., B.A. Baum, K.I. Strabala, and R.A. Frey (2002). *Cloud Top Properties and Cloud Phase Algorithm Theoretical Basis Document*, MODIS Algorithm Theoretical Basis Document No. ATBD-MOD-04. NASA. Available at http://modis.gsfc.nasa.gov/data/atbd/atbd_mod04.pdf.

Minnis, P., D.F. Young, D.P. Kratz, *et al.* (1997). *Clouds and the Earth's Radiant Energy System (CERES) Algorithm Theoretical Basis Document: Cloud Optical Property Retrieval*, Subsystem 4.3. Pasadena, CA: NASA. Available at http://asd-www.larc.nasa.gov/ATBD/pdf_docs/r2_2/ceres-atbd2.2-s4.3.pdf.

Mulligan, M., and S.M. Burke (2005). *FIESTA: Fog Interception for the Enhancement of Streamflow in Tropical Areas*, Final Technical Report for AMBIOTEK contribution to DfID-FRP project R7991. Leigh-on-Sea, UK: AMBIOTEK. Also available at www.ambiotek.com/fiesta.

Myers, C.W. (1969). The ecological geography of cloud forest in Panamá. *American Museum Novitates* **396**: 1–52.

Nadkarni, N., and N. Wheelwright (eds.) (2000). *Monteverde: Ecology and Conservation of a Tropical Cloud Forest*. New York: Oxford University Press.

Nair, U.S., J.A. Rushing, R. Ramachandran, *et al.* (1999). Detection of cumulus cloud fields in satellite imagery. *Earth Observing Systems IV, Proceedings of SPIE* **3**(750): 345–355.

Nair, U.S., R.O. Lawton, R.M. Welch, and R.A. Pielke Sr. (2003). Impact of land use on tropical montane cloud forests: sensitivity of cumulus cloud field characteristics to lowland deforestation. *Journal of Geophysical Research* **108**: 4206–4218.

Nair, U.S., S. Asefi, R.M. Welch, *et al.* (2008). Biogeography of tropical montane cloud forests. II. Mapping of orographic cloud immersion. *Journal of Applied Meteorology and Climatology* **47**: 2183–2197.

Pielke, R.A., W.R. Cotton, R.L. Walko, *et al.* (1992). A comprehensive meteorological modeling system: RAMS. *Meteorological and Atmospheric Physics* **49**: 69–91.

Pounds, J.A., M.P.L. Fogden, and J.H. Campbell (1999). Biological response to climate change on tropical mountain. *Nature* **389**: 611–614.

Pounds, J.A., M. Bustamante, L.A. Coloma, *et al.* (2006). Widespread amphibian extinctions from epidemic disease driven by global warming. *Nature* **439**: 161–167.

Ray, D.K., U.S. Nair, R.O. Lawton, R.M. Welch, and R.A. Pielke Sr. (2006). Impact of land use on Costa Rican tropical montane cloud forests: sensitivity of orographic cloud formation to deforestation in the plains. *J. Geophysical Research* **111**: D02108, doi: 10.1029/2005JD006096.

Rhodes, A.L., A.J. Guswa, and S.E. Newell (2006). Seasonal variation in the stable isotopic composition of precipitation in the tropical montane forests of Monteverde, Costa Rica. *Water Resources Research* **42**, W11402, doi:10.1029/2005WR004535.

Richardson, A.D., E.G. Denny, T.G. Siccama, and X. Lee (2002). Evidence for a rising cloud ceiling in eastern North America. *Journal of Climate* **16**: 2093–2098.

Rocchio, L. (2007). *The Landsat Program*. Available at http://landsat.gsfc.nasa.gov.

Silver, W.L., E. Marin-Spiotta, and A.E. Lugo (2001). El Caribe. In *Bosques nublados del neotropico*, eds. M. Kappelle and A.D. Brown, pp. 155–181. Santo Domingo de Heredia, Costa Rica: INBio.

Stadtmüller, T. (1987). *Cloud Forests in the Humid Tropics: A Bibliographic Review*. Tokyo: The United Nations University, and Turrialba, Costa Rica: Centro Agronomico Tropical de Investigacion y Ensenanza.

Still, C.J., P.N. Foster, and S.H. Schneider (1999). Simulating the effects of climate change on tropical montane cloud forests. *Nature* **389**: 608–610.

Sugden, A. (1982a). The vegetation of the Serranía de Macuira, Guajira, Colombia: a contrast of arid lowlands and an isolated cloud forest. *Journal of the Arnold Arboretum* **63**: 1–30.

Sugden, A. (1982b). The ecological, geographic and taxonomic relationships of an isolated Colombian cloud forest, with implications for island biogeography. *Journal of the Arnold Arboretum* **63**: 31–61.

Tremback, C.J., and R. Kessler (1985). A surface temperature and moisture parameterization for use in mesoscale numerical models. In *Proceedings of 7th AMS Conference on Numerical Weather Prediction*, June 17–20, Montreal, Quebec, Canada, pp. 355–358. Boston, MA: American Meteorological Society.

Van der Molen, M.K., A.J. Dolman, M.J. Waterloo, and L.A. Bruijnzeel (2006). Climate is affected more by maritime than by continental land use change: a multiple-scale analysis. *Global and Planetary Change* **54**: 128–149.

Van Steenis, C.G.G.J. (1972). *The Mountain Flora of Java*. Leiden, the Netherlands: E.J. Brill.

Vázquez-García, J.A. (1995). Cloud forest archipelagos: preservation of fragmented montane ecosystems in tropical America. In *Tropical Montane Cloud Forests*, eds. L.S. Hamilton, J.O. Juvik, and F.N. Scatena, pp. 314–332. New York: Springer-Verlag.

Welch, R.M., S. Asefi, J. Zeng, *et al.* (2008). Biogeography of tropical montane cloud forests. I. Remote sensing of cloud base heights. *Journal of Applied Meterology and Climatology* **47**: 960–975.

23 Understanding the role of fog in forest hydrology: stable isotopes as tools for determining input and partitioning of cloud water in montane forests

M. Scholl
US Geological Survey, Reston, Virginia, USA

W. Eugster
Swiss Federal Institute of Technology ETH, Zürich, Switzerland

R. Burkard
University of Bern, Bern, Switzerland

ABSTRACT

Understanding the hydrology of tropical montane cloud forests (TMCF) has become essential as deforestation of mountain areas proceeds at an increased rate worldwide. Passive and active cloud water collectors, throughfall and stemflow collectors, visibility or droplet size measurements, and micrometeorological sensors are typically used to measure fog water inputs to ecosystems. In addition, stable isotopes may be used as a natural tracer for fog and rain. Previous studies have shown that the isotopic signature of fog tends to be more enriched in the heavier isotopes ^{2}H and ^{18}O than that of rain, due to differences in condensation temperature and history. Differences between fog and rain isotopes are largest for synoptic-scale rain storms vs. local fogs or orographic clouds. Isotopic differences have also been observed between locally generated rain and fog on mountains with orographic clouds, but only a few studies have been conducted. Quantifying fog deposition using isotope methods is more difficult in forests receiving mixed precipitation, due to limitations in the ability of sampling equipment to separate fog from rain.

This chapter describes the various types of fog most relevant to MCF and the importance of fog water deposition in the hydrological budget. A brief overview of isotope hydrology provides the background needed to understand isotope applications in cloud forests. A summary of previous work explains isotopic differences between rain and fog in different environments, and how monitoring the isotopic signature of surface water, soil water, and tree sap can yield estimates of the contribution of fog water to streamflow, recharge, and transpiration. Next, instrumentation to measure fog and rain, and methods to determine isotopic concentrations in plant and soil water are discussed. The chapter concludes with the identification of some of the more pressing research questions in this field and offers various suggestions for future research.

INTRODUCTION

The importance of fog in the water budget of montane cloud forests (MCF) has been recognized for many years (Kerfoot, 1968; Zadroga, 1981; cf. Delay and Giambelluca, this volume; Marzol-Jaén, this volume), but there are still unresolved questions as to how cloud-affected ecosystems actually function hydrologically (Bruijnzeel *et al.*, 2005; Giambelluca *et al.*, this volume). Accurate water balances that include both fog and horizontal precipitation (precipitation not captured by a standard rain gage) are needed to quantify fog contributions to groundwater recharge and streamflow, to understand the role of fog water inputs in terms of nutrient and pollutant loadings, and

Tropical Montane Cloud Forests: Science for Conservation and Management, eds. L. A. Bruijnzeel, F. N. Scatena, and L. S. Hamilton. Published by Cambridge University Press.

to quantify fog as a moisture source during rainless periods. Clouds with low liquid-water content or events with low wind speeds may not deposit measurable amounts of water but may still have an effect on forest functioning via the reduction of transpiration from the plants (García Santos, 2007; Fischer *et al.*, 2009; McJannet *et al.*, this volume; cf. Schawe *et al.*, this volume). Distinguishing fog events, events with fog plus rain, and rain events aids in determining rates of cloud water interception by a forest, and how much water input would be missed by using conventional rain-gage-based water-balance techniques only (Holwerda *et al.*, 2006; García Santos, 2007; Giambelluca *et al.*, this volume).

The stable isotopes of water (^{2}H and ^{18}O) are a potentially useful tool for determining the amount and transport pathways of cloud water in MCF. To use these tools effectively requires an understanding of the underlying processes controlling stable isotope composition of natural waters. This chapter summarizes the results of previous studies involving stable isotopes of fog or cloud water, and gives an overview of how isotope ratios may be used to provide insight into water balances and water transport in cloud forests.

A necessary first step is to define what is meant by cloud water, fog, horizontal precipitation, etc., and what one is trying to quantify in terms of the total water input received by a cloud-affected forest. In meteorology, a cloud touching the ground with a horizontal visibility less than 1 km is defined as fog. This definition does not serve well in forest hydrology, because it does not include the aspect of whether or not there is measurable precipitation from fog. The most precise way to distinguish fog from rain is by droplet size. The American Meteorological Society *Glossary of Meteorology* (Glickman, 2000) states: "Fog differs from cloud only in that the base of fog is at the earth's surface while clouds are above the surface" and "A diameter of 0.2 mm has been suggested as an upper limit to the size of drops that shall be regarded as cloud drops; larger drops fall rapidly enough so that only very strong updrafts can sustain them. Any such division is somewhat arbitrary, and active cumulus clouds sometimes contain cloud drops much larger than this." In this chapter, the term "fog" is used to refer to cloud droplets that are small enough to remain suspended in the air without falling. "Cloud water" refers to the entire size spectrum of fog, drizzle, and rain present in a cloud. Rain is defined as having a droplet size greater than 0.5 mm (Glickman, 2000); in this chapter, the term "rain" refers to droplets large enough to fall into a standard rain gage, with the understanding that drop size will depend on the prevailing wind conditions (cf. Nespor and Sevruk 1999; Holwerda *et al.*, 2006; Giambelluca *et al.*, this volume).

Climatologically, there are several types of fog that may occur in MCF, including advection fog, sea fog, steam fog, radiation fog, and orographic clouds (see Bruijnzeel *et al.* (2005) for more detail). Advection fog is fog that is formed elsewhere and then transported horizontally, e.g. to a mountain range. In coastal regions, such advection fog may have originated as sea fog or steam fog. Sea fog forms when warm air masses move over cold ocean water, and the fog droplets represent mostly the condensate of the water vapor present in the air before cooling below its dew point temperature. This type of fog occurs frequently off the Pacific coast of North and South America and affects coastal mountain ranges in those areas. Conversely, steam fog commonly forms over warm water bodies when evaporation from the water surface exceeds the capacity of the colder air above it to take up moisture, after which the excess vapor condenses to liquid water drops. Radiation fogs occur when moist air cools below its dew point temperature. Typically, radiation fog is formed overnight in mountain valleys and basins with little or no drainage of cold air (cf. Liu *et al.*, this volume). It rarely occurs in mountain valleys where a strong down-valley wind occurs at night. Orographic clouds (upslope or mountain fogs) form when moist air is topographically forced to rise. This may occur on a large scale where the prevailing winds blow against entire mountain ranges, or on a smaller scale where thermally forced winds blow upslope. The small droplets typical of fog may be mixed with larger droplets; either rain, or droplets that, over flat terrain, would be large enough to precipitate as rain, but which are now kept in the orographic clouds due to strong updrafts. Thus, orographic clouds may be accompanied by significant wind and rainfall (cf. Giambelluca *et al.*, this volume). Because raindrops are typically several orders of magnitude larger in volume and amount of liquid water, orographic clouds may deposit much higher amounts of water than do advective fogs with their small droplets (Bruijnzeel *et al.*, 2005).

Droplet size measurements have only been made infrequently at remote mountain sites and what has been obtained in most studies is an operational definition of fog deposition, based on the type of instrument or sampler used. From a water balance perspective, it is important to quantify all incoming precipitation types. This involves measuring the size fraction of precipitation that is not measured in a standard rain gage and the amount that a rain gage may miss due to differences in collection efficiency compared to a vegetation surface. Several other chapters in this volume (notably Frumau *et al.*; García Santos and Bruijnzeel; Giambelluca *et al.*; Holwerda *et al.* #28; Mulligan *et al.* #25; cf. Tobón *et al.* #26) discuss the latter in some detail.

ISOTOPE FRACTIONATION PROCESSES RELEVANT TO CLOUD FOREST PRECIPITATION

The chemical composition of water is H_2O, a combination of two hydrogen (^{1}H) and one oxygen (^{16}O) atoms. A very small fraction of this water, however, is not composed of molecules containing the most abundant ^{1}H and ^{16}O atoms, but contains

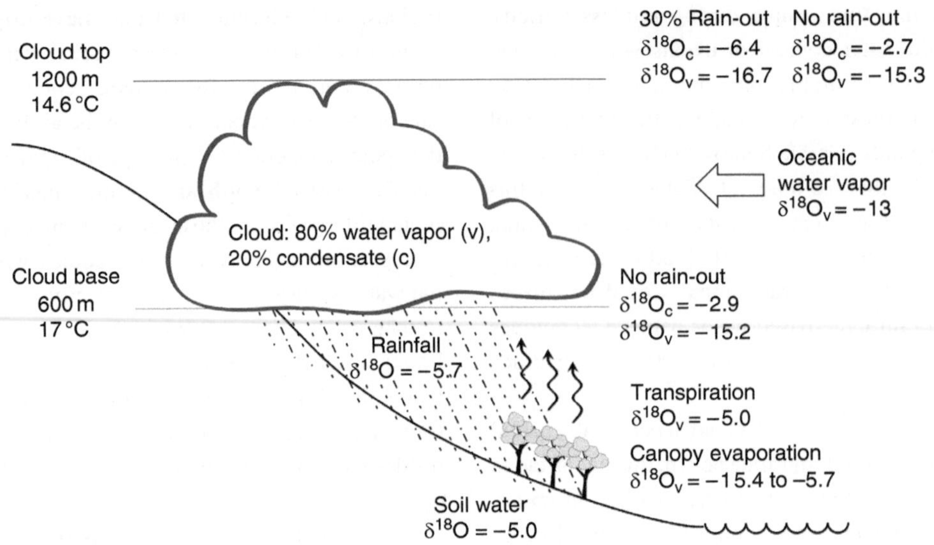

Figure 23.1. Isotopic transformations in the water cycle in a mountain cloud forest, illustrated using hypothetical $\delta^{18}O$ values in ‰. The example is an orographic cloud of 600 m vertical extent that originated from oceanic water vapor. The isotopic compositions of vapor ($\delta^{18}O_v$) and fog condensate ($\delta^{18}O_c$) are calculated using Rayleigh distillation equations and fractionation factors as given in Clark and Fritz (1997). Isotopic fractionation at cloud top and cloud base temperatures is shown in the figure. Isotopic values for two situations are illustrated, where the top of the cloud is raining (30% rain-out) and not raining (no rain-out), while the bottom of the cloud does not produce rain. With this small range in temperatures, loss of rain from the cloud top is necessary for isotopic composition to differ significantly with altitude within the cloud.

molecules with either ^2H (deuterium, often also abbreviated with D) substituted in place of ^1H, or ^{18}O in place of ^{16}O. The different isotopic variations of the molecules are called isotopologs. These isotopes are stable, that is, they are not subject to radioactive decay. Because concentrations of ^{18}O and ^2H are typically very low, (i) the concentrations are expressed as the difference in $^{18}O/^{16}O$ ratio between a sample and a known standard, divided by the ratio of the standard; this yields values that are labeled with δ (e.g. $\delta^{18}O$); and (ii) these δ values are expressed as per mil (‰) difference from the known standard. For water samples, the Vienna Standard Mean Ocean Water (V-SMOW) is typically used; its value is 0‰ for both $\delta^{18}O$ and δ^2H. Most precipitation and natural water samples have $\delta^{18}O$ and $\delta\ ^2H$ values that are more negative than ocean water. Because ^2H is heavier than ^1H, and ^{18}O is heavier than ^{16}O, the terms "light" and "heavy" have come into use when describing waters with different isotopic composition. "Lighter" water (i.e. depleted in the heavier isotopes) has a more negative δ value, and "heavier" water (enriched with the heavier isotopes) has a more positive δ value. The precision of laboratory analyses for ^{18}O is generally 0.2 ‰ or better, whereas that for ^2H is 2 ‰ or better. Textbooks such as Mook and de Vries (2001) or Clark and Fritz (1997) provide more details for the interested reader.

Isotopic fractionation between vapor and liquid water involves temperature-dependent proportions of the heavier isotopes of hydrogen and oxygen partitioning into the liquid phase during either condensation or evaporation (Clark and Fritz, 1997; and references therein). The heavier isotopes have a lower diffusion velocity and are less reactive than the lighter ones because their bonds with other molecules are stronger than those of the lighter isotopes. These characteristics lead to repeatable patterns of isotopic composition for water in different parts of the hydrological cycle.

The condensation process in clouds is similar enough worldwide for the $\delta^{18}O$ and δD composition of precipitation to exhibit a linear relationship: $\delta D = 8.20\ \delta^{18}O + 11.27$, known as the global meteoric water line (GMWL) (Rozanski *et al.* (1993), after Craig, 1961). Differences in climate and humidity between different areas cause local meteoric water lines to differ in slope and intercept from the global average. Therefore, ideally, a local meteoric water line should be used in the interpretation of the isotope hydrology of an area. The stable isotope composition of precipitation varies with temperature and the history of the water vapor mass producing the precipitation. The classic conceptual model for progressive isotopic depletion of rainfall, that is, a preferential rain-out of the heavier water, involves a moist air mass rising and cooling. As condensation occurs and rain enriched in ^2H and ^{18}O falls to the ground, the source vapor in the cloud becomes increasingly depleted in these heavier isotopes, causing subsequent rainfall to be more isotopically depleted (although the rain at any time is enriched relative to the vapor it condensed from). The end result of these processes is that rainfall sampled from the same vapor mass, over time or along an altitudinal transect, will become progressively depleted isotopically (Dansgaard, 1964; Siegenthaler and Oeschger, 1980; and many others). Complicating this conceptual model is the situation where the cloud intersects the land surface, and precipitation is sampled within the cloud (Figure 23.1). In this case, samples may have an overall isotopic composition resulting from differently sized water droplets, water droplets from different altitudes within the cloud, and evaporation and re-equilibration of falling raindrops within the cloud. There may also be additional vapor from evaporation or transpiration incorporated into the cloud. Isotopic compositions under such conditions are not as predictable, and only a few studies are available. Some clouds have stratified isotopic composition, as evidenced by isotopic analysis of concentric layers in hailstones (Jouzel *et al.*, 1975) and by modeling of isotopic distributions within clouds (Federer *et al.*, 1982; Gedzelman and Arnold, 1994).

Fog has been found to be isotopically enriched compared to rainfall in the same area, when fog and rainfall are generated by separate climatic processes, for example, in areas with distinct wet and dry seasons. Advective ocean-derived fog (Aravena *et al.*, 1989; Ingraham and Matthews, 1995; Dawson, 1998) condenses at ambient air temperature and may even equilibrate with warmer temperatures as it moves onshore. Orographic or upslope fogs may also represent early-stage condensation from a moist air mass, at temperatures near those of the land surface (Ingraham and Matthews, 1988). In contrast, rains from frontal or low-pressure systems typically condense higher in the atmosphere at much colder temperatures, and the cloud as a whole becomes progressively isotopically lighter during rain-out (Scholl *et al.*, 2002, 2007).

When fog and rain occur as part of the same climatic process, fog droplets that condensed at (or equilibrated with) air temperature near the land surface could be more isotopically enriched than raindrops in the same cloud, as long as the raindrops formed near the top of the cloud and progressive vapor depletion is occurring due to rain-out processes and lower temperatures (Figure 23.1). This situation was postulated for the Otago uplands of New Zealand, where fog and rain often occurred simultaneously and the respective isotopic compositions fell along the same local meteoric water line (Ingraham and Mark, 2000). For mountains frequently immersed in raining orographic clouds, such as in Hawai'i (Scholl *et al.*, 2002, 2007; Table 23.1), Puerto Rico (Te Linde *et al.*, 2001; Burkard and Eugster, unpublished data in Table 23.1; Scholl *et al.*, 2006) or Costa Rica (Burkard *et al.*, 2003, unpublished data in Table 23.1; Rhodes *et al.*, this volume #24;), isotopic signatures of rain and fog may not be as distinct. Many measurements in these areas showed fog to be isotopically enriched compared to rain, but in some cases, isotopic signatures were similar, and measurements of individual precipitation events in both Puerto Rico and Costa Rica included instances when fog isotopic composition was more depleted than rain isotopic composition (Table 23.1). The reasons for the latter observations are unknown, and further studies need to be done.

Throughfall (TF) is precipitation collected underneath a vegetation canopy. The definition of TF generally includes both precipitation that has been intercepted by the canopy and then drips to the ground, and precipitation falling directly through gaps within the canopy without interception. The intercepted precipitation may include rain and/or fog, and the canopy has a storage capacity that must be filled before the precipitation begins dripping off to the ground (Bruijnzeel, 2000). Studies of isotopic composition of TF generally reported TF to be enriched relative to open-sky rainfall, although depletion has also been observed in some samples (Scholl *et al.*, 1995). The difference in isotopic composition between TF and rain is thought to be due to evaporation and exchange with atmospheric vapor during storage in the canopy, with the magnitude of the change depending on temperature, humidity, and the residence time of the water in the canopy. Published isotopic enrichments of ^{18}O in TF range from an average of 0.3 ‰ in a boreal pine forest in Sweden (Saxena, 1986), to 0.38 ‰ in spruce and 0.36 ‰ in beech forests in Germany (Brodersen *et al.*, 2000), to as much as 1 ‰ in coniferous and deciduous forests in Georgia, USA (Kendall, 1992). Isotopic composition of TF can also be affected by placement of the collector; for example, Brodersen *et al.* (2000) found higher isotopic enrichment under tree crown centers than under the crown periphery. In a montane forest in Hawai'i, isotopic depletion in TF samples was thought to result from the collectors receiving water predominantly from isotopically depleted intense rainfall events, while the canopy blocked much of the locally generated, isotopically enriched orographic rainfall and fog drip from falling into the collectors (Scholl *et al.*, 1995).

Isotopic fractionation of water occurs during evaporation from soils, but not during the root uptake process in transpiration (Thorburn *et al.*, 1993a; Dawson and Ehleringer, 1998). Plant leaves contain water that is isotopically enriched due to evaporation, but it is generally thought that there is no significant isotopic fractionation of the vapor in the process of transpiration outflux from leaves, so that under steady-state conditions, transpired vapor is assumed to have the same isotopic composition as the water taken up by the roots (Dawson and Ehleringer, 1998; Wang and Yakir, 2000).

Gat (2000) reviewed the different sources of atmospheric vapor and how isotopes may be used to trace water sources in the atmospheric water balance. The deuterium excess (*d*) is defined as $d = (\delta^2H - 8\ \delta^{18}O)$. Because of differences in the fractionation of oxygen and hydrogen during evaporation at different humidity levels, precipitation from water vapor that originates from re-evaporated rainfall (vapor from canopy interception, lakes and wetlands, or falling rain) may have a larger deuterium excess value than precipitation originating from water vapor evaporated from the oceans. This parameter has been used to assess the contribution to precipitation at a particular location by recycled moisture. Examples include the Amazon

Table 23.1 *Stable isotope values ($\delta^{18}O$, $\delta^{2}H$) of fog, cloud water, and rain*

	Study	Location	Altitude (m)	Fog type	Sampling interval	Number of samples	Fog $\delta^{18}O$ (‰)	Fog $\delta^{2}H$ (‰)	Rain $\delta^{18}O$ (‰)	Rain $\delta^{2}H$ (‰)
1	Gonfiantini & Longinelli (1962)	Shipboard, off Newfound land	Sea level	Advective (oceanic)	≤ 4 hours	Avg. of 14 fog, 18 rain	–2.4	–	–5.8	–
2	Clark *et al.* (1987)[a]	Oman		Advective orographic (monsoon)	–		–0.43 to +1.13	–2.1 to +10.5	–	–
3	Ingraham & Matthews (1988)	Kenya	1700	Advective orographic	Grab: 10–12 days:	Avg. of 4 Avg. of 2	–75 –1.6	+13 to +6	–	–
4	Aravena *et al.* (1989)	Chile		Advective (oceanic)	–	Avg. of 13 fog, 4 rain	–1.86	–3.2	–5.6	–40
5	Ingraham & Matthews (1990, 1995)	California	–	Advective (oceanic)	Cumulative samples, intervals not given	4-year average, 3 sites	–2.3 –2.8 –2.4	–10 –15 –10	–6.5 –7.0 –6.3	–38 –40 –36
6	Dawson (1998) and ongoing observations[b]	California		Advective (oceanic)	Monthly averages, 3-year study	17 fog collectors, rain not given	+2.7 to –4.9	+6.4 to –8.9	–5.5 to –15.6	–28.7 to –84.6
7	Feild and Dawson (1998)[c]	Costa Rica	1500	Orographic	Event	Averages, 82 total	–2.9	–6.6	–2.8	–24.9
8	Dawson and Vidiella (1998)	Chile	68	Advective oceanic	Monthly, per event in summer	Averages, 212 total	–1.3	–4.6	–3.7	–32.8
9	Dawson, 2000–2004 (pers. comm.)	California	140	Advective oceanic	Monthly since 1/00	Averages, 97 samples	–2.3	–8.1	–8.4	–46.2
10	Dawson, 2000–2004 (pers. comm.)	California	128	Advective oceanic	Monthly since 1/00	Averages, 155 samples	–3.3	–7.1	–8.9	–55.3
11	Corbin *et al.* (2005)	California	9	Advective oceanic	Monthly, and per event in summer	Averages, 38 total	–2.7	–6.4	–5.3	–40.8
12	Ingraham and Mark (2000)	New Zealand	736, 870, & 1140 m	Advective	Approx. monthly, 2 summers	3 sites, 9 to 11 samples per site	–5.3 to –10.4	–30 to –71	–7.9 to –12	–50 to –87
13a	te Linde *et al.* (2001)[d]	Puerto Rico	1018	Orographic cloud (trade wind)	Daily for 5 weeks	Avg. of 4 fog, 4 rain:	–3.8	–17	–4.1	–17
13b	te Linde *et al.* (2001)[d]	Puerto Rico	1018	Orographic cloud (trade wind)	Daily for 5 weeks	Composite (bulk) sample:	–1.9		–2.4	
14a	Burkard, Eugster & Holwerda, unpublished data[e]	Puerto Rico	1015	Orographic cloud (trade wind)	1 sample per event, 12-hr avg.	fog>rain: 36	–1.21	–1.70	–1.62	–6.2

14b	Burkard, Eugster & Holwerda, unpublished data[e]	Puerto Rico	1015	Orographic cloud (trade wind)	1 sample per event, 12-hr avg.	fog<rain: 8	−1.32	−4.70	−0.47	−2.0
15a	Scholl *et al.* (2007)[f]	Leeward Maui	1220	Orographic cloud (thermal)	2 year vwa of monthly samples	19	−2.7	−6	−5.5	−31
15b	Scholl *et al.* (2007)	Windward Maui	1950	Orographic cloud (trade wind)	2 year vwa of monthly samples	22	−4.1	−16	−5.5	−28
16	Still *et al.* (2003)	Costa Rica	–	–	Not given	1 sample each	−4.9	−21	−8.1	−45
17a	Rhodes *et al.* (this volume #24)[g]	Costa Rica	1460	Bulk precipitation incl. fog	Variable; avg. for each season	10 wet season			−7.5	−49
17b	Rhodes *et al.* (this volume #24)[g]	Costa Rica	1460	Bulk precipitation incl. fog	Variable; avg. for each season	3 transitional season			−1.9	+4
18a	Burkard, Schmid & Eugster, unpublished data[e]	Costa Rica	1460	Orographic cloud	1 sample per event, 24-hr avg.	fog>rain: 31	−1.99	−6.46	−2.39	−12.3
18b	Burkard, Schmid & Eugster, unpublished data[e]	Costa Rica	1460	Orographic cloud	1 sample per event, 24-hr avg.	fog<rain: 6	−3.14	−15.40	−2.23	−10.2
19	[8]Fischer and Still, (2007)[h]	Santa Cruz I., California	296	Advective oceanic	Monthly for 34 months	37 fog, 17 rain	−2.52	−11.4	−5.16	−31
20	Liu *et al.* (2007)	SW China	750	Radiation	Fog: biweekly, rain: monthly and event	116 rain, 59 fog	−3.3	−11.5	−7.9	−56.8
21	Scholl *et al.* (2006) and unpublished	Puerto Rico	1050	Orographic cloud (trade wind)	monthly	36 rain, 36 fog	−2.17	−1.3	−3.19	−10.1

[a] Data from I. Clark, written communication, 2005; data were shown in Figure 8 in Clark *et al.*, 1987.

[b] Research is ongoing, data reported here are for the entire period of data collection at the site to date (T.E. Dawson, written comm., 2004).

[c] Values reported here are more comprehensive than those reported in the publication (T.E. Dawson, written comm., 2004).

[d] Values in 13a are the average of 4 selected events; 13b is the volume-weighted average of all the daily samples.

[e] Values are averaged separately for conditions where δD or $\delta^{18}O$ in fog was enriched compared to rain (fog>rain; 14a, 18a) and conditions with both isotopes being depleted in fog compared to rain (fog<rain; 14b, 18b).

[f] Fog averages for Maui exclude sampling periods with large storms, because rain dominated the fog collector sample. vwa, volume-weighted average.

[g] Samples are bulk precipitation; averages are for wet season (17a, convective precipitation) and transitional season (17b, orographic precipitation).

[h] Average values are from D. Fischer, written comm. Their publication gives values in graph format only.

Basin (Salati *et al.*, 1979; Martinelli *et al.*, 1996), Hawai'i (Scholl *et al.*, 2007), and Costa Rica (Rhodes *et al.*, 2006; Rhodes *et al.*, this volume #24; cf. Nair *et al.*, this volume).

ISOTOPES IN RAINFALL AND FOG: A REVIEW

Published studies including measurements of the stable isotope content of fog span about 40 years (Table 23.1), but are surprisingly few considering the large number of isotope hydrology studies conducted during that time. Reported isotope values for fog water cover a large range, from –71 to +13‰ for δ^2H and –10.4 to + 2.7‰ for δ^{18}O, depending on the temperature and vapor sources in the study areas (Table 23.1).

A scatter-plot of δ^{18}O vs. δ^2H values in fog or cloud water and in rainfall from the studies listed in Table 23.1 is shown in Figure 23.2. Fog values plot both above and below the GMWL, but as a group, the values are parallel to the meteoric water line as is expected for precipitation samples. As also observed for fog and rain samples from individual sites, the fog values as a group are generally enriched in the heavier isotopes compared to the rain samples, although the two groups overlap because of the large range of temperatures for the sites.

Gonfiantini and Longinelli (1962) published δ^{18}O analyses of fog and rain collected on a ship near the fishing banks off Newfoundland and Labrador. The fog was formed at the sea

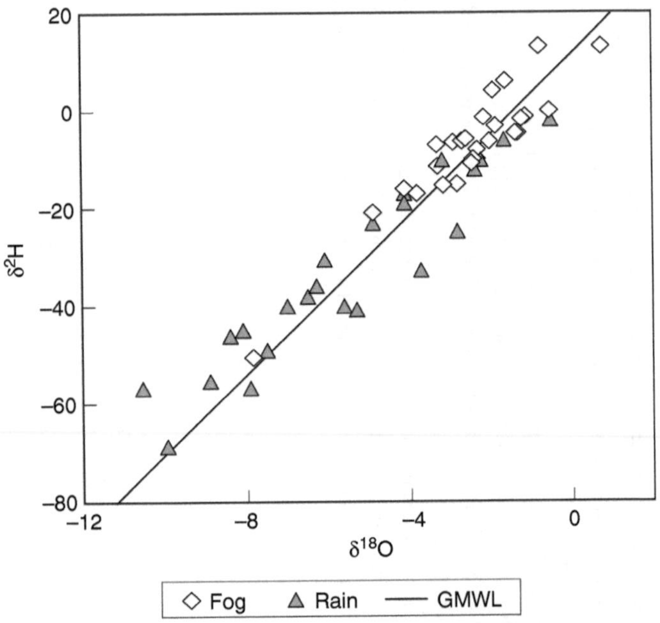

Figure 23.2. Fog isotopic values from around the world based on the studies listed in Table 23.1, compared to rain isotope values from the same site (if given) and the global meteoric water line (GMWL). All values are in ‰. Single symbols are individual values, or the average value when a range was reported in the publication.

surface due to warmer air moving over cold water (sea fog). The authors noted that the δ^{18}O values ranged from that of sea water to water vapor in equilibrium with sea water. Although the range of δ^{18}O in fog overlapped the range of rain δ^{18}O values, the fog samples on average were isotopically enriched compared to rain in the same area.

Clark *et al.* (1987) and Ingraham and Matthews (1988), in regional isotope hydrology studies conducted in arid climates, found fog to be an important component of the local hydrology. In northern Kenya, Ingraham and Matthews (1988) sampled fog, groundwater, and a river in a desert basin where isolated high mountains received orographic fog precipitation. They found the fog to be substantially isotopically enriched compared to a groundwater sample derived from rain, and groundwater in three other areas appeared to be a mixture of fog and rain. Clark *et al.* (1987) found that monsoon fog and light rain in the mountains of southern Oman constituted a source of recharge for groundwater in the mountains and in the adjacent coastal plain. These monsoon-cloud precipitation samples had isotopic signatures similar to sea water, and the precipitation was assumed to originate from a single-stage evaporation–condensation cycle near the coast.

In an isotope hydrology study of the windward side of East Maui, Hawai'i, Scholl *et al.* (2002) found isotopic values of mountain streams and springs to be unexpectedly enriched in the heavier isotopes for their altitude, compared to volume-weighted average rainfall measured in a transect up the mountain side. The best explanation for the observed isotopic composition was that fog drip contributed substantially to streamflow and shallow groundwater. This work suggested that water balances based only on streamflow and rain gage precipitation measurements could have a large degree of error. A subsequent study was done (Scholl *et al.*, 2007; Table 23.1) on both the windward and leeward sides of East Maui, with cloud water samples collected from passive collectors at monthly intervals for 2 years. On the windward side, fog occurred as part of the trade wind orographic clouds, which had droplet sizes ranging from fog to rain. On the leeward side, thermal heating over the land generated an upslope fog with relatively lower liquid-water content. Fog was isotopically enriched compared to rain at both sites (Table 23.1), but isotopic differences between fog and rain were larger on the leeward side, where synoptic-scale storms accounted for most of the rainfall. With few exceptions, each month's fog collector sample was more enriched than the rain collector sample, though the two samples were frequently similar in composition (Scholl *et al.*, 2007).

Near Pico del Este in the Luquillo Mountains of eastern Puerto Rico, Burkard and Eugster (Table 23.1) sampled fog, rain, throughfall, and stemflow in a follow-up study after the preliminary work of te Linde *et al.* (2001). Fog and rain from each precipitation event were sampled for approximately 1 month. Using a Caltech-type active strand cloud-water collector (Daube

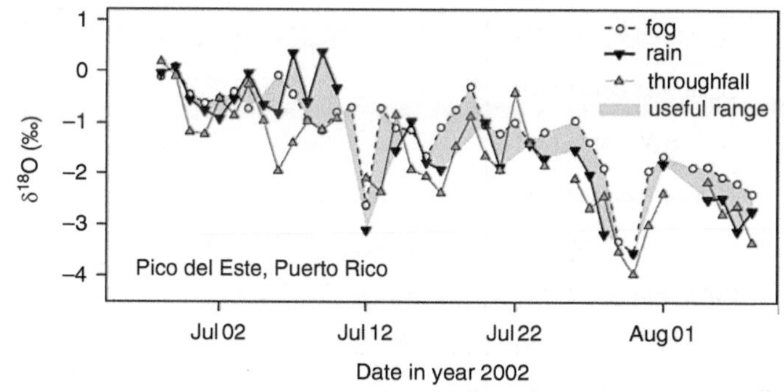

Figure 23.3. Isotopic signatures of $\delta^{18}O$ in rain, fog, and throughfall measured in an elfin cloud forest in Puerto Rico (daily means). The time traces reveal the importance of synoptic-scale changes in weather. If throughfall concentrations are in the gray-shaded area, then a two end-member mixing model can be used. Frequently, however, throughfall $\delta^{18}O$ was below fog and rain (W. Eugster *et al.*, unpublished data).

et al., 1986), fog was efficiently separated from rain even when the two occurred simultaneously. The fog collector was switched on whenever visibility dropped to less than 500 m while sampling continued until the 10-min running mean visibility exceeded 500 m for several minutes. This threshold was chosen to minimize contamination by insects and dry aerosols; fog with a visibility in excess of 500 m tends to contribute very little to total fog deposition (Thalmann *et al.*, 2002; Burkard *et al.*, 2003). Rain and fog water samples for isotopic analysis were always taken simultaneously, such that the durations of sampling of fog and rain matched. The local type of fog is an advected cloud formed over a warm ocean, with the cloud base below the mountain top for almost 75% of the time (cf. Holwerda *et al.*, 2006; Holwerda *et al.*, this volume #28). Isotopic concentrations of ^{18}O and ^{2}H were usually (but not always) enriched in fog compared to rain (Figure 23.3). The difference in the isotopic signal, however, was quite small (Figure 23.3 and Table 23.1) compared to the temporal variability caused by weather conditions and prevailing wind directions. Conditions during which fog was depleted in ^{18}O and ^{2}H may have been partly due to time lags between fog and rain (Figure 23.3). During some events, a change in isotopic signature occurred 12–24 hours (the minimum sampling interval used in the study) earlier in fog than in rain (Eugster, 2007). This result is consistent with the potential explanation that coalescence of rain consisting of relatively large droplets, although originating from the same air mass as the fog, takes longer than the coalescence of fog alone. The fact that isotopic signatures at Pico del Este did not differ much between fog and rain after considering such a time lag, suggests that under these humid tropic conditions, condensation and re-evaporation processes occur in an atmospheric system saturated with water vapor. Thus, the coalescence of droplets as the air mass ages determines how much of the water is available as large droplets (rain) and what fraction remains as small droplets (fog).

The isotopic signature of composite throughfall at Pico del Este was often depleted in ^{18}O and ^{2}H relative to both rain and fog (Figure 23.3), which may indicate that additional processes (not fully understood as yet) could affect throughfall isotopic composition. Twenty roving collectors were randomly placed at 80 pre-defined positions, and isotopic analysis was performed on a single sub-sample after bulking the water from all collectors. Although this sampling design should capture the spatial variability of throughfall amounts, it may be sensitive to the influence of isotopic "hot spots" where the influence of rainfall contributions is disproportionately large (so-called "drip points"; Holwerda *et al.*, 2006).

Similar conditions to those described above were found in a windward cloud forest in the Monteverde area in northern Costa Rica (Burkard *et al.*, 2003 in Table 23.1; cf. Schmid *et al.*, this volume), where a few events did not correspond to the typical pattern that stable isotopes are enriched in fog compared to rain. This site, which was not as close to the ocean as the example from Puerto Rico, experienced an even wider range of isotopic concentrations, depending on weather conditions. As in Puerto Rico, isotopic signatures in fog and rain followed each other closely (data not shown).

Rhodes *et al.* (this volume #24) analyzed samples of bulk precipitation and throughfall for stable isotopes in a leeward cloud forest at Monteverde. The bulk precipitation samples spanned the wet and transitional seasons, when the precipitation source changed from convection associated with the Intertropical Convergence Zone (ITCZ) to orographic uplift. Samples from orographic precipitation were assumed to include (wind-driven) rain and fog, and were isotopically enriched compared to samples collected from convective rainfall (Table 23.1). Concurrent fog and rain samplers deployed for a period in January had similar isotopic composition.

A few isotopic studies involving fog, rain, and groundwater included analyses of leaf water (Aravena *et al.*, 1989) or tree sap and soil water (Ingraham and Matthews, 1995; Scholl *et al.*, 2002); others used isotopes to examine the role of fog in ecosystems (e.g. Dawson, 1998 and subsequent work; Feild and Dawson, 1998). These studies found fog water to be isotopically enriched compared to rainwater, and each study showed that the vegetation utilized fog water under some conditions. The investigation by Dawson (1998) in the redwood forests of California was the most extensive, showing conclusively that the ecosystem depended on fog as a water source as the soil dried out during the rainless summer period.

The differences in isotopic values for rain and fog in each of the studies listed in Table 23.1 are shown in Figure 23.4. When a range of values was given in the original paper, the mean values were used to calculate the difference. Both deuterium and oxygen-18 plots are shown because some studies involved only one of the isotopes. The graphs show differences in isotopic

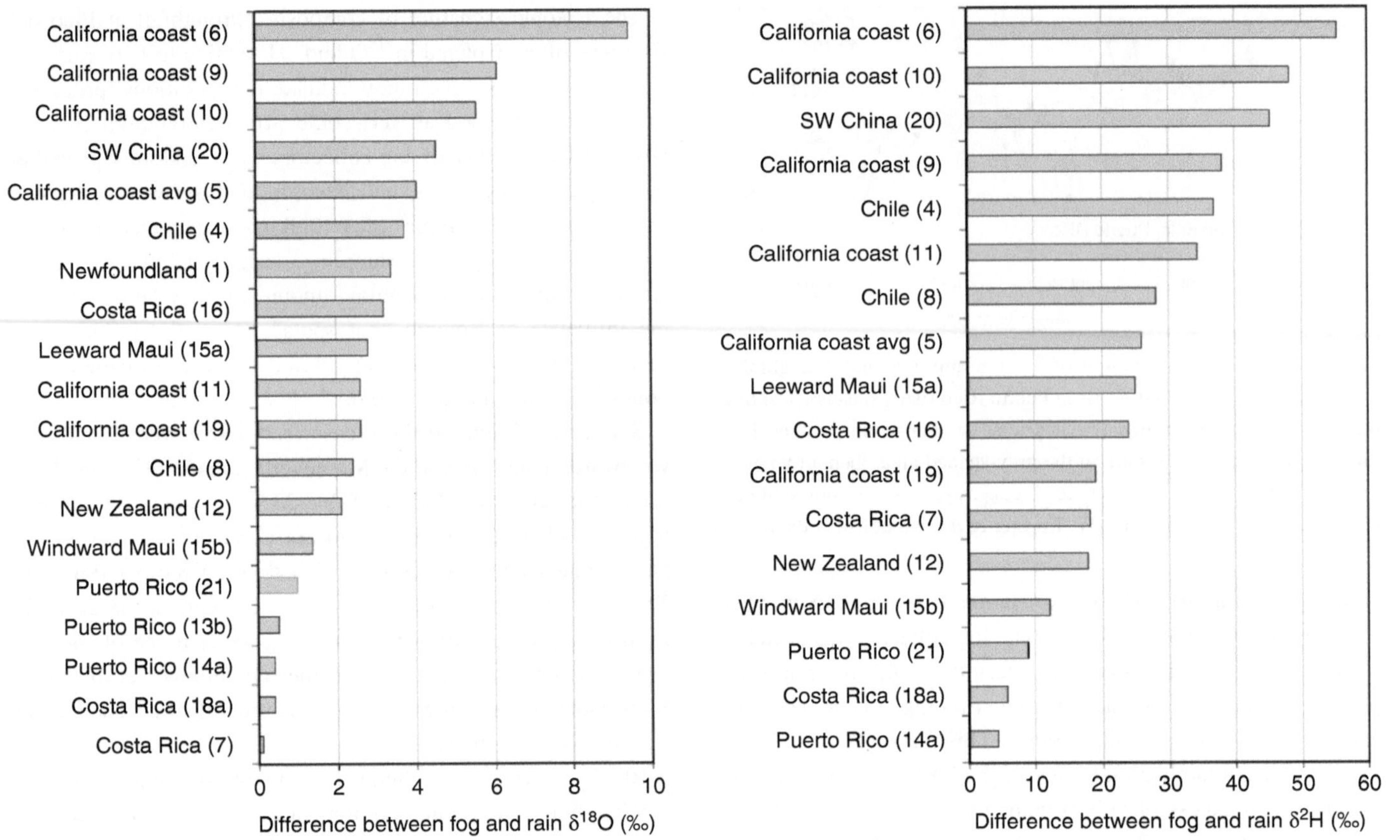

Figure 23.4. Magnitude of the difference between values for ^{18}O and ^{2}H in fog and rain. Numbers correspond with numbered studies in Table 23.1. For study 13a (Puerto Rico) there was no measurable difference between $\delta^{2}H$ values in fog and rain.

values of fog and rain in different localities, but these should not be considered definitive as different sampling frequencies, methods, and numbers of samples were associated with the different studies. The graphs indicate that the smallest differences between fog and rain (<2‰ in ^{18}O and <20‰ in ^{2}H) are associated with orographic clouds on mountains in Puerto Rico, windward Hawai'i, and Costa Rica. The largest differences (>4‰ in ^{18}O and >30‰ in ^{2}H) involved advected oceanic fog, with rain as a separate climatic process (California and Chile).

To summarize previous work, relatively few studies have examined the isotopic composition of fog, and of these, even fewer constituted more than reconnaissance-level surveys. Of the studies listed in Table 23.1, only four involved long-term sampling, yielding average isotopic values in fog and rain over 2 to 4 years, and only three studies investigated stable isotopes in fog and rain in detail for every precipitation event during at least 1 month. Most of the studies found average fog isotopic composition to be more enriched than average rain; however, sampling in Puerto Rico and Costa Rica revealed situations where the reverse was true (Figure 23.3, Table 23.1). In some studies, collection methods did not entirely separate fog and rain, so that average values may not reflect true isotopic differences between fog and rain. Tracing fog water using isotopic analysis of plant water has been successful in some environments, particularly where fog and rain are governed by different climatic processes.

SAMPLING METHODS FOR STABLE ISOTOPES

Precipitation

GENERAL CONSIDERATIONS

The main goal when collecting a precipitation sample for stable isotope analysis is to avoid changes in isotopic composition between collection and analysis. Water in a container open to the atmosphere immediately begins to change in isotopic composition due to evaporation and exchange with ambient vapor (Ingraham and Criss, 1993). If evaporation takes place in an atmosphere of high humidity, the altered isotopic composition will be close to the meteoric water line (Gonfiantini, 1986; reprinted in Clark and Fritz, 1997), so that the evaporation is not detectable but the sample values are incorrect (Scholl *et al.*, 1995). Evaporation and exchange are minimized if the sampler is designed to inhibit evaporation, collects relatively large samples with a small surface area of sample exposed to the atmosphere, and if the collection reservoir is shielded against direct insolation. Water samples for stable isotope analysis should be collected in glass bottles with "poly-seal" caps (caps with conical plastic insert). Mook and de Vries (2001) give a detailed discussion of general water sampling and isotopic analytical methods.

FOG, RAIN, AND THROUGHFALL COLLECTORS

Most collectors used for precipitation sampling on a timescale longer than a few hours involve some kind of system whereby the rain is collected in a funnel, or fog is collected on a vertical screen or strings, after which the water is routed to a collection bottle. There are two ways to preserve the isotopic composition of the water in a precipitation collector:

(i) a layer of mineral oil is added to serve as a barrier against evaporation. Because of its greater density, precipitation entering the collector will sink below the protective oil layer. If the sampling interval is longer than a week, the oil layer needs to be at least 1-cm thick, as water will evaporate through thinner layers (Scholl *et al.*, 1995, 2002).

(ii) The sampling container has the smallest possible area open to the atmosphere, so that the degree of evaporation or exchange is small relative to sample volume. There are many ways to accomplish this, ranging from the use of tubing extending from the stem of the funnel to the bottom of the sampling container, adding a loop in the tubing that will then remain full of water after precipitation and so provide a barrier against sample evaporation (though this does not work well with small samples), collecting the sample in a bag to minimize headspace, or using a plastic ball that allows the water to flow underneath into the container.

For any precipitation collector design, a control container with known volume and isotopic composition should be placed at the collection location to test the selected sampling method.

To collect the small fog droplets separately from larger raindrops, an active strand cloud water collector (known as a CASCC; Daube *et al.*, 1986, 1987) may be used. This device has a fan that pulls the air from under a rain-shed across an array of strings where the fog droplets impact on Teflon or stainless steel strings and are guided to a collection bottle. The size and spacing of the strings and the air velocity generated by the fan determine the size of the droplets captured. The collector inlet must face the wind. A drawback of this collector for remote sites is the high power consumption of the fan. A test in the Adirondack Mountains of New York, USA (M. Scholl, unpublished data) showed that isotopic samples of cloud water collected with a small CASCC (18 × 18 cm inlet) had the same composition as samples passively collected on wire mesh, suggesting that the fan's airflow velocity does not cause significant evaporation and isotopic fractionation of the fog droplets as they impact on the strings. Further tests, however, should be done with the larger, higher-velocity CASCCS. Because CASCCs are extremely efficient, they produce the amount of water required for isotopic analysis within a short time.

Tree cores, twigs, and soil water

The ^{18}O or ^{2}H isotopic composition of sap can indicate the transpiration water source for the plants in a forest (White *et al.*, 1985; Thorburn *et al.*, 1993b; Dawson and Ehleringer, 1998; Meinzer *et al.*, 1999; and many others). Dawson and Ehleringer (1998) and Dawson (1993) provided thorough reviews of the use of plant isotope tracers in watershed hydrology and may be consulted for additional detail. Tree sap for isotopic analysis may be collected from cores (White *et al.*, 1985) or twigs (Dawson, 1993; Thorburn *et al.*, 1993b). The water in the xylem is the transpiration water, so when taking core samples, sapwood should be separated from heartwood as the isotopic composition of the heartwood may represent stored water from other sources or time periods. For twig samples, twigs with mature bark and measuring at least 1 cm diameter are removed from the plant, the bark is scraped off, and twigs cut into short pieces to fit into an airtight glass jar. Samples are frozen until analysis. The most commonly used methods for extracting the water from plant tissues are azeotropic distillation (Revesz and Woods, 1990; Thorburn *et al.*, 1993b) and cryogenic vacuum distillation (West *et al.*, 2006), after which the water is analyzed for isotopic composition using standard methods.

Sampling the isotopic composition of soil water in forests involves more uncertainty than precipitation and plant water, as water infiltrating into the soil is subject to evaporation in the unsaturated zone, particularly in the top of the soil profile (Allison *et al.*, 1983; Barnes and Allison, 1983; Barnes and Turner, 1998). Knowledge of the location of the root zone for the plants of interest is helpful to make optimum use of both soil and sap isotopic data from a site. Samples for isotopic composition of soil water are obtained by extraction from soil cores or from the outflow of lysimeters. Suction-type lysimeters introduce the possibility of isotopic fractionation, although Landon *et al.* (1999) found in laboratory tests that lysimeter samples were not significantly fractionated. Soil core samples are placed in airtight glass or metal containers immediately after collection and kept refrigerated until analysis to inhibit microbial activity and evaporation. Extraction techniques include vacuum distillation, azeotropic distillation, microdistillation with zinc, and immiscible displacement using centrifugation, with azeotropic distillation generally found to be the best method (Barnes and Turner, 1998).

TRACING FOG THROUGH THE FOREST WATER CYCLE

Sampling interval considerations

Both long-term, infrequent sampling and short-term, intensive sampling strategies have been used to study the isotope hydrology of cloud forests. There are advantages and disadvantages to each approach, and the selected approach also depends on the research questions, site accessibility, and funding. Short-term, event-based sampling of isotopes is important to investigate processes that may vary over short time intervals, such as interception of rainfall or fog, crown drip, transpiration, and storm runoff. Some simplifying assumptions about fractionation processes within the canopy are necessary, but short-term event-based sampling may help to resolve this. Longer-term cumulative-sample studies integrate short-term variations in samples and yield a volume-weighted average isotopic value. This strategy can work well for watershed processes that involve longer timescales or for understanding spatial variability in mountainous areas. For example, the pathway from fog and rainwater inputs to groundwater may in some areas be on the scale of months to years, and for such an application the variation in isotopic signature with each event can be considered negligible. Careful consideration should be taken concerning the relevant processes and their respective timescales, from which an optimum sampling strategy may be derived.

The sampling interval for precipitation also needs to be determined with consideration of the study objectives, and the climatology and precipitation patterns at the site. The stable isotope composition of rainfall commonly changes during the course of a precipitation event (McDonnell *et al.*, 1990; Clark and Fritz, 1997, p. 50; Barnes and Turner, 1998), and these changes are not predictable because of variations in cloud processes producing the rain, and because precipitation is normally sampled at a fixed location while cloud systems move past. Several convective precipitation cells with individual isotopic signatures may contribute to one single precipitation event observed at a fixed location. To resolve this variation, samples must be taken over the course of a precipitation event. At the opposite end of the sampling time scale, for regional hydrological studies, 6-month cumulative samples of precipitation may be an adequate resolution, provided that evaporation between rain events is prevented. Most of the studies discussed in this chapter had sampling intervals ranging from daily to monthly.

Mixing models and stream hydrograph separation techniques

In locations where fog and rain have distinct isotopic signatures, a simple linear mixing model may be applicable for evaluating possible contributions of the two source waters to streamflow, plant transpiration, or groundwater recharge. For example, fog and rain are considered the two independent source water end members, which are mixed in a certain volumetric proportion to yield the resulting component (e.g. transpiration water). The fraction of transpiration that comes from fog water, f_F, is

$$f_F = \frac{\delta_T - \delta_R}{\delta_F - \delta_R}$$

where the isotopic ratios of ^{18}O or ^{2}H are denoted with δ, and the indexes T, F, and R, indicate the sampled transpiration water (sap), fog, and rain, respectively. Mixing models work best in situations where either: (i) there is no fractionation of the source water isotopes in the process of transport through the hydrological cycle, or (ii) the fractionation is known and constant. In the case of streamflow studies or recharge over short time intervals, precipitation passes through the unsaturated zone, and its isotopic composition changes by mixing with soil pore water already present. Similarly, for water intercepted by the canopy, evaporative isotopic fractionation may be an issue. In such cases, the problem might require a three or even four end-member mixing model.

Stable isotope techniques have been used for more than 30 years in watershed hydrograph separation studies. The recognition that rain events often have significantly different isotopic composition than water stored in the subsurface led to the finding that streamflow generally includes a significant proportion of pre-event water stored in the watershed, rather than just runoff from the current storm event (Sklash *et al.*, 1976; Sklash and Farvolden, 1979; Buttle, 1998). These techniques may have application in determining the amount of cloud water contributing to streamflow in mountain forests. Another technique involves using the seasonal fluctuations in precipitation isotopes to determine residence time for water in a watershed (Maloszewski and Zuber, 1982; McGuire *et al.*, 2002). Seasonal differences in rain isotopes tend to be smaller in the tropics than at higher latitudes, but the required differences in precipitation isotopes for the application of such techniques can occur in mountain areas having distinct rainy and dry seasons, with fog providing significant moisture during the dry season (Rhodes *et al.*, 2006; Liu *et al.*, 2007; Liu *et al.*, this volume; cf. Bruijnzeel 2001), or in areas having orographically generated local rainfall alternating with synoptic-scale storms (Scholl *et al.*, 2006, 2007). Evaporative fractionation may also be useful as a check on other methods for determining the evaporation rate of precipitation intercepted by the canopy, provided throughfall has a significant evaporative signature (Saxena, 1986; Brodersen *et al.*, 2000).

In principle, stable isotopes constitute a powerful tool to determine the relative influences of fog and rain in the local water budget (Dawson, 1998; Liu *et al.*, 2005), and may also yield insights into water dynamics in the plant canopy. In reality, conditions are sometimes more complex than this, and simple mixing models may not provide the desired information. However, these models can be a valuable approach for testing conceptual models of water flow in mountain areas affected by fog, yielding information on previously unexpected or unknown processes that prove to be relevant at a certain site.

DIRECTIONS FOR FUTURE RESEARCH

Previous studies have suggested that stable isotopes can be a useful tool in identifying the role of fog water in the hydrological and ecological functioning of cloud-affected forests, but comparatively few studies have been conducted and several important questions remain. In areas where fog and rain are generated by separate climate processes, fog is isotopically enriched compared to rainfall, and this contrast has been used effectively to demonstrate the importance of fog water to, for example, the redwood forest of coastal California (Dawson, 1998). Tracing fog water inputs to an ecosystem is least ambiguous where fog is the only source of water for all or part of the year (Aravena *et al.*, 1989; Ingraham and Matthews, 1995; Dawson, 1998; Liu *et al.*, 2005). However, in areas receiving mixed fog and rain, like many montane cloud forests experiencing significant orographic cloud water input (Holwerda *et al.*, 2006; Giambelluca *et al.*, this volume), the isotopic composition of rain and fog has not been as well documented. The data from the elfin cloud forest at Pico del Este in Puerto Rico (Table 23.1, Figure 23.3) show conclusively that fog and rain sampled from the same cloud are often isotopically different. Other studies in climates with locally generated orographic rainfall have also found fog to be isotopically enriched compared to rain, although such differences were relatively small (Scholl *et al.*, 2007). Larger differences have been found between different types of precipitation events (local orographic vs. synoptic scale). There is a need for more studies of throughfall isotopic composition, as well as of rain and fog water, particularly in relation to the determination of the water dynamics and sources of precipitation-dependent parts of the ecosystem such as epiphytes (cf. Hietz, this volume; Tobón *et al.*, this volume #26).

Because much of the precipitation input to forest soils consists of throughfall, it is important to understand the isotopic composition of throughfall if isotope techniques are to be used to understand streamflow sources, ecosystem functioning, or recharge patterns (cf. Goller *et al.*, 2005; Liu *et al.*, 2005, 2007). There is some disagreement as to whether the isotopic composition of throughfall is always enriched relative to rainfall. Brodersen *et al.* (2000) shed some light on this issue, showing that the placement of the collector is important. No comparative work is available for montane cloud forests. Both throughfall and fog water may be isotopically enriched relative to rain, which may render it difficult to separate contributions by rain and fog to overall canopy interception using isotope mass-balance techniques (cf. Schmid *et al.*, this volume). Similarly, large inputs of evaporatively enriched throughfall to a forest may make differentiating between fog and rain inputs more difficult in longer-term studies. It may also be possible for throughfall to become depleted by vapor exchange at high humidities (Brodersen *et al.*, 2000), which may be relevant in cloud forest situations.

The design of fog collectors for stable isotope sampling also needs further work. In areas with mixed fog and rain precipitation, passive fog collectors generally collect some proportion of wind-blown rainfall, even if they have a cover to keep out the rain (Fischer and Still, 2007; McJannet *et al.*, 2007; Scholl *et al.*, 2007; Frumau *et al.*, this volume; Giambelluca *et al.*, this volume). Active collectors require a power source which makes them less practical for use in remote areas or for long periods of time. Such problems in the sampling of fog and rain have led to some uncertainty in measured isotopic compositions. Measurements that would help to resolve these questions include analysis of isotopic compositions across the range of droplet sizes within a raining cloud, and at several altitudes within a single cloud. Despite the present sampling limitations, it is clear that in hydrological studies involving isotope techniques in mountain forests, relying on the isotopic composition of precipitation as collected in a standard funnel-type collector under open sky could lead to misinterpretation. Given the magnitude of potential fog capture by the canopy of many cloud-affected forests, isotope samples should be analyzed from rain, throughfall, and cloud water collectors alike to ascertain the relative contributions by rainfall and fog to overall ecosystem inputs.

ACKNOWLEDGEMENTS

We thank Dr. Douglas Burns and Dr. Karen Rice of the U. S. Geological Survey, and an anonymous reviewer, for helpful comments on the manuscript.

REFERENCES

Allison, G.B., C.J. Barnes, and M.W. Hughes (1983). Distribution of deuterium and oxygen-18 in dry soils. II. Experimental. *Journal of Hydrology* **64**: 377–397.

Aravena, R., O. Suzuki, and A. Pollastri (1989). Coastal fog and its relation to groundwater in the IV region of northern Chile. *Chemical Geology (Isotope Geoscience Section)* **79**: 83–91.

Barnes, C.J., and G.B. Allison (1983). The distribution of oxygen-18 and deuterium in dry soils. I. Theory. *Journal of Hydrology* **60**: 141–156.

Barnes, C.J., and J.V. Turner (1998). Isotopic exchange in soil water. In *Isotope Tracers in Catchment Hydrology*, eds. C. Kendall and J.J. McDonnell, pp. 137–163. Amsterdam: Elsevier.

Brodersen, C., S. Pohl, M. Lindenlaub, C. Leibundgut, and K. von Wilpert (2000). Influence of vegetation structure on isotope content of throughfall and soil water. *Hydrological Processes* **14**: 1439–1448.

Bruijnzeel, L.A. (2000). Forest hydrology. In *The Forest Handbook*, ed. J.C. Evans, pp. 301–343. Oxford, UK: Blackwell Scientific.

Bruijnzeel, L.A. (2001). Hydrology of tropical montane cloud forests: a reassessment. *Land Use and Water Resources Research* **1**: 1–8.

Bruijnzeel, L.A., W. Eugster, and R. Burkard (2005). Fog as an input to the hydrological cycle. In *Encyclopaedia of Hydrological Sciences*, eds. M.G. Anderson and J.J. McDonnell, pp. 559–582. Chichester, UK: John Wiley.

Burkard, R. (2003). *Fogwater flux measurements above different vegetation canopies*. Ph.D. thesis, University of Bern, Bern, Switzerland.

Burkard, R., P. Bützberger, and W. Eugster (2003). Vertical fogwater flux measurements above an elevated forest canopy at the Lägeren research site, Switzerland. *Atmospheric Environment* **37**: 2979–2990.

Buttle, J.M. (1998). Fundamentals of small catchment hydrology. In *Isotope Tracers in Catchment Hydrology*, eds. C. Kendall and J.J. McDonnell, pp. 1–49. Amsterdam: Elsevier.

Clark, I.D., P. Fritz, O.P. Quinn, *et al.* (1987). Modern and fossil groundwater in an arid environment: a look at the hydrogeology of southern Oman. In *Symposium on Isotope Techniques in Water Resources Development*, pp. 167–187. Vienna: International Atomic Energy Agency.

Clark, I.D., and P. Fritz (1997). *Environmental Isotopes in Hydrogeology*. New York: CRC Press.

Corbin, J.D., M.A. Thomsen, T.E. Dawson, and C.M. D'Antonio (2005). Summer water use by California coastal prairie grasses: fog, drought, and community composition. *Oecologia* **145**: 511–521.

Craig, H. (1961). Isotopic variations in meteoric waters. *Science* **133**: 1702–1703.

Dansgaard, W. (1964). Stable isotopes in precipitation. *Tellus* **16**: 436–468.

Daube B.C. Jr., R.C. Flagan, and M.R. Hoffmann (1986). *Active Cloudwater Collector*. US Patent No. 4697462.

Daube, B., K.D. Kimball, P.A. Lamar, and K.C. Weathers (1987). Two new ground-level cloud water sampler designs which reduce rain contamination. *Atmospheric Environment* **21**: 893–900.

Dawson, T.E. (1993). Water sources of plants as determined from xylem-water isotopic composition: perspectives on plant competition, distribution, and water relations. In *Stable Isotopes and Plant Carbon–Water Relations*, eds. J.R. Ehleringer, A.E. Hall, and G.D. Farquhar, pp. 465–496. San Diego, CA: Academic Press.

Dawson, T.E. (1998). Fog in the California redwood forest: ecosystem inputs and use by plants. *Oecologia* **117**: 476–485.

Dawson, T.E., and J.R. Ehleringer (1998). Plants, isotopes and water use: a catchment-scale perspective. In *Isotope Tracers in Catchment Hydrology*, eds. C. Kendall and J.J. McDonnell, pp. 165–202. Amsterdam, Elsevier.

Dawson, T.E., and P.E. Vidiella (1998). Plant–fog interactions in California and Chile. In *Proceedings of the 1st International Conference on Fog and Fog Collection*, eds. R.S. Schemenauer and H. Bridgman, pp. 225–228. Ottawa, Canada: IDRC.

Eugster, W. (2007). The relevance of fog for the vegetation: is it the water or the nutrients that matter? In *Proceedings of the 4th International Conference on Fog, Fog Collection and Dew*, eds. A. Biggs and P. Cereceda, pp. 359–362. Santiago de Chile: Catholic University.

Federer, B., N. Brichet, and J. Jouzel (1982). Stable isotopes in hailstones. I. The isotopic cloud model. *Journal of the Atmospheric Sciences* **39**: 1323–1335.

Feild, T.S., and T.E. Dawson (1998). Water sources used by *Didymopanax pittieri* at different life stages in a tropical cloud forest. *Ecology* **79**: 1448–1452.

Fischer, D.T., and C.J. Still (2007). Evaluating patterns of fog water deposition and isotopic composition on the California Channel Islands. *Water Resources Research* **43**: W04420, doi:101029/2006WR005124.

Fischer, D.T., C.J. Still, and A.P. Williams (2009). Significance of summer fog and overcast for drought stress and ecological functioning of coastal California endemic plant species, *Journal of Biogeography* **36**: 783–799.

García Santos, G. (2007). *An ecohydrological and soils study in a subtropical montane cloud forest in the National Park of Garajonay, La Gomera, (Canary Island, spain)*. Ph.D. thesis, VU University Amsterdam, Amsterdam, The Netherlands. Available at www.falw.vu.nl/nl/onderzoek/earth-sciences/geo-environmental-science-and-hydrology/hydrology-dissertations/index.asp.

Gat, J.R. (2000). Atmospheric water balance: the isotopic perspective. *Hydrological Processes* **14**: 1357–1369.

Gedzelman, S.D., and R. Arnold (1994). Modeling the isotopic composition of precipitation. *Journal of Geophysical Research* **99**: 10455–10471.

Glickman, T.S. (2000). *Glossary of Meteorology*, 2nd edn. Boston, MA: American Meteorological Society. Also available at http://amsglossary.allenpress.com/glossary.

Goller, R., W. Wilcke, M. Leng, *et al.* (2005). Tracing water paths through small catchments under a tropical montane rain forest in south Ecuador by an oxygen isotope approach. *Journal of Hydrology* **308**: 67–80.

Gonfiantini, R., and A. Longinelli (1962). Oxygen isotopic composition of fogs and rains from the North Atlantic. *Experientia* **18**: 222–223.

Holwerda, F., R. Burkard, W. Eugster, *et al.* (2006). Estimating fog deposition at a Puerto Rican elfin cloud forest site: comparison of the water-budget and eddy covariance methods. *Hydrological Processes* **20**: 2669–2692.

Ingraham, N.L., and R.E. Criss (1993). Effects of surface area and volume on the rate of isotopic exchange between water and water vapor. *Journal of Geophysical Research (Atmospheres)* **98**: 20547–20553.

Ingraham, N.L., and A.F. Mark (2000). Isotopic assessment of the hydrological importance of fog deposition on tall snow tussock grass on southern New Zealand uplands. *Austral Ecology* **25**: 402–408.

Ingraham, N.L., and R.A. Matthews (1988). Fog drip as a source of groundwater recharge in northern Kenya. *Water Resources Research* **24**: 1406–1410.

Ingraham, N.L., and R.A. Matthews (1990). A stable isotope study of fog: the Point Reyes Peninsula, California, U.S.A. *Chemical Geology (Isotope Geoscience Section)* **80**: 281–290.

Ingraham, N.L., and R.A. Matthews. (1995). The importance of fog drip water to vegetation: Point Reyes Peninsula, California. *Journal of Hydrology* **164**: 269–285.

Jouzel, J., L. Merlivat, and E. Roth (1975). Isotopic study of hail. *Journal of Geophysical Research* **80**: 5015–5030.

Kendall, C. (1992). Temporal, spatial and species effects on the oxygen and hydrogen isotopic compositions of throughfall. *EOS, Transactions of the American Geophysical Union* **73**: 160.

Kerfoot, O. (1968). Mist precipitation on vegetation. *Forestry Abstracts* **29**: 8–20.

Landon, M.K., G.N. Delin, S.C. Komor, and C.P. Regan (1999). Comparison of the stable-isotopic composition of soil water collected from suction lysimeters, wick samplers, and cores in a sandy unsaturated zone, *Journal of Hydrology* **224**: 45–54.

Liu, W.J., Y.P. Zhang, H.M. Li, and H.M. Liu (2005). Fog drip and its relation to groundwater in the tropical seasonal rain forest of Xishuangbanna, Southwest China. *Water Research* **39**: 787–794.

Liu, W.J., W.Y. Liu, P.J. Li, *et al.* (2007). Using stable isotopes to determine sources of fog drip in a tropical seasonal rain forest of Xishuangbanna, SW China. *Agricultural and Forest Meteorology* **143**: 80–91.

Maloszewski, P., and A. Zuber (1982). Determining the turnover time of groundwater systems with the aid of environmental tracers. I. Models and their applicability. *Journal of Hydrology* **57**: 207–231.

Martinelli, L.A., R.L. Victoria, L.S.L. Sternberg, A. Ribeiro, and M.Z. Moreira (1996). Using stable isotopes to determine sources of evaporated water to the atmosphere in the Amazon basin. *Journal of Hydrology* **183**: 191–204.

McDonnell, J.J., M. Bonell, M.K. Stewart, and A.J. Pearce (1990). Deuterium variations in storm rainfall: implications for stream hydrograph separation. *Water Resources Research* **26**: 455–458.

McGuire, K.J., D.R. DeWalle, and W.J. Gburek (2002). Evaluation of mean residence time in subsurface waters using oxygen-18 fluctuations during drought conditions in the mid-Appalachians. *Journal of Hydrology* **261**: 132–149.

McJannet, D.L., J.S. Wallace, and P. Reddell (2007). Precipitation interception in Australian tropical rainforests. II. Altitudinal gradient of cloud interception, stemflow, throughfall and interception. *Hydrological Processes* **21**: 1703–1718.

Meinzer, F.C., J.L. Andrade, G. Goldstein, *et al.* (1999). Partitioning of soil water among canopy trees in a seasonally dry tropical forest. *Oecologia* **121**: 293–301.

Mook, W.G., and J.J. de Vries (2001). *Environmental Isotopes in the Hydrological Cycle: Principles and Applications, Vol. I, Introduction: Theory, Methods, Review*. Paris: UNESCO, and Vienna: IAEA. Also available at www.iaea.org/programmes/ripc/ih/volumes/volume1.htm.

Nespor, V., and B. Sevruk (1999). Estimation of wind-induced error of rainfall gauge measurements using a numerical simulation. *Journal of Atmospheric and Oceanic Technology* **16**: 450–464.

Revesz, K., and P.H. Woods (1990). A method to extract soil water for stable isotope analysis. *Journal of Hydrology* **115**: 397–406.

Rhodes, A.L., A.J. Guswa, and S.E. Newell (2006). Seasonal variation in the stable isotopic composition of precipitation in the tropical montane forests of Monteverde, Costa Rica. *Water Resources Research* **42**, W11402, doi:10.1029/2005WR004535.

Rozanski, K., Araguas-Araguas, L., and R. Gonfiantini (1993). Isotopic patterns in modern global precipitation. In *Climate Change in Continental Isotopic Records*, eds. P.K. Swart, K.C. Lohman, J. McKenzie, and S. Savin, pp. 1–36. Washington, DC: American Geophysical Union.

Salati, E., A. Dall'Olio, E. Matsui, and J.R. Gat (1979). Recycling of water in the Amazon basin: an isotopic study. *Water Resources Research* **15**: 1250–1258.

Saxena, R.K. (1986). Estimation of canopy reservoir capacity and oxygen-18 fractionation in throughfall in a pine forest. *Nordic Hydrology* **17**: 251–260.

Scholl, M.A., S.E. Ingebritsen, C.J. Janik, and J.P. Kauahikaua (1995). *An isotope hydrology study of the Kilauea Volcano area, Hawaii* U.S. Geological Survey Water Resources Investigations Report No. 95–4213. Washington, DC: U.S. Government Printing Office.

Scholl, M.A., S.B. Gingerich, and G.W. Tribble (2002). The influence of microclimates and fog on stable isotope signatures used in interpretation of regional hydrology: East Maui, Hawaii. *Journal of Hydrology* **264**: 170–184.

Scholl, M.A., J.B. Shanley, and J.W. Troester (2006). Stable isotope measurements of rain, cloud water, and streams in the Luquillo mountains, Puerto Rico, *Geological Society of America, Abstracts with Programs* **38**(7): 97.

Scholl, M.A., T.W. Giambelluca, S.B. Gingerich, M.A. Nullet, and L.L. Loope (2007). Cloud water in windward and leeward mountain forests: the stable isotope signature of orographic cloud water. *Water Resources Research* **43**: W12411, doi:101029/2007WR006011.

Siegenthaler, U., and H. Oeschger (1980). Correlation of ^{18}O in precipitation with temperature and altitude. *Nature* **285**: 314–317.

Sklash, M.G., and R.N. Farvolden (1979). The role of groundwater in storm runoff. *Journal of Hydrology* **43**: 45–65.

Sklash, M.G., R.N. Farvolden, and P. Fritz (1976). A conceptual model of watershed response to rainfall, developed through the use of oxygen-18 as a natural tracer. *Canadian Journal of Earth Sciences* **13**: 271–283.

Still, C.J., P.N. Foster, A. Pounds, and A. Williams (2003). Preliminary measurements of oxygen-18 and hydrogen-2 in water samples collected from a cloud forest in Monteverde, Costa Rica. *EOS, Transactions of the American Geophysical Union, Fall Meeting Supplement* **84**: Abstract B31E-0348.

Te Linde, A.H., L.A. Bruijnzeel, J. Groen, F.N. Scatena, and H.A.J. Meijer (2001). Stable isotopes in rainfall and fog in the Luquillo Mountains, eastern Puerto Rico: a preliminary study. In *Proceedings of the 2nd International Conference on Fog and Fog Collection*, eds. R.S. Schemenauer and H.A. Puxbaum, pp. 181–184. Ottawa, Canada: IDRC.

Thalmann, E., R. Burkard, T. Wrzesinsky, W. Eugster, and O. Klemm (2002). Ion fluxes from fog and rain to an agricultural and a forest ecosystem in Europe. *Atmospheric Research* **64**: 147–158.

Thorburn, P.J., T.J. Hatton, and G.R. Walker (1993a). Combining measurements of transpiration and stable isotopes of water to determine groundwater discharge from forests. *Journal of Hydrology* **150**: 563–587.

Thorburn, P.J., G.R. Walker, and J.-P. Brunel (1993b). Extraction of water from *Eucalyptus* trees for analysis of deuterium and oxygen-18: laboratory and field techniques. *Plant, Cell, and Environment* **16**: 269–277.

Wang, X.-F., and D. Yakir (2000). Using stable isotopes of water in evapotranspiration studies. *Hydrological Processes* **14**: 1407–1421.

West, A.G., S.J. Patrickson, and J.R. Ehleringer (2006). Water extraction times for plant and soil materials used in stable isotope analysis. *Rapid Commmunications in Mass Spectrometry* **20**: 1317–1321.

White, J.W.C., E.R. Cook, J.R. Lawrence, and W.S. Broecker (1985). The D/H ratios of sap in trees: implications for water sources and tree ring D/H ratios. *Geochimica et Cosmochimica Acta* **49**: 237–246.

Zadroga, F. (1981). The hydrological importance of a montane cloud forest area of Costa Rica. In *Tropical Agricultural Hydrology*, eds. R. Lal and E. W. Russell, pp. 59–73. New York: John Wiley.

24 Using stable isotopes to identify orographic precipitation events at Monteverde, Costa Rica

A. L. Rhodes, A. J. Guswa and S. E. Newell
Smith College, Northampton, Massachusetts, USA

ABSTRACT

Monteverde, in north-western Costa Rica, is located on the leeward side of the Continental Divide, and experiences strong seasonal variations in precipitation. The majority of rainfall to the region occurs during the wet season (May–October) when the Intertropical Convergence Zone (ITCZ) brings convective rainfall. In addition, trade winds produce orographic uplift and condensation, and this hydrological input is dominant during the transitional (November–January) and dry (February–April) seasons. This chapter assesses the viability of using stable isotopes of oxygen and hydrogen to identify and distinguish waters arisen from these two condensation mechanisms, with the future goal of using these signatures to trace precipitation through the hydrological cycle.

Open-field precipitation and throughfall were sampled at Monteverde at an elevation of 1460 m.a.s.l. from mid-June 2003 to mid-January 2004. Cumulative precipitation was collected over a variable sampling interval that ranged from 1 to 48 days. The sampling interval averaged 4 days during June–July 2003 and January 2004; otherwise, samples were collected on average every 26 days. The wet season yielded a wide range of isotopic values, from −13.1 to −3.0‰ ($\delta^{18}O$) and −96 to −15‰ ($\delta^{2}H$). The isotopically lightest samples were collected during June and September/October, when the ITCZ is over Costa Rica. Rain and fog samples collected during the transitional season were isotopically heavier, ranging from −4.3 to −1.9‰ ($\delta^{18}O$) and −16 to + 6‰ ($\delta^{2}H$). January precipitation samples all yielded positive $\delta^{2}H$ values and had isotopic compositions similar to those of fog water samples reported elsewhere. The observed variation over the study period is consistent with what is expected for rain events having different condensation histories. Higher (heavier) isotopic values resulted from orographic precipitation events whereas lower (lighter) isotopic values are produced by convective and other precipitation events where the vapor mass has become progressively depleted by rain-out processes. Precipitation samples collected between September and January had *d*-excess values that ranged from 16% to 21‰. The highest *d*-excess values were found in January when orographic precipitation dominates, indicating the possibility of a recycled water component to the regional hydrological cycle.

INTRODUCTION

The montane cloud forests of Monteverde are located on the Cordillera de Tilarán, the volcanic range that divides north-western Costa Rica (Figure 24.1). The climate of this region is marked by three distinct seasons: viz. wet, transitional, and dry (Clark *et al.*, 2000). The wet season (May–October) corresponds to the time when the Intertropical Convergence Zone (ITCZ) is located over Costa Rica, and precipitation is dominated by intensive convective events. The transitional (November–January) and dry (February–April) seasons correspond to the months when the ITCZ is located to the south of Costa Rica. During this time, fog is generated via orographic uplift. The trade winds force moist air from the Atlantic Ocean up the Tilarán Mountains, and the region becomes enshrouded in clouds and drizzle (cf. Lawton *et al.*, this volume). Schmid *et al.* (this volume) concluded that 4–7% of water inputs to a windward cloud forest in the Monteverde region during the 2003 dry season came from fog.

Whilst much of the annual precipitation can be attributed to the presence of the ITCZ during the wet season, the importance

Tropical Montane Cloud Forests: Science for Conservation and Management, eds. L. A. Bruijnzeel, F. N. Scatena, and L. S. Hamilton. Published by Cambridge University Press.

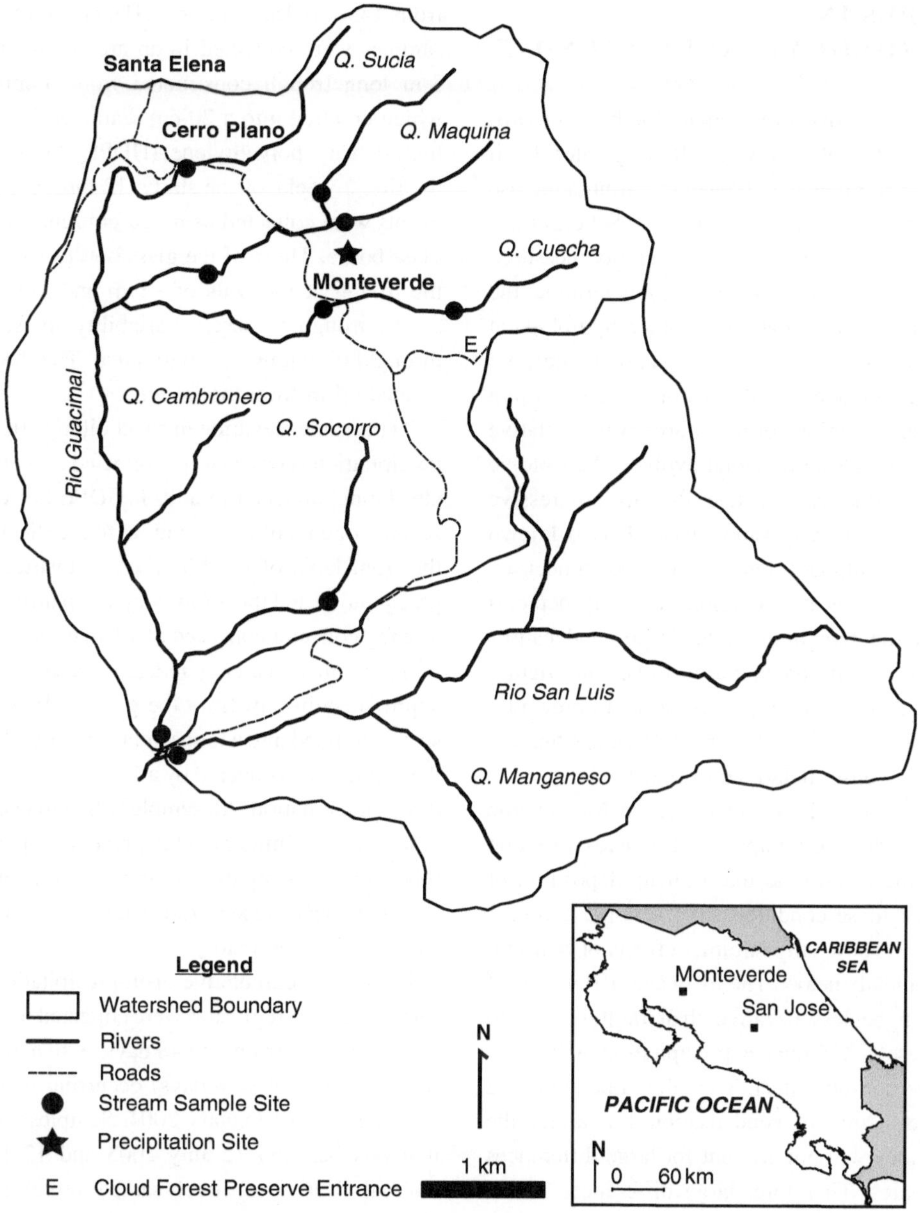

Figure 24.1. Rio Guacimal and Rio San Luis watersheds, showing sites sampled for stable isotope composition.

of orographic precipitation during the dry season should not be underestimated. This is especially true on the leeward (Pacific) side of the Continental Divide. Orographic precipitation is limited to the upper elevations where clouds spill over the mountains before dispersing as the air masses warm during their descent. Despite the limited availability of water during the dry season, this region is home to nearly all of the development associated with Monteverde. The recent boom in ecotourism has put significant stress on water resources, especially during the dry season when visitors increase in number and overall water demands are greater. In addition to these community impacts, recent work has shown evidence that regional land-use and climate change may lead to increased cloud base elevations (Still *et al.*, 1999; Lawton *et al.*, 2001; Karmalkar *et al.*, 2008; Nair *et al.*, this volume). This would reduce the orographic component of precipitation to the hydrological cycle, affecting both ecology and water resources (cf. Zadroga, 1981; Clark *et al.*, 1998). Because of the potential for reduced water supply and the reality of increased demand, a better understanding of the role of orographic precipitation is essential for sustaining the vibrant Monteverde region.

This chapter presents results from a preliminary investigation of precipitation patterns observed in Monteverde. The completion of this study can be found in Rhodes *et al.* (2006), which presents a longer data-set from 2003 to 2005 and expanded discussion (see also Guswa *et al.*, 2007).

STABLE ISOTOPES IN HYDROMETEOROLOGY: BACKGROUND

Whilst seasonality provides a first-order separation between orographic and ITCZ-derived rainfall in Costa Rica, greater detail regarding the hydrological inputs may be gained through the use of stable isotopes (cf. Schmid *et al.*, this volume; Scholl *et al.*, this volume). The isotopic composition of water (as measured through $\delta^{18}O$ and $\delta^{2}H$) may also be used to determine the contribution of orographic precipitation to other hydrological fluxes within the watershed (e.g. plant uptake and streamflow; Scholl *et al.*, 2002). The relative abundances of ^{18}O and ^{2}H in a first-stage condensate are functions of the source water, relative humidity, and temperature. As a rain cloud evolves, the isotopic composition of precipitation can change due to progressive condensation from the surrounding vapor mass. This is known as the rain-out effect; continued removal of the condensed phase from the vapor mass by precipitation gradually depletes the heavy isotopes in the remaining water vapor and causes the isotopic composition of the precipitation to become lighter as the storm progresses. The rain-out process leads to precipitation at the end of a rain event with $\delta^{18}O$ and $\delta^{2}H$ values that are more negative than at the onset (Clark and Fritz, 1997).

Consistent with their name, the cloud forests of Monteverde are found in areas for which orographic uplift leads to cloud condensation at the same location as the ultimate deposition of the liquid water. Under these conditions, the isotopic composition of precipitation will show only minimal effects of rain-out, and water will be isotopically heavy. The $\delta^{18}O$ and $\delta^{2}H$ values of orographic rainfall will be less negative than rainfall derived from the ITCZ, because ITCZ-related precipitation will show greater evidence of the rain-out effect. Though temperature and relative humidity at the time of condensation can affect the isotopic composition, they will not account for large differences in δ-values in the tropics. Therefore, large differences in the isotopic composition of precipitation may be used to identify the condensation mechanism of liquid water at Monteverde (Rhodes *et al.*, 2006), as well as in other mountainous regions in the tropics (Scholl *et al.*, this volume).

METHODS

During the period from 26 June 2003 to 18 January 2004 covering the wet and transitional seasons, precipitation samples were collected at a forested site in the Rio Guacimal watershed (discussed in Rhodes *et al.*, this volume #45) at an elevation of 1460 m.a.s.l. where two Holdridge Life Zones overlap (Figure 24.1). On this leeward Pacific slope, so-called "premontane wet forest" extends from 800 to 1500 m.a.s.l. whereas "lower montane wet forest" (cloud forest) encompasses higher elevations from 1450 to 1600 m.a.s.l. (Haber, 2000). Fourteen throughfall samples were collected in an area of secondary growth using a 3-m long trough constructed from 5-cm diameter PVC pipe, which emptied into a 20-cm diameter funnel, attached to a 20-l, high-density polyethylene (HDPE) bucket. In addition, during the first 2 weeks of the study, 12 throughfall samples from three events were collected using 20-cm funnels, each attached to a 4-l glass bottle. Three of the glass bottles were positioned alongside the trough at intervals of ~1 m and sampled at the same time as the trough to assess variability in isotopic composition of throughfall within a small area. The fourth glass bottle was located ~5 m from the trough.

At the same elevation in an adjoining 100-m^2 clearing, 13 open-precipitation samples were collected as well, using a 20-cm diameter funnel attached to a 20-l, HDPE bucket. In January 2004, a second open collector and a fog collector were installed on the front lawn of the Monteverde Institute (yielding two open-precipitation and three fog samples) at an elevation of 1400 m.a.s.l. The fog collector consisted of a 1-m high, fine-mesh screen formed into a 20-cm diameter cylinder; the base of the cylinder was erected approximately 2 m from the ground. Moisture scavenged by the screen emptied into a funnel attached to a 4-l glass bottle. The top of the cylinder was covered by a 1 × 1 m roof to minimize the input of direct precipitation. All sample collection vessels were coated with at least 1 cm of mineral oil to prevent evaporation (cf. Scholl *et al.*, this volume). Samples were transferred to 30-ml glass bottles with poly-seal caps and kept refrigerated (~10 °C) until transport to the stable isotope laboratory.

In total, 44 cumulative open-precipitation, fog, and throughfall samples and 18 replicates were collected on 14 dates. The sampling interval ranged from 1 to 48 days, with a high sampling frequency (average interval = 4 days) occurring between 28 June–12 July 2003 and 12–18 January 2004. Samples were collected at longer intervals between 12 July 2003 and 12 January 2004 (average interval = 26 days). Each sample represents an accumulation of the rain events that occurred over the sampling interval, except for the October collection when the bucket overflowed. In addition to precipitation, on 11 January 2004 ten streamwater samples were collected from major tributaries of the upper Rio Guacimal watershed and at the mouth of the Rio San Luis (Figure 24.1).

Stable isotope compositions of precipitation and streamwater samples were measured at the University of Nevada at Reno. Water $\delta^{18}O$ analyses were performed using a Micromass Multi-Prep-IsoPrime system, according to the CO_2–H_2O equilibration method of Epstein and Mayeda (1953). Water $\delta^{2}H$ analyses were performed using a Eurovector elemental analyzer–Micromass IsoPrime system, after the method of Morrison *et al.* (2001). Results of $\delta^{18}O$ and $\delta^{2}H$ are analyses reported in the customary units of per mil (‰) vs. V-SMOW (Morrison *et al.*, 2001). The $\delta^{18}O$ and $\delta^{2}H$ measurements have an uncertainty of 0.1‰ and 1‰, respectively.

RESULTS AND DISCUSSION

Isotope compositions for all precipitation samples ranged from −12.8 to −1.3‰ ($\delta^{18}O$) and −94 to + 8‰ (δ^2H). No differences in stable isotope compositions were observed between samples from the two open-precipitation locations and between open-precipitation and fog samples. However, the isotopic composition differed among the four throughfall (*TF*) collectors, despite the fact that these collectors were positioned within close proximity of each other. The isotopic composition of *TF* varied up to 0.8‰ for $\delta^{18}O$ and up to 5‰ for δ^2H among the different collectors over the same sampling interval. In addition, differences between open-field precipitation and *TF* collected over the same period ranged from −0.9 to + 1.4‰ for $\delta^{18}O_{OPEN\text{-}TF}$ and from −8 to + 8‰ for $\delta^2H_{OPEN\text{-}TF}$. The direction and magnitude of these differences varied between sampling periods, and they showed no consistent temporal pattern. Causes for the isotopic differences between *TF* collectors and between open-field precipitation and *TF* are unclear from this limited data-set; the differences may have resulted from incomplete mixing of precipitation within the forest canopy – either during a storm or among different events (Saxena, 1986). Volume-weighted averages of $\delta^{18}O$ and δ^2H for both open-field precipitation and *TF* calculated over the period 26 June–12 July 2004 were within analytical error.

Whilst leeward-slope transect data are not available to directly test the effect of progressive rain-out and variations in temperature with elevation on the $\delta^{18}O$ values of precipitation, these processes are not thought to explain the observed differences in the composition of precipitation collected at 1400 and 1460 m.a.s.l. (see discussion in Guswa *et al.*, 2007). Scholl *et al.* (1996, 2002) found that $\delta^{18}O$ values varied with elevation by ~ −1.4 ‰ km^{-1} on the leeward slopes of the islands of Maui and Hawai'i. This change is similar to the presently observed differences between precipitation collectors located at the same elevation, so any effect of rain-out on isotopic composition due to slight elevation differences is within the observed variability, and very small compared to the seasonal changes in $\delta^{18}O$.

Figure 24.2 shows that all precipitation and stream data plot near the global meteoric water line (GMWL) defined by Craig (1961), whereas Figure 24.3 illustrates the strong temporal variation in isotopic composition of precipitation that is consistent with the seasonal climate in Monteverde. The lightest isotopic composition, marked by the lowest $\delta^{18}O$ and δ^2H values, occurred in samples collected during the rainiest periods of the wet season in June 2003 ($\delta^{18}O_{Open}$ = −11.1‰ and δ^2H_{Open} = −83‰) and October 2003 ($\delta^{18}O_{TF}$ = −12.4‰ and δ^2H_{TF} = −82‰), when the ITCZ was positioned directly over Costa Rica (cf. Clark *et al.*, 2000). Precipitation collected during the transitional season was much heavier and ranged from −3.9 to −1.3‰ ($\delta^{18}O$) and −14 to + 8‰ (δ^2H). In January, at the end of the transitional season in Monteverde when orographic precipitation dominates, all samples yielded positive δ^2H values and were similar to stable isotope compositions of fog water reported elsewhere (Ingraham and Matthews, 1988, 1995; Aravena *et al.*, 1989; Dawson, 1998; cf. Scholl *et al.*, this volume).

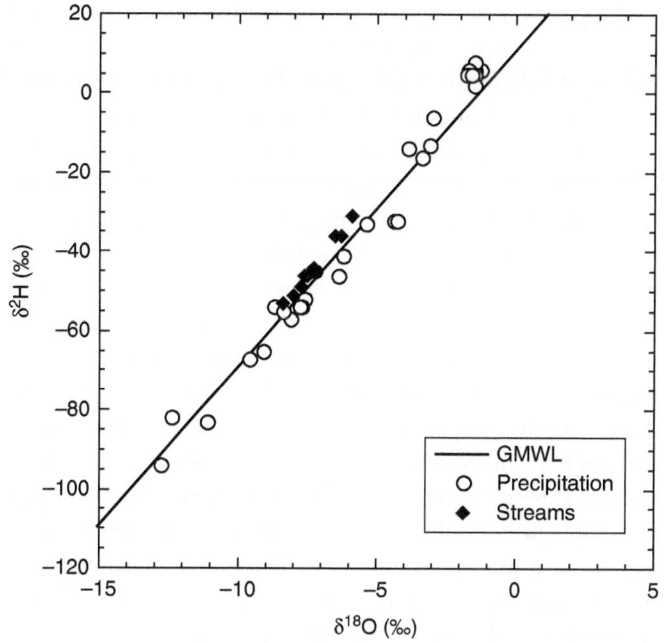

Figure 24.2. Stable isotope compositions of precipitation and stream samples collected between 24 June 2003 and 18 January 2004.

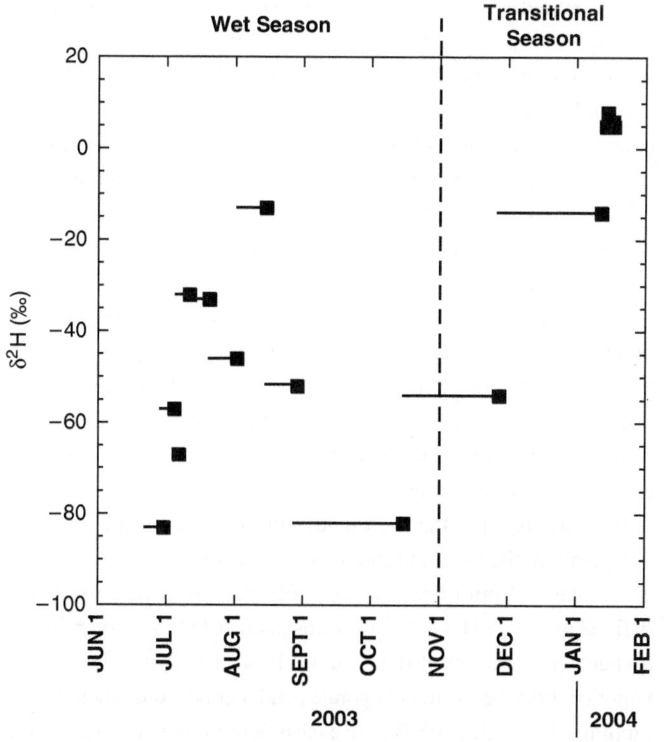

Figure 24.3. Temporal variation of δ^2H in precipitation at Monteverde. Bars show duration of sample collection period. All samples are open-field precipitation, except October, which is throughfall.

The present data suggest that rain-out has a strong control on the stable isotope composition of precipitation, and that the degree of rain-out decreases from the wet to the transitional season. This coincides with the seasonal occurrence of convective storms during the wet season and a dominance of orographic precipitation during the transitional season. Rain events associated with the ITCZ have long travel paths before reaching Monteverde, and therefore have light stable isotope compositions. During the transitional season, however, the rain-out effect on stable isotopic composition is small, as exhibited by the heavier $\delta^{18}O$ and $\delta^{2}H$ values. Differences in condensation temperatures alone – without a significant rain-out process – cannot account for the observed differences in stable isotopic compositions of precipitation between the wet and transitional seasons at Monteverde. Even though convective clouds can condense at higher elevations (and thus lower temperatures), than do near-surface orographic clouds, equilibrium fractionation factors for ^{18}O and ^{2}H in vapor-water reactions are larger at lower temperatures (Majoube, 1971). Thus, precipitation condensed at lower temperatures will be isotopically enriched relative to precipitation condensed at higher temperatures. Because air masses continually cool by adiabatic expansion as a storm progresses, progressive distillation of the heavy isotopes from vapor during the rain-out process is necessary for precipitation to evolve toward lighter isotopic compositions (Clark and Fritz, 1997).

Scholl *et al.* (2002) observed similar results in the Hawaiian Islands; storm systems that progressed at sea before reaching the islands were isotopically depleted relative to precipitation that originated when warm, trade-wind-driven air rose over topography. Further work is needed to explain the large variation in isotopic composition during the wet season ($\delta^{18}O = -12.8$ to -3.1‰ and $\delta^{2}H = -94$ to -13‰). Measured isotope values in precipitation increased from July through early August, when the ITCZ is north of Monteverde. The observed variation could reflect differences in storm travel distance, variation in condensation temperature, the mixing of convective and orographic events, or a combination of these effects.

In addition to the seasonal controls on stable isotopic composition, the precipitation data show evidence of increasing *d*-excess values from June to January (Figure 24.4), where *d*-excess is defined as $d = \delta^{2}H - 8 * \delta^{18}O$ (Dansgaard, 1964). A proportionally higher concentration of deuterium relative to oxygen-18 in water vapor occurs when water evaporates under conditions when relative humidity is less than 100%; the excess deuterium is incorporated in moisture when the vapor condenses. The global meteoric water line has a *d*-excess value of ~10‰, due to evaporation from the oceans at an average relative humidity of ~85% (Clark and Fritz, 1997). A second stage of evaporation will introduce proportionally more deuterium into the vapor phase, leading to precipitation with even larger *d*-excess values. Therefore, a *d*-excess value greater than ~10‰ indicates that the precipitation contains water either recycled

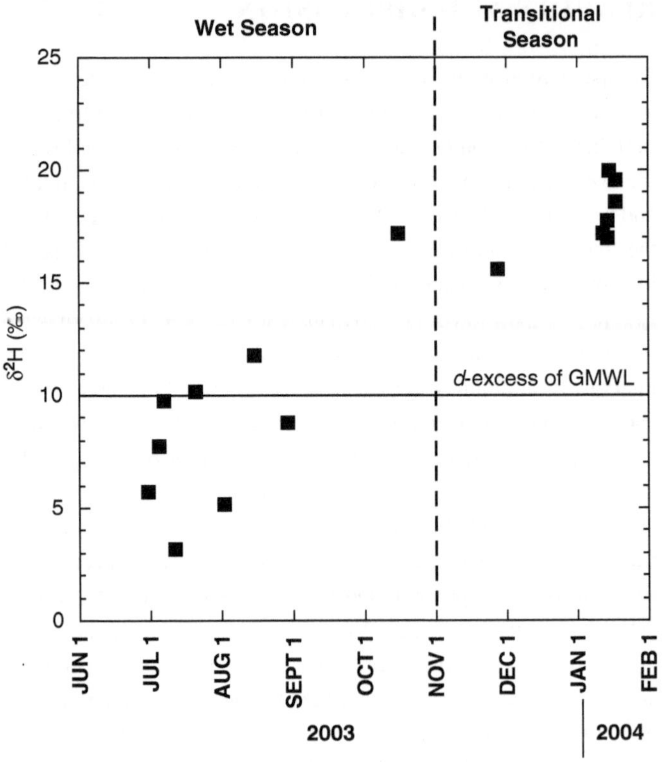

Figure 24.4. Temporal variation of *d*-excess values in precipitation at Monteverde.

from land or condensed from vapor that formed when relative humidity was less than ~85%.

In a survey of streamwater samples throughout Costa Rica, Lachniet and Patterson (2002) observed elevated and increasing *d*-excess values inland from the Caribbean Sea that were up to 20‰, suggesting that recycled water from evaporation was an important contributor to the inland water budget. This phenomenon was also observed by Gat and Matsui (1991) in the Amazon Basin, who attributed a 3‰ increase in the *d*-excess value of precipitation to an evaporative flux. The present data show that between September and January, *d*-excess values of precipitation were higher than 15.6‰ and reached values as large as 22.6‰ (Figure 24.4). Similar to these other studies, the present data may thus be interpreted as indicating a contribution from recycled water to vapor masses moving inland. This phenomenon appears most significant during the transitional season when orographic precipitation tends to be the dominant source of water, and moisture derived from the Caribbean mixes with terrestrial water added via evaporation. One would expect to observe high *d*-excess values throughout the dry season as well.

The high *d*-excess value (17.2‰) observed in bulk precipitation collected from September to October suggests that precipitation at Monteverde also contains recycled water during the rainiest months of Costa Rica's wet season, whereas a recycled water signal may be absent in the beginning of the wet season.

Comparison of precipitation depths measured at La Selva (lowland Caribbean side) and Palo Verde (Pacific side) (Organization for Tropical Studies, 2004) shows that the Pacific slope receives more rain than the Caribbean side of the continental divide during September and October. Furthermore, streamwater samples collected on 11 January 2004 from major tributaries in the Rio Guacimal watershed showed elevated *d*-excess values (Figure 24.4). This is consistent with the precipitation signature during the transitional season, but also may suggest that precipitation at the end of the wet season contributes to transitional season base flow. Further work investigates the relationship between changing wind and precipitation patterns associated with the seasonal strengthening and weakening of the ITCZ in the Pacific equatorial region and their connection to observed stable isotopic compositions of precipitation at Monteverde (Rhodes *et al.*, 2006). Guswa *et al.* (2007) show an attenuated version of the seasonal isotopic signal in precipitation in stream water, illustrating the importance of trade-wind-derived, orographic precipitation to water resources of the Monteverde region.

CONCLUSIONS

Stable isotopes may serve as an important tool in tracing water with different formation histories through cloud-forest watersheds. The isotopic composition of precipitation collected in the Monteverde area in north-western Costa Rica shows strong temporal variations consistent with differences in condensation history of rain events that occur in the wet and transitional seasons. Compositions of ITCZ-related convective precipitation during the wet season are isotopically light, whereas heavier values are observed for orographic precipitation during the transitional season. High *d*-excess values for samples collected between September and January imply a recycled component and/or a change in source from the Caribbean to the Pacific Ocean for precipitation that falls in the Monteverde region. The presence of a recycled component indicates that lowland deforestation, and the associated reduction in evaporation, would reduce precipitation in Monteverde, as suggested by Lawton *et al.* (2001) and Nair *et al.* (this volume), and this impact would be most severe in the dry season. Further work will trace the different precipitation types through the Rio Guacimal watershed and will investigate possible topographic controls on the spatial variability of orographic water in the Monteverde region.

ACKNOWLEDGEMENTS

We are grateful to Nat Scrimshaw, The Monteverde Institute, and the Monteverde Community Arts Center, who provided logistical support and staff resources for this research. S. Poulson advised the stable isotope analyses at University of Nevada, Reno. M. Marin, A. Sarma, I. Johnson, and J. Neibler assisted with field work. Funding was provided by Smith College faculty development grants and the Smith College Summer Science Internship Program.

REFERENCES

Aravena, R., O. Suzuki, and A. Pollastri (1989). Coastal fog and its relation to groundwater in the IV region of northern Chile. *Chemical Geology (Isotope Geoscience Section)* **79**: 83–91.

Clark, I. D., and P. Fritz (1997). *Environmental Isotopes in Hydrogeology*. Boca Raton, FL: Lewis Publishers.

Clark, K. L., N. M. Nadkarni, D. Schaefer, and H. L. Gholz (1998). Atmospheric deposition and net retention of ions by the canopy in a tropical montane forest, Monteverde, Costa Rica. *Journal of Tropical Ecology* **14**: 27–45.

Clark, K., R. Lawton, and P. Butler (2000). The physical environment. In *Monteverde: Ecology and Conservation of a Tropical Cloud Forest*, eds. N. M. Nadkarni and N. T. Wheelwright, pp. 15–34. Oxford, UK: Oxford University Press.

Craig, H. (1961). Isotope variations in meteoric water. *Science* **133**: 1702–1703.

Dansgaard, W. (1964). Stable isotopes in precipitation. *Tellus* **16**: 436–468.

Dawson, T. (1998). Fog in the California redwood forest: ecosystem inputs and use by plants. *Oecologia* **117**: 476–485.

Epstein S., and T. Mayeda (1953). Variation of O^{18} content of waters from natural sources. *Geochimica et Cosmochimica Acta* **4**: 213–224.

Gat, J., and E. Matsui (1991). Atmospheric water balance in the Amazon Basin: an isotopic evapotranspiration model. *Journal of Geophysical Research* **96**:13 179–13 188.

Guswa, A. J., A. L. Rhodes, and S. E. Newell (2007). Importance of orographic precipitation to the water resources of Monteverde, Costa Rica. *Advances in Water Resources* **30**: 2098–2112.

Haber, W. (2000). Plants and vegetation. In *Monteverde: Ecology and Conservation of A tropical Cloud Forest*, eds. N. M. Nadkarni and N. T. Wheelwright, pp. 39–94. Oxford, UK: Oxford University Press.

Ingraham, N., and R. Matthews (1988). Fog drip as a source of ground water recharge in northern Kenya. *Water Resources Research* **24**: 1406–1410.

Ingraham, N., and R. Matthews (1995). The importance of fog-drip water to vegetation: Point Reyes Peninsula, California. *Journal of Hydrology* **164**: 269–285.

Karmalkar, A. V., R. S. Bradley, and H. F. Diaz (2008). Climate change scenario for Costa Rican montane forests. *Geophysical Research Letters* **25**, L11702, doi: 10.1029/2008GL033940.

Lachniet, M., and W. Patterson (2002). Stable isotope values of Costa Rican surface waters. *Journal of Hydrology* **260**: 135–150.

Lawton, R. O., U. S. Nair, R. A. Pielke Sr., and R. M. Welch (2001). Climatic impact of tropical lowland deforestation on nearby montane cloud forests. *Science* **294**: 584–587.

Majoube, M. (1971). Fractionnement en oxygène-18 et en deutérium entre l'eau et sa vapeur. *Journal of Chemical Physics* **197**: 1423–1436.

Morrison J., T. Brockwell, T. Merren, F. Fourel, and A. M. Phillps (2001). On-line high-precision stable hydrogen isotopic analyses on nanoliter water samples. *Analytical Chemistry* **73**: 3570–3575.

Organization for Tropical Studies (2004). *Meteorological Data at La Selva Biological Station (1957–2003) and Palo Verde Meteorological Station (1996–2004)*. Available at www.ots.ac.cr.

Rhodes, A. L., A. J. Guswa, and S. E. Newell (2006). Seasonal variation in the stable isotopic composition of precipitation in the tropical montane forests of Monteverde, Costa Rica. *Water Resources Research* **42**, W11402, doi:10.1029/2005WR004535.

Saxena, R. (1986). Estimation of canopy reservoir and O-18 fractionation in throughfall in a pine forest. *Nordic Hydrology* **17**: 251–260.

Scholl, M., S. Ingebritsen, C. Janik, and J. Kauahikaua (1996). Use of precipitation and groundwater isotopes to interpret regional hydrology on a tropical volcanic island: Kilauea volcano area, Hawaii. *Water Resources Research* **32**: 3525–3537.

Scholl, M., S. Gingerich, and G. Tribble (2002). The influence of microclimates and fog on stable isotope signatures used in interpretation of regional hydrology: East Maui, Hawaii. *Journal of Hydrology* **264**: 170–184.

Still, C., P. Foster, and S. Schneider (1999). Simulating the effects of climate change on tropical montane cloud forests. *Nature* **389**: 608–610.

Zadroga, F. (1981). The hydrological importance of a montane cloud forest area of Costa Rica. In *Tropical Agricultural Hydrology*, eds. R. Lal and E. W. Russell, pp. 59–73. New York: John Wiley.

25 Using "biosensors" to elucidate rates and mechanisms of cloud water interception by epiphytes, leaves, and branches in a sheltered Colombian cloud forest

M. Mulligan
King's College London, London, UK
A. Jarvis
International Centre for Tropical Agriculture, Cali, Colombia
J. González
King's College London, London, UK
L. A. Bruijnzeel
VU University, Amsterdam, the Netherlands

ABSTRACT

Cloud water interception (*CWI*) by vegetation is a characteristic hydrological process in tropical montane cloud forests (TMCF). Its magnitude is difficult to measure directly and so artificial collectors tend to be used for comparisons between sites. However, artificial collectors say little about the actual inputs of cloud water (fog) to the vegetation. Scaling inputs via *CWI* from point measurements up to the watershed scale remains an important challenge. This chapter uses a combination of novel field monitoring techniques and epiphyte biomass measurements to quantify the magnitude and better understand the mechanisms of *CWI* in a sheltered, continental TMCF in south-western Colombia. Relationships were sought between amounts of fog water captured by complex vegetation structures including branches, leaves, and bryophytic epiphytes, for use in scaling up point measurements of fog water to the watershed scale using a simple GIS-based extrapolation.

INTRODUCTION

Tropical montane cloud forests (TMCF) often occur in hydrologically important headwater areas, and there is a growing recognition of their significance as reliable suppliers of high-quality water to downstream areas, particularly during prolonged rainless periods or where those areas are (seasonally) dry (Zadroga, 1981; Brown *et al.*, 1996; Mulligan and Burke, 2005a, b). Nevertheless, TMCF are, hydrologically speaking, amongst the least known of tropical forest ecosystems and the precise hydrological role of the ubiquitous ground-level cloud (fog) has remained elusive (Bruijnzeel and Proctor, 1995; Bruijnzeel, 2005; cf. Part III, this volume).

Many TMCF are characterized by a high density, biomass, and diversity of epiphytes, including angiosperms, pteridophytes, bryophytes, lichens, and algae (Gentry, 1988; Wolf, 1993). Epiphytic communities may produce an extensive cover on the trunks and branches of trees in TMCF (Frahm and Gradstein, 1991) whereas epiphyllic communities have been found to colonize leaves (Kiew, 1982). Knowledge of the hydrological properties and role of epiphytic vegetation in TMCF is on the increase (Pócs, 1980; Nadkarni, 1984; Frahm, 1990; Veneklaas *et al.*, 1990; Hölscher *et al.*, 2004; Köhler *et al.*, 2007; Köhler *et al.*, this volume; Tobón *et al.*, this volume #26). For example, the saturated storage capacity of (mostly bryophytic) epiphytes is often two to five times their dry biomass, with reported maximum values of 7.2 mm (or 72 000 l ha^{-1}) for an upper montane cloud forest in Colombia (Veneklaas *et al.*, 1990) and *c.* 8.0 mm for a leeward lower montane cloud forest in Costa Rica (Clark, 1994, cf. Häger and Dohrenbusch, this volume).

Tropical Montane Cloud Forests: Science for Conservation and Management, eds. L. A. Bruijnzeel, F. N. Scatena, and L. S. Hamilton. Published by Cambridge University Press.

Although epiphytes constitute only one component of the forest's intercepting surface (which also includes leaves, stems, branches, and leaf litter), overall canopy storage values for lowland rain forests with low epiphyte mass are typically much lower (in the order of 1.0 ± 0.25 mm; Jetten, 1996) than those derived for epiphyte-rich TMCF (2–5 mm; Köhler *et al.*, 2007). Whilst epiphytes thus appear to be important hydrologically, Hölscher *et al.* (2004) presented evidence that the *actual* contribution by non-vascular epiphytes to overall rainfall interception may be relatively low, despite their substantial *potential* water storage capacity. They attributed this to the fact that during the rainy season the mosses are frequently wetted by rain and hence only a fraction of the potential storage is actually available for the absorption of additional water at the beginning of the next rainfall. Similar information on the interception of cloud water (*CWI*) by epiphytes is almost non-existent (e.g. Chang *et al.*, 2002; Tobón *et al.*, this volume #26).

This chapter combines long-term field monitoring of *CWI* as indicated by "biosensors" (living branches and epiphyte mats) in the Tambito experimental watersheds of south-western Colombia with field measurements of epiphyte biomass and surface area. The aim of the study is to better understand the magnitude, mechanisms and underlying processes of *CWI* by epiphytes, leaves, and branches under the typically sheltered (i.e. low wind speeds) conditions prevailing at Tambito. The results are compared with the *CWI* by various types of passive fog gages as determined in short-term laboratory studies in an experimental cloud chamber (M. Mulligan *et al.*, unpublished data) to better assess the representativity of fog gages as indicators of *CWI* by live forests and to assist in scaling up point measurements of *CWI* to the watershed scale.

THE TAMBITO EXPERIMENTAL WATERSHEDS

The Tambito experimental watersheds are located on the Pacific slopes of the Western (Occidental) Cordillera of the Colombian Andes in the Department of Cauca, Colombia (2° 30.4′ N, 76° 59.9′ W). The watersheds above the confluence of the Tambito and Palo Verde rivers range in elevation between 1374 and 2894 m.a.s.l. and jointly cover an area of 1424 ha. The dominant forest has been classified as Andean tropical sub-montane cloud forest (corresponding with the lower montane cloud forest (LMCF) formation proposed by Bruijnzeel, 2005). Important plant families include the Cecropiaceae, Lauraceae, Melastomataceae, and Palmae, with smaller numbers of Clusiaceae, Urticaceae, Rubiaceae, and others (Jarvis, 2005; Letts *et al.*, this volume). Upper montane cloud forest (UMCF) is found in the upper reaches of the watersheds above 2500 m.a.s.l. Canopy height and basal area of the LMCF vary between 6.9 and 10.2 m, and 27 and 51 m^2 ha^{-1} (Jarvis, 2005); epiphytes are abundant and dominated by mosses.

Annual rainfall in the area is very high (3600–7000 mm) and increases by *c.* 400 mm 100 m^{-1} rise in elevation; the highest monthly totals occur between October and May (>500 mm $month^{-1}$) but there are regular rain-free periods between June and mid September (Letts *et al.*, this volume). The mean annual temperature at 1450 m.a.s.l. is 18 °C with an average diurnal range of 7 °C. Solar radiation is low (intensities of 264 W m^{-2} on average in the wet season and 335 W m^{-2} in the dry season) because of frequent cloud shadow or cloud immersion (Letts *et al.*, this volume). Humidity is >95% for 16 h day^{-1} on average in the open vs. 22 h day^{-1} in the interior of closed forest. Wind speeds are typically very low (<2 m s^{-1} in the open, <0.6 m s^{-1} inside the forest) due to the sheltered setting of the watersheds. An analysis by Mulligan and Burke (2005a) indicated that more than half (53%) of the 529 UNEP–WCMC known cloud forest points throughout the tropics occur on mountain sides sheltered from the dominant local wind direction. As such, sheltered TMCF like the forest at Tambito are not uncommon. Fog frequency varies seasonally, with *CWI* being greatest between September and February, though the pattern is somewhat spatially variable (González, this volume). The diurnal pattern of *CWI* shows greater fog frequency at night, especially at high elevations.

QUANTIFYING CLOUD WATER INTERCEPTION

Theory

The rate of *CWI* depends upon the volume (and drop-size distribution) of suspended water droplets passing by, and upon the effectiveness of the vegetation at intercepting these. This effectiveness is predominantly a function of surface area, vegetation structure and aerodynamic smoothness (Merriam, 1973). Based on wind-tunnel experiments the following equation derived by Merriam (1973) for the amount of fog that would impinge on a unit of (leaf) surface area was found to describe the process adequately:

$$F = AH \cdot u \cdot t \cdot S_c \quad (25.1)$$

where

F = fog drip (ml s^{-1} m^{-2})

AH = absolute atmospheric humidity (%)

u = wind speed (m s^{-1})

t = time (s)

S_c = empirical scavenging coefficient (dimensionless)

and

$$AH = \frac{e}{T_a R_v} \quad (25.2)$$

with

e = ambient vapor pressure (hPa)

T_a = air temperature (K)

R_v = gas constant for water vapor (0.00463, dimensionless).

The scavenging coefficient depends on the drop-size distribution and velocity of the fog as well as on the area, size and spatial arrangement of the catching surface. As such, S_c is difficult to predict in a diverse tropical forest canopy made up of complex leaf, branch, epiphyll and epiphyte structures, although it may be estimated for a particular vegetation type if all other terms in Eq. (25.1) can be measured (Merriam, 1973).

Methods

GENERAL APPROACH

The magnitudes of *CWI*, storage of water, and subsequent evaporation and drainage of intercepted water from epiphytes were inferred through field monitoring of the wetting and drying patterns of "biosensors" (epiphyte mats, leafy branches). In a companion study (M. Mulligan *et al.*, unpublished data), controlled laboratory experiments were conducted to quantify the respective processes for individual epiphytes, branches, and leaves as well as for three commonly used types of passive fog gages. Next, the watershed-wide biomass and surface area of mossy epiphytes and the leaf area index (*LAI*) of the tree stratum were estimated. In combination, this information was used with climatic and other terrain and vegetation characteristics within a Geographic Information System (GIS) to produce a preliminary estimate of watershed-wide *CWI* by the LMCF at Tambito based on Eq. (25.1).

CLIMATIC MEASUREMENTS AND FOG DYNAMICS

Two automatic weather stations (AWS) were installed in June 1997 and operated until March 2001. The Campo station was located in a cleared pasture area at 1450 m.a.s.l. (facing north-east and 30 m from the nearest forest edge) and monitored long-term meteorological conditions (solar radiation, cloud cover, air temperature, relative humidity, wind speed, and leaf wetness, all measured at 1.5 m). The Bosque station was installed at the same time at 1620 m.a.s.l. inside primary LMCF on a south-east-facing planar slope and monitored the same variables as the Campo station, but below the forest canopy (with relative humidity and temperature measured at 10 m and all other variables at 1.5 m). Due to problems with power supply and extreme humidity, both stations have a broken record. The available data are in some cases at 5-min intervals and on other occasions as hourly means of 10-min interval data. Additional instruments were sited at five elevations between 1375 and 2340 m.a.s.l. to measure altitudinal gradients in fog occurrence (wire harps; Gonzàlez, this volume), temperature, humidity, and rainfall within the LMCF belt.

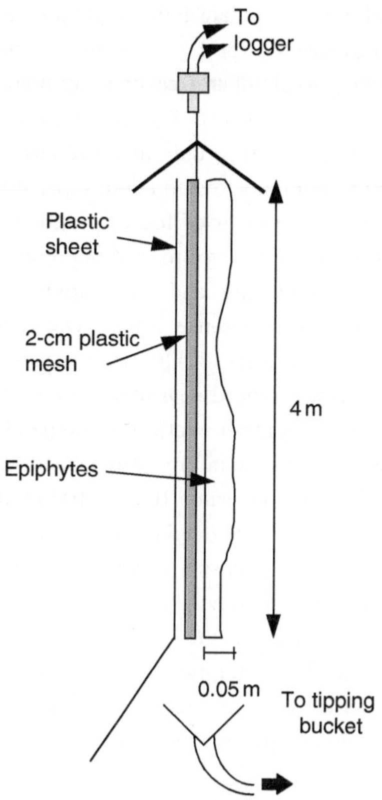

Figure 25.1. The weighing system for monitoring the water dynamics of mossy epiphytes under field conditions with rainfall excluded.

FIELD DETERMINATION OF EPIPHYTE WATER DYNAMICS

From July 1997 to April 2001 a field monitoring experiment to monitor *CWI* by mossy epiphytes was operated at the Bosque AWS site. The experimental design as described in Figure 25.1 consisted of 2.39 kg (dry weight) of locally collected mossy epiphytes contained within a coarse (4-cm mesh width) net (4.5 m tall by 0.5 m wide by 0.05 m thick) suspended from an electronic weighing system accurate to 1 g and connected to a DataTaker logger. The weighing system was calibrated *in situ* and allowed continuous monitoring of epiphyte water storage. This "biosensor" had an area of 2.25 m^2, and an estimated epiphyte leaf surface area of 885.3 m^2 according to the allometric relationship described in this chapter (see below). The entire apparatus was attached to a horizontal branch at 5 m and protected from incident rainfall by a sheet-metal hood. Fog drip was collected in a gutter beneath the epiphytes and channeled into an Environmental Measurements (Newcastle, UK) tipping bucket rain gage (resolution 0.2 mm). A polythene backing acted to produce a barrier to *CWI* on one side, thereby mimicking the condition of mossy epiphytes growing on a tree trunk. The epiphyte mass did not include any "canopy humus" (Nadkarni, 1984) since this is not a significant feature in the LMCF at Tambito. The device was operated for 2 years with rain exclusion and plastic backing to simulate bole epiphytes (as used in this chapter) and subsequently for 1 year without rainfall exclusion and plastic backing to simulate free-standing branch and crown epiphytes (not discussed here). Weight-based systems cannot measure *CWI* directly, as they only measure changes in storage which represent *CWI* minus (evaporation + fog drip). However, the independent measurement of fog drip allows the weighing system to measure *CWI* during times of negligibly low evaporation (i.e. during fog events), and evaporation during times of negligible *CWI* (i.e. in the absence of fog). The system cannot be used effectively during times of high wind speeds.

A similar set-up was placed nearby using a standard net only (2 m × 2.5 m, 0.25 cm mesh width) for comparability with the epiphyte-containing version and to further understand structural controls on *CWI*. Within the same 10 m × 10 m plot, a large tree (20-m tall, diameter at breast height, DBH = 42 cm) was instrumented with leaf wetness sensors (Campbell Scientific 237F) at 2, 2.5, 4, 5, 6, 8, 9, and 10 m height (cf. Letts *et al.*, this volume) to provide an estimate of the leaf surface wetness profile resulting from cloud and rainfall interception. An SKH 2013 Skye Instruments temperature and humidity probe on the same tree at 10 m height completed the set-up.

WHOLE-BRANCH EXPERIMENTS

To better understand *CWI* by leaves and branches a separate system was designed. Unlike epiphytes, branches do not survive for very long once cut so this system was used for short-term (overnight to a few nights) experiments only. The branch-weighing system was installed in the open at the Campo AWS site to typify the increased exposure of branches compared with bole epiphytes. The structure comprised a large wooden frame supporting a strain-gage-based continuous weighing device connected to the data-logger of the AWS. The weighing system was calibrated *in situ*. On a number of still evenings epiphyte-free branches of representative genera (*Miconia, Psychotria, Clusia, Welfia regia*, and Gesneriaceae spp.) were cut, the cut sealed to reduce water loss, and the branch suspended overnight or over a series of nights. The logger recorded weight every 5 min as an average of readings taken every minute. Because the change in weight between two readings reflects water gain (*CWI*) minus water loss (fog drip + evaporation) and no independent measurements of fog drip were possible for branches, this system did not allow the monitoring of rates of *CWI* once fog drip started. However, before the commencement of fog drip the rate of change of weight may be taken to approximate *CWI* (assuming that evaporation is zero under foggy conditions). The system also allowed an assessment of leaf water storage capacity and post-fog evaporation. For each sampled branch, leaf area was measured after the experiment by scanning and histogram stretching (contrast enhancement) to separate leaf from background (O'Neal *et al.*, 2002).

Table 25.1 *The epiphyte biomass index (*EBI*) as used in this study*

EBI	Description
0	No epiphyte coverage
1	1–2 cm thick, 33% partial cover on trunk only
2	1–2 cm thick, 66% partial cover on trunk only
3	1–2 cm thick, full cover on trunk only
4	1–2 cm thick, full cover on trunk and crown only
5	1–2 cm thick, full cover on trunks and major branches
6	2–3 cm thick, full cover on trunks only
7	2–3 cm thick, full cover on trunks and major branches
8	>3 cm thick, full cover on trunks and major branches
9	>3 cm thick, full cover on trunks and major branches, plus cover on lianas
10	>3 cm thick, full cover on trunks, major branches, and lianas with hanging growth forms

ESTIMATION OF EPIPHYTE BIOMASS

To scale up point measurements of *CWI* by epiphytes to the watershed scale in a mechanistic manner, epiphyte biomass needs to be assessed. The biomass of epiphytes in TMCF is notoriously difficult to quantify because of differences in micro-climate and aspect within the canopy, tree age, and tree species (Wolf, 1993; Hietz and Hietz-Seifert, 1995; Wolf *et al.*, 2009; Köhler *et al.*, this volume). Estimation of epiphyte biomass (and distribution) is limited further by poorly developed allometric techniques for these groups (cf. McCune, 1994).

Given the dominance of mosses in the Tambito LMCF (cf. Jarvis, 2005), the following refers specifically to bryophytes. The hydrological properties of bromeliads and other epiphytic vegetation were not investigated, though they may contribute to canopy water dynamics in TMCF. Köhler *et al.* (this volume) derived a water storage value for bromeliads of 0.25 mm (*c.* 27% of the corresponding value for bryophytes) in an upper montane forest in Costa Rica. A two-step approach was followed here. First, a subjective epiphyte biomass index (*EBI*, Table 25.1) was developed to estimate the degree of epiphytism on tree trunks (boles) and stems (branches). This index was then calibrated against measured epiphytic biomass using three small (2 m × 3 m) plots at 1450 m.a.s.l. in which all trees were felled and stripped of all epiphyte material at progressive 1-m height intervals. In addition, for a series of individual randomly sampled trees around these plots only the first 2 m of bole were sampled (i.e. no felling). Field-moist weights of harvested epiphytes were determined and randomly chosen control samples were kept for oven drying in order to derive dry weights. Tree height, *EBI*, and DBH were measured as well. In total, data from 42 individual trees were available to derive a relationship between *EBI* and measured epiphyte biomass per square meter of stem.

Next, this relationship was applied to estimate epiphyte biomass at a further five (10 m × 10 m) plots at 1300, 1400, 1650, 1700, and 1900 m.a.s.l. located on the east-facing planar slope of Cerro el Perro – the largest block of primary LMCF in the watershed. Values of *EBI*, DBH, and tree height *h* (in m) were determined for a random sample of trees in each of these plots and overall stand density per plot was measured. Trunks were assumed to be cylindrical in shape given the form of the trees and thus trunk surface area (*TSA*, m^2) is simply given by:

$$TSA = \pi \cdot \text{DBH} \cdot h. \qquad (25.3)$$

Epiphyte biomass (kg dry weight per sample of *n* trees) then becomes:

$$\sum\nolimits_{i=1}^{n} TSA \cdot EB_{[EBI]} \qquad (25.4)$$

where

n = number of sampled trees per plot

$EB_{[EBI]}$ = estimated epiphyte biomass per m^2 of bole for a given *EBI*.

Hence, epiphyte biomass per plot (kg) is given by:

$$(EB_{\text{sample}} \cdot sd)/n \qquad (25.5)$$

where

sd = stand density (number of trees per plot).

Given the difficulties in movement and working in primary LMCF on very steep slopes, this methodology for epiphyte biomass estimation, although considered approximate at best, was both feasible and rapid. The method does not account properly for extensive crown epiphytism which, though not as widespread as bole epiphytism in the Tambito forests, is important in exposed (emergent) crowns (Wolf, 1993; Köhler *et al.*, 2007; Köhler *et al.*, this volume).

ESTIMATING EPIPHYTE SURFACE AREA

Most models of canopy interception incorporate some measurement of the surface area of foliage, such as *LAI*, since this is a major control on the canopy storage capacity for water. A total of 47 mossy epiphyte samples (15–190 g each) were collected randomly from around the Tambito watershed. Surface area was calculated by breaking apart each sample to expose individual leaflets and scanning at 600 dpi (65 images). Histogram stretching (contrast enhancement) was used to separate epiphyte material from background and the surface area of leaflets was calculated electronically using raster GIS. The surface measures were then correlated with measured dry biomass for the same samples for upscaling of surface area. It should be noted that the surface area will inevitably be underestimated by this method, due to the very small and intricate leaf pattern of epiphytes.

RESULTS AND DISCUSSION

Patterns of humidity and cloud cover

Ground-level cloud (i.e. fog) is a daily occurrence in Tambito except during the driest periods of the dry season. Average relative humidity (*RH*) at the Campo AWS (1450 m.a.s.l.) during the interrupted (8-month) record between July 1998 and August 1999 was >95% from 17.00 h to 08.00 h daily. The minimum daily *RH* during the same period was, on average, 79% (usually attained around noon time). The forest interior Bosque site (1650 m.a.s.l., 10 m from the ground) had an $RH > 95\%$ from 13.00 h through to 11.00 h daily on average. Minimum *RH* at this site was 94% on average (reached around 11.00 h). Thus, on average, the air is close to saturation for 16 out of 24 h in the open at 1450 m.a.s.l. and for as many as 22 out of 24 h beneath the forest canopy at 1650 m.a.s.l. The leaf wetness sensors at the Campo site also indicated the grass to be wet throughout the night although the leaves dried up rapidly between 06.00 h and 09.00 h daily. Conversely, leaves at the Bosque site (2.5–10 m from the ground) were, on average, around 10% wet for most of the photosynthetic day (Letts *et al.*, this volume). Moisture interception by the leaves typically occurred slowly from 12.00 to 19.00 h but increased rapidly from 19.00 h onward as the air became saturated. Also, *RH* exhibited a strong seasonal pattern with average monthly values at the Campo site ranging from a minimum of 91% in July to ~100% in November.

Although direct measurements of visibility are lacking, the observed patterns for *RH* and leaf wetness suggest that some fog interception is likely to occur from 11.00 to 06.00 h daily. Where intercepting surfaces are exposed to sunlight, intercepted water may be re-evaporated during a short window of opportunity between 06.00 and 10.00 h. González (this volume) used wire harps at different elevations within the watershed to obtain a first estimate of fog incidence. Average daily values were 1.15 ± 0.90 mm at the Campo AWS (1450 m.a.s.l.) vs. 2.2 ± 1.3 mm at 2340 m.a.s.l. within the zone of most frequent cloud (>2200 m.a.s.l.; González, this volume).

Field determination of cloud water interception by epiphytes

METEOROLOGICAL CONDITIONS

Two periods of hourly data (averaged from 10-min readings) from the biosensor with contrasting meteorological conditions are examined here. The first concerns a dry period (15–22 August 1999) and the second a wet period (7–14 August 1999). Average meteorological conditions for the two periods are summarized in Table 25.2. There are clear differences in rainfall, temperature and solar radiation between the two periods but only the relative humidity was significantly different ($p < 0.001$).

Table 25.2 *Average meteorological conditions during the epiphyte water dynamics field experiment in the forest at 1650 m.a.s.l. in August 1999*

Period	Rainfall (mm)	Mean daytime solar radiation (Wm^{-2})	Mean air temperature at 10 m (°C)	Mean relative humidity at 10 m (%)
Dry	30.1	39.1 (6.25)	16.71 (0.17)	96.2 (0.42)
Wet	46.8	30.1 (2.36)	16.54 (0.11)	100 (0.18)
Significantly different at $p < 0.001$	No	No	No	Yes

The figures in brackets are standard errors.

Hourly meteorological conditions for the two periods are shown in Figures 25.2a and 25.2b, respectively. It is clear that high humidity levels tend to occur within the canopy at night and these are often, but not always, associated with rainfall events. During wet periods, within-canopy humidity can remain at 100% for days on end, whilst dry periods, with cloudless conditions, lead to high air temperatures and low humidity during the day.

WATER STORAGE DYNAMICS

Figures 25.3a and 25.3b show the patterns of changes in water storage of the epiphyte biosensor and the nylon net in relation to ambient humidity, rainfall and wind speed measured at 1.5 m during the wet and dry period, respectively. Relative humidity remained consistently close to 100% at night. However, wind speeds in the forest interior were consistently close to zero. Hence, according to Eq. (25.1) cloud water availability would be low. Using Eq. (25.1), for the part of the wet period between 9 and 14 August 1999 for which wind data were available (wind data are missing for the first two days, Figure 25.3a), the calculated horizontal water vapor flux was, on average, 10 872 ml m^{-2} h^{-1} between sunset and sunrise (average wind speed $u = 0.25$ m s^{-1}) compared to 59 400 ml m^{-2} h^{-1} between sunrise and sunset ($u = 1.5$ m s^{-1}). High humidity combined with the higher wind speeds during daylight hours led to increases in epiphyte water storage as long as solar radiation was low. Similarly, decreases in water storage occurred when solar radiation was high, indicating that complex evaporation/interception cycles may exist.

Average water absorption rates were calculated using the same Eq. (25.1) for rainless hours with $RH > 95\%$ and $u < 0.5$ m s^{-1} and where the epiphyte trap indicated net wetting, during the period 9–14 August 1999 ($n = 15$). These periods were considered to be less prone to potential errors induced by raindrop impact or buffeting of the apparatus by wind. For a unit area with $S_c = 1$ this gives a potential water uptake rate of 8611 ml m^{-2} h^{-1} (range: 0–15 194). Since the average measured rate of uptake for the same period and the same conditions ($n = 15$) was 4.5 ml m^{-2} $hour^{-1}$ (range: 0.002–38.9), the scavenging coefficient derived for these epiphytes was a very low 0.05% on average (range: 2.3×10^{-5}–0.45%). The pattern of water storage for the plain nylon net was simpler than that for the epiphyte mass and often showed rapid water capture followed by progressive drainage and evaporative loss, with the rate of loss changing in proportion with the amount of water stored (Figure 25.3). The pattern of water storage for the epiphytes is much more complex, perhaps because both the rate of uptake and the rate of drainage are controlled by the volume of water stored as well as by ambient conditions. Low *RH* conditions appeared to be associated with rapid drying of both the epiphytes and the nylon net.

RATES OF WETTING AND DRYING

By separating the epiphyte water storage data for the 2 weeks under consideration into drying hours ($n = 228$) and wetting hours ($n = 157$), average rates of wetting and drying could be calculated. Since no fog drip was recorded by the tipping bucket during this entire period, wetting should occur whenever *CWI* exceeded evaporation, and drying whenever evaporation exceeded *CWI*. The inferred average wetting rate was 19.4 ml h^{-1} (equivalent to 8.1 ml h^{-1} kg dry $biomass^{-1}$) whereas the average drying rate was 11.0 ml h^{-1} (4.6 ml h^{-1} kg dry $biomass^{-1}$). Since no fog drip from the epiphytes was measured, drying must have occurred through evaporation alone. The observed lack of fog drip is not surprising given the very small fraction of the epiphyte storage capacity that was filled under conditions of rainfall exclusion. Storage varied between 0.004% and 7.93%, with an average of 2.44% for this 16-day period (384 hours). Köhler *et al.* (2007) and Köhler *et al.* (this volume) estimated moisture contents of 100–400% for rain-exposed mosses, indicating the dominant contribution of rainfall to the water storage dynamics of mosses (cf. García-Santos, 2007).

EVAPORATION AND FOG DRIP

Analysis of all data for hours without rain between 7 and 22 August 1999 indicated no relationship between the rate of weight loss (evaporation) and epiphyte water storage. This is not surprising given that the range of storage achieved was always less than 8%. Analysis of inferred cumulative *CWI* and evaporation for the entire period indicated an approximate balance between the two, with 3040 ml of *CWI* and 2934 ml of evaporation (hence the small increment in storage, Figure 25.4). Evaporation exceeded *CWI* for much of the period but *CWI*

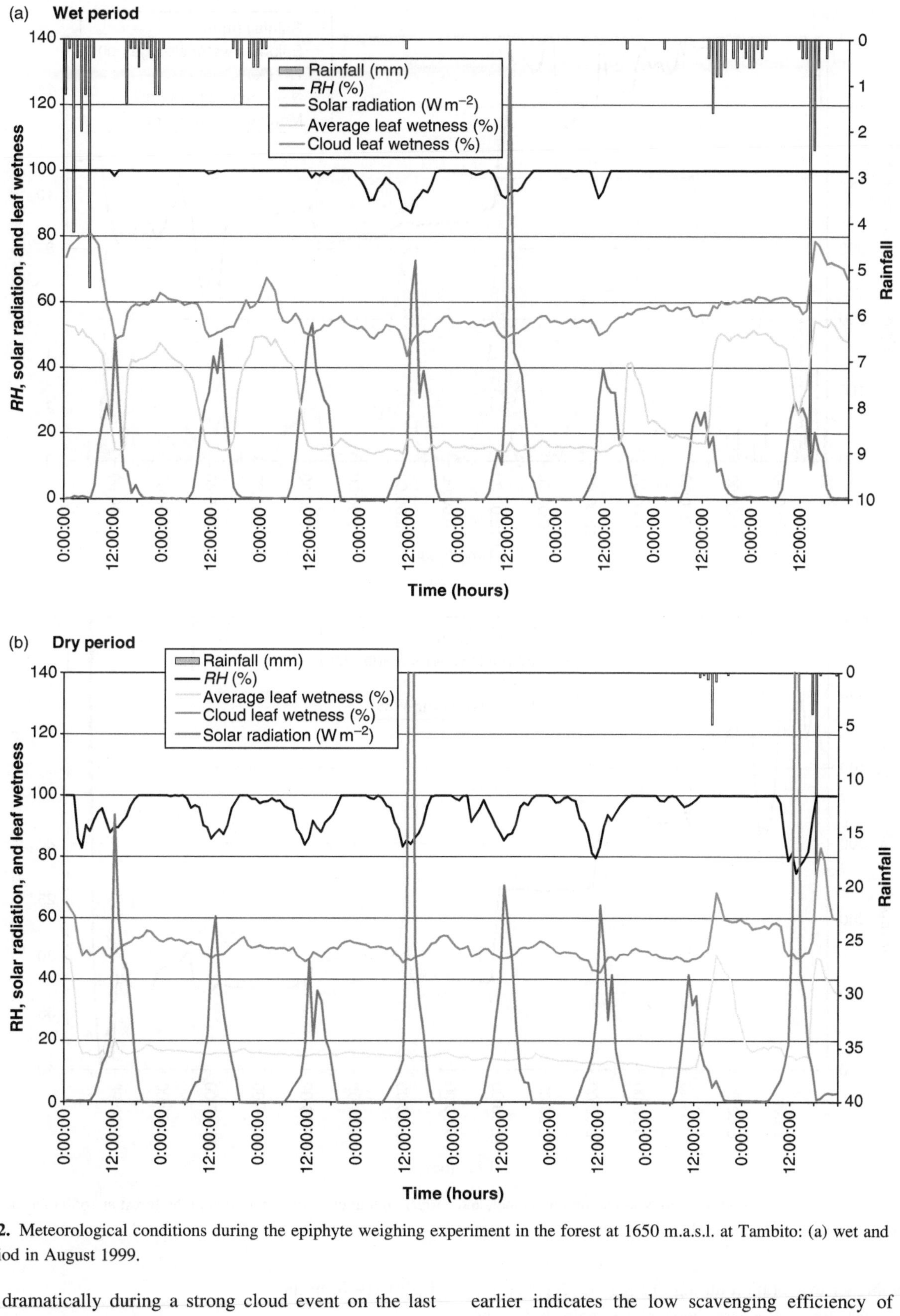

Figure 25.2. Meteorological conditions during the epiphyte weighing experiment in the forest at 1650 m.a.s.l. at Tambito: (a) wet and (b) dry period in August 1999.

increased dramatically during a strong cloud event on the last day. The fact that the inferred *CWI* values constitute only a fraction of the calculated horizontal water vapor fluxes cited earlier indicates the low scavenging efficiency of the bryophytes. Figure 25.4 illustrates further that cloud water absorption and evaporation were in balance during wet periods

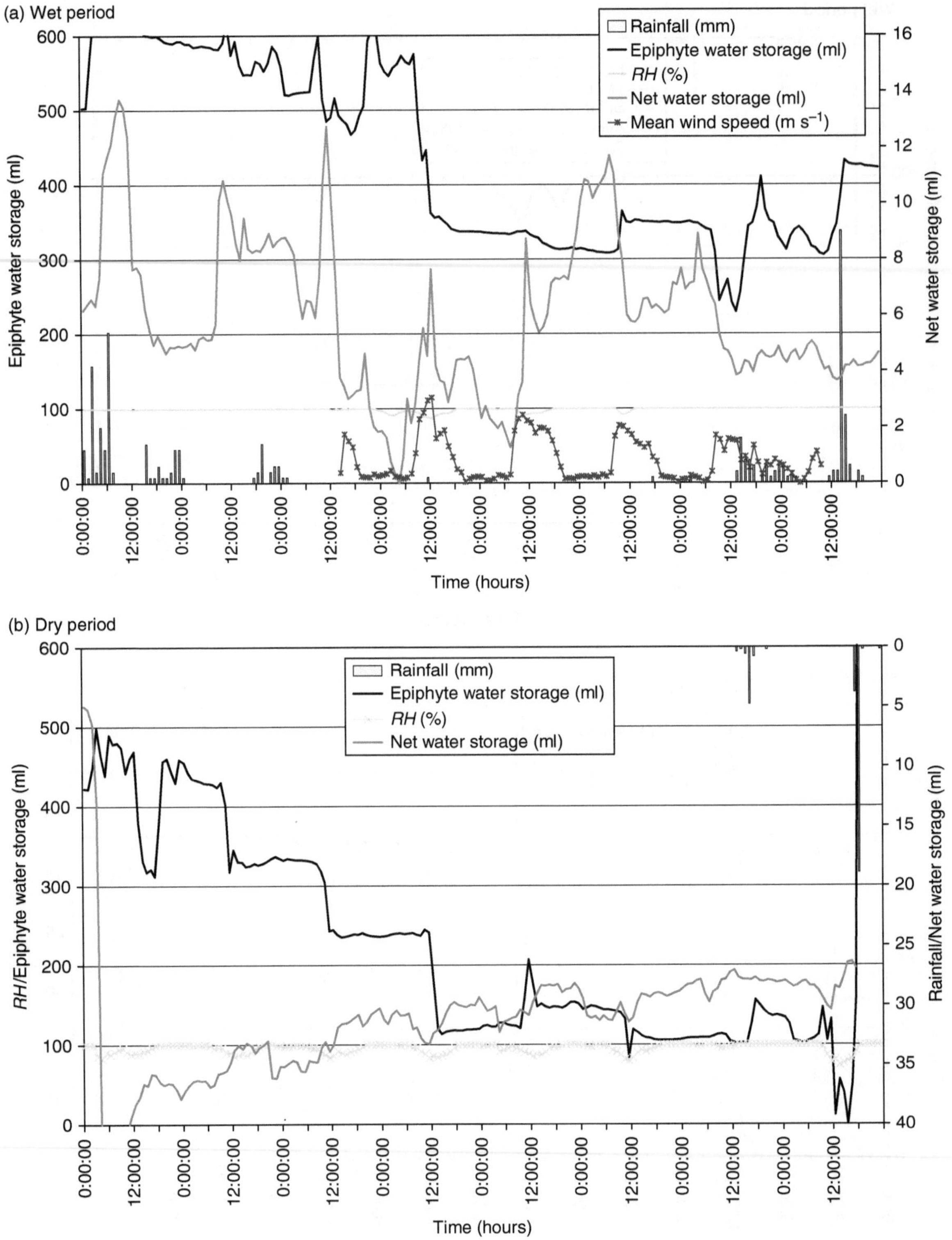

Figure 25.3. Water storage dynamics of epiphytes and of a nylon net, and corresponding climatic conditions in the forest at 1650 m.a.s.l. at Tambito: (a) wet and (b) dry period in August 1999.

whereas the higher evaporation during dry periods depleted the amount of water stored. Recharge took place again during the next wet period.

Epiphyte biomass

Measured epiphyte biomass at the three 3 m × 2m plots at 1450 m.a.s.l. averaged 0.32 t dry matter ha^{-1} for understory

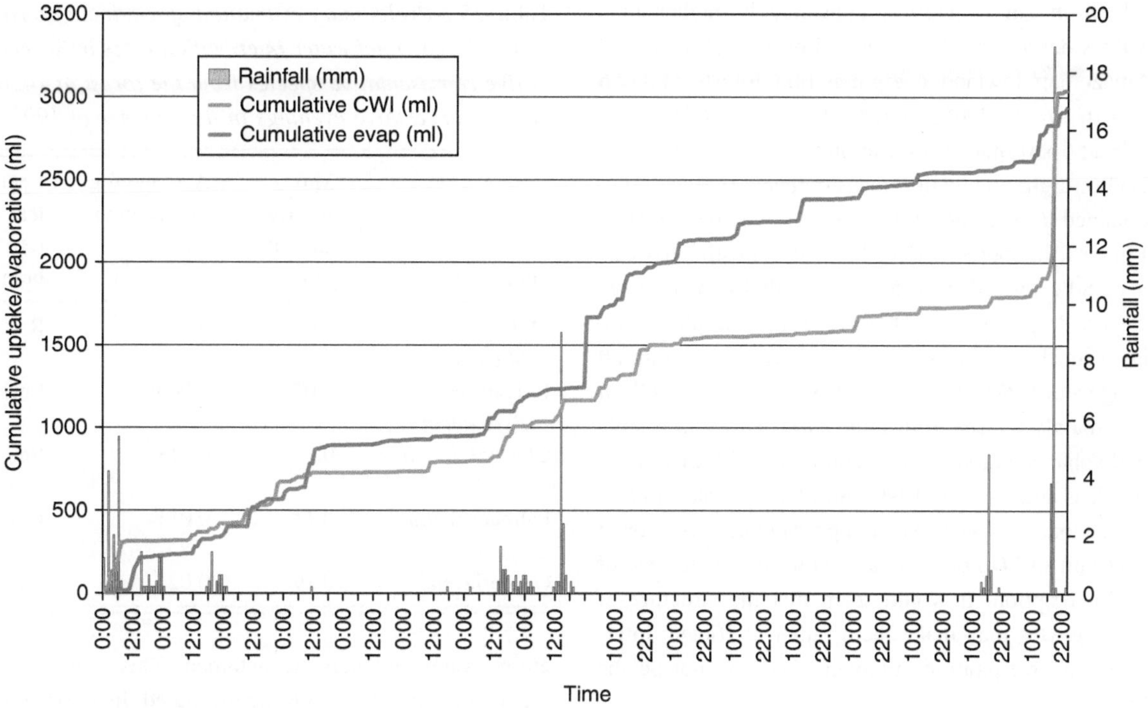

Figure 25.4. Rainfall, cumulative cloud water absorption, and cumulative evaporation from epiphytes in the forest at 1650 m.a.s.l. at Tambito between 7 and 22 August 1999.

epiphytes and a further 2.28 t ha^{-1} for canopy epiphytes, giving a total of 2.6 t ha^{-1}. Reported values of epiphytic biomass in LMCF (reviewed by Köhler *et al.*, 2007) range from 2.1 t ha^{-1} for a stunted ridge-top forest in Jamaica to 33.1 t ha^{-1} for a tall leeward LMCF on the Pacific slope near Monteverde in Costa Rica (cf. Häger and Dohrenbusch, this volume), with much variability between studies and sites. The low value derived for the LMCF at Tambito may reflect a combination of limitations in the sampling procedure (which tended to underestimate contributions by epiphytes higher up in the canopy) and the fact that the study site was situated below the level of the most pronounced cloud development (>2200 m.a.s.l.; Gonzalez, this volume).

The relationship between measured epiphyte dry biomass (*EDB*, in kg m^{-2} per trunk) and the epiphyte biomass index of Table 25.1 (*EBI*) for the three plots and surrounding trees was determined as:

$$EDB = 0.0992 \cdot EBI \quad (r^2 = 0.75). \tag{25.6}$$

Application of Eq. (25.6) to five additionally sampled plots between 1300 and 1900 m.a.s.l. indicated that epiphyte biomass might vary from 1.4 t ha^{-1} at 1400 m.a.s.l. to 7.1 t ha^{-1} at 1900 m.a.s.l. Indeed, *EDB* (in t ha^{-1}) increased systematically by about 1 t ha^{-1} 100 m^{-1} according to:

$$EDB = 0.0109 \cdot H - 13.575 \quad (r^2 = 0.99, n = 5) \tag{25.7}$$

where elevation is given in m.

Combining Eq. (25.7) with the digital elevation model for the Tambito watersheds gave an estimated areal average epiphyte biomass of 5.6 t ha^{-1}. This value must be considered too low for the entire catchment because the elevational range for which Eq. (25.7) was derived did not extend into the cloud belt proper (González, this volume) and *EDB* is known to increase dramatically upon entering the zone of heaviest condensation (Wolf, 1993; cf. Köhler *et al.*, 2007). However, the area above 2200 m.a.s.l. makes up about 30% of the watershed's area above 1000 m.a.s.l., and therefore the values derived for the part below 2200 m.a.s.l. is likely to hold for the majority of the watershed. Primary forest at Tambito appeared to contain higher epiphyte biomass than did secondary forest (cf. Köhler *et al.*, this volume), although no quantification of this was made.

Surface area of epiphytes

The relationship between individual epiphyte biomass (*EDB*, in kg) and measured surface area (*ESA*, in m^2) was:

$$ESA = 37 \cdot EDB \quad (r^2 = 0.81, n = 47). \tag{25.8}$$

This means that the specific leaf area of epiphytes is extremely high at 370 $cm^2\ g^{-1}$ (37 $m^2\ kg^{-1}$), and this is likely to still be an underestimate given the difficulty of scanning leaves with such a complex fractal form as the samples under consideration. Using the *EDB* value derived for the plots at 1450 m.a.s.l. gave an epiphyte surface area per unit ground area (the "epiphyte area index") of 9.88 m^2 m^{-2}. This is considerably higher than foliar *LAI* values derived by destructive biomass measurements for most Amazonian lowland evergreen rain forests (4.3–7.5 m^2 m^{-2}; Roberts *et al.*, 2005) or indeed (canopy) *LAI* values for the LMCF at 1650 m.a.s.l. in Tambito (4.0 ± 1.2 m^2 m^{-2}, $n = 25$) based on light attenuation measurements (Letts, 2003). Similarly obtained *LAI* values for a lower montane forest at 1900–2200 m.a.s.l. in southern Ecuador with an estimated epiphyte cover of 80% ranged from 5.2 to 9.3 (Fleischbein *et al.*, 2005). However, despite their high surface area, epiphytes are likely to play a much smaller role in the interception of rainfall than the foliage does. Because the epiphytes in the LMCF at Tambito grow mainly on tree trunks, the surface area accessible to rainfall and solar radiation is small compared with that of the foliage (Veneklaas and Van Ek, 1990). Similarly, Hölscher *et al.* (2004) estimated that epiphytes in a tall upper montane oak forest in Costa Rica (with an *EDB* of 3.4 t ha^{-1}) contributed only 8% of total canopy interception. It would seem that the structure of (stem-based) epiphytes may make them rather inefficient interceptors of vertical precipitation, although this may not be the case for *CWI*.

Whole-branch experiments

The continuous weighing of whole branches produced a data-set of 4906 5-min readings distributed over 17 days between 12 July and 14 August 1999. Rates of *CWI*, evaporation, and the magnitude of the water storage capacity were determined for branches from five species representing five major families (Table 25.3). A comparison of weight gain against measured *RH* at the Campo AWS indicated that net gains only occurred when *RH* exceeded 95% and increased with increasing *RH* beyond this threshold. Because climatic conditions differed between successive weighings, direct comparisons of rates of evaporation and interception for the respective taxa were not possible. Table 25.3 further shows that maximum amounts of cloud water held on the leaves, measured for the taxa were mainly in the order of 0.1–0.2 mm, whilst *CWI* rates (including evaporation) were generally 0.01–0.02 mm h^{-1}. For an *LAI* of 4.0 (Letts, 2003) this would be equivalent to foliar *CWI* rates of 0.04–0.08 mm h^{-1} (0.65–1.3 mm day^{-1} or 237–474 mm $year^{-1}$), given the estimated fog frequency at this elevation. González (this volume) obtained an average daily *CWI* of 1.15 mm using a modified wire harp at the same site.

Upscaling of *CWI* by epiphytes to the watershed scale

Combining the average wetting and drying rates derived from the biosensor at 1650 m.a.s.l. with climatic observations (wet and dry periods) and the extrapolated areal epiphytic biomass of 5.6 t ha^{-1}, a first areal estimate of *CWI* by epiphytes for the entire watershed may be obtained. This scaling is admittedly crude since it is based on the estimated distribution of epiphytic vegetation and the assumption that the atmospheric conditions measured at the Bosque site are applicable throughout the watershed. In reality, cloud cover is likely to increase somewhat in density and frequency with elevation (González, this volume) whereas evaporation will decrease with elevation. A series of *T*/*RH* stations at altitudes from 1431 to 2496 m.a.s.l. showed a fairly constant mean *RH* but a significant decrease in temperature and absolute humidity with elevation (Gonzalez, 2007). It is also pertinent to remember that, for simplicity, the wetting and drying behavior of the mossy epiphytes was studied *with rainfall excluded* and so the results are most applicable to rainless periods. Data are available without rainfall exclusion for future analysis.

Table 25.3 *Preliminary estimation of maximum storage capacity and average cloud water interception rates by severed branches of five representative species from the forest at 1450 m.a.s.l. during consecutive evenings in the summer of 1999*

Specimen	Max storage capacity (mm)	Average fog interception rate (mm h^{-1})	Leaf texture, leaf surface area (cm^2 per branch)
Melastomataceae: *Miconia*	>0.02	0.0015	Rough, 3462
Rubiaceae: *Psychotria*	0.07	0.0058	Smooth, 3237
Clusiaceae: *Clusia*	0.17	0.018	Smooth and waxy, 2213
Palmae: *Welfia regia*	0.11	0.011	Rough, 1534
Gesneriaceae	0.19	0.021	Pubescent, 2512

The average rate of *CWI* by epiphytes (during wetting periods) indicated by the biosensor was 8.1 ml h^{-1} kg dry $biomass^{-1}$). This would equate to an areally averaged rate of 0.0077 mm h^{-1} (range: 0.0011–0.014 mm h^{-1}) over the watershed using the observed relationship between *EDB* (t ha^{-1}) and elevation (m): $EDB = 0.010 \cdot \text{elevation} - 13.57$ ($r^2 = 0.99$, $n = 5$; Jarvis, 2000) or 0.0045 mm h^{-1} using the mean value for an *EDB* of 5.6 t ha^{-1}. Assuming a minimum of 16 and a maximum of 22 h day^{-1} with *RH* in excess of 95% (taken as a proxy for the occurrence of fog), this would represent an extra input of water via *CWI* by epiphytes of 45–62 mm $year^{-1}$ with *EDB* varying according to elevation, or 26–36 mm $year^{-1}$ using the watershed-averaged *EDB*. Such amounts are negligibly small compared to the altitudinally varying 3600–7000 mm of rainfall received annually at Tambito. In addition, where epiphytes are not subject to saturation by intercepted rainfall, part of this water is re-evaporated again at an average rate of 4.6 ml h^{-1} kg dry $biomass^{-1}$, or 0.0044 mm h^{-1}

using the altitudinally varying *EDB* and 0.0026 mm h^{-1} for the watershed-averaged *EDB*. Evaporation occurs, on average, for between 2 and 8 h day^{-1}, producing an estimated annual areal evaporation loss from epiphytes of only 3–13 mm (for altitudinally varying *EDB*) or 1.9–7.6 mm $year^{-1}$ (for watershed-averaged *EDB*). The remainder (32–59 mm $year^{-1}$ with altitudinally varying *EDB* or 24–28 mm $year^{-1}$ with watershed-averaged *EDB*) would be fog drip but this occurs at such a slow rate (0.0055–0.0100 mm h^{-1}, equivalent to 1 bucket-tip per 20–35 h of sustained drip) that its measurement would be impossible using conventional sensors.

CONCLUSIONS

This chapter has described some alternate procedures for, and results from, a series of field experiments designed to better understand the dynamics of cloud water interception (*CWI*) by epiphytic mosses and leaves (acting as "biosensors") under calm atmospheric conditions with the intention of facilitating upscaling of point measurements of *CWI* to the watershed scale. Estimated epiphyte biomass in the Tambito experimental watershed averaged 2.6 t ha^{-1} at 1450 m.a.s.l. and increased with elevation to give an estimated areal average of 5.6 t ha^{-1} for the lower montane cloud forest belt between 1300 and 2200 m.a.s.l. The epiphytic leaf area index was calculated as 9.88 m^2 m^{-2} which is more than double the estimated foliar *LAI* for the LMCF at 1650 m.a.s.l. Field results for the absorption, retention, and evaporation of cloud water by mossy epiphytes protected against wetting by rainfall indicated: (i) very low levels of saturation in the absence of rainfall (<8% moisture content), (ii) an average *CWI* rate of 8.1 ml h^{-1} kg dry $biomass^{-1}$during wetting, (iii) an average rate of drying of 4.6 ml h^{-1} kg dry $biomass^{-1}$, and (iv) a very low scavenging coefficient (0.05%) for the moss which also proved to be a function of water storage. Epiphytes were shown to accumulate fog inputs very slowly but also to have a high storage capacity and to release moisture slowly via evaporation. No fog drip was recorded, presumably due to the fact that the biosensor was sheltered against direct inputs from rainfall and crown drip. Upscaling of these point observations to the watershed scale demonstrated that *CWI* by epiphytes at Tambito is likely to constitute less than 1% of regular rainfall inputs. Values of *CWI* derived for entire branches (but with their epiphytes removed) exposed to fog in open pasture ranged from 237 to 474 mm $year^{-1}$, an order of magnitude greater than the *CWI* inferred from the biosensor for epiphytes within the forest interior. These very low values for both *CWI* and re-evaporation from epiphytes are likely to reflect the position of the biosensor within the forest and the limited exchange with the atmosphere above the canopy. Both radiation and wind speeds are much reduced near the ground, whereas humidities remain high (cf. Motzer, 2005), thereby reducing evaporative losses. Tobón *et al.* (this volume #26) showed that *net* fog absorption (i.e. absorption minus concurrent evaporation) increased as one descends within the canopy whereas evaporative losses decreased strongly. Because results obtained with the biosensor are sensitive to wind speed, this approach is less suitable for use higher up in the canopy. However, the results might perhaps be scaled using a series of passive fog catchers positioned at various heights above the ground.

ACKNOWLEDGEMENTS

The authors would like to thank all of the Colombian institutions that have provided data and logistic and field support for this research. Particular thanks go to the late Dr. Alvaro Jose Negret, Fundacion Proselva and the University of Cauca for permission to carry out research at Tambito and to Quintin and family for taking care of us in the field. Thanks go to Ron Knapp, William Bell, and others at CIAT in Cali for their logistic support. Mauricio Rincon-Romero digitized the Tambito DEM from IGAC 1 : 25 000 base cartography. MM in particular would like to thank all undergraduate, masters and Ph.D. students and friends that have helped in the construction of field and laboratory equipment, often under arduous conditions. This research was funded by the Royal Society research grants (RSRG 18/19443), the University of London central research fund, the British Council in Colombia, and the Royal Geographical Society of the UK.

REFERENCES

Brown, M. B., I. de la Roca, A. Vallejo, *et al.* (1996). *A Valuation Analysis of the Role of Cloud Forests in Watershed Protection, Sierra de las Minas Biosphere Reserve, Guatemala, and Cusuco N.P., Honduras.* Philadelphia, PA: RARE Center for Tropical Conservation.

Bruijnzeel, L. A. (2005). Tropical montane cloud forest: a unique hydrological case. In *Forests, Water and People in the Humid Tropics*, eds. M. Bonell and L. A. Bruijnzeel, pp. 462–483. Cambridge, UK: Cambridge University Press.

Bruijnzeel, L. A., and J. Proctor (1995). Hydrology and biochemistry of tropical montane cloud forests: what do we really know? In *Tropical Montane Cloud Forests*, eds. L. S. Hamilton, J. O. Juvik, and F. N. Scatena, pp. 38–78. New York: Springer-Verlag.

Chang, S. C., I. L. Lai, and J. T. Wu (2002). Estimation of fog deposition on epiphytic bryophytes in a subtropical montane forest ecosystem in northeastern Taiwan. *Atmospheric Research* **64**: 159–167.

Clark, K. L. (1994). The role of epiphytic bryophytes in the net accumulation and cycling of nitrogen in a tropical montane cloud forest. Ph.D. thesis, University of Florida, Gainesville, FL.

Fleischbein, K., W. Wilcke, C. Valarezo, W. Zech, and K. Knoblich (2005). Water budgets of three small catchments under montane forest in Ecuador: experimental and modelling approach. *Hydrological Processes* **20**: 2491–2507.

Frahm, J. P. (1990). The ecology of epiphytic bryophytes on Mt. Kinabalu, Sabah (Malaysia). *Nova Hedwigia* **51**: 121–132.

Frahm, J. P., and S. R. Gradstein (1991). An altitudinal zonation of tropical rain forests using bryophytes. *Journal of Biogeography* **18**: 669–676.

García-Santos,, G. (2007). An ecohydrological and soils study in a montane cloud forest in the National Park of Garajonay, La Gomera (Canary Islands, Spain). PhD Thesis, VU University Ámsterdam, Ámsterdam, The Netherlands. [http://www.falw.vu.nl/nl/onderzoek/earth-sciences/geo-environmental-science-and-hydrology/hydrology-dissertations/index.asp].

Gentry, A. H. (1988). Changes in plant community diversity and floristic composition along environmental and geographical gradients. *Annals of the Missouri Botanical Garden* **75**: 1–34.

Gonzalez, J. (2007). Cloud interception by trees in a tropical montane cloud forest of south-western Colombia. Ph.D. thesis. King's College London, University of London, UK.

Hietz, P., and U. Hietz-Seifert (1995). Structure and ecology of epiphyte communities of a cloud forest in central Veracruz, Mexico. *Journal of Vegetation Science* **6**: 719–728.

Hölscher, D., L. Köhler, A. I. J. M. Van Dijk, and L. A. Bruijnzeel (2004). The importance of epiphytes to total rainfall interception by a tropical montane rain forest in Costa Rica. *Journal of Hydrology* **292**: 308–322.

Jarvis, A. (2000). Quantifying the hydrological role of cloud deposition onto epiphytes in a tropical montane cloud forest, Colombia. B.Sc. thesis, King's College London, University of London, UK. Also available at www.ambiotek.com/herb.

Jarvis, A. (2005). Terrain controls on the distribution of tree species diversity and structure in tropical lowland and tropical montane forest. Ph.D. thesis, King's College London, University of London, UK. Also available at www.ambiotek.com/theses.

Jetten, V. G. (1996). Interception of tropical rain forest: performance of a canopy water balance model. *Hydrological Processes* **10**: 671–685.

Kiew, R. (1982). Observations on leaf color, epiphyll cover, and damage on Malayan *Iguanura wallichiana*. *Principes* **26**: 200–204.

Köhler, L., C. Tobón, K. F. A. Frumau, and L. A. Bruijnzeel (2007). Biomass and water storage dynamics of epiphytes in old-growth and secondary montane cloud forest stands in Costa Rica. *Plant Ecology* **193**: 171–184.

Letts, M. J. (2003). Carbon assimilation and productivity in a north-west Andean tropical montane cloud forest. Ph.D. thesis, King's College London, University of London, UK. Also available at www.ambiotek.com/theses.

McCune, B. (1994). Using epiphyte litter to estimate epiphyte biomass. *Bryologist* **97**: 396–401.

Merriam, R. A. (1973). Fog drip from artificial leaves in a fog wind tunnel. *Water Resources Research* **9**: 1591–1597.

Motzer, T. (2005). Micrometeorological aspects of a tropical mountain forest. *Agricultural and Forest Meteorology* **135**: 230–240.

Mulligan, M., and S. M. Burke (2005a). *FIESTA: Fog Interception for the Enhancement of Streamflow in Tropical Areas*, Appendix 4a to Final Technical Report of DFID-FRP Project no. R7991. Leigh-on-Sea, UK: AMBIOTEK. Also available at www.ambiotek.com/fiesta.

Mulligan, M., and S. M. Burke (2005b). *Global Cloud Forests and Environmental Change in a Hydrological Context*. Leigh-on-Sea, UK: AMBIOTEK. Also available at www.ambiotek.com/cloudforests.

Nadkarni, N. M. (1984). Epiphyte biomass and nutrient capital of a neotropical elfin forest. *Biotropica* **16**: 249–256.

O'Neal, M. E., D. A. Landis, and R. Isaacs (2002). An inexpensive, accurate method for measuring leaf area and defoliation through digital image analysis. *Journal of Economic Entomology* **95**: 1190–1194.

Pócs, T. (1980). The epiphytic biomass and its effect on the water balance of two rain forest types in the Uluguru Mountains (Tanzania, East Africa). *Acta Botanica Academiae Scientiarum Hungaricae* **26**: 143–167.

Roberts, J. M., J. H. C. Gash, M. Tani, and L. A. Bruijnzeel (2005). Controls on evaporation in lowland tropical rainforest. In *Forests, Water and People in the Humid Tropics*, eds. M. Bonell and L. A. Bruijnzeel, pp. 287–313. Cambridge, UK: Cambridge University Press.

Veneklaas, E. J., and R. van Ek (1990). Rainfall interception in two tropical montane cloud forests, Colombia. *Hydrological Processes* **4**: 11–326.

Veneklaas, E. J., R. J. Zagt, A. van Leerdam, *et al.* (1990). Hydrological properties of the epiphyte mass of a montane tropical rain forest, Colombia. *Vegetatio* **89**: 183–192.

Wolf, J. H. D. (1993). Ecology of epiphytes and epiphyte communities in montane rainforests, Colombia. Ph.D. thesis, University of Amsterdam, Amsterdam, The Netherlands.

Wolf, J. H. D., S. R. Gradstein, and N. M. Nadkarni (2009). A protocol for sampling vascular epiphyte richness and abundance. *Journal of Tropical Ecology* **25**: 107–121.

Zadroga, F. (1981). The hydrological importance of a montane cloud forest area of Costa Rica. In *Tropical Agicultural Hydrology*, eds. R. Lal and E. W. Russell, pp. 59–73. New York: John Wiley.

26 Water dynamics of epiphytic vegetation in a lower montane cloud forest: fog interception, storage, and evaporation

C. Tobón
Universidad Nacional de Colombia, Medellin, Colombia

L. Köhler
University of Göttingen, Göttingen, Germany

K. F. A. Frumau and L. A. Bruijnzeel
VU University, Amsterdam, the Netherlands

R. Burkard and S. Schmid
University of Bern, Bern, Switzerland

ABSTRACT

Epiphytic vascular plants and bryophytes constitute an important component of cloud forest canopies. Because of their different characteristics compared with leaves and other tree structural elements, epiphytes can be expected to behave differently in terms of their ability to intercept and store rain and cloud water, whereas losses through evaporation and drip may also occur at different rates. The water dynamics of epiphytes were studied in a windward lower montane cloud forest in northern Costa Rica. The exposed site experienced frequent horizontal precipitation (fog and wind-driven rain) as well as strong winds. *In situ* epiphyte wetting experiments were conducted at different levels within the 20-m canopy during a series of fog events using pre-weighed branches with known epiphyte biomass, while making simultaneous measurements of fog density and drop-size spectrum on a tower extending above the canopy. Rates of water loss via evaporation from pre-wetted epiphyte-laden branches suspended at different heights within the canopy were determined on dry days. Storage capacities were determined by gravimetric means, both in the field and under controlled conditions. Total epiphyte biomass of the forest was estimated through systematic sampling of three emergent trees and five sub-canopy trees in combination with a diameter survey of four plots of 1000 m^2 each. Fog interception rates by epiphyte-laden branches differed with position in the canopy, with an average rate of 54.7 ml $hour^{-1}$ kg^{-1} of oven-dry biomass. Absorption rates were correlated with fog liquid water content and initial moisture content of the sample. Rates of water loss during dry conditions differed considerably – depending on sample exposition, ambient climatic conditions, and sample moisture content – but followed an exponential decline with time. Maximum water storage capacity values differed also, with higher values after artificial spraying than after exposure to natural fog, suggesting that the samples were not completely saturated under field conditions with fog only.

INTRODUCTION

Cloud-forested headwater areas on the Atlantic (windward) slopes of northern Costa Rica exhibit very high runoff coefficients. This has been attributed to, *inter alia*, a combination of reduced evaporative losses in the cloud belt and extra ("occult") inputs of wind-driven cloud water captured by the vegetation (Zadroga, 1981). On the Pacific side of the Continental Divide,

Tropical Montane Cloud Forests: Science for Conservation and Management, eds. L. A. Bruijnzeel, F. N. Scatena, and L. S. Hamilton. Published by Cambridge University Press.

throughfall in a leeward cloud forest in the same area (Monteverde) was reported to be low (65% of incident rainfall) despite an estimated annual extra input of "fog water" of 886 mm (Clark *et al.*, 1998). Epiphyte biomass in this forest was estimated at 33.1 t ha^{-1} (Nadkarni *et al.*, 2004), more than twice the value derived by Köhler *et al.* (2007) for a nearby windward cloud forest at a similar elevation (see also Häger and Dohrenbusch, this volume).

The frequent fog incidence commonly associated with montane cloud forests is widely considered to be responsible for the abundance of epiphytes in these forests (Frahm and Gradstein, 1991; Wolf, 1993; Bohlman *et al.*, 1995; Hemp, 2002; Gradstein *et al.*, this volume; Mulligan, this volume). Because of their different characteristics compared with leaves, vascular epiphytes such as bromeliads, and non-vascular epiphytes such as mosses and filmy ferns, can be expected to behave differently in terms of their ability to intercept and store cloud water, whereas losses through evaporation may occur at a different rate as well (Veneklaas *et al.*, 1990; Richardson *et al.*, 2000; Köhler *et al.*, this volume; Mulligan *et al.*, this volume #25).

The exact role of epiphytes in the canopy water balance is still incompletely understood but arguably assumes extra importance given the major difficulties associated with the accurate quantification of total water inputs to these forests. Under the prevailing windy conditions the separation of "ordinary" precipitation from wind-driven (horizontal) rain and fog is particularly difficult (cf. chapters by Frumau *et al.*, Giambelluca *et al.*, Häger and Dohrenbusch, Holwerda *et al.* #29, and Schmid *et al.*, this volume). This calls for novel approaches, including process-based modeling (Murakami, 2006) and stable isotope techniques (Scholl *et al.* this volume; cf. Rhodes *et al.*, 2006). Hölscher *et al.* (2004) used a running-water balance model to assess the contribution by epiphytes to total rainfall interception in an upper montane forest not subjected to significant fog inputs in the Cordillera de Talamanca of Costa Rica. To apply a similar approach under conditions of mixed precipitation (i.e. rain and fog), separate information is required on the wetting and drying characteristics as well as the storage capacities of leaves and epiphytes.

Previous studies of fog interception by epiphytes have mostly used disturbed samples. Epiphytes have been collected from branches and placed in plastic nets or racks inside the forest or in the open and have been exposed to rainfall and fog, or sprayed with water to determine the maximum storage capacity (e.g. Veneklaas *et al.*, 1990; Chang *et al.*, 2002; Hölscher *et al.*, 2004; Mulligan *et al.*, this volume #25). However, such approaches are likely to change the structure and composition of epiphyte arrangements on individual branches (Van Leerdam and Zagt, 1989). Moreover, upon entering a forest, fog is typically moving in a spatially heterogeneous manner according to spatial changes in wind speed and canopy density (Lovett, 1984). As a result, fog interception and evaporation rates will vary depending on position in the canopy and exposure to the prevailing winds and radiation. To take these considerations into account the present study used suspended branches with their epiphytes still in place at different heights within the canopy. The samples were exposed to the elements during foggy or dry conditions whilst determining the corresponding changes in weight and recording ambient weather variables.

This work was carried out as part of a larger project aiming to evaluate the hydrological impacts of cloud forest conversion to pasture in the Monteverde area of Costa Rica (Bruijnzeel, 2006).

STUDY SITE

The studied windward lower montane cloud forest was situated on the Atlantic slope in the Tilarán range at *c.* 1460 m.a.s.l., near Santa Elena, Monteverde, north-western Costa Rica (10° 21′ 33″ N, 84° 48′ 5″ W). Rainfall input between 1 July 2003 and 30 June 2004 was about 6000 mm with a somewhat less rainy period between February and April. Due to the prevailing strong winds, horizontal precipitation occurred frequently (both in the form of wind-driven rain and fog). Measured horizontal precipitation for the July 2003–June 2004 period was 2740 mm. Average temperature in 2003 was 17.0 °C (monthly range 15.4–17.7 °C). Relative humidity was generally above 90% and foggy conditions prevailed for 50% (at night) to 60% (during the day) of the time (K. F. A. Frumau, unpublished data). The main canopy reached a height of *c.* 20 m, with epiphyte-laden emergents extending from the canopy by several meters. Common trees in the upper canopy included *Ficus crassiuscula, Elaeagia auriculata, Weinmannia wercklei*, and various Myrtaceae (Lawton and Dryer, 1980). Tree ferns and palms of ~4 m height dominated the understorey. Epiphytic vascular plants and bryophytes were abundant and represented an estimated biomass of 16.2 t ha^{-1} (Köhler *et al.*, 2007).

METHODOLOGY

Branch sections of *c.* 0.5 m length and diameters of 1.0–5.0 cm were collected from several trees using rope climbing techniques. The branches were covered with bryophytes, lichens, and small ferns, which were left in place. To measure epiphyte fog interception or evaporation rates, sample branches were placed on open trays at three different heights within the canopy (6, 15, and 22 m) using arms extending from a 25-m tall canopy access tower. Wind speed (A100 cup anemometers, Vector Instruments, UK), temperature and humidity (HMP45C sensors, Vaisala, Finland) were measured at these same heights, and solar radiation (CM7, Kipp and Zonen, the Netherlands) above the canopy. Occurrence of horizontal precipitation was recorded at all four heights as well using wire harps and cylindrical louvered screen gages (Frumau *et al.*, this volume) whereas fog liquid water content (*LWC*) was derived from above-canopy measurements of fog drop-size spectrum made with a cloud-particle spectrometer (Burkard *et al.*, 2003; cf. Schmid *et al.*, this volume).

Rates of fog water absorption by the epiphytes were determined by exposing three pre-weighed and air-dried sample branches per measurement height and sampling occasion to passing fog, and recording changes in weight on an hourly basis using portable weighing equipment. Samples were exposed throughout the daytime (07:00–17:00 h) as long as conditions with fog (but not rain) lasted. Care was taken to avoid that samples received water dripping from other samples in higher positions. In total, 67 sample branches were used during 11 fog events. On dry days, rates of water loss via evaporation were determined using three pre-wetted samples per measurement height and run. Drying experiments were carried out during daytime conditions on 19 days between February 2003 and February 2004 without fog or rainfall, involving a total of 102 samples. After each absorption or drying experiment, all epiphytes were collected per branch, weighed to the nearest 0.1 g and dried at 70 °C for 48 hours. Branch volume, surface area, and length were also determined. The observed hourly differences in sample weight and oven-dry weights allowed computation of the corresponding (net) fog absorption or evaporation rates. Water contents were expressed as percentages of dry weights.

Similar epiphyte-covered branches ($n = 53$) as used in the field experiments were used for the laboratory determination of the maximum water storage capacity. Samples were sprayed with water until saturation, covered with plastic against evaporation and suspended for about 30 min until drainage finished. Next, the epiphytes were separated from the branch and gravimetric moisture contents determined as described above. The mean gravimetric water content for all samples was taken to represent the maximum epiphyte water storage capacity. The value of total epiphyte biomass for the forest as derived by Köhler *et al.* (2007) was used to extrapolate fog water absorption and evaporation rates (ml hour^{-1} kg^{-1} of dry biomass) and storage capacity values (%) to the stand level (in mm hour^{-1} and mm, respectively).

Between March and July 2003, bryophytes and crown humus were also sampled frequently (though irregularly) from 0.5-m branch sections located in the inner crowns at 15–20 m height to determine day-to-day variations of *in situ* gravimetric water content ($n = 600$; Köhler *et al.*, 2007). The observed difference in maximum and minimum water contents during this period was multiplied times the corresponding biomass to obtain an estimate of the *in situ* water storage capacity associated with dry biomass of epiphytes and canopy humus at the stand level (Köhler *et al.*, 2007).

Simple and multiple regressions were used to investigate to what extent fog absorption rates, epiphyte storage capacity, and fog *LWC* were correlated, and to assess the influence of different climatic parameters on fog absorption and evaporation rates (Pearson correlation coefficients).

Epiphyte absorption and evaporation rates at different measurement heights were compared on the basis of geometric means, minimum and maximum values, Pearson correlation coefficients as well as coefficients of variation and an ANOVA test.

RESULTS AND DISCUSSION

Fog interception by epiphytes

Fog events occurred throughout the year, mostly together with wind-driven rain. However, during the drier February–April period, fog events often occurred separately from rainfall. During 11 such events – which represented a wide range of intensities – epiphyte-covered branches were exposed to passing fog. Fog absorption increased asymptotically with time (Figure 26.1), and in individual events reached 88% of sample oven-dry weight. Overall average fog absorption was 5.6% (± 3.0) of dry weight hour^{-1} or 54.7 ml hour^{-1} kg^{-1}. *Net* absorption rates (i.e. gross absorption minus concurrent evaporation, see below) at 15 and 22 m were not significantly different at *c.* 5% of dry epiphyte weight hour^{-1}, but the average rate of 6.8% at 6 m height was significantly higher ($p < 0.03$) (Table 26.1, Figure 26.1). Therefore, net fog absorption efficiency increased from the top of the canopy downward. In addition, during some hours with low-intensity fog, samples were observed to lose weight

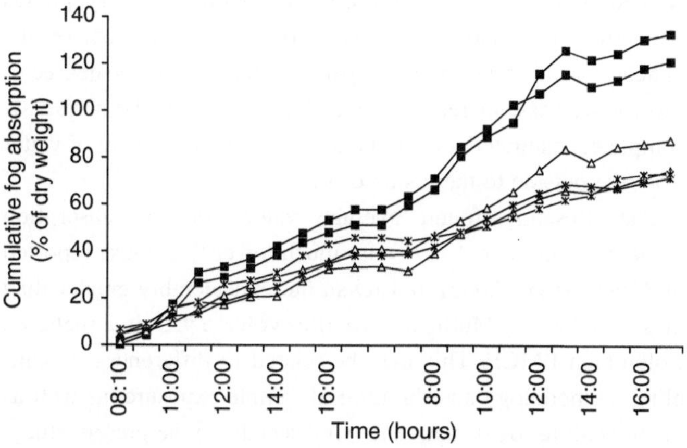

Figure 26.1. Cumulative daytime absorption (% of dry weight) of fog water by epiphytes exposed to foggy conditions in a windward lower three different montane cloud forest, Monteverde. Paired samples were positioned at heights: 6(■), 15(Δ), and 22 m(x) above ground level.

Table 26.1 *Average and maximum values of fog absorption and evaporation rates (standard deviations between brackets) of epiphytes at three different heights within a windward lower montane cloud forest, Monteverde*

	22 m	15 m	6 m
Net fog absorption rate			
(gain as % of dry weight hour^{-1})			
Average	4.8 (± 3.0)	5.2 (±2.4)	6.8 (±3.2)
Maximum	13.4	13.2	16.2
Interception rate (ml hour^{-1} kg^{-1})	49.4 (± 31.4)	52.6 (± 27.1)	62.3 (± 34.7)
Evaporation rate			
(loss as % of dry weight hour^{-1})			
Average	13.3 (± 10.3)	12.7 (± 10.8)	10.0 (± 8.8)
Maximum	55.4	45.1	36.5

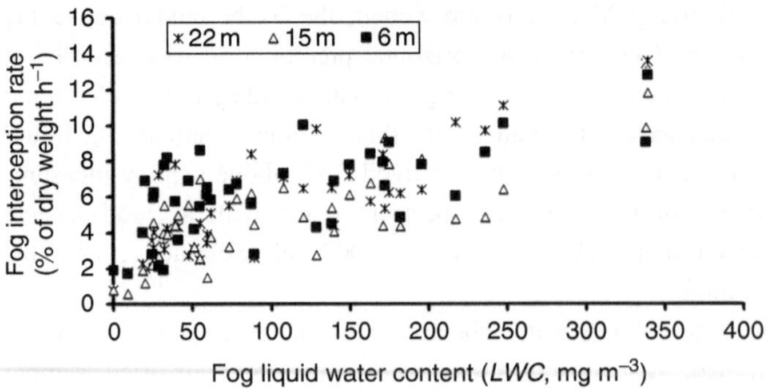

Figure 26.2. Relationship between fog liquid water content (mg m^{-3}) and hourly absorption of fog water by epiphytes (% of dry weight) exposed at three different heights in a windward lower montane cloud forest, Monteverde.

again – especially in the more exposed and ventilated upper third of the canopy. Conversely, at 6 m (i.e. below the main canopy) the air remained near-saturated, whereas radiation levels were low and ventilation poor (Table 26.1).

When exposed to consecutive fog events, samples accumulated on average up to 133% of their dry weight (Figure 26.1), with the largest increases noted at the lowermost position. Of the samples that were exposed to continuous fog for three days in a row, some began to drip on the second day after their weight had reached about 180% of the dry weight. However, consecutive measurement showed that all samples continued to increase their weight throughout the duration of the fog event (Figure 26.1). Visual inspection revealed that even when the sample surface was almost saturated, the inside was still partially dry. Thus, part of the surface excess water dripped to the forest floor while wetting of the inner part of the moss sample continued. According to Chang *et al.* (2002), fog absorption by bryophytes is controlled by surface wetness. However, the present results indicate that absorption is only partly influenced by the degree of surface wetness; increases in weight, even when the surface is completely saturated, occur due to continued movement of water from the outside to the inside of the sample.

The presently found average values of fog absorption (49.4–62.3 ml hour^{-1} kg^{-1}) are much lower than those reported by Chang *et al.* (2002) in Taiwan but considerably greater than those obtained by Mulligan *et al.* (this volume #25) in a sheltered Colombian LMCF. This may be related to differences in sampling methodology and duration of sample exposure, as well as the prevailing fog density and wind speeds. In the present study, the samples were undisturbed and may have mimicked actual conditions rather better than disturbed and repacked samples. Also, the largest increases in sample weight were observed at the beginning of the experiments, which further explains the larger values found by Chang *et al.* (2002) who exposed completely dry samples during very short periods only (up to 1.5 hours). The very low rates of fog absorption reported for the Colombian forest may reflect the very low wind speeds and reduced fog density prevailing beneath the canopy at Tambito (see discussion in Mulligan *et al.*, this volume #25).

Amounts of fog absorption by the epiphytes correlated rather poorly with climatic parameters that would be expected to be of importance, such as atmospheric humidity or wind speed ($r^2 < 0.24$). This is due to the fact that during fog events relative humidity at all heights was close to saturation anyway, whereas wind speeds were highly variable. No relationship with average wind speeds was found either. However, absorbed amounts of fog water correlated better ($r^2 = 0.56$) with fog *LWC* (Figure 26.2). Furthermore, the multi-factor ANOVA revealed a significant relation ($p < 0.05$) between epiphyte fog absorption rate, fog *LWC* and (to a much lesser degree) wind speed (Figure 26.3).

Generally, the rate of fog absorption was somewhat larger at the beginning of the experiments – when the samples were still air-dry – implying an inverse correlation between the rate of fog absorption (ml hour^{-1} kg^{-1}) and sample gravimetric moisture content (%, Figure 26.4). The epiphytes were capable of absorbing more than twice their own dry weight below the main canopy (at 6 m) though smaller values were attained at higher positions in the canopy (Figure 26.4).

Evaporation from epiphytes

Figure 26.5 shows the exponential decreases in epiphyte water content during four consecutive dry days (1–4 February 2004) at 22, 15, and 6 m height, respectively. The surfaces of the samples generally dried within 4 hours but most samples remained moist inside even after several days of exposition to dry conditions (e.g. >100% gravimetric moisture; Figure 26.5). The continued supply of moisture from the inner to the outer parts is thought to

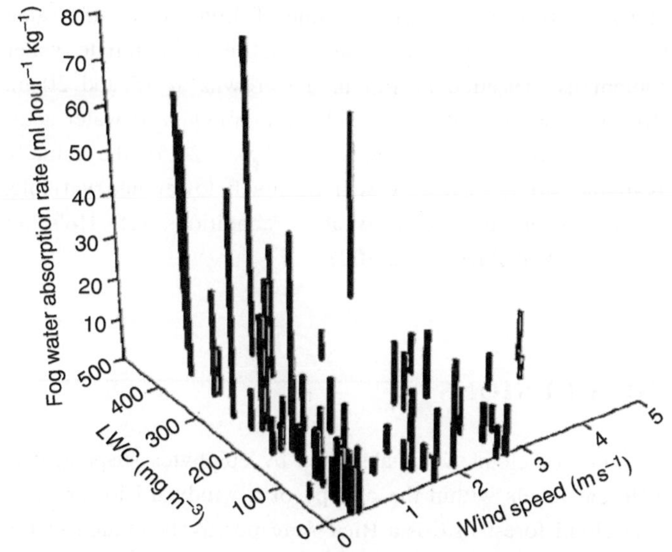

Figure 26.3. Dependence of fog absorption (ml hour^{-1} kg^{-1}) by the epiphyte surface layer on fog liquid water content (*LWC*, mg m^{-3}) and wind speed (m s^{-1}) in a windward lower montane cloud forest, Monteverde.

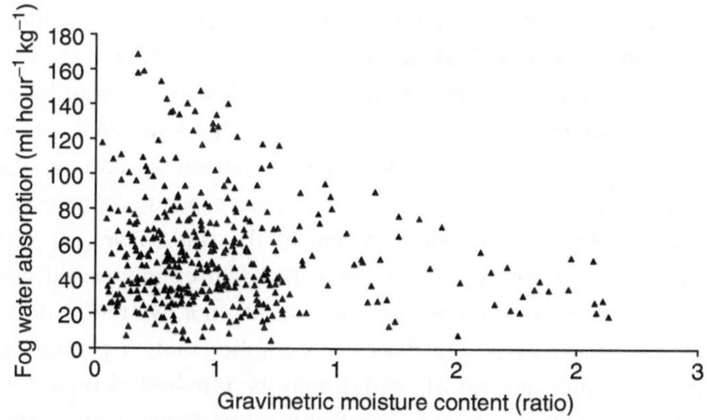

Figure 26.4. Fog water absorption by epiphytes (ml hour^{-1} kg^{-1}) vs. epiphyte gravimetric water content (% of dry weight) in a windward lower montane cloud forest, Monteverde.

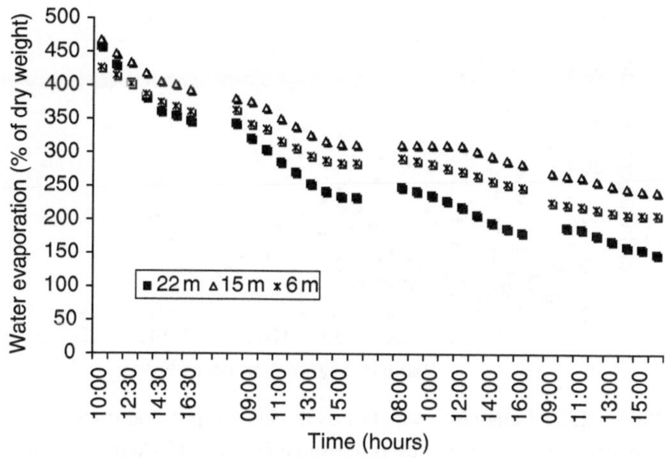

Figure 26.5. Decrease in daytime water content (% of dry weight) of nine epiphyte samples at three heights each ($n = 27$) during four consecutive dry days, in a windward lower montane cloud forest, Monteverde.

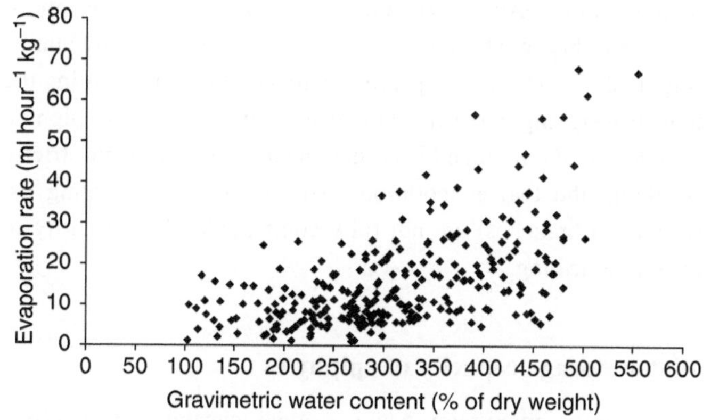

Figure 26.6. Relationship between daytime weight loss through evaporation from epiphytes (ml hour^{-1} kg^{-1}) and epiphyte water content (% dry weight) in a windward lower montane cloud forest, Monteverde. Data from three different heights were pooled.

sustain epiphyte physiological processes and to ensure survival during dry conditions (cf. Hietz, this volume).

Average evaporation rates and total loss of water by epiphytes located at the top of the tree crowns were much higher than inside the forest ($p < 0.05$; Table 26.1). Evaporation was related ($r^2 = 0.47$) to gravimetric water content of the sample (Figure 26.6) and to net radiation as measured at the top of the canopy ($r^2 = 0.43$; epiphytes at 22 m only), but not to the Penman reference evaporation equivalent ($r^2 = 0.18$) as determined for above-canopy conditions.

On average, rates of water lost by evaporation were 1.5–2.8 times higher than rates of net fog water absorption (Table 26.1). As such, the large losses of water from thoroughly wetted epiphytes were not compensated by fog inputs, and additional water must be derived from rainfall. The situation is less critical at lower levels in the canopy where epiphytes retained greater quantities of water in between fog or rainfall events (Figure 26.5) and where net absorption rates were higher (Figure 26.1). These wetting and drying patterns have important implications for the throughfall dynamics of cloud forests. Often, the epiphytes will still be partially wet upon the start of a rainfall event and less water will be required to replenish epiphyte water storage capacity. This, in turn, is expected to result in more water passing through the canopy and reaching the soil surface. The different wetting and drying characteristics of epiphytes compared to those of leaves also render the modeling of overall canopy interception much more complex (cf. Hölscher *et al.*, 2004; Murakami, 2006).

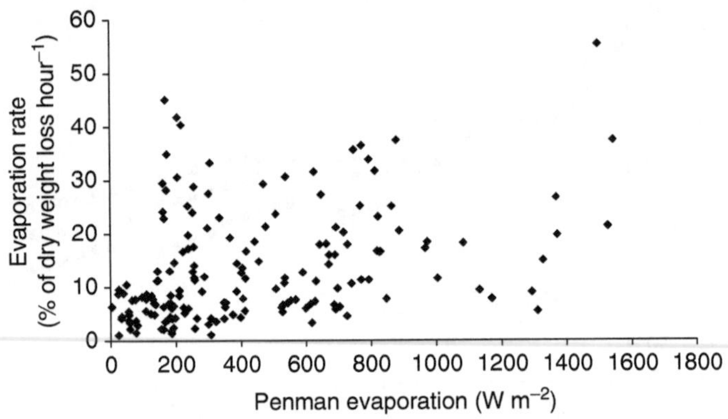

Figure 26.7. Cumulative weight loss through evaporation from epiphytes and canopy humus as measured *in situ* at 15–20 m height in a windward lower montane cloud forest at Monteverde, during three consecutive rainless days in April 2003.

Evaporation from epiphytes sampled *in situ* from the inner parts of tree crowns at 15–20 m height during three consecutive rainless days in April 2003 amounted to 251% of dry weight in the case of bryophytes and 117% for less exposed canopy humus (Figure 26.7). This finding agrees with the observation during the branch weighing experiments that the outside parts of epiphyte samples dried out much faster than the insides. Although, strictly speaking, the two evaporation data-sets (*in situ* sampling vs. suspended branches) are not fully comparable, the similarity in results is striking.

Water storage capacity of epiphytes

The average maximum water storage capacity of epiphyte samples (laboratory wetting tests) was determined at 323 ± 106% of dry weight. The high standard deviation reflects the variable composition of epiphytic species on the sample branches. Although the bulk of the epiphytes consisted of mosses with an intrinsically high storage capacity, the presently found storage value falls in the lower part of the range reported for various cloud forests (Kershaw, 1985; Shaw and Goffinet, 2000) and is much lower than the value derived by Pypker *et al.* (2002) for a Douglas fir forest in the Pacific North-West using very small (3 g) undisturbed epiphyte samples. Such differences may be related to differences in size, composition, and shape of the samples used in the various studies. The presently used samples were much larger and consisted of a mixture of bryophytes, ferns, and bromeliads rather than bryophytes only.

Combining the maximum water storage capacity of 323% with the total epiphyte biomass of 16 215 kg ha^{-1} derived for this mature cloud forest by Köhler *et al.* (2007) allowed an estimation of the (maximum) amount of water that can be held by the epiphytes at the stand scale as 5.24 mm. Kőhler *et al.* (2007) derived a very similar value of 4.95 mm for the same forest based on the maximum variation in epiphyte water content as measured *in situ* in the crowns at 15 and 20 m. However, because of the considerable amounts of water actually retained by the epiphytes (cf. Figure 26.5), the actually available storage capacity will be much lower and variable, depending on antecedent weather conditions (cf. Hölscher *et al.*, 2004; Köhler *et al.*, 2007).

CONCLUSIONS

Amounts of cloud water absorbed by epiphytes suspended at different levels within the canopy of a windward lower montane cloud forest in Costa Rica were positively related to fog liquid water content, whereas absorption rates decreased as epiphyte water content increased. Net fog absorption rates were always higher below the main canopy (6 m) than at the canopy level (22 m), whereas evaporation rates were persistently lower at 6 m. Evaporation losses on dry days were typically 150–280% of average fog absorption rates, depending on canopy position. Therefore, epiphyte water content within the canopy was both higher and varied less with time than in the top of the canopy. Evaporation followed a logarithmic decay pattern and was inversely related to remaining water content of the sample. Although both laboratory and *in situ* tests suggested a maximum water storage capacity value associated with epiphytes of *c.* 5 mm at the stand level, *in situ* evaporation experiments also indicated that epiphytes did not fully dry out between precipitation events, not even after three to four consecutive dry days. This implies that in practice a much smaller amount of rainfall may be required to replenish epiphyte storage capacity depending on previous wetting and drying cycles. As a result, amounts of throughfall will be enhanced accordingly, also because epiphytes were observed to start dripping during dense fog events before full saturation was reached.

ACKNOWLEDGEMENTS

Financial support from the Forestry Research Programme of the Department for International Development of the UK (FRP-DFID project R7991) is gratefully acknowledged. The views expressed here are not necessarily those of DFID. Dr. Peter Hietz is thanked for reviewing the draft manuscript. Special thanks go to Eulogio Jimenez of the San Gerardo Lodge for site permission and help, and to the FIESTA field team for their assistance.

REFERENCES

Bohlman, S. A., T. J. Matelson, and N. M. Nadkarni (1995). Moisture and temperature patterns of canopy humus and forest floor soil of a montane cloud forest, Costa Rica. *Biotropica* **27**: 13–19.

Bruijnzeel, L. A. (2001). Hydrology of tropical montane cloud forests: a reassessment. *Land Use and Water Resources Research* **1**: 1.1–1.18.

Bruijnzeel, L. A. (compiler) (2006). *Hydrological Impacts of Converting Tropical Montane Cloud Forest to Pasture, with Initial Reference to Northern Costa Rica*, Final Technical Report DFID-FRP Project no. R7991. Amsterdam: VU University Amsterdam, and Aylesford, UK: Forestry Research Progamme of the UK, Department for International Development. Also available at www.geo.vu.nl/~fiesta.

Burgess, S. S. O, and T. E. Dawson (2004). The contribution of fog to water relations of *Sequoia sempervirens* (D. Don): foliar uptake and prevention of dehydration. *Plant, Cell and Environment* **27**: 1023–1034.

Burkard, R., P. Bützberger, and W. Eugster (2003). Vertical fogwater flux measurement above an elevated forest canopy at the Lägeren research site, Switzerland. *Atmospheric Environment* **37**: 2979–2990.

Chang, S. -C., I. L. Lai, and J. Wu (2002). Estimation of fog on epiphytic bryophytes in a subtropical montane forest ecosystem in north-eastern Taiwan. *Atmospheric Research* **64**: 159–167.

Clark, K. L., N. M. Nadkarni, D. Schaefer, and H. L. Gholz (1998). Atmospheric deposition and net retention of ions by the canopy in a tropical montane forest, Monteverde, Costa Rica. *Journal of Tropical Ecology* **14**: 27–45.

Frahm, J. -P., and S. R. Gradstein (1991). An altitudinal zonation of the tropical rain forest using bryophytes. *Journal of Biogeography* **18**: 669–678.

Hemp, A. (2002). Ecology of the pteridophytes on the southern slopes of Mt. Kilimanjaro. I. Altitudinal distribution. *Plant Ecology* **159**: 211–239.

Hölscher, D., L. Köhler, A. I. J. M. van Dijk, and L. A. Bruijnzeel (2004). The importance of epiphytes in rainfall interception by a tropical montane rainforest in Costa Rica. *Journal of Hydrology* **292**: 308–322.

Kershaw, K. A. (1985). *Physiological Ecology of Lichens*. Cambridge, UK: Cambridge University Press.

Köhler, L., C. Tobón, K. F. A. Frumau, and L. A. Bruijnzeel (2007). Biomass and water storage of epiphytes in old-growth and secondary montane rain forest stands in Costa Rica. *Plant Ecology* **193**: 171–184.

Lawton, R., and V. Dryer (1980). The vegetation of the Monteverde Cloud Forest Reserve. *Brenesia* **18**: 101–116.

Lovett, G. M. (1984). Rates and mechanisms of cloud water deposition to a subalpine balsam fir forest. *Atmospheric Environment* **18**: 361–371.

Murakami, S. (2006). A proposal for a new forest canopy interception mechanism: splash droplet evaporation. *Journal of Hydrology* **319**: 72–82.

Nadkarni, N. M., D. Schaefer, T. J. Matelson, and R. Solano (2004). Biomass and nutrient pools of canopy and terrestrial components in a primary and a secondary montane cloud forest, Costa Rica. *Forest Ecology and Management* **198**: 223–236.

Pócs, T. (1980). The epiphytic biomass and its effect on the water balance of two rain forest types in the Uluguru Mountains (Tanzania, East Africa). *Acta Botanica Academiae Scientiarum Hungaricae* **26**: 143–167.

Pypker, T. G., B. J. Bond, and M. H. Unsworth (2002). *The Role of Epiphytes in the Interception and Evaporation of Rainfall in Old-Growth Douglas-Fir Forests in the Pacific Northwest*. Corvalis, OR: Oregon State University.

Rhodes, A. L., A. J. Guswa, and S. E. Newell (2006). Seasonal variation in the stable isotopic composition of precipitation in the tropical montane forests of Monteverde, Costa Rica. *Water Resources Research* **42**, W11402, doi:10.1029/2005WR004535.

Richardson, B. A., M. J. Richardson, F. N. Scatena, and W. H. McDowell (2000). Effects of nutrient availability and other elevational changes on bromeliad populations and their invertebrate communities in a humid tropical forest in Puerto Rico. *Journal of Tropical Ecology* **16**: 167–188.

Shaw, A. J., and B. Goffinet (2000). *Bryophyte Biology*. Cambridge, UK: Cambridge University Press.

Van Leerdam, A., and R. J. Zagt (1989). The epiphyte vegetation of an Andean forest in Colombia: aspects of its hydrology and distribution in the canopy. M.Sc. thesis, University of Utrecht, Utrecht, the Netherlands.

Veneklaas, E. J., R. J. Zagt, A. van Leerdam, *et al.* (1990). Hydrological properties of epiphyte mass of a montane tropical rain forest, Colombia. *Vegetatio* **89**: 183–192.

Wolf, J. (1993). Diversity patterns and biomass of epiphytic bryophytes and lichens along an altitudinal gradient in the northern Andes. *Annals of the Missouri Botanical Garden* **80**: 928–960.

Zadroga, F. (1981). The hydrological importance of a montane cloud forest area of Costa Rica. In *Tropical Agricultural Hydrology*, eds. R. Lal and E. W. Russell, pp. 59–73. New York: John Wiley.

27 Epiphyte biomass in Costa Rican old-growth and secondary montane rain forests and its hydrological significance

L. Köhler and D. Hölscher
University of Göttingen, Göttingen, Germany
L. A. Bruijnzeel
VU University, Amsterdam, the Netherlands
C. Leuschner
University of Göttingen, Göttingen, Germany

ABSTRACT

Epiphyte biomass and associated canopy water storage capacity may vary greatly in tropical montane forests depending on climate, forest structure, and stand age. This study compares old-growth and secondary forests in the upper montane belt of the Cordillera de Talamanca (Costa Rica) with respect to biomass of non-vascular and vascular epiphytes and their effect on water fluxes in the canopies of an old-growth forest, an early-successional stand (10–15 years of age), and a mid-successional stand (*c*. 40 years). Irrespective of stand age, epiphyte communities were strongly dominated by non-vascular plants (70–99% of total epiphytic biomass). Epiphyte biomass in the old-growth forest (3400 kg ha^{-1}) was more than 20 times that of the youngest stand (160 kg ha^{-1}) and more than six times that of the intermediate stand (520 kg ha^{-1}). Consequently, the water storage capacity of non-vascular epiphytes and canopy humus increased from 0.06 mm in the early-successional, via 0.18 mm in the mid-successional, to 0.97 mm in the old-growth stand. Thus, the recolonization by epiphytes of tropical successional forests after clear-cutting, and the restoration of epiphytic water storage capacity will require many decades if not centuries.

INTRODUCTION

Tropical montane cloud forests (TMCF) are among the most epiphyte-rich forest types of the world. This epiphyte abundance is widely believed to reflect the prevailing microclimatic conditions such that forests in regions with high rainfall and frequent fog typically harbor more epiphyte biomass than stands at drier locations (Frahm and Gradstein, 1991; Wolf, 1993). This, in turn, may influence the interception of rainfall and cloud water (e.g. Pócs, 1980; Nadkarni, 1984; Veneklaas and Van Ek, 1990; Ataroff and Rada, 2000; Tobón *et al*., this volume #26). Bryophytes and other poikilohydric epiphytes will affect canopy water fluxes mainly through the enhancement of overall canopy water storage capacity. Therefore, information on epiphyte biomass and composition and its hydrological properties is a prerequisite for better understanding cloud and rain water interception in TMCF, and for building more realistic process models (Hölscher *et al*., 2004).

Published estimates of epiphytic biomass at the stand level for old-growth TMCF range from 370 kg ha^{-1} in a cyclone-ridden stunted ridge-top forest in Jamaica (Tanner, 1980, 1985) to 44 000 kg ha^{-1} in an upper montane cloud forest in Colombia (Hofstede *et al*., 1993). Much more information is available in this respect for Neotropical than for Paleotropical forests. As a consequence of rapid deforestation in the past few decades, land-use types other than old-growth forest already cover large areas in the montane tropics, and this area is on the increase (Mulligan, this volume). When agricultural activities are

Tropical Montane Cloud Forests: Science for Conservation and Management, eds. L. A. Bruijnzeel, F. N. Scatena, and L. S. Hamilton. Published by Cambridge University Press.

abandoned, secondary forest may establish with consequences for biodiversity, forest hydrological processes and biogeochemical cycling (e.g. Kappelle *et al.*, 1995; Helmer, 2000; Giambelluca, 2002). Epiphyte colonization of secondary forests is a process that may well take decades. Therefore, old-growth and secondary forests are likely to differ in epiphyte biomass and canopy hydrological functioning. However, information on epiphyte biomass in secondary TMCF is scarce (e.g. Nadkarni *et al.*, 2004) and data on their hydrological properties even scarcer (Köhler *et al.*, 2007; Tobón *et al.*, this volume #26).

To better understand canopy water fluxes in TMCF, data on epiphyte biomass and associated water storage capacity alone are not sufficient. Information on the temporal changes in epiphytic water storage is arguably more important because this is what controls the actual interception capacity (Hölscher *et al.*, 2004; Tobón *et al.*, this volume #26). The present study was conducted in three leeward upper montane old-growth and secondary forest stands in the Cordillera de Talamanca in Costa Rica. The objectives were to: (i) describe the composition and distribution of the epiphytic vegetation and canopy humus in the respective stands, (ii) estimate epiphyte biomass at the stand level, and (iii) analyze *in situ* water storage dynamics of epiphytic bryophytes and canopy humus.

STUDY SITES

The study was carried out between November 1999 and August 2000 in the headwater area of the Rio Savegre watershed on the Pacific slopes of the Cordillera de Talamanca at an elevation of *c.* 2900 m.a.s.l. The studied stands were a 10–15-year-old early-successional forest (ESF), a 40-year-old mid-successional forest (MSF), and an old-growth upper montane rain forest (OGF) without major visible human impact. All stands were located in the Los Santos Forest Reserve (9° 35′ 40″ N, 83° 44′ 30″ W) where studies on water and nutrient dynamics have also been carried out (Köhler *et al.*, 2006; Hölscher *et al.*, this volume). The ESF and MSF were located at a distance to each other of ~500 m whereas the OGF was about 2 km away. Slopes were steep throughout (ESF 30°, MSF 25°, OGF 31°) and aspect was south at the ESF and MSF sites vs. north-east at the OGF. The OGF and MSF sites were previously the subject of detailed investigations of the diversity of vascular plants and forest structure (Kappelle *et al.*, 1995, 1996). Both stands were dominated by *Quercus copeyensis* C.H. Müll. Besides *Q. copeyensis*, a number of early-successional tree species were present in the ESF. In the OGF, a second oak species, *Q. costaricensis* Liebm., was co-dominant. The height of the upper tree layer was 5–9 m in the ESF, 11–15 m in the MSF, and 30–35 m in the OGF. The number of stems with a diameter at breast height (DBH) $\geq$ 3 cm was 3460 ha^{-1} (OGF), 5730 ha^{-1} (ESF), and 4800 ha^{-1} (MSF). Average annual temperature at a station 15 km east of the study site (Villa Mills, 3000 m.a.s.l.) is 10.9 °C and average annual rainfall is 2812 mm, with a dry season from December or January to April when average monthly totals range between 20 and 100 mm (Instituto Meteorológico Nacional, 1988). Additional water inputs through horizontal precipitation were considered to be small at these relatively sheltered sites (Köhler, 2002).

METHODS

Epiphyte sampling

DISTRIBUTION AND COMPOSITION OF EPIPHYTES

Due to the dominance of oaks in the studied stands, only *Q. copeyensis* trees were considered for epiphyte sampling. In each stand, six dominant oaks were sampled, yielding 344 samples in total. Selection of sample trees was based on accessibility of the crown for climbing and on general visual representativity in terms of epiphyte biomass. Trees were climbed using either single-rope techniques (Perry, 1978) or a ladder in the case of smaller trees. Trees in the ESF were cut and epiphytes sampled on the ground at 1-m intervals along the vertical axis, and from the outer parts of the smaller branches. In the other stands, the sampled trees were stratified into main sections (trunk, inner branches, middle branches, and outer branches) following Johansson (1974). The ratio of epiphyte dry mass to substrate surface was determined by collecting a varying number of samples per branch section (5–10). Samples were collected from areas including the upper and lower sides of a branch. Inner branch samples were taken at sections where the stems forked into major branches. Samples from the outer branches were collected by cutting off the branches as they were inaccessible through climbing. Tree trunks were sampled by stripping all epiphytes within bands encircling the trunk at different heights. For trees of larger DBH, rectangular areas of 20 × 30 cm (exposition N, S, W, E) were sampled at different heights on the trunk. Samples were separated in the laboratory into bryophytes and lichens, vascular plants, and canopy humus (partly or highly decomposed organic material). The fractionated samples were oven-dried at 70 °C for 48 hours and weighed to the nearest 0.1 g. Species identification was done by Holz *et al.* (2002).

TOTAL EPIPHYTE BIOMASS

Epiphyte biomass at the stand level in the OGF and MSF was calculated from data on biomass per unit bark area, measured stem and branch circumference, and, for branches, distance from the stem to the sampling position, and the number of first-order branches. Based on the assumption that total sapwood area of a tree varies only little with height in the canopy (West *et al.*, 1999), cumulative cross-sectional areas and branch surface areas were estimated for different height sections of a tree. This allowed the calculation of epiphyte biomass at the stand level from epiphyte biomass per unit bark area and stem density (Köhler, 2002). For the shrub layer of the OGF and MSF, epiphyte biomass estimation was based on data collected by Kunz (2000) who determined epiphyte biomass on more than 600 shrubs in the two stands. In the ESF, total branch surface area per tree was obtained from length and width measurements on 20 branches distributed among six oak trees. To calculate trunk surface area, stems were divided into 1-m segments which were assumed to have a conical shape. The total surface area of a tree obtained by this procedure was then multiplied times epiphyte biomass per unit bark area and total stem number (DBH $\geq$ 3 cm) to obtain epiphyte biomass at the stand level.

The number of tank bromeliads was counted on six dominant oaks per stand and assigned to two size classes (diameters of 10–40 cm and >40 cm) to estimate their biomass at stand level. Further details on epiphyte sampling techniques are given in Köhler (2002).

Water content of epiphytes

In the OGF, the mat-building moss *Leptodontium exasperatum* Cardot dominated the epiphyte vegetation in terms of biomass. Therefore, *in situ* water content monitoring was concentrated on this species for a period of 10 months (November 1999 – August 2000). Seven to ten moss samples were taken every 4 weeks with a small core sampler (8.4 cm diameter, 10 cm length) from three inner branches of a single tree at 20–25 m height. Occasionally, moss samples were also taken from other trees for comparison. A total of 76 moss samples were collected. Sampling was carried out between 09.00 and 13.00 h, and covered both the dry and the rainy season. Samples were stored immediately in plastic bags and transported to the laboratory for gravimetric determination of their water content (drying at 70 °C for 48 hours).

The difference between the highest and lowest *in situ* moss water content values (percentage of dry weight) during the 10-month sampling period was taken to represent maximum water storage. Multiplying the latter times stand biomass of non-vascular epiphytes gave the associated maximum water storage at the stand level.

The hydrological properties of canopy humus were not investigated at San Gerardo but were approximated from measurements of maximum water storage in canopy humus in a lower montane cloud forest at Monteverde, Costa Rica (Köhler *et al.*, 2007) and biomass data of canopy humus from San Gerardo. Maximum water storage of tank bromeliads at the stand level was estimated by experimentally determining the storage capacity of 13 individuals with rosette diameters between 18 and 46 cm and field counts of the number of bromeliads in the two size classes defined above.

Drying experiments

Drying experiments with non-vascular epiphytes were carried out in the rainy and the dry season to estimate rates of evaporation of intercepted water under near-natural conditions. Samples of the moss *Leptodontium* ($n = 7$), the pendant bryophyte *Pilotrichella flexilis* (Hedw.) Ångstr. ($n = 7$), and the macro-lichen *Sticta ferax* Müll. Arg. ($n = 8$) were saturated in water for 5 min. After a drainage phase of 10 min the samples were exposed to the atmosphere at 1 m above the forest floor within the OGF. Subsequent water loss through evaporation was determined at regular intervals by weighing (Köhler, 2002).

RESULTS

Spatial distribution and composition of epiphyte communities

On the canopy trees of the OGF, epiphyte vegetation was dominated by bryophytes and lichens. Biomass of non-vascular epiphytes was rather low on the inner branches (41% of total epiphytic biomass), but the proportion of bryophytes and lichens increased toward the middle (71%) and outer branches (98%). Separation between bryophytes and lichens was not possible for all branch samples; nevertheless, bryophytes were by far the dominant group. On the inner branches, they made up 95% of non-vascular epiphyte biomass. The relative importance of canopy humus in the OGF declined from inner to outer branches. However, the high mass of canopy humus on the inner branches (53%) may have been caused in part by the mode of sampling. Samples were collected where stems forked into major branches and this situation may not have been entirely representative. Vascular epiphytes contributed 2–11% of total epiphyte biomass, with no clear distribution pattern within the crown sections (Figure 27.1). The absolute biomass of bryophytes and lichens per unit branch area was 466 $\pm$ 584 g m^{-2} on the middle branches and only 174 $\pm$ 219 g m^{-2} on the outer branches. On trunk surfaces, the dominant epiphytic life forms were also bryophytes and lichens (94%). Canopy humus was negligible and vascular plants on trunks contributed only 6% to total epiphyte mass.

In the MSF, nearly the entire epiphyte biomass consisted of bryophytes whereas the percentage of lichens was very low. Tank bromeliads were not included in Figure 27.1 since these were quantified separately by direct counting. In the ESF, the dominance of bryophytes was even greater (98–100%).

Epiphyte biomass at the stand level

In the OGF, dominant oak trees held an estimated average 12.1 kg of epiphyte biomass per tree. At the stand level, total epiphytic

Table 27.1 *Epiphytic biomass at the stand level in an old-growth forest (OGF) and in early- and mid-successional (secondary) forests (ESF and MSF) at San Gerardo, Talamanca range, Costa Rica. Bromeliads included in vascular epiphyte numbers*

	Epiphytic component (kg ha^{-1})								
	Total	Non-vascular		Vascular		*Bromeliads*		Canopy humus	
OGF	3400	2385	(70%)	770	(23%)	*316*	*(9%)*	245	(7%)
ESF	160	159	(99%)	1	(1%)	0		0	
MSF	520	463	(89%)	57	(11%)	*47*	*(9%)*	0	

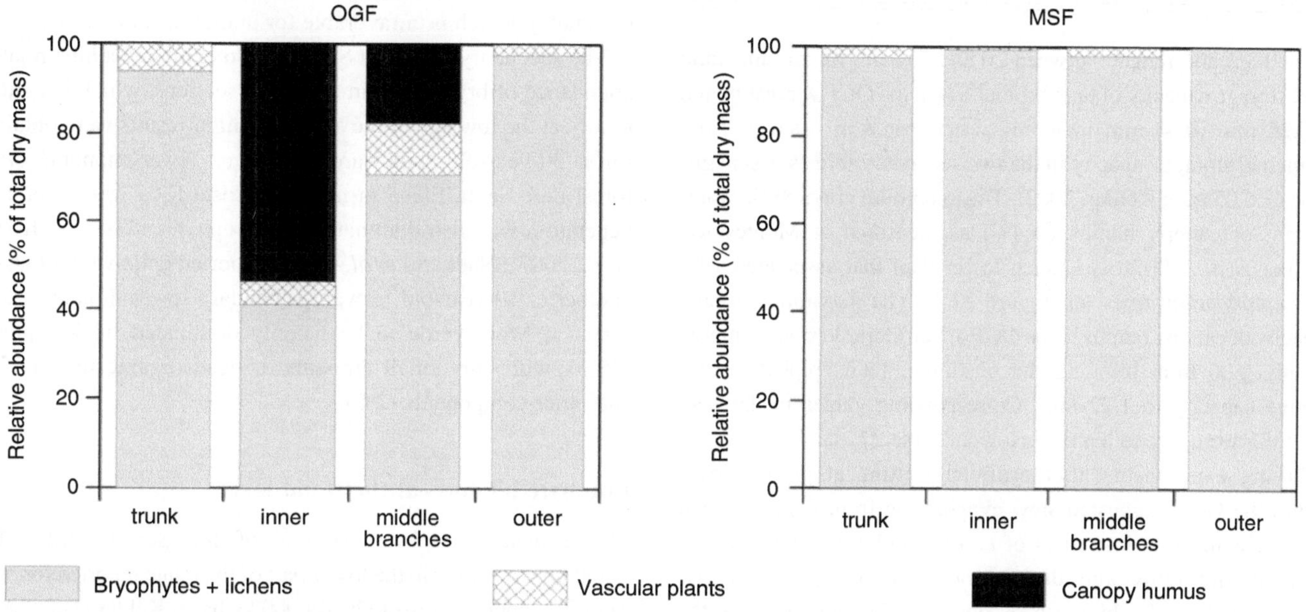

Figure 27.1. Spatial distribution and composition of epiphyte biomass and canopy humus in an old-growth forest (OGF) and a mid-successional forest (MSF) at San Gerardo, Talamanca range, Costa Rica.

biomass on all upper canopy trees was estimated at 2820 kg ha^{-1} of which 1920 kg ha^{-1} were non-vascular epiphytes. *Leptodontium* mats contributed more than 30% of the latter. Tank bromeliads were particularly abundant in the upper canopy layer and had an estimated biomass of 316 kg ha^{-1}. The epiphyte biomass of the mid-canopy layer was estimated at about 500 kg ha^{-1} whereas that associated with the shrub layer was only 80 kg ha^{-1} and consisted almost entirely of non-vascular epiphytes. Thus, total epiphyte biomass at the stand level equalled about 3400 kg ha^{-1} (Table 27.1).

In the secondary forests epiphyte biomass was estimated at only 160 kg ha^{-1} in the ESF vs. 520 kg ha^{-1} in the MSF (Table 27.1). Amounts of canopy humus in the secondary stands were negligibly small.

Water content of epiphytes

The water content of *Leptodontium* mats in the OGF varied between 24 ± 1 (SD) and 406 ± 31% of dry weight (Figure 27.2), suggesting a maximum water storage capacity of 382%. Estimated capacities of non-vascular epiphytes at the stand level were 0.06 mm for the ESF, 0.18 mm for the MSF, and 0.91 mm for the OGF (Table 27.2). The water storage capacity of individual bromeliads increased significantly with diameter of the rosettes

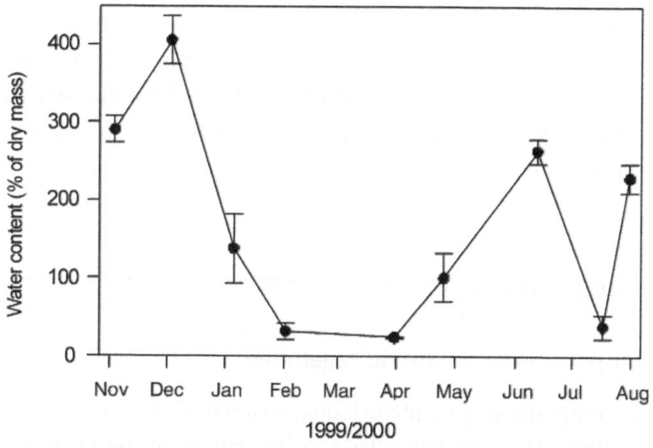

Figure 27.2. Water content of the epiphytic moss *Leptodontium* as measured *in situ* within the tree crowns of an old-growth forest at San Gerardo, Talamanca range, Costa Rica (mean ± SD, n = 6–10).

Table 27.2 *Water storage capacities of non-vascular epiphyte biomass and canopy humus at the stand level in three forests at San Gerardo, Talamanca range, Costa Rica*

	Maximum water storage (mm)		
	Non-vascular epiphytes	Canopy humus	Total
OGF	0.91	0.06	0.97
ESF	0.06	0	0.06
MSF	0.18	0	0.18

(r^2 = 0.86) and ranged between 20 and 750 ml. Maximum stand level storage capacity of tank bromeliads in the OGF was estimated at 0.25 mm (0.34 mm including water stored in tissues). Tank bromeliad storage capacity in the two secondary forests was negligible (<0.05 mm; Köhler, 2002). The maximum gravimetric water content of canopy humus (264%) as measured at Monteverde (Köhler *et al.*, 2007) was much lower than that associated with the *Leptodontium* mats (cf. Figure 27.2). The maximum storage capacity of canopy humus in the OGF at San Gerardo was estimated at only 0.06 mm, bringing the maximum total epiphyte water storage capacity to 1.22 mm. Corresponding values in the ESF and MSF were five to ten times lower (Table 27.2).

Drying experiments after artificial wetting at ground level within the OGF suggested slow evaporation from *Leptodontium* mosses, with daily loss rates of *c.* 12% and 15% of the stored water during rainy and dry season conditions, respectively. Extrapolated to the biomass of non-vascular epiphytes at the stand level, this would equal a daily water loss of 0.03 mm (rainy season) and 0.04 mm (dry season). Hanging bryophytes such as *Pilotrichella flexilis* showed an initially fast, but exponentially decreasing, evaporation rate during the dry season (50% within 65 min) and more moderate rates during the rainy season (50% within 220 min). Evaporation curves derived for the macrolichen *Sticta ferax* were comparable to those for *Pilotrichella* in both seasons (Köhler, 2002).

DISCUSSION

Composition of epiphytic vegetation

The composition of epiphyte biomass varied with crown location and stand age. The percentage of bryophytes in the OGF was generally lower on the inner branches and increased toward the middle and outer branches. Canopy humus showed a reverse distribution with a high percentage on the inner branches and lower values on the middle and outer branches. Similar patterns of epiphytic composition and distribution have been reported for dominant trees in several other Neotropical montane forests (Nadkarni, 1984; Ingram and Nadkarni, 1993; Wolf, 1993; Freiberg and Freiberg, 2000). The increase in relative abundance of bryophytes toward the outer crown reflects the fact that cryptogams are typically the first to colonize small (young) branches, gradually giving way to vascular epiphytes. The latter may require dead organic matter and the presence of bryophytes, as well as more time to establish (Ingram and Nadkarni, 1993; Freiberg and Freiberg, 2000). Canopy humus also needs much more time to accumulate and consequently its percentage is highest in the inner crowns where branches are oldest. Stem surfaces only showed trace amounts of canopy humus because of the vertical surface orientation which is unfavorable for humus accumulation.

The secondary forests at San Gerardo exhibited a high relative abundance of bryophytes in all crown sections, which is thought to reflect the low age of the stands. Similar results were obtained for a 30-year-old windward secondary lower montane cloud forest near Santa Elena in northern Costa Rica where epiphyte vegetation was also dominated by bryophytes (83–97%; Köhler *et al.*, 2007). Nadkarni *et al.* (2004) reported epiphyte biomass in another *c.* 50-year-old leeward secondary lower montane cloud forest at Monteverde to be heavily dominated by bryophytes (95%), with only small amounts of dead organic matter (3%) and other components (2%).

Epiphyte biomass at the stand level

The estimated epiphyte biomass of the San Gerardo OGF (3400 kg ha^{-1}) falls in the lower part of the range reported for other tropical montane forests (370–44 000 kg ha^{-1}; Köhler *et al.*, 2007). This is almost certainly due to the unfavorable distribution of rainfall at San Gerardo. Site factors that typically promote epiphyte abundance in TMCF include year-round high atmospheric humidity and fog incidence, moderate temperatures and high precipitation (Jarvis and Mulligan, this volume; cf. Häger and Dohrenbusch, this volume). Poikilohydric non-vascular epiphytes are particularly sensitive to prolonged dry periods and depend on sufficiently frequent wet periods for their survival (Benzing, 1998; Hietz, this volume). Long-term climatic data for the San Gerardo area indicate that 3-month periods with less than 20 mm of rainfall do occur (Instituto Meteorológico Nacional, 1988). Between 1 January and 31 March 2000, 68% of the days were without rain (or fog). The duration of the longest dry spell in 2000 was 32 days. Measurable horizontal precipitation occurred only on a few occasions (Köhler, 2002) and the associated additional water input can be safely assumed to be low. These seasonally dry conditions are also reflected in the composition of the bryophyte community at San Gerardo, such as the high percentage of mats of mosses (e.g. *Leptodontium*) and the nearly complete absence of bryophyte feathers, tails, or pendants in the canopy (Holz *et al.*, 2002).

Few comparative data exist for secondary TMCF. Nadkarni *et al.* (2004) obtained a value of 200 kg ha^{-1} for a *c.* 50-year-old forest at Monteverde. Although this value is dwarfed by the

33 100 kg ha^{-1} obtained by the same authors for a nearby OGF, it is intermediate between the epiphyte biomass values derived for the 10–15- and 40-year-old secondary stands at San Gerardo. The 30-year-old secondary forest near Santa Elena referred to earlier had an estimated epiphyte biomass of 1035 kg ha^{-1} compared to 16 215 kg ha^{-1} in nearby OGF (Köhler *et al.*, 2007). These relatively high values probably reflect the very high rainfall and fog incidence at the site (Frumau *et al.*, this volume; Tobón *et al.*, this volume #26). Although the data-set is still small, epiphyte biomass in all three Costa Rican studies was much lower in secondary stands than in nearby OGF. Furthermore, whilst epiphyte biomass generally increases during succession (Hale, 1967, in Coxson and Nadkarni, 1995) it clearly takes many decades, if not centuries to attain the biomass levels observed in the respective Costa Rican OGF.

Water storage in epiphytes

The maximum water content of bryophytes as measured *in situ* in the canopy of the San Gerardo OGF (406% of dry weight) falls within the range of values reported by other studies (200–500%; Pócs, 1980; Nadkarni, 1984; Frahm, 1990; Köhler *et al.*, 2007). Thus, bryophytes can significantly increase the overall water storage capacity of TMCF canopies. Whilst canopy storage values for lowland rain forests with low epiphyte mass typically are in the order of 1.0 ± 0.25 mm (Jetten, 1996), values derived for epiphyte-rich TMCF are generally much higher (2–5 mm; Pócs, 1980; Veneklaas and van Ek, 1990; Köhler *et al.*, 2007; Oesker *et al.*, this volume; Tobón *et al.*, this volume #26). The presently found potential water storage capacity associated with non-vascular epiphytes alone in the OGF at San Gerardo was estimated at 0.9 mm. Even after adding the contributions by tank bromeliads and canopy humus, this value did not increase to more than 1.2 mm due to the rather low overall epiphyte biomass. Corresponding values for the secondary forests at San Gerardo were several times lower (Tables 27.1 and 27.2).

Actual epiphyte water storage capacities in the San Gerardo forests may be even smaller, depending on the balance between previous precipitation inputs and losses via evaporation and drip (Veneklaas and Ek, 1990). Application of an analytical interception model to the San Gerardo OGF suggested that the contribution by non-vascular epiphytes to overall rainfall interception was relatively low (6%) despite their substantial water storage capacity. The main reason for this finding is that during the rainy season the mosses are frequently wetted by rain and hence only a fraction of the potential storage capacity is actually available for absorbing additional moisture at the beginning of the next rainfall (cf. García-Santos, 2007; Tobón *et al.*, this volume #26). Although the epiphytes dried out severely during the dry season (Hölscher *et al.*, 2004), rainfall interception during this period was small as there was little rain to be captured anyway.

CONCLUSIONS

Epiphyte biomass in old-growth and secondary montane forests on the leeward slope of the Talamanca range in Costa Rica was largely dominated by bryophytes, whereas amounts were low compared to other montane tropical locations, most probably because of a seasonal rainfall regime. Epiphyte biomass in 10–40-year-old stands was very much smaller than that in nearby old-growth forest. The re-establishment of epiphyte communities requires much more time than the restoration of forest structure during succession after forest clear-cutting. As a result, canopy water storage dynamics may only approach prior conditions after many decades or even centuries.

ACKNOWLEDGEMENT

The present study was funded by a grant from the German Science Foundation (DFG; HO 2119/1–3) which is gratefully acknowledged.

REFERENCES

Ataroff, M., and F. Rada (2000). Deforestation impact on water dynamics in a Venezuelan Andean cloud forest. *Ambio* **29**: 440–444.

Benzing, D. H. (1998). Vulnerabilities of tropical forests to climate change: the significance of resident epiphytes. *Climatic Change* **39**: 519–540.

Coxson, D. S., and N. M. Nadkarni (1995). Ecological roles of epiphytes in nutrient cycles of forest ecosystems. In *Forest Canopies*, eds. M. D. Lowman and N. M. Nadkarni, pp. 495–543. San Diego, CA: Academic Press.

Edwards, P. J., and P. J. Grubb (1977). Studies of mineral cycling in a montane rain forest in New Guinea. I. The distribution of organic matter in the vegetation and soil. *Journal of Ecology* **65**: 943–969.

Frahm, J. P. (1987). Which factors control the growth of epiphytic bryophytes in tropical rainforest? *Symposia Biologica Hungarica* **35**: 639–648.

Frahm, J. P. (1990). The ecology of epiphytic bryophytes on Mt. Kinabalu, Sabah (Malaysia). *Nova Hedwigia* **51**: 121–132.

Frahm, J. P., and S. R. Gradstein (1991). An altitudinal zonation of tropical rain forests using bryophytes. *Journal of Biogeography* **18**: 669–676.

Freiberg, M., and E. Freiberg (2000). Epiphyte diversity and biomass in the canopy of lowland and montane forests in Ecuador. *Journal of Tropical Ecology* **16**: 673–688.

García-Santos, G. (2007). *An ecohydrological and soils study in a montane cloud forest in the National Park of Garajonay, La Gomera (Canary Islands, Spain).* PhD Thesis, VU University Ámsterdam, Ámsterdam, The Netherlands. [http://www.falw.vu.nl/nl/onderzoek/earth-sciences/geo-environmental-science-and-hydrology/hydrology-dissertations/index.asp].

Giambelluca, T. W. (2002). Hydrology of altered tropical forest. *Hydrological Processes* **16**: 1665–1669.

Helmer, E. H. (2000). The landscape ecology of tropical secondary forest in montane Costa Rica. *Ecosystems* **3**: 98–114.

Hofstede, R. G. M., J. H. D. Wolf, and D. H. Benzing (1993). Epiphytic biomass and nutrient status of a Colombian upper montane rain forest. *Selbyana* **14**: 37–45.

Hölscher, D., L. Köhler, A. I. J. M. Van Dijk, and L. A. Bruijnzeel (2004). The importance of epiphytes to total rainfall interception by a tropical montane rain forest in Costa Rica. *Journal of Hydrology* **292**: 308–322.

Holz, I., S. R. Gradstein, J. Heinrichs, and M. Kappelle (2002). Bryophyte diversity, microhabitat differentiation and distribution of life forms in Costa Rican upper montane *Quercus* forest. *The Bryologist* **105**: 334–348.

Ingram, S.W., and N.M. Nadkarni (1993). Composition and distribution of epiphytic organic matter in a neotropical cloud forest, Costa Rica. *Biotropica* **25**: 370–383.

Instituto Meteórologico Nacional (1988). *Catastro de las series de precipitaciones medidas en Costa Rica.* San José, Costa Rica: Ministerio de Recursos Naturales, Energía y Minas.

Jacobsen, N.H.G. (1978). An investigation into the ecology and productivity of epiphytic mosses. *South African Journal of Botany* **44**: 297–312.

Jetten, V.G. (1996). Interception of tropical rain forest: performance of a canopy water balance model. *Hydrological Processes* **10**: 671–685.

Johansson, D. (1974). Ecology of vascular epiphytes in West African rain forest. *Acta Phytogeographica Suedica* **59**: 1–136.

Kappelle, M., P.A. Kennis, and R.A.J. Vries (1995). Changes in diversity along a successional gradient in a Costa Rican upper montane *Quercus* forest. *Biodiversity and Conservation* **4**: 10–34.

Kappelle, M., T. Geuze, M.E. Leal, and A.M. Cleef (1996). Successional age and forest structure in a Costa Rican upper montane *Quercus* forest. *Journal of Tropical Ecology* **12**: 681–689.

Köhler, L. (2002). *Die Bedeutung der Epiphyten im ökosystemaren Wasser- und Nährstoffumsatz verschiedener Altersstadien eines Bergregenwaldes in Costa Rica,* Berichte des Forschungszentrums Waldökosysteme, Series A, Vol. 181. Göttingen, Germany: University of Göttingen.

Köhler, L., D. Hölscher, and Ch. Leuschner (2006). Above-ground water and nutrient fluxes in three successional stages of Costa Rican montane oak forest with contrasting of epiphyte abundance. In *Ecology and Conservation of Neotropical Montane Oak Forests*, ed. M. Kappelle, pp. 271–282. Berlin: Springer-Verlag.

Köhler, L., C. Tobón, K.F.A. Frumau, and L.A. Bruijnzeel (2007). Biomass and water storage of epiphytes in old-growth and secondary montane rain forest stands in Costa Rica. *Plant Ecology* **193**: 171–184.

Kunz, M. (2000). Diversität epiphytischer Kryptogamen in der Strauchschicht montaner Eichenwälder Costa Ricas. M.Sc. thesis, University of Göttingen, Göttingen, Germany.

Nadkarni, N.M. (1984). Epiphyte biomass and nutrient capital of a neotropical elfin forest. *Biotropica* **16**: 249–256.

Nadkarni, N.M., D. Schaefer, T.J. Matelson, and R. Solano (2004). Biomass and nutrient pools of canopy and terrestrial components in a primary and a secondary montane cloud forest, Costa Rica. *Forest Ecology and Management* **198**: 223–236.

Perry, D.R. (1978). A method of access into the crowns of emergent and canopy trees. *Biotropica* **10**: 155–157.

Pócs, T. (1980). The epiphytic biomass and its effect on the water balance of two rain forest types in the Uluguru Mountains (Tanzania, East Africa). *Acta Botanica Academiae Scientiarum Hungaricae* **26**: 143–167.

Tanner, E.V.J. (1980). Studies on the biomass and productivity in a series of montane rain forest in Jamaica. *Journal of Ecology* **68**: 573–588.

Tanner, E.V.J. (1985). Jamaican montane forests: nutrient capital and cost of growth. *Journal of Ecology* **73**: 553–568.

Van Reenen, G.B.A., and S.R. Gradstein (1983). A transect analysis of the bryophyte vegetation along an altitudinal gradient on the Sierra Nevada de Santa Marta, Colombia. *Acta Botanica Neerlandica* **32**: 163–175.

Van Reenen, G.B.A., and S.R. Gradstein (1984). An investigation of bryophyte distribution and ecology along an altitudinal gradient in the Andes of Colombia. *Journal of Hattori Botanical Laboratory* **56**: 79–84.

Veneklaas, E., and R. van Ek (1990). Rainfall interception in two tropical montane rain forests, Colombia. *Hydrological Processes* **4**: 311–326.

Veneklaas, E.J., R.J. Zagt, A. van Leerdam, *et al.* (1990). Hydrological properties of the epiphyte mass of a montane tropical rain forest, Colombia. *Vegetatio* **89**: 183–192.

Weaver, P.L. (1972). Cloud moisture interception in the Luquillo Mountains of Puerto Rico. *Caribbean Journal of Science* **12**: 129–144.

West, G.B., J.H. Brown, and B.J. Enquist (1999). A general model for the structure and allometry of plant vascular systems. *Nature* **400**: 664–667.

Wolf, J.H.D. (1993). Diversity patterns and biomass of epiphytic bryophytes and lichens along an altitudinal gradient in the northern Andes. *Annals of the Missouri Botanical Garden* **80**: 928–960.

28 Comparison of passive fog gages for determining fog duration and fog interception by a Puerto Rican elfin cloud forest

F. Holwerda and L. A. Bruijnzeel
VU University, Amsterdam, the Netherlands

F. N. Scatena
University of Pennsylvania, Philadelphia, Pennsylvania, USA

ABSTRACT

Between 5 March and 10 May 2001, the performance of three types of passive fog gages (wire harp *WH*, standard fog collector *SC*, and Juvik gage *JU*) was compared at a wind-exposed Puerto Rican elfin cloud forest site. The gages were used to determine the timing and duration of fog and of fog-induced crown drip. Gage to canopy conversion factors were derived from the ratio between throughfall, and fog collected by the respective gages during periods with fog-only and negligible evaporation loss. The fog gages indicated very similar timing and fog duration, presumably because dense fog prevailed for 75–80% of the time and winds were generally sufficiently strong (4–5 m s^{-1}). Resulting gage-to-canopy factors were 0.12 for *WH* and *SC*, and 0.10 for *JU*. Estimating fog interception by the canopy using these conversion factors gave very similar values for the three gage types (mean rates of 0.15 ± 0.01 mm $hour^{-1}$). Throughfall typically started about 5 hours after the fog gages indicated the beginning of a fog event. This time lag probably reflects the filling of the storage capacity of the canopy, because wet canopy evaporation was negligible. Integrating the estimated fog interception rates over the time lag suggested an average value of 0.4 mm for the canopy storage capacity during events with fog-only.

INTRODUCTION

Rates and amounts of fog interception by vegetation depend on wind speed, fog liquid water content (*LWC*) and duration, as well as surface area and geometry of the vegetation (Schemenauer, 1986). Information on the timing and duration of fog can be obtained with passive fog gages, provided these are protected from rainfall and equipped with a recording device (Bruijnzeel *et al.*, 2005). Fog *LWC* may also be evaluated from collections by passive gages when information on their collection efficiency and prevailing wind speeds is available (e.g. Schemenauer and Joe, 1989). A variety of passive gages is available, and there has been some discussion as to what is the most suitable type of gage to characterize local fog conditions (Juvik and Nullet, 1995a; Schemenauer and Cereceda, 1995; cf. Delay and Giambelluca, this volume; Frumau *et al.*, this volume). For example, a cylindrical gage is considered superior to a flat screen, because it has uniform exposure to all wind directions (Juvik and Nullet, 1995a; cf. García Santos, 2007; Giambelluca *et al.*, this volume). On the other hand, a flat screen generally has a much larger collection area than a cylindrical gage, and may thus measure fog when *LWC* or wind speeds are low (Schemenauer and Cereceda, 1995).

Amounts of fog water collected by a fog gage are not necessarily representative of those actually intercepted by a forest canopy. However, good relationships have been found between collection rates of passive gages and fog-induced throughfall, *TF* (Joslin *et al.*, 1990; Vong *et al.*, 1991; Juvik and Nullet, 1995b;

Tropical Montane Cloud Forests: Science for Conservation and Management, eds. L. A. Bruijnzeel, F. N. Scatena, and L. S. Hamilton. Published by Cambridge University Press.

Holwerda *et al.*, this volume #29). Dividing amounts of fog-induced *TF* by the corresponding volumes collected by a fog gage yields a so-called gage-to-canopy conversion factor, which theoretically integrates the difference in fog-catching efficiency between gage and canopy (Schellekens *et al.*, 1998). This factor may then be used to estimate fog interception from the fog gage record over an extended period.

This paper reports on a comparison of three types of passive fog gages (wire harp *WH*, "standard" fog collector *SC*, and cylindrical aluminum louvered screen *JU*) as used between 5 March and 10 May 2001 at a wind-exposed elfin cloud forest site near Pico del Este, Puerto Rico. The fog gages were equipped with a tipping bucket and data-logging system to obtain continuous records. Throughfall and above-canopy climatic parameters (including wind speed and wind direction) were measured as well. The following questions were addressed: (i) what is the most suitable type of gage to characterize fog conditions (timing, duration, intensity) at wind-exposed sites such as Pico del Este?; (ii) how do estimates of fog interception based on gage to canopy factors differ between the three types of gages?; and (iii) how does the timing of canopy drip relate to fog occurrence according to the respective gages?

STUDY AREA

The study site was located at *c.* 1010 m.a.s.l. near Pico del Este (1052 m.a.s.l.), the easternmost summit in the Luquillo Mountains, Puerto Rico. Fog measurements were made at a local ridge top whereas additional climatic and *TF* measurements were made about 30 m below the ridge top on a 17° slope facing N40E. The prevailing trade winds blow mainly from the east, and wind speeds are typically 4–5 m s^{-1} with little seasonal variation (Briscoe, 1966; Brown *et al.*, 1983). Mean annual rainfall at Pico del Este is 4435 mm $year^{-1}$. On average, most rainfall occurs during the months of May, October, and November (450–500 mm each), whereas January, February, and March typically receive 250–300 mm each (García-Martinó *et al.*, 1996). Mean monthly relative humidity below the forest canopy at Pico del Este is close to saturation (99.7%; Weaver, 1972). Fog occurs for about 60–75% of the time during the day and for about 95–100% of the time at night (Baynton, 1968; Holwerda *et al.*, 2006). Mean monthly temperatures range between 17 °C and 20 °C (Brown *et al.*, 1983). The vegetation at Pico del Este is typical of the elfin cloud forest type found on peaks and ridges above 900 m.a.s.l. in the Luquillo Mountains, and is dominated by *Tabebuia rigida*, *Ocotea spathulata*, and *Micropholis garcinaefolia* having a mean height of ~3 m. The forest is rich in epiphytes, with mosses and bromeliads covering branches, stems, and part of the soil surface (Weaver, 1995). Leaf area index, as estimated from light extinction measurements (F. Holwerda, unpublished data), was 2.1 ± 0.7, close to the value of 2.0 found by Weaver *et al.* (1986) for the same forest using destructive sampling methods.

METHODS

General climatic parameters

Above-canopy climatic measurements were made at a measuring interval of 30 s in a 7.5-m tall mast placed in a small, natural clearing. The measurements were recorded by a Campbell Scientific 21X data-logger that stored the averaged data every 5 min in an external storage module. Incoming and outgoing short- and long-wave radiation components were measured separately above the canopy with Kipp and Zonen albedometers and Eppley Pyrgeometers, respectively. Net radiation was determined by adding the respective components. Air temperature and relative humidity were measured with a combined sensor (Vaisala, HMP45C) housed in a ventilated radiation screen. Wind speed and direction were measured using an A100R switching anemometer (at 5 m) and a W200P potentiometer wind vane (at 7.4 m), respectively (Vector Instruments, UK). Rainfall was measured with a tipping bucket rain gage (Campbell Scientific Ltd., ARG100, 500 cm^2 orifice, 0.20 mm per tip) placed in a small clearing about 10 m from the mast. Evaporation from the wet canopy (E_w) was calculated from the above-canopy climatic data using the wet variant of the Penman–Monteith equation (Monteith, 1965).

Fog

Fog occurrence was measured using three types of passive gages: a wire harp *WH* (Goodman, 1985), a modification of the "standard" fog collector *SC* (Schemenauer and Cereceda, 1994), and a cylindrical aluminum louvered screen *JU* (Juvik and Ekern, 1978). Two gages of each type were installed next to each other on the ridge top with the center of their collection area at 2.5 m above the ground. The gages were freely exposed to the prevailing winds and not protected against rainfall. The *WH* consisted of a 0.5 × 0.5 m flat screen with 0.8 mm vertical nylon strings spaced 2 mm apart. To achieve better comparability with the *WH*, the dimensions of the *SC* were adjusted to 0.5 × 0.5 m instead of the 1.0 × 1.0 m proposed originally by Schemenauer and Cereceda (1994) (see also García-Santos and Bruijnzeel, this volume). The polypropylene mesh of the *SC* consists of a triangular weave of a flat fiber about 1 mm wide and 0.1 mm thick (Schemenauer and Cereceda, 1994). Both the *WH* and *SC* were positioned so as to face the easterly trade winds. The *JU* gage consists of a cylindrical aluminum louvered screen (13 cm in diameter by 40 cm in height) (see Juvik and Ekern, 1978 or Juvik and Nullet, 1995a for details). Tubing from each of the gages led into tipping bucket systems (5 ml per tip) connected to custom-built data-loggers (VU

University Amsterdam). Volumes of fog water collected by the gages (l) were converted to water depth equivalent (mm) by dividing by the collecting area of the respective gages (i.e. 0.25 m^2 for *WH* and *SC*, and 0.052 m^2 for *JU*).

Throughfall

Throughfall (*TF*) was measured with two large steel gutters (projected surface areas of 1.11 and 1.14 m^2) placed near the meteorological mast. Each gutter had a tipping bucket system (50 ml per tip) connected to a custom-built data-logger (VU University Amsterdam).

Derivation of fog gage to canopy conversion factors

For a fog-only event, the wet canopy water budget of a forest reads:

$$CWI = TF + SF + E_i \tag{28.1}$$

in which *CWI* is gross cloud water (fog) interception, *TF* throughfall, *SF* stemflow, and E_i evaporation from the wet canopy (interception loss); all components are expressed in mm of water per time period. Because *SF* during events with fog-only is probably negligible, it was not considered separately.

For saturated canopy conditions and negligible wet canopy evaporation rate (E_w), net fog interception (CWI_{net}) more or less equals *TF*. Dividing gross or net values of *CWI* by the amount of fog collected by a fog gage of type x (CWI_x) yields the corresponding gage to canopy conversion factor (f_x) (Schellekens *et al.*, 1998).

The simplified conditions under which *TF* equals (net) *CWI* were met by using all 30-min measuring intervals having $TF > 0.1$ mm and $E_w < 0.01$ mm. The first 3 hours after rainfall had ceased were excluded from the analysis to avoid any contributions by rain-induced *TF*.

RESULTS AND DISCUSSION

Comparison of fog collection rates by the different gages

After defining fog events as being separated by a fog-free period of at least 3 hours to ensure a fully dry canopy at the start of an event (Schellekens *et al.*, 1998) 40 events were extracted from the record for the period between 5 March and 10 May 2001. Mean event duration was about 34 hours. However, median duration was only about 19 hours because of the strongly positively skewed distribution of event durations. Comparisons between fog collection rates by the three types of gages were made using data from eight fog-only events (Figure 28.1). Collection rates by the *SC* and the *WH* were equal (Figure 28.1a) whereas those by the *JU* were 24% higher (Figure 28.1b). The *JU* showed a

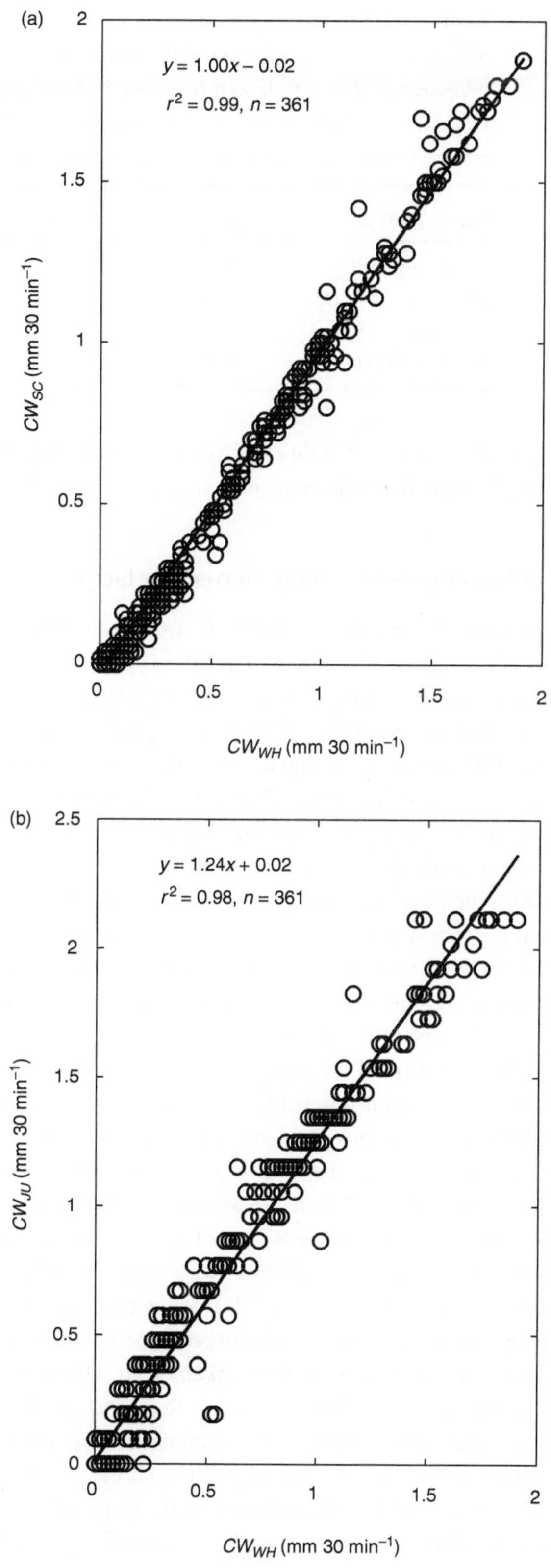

Figure 28.1. Relationships between fog collection rates (CWI_x, mm 30 min^{-1}) by (a) the wire harp *WH* and standard gauge *SC* and (b) *WH* and the Juvik gauge *JU* for eight fog-only events between 5 March and 10 May 2001 at Pico del Este.

Table 28.1 *Average gage-to-canopy factors* f_x *as derived for a wire harp* WH, *a modified standard fog collector* SC, *and a Juvik-type fog gage* JU *based on 120 30-min intervals with throughfall* TF > *0.1 mm and wet canopy evaporation* E_w < *0.01 mm between 5 March and 10 May 2001 at Pico del Este, Puerto Rico. Also shown are coefficients of variation CV of* f_x *and relationships between above-canopy wind speed* U *and* f_x

Fog gage type	f_x	CV (%)	n	Wind-dependency $f_x = aU + b$	r^2	p value
Wire harp (*WH*)	0.12	20	120	$f_x = 0.008U + 0.073$	0.10	<0.01
Modified standard collector (*SC*)	0.12	17	120	$f_x = 0.005U + 0.094$	0.05	0.01
Juvik-type (*JU*)	0.10	20	120	$f_x = 0.001U + 0.093$	0.00	0.68

slightly delayed response at times of low fog incidence, probably because of its smaller collection area.

Derivation of gage to canopy conversion factors

Gage-to-canopy conversion factors f_x were calculated using 120 30-min intervals with throughfall $TF > 0.1$ mm, wet canopy evaporation $E_w < 0.01$ mm, and excluding the first 3 hours after rainfall P had ceased (Table 28.1). Because *CWI* rates obtained with the *WH* and *SC* were equal, the corresponding conversion factors were also equal (0.12). Because of the higher collection rates of the *JU* (Figure 28.1), f_{JU} was slightly lower (Table 28.1; cf. Frumau *et al.*, this volume). The variability of f_x was very similar for the three types of gages (17–20%, Table 28.1). Whilst f_{WH} and f_{SC} were weakly but significantly correlated (at the 0.05 level) with wind speed (U), no such relationship was found for f_{JU} (Table 28.1). The strings of the harp started to vibrate at high values of U, causing already intercepted drops to fall off again and be blown away (cf. Frumau *et al.*, this volume). However, there was no relationship between f_{WH} and U when 30-min intervals with $U > 8$ m s^{-1} were excluded from the analysis. Because wind speeds greater than 8 m s^{-1} do not often occur at the study site (only for 2% of the time during the 9-week study period), effects of wind-dependency of f_{WH} on calculated *CWI* totals were unimportant. Neither was there a relationship between f_{SC} and U for $U < 8$ m s^{-1}. Conversely, f_{JU} showed no wind-dependency at high U, probably because its rigid aluminum screen is not easily shaken. It is unknown to what extent the calculated f_x values are also valid at very low wind speeds ($U < 2$ m s^{-1}). Under such conditions, gravitational settling rather than turbulent diffusion governs fog deposition on the forest canopy (Beswick *et al.*, 1991), and collection efficiencies of fog gages often drop under such conditions (Schemenauer and Joe, 1989). At the Pico del Este site, however, winds are typically 4–5 m s^{-1}, and they do not often fall below 2 m s^{-1} (14% of the time during the study period).

In addition, the collection efficiencies of the flat *WH* and *SC* depend on wind direction because of their fixed orientation, while the cylindrical *JU* offers uniform exposure to winds from all directions (Juvik and Nullet, 1995a; cf. Schemenauer and Cereceda, 1995). However, winds at Pico del Este came from easterly directions for 77% of the time during the study period, rendering effects of gage orientation on f_x relatively unimportant (cf. Giambelluca *et al.*, this volume). Conversely, wind direction effects on the catch of a similarly sized *SC* at an exposed ridge top in La Gomera (Canary Islands) appeared substantial (García-Santos, 2007).

Fog duration and interception

Eight events with fog-only occurring between 5 March and 10 May 2001 were analyzed in detail (Table 28.2). The *WH* and *SC* indicated almost equal event durations (means of 16 hours), while the *JU* gave somewhat lower values (mean of 15 hours, Table 28.2). The difference was most pronounced for event no. 31, when event duration was 6 hours according to the JU, vs. 13 and 12 hours according to the *WH* and *SC*, respectively (Table 28.2). However, the fog was very light at the start of this particular event (with collection rates by the *WH* < 0.1 mm 30 min^{-1}). Under such conditions, the collection area of the *JU* was probably too small to yield measurable amounts of fog (Figure 28.1). Between 25 June and 6 August 2002, Holwerda *et al.* (2006) measured visibility (*VIS*, m) at the same site using a Present Weather Detector. Dense fog (*VIS* < 250 m) occurred for about 77% of the time, while fog with *VIS* between 250 and 1000 m occurred for only about 8% of the time. Therefore, and because of the sufficiently strong winds prevailing at the site (4–5 m s^{-1}), the variability in event duration as indicated by the respective gage types was generally less than 5% (Table 28.2). The variability in estimates of event *CWI* by the forest as obtained by the use of the respective gage to canopy factors f_x, was generally small as well (≤10%, Table 28.2). Mean hourly net *CWI* rates were calculated for each gage type using all hours without rainfall and excluding the first 3 hours after rainfall had ceased. This gave values of 0.15 ± 0.10 mm hour^{-1} (SD) for the *WH*, 0.14 ± 0.11 mm hour^{-1} for the *SC*, and 0.15 ± 0.11 mm hour^{-1} for the *JU*. The presently found average rate of

Table 28.2 *Fog interception* (CWI_x, *mm) and event duration (hours) as indicated by a wire harp* WH, *a modified standard fog collector* SC, *and a Juvik-type gage* JU *for eight fog-only events between 5 March and 10 May 2001 at Pico del Este, Puerto Rico.* CWI *by the forest canopy based on the respective average gage-to-canopy factors* f_x. *Amounts of fog drip* (TF) *added for comparison*

	Duration (hours)					CWI_x (mm)			CWI (mm)					TF
Event	*WH*	*SC*	*JU*	Mean	CV (%)	*WH*	*SC*	*JU*	*WH*	*SC*	*JU*	Mean	CV (%)	(mm)
5	16	17	16	16	2	26.7	26.5	35.3	3.3	3.3	3.4	3.3	3	2.1
6	18	18	17	17	2	15.0	13.4	20.8	1.8	1.7	2.0	1.8	10	1.1
7	16	16	15	16	3	8.1	6.0	10.9	1.0	0.7	1.1	0.9	18	0.1
8	36	35	35	35	2	66.4	64.7	82.4	8.1	8.0	8.0	8.1	1	6.0
10	15	15	14	15	4	31.3	30.8	38.8	3.8	3.8	3.8	3.8	1	3.2
16	5	3	3	4	25	1.1	0.9	1.4	0.1	0.1	0.1	0.1	11	0.0
29	14	14	11	13	11	5.6	5.0	7.4	0.7	0.6	0.7	0.7	8	0.0
31	13	12	6	10	39	1.3	0.8	1.1	0.2	0.1	0.1	0.1	29	0.0
Mean	16	16	15	16	11	19.4	18.5	24.8	2.4	2.3	2.4	2.4	10	1.6

0.15 ± 0.11 mm hour^{-1} is slightly higher than the 0.11 ± 0.05 mm hour^{-1} derived for conditions of fog-only by Holwerda *et al.* (2006) for the same site during a 44-day observation period in the summer of 2002 using the wet canopy water-budget approach. Holwerda *et al.* (2006) acknowledged that the latter value might be an underestimate, because it was based on observations made during daytime periods during which wind speeds and cloud *LWC* were lower than at night. On the other hand, Holwerda *et al.* (2006) minimized effects of spatial variability in *TF* by the use of 20 roving gages, while in the present study *TF* may have been overestimated because of the use of only two fixed gutters.

Timing of fog-induced throughfall relative to fog interception

Figure 28.2 shows estimated net *CWI* rates together with *TF* and E_w rates during four of the eight events listed in Table 28.2. Half-hourly *CWI* rates as derived from the three types of gages were very similar. All fog events listed generally started by late afternoon or early evening (Figure 28.2). As soon as an event started, E_w became very small, because relative humidity approached 100% and net radiation was close to zero. Figure 28.2 also shows the considerable time lag between the start of a fog event and the start of crown drip. This most probably reflects the filling of the canopy storage *S* (including mosses and epiphytes) since E_w was negligible (cf. Joslin *et al.*, 1990). Under conditions of negligible E_w, the magnitude of the time lag depends on the rate of *CWI* and the effective value of *S*. Whilst the applied event separation period of 3 hours was sufficient to dry up the leaves, a longer period must be required for the drying out of mosses on branches and stems. Hence, the instantaneous (effective) value of *S* will depend on previous wetting and drying cycles (Hölscher *et al.*, 2004; cf. Köhler *et al.*, *this volume*; Mulligan *et al.*, this volume; Tobón *et al.*, this volume #26). The average value of *S* was estimated by integrating calculated *CWI* rates over the duration of the time lag for each *TF*-generating fog event between 5 March and 10 May 2001 ($n = 29$). The mean time lag was about 5 hours (range: 2–12 hours), yielding a mean storage value of about 0.4 mm (range: 0.0–0.7 mm). However, it often took another few hours before *TF* rates became similar to estimated *CWI* rates (Figure 28.2), suggesting that the actual storage capacity was higher than 0.4 mm. A similar process was observed by Chang *et al.* (2002), who measured the increase in weight of (air-dried) bryophytes under foggy conditions in a cypress forest in Taiwan. Fog was seen to begin to drip from the mosses while they were still increasing in weight, i.e. before full saturation was reached (see also discussion in Tobón *et al.*, this volume #26).

CONCLUSIONS

Between 5 March and 10 May 2001, the performance of three passive fog gages (wire harp *WH*, a modified standard fog collector *SC*, and a Juvik-type gage *JU*) was compared at a wind-exposed Puerto Rican elfin cloud forest site.

At low rates of fog incidence, the *JU* indicated shorter fog duration than did either the *WH* or *SC*, probably because of its much smaller collection area (0.052 m^2 vs. 0.25 m^2 for *WH* and *SC*). In general, however, the difference in inferred fog duration according to the three gage types was less than 5%, because dense fog prevailed for 75–80% of the time and winds were sufficiently strong. Gage-to-canopy conversion factors, calculated as the ratio between throughfall (fog drip) and fog collection by the gage during periods of fog-only and negligible evaporation loss, were 0.12 for the *WH* and *SC*, and 0.10 for the *JU*. As a result, estimated mean fog interception rates by the forest using the respective factors were effectively equal at 0.15 ± 0.10 mm hour^{-1} for

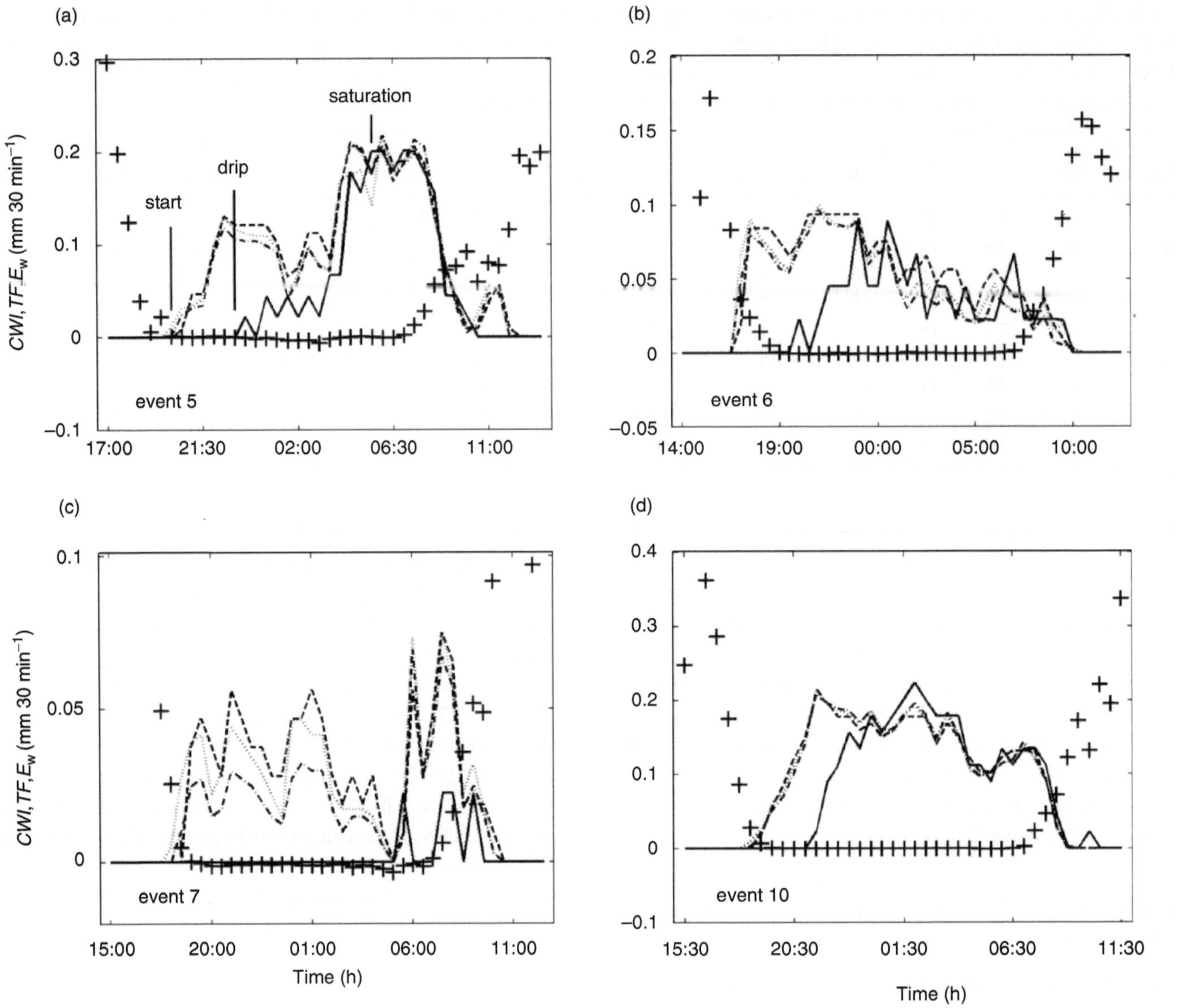

Figure 28.2. Fog deposition rates (*CWI*), as calculated using gauge to canopy conversion factors f_x (Table 28.1), according to the wire harp *WH* (dotted line), standard gage *SC* (dashed dotted line), and Juvik gage *JU* (dashed line) for four selected fog-only events (Table 28.2) between 5 March and 10 May 2001 at Pico del Este. Throughfall *TF* (solid line) and wet canopy evaporation rates E_w (plusses) have been added.

the *WH*, 0.14 ± 0.11 mm hour^{-1} for the *SC*, and 0.15 mm ± 0.11 hour^{-1} for the *JU*.

Throughfall typically started about 5 hours after the fog gages had indicated the start of a fog event. Integrating the estimated fog interception rates over the time lag gave an average value of 0.4 mm for the effective canopy storage capacity.

ACKNOWLEDGEMENTS

This work was supported by a grant from the Netherlands Foundation for the Advancement of Tropical Research (WOTRO, grant no. W76–206). The authors thank C. Estrada, G. Guzman, S. Moya, and C. Torrens (U.S. Department of Agriculture Forest Service) for help during the fieldwork and Dr. Robert Schemenauer for helpful comments on an earlier draft.

REFERENCES

Baynton, H. W. (1968). Ecology of an elfin forest in Puerto Rico. II. Microclimate of Pico Del Oeste. *Journal of the Arnold Arboretum* **49**: 419–430.

Beswick, K. M., K. J. Hargreaves, M. W. Gallagher, T. W. Choularton, and D. Fowler (1991). Size-resolved measurements of cloud droplet deposition velocity to a forest canopy using an eddy-correlation technique. *Quarterly Journal of the Royal Meteorological Society* **117**: 623–645.

Briscoe, C. B. (1966). *Weather in the Luquillo Mountains of Puerto Rico.* Rio Piedras, Puerto Rico: Institute of Tropical Forestry.

Brown, S., A. E. Lugo, S. Silander, and L. Liegel (1983). *Research History and Opportunities in the Luquillo Experimental Forest*. New Orleans, LA: U.S. Department of Agriculture Forest Service, Southern Forest Experiment Station.

Bruijnzeel, L. A., W. Eugster, and R. Burkard (2005). Fog as an input to the Hydrological cycle. In *Encyclopaedia of Hydrological Sciences*, eds. M. G. Anderson and J. J. McDonnell, pp. 559–582. Chichester, UK: John Wiley.

Chang, S. C., I. L. Lai, and J. T. Wu (2002). Estimation of fog deposition on epiphytic bryophytes in a subtropical montane forest ecosystem in north-eastern Taiwan. *Atmospheric Research* **64**: 159–167.

García-Martinó, A. R., G. S. Warner, F. N. Scatena, and D. L. Civco (1996). Rainfall, runoff and elevation relationships in the Luquillo Mountains of Puerto Rico. *Caribbean Journal of Science* **32**: 413–424.

García-Santos, G. (2007). *An ecohydrological and soils study in a montane cloud forest in the National Park of Garajonay, La Gomera (Canary Islands, Spain)*. PhD Thesis, VU University Ámsterdam, Ámsterdam, The Netherlands. [http://www.falw.vu.nl/nl/onderzoek/earth-sciences/geo-environmental-science-and-hydrology/hydrology-dissertations/index.asp].

Goodman, J. (1985). The collection of fog drip. *Water Resources Research* **21**: 392–394.

Hölscher, D., L. Köhler, A. I. J. M. Van Dijk, and L. A. Bruijnzeel (2004). The importance of epiphytes to total rainfall interception by a tropical montane rain forest in Costa Rica. *Journal of Hydrology* **292**: 308–322.

Holwerda, F., R. Burkard, W. E. Eugster, *et al.* (2006). Estimating fog deposition at a Puerto Rican elfin cloud forest site: comparison of the water budget and eddy covariance methods. *Hydrological Processes* **20**: 2669–2692.

Joslin, J. D., S. F. Mueller, and M. H. Wolfe (1990). Test of models of cloud-water deposition to forest canopies using artificial and living collectors. *Atmospheric Environment* **24A**: 2893–2903.

Juvik, J. O. and P. C. Ekern. (1978). *A Climatology of Mountain Fog on Mauna Loa, Hawai'i Island, Technical Report No. 118*. Honolulu, HI: Water Resources Research Center, University of Hawai'i.

Juvik, J. O., and D. Nullet (1995a). Comments on "a proposed standard fog collector for use in high elevation regions." *Journal of Applied Meteorology* **34**: 2108–2110.

Juvik, J. O., and D. Nullet (1995b). Relationships between rainfall, cloud-water interception and canopy throughfall in a Hawaiian montane forest. In *Tropical Montane Cloud Forests*, eds. L. S. Hamilton, J. O. Juvik, and F. N. Scatena, pp. 165–182. New York: Springer-Verlag.

Monteith, J. L. (1965). Evaporation and the environment. *Symposia of the Society for Experimental Biology* **19**: 245–269.

Schellekens, J., L. A. Bruijnzeel, A. J. Wickel, F. N. Scatena, and W. L. Silver (1998). Interception of horizontal precipitation by elfin cloud forest in the Luquillo Mountains, eastern Puerto Rico. In *Proceedings of the 1st International Conference on Fog and Fog Collection*, eds. R. S. Schemenauer and H. A. Bridgman, pp. 29–32. Ottawa, Canada: IDRC.

Schemenauer, R. S. (1986). Acidic deposition to forests: the 1985 Chemistry of High Elevation Fog (CHEF) Project. *Atmosphere–Ocean* **24**: 303–328.

Schemenauer, R. S., and P. Cereceda (1994). A proposed standard fog collector for use in high-elevation regions. *Journal of Applied Meteorology* **33**: 1313–1322.

Schemenauer, R. S., and P. Cereceda (1995). Reply to comments by Juvik and Nullet (1995). *Journal of Applied Meteorology* **34**: 2111–2112.

Schemenauer, R. S., and P. I. Joe (1989). The collection efficiency of a large fog collector. *Atmospheric Research* **24**: 53–69.

Vong, R. J., J. T. Sigmon, and S. F. Mueller (1991). Cloud water deposition to Appalachian forests. *Environmental Science and Technology* **25**: 1014–1021.

Weaver, P. L. (1972). Cloud moisture interception in the Luquillo Mountains in Puerto Rico. *Caribbean Journal of Science* **12**: 129–144.

Weaver, P. L. (1995). The colorado and dwarf forests of Puerto Rico's Luquillo Mountains. In *Tropical Forests: Management and Ecology*, eds. A. E. Lugo and C. Lowe, pp. 109–141. New York: Springer-Verlag.

29 Fog interception in a Puerto Rican elfin cloud forest: a wet-canopy water budget approach

F. Holwerda, L. A. Bruijnzeel, A. L. Oord
VU University, Amsterdam, the Netherlands

F. N. Scatena
University of Pennsylvania, Philadelphia, Pennsylvania, USA

ABSTRACT

Between 10 July and 25 October 2001, fog interception (*CWI*) by a Puerto Rican elfin cloud forest at 970 m.a.s.l. was studied. Values of *CWI* were estimated from the wet-canopy water budget as the sum of throughfall, stemflow, and interception loss, minus rainfall corrected for the effect of slope. *CWI* was also estimated from throughfall measurements during periods of fog-only. Timing and duration of fog were measured using a wire harp that was protected against rainfall. Estimated rates of *CWI* correlated significantly ($p < 0.01$) with collections by the fog gage. Gage to canopy factors were determined from the slopes of these relationships and were considered to integrate differences in fog-catching efficiency of the wire harp and the forest canopy. Applying these factors to the fog gage record yielded an estimated 1.1–1.4 mm day^{-1} of fog *CWI*, equivalent to 10–12% of mean daily corrected rainfall (11.3 mm day^{-1}). The wet-canopy water budget method gave the best results when rainfall amounts were small compared to fog, but error bands of estimated *CWI* became too wide to give meaningful results when rainfall was equal to or larger than the amount of fog. Furthermore, solving the water budget for large rainstorms gave negative values of *CWI*, possibly because of underestimated stemflow.

INTRODUCTION

Although the hydrological importance of tropical montane cloud forest (TMCF) has become increasingly recognized during the last two decades (Zadroga, 1981; Bruijnzeel, 2005), reliable estimates of cloud water (fog) interception (*CWI*) by TMCF are still relatively scarce (reviewed by Bruijnzeel, 2005). A common approach is to compare amounts of throughfall (*TF*) plus stemflow (*SF*) with amounts of incident rainfall (*P*) during periods with and without fog (Harr, 1982; Schellekens *et al.*, 1998; Holder, 2003; cf. McJannet *et al.*, 2007). Because wet-canopy evaporation losses are typically neglected, this technique provides a minimum estimate of *CWI* (see Bruijnzeel, 2005 for discussion). Furthermore, because TMCF often occurs on wind-exposed ridges or mountain peaks where rain tends to fall at an angle for much of the time, the effective depth of rainfall incident to sloping ground differs from that indicated by measurements made with a conventional rain gage (Sharon, 1980; cf. García-Santos and Bruijnzeel, this volume; Giambelluca *et al.*, this volume). Also, rainfall measurements by gages raised above ground level are subject to systematic error because of the distorted wind field above the gage orifice (Sevruk, 1982). As such, rainfall measurements at TMCF sites may be subject to large uncertainty, which further complicates quantification of *CWI* via comparative measurements (cf. Blocken *et al.*, 2005; Frumau *et al.*, 2006; Holwerda *et al.*, 2006a; Mulligan *et al.*, 2006). The proper quantification of *TF* requires large numbers of gages to account for the high spatial variability of tropical forest canopies and the use of "roving" gages has been recommended to increase the sample area and reduce the error (Lloyd and Marques, 1988; Holwerda *et al.*, 2006b). However, the roving gage method has been little used so far in TMCF (Bruijnzeel, 2005; cf. Fleischbein *et al.*, this volume), and errors in published estimates of *CWI* based on conventionally measured *TF* are likely to be substantial (cf. Hafkenscheid *et al.*, 2002; Holwerda *et al.*, 2006b; Prada *et al.*, 2009).

Tropical Montane Cloud Forests: Science for Conservation and Management, eds. L. A. Bruijnzeel, F. N. Scatena, and L. S. Hamilton. Published by Cambridge University Press.

Modeled estimates of fog deposition in complex terrain show pronounced spatial variability (Walmsley *et al.*, 1996; Mulligan and Burke, 2005), as do comparative studies using differently exposed fog gages (e.g. Juvik and Ekern, 1978; Cavelier *et al.*, 1996; cf. González, this volume). However, field studies of small-scale spatial variability of *CWI* inferred from net precipitation (i.e. *TF* + *SF*) measurements in TMCF of contrasting exposition are rare (e.g. Weaver, 1972; Stadtmüller and Agudelo, 1990; García-Santos, 2007; Prada *et al.*, 2009; Giambelluca *et al.*, this volume; Häger and Dohrenbusch, this volume). This chapter reports on a study of *CWI* in an elfin cloud forest at 970 m.a.s.l. in the Luquillo Mountains of Puerto Rico between 10 July and 25 October 2001. The present data complement an earlier study at the nearby but more exposed Pico del Este (Holwerda *et al.*, 2006a).

STUDY AREA

The measurements were made in a 0.54-ha watershed situated *c.* 200 m east-northeast of Pico del Oeste (1020 m.a.s.l.) in the eastern Luquillo Mountains of Puerto Rico (Figure 29.1). The watershed is located between 950 and 990 m.a.s.l. in the headwaters of the Rio Abajo. It has an easterly aspect, and a mean slope of ~11°. The area is mainly covered with 2–5 m tall (on average ~3 m) elfin cloud forest dominated by *Tabebuia rigida* and *Ocotea spathulata* with an estimated leaf area index of 2.0–3.5 (Weaver, 1995). The forest is rich in epiphytes, with mosses and bromeliads covering branches, stems, and part of the soil surface (Weaver, 1972). Small fragments of palm forest occur along the lower reaches of the stream channel.

Meteorological parameters were measured above the canopy at nearby palm forest (900 m.a.s.l.) and elfin cloud forest (1010 m.a.s.l.) sites (see Figure 29.1 for locations). The palm forest site is dominated by Sierra palm (*Prestoa montana*) with a mean height of ~6 m (Holwerda, 2005) whereas the forest near Pico del Este (Weaver, 1995; Holwerda *et al.*, 2006a) is very similar to that of the study watershed. Mean annual rainfall is about 4500 mm (Baynton, 1968; García-Martinó *et al.*, 1996). On average, most rainfall occurs in May, October, and November (450–500 mm each), whereas January, February, and March typically receive 250–300 mm each (García-Martinó *et al.*, 1996). Mean monthly temperatures vary from *c.* 17 °C in January–February to *c.* 20 °C in September–October (Baynton, 1968; Brown *et al.*, 1983) whereas the mean relative humidity is very high at about 98% (Weaver, 1995). The prevailing trade winds blow mainly from the east; the mean monthly wind speed at Pico del Oeste is 4–5 m s^{-1} (Baynton, 1968). Fog is very common. On Pico del Oeste, Baynton (1968) reported fog to persist for *c.* 60% of the time during the day and for nearly 100% of the time at night. At Pico del Este, fog was observed for about 75% and 95% of the time during the day and at night, respectively, between 25 June

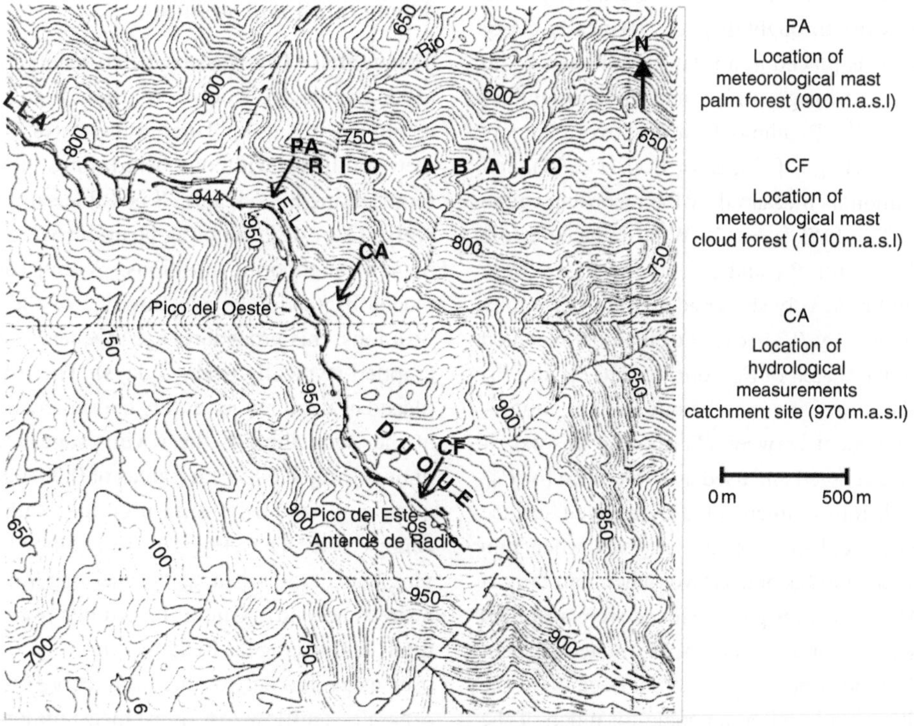

Figure 29.1. Location of climatic masts in the palm forest (900 m.a.s.l.) and the elfin cloud forest (1010 m.a.s.l.), and location of hydrological measurements (970 m.a.s.l.) near Pico del Oeste, Luquillo Mountains, Puerto Rico. Spacing of contour lines is 10 m.

and 7 August 2005 (Holwerda *et al.*, 2006a; cf. Holwerda *et al.*, this volume #28).

METHODS

General climatic measurements

Continuous climatic measurements were made in 7.5 m and 12 m tall masts placed at the nearby palm and cloud forest sites (Figure 29.1), respectively, using identical instrumental set-ups. Measurements included incoming and outgoing short- and long-wave radiation to give net radiation, wind speed and direction, as well as temperature and relative humidity. For instrumental details see Holwerda (2005). Climatic parameters for the *CWI* study site (970 m.a.s.l.) were calculated as the means of those measured at the palm (900 m) and cloud forest (1010 m) sites.

Rainfall

Rainfall (P, mm) was measured at the center of the watershed with a tipping bucket rain gage (P_{tb}, manufactured at the Vrije Universiteit Amsterdam, 500 cm^2 orifice at 1.5 m above the ground, 0.10 mm per tip) connected to a data-logger that registered rainfall to the nearest second; recordings were converted later to the hourly totals used in this analysis. Rainfall was also measured with two totalizing rain gages (P_{t1}, P_{t2}, 100 cm^2 orifice at 0.3 m, and spaced about 100 m apart). The two totalizers were read at the same time as the throughfall gages (typically every 3 days). Each rain gage was placed at a relatively sheltered site with the mean angle between gage orifice and the surrounding tree tops being less than 20°. Readings from the two totalizers agreed very well ($P_{t2} = 1.01P_{t1} - 0.28$, r2 = 1.00, $n = 33$), and were c. 3% lower than amounts measured by the recording gauge ($P_{t1} = 0.97P_{tb} - 0.90$, r2 = 1.00, $n = 37$). Wind-induced errors were considered negligible for P_{t1} and P_{t2}, as the gages were close to ground level and relatively sheltered. Wind effects may have affected measurements by P_{tb} because of its more elevated position. The slightly higher catch of P_{tb} compared to P_{t1} and P_{t2} is somewhat suspect, therefore, and may be due to gage calibration errors. Because agreement between P_{t1} and P_{t2} was excellent, their readings were averaged and used as a reference for P_{tb}.

The intensity at which rain is intercepted by a given surface depends on the angle of incidence, i.e. it is highest when rain falls normal to the surface, and becomes lower when rain falls at an angle (Sharon, 1980). The rain gage orifices were installed horizontally at a slope of about 11° that was orientated perpendicularly to the easterly trade winds (Figure 29.1). Because of the relatively windy conditions, rain fell at an angle for much of the time, and the actual rainfall incident to the forest canopy must therefore have been higher than that measured by the gages. A trigonometric model was used to compute the rainfall normal to a slope (P_a) from conventional (horizontal orifice) rainfall (P) measurements (Sharon, 1980). For each event, the median rainfall inclination was calculated using relationships between rainfall intensity, raindrop size, terminal fall velocity of the rain drops, and wind speed (Holwerda *et al.*, 2006a). Rainfall events were defined as being separated by a dry period of at least 3 hours (Schellekens *et al.*, 1998). In applying the Sharon (1980) model, it was assumed that the canopy surface runs parallel to the sloping ground, and that the proportion of the rain reaching the ground without touching the vegetation is negligible (cf. Holwerda *et al.*, 2006a).

Fog

Fog occurrence was measured in the center of the watershed with a 0.5 × 0.5 m wire harp (Goodman, 1985) orientated perpendicularly to the prevailing winds at a height of 3 m above the ground. The fog gage was protected from rainfall by a 2 × 2 m cover placed ~3 cm above the gage and slanting slightly in a N240E direction. A 50 × 6 cm gutter at the bottom of the gage led the collected cloud water to a tipping bucket (5 ml capacity) and logger device. Hourly amounts of fog collected by the gage (litres) were converted to depth equivalent (CWI_g, mm) by dividing through the gage collecting area.

Throughfall and stemflow

Throughfall (*TF*) was measured using 67 totalizing rain gages (100 cm^2 each). For 45 of the 67 gages, a roving sampling technique was used. Three transects of 79, 63, and 37 m length were outlined with numbered flags placed at 1-m intervals, representing 80, 64, and 38 possible sampling positions, respectively. These 182 sampling positions were divided into 45 groups of 4–5 neighboring positions. Within each group, a gage was placed by randomly selecting one of the 4–5 possible sampling positions; this procedure was repeated each time the gages were emptied (typically every 3 days). The remaining 22 gages were placed at fixed but randomly chosen positions, and were emptied at the same time as the roving gages. In addition, three steel gutters (projected surface areas of 0.98, 0.90, and 1.03 m^2, placed at an angle of ~18°) were used to obtain continuous records of canopy drip. The gutters were equipped with custom-built tipping bucket (50 ml capacity) and logger devices (recording to the nearest second). Gutter recordings were converted to hourly totals and averaged. Readings from the 67 totalizing gages were used as a reference for the mean record of the gutters. Because no separate stemflow (*SF*, mm) measurements were made, the previously recorded value of 5% obtained by Weaver (1972) for this forest type was used throughout.

Rainfall interception loss

Interception loss (E_i, mm) was calculated using the analytical model of Gash (1979) for corrected rainfall (P_a). Application of the model required the determination of the following canopy parameters: (i) canopy saturation value (S, mm), i.e. the amount of water left on the canopy when TF has ceased; (ii) "direct" TF coefficient (p), i.e. the proportion of the rain which falls to the ground without touching the vegetation; (iii) trunk water capacity (S_t, mm); and (iv) the proportion of P_a diverted to stemflow SF (p_t). Values of S were determined using the method of Leyton *et al.* (1967) for storms $\geq$ 1.5 mm. Because under foggy conditions, amounts of TF will be raised, only (nearly) fog-free rainfall events were selected for this analysis. The same constraint applied to events used for the determination of p using the method of Jackson (1975) for storms $\leq$ 1 mm. Evaporation from the trunks was not considered, i.e. S_t was assumed negligible. The mean SF fraction of 0.05 was used for p_t. The Gash (1979) model also requires information on the average rainfall intensity falling onto a saturated canopy ($\overline{R_a}$, mm hour^{-1}, taken from the recording gage and corrected for the effect of slope), and the average evaporation rate from the wet canopy ($\overline{E_w}$, mm hour^{-1}) which was calculated using the wet variant of the Penman–Monteith equation (Monteith, 1965). The condition of a saturated canopy is usually maintained when $\overline{R_a}$ exceeds 0.5 mm hour^{-1} (Gash, 1979).

Cloud water interception

Cloud water interception (CWI) was calculated from the wet-canopy water budget for each rainfall event j:

$$CWI_{\mathrm{wb},j} = TF_j + SF_j + E_{\mathrm{i},j} - P_{a,j} \tag{29.1}$$

with the remaining terms as defined previously. In addition, for each period j with fog-only, fog interception ($CWI_{\mathrm{tf,j}}$) was equated to the corresponding amount of throughfall TF_j. Fog events were defined as being separated by 3 hours without fog to allow canopy drying before the next event. In case a fog-only period was preceded by rainfall, the first 3 hours were excluded to avoid contributions by rain-induced TF. Because SF was not measured, it was neglected in the calculations of CWI_{tf}.

STANDARD ERRORS

The standard error (SE) of $CWI_{\mathrm{wb},j}$ was determined from the errors in the respective budget components using:

$$\mathrm{SE}_{CWI_{\mathrm{wb},j}} = \sqrt{\mathrm{SE}^2_{TF,j} + \mathrm{SE}^2_{SF,j} + \mathrm{SE}^2_{\mathrm{E}_{i,j}} + \mathrm{SE}^2_{\mathrm{P}_{a,j}}} \tag{29.2}$$

in which $\mathrm{SE}_{x,j}$ is the standard error in water equivalent (mm) for a water budget component x per rainfall event j. For those components for which no SE could be calculated (SF and E_i), the error equaled the estimated or calculated value for that component. The error in P_a was estimated as the difference between the catch of totalizers P_{t1} and P_{t2} (i.e. 1% of P_a) plus the slope correction itself. The SE of mean TF was calculated from the distribution of the catch of the 67 individual gages per sampling interval.

RESULTS

Fog occurrence in relation to climatic conditions

Rates of cloud water collection by the wire harp showed a distinct diurnal pattern (CWI_g, Figure 29.2f). Collection rates were usually highest at night (0.5–0.6 mm hour^{-1}) and decreased rapidly after sunrise to less than 0.2 mm hour^{-1} during the day. Although this pattern is partly related to that of wind speed U (Figure 29.2c), it also indicates that fog was almost continuously present at night due to the lowering of the cloud base caused by nocturnal cooling. During the day, however, the cloud base would sometimes rise above the elevation of the site due to daytime heating. Between 10 July and 25 October 2001, the mean amount of fog collected by the wire harp was 9.0 mm day^{-1} (Table 29.1). Out of the 100 days with data, 99 days had fog and only one day was entirely fog-free. Overall fog occurrence was 73% of the time (55% and 89% during day- and night-time conditions, respectively). Defining fog events as being separated by a fog-free period of at least 3 hours yielded 64 events with a mean duration of 27 hours (Table 29.1). The mean amount of fog intercepted by the wire harp per event was 14.1 mm. The most prolonged event lasted 188 hours (*c.* 8 days, 14–22 October 2001), during which time the wire harp intercepted a total of ~135 mm of fog.

Mean diurnal patterns of net radiation (R_n), temperature (T), wind speed (U), wind direction (U_{dir}), and relative humidity (RH) between 10 July and 25 October 2001, as calculated from the data collected at the two nearby climatic stations, are shown in Figure 29.2. The daily pattern of R_n (Figure 29.2a) shows that cooling at night by eradiation was hampered by the persistence of fog (Figure 29.2f). The low value of mean midday R_n (*c.* 300 W m^{-2}, Figure 29.2a) indicates that fog or clouds also attenuated much of the incident radiation during the day. This resulted in a small difference (*c.* 1.5 °C) between mean maximum and minimum T (Figure 29.2b). At night RH was close to 100%, while the mean midday value was about 95% (Figure 29.2e), again indicating that fog was usually heavier and more persistent at night than during the day. Winds were predominantly from the east (Figure 29.2d) and somewhat stronger at night than during the day (2.8–3.0 m s^{-1}, Figure 29.2c).

Rainfall

Mean daily rainfall between 10 July and 25 October 2001 was 10.9 mm day^{-1} (Table 29.2), and from the 97 days with data,

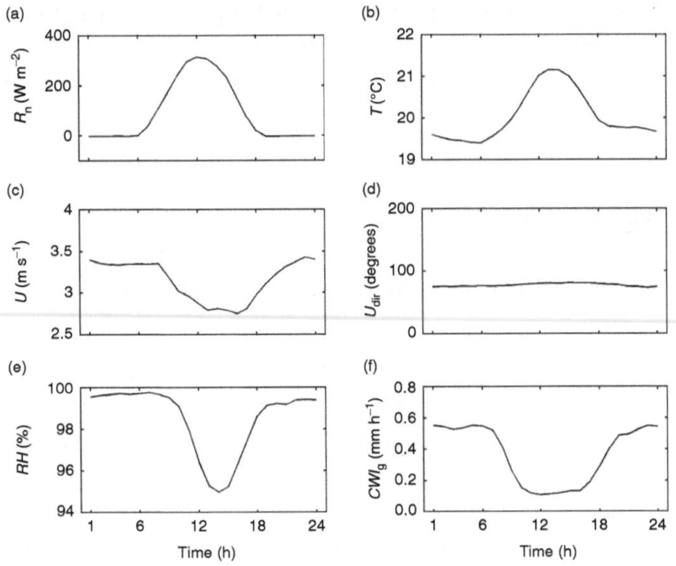

Figure 29.2. Mean diurnal patterns of (a) net radiation (R_n), (b) temperature (T), (c) wind speed (U), (d) wind direction (U_{dir}), and (e) relative humidity (RH), as calculated from data of two nearby climatic stations (see Figure 29.1), and (f) daily pattern of fog collection (CWI_g) by a wire harp placed in the center of the study watershed (970 m.a.s.l.) between 10 July and 25, October 2001.

only three days did not have rain. Overall, 112 rainfall events were distinguished, with a mean duration of 10 hours, a mean size of 9.4 mm, and a mean intensity of ~0.7 mm hour^{-1} (Table 29.2). Median storm duration, size, and intensity were 8.5 hours, 5.2 mm, and ~0.4 mm hour^{-1}, respectively. Average rainfall after correction for the effect of slope (P_a, 11.3 mm day^{-1}) was 0.4 mm (4%) above measured rainfall (P, 10.9 mm day^{-1}, Table 29.2). Rainfall events were nearly always accompanied by fog; on average, fog occurred for 78% of the time during a rainfall event, and only nine out of the 112 events (8%) were completely fog-free.

Throughfall and stemflow

Between 10 July and 25 October 2001, the 67 totalizing TF gages were sampled 40 times. Mean TF per sampling occasion correlated well with P_a ($TF = 0.90P_a$, r2 = 0.98, $n = 40$). Cumulative TF per sampling position was divided by the corresponding P_a and multiplied by 100 to give relative throughfall (TF/P) at the 204 sampling positions (Figure 29.3a). The variation in point TF was considerable, ranging from 16% to 261% of P_a. Mean point TF was 90% of P_a, whereas median point TF was 83%; the standard deviation of the distribution shown in Figure 29.3a is 45%. Values of TF exceeded P_a in 57 out of 204 cases (28%). The coefficient of variation (CV, %) of mean TF per sampling occasion increased with decreasing P_a (Figure 29.3b). The highest CVs (>100%) were observed for $P_a < 10$ mm. In the range $P_a = 10$–50 mm, the CV generally varied between 50% and 100%, whereas for $P_a > 50$ mm, it was generally between 50% and 60%. Mean TF for the 97 days with available rainfall data was 10.2 mm day^{-1}, corresponding to 90% of P_a (Table 29.2). Mean stemflow (SF) calculated using the fixed fraction of 0.05 was 0.6 mm day^{-1}. The mean sum of mean TF and SF (i.e. net precipitation P_{net}) was 10.8 mm day^{-1}, corresponding to 95% of P_a.

Table 29.1 *Statistical parameters of 64 fog* (CWI_g) *events, as measured with a wire harp between 10 July and 25 October 2001 at the Pico del Oeste watershed site, Puerto Rico*

Fog intercepted by the wire harp (CWI_g)	9.0 mm day^{-1} ($n = 100$, range: 0–26.6 mm day^{-1})
No. of fog events	64
Mean duration of fog event	27 hours (range: 1–188 hours)
Mean fog intercepted per event	14.1 mm (range: 0.0–134.5 mm)

Table 29.2 *Statistical parameters of 112 rainfall* (P) *events between 10 July and 25 October 2001 at the Pico del Oeste watershed site, and summarized corrections for the effect of slope. Mean daily amounts of throughfall* (TF) *and stemflow* (SF) *added for comparison*

Mean measured rainfall (P)	10.9 mm day^{-1} ($n = 97$; range: 0–76 mm day^{-1})
No. of rainfall events	112
Mean duration of rainfall event	10 hours (range: 1–43 hours)
Mean rainfall depth per event	9.4 mm (range: 0.1–77.1 mm)
Mean rainfall intensity (R)	0.7 mm hour^{-1} (range: 0.0–6.3 mm hour^{-1})
Mean slope correction factor	1.04 (range: 0.99–1.07)
Mean rainfall after slope correction (P_a)	11.3 mm day^{-1}
Throughfall (TF)	10.2 mm day^{-1} ($n = 97$, 90% of P_a)
Stemflow (SF)	0.6 mm day^{-1} ($n = 97$, 5% of P_a)

Modeled rainfall interception loss

Regressing TF against P_a for nine events with $P_a < 1$ mm and negligible fog ($F_g < 0.2$ mm) yielded a value for the free TF coefficient (p) of 0.23 (Figure 29.4a). Fitting a line of slope 0.95 ($1 - p_t$) through the upper points of the TF vs. P_a graph for eight storms with $P_a > 1.5$ mm and $F_g < 0.5$ mm gave an estimated canopy saturation value (S) of 0.50 mm, whereas the directly fitted line suggested a value of 0.46 mm (Figure 29.4b).

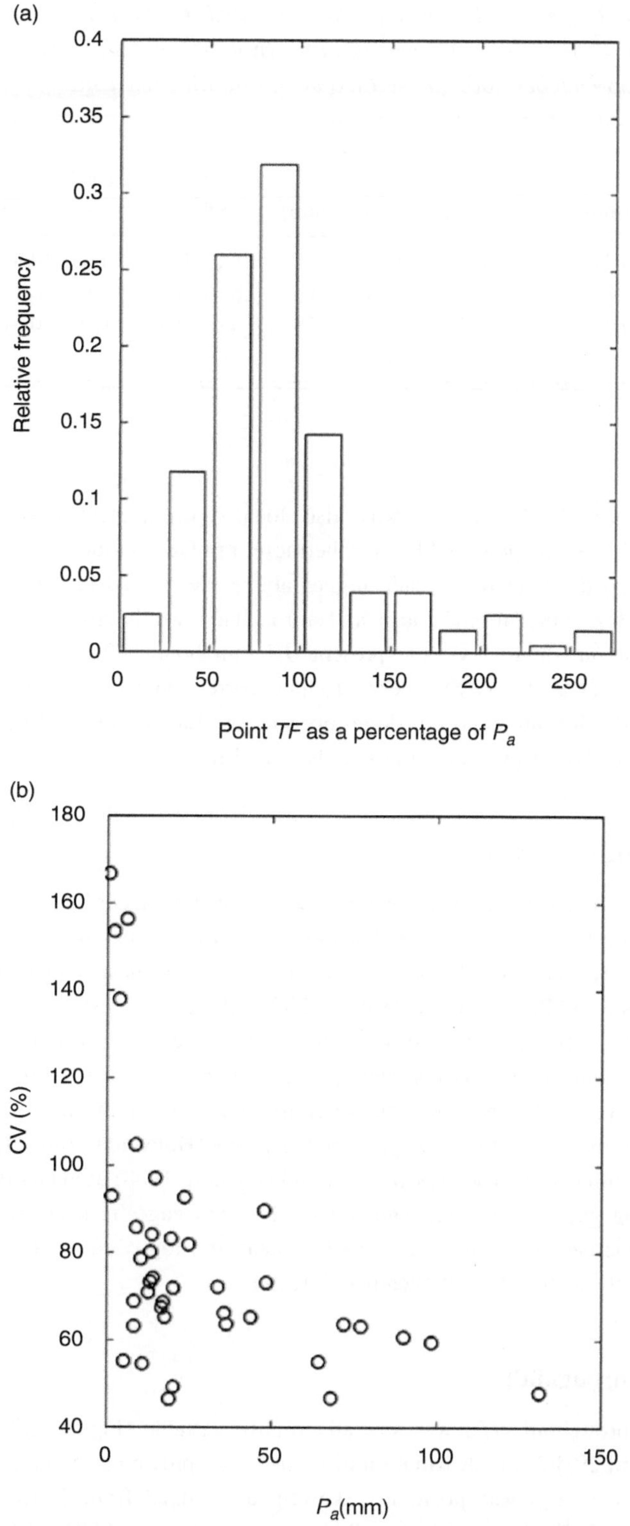

Figure 29.3. (a) Frequency distribution of point throughfall (*TF*) expressed as a percentage of corrected rainfall (P_a) at 204 sampling positions in the Pico del Oeste elfin cloud forest between 10 July and 25 October 2001. (b) Relationship between P_a and the coefficient of variation CV (100 × SD/mean) of mean *TF* for 40 sampling intervals (using 67 gages) for the same period.

The mean wet-canopy evaporation rate ($\overline{E_w}$) calculated for all hours with rainfall intensity $R_a \geq 0.50$ mm hour^{-1} was 0.06 mm hour^{-1}. The corresponding mean rainfall intensity $\overline{R_a}$ was 3.50 mm hour^{-1}, yielding the very low mean relative evaporation rate $\overline{E_w}/\overline{R_a}$ of 0.02 and model calculations were performed using this value. The mean amount of rainfall necessary to saturate the canopy P_a' (Gash, 1979) was 0.70 mm, and slightly higher than S (0.50 mm). Components of overall interception loss E_i as calculated with the analytical model for the 112 rainfall events were as follows: evaporation from the wetted canopy contributed only 28% (18.4 mm) of the total calculated E_i of 66.5 mm, the bulk of which (65%, or 43 mm) was associated with drying up of the canopy in between storms. Wetting up of the canopy contributed the remaining 8% (5.1 mm). Total calculated E_i was 6% of P_a (1099 mm).

Fog interception

The 112 rainfall (P_a) events of the study period gave as many (potential) estimates of fog interception (CWI_{wb}) based on the wet-canopy water budget equation. However, events for which the calculated error of CWI_{wb} ($SE_{CWI_{wb}}$) exceeded the estimate itself ($n = 30$), or for which the resulting CWI_{wb} value was negative ($n = 45$) were left out of the analysis (Table 29.3). The remaining 37 events (33%) that passed these selection criteria received less P_a (mean 4.6 mm event^{-1}) than fog (CWI_g as measured by the wire harp; mean 9.3 mm event^{-1}), while events with $SE_{CWI_{wb}} > CWI_{wb}$ had comparable amounts of P_a (mean 6.5 mm event^{-1}) and CWI_g (mean 6.4 mm event^{-1}). Events for which CWI_{wb} was negative, had much greater P_a (mean 16.3 mm) than CWI_g (mean 4.9 mm) (Table 29.3).

Mean measured event rainfall (P) for the 37 events that passed the criteria was 4.4 mm (not shown), whereas mean corrected event rainfall (P_a) was 5% higher (4.6 mm, Table 29.3). Mean event duration was about 14 hours, and mean fog occurrence was nearly 100%. Mean P_{net} ($TF + SF$, 5.7 mm) was about 24% higher than P_a. The estimated mean fog interception per event (CWI_{wb}) was 1.6 ± 0.8 mm. *CWI* rates (mm hour^{-1}) were estimated by dividing event CWI_{wb} by event duration, and these rates are plotted against corresponding collection rates CWI_g by the wire harp in Figure 29.5a. Despite considerable variation, estimated CWI_{wb} rates correlated significantly (at the 0.01 level) with CWI_g. The slope of the regression of CWI_{wb} against CWI_g (0.16, Figure 29.5a) can be taken to represent a "gage-to-canopy" conversion factor, which theoretically integrates differences in the fog-catching efficiencies of the wire harp and the forest canopy (Schellekens *et al.*, 1998; cf. García-Santos, 2007). Applying the latter factor to the record of the wire harp would suggest an average fog interception of about 1.4 mm day^{-1} (range: 0.0–4.2 mm day^{-1}, $n = 100$).

Table 29.3 *Summary of wet-canopy water budget components for rainfall* (P_a) *events for which (i) the calculated standard error in fog interception* ($SE_{CWI_{wb}}$, *Eq. 29.2) was larger than the estimate itself* ($n = 30$)*; (ii)* CWI_{wb} *was negative* ($n = 45$)*, or (iii)* CWI_{wb} *was positive and larger than* $SE_{CWI_{wb}}$ ($n = 37$) *between 10 July and 25 October 2001 at the Pico del Oeste watershed site*

Event category	No. of events	Duration (hours)	Fog occurrence (%)	CWI_g (mm)	P_a (± SE) (mm)	TF (± SE) (mm)	SF (± SE) (mm)	E_i (± SE) (mm)	CWI_{wb} (± SE) (mm)
$SE_{CWI_{wb}} > CWI_{wb}$	30	13	68	6.4	6.5 (± 0.4)	6.1 (± 0.5)	0.3 (± 0.3)	0.5 (± 0.5)	0.4 (± 0.9)
$CWI_{wb} < 0$	45	12	71	4.9	16.3 (± 0.8)	13.2 (± 1.0)	0.8 (± 0.8)	0.8 (± 0.8)	−1.5 (± 1.7)
$SE_{CWI_{wb}} < CWI_{wb}$ and $CWI_{wb} > 0$	37	14	95	9.3	4.6 (± 0.3)	5.5 (± 0.5)	0.2 (± 0.2)	0.5 (± 0.5)	1.6 (± 0.8)

Throughfall-based fog interception during fog-only periods (CWI_{tf}, excluding the first 3 hours after rainfall) were summed per fog event, yielding 37 estimates of CWI_{tf}. Estimated hourly rates correlated significantly (at the 0.01 level) with CWI_g (Figure 29.5b). The slope of the regression (0.12) again represents a gage-to-canopy conversion factor; applying the latter to the record of the wire harp yielded an average fog interception of about 1.1 mm day^{-1} (range: 0.0–3.3 mm day^{-1}, $n = 100$), i.e. slightly smaller than the water-budget based estimate (Figure 29.5a).

DISCUSSION

Rainfall

The presently measured mean daily rainfall (P) of 11 mm (Table 29.2) is close to the (long-term) value of 120–13 mm for nearby Pico del Este (García-Martinó *et al.*, 1996). Rainfall incident to the sloping canopy (P_a) was calculated from P using a trigonometric model. In doing so, it was assumed that the proportion of direct TF (p) was negligibly small. Arguably, direct TF would need to be corrected in a similar manner as rainfall (i.e. plus ~4%, Table 29.2) but this would increase the mean water-budget-based fog interception estimate (CWI_{wb}) by only 20–3%. The use of mean rainfall intensities and wind speeds per event in the model calculations introduced further simplifications. In reality, rainfall intensity, wind speed, and the angle of incidence of rain drops varied during events. In mountainous terrain, winds are greatly influenced by local topography; therefore, an unknown additional error was introduced by using the mean of the wind conditions as measured at the palm and cloud forest sites.

The mean correction factor for slope effects (+4%, Table 29.2) was much lower than that calculated by Holwerda *et al.* (2006a) for nearby Pico del Este (+28%). However, the Pico del Este site was not only windier (mean wind speed 50–6 m s^{-1}; Holwerda *et al.*, 2006a) than the present site (~3 m s^{-1}, Figure 29.2c), but it was also located on a steeper slope (17° vs. the present 11°). Furthermore, rainfall measurements at Pico del Este were made higher above the ground (1.8 m vs. 0.3 m in the present study), and storms had lower intensity (mean 0.4 mm hour^{-1} vs. the present 0.7 mm hour^{-1}, Table 29.2). García-Santos (2007) derived slope correction values of 130–18% for an exposed ridge-top site on La Gomera (Canary Islands) for average wind speeds of *c.* 3 m s^{-1}.

Fog occurrence

Fog occurred on average for 73% of the time according to the wire harp. This is somewhat lower than the values obtained by Holwerda *et al.* (2006a) and Baynton (1968) for nearby Pico del Este and Pico del Oeste (85% and 80% of the time, respectively). The difference may reflect the lower elevation at which the present study was conducted (i.e. 970 m.a.s.l. vs. 1010–1020 m.a.s.l.). Differences in measured fog occurrence may also relate to the use of different types of fog gages (Holwerda *et al.*, this volume #28). The present study and Baynton (1968) used passive fog gages (wire harp and louvered screen gage, respectively), whereas Holwerda *et al.* (2006a) measured fog occurrence directly with a Present Weather Detector.

Throughfall

Throughfall (TF) was spatially highly variable (Figures 29.3a and 29.3b); the distribution of point TF (expressed as a percentage of P_a) was positively skewed and ranged from 160% to 261%. Further, at 57 of the 204 sampling positions (28%) cumulative TF exceeded P_a (Figure 29.3a). Measuring TF at 80 sampling positions during 6 weeks in a similar, 3-m tall elfin cloud forest at nearby Pico del Este (using 20 roving gages), Holwerda *et al.* (2006a) found that at 34 of the 80 positions (43%) TF exceeded P_a. The distribution shown in Figure 29.3a has a mean relative TF and SD of 90% and 45% (of P_a), respectively, and

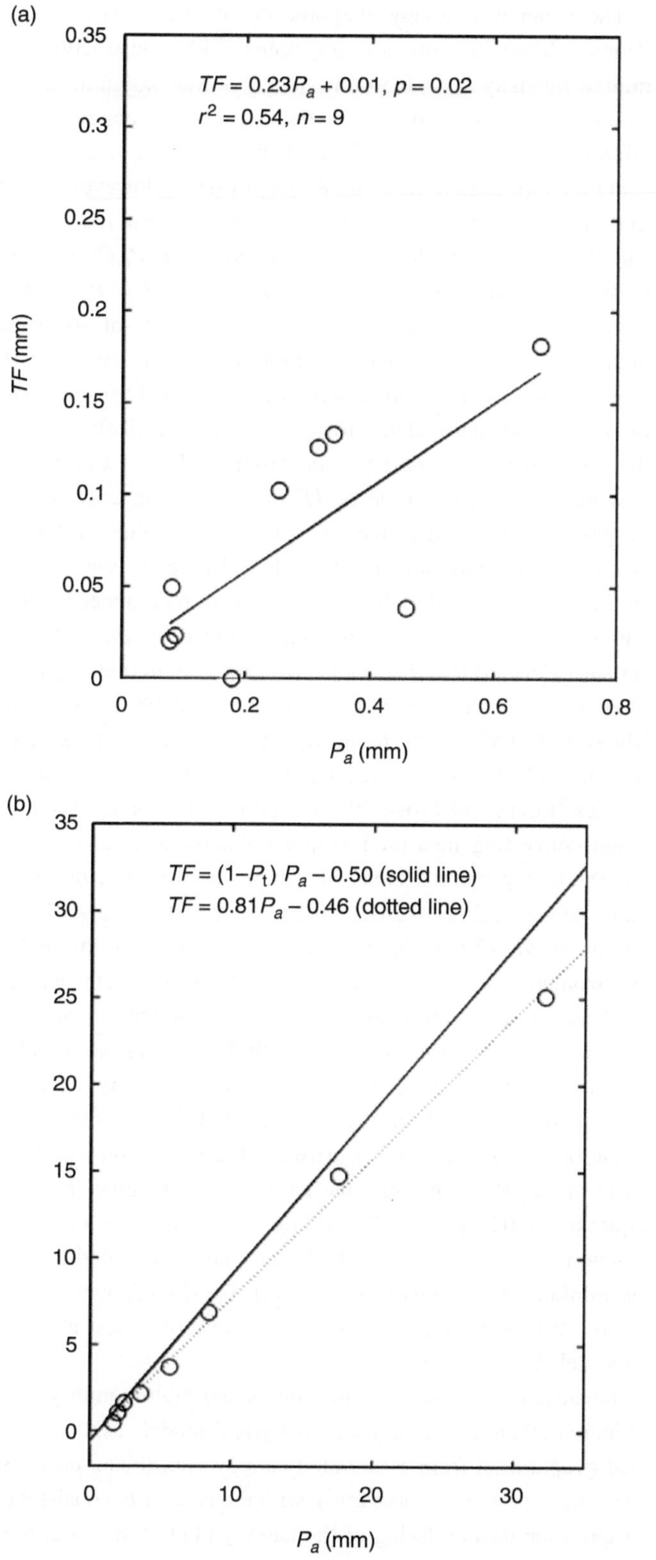

Figure 29.4. (a) Derivation of the direct throughfall coefficient (p, 0.23) at the Pico del Oeste watershed site following Jackson (1975). (b) Estimation of the canopy saturation value (S, 0.50 mm) by the method of Leyton *et al.* (1967). Stemflow fraction (p_t) taken as 0.05; the dotted line represents the best fit to the TF–P_a data pairs.

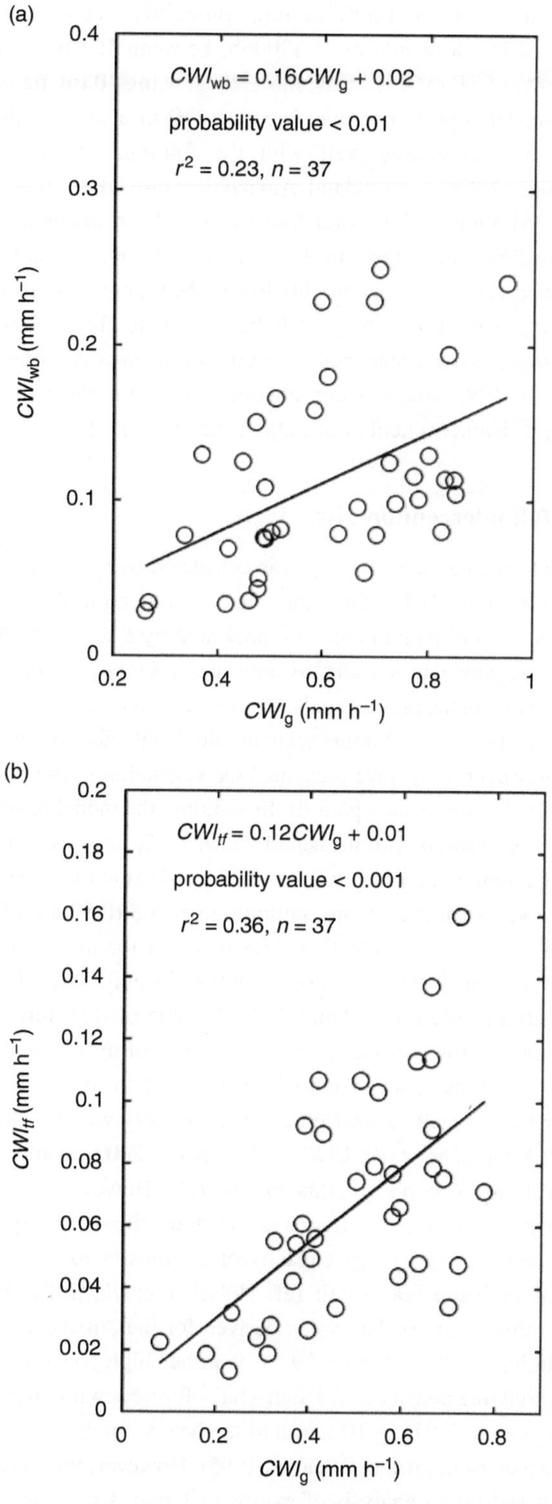

Figure 29.5. (a) Relationship between fog collection rates as measured by the wire harp (CWI_g, mm hour^{-1}) and water-budget-based fog interception rates (CWI_{wb}, mm hour^{-1}) for 37 rainfall (P_a) events between 10 July and 25 October 2001 at the Pico del Oeste watershed site. (b) Relationship between CWI_g and throughfall-based fog interception rates (CWI_{tf}, mm hour^{-1}) for 37 fog events during the same period.

hence, a CV of about 50%. Similar variability (about 48%) was observed by Holwerda *et al.* (2006b) between 160 point measurements of *TF* (March–November 2001, using 30 roving gages) in 20-m tall lower montane forest at 340 m a.s.l. The present result also compares well with the findings of Lloyd and Marques (1988) for lowland Amazonian rain forest (CV 500–55%). Although elfin cloud forest differs from lower montane and lowland rain forests in that the trees are much smaller and species diversity is considerably lower, the high variability of *TF* in the present forest may well be related to the presence of numerous, twisted branches covered with mosses and epiphytes (Weaver, 1995) and has consequences for the number of gages used (cf. Hafkenscheid *et al.*, 2002; Prada *et al.*, 2009).

Rainfall interception loss

Rainfall interception loss (E_i) was calculated using the analytical model of Gash (1979). One important assumption in the model is that the rainfall pattern can be represented by a series of discrete storms, separated by sufficiently long intervals for the canopy to dry completely. In the present study, storms were considered separate if preceded by at least 3 hours without rain. Continued fog interception, however, may have prevented the vegetation from drying up completely during such periods. In addition, the model results can only be compared with measured values of E_i for storms without fog, the number of which was small (only nine out of 112 storms).

The value for the canopy saturation value S (0.50 mm, Figure 29.4b) compares well with values reported for the elfin cloud forest at nearby Pico del Este (0.400–0.50 mm; Wickel, 1997; Holwerda *et al.*, this volume #28). A value of 0.50 mm would seem rather low considering the abundance of mosses and other epiphytes. Mosses are potentially capable of storing amounts of water that exceed several times their own dry weight (Weaver, 1972; Veneklaas *et al.*, 1990; Köhler *et al.*, 2007; Köhler *et al.*, this volume; Tobón *et al.*, this volume #26). However, because of the persistently wet conditions at the study site, it is very likely that the effective storage capacity of the mosses to absorb additional moisture was small (cf. Hölscher *et al.*, 2004; Köhler *et al.*, this volume). The value derived for the direct throughfall coefficient p (0.23, Figure 29.4a) is rather high considering the very high tree density (3671 stems ha^{-1} for trees with diameter at breast height DBH $\geq$ 10 cm) and the heavy epiphyte and bryophyte load of the forest (Weaver, 1995). However, this value of p was based on an analysis of storms < 1 mm, while the canopy saturation value was already estimated at 0.50 mm (Figures 29.4a and 29.4b). Therefore, it is possible that the derived p value includes some indirect *TF* occurring after canopy saturation. Removal of intercepted rain drops from the vegetation by wind may also have increased the value of p, whereas the possibility of drops splashing off the hard, leathery leaves cannot be excluded either (cf. Schellekens *et al.*, 2000).

The mean wet-canopy evaporation rate $\overline{E_w}$ – based on the Penman–Monteith equation for hours with (slope corrected) rainfall intensity $R_a \geq 0.50$ mm hour^{-1} – was 0.06 mm hour^{-1}, which is equivalent to ~40 W m^{-2}. The corresponding mean rainfall intensity $\overline{R_a}$ was 3.50 mm hour^{-1}, yielding the very low mean relative evaporation rate $\overline{E_w}/\overline{R_a}$ of 0.02. However, $\overline{E_w}/\overline{R_a}$ as derived from the slope of the regression between measured E_i and P_a (Gash, 1979) for storms with little fog ($P_a/CWI_g > 10$) yielded the much higher value of 0.09 ($E_i = 0.09P_a + 0.79$, $r^2 = 0.37$, $n = 13$), implying an evaporation rate of ~0.27 mm hour^{-1}. Using the Penman–Monteith-based rate of $\overline{E_w}$, the model underestimated "measured" E_i losses for storms in excess of 10 mm in particular. However, it is not likely that the difference between measured and observed E_i was caused by a systematic underestimation of *TF* in view of the large number of gages used. Rather, the use of a fixed value of 5% for stemflow (*SF*) may have been too low. In recent years, several studies have reported high *SF* values for stunted upper montane and elfin cloud forests, e.g. Hafkenscheid *et al.* (2002) in Jamaica (18% of P), McJannet *et al.* (2007) in northern Queensland (14% of P, 11% of P_a), and Kumaran (2008) in Peninsular Malaysia (30.5% of P). Measuring *SF* in rain forests is generally difficult due to the large variation within and between tree species (Levia and Frost, 2003). Additional error is introduced when converting measured volumes (liters) to areal estimates (mm) using projected areas of sample tree crowns or overall stem density (cf. Hafkenscheid *et al.*, 2002). Finally, Tóbon-Marin *et al.* (2000) found that the *SF* fraction in lowland Amazonian rain forest increased with storm size whereas McJannet *et al.* (2007) observed relative amounts of *SF* to be enhanced during rainfall events with fog in various montane forests in northern Queensland. Therefore, *SF* may well have been underestimated in the present study by using a fixed value of 5%. The difference between the regression-based estimate of $\overline{E_w}/\overline{R_a}$ (0.09) and that based on the Penman–Monteith equation (0.02) was 0.07. Adding this difference to the *SF* fraction of 0.05 suggests that *SF* may have accounted for 12% of rainfall. By comparison, Weaver (1972) estimated *SF* at about 10% of the rainfall for a ridge-top elfin forest at nearby Pico del Este.

Because of the low radiation inputs and high humidity levels (Figures 29.2a and 29.2e), the analytical model results suggest that evaporation from a saturated canopy contributed only 28% (18.4 mm) of total E_i (66.5 mm), with the bulk of E_i consisting of evaporation during drying of the canopy (43.0 mm, about 65%) after rainfall ceased. However, the frequently foggy conditions at the site may have invalidated the assumption of complete canopy drying between storms. To assess the (approximate) effect of incomplete canopy drying between storms on the water-budget-based fog gage to canopy conversion factor (0.16, Figure 29.5a), the (effective) canopy saturation value (S) was decreased from 0.50 to 0.25 mm. This had a comparatively small effect on the magnitude of the gage to canopy factor (0.15, i.e. −6%).

Fog interception

Solving the wet-canopy water budget (Eq. 29.1) for each of the 112 rainfall events (Table 29.2) yielded 37 estimates of fog interception (CWI_{wb}) that were both positive and larger than their error estimates $SE_{CWI_{wb}}$. However, for the majority of events, estimates of CWI_{wb} were either smaller than $SE_{CWI_{wb}}$ ($n = 30$) or even negative ($n = 45$) (Table 29.3). As discussed previously, the difference between modeled and observed interception loss (E_i) may have been caused by an underestimation of *SF*. For the same reason, estimates of CWI_{wb} may become negative when solving the budget equation for large rainfalls. In addition, errors in the individual water budget terms became large compared to the magnitude of CWI_{wb} whenever P_a exceeded the catch of the wire harp (CWI_g). This also applied to events with comparable amounts of P_a and CWI_g. Thus, solving the wet-canopy water budget for CWI_{wb} gave the best results when P_a was small compared to CWI_g (Table 29.3).

Although the variation in estimates of CWI_{wb} was large due to accumulation of errors in individual water budget terms (Table 29.3), estimated fog interception correlated significantly with collection rates by the wire harp (Figure 29.5a). Also, throughfall-based fog interception rates (CWI_{tf}) during periods with fog-only correlated slightly better with CWI_g (Figure 29.5b), probably because the error in CWI_{tf} was determined solely by variation in *TF*. The gage to canopy conversion factor based on CWI_{tf} (0.12, Figure 29.5b) was somewhat lower than that based on CWI_{wb} (0.16, Figure 29.5a), probably because both *SF* and E_i were neglected in the calculations of CWI_{tf}. However, considering the wide error bands associated with CWI_{wb} (Table 29.3), this difference may also reflect the error in CWI_{wb}.

Applying the gage-to-canopy conversion factors to the record of the wire harp suggested a fog interception rate of 1.10–1.4 mm day^{-1}, equivalent to 10–12% of mean P_a. Interestingly, the presently derived estimate is 30–50% lower than the 2.1 mm day^{-1} value derived by Holwerda *et al.* (2006a) for very similar forest at nearby Pico del Este. This difference is thought to largely reflect contrasts in site exposure. The Pico del Este site was located just below a ridge top and more exposed to winds than the present site.

CONCLUSIONS

Fog interception (*CWI*) by a Puerto Rican elfin cloud forest at 970 m.a.s.l. was studied between 10 July and 25 October 2001. Fog, as indicated by a shielded wire harp, occurred on average for 73% of the time (55% during the day and 89% at night). Mean daily *CWI* rates as estimated from the wet-canopy water budget (1.4 mm) and from throughfall measurements during periods with fog-only (1.1 mm) were 12% and 10% of mean corrected daily rainfall (11.3 mm), respectively. At 1.1–1.4 mm day^{-1}, *CWI* was 30–50% lower than the water-budget-based estimate of 2.1 mm day^{-1} for a similar, but more exposed elfin cloud forest in the same area.

ACKNOWLEDGEMENTS

This work was supported by a grant from the Netherlands Foundation for the Advancement of Tropical Research (WOTRO, grant no. W76–206). The authors thank C. Estrada, G. Guzman, S. Moya, and C. Torrens (U.S. Department of Agriculture Forest Service) for their help during the field work and Dr. Werner Eugster for helpful comments on an earlier draft.

REFERENCES

Baynton, H. W. (1968). Ecology of an elfin forest in Puerto Rico. II. Microclimate of Pico Del Oeste. *Journal of the Arnold Arboretum* **49**: 419–430.

Blocken, B., J. Carmeliet, and J. Poesen (2005). Numerical simulation of the wind-driven rainfall distribution over small-scale topography in space and time. *Journal of Hydrology* **315**: 252–273.

Brown, S., A. E. Lugo, S. Silander, and L. Liegel (1983). *Research History and Opportunities in the Luquillo Experimental Forest.* New Orleans, LA: U.S. Department of Agriculture Forest Service, Southern Forest Experiment Station.

Bruijnzeel, L. A. (2005). Tropical montane cloud forests: a unique hydrological case. In *Forests, Water and People in the Humid Tropics*, eds. M. Bonell and L. A. Bruijnzeel, pp. 462–483. Cambridge, UK: Cambridge University Press.

Cavelier, J., D. Solis, and M. A. Jaramillo (1996). Fog interception in montane forests across the Central Cordillera of Panama. *Journal of Tropical Ecology* **12**: 357–369.

Frumau, K. F. A., L. A. Bruijnzeel, and C. Tobón (2006). Measurement of precipitation in montane tropical catchments: comparative performance of conventional, spherical and "potential" rain gages. In *Forest and Water in a Changing Environment*, eds. S. R. Liu, G. Sun, and P. S. Sun, pp. 104–108. Vienna: IUFRO, and Beijing: Chinese Academy of Forestry.

García-Martinó, A. R., G. S. Warner, F. N. Scatena, and D. L. Civco (1996). Rainfall, runoff and elevation relationships in the Luquillo Mountains of Puerto Rico. *Caribbean Journal of Science* **32**: 413–424.

García-Santos, G. (2007). *An ecohydrological and soils study in a montane cloud forest in the National Pak of Garajonay, La Gomera (Canary Islands, Spain).* Ph.D. thesis VU University, Amsterdam, the Netherlands. Also available at www.falw.vu.nl/nl/onderzoek/earth-sciences/geo-environmental-science-and-hydrology/hydrology-dissertations/index.asp.

Gash, J. H. C. (1979). An analytical model of rainfall interception by forests. *Quarterly Journal of the Royal Meteorological Society* **105**: 43–55.

Goodman, J. (1985). The collection of fog drip. *Water Resources Research* **21**: 392–394.

Hafkenscheid, R. L. L. J., L. A. Bruijnzeel, R. A. M. de Jeu, and N. J. Bink (2002). Water budgets of two upper montane rain forests of contrasting stature in the Blue Mountains, Jamaica. In *Proceedings of the 2nd International Colloquium on Hydrology and Water Management in the Humid Tropics*, Technical Documents in Hydrology No. **52**, ed. J. S. Gladwell, pp. 399–424. Paris: IHP–UNESCO.

Harr, R. D. (1982). Fog-drip in the Bull Run municipal watershed, Oregon. *Water Resources Bulletin* **18**: 785–789.

Holder, C. D. (2003). Fog precipitation in the Sierra de las Minas Biosphere Reserve, Guatemala. *Hydrological Processes* **17**: 2001–2010.

Hölscher, D., L. Köhler, A. I. J. M. Van Dijk, and L. A. Bruijnzeel (2004). The importance of epiphytes to total rainfall interception by a tropical montane rain forest in Costa Rica. *Journal of Hydrology* **292**: 308–322.

Holwerda, F. (2005). Water and energy budgets of rain forests along an elevation gradient under maritime tropical conditions. Ph.D. thesis VU University, Amsterdam, the Netherlands. Also available at www.falw.vu.nl/nl/onderzoek/earth-sciences/geo-environmental-science-and-hydrology/hydrology-dissertations/index.asp.

Holwerda, F., R. Burkard, W. E. Eugster, *et al.* (2006a). Estimating fog deposition at a Puerto Rican elfin cloud forest site: comparison of the water budget and eddy covariance methods. *Hydrological Processes* **20**: 2669–2692.

Holwerda, F., F. N. Scatena, and L. A. Bruijnzeel (2006b). Throughfall in a Puerto Rican lower montane rain forest: a comparison of sampling strategies. *Journal of Hydrology* **327**: 592–602.

Jackson, I. J. (1975). Relationships between rainfall parameters and interception by tropical forest. *Journal of Hydrology* **24**: 215–238.

Juvik, J. O. and P. C. Ekern (1978). *A Climatology of Mountain Fog on Mauna Loa, Hawai'i Island*, Technical Report No. 118. Honolulu, HI: Water Resources Research Center, University of Hawai'i.

Köhler, L., C. Tobón, K. F. A. Fruman, and L. A. Bruijnzeel (2007). Biomass and water storage of epiphytes in old-growth and secondary montane rain forest stands in Costa Rica. *Plant Ecology* **193**: 171–184.

Kumaran, S., (2008). Hydrometeorology of Tropical montane rain forests of Gunung Brinchang, Pahang Darul Makmur, Malaysia. Ph.D. thesis, University Putra Malaysia, Serdang, Malaysia.

Levia, D. R. Jr., and E. E. Frost (2003). A review and evaluation of stemflow literature in the hydrological and biogeochemical cycles of forested and agricultural ecosystems. *Journal of Hydrology* **274**: 1–29.

Leyton, L., E. R. C. Reynolds, and F. B. Thompson (1967). Rainfall interception in forest and moorland. In *International Symposium on Forest Hydrology*, eds. W. E. Sopper and H. W. Lull, pp. 163–168. Oxford, UK: Pergamon Press.

Lloyd, C. R., and A. O. Marques Filho (1988). Spatial variability of throughfall and stemflow measurements in Amazonian rain forest. *Agriculture and Forest Meteorology* **42**: 63–73.

McJannet, D., J. Wallace, and P. Reddell (2007). Precipitation interception in Australian tropical rainforests. II. Altitudinal gradients of cloud interception, stemflow, throughfall and interception. *Hydrological Processes* **21**: 1703–1718.

Monteith, J. L. (1965). Evaporation and the environment. *Symposia of the Society for Experimental Biology* **19**: 245–269.

Mulligan, M., and S. M. Burke (2005). *DFID FRP Project R7991, FIESTA: Fog Interception for the Enhancement of Streamflow in Tropical Areas*, Final Technical Report of AMBIOTEK/King's College London contributions. Leigh-on-Sea, UK: AMBIOTEK. Also available at www.ambiotek.com/fiesta.

Mulligan, M., K. F. A. Frumau, and L. A. Bruijnzeel (2006). Falling at the first hurdle: spatial rainfall variability and the problem of closing catchment water budgets in tropical montane environments. In *Forest and Water in a Changing Environment*, eds. S. R. Liu, G. Sun, and P. S. Sun, pp. 114–117. Vienna: IUFRO, and Beijing: Chinese Academy of Forestry.

Prada, S., M. Menezes de Sequeiro, C. Figueira, and M. Oliveira da Silva (2009). Fog precipitation and rainfall interception in the natural forests of Madeira Island (Portugal). *Agricultural and Forest Meteorology* **149**: 1179–1187.

Schellekens, J., L. A. Bruijnzeel, A. J. Wickel, F. N. Scatena, and W. L. Silver (1998). Interception of horizontal precipitation by elfin cloud forest in the Luquillo Mountains, eastern Puerto Rico. In *Proceedings of the 1st International Conference on Fog and Fog Collection*, eds. R. S. Schemenauer and H. A. Bridgman, pp. 29–32. Ottawa, Canada: IDRC.

Schellekens, J., L. A. Bruijnzeel, F. N. Scatena, N. J. Bink, and F. Holwerda (2000). Evaporation from a tropical rain forest, Luquillo Experimental Forest, eastern Puerto Rico. *Water Resources Research* **36**: 2183–2196.

Sevruk, B. (1982). *Methods of Correction for Systematic Error in Point Precipitation Measurement*. Geneva, Switzerland: World Meteorological Organization.

Sharon, D. (1980). Distribution of hydrologically effective rainfall incident on sloping ground. *Journal of Hydrology* **46**: 165–188.

Stadtmüller, T., and N. Agudelo (1990). Amount and variability of cloud moisture input in a tropical cloud forest. *International Association of Hydrological Sciences Publication* **193**: 25–32.

Tóbon-Marin, C., W. Bouten, and J. Sevink (2000). Gross rainfall and its partitioning into throughfall, stemflow and evaporation of intercepted water in four ecosystems in western Amazonia. *Journal of Hydrology* **237**: 40–57.

Veneklaas, E. J., R. J. Zagt, A. van Leerdam, *et al.* (1990). Hydrological properties of the epiphyte mass of a montane tropical rain forest, Colombia. *Vegetatio* **89**: 183–192.

Walmsley, J. L., R. S. Schemenauer, and H. A. Bridgman (1996). A method for estimating the hydrological input from fog in mountainous terrain. *Journal of Applied Meteorology* **35**: 2237–2249.

Weaver, P. L. (1972). Cloud moisture interception in the Luquillo Mountains in Puerto Rico. *Caribbean Journal of Science* **12**: 129–144.

Weaver, P. L. (1995). The colorado and dwarf forests of Puerto Rico's Luquillo Mountains. In *Tropical Forests: Management and Ecology*, eds. A. E. Lugo and C. Lowe, pp. 109–141. New York: Springer-Verlag.

Wickel, A. J. (1997). Rainfall interception modelling for two tropical forest types in the Luquillo Experimental Forest, *Puerto Rico*. M.Sc. thesis, VU University, Amsterdam, the Netherlands.

Zadroga, F. (1981). The hydrological importance of a montane cloud forest area of Costa Rica. In *Tropical Agricultural Hydrology*, eds. R. Lal and E. W. Russell, pp. 59–73. New York: John Wiley.

30 Fog gage performance under conditions of fog and wind-driven rain

K. F. A. Frumau
VU University, Amsterdam, the Netherlands

R. Burkard and S. Schmid
University of Bern, Bern, Switzerland

L. A. Bruijnzeel
VU University, Amsterdam, the Netherlands

C. Tobón
Universidad Nacional de Colombia, Medellin, Colombia

J. C. Calvo-Alvarado
Instituto Tecnológico de Costa Rica, Cartago, Costa Rica

ABSTRACT

Fog and wind-driven rain (*WDR*) are difficult to measure separately and reported measurements of "fog" often represent a combination of fog and *WDR*. In this chapter the term "horizontal precipitation" (*HP*) is used instead. Understanding of "typical" amounts of *HP* intercepted by different types of cloud forest is hampered by a lack of comparative information on local fog climatology. Usually some kind of "fog gage" is used to characterize fog occurrence and amounts. Collection efficiencies of three passive fog gages, viz. a wire harp, a modified cylindrical gage (Juvik-type), and a tunnel gage (Daube-type), were derived by comparing the volumes of water collected by the respective gages with cloud water fluxes derived from fog liquid water content (*LWC*) as measured by a cloud particle spectrometer during conditions of fog at a windward cloud forest site in northern Costa Rica. The collection efficiency of the three gages proved linearly related to the horizontal cloud water flux as measured by the gages themselves. Therefore, no additional information on wind speed, droplet size, and fog *LWC* was needed. During conditions of *HP*, relative collection efficiencies were derived by comparing the volumes collected by the respective gages. The modified Juvik gage had an efficiency close to 100%, independently of wind speed and direction. The efficiency of the wire harp depended critically on wind speed, whereas the tunnel gage collected additional precipitation at small precipitation angles and low wind speeds.

INTRODUCTION

In montane tropical areas below the snowline, precipitation (*P*) is normally composed of drizzle and rain. Precipitation may fall under an angle due to wind, with drizzle (drop size range 100–500 μm, fall velocities 25–200 cm s^{-1}) being more susceptible to wind than rain (drop sizes > 500 μm, fall velocities > 200 cm s^{-1}; Pruppacher and Klett, 1978). Fog consists of small droplets (<100 μm) suspended in the air that are easily transported by wind and atmospheric turbulence. As such, horizontally transported wind-driven rain (*WDR*) and fog (*F*) are typically not included in conventional rainfall measurements which mostly represent the vertical (downward) component of precipitation (*VP*) only. The sum of *WDR* and *F* may be called "horizontal precipitation" (*HP*) and is usually measured using some kind of vertical trapping surface (Bruijnzeel *et al.*, 2005). Amounts of fog settling onto a surface through sedimentation are usually very small compared to turbulence-driven fluxes, with the relative importance of the two processes depending

Tropical Montane Cloud Forests: Science for Conservation and Management, eds. L. A. Bruijnzeel, F. N. Scatena, and L. S. Hamilton. Published by Cambridge University Press.

on the size distribution of the fog droplets and the strength of turbulent mixing (Beswick *et al.*, 1991; Vermeulen *et al.*, 1997).

Tropical montane cloud forest (TMCF) and other areas frequently immersed in fog have been reported to intercept significant amounts of fog, particularly during periods of low rainfall (see successive reviews by Bruijnzeel and Proctor, 1995; Bruijnzeel, 2005; cf. Bruijnzeel *et al.*, this volume). Quantification of cloud water interception (CWI) by tall vegetation in complex mountain terrain is not straightforward and several approaches – ranging from the use of simple passive fog gages (Cavelier *et al.*, 1996; Ataroff, 1998) to the comparison of gross and net precipitation (Holder, 2003) – have been attempted, along with modeling (Slinn, 1982; Lovett, 1984; Walmsley *et al.*, 1996) and, more recently, direct measurement of fog deposition using eddy covariance methods (Beswick *et al.*, 1991; Kowalski and Vong, 1999; Burkard *et al.*, 2002; Eugster *et al.*, 2006; Holwerda *et al.*, 2006; Beiderwieden *et al.*, 2008; cf. Schmid *et al.*, this volume).

Information on fog liquid water content (*LWC*) in areas with TMCF is scarce, although mean values of standard variables like wind speed, temperature, and humidity are becoming more widely available (Jarvis and Mulligan, this volume). Therefore, amounts of fog captured by passive fog gages and expressed as horizontal cloud water fluxes (*CWF*; mm m^{-2} $hour^{-1}$) may serve as a relatively simple measure for site comparisons and model input. With the exception of a few areas such as the Canary Islands (García-Santos, 2007; Ritter *et al.*, 2008; cf. Marzol-Jaén, this volume; Marzol-Jaén *et al.*, this volume), Puerto Rico (Baynton, 1969; Eugster *et al.*, 2006; Holwerda *et al.*, 2006; Holwerda *et al.*, this volume #28 and #29), the Hawaiian Islands (Juvik and Ekern, 1978; Delay and Giambelluca, this volume; Giambelluca *et al.*, this volume), and the Monteverde area of northern Costa Rica (Clark *et al.*, 1998; Häger and Dohrenbusch, this volume; Lawton *et al.*, this volume), there is a distinct lack of comparative information on spatial variations in local fog climatology. This hampers our understanding of the "typical" amounts of fog intercepted by different types of cloud forest (cf. Bruijnzeel, 2005).

More fundamentally, comparatively little is known of the (relative) performance of such widely used passive fog gages as the wire harp (Goodman, 1985), the cylindrical louvered shade screen (Juvik and Ekern, 1978), or the "standard fog collector" (Schemenauer and Cereceda, 1994; cf. Juvik and Nullet, 1995) and modifications thereof (Marzol, 2002), particularly under the conditions of low-intensity rain and strong winds observed on many tropical mountain slopes and ridges (cf. Häger and Dohrenbusch, this volume; Holwerda *et al.*, this volume #28; Giambelluca *et al.*, this volume; Tanaka *et al.*, this volume).

Schemenauer and Joe (1989) inferred the collection efficiency of large (48-m^2) polypropylene screens from the difference in atmospheric *LWC* measured in front of and behind the screen and found an average efficiency of 66% at wind speeds between 3.5 and 6.5 m s^{-1}, dropping to 26% at wind speeds <2 m s^{-1}. The volume of fog water actually collected by the screen was 2.9 times lower than predicted from these efficiencies, implying that the actual efficiency was closer to 20%. Schemenauer and Cereceda (1994) reported the same efficiency for a 1-m^2 screen made of the same mesh material (two layers) as for the large 48-m^2 screens, whereas a quarter-sized screen (0.25-m^2) collected less due to the use of a single mesh layer. Holwerda *et al.* (this volume #28) found comparable amounts of water were collected in Puerto Rico by a single-layered quarter-sized polypropylene screen and a quarter-sized wire harp (cf. Goodman, 1985).

This chapter investigates the collection efficiencies of a quarter-sized wire harp, a modified Juvik-type gage (see below), and a tunnel-type gage (Daube *et al.*, 1987) under the wet and windy conditions prevailing on the Atlantic side of the Continental Divide in northern Costa Rica. The performance of the respective gages was determined both during conditions of fog-only, and during times of low-intensity wind-driven rain and strong wind (i.e. *HP*). Measurements per gage type were not replicated.

MATERIALS AND METHODS

Study site

The experiment was conducted at (or just below, depending on the instrument under consideration) the top of a 24-m high tower extending above the canopy of a *c.* 20 ± 2 m tall windward lower montane cloud forest at 1460 m.a.s.l., ~10 km NE of the town of Santa Elena in northern Costa Rica (10° 21′ 33″ N, 84° 48′ 5″ W). The tower was located on the east-facing slope of a fully forested north–south orientated ridge, halfway between the ridge top (40 m) and the valley. The tower location was subject to the predominantly north-easterly winds, although the flow was locally deflected southward as an upslope valley wind. A general description of the climatic conditions in the study area has been given by Clark *et al.* (2000), whereas additional climatic and site information can be found in Tobón *et al.* (this volume #26) and Schmid *et al.* (this volume). Fog gage measurements were made between September 2002 and August 2004.

Instrumentation

A cloud droplet spectrometer (FM-100, Droplet Measurement Technologies Inc., Boulder, CO, USA) was in operation at 26.5 m between 20 February and 13 May 2003 and was used for the evaluation of fog *LWC* (mg m^{-3}). The FM-100 allows the counting and characterization of fog droplets with diameters of 2–50 μm in up to 40 size classes (Burkard *et al.*, 2002). Values of *LWC* for each size class were computed, based on the geometric mean volume of spherical droplets derived from the log-normal distribution of size counts.

Figure 30.1. (a) Wire harp and modified Juvik-type fog gage, and (b) Daube-type tunnel gage as used in the comparative fog gage experiment at the San Gerardo cloud forest site, northern Costa Rica. Note anemometer in the background.

A modified Juvik-type fog gage (*MJU*), consisting of a louvered cylindrical aluminum shade screen of 40.5 cm height and a cross-sectional area of 540 cm^2 (Juvik and Ekern, 1978), was installed at 22 m (i.e. 3 m above the top of the adjacent canopy) for the measurement of *HP*. The modification consisted of an added funnel at the top of the gage to allow for the separate measurement of vertical and horizontal fluxes (Figure 30.1a). No attempt was made to shield the gage as has been done by others to prevent precipitation from entering (e.g. McJannet *et al.*, 2007), in view of the large inclination of the prevailing low-intensity rain and strong winds (cf. Clark *et al.*, 1998; Häger and Dohrenbusch, this volume). The resolution of the gage was 0.07 mm.

A 50 × 50 cm wire harp (*WH*) with 0.8-mm diameter nylon strings spaced 2 mm apart was placed next to the *MJU*, with its orientation about normal to the main fog-carrying wind direction (Figure 30.1a). Both effective cross-sectional area and collection efficiency of the *WH* depend on wind direction (cf. Friedlander, 1977). As a result, measurement resolution varied somewhat, but was typically 0.015 mm for winds normal to the screen.

A two-stage tunnel-type gage (*TTG*; Daube *et al.*, 1987) was installed at 26 m and operated from October 2003 onward for the measurement of *HP*. The *HP* passing the 610-cm^2 intake was separated into *WDR* (collected in the first stage by impaction on a 45° ramp) whereas the smaller fog droplets were led on to the second stage where they impacted on two wire harps (having two rows of nylon strings and 625 cm^2 each) placed in series at an angle to facilitate droplet drainage by wind drag along the strings (Figure 30.1b).

Derivation of fog collection efficiencies

Collection efficiencies of the respective types of fog gages were derived separately for conditions of fog-only and for conditions of *HP* (*WDR* plus fog). Derivation of the collection efficiency of a fog gage depends critically on the calibration of the tipping buckets used to record the catch. Separate calibrations were carried out in the laboratory as well as in the field for flow rates of 1–11 ml min^{-1} which converted to 1.1–12.2 mm $hour^{-1}$ for the *MJU* and 1.2–12.9 mm $hour^{-1}$ for the *TTG*. It was assumed that the calibrations were also valid for flow rates outside the specified range because no significant differences were found between the extremes of the tested flow rates.

Collection efficiencies of the respective gages were determined for 30-min sampling periods, with collected volumes converted to horizontal cloud water flux (*CWF*, in mm m^{-2} 30 min^{-1}) by normalizing the collected water by the corresponding cross-sectional area (m^2) of the gage. The spectrometer-based fog *LWC* data (mg m^{-3}), which served as the reference, were converted to *CWF* by multiplying them times wind speed (m s^{-1}) over the duration of the sampling period (30 min). Efficiencies of the *WH* and *MJU* for conditions of fog-only were derived from corresponding *LWC* data as measured by the FM-100, whereas that for the *TTG* was derived indirectly by comparison with the *MJU* (corrected for efficiency), because the *TTG* was installed only after termination of the experiment with the FM-100. Relative collection efficiencies for conditions with *HP* were derived by comparing the volumes collected by the three gage types under varying climatic conditions. Although the FM-100 was installed 4.5 m above the *WH* and the *MJU*, it was assumed that, on average, the difference in *LWC* between the two heights was negligible (cf. Burkard *et al.*, 2002).

Fog *LWC* values as measured with the FM-100 were selected on the basis of the steady cloud criterion introduced by Vong and Kowalski (1995), using a value of 0.5 instead of 0.3. This

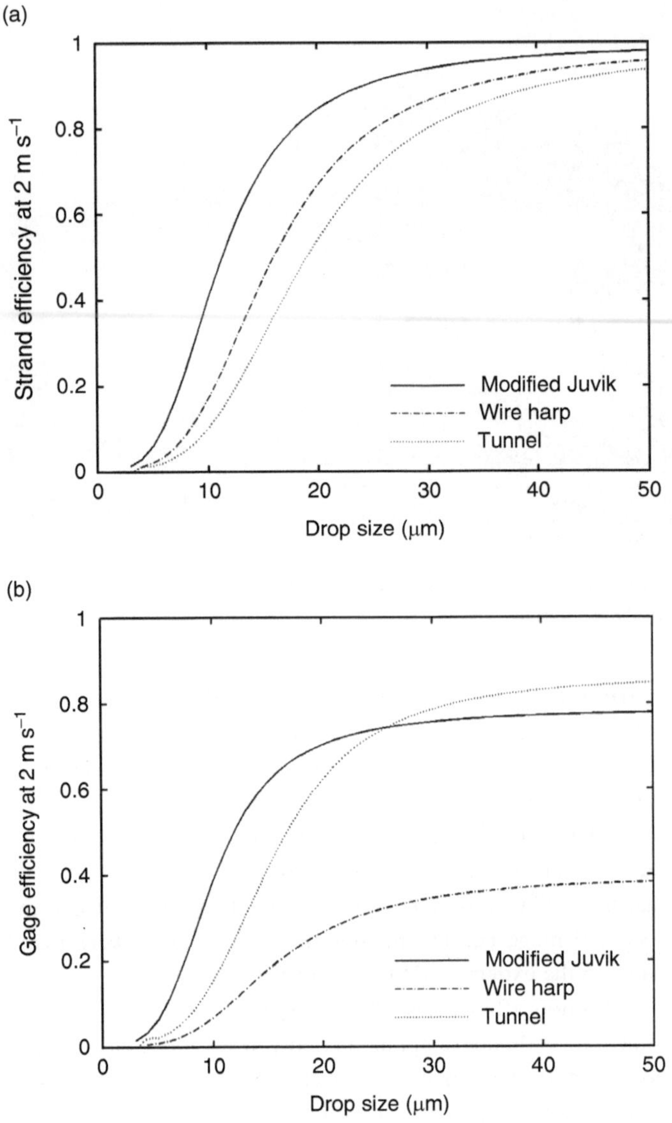

Figure 30.2. Theoretical catch efficiency as a function of drop size for the strands of the three types of fog gages used in the experiment (a); (b) the same extrapolated to the entire gage. All values calculated for a wind speed of 2 m s^{-1}.

criterion was also applied to the fog gage records to minimize gage wetting and drying effects during short, intermittent fog events. Sampling periods for fog-only were selected from fog events exceeding 1.5 hours duration and having no precipitation (as determined by an adjacent rainfall recorder, 0.2 mm resolution). Fog was assumed to occur whenever the catch by the *MJU* exceeded 0.14 mm $hour^{-1}$ (i.e. twice the resolution of the gage). This value corresponds with the detection limit of the device for a fog having an *LWC* of 100 mg m^{-3} and passing at a speed of 1 m s^{-1} (assuming a collection efficiency of 40%). When the rain gage record showed a 0.2 mm tip at more or less regular intervals that exceeded 1 hour, it was taken to represent fog deposition rather than rain. Sampling periods with *HP* were selected using a lower limit of 1 mm $hour^{-1}$ (definition of light rain) as measured for *HP* and *VP*. Finally, an additional selection criterion was applied for wind direction (N310E–N030E) and wind speed (>1 m s^{-1}) to ensure a minimum amount of unobstructed flow over a relatively homogeneous canopy to the fog gages and the spectrometer.

RESULTS AND DISCUSSION

Theoretical collection efficiencies of different fog gages

Based on theoretical considerations (Langmuir and Blodgett, 1946), a single strand of the *MJU* material was predicted to be more efficient for fog droplet impaction than the *WH* or the *TTG* at low wind speed (2 m s^{-1}) and small drop sizes (Figure 30.2a). The efficiency curves shifted toward the left at higher wind speeds (not shown), reflecting increased impaction efficiency for the same drop size due to the associated increase in momentum. In reality, some of the examined fog gages rather represented multiple strands. The four rows of strands of the *TTG* increased the actual fraction of air sampled compared to the single row of the *WH*. Also, the *MJU* presented a more complicated case due to its cylindrical shape, but for the purpose of comparison it was assumed to consist of two rows of strands having a surface area equal to half that of the cylinder. Following Friedlander (1977) the fraction of air sampled by the *WH*, *MJU*, and *TTG* would then be 40%, 79%, and 87%, respectively. The theoretical collection efficiencies (η) of the respective gages are compared in Figure 30.2b as a function of fog droplet size for a wind speed of 2 m s^{-1}. These were calculated from $\eta = 1 - (1 - \eta_s * fr)^r$, where η_s is the efficiency of a single strand (Figure 30.2a), *r* the number of intercepting rows, and *fr* the fractional coverage of each row.

Figure 30.3 shows the theoretical dependence of the efficiency of the *MJU* on wind speed for three selected drop sizes. Gage efficiency increased with wind speed but leveled off quickly for winds speeds between 2 and 4 m s^{-1}. The drop-size spectrum of the fog at the study site was nearly normally distributed (mean kurtosis 0.256 ± 0.007, mean skewness 0.064 ± 0.035), with an average median droplet diameter of 15.1 ± 1.35 μm. Therefore, much of the fog's *LWC* had a droplet size larger than the ~10 μm that marked the 50% collection efficiency level for the *MJU*, even at a low wind speed of 2 m s^{-1} (Figure 30.3).

Fog gage efficiency during conditions of fog-only

MODIFIED JUVIK GAGE AND WIRE HARP

Application of the selection criteria described above allowed the collection efficiencies of the *MJU* and *WH* to be examined for 66 and 16 half-hourly records, respectively. The cloud water flux

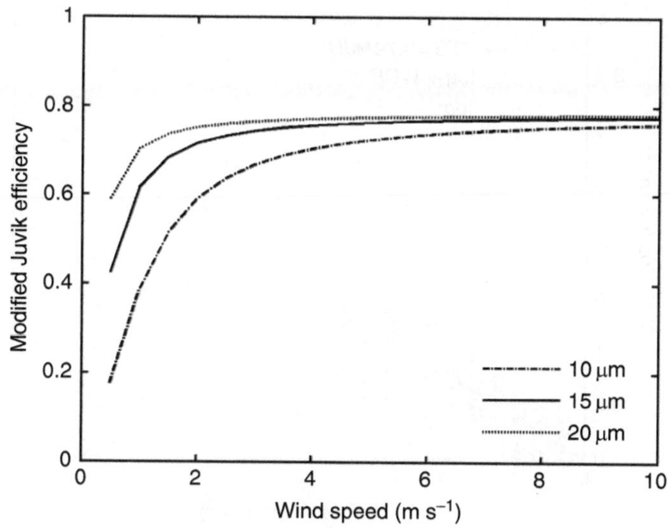

Figure 30.3. Theoretical catch efficiency curves derived for the modified Juvik-type gage as a function of wind speed and droplet size.

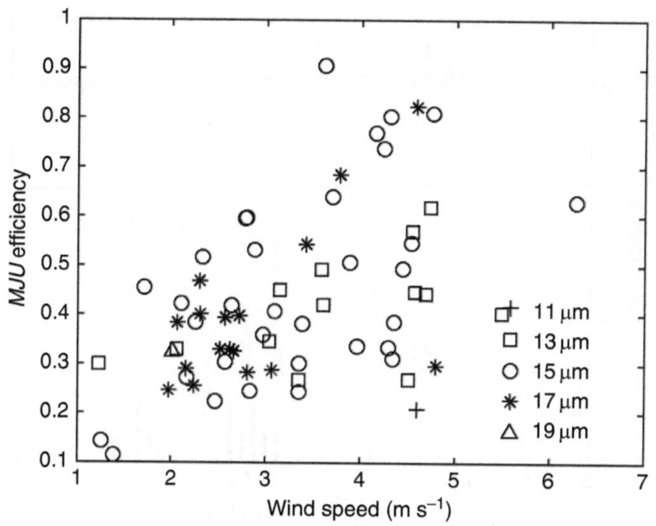

Figure 30.5. Collection efficiency (C_{eff}) of the modified Juvik-type gage as a function of wind speed and droplet size.

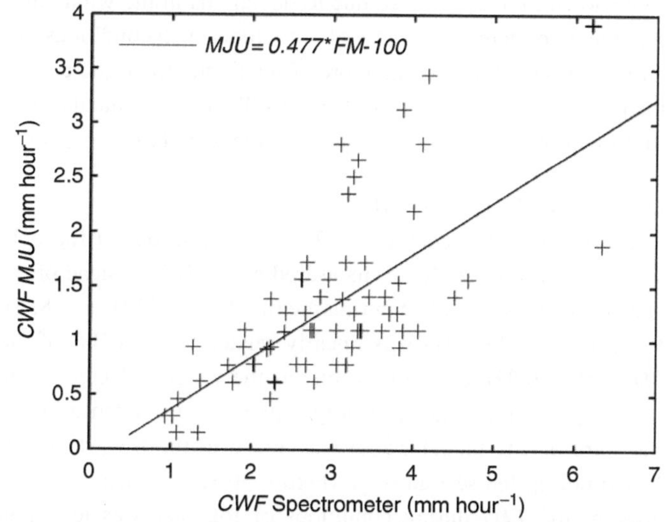

Figure 30.4. Cloud water flux (*CWF*, in mm m^{-2}) collected by the modified Juvik-type gage vs. that of the spectrometer.

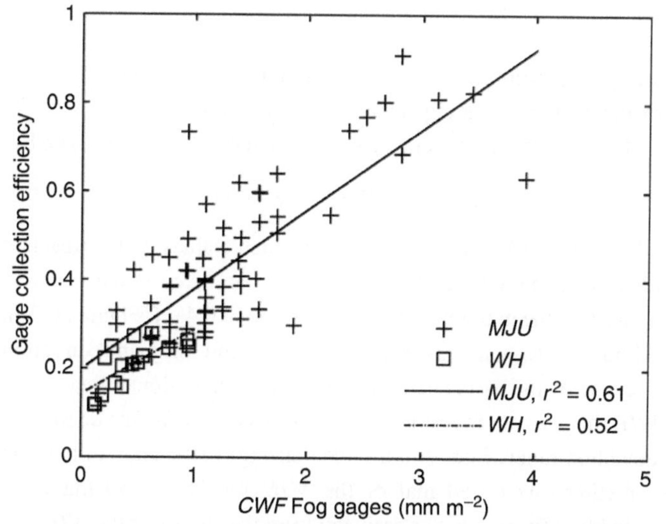

Figure 30.6. Collection efficiencies (C_{eff}) of the modified Juvik-type and wire harp gages as a function of the cloud water flux (*CWF*, in mm m^{-2})) measured by the respective gages.

CWF of the *MJU*HP proved to be linearly related with that based on the spectrometer (Figure 30.4). An average collection efficiency of 47.7% was found for the *MJU*, although efficiency varied widely (Figure 30.4), depending on wind speed and fog droplet size as illustrated further by Figure 30.5. The remaining scatter is probably due to variations in *LWC* because events with a low *LWC* tend to produce lower collection efficiency values because of the larger relative losses by evaporation from the gage surface as well as delayed drainage during wetting up of the gage.

Naturally, the *CWF* as measured by the gage itself integrates the combined effects of variations in wind speed, fog droplet size, and *LWC* on collection efficiency, and simple linear relationships between *CWF* and gage efficiency were found, both for the *MJU* and the *WH* (Figure 30.6).

The collection efficiency of the *MJU* (C_{effMJU}) is given by:

$$C_{eff_{MJU}} = 0.36 * C_t * CWF_{MJU} + 0.2 \qquad (30.1)$$

where C_t is a correction factor equal to 30/sampling period duration (min) and *CWF* is expressed in mm m^{-2} per sampling period (in minutes). The slope of Eq. (30.1) is 0.36 ± 0.07 (at 95% confidence) and the offset 0.2 ± 0.06, with $r^2 = 0.61$. The maximum collection efficiency value derived from the selected data was ~0.9, although values appeared to plateau at a value of 0.8 (Figure 30.6), i.e. similar to the theoretical value predicted in Figure 30.2b. Thus, for conditions of fog-only, Eq. (30.1) is valid up to *MJU* (C_{effMJU} = 0.8. However, for higher flux values, fog

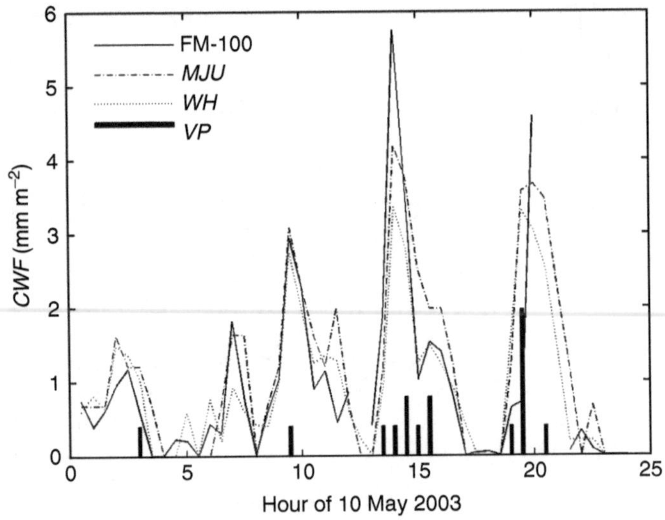

Figure 30.7. Comparison of the cloud water flux (*CWF*, in mm m^{-2}) as measured by the spectrometer vs. those measured by the modified Juvik-type gage and wire harp gages (values corrected for gage collection efficiency). Data are for 10 May 2003.

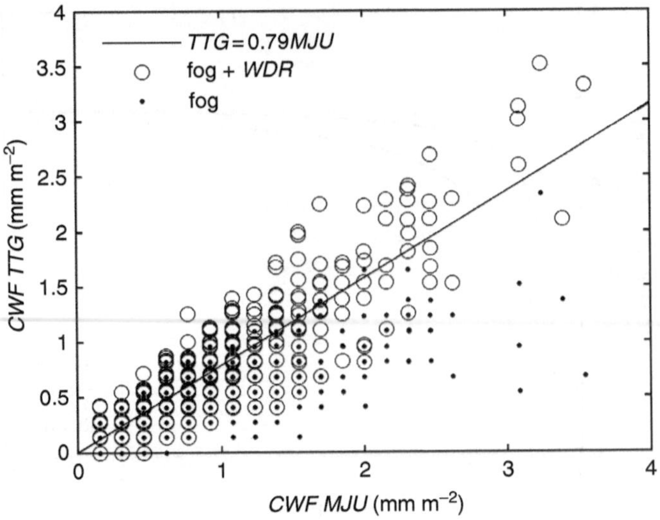

Figure 30.8. Cloud water flux (*CWF*, in mm m^{-2}) as measured by the Daube tunnel gage vs. that measured by the modified Juvik-type gage.

and precipitation tend to be mixed and the collection efficiency of the *MJU* may be expected to approach unity.

The collection efficiency for the wire harp (C_{effWH}) is given by:

$$C_{eff_{WH}} = 0.29{*}C_t{*}CWF_{WH} + 0.14 \quad (30.2)$$

where Eq. (30.2) is valid for efficiency values ≤ 0.3 (reached already at low *CWF*; Figure 30.6). The observed maximum was again near the maximum theoretical value of 0.4 (cf. Figure 30.2b). Although (C_{effWH} was expected to depend on wind direction because of the associated reduction in cross-sectional area of the *WH*, this was not investigated due to a lack of data outside the 20° deviation range. Furthermore, the collection efficiency of the *WH* sometimes exceeded that of the *MJU* for $LWC < 50$ mg m^{-3}, possibly reflecting better drainage from the strings of the *WH*.

The fluxes measured with the *MJU* and *WH* after correction for gage efficiency closely matched the *CWF* derived with the FM-100, as shown in Figure 30.7 for 10 May 2003.

As a further check of the two collection efficiency equations, the dependence of the ratio of the corrected *CWFs* for the *WH* and the *MJU* on wind speed was examined for conditions of fog-only (n = 883) between 1 January and 1 July 2004. No such dependence was found, suggesting that any dependencies of gage efficiency on wind speed were corrected for in the same manner for the two gages. Neither did the ratios of the corrected *CWFs* depend on the magnitude of the *CWF* itself, the average ratios being 1.11 ± 0.57 for low *CWF* values ($\leq$0.5 mm m^{-2} per 30 min), 0.98 ± 0.29 for intermediate *CWF* values (0.5–1 mm m^{-2} per 30 min), and 1.06 ± 0.16 for high *CWF* values (1–1.5 mm m^{-2} per 30 min). Occasionally, high ratios were found during periods of high *CWF*, making the *WH* seemingly more efficient (not shown), but these were due to deviations in the wind angle relative to the screen in excess of 20°. Under such conditions, the harp is expected to become more efficient due to an increase in the percentage of air sampled (from *c.* 40% at perpendicular flow to 46% at 30° angle, for example; Friedlander, 1977).

DAUBE TUNNEL GAGE

The collection efficiency of the *TTG* for conditions of fog-only was derived from the fluxes measured by the *MJU* instead of the spectrometer as measured during the first half of 2004 (n = 885). Collection by the *TTG* was linearly related to the *CWF* of the *MJU* (r^2 = 0.82), but only as long as the water collected in the first stage of the gage (probably representing condensation of moisture on the tunnel walls) was added to the fog water collected during the second stage (Figure 30.8). As such, performance of the *TTG* during conditions of fog-only was less than optimal. As found earlier for the *MJU* and *WH*, the collection efficiency of the *TTG* appeared to be linearly related to its own *CWF* (Figure 30.9).

The collection efficiency of the *TTG* (C_{effTTG}) is given by:

$$C_{eff_{TTG}} = 0.40{*}C_t{*}CWF_{TTG} + 0.16. \quad (30.3)$$

The slope (at 95% confidence) is 0.40 ± 0.02 and the offset 0.16 ± 0.01, with r^2 = 0.70. As found earlier for the *MJU*, maximum theoretical and experimental collection efficiencies of the *TTG* were about 0.8 (cf. Figures 30.2b and 30.9), suggesting that Eq. (30.3) is valid up to this value.

Fog gage efficiency during conditions of wind-driven rain

Because *WDR* made up 86% of total *HP* at the study site (K. F. A. Frumau, unpublished data), fog gage efficiencies were also

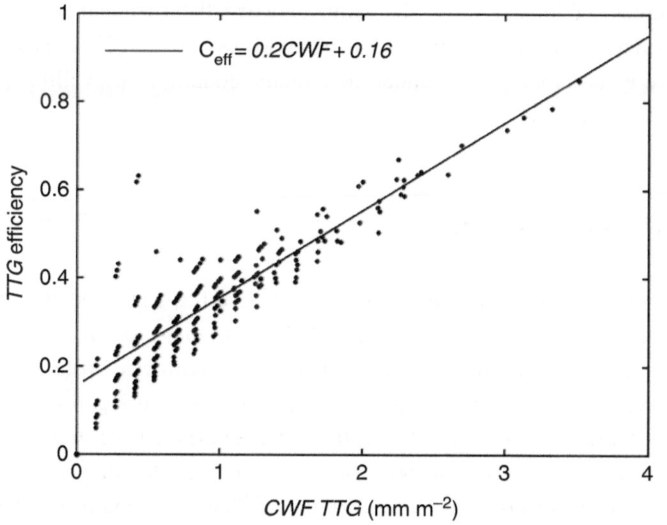

Figure 30.9. Collection efficiency (C_{eff}) of the Daube tunnel gage as a function of the cloud water flux (*CWF*, in mm m^{-2}) as measured by the gage itself.

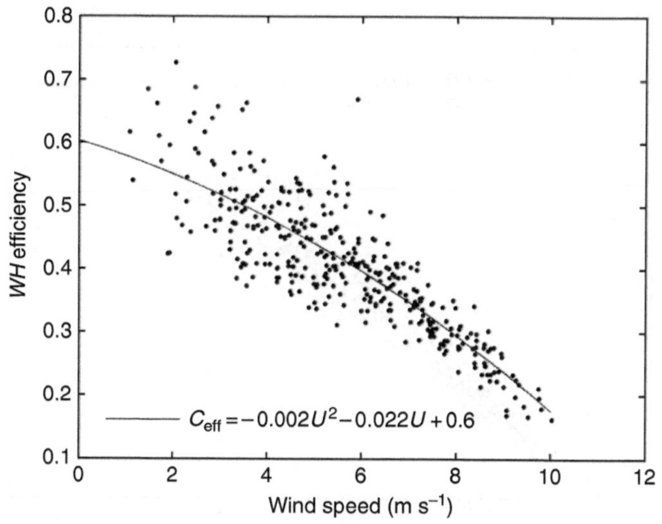

Figure 30.10. Collection efficiency of the wire harp gage as a function of wind speed.

examined under conditions of *HP*. However, because no independent means of evaluating collection efficiencies were available for these conditions, only *relative* efficiencies could be derived. The *MJU* was taken as the reference because of its high efficiency at high flux values (Figure 30.6). To start with, the *MJU* was assumed to represent *HP* without deflecting drizzle- and rain-sized drops due to its permeable surface. Furthermore, the catch of the *MJU* and the *TTG* was supposed to be independent of wind direction because of the cylindrical shape and rotating ability of the respective gages.

MODIFIED JUVIK GAGE

To test for potentially reduced efficiency because of the *MJU*'s permeability and flow deflection, the performance of the *MJU* was compared with that of another modification equipped with a second cylindrical screen of 6.5-cm diameter placed inside the outer cylinder. In this way, gage permeability was reduced, and the trapping surface increased by almost 50%, while maintaining the outward cross-sectional area. The data record was split according to prevailing wind speeds because the doubly modified gage might be subject to potentially larger flow deflection at relatively low wind speeds (1–5 m s^{-1}), whereas the *MJU* might suffer increased losses at higher wind speeds (>5 m s^{-1}) due to its greater permeability and the possibility of droplets being blown off again. Splash losses were assumed negligible for both gages or otherwise to be similar.

The catch ratio of the two gages (*MJU*/doubly modified gage) was essentially unity (0.999, $r^2 = 1.00$, $n = 73$) for the observed intensity range for *HP* of 1–22 mm hour^{-1}. At wind speeds exceeding 5 m s^{-1}, the doubly modified gage caught marginally more water than did the *MJU* (ratio 0.994, $r^2 = 0.99$, $n = 171$). Similar results were obtained for slightly different wind speed criteria. Apparently, the potentially larger flow deflection effects of the doubly modified gage were either offset by its higher surface collection efficiency, or deflection was insignificant for *WDR* due to the latter's high momentum. The very small difference with the *MJU* lends further confidence to the assumption of 100% efficiency for this gage during times of *HP*, independently of wind speed and direction.

WIRE HARP

The collection efficiency of the *WH* (taking the *MJU* as a reference) during times of *HP* was found to decrease strongly with increasing wind speed (Figure 30.10). Droplets intercepted by the *WH* were blown off again because of drag by wind and vibration of the strands (cf. Joslin *et al.*, 1990; Holwerda *et al.*, this volume #28)). The higher collection efficiency observed at low wind speeds compared with the results for conditions of fog-only are considered to have been caused by additional catch of *WDR* by the collection trough of the harp.

TUNNEL GAGE

The fluxes measured with the *TTG* and *MJU* under conditions of *HP* appeared to be linearly related and almost equal (Figure 30.11a), confirming the earlier assumption of 100% efficiency for the two gages (unless environmental effects affect both gages similarly). However, the flux indicated by the *TTG* was slightly larger than that of the *MJU* for *CWF* values between 2 and 8 mm m^{-2}. These larger *CWF* values were only found during times of low wind speeds and the two *CWF*s converged again to similar values at higher wind speeds (Figure 30.11a). Close inspection of the *TTG* frontal opening revealed it to be inclined slightly backward and this allowed some of the vertical component of precipitation to enter as well. This

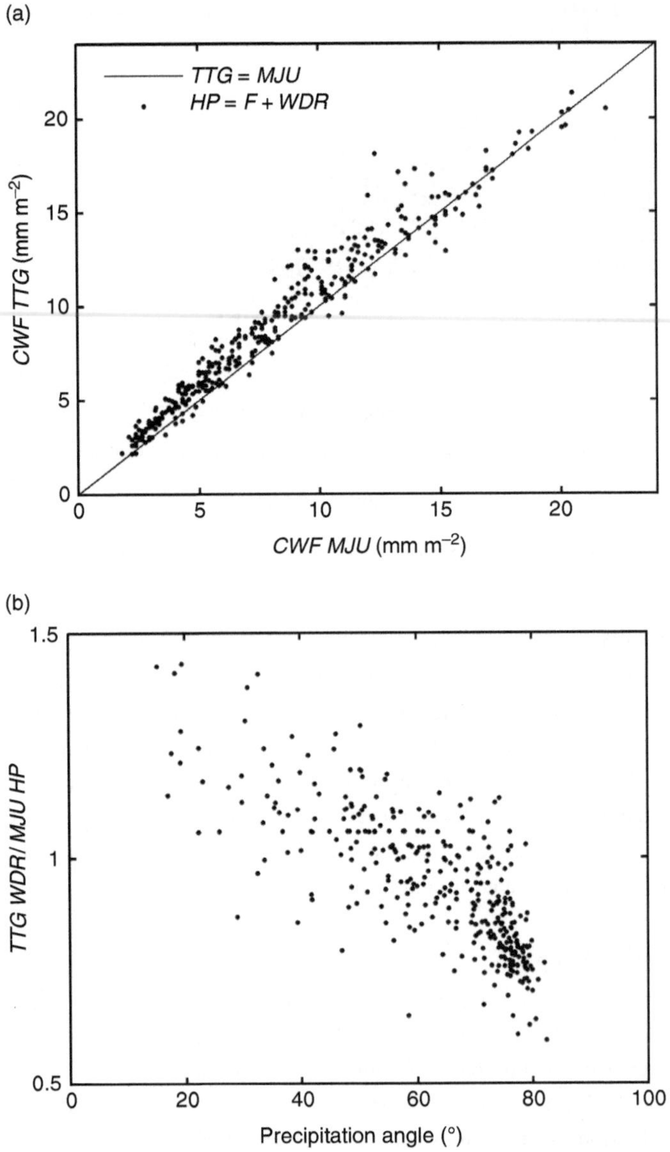

Figure 30.11. (a) Cloud water flux (*CWF*, in mm m^{-2}) as measured by the Daube tunnel gage vs. that measured by the modified Juvik-type gage; (b) collection efficiency of the tunnel gage as a function of precipitation angle.

effect should be reflected in the catch of *WDR* by the *TTG* as illustrated in Figure 30.11b.

The precipitation angle may be derived using simple trigonometry from the vertical and horizontal components of precipitation as measured separately with the *MJU* (Frumau *et al.*, 2006). The collection efficiency of the *TTG* decreased with increase in precipitation angle from the vertical, reflecting the smaller additional catch of the vertical component. The strong decrease in gage efficiency near rain angles of 80° is caused by increased amounts of *WDR* being forced up the ramp toward the second stage at high wind speeds. This was facilitated also by the internal enhancement of the flow due to a 25% reduction in cross-sectional area at the ramp between the first and second stages. Thus, the separation of fog and *WDR* by the *TTG* appears to be less than perfect under the climatic conditions prevailing at the study site.

CONCLUSIONS

The collection efficiencies of three passive fog gages (Modified Juvik-type gage *MJU*, wire harp *WH*, and tunnel-type gage *TTG*) as used at a windward cloud forest site in northern Costa Rica were evaluated during times of fog-only, and during times of wind-driven rain plus fog (*HP*). Efficiencies during fog-only were shown to depend on wind speed, fog droplet size and fog liquid water content. However, gage efficiencies could also be approximated using linear relationships with the horizontal cloud water flux (*CWF*) as measured by the respective collector types themselves. Collection efficiencies of the *MJU* and *TTG* gradually increased to 0.8 with increases in *CWF*, in accordance with theoretical expectations for larger drop sizes and higher wind speeds. The maximum collection efficiency of the *WH* was only about 0.30, again in accordance with theoretical expectation.

Collection efficiency during conditions of *HP* by the *MJU* proved to be independent of wind speed or direction; also, collection efficiency was very similar to the catch of a similar gage with artificially lowered permeability to air flow. The collection by the *TTG* was linearly related (1 : 1) to that of the *MJU*, suggesting that both gages are 100% efficient under conditions of *HP*, regardless of wind direction. However, the catch of the *TTG* depended indirectly on wind speed via the latter's effect on precipitation angle, due to a slightly reclined frontal opening that allowed some vertical precipitation to enter. The collection efficiency of the *WH* decreased rapidly with increased wind speed, efficiencies being only 20–60% of those derived for the *MJU*.

Summarizing, the *MJU*, *TTG*, and *WH* are all useful for measurement of fog during conditions of fog-only, although the presently derived efficiency for the *WH* is valid only for wind directions normal (±20°) to the collector surface. Measurement of *HP* (fog in combination with *WDR*) can be achieved with the *MJU* and *TTG* but not with the *WH*.

ACKNOWLEDGEMENTS

This work was funded by the British Department for International Development (DFID) as part of the Forestry Research Programme of the Renewable Natural Resources Research Strategy (RNRRS), project no. R7991. The views expressed here are not necessarily those of DFID. Fog liquid water content measurements were made as part of a project funded by the Swiss National Science Foundation under grant no. 2100–068051.02.

REFERENCES

Ataroff, M. (1998). Importance of cloud water in Venezuelan Andean cloud forest water dynamics. In *Proceedings of the 1st International Conference on Fog and Fog Collection*, eds. R. S. Schemenauer and H. A. Bridgman, pp. 25–28. Ottawa, Canada: IDRC.

Baynton, H. W. (1969). The ecology of an elfin forest in Puerto Rico. III. Hilltop and forest influences on the microclimate of Pico del Oeste. *Journal of the Arnold Arboretum* **50**: 80–92.

Beiderwieden, E., V. Wolff, Y. J. Hsia, and O. Klemm (2008). It goes both ways: measurements of simultaneous evapotranspiration and fog droplet deposition at a montane cloud forest. *Hydrological Processes* **22**: 4181–4189.

Beswick, K. M., K. J. Hargreaves, M. W. Gallagher, T. W. Choularton, and D. Fowler (1991). Size-resolved measurements of cloud droplet deposition velocity to a forest canopy using an eddy-correlation technique. *Quarterly Journal of the Royal Meteorological Society* **117**: 623–645.

Bruijnzeel, L. A. (2005). Tropical montane cloud forests: a unique hydrological case. In *Forests, Water and People in the Humid Tropics*, eds. M. Bonell and L. A. Bruijnzeel, pp. 462–483. Cambridge, UK: Cambridge University Press.

Bruijnzeel, L. A., and J. Proctor (1995). Hydrology and biogeochemistry of tropical montane cloud forests: what do we really know? In Tropical Montane Cloud Forests, eds. L. S. Hamilton, J. O. Juvik, and F. N. Scatena, pp. 38–78. New York: Springer-Verlag.

Bruijnzeel, L. A., W. Eugster, and R. Burkard (2005). Fog as an input to the hydrological cycle. In *Encyclopaedia of Hydrological Sciences*, eds. M. G. Anderson and J. J. McDonnell, pp. 559–582. Chichester, UK: John Wiley.

Burkard, R., W. Eugster, T. Wrzesinsky, and O. Klemm (2002). Vertical divergence of fogwater fluxes above a spruce forest. *Atmospheric Research* **64**: 133–145.

Cavelier, J., D. Solis, and M. A. Jaramillo (1996). Fog interception in montane forest across the Central Cordillera of Panama. *Journal of Tropical Ecology* **12**: 357–369.

Clark, K. L., N. M. Nadkarni, D. Schaeffer, and H. L. Gholz (1998). Atmospheric deposition and net retention of ions by the canopy in a tropical montane forest, Monteverde, Costa Rica. *Journal of Tropical Ecology* **14**: 27–45.

Clark, K. L., R. O. Lawton, and P. Butler (2000). The physical environment. In *Monteverde: Ecology and Conservation of a Tropical Cloud Forest*, eds. N. M. Nadkarni and N. T. Wheelwright, pp. 15–34. Oxford, UK: Oxford University Press.

Daube, B., K. D. Kimball, P. A. Lamar, and K. C. Weathers (1987). Two new ground-level cloud water sampler designs which reduce rain contamination. *Atmospheric Environment* **21**: 893–900.

Demoz, B. B., J. L. Collett, and B. C. Daube (1996). On the Caltech Active Strand Cloudwater Collectors. *Atmospheric Research* **41**: 47–62.

Eugster, W. R.Burkard, F. Holwerda, F. N. Scatena, and L. A. Bruijnzeel (2006). Characteristics of fog and fog-water fluxes in a Puerto Rican elfin cloud forest. *Agricultural and Forest Meteorology* **139**: 288–306.

Friedlander, S. K., (1977). *Smoke, Dust, and Haze: Fundamentals of Aerosol Behavior*. New York: John Wiley.

Frumau, K. F. A., L. A. Bruijnzeel, and C. Tobón (2006). *Hydrological Measurement Protocol for Montane Cloud Forests*, Annex 2, Final Technical Report on DFID-FRP Project No. R7991. Amsterdam: VU University, and Aylesford, UK: Forestry Research Program of the UK Department for International Development.

García-Santos, G. (2007). *An ecohydrological and soils study in a subtropical montane cloud forest in the National Park of Garajonay, La Gomera, (Canary Islands, Spain)*. Ph.D. thesis, VU University Amsterdam, Amsterdam, the Netherlands. Also available at www.falw.vu.nl/nl/onderzoek/earth-sciences/geo-environmental-science-and-hydrology/hydrology-dissertations/index.asp.

Goodman, J. (1985). The collection of fog drip. *Water Resources Research* **21**: 392–394.

Holder, C. D. (2003). Fog precipitation in the Sierra de las Minas Biosphere Reserve, Guatemala. *Hydrological Processes* **17**: 2001–2010.

Holwerda, F., R. Burkard, W. Eugster, *et al.* (2006). Estimating fog deposition at a Puerto Rican elfin cloud forest site: comparison of the water-budget and eddy covariance methods. *Hydrological Processes* **20**: 2669–2692.

Joslin, J. D., S. F. Mueller, and M. H. Wolfe (1990). Tests of models of cloudwater deposition to forest canopies using artificial and living collectors. *Atmospheric Environment* **24**: 3007–3019.

Juvik, J. O., and P. C. Ekern (1978). *A Climatology of Mountain Fog on Mauna Loa, Hawai'i Island*, Technical Report No. 118. Honolulu, HI: Water Resources Research Center, University of Hawai'i.

Juvik, J. O., and D. Nullet (1995). Comments on "a proposed standard fog collector for use in high elevation regions." *Journal of Applied Meteorology* **34**: 2108–2110.

Kowalski, A. S., and R. J. Vong (1999). Near-surface fluxes of cloud water evolve vertically. *Quarterly Journal of the Royal Meteorological Society* **125**: 2663–2684.

Langmuir, I., and K. B. Blodgett (1946). *A Mathematical Investigation of Water Droplet Trajectories*, Army Air Forces Technical Report No. 5418. Washington, DC: U.S. Army Air Forces Headquarters, Air Technical Service Command.

Lovett, G. M. (1984). Rates and mechanisms of cloud water deposition to a subalpine balsam fir forest. *Atmospheric Environment* **18**: 361–371.

Marzol, M. V. (2002). Fog water collection in a Rural Park in the Canary Islands (Spain). *Atmospheric Research* **64**: 239–250.

McJannet, D., J. Wallace, and P. Reddell (2007). Precipitation interception in Australian tropical rainforests. II. Altitudinal gradients of cloud interception, stemflow, throughfall and interception. *Hydrological Processes* **21**: 1703–1718.

Pruppacher, H. R., and J. D. Klett. (1978). *Microphysics of Clouds and Precipitation*. Dordrecht, the Netherlands: D. Reidel.

Ritter, A., C. M. Regalado, and G. Aschan (2008). Fog water collection in a subtropical elfin laurel forest of the Garajonay National Park (Canary Islands): a combined approach using artificial fog catchers and a physically based impaction model. *Journal of Hydrometeorology* **9**: 920–935.

Schemenauer, R. S., and P. Cereceda (1994). A proposed standard fog collector for use in high-elevation regions. *Journal of Applied Meteorology* **33**: 1313–1322.

Schemenauer, R. S., and P. I. Joe (1989). The collection efficiency of a large fog collector. *Atmospheric Research* **24**: 53–69.

Slinn, W. G. N. (1982). Predictions for particle deposition to vegetative canopies. *Atmospheric Environment* **16**: 1785–1794.

Vermeulen, A. T., G. P. Wyers, F. G. Romer, *et al.* (1997). Fog deposition on a coniferous forest in the Netherlands. *Atmospheric Environment* **31**: 375–386.

Vong, R. J., and A. S. Kowalski (1995). Eddy-correlation measurements of size-dependent cloud droplet turbulent fluxes to complex terrain. *Tellus Series B* **47**: 331–352.

Walmsley, J. L., R. S. Schemenauer, and H. A. Bridgman (1996). A method for estimating the hydrologic input from fog in mountainous terrain. *Journal of Applied Meteorology* **35**: 2237–2249.

31 The wet-canopy water balance of a Costa Rican cloud forest during the dry season

S. Schmid, R. Burkard
University of Bern, Bern, Switzerland
K. F. A. Frumau, C. Tobón, L. A. Bruijnzeel
VU University, Amsterdam, the Netherlands
R. Siegwolf
Paul Scherrer Institute, Villingen, Switzerland
W. Eugster
ETH Zurich, Zurich, Switzerland

ABSTRACT

Fog deposition, precipitation, throughfall, and stemflow were measured in a windward tropical montane cloud forest near Monteverde, Costa Rica, for a 65-day period during the dry season of 2003. Net fog deposition was measured directly with the eddy covariance method and amounted to 1.2 ± 0.1 mm day^{-1} (mean $\pm$ standard error). Fog water deposition was 4–7% of incident rainfall for the entire period. Stable isotope concentrations ($\delta^{18}O$ and $\delta^{2}H$) were determined in a large number of samples of each water component. Comparisons between direct fog deposition measurements and the results of a mass-balance model using the stable isotopes as tracers indicated that the latter might be a good tool to estimate fog deposition in the absence of direct measurement under many (but not all) conditions. At 506 mm, measured water inputs over the 65 days (fog plus rain) fell short by 46 mm compared to the canopy output of 552 mm (throughfall, stemflow, and evaporation). The discrepancy is attributed to underestimation of rainfall during conditions of high wind.

INTRODUCTION

Montane cloud forests are widely believed to receive significant extra amounts of water to the water budget by the capture of water from passing fog and low cloud. The quantification of this additional water input is of great practical importance where upland watersheds with cloud forests supply water to local and downstream populations (Zadroga, 1981; Brown *et al.*, 1996; Rhodes *et al.*, this volume #45). Fog water inputs are usually quantified indirectly, be it through comparison of rainfall and crown drip during times with and without fog (e.g. Harr, 1982; Sigmon *et al.*, 1989; Hafkenscheid *et al.*, 2002; Holder, 2003), modeling (e.g. Lovett, 1984; Yin and Arp, 1994; Walmsley *et al.*, 1996; Ritter *et al.*, 2008), or, most frequently, by simply using some kind of passive "fog" gage (e.g. Juvik and Ekern, 1978; Goodman, 1985; cf. Bruijnzeel *et al.*, 2005). The drawback of indirect methods is that the quantification of fog deposition is dependent on correct rainfall and net precipitation measurements and by necessity represents net amounts because of the inclusion of unmeasured amounts of water evaporated from the wetted canopy. Reliance on fog gages introduces the problem of separating contributions by fog and (wind-driven or inclined) rain to the overall catch (McJannet *et al.*, 2007; cf. Frumau *et al.*, this volume; Giambelluca *et al.*, this volume; Tanaka *et al.*, this volume). An additional problem is the translation of collected water amounts (by a vertical surface) to effective hydrological input (onto a horizontal plane), and the impossibility of any type of gage to represent a living natural vegetation. The distinction between fog and rain water inputs is crucial, because the latter are – due to the larger droplet sizes involved – deposited

Tropical Montane Cloud Forests: Science for Conservation and Management, eds. L. A. Bruijnzeel, F. N. Scatena, and L. S. Hamilton. Published by Cambridge University Press.

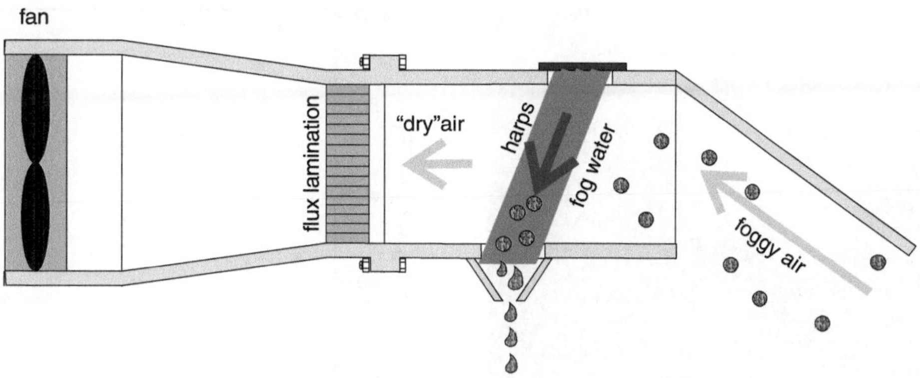

Figure 31.1. Schematic representation of the CASCC fog collector.

independently of the surface cover. Fog water, however, is strongly influenced by turbulence. Because of the rougher and larger surface area of forest canopies, fog water deposition is much larger over forests than over, for example, pasture (e.g. Thalmann *et al.*, 2002).

This chapter presents the results of direct net fog water deposition measurements by means of the eddy covariance method (Beswick *et al.*, 1991; Beiderwieden *et al.*, 2008) as made above a windward montane cloud forest in northern Costa Rica. The results are compared with those obtained with a mass-balance model using the stable isotopes ^{18}O and ^{2}H as tracers (cf. Guswa *et al.*, 2007; Scholl *et al.*, this volume). To further validate the direct fog deposition measurements, the remaining components of the wet-canopy water budget equation (Holwerda *et al.*, 2006a) were measured or calculated and the different approaches evaluated.

MATERIALS AND METHODS

Site description

Hydrological and micro-meteorological measurements were made between 10 February and 13 May 2003 at 1460 m.a.s.l. in the San Gerardo headwater area within the Caño Negro drainage basin on the exposed Atlantic slopes of the Cordillera de Tilarán in northern Costra Rica, *c.* 10 km NE of the town of Santa Elena (10° 21′ 33″ N, 84° 48′ 5″ W). The meteorological measurements were conducted on top of a 24-m tall tower situated about 200 m from the crest of a fully forested 10° slope of N350E exposure. Measurements of net precipitation were made on a nearby slope with a somewhat different exposure and gradient (slope 30°, N270E). The forest was a windward lower montane cloud forest with its main canopy surface at 21–22 m from which epiphyte-laden emergents were sticking out by several meters. Overall epiphyte biomass was determined at 16.2 t ha^{-1} (Köhler *et al.*, 2007). Tree ferns and palms were common in the understory. Long-term rainfall data for the upper Atlantic slopes are not available, but conventionally measured rainfall input between 1 July 2003 and 30 June 2004 at the site was *c.* 6000 mm (K.F.A. Frumau, unpublished data). A somewhat drier period tends to occur in the area between February and April, with monthly rainfall of generally less than 150 mm, vs. 500 mm or more during the remainder of the year (Clark *et al.*, 2000). Due to the prevailing high wind speeds there is a major horizontal precipitation component in the form of wind-driven rain and fog (cf. Clark *et al.*, 1998; Frumau *et al.*, this volume; Häger and Dohrenbusch, this volume). Tobón *et al.* (this volume #26) provide further basic information on site climate.

Methods and instrumentation

Net fog water deposition was measured directly by the eddy covariance technique. The measurements were performed with a three-dimensional ultrasonic anemometer (model 1199 HSE, Gill Ltd., Solent, UK) coupled with an active high-speed FM-100 cloud particle spectrometer (Droplet Measurement Technologies Inc., Boulder, CO, USA). Fog droplets were continuously measured in 40 size classes between 2 and 50 μm diameter and recorded together with the three-dimensional wind speed information at a frequency of 12.5 times per second. For further details on the eddy covariance set-up the reader is refered to Burkard *et al.* (2003). Vertical rainfall was measured by several standard rain gages (100-cm^2 orifice). The precipitation measurements were corrected for aerodynamic losses around the gage due to wind following the procedure of Førland *et al.* (1996), and for the effects of sloping ground according to Sharon (1980). Horizontal precipitation (i.e. fog and wind-driven rain) was measured using a modification of the cylindrical louvered gage of Juvik and Ekern (1978) (cf. Frumau *et al.*, this volume). Throughfall (*TF*) was measured with the roving gage technique using 60 totalizers (cf. Lloyd and Marques, 1988). Stemflow (*SF*) was measured on 30 trees using spiral gutters and made up <1% of uncorrected rainfall (K.F.A. Frumau and C. Tobón, unpublished data). Wet-canopy evaporation was calculated with the wet-canopy version (i.e.

Figure 31.2. Locally built rotating gage with vertical orifice (RGVO) to capture horizontal precipitation.

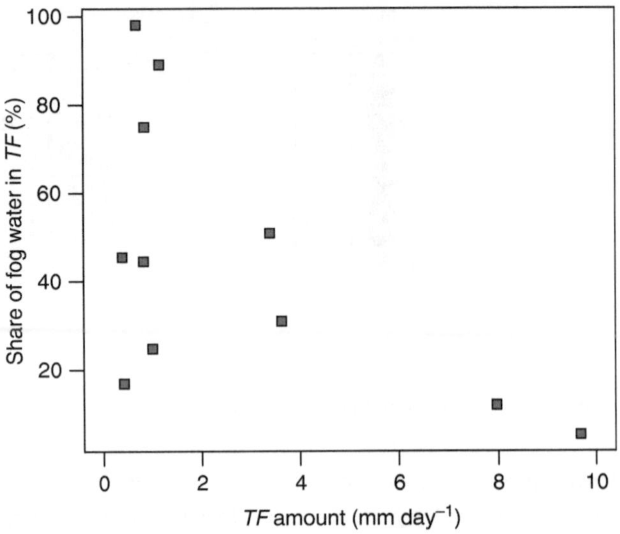

Figure 31.3. Measured throughfall (*TF*) amounts and corresponding fractions of fog water in throughfall as calculated with the mixing model of Eq. (31.1) using ^{18}O as the tracer at the San Gerardo cloud forest site.

surface resistance $r_s = 0$) of the Penman–Monteith equation (Monteith, 1965).

In addition to the eddy covariance measurements, fog water deposition was also estimated with a mixing model as described by Brunel *et al.* (1995), using the stable isotopes ^{18}O and ^{2}H as tracers. Samples for the determination of isotope concentrations were taken on a nominally daily basis from rain, fog, *TF*, and *SF*. Fog was collected with a modified Caltech Active Strand Cloud-water Collector (CASCC; Demoz *et al.*, 1996) with a shielded inlet to avoid raindrops (>100 m) from being collected (Figure 31.1). The teflon strands collect all fog droplets larger than approximately 5–7 μm. Rain water was collected with a sampler built according to the specifications of IAEA (2002) whereas a locally constructed, rotating gage was used to sample wind-driven rain (horizontal precipitation) (Figure 31.2). A representative sub-sample of the water collected by each individual *TF* and *SF* collector was used for isotope analysis. For days on which samples were available from all water types, the share of fog and rain water in *TF* was calculated using the following mixing model containing two end members:

$$f = (\delta_{TF} - \delta_F)/(\delta_P - \delta_F), \quad (31.1)$$

where f is the fraction of rain water in *TF* and the various subscripts of δ denote the respective concentrations of either ^{18}O or ^{2}H in *TF*, fog (*F*) and rain (*P*).

RESULTS AND DISCUSSION

Fog deposition

Average net fog deposition measured by the eddy covariance system for conditions with visibility below 1000 m was 0.05 mm h^{-1} or 14.2 mg m^{-2} s^{-1}. This rate is similar to the fog deposition measured with the same equipment in an elfin cloud forest near Pico del Este in Puerto Rico (0.04 mm h^{-1} or 10.2 mg m^{-2} s^{-1}; Eugster *et al.*, 2006; Holwerda *et al.*, 2006a) and lies within the range of reported values measured with this technique outside the tropics (Beswick *et al.*, 1991; Vong and Kowalski, 1995; Vermeulen *et al.*, 1997; Burkard *et al.*, 2003). The average daily deposition rate of 1.2 mm also lies within the range of 0.2–4.0 mm day^{-1} of reported cloud water interception rates in tropical montane areas as obtained by various indirect methods (Bruijnzeel and Proctor, 1995; Bruijnzeel, 2005). Fog deposition expressed as a percentage of rainfall (4%) was at the low end of the spectrum reported by Bruijnzeel and Proctor (1995) and Bruijnzeel (2005) (4–281% of corresponding rainfall). Whilst the observed daily deposition rates are plausible, the contribution of fog water to total water input proved very small. Based on the direct measurements, fog water added a mere 19 mm of water to the forest water budget during the 65-day measurement period during which an unusually low fog frequency of 26% was observed. The higher than normal position of the cloud base, and the associated lack of fog, were caused by the occurrence of the *temporales del pacifico* weather pattern (cf. Clark *et al.*, 2000). During several days, the field site was climatically speaking situated on the leeward side of the Continental Divide due to the occurrence of the associated warm and dry western winds. According to Clark *et al.* (2000), this type of weather system tends to occur mostly during the hurricane season (August–October) and its occurrence in March 2003 may have led to non-representative conditions. However, Clark *et al.* (2000) also report an average of 20–25% cloud immersion for the upper slopes and ridges along the Continental Divide in the (leeward) area of nearby Monteverde during the

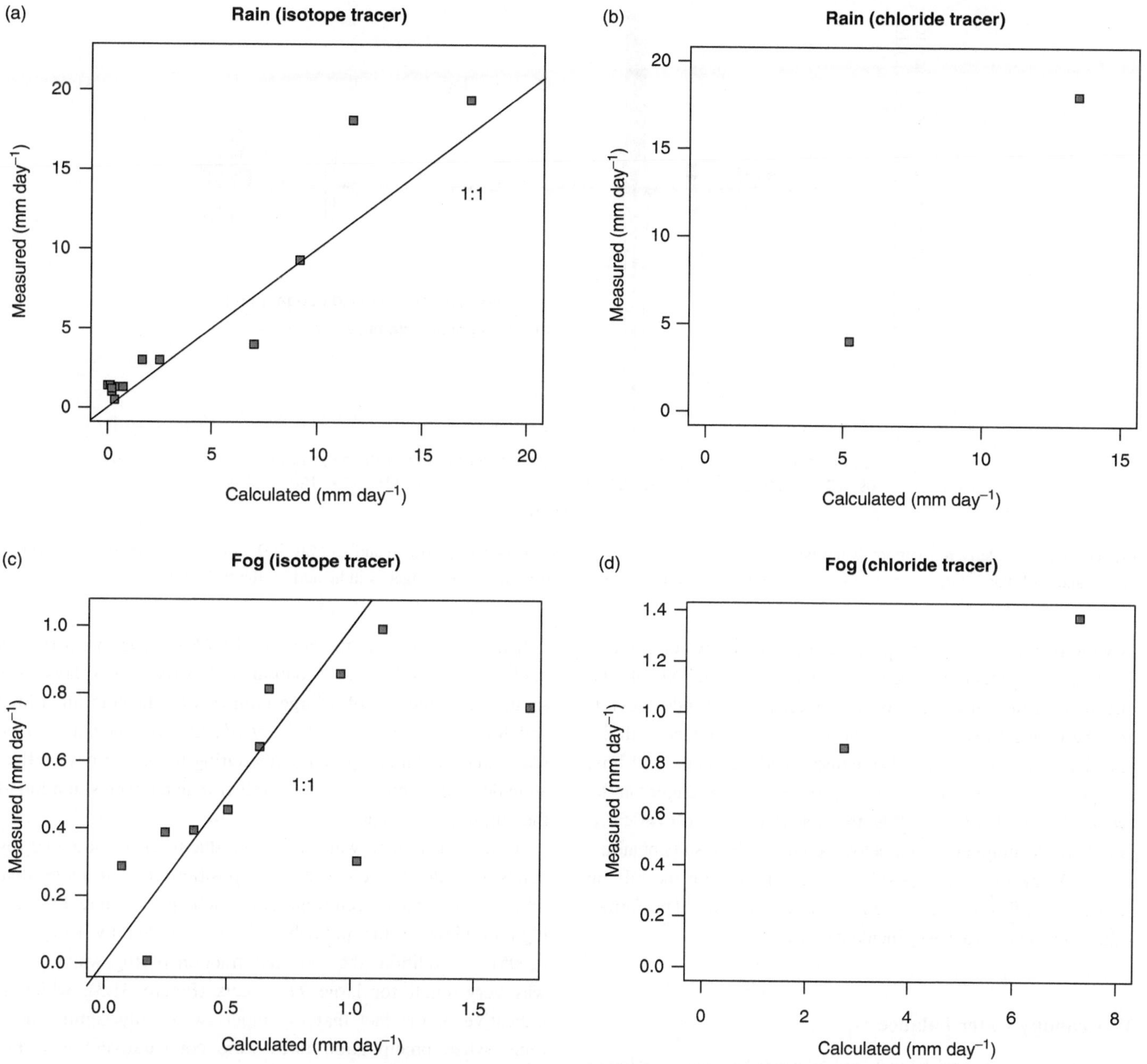

Figure 31.4. Directly measured (eddy covariance-based) daily net fog water deposition rates vs. rates calculated with the mixing model of Eq. (31.1) using ^{18}O as the tracer ($r^2 = 0.52$, $p = 0.009$) at the San Gerardo cloud forest site.

dry season. These percentages compare well with the presently observed 26% of fog occurrence. Therefore, the inferred 4% fog contribution during the study period may well be typical for the dry season after all, although longer-term observations would need to confirm this (cf. Guswa *et al.*, 2007; Lawton *et al.*, this volume).

Isotope concentrations of all water types were available for 21 sampling days. The fractions of fog water in throughfall as calculated with the mixing model of Eq. (31.1) were outside the acceptable range of 0–100% on seven and eight occasions for ^{2}H and ^{18}O, respectively. Therefore the model failed on 33–38% of the available data. The fractions obtained for the remaining days are shown in Figure 31.3. The calculated fractions of fog water in *SF* were all outside the acceptable range, whereas values of δ^2H and δ^{18}O in evaporated water were not determined. Therefore, the daily fog deposition according to the mixing model approach was estimated by assuming equal fractions of fog water in *TF*, *SF*, and evaporated water, and multiplying this times the respective amounts of water. On two days, the inferred fog deposition exceeded 4 mm day^{-1}, which was considered unrealistically high

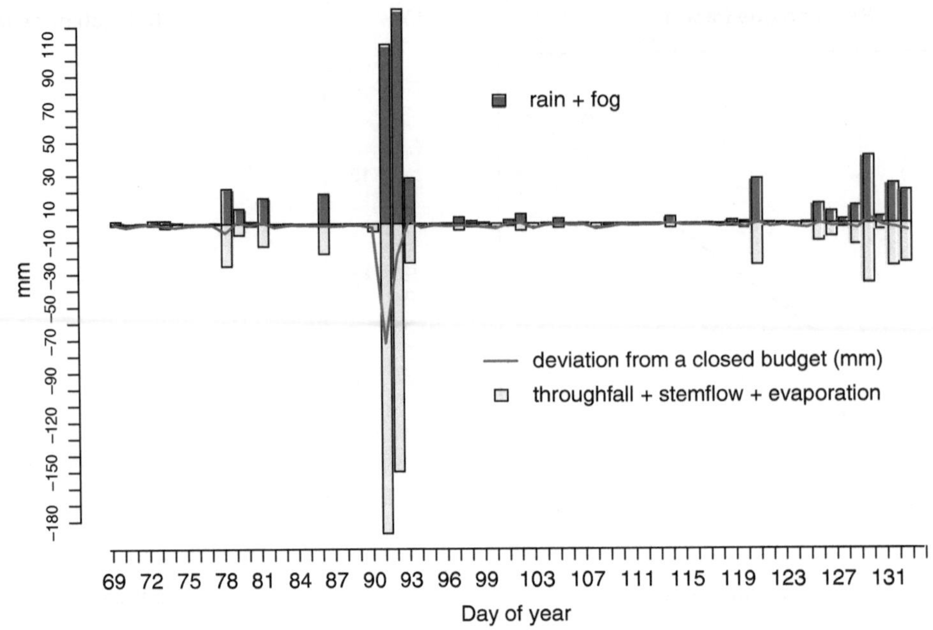

Figure 31.5. The daily wet-canopy water balance at the San Gerardo forest site between 19 March and 13 May 2003. Inputs are displayed by positive values and dark bars, outputs are negative with gray bars, and the deviation from a closed budget is indicated by the bold line.

in view of measured wind speeds and fog liquid water content. These data were therefore excluded from the comparison with the directly measured amounts shown in Figure 31.4. The eddy covariance-based fog fluxes and the deposition rates estimated with the mass balance approach for ^{18}O are distributed around the 1 : 1 line, except for two events. Because the correlations were higher in the case of ^{18}O ($r = 0.72, p = 0.009$) than for ^{2}H as a tracer ($r = 0.54, p = 0.045$), the numbers given below pertain to the results obtained with ^{18}O. Overall, fog deposition as calculated with the mixing model was 1.5 times the directly measured amount. The former value corresponds to 7% of incident rainfall.

Wet-canopy water balance

For the computation of the wet-canopy water balance, a period of 65 days with uninterrupted measurements of all components was selected (9 March – 13 May 2003). For this period, the following amounts (all in mm) were determined:

$$487_{(\text{corrected rainfall})} + 19_{(\text{direct fog})} \approx 497_{(TF)} + 50_{(\text{wet-canopy evaporation})} + 5_{(SF)}. \qquad (31.2)$$

For these 65 days, the difference between canopy inputs (rain and fog) and outputs (*TF*, *SF*, evaporation) was not statistically significant (paired Wilcoxon rank sum test, 95% confidence level, $p = 0.30$). However, on days with high wind speeds and inclined rainfall, the difference between inputs and outputs became quite large (Figure 31.5: days 91 and 92, or 1 and 2 April 2003).

If it is assumed that the error in the *TF* measurements (which made up 97% of the total output during these two days) was small due to the use of a large number of collectors (cf. Lloyd and Marques, 1988; Holwerda *et al.*, 2006b), then it is most likely that the "missing" amounts during these two days reflect an underestimation of the inputs rather than an overestimation of the output components.

If the unaccounted water had been due to underestimated fog deposition alone, then a daily deposition rate of more than 18 mm would have been required, which appears unrealistically high in view of the published range reported by Bruijnzeel (2005). In addition, the modeled fraction of fog water in *TF* was very small for large *TF* events (Figure 31.3), which is indicative of the fact that fog inputs were only significant at times when precipitation rates were comparatively low. For example, a daily *TF* amount of 10 mm would yield a contribution of ~5% by fog according to the mixing model (Eq. 31.1). Measured daily *TF* totals for 1 and 2 April 2003 were very much larger (165 and 134 mm, respectively, with corresponding rainfall amounts of 134 and 44 mm, respectively). Whilst no isotopic information was available for these extreme events, based on Figure 31.3 it would be expected that the associated fraction of fog water in *TF* for these two events would be well below 5%. This suggests that total fog deposition must have been less than 15 mm during these two days (the eddy covariance system measured a total amount of 5.5 mm). Therefore, it is concluded that the bulk of the unexplained gap of 46 mm in the wet-canopy water balance (Eq. 31.2) was due to underestimation of rainfall, despite the application of corrections for aerodynamic losses

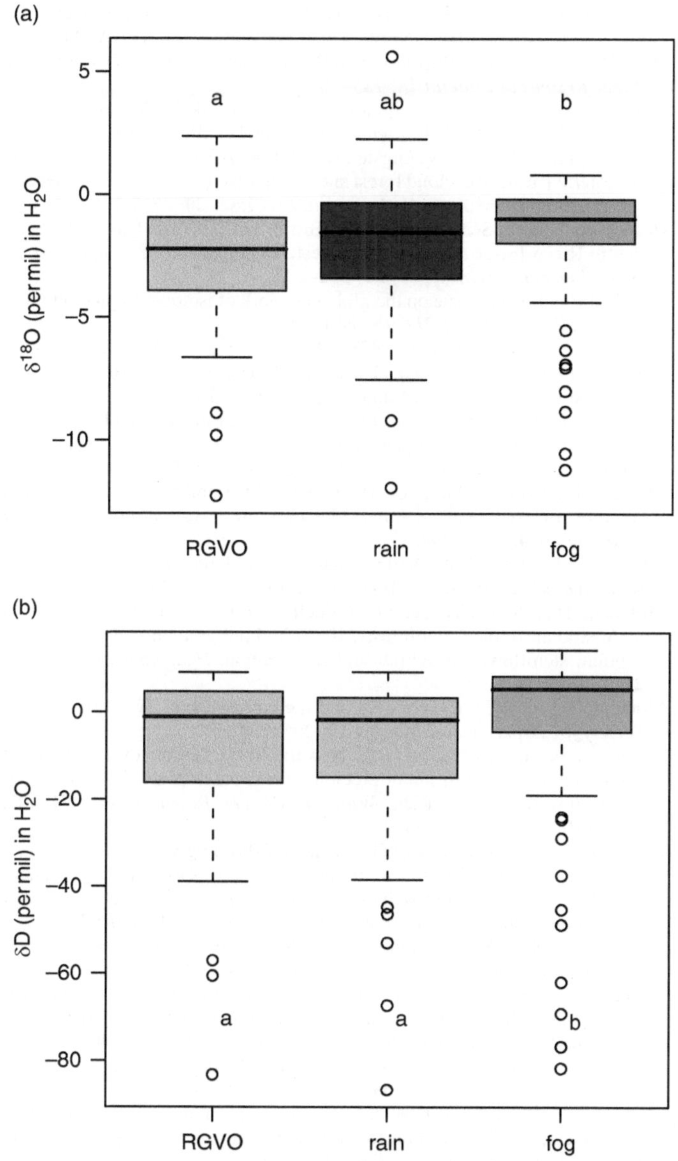

Figure 31.6. Isotopic signatures of source waters: (a) oxygen isotopic ratios ($\delta^{18}O$), (b) hydrogen isotopic ratios (δD).

around the gage and for slope effects. This argumentation is supported by measurements made with the louvered cylindrical gage (*MJU*) and the rotating vertical orifice gage, both of which collected large volumes of water during these two days. The *MJU* recorded 459 and 100 mm of horizontal precipitation for 1 and 2 April, respectively. Furthermore, isotopic values of the water collected with the rotating vertical orifice gage showed a clear difference with those of fog water (as collected by the modified CASCC) throughout the measuring campaign (Figure 31.6), whereas no significant difference was found with δ-values in rain water. It is most likely, therefore, that the water collected by the rotating gage during the storm events of 1 and 2 April was mostly (wind-driven) rain.

CONCLUSIONS

The components of the wet-canopy water balance of a windward montane cloud forest in north-western Costa Rica were measured or calculated for a period of 65 days during the dry season of 2003. There was no statistically significant difference between totals of canopy inputs (fog and rain) and outputs (throughfall, stemflow, and wet-canopy evaporation) over this period. Fog inputs ranged between 4% (eddy covariance-based net flux measurements) and 7% (tracer-based mixing model using ^{18}O) of incident rainfall. Rainfall was corrected for the effects of sloping ground and aerodynamic losses around the gage. On average, the corrected amounts were 27% higher. However, for an extreme composite storm event with strong winds and therefore substantial inclined or near-horizontal rainfall, the applied corrections were insufficient and rainfall was still underestimated. During the same event, throughfall greatly exceeded incoming rainfall and a cylindrical louvered screen gage collected large amounts of water, thereby confirming the occurrence of near-horizontal precipitation.

The isotope-based mixing model was found to be a good tool to assess the fraction of fog water in throughfall during many precipitation events. However, this method needs further refinement as several relevant questions remain still unsolved, including: (i) why did the inferred fraction for almost half of the events lie outside the valid range of 0–100%?, (ii) why did the method not work for stemflow samples?, (iii) does isotopic fractionation occur while fog water is sampled with an Active Strand Cloud Water Collector?, and (iv) how large is the effect of evaporation on isotopic concentrations in throughfall water?

An extension of the available data-set for both directly measured net fog deposition and isotopic composition is expected to increase confidence in the idea that the isotope mixing approach could usefully replace the expensive and sophisticated eddy covariance instrumentation. However, it would first be necessary to understand why the results obtained with ^{2}H as a tracer differed so much from those based on ^{18}O (see also Scholl *et al.*, this volume) at this wind-exposed location, when Rhodes *et al.* (2006) and Rhodes *et al.* (this volume) found isotopic values in precipitation at nearby (leeward) Monteverde followed the global meteoric water line.

ACKNOWLEDGEMENTS

This study was supported by the Swiss National Science Foundation (grant no. 2100–068051) and the Forestry Research Programme of the UK Department for International Development (DFID) (project R 7991). The views expressed here are not necessarily those of DFID. We thank Professor Jeff Collett

(Colorado State University) for the loan of two CASCC collectors, and Jürg Schenk (University of Bern), for technical assistance with the eddy covariance set-up.

REFERENCES

Beiderwieden, E., V. Wolff, Y. J. Hsia, and O. Klemm (2008). It goes both ways: measurements of simultaneous evapotranspiration and fog droplet deposition at a montane cloud forest. *Hydrological Processes* **22**: 4181–4189.

Beswick, K. M., K. J. Hargreaves, M. W. Gallagher, T. Choularton, and D. Fowler (1991). Size-resolved measurements of cloud droplet deposition velocity to a forest canopy using an eddy correlation technique. *Quarterly Journal of the Royal Meteorological Society* **117**: 623–645.

Brown, M. B., I. de la Roca, A. Vallejo, *et al.* (1996). *A Valuation Analysis of the Role of Cloud Forests in Watershed Protection: Sierra de las Minas Biosphere Reserve, Guatemala and Cusuco N.P., Honduras.* Philadelphia, PA: RARE Center for Tropical Conservation.

Bruijnzeel, L. A. (2005). Tropical montane cloud forests: a unique hydrological case. In *Forests, Water and People in the Humid Tropics*, eds. M. Bonell and L. A. Bruijnzeel, pp. 462–483. Cambridge, UK: Cambridge University Press.

Bruijnzeel, L. A., and J. Proctor (1995). Hydrology and biogeochemistry of tropical montane cloud forests: what do we really know? In *Tropical Montane Cloud Forests*, eds. L. S. Hamilton, J. O. Juvik, and F. N. Scatena, pp. 38–78. New York: Springer-Verlag.

Bruijnzeel, L. A., W. Eugster, and R. Burkard (2005). Fog as a hydrological input. In *Encyclopaedia of Hydrological Sciences*, ed. M. G. Anderson, pp. 559–582. Chichester, UK: John Wiley.

Brunel, J. -P., G. K. Walker, and A. K. Kenneth-Smith (1995). Field validation of isotopic procedures for determining sources of water used by plants in a semi-arid environment. *Journal of Hydrology* **167**: 351–368.

Burkard, R., P. Bützberger, and W. Eugster (2003). Vertical fogwater flux measurement above an elevated forest canopy at the Lägeren research site, Switzerland. *Atmospheric Environment* **37**: 2979–2990.

Clark, K. L., N. M. Nadkarni, D. Schaeffer, and H. L. Gholz (1998). Atmospheric deposition and net retention of ions by the canopy in a tropical montane forest, Monteverde, Costa Rica. *Journal of Tropical Ecology* **14**: 27–45.

Clark, K. L., R. O. Lawton, and P. R. Butler (2000). The physical environment. In *Monteverde: Ecology and Conservation of a Tropical Cloud Forest*, eds. N. M. Nadkarni and N. T. Wheelwright, pp. 15–34. Oxford, UK: Oxford University Press.

Demoz, B. B., J. L. Collett, and B. C. Daube (1996). On the Caltech Active Strand Cloudwater Collectors. *Atmospheric Research* **41**: 47–62.

Eugster, W., R. Burkard, F. Holwerda, F. N. Scatena, and L. A. Bruijnzeel (2006). Characteristics of fog and fog-water fluxes in a Puerto Rican elfin cloud forest. *Agricultural and Forest Meteorology* **139**: 288–306.

Førland, E. J., P. Allerup, B. Dahlström, *et al.* (1996). *Manual for Operational Correction of Nordic Precipitation Data.* Oslo, Norway: Norwegian Meteorological Institute.

Goodman, J. (1985). The collection of fog drip. *Water Resources Research* **21**: 392–394.

Guswa, A. J., A. L. Rhodes, and S. E. Newell (2007). Importance of orographic precipitation to the water resources of Monteverde, Costa Rica. *Advances in Water Resources* **30**: 2098–2112.

Hafkenscheid, R. L. L. J., L. A. Bruijnzeel, R. A. M. de Jeu, and N. J. Bink (2002). Water budgets of two upper montane rain forests of contrasting stature in the Blue Mountains, Jamaica. In *Proceedings of the 2nd International Colloquium on Hydrology and Water Management in the Humid Tropics*, Technical Documents in Hydrology No. 52, ed. J. S. Gladwell, pp. 399–424. Paris: IHP-UNESCO, and Panama City: CATHALAC.

Harr, R. D. (1982). Fog drip in the Bull Run municipal watershed, Oregon. *Water Resources Bulletin* **18**: 785–789.

Holder, C. D. (2003). Fog precipitation in the Sierra de las Minas Biosphere Reserve, Guatemala. *Hydrological Processes* **17**: 2001–2010.

Holwerda, F., R. Burkard, W. Eugster, *et al.* (2006a). Estimating fog deposition at a Puerto Rican elfin cloud forest site: comparison of the water-budget and eddy covariance methods. *Hydrological Processes* **20**: 2669–2692.

Holwerda, F., F. N. Scatena, and L. A. Bruijnzeel (2006b). Throughfall in a Puerto Rican lower montane rain forest: a comparison of sampling strategies. *Journal of Hydrology* **327**: 592–602.

IAEA (2002). Special issue on the global network of isotopes in precipitation. *Water and Environment Newsletter* 16: 5.

Juvik, J. O., and P. C. Ekern (1978). *A Climatology of Mountain Fog on Mauna Loa, Hawaii Island*, Technical Report No. 118. Honolulu, HI: Water Resources Research Center, University of Hawai'i.

Köhler, L., C. Tobón, K. F. A. Frumau, and L. A. Bruijnzeel (2007). Biomass and water storage of epiphytes in old-growth and secondary montane rain forest stands in Costa Rica. *Plant Ecology* **193**: 171–184.

Lloyd, C. R., and A. O. Marques (1988). Spatial variability of throughfall and stemflow measurements in Amazonian rain forest. *Agricultural and Forest Meteorology* **43**: 277–294.

Lovett, G. M. (1984). Rates and mechanisms of cloud water deposition to a subalpine balsam fir forest. *Atmospheric Environment* **18**: 361–371.

McJannet, D., J. S. Wallace, and P. Reddell (2007). Precipitation interception in Australian tropical rainforests. II. Altitudinal gradients of cloud interception, stemflow, throughfall and interception. *Hydrological Processes* **21**: 1703–1718.

Monteith, J. L. (1965). Evaporation and the environment. *Symposium of the Society for Experimental Biology* **19**: 205–234.

Rhodes, A. L., A. J. Guswa, and S. E. Newell (2006). Seasonal variation in the stable isotopic composition of precipitation in the tropical montane forests of Monteverde, Costa Rica. *Water Resources Research* **42**, W11402, doi:10.1029/2005WR004535.

Ritter, A., C. M. Regalado, and G. Aschan (2008). Fog water collection in a subtropical elfin laurel forest of the Garajonay National Park (Canary Islands): a combined approach using artificial fog catchers and a physically based impaction model. *Journal of Hydrometeorology* **9**: 920–935.

Sharon, D. (1980). The distribution of hydrologically effective rainfall incident on sloping ground. *Journal of Hydrology* **46**: 165–188.

Sigmon, J. T., F. G. Gilliam, and M. E. Partin (1989). Precipitation and throughfall chemistry for a montane hardwood ecosystem: potential contributions from cloud water. *Canadian Journal of Forest Research* **19**: 1240–1247.

Slinn, W. G. N. (1982). Predictions for particle deposition to vegetative canopies. *Atmospheric Environment* **16**: 1785–1794.

Thalmann, E., R. Burkard, T. Wrzesinsky, W. Eugster, and O. Klemm (2002). Ion fluxes from fog and rain to an agricultural and a forest ecosystem in Europe. *Atmospheric Research* **64**: 147–158.

Vermeulen, A. T., G. P. Wyers, F. G. Römer, *et al.* (1997). Fog deposition on a coniferous forest in the Netherlands. *Atmospheric Environment* **31**: 375–386.

Vong, R. J., and A. S. Kowalski (1995). Eddy correlation measurements of size dependent cloud droplet turbulent fluxes to complex terrain. *Tellus Series B* **47**: 331–352.

Walmsley, J. L., R. S. Schemenauer, and H. A. Bridgman (1996). A method for estimating the hydrological input from fog in mountainous terrain. *Journal of Applied Meteorology* **35**: 2237–2249.

Yin, X. W., and P. A. Arp (1994). Fog contributions to the water budget of forested watersheds in the Canadian Maritime Provinces: a generalized algorithm for low elevations. *Atmosphere–Ocean* **32**: 553–566.

Zadroga, F. (1981). The hydrological importance of a montane cloud forest area of Costa Rica. In *Tropical Agricultural Hydrology*, eds. R. Lal and E. W. Russell, pp. 59–73. New York: John Wiley.

32 Measured and modeled rainfall interception in a lower montane forest, Ecuador

K. Fleischbein
University of Giessen, Giessen, Germany
W. Wilcke
Johannes Gutenberg University of Mainz, Mainz, Germany
R. Goller
University of Bayreuth, Bayreuth, Germany
C. Valarezo
Universidad Nacional de Loja, Loja, Ecuador
W. Zech
University of Bayreuth, Bayreuth, Germany
K. Knoblich
University of Giessen, Giessen, Germany

ABSTRACT

The evaporative loss of intercepted water from the canopy constitutes an important element of the water budget of forests. Starting April 1998, incident precipitation (*P*), throughfall (*TF*), and stemflow (*SF*) were measured in five transects laid out in three small watersheds (~10 ha each) with lower montane rain forest at 1900–2200 m.a.s.l. in South Ecuador. Interception loss (*I*) was also modeled using the analytical model of Gash (1979). The storage capacity of the leaves and of the trunks and branches, as well as the direct throughfall, and stemflow fractions were determined using conventional regression approaches. In addition, apparent total evaporation (*ET*) was determined from the water budget for the three watersheds. Mean annual *P* in the first 4 years ranged between 2363 and 2592 mm among the three watersheds. Average *I* derived from weekly measurements of *P*, *TF*, and *SF* ranged between 2.0 and 3.5 mm day^{-1} (i.e. 32–50% of *P*). Modeled average *I* was similar to measured values at 2.1–3.4 mm day^{-1} (32–49% of *P*). We found that *I* constituted an important part of the average estimated watershed *ET* of 3.5–4.3 mm day^{-1}. The high evaporative losses are attributed to a combination of low rainfall intensities, the usual absence of fog, high canopy density, abundant epiphytes, and advected energy from lower elevations.

INTRODUCTION

Rainfall interception and its subsequent evaporation (interception loss, *I*) is an important control of the water yield of forested watersheds (Sopper and Lull, 1967; Bruijnzeel, 2000). Interception loss (*I*) depends on the available evaporative energy – which can be particularly high near the equator – and on the evaporative demand of the atmosphere, which is mainly driven by prevailing wind speed and vapor pressure deficits. In species-rich tropical rain forests, *I* tends to be highly variable even at a scale of a few hectares and less (Lloyd and Marques, 1988; Wilcke *et al.*, 2001; Holwerda *et al.*, 2006). This is generally thought to reflect high spatial variability in vegetation density, composition, tree crown exposure, and architecture, which all affect the aerodynamic roughness of the canopy and its capacity to store water (Roberts

Tropical Montane Cloud Forests: Science for Conservation and Management, eds. L. A. Bruijnzeel, F. N. Scatena, and L. S. Hamilton. Published by Cambridge University Press.

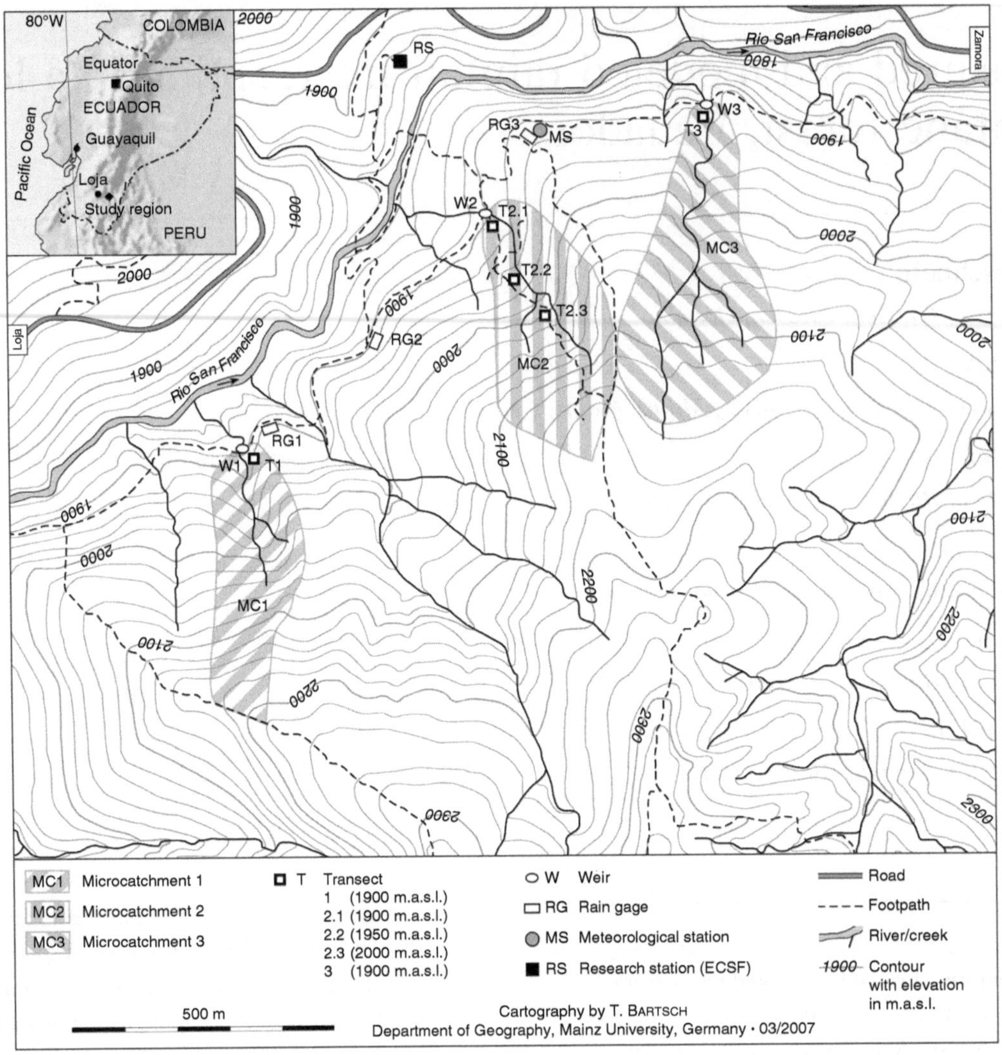

Figure 32.1. Location of the studied micro-watersheds (MC), rain gages, climate station, and the five transects for the measurement of rainfall interception in a lower montane forest in southern Ecuador.

et al., 2005). It has also been suggested that the water cascading through a multiple-layered canopy creates an added variability of its own (Jetten, 1996), whereas in wet montane forests the presence of abundant epiphytes and bryophytes creates additional complexity (Hölscher *et al.*, 2004; cf. Köhler *et al.*, this volume; Tobón *et al.*, this volume #26).

To obtain reliable estimates of throughfall (*TF*) and *I* in (montane) rain forests, an adequate sampling strategy is required which usually involves the use of numerous roving gages and tends to be labor-intensive (Lloyd and Marques, 1988; Holwerda *et al.*, 2006). Therefore, *I* has also been predicted by rainfall interception models. The analytical model of Gash (1979), a simplification of the data-demanding running water balance model of Rutter *et al.* (1971) is frequently used for this purpose.

The aim of this chapter is to model the interception loss from an epiphyte-rich lower montane rain forest not subject to much fog and low cloud (Bendix *et al.*, 2008) in South Ecuador using the Gash model. The reliability of the model results were checked by comparing with measured weekly interception losses.

MATERIAL AND METHODS

Study site

Three forested micro-watersheds (MC1–MC3) located between the cities of Loja (04° 00′ S, 79° 12′ W) and Zamora (04° 05′ S, 78° 58′ W) in the province of Zamora–Chinchipe in South Ecuador were selected for hydrological studies (Wilcke *et al.*, 2001; Figure 32.1). Watershed MC1 was *c.* 8 ha, MC2 *c.* 10 ha, and MC3 *c.* 13 ha. The watersheds were situated at 1900–2200 m.a.s.l. and drained into the Rio San Francisco which in turn flows toward the Amazon Basin.

Weather data were recorded by a meteorological station in operation since April 1998 and positioned on a ridge top within a

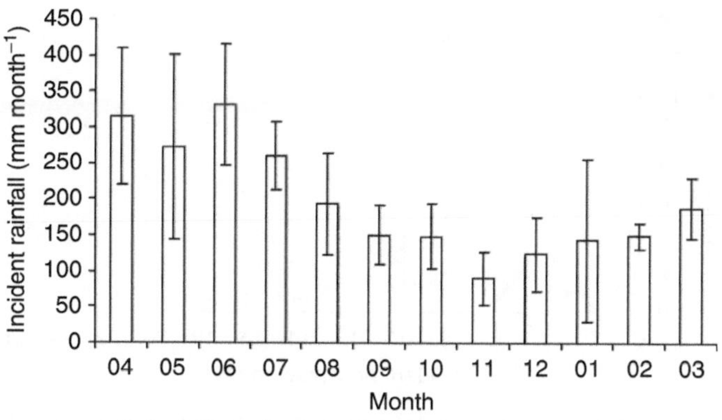

Figure 32.2. Mean monthly precipitation totals and their standard deviations between April 1998 and April 2002 as measured at the three rain gage stations in Figure 32.1.

clear-cut area between MC2 and MC3 at 1952 m.a.s.l. (Figure 32.1). Between April 1998 and April 2000, the average annual temperature was 16.2 °C, with an average annual temperature gradient of 0.7 °C 100 m^{-1} increase in altitude as derived from additional temperature measurements at 2927 m.a.s.l. (Bendix *et al.*, 2008). Although some months have higher rainfall than others (notably March–July), there is no clear rainy or dry season (Figure 32.2).

The forest at the study site has been classified as *bosque siempreverde montano*, or evergreen montane forest (Balslev and Øllgaard, 2002). In the classification of Bruijnzeel and Hamilton (2000) it would represent a lower montane rain forest, although the abundance of epiphytes and bryophytes (Fleischbein *et al.*, 2005) suggests a transition to lower montane cloud forest. Depending on relief and altitudinal range, Paulsch (2002) divided the forests of the study area into ravine forest and ridge forest, on the basis of structural properties. The forest in MC1 probably represents advanced natural regrowth and in the lower part of MC2 there are indications of human disturbance. Lauraceae, Melastomataceae, and Euphorbiaceae are the most important tree families present. The tallest and most species-rich forest is found on footslopes and in ravines, where the canopy reaches 25 m with some emergents reaching up to 35 m. The present study sites were located in the primary ravine forest zone at lower altitudes (Figure 32.1). The forest has a stem density of 500–1250 ha^{-1} (for diameter at breast height DBH $\geq$ 0.1 m) and of 1100–3100 ha^{-1} (DBH $\geq$ 0.05 m; Homeier, 2004). Tree ferns (*Cyathea* sp.) were numerous in the lower stratum whereas epiphyte and bryophyte cover was estimated at 80%. The cover fraction of the canopy stratum was estimated at 40–50%, vs. 60% and 50% for the middle and lower canopy strata, respectively (Paulsch, 2002; Fleischbein *et al.*, 2005). Leaf area index (*LAI*) based on light transmission measurements ranged from 5.19–9.32 (Fleischbein *et al.*, 2005). Steep slopes (mean inclination 38°, locally up to 70°) characterize the study area (cf. Wilcke *et al.*, this volume).

Instrumentation

A total of five transects were established for the measurement of *I* (Figure 32.1). Three transects were located at 1900–1910 m.a.s.l. in the respective watersheds (T1, T2.1, and T3), and the remaining two were established higher up in MC2; one at 1950–1960 m (T2.2) and the other at 2000–2010 m (T2.3). Along each transect, five *TF* gages and – at T1, T2.1, and T3 (in the lower part of each watershed) – also five stemflow (*SF*) collectors were installed in April 1998. In May 2000, three additional *TF* collectors were added to each transect. To measure rainfall input (*P*) for each watershed, three stations consisting of five rain gages each were established at adjacent deforested sites (Figure 32.1). All *P* and *TF* collectors consisted of standard Hellmann-type gages with a fixed position to minimize forest and soil disturbance in the very steep terrain. The 2-l sampling bottles were equipped with a funnel of 115-mm diameter at 0.3 m above the soil surface. A table-tennis ball was placed in each funnel to reduce evaporation. In addition, rainfall collecting bottles were wrapped in aluminum foil to shield them from radiation. Any litter fallen into the collectors was removed at least twice per week. Stemflow (*SF*) was collected using a polyurethane spiral around the bole of each sample tree, following Likens and Eaton (1970). A funnel was attached to the bottom end of the spiral and fed into a 10-l vessel. The size of the latter was sufficient to prevent overflow in all but a few exceptions. The 15 sampled trees represented seven species (including the most common tree fern) with DBH values ranging from 33 to 123 cm (Fleischbein *et al.*, 2005).

Values of *P*, *TF*, and *SF* were measured between April 1998 and April 2002 on a weekly basis. After determination of the water volume, an aliquot of 100 ml was collected for chemical analysis and the remaining water was used to clean the samplers (see also Zimmermann *et al.*, 2007). All collectors were replaced at 6-month intervals.

Hydrological parameters

Interception losses were calculated from weekly measurements of *TF*, *SF*, and *P* as:

$$I = P - (TF + SF) \tag{32.1}$$

where all terms are as defined previously and expressed in mm week^{-1}.

The water-balance equation (Ward and Robinson, 1990) was used to calculate total evapotranspiration (*ET*) at the watershed scale:

$$P = Q + ET + \Delta S + \Delta G + \Delta L \tag{32.2}$$

where Q is the streamflow leaving the watershed, ΔS and ΔG the changes in soil moisture and groundwater storage over the period of measurement, respectively, and ΔL any unrecorded leakage into or out of the watershed. The contributions of ΔS and ΔG could be neglected safely because of the long duration of the present study (4 years). Similarly, preliminary seismic surveys indicated the presence of dense layers at the base of each watershed that are likely to impede deep seepage into or out of the watersheds (S. Hecht, personal communication). As such, Eq. (32.2) was simplified to give apparent ET as:

$$ET = P - Q. \quad (32.3)$$

To quantify surface flows (Q), 90° V-notch weirs with stilling basins were installed at the outlet of each of the three watersheds in April 1998. Initially, water levels were recorded manually and only once per week. After January 1999, water levels were measured continuously using pressure transducers and recorded at hourly intervals by data-loggers. Water levels were converted to discharge (l s^{-1}) using empirical equations (Fleischbein *et al.*, 2006). Because the loggers suffered frequent breakdowns, surface flows were also modeled with TOPMODEL (Beven *et al.*, 1995). Further details have been given by Fleischbein *et al.* (2006) who also provide estimates of the errors associated with each water budget component.

Derivation of canopy parameters

The analytical model of interception requires knowledge of the canopy structure as described by the following parameters: the storage capacity of the leaves (S) and of the trunks (S_t); the "direct" throughfall fraction (p); and the stemflow fraction (p_t) (Gash and Morton, 1978; Gash, 1979). The values of these four parameters are usually derived by regression analysis using measured daily or event-based values of P and TF or SF (Jackson, 1975; Gash and Morton, 1978). Because only weekly totals of TF and SF were available, the regression analyses were applied to the 19 occasions within the entire monitoring period during which only a single rainfall event occurred in a particular week as evidenced by the daily rainfall record of the main weather station. Rainfalls with less than 2 mm week^{-1} did not produce measurable TF (Fleischbein *et al.*, 2005). Among the 19 weeks with $P \geq 2$ mm, seven events had rainfall in excess of 10 mm. The latter were used for the determination of S, p_t, and S_t (using the methods of Leyton *et al.* (1967) and Gash and Morton (1978), respectively) as canopy saturation was guaranteed under these conditions (Figure 32.3). Based on visual inspection of Figure 32.3, rainfalls between 2 and 10 mm week^{-1} were considered insufficient to fully saturate the dense and epiphyte-laden canopy. This is more than the canopy saturation value derived by Fleischbein *et al.* (2005) using the regression of TF on P, possibly because of interception losses during the rain events. To derive p, the 12 rainfall events <10 mm were used following the method of Jackson (1975) (see Fleischbein *et al.* (2005) for details).

The analytical model of rainfall interception

The analytical model of rainfall interception is a simplification of the numerical Rutter model and can be applied on a daily basis or on an event basis (Gash, 1979). Briefly, storms are divided into those large enough to fully saturate the canopy and those that do not. The amount of water needed to saturate the canopy ($P_G{}'$) is defined as:

$$P_G{}' = \frac{-\overline{R}S}{\overline{E}_w} \ln\left[1 - \frac{\overline{E}_w}{\overline{R}}(1 - p - p_t)^{-1}\right] \quad (32.4)$$

where $\overline{R}$ (mm h^{-1}) is the average precipitation intensity falling on a saturated canopy and $\overline{E}_w$ (mm h^{-1}) the average evaporation from the wetted canopy. Next, the model uses a series of simple expressions to calculate the interception loss associated with different phases within a storm (notably wetting up of the canopy, full saturation, and drying-up after rainfall ceases; Gash, 1979). In applying the analytical model, after $P_G{}'$ has been filled, saturated canopy conditions are generally assumed to be maintained as long as hourly rainfall exceeds a threshold. $\overline{E}_w$ is usually calculated using the wet-canopy version of the Penman–Monteith evaporation equation for the hours in which rainfall exceeds the threshold and (ideally) using ambient weather variables as measured above the canopy (Monteith, 1965; Gash, 1979; Schellekens *et al.*, 1999).

The analytical model was parameterized with the independently determined canopy parameters S, S_t, p, and p_t for the five transects and run using hourly precipitation and $\overline{E}_w$ values computed from weather variables measured at the meteorological station (i.e. above grassland) for the period April 1998 – April 2002.

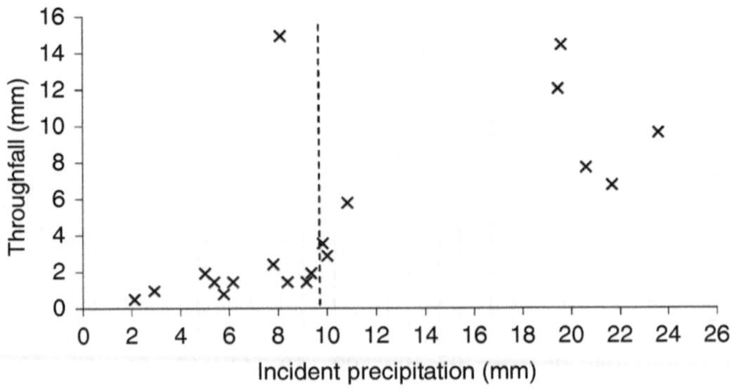

Figure 32.3. Example of incident rainfall vs. throughfall as measured by a selected rain gage in transect T3. Only the 19 weeks with one single rain event >2 mm between April 1998 and April 2002 are shown. The dotted line at an incident rainfall of 10 mm indicates full canopy saturation as derived by visual impression.

RESULTS

Mean annual P between April 1998 and April 2002 for the three watersheds ranged between 2363 and 2592 mm, with an overall mean of 2504 ± 123 mm. During the 1468 days of observation, 2971 rainfall events were registered at the climate station. Events were considered separate after a dry period of at least 2 hours.

Table 32.1 *Average values of the storage capacity of the canopy* (S) *and trunks* (S_t)*, the direct throughfall fraction* (p)*, and the stemflow fraction* (p_t) *and their standard deviations for the forest in the five study transects (T1, T2.1, T2.2, T2.3, and T3); selected results for other montane tropical forests added for comparison*

S (mm)	*p*	S_t (mm)	p_t	Country	Vegetation[a]	Reference
1.91 ± 1.94	0.42 ± 0.2	0.041 ± 0.019	0.003 ± 0.001	Ecuador	LMRF	This study
2.71 ± 1.27	0.31 ± 0.13	0.040 ± 0.015	0.003 ± 0.001	Ecuador	T1, LMRF	This study
1.53 ± 2.91	0.54 ± 0.16	0.047 ± 0.016	0.004 ± 0.001	Ecuador	T2.1, LMRF	This study
2.16 ± 1.76	0.39 ± 0.13			Ecuador	T2.2, LMRF	This study
0.99 ± 1.41	0.36 ± 0.12			Ecuador	T2.3, LMRF	This study
2.02 ± 2.08	0.52 ± 0.31	0.037 ± 0.024	0.002 ± 0.001	Ecuador	T3, LMRF	This study
0.89	0.25			Tanzania	LMRF	Jackson (1975)
2.5				Panamá	LMRF	Cavelier *et al.* (1997)
1.15	0.23	0.023		Puerto Rico	LMRF	Schellekens *et al.* (1999)
1.89 (1.08 leaves 0.81 mosses)				Costa Rica	UMRF	Hölscher *et al.* (2004)
5.0	0.52–0.54			Colombia	UMCF	Van Leerdam and Zagt (1989); Veneklaas and Ek (1990)
1.30–1.57	0.05–0.13	0.15–0.20	0.20–0.39	Jamaica	UMCF	Hafkenscheid *et al.* (2002)
0.43–0.50	0.23	0.05		Puerto Rico	ECF	Holwerda *et al.* this volume, #29

[a] LMRF, lower montane rain forest; UMRF, upper montane rain forest; UMCF, upper montane cloud forest; ECF, elfin cloud forest.

Overall mean intensity was 0.41 mm hour^{-1} and mean duration was 9 hours and 18 min. About 73% of events received 0–2 mm of rain and 20% were larger than 2 mm but smaller than 10 mm. Only 7% of storms were larger than 10 mm. The largest registered event was 215 mm and lasted for 12 hours.

Mean annual *TF* totals at the five study transects ranged between 1246 and 1589 mm (49–67% of *P*), with an overall mean of 1473 ± 197 (59% of *P*). Mean annual *SF* totals during the same period at the three lowermost transects were only 23–26 mm and accounted for a mere 0.9–1.1% of *P* (Fleischbein *et al.*, 2006). Therefore, corresponding inferred values of *I* ranged from 748 to 1286 mm year^{-1} (32–50% of *P*), with an overall annual mean of 1006 ± 197 mm (40 ± 7.9% of *P*).

The values of the respective canopy parameters are summarized in Table 32.1, both for the forest as a whole and for the individual transects. Derived canopy saturation values (*S*) and free throughfall coefficients (*p*) varied considerably between transects, with overall means of 1.91 mm and 0.42, respectively. Stemflow-related parameters varied less between sites (Table 32.1).

Mean interception loss calculated with Eq. (32.1) using weekly measurements of *P*, *TF*, and *SF* varied between 32% and 50% of *P* (equivalent to 2.0–3.5 mm day^{-1}) among the five study transects (Figure 32.4). Corresponding modeled values ranged between 32% and 49% of *P* (2.1–3.4 mm day^{-1}) and were thus close to measured values although a major discrepancy (>10%) was observed for transect T2.2. The goodness of fit (*r*)

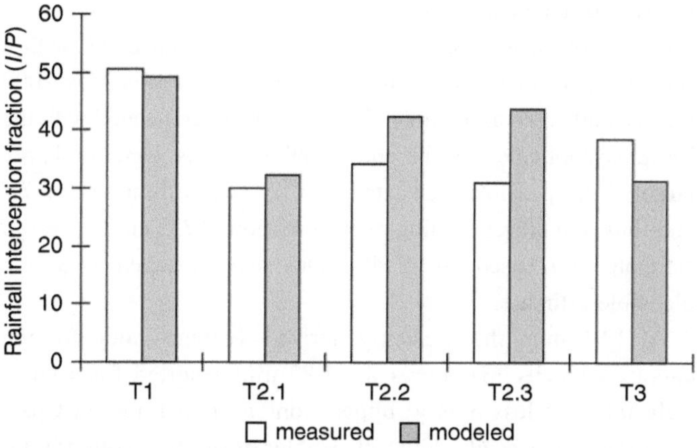

Figure 32.4. Measured and modeled rainfall interception fractions (*I*/*P*) for the five study transects in lower montane forest in southern Ecuador between April 1998 and April 2002.

for regressions linking modeled and measured weekly *I* totals for the five transects varied between 0.58 and 0.72.

The average rainfall intensity onto a saturated canopy $\overline{R}$ was 5.7 mm day^{-1} whereas mean evaporation from the wetted canopy E_w was 3.5 mm day^{-1}. The contribution to overall modeled *I* by wetting up of the canopy during large storms was 37%, vs. 41% during saturated canopy conditions, and 22% during canopy drying. The large storms contributed 51% to total modeled *I*.

Average watershed-based annual evapotranspiration totals (*ET*) were 1570 mm for watershed MC1, 1281 mm for MC2, and 1546 mm for MC3 (equivalent to 54–61% of *P*). These annual estimates would correspond with average daily *ET* rates of 3.5–4.3 mm. Taking the derived *ET* and *I* estimates at face value would imply the latter to make up a high (58–82%) fraction of overall *ET* in this lower montane forest.

DISCUSSION

The percentage of *P* reaching the soil as *TF* at the study site is at the lower end (or even lower in the case of transect T1) of the range of 55–101% compiled by Bruijnzeel and Proctor (1995) and Bruijnzeel (2001) for tropical montane (cloud) forests. Oesker *et al.* (this volume) reported an *I* value of 29% of *P* for one selected year (November 2001–November 2002) at a site near transect T2.1. The currently derived overall average *I* of 61% is comparable to the 63–65% reported by Cavelier *et al.* (1997) and Clark *et al.* (1998) for similarly epiphyte-laden lower montane (cloud) forests in Panamá and Costa Rica, respectively. This similarity is all the more remarkable in view of the difficulty to accurately determine *TF* in species-rich forest. If the results obtained by Lloyd and Marques (1988) for a tropical lowland forest in Brazil also applied to the present montane forest, at least 40 collectors would be needed to estimate the mean *TF* with a standard error of *c.* 10%. This was only realized after May 2000 for the study area as a whole. Thus, the errors associated with the *TF* measurements for the individual transects were probably larger. This possibly translates into an underestimation of *TF* and thus in a corresponding overestimation of *I*. Yet, the overall estimate of *I* based on all 40 gages may be taken as a first plausible estimate.

At 1.91 mm, the presently derived average value for the canopy capacity (*S*) equals the 1.89 mm reported for a similarly tall (but less mossy) upper montane rain forest in Costa Rica (Hölscher *et al.*, 2004; cf. Köhler *et al.*, this volume). Of this 1.91 mm, about 0.38 mm can be attributed to the foliar component of the canopy according to the laboratory measurements of Oesker *et al.* (this volume). These storage values are higher than found for forests under less wet conditions (e.g. Jackson, 1975) but true cloud forests having very high epiphyte and bryophyte biomass typically exhibit much higher values (up to 5.0–6.0 mm; Veneklaas and Ek, 1990; Köhler *et al.*, 2007). However, Hölscher *et al.* (2004) and Köhler *et al.* (2007) have drawn attention to the fact that the actually available capacity to store additional moisture tends to be much smaller than these potential values because the mosses are often wetted by previous rain and fog (cf. Tobón *et al.*, this volume #26).

The average value derived for the direct throughfall fraction ($p = 0.42$) must be considered as very high given the high *LAI* of the study forest (5.2–9.3; Fleischbein *et al.*, 2005). Schellekens *et al.* (1999) suggested that values based on canopy photography or light transmission measurements may be more reliable, because rain drops splashing off the leaves will contribute to direct *TF* and so enhance the apparent value of *p*. Nevertheless, *p* correlated significantly with *LAI* as measured above each *TF* gauge ($r = -0.49$, $\alpha < 0.02$; Fleischbein *et al.*, 2005). However, these estimates of *p* were based on an analysis of small storms occurring under dry conditions with only one rainfall event per week. As such, they may be less representative for wetter periods when the foliage may be more developed (cf. Brouwer, 1996). Peaks in litterfall were observed during the dry period in December but also during a wet and stormy period in July (Wilcke *et al.*, 2002).

Finally, at 1281–1570 mm, apparent annual *ET* totals (including interception evaporation *I*) were at the upper end or higher than the range of 1155–1380 mm compiled by Bruijnzeel (2001) for equatorial lower montane forest with negligible fog incidence. It is well known that estimates of *ET* obtained with the water budget approach may be (much) too high if subterranean losses from the watershed are not taken into account (Ward and Robinson, 1990). To check whether the study watersheds were watertight, additional experiments were undertaken. In a study of the ^{18}O signal of the water in various ecosystem compartments and fluxes, Goller *et al.* (2005) were unable to detect any indications of additional water sources contributing to streamflow other than the subsoil. Furthermore, preliminary seismic surveys indicated the presence of a dense layer underneath the watersheds that is likely to impede deep-seated seepage (S. Hecht, personal communication). However, it cannot be excluded that peak flows were underestimated (Fleischbein, 2004). Fleischbein *et al.* (2006) discussed the magnitude of the various possible errors in the water balance in some detail and drew attention to the fact that only the surface area of watershed MC2 was known with sufficient precision. As such, the corresponding *ET* value of 1281 mm year^{-1} calculated for watershed MC2 may be considered the most reliable of the three estimates. It is also close to the average value (1265 mm, $n = 7$) derived for similar forests elsewhere in the tropics (Bruijnzeel and Proctor, 1995).

Rainfall interception losses made up a large proportion of total *ET* (possibly up to 81%). The high *I* fraction may be attributed to the high density of the canopy of the study forest as indicated by its high *LAI* (5.2–9.3 as measured above the 40 throughfall collectors) and a coverage of up to 80% of the tree surfaces by epiphytes and mosses (Fleischbein *et al.*, 2005). In addition, wet canopy evaporation may have been enhanced by advected energy provided by the influx of warmer and drier air moving upslope (cf. Schellekens *et al.*, 1999). Motzer *et al.* (this volume) determined an annual evaporation equivalent of 800 mm based on net radiation measurements for this area, whereas transpiration according to the Penman–Monteith equation was *c.* 560 mm. The latter value is similar to the presently derived estimate of 459 ± 152 mm obtained by subtracting *I* from watershed-budget-based *ET* totals (Fleischbein *et al.*, 2006). However, the theoretically remaining 340 mm available for *I* (i.e. 800 minus 460 mm and representing <14% of the annual rainfall of 2500 mm) must be considered unrealistically low in view of the measured rainfall interception being >40%, presumably because contributions by advected energy were not included in Motzer's net radiation equivalent.

CONCLUSION

Measured long-term interception losses from a lower montane rain forest not subjected to fog or low cloud and based on 40 throughfall gages distributed among five transects in South Ecuador indicated wet-canopy evaporation totals (32–50% of rainfall) that were at the upper end or even higher than the range of values reported for similar forests elsewhere in the tropics.

Annual evapotranspiration totals obtained with the watershed water-budget technique gave rather high values for two out of three nearly adjacent (and allegedly watertight) watersheds. Combining average annual interception loss with the lowest – and possibly most plausible – estimate of total *ET* suggested the latter to be strongly dominated by the interception component.

Factors contributing to this are thought to include the high density of the canopy, a high incidence of epiphytes and mosses on branches and stems, relatively low rainfall intensities, and possibly advected energy in the form of warmer and drier air moving upslope.

ACKNOWLEDGEMENTS

We thank the Ministerio del Ambiente of the Republic of Ecuador for permitting the research (No. 002–IC-FLO-DFZ-MA), the Fundación Científica de San Francisco (FCSF, now Nature and Culture International, NCI) for access to the study area and the research station, P. Emck for providing climatic data, and the Deutsche Forschungsgemeinschaft (FOR 402, Wi 1601/5-1, -3) for funding this study. Wolfgang Wilcke acknowledges the receipt of a Heisenberg grant from the Deutsche Forschungsgemeinschaft (Wi 1601/3-1, -2). We furthermore are indebted to Sampurno Bruijnzeel for his valuable suggestions to improve this chapter and for general support.

REFERENCES

Balslev, H., and B. Øllgaard (2002). Mapa de vegetación del sur de Ecuador. In *Botánica Austroecuatoriana: estudios sobre los recursos vegetales en las Provincias de El Oro, Loja y Zamora-Chinchipe*, eds. M. Z. Aguirre, J. E. Madsen, E. Cotton, and H. Balslev, pp. 51–64. Quito, Ecuador: Ediciones Abya-Yala.

Bendix, J., R. Rollenbeck, M. Richter, P. Fabian, and P. Emck (2008). Climate. In *Gradients in a Tropical Mountain Ecosystem of Ecuador*, eds. E. Beck, J. Bendix, I. Kottke, F. Makeschin, and R. Mosandl, pp. 63–74. New York: Springer-Verlag.

Beven, K. J., R. Lamb, P. F. Quinn, R. Romanowicz, and J. Freer (1995). TOPMODEL. In *Computer Models of Watershed Hydrology*, ed. V. P. Singh, pp. 627–668. Highlands Ranch, CO: Water Resources Publications.

Brouwer, L. C. (1996). *Nutrient Cycling in Pristine and Logged Tropical Rain Forest, Guyana*, Tropenbos Guyana Series No. 1. Georgetown, Guyana: Tropenbos Foundation.

Bruijnzeel, L. A. (2000). Forest hydrology. In *The Forests Handbook*, ed. J. C. Evans, pp. 301–343. Oxford, UK: Blackwell Scientific.

Bruijnzeel, L. A. (2001). Hydrology of tropical montane cloud forests: a reassessment. *Land Use and Water Resources Research* **1**: 1–18.

Bruijnzeel, L. A., and L. S. Hamilton (2000). *Decision Time for Cloud Forests*, IHP Humid Tropics Program Series No. 13. Paris: UNESCO, Amsterdam: IUCN-NL and Gland, Switzerland: WWF.

Bruijnzeel, L. A., and J. Proctor (1995). Hydrology and biogeochemistry of tropical montane cloud forests: what do we really know? In *Tropical Montane Cloud Forests*, eds. L. S. Hamilton, J. O. Juvik, and F. N. Scatena, pp. 38–78. New York: Springer-Verlag.

Cavelier, J., M. Jaramillo, D. Solis, and D. De Leon (1997). Water balance and nutrient inputs in bulk precipitation in tropical montane cloud forest in Panama. *Journal of Hydrology* **193**: 83–96.

Clark, D. L., N. M. Nadkarni, and H. L. Gholz (1998). Growth, net production, litter decomposition, and net nitrogen accumulation by epiphytic bryophytes in a tropical montane forest. *Biotropica* **30**: 12–23.

Fleischbein, K. (2004). *Wasserhaushalt eines Bergwaldes in Ecuador: Experimenteller und modellhafter Ansatz auf Einzugsgebietsebene*. Giessen, Germany: Lenz-Verlag.

Fleischbein, K., W. Wilcke, R. Goller, *et al.* (2005). Rainfall interception in a lower montane forest in Ecuador: effects of canopy properties. *Hydrological Processes* **19**: 1355–1371.

Fleischbein, K., W. Wilcke, C. Valarezo, W. Zech, and K. Knoblich (2006). Water budgets of three small catchments under montane forest in Ecuador: experimental and modelling approach. *Hydrological Processes* **20**: 2491–2507.

Gash, J. H. C. (1979). An analytical model of rainfall interception in forests. *Quarterly Journal of the Royal Meteorological Society* **105**: 43–55.

Gash, J. H. C., and A. J. Morton (1978). An application of the Rutter Model to the estimation of the interception loss from Thetford Forest. *Journal of Hydrology* **38**: 49–58.

Goller, G., W. Wilcke, M. J. Leng, *et al.* (2005). Tracing water paths through small catchments under a tropical montane rain forest in south Ecuador by an oxygen isotope approach. *Journal of Hydrology* **308**: 67–80.

Hafkenscheid, R. L. L. J., L. A. Bruijnzeel, R. A. M. de Jeu, and N. J. Bink (2002). Water budgets of two upper montane rain forests of contrasting stature in the Blue Mountains, Jamaica. In *Proceedings of the 2nd International Colloquium on Hydrology and Water Management in the Humid Tropics*, Technical Documents in Hydrology No. 52, ed. J. S. Gladwell, pp. 399–424. Paris: IHP-UNESCO, and Panama City: CATHALAC.

Hölscher, D., L. Köhler, A. I. J. M. Van Dijk, and L. A. Bruijnzeel (2004). The importance of epiphytes to total rainfall interception by a tropical montane rain forest in Costa Rica. *Journal of Hydrology* **292**: 308–322.

Holwerda, F., F. N. Scatena, and L. A. Bruijnzeel (2006). Throughfall in a Puerto Rican lower montane rain forest: a comparison of sampling strategies. *Journal of Hydrology* **327**: 592–602.

Homeier, J. (2004). *Baumdiversität, Waldstruktur und Wachstumsdynamik zweier tropischer Bergregenwälder in Ecuador und Costa Rica*. Berlin, Germany: J. Cramer-Verlag.

Homeier, J., H. Dalitz, and S.-W. Breckle (2002). Waldstruktur und Baumartendiversität im montanen Regenwald der Estación Cientifica San Francisco in Südecuador. *Berichte der Reinhold-Tüxen-Gesellschaft* **14**: 109–118.

Jackson, I. J. (1975). Relationships between rainfall parameters and interception by tropical forest. *Journal of Hydrology* **24**: 215–238.

Jetten, V. (1996). Interception of tropical rainforest: performance of a canopy water balance model. *Hydrological Processes* **10**: 671–685.

Köhler, L., C. Tobón, K. F. A. Frumau, and L. A. Bruijnzeel (2007). Biomass and water storage of epiphytes in old-growth and secondary montane rain forest stands in Costa Rica. *Plant Ecology* **193**: 171–184.

Leyton, L., E. R. C. Reynolds, and F. B. Thompson (1967). Rainfall interception in forest and moorland. In *International Symposium on Forest Hydrology*, eds. W. E. Sopper and H. W. Lull, pp. 163–168. Oxford, UK: Pergamon Press.

Likens, G. E., and J. S. Eaton (1970). A polyurethane stemflow collector for trees and shrubs. *Ecology* **51**: 937–939.

Lloyd, C. R., and A. O. Marques Jr. (1988). Spatial variability of throughfall and stemflow measurements in Amazonian rain forest. *Agriculture and Forest Meteorology* **42**: 63–73.

Monteith, J. L. (1965). Evaporation and the environment. *Symposia of the Society for Experimental Biology* **19**: 245–269.

Paulsch, A. (2002). Development and application of a classification system for undisturbed and disturbed tropical montane forests based on vegetation structure. Ph.D. thesis, University of Bayreuth, Bayreuth, Germany.

Roberts, J. M., J. H. C. Gash, M. Tani, and L. A. Bruijnzeel (2005). Controls on evaporation in lowland tropical rainforest. In *Forests, Water and People in the Humid Tropics*, eds. M. Bonell and L. A. Bruijnzeel, pp. 287–313. Cambridge, UK: Cambridge University Press.

Rutter, A. J., K. A. Kershaw, P. C. Robins, and A. J. Morton (1971). A predictive model of rainfall interception in forests. I. Derivation of the model from observations in a plantation of corsican pine. *Agricultural Meteorology* **9**: 367–384.

Schellekens, J., F. N. Scatena, L. A. Bruijnzeel, and A. J. Wickel (1999). Modelling rainfall interception by a lowland tropical rain forest in northeastern Puerto Rico. *Journal of Hydrology* **225**: 68–184.

Sopper, W. E., and H. W. Lull (eds.) (1967). *Forest Hydrology: Proceedings of a National Science Foundation Advanced Science Seminar*, Pennsylvania State University, University Park, PA, 29 Aug – 10 Sept 1965. Oxford, UK: Pergamon Press.

Van Leerdam, A., and R. J. Zagt (1989). The epiphyte vegetation of an Andean forest in Colombia: aspects of its hydrology and distribution in the canopy. M.Sc. thesis, University of Utrecht, Utrecht, the Netherlands.

Veneklaas, E. J., and R. van Ek (1990). Rainfall interception in two tropical montane rain forests, Colombia. *Hydrological Processes* **4**: 311–326.

Ward, R. C., and M. Robinson (1990). *Principles of Hydrology*. London: McGraw-Hill.

Wilcke, W., S. Yasin, C. Valarezo, and W. Zech (2001). Change in water quality during the passage through a tropical montane rain forest in Ecuador. *Biogeochemistry* **55**: 45–72.

Wilcke, W., S. Yasin, U. Abramowski, C. Valarezo, and W. Zech (2002). Nutrient storage and turnover in organic layers under tropical montane rain forest in Ecuador. *European Journal of Soil Science* **53**: 15–27.

Zimmermann, A., W. Wilcke, and H. Elsenbeer (2007). Spatial and temporal patterns of throughfall quantity and quality in a tropical montane forest in Ecuador. *Journal of Hydrology* **343**: 80–96.

33 Measuring cloud water interception in the Tambito forests of southern Colombia

J. González

King's College London, London, UK

ABSTRACT

This chapter discusses the results obtained with passive cloud water collectors installed at five elevations between 1400 and 2400 m.a.s.l. in the Tambito area of southern Colombia. Harp-based cloud water interception (*CWI*) proved significantly higher in the upper montane cloud forest (UMCF) zone (>2200 m.a.s.l.) compared to the lower montane cloud forest (LMCF) zone. Average *CWI* values ranged from 0.1 to 1.3 mm day^{-1} but there was no significant correlation between altitude and average *CWI* below the main cloud belt, presumably due to differences in exposure between sites. Furthermore, *CWI* exhibited differences in seasonality with elevation, with maximum *CWI*/rainfall ratios in the UMCF zone during the dry season and earlier in the year at lower elevations. In addition, a comparison of the wire harp to the more widely use method of measuring the excess throughfall minus rainfall was explored in the LMCF.

INTRODUCTION

Despite the importance of cloud water interception (*CWI*) to the water budget of tropical montane cloud forest (TMCF), its quantification remains difficult. Various types of artificial cloud water collectors have been used to monitor fog incidence at many TMCF sites (Bruijnzeel and Proctor, 1995; Bruijnzeel 2001), but there is no standard device to monitor *CWI*. Interpreting the results obtained with fog collectors is confounded by the difficulty to distinguish rainfall from fog (cf. Frumau *et al.*, this volume; Giambelluca *et al.*, this volume; Tanaka *et al.*, this volume), whereas the fog catch of some types of gages is also affected by wind direction (Juvik and Nullet, 1995a; García-Santos, 2007). More importantly, there is the intrinsic problem that any relationships between the catch of these artificial interceptors and forest vegetation are location- and instrument-specific (Bruijnzeel and Proctor, 1995; Juvik and Nullet, 1995b; Giambelluca *et al.*, this volume).

The Centre of Pacific Environmental Studies Tambito (CEAT) is one of the few sites with cloud forest in Colombia where fog incidence has been monitored in both upper and lower montane cloud forest. The Tambito Forest Reserve is located on the steep Pacific slopes of the Western Cordillera of the Colombian Andes in the municipality of El Tambo, Cauca province (2° 30′ N, 76° 60′ W). Rainfall in the area is very high (3600–7000 mm year^{-1} depending on elevation) with the highest monthly totals occurring between October and May (>500 mm month^{-1}) but with regular rain-free periods between June and mid-September (Letts *et al.*, this volume). Wind speeds are typically very low (<2 m s^{-1}).

This chapter presents amounts of *CWI* as obtained with two types of fog collectors in the Tambito area, viz. a modified Grunow-type collector and a wire harp collector that was shielded against direct precipitation. Variations in catch with elevation and season were explored using harps placed at five different elevations between 1375 and 2340 m.a.s.l. In addition, the comparative performance of the two collector types under the prevailing climatic conditions was examined at two locations in the lower reaches of the study watershed. Finally, *CWI* was estimated as the differential between measured throughfall and rainfall at one site in the lower reaches of the watershed (González, 2005).

METHODS

Passive fog gages

The modified Grunow-type collector (*MGC*; cf. Grunow, 1952; Cavelier *et al.*, 1996) consisted of a 30-cm high cylinder of 10-cm diameter made out of plastic wire mesh (#5 mm, thickness

Tropical Montane Cloud Forests: Science for Conservation and Management, eds. L. A. Bruijnzeel, F. N. Scatena, and L. S. Hamilton. Published by Cambridge University Press.

0.5 mm). The cylinder was mounted on top of a funnel that drained into a collecting bottle through a plastic tube. No shielding against rainfall was applied. The volume of rainfall captured by another funnel of the same dimensions placed nearby was subtracted from the volume intercepted by the *MGC* and the difference was taken to represent *CWI*. No correction was made for any extra rain water captured by the cylinder because of the prevailing low wind speeds. Two *MGC* were used, one placed in a pasture area close to the CEAT accommodation facilities (Campo station, 1450 m.a.s.l.) and one in a secondary forest (Rio station, 1374 m.a.s.l.). Both gages were suspended at canopy height (*c*. 2 m above ground level). Measurements were taken every 3 days from August 1999 until May 2001.

The modified wire harp collector (*MWH*) (cf. Nagel, 1956; Goodman, 1985) consisted of a 1 × 1 m PVC frame equipped with 48 vertical nylon strings of 1-mm diameter each and spaced at 2-cm intervals. A plastic roof (2 m^2) was placed above the harp to prevent contamination by rainfall. Cloud water coalescing on the wires drained into slightly sloping near-horizontal plastic tubing through 5-mm holes, to minimize evaporation and lateral contamination by rainfall. The water was ultimately collected in a bottle (1 l). Up to six *MWH* were installed at each of five sites and suspended next to each other at 2 m from the ground using ropes and poles placed between trees (Figure 33.1). Table 33.1 summarizes the locations and general characteristics of each collector site. Volumes of fog water captured by the *MWH* were measured only fortnightly due to accessibility limitations in the very steep terrain. Measurements were made between July 2000 and May 2001.

Cloud water interception estimated as the differential between throughfall and rainfall

Throughfall (*TF*) was collected manually at the Rio site using 30 funnels (1265 cm^2 total surface area). The funnels were attached to small bottles which were suspended to a net (6.5 m^2) placed ~1 m above the ground in the understory of a secondary forest with a well-developed canopy (leaf area index, $LAI = 2.77$). Throughfall typically has a high spatial variability and therefore the gages were relocated randomly after each reading (Lloyd and Marques, 1988; Holwerda *et al.*, 2006). The net facilitated this

Table 33.1 *Locations and general description of the cloud water interception stations at Tambito*

Station	Elevation	Forest type	Aspect	Coordinates
Rio	1,374 m.a.s.l.	Secondary forest	SW	2.50555 N
	river junction	LMCF		76.99913 W
Campo	1,450 m.a.s.l.	Pastures	SW	2.50634 N
	Tambito slope	LMCF		76.99911 W
Bosque	1,650 m.a.s.l.	Primary forest	NE	2.5139 N
	Palo Verde slope	LMCF		76.99478 W
Pela Huevos	2050 m.a.s.l.	Secondary forest/ pastures	SW	2.50556 N
	Tambito slope	LMCF		76.98754 W
Asomadero	2,340 m.a.s.l.	Secondary forest	SW	2.5047 N
	Tambito slope	UMCF		76.9753 W

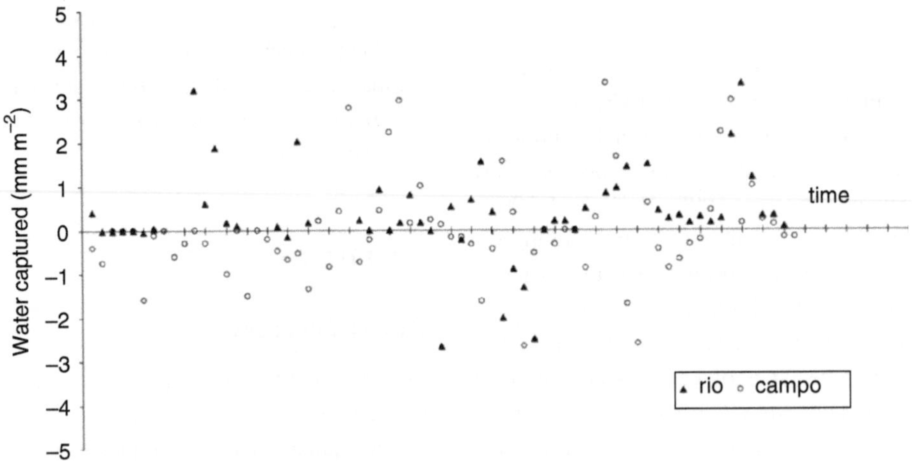

Figure 33.1. Scatter-plot of apparent cloud water interception based on the difference in amounts of water captured by a Grunow-type collector and an ordinary rain gage.

since gage location could be determined easily on the net and the net-suspended bottles were self-leveling. This design provided an estimate of *CWI* only during events when *TF* exceeded rainfall (cf. Holder, 2003). The advantage of estimating *CWI* with this method is that it reflects the efficiency of interception of native vegetation, rather than an instrumental standard. Data on *TF* were collected during 10 days in the summer of 1999, 15 days in the summer of 2000, and 5 days during each of the first 6 months of 2000.

RESULTS AND DISCUSSION

Performance of the modified Grunow-type collector

The *MGC* and rain gages at the Rio and Campo stations yielded 164 data pairs but on as many as 43% of all sampling occasions the rain gage gave higher values than the *MGC*, thereby making it impossible to estimate amounts of *CWI* (Figure 33.2). Normally, Grunow-type collectors tend to overestimate *CWI* because of extra amounts of rainfall being trapped by the wire mesh as compared to an ordinary rain gage, particularly under windy conditions (Cavelier *et al.*, 1996; cf. Bruijnzeel, 2001). Also, saturation of the mesh by rain drops may occur, thereby increasing the effective area of interception (Schemenauer and Cereceda, 1994). However, in view of the low wind speeds in the study area and the use of a rather coarse mesh (5 mm vs. the customary 1 mm), such explanations are less than satisfactory. Evaporation of intercepted droplets from the wire mesh before draining into the container may lead to underestimation of *CWI* (cf. Dawson, 1998; Frumau *et al.*, this volume). Yet it is unlikely that this process would be sufficiently important to mask the entire catch of fog water in view of the high levels of cloudiness and humidity in the area (González, 2000; Letts *et al.*, this volume) and the coarseness of the wire mesh used.

Despite these limitations, the mean estimated fog incidence at Tambito on all days when apparent *CWI* values were positive was within the reported range of 0.2–2.2 mm day^{-1} for TMCF environments using similar types of fog gages (Bruijnzeel and Proctor, 1995). Average daily *MGC*-based *CWI* at the Rio station was 0.94 ± 1.20 mm (range 0.05–6.40 mm; $n = 42$) vs. 1.15 ± 0.90 (range 0.03–3.64 mm, $n = 37$) at the Campo station. Although the difference is not statistically significant, the lower value derived for the Rio station may reflect a lesser exposure to wind compared to the more open Campo site.

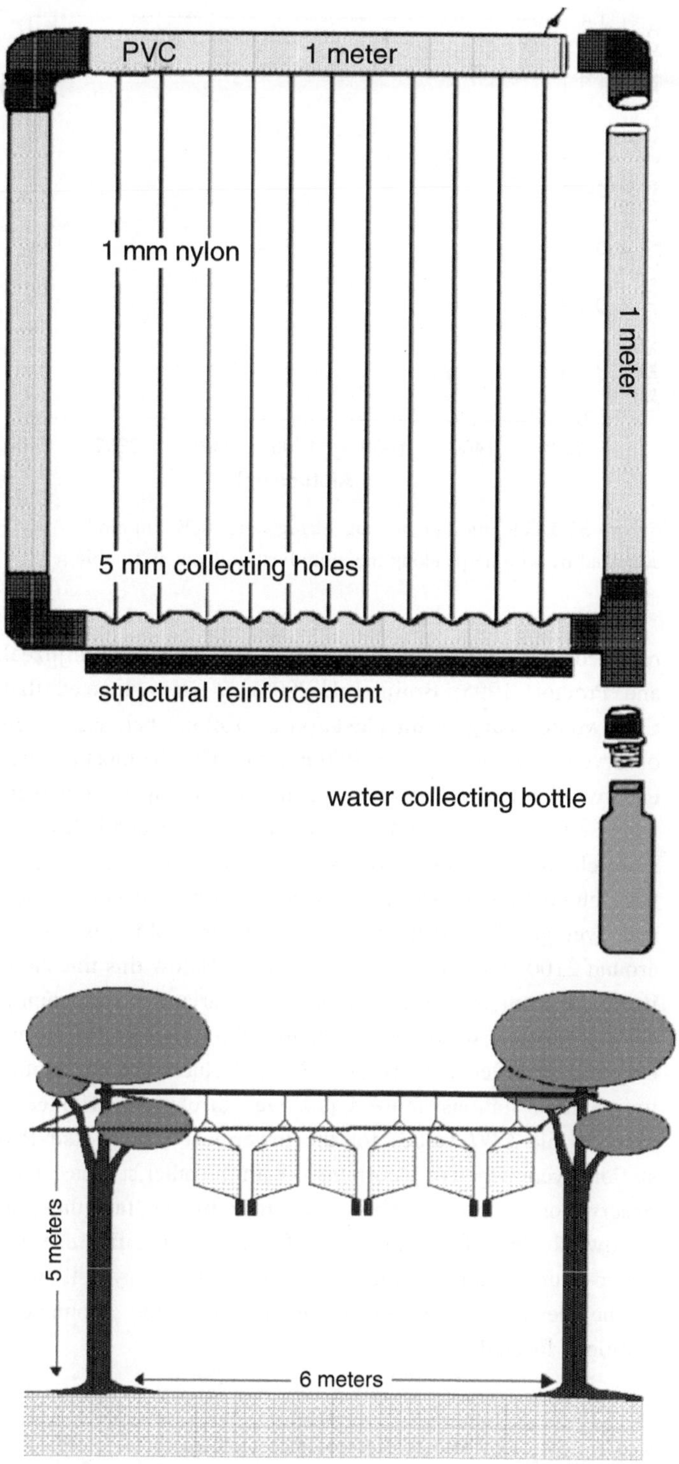

Figure 33.2. Schematic diagram of the modified wire harp (as used in the year 2000) and its placement in relation to the forest canopy at Tambito.

Variations in wire-harp-based *CWI* estimates with elevation

The data collected between July 2000 and May 2001 using *MWH* gages at five different elevations (Table 33.1) suggested average daily *CWI* values to range from 0.12 ± 0.06 mm in lower Tambito (*c.* 1400 m.a.s.l.) to 2.22 ± 1.29 mm in the upper parts (2340 m.a.s.l.). As such, with the exception of the uppermost site, the Tambito data fall mostly in the lower part of the range

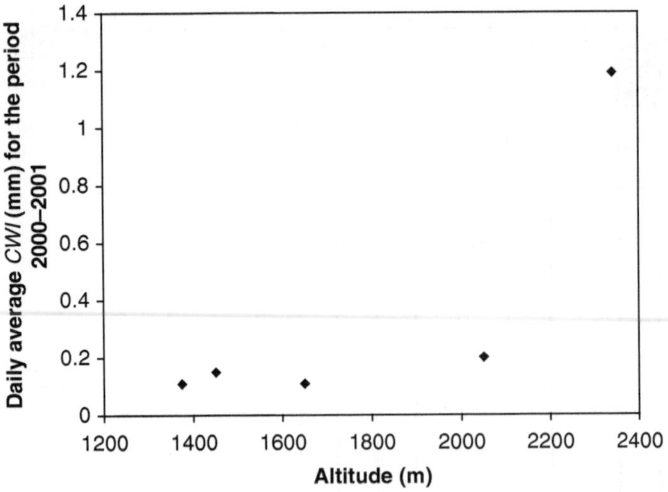

Figure 33.3. Variation in amounts of fog water (*CWI*, in mm) as collected by wire harps along the altitudinal gradient at Tambito.

reported for similar instrumentation in TMCF (Bruijnzeel and Proctor, 1995; Bruijnzeel, 2005). It was expected that *CWI* would change with elevation as a cloud belt was often observed at altitudes above 2200 m.a.s.l., with the most persistent low cloud being observed on the upper ridges. However, because four out of five *MWH* stations were located below the main cloud belt and only one station (Asomadero) was within the cloud belt, there was no linear relationship between altitude and average *CWI*, although a distinct threshold was visible around 2100–2200 m.a.s.l. (Figure 33.3). Below this threshold it would seem that micro-topographic variations influencing site exposure to wind and fog caused the rather minor differences in observed *CWI* (Figure 33.3). Values of *CWI* for the upper three stations showed positive correlations with each other, while *CWI* values for the lowermost Campo and Rio stations were negatively correlated with the other stations. This observation may be partially explained by the fact that the windward side of the mountain facing the Pacific receives water-saturated air flowing up the slopes that may spill over to the leeward part of the mountain where the Asomadero station is located.

Seasonality of *CWI* at different elevations

The development of low cloud in the rugged terrain of the study area may show a complex spatial pattern that may vary seasonally (cf. Lawton *et al.*, this volume). In particular, smaller changes in cloud incidence would be expected in the UMCF zone above 2200 m.a.s.l. compared to the LMCF zone. Figure 33.4 shows the seasonal fog incidence based on the *MWH* measurements at the five stations. Values of *CWI* within the upper MCF zone (Asomadero station, 2340 m.a.s.l.) seems to exhibit a bimodal seasonal distribution with the highest values coinciding with the season of maximum rainfall (October–December), and somewhat elevated fog incidence in early summer as well. Interestingly, the average value of 3.7 mm day^{-1} as registered at the Asomadero station between October 2000 and January 2001 is close to the highest fog-gage based *CWI* values reported in the literature for TMCF sites (4.0–6.3 mm day^{-1}; Bruijnzeel and Proctor, 1995; Bruijnzeel, 2005), although in the latter cases contributions by wind-driven rain could not be excluded (Bruijnzeel, 2005). It is also close to the 4.06 mm day^{-1} obtained for the last 3 months of 1999 at the 20 de Julio station (2200 m.a.s.l.) during the early stages of this study (González, 2000). Variable but lower values of *CWI* were observed during the rest of the year at Asomadero. A somewhat similar but much more subdued pattern was observed at the Bosque station (1650 m.a.s.l., and well below the main cloud belt), but not at the Pela Huevos station (2050 m.a.s.l.), despite the fact that the latter was situated much closer to the average cloud condensation level. Indeed, at Pela Huevos and the two lowermost stations there was no marked seasonal cycle in *CWI*, although values increased moderately in December and February (Figure 33.4). Minimum values of *CWI* for most stations were observed in August, which corresponds with the last month of the "dry" season in terms of rainfall (González, 2000). The smallest variability between stations was observed in February, when all stations gave *CWI* values of less than 1 mm day^{-1} whereas the largest variability occurred in November (Figure 33.4).

Seasonality in *CWI* compared to rainfall

Given the great differences in water inputs in the form of rainfall (*P*) and fog (as represented by the biweekly *MHW*-based measurements of *CWI*) in the wet and dry seasons (Figure 33.4), it is of interest to explore during what time of year *CWI* is potentially the most important at the respective elevations. Whilst the results should be considered preliminary, given that average daily *CWI* values for each month are based on only two biweekly measurements and not necessarily representative of fog capture by the vegetation, the monthly *CWI*/*P* ratios (%) do provide a first indication of differences in the seasonal contribution of fog to overall precipitation at the different stations (Table 33.2).

At the highest station, the relative average monthly contribution of *CWI* was greatest (7.9%), despite the fact that rainfall was more abundant than at lower altitudes (9890 vs. 5167 mm). The rest of the stations gave *CWI*/*P* values of 1.1–1.6%, suggesting that contributions by fog to total water inputs below the main cloud belt at Tambito are rather minor. It is pertinent to highlight that Tambito is situated on the leeward side of the mountain and experiences very light wind conditions,

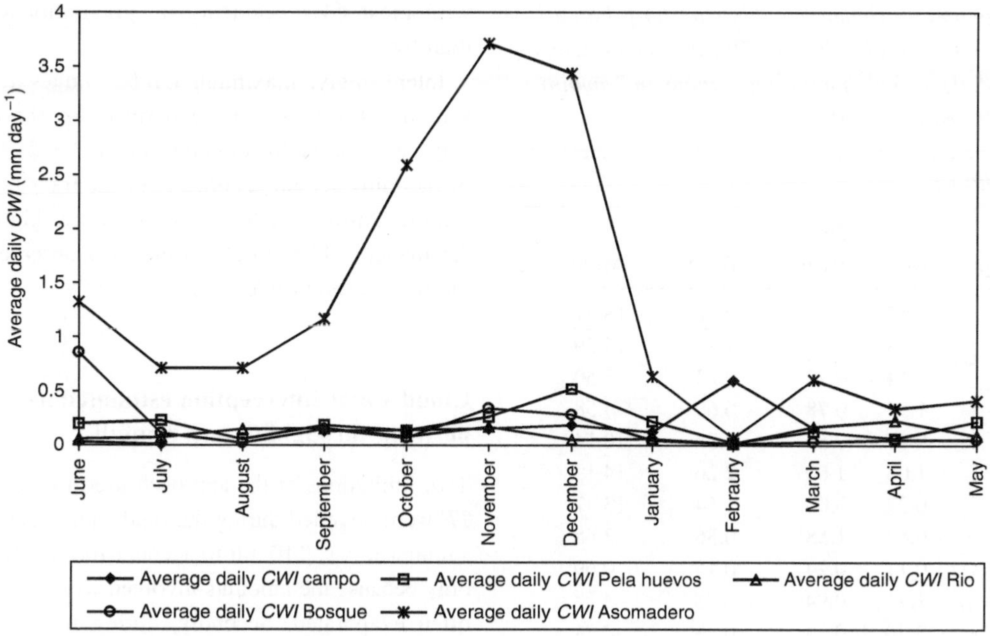

Figure 33.4. Seasonality of amounts of fog water (*CWI*, in mm) as collected by wire harps along the altitudinal gradient at Tambito.

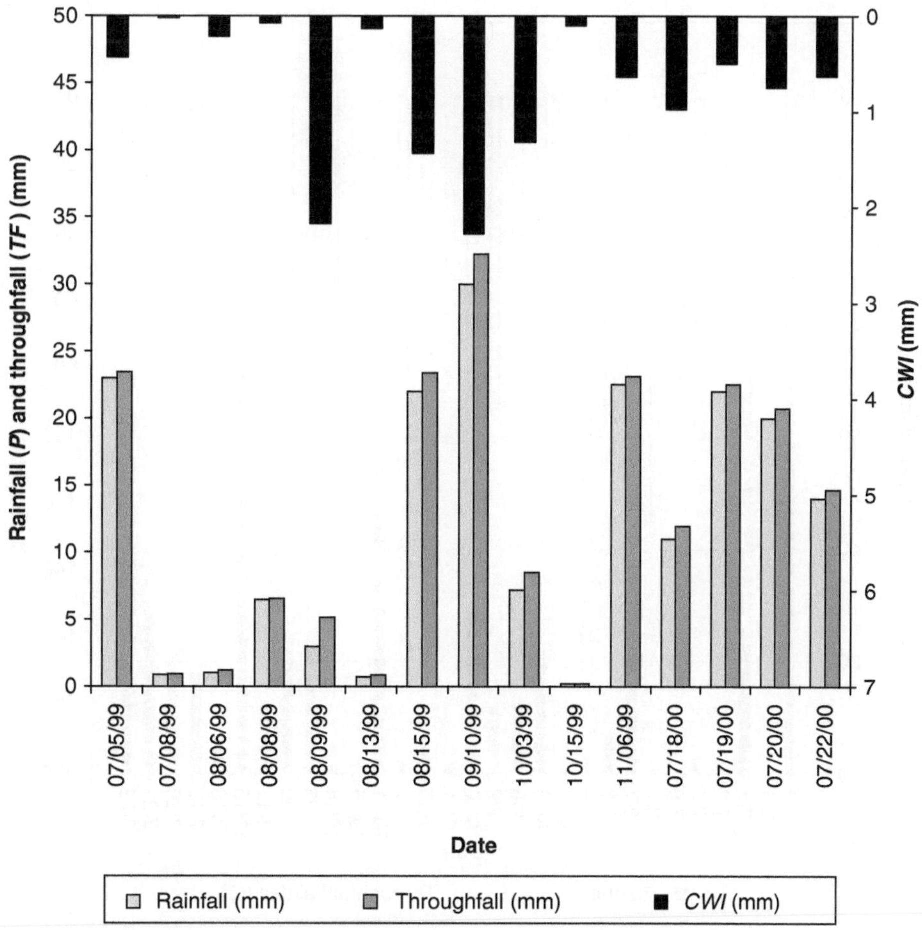

Figure 33.5. Amounts of cloud water interception (mm) in a secondary lower montane cloud forest at Tambito (1375 m.a.s.l.). Estimates are based on excess amounts of throughfall over incident rainfall for 15 storms between July 1999 and July 2000.

Table 33.2 *Average and maximum monthly wire-harp-based cloud water interception to rainfall* (CWI/P) *ratios (expressed as percentage of rainfall) at five monitoring stations in Tambito between June 2000 and May 2001*

	CWI/P(%)				
Month	Campo	Rio	Pela Huevos	Bosque	Asomadero
June	0.26	0.55	2.50	9.32	18.91
July	0.41	2.38	4.57	1.80	14.29
August	0.21	3.14	0.72	0.23	5.50
September	1.26	1.52	0.78	0.60	4.58
October	0.96	0.57	0.84	0.38	10.14
November	0.94	1.05	1.60	2.26	14.19
December	0.97	0.27	3.06	1.54	13.47
January	1.17	0.58	1.58	0.36	3.48
February	5.71	0.11	0.24	0.15	0.69
March	1.59	1.71	0.94	0.25	3.58
April	0.00	3.30	0.53	0.38	2.74
May	0.00	0.81	1.58	0.37	2.73
Annual	1.12	1.33	1.58	1.47	7.86
Max	5.71	3.30	4.57	9.32	18.91

with most *CWI* occurring by gravitational deposition rather than by impaction.

Interestingly, maximum *CWI/P* values for the upper three stations (4.6–18.9%) were derived for the dry season (June–July), and for the lower two stations (3.3–5.7%) in the latter half of the rainy season (February) or the transition toward the drier season (April) (Table 33.2). However, high *CWI/P* values at Asomadero (13.5–14.2%) were also observed at the height of the wet season (Table 33.2).

Cloud water interception estimated as the differential between throughfall and rainfall

Two limitations of the approach used to measure *CWI* based on *TF* were unveiled during the field campaigns. Firstly, during the summer it was difficult to account for any *TF* generated by fog-only because the amounts involved were so small that they were often evaporated. Secondly, during rain events exceeding 40 mm, the capacity of the collection bottles was often exceeded. Fortunately, 90% of the days sampled in 1999 and 2000 had less than 40 mm of daily rainfall.

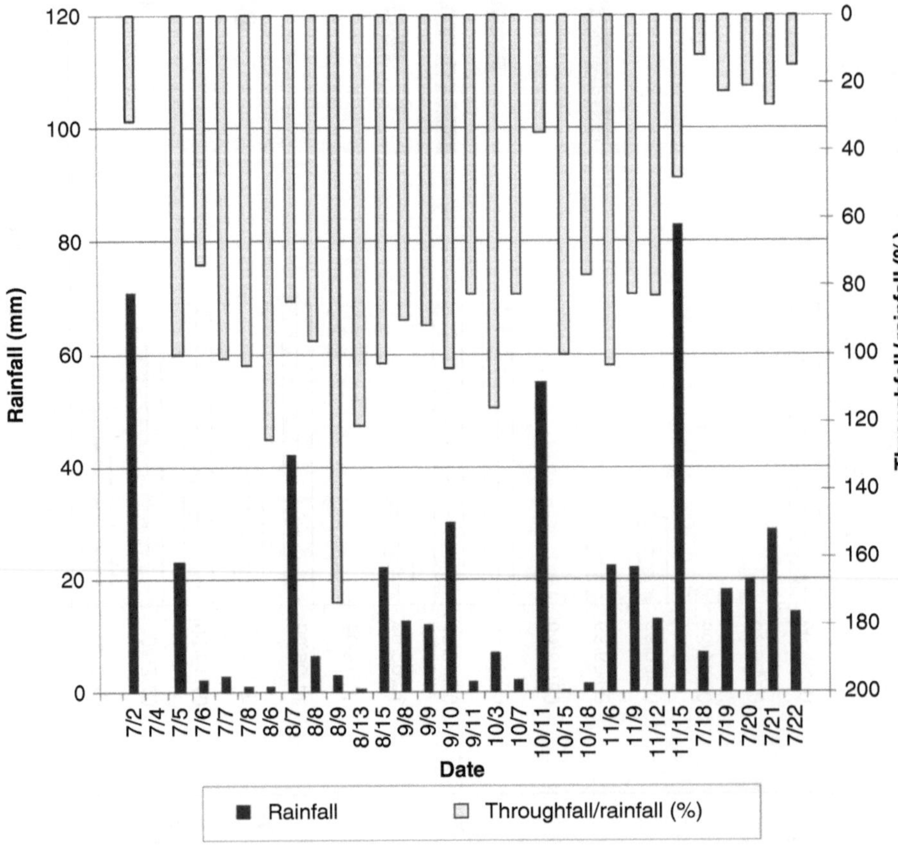

Figure 33.6. Throughfall as a percentage of rainfall in a secondary lower montane cloud forest at Tambito (1375 m.a.s.l.) between July 1999 and July 2000.

Average daily P during the 40 days sampled was 13.7 mm whereas average TF was 9 mm (65.7% of P). However, on 15 out of these 40 days, TF exceeded P, which was attributed to the capturing of fog (Figure 33.5). It should be noted that the spatial variability (standard error) of TF was sometimes greater than the inferred CWI during the summer months which brings into question the certainty of the results. However, even after removing those days from the analysis the average was not greatly affected.

The highest CWI was recorded on 9 August and 10 September 1999, with values of 2.2 mm and 2.3 mm, respectively. Average daily CWI estimated using this method was 0.8 mm. Since TF/P ratios were determined only for 40 days, it is difficult to compare these results with those of the wire harps which were monitored over a full year and at biweekly intervals only. The MWH gage at the Rio site where TF was measured (1375 m.a.s.l.) had an average daily catch of 1.1 mm between June 2000 and May 2001. Extrapolating the results for the 15 days with positive TF-based CWI to a whole year gave an estimated 280 mm of additional input (9.3% of the recorded annual rainfall total at this location). Values of CWI did not show a significant correlation with rainfall (Figure 33.6).

CONCLUSION

Under the high-rainfall conditions prevailing at Tambito, modified Grunow-type fog gages did not perform satisfactorily, despite very low wind speeds. Wire-harp-based cloud water interception (CWI) rates were significantly higher in the UMCF zone above 2200 m.a.s.l. and showed a different seasonality than at lower elevations. Because of the relatively sheltered location of the area and the low wind speeds, fog interception occurred mainly via deposition rather than impaction. This, together with the high rainfall, caused average monthly CWI/rainfall ratios to be very low (1.1–1.6%) except in the UMCF zone (7.9%). Maximum relative fog incidence tended to occur during the dry season at higher elevations, and earlier in the year at lower elevations. Values of CWI derived from short-term ($n = 40$) measurements of rainfall and throughfall at 1375 m.a.s.l. during the summer season suggested an average value of 0.8 mm day^{-1}.

ACKNOWLEDGEMENT

This project was carried out within the framework of the Hydrology, Ecology, and Regional Biodiversity (HERB) project of King's College London coordinated by Dr. Mark Mulligan.

REFERENCES

Bruijnzeel, L. A. (2001). Hydrology of tropical montane cloud forest: a reassessment. *Land Use and Water Resources Research*: **1**: 1–18.

Bruijnzeel, L. A. (2005). Tropical montane cloud forests: a unique hydrological case. In *Forests, Water and People in the Humid Tropics*, eds. M. Bonell and L. A. Bruijnzeel, pp. 462–483. Cambridge, UK: Cambridge University Press.

Bruijnzeel, L. A., and J. Proctor (1995). Hydrology and biogeochemistry of tropical montane cloud forests: what do we really know? In *Tropical Montane Cloud Forests*, eds. L. S. Hamilton, J. O. Juvik, and F. N. Scatena, pp. 38–78. New York: Springer-Verlag.

Cavelier, J., D., Solis, and M. A. Jaramillo (1996). Fog interception in montane forests across the Central Cordillera of Panama. *Journal of Tropical Ecology* **12**: 357–369.

Dawson, T. E. (1998). Fog in the California redwood forest: ecosystem inputs and use by plants. *Oecologia* **117**: 476–485.

García-Santos, G. (2007). *An ecohydrological and soils study in a montane cloud forest in the National Park of Garajonay, La Gomera (Canary Islands, Spain)*. PhD Thesis, VU University Ámsterdam, Ámsterdam, The Netherlands. [http://www.falw.vu.nl/nl/onderzoek/earth-sciences/geo-environmental-science-and-hydrology/hydrology-dissertations/index.asp].

González, J. (2000). Monitoring cloud interception in a tropical montane cloud forest of the Southwestern Colombian Andes. *Advances in Environmental Monitoring and Modelling* **1**: 97–117. Also available at www.kcl.ac.uk/advances.

González, J. (2005). Cloud interception by trees in a tropical montane cloud forest of southwestern Colombia. Ph.D. thesis, King's College London, University of London, UK.

Goodman, J. (1985). The collection of fog drip. *Water Resources Research* **21**: 392–394.

Grunow, H. (1952). Fog precipitation. *Berichte des Deutschen Wetterdienstes in der U.S. Zone, Bad Kissingen* **35**(42): 30–34.

Holder, C. D. (2003). Fog precipitation in the Sierra de las Minas Biosphere Reserve, Guatemala. *Hydrological Processes* **17**: 2001–2010.

Holwerda, F., F. N. Scatena, and L. A. Bruijnzeel (2006). Throughfall in a Puerto Rican lower montane rain forest: a comparison of sampling strategies. *Journal of Hydrology* **327**: 592–602.

Juvik, J. O., and D. Nullet (1995a). Comments on "a proposed standard fog collector for use in high elevation regions." *Journal of Applied Meteorology* **34**: 2108–2110.

Juvik, J. O., and D. Nullet (1995b). Relationships between rainfall, cloud-water interception and canopy throughfall in a Hawaiian montane forest. In *Tropical Montane Cloud Forests*, eds. L. S. Hamilton, J. O. Juvik, and F. N. Scatena, pp. 165–182. New York: Springer-Verlag.

Lloyd, C. R., and A. O. Marques Jr. (1988). Spatial variability of throughfall and stemflow measurements in Amazonian rain forest. *Agriculture and Forest Meteorology* **42**: 63–73.

Nagel, A. (1956). Fog precipitation on Table mountain. *Quarterly Journal of the Royal Meteorological Society* **82**: 452–460.

Schemenauer, R. S., and P. Cereceda (1994). A proposed standard fog collector for use in high elevation regions. *Journal of Applied Meteorology* **33**: 1313–1322.

34 Relationships between rainfall, fog, and throughfall at a hill evergreen forest site in northern Thailand

N. Tanaka
The University of Tokyo, Bunkyo-ku, Tokyo, Japan

K. Kuraji
The University of Tokyo, Seto, Aichi, Japan

C. Tantasirin
Kasetsart University, Bangkok, Thailand

H. Takizawa
Nihon University, Fujisawa, Kanagawa, Japan

N. Tangtham
Kasetsart University, Bangkok, Thailand

M. Suzuki
The University of Tokyo, Bunkyo-ku, Tokyo, Japan

ABSTRACT

This study used hourly data of rainfall, water captured by a sheltered fog gage, and wind speed as collected at a montane forest site in northern Thailand during nearly 3 years, to test the efficiency of the rain-protected passive fog gage as a predictor of fog occurrence. To separate possible contributions by wind-driven rain (*WDR*) from fog, the maximum rate of water input to the fog gage during rainless conditions (Fog_{max}) was derived as a function of wind speed. During periods with rain and fog, the fog gage often produced values above the Fog_{max} line, suggesting contributions by *WDR*. The specific conditions of rainfall intensity and wind speed for which this happened were identified and the corresponding data were excluded from the fog data-set for subsequent reanalysis. Based on the recalculated data-set, inter-annual and seasonal variations as well as the diurnal pattern of fog occurrence at the studied forest are described. Fog-induced canopy drip during rainless periods was only 19.3 mm over the 3 years, being less than 0.5% of total throughfall and *c*. 33% of the corresponding catch by the fog gage (58 mm). However, the fog gage captured nearly 18 times more water (1033 mm) during all times when *WDR* could reasonably be excluded.

INTRODUCTION

Although tropical montane forests are widely distributed over South-East Asia (Doumenge *et al*., 1995; Bubb *et al*., 2004; Mulligan, this volume), little is known about the occurrence and associated hydrological impacts of fog in these forests (Bruijnzeel and Proctor, 1995; Bruijnzeel, 2005). Forests in northern Thailand at elevations above 1000 m.a.s.l. are classified as tropical montane forests and these forests are occasionally subjected to fog (e.g. Werner and Sansitsuk, 1993). Concerns over the hydrological role of the forests are increasing, because populations living downstream expect them to maintain a steady supply of water for agricultural and other activities. There have been a number of hydrological studies in the montane forests of northern Thailand (e.g. Tangtham, 1974; Giambelluca *et al*., 1996; Tanaka *et al*., 2003), but few have described the role of fog (cf. Tanaka

Tropical Montane Cloud Forests: Science for Conservation and Management, eds. L. A. Bruijnzeel, F. N. Scatena, and L. S. Hamilton. Published by Cambridge University Press.

et al., 2005). Fog may not only constitute an extra input to the forest water budget but also result in lowered evaporation totals (Bruijnzeel and Proctor, 1995; Bruijnzeel 2005; cf. García-Santos, 2007) and produces wet canopy conditions that may affect both photosynthetic (Letts and Mulligan, 2005) and forest productivity rates (Letts *et al.*, this volume).

To monitor fog occurrence, often a passive fog collector of some kind is installed in the open (e.g. Juvik and Nullet, 1995; Cavelier *et al.*, 1996; Hutley *et al.*, 1997; González, 2000; McJannet *et al.*, 2007; Marzol-Jaén *et al.*, this volume). The present study employed a cylindrical gage of the type described by Juvik and Nullet (1995) which was equipped with a "hat" to shield it against direct rainfall inputs. This chapter examines the performance of the sheltered gage as a fog sensor under the climatic conditions prevailing in upland northern Thailand. In addition, allowing for the performance of the gage, patterns of fog occurrence and crown drip resulting from fog during rainless periods were investigated.

MATERIALS AND METHODS

Study site

The present study was conducted at Kog-Ma watershed (18° 48′ N, 98° 54′ E, 1265–1450 m.a.s.l.) on the east-facing slope of Mount Pui (1685 m.a.s.l.), 10 km west of Chiang Mai, northern Thailand (Figure 34.1). According to the Preliminary Forest Land Use Assessment, conducted in 2000 by the Wildlife and Plant Conservation Department of Thailand, approximately 56% of the land in northern Thailand is classified as forest, with deciduous forest mainly occurring in the lowlands (73000 km^2) and evergreen forest in montane areas (defined as lying above 1000 m.a.s.l.; 20000 km^2). The vegetation at Kog-Ma watershed is an undisturbed hill evergreen forest with canopy heights ranging from 25 to 40 m. Dominant tree species include *Lithocarpus*, *Quercus*, and *Castanopsis* spp., whereas undergrowth, shrubs, and various epiphytes are present as well (Tangtham, 1974).

Mean annual rainfall at Kog-Ma watershed between 1966 and 1978 was 2084 mm (Chunkao *et al.*, 1981) and rainfall is highly seasonal. Most of the annual total occurs in the wet season from April to October whereas no significant rain falls during the dry season from November to March. Mean annual temperature during the study period (December 1999 – November 2002) was 19.7 °C. April and December were the warmest and coldest months, with monthly mean temperature of 23.3 and 16.0 °C, respectively. The seasonal change in atmospheric humidity is considerable, with a mean relative humidity of *c.* 86% during the 7-month wet season vs. 68% during the 5-month dry season. Diurnal and seasonal trends in wind speed and direction are also variable as described by Komatsu *et al.* (2003).

Instrumentation

Rainfall above the canopy was measured hourly using a tipping-bucket rain gage (No. 34-T, Ohta Keiki Corporation; resolution 0.5 mm) located at the top of a 50-m tall observation tower (see Figure 34.1 for location). Wind speed was recorded at the same height as the rain gage using an AC750, Makino Ohyousokki anemometer. To monitor fog incidence, a cylindrical louvered aluminum screen (height 40.6 cm, diameter 12.7 cm; Juvik and Nullet, 1995) was also installed at the top of the tower. The apparatus was covered by a circular hat of 57 cm diameter in

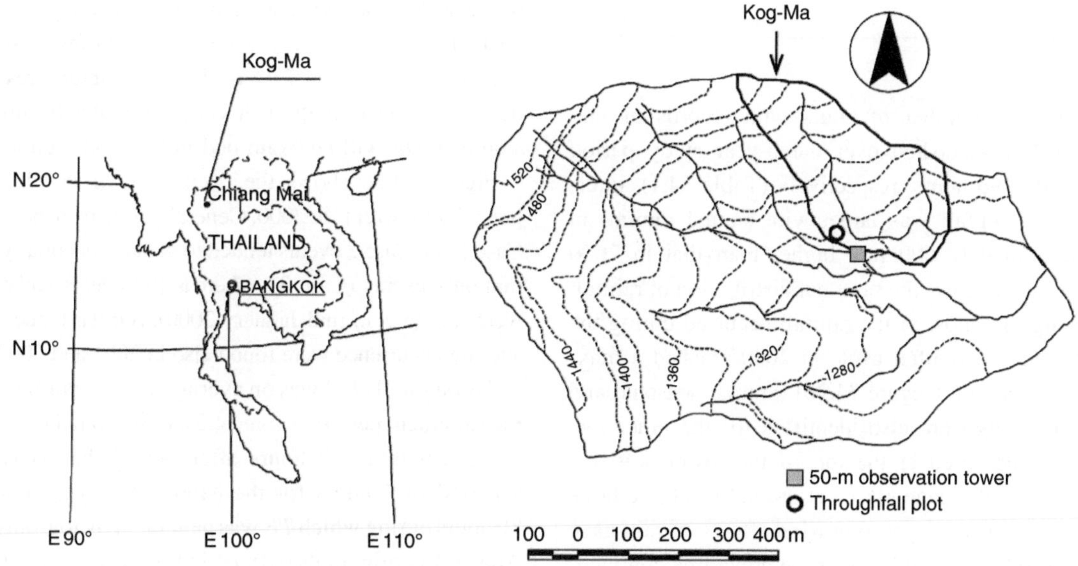

Figure 34.1. Location and topography of Kog-Ma watershed plus locations of observation tower and throughfall measurement plot.

Table 34.1 *Rainfall and fog gage capture observed at Kog-Ma watershed in northern Thailand during the 3-year study period. Number of hours with occurrence of rainfall and fog capture are also indicated*

Year[a]	Rainfall		Fog gage capture			
	Amount	Number of hours	Amount	Number of hours		
				Total	Without rain (Case 1)	With simultaneous rain (Case 2)
	(mm)	(hours)	(ml)	(hours)	(hours)	(hours)
2000	1610	565	26078	560	282	278
2001	1660[b] (1595)[d]	592[b] (527)[d]	16454[c]	423[c]	189[c]	234[c]
2002	2244	639	33363	539	219	320
Three-year total	5513 (5448)[d]	1796 (1731)[d]	75894	1522	690	832

[a] Year 2000: 1 December 1999 to 30 November 2000; Year 2001: 1 December 2000 to 30 November 2001; Year 2002: 1 December 2001 to 30 November 2002.
[b] Rainfall and number of hours with rain, those were observed during data gap period of the fog gage (see footnote *c* to this table), are included in these values.
[c] Note that, from March 2001 to May 2001, the fog gage was out of operation.
[d] Rainfall and number of hours with rain, excluding those observed in the data gap of the fog gage, are given in parentheses.

an attempt to prevent rain drops from entering the gage. Collected water volumes were measured hourly using a covered tipping-bucket rain gage (No. 34-T, Ohta Keiki Co.; with a resolution of 15.7 ml or 0.3 mm of water on the cross-sectional area of gage screen) connected to a HOBO Event logger (Onset Computer Corporation). Throughfall (*TF*) was observed hourly in a plot situated 50 m north-west of the tower using four tipping-bucket rain gages of the same type as that installed on the tower (orifice area 314 cm^2 per gage). The gages were placed randomly in a fixed spatial arrangement. Observations were made from 1 December 1999 to 4 November 2002.

RESULTS

The total rainfall and number of hours with records of rain observed at Kog-Ma watershed over the 3-year study period was 5513 mm and 1796 hours, respectively (Table 34.1). Interannual variation in rainfall was large, with annual rainfall in 2002 being approximately 600 mm higher than that in 2000 and 2001. Figure 34.2a shows the seasonal distribution of rainfall in 2000, indicating that most of the rainfall occurred during the wet season. The diurnal rainfall cycle in 2000 reached a maximum during the afternoon (Figure 34.2b). Similar seasonal and diurnal rainfall patterns were also identified for the other two years. Rainfall as measured at the top of the tower was 1% greater than the catch of a gage placed at ground level in a large clearing some 250 m away, suggesting wind effects on rain gage catch at the tower to be negligible under the prevailing continental tropical conditions (Tanaka *et al.*, 2005).

The total volume of water collected by the fog gage was 75/894 ml over the 35-month study period (Table 34.1) or 1472 mm of water depth per unit vertical cross-sectional area of the gage. The fog gage recorded water inputs for a total of 1522 hours, which was only 274 hours less than the total period with rainfall being recorded (Table 34.1). Taken at face value, this would suggest that fog interception by the forest canopy should constitute an important hydrological process in the Kog-Ma watershed. However, it cannot be excluded that the fog gage captured some wind-driven rain (*WDR*) during particularly windy events. Therefore, precipitation events were classified into three cases, i.e. fog-only (Case 1; 690 hours), fog plus simultaneous rainfall (Case 2; 832 hours), and rain-only (Case 3, 899 hours) (Table 34.1). Percentages of fog gage capture during Case 1 and Case 2 events were 38% and 72% of the total capture, respectively. The degree of contamination of captured water by unwanted inclusion of *WDR* will be examined in the next section.

Figure 34.2c shows the occurrence of the different cases of precipitation events in 2000. Generally, fog incidence associated with Case 1 and Case 2 events tended to occur sequentially before or after rain-only events (Case 3), whereas fog events without any rainfall were rare (e.g. in mid-January 2000). Similar tendencies of rainfall and fog occurrence were found also in 2001 and 2002 (not shown).

Throughfall (*TF*) was on average 92% of rainfall over the entire measurement period (Table 34.2). Drip from the canopy generally stopped within 1–2 hours after rainfall had ceased. Assuming a period of 3 hours for the canopy to dry up completely gave 80 hours during which *TF* was generated by fog-only (Table 34.2). Annual fog drip totals defined in this way were only 7.6, 5.6, and 6.1 mm in 2000, 2001, and 2002, respectively. Thus, fog drip

Table 34.2 *Throughfall and fog-induced canopy drip during rainless period observed at Kog-Ma watershed during the 3-year study period*

Year[a]	Throughfall		Fog-induced canopy drip during rainless period[b]			
	Amount (mm)	Fraction of annual rainfall (%)	Amount (mm)	Fraction of annual throughfall (%)	Number of hours (hours)	Fog gage capture (ml)
2000	1500	93	7.6	0.5	29	1319
2001[c]	1561	94	5.6	0.3	26	848
2002	2031	91	6.1	0.3	25	832
Three-year total	5092	92	19.3	0.4	80	2999

[a] See footnote *a* to Table 34.1.
[b] Fog-induced canopy drip during rain-free hours indicates throughfall during hours when throughfall occurred with simultaneous records of fog gage capture, despite absence of rainfall during the past 3 hours (see text).
[c] See footnote *c* to Table 34.1.

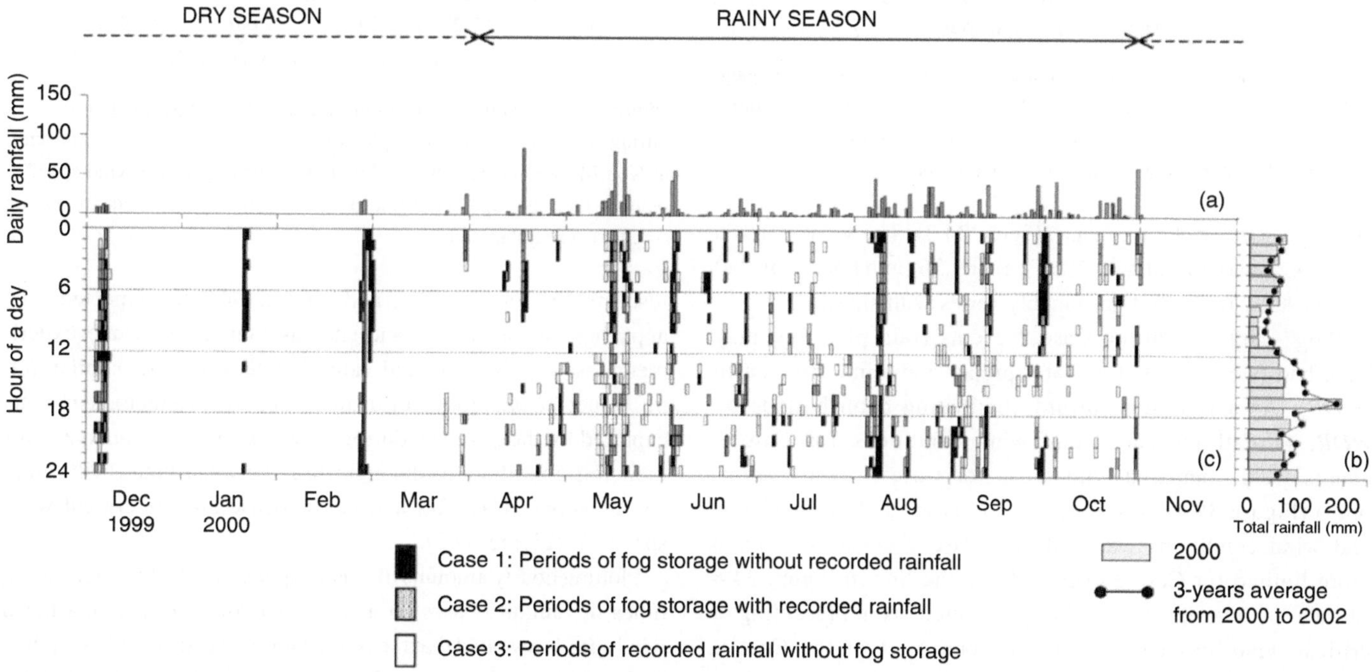

Figure 34.2. (a) Seasonal variation in rainfall, (b) diurnal cycle of rainfall, and (c) occurrence of fog incidence and rainfall at Kog-Ma watershed in the year 2000.

proper contributed <0.5% to total annual *TF*, and occurred during 25–29 hours per year (Table 34.2).

DISCUSSION

Fog gage performance

To estimate "true" fog occurrence, the performance of the fog gage was investigated in relation to meteorological conditions. Figure 34.3 shows the relationship between fog capture by the gage during Case 1 events (fog-only) and corresponding hourly mean wind speeds. The (hand-drawn) line represents the upper limit of measured fog capture, which increases with wind speed. Fog capture amounts plotting below this line presumably resulted from episodes with lower fog liquid water content. Fog capture (F, ml hour^{-1}) during Case 1 events can be expressed as:

$$F = Fog_{\max} \cdot f \cdot U, \quad (34.1)$$

where $Fog_{\max}$ is the slope of the limiting line, f relative fog density (ranging between 0 in the absence of fog and 1 at maximum density), and U the hourly mean wind speed (m s^{-1})

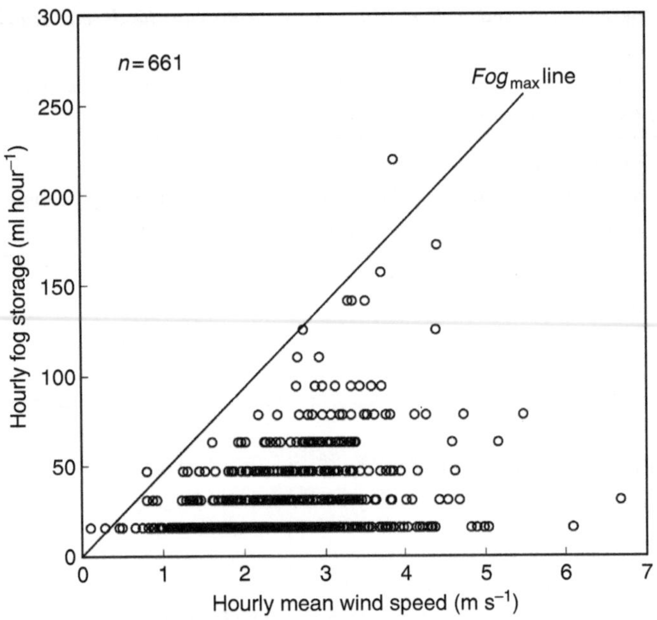

Figure 34.3. Relationship between hourly fog captured by the fog gage (ml) during times without rain (Case 1 events) and hourly mean wind speed (m s^{-1}) at Kog-Ma watershed. The solid line shows the upper envelope of captured fog vs. the mean wind speed.

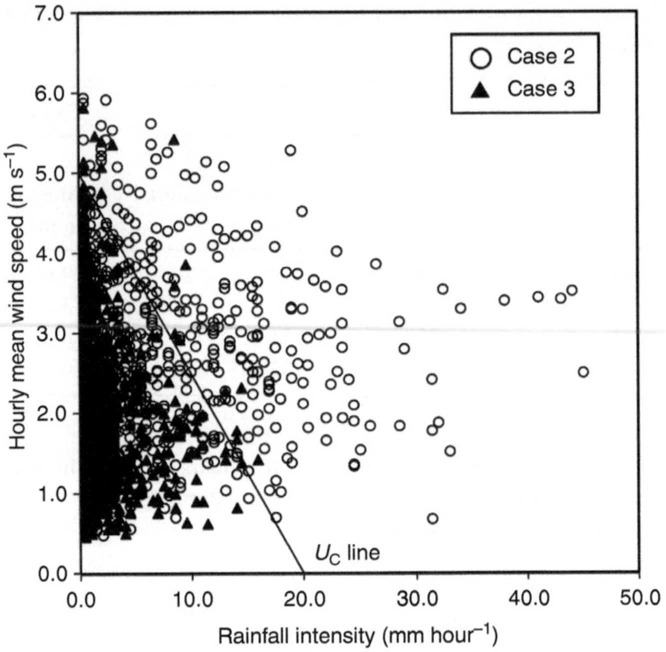

Figure 34.4. Comparison of rainfall and meteorological conditions during Case 2 events (○, rain plus fog) and Case 3 events (▲, rain-only) at Kog-Ma watershed. The U_C line represents the critical wind speed below which the fog gage is not likely to capture much wind-driven rain (see text for explanation).

at the position of the fog gage (Figure 34.3). The value derived for Fog_{max} was 70 (ml $hour^{-1}$ m^{-1} s) or 1.36 (mm $hour^{-1}$ m^{-1} s) for the study tower and presumably varies with sites.

"Fog" captured during Case 2 events (rain plus fog) may have had two sources: fog water proper and wind-driven rain (*WDR*). As a first step toward eliminating contributions by *WDR*, rainfall intensities and wind conditions for Case 2 events were compared with those for Case 3 events (rain-only) (Figure 34.4). The range in observed rainfall intensities and wind conditions was wide for Case 2 events but much more limited for Case 3 events. Thus, the line in Figure 34.4 (drawn by eyesight) may be interpreted as representing a critical wind speed (U_C) that envelops most of the Case 3 events. Assuming that the process by which rain water reaches the fog gage screen is determined mainly by raindrop size (in turn closely related to rainfall intensity; Marshall and Palmer, 1948), and wind speed (cf. Frumau *et al.*, this volume; Giambelluca *et al.*, this volume), values below the U_C line would indicate conditions during which the fog gage was unlikely to collect significant amounts of *WDR*. As shown in Figure 34.4, U_C varied with rainfall intensity, with small drops (low intensities) requiring lower wind speeds to become inclined and vice versa (cf. Sharon, 1980; Frumau *et al.*, this volume):

$$U_C = U_{C0} - a \cdot r, \quad (34.2)$$

where r (mm $hour^{-1}$) is rainfall intensity, and U_{C0} (5.0 m s^{-1}) and a (about 0.25) are the y-intercept and the slope of the line, respectively. Values of U_{C0} and a in Eq. (34.2) are expected to be applicable to other montane sites as long as the same experimental set-up is used and rainfall intensities are similar (cf. McJannet *et al.*, 2007; Frumau *et al.*, this volume). Amounts captured by the fog gage during Case 2 events were analyzed for their fog contribution, distinguishing between hours with wind speeds above (Case 2a) and below (Case 2b) the critical wind speed U_C (Figure 34.5).

Plotting hourly amounts of water captured by the fog gage during times of rainfall (Cases 2a and 2b) against corresponding mean wind speeds revealed that measured "fog" exceeded the Fog_{max} line during 84 hours (Figure 34.5). Assuming similar fog densities for periods with and without rainfall, these excess amounts must include *WDR*. Indeed, most of the values exceeding the Fog_{max} line occurred during Case 2b conditions, i.e. when wind speeds exceeded U_C (Figure 34.5). Therefore, the hourly fog gage capture data-set was recalculated, this time with all fog capture measured during Case 2b occurrences excluded. It is recognized that this approach leaves out any fog that may have occurred during these windy periods and as such it will underestimate the total contribution of fog during periods of rainfall somewhat.

"True" fog occurrence

The recalculated hourly fog data-set gave a 30% reduction in the total amount of water captured by the fog gage during the entire

Table 34.3 *Fog gage capture observed during Case 1, Case 2a, and Case 2b events (see text for details). Recalculated fog gage capture is also indicated*

Year[a]	Observed fog gage capture						Recalculated fog gage capture	
	Case 1		Case 2a		Case 2b		Case 1 plus Case 2a	
	Number of hours (hours)	Amount (ml)	Number of hours (hours)	Amount (ml)	Number of hours (hours)	Amount (ml)	Number of hours (hours)	Amount (ml)
2000	282	9498	223	8682	55	7897	505	18181
2001[b]	189	4773	180	6516	54	5165	369	11288
2002	219	6829	253	16956	67	9577	472	23786
Three-year total	690	21101	656	32154	176	22639	1346	53254

[a] See footnote *a* to Table 34.1.
[b] See footnote *c* to Table 34.1.

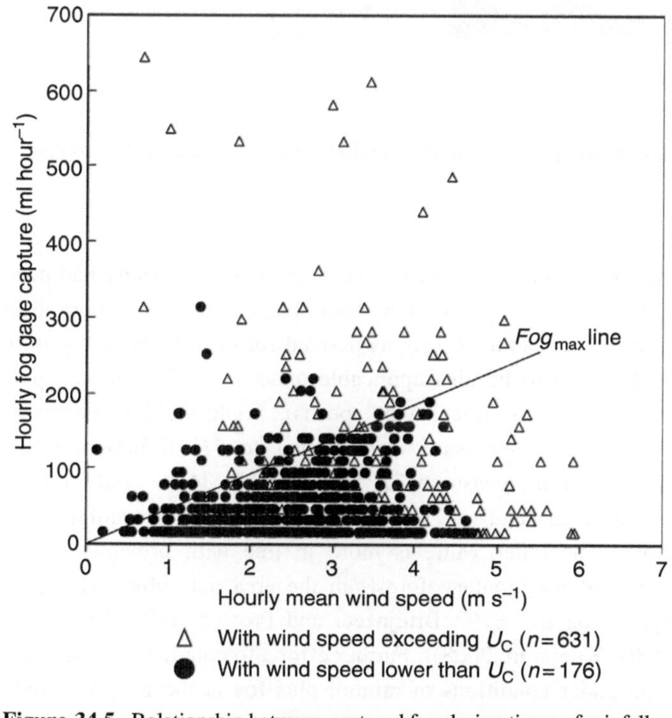

Figure 34.5. Relationship between captured fog during times of rainfall (Case 2 events), and corresponding mean wind speed at Kog-Ma watershed. Captured fog amounts are distinguished according to wind speed: (●) wind speeds < critical wind speed U_C and (Δ) wind speeds ≥ U_C.

observation period, from 75.9 to 53.2 l, or from 1472 to 1033 mm. Also, the total duration of fog gage capture decreased by 11.6%, from 1522 to 1346 hours. Table 34.3 summarizes (recalculated) fog incidence and duration over the 3-year study period.

Distinct inter-annual variation in the occurrence of fog was observed. For example, annual (recalculated) fog capture in 2000, 2001, and 2002 varied roughly twofold, with annual totals of 18.1, 11.3, and 23.8 l, respectively (Table 34.3). Inter-annual variation in the number of hours with fog was less pronounced, ranging from 376 to 507 hours. Interestingly, the variation in annual fog duration seemed to vary in a non-linear fashion with the variation in annual fog capture (not shown), presumably because of variations in fog density (cf. f in Eq. 34.1) between years.

The average diurnal variation in total "fog" capture by the tower gage during Case 1 and Case 2 events for the entire observation period shown in Figure 34.6 indicates a clear peak between 16.00 and 17.00 h that corresponds with the diurnal peak in rainfall. However, the similarity in the two patterns disappeared after excluding the capture during hours with high winds (Case 2b events), revealing higher fog capture during the night (22.00–06.00 h) and the early morning hours (06.00–10.00 h).

Throughfall and fog drip

Relative amounts of total *TF* (*c.* 92% of rainfall, Table 34.2) were rather high for South-East Asian montane forests and are reminiscent of the values normally reported for wetter and mossier forests subject to substantial amounts of horizontal precipitation (Bruijnzeel and Proctor, 1995; Bruijnzeel, 2005). As such, this high value might reflect an as yet unquantified contribution by fog water inputs. However, only four gages in fixed position were used for estimating *TF*, although high spatial variability is normally reported for these types of forests (Lloyd and Marques, 1988; Holwerda *et al.*, 2006). Tanaka *et al.* (2005) conducted more extensive *TF* measurements using 30 totalizing fixed gages (with an orifice area of 346 cm^2 per

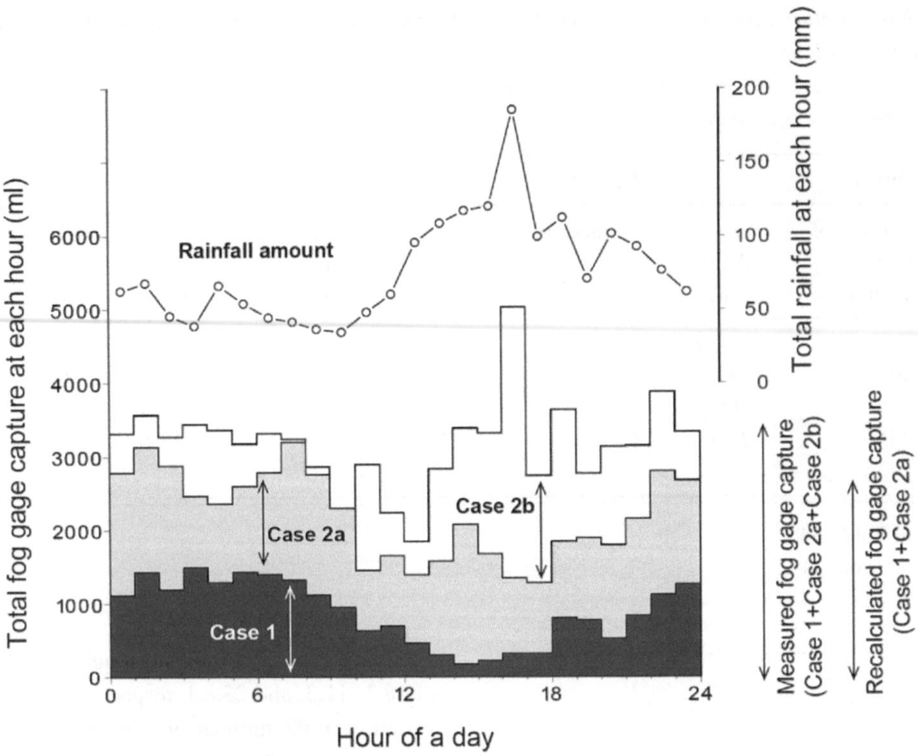

Figure 34.6. Averaged diurnal variations in measured and recalculated fog capture during the 3-year study period at Kog-Ma watershed as compared with the diurnal pattern of rainfall.

gage) in the same stand for 4 years and reported the coefficients of variation (CV) associated with each of the gages to range from 0.132 to 0.146. Combining these CVs and a *t*-value of 2.05 ($p = 0.05$, 29 degrees of freedom) yields a required number of gages of 29–36 for the estimation of mean annual *TF* with an accuracy of 5% (cf. Lloyd and Marques, 1988). This number is considerably greater than the four gages used in the present study, suggesting potentially large uncertainty in the amounts of *TF* reported here.

Amounts of crown drip generated by fog-only were very small indeed (19.3 mm for the entire study period) and constituted less than 0.5% of total *TF* (Table 34.2). The corresponding catch of the fog gage during these 80 hours of fog-only was 2.99 l or 58 mm, suggesting that 33% of the gage catch was turned into crown drip. However, hours when fog drip occurred made up only a small part of total fog occurrence while total fog gage catch during Case 1 (fog-only) and Case 2a (rain plus fog at sub-critical wind speeds, cf. Figure 34.5) events was nearly 18 times that during times of fog drip (Table 34.3). García-Santos (2007) also found *TF* during times of fog-only to be very low (6% of the catch of a quarter-sized standard fog gage placed above the canopy) at a ridge-top site on La Gomera (Canary Islands), but their *TF* exceeded incident rainfall during days with fog plus rain. These authors hypothesized that a greater proportion of intercepted fog was able to reach the forest floor once the canopy storage capacity had been filled by rainfall. As such, assuming the gage-to-canopy-drip conversion factor of 33% as derived for conditions of fog-only at Kog-Ma to be also applicable under conditions of fog plus rainfall (at sub-critical wind speeds), would yield an estimated 344 mm of fog water contributing to *TF*. Subtracting this amount from measured *TF* totals (Table 34.2) would reduce the average relative *TF* amount from *c.* 92% of rainfall to *c.* 86%. The latter value is more in line with previous results obtained for montane forests in the area not subjected to frequent fog (62–87%; Bruijnzeel and Proctor, 1995; Liu *et al.*, 2002; Kumaran, 2008)). Further efforts to quantify fog interception under conditions of rainfall plus fog at the Kog-Ma study site are needed to test this contention.

CONCLUSIONS

An hourly data-set of rainfall, water captured by a sheltered fog gage, and wind speed as measured above the canopy of a montane forest in northern Thailand during nearly 3 years was used to test the efficiency of a rain-protected passive fog gage as a fog sensor. During times of rainfall plus fog, the fog gage often captured more water than expected on the basis of a relationship between gage catch and wind speed derived during times of fog-only, suggesting contributions by wind-driven rainfall. After identifying the specific conditions of rainfall intensity and wind speed under which this contamination occurred, the corresponding amounts were excluded from the data-set, lowering the formerly derived total gage capture of 1472 mm to 1033 mm. Only 55 mm of the latter total was recorded during times of rain-free fog drip, producing 19.3 mm of drip from the canopy (33% of corresponding gage catch). Fog drip constituted less than 0.5% of total throughfall (*TF*). Whilst additional contributions to *TF* by fog interception during times of other fog events (e.g. times of fog plus rain) are as yet unknown, total fog contribution to *TF* is likely to exceed 344 mm for the entire study period.

ACKNOWLEDGEMENTS

We thank Drs. Natsuko Yoshifuji and Tomonori Kume at the University of Tokyo for providing meteorological data used in this study, the local assistants for help with the field work and logistic support, and Professor James O. Juvik for providing the fog gage. We sincerely thank Professors Thomas W. Giambelluca and Sampurno Bruijnzeel for their helpful comments on an earlier draft of this paper. This study was funded by the Global Energy and Water Cycle Experiment, Asian Monsoon Experiment-Tropics (GEWEX/GAME-Tropics) project and the Core Research for Evolutional Science and Technology (CREST) program of the Japan Science and Technology (JST) Agency.

REFERENCES

Bruijnzeel, L. A. (2001). Hydrology of tropical montane cloud forest: a reassessment. *Land Use and Water Resources Research* **1**: 1–18.

Bruijnzeel, L. A. (2005). Tropical montane cloud forests: a unique hydrological case. In *Forests, Water and People in the Humid Tropics*, eds. M. Bonell and L. A. Bruijnzeel, pp. 462–483. Cambridge, UK: Cambridge University Press.

Bruijnzeel, L. A., and J. Proctor (1995). Hydrology and biogeochemistry of tropical montane cloud forest: what do we really know? In *Tropical Montane Cloud Forests*, eds. L. S. Hamilton, J. O. Juvik, and F. N. Scatena, pp. 38–78. New York: Springer-Verlag.

Bubb, P., I. May, L. Miles, and J. Sayer (2004). *Cloud Forest Agenda*, UNEP–WCMC Biodiversity Series No. 20. Cambridge, UK: UNEP–WCMC.

Cavelier, J., D. Solis, and M. A. Jaramillo (1996). Fog interception in montane forests across the Central Cordillera of Panamá. *Journal of Tropical Ecology* **12**: 357–369.

Chunkao, K., N. Tangtham, S. Boonyawat, and W. Niyom (1981). Watershed management research on mountainous land. In *15-Year Tentative Report 1966–1980*. Bangkok, Thailand: Department of Conservation, Faculty of Forestry, Kasetsart University (in Thai with an English abstract).

Doumenge, C., D. Gilmour, M. R. Pérez, and J. Blockhus. (1995). Tropical montane cloud forests: conservation status and management issues. In *Tropical Montane Cloud Forests*, eds. L. S. Hamilton, J. O. Juvik, and F. N. Scatena, pp. 24–37. New York: Springer-Verlag.

García-Santos, G. (2007). *An ecohydrological and soils study in a montane cloud forest in the National Park of Garajonay, La Gomera (Canary Islands, Spain)*. PhD Thesis, VU University Ámsterdam, Ámsterdam, The Netherlands. [http://www.falw.vu.nl/nl/onderzoek/earth-sciences/geo-environmental-science-and-hydrology/hydrology-dissertations/index.asp].

Giambelluca, T. W., L. T. Tran, A. D. Ziegler, T. P. Menard, and M. A. Nullet (1996). Soil–vegetation–atmosphere processes: simulation and field measurement for deforested sites in northern Thailand. *Journal of Geophysical Research* **101**: 25 867–25 885.

González, J. (2000). Monitoring cloud interception in a tropical montane cloud forest of the southwestern Colombian Andes. *Advances in Environmental Monitoring and Modelling* **1**: 97–117.

Holwerda, F., F. N. Scatena, and L. A. Bruijnzeel (2006). Throughfall in a Puerto Rican lower montane rain forest: a comparison of sampling strategies. *Journal of Hydrology* **327**: 592–602.

Hutley, L. B., D. Doley, D. J. Yates, and A. Boonsaner (1997). Water balance of an Australian subtropical rainforest at altitude: the ecological and physiological significance of intercepted cloud and fog. *Australian Journal of Botany* **45**: 311–329.

Juvik, J. O., and D. Nullet (1995). Comments on "a proposed standard fog collector for use in high elevation regions." *Journal of Applied Meteorology* **34**: 2108–2110.

Komatsu, H., N. Yoshida, H. Takizawa, *et al.* (2003). Seasonal trend in the occurrence of nocturnal drainage flow on a forested slope under a tropical monsoon climate. *Boundary–Layer Meteorology* **106**: 573–592.

Kumaran, S. (2008). Hydrometeorology of tropical montane rain forests of Gunung Brinchang, Pahang Darul Makmur, Malaysia. Ph.D. thesis, University Putra Malaysia, Serdang, Malaysia.

Letts, M. G., and M. Mulligan (2005). The impact of light quality and leaf wetness on photosynthesis in north-west Andean tropical montane cloud forest. *Journal of Tropical Ecology* **21**: 549–557.

Liu, W., J. E. D. Fox, and Z. Xu (2002). Nutrient fluxes in bulk precipitation, throughfall and stemflow in montane subtropical moist forest on Ailao Mountain in Yunnan, south-west China. *Journal of Tropical Ecology* **18**: 527–541.

Lloyd, C. R., and F.A. de O. Marques (1988). Spatial variability of throughfall and stemflow measurement in Amazonian rainforest. *Agriculture and Forest Meteorology* **42**: 63–73.

Marshall, J. S., and W. M. Palmer (1948). The distribution of raindrops with size. *Journal of Meteorology* **5**: 165–166.

McJannet, D., J. S. Wallace, and P. Reddell (2007). Precipitation interception in Australian tropical rainforests. II. Altitudinal gradients of cloud interception, stemflow, throughfall and interception. *Hydrological Processes* **21**: 1703–1718.

Sharon, D. (1980). The distribution of hydrologically effective rainfall incident on sloping ground. *Journal of Hydrology* **46**: 165–188.

Tanaka, K., H. Takizawa, N. Tanaka, *et al.* (2003). Transpiration peak over a hill evergreen forest in northern Thailand in the late dry season: assessing the seasonal changes in evapotranspiration using a multilayer model. *Journal of Geophysical Research* **108**: 4533, doi:10.1029/2002JD 003028.

Tanaka, N., C. Tantasirin, K. Kuraji, M. Suzuki, and N. Tangtham (2005). Inter-annual variation in rainfall interception at a hill evergreen forest in northern Thailand. *Bulletin of the Tokyo University Forest* **113**: 11–44.

Tangtham, N. (1974). Preliminary study of the ecosystem of hill evergreen forest in northern Thailand. Ph.D. thesis, Pennsylvania State University, University Park, PA, USA.

Veneklaas, E. J., and R. van Ek (1990). Rainfall interception in two tropical montane rain forests, Colombia. *Hydrological Processes* **4**: 311–326.

Werner, W. L., and T. Santisuk (1993). Conservation and restoration of montane forest communities in Thailand. In *Restoration of Tropical Forest Ecosystems*, eds. H. Lieth and M. Lohmann, pp. 193–202. Dordrecht, the Netherlands: Kluwer.

35 History of fog and cloud water interception research in Hawai'i

J. K. DeLay and T. W. Giambelluca
University of Hawai'i at Mānoa, Honolulu, Hawai'i, USA

ABSTRACT

Research on fog climatology and cloud water interception (*CWI*) in the montane cloud forests of Hawai'i spans nearly 50 years, from the pioneering work of Wendell Mordy and Paul Ekern on Lâna'i in the 1950s, through the continuing efforts of James Juvik since 1972. This work has helped to improve understanding of the spatial patterns of fog occurrence and to quantify *CWI* in forests and other vegetation. Reported *CWI* or fog incidence estimates are as high as 4982 mm year^{-1} at particularly exposed locations, although most windward sites within the cloud zone are in the range between 280 and 1130 mm year^{-1}, with leeward sites receiving between 100 and 500 mm year^{-1}, and less than 250 mm year^{-1} in high-elevation areas above the trade-wind inversion. Most of the early work was based on mechanical fog gage measurements whose well-known limitations make accurate estimation of actual *CWI* by a forest canopy difficult. Advancing the current level of understanding will have to come from studies incorporating other methods, such as the wet-canopy water budget and stable isotope mass balance approaches, in addition to the continued use of recording fog and throughfall gages.

INTRODUCTION

The summit areas of 12 of the volcanoes comprising the main Hawaiian Islands lie at or above the usual cloud base height, and are frequently immersed in or ringed by ground-level orographic cloud or fog (Figure 35.1). In Hawai'i, as elsewhere, the presence of fog has profound effects on the biota, producing distinct ecological zones known collectively as tropical montane cloud forest (TMCF). The movement of fog by wind results in deposition of cloud droplets on the leaves and stems of plants, and other obstacles. This cloud water interception (*CWI*),[1] locally called "fog drip," has long been recognized as a distinct supplemental hydrological input, contributing to soil water reserves, groundwater recharge, and streamflow in windward, high-rainfall areas, and providing life-sustaining moisture on otherwise dry, leeward and high-elevation slopes. Quantifying *CWI* in Hawai'i has proven difficult, despite research spanning almost 50 years. In this chapter, prior efforts to estimate *CWI* are reviewed and the limitations of current knowledge of *CWI* in Hawai'i are identified.

CLOUD WATER IN HAWAIIAN CULTURE

Long before western contact, Hawaiians had a deep appreciation for the natural processes affecting the islands, as reflected in their language. The clouds, fog, and mists that develop on the islands' mountain slopes have many Hawaiian names, each of which can have multiple meanings according to Pukui and Elbert (1986), who also provide some literal translations of Hawaiian phrases related to these phenomena as follows. *'Ohu* describes the "fog [or] light cloud on a mountain", and also means "to be adorned as with *leis*," perhaps in reference to the ring of clouds that often encircle Hawai'i's peaks. Hawaiians even classified fog or mist according to its density, movement, and color. *Uhiwai* is a "dense mist," *kili 'ohu* refers to "fine mist," and *ko'i 'ula* is a "rainbow colored mist." *Lilinoe* can also refer to fine mist and to the Hawaiian mist goddess of that name. Mist of the type that leads to *CWI* is called *noe kolo* or "mist creeping." *Kawaû* means "moist with fog" and the phrase *he 'ohu ke aloha 'a 'ohe kuahiwi*

[1] Strictly speaking, the term "cloud water interception" (*CWI*) refers to deposition of cloud water (fog) on vegetation and other obstacles. This can be estimated from canopy water balance measurements. Many studies instead use passive fog gages and report the results as fog (cloud water) interception, fog deposition, or fog incidence. Herein, we refer to the results of previous fog measurements using fog gages or screens as *FS*, and those based on canopy water balance as *CWI*.

Tropical Montane Cloud Forests: Science for Conservation and Management, eds. L. A. Bruijnzeel, F. N. Scatena, and L. S. Hamilton. Published by Cambridge University Press.

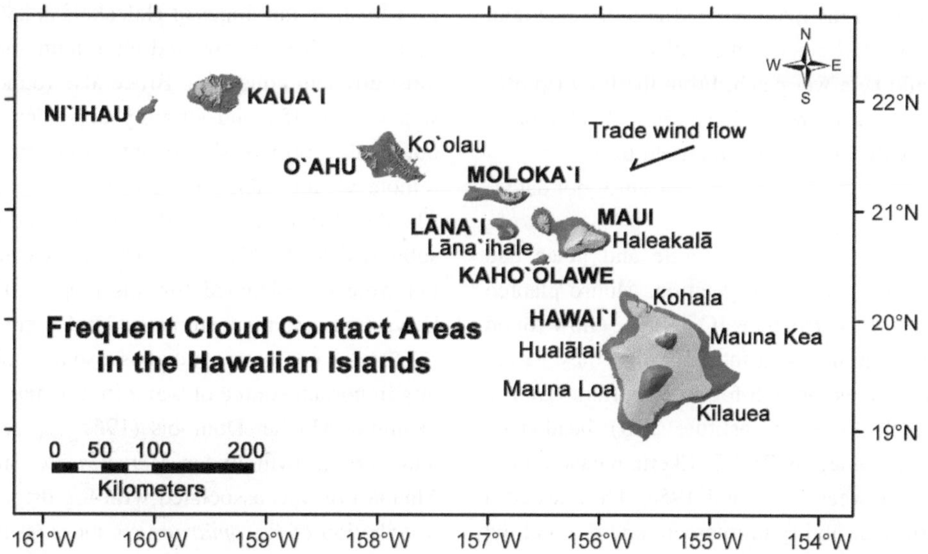

Figure 35.1. Map showing areas of frequent cloud contact (lighter shades) corresponding to elevations between 600 and 2500 m.a.s.l. Cloud contact can also occur above 2500 m.a.s.l. on larger mountains and tends to occur at the lowest elevations on steep, smaller mountains such as the Ko'olau Range on O'ahu. Some areas of potential TMCF have been converted to pasture or are covered by recent lava flows.

kau 'ole... "Love is mist, settles on all the hills" is indicative of the romantic association with fog. The remote and uninhabited mountain forest areas of Hawai'i including its TMCF are sacred Hawaiian areas called *wao akua*, in reference to the spirits (*akua*) that dwell there (Pukui and Elbert, 1986).

In leeward areas with low rainfall, aboriginal Hawaiian agriculture and settlement may have been at least partially facilitated by *CWI* in the adjacent montane dryland forest areas which have since been deforested. The Lapakahi region on the leeward side of Kohala Mountain on Hawai'i is the site of archeological remains associated with rain-fed agriculture. Artifacts include water cisterns (Rosendahl, 1972), some of which may have been used to collect cloud water (P.M. Vitousek, personal communication, 2003). At higher elevations on the arid leeward side of East Maui, archaeologic evidence indicates pre-contact Hawaiian settlements where there is now minimal rainfall and currently no significant surface water. Stock *et al.* (2003) postulated that extensive removal of leeward cloud forests dried up springs and streams that once supported habitation in those areas. Current resettlement efforts by native Hawaiians in the dry Kahikinui area may be facilitated by plans for harvesting of fog water with large-scale mechanical collectors (Juvik, 1998).

CHRONOLOGY OF CLOUD WATER INTERCEPTION RESEARCH

He ala ehu aku kênâ

A misty pathway, that!

(Pukui and Elbert, 1986)

Scientific observations of fog and *CWI* in the Hawaiian Islands began appearing in the literature around the middle of the twentieth century. Powers and Wentworth (1941) made numerous observations of the orographic cloud in the Saddle area between Mauna Loa and Mauna Kea during the summer of 1939. In addition to estimating the depth (vertical thickness) and frequency of the cloud during this period, they observed *CWI* occurring in open *mâmane* (*Sophora chrysophylla*) forest noting, "This mist accumulates on grasses and particularly on the foliage and branches of trees... dripping to the ground leads to a much more luxuriant growth." Squires and Warner (1957) later made airborne observations of cloud liquid water content, drop size, and cloud morphology in orographic clouds over the Saddle area.

While fog was recognized as a possible water source on various islands, efforts to measure and manage *CWI* in Hawai'i trace their roots to the Island of Lâna'i. The first modern recognition of fog as a water source in Hawai'i is credited to George Munro (1866–1963). Munro was a prolific naturalist who traveled the Hawaiian Islands and lived on Lâna'i in the early 1900s, recording his observations and taking conservation actions as a land manager. In his nineties he wrote a collection of his general observations on Lâna'i, including the following excerpt:

> A 125 foot high Norfolk Island pine tree [*Araucaria heterophylla*] at Kô'ele, [his upland Lanâ'i home] would condense fog when there was a brisk wind driving through its foliage. The drops falling on the iron roof under it made quite a tapping noise... On the mountain there is a certain amount of condensation of fog on leaves of the trees; and this fog-drop, as it is called, is now being measured on Lâna'i by the Hawaiian Pineapple Company with interesting

> results. I have traversed the main ridge after a dry spell and although the trail would be dry in most places, under every tall tree there would be a wet patch. From the fog drip off two or three low half-dead trees, I have seen a trickle of water or a little stream running down an old trail.
>
> (Munro, not dated)

By the time Munro arrived on Lâna'i, cattle and goats had denuded most of the island's native vegetation. Munro planted many of the *Araucaria columnaris* trees (Cook pine) now found along the trail near the summit, with the hope that they would increase local water resources through fog interception. Later, a group from the Pineapple Research Institute (PRI), headed by Wendell Mordy and later joined by Paul C. Ekern, measured fog interception on Lâna'i between 1955 and 1958. They tested a variety of materials for their ability to intercept and drain cloud water. They also took several approaches to measuring crown drip beneath some of the isolated trees probably planted earlier by Munro. Their results were released in a series of reports (Mordy, 1955; Peperfale, 1956; Ekern, 1958, 1960) and a classic summary paper (Ekern, 1964). In the reports, the authors acknowledged that the mechanical fog collectors were not necessarily representative of the *CWI* by vegetation and indicated that measuring crown drip during periods of low rainfall was a more appropriate approach to determine *CWI*. They collected crown drip (or throughfall, *TF*) beneath an *Araucaria* tree using eight rain gages and sheets of galvanized roofing. During light rains, *TF* collected in this manner amounted to 13 times the rainfall. Ekern noted that the vertical silhouette of the tree was roughly equal to the horizontal projected area of the crown. He assumed that the tree intercepted twice the amount of rainfall measured in the open and estimated annual *CWI* as 762 mm (Ekern, 1964). In the summer of 1958, louvered screen cylinders were deployed as fog collectors at about 2 m above the ground and at canopy height (~9 m). The collectors showed increased fog interception with height at a rate greater than that expected from the increase in wind speed, suggesting that interception occurred in the canopy during periods when the orographic cloud failed to contact the ground. Ekern emphasized the significant contribution of *CWI* during periods of low rainfall as indicated by fog gage measurements. For example, during a 3-year period, the unit vertical catch of water by a wire harp fog gage was three times as much as the standard rain gage measurement on days when rainfall was less than 0.254 mm (Ekern, 1964). Recently, *FS*- and *TF*-based observations on Lâna'i were started again to further clarify the amount and distribution of *CWI* on the island (see below; Juvik, 2009).

Elsewhere, other studies were carried out to assess the feasibility of harvesting fog as a water source. For example, Bruce (1966) tested the capture efficiency of 13 types of flat screen panels using small-scale installations in the Ko'olau Gap area of East Maui on the slopes of Haleakalâ at 2300 m.a.s.l. A wire harp design and Kaiser louvered aluminum shade screen proved the most efficient collectors. Bruce also found that fog interception was site-specific and strongly dependent on local topography, but could form a significant supplemental water source for remote visitor cabins if collectors were appropriately located. The detailed accounts of Bruce's methodology and results were submitted to the National Park Service office in San Francisco but were not obtained for this review. The recommended fog harvesting systems were never implemented.

Cloud water interception has also been identified as a biologically important source of water in mountain areas in Hawai'i. For example, Mueller-Dombois (1967) suggested that dense shrub and herb growth under *Metrosideros* and *Sophora* forest on Mauna Loa was associated with fog drip. He also attributed the distribution of *Racomitrium sp.* moss on the northern aspects of lava flows in the alpine desert of Mauna Loa (3050–3350 m.a.s.l.) to *CWI* from moist upslope winds. The mosses proved to be efficient cloud water collectors, absorbing 300% of their own dry weight after a half day of cloud contact. Using Grunow-type wire mesh fog interceptors (Grunow, 1952 as cited by Nagel, 1956), Smathers and Mueller-Dombois (1972) measured interception on the floor of Kîlauea caldera and in the Ka'u Desert to examine the influence of cloud water on primary succession. The authors did not find a significant difference between habitat types within the caldera and surmised that at the exposed Ka'u Desert location, wind-blown rainfall accounted for increased interception by the Grunow collectors. However, they attributed some of the additional interception from the caldera measurements made in sheltered locations, to steam clouds. They further identified the interception of steam cloud water by rock surfaces and the concentration of rainfall and cloud water in cracks as a possible, biologically important source of moisture.

Beginning in the early 1970s and continuing to the present, James Juvik has continued research on *CWI* in Hawai'i using fog screen (*FS*) measurements. Much of his research has focused on the altitudinal distribution of fog occurrence, using transects of fog gage stations across elevation gradients. Juvik tried several methods, of varying levels of sophistication, to separate wind-blown rainfall from fog collected by the gage, based on drop-size distribution and wind speed (Juvik and Perreira, 1974; McKnight and Juvik, 1975; Juvik, 1977; Juvik and Ekern, 1978). Juvik and Perreira (1974) established a transect of louvered-screen fog collectors on windward Mauna Loa between 610 m and 3415 m.a.s.l. and reported an increase in the fog/rainfall ratio with elevation, reaching 0.65 at the top of the transect. The highest absolute fog interception occurred at 1580 m.a.s.l., totaling 638 mm over 7 months. The summer months of that study were affected by a severe drought, during which fog at 1580 m.a.s.l. was recorded in twice as many weeks as rainfall whereas 871 mm of *FS* (135% of rainfall) was observed vs. 65% during the winter

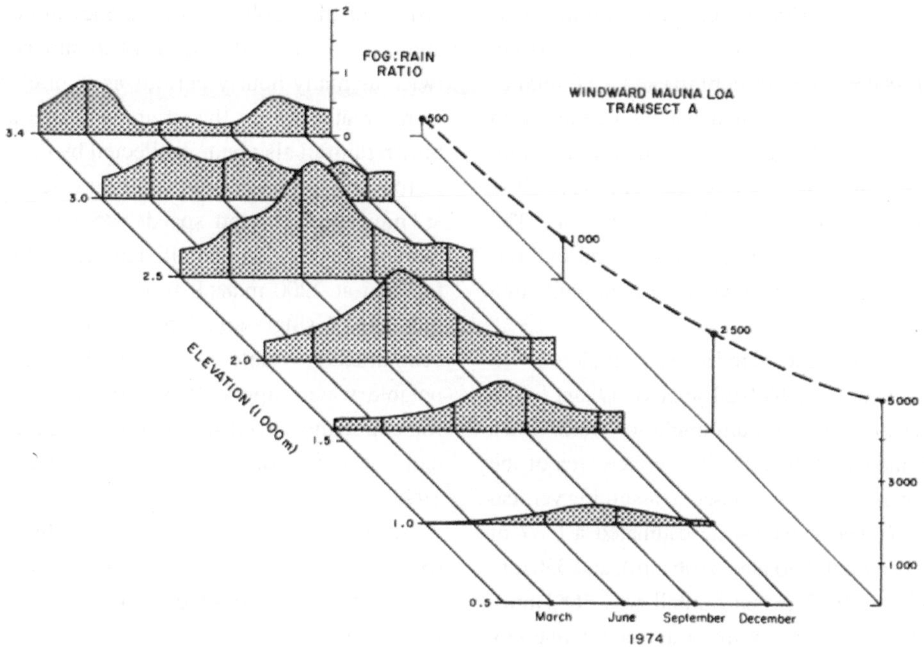

Figure 35.2. Seasonal fog-to-rain ratios on windward Mauna Loa (from Juvik and Ekern, 1978).

months. Although the highest absolute *FS* interception occurred at 1580 m.a.s.l., the highest *FS* to rain ratio occurred at 2350 m.a.s.l. (Figure 35.2).

To characterize the fog and estimate the effect of drop size on collection efficiency, McKnight and Juvik (1975) sampled drop-size distributions in ground-level fog on Mauna Loa using glass slides with a magnesium oxide coating. The most frequently occurring drop sizes were less than 20 μm. They also developed a method to separate wind-blown rainfall collected by the fog gage based on drop size and wind speed. Blanchard (1953) made ground-based observations of the drop-size distribution of Hawaiian rainfall of various intensities on Mauna Loa. In general, drop size increased with rainfall intensity, but not in a linear fashion (see Giambelluca *et al.*, this volume for further discussion). Juvik did not employ the model because drop-size distributions were not recorded during his investigations.

Juvik (1977) and Juvik and Ekern (1978) continued to maintain a network of climate stations in elevation transects on Mauna Loa and Hualâlai on the island of Hawai'i from 1974 to 1976. Fog proved to be concentrated in a belt from 1500 to 2500 m.a.s.l. on the windward slope of Mauna Loa. Furthermore, cloud depth, density, and wind speeds were all lower on the leeward side of the mountain, whereas the distribution of fog on the mountain's slopes was shown to be influenced by the trade-wind inversion and local convection. On the windward side of Mauna Loa, much lower fog interception rates were measured at sites sheltered by local topography. For example, an annualized total *FS* of 859 mm was recorded at an exposed site at 1580 m.a.s.l., but only 28 mm was received at 1692 m.a.s.l. in the lee of Mauna Loa's south-east ridge. Juvik and Ekern (1978) concluded that fog plays a significant role in the water balance of Mauna Loa.

On Hualâlai (in the western part of Hawai'i Island, Figure 35.1) Juvik (1977) found measurable *FS* at a pasture site, but no *FS* in a nearby 60-m clearing within 12- to 15-m tall continuous *'ôhi'a* (*Metrosideros* sp) forest. This finding demonstrated that the horizontal flux of fog water was higher in the open than within the forest. Studies of *CWI* on Hualâlai have recently been resumed. Preliminary results (Juvik and Tango, 2003) indicate that fog is a significant contributor to a water budget surplus at mid-slope elevations (1200–1800 m.a.s.l.). The Hualâlai study includes ongoing *TF* measurements that should better quantify the contribution of *CWI* to the forest water budget.

Juvik (1977) recorded high rates of *FS* with a louvered screen collector that were positively correlated ($r = 0.92$) with increases in soil moisture beneath the canopy of *mâmane* (*Sophora chrysophylla*) forest, supporting the earlier casual observations by Powers and Wentworth (1941). Later, Juvik and Nullet (1995b) provided additional evidence for the significance of *CWI* in *mâmane* forest in a 1-year study of the canopy water balance of high-elevation *mâmane* forest on leeward Mauna Kea. They deployed both a louvered screen cylinder and Grunow-type fog interceptors (cf. Nagel, 1956) in an open site in conjunction with two 3.8-m *TF* troughs beneath an adjacent 7.6-m tall *mâmane* tree. The study was conducted at 2600 m.a.s.l., above the usual height of the trade-wind inversion (which tends to form the upper limit of the moist cloud belt), during a year when rainfall was only 257 mm, roughly half the long-term average of 500 mm. They reported an *FS* of 98 mm (38% of rainfall) and a *TF* total of 194 mm (76% of rainfall), most of which was associated with upslope, afternoon winds and displayed a summer maximum. They also noted that during a 5-month period with very little rainfall, the proportion of *TF* increased to 147% of rainfall. The authors attributed the increase in *TF* during this period to *CWI*, reporting that the fog collector recorded a vertical unit area equivalent totaling 197% of rainfall.

Ekern (1983) measured *FS* using louvered screen cylinders at 625 m.a.s.l. on the leeward side of the Ko'olau Range on O'ahu for an 18-month period. He assumed a maximum angle for wind-blown raindrops of 45° which increased the effective surface area of the gage collecting wind-blown rainfall as opposed to assuming vertical rainfall. Based on the *FS* measurements, he estimated a *CWI* of 237 mm during 1981 (when rainfall totaled 3806 mm), and 330 mm of *CWI* during 8 months in 1982 (when rainfall was 3682 mm). Extrapolating *CWI* values to the 26% of the watershed defined by the cloud zone, Ekern estimated that the supplemental cloud water contributed 62 mm to the watershed as a whole during 1981.

Juvik and Nullet (1995b) recorded rainfall, *FS*, and estimated *CWI* at 1160 m.a.s.l. in the Kohala Mountains (Island of Hawai'i) from an open site near the steep headwall of the Alakahi Valley, and recorded canopy *TF* with troughs beneath a nearby patch of *'ôhi'a* (*Metrosideros polymorpha*) trees. They determined *FS* from a louvered screen as 1365 mm and *TF* as 606 mm during a relatively dry 100-day period when rainfall was only 336 mm. These open-site conditions can be considered representative of a ridge top or forest edge in the area, but not of continuous forest. More recent work near that site suggests that *FS* and *CWI* at Kohala are lower within continuous canopy than at the previously investigated exposed (leading edge) sites. DeLay (2005) measured *FS*, *TF*, stemflow, and variables used to estimate evaporation, in a continuous canopy site at Alakahi during two 4-month periods with contrasting rainfall receipts. An interception model was calibrated to predict *TF* during events when fog was absent and *CWI* was determined as the difference between predicted and estimated *TF* during fog events. The model results suggested a *CWI* contribution to *TF* of about 10% of the total *TF* for both periods combined. Rainfall was 1486 mm, *TF* 1288 mm, *FS* 154 mm, potential evaporation 650 mm, and *CWI* 131 mm. During the low-rainfall period, however, *FS* was much greater and the fraction of *TF* increased from 84% to 99% of rainfall.

Juvik (1998) measured cloud water inputs at Kahikinui on the leeward slope of Haleakalâ, Maui along an elevational transect to assess the potential for fog collection as a water supply to facilitate resettlement of the area by Native Hawaiians. Fog gages were placed at a height of 3 m to measure input, and several types of commercial, plastic shade-screen material were tested for their collection efficiency at elevations between 1350 and 2200 m.a.s.l. during a 14-month period. Fog occurred in twice as many hourly periods as rainfall, and both fog and rain were greater during the summer. To reduce the effect of wind-blown rainfall also being collected by the fog gage, the definition of fog events was limited to times when rainfall intensities were <3 mm hour^{-1}, wind speeds <5 m s^{-1}, and fog gage output exceeded twice the rainfall. This correction changed the ratio of fog/rain at 2200 m.a.s.l. from 3.36 to a still significant 1.97 in the case of single-layer, 50% shade screen. It was determined economically viable to attempt large-scale fog collection for potable water supply. It was further estimated that a 1000-m^2 fog collecting screen would, on average, harvest 4–5 m^3 of water daily at a cost of US$1–2 m^{-3} over a 20-year period (Juvik, 1998).

At Kîlauea Volcano, Heath and Huebert (1999) studied the role of nitrogen-enriched cloud water from nearby volcanic emissions, as a possible source of fixed nitrogen for forest plants. As part of their study, they estimated *CWI* using a canopy water balance approach incorporating rain, *TF*, and stemflow measurements in 12–16-m tall *Metrosideros* forest over a 1-year period. They estimated that *CWI* added 1130 mm year^{-1}, about 60% of rainfall measured in the open. Heath (2001) refined the canopy water balance approach and presented results from an extended measurement period of 5 years. The estimated *CWI* for this period was much higher than the initial figure, with an annual average of 2050 mm. A subsequent reanalysis involving correction of the evaporation estimate resulted in a *CWI* estimate of 1640 mm year^{-1} (J. Carrillo, personal communication, 2007). A particle volume monitor employed for a 3-year period to characterize fog incidence based on liquid water content, indicated that fog was present about 20% of the time.

The process of orographic cloud formation and rainfall in Hawai'i tends to concentrate ^{18}O isotopes in cloud water as compared to rainfall with increasing elevation, and Scholl *et al.* (2002) concluded that ^{18}O enrichment in the water of windward East Maui streams and springs was indicative of significant input of fog water to the hydrological system.

Giambelluca *et al.* (this volume) estimated *CWI* using the canopy water balance technique for both leeward and windward forest sites on Haleakalâ, Maui. At Auwahi (1219 m.a.s.l.), the leeward, dry forest site, annualized *CWI* was 151 mm, or 37% of rainfall. For Waikamoi (1951 m.a.s.l.), on the windward slope, *CWI* was 1073 mm, or 40% of rainfall.

A comprehensive survey of *CWI* by Cook pines across the upper watershed of Lâna'i Island (650–1066 m.a.s.l.) was conducted recently by Juvik (2009). This study paired a series of six automatic weather stations (including passive fog gages) placed in the open along the elevational transect with measurements of *TF* and stemflow (as well as soil moisture) at hourly intervals.

For several sites higher up on the ridge (>900 m.a.s.l.), the results indicated that *TF* beneath the 20–25-m tall Cook pines was more than an order of magnitude higher than that derived by Ekern (1964) for his ~10-m tall *Araucaria* at 838 m.a.s.l. in the 1950s. Specifically, the Cook pines in the frequently cloud-immersed summit area collected more than 9500 mm of *TF* annually, compared to less than 1000 mm of rain measured nearby in the open. Spatially weighted extrapolations were made of *TF* measured beneath individual trees using a digital terrain model characterizing exposure, slope angle, and slope curvature for all 1875 mature *Araucaria* trees in the upper watershed to produce an approximation of the extra groundwater recharge attributable to fog drip (Juvik, 2009).

Summarizing past studies in Hawai'i, Table 35.1 lists selected prior studies of *CWI* in Hawai'i, beginning with the pioneering work of Ekern (1964) until the recent studies by Juvik (2009) and Giambelluca *et al.* (this volume). Observed totals of rainfall (*P*), throughfall (*TF*), *FS*, and *CWI* are given for each study period. However, to facilitate comparison among different studies, daily values are also given. Figure 35.3 also presents the results of these studies as annual equivalents. On an annual basis, inferred *CWI* totals range from as little as 22–28 mm on leeward Mauna Loa (Juvik and Ekern, 1978) to as much as 4982 mm year^{-1} at a particularly exposed site in the Kohala Mountains (Juvik and Nullet, 1995b) and 3243 mm year^{-1} in the summit region of Lâna'i Island (Juvik, 2009). Excluding these exceptionally high values, which are almost certain to include as yet unknown amounts of wind-blown rainfall not recorded by an ordinary gage, *CWI* and *FS* are generally between 280 and 1130 mm year^{-1} for windward sites within the cloud zone, between 100 and 500 mm year^{-1} at leeward montane sites, and less than 250 mm year^{-1} at elevations above the usual cloud zone (Figure 35.3). The wide range of estimates is probably indicative of both the spatial diversity of *CWI* and of the uncertainty in the estimates (see below).

SOURCES OF UNCERTAINTY

Most of the *CWI* estimates summarized in this chapter were derived from open-site fog gage (*FS*) measurements. Researchers have long acknowledged that the use of fog gages for estimating *CWI* is problematic, but is advantageous for reasons of convenience and comparability among sites (Bruijnzeel *et al.*, 2005). The most obvious problem with most fog gages is the difficulty in distinguishing fog from rain, particularly under windy conditions. Nagel (1956) and Grunow (1960) investigated the effects of drop size and wind speed on collection by Grunow-type gages and subsequent researchers have made similar efforts for other gage designs. Some designs, such as that used by Juvik in some locations, employ a cover or "hat" to prevent most rain from entering the gage (Juvik and Nullet, 1995a; cf. McJannet *et al.*, 2007a,b; Tanaka *et al.*, this volume). The efficiency of such a protective cover will depend on prevailing raindrop sizes and wind speeds which together determine the angle of incidence of the rain (Sharon, 1980; Holwerda *et al.*, 2006). Blanchard's 1953 study of the drop-size distribution in Hawaiian rainfall on Mauna Loa revealed that drop size increases with rainfall intensity but not in a linear fashion (see Giambelluca *et al.*, this volume for further discussion). This implies that, given the same wind speed, fog gages will intercept a smaller proportion of incident rainfall during higher intensity events. However, this does not necessarily imply that the total rainfall intercepted by the fog gage during high-intensity events is less than that collected during light rains. With other gages, such as the "standard fog collector" (or *SFC*) proposed by Schemenauer and Cereceda (1994), or wire harps (Goodman 1985), measurements unquestionably represent a mixture of fog and wind-blown rain. The *SFC* is a vertically orientated planar screen, with a fixed orientation, so that changes in wind direction will also affect measurements (Juvik and Nullet, 1995a; García-Santos, 2007). As shown by Frumau *et al.* (this volume), water drops captured by a wire harp may be blown off again at higher wind speeds. In addition, the collection efficiency of the two types of gages depends on the prevailing wind speed. The louvered cylindrical gage used by Juvik and Ekern (1978) and in many other subsequent studies in Hawai'i suffers much less from these disadvantages (Frumau *et al.*, this volume). The Juvik-style gage, like any other, has a collection efficiency unique to the design, regardless of whether it is exposed to wind-blown rainfall. This efficiency may not be equal for cloud water and rainfall, and may change with wind speed and drop size.

Beyond these mechanical issues, a fundamental question is how such measurements can be related to *CWI* by structurally complex natural vegetation stands. The first published record of *FS* measurements using a fog gage was by Marloth (1904) as cited by Nagel (1956). Marloth's gage employed grass mounted over a rain gage and this simple pioneering design was probably more representative of *CWI* by natural vegetation than subsequent attempts using mechanical screens. Even if rainfall can be physically or mathematically excluded from mechanical collector measurements, no *a priori* relationship can be expected between interception by a vertical collector and water depth beneath the canopy. The volume of water collected in fog gages is routinely reported as a depth (volume per unit area of the vertically oriented collector) and then compared with rainfall measurements, conventionally expressed as a depth (volume per horizontal area). As acknowledged by previous researchers in Hawai'i, it is clearly not correct to assume that the fog gage depth (over a vertically oriented area) can be equated with the depth of *CWI* by vegetation over a horizontal area. However, fog gage measurements are frequently interpreted that way.

Table 35.1 *Summary of selected cloud water interception studies in Hawai'i*

Site			Observations			Method[a]	Study period				Daily				Ref[f]
Location	Aspect	Elev. (m)	Start	End	Period (mos)		*TF*[b]	*P*[c]	*CWI*[d]	*FS*[e]	*TF*(mm)	*P*[c]	*CWI*[d]	*FS*[e]	
Lâna'ihale	Wind	838	Jul-55	Jul-58	36	1,3	9931	3785	2286		9.07	3.46	2.09		1
Lâna'ihale	Wind	838	Aug-55	Jul-58	36	5		2642		1803		2.41		1.65	1
Lâna'ihale	Wind	921	May-07	Aug-08	10	4	4922	401		2637	16.6	1.6		10.8	13
Kîlauea	ENE	1210	Jun-74	Jun-75	9	4		1794		381		6.55		1.39	2
Kîlauea	Wind	1190	Jun-95	Jun-96	12	1	NR[g]	1900	1130			5.21		3.10	3
Kîlauea	Wind	1190	Jun-95	Jun-00	60	1,2	NR[g]	14780	12320		7.97	6.80	5.62		10
Kîlauea	Wind	1190	Jun-95	Jun-00	60	1,2	NR[g]	NR[g]	NR[g]		8.82	6.80	4.49		11
M. Loa	NE	610	Oct-72	Apr-73	7	4		2999		168		14.1		0.79	4
M. Loa	ENE	1580	Oct-72	Apr-73	7	4		1285		638		6.04		3	4
M. Loa	NE	2530	Oct-72	Apr-73	7	4		450		293		2.11		1.38	4
M. Loa	N	3415	Oct-72	Apr-73	7	4		184		40		0.86		0.19	4
M. Loa	NE	2530	Jun-73	Aug-73	3	4		65		87		0.71		0.95	4
M. Loa	ENE	1580	Jan-74	Dec-74	12	4		2449		681		6.71		1.87	2
M. Loa	ENE	1580	Jan-75	Dec-75	12	4		2679		859		7.34		2.35	2
M. Loa	NE	2530	Jan-74	Dec-74	12	4		1031		734		2.82		2.01	2
M. Loa	NE	2530	Jan-75	Dec-75	12	4		1046		678		2.87		1.87	2
M. Loa	N	3415	Jan-74	Dec-74	12	4		555		233		1.52		0.64	2
M. Loa	N	3415	Jan-75	Dec-75	12	4		300		138		0.82		0.38	2
M. Loa	SE	1692	Jun-74	Jun-75	9	4		588		21		2.15		0.08	2
M. Loa	SE	2012	Jun-74	Jun-75	9	4		644		17		2.35		0.06	2
Hualâlai	WSW	981	Aug-74	Dec-75	14	4		1857		134		4.36		0.32	2
Hualâlai	WSW	1905	Aug-74	Dec-75	17	4		1192		304		2.30		0.59	2
Hualâlai	Sum.	2496	Aug-74	Dec-75	17	4		528		832		1.02		1.61	2
Hualâlai	Lee	1400	Feb-02	Jan-03	12	4		689		98		1.89		0.27	5
Ko'olau	Lee	625	Jan-81	Dec-81	12	4		3806		237		10.4		0.65	6
Ko'olau	Lee	625	Jan-82	Aug-82	8	4		3682		330		15.1		1.36	6
Kohala	NE	1160	Aug-84	Nov-84	3	1,4	606	336		1365	6.06	3.36		13.6	7
Kohala	NE	1180	Mar-03	Jul-04	8	1,2,4	1288	1486	131	154	5.41	6.24	0.55	0.65	12
M. Kea	SW	2600	Oct-91	Sep-92	12	1,4	194	257		98	0.53	0.70		0.27	7
Haleakalâ	Lee	2200	May-96	Jun-97	14	6		567		839		1.33		1.97	8
Haleakalâ	Lee	1219	Feb-02	Aug-02	7	1,2,3,6	100	155	57	60	0.72	1.12	0.73	0.43	9
Haleakalâ	Wind	1951	Nov-01	Aug-02	9	1,2,3,6	2298	1936	773	1856	8.74	7.36	2.98	7.06	9

[a] Methods: 1, canopy water balance based on *TF* and rainfall measurements; 2, evaporation estimate included in canopy water balance estimate; 3, accounted for windblown rainfall; 4, fog gage, louvered screen; 5, fog gage, harp-type; 6, fog gage, mesh screen-type; 7, fog gage, Grunow-type (Grunow, 1952 as cited by Nagel 1956).

[b] *TF*: measured *TF*.

[c] *P*: measured rainfall.

[d] *CWI*: cloud water interception derived using various methods (see Methods column).

[e] *FS*: fog screen measurements.

[f] References: 1, Ekern (1964); 2, Juvik and Ekern (1978); 3, Heath and Huebert (1999); 4, Juvik and Perreira (1974); 5, Juvik and Tango (2003); 6, Ekern (1983); 7, Juvik and Nullet (1995b); 8, Juvik (1998); 9, Giambelluca *et al.* (this volume); 10, Heath (2001); 11, Carillo (personal communication 2007); 12, DeLay (2005); 13: Juvik (2009).

[g] NR: *TF* measured but not reported.

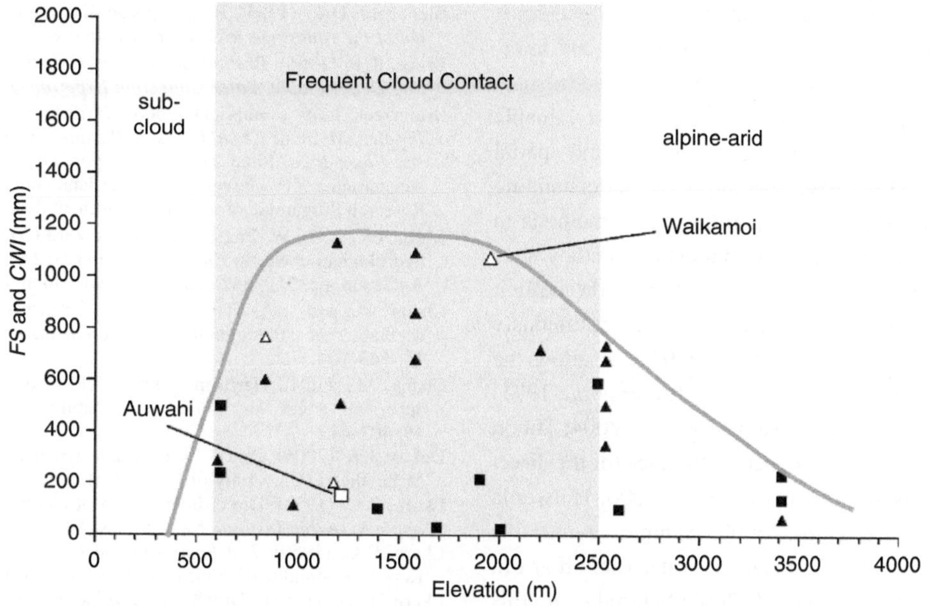

Figure 35.3. Annualized *FS* (black) and *CWI* (white) totals from selected studies presented in Table 35.1 shown in relation to elevation. Windward sites (triangles) and leeward sites (squares) are also distinguished. Cloud contact zone corresponds to shaded area in Figure 35.1. NB: the figure does not include the exceptionally high annualized *FS* total of 4982, mm reported for 1190 m.a.s.l. on windward Kohala (Juvik and Nullet, 1995b).

A related problem concerns the fact that each forest or collection of trees represents a unique situation in terms of its effective intercepting surface area whereas the storage capacity of the abundant epiphytes and bryophytes that characterize many cloud forests tends to vary in time as well, thus adding further complexity (Hölscher *et al.*, 2004; Tobón *et al.*, this volume #26).

With Hawai'i's persistent winds, rain seldom falls vertically. Wind-blown rain complicates the interpretation not only of fog gage observations, as described above, but also canopy water balance measurements in general. Rain gages, with their horizontally oriented orifices, do not accurately represent rainfall incident on sloping surfaces, isolated trees, or upwind edges, especially during conditions of strong wind (Sharon, 1980; Nespor and Sevruk, 1999; Holwerda *et al.*, 2006). Rainfall tends to be enhanced on windward-facing slopes and reduced on leeward-facing slopes. Failure to account for this effect can create large errors in estimates of *CWI* derived from comparisons of *TF* and rainfall. Previous research in Hawai'i has emphasized that a greater proportion of fog interception occurs during low-rainfall periods (e.g. Ekern, 1964; Juvik and Nullet, 1995b). However, this finding may also result from the use of vertical fog gages, which tend to collect a greater proportion of wind-blown rainfall during lighter rains when drops are smaller. A standard rain gage is likely to miss such near-horizontal precipitation. The recently introduced spherical rain gage (Chang and Flannery, 2001; Chang and Harrison, 2005) has been designed specifically to present the same effective gage orifice regardless of the inclination of the rainfall and so should allow measurement of what may be termed the "potential" or "true" precipitation. Frumau *et al.* (2006) compared the performance of a standard rain gage, a spherical gage, and a modification of the louvered cylindrical gage of Juvik and Ekern (1978) with which the angle of incident rain can be evaluated, under wet and windy conditions in Costa Rica. Although the spherical gage was clearly superior to the standard gage, its catch lagged behind that of the modified louvered cylinder during conditions of low rainfall intensity and even more so during times of fog. Further testing of both new devices under a range of conditions is desirable.

An important goal of fog research in Hawai'i is to determine the hydrological input to Hawai'i's watersheds from *CWI*. In this regard, it is the comparatively large areas of continuous forest, rather than forest edges or exposed vegetation on ridge tops that are the spatially significant areas of concern. However, many of the previous studies are based on fog gage measurements taken in large clearings or along topographic ridges, or from canopy water balance observations in isolated patches of vegetation or at forest edges (cf. Juvik, 2009). Both wind velocity and fog density is much higher at the forest edge or in the open than within the canopy of a closed-canopy forest. Measurements may not, therefore, be representative of the majority of the cloud forest area. At a forest-edge site in the Catskill Mountains, for example, Weathers *et al.* (1995) found that *CWI* decreased to half its edge value at a point only 28 m into the forest. Similarly, relative amounts of *TF* as measured at 2 m from the edge of a regenerating cloud forest in Costa Rica were more than twice those at 10 m from the edge at high wind speeds although amounts were similar at low wind speeds (Bruijnzeel, 2006). Such effects must be taken into consideration when selecting field sites and interpreting results.

The years of fog research in Hawai'i have provided valuable insights into the spatial patterns of fog occurrence and spatial differences in (potential) fog fluxes. But improved understanding will have to come from studies incorporating other methods in addition to fog gage measurements, and focusing on study sites within continuous forest. Alternative methods include explicit canopy water balance observations using robust continuously recording equipment (cf. McJannet *et al.*, 2007a) to allow the testing of running water balance models (e.g. Rutter *et al.*, 1971) for canopy and bryophytes alike (cf. Hölscher *et al.*, 2004; Tobón *et al.*, this volume #26); eddy covariance techniques for the direct measurement of fog deposition (Eugster *et al.*, 2006; Holwerda *et al.*, 2006; Beiderwieden *et al.*, 2008; Schmid *et al.*, this volume), and analysis of stable isotopes of water (Scholl *et al.*, 2002; Rhodes *et al.*, 2006; Guswa *et al.*, 2007; Schmid *et al.*, this volume; Scholl *et al.*, this volume).

CONCLUSION

After half a century of fog research in Hawai'i, patterns are beginning to emerge regarding the spatial extent and potential amounts of water derived directly from deposition of cloud drops on (tall) vegetation. Much of the credit for this improved understanding goes to the pioneers in this field, especially Wendell Mordy, Paul C. Ekern, and James O. Juvik, whose efforts have provided much of what is now known about cloud water interception in Hawai'i. But because of the complexities of the various interactions between the atmosphere and the vegetation, and the spatial variability of both fog and vegetation structure, much remains to be done.

ACKNOWLEDGEMENTS

This material is based upon work supported by the National Science Foundation under Grant no. EAR-0309731 and by the US Geological Survey, Biological Resources Discipline, Global Change Program. Special thanks are due to James Juvik and Paul Ekern for providing documentation and critical comments. Lan Cuo assisted with the spatial data used to develop Figure 35.1.

REFERENCES

Beiderwieden, E., V. Wolff, Y. J. Hsia, and O. Klemm (2008). It goes both ways: measurements of simultaneous evapotranspiration and fog droplet deposition at a montane cloud forest. *Hydrological Processes* **22**: 4181–4189.

Blanchard, D. C. (1953). Raindrop size-distribution in Hawaiian rains. *Journal of the Atmospheric Sciences* **10**: 457–473

Bruce, R. P. (1966). *Preliminary Report, Haleakalâ National Park, Ko'olau Fog Drip Potable Water Collection Experiment.* Mimeographed, 8 p.)

Bruijnzeel, L. A. (compiler) (2006). *Hydrological Impacts of Converting Tropical Montane Cloud Forest to Pasture,with Initial Reference to Northern Costa Rica*, Final Technical Report DFID-FRP Project No. R7991. Amsterdam: VU University Amsterdam, and Aylesford, UK: Forestry Research Progamme of the UK Department for International Development.

Bruijnzeel, L. A., W. Eugster, and R. Burkard (2005). Fog as an input to the hydrological cycle. In *Encyclopaedia of Hydrological Sciences*, ed. M. G. Anderson, pp. 559–582. Chichester, UK: John Wiley.

Chang, M., and L. A. Flannery (2001). Spherical gages for improving the accuracy of precipitation measurements. *Hydrological Processes* **15**: 643–654.

Chang, M., and L. Harrison (2005). Field assessments on the accuracy of spherical gages in rainfall measurements. *Hydrological Processes* **19**: 403–412.

DeLay, J. K. (2005). Canopy water balance of an elfin cloud forest at Alakahi. M.Sc. thesis, University of Hawai I, Honolulu, HI.

Ekern, P. C. (1958). Direct interception of cloud water on Lâna'ihale. *Pineapple Research Institute News* **6**: 129–133.

Ekern, P. C. (1960). *The Importance of Fog Drip on the Watershed*, Final Report. Honolulu, HI: Pineapple Research Institute.

Ekern, P. C. (1964). Direct interception of cloud water on Lâna'ihale, Hawai'i. *Soil Science Society of America Proceedings* **28**: 419–421.

Ekern, P. C. (1983). *Measured Evaporation in High Rainfall Areas, Leeward Ko'olau Range, O'ahu, Hawai'i*, Technical Report No. 156. Honolulu, HI: Water Resources Research Center, University of Hawai'i.

Eugster, W., R. Burkard, F. Holwerda, F. N. Scatena, and L. A. Bruijnzeel (2006). Characteristics of fog and fog-water fluxes in a Puerto Rican elfin cloud forest. *Agricultural and Forest Meteorology* **139**: 288–306.

Frumau, K. F. A., L. A. Bruijnzeel, and C. Tobón (2006). Measurement of precipitation in montane tropical catchments: comparative performance of conventional, spherical and "potential" rain gages. In *Forest and Water in a Changing Environment*, eds. S. R. Liu, G. Sun, and P. S. Sun, pp. 104–108. Vienna: IUFRO, and Beijing: Chinese Academy of Forestry.

García-Santos, G. (2007). *An ecohydrological and soils study in a montane cloud forest in the National Park of Garajonay, La Gomera (Canary Islands, Spain).* PhD Thesis, VU University Ámsterdam, Ámsterdam, The Netherlands. [http://www.falw.vu.nl/nl/onderzoek/earth-sciences/geo-environmental-science-and-hydrology/hydrology-dissertations/index.asp].

Goodman, J. (1985). The collection of fog drip. *Water Resources Research* **21**: 392–394.

Grunow, J. (1952). Nebelniederschlag. *Sonderdruck aus Berichte des Deutschen Wetterdienstes in der U.S. Zone, Bad Kissingen* **42**: 30–34.

Grunow, J. (1960). The productiveness of fog precipitation in relation to the cloud droplet spectrum. *Proceedings of the Cloud Physics Conference*, Woods Hole, MA, *Geophysical National Research Council* **746**: 110–117.

Guswa, A. J., A. L. Rhodes, and S. E. Newell (2007). Importance of orographic precipitation to the water resources of Monteverde, Costa Rica. *Advances in Water Resources* **30**: 2098–2112.

Heath J. A. (2001). Atmospheric nutrient deposition in Hawai'i: methods, rates and sources. Ph.D. thesis, University of Hawai'i, Honolulu, HI.

Heath, J. A., and B. Huebert (1999). Cloudwater deposition as a source of fixed nitrogen in a Hawaiian montane forest. *Biogeochemistry* **44**: 119–134.

Hölscher, D., L. Köhler, A. I. J. M. van Dijk, and L. A. Bruijnzeel (2004). The importance of epiphytes to total rainfall interception by a tropical montane rain forest in Costa Rica. *Journal of Hydrology* **292**: 308–322.

Holwerda, F., R. Burkard, W. E. Eugster, *et al.* (2006). Estimating fog deposition at a Puerto Rican elfin cloud forest site: comparison of the water budget and eddy covariance methods. *Hydrological Processes* **20**: 2669–2692.

Juvik, J. O. (1977). Mountain fog on Mauna Loa, Hawai'i. Ph.D. thesis, University of Hawai'i, Honolulu, HI.

Juvik, J. O. (1998). Mountain climatology and large scale cloud water recovery at Kahikinui, Maui, Hawaiian Islands. In *Proceedings of the 1st International Conference on Fog and Fog Collection*, eds. R. S. Schemenauer and H. A. Bridgman, pp. 437–440. Ottawa, Canada: International Development Research Center.

Juvik, J. O. (2009). The seminal Hawaiian cloud forest "Lâna'i fog drip study": 50th anniversary reassessment. In *Tropical Hydrology Symposium,* eds. F. L. Ogden, R. S. Harmon, and R. F. Stallard, p. 229. Panama City: Smithsonian Institution.

Juvik, J. O., and P. C. Ekern (1978). *A Climatology of Mountain Fog on Mauna Loa, Hawai'i Island*, Technical Report No. 118. Honolulu, HI: Water Resources Research Center, University of Hawai'i.

Juvik, J. O., and D. Nullet (1995a). Comments on a proposed standard fog collector for use in high elevation regions. *Journal of Applied Meteorology* **34**: 2108–2110.

Juvik, J. O., and D. Nullet (1995b). Relationships between rainfall, cloud-water interception, and canopy throughfall in a Hawaiian montane forest. In *Tropical Montane Cloud Forests*, eds. L. S. Hamilton, J. O. Juvik, and F. N. Scatena, pp. 165–182. New York: Springer-Verlag.

Juvik, J. O., and D. J. Perreira (1974). Fog interception on Mauna Loa, Hawai'i. *Proceedings of the Association of American Geographers* **6**: 22–24.

Juvik, J. O., and L. K. Tango (2003). *Climate and Water Resources of Pu'u Wa'a Wa'a, North Kona*. Hilo, HI: Department of Land and Natural Resources, Division of Forestry and Wildlife.

Marloth, R. (1904). Results of experiments on Table Mountain for ascertaining the amount of moisture deposited from the south-east clouds. *Transactions of the South African Philosophical Society* **14**: 403–408.

McJannet, D., J. S. Wallace, and P. Reddell (2007a). Precipitation interception in Australian tropical rainforests. I. Measurement of stemflow, throughfall and cloud interception. *Hydrological Processes* **21**: 1692–1702.

McJannet, D., J. S. Wallace, and P. Reddell (2007b). Precipitation interception in Australian tropical rainforests. II. Altitudinal gradients of cloud interception, stemflow, throughfall and interception. *Hydrological Processes* **21**: 1703–1718.

McKnight, J. H., and J. O. Juvik (1975). *Methodological Approaches in Hawaiian Fog Research*, Technical Report No. 85. Honolulu, HI: Water Resources Research Center, University of Hawai'i.

Mordy, W. (1955). *Cloud Seeding Project*, Final Report. Honolulu, HI: Pineapple Research Institute of Hawai'i and Experiment Station of the Hawaiian Sugar Planters Association Meteorology Department.

Mueller-Dombois, D. (1967). Ecological relations in the alpine and subalpine vegetation on Mauna Loa, Hawai'i. *Journal of the Indian Botanical Society* **46**: 403–411.

Munro, G. C. Not dated. *The Story of Lâna'i*. Typed copy. Donated to University of Hawai'i at Mânoa in 1989.

Nagel, J. F. (1956). Fog precipitation on Table Mountain. *Quarterly Journal of the Royal Meteorological Society* **82**: 452–460.

Nespor, V., and B. Sevruk (1999). Estimation of wind-induced error of rainfall gage measurements using a numerical simulation. *Journal of Atmospheric and Oceanic Technology* **16**: 450–464.

Peperfale, P. (1956). Evaluation of the data collected at the fog drip station from August 1, 1955 to May 6, 1956. Honolulu, H: Hawaiian Pineapple Company, Ltd. (unpublished).

Powers, H. A., and C. K. Wentworth (1941). Air movement and weather in the Mauna Kea-Mauna Loa Saddle, Hawai'i. *Bulletin of the American Meteorological Society* **22**: 6–13.

Pukui, M. K., and S. H. Elbert (1986). *Hawaiian Dictionary*. Honolulu, HI: University of Hawai'i Press.

Rhodes, A. L., A. J. Guswa, and S. E. Newell (2006). Seasonal variation in the stable isotopic composition of precipitation in the tropical montane forests of Monteverde, Costa Rica. *Water Resources Research* **42**, W11402, doi:10.1029/2005WR004535.

Rosendahl, P. H. (1972). Aboriginal agriculture and residence patterns in upland Lapakahi, island of Hawai'i. Ph.D. thesis, University of Hawai'i at Mânoa, Honolulu, HI.

Rutter, A. J., K. A. Kershaw, P. C. Robins, and A. J. Morton (1971). A predictive model of rainfall interception in forests. Derivation of the model from observations in a plantation of Corsican pine. *Agricultural Meteorology* **9**: 367–384.

Schemenauer, R. S., and P. Cereceda (1994). A proposed standard fog collector for use in high elevation regions. *Journal of Applied Meteorology* **33**: 1313–1322.

Scholl, M., S. B. Gingerich, and G. Tribble (2002). The influence of microclimates and fog on stable isotope signatures used in the interpretation of regional hydrology: East Maui, Hawai'i. *Journal of Hydrology* **264**: 170–184.

Sharon, D. (1980). Distribution of hydrologically effective rainfall incident on sloping ground. *Journal of Hydrology* **46**: 165–188.

Smathers, G. A., and D. Mueller-Dombois (1972). *Invasion and Recovery of Vegetation after a Volcanic Eruption in Hawaii*, Technical Report No. 10. Honolulu, HI: Water Resources Research Center, University of Hawai'i.

Squires, P., and J. Warner (1957). Some measurements in the orographic cloud on the island of Hawai'i and in trade wind cumuli. *Tellus* **9**: 477–494.

Stock, J., J. Coil, and P. V. Kirch (2003). Paleohydrology of arid southeastern Maui, Hawaiian Islands, and its implications for prehistoric human settlement. *Quaternary Research* **59**: 12–24.

Weathers, K. C., G. M. Lovett, and G. E. Likens (1995). Cloud deposition to a spruce forest edge. *Atmospheric Environment* **29**: 665–672.

36 Interpreting canopy water balance and fog screen observations: separating cloud water from wind-blown rainfall at two contrasting forest sites in Hawai'i

T. W. Giambelluca, J. K. DeLay, and M. A. Nullet
University of Hawai'i at Mānoa, Honolulu, Hawai'i, USA

M. Scholl
U.S. Geological Survey, Reston, Virginia, USA

S. B. Gingerich
U.S. Geological Survey, Honolulu, Hawai'i, USA

ABSTRACT

Based on field measurements made at dry (Auwahi) and wet (Waikamoi) cloud forest sites on the island of Maui, a preliminary analysis of fog gage measurements and wet-canopy water balance estimates was made. Accounting for effects of wind-blown rainfall and varying wind direction, estimates of cloud water flux were derived based on fog gage observations. Throughfall (*TF*) measurements, incident rainfall estimates, and calculated amounts of wet-canopy evaporation were used to estimate event totals of cloud water interception (*CWI*) by the vegetation at each site. Measured *TF* was about 37% of incident rainfall at Auwahi, and 119% at Waikamoi. At both sites *TF* was dominated by rainfall, but was significantly influenced by fog at Waikamoi only. Fog contributed at an average frequency of once every two days at Auwahi and about twice in three days at Waikamoi. Derived *CWI* totals were equivalent to 151 mm year^{-1} at Auwahi and 1073 mm year^{-1} at Waikamoi. At Auwahi, however, the majority of intercepted water was re-evaporated from the wet vegetation, and never reached the ground. Total *CWI* was related to fog screen catch and cloud water flux at Waikamoi, but not at Auwahi.

INTRODUCTION

In Hawai'i, as in other mountainous tropical areas, the presence of fog has profound effects on the biota, producing distinct ecological zones known collectively as tropical montane cloud forests (TMCF). The movement of fog by wind results in deposition of cloud droplets on the leaves and stems of plants, and other obstacles. This cloud water interception (*CWI*), locally called "fog drip," has long been recognized as a hydrological input, contributing to soil water, groundwater, and streamflow in windward, high-rainfall areas, and providing life-sustaining moisture on otherwise dry, leeward, and high-elevation slopes. Quantifying *CWI* in Hawai'i has proven difficult, though, despite research efforts spanning more than 50 years (DeLay and Giambelluca, this volume).

Much of the previous *CWI* research in Hawai'i has been based on artificial fog collectors (Delay and Giambelluca, this volume). For most types of fog gages, interpreting measurements is confounded by the difficulty to distinguish rainfall from fog whereas the fog catch of some gages is also affected by wind direction (Juvik and Nullet, 1995a; García-Santos, 2007). But even when corrected for rainfall and wind direction, fog gages are usually not representative of *CWI* occurring in the natural vegetation (Bruijnzeel *et al.*, 2005). Wet-canopy water balance methods are often used to assess *CWI*, but these are themselves subject to error, especially in sloping and windy terrain where incident rainfall differs from the amount measured (Holwerda *et al.*, 2006; cf. García-Santos, 2007; Holwerda *et al.*, this volume #29).

Tropical Montane Cloud Forests: Science for Conservation and Management, eds. L. A. Bruijnzeel, F. N. Scatena, and L. S. Hamilton. Published by Cambridge University Press.

This chapter presents preliminary results of a field study aimed at overcoming some of the limitations of prior *CWI* studies in Hawai'i. The objectives were to: (i) make field measurements pertaining to *CWI* at a dry and at a wet cloud forest site on Mt. Haleakala, Maui; (ii) develop and apply methods for interpreting fog gage measurements in the presence of wind-blown rainfall and varying wind direction; and (iii) account for the effect of terrain slope on incident rainfall when computing *CWI* via the canopy water balance approach.

MATERIALS AND METHODS

Sites were chosen so as to represent the extremes of dry and wet cloud forest environments, and field measurements aimed at evaluating fog water inputs to the respective ecosystems based on fog gage measurements, the canopy water balance approach, and the use of stable isotopes of water to identify the proportions of fog and rain water in surface water, soil water, groundwater, and plant tissue. Here, a preliminary analysis is presented of fog screen observations and canopy water balance based on data for the first 7 months of observations at Auwahi (a dry site) versus 9 months at Waikamoi (a wet site). The use of stable isotopes to assess fog water inputs is discussed in some detail by Scholl *et al.* (this volume).

Field sites

Field instrumentation was set up at two contrasting forest sites on Mt. Haleakala, viz. at a leeward dry forest site at 1219 m.a.s.l. in Ulupalakua Ranch, called "Auwahi," and at a wet windward site at 1951 m.a.s.l. within the Nature Conservancy's Waikamoi Forest Reserve, called "Waikamoi." Both Waikamoi (Medeiros *et al.*, 1995) and Auwahi (Wagner *et al.*, 1999) have been identified as "hot spots" of endemic biodiversity. Hawaiian dry forests were the most diverse of Hawaiian ecosystems and Auwahi is the most diverse dry forest location left in Hawai'i with more than 40 genera present. At Auwahi, instruments for canopy water balance measurements were placed in low shrubs and trees of up to ~2.5 m in height (estimated leaf area index, $LAI = 0.82$), located just upslope of a pasture. Wind direction is often upslope, hence the vegetation is exposed to air moving across the open pasture land. Because of this exposure, interception of fog and wind-blown rainfall is much higher than would be the case with continuous vegetation of the same height (e.g. Weathers *et al.*, 1995). However, the study site is typical of the shrub and tree cover in the area, which occurs exclusively in small isolated patches. The wet windward forest site at Waikamoi is more typical of other Hawaiian TMCF (estimated *LAI* 2.0), with the dominant tree genera including *Metrosideros*, *Acacia*, *Cheirodendron*, *Hedyotis*, *Meliocope*, and *Ilex* (A. Medeiros, personal communication; Medeiros *et al.*, 1998). Here, the site was at the boundary between continuous 4–6-m high forest and 0.5–1.5-m high shrub, with the predominant wind direction bringing air from low to high vegetation at night and from high to low vegetation during the day. Hence, there may be some edge-related enhancement of fog and wind-blown rain interception at this site. Wind speeds as measured at approximately the height of the tallest vegetation averaged 2.8 ± 1.3 m s^{-1} at Auwahi and 2.7 ± 0.9 m s^{-1} at Waikamoi. Epiphytes, including alien "Spanish moss" (*Tillandsia usneoides*), are found at both sites, adding to above-ground water storage capacity.

Field measurements

Field measurements at each site consisted of the following: solar radiation (model 8–48, Eppley, Newport, RI, USA), net radiation (model Q*7.1, REBS, Seattle, WA, USA), canopy (model 4000.4ZL, Everest, Tucson, AZ, USA), soil (model TCAV, Campbell, Logan, UT, USA), and air (model HMP35C, Vaisala, Helsinki, Finland) temperatures, relative humidity (model HMP35C, Vaisala), wind speed and direction (models 014A and 024A, respectively, MetOne, Grants Pass, OR, USA), soil heat flux (model HFT-3, REBS) and soil moisture (model CS615, Campbell), rainfall (model TE525, Texas Electronics, Dallas, TX, USA), fog and throughfall. For fog observations, a visibility sensor (results not reported here) was used plus a vertical, 1-m^2 fog screen of the Schemenauer and Cereceda (1994) design, the catch of which was recorded using a tipping-bucket rain gage. Throughfall (*TF*) was measured with four collectors (only two collectors reported here for Waikamoi), each consisting of three 600×4 cm troughs giving a total collection area per gage of approximately 0.7 m^2 (cf. McJannet *et al.*, 2007a). Each collector had a large tipping bucket of our own design (nominal calibration 0.15 l per tip, equivalent to ~0.21 mm of *TF* per tip). Data-loggers recorded the times of occurrence of each tip for rainfall, fog screen, and *TF* gages. Other meteorological sensors were sampled at 10-s intervals and recorded hourly.

Interpretation of fog screen observations

Fog gages of various designs have been proposed for field monitoring of the horizontal flux of fog. As mentioned, a common problem with such gages is the difficulty in distinguishing fog from rainfall. Some designs, such as the louvered aluminum cylindrical gage of Juvik and Nullet (1995a), use a cover in an attempt to prevent rain from entering the gage (cf. McJannet *et al.*, 2007a). With other designs, the amount of rainfall captured by the gage should be estimated and deducted from the observed total. To correct for rain capture, the volume of rainfall incident on the 1-m^2 frame of fog screen used in the present study was estimated as follows. Under anything but calm conditions, rain drops will fall diagonally, rather than vertically. The angle of a drop's path, with respect to the vertical, depends on the wind speed and the terminal velocity of the drop, the latter being a function of drop size (Sharon, 1980). Because rain drop size is not uniform, drops fall at a range of angles at any given time, with small drops falling at a larger angle from the vertical. The first task, then, is to estimate the drop-size distribution of incident rainfall.

The data obtained by Blanchard (1953) for orographic rain on the island of Hawai'i at similar elevations to those of the study sites were used to characterize the drop-size distribution. It was found that the total number of drops was a function of rainfall intensity, and could be reasonably well described by:

$$N = 7872.9R^{-0.4551} \quad (36.1)$$

where N = the total drop count (number of drops m^{-3}), and R = rainfall intensity (mm hour^{-1}). It was also found that a gamma distribution can be used to predict the shape of the drop-size distribution for a given rain rate:

$$n_i = N\lambda^\eta d_i^{\eta-1} \frac{e^{-\lambda d_i}}{e^{\ln(\Gamma(\eta))}} \quad (36.2)$$

where n_i = the drop count in drop-size class i (drops m^{-3}), d_i = average drop diameter (m) for drop-size class i, e = the base of the natural logarithm (2.718), Γ = the gamma function, and where the parameters λ and η can be adjusted to obtain the best fit against the observed drop-size distribution. Eleven drop-size classes were used with central values of 0.1 through 2.1 mm with a 0.2-mm interval. The shape of the distribution was best preserved by setting λ to a constant value of 8.0 and defining η as a function of the rainfall intensity, as:

$$\eta = 1.8498R^{0.4774}. \quad (36.3)$$

The results yielded drop counts in each drop-size class that were proportionally correct but must be scaled to reproduce the rain rate.

Next, the drop count per unit volume needs to be converted to a rain rate, i.e. the amount of water falling through a horizontal plane per unit time. To do this, drop terminal velocity was estimated as a function of drop diameter. Using data from Gunn and Kinzer (1948) the following function for estimating drop terminal velocity was developed:

$$vel_i = -0.88171d_i^2 + 0.051316d_i - 0.27019 \quad (36.4)$$

where vel_i = terminal velocity of drops (m s^{-1}) with diameter d_i as defined previously. The number of drops in each size class passing through a horizontal plane per unit time can be estimated as:

$$m_i = n_i vel_i \quad (36.5)$$

where m_i = the number of drops in size class i passing through a 1-m^2 horizontal area per second (drops s^{-1}). The volume of water passing through a 1-m^2 plane is the product of the drop count rate (m_i) and individual drop volume, which, if drops are assumed to be spherical, is defined as:

$$vol_i = \tfrac{4}{3}\pi\left(\frac{d_i}{2}\right)^3 \quad (36.6)$$

where vol_i = the volume of drop (m^3) of diameter d_i. An estimate of the rainfall intensity in each drop-size class can be derived as:

$$r'_i = 3600 m_i vol_i \quad (36.7)$$

where r'_i = the unscaled rainfall intensity for drop-size class i (m hour^{-1}) and the coefficient 3600 converts units from m s^{-1} to m hour^{-1}. The total unscaled rain rate is then obtained by summing the volume in the 11 individual drop-size classes:

$$R' = \sum_{i=1}^{11} r'_i \quad (36.8)$$

where R' = unscaled rain rate (m hour^{-1}). The ratio of this value to the measured rainfall intensity, gives a dimensionless scaling factor (S):

$$S = \frac{R}{1000R'} \quad (36.9)$$

where the coefficient 1000 converts R' into mm hour^{-1} units compatible with R. Applying this scaling factor to each drop-size class, the rainfall intensity in each drop-size class (r_i, m hour^{-1}) can be estimated:

$$r_i = Sr'_i. \quad (36.10)$$

The vertical 1-m^2 fog screen and 10-cm wide collecting trough projected (via the rain-drop trajectory) onto a horizontal plane forms a parallelogram, whose length in the downwind direction can be derived as:

$$X_i = \frac{U}{vel_i} + 0.05 for \frac{U}{vel_i} > 0.05 \quad (36.11)$$

$$X_i = 0.1 \text{ for } \frac{U}{vel_i} \leq 0.05 \quad (36.12)$$

where X_i = the downwind length of the horizontally projected area of the fog screen (m), U = the wind velocity (m s^{-1}), and 0.05 m the width of the upwind half of the collecting trough. Under very light winds or calm conditions, with drops falling nearly straight down, X_i is reduced to a minimum value of 0.1 m, the full width of the collecting trough. The height of the parallelogram h (m) can be estimated as:

$$h = |\sin(\theta - \varphi)| \quad (36.13)$$

where θ = the wind direction (°), and ϕ = orientation of the screen (°). The projected area for each drop-size class (A_i, m^2) is:

$$A_i = hX_i. \tag{36.14}$$

The volume of rain incident on the fog screen can now be estimated for each drop-size class as:

$$y_i = r_i A_i \tag{36.15}$$

where y_i = the volume of rain (m^3 hour^{-1}), in drop-size class i, incident on the fog screen. The sum (Y):

$$Y = \sum_{i=1}^{11} y_i \tag{36.16}$$

gives the total amount of rain incident on the fog screen (m^3 hour^{-1}).

The net fog catch (FC) by the screen (m^3 hour^{-1}) was computed as:

$$FC = FS - kY \tag{36.17}$$

where FS = measured amount of fog plus rainfall caught by the fog screen (m^3 hour^{-1}), and k = rain-catch efficiency of the screen (ratio of rainfall caught to incident rainfall). To estimate k, values were iteratively adjusted downward until most of the resulting negative FC values were eliminated. Doing so with data from both field sites resulted in a value of 0.30 for k, which eliminated all but 3% and 0.9% of negative instances of FC at Auwahi and Waikamoi, respectively. Subsequently, all negative values of FC were reassigned as zero.

According to Schemenauer and Cereceda (1994), the fog-catch efficiency of their screen is 0.66. Catch efficiency is probably sensitive to wind direction relative to screen orientation, and possibly also to wind speed, but insufficient data were available to test this quantitatively. Therefore, the Schemenauer and Cereceda (1994) value was used to simply estimate the amount of fog flowing through the 1-m^2 screen ($FLUX_{screen}$ in m^3 hour^{-1}) as:

$$FLUX_{screen} = \frac{FC}{0.66}. \tag{36.18}$$

However, a more useful variable is the flow of fog water through a vertical plane normal to the wind direction ($FLUX_{norm}$, in m^3 hour^{-1}) which can be estimated from:

$$FLUX_{norm} = \frac{FLUX_{screen}}{\sin(\theta - \varphi)} \tag{36.19}$$

where θ = wind direction (° from N), and ϕ = orientation of the screen (°). Because unreasonable amplification of $FLUX_{norm}$ occurred for small relative wind angles, a minimum value of 30° was set for $(\theta - \phi)$. Values of $FLUX_{norm}$ may be expressed in depth units relative to the vertical catch area of the screen, as:

$$CWF = 1000\frac{FLUX_{norm}}{B} \tag{36.20}$$

where CWF = cloud water flux per unit vertical area (mm hour^{-1}), B = the vertical catch area of the screen (m^2), and the coefficient 1000 converts units from m hour^{-1} to mm hour^{-1}.

Wet-canopy water balance

Cloud water interception by the vegetation (CWI) can be computed from measurements of water inputs and outputs to and from the wetted canopy. A simple conservation of mass approach to canopy water exchange can be expressed as:

$$RI + CWI = E + TF + SF \tag{36.21}$$

where RI = rainfall incident on the canopy, E = evaporation of water from the wetted canopy, TF = throughfall, and SF = stemflow. The units for all terms are depth per unit time (e.g. mm hour^{-1}) but canopy water balance calculations are best done on an event basis. By rearranging Eq. (36.21) and measuring or estimating RI, E, TF, and SF, CWI can be derived (Holwerda *et al.*, 2006).

Here, events are defined as periods of measured RI, TF, and/or FC, with a maximum allowable break (period of no recorded RI, TF, or FC) of 3 hours to allow the canopy to dry up (cf. Schellekens *et al.*, 1999; Tanaka *et al.*, this volume). Although it is assumed that water storage in the canopy is zero at the start of each event, this is certainly not the case for successive events occurring at night, nor is it likely that water stored in epiphytes is completely evaporated within this period (cf. Köhler *et al.*, 2007; Köhler *et al.*, this volume; Tóbon *et al.*, this volume #26). In subsequent analyses, the present procedure to account for storage carry-over from the previous event will be adjusted.

A major source of error in this approach derives from the use of uncorrected rain gage measurements to estimate RI. Herwitz and Slye (1992) have shown that with wind-driven rainfall (WDR), variations in canopy height lead to an uneven distribution of incident rainfall over a forest canopy. In addition, WDR incident on sloping land surfaces can differ significantly from that measured by a horizontally oriented gauge (Sharon, 1980; cf. Blocken *et al.*, 2005). Undulations in the canopy surface are complex and difficult to account for. However, one would expect the effects of such variations to offset one another as the sample area increases. A large enough sampling area for TF measurements would allow one to neglect this issue. While the TF measurements reported in this chapter (average of two gages of approximately 0.72-m^2 collecting area each) are probably not adequate in this respect, effects of canopy variability on rain catch are not treated here. However, effects of slope are addressed by calculating the incident rainfall RI relative to measured rainfall R as:

$$\frac{RI}{R} = \frac{\cos(\beta)}{\cos(\gamma)} \quad \text{for} \ \cos(\beta) \geq 0 \tag{36.22}$$

$$\frac{RI}{R} = 0 \text{ for } \cos(\beta) \geq 0 \quad (36.23)$$

where γ = rainfall zenith angle, and β = angle of incidence of rainfall on slope:

$$\cos(\beta) = \cos(\delta)\cos(\gamma) + \sin(\delta)\sin(\gamma)\cos(\theta - \alpha) \quad (36.24)$$

where δ = slope, θ = wind direction, and α = slope aspect (Sharon, 1980).

With small drop size and persistently high wind velocity prevailing in Hawai'i, rain gages can be expected to significantly underestimate rainfall. Corrections can be made using estimates of drop size and wind velocity (Holwerda *et al.*, 2006; cf. García-Santos, 2007) and this will be done in a subsequent analysis of the present data. Throughfall was measured as described under field methods whereas stemflow was assumed to be negligible. This simplifying assumption is another source of error in the present analysis which will need to be rectified in later work. Evaporation of water from the wetted canopy is often assumed to occur at the potential evaporation (*PE*) rate, estimated as a function of meteorological variables. Here, the Penman–Monteith equation (Monteith, 1973) was used for this. However, it should be noted that some researchers have found evidence of wet canopy evaporation rates to be much greater than that estimated by the Penman–Monteith equation for tropical island and coastal sites outside Hawai'i (Waterloo *et al.*, 1999; Schellekens *et al.*, 2000; McJannet *et al.*, 2007b; McJannet *et al.*, this volume). Meteorological measurements at each site were used to calculate *PE* during each rainfall/fog event.

In addition to evaporation during the event, one must account for stored water remaining in and on the canopy at the end of an event. This water, also lost by evaporation, is more difficult to estimate, particularly that held in epiphytes and bryophytes (Köhler *et al.*, 2007; Tóbon *et al.*, this volume #26). Here, it was assumed that during any event producing *TF*, storage at the end of the event equaled the canopy water storage capacity. For events not producing *TF*, a conservative estimate of *CWI* was obtained by assuming final canopy storage to be zero. This assumption may lead to significant underestimation of total *CWI* if the climate is characterized by a large proportion of small events. In this preliminary study, water storage values were assumed of 1 and 2 mm, respectively, for the Auwahi and Waikamoi sites. Negative instances of event *CWI* were reset to zero as these were considered indicative of errors in water budget component measurements or estimates. In such cases, the amount of the adjustment is reported as the *CWI error*.

RESULTS

Table 36.1 gives site and study period characteristics for the two stations. It should be noted that periods during which any of the key instruments malfunctioned were removed from the analysis, resulting in significant missing data periods, particularly at Auwahi (24%, Table 36.1).

Table 36.1 *Study site and study period characteristics*

	Auwahi	Waikamoi
Elevation (m.a.s.l.)	1219	1951
Aspect	leeward	windward
Start of study period	12 Feb 02	4 Nov 01
End of study period	14 Aug 02	14 Aug 02
Period total (days)	183	283
Missing data (days)	44	20
Missing data (%)	24	7
Observation period total (days)	139	263

Using the methods outlined above, the quantity of rainfall caught by the fog screen was estimated for hourly periods at the two sites, and subtracted from the total amount of water captured by the screen (*FS*). Subsequently, the amount of cloud water flux through a vertical plane normal to the wind direction (*CWF*) was estimated. Events were identified and totals summed for each event period. Table 36.2 gives the frequency and duration of water input (rain and/or fog) events, raw observations of *R*, *TF*, and *FS*, and calculated *CWF*. For comparison, annual totals were derived by multiplying the per day values by 365. Annual rainfall totals given in Table 36.2 differed from the mean annual rainfall at the two sites; annualized study-period rainfall at Auwahi (397 mm) was much lower than the mean annual value of 890 mm (based on 3 years of data at the site), while Waikamoi's annualized study-period rainfall (2573 mm) was higher than its annual mean of 2398 mm (based on 2 years of data). Canopy water balance results are summarized in Table 36.3 and discussed below.

DISCUSSION

Correcting the fog screen values attributable to cloud water (*FC*, not shown) for wind-driven rainfall (*WDR*) resulted in significantly lower amounts than measured values (*FS*). However, subsequent correction for wind direction resulted in a significant increase again. The net effect of the two corrections was to produce cloud water flux values (*CWF*) that were 10% and 24% higher than *FS* at Auwahi and Waikamoi, respectively (Table 36.2). This result is certainly no more than a coincidence, as the amounts by which *FS* is inflated by *WDR*, and reduced by changing wind direction relative to the fixed screen orientation are dependent on numerous site- and installation-specific variables, which do not necessarily have self-canceling effects on the measurements. One can reasonably expect *FS* to diverge from

Table 36.2 *Event statistics, raw observation totals, and cloud water flux estimates for Auwahi and Waikamoi*

	Auwahi				Waikamoi			
	Study period	Per day	Per rain/ fog day	Annual	Study period	Per day	Per rain/ fog day	Annual
Frequency and duration								
No. of days with rain and/or fog	71	0.80	1.00	291	187	0.71	1.00	260
No. of rain and/or fog events	89	0.64	1.25	234	211	0.80	1.13	293
Event duration (hr)	329	2.4	4.6	865	1756	6.7	9.4	2439
Mean event duration (hr:min)	3:42			8:19				
Raw observations								
R: Rainfall (mm)	151	1.09	2.13	397	1853	7.04	9.91	2573
TF: Throughfall (mm)	100	0.72	1.41	263	2298	8.74	12.29	3192
FS: Fog screen (mm)	60	0.43	0.85	158	1856	7.06	9.93	2578
Cloud water flux								
CWF: Cloud water flux (mm)	66	0.47	0.93	173	2300	8.75	12.30	3194

Table 36.3 *Canopy water balance results for Auwahi and Waikamoi*

	Auwahi				Waikamoi			
	Study period	Per day	Per rain/ fog day	Annual	Study period	Per day	Per rain/ fog day	Annual
RI: Incident rainfall (mm)	155	1.12	2.18	407	1936	7.36	10.35	2688
TF: Throughfall (mm)	100	0.72	1.41	263	2298	8.74	12.29	3192
E_{event}: Evaporation during events (mm)	34	0.24	0.48	90	133	0.51	0.71	185
E_{after}: Evaporation after events (mm)	57	0.81	1.58	150	264	1.00	1.41	367
CWI: Cloud water interception (mm)	57	0.73	1.44	151	773	2.98	4.13	1073
CWI error (mm)	21	0.08	0.15	54	13	0.05	0.07	18
RI+*CWI*: Total water input (mm)	212	1.85	3.62	557	2708	10.30	14.48	3761
Number of *CWI* events	66	0.47	0.93	173	174	0.66	0.93	242

CWF by larger amounts at other sites. For example, García-Santos (2007) derived an underestimation of 63% for *CWF* based on a quarter-sized fog planar fog gage at a windy site on La Gomera (Canary Islands). Hence, as an estimate of *CWF*, it is not recommended to use uncorrected *FS* values from a planar fog screen such as the Schemenauer and Cereceda (1994) gage (cf. Juvik and Nullet, 1995a).

Comparison of daily *R*, *TF*, and *CWF* values for the two sites (Table 36.2) underscores the extreme contrast between windward and leeward exposures on mountain slopes in Hawai'i, although some of the difference may be explained by the difference in elevation (1219 m.a.s.l. at Auwahi vs. 1951 m.a.s.l. at Waikamoi). Water input events were more frequent, longer in duration, and characterized by higher intensities of *R*, *TF*, and *CWF* at Waikamoi as compared with Auwahi. The contrast is most striking in terms of *CWF*, which was more than 18 times greater, on average, at Waikamoi than Auwahi. Even as a percentage of rainfall, *CWF* was much greater at Waikamoi (124%) than at Auwahi (44%). The *CWF* values at Waikamoi suggest that fog may be a major input of water, adding significantly to the already high amount derived from rainfall (Table 36.2).

Accounting for the effects of *WDR* in sloping terrain can reduce or increase the estimate of incident rainfall compared to uncorrected measured rainfall depending on wind direction and slope aspect (cf. Bradley *et al.*, 1997; Blocken *et al.*, 2006). For both stations, estimated incident rainfall was lower than measured rainfall at times, but was higher on average, indicating predominance of upslope wind during periods of rain at both sites. Rain catch by the vegetation above the *TF* gages at both sites was, undoubtedly, also enhanced at certain times – and

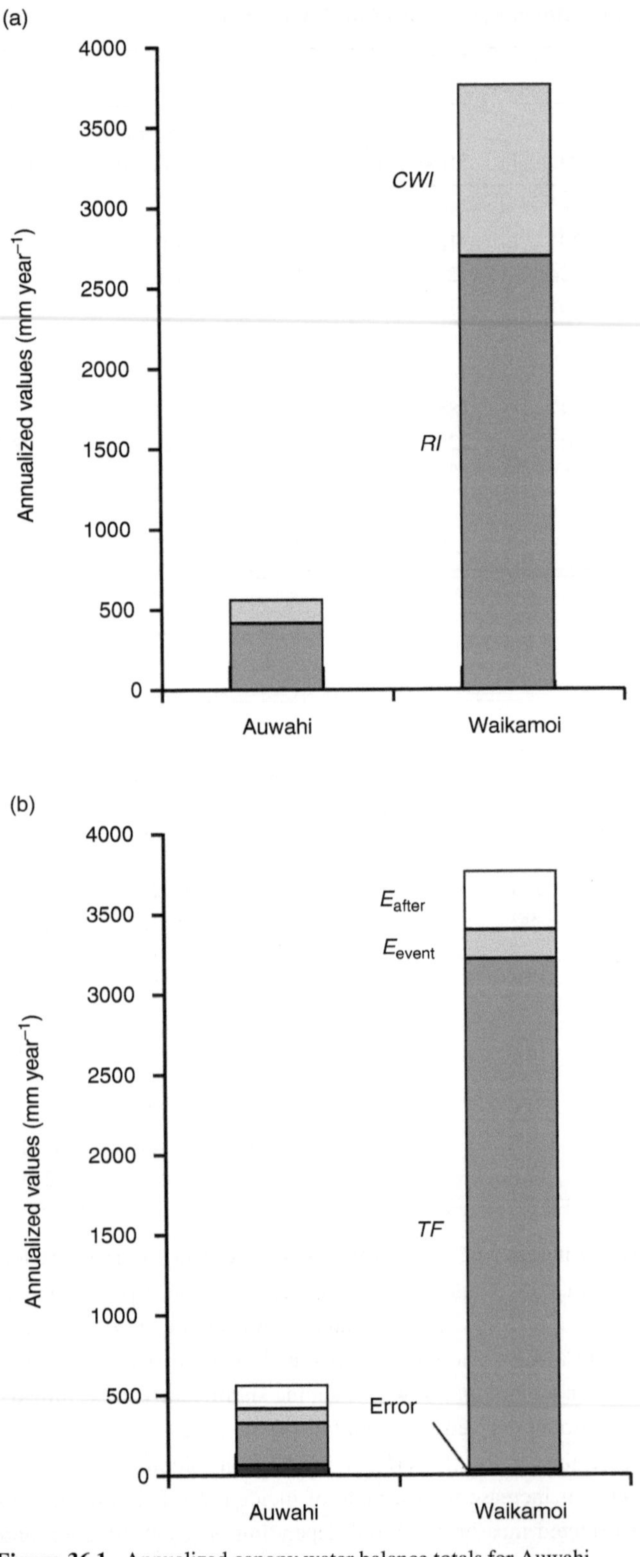

Figure 36.1. Annualized canopy water balance totals for Auwahi and Waikamoi, Haleakala, Maui, Hawai'i: (a) water input equal to incident rainfall (*RI*) plus cloud water interception (*CWI*); and (b) disposition of water input equal to the sum of throughfall (*TF*), evaporation of intercepted water during events (E_{event}), evaporation of stored water following events (E_{after}), and error revealed when $TF - RI + E_{\text{event}} + E_{\text{after}} < 0$.

reduced at other times – as changing wind direction and speed either improved or reduced the exposure of parts of the canopy to wind-blown rainfall (cf. Schmid *et al.*, this volume). No attempt was made to account for these effects in this preliminary analysis (cf. Blocken *et al.*, 2005, 2006).

Measured *TF* was about 65% of *RI* at Auwahi, and 119% of *RI* at Waikamoi (Table 36.3). This difference is indicative of the much greater amount of *CWF* at Waikamoi. Based on the wet-canopy water balance estimates, fog contributed to water input events at an average frequency of once every 2 days at Auwahi and about twice in 3 days at Waikamoi. Derived estimates of *CWI* were equivalent to 151 mm year^{-1} (adding 37% to the water input derived from *RI*) at Auwahi and 1073 mm year^{-1} (adding 40% to *RI*) at Waikamoi (Table 36.3). The total water inputs for the two sites are compared in Figure 36.1a, illustrating the large contribution of *CWI* at the wet, windward site. At Auwahi, *CWI*, although much smaller in absolute terms, is significant in relation to rainfall. The disposition of water input at the two sites is illustrated in Figure 36.1b. Considering that in the absence of *CWI* (and ignoring stemflow), *TF* would be equal to *RI* minus evaporative losses, comparing Figures 36.1a and 36.1b reveals that *TF* is certainly enhanced significantly at Waikamoi, but not at Auwahi. This stems from the fact that, at Auwahi, *TF* was greater than *RI* for only 15 out of 89 events. Hence, the bulk of *CWI* was evaporated again from the wetted vegetation, and never reached the ground. These results suggest that direct *CWI* contributions to the net hydrological input at the soil surface at Auwahi are minimal. The evaporation of intercepted cloud water can be expected to reduce transpiration (cf. Beiderwieden *et al.*, 2008; Reinhardt and Smith, 2008), which can be interpreted as representing a net hydrological gain. However, under the perennially dry soil conditions at Auwahi, transpiration is likely to be low in any case, limiting this type of indirect contribution as well.

Whilst *CWF* in relation to rainfall was much greater at Waikamoi than Auwahi (Table 36.2), the reverse is true for *CWI* (Table 36.3). Comparing the two cloud-water variables, *CWI* was 86% of *CWF* at Auwahi and only one-third of *CWF* at Waikamoi. Taken at face value, this implies that the vegetation at the Auwahi site is more efficient at catching cloud water than the vegetation at Waikamoi, which is reasonable given the exposure of the Auwahi site to unimpeded flow of fog across the adjacent pasture. For this same reason, however, the Auwahi site may be more susceptible to error from unaccounted enhancement of incident rainfall. It should also be emphasized that both sites, but especially Auwahi, are exposed such that cloud water and rain deposition are enhanced by air entering the vegetation from the side. Measurements made further away from the edge are likely to have given much lower values (Weathers *et al.*, 1995).

At both sites, the occurrence of rain without fog is infrequent (Table 36.4). While fog without recorded rainfall is fairly common at both sites, water input during these fog-only events

Table 36.4 *Percentage of each water balance component and event occurrence for events with rain only, fog only, and rain and fog at Auwahi and Waikamoi*

	Auwahi			Waikamoi		
	Rain only	Fog only	Rain and fog	Rain only	Fog only	Rain and fog
R: Rainfall	2	0	98	2	0	98
RI: Incident rainfall	2	0	98	2	0	98
TF: Throughfall	1	1	99	1	0	99
E_{event}: Evaporation during events	5	34	61	6	14	80
E_{after}: Evaporation after events	9	19	72	7	13	80
CWI: Cloud water interception	7	41	53	2	7	90
CWI error	1	0	99	24	0	76
RI+CWI: Total water input	4	11	85	2	2	96
Percentage of CWI events[a]	7	39	49	12	31	53

[a] Note that an additional 4% of CWI events at both Auwahi and Waikamoi had recorded throughfall with no recorded rainfall or fog.

Table 36.5 *Coefficients of determination* (r^2) *of regression equations relating various measured and derived variables at the two study sites*

	Auwahi	Waikamoi
$TF = f(R)$	0.965	0.966
$TF = f(RI)$	0.945	0.979
$TF = f(FS)$	0.387	0.778
$TF = f(CWF)$	0.174	0.700
$TF = f(CWI)$	0.017[a]	0.771
$CWI = f(FS)$	0.002[a]	0.650
$CWI = f(CWF)$	0.011[a]	0.610
n	89	211

TF = throughfall; *R* = rainfall; *RI* = rainfall incident on sloping surface; *FS* = fog screen capture; *CWF* = cloud water flux according to screen; *CWI* = cloud water interception according to canopy water budget.

[a] Not significant at 0.001.

constitutes a significant portion of the total only for the Auwahi site, where 41% of *CWI* and 11% of total water input occurred during fog-only events. Events in which both rain and fog were recorded contributed 85 and 96% of total water input at Auwahi and Waikamoi, respectively.

Table 36.5 gives the results of a regression analysis conducted to elucidate the relationships between some of the measured and calculated variables. Regressions were done for event totals. Event *TF* was related to both rainfall and cloud water variables at Waikamoi, although much more strongly to rainfall. However, at Auwahi, *TF* was only weakly related to *FS* and *CWF*, and unrelated to *CWI*. Similarly, *CWI* was related to *FS* and *CWF* at Waikamoi, but not at Auwahi, i.e. the fog screen gave a reasonable good indication of variations in *CWI* at Waikamoi, but was of no value in that regard at Auwahi. These statistics confirm that *TF* is dominated by rainfall at both sites, and is significantly influenced by fog only at Waikamoi. The present efforts to improve the utility of fog screen measurements by removing effects of wind-blown rainfall and varying wind direction did not result in improved prediction of *CWI* at either station.

In Figure 36.2, the presently derived *CWI* estimates have been plotted against elevation along with previous estimates, obtained mostly from fog screens (i.e. *FS*). The exceptionally high estimates given by Juvik and Nullet (1995b) for Kohala, which were based on only 100 days of observation, have been left out from the diagram. The zone of frequent cloud contact is also shown by shading. The Auwahi estimate appears somewhat below the average for its elevation, while Waikamoi appears to be near the upper end of estimates. Overall, the maximum values as a function of elevation (Figure 36.2, gray line) match reasonably well with the qualitative fog zones, with the highest values observed between about 600 and 2600 m.a.s.l. However, the range of values is almost certainly indicative of more than the true variability. Hence, there is still work to do.

CONCLUSION

The results presented here indicate that cloud water interception (*CWI*) adds significantly to rainfall at both dry and wet cloud forest sites in Hawai'i, although most water derived from fog was evaporated again before reaching the ground at the dry site. Methods were developed and applied to improve the utility of fog screen observations. Removing the effects of wind-blown rainfall and varying wind direction, was expected to yield a better indicator of *CWI*. However, corrected cloud water fluxes (*CWF*) did not do a better job of predicting *CWI* as determined

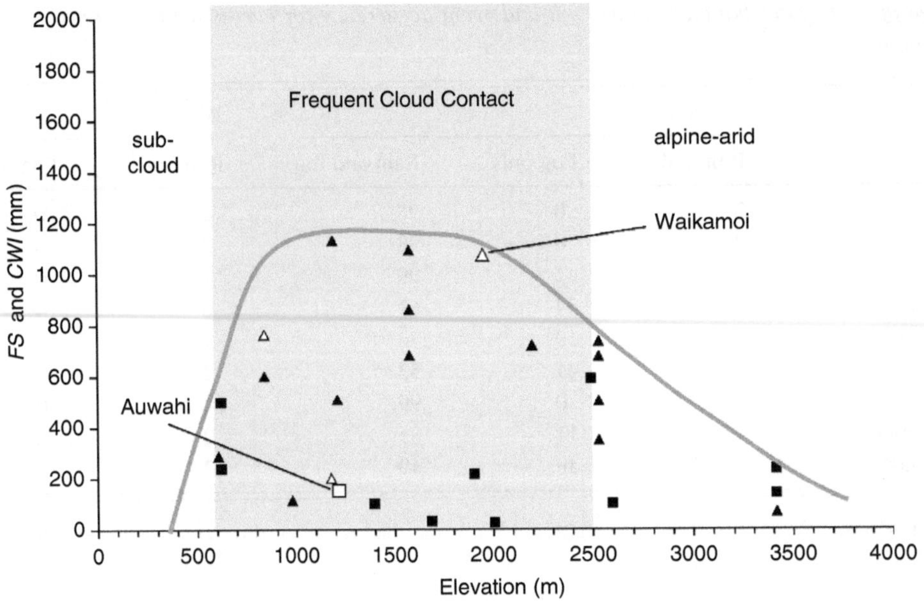

Figure 36.2. Annualized estimates of cloud water interception at Waikamoi and Auwahi, and prior fog screen (*FS*, black symbols) and canopy cloud water interception (*CWI*, white symbols) estimates, shown as a function of elevation; prior studies summarized by DeLay and Giambelluca (this volume). Windward sites (triangles) and leeward sites (squares) are distinguished. Shading corresponds to zone of frequent cloud contact as depicted in DeLay and Giambelluca (this volume). Gray line shows approximate *CWI* vs. elevation envelope.

by canopy water balance. At the dry site, neither the raw fog screen observations nor the calculated *CWF* had any predictive value for *CWI* by the vegetation.

Throughfall (*TF*) was dominated by rainfall at both sites, and was significantly influenced by fog at the windward site only. Very high rates of screen-based *CWF* and water-budget-based *CWI* were estimated for the windward site. At the leeward site, a high percentage of *CWF* was intercepted by the vegetation, but, again, most of the intercepted cloud water was evaporated before reaching the ground. It should be noted, however, that the estimated evaporation is strongly dependent on the assumed (and as yet unmeasured) storage capacity, and, hence, both evaporation and *CWI* are uncertain.

The results presented here should be regarded as preliminary. The Auwahi site and to a lesser extent Waikamoi probably received cloud water inputs that were enhanced by the transition from low to high stature vegetation. Cloud water input is probably lower for comparable locations within continuous vegetation of the same height. The data presented here constitute a small sample and may not represent average conditions, especially in the case of Auwahi where extremely dry conditions prevailed. Furthermore, several simplifying assumptions were made, e.g. with respect to canopy storage capacities, and these must be tested. Additional data from the two sites are currently being analyzed and methods of analysis refined. The canopy water balance estimates will also be integrated with estimates of cloud water input based on stable isotope analysis (cf. Scholl *et al.*, this volume).

ACKNOWLEDGEMENTS

The work presented in this paper was supported by the U.S. Geological Survey, Water Resources Division, Venture Capital Fund; the National Science Foundation under Grant No. EAR-0309731; and by the U.S. Geological Survey, Biological Resources Discipline, Global Change Program. Any use of trade, product, or firm names is for descriptive purposes only and does not imply endorsement by the U.S. Government. The authors thank Veronica Hutton and Corina Niemand for contributions to early stages of the data analysis.

REFERENCES

Beiderwieden, E., V. Wolff, Y. J. Hsia, and O. Klemm (2008). It goes both ways: measurements of simultaneous evapotranspiration and fog droplet deposition at a montane cloud forest. *Hydrological Processes* **22**: 4181–4189.

Blanchard, D. C. (1953). Raindrop size-distribution in Hawaiian rains. *Journal of Meteorology* **10**: 457–473.

Blocken, B., J. Carmeliet, and J. Poesen (2005). Numerical simulation of the wind-driven rainfall distribution over small-scale topography in space and time. *Journal of Hydrology* **315**: 252–273.

Blocken, B., J. Poesen, and J. Carmeliet (2006). Impact of wind on the spatial distribution of rain over micro-scale topography: numerical modeling and experimental verification. *Hydrological Processes* **20**: 345–368.

Bradley, S. G., W. R. Gray, L. D. Pigott, *et al.* (1997). Rainfall redistribution over low hills due to flow perturbation. *Journal of Hydrology* **202**: 33–47.

Bruijnzeel, L. A., W. Eugster, and R. Burkard (2005). Fog as an input to the hydrological cycle. In *Encyclopaedia of Hydrological Sciences*, eds. M. G. Anderson and J. J. McDonnell, pp. 559–582. Chichester, UK: John Wiley.

García-Santos, G. (2007). *An ecohydrological and soils study in a montane cloud forest in the National Park of Garajonay, La Gomera (Canary Islands, Spain).* PhD Thesis, VU University Ámsterdam, Ámsterdam, The Netherlands. [http://www.falw.vu.nl/nl/onderzoek/earth-sciences/geo-environmental-science-and-hydrology/hydrology-dissertations/index.asp].

Gunn, R., and G. D. Kinzer (1948). The terminal velocity of fall for water droplets in stagnant air. *Journal of Meteorology* **6**: 243–248.

Herwitz, S. R., and R. E. Slye (1992). Spatial variation in the interception of inclined rainfall by a tropical rainforest canopy. *Selbyana* **13**: 62–71.

Holwerda, F., R. Burkard, W. E. Eugster, *et al.* (2006). Estimating fog deposition at a Puerto Rican elfin cloud forest site: comparison of the water budget and eddy covariance methods. *Hydrological Processes* **20**: 2669–2692.

Juvik, J. O., and D. Nullet (1995a). Comments on a proposed standard fog collector for use in high elevation regions. *Journal of Applied Meteorology* **34**: 2108–2110.

Juvik, J. O., and D. Nullet (1995b). Relationships between rainfall, cloud-water interception, and canopy throughfall in a Hawaiian montane forest. In *Tropical Montane Cloud Forests*, eds. L. S. Hamilton, J. O. Juvik, and F. N. Scatena, pp. 165–182. New York: Springer-Verlag.

Köhler, L., C. Tobón, K. F. A. Frumau, and L. A. Bruijnzeel (2007). Biomass and water storage dynamics of epiphytes in old-growth and secondary montane rain forest stands in Costa Rica. *Plant Ecology* **193**: 171–184.

McJannet, D., J. S. Wallace, and P. Reddell (2007a). Precipitation interception in Australian tropical rainforests. I. Measurement of stemflow, throughfall and cloud interception. *Hydrological Processes* **21**: 1692–1702.

McJannet, D., J. S. Wallace, and P. Reddell (2007b). Precipitation interception in Australian tropical rainforests. II. Altitudinal gradients of cloud interception, stemflow, throughfall and interception. *Hydrological Processes* **21**: 1703–1718.

Medeiros, A. C., L. L. Loope, and B. W. Hobdy (1995). Conservation of Cloud Forests in Maui County (Maui, Molokai, and Lanai), Hawaiian Islands. In *Tropical Montane Cloud Forests*, eds. L. S. Hamilton, J. O. Juvik, and F. N. Scatena, pp. 223–233. New York: Springer-Verlag.

Medeiros, A. C., C. F. Davenport, and C. G. Chimera (1998). *Auwahi: Ethnobotany of a Hawaiian Dryland Forest*, Technical Report No. 114. Honolulu, HI: Cooperative National Park Resource Studies Unit.

Monteith, J. L. (1973). *Principles of Environmental Physics.* Amsterdam: Elsevier.

Reinhardt, K., and W. K. Smith (2008). Impacts of cloud immersion on microclimate, photosynthesis and water relations of *Abies fraseri* (Pursh.) Poiret in a temperate mountain cloud forest. *Oecologia* **158**: 229–238.

Schellekens, J., F. N. Scatena, L. A. Bruijnzeel, and A. J. Wickel (1999). Modelling rainfall interception by a lowland tropical rainforest in north-eastern Puerto Rico. *Journal of Hydrology* **225**: 168–184.

Schellekens J., L. A. Bruijnzeel, F. N. Scatena, N. J. Bink, and F. Holwerda (2000). Evaporation from a tropical rain forest, Luquillo Experimental Forest, eastern Puerto Rico. *Water Resources Research* **36**: 2183–2196.

Schemenauer, R. S., and P. Cereceda (1994). A proposed standard fog collector for use in high elevation regions. *Journal of Applied Meteorology* **33**: 1313–1322.

Sharon, D. (1980). The distribution of effective rainfall incident on sloping ground. *Journal of Hydrology* **46**: 165–188.

Wagner, W. L., D. R. Herbst, and S. H. Sohmer (1999). *Manual of the Flowering Plants of Hawai'i.* Honolulu, HI: University of Hawai'i Press.

Waterloo, M. J., L. A. Bruijnzeel, H. F. Vugts, and T. T. Rawaqa (1999). Evaporation from *Pinus caribaea* plantations on former grassland soil under maritime tropical conditions. *Water Resources Research* **35**: 2133–2144.

Weathers, K. C., G. M. Lovett, and G. E. Likens (1995). Cloud deposition to a spruce forest edge. *Atmospheric Environment* **29**: 665–672.

37 Historical background of fog water collection studies in the Canary Islands

M. V. Marzol-Jaén
University of La Laguna, La Laguna, Tenerife, Canary Islands, Spain

ABSTRACT

This chapter constitutes a bibliographic compilation regarding cloud occurrence and fog water collection in the Canary Islands, ranging from legendary stories of water collection from the famous "fountain tree" on the island of El Hierro, to current research on the role of fog in the water budget of *laurisilva* and *fayal-brezal* forests. Fog water was a very important resource for the early inhabitants of the islands and allows the presence of evergreen forests in areas with low annual rainfall and a long dry summer season.

INTRODUCTION

The Canary Islands have aroused interest from the classical age onward for various reasons (Martínez, 1992; Hernández, 1998). First of all because of their insularity; then because of their main mountain (the Teide volcano on Tenerife, 3718 m.a.s.l.); from the fifteenth to the seventeenth century because of their "fountain tree"; and nowadays because of their suitable weather conditions and special flora. One of the stories that generated the most curiosity and led to many descriptions is that of the Garoé tree – also known as the "holy tree," the "fountain tree," or the "tree of life," as Saint Isidore of Seville called it. The legend is based on the fact that on El Hierro – the westernmost island in the Canaries – there was a large tree, possibly a *til* (*Ocotea foetens*) that provided enough water for the entire Bimbache people (as the former dwellers of El Hierro are called). El Hierro (28° N, 18° W) is part of an ancient volcano with a surface area of 268.7 km^2 and the highest summit just below 1400 m.a.s.l. (Figure 37.1). From the fifteenth century onward, the persistent lack of water on the island and the occurrence of long drought periods have forced its inhabitants to develop ingenious procedures to collect and use rain and fog water.

HISTORIC REFERENCES TO THE GAROÉ FOUNTAIN TREE

The most detailed references to the phenomenon of the "tree that weeps" date back to the fifteenth century. In *Historia de las Indias*, Fray Bartolomé de las Casas (1550) wrote about the volume of water collected; in *Descripción de las Islas Canarias* (1590), Torriani described the fog phenomenon; in *Historia de la Conquista de las Siete Islas Canarias*, Fray Juan de Abreu y Galindo (1632) described the tree in great detail; and in *Historia General y Natural de las Indias* (1535–1548), Gonzalo Fernández de Oviedo talked about the frequency with which fog appeared. Here are some of these interesting descriptions (cf. Figure 37.2):

> *More than 1000, pipes, that would be 25 000 or 30 000 pitchers, can be held in the reservoir or tank . . . so that next to the tree and tank there is a house and guard.*
>
> De Las Casas (1550)

> *This tree stops the fog because of its numerous branches and leaves, which bulge wet . . . and unable to conserve it in the form of vapor, converts it into extremely thick droplets that fall into the ditch.*
>
> Torriani (1582)

> *It is 12 hands thick and four wide, and 40 high from the base to the top, and the surrounding canopy 120 feet. The fruit is like acorns . . . with laurel leaves . . . which are always green. There are two tanks or large troughs, each of them 20 square feet and 16 hands deep, made of rough stone. Each of them gathers 20 leather bottles of water. Next to the tree stands a guard appointed by the Council, with his house and wage, who gives each inhabitant seven jugs of water.*
>
> Abreu y Galindo (1632).

> *Every day God provides the indigenous inhabitants of El Hierro with "heavenly water". Two hours before sunrise the tree begins to sweat thanks to a cloud that stands above it and*

Tropical Montane Cloud Forests: Science for Conservation and Management, eds. L. A. Bruijnzeel, F. N. Scatena, and L. S. Hamilton. Published by Cambridge University Press.

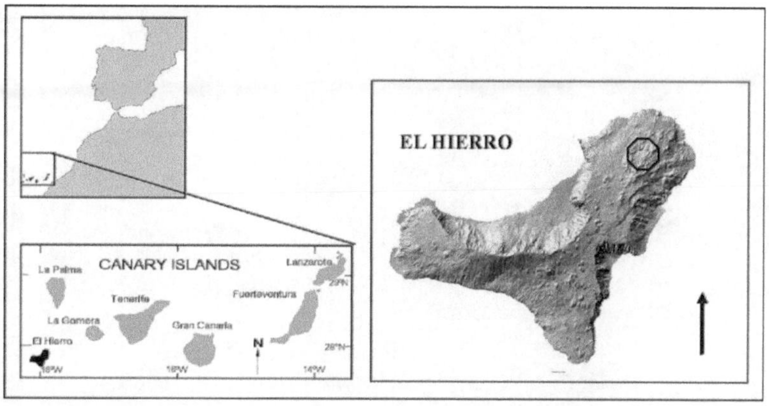

Figure 37.1. Geographic situation of El Hierro (Canary Islands) and location of the former Garoé tree.

Figure 37.2. Engraving of the Garoé tree by J. Hilton, 1748 (*The Universal Magazine of Knowledge and Pleasure*).

> *continues thus until two hours after the sun rises. Fours hours, hence, does this process last, allowing the water that is excellent and healthy for both men and cattle to fall into a pool.*
>
> Fernández de Oviedo (1535–1548)

Theodor de Bry (1597), an explorer of the Americas from a Belgian family of engravers, wrote a fictional narrative concerning the Garoé tree phenomenon and following his return from the Americas made an engraving of the Garoé, which was copied throughout Europe during the seventeenth and eighteenth centuries (Figure 37.2).

> *In the island of El Hierro, one of the Canaries, also called the Fortunate Isles, can be found a strange and very scarce marvel of nature, seen there on the voyage to the Americas . . . They describe the only tree on that island from which drips sufficient soft water to quench not only the thirst of the men of that island, but also all the animals in general. For there are no other streams of water in all the island, nor pools, nor springs, nor other sources or reservoirs, from which water to drink can be drawn. The said tree, as many write, is quite tall and has many leaves like those of a walnut tree, though a little bigger, and is surrounded by a wall in the form of a well or fountain so that the water might drop into it. Every day the tree is enveloped by a dense fog for two hours before sunrise and until the sun comes out and eats up the fog. While the fog covers the tree, the tree sweats and from the trunk and leaves drops sufficient water which is praised for its virtues and good taste.*
>
> De Bry (1597)

Figure 37.3. Where the Garoé tree once stood, there now stands a 7-m high *til* (*Ocotea foetens*), which captures fog water like the former Garoé tree. See Figure 37.1 for location. (See also color plate.)

The fountain tree was so important that it is depicted on the island's municipal coat of arms until today, but it was uprooted during a hurricane in 1610, leaving the inhabitants in distress (Gioda *et al.*, 1992). Today, a 7-m high *Ocotea* tree stands at the location of the former fountain tree near the top of the Hoya de Tejis cliffs, in the north-eastern part of the island at an elevation of 1004 m.a.s.l., about 9 km from the coastline (Figure 37.3; see Figure 37.1 for location).

HISTORIC DESCRIPTIONS OF THE "SEA OF CLOUDS"

The first scientific studies and meteorological measurements of the cloud formation causing fog in the Canary Islands date from the 1930s. Before then, there were only physical descriptions made by people who climbed through this "sea of clouds" when scaling Mt. Teide (Figure 37.4). Many of these descriptions were made from the second half of the eighteenth century onward, coinciding with the development of science, a burgeoning interest in natural phenomena, and the extraordinary descriptive undertakings of naturalists. What is more, the Canary Islands were a necessary transit point for expeditions to the Americas at the time (von Humboldt, 1799; De Buch, 1836; Wilde, 1837; Coquet, 1884) and there is a wealth of bibliographic material about these travels (García Pérez, 1988; González Lemus, 1995; Pico and Corbella, 2000; Ory, 2004, etc.).

Many visitors who passed through Tenerife climbed to the summit of Mt. Teide and described the morphology of the sea of clouds, its frequency, temperature, height, and the atmosphere inside the fog-affected forest, etc. (e.g. Glas, 1761; Pruneda, 1848; Verneau, 1891; Ossuna and Saviñon, 1893). Here is one example of such descriptions, written by George Glas in September 1761:

Figure 37.4. The "sea of clouds" covering the northern side of Tenerife. The Teide volcano can be seen in the background (3717 m.a.s.l.). Photo by Mª V. Marzol. (See also color plate.)

> *... but what was most noteworthy for my travelling companion was the extraordinary and highly unusual appearance of the clouds below us; they looked like an ocean, only the surface was not so blue or smooth, but looked like very white cotton, and where this ocean of clouds, as I might call it, touched the shore, it seemed to froth like waves breaking on the beach . . . When we descended towards the clouds on our way back from the summit, and we entered it, it seemed like thick fog or vapor, of the same consistency that we frequently see in England. All the trees in the forest and our clothes were soaked.*
>
> Glas (1761)

Table 37.1 *Amounts of water (mm) collected by a rain gage placed in the open (I) and underneath a pine forest canopy (II) at El Realejo Bajo (Tenerife) in 1951 (after Ceballos and Ortuño, 1952), and the ratio between the two (III)*

	J	F	M	A	M	J	J	A	S	O	N	D	Year
I	88	267	77	47	35	6	0	0	43	58	117	158	856
II	314	698	109	252	165	157	43	193	170	205	277	457	3040
III	3.5	2.6	1.4	5.4	4.7	26.2	–	–	4.0	3.5	2.4	2.9	3.6

Data obtained and modified from Ceballos and Ortuño (1952, p. 422).

Table 37.2 *Observations of the "sea of clouds" as made at the meteorological observatory of Izaña (Tenerife, Canary Islands) between 1939 and 1950 (after Font Tullot, 1951)*

Month	Mean of 3[a] readings day^{-1} (%)	Mean of 1 reading day^{-1} (%)	Days with fog (%)	Days without fog (%)	Height of inversion (m.a.s.l.)
January	29.0	58.1	25.8	16.1	1570
February	35.7	57.2	28.5	14.3	1600
March	41.9	58.1	29.0	12.9	1650
April	60.0	70.0	23.3	6.7	1540
May	71.0	80.6	16.1	3.2	1660
June	83.3	96.7	3.3	0.0	1400
July	90.3	93.5	0.0	6.4	1240
August	70.9	90.3	2.0	7.6	1230
September	60.0	83.3	10.0	6.6	1470
October	48.1	61.2	29.0	9.7	1590
November	30.0	53.3	33.3	13.3	1620
December	32.2	54.8	29.0	16.1	1670

[a] At 6.00, 12.00, and 17.00 h.

SCIENTIFIC STUDIES OF THE "SEA OF CLOUDS"

The first studies concerning the sea of clouds in Tenerife are those of Von Ficker (1930) and Barasoaín (1943). While the former put forward the idea of two superimposed layers of trade wind separated by a thermal inversion, the latter proposed that the height of the sea of clouds depends on the difference in pressure between Santa Cruz de Tenerife (36 m.a.s.l.) and Izaña (2367 m.a.s.l.).

The forest engineers Ceballos and Ortuño (1952) were the first to carry out experimental collections of fog water captured from the sea of clouds by the pine forests of Tamadaba (on the island of Gran Canaria) and El Realejo Bajo (Tenerife) (Table 37.1). They used two rain gages throughout the year 1951, one installed underneath the trees and the other placed outside the range of the trees, and collected on average three times more water from the gage in the forest than from the one in the clearing. The results of these early trials were subsequently published in many different works (Dansereau, 1968; Fernandopullé *et al.*, 1975). Although based on an absolute minimum number of gages and therefore only indicative at best in the absence of information on the degree of spatial variability within the stands (cf. Holwerda *et al.*, this volume #29; Tanaka *et al.*, this volume), the fog drip results listed in Table 37.1 illustrate the importance of cloud water interception, particularly during the dry summer months.

Font Tullot (1951) described the seasonal variation of the sea of clouds, its maximum frequency of occurrence in summer, and its minimum occurrence in winter. He also established the daily variation in cloud height: lower in the early morning hours (due to compression by downward winds at night), and higher after midday (because of warming up of the slopes and convective movement of the air). Table 37.2 summarizes Font Tullot's significant contribution to our knowledge of the behavior of the sea of clouds at the observatory of Izaña (Tenerife, 2367 m.a.s.l.) between 1939 and 1950. It is pertinent to note that the cloud deck becomes more frequent in summer, particularly in July (present on 90% of the days).

Valladares (1996) used an alternative methodology (advanced by Marzol *et al.*, 1988) to locate the altitudinal range of the sea of clouds on the northern slopes of Tenerife, recording the hours

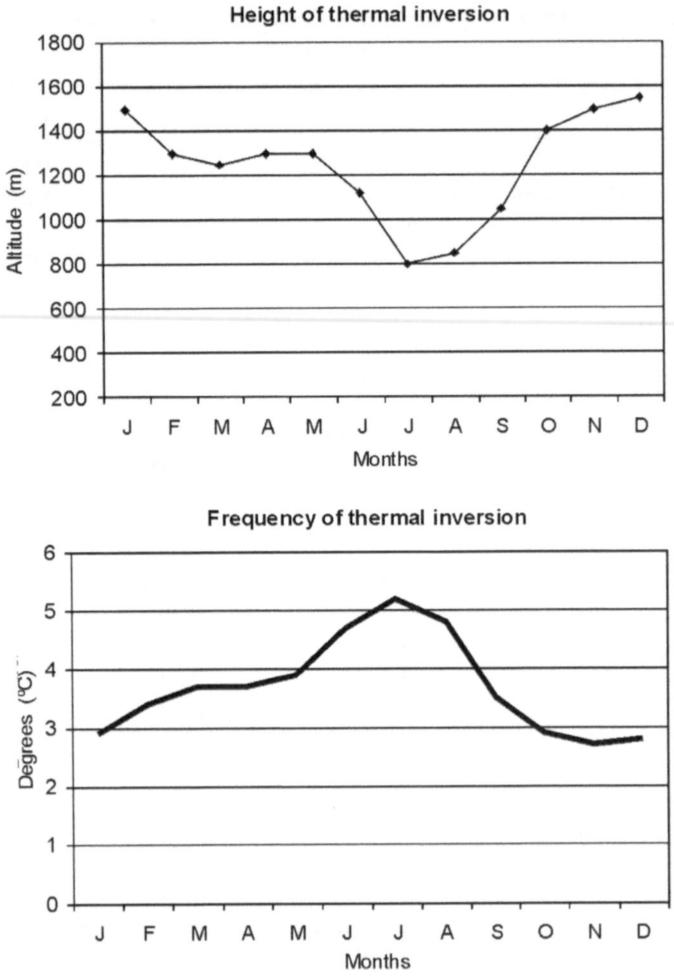

Figure 37.5. Seasonal variation of the height and frequency of thermal inversion in Canary Islands (after Dorta, 1996).

Figure 37.6. Present-day fog harvesting experiments near the summits of El Hierro, above 1000 m.a.s.l. (a) An isolated 5-m high *Erica arborea* tree, located at the top of the Jinama cliffs, drips fog water onto the cemented ground beneath down to a water tank. (b) Eight *Erica* bushes capturing fog at El Cres. Photos by Mª V. Marzol. (See also color plate.)

when relative humidity exceeded 95%, as indicated by hygrometers installed above, below, and within the sea of clouds. Dorta (1996) investigated the annual variation in the position of the thermal inversion layer in the Canary Islands using radiosonde data (1983–1992) which confirmed the high frequency of occurrence of the phenomenon. In summer, there is subsidence inversion on 95% of the days vs. only 75% in winter. The altitude of the inversion also varies from *c.* 800 m.a.s.l. in summer to *c.* 1200 m.a.s.l. in winter (Figure 37.5).

WATER COLLECTION FROM THE SEA OF CLOUDS

There are numerous examples in the Canarian archipelago of the use of cloud water for human consumption based on a variety of capturing systems. In 1948, after a prolonged drought on El Hierro, the farmer Tadeo Casañas came up with the idea for a system to collect the fog water dripping from 14 *brezos* (*Erica arborea* or tree heath), which was then funneled through a rudimentary system of agave leaves (*Agave americana*) to a collection drum. This system managed to collect 14 l min^{-1} during the night when the fog is denser. Nowadays, many of the water sources located around the summits of the island at an altitude of *c.* 1200 m.a.s.l. obtain their water from the interception of fog by ~*c.* 4-m high *Erica* trees and bushes (Figure 37.6).

In recent years there has been a lot of interest in measuring the amount of fog water deposited on vegetation in the Canaries and many experiments have been carried out. Some of these experiments are incorrect from a methodologic point of view, but the use of different types of instruments prevents a proper comparison of the respective results (cf. García Prieto *et al.*, 1960; Morey and González, 1966; Lösch, 1993; Marzol *et al.*, 1994).

Kämmer (1974) undertook several projects in Tenerife to measure fog water capture by different vegetation types. He

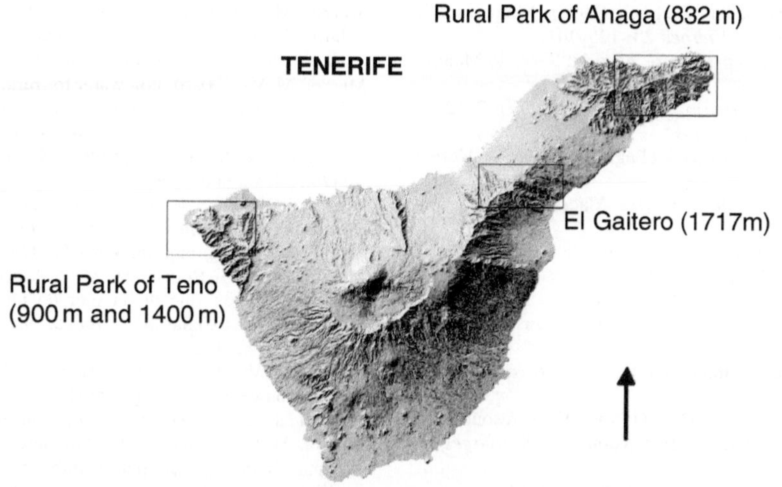

Figure 37.7. Location of study areas for fog water collection potential on the island of Tenerife, Canary Islands, by Mª V. Marzol and co-workers.

installed 25 Grunow-type fog collectors (50-cm^2 cross-sectional area; Grunow, 1952, 1965) and 25 rain gages (Hellman-type) close to the latter, at 150 cm above ground level at different elevations and on slopes of different orientations, both in clearings. He also measured fog water beneath trees and experimented with various species (*Erica arborea, E. scoparia, Persea indica, Laurus azorica, Pinus canariensis*, and *Myrica faya*, etc.), which led him to favor the greater efficiency of needle-like leaves over lanceolate ones, particularly in conditions of gentle winds and dense fog. Kämmer also noted that when the winds were stronger and the fog droplets bigger, both types of leaf tended to collect similar amounts of fog water. The largest difference in amounts of water collected by a rain gage placed in the open and a nearby fog collector (receiving both rain and fog) was found at an altitude of *c*. 1000 m.a.s.l., with the latter collecting 7.8 times more. Because of the difficulty to differentiate between fog and rain, Kämmer used the expression "difference of measure" (i.e. the differential). At higher altitudes, this difference was reduced to 4.8 times more water being collected by the fog gage, whereas at lower altitudes it only amounted to 1.2 times more (Kämmer, 1969–1972).

Aboal (1998) and Aboal *et al.* (1999a,b), in collaboration with the Ecological and Geobotanical Institute of the University of Düsseldorf (Germany), investigated the role of fog in the evergreen *laurisilva* forest in northern Tenerife. In an experimental 400-m^2 plot within the Agua García forest at an altitude of 775 m.a.s.l., throughfall was measured with 40 rain gages placed beneath the trees and the amounts were compared with rainfall measured at a meteorological station 40 m away in a forest clearing. It was concluded that fog only supplied 2% of the water reaching the forest floor in this sheltered *laurisilva* forest.

Marzol *et al.* (1996, 1997, 1998, 2005) researched the sea of clouds and its water supplying potential at various locations and at different altitudes in Tenerife, mainly at El Gaitero and in the Rural Parks of Anaga and Teno (Figure 37.7). These studies used Standard Fog Collectors (*SFC*, 1-m^2 collection screen; Schemenauer and Cereceda, 1994) and Quarter Fog Collectors (*QFC*, 0.25-m^2 screen; Marzol, 2002a) as part of automatic weather stations to characterize local fog climates. The fog water collected by an *SFC* was compared to that of a *QFC* for a period of 13 months at two sites (Teno and Anaga, both on Tenerife), to derive the correction factor linking the two gages; the obtained correction factor was 3.6 at Teno and 3.5 at Anaga (Marzol, 2002b).

The results suggested that ridge and summit areas at 800–900 m.a.s.l. have the greatest potential for fog water collection, with amounts collected by the fog screens often exceeding those in the rain gages. These summit areas are frequently affected by fog and occupied by a stunted variant of *laurisilva* known locally as *fayal–brezal* forest dominated by *Erica arborea* and *Myrica faya* (see Marzol-Jaén *et al.*, this volume for details). A similar conclusion was drawn by García-Santos (2007) for a ridge-top forest at an altitude of 1270 m.a.s.l. in central La Gomera, using similar methodology (Ritter *et al.*, 2008), and by Prada *et al.* (2009) for an exposed *Erica* stand on Madeira.

REFERENCES

Aboal, J. (1998). Los flujos netos hidrológicos y químicos asociados a un bosque de laurisilva en Tenerife. Ph.D. thesis, Department of Vegetation Ecology, Universidad de La Laguna, La Laguna, Tenerife, Spain.

Aboal, J., S. Jiménez, D. Morales, and J. Hernández (1999a). Rainfall interception in laurel forest in the Canary Islands. *Agricultural and Forest Meteorology* **97**: 73–86.

Aboal, J., D. Morales, J. Hernández, and S. Jiménez (1999b). The measurement and modelling of the variation of stemflow in a laurel forest in Tenerife, Canary Islands. *Journal of Hydrology* **221**: 161–175.

Abreu y Galindo, J. de (1632). *Historia de la Conquista de las Siete Islas Canarias*

Aschan, G., S. Jiménez, D. Morales, and R. Löch (1994). Microclimatic aspects of a laurel forest in Tenerife. *Vieraea* **23**: 125–141.

Barasoaín, J. (1943). *El mar de nubes en Tenerife*. Madrid: Servicio Meteorológico Nacional.

Bello Espinosa, D. (1931). Un jardín canario. *El Campo* **XVI** (79): 7–12.

Casas, F. de las (1550). *Historia de las Indias*.

Ceballos, L., and F. Ortuño (1952). El bosque y el agua en Canarias. *Montes* **48**: 418–423.

Coquet, A. (1884). *Une excursión aux iles Canaries*. Madrid: Edición española de 1982.

Dansereau, P. (1968). Estructura y funciones del bosque de Laurisilva en las Canarias. *Collectanea Botánica*, Vol. **II**, part 1.

De Bry, Th. (1597).

De Buch, L. (1836). *Description physique des Iles Canaries*. Paris: F.G. Levrault.

Dorta, P. (1996). Las inversiones térmicas en Canarias. *Investigaciones Geográficas* **15**: 109–126.

Fernandopullé, D., R. Heras, and J. Saenz Oiza (1975). Water resources of volcanic islands. A pilot project in the Canaries. *Nature and Resources* **11**: 8–12.

Ficker, H. Von. (1930). *Die meteorologischen Verhältnisse der Insel Teneriffa*. Berlin: Presse der Akademie.

Font Tullot, I. (1951). El espesor de la capa superficial de aire marítimo en la región de las Islas Canarias. *Revista de Geofísica* **X** (40): 281–291.

García Pérez, J. (1988). *Viajeros ingleses en las Islas Canarias durante el siglo XIX*. Santa Cruz de Tenerife: Caja General de Ahorros.

García Prieto, P., F. Ludlam, and P. Saunders (1960). The possibility of artificially increasing rainfall on Tenerife in the Canary Islands. *Weather* **5**: 39–51.

García-Santos, G. (2007). *An ecohydrological and soils study in a montane cloud forest in the National Park of Garajonay, La Gomera (Canary Islands, Spain)*. PhD Thesis, VU University Ámsterdam, Ámsterdam, The Netherlands. [http://www.falw.vu.nl/nl/onderzoek/earth-sciences/geo-environmental-science-and-hydrology/hydrology-dissertations/index.asp].

Gioda, A., A. Acosta Baladón, P. Fontanel, Z. Hernández, and A. Santos (1992). L'arbre fontaine. *La Recherche* **249**: 1400–1408.

Glas, G. (1761). *Descripción de las Islas Canarias*. Reprinted 1982, Tenerife: Ediciones Goya.

Gómez, L., and A. Fernández (2003). Primeros resultados del seguimiento de la precipitación de la niebla en el Parque Nacional de Garajonay. In *La biogeografía: Ciencia geográfica y ciencia biológica*, eds. M. E. Arozena, E. Beltrán, and P. Dorta, pp. 233–247. La Laguna, Tenerife: Universidad de La Laguna.

González Lemus, N. (1995). *Las islas de la ilusión (británicos en Tenerife. 1850–1900)*. Las Palmas de Gran Canaria: Cabildo Insular de Gran Canaria.

Grunow, J. (1952). Nebelniederschlag. Sonderdruck aus Berichte des Deutschen Wetterdienstes in der U.S. Zone, *Bad Kissingen* **42**: 30–34.

Grunow, J. (1965). Interception in a Spruce stand on the Hopenpeissenberg and methods of measuring it. *Fortwissenschafftlich Zentralblatt* **84**: 212–229.

Hernández, A. (1998). *Garoé: Iconografía del Árbol del Agua*. Santa Cruz de Tenerife: Dirección General de Patrimonio Histórico.

Kämmer, F. (1974). Klima und Vegetation auf Tenerife besonders im Hinblick auf den Nebelniederschlag. *Scripta Geobotanica* **7**: 1–78.

Lösch, R. (1993). Water relations of Canarian laurel forest trees. In *Water Transport in Plants under Climatic Stress*, eds. J. Borghetti and A. Raschi, pp. 243–246. Cambridge, UK: Cambridge University Press.

Martínez, M. (1992). *Canarias en la Mitología*. Santa Cruz de Tenerife: Cabildo Insular de Tenerife y Centro de la Cultura Popular Canaria.

Marzol, M. V. (2002a). Un système de captation passive de l'eau du brouillard: Application et résultats obtenus aux Îles Canaries (1992–2001). *Publications de l'Association Internationale de Climatologie* **14**: 87–94.

Marzol, M. V. (2002b). Fog water for rural zones: experiences in a Rural Park in the Canary Islands (Spain). *Atmospheric Research* **64**: 239–250.

Marzol, M. V. (2005). *La captación del agua de la niebla en la isla de Tenerife*. Santa Cruz de Tenerife: Caja General de Ahorros.

Marzol, M. V., and P. Valladares (1998). Evaluation of fog water collection in Anaga (Tenerife, Canary Islands). In *Proceedings of the 1st International Conference on Fog and Fog Collection*, eds. R. Schemenauer and H. A. Bridgman, pp. 449–452. Ottowa, Canada: IDRC.

Marzol, M. V., J. Rodríguez, M. E. Arozena, and M. Luis (1988). Rapport entre la dynamique de la mer de nuages et la vegétation au Nord de Tenerife (Iles Canaries). *Publications de l'Association Internationale de Climatologie* **4**: 273–283.

Marzol, M. V., P. Dorta, P. Valladares, P. Morin, and M. Abreu (1994). La captation de l'eau à Tenerife (Iles Canaries): l'utilisation des brouillards. Publications de *l'Association Internationale de Climatologie* **7**: 83–91.

Marzol, M. V., J. Sánchez, P. Valladares, R. Pérez, and P. Dorta (1996). La captación del agua del mar de nubes en Tenerife: método e instrumental. In *Clima y agua: La gestión de un recurso climático*, eds. M V. Marzol, P. Dorta, and P. Valladares, pp. 333–350. Madrid: Tabapress.

Marzol, M. V., P. Dorta, P. Valladares, and R. Pérez (1997). Le potentiel hydrique de la base de la mer de nuages à Tenerife (Iles Canaries). *Publications de l'Association Internationale de Climatologie* **10**: 93–101.

Morey, M., and F. González (1966). *Aspectos de la utilización del agua por las plantas en las Islas Canarias*. Madrid: Boletín de la Real Sociedad Española de Historia Natural.

Ory Ajamil, F. (2004). *Ciencia y presencia extranjera en las Islas Canarias: El aire*. Santa Cruz de Tenerife: Thesaurus.

Ossuna y Saviñon, M. (1893). *Viaje al Pico de Tenerife: Descripción geológica de este monte volcánico*. La Laguna, Tenerife: Nueva Gráfica.

Oviedo, G. F. de (1535–1548). *Historia Generaly Natural de las Indias*.

Pico, B., and D. Corbella (2000). *Viajeros franceses a las Islas Canarias*. Santa Cruz de Tenerife: Instituto de Estudios Canarios.

Prada, S., M. Menezes de Sequeiro, C. Figueira, and M. Oliveira da Silva (2009). Fog precipitation and rainfall interception in the natural forests of Madeira Island (Portugal). *Agricultural and Forest Meteorology* **149**: 1179–1187.

Pruneda, V. (1848). *Un viaje a las Islas Canarias (confinado á ellas seis años en el de 1845)*. Teruel: Imprenta de Anselmo Zarzoso y Cía.

Ritter, A., C. M. Regalado, and G. Aschan (2008). Fog water collection in a subtropical elfin laurel forest of the Garajonay National Park (Canary Islands): a combined approach using artificial fog catchers and a physically based impaction model. *Journal of Hydrometeorology* **9**: 920–935.

Schemenauer, R., and P. Cereceda (1994). A proposed standard fog collector for use in high elevation regions. *Journal of Applied Meteorology* **33**: 1313–1322.

Torriani, L. (1582). *Descripción de las Islas Canarias*. Reprinted 1978, Santa Cruz de Tenerife: Ediciones Goya.

Valladares, P. (1996). Propuesta metodológica para el estudio del Mar de nubes: el caso del año 1989 en el norte de Tenerife. *Investigaciones Geográficas* **15**: 99–108.

Verneau, R. (1891). *Cinco años de estancia en las islas Canarias*. La Orotava: Ediciones JADL.

Von Humboldt, A. (1799). *Viaje a las Islas Canarias*. Reprinted 1995, Santa Cruz de Tenerife: Lemus.

Wilde, W. (1837). *Viaje a Tenerife*. Madrid: Tauro.

38 Effects of fog on climatic conditions at a sub-tropical montane cloud forest site in northern Tenerife (Canary Islands, Spain)

M. V. Marzol-Jaén
University of La Laguna, La Laguna, Tenerife, Canary Islands, Spain
J. Sànchez-Megía
National Meteorological Institute, Tenerife, Canary Islands, Spain
G. García-Santos
VU University, Amsterdam, the Netherlands

ABSTRACT

This chapter discusses microclimatic conditions at a ridge crest site in northern Tenerife carrying an impoverished and stunted forest, locally called *fayal–brezal*. The site is characterized by persistent fog and low cloud brought in by the prevailing trade winds. Temperatures, relative humidity, and wind speed differed between times with and without fog, with lower temperatures and wind speeds, and increased humidity occurring during times of fog. Average rainfall was low (474 mm year^{-1}) with a long dry period in summer. Fog incidence as measured with "standard fog collectors" placed on the windward edge of the forest suggested potentially large amounts of fog water may be available for capturing by the vegetation during rain-free periods. This extra water is thought to explain the survival of an evergreen forest despite the low and strongly seasonal rainfall.

INTRODUCTION

Tenerife is the largest island within the Canaries with a surface area of 2065 km^2 (Figure 38.1). Despite an annual influx of 5 million tourists, nearly half of the island's area is protected as belonging to one of the six categories of environmental protection established under Spanish law (Martín *et al.*, 1995). The steep relief and particular atmospheric conditions prevailing in this part of the Atlantic have produced a distinct altitudinal vegetation zonation that closely corresponds with changes in climatic conditions (Kämmer, 1974; Arozena, 1991; Fernández-Palacios, 1999). Between sea level and 400 m.a.s.l., rainfall is less than 400 mm year^{-1} and the natural vegetation consists of coastal bushes dominated by *Euphorbia* species. There is a transitional vegetation formation with *Juniperus phoenica* and *Euphorbia*s between 400 and 700 m.a.s.l. whereas between 700 and 1100 m.a.s.l., so-called *laurisilva* is found, a tall evergreen forest dominated by various laurel species. The *laurisilva* zone receives 600–900 mm of rainfall year^{-1} and roughly coincides with the cloud belt locally known as the "sea of clouds" (cf. Marzol-Jaén, this volume). Above the *laurisilva*, rainfall and cloud incidence decrease again and pine forests (*Pinus canariensis* and *P. alephensis*) take over, whereas between 2000 and 2600 m.a.s.l. a low mountain scrub prevails.

The *laurisilva* is a particularly important ecosystem representing the vegetation that existed on Earth during the Tertiary era, especially in southern Europe and northern Africa. Nowadays, it is nearly exclusively preserved in humid insular areas above 700 m.a.s.l. that face the trade winds bringing orographic precipitation and clouds (Fernández-Palacios, 1999). It contains numerous endemic species of flora and fauna, as well as other species in danger of extinction (Arozena, 1987; Ohsawa *et al.*, 1999; cf. Ohsawa *et al.*, this volume). On exposed upper slopes, ridges, and summits within the *laurisilva* zone, an impoverished forest occurs (locally called *fayal–brezal*) which is dominated by tree

Tropical Montane Cloud Forests: Science for Conservation and Management: eds. L. A. Bruijnzeel, F. N. Scatena, and L. S. Hamilton. Published by Cambridge University Press.

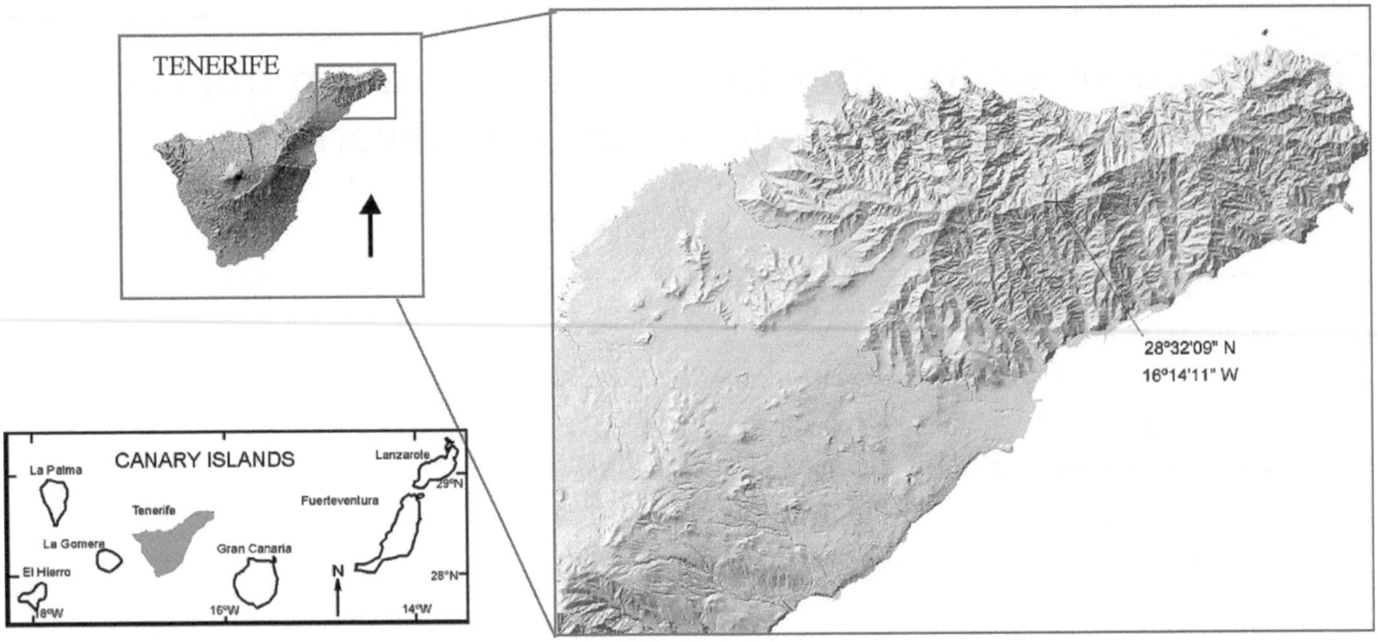

Figure 38.1. Location of study site within the Anaga mountains of northern Tenerife, and general geographic location of Tenerife within the Canary Islands.

heath (*Erica arborea*) and indigenous beech (*Myrica faya*). The *fayal–brezal* forest is essentially a kind of sub-tropical elfin cloud forest, with abundant mosses hanging from the branches (mostly *Usnea barbata*) and on the stems (Ohsawa *et al.*, 1999). In places it forms a transition between the *laurisilva* and the pine forest (Arozena, 1991; Fernández-Palacios *et al.*, 2001). *Fayal–brezal* trees are often stunted (3–4 m high) and, at very wind-exposed locations, also wind-pruned (Ohsawa *et al.*, 1999; cf. García-Santos, 2007).

The foliage of the dominant *Erica* species in *fayal–brezal* is needle-like and highly suitable for the capturing of passing fog droplets (Kämmer, 1974; cf. Prada *et al.*, 2009). Fog is considered to play an important hydrological role in the survival of this type of forest, both in terms of the supply of extra inputs of water during prolonged rainless periods (Marzol, 1988, 2002a; Marzol and Valladares, 1998; Prada *et al.*, 2009; García-Santos, 2007) and through its moderating effect on evaporation (García-Santos, 2007). This chapter discusses the fog climate of a ridge-top *fayal–brezal* forest in northern Tenerife, and the effects of fog on temperature and humidity.

MATERIALS AND METHODS

The study site is located at 842 m.a.s.l. near the summit of the 14 419-ha Anaga Rural Park (28° 32′ 09″ N; 16° 14′ 11″ W), in the north-eastern part of Tenerife (Figure 38.1). The Anaga mountain range is the first significant orographic obstacle encountered by the trade winds, which must ascend and accelerate up the northern slopes and thus change their direction from NE to NNW. This process induces adiabatic cooling of the air and the subsequent condensation of water vapor in the form of orographic fog, to which advection fog is added as the "sea of clouds" drifting in from the Atlantic reaches the island (Dorta, 1996; Marzol, 2002b). As a result, the summit of the mountain range is humid all year (Marzol and Valladares, 1998).

The climate at the study site was monitored by an SEAC (Spanish Society for Cybernetic Applications) automated weather station (AWS) from August 1996 to June 2005. Averaged temperature, humidity, rainfall, wind speed, and wind direction data were stored at 10-min intervals (see Marzol, 2005 for instrumental details). Fog water potential was studied using a standard fog collector (*SFC*) as per the design of Schemenauer and Cereceda (1994) from August 1996 to August 2000, and subsequently using one quarter-sized fog collector (*QFC*) as designed by Marzol (2002b, 2005). Double-layered Raschel mesh was used in both screen types, with a cover coefficient of 60%. Overall, collections by the two screen types differed by a factor of 3.5 on average instead of 4.0 as would be expected on the basis of the difference in screen size alone (1 m^2 vs. 0.25 m^2) (Marzol, 2002). Collection efficiency of the *SFC* was 15% higher than that of the *QFC*. The data used for this analysis correspond to the standard mesh.

The two screens were placed next to each other in the open with their center at 2.0 m above ground level and facing the prevailing NNW winds at a distance of 4 km from the coast and roughly 3 m above the surrounding *fayal–brezal* slope forest. Fog water volumes were recorded every 10 min (tipping bucket,

0.1 mm per tip), at the same time as the other climatic variables. All variables were analyzed on hourly and daily timescales. Amounts of water captured by the two fog collectors and the rain gage were compared on a daily scale ($n = 2991$) whereas hourly information (71 784 hours or 92% of the total time between August 1996 and June 2005) was used to derive average seasonal and diurnal patterns for temperature, humidity, and wind speed. In doing so, a distinction was made between periods with fog-only, and periods without fog or rain in an attempt to evaluate the effects of fog on the magnitude of the respective climatic variables. No corrections were applied for the effect on fog water capture by a screen of fixed orientation to account for variations in wind direction (cf. Juvik and Nullet, 1995; García-Santos, 2007; Giambelluca *et al.*, this volume).

RESULTS AND DISCUSSION

Climatic conditions

The mean annual temperature between 1996 and 2005 at the study site was mild: 13.4 ± 0.6 °C. In the summer, average mean monthly temperatures went up to only 16.5 ± 0.9 °C, because the frequent cloudiness prevailing in June, July, and August prevented the sun's rays from heating the surface much; in winter, mean temperatures decreased to an average of 10.7 ± 0.9 °C. The absolute maximum and minimum temperatures recorded were 44.8 °C (in August 1999), and 4.4 °C (in January 1997), respectively. The Anaga area is humid throughout the year (mean annual relative humidity 83 ± 3%) but, strikingly enough, more so in summer than in winter due to the frequent presence of fog in summer (see below). The prevailing winds come from the NNW, as they are channeled through the ravines crossing the mountain range, and had an average speed of 6.0 ± 0.5 m s^{-1}. The strongest winds occurred at night. During the 9-year observation period, wind speeds exceeded 25 m s^{-1} on 160 days. Average annual rainfall was 474 ± 28 mm, but this must be considered an underestimate in view of the prevailing wind regime which not only causes aerodynamic wind losses around the gage (Sevruk, 1982; Førland *et al.*, 1996) but also make the rain come in at variable angles. For example, García-Santos (2007) estimated wind losses around the gage to be about 20% at an exposed crest site of similarly low rainfall in central La Gomera with an average wind speed (as measured above the canopy of the *fayal–brezal*) of 3.0 ± 1.9 m s^{-1}. Holwerda *et al.* (2006) derived an average wind loss of *c.* 8% at a high-rainfall site with elfin cloud forest in Puerto Rico where wind speeds were mostly between 5 and 6 m s^{-1} (cf. Holwerda *et al.*, this volume #29). Depending on the angle of the slope vs. that of the incident rain, a further correction is required (Sharon, 1980) that varied between 18% and 28% in the cited studies from La Gomera and Puerto Rico. The corresponding average annual amount of water captured by the fog collectors during times of fog-only was 1214 ± 86 mm, suggesting high fog incidence at this exposed site. The latter value is intermediate between the 468–594 mm measured with a *QFC* at the La Gomera ridge-top site cited earlier (before correction for gage efficiency or wind direction; García-Santos, 2007) and the estimated 5110 mm of fog-induced drip observed in an *Erica arborea* stand in nearby Madeira (Prada *et al.*, 2009). Given the exposed nature of the latter site, however, it is likely that much of this "fog" capture consisted of wind-driven rain not recorded by the rain gage placed in the open (cf. Giambelluca *et al.*, this volume; Schmid *et al.*, this volume).

Changes in climatic conditions during times of fog

During times of fog, microclimatic conditions at the Anaga summit site were distinctly different from those associated with non-fog conditions (Figure 38.2). Temperatures decreased on average by 1.6 °C (2.6 °C in summer, and 1.0 °C in winter), whereas relative humidity was raised on average by 22% (range: +17% in September to +31% in March). Also, wind speeds during times of fog were higher by 1.9 m s^{-1} on average, with only slight seasonal variation (Figure 38.2). During dry weather (i.e. neither rain nor fog present), temperatures were clearly higher (especially in summer), while humidity dropped (sometimes to 70%) and winds were lighter than during times of fog. The increase in temperature and the sharp decrease in atmospheric humidity in February (Figure 38.2) were caused by a predominance of warm, dry winds from the Sahara during this month between 1997 and 2001, but this is considered statistically anomalous from a long-term perspective.

Fog water collection

For a given collector type, the main factors influencing amounts of fog water capture relate to geographic features, such as altitude, distance to the ocean, and slope angle and exposition (Cereceda *et al.*, 1988; Marzol, 2005) as well as atmospheric factors, notably wind speed, fog liquid water content (*LWC*), and duration of the fog event (Schemenauer and Cereceda, 1994; Walmsley *et al.*, 1996, 1999). Figure 38.3 shows the seasonal pattern of rainfall and fog water volumes as collected by the fog screens. Although no information was collected about fog *LWC* or droplet size distribution (cf. Holwerda *et al.*, 2006; Schmid *et al.*, this volume), and although no measurements were made of fog drip beneath the forest canopy (cf. Prada *et al.*, 2009), it appears safe to state that the ecological value of fog is highly significant, given the fact that fog collection rates by the screens were highest during the long rainless summer months (Figure 38.3; cf. García-Santos, 2007). Conversely, most rain falls in

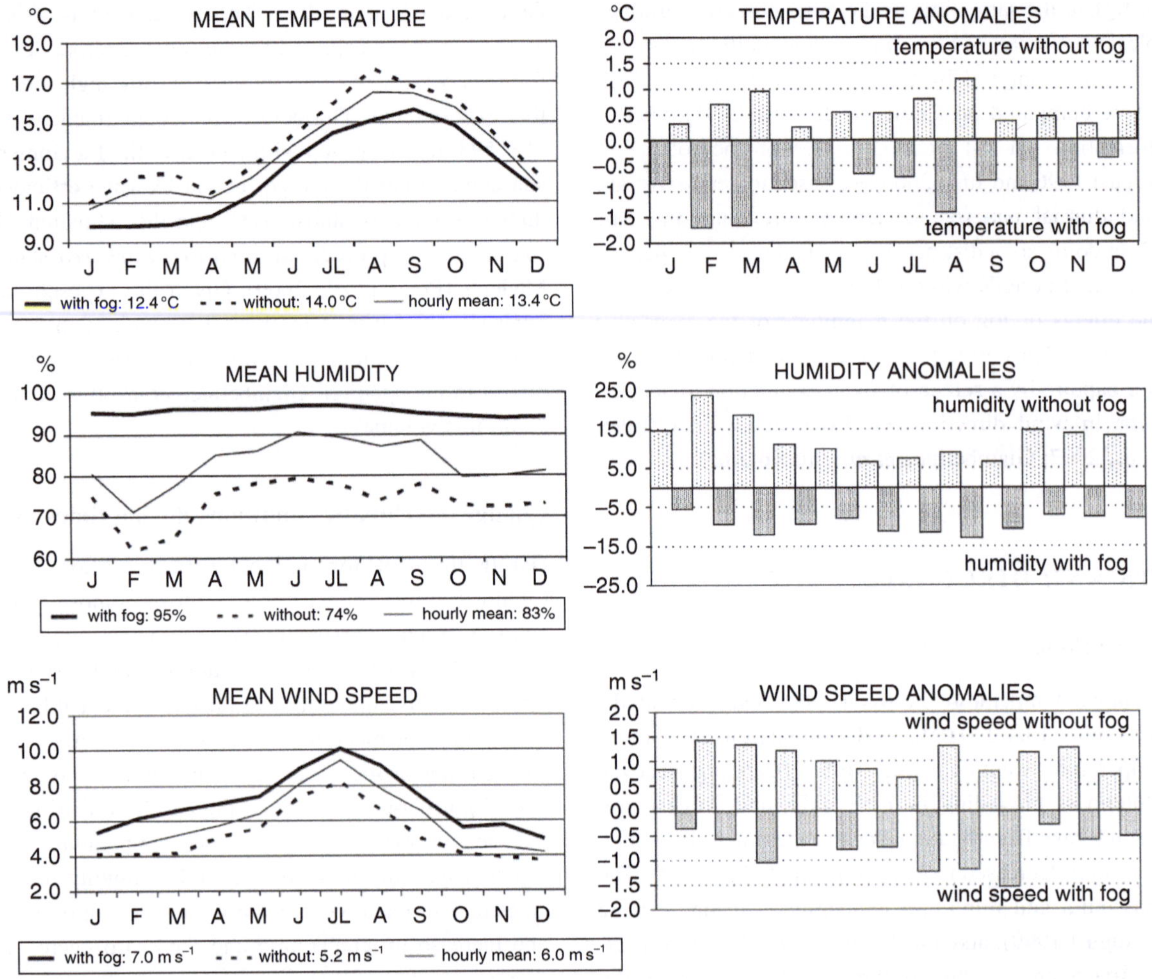

Figure 38.2. Average mean monthly temperature, relative humidity, and wind speed at the Anaga summit site (August 1996–June 2005) in northern Tenerife and their variations or anomalies during times of fog or non-fog.

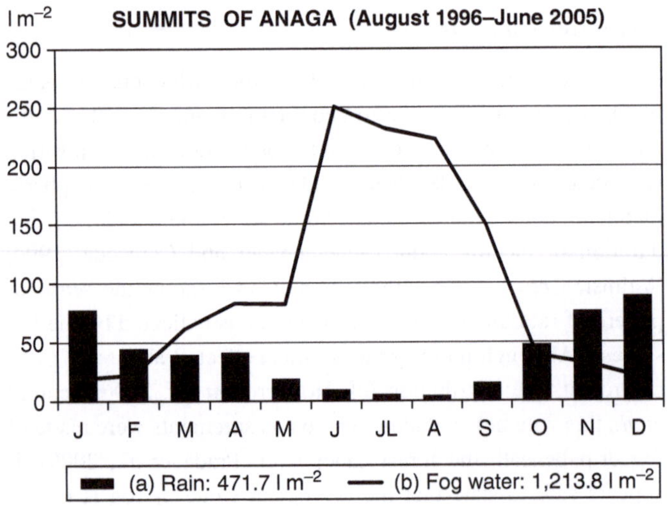

Figure 38.3. Average monthly distribution of (a) rainfall and (b) fog water inputs (both in l m^{-2}) at the Anaga summit site in northern Tenerife (August 1996–June 2005).

winter (Figure 38.3). During the 9-year observation period, the maximum monthly collection derived solely from fog was 493 mm water equivalent (July 2001); the daily maximum was 63 mm (14 August 2000), and the hourly maximum rate 6.7 mm (17 May 2002, 08.00–09.00 h) (Figure 38.3).

The amounts of fog and rain water collected in the months of July 2002 (summer conditions) and February 2005 (winter conditions) are summarized in Table 38.1. In July 2002 the screens collected water (be it rain or fog) during 81% of all hours vs. 28% in February 2005. It did not rain much in July (9.5 mm) but a significant amount of fog was collected (864 mm depth equivalent), whereas in February it rained a lot (150.4 mm) while very little fog water was collected (42 mm).

In the absence of actual measurements one can only speculate about the amount of fog captured by the Anaga *fayal–brezal* forest. The prevailing wind speeds were within the limits for efficient fog collection (Schemenauer and Joe, 1989) whereas fog water volumes collected by the screens were more than twice those reported by García-Santos (2007) for a ridge top in La Gomera.

Table 38.1 *Water volumes collected by the quarter-sized fog collector* (QFC) *and the rain gage in July 2002 and February 2005 at the Anaga summit site, northern Tenerife*

	July 2002	February 2005
Hourly-scale information	744 hours	672 hours
Number of rain hours	48	120
Hours when *QFC* collected water (rain and/or fog)	601	191
Hours when *QFC* collected only fog	553	71
Rainfall (mm) as measured by the rain gage	9.5	150.4
Total water collected by *QFC* (rain plus fog)	998 L m^{-2}	206.0 L m^{-2}
Fog water collected by *QFC* during rain-free hours[a]	864 L m^{-2}	42 L m^{-2}

[a] Excluding the information of hours with simultaneous fog and rain.

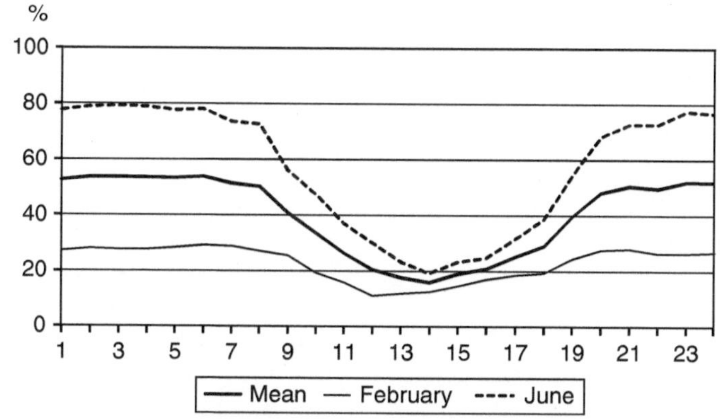

Figure 38.4. Average diurnal pattern of fog occurrence at the Anaga summit site in northern Tenerife, and its seasonal variation.

On the other hand, in the latter study it was observed that amounts of fog drip during times of fog-only were only 5.9% of incident fog as measured by a *QFC*. Interestingly, crown drip on days with both rainfall and fog (though not occurring simultaneously) was 108% of wind- and slope-corrected rainfall. Perhaps the wetting of mosses and leaves by rain allows more of the captured fog water to reach the forest floor than during times of fog-only (García-Santos, 2007).

Fog frequency at the Anaga summit

Fog occurred during 41% of the time at the Anaga summit site. Fog frequency varied significantly, both seasonally and diurnally (Figure 38.4). On a seasonal scale, fog was much more frequent in the summer (60%) than in the winter (30%) whereas on the diurnal scale, fog was much more frequent at night (50%) than during the day (20%). Moreover, the diurnal variation was more pronounced in summer than in winter (Figure 38.4).

CONCLUSION

Although actual amounts of fog water captured by the *fayal–brezal* ridge crest forest under study in the Anaga mountains of northern Tenerife were not measured, there can be little doubt that fog is a major factor in the maintenance of this evergreen forest under conditions of low and highly seasonal rainfall. The presence of fog reduced temperatures and raised atmospheric humidity, but mostly represented the dominant source of water during the long dry summer.

ACKNOWLEDGEMENT

We would like to express our gratitude to Professor L. A. Bruijnzeel for his interest and very relevant comments on the chapter.

REFERENCES

Arozena, M. E. (1987). *Estudio geográfico del Monte de El Cedro.* Santa Cruz de Tenerife: Excelentísimo Cabildo Insular de La Gomera.

Arozena, M. E. (1991). *Los paisajes naturales de la Gomera.* Santa Cruz de Tenerife: Excelentísimo Cabildo Insular de La Gomera.

Cereceda, P., R. Schemenauer, and N. Carvajal (1988). Factores tipográficos que determinan la distribución de las neblinas costeras en el Tofo (IV Región de Coquimbo-Chile). *X Congreso Nacional de Geografía y I Jornadas de Cartografía temática, Santiago de Chile:* B1–B4.

Dorta, P. (1996). Las inversiones térmicas en Canarias. *Investigaciones Geográficas* **15**: 109–126.

Fernández-Palacios, J. (1999). Marco ecológico de las Islas Canarias. In *Ecología y cultura en Canarias*, eds. A. Machado, F. García-Talavera, E. Villalba, *et al.*, pp. 83–105. Santa Cruz de Tenerife: Museo de la Ciencia y el Cosmos.

Fernández-Palacios, J., Vera, A. and A. Brito (2001). Los Ecosistemas. In *Naturaleza de las Islas Canarias*, eds. J. Fernández-Palacios and J. Martín Esquivel, pp. 157–179. Santa Cruz de Tenerife: Turquesa.

Førland, E. J., P. Allerup, B. Dahlström, *et al.* (1996). *Manual for Operational Correction of Nordic Precipitation Data.* Oslo, Norway: Norwegian Meteorological Institute.

García-Santos, G. (2007). *An ecohydrological and soils study in a montane cloud forest in the National Park of Garajonay, La Gomera (Canary Islands, Spain).* Ph.D. thesis, VU University Amsterdam, Amsterdam, The Netherlands. Also available at www.falw.vu.nl/nl/onderzoek/earth-sciences/geo-environmental-science-and-hydrology/hydrology-dissertations/index.asp.

Holwerda, F., R. Burkard, W. E. Eugster, *et al.* (2006). Estimating fog deposition at a Puerto Rican elfin cloud forest site: comparison of the water budget and eddy covariance methods. *Hydrological Processes* **20**: 2669–2692.

Juvik, J. O., and D. Nullet (1995). Comments on "a proposed standard fog collector for use in high elevation regions." *Journal of Applied Meteorology* **34**: 2108–2110.

Kämmer, F. (1974). Klima und Vegetation auf Tenerife Besonders im Hinblick auf den Nebelniederschlag. *Scripta Geobotanica* **7**: 1–78.

Martín, J., H. García, C. Redondo, I. García, and I. Carralero (1995). *La red canaria de espacios naturales protegidos.* Santa Cruz de Tenerife: Gobierno de Canarias.

Marzol, M. V. (1988). *La lluvia, un recurso natural para Canarias.* Santa Cruz de Tenerife: Caja General de Ahorros.

Marzol, M. V. (2002a). Fog water collection in a rural park in the Canary Islands. *Atmospheric Research* **64**: 239–250.

Marzol, M. V. (2002b). Un système de captation passive de l'eau du brouillard: application et résultat obtenus aux Îles Canaries (1992–2001). *Publications de l'Association Internationale de Climatologie* **14**: 87–94.

Marzol, M. V. (2005). *La captación del agua de la niebla en la isla de Tenerife*. Santa Cruz de Tenerife. Caja General de Ahorros.

Marzol, M. V., and P. Valladares (1998). Evaluation of fog water collection in Anaga (Tenerife, Canary Islands). In *Proceedings of the 1st International Conference on Fog and Fog Collection*, eds. R. Schemenauer and H. Bridgman, pp. 449–452. Ottawa, Canada: IDRC.

Ohsawa, M., W. Wildpret, and M. Del Arco (1999). *Anaga Cloud Forest: A Comparative Study on Evergreen Broad-Leaved Forest and Trees of the Canary Islands and Japan*. Chiba, Japan: Laboratory of Ecology, Chiba University.

Prada, S., M. Menezes de Sequeiro, C. Figueira, and M. Oliveira da Silva (2009). Fog precipitation and rainfall interception in the natural forests of Madeira Island (Portugal). *Agricultural and Forest Meteorology* **149**: 1179–1187.

Schemenauer, J., and P. Cereceda (1994). A proposed standard fog collector for use in high-elevation regions. *Journal of Applied Meteorology* **33**: 1313–1322.

Schemenauer, J., and P. Joe (1989). The collection efficiency of a massive fog collector. *Atmospheric Research* **24**: 53–69.

Sevruk, B. (1982). *Methods of Correction for Systematic Error in Point Precipitation Measurement*, Report No. 589. Geneva, Switzerland: World Meteorological Organization.

Sharon, D. (1980). Distribution of hydrological effective rainfall incident on sloping ground. *Journal of Hydrology* **46**: 165–188.

Walmsley, J., and R. Schemenauer (1996). A method for estimating the hydrologic input from fog in mountainous terrain. *Journal of Applied Meteorology* **35**: 2237–2248.

Walmsley, J., W. R. Burrows, and R. Schemenauer (1999). The use of routine weather observations to calculate liquid water content in summertime high-elevation fog. *Journal of Applied Meteorology* **38**: 369–384.

Part IV
Nutrient dynamics in tropical montane cloud forests

L. A. Bruijnzeel and F. N. Scatena

39 Spatial and temporal dynamics of atmospheric water and nutrient inputs in tropical mountain forests of southern Ecuador

R. Rollenbeck and J. Bendix
University of Marburg, Marburg, Germany

P. Fabian
Technical University of Munich, Munich, Germany

ABSTRACT

As part of an interdisciplinary research program, the spatial and temporal variability of precipitation and associated nutrient inputs in southern Ecuador have been investigated since January 2002. The study site is located at the northern margin of the Podocarpus National Park in the vicinity of Loja, about 500 km south of Quito, at altitudes ranging from 1800 to 3180 m.a.s.l. Due to its low density, the conventional rainfall station network fails to register the highly variable distribution of rain whereas fog is not accounted for at all. Hence, a new measurement infrastructure had to be installed. For the first time in a tropical montane forest setting, a Weather Radar was used, covering a radius of 60 km and reaching from the Amazon Basin to the coastal plains of the region. Furthermore, a dense network of sampling stations provided data about the altitudinal gradient of fog water inputs and the chemical properties of the different precipitation types. This combined approach provided important information on the formative processes of rain events on the eastern escarpment of the Andes. Rainfall distribution proved far more variable than previously known and strongly coupled to the orographic characteristics of the landscape. Maxima occurred especially on the exposed mountain slopes in the eastern parts of the Radar range, whereas the highest crests of the Andes received less precipitation. The study area has two cloud condensation levels, occurring at 1500–2000 m and 2500–3500 m.a.s.l., respectively. Fog was estimated to provide an additional 5–35% of water to conventionally measured rainfall. As with rainfall, fog capture exhibited a marked altitudinal gradient. Precipitation samples were analyzed for their main chemical properties to calculate spatial gradients as well as temporal trends of nutrient inputs. Although average matter fluxes were fairly low, episodic events contributed relevant amounts to the overall ecosystem nutrient budget.

INTRODUCTION

The biodiversity of tropical montane cloud forests (TMCF) is strongly influenced by climatic conditions (Frahm and Gradstein, 1991; Barthlott *et al.*, 1996; cf. Jarvis and Mulligan, this volume; Hemp, this volume #12). Amount and frequency of precipitation (both rainfall and cloud water interception) is one of the key factors controlling the development of these ecosystems, with upper montane cloud forests (UMCF) generally experiencing wetter conditions than lower montane cloud forests (LMCF) or lower montane rain forests (LMRF) (Grubb, 1977; Bruijnzeel, 2001; Schawe *et al.*, this volume). The spatial distribution and temporal dynamics of precipitation also affect the magnitude of atmospheric contributions to an ecosystem's nutrient budget. Although various studies on nutrient inputs and balances in LMRF and TMCF have been conducted in the last decades (e.g. Bruijnzeel, 1991; Asbury *et al.*, 1994; Cavelier *et al.*, 1997; Clark *et al.*, 1998; Hafkenscheid, 2000; Boy and Wilcke, 2008; Boy *et al.*, 2008a,b), current knowledge of the underlying processes is still insufficient, mostly due to the great complexity

Tropical Montane Cloud Forests: Science for Conservation and Management, eds. L. A. Bruijnzeel, F. N. Scatena, and L. S. Hamilton. Published by Cambridge University Press.

and heterogeneity of precipitation and cloud water inputs in steep montane terrain (cf. Daly *et al.*, 1994; Walmsley and Schemenauer, 1996; Blocken *et al.*, 2005; Mulligan *et al.*, 2006).

The PREDICT (Precipitation Dynamics and Chemical Properties in Tropical Montane Forest) project initiated in 2002 aims to provide an in-depth analysis of the dynamics of atmospheric water and matter fluxes in tropical montane forest in southern Ecuador. It is part of a comprehensive ecological research program ("Functionality in a tropical mountain forest of Southern Ecuador") having its central study area within the Parque Nacional Podocarpus (4° S/79° W) between 1800 m and 3200 m.a.s.l. (Beck and Müller-Hohenstein, 2001; Beck *et al.*, 2008). The research area is covered by pasture and various types of primary and secondary tropical montane forests giving way to *paramó* vegetation at higher elevations. Climatic conditions in the area have been investigated since 1998 by means of three automatic weather stations (AWS; Richter, 2003). However, the high spatio-temporal variability of cloudiness and rainfall in the area was not covered adequately by these few stations. In addition, the collection of information about the potentially important contribution of fog to the water and nutrient budgets of the forests was not part of the original station design.

This study combines meteorological measurements, remote sensing techniques and modeling approaches to quantify rainfall distribution, fog incidence, and associated atmospheric nutrient inputs. This information is to serve as the climatologic basis for sustainable management of the vulnerable forests and to aid in the preservation of the high biodiversity found in the Park (Werner *et al.*, 2005; Hilt *et al.*, 2006). Data analysis includes information on meso- and larger-scale atmospheric processes to obtain a better understanding of synoptic influences on local meteorological conditions.

MATERIALS AND METHODS

Study area, experimental set-up, and types of data collected

The wider surroundings of the study area, the so-called *Nudo de Loja* (knot of Loja) possess a rather special topography: whilst most of Colombia, Ecuador, and Perú are characterized by two distinct mountain chains, here the mountains are grouped into various small chains and intersecting valleys. Also, the average elevation of the mountains in this part of the Andes is some 1000 m lower than in other parts (Figure 39.1). Hence, the part extending from roughly 4–7° S is called the "depression of Huancabamba" after a small Peruvian village in the vicinity. Although the *Nudo de Loja* forms the lowest traverse for air masses carried by the persistent trade winds from the Amazon Basin to the Pacific coast, it is characterized by an extreme spatial variability of climate. For example, the town of Zamora (20 km to the east of the study area) receives about 2000 mm of rain per year in 12 humid months, whereas the valley of Catamayo 70 km further west is highly arid (12 arid months) and receives only 300 mm rainfall on average.

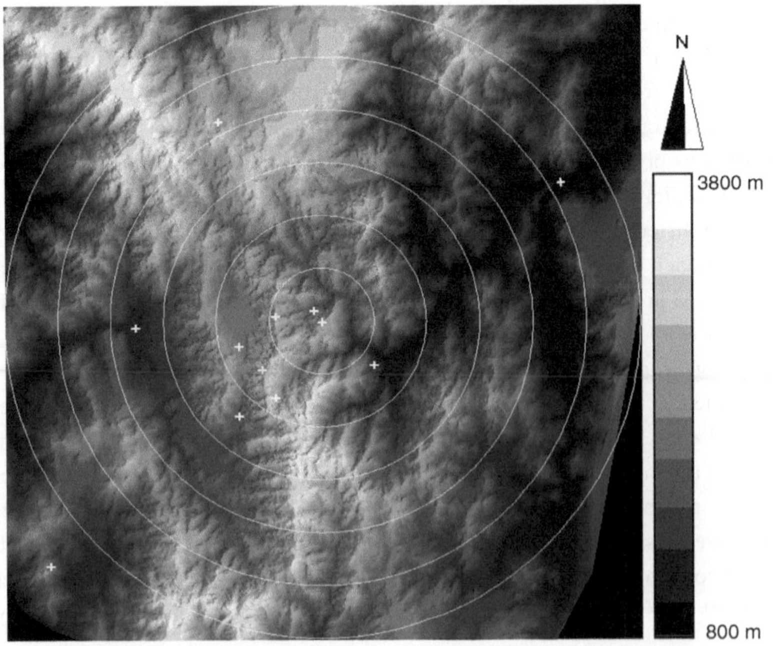

Figure 39.1. Elevation model of the *Nudo de Loja* area, southern Ecuador. Weather Radar ring ranges shown are at 10-km intervals. Crosses indicate the locations of climate stations.

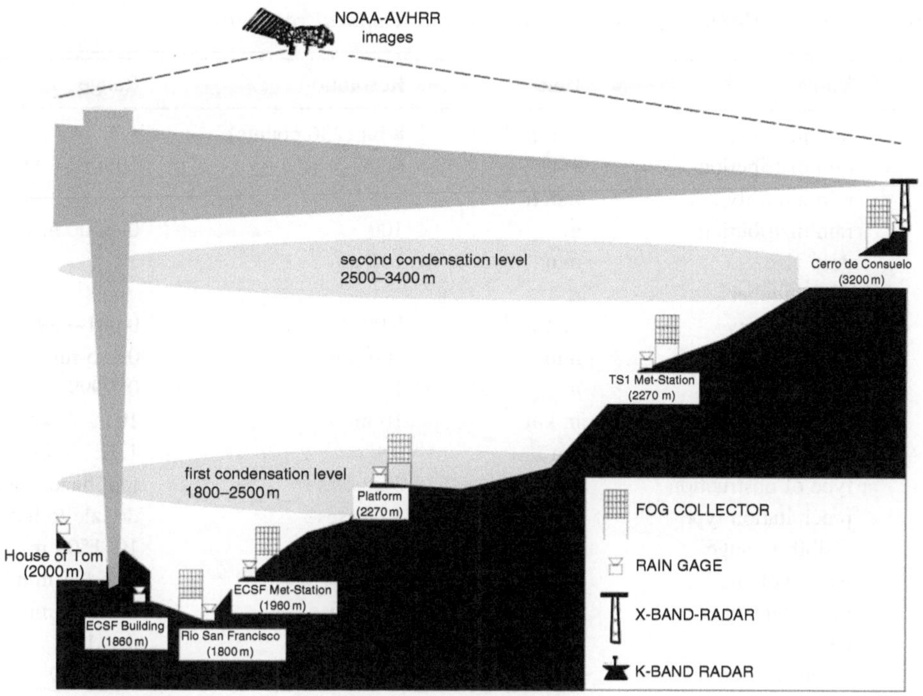

Figure 39.2. Experimental set-up of the PREDICT Project. Also shown are the two cloud condensation levels typical for the study area and resulting from cloudiness arising in the Amazon Basin (lower level) and local convection (higher level), respectively.

The general climate, forest, and soils in the vicinity of the main research station (ECSF, Estacion Cientifica de San Francisco, 3° 58' 18" S/79° 04' 45" W, 1860 m.a.s.l.; Figure 39.2) are described by Fleischbein *et al.* (this volume), Oesker *et al.* (this volume), and Wilcke *et al.* (this volume). A more detailed climatic characterization has been given by Rollenbeck *et al.* (2007) and Bendix *et al.* (2008b).

The experimental set-up established for the PREDICT project included instruments that work on various scales and at different temporal resolutions (Figure 39.2, Table 39.1). The spatio-temporal dynamics of rainfall were analyzed using weather radar images of an X-band local area Weather Radar (LAWR, Danish Hydraulic Institute), a low-cost alternative to larger weather radar systems. It provides the typical range of X-band systems (60-km radius) but lacks some of the more sophisticated capabilities of operational systems, such as vertical scans or Doppler technology (Jensen, 2002; Rollenbeck and Bendix, 2006). The system, which is the first meteorological radar in the Andes, was installed on the highest peak in the study area, the Cerro de Consuelo at 3180 m.a.s.l. (Figure 39.2). Its measurement range extended far beyond the central study area and thus enabled insight into larger-scale rain-forming processes which helped to better understand the climatologic and orographic influence of the mountains on rainfall distribution. The images were generated every 5 min and had a resolution of 500×500 m. Hence, the central study area was covered by about 25 radar pixels, which is a major improvement compared to the three point observations provided previously by the climatic stations. The three existing AWS plus nine additional totalizing rain gages placed along the altitudinal gradient were used to calibrate the LAWR data.

Fog water inputs were determined by six square (1-m^2) polypropylene mesh collectors placed along the altitudinal gradient, following the design proposed by Schemenauer and Cereceda (1994). These were chosen for their simple set-up and were exposed to the main easterly wind direction. Whilst it is recognized that the use of square fog screens of fixed orientation may lead to underestimation of fog water inputs due to variations in wind direction (Juvik and Nullet, 1995; García-Santos, 2007), the effect is considered small in the present case because the frequency of winds from the easterly sector was *c.* 90%. Water samples from each site were collected weekly to determine fog capture totals and their chemical properties.

To evaluate the performance of the non-recording fog collectors under conditions of drizzle and wind-driven rain, a set of more sophisticated automatic instruments that measured various properties of precipitation every 5 min was installed close to the main research station. These included: (i) A BIRAL VPF-730 scatterometer measuring atmospheric extinction and horizontal visibility, rain rate, particle number, droplet spectra, and rain type; (ii) an automatic impact collector (Eigenbrodt NES210 coupled to an ONED 250 fog detector) to sample fog water; and (iii) an automatic rain-water sampler (Eigenbrodt NMO 191/s) that determined rain rate and measured the values of pH

Table 39.1 *Instruments used in the PREDICT project and their spatio-temporal resolution*

Device	Value	Unit	Resolution	Range	Interval
X-band Radar	rain intensity	mm h^{-1}	8 bit (256 counts)		5 min
	rain distribution	km^2	0.25	60 km = 11300 km^2	5 min
K-band Radar	rain intensity	mm h^{-1}			1 min
METEK MRR-2	rain distribution	m	100	0–6000 m vertical	1 min
	drop size	mm	100 μm		1 min
	drop number	*n*			1 min
Scatterometer	rain intensity	mm h^{-1}	0.0001	0.001–999	5 min
BIRAL VPF 730	drop size	mm	100 μm	0.1–5 mm	5 min
	drop number	*n*	1	0–9999	5 min
	visibility range	m km^{-1}	10 m	10 m–75 km	5 min
	atm. extinction	km^{-1}	0.1	1.3E^{-5}–0.1	5 min
	type of obstruction	–	3 classes	fog, haze, smoke	5 min
	precipitation type	–	8 classes	drizzle to heavy rain	5 min
Fog detector	visibility range	m	10 m	10–1500 m	5 min
Fog collector	water volume	mm^3	0,1	0.1–475 mm^3	7 d
Rain monitor	rain intensity	mm	0.2 mm	0.2–100 mm	5 min
	pH value	log *n* H$^+$	0.1	0.1–14	1 h
	conductivity	μS	0.1	0.1–1999	1 h
Rain gage	rain intensity	mm	0.1	0.2–100 mm	7 d
Fog collector (mesh)	water volume	mm	0.1	0.2–100 mm	7 d
NOAA-AVHRR	images	–	1 km	–	0.5–1 d
Weather cam	images	–	640 × 480 pixels	5 km	5 min

and electric conductivity in real time. Sample bottles were automatically changed every day. Operation time was recorded on a standard PC in order to calculate average liquid water content (*LWC*) of the fog for the sample intervals. An AWS provided hourly data of temperature, relative humidity, wind field, and global radiation.

The measurements inside the central research area were supplemented by an additional station on the El Tiro ridge, *c*. 10 km to the west of the ECSF. This site was equipped with two fog collectors, a rain gage, and an AWS. The second fog collector was of the cylindrical type using Teflon wires as collecting surface (Kroneis, Austria). The El Tiro ridge constitutes the main division between the Rio San Francisco valley and the basin of Loja (Figure 39.1). Strong funneling of easterly winds along the valley that converged with the thermal upslope breeze from Loja caused a high frequency of rain and fog throughout the year at this site.

A further five AWS plus five official principal stations of the Ecuadorian Weather Service (INAMHI) supplying daily totals of rainfall operating within the range of the LAWR (cf. Figure 39.1) were used for calibration of the measured reflectivities (conversion to rain rate). Weather satellite data for the entire area of Ecuador are obtained by the project's High Resolution Picture Transmission (HRPT) receiving station operated in cooperation with INAMHI at their headquarters in Quito. Two daily passes of the NOAA satellite series Advanced Very High Resolution Radiometer (AVHRR) imagery were captured and transferred via Internet to Marburg (Germany) for further analysis (Bendix *et al*., 2004).

During an intensive field campaign from December 2001 to April 2003, additional measurements of vertical rainfall profiles were made using a vertical K-band Doppler rain radar (METEK MRR-2; Klugmann *et al*., 1996). These measurements were mostly made at the ECSF station building in the Rio San Francisco valley at 1850 m.a.s.l. (cf. Figure 39.2), with supplementary mobile measurements within the range of the LAWR (up to 45 km distance). To ensure continuous observation of the synoptic situation in the San Francisco valley, a simple digital weather cam was installed at the ECSF that took an image every 5 min. These data allowed monitoring and detection of the cloud base altitude (local condensation level) within the ECSF research area (cf. Lawton *et al*., this volume).

Data preprocessing methods and integration with numeric models

Most of the instruments used here have an experimental or semi-experimental character. Hence, preprocessing schemes were developed to allow use of the data for climatologic analysis. For the LAWR, a specific scheme for correction

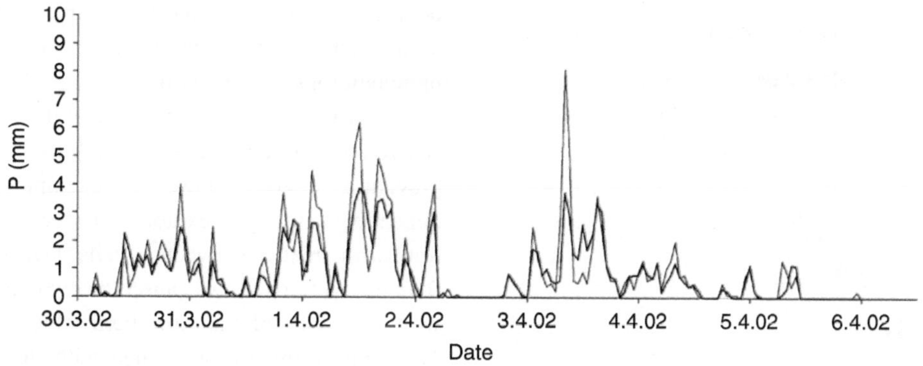

Figure 39.3. Example of LAWR calibration against rainfall recorders. Black line: rain-gage data (hourly totals) for site TS1 (2 km from radar station), gray line: radar-based rainfall.

and calibration of rain rate was developed, because this type of radar differs strongly from more commonly used weather radars (see Rollenbeck and Bendix (2006) for details). The calibrated images give reliable results up to a range of 40 km (Figures 39.3 and 39.4).

To assess the influence of the regional atmospheric circulation on locally observed weather variability, data of the NOAA–AVHRR sensor were used to derive maps of cloud type and frequency, cloud top altitude, liquid water path, and optical depth (Bendix *et al.*, 2004). Together with the information on cloud base height as obtained from the webcam, values of cloud *LWC* were deduced (Bendix *et al.*, 2008a; cf. Welch *et al.*, 2008).

For the polypropylene mesh fog collectors, characteristics of the collecting surface and possible sources of error were assessed by referencing the collected volumes against the data obtained by the automated measurement systems, notably the automatic fog collector and the scatterometer (Table 39.1). The polypropylene mesh (mesh width 1.5 mm) proved capable of stripping an average of 50% of the actual *LWC* of the air passing through the 1-m^2 frame. Errors due to varying wind velocities, directions and additional contribution by wind-driven rain and drizzle were accounted for. Comparison of the fog gage records and the totalizing rain gages showed that the greater the wind-loss error of the conventional rainfall measurement, the higher the catch of the mesh collector. Thus, the collected volumes should not be treated primarily as fog water but rather as a method to compensate the known errors of the standard rain gages (Sevruk, 1981; Rollenbeck *et al.*, 2007, cf. Frumau *et al.*, this volume; Giambelluca *et al.*, this volume). Such errors can be especially large under the windy conditions prevailing in exposed mountainous terrain (cf. Cavelier *et al.*, 1996; Bruijnzeel, 2001; Holwerda *et al.*, 2006; García-Santos, 2007; Häger and Dohrenbusch, this volume; Marzol-Jaén *et al.*, this volume; McJannet *et al.*, this volume). The additional amounts collected by the fog screens were consequently treated as representing various forms of horizontal precipitation (*HP*) including fog, drizzle, and other wind-driven rain (cf. Frumau *et al.*, this volume; Giambelluca *et al.*, this volume). Although the true water inputs to TMCF depend on vegetation type and surface structure as well (Bruijnzeel *et al.*, 2005; Prada *et al.*, 2009; Mulligan *et al.*, this volume #25), the combined use of conventional rain gages and fog collectors should give more representative input values than rain gages only.

Water samples for chemical analysis were collected along the altitudinal gradient from all collector types. The samples were analyzed for major cations and anions, pH, and electric conductivity (*EC*) using methods specified by Boy *et al.* (2008a,b). A total of 17 samples per week were retrieved from the different collectors. The analyses were evaluated to determine the seasonal variability of the atmospheric contribution to the forest nutrient budget as well as spatial variations along the altitudinal gradient. Certain episodes and events of high nutrient input were detected and to determine potential sources, backward trajectories were computed using the FLEXTRA model based on ECMWF reanalysis data and a special post-processing routine (Stohl and Trickl, 1999; Fabian *et al.*, 2005; Rollenbeck, 2005).

RESULTS

Rainfall and cloud water patterns and totals

The extreme climatologic variability of the study area is reflected well by the observed distribution of cloud frequency (Figure 39.5). The eastern slopes have higher cloud frequencies than the lowlands of the Amazon Basin, whereas to the west of the Loja region low cloud frequency and arid conditions prevail. The central study area is characterized by a "tongue" of high cloud frequency, entering from the east and continuing toward the north-west (Figure 39.5).

The spatial distribution of rainfall as seen by the radar (Figure 39.6) is similar to that observed for cloud frequency (Figure 39.5). Annual rainfall totals according to the LAWR indicate that some exposed eastern slopes receive up to

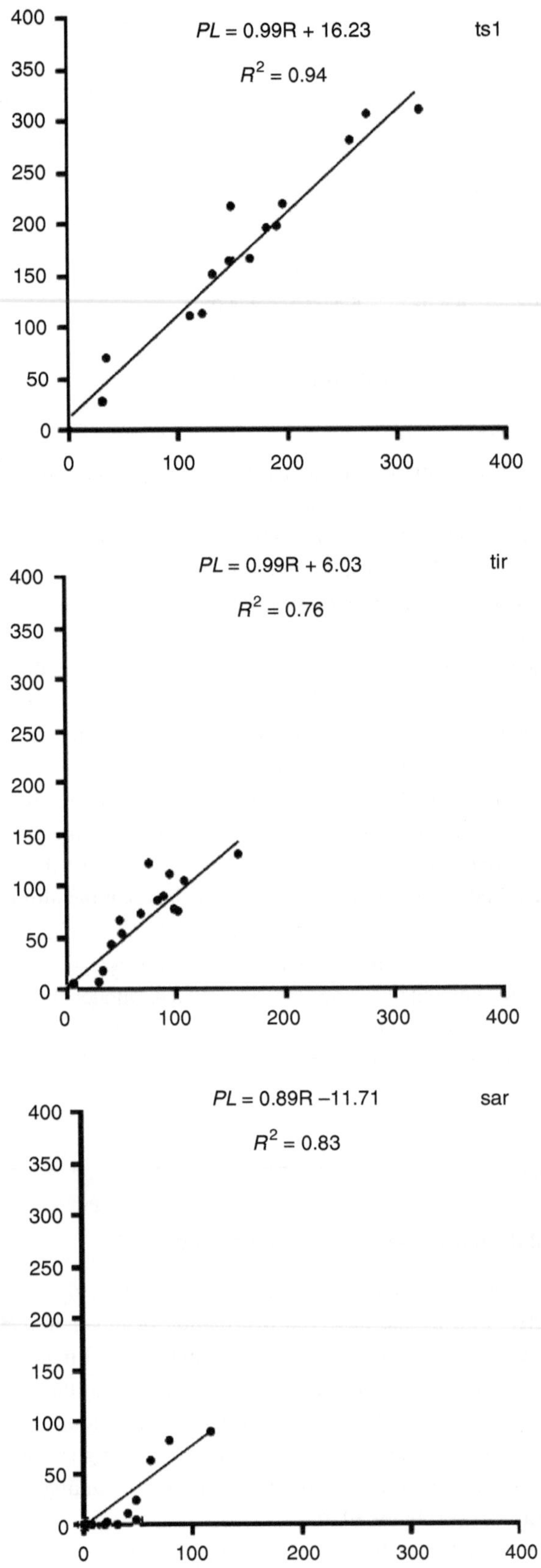

Figure 39.4. Examples of conventionally measured monthly rainfall totals (*R*, *x*-axis) against radar-based rainfall totals (*PL*, *y*-axis).

6000 mm of precipitation per year. In contrast, in the western half of the radar range, rainfall rarely exceeded 600 mm, even on mountain tops (Figure 39.6).

The seasonal variability of rainfall distribution in the area showed distinct patterns that could not be attributed only to the prevailing easterly flow. For example, the period between September and April exhibited a more heterogeneous distribution, with smaller convective cells. During this period, the less frequent westerly flows caused heavy convective rainfall on the western slopes of the Cordillera which are normally under a *Föhn*-type influence associated with the trade winds (Richter, 2003). Conversely, the season from May to September showed a clear bimodal pattern of rainfall formation, with maxima occurring on the eastern slopes, while the western part remained much drier (Bendix *et al.*, 2008b). Very frequent but less intense precipitation was responsible for the high annual totals on the eastern slopes, with rainfall amounts ranging from 1000 mm to almost 6000 mm (Figure 39.6).

The LAWR data indicated that the heaviest rain is associated with steep slopes facing the airflow coming from the Amazon Basin. The actual amount of rainfall depends on altitude, but the vertical gradients themselves also showed considerable spatial variability. Some areas probably received annual rainfall totals in excess of 7000 mm.

From data obtained with the vertically pointing MRR-radar, a vertical increase of 35% was determined in rain rates in the free atmosphere going from the valley (1860 m.a.s.l.) to the Cerro de Consuelo (3180 m.a.s.l.). A similar increase with elevation was observed by the recording rain gages and fog collectors (Figure 39.7). According to the longest available data-set (1998–2005) the average annual precipitation total (rainfall only) for the research area was calculated to be 2176 mm at 1960 m.a.s.l., increasing to *c.* 4740 mm at 3180 m.a.s.l. This corresponds to a vertical gradient in rainfall input of *c.* 210 mm 100 m^{-1} elevation increase (Bendix *et al.*, 2008a; cf. Oesker *et al.* this volume), which is slightly less than the 235 mm 100 m^{-1} observed in nearby Bolivia (Schawe *et al.*, this volume).

The vertical gradients in rainfall and *HP* illustrated in Figure 39.7 show the belt between 2300 and 2700 m.a.s.l. to have the highest totals for both rainfall and *HP*, while the highest elevations have maximum *HP* but less rain. The intermediate zone between 2800 and 3000 m.a.s.l. shows a local minimum for both precipitation types. The average vertical gradient for the sum of the two precipitation types was about 360 mm 100 m^{-1}. Amounts of *HP* increased with elevation from a mere 2% at 1800 m.a.s.l., via 5.6% (121 mm) at 1960 m.a.s.l. (ECSF) and 9.6% (527 mm) at 2660 m.a.s.l., to 41.2% (1958 mm) at the Cerro de Consuelo (3180 m.a.s.l.) (Bendix *et al.*, 2008a). Thus, with increasing altitude, fog water and horizontal rain became increasingly important contributors to total atmospheric water input. Areal total precipitation was estimated at *c.* 3600 mm per year,

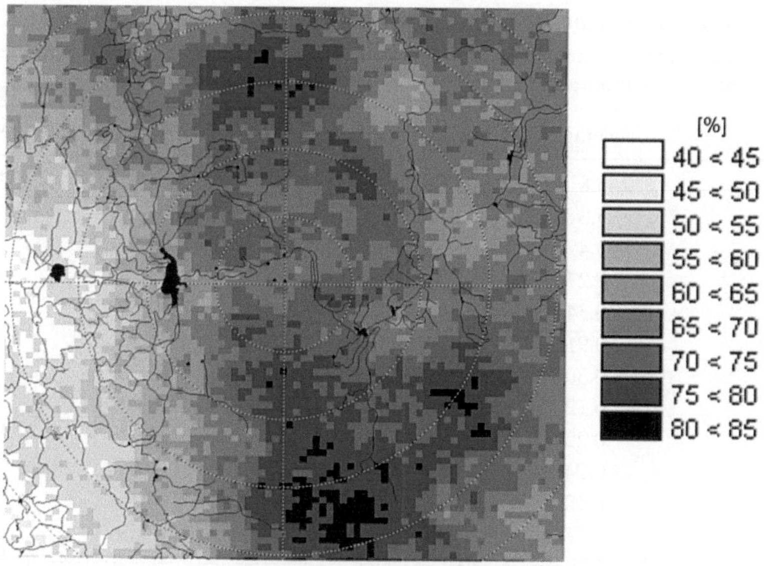

Figure 39.5. Average cloud frequency (2002–2003) as derived from NOAA–AVHRR data.

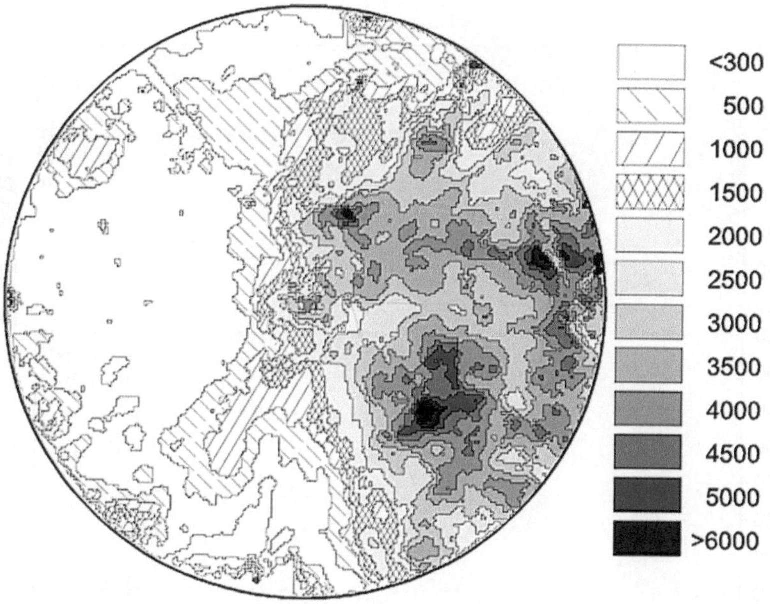

Figure 39.6. Average annual radar-based rainfall totals for 2002–2003. The figure covers the same region as Figure 39.5.

which is 80% higher than indicated by the conventional rainfall measurement by the primary climate station at 1960 m.a.s.l.

Figure 39.7 further shows the distinct seasonality for rainfall, especially at lower elevations, in contrast to fog precipitation which is more uniformly distributed over the year.

Rainfall and cloud water chemistry

Altitudinal gradients and temporal variations were also examined for the chemical composition of the collected rain and fog water. To compare different sampling intervals, all values were normalized to volume-weighted monthly averages of the respective parameters. In general, the sampled water had very low ionic concentrations, and electric conductivity rarely exceeded values of 20 $\mu S\ cm^{-1}$. Values were even lower in the rainy season because of dilution. Fog, as expected, showed slightly higher conductivity, because its smaller drop sizes offer a larger reactive surface for interaction with atmospheric aerosols (Asbury *et al.*, 1994). The same applied to pH: whilst values for rain ranged from 3.8–6.7, fog was slightly more acid (range 3.5–6.1). The observed ranges for both variables agreed well with data from other groups working in the same area (Boy *et al.*, 2008a; Oesker *et al.*, this volume). For other studies, both higher (e.g. Asbury *et al.*, 1994; Liu *et al.*, 2003) and lower values (e.g.

Table 39.2 *Average monthly nutrient deposition at the Reserva Biologica San Francisco (2002–2004)*

	Rain			Fog		
	Average	Max	Min	Average	Max	Min
Precipitation (mm)	246.8	1006.7	18.0	62.0	247.3	2.0
pH (*n*)	5.3	6.7	3.8	4.9	6.1	3.5
Conductivity ($\mu S\ cm^{-1}$)	3.4	38.6	0.0[a]	11.5	204.8	0.8
NH_4^+ ($kg\ ha^{-1}$)	0.27	3.34	0.0[a]	0.26	3.99	0.0*
Ca^{2+} ($kg\ ha^{-1}$)	0.34	3.28	0.0[a]	0.09	1.17	0.0*
Cl^- ($kg\ ha^{-1}$)	0.91	9.32	0.04	0.26	2.76	0.01
PO_4^{3-} ($kg\ ha^{-1}$)	0.30	2.52	0.0[a]	0.08	1.09	0.0[a]
Mg^{2+} ($kg\ ha^{-1}$)	0.12	1.49	0.0[a]	0.03	0.69	0.0[a]
NO_3^- ($kg\ ha^{-1}$)	0.22	2.08	0.0[a]	0.41	6.57	0.0[a]
K^+ ($kg\ ha^{-1}$)	0.24	1.96	0.0[a]	0.09	0.85	0.0[a]
Na^+ ($kg\ ha^{-1}$)	0.52	9.14	0.0[a]	0.15	3.08	0.0[a]
SO_4^{2-} ($kg\ ha^{-1}$)	0.77	3.73	0.0[a]	0.59	4.15	0.0[a]

[a] Below detection limit.

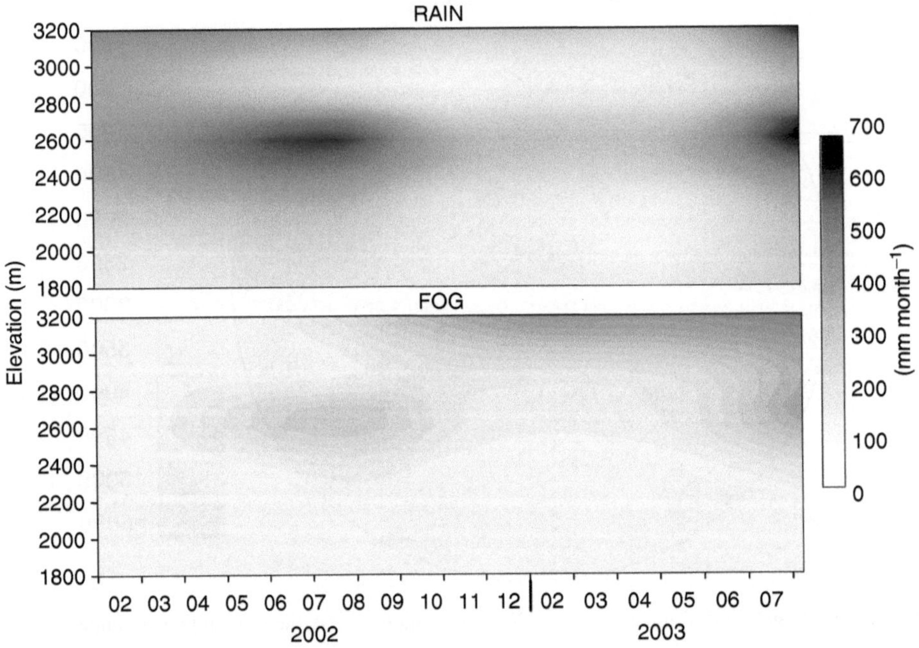

Figure 39.7. Monthly totals of rain and fog/drizzle in the Reserva Biologica San Francisco: *y*-axis shows elevation, *x*-axis the monthly timescale, while the degree of shading represents the observed monthly precipitation.

Veneklaas *et al.*, 1990; Clark *et al.*, 1998) have been reported, depending on local site conditions (cf. Hafkenscheid, 2000).

The changes in pH and conductivity with elevation were also more pronounced for fog/drizzle than for rain water. Conductivity increased but pH decreased with altitude. Conversely, ionic concentrations were similar across elevations and significant variations in rain water chemistry were visible only during specific events or periods. The highest ionic loads were observed at the El Tiro station which experiences extreme convergence due to its position at the end of the San Francisco valley (cf. Figure 39.1). Whilst marine air masses from the Pacific seemed to be the major sources of sodium and chloride, sulfate, nitrate, and ammonia were more associated with easterly wind directions.

Generally, atmospheric inputs to the ecosystem nutrient budget were low (Table 39.2). However, according to Wilcke *et al.* (2001) and Boy *et al.* (2008b) the study area is characterized by high inter-annual variability, with certain periods

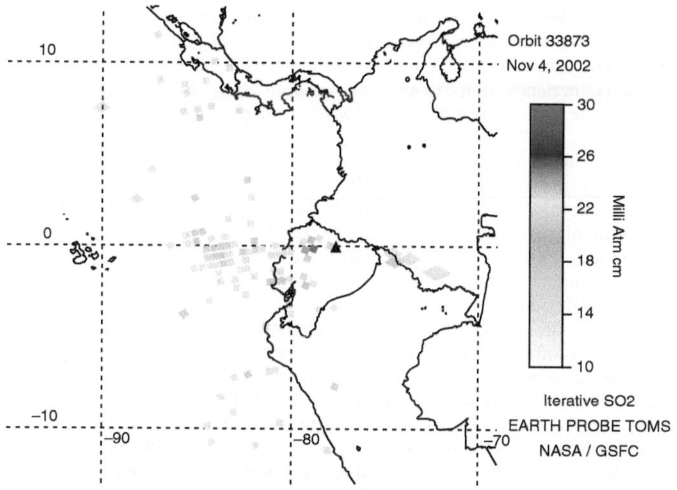

Figure 39.8. Sulfate emissions by the El Reventador volcano (black triangle) as observed by the TOMS satellite (diamonds, size proportional to atmospheric concentration). The gray lines show 10-day-backward trajectories, indicating the typical north-easterly trade winds that transported the emissions to the study area. (See also color plate.)

showing a net loss of nutrients and others a net gain that most likely originates from variable atmospheric inputs (cf. Boy and Wilcke, 2008).

Even short episodes or single events can make a significant contribution in terms of atmospheric nutrient loading. For example, on 3 November 2002, the Reventador volcano erupted in northern Ecuador and emitted a huge cloud of volcanic ash that (according to the authors' own observations) mainly affected the capital, Quito. Four days later, water samples taken from the fog collectors showed concentrations of sulfate that were seven times higher than normal, with maximum values reaching 14 mg l^{-1}. This event alone deposited some 13 kg ha^{-1} of sulfate in the research area, representing a relevant contribution to the ecosystem (cf. Veneklaas *et al.*, 1990). Inspection of local wind data associated with this event gave no hint of a northerly component but the FLEXTRA trajectories and comparison with TOMS satellite data clearly showed a connection to the eruption (Figure 39.8).

Similar observations of elevated concentrations of atmospheric nitrogen inputs pointed to a connection with biomass burning episodes in the Amazon Basin (Fabian *et al.*, 2005; cf. Nepstad *et al.*, 1999). During the dry season in Brazil (December to March) higher values of nitrate and ammonia input in fog and rain were observed in the study area, reaching monthly average concentrations of up to 2 mg l^{-1} and a deposition rate of *c.* 6 kg of N ha^{-1} $month^{-1}$, a very high value for TMCF (Hafkenscheid, 2000).

CONCLUSIONS AND OUTLOOK

Influences of topographic configuration on rainfall and cloud water distribution in the region were clearly visible in the radar and satellite images (Figures 39.5 and 39.6). The extraordinary positioning of the *Nudo de Loja* caused extreme spatial variability in rainfall that would be impossible to assess with a conventional rain gage network. Hence, remote sensing approaches constitute the only way to deal with this problem in an adequate manner.

The strong altitudinal gradient in precipitation observed in many mountain regions (cf. Häger and Dohrenbush, this volume; Schawe *et al.*, this volume; McJannet *et al.*, this volume) is confirmed by the present study, although the gradients showed considerable spatial variability. In addition, the considerable horizontal precipitation inputs (fog, drizzle, and wind-driven rain) recorded at higher elevations indicate that ecosystem water and nutrient inputs have been significantly underestimated in the past. Further investigations will focus on structural parameters of the vegetation to enable the development of a fog capturing model (cf. Bruijnzeel *et al.*, 2005) to convert potential *HP* incidence as measured by the fog screens to actually captured inputs (cf. Giambelluca *et al.*, this volume; Tanaka *et al.*, this volume).

The existence of two condensation levels in the study area was confirmed by the rainfall and *HP* measurements (Figure 39.7) and shows a certain coincidence with vegetation zonation (Homeier *et al.*, 2008). Probably, the high cloudiness and the rather high precipitation totals, along with high wind speeds at higher elevations act as a limiting factor for tree growth (Bader *et al.*, 2007; cf. Leuschner *et al.*, 2007; Moser *et al.*, 2007).

Some frequently stated assumptions about the characteristics of tropical rainfall climates have to be revised on the basis of the present results. Whilst in the lowlands the majority of rainfall events were indeed convective, advective precipitation was the dominant form of precipitation in the mountains. Moreover, meso-scale processes and local processes interacted, to cause rainy episodes with durations ranging from a few hours to several days.

Atmospheric contributions to the ecosystem nutrient budget were generally low, but inputs were periodically higher. Geogenic dust and volcanic eruptions were the main sources of atmospheric nutrients but anthropogenic sources (mainly biomass burning in the Amazon Basin) were important at times as well. Whilst the overall influence of these fluctuations on the TMCF's long-term nutrient balance is not clear yet, integration of the present data with nutrient losses obtained by other groups (Boy *et al.*, 2008b) is expected to shed more light on the matter.

ACKNOWLEDGEMENT

The authors are indebted to the German Research Foundation (DFG) for the funding of this work within the scope of the Research Unit RU402 "Biodiversity and Sustainable Management of a Megadiverse Mountain Ecosystem in South Ecuador" (BE 1780/15–1, FA 62/18–1, NA 783/1–1).

REFERENCES

Asbury, C. E., W. H. McDowell, R. Trinidad-Pizarro, and S. Berrios (1994). Solute deposition from cloud water to the canopy of a Puerto Rican montane forest. *Atmospheric Environment* **28**: 1773–1780.

Bader M. Y., I. van Geloof, and M. Rietkerk (2007). High solar radiation hinders tree establishment above the alpine treeline in northern Ecuador. *Plant Ecology* **191**: 33–45.

Barthlott, W., W. Lauer, and A. W. Placke (1996). Global distribution of species diversity in vascular plants: towards a world map of phytodiversity. *Erdkunde* **50**: 317–327.

Beck E., and K. Müller-Hohenstein (2001). Analysis of undisturbed and disturbed tropical mountain forest ecosystems in Southern Ecuador. *Die Erde* **132**: 1–8.

Beck, E., J. Bendix, I. Kottke, F. Makeschin, and R. Mosandl (eds.) (2008). *Gradients in a Tropical Mountain Ecosystem of Ecuador*. New York: Springer-Verlag.

Bendix J., R. Rollenbeck, and E. Palacios (2004). Cloud classification in the Tropics: a suitable tool for climate ecological studies in the high mountains of Ecuador. *International. Journal of Remote Sensing* **25**: 4521–4540.

Bendix J., R. Rollenbeck, and C. Reudenbach (2006). Diurnal patterns of rainfall in a tropical Andean vally of southern Ecuador as seen by a vertically pointing K-Band Doppler Radar. *International Journal of Climatology* **26**: 829–846.

Bendix J., R. Rollenbeck, D. Göttlicher, T. Nauss, and P. Fabian (2008a). Seasonality and diurnal course of very low clouds in a deeply incised valley of the eastern tropical Andes (South Ecuador) as observed by a cost effective WebCam system. *Meteorological Applications* **15**: 281–291.

Bendix, J., R. Rollenbeck, M. Richter, P. Fabian, and P. Emck (2008b). Climate. In *Gradients in a Tropical Mountain Ecosystem of Ecuador*, eds. E. Beck, J. Bendix, I. Kottke, F. Makeschin, and R. Mosandl, pp. 63–74. New York: Springer-Verlag.

Blocken, B., J. Carmeliet, and J. Poesen (2005). Numerical simulation of the wind-driven rainfall distribution over small-scale topography in space and time. *Journal of Hydrology* **315**: 252–273.

Boy, J., and W. Wilcke (2008). Tropical Andean forest derives calcium and magnesium from Saharan dust. *Global Biogeochemical Cycles* **22**, GB1027, doi: 10.1029/2007GB002960.

Boy, J., R. Rollenbeck, C. Valarezo, and W. Wilcke (2008a). Amazonian biomass burning-derived acid and nutrient deposition in the north Andean montane forest of Ecuador. *Global Biogeochemical Cycles* **22**, GB4011, doi:10.1029/2007GB003158.

Boy, J., C. Valarezo, and W. Wilcke (2008b). Water flow paths in soil control element exports in an Andean tropical montane forest. *European Journal of Soil Science* **59**: 1209–1227.

Bruijnzeel, L. A. (1991). Nutrient input–output budgets of moist tropical forest ecosystems: a review. *Journal of Tropical Ecology* **7**: 1–24.

Bruijnzeel, L. A. (2001). Hydrology of tropical montane cloud forests: a reassessment. *Land Use and Water Resources Research* **1**: 1–18.

Bruijnzeel, L. A., W. Eugster, and R. Burkard (2005). Fog as an input to the hydrological cycle. In *Encyclopaedia of Hydrological Sciences*, eds. M. G. Anderson and J.J. McDonnell, pp. 559–582. Chichester, UK: John Wiley.

Cavelier, J., D. Solis, and M. A. Jaramillo (1996). Fog interception in montane forest across the Central Cordillera of Panama. *Journal of Tropical Ecology* **12**: 357–369.

Cavelier, J., M. A. Jaramillo, D. Solis, and D. de León (1997). Water balance and nutrient inputs in bulk precipitation in tropical montane cloud forest in Panama. *Journal of Hydrology* **193**: 83–96.

Clark, K. L., N. M. Nadkarni, D. Schaefer, and H. L. Gholz (1998). Cloud water and precipitation chemistry in a tropical montane forest, Monteverde, Costa Rica. *Atmospheric Environment* **32**: 1595–1603.

Daly, C., R. P. Neilson, and D. L. Phillips (1994). A statistical–topographical model for mapping climatological precipitation over mountainous terrain. *Journal of Applied Meteorology* **33**: 140–158.

Fabian P., M. Kohlpaintner, and R. Rollenbeck (2005). Biomass burning in the Amazon: fertilizer for the mountaineous rain forest in Ecuador. *Environmental Science and Pollution Research* **12**: 290–296.

Frahm, J. P., and S. R. Gradstein (1991). An altitudinal zonation of tropical rain forests using bryophytes. *Journal of Biogeography* **18**: 669–676.

García-Santos, G. (2007). *An ecohydrological and soils study in a montane cloud forest in the National Park of Garajonay, La Gomera (Canary Islands, Spain)*. PhD Thesis, VU University Ámsterdam, Ámsterdam, The Netherlands. [http://www.falw.vu.nl/nl/onderzoek/earth-sciences/geo-environmental-science-and-hydrology/hydrology-dissertations/index.asp].

Grubb, P. J. (1977). Control of forest growth and distribution on wet tropical mountains: with special reference to mineral nutrition. *Annual Review of Ecology and Systematics* **8**: 83–107.

Hafkenscheid, R. L. L. J. (2000). Hydrology and biogeochemistry of tropical montane rain forests of contrasting stature in the Blue Mountains, Jamaica. Ph.D. thesis, VU University Amsterdam, Amsterdam, the Netherlands. Also available at http://dare.ubvu.vu.nl/bitstream/1871/12734/1/tekst.pdf.

Hilt, N., G. Brehm, and K. Fiedler (2006). Diversity and ensemble composition of geometrid moths along a successional gradient in the Ecuadorian Andes. *Journal of Tropical Ecology* **22**: 155–166.

Holwerda, F., R. Burkard, W. Eugster, *et al.* (2006). stimating fog deposition at a Puerto Rican elfin cloud forest site: comparison of the water-budget and eddy covariance methods. *Hydrological Processes* **20**: 2669–2692.

Homeier, J., F. A Werner, S. R. Gradstein, S. W. Breckle, and M. Richter (2008) Potential vegetation and floristic composition of Andean forests in South Ecuador, with a focus on the RBSF. In *Gradients in a Tropical Mountain Ecosystem of Ecuador*, eds. E. Beck, J. Bendix, I. Kottke, F. Makeschin, and R. Mosandl, pp. 87–100. Berlin: Springer-Verlag.

Jensen, N. E. (2002). X-Band local area weather radar: preliminary calibration results. *Water Science Technology* **45**: 135–138.

Juvik, J. O., and D. Nullet (1995). Comments on "a proposed standard fog collector for use in high elevation regions." *Journal of Applied Meteorology* **34**: 2108–2110.

Klugmann, D., K. Heinsohn, and H. -J. Kirtzel (1996). A low cost 24 GHz FM-CW-Doppler radar rain profiler. *Contributions to Atmospheric Physics* **69**: 247–253.

Leuschner, Ch., G. Moser, C. Bertsch, M. Röderstein, and D. Hertel (2007). Large altitudinal increase in tree root/shoot ratio in tropical mountain forests of Ecuador. *Basic and Applied Ecology* **8**: 219–230.

Liu, H., D.J. Jacob, I. Bey, *et al.* (2003). Transport pathways for Asian pollution outflow over the Pacific: interannual and seasonal variations. *Journal of Geophysical Research* **108** (D20): 8786.

Moser, G., D. Hertel, and Ch. Leuschner (2007). Altitudinal change in LAI and stand leaf biomass in tropical montane forests: a transect study in Ecuador and a pan-tropical meta-analysis. *Ecosystems* **10**: 924–935.

Mulligan, M., K. F. A. Frumau, and L. A. Bruijnzeel (2006). Falling at the first hurdle: spatial rainfall variability and the problem of closing catchment water budgets in tropical montane environments. In *Forest and Water in a Changing Environment*, eds. S. R. Liu, G. Sun, and P. S. Sun, pp. 114–117. Vienna: IUFRO, and Beijing: Chinese Academy of Forestry.

Nepstad, D. C., A. Veríssimo, A. Alencar, *et al.* (1999). Large-scale impoverishment of Amazonian forests by logging and fire. *Nature* **398**: 505–508.

Prada, S., M. Menezes de Sequeiro, C. Figueira, and M. Oliveira da Silva (2009). Fog precipitation and rainfall interception in the natural forests of Madeira Island (Portugal). *Agricultural and Forest Meteorology* **149**: 1179–1187.

Richter, M. (2003). Using epiphytes and soil temperatures for eco-climatic interpretations in Southern Ecuador. *Erdkunde* **57**: 161–181.

Rollenbeck, R., and J. Bendix (2006). Experimental calibration of a cost-effective X-band weather radar for climate ecological studies in southern Ecuador. *Atmospheric Research* **79**: 296–316.

Rollenbeck, R., P. Fabian, and J. Bendix (2005). Precipitation dynamics and chemical properties in tropical mountain forests of Ecuador. *Advances in Geosciences* **6**: 1–4.

Rollenbeck, R., J. Bendix., P. Fabian, *et al.* (2007). Comparison of different techniques for the measurement of precipitation in tropical montane rain forest regions. *Journal of Atmospheric and Oceanic Technology* **24**: 156–168.

Schemenauer, R., and P. Cereceda (1994). A proposed standard fog collector for use in high-elevation regions. *Journal of Applied Meteorology* **33**: 1313–1322.

Sevruk, B. (1981). Methodische Untersuchung des systematischen Messfehlers des Regenmessers nach Hellmann. Ph.D. thesis, ETH Zürich, Zürich, Switzerland.

Stohl, A., and T. Trickl (1999). A textbook example of long-range transport: simultaneous observation of ozone maxima of stratospheric and North American origin in the free troposphere over Europe. *Journal of Geophysical Research* **104** (D23): 30 445–30 462.

Veneklaas, E. J., R. J. Zagt, A. R. J. Van Leerdam, *et al.* (1990). Hydrological properties of the epiphyte mass of a montaine tropical rain forest. *Vegetatio* **89**: 183–192.

Walmsley, J. L., and R. S. Schemenauer (1996). A method for estimating the hydrologic input from fog in mountainous terrain. *Journal of Applied Meteorology* **35**: 2237–2249.

Welch, R. M., S. Asefi, J. Zeng, *et al.* (2008). Biogeography of tropical montane cloud forests. I. Remote sensing of cloud base heights. *Journal of Applied Meterology and Climatology* **47**: 960–975.

Werner, F. A., J. Homeier, and S. R. Gradstein (2005). Diversity of vascular epiphytes on isolated remnant trees in the montane forest belt of southern Ecuador. *Ecotropica* **11**: 21–40.

Wilcke, W., S. Yasin, C. Valarezo, and W. Zech (2001). Nutrient budget of three microcatchments under tropical montane forest in Ecuador: preliminary results. *Die Erde* **132**: 61–74.

40 Fog deposition and chemistry in a sub-tropical montane cloud forest in Taiwan

S. C. Chang, C. F. Yeh, M. J. Wu, Y. T. Chen, Y. J. Hsia
National Dong Hwa University, Hualien, Taiwan

C. P. Wang
Taiwan Forestry Research Institute, Taipei, Taiwan

J. T. Wu
Academia Sinica, Taipei, Taiwan

ABSTRACT

Annual fog deposition and atmospheric chemical deposition were evaluated for a strongly fog-affected coniferous forest at Chi-Lan, Taiwan. Fog capture efficiencies of *Chamaecyparis obtusa* var. *formosana* leaves were measured at three heights within the canopy using *in situ* exposure experiments under contrasting climatic conditions. The efficiencies obtained in this way were multiplied times leaf biomass to calculate stand-level fog deposition rates. Furthermore, a statistical model was developed, linking fog deposition rate to visibility. Using the latter model, annual fog deposition from March 2003 to February 2004 was calculated to be 297 mm, or ~9% of the total atmospheric water input. Fog contributions exhibited a highly seasonal pattern that depended mainly on the amount of precipitation. Due to the higher chemical concentrations of fog compared to precipitation, nutrient deposition via fog played a significant role. Inorganic nitrogen was absorbed by the canopy whereas potassium was leached. It is concluded that fog constitutes an important factor influencing the water and nutrient dynamics of the montane forest ecosystem under study.

INTRODUCTION

The concept of a "montane cloud forest belt" has been recognized in Taiwan for decades (Su, 1984). Although there were few meteorological records on cloud heights at the time, the elevational range of the cloud belt was inferred from temperature lapse rates as calculated from data of more than 150 weather stations (Su, 1984). In central Taiwan, where the mountain range exceeds an elevation of 3000 m.a.s.l., the cloud forest belt is found between 1500 and 2500 m.a.s.l. whereas in northern Taiwan, the cloud base may be as low as 700 m.a.s.l. Within this cloud belt the equivalent climatic zone is warm-temperate to temperate. The associated vegetation zone is dominated by *Quercus*, with a special sub-type dominated by *Chamaecyparis obtusa* var. *formosana* (Su, 1984).

Approximately 59% of the total 36 000 km^2 of the country is forested (Cheng *et al.*, 2002). If cloud forest is defined as the forest located between 1000 and 2500 m elevation, then the cloud forest belt may account for 27% of the total area of the island. Since this area is almost entirely covered by forests, approximately 45% of the forest in Taiwan may be classified as being cloud-affected forest.

Although nearly half the forest in Taiwan may thus be subject to some degree of fog incidence, until recently little was known about the physical environment characterizing this region or about the effects of fog on the ecosystem (Lin and Peng, 1998; Chang *et al.*, 2002). A preliminary hydrometeorological study in an old-growth cloud-affected *Chamaecyparis obtusa* var. *formosana* forest conducted in the Yuanyang Lake Nature Reserve (YYL site) in north-eastern Taiwan revealed that fog duration at the site was *c.* 3400 hours $year^{-1}$ (Chang *et al.*, 2002). During one month (February 2001), 57 mm of cloud water was calculated to be absorbed by the epiphytic bryophytes of the forest. This was estimated to account for 37% of the total atmospheric

Tropical Montane Cloud Forests: Science for Conservation and Management, eds. L. A. Bruijnzeel, F. N. Scatena, and L. S. Hamilton. Published by Cambridge University Press.

water input and for more than 50% of the associated nutrient inputs (Chang *et al.*, 2002).

Because of the potential significance of fog in this region, a more comprehensive fog research program was established in 2001 in a *Chamaecyparis* stand about 2.5 km from the YYL site. This program consists of the long-term monitoring of nutrient fluxes into, within, and out of the forest ecosystem. One of the main objectives of the program is to quantify fog deposition rates using various complementary methods, including *in situ* exposure (Chang *et al.*, 2002; cf. Tobón *et al.*, this volume #26), inferential micrometeorological methods (Lovett, 1994), and eddy covariance measurements (Vong and Kowalski, 1995; Beiderwieden *et al.*, 2008; cf. Schmid *et al.*, this volume). The program is supported by several ongoing research projects of the National Science Council (NSC) of Taiwan and by cooperative projects between NSC and the Deutsche Forschungsgemeinschaft (DFG) of Germany. The central goal of these projects is to clarify the role of fog in the hydrological and biogeochemical cycles of montane ecosystems in Taiwan, and to extend the knowledge of cloud forests in the sub-tropical region. This paper presents the results of the *in situ* fog exposure experiment and the fluxes of water and nutrients in bulk precipitation, fog, throughfall, and stemflow for a 1-year period.

MATERIALS AND METHODS

Study site

The Chi-Lan Mountain (CLM) study site is approximately 300 ha in area and ranges in elevation from 1400 to 1800 m.a.s.l. The site is situated on the eastern slope of the Syue-Shan mountain range (highest peak 3886 m.a.s.l.) at 24° 35′ N, 121° 25′ E (Figure 40.1). The site was chosen because of its even-aged forest structure and relatively uniform topography, which is rare at this altitude in Taiwan. The *Chamaecyparis* forest extends continuously downhill for more than 1 km at an average slope of 14° toward the south-east (Figure 40.1), thereby providing an excellent opportunity for the study of atmosphere–canopy interactions using micrometeorological methods.

In 2002 a 1-ha area was selected for intensive long-term research on cloud water dynamics. Meteorological instruments on a 23-m walk-up tower recorded net radiation (Kipp and Zonen CNR1, the Netherlands), wind speed and wind direction (Wind monitor 05103, R.M.Young, USA), air temperature, relative humidity (RHT2nl, Delta-T, UK and R.M.Young 41372VC aspired T/RH, USA), and visibility (MIRA visibility sensor 3544, Norway). Meteorological data were stored as 10-min averages (CR5000, Campbell, USA and DL2c, Delta-T, UK data-loggers) and automatically transferred to a personal computer. Visibility data were used to measure fog occurrence and for switching on a "Caltech"-type active fog collector (Demoz *et al.*, 1996). From the initiation of the study in 2002 until March 2005, there was no AC power supply at the site. To save battery power, fog water was only collected when visibility was less than 500 m, even though the occurrence of fog was defined for visibilities of less than 1000 m. After March 2005, fog water was collected when visibility was less than 1000 m.

According to the visibility data collected between 2002 and 2004, the site is subjected to frequent fog of long duration. From March 2003 to February 2004, fog was present for 38% of the time (3300 hours). Daily fog duration shows a clear seasonal variation with a maximum of 14.2 hour day^{-1} in November and a minimum of 2.7 hour day^{-1} in July (Figure 40.2). A similar seasonal pattern had been observed previously at the nearby YYL site (Chang *et al.*, 2002). The CLM site is also subjected to regular south-easterly winds moving upslope during the day and to downhill wind during the night (Figure 40.3).

In contrast to the old-growth *C. obtusa* var. *formosana* and *C. formosensis* forests of the surrounding area and the nearby Yuanyang Lake Nature Reserve, the most valuable trees had been clear-cut from the CLM site around 1960. The current stand has regenerated naturally from seed. A vegetation survey conducted in 2002 and 2003 showed that trees with a diameter at breast height (DBH) > 1 cm were mostly *C. obtusa* var. *formosana*. This species had a density of 1820 stems ha^{-1} and comprised 81.8% (41.5 m^2 ha^{-1}) of the total basal area of the stand. Thirty-two deciduous tree species comprising the over- and understorys contributed the remaining 18.2% of basal area and had a combined density of 4035 stems ha^{-1}. The *C. obtusa* trees had an average DBH of 14.7 cm and an average height of 10.3 m.

The above-ground biomass of the *C. obtusa* trees was estimated by systematically sampling stems and branches of 11 differently sized trees. Stem biomass was estimated from average dry wood density and diameter measurements made every 50-cm along the stems. Branch and leaf biomass was measured on every fourth branch counted from the lowest live branch. Samples were cut, oven-dried at 60 °C until constant weight, and weighed. Regressions between dry weights of wood and leaves (kg), and branch basal diameter (cm) were developed and applied to all branches of the sample trees to derive allometric equations between tree DBH and total leaf, branch, and stem dry weights per tree. All of the following allometric equations had coefficients of determination (r^2) greater than 0.95:

$$\ln(\text{leaves}) = -5.3 + 2.6 * \ln(\text{DBH}) \qquad (40.1)$$

$$\ln(\text{branches}) = -5.8 + 2.8 * \ln(\text{DBH}), \text{ and} \qquad (40.2)$$

$$\ln(\text{stems}) = -2.8 + 2.4 * \ln(\text{DBH}). \qquad (40.3)$$

Using these equations, the stem, branch, and leaf biomass of the stand were estimated at 26.6, 17.0, and 17.4 t ha^{-1},

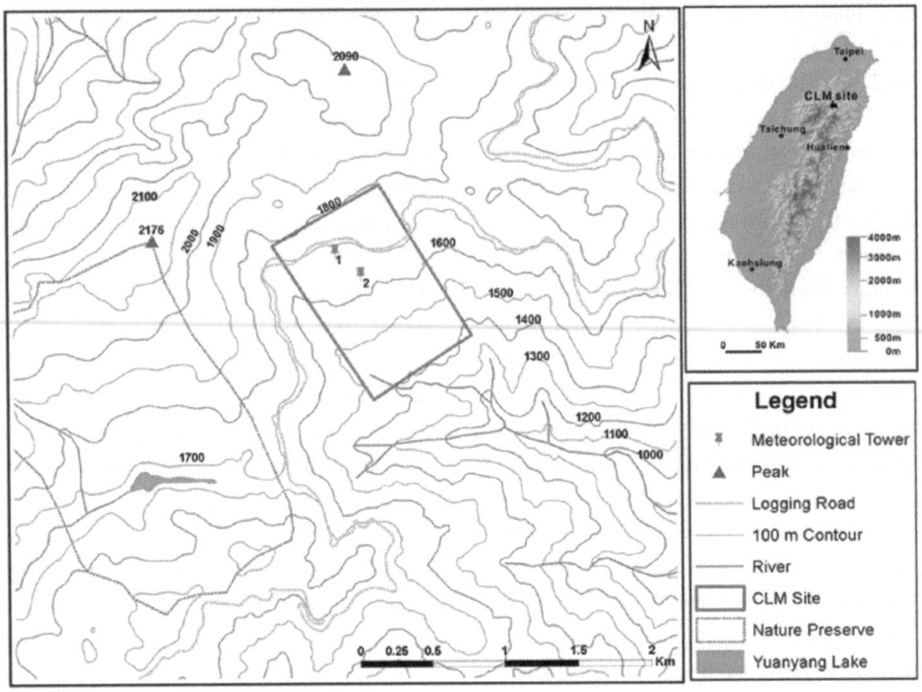

Figure 40.1. Topographic map of Taiwan showing the main cities and detailed contours of the Chi-Lan Mountain (CLM) study site. (See also color plate.)

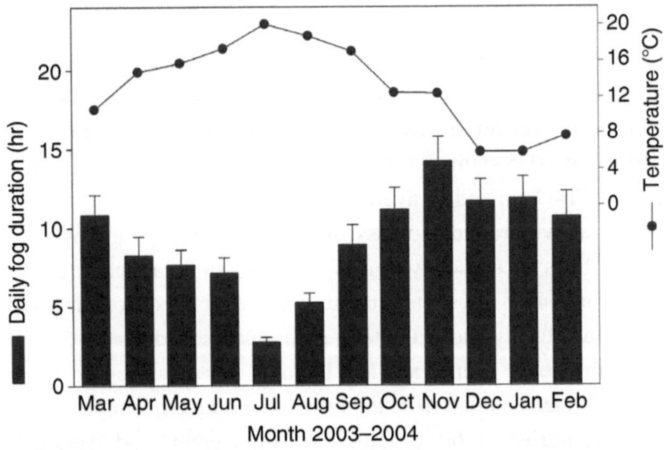

Figure 40.2. Average daily fog duration (solid bars) and average monthly air temperature (solid circles) from March 2003 to February 2004 at the Chi-Lan Mountain site. Fog duration of 5 min was added to daily fog duration whenever average visibility of the 5-min period was less than 1000 m (error bar = standard error of the mean).

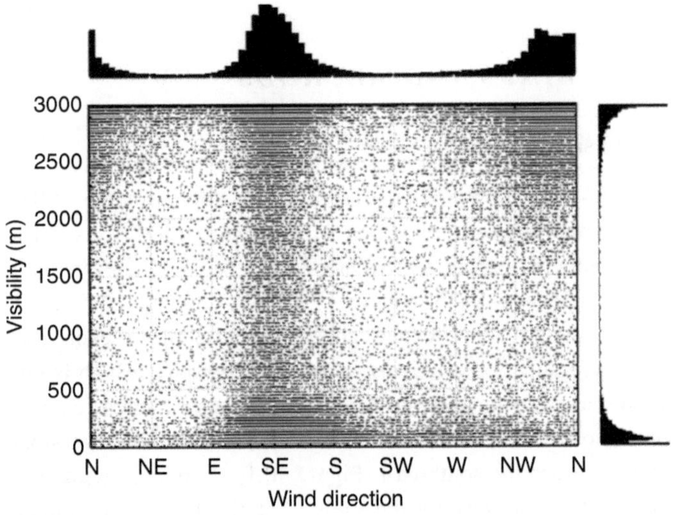

Figure 40.3. Patterns of fog occurrence and wind regime at the Chi-Lan Mountain site in 2003. The dots in the large diagram represent 5-min records of visibility and wind direction. The upper and right-hand diagrams show the frequency distributions of wind direction and visibility, respectively.

respectively. The vertical distribution of leaf weights of the 11 sample trees was also recorded and used to develop allometric equations to estimate the leaf weight of every vertical meter of canopy. Thus, the vertical distribution of leaf biomass of the stand could be calculated and used for the estimation of fog deposition rates at different levels within the canopy (see next section).

Determining in situ fog deposition

In situ exposure of *C. obtusa* leaves was used to determine the amount of water gained by the leaves via fog deposition and absorption. Before exposure, healthy twigs were cut and air-dried for 24 hours to remove surface wetness. No visually perceptible wilting of the twigs occurred during this process due to the high humidity of the site. The twigs were then hung from the tower at three heights, representing the top, middle, and lower layers of the canopy. Beneath the twigs plastic plates were placed to catch any fog water dripping from the saturated leaves. During the exposure, which lasted for at least 60 min, changes in the weights of twigs and plates were recorded every 10 min. After the experiment the leaves were brought to the laboratory and their dry weights were measured after oven-drying at 60 °C until constant weight (usually within 48 hours). The fog capturing efficiency (*FCE*) was then calculated as the average increase in weight during the exposure per unit time and unit dry weight of leaves. In total, 32 exposure experiments were conducted from December 2002 to October 2003 under various weather conditions (cf. Tobón *et al.*, this volume #26).

Upscaling fog deposition to the stand scale

The average depth of the canopy at the tower site was 6.2 m. Three layers were distinguished, viz.: (i) the top layer L_1 from tree top (10.3 m) to the intersecting line of the adjacent trees (8.5 m); (ii) the middle layer L_2 from 8.5 m to 6.3 m, and (iii) the lower layer L_3 from 6.3 m to the lowest branch (4.1 m). The L_2 and L_3 layers were divided such that the two layers had the same thickness.

The fog deposition rate at the stand scale for each exposure experiment was calculated as:

$$F = FCE_1 * B_1 + FCE_2 * B_2 + FCE_3 * B_3 \qquad (40.4)$$

where F (kg H_2O ha^{-1} 5 min^{-1}) is the fog deposition rate at the stand scale, FCE_1 to FCE_3 (kg H_2O kg dry leaf weight^{-1} 5 min^{-1}) the respective fog capturing efficiencies of layers L_1 to L_3, and B_1 to B_3 (kg ha^{-1}) the oven-dry leaf biomass of the respective layers (estimated at 5700, 6900, and 4100 kg ha^{-1}, respectively). The *C. obtusa* leaves are scale-like and densely imbricate. Since the mass and surface area of the leaves exhibited high linear correlation (leaf area [m^2] = 7.76*10^{-3} * leaf dry weight [g], $r^2 = 0.88$, $n = 18$), the *FCEs* were calculated on a mass basis for convenience.

Nutrient fluxes

As part of the long-term monitoring program at the CLM site, bulk precipitation, throughfall, stemflow, and fog water were collected biweekly and their chemical composition was analyzed. A simplified version of the Caltech Active Strand Cloudwater Collector (Demoz *et al.*, 1996) was installed 10 m above the canopy on the instrument tower. The collecting unit contained two rows of 75 Teflon strands each, which were 17 cm long and 0.4 mm in diameter, spaced 1.0 mm apart. The collector was run automatically at a 5-min-on/5-min-off rhythm whenever visibility was less than 500 m. The inlet of the collector was protected against rain and faced the prevailing daytime wind direction. Bulk precipitation was collected in a clearing 500 m away from the stand using a polyethylene (PE) funnel (20-cm diameter) and a 20-l PE collecting bottle. Three replicates were installed at 3.5 m height, at the same level as two adjacent rain gages. Throughfall was collected using 12 funnel-bottle systems placed systematically within the stand. Stemflow was collected using polyurethane collars attached to five *C. obtusa* stems and the water was conducted to 150-l containers. All of the water samples were analyzed for pH (WTW pH 340i, Germany) and electric conductivity (WTW Con. 340i, Germany) in the field and brought to the laboratory for determining concentrations of sodium, potassium, magnesium, calcium, manganese, iron, aluminum, sulfate (ICP-OES, Jobin-Yvon Horiba Group, JY2000, USA), chloride (Zall *et al.*, 1956), nitrate, ammonium, and total nitrogen (Flow Injection Analyzer, Lachat QuikChem 8000 series, USA).

RESULTS

Stand-level fog deposition during individual events

After upscaling of the fog capturing efficiencies to the stand level, corresponding fog deposition rates during the 32 exposure experiments ranged from 0 to 238 kg H_2O ha^{-1} 5 min^{-1}. Figure 40.4 shows the results for 22 exposure experiments that were conducted under foggy conditions (visibility in the range 21–791 m). The highest deposition rate (238 kg H_2O ha^{-1} 5 min^{-1}) was observed on 31 May 2003 when average visibility was only 21 m. During this event, more than half of the total fog water flux was contributed by the top layer of the canopy (53%), followed by the middle (32%), and lower layers (15%).

ANNUAL AND SEASONAL FOG DEPOSITION

Fog deposition rates derived from the exposure experiments were related to visibility data (Figure 40.4). As such, annual fog deposition for the site might be estimated from visibility. However, before doing this, the relationship between fog deposition rate and key parameters of the widely used process-based fog deposition model developed by Lovett (1984) was analyzed. The Lovett model calculates the one-dimensional movement of fog droplets across each 1-m layer of canopy using the equation:

$$F = \Delta C/R \qquad (40.5)$$

where ΔC is the gradient of fog droplet concentration, and R the resistance of the air against deposition. The fog droplet spectrum

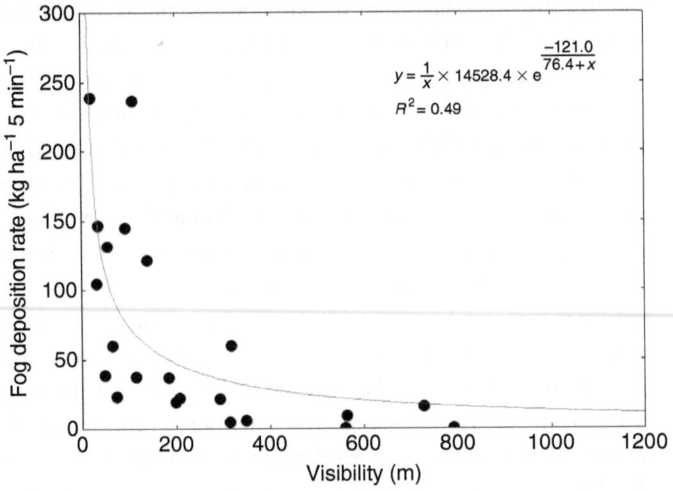

Figure 40.4. Measured (dots) and predicted (line) fog deposition rates at the Chi-Lan Mountain site. Measured values are results of the *in situ* exposure experiments, whereas predicted values are based on the statistical model of Eq. (41.7).

can be measured using a commercial droplet monitoring system (e.g. the FM-100 manufactured by Droplet Measurement Technologies, USA; Beiderwieden *et al.*, 2008; cf. Frumau *et al.*, this volume; Schmid *et al.*, this volume). As a more affordable substitute, Klemm *et al.* (2005) showed that appropriate parameterizations of fog liquid water content and droplet size distribution could also be obtained from visibility data, which, together with wind speed, are the most important meteorological factors in the Lovett model. The Klemm *et al.* (2005) parameterizations were used together with measured stand characteristics to calibrate the model for the CLM study site. Fog deposition rates during visibilities ranging from 100 to 1000 m were calculated for different wind speeds (0.5, 1.18, 1.5, 2.0, and 3.0 m s^{-1}). The vegetation parameters in the model were kept constant.

The obtained relations between deposition rates and visibility for three different wind speeds are shown in Figure 40.5a–c. The data points were fitted using the following type of equation:

$$F = a * e^{(b/(V+c))} \Big/ V \qquad (40.6)$$

where V is the visibility (m), and a, b, and c the respective regression constants. Under constant wind speeds, Eq. (40.6) gave very good predictions of fog deposition (Figure 40.5a–c),

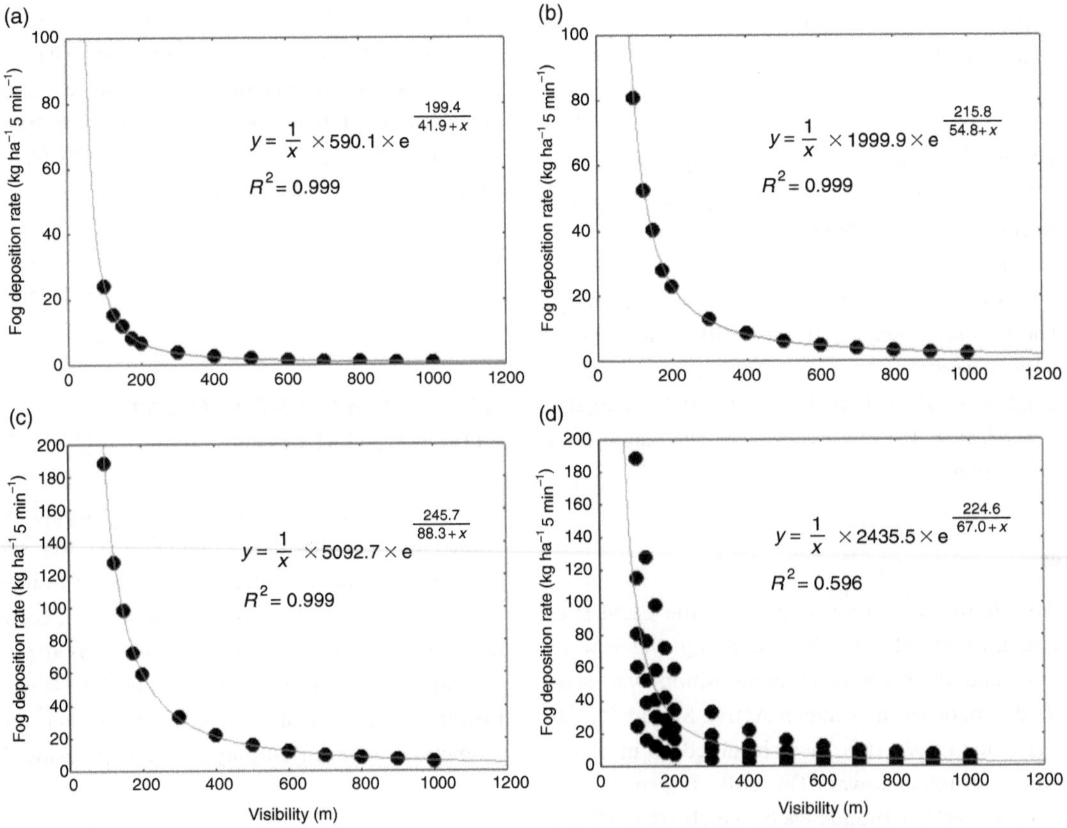

Figure 40.5. Modeled fog deposition rates (after Lovett, 1984) vs. visibility at the Chi-Lan Mountain site. Model constants varied with wind speed: (a) 0.5 m s^{-1}, (b) 1.5 m s^{-1}, and (c) 3.0 m s^{-1}. (d) Simulations for wind speeds of 0.5, 1.18, 1.5, 2.0, and 3.0 m s^{-1}.

Table 40.1 *Total fluxes of water and nutrient in bulk precipitation* (BP), *fog deposition* (FG), *throughfall* (TF), *and stemflow* (SF) *at the Chi-Lan Mountain site between 27 February 2003 and 4 March 2004; relative contributions by fog deposition also indicated*

	BP	*FG*	*TF*	*SF*	*BP + FG*	*FG/(BP + FG)* (%)
Water (mm $year^{-1}$)	2963	303	2606	122	3266	9
H (kg ha^{-1} $year^{-1}$)	0.17	0.26	0.08	0.07	0.43	60
Na (kg ha^{-1} $year^{-1}$)	9.7	1.9	8.5	0.5	11.5	16
K (kg ha^{-1} $year^{-1}$)	6.7	1.2	11.7	0.7	7.9	15
Mg (kg ha^{-1} $year^{-1}$)	3.7	0.4	3.6	0.1	4.1	11
Ca (kg ha^{-1} $year^{-1}$)	6.8	1.1	7.2	0.4	7.9	14
Mn (kg ha^{-1} $year^{-1}$)	0.2	0.1	0.7	0.0	0.3	23
Fe (kg ha^{-1} $year^{-1}$)	1.0	0.3	0.8	0.0	1.3	21
Al (kg ha^{-1} $year^{-1}$)	1.2	0.1	1.0	0.0	1.4	10
SO_4-S (kg ha^{-1} $year^{-1}$)	8.5	5.0	9.1	0.5	13.6	37
Cl (kg ha^{-1} $year^{-1}$)	11.7	5.1	13.6	1.0	16.8	31
NO_3-N (kg ha^{-1} $year^{-1}$)	2.8	3.0	1.6	0.0	5.8	51
NH_4-N (kg ha^{-1} $year^{-1}$)	2.6	3.4	1.9	0.1	6.0	57
DON (kg ha^{-1} $year^{-1}$)	0.8	0.5	1.7	0.1	1.4	39

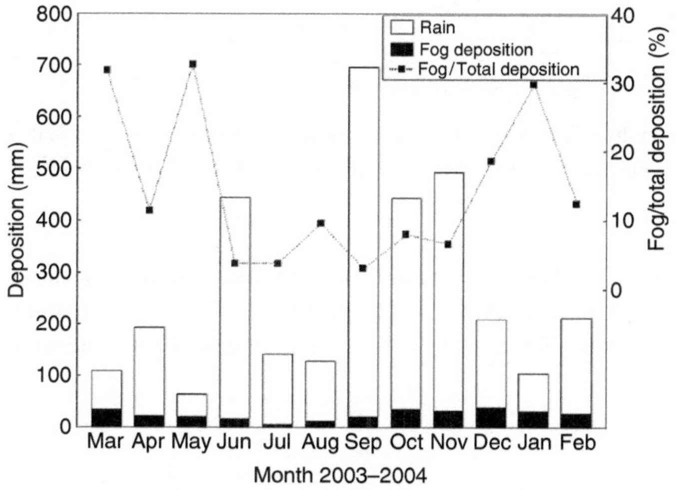

Figure 40.6. Seasonal variation of fog and rainfall inputs at the Chi-Lan Mountain site between March 2003 and February 2004. Solid squares represent the percentages of fog to total input.

whereas under conditions of variable wind speed the predictions were reasonable (Figure 40.5d).

Therefore, the statistical model of Eq. (40.6) was adopted and fitted to the fog deposition rates (*F*) and associated visibility values (*V*) of the exposure experiments, resulting in the following equation for the CLM site (Figure 40.4):

$$F = 14528.4 * e^{(-121.0/(V+76.4))}/V \quad (40.7)$$

where *F* is in kg H_2O ha^{-1} 5 min^{-1} and *V* in m.

Annual fog deposition was estimated using Eq. (40.7) from visibility data recorded between March 2003 and February 2004, using only visibility values of less than 1000 m to denote foggy conditions. A total of 297 mm of fog water was predicted to be captured by the canopy during this period, accounting for 9.2% of the total water input of 3237 mm (rain + fog). The contribution of fog varied seasonally: in winter and spring, when amounts of rainfall are low, fog may contribute up to 30% of the total water input (Figure 40.6). Conversely, in the wet summer months the water input is dominated by rain and the contribution of fog is even less than 4% in June, July, and September 2003 (Figure 40.6).

ATMOSPHERIC NUTRIENT INPUTS

Fluxes of water and nutrients in bulk precipitation, fog, throughfall, and stemflow were calculated for the period from 27 February 2003 to 4 March 2004. Table 40.1 lists the respective fluxes as well as the percentages of water and nutrients contributed by fog deposition to total atmospheric inputs.

Due to the higher nutrient concentrations in fog water compared to rain, the fog-to-total deposition ratios for all nutrients exceeded the ratio for water volume (9%). Fluxes of hydrogen (H^+), nitrate-N, and ammonium-N in fog were even higher than those in bulk precipitation (Table 40.1).

The biweekly mass balances of water and nutrients in the canopy are analyzed further in Figure 40.7. Total atmospheric water deposition (bulk precipitation + fog) exceeded net

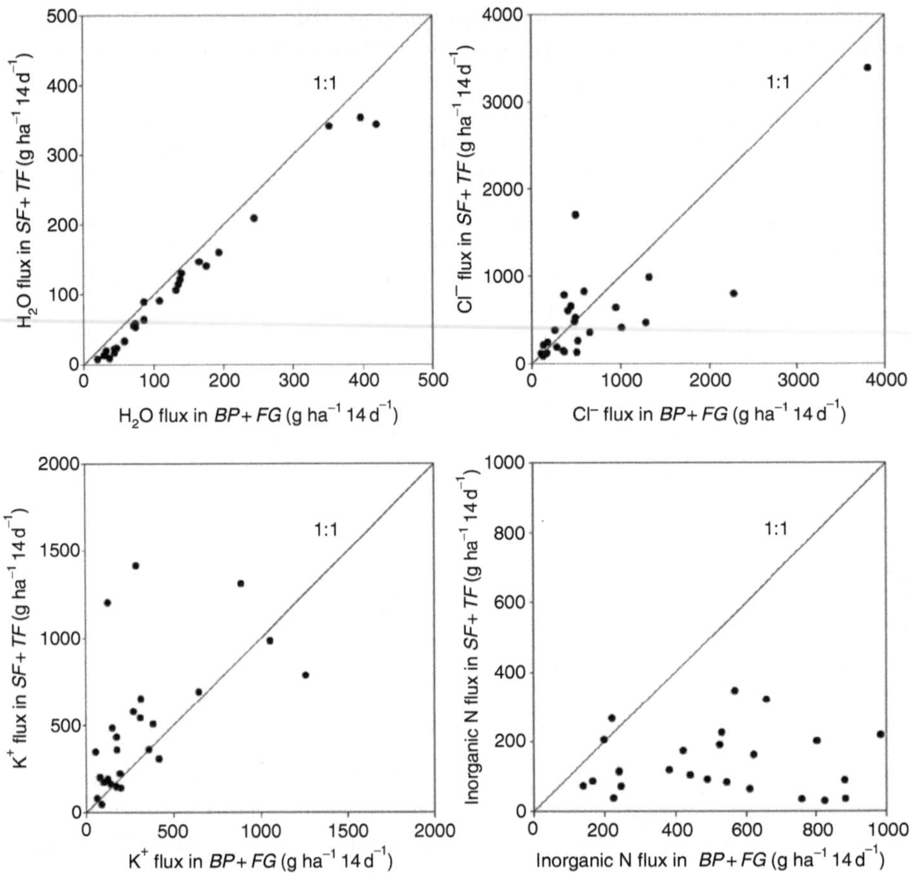

Figure 40.7. Biweekly canopy mass balances of water and selected nutrients at the Chi-Lan Mountain site between 27 February 2003 and 4 March 2004. Atmospheric inputs (bulk precipitation + fog, *BP* + *FG*) are plotted against fluxes in net precipitation (throughfall + stemflow, *SF* + *TF*).

precipitation (throughfall + stemflow) for almost all of the 25 sampled periods (Figure 40.7a). The only exception was the period 11–24 July 2003 when total deposition (86 mm) was slightly lower than net precipitation (89 mm).

For physiologically "neutral" ions like chloride, atmospheric inputs and net outputs from the canopy clustered around the 1 : 1 line (Figure 40.7b), although the annual deposition was about 3 kg ha^{-1} year^{-1} higher than the chloride flux in net precipitation (Table 40.1). Conversely, the biweekly mass balances showed a clear pattern of leaching for potassium (Figure 40.7c) and of absorption by the canopy for inorganic nitrogen (NO_3-N and NH_4-N; Figure 40.7d). Whilst potassium leaching has been widely reported for a variety of forest types (Langusch *et al.*, 2003), absorption and retention of nitrogen during the passage of rainfall and fog through the canopy seems typical of cloud forests having high bryophyte biomass (Asbury *et al.*, 1994; Clark *et al.*, 1998; Hafkenscheid, 2000).

Fog deposition is thus seen to contribute significantly to total atmospheric nutrient inputs at the CLM site. To evaluate the importance of these contributions to the overall nutritional status of the *C. obtusa* forest, additional information on nutrient stocks in soil and biomass, as well as on internal nutrient cycling rates, is needed. Some of these issues are currently being studied at the site.

DISCUSSION AND OUTLOOK

The *in situ* exposure experiment is only one of the methods employed at the CLM site to quantify fog deposition rates in the *C. obtusa* forest but the present results support the general belief that fog is an important environmental factor in this montane ecosystem. In addition, this "direct" measurement of fog deposition may serve as a validation of the results obtained with other, more indirect methods. For example, an annual fog deposition of 313 mm was obtained by applying the Lovett (1984) deposition model for the same time period (S. C. Chang, unpublished data) which agrees closely with the 297 mm derived with the exposure experiments.

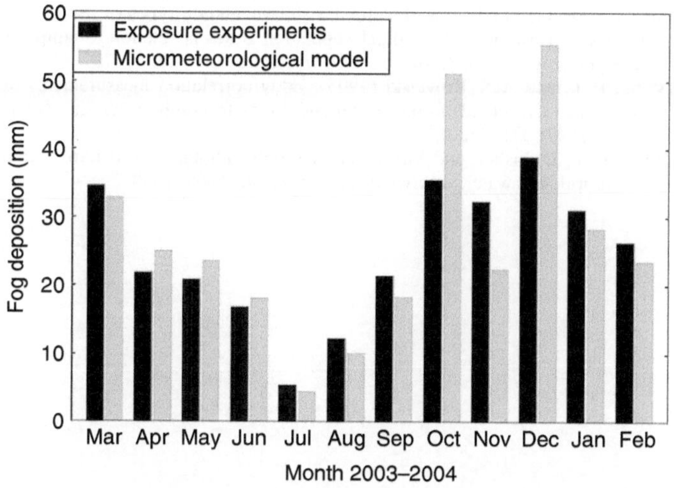

Figure 40.8. Comparison of methods for estimating fog deposition at the Chi-Lan Mountain site. The micrometeorological (Lovett) model predicted an annual deposition of 313 mm vs. 297 mm according to the *in situ* exposure experiment.

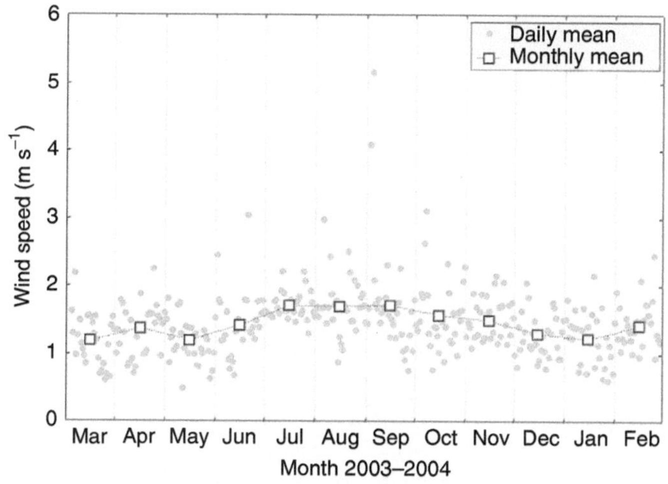

Figure 40.9. Seasonal variation of wind speeds at the Chi-Lan Mountain site from March 2003 to February 2004. Both daily and monthly means were calculated from 5-min records.

The associated monthly fog deposition totals were compared with those of the exposure experiment, which revealed two distinctive patterns (Figure 40.8): from January to September the two methods generated about the same amount of deposition, whereas in October and December 2003 differences between the two methods were large. The scaling-up of fog deposition rates obtained during a limited number of exposure episodes to much longer periods may well introduce substantial error, whereas the effect of wind speed was not taken into account in the statistical model (cf. Figure 40.4). However, the apparent discrepancy in fog deposition rates for these two months is not necessarily a result of the exclusion of wind speed from the statistical model because average wind speeds were similar to those observed in the preceding and subsequent months (Figure 40.9). Furthermore, low fog deposition rates were calculated both by the statistical model and the Lovett model for the July to September period that is characterized by slight winds and short fog duration (cf. Figure 40.2).

Work is ongoing to also measure fog deposition rates using the eddy covariance method (Klemm *et al.*, 2006; Beiderwieden *et al.*, 2008). The comparison of different methods at the same site will help to better understand the fog deposition and capturing process in general, and to further clarify the role of fog in the sub-tropical montane cloud forest of Taiwan.

ACKNOWLEDGEMENTS

We thank the Forest Conservation and Management Administration, Veterans Affairs Commission for permission of using the study site. The project is funded by the National Science Council, Taiwan (grant no. NSC-92-2313-B-259-002-) and partly by the Deutscher Akademischer Austauschdienst (DAAD), Germany.

REFERENCES

Asbury, C. E., W. H. McDowell, R. Trinidad-Pizarro, and S. Berrios (1994). Solute deposition from cloud water to the canopy of a Puerto Rican montane forest. *Atmospheric Environment* **28**: 1773–1780.

Beiderwieden, E., V. Wolff, Y. J. Hsia, and O. Klemm (2008). It goes both ways: measurements of simultaneous evapotranspiration and fog droplet deposition at a montane cloud forest. *Hydrological Processes* **22**: 4181–4189.

Chang, S. C., I. L. Lai, and J. T. Wu (2002). Estimation of fog deposition on epiphytic bryophytes in a subtropical montane forest ecosystem in northeastern Taiwan. *Atmospheric Research* **64**: 159–167.

Cheng, J. D., L. L. Lin, and H. S. Lu (2002). Influences of forests on water flows from headwater watersheds in Taiwan. *Forest Ecology and Management* **165**: 11–28.

Clark, K. L., N. M. Nadkarni, D. Schaefer, and H. L. Gholz (1998). Cloud water and precipitation chemistry in a tropical montane forest, Monteverde, Costa Rica. *Atmospheric Environment* **32**: 595–1603.

Demoz, B. B., J. L. Collett, and B. C. Daube (1996). On the Caltech Active Strand Cloudwater Collectors. *Atmospheric Research* **41**: 47–62.

Hafkenscheid, R. L. L. J. (2000). *Hydrology and biogeochemistry of tropical montane rain forests of contrasting stature in the Blue Mountains, Jamaica*. Ph.D. thesis, VU University Amsterdam, Amsterdam, the Netherlands. Also available at http://dare.ubvu.vu.nl/bitstream/1871/12734/1/tekst.pdf.

Klemm, O., T. Wrzesinsky, and C. Scheer (2005). Fog water flux at a canopy top: direct measurement versus one-dimensional model. *Atmospheric Environment* **39**: 5375–5386.

Klemm, O., S. C. Chang, and Y. J. Hsia (2006). Energy fluxes at a subtropical mountain cloud forest. *Forest Ecology and Management* **224**: 5–10.

Langusch, J. J., W. Borken, M. Armbruster, N. B. Dise, and E. Matzner (2003). Canopy leaching of cations in Central European forest ecosystems: a regional assessment. *Journal of Plant Nutrition and Soil Science* **166**: 168–174.

Lin, N. H., and C. M. Peng (1998). Chemistry of mountain clouds observed in the northern Taiwan. In *Proceedings of the 1st International Conference on Fog and Fog Collection*, eds. R. S. Schemenauer and H. A. Bridgman, pp. 117–120. Ottawa, Canada: IDRC.

Lovett, G. M. (1984). Rates and mechanisms of cloud water deposition to a subalpine balsam fir forest. *Atmospheric Environment* **18**: 361–371.

Lovett, G. M. (1994). Atmospheric deposition of nutrients and pollutants in North America: an ecological perspective. *Ecological Applications* **4**: 629–650.

Su, H. J. (1984). Studies on the climate and vegetation types of the natural forests in Taiwan. II. Altitudinal vegetation zones in relation to temperature gradient. *Quarterly Journal of Chinese Forestry* **17**: 57–73.

Vong, R. J., and A. S. Kowalski (1995). Eddy correlation measurements of size-dependent cloud droplet turbulent fluxes to complex terrain. *Tellus Series B* **47**: 331–352.

Zall, D. M., D. Fisher, and M. Q. Garner (1956). Photometric determination of chlorides in water. *Analytical Chemistry* **28**: 1665–1668.

41 Fog and rain water chemistry in the seasonal tropical rain forest of Xishuangbanna, south-west China

W. J. Liu, H. M. Li, Y. P. Zhang

Xishuangbanna Tropical Botanical Garden, Chinese Academy of Sciences, Kunming, Yunnan, P. R. China

C. M. Wang

Southwest Forestry College, Kunming, Yunnan, P. R. China

F. R. Meng

University of New Brunswick, Fredericton, Canada

ABSTRACT

Fog water, fog drip, and rainfall chemistry were examined at a seasonal tropical rain forest site in Xishuangbanna, south-west China between November 2001 and October 2002. During this period, radiation fog occurred on 204 days, with a total duration of 1949 hours of which 1618 hours (37% of the total time) occurred during the dry season (November to April). Mean pH values of fog water, fog drip, and rain were 6.78, 7.30, and 6.13, respectively. The ion with the highest concentration in both fog and rain water was bicarbonate (HCO_3^-), followed by calcium, magnesium, and ammonium. Concentrations of nitrate, HCO_3^-, ammonium, calcium, and potassium in fog water collected in the latter half of the dry season were significantly higher ($p < 0.05$) than earlier in the dry season. Ionic concentrations in fog drip were higher than those in fog water, except for ammonium and hydrogen. This is attributed to the washing-off of dust and ash-derived nutrients deposited on the leaves and by the leaching of alkaline ions from the leaves. Dry deposition of ash and dust is most probably related to biomass burning and road construction activity.

INTRODUCTION

Xishuangbanna Forest Reserve (XFR) is situated on the northern edge of the tropical zone of South-East Asia and is contiguous to Laos to the south and Myanmar to the south-west. This area constitutes an intermediate biogeographic zone between continental mainland Asia and the South-East Asian Peninsula, and thus has floristic elements from both the south and the north (Cao *et al.*, 1996).

Being under the direct impact of the East Asian monsoon, Xishuangbanna is dominated by warm and humid air masses from the Indian Ocean in summer and by continental air masses of sub-tropical origin in winter. This results in a rainy season between May and October, and a dry season between November and April. Furthermore, the Hengdwan Mountains to the north of the region act as a major barrier keeping out cold air from the north in winter. Mean annual rainfall (1487 mm) is considerably lower than observed for most rain forests in other parts of the world (Richards, 1996). Due to its unique geographic location and climatic features, the area supports a tropical rain forest with a small proportion of deciduous tree species (Cao *et al.*, 1996).

According to the classification system of vegetation in China, a large proportion of the forest in this region is seasonal tropical rain forest, found primarily in wet valleys, lowlands and on low hills (less than 1000 m.a.s.l.) where heavy radiation fogs occur frequently. Radiation fogs are typically formed during the night when the cooling effect of outgoing long-wave radiation reduces the air temperature to or below its dew point under conditions of very low wind speeds (cf. Gradstein *et al.*, this volume).

Deposition of fog water onto vegetation represents an important hydrological and chemical input in some tropical montane and coastal ecosystems (cf. Asbury *et al.*, 1994; Bruijnzeel and Proctor, 1995; Clark *et al.*, 1998; Hafkenscheid, 2000; Chang *et al.*, this volume; Rollenbeck *et al.*, this volume). However,

Tropical Montane Cloud Forests: Science for Conservation and Management, eds. L. A. Bruijnzeel, F. N. Scatena, and L. S. Hamilton. Published by Cambridge University Press.

comparatively little attention has been paid to the fog-inundated seasonal rain forests of Xishuangbanna, and no general picture about the importance of radiation fog as a contributor of nutrients to this type of forest has been developed. Therefore, rainfall, fog water, and fog drip were collected from November 2001 to October 2002.

STUDY AREA

The present study was conducted at a seasonal tropical rain forest site (21° 55′ 39″ N, 101° 15′ 55″ E, 750 m.a.s.l.) within the XFR. The site is located about 800 km from the Bay of Bengal toward the south-west and about 600 km from the Bay of Beibu toward the east. The forest is surrounded by agricultural areas with low residential housing densities. A national road that was being widened during the sampling period lies about 1.5 km toward the south-west.

Average canopy height of this forest is about 35 m, and the canopy trees usually develop strong buttresses (for example, *Terminalia myriocarpa*, *Pometia tomentosa*). Leaf area index is on average 6.34. The tree stratum is subdivided into three layers represented by different species. More than 70% of the trees occur beneath the main canopy in the layer below 16 m. Epiphytes comprise many species of algae, lichens, mosses, and ferns. Woody climbers such as *Byttneria integrifolia* and *Gnetum montanum* are very common (Cao *et al.*, 1996).

Long-term climate records indicate a mean annual air temperature of 21.7 °C. June is the hottest month with a maximum monthly temperature of 25.7 °C. The lowest monthly minimum of 15.9 °C occurs in January. Mean annual rainfall is 1487 mm, of which 1294 mm (87%) occurs in the rainy season vs. 193 mm (13%) in the dry season. Radiation fog occurs almost every day from November to April and is heaviest from midnight (23:00–02:00 h) until mid-morning (09:00–11:00 h) (Liu *et al.*, 2004).

METHODS

Sample collection

Twelve plastic funnel collectors (80 cm in diameter), each connected to a 1-l plastic bottle, were placed at 0.7 m above the forest floor in a fixed but random pattern to collect fog dripping from the canopy. Litterfall was excluded from the bottle by a nylon mesh (1-mm mesh). Each 503 ml of water collected was equivalent to 1 mm of crown drip. The collectors were read in the morning after fog drip had stopped, and the funnels were leveled and cleaned of any litter present. After each fog drip sampling for chemical analysis, the funnels and bottles were replaced with clean ones. To reduce their number, samples were volume-weighted for each collection event. Stemflow was not collected because no fog water was seen to move to the forest floor along the stems. Visual observations indicated that only a small part of the bark of the upper stems was soaked occasionally.

Fog samples were collected on an event basis using passive cylindrical string collectors (Schmitt, 1987). The collectors consisted of two 100-mm diameter Teflon rings with a vertical distance of 30 cm in between and an 80-mm diameter hole in each ring. The lower sampling limit (50% collection efficiency) of the collector was at fog droplet radii of 6 and 3 μm at wind speeds higher than 1 and 5 m s^{-1}, respectively, with a shift toward higher collection efficiencies for smaller droplets at increased wind velocity (Schmitt, 1987). Two of these collectors were mounted at 38 m (0.5 m above the canopy) on a 72-m high meteorological tower. All captured fog water drained into a 2-l watertight plastic container via plastic tubing. Collectors were covered by a plastic bag when not in use and cleaned with distilled water before each sample collection period.

Rain water samples were collected with a stainless steel funnel (collecting area of 314 cm^2) connected to a 2-l plastic bottle. The funnel was mounted on top of the tower to avoid interference from trees and the tower itself. The duration of each sampling period was 24 hours. The funnel was washed twice daily in the morning and evening to avoid contamination by dry deposition. A tipping-bucket rain gage was mounted on top of the tower to determine rainfall intensity and event duration at hourly intervals in combination with an automated data-logger throughout 2001–2002 (Liu *et al.*, 2004).

A visual record of the frequency and duration of fog throughout the day was maintained. A fog day was defined as a day when visibility was 1000 m or less for more than 15 min. Between November 2001 and October 2002, 17 samples were collected of fog and fog drip each, vs. nine samples of rain water.

Chemical analysis

Samples of fog water, fog drip, and rain water were collected in clean 250-ml polyethylene bottles and transported to the Xishuangbanna Water Extension Laboratory. Measurements of pH were made immediately with a digital pH-meter. Next, the samples were filtered (0.45 μm) and refrigerated at 4 °C until chemical analysis, approximately 7 days later, for major ionic components. Concentrations of sodium (Na^+), potassium (K^+), calcium (Ca^{2+}), and magnesium (Mg^{2+}) were analyzed using atomic absorption spectrometry, and chloride (Cl^-), sulfate (SO_4^{2-}), and nitrate (NO_3^-) using ion chromatography. Bicarbonate (HCO_3^-) was also measured by ion chromatography, using distilled, deionized water as the eluent. Ammonium (NH_4^+) was determined using a spectrophotometer (at 655 nm) after chemical reactions (for methodological details see Xie and Wang, 1998).

Table 41.1 *Volume-weighted mean ion concentrations (μmol l^{-1}) with minimum and maximum values, along with pH values in fog, fog drip, and rain water in the seasonal tropical rain forest at Xishuangbanna, SW China during 2001–2002*

	Fog water (n = 17)			Fog drip (n = 17)			Rain water (n = 9)		
Ion	Mean	Min.	Max.	Mean	Min.	Max.	Mean	Min.	Max.
pH	6.78	5.71	7.92	7.30	6.14	8.18	6.13	5.69	7.25
Cl^-	22.7	n.d.	59.4	35.4	5.8	85.1	5.5	n.d.	55.5
SO_4^{2-}	27.2	n.d.	59.3	31.9	9.3	81.6	6.1	5.2	48.1
NO_3^-	30.7	6.7	68.7	79.1	6.6	98.4	8.3	6.3	62.5
HCO_3^-	85.2	22.3	172.7	149.9	26.9	192.7	37.3	16.4	79.6
Na^+	16.8	11.9	57.4	40.7	14.1	112.6	9.3	3.6	45.2
K^+	29.7	8.2	69.8	118.2	16.3	154.3	7.5	2.1	58.2
Ca^{2+}	66.4	9.4	107.1	97.2	12.4	136.4	18.6	4.4	113.7
Mg^{2+}	54.8	7.5	97.6	126.8	8.2	201.8	8.3	3.8	92.5
NH_4^+	52.7	n.d.	74.2	27.2	n.d.	59.0	13.2	n.d.	54.8

n.d.: not detectable.

Data analysis

Volume-weighted mean ion concentrations were calculated for fog, fog drip, and rain events. Since very large variances were common, ranges rather than standard deviations are given. Wilcoxon's two-sample non-parametric tests were used to detect differences in ion concentrations in fog, fog drip, and rain between early dry season and middle to late dry season sampling periods. SPSS 10.0 for Windows was used for all statistical calculations.

RESULTS

Fog frequency

The frequency of fog events between November 2001 and October 2002 was high (Figure 41.1). Fog events were more frequent and lasted longer during the cool–dry season (November–February) and hot–dry season (March–April) than during the rainy season months (May–October). This seasonal variation in fog frequency reflects the general meteorological conditions, with low air temperatures resulting from strong night-time radiative cooling, and region-wide stagnation resulting from an overlying subsidence inversion favoring the formation of radiation fog during the dry season. In total, fog was observed on 204 days. The total duration of fog was 1949 hours (about 22% of the total time), of which 1618 hours (37% of total time) occurred in the dry season (November–April). The cool–dry and hot–dry season months had average daily fog durations of 11.4 and 9.2 hours, respectively, while the rainy season months had 6.0 hours of fog on average. However, on many occasions fog during the rainy season only occurred at the very summit of the study site.

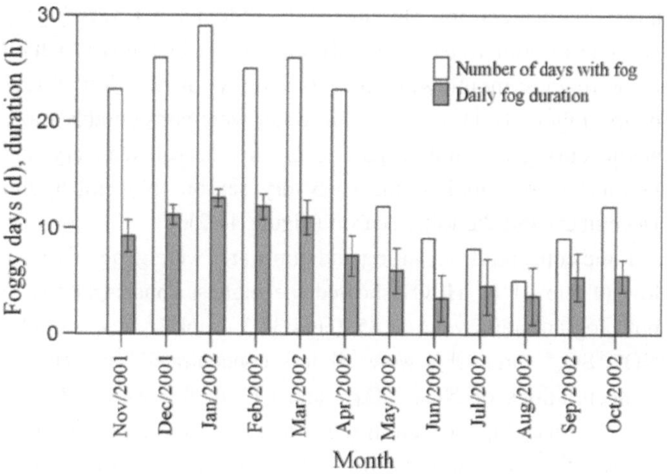

Figure 41.1. Monthly distribution of foggy days and average daily fog duration in the seasonal tropical rain forest at Xishuangbanna, SW China during 2001–2002. Error bars represent standard deviations.

Fog and rain water chemistry

Seventeen fog episodes were sampled for chemical analysis. Eleven episodes were sampled in the cool–dry season, four in the hot–dry season, and two in the rainy season. The fog episodes occurred predominantly from the middle of the night through mid-morning and lasted 8–13 hours. In addition, nine rain water samples were collected, two in the cool–dry season, two in the hot–dry season, and five in the rainy season.

Great variability was found in the volume-weighted mean ion concentrations and pH of fog, fog drip, and rain water (Table 41.1). The mean pH values of fog, fog drip and rain water were 6.78, 7.30, and 6.13, respectively. Therefore, fog drip was alkaline on average whereas both fog and rain water were less acidic than pure water in equilibrium with atmospheric carbon dioxide

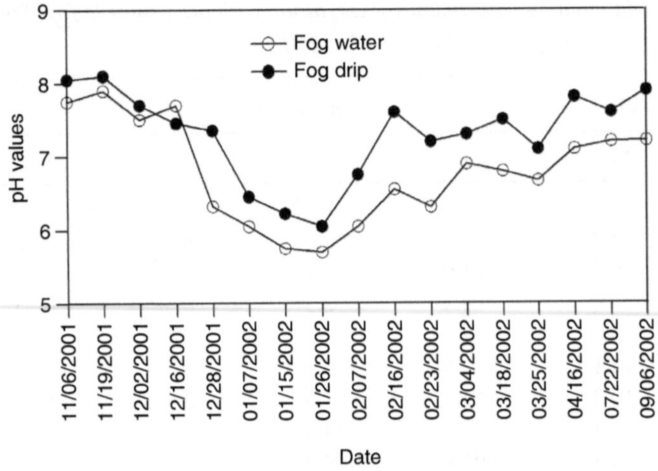

Figure 41.2. Values of pH incident fog and fog drip in the seasonal tropical rain forest at Xishuangbanna, SW China during 2001–2002.

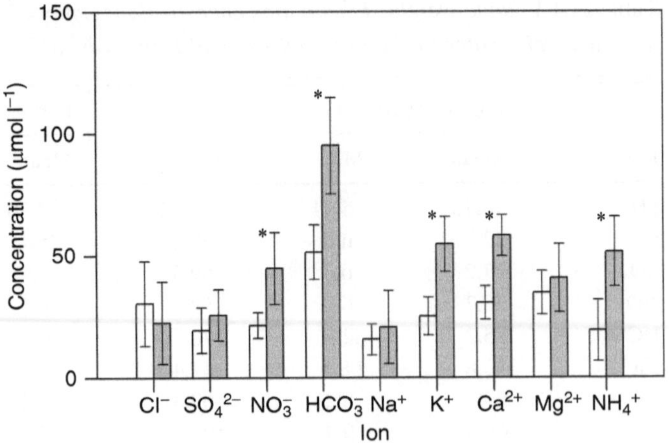

Figure 41.3. Ion concentrations (mean ± 1 SD) in fog water collected in the early (November and December, open bars, $n = 5$) and middle to late dry season (February–April, closed bars, $n = 6$) in the seasonal tropical rain forest at Xishuangbanna, SW China during 2001–2002. * represent $p < 0.05$, Wilcoxon two-sample non-parametric tests.

(pH = 5.65). In general, pH values in fog drip were higher than those in incident fog (Figure 41.2) although pH was highly variable in both water types. pH was lowest in January, which is the foggiest month with an average daily fog duration of 12.8 hours (Figure 41.1). Fog and rain water were considerably more acidic during the middle part of the dry season (January and February) compared to the early dry season (November and December) and the rainy season (Figure 41.2).

Ionic concentrations in fog and rain water were generally very low (Table 41.1). HCO_3^- showed the highest concentrations in both fog and rain water at 85.2 and 37.3 μmol l^{-1}, respectively. NO_3^-, SO_4^{2-} and Cl^-, were all less concentrated than HCO_3^-. Concentrations of NO_3^-, SO_4^{2-} and Cl^- in fog were 2.1, 2.7, and 1.6 times higher than in rain water. The dominant cation in fog was Ca^{2+} followed by Mg^{2+}, NH_4^+, K^+, Na^+, and H^+ with a more or less similar sequence in rain water.

The mean concentrations of all ionic components in fog drip were usually higher than those in incident fog, except for NH_4^+, with K^+ and Mg^{2+} concentrations being increased the most (Table 41.1). Fog samples collected during the middle to late dry season (February to April) had higher concentrations of NO_3^-, HCO_3^-, NH_4^+, Ca^{2+}, and K^+ compared to those collected in the early dry season (November–December) ($p < 0.05$; Figure 41.3). Ion concentrations in fog drip and rain water also showed a similar pattern during these two sampling periods that coincided with visible atmospheric haze, presumably due to agricultural activities and biomass burning.

DISCUSSION

Hydrological importance of fog drip

No comparative data on drip generated by radiation fog in other tropical rain forests seem to be available (Bruijnzeel and Proctor, 1995; Bruijnzeel, 2005; cf. Gradstein *et al.*, this volume; Tanaka *et al.*, this volume). The presently obtained annual and dry season fog drip totals (89.4 and 76.8 mm) constituted only 5% of annual rainfall but as much as 49% of incident precipitation during the dry season (Liu *et al.*, 2004). Fog drip represents an underestimate of the total fog water input into this forest because the direct evaporation of fog water from the wetted canopy was not included (cf. Holder, 2003; Tobón *et al.*, this volume). On some light fog days, all fog water intercepted by the canopy was evaporated again and did not reach the ground (cf. Giambelluca *et al.*, this volume). Fog drip was more frequent and constituted a larger proportion of total input during the dry season, a pattern observed in many other montane cloud forests with a seasonal rainfall regime (e.g. Clark *et al.*, 1998; Holder, 2003). At 89 mm, the annual fog drip total at Xishuangbanna falls within the lower range of values reported for other tropical montane forests (Bruijnzeel, 2005; cf. Bruijnzeel *et al.*, this volume). The low fog drip observed in the study compared may be partially attributed to the low wind speeds prevailing at our site (<2 m s^{-1}; Liu *et al.*, 2004; cf. Tanaka *et al.*, this volume). Isotopic studies in the same forest demonstrated that fog drip did not contribute to groundwater recharge (Liu *et al.*, 2005, 2007).

Chemical composition of water

The study site had generally low ion concentrations in both fog and rain water (Table 41.1), illustrating its remoteness from industry, oceans, large population centers, or volcanoes (cf. Hafkenscheid, 2000; Rollenbeck *et al.*, this volume) as well as its comparatively sheltered location (cf. Boy and Wilcke, 2008). The near-neutral to mildly alkaline pH values in fog and rain at Xishuangbanna indicate neutralization of acids by Ca^{2+} and Mg^{2+} and ammonia derived from soil dust and agricultural activity

(Kumar *et al.*, 2003). NH_4^+ concentrations were higher than NO_3^- and SO_4^{2-} concentrations (Table 42.1), resulting in sufficient acid neutralization and pH values typically much higher than found in many other unpolluted regions (Collett *et al.*, 2002). Similar pH values in fog (5.4–7.5) have been reported elsewhere in rural south-west China (Huang *et al.*, 1992). Sources of NH_3 include agricultural activities (livestock and fertilizing) (Yamasoe *et al.*, 2000) and the large smallholder rubber processing industry in the Xishuangbanna region. In addition, during the dry season, NH_3 may have been emitted through biomass burning, which commonly occurs in the area during this time of year (cf. Crutzen and Andreae, 1990). Potassium is a well-recognized atmospheric tracer for biomass burning, and its concentration was elevated significantly during the burning period (cf. Yamasoe *et al.*, 2000; Fabian *et al.*, 2005; Boy *et al.*, 2008).

Studies on atmospheric aerosols in the regional boundary layer of Xishuangbanna also indicated an excess of soil-originated aerosols which may be blown into the atmosphere during and after cultivation (Huang *et al.*, 1992). Cation-rich dust may also have been derived from construction activities associated with the widening of a dirt road located about 1.5 km upwind from the study site.

The anion with the highest concentration in both fog and rain water was HCO_3^-. Eckardt and Schemenauer (1998) and Schemenauer and Cereceda (1992) also found HCO_3^- to be a very important anion in fog at remote tropical desert locations. In the present case, local soil-derived dust containing fertilizer salts and rock-derived carbonates may well be the prime source of the HCO_3^- over absorption of atmospheric CO_2.

The higher ionic concentrations in incident fog compared to rain water at Xishuangbanna agree well with many other studies, both in temperate (Schemenauer and Cereceda, 1992) and tropical fog-affected areas (Hafkenscheid, 2000; cf. Chang *et al.*, this volume; Rollenbeck *et al.*, this volume). During the middle part of the dry season, the local topography favors strong temperature inversions with low wind speeds and the formation of widespread dense radiation fog (Liu *et al.*, 2004). When fog forms, air pollutants are trapped within the stable inversion layer with its long residence time, and gases and particles become incorporated within the droplets (Fuzzi *et al.*, 1992). This process may also explain why fog was considerably more acidic during the middle of the dry season compared to the early dry season and rainy season when such inversions were less pronounced or lacking entirely (Figure 41.2).

The pH increased significantly from incident fog to fog drip (Table 41.1). This may be caused by the wash-off of soil- and ash-derived nutrients deposited on the canopy during rain- and fog-free periods, and by leaching of alkaline ions (notably K^+) from the leaves. Loss of H^+ ions and pH increase in throughfall is generally believed to be due to exchange of H^+ with other cations on canopy exchange sites (Rao *et al.*, 1995). Leaching may contribute significantly to enhanced fluxes of Ca^{2+} and Mg^{2+} in throughfall, especially where NH_4^+ inputs are substantial, leading to exchange between base cations and NH_4^+ (Beier *et al.*, 1992). Interestingly, NO_3^- was also much higher in fog drip than in incident fog at Xishuangbanna (Table 41.1) whereas in many cloud forests the reverse is the case due to retention of NO_3^- and NH_4^+ by epiphytes and bryophytes (Clark *et al.*, 1998; Hafkenscheid, 2000; Chang *et al.*, this volume).

CONCLUSIONS

This study is the first to describe fog and fog drip chemistry in the seasonal tropical rain forest of Xishuangbanna. High radiation fog occurrence was recorded during 2001–2002, with January having the highest overall fog frequency. Annual fog drip total was 89.4 mm, contributing an estimated 5% of annual rainfall but 49% to dry-season rainfall.

Wide variation was observed in ion concentrations between individual events. Mean pH values of incident fog, fog drip, and rain water were 6.78, 7.30, and 6.13, respectively, reflecting high concentrations of NH_4^+, Ca^{2+}, and Mg^{2+} and low concentrations of SO_4^{2-} and NO_3^- at this remote site. Fog and rain water were considerably more acidic during the middle part of the dry season compared to the early dry season and the rainy season. Fog consistently showed much higher ion concentrations than rain. Both pH and most ion concentrations were significantly higher in fog drip than in incident fog, suggesting an effect of dry deposition or canopy leaching. Dry deposition was probably enhanced by frequently occurring biomass burning, and construction works associated with the widening of a nearby road. Whilst fog deposition may not contribute to total hydrological inputs as much as precipitation does during certain times of year, the high frequency and long immersion time of fog during the dry season substantially increase nutrient inputs delivered to the forest.

ACKNOWLEDGEMENTS

This research was supported by the Education Program of "Western Light" of CAS, the NSFC (30770368, 30570308) and NSFY (2006C0057M). The authors are grateful to Professor L. A. Bruijnzeel of the VU University, Amsterdam, the Netherlands, for editorial comments.

REFERENCES

Asbury, C. E., W. H. McDowell, R. Trinidad-Pizarro, and S. Berrios (1994). Solute deposition from cloud water to the canopy of a Puerto Rican montane forest. *Atmospheric Environment* **28**: 1773–1780.

Beier, C., P. Gundersen, and L. Rasmussen (1992). A new method for estimation of dry deposition of particles based on throughfall measurements in a forest edge. *Atmospheric Environment* **26**A: 1553–1559.

Boy, J., and W. Wilcke (2008). Tropical Andean forest derives calcium and magnesium from Saharan dust. *Global Biogeochemical Cycles* **22**, GB1027, doi: 10.1029/2007GB002960.

Boy, J., R. Rollenbeck, C. Valarezo, and W. Wilcke (2008). Amazonian biomass burning-derived acid and nutrient deposition in the north Andean montane forest of Ecuador. *Global Biogeochemical Cycles* **22**, GB4011, doi:10.1029/2007GB003158.

Bruijnzeel, L. A. (2005). Tropical montane cloud forests: a unique hydrological case. In *Forests, Water and People in the Humid Tropics*, eds. M. Bonell and L. A. Bruijnzeel, pp. 462–483. Cambridge, UK: Cambridge University Press.

Bruijnzeel, L. A., and J. Proctor (1995). Hydrology and biogeochemistry of tropical montane cloud forests: what do we really know? In *Tropical Montane Cloud Forests*, eds. L. S. Hamilton, J. O. Juvik, and F. N. Scatena, pp. 25–46. New York: Springer-Verlag.

Cao, M., J. H. Zhang, and Z. L. Feng (1996). Tree species composition of a seasonal rain forest in Xishuangbanna, SW China. *Tropical Ecology* **37**: 183–192.

Clark, K. L., N. M. Nadkarni, D. Schaefer, and H. L. Gholz (1998). Cloud water and precipitation chemistry in a tropical montane forest, Monteverde, Costa Rica. *Atmospheric Environment* **32**: 1595–1603.

Collett, J.Jr., A. Bator, D. E. Sherman, *et al.* (2002). The chemical composition of fogs and intercepted clouds in the United States. *Atmospheric Environment* **64**: 29–40.

Crutzen, P. J., and M. O. Andreae (1990). Biomass burning in the tropics: impact on the atmospheric chemistry and biogeochemical cycles. *Science* **250**: 1669–1678.

Eckardt, E. D., and R. S. Schemenauer (1998). Fog water chemistry in the Namib Desert, Namibia. *Atmospheric Environment* **32**: 2595–2599.

Fabian P., M. Kohlpaintner, and R. Rollenbeck (2005). Biomass burning in the Amazon: fertilizer for the mountaineous rain forest in Ecuador. *Environmental Science and Pollution Research* **12**: 290–296.

Fuzzi, S., M. C. Facchini, G. Orsi, and D. Ferri (1992). Seasonal trend of fog water chemical composition in the Po Valley. *Environmental Pollution* **75**: 75–80.

Hafkenscheid, R. L. L. J. (2000). Hydrology and biogeochemistry of tropical montane rain forests of contrasting stature in the Blue Mountains, Jamaica. Ph.D. thesis, VU University Amsterdam, Amsterdam, the Netherlands. Also available at http://dare.ubvu.vu.nl/bitstream/1871/12734/1/tekst.pdf.

Holder, C. D. (2003). Fog precipitation in the Sierra de las Minas Biosphere Reserve, Guatemala. *Hydrological Processes* **17**: 2001–2010.

Huang, Y. S., H. G. Guo, and F. X. Liu (1992). The chemical composition of fog water in industrial and non-industrial districts. *Acta Geographica Sinica* **47**: 66–73.

Kumar, R., A. Rani, K. M. Kumari, and S. S. Srivastava (2003). Direct measurement of atmospheric dry deposition to natural surfaces in a semi-arid region of north central India. *Journal of Geophysical Research* **108**: 4625.

Liu, W. J., F. R. Meng, Y. P. Zhang, and H. M. Li (2004). Water input from fog drip in the tropical rain forest of Xishuangbanna, SW China. *Journal of Tropical Ecology* **20**: 517–524.

Liu, W. J., Y. P. Zhang, H. M. Li, and H. M. Liu (2005). Fog drip and its relation to groundwater in the tropical seasonal rain forest of Xishuangbanna, Southwest China. *Water Research* **39**: 787–794.

Liu, W. J., W. Y. Liu, P. J. Li, *et al.* (2007). Using stable isotopes to determine sources of fog drip in a tropical seasonal rain forest of Xishuangbanna, SW China. *Agricultural and Forest Meteorology* **143**: 80–91.

Rao, P. S. P., G. A. Momin, P. D. Safai, A. G. Pillai, and L. T. Khemani (1995). Rain water and throughfall chemistry in the Silent Valley forest in South India. *Atmospheric Environment* **29**: 2025–2029.

Richards, P. W. (1996). *The Tropical Rain Forest*, 2nd edn. Cambridge, UK: Cambridge University Press.

Schemenauer, R. S., and P. Cereceda (1992). Monsoon cloud water chemistry on the Arabian Peninsula. *Atmospheric Environment* **26**: 1583–1587.

Schmitt, G. (1987). *Methoden und Ergebnisse der Nebelanalyse*, Report No. 72. Frankfurt/Main, Germany: Institut für Meteorologie und Geophysik der Universität Frankfurt.

Xie, X. Q., and L. J. Wang (1998). *Observation and Analysis of Water Environment Factors*. Beijing: Standard Press of China.

Yamasoe, M. A., P. Artaxo, A. H. Miguel, and A. G. Allen (2000). Chemical composition of aerosol particles from direct emissions of vegetation fires in the Amazon basin: water-soluble species and trace elements. *Atmospheric Environment* **34**: 1641–1653.

42 Spatial heterogeneity of throughfall quantity and quality in tropical montane forests in southern Ecuador

M. Oesker
University of Hohenheim, Stuttgart, Germany

J. Homeier
University of Göttingen, Göttingen, Germany

H. Dalitz
University of Hohenheim, Stuttgart, Germany

L. A. Bruijnzeel
VU University, Amsterdam, the Netherlands

ABSTRACT

Canopy structure and throughfall (*TF*) were determined in three different forest types within the tropical montane rain forest belt in southern Ecuador. Heterogeneity of *TF* amounts and selected nutrient concentrations were compared to heterogeneity of canopy structure and tree species diversity. Canopy structure was characterized using hemispheric images and software calculating radiation beneath the canopy, mean leaf angle, and canopy openness. *TF* was sampled over a 1-year period (November 2001–November 2002), and analyzed for pH, electric conductivity, potassium, calcium, and magnesium. Radiation penetrating through the canopy ranged between 9.7% and 17.2% and gap fractions between 6.1% and 9.5% in the respective forests. At 71%, 85%, and 91% of incident precipitation, *TF* differed significantly between the three forest types, although standard deviations (SD) were high. The highest heterogeneity in *TF* (as represented by SD) was found for the forest type with the greatest heterogeneity in canopy structure, and vice versa. Heterogeneity of element concentrations in *TF* (again represented by their SD) exhibited strong correlations ($r^2 = 0.912$–0.987) with tree species diversity per forest as expressed by the Shannon–Wiener diversity index. Rates of nutrient leaching from seven tree species were determined experimentally. Amounts of elements leached differed between species, and specific patterns were observed per species. These findings suggest that higher tree diversity leads to greater complexity in leaching patterns and to greater heterogeneity in *TF* nutrient composition. Water storage capacities of single leaves (S_L) of 11 tree species were determined experimentally and ranged from <30 ml m^{-2} of leaf surface area to >200 ml m^{-2}, depending on species and leaf angle. Combining the mean forest leaf area index (5.37 m^2 m^{-2}) with the average S_L (70.7 ml m^{-2}) and an overall mean leaf angle of 46° gave a theoretical maximum water storage capacity of 0.38 mm for the foliar component of the canopy.

INTRODUCTION

Throughfall (*TF*) is a major pathway in forest ecosystems for the transfer of nutrients from the canopy to the forest floor. Nutrients can be washed off, or leached out of leaves, by rain or intercepted cloud water (fog) and transported to the soil via crown drip (Tukey, 1970; Parker, 1983; Chang *et al.*, this volume; Liu *et al.*, this volume). Information about quality and quantity of *TF* is often used to compare forests dominated by different species (e.g. Ulrich and Pankrath, 1983; Ahmad-Shah and

Tropical Montane Cloud Forests: Science for Conservation and Management, eds. L. A. Bruijnzeel, F. N. Scatena, and L. S. Hamilton. Published by Cambridge University Press.

Rieley, 1989; Houle *et al.*, 1999), or to describe differences in forest type and associated ecological functioning along broad altitudinal (Veneklaas, 1990; Balestrini *et al.*, 1998), age (Hölscher *et al.*, 2003) or local topographic gradients (Oesker *et al.*, 2008). In tropical montane rain forests (including cloud forests) *TF* is often characterized by high spatial variability (Clark *et al.*, 1998; Hafkenscheid *et al.*, 2002; Holwerda *et al.*, 2006a,b; Zimmermann *et al.*, 2007; Oesker, 2008). Although wind affects *TF* variability in some cases through its effect on rainfall inclination (Herwitz and Slye, 1992; Lin *et al.* 1997; Clark *et al.*, 1998; Holwerda *et al.*, 2006b; cf. Tanaka *et al.*, this volume), the prime influence is generally thought to be exerted by canopy structure and geometry (Lloyd and Marques, 1988; Loescher *et al.*, 2002), including the degree of epiphyte loading (Cavelier *et al.*, 1997; Clark *et al.*, 1998; Hölscher *et al.*, 2004; Fleischbein *et al.*, 2006). Ecological consequences of *TF* heterogeneity for plant diversity (e.g. in terms of seedling establishment and therefore coexistence of species) have been discussed by Dalitz *et al.* (2004) and Oesker (2008), and for micro-scale variation in soil fertility conditions by Wilcke *et al.* (2001), Musila *et al.* (2005), and (Musila 2007).

This chapter investigates the hypothesis that quality and quantity of *TF* differ significantly between forests of contrasting diversity and different canopy structure because of differences in their leaching properties. Quantity and quality of *TF* were determined in three montane rain forest types with different canopy structure between 1950 and 2275 m.a.s.l. in southern Ecuador. In addition, nutrient leaching rates for seven tree species and the water storage capacities of the leaves of 11 different tree species were compared experimentally. The current work was conducted as part of a large interdisciplinary ecological research program in the tropical high mountains of southern Ecuador (Beck and Müller-Hohenstein, 2001; Beck *et al.*, 2008).

MATERIALS AND METHODS

Study site

The study site was situated in the reserve Estación Científica San Francisco (ECSF) (03° 58′ S, 79° 04′ W), close to the Podocarpus National Park in South Ecuador. The ECSF is situated on the eastern slopes of the Andes and its forests can be characterized as tropical montane rain forest (Balslev and Øllgaard, 2002; *bosque siempreverde montano*). Steep slopes and deep ravines characterize the study area (Wilcke *et al.*, 2001; cf. Wilcke *et al.*, this volume). Mean annual precipitation (*P*) (1998–2005) ranges from 2176 mm at 1950 m.a.s.l. (ECSF) via 2193 mm at 2270 m.a.s.l. (Plataforma station) to 4480 mm at 2660 m.a.s.l. (Station TS1) (Bendix *et al.*, 2008). The main rainy season is from April to June, but all months receive more than 100 mm (Emck 2007; Rollenbeck *et al.*, 2007). The greatest variability in rainfall occurs during the weakly developed dry season in November (Rollenbeck *et al.*, 2007). Average temperature at the ECSF is 16.2 °C with an average lapse rate of 0.61 °C 100 m^{-1} (Emck, 2007; Bendix *et al.*, 2008).

More than 280 tree species have been identified in the ECSF research area (Homeier *et al.*, 2008). Most abundant are *Graffenrieda emarginata*, *Purdiaea nutans*, and *Alchornea grandiflora*. The present study was conducted in three of the five local forest types defined by Homeier *et al.* (2002). The three forests differed in terms of their tree species composition and tree density (Homeier, 2004). The first forest type (I, valley) is found in gorges between 1960 and 2070 m.a.s.l. With its main canopy at 25–30 m and some trees reaching up to 35 m, it is the tallest forest in the study area. The average number of stems with diameter at breast height (DBH) $\geq$ 5 cm is 1825 ha^{-1} and the mean basal area 42.4 m^2 ha^{-1}. It is also the most species-rich forest, with Lauraceae, Mimosaceae, and Melastomataceae as the most abundant tree families (see Table 42.1 below). The second forest type (II, lower ridge) is situated on ridges and upper slopes between 1980 and 2090 m.a.s.l. It has a canopy height of 15–20 m and a mean basal area of 35.6 m^2 ha^{-1}. Stem density is 2542 ha^{-1} and the most abundant tree families are the Melastomataceae, Lauraceae, and Euphorbiaceae. The third forest type (III, upper ridge) occurs on ridges between 2140 and 2210 m.a.s.l., has a more open canopy at a height of 10–15 m, and a mean basal area of 27.2 m^2 ha^{-1}. Average stem density is 2250 ha^{-1} and most species belong to the Melastomataceae, Rubiaceae, Clusiaceae, and Euphorbiaceae. Up to 35% of all identified tree individuals consist of *Graffenrieda emarginata* (Melastomataceae), *Alchornea grandiflora* (Euphorbiaceae), *Purdiaea nutans* (Cyrillaceae), and *Podocarpus oleifolius* (Podocarpaceae), and overall tree diversity in forest type III is lowest (Homeier, 2004; see also Table 42.1).

Epiphyte loading (given as bryophyte coverage on stems and branches) of the respective forest types according to Kürschner and Parolly (2004) varied considerably, ranging between 3.8 t ha^{-1} in the lower reaches of forest type II to 13.4 t ha^{-1} in the valley forest (Table 42.1).

Throughfall determination

Totals of *TF* were measured weekly in each forest type in three plots of 20 × 20 m. The plots were chosen for their representativity and equipped with nine *TF* collectors placed in a 5-m grid. Four additional collectors were placed in one plot per forest type (see Figure 42.1 for gage lay-out), bringing the total of *TF* collectors per forest to 31. Incident precipitation (*P*) was measured above (at 2275 m.a.s.l.) and below (at 1950 m.a.s.l.) the study area with five collectors at each site. The catch of the *TF* collectors below 2100 m.a.s.l. was compared to that of the *P* collectors at 1950 m and the *TF* above 2100 m to *P* at 2275 m.

Table 42.1 *Characteristics of forest types I, II, and III*

Parameter				Forest type I	Forest type II	Forest type III
Plot altitude	(range)	(m.a.s.l.)		1960–2070	1980–2090	2140–2210
Canopy height	(range)	(m)		25–30	15–20	10–15
Tree density (tree DBH > 5 cm)	(SD)	(trees ha^{-1})	($n = 3$)	1825 (± 143)	2542 (± 470)	2250 (± 391)
Tree basal area (tree DBH > 5 cm)	(SD)	($m^2\ ha^{-1}$)	($n = 3$)	42.4 (± 6.9)	35.6 (± 7.1)	27.2 (± 2.9)
No. of tree species per 400 m^2	(SD)	(n)	($n = 3$)	40.3 (± 5.6)	33.0 (± 4.3)	26.3 (± 0.9)
Shannon–Wiener index	(SD)	N_1	($n = 3$)	32.4 (± 5.4)	23.1 (± 4.1)	18.2 (± 1.9)
Mean temperature	(min/max)	(C°)	(n=2/n=1)	14.9 (9.9/25.6)	14.9 (9.8/25.6)	14.2 (9.0/25.5)
Biomass of epiphytic bryophytes (after Kürschner and Parolly, 2004)	(range)	(t ha^{-1})	(n = 3–8)	13.41 (5.7–26.4)	3.80 (3.3–5.2)	11.58 (6.5–14.8)
Bryophyte water storage (after Kürschner and Parolly, 2004)	(range)	(mm)	(n = 3–8)	9.15 (4.0–18.9)	2.66 (1.9–2.9)	4.65 (3.3–5.7)
Light below canopy	(SD)	(%)	($n = 31$)	9.7 (± 2.3)	14.2 (± 2.3)	17.2 (± 4.8)
Visible sky	(SD)	(%)	($n = 31$)	6.10 (± 1.29)	7.48 (± 0.94)	9.54 (± 1.94)
Leaf area index (*LAI*)	(SD)	($m^2\ m^{-2}$)	($n = 31$)	5.51 (± 1.42)	5.75 (± 1.43)	4.88 (± 1.13)
Mean leaf angle	(SD)	(°)	($n = 31$)	34.90 (± 17.86)	52.54 (± 19.00)	50.88 (± 17.76)
Throughfall (*TF*)	(SD)	(% rainfall)	($n = 31$)	70.80 (± 22.34)	85.30 (± 25.12)	90.95 (± 36.45)

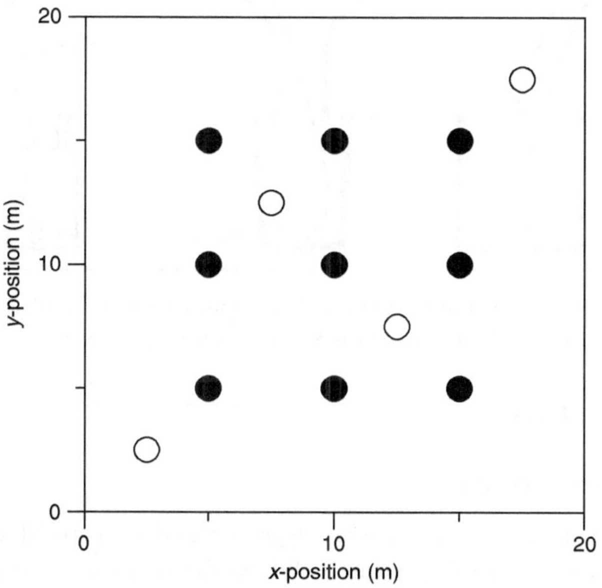

Figure 42.1. Throughfall (*TF*) measurement design: nine collectors placed in a 5-m grid for all 400-m^2 plots (solid circles), plus four additional collectors (open circles) in one plot per forest type.

All collectors were constructed from polyethylene funnels (15 cm diameter) with a vertical 9-cm PVC ring giving a collecting area of 198.5 cm^2, placed at 1.2 m above ground level. The funnels were connected via polyethylene tubing to a PVC bottle (3.785 l capacity). Nylon nets with 0.5-mm mesh were mounted between the funnel and the tubing to avoid contamination by organic debris. All equipment was cleaned with 1% nitric acid and distilled water before the experiment. After each sampling occasion the gages were cleaned with distilled water, and gages affected by algal growth were replaced. Evaporation from the gages was negligible (<0.5% per week, $n = 36$). The catch obtained with this type of collector showed a very high correlation with that of standard rain gages (Hellmann type; Rollenbeck *et al.*, 2007).

Measurements of *P* and *TF* were made between November 2001 and November 2002 on a weekly basis, but during times of intensive rainfall, *P* and *TF* were sampled more frequently. Stemflow (*SF*) was not determined but observations by Wilcke *et al.* (2001) in nearby tall forest at 1910 m.a.s.l. showed *SF* to be negligible at ~1% of *P*. Each sampling event involved determination of *TF* volume, collection of a sample for nutrient analysis, and cleaning of the nets and funnels. Electric conductivity (*EC*) and pH were measured the same day for general characterization of the samples. All samples were subsequently filtered (Schleicher & Schnüll folded filter type 595$^1/_2$) and sent to Germany for analysis of their potassium (K^+), calcium (Ca^{2+}), and magnesium (Mg^{2+}) concentrations (flame photometry and AAS, Perkin–Elmer 5100). These elements were chosen because of their contrasting mobility and electric exchange capacity (cf. Zimmermann *et al.*, 2007; Chang *et al.*, this volume; Liu *et al.*, this volume).

Hemispheric images were taken above each *TF* collector using a 180° fish-eye lens (Nikon F8-E8). Images were analyzed with HemiView Canopy Analysis Software (Version 2.1 SR1, DeltaT Devices Ltd., UK) to compute leaf area index (*LAI*), mean leaf angle, canopy openness, and radiation conditions beneath the canopy. The *LAI* values produced by HemiView required a correction factor (Chen *et al.*, 1991) which was determined at 1.8 for forest type II by Ohlemacher (2001) and was used throughout this study.

Leaf water storage capacity

Water storage capacities of individual leaves (S_L) were determined for 11 tree species: *Alchornea grandiflora* (Euphorbiaceae), *Clethra revoluta* (Clethraceae), *Clusia* sp. (Clusiaceae), *Graffenrieda emarginata* (Melastomataceae), *Heliocarpus americanus* (Tiliaceae), *Isertia laevis* (Rubiaceae), *Piptocoma discolor* (Asteraceae), *Podocarpus oleifolius* (Podocarpaceae), *Purdiaea nutans* (Cyrillaceae), *Tabebuia chrysantha* (Bignoniaceae), and *Vismia tomentosa* (Clusiaceae). A predetermined volume of 0.176 ml of distilled water was sprayed from a distance of 15 cm onto the leaf surface. This was repeated for three different leaf angles (20°, 45°, and 80°), using 20 randomly chosen leaves from 20 different individuals per species. The number of sprayings required for the first droplet to form and leave the leaf was determined. The spraying area for the system was determined at 290 cm^2. Because the spraying area exceeded the leaf area, the relative volume of water reaching the leaf was used for the calculation of S_L in ml m^{-2} of leaf area.

Nutrient leaching experiment

Species-specific nutrient leaching was investigated experimentally on the following seven out of the 11 species used for water storage capacities: *Graffenrieda emarginata* (Melastomataceae), *Heliocarpus americanus* (Tiliaceae), *Isertia laevis* (Rubiaceae), *Piptocoma discolor* (Asteraceae), *Podocarpus oleifolius* (Podocarpaceae), *Purdiaea nutans* (Cyrillaceae), and *Tabebuia chrysantha* (Bignoniaceae). These species were chosen because they represent a wide spectrum of families and leaf types while belonging to the more abundant tree species. Twenty leaves (10 young and 10 mature leaves) per species were selected from different individuals, representing the entire study area and the altitudinal range of each species' occurrence. Nutrient leaching rates were determined using a specially developed semi-automated leaching simulation analysis system (SALSA) (Figure 42.2). A leaf from a freshly cut branch was clamped into the SALSA system and 2000 ml of distilled water was sprayed onto the leaf. Samples were taken of the leachate and the distilled water, and their electric conductivity (*EC*), pH, and nutrient concentrations were determined. The treated leaves were scanned for their surface area directly after the experiment.

Statistical analysis

Standard deviations (SD) were used as a measure of parameter heterogeneity. ANOVA and least significant difference (LSD) test were used to test for significant differences. Statistical analysis was performed using STATISTICA (Version 5.1 M, StatSoft Inc., Tulsa, Oklahoma, USA) for calculating the ANOVA, and XACT (Version 8.05c, SciLab GmbH, Germany) for calculating correlation coefficients.

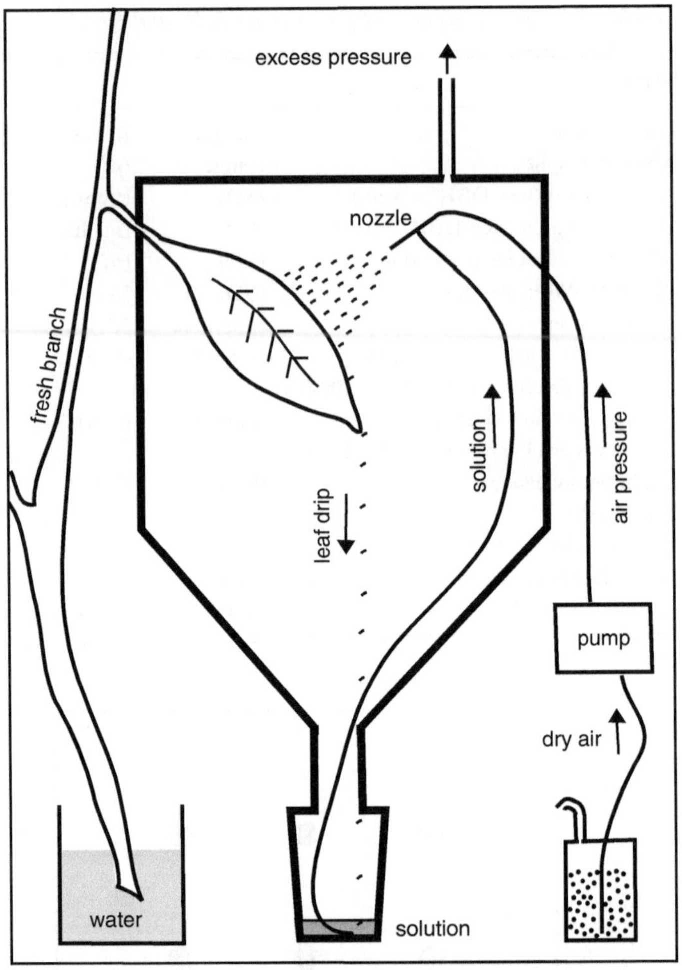

Figure 42.2. Schematic diagram of the semi-automatic leaching simulation analysis system (SALSA) used in the present study.

RESULTS

Forest structure

The three investigated forest types differed in terms of tree species composition (Shannon–Wiener diversity index, N_1) and various structural parameters (Table 42.1). The highest tree diversity was found in forest I, followed by forests II and III. The canopy of the poorest and shortest forest (III) was comparatively open and, as a result, had higher light penetration and the lowest *LAI* (ANOVA, post-hoc LSD test, $p \leq 0.05$). The second forest was intermediate in terms of light penetration and openness but its *LAI* did not differ significantly from that derived for forest I (Table 42.1). However, average leaf angles were significantly different.

Throughfall quantity

Over the 1-year observation period 1948 mm of *P* were collected at 1950 m.a.s.l. vs. 2737 mm at 2275 m.a.s.l., giving an overall mean P value of 2343 mm. The quantity of TF differed significantly between the three forests (ANOVA, post-hoc LSD test, $p \leq 0.05$; Table 42.1). The highest TF/P fraction was found for forest III (91%), followed by forest II (85%) and forest I (71%). TF in forest III also exhibited the highest SD but differences between forests II and I were not that distinct (Table 42.1).

The annual wet-canopy nutrient input/output balance – calculated by subtracting the amounts of nutrients carried in TF from those brought in by bulk rainfall P – is shown for each forest in Figure 42.3. The greatest difference between inputs and outputs for the three elements considered (K, Mg, and Ca) and pH (calculated as H^+) was found for forest I (valley), and the smallest for forest III (upper ridge). The SDs of Mg, K, and Ca concentrations in TF were largest for forest I (valley), suggesting it to be the most heterogeneous, followed by forest II (lower ridge) and forest III (upper ridge). Net retention of Ca and H occurred upon passage through the canopy (negative balance), whereas amounts of K and Mg were increased; K was leached more than Mg in all three forest types (Figure 42.3).

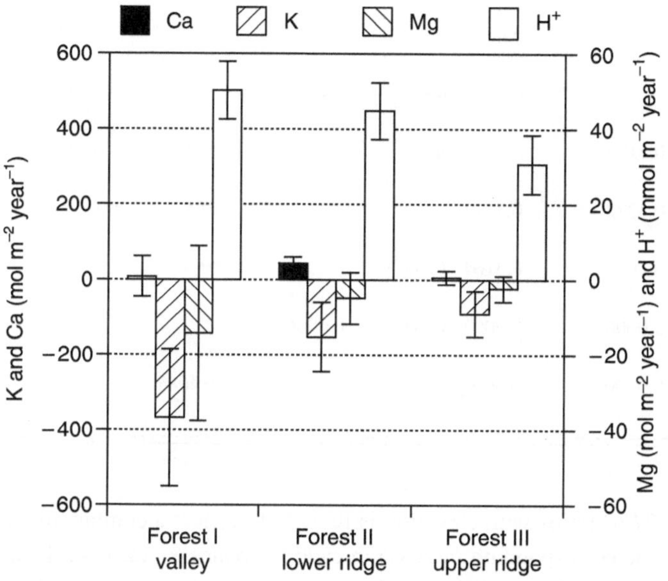

Figure 42.3. Net input–output budgets of nutrient fluxes associated with passage of the canopy in the respective forest types. Fluxes of K, Ca, and Mg given in mol m^{-2} $year^{-1}$ ($n = 31$), those of H^+ in mmol m^{-2} $year^{-1}$ (excluding extreme values; forest I: $n = 30$, forest II: $n = 31$; forest III: $n = 28$). Negative values imply ion retention and positive values net ion release.

Leaf water storage capacity

Experimentally inferred leaf water storage capacity values (S_L) differed between tree species and leaf angles (Figure 42.4, Table 42.2). The highest storage capacities were determined at a leaf angle of 20°, with values ranging from 29.3 ml m^{-2} of leaf area for *Isertia laevis* to 212.7 ml m^{-2} of leaf area for *Alchornea grandiflora*. Seven out of the 11 investigated species had average storage capacities in excess of 100 ml m^{-2} of leaf area and two in excess of 200 ml m^{-2} of leaf area at this low angle. Values of S_L at an angle of 45° or 80° were significantly lower (ANOVA, post-hoc Scheffé test, $p \leq 0.001$). At an angle of 45°, *Heliocarpus americanus* and *Clethra revoluta* had the highest storage capacity (differing from all other species), whereas *Isertia laevis* leaves represented the lowest storage (Table 42.2).

Based on hemispherical photography, an average leaf angle of 46° (SD 20°) was derived for all three forests combined (Table 42.1). Therefore, the average S_L of single leaves determined for an angle of 45° across all experimental species (70.74 ml m^{-2} of leaf area) was used to extrapolate to stand level. Combining this figure with the average LAI for the three forests (5.37, Table 42.1) gave an estimated leaf water storage capacity at the stand level of 0.38 mm (Oesker, 2008).

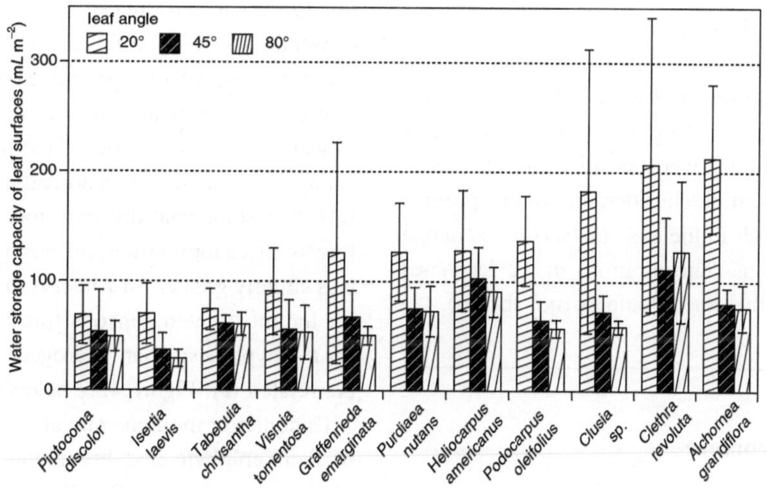

Figure 42.4. Comparative leaf water storage capacities for different tree species and leaf angles (20°, 45°, and 80°) expressed as ml $m^{-2}{}_{\text{leaf area}}$ ($n = 20$).

Table 42.2 *Significant differences in leaf water storage capacities of 11 species at an angle of inclination of 45°. Shown are the p-values and the level of significance (ANOVA, post-hoc LSD test; n.s., not significant; *, $p \leq 0.05$; **, $p \leq 0.01$; ***, $p \leq 0.001$; $n = 20$)*

	Isertia laevis	*Piptocoma discolor*	*Vismia tomentosa*	*Tabebuia chrysantha*	*Podocarpus oleifolius*	*Graffenrieda emarginata*	*Clusia* spc.	*Purdiaea nutans*	*Alchornea grandiflora*	*Heliocarpus americanus*	*Clethra revoluta*
Isertia laevis		*	*	**	***	***	***	***	***	***	***
Piptocoma discolor	0.0400		n.s.	n.s.	n.s.	n.s.	*	**	***	***	***
Vismia tomentosa	0.0153	0.7236		n.s.	n.s.	n.s.	*	*	**	***	***
Tabebuia chrysantha	0.0015	0.2543	0.4230		n.s.	n.s.	n.s.	n.s.	*	***	***
Podocarpus oleifolius	0.0004	0.1335	0.2431	0.7166		n.s.	n.s.	n.s.	*	***	***
Graffenrieda emarginata	0.0001	0.0658	0.1296	0.4717	0.7175		n.s.	n.s.	n.s.	***	***
Clusia spc.	0.0000	0.0122	0.0286	0.1675	0.3082	0.5189		n.s.	n.s.	***	***
Purdiaea nutans	0.0000	0.0047	0.0119	0.0878	0.1780	0.3322	0.7417		n.s.	***	***
Alchornea grandiflora	0.0000	0.0005	0.0015	0.0178	0.0441	0.1031	0.3164	0.5009		**	***
Heliocarpus americanus	0.0000	0.0000	0.0000	0.0000	0.0000	0.0000	0.0002	0.0008	0.0066		n.s.
Clethra revoluta	0.0000	0.0000	0.0000	0.0000	0.0000	0.0000	0.0000	0.0000	0.0004	0.4084	

Nutrient leaching experiment

Experimentally obtained nutrient leaching rates per unit water and leaf area of the investigated species showed large variation between species (Figure 42.5). Positive bars in Figure 42.5 imply a net loss of elements leached out of the leaves. For *EC*, two groups could be distinguished: (i) leachates from *Graffenrieda emarginata*, *Podocarpus oleifolius*, and *Piptocoma discolor* had relatively high *EC* values, whereas *Isertia laevis*, *Purdiaea nutans*, and *Heliocarpus americanus* exhibited much lower *EC* values (ANOVA, post-hoc – LSD, $p \leq 0.05$). *Tabebuia chrysantha* showed an intermediate value. Of the various cations, K was leached the most, followed by Ca and Mg. The highest amount of leached K was found for the gymnosperm *Podocarpus oleifolius*. Proton concentrations were positive in all cases, except for *Isertia laevis*, indicating considerable loss of protons from the leaves in general and, in the case of *Isertia*, exchange processes with cations. Protons concentrations in the leachates were more variable than for the other elements (not shown).

DISCUSSION

Throughfall and forest structure

The three forest types differed in tree species composition, canopy structure, and *TF* fraction (Table 42.1). The observed increase in *TF*/*P* ratios with elevation is thought to reflect a combination of the corresponding increase in canopy openness (Table 42.1) and (to an unknown degree) of contributions by horizontal precipitation (*HP*, fog, drizzle, and wind-driven rain) missed by the conventional rain gages but intercepted by the trees (Herwitz and Slye, 1992; Holwerda *et al.*, 2006b; McJannet *et al.*, 2007; cf. Giambelluca *et al.*, this volume; Holwerda *et al.*, this volume #29). Bendix *et al.* (2008) reported an extra 121 mm year^{-1} of *HP* (5.6% of *P*) at 1960 m.a.s.l. (ECSF station), increasing to 210 mm (9.6%) at 2270 m.a.s.l. (Plataforma station).

Whilst forest I (valley) exhibited the highest tree species diversity, forest III (upper ridge) was the most heterogeneous in terms of light penetration and canopy openness. Forests I and II (lower ridge) had significantly different below-canopy radiation, canopy openness, and mean leaf angle, but they did not differ in *LAI*. It is somewhat difficult, therefore, to distinguish these two forests in canopy structural terms (Table 42.1). The SD of *TF* was slightly greater for forest II than for forest I, and was highest for the most heterogeneous forest (III), providing some support for the hypothesis that heterogeneity of *TF* and canopy structure are related (cf. Figure 42.7 below).

Canopies of montane rain and cloud forests differ also in terms of their epiphyte and bryophyte loads (Frahm and Gradstein, 1991; Kürschner and Parolly, 2004; Köhler *et al.*, 2007). Both *TF* fractions and bryophyte levels typically increase from lower

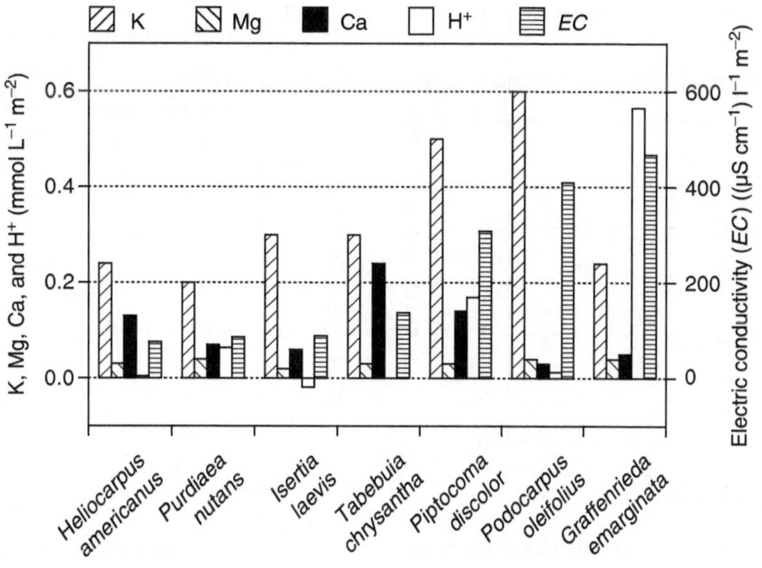

Figure 42.5. Leaf nutrient leaching rates for seven tree species in mmol l^{-1} of simulated rain and m^{-2}leaf area; electric conductivity (*EC*) in µS cm^{-1} l^{-1} of simulated rain and m^{-2}leaf area ($n = 20$).

montane rain forest (LMRF) via lower montane cloud forest (LMCF) to upper montane cloud forest (UMCF) (Frahm and Gradstein, 1991; Bruijnzeel, 2001). The presently found *TF* values for forests I and III are very close to the respective overall averages derived for LMRF and LMCF, respectively, whereas *TF* for forest II is intermediate (cf. Bruijnzeel, 2001; Bruijnzeel *et al.*, this volume). The average *TF* of 70.8% observed for the valley forest is distinctly higher than the 59% reported for nearby valley forest by Fleischbein *et al.* (this volume) using five to eight *TF* gages per plot instead of the presently used 31 gages. Measured *TF* fractions normally tend to be raised by the use of a larger number of gages, particularly if these are regularly relocated (Lloyd and Marques, 1988; Holwerda *et al.*, 2006a). As such, the presently obtained *TF* fractions may be considered more realistic than the ones reported by Fleischbein *et al.* (2006) and Fleischbein *et al.* (this volume). Some support for this contention comes from the fact that Zimmermann *et al.* (2007) obtained a value of 79% for a forest at 1950 m.a.s.l. in the same area using 25 gages. However, because the latter observations only pertained to a period of 3 months, the results are not directly comparable to the annual values presented here and by Fleischbein *et al.* (this volume).

Leaf and canopy storage capacity

Differences in leaf water storage capacities S_L between species as found in the laboratory experiment were large, ranging from <30 ml m^{-2} of leaf area to >200 ml m^{-2}, depending on tree species and leaf angle (Figure 42.4). As expected, the presently derived maximum stand-level water storage capacity value of 0.38 mm for the *foliar* component of the canopy is much smaller than the 1.91 mm determined by Fleischbein *et al.* (this volume) for the canopy as a whole, based on *P* and *TF* data measured in the field. The latter value includes the storage capacity afforded by epiphytes and bryophytes (S_{EPI}). Kürschner and Parolly (2004) estimated S_{EPI} for the lower part of forest I (valley) at 9.5 mm vs. 2.7 and 4.6 mm for the lower and upper parts of forest II, respectively; the epiphytes in the upper part of forest III were estimated to represent a potential storage of 4.7 mm. Figure 42.6 compares the data of Kürschner and Parolly (2004) on bryophyte biomass and S_{EPI} with the presently measured interception loss (*P* minus *TF*) for directly comparable plots. Interception loss appears to be only poorly correlated with bryophyte biomass (as also found by Fleischbein *et al.*, 2006), but interception exhibits a surprisingly strong correlation with S_{EPI} (Figure 42.6). This finding is all the more remarkable because the S_{EPI} values of Kürschner and Parolly (2004) represent *potential* rather than effective storage values. As pointed out by Hölscher *et al.* (2004) and Köhler *et al.* (2007) – and as demonstrated further by the much lower storage value derived by Fleischbein *et al.* (this volume) for the canopy as a whole under field conditions – the effective capacity of bryophytes to accommodate additional rainfall tends to be much smaller than the amount represented by S_{EPI} because of wetting by previous rainfall (cf. Köhler *et al.*, this volume; Tobón *et al.*, this volume #26).

Forest diversity and nutrient leaching

The second hypothesis – that a forest with higher tree species diversity should have higher heterogeneity in *TF* nutrient fluxes – is confirmed by the data shown in Figure 42.7. Forest tree diversity as represented by the Shannon–Wiener diversity index (N_1) was

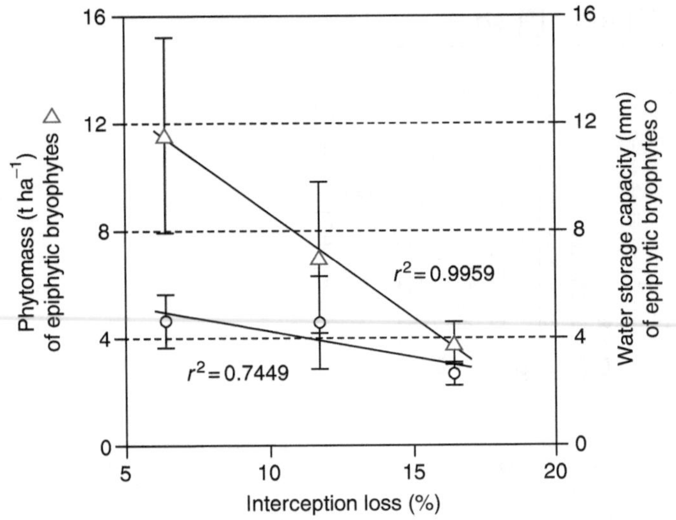

Figure 42.6. Biomass (Δ) and water storage capacity (○) of epiphytic bryophytes in the investigated types of forests (data from Kürschner and Parolly, 2004) vs. presently determined interception losses (%) for directly comparable forest types.

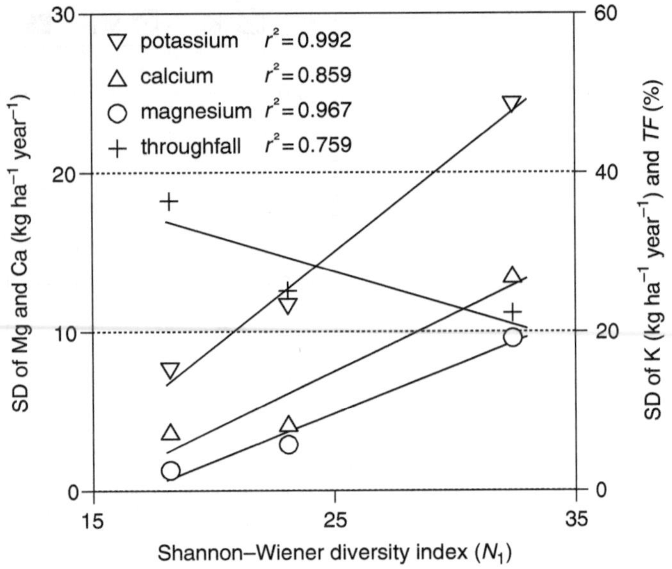

Figure 42.7. Standard deviation (SD) of magnesium, calcium, and potassium fluxes in throughfall (kg ha^{-1} year^{-1}) and standard deviation of throughfall (%; $n = 31$) vs. the Shannon–Wiener diversity index (N_1) of the three investigated forests.

strongly correlated with the SD of magnesium, calcium, and potassium fluxes in *TF* ($r > 0.95$). The highest values of both tree species diversity and heterogeneity in *TF* quality were obtained for forest I (valley), and the lowest for forest III (upper ridge) (Figure 42.7). A reverse pattern was found for the SD of *TF* quantity, i.e. the highest spatial variability occurred in the least diverse forest.

It is important to note that differences in experimentally determined species-specific leaching rates were even greater (Figure 42.5) than those observed in the field, suggesting that tree diversity does contribute to spatial heterogeneity of *TF* quality (cf. Hansen, 1996). However, leaching rates as obtained with the SALSA system are not fully comparable with those incurred by natural rain, which normally has lower pH values than the distilled water used in the experiments (cf. Boy *et al.*, 2008; Oesker, 2008; Rollenbeck *et al.*, this volume). Calcium provides a case in point as it was leached in the SALSA experiment but retained in the canopy. As such, the SALSA approach seems to be useful mostly for comparative purposes. Also, to obtain a more complete picture the nutrient dynamics of epiphytes, bryophytes and canopy humus would need to be included in the experiments (Oesker, 2008). Throughfall in cloud forests is often much reduced in nitrate and phosphate compared to concentrations in incident rainfall and fog water (Clark *et al.*, 1998; Hafkenscheid, 2000; cf. Chang *et al.*, this volume).

CONCLUSION

In conclusion, the present data suggest that heterogeneity in *TF* amounts in the studied montane rain forests is influenced mostly by heterogeneity in canopy structure, whereas the heterogeneity in *TF* quality is influenced by tree species diversity, presumably because of species-specific differences in leaf leachability. Effects of differences in epiphyte biomass and associated nutrient absorption and release would need to be included in future experiments.

ACKNOWLEDGMENTS

Special thanks are due to H. Lucero (Universidad Tecnica Particular de Loja, Ecuador) for collaboration. We are grateful to Dr. P. Emck and Dr. R. Rollenbeck for the use of meteorological data and to W. Musila for help with editing the draft manuscript. Field assistance was received from the following individuals: R. Armijos, E. Klingelmann, M. Lara, C. Morchner, K. Ristau, A. Schaper, A. Scheffer, A. Steinfelder, D. Valez, T. Walkenhorst, and from others at the ECSF. Thanks are expressed to the Deutsche Forschungsgemeinschaft (FOR 402) for financing this research, and to the Ministerio del Medio Ambiente de Loja for permitting the investigation.

REFERENCES

Ahmad-Shah, A., and J. O. Rieley (1989). Influence of tree canopies on the quality of water and amount of chemical elements reaching the peat surface of a basin mire in the Midlands of England. *Journal of Ecology* **77**: 357–386.

Balestrini, R., L. Galli, A. Tagliaferri, and G. Tartari (1998). Study on throughfall deposition in two North Italian forest sites (Valtellina, Lombardy). *Chemosphere* **36**: 1095–1100.

Balslev, H., and B. Øllgaard (2002). Mapa de vegetación del sur de Ecuador. In *Botánica Austroecuatoriana: Estudios sobre los recursos vegetales en las provincias de El Oro, Loja y Zamora-Chinchipe*, eds. M. Z. Aguirre, E. J. Madsen, E. Cotton, and H. Balslev, pp. 51–64. Quito, Ecuador: Ediciones Abya-Yala.

Beck, E., and K. Müller-Hohenstein (2001). Analysis of undisturbed and disturbed tropical mountain forest ecosystems in Southern Ecuador. *Die Erde* **132**: 1–8.

Beck, E., J. Bendix, I. Kottke, F. Makeschin, and R. Mosandl (eds.) (2008). *Gradients in a Tropical Mountain Ecosystem of Ecuador*. Berlin: Springer-Verlag.

Bendix, J., R. Rollenbeck, M. Richter, P. Fabian, and P. Emck (2008). Climate. In *Gradients in a Tropical Mountain Ecosystem of Ecuador*, eds. E. Beck, J. Bendix, I. Kottke, F. Makeschin, and R. Mosandl, pp. 63–74. Berlin: Springer-Verlag.

Boy, J., R. Rollenbeck, C. Valarezo, and W. Wilcke (2008). Amazonian biomass burning-derived acid and nutrient deposition in the north Andean montane forest of Ecuador. *Global Biogeochemical Cycles* **22**, GB4011, doi:10.1029/2007GB003158.

Bruijnzeel, L. A. (2001). Hydrology of tropical montane cloud forests: a reassessment. *Land Use and Water Resources Research* **1**: 1–18.

Cavelier, J., M. Jaramillo, D. Solis, and D. DeLeon (1997). Water balance and nutrient inputs in bulk precipitation in tropical montane cloud forest, Panama. *Journal of Hydrology* **193**: 83–96.

Chen, J. M., T. A. Black, and R. S. Adams (1991). Evaluation of hemispherical photography for determining plant area index and geometry of a forest stand. *Agricultural and Forest Meteorology* **56**: 129–143.

Clark, K. L., N. M. Nadkarni, D. Schaeffer, and H. L. Gholz (1998). Atmospheric deposition and net retention of ions by the canopy in a tropical montane forest, Monteverde, Costa Rica. *Journal of Tropical Ecology* **14**: 27–45.

Dalitz, H., J. Homeier, H. R. Salazar, and A. Wolter. (2004). Spatial heterogeneity generating plant diversity? In *Proceedings of the 2nd Symposium of the A.F.W. Schimper-Foundation*, eds. S. W. Breckle, B. Schweizer, and A. Fangmeier, pp. 199–213. Stuttgart, Germany: Verlag Günter Heimbach.

Emck, P. (2007). A climatology of South Ecuador: with special focus on the major Andean ridge as Atlantic–Pacific climate divide. Ph.D. thesis, University of Erlangen, Erlangen, Germany. Also available at www.opus.ub.uni-erlangen.de/opus/volltexte/2007/656/.

Fleischbein, K., W. Wilcke, R. Goller, *et al.* (2006). Rainfall interception in a lower montane forest in Ecuador: effects of canopy properties. *Hydrological Processes* **19**: 1355–1371.

Frahm, J. -P., and S. R. Gradstein (1991). An altitudinal zonation of the tropical rain forest using bryophytes. *Journal of Biogeography* **18**: 669–678.

Hafkenscheid, R. L. L. J. (2000). Hydrology and biogeochemistry of tropical montane rain forests of contrasting stature in the Blue Mountains, Jamaica. Ph.D. thesis, VU University Amsterdam, Amsterdam, the Netherlands. Also available at http://dare.ubvu.vu.nl/bitstream/1871/12734/1/tekst.pdf.

Hafkenscheid, R. L. L. J., L. A. Bruijnzeel, R. A. M. de Jeu, and N. J. Bink (2002). Water budgets of two upper montane rain forests of contrasting stature in the Blue Mountains, Jamaica. In *Hydrology and Water Management in the Humid Tropics*, ed. J. S. Gladwell, pp. 399–424. Paris: UNESCO, and Panama City: CATHALAC.

Hansen, K. (1996). In-canopy throughfall measurements of ion fluxes in Norway spruce. *Atmospheric Environment* **30**: 4065–4076.

Herwitz, S. R., and R. E. Slye (1992). Spatial variability in the interception of inclined rainfall by a tropical rainforest canopy. *Selbyana* **13**: 62–71.

Hölscher, D., L. Köhler, C. Leuschner, and M. Kappelle (2003). Nutrient fluxes in stemflow and throughfall in three successional stages of an upper montane rain forest in Costa Rica. *Journal of Tropical Ecology* **19**: 557–565.

Hölscher, D., L. Köhler, A. I. J. M. van Dijk, and L. A. Bruijnzeel (2004). The importance of epiphytes to total rainfall interception by a tropical montane rain forest in Costa Rica. *Journal of Hydrology* **292**: 308–322.

Holwerda, F., F. N. Scatena, and L. A. Bruijnzeel (2006a). Throughfall in a Puerto Rican lower montane rain forest: a comparison of sampling strategies. *Journal of Hydrology* **327**: 592–602.

Holwerda, F., R. Burkard, W. E. Eugster, *et al.* (2006b). Estimating fog deposition at a Puerto Rican elfin cloud forest site: comparison of the water budget and eddy covariance methods. *Hydrological Processes* **20**: 2669–2692.

Homeier, J. (2004). *Baumdiversität, Waldstruktur und Wuchsdynamik zweier tropischer Bergregenwälder in Ecuador und Costa Rica*, Dissertationes Botanicae No. 391. Stuttgart, Germany: J. Cramer.

Homeier, J., H. Dalitz, and S. -W. Breckle (2002). Waldstruktur und Baumartendiversität im montanen Regenwald der Estación Científica San Francisco in Südecuador. *Berichte der Reinhold-Tüxen-Gesellschaft* **14**: 109–118.

Homeier, J., F. A. Werner, S. -W. Breckle, S. R. Gradstein, and M. Richter (2008). Potential vegetation and floristic composition of andean forests in south ecuador, with a focus on the RBSF. In *Gradients in a Tropical Mountain Ecosystem of Ecuador*, eds. E. Beck, J. Bendix, I. Kottke, F. Makeschin, and R. Mosandl, pp. 87–100. Berlin: Springer-Verlag.

Houle, D., R. Quimet, R. Paquin, and J. -G. Laflamme (1999). Interaction of atmospheric deposition with a mixed hardwood and a coniferous forest canopy at the Lake Clair Watershed (Duchesnay, Quebec). *Canadian Journal of Forest Research* **29**: 1944–1957.

Köhler, L., C. Tobón, K. F. A. Frumau, and L. A. Bruijnzeel (2007). Biomass and water storage of epiphytes in old-growth and secondary montane rain forest stands in Costa Rica. *Plant Ecology* **193**: 171–184.

Krebs, C. J. (1999). *Ecological Methodology*, 2nd edn. Menlo Park, CA: Benjamin/Cummings.

Kürschner, H., and G. Parolly (2004). Phytomass and water-storing capacity of epiphytic rain forest bryophyte communities in S. Ecuador. *Botanische Jahrbücher für Systematik, Pflanzengeschichte und Pflanzengeographie* **125**: 489–504.

Lin, T. C., S. P. Hamburg, H. B. King, and Y. J. Hsia (1997). Spatial variability of throughfall in a subtropical rain forest in Taiwan. *Journal of Environmental Quality* **26**: 172–180.

Lloyd, M., and A. Marques (1988). Spatial variability of throughfall and stemflow measurements in Amazonian rainforest. *Agricultural and Forest Meteorology* **42**: 63–73.

Loescher, H. W., J. S. Powers, and S. F. Oberbauer (2002). Spatial variation of throughfall volume in an old-growth tropical wet forest, Costa Rica. *Journal of Tropical Ecology* **18**: 397–407.

McJannet, D., J. S. Wallace, and P. Reddell (2007). Precipitation interception in Australian tropical rainforests. II. Altitudinal gradients of cloud interception, stemflow, throughfall and interception. *Hydrological Processes* **21**: 1703–1718.

Musila, W. (2007). Multi-scale analysis of spatial heterogeneity of Kakamega tropical forest soils: role of disturbance, succession, soil depth, trees and NIRS. Ph.D. thesis, Institute of Botany, University of Hohenheim, Hohenheim, Germany.

Musila, W., H. Todt, D. Uster, and H. Dalitz (2005). Is geodiversity correlated to biodiversity? A case study of the relationship between spatial heterogeneity of soil resources and tree species diversity in a western Kenyan rainforest. In *African Biodiversity*, eds. B. A. Huber, B. J. Sinclair, and K. H. Lampe, pp. 405–414. Berlin: Springer-Verlag.

Oesker, M. (2008). Untersuchungen zur räumlichen Heterogenität von Kronenstruktur und Bestandesniederschlag in einem tropischen Bergregenwald. Ph.D. thesis, Institute of Botany, University of Hohenheim, Hohenheim, Germany.

Oesker, M., H. Dalitz, S. Günther, J. Homeier, and S. Matezki (2008). Spatial heterogeneity patterns: gorges and crests. In *Gradients in a Tropical Mountain Ecosystem of Ecuador*, eds. E. Beck, J. Bendix, I. Kottke, F. Makeschin, and R. Mosandl, pp. 267–274. Berlin: Springer-Verlag.

Ohlemacher C. (2001). Untersuchungen zum Lichtklima und zur Struktur eines Bergregenwaldes im Süden von Ecuador. Diploma thesis, Institute of Botany, University of Hohenheim, Hohenheim, Germany.

Parker, G. G. (1983). Throughfall and stemflow in the forest nutrient cycle. *Advances in Ecological Research* **13**: 57–125.

Rollenbeck, R., J. Bendix, P. Fabian, *et al.* (2007). Comparison of different techniques for the measurement of precipitation in tropical montane rain forest regions. *Journal of Atmospheric and Oceanic Technology* **24**: 156–168.

Tukey, H. B. (1970). The leaching of substances from plants. *Annual Review of Plant Physiology* **21**: 305–324.

Ulrich, B., and J. Pankrath (1983). *Effects of Accumulation of Air Pollution in Forest Ecosystems*. Dordrecht, the Netherlands: D. Reidel.

Veneklaas, E. J. (1990). Nutrient fluxes in bulk precipitation and throughfall in two montane tropical rain forests, Colombia. *Journal of Ecology* **78**: 974–992.

Wilcke, W., S. Yasin, C. Valarezo, and W. Zech (2001). Change in water quality during the passage through a tropical montane rain forest in Ecuador. *Biogeochemistry* **55**: 45–72.

Zimmermann, A., W. Wilcke, and H. Elsenbeer (2007). Spatial and temporal patterns of throughfall quantity and quality in a tropical montane forest in Ecuador. *Journal of Hydrology* **343**: 80–96.

43 Effect of topography on soil fertility and water flow in an Ecuadorian lower montane forest

W. Wilcke, J. Boy
Johannes Gutenberg University of Mainz, Mainz, Germany

R. Goller
University of Bayreuth, Bayreuth, Germany

K. Fleischbein
University of Potsdam, Potsdam, Germany

C. Valarezo
Universidad Nacional de Loja, Loja, Ecuador

W. Zech
University of Bayreuth, Bayreuth, Germany

ABSTRACT

Tropical montane forests are frequently located on steep slopes with pronounced differences in topographic exposure, related microclimatic conditions and hence in composition and structure of the vegetation over small distances. The objective of this work was to test the hypothesis that topographic position significantly influences soil fertility and water flow in these forests. Soil properties were determined at various topographic positions and water samples of selected ecosystem fluxes analyzed over a 1-year period for oxygen isotopes in three small, steep watersheds under lower montane forest in the Eastern Cordillera of the Andes in southern Ecuador. The soils are subject to lateral material movement (landsliding and solifluction). This, together with the pronounced variation in climatic conditions and vegetation over small distances, resulted in high heterogeneity of soil properties. The pH of the A-horizon ranged between 3.7 and 6.4; concentrations of base metals (calcium, magnesium), sulfur and phosphorus, and trace metals (manganese, zinc) showed enormous spatial variation (coefficient of variation: 358–680% over a surface area of <30 ha). The steepness of the study area and the large contrast in hydraulic conductivities of the organic layer and the mineral soil resulted in a hillslope flow regime dominated by fast lateral flow. During baseflow conditions, $\delta^{18}O$ values were similar to that of the sub-soil solution, but rapidly became similar to values in the top-soil solution during rain storms. The chemical composition of stormflows resembled that of the litter leachate. Stormflow had lower pH and higher organic carbon and metal concentrations than did baseflow. It is concluded that topographic position and lateral transport of water and matter (as a consequence of the pronounced inclination) are important controls of the water and nutrient cycles of the study forest.

INTRODUCTION

The frequent location of montane tropical forests on steep slopes results in high variability of abiotic environmental conditions over small distances. For example, Cox *et al.* (2002) found that topographic position (ridges vs. valleys) explained 11–60% of the total variation in soil properties at the Luquillo Experimental Forest in Puerto Rico (cf. Roman *et al*, this volume). Such large variability creates a multitude of ecological niches. Lateral movement of material via erosion, solifluction, and landslides

Tropical Montane Cloud Forests: Science for Conservation and Management, eds. L. A. Bruijnzeel, F. N. Scatena, and L. S. Hamilton. Published by Cambridge University Press.

further increases the heterogeneity of soil properties, resulting in highly variable site conditions even when climatic factors and topographic position are similar. The resulting heterogeneous substrates and soil-forming conditions, together with a spatially variable water regime, explain the enormous variation in forest soil properties at the scale of even a few meters (Silver *et al.*, 1994; Wilcke *et al.*, 2002; Roman *et al.*, this volume). This high variability of site conditions is likely to be an important driver of the high biodiversity reported for the northern Andes, the most plant species-rich zone of the Neotropics (Henderson *et al.*, 1991; Kessler, 2002; Barthlott *et al.*, 2005). Once established, plant diversity probably further enhances the variability of soil properties (cf. Oesker *et al.*, this volume).

Landsliding is part of the natural dynamics of montane forest on steep slopes in Ecuador as indicated by the presence of numerous truncated soil profiles (Frei, 1958; Wilcke *et al.*, 2003). Wilcke *et al.* (2003) reported that 3.7% of a 1117-ha study area was covered with visible landslides, most of which were considered to have been triggered by natural causes. Landslides play an important role in forest regeneration because some tree species are light-demanding and thus benefit from openings created by landslides (Guariguata, 1990; Bussmann *et al.*, 2008; Homeier *et al.*, 2008b). In addition, landslides (and probably also solifluction) produce a layering of the substrate that is likely to influence the direction of water flow. Furthermore, through the accumulation of organic matter on top of the mineral soil an extra layer is formed that has both lower bulk density and higher hydraulic conductivity than the underlying mineral layers (Huwe *et al.*, 2008). Combined with the steepness, it is reasonable to assume that shallow lateral water flow plays an important role in montane steepland hydrological and biogeochemical functioning. Indeed, rapid subsurface stormflow is a frequent phenomenon on steep forested hillslopes under conditions of high rainfall (Casper, 2002; Schellekens *et al.*, 2004; Bonell, 2005; Boy *et al.*, 2008). This is largely attributable to a combination of the abundant presence of macropores in forest top-soils (Buttle and McDonald, 2000) – created in turn by the high biological activity and stone content of such layers (cf. Pritchett, 1979) – and the abrupt changes in soil bulk density and hydraulic conductivity going from the organic layer to the underlying mineral soil (Gilmour *et al.*, 1987; Wilcke *et al.*, 2002; Bonell, 2005; Huwe *et al.*, 2008). Near-surface flow paths in montane forests have also been inferred from hydrochemical observations. For example, Wilcke *et al.* (2001) reporterd an increase in the concentrations of aluminum (Al), copper (Cu), zinc (Zn), and total organic carbon (TOC) as well as a decrease in pH in streamflow during and shortly after rain storms in a south Ecuadorian montane forest. Concentrations of all these elements (except Al) were highest in the organic layer (Wilcke *et al.*, 2002), suggesting that a significant portion of the stormflow came from near-surface soil horizons.

The application of stable isotopes as conservative tracers in investigations of runoff generation and other hillslope hydrological processes has become a common tool in the last two decades (Kendall and Caldwell, 1998) and applications in the humid tropics are on the increase (Buttle and McDonnell, 2005; cf. Guswa *et al.*, 2007; Rhodes *et al.*, this volume #24). The rationale for using $\delta^{18}O$ values in tracing water paths lies in their high temporal variation in precipitation (Genereux and Hooper, 1998;, Rhode, 1998) which is related to isotope fractionation in the atmosphere during evaporation/condensation cycles (Kendall *et al.*, 1999). However, for maximum elucidation of flowpaths, isotope data should be combined with hydrological information (Kennedy *et al.*, 1986; Bonell *et al.*, 1998; Kendall *et al.*, 1999; Buttle and McDonnell, 2005).

The objectives of the present work are to: (i) quantify the heterogeneity of soil properties contributing to soil fertility, and to explore the role of systematic controls such as topographic position and landslides, and (ii) assess the consequences of the steepness and small-scale topographic variability for water transport in small forested watersheds in lower montane forest in south Ecuador.

MATERIALS AND METHODS

Study area

The study area is located on the eastern slopes of the Eastern Cordillera in southern Ecuador between the cities of Loja and Zamora at 04° 00′ S and 79° 05′ W (see Figure 32.1 in Fleischbein *et al.*, this volume). Measurements were conducted in three small (8–13 ha) and steep (30–50°) watersheds under montane forest between altitudes of 1900–2200 m.a.s.l. Between May 1999 and April 2002, on average 2448 mm of precipitation was recorded, with June being the wettest month and the November–January period the driest. The mean temperature at 1950 m.a.s.l. was 15.5 °C with a seasonal amplitude of ~2 °C. Further climatic information can be found in Bendix *et al.* (2008) and Motzer *et al.* (this volume). Soils are mostly young and have developed in postglacial landslide material or possibly in periglacial cover beds. The underlying bedrock consists of interbedded Palaeozoic phyllites, quartzites, and meta-sandstones (Wilcke *et al.*, 2001).

Micro-watersheds MC2 and MC3 are entirely forested, whereas the upper part of micro-watershed MC1 was used for agriculture until about 10 years ago and is now covered by a mixture of grasses and scrub. The study forest has been classified as evergreen montane forest (Balslev and Øllgaard, 2002) and as montane cloud forest (Valencia *et al.*, 1999), corresponding to the lower montane rain forest formation of Grubb (1977) grading into lower montane cloud forest at higher elevations (*sensu* Bruijnzeel and Hamilton, 2000). Further information on the floristics and structure of the forest can be found in Homeier (2004), Homeier *et al.* (2008a), and Oesker *et al.* (this volume).

Hydrological measurements and water sampling

Incident rainfall was measured weekly between 16 August 2000 and 15 August 2001 using five bulk precipitation collectors at each of three cleared sites in the vicinity of the forest (see Figure 32.1 in Fleischbein *et al.* (this volume) for locations). After determination of rainwater volumes an aliquot of the bulked sample was taken for chemical analysis. Measurements were made of throughfall, litter leachate, and the soil solution at a total of five transects (MC1, MC2.1, MC2.2, MC2.3, and MC3). The transects were about 20 m long and situated on the lower part of the slope at 1900–1910 m.a.s.l. in each of the three watersheds (MC1, MC2.1, and MC3), and at 1950–1960 m.a.s.l. (MC2.2) and 2000–2010 m.a.s.l. (MC2.3) in MC2 (Figure 32.1). Each transect had five (eight after May 2000) evenly distributed throughfall collectors, plus three collectors for sampling (but not measuring) lateral near-surface flow, and three suction lysimeters for soil solution sampling at 0.15 m and 0.30 m depth in the mineral soil. The precipitation and throughfall samplers have been described in detail by Fleischbein *et al.* (this volume). The lateral-flow collectors consisted of a halved plastic pipe covered with a polyethylene net (0.5-mm mesh width) connected to a 1-l polyethylene sampling bottle. The collectors were installed along the contour within the organic layer with the open side facing upslope to catch water moving through the organic layer parallel to the soil surface. Collectors were positioned in the upper, middle, and lower part of each transect (Goller *et al.*, 2005). Suction lysimeters (sintered glass cups, 1 ± 0.1 m pore size, UMS, Munich, Germany) served to extract soil solution with the help of a vacuum pump (pressures were adjusted weekly in accordance with prevailing matric potentials as measured with tensiometers). All solution samples were collected weekly (Goller *et al.*, 2005).

To determine the saturated hydraulic conductivity (K^*) of the soil a Guelph constant-head well permeameter was used (Elrick and Reynolds, 1992). The hydraulically effective soil was divided into three layers: organic layer, mineral top-soil, and mineral sub-soil. At 20 locations in MC2, K^* was determined separately by consecutively removing the layers. Streamwater levels were recorded hourly using a custom-made pressure transducer at the outlet of each watershed. Due to frequent malfunctioning of the data-loggers (70–80% of the total time), extensive gaps in the streamflow record occurred. Streamflow during the gaps was estimated using TOPMODEL after validation against measured data as described in detail by Fleischbein *et al.* (2006). Samples of streamflow during baseflow conditions were collected weekly between 16 August 2000 and 15 August 2001. A single stormflow sample was taken immediately after the arbitrarily selected rainstorm of 19 September 2000 for stable isotope analysis. Water samples for $^{18}O/^{16}O$ analysis were immediately stored in tightly sealed 10- and 100-ml glass bottles and kept at 7 °C until analysis after 6–12 months after sampling.

Soil sampling

All horizons (Oi, Oe, Oa, A, B, and C) of 47 soils in the three watersheds were sampled. Sampling included 10 soil pits distributed along each of two transects, one running at *c.* 1 m distance along the stream in MC2 between 1880 and 2100 m.a.s.l. ("valley bottom"), the other on the ridge of the same watershed between 1890 and 2110 m.a.s.l. ("ridge top"). Samples were taken from the walls of pits and air-dried. Organic horizons were separated from mineral soil at the point of an abrupt increase in density from about 0.1 to 0.2 g cm^{-3} in the organic layer to *c.* 1 g cm^{-3} in the mineral soil. The organic horizons were ground and the mineral horizons sieved to <2 mm. All samples were stored in closed plastic bags at room temperature until analysis.

Chemical analyses

The following parameters were determined in the solid soil samples: soil pH in water (soil : solution ratio 1 : 2.5) with a standard pH electrode (Orion U402-S7), total carbon (C), nitrogen (N), and sulfur (S) concentrations with a CHNS-analyzer (Vario EL, Elementar Analysensysteme), effective cation-exchange capacity (ECEC) by extraction with 1M NH_4NO_3 (Zeien and Brümmer, 1989), base saturation (BS) by calculating the proportion of charge-equivalent of extractable base cations (calcium (Ca) + potassium (K) + magnesium (Mg) + sodium (Na)) of the ECEC; as well as concentrations of total Al, Ca, K, Mg, Mn, Na, P, and Zn by digestion with concentrated HNO_3/HF (mineral soil samples; Zeien and Brümmer, 1989) or concentrated HNO_3 under pressure (organic samples; Heinrichs *et al.*, 1986). Concentrations of P in the extracts were determined with ICP atomic emission spectrometry (GBC Integra XMP, GBC Scientific Equipment Pty, Australia), those of the metals with flame atomic absorption spectrometry (Varian AA400).

Stable isotope analysis ($\delta^{18}O$) was performed on all water samples of every second sampling week between 16 August 2000 and 15 August 2001 using mass spectrometric methods (Goller *et al.*, 2005).

Calculations and statistical evaluations

Correlation analyses followed the least-squares method. Soil properties along the valley-bottom and ridge-top transects were compared pair-wise using the paired-differences test for dependent data rows (Hartung and Elpelt, 1989, p. 534) to see whether the mean difference between pairs differed significantly from zero. Significance levels were generally set at $p < 0.05$.

RESULTS AND DISCUSSION

Soil heterogeneity

Overall, the predominant soils were humic Dystrudepts (USDA–NRCS, 2003) which occupied 28%, 60%, and 28% of the area in watersheds 1, 2, and 3 (Yasin, 2001). The valley-bottom (VB) transect was dominated by oxyaquic Dystrudepts and the ridge-top (RT) transect by humic Dystrudepts. All soils were shallow, loamy-skeletal, and had high mica contents (Yasin, 2001).

Many of the properties of the A-horizons were highly variable (Table 43.1). This is particularly true for the base metals (Ca, Mg), for P, S, and trace metals (Mn, Zn) which had coefficients of variation (CV) as high as 358–680%. A similarly high heterogeneity has been reported by Wilcke *et al.* (2002) for the organic layers in the study watersheds. Wilcke *et al.* (2001) demonstrated a close correlation between Ca and Mg concentrations in the soil organic layer and in throughfall and litter leachate. Thus, the cycles of some nutrients in the study forest appear closely linked to the very high soil chemical heterogeneity, although arguably it is debatable what comes first: soil or floristic heterogeneity (cf. Oesker *et al.*, this volume).

Part of the observed variation in soil properties is related to systematic factors such as topographic position. Soils from the RT had, on average, significantly lower pH (organic layer: 4.0, A-horizons: 4.2) than did VB soils (4.6, 4.9, paired-differences test, $p < 0.05$). The mean base saturation in the A-horizons was significantly lower in RT soils than in VB soils. Also, the mean ECEC in the A-horizons was significantly higher in RT soils than in VB soils because of the higher organic matter concentrations of the former (Figure 43.1). In O- and A-horizons of VB soils, mean total concentrations of Ca, Mg, Mn, P, and Zn were higher than in RT soils; many of these differences were significant (Table 43.2). The enhanced acidification, lower base saturation, and lower total element concentrations in RT soils compared to VB soils are considered the result of element leaching from RT to VB positions, partly enhanced by the higher water inputs and lower temperatures at higher elevations (Bendix *et al.*, 2008; Rollenbeck *et al.*, this volume; cf. Oesker *et al.*, this volume). Similar results were reported by Scatena and Lugo (1995) who found lower exchangeable base metal concentrations, higher soil organic matter, and acidity at ridge-top than at valley positions in two steep watersheds in Puerto Rico (cf. Roman *et al.*, this volume).

In the A-horizons of the VB transect (which represented 10 soils with similar topographic position and exposure, but variable altitude and therefore variable water inputs and temperatures), pH ($r = -0.62$), total concentrations of Ca (−0.59), Mg (−0.72), Na (−0.81), P (−0.61), and Zn (−0.69), as well as exchangeable concentrations of Ca (−0.65), K (−0.59), Mg (−0.58), and Mn (−0.63) all exhibited significant negative correlations with

Table 43.1 *Descriptive statistics of selected properties of 47 A-horizons collected in three micro-watersheds under lower montane forest in southern Ecuador*

Property	Unit	Mean	Median	Standard deviation	Coefficient of variation	Minimum	Maximum
pH (H_2O)		4.3	4.6	4.2	99	3.7	6.4
CEC	$mmol_c\ kg^{-1}$	72	67	37	51	15	152
BS	%	29	19	30	104	1.3	103
Total concentrations							
C	$g\ kg^{-1}$	35	34	20	59	2.8	92
N	$g\ kg^{-1}$	4.5	3.2	8.0	176	0.52	58
P	$g\ kg^{-1}$	3.0	0.68	14	469	not detected	102
S	$g\ kg^{-1}$	1.2	0.38	5.5	476	0.12	40
Ca	$g\ kg^{-1}$	2.5	0.39	14	557	0.02	99
Mg	$g\ kg^{-1}$	1.5	0.73	5.5	358	0.05	40
K	$g\ kg^{-1}$	17	17	6.1	37	0.38	32
Al	$g\ kg^{-1}$	47	47	15	31	5.2	84
Fe	$g\ kg^{-1}$	21	22	8.8	42	0.42	39
Mn	$g\ kg^{-1}$	1.5	0.46	7.3	492	0.14	52
Zn	$g\ kg^{-1}$	2.9	0.07	20	680	0.02	139
Cu	$mg\ kg^{-1}$	7.8	5.1	15	195	not detected	92
Element ratios							
C/N	–	11	12	5.4	48	1.7	36
C/P	–	50	44	37	74	7.0	141
C/S	–	86	82	34	40	7.8	160

Table 43.2 *Selected total element concentrations of the O- and A-horizons along the valley bottom (VB) and ridge top (RT) transects under montane rain forest in Ecuador ($n = 10$). Different letters indicate significant differences between the VB and the RT soils at $p < 0.05$ according to the paired-differences test.*

		O-horizons		A-horizons	
		VB	RT	VB	RT
Al	(g kg^{-1})	4.4	7.6	50	48
Ca	(g kg^{-1})	8.7a	4.3b	0.67a	0.37b
Fe	(g kg^{-1})	1.6	2.1	24	21
K	(g kg^{-1})	3.2	3.9	19	18
Mg	(g kg^{-1})	2.5a	1.5b	0.94a	0.70b
Mn	(g kg^{-1})	0.28	0.22	0.49a	0.29b
N	(g kg^{-1})	23	22	2.7	4.2
Na	(g kg^{-1})	0.11	0.13	1.7a	0.88b
P	(g kg^{-1})	1.0	0.96	0.61a	0.50b
S	(g kg^{-1})	2.5	2.3	0.32	0.35
Zn	(g kg^{-1})	0.02	0.02	0.08a	0.07b

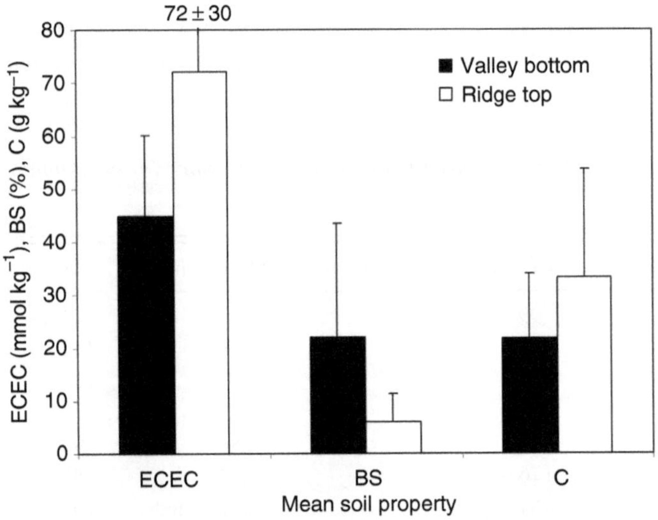

Figure 43.1. Mean effective cation-exchange capacity (ECEC), base saturation (BS), and carbon (C) concentrations of 10 soils each in valley-bottom (VB) and ridge-top (RT) positions. Error bars indicate standard deviations.

altitude, while exchangeable iron (Fe) concentrations (0.61), C/P ratios (0.61), and C storages (0.57) correlated positively. Thus, acidification and leaching in soils of similar topographic position were enhanced, and organic matter decomposition reduced, with elevation over comparatively short distances. Again, this is probably related to systematically decreasing soil temperatures and increasing precipitation with altitude even over several hundreds of meters. However, in the RT transect, only few correlations between soil properties and altitude were found, suggesting that lateral leaching of nutrients to the valley was consistently more dominant than the variation in temperature and precipitation with altitude along the studied transect (not shown).

Another systematic reason for the observed high soil heterogeneity relates to the effect of landsliding on soil fertility. In 1998, aerial photographs and field studies indicated that 3.7% of a 1117-ha study area was covered with visible landslides (Wilcke *et al.*, 2003). Most of these were shallow translational movements in a 2–5-m thick weathered layer and were classified as translational debris slides and earth flows (Crozier, 1986). The most important effect of landsliding was the removal of the organic layer. The mean mass of all O-horizons on four landslides increased in the order: landslide head (46 t ha^{-1}) < central part (101 t ha^{-1}) < toe (102 t ha^{-1}) < reference soils (291 t ha^{-1}), implying a trend in nutrient storage of the same order. However, there were no consistent differences in pH, ECEC and base saturation of the mineral soils on and adjacent to the studied landslides (Wilcke *et al.*, 2003).

Table 43.3 *Ranges and means of saturated hydraulic conductivities (K^*) in the soil horizons of the study area as determined with the Guelph permeameter*

Horizon (m s^{-1})	Min.	Max.	Geometric mean	*n*
O	8.8×10^{-5}	3.0×10^{-3}	5.0×10^{-4}	15
A	5.6×10^{-8}	8.2×10^{-4}	2.1×10^{-6}	23
B	1.1×10^{-9}	1.1×10^{-4}	4.6×10^{-7}	25

Hillslope hydrological flow paths

Near-surface lateral fluxes, explaining the observed distribution of nutrients in the landscape, would be favored by abrupt decreases in K^* with soil depth. The depth distribution of K^* in selected individual soils and its spatial distribution were investigated in MC2 (Table 43.3). Overall geometric mean K^* of the entire soil profile was in the order of 10^{-7} m s^{-1} but that for the organic layer was 1000-fold higher (in the order of 10^{-4} m s^{-1}) and still 10-fold higher in the near-surface mineral soil horizons (in the order of 10^{-6} s^{-1}). The large difference in K^* values of the organic layer and the mineral soil possibly favors ponding of water on top of the mineral soil and subsequent lateral flow in the organic layer (Huwe *et al.*, 2008; cf. Elsenbeer, 2001; Bonell, 2005).

To reduce the scatter in the K^* data, the study watershed was divided into four sectors according to topographic position and similarity in K^*, viz. (i) the uppermost part, (ii) the west-exposed slope, (iii) the east-exposed slope, and (iv) the riparian zone. Mineral soils in the uppermost part of the watershed showed an

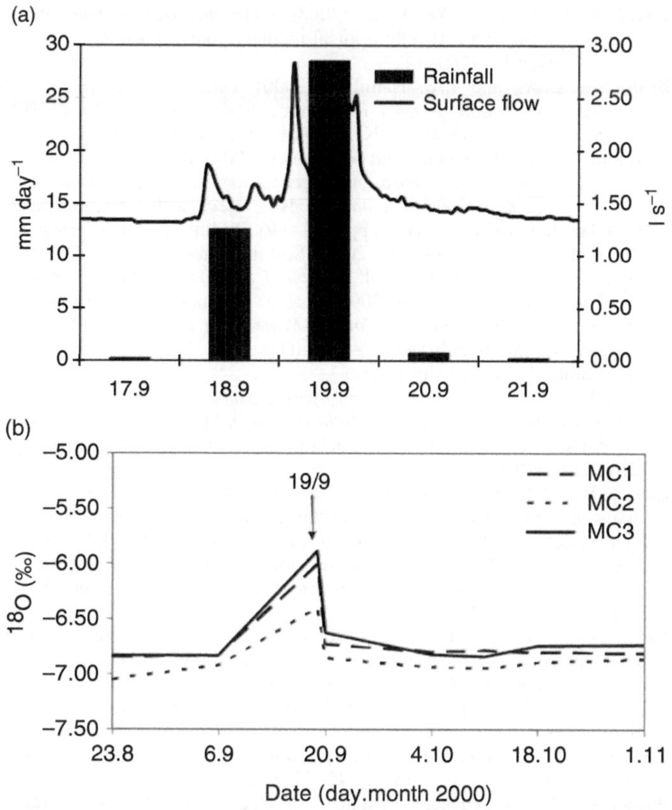

Figure 43.2. Temporal course of (a) surface discharge and daily rainfall between 17 and 21 September 2000, and (b) $\delta^{18}O$ values in stream water between 23 August and 1 November 2000 in a steep watershed under lower montane forest in southern Ecuador.

elevated content of fine particles, and K^* was low (in the order of 10^{-9} m s^{-1}). Conversely, riparian soils showed a high content of coarse material, producing the highest K^* value within the watershed (in the order of 10^{-5} m s^{-1}) (cf. Huwe *et al.*, 2008).

The vertical and lateral distribution of K^* thus supports the hypothesis that lateral flow is an important control of element distribution within, and outputs from the watershed (cf. Boy *et al.*, 2008). As a further test of this, the $\delta^{18}O$ signatures of ecosystem water fluxes collected on the day of an arbitrarily selected rainstorm (19 September 2000) were determined when the water levels of the streams were clearly increased. During the study year, $\delta^{18}O$ values of rainfall varied between −12.6% and +2.1%. There was no correlation between $\delta\ ^{18}O$ values in rainfall and temperature or rainfall volume, indicating that $\delta^{18}O$ in rainfall was attributable to air masses of different provenances. The $\delta^{18}O$ values of throughfall (−12.4% to +0.6%) and lateral flow through the organic layer were similar to those in rainfall (−10.5% to + 2.8%). However, variations in $\delta\ ^{18}O$ values of soil water and stream baseflow were smaller (−9.1% to −3.0% and −5.8% to −8.7%, respectively) than those in rainfall, throughfall, and lateral flow (Goller *et al.*, 2005). This dampening of the $\delta\ ^{18}O$ signal between rainfall and stream water is typical of forested ecosystems (Förstel, 1996). The contrasting $\delta^{18}O$ values of rainfall and soil water enable the distinction of different water sources to stream discharge during rainstorms (e.g. Bonell *et al.*, 1998; Genereux and Hooper, 1998).

On 19 September 2000, $\delta\ ^{18}O$ values in stormflow increased immediately from a value of *c.* −7% (baseflow conditions) to *c.* −6.0% in all three watersheds (Figure 43.2), suggesting a contribution of water from a different source having a $\delta\ ^{18}O$ value similar to those of rainfall or lateral flow in the organic layer. This contention was confirmed by higher moisture contents in the organic layer than in the upper mineral soil during this particular event (data not shown). Thus, whilst water flow paths through the soil profile were largely vertical during non-rainfall (baseflow) conditions before being deflected laterally upon meeting an impervious layer, short-term lateral flow, mainly in the organic layers, occurred during intense storm events, particularly in pre-saturated soil (Goller *et al.*, 2005; cf. Bogner *et al.*, 2008). Additional evidence for the importance of near-surface flows in the study watersheds has been provided recently by Boy *et al.* (2008). Lateral flows through the carbon- and nutrient-rich top layers during rain events contributed more than 50% of the total export of C, N, Al, and Mn, and about 33% of all base cation export from the watershed (Boy *et al.*, 2008).

CONCLUSIONS

The present results demonstrate that soil properties in the investigated lower montane forest are highly variable, particularly for Ca, Mg, S, P, Mn, and Zn concentrations. Part of this variation can be explained by topographic position and the influence of landslides. Ridge-top soils had higher humus concentrations, were more acid, and were poorer in nutrients than valley-bottom soils, because of stronger acidification and nutrient leaching on the more exposed ridges and upper slopes. Furthermore, for valley-bottom soils, relatively minor altitudinal differences (a few tens of meters) exerted additional control on chemical properties, whereas for ridge-top soils such influences were minor. Generally, soils became more acid, richer in organic matter (but less well decomposed), and poorer in both exchangeable and total base metal concentrations with elevation, probably because of the associated decreases in temperature and increases in precipitation. Landslides removed the organic layer and the uppermost part of the A-horizons at the top end of the landslide, and redeposited the material in the toe area, resulting in considerably lowered nutrient contents in O- and A-horizons on landslides and producing additional spatial heterogeneity.

However, topographic position and landslide influences only explained part of the overall variability in soil chemical properties, because considerable variation was observed among soils in similar topographic positions or between different landslides.

The distribution of nutrients in the landscape, with richer soils in lower positions and poorer soils at the higher locations, suggests that lateral redistribution of material and nutrients plays an important role. Lateral redistribution of soil nutrients is favored by strongly decreasing soil hydraulic conductivity (K^*) values with depth and by increasing K^* values when going from high to low positions in the watershed.

Average $\delta^{18}O$ values of water traveling from the canopy through the litter layer and soil to the stream showed a systematic decrease, suggesting that – during baseflow conditions – flows above the impervious layer at which the flow was deflected laterally, were predominantly vertical. During storm events, transport conditions switched to rapid lateral flow, mostly at the interface between the organic layers and the mineral soil, as evidenced by the $\delta^{18}O$ signal of stormflow water which became similar to values encountered in rain water and the near-surface soil solution.

ACKNOWLEDGEMENTS

We are grateful to Nature and Culture International (NCI) in Loja, Ecuador, for access to the study area and the research station, and to the Instituto Ecuatoriano Forestal de Areas Naturales y Vida Silvestre (INEFAN) for permission to conduct this work. Special thanks are due to Katrin Wagner, Hans-Jürgen Tobschall, and Melanie Leng for their contributions to the stable isotope study. This investigation was funded by the Deutsche Forschungsgemeinschaft (DFG) under grant no. FOR 402, Wi1601/5-2,-3.

REFERENCES

Balslev, H., and B. Øllgaard (2002). Mapa de vegetación del sur de Ecuador. In *Botánica Austroecuatoriana: Estudios sobre los recursos vegetales en las Provincias de El Oro, Loja y Zamora-Chinchipe*, eds. M. Z. Aguirre, J. E. Madsen, E. Cotton, and H. Balslev, pp. 51–64. Quito, Ecuador: Ediciones Abya-Yala.

Barthlott, W., J. Mutke, M. D. Rafiqpoor, G. Kier, and H. Kreft (2005). Global centers of vascular plant diversity. *Nova Acta Leopoldina NF* **92** (342): 61–83.

Bendix, J., R. Rollenbeck, M. Richter, P. Fabian, and P. Emck (2008). Climate. In *Gradients in a Tropical Mountain Ecosystem of Ecuador*, eds. E. Beck, J. Bendix, I. Kottke, F. Makeschin, and R. Mosandl, pp. 63–74. Berlin: Springer-Verlag.

Bogner, C., S. Engelhardt, J. Zeilinger, and B. Huwe (2008). Visualization and analysis of flow patterns and water flow simulations in disturbed and undisturbed tropical soils. In *Gradients in a Tropical Mountain Ecosystem of Ecuador*, eds. E. Beck, J. Bendix, I. Kottke, F. Makeschin, and R. Mosandl, pp. 403–412. Berlin: Springer-Verlag.

Bonell, M. (2005). Runoff generation in tropical forests. In *Forests, Water and People in the Humid Tropics*, eds. M. Bonell and L. A. Bruijnzeel, pp. 314–406. Cambridge, UK: Cambridge University Press.

Bonell, M., C. J. Barnes, C. R. Grant, A. Howard, and J. Burns (1998). Oxygen and hydrogen isotopes in rainfall-runoff studies. In *Isotope Tracers in Catchment Hydrology*, eds. C. Kendall and J. J. McDonnell, pp. 347–390. Amsterdam: Elsevier.

Boy, J., C. Valarezo, and W. Wilcke (2008). Water flow paths in soil control element exports in an Andean tropical montane forest. *European Journal of Soil Science* **59**: 1209–1227.

Bruijnzeel, L. A., and L. S. Hamilton (2000). *Decision Time for Cloud Forests*, IHP Humid Tropics Programme Series No. 13. Paris: IHP-UNESCO, Amsterdam: IUCN-NL, and Gland, Switcerland: WWF.

Bussmann, R. W., W. Wilcke, and M. Richter (2008). Landslides as important disturbance regimes: causes and regeneration. In *Gradients in a Tropical Mountain Ecosystem of Ecuador*, eds. E. Beck, J. Bendix, I. Kottke, F. Makeschin, and R. Mosandl, pp. 319–346. Berlin: Springer-Verlag.

Buttle, J. M., and D. J. McDonald (2000). Soil macroporosity and infiltration characteristics of a forest podzol. *Hydrological Processes* **14**: 831–848.

Buttle, J. M., and J. McDonnell (2005). Isotope tracers in catchment hydrology in the humid tropics. In *Forests, Water and People in the Humid Tropics*, eds. M. Bonell and L. A. Bruijnzeel, pp. 770–789. Cambridge, UK: Cambridge University Press.

Casper, M. (2002). *Die Identifikation hydrologischer Prozesse im Einzugsgebiet des Dür-reychbaches (Nordschwarzwald)*, Mitteilungen des Instituts für Wasserwirtschaft und Kulturtechnik No 210. Karlsruhe, Germany: University of Karlsruhe.

Cox, S. B., M. R. Willig, and F. N. Scatena (2002). Variation in nutrient characteristics of surface soils from the Luquillo Experimental Forest of Puerto Rico: a multivariate perspective. *Plant and Soil* **247**: 189–198.

Crozier, M. J. (1986). Classification of slope movements. In *Landslides: Causes, Consequences, and Environment*, ed. M. J. Crozier, pp. 3–31. London: Croom Helm.

Elrick, D. E., and W. D. Reynolds (1992). Infiltration from constant-head well permeameters and infiltrometers. In *Advances in Measurement of Soil Physical Properties: Bringing Theory into Practice*, Special Publication No. 30, eds. C. G. Topp *et al.*, pp. 1–24. Madison, WI: Soil Science Society of America.

Elsenbeer, H. (2001). Hydrologic flowpaths in tropical rainforest soilscapes: a review. *Hydrological Processes* **15**: 1751–1759.

Fleischbein, K., W. Wilcke, C. Valarezo, W. Zech, and K. Knoblich (2006). Water budgets of three small catchments under montane forest in Ecuador: experimental and modelling approach. *Hydrological Processes* **20**: 2491–2507.

Förstel, H. (1996). Hydrogen and oxygen isotopes in soil water: use of ^{18}O and D in soil water to study the soil–plant–water system. In *Mass Spectrometry of Soils*, ed. T. W. Boutton, pp. 285–310. New York: Marcel Dekker.

Frei, E. (1958). Eine Studie über den Zusammenhang zwischen Bodentyp, Klima und Vegetation in Ecuador. *Plant and Soil* **9**: 215–236.

Genereux, D. P., and R. P. Hooper (1998). Oxygen and hydrogen isotopes in rainfall-runoff studies. In *Isotope Tracers in Catchment Hydrology*, eds. C. Kendall and J. J. McDonnell, pp. 319–346. Amsterdam: Elsevier.

Gilmour, D. A., M. Bonell, and D. S. Cassells (1987). The effects of forestation on soil hydraulic properties in the Middle Hills of Nepal: a preliminary assessment. *Mountain Research and Development* **7**: 239–249.

Goller, R., W. Wilcke, M. Leng, *et al.* (2005). Tracing water paths through small catchments under a tropical montane rain forest in south Ecuador by an oxygen isotope approach. *Journal of Hydrology* **308**: 67–80.

Grubb, P. J. (1977). Control of forest growth and distribution on wet tropical mountains: with special reference to mineral nutrition. *Annual Review of Ecology and Systematics* **8**: 83–107.

Guariguata, M. R. (1990). Landslide disturbance and forest regeneration in the upper Luquillo mountains of Puerto Rico. *Journal of Ecology* **78**: 814–832.

Guswa, A. J., A. L. Rhodes, and S. E. Newell (2007). Importance of orographic precipitation to the water resources of Monteverde, Costa Rica. *Advances in Water Resources* **30**: 2098–2112.

Hartung, J., and B. Elpelt (1989). *Multivariate Statistik.* Munich, Germany: Oldenbourg-Verlag.

Henderson, A., S. P. Churchill, and J. L. Luteyn (1991). Neotropical plant diversity. *Nature* **351**: 21–22.

Heinrichs, H., H.-J. Brumsack, N. Loftfield, and N. König (1986). Verbessertes Druckaufschlußsystem für biologische und anorganische Materialien. *Zeitschrift für Pflanzenernährung und Bodenkunde* **149**: 350–353.

Homeier, J. (2004). *Baumdiversität, Waldstruktur und Wachstumsdynamik zweier tropischer Bergregenwälder in Ecuador und Costa Rica*, Dissertationes Botanicae No. 391. Berlin: J. Cramer.

Homeier, J., F. A. Werner, S.-W. Breckle, S. R. Gradstein, and M. Richter. (2008a). Potential vegetation and floristic composition of Andean forests in south Ecuador, with a focus on the RBSF. In *Gradients in a Tropical*

Mountain Ecosystem of Ecuador, eds. E. Beck, J. Bendix, I. Kottke, F. Makeschin, and R. Mosandl, pp. 87–100. Berlin: Springer-Verlag.

Homeier, J., S.-W. Breckle, and M. Richter (2008b). Gap dynamics in a tropical montane rain forest. In *Gradients in a Tropical Mountain Ecosystem of Ecuador*, eds. E. Beck, J. Bendix, I. Kottke, F. Makeschin, and R. Mosandl, pp. 311–318. Berlin: Springer-Verlag.

Huwe, B., B. Zimmermann, J. Zeilinger, M. Quizhpe, and H. Elsenbeer (2008). Gradients and patterns at local, field and catchment scales. In *Gradients in a Tropical Mountain Ecosystem of Ecuador*, eds. E. Beck, J. Bendix, I. Kottke, F. Makeschin, and R. Mosandl, pp. 391–402. Berlin: Springer-Verlag.

Kendall, C., and E. A. Caldwell (1998). Fundamentals of isotope geochemistry. In *Isotope Tracers in Catchment Hydrology*, eds. C. Kendall and J. J. McDonnell, pp. 51–86. Amsterdam: Elsevier.

Kendall, C., J. B. Shanley, and J. J. McDonnell (1999). A hydrometric and geochemical approach to test the transmissivity feedback hypothesis during snowmelt. *Journal of Hydrology* **219**: 188–205.

Kennedy, V. C., C. Kendall, G. W. Zellweger, T. A. Wyerman, and R. J. Avanzino (1986). Determination of the components of stormflow using water chemistry and environmental isotopes, Mattole River basin, California. *Journal of Hydrology* **84**: 107–140.

Kessler, M. (2002). The elevational gradient of Andean plant endemism: varying influences of taxon-specific traits andtopography at different taxonomic levels. *Journal of Biogeography* **29**: 1159–1166.

Pritchett, W. L. (1979). *Properties and Management of Forest Soils*. New York: John Wiley.

Rhode, A. (1998). Snowmelt-dominated systems. In *Isotope Tracers in Catchment Hydrology*, eds. C. Kendall and J. J. McDonnell, pp. 391–434. Amsterdam: Elsevier.

Scatena, F. N., and A. E. Lugo (1995). Geomorphology, disturbance, and the soil and vegetation of two subtropical wet steepland watersheds of Puerto Rico. *Geomorphology* **13**: 199–213.

Schellekens, J., F. N. Scatena, L. A. Bruijnzeel, *et al.* (2004). Stormflow generation in a small rainforest catchment in the Luquillo Experimental Forest, Puerto Rico. *Hydrological Processes* **18**: 505–530.

Silver, W. L., F. N. Scatena, A. H. Johnson, T. G. Siccama, and M. J. Sanchez (1994). Nutrient availability in montane wet tropical forest: spatial patterns and methodological considerations. *Plant and Soil* **164**: 129–145.

USDA–NCRS 2003. *Keys to Soil Taxonomy*, 9th edn. Available at http://soils.usda.gov./technical/classification/tax_keys/keysweb.pdf.

Valencia, R., C. Cerón, W. Palacios, and R. Sierra (1999). Las formaciones naturales de la Sierra del Ecuador. In *Propuesta preliminar de un sistema de clasificación de vegetación para el Ecuador continental*, ed. R. Sierra, pp. 79–108. Quito, Ecuador: Proyecto INEFAN/GEF-BIRF.

Wilcke, W., S. Yasin, C. Valarezo, and W. Zech (2001). Change in water quality during the passage through a tropical montane rain forest in Ecuador. *Biogeochemistry* **55**: 45–72.

Wilcke, W., S. Yasin, U. Abramowski, C. Valarezo, and W. Zech (2002). Nutrient storage and turnover in organic layers under tropical montane rain forest in Ecuador. *European Journal of Soil Science* **53**: 15–27.

Wilcke, W., H. Valladarez, R. Stoyan, *et al.* (2003). Soil properties on a chronosequence of landslides in montane rain forest, Ecuador. *Catena* **53**: 79–95.

Yasin, S. (2001). *Water and Nutrient Dynamics in Microcatchments under Montane Forest in the South Ecuadorian Andes*. Bayreuth, Germany: University of Bayreuth.

Zeien, H., and G. W. Brümmer (1989). Chemische Extraktion zur Bestimmung von Schwermetallbindungsformen in Böden. *Mitteilungen der Deutschen Bodenkundlichen Gesellschaft* **59**: 505–510.

44 Human impacts on stream-water chemistry in a tropical montane cloud forest watershed, Monteverde, Costa Rica

A. L. Rhodes and A. J. Guswa
Smith College, Northampton, Massachusetts, USA

S. Dallas
Monteverde Institute, Monteverde, Puntarenas, Costa Rica

E. M. Kim, S. Katchpole, and A. Pufall
Smith College, Northampton, Massachusetts, USA

ABSTRACT

Monteverde, Costa Rica is home to a tropical montane cloud forest that illustrates the balances between habitat and development inherent to ecotourism. The Monteverde Cloud Forest Reserve, located on the leeward side of the Continental Divide, has experienced a 100-fold increase in visitors since its inception 20 years ago. The associated growth in population and commercial development has the potential to impact water resources. Over three years, more than 400 stream-water samples were collected from eight sites above and below the main road within the Rio Guacimal watershed to assess the effect of development on water quality. The chemistry of upstream samples reflects mineral weathering and cation exchange reactions in the forest soils. Comparisons of downstream samples to these baseline data showed evidence of anthropogenic impacts: chemical concentrations were two to five times higher at downstream locations. The highest concentrations were observed at the site with the highest population density. These results point to the value of forest preserves, specifically the Monteverde Cloud Forest Reserve and the Bosque Eterno de Los Niños (or the Children's Eternal Rainforest), in limiting growth in riparian areas, which in turn helps to protect the quality of water resources for downstream communities.

INTRODUCTION

Tropical montane cloud forests (TMCFs) are ecosystems of extraordinary biological diversity whose existence depends on frequent immersion in clouds and mist. Monteverde, in northern Costa Rica, harbors highland TMCFs that exemplify the delicate balances between climate, hydrology, habitat, and development. This region has received significant research attention due to its high biodiversity (cf. Nadkarni and Wheelwright, 2000), and is undergoing rapid development as an ecotourist destination. The success of ecotourism has spurred the construction of new hotels, restaurants, businesses, and homes in Monteverde and the neighboring communities of Cerro Plano and Santa Elena (cf. Calvo-Alvarado *et al.*, this volume).

Water plays a prominent role in Monteverde; it is essential for the local ecosystem as well as the community. Drinking water is taken from springs and streams in the Monteverde Cloud Forest Reserve, and increased demands for water are expected to reduce stream baseflows in the future. This could affect aquatic habitats and limit the ability of these streams to dilute household and agricultural wastewater discharge and run-off from agricultural land. Decreased cloud formation in the region brought on by climate change (Still *et al.*, 1999; Lawton *et al.*, 2001; Karmalkar *et al.*, 2008; Foster, this volume) could further stress the quantity and quality of the water. This study characterizes the water chemistry of a 19.7-km^2 headwater watershed of the Rio Guacimal to illustrate the impacts of land use on water chemistry.

Tropical Montane Cloud Forests: Science for Conservation and Management, eds. L. A. Bruijnzeel, F. N. Scatena, and L. S. Hamilton. Published by Cambridge University Press.

STUDY AREA

Monteverde (10° 08′ N, 84° 48′ W; 1500, m.a.s.l.) is located on the leeward (western and Pacific) side of the Cordillera de Tilarán, a stretch of the volcanic range that defines the Continental Divide of Costa Rica (Figure 44.1). As described by Cignolini and Chaves (1986), the regional bedrock geology consists of calc-alkalic rocks of the Aguacate volcanic series (Miocene age, composed of basaltic andesite and andesite lava flows) that is associated with the Guacimal Pluton (late Miocene age, composed of quartz-diorite, quartz-monzonite, and granite). Located near the Costa Rican Gold Belt, the Aguacate Formation in this region is commonly hydrothermally altered. The Aguacate Formation and Guacimal Pluton are discordantly overlain by the Monteverde Formation (Pliocene), which consists of basaltic andesite and andesite lava flows and thick deposits of heavily weathered rhyodacitic ignimbrites. Mineralogy and major element chemistry of these rock formations in the gold-belt region near Monteverde have been presented by Cignolini and Chaves (1986). The headwater streams contributing to the Rio Guacimal (Quebrada Cuecha, Q. Máquina, and Q. Sucia) described in this study are incised into basaltic andesite flows and into rhyodacitic ignimbrite deposits, whereas the lower-elevation tributaries (Q. Socorro and Q. Cambranero) predominately drain the Aguacate Formation (cf. Figure 44.1). Hydrothermal alteration and associated calcite veining is visible in bedrock along the Rio Guacimal, ~500 m upstream of its confluence with the Q. Cambranero.

Soils in the study area are reported as Andisols (Clark *et al.*, 2000; cf. Schaetzl and Anderson, 2005), in line with the humid climate and volcanic substrate. Unpublished soil chemistry data by Kim (2002), from two forested locations in the Q. Cuecha sub-watershed, show the organic and A-horizons to have low cation exchange capacities (5.3–28 meq 100 g^{-1}), high base saturation (>95%; dominated by exchangeable calcium (Ca) and magnesium (Mg)), and very low exchangeable aluminum (<0.3 meq 100 g^{-1}); accordingly, these horizons are slightly acidic (pH range: 5.9–6.6).

The climate of the Monteverde region is marked by three distinct seasons that are related to the north–south migration of the Intertropical Convergence Zone (ITCZ) across Costa Rica. Annually, Monteverde receives ~2500, mm of precipitation, with ~70% occurring during the wet season (May–October) when the ITCZ is positioned directly over Monteverde. The transitional (November–January) and dry (February–April) seasons correspond to the months when the ITCZ is located south of Costa Rica. Average precipitation during the dry season is less than 50 mm per month (Clark *et al.*, 2000). Amounts of orographic precipitation on the leeward (i.e. Pacific) slope vary with elevation (Guswa *et al.*, 2007; Häger and Dohrenbusch, this volume), and annual precipitation as collected with standard rain gages is likely to be underestimated because of wind effects (Holwerda *et al.*, 2006; Giambelluca *et al.*, this volume; Tanaka *et al.*, this volume), particularly on the higher slopes close to the Continental Divide where both strong winds and horizontal precipitation (fog and wind-driven rain) are prevalent (Clark *et al.*, 1998a; Häger and Dohrenbusch, this volume). In comparison, average annual rainfall at the Pacific coast, as measured at the Palo Verde Biological Station (Organization for Tropical Studies, 2005), is 2.5 times less than that at Monteverde, suggesting that the headwater areas of the Monteverde region are important suppliers of water to downstream communities on the leeward slope (cf. Zadroga, 1981). Much work has been initiated in recent years to better understand (forest) hydrological processes operating in the Monteverde region (e.g. Fallas, 2002; Frumau *et al.*, 2006; Mulligan *et al.*, 2006; Rhodes *et al.*, 2006; Guswa *et al.*, 2007; Köhler *et al.*, 2007; Frumau *et al.*, this volume; Rhodes *et al.*, this volume #24; Schmid *et al.*, this volume; Tobón *et al.*, this volume #26 and #52). Others have focused on the hydroclimatic impacts of land-cover change in the area (Lawton *et al.*, 2001; Nair *et al.*, 2003; Ray *et al.*, 2006; Lawton *et al.*, this volume; Nair *et al.*, this volume).

This study focuses on the headwater area (1970, ha) of the Rio Guacimal (Figure 44.1), which extends from 1842, m.a.s.l. on the Continental Divide to its confluence with the San Luis River (700 m.a.s.l.). The upper elevations of the watershed are forested and largely undisturbed by human activity; they include primary and secondary forest as well as forest preserves (Monteverde Cloud Forest Reserve and the Bosque Eterno de Los Niños/Children's Eternal Rainforest). Development is concentrated in the towns of Santa Elena and Cerro Plano, and continues along the main road that leads to the Monteverde Cloud Forest Reserve. Agricultural activity – farms for dairy cows, pigs, coffee, and other crops – is concentrated below the main road.

APPROACH AND METHODS

Since 2000, stream-water samples have been collected throughout the watershed and analyzed for major ion chemistry. This chapter presents the results of 432 stream-water samples collected at eight sites representing remote and populated areas (see Figure 44.1 for sampling locations). Three sites (QC-300, QM-300, and HB-100) are located on first-order streams that drain forested, upland regions, and serve as reference locations for assessing human influences on water quality further downstream. Three additional sites (RG-200, QM-200, QS-200) are located just downstream of the main road to the Monteverde Cloud Forest Preserve that bisects the watershed. Of these, QS-200 drains the most commercial and densely populated area; all three have similar climate and topography. At lower elevations, the watershed becomes warmer and drier. QSO-100 (~940

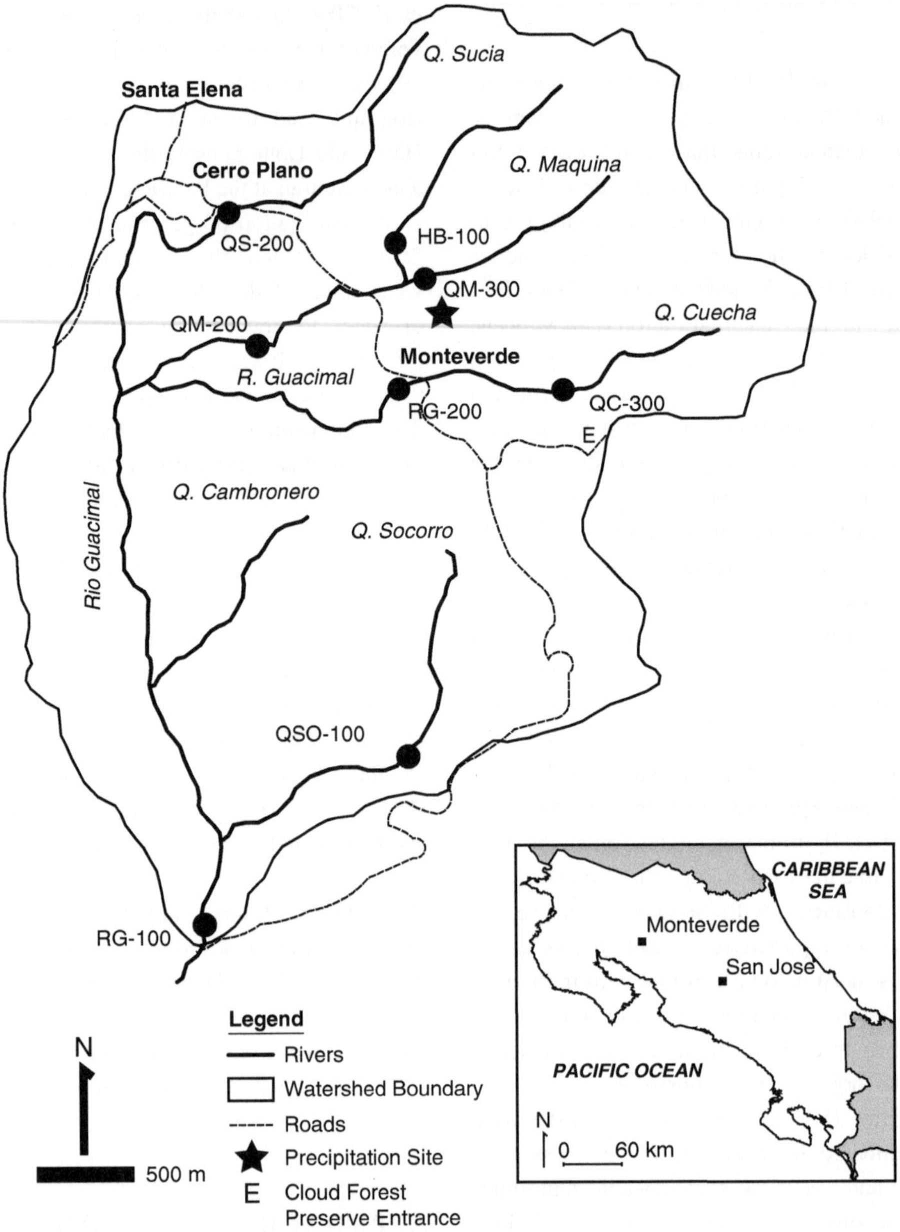

Figure 44.1. The Rio Guacimal watershed in northern Costa Rica and stream-water sampling locations.

m.a.s.l.) receives runoff from forested, pasture, and low-density residential areas. The lowermost (700 m.a.s.l.) sampling point (RG-100) receives water from all other sites and runoff from larger-scale agriculture, including a commercial pig farm. Stream-water samples were collected, on average, every 24 days, although the number of days between collections varied between 2 and 74 days. Rainfall and throughfall samples were collected in a clearing and beneath the canopy of secondary, tropical montane forest at three locations during July–October 2003; two of these sites were at 1400, m.a.s.l. and one site was at 1460, m.a.s.l. The upper regions of the Rio Guacimal watershed, near the Continental Divide, are occupied by various types of tropical montane cloud forest (Lawton and Dryer, 1980; Häger and Dohrenbusch, this volume). However, the higher peaks along the Continental Divide appear to limit the extent of orographic cloud water inputs to lower elevations on the leeward slope, particularly in the western region of the watershed (Guswa *et al.*, 2007). The precipitation collectors in this study are located in tropical montane forests, as opposed to tropical montane cloud forest, and they are likely to receive less cloud water input than what is added to the upper portion of the watershed (Häger and Dohrenbusch, this volume).

All water samples were collected in acid-washed polyethylene bottles, after rinsing three times with the sample, and then refrigerated. Samples were analyzed at the Aqueous Geochemistry laboratory at Smith College (Northampton, MA, USA) for pH, electric conductance (EC), acid-neutralizing capacity (ANC), major cation (calcium (Ca^{2+}), magnesium (Mg^{2+}), sodium (Na^{+}), potassium (K^{+}), and ammonium (NH_4^{+})) and anion concentrations (fluoride (F^{-}), chloride (Cl^{-}), nitrate (NO_3^{-}), phosphate (PO_4^{3-}), and sulfate (SO_4^{2-})), as well as dissolved silica. ANC was measured by Gran titration on unfiltered samples. Samples then were vacuum filtered through a 0.45-μm Millipore membrane. Cation concentrations of acidified splits and anion concentrations of non-acidified splits were determined by ion chromatography, using a Dionex 500 ion chromatograph. Instruments were standardized to solutions of known concentrations; errors in accuracy are <0.13 mg l^{-1} for cations and <0.18 mg l^{-1} for anions. Dissolved silica concentrations were determined by colorimetry using a Bauch and Lomb Spectronic 21 spectrometer. For each sample, charge balances were calculated in equivalents as

$$(\Sigma \text{ cations} - \Sigma \text{ anions})/(\Sigma \text{ cations} + \Sigma \text{ anions}) * 100\%.$$

Samples having imbalances greater than 5% were either reanalyzed or assessed for cause of the imbalance. An average ANC error of 20 μeq l^{-1} was estimated from the charge balance calculations. Results are reported in molar (μmol l^{-1}) units; equivalents (μeq l^{-1}) are used when data interpretation requires charge balances. Due to the scope of this chapter, results for dissolved silica are not presented.

Impacts of land use on stream-water chemistry are evaluated by comparing samples collected from the upland, forested streams to those suspected to be impacted by development. In an undisturbed region, one would expect the dominant controls on stream-water chemistry to be mineral weathering and cation exchange reactions between precipitation and soil and rock materials. Base cations (Ca^{2+}, Mg^{2+}, Na^{+}, K^{+}) and alkalinity (measured as ANC) are added in equal amounts to stream water by these processes. Point- and non-point-source pollution may add strong acids (indicated by enhanced SO_4^{2-} and NO_3^{-} levels) and salts (identified by Cl^{-}), although natural sources for these anions exist as well. For example, release of volcanic gasses from Volcan Arenal, located ~25km upwind, may contribute sulfate to Monteverde. In general, atmospheric deposition of SO_4^{2-}, NO_3^{-}, and Cl^{-} might be significant in montane cloud forests because ion concentrations in cloud water typically are three to ten times greater than in precipitation (Clark *et al.*, 1998b; cf. Chang *et al.*, this volume; Liu *et al.*, this volume). Wastewater discharges and surface runoff containing detergents or fertilizers may also contribute alkalinity, which increases the ANC and base cation concentrations of the streams (Rhodes *et al.*, 2001). Dust generated from dirt roads could affect stream-water chemistry in a similar manner as mineral weathering processes by adding base cations and ANC.

To evaluate whether anthropogenic compounds from the atmosphere are affecting stream chemistry in the Monteverde area, the chemistry of throughfall and bulk precipitation was compared to the chemistry of stream water in the cloud forest zone. The impact of pollution added by local land-use activities was determined by comparing sites below the main road against these remote forested sites. Possible correlations were investigated between the sum of base cations (SBC = Ca^{2+}+ Mg^{2+}+ Na^{+}+ K^{+}) and the sum of acid and salt anions (SAA = SO_4^{2-} + NO_3^{-} + Cl^{-}) for each stream sampling site. A weak correlation between SBC and SAA would be consistent with natural processes having the strongest control on stream-water chemistry since weathering of minerals adds far more base cations than acid anions. Conversely, it is expected that pollution from multiple land uses will cause correlation between SBC and SAA to become stronger, with pollution signatures exhibiting greater variability as the intensity of human activity is not constant throughout the year (cf. Fabian *et al.*, 2005; Boy and Wilcke, 2008; Boy *et al.*, 2008a). The seasonality of precipitation also plays a role in either adding (through the production of surface runoff; cf. Tobón *et al.*, this volume #52) or diluting pollution.

Mineral weathering by carbonic acid and cation exchange reactions in soils leads to a one-to-one increase in SBC and ANC (cf. Langmuir, 1997). SBC exceeds ANC when strong acids reduce alkalinity or salts add base cations. At the forested sites, SBC is expected to be slightly higher than ANC due to (mildly) acid rain, acid cloud water, and naturally occurring organic acids from the environment (cf. Langmuir, 1997; Rhodes *et al.*, 2001). Larger differences between SBC and ANC than observed at the forested sites are considered to represent the impact of pollution on the other streams.

RESULTS AND DISCUSSION

Precipitation chemistry

The chemical composition of bulk precipitation as measured in the open at 1400, m.a.s.l. (Figure 44.2) is consistent with the range of results reported for Monteverde cloud forests by Clark *et al.* (1998b), but presently determined values are more concentrated than the annual average results for Ca, Na, and K presented in Clark *et al.* (1998a). Similar average concentrations of Na^{+} and Cl^{-} (~17 μmol l^{-1} each) suggest that marine aerosols have a strong influence on precipitation chemistry at Monteverde. Average SO_4^{2-} and NO_3^{-} concentrations (7.1 and 2.2 μmol l^{-1}, respectively) are lower than in the north-eastern USA (NADP 2002 data, accessed in 2004), where rainfall is affected by industrial air pollution. In addition to marine aerosols, sulfate in rain that reaches Monteverde could be derived partly from volcanic emissions; Volcan Arenal is located ~25 km upwind. Interestingly, the precipitation collected was not acidic (pH = 5.5–6.1; ANC = 15–47 μeq l^{-1}). In the Monteverde Cloud Forest

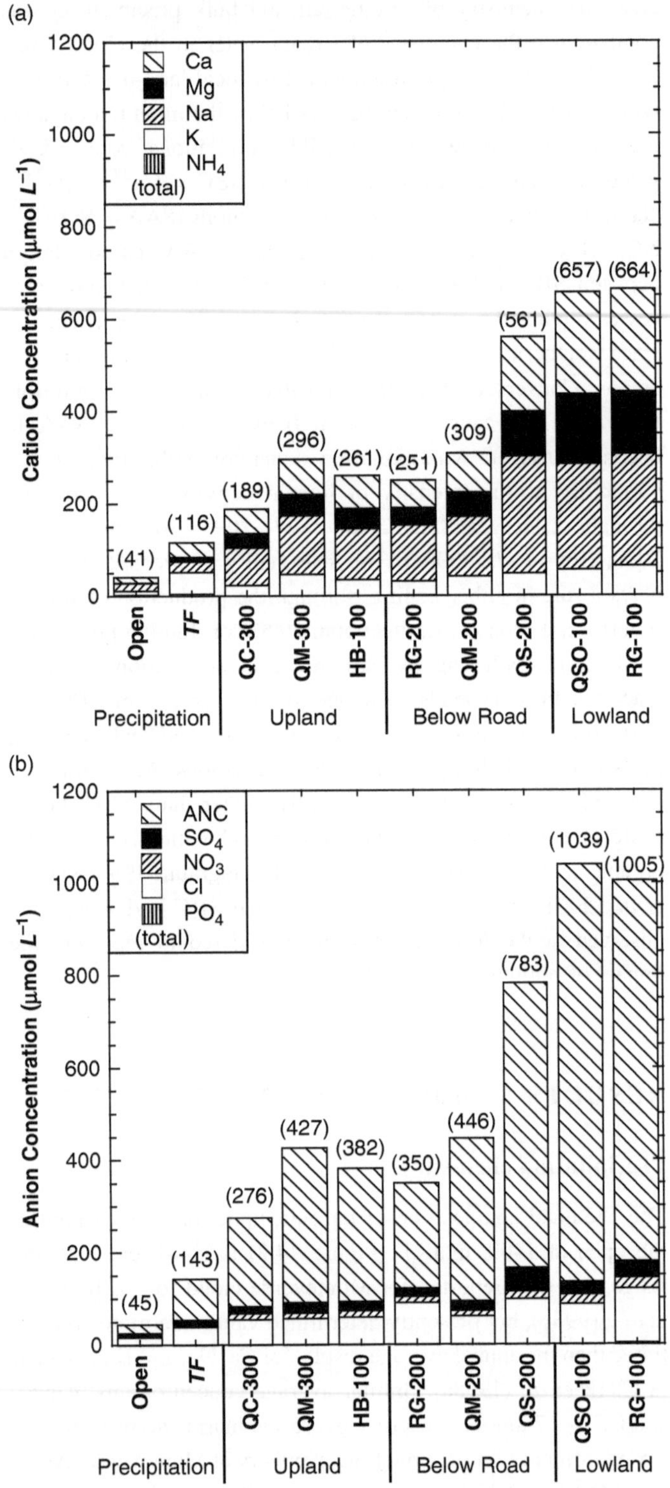

Figure 44.2. Average (a) cation and (b) anion concentrations of precipitation and stream-water samples in the Rio Guacimal watershed. Open, open-field precitation; *TF*, throughfall.

Reserve, Clark *et al.* (1998a) measured lower average annual pH for precipitation (5.0) and mixed cloud water and wind-driven rain (pH = 4.88). The difference with the presently obtained pH value might reflect differences in the catch of fog and wind-driven rain by the respective collector types (cf. Clark *et al.*, 1998a). Except for magnesium, the average concentrations of the non-salt base cations (Ca^{2+} ~11.8, Mg^{2+} ~ 1.1, and K^+ ~ 8.4 µmol l^{-1}) are notably higher than the annual average base cation concentrations reported by Clark *et al.* (1998a) for Monteverde and by Hendry *et al.* (1984) for Turrialba, in central Costa Rica. The difference may help to explain the near-neutral pH and positive ANC values found by the present study, although the underlying cause is difficult to infer from the limited precipitation sampling of this study. In the Monteverde Cloud Forest Preserve, Clark *et al.* (1998a,b) observed much higher concentrations for all ions in cloud water than in rain, attributing enhanced cation and nitrate concentrations in rain to burning of agricultural and forest biomass in lowland areas, particularly during the dry season (cf. Fabian *et al.*, 2005; Boy *et al.*, 2008a; Rollenbeck *et al.*, this volume). These observations, in general, should extend to the upper regions of the Rio Guacimal watershed as well. Atmospheric mixing of precipitation with dust and ash – whether from biomass burning, volcanoes, or dirt roads – could also neutralize the acidity normally observed in rain.

Differences in chemistry between precipitation measured in the open and beneath vegetation (throughfall) typically reflect a combination of effects. Wash-off of atmospheric material accumulated on plant surfaces by dry deposition, leaf exudates, and direct evaporation of intercepted water would increase ion concentrations, whereas further exchange processes between precipitation and vegetation (mostly leaching) usually add substantial amounts of K^+ to throughfall (Parker, 1983; Cavelier *et al.*, 1997; Hafkenscheid, 2000; Zimmermann *et al.*, 2007; cf. Chang *et al.*, this volume; Oesker *et al.*, this volume). Results for Ca^{2+} and Mg^{2+} vary between study areas as a function of rainfall regime and geological substrate (Clark *et al.*, 1998a; Hafkenscheid, 2000; Boy *et al.*, 2008a; cf. Chang *et al.*, this volume; Liu *et al.*, this volume; Oesker *et al.*, this volume). In TMCFs, NH_4^+, NO_3^-, and PO_4^{3-} levels and fluxes in throughfall are often distinctly lower than corresponding values in precipitation, presumably due to absorption by the abundantly present epiphytes (Clark *et al.*, 1998a; Hafkenscheid, 2000; Chang *et al.*, this volume).

Stream-water chemistry

As indicated further in Figure 44.2, average concentrations of major ions in stream water increased with decreasing elevation and increasing land use intensity. Chemical concentrations in upland forest streams were 1.5–3 times higher than in open-field precipitation and throughfall due to contributions by mineral weathering and cation exchange processes in the soil. The solute loads of streams receiving surface runoff from populated areas were 2–5 times higher than those for the upland stream-water sites. Among the upland sites (HB-100, QM-300, and QC-300), cationic concentrations decreased in the order $Na^+ > Ca^{2+} >$

$Mg^{2+} > K^+ >> NH_4^+$. The ratios between cations were the same at these sites, suggesting that bedrock mineralogy and soil chemistry in the Quebrada Maquina and Q. Cuecha watersheds are similar. The cation results are consistent with the andesitic bedrock and dacitic ignimbrite deposits of the Monteverde Formation and the high base saturation levels (48–99%) of organic and mineral soil horizons (Kim *et al.*, 2002). The upland streams sampled at QM-300 and HB-100 had higher overall base cation and ANC concentrations than the water at QC-300, suggesting that sub-surface water in the Q. Maquina watershed has a deeper flowpath and a longer contact time with soil and rock materials before entering the stream channel than the water in the Q. Cuecha (Figures 44.1 and 44.2).

For acid and salt anions, chloride concentration was highest ($Cl^- > SO_4^{2-} \cong NO_3^- >> PO_4^{3-}$) and varied little among the three upland sites (Figure 44.2). $SO_4^{2-} : Cl^-$ ratios in these streams were higher than in precipitation, and, like base cations and ANC, SO_4^{2-} concentrations were greater at QM-300 and HB-100 than at QC-300. This indicates that mineral weathering, for instance of accessory sulfide minerals commonly found in volcanic rocks, adds sulfate to the streams.

Strong acid and salt anion concentrations of the Q. Maquina and Q. Cuecha waters increased only slightly downstream of the main road (Figure 44.2). Elevated concentrations of chloride and sodium at RG-200 (Q. Cuecha is renamed Rio Guacimal where it intersects the road) were observed occasionally, presumably due to addition of cheese-processing wastewater discharged by the Monteverde Cheese Factory, located 100 m upstream of the RG-200 sampling site. Overall, the Monteverde Cloud Forest Preserve, the Bosque Eterno de los Niños, and other adjacent forests have limited development along the riparian zone of the upper reaches of these streams, which minimizes non-point source pollution.

In stark contrast, the Q. Sucia (site QS-200), which drains the most densely populated area in the watershed (Figure 44.1), showed much higher concentrations for all ions (Figure 44.2). Plotting the sum of base cations (SBC) against the sum of acid and salt anions (SAA) shows that pollution strongly controls the water chemistry of the Q. Sucia (Figure 44.3). Acid anion concentrations (SAA = 150–400 μeq l^{-1}) were more than twice those observed at the upland sites (SAA = 75–150 μeq l^{-1}). Furthermore, the wide range in SBC values (550–1350 μeq l^{-1}) combined with higher average ANC (616 μeq l^{-1}) indicate that alkaline pollution is added to the water of the Q. Sucia as well. Sources of alkaline pollution included sodium bicarbonate soaps and detergents in wastewater commonly observed in discharge to Q. Sucia from houses and laundromats. Mineral weathering of cements used in building construction may be an additional source of base cations and alkalinity to stream water from the developed areas.

In contrast to the elevated concentrations observed at QS-200, those observed in the Q. Socorro further downstream in the

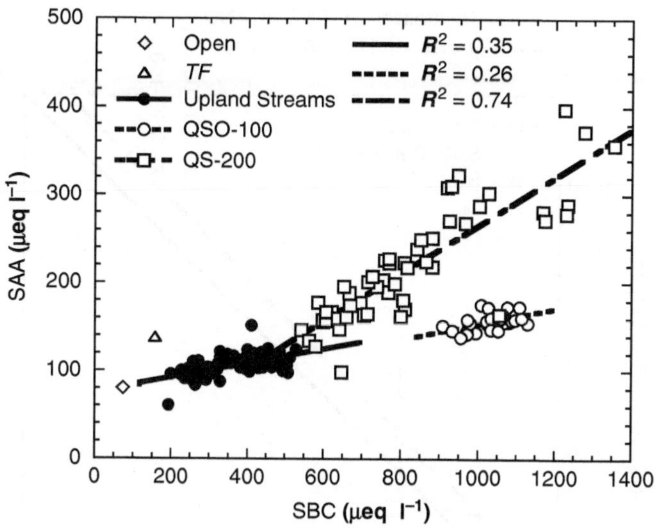

Figure 44.3. Sum of acid and salt anions (SAA) vs. sum of base cations (SBC) in waters collected in the Rio Guacimal watershed.

watershed cannot be attributed to high inputs of pollution. The range of SAA (125–200 μeq l^{-1}) is slightly elevated but comparable to what is seen in the upland sites (SAA = 75–150 μeq l^{-1}). Also, whilst SBC is high (900–1150 μeq l^{-1}), its variability is comparable to that in the samples from the upland sites (Figure 44.3). These observations suggest that mineral weathering and cation exchange are the major processes controlling the chemistry of the Q. Socorro, and agricultural activites and development has a minimal impact on this stream. Changes in bedrock composition, higher temperatures, and longer residence time of sub-surface flow at this lower elevation site could all explain the higher concentrations. Not far from QSO-100, calcite veins cross-cut hydrothermally altered andesite bedrock of the Aguacate Formation near the Rio Guacimal, and gold mineralization in the region (Cigolini and Chaves, 1986) suggests that hydrothermal veining and alteration may be extensive in the lower region of the watershed. Calcite chemically weathers much more rapidly than most silicate minerals (Langmuir, 1997), and even low percentages of hydrothermal calcite veins in bedrock can contribute significantly to the total calcium and alkalinity of stream water (cf. Mast *et al.*, 1990).

The net effect of adding acids, salts, and alkalinity to streams is shown in Figure 44.4. If base mineral weathering and cation exchange reactions were the only controls on stream-water chemistry, a theoretical 1 : 1 relationship between SBC and ANC, with an intercept of zero μeq l^{-1}, would exist. Whilst alkaline pollution does not change the observed 1 : 1 relationship, acids and salts (which remove ANC and add base cations, respectively) shift values away from this theoretical line (Rhodes *et al.*, 2001). On average, atmospheric effects on precipitation and sulphide weathering account for approximately an 80 μeq l^{-1} shift in the intercept of SBC versus ANC at the remote forested

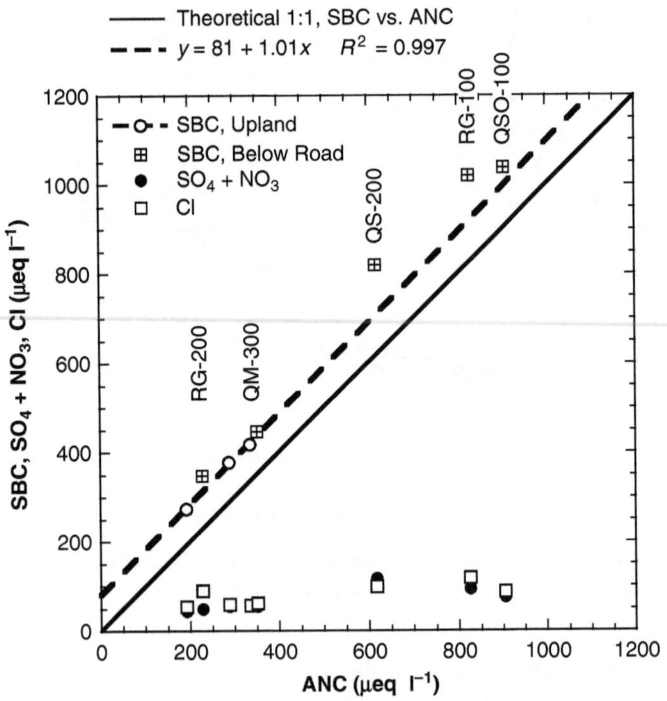

Figure 44.4. Sum of base cations (SBC) vs. acid-neutralizing capacity (ANC) quantifying the impact of pollution at each sampling location within the Rio Guacimal watershed. Acidic pollution (SO_4^{2-} + NO_3^-) is approximately equal to salt (Cl^-) additions.

sites (Figure 44.4). Relative to this baseline, further deviations from the theoretical relationship between SBC and ANC quantify the chemical impact of pollution. Figure 44.4 shows clearly that the Q. Sucia and the lower Rio Guacimal are the most impacted. In all eight sites, acidic pollution (SO_4^{2-} + NO_3^-) is approximately equal to salt additions (Cl^-), and these concentrations are greatest at sites QS-200 and RG-100.

CONCLUSIONS

Comparison of stream-water chemistry upstream and downstream of development areas in the Monteverde area of northern Costa Rica shows clear evidence of impacts on water quality arising from urban and agricultural development. Among three sampling sites located just downstream of the main road, the anthropogenic signature was strongest at the site representing the highest degree of development (QS-200). The chemistry of the streams at the other two sites (RG-200 and QM-200, situated at comparable positions in the landscape as site QS-200) was quite similar to that of the upland forest sites. Although QM-200 is downstream of several restaurants and hotels, and RG-200 is downstream of the Monteverde Cheese Factory, the area of forest cover upstream of these sites is much greater than the densely populated catchment upstream of QS-200. These results illustrate the value of the forest preserves, specifically the Monteverde Cloud Forest Preserve and the Bosque Eterno de Los Niños (Childen's Eternal Rainforest), in limiting the impacts on water quality of development and associated point- and non-point-source pollution in recharge regions of watersheds. Future work will seek to elucidate details of mineral weathering, causes behind temporal variations in stream chemistry (cf. Boy *et al.*, 2008b), and links between hydrological variation and geochemical variability.

ACKNOWLEDGEMENTS

We thank Nat Scrimshaw, The Monteverde Institute for providing logistic support and staff resources. We also thank B. Scheffe, A. Harmon, M. Marin, J. Torres, and J. Neibler, all of MVI, who assisted with stream sampling; S. Newell, C. Chazen, and M. Mick assisted with chemical analysis. Funding was provided by Smith College faculty development grants and the Smith Summer Science Internship Program.

REFERENCES

Boy, J., and W. Wilcke (2008). Tropical Andean forest derives calcium and magnesium from Saharan dust. *Global Biogeochemical Cycles* **22**, GB1027, doi: 10.1029/2007GB002960.

Boy, J., R. Rollenbeck, C. Valarezo, and W. Wilcke (2008a). Amazonian biomass burning-derived acid and nutrient deposition in the north Andean montane forest of Ecuador. *Global Biogeochemical Cycles* **22**, GB4011, doi:10.1029/2007GB003158.

Boy, J., C. Valarezo, and W. Wilcke (2008b). Water flow paths in soil control element exports in an Andean tropical montane forest. *European Journal of Soil Science* **59**: 1209–1227.

Cavelier, J., M. A. Jaramillo, D. Solis, and D. de León (1997). Water balance and nutrient inputs in bulk precipitation in tropical montane cloud forest in Panama. *Journal of Hydrology* **193**: 83–96.

Cignolini, C., and R. Chaves (1986). Geological, petrochemical and metallogenic characteristics of the Costa Rican gold belt: contribution to new explorations. *Geologische Rundschau* **75**: 737–754.

Clark, K. L., N. M. Nadkarni, D. Schaefer, and H. L. Gholz (1998a). Atmospheric deposition and net retention of ions by the canopy in a tropical montane forest, Monteverde, Costa Rica. *Journal of Tropical Ecology* **14**: 27–45.

Clark, K., N. Nadkarni, D. Schaefer and H. Gholz (1998b). Cloud water and precipitation chemistry in a tropical montane forest, Monteverde, Costa Rica. *Atmospheric Environment* **32**: 1595–1603.

Clark, K. L., R. O. Lawton, and P. Butler (2000). The physical environment. In *Monteverde: Ecology and Conservation of a Tropical Cloud Forest*, eds. N. M. Nadkarni and N. T. Wheelwright, pp. 15–34. New York: Oxford University Press

Fabian P., M.Kohlpaintner, and R. Rollenbeck. (2005). Biomass burning in the Amazon: fertilizer for the mountainous rain forest in Ecuador. *Environmental Science and Pollution Research* **12**: 290–296.

Fallas, J. (2002). Net precipitation patterns in undisturd and fragmented Costa Rican cloud forests. In *Proceedings of the 2 International Colloquium on Hydrology and Water Management*, ed. J. S. Gladwell, 389–398. Paris: UNESCO, and Panamá City: CATHALAC.

Frumau, K. F. A., L. A. Bruijnzeel, and C. Tobón (2006). Measurement of precipitation in montane tropical catchments: comparative performance of conventional, spherical and "potential" rain gages. In *Forest and Water in a Changing Environment*, eds. S. R. Liu, G. Sun, and P. S. Sun, pp. 104–108. Vienna: IUFRO, and Beijing: Chinese Academy of Forestry

Guswa, A. J., A. L. Rhodes, and S. E. Newell (2007). Importance of orographic precipitation to the water resources of Monteverde, Costa Rica. *Advances in Water Resources*, **30**: 2098–2112.

Hafkenscheid, R. L. L. J. (2000). Hydrology and biogeochemistry of tropical montane rain forests of contrasting stature in the Blue Mountains, Jamaica. Ph.D. thesis, VU University Amsterdam, Amsterdam, the Netherlands. Also available at http://dare.ubvu.vu.nl/bitstream/1871/12734/1/tekst.pdf.

Hendry, C. D., C. W. Barrish, and E. S. Edgerton (1984). Precipitation chemistry at Turrialba, Costa Rica. *Water Resources Research* **20**: 1677–1684.

Holwerda, F., R. Burkard, W. Eugster, *et al.* (2006). Estimating fog deposition at a Puerto Rican elfin cloud forest site: comparison of the water-budget and eddy covariance methods. *Hydrological Processes* **20**: 2669–2692.

Karmalkar, A. V., R. S. Bradley, and H. F. Diaz (2008). Climate change scenario for Costa Rican montane forests. *Geophysical Research Letters* **25**, L11702, doi: 10.1029/2008GL033940.

Kim, E. (2002). A water quality, hydrology, and policy study of a tropical cloudforest watershed in Monteverde, Costa Rica. B.A. thesis, Smith College, Northampton, MA, USA.

Kim, E., A. Rhodes, S. Katchpole, *et al.* (2002). Water quality study of a cloudforest watershed in Monteverde, Costa Rica. *The Geological Society of America, 2002 Northeastern Section Annual Meeting, Abstracts with Programs*, **34**: 18.

Köhler, L., C. Tobón, K. F. A. Frumau, and L. A. Bruijnzeel (2007). Biomass and water storage of epiphytes in old-growth and secondary montane rain forest stands in Costa Rica. *Plant Ecology* **193**: 171–184.

Langmuir, D. (1997). *Aqueous Environmental Geochemistry*. Upper Saddle River, NJ: Prentice Hall.

Lawton, R. O., and V. Dryer (1980). The vegetation of the Monteverde Cloud Forest Preserve. *Brenesia* **18**: 101–116.

Lawton, R., U. Nair, R. Pielke, and R. Welch (2001). Climatic impact of tropical lowland deforestation on nearby montane cloud forests. *Science* **294**: 584–587.

Mast, M. A., J. I. Drever, and J. Baron (1990). Chemical-weathering in the Loch Vale Watershed, Rocky Mountain National Park, Colorado. *Water Resources Research* **26**: 2971–2978.

Mulligan, M., K. F. A. Frumau, and L. A. Bruijnzeel (2006). Falling at the first hurdle: spatial rainfall variability and the problem of closing catchment water budgets in tropical montane environments. In *Forest and Water in a Changing Environment*, eds. S. R. Liu, G. Sun, and P. S. Sun, pp. 104–108. Vienna: IUFRO, and Beijing: Chinese Academy of Forestry.

Nadkarni, N. M., and N. T. Wheelwright (2000). Introduction. In *Monteverde: Ecology and Conservation of a Tropical Cloud Forest*, eds. N. M. Nadkarni and N. T. Wheelwright, pp. 3–13. New York: Oxford University Press.

Nair, U. S., R. O. Lawton, R. M. Welch, and R. A. Pielke Sr. (2003). Impact of land use on tropical montane cloud forests: sensitivity of cumulus cloud field characteristics to lowland deforestation. *Journal of Geophysical Research* **108** (D7): 4206–4218, doi:10.1029/2001JD001135.

National Atmospheric Deposition Program (NADP) (2002). *Annual Isopleth Map*. Available at http://nadp.sws.uiuc.edu/.

Organization for Tropical Studies (2005). *Palo Verde Meteorological Data*. Available at www.ots.ac.cr/en/.

Parker, G. G. (1983). Throughfall and stemflow in the forest nutrient cycle. In *Advances in Ecological Research 13*, eds. A. Macfadyen and E. D. Ford, pp. 58–136. New York: Academic Press.

Ray, D. K., U. S. Nair, R. O. Lawton, R. M. Welch, and R. A. Pielke Sr. (2006). Impact of land use on Costa Rican tropical montane cloud forests: sensitivity of orographic cloud formation to deforestation in the plains. *Journal of Geophysical Research* **111**: D02108, doi: 10.1029/2005JD006096

Rhodes, A., R. Newton, and A. Pufall (2001). Influences of land use on water quality of a diverse New England watershed. *Environmental Science and Technology* **35**: 3640–3645.

Rhodes, A. L., A. J. Guswa, and S. E. Newell (2006). Seasonal variation in the stable isotopic composition of precipitation in the tropical montane forests of Monteverde, Costa Rica. *Water Resources Research* **42**: W11402, doi:10.1029/2005WR004535.

Schaetzl, R., and S. Anderson (2005) *Soils: Genesis and Morphology*. Cambridge, UK: Cambridge University Press.

Still, C., P. Foster, and S.H. Schneider (1999). Simulating the effects of climate change on tropical montane cloud forests. *Nature* **389**: 608–610.

Zadroga, F. (1981). The hydrological importance of a montane cloud forest area of Costa Rica. In *Tropical Agricultural Hydrology*, eds. R. Lal and E. W. Russell, pp. 59–73. New York: John Wiley.

Zimmermann, A., W. Wilcke, and H. Elsenbeer (2007). Spatial and temporal patterns of throughfall quantity and quality in a tropical montane forest in Ecuador. *Journal of Hydrology* **343**: 80–96.

45 Is there evidence for limitations to nitrogen mineralization in upper montane tropical forests?

W. L. Silver, A. W. Thompson, D. J. Herman, and M. K. Firestone
University of California, Berkeley, California, USA

ABSTRACT

The structure and function of forest ecosystems often change along altitudinal gradients in the tropics, culminating in short-stature, low-productivity cloud forests at the uppermost elevations. Field data and literature values were used to examine patterns in nitrogen mineralization on tropical mountains and to discuss the potential for nitrogen limitation to net primary productivity. Few trends in net nitrogen mineralization within and across elevation gradients in the tropics were found, and rates were generally comparable to those found in tropical forests at low elevations. Gross nitrogen mineralization rates were much higher than net rates, and in Puerto Rico upper montane forests exhibited higher gross nitrogen mineralization than lower elevation forests. Work from Puerto Rico found no effect of short-term anaerobic conditions. In Hawai'i gross nitrogen mineralization increased with substrate age. Ammonium availability was augmented by dissimilatory nitrate reduction to ammonium in montane forests; nitrogen conservation via this pathway exceeded losses via N_2O production. Patterns in nitrogen circulation in upper montane forests in Puerto Rico showed that elfin and palm forests had lower nitrogen use efficiency and a higher proportion of nitrogen mineralized from decomposing litter relative to other forest types, further indicating that rates of nitrogen supply in these forests are considerable. In summary, the data reviewed in this chapter suggest that nitrogen limitation alone cannot explain patterns in the structure and function of tropical montane forest vegetation. Alternative factors are offered that warrant further investigation.

INTRODUCTION

Humid montane tropical forests in general, and montane cloud forests in particular, are often characterized by low net primary productivity, low standing biomass, and short stature relative to lowland tropical forests (Grubb, 1977). Whilst there are many potential mechanisms that can contribute to this, nitrogen limitation has frequently been suggested as an important and probable cause (Tanner *et al.*, 1990, 1992, 1998; Vitousek *et al.*, 1993; Raich *et al.*, 1997; Leuschner *et al.*, 2007; Moser *et al.*, 2007; cf. Soethe *et al.*, 2008a; Benner *et al.*, this volume). The basis for this lies in characteristics of the nitrogen cycle that differ fundamentally from other biogeochemical cycles. Nitrogen inputs to terrestrial ecosystems come primarily from the fixation of atmospheric nitrogen (as opposed to mineral weathering), and its cycling is predominantly controlled by biota that are thought to be highly sensitive to soil temperature, moisture availability, aeration, and nutrient supply rates over short temporal scales. Montane tropical forests are generally cooler than their lower elevation counterparts and often receive high rainfall (Grubb, 1977). Combined, these two factors typically lead to wetter soil conditions, particularly in upper montane forests (Bruijnzeel and Proctor, 1995; Santiago *et al.*, 2000 cf. Schawe *et al.*, this volume), where surface soils can experience periods of oxygen (O_2) depletion (Silver *et al.*, 1999). Depletion of O_2 is linked with temperature and rainfall, with lower temperatures often resulting in slower decomposition rates, higher soil organic matter pools (Silver *et al.*, 1999 Schuur, 2001 Schuur *et al.*, 2001 cf. Roman *et al.*, this volume), and increased water retention, leading in turn to conditions where O_2 consumption can exceed resupply via diffusive transport from the atmosphere (Silver *et al.*, 1999).

Nitrogen cycling is highly dynamic in humid, organic-rich tropical soils, making accurate characterization of patterns in nitrogen availability difficult. Most studies have only measured net rates of soil nitrogen cycling, using a simple analytical

Tropical Montane Cloud Forests: Science for Conservation and Management, eds. L. A. Bruijnzeel, F. N. Scatena, and L. S. Hamilton. Published by Cambridge University Press.

technique that derives nitrogen availability from the net change in mineral nitrogen (i.e. ammoniumn (NH_4^+) and nitrate (NO_3^-)) pools at two time points, before and after a field or laboratory incubation. A wide range of incubation lengths has been used (24 hours to 40 days). This complicates direct comparisons because nitrogen is likely to be produced or consumed throughout the incubation period, and the termination point is arbitrary with regard to microbial process rates.

Gross rates of soil nitrogen cycling provide a more realistic assessment of the dynamic nature of inorganc nitrogen pools. However, the assumptions of the procedure (principally the lack of internal recycling) are only valid over short time intervals, usually from 6 to 48 hours. Gross nitrogen mineralization and nitrification are determined through isotope pool dilution techniques following ^{15}N additions to the end-product pool (NH_4^+ for gross mineralization and NO_3^- for gross nitrification). The NH_4^+ and NO_3^- pools assessed by this method are those into which the added ^{15}N compounds can diffuse; consequently, these are the pools accessible to plant roots through diffusion (Hart *et al.*, 1994). Comparing rates of carbon mineralization to gross nitrogen transformations has led to the conclusion that nitrogen is being mineralized from labile, nitrogen-rich compounds (Schimel and Bennett, 2004) which is consistent with the enzymatic processes known to directly yield NH_4^+ (Myrold, 1998). The use of isotope tracers can also provide a measure of the fates of nitrogen within the soil (e.g. microbial uptake – termed immobilization, dissimilatory or assimilatory reduction, or storage as organic nitrogen), and when coupled with measurements of ^{15}N uptake by plants and ^{15}N gas emissions it can give a relatively complete picture of internal nitrogen supply and some of the factors affecting supply and loss rates over short time periods (Hart *et al.*, 1994; cf. Soethe *et al.*, 2006).

This study reviews patterns in net and gross soil nitrogen cycling for a series of montane tropical forest elevation gradients. The primary aim was to determine whether patterns of mineral nitrogen supply in soils vary systematically with elevation, redox conditions, or patterns in soil carbon and nitrogen pools. Both net and gross nitrogen mineralization and nitrification measurements are used as indices of nitrogen cycling rates, and the values obtained are compared with other data from montane and lowland tropical forests.

MATERIALS AND METHODS

Study area

This analysis uses a combination of new data and values from the literature. The new data come from a study conducted along a 700-m elevation gradient (350 to 1050 m.a.s.l.) in the Luquillo Experimental Forest (LEF), as part of the Long Term Ecological Research program in Puerto Rico, USA (18° 30′ N, 65° 80′ W). Mean annual temperature ranges from *c.* 26 °C at the lowest elevation to *c.* 19 °C at upper elevations. Annual precipitation ranges from approximately 3500 mm at 350 m elevation to >5000 mm at 1050 m.a.s.l., and is evenly distributed throughout the year (Brown *et al.*, 1983; Weaver, 1994; García-Martino *et al.*, 1996). The cloud-affected zone, which includes tropical montane wet forest (TMW, 650–750 m.a.s.l.), palm forest (750–920 m.a.s.l.), taller cloud forest (820–930 m.a.s.l.), and elfin forest (>900 m.a.s.l.), receives approximately 4500 mm of rainfall annually (García-Martino *et al.*, 1996). Elfin forests are most common on exposed ridges and slopes, and are shorter in stature than the other forests in the research area. The taller cloud forest occurs in swales and on leeward slopes, where the vegetation is taller than in elfin forests (4–9 m vs. 2–3 m, respectively), and has different plant community composition (Brown *et al.*, 1983; Weaver, 1995). The elfin forest receives an estimated additional input via cloud water interception of *c.* 770 mm year^{-1} (Holwerda *et al.*, 2006; cf. Holwerda *et al.*, this volume #29). Soils are classified as clay-rich ultisols (Brown *et al.*, 1983; Roman *et al.*, this volume). Soils are relatively deep and highly weathered, and are rich in exchangeable iron, soil organic matter content, total nitrogen, and extractable phosphorus relative to lower elevation forests. Soil pH (0–10 cm depth) in the elfin zone averages 5.41 (± 0.15) and is significantly higher (less acidic) than in lower elevation forests (Silver *et al.*, 1999). Tree height and some soil nutrient pools (magnesium, potassium) appear to co-vary along the gradient (Roman *et al.*, this volume).

Experimental approach

Twelve 10 × 30 m permanent plots were established in 1994 between ~650 and 970 m.a.s.l., with three replicate plots in each of the four main forest types. Plots were located on ridges and slope positions, where the respective forest types reach their best development (Weaver, 1994). The sites are part of a long-term study to estimate the effects of climate on forest net primary productivity and biogeochemical cycling.

For the determination of net nitrogen mineralization rates and nitrogen pool sizes, three soil samples were collected at 0–10 cm depth in each plot. Samples were split in the laboratory, with one half of each sample being extracted the same day with 2M KCl (Hart *et al.*, 1994). The other half were covered with an air-permeable lid and incubated in the dark at approximately 23 °C for 7 days and then extracted as above. All extracts were frozen for later colorometric analysis of NH_4^+ and $NO_3^- + NO_2$ at the International Institute of Tropical Forestry (IITF) using an Alpkem auto analyzer. Net nitrogen mineralization and nitrification rates were calculated according to Hart *et al.* (1994).

For gross nitrogen cycling rates, five soil samples (0–10 cm depth) were collected from two plots each in the TMW, palm forest, and elfin forest. Soil samples ($n = 5$ per plot and two replicate 10×30 m plots) were also collected from ridge, slope, and valley positions at 350 m elevation in the Bisley Research Watersheds. Samples were express-mailed within 12 hours to the University of California at Berkeley. Upon arrival, samples were split such that half of the samples received $^{15}NH_4SO_4$ at a soil concentration of 0.023 μg g^{-1} (final soil enrichment of 2.4 atom % $^{15}NH_4$), whereas the other half received $^{15}KNO_3$ added at a soil concentration 0.12 μg g^{-1} (final soil enrichment of 12.4 atom % $^{15}NO_3$). Thirty grams of oven-dry equivalent sample from each set of labeled soils were incubated in 225-ml jars under ambient conditions for 0, 6, and 24 hours, and under a dinitrogen (N_2) headspace for 0, 3, 6, and 24 hours. At the end of incubations, the jars were extracted with 150 ml of 2M KCl. Incubations were done at 25 °C. Extracts were prepared for isotope analysis by diffusion (Herman *et al.*, 1995), and nitrogen-isotope ratios were measured using an automated nitrogen-carbon analyzer coupled to an isotope ratio mass spectrometer (ANCA–IRMS). Nitrous oxide was determined by gas chromatography using a ^{63}Ni detector, and nitrogen-gas isotope ratios using a trace gas module coupled to an IRMS. Rates of dissimilatory NO_3^- reduction to NH_4^+ (DNRA) – an anaerobic microbial process that has been shown to rapidly convert NO_3^- to NH_4^+ in some upland soils (Silver *et al.*, 2001, 2005; Pett-Ridge and Firestone, 2005; Pett-Ridge *et al.*, 2006; Templer *et al.*, 2008) – were also measured. DNRA has the potential to conserve nitrogen in ecosystems with high potential NO_3^- losses from denitrification and leaching (Silver *et al.*, 2001). DNRA rates were calculated as the difference in $^{15}NH_4^+$ atom % between sampling periods, multiplied by the mean NH_4^+ pool size during the interval, and corrected for the mean residence time (MRT) of the NH_4^+ pool. This was then divided by the mean $^{15}NO_3^-$ atom % during the interval to account for the isotopic composition of the source pool. To estimate MRT of the $^{15}NH_4^+$ pool, the initial NH_4^+ pool (μg g^{-1}) was divided by the rate of gross consumption using the 0–24-hour interval data from the simultaneous laboratory experiment. The MRT value from the 0–24-hour interval was chosen because it reduced the impact of short-term soil disturbance, and because it was the largest MRT value measured, its use in the DNRA calculation provided the most conservative estimates. Gross mineralization, nitrification, and NH_4^+ and NO_3^- consumption were calculated according to Kirkham and Bartholomew (1954), with means of the 6- and 24-hour incubations reported.

Field litterfall and decomposition experiments were used to estimate patterns in nitrogen cycling in forests along the elevational gradient. Litterfall was collected in five baskets (approximately 0.16 m^2 each) per plot. Baskets were lined with fine mesh and suspended off the ground on 50-cm stakes to avoid contact with the soil surface and perforated at the bottom to allow drainage. Litter was collected every 2 weeks from 1994 to 1999, dried at 65 °C, and sorted into leaves, fruits and flowers, fine wood (<5 cm diameter), and miscellaneous material. Each litter fraction was then redried and weighed. For nutrient analyses, samples were bulked by plot, forest type, and month yielding one nutrient sample per sorting category per forest type per month ($n = 16$ samples per month). Dried and ground samples were analyzed for total carbon and nitrogen using a CE Elantech NC2100 elemental analyzer (Lakewood, NJ). Decomposition of leaf litter was measured over a 2-year period at all the upper montane sites ($n = 12$ plots). Approximately 5 g of dried leaf litter was placed in mesh bags with <1 mm openings. Five bags were collected from each plot after 3, 9, 16, and 24 months. Samples were dried at 65 °C, weighed to determine mass, and redried prior to ashing at 550 °C for 12 hours to determine the amount of mineral contamination. Dried and ground samples were analyzed for total carbon and nitrogen as described above. Examination of the data determined that an exponential relationship best explained patterns in mass loss over time (Wieder and Lang, 1982), and this was used to estimate a decomposition coefficient (k).

Statistical procedures

Statistical analyses were performed using Systat (Wilkinson, 1984). Analysis of variance (ANOVA) was used to determine whether changes in pool sizes or rates occurred among treatments. Data were log-transformed when necessary to meet assumptions for ANOVA. Significant differences were determined as $p < 0.05$ unless noted otherwise.

RESULTS AND DISCUSSION

Net nitrogen transformations in montane and lowland tropical forests

Results from the Puerto Rico elevational gradient and other tropical montane forest transect studies suggest that potential net nitrogen mineralization and nitrification do not generally vary systematically along elevation gradients within or among study sites. In the Luquillo Mountains of Puerto Rico, net nitrogen mineralization rates ranged from −2.1 (± 0.7) to 6.0 (± 5.5) μg g^{-1} day^{-1} and averaged 0.2 (± 0.6) μg g^{-1} day^{-1}. Net nitrification rates exhibited a smaller range (0.0 ± 0.0 to 2.0 ± 1.0 μg g^{-1} day^{-1}) and averaged 0.5 (± 0.2) μg g^{-1} day^{-1}. Neither net nitrogen mineralization nor net nitrification rates followed statistically significant patterns with elevation or forest type (Table 45.1). Although all samples were incubated at laboratory temperatures, rates varied significantly among sites, suggesting that other factors (e.g. organic nitrogen pools, redox potential, pH) are controlling net nitrogen fluxes.

Table 45.1 *Patterns in net nitrogen cycling along an elevation gradient in the Luquillo Experimental Forest, Puerto Rico. Net nitrogen mineralization from field measurements is compared with estimates from the difference between gross nitrogen mineralization and gross ammonium consumption. Upper-case letters identify statistically significant differences among elevations. Lower-case letters identify statistically significant differences between measured and estimated rates. Values are means and standard errors.* $p < 0.05$ *unless otherwise noted. Short cloud forest is referred to as elfin forest in the text*

Elevation (m)	Forest	Net N mineralization ($\mu g\ g^{-1}\ day^{-1}$)	Net nitrification ($\mu g\ g^{-1}\ day^{-1}$)	Estimated net N mineralization ($\mu g\ g^{-1}\ day^{-1}$)
635	TMW	0.44 ± 0.16	0.84 ± 0.13 AC	
731	TMW	−0.56 ± 0.15	0.23 ± 0.18 B	0.06 ± 0.28 [a]
741	TMW	6.03 ± 5.48	1.98 ± 1.05 A	0.63 ± 0.30
806	Palm	0.37 ± 0.70	1.20 ± 0.53 C	
858	Palm	0.45 ± 0.40	0.75 ± 0.07 D	0.09 ± 0.10
871	Taller cloud	−2.07 ± 0.72	0.00 ± 0.00 BE	
915	Taller cloud	−0.33 ± 0.22	0.00 ± 0.00 BE	
923	Palm	−0.88 ± 0.48	0.25 ± 0.25 BD	0.20 ± 0.06 a
930	Taller cloud	−0.09 ± 0.41	0.28 ± 0.10 DE	
949	Short cloud	−0.57 ± 0.18	0.22 ± 0.11 DE	0.07 ± 0.04 a
967	Short cloud	−0.29 ± 0.28	0.00 ± 0.00 B	
968	Short cloud	−0.22 ± 0.27	0.59 ± 0.22 CD	0.54 ± 0.15 a

[a] $p = 0.085$.

Net nitrogen transformation rates in the LEF were similar to those from other montane systems (Figure 45.1). Few patterns in net nitrogen mineralization or nitrification are discernable with elevation or temperature. Montane forest soils in Costa Rica (Marrs *et al.*, 1988) and Borneo (ultrabasic soils only, Kitayama and Aiba, 2002) showed a linear decrease in net nitrogen mineralization and nitrification with elevation, but in Costa Rica rates were not sensitive to temperature in a laboratory experiment, where high soil moisture and presumably low soil aeration were the primary limiting factors to net nitrogen transformations (Marrs *et al.*, 1988). Similarly, net nitrogen transformations in soils from Colombian cloud forests were insensitive to temperature, but responded positively to drying, again suggesting aeration as a potential controlling mechanism (Cavelier *et al.*, 2000; cf. Leuschner *et al.*, 2007). Vitousek and Matson (1988) reported a range of net nitrogen mineralization rates from 0.8–6.8 $\mu g\ g^{-1}\ day^{-1}$ for lowland tropical forest soils; montane forest ecosystems had somewhat greater net nitrogen immobilization rates (i.e. uptake in microbial biomass), but similar net mineralization potential, with rates ranging from −2.6–6.0 $\mu g\ g^{-1}\ day^{-1}$.

Gross nitrogen mineralization, retention, and loss

In the upper elevations of the LEF, gross nitrogen mineralization rates were considerably higher than net rates and ranged from 19–21 (both ± 5) $\mu g\ g^{-1}\ day^{-1}$ whereas gross nitrification rates ranged from 16 (± 5) to 19 (± 6) $\mu g\ g^{-1}\ day^{-1}$ (Figure 45.2). Gross mineralization and nitrification rates were significantly higher in the upper montane forests (>600 m.a.s.l.) than in lower elevation forest types (*c.* 350 m.a.s.l.) and much of the mineralized nitrogen was consumed, primarily through high gross nitrification rates (Figure 45.2).

Rates of gross nitrogen mineralization in the LEF were similar to rates reported for Hawai'i, the only other tropical montane site having comparable data (Figure 45.3). The Hawaiian data suggest that gross nitrogen mineralization increases with increasing substrate age (Riley and Vitousek, 1995; Hall and Matson, 2003; Figure 45.3). Sites on substrates older than 1 million years had similar gross nitrogen mineralization rates as the Puerto Rican montane soils, which are derived from highly weathered saprolite (Buss *et al.*, 2008). Gross nitrification rates were significantly higher in the montane forests of the LEF than in lower elevation LEF soils, or montane forests in Hawai'i or tropical Australia (Figure 45.4).

Upper montane tropical forest soils often experience periodic reducing conditions (Silver *et al.*, 1999) that can significantly affect microbial nitrogen cycling processes (Silver *et al.*, 2001, 2005; Pett-Ridge *et al.*, 2006). Gross nitrogen mineralization rates were insensitive to short-term anaerobic conditions in both upper and lower elevation Puerto Rican forests. In contrast, rates of dissimilatory NO_3^- reduction to NH_4^+ (DNRA) increased when soils became anoxic in the normally well-drained lower-elevation ridges and in the tropical montane wet forest. In nature, these soils are generally well aerated and experience reducing conditions infrequently. Slope and valley soils at the lower elevation site showed no significant increase in DNRA with

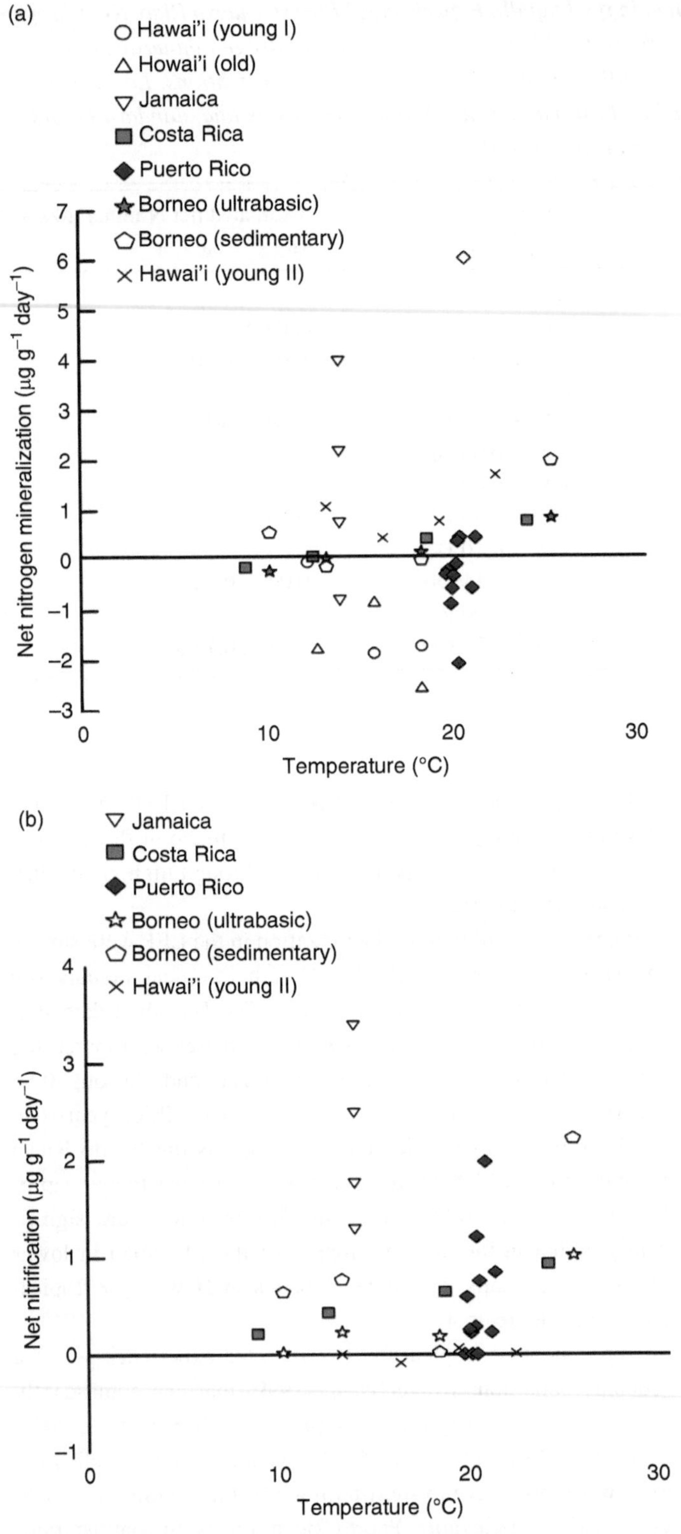

Figure 45.1. Rates of (a) net nitrogen mineralization and (b) nitrification along elevation and temperature gradients on tropical mountains. Data are from: young and old Hawaiian soils I (Vitousek *et al.*, 1988); Jamaica (Tanner, 1977); Costa Rica (Marrs *et al.*, 1988); Puerto Rico (this study); Borneo (Kitayama and Aiba, 2002); and young Hawaiian soil II (Raich *et al.*, 1997).

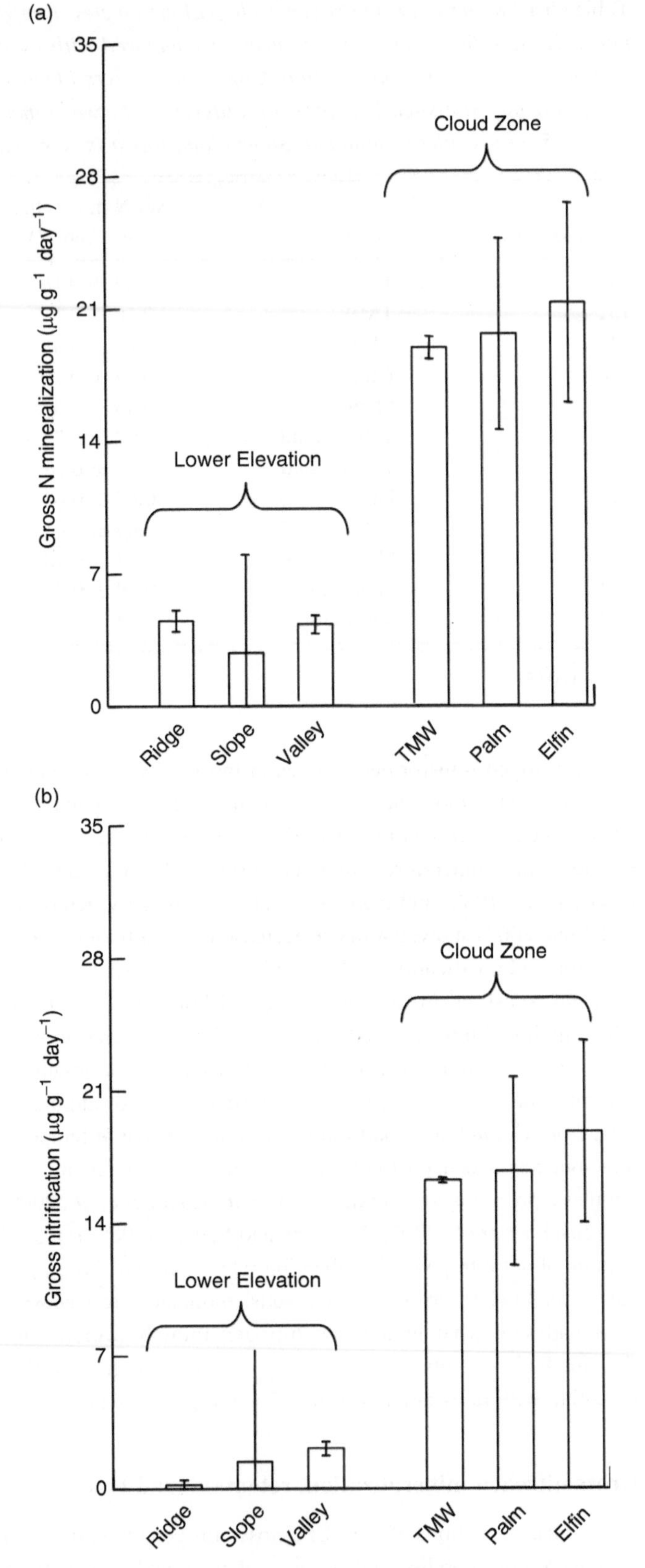

Figure 45.2. Rates of (a) gross mineralization and (b) gross nitrification in low-elevation forests and montane cloud forests in the Luquillo Experimental Forest, Puerto Rico. Values are means and standard errors.

anaerobic conditions, nor did the palm or elfin forest soils (Figure 45.5). These soils are likely to experience frequent low redox potential conditions and may not have been significantly affected by the short-term anaerobiosis imposed during the incubations. Rates of DNRA accounted for approximately 10% of gross mineralization rates. DNRA is also likely to play an important role in decreasing the amount of nitrogen available for N_2O production. Humid tropical forests are the largest natural source of N_2O, a potent greenhouse gas (Lashof and Ahuja, 1990) and catalyst for stratospheric ozone depletion (Cicerone, 1987). Nitrous oxide is produced both via nitrification and denitrification, although denitrification is thought to dominate in perhumid soils (Groffman and Tiedje, 1989). Like DNRA, denitrification is generally thought to be controlled by soil redox conditions, nitrification rates, NO_3^- availability, and labile carbon pools (Tiedje *et al.*, 1982). Nitrous oxide fluxes from LEF soils are high, averaging 34.0 ± 9.6 ng cm^{-2} hour^{-1} (Silver *et al.*, 2001). Rates of DNRA significantly exceeded N_2O fluxes (Figure 45.6), suggesting that DNRA may limit N_2O production by decreasing the NO_3^- available for denitrification (Templer *et al.*, 2008). Denitrification to N_2 is another potential pathway for nitrogen loss under anaerobic conditions (Tiedje *et al.*, 1982). The relative importance of nitrogen loss via N_2 emissions is not well known for montane tropical forests, primarily due to methodological constraints.

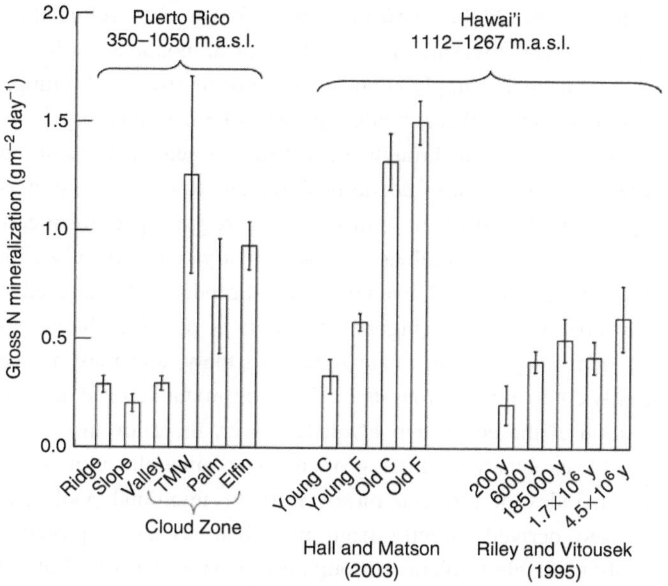

Figure 45.3. Gross mineralization rates in Puerto Rican and Hawaiian montane forest soils. Values are means and standard errors. C, Control; F, fertilized with nitrogen for the Hall and Matson (2003) study; the *x*-axis for Riley and Vitousek (1995) is substrate age.

Patterns of nitrogen circulation

Patterns of nutrient circulation via plant uptake, nutrient use efficiency, and decomposition can provide an indirect index of nutrient availability. Theory would suggest that plants growing in a nutrient-poor soil would cycle those nutrients conservatively and efficiently by maximizing production per unit of nutrient uptake. At the stand level, this can lead to the production of litter with low nutrient quality (due to nutrient resorption prior to senescence), slow decomposition rates, and a positive feedback to low nutrient availability (Vitousek, 1982, 1984; Silver, 1994). In the LEF, leaf litterfall mass, nutrient and chemical content, and the rate of leaf litter decay were measured to determine

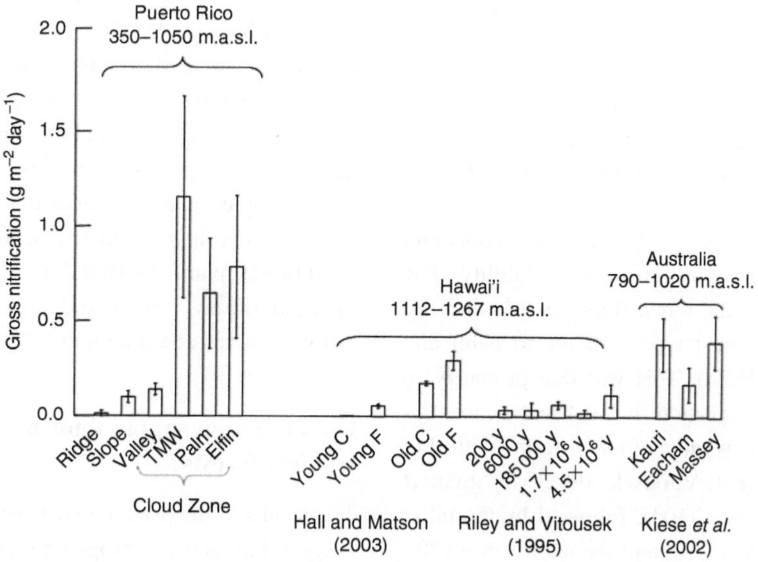

Figure 45.4. Gross nitrification rates in Puerto Rican, Hawaiian, and Australian montane forest soils. Values are means and standard errors. C, Control; F, fertilized with nitrogen for the Hall and Matson (2003) study; the *x*-axis for Riley and Vitousek (1995) is substrate age. The *x*-axis for Kiese *et al.* (2002) refers to site names.

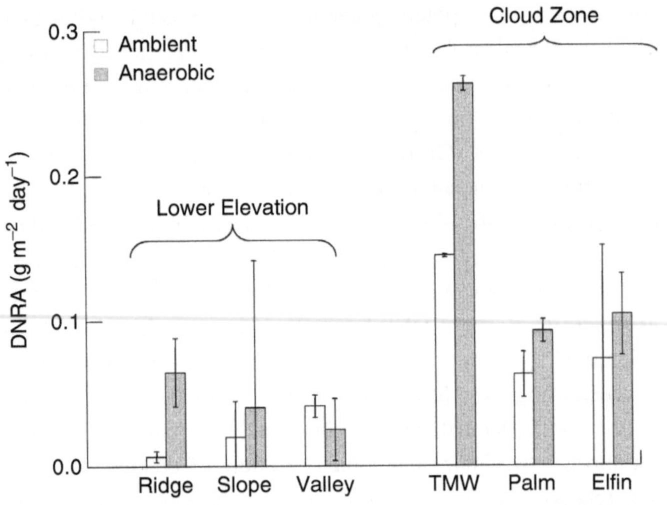

Figure 45.5. The effects of an N_2 headspace on dissimilatory nitrate reduction to ammonium in low-elevation forests and montane cloud forests in the Luquillo Experimental Forest, Puerto Rico. Values are means and standard errors.

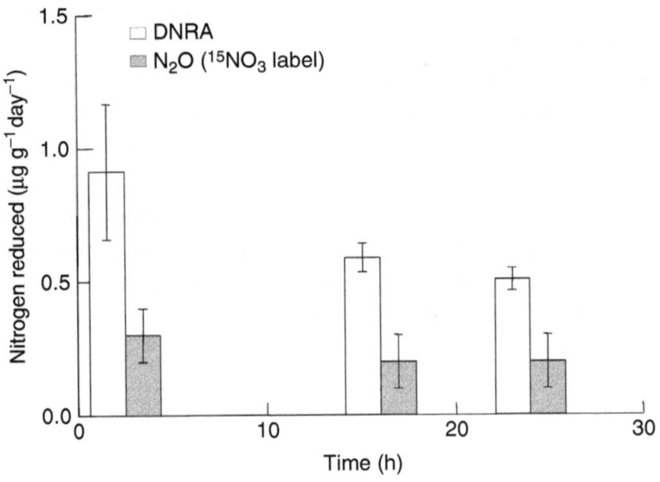

Figure 45.6. Patterns in dissimilatory nitrate reduction to ammonium and nitrous oxide fluxes from montane cloud forests in the Luquillo Experimental Forest, Puerto Rico. Values are means and standard errors.

whether patterns of nutrient circulation through these pools and fluxes appeared to be influenced by nitrogen availability. The amount of nitrogen returned to the forest floor via leaf litterfall was lower in the two cloud forest types relative to palm and montane wet forest (Figure 45.7a). This was due primarily to declining litterfall mass with elevation, not nutrient concentrations. The nitrogen use efficiency value of leaf litterfall (litterfall mass divided by nitrogen content; Vitousek, 1982) was greatest in the tropical montane wet forest (104), followed by the taller cloud forest (92), the elfin forest (82), and the palm forest (78). The low nitrogen use efficiency in the palm forest likely derives from high nitrogen concentrations in leaf tissue of the dominant palm species.

Litter decay rates did not follow the same trend as litter production (Figure 45.7b), and appeared to be related to carbon : nitrogen and lignin : nitrogen ratios in the decomposing material (Figure 45.7c,d). When comparing amounts of nitrogen produced in leaf litterfall annually with amounts released from decomposing leaf tissues, the pattern follows that for nitrogen use efficiency, with the lowest proportion of recycling occurring in the tropical wet forest and the greatest proportion occurring in the palm and elfin forests. This would suggest that in the Luquillo Mountains the higher elevation cloud forests are cycling their nitrogen less conservatively than the tropical montane wet forest, and that nitrogen supply is not likely to be as strongly limiting.

At first glance these results appear to be contradictory. Low nitrogen circulation through the vegetation generally results in poor litter quality, slow decomposition, and low nitrogen mineralization rates. However, in this study, nitrogen supply (as measured by gross N mineralization) did not vary systematically with elevation (Figure 45.3), even though leaf productivity declined at the upper elevations (Figure 45.7a). It is possible that gross nitrogen assays overestimate nitrogen supply, and may not be readily comparable to long-term fluxes of production and decomposition at the ecosystem level. However, there are also other possible explanations. First, gross nitrogen mineralization assays are an index of nitrogen turnover in the microbial pool. The nitrogen derived directly from microbial bodies is probably highly available to plants (through diffusion in the soil), but not directly correlated with nitrogen inputs via leaf litter. Second, not all nitrogen is retained within the ecosystem. Some montane tropical forests have been shown to exhibit high rates of N_2O efflux and denitrification (Keller *et al.*, 1986; Silver *et al.*, 1999; Hall and Matson, 2003) due to high gross nitrogen transformations in soils, fluctuating redox conditions, and abundant labile carbon pools. The humid conditions of the LEF are also likely to lead to considerable nitrogen loss in leachate (McDowell and Asbury, 1994). Nitrogen export via these pathways undoubtedly accounts for some of the mineralized nitrogen not taken up by plants and microbes. Finally, if plants are limited by some other factor that does not directly affect nitrogen mineralization (e.g. high wind stress, low light levels, anaerobic soil conditions), plant demand for nitrogen may be low, inorganic nitrogen supply rates could exceed plant uptake, and overall nitrogen losses could be high.

Does nitrogen supply limit productivity of montane tropical forests?

The results of the present experiments and review of the literature suggest that rates of nitrogen mineralization and nitrification vary within and across tropical mountains, but generally do not decrease systematically with elevation. Rates of mineral nitrogen supply were generally high and not likely to be a limiting factor to

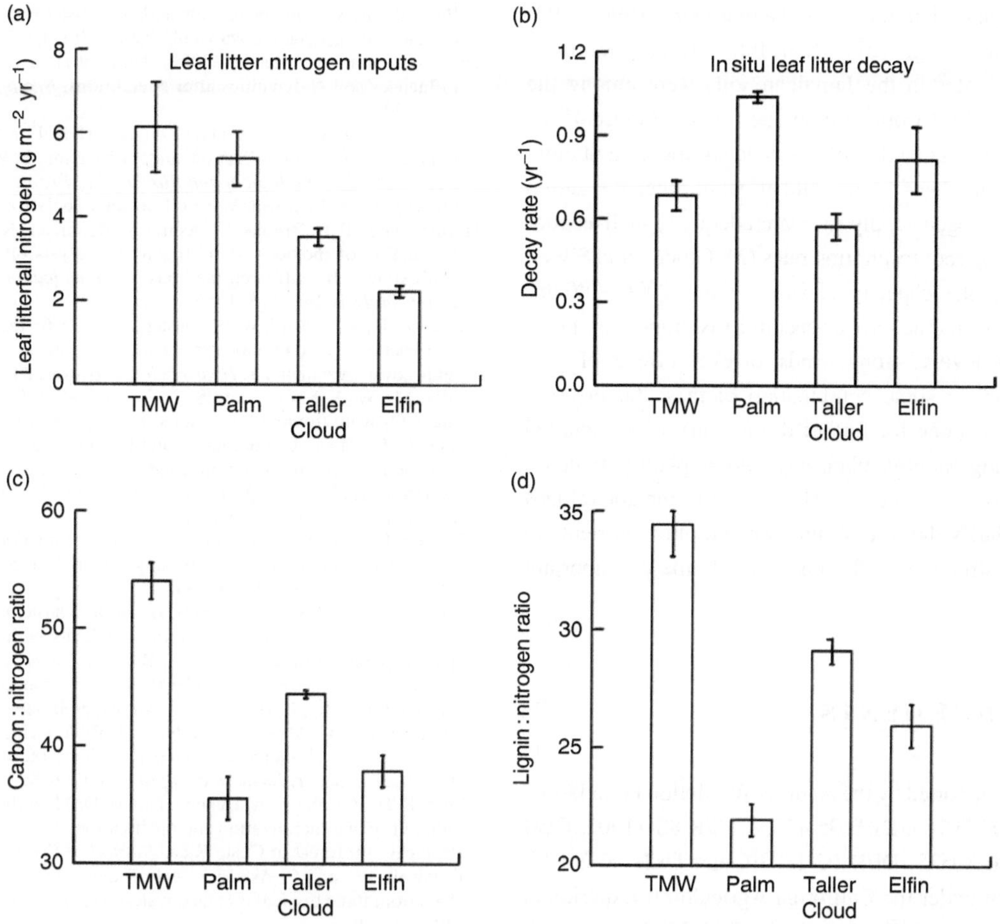

Figure 45.7. (a) Nitrogen content, (b) decomposition rates, (c) carbon-to-nitrogen ratios, and (d) lignin-to-nitrogen ratios for leaf litter from four upper montane forest types in the Luquillo Experimental Forest, Puerto Rico. Values are means and standard errors.

net primary productivity (NPP) in an altitudinal sequence of montane forests in Puerto Rico. Rates of gross nitrogen mineralization in Puerto Rico and Hawai'i were much higher than net rates. Periods of poor aeration did not appear to limit gross nitrogen mineralization, and in some cases stimulated nitrogen conservation via dissimilatory nitrate reduction to ammonium (DNRA). In the Luquillo Experimental Forest, litter quality did not follow a trend with elevation, nor did litter decomposition. The most productive forest along the elevation gradient (tropical montane wet forest) had the highest carbon: nitrogen and lignin: nitrogen ratios in leaf litter (cf. Figure 47.7c, d), exhibited a slower decay rate than the elfin forest, as well as the highest nitrogen use efficiency, and the lowest degree of nitrogen recycling from litter.

These findings suggest that mineral nitrogen supply is unlikely to result in strong nitrogen limitation to NPP in montane tropical forests. However, soil nitrogen mineralization assays are only one index of nitrogen availability and potential limitation. Forest fertilization studies measure growth responses to increased nutrient availability, and constitute a widely accepted approach to assessing nutrient limitation. However, these studies are not without problems either. Most fertilization studies focus only on above-ground tissues (e.g. litterfall and/or stem growth) and do not account for potential shifts in allocation from below- to above-ground processes (cf. Leuschner *et al.*, 2007; Moser *et al.*, 2007; Hertel and Leuschner, this volume), thereby overestimating the degree of nutrient limitation to the biota. Furthermore, the diversity of most tropical forests makes it difficult to determine patterns in species-specific responses to fertilization. Different species may respond differently to nutrient additions, making it difficult to estimate whole ecosystem responses if small plots are used (Tanner *et al.*, 1998).

Nitrogen fertilization studies have been conducted in montane forests in Hawai'i, Jamaica, and Venezuela (summarized in Tanner *et al.*, 1998; Benner *et al.*, this volume). Stem growth increased after nitrogen additions on young volcanic soils in Hawai'i (Vitousek *et al.*, 1993; Raich *et al.*, 1997), but not on older soils (Vitousek and Farrington, 1997), thereby matching the pattern in gross nitrogen supply discussed above. In Jamaica and Venezuela, plant responses to fertilization occurred only after a significant lag time (approximately 2 years). Stem growth and/or litterfall increased at both sites (Tanner *et al.*, 1990, 1992) suggesting potential nitrogen limitation. Interestingly, net nitrogen mineralization rates in the Jamaican soils were among the highest rates reported for tropical montane forests (Figure 45.1).

Some montane forests clearly do experience nitrogen limitation. It is also possible that other factors could affect nitrogen utilization by plants, even if nitrogen supply rates are adequate, such as poor soil aeration affecting root respiration rates (McGroddy and Silver, 2000) and root uptake capacity (Soethe *et al.*, 2006, 2008b), unfavorable element interactions or specific toxicities (e.g. polyphenols), low light levels, strong winds, or even effects of high ultraviolet radiation, or some combination of these factors (see Benner *et al.*, this volume for further discussion). More detailed studies of the rooting patterns, plant physiology, plant population and community ecology, and biogeochemistry of montane (cloud) forests will help clarify the role of nitrogen and other nutrients in determining the structure and function of these important ecosystems.

ACKNOWLEDGEMENTS

This research was supported by the Andrew W. Mellon Foundation, NSF grants DEB 00-89783, DEB 05-42558, BSR 88-11902, DEB 94-11973, DEB 00-80538, DEB 02-18039, and DEB 06-20910, and was carried out under the California Agricultural Experiment Station projects # 7069-MS (WLS) and # 6117-H (MKF). We thank Peter Vitousek for helpful feedback on an earlier draft.

REFERENCES

Brown, S., A. E. Lugo, S. Silander, and L. Liegal (1983). *Research History and Opportunities in the Luquillo Experimental Forest*, General Technical Report No. SO-441. New Orleans, LA: U.S. Department of Agriculture-Forest Service Southern Forest Experimental Station.

Bruijnzeel, L. A., and J. Proctor (1995). Hydrology and biogeochemistry of tropical montane cloud forests: what do we really know? In *Tropical Montane Cloud Forests*, eds. L. S. Hamilton, J. O. Jarvik, and F. N. Scatena, pp. 38–78. New York: Springer-Verlag.

Buss, H. L., A. F. White, D. Vivit, *et al.* (2008). Mineral weathering in a deep volcaniclastic saprolite, Luquillo Mountains, Puerto Rico. *Geochimica et Cosmochimica Acta* **72**: 124–133.

Cavelier, J., E. Tanner, and J. Santamaria (2000). Effect of water, temperature and fertilizers on soil nitrogen net transformations and tree growth in an elfin cloud forest of Columbia. *Journal of Tropical Ecology* **16**: 83–99.

Cicerone, R. J. (1987). Changes in stratospheric ozone. *Science* **237**: 35–42.

García-Martinó, A. R., G. S. Warner, F. N. Scatena, and D. L. Civco (1996). Rainfall, runoff and elevation relationships in the Luquillo Mountains of Puerto Rico. *Caribbean Journal of Science* **32**: 413–424.

Groffman, P. M., and J. M. Tiedje (1989). Denitrification in north temperate forest soils: spatial and temporal patterns at the landscape and seasonal scales. *Soil Biology and Biochemistry* **21**: 613–620.

Grubb, P. J. (1977). Control of forest growth and distribution on wet tropical mountains: with special reference to mineral nutrition. *Annual Review of Ecology and Systematics* **8**: 83–107.

Hafkenscheid, R. L. L. J. (2000). Hydrology and biogeochemistry of tropical montane rain forests of contrasting stature in the Blue Mountains, Jamaica. Ph.D. thesis, VU University Amsterdam, Amsterdam, the Netherlands. Also available at http://dare.ubvu.vu.nl/bitstream/1871/12734/1/tekst.pdf.

Hall, S. J., and P. A. Matson (2003). Nutrient status of tropical rain forests influences soil N dynamics after N additions. *Ecological Monographs* **73**: 107–129.

Hart, S. C., J. M. Stark, E. A. Davidson, and M. K. Firestone (1994). Nitrogen mineralization, immobilization, and nitrification. In *Methods of Soil Analysis*, Part 2, *Microbiological and Biochemical Properties*, ed. R. W. Weaver, pp. 985–1018. Madison, WI: Soil Science Society of America.

Herman, D. J., P. D. Brooks, M. Ashraf, F. Azam, and R. L. Mulvaney (1995). Evaluation of methods for nitrogen-15 analysis of inorganic nitrogen in soil extracts. II. Diffusion methods. *Communications in Soil Science and Plant Analysis* **26**: 1675–1685.

Holwerda, F., R. Burkard, W. E. Eugster, *et al.* (2006). Estimating fog deposition at a Puerto Rican elfin cloud forest site: comparison of the water budget and eddy covariance methods. *Hydrological Processes* **20**: 2669–2692.

Keller, M., W. A. Kaplan, and S. C. Wofsy (1986). Emissions of N_2O, CH_4, and CO_2 from tropical forest soils. *J. Geophys. Res.* **91**: 11 791–11 802.

Kiese, R., H. Papen, E. Zumbusch, and K. Butterbach-Bahl (2002). Nitrification activity in tropical rain forest soils of the Coastal Lowlands and Atherton Tablelands, Queensland, Australia. *Journal of Plant Nutrition and Soil Science* **165**: 682–685.

Kirkham, D., and W. V. Bartholomew (1954). Equations for following nutrient transformations in soil, utilizing tracer data. *Soil Science Society of America Proceedings* **18**: 33–34.

Kitayama, K., and S. Aiba (2002). Ecosystem structure and productivity of tropical rain forests along altitudinal gradients with contrasting soil phosphorus pools on Mount Kinabalu, Borneo. *Journal of Ecology* **90**: 37–51.

Lashof, D. A., and D. R. Ahuja (1990). Relative contributions of greenhouse gas emissions to global warming. *Nature* **344**: 529–531.

Leuschner, Ch., G. Moser, C. Bertsch, M. Röderstein, and D. Hertel (2007). Large altitudinal increase in tree root/shoot ratio in tropical mountain forests of Ecuador. *Basic and Applied Ecology* **8**: 219–230.

Marrs, R. H., J. Proctor, A. Heaney, and M. D. Mountford (1988). Changes in soil nitrogen mineralization and nitrification along an altitudinal transect in tropical rain forest in Costa Rica. *Journal of Ecology* **76**: 466–482.

McDowell, W., and C. Asbury (1994). Export of carbon, nitrogen, and major ions from three tropical montane watersheds. *Limnology and Oceanography* **39**: 111–125.

McGroddy, M. and W. L. Silver (2000). Variations in belowground carbon storage and soil CO_2 flux rates along a wet tropical climate gradient. *Biotropica* **32**: 614–624.

Moser, G., D. Hertel, and Ch. Leuschner (2007). Altitudinal change in LAI and stand leaf biomass in tropical montane forests: a transect study in Ecuador and a pan-tropical meta-analysis. *Ecosystems* **10**: 924–935.

Myrold, D. D. (1998). Microbial nitrogen transformations. In *Principles and Applications of Soil Microbiology*, eds. D. M. Sylvia, J. J. Fuhrmann, P. G. Hartel, and D. A. Zuberer, pp. 259–294. Upper Saddle River, NJ: Prentice Hall.

Pett-Ridge, J., and M. K. Firestone (2005). Redox fluctuation structures microbial community in a wet tropical soil. *Applied and Environmental Microbiology*, **71**: 6998–7007.

Pett-Ridge, J., W. L. Silver, and M. K. Firestone (2006). Redox fluctuations in a humid tropical forest soil impact N-cycling rates by framing the composition of the soil microbial community. *Biogeochemistry* **81**: 95–110.

Píco, R. (1974). *The Geography of Puerto Rico*. Chicago, IL: Aldine.

Raich, J. W., A. E. Russell, and P. M. Vitousek (1997). Primary productivity and ecosystem development along an elevational gradient on Mauna Loa, Hawai'i. *Ecology* **78**: 707–721.

Riley, R. H., and P. M. Vitousek (1995). Nutrient dynamics and nitrogen trace gas flux during ecosystem development in montane rain forest. *Ecology* **76**: 292–304.

Santiago, L. S., G. Goldstein, F. C. Meinzer, J. H. Fownes, and D. Mueller-Dombois (2000). Transpiration and forest structure in relation to soil waterlogging in a Hawaiian montane cloud forest. *Tree Physiology* **20**: 673–681.

Schimel, J. P., and J. Bennett (2004). Nitrogen mineralization: challenges of a changing paradigm. *Ecology* **85**: 591–602.

Schuur, E. (2001). The effect of water on decomposition dynamics in mesic to wet Hawaiian montane forests. *Ecosystems* **4**: 259–273.

Schuur, E. A. G., O. A. Chadwick, and P. A. Matson (2001). Carbon cycling and soil carbon storage in mesic to wet Hawaiian montane forests. *Ecology* **82**: 3182–3196.

Silver, W. (1994). Is nutrient availability related to plant nutrient use in humid tropical forests? *Oecologia* **98**: 336–343.

Silver, W. L., M. Keller, and A. E. Lugo (1999). Soil oxygen availability and biogeochemical cycling along elevation and topographic gradients in Puerto Rico. *Biogeochemistry* **44**: 301–328.

Silver, W. L., D. J. Herman, and M. K. Firestone (2001). Dissimilatory nitrate reduction to ammonium in upland tropical forest soils. *Ecology* **82**: 2410–2416.

Silver, W. L., A. W. Thompson, M. K. Firestone, A. Reich, and J. J. Ewel (2005). Nitrogen retention and loss in tropical plantations and old growth forests. *Ecological Applications* **15**: 1604–1614.

Soethe, N., J. Lehmann, and C. Engels (2006). The vertical pattern of rooting and nutrient uptake at different altitudes of a south Ecuadorian montane forest. *Plant and Soil* **286**: 287–299.

Soethe, N., J. Lehmann, and C. Engels (2008a). Nutrient availability at different altitudes in a tropical montane forest in Ecuador. *Journal of Tropical Ecology* **24**: 397–406.

Soethe, N., W. Wilcke, J. Homeier, J. Lehmann, and C. Engels (2008b). Plant growth along the altitudinal gradient: role of plant nutritional status, fine root activity, and soil properties. In *Gradients in a Tropical Mountain Ecosystem of Ecuador*, eds. E. Beck, J. Bendix, I. Kottke, F. Makeschin, and R. Mosandl, pp. 259–266. Berlin: Springer-Verlag.

Tanner, E. V. J. (1977). Four montane rain forests of Jamaica: a quantitiative characterization of the floristics, the soils and the foliar mineral levels, and a discussion of the interrelations. *Journal of Ecology* **65**: 883–918.

Tanner, E. V. J., V. Kapos, S. Freskos, J. R. Healey, and A. M. Theobald (1990). Nitrogen and phophorus fertilization of Jamaican montane forest trees. *Journal of Tropical Ecology* **6**: 231–238.

Tanner, E. V. J., V. Kapos, and W. Franco (1992). Nitrogen and phosphorus fertilization effects on Venezuelan montane forest trunk growth and litterfall. *Ecology* **73**: 78–86.

Tanner, E. V. J., P. M. Vitousek, and E. Cuevas (1998). Experimental investigation of nutrient limitation of forest growth on wet tropical mountains. *Ecology* **79**: 10–22.

Templer, P. H., W. L. Silver, J. Pett-Ridge, K. DeAngelis, and M. K. Firestone (2008). Plant and microbial controls on nitrogen retention and loss in a humid tropical forest. *Ecology* **89**: 3030–3040.

Tiedje, J. M., A. J. Sexton, D. D. Myrold, and J. A. Robinson (1982). Denitrification: ecological niches, competition and survival. *Antonie van Leeuwenhoeck* **48**: 569–583.

Vitousek, P. (1982). Nutrient cycling and nutrient use efficiency. *American Naturalist* **119**: 553–572.

Vitousek, P. (1984). Litterfall, nutrient cycling, and nutrient limitation in tropical forests. *Ecology* **65**: 285–298.

Vitousek, P. M., and H. Farrington (1997). Nutrient limitation and soil development: experimental test of a biogeochemical theory. *Biogeochemistry* **37**: 63–75.

Vitousek, P. M., and P. A. Matson (1988). Nitrogen transformations in a range of tropical forest soils. *Soil Biology and Biochemistry* **20**: 361–367.

Vitousek, P. M., P. A. Matson, and D. R. Turner (1988). Elevational and age gradients in Hawaiian montane rainforest: foliar and soil nutrients. *Oecologia* **84**: 362–370.

Vitousek, P. M., L. R. Walker, L. D. Whiteaker, and P. A. Matson (1993). Nutrient limitations to plant growth during primary succession in Hawaii Volcanoes National Park. *Biogeochemistry* **23**: 197–215.

Weaver, P. L. (1994). *Bano de Oro Natural Area Luquillo Mountains, Puerto Rico*. New Orleans, LA: U.S. Department of Agriculture Forest Service Southern Forest Experiment Station.

Weaver, P. L. (1995). The Colorado and dwarf forests of Puerto Rico's Luquillo Mountains. In *Tropical Forests: Management and Ecology*, eds. A. E. Lugo and C. Lowe, pp. 109–141. New York: Springer-Verlag.

Wieder, R. K., and G. E. Lang (1982). A critique of the analytical methods used in examining decomposition data obtained from litter bags. *Ecology* **63**: 1636–1642.

Wilkinson, L. (1984). *Systat, Version 1*. Evanston, IL: Systat Inc.

46 Fine root mass and fine root production in tropical moist forests as dependent on soil, climate, and elevation

D. Hertel and Ch. Leuschner
University of Göttingen, Göttingen, Germany

ABSTRACT

This chapter presents a meta-analysis of fine root mass and productivity in tropical moist forests in terms of the dependence on various environmental factors, using 87 data-sets from both Paleo- and Neotropical forests. The present review differs from earlier analyses in that it focuses strictly on the fine root fraction (<2 mm in diameter) and applies relatively rigid criteria with respect to the selection of data (a.o. to prevent the merging of data on live and dead root mass). Forests in the upper montane belt (>2000 m.a.s.l.) have markedly higher live fine root biomass compared to mid-elevation and lowland forests, both in the Paleotropics/Australia and the Neotropics. Hence, the ratio of shoot to fine root biomass decreases significantly with elevation. Fine root production is negatively related to above-ground biomass. These findings highlight the increasing ecological importance of the fine root system of tropical moist forests with increasing elevation.

INTRODUCTION

Fine roots play an important role in the functioning of trees because they are the organs of water and nutrient acquisition. Although representing a relatively small part of total tree biomass, fine roots often consume a large portion of the annual carbon gain (Grier *et al.*, 1981; Vogt *et al.*, 1996). Decaying fine roots are a major source of carbon addition to the soil organic matter pool. The rising interest in the below-ground compartment of forests in the last decades has led to an increasing number of studies on fine root biomass and turnover, mainly in temperate and boreal forests (see global reviews by Vogt *et al.*, 1996; Cairns *et al.*, 1997; Gill and Jackson, 2000). Despite the widely recognized importance of tropical forests in the global carbon budget (Malhi *et al.*, 1999; Post and Kwon, 2000; Lal, 2005), current knowledge about structure and function of the fine root system in tropical moist forests is limited, particularly for tropical montane (cloud) forests (Bruijnzeel and Veneklaas, 1998).

For this review the available information on fine root mass and fine root production of all types of moist old-growth forests (ranging from lowland forests to upper montane (cloud) forests) in the American, African, and Australasian tropics was assembled to investigate relations between fine root mass or fine root production on the one hand, and temperature, rainfall, elevation, and soil acidity (pH) on the other. The key question addressed here is: "Is there a principal difference in fine root mass and productivity between lowland and montane tropical forests?"

Moist tropical lowland forests are known to differ significantly from montane forests in above-ground structure and species composition, but also in soil chemical properties and nutrient availability (Vitousek and Sanford, 1986; Marrs *et al.*, 1988; Richards, 1996; Soethe *et al.*, 2008a,b; cf. Benner *et al.*, this volume; Roman *et al.*, this volume; Silver *et al.*, this volume). In the majority of lowland forests, tree nutrition is characterized by fast litter decay and high mineralization rates (Jenny *et al.*, 1949; Meentemeyer, 1977; Anderson and Swift, 1983). Montane forests, on the other hand, experience lower temperatures (Grubb, 1977; Hemp, this volume #12; Jarvis and Mulligan, this volume) and, often, adverse soil conditions such as partial anoxia and low pH (Schrumpf *et al.*, 2001; cf. Roman *et al.*, this volume) leading, in turn, to slower decomposition rates (Tanner, 1981; Schuur 2001; Takyu *et al.*, 2003; cf. Benner *et al.*, this volume). Therefore, major differences in structure and dynamics of the fine root

Tropical Montane Cloud Forests: Science for Conservation and Management, eds. L. A. Bruijnzeel, F. N. Scatena, and L. S. Hamilton. Published by Cambridge University Press.

system between lowland and montane forests are to be expected. Recently, Moser *et al.* (2007) examined the altitudinal changes in forest leaf area index (*LAI*), specific leaf area (*SLA*), and leaf biomass using pan-tropical data from 39 montane forest stands. *LAI* decreased with elevation by about 1 $m^2\ m^{-2}\ km^{-1}$, implying a reduction of 40–60% between 1000 and 3000 m.a.s.l. Such a strong reduction in leaf surface area with elevation can be expected to have important implications for forest productivity (i.e. canopy carbon gain; Moser *et al.*, 2007).

The present meta-analysis differs from earlier reviews in several ways: (i) only records of fine root data for forests in the humid tropics were considered, in contrast to other more global studies (Vogt *et al.*, 1996; Cairns *et al.*, 1997; Gill and Jackson, 2000); (ii) unlike these earlier studies, data on live (biomass), dead (necromass), and total (live + dead) fine root mass were not combined; (iii) only fine root data in the strict sense (mostly <2 mm) were considered; and (iv) only production data obtained by direct measurements of root mass change over time or by calculations based on fine root respiration measurements in combination with records of fine root mass standing stock were included. Despite these comparatively rigid criteria, more than 80 data-sets were compiled and analyzed.

THE DATABASE

The database used for this analysis was compiled from 42 available literature records of fine root mass or production as obtained in 87 different stands in the humid tropics of South America, Africa, Asia, Australia and Oceania (incl. Hawai'i) (tabulated in Appendix 46.1). Stand selection was done mainly according to the global distribution limits of tropical moist forests as defined by Richards (1996), using a rainfall limit of 1700 mm $year^{-1}$ to separate tropical wet and dry forests (Richards, 1996). Only studies giving mass or production data of fine roots <3 mm in diameter were selected; the majority of the data refer to fine root data <2 mm, although in one single case data of fine root mass <4 mm were admitted. In most studies, fine root mass was recorded on a single sampling occasion; where data from repeated sampling were provided, mean values were admitted. Care was taken to separate live (biomass), dead (necromass), and total (live + dead) fine root mass records. Sampling depths differed markedly between investigations; although most studies recorded fine root mass only within the upper 10–50 cm of soil, some investigators sampled to a depth of 400 or even 600 cm. However, no standardization of root mass values for a defined soil depth was applied in this analysis, because many studies showed the fine root systems to be concentrated in the uppermost soil horizons, even where very deep profiles of several meters were investigated (e.g. Nepstad *et al.*, 2002). Nevertheless, this inconsistency in sampling depth between studies may have contributed to the variance in the overall data-set.

Methodologic differences made it difficult to select adequate fine root production data. Only data recorded by direct analysis of the existing fine root system were admitted: e.g. root mass changes over time applying sequential soil coring (e.g. Sánchez-Gallén and Alvarez-Sánchez, 1996; Sundarapandian and Swamy, 1996), in-growth techniques (e.g. Cuevas and Medina, 1988; Cavelier *et al.*, 1999; Stewart, 2000), or approximations of fine root productivity based on fine root respiration measurements in combination with records of the fine root mass standing stock (Herbert and Fownes, 1999). All indirect estimates, including those derived from above-ground structural parameters, were excluded from the analysis.

Linear and simple non-linear regression analyses of the dependence of fine root mass or production on soil, climate or elevation were done using Xact software (SciLab, Hamburg, Germany, version 7.12). All other statistical analyses were conducted with the package SAS (SAS Institute, Cary, NC, USA, version 6.12).

FINE ROOT BIOMASS AND NECROMASS OF MATURE MOIST TROPICAL FORESTS

Fine root biomass

In 55 observations, fine root material was separated into live and dead roots. Standing stocks of fine root biomass in these studies varied widely between 8 and 1440 g m^{-2} (see Appendix 46.1 for details), with an overall average fine root biomass of 451 ± 45 g m^{-2}. Leuschner and Hertel (2003) derived a mean fine root biomass of 442 g m^{-2} (range: 69–1500 g m^{-2}) in an analysis of 129 studies on temperate coniferous and broad-leaved forests. Thus, the mass of living fine roots in mature tropical forests does not differ significantly from that in temperate forests. This finding is surprising, because tropical moist forests have a much larger above-ground biomass compared to forests in other climatic regions (Dixon *et al.*, 1994; Malhi *et al.*, 1999, 2006b). Assuming that a larger above-ground biomass requires higher nutrient and water uptake by the fine root system (Soethe *et al.*, 2008a,b), one might be tempted to conclude that the fine roots of tropical forests are more efficient in terms of surface-specific nutrient and water uptake rates. On the other hand, it has to be considered that water and nutrient uptake in the humid tropics takes place throughout the year, compared to a growing season of 5–8 months in temperate regions.

Grouping the data by elevation showed that fine root biomass of mature lowland forests (<400 m.a.s.l.) does not differ significantly from that of lower montane forests (400–2000 m.a.s.l.), but the fine root biomass of upper montane forests (above 2000 m.a.s.l.) is about three times larger (Figure 46.1). These three elevational categories were chosen to separate typical lowland

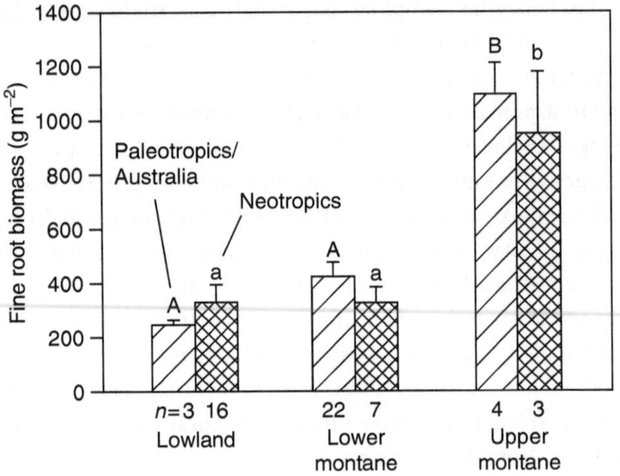

Figure 46.1. Fine root biomass of old-growth forests in tropical humid lowland (<400 m.a.s.l.), lower montane (400–2000 m.a.s.l.), and upper montane (>2000 m.a.s.l.) regions. Both means and standard errors are shown. Different capital or lower-case letters indicate significant differences among the elevation groups of the two provenances; $p < 0.05$, Scheffé's test for unbalanced, parametrically distributed samples.

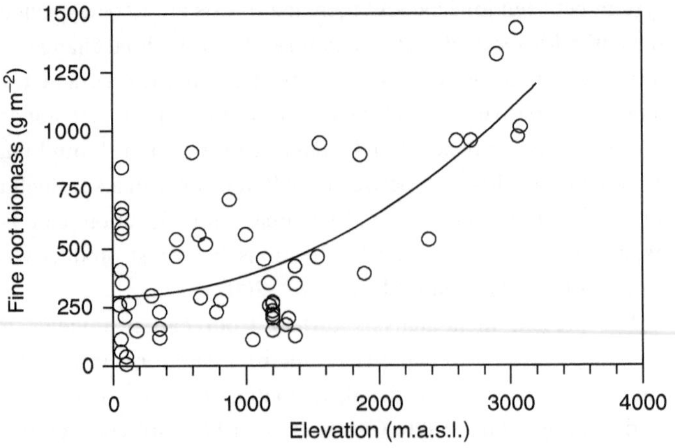

Figure 46.2. Relationship between fine root biomass and elevation for old-growth moist tropical forests ($r^2_{adj} = 0.48$, $p < 0.001$, $n = 55$). See Table 46.1 for more information on the regression analysis.

forests from better-drained hill and lower montane forest stands, and to combine all high-elevation forests with less favorable temperature and soil water regimes in a separate class (cf. Schrumpf *et al.*, 2001; Schawe *et al.*, 2007; Moser *et al.*, 2008; Roman *et al.*, this volume). It is important to note that this grouping does not reflect floristic or physiognomic vegetation zones at different elevations as introduced e.g. by Grubb (1977) or Richards (1996). In none of the three elevation groups could a significant difference in fine root biomass be found between Paleotropical (including Australian) and Neotropical forests (Figure 46.1). Whilst the majority of studies on fine root biomass in the Paleotropics pertain to lower montane forests, corresponding investigations in the Neotropics were conducted mainly in the lowlands. So far, only four studies of fine root biomass have been conducted in Paleotropical upper montane forests, and even less (three) in the Neotropics (Figure 46.1 and Appendix 46.1), although new work has recently been completed in eastern Bolivia (Hertel and Wesche, 2008) or is under way in the Eastern Cordillera in Perú (Malhi *et al.*, 2006a).[1]

An analysis of the relationships between fine root biomass and various environmental factors showed that live fine root mass increases exponentially with elevation (Figure 46.2). Although fine root biomass of old-growth forests at lower altitudes varies widely, elevation explains a major portion of the variance of the data-set (Table 46.1). Due to its close relationship with elevation, air temperature correlated negatively to fine root biomass. Interestingly, no other factor considered was found to influence fine root biomass significantly for the data-set as a whole, including mean annual precipitation, soil acidity (pH), or above-ground biomass (Table 46.1), although high correlations between fine root biomass and C/N ratio of the organic layer or topsoil pH have been reported by individual transect studies (e.g. Moser *et al.*, 2008). The generally observed increase in fine root biomass with elevation is all the more interesting because above-ground biomass did decrease linearly with elevation and, therefore, with decreasing temperatures (Table 46.1). Together, the linear decrease in above-ground biomass and the non-linear increase in fine root biomass led to a significant decrease in the shoot/fine root biomass ratio with elevation (Figure 46.3, Table 46.1).

Which factors cause the observed increase in fine root biomass with elevation? Tropical forests at higher elevations are growing under conditions of lower temperatures and, in many regions, high precipitation (Grubb, 1977; Bruijnzeel and Veneklaas, 1998; Jarvis and Mulligan, this volume). Upper montane cloud forest soils are often waterlogged, oxygen-depleted, and acid (Tanner *et al.*, 1998; Silver *et al.*, 1999; Santiago *et al.*, 2000; Schrumpf *et al.*, 2001; Schawe *et al.*, 2007; cf. Roman *et al.*, this volume; Schawe *et al.*, this volume). Under such conditions, nutrient cycling rates and nutrient availability are often reduced (Meentemeyer, 1977; Schuur, 2001; Benner *et al.*, this volume). However, high soil acidity and nutrient limitation (e.g. of phosphorus) occur in tropical lowlands as well (Vitousek and Sanford, 1986; Tanner *et al.*, 1998; Malhi *et al.*, 2004; cf. Roman *et al.*, this volume). Interestingly, for a number of lowland forests comparatively high fine root biomass values (500–850 g m^{-2}) have been reported, whereas values <500 g m^{-2} are absent for high-elevation sites (Figure 46.2).

There is evidence that tropical montane forest growth is negatively affected by low nutrient availability (notably nitrogen). Studies by Marrs *et al.* (1988) and Kitayama and Aiba (2002) showed that N mineralization in tropical forest soils decreases with elevation (see also Benner *et al.*, this volume; Silver *et al.*, this

[1] Results not included in the present analysis.

Table 46.1 *Relationships between fine root biomass, necromass, total mass (live + dead), above-ground biomass, and log (shoot/fine root) biomass ratio of old-growth moist tropical forests, and various environmental variables*

Parameter	Source	Regression equation	r^2	r^2_{adj}	p	n
Fine root biomass	Elevation	$y = 296 + 8.85 \cdot 10^{-5} \cdot x^2$	**0.49**	**0.48**	**0.001**	55
	Air temperature	$y = 185 + 81.1 \cdot x^{-2}$	**0.37**	**0.36**	**0.001**	46
	Precipitation	$y = 474 - 0.0079 \cdot x$	0.00	−0.02	0.877	55
	Soil proton concentration	$y = 441 + 1.79 \cdot 10^5 \cdot x$	0.01	−0.02	0.639	33
	Above-ground biomass	$y = 452 + 0.36 \cdot x$	0.02	−0.03	0.580	23
Fine root necromass	Elevation	$y = 26.4 + 0.41 \cdot x$	**0.46**	**0.44**	**0.001**	21
	Air temperature	$y = -68 + 1.60 \cdot 10^5 \cdot x^{-2}$	**0.64**	**0.62**	**0.001**	18
	Precipitation	$y = -132 + 0.23 \cdot x$	0.15	0.10	0.085	21
	Soil proton concentration	$y = 359 + 2.71 \cdot 10^9 \cdot x^2$	**0.73**	**0.71**	**0.001**	20
Total fine root mass	Elevation	$y = 941 + 0.20 \cdot x$	0.01	−0.01	0.433	49
	Air temperature	$y = 1328 - 6.84 \cdot x$	0.00	−0.02	0.863	44
	Precipitation	$y = 1752 - 0.22 \cdot x$	0.03	0.00	0.276	49
	Soil proton concentration	$y = 548 + 2.65 \cdot 10 \cdot x^6$	**0.32**	**0.30**	**0.001**	35
Above-ground biomass	Elevation	$y = 344 - 0.0613 \cdot x$	**0.19**	**0.17**	**0.005**	31
	Air temperature	$y = -20.1 + 16.1 \cdot x$	**0.39**	**0.36**	**0.001**	30
Log(shoot biomass/fine root biomass)	Elevation	$y = 2.41 - 0.0004 \cdot x$	**0.64**	**0.62**	**0.001**	23

Notes: r^2 and adjusted r^2 refer to linear or simple non-linear regression models. Regression results with $p < 0.05$ are printed in bold. Fine root mass in g m^{-2}, above-ground biomass in Mg ha^{-1}, log (shoot/fine root) biomass ratio in g m^{-2} g^{-1} m^{-2}; other variable units: m.a.s.l. (elevation), °C (mean annual air temperature), mm $year^{-1}$ (annual precipitation), mol l^{-1} (soil proton concentration, derived from pH(H_2O) data).

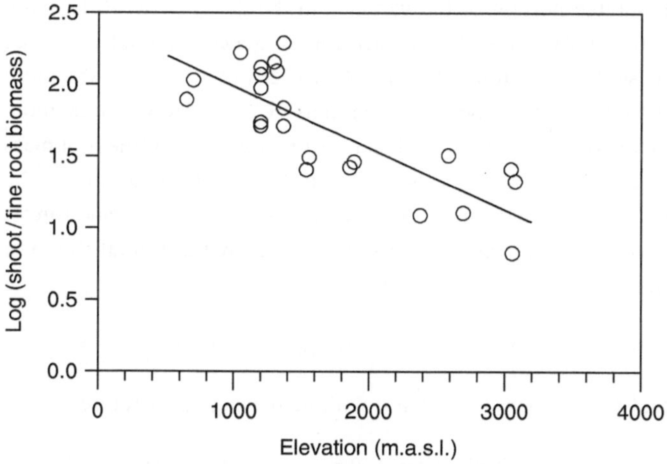

Figure 46.3. Relationship between log(shoot/fine root) biomass ratio and elevation for old-growth moist tropical forests ($r^2_{adj} = 0.51$, $p < 0.001$, $n = 23$). See Table 46.1 for more information on the regression analysis.

volume). However, mineralization rates in the Costa Rican study by Marrs *et al.* (1988) were not found to be sensitive to temperature in a laboratory experiment and the authors concluded that the observed reduction in mineralization rates with elevation rather reflected the increasingly wetter soil conditions at higher altitudes (cf. Bruijnzeel and Proctor, 1995; Schawe *et al.*, this volume). Nevertheless, since neither high water content nor high soil acidity were found to influence fine root biomass of tropical forests directly (Table 46.1), it could be speculated that retarded decomposition and mineralization, but also reduced nutrient uptake rates caused by lower temperatures, might lead to increased nutrient limitation at higher elevations (see also the discussion in Soethe *et al.*, 2008a,b). A temperature effect on nutrient supply and uptake, which is absent in lowland forests, might be one possible explanation for the present finding that tropical forests at higher altitudes invest more carbon into the fine root system compared to lowland forests (cf. Leuschner *et al.*, 2007).

Necromass and total fine root mass

The majority of fine root studies in old-growth moist tropical forests recorded total fine root mass (live + dead roots) or fine root biomass only (Appendix 46.1). Only 21 observations comprised separate data on fine root necromass. Fine root necromass varied to a greater extent than fine root biomass: an average of 525 g m^{-2} (range: 34–2400 g m^{-2}) was estimated (Appendix 46.1). Thus, not only the mean, but also the maximum value of fine root necromass of tropical moist forests exceeds the corresponding fine root biomass values. However, part of this large variability is likely to be caused by methodologic differences between studies and uncertainty in the determination of dead fine roots.

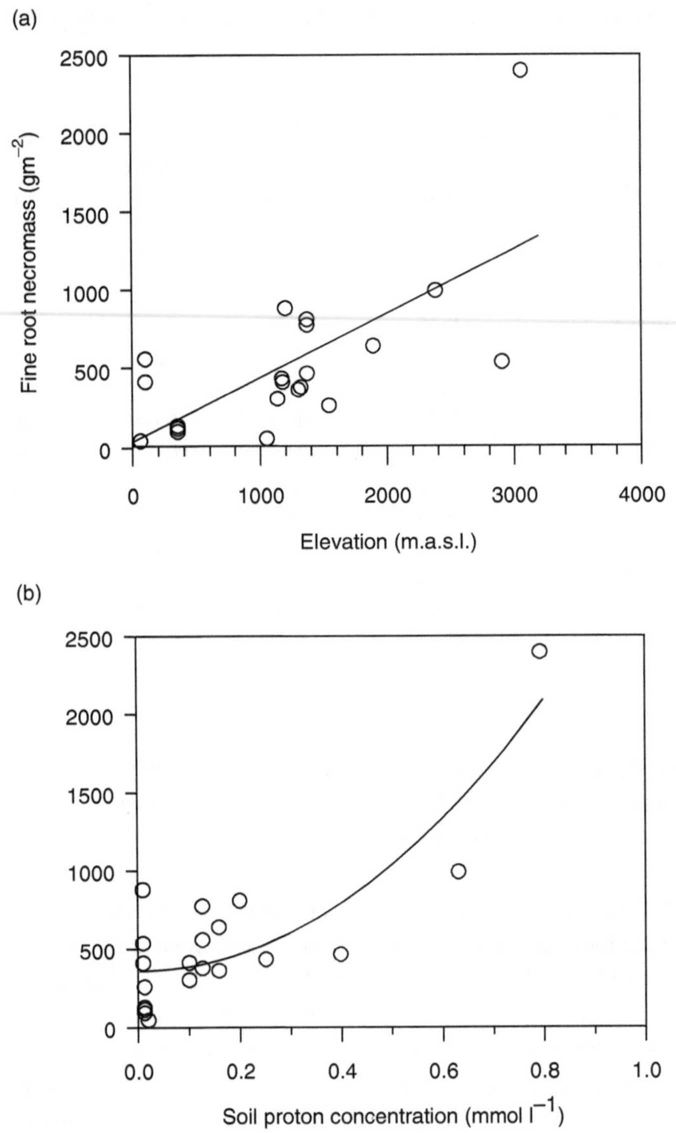

Figure 46.4. Relationships between fine root necromass of old-growth moist tropical forests and (a) elevation ($r^2_{adj} = 0.44, p < 0.001, n = 21$), and (b) top-soil proton concentration ($r^2_{adj} = 0.71, p < 0.001, n = 20$). See Table 46.1 for more information on the regression analysis.

As did fine root biomass, fine root necromass was also found to increase with elevation (Figure 46.4a). Air temperature correlated well with fine root necromass, but precipitation showed only a weak and non-significant influence. Soil acidity explained the largest proportion of the variance in the necromass data (Table 46.1). Even though site¯s with high soil acidity were relatively few, the derived exponential relationship was highly significant (Figure 46.4b).

Different reasons may underlie the high fine root necromass values of tropical moist forests, viz. (i) a low decomposition rate of the dead roots, (ii) a high root mortality rate, or (iii) both. Both decreasing temperatures (with altitude) and increasing soil acidity are known to reduce microbial activity and hence decomposition of organic material (Meentemeyer, 1977; Berendse *et al.*, 1989). On the other hand, the lower temperatures, high soil water content, high proton and aluminum concentrations, and often low nutrient availability associated with upper montane cloud forests (Hafkenscheid, 2000; Soethe *et al.*, 2006, 2008a,b; Schawe *et al.*, 2007; Roman *et al.* this volume) are likely to be unfavorable for root growth; studies by Pregitzer *et al.* (1993) and Godbold *et al.* (2003) in temperate forests showed that fine root longevity is shorter under adverse soil chemical conditions; Soethe *et al.* (2008b) demonstrated a similar effect in upper montane and high-elevation elfin cloud forests in Ecuador. Therefore, it is most probable that both reduced root decay rates and enhanced root mortality rates are responsible for the observed increase in fine root necromass at higher altitudes (Figure 46.4).

Forty percent of the studies considered did not distinguish between live and dead root material and thus gave total fine root mass values only (Appendix 47.1). Adding the results of the investigations that recorded both live fine root mass and necromass, total fine root mass displayed a remarkable range: 51–7430 g m^{-2}. The maximum values in the present data-set exceed that for total fine root mass (2805 g m^{-2}) in the literature survey of Vogt *et al.* (1996) and even the maximum value of 4900 g m^{-2} for fine plus coarse root mass as reported by Jackson *et al.* (1996) on a global scale.

Regression analysis revealed only poor correlations between total fine root mass and various environmental factors. Neither elevation, nor air temperature, or annual precipitation explained a notable portion of the variance in the data (Table 46.1). Only soil acidity showed a significant influence on total fine root mass, but to a much lower extent (32% of variance explained) than found for fine root necromass (73%). However, a more detailed analysis indicated that the relation of total fine root mass to different parameters was obscured by high variability in the data for lowland forests. A separate analysis for montane forests (>400 m.a.s.l.) did show a significant increase in total fine root mass with increasing elevation:

$$y = -139 + 0.74x, r^2 = 0.47, r^2_{adj} = 0.44, p<0.001, n = 24$$

and an exponential decrease with increasing temperature:

$$y = 52.6 + 2.3\,10^5 x^{-2}, r^2 = 0.67, r^2_{adj} = 0.65, p<0.001, n = 23.$$

Since the number and origin of the data on fine root biomass ($n = 55$) and necromass ($n = 21$) are different, it is difficult to clarify which factors caused the observed large variation in total fine root mass. High acidity, which may occur in lowland as well as in montane soils, seems to influence primarily fine root necromass, but not fine live root mass (Table 46.1). On the other hand, both fine root necromass (Figure 46.4a) and biomass (Figure 46.2) increased with elevation. Thus, the large total fine root masses of high-elevation forests are the consequence of both high amounts of dead and living fine roots.

FINE ROOT PRODUCTION OF MATURE TROPICAL MOIST FORESTS

Studies on fine root production in tropical moist forests have been conducted less frequently than those on fine root mass (9 vs. 38 studies, respectively). Reported annual fine root production values range between 75 and 2193 g m^{-2} year^{-1} (Appendix 46.1). This span is somewhat wider than the range of fine root production estimates of 43–1317 g m^{-2} year^{-1} derived in the global survey by Vogt *et al.* (1996) and considerably broader than the range of 350–1100 g m^{-2} year^{-1} given for temperate forests by Fogel (1985). Also, the presently derived average fine root production for tropical moist forests of 557 g m^{-2} year^{-1} ($n = 22$) exceeds the global average of 392 g m^{-2} year^{-1} inferred by Vogt *et al.* (1996).

Whilst Neotropical lowland forests do not differ significantly in average fine root production from Neotropical lower montane forests (405 and 540 g m^{-2} year^{-1}, respectively), that in Neotropical upper montane forests is significantly higher (1721 g m^{-2} year^{-1} on average; Figure 46.5). Unfortunately, a generalization of this finding is hampered by the fact that the current data-set contains only two data points from forests

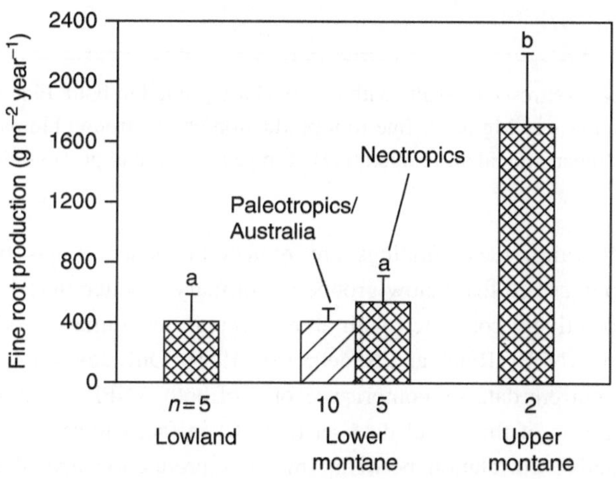

Figure 46.5. Annual fine root production of old-growth forests in tropical humid lowland (<400 m.a.s.l.), lower montane (400–2000 m.a.s.l.), and upper montane (>2000 m.a.s.l.) regions. Both means and standard errors are given. Different letters indicate significant differences among the elevation groups of Neotropical forests; $p < 0.05$, Scheffé's test for unbalanced, parametrically distributed samples.

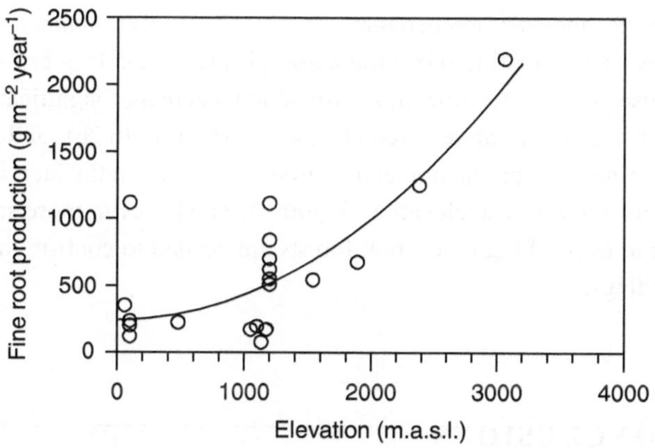

Figure 46.6. Relationship between annual fine root production of old-growth moist tropical forests and elevation ($r^2_{adj} = 0.47$, $p < 0.001$, $n = 22$). See Table 46.2 for more information on the regression analysis.

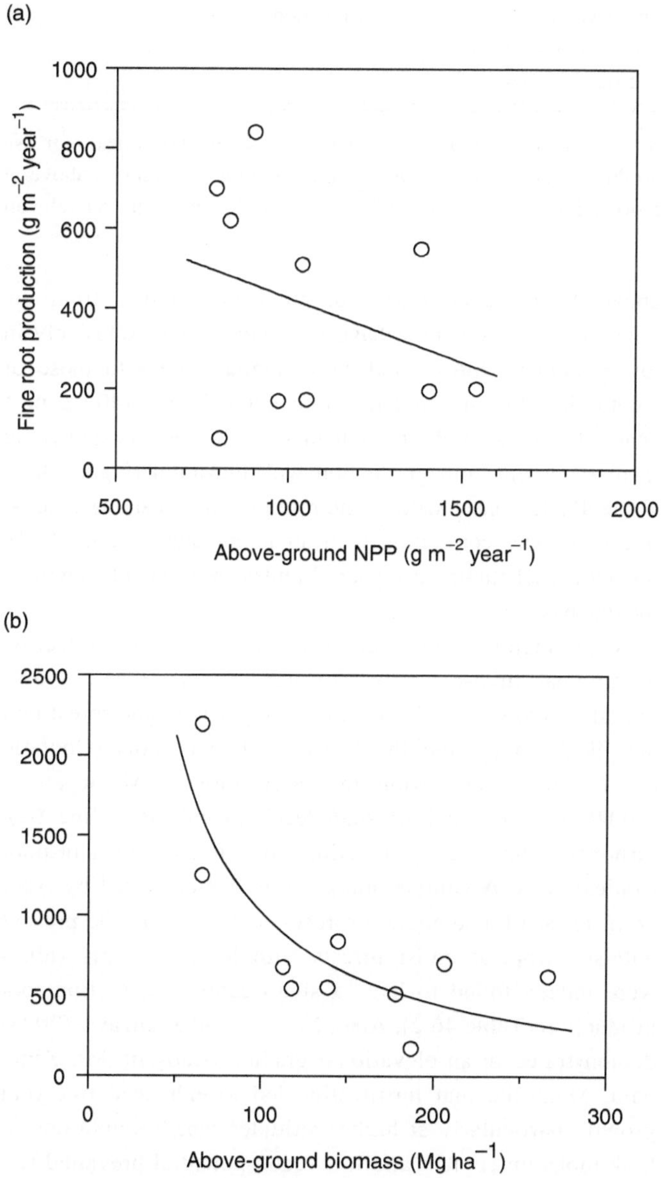

Figure 46.7. (a) Relationship between annual fine root production and above-ground net primary productivity of old-growth moist tropical forests ($r^2_{adj} = -0.01$, $p = 0.36$, $n = 10$). (b) The same with above-ground biomass ($r^2_{adj} = 0.86$, $p < 0.001$, $n = 10$). See Table 46.2 for more information on the regression analysis.

Table 46.2 *Relationships between fine root production, above-ground production, and fine root production/above-ground biomass ratio of old-growth moist tropical forests, and various site variables*

Parameter	Source	Regression equation	r^2	r^2_{adj}	p	n
Fine root production	Elevation	$y = 248 + 0.0002{\cdot}x^2$	**0.64**	**0.62**	**0.001**	22
	Air temperature	$y = 21.9 + 1.61{\cdot}10^5\ x^{-2}$	**0.58**	**0.56**	**0.001**	22
	Precipitation	$y = -51.4 + 0.202{\cdot}x$	0.12	0.07	0.119	22
	Soil pH(H_2O)	$y = 120 + 1.28{\cdot}10^5{\cdot}x^{-4}$	**0.45**	**0.40**	**0.009**	14
	Above-ground biomass	$y = -118 + 11362{\cdot}x^{-1}$	**0.67**	**0.63**	**0.010**	10
	Above-ground production	$y = 745 - 0.32{\cdot}x$	0.10	−0.01	0.364	10
Above-ground production	Elevation	$y = 1538 - 0.353{\cdot}x$	**0.50**	**0.47**	**0.001**	23
	Air temperature	$y = -163 + 70.3{\cdot}x$	**0.70**	**0.69**	**0.001**	23
Fine root production · above-ground biomass^{-1}	Elevation	$y = 0.028 + 3.61{\cdot}10^{-15}{\cdot}x^4$	**0.96**	**0.95**	**0.001**	10

Notes: r^2 and adjusted r^2 refer to linear or simple non-linear regression models. Regression results with $p < 0.05$ are printed in bold. Fine root production and above-ground production in g m^{-2} $year^{-1}$, above-ground biomass in Mg ha^{-1}, fine root production/above-ground biomass ratio in g m^{-2} $year^{-1}$ · g^{-1} m^{-2}; other variable units: m.a.s.l. (elevation), °C (mean annual air temperature), mm $year^{-1}$ (annual precipitation).

above 2000 m.a.s.l. Fine root production studies from the Paleotropics or Australia have been conducted exclusively in lower montane forests and show similar values to those of Neotropical lowland and lower montane forests (409 g m^{-2} $year^{-1}$ on average). Regression analysis showed an exponential increase in fine root production with elevation (Figure 46.6, Table 46.2). Consequently, fine root production shows a negative non-linear correlation with air temperature (Table 46.2), but additional studies at higher altitudes are needed to augment the database.

Which factors could explain the high fine root production of tropical high-elevation forests? As mentioned above, prevalent growth conditions in the upper montane forest belt are likely to shorten the lifespan of fine roots (Godbold *et al.*, 2003). Following the conjecture of Malhi *et al.* (2004), reduced root lifespan leads to enhanced fine root turnover (and thus production) on acidic, nutrient-poor tropical sites. A similar relation was demonstrated by Aber *et al.* (1985) for temperate forest ecosystems. In the present data-set, tropical moist forests with lower soil pH values were indeed found to have a significantly higher fine root production (Table 46.2). Also, Nomura and Kikuzawa (2003) demonstrated in an elevational gradient study on Mt. Kinabalu, Malaysia, that fertilization led to enhanced fine root growth, particularly at higher altitudes where conditions of high moisture content and low redox potential prevailed (cf. Schawe *et al.*, this volume). Apparently, low nutrient supply reduces fine root longevity instead of increasing it. Trees at high elevations may compensate for this loss in root mass by increasing fine root production (cf. Moser *et al.*, 2008; Soethe *et al.*, 2008b).

If valid, these findings are remarkable since it has been hypothesized that below-ground net primary production (BNPP) is positively correlated with above-ground net primary production (ANPP) (Raich and Nadelhoffer, 1989). Only few studies in the current data-set comprise records of both ANPP and BNPP. Analysis of this small data-set ($n = 10$) in fact showed a weak negative correlation between fine root production and above-ground productivity (Figure 46.7a, Table 46.2), not a positive one as predicted by the Raich and Nadelhoffer (1989) hypothesis. Although the relationship in Figure 46.7a is not significant, it is noticeable that sites with low ANPP can have high or low fine root productivity, whereas forests with high ANPP have comparably low fine root productivity (Figure 46.7a). The contrasting elevational trends derived for below- and above-ground production result from the strong positive correlation between ANPP and air temperature, and the negative temperature dependence of fine root production (Table 46.2). This has two consequences: (i) fine root production decreases significantly with increasing above-ground biomass (Figure 46.7b), and (ii) the fine root production/above-ground biomass ratio increases exponentially with elevation (Figure 46.8). However, more studies in tropical high-elevation forests are needed to confirm these findings.

CONCLUSIONS

The present meta-analysis is based on fine root data from more than 80 tropical moist forest stands, both from the lowlands and the montane belt. Compared to earlier reviews of fine root mass

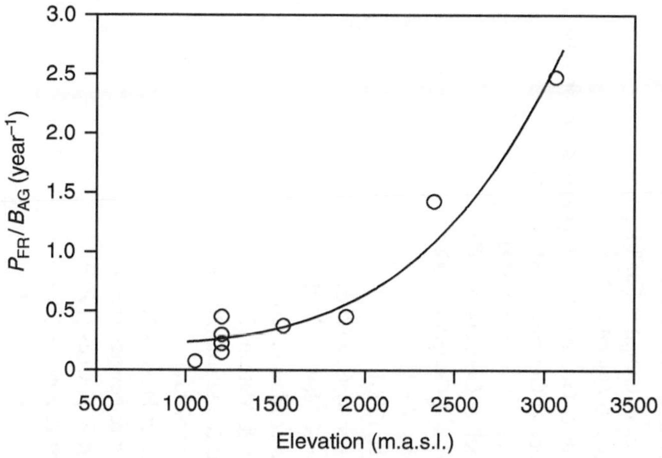

Figure 46.8. Relationship between the ratio of annual fine root production P_{FR} to above-ground biomass B_{AG}, and elevation for old-growth moist tropical forests ($r^2_{adj} = 1.00$, $p < 0.001$, $n = 10$). See Table 46.2 for more information on the regression analysis.

in forests, the presently used data selection criteria that were more rigid with respect to root diameter, root vitality status, and technique of production estimation.

Despite large differences in above-ground biomass, old-growth tropical moist forests have remarkably similar values of fine root biomass compared to temperate forests, whereas annual fine root production rates appear to be somewhat higher than in temperate or boreal forests.

For fine root biomass (i.e. live roots), a negative relation was found with temperature, whereas soil proton concentration (acidity) had no influence. Fine root necromass is negatively related to temperature as well, but at the same time it shows a positive relationship with soil acidity. Although large fine root systems and high fine root productivity may be found in tropical lowland forests at both nutrient-poor and acidic sites, tropical montane forests were found to differ from lowland forests by their consistently higher live and dead fine root mass, as well as by their higher fine root productivity. Since both above-ground biomass and above-ground production typically decrease with elevation, a marked shift in carbon allocation from shoot to root occurs with increasing elevation in old-growth tropical moist forests (cf. Leuschner *et al.*, 2007; Moser *et al.*, 2008). These findings suggest an enhanced ecological importance of the fine root system in these forests with elevation.

Appendix 46.1 *List of study sites used in the present review*

Cit. No.	Location	Country	Longitude	Latitude	Forest type	References
(a) Location of sites						
1	Yapo Forest	Ivory Coast	04°04′W	05°27′N	lowland	Huttel (1975), Huttel & Bernhard-Reversat (1975)
2	Banco Plateau	Ivory Coast	04°04′W	05°27′N	lowland	Huttel (1975), Huttel & Bernhard-Reversat (1975)
3	Banco Thalweg	Ivory Coast	04°04′W	05°27′N	lowland	Huttel (1975), Huttel & Bernhard-Reversat (1975)
4	Korup National Park	Cameroon	09°00′E	05°00′N	lowland	Newbery *et al.* (1997), Chuyong *et al.* (2000)
5	Korup National Park	Cameroon	09°00′E	05°00′N	lowland	Newbery *et al.* (1997), Chuyong *et al.* (2000)
6	SW Cameroon	Cameroon	10°27′E	03°03′N	lowland	Ibrahim *et al.* (2002)
7	SW Cameroon	Cameroon	10°45′E	02°57′N	lower montane	Ibrahim *et al.* (2002)
8	SW Cameroon	Cameroon	10°45′E	03°06′N	lower montane	Ibrahim *et al.* (2002)
9	Kodayar	India	77°15′E	08°29′N	lower montane	Sundarapandian & Swamy (1996)
10	Kodayar	India	77°15′E	08°29′N	lower montane	Sundarapandian & Swamy (1996)
11	Pasoh Forest	Malaysia	102°35′E	00°34′S	lowland	Kira (1978)
12	Mt. Kinabalu	Malaysia	116°33′E	06°05′N	lower montane	Kitayama & Aiba (2002)
13	Mt. Kinabalu	Malaysia	116°33′E	06°05′N	upper montane	Kitayama & Aiba (2002)
14	Mt. Kinabalu	Malaysia	116°33′E	06°05′N	upper montane	Kitayama & Aiba (2002)
15	Mt. Kinabalu	Malaysia	116°33′E	06°05′N	lower montane	Kitayama & Aiba (2002)
16	Mt. Kinabalu	Malaysia	116°33′E	06°05′N	upper montane	Kitayama & Aiba (2002)
17	Mt. Kinabalu	Malaysia	116°33′E	06°05′N	upper montane	Kitayama & Aiba (2002)
18	Mt. Kinabalu	Malaysia	116°33′E	06°05′N	lower montane	Kitayama & Aiba (2002)
19	Mt. Kinabalu	Malaysia	116°33′E	06°05′N	lower montane	Kitayama & Aiba (2002)
20	N Queensland	Australia	145°04′E	15°05′S	lowland	Maycock & Gongdon (2000)
21	N Queensland	Australia	145°08′E	15°04′S	lowland	Maycock & Gongdon (2000)
22	Mt. Kerigomna	New Guinea	145°11′E	06°00′S	upper montane	Edwards & Grubb (1977), Edwards & Grubb (1982)
23	N Queensland	Australia	145°30′E	18°51′S	lower montane	Maycock & Gongdon (2000)
24	N Queensland	Australia	145°44′E	18°11′S	lower montane	Maycock & Gongdon (2000)
25	N Queensland	Australia	145°57′E	18°25′S	lowland	Maycock & Gongdon (2000)
26	N Queensland	Australia	146°10′E	19°00′S	lower montane	Maycock & Gongdon (2000)
27	Big Island, Hawai'i	USA	155°16′W	19°20′N	lower montane	Ostertag (2001)
28	Big Island, Hawai'i	USA	155°21′W	19°44′N	lower montane	Ostertag (2001)
29	Maui, Hawai'i	USA	156°13′W	20°46′N	lower montane	Schuur & Matson (2001)
30	Maui, Hawai'i	USA	156°13′W	20°46′N	lower montane	Schuur & Matson (2001)
31	Maui, Hawai'i	USA	156°13′W	20°46′N	lower montane	Schuur & Matson (2001)
32	Maui, Hawai'i	USA	156°13′W	20°46′N	lower montane	Schuur & Matson (2001)
33	Maui, Hawai'i	USA	156°13′W	20°46′N	lower montane	Schuur & Matson (2001)
34	Kauai, Hawai'i	USA	159°38′W	22°08′N	lower montane	Ostertag (2001)
35	Hawai'i	USA	155–160°W	19–22°N	lower montane	Herbert & Fownes (1999)
36	Hawai'i	USA	155–160°W	19–22°N	lower montane	Herbert & Fownes (1999)
37	Hawai'i	USA	155–160°W	19–22°N	lower montane	Herbert & Fownes (1999)
38	Hawai'i	USA	155–160°W	19–22°N	lower montane	Herbert & Fownes (1999)
39	Hawai'i	USA	155–160°W	19–22°N	lower montane	Herbert & Fownes (1999)
40	Los Tuxtlas, Veracruz	Mexico	95°05′W	18°35′N	lower montane	Jaramillo *et al.* (2003)
41	Los Tuxtlas, Veracruz	Mexico	95°05′W	18°35′N	lower montane	Jaramillo *et al.* (2003)

42	Los Tuxtlas, Veracruz	Mexico	95°05′W	18°35′N	lower montane	Jaramillo *et al.* (2003)
43	Los Tuxtlas, Veracruz	Mexico	95°04–09′W	18°34–36′N	lower montane	Sánchez-Gallén & Alvarez-Sánchez (1996)
44	Monteverde	Costa Rica	84°48′W	10°18′N	lower montane	Vance & Nadkarni (1992)
45	La Selva	Costa Rica	83°59′W	02°54′N	lowland	Sanford (1989)
46	La Selva	Costa Rica	83°59′W	10°26′N	lowland	Gower (1987)
47	La Selva	Costa Rica	83°59′W	10°26′N	lowland	Gower (1987)
48	La Selva	Costa Rica	83°50′W	09°53′N	lowland	Denslow *et al.* (1998)
49	La Selva	Costa Rica	83°50′W	09°53′N	lowland	Denslow *et al.* (1998)
50	La Selva	Costa Rica	83°50′W	09°53′N	lowland	Denslow *et al.* (1998)
51	La Selva	Costa Rica	83°50′W	09°53′N	lowland	Denslow *et al.* (1998)
52	La Selva	Costa Rica	83°50′W	09°53′N	lowland	Denslow *et al.* (1998)
53	La Selva	Costa Rica	83°50′W	09°53′N	lowland	Denslow *et al.* (1998)
54	La Selva	Costa Rica	83°50′W	10°26′N	lowland	Raich (1980), Raich (1983)
55	Cordillera de Talamanca	Costa Rica	83°47′W	09°35′N	upper montane	Hertel *et al.* 2003
56	Barro Colorado Island	Panama	80°14′W	08°43′N	lower montane	Cavelier (1992)
57	Barro Colorado Island	Panama	80°09′W	09°07′N	lowland	Cavelier (1992)
58	Barro Colorado Island	Panama	79°51′W	09°09′N	lowland	Cavelier *et al.* (1999)
59	Barro Colorado Island	Panama	79°51′W	09°09′N	lowland	Yavitt & Wright (2001)
60	Podocarpus National Park	Ecuador	79°10′W	04°06′S	upper montane	Röderstein *et al.* (2005), Moser *et al.*, unpubl.
61	Podocarpus National Park	Ecuador	79°04′W	03°58′S	lower montane	Röderstein *et al.* (2005), Moser *et al.*, unpubl.
62	Podocarpus National Park	Ecuador	79°04′W	03°59′S	upper montane	Röderstein *et al.* (2005), Moser *et al.*, unpubl.
63	Podocarpus National Park	Ecuador	78°58′W	04°07′S	lower montane	Röderstein *et al.* (2005), Moser *et al.*, unpubl.
64	Podocarpus National Park	Ecuador	78°58′W	04°07′S	lower montane	Röderstein *et al.* (2005), Moser *et al.*, unpubl.
65	Blue Mountains	Jamaica	77°W	18°N	lower montane	Tanner (1980), Tanner (1977)
66	La Planada	Colombia	77°24′W	01°05′N	lower montane	Cavelier *et al.* (1996)
67	Araracuara, Amazonia	Colombia	72°27′W	00°37′S	lowland	Pavlis & Jenik (2000)
68	San Carlos de Rio Negro	Venezuela	67°03′W	01°54′N	lowland	Jourdan & Escalante (1980), Stark & Spratt (1977)
69	San Carlos de Rio Negro	Venezuela	67°03′W	01°54′N	lowland	Cuevas & Medina (1988)
70	San Carlos de Rio Negro	Venezuela	67°03′W	01°54′N	lowland	Cuevas & Medina (1988)
71	San Carlos de Rio Negro	Venezuela	67°03′W	01°54′N	lowland	Cuevas & Medina (1988)
72	Luquillo Exp. For./UEP	Puerto Rico	65°50′W	18°18′N	upper lowland	Silver & Vogt (1993)
73	Luquillo Exp. For.	Puerto Rico	65°50′W	18°18′N	lower montane	McGroddy & Silver (2000)
74	Luquillo Exp. For./LEP	Puerto Rico	65°50′W	18°18′N	upper lowland	Silver *et al.* (1996), Silver & Vogt (1993)
75	Luquillo Exp. For.	Puerto Rico	65°50′W	18°18′N	lower montane	McGroddy & Silver (2000)
76	Luquillo Exp. For./C	Puerto Rico	65°50′W	18°18′N	upper lowland	Silver *et al.* (1996), Silver & Vogt (1993)
77	Luquillo Exp. For.	Puerto Rico	65°50′W	18°18′N	lowland	McGroddy & Silver (2000)
78	Luquillo Exp. For.	Puerto Rico	65°50′W	18°18′N	upper lowland	McGroddy & Silver (2000)
79	Luquillo Exp. For.	Puerto Rico	65°50′W	18°18′N	lower montane	McGroddy & Silver (2000)
80	Gran Sabana	Venezuela	61°01′W	05°00′N	lower montane	Priess *et al.* (1999)
81	Tapajós, Amazonia	Brazil	54°57′W	02°54′S	lowland	Davidson *et al.* (2004)
82	Tapajós Nat. For.	Brazil	54°57′W	02°54′S	lowland	Nepstad *et al.* (2002)
83	Tapajós Nat. For.	Brazil	54°57′W	02°54′S	lowland	Silver *et al.* (2000)
84	Tapajós Nat. For.	Brazil	54°57′W	02°54′S	lowland	Nepstad *et al.* (2002)
85	Tapajós Nat. For.	Brazil	54°57′W	02°54′S	lowland	Silver *et al.* (2000)
86	Manaus, Amazonia	Brazil	50°06′W	02°35′S	lowland	Malhi *et al.* (1999)
87	Manaus, Amazonia	Brazil	50°05′W	02°35′S	lowland	Klinge (1973)

Cit. No.	Elevation (m.a.s.l.)	Precipitation (mm)	Temperature (°C)	Soil type	Top-soil pH (H_2O)	Above-ground biomass ($t\ ha^{-1}$)	Above-ground produc-tion ($g\ m^{-2}\ yr^{-1}$)	Sampling depth (cm)	Fine root diameter (mm)	Fine root biomass ($g\ m^{-2}$)	Fine root necro-mass (gm^{-2})	Total fine root mass (gm^{-2})	Fine root produc-tion ($g\ m^{-2}\ yr^{-1}$)
(b) Parameters													
1	100	1800	26.5	–	–	–	–	130	2	–	–	73	–
2	100	1800	26.5	–	–	–	–	130	2	–	–	54	–
3	100	1800	26.5	–	–	–	–	130	2	–	–	51	–
4	50	5180	26.0	Oxisols/Ultisols	4.4	–	–	50	2	–	–	307	–
5	50	5180	26.0	Oxisols/Ultisols	4.2	–	–	50	2	–	–	363	–
6	110	1800	25.0	Oxisols	–	–	–	25	2	–	–	4700	–
7	760	1750	25.0	Oxisols	5.5	–	–	25	2	–	–	7430	–
8	560	3100	25.0	Oxisols	4.0	600	–	25	2	–	–	1740	–
9	480	3146	27.0	–	–	–	–	25	3	468	–	–	224
10	480	3146	27.0	–	–	–	–	25	3	540	–	–	226
11	200	1807	25.5	–	–	425	–	–	1	–	–	2000	–
12	700	2300	24.0	–	4.5	554	1715	30	2	520	–	–	–
13	3050	2100	10.0	–	5.3	370	199	30	2	1440	–	–	–
14	2700	2200	12.0	–	5.1	122	725	30	2	960	–	–	–
15	650	2300	24.0	–	4.1	437	1913	30	2	560	–	–	–
16	2590	2300	12.0	–	3.4	308	780	30	2	960	–	–	–
17	3080	2100	10.0	–	4.9	215	816	30	2	1020	–	–	–
18	1560	2300	18.0	–	4.0	294	1222	30	2	950	–	–	–
19	1860	2300	18.0	–	5.4	238	813	30	2	900	–	–	–
20	90	2000	–	Entisol	–	–	–	40	2	210	–	–	–
21	120	2000	–	Alfisol	–	–	–	40	2	270	–	–	–
22	2500	4000	13.0	–	6.0	295	–	25	3	–	–	950	–
23	813	2000	–	Alfisol	–	–	–	40	2	280	–	–	–
24	600	2000	–	Oxisol	–	–	–	40	2	910	–	–	–
25	50	2000	–	Entisol	–	–	–	40	2	260	–	–	–
26	880	2000	–	Oxisol	–	–	–	40	2	710	–	–	–
27	1176	2500	16.0	Hydric Dystrandepts	5.0	–	1053	20	2	257	410	667	173
28	1170	2500	16.0	Typic Hydrandepts	3.6	–	970	30	2	354	433	787	169
29	1320	3350	16.0	Inceptisols/ Andisols	3.9	250	930	51	2	202	377	579	–

30	1300	4050	16.0	Inceptisols/ Andisols	3.8	250	740	51	2	175	361	536	–
31	1370	2750	16.0	Inceptisols/ Andisols	3.9	239	970	51	2	349	774	1123	–
32	1370	2200	16.0	Inceptisols/ Andisols	3.4	250	940	51	2	128	465	593	–
33	1370	2450	16.0	Inceptisols/ Andisols	3.7	217	1040	51	2	425	809	1234	–
34	1134	2500	16.0	Plinthic Acrudox	4.0	–	801	15	2	457	302	759	75
35	1200	2500	16.0	–	5.0	178	1040	30	2	152	–	–	510
36	1200	2500	16.0	–	4.1	145	900	30	2	266	–	–	840
37	1200	2500	16.0	–	3.6	266	830	50	2	203	–	–	620
38	1200	2500	16.0	–	4.0	206	790	10	2	218	–	–	700
39	1200	2500	16.0	–	3.8	139	1380	20	2	272	–	–	550
40	1000	4500	27.0	Vitric Andosols	6.3	–	–	100	4	–	–	480	–
41	1000	4500	27.0	Vitric Andosols	6.3	–	–	100	4	–	–	300	–
42	1000	4500	27.0	Vitric Andosols	6.3	–	–	100	4	–	–	270	–
43	1100	4700	24.0	–	–	–	1405	20	3	–	–	–	195
44	1550	2000	–	Typic Dystrandept	–	–	–	–	2	–	–	881	–
45	100	3900	25.0	Andic Humitropept	–	–	–	15	2	–	–	140	–
46	60	3800	20.0	Fluvaquentic Hapludoll	–	–	–	50	2	61	–	–	–
47	60	3800	20.0	Oxic Dystrandept	–	–	–	50	2	114	–	–	–
48	70	4000	24.0	Ultisols (Tropohumult)	–	–	–	10	2	675	–	–	–
49	70	4000	24.0	Ultisols (Tropohumult)	–	–	–	10	2	566	–	–	–
50	70	4000	24.0	Ultisols (Tropohumult)	–	–	–	10	2	848	–	–	–
51	70	4000	24.0	Ultisols (Tropohumult)	–	–	–	10	2	647	–	–	–
52	70	4000	24.0	Ultisols (Tropohumult)	–	–	–	10	2	590	–	–	–
53	70	4000	24.0	Ultisols (Tropohumult)	–	–	–	10	2	355	–	–	–

(*cont.*)

Cit. No.	Elevation (m.a.s.l.)	Preci-pitation (mm)	Tem-pera-ture (°C)	Soil type	Top-soil pH (H_2O)	Above-ground biomass (t ha^{-1})	Above-ground produc-tion (g m^{-2} yr^{-1})	Sampling depth (cm)	Fine root diameter (mm)	Fine root biomass (g m^{-2})	Fine root necro-mass (gm^{-2})	Total fine root mass (gm^{-2})	Fine root produc-tion (g m^{-2} yr^{-1})
54	70	3980	24.1	Andic Humitropepts	–	–	–	50	2	–	–	289	–
55	2900	2700	11.0	Humic Andosols	5.0	–	–	28	2	1329	537	1866	–
56	1200	3900	19.0	Inceptisols	4.0	–	–	25	2	–	–	400	–
57	60	2567	27.0	Oxisols	5.5	–	–	25	2	–	–	280	–
58	60	2612	26.0	Alfisol	–	–	–	10	2	–	–	130	352
59	60	2620	26.0	Alfisol	–	–	–	100	2	410	34	444	–
60	3060	4500	9.0	Humic Cambisols	3.1	8	–	20	2	978	2398	3376	2193
61	1890	1950	15.7	Humic Cambisols	3.8	50	–	20	2	393	640	1033	676
62	2380	5000	12.9	Humic Cambisols	3.2	24	–	20	2	539	992	1531	1249
63	1050	2205	20.6	Cambisols	4.7	190	–	20	2	112	47	159	171
64	1540	2500	17.8	Cambisols	4.9	45	–	20	2	464	259	723	544
65	1720	3000	15.0	Oxisols	3.5	223	–	50	2	–	–	1120	–
66	1800	3750	15.0	Dystrandepts/ Dystropepts	4.2	–	–	45	2	–	–	1169	–
67	160	3059	25.7	Alisols/Acrisols/ Ferrasols	3.7	–	–	50	2	–	–	3924	–
68	100	3500	28.0	Oxisols	–	–	1540	40	2	–	–	2692	201
69	100	3565	26.0	Oxisols	–	–	–	10	2	–	–	–	1117
70	100	3565	26.0	Sandy spodosols	–	–	–	10	2	–	–	–	120
71	100	3565	26.0	Sandy spodosols	–	–	–	10	2	–	–	–	235
72	350	3500	–	Ultisols	4.9	–	–	35	2	120	93	213	–
73	780	4000	26.1	Clayey Ultisols	–	–	–	10	2	230	–	–	–
74	350	3500	–	Ultisols	4.9	–	–	35	2	158	128	286	–
75	1000	5000	25.0	Clayey Ultisols	–	–	–	10	2	560	–	–	–
76	350	3500	–	Ultisols	4.9	–	–	35	2	230	114	344	–
77	180	2900	33.9	Clayey Ultisols	–	–	–	10	2	150	–	–	–
78	290	2900	30.7	Clayey Ultisols	–	–	–	10	2	300	–	–	–
79	660	4000	27.7	Clayey Ultisols	–	–	–	10	2	290	–	–	–

80	1200	2200	20.0	Oxisols (Acrohumox)	5.0	–	–	30	2	234	880	1114	1113
81	100	2000	–	Oxisols/ Haplustox	4.5	235	–	600	2	–	–	420	–
82	100	2000	25.0	Oxisols (Haplustox)	4.2	291	–	600	2	–	–	340	–
83	100	2000	25.0	Oxisols/Ultisols (Sand)	3.9	–	–	40	2	8	558	566	–
84	100	2000	25.0	Oxisols (Haplustox)	4.2	305	–	600	2	–	–	420	–
85	100	2000	25.0	Oxisols/Ultisols (Clay)	4.0	–	–	40	2	40	412	452	–
86	100	2200	25.5	Clay Oxisols	–	350	–	–	–	–	–	3800	–
87	100	2200	25.5	Latosol	–	–	–	107	2	–	–	843	–

REFERENCES

Aber, J. D., J. M. Melillo, K. J. Nadelhoffer, C. A. McClaugherty, and J. Pastor (1985). Fine root turnover in forest ecosystems in relation to quantity and form of nitrogen availability: a comparison of two methods. *Oecologia* **66**: 317–321.

Anderson, J. M., and M. J. Swift (1983). Decomposition in tropical forests. In *Tropical Rain Forest: Ecology and Management*, eds. S. L. Sutton, T. C. Whitmore, and A. C. Chadwick, pp. 287–309. Oxford, UK: Blackwell Scientific.

Berendse, F., R. Bobbink, and G. Rouwenhorst (1989). A comparative study on nutrient mineralization. *Oecologia* **78**: 338–348.

Bruijnzeel, L. A., and J. Proctor (1995). Hydrology and biogeochemistry of tropical montane cloud forests: what do we really know? In *Tropical Montane Cloud Forests*, eds. L. S. Hamilton, J. O. Juvik, and F. N. Scatena, pp. 38–78. New York: Springer-Verlag.

Bruijnzeel, L. A., and E. J. Veneklaas (1998). Climatic conditions and tropical montane forest productivity: the fog has not lifted yet. *Ecology* **79**: 3–9.

Cairns, M. A., S. Brown, E. H. Helmer, and G. A. Baumgardner (1997). Root biomass allocation in the world's upland forests. *Oecologia* **111**: 1–11.

Cavelier, J. (1992). Fine-root biomass and soil properties in a semideciduous and a lower montane rain forest in Panama. *Plant and Soil* **142**: 187–201.

Cavelier, J., J. Estevez, and B. Arjona (1996). Fine root biomass in three successional stages of an Andean cloud forest in Colombia. *Biotropica* **28**: 728–736.

Cavelier, J., S. J. Wright, and J. Santamaria (1999). Effects of irrigation on litterfall, fine root biomass and production in a semideciduous lowland forest in Panama. *Plant and Soil* **211**: 207–213.

Chuyong, G. B., D. M. Newbery, and N. C. Songwe (2000). Litter nutrients and retranslocation in a central African rain forest dominated by ectomycorrhizal trees. *New Phytologist* **148**: 493–510.

Cuevas, E., and E. Medina (1988). Nutrient dynamics within Amazonian forests. II. Fine root growth, nutrient availability and leaf litter decomposition. *Oecologia* **76**: 222–235.

Davidson, E. A., F. Y. Ishida, and D. C. Nepstad (2004). Effects of an experimental drought on soil emissions of carbon dioxide, methane, nitrous oxide, and nitric oxide in a moist tropical forest. *Global Change Biology* **10**: 718–730.

Denslow, J. S., A. M. Ellison, and R. E. Sanford (1998). Treefall gap size effects on above- and below-ground processes in a tropical wet forest. *Journal of Ecology* **86**: 597–609.

Dixon, R. K., S. Brown, R. A. Houghton, *et al.* (1994). Carbon pools and flux of global forest ecosystems. *Science* **263**: 185–190.

Edwards, P. J., and P. J. Grubb (1977). Studies of mineral cycling in a montane rainforest in New Guinea. I. The distribution of organic matter in the vegetation and soil. *Journal of Ecology* **65**: 943–969.

Edwards, P. J., and P. J. Grubb (1982). Studies of mineral cycling in a montane rainforest in New Guinea. IV. Soil characteristics and the division of mineral elements between the vegetation and soil. *Journal of Ecology* **70**: 649–666.

Fogel, R. (1985). Roots as primary producers in below-ground ecosystems. In *Ecological Interactions in Soil: Plants, Microbes and Animals*, eds. A. H. Fitter, D. Atkinson and D. J. Read, pp. 23–35. Oxford, UK: Blackwell Scientific.

Gill, R. A., and R. B. Jackson (2000). Global patterns of root turnover for terrestrial ecosystems. *New Phytologist* **147**: 13–31.

Godbold, D. L., H. W. Fritz, G. Jentschke, H. Meesenburg, and P. Rademacher (2003). Root turnover and root necromass accumulation of Norway spruce (*Picea abies*) are affected by soil acidity. *Tree Physiology* **23**: 915–921.

Gower, S. T. (1987). Relations between mineral nutrient availability and fine root biomass in two Costa Rican tropical forests: a hypothesis. *Biotropica* **19**: 171–175.

Grier, C. C., K. A. Vogt, M. L. Keyes, and R. L. Edmonds (1981). Biomass distribution and above- and below-ground production in young and mature *Abies amabilis* zone ecosystems of the Washington Cascades. *Canadian Journal of Forest Research* **11**: 155–167.

Grubb, P. J. (1977). Control of forest growth and distribution on wet tropical mountains, with special reference to mineral nutrition. *Annual Review of Ecology and Systematics* **8**: 83–107.

Hafkenscheid, R. L. L. J. (2000). Hydrology and biogeochemistry of tropical montane rain forests of contrasting stature in the Blue Mountains, Jamaica. Ph.D. thesis, VU University Amsterdam, Amsterdam, the Netherlands. Also available at http://dare.ubvu.vu.nl/ bitstream/1871/12734/1/tekst.pdf.

Herbert, D. A., and J. H. Fownes (1999). Forest productivity and efficiency of resource use across a chronosequence of tropical montane soils. *Ecosystems* **2**: 242–254.

Hertel, D, and K. Wesche (2008). Tropical moist *Polylepis* stands at the treeline in East Bolivia: the effect of elevation on stand microclimate, above- and below-ground structure, and regeneration. *Trees* **22**: 303–315.

Hertel, D., C. Leuschner, and D. Hölscher (2003). Size and structure of fine root systems in old-growth and secondary tropical montane forests (Costa Rica). *Biotropica* **35**: 143–153.

Huttel, C. (1975). Root distribution and biomass in three Ivory Coast rain forest plots. In *Tropical Ecological Systems*, eds. F. B. Golley and E. Medina, pp. 123–130. New York: Springer-Verlag.

Huttel, Ch., and F. Bernhard-Reversat (1975). Recherches sur l'écosystème de la forêt subéquatoriale de basse Côte-d'Ivoire: cycle de la matière organique. *Terre et Vie* **29**: 203–228.

Ibrahima, A., P. Schmidt, P. Ketner, and G. J. M. Mohren (2002). *Phytomasse et cycle des nutrients dans la forêt tropicale dense humide du Sud Cameroun*, Tropenbos-Cameroon Documents No. 9. Kribi, Cameroon: The Tropenbos–Cameroon Programme.

Jackson, R. B., J. Canadell, J. R. Ehleringer, *et al.* (1996). A global analysis of root distribution for terrestrial biomes. *Oecologia* **108**: 389–411.

Jaramillo, V. J., R. Ahedo-Hernández, and J. B. Kauffman (2003). Root biomass and carbon in a tropical evergreen forest of Mexico: changes with secondary succession and forest conversion to pasture. *Journal of Tropical Ecology* **19**: 457–464.

Jenny, H., S. P. Gessel, and F. T. Bingham (1949). Comparative study of decomposition rates of organic matter in temperate and tropical regions. *Soil Science* **68**: 419–432.

Jourdan, C. F., and G. Escalante (1980). Root productivity in an Amazonian rain forest. *Ecology* **61**: 14–18.

Kira, T. (1978). Community architecture and organic matter dynamics in tropical lowland rain forest of Southeast Asia with special reference to Pasoh Forest, West Malaysia. In *Tropical Trees as Living Systems*, eds. P. B. Tomlinson and M. H. Zimmerman, pp. 561–590. New York: Cambridge University Press.

Kitayama, K., and S. -I. Aiba (2002). Ecosystem structure and productivity of tropical rain forests along altitudinal gradients with contrasting soil phosphorus pools on Mount Kinabalu, Borneo. *Journal of Ecology* **90**: 37–51.

Klinge, H. (1973). Root mass estimation in lowland tropical rain forests of central Amazonia, Brazil. I. Fine root masses of a pale yellow latosol and a giant humus podzol. *Tropical Ecology* **14**: 29–38.

Lal, R. (2005). Forest soils and carbon sequestration. *Forest Ecology and Management* **220**: 242–258.

Leuschner, Ch., and D. Hertel (2003). Fine root biomass of temperate forests in relation to soil acidity and fertility, climate, age and species. *Progress in Botany* **64**: 405–438.

Leuschner, Ch., G. Moser, C. Bertsch, M. Röderstein, and D. Hertel (2007). Large altitudinal increase in tree root/shoot ratio in tropical mountain forests of Ecuador. *Basic and Applied Ecology* **8**: 219–230.

Malhi, Y., D. D. Baldocchi, and P. G. Jarvis (1999). The carbon balance of tropical, temperate and boreal forests. *Plant, Cell and Environment* **22**: 715–740.

Malhi, Y., T. R. Baker, O. L. Phillips, *et al.* (2004). The above-ground coarse wood productivity of 104 Neotropical forest plots. *Global Change Biology* **10**: 563–591.

Malhi, Y., P. Meir, M. Bird, N. Salinas, and M. Silman (2006a). *Detailed Assessment of Ecosystem Carbon Dynamics along a Tropical Forest Altitudinal Gradient*, research proposal to the National Environment Research Council of the UK. Oxford, UK: University of Oxford.

Malhi, Y., D. Wood, T. R. Baker, *et al.* (2006b). The regional variation of aboveground live biomass in old-growth Amazonian forests. *Global Change Biology* **12**: 1107–1138.

Marrs, R., J. Proctor, A. Heaney, and M. Mountford (1988). Changes in soil nitrogen mineralization and nitrification along an altitudinal transect in tropical rain forest in Costa Rica. *Journal of Ecology* **76**: 466–482.

Maycock, C. R., and R. A. Congdon (2000). Fine root biomass and soil N and P in north Queensland rain forests. *Biotropica* **32**: 185–190.

McGroddy, M., and W. L. Silver (2000). Variations in belowground carbon storage and soil CO_2 flux rates along a wet tropical climate gradient. *Biotropica* **32**: 614–624.

Meentemeyer, V. (1977). Climatic regulation of decomposition rates of organic matter in terrestrial ecosystems. In *Environmental Chemistry and Cycling Processes*, eds. D. C. Adrians and I. L. Brisbin, pp. 779–789. Washington, DC: U.S. Department of Energy.

Moser, G., D. Hertel, and Ch. Leuschner (2007). Altitudinal change in LAI and stand leaf biomass in tropical montane forests: a transect study in Ecuador and a pan-tropical meta-analysis. *Ecosystems* **10**: 924–935.

Moser, G., M. Röderstein, N. Soethe, D. Hertel, and Ch. Leuschner (2008). Altitudinal changes in stand structure and biomass allocation of tropical mountain forest in relation to microclimate and soil chemistry. In *Gradients in a Tropical Mountain Ecosystem of Ecuador*, eds. E. Beck, J. Bendix, I. Kottke, F. Makeschin, and R. Mosandl, pp. 229–242. Berlin: Springer-Verlag.

Nepstad, D. C., Moutinho, P., Dias, M. B., *et al.* (2002). The effects of partial throughfall exclusion on canopy processes, aboveground production, and biogeochemistry of an Amazon forest. *Journal of Geophysical Research* **107**: 1–18.

Newbery, D. M., I. J. Alexander, and J. A. Rother (1997). Phosphorus dynamics in a lowland African rain forest: the influence of ectomycorrhizal trees. *Ecological Monographs* **67**: 367–409.

Nomura, N., and K. Kikuzawa (2003). Productive phenology of tropical montane forests: fertilization experiments along a moisture gradient. *Ecological Research* **18**: 573–586.

Ostertag, R. (2001). Effects of nitrogen and phosphorus availability on fine-root dynamics in Hawaiian montane forests. *Ecology* **82**: 485–499.

Pavlis, J., and J. Jenik (2000). Roots of pioneer trees in the Amazonian rain forest. *Trees* **14**: 442–455.

Post, W. M., and K. C. Kwon (2000). Soil carbon sequestration and land use change, processes and potential. *Global Change Biology* **6**: 317–327.

Pregitzer, K. S., R. L. Hendrick, and R. Fogel (1993). The demography of fine roots in response to patches of water and nitrogen. *New Phytologist* **125**: 575–580.

Priess, T., C. Then, and H. Fölster (1999). Litter and fine-root production in three types of tropical premontane rain forest in Venezuela. *Plant Ecology* **143**: 171–187.

Raich, J. W. (1980). Fine root regrow rapidly after forest felling. *Biotropica* **12**: 231–232.

Raich, J. W. (1983). Effects of forest conversion on the carbon budget of a tropical soil. *Biotropica* **15**: 177–184.

Raich, J. W., and K. J. Nadelhoffer (1989). Belowground carbon allocation in forest ecosystems: global trends. *Ecology* **70**: 1346–1354.

Richards, P. W. (1996). *The Tropical Rain Forest*, 2nd edn. Cambridge, UK: Cambridge University Press.

Röderstein, M., D. Hertel, and C. Leuschner (2005). Above- and belowground litter production in three tropical mountain forests (Ecuador). *Journal of Tropical Ecology* **21**: 483–492.

Sánchez-Gallén, I., and J. Alvarez-Sánchez (1996). Root productivity in a lowland tropical rain forest in Mexico. *Vegetatio* **123**: 109–115.

Sanford, R. L. (1989). Fine root biomass under a tropical forest light opening in Costa Rica. *Journal of Tropical Ecology* **5**: 251–256.

Santiago, L. S., G. Goldstein, F. C. Meinzer, J. H. Fownes, and D. Mueller-Dombois (2000). Transpiration and forest structure in relation to soil waterlogging in a Hawaiian montane cloud forest. *Tree Physiology* **20**: 673–681.

Schawe, M., S. Glatzel, and G. Gerold (2007). Soil development along an altitudinal transect in a Bolivian tropical montane rainforest: Podzolization vs. hydromorphy. *Catena* **69**: 83–90.

Schrumpf, M., M. Guggenberger, G. Schubert, C. Valarezo, and W. Zech (2001). Tropical montane rainforest soils: development and nutritional status along an altitudinal gradient in the south Ecuadorian Andes. *Die Erde* **132**: 43–59.

Schuur, E. (2001). The effect of water on decomposition dynamics in mesic to wet Hawaiian montane forests. *Ecosystems* **4**: 259–273.

Schuur, E. A. G., and P. A. Matson (2001). Net primary productivity and nutrient cycling across a mesic to wet precipitation gradient in Hawaiian montane forest. *Oecologia* **128**: 431–442.

Silver, W. L., and K. A. Vogt (1993). Fine root dynamics following single and multiple disturbance in a subtropical wet forest ecosystem. *Journal of Ecology* **81**: 729–738.

Silver, W. L., F. N. Scatena, A. H. Johnson, T. G. Siccama, and F. Watt (1996). At what temporal scales does disturbance affect belowground nutrient pools? *Biotropica* **28**: 441–457.

Silver, W. L., M. Keller, and A. E. Lugo (1999). Soil oxygen availability and biogeochemical cycling along elevation and topographic gradients in Puerto Rico. *Biogeochemistry* **44**: 301–328.

Silver, W. L., J. Neff, M. McGroddy, *et al.* (2000). Effects of soil texture on belowground carbon and nutrient storage in a lowland Amazonian forest ecosystem. *Ecosystems* **3**: 193–209.

Soethe, N., J. Lehmann, and C. Engels (2006). The vertical pattern of rooting and nutrient uptake at different altitudes of a south Ecuadorian montane forest. *Plant and Soil* **286**: 287–299.

Soethe, N., J. Lehmann, and C. Engels (2008a). Nutrient availability at different altitudes in a tropical montane forest in Ecuador. *Journal of Tropical Ecology* **24**: 397–406.

Soethe, N., W. Wilcke, J. Homeier, J. Lehmann, and C. Engels (2008b). Plant growth along the altitudinal gradient: role of plant nutritional status, fine root activity, and soil properties. In *Gradients in a Tropical Mountain Ecosystem of Ecuador*, eds. E. Beck, J. Bendix, I. Kottke, F. Makeschin, and R. Mosandl, pp. 259–266. Berlin: Springer-Verlag.

Stark, N., and M. Spratt (1977). Root biomass and nutrient storage in rain forest oxisols near San Carlos de Rio Negro. *Tropical Ecology* **18**: 1–9.

Stewart, C. G. (2000). Fine root growth in response to local nutrient enrichment in two tropical montane forests. *Biotropica* **32**: 369–373.

Sundarapandian, S. M., and P. S. Swamy (1996). Fine root biomass distribution and productivity patterns under open and closed canopies of tropical forest ecosystems at Kodayar in Western Ghats, South India. *Forest Ecology and Management* **86**: 181–192.

Takyu, M., S. I. Aiba, and K. Kitayama (2003). Changes in biomass, productivity and decomposition along topographical gradients under different geological conditions in tropical lower montane forests on Mount Kinabalu, Borneo. *Oecologia* **134**: 397–404.

Tanner, E. V. J. (1977). Four montane rain forests of Jamaica: a quantitative characterization of the floristics, the soils and the foliar mineral levels, and a discussion of the interrelations. *Journal of Ecology* **65**: 883–918.

Tanner, E. V. J. (1980). Studies on the biomass and productivity in a series of montane rain forests in Jamaica. *Journal of Ecology* **68**: 573–588.

Tanner, E. V. J. (1981). The decomposition of leaf litter in Jamaican montane rain forests. *Journal of Ecology* **69**: 263–273.

Tanner, E. V. J., P. M. Vitousek, and E. Cuevas (1998). Experimental investigation of nutrient limitation of forest growth on wet tropical mountains. *Ecology* **79**: 10–22.

Vance, E. D., and N. M. Nadkarni (1992). Root biomass distribution in a moist tropical montane forest. *Plant and Soil* **142**: 31–39.

Vitousek, P. M., and R. L. Sanford (1986). Nutrient cycling in moist tropical forest. *Annual Review of Ecology and Systematics* **17**: 137–167.

Vogt, K. A., D. J. Vogt, *et al.* (1996). Review of root dynamics in forest ecosystems grouped by climate, climatic forest type and species. *Plant and Soil* **187**: 159–219.

Yavitt, J. B., and S. J. Wright (2001). Drought and irrigation effects on fine root dynamics in a tropical moist forest, Panama. *Biotropica* **33**: 421–434.

Part V
Cloud forest water use, photosynthesis, and effects of forest conversion

L. A. Bruijnzeel and F. N. Scatena

47 Transpiration and microclimate of a tropical montane rain forest, southern Ecuador

T. Motzer, N. Munz
University of Mannheim, Mannheim, Germany
D. Anhuf
University of Passau, Passau, Germany
M. Küppers
University of Hohenheim, Stuttgart, Germany

ABSTRACT

Climatic measurements were made within and above a tropical lower montane (cloud) forest at 1975 m.a.s.l. in the southern Ecuadorian Andes to assess micro-meteorological conditions and calculate overall forest transpiration (using the Penman–Monteith model). Transpiration by individual trees was determined by means of sapflow measurements (Granier-type gages), supplemented by porometric measurements of leaf gas exchange. The light environment of the forest was characterized by high spatio-temporal variability. Modeled forest-floor photosynthetic photon flux density (*PPFD*) varied on average between 5.5% and 10.5% of incident radiation. Thermo-hygric gradients within the forest were weak, and the understory stratum was aerodynamically well coupled to the atmosphere above the forest, suggesting efficient turbulent mixing of the forest air. Daily sapflow totals ranged from 2–165 l day^{-1} and increased dramatically with tree height, trunk diameter, and crown dominance. Stand-scaled sapflow E_c was modeled based on stand structural parameters in a two-layer approach. The contribution of sub-canopy and understorey individuals to total stand transpiration was only about 20%, illustrating the dominance of upper canopy trees in the process. Penman–Monteith-based estimates of forest transpiration E_a showed a high energy efficiency (average E_a ~ 70% of net radiation), but due to low solar inputs, E_a remained low at 560 mm year^{-1} (representing 26.4% of the annual precipitation of 2120 mm during the measurement period). Stand-scaled sapflow E_c accounted for 85% to more than 90% of daily E_a and amounted to 1.8 mm day^{-1} on average (range: 0.7–2.8 mm day^{-1}) during the drier season.

INTRODUCTION

Tropical montane forests are considered to play a crucial role in local and regional water budgets and atmospheric circulation (e.g. Lawton *et al.*, 2001; Nair *et al.*, 2003; Ray *et al.*, 2006; Van der Molen *et al.*, 2006; cf. Foster, this volume). Generally, knowledge of biophysical interactions between forest canopies and the overlying atmosphere is essential for sound interpretation of hydrological and microclimatic processes in forested areas. These processes comprise the conversion of solar energy to latent and sensible heat at the canopy surface and its transmission into the forest interior, thereby influencing forest and soil energy budgets, as well as plant physiological activity, including transpiration (Roberts *et al.*, 1993, 2005). High-resolution eco-physiological and meteorological field data are indispensable for the quantification of site-specific energy and water budgets.

There have been some studies of the water budget of tropical montane forests (reviewed by Bruijnzeel 2001, 2005), but most of these have calculated evapotranspiration (*ET*) using a water balance approach at the watershed scale (e.g. Edwards, 1979; Schellekens *et al.*, 2000; Fleischbein *et al.*, 2006; Wilcke *et al.*, 2008; Fleischbein *et al.*, this volume) from which the transpirational component of *ET* was derived only indirectly by subtracting

Tropical Montane Cloud Forests: Science for Conservation and Management, eds. L. A. Bruijnzeel, F. N. Scatena, and L. S. Hamilton. Published by Cambridge University Press.

wet-canopy evaporation (interception) from total *ET*. Whilst this approach is already crude in itself because of the possibility of error accumulation in estimated *ET* (Ward and Robinson, 2000), the method greatly underestimates *ET* in the case of cloud-affected forests where occult cloud water inputs are not explicitly taken into account in the water budget (Bruijnzeel and Proctor, 1995). Transpiration (i.e. evaporation from a dry canopy, E_t) in montane tropical forests has been determined occasionally using (destructive) potometric methods (e.g. Weaver, 1975; Zotz *et al.*, 1998) or via *in situ* measurements of sapflow (Santiago *et al.*, 2000; García-Santos, 2007; McJannet *et al.*, 2007). However, such tree-based approaches require upscaling to the stand level which is not always easy in species-rich rain forests (McJannet *et al.*, 2007; Küppers *et al.*, 2008). Similarly, applying micro-meteorological approaches such as the eddy covariance technique has its own problems in steep and complex topography (Bruijnzeel and Proctor, 1995; Turnipseed *et al.*, 2003; Holwerda, 2005; Beiderwieden *et al.*, 2008). In short, quantitative information on transpirational water loss from tropical montane (cloud) forests has remained scarce (Bruijnzeel, 2001, 2005; McJannet *et al.*, 2007; cf. García-Santos, 2007; McJannet *et al.*, this volume).

This scarcity of information is equally true for intra-forest climatic data in montane tropical forests. This chapter describes the micro-meteorological conditions prevailing within a lower montane forest situated just below the main cloud belt in the Ecuadorian Andes, and their implications for tree and stand transpiration. Moreover, gage-based transpiration rates and rates calculated with the Penman–Monteith equation are compared.

MATERIALS AND METHODS

Field site

The study site is located in the lower montane rain forest belt of the Reserva Biológica San Francisco in the southern Ecuadorian Andes (03° 58′ S, 79° 04′ W). Measurements were carried out at 1975 m.a.s.l. within a 480-m^2 forest plot in steep terrain (up to 40–45°, NNW-exposed). Average canopy height was 18 m. The forest was stratified into an overstorey layer (>14 m), a mid-/sub-canopy stratum (2.5–14 m; arguably with a non-continuous boundary between mid- and sub-canopy at about 6 m height), and an understory layer (<2–2.5 m; for details see Motzer, 2003). Stand density was 812 stems ha^{-1} (all trees with diameter at breast height DBH > 10 cm). Leaf area index (*LAI*) averaged 6.4 (range 3.7–9.8; Motzer, 2003; cf. Oesker *et al.*, this volume) and the frequency of epiphytes and lianas was moderate compared to forest at higher altitudes in the area (Kürschner and Parolly, 2004; Motzer *et al.*, 2005).

Regional weather data were provided by a standard automatic weather station (AWS) at 1952 m.a.s.l. (see Figure 32.1 in Fleischbein *et al.*, this volume). Average annual precipitation between 1998 and 2002 was 2067 mm, with a marked rainy season from April to June (with maximum rainfall in May, up to 257 mm) and a drier period between November and January (minimum in November 90 mm). Mean annual air temperature was 15.5 °C (1999–2002), varying from 14.5 °C (July) to 16.2 °C (November), although under dry conditions maximum day-time values sometimes exceeded 25 °C. Atmospheric humidity was generally high and night-time saturation often occurred, with dew upon the plant surfaces being observed most early mornings. Low cloud or dense fog was infrequent and horizontal precipitation was estimated to be less than 10% of annual precipitation at altitudes between 1800 and 2000 m. a.s.l. (Bendix *et al.*, 2008; cf. Rollenbeck *et al.*, this volume). As such, and because of its epiphyte and bryophyte load (Fleischbein *et al.*, this volume) the study forest may be tentatively classified as representing a transition between lower montane rain forest (LMRF) *sensu* Grubb (1977) and the lower montane cloud forest (LMCF) of Bruijnzeel and Hamilton (2000). Fieldwork was conducted during a pilot study in a comparable plot nearby (November 1999–February 2000) and during the main phase in the plot described above (transpiration: September 2000–March 2001, microclimate until February 2002).

Measurements

STAND STRUCTURE

Relevant stand structural parameters such as stem diameter, tree height, projected crown area, and crown depth were determined and incorporated into a GIS-system (ArcView 3.2a). Canopy openness and *LAI* were inferred from optical measurements using an LAI2000 plant canopy analyzer (Li-Cor, USA).

FOREST MICROCLIMATE

Microclimatic parameters within and below the canopy were measured using a custom-made steel wire suspension system with aluminum frames carrying the sensors. The system was attached to a branch of *Trichilia guianense*, a dominant canopy tree of 18 m height, and secured with guy wires tied to the ground. Air temperature and humidity (Vaisala HMP35 probes), photosynthetic photon flux density (*PPFD*; SKP215, Skye Instruments) and wind speed (A100R cup anemometer, Vector Instruments) were all measured at 4 m, 8 m, and 13 m above the ground. A telescoping extension was mounted in the upper crown of the same tree to allow measurements of climatic conditions at about 2 m above the vegetation surface (20 m above the ground). In addition to the wind speed and *PPFD* sensors described above, lightweight instruments were used for temperature and humidity (Vaisala 50Y probe), short-wave radiation (Li-Cor LI 200SA), and net radiation (Kipp and Zonen, NRLite). All data were sampled at 30-s intervals and averaged over 5-min periods, using a Campbell Scientific CR10X data-logger with an upstream

AM416 Relay Multiplexer, and stored in a Cambell Scientific SM716 storage module.

Values of *PPFD* at the forest floor were modeled for the entire plot by analyzing 52 hemispheric photographs using the Software HemiView 2.1 (Delta-T Devices, UK). The shading influence of the narrowed horizon was taken into account, and the light extinction coefficient in the model was optimized using the continuously monitored *PPFD* in the understory. This enhanced model performance considerably compared to the HemiView default radiation model (Motzer, 2005).

SAPFLOW IN INDIVIDUAL TREES

To estimate individual tree water use, xylem sapflow was measured using the thermal dissipation method developed by Granier (1985). These sapflow gages consist of two 20-mm long needle-probes (manufactured by UP, Germany) inserted horizontally above each other into the sapwood of the tree trunk. The upper probe is heated continually, creating a temperature difference with the lower, non-heated probe. This temperature difference is inversely correlated with sapflux density J_s (ml $cm^{-2}_{(sapwood)}$ min^{-1}) as calculated using an empirical relationship proposed by Granier (1985). Total sapflow F (ml min^{-1}) is calculated conventionally by multiplying J_s times the conducting sapwood area A_s, which can be estimated with sufficient accuracy from wood cores (Diawara *et al.*, 1991). Specific decay functions were applied to account for radial changes in J_s (Motzer, 2003). Sapflow was measured on 10 trees (plus seven in the pilot plot) of different taxonomy and structural position, reflecting the floristic and structural composition of the plot. In order to account for sapflow variations due to inhomogeneities in sapwood, extra sensors were inserted in opposite stem segments and at 20–40 mm depth on three trees (see Motzer (2003) for details).

Sapflow data for individual trees were scaled to the stand level (E_c) using a two-layer model approach. Due to the strong correlation between F and conducting sapwood area A_s, stem cross-sectional area was chosen as the scaling variable:

$$E_c = \frac{\left(A_{b,OV}\,\overline{F_{b,OV}}\right) + \left(A_{b,SC}\,\overline{F_{b,SC}}\right)}{A_{plot}} \tag{47.1}$$

where A_{plot} is the surface area of the plot (m^2), A_b the basal area of the tree stems (m^2), $\overline{F_b}$ the average sapflow per unit basal area for the measured trees (l m^{-2} min^{-1}), while the suffixes *OV* and *SC* denote the overstory (14–18 m) and mid- to sub-canopy trees (2.5–14 m), respectively. Using a GIS-based crown map, some individuals smaller than 14 m were assigned to the overstory group, when at least 50% of their crown area contributed to the upper canopy. Understory transpiration was estimated using porometric data. Its share of E_c remained below 2–4% and therefore was neglected (see Motzer *et al.*, 2005 for details).

STAND TRANSPIRATION

Actual transpiration of the forest stand was also determined with the well-known Penman–Monteith equation (Monteith, 1965), with the physiologically controlled crown or surface conductance g_c calculated according to the non-linear multifactorial model of Jarvis (1976) as adapted by Stewart (1988):

$$g_c = g_{c,max} \cdot f(PPFD)\,f(VPD_c)\,f(T_c)\,f(\Psi_s)\,f(LAI). \tag{47.2}$$

In Eq. (47.2), the maximum crown conductance $g_{c,max}$ is reduced by stress functions that incorporate *PPFD*, vapor pressure deficit at crown level VPD_c, crown temperature T_c, soil water potential Ψ_s and *LAI* (Jarvis, 1976; Jones, 1992). The respective coefficients were calculated from non-linear regression analysis using field data of leaf conductance and transpiration at leaf level using a Li-Cor LI 1600 steady-state porometer (see Motzer *et al.*, 2005 for details). Access to the tree crowns was provided by rope-climbing techniques and a rope ladder. Therefore, gas exchange measurements could be realized only within the central crowns of securely accessible canopy trees. The porometry data-set consisted of about 2000 measurements on five canopy trees and 10 understory individuals, representing 21 days at the end of the dry season between January and March 2001.

RESULTS

The annual short-wave radiation total above the forest stand was 3937 MJ m^{-2} (October 2000–September 2001). The upper canopy layer accounted for the interception of more than 90% of incident *PPFD*. In the understory, measured *PPFD* transmission at 4 m above ground averaged 5.5%, but depending on the direct to diffuse radiation ratio (as deduced from HemiView modeling), fluctuations were high (Figure 47.1). Spatial variation of modeled *PPFD* at the forest floor was considerable as well. Compared to point quantum sensor measurements made at 4 m, the spatial model yielded slightly higher *PPFD* values (5.5–10.5% of incident *PPFD*), depending on spatial position (Figure 47.2).

Figure 47.3 shows the thermo-hygric stratification of the forest air under (a) more or less standard day-time meteorological conditions (characterized by sufficient radiant energy influx (11.6 MJ m^{-2}) and moderate but gusty valley winds; 13 December 2000), and (b) during an overcast and rainy day (15 April 2001). As a result of the topographically conditioned narrowed horizon (proportion of visible sky 0.66), direct radiation did not reach the forest canopy until 08.00 h. Heating of the vegetation surface then led to rapid increases in air temperature T_a and vapor pressure deficit *VPD* above the canopy, but within the forest, both T_a and *VPD* increased only after a marked delay, particularly under sunny conditions (Figure 47.3a). Interestingly,

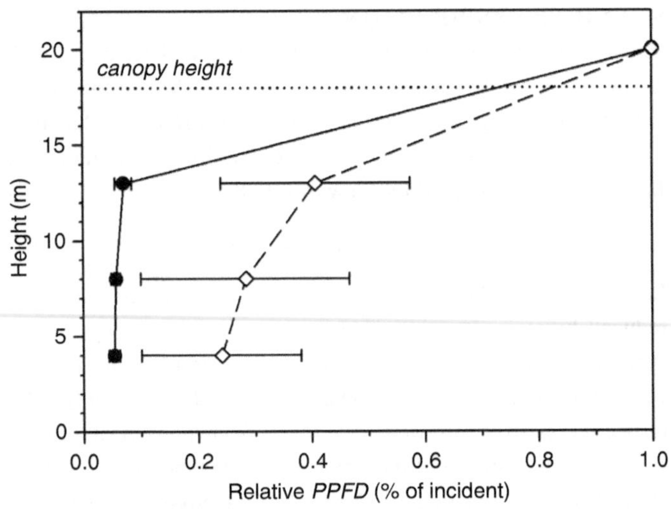

Figure 47.1. Light climate of the forest: transmission of photosynthetic photon flux density *PPFD* as measured with quantum sensors at the meteorological station; (●) daily mean and (◇) maximum values during the study period.

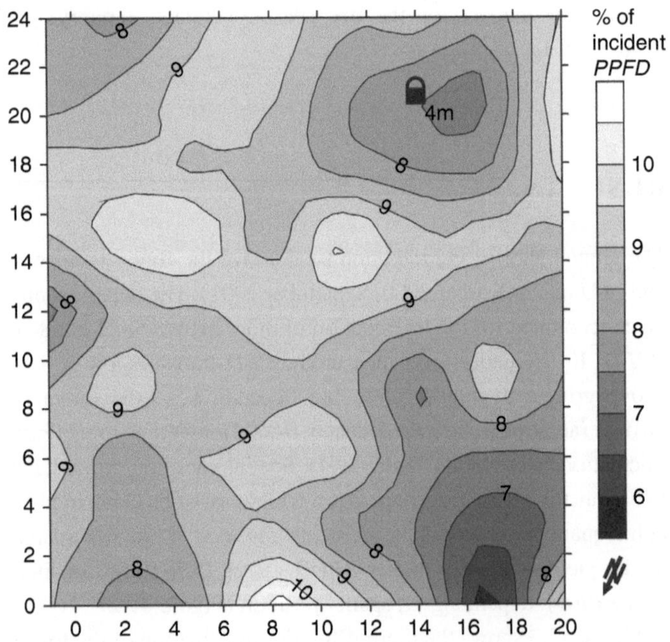

Figure 47.2. Light climate of the forest: average light environment at the forest floor as modeled with HemiView 2.1. Symbol indicates position of quantum sensor.

all curves proceeded almost in phase throughout the day. Day-time T_a (and *VPD*) were almost identical within the mid-canopy stratum (8–13 m), and ground-level T_a was only slightly lower. The onset of rain on 15 April interrupted the normal increase in both variables during the morning, causing a rapid decline of all curves and a collapse of the preceding stratification (Figure 47.3b). The air in the forest interior became saturated. When rainfall intensity decreased (around 15.00–17.00 h), the profile re-established under almost rainless conditions, except in the understory. During the entire study period, day-time maximum T_a above the canopy (20 m) rarely exceeded 28 °C, whereas within the forest mean temperatures were about 20% lower (Motzer, 2005).

Sapflow rate increased drastically with tree height and stem DBH (Figure 47.4). Average daily totals of F amounted to 80–120 l day^{-1} (with an absolute maximum of 165 l day^{-1}) for dominant canopy trees, primarily depending on day-time climatic conditions. Daily values of F for intermediate and suppressed mid- or sub-canopy trees were lower by an order of magnitude (18.9 l day^{-1}, range 1.3–9.7).

A time series of stand-scaled sapflow E_c and Penman–Monteith-based actual transpiration E_a demonstrates that both were in phase during daylight conditions, but not in the late evening and during the night (Figure 47.5). Whilst E_a dropped immediately after sunset, E_c decreased only gradually and zero flow usually did not occur until the early morning (if at all). For this reason, late-evening values of E_c had to be neglected when investigating the meteorological triggers of E_c (see Figure 47.8 below). Night-time gain of latent heat (i.e. condensation, negative E_a) was frequent but remained usually well below 5% of day-time latent heat totals, although this depended on night-time atmospheric conditions. When separating transpiration for the overstorey and mid- or sub-canopy categories, the latter showed a contracted daily course (Figure 47.5). However, the contribution by mid- and sub-canopy trees to daily stand transpiration totals was only about 20% of E_c on average, with short-term peaks of 35–40% of E_c under well-ventilated sunny conditions (e.g. on 16 December in Figure 47.5). Maximum E_a exceeded 0.4 mm hour^{-1}, whereas E_c was well below 0.3 mm hour^{-1} even under favorable conditions. Under certain conditions, sapflow-based stand transpiration E_c exceeded Penman–Monteith-based estimates E_a, both in terms of day-time hourly values (e.g. on 22 December in Figure 47.5) and daily sums (Figure 47.6).

On a daily basis, E_c was linearly correlated with summed *PPFD* (Figure 47.7). Multiple regression analysis revealed that 75% of the variance in E_c could be explained by *PPFD* alone, whereas incorporation of additional meteorological parameters such as *VPD*, wind speed, and T_a enhanced the explained variance by only 6% (and chiefly attributable to *VPD*). When plotting hourly values of E_c against *VPD*, on prevalent sunny days E_c increased initially with *VPD* in a quasi-linear manner (Figure 47.8). Under conditions of higher atmospheric evaporative demand (*VPD* > 12 hPa), *PPFD* was usually above 1200 mol m^{-2} s^{-1}. In this range, the relationship between E_c and *VPD* tended to level off and sapflow became relatively insensitive to changes in *PPFD* or *VPD* (Figure 47.8). During the daylight period (07.00–18.00 h) this relationship was strongly exponential ($r^2 = 0.79$, $p < 0.001$).

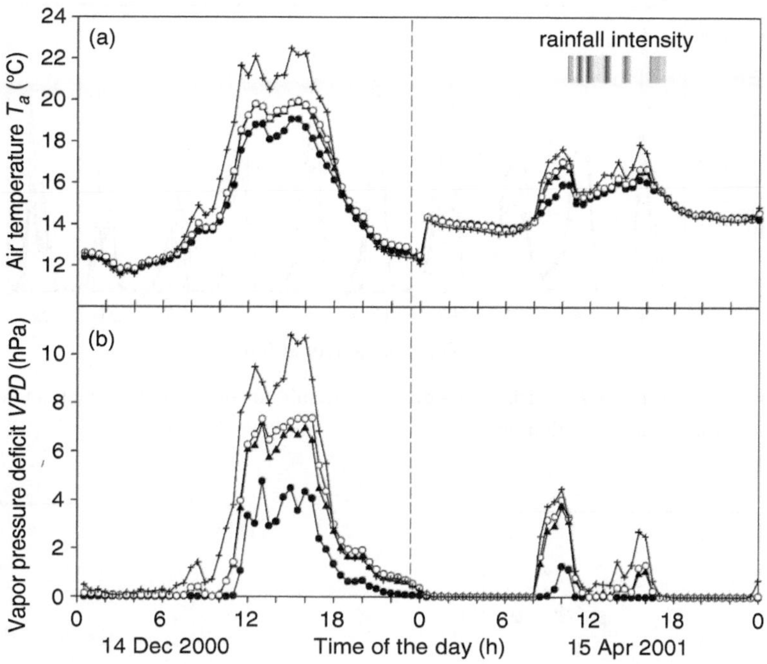

Figure 47.3. Daily cycles of (a) air temperature T_a and (b) water vapor pressure deficit *VPD* along the measured vertical profile on a sunny day (left-hand diagram) and on a rainy day (right-hand diagram; rainfall intensity schematic). Symbols indicate measured data at heights of 4 m (●), 8 m (▲), 13 m (○) within the forest and above the canopy at 20 m (+).

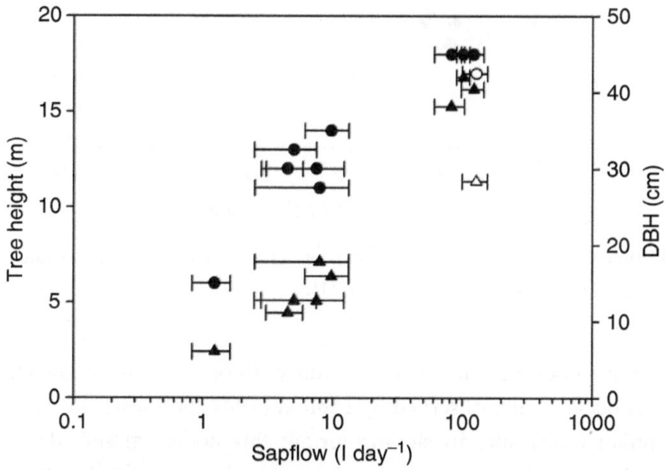

Figure 47.4. Daily sapflow totals of the investigated tree species as a function of (●) stem diameter (DBH) and (▲) tree height. Open symbols: *Cecropia montana*. Sapflow plotted on a logarithmic scale to linearize the relationship with DBH ($r^2 = 0.75$, $p < 0.001$) and tree height (dashed line, $r^2 = 0.8$, $p < 0.001$), both excluding *Cecropia*.

DISCUSSION

Local climatic conditions, topographic characteristics, and forest structure all influence the microclimate above and within the forest, and this, in turn, controls the eco-physiology of the plants. Micro-meteorological gradients within the forest may provoke gradients of physiological activity as well (Roberts *et al.*, 1993; cf. Küppers *et al.*, 2008). Therefore, thermo-hygric and light conditions in the vertical stand profile are crucial for the interpretation and simulation of eco-physiological processes such as photosynthesis and transpiration. The attenuation of light penetrating into the studied forest, and the spatio-temporal variability of the light transmission were considerable (Figures 47.1 and 47.2; cf. Oesker *et al.*, this volume). Average *PPFD* at the forest floor was at the upper limit of the range reported for other tropical (lowland) forest sites (Shuttleworth *et al.*, 1984; Walter and Breckle, 1991; Anhuf and Rollenbeck, 2001), which may partly be attributed to the high proportion of diffuse light at this montane site. Even though modeled *PPFD* levels at the forest floor (Figure 47.2) were slightly higher than measured ones, HemiView modeling provides valuable insights into the radiation regime of a forest, and may also support other ecological studies (e.g. seedling dynamics). However, in mountainous terrain it is essential to apply site-specific correction factors to account for the narrowed horizon (proportion of visible sky, which amounted to 0.66 in this particular case) and the lower general atmospheric extinction coefficient (Motzer, 2005; cf. Oke, 1987).

Generally, the absorption of solar radiation in the upper canopy of dense, multi-layered forest stands generates a stable stratification of air density, with warm, light air in the upper part of the forest and a cooler understory stratum below (Dolman *et al.*, 1991; Szarzynski and Anhuf, 2001). Wind gusts may penetrate the upper canopy, but the understory usually remains unaffected by turbulent air movement, leading to a decoupled

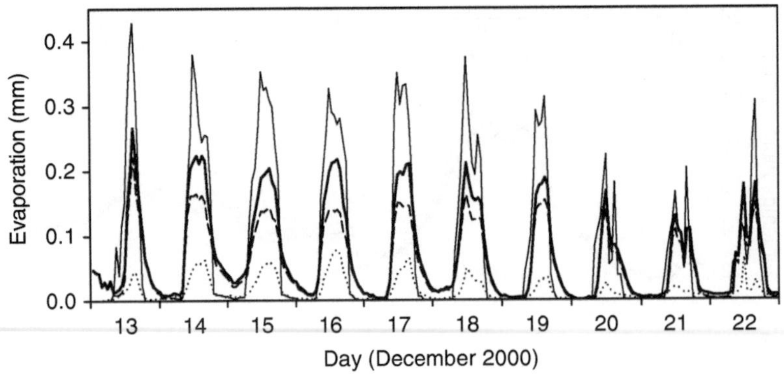

Figure 47.5. Daily courses of stand-scaled sapflow E_c (thick line) – divided into mid- or sub-canopy (dotted) and overstory (dashed) portion – and Penman–Monteith estimates of forest transpiration E_a (thin line).

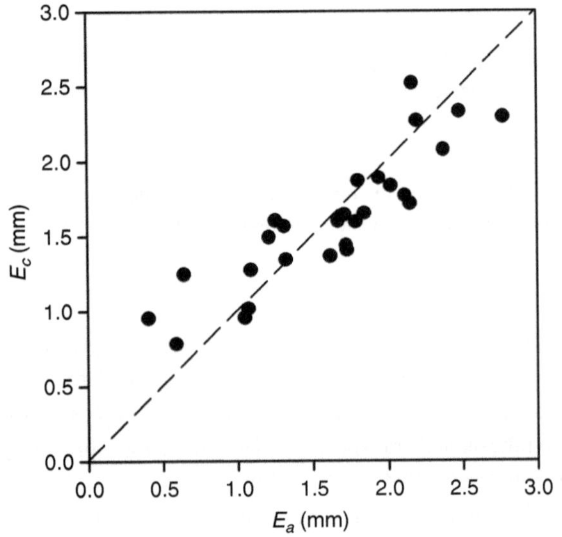

Figure 47.6. Canopy transpiration E_c in relation to actual transpiration E_a (Penman–Monteith estimate). Rainy days were excluded.

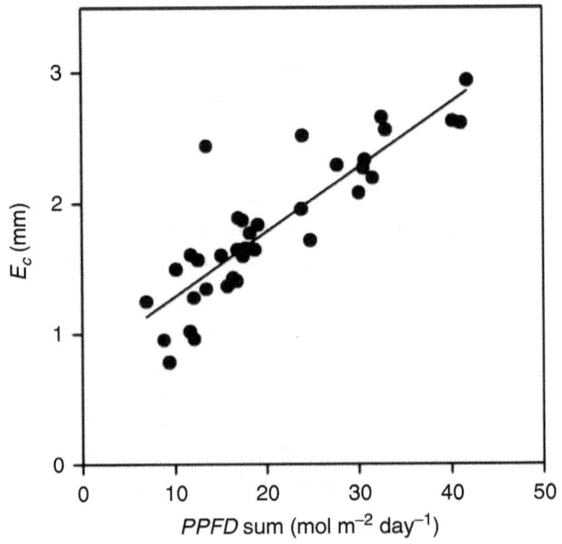

Figure 47.7. Canopy transpiration E_c as a function of daily *PPFD* sums. Linear regression $r^2 = 0.75$, $p < 0.001$.

course of temperature and humidity. However, the studied forest exhibited comparatively weak thermo-hygric gradients and a more or less in-phase progression of T_a and *VPD* throughout the measured profile (Figure 47.3). Together with the observation of sometimes higher wind speeds within the forest than slightly above the canopy (Motzer, 2005), this may be taken as evidence of relatively efficient turbulent mixing of the forest air volume – possibly as a result of the typically gusty valley winds, that invoke intensive leaf and crown movement, thereby allowing deep penetration of eddies. Similarly, the unusual night-time situation, characterized by the absence of a temperature inversion along the forest profile (e.g. Szarzynski and Anhuf, 2001), seems to be linked to the local wind patterns (cf. Bendix *et al.*, 2008). After sunset, radiative cooling of the upper canopy layer is accompanied by the establishment of typically calm downslope winds. The cool air masses, sliding gently downslope upon the canopy surface, may sink through local canopy openings down to the forest floor. This phenomenon was occasionally observed by Molion (1987) at a lowland Amazonian forest site. In sloping terrain this nocturnal subsidence leads to the development of two more or less separate katabatic flow systems of cold air: one above the canopy surface, and a second near the forest floor. This contention was supported partly by anemometer measurements (Motzer, 2005), but further evidence, e.g. through stratified eddy correlation measurements, would be desirable.

The current analysis of meteorological influences on canopy transpiration indicates that diurnal radiation levels are the primary control of overall stomatal status and daily transpiration totals of the canopy trees (Figure 47.7). Within this *PPFD*-imposed range, changes in *VPD* appear to act as the main short-term trigger of transpiration during the course of the day. The stand-scaled sapflow data gave strong evidence for an upper limit to forest transpiration. This tendency of sapflow to level off

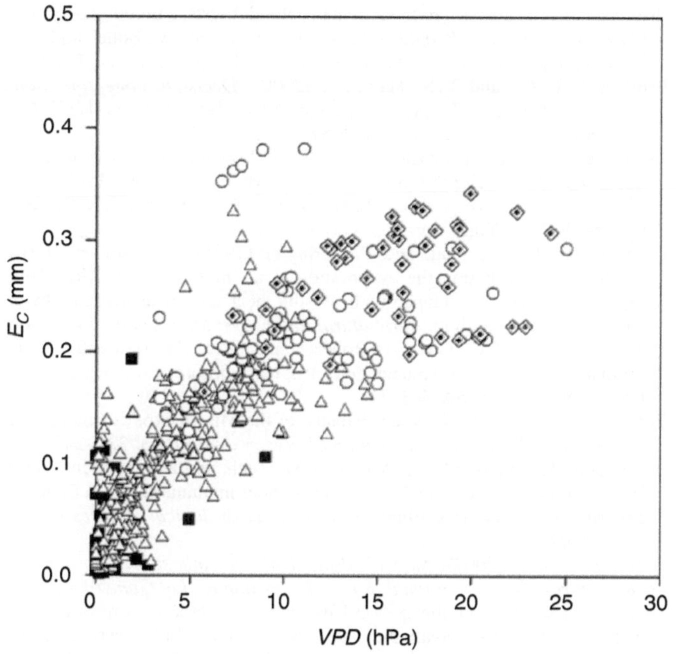

Figure 47.8. Canopy transpiration E_c in relation to above-canopy vapor pressure deficit *VPD* at different light levels. Hourly data between 07.00 and 18.00 h; different symbols indicate differences in *PPFD* over the previous 1-hour period (*PPFD* in μmol m^{-2} s^{-1}): (■) <50, (△) 50–600, (○) 600–1200, (◈) >1200. Non-linear regression: $y = 0.2767$ $(1-e^{-0.1533x})$, $r^2 = 0.79$, $p < 0.001$.

at higher *VPD* levels (Figure 47.8) has been reported for tropical lowland trees as well (Meinzer *et al.*, 1993; Granier *et al.*, 1996b; Oren *et al.*, 1996; Anhuf *et al.*, 1999; cf. García-Santos, 2007), and can be attributed to stomatal or hydraulic limitations. Measured tree transpiration rates were highest for dominant overstory trees with extensive crowns exposed to the turbulent atmospheric environment above the forest, and decreased rapidly with tree height and successive shading of the crowns (Figure 47.4). Differences in transpirational behavior were more pronounced between trees of different structural positions than between species (cf. Küppers *et al.*, 2008). Therefore, incorporating several size- and structure-related variables into the upscaling procedure is of major importance for the reliability of sapflow-based stand transpiration (Granier *et al.*, 1996a,b; Oren *et al.*, 1998; cf. McJannet *et al.*, 2007). In this context, differing physiological traits or successional status of the measured species have to be considered as well. For example, *Cecropia montana*, a light-demanding, fast-growing, late-stage pioneer species that possesses large xylem vessels and a high conductivity (Richards, 1996), exhibited above-average transpiration rates for its stem diameter (Figure 47.4; cf. Küppers *et al.*, 2008). However, due to its exceptional status, and the fact that gaps made up only a small proportion of the forest, *Cecropia* was not included in the overall stand transpiration model.

At values of 2.0 (Penman–Monteith estimate) and 1.8 (sapflow gages) mm day^{-1}, the average short-term transpiration rates derived for the present LMCF during the (relatively drier) period between January and March 2001 fall within the range reported for this type of forest (1.0–2.4 mm day^{-1}; Holwerda, 2005; Gomez-Cardenas, 2009; Bruijnzeel *et al.*, this volume) and are comparable to the 1.7 mm day^{-1} (591 mm $year^{-1}$) obtained by McJannet *et al.* (this volume) for a similar forest type in northern Australia using heat-pulse velocity gages to measure sapflow. The present Penman–Monteith-based estimate for the year October 2000–September 2001 of 561 mm (1.54 mm day^{-1}; Motzer, 2003) is also of the same order of magnitude as the crude average estimate derived by Fleischbein *et al.* (this volume) for the present forest between 1900 and 2200 m.a.s.l. by subtracting measured annual interception loss from overall watershed-budget-based annual *ET* (460 ± 150 mm $year^{-1}$ or 1.26 ± 0.41 mm day^{-1}).

Transpiration studies in tropical montane forest have either employed bulk micro-meteorological approaches such as eddy covariance measurements (Holwerda, 2005; Beiderwieden *et al.*, 2008; Giambelluca *et al.*, 2009) or (increasingly) a large number of sapflow gages and scaling up to stand level (Santiago *et al.*, 2000; García-Santos, 2007; McJannet *et al.*, 2007; Küppers *et al.*, 2008; Gomez-Cardenas, 2009). When making such comparisons it is important to bear in mind the sometimes incomplete compatibility of different approaches (Bruijnzeel and Proctor, 1995). This is also illustrated when comparing the present sapflow-based stand transpiration E_c to the Penman–Monteith-based estimates of forest transpiration E_a. On an hourly basis, E_c frequently exceeded E_a after sunset, when E_a rapidly dropped with fading solar energy input whereas E_c decreased more gradually (Figure 47.5). This phenomenon can be attributed to the recharge of the plant water storage when leaf transpiration ceases (Schulze *et al.*, 1985; Goldstein *et al.*, 1998; Phillips *et al.*, 2003) although night-time transpiration driven by continued atmospheric evaporative demand has been reported occasionally as well (García-Santos, 2007). Whilst long-term stand-scale sapflow usually equates total transpirational water loss from a forest (Granier *et al.*, 1996a; Köstner *et al.*, 1998; Wullschleger *et al.*, 1998), water storage dynamics may influence transpiration estimates when sunny days with high evaporative demand are followed by overcast and rainy days. This could explain the non-conformity of E_c and E_a under conditions of lower evaporative demand ($E_c > E_a$ in Figure 47.6), when leaf transpiration is suppressed and measured sap flux is amplified by the recharge flow through the stem.

In conclusion, a sound understanding of baseline biophysical interactions at the vegetation–atmosphere interface is essential for the assessment of possible impacts of human interference and global change-related processes on tropical montane ecosystems, including their stability and biodiversity (Pounds *et al.*,

1999, 2006), as well as effects on regional climatic circulation and water budgets (Foster, this volume). Additional reliable field data on the micro-meteorology, energy balance, and water fluxes associated with different types of tropical montane (cloud) forests remain needed to improve parameterization of regional and global climate models (cf. Foster, this volume). In view of the non-ideal conditions for meteorological top–down approaches in complex, rough terrain with heterogeneous surface characteristics (cf. Turnipseed *et al.*, 2003; Holwerda, 2005; Beiderwieden *et al.*, 2008), bottom–up approaches like sapflow measurements scaled up to the forest stand seem to be advantageous, particularly in long-term studies (Santiago *et al.*, 2000; McJannet *et al.*, 2007; Gomez-Cardenas, 2009). Needless to say, this requires a sufficient number of sapflow sensors to account for spatial variability in species-rich forests (Granier *et al.*, 1996b; Köstner *et al.*, 1996; Oren *et al.*, 1998). However, according to McJannet *et al.* (2007) and Küppers *et al.* (2008), transpiration in montane forests is governed primarily by tree size and exposure or position in the canopy rather than by species, thereby paving the way for robust procedures to scale up sapflow measurements on individual trees to the stand level.

ACKNOWLEDGEMENTS

We are grateful to Dieter Schmitt and Jörn Lippert for essential assistance during fieldwork and data analysis. We thank Professor Michael Richter and Paul Emck for providing standard meteorological data, and the Fundación Científica San Francisco for providing the research facilities. Further, we like to thank Sampurno Bruijnzeel for his tireless support, helpful comments, and his considerable contributions during the finalization of this chapter. This study was partly supported by the Deutsche Forschungsgemeinschaft DFG (grant nos. Ku592/21–1, Ku592/30–1).

REFERENCES

Anhuf, D., and R. Rollenbeck (2001). Canopy structure of the Rio Surumoni rain forest (Venezuela) and its influence on microclimate. *Ecotropica* **7**: 21–32.

Anhuf, D., T. Motzer, R. Rollenbeck, B. Schröder, and J. Szarzynski (1999). Water budget of the Surumoni crane site (Venezuela). *Selbyana* **20**: 179–185.

Beiderwieden, E., V. Wolff, Y. J. Hsia, and O. Klemm (2008). It goes both ways: measurements of simultaneous evapotranspiration and fog droplet deposition at a montane cloud forest. *Hydrological Processes* **22**: 4181–4189.

Bendix, J., R. Rollenbeck, M. Richter, P. Fabian, and P. Emck (2008). Climate. In *Gradients in a Tropical Mountain Ecosystem of Ecuador*, eds. E. Beck, J. Bendix, I. Kottke, F. Makeschin and R. Mosandl, pp. 63–74. Berlin: Springer-Verlag.

Bruijnzeel, L. A. (2001). Hydrology of tropical montane cloud forests: a reassessment. *Land Use and Water Resources Research* **1**: 1–18.

Bruijnzeel, L. A. (2005). Tropical montane cloud forests: a unique ecosystem. In *Forests, Water and People in the Humid Tropics*, eds. M. Bonell and L. A. Bruijnzeel, pp. 463–482. Cambridge, UK: Cambridge University Press.

Bruijnzeel, L. A., and L. S. Hamilton (2000). *Decision Time for Cloud Forests*, IHP Humid Tropics Programme Series No. 13. Paris: UNESCO, Amsterdam: IUCN-NL, and Gland, Switzerland: WWF.

Bruijnzeel, L. A., and J. Proctor (1995). Hydrology and biogeochemistry of tropical montane cloud forests: what do we really know? In *Tropical Montane Cloud Forests*, eds. L. S. Hamilton, J. O. Juvik, and F. N. Scatena, pp. 38–78. New York: Springer-Verlag.

Diawara, A., D. Lousteau, and P. Berbigier (1991). Comparison of two methods for estimating the evaporation of a *Pinus pinaster* (Ait.) stand: sap flow and energy balance with sensible heat flux measurements by an eddy covariance method. *Agricultural and Forest Meteorology* **54**: 49–66.

Dolman, A. J., J. H. C. Gash, J. Roberts, and W. J. Shuttleworth (1991). Stomatal and surface conductance of tropical rainforest. *Agricultural and Forest Meteorology* **54**: 303–318.

Edwards, K. A. (1979). The water balance of the Mbeya experimental catchments. *East African Agricultural and Forestry Journal* **43**: 231–247.

Fleischbein, K., W. Wilcke, C. Valarezo, W. Zech, and K. Knoblich (2006). Water budgets of three small catchments under montane forest in Ecuador: experimental and modelling approach. *Hydrological Processes* **20**: 2491–2507.

García-Santos, G. (2007). *An ecohydrological and soils study in a montane cloud forest in the National Park of Garajonay, La Gomera (Canary Islands, Spain)*. Ph.D. thesis, VU University Amsterdam, Amsterdam, the Netherlands. Also available at www.falw.vu.nl/nl/onderzoek/earth-sciences/geo-environmental-science-and-hydrology/hydrology-dissertations/index.asp.

Giambelluca, T. W., R. E. Martin, G. P. Asner, *et al.* (2009). Evapotranspiration and energy balance of native wet montane cloud forest in Hawai'i. *Agricultural and Forest Meteorology* **149**: 230–243.

Goldstein, G., J. L. Andrade, F. C. Meinzer, *et al.* (1998). Stem water storage and diurnal patterns of water use in tropical forest canopy trees. *Plant, Cell and Environment* **21**: 397–406.

Gomez-Cardenas, M. (2009). Transpiration by contrasting vegetation cover types in the montane cloud forest belt of eastern Mexico. Ph.D. thesis, Iowa State University, Ames, IA, USA.

Granier, A. (1985). Une nouvelle méthode pour la mesure du flux de sève brute dans le tronc des arbres. *Annales des Sciences Forestières* **42**: 193–200.

Granier, A., P. Biron, N. Bréda, J. Y. Pontailler, and B. Saugier (1996a). Transpiration of trees and forest stands: short- and long-term monitoring using sapflow methods. *Global Change Biology* **2**: 265–274.

Granier, A., R. Huc, and S. T. Barigah (1996b). Transpiration of natural rainforest and its dependence on climatic factors. *Agricultural and Forest Meteorology* **78**: 19–29.

Grubb, P. J. (1977). Control of forest growth and distribution on wet tropical mountains: with special reference to mineral nutrition. *Annual Review of Ecology and Systematics* **8**: 83–107.

Holwerda, F. (2005). Water and energy budgets of rain forests along an elevational gradient under maritime tropical conditions. Ph.D. thesis, VU University Amsterdam, Amsterdam, the Netherlands. Also available at www.falw.vu.nl/images_upload/B321FA03–1279-D040–8B25E94A8EBDB962.pdf.

Jarvis, P. G. (1976). The interpretation of the variation in leaf water potentials and stomatal conductance found in canopies in the field. *Philosophical Transactions of the Royal Society of London Series B* **273**: 593–610.

Jones, H. G. (1992). *Plants and Microclimate*. Cambridge, UK: Cambridge University Press.

Köstner, B. M. M., P. Biron, R. Siegwolf, and A. Granier (1996). Estimates of water vapour flux and canopy conductance of Scots Pine at the tree level utilizing different xylem sapflow methods. *Theoretical and Applied Climatology* **53**: 105–113.

Köstner, B. M. M., A. Granier, and J. Čermák (1998). Sapflow measurements in forest stands: methods and uncertainties. *Annales des Sciences Forestières* **55**: 13–27.

Küppers, M., T. Motzer, D. Schmitt, *et al.* (2008). Stand structure, transpiration responses in trees and vines, and stand transpiration of different forest types within the mountain rainforest. In *Gradients in a Tropical Mountain ecosystem of Ecuador*, eds. E. Beck, J. Bendix, I. Kottke, F. Makeschin, and R. Mosandl, pp. 243–258. Berlin: Springer-Verlag.

Kürschner, H., and G. Parolly (2004). Phytomass and water-storing capacity of epiphytic rain forest bryophyte communities in S. Ecuador. *Botanische*

Jahrbücher für Systematik, Pflanzengeschichte und Pflanzengeographie **125**: 489–504.

Lawton, R. O., U. S. Nair, R. A. Pielke Sr., and R. M. Welch (2001). Climatic impact of tropical lowland deforestation on nearby montane cloud forests. *Science* **294**: 584–587.

McJannet, D. L., P. G. Fitch, M. G Disher, and J. Wallace (2007). Measurements of transpiration in four tropical rainforest types of north Queensland, *Australia. Hydrological Processes* **21**: 3549–3564.

Meinzer, F. C., G. Goldstein, N. M. Holbrook, P. Jackson, and J. Cavelier (1993). Stomatal and environmental control of transpiration in a lowland tropical forest tree. *Plant, Cell and Environment* **16**: 429–436.

Molion, L. C. B. (1987). Micrometeorology of an amazonian rain forest. In *The Geophysiology of Amazonia: Vegetation and Climate Interactions*, ed. R. E. Dickinson, pp. 255–272. New York: John Wiley.

Monteith, J. L. (1965). Evaporation and environment. *Symposium of the Society of Experimental Biology* **19**: 206–234.

Motzer, T. (2003). *Bestandesklima, Energiehaushalt und Evapotranspiration eines neotropischen Bergregenwaldes. Forstmeteorologische und ökophysiologische Untersuchungen in den Anden Süd-Ecuadors.* Mannheim, Germany: Department of Geography, University Mannheim.

Motzer, T. (2005). Micrometeorological aspects of a tropical mountain forest. *Agricultural and Forest Meteorology* **135**: 230–240.

Motzer, T., N. Munz, M. Küppers, D. Schmitt, and D. Anhuf (2005). Stomatal conductance, transpiration and sap flow of tropical montane rain forest trees in the southern Ecuadorian Andes. *Tree Physiology* **25**: 1283–1293.

Nair, U. S., R. O. Lawton, R. M. Welch, and R. A. Pielke Sr. (2003). Impact of land use on tropical montane cloud forests: sensitivity of cumulus cloud field characteristics to lowland deforestation. *Journal of Geophysical Research* **108**(D7): 4206–4218.

Oke, T. R. (1987). *Boundary Layer Climates*, 2nd edn. London: Methuen.

Oren, R., R. Zimmermann, and J. Terborgh (1996). Transpiration in upper Amazonia floodplain and upland forests in response to drought-breaking rains. *Ecology* **77**: 968–973.

Oren, R., N. Phillips, G. Katul, B. E. Ewers, and D. E. Pataki (1998). Scaling xylem sap flux and soil water balance and calculating variance: a method for partitioning water flux in forests. *Annales des Sciences Forestières* **55**: 191–216.

Phillips, N., M. G. Ryan, B. J. Bond, *et al.* (2003). Reliance on stored water increases with tree size. *Tree Physiology* **23**: 237–245.

Pounds, J. A., M. P. L. Fogden, and J. H. Campbell (1999). Biological response to climate change on a tropical mountain. *Nature* **398**: 611–615.

Pounds, J. A., M. Bustamante, L. A. Coloma, *et al.* (2006). Widespread amphibian extinctions from epidemic disease driven by global warming. *Nature* **439**: 161–167.

Ray, D. K., U. S. Nair, R. O. Lawton, R. M. Welch, and R. A. Pielke **Sr**. (2006). Impact of land use on Costa Rican tropical montane cloud forests: sensitivity of orographic cloud formation to deforestation in the plains. *Journal of Geophysical Research* **111**: D02108, doi:10.1029/2005JD006096.

Richards, P. W. (1996). *The Tropical Rainforest*, 2nd edn. Cambridge, UK: Cambridge University Press.

Roberts, J., O. M. R. Cabral, G. Fisch, *et al.* (1993). Transpiration from an Amazonian rainforest calculated from stomatal conductance measurements. *Agricultural and Forest Meteorology* **65**: 175–196.

Roberts, J. M., J. H. C. Gash, M. Tani, and L. A. Bruijnzeel (2005). Controls on evaporation in lowland tropical rainforest. In *Forests, Water and People in the Humid Tropics*, eds. M. Bonell, and L. A. Bruijnzeel pp. 287–313. Cambridge, UK: Cambridge University Press.

Santiago, L. S., G. Goldstein, F. C. Meinzer, J. H. Fownes, and D. Müller-Dombois (2000). Transpiration and forest structure in relation to soil waterlogging in a Hawaiian montane cloud forest. *Tree Physiology* **20**: 673–681.

Schellekens, J., L. A. Bruijnzeel, F. N. Scatena, N. J. Bink, and F. Holwerda (2000). Evaporation from a tropical rain forest, Luquillo Experimental Forest, eastern Puerto Rico. *Water Resources Research* **36**: 2183–2196.

Schulze, E. -D., J. Čermák, R. Matyssek, *et al.* (1985). Canopy transpiration and water fluxes in the xylem of the trunk of *Larix* and *Picea* trees: a comparison of xylem flow, porometer and cuvette measurements. *Oecologia* **66**: 475–483.

Shuttleworth, W. J., J. H. C. Gash, C. R. Lloyd, *et al.* (1984). Observations of radiation changes above and below the Amazonian forest. *Quarterly Journal of the Royal Meteorological Society* **110**: 1163–1169.

Stewart, J. B. (1988). Modelling surface conductance of pine forest. *Agricultural and Forest Meteorology* **43**: 19–35.

Szarzynski, J., and D. Anhuf (2001). Micrometeorological conditions and canopy energy exchange of a neotropical rain forest (Surumoni–Crane Project, Venezuela). *Plant Ecology* **153**: 231–239.

Turnipseed, A. A., D. E. Anderson, P. D. Blanken, W. M. Baugh, and R. K. Monson (2003). Airflows and turbulent flux measurements in mountainous terrain. I. Canopy and local effects. *Agricultural and Forest Meteorology* **119**: 1–21.

Van der Molen, M. K., A. J. Dolman, M. J. Waterloo, and L. A. Bruijnzeel (2006). Climate is affected more by maritime than by continental land use change: a multiple-scale analysis. *Global and Planetary Change* **54**: 128–149.

Walter, H., and S. W. Breckle. (1991). Ökologie der Erde: Ökologische Grundlagen in globaler Sicht, Vol. 1. Heidelberg, Germany: Spektrum Akademischer Verlag.

Ward, R. C. and M. Robinson (2000) *Principles of Hydrology*, 4th edn. New York: McGraw-Hill.

Weaver, P. L. (1975). Transpiration rates in the elfin forest of the Luquillo Mountains of Puerto Rico. *Caribbean Journal of Science* **15**: 21–30.

Whitehead, D., and T. M. Hinckley (1991). Models of water flux through forest stands: critical leaf and stand parameters. *Tree Physiology* **9**: 35–57.

Wilcke, W., S. Yasin, K. Fleischbein, *et al.* (2008). Water relations. In *Gradients in a Tropical Mountain Ecosystem of Ecuador*, eds. E. Beck, J. Bendix, I. Kottke, F. Makeschin, and R. Mosandl, pp. 193–202. Berlin: Springer-Verlag.

Wullschleger, S. D., F. C. Meinzer, and R. A. Vertessy (1998). A review of whole-plant water use studies in trees. *Tree Physiology* **18**: 499–512.

Zotz, G., M. T. Tyree, S. Patiño, and M. R. Carlton (1998). Hydraulic architecture and water use of selected species from a lower montane forest in Panama. *Trees* **12**: 302–309.

48 Physiological variation in Hawaiian *Metrosideros polymorpha* across a range of habitats: from dry forests to cloud forests

L. S. Santiago
University of California, Riverside, California, USA

T. J. Jones and G. Goldstein
University of Miami, Coral Gables, Florida, USA

ABSTRACT

Hydraulic characteristics of the common Hawaiian tree species, *Metrosideros polymorpha*, were compared in cloud forest, dry forest, bogs, and along an altitudinal gradient to understand how habitat plays a role in the evolution of plant hydraulic features. Plants of intermediate altitudes within the cloud forest zone suffered 50% reduction of hydraulic conductivity at higher water potentials than did low-and high-altitude plants, indicating that plants from cloud forest habitats are more susceptible to cavitation. Xylem area per unit leaf area increased with altitude, and was relatively high in dry forest and bogs, suggesting that more xylem is necessary to support leaf gas exchange in plants stressed by drought, waterlogging, or high altitude. Further, transpiration, leaf traits, and forest structure were examined at an extremely wet cloud forest site (>5000 mm of precipitation per year) to evaluate physiological limitations associated with waterlogging as a mechanism for reduced canopy leaf area. Leaf area index (*LAI*) and stand basal area were lower on level, waterlogged sites than on moderately sloped, well-drained sites. Stand transpiration varied from 79–89% of potential evapotranspiration (*PET*) for sloping sites and from 28–51% of *PET* for level sites. Leaf area index was a good predictor of stand transpiration. Whole-tree transpiration was lower at level sites with waterlogged soils, but was similar to that for trees on level sites when normalized by leaf area. Whole-tree leaf area per unit stem diameter was lower at level sites, suggesting that soil oxygen deficiency may reduce leaf area, but transpiration on a leaf area basis did not differ substantially, thereby maintaining leaf physiological behavior. A high degree of above-ground rooting was observed in cloud forest habitats with flooded soils, suggesting that soil waterlogging is a major limitation to plant growth in some Hawaiian cloud forests.

INTRODUCTION

Tropical forests often exhibit substantial variation in vegetation structure in relation to gradients in moisture availability (Gentry, 1988; Wright, 1992; Vázquez and Givnish, 1998; Bach, 2004; cf. Häger and Dohrenbusch, this volume; Schawe *et al.*, this volume). Forests on the dry side of this gradient commonly have a greater proportion of species that lose their leaves during the dry season. Deciduous phenology appears to reduce plant water use during the dry season when water availability may limit physiological activity, whereas forests on the wet side of this gradient are usually dominated by species with evergreen leaves that exhibit lower rates of gas exchange and reduced demand on soil nutrients to replace leaves (Santiago *et al.*, 2004b). Moisture availability usually increases with increasing altitude on tropical mountains, and is highest where low clouds, fog, or horizontal precipitation is common (Bruijnzeel and Proctor, 1995; Cavelier, 1996; Schawe *et al.*, this volume). Thus, upper montane cloud forests (UMCF) subjected to high rainfall and high cloud incidence represent an extreme in moisture availability in tropical habitats and tend to be dominated by species that exhibit thicker and smaller (sclerophyllous) leaves than do lowland forests or

Tropical Montane Cloud Forests: Science for Conservation and Management, eds. L. A. Bruijnzeel, F. N. Scatena, and L. S. Hamilton. Published by Cambridge University Press.

lower montane rain forests not affected by frequent fog (Grubb *et al.*, 1963; Grubb, 1977; Tanner and Kapos, 1982; Moser *et al.*, 2008). They also have an evergreen phenology shared by species of wet lowland forest (Vázquez and Givnish, 1998). Information on plant hydraulic architecture has the potential to provide added insights to comparisons of leaf-level measurements and links leaf-level observations with physiological processes at the branch and whole-organism levels of organization (Santiago *et al.*, 2004a). Therefore, this chapter focuses on comparative hydraulic architecture and xylem vulnerability to cavitation of trees across a broad array of tropical habitats in order to place physiological observations of UMCF species in a broader context and explore the possibility of scaling plant water use to gain information on ecosystem transpiration fluxes.

Although cloud forests encompass a diversity of forest types that exhibit contrasting hydrological behavior (Bruijnzeel, 2005) and nutrient dynamics (Soethe *et al.*, 2008; Benner *et al.*, this volume; Roman *et al.*, this volume), several general trends in cloud forest eco-physiology have been observed compared to lowland forests or montane forests not affected by frequent fog and low cloud. Species in UMCF tend to exhibit lower foliar nutrient concentrations of nitrogen and phosphorus (Tanner, 1977; Vitousek and Sanford, 1986; Vitousek *et al.*, 1992; Soethe *et al.*, 2008). Reduced nutrient concentrations coupled with thicker and tougher leaves are consistent with evidence that nutrient limitation has a strong effect on the physiognomy of cloud forest ecosystems (Hafkenscheid, 2000; Schuur, 2003; Moser *et al.*, 2007, 2008). In addition, UMCF plants appear sensitive to water shortage. Stomatal conductance of UMCF plants during clear weather has been shown to decrease with increasing vapor pressure deficit (*VPD*) of the air (Körner *et al.*, 1983; Meinzer *et al.*, 1984; Jane and Green, 1985; Cavelier, 1990; García-Santos, 2007; Küppers *et al.*, 2008), indicating that adaptation to humid conditions may come at the cost of desiccation resistance, relative to plants from lower montane cloud forest (LMCF) or lowland forest. Montane forest species exhibit relatively high minimum tissue water potential compared to lowland forest species (Cavelier, 1996), supporting the notion that stomata close at higher water potentials than plants acclimated or adapted to drier habitats. Therefore, cloud forest species operate under constraints of relatively low nutrient availability, yet relatively high water availability (Soethe *et al.*, 2008; Schawe *et al.*, this volume), suggesting that intra-specific comparisons across a range of habitats may help to clarify effects of environmental factors on hydraulic architecture and plant water use.

Metrosideros polymorpha probably occurs over as broad a habitat range as any known woody plant species (Dawson and Stemmerman, 1990). *Metrosideros polymorpha* is the dominant species on new lava flows in Hawai'i and persists as a community dominant throughout ecosystem development to sites with mature, well-developed soils. It also spans a broad water availability gradient from lowland dry forest (<400 mm precipitation year^{-1}) to some of the wettest places on Earth, and the elevation gradient from sea level to the tree-line. Because of this broad environmental range, *M. polymorpha* represents a model organism for understanding intra-specific adaptation to environmental constraints. In addition, *M. polymorpha* facilitates scaling of water fluxes from individual trees to ecosystem levels because of its mono-dominance in many Hawaiian forest stands (cf. Gerrish and Mueller-Dombois, 1999). Cordell *et al.* (1998) demonstrated that ecophysiological features and anatomic characteristics of *M. polymorpha* on the island of Hawai'i are largely induced by the environment, whereas, leaf size is determined by genetic background. Field populations of *M. polymorpha* exhibit increasingly thicker leaves with increasing altitude, although photosynthesis per unit leaf area remains constant (Cordell *et al.*, 1998), implying that photosynthetic rate per unit mass decreases with increasing altitude. In extremely wet forest on the island of Maui, *M. polymorpha* trees exhibited lower leaf area in anoxic soils, but transpiration per unit leaf area was constant, hence changes in hydraulic architecture resulted in homeostasis of leaf physiological activity (Santiago *et al.*, 2000). Therefore, *M. polymorpha* appears to exhibit variation in leaf anatomy, especially thickness, and overall plant form at sites along environmental gradients, and variation in hydraulic architecture (leaf area per sapwood area) among microhabitats within a single site. Here, an analysis is presented of hydraulic architecture and xylem vulnerability of *M. polymorpha* across an environmental gradient to determine the contribution of the hydraulic system to variation in plant form and function with special emphasis on the constraints of plant water use in cloud forest environments.

MATERIALS AND METHODS

Study sites

The present study was conducted across a range of sites that varied in elevation, climate, and vegetation structure at four sites on the island of Hawai'i and one site on the island of Oahu (Table 48.1). The old lava flow at the University of Hawai'i at Hilo, approximately 1 km west of Komohana Street, was used as the lowland wet forest site. The UMCF site was located on the north side of Mauna Loa, Hawai'i, whereas a site along the saddle road on the same island was used to represent a dry sub-alpine scrub habitat located above the thermal inversion. Further information on the lowland wet forest, UMCF, and sub-alpine scrub sites can be found in Cordell *et al.* (1998). Furthermore, the Kaupulehu dry forest on the south side of the island of Hawai'i was used to represent a lowland dry forest, and the stunted vegetation in Ka'au crater on the south side of the island of Oahu to represent a lowland bog.

Table 48.1 *Site descriptions, location, elevation, mean annual temperature* (MAT), *and mean annual precipitation* (MAP) *for the present study areas on Hawai'i and Oahu*

Site		Elevation (m.a.s.l.)	*MAT* (°C)	*MAP* (mm)
Lowland dry forest	Kaupulehu, Hawai'i	600	19.4	400
Lowland wet forest	Hilo, Hawai'i	107	23.0	3600
Lowland bog	Ka'au crater, Oahu	418	20.7	2000
Upper montane cloud forest	Mauna Loa, Hawai'i	1981	12.0	2000
Sub-alpine scrub	Mauna Loa, Hawai'i	2458	10.0	1500

Percent loss of hydraulic conductivity

In order to measure vulnerability to drought-induced hydraulic failure, percent loss of hydraulic conductivity was measured on excised branches of *M. polymorpha*, based on the methods of Sperry *et al.* (1988) as modified by Jones (2001). Upper crown branches were collected from mature *M. polymorpha* individuals before sunrise and transported to the laboratory. The first fully developed stems were cut from the branch and connected to the hydraulic conductivity apparatus (Sperry *et al.*, 1988) in filtered water (0.2 μm) to prevent xylem embolism. Cut ends were shaved with Teflon-coated razor blades. The hydraulic conductivity apparatus consisted of a beaker supplying filtered water under gravitational pressure to the stem. The apparatus was frequently flushed with 10% bleach solution to avoid microbial growth. To prevent removal of existing embolisms, an appropriate hydraulic head height was determined for *M. polymorpha* by recording changes in hydraulic conductivity with different gravitational pressures (Kolb and Davis, 1994). The optimal gravitational pressure was determined to be approximately 1.4 kPa. Flow rates were determined volumetrically 5–15 min after connection when they became steady. Initial hydraulic conductivity (k_H) was estimated as the rate of water flux (J, mmol s^{-1}) in a stem when a pressure gradient (dP/dx, MPa m^{-1}) was applied across the stem:

$$k_H = \frac{J}{(dP/dx)}. \quad (48.1)$$

Maximum k_H was measured after a 20-min high-pressure (175 kPa) flush from a captive air tank to dissolve and remove air bubbles from embolized vessels. After restoring conductivity, k_H was measured as described above to obtain $k_{H\,max}$. Percent loss of hydraulic conductivity (*PLC*) was determined as:

$$PLC = \frac{100 \cdot (k_{H\,max} - k_H)}{k_{H\,max}} \quad (48.2)$$

where $k_{H\,max}$ is maximum hydraulic conductivity (mmol m s^{-1} MPa^{-1}) and k_H represents native hydraulic conductivity as defined above.

Vulnerability curves

The percentage of embolized vessels in a stem segment in relation to water potential was used to construct vulnerability curves for each species using the bench dehydration method (Sperry *et al.*, 1988; Jones, 2001). Plant material was air-dried for up to 6 days to allow the development of leaf water potentials ranging from −0.09 to −6.9 MPa. Before measuring k_H, plants were placed in double-sealed plastic bags (damp paper towels between bags) for up to 48 hours to allow equilibration between leaf and xylem water potentials. Leaf water potential (Ψ_L) was measured using a pressure chamber (PMS Instruments, Corvallis, Oregon, USA), concurrently with stem k_H. The relationship between Ψ_L and k_H was fitted using the exponential sigmoid equation (Pammenter and Wiligen, 1998):

$$PLC = \frac{100}{(1 + e^{(a(\Psi_L - b))})} \quad (48.3)$$

where *PLC* is the percent loss in k_H as defined above, Ψ_L the leaf water potential corresponding to k_H measurements, constant a the slope of the line, and b the estimated Ψ_L at 50% loss in conductivity (PLC_{50}). The Ψ_L was also recorded at 100% loss in conductivity (PLC_{100}). Non-linear regressions were conducted with SAS (SAS Institute Inc., 1992).

Hydraulic architecture

Following vulnerability curve measurements, the xylem area and all of the leaf area distal to the excised stem segment were measured for calculation of xylem area per unit leaf area (*XA/LA*, m^2 m^{-2}), as an index of potential difference in water transport capacity relative to transpirational demand (Meinzer *et al.*, 1997). This index represents the amount of xylem area supporting a given foliar area. Higher values of this index are assumed to represent greater integrated environmental stress on the plant hydraulic system. In addition, the mass and area of 15–30 leaves were measured per site for the calculation of leaf mass per area (*LMA*, g m^{-2}).

RESULTS

Percentage loss of hydraulic conductivity (*PLC*) as a function of water potential (Ψ_L) showed a similar pattern across stems from all sites (Figure 48.1). The Ψ_L at which PLC_{50} occurred was lowest in sub-alpine scrub, indicating ability to withstand low Ψ_L in individuals from high altitude populations (Table 48.2). Species

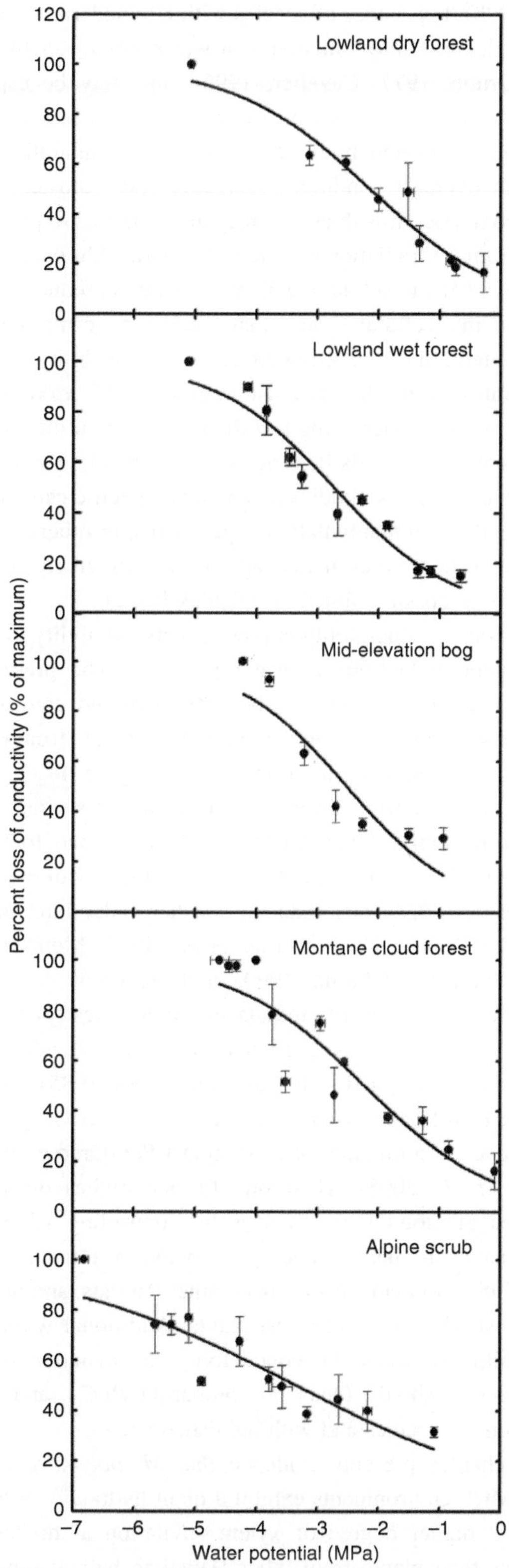

Figure 48.1. Percent loss of hydraulic conductivity as a function of stem water potential for Hawaiian *Metrosideros polymorpha* populations from five contrasting habitat types. Each point represents the mean of three branches (±1 SE). Lines represent best fit non-linear regressions for Eq. (49.3). See Table 49.2 for regression parameters.

Table 48.2 *The xylem water potential at which 50%* (PLC_{50}) *and 100%* (PLC_{100}) *loss of hydraulic conductivity occurs in stems, and the slope of the percent loss of stem hydraulic conductivity–water potential relationship for* Metrosideros polymorpha *trees from five contrasting habitats on the islands of Hawai'i and Oahu*

Site	PLC_{50}	PLC_{100}	Slope
Lowland dry forest	−2.1	−5.1	0.898
Lowland wet forest	−2.7	−5.1	1.038
Lowland bog	−2.5	−4.3	1.110
Montane cloud forest	−2.2	−4.3	0.943
Sub-alpine scrub	−3.3	−6.8	0.494

from relatively moist sites such as montane cloud forest, lowland wet forest, and the bog exhibited PLC_{50} values at relatively higher Ψ_L than did sub-alpine scrub, indicating greater susceptibility to cavitation in plants from these populations (Table 48.2). Lowland dry forest exhibited the highest PLC_{50} values, but also relatively low PLC_{100}, indicating that although dry-forest species reached PLC_{50} at a high Ψ_L, loss of conductivity above PLC_{50} occurred gradually. Lowland wet forest plants had relatively low PLC_{100} values that were similar to the lowland dry forest population (Table 48.2). Montane cloud forest and bog plants exhibited the highest PLC_{100} values, indicating the greatest vulnerability to cavitation at relatively high Ψ_L. The slope of the vulnerability curve was greatest in bog plants which experience the highest soil moisture levels, whereas plants from the sub-alpine scrub had the most gradual slope, consistent with greater ability to withstand desiccation in high-elevation populations (Table 48.2).

Mean leaf mass per unit area (*LMA*) ranged from a minimum of 206.5 g m^{-2} in lowland wet forest to a maximum of 419.8 g m^{-2} in sub-alpine scrub. Upper montane cloud forest, lowland dry forest, and the bog showed moderate values around 260 g m^{-2} (Figure 48.2). Stem xylem area per unit leaf area (*XA*/*LA*) ranged from a minimum of $2.0 \cdot 10^5$ m^2 m^{-2} in lowland wet forest to a maximum of $3.8 \cdot 10^5$ m^2 m^{-2} in sub-alpine scrub. This variation in *XA*/*LA* indicates an increase in the amount of xylem area needed to support a given leaf area, and therefore increasing integrated environmental stress in the order: lowland wet forest < montane cloud forest < lowland dry forest < bog < sub-alpine scrub (Figure 48.2).

DISCUSSION

Changes in hydraulic architecture across environmental gradients

Plant hydraulic architecture and vulnerability to cavitation in Hawaiian *Metrosideros polymorpha* varied across a range of habitats in a manner that is consistent with a range of

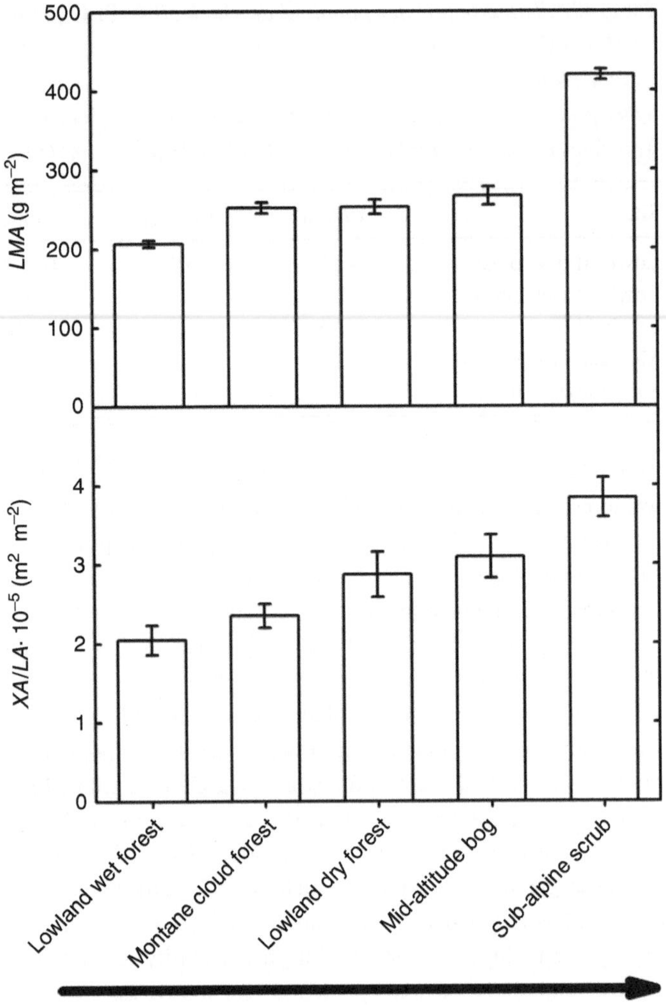

Figure 48.2. Mean leaf mass per unit area (*LMA*) and xylem area per unit leaf area (*XA*/*LA*) for terminal branches of Hawaiian *Metrosideros polymorpha* populations from five contrasting habitat types (n = 15–30).

environmental stresses among habitat types. The xylem area needed to support physiological activity of a given leaf area increased from lowland wet forest to sub-alpine scrub, indicating increasing environmental stress along this axis of habitat types (Figure 48.2). The data support the notion that plants of moist environments generally exhibit greater susceptibility to xylem cavitation, whereas plants above or below zones of frequent waterlogging appear able to withstand lower water potential with less cavitation of the hydraulic system.

The increase in *LMA* with increasing environmental stress and altitude is in line with previous reports. Several studies of variation in leaf morphology along tropical altitudinal gradients have documented similar trends of increasing leaf thickness (Grubb, 1977; Körner *et al.*, 1983; Cordell *et al.*, 1998; Moser *et al.*, 2007), and suggest that increased *LMA* may serve various functions. Leaves of *M. polymorpha* have a hypodermis which increases substantially with increasing leaf thickness (Cordell *et al.*, 1998). For leaves of species with a hypodermis, increases in leaf thickness may function as a water supply for the mesophyll (Grubb, 1977; Cavelier, 1996), and may be especially important in situations where leaf water is used to support leaf gas exchange during brief periods of light availability during breaks in overcast conditions (Cavelier and Goldstein, 1989). Increased hypodermis thickness may also function to protect leaf organs from UV radiation (Flenley, 1995). In addition, the strong increase in *LMA* in sub-alpine plants may reflect reduction of cell sizes and intercellular spaces that could reduce the apoplastic water content, thus increasing the supercooling ability and freezing tolerance of the leaf (Goldstein *et al.*, 1985; Cordell *et al.*, 2000). Therefore, increasing leaf thickness with altitude on tropical mountains potentially reflects a variety of adaptations to environmental stress. Studies that address specific causes of leaf thickness through manipulation experiments, or otherwise rigorously controlled experimental approaches are likely to distinguish among possible functions of thick leaves.

Variation in vulnerability curves reflects the ability of a plant to withstand reductions in water potential. The present data suggest that *M. polymorpha* plants from humid environments are more sensitive to xylem cavitation than plants from extreme environments, such as sub-alpine scrub (Figure 48.1). Several lines of evidence from a variety of tropical forests suggest that plants from humid environments, especially cloud forests, are sensitive to desiccation and close their stomata with even slight increases in *VPD*, thus reducing further reductions in water potential (Weaver, 1975; Körner *et al.*, 1983; Meinzer *et al.*, 1984; Cavelier and Mejia, 1990; cf. García-Santos, 2007). In New Zealand, closure of stomata by cloud forest plants under slight water stress was attributed to poorly developed root systems in waterlogged soils (Jane and Green, 1985). However, it is a general trend that most species show decreasing stomatal conductance as a function of increasing *VPD* (Oren *et al.*, 1999; Roberts *et al.*, 2005). Therefore, further studies on stomatal responses of cloud forest species by controlling *VPD* during gas exchange are likely to distinguish between stomatal behavior that is similar to plant species from other habitats, and purported extra sensitivity of cloud forest plants. Additional work on the vulnerability of hydraulic conductivity to cavitation of cloud forest species should focus on comparing PLC_{50} and PLC_{100} values among species and with published values.

This chapter presents evidence that *M. polymorpha* plants from UMCF environments exhibit a plant hydraulic system that reaches a higher degree of xylem cavitation at higher water potentials than plants from other Hawaiian habitat types. It is emphasized that cloud forests are part of a continuum of tropical habitats in which resources are governed by a constellation of environmental factors including water, light, temperature, and nutrient availability. The present results suggest that further comparative studies, especially inter-specific investigations of comparative plant physiology, will augment our knowledge of cloud forest plant function.

Scaling plant transpiration in cloud forest

Until recently, most estimates of transpiration from montane cloud forests were based on water budget calculations for small watersheds in which apparent interception evaporation is subtracted from total evapotranspiration to give a crude estimate of transpiration (Bruijnzeel and Proctor, 1995; Bruijnzeel, 2005). More recently, micrometeorological approaches such as eddy covariance and related techniques have yielded plausible values for cloud forest transpiration (Holwerda, 2005; Beiderwieden *et al.*, 2008). However, such top–down approaches do not take into account the contribution of plant species or growth forms, which can be important for predicting effects of vegetation on the hydrological cycle (Santiago, 1998a; Küppers *et al.*, 2008; cf. Motzer *et al.*, this volume). Developments in sapflow methodology now allow robust measurement of whole-tree transpiration in cloud forest environments (McJannet *et al.*, 2007; Küppers *et al.*, 2008; Gomez-Cardenas, 2009; McJannet *et al.*, this volume; Motzer *et al.*, this volume), permitting the use of bottom–up models that predict how systems function at larger scales based on extrapolations of measurements at small spatial and short temporal scales (Jarvis, 1993). Hawaiian montane forests are optimal systems for applying bottom–up models because the dominance of a single canopy species facilitates scaling from leaves or whole-tree measurements to the canopy scale. Results of bottom–up models in Hawaiian montane forest suggest that a combination of whole-tree transpiration and microclimatic measurements can be successfully employed to estimate the contribution of plant transpiration fluxes to the hydrological cycle (Santiago, 1998a).

In montane cloud forest on Maui, Santiago *et al.* (2000) reported that waterlogging due to local topography had major effects on plant hydraulic architecture and transpiration. As basal xylem area increased, whole-tree leaf area increased on moderately sloped, well-drained sites, but remained relatively constant on level, waterlogged sites (Figure 48.3). Such variation in *XA*/*LA* indicates that trees at waterlogged micro-sites required more xylem to meet the physiological needs of their leaves. In addition, trees on waterlogged substrates exhibited a greater degree of above-ground rooting (Santiago, 1998b, 2000), which may increase susceptibility to drought (Holbrook and Putz, 1996). Leaf area index (*LAI*, $m^2\ m^{-2}$) and stand basal area were also lower on level sites than on sloping sites, and because *LAI* was a positive predictor of stand transpiration (Figure 48.4), the contribution of tree transpiration to total evapotranspiration was lower on level sites than on sloping sites. However, when transpiration was expressed on a leaf area basis, comparisons between level and sloping sites did not differ substantially, indicating that leaf-level physiological behavior was essentially conserved (Santiago *et al.*, 2000). Therefore, plant hydraulic architecture has the potential to shift in response to local resource availability within a site, similar to responses observed among contrasting habitat types (cf. Leuschner *et al.*, 2007; Moser *et al.*, 2007; Hertel and Leuschner, this volume). In the case of *M. polymorpha*, shifts in hydraulic architecture resulted in variation in whole-plant transpiration.

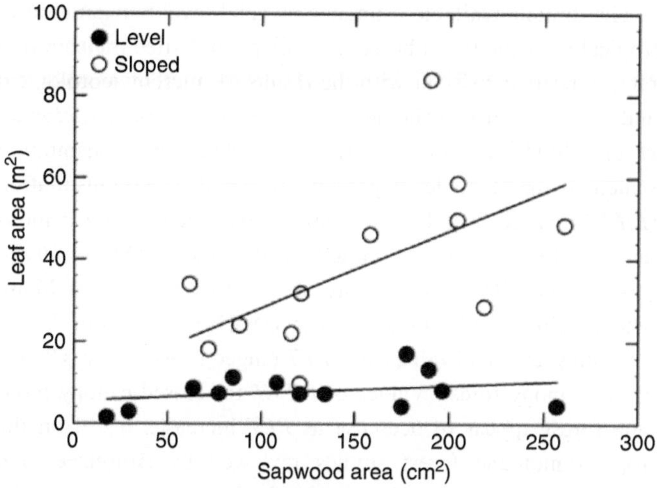

Figure 48.3. Total tree leaf surface area (*LA*) as a function of basal stem xylem area (*XA*) for *Metrosideros polymorpha* trees at Waikamoi, Maui. Lines represent best-fit linear regressions for sloped ($LA = 0.19{\cdot}XA+9.38$; $r^2 = 0.36$; $p = 0.045$) and level sites ($LA = 0.019{\cdot}XA+5.57$; $r^2 = 0.10$; $p = 0.292$) (modified from Santiago *et al.*, 2000).

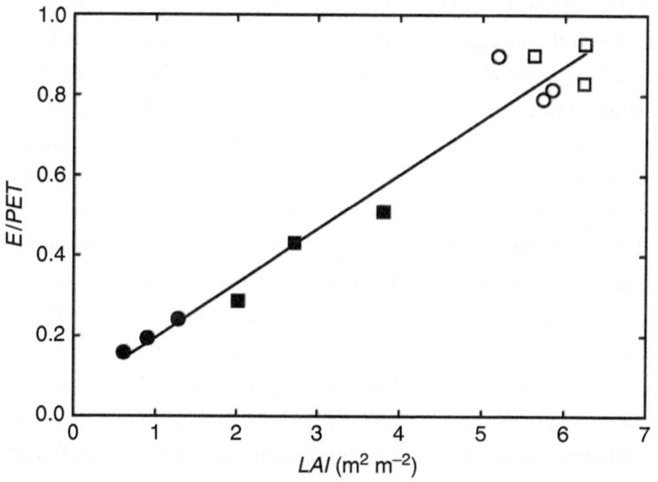

Figure 48.4. The ratio of stand transpiration to potential evapotranspiration (*E*/*PET*) (Penman, 1948) as a function of canopy leaf area index (*LAI*) for sloped, closed-canopy (○) and level, open-canopy (●) sites, and total stand *LAI*, including understory vegetation, for sloped (□) and level (■) sites. The line represents best-fit linear regression for all sites ($E/PET = 0.136{\cdot}LAI+0.059$; $r^2 = 0.96$; $p < 0.001$) (modified from Santiago *et al.*, 2000).

The limited available estimates of stand-level transpiration in tropical montane forest based on scaling whole-tree transpiration rates appears consistent with the results of micrometeorological and water balance techniques (Motzer *et al.*, 2005; McJannet *et al.*, 2007). In southern Ecuador, whole-tree transpiration scaled to the stand level per unit potential evapotranspiration (*E/PET*) varied from 0.55 to 0.90 in closed-canopy lower montane rain forest subject to limited cloud incidence (Motzer *et al.*, this volume). These values are similar to reports from Maui, where estimates of transpiration in UMCF ranged from 258 to 291 mm $year^{-1}$ and values of *E/PET* ranged from 0.79 to 0.89 for closed canopy forest. Values of *E/PET* for closed-canopy montane forest appear to decrease as *PET* increases based on the tropical montane forest studies reviewed by Bruijnzeel and Proctor (1995) and Bruijnzeel (2005). Leaf area index can be useful for predicting evaporative fluxes from some types of UMCF. On Maui, *LAI* was a strong predictor of transpiration when values were compared across the range of *LAI* in UMCF (Figure 48.4), supporting the idea that *LAI* is the most important variable for relating vegetation structure to energy and mass exchange because it is a measure of the evaporative surface area (Running and Coughlan, 1988) as well as a strong determinant of photosynthetic capacity (Moser *et al.*, 2007). In southern Ecuador, Motzer *et al.* (this volume) reported that although trees with greater projected crown area had higher rates of transpiration, crown exposure was also an important predictor of transpiration of individual trees, as upper canopy trees accounted for 80% of stand transpiration. In northern Queensland, McJannet *et al.* (2007) determined that total stand transpiration was controlled by forest attributes such as stem density, tree size, and sapwood area (cf. McJannet *et al.*, this volume). Future attempts that employ measurements of plant transpiration to estimate the contribution of vegetation to the hydrological cycle should focus on *LAI*, crown exposure, and stand density, in addition to including sufficient variation in micro-topography to represent local variation if applicable.

CONCLUSIONS AND RECOMMENDATIONS

Based on the available physiological data it appears that cloud forest plants are more susceptible to xylem cavitation than plants from habitats that experience dry conditions with greater frequency. It is also evident that vegetation in cloud forests plays an important role in the hydrological cycle and that measurement of plant hydraulics allows quantification of this role at the scale of individual trees, thus improving our understanding of the contribution of vegetation in evaporative fluxes to the atmosphere. Many previous studies in cloud forests have focused on why cloud forest plants differ from lowland forest species. It is apparent that a variety of direct and indirect environmental factors, including light, nutrient availability, waterlogging, atmospheric humidity, and temperature, to name a few, contribute to the reduced stature of certain types of cloud forest (cf. Benner *et al.*, this volume; Roman *et al.*, this volume). However, to differentiate among the multitude of factors that influence vegetation structure, future studies must also employ manipulation experiments, such as fertilization experiments which unequivocally demonstrate nutrient limitation (Tanner *et al.*, 1990, 1998; Vitousek *et al.*, 1993; Herbert and Fownes, 1995; cf. Soethe *et al.*, 2008). New forms of manipulation, such as experimental light enrichment offer potentially powerful tests of light limitation (Graham *et al.*, 2003; cf. Brodersen *et al.*, 2008). Studies along environmental gradients where most ecosystem state factors can be held constant, thus forming a natural experiment, can identify important factors controlling ecosystem function (Austin and Vitousek, 1998; Schuur and Matson, 2001; Santiago *et al.*, 2004b). Where field studies may be too complicated to isolate specific mechanisms, highly controlled greenhouse or common garden experiments may be the answer. Therefore, with imagination as the only limitation, rigorous experimental design represents the only format for separating myth from mechanism in cloud forest ecosystem function.

ACKNOWLEDGEMENTS

We thank Susan Cordell for local information regarding research sites, Rick Meinzer for advice on hydraulic conductivity measurements, and the National Science Foundation and the Moore Foundation for partial financial support.

REFERENCES

Austin, A. T., and P. M. Vitousek (1998). Nutrient dynamics on a precipitation gradient in Hawai'i. *Oecologia* **113**: 519–529.

Bach, K. (2004). Vegetationskundliche Untersuchungen zur Höhenzonierung tropischer Bergregenwälder in den Anden Boliviens. Ph.D. thesis, University of Göttingen, Göttingen, Germany.

Beiderwieden, E., V. Wolff, Y. J. Hsia, and O. Klemm (2008). It goes both ways: measurements of simultaneous evapotranspiration and fog droplet deposition at a montane cloud forest. *Hydrological Processes* **22**: 4181–4189.

Brodersen, C. R., T. C. Vogelmann, W. E. Williams, and H. L. Gorton (2008). A new paradigm in leaf-level photosynthesis: direct and diffuse lights are not equal. *Plant, Cell and Environment* **31**: 159–164.

Bruijnzeel, L. A. (2005). Tropical montane cloud forest: a unique hydrological case. In *Forests, Water and People in the Humid Tropics*, eds. M. Bonell and L. A. Bruijnzeel, pp. 462–483. Cambridge, UK: Cambridge University Press.

Bruijnzeel, L. A., and J. Proctor (1995). Hydrology and biogeochemistry of tropical montane cloud forests: what do we really know? In *Tropical Montane Cloud Forests*, eds. L. S. Hamilton, J. O. Juvik, and F. N. Scatena, pp. 38–78. New York: Springer-Verlag.

Cavelier, J. (1990). Tissue water relations in elfin cloud forest tree species of Serranía de Macuira, Guajira, Colombia. *Trees: Structure and Function* 4: 155–163.

Cavelier, J. (1996). Environmental factors and ecophysiological processes along altitudinal gradients in wet tropical mountains. In *Tropical Forest Plant Ecophysiology*, eds. S. S. Mulkey, R. L. Chazdon, and A. P. Smith, pp. 399–439. New York: Chapman and Hall.

Cavelier, J., and G. Goldstein (1989). Leaf anatomy and water relations in tropical elfin cloud forest tree species. In *Structural and Functional Responses to Environmental Stresses*, eds. K. H. Krebb, H. Richter, and T. M. Hinckley, pp. 243–253. The Hague: SPB Academic Publishing.

Cavelier, J., and C. A. Mejia (1990). Climatic factors and tree stature in the elfin cloud forest of Serranía de Macuira, Columbia. *Agricultural and Forest Meteorology* **53**: 105–123.

Cordell, S., G. Goldstein, D. Mueller-Dombois, D. Webb, and P. M. Vitousek (1998). Physiological and morphological variation in *Metrosideros polymorpha*, a dominant Hawaiian tree species, along an altitudinal gradient: the role of phenotypic plasticity. *Oecologia* **113**: 188–196.

Cordell, S. G., G. Goldstein, P. J. Melcher, and F. C. Meinzer (2000). Photosynthetic and freezing avoidance in *Metrosideros polymorpha* at treeline in Hawaii. *Arctic, Antarctic and Alpine Research* **32**: 381–387.

Dawson, J. W., and L. Stemmerman (1990). *Metrosideros* (Myrtaceae). In *Manual of the Flowering Plants of Hawai'i*, eds. W. L. Wagner, D. R. Herbst, and S. H. Sohmer, pp. 964–970. Honolulu, HI: Bernice P. Bishop Museum.

Flenley, J. R. (1995). Cloud forest, the *Massenerhebung* effect, and ultraviolet insolation. In *Tropical Montane Cloud Forests*, eds. L. S. Hamilton, J. O. Juvik, and F. N. Scatena, pp. 150–155. New York: Springer-Verlag.

García-Santos, G. (2007). *An ecohydrological and soils study in a montane cloud forest in the National Park of Garajonay, La Gomera (Canary Islands, Spain)*. Ph.D. thesis, VU University Amsterdam, Amsterdam, the Netherlands. Also available at www.falw.vu.nl/nl/onderzoek/earth-sciences/geo-environmental-science-and-hydrology/hydrology-dissertations/index.asp.

Gentry, A. H. (1988). Changes in plant community diversity and floristic composition on environmental and geographical gradients. *Annals of the Missouri Botanical Garden* **75**: 1–34.

Gerrish, G. C., and D. Mueller-Dombois (1999). Measuring stem growth rates for determining age and cohort analysis of a tropical evergreen tree. *Pacific Science* **53**: 418–429.

Giambelluca, T. W., R. E. Martin, G. P. Asner, *et al.* (2009). Evapotranspiration and energy balance of native wet montane cloud forest in Hawai'i. *Agricultural and Forest Meteorology* **149**: 230–243.

Goldstein, G., F. Rada, and A. Azocar (1985). Cold hardiness and supercooling along an altitudinal gradient in Andean giant rosette species. *Oecologia* **68**: 147–152.

Gomez-Cardenas, M. (2009). Transpiration by contrasting vegetation cover types in the montane cloud forest belt of eastern Mexico. Ph.D. thesis, Iowa State University, Ames, IA, USA.

Graham, E. A., S. S. Mulkey, K. Kitajima, N. G. Phillips, and S. J. Wright (2003). Cloud cover limits net CO_2 uptake and growth of a rainforest tree during tropical rainy seasons. *Proceedings of the National Academy of Sciences USA* **100**: 572–576.

Grubb, P. J. (1977). Control of forest growth and distribution on wet tropical mountains: with special reference to mineral nutrition. *Annual Review of Ecology and Systematics* **8**: 83–107.

Grubb, P. J., J. R. Lloyd, T. D. Pennington, and T. C. Whitmore (1963). A comparison of montane and lowland rainforest in Ecuador. I. The forest structure, physiognomy, and floristics. *Journal of Ecology* **51**: 567–601.

Hafkenscheid, R. L., (2000). Hydrology and biogeochemistry of tropical montane rain forests of contrasting stature in the Blue Mountains, Jamaica. Ph.D. thesis, VU University Amsterdam, Amsterdam, the Netherlands. Also available at http://dare.ubvu.vu.nl/bitstream/1871/12734/1/tekst.pdf.

Herbert, D. A., and J. H. Fownes (1995). Phosphorus limitation of forest leaf area and net primary production on a highly weathered soil. *Biogeochemistry* **29**: 223–235.

Holbrook, N. M., and F. E. Putz (1996). Water relations of epiphytic and terrestrially rooted strangler figs in a Venezuelan palm savanna. *Oecologia* **106**: 424–431.

Holwerda, F. (2005). Water and energy budgets of rain forests along an elevational gradient under maritime tropical conditions. Ph.D. thesis, VU University Amsterdam, Amsterdam, the Netherlands. Also available at www.falw.vu.nl/images_upload/B321FA03-1279-D040-8B2SE94A8EBDB962.pdf.

Jane, G. T., and T. G. A. Green (1985). Patterns of stomatal conductance in six evergreen tree species from a New Zealand cloud forest. *Botanical Gazette* **146**: 413–420.

Jarvis, P. G. (1993). Prospects for bottom-up models. In *Scaling Physiological Process*, eds. J. R. Ehleringer and C. B. Field, pp. 117–126. New York: Academic Press.

Jones, T. J., (2001). *Hydraulic architecture, vulnerability to cavitation, and embolism repair in a eucalypt and coffee species.* M.Sc. thesis, University of Hawaii, Honolulu, HI, USA.

Kolb, K. J., and S. D. Davis (1994). Drought tolerance and xylem embolism in co-occurring species of coastal sage and chaparral. *Ecology* **75**: 648–659.

Körner, C., A. Allison, and H. Hilscher (1983). Altitudinal variation of leaf diffuse conductance and leaf anatomy in heliophytes of montane New Guinea and their interrelation with microclimate. *Flora* **174**: 91–135.

Küppers, M., T. Motzer, D. Schmitt, *et al.* (2008). Stand structure, transpiration responses in trees and vines, and stand transpiration of different forest types within the mountain rainforest. In *Gradients in a Tropical Mountain Ecosystem of Ecuador*, eds. E. Beck, J. Bendix, I. Kottke, F. Makeschin, and R. Mosandl, pp. 243–258. Berlin: Springer-Verlag.

Leuschner, Ch., G. Moser, C. Bertsch, M. Röderstein, and D. Hertel (2007). Large altitudinal increase in tree root/shoot ratio in tropical mountain forests of Ecuador. *Basic and Applied Ecology* **8**: 219–230.

McJannet, D., P. Fitch, M. Disher, and J. Wallace (2007). Measurements of transpiration in four tropical rainforest types of north Queensland, Australia. *Hydrological Processes* **21**: 3549–3564.

Meinzer, F. C., G. Goldstein, and M. Jaimes (1984). The effect of atmospheric humidity on stomatal control of gas exchange in two tropical coniferous species. *Canadian Journal of Botany* **62**: 591–595.

Meinzer, F. C., J. L. Andrade, G. Goldstein, *et al.* (1997). Control of transpiration from the upper canopy of a tropical forest: the role of stomatal, boundary layer and hydraulic architecture components. *Plant, Cell and Environment* **20**: 1242–1253.

Moser, G., D. Hertel, and Ch. Leuschner (2007). Altitudinal change in LAI and stand leaf biomass in tropical montane forests: a transect study in Ecuador and a pan-tropical meta-analysis. *Ecosystems* **10**: 924–935.

Moser, G., M. Röderstein, N. Soethe, D. Hertel, and Ch. Leuschner (2008). Altitudinal changes in stand structure and biomass allocation of tropical mountain forest in relation to microclimate and soil chemistry. In *Gradients in a Tropical Mountain Ecosystem of Ecuador*, eds. E. Beck, J. Bendix, I. Kottke, F. Makeschin, and R. Mosandl, pp. 229–242. Berlin: Springer-Verlag.

Motzer, T., N. Munz, M. Küppers, D. Schmitt, and D. Anhuf (2005). Stomatal conductance, transpiration and sap flow of tropical montane rain forest trees in the southern Ecuadorian Andes. *Tree Physiology* **25**: 1283–1293.

Oren, R., J. S. Sperry, G. Katul, *et al.* (1999). Survey and synthesis of intra- and interspecific variation in stomatal sensitivity to vapor pressure deficit. *Plant, Cell and Environment* **22**: 1515–1526.

Pammenter, N. W., and C. V. Wiligen (1998). A mathematical and statistical analysis of the curves illustrating vulnerability of xylem to cavitation. *Tree Physiology* **18**: 589–593.

Penman, H. L. (1948). Natural evaporation from open water, bare soil and grass. *Proceedings of the Royal Society of London Series A* **193**: 120–146.

Roberts, J. M., J. H. C. Gash, M. Tani, and L. A. Bruijnzeel (2005). Controls on evaporation in lowland tropical rainforest. In *Forests, Water and People in the Humid Tropics*, eds. M. Bonell and L. A. Bruijnzeel, pp. 287–313. Cambridge, UK: Cambridge University Press.

Running, S. W., and J. C. Coughlan (1988). A general model for forest ecosystem processes for regional application. I. Hydrologic balance, canopy gas exchange, and primary production processes. *Ecological Modeling* **42**: 125–154.

Santiago, L. S. (1998a). A bottom-up approach to scaling canopy transpiration in a Hawaiian montane cloud forest. *Selbyana* **19**: 297.

Santiago, L. S. (1998b). Transpiration, forest structure and regeneration in relation to waterlogged soils in the cloud forest of Waikamoi, Maui. M.Sc. thesis, University of Hawaii, Honolulu, HI, USA.

Santiago, L. S. (2000). Use of coarse woody debris by the plant community of a Hawaiian montane cloud forest. *Biotropica* **32**: 633–641.

Santiago, L. S., G. Goldstein, F. C. Meinzer, J. Fownes, and D. Mueller-Dombois (2000). Transpiration and forest structure in relation to soil waterlogging in a Hawaiian montane cloud forest. *Tree Physiology* **20**: 673–681.

Santiago, L. S., G. Goldstein, F. C. Meinzer, *et al.* (2004a). Leaf photosynthetic traits scale with hydraulic conductivity and wood density in Panamanian forest canopy trees. *Oecologia* **140**: 543–550.

Santiago, L. S., K. Kitajima, S. J. Wright, and S. S. Mulkey (2004b). Coordinated changes in photosynthesis, water relations and leaf nutritional traits of canopy trees along a precipitation gradient in lowland tropical forest. *Oecologia* **139**: 495–502.

Schuur, E. A. G. (2003). Productivity and global climate revisited: the sensitivity of tropical forest growth to precipitation. *Ecology* **84**: 1165–1170.

Schuur, E. A. G., and P. A. Matson (2001). Net primary productivity and nutrient cycling across a mesic to wet precipitation gradient in Hawaiian montane forest. *Oecologia* **128**: 431–442.

Soethe, N., W. Wilcke, J. Homeier, J. Lehmann, and C. Engels (2008). Plant growth along the altitudinal gradient: role of plant nutritional status, fine root activity, and soil properties. In *Gradients in a Tropical Mountain Ecosystem of Ecuador*, eds. E. Beck, J. Bendix, I. Kottke, F. Makeschin, and R. Mosandl, pp. 259–266. Berlin: Springer-Verlag.

Sperry, J. S., J. R. Donnelly, and M. T. Tyree (1988). A method for measuring hydraulic conductivity and embolism in xylem. *Plant, Cell and Environment* **11**: 35–40.

Tanner, E. V. J. (1977). Four montane rain forests of Jamaica: a quantitative characterization of the floristics, the soils and the foliar mineral levels, and a discussion of the interrelations. *Journal of Ecology* **65**: 883–918.

Tanner, E. V. J., and V. Kapos (1982). Leaf structure of Jamaican upper montane rainforest trees. *Biotropica* **14**: 16–24.

Tanner, E. V. J., V. Kapos, S. Frescos, J. R. Healey, and A. M. Theobald (1990). Nitrogen and phosphorus fertilization effects on Venezuelan montane forest trunk growth and litterfall. *Ecology* **73**: 78–86.

Tanner, E., P. Vitousek, and E. Cuevas (1998). Experimental investigation of nutrient limitation of forest growth on wet tropical mountains. *Ecology* **79**: 10–22.

Vázquez, J. A., and T. Givnish (1998). Altitudinal gradients in tropical forest composition, structure, and diversity in the Sierra de Manantlán. *Journal of Ecology* **86**: 999–1020.

Vitousek, P. M., and R. L. Sanford (1986). Nutrient cycling in moist tropical forest. *Annual Review of Ecology and Systematics* **17**: 137–167.

Vitousek, P. M., G. H. Aplet, D. R. Turner, and J. J. Lockwood (1992). The Mauna Loa environmental matrix: foliar and soil nutrients. *Oecologia* **89**: 372–382.

Vitousek, P. M., L. R. Walker, L. D. Whiteaker, and P. A. Matson (1993). Nutrient limitations to plant growth during primary succession in Hawai'i Volcanoes National Park. *Biogeochemistry* **23**: 197–215.

Weaver, P. L. (1975). Transpiration rates in the elfin forest of the Luquillo Mountains of Puerto Rico. *Caribbean Journal of Science* **15**: 21–30.

Wright, S. J. (1992). Seasonal drought, soil fertility and the species density of tropical forest plant communities. *Trends in Ecology and Evolution* **7**: 260–263.

49 Environmental controls on photosynthetic rates of lower montane cloud forest vegetation in south-western Colombia

M. G. Letts
University of Lethbridge, Lethbridge, Alberta, Canada

M. Mulligan
King's College London, London, UK

M. E. Rincón-Romero
Universidad del Valle, Cali, Colombia

L. A. Bruijnzeel
VU University, Amsterdam, the Netherlands

ABSTRACT

A variety of microclimatic and edaphic factors have been shown to limit photosynthetic productivity in tropical montane cloud forest (TMCF) ecosystems. It is now understood that multiple controls may limit photosynthesis within individual TMCFs, and that the relative importance of each control varies between sites. This chapter describes the nutrient status, microclimate, leaf structural traits and photosynthetic gas-exchange characteristics of lower montane cloud forest (LMCF) vegetation at the Centro de Estudios Ambientales Tambito site, a wet LMCF reserve located on the Pacific slopes of Colombia's Western Cordillera. Neither periodic water shortage nor prolonged waterlogging was observed at Tambito. Total soil nitrogen and available phosphorus were above the range typically observed in lowland evergreen rain forests (LERF), while exchangeable calcium levels were lower. In terms of soil nutrient status, total nitrogen and available phosphorus at Tambito were above the range typically observed in more productive LERF, while exchangeable calcium levels were lower. Leaf nutrient contents observed at Tambito were broadly similar to values observed in LERF. Photosynthetic photon flux density (*PPFD*) remained well below the light-saturation level for leaf-scale photosynthesis (*A*) throughout the day during the wet season and for 21 hours day^{-1} during the dry season. Cloudiness may reduce the competitive advantage of high A_{max} in canopy leaves, thereby increasing the fitness of alternative traits conferred by low specific leaf area, including nutrient-use efficiency and leaf longevity. Therefore, it was concluded that persistent *PPFD* limitation of photosynthesis may help to explain the prevalence of sclerophylly in tropical montane cloud forests.

INTRODUCTION

In a review of tropical montane cloud forest (TMCF) productivity and stature, Bruijnzeel and Veneklaas (1998) outlined a number of environmental controls on productivity, while highlighting the need for greater understanding of the relative importance of each factor. Some proposed controls were microclimatic, such as low minimum temperatures (Kitayama and Aiba, 2002; Hemp, this volume #12), and persistently high humidity levels associated with fog and low cloud (Letts and Mulligan, 2005; cf. Reinhardt and Smith, 2008; Schawe *et al.*, this volume). Other controls were pedological, including effects of high soil acidity, persistent waterlogging, and ensuing nutrient limitations

Tropical Montane Cloud Forests: Science for Conservation and Management, eds. L. A. Bruijnzeel, F. N. Scatena, and L. S. Hamilton. Published by Cambridge University Press.

(Leuschner *et al.*, 2007; Moser *et al.*, 2007; Benner *et al.*, this volume; Roman *et al.*, this volume; Santiago *et al.*, this volume). High soil acidity may be associated with aluminum toxicity and with high concentrations of phenolic compounds, which may contribute to reduced nutrient uptake and above-ground productivity in certain TMCFs (Hafkenscheid, 2000; cf. Hertel and Leuschner, this volume).

Leaf nutrient concentration and specific leaf area (*SLA*, $cm^2\ g^{-1}$) are known to scale well with maximum photosynthesis rate (A_{max}) across a range of climates, but A_{max} is inversely related to leaf longevity (Reich *et al.*, 1991, 1999; Wright *et al.*, 2004) and nutrient-use efficiency (*NUE*; Poorter and Evans, 1998). Leaves of TMCF vegetation are often long-lived (Grubb, 1977; Moser *et al.*, 2007) and sclerophyllous (Cavelier and Goldstein, 1989; Velázquez-Rosas *et al.*, 2002), compared to those of lowland evergreen rain forest (LERF). Leaves with low *SLA* generally exhibit lower A_{max}, because of their investment in chloroplast-free cell wall materials (Shipley *et al.*, 2006), presumably to promote longevity and *NUE* while reducing herbivory. Lower A_{max} would, thus, be expected in TMCF species with sclerophyllous leaves, especially when leaf nutrient concentrations are low, as has been observed in TMCFs on Hawai'i (Gerrish *et al.*, 1988; cf. Santiago *et al.*, this volume), Jamaica (Hafkenscheid, 2000), and at higher elevations in Borneo (Kitayama and Aiba, 2002). The reason for thick leaves in TMCF is not completely understood, but hypotheses include low nutrient availability (Tanner *et al.*, 1998), high UV-radiation intensity (Flenley, 1995) and the need for leaf longevity to ensure: (i) adequate lifetime carbon gain despite low photosynthetic photon flux density (*PPFD*) (Williams *et al.*, 1989) and (ii) higher leaf nutrient residence time (Aerts and Van der Peijl, 1993).

Air temperature, vapor pressure deficit (*VPD*) and mean solar irradiance are lower in TMCF than in LERF (Grubb, 1977; Jarvis and Mulligan, this volume). Low *VPD* reduces the transpiration rate for any given leaf temperature and stomatal conductance, while ground-level cloud decreases incident *PPFD* (Cavelier and Mejía, 1990; Reinhardt and Smith, 2008; Mildenberger *et al.*, 2009) and promotes leaf wetness (Letts and Mulligan, 2005). Typically *PPFD* is 50% lower at sites with stunted TMCF than in LERF (Bruijnzeel and Veneklaas, 1998; Holwerda, 2005), and bright sunshine typically occurs for 3 hours day^{-1} or less, compared to 5.3–5.8 hour day^{-1} in LERF (Grubb, 1977; Letts, 2003). Limitation of canopy photosynthesis by *PPFD* has been demonstrated in some cloud-affected montane forests (Reinhardt and Smith, 2008; Mildenberger *et al.*, 2009) and is known to occur seasonally, even in lowland rain forest (Graham *et al.*, 2003). Cloud water interception can cause both adaxial (upper surface) and abaxial (underside) leaf wetness. Adaxial leaf wetness, which occurs during rainfall, does not inhibit photosynthesis directly, as most TMCF leaves are hypostomatous (Dietz *et al.*, 2007). Abaxial wetness during fog may reduce photosynthetic gas-exchange directly, by covering open stomata with a film of water (Fogg, 1947; Smith and MacLean, 1989). This effect has been demonstrated experimentally (Letts and Mulligan, 2005), but is thought to be negligible where water repulsion, through beading and subsequent shedding of drops, prevents the formation of a surface film (Smith and McClean, 1989; Boyce *et al.*, 1991; Reinhardt and Smith, 2008) or where the stomata are protected against wetting (Lai *et al.*, 2007; Mildenberger *et al.*, 2009).

Although TMCFs are extensive throughout the 700-km north–south axis of the Western Cordillera in the Chocó climate region of Colombia, very little consideration has been given to the pedological and climatic factors potentially limiting photosynthetic productivity in the region. This study first presents an assessment of nutrient concentrations in soils and vegetation at Centro de Estudios Ambientales Tambito (Tambito), a 5.5 × 4.1 km cloud forest reserve at 1374–2894 m.a.s.l. in the Western Cordillera. Next, the forest microclimate is characterized, with particular emphasis on *PPFD* and leaf wetness, because of their potential importance as controls on photosynthetic gas exchange.

STUDY SITE

The present study was carried out in lower montane cloud forest (LMCF) at two elevations (1450–1600 m.a.s.l. and 2150–2200 m.a.s.l.) at Tambito, Cauca, Colombia (2° 30′ N, 77° 00′ W; Figure 49.1). About two-thirds of the study site consisted of primary LMCF, while the remainder was secondary forest of about 16 years of age in 1999. Canopy heights ranged from 10 to 15 m at 1450 m.a.s.l. and from 8 to 12 m at 2150 m.a.s.l. Based on 6 years of data, daily mean air temperature (T_a) and rainfall were 18 °C and 3600 mm $year^{-1}$ at 1450 m.a.s.l. vs. 12 °C and 7000 mm $year^{-1}$ at 2150 m.a.s.l. Rainfall exceeded 500 mm $month^{-1}$ at 1450 m from October to May. August was the driest month, with a range of just 100–220 mm $month^{-1}$. However, extended rain-free periods were not uncommon from June to mid-September, especially during the 1997 El Niño event, when July was rain-free. Cloud water interception (*CWI*) occurred throughout the year, but ratios of *CWI* to rainfall were generally low, especially during the rainy season (González, this volume; Mulligan *et al.*, this volume #25).

In terms of its geological setting, the Pacific-facing hillslopes of the Western Cordillera are underlain mostly by Cretaceous sedimentary and volcanic rocks (Millward *et al.*, 1984; Kerr *et al.*, 1998). In the Tambito area, tholeiitic basalts were prevalent and interspersed by fractured sedimentary blocks, but phyllites and schists were also present. Tambito is located 95 km to

Figure 49.1. Map of study sites: (i) Tambito, Cauca, Colombia; (ii) Auca, Orellana, Ecuador.

the west of Volcán Purace (4780 m.a.s.l.), an active, composite volcano of the Central Cordillera, but no evidence of ash deposits was noted at Tambito.

METHODS

Soil nutrient concentrations and pH

Soils were sampled for nutrient analysis at 18 sites, ranging in elevation from 1400 to 2250 m.a.s.l. Seven samples were taken in primary and secondary forest each, while the remaining four were sampled in pasture at 1400 m.a.s.l. The samples were analyzed for soil organic matter (SOM), available phosphorus (P_{av}), total nitrogen (N_t), extractable potassium (K), calcium (Ca), and magnesium (Mg), as well as total boron (B_t) and aluminum (Al) concentrations. Boron was measured because agricultural studies have demonstrated that soils with low pH, high sand content, and low clay content may be B-deficient (Goldberg, 1997; Shorrocks, 1997). Aluminum was measured because previous studies have suggested that Al toxicity may limit the productivity of TMCF under conditions of low pH (Hafkenscheid, 2000).

After clearing litter, soils were sampled at 10-cm intervals from 5 to 65 cm depth, using a soil corer. Samples in the primary and secondary forests were taken from paired sites with similar elevation and aspect. All analyses were carried out by the Soils Department of the International Center for Tropical Agriculture (CIAT) at Cali, according to procedures outlined in Mackean (1993). Soil organic matter was determined by the Walkey–Black method, and multiplied by 1.72 to obtain percentage OM content. Available phosphorus was determined using the Bray II procedure. Total nitrogen was determined by sulfuric acid digestion. Exchangeable Ca, K, and Mg were determined by ammonium acetate extraction (1M, pH 7). Total B was determined by hot water extraction, while exchangeable Al was extracted in 1M KCl. One-way analysis of covariance (ANCOVA) was carried out to compare soil nutrient concentrations in primary forest, secondary forest and pasture, using log-transformed soil nutrient levels as dependent variables and depth as the concomitant variable for each of the three independent sample populations. The process was repeated for each nutrient. Soil pH was measured separately at 19 primary forest sites ranging in elevation from 1150 to 2250 m.a.s.l.; pH was determined from water extracts of soil samples, using a glass-electrode pH meter (Hach). Measurements were recorded as the average of three readings for each sample.

Foliar nutrient levels

Leaf nutrient concentrations were assessed on leaves of seven genera of plants, including four trees and three understory plants found in two altitude ranges: 1450–1600 m.a.s.l. and 2150–2200 m.a.s.l. Structurally and morphologically similar plants were chosen for sampling within the two elevation zones. At 1450–1600 m.a.s.l., the canopy trees examined were *Clusia* sp.,

Cecropia garciae, *Miconia* sp., and *Psychotria racemosa*, while the understory plants were *Alloplectus teuscheri*, *Anthurium* sp., and unidentified palms. At 2150–2200 m.a.s.l., canopy trees included *Clusia pentandra*, *Cecropia bullata*, *Miconia* sp., and *Psychotria cuatrecasasii*, while the understory plants sampled were *Alloplectus schultzei*, *Anthurium* sp., and palms. In August 1999, 24 leaves were removed in the early afternoon from the sunlit portions of the crown of each tree, using a leaf cutter. All of the trees had a diameter at breast height (DBH) of at least 5 cm and a height of at least 3 m. At least one sample was obtained for each tree species.

The leaf samples were oven-dried at 70 °C for two days and weighed and measured for *SLA* determination. Following overnight pre-digestion, the samples were prepared for nutrient analysis by double acid digestion (Miyazawa *et al.*, 1999). Total N was determined according to the Kjeldahl method (Bremmer and Mulvaney, 1982). Spectrophotometry was used to determine P concentrations (Murphy and Riley, 1965), while K, Ca, and Mg levels were analyzed through nitric-perchloric (2 : 1) digestion and spectrophotometric atomic absorption (Benton and Jones, 1989). Boron levels were analyzed through calcination and molecular spectrophotometry (Mackean, 1993). Mass-based concentrations were converted to area-based levels, according to *SLA*.

To compare leaf nutrient content in trees differing by successional stage, additional samples were taken from secondary forest specimens of *Cecropia* sp., *Clusia* sp., *Miconia* sp., and *Psychotria* sp. at both altitudes. In the understory species, samples were composed of leaves from several individuals. Each plant selected for photosynthesis measurements was sampled for foliar nutrient analysis, to facilitate comparisons of nutrient content with A_{max}. Samples of each tree were deposited in the CAUT Herbarium of the Popayan Museum of Natural History.

Climate monitoring

Two automatic weather stations were installed within the LMCF zone, one in open pasture at 1450 m.a.s.l. (2° 30′ 25″ N, 77° 0′ 2″ W) and one within primary forest at 1600 m.a.s.l. (2° 30′ 45″ N, 76° 59′ 42″ W). The following variables were monitored: global radiation (*K*↓), *PPFD*, ultraviolet-A (UV-A) radiation, T_a, relative humidity (*RH*), and rainfall. Values were recorded every 10 min, averaged over hourly intervals, and stored using Datataker© loggers (Data Electronics, Australia). Measurements of *K*↓ and UV-A were made with an SKS 1110 pyranometer and SKU 420 UV-A sensor (Skye Instruments, UK); T_a was measured with a thermistor (SKH 2021), while *RH* was measured with a capacitance-based probe (SKH 2013, Skye Instruments, UK). These sensors were placed within a miniature Stevenson radiation shield at a height of 1.5 m. Data were obtained from 28 July to 30 November 1998 and from 23 June to 14 August 1999. Here, the focus will be on a comparison between August and November 1998, representing the dry and wet seasons, respectively. November 1998 was wetter than average, with rainfall in excess of 700 mm at 1450 m.a.s.l.

Soil matric potential was monitored at the pasture site from June 1997 to November 1999, at depths of 0.4 and 1.2 m, using single gypsum blocks (Soil Moisture Corporation, USA). Only data from the dry season of 1997 are reported here. Soil moisture was also monitored using a capacitive soil moisture probe (Theta Probe, Delta-T Devices, UK), inserted horizontally at a depth of 30 cm. Data were recorded at 10-min intervals and averaged hourly. Data are reported for the period 22 September to 12 November 2000. The voltage returned by the Theta Probe had a near-linear relationship with volumetric soil moisture (Delta-T Devices, 1999).

To compare LMCF light intensity and quality with lowland rain forest (LERF) conditions, identical radiation measurements were made at Auca, Orellana, Ecuador (0.7° S, 76.9° W, 300 m.a.s.l.), 300 km south of the Tambito site. Mean annual T_a and rainfall at Auca are 26 °C and 2860 mm. Measurements of *K*↓ and *PPFD* were taken from 4 October to 31 December 2000.

Photosynthetic gas-exchange measurements

Net photosynthesis (*A*) measurements were made *in situ*, using a CIRAS-1 Portable Differential CO_2/H_2O infra-red gas analyzer, with a 2.5-cm^2 PLC(B) broad-leaf cuvette (PP Systems, Hitchin, UK). Readings were obtained from healthy, fully developed leaves of the same specimens described for leaf nutrient analysis. Measurements were made between June and August 1999, with the exception of *Clusia garciae*, for which measurements were made in February 1999. To determine mean A_{max} for each species, at least 20 fully lit leaves were sampled from each tree specimen at heights of 2–5 m. The CO_2 concentration was set to 375 ppm.

Curves of *PPFD* vs. *A* were produced using the Thornley and Johnson (1990) formulation, to determine light saturation intensity ($PPFD_{sat}$), defined as the *PPFD* at which modeled *A* reached 90% of its asymptotic maximum rate. Data were removed when they did not satisfy this criterion, resulting in a total of 595 *PPFD*-saturated measurements for the 14 species, out of an original sample of 2094. To test the hypothesis that photosynthetic gas exchange is inhibited when fog interception wets the abaxial, or lower, surface of LMCF leaves, A_{max} was first measured with only the adaxial, or upper, surface wetted, and then with both surfaces wetted by a fine spray. An atomizer was used to wet the abaxial surface for 2 min before the measurements were taken. Readings were also obtained with both surfaces dry, as a control. The presence of water on leaf surfaces did not affect the accuracy of the A_{max} measurements, because the CIRAS-1 system corrects for the presence of water vapor and for volume

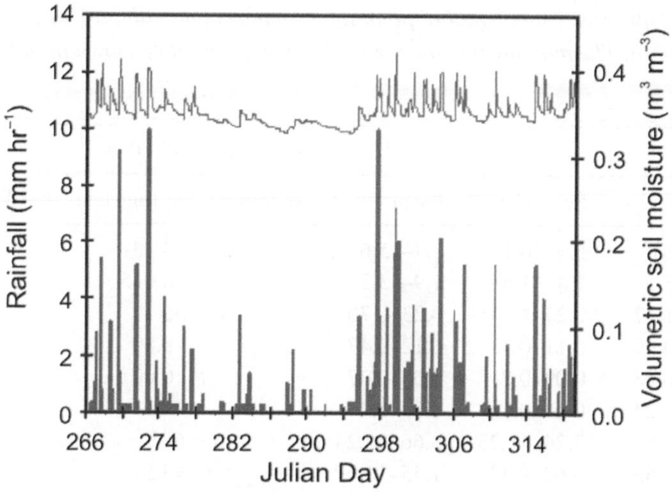

Figure 49.2. Hourly volumetric soil moisture content (θ_v) and rainfall intensity at Tambito, Cauca, Colombia under pasture at 1450 m.a.s.l., from 22 September to 12 November 2000.

changes due to increases in air volume as water vapor content increases (K. J. Parkinson, personal communication). For the leaf wetness experiment, photosynthesis measurements were taken between 13 and 19 February 2000.

RESULTS

Soil moisture

Tambito soils are well-drained sandy loams with an average composition of 57% sand, 21% silt, and 22% clay (Rincón-Romero, 2000). The soil moisture data obtained at the pasture site from 22 September to 12 November 2000 (Figure 49.2) are typical of the soil moisture regime at Tambito. Rainfall events briefly increased volumetric moisture content (θ_v) at a depth of 0.3 m, but soils quickly returned to field capacity ($\theta_v \approx 0.37$), as a result of drainage and water use by vegetation. The steep topography and well-drained soils limited the duration of water-logging, except at small, isolated patches of shallow sloping terrain. During dry periods, the absence of rainfall led to sustained, though modest, reductions in soil moisture (Figure 49.2). During the 1997 El Niño event, there was no rainfall at 1450 m.a.s.l. for 64 consecutive days beginning in late June. During this exceptional drought, the soils of the pasture site dried to 0.31 m^3 m^{-3} at a depth of 0.4 m, and to 0.36 m^3 m^{-3} at 1.2 m. These values are above the wilting point of 0.23 m^3 m^{-3} for the pasture site, as estimated using soil textural properties (Brady and Weil, 1999).

Soil nutrient status, acidity, and aluminum levels

At shallow depths, the soils in primary cloud forest had far higher levels of N_t, P_{av}, K, Ca, and B_t than those of secondary forest and pasture (Table 49.1). Deeper in the soil profile, levels did not differ significantly. Average extractable Al content was higher in secondary forest (1.5–4.7 meq 100 g^{-1}) than in primary forest (1.2–2.4 meq 100 g^{-1}), with the highest values recorded in the shallower layers. Mean Al : Ca ratios increased from 0.30 to 1.95 from 5 cm to 45 cm depth in primary forest soils, but were much higher in secondary forest soils, increasing from 5.89 to 8.28 at the same depths. In primary forest, soil pH varied from 3.6–5.7 at a depth of 10 cm (n = 19; mean pH = 4.6 ± 0.6), and from pH 4.7–5.6 at 50 cm (n = 19; mean pH = 5.1 ± 0.3). Average pH was not significantly lower above 2000 m.a.s.l. than below 1600 m.a.s.l., although the two lowest individual values were recorded above 2000 m.a.s.l.

Foliar nutrient concentrations

Despite the much lower nutrient concentrations of secondary forest soils (Table 49.1), leaf nutrient concentrations were not significantly lower in secondary forest canopy leaves than in those of primary forest (Table 49.2), except in the case of B at 2150–2200 m.a.s.l. (t-test, $p < 0.05$). Leaf N, P, K, and Mg concentrations were similar to those of an LMCF at the Reserva Biológica San Francisco (RBSF), in southern Ecuador (Soethe *et al.*, 2008), and fell within the upper portion of the ranges observed for upper montane cloud forests in the Blue Mountains of Jamaica (Hafkenscheid, 2000). However, leaf Ca concentrations were higher in the species observed at Tambito than in these TMCFs. With data for all canopy tree leaves in primary or secondary forest stages combined, leaf N and K levels were not significantly lower at 2150–2200 m.a.s.l. than at 1450–1600 m.a.s.l. ($p = 0.08$ for both elements), but leaf B concentrations were significantly lower at 2150 m.a.s.l (t-test; $p < 0.05$). In percentage terms, *Clusia* sp. leaves had much lower N, P, and B concentrations than those of all other tree species, as well as understory herbs and shrubs (Table 49.3). However, the specific leaf area (*SLA*) of its sclerophyllous leaves (40 cm^2 g^{-1}) at 1450–1600 m.a.s.l. was much lower than that for *Cecropia* sp. (102 cm^2 g^{-1}), *Miconia* sp. (72 cm^2 g^{-1}), and *Psychotria* sp. (62 cm^2 g^{-1}), so that values were less distinct on an area basis. To determine the relationship between leaf nutrient concentration and A_{max}, the mean leaf nutrient content obtained from each plant was paired with its corresponding mean A_{max} value. In canopy tree species A_{max} was correlated with leaf N ($r^2 = 0.60$) and K ($r^2 = 0.79$) (Figure 49.3), but not in understory species; A_{max} was not correlated with foliar P, Ca, Mg, or B levels.

Short-wave radiation, *PPFD*, and UV radiation

Mean short-wave radiation ($K\downarrow$) totals in the LCMF belt of Tambito were 12.0 ± 3.7 MJ m^{-2} day^{-1} during the dry season (August) and 7.6 ± 2.1 MJ m^{-2} day^{-1} during the wet season

Table 49.1 *Soil organic matter (SOM), total nitrogen, available phosphorus, exchangeable potassium, magnesium, calcium, and aluminum, and total boron concentrations at Tambito, Cauca, Colombia. The maximum value for each depth interval is shown in bold italics*

Property and depth (cm)		Mean			Range			One-way ANCOVA
		Primary	Secondary	Pasture	Primary	Secondary	Pasture	(depth vs. log [nutrient], $k=3$)
SOM (%)	5	***28***	11.7	8.9	9.1–70.8	5.3–26.1	3.7–15.6	df = 32, $F=4.83$
	15	***9.3***	7.6	6.7	5.4–16.1	2.3–11.6	1.9–13.2	$P=0.015$
N_t (%)	5	***1.26***	0.50	0.47	0.71–1.77	0.27–1.03	0.23–0.70	df = 55
	15	***0.52***	0.28	0.31	0.23–0.99	0.11–0.46	0.12–0.47	$F=9.89$
	35	***0.27***	0.16	0.20	0.08–0.48	0.09–0.21	0.06–0.37	$P<0.001$
	45	***0.19***	0.11	0.15	0.09–0.24	0.04–0.14	0.06–0.24	
P_{av} (ppm)	5	***30.9***	11.5	8.40	13.3–39.8	3.24–18.23	3.66–13.24	df = 64
	15	***5.13***	3.64	4.54	1.69–9.53	0.67–8.43	1.45–10.69	$F=3.51$
	35	***3.03***	2.73	1.09	0.34–9.32	0.60–7.58	0.87–1.25	$P=0.036$
	45	1.40	***1.47***	0.88	0.56–2.83	0.59–3.19	0.46–1.37	
K (meq 100 g^{-1})	5	***0.88***	0.34	0.25	0.38–1.53	0.18–0.69	0.16–0.36	df = 62
	15	***0.37***	0.13	0.18	0.12–0.68	0.07–0.26	0.07–0.28	$F=21.55$
	35	***0.27***	0.05	0.18	0.06–0.79	0.03–0.11	0.07–0.28	$P<0.001$
	45	***0.21***	0.06	0.09	0.05–0.70	0.04–0.10	0.05–0.12	
Mg (meq 100 g^{-1})	5	***2.61***	0.48	0.73	1.53–5.10	0.35–0.72	0.31–1.15	df = 59
	15	***1.00***	0.29	0.58	0.38–2.06	0.17–0.53	0.20–0.84	F = 29.36
	35	***0.51***	0.19	0.34	0.26–0.86	0.12–0.33	0.15–0.46	P<0.001
	45	***0.37***	0.15	0.34	0.20–0.70	0.13–0.16	0.15–0.46	
Ca (meq 100 g^{-1})	5	***7.91***	0.81	1.96	3.17–12.10	0.20–1.97	0.69–3.69	df = 57
	15	***2.48***	0.39	1.55	0.88–6.82	0.11–1.23	0.42–2.83	F = 35.33
	35	***1.08***	0.24	1.05	0.51–2.63	0.11–0.51	0.36–1.97	P<0.001
	45	0.64	0.23	***0.81***	0.30–0.96	0.10–0.36	0.29–1.73	
B_t (ppm)	5	***9.08***	1.33	0.55	2.00–24.80	0.40–2.45	0.31–0.95	df = 55
	15	***0.67***	0.39	0.29	0.36–1.21	0.12–0.68	0.14–0.46	F = 7.73
	35	***0.26***	0.16	0.17	0.11–0.51	0.06–0.21	0.12–0.26	P = 0.001
	45	***0.26***	0.11	0.17	0.11–0.63	0.07–0.15	0.12–0.26	
Al (meq 100 g^{-1})	5	2.38	***4.77***	1.86	0.87–7.80	1.56–7.58	1.56–2.18	df = 57
	15	1.86	***3.23***	1.80	0.76–4.37	2.29–4.99	1.14–2.60	F = 10.65
	35	1.18	***1.72***	1.18	0.52–2.60	1.56–2.39	0.52–1.77	P = 0.001
	45	1.25	***1.49***	0.90	0.52–3.33	1.25–1.66	0.31–1.56	

(November) of 1998. By comparison, mean insolation at Auca, Ecuador, was 15.6 MJ m^{-2} day^{-1}. During the dry season, $K\downarrow$ was similar at Tambito and Auca in the mornings, but significantly higher at Auca from 13.00 to 16.00 h (Table 49.4). However, maximum-recorded $K\downarrow$ values were higher at Tambito, except at 15.00 and 16.00 h. During the wet season, $K\downarrow$ was significantly lower in the LCMF belt of Tambito than at Auca at all hours. Wet-season radiation levels at Tambito were also significantly lower than dry-season levels, except in the late afternoon. Values of *PPFD* were significantly lower at Tambito than at Auca for all hours of the day (Table 49.4). Maximum-recorded *PPFD* levels were also higher at Auca than at Tambito. As observed for $K\downarrow$, *PPFD* intensity at Tambito was higher during the dry season than during the wet season, from 08.00 to 14.00 h. Mean UV-A levels, which are proportional to UV-B intensity, were significantly lower at Tambito than at Auca at all hours. During the dry season, however, maximum recorded UV-A radiation was higher at Tambito. The result is that the forest canopy receives low intensity UV-A more frequently in LMCF than in LERF, but UV-A intensity is at least as high in LMCF during clear-sky conditions (Figure 49.4).

Photosynthesis

Under *PPFD*-saturated conditions, mean leaf chamber temperature (T_l) was 25.4 °C at 1450–1600 m.a.s.l. and 21.9 °C at 2150–2200 m.a.s.l., with corresponding mean *VPD* values of 1.08 kPa and 0.41 kPa, respectively. Under these conditions, A_{max} ranged from 5.4 to 10.6 μmol m^{-2} s^{-1} in leaves of canopy tree species (Table 49.5) although rates as high as 13.1 μmol m^{-2} s^{-1} were

Table 49.2 *Nutrient concentrations in leaves of (i) primary and secondary montane cloud forest trees at Tambito, Colombia, (ii) primary forest trees at Reserva Biologica San Francisco, Ecuador, and (iii) primary and secondary forest trees of the Blue Mountains, Jamaica*

	Tambito, Colombia (1450–1600 m)		Tambito, Colombia (2150–2200 m)		RBSF, Ecuador[a] (1900 m)	(2400 m)	Blue Mountains[b] (1600–1900 m)
Leaf nutrient	Primary	Secondary	Primary	Secondary	Primary	Primary	All types (range)
N (%)	2.40 ± 0.82	2.04 ± 1.33	1.33 ± 0.74	1.31 ± 0.27	2.17 ± 1.30	1.35 ± 1.20	0.49–1.57
P (%)	0.12 ± 0.03	0.09 ± 0.04	0.14 ± 0.10	0.13 ± 0.07	0.22 ± 0.03	0.12 ± 0.01	0.04–0.13
K (%)	1.12 ± 0.48	0.88 ± 0.11	0.72 ± 0.37	0.55 ± 0.13	1.00 ± 0.12	0.70 ± 0.04	0.18–1.60
Ca (%)	1.36 ± 0.67	1.38 ± 1.12	1.27 ± 0.70	1.00 ± 1.02	0.62 ± 0.17	0.74 ± 0.10	0.17–1.57
Mg (%)	0.39 ± 0.28	0.16 ± 0.07	0.33 ± 0.08	0.24 ± 0.09	0.34 ± 0.06	0.27 ± 0.20	0.10–0.63
B (ppm)	18.8 ± 11.7	15.2 ± 5.1	9.4 ± 1.7	5.2 ± 0.9			

[a] Reserva Biologica San Francisco, Ecuador (RBSF); data from Soethe *et al.* (2008).
[b] Blue Mountains, Jamaica; data from Hafkenscheid (2000).

Table 49.3 *Leaf nutrient concentrations in seven genera of plants at Tambito, Cauca, Colombia; the range of leaf nutrient concentrations observed in Lowland Evergreen Rain Forest (LERF; Jordan, 1985) is shown for comparison*

Nitrogen (%)		Phosphorus (%)		Potassium (%)	
Anthurium sp.	2.28	*Anthurium* sp.	0.195	*Anthurium* sp.	2.56
Psychotria sp.	2.24	*Cecropia* sp.	0.177	Palmae	1.28
Alloplectus sp.	2.11	*Psychotia* sp.	0.131	*Psychotria* sp.	1.16
Cecropia sp.	2.07	*Miconia* sp.	0.114	*Alloplectus* sp.	1.15
Miconia sp.	2.07	Palmae	0.110	*Cecropia* sp.	1.10
Palmae	1.96	*Alloplectus* sp.	0.105	*Clusia* sp.	0.69
Clusia sp.	0.96	*Clusia* sp.	0.065	*Miconia* sp.	0.58
LERF range	1.1–4.2	LERF range	0.028–0.28	LERF range	0.2–6.73
Calcium (%)		**Magnesium (%)**		**Boron (ppm)**	
Anthurium sp.	2.22	*Anthurium* sp.	0.818	*Anthurium* sp.	23.9
Clusia sp.	1.87	*Alloplectus* sp.	0.683	Palmae	20.3
Miconia sp.	1.82	Palmae	0.585	*Psychotria* sp.	18.8
Alloplectus sp.	1.61	*Psychotria* sp.	0.378	*Alloplectus* sp.	15.7
Palmae	1.49	*Cecropia* sp.	0.297	*Miconia* sp.	12.4
Cecropia sp.	0.88	*Clusia* sp.	0.250	*Cecropia* sp.	11.6
Psychotria sp.	0.79	*Miconia* sp.	0.242	*Clusia* sp.	8.0
LERF range	0.1–4.4	LERF range	0.1–0.843		

observed in February 2000, during control (dry-leaf) measurements of the leaf wetting experiment. Mean A_{max} ranged from 3.5 to 9.3 μmol m^{-2} s^{-1} in understory plants (Table 49.5). Overall, A_{max} fell within the lower portion of the 5–25 mol m^{-2} s^{-1} range observed in tree species of LERF. Rates were numerically higher within the lower elevation range for six out of seven canopy and understory species examined in 1999, but mean A_{max} was not significantly higher at 1450–1600 m.a.s.l. (8.1 ± 2.0 μmol m^{-2} s^{-1}) than at 2150–2200 m.a.s.l. (6.4 ± 2.7 μmol m^{-2} s^{-1}) at the 95% confidence level (paired, two-tailed *t*-test, df = 6, $p = 0.08$). For canopy tree species only, A_{max} also did not differ between the two elevation ranges (9.0 ± 1.1 μmol m^{-2} s^{-1} vs. 8.2 ± 2.2 μmol m^{-2} s^{-1}; df = 3, $p = 0.42$). Mean stomatal conductance (g_s) was usually in the 150–550 mmol m^{-2} s^{-1} range, with a tendency for higher values in sunlit tree leaves than in understory vegetation. However, low *VPD* suppressed transpiration rates (*E*), so that photosynthetic water-use efficiency (*WUE*, i.e. A_{max}/E) remained moderately high (Table 49.5). The mean ratios of internal to atmospheric CO_2 concentration (c_i/c_a) ranged from 0.77 to 0.89 under *PPFD*-saturated conditions.

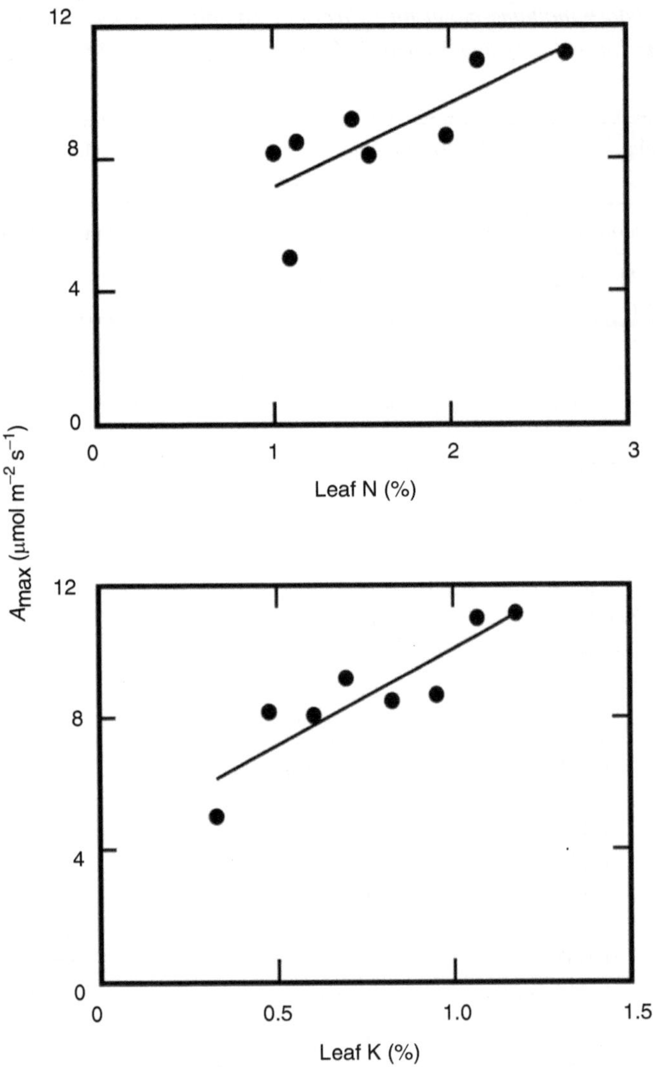

Figure 49.3. Relationship between leaf nutrient concentrations (above: nitrogen, below: potassium) and maximum photosynthetic rates at *PPFD* saturation (A_{max}) in lower montane cloud forest plants at Tambito.

As a result of the low *SLA* of LMCF trees at 1450 m.a.s.l. (69 ± 26 $cm^2\ g^{-1}$), mass-based photosynthesis rates (A_{mass}) were lower at Tambito (62 ± 22 nmol $g^{-1}\ s^{-1}$) than in LERF trees of the same families (143 ± 74 nmol $g^{-1}\ s^{-1}$; Poorter and Remkes, 1990; Reich *et al.*, 1991; Valladares *et al.*, 2000). This was especially true of the most sclerophyllous tree (*Clusia* sp.), for which A_{mass} was only 35 nmol $g^{-1}\ s^{-1}$. Relative to the worldwide leaf economics spectrum (Wright *et al.*, 2004), canopy leaves of *Clusia* sp. exhibited one of the lowest A_{mass} values among tropical plants, whereas A_{mass} was closer to the mean in *Cecropia* sp., *Miconia* sp., and *Psychotria* sp. (Figure 49.5). All four tree species exhibited lower than average A_{mass}: N ratios for their respective N contents.

Upon experimental wetting of both the adaxial and abaxial leaf surfaces, A_{max} decreased from 7.7 ± 1.3 to 5.6 ± 0.7 mol $m^{-2}\ s^{-1}$ in *Anthurium* sp., from 8.1 ± 0.7 to 4.4 ± 1.1 mol $m^{-2}\ s^{-1}$ in *Cecropia garciae*, from 13.1 ± 1.0 to 2.1 ± 0.7 mol $m^{-2}\ s^{-1}$ in *Miconia* sp., and from 11.6 ± 0.2 to −0.5 ± 0.3 mol $m^{-2}\ s^{-1}$ in *Psychotria racemosa* (Letts and Mulligan, 2005). Full wetting also caused a 100% reduction in A_{max} in *Clusia* sp., though measurements were only obtained to a *PPFD* intensity of 700 µmol $m^{-2}\ s^{-1}$. The 4% increase in A_{max} observed upon adaxial wetting of *Anthurium* sp. and *Cecropia garciae* was not statistically significant (*t*-test, $p > 0.05$).

DISCUSSION

The above-ground productivity of TMCF ecosystems is usually lower than in LERF, ranging from ~200 to 2000 g m^{-2} $year^{-1}$ globally (Weaver and Murphy, 1990; Herbert and Fownes, 1999; Kitayama and Aiba, 2002). The specific soil and microclimatic factors that interact to cause low productivity vary from site to site (Waide *et al.*, 1998; Soethe *et al.*, 2008; cf. Roman *et al.*, this volume). In this section, an initial assessment of the nutrient status and microclimate of the Tambito LMCF is provided, with an emphasis on implications for forest productivity.

Soil characteristics and leaf nutrient status

Soil moisture under pasture remained mostly near field capacity, due to frequent rainfall input and modest irradiation levels, combined with rapid drainage of porous ($\theta_{sat} = 0.66\ m^3\ m^{-3}$), sandy soils on steeply sloping terrain. Whilst this suggests that neither periodic dryness nor waterlogging are likely stressors (Figure 49.2), these results would need to be replicated at a range of elevations in view of the general increase in rainfall and cloud incidence with elevation in the area (Mulligan *et al.*, this volume #25; González, this volume). Soils were acidic (pH 3.6–5.7) but not excessively so (cf. Cavelier, 1992; Bautista-Cruz and del Castillo, 2005; Schawe *et al.*, 2007, Roman *et al.*, this volume).

Nitrogen and phosphorus are usually the chief limiting elements in TMCF (Vitousek and Sanford, 1986; Benner *et al.*, this volume). However, Kitayama and Aiba (2002) found that soil N and P peaked at intermediate elevation in a transect study in Borneo, while P was strongly affected by substrate quality. Consistent with these findings, Cavelier (1992) found that N values were higher in a 1200 m.a.s.l. LMCF in Panama than in the adjacent lowlands while Wilcke *et al.* (2001) observed relatively high N_t (1.2–2.1%) and total P (P_t; 0.3–0.9%) at a 1960–2450 m.a.s.l. site in southern Ecuador, though concentrations decreased with elevation. At Tambito, soil nutrient characteristics were similar to those in Panamanian (Cavelier, 1992) and Ecuadorian TMCF (Wilcke *et al.*, 2008), with relatively high soil N_t and a similar range of P_{av} (Table 49.1) compared to a set of lowland rainforest soils of varying substrate quality (Malhi *et al.*, 2009). Soil K and Mg were found in concentrations consistent with the range observed by Cavelier (1992) in LERF.

Table 49.4 *Mean diurnal pattern of (i) incoming short-wave radiation (K↓), (ii)* PPFD, *and (iii) UV-A radiation in Lowland Evergreen Rain Forest (LERF; Auca, Ecuador, 300 m.a.s.l.) and in Lower Montane Cloud Forest (LMCF; Tambito, 1450 m.a.s.l.) during the dry and wet seasons*

(i) Incoming short-wave radiation, *K*↓

Hour	Mean ($W\ m^{-2}$) and SD			*t*-value (df)		*p* value	
	LERF (Auca)	LMCF (Dry)	LMCF (Wet)	LERF> LMCF (Dry)	LMCF(Dry)>LMCF (Wet)	LERF> LMCF (Dry)	LMCF(Dry)>LMCF (Wet)
0800 h	236 ± 121	202 ± 145	130 ± 42	1.22 (112)	2.88 (66)	>0.8	0.005
0900 h	343 ± 148	387 ± 240	280 ± 154	1.04 (113)	2.21 (67)	>0.8	0.02
1000 h	528 ± 265	575 ± 324	331 ± 195	0.77 (112)	3.78 (65)	>0.7	0.0005
1100 h	590 ± 288	588 ± 388	335 ± 170	0.03 (112)	3.55 (65)	>0.2	0.0005
1200 h	610 ± 279	453 ± 287	264 ± 111	2.42 (108)	3.15 (62)	0.01	0.0025
1300 h	607 ± 266	388 ± 287	208 ± 85	3.77 (109)	3.49 (62)	0.0001	0.0005
1400 h	532 ± 225	250 ± 180	181 ± 78	7.08 (110)	2.06 (64)	0.0001	0.025
1500 h	370 ± 141	155 ± 123	158 ± 55	8.28 (112)	0.10 (66)	0.0001	>0.25
1600 h	232 ± 90	96 ± 69	97 ± 54	8.82 (111)	0.08 (65)	0.0001	>0.25

(ii) Photosynthetic photon flux density, *PPFD*

Hour	Mean ($mol\ m^{-2}\ s^{-1}$) and SD			*t*-value (df)		*p* value	
	LERF (Auca)	LMCF (Dry)	LMCF (Wet)	LERF> LMCF (Dry)	LMCF(Dry)>LMCF (Wet)	LERF> LMCF (Dry)	LMCF(Dry)>LMCF (Wet)
0800 h	444 ± 236	258 ± 141	80 ± 61	5.19 (112)	6.90 (66)	0.0001	0.0001
0900 h	631 ± 276	494 ± 224	244 ± 148	2.84 (113)	5.49 (67)	0.0005	0.0001
1000 h	936 ± 462	748 ± 294	330 ± 188	2.62 (112)	6.97 (65)	0.005	0.0001
1100 h	1035 ± 493	755 ± 367	355 ± 190	3.38 (112)	5.70 (65)	0.001	0.0001
1200 h	1067 ± 477	599 ± 369	279 ± 145	5.58 (108)	4.65 (62)	0.0001	0.0001
1300 h	1062 ± 460	514 ± 358	205 ± 105	6.75 (109)	4.79 (62)	0.0001	0.0001
1400 h	939 ± 389	338 ± 228	170 ± 85	10.18 (110)	4.05 (64)	0.0001	0.0001
1500 h	666 ± 249	203 ± 140	134 ± 61	12.58 (112)	2.70 (66)	0.0001	0.005
1600 h	436 ± 170	126 ± 83	77 ± 39	12.89 (111)	3.14 (65)	0.0001	0.0025

(iii) Ultraviolet radiation, UV-A

Hour	Mean ($W\ m^{-2}$) and SD			*t*-value (df)		*p* value	
	LERF (Auca)	LMCF (Dry)	LMCF (Wet)	LERF> LMCF (Dry)	LMCF(Dry)>LMCF (Wet)	LERF> LMCF (Dry)	LMCF(Dry)>LMCF (Wet)
0800 h	10.4 ± 4.1	4.2 ± 1.4	2.4 ± 0.9	11.66 (112)	6.23 (66)	0.0001	0.0001
0900 h	14.6 ± 4.7	8.5 ± 2.8	5.3 ± 2.5	8.46 (113)	4.72 (67)	0.0001	0.0001
1000 h	17.5 ± 5.2	13.8 ± 4.4	7.3 ± 4.1	3.85 (112)	5.93 (65)	0.0005	0.0001
1100 h	18.5 ± 4.8	15.7 ± 6.0	7.9 ± 4.0	2.28 (112)	6.00 (65)	0.02	0.0001
1200 h	18.6 ± 4.6	12.8 ± 6.4	6.5 ± 2.7	4.47 (108)	4.83 (62)	0.0001	0.0001
1300 h	18.8 ± 4.6	11.2 ± 6.2	5.0 ± 2.1	6.03 (109)	5.05 (62)	0.0001	0.0001
1400 h	17.9 ± 4.3	7.6 ± 4.1	4.1 ± 1.8	11.51 (110)	4.27 (64)	0.0001	0.0001
1500 h	14.8 ± 3.7	4.9 ± 3.0	3.2 ± 1.2	14.69 (112)	2.92 (66)	0.0001	0.0025
1600 h	9.8 ± 2.6	3.0 ± 1.6	1.8 ± 0.8	16.72 (111)	0.67 (65)	0.0001	>0.25

K↓ and *PPFD* data reproduced from Letts and Mulligan (2005), with editorial permission. SD, standard deviation. df, degrees of freedom.

Higher soil Ca was observed at Tambito than in either the Panamanian or Ecuadorian sites described above, but was lower than typically observed in LERF. Mean soil B content was 9.08 ppm at 5 cm depth in the primary forest, but only 0.55–1.33 ppm after disturbance (Table 49.1). The former value is nearly an order of magnitude higher than reported for a LERF on Barro Colorado Island, Panamá (0.94 ppm; John *et al.*, 2007).

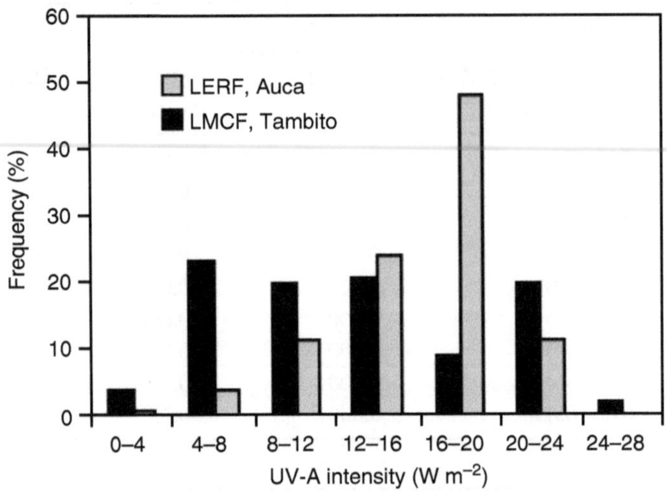

Figure 49.4. Frequency of occurrence of UV-A radiation intensity ranges in the lower montane cloud forest belt at Tambito, Cauca, Colombia (1450 m.a.s.l., LMCF) and in lowland evergreen rain forest at Auca, Ecuador (300 m.a.s.l., LERF).

A comparison of mass-based leaf nutrient concentrations between TMCF (Table 49.3) and LERF (Malhi *et al.*, 2009) reveals a slightly different pattern, with TMCF leaves showing similar or somewhat lower N, higher K, and similar Mg concentrations. Foliar P is highly variable in TMCF, ranging from less than 0.01% in nutrient-poor soils of Borneo (Kitayama and Aiba, 2002) to 0.22% in southern Ecuador (Soethe *et al.*, 2008). The variability of leaf nutrient content observed even within individual MCF sites (Hafkenscheid, 2000: Kitayama and Aiba, 2002) highlights the need for further collection and synthesis of nutrient and photosynthesis data from a range of MCF ecosystems differing in elevation, wetness, seasonality and substrate quality, to develop an improved understanding of photosynthetic nutrient-use strategy in response to environmental conditions.

Photosynthesis and leaf structure in a low-*PPFD* environment

The linear relationship between *PPFD* and net primary productivity in moist environments is well documented at the ecosystem

Table 49.5 *Photosynthetic gas-exchange characteristics of lower montane cloud forest vegetation at Tambito, including the* PPFD *intensity at which photosynthesis rates reached their asymptotic peak* ($PPFD_{sat}$)*, maximum photosynthetic rate* (A_{max})*, photosynthetic water-use efficiency* (WUE)*, stomatal conductance* (g_s)*, the ratio between internal and atmospheric* CO_2 *concentrations* (c_i/c_a)*, and the number of leaves sampled,* N

Plant type	Specimen and altitude	$PPFD_{sat}$ (μmol m^{-2}s^{-1})	A_{max} (μmol m^{-2}s^{-1})	WUE^a (mmol mol^{-1})	g_s (mmol m^{-2}s^{-1})	c_i/c_a	N
Tree	*Clusia* sp. 1480 m	600	8.8 ± 3.4	2.43 ± 1.43	323 ± 269	0.80 ± 0.13	30
	Clusia pentandra 2160 m	670	8.2 ± 0.5	5.88 ± 2.08	537 ± 394	0.80 ± 0.10	11
	Cecropia garciae 1445 m	950	8.7 ± 2.2	2.88 ± 1.08	296 ± 99	0.85 ± 0.02	30
	Cecropia bullata 2145 m	950	10.2 ± 2.2	6.13 ± 0.80	438 ± 197	0.83 ± 0.16	24
	Miconia sp. 1445 m	560	8.1 ± 2.0	3.89 ± 1.89	300 ± 341	0.77 ± 0.07	33
	Miconia sp. 2160 m	420	5.4 ± 1.2	3.89 ± 1.77	341 ± 32	0.86 ± 0.08	63
	Psychotria racemosa 1450 m	610	10.6 ± 1.8	4.63 ± 2.01	1327 ± 1187	0.89 ± 0.07	15
	Psychotria cuatrecasasii 2160 m	910	9.2 ± 2.2	5.06 ± 2.42	520 ± 438	0.81 ± 0.09	75
Understory	*Alloplectus teuscheri* 1600 m	370	6.6 ± 0.3	3.30 ± 1.65	421 ± 245	0.85 ± 0.10	8
	Alloplectus schultzei 2160 m	230	3.5 ± 0.5	3.47 ± 1.03	218 ± 119	0.89 ± 0.05	51
	Anthurium sp. 1450 m	600	9.3 ± 2.7	3.60 ± 1.37	442 ± 208	0.84 ± 0.04	178
	Anthurium sp. 2160 m	240	4.5 ± 1.1	6.63 ± 1.68	157 ± 100	0.81 ± 0.08	22
	Palm 1600 m	220	4.4 ± 0.5	5.53 ± 3.35	300 ± 141	0.88 ± 0.06	15
	Palm 2160 m	210	4.2 ± 1.5	5.34 ± 1.24	175 ± 168	0.85 ± 0.08	40

[a] $WUE = A_{max}/E$.

Source: Data from Letts and Mulligan (2005), with editorial permission.

level (e.g. Linder, 1985; Landsberg, 1986; Prince *et al.*, 1994). Graham *et al.* (2003) found *PPFD* limitation to be the main cause of reduced carbon uptake in the Panamanian LERF canopy tree *Luehea seemannii* during the rainy season, despite *PPFD* intensities that were 42–62% higher than at Tambito. Compared to the LERF at Auca, *PPFD* intensity was 44% lower at Tambito in the dry season and 74% lower during the wet season of November, 1998. Mean *PPFD* did not exceed saturating intensities ($PPFD_{sat}$) during the wet season and only exceeded $PPFD_{sat}$ from 10.00–12.00 h during the dry season (Tables 49.4 and 49.5). This suggests that all leaves at Tambito should experience *PPFD* limitation of photosynthesis, regardless of canopy position. Future studies of photosynthesis (*A*) in TMCF should be designed to compare the negative impacts of low *PPFD* and abaxial leaf wetness on *A*, against the benefits provided by diffuse radiation, such as improved light penetration into the canopy and a reduction of photo-inhibitory effects (Goudriaan, 1977; Roderick *et al.*, 2001; Reinhardt and Smith, 2008).

Recent work in temperate and sub-tropical cloud forests has revealed that the degree to which shoot- and canopy-scale photosynthesis is reduced by low *PPFD* depends on the nature of the cloud cover and the photosynthetic strategy of the dominant vegetation type. In Fraser fir (*Picea rubens*) of the southern Appalachians, daily mean *A* was only slightly lower at the shoot level on cloudy days than on sunny days, whereas *A* was 35% lower on foggy days. In a sub-tropical cloud forest in Taiwan, Mildenberger *et al.* (2009) found that a 64% reduction of *PPFD*, due to fog, resulted in only a slight decrease in net ecosystem exchange of CO_2 (*NEE*) in a native *Chamaecyparis* forest. However, *NEE* was reduced by up to 75% in a nearby plantation of the introduced species *Cryptomeria japonica* (Mildenberger *et al.*, 2009). The difference was attributed to the lower $PPFD_{sat}$ of *Chamaecyparis* and to its sunken stomata, which prevented wetting of the abaxial side of leaves during prolonged and dense fog (Mildenberger *et al.*, 2009).

In addition to the negative effect of persistent cloudiness on carbon gain, Williams *et al.* (1989) hypothesized that such conditions might create selective pressure promoting leaf longevity. Long-lived foliage is usually thicker and harder, with lower specific leaf area (*SLA*) due to the higher construction cost required to produce such leaves (Poorter, 1994). Leaves with low *SLA* tend to have lower mass-based leaf N concentration (Moser *et al.*, 2007) and lower A_{max}, because cell wall materials do not photosynthesize and tend to make leaves less permeable to gases (Shipley *et al.*, 2006). Overall, the vegetation at Tambito was characterized by relatively low *SLA* and A_{mass}, low A_{mass}: leaf N ratios (Figure 49.5), and relatively high water use efficiency (*WUE*), partly in response to low *VPD*. However, this does not preclude high *NUE* in terms of net lifetime carbon gain for a given N content, because of the higher leaf longevity of leaves with a low *SLA* (Reich *et al.*, 1991).

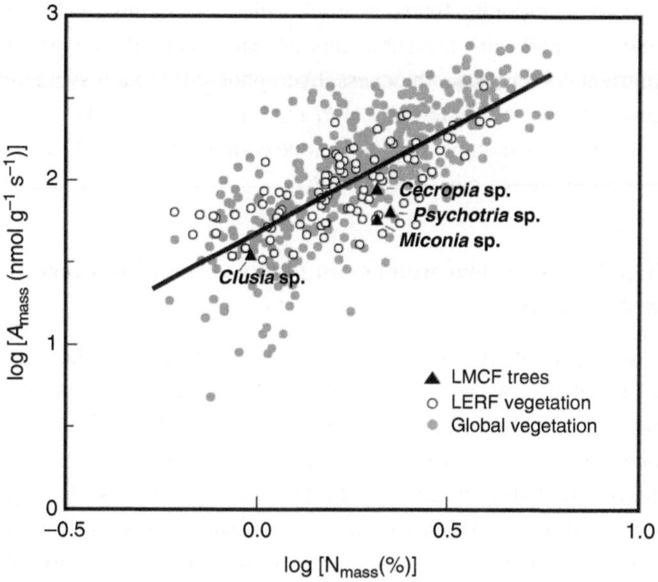

Figure 49.5. Relationship between mass-based leaf nitrogen concentration and light-saturated photosynthetic gas-exchange (A_{mass}) in a lower montane cloud forest at Tambito (LMCF), as compared with lowland evergreen rain forest (LERF) and global vegetation. Global data obtained with permission from GLOPNET (Wright *et al.*, 2004).

The prevalence of leaves with high longevity and low *SLA* in TMCF is consistent with the theory that long-lived, sclerophyllous leaves have evolved to maximize lifetime carbon gain and *NUE* in an environment strongly limited by low *PPFD*. However, low *SLA* may also reduce leaf area index (*LAI*), thereby limiting whole-plant carbon gain and asymptotic maximum tree stature (cf. Thomas and Bazzaz, 1999).

Consistent with the tendency toward greater sclerophylly at higher elevations, only a small decrease in leaf biomass was observed at the stand level in a transect from lower montane forest at 1000 m.a.s.l. to stunted elfin cloud forest at 3060 m.a.s.l. in southern Ecuador (Moser *et al.*, 2007). A 30% reduction in *SLA* partly compensated for lowered *LAI*. Along with the tendency toward more sclerophyllous foliage, leaf lifespan increased from 16 to 25 months with elevation (Moser *et al.*, 2007; cf. Velázquez-Rosas *et al.*, 2002). However, in contrast with the conditions at Tambito, mean *PPFD* levels in Bolivian cloud forests between 1850 and 3050 m.a.s.l. exceeded the 500 μmol m^{-2} s^{-1} $PPFD_{sat}$ required for canopy leaves for a majority of daytime hours (Gerold *et al.*, 2008), and large reductions in both forest stature and *SLA* were observed with altitude despite only a 17% reduction in *PPFD*. To develop a more complete understanding of MCF photosynthesis strategy, nutrient cycling, water use, and forest stature, there exists a need for integration of ecological (Moser *et al.*, 2007), soil (Schawe *et al.*, 2007), and hydrometeorological (Gerold *et al.*, 2008; Schawe *et al.*, this volume) transect studies (cf. Malhi *et al.*, 2006). Such studies should incorporate forest microclimatic monitoring (including leaf wetness) with measurements of leaf structural traits (*SLA*, nutrient content, size, thickness, hydrophobicity), photosynthetic gas-exchange characteristics ($PPFD_{sat}$, A_{max}, A, g_s, E, WUE, and c_i/c_a) and foliar stable carbon isotope composition, which serves as an integrative measure of *WUE*.

The impact of leaf wetness on photosynthesis and forest productivity

The impact of wetting on photosynthesis depends on which side of the leaf is wetted, because most TMCF leaves are hypostomatous (Dietz *et al.*, 2007). Brewer and Smith (1995) found that adaxial wetness stimulated photosynthesis in water-repellent leaves, probably by cooling the leaf surface. The rise in A_{max} observed upon adaxial wetting was statistically insignificant at Tambito, whereas A_{max} decreased strongly in response to full (abaxial and adaxial) leaf wetting. The reduction or elimination of photosynthesis in wettable species is likely due to the coverage of stomata by a thin film of water (e.g. Fogg, 1947; Smith and McClean, 1989), as may occur in species lacking hydrophobicity during prolonged fog interception events.

The degree to which leaf wetness may decrease productivity in MCF depends on: (i) the surface wetted (adaxial or abaxial), (ii) the frequency of wetness, and (iii) the proportion of the leaf surface wetted (Reinhardt and Smith, 2008; Mildenberger *et al.*, 2009). At Tambito, from 9 August to 9 September 1999, resistance-based wetness sensors recorded mean midday adaxial wetness near 12% (proportion of sensor wetted) within the upper canopy (10 m) at 1600 m.a.s.l., but over 50% at a height of 2.5–5.0 m above the ground (Letts and Mulligan, 2005). Resistance-based wetness sensors placed within the LMCF at 2.5 m above the ground recorded abaxial wetness above 40% at midday on days with cloud interception, although these sensors did not possess the hydrophobic water-shedding mechanisms or tilt angles of leaf surfaces and likely overestimated the degree of leaf wetness (cf. Sentelhas *et al.*, 2007; Reinhardt and Smith, 2008). Some leaves, such as those of *Clusia bullata*, were quite susceptible to abaxial wetting by wind-driven cloud at Tambito, whereas most others remained abaxially dry, regardless of the duration of the fog (M. G. Letts, personal observation; cf. Mildenberger *et al.*, 2009). On Sulawesi, Indonesia, Dietz *et al.* (2007) observed that abaxial leaf surfaces generally remained dry during rainfall in a tropical montane forest with a closed canopy at 32 m. However, in temperate montane ecosystems of the United States, Brewer and Smith (1995) found abaxial wetness after both rainfall and dewfall.

Frequent adaxial wetness can lead to the establishment of epiphyllic organisms on long-lived foliage. These organisms compete for light with the leaf (Dietz *et al.*, 2007), but are often capable of photosynthesis themselves, thereby reducing the impact of leaf wetness on tree productivity. Furthermore, the higher proportional wetness of adaxial surfaces indicates that only a small increase in A_{max} from leaf cooling (Brewer and Smith, 1995) would be required to offset stomatal limitation resulting from abaxial wetness. Given the remaining uncertainty regarding the impact of leaf wetness on leaf, shoot and stand-level photosynthesis, there is a need for: (i) an improved understanding of the process of adaxial and abaxial leaf wetting by cloud interception and its subsequent evaporation, and (ii) the development of improved techniques to monitor leaf wetness in MCF (cf. Reinhardt and Smith, 2008; Mulligan *et al.*, this volume #25).

CONCLUSIONS

Of the major soil nutrients (N, P, K, Ca, and Mg), only Ca was found at lower concentrations at Tambito than in LERF. No evidence was found of soil dryness or waterlogging, and soils were only moderately acidic, in the pH 3.6–5.7 range. Although fertilization studies have not been carried out in the study region, leaf nutrient concentrations were similar to those found in other MCFs and were within the middle portion of the range observed in LERF. This suggests that the relative importance of nutrient deficiency as a control on photosynthetic productivity may be comparatively low compared to other environmental controls in the LMCF belt at Tambito. Whereas evidence of greater nutrient limitation than LERF was sparse, the persistence of low cloud was shown to be limiting to photosynthesis. Experimental wetting of the abaxial surface of LMCF leaves, as may occur in hydrophilic foliage during prolonged *CWI* events, was shown to reduce A_{max}. More importantly, *PPFD* intensity did not reach saturating levels for photosynthesis during the wet season, and only reached $PPFD_{sat}$ for about 3 hours day^{-1} during the dry season. This may contribute to lower forest productivity and to selective pressures favoring leaf longevity and sclerophylly. Where *PPFD* is frequently below saturating levels during daylight hours throughout the year, the benefit of increasing leaf lifespan conferred by low *SLA* may outweigh the cost of lower A_{max}, thereby maximizing net leaf carbon gain. The results of this study highlight the importance of cloudiness as a key factor contributing to lower photosynthetic rate in TMCFs of the Chocó climate region.

ACKNOWLEDGEMENTS

The authors thank Fundación Proselva for permission to carry out research at Tambito. Funding was provided by the Fonds pour la formation des chercheurs et l'aide à la recherche (FCAR

Québec, MGL), the University of London Central Research Fund (MGL), the Royal Society (MM), King's College London (MGL, MM), and the British Council LINK programme (MM). Universidad del Cauca and Corporación Autónoma del Cauca provided logistic support in the field. Nutrient analyses were performed by Dr. Octavio Mosquera (CIAT). Mauricio Larrea maintained the weather station at Auca. We are also indebted to forest rangers of the Ministerio del Medio Ambiente for their expertise, and to the people of Juntas and Huisitó, Municipality of El Tambo for their hospitality. Two anonymous reviewers are thanked for useful comments on the manuscript.

REFERENCES

Aerts, R., and M. J. Van der Peijl (1993). A simple model to explain the dominance of low productive perennials in nutrient-poor habitats. *Oikos* **66**: 144–147.

Bautista-Cruz, A., and R. F. del Castillo (2005). Soil changes during secondary succession in a tropical montane cloud forest area. *Soil Science Society of America Journal* **69**: 906–914.

Benton, J. B., and J. Jones (1989). *Plant Analysis Techniques*. Athens, GA: Benton–Jones Laboratories.

Boyce, R. L., D. C. McCune, and G. P. Berlyn (1991). A comparison of foliar wettability of red spruce and balsam fir growing at high elevation. *New Phytologist* **117**: 543–555.

Brady, N. C., and R. R. Weil (1999). *The Nature and Properties of Soils*, 12th edn. Upper Saddle River, NJ: Prentice Hall.

Bremmer, J. M., and C. S. Mulvaney. (1982). Nitrogen-total. In *Methods of Soil Analysis, Part 2, Chemical and Microbiological Properties*, eds. A. L. Page, R. H. Miller, and D. R. Keeney, pp. 595–624. Madison, WI: American Society of Agronomy and Soil Science Society of America.

Brewer, C. A., and W. K. Smith (1995). Leaf surface wetness and gas exchange in the pond lily *Nuphar polysepalum* (Nymphaeaceae). *American Journal of Botany* **82**: 1271–1277.

Bruijnzeel, L. A., and J. Proctor (1995). Hydrology and biogeochemistry of tropical montane cloud forests: what do we really know? In *Tropical Montane Cloud Forests*, eds. L. S. Hamilton, J. O. Juvik, and F. N. Scatena, pp. 38–78. New York: Springer-Verlag.

Bruijnzeel, L. A., and E. J. Veneklaas (1998). Climatic conditions and tropical montane forest productivity: the fog has not lifted yet. *Ecology* **79**: 3–9.

Bruijnzeel, L. A., M. J. Waterloo, J. Proctor, A. T. Kuiters, and B. Kotterink (1993). Hydrological observations in montane rain forests on Gunung Silam, Sabah, Malaysia, with special reference to the '*Massenerhebung*' effect. *Journal of Ecology* **81**: 145–167.

Cavelier, J. (1992). Fine-root biomass and soil properties in a semideciduous and a lower montane rain forest in Panama. *Plant and Soil* **142**: 187–201.

Cavelier, J., and G. Goldstein (1989). Leaf anatomy and water relations in tropical elfin cloud forest tree species. In *Structural and Functional Responses to Environmental Stresses*, eds. K. H. Kreeb, H. Richter, and T. M. Hinckley, pp. 243–253. The Hague: SPB Academic Publishing.

Cavelier, J., and C. Mejía (1990). Climatic factors and tree stature in the elfin cloud forest of Serrania de Macuira, Colombia. *Agricultural and Forest Meteorology* **53**: 105–123.

Delta-T Devices (1999). *Theta Probe Soil Moisture Sensor Type ML2x: User's Manual*. Cambridge, UK: Delta-T Devices.

Dietz, J., Ch. Leuschner, D. Hölscher, and H. Kreilein (2007). Vertical patterns and duration of surface wetness in an old-growth tropical montane forest, Indonesia. *Flora* **202**: 111–117.

Flenley, J. R. (1995). Cloud forest, the *Massenerhebung* effect, and ultraviolet insolation. In *Tropical Montane Cloud Forests*, eds. L. S. Hamilton, J. O. Juvik, and F. N. Scatena, pp. 150–155. New York: Springer-Verlag.

Fogg, G. E. (1947). Quantitative studies on the wetting of leaves by water. *Proceedings of the Royal Society of London Series B* **134**: 503–522.

Gerold, G., M. Schawe, and K. Bach (2008). Hydrometeorologic, pedologic, and vegetation patterns along an elevational transect in the montane forest of the Bolivian Yungas. *Die Erde* **139**: 141–168.

Gerrish G., D. Mueller-Dombois, and K. W. Bridges (1988). Nutrient limitation and *Metrosideros* forest dieback in Hawaii. *Ecology* **69**: 723–727.

Goldberg, S. (1997). Reactions of boron with soils. *Plant and Soil* **193**: 35–48.

Goudriaan, J. (1977). *Crop Micrometeorology: A Simulation Study*. Wageningen, the Netherlands: PUDOC.

Graham, E. A., S. S. Mulkey, K. Kitajima, N. G. Phillips, and S. J. Wright (2003). Cloud cover limits net CO_2 uptake and growth of a rainforest tree during tropical rainy seasons. *Proceedings of the National Academy of Sciences USA* **100**: 572–576.

Grubb, P. J. (1977). Control of forest growth and distribution on wet tropical mountains: with special reference to mineral nutrition. *Annual Review of Ecology and Systematics* **8**: 83–107.

Hafkenscheid, R. L. L. J. (2000). Hydrology and biogeochemistry of tropical montane rain forests of contrasting stature in the Blue Mountains, Jamaica. Ph.D. thesis, VU University Amsterdam, Amsterdam, the Netherlands. Also available at http://dare.ubvu.vu.nl/bitstream/1871/12734 /1/tekst.pdf.

Herbert, D. A., and J. H. Fownes (1999). Forest productivity and efficiency of resource use across a chronosequence of tropical montane soils. *Ecosystems* **2**: 242–254.

Holwerda, F. (2005). Water and energy budgets of rain forests along an elevation gradient under maritime tropical conditions. Ph.D. thesis, VU University Amsterdam, Amsterdam, the Netherlands. Also available at www.falw.vu.nl/nl/onderzoek/earth-sciences/geo-environmental-science-and-hydrology/hydrology-dissertations/index.asp.

John, R., J. W. Dalling, K. E. Harms, *et al.* (2007). Soil nutrients influence spatial distributions of tropical tree species. *Proceedings of the National Academy of Sciences of the USA* **104**: 864–869.

Jordan, C. F. (1985). *Nutrient Cycling in Tropical Forest Ecosystems*. Chichester, UK: John Wiley.

Kerr, A. C., J. Tarney, A. Nivia, G. F. Marriner, and A. D. Saunders (1998). The internal structure of oceanic plateaus: Inferences from obducted Cretaceous terranes in western Colombia and the Caribbean. *Tectonophysics* **292**: 173–188.

Kitayama, K., and S. I. Aiba (2002). Ecosystem structure and productivity of tropical rain forests along altitudinal gradients with contrasting soil phosphorus pools on Mount Kinabalu, Borneo. *Journal of Ecology* **90**: 37–51.

Lai, I. L., W. H. Schröder, J. T. Wu, *et al.* (2007). Can fog contribute to the nutrition of *Chamaecyparis obtusa* var. *formosana*? Uptake of a fog solute tracer into foliage and transport to roots. *Tree Physiology* **27**: 1001–1009.

Landsberg, J. J. (1986). *Physiological Ecology of Forest Production*. New York: Academic Press.

Letts, M. G. (2003). Carbon assimilation and productivity in a north-west Andean tropical montane cloud forest. Ph.D. thesis, King's College London, University of London, London, UK. Also available at http://people.uleth.ca/~matthew.letts/letts thesis 2003.pdf.

Letts, M. G., and M. Mulligan (2005). The impact of light quality and leaf wetness on photosynthesis in north-west Andean tropical montane cloud forest. *Journal of Tropical Ecology* **21**: 549–557.

Leuschner, Ch., G. Moser, C. Bertsch, M. Röderstein, and D. Hertel (2007). Large altitudinal increase in tree root/shoot ratio in tropical mountain forests of Ecuador. *Basic and Applied Ecology* **8**: 219–230.

Linder, S. (1985). Potential and actual production in Australian forest stands. In *Research for Forest Management*, eds. J. J. Lansbery and W. Parsons, pp. 11–35. Melbourne, Australia: CSIRO.

Mackean, S. (1993). *Manual de análisis de suelos y plantas*. Santiago de Cali, Colombia: Centro Internacional de Agricultura Tropical (CIAT).

Malhi, Y., P. Meir, M. Bird, N. Salinas, and M. Silman (2006). *Detailed Assessment of Ecosystem Carbon Dynamics along a Tropical Forest Altitudinal Gradient*, research proposal to the National Environment Research Council of the UK. Oxford, UK: University of Oxford.

Malhi, Y., L. E. O. C. Aragão, D. B. Metcalfe, *et al.* (2009). Comprehensive assessment of carbon productivity, allocation and storage in three Amazonian forests. *Global Change Biology* **15**: 1255–1274.

Medina, E., E. Cuevas, and P. L. Weaver (1981). Composición foliar y transpiración de especies lenosas de Pico del Este, Sierra de Luquillo, Puerto Rico. *Acta Científica Venezolana* **32**: 159–165.

Mildenberger, K., E. Beiderwieden, Y. J. Hsia, and O. Klemm (2009). CO_2 and water vapor fluxes above a subtropical mountain cloud forest: the effect of light conditions and fog. *Agricultural and Forest Meteorology*, doi:10.1016/j.agrformet.2009.06.004

Millward, D., G. F. Marriner, and A. D. Saunders (1984). Cretaceous tholeiitic volcanic rocks form the western Cordillera of Colombia. *Journal of the Geological Society* **141**: 847–860.

Miyazawa, M., M. A. Pavan, T. Muraoka, C. A. F. S. Carmo, and W. J. Mello (1999). Análise química de tecidos vegetais. In *Manual de análise química de solos, plantas e fertilizantes*, ed. F. C. Silva, pp. 172–223. Brasília, Brasil: EMBRAPA.

Moser, G., D. Hertel, and Ch. Leuschner (2007). Altitudinal change in LAI and stand leaf biomass in tropical montane forests: a transect study in Ecuador and a pan-tropical meta-analysis. *Ecosystems* **10**: 924–935.

Murphy, J., and J. P. Riley (1965). A modified single solution method for estimation of phosphate in natural maters, *Analitica Chimica Acta* **27**: 3–36.

Poorter, H. (1994). Construction costs of biomass: a whole plant perspective. In *A Whole-Plant Perspective on Carbon–Nitrogen Interactions*, eds. J. Roy and E. Garnier, pp. 111–127. The Hague: SPB Academic Publishing.

Poorter, H., and J. R. Evans (1998). Photosynthetic nitrogen-use efficiency of species that differ inherently in specific leaf area. *Oecologia* **116**: 26–37.

Poorter, H., and C. Remkes (1990). Leaf area ratio and net assimilation rate of 24 wild species differing in relative growth rate. *Oecologia* **83**: 553–559.

Prince, S. D., C. O. Justice, and B. Moore III (1994). *Monitoring and Modeling of Terrestrial Net and Gross Primary Production*, IGBP-DIS Working Paper No. 8. College Park, MD: University of Maryland.

Reich, P. B., C. Uhl, M. B. Walters, and D. S. Ellsworth (1991). Leaf life-span as a determinant of leaf structure and function among 23 Amazonian tree species. *Oecologia* **86**: 16–24.

Reich, P. B., D. S. Ellsworth, M. B. Walters, *et al.* (1999). Generality of leaf trait relationships: a test across six biomes. *Ecology* **80**: 1955–1969.

Reinhardt, K., and W. K. Smith (2008). Impacts of cloud immersion on microclimate, photosynthesis and water relations of *Abies fraseri* Pursh. Poiret in a temperate mountain cloud forest. *Oecologia* **158**: 229–238.

Rincón-Romero, M. E. (2000). Modelling hydrological sensitivity to land-use change in tropical mountainous environments. Ph.D. thesis, King's College, University of London, London, UK.

Roderick, M. L., G. D. Farquhar, S. Berry, and I. R. Noble (2001). On the direct effect of clouds and atmospheric particles on the productivity and structure of vegetation. *Oecologia* **129**: 21–30

Schawe, M., S. Glatzel, and G. Gerold (2007). Soil development along an altitudinal transect in a Bolivian tropical montane rainforest: podzolization vs. hydromorphy. *Catena* **69**: 83–90.

Sentelhas, P. C., T. J. Gillespie, and E. A. Santos (2007). Leaf wetness duration measurement: comparison of cylindrical and flat plate sensors under different field conditions. *International Journal of Biometeorology* **51**: 265–273.

Shipley B., M. J. Lechowicz, I. Wright, and P. B. Reich (2006). Fundamental trade-offs generating the worldwide leaf economics spectrum. *Ecology* **87**: 535–541.

Shorrocks, V. M. (1997). The occurrence and correction of boron deficiency. *Plant and Soil* **193**: 121–148.

Smith, W. K., and T. M. McClean (1989). Adaptive relationship between leaf water repellency, stomatal distribution and gas exchange. *American Journal of Botany* **76**: 465–469.

Soethe, N., W. Wilcke, J. Homeier, J. Lehmann, and C. Engels (2008). Plant growth along the altitudinal gradient: role of plant nutritional status, fine root activity, and soil properties. In *Gradients in a Tropical Mountain Ecosystem of Ecuador*, eds. E. Beck, J. Bendix, I. Kottke, F. Makeschin, and R. Mosandl, pp. 259–266. Berlin: Springer-Verlag.

Tanner, E. V. J., P. M. Vitousek, and E. Cuevas (1998). Experimental investigation of nutrient limitation of forest growth on wet tropical mountains. *Ecology* **79**: 10–22.

Thomas, S. C., and F. A. Bazzaz (1999). Asymptotic height as a predictor of photosynthetic characteristics in Malaysian rain forest trees. *Ecology* **80**: 1607–1622.

Thornley, J. H. M., and I. R. Johnson (1990). *Plant and Crop Modelling: A Mathematical Approach to Plant and Crop Physiology*. Oxford, UK: Oxford University Press.

Valladares, F., S. J. Wright, E. Lasso, K. Kitajima, and R. W. Pearcy (2000). Plastic phenotypic response to light of 16 congeneric shrubs from a Panamanian rainforest. *Ecology* **81**: 1925–1936.

Vázquez, J. A., and T. J. Givnish (1998). Altitudinal gradients in tropical forest composition, structure and diversity in the Sierra de Manantlan. *Journal of Ecology* **86**: 999–1020.

Velázquez-Rosas, N., J. Meave, and S. Vázquez-Santana (2002). Elevational variation of leaf traits in montane rain forest tree species at La Chinantla, southern Mexico. *Biotropica* **34**: 534–546.

Vitousek, P. M., and R. L. Sanford (1986). Nutrient cycling in moist tropical forest. *Annual Review of Ecology and Systematics* **17**: 137–167.

Waide, R. B., J. K. Zimmerman, and F. N. Scatena (1998). Controls of primary productivity: lessons from the Luquillo Mountains in Puerto Rico. *Ecology* **79**: 31–37.

Weaver, P. L., and P. G. Murphy (1990). Structure and productivity in Puerto Rico's Luquillo Mountains. *Biotropica* **22**: 69–82.

Wilcke, W., S. Yasin, C. Valarezo, and W. Zech (2001). Change in water quality during the passage through a tropical montane rain forest in Ecuador. *Biogeochemistry* **55**: 45–72.

Williams, K., C. B. Field, and H. A. Mooney (1989). Relationships among leaf construction cost, leaf longevity and light environment in rain forest plants of the genus *Piper*. *American Naturalist* **133**: 198–211.

Wright, I. J., P. B. Reich, W. Westoby, *et al.* (2004). The worldwide leaf economics spectrum. *Nature* **428**: 821–827.

50 Comparative water budgets of a lower and an upper montane cloud forest in the Wet Tropics of northern Australia

D. L. McJannet
CSIRO Land and Water, Indooroopilly, Queensland, Australia

J. S. Wallace
CSIRO Land and Water, Townsville, Queensland, Australia

P. Reddell
CSIRO Land and Water, Atherton, Queensland, Australia

ABSTRACT

This chapter presents a comparison of the water budgets of a lower and an upper montane cloud forest in Australia's Wet Tropics region based on field measurements of rainfall, throughfall (*TF*), stemflow (*SF*), transpiration, and cloud water interception (*CWI*). The proportions of total precipitation measured as *TF* and *SF* varied between the two sites. At the lower montane cloud forest site (Upper Barron), *SF* was 7% while *TF* was just 64%. The upper montane cloud forest site (Bellenden Ker) showed higher *SF* at 10% while *TF* was also high at 83%. *CWI* at the two sites was quantified using a wet-canopy water balance methodology and was found to contribute up to 65% of the monthly water input during dry season months. At Upper Barron, *CWI* was 19% of average annual precipitation while at Bellenden Ker it was as much as 29%. These measurements resulted in an overall canopy interception evaporation loss of 29% at Upper Barron and just 7% at Bellenden Ker. About 20% of total precipitation was lost through transpiration at Upper Barron and just 5% at Bellenden Ker. Transpiration losses were less than water losses through wet-canopy evaporation at both sites. Both sites have a large annual net surplus of water to sustain streamflow and groundwater recharge. Total evaporation losses at Upper Barron accounted for about 50% of total precipitation input, leaving 50% for runoff and groundwater recharge. At Bellenden Ker the exceptionally wet conditions resulted in evaporation totalling just 13% of total precipitation, with 87% for runoff and recharge. The seasonality of water balance components is examined and the results are discussed in light of future climate change issues.

INTRODUCTION

The vast majority of the tropical rain forests of Australia are found in the Wet Tropics World Heritage Area, which is located in north Queensland. The Wet Tropics region covers an area of about 9000 km^2 on the coastal strip adjacent to the Great Barrier Reef. This region has high economic importance (AUD \$2 billion per year from tourism alone; OESR, 2002) and exceptional environmental value, and 64% of the area is listed as World Heritage National Park (WTMA, 2007). Most watersheds in the region comprise complex mosaics of different land uses and vegetation, with rain forests frequently making up a significant proportion of the vegetation cover.

Despite the economic and environmental importance of this region, there is little quantitative information on the hydrology of the region's different forest types, or of their hydrological characteristics compared to similar areas under agricultural forms of land use. Consequently, the effects of changes in vegetation type and cover on watershed hydrological response

Tropical Montane Cloud Forests: Science for Conservation and Management, eds. L. A. Bruijnzeel, F. N. Scatena, and L. S. Hamilton. Published by Cambridge University Press.

Table 50.1 *Site characteristics for the Upper Barron (lower montane cloud forest) and Bellenden Ker (upper montane cloud forest) field sites*

	Upper Barron	Bellenden Ker
Measurement duration	13/09/2003–30/06/2005	05/06/2004–30/06/2005
Altitude	1050 m	1560 m
Latitude	17°27.1	17°16.0 S
Longitude	145°29.7	145°51.0 E
Geology	Basalt	Granite
Forest type[a]	CNVF	SMVFT
LAI	4.1 m^2 m^{-2}	3.3 m^2 m^{-2}
Canopy gap	4.0%	3.5%
Stem density	925 stems ha^{-1}	2019 stems ha^{-1}
Basal area	69 m^2 ha^{-1}	74 m^2 ha^{-1}
Annual average precipitation (mm)[b]	2983	6434
Long-term annual precipitation (mm)	–	8100
Annual average ET_0 (FAO-56) (mm)[b]	1087	910
Annual average E_0 (Penman) (mm)[b]	1237	1,073
Average daily solar radiation (MJ m^{-2})	17.2	12.6
Average daily temperature (°C)	18.3	14.2
Average daily relative humidity (%)	93.9	87.8
Average daily wind speed (m s^{-1})	2.9	3.2

[a] CNVF, Complex Notophyll Vine Forest; SMVFT, Simple Microphyll Vine Fern Thicket.
[b] Annual precipitation and potential evaporation are calculated from the total of monthly averages for the observation period at each site.

to rainfall, water yield, and water quality from these humid tropical landscapes are poorly understood. Such knowledge is vital in developing a predictive understanding of the effects of possible land-use and climate changes on water resources in the region.

Information on the water budget of Australian tropical rain forests is limited to only a handful of studies, notably that of Hutley *et al.* (1997) who measured water balance components for one year in a sub-tropical upland forest subject to fog in south-east Queensland, and a study by Gilmour (1975) who undertook measurement of water balance components in a coastal lowland rain forest watershed in north Queensland over a 4-year period. These studies form a good basis for understanding the water balance of Australian tropical rain forests. However, it is likely that different rain forest types will exhibit different water budgets due to the strong rainfall gradients and distinct forest structural differences (Tracey, 1982). Therefore, further water balance studies are needed.

In response to these knowledge gaps, a series of measurement sites was established across the Wet Tropics for monitoring their water balance components (McJannet *et al.*, 2007a,b). Sites were selected to represent different forest types subjected to different climatic, altitudinal and geological influences. This chapter presents annual and seasonal water budget results for two of these sites; a lower montane cloud forest (LMCF) and an upper montane cloud forest (UMCF).

METHODS AND MATERIALS

Site characteristics

Two field locations were chosen for this study; an LMCF site called Upper Barron (UB) and an UMCF site called Bellenden Ker (BK, Figure 50.1). Both locations experience distinct wet and dry season rainfall patterns and receive more that 50% of their annual rainfall in January, February, and March (Table 50.1). The UB site is located in the Longlands Gap State Forest on the Atherton Tablelands at an elevation of 1050 m.a.s.l. Further basic climatic information is given in Table 50.1. The BK site in the Wooroonooran National Park is positioned on top of the second highest mountain in Queensland at an elevation of 1560 m.a.s.l. UB has an *LAI* of 4.1 and a canopy height of 25 m while BK has a leaf area index (*LAI*) of 3.3 and a canopy at 8 m. According to the Australian tropical forest classification system of Tracey (1982), UB is a Complex Notophyll Vine Forest and BK a Simple Microphyll Vine Fern Thicket. Both sites have similar basal areas (Table 50.1), although stem density at UB (925 stems ha^{-1}) is much lower than at BK (2019 stems ha^{-1}).

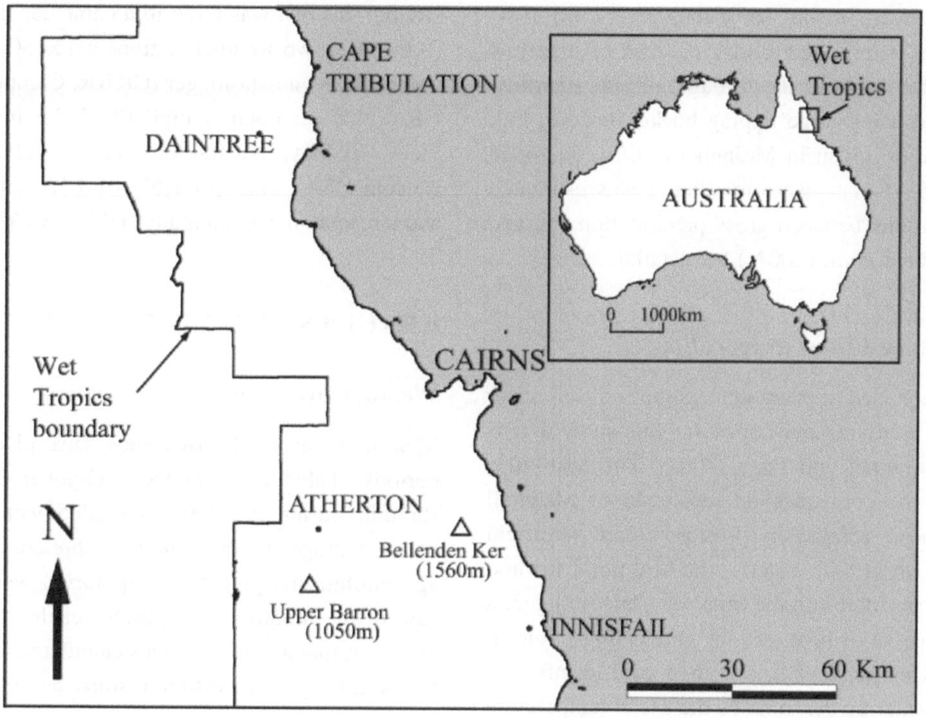

Figure 50.1. The Australian Wet Tropics and present field site locations (open triangles).

The soils at UB are derived from basalt and are quite deep (> 12m) whereas at BK, soils are derived from granite and much rockier.

Cloud forest water balance

The basic water balance of a cloud forest can be expressed as:

$$(P_g + P_c) = I + T + E_s + Q + D + \Delta S \tag{50.1}$$

where ($P_g + P_c$) represent precipitation inputs in the form of "ordinary" rainfall (P_g) and "cloud interception" (P_c), respectively. In this chapter, P_c includes horizontal (wind-driven) rain. I is canopy interception (wet-canopy evaporation) which is calculated as the difference between precipitation inputs ($P_g + P_c$, see next paragraph for details) and measurements of throughfall (*TF*) and stemflow (*SF*) combined. T is transpiration of the forest and E_s is forest floor evaporation. Q and D represent streamflow and groundwater recharge, whereas over annual periods the change in soil water storage (ΔS) is considered to be small. The sum of Q and D is then given by the difference between the amount of water entering the soil ($SF + TF$) and evapotranspiration ($T + E_s$).

Rainfall and cloud water interception

Total precipitation inputs at each site were determined from the combination of rainfall measurements (P_g) and cloud water interception (P_c) estimates. Half-hourly rainfall (P_g) was first corrected for wind losses due to turbulence around the gage using concurrent wind speed data as described by Førland *et al.* (1996). Allowances were then made for wind-blown (near-horizontal) rain being intercepted on sloping land (Sharon, 1980). This correction utilized half-hourly rainfall, wind speed, and wind direction data along with the slope and aspect of each site in the trigonometric model described by Holwerda (2005). Wind loss corrections added approximately 3% and 2% to the measured rainfall values at UB and BK, respectively. Corrections for slope interception of near-horizontal rain added 6% and 8%, respectively (McJannet *et al.*, 2007c). Rainfall totals that had both these corrections applied are referred to in the text as P_{ga}.

Canopy interception

Estimates of P_c at each site were made using the wet-canopy water balance methodology as described in McJannet *et al.* (2007b). This uses measurements of P_{ga}, throughfall (*TF*), and stemflow (*SF*) (which are described in the section below) and only relies on a fog gage (Juvik and Ekern, 1978; cf. Frumau *et al.*, this volume) to define the occurrence of fog rather than the absolute amount (see McJannet *et al.*, 2007b for details). This indirect use of the fog gage overcomes any problems associated with how well the gage mimics the cloud interception capacity of the forest (cf. Giambelluca *et al.*, this volume; Tanaka *et al.*, this volume).

Wet-canopy evaporation or canopy interception (I) in this study was determined as the difference between precipitation ($P_{ga} + P_c$) and throughfall (*TF*) plus stemflow (*SF*).

Measurements of *TF* were made using long (6–12 m) PVC troughs while (*SF*) was determined using a network of stemflow collars. Water collected by both the throughfall and stemflow systems was measured using large tipping-bucket devices. Full technical descriptions are given in McJannet *et al.* (2007b). In periods when *TF* and *SF* data were missing, *I* was estimated using regression equations between gross precipitation P_{ga} and ($SF + TF$) (see McJannet *et al.* (2007c) for details).

Transpiration and forest floor evaporation

Heat-pulse sensors were used to measure transpiration (soil water uptake). The system used was developed for this study and is fully described in McJannet and Fitch (2004). The heat-pulse method is based on the compensation technique of Marshall (1958) which has been refined by Swanson and Whitfield (1981) and Hatton *et al.* (1990). Briefly, the heat-pulse method uses a heating probe, inserted into the trunk of a tree, to inject a short (usually <2 s) pulse of heat into the conducting tissue of the tree (sapwood). This pulse of heat is then used to infer the rate of movement of water up the trunk by the use of temperature sensing probes inserted above and below the heating probe. This monitoring arrangement compensates for the process of heat diffusion and allows heat convection to be isolated and measured.

Heat-pulse sensors were installed in nine trees at UB and in eight trees at BK. The sapwood depth of each tree was determined from increment cores, and heat-pulse sensors were spaced at approximately even depth intervals in the sapwood to account for radial variation in sap velocity (cf. García-Santos, 2007). Trees larger than 50 cm diameter at breast height (DBH) were fitted with four probe sets whereas smaller trees were given two probe sets. This procedure was followed to ensure more accurate estimates of transpiration from the larger trees which dominate overall forest water use (McJannet *et al.*, 2007a).

For subsequent monthly and seasonal transpiration analysis, missing daily transpiration values were estimated from the relationship between reference evaporation (ET_0; Allen *et al.*, 1998) (or total solar radiation when ET_0 data were not available), and transpiration (see McJannet *et al.* (2007a) for details).

Evaporation from the forest floor (E_s) was not measured because of its small contribution. A value of 0.2 mm day^{-1}, which is the average of results from other studies (Jordan and Heuveldop, 1981; Kelliher *et al.*, 1992) was used instead.

Weather Measurements

At UB an Automatic Weather Station (AWS) (Environdata, Warwick, Australia) was installed at a distance of 50 m from the forest edge at a height of 2 m. A similar AWS was installed at BK but this one was fixed to a cable car tower at canopy height (8 m). The two weather stations made 30-min measurements as controlled by a data-logger (CR10x, Campbell Scientific, Logan, USA) that recorded rainfall (TB3 0.5 mm, Hydrological Services, Sydney, Australia), solar radiation (Li 200x, LiCor, Lincoln, USA), wind speed (6504-FS, Unidata, Willetton, Australia) and temperature and humidity (HMP45C, Vaisala, Finland).

RESULTS

Weather overview

Measurements at the two sites took place over varying time periods (Table 50.1) between October 2003 and June 2005. Rainfall during the 2004 and 2005 wet seasons was slightly below average. To facilitate inter-site comparisons, annual average rainfall and potential evaporation were estimated from the sum of the monthly averages for each site. It should be noted, however, that annual averages calculated in this way are affected by the timing of the different study periods for each site.

Stemflow and throughfall

Figure 50.2 shows the relationship between P_{ga} and *SF* and *TF* for UB and BK, distinguishing between days with and without cloud water interception P_c. The power curves fitted to the data in Figure 50.2 illustrate the strength of the relationships. For example, at UB, the coefficient of determination (r^2) for the relationship between between P_{ga} and *SF* on days without P_c was 0.94 (Figure 50.2a) vs. 0.99 for that linking P_{ga} and *TF* (Figure 50.2c). Similarly high correlations between P_{ga} and *TF* ($r^2 = 0.99$) and between P_{ga} and *SF* ($r^2 = 0.97$) were observed for BK on days without P_c. However, for days where P_c did occur, the relationships between P_{ga} and *TF* and *SF* were weaker ($r^2 = 0.78$–0.93) because measured rainfall did not account for all of the water input (Figure 50.2).

Cloud water interception

Over the study period 19% of total precipitation input ($P_{ga} + P_c$) at the UB site was from P_c vs. 29% at BK (Figure 50.3). The percentage contribution of P_c during each month showed a distinct seasonal trend, and reached 65% in the dry season months (May–October) at UB and 55% at BK. This observation is linked to the variation in rainfall quantity and intensity across the seasons. During the early wet season (November–December), rain falls in short high-intensity "build up" storms. Later in the wet season (January–March) large-scale monsoonal depressions dominate and rain falls in much larger quantities. During the drier months rainfall is less intensive and often characterized by long periods of light rain and drizzle (Gilmour, 1975).

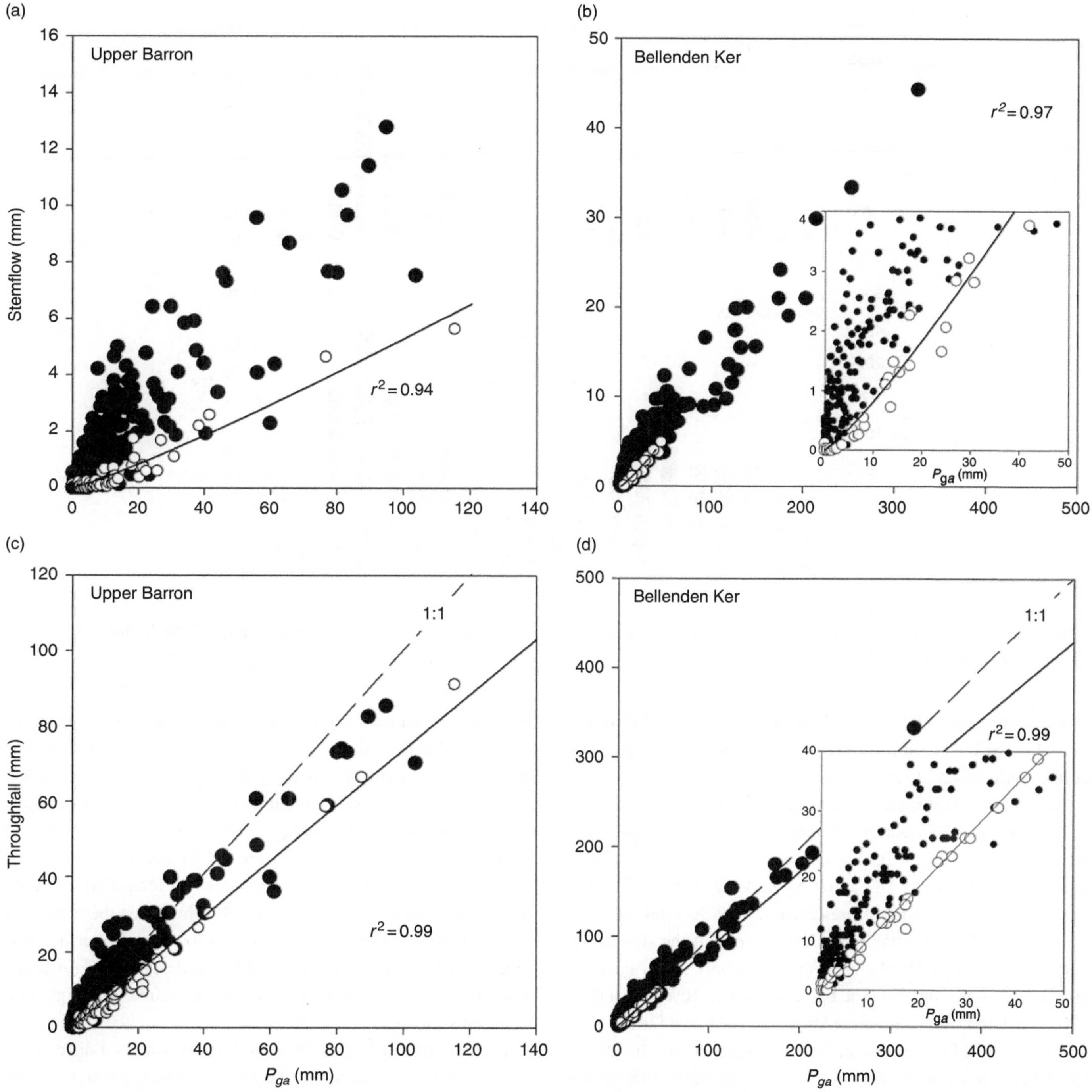

Figure 50.2. Relationships between stemflow and rainfall (corrected for wind losses and topographic effects, P_{ga}), and throughfall and rainfall at Upper Barron and Bellenden Ker. Days with and without cloud water interception shown as solid and open circles, respectively. Coefficients of determination pertain to days without P_c. Inset for Bellenden Ker figures shows details for low rainfall region of graph.

Transpiration

Daily transpiration (T) at each of the two sites was determined by scaling measurements from sample trees to the plot level using the strong relationships exhibited between tree diameter and daily water use (Figure 50.4). Relationships such as that shown in Figure 50.4 were developed on a daily basis for each site and used to estimate the daily water use of all trees on a 1600 m^2 plot. It is interesting to note that such a strong relationship between tree size and daily water use exists despite the fact that nearly all sampled trees belonged to different species.

Both sites showed distinct seasonal trends in T (Figure 50.5). At UB the highest rates tended to occur around October and November vs. September at BK. Over the entire study period

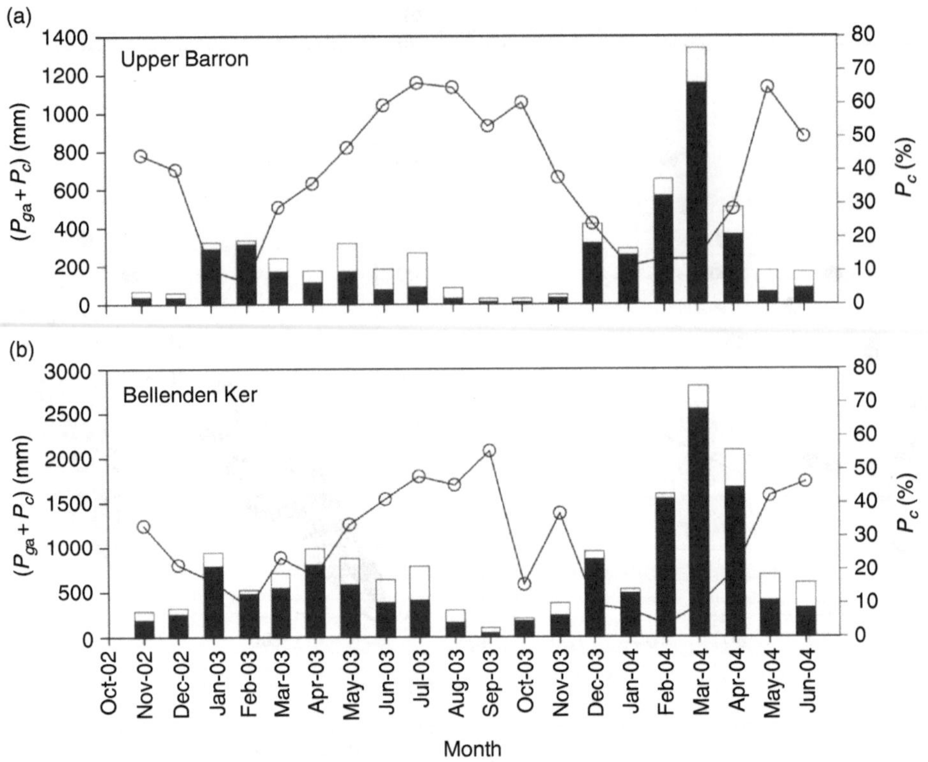

Figure 50.3. Monthly rainfall (corrected for wind and topographic effects, P_{ga}) and cloud water interception (P_c) at (a) Upper Barron and (b) Bellenden Ker. Corresponding contribution of cloud water interception to total water input is shown on the secondary *y*-axis.

average daily rates of T were 1.7 mm at UB and only 1.0 mm at BK. The limiting factor for transpiration at BK is the low number of days with favorable conditions.

To explore the influence of stem size distribution on total T, an analysis of the distribution of stem size, sapwood area and transpiration across DBH size classes was undertaken at both sites (Figure 50.6). At UB about 55% of all stems fell in the 10–20-cm DBH class; this represented 32% of the total sapwood area but only 22% of total T. At the other end of the size range, trees in the >70-cm DBH class represented only 2% of the total number of stems on the plot but contributed 10% of total T. A very different case is presented by the dense forest (2019 stems ha^{-1}) at BK which is dominated by stems in the 10–20-cm DBH size class (65%). This size class represented 45% of the total sapwood area and 50% of the total plot water use. The extreme environmental conditions experienced by the UMCF at BK have apparently resulted in a stunted forest dominated by small-diameter trees. Whilst these small trees use very little water, their number enables them to dominate total stand water use (Figure 50.6).

Water balance

Both cloud forest sites showed distinct seasonal variation in the components of their water balance, reflecting the variation in precipitation and energy inputs between wet and dry seasons. Figure 50.7 demonstrates how evaporative losses via T, I, and E_s affect the seasonality of the net water balance ($Q + D + \Delta S$) at each site. With the huge precipitation inputs at BK, all months showed a net production of ($Q + D + \Delta S$). The occurrence of negative net water balance totals around the end of the dry season at UB indicates that evaporative losses can temporarily exceed precipitation inputs in LMCF. The UB forest must therefore rely on soil moisture stores or have access to groundwater at these times. Whilst T was reduced during the wettest months, I reached a maximum at this time of year, therefore maintaining high total evaporative losses (Figure 50.7).

The data from UB also illustrate how inter-annual differences in wet-season rainfall affect the water balance. Rainfall during the 2004/05 wet season was only 47% (1586 mm) of that during the 2003/04 wet season (3384 mm). However, in 2004/05 ($Q + D + \Delta S$) decreased by an even greater proportion, being only 31% (709 mm) of its value in 2003/04 (2299 mm). This illustrates how there can be a two-fold effect of lower precipitation on forest water yield; firstly, precipitation inputs are reduced, and secondly, evaporative losses (as a percentage of inputs) increase.

Water balance component measurements at each of the sites were undertaken over different periods, thus complicating comparisons between sites and with other studies. To help overcome

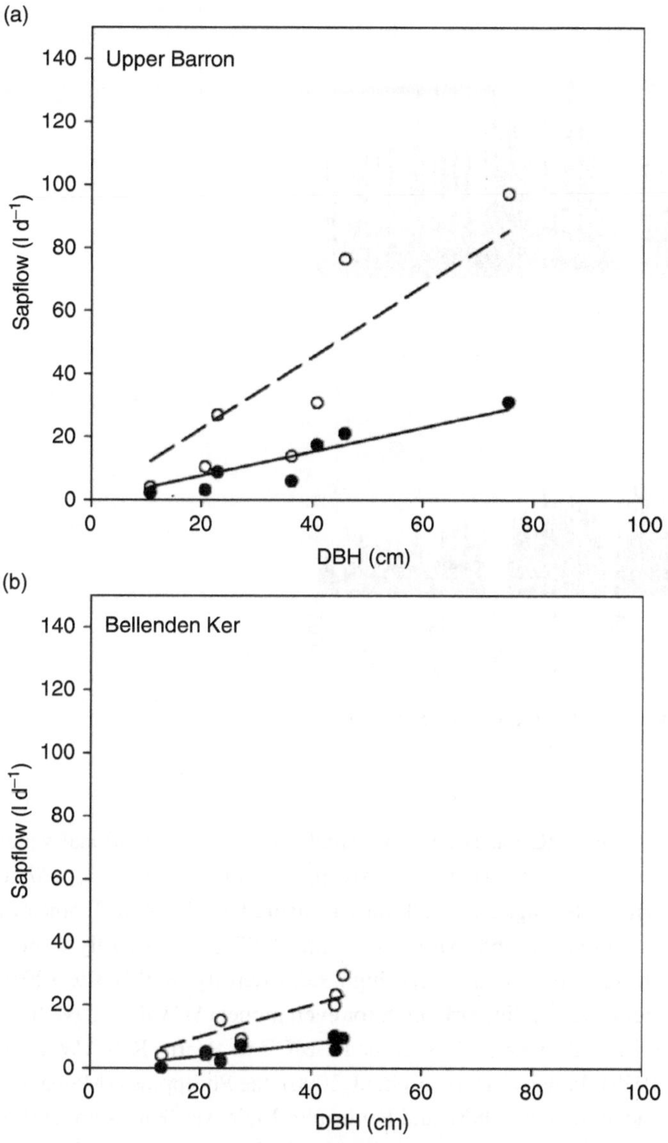

Figure 50.4. Relationship between diameter at breast height (DBH) of individual sample trees and daily sapflow (*T*) at (a) Upper Barron and (b) Bellenden Ker for days with high (dashed line) and low (solid line) transpiration. UB High $T = 1.13$ DBH($r^2 = 0.76$), UB Low $T = 0.38$ DBH ($r^2 = 0.84$). BK High $T = 0.50$ DBH ($r^2 = 0.78$), BK Low $T = 0.19$ DBH$^{1.15}$ ($r^2 = 0.72$).

this, monthly averages were calculated of all water balance components throughout the observation period for each site. Annual averages of each water balance component were then calculated from the sum of these monthly averages.

Figure 50.8 provides an overview of the normalized water balance components at the two sites for the duration of the measurements. These figures are based on the water balance components described in Eq. (50.1), with precipitation input pathways shown in ellipses, evaporative losses in rectangles, and streamflow/groundwater recharge in hexagons. Percentages of gross precipitation are shown in brackets.

At UB, cloud water interception (P_c) added 19% to gross precipitation. *SF* was 7% of gross precipitation (9% of P_{ga}) whereas *TF* was 64% (79% of P_{ga}), resulting in an overall canopy interception *I* of 29% (35% of P_{ga}). Evaporative losses through *I* were much greater than those via *T* (20% of gross precipitation input), indicating that evaporation from the wetted canopy dominates overall *ET* at UB. The value of ($Q + D$) was equivalent to 49% of gross precipitation.

The water balance at BK is a very different story in that not only are the overall precipitation inputs much greater, but evaporative losses are much less – both in relative and absolute terms (Figure 50.8). Of the 7471 mm of gross precipitation a mere 4% was lost through *T*, and only 7% through *I*. As a result, *ET* is equivalent to only 12% of gross precipitation, leaving 88% for ($Q + D$). This water balance analysis clearly illustrates the hydrological difference between LMCF and UMCF in the study area, thereby confirming the tentative distinction between the two forest types proposed by Bruijnzeel (2001, 2005) based on more limited data.

DISCUSSION AND CONCLUSIONS

Values of *SF* determined for the lower montane cloud forest at UB were quite high at 7% of gross precipitation and nearly 9% of P_{ga}, compared to values of 2% or less commonly reported for this type of forest (Bruijnzeel and Proctor, 1995; Bruijnzeel, 2005). At UB *TF* was 64% of gross precipitation and 79% of P_{ga}which is somewhat lower than the average value for LMCF (~ 90%; Bruijnzeel, 2005); however, the lower *TF* at this site is offset somewhat by the high *SF* and at 88% of P_{ga}, overall net precipitation ($TF + SF$) is similar to the 92% derived on average for this vegetation type by Bruijnzeel (2005). At 29% of gross precipitation and 35% of P_{ga}, the apparent wet-canopy evaporation (interception loss *I*) obtained for the forest at UB is very high compared to the overall average for LMCF (8%), although it is close to results reported for comparable conditions in Costa Rica (Clark *et al.*, 1998). LMCFs that are subject to less rainfall and cloud water incidence than experienced by the forest at UB generally exhibit even lower *TF* percentages (Cavelier *et al.*, 1997; Ataroff, 1998; Wilcke *et al.*, 2001; Fleischbein *et al.*, 2005; cf. Giambelluca *et al.*, this volume), although underestimation of TF by a less than optimal sampling design cannot always be excluded (cf. Holwerda *et al.*, 2006a).

On an annual basis, the residual water yield (streamflow plus groundwater recharge) at UB is roughly equivalent to 50% of gross precipitation and similar to the 60% reported for nearby lowland rain forest by Gilmour (1975). Transpiration for the LMCF of the UB site was 591 mm year^{-1} which is comparable to the estimates of 560 mm derived for a similar forest in Venezuela using porometry (Ataroff and Rada, 2000) and

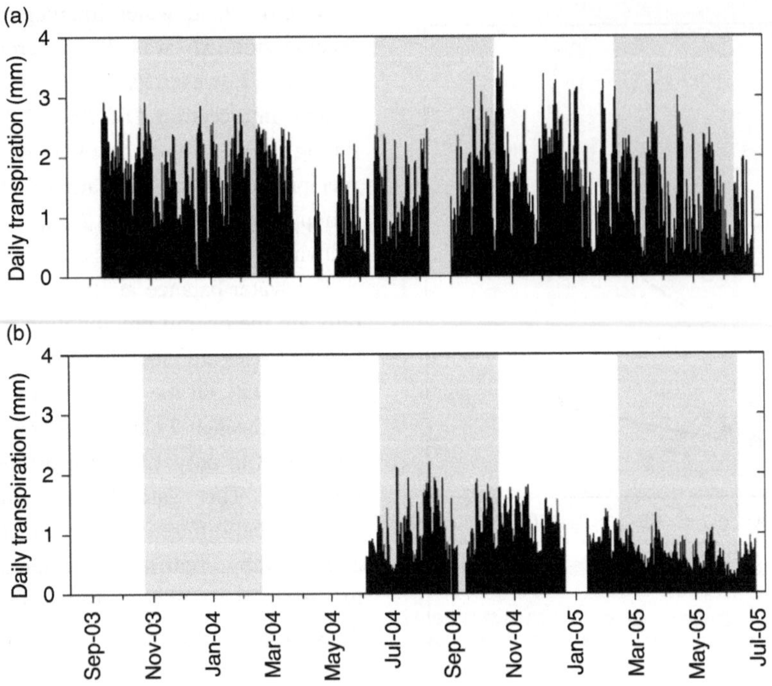

Figure 50.5. Daily transpiration (mm) for (a) Upper Barron and (b) Bellenden Ker. Wet season shaded in gray.

513 mm reported for a sub-tropical LMCF in south-east Queensland (Hutley *et al.*, 1997), but higher than the 365 mm reported for an LMCF in Costa Rica subject to higher rainfall (Bruijnzeel, 2006). McJannet *et al.* (2007a) suggested that the lower transpiration of Australian rain forests might be related to a combination of forest structural differences (i.e. fewer large trees), climatic differences (frequent canopy wetting and coastal proximity; cf. Gilmour, 1975; Wallace and McJannet, 2006), or possibly differences in tree physiology. Forest structural differences, particularly the possibility of Australian rain forest containing fewer larger trees, could be attributed to the frequent occurrence of cyclones on the Wet Tropics coast. Such cyclones often result in destruction of larger trees that could, in turn, be preventing the formation of tall canopies with large-crowned emergent trees (Gilmour, 1975). The overall *ET* total derived for the UB forest (1518 mm year^{-1}) not only exceeds by far the average value for LMCF (<1000 mm; Bruijnzeel, 2005), but also that for lower montane rain forests not affected by fog (1265 mm; Bruijnzeel and Proctor, 1995). More than half of the total *ET* at UB is contributed by wet-canopy evaporation. Interception studies from rain forests around the world suggest that the amount of water evaporated directly from the wetted canopy can be very high under "maritime" conditions experienced at sites, such as Upper Barron, which are located near continental edges or on islands (Schellekens *et al.*, 2000; Giambelluca *et al.*, 2009). This enhanced evaporation is believed to be at least partly explained by advection of energy from the nearby oceans (Schellekens *et al.*, 2000).

The UMCF at BK has revealed some quite exceptional water balance results. At 10% of gross precipitation (14% of P_{ga}), *SF* at BK is the highest of all sites measured in the Wet Tropics of north Queensland (McJannet *et al.*, 2007c) and is likely to be a consequence of the very high stem density at this site (2019 stems ha^{-1}). Similarly high (or even higher) *SF* values have been observed in multiple-stemmed UMCF in Puerto Rico (Weaver, 1972), Jamaica (Hafkenscheid, 2000), the Philippines (Mamanteo and Veracion, 1985), and Peninsular Malaysia (Kumaran, 2008). At Bellenden Ker *TF* was also high at 83% of gross precipitation (116% of P_{ga}) and is likely to be the combined result of a lower *LAI* (3.3 m^2 m^{-2}), the reduced evaporation at this site, as well as the extra water inputs afforded by cloud water interception and horizontal precipitation (see also Juvik, 2009 and Prada *et al.*, 2009 for other examples of extremely high *TF* due to predominantly wind-driven inputs). Expressed as a percentage of P_{ga}, the observed *TF* is close to the overall average (109%) derived for this forest type by Bruijnzeel (2001, 2005). Despite the availability of very large amounts of gross precipitation, both *T* and *I* were relatively small at 5% and 7% of gross precipitation, respectively. Apparently, the enhanced evaporation believed to be occurring at UB due to the proximity to the ocean is not obvious at BK (despite being situated even closer to the Pacific; Figure 50.1), presumably due to prolonged periods of immersion in cloud and a sparser canopy. The large inputs and low evaporative losses at BK result in the availability of 87% (6500 mm) of precipitation for streamflow and groundwater recharge ($Q + D$).

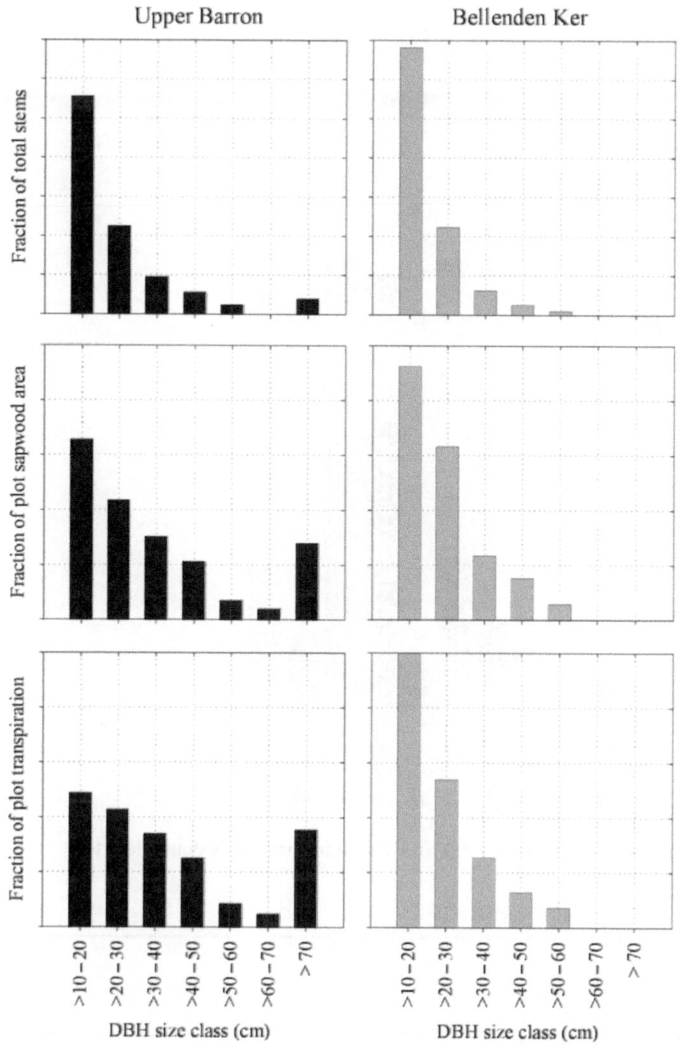

Figure 50.6. Distribution of stems and sapwood area for different size classes at Upper Barron (left-hand column) and Bellenden Ker (right-hand column). Relationship between DBH size class at each site and fraction of total stems (top row), fraction of total sapwood area (middle row), and fraction of total plot transpiration for the study duration (bottom row).

The only other cloud forest sites for which E_t values have been reported under similarly extreme climatic conditions are the San Gerardo LMCF in northern Costa Rica (Bruijnzeel, 2006; cf. Frumau *et al.*, this volume) and the Pico del Este elfin cloud forest (ECF) in Puerto Rico (Holwerda, 2005; cf. Holwerda *et al.*, this volume #29). The San Gerardo forest receives ~6000 mm of rainfall per year plus an estimated 2740 mm of horizontal precipitation (cloud water plus wind-driven rain at angles >85° from the vertical), whereas the Pico del Este forest is subjected to 4435 mm of rainfall plus *c.* 780 mm of cloud water interception (Holwerda *et al.*, 2006b). Interestingly, despite major contrasts in forest stature or *LAI* between the San Gerardo LMCF (20–22 m tall, *LAI* = 3.0), the Bellenden Ker UMCF (8 m, *LAI* 3.3), and the Pico del Este ECF (2.5–3 m, *LAI* 2.1), their annual transpiration totals are comparable at 365 mm, 349 mm, and 296 mm, respectively. However, whilst the trend in T follows that for forest stature, T values normalized for site evaporative conditions (represented by Penman's E_0) do not exhibit a similar trend, with E_t/E_0 for BK (0.33) being distinctly lower than the values for the San Gerardo LMCF (0.40) and the Pico del Este ECF (0.43) (cf. Bruijnzeel *et al.*, this volume).

Results showed that there can be a two-fold effect of El Niño conditions on forest water yield through the combination of reduced precipitation inputs and increased evaporative losses. Similar findings have been reported by Chiew and McMahon (2002) when modeling the impacts of climate change on Australian streamflow. They found that in wet and temperate watersheds the percentage change in streamflow was about twice the percentage change in rainfall. The potential impact of this finding is further amplified by the report of Albritton *et al.* (2001) which summarizes climate change projections according to a number of models and states that the recent trend for sea surface temperatures in the tropical Pacific to become more El Niño-like are projected to continue.

In his review of the hydrology of tropical montane cloud forests, Bruijnzeel (2005) noted that there is increasing evidence that these forests and their water balances are threatened by global warming. For example, with higher temperature the average cloud condensation level will tend to rise (Still *et al.*, 1999; cf. Lawton *et al.*, 2001), removing the capacity for intercepting cloud water and potentially increasing evaporation (Reinhardt and Smith, 2008; Mildenberger *et al.*, 2009; Foster, this volume). The net effect of this would be to reduce water yield (assuming rainfall inputs to remain similar). Ecologically, such changes in canopy moisture conditions could also have a serious impact on plant (especially epiphytes; Zotz and Bader, 2009; Nadkarni, this volume) and animal species that are dependent on persistently wet, humid conditions (Pounds *et al.*, 1999, 2006). Williams *et al.* (2003) used a modeling approach to investigate the potential impacts of climate change on the spatial distribution of endemic rain forest vertebrates in north Queensland, Australia. They concluded that increases in temperature, raised orographic cloud base and reduced dry season precipitation would have the potential to result in the extinction of many species currently found in mountain areas due to significant reduction or complete loss of core environment.

Overall, the present research has highlighted the distinctly different hydrological characteristics of two montane cloud forest types in northern Australia. It is clear from this research that montane forest types with a high degree of exposure to prevailing winds and frequent immersion in cloud receive significant additional inputs of water through cloud interception. Long-term measurements of such inputs are crucial for water balance studies in these types of forest. With a fuller understanding of how the

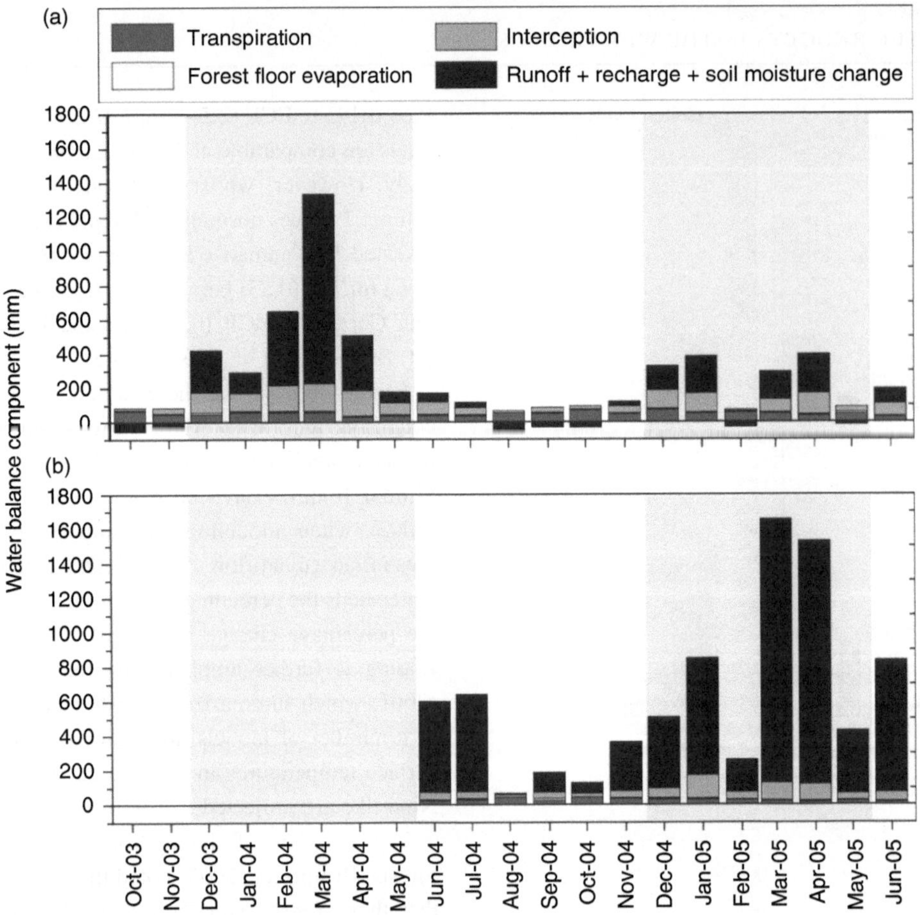

Figure 50.7. Monthly water balance components (mm) for (a) Upper Barron and (b) Bellenden Ker (b). Evapotranspiration is equivalent to transpiration plus forest floor evaporation. Wet season months are shaded gray.

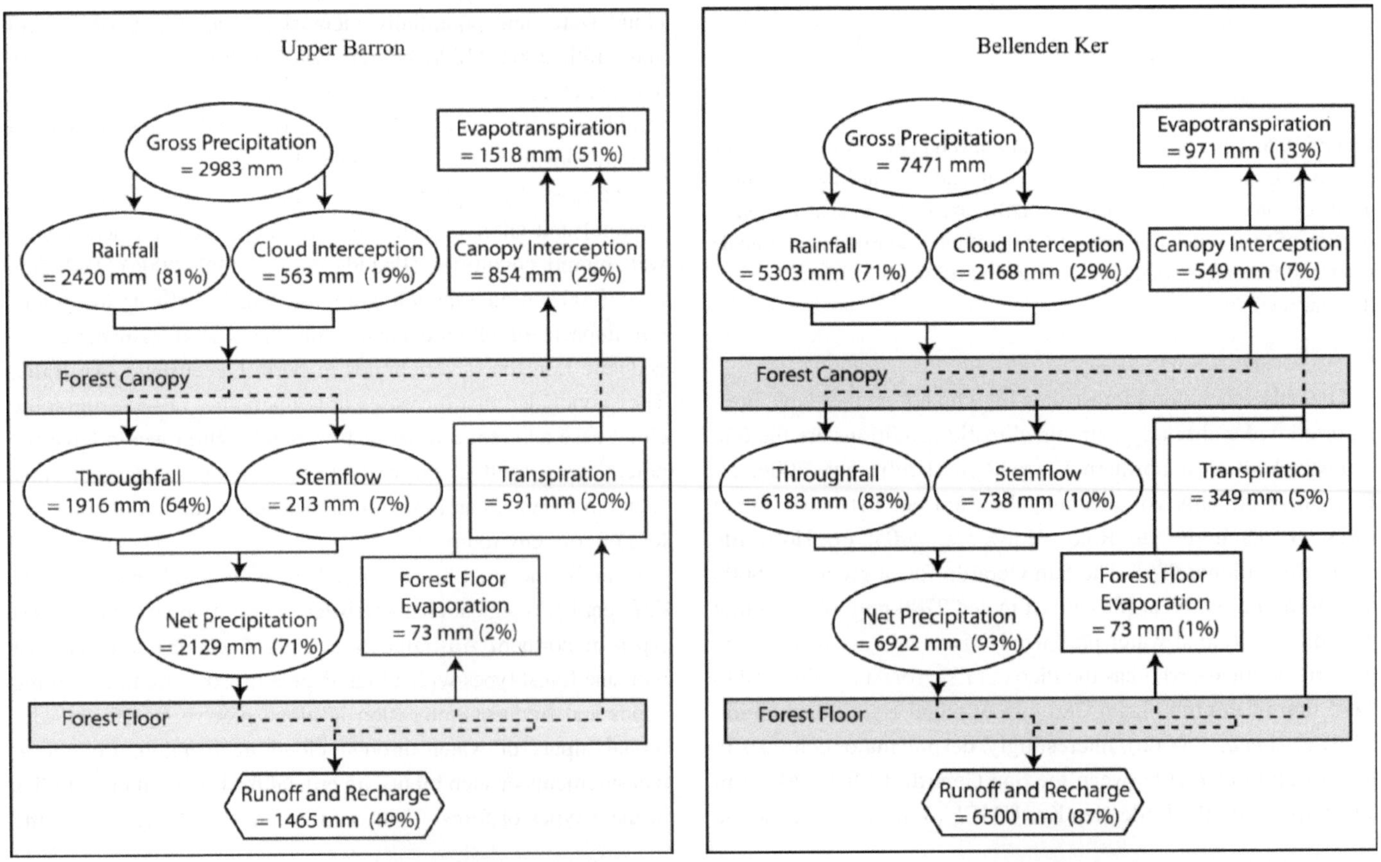

Figure 50.8. Annual average water balance components for Upper Barron (lower montane cloud forest) and Bellenden Ker (upper montane cloud forest). Precipitation input pathways are shown in ellipses, evaporative losses are in rectangles and residual streamflow/groundwater recharge totals are shown in a hexagon. Percentages of gross precipitation are shown in brackets.

natural systems of the tropics function and the services these forests provide one can make better informed decisions about how to utilize these landscapes, taking into account the pros and cons of alternative land uses and potential changes to climate.

ACKNOWLEDGEMENTS

The authors would like to thank Mark Disher, Peter Fitch, Andrew Ford, Peter Richardson, Trevor Parker, Adam McKeown, Trudi Prideaux, Jenny Holmes, and Pepper Brown for their help with field installation and maintenance. Fred Scatena, Sampurno Bruijnzeel, and Tom Giambelluca provided valuable feedback for improving this paper. Funding for this research was provided by the Cooperative Research Centre for Tropical Rainforest Ecology and Management.

REFERENCES

Albritton, D. L., M. R. Allen, A. P. M. Baede, *et al.* (2001). Summary for policy-makers. In *Climate Change 2001: The Scientific Basis. Contribution of Working Group I to the Third Assessment Report of the Intergovernmental Panel on Climate Change*, eds. J. T. Houghton, Y. Ding, D. J. Griggs, *et al.*, pp. 21–98. Cambridge, UK: Cambridge University Press.

Allen, R. G., L. S. Pereira, D. Raes, and M. Smith (1998). *Crop Evapotranspiration: Guidelines for Computing Crop Water Requirements.* Rome: FAO.

Ataroff, M. (1998). Importance of cloud water in Venezuelan Andean cloud forest water dynamics. In *Proceedings of the 1st International Conference on Fog and Fog Collection*, eds. R. S. Schemenauer and H. A. Bridgman, pp. 25–28. Ottawa, Canada: IDRC.

Ataroff, M., and F. Rada (2000). Deforestation impact on water dynamics in a Venezuelan Andean cloud forest. *Ambio* **29**: 440–444.

Bruijnzeel, L. A. (2001). Hydrology of tropical montane cloud forests: a reassessment. *Land Use and Water Resources Research* **1**: 1–18.

Bruijnzeel, L. A. (2005). Tropical montane cloud forest: a unique hydrological case. In *Forests, Water and People in the Humid Tropics*, eds. M. Bonell and L. A. Bruijnzeel, pp. 462–483. Cambridge, UK: Cambridge University Press.

Bruijnzeel, L. A. (ed.) (2006). *Hydrological Impacts of Converting Tropical Montane Cloud Forest to Pasture, With Initial Reference to Northern Costa Rica*, Final Technical Report for Project R7991, DFID Forestry Research Programme. Amsterdam, the Netherlands: VU University Amsterdam.

Bruijnzeel, L. A., and J. Proctor (1995). Hydrology and biogeochemistry of tropical montane cloud forests: what do we really know? In *Tropical Montane Cloud Forests*, eds. L. S. Hamilton, J. O. Juvik, and F. N. Scatena, pp. 38–78. New York: Springer-Verlag.

Cavelier, J., M. Jaramillo, D. Solis, and D. de Leon (1997). Water balance and nutrient inputs in bulk precipitation in tropical montane cloud forest in Panama. *Journal of Hydrology* **193**: 83–96.

Chiew, F. H. S., and T. A. McMahon (2002). Modelling the impacts of climate change on Australian streamflow. *Hydrological Processes* **16**: 1235–1245.

Clark, K. L., N. M. Nadkarni, D. Schaefer, and H. L. Gholz (1998). Atmospheric deposition and net retention of ions by the canopy in a tropical montane forest, Monteverde, Costa Rica. *Journal of Tropical Ecology* **14**: 27–45.

Fleischbein, K., W. Wilcke, R. Goller, *et al.* (2005). Rainfall interception in a lower montane forest in Ecuador: effects of canopy properties. *Hydrological Processes* **19**: 1355–1371.

Førland, E. J., P. Allerup, B. Dahlström, *et al.* (1996). *Manual for Operational Correction of Nordic Precipitation Data.* Oslo, Norway: Norwegian Meteorological Institute.

García-Santos, G. (2007). *An ecohydrological and soils study in a montane cloud forest in the National Park of Garajonay, La Gomera (Canary Islands, Spain).* PhD Thesis, VU University Ámsterdam, Ámsterdam, The Netherlands. [http://www.falw.vu.nl/nl/onderzoek/earth-sciences/geo-environmental-science-and-hydrology/hydrology-dissertations/index.asp].

Giambelluca, T. W., R. E. Martin, G. P. Asner, *et al.* (2009). Evapotranspiration and energy balance of native wet montane cloud forest in Hawai'i. *Agricultural and Forest Meteorology* **149**: 230–243.

Gilmour, D. A. (1975). Catchment water balance studies on the wet tropical coast of north Queensland. Ph.D. thesis, James Cook University, Townsville, Queensland, Australia.

Hafkenscheid, R. R. L. J. (2000). Hydrology and biogeochemistry of tropical montane rain forests of contrasting stature in the Blue Mountains, Jamaica. Ph.D. thesis. VU University Amsterdam, Amsterdam, the Netherlands. Also available at http://dare.ubvu.vu.nl/bitstream/1871/12734/1/tekst.pdf.

Hatton, T. J., E. A. Catchpole, and R. A. Vertessy (1990). Integration of sapflow velocity to estimate plant water use. *Tree Physiology* **6**: 201–209.

Holwerda, F. (2005). Water and energy budgets of rain forests along an elevation gradient under maritime tropical conditions. Ph.D. thesis, VU University Amsterdam, Amsterdam, the Netherlands. Also available at www.falw.vu.nl/nl/onderzoek/earth-sciences/geo-environmental-science-and-hydrology/hydrology-dissertations/index.asp.

Holwerda, F., F. N. Scatena, and L. A. Bruijnzeel (2006a). Throughfall in a Puerto Rican lower montane rain forest: a comparison of sampling strategies. *Journal of Hydrology* **327**: 592–602.

Holwerda, F., R. Burkard, W. E. Eugster, *et al.* (2006b). Estimating fog deposition at a Puerto Rican elfin cloud forest site: comparison of the water budget and eddy covariance methods. *Hydrological Processes* **20**: 2669–2692.

Hutley, L. B., D. Doley, D. Yates, and A. Boonsaner (1997). Water balance of an Australian subtropical rainforest at altitude: the ecological and physiological significance of intercepted cloud and fog. *Australian Journal of Botany* **45**: 311–329.

Jordan, C. F., and J. Heuveldop (1981). The water budget of an Amazonian rain forest. *Acta Amazonica* **11**: 87–92.

Juvik, J. O. (2009). The seminal Hawaiian cloud forest "Lâna'i fog drip study": 50th anniversary reassessment. In *Tropical Hydrology Symposium*, eds. F. L. Ogden, R. S. Harmon, and R. F. Stallard, p. 229. Panama City: Smithsonian Tropical Research Institute.

Juvik, J. O., and P. C. Ekern (1978). *A Climatology of Mountain Fog on Mauna Loa Hawaii Island*, Technical Report No. 118. Honolulu, HI: Water Resources Research Center, University of Hawai'i.

Kelliher, F. M., B. M. Kostner, D. Y. Hollinger, *et al.* (1992). Evaporation, xylem sapflow, and tree transpiration in a New Zealand broad leaved forest. *Agricultural and Forest Meteorology* **62**: 53–73.

Kumaran, S. (2008). Hydrometeorology of tropical montane rain forests of Gunung Brinchang, Pahang Darul Makmur, Malaysia. Ph.D. thesis, University of Putra Malaysia, Serdang, Malaysia.

Lawton, R. O., U. S. Nair, R. A. Pielke Sr., and R. M. Welch (2001). Climatic impact of tropical lowland deforestation on nearby montane cloud forests. *Science* **294**: 584–587.

Mamanteo, B. P., and V. P. Veracion (1985). Measurements of fog drip, throughfall and stemflow in the mossy and Benguet pine (*Pinus kesiya* Royle ex Gordon) forests in the upper Agno river basin. *Sylvatrop (Philippines Forestry Research Journal)* **10**: 271–282.

Marshall, D. C. (1958). Measurement of sapflow in conifers by heat transport. *Plant Physiology* **6**: 385–396.

McJannet, D. L, and P. G. Fitch (2004). *A Flexible and Easily Constructed Heat Pulse System for Monitoring Sapflow in Trees*, Technical Report No. 39/04. Canberra: CSIRO Land and Water. Also available at www.clw.csiro.au/publications/technical2004/tr39–04.pdf.

McJannet, D. L., P. G. Fitch, M. G. Disher, and J. S. Wallace (2007a). Measurements of transpiration in four tropical rainforest types of north Queensland, Australia. *Hydrological Processes* **21**: 3549–3564.

McJannet, D. L., J. S. Wallace, and P. Reddell (2007b). Precipitation interception in Australian tropical rainforests. I. Measurement of stemflow, throughfall and cloud interception. *Hydrological Processes* **21**: 1692–1702.

McJannet, D. L., J. S. Wallace, and P. Reddell (2007c). Precipitation interception in Australian tropical rainforests. II. Altitudinal gradient of cloud interception, stemflow, throughfall and interception. *Hydrological Processes* **21**: 1703–1718.

Mildenberger, K., E. Beiderwieden, Y. J. Hsia, and O. Klemm (2009). CO_2 and water vapor fluxes above a subrtropical mountain cloud forest: the effect of light conditions and fog. *Agricultural and Forest Meteorology* **149**: 1730–1736.

OESR (2002). *The Contribution of International and Domestic Visitor Expenditure to the Queensland Regional Economies 1998–99*. Brisbane, Queensland: Office of Economic and Statistical Research.

Pounds, A. J., M. P. L. Fogden, and J. H. Campbell. (1999). Biological response to climate change on tropical mountain. *Nature* **389**: 611–614.

Pounds, J. A., M. R. Bustamante, L. A. Coloma, *et al.* (2006). Widespread amphibian extinctions from epidemic disease driven by global warming. *Nature* **439**: 161–167.

Prada, S., M. Menezes de Sequeiro, C. Figueira, and M. Oliveira da Silva (2009). Fog precipitation and rainfall interception in the natural forests of Madeira Island (Portugal). *Agricultural and Forest Meteorology* **149**: 1179–1187.

Reinhardt, K., and W. K. Smith (2008). Impacts of cloud immersion on microclimate, photosynthesis and water relations of *Abies fraseri* Pursh. Poiret in a temperate mountain cloud forest. *Oecologia* **158**: 229–238.

Schellekens, J., L. A. Bruijnzeel, F. N. Scatena, N. J. Bink, and F. Holwerda (2000). Evaporation from a tropical rain forest, Luquillo Experimental Forest, Eastern Puerto Rico. *Water Resources Research* **36**: 2183–2196.

Sharon, D. (1980). The distribution of hydrologically effective rainfall incident on sloping ground. *Journal of Hydrology* **46**: 165–188.

Still, C. J., P. N. Foster, and S. H. Schneider (1999). Simulating the effects of climate change on tropical montane cloud forests. *Nature* **398**: 608–610.

Swanson, R. H., and D. W. A. Whitfield (1981). A numerical analysis of heat pulse velocity theory and practice. *Journal of Experimental Botany* **32**: 221–239.

Tracey, J. G. (1982). *The Vegetation of the Humid Tropical Region of North Queensland*. Melbourne, Victoria: CSIRO.

Wallace, J. S, and D. L. McJannet (2006). On interception modelling of a lowland coastal rainforest in northern Queensland, Australia. *Journal of Hydrology* **329**: 477–488.

Weaver, P. L. (1972). Cloud moisture interception in the Luquillo mountains of Puerto Rico. *Caribbean Journal of Science* **12**: 129–144.

Wilcke, W., S. Yasin, C. Valarezo, and W. Zech (2001). Change in water quality during the passage through a tropical montane rain forest in Ecuador. *Biogeochemistry* **55**: 45–72.

Williams, S., E. Bolitho, and S. Fox (2003). Climate change in Australian tropical rainforests: an impending environmental catastrophe. *Proceedings of the Royal Society of London Series B* **270**: 1887–1893.

WTMA (2007). *Wet Tropics World Heritage Area 2007*. Cairns, Queensland: Wet Tropics Management Authority. Also available at www.wettropics.gov.au/mwha/mwha_pdf/maps/WT_tenure.pdf.

Zotz, G., and M. Y. Bader (2009). Epiphytic plants in a changing world: global change effects on vascular and non-vascular epiphytes. *Progress in Botany* **70**: 147–170.

51 Effects of forest disturbance and regeneration on net precipitation and soil water dynamics in tropical montane rain forest on Mount Kilimanjaro, Tanzania

M. Schrumpf
Max Planck Institute for Biogeochemistry, Jena, Germany

H. V. M. Lyaruu
University of Dar es Salaam, Dar es Salaam, Tanzania

J. C. Axmacher
University College London, London, UK

W. Zech
University of Bayreuth, Bayreuth, Germany

L. A. Bruijnzeel
VU University, Amsterdam, the Netherlands

ABSTRACT

The montane rain forest belt on Mt. Kilimanjaro forms an important water source for northern Tanzania that is threatened by both logging and fire. The aim of this study was to investigate consequences of forest fragmentation on various aspects of the water cycle. Soil properties, rainfall, throughfall, and soil water suction were analyzed for mature forest, secondary forest patches, and large clearings. A total of 10 plots located on the south-western slopes of the mountain between 2100 and 2300 m.a.s.l. were monitored from May 2000 until June 2002. Annual rainfall amounts ranged from 2000–2600 mm and showed high spatial and inter-annual variability. Rainfall interception ranged from 3% to 9% of incident rainfall in clearings to a maximum of 32% in forests. In general, soils under mature forest were wettest and showed only minor moisture fluctuations through the year. Soils of secondary forest sites were driest and soil water suction exhibited the largest fluctuations. The difference between the two forest types may reflect a combination of differences in interception, evaporation from the forest floor, and transpiration, because ventilation and radiation penetration can be expected to be enhanced in fragmented secondary forest. In clearings the higher throughfall and presumably lower transpiration rates led to moister conditions compared to adjacent secondary forest sites. Top-soil sand contents of the Andisols differed between sites, with disturbed sites having higher sand contents and consequently lower water contents at similar soil water suctions than did mature forest sites. Hence, soil water storage at disturbed sites was significantly lower than under mature forest during dry conditions, e.g. the secondary forest soil only stored 69% of the water stored under the mature forest in the upper 45 cm of the mineral soil at a suction of 55 kPa. The results indicate that on Kilimanjaro, soil conditions under mature forest are most suitable for optimum water storage while the transformation of closed mature forest to a mosaic of forests in different regeneration stages leads to high spatio-temporal variability in soil moisture and drier conditions in secondary forest patches.

Tropical Montane Cloud Forests: Science for Conservation and Management, eds. L. A. Bruijnzeel, F. N. Scatena, and L. S. Hamilton. Published by Cambridge University Press.

INTRODUCTION

Although Tanzanian rain forests cover less than 2% of the country, they form important watershed areas ensuring a stable water supply (Bjørndalen, 1992). The most important watershed for northern Tanzania and southern Kenya (the Pangani River Basin) is fed largely by the forest belt of Mt. Kilimanjaro, and supports highly productive agriculture around the mountain (Røhr, 2003; Hemp, 2005a). Because the hydro-electric power stations along the Pangani River rely on a constant water flow, the conservation of Kilimanjaro's water resources is not only of local, but also of regional interest (IUCN, 2003). Sarmett and Faraji (1991) studied the dry-season discharge of rivers along the lower slopes of Mt. Kilimanjaro between the 1950s and 1970s. They observed a decrease in the flow of streams that were not spring-fed and ascribed this to increased water diversion for irrigation and changes in land use accompanied by loss of forest cover as well as soil degradation. Since the beginning of the last century, the forests of Mt. Kilimanjaro have experienced major changes due to logging and the establishment of forest plantations on the western and north-eastern slopes, resulting in fragmentation of the natural forest (Lamprey *et al*., 1991). During the past few decades, the forests on the lower southern slopes have been affected by selective logging which has led to an opening of the forest and a mosaic of secondary vegetation in different regeneration stages marked by shifts in plant species composition (Mwasaga, 1991). Recently, illegal logging has also expanded to higher elevations, after the lower slopes had been mostly depleted of valuable timber species such as *Ocotea usambarensis* (Lambrechts *et al*., 2002) whereas fire is a problem at higher elevations around the timber-line (Hemp, 2005b; Hemp, this volume #58).

Bruijnzeel (2004) summarized current knowledge on the hydrological effects of tropical forest conversion and concluded that forest clearing tends to reduce interception and transpiration losses and therefore leads to an increase in soil moisture and stream water yields. However, this contention only holds as long as key soil properties such as infiltration capacity and water retention do not become degraded by accelerated surface erosion. Consequences of such deterioration could be a reduction of dry-season flows and an increase in storm flows (Bruen, 1989; Bruijnzeel, 2004). Once the total water-holding capacity of the soil is severely reduced by prolonged erosion, it is doubtful whether reforestation is capable of fully restoring the original hydrological condition (Bruijnzeel, 2004; Scott *et al*., 2005). Furthermore, the increased water use of fast-growing secondary vegetation could lead to decreases in soil moisture reserves and stream baseflows (Parker, 1985; Giambelluca, 2002). The duration of the period after which stream discharges return to pre-disturbance levels is essentially unknown but will depend amongst others on the growth rate of the vegetation (Giambelluca, 2002).

Amongst tropical forests, cloud forests constitute a special case due to the added inputs afforded by the interception of fog water by the forest vegetation (Stadtmüller, 1987). Such additional inputs are largely lost after cloud forest conversion to shorter vegetation types (e.g. pasture, annual cropping) and may even result in reduced streamflow and groundwater recharge (Zadroga, 1981; Ingwersen, 1985; Bruijnzeel, 2005).

Despite the importance of Mt. Kilimanjaro's forests, information on their hydrological behavior is mostly lacking (Røhr, 2003; cf. Hemp, 2005b; Hemp, this volume #58). The aim of the present study is to characterize specific physical properties of the forest's soils and to analyze the impacts of forest disturbance and regrowth on rainfall, throughfall, soil water suction, and water content. The investigation was restricted to a relatively small area containing mature forest, 55–60-year-old secondary forest, and clearings with low scrub, ferns, and herbs between 2100 and 2250 m.a.s.l. on the southern slopes of the mountain.

MATERIAL AND METHODS

Study area

The study was conducted between 2100 and 2300 m.a.s.l. on the south-western slopes of Mt. Kilimanjaro in the forest belt above the village of Machame. On the most humid southern slopes, a maximum annual rainfall of about 3000 mm has been measured between altitudes of 2200 and 2400 m.a.s.l. (Røhr and Killingtveit, 2003; Hemp, 2005b; cf. Hemp, this volume #12). According to the floristic classification of Hemp (this volume #12), the study area is located within the *Ocotea* (camphor tree) belt of which the lower part (1800–2200 m.a.s.l.) can be classified as lower montane rain forest (LMRF) and the upper part (2200–2500 m.a.s.l.) as lower montane cloud forest (LMCF). Soils have developed in layered volcanic ashes that overlay phonolites and trachytes of the Lent group (Downie and Wilkinson, 1972; Schrumpf, 2004).

Vegetation characteristics

Complete species lists for the study sites and the structural characteristics of the vegetation have been presented by Axmacher (2003). Trees in the mature forest reach heights of more than 40 m and may have a diameter at breast height of as much as 1.8 m. The dominant tree species is *Ocotea usambarensis*. Tree ferns (*Cyathea manniana*) typically occur in the understory, and there is a high density and diversity of epiphytes, among them many pteridophytes and bryophytes (cf. Hemp, 2002; Hemp, this volume #12).

Older secondary forests in the area have resulted from logging during the Second World War (Wood, 1964; DOS, 1968).

The canopy layer in these forests consists mainly of the pioneer *Macaranga kilimanjarica* and some young *Ocotea* trees. With a maximum of 0.5–1.0 m, stem diameters in the secondary forest are smaller compared to mature forest and canopy height is lower, reaching a maximum of 35 m. Abundance and diversity of epiphytes is also reduced and the moss layer on tree trunks is markedly thinner (Axmacher, 2003). Clearings (>500 m^2) created by more recent logging are dominated by *Pteridium aquilinum*, *Rubus steudneri*, and *Begonia meyeri-johannis*. According to regional foresters, some of these clearings have remained in the present state for over a decade without any invasion of tree species (cf. Hartig and Beck, 2003; Aide *et al.*, this volume).

Study design

Four plots in mature forest, three plots in secondary forest, and three clearings of 400 m^2 each were selected. Measurements started in May 2000 and were continued until June 2002. All plots were on slope gradients of less than 10°. One plot in a clearing had to be relocated in May 2001 due to a fire experiment. Since the area below 2200 m.a.s.l. was mostly depleted of mature forest, the corresponding plots had to be selected at slightly higher elevations (2250–2335 m.a.s.l.) than the disturbed sites (2075–2165 m.a.s.l.). Close to each plot, a soil pit was dug to a maximum depth of 2.3 m. Soil description and classification were done in accordance with the US Soil Taxonomy (Soil Survey Staff, 2003). Samples were taken as composite samples of three profile walls from each soil horizon, including the litter layer. Five additional undisturbed samples for the analysis of soil water retention characteristics and bulk density were extracted using 100-cm^3 steel rings at each of four soil depths (0.05–0.15, 0.20–0.30, 0.55–0.65, and 1.00–1.10 m). The homogeneity of the soils on the plots was also tested using a soil auger (Schrumpf, 2004).

Rainfall (*P*) and throughfall (*TF*) were collected with identical funnel-type gages that had a diameter of 115 mm. A table-tennis ball was placed in each funnel to reduce evaporation losses, and a 0.5-mm mesh net was fitted between the collector and funnel junction. As the mature forests and the regenerating sites were located at some distance from each other, *P* was measured in two clearings at 2100 m.a.s.l. and in a third clearing at 2250 m.a.s.l. close to the mature forest sites. Five *P* collectors were installed per clearing, placed on poles 1.5 m above the ground after removal of the surrounding shrub vegetation. To measure *TF*, every plot was equipped with 10 randomly distributed collectors whose positions remained fixed throughout the duration of the study. The collection bottles were partly buried so that the rim of the funnel was 0.3 m above the ground. According to studies summarized in Thimonier (1998) and Bruijnzeel (1989), the 10 collectors used to determine average *TF* per plot are at the lower end for sound quantitative results. However, if means per vegetation type are considered, the 40 collectors for the mature forest and the 30 collectors each for the secondary forest and clearings come close to the required number of collectors suggested by Bruijnzeel (1989) and Holwerda *et al.* (2006) for *TF* studies in tropical rain forests.

In a preliminary study, stemflow (*SF*) was collected on eight mature trees using gutter-like collectors made of cellular rubber coiled around the tree trunk. Initial analysis of *SF* in these forests revealed that it accounted for <1% of total *TF* in both forest types. This is in accordance with studies in other LMRF with similar structure, where *SF* is usually 1–2% of *P* (Bruijnzeel and Proctor, 1995). Therefore, measurements of *SF* were terminated after one year and will not be discussed further.

Soil water suction (*h*, hPa) was determined using pressure transducer tensiometers (Stitch Tensiometer T1, UMS, Munich) installed at four depths (0.10–0.15, 0.20–0.25, 0.60–0.65, and 1.10–1.15 m). Two tensiometers were used per soil depth and plot. Thus, it was not possible to record the heterogeneity of soil moisture conditions at single sites. Nevertheless, the results are considered useful in giving a general idea of differences in soil moisture conditions under the respective vegetation types. Because the tensiometer at 0.1 m depth was positioned in a loose soil layer of low bulk density, suction values were not always reliable during dry conditions.

Starting in May 2000, tensiometer readings and measurements of *P* and *TF* were recorded twice a week on the same day for all plots during the first year. Sampling continued on a weekly basis in the second year until November 2001 although measurements in the mature forest were made 2 days after the measurements in the other plots for logistic reasons (Schrumpf, 2004). From December 2001 until June 2002, *P*, *TF*, and soil water suction at the lower plots (clearings, secondary forests) were recorded twice a week again, whilst readings at the mature forest plots continued to be taken once a week.

SOIL PHYSICAL CHARACTERIZATION

Particle-size distribution was determined on field-moist samples. For dispersion, 5 g of field-moist soil were shaken overnight in 500 ml of de-ionized water after adding 20 ml of 0.1M $Na_4P_2O_7$ solution. Sand-sized particles were recorded via sieving, while silt and clay contents were determined using the pipette method (Gee and Bauder, 1986). Soil moisture retention curves were determined on undisturbed soil cores (100 cm^3) that had been stored at 4 °C in field-moist condition. The retention curves were determined using ceramic plates, starting with water-saturated samples and subsequent successive dehydration to suctions of 0.316, 1.0, 3.16, 10.0, and 31.6 kPa. Corresponding moisture contents were determined gravimetrically after equilibration. Finally, samples were dried at 110 °C to determine their dry bulk densities.

Data analysis

The Soil Hydraulic Properties Fitting (SHYPFIT) program (Durner, 1994) and the soil water retention model of Van Genuchten (1980) were used to fit measured suction and water content data pairs to derive continuous soil moisture retention curves for individual horizons below each vegetation type. The respective curves were then used to estimate mean soil water contents from measured suctions, both per horizon and for the entire soil profile. Calculations were made separately for rainy and dry seasons using the equations given in Lilienfein *et al.* (1999). Because the deepest soil horizon studied was only at 1.15 m, estimation of soil water storage for the whole pedon was limited by definition to a depth of 1.25 m. For estimation of profile water storage, the profile was divided into four zones: 0–0.15, 0.15–0.45, 0.45–0.75, and 0.75–1.25 m. In doing so, it was assumed that bulk densities and soil water characteristics as measured within the respective zones were valid for the entire zone. Thus, the derived water storage values will only represent approximations.

Statistical analyses were conducted using the statistical work package Statistica 5.0 (StatSoft, Tulsa, OK). If the analysis of variance (ANOVA) indicated significant ($p < 0.05$) differences between treatments, Scheffé tests were used to analyze individual differences among treatments.

RESULTS AND DISCUSSION

Soil properties

All soil profiles analyzed had developed in layered volcanic ashes and contained buried soils. Depending on their individual make-up, soils were classified as Fulvudands, Epiaquands, or Placaquands (Schrumpf, 2004). At the mature forest sites, cemented horizons of varying thickness and expression were found at soil depths between 0.85 m and 1.35 m. Whilst these were completely absent in some profiles, they reached a thickness of up to 0.2 m in others. Judging by the degree of mottling above them, the cemented horizons served as effective barriers for percolating water (Schrumpf, 2004).

Bulk density was not significantly different between treatments. However, top-soil sand content (>50 μm) under secondary forest was significantly higher than under mature forest (Table 51.1). Sand contents in the clearings were intermediate whereas the highest clay contents were observed at the mature forest sites.

Table 51.1 *Bulk density part to be merged with particle size distribution part*

	Bulk density[a] (g cm^{-3})
Mature forest	
0–0.15	0.32 ± 0.06
0.15–0.3	0.40 ± 0.04
0.30–0.60	
0.60–1.00	0.61 ± 0.10
1.00–1.50	0.56 ± 0.05
Secondary forest	
0–0.15	0.37 ± 0.03
0.15–0.3	0.38 ± 0.03
0.30–0.60	
0.60–1.00	0.61 ± 0.05
1.00–1.50	0.61 ± 0.02
Clearings	
0–0.15	0.30 ± 0.02
0.15–0.30	0.35 ± 0.06
0.30–0.60	
0.60–1.00	0.63 ± 0.07
1.00–1.50	0.53 ± 0.01

[a] Only 0.05–0.15, 0.20–0.30, 0.60–0.70, 1.05–1.15.

Andisols are known to contain large amounts of plant-available water and hygroscopic water (Shoji *et al.*, 1993; cf. Tobón *et al.*, this volume #52). This is also the case at Mt. Kilimanjaro where comparatively high water contents were observed at the water suctions analyzed (Figure 51.1). All soils were extremely porous, with porosities (defined as the water content at saturation or $h = 0$) almost invariably exceeding 70%. For secondary forest sites and clearings, the shape of the soil water characteristic changed with soil depth, in accordance with the more coarsely textured top-soil and higher clay contents of deeper soil horizons (Table 51.1). Differences within the profile were much less pronounced at the mature forest sites whose soil water characteristic curves much resembled those presented by Moldrup *et al.* (2003) for 18 Japanese Andosols (see also Tobón *et al.*, this volume #52).

Top-soil (0.05–0.15 m) water content at high suction values (≥100 hPa or pF 2.0) at the mature forest sites was significantly higher (Scheffé test, $p < 0.05$) than at the secondary forest sites and clearings. At a depth of 0.20–0.25 m, water content in the mature forest soil was also significantly higher than that of the secondary forest at a suction of 31.6 hPa and higher than that of both secondary forest and clearings at a suction of 100 hPa. No significant differences in soil water contents at any suction were obtained for the sub-soil (0.6 and 1.1 m; Figure 51.1).

The observed differences in top-soil water characteristic curves among sites indicate contrasts in pore size distribution and probably reflect the observed differences in the content of sand-sized

Table 51.2 *Average bulk density and particle-size distribution sat four soil depths under mature forest, secondary forest and clearings (SE: standard error,* n = *3)*

	Particle size							
Bulk density (g kg^{-1})	<2 μm	SE	2–20 μm	SE	20–100 μm	SE	>100 μm	SE
Mature forest								
0–0.10	689	28	185	12	52	10	75	17
0.15–0.25	747	71	99	78	57	13	97	26
0.60–0.70	634	126	161	86	62	23	142	21
1.05–1.15	494	70	262	27	64	16	179	71
Secondary forest								
0–0.10	420	46	217	42	96	5	267	6
0.15–0.25	514	62	205	24	68	7	213	82
0.60–0.70	493	52	295	27	70	6	142	31
1.05–1.15	563	9	248	8	43	7	146	21
Clearings								
0–0.10	638	67	150	39	49	17	163	32
0.15–0.25	585	69	261	87	50	10	104	28
0.60–0.70	513	70	261	24	57	14	169	70
1.05–1.15	594	12	218	28	53	10	134	35

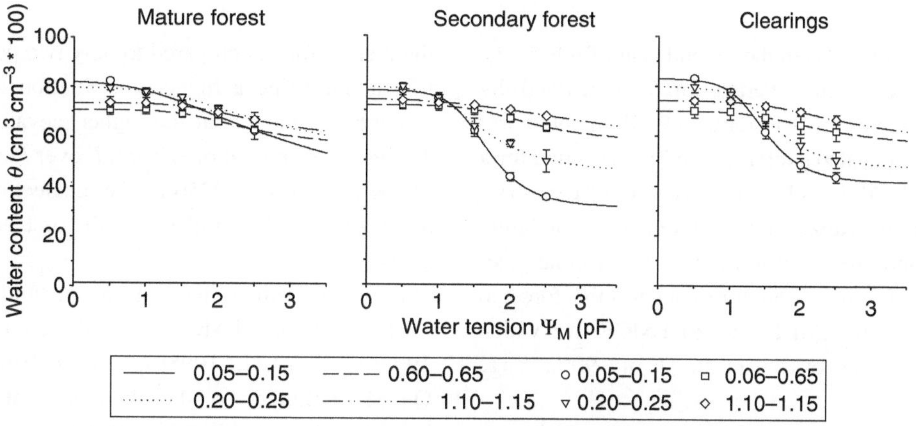

Figure 51.1. Measured data points (± standard error, $n = 3$) and fitted (lines) soil-water characteristic curves at four soil depths below mature forest, secondary forest, and clearings on the southern slopes of Mt. Kilimanjaro.

particles because bulk densities (and therefore compaction) did not differ between sites (Table 51.2). Andisols are known to irreversibly form stable, sand-sized aggregates after drying beyond a critical suction value between pF 3 and 4 (Wada, 1989; Shoji *et al.*, 1993). Because the uppermost ash layers of the respective sites are supposedly of similar age (Schrumpf, 2004), the higher content of sand-size particles at the disturbed sites may be the result of more intensive drying of the top-soil following opening up of the forest.

Rainfall, throughfall and interception

Annual amounts of rainfall (*P*) ranged between 1960 and 2600 mm and were distinctly higher in the first year as compared to the second. About 70–80% of *P* reached the soil as throughfall (*TF*) in the forest sites, whereas it was more than 90% in the clearings (Table 51.3). In the first year, derived average interception loss (neglecting stemflow) for the secondary forest sites was higher than that for the mature forest sites (27% and 18% of *P*, respectively), whereas it was similar for both forest types during the drier, second year (32% and 30%).

Because of the difference in site elevation – and therefore rainfall and fog incidence – *TF* amounts in the secondary and mature forests cannot be compared directly. As such, it remains unclear to what extent the secondary forest at 2100 m.a.s.l. exhibited a similar interception loss as old-growth forest would at that elevation. Average *TF* in the secondary forest ranged from

Table 51.3 *Amounts of rainfall, throughfall, and intercepted water during the two years of study for the different vegetation types considered*

	Rainfall (mm)		Throughfall (mm)			Interception (mm)		
	2,100 m.a.s.l.	2,250 m.a.s.l.	Clearings	Secondary forest	Mature forest	Clearings	Secondary forest	Mature forest
Year 1	2600	2480	2490 ± 80	1890 ± 100	2040 ± 27	110 ± 80 (3%)	712 ± 100 (27%)	438 ± 33 (18%)
Year 2	2210	1960	2010 ± 40	1500 ± 20	1370 ± 20	210 ± 40 (9%)	710 ± 20 (32%)	590 ± 20 (30%)

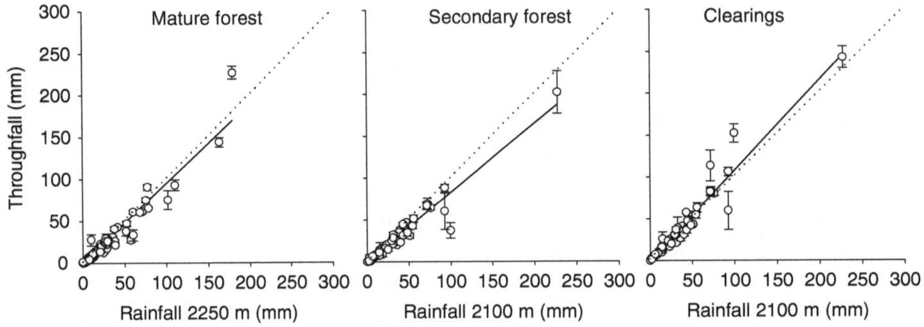

Figure 51.2. Relationships between rainfall and throughfall (± standard error) for the three studied vegetation types on the southern slopes of Mt. Kilimanjaro from June 2000 to May 2001 (first year). Dotted lines give the 1 : 1 relation and full lines show linear regressions.

73% of *P* in the first year to 77% in the second year (Table 51.3). These values are close to the average of 75% derived by Bruijnzeel (2005) for mature LMRF (range: 67–81%) not appreciably affected by fog. It is also similar to the 78% reported for a sub-montane forest at 1500 m.a.s.l. in the (drier) West Usambara Mountains, elsewhere in Tanzania (Lundgren and Lundgren, 1979). Also, at 82% and 70% of *P* in the first and second year, respectively, *TF* values for the mature forest suggest this forest to be hydrologically closer to LMRF than to LMCF (for which Bruijnzeel (2005) derived an overall mean *TF* of ~90%; range 55–100%).

In the absence of detailed records of *P*, *TF*, and fog incidence it is not possible to quantify the contribution of fog to *TF* at the mature forest site. Duane *et al.* (2008) measured the relative humidity at seven locations between 2340 and 5800 m a.s.l. on the south-western slopes of Mt. Kilimanjaro. With an average value of 97.7%, humidity was highest at their lowermost forest station which corresponds to the highest mature forest plot of the present study. Based on automated camera images, Duane *et al.* (2008) assumed the presence of fog and clouds in the forest at relative humidity values of 95% or higher, which occurred almost every night at 2340 m.a.s.l. Thus, cloud water is likely to frequently wet the mature forest of the present study. The additional humidity afforded by the fog may result in enhanced *TF* if fog capture occurs prior to rainfall events so that the amount necessary for canopy saturation is reduced due to an already moist canopy (García-Santos, 2007). At eight out of 100 half-weekly periods during the first year, *TF* exceeded *P* in the mature forest compared to only two periods in the secondary forest, indicating a higher contribution of canopy stripping to throughfall amounts at the higher elevation sites (Figure 51.2). Taking the sum total of excess *TF* over *P* as a first approximation of fog incidence at 2250 m (i.e. neglecting wet-canopy evaporation) gave a value of 65 mm in the first and 40 mm in the second study year.

Low *TF* values have been attributed to high epiphyte loading in some LMCFs (e.g. Cavelier *et al.*, 1997; Ataroff 1998; Clark *et al.*, 1998; cf. Fleischbein *et al.*, this volume). On Mt. Kilimanjaro, the abundance of epiphytic bryophytes known to be capable of intercepting large quantities of water (cf. Pócs, 1980) was greater at the mature forest than at the secondary forest sites, but rainfall interception by the latter was higher. This could be partly explained by the overall higher density of the tree layer in the secondary as compared to the mature forest (Axmacher, 2003) and the presumably higher frequency of clouds around 2300 m.a.s.l. (cf. Duane *et al.*, 2008) – adding to the amount of water stored in the epiphyte layer and so reducing their intercepting capacity (Hölscher *et al.*, 2004; cf. Köhler *et al.*, this volume; Oesker *et al.*, this volume).

Rainfall variability in time and space

Northern Tanzania has a bimodal rainfall distribution with two rainy seasons, a longer and a shorter one. The first usually occurs from March to June and the latter around December and January,

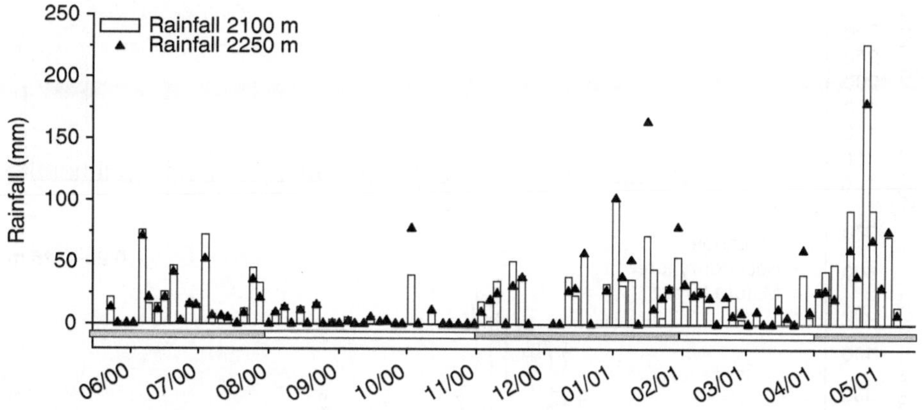

Figure 51.3. Variability of rainfall during the first study year at 2100 and 2250 m.a.s.l. on the southern slopes of Mt. Kilimanjaro. Each bar or data point represents the cumulative rainfall of one sampling period (of 3 and 4 days alternating). Gray areas below the graph mark rainy periods, white ones dry periods.

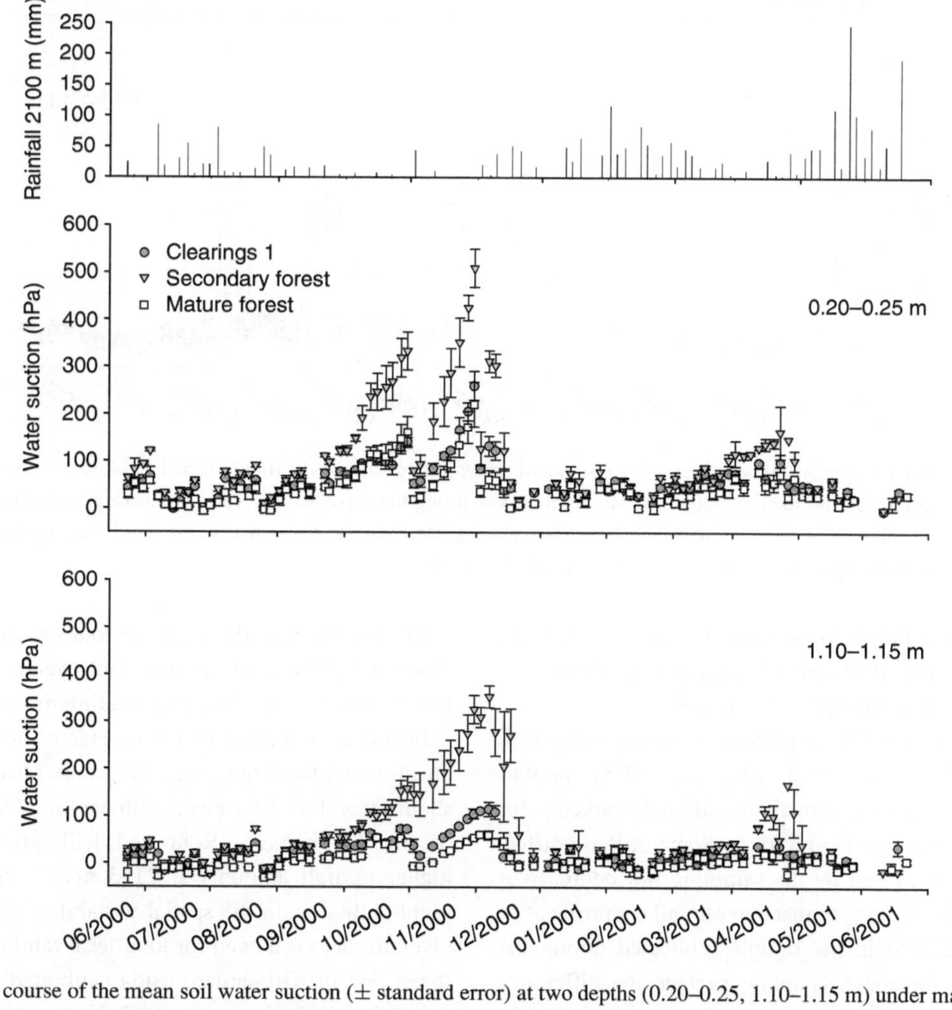

Figure 51.4. Seasonal course of the mean soil water suction (± standard error) at two depths (0.20–0.25, 1.10–1.15 m) under mature forest, secondary forest, and clearings on the southern slopes of Mt. Kilimanjaro during the first study year. The top graph shows cumulative half-weekly rainfall amounts collected prior to the tensiometer readings for comparison.

as was the case in the first study year (Figure 51.3). In the second year, there was no minor rainy season, which led to overall lower annual rainfall amounts (Table 51.3) and drier soil conditions (Figures 51.4 and 51.5).

High variability in both annual and monthly rainfall amounts has often been reported for Tanzania (Nieuwolt, 1974; Lundgren and Lundgren, 1979; Sarmett and Faraji, 1991). Long-term observations made at an altitude of 2100 m on Mt. Kilimanjaro

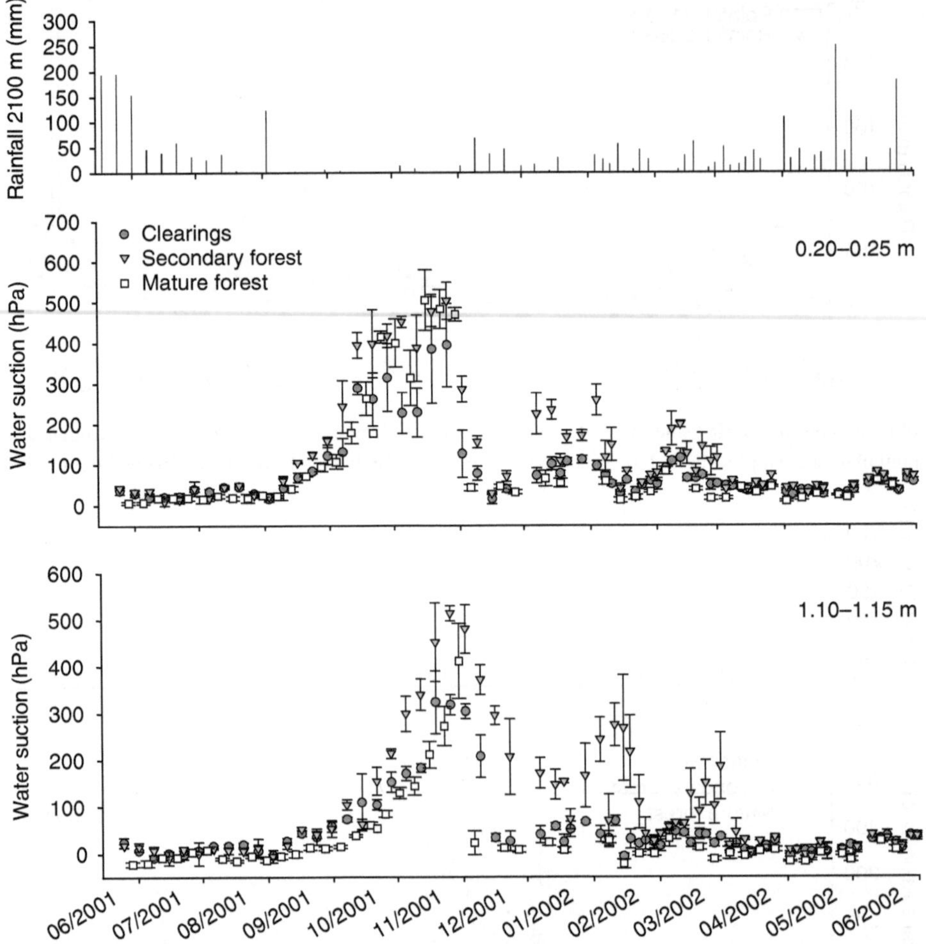

Figure 51.5. Seasonal course of mean soil water suction (± standard error) at two soil depths (0.20–0.25, 1.10–1.15 m) under mature forest, secondary forest, and clearings on the southern slopes of Mt. Kilimanjaro during the second study year. In clearings and secondary forests, readings were first made weekly (June to November) and later twice a week. Readings in the mature forest were done weekly but not on the same day as the other sites. The top graph shows cumulative weekly rainfall amounts.

gave a mean annual rainfall of 1840 mm with totals varying from 1200–3815 mm between 1945 and 1958 (Hedberg, 1964). The presently measured totals are within this range.

Although overall seasonal distribution and annual totals were similar (Table 51.3, Figures 51.3, 51.4, and 51.5), rainfall amounts at 2100 and 2250 m sometimes differed markedly for individual sampling periods (Figure 51.3). Some of these differences may have been caused by the sampling procedure, as it took a full day to visit all sites, so that e.g. rainfall occurring after midday was not included in the samples collected during the morning. However, for some sampling periods the difference was not compensated upon the next sampling, suggesting individual storm events to affect only some sites, but not the whole area. According to Nieuwolt (1974), Mt. Kilimanjaro is among the regions with the highest frequency of convectional rainstorms in Tanzania. Such storms – which tend to have limited temporal and spatial distribution (Nieuwolt, 1974) – will contribute to higher rainfall variability.

In the present study, annual *P* totals at 2100 m.a.s.l. exceeded those at 2250 m.a.s.l. in both study years, despite the claim of Hemp (this volume #12) that maximum rainfall occurs between 2200 and 2400 m.a.s.l. However, Hemp's totals are inferred from uncalibrated totalizing gages read only once or twice a year and should therefore be viewed with caution. On the southern slopes of Mt. Kilimanjaro, Røhr and Killingtveit (2003) measured higher rainfall amounts at 2153 m.a.s.l than at 2367 m.a.s.l., highlighting again the spatial variability of rainfall distribution. Overall, there is a need for long-term rainfall observations in the forest belt of Kilimanjaro using calibrated recording equipment (cf. Røhr and Killingtveit, 2003; Nauss *et al.*, 2009).

Annual course of soil water suction

As shown in Figure 51.4, the secondary forest sites typically exhibited the highest soil water suctions as compared to mature forest and clearings during the first year, which is in accordance with the lower *TF* observed in the secondary forest (Table 51.3). Although the largest amounts of TF were measured in the clearings, soil water suction was usually lowest (i.e. soils were wettest) under mature forest. Differences between vegetation types were most pronounced during the dry seasons. Even during short dry periods, suction levels in the secondary forest soils responded rapidly, whereas in the mature forest, soil water suction only exhibited significant changes during the main dry season (Figure 51.4).

During the second (and generally drier) year, tensiometer readings at the mature forest sites were not conducted on the same day as for the secondary forest sites. This has to be taken into account when interpreting Figure 51.5. Whilst the mature forest soil was wetter during the long dry season in 2000, suctions in 2001 were as high as in the secondary forest. However, during the following minor dry periods, suction levels in the mature forest soil remained low, whereas fluctuations at the secondary forest sites were again much greater. Clearings were on average wetter than secondary forest sites, especially during drier periods.

Daws *et al.* (2002) observed significantly lower soil water suctions on slopes compared with plateau sites in a semi-deciduous tropical forest in Panamá. On Mt. Kilimanjaro, current logging activities concentrate on valleys rather than ridges. As a consequence, clearings are usually found in valley or downslope positions, whereas older secondary forest sites are concentrated on ridges. Nevertheless, analysis of soil water suction values in six clearings in different relief positions did not result in significant differences. Therefore, and because the mature forest sites were located in upslope, downslope, and plateau positions and exhibited the lowest and least variable suction values, relief position is not supposed to be an important factor controlling differences in soil moisture between study sites.

The higher soil water levels in the clearings compared to adjacent secondary forest sites (Figures 51.4 and 51.5) can be explained by the lower interception (Table 51.3) and presumably lower transpiration rates of the scrubs and herbs of the clearings. Similar results have been obtained in natural forest gaps and in recent clearings created by tropical forest cutting and burning (Parker, 1985; Klinge, 1997). Accordingly, one would also have expected the soil of the mature forest to be drier than the clearings, but this was not the case. One reason for this might be the occurrence of cemented horizons at the mature forest sites hindering water percolation. However, the crusts were not continuous and the depth of their occurrence varied. In one soil profile, no cemented horizons were observed down to a depth of over 2 m, and yet water was seen standing close to the profile rim during the rainy season. This indicates a generally high water table during that time of the year and the crusts are therefore not solely responsible for the high water content of the mature forest sites (Schrumpf, 2004).

Although neither fog incidence nor soil water uptake were measured in the Kilimanjaro LMCF, the frequent visual observation of fog and low cloud (Duane *et al.*, 2008), the high humidity within the stand, and the high density of epiphytic bryophytes and filmy ferns (cf. Hemp, 2002; Richter, 2003) are all suggestive of a comparatively low soil water uptake, whereas the secondary forests did not exhibit these characteristics and were still dominated by early successional species that may well have a higher water use (cf. Giambelluca, 2002). This might go some way toward explaining the observed difference in soil water status between the old-growth and secondary stands (see also below). There is a clear need for further ecohydrological work on Kilimanjaro.[1]

Furthermore, forest fragmentation influences water relations as edge effects become effective. Kapos (1989) found that due to higher temperatures and vapor pressure deficits at forest edges, the water demand of plants inside the forest increased up to 40 m from the forest edge. Also, both ventilation and light penetration are higher at forest edges, further increasing atmospheric water demand. Because the forest at lower elevations on Mt. Kilimanjaro is currently strongly fragmented (Lambrechts *et al.*, 2002), the associated increase in forest edges may have led to an overall increase in evaporation from secondary forest, leading in turn to drier soil conditions.

Soil water storage

As a consequence of the shape of the respective soil water characteristic curves (Figure 51.1), soil water content and storage was similar among the studied vegetation types at soil water suctions close to saturation. However, at a higher suction of 550 hPa (which was among the highest values measured in the top-soil; Figures 51.4 and 51.5), the water content at the secondary forest sites was only 48% of that estimated for the mature forest soil at a depth of 0.1 m and 80% at 0.2 m, respectively. Assuming that the measured properties at 0.1 m depth were representative for the top 0.15 m, and the values measured at 0.25 m for the layer between 0.15 and 0.45 m, the top 45 cm of the secondary forest soil would store only 69% of the amount held by the mature forest soil at a suction of 550 hPa.

Soil water suctions were usually lowest under mature forest. At the same time, water contents at a given suction were higher in the upper horizons of the mature forest soil compared to the other sites. Hence, the estimated average water storage of the soils was also highest under mature forest. Mean estimated soil water storage totals per vegetation type down to a depth of 1.25 m

[1] Planning of new research on the climate dynamics (including fog incidence), and the chief controls on water and carbon fluxes along the elevational gradient in the Kilimanjaro region is in an advanced stage (Nauss *et al.*, 2009; Tenhunen *et al.*, 2009).

Table 51.4 *Median of the estimated soil water storage down to a depth of 1.25 m under different vegetation types (± standard error, n = 3, 4); different letters indicate significant differences (Scheffé test,* $p < 0.05$*)*

	Σ rainfall (mm)		Estimated mean soil water storage, 0–1.25 m (mm)		
	2100 m.a.s.l.	2250 m.a.s.l.	Clearings	Secondary forest	Mature forest
Long rainy season 05/00–07/00	460	370	$928^{a} \pm 4$	$856^{b} \pm 10$	$875^{b} \pm 6$
Long dry season 8/00–10/00	130	160	$893^{a} \pm 4$	$769^{b} \pm 3$	$804^{c} \pm 4$
Short rainy season 11/00–01/01	780	830	$921^{a} \pm 3$	$844^{b} \pm 8$	$848^{b} \pm 3$
Short dry season 02/01–03/01	250	260	$909^{a} \pm 4$	$798^{b} \pm 6$	$830^{c} \pm 2$
Long rainy season 04/01–07/01	1570	1500	$925^{a} \pm 2$	$869^{b} \pm 10$	$868^{b} \pm 14$
Long dry season 08/01–10/01	150	100	$851^{a} \pm 6$	$747^{b} \pm 2$	$777^{b} \pm 10$
Short rainy season 11/00–01/02	430	520	$908^{a} \pm 4$	$774^{b} \pm 9$	$830^{c} \pm 7$
Short dry season 02/02–03/02	300	380	$913^{a} \pm 2$	$793^{b} \pm 14$	$838^{c} \pm 12$
Long rainy season 04/02–07/02	1400	560	$909^{a} \pm 2$	$829^{b} \pm 2$	$850^{b} \pm 9$

are provided in Table 51.4, distinguishing between rainy and dry seasons. Although rainfall amounts were similar between treatments, soil water storage differed significantly at all times. Particularly during drier periods, soil water storage at the secondary forest sites was significantly lower than in the clearings. Overall, soil conditions under the mature forest were more suitable for high water storage than below secondary forest, although differences were not always significant at different times of the year (Table 51.4).

CONCLUSION

Since water is a most valuable resource in Tanzania, the maintenance of water yields from forested headwater areas is of great importance. The high inter-annual and spatial variability of rainfall underlines the necessity for the buffering of rainfall peaks and droughts for a reliable water supply, which is supported by the presence of undisturbed natural forest. The results of the present study reveal that opening of the lower montane cloud forest and conversion to secondary vegetation leads to higher spatial and seasonal variation of soil water suctions and soil water contents as compared to old-growth forest. However, the observed differences in soil water suction between sites should not be attributed purely to contrasts in vegetation cover, because effects of differences in site conditions (notably top-soil sand content and presence of cemented horizons) – although supposed to be minor – cannot be ruled out completely.

The overall effect of forest fragmentation on watershed-scale water yields cannot be predicted from the present data, but the results do indicate a change in the water-holding capacity of the soil after forest disturbance. In a waterhed study in southern Tanzania, Edwards (1979) found that infiltration rates and erosion did not change following montane forest clearing, and water yields increased without adverse effect on seasonal flow distribution. Streamflow measurements at the watershed scale are required on Mt. Kilimanjaro to quantify the effects of forest fragmentation and conversion of old-growth forest to young secondary growth on stream discharge. Also, the small-scale spatial variability of top-soil texture and the possible relation to forest disturbance deserves more attention, since this has been shown to have a high impact on soil water storage.

ACKNOWLEDGEMENTS

This study was supported by a grant from the German Science Foundation (DFG, Ze 154/1–4). Furthermore we are greatly indebted to the following organizations: Department of Mines and Minerals Tanzania, Forestry and Beekeeping Division Tanzania (FBD), Kilimanjaro National Park Authority (KINAPA), Tanzania National Park Authority (TANAPA), and the Tanzanian Commission for Science and Technology (COSTECH).

REFERENCES

Ataroff, M. (1998). Importance of cloud water in Venezuelan Andean cloud forest water dynamics. In *Proceedings of the 1st International Conference on Fog and Fog Collection*, eds. R. S. Schemenauer and H. A. Bridgman, pp. 25–28. Ottawa, Canada: IDRC.

Axmacher, J. (2003). Diversität von Geometriden (Lepidoptera) und Gefäßpflanzen entlang von Habitatgradienten am Südwest-Kilimanjaro. Ph.D. thesis, University of Bayreuth, Bayreuth, Germany.

Bjørndalen, J. E. (1992). Tanzania's vanishing rain forests: assessment of nature conservation values, biodiversity and importance for water catchment. *Agriculture, Ecosystems and Environment* **40**: 313–334.

Bruen, M. (1989). Hydrological considerations for development in the Eastern Usambara Mountains. In *Forest Conservation in the East Usambara Mountains, Tanzania*, eds. A. C. Hamilton and R. Bernsted-Smith, pp. 117–140. Gland, Switzerland: IUCN.

Bruijnzeel, L. A. (1989). Nutrient cycling in moist tropical forests: the hydrological framework. In *Mineral Nutrients in Tropical Forest and Savanna Ecosystems*, ed. J. Proctor, pp. 383–415. Oxford, UK: Blackwell Scientific.

Bruijnzeel, L. A. (2004). Hydrological functions of tropical forests: not seeing the soil for the trees? *Agriculture, Ecosystems and Environment* **104**: 185–228.

Bruijnzeel, L. A. (2005). Tropical montane cloud forest: a unique hydrological case. In *Forests, Water and People in the Humid Tropics*, eds. M. Bonell and L. A. Bruijnzeel, pp. 462–483. Cambridge, UK: Cambridge University Press.

Bruijnzeel, L. A., and J. Proctor (1995). Hydrology and biogeochemistry of tropical montane cloud forests: what do we really know? In *Tropical Montane Cloud Forests*, eds. L. S. Hamilton, J. O. Juvik, and F. N. Scatena, pp. 38–78. New York: Springer-Verlag.

Cavelier, J., M. A. Jaramillo, D. Solis, and D. de León (1997). Water balance and nutrient inputs in bulk precipitation in tropical montane cloud forests in Panama. *Journal of Hydrology* **193**: 83–96.

Clark, K. L., N. M. Nadkarni, D. Schaefer, and H. L. Gholz (1998). Atmospheric deposition and net retention of ions by the canopy in a tropical montane forest, Monteverde, Costa Rica. *Journal of Tropical Ecology* **14**: 27–45.

Daws, M. I., C. E. Mullins, D. F. R. P. Burslem, S. R. Paton, and J. W. Dalling (2002). Topographic position affects the water regime in a semideciduous tropical forest in Panamá. *Plant and Soil* **238**: 79–90.

DOS (1968). *Y 742 Series, Sheet Nos. 42/3, /4, 56/1, /2, /4, 57/1*. London: Directorate of Overseas Surveys.

Downie, C., and P. Wilkinson (1972). *The Geology of Kilimanjaro*. Sheffield, UK: University of Sheffield.

Duane, W. J., N. C. Pepin, M. L. Losleben, and D. R. Hardy (2008). General characteristics of temperature and humidity variability on Kilimanjaro, Tanzania. *Arctic, Antarctic, and Alpine Research* **40**: 323–334.

Durner, W. (1994). *SHYPFIT User's Manual*, Research Report No. 94.1. Bayreuth, Germany: Institute of Hydrology, University of Bayreuth.

Edwards, K. A. (1979). The water balance of the Mbeya experimental catchments. *East African Agricultural and Forestry Journal* **1**: 231–247.

Gee, G. W., and J. W. Bauder (1986). Particle-size analysis. In *Methods of Soil Analysis*, Part 1, *Physical and Mineralogical Methods*, ed. A. K Klute, pp. 383–411. Madison, WI: Soil Science Society of America.

Giambelluca, T. W. (2002). Hydrology of altered tropical forest. *Hydrological Processes* **16**: 1665–1669.

Hartig, K., and E. Beck (2003). The Bracken fern (*Pteridium arachnoideum* Kaulf.) dilemma in the Andes of Southern Ecuador. *Ecotropica* **9**: 3–13.

Hedberg, O. (1964). Features of afroalpine plant ecology. *Acta Phytogeographica Suedica* **49**: 1–44.

Hemp, A. (2002). Ecology of the pteridophytes on the southern slopes of Mt. Kilimanjaro. I. Altitudinal distribution. *Plant Ecology* **159**: 211–239.

Hemp, A. (2005a). The banana forests of Kilimanjaro: biodiversity and conservation of the Chagga homegardens. Biodiversity and Conservation, doi: 10.1007/s10531-004-8230-8.

Hemp, A. (2005b). Climate change driven forest fires marginalize the impact of ice cap wasting on Kilimanjaro. *Global Change Biology* **11**: 1013–1023.

Hölscher, D., L. Köhler, A. I. J. M. van Dijk, and L. A. Bruijnzeel (2004). The importance of epiphytes epiphytes to total rainfall interception by a tropical montane rain forest in Costa Rica. *Journal of Hydrology* **292**: 308–322.

Holwerda, F., F. N. Scatena, and L. A. Bruijnzeel (2006). Throughfall in a Puerto Rican lower montane rain forest: a comparison of sampling strategies. *Journal of Hydrology* **327**: 592–602.

Ingwersen, J. B. (1985). Fog drip, water yield and timber harvesting in the Bull Run Municipal Watershed, Oregon. *Water Resources Bulletin* **21**: 469–473.

IUCN (2003). *Pangani Situation Analysis*. Nairobi, Kenya: IUCN Eastern Africa Programme.

Kapos, V. (1989). Effects of isolation on the water status of forest patches in the Brazilian Amazon. *Journal of Tropical Ecology* **5**: 173–185.

Klinge, R. (1997). Wasser- und Nährstoffdynamik im Boden und Bestand beim Auffbau einer Holzplantage im östlichen Amazonasgebiet. Ph.D. thesis, Georg-August-University Göttingen, Göttingen, Germany.

Lambrechts, C., B. Woodley, A. Hemp, C. Hemp, and P. Nnyiti (2002). *Aerial Survey of the Threats to Mt. Kilimanjaro Forests*. Dar es Salaam, Tanzania: UNDP, UNOPS, UNF, UNEP, Kenya Wildlife Service, and University of Bayreuth.

Lamprey, R. H., F. Michelmore, and H. F. Lamprey (1991). Changes in the boundary of the montane rainforest on Mt. Kilimanjaro between 1958–1987. In *The Conservation of Mount Kilimanjaro*, ed. W. D. Newmark, pp. 9–16. Gland, Switzerland: IUCN.

Lilienfein, J., W. Wilcke, M. A. Ayarza, *et al.* (1999). Annual course of matric potential in different used savanna oxisols in Brazil. *Soil Science Society of America Journal* **63**: 1778–1785.

Lundgren, L., and B. Lundgren (1979). Rainfall, interception and evaporation in the Mazumbai forest reserve, West Usambara Mts., Tanzania and their importance in the assessment of land potential. *Geografiska Annaler* **61**: 157–178.

Moldrup, P., S. Yoshikawa, T. Olesen, T. Komatsu, and D. E. Rolston (2003). Gas diffusivity in undisturbed volcanic ash soils: test of soil-water-characteristic-based prediction models. *Soil Science Society of America Journal* **67**: 41–51.

Mwasaga, B. C. (1991). The natural forest of Mount Kilimanjaro. In *The Conservation of Mount Kilimanjaro*, ed. W. D. Newmark, pp. 136–145. Gland, Switzerland: IUCN.

Nauss, T., J. Bendix, and L. A. Bruijnzeel (2009). Climate dynamics of the Kilimanjaro region. In *Kilimanjaro Ecosystems under Global Change: Linking Biodiversity, Biotic Interactions and Biogeochemical Ecosystem Processes*, ed. I. Steffan-Dewenter. Proposal for the establishment of a Research Unit submitted to the Deutsche Forschungsgemeinschaft.

Nieuwolt, S. (1974). Rainstorm distributions in Tanzania. *Geografiska Annaler* **56A**: 241–250.

Parker, G. G. (1985). The effect of disturbance on water and solute budgets of hillslope tropical rainforest in Northeastern Costa Rica. Ph.D. thesis, University of Georgia, Athens, GA, USA.

Pócs, T. (1980). The epiphytic biomass and its effect on the water balance of two rain forest types in the Uluguru Mountains (Tanzania, East Africa). *Acta Botanica Academiae Scientiarum Hungariae* **26**: 143–167.

Richter, M. (2003). Using plant functional types and soil temperatures for ecoclimatic interpretations in southern Ecuador. *Erdkunde* **57**: 161–181.

Røhr, P. C. (2003). A hydrological study concerning the southern slopes of Mt. Kilimanjaro, Tanzania. Ph.D. thesis, Norwegian University of Science and Technology, Trondheim, Norway.

Røhr, P. C., and A. Killingtveit (2003). Rainfall distribution on the slopes of Mt. Kilimanjaro. *Hydrological Sciences Journal* **48**: 65–77.

Sarmett, J. D., and S. A. Faraji (1991). The hydrology of Mount Kilimanjaro: an examination of dry season runoff and possible factors leading to its decrease. In *The Conservation of Mount Kilimanjaro*, ed. W. D. Newmark, pp. 53–70. Gland, Switzerland: IUCN.

Schrumpf, M. (2004). Biogeochemical investigations in old growth and disturbed forest sites at Mount Kilimanjaro. Ph.D. thesis, University of Bayreuth, Bayreuth, Germany.

Scott, D. F., L. A. Bruijnzeel, and J. Mackensen (2005). The hydrological and soil impacts of forestation in the tropics. In *Forests, Water and People in the Humid Tropics*, eds. M. Bonell and L. A. Bruijnzeel, pp. 622–651. Cambridge, UK: Cambridge University Press.

Shoji, S., M. Nanzyo, and R. A. Dahlgren (1993). *Volcanic Ash Soils*. Amsterdam, the Netherlands: Elsevier.

Soil Survey Staff (2003). *Keys to Soil Taxonomy*. Washington, DC: Natural Resource Conservation Service of the U.S. Department of Agriculture.

Stadtmüller, T. (1987). *Cloud Forests in the Humid Tropics: A Bibliographic Review*. Turrialba, Costa Rica: Centro Agronomico Tropical de Investigación y Ensenanza, and Tokyo: The United Nations University.

Tenhunen, J., D. Otieno, B. Huwe, and B. Glaser (2009). Patterns in water and carbon flux controls along land use and climate gradients. In *Kilimanjaro Ecosystems under Global Change: Linking Biodiversity, Biotic Interactions and Biogeochemical Ecosystem Processes*, ed. I. Steffan-Dewenter. Proposal for the establishment of a Research Unit submitted to the Deutsche Forschungsgemeinschaft.

Thimonier, A. (1998). Measurement of atmospheric deposition under forest canopies: Some recommendations for equipment and sampling design. *Environmental Monitoring and Assessment* **52**: 353–387.

Van Genuchten, M. Th. (1980). A closed-form equation for predicting the hydraulic conductivity of unsaturated soils. *Soil Science Society of America Journal* **44**: 892–898.

Wada, K. (1989). Allophane and imogolite. In *Minerals in Soil Environments*, ed. J. B. Dixon, pp. 603–638. Madison, WI: Soil Science Society of America.

Wood, P. J. (1964). A note on forestry on Kilimanjaro. *Tanganyika Notes and Records* **64**: 111–114.

Zadroga, F. (1981). The hydrological importance of a montane cloud forest area of Costa Rica. In *Tropical Agricultural Hydrology*, eds. R. Lal and E. W. Russell, pp. 59–73. New York: John Wiley.

52 Changes in soil physical properties after conversion of tropical montane cloud forest to pasture in northern Costa Rica

C. Tobón
Universidad Nacional de Colombia, Medellín, Colombia

L. A. Bruijnzeel and K. F. A. Frumau
VU University, Amsterdam, the Netherlands

J. C. Calvo-Alvarado
Instituto Tecnológico de Costa Rica, Cartago, Costa Rica

ABSTRACT

Within the framework of a larger project studying the hydrological impacts of converting tropical montane cloud forest to pasture in the Tilarán range of northern Costa Rica, physical and hydraulic properties of various volcanic soils were compared in two small watersheds covered with mature lower montane cloud forest and pasture, respectively. *In situ* and laboratory experiments were conducted to determine trends in soil texture, bulk density, porosity, water retention characteristics, infiltration, and (un)saturated hydraulic conductivities with depth under the two types of land cover. Despite their predominantly sandy texture, the soils were rich in organic matter and non-crystalline material such as allophane. Bulk densities were very low and similar between sites for corresponding soil horizons, except for the pasture top-soil which was more compacted, particularly on cow trails. Soil porosity was very high throughout the profile and dominated by macro- and mesopores, again with the exception of the pasture top-soil and the cow trails. Water retention at a suction of 1500 kPa (permanent wilting point) was very high, except in gravelly C-horizons which had low retention capacity. Amounts of plant-available water (i.e. held at suctions between 10 and 1500 kPa) were also high. Surface infiltration rates were relatively high and dominated by "bypass" flow via macropores in the gravelly horizon (at 20–30 cm depth). Spatial variability in infiltration rates was high in the pasture but less in the forest. Saturated hydraulic conductivity at the soil surface was high in general but considerably reduced in the pasture. Values obtained by well permeametry (Guelph permeameter) were higher and considered more representative than those obtained by tension infiltrometry. Unsaturated hydraulic conductivities were also quite high, even at high soil water suctions, except for the gravelly C-horizon, where conductivity decreased almost linearly with decreasing water content. Overall, the results indicate that converting cloud forest to grazed pasture diminishes infiltration of water into the soil due to increased top-soil density and reduced porosity, potentially resulting in increased overland flow and runoff response to rainfall.

INTRODUCTION

Montane cloud forests are important for their high water yield due to a combination of reduced evaporation losses and extra inputs afforded by the interception of wind-driven rain and fog (Zadroga, 1981; McJannet *et al.*, this volume), for erosion control on steep slopes, and for their unique biodiversity (Hamilton *et al.*, 1995; Kappelle and Brown, 2001) As populations depending on the clean water emanating from these

Tropical Montane Cloud Forests: Science for Conservation and Management, eds. L. A. Bruijnzeel, F. N. Scatena, and L.S. Hamilton. Published by Cambridge University Press.

forests increase, so does the importance of cloud forest conservation and understanding the processes underlying their hydrological behavior. Whilst progress has been made in recent years with the quantification of horizontal precipitation inputs to cloud forests (Juvik and Nullet, 1995; Holwerda *et al.*, 2006; McJannet *et al.*, 2007b; Giambelluca *et al.*, this volume; Schmid *et al.*, this volume), and evaporative losses from cloud forests (Santiago *et al.*, 2000; Holwerda, 2005; Motzer *et al.*, 2005; McJannet *et al.*, 2007a; Giambelluca *et al.*, 2009), their soil water dynamics have remained rather poorly documented (Herrmann, 1971; Hafkenscheid *et al.*, 2002; García-Santos, 2007; Bogner *et al.*, 2008; Schrumpf *et al.*, this volume).

Soils under mature tropical forests are generally well aggregated and typically have high infiltration capacity and low (if any) infiltration-excess overland flow (Bonell, 2005) Upon conversion to pasture or annual cropping, considerable changes in soil properties usually occur, particularly after mechanized logging and clearing (Lal, 1987; Schwartz *et al.*, 2000; Grip *et al.*, 2005; Huwe *et al.*, 2008; Zimmermann and Elsenbeer, 2008) Cloud forest clearing on steep slopes often does not involve heavy machinery, whereas rainfall intensities tend to be lower at higher elevations, whereas, in addition, bracken ferns often invade freshly burned montane areas, thereby affording some surface protection against erosion (Aide *et al.*, this volume). As such, the surface impacts of cloud forest clearing may not be all that pronounced (e.g. Edwards, 1979; Duisberg-Waldenberg, 1980), although compaction of pasture top-soils by roaming cattle can be substantial, as demonstrated by the work of Zimmermann and Elsenbeer (2008) on soils derived from metamorphic rocks in southern Ecuador. Volcanic soils represent a special case in that these may show irreversible physical changes upon drying (Kubota, 1976) However, there is comparatively little information on the physical and hydraulic properties of volcanic soils under pristine cloud forest conditions and changes therein after forest conversion (Regalado and Ritter, 2006; García-Santos, 2007; Podwojewski *et al.*, 2008; Schrumpf *et al.*, this volume).

Applications of watershed hydrological response models have increased steadily and these require spatially representative parameters describing the physical and hydraulic properties of the soils (Saghafian *et al.*, 1995; Woolhiser *et al.*, 1996; Parasuraman *et al.*, 2006; Lazarovitch *et al.*, 2007; cf. Huwe *et al.*, 2008) To address this lack of information regarding the soil hydrological impacts of tropical montane cloud forest conversion to pasture, detailed field and laboratory measurements were made of the texture, bulk density, porosity, water retention, infiltration, and hydraulic conductivity of volcanic soils in two small watersheds with mature cloud forest and pasture in the Tilarán range of northern Costa Rica.

STUDY AREA

This study was carried out in two small first-order watersheds in the San Gerardo area on the wet Atlantic slopes of the Tilarán range near Santa Elena, northern Costa Rica (10° 21′ 33″ N, 84° 48′ 05″ W). The cloud forest watershed (3.5 ha) was situated between 1450 and 1600 m.a.s.l. and the pasture watershed (8.7 ha) between 1520 and 1620 m.a.s.l. The area is characterized by high rainfall (4400–6000 mm year^{-1}) and strong winds, and receives considerable amounts of horizontal precipitation, bringing the total annual input to ~9000 mm (Frumau *et al.*, 2006) Whilst the drier Pacific slopes experience a well-defined dry season from January until May, the dry season is much less pronounced on the wet Atlantic slopes (Zadroga, 1981; Clark *et al.*, 2000) Average temperature in 2003 was 17.0 °C (monthly range 15.4–17.7 °C). Relative humidity was generally above 90% and foggy conditions prevailed for 50% (at night) to 60% (during the day) of the time (K. F. A. Frumau, unpublished data).

The mature cloud forest was about 20–22 m tall, with *Ficus crassiuscula, Elaeagia auriculata, Weinmannia wercklei*, and several Myrtaceae being the most common tree species (Guariguata and Kattan, 2003) Tree ferns and palms (<4 m) dominated the understory. Epiphytes were abundant, reaching an estimated total biomass of 16.2 t ha^{-1} (Köhler *et al.*, 2007) The *Pennisetum clandestinum* pasture site was dominated by herbaceous vegetation (mostly bracken) and had numerous cow trails that tended to follow the contours and which occupied, on average, about 30% of the watershed area.

Soils have developed in slightly to moderately weathered lahar deposits of the Quaternary Monteverde formation and in volcanic ashes deposited during a series of eruptive events since the late Miocene (Vargas, 2001) Soil profiles were multiple-layered and soil age increased with depth (Saigusa *et al.*, 1987) The majority of the soils in the study area had similar upper horizons (A, A/B, Bw, and C) but their thickness varied considerably. They were classified as Udands, notably thaptic Hapludands, typic Hapludands, and typic Hydrudands (C. Tobón, unpublished data).

METHODOLOGY

General approach

To define the soil sampling scheme and the number of field sites required to evaluate the various soil physical and hydraulic properties, a detailed soil survey was carried out first in which the soils were classified at the great group level (Soil Survey Staff, 1996) Soil sampling and subsequent field and laboratory measurements were carried out between August 2003 and July 2004. The approximate effects of land use on soil properties were evaluated by comparing the specific properties of paired plots under undisturbed cloud forest and pasture.

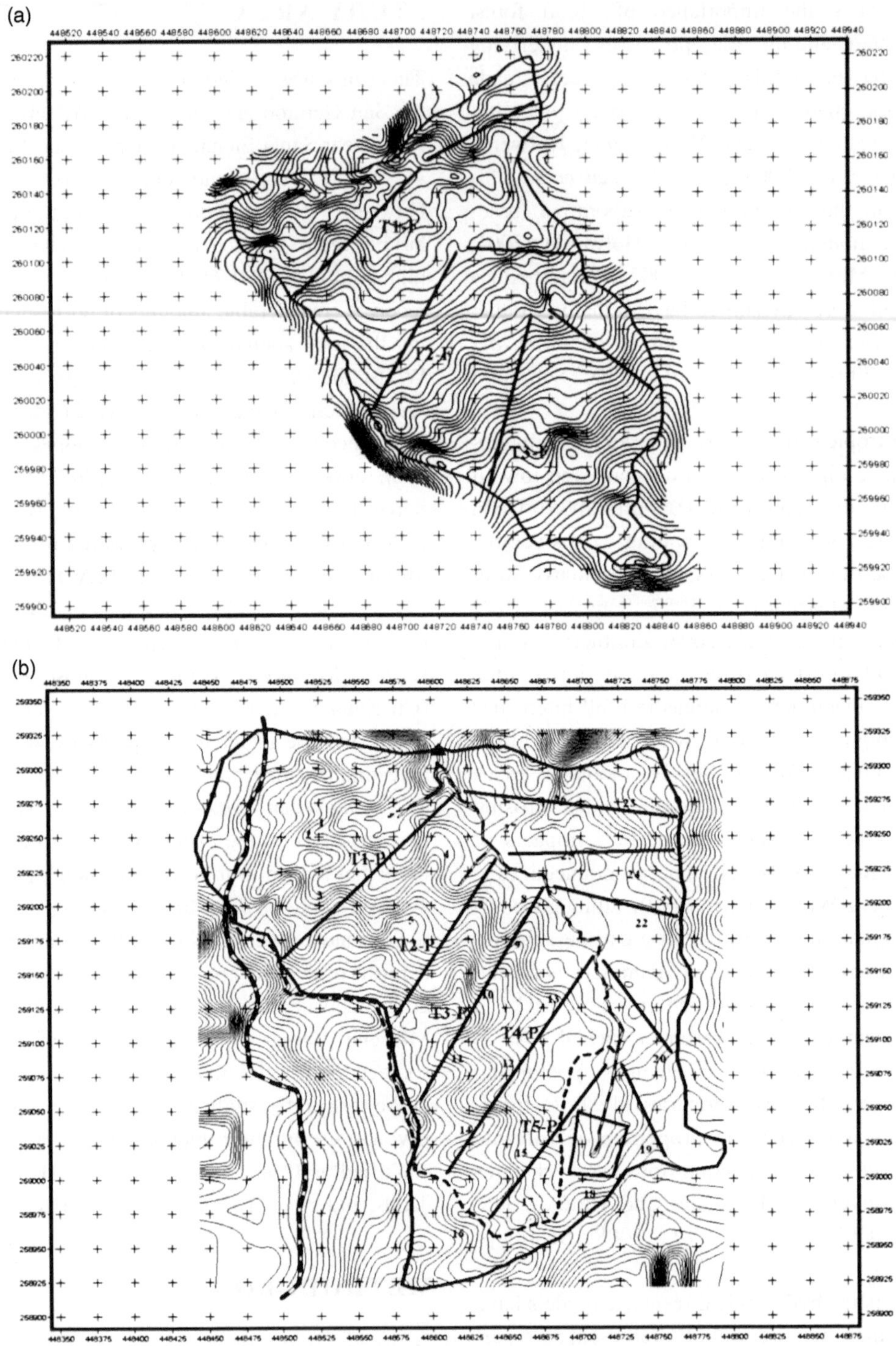

Figure 52.1. Soil sampling transects within (a) mature cloud forest (n = 19 sampling sites) and (b) grazed pasture (n = 27 sampling sites) near San Gerardo, Tilarán range, northern Costa Rica.

In the forest, three transects were laid out which ran from the nearest low ridge on either side of the stream to the corresponding water divides (Figure 52.1a), with a total of 19 sampling sites. In the pasture, five transects containing 27 sampling sites were selected (Figure 52.1b). At each site, soil samples for the laboratory determination of soil physical and hydraulic properties were collected from the different soil horizons (A, Bw, C, 2Ab, 2Bwb, 2C, A/Bb, 3Bwb, and 3C, where b denotes "buried") down to a depth of ~1.5 m (containing the majority of fine roots; Carvajal, 2003) In an attempt to relate and compare field and laboratory results, samples for laboratory analysis were taken close (0.5 m) to the points and at the same depths where the

Table 52.1 *Physical properties and soil organic matter of volcanic soils under cloud forest and grazed pasture near San Gerardo, Monteverde, northern Costa Rica. Average values, standard deviations, and the number of samples are presented, respectively, in each column*

Soil horizon	Texture		Bulk density (kg m^{-3})		Porosity		Soil organic matter (%)	
	Forest	Pasture	Forest	Pasture	Forest	Pasture	Forest	Pasture
A	Loam	Loam	406.5_19_100	527.6_165_123	0.85_0.01_100	0.79_0.07_123	16.2_2.3_10	7.8_3.6_14
A/B	Loam		484.2_21_100		0.81_0.01_34		13.2_3.4_10	9.6_3.3_14
Bw	Sandy loam	Sandy loam	611.8_38_49	626.7_117_103	0.77_0.02_49	0.76_0.04_103	7.8_1.7_10	9.2_2.9_14
C	Coarse to gravelly sand	Sand	712.4_21_64	689.6_100_81	0.73_0.02_64	0.73_0.04_81	2.7_0.78_10	4.6_2.1_14
Ab	Sandy clay	Sandy clay	639.0_159_47	633.7_153_97	0.77_0.06_47	0.76_0.06_97	9.9_3.1_5	10.2_2.9_5
Bw1	Loam	Sandy loam	667.4_191_54	639.5_110_77	0.74_0.06_54	0.75_0.04_77	3.8_0.98_5	6.2_1.8_5
2C1	Fine sand	Fine sand	811.1_80_48	889.5_158_91	0.69_0.04_48	0.66_0.06_91	2.45_0.56_5	2.5_0.87_5
A/Bb	Sandy clay	Sandy loam	693.7_132_66	647.8_141_90	0.73_0.05_66	0.75_0.05_90		
Bw2	Sandy clay	Sandy loam	723.4_121_51	773.9_136_69	0.72_0.05_51	0.70_0.05_69		
2C2	Loamy sand	Fine sand	778.0_76_50	791.0_177_84	0.70_0.05_50	0.70_0.07_84		

field measurements were carried out. All laboratory measurements were conducted at the Agricultural Research Center of the University of Costa Rica (CIA-UCR) in San José.

Soil texture, bulk density, soil organic matter, total porosity, and water retention characteristics

To determine soil bulk density, undisturbed samples were collected with stainless steel rings of 100 cm^3 at each site and depth. The samples were weighed and oven-dried at 105 °C for 48 hours. Bulk density was calculated as the ratio of the mass of the oven-dry sample to its bulk volume. In total, 456 samples were collected from the forest and 548 from the pasture.

For soil texture and particle size distribution, samples of 200 g were collected from each site and depth and subjected to a modified sieve–pipette sedimentation method using 1N NaOH as a dispergent and applying mechanical agitation (CIA-UCR, 1984) The USDA system was used for particle size classification and size limits (Soil Survey Staff, 1996) For soil organic matter (SOM) determination, three samples were collected from each soil horizon and at each forest and pasture site. Samples were mixed per soil horizon, and mean SOM content was determined for each soil horizon using the Walkey–Black method (Black, 1967) down to a depth of 100 cm (containing the first seven horizons; see Table 52.1). Total porosity, defined as total pore volume divided by sample volume, was evaluated from bulk density (kg m^{-3}) and average particle density (taken as 2600 kg m^{-3}; Baver, 1956)

For the characterization of soil water retention curves (WRCs), triplicate samples of 5 cm diameter and 1 cm height were collected at each site and depth. In total, 857 samples were collected, 392 in the forest and 465 in the pasture. Each core was wrapped in aluminum foil and placed in plastic bags to avoid disturbance during transportation to the laboratory. A complete retention curve was determined in order to allow evaluation of the amount of water lost at low suctions. First, the samples were weighed after reaching saturation on a sand box. For low levels of soil water suction (1, 3, 5, 10, and 20 kPa) a sand-box apparatus (Stakman and Van der Harst, 1969) was used. Samples were weighed after equilibration (normally within 48 hours). A pressure-plate apparatus was used to determine moisture contents associated with pressures of 30, 100, 300, 500, 1000, and 1500 kPa, with weighing after equilibration (Klute and Dirksen, 1986) The resulting experimental data pairs were fitted with the RECT 6.0 program (Riverside Salinity Laboratory, 2001) to obtain continuous mathematic descriptions for the WRCs. Because the shape of the WRCs indicated that many of the soils had a bimodal pore size distribution, the equations of Durner (1994) rather than those of Van Genuchten (1980) were used.

Infiltration and unsaturated hydraulic conductivity

An SMS, SW-080B tension infiltrometer (TI) was used to measure infiltration and derive the unsaturated hydraulic conductivity of surface soils through successive measurements at pressure heads of −15, −10, and −5 cm (Reynolds and Elrick, 1991) By adding a final measurement of steady-state infiltration at a tension close to zero, the saturated hydraulic conductivity could be estimated for comparison with the results obtained with other methods (see below). Generally, the time required to approach steady-state conditions varied from ~50 to 140 min. In total, 21 measurements were made in the cloud forest vs. 27 in the pasture, plus seven sites on cow trails. It was found that after the initial steady state was reached, the infiltration rate decreased for a while, after which it increased again to initial steady-state values. To better understand the infiltration dynamics and the underlying two- and three-dimensional soil water fluxes, soil water content was measured simultaneously at selected sites using small TDR probes (Campbell Scientific, 10-cm length) placed underneath the infiltrometer disk, and at specific depths (i.e. at 5, 10, and 15 cm in the A- and Bw-horizons, and at 22 and 25 cm in the top and middle part of the coarse C-horizon). Infiltration rate and soil water content were measured again after 1 week (Vogeler *et al.*, 1996; Wang *et al.*, 1998).

Saturated hydraulic conductivity

Values of saturated hydraulic conductivity (K_s) appear to be influenced by the method used. In addition, values are highly sensitive to sample size, flow geometry and sample collection procedures (Bouma, 1983; Reynolds and Elrick, 1991; Stolte *et al.*, 1994; Davis *et al.*, 1999) Therefore, K_s was characterized using both field and laboratory methods and differently sized samples, viz. (i) the constant-head method using small soil cores (7.5. × 7.5 cm) in the laboratory (Klute and Dirksen, 1986), (ii) field tension infiltrometry (with 20-cm diameter disk), and (iii) the constant-head well permeameter method in uncased boreholes of 6-cm diameter (Guelph permeameter; Reynolds and Elrick, 1985, 1991).

Soil macropores, when filled with water, permit rapid turbulent flow (Beven and Germann, 1982) which may disturb core samples during laboratory experiments. Therefore, a fine nylon cloth (250-μm mesh width) was placed at the base of the core samples to avoid loss of material. In the tension infiltrometer approach the algebraic approximation for steady-state unconfined saturated infiltration rates as proposed by Wooding (1968) was used to calculate K_s from measured infiltration. Determination of field K_s in uncased boreholes involves the use of analytical solutions which are based on the assumption that the soil is homogenous throughout the profile (Reynolds and Elrick, 1985; Xiang, 1994) This condition was not met in the layered soils under consideration. According to Xiang *et al.* (1997) it is important to estimate the conductivity for each soil layer separately in such cases. Therefore, measurements of steady-state rates were made at two different depths, usually at 0.1 m and at 0.2–0.3 m (representing the A- and Bw-horizons and the C-horizon, respectively) but occasionally at greater depth (0.5 m) if required by horizon distribution. Successive measurements were made in separate boreholes placed at a distance of *c.* 0.5 m. In this way, effects of soil heterogeneity were minimized while avoiding side effects of the previous measurement.

Statistical analysis

In general, the experimental design for the statistical analysis was randomized per soil horizon. Soil properties were compared between soil depths and sites within the forest or pasture, and between forest and pasture, on the basis of geometric means, minimum and maximum values, the coefficient of variation, and the Pearson correlation coefficient. Data were analyzed using analysis of variance (ANOVA). Differences were considered significant at 0.05 probability level. For the spatial distribution of K_s, a log-normal statistical distribution was assumed, a common procedure for *in situ* and undisturbed core measurements (Warrick and Nielsen, 1980; Reynolds *et al.*, 2000; see also Zimmermann and Elsenbeer (2008) for in-depth discussion). Therefore, the data were log-transformed (ln) prior to analysis and the antilog taken before interpretation of the results. Correlations between results obtained with different methods were conducted on a site-to-site basis within each land-cover type and soil layer using the ln-transformed K_s values.

RESULTS AND DISCUSSION

Soil texture

Table 52.1 gives the average values and standard deviations for texture, bulk density, and porosity of each soil horizon down to a depth of 1.5 m. In both forest and pasture, sandy textures were predominant throughout the soil profiles and along transects. The uppermost C-horizon consisted mainly of sandy and gravelly particles of about 2 mm (lapilli) that were deposited throughout the study area during the last major volcanic event (Vargas, 2001) Differences in particle size distribution of corresponding horizons beneath forest and pasture were not significant ($p <$ 0.05). However, in the forest, the first C-horizon contained a larger amount of organic material than in the pasture. This organic matter is thought to have derived from the surface layer (Ao) and to have accumulated at the boundary between the highly permeable C-horizon and the denser buried 2Ab-horizon below, presumably due to the sudden decrease in hydraulic conductivity (cf. Table 52.2 below). In the pasture, the sandy C-horizon material was partly mixed with finer material eluviated from the upper Ap-horizon, producing a more loamy overall texture (Table 52.1).

The predominantly sandy texture of the soils is mainly due to the presence of the volcanic ashes, and partly the result of wetting and drying cycles producing the formation of micro-aggregates (Kubota, 1976; Shoji *et al.*, 1993; cf. Schrumpf *et al.*, this volume). Moreover, non-crystalline materials present in volcanic soils act as cementing agents and hamper complete dispersion of mineral particles during textural analysis (Ping *et al.*, 1988), typically resulting in an underestimation of clay content.

Bulk density

Bulk densities were very low throughout the profiles, except for the gravelly C-horizon which showed the highest values (Table 52.1). Low density is a special feature of volcanic soils and related to their highly porous structure, reflecting an abundance of non-crystalline materials and organic matter (Shoji *et al.*, 1993). The bulk density values obtained for the San Gerardo cloud forest soils are among the lowest reported for volcanic soils, but they are comparable to other young volcanic soils (Shoji *et al.*, 1993), with similar amounts of organic matter (Table 52.1). Comparing bulk density values within the same profile, there were no significant differences ($p < 0.05$) between different soil layers, both under forest and pasture, except for the top-soil (A- and A/B-horizons) which exhibited significantly lower values under both land-cover types. However, the difference was considerably less in the case of pasture (Table 52.1).

Bulk densities of each soil layer were also compared (horizontally) between sampling sites (separately for forest and pasture). In the forest, surface soil layers showed no significant differences ($p < 0.05$) between sites but there were large differences for subsurface layers as evidenced by their large standard deviations (Table 52.1). This reflects differences in parent material which varied from sandier volcanic ashes, to lahar deposits producing more clayey sub-soil horizons.

As for the pasture, differences in soil bulk density between sites were much larger, notably for the first 25 cm (Table 52.1). This was mainly due to the deviating values obtained for profiles positioned on slopes where the original A-horizon had become mixed with the underlying A/B- or Bw-horizons. At some sites, the entire top-soil had become mixed with C-horizon material, producing a new top-layer with a coarser texture (sandy loam), possibly due to soil disturbance caused by cattle activity. Moreover, there was a trend toward increasing (top-soil) density values going from the ridges to riparian footslopes. This seems related to erosion of soil particles from the upper parts of the slopes and from cow trails by overland flow (C. Tobón, unpublished data) followed by deposition on the gentler footslopes (Harden, 1992) Comparing soil bulk densities of the first 25 cm below cow trails and non-trail surfaces showed even larger differences ($p < 0.01$). The regular passage of cows and people along the paths produced much higher density values (Figure 52.2a), as has also been found for coarser-textured soils elsewhere (Patel and Singh, 1981; Harden, 1992; Ziegler *et al.*, 2004).

Comparisons of mean bulk density values between forest and pasture show these to differ significantly ($p < 0.05$) in the first 25 cm, with values being larger in the pasture (Figure 52.2a) However, for deeper layers, densities did not differ much between treatments (Pearson $r = 0.89$, $p = 0.0014$) (Figure 52.2). Spatial variability in C-horizon bulk density under pasture was much higher than under forest (Table 52.1), although the degree of mixing varied between sites.

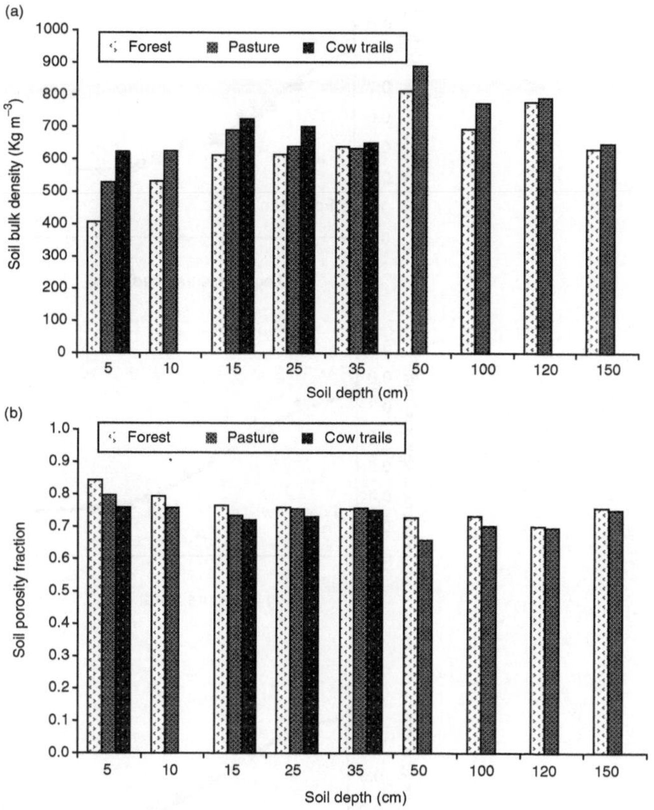

Figure 52.2. (a) Bulk density (kg m^{-3}) and (b) porosity (fraction) of the different soil horizons beneath mature cloud forest and grazed pasture sites near San Gerardo, Tilarán range, northern Costa Rica.

Porosity

In line with the low density and well-developed structure of the soils, total porosity values were very high at 0.69–0.85 in the forest and 0.66–0.79 in the pasture (Table 52.1; cf. Furuhata and Hayashi, 1980) For most soil horizons significant differences occurred between forest and pasture ($p < 0.05$), whereas within the pasture significant differences ($p < 0.05$) were observed between top-soil porosities associated with cow trails and non-trail surfaces (Figure 52.2b).

The average 30% increase in A-horizon bulk density in the pasture compared to the forest produced an average reduction in top-soil porosity of only 7% (Table 52.1). As will be shown below, this change was primarily effected by a reduction in pore volume associated with macropores.

Water retention characteristics

The water retention curves (WRCs) for the respective layers of the forest soil showed broadly similar tendencies indicating broadly comparable soil properties (Figure 52.3a) Similar trends were observed for the pasture soils (Figure 52.3b), except below the cow trails which showed completely different behavior

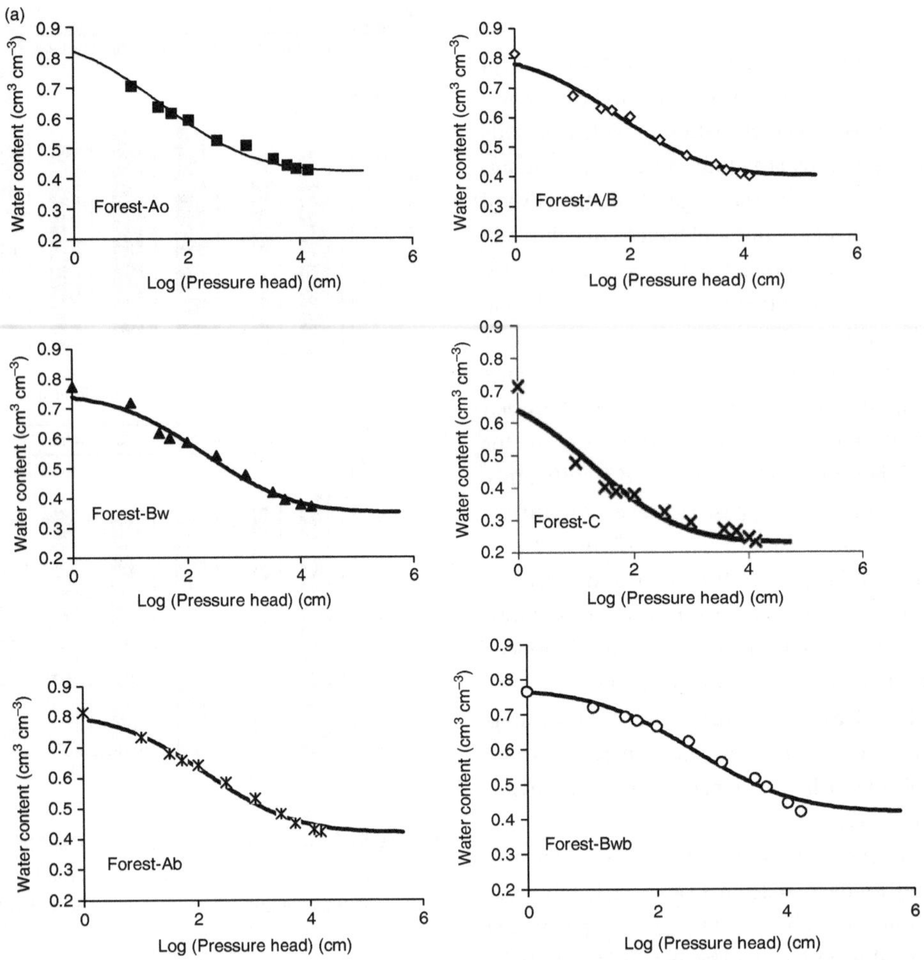

Figure 52.3. Measured and fitted water retention curves for soils under (a) mature cloud forest, (b) grazed pasture, and (c) compacted cow trails near San Gerardo, Tilarán range, northern Costa Rica.

(rather resembling that of clayey soils; Figure 52.3c) because of their compacted nature. Most WRCs resembled the curves normally derived for coarse-textured soils, despite the very low bulk densities of the present soils. According to Luxmoore (1981), macropores empty at pressures between 0 and 0.3 kPa and mesopores between 0.3 and 30 kPa. Thus, a slight trend was discernible toward increased soil water retention with depth, reflecting the larger numbers of mesopores (2–200 μm) and micropores (<2 μm) at depth compared to a dominance of macropores (>200 μm) and micropores near the surface (Luxmoore, 1981) As a result, water content close to saturation was larger in surface horizons than in deeper layers (Figure 52.3a,b).

Generally, plant-available water (*PAW*) is defined as the amount of water held between field capacity and permanent wilting point, estimated as the difference in water contents between soil matric potentials of 33 and 1500 kPa (Marshall *et al.*, 1996). In the forest, values of *PAW* defined in this way, ranged from 0.09 to 0.17 $cm^3\ cm^{-3}$ in the upper layers and from 0.11 to 0.20 $cm^3\ cm^{-3}$ in deeper layers. In the pasture, values varied from 0.13–0.19 $cm^3\ cm^{-3}$ in top-soil vs. 0.12–0.18 in deeper layers. Whilst a suction value of 1500 kPa is widely accepted to characterize the permanent wilting point for many plants growing in a wide range of soils, the generally used value for field capacity (33 kPa) is not appropriate for volcanic soils. Instead, for Andisols formed under humid climates, a value of 10 kPa is more suitable (Saigussa *et al.*, 1987) Taking 10 kPa to denote field capacity, *PAW* values increased by up to 80% for the forest soils and up to 35% for the pasture soils, with the largest increases noted for top-soils.

The studied soils appeared to lose a large part of their water already at comparatively low suctions (*c.* 36% at suctions <33 kPa). At the same time, the shape of the WRCs indicated that most soil horizons were capable of retaining large amounts of water at wilting point (1500 kPa). Indeed, the present soils retained roughly twice the amount of water held by highly clayey

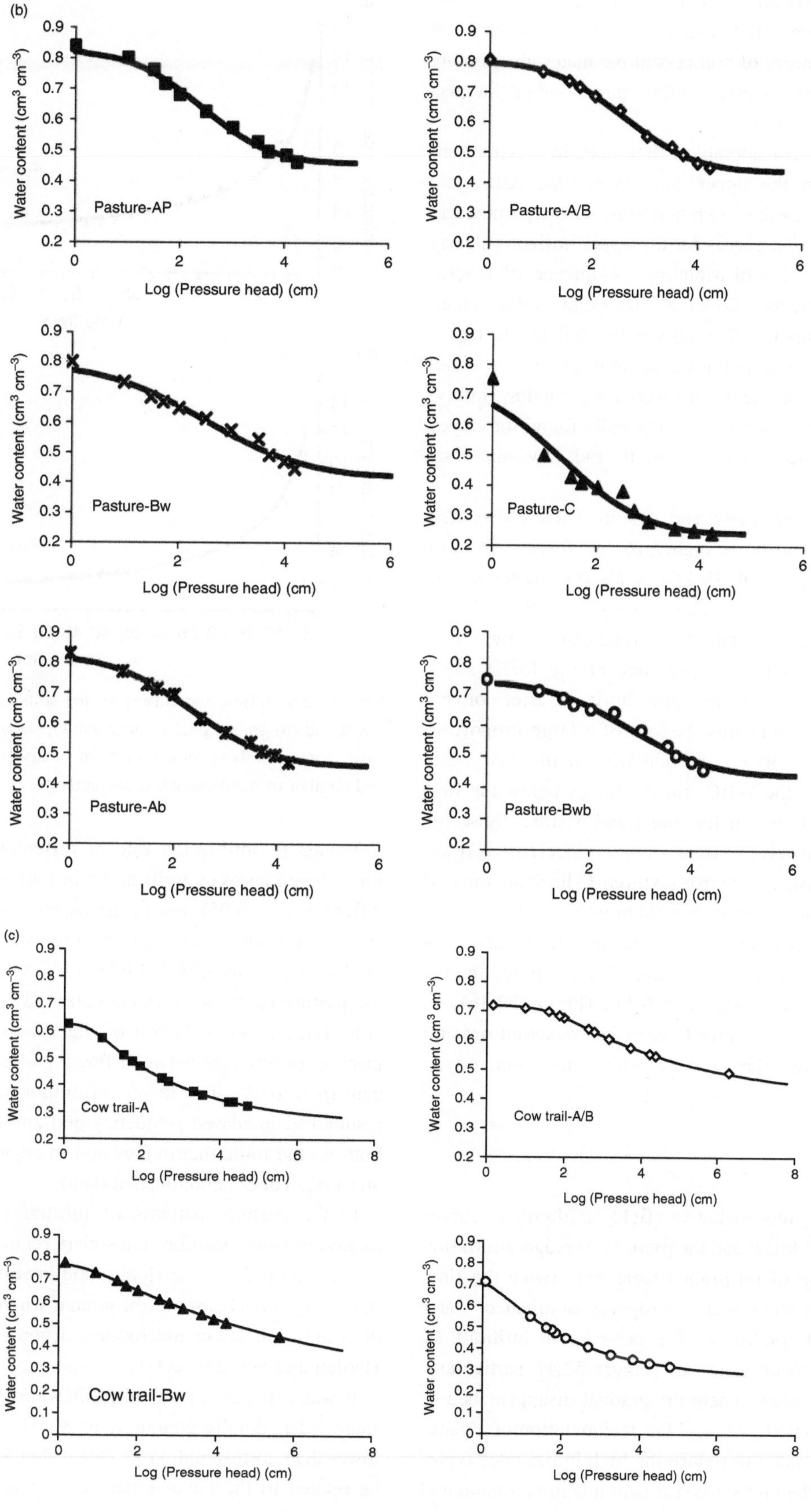

Figure 52.3. *(cont.)*

soils in the Amazon (Tobón, 1999; Tomasella *et al.*, 2000) This high residual water retention of volcanic soils has been attributed to their high concentrations of non-crystalline materials (notably allophanic clays) intimately mixed with organic matter (Saigusa *et al.*, 1987; Ito *et al.*, 1991).

The WRCs suggest considerable differences in water retention capacity between the upper soil layers (A-, A/B-, and Bw-horizons) and the C- and Bwb-horizons below, particularly in the cloud forest. Whilst curves for the upper horizons clearly show a bimodal pore size distribution (dominance of macro- and micropores), in deeper horizons this trend is less clear, implying fewer macropores. The curves for soil layers below 1 m depth generally (but not always) showed only a small reduction in soil water content with increasing suction (implying a dominance of micropores), as is normally found for clayey soils with a low concentration of roots and low microbial activity.

Comparing WRCs for forest and pasture soils shows that although water availability is higher in the forest for most layers, the water retained at 33 kPa is always higher in the pasture. This is thought to reflect changes in the relative abundances of macropores and their replacement by either meso- or micropores. The disappearance of the forest soil's dual pore size distribution in the upper horizons after conversion to grazed pasture indicates the loss of a large proportion of the macro- and mesopores, particularly on the cow trails (Figure 52.3c). Whilst the WRCs for the layers below the first C-horizon were rather similar for forest and pasture, those for the C-horizons themselves were rather different (Figure 52.3a, b). In the forest, the coarse-textured C-horizon showed very fast water release even at low suctions (<1 kPa), whilst for the more loamy C-horizon of the pasture the release was more gradual. In addition, the pasture C-horizon retained a very large amount of water was at 1500 kPa (Figure 52.3b). As indicated previously, the pasture C-layer has received organic and fine material from the top-soil, which has rendered its texture more loamy.

Infiltration

Figure 52.4 shows representative field infiltration curves obtained for the cloud forest and the pasture. Average infiltration rates at the beginning of infiltration were only twice the final ponded values, suggesting that macropores dominated water flow under saturated conditions. The decrease in infiltrability with time seems to occur quite fast (Figure 52.4), particularly after conversion to pasture where the gradual disappearance of macropores due to compaction, and the accumulation of a dense mass of grass roots tended to reduce the high initial rates typically associated with the undisturbed situation (Zimmermann and Elsenbeer, 2008).

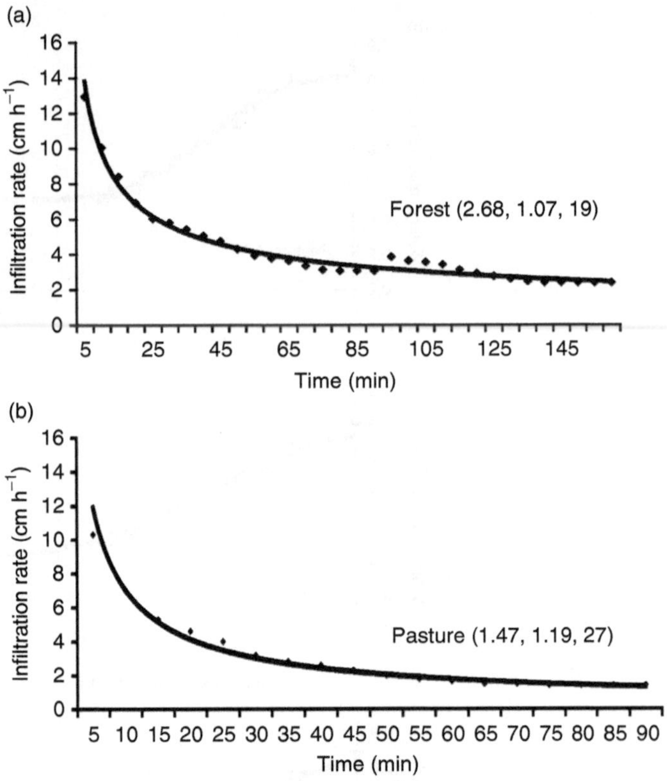

Figure 52.4. Infiltration curves for top-soils under (a) mature cloud forest and (b) grazed pasture near San Gerardo, Tilarán range, northern Costa Rica. Numbers shown are final steady-state rate, standard error, and number of measurements, respectively.

Values of infiltration rate at different locations within the forest (even between trails and non-trails) were not significantly different ($p < 0.05$), but in the pasture (also on the cow trails), spatial variability was high. Average surface infiltration rate in the forest (f_c) was 2.68 ± 1.07 cm h^{-1} vs. 1.47 ± 1.19 cm h^{-1} in the pasture (non-trail surfaces) and 0.87 ± 0.34 cm h^{-1} on the cow trails. After inclusion of the values for cow trails in the pasture, differences between forest and pasture became significant ($p < 0.05$). The lower infiltration rates on the cow paths resulted in increased frequency and amounts of overland flow, both on the trails themselves and in adjacent downslope grassy areas (C. Tobón, unpublished data).

In the pasture, patterns of infiltration were systematically related to slope position; rates were relatively high on the upper and middle parts of the slopes, and lowest on the footslopes and in concave areas close to the stream where soils had higher bulk densities, and fewer macropores according to their WRCs (cf. Harden and Scruggs, 2003).

It was expected that final infiltration rates f_c would approach saturated hydraulic conductivity K_s. However, f_c was typically lower than corresponding K_s (see Table 52.2 below), which may be related to the development of a perched water table above the gravelly C-horizon during the infiltration experiments. As

Table 52.2 *Values of geometric means, maxima, minima, coefficient of variation (CV), and number of samples* (n) *of saturated hydraulic conductivity (cm day^{-1}) of different layers of volcanic soils under cloud forest, grazed pasture, and cow trails near San Gerardo, Monteverde, northern Costa Rica*

Soil	Guelph permeameter					Tension infiltrometer					Small-cores method				
horizon	Mean	Max.	Min.	CV%	*n*	Mean	Max.	Min.	CV%	*n*	Mean	Max.	Min.	CV%	*n*
Forest															
A	71.4	183.4	34.5	54.4	16	37.1	62.3	9.2	10.4	21	47.2	90.9	15.9	53.4	30
Bw	18.1	21.4	15.7	16.0	5						15.7	20.7	9.6	23.7	19
C	1204	2563	658.1	83.8	5						3171	4827	2235	31.2	12
2Ab	37.6	57	28.4	33.4	5						29.6	63.4	15.9	60.8	12
2Bwb											13.3	51.0	4.5	146	7
Pasture															
A	24.3	48.4	2.9	88.5	17	17.7	38.2	2.2	10.5	27	16.9	37.4	7.2	53.6	25
Bw	29.5	56.7	17.3	51.0	5						24.1	41.1	6.5	37.5	14
C	580	1025	193	62.6	5						1414	1761	1143	15.1	14
2Ab	14.2	45.7	4.6	127.1	5						14.1	24.0	7.4	29.5	11
2Bwb											8.5	15.9	2.3	52.7	9
Cow trails															
A-Bw						3.2	8.3	1.3	2.5	7	3.3	105.4	0.3	1401	9

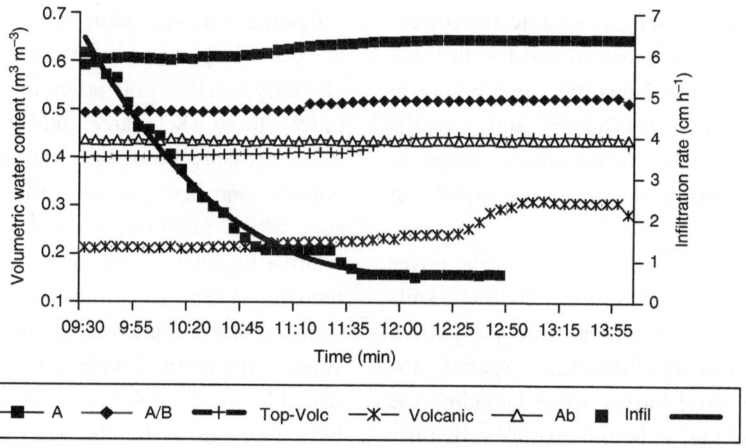

Figure 52.5. Trends in infiltration rate as a function of soil water content in a layered volcanic soil beneath mature cloud forest, San Gerardo, Tilarán range, northern Costa Rica.

illustrated in Figure 52.5, soil water content in the top (A) layer changed moderately after the start of infiltration, while underneath this layer (A/B) a response was only observed later (although before attaining steady-state conditions). This indicates that water moved from the less permeable A- and Bw-horizons to the more permeable C-horizon below. Once the wetting front reached the top of this coarse layer, ponding occurred because of the poor contact between the two layers. Only after saturation was reached at the top of the coarse layer, water began to move through it, albeit at a rate controlled by the conductivity of the less permeable horizons above it, and until the wetting front reached the 2Ab- or 2A/Bb-horizons found below the C-horizon. These deeper layers had considerably lower permeabilities than either the gravelly C-horizon or the top-layers (Table 52.2 below).

Saturated hydraulic conductivity

Table 52.2 presents average, maximum, and minimum values as well as the coefficient of variation of K_s for the different profiles

Table 52.3 *Site-wise Pearson correlation coefficient (r) and statistical significance* (p-*value*) *of differences in saturated hydraulic conductivity as obtained using three different methods: Tension infiltrometer (TI), Guelph permeameter (GP), and intact soil cores (SC). The analysis was conducted using ln-transformed data from cloud forest and grazed pasture near San Gerardo, Monteverde, Costa Rica*

Land Cover	Methods compared	r-value	p-value
Cloud forest	GP vs. TI	0.2876	0.3768
	GP vs. SC	0.2121	0.4563
	TI vs. SC	0.0835	0.7821
Pasture	GP vs. TI	0.4722	0.2457
	GP vs. SC	0.3864	0.3123
	TI vs. SC	0.2156	0.4465

as obtained with the three methods used (i.e. constant-head permeametry in the laboratory, tension infiltrometry, and Guelph permeameter in the field). Significant differences were found between the respective soil layers (vertical comparison) per land cover, the differences being largest between the surface layers (A–Bw) and the C-horizon, both under forest and under pasture (Table 52.2). Values of K_s under forest were generally significantly higher ($p < 0.05$) than beneath pasture for corresponding soil horizons (horizontal comparison) and for all three methods. The higher values for the forest sites are primarily related to the presence of large macropores and a well-developed soil structure. K_s reached its maximum values in the coarse-textured upper C-horizon but decreased rapidly in the layers underneath.

Comparing corresponding soil layers for the two land-cover types, spatial variability in K_s was much lower in the forest soils regardless of the method used, in agreement with the pattern described earlier for soil bulk density (Table 52.1). Spatial variability in pasture K_s was increased further when including the cow trails (Table 52.2). However, although spatial variability was larger in the pasture, the CV of K_s in the C-layer was larger in the forest, possibly because of spatial variations in amounts of illuviated organic matter. The high CV values obtained with the small-core method for the 2Ab- and 2Bwb-layers underneath the C-horizon mainly reflect variability in soil texture resulting, in turn, from differences in parent material (Table 52.1).

Top-soil K_s values as measured with the Guelph permeameter and on small cores were consistently lower close to the stream (narrow riparian area) than at upslope locations, particularly in the pasture. This pattern is the inverse of that found earlier for soil bulk density and caused a decrease in the time to ponding in a downslope direction, such that overland flow occurrence was increased even during small rainfall events (cf. Harden and Scruggs, 2003).

Table 52.2 also shows that K_s values differed between methods. Generally, for all soil layers and land covers, values of field K_s obtained with the Guelph permeameter were the highest, except for the C-layer, for which the small-cores method gave higher values. Although comparisons between different techniques for measuring K_s suffer from the lack of independent benchmark values, an attempt is made below to explain some of the observed contrasts. The higher K_s values found generally with the Guelph permeameter can be explained by the larger sample size (i.e. vertical area) employed by this method, thereby allowing more representative inclusion of large macropores (compared to the small-cores method; Davis *et al.*, 1999) and multidimensional flow (compared to the one-dimensional flow imposed by tension infiltrometry; Reynolds *et al.*, 1985, 2000). Also, although most 15 cm deep auger holes included only the A-horizon, at some pasture sites they also included the A/B- or Bw-horizon. Because K_s values of the latter were generally higher than that of the A-horizon (Table 52.2), their inclusion would positively influence the result.

Tension infiltrometry yielded significantly ($p < 0.05$) lower mean, maximum, and minimum values of top-soil K_s values compared to those obtained with the Guelph permeameter or the small-cores method (with the exception of the pasture in the latter case). This may be related to one or both of the following reasons: while smoothing the soil surface to install the ceramic plate prior to the measurement, even if done very carefully, surface soil pores may be destroyed or become partially closed (Schwärzel and Punzel, 2007). Alternatively, a lower pressure head than imposed may develop in part of the contact sand between the infiltrometer disk and the underlying soil, which produces a decrease in flow through macropores relative to what would have occurred without the presence of the contact sand (Reynolds *et al.*, 2000).

In the sandy to gravelly C-layer, values of K_s obtained with the small-cores method were 2.4–2.6 times higher than those produced by the Guelph permeameter, whereas values were similar between methods for the other soil horizons (Table 52.2). This may reflect rapid preferential flow through the large macropores of this coarse layer which may have extended along the entire length of the sample cores. Moreover, it was observed that at the beginning of the experiments water passing through the samples flushed out organic matter and other fine material, thereby leaving only gravel and coarse sand. Indeed, steady-state K_s values were higher at the end of the experiments although the samples were completely saturated from the start.

The lower K_s values obtained with the small-cores method for the other, less coarse layers might theoretically be explained by compaction during coring or simply to inadequate sample size (Bouma, 1983; Davis *et al.*, 1999; Reynolds *et al.*, 2000). However, compaction of the present samples is unlikely because of their sandy loam texture and careful collection

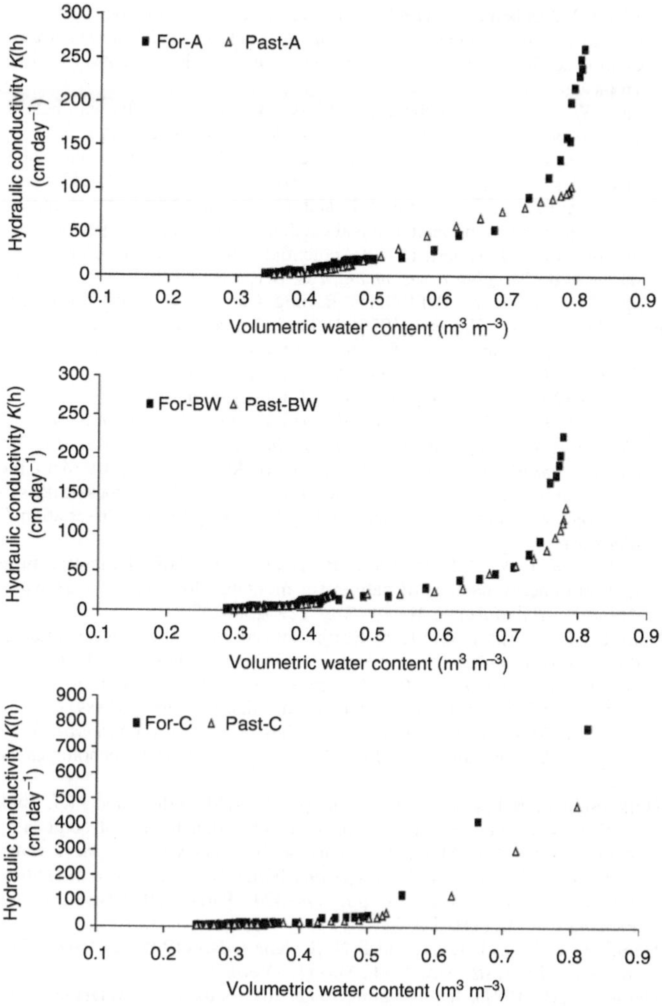

Figure 52.6. Relationships between unsaturated hydraulic conductivity (cm h^{-1}) and volumetric water content (cm^3 cm^{-3}) for three main horizons within layered volcanic soils beneath mature cloud forest and grazed pasture near San Gerardo, Tilarán range, northern Costa Rica.

and transport. Moreover, large samples are more difficult to collect and transport without disturbance. Rather, the present results can be explained by the fact that the soils had visible pores in all directions that were easily closed off by the core walls upon sampling, thereby restricting preferential flow.

The value of K_s measured in the field includes any effects of entrapped air and may be twice as small as the true saturated value (Stephens *et al.*, 1983; Reynolds and Elrick, 1985). However, this is unlikely to be the case in the present soils which remained wet throughout the study period, thereby impeding air entry to most soil pores. Therefore, the values obtained with the Guelph permeameter are considered the most representative.

Unsaturated hydraulic conductivity

Figure 52.6 presents the unsaturated hydraulic conductivity K_h as a function of volumetric water content (θ) for the three main soil layers (A, Bw, and C) in the forest and pasture. As found earlier for the WRCs (Figure 52.3), the conductivities for the upper two soil layers in the forest show the dominant influence of macropores, in that conductivity at high water content ($\theta > 0.75$) decreased rapidly for only small decreases in water content. At low water contents ($\theta < 0.40$) the changes in K_h were very small. Conversely, K_h in the sandy C-layer decreased to a very low value already at a water content of about 0.50 cm^3 cm^{-3} (Figure 52.6). Thus, whilst K_s of the C-horizon greatly exceeded that of the A- or Bw-horizon (Table 52.2), the opposite was true for K_h at relatively minor suction values (10–30 kPa) corresponding to θ values of 0.47–0.42.

Relationships of K_h and θ differed considerably between forest and pasture, particularly for the A-horizon and to a lesser extent the C-horizon (Figure 52.6). For example, at a suction of 10 kPa (corresponding to a θ of 0.59 cm^3 cm^{-3}), K_h in the A-horizon of the forest was about three orders of magnitude higher than in the pasture, while in the Ab-horizon below the C-layer it was still twice as high (C. Tobón, unpublished data). However, at relatively low water contents ($\theta < 0.45$ cm^3 cm^{-3}) K_h was similar in forest and pasture soils, except for the cow trails where K_h was very low throughout the moisture range (not shown). Such contrasts are associated with the reduction in the number of macropores in the pasture and their replacement by meso- and micropores in the top layers (Ap and Bw). The present results agree with the observations of Buytaert (2004) and Podwojewski *et al.* (2008) on volcanic soils in Ecuador and illustrate the fragility of these soils.

CONCLUSIONS

The present results demonstrate that the soil physical and hydraulic properties of volcanic soils under old-growth montane cloud forest are rather unique in that they have very low bulk densities throughout the soil profile, high porosity (provided mainly by macropores), high water retention at 1500 kPa, high plant-water availability (between suctions of 10 and 1500 kPa) and high infiltration rate and hydraulic conductivity. Conversion to grazed pasture produced important changes in these properties, particularly in the top 25 cm. Specifically, top-soil infiltration and hydraulic conductivity were strongly reduced, as was soil porosity, whereas bulk densities were increased. These changes reflect changes in pore size distributions, notably a decrease in the number of macropores and their replacement by meso- and micropores. The greatest changes were observed for the soils underneath cow trails in the pasture, where infiltration was greatly diminished and runoff production increased.

The large difference in K_s of the top-soil and the C-horizon under forest is likely to result in a stormflow regime dominated by fast lateral sub-surface flow (Bonell, 2005) However, on the steep and upper slopes of the pasture watershed, the gravelly C-horizon had become mixed with material from the layers above, resulting in a more loamy texture, slightly lower bulk density and considerably reduced K_s, as well as in improved water retention compared to the original C-horizon. Thus, contrasts between the altered C-horizon and the layers above and below were less pronounced and flows to deeper layers more continuous. This may lead, in turn, to a less responsive lateral sub-surface flow regime.

The results of this study indicate that land cover had a major effect on key soil properties, and consequently on water movement in these layered volcanic soils. As such, parameterization of soil hydraulic properties in watershed models based on soil type or texture alone is likely to lead to poor predictions of the hydrological impact of cloud forest conversion to pasture.

ACKNOWLEDGEMENTS

Financial support for this study by the Forestry Research Programme of the Department for International Development (DFID) of the UK (project R7991) is gratefully acknowledged. The views expressed here are not necessarily those of DFID. Eulogio Jimenez and the owners of the forest and pasture catchments are thanked for their permission to use the sites.

REFERENCES

Baver, L. D. (1956). *Soil Physics*. New York: John Wiley.

Beven, K., and P. Germann (1982). Macropores and water flow in soils. *Water Resources Research* **18**: 1311–1325.

Black, C. E. (1967). *Method of Soil Analysis*, American Society of Agronomists Monograph No. 9. Madison, WI: American Society of Agronomists.

Bogner, C., S. Engelhardt, J. Zeilinger, and B. Huwe (2008). Visualization and analysis of flow patterns and water flow simulations in disturbed and undisturbed tropical soils. In *Gradients in a Tropical Mountain Ecosystem of Ecuador*, eds. E. Beck, J. Bendix, I. Kottke, F. Makeschin, and R. Mosandl, pp. 403–412. Berlin: Springer-Verlag.

Bonell, M. (2005). Runoff generation in tropical forests. In *Forests, Water and People in the Humid Tropics*, eds. M. Bonell and L. A. Bruijnzeel, pp. 314–406. Cambridge, UK: Cambridge University Press.

Bouma, J. (1983). Use of soil survey data to select measurement techniques for hydraulic conductivity. *Agricultural Water Management* **6**: 177–190.

Buytaert, W. (2004). The properties of the soils of the south Ecuadorian *páramo* and the impact of land use changes on their hydrology. Ph.D. thesis, Catholic University Leuven, Leuven, Belgium.

Carvajal, A. (2003). Distribución de raíces finas en suelos del bosque nuboso y pastos, en Monteverde, Costa Rica. B.Sc. thesis, Instituto Tecnológico de Costa Rica, Cartago, Costa Rica.

Centro de Investigaciones Agronómicas (CIA-UCR) (1984). *Manual de laboratorio de edafología*. San José, Costa Rica: CIA-UCR.

Clark, K. L., R. O. Lawton, and P. R. Butler (2000). The physical environment. In *Monteverde: Ecology and Conservation of a Tropical Cloud Forest*, eds. N. M. Nadkarni and N. T. Wheelwright, pp. 15–34. Oxford, UK: Oxford University Press.

Davis, S. H., R. A. Vertéis, and R. P. Silberstein (1999). The sensitivity of a catchment model to soil hydraulic properties obtained by using different measuring techniques. *Hydrological Processes* **13**: 677–688.

Duisberg-Waldenberg, P. (1980). Erosión y conservación de suelos. In *Estudio ecológico integral de las zonas de afectación del Proyecto Arenal*, Centro Científica Tropical, 2–1–2–65. San José, Costa Rica: Centro Científica Tropical.

Durner, W. (1994). Hydraulic conductivity estimation for soils with heterogeneous pore structure. *Water Resources Research* **30**: 211–223.

Edwards, K. A. (1979). The water balance of the Mbeya experimental catchments. *East African Agricultural and Forestry Journal* **1**: 231–247.

Frumau, K. F. A., L. A. Bruijnzeel, and C. Tobón (2006). Measurement of precipitation in montane tropical catchments: comparative performance of conventional, spherical and "potential" rain gages. In *Forest and Water in a Changing Environment*, eds. S. R. Liu, G. Sun, and P. S. Sun, pp. 104–108. Vienna: IUFRO, and Beijing: Chinese Academy of Forestry.

Furuhata, A., and S. Hayashi (1980). Relation between soil structure and soil pore composition: case of volcanogenous soils in Tokachi district. *Research Bulletin o the Hokkaido National Agricultural Experimental Station* **126**: 53–58 (in Japanese, with English summary).

García-Santos, G. (2007). *An ecohydrological and soils study in a montane cloud forest in the National Park of Garajonay, La Gomera (Canary Islands, Spain)*. Ph.D. thesis, VU University Amsterdam, Amsterdam, the Netherlands. Also available at www.falw.vu.nl/nl/onderzoek/earth-sciences/geo-environmental-science-and-hydrology/hydrology-dissertations/index.asp.

Giambelluca, T. W., R. E. Martin, G. P. Asner, *et al.* (2009). Evapotranspiration and energy balance of native wet montane cloud forest in Hawai'i. *Agricultural and Forest Meteorology* **149**: 230–243.

Grip, H., J. M. Fritsch, and L. A. Bruijnzeel (2005). Soil and water impacts during forest conversion and stabilization to new land use. In *Forests, Water and People in the Humid Tropics*, eds. M. Bonell and L. A. Bruijnzeel, pp. 563–589. Cambridge, UK: Cambridge University Press.

Guariguata, M. R., and G. H. Kattan (2003). *Ecología y conservación de bosques Neotropicales*. San José, Costa Rica: Editorial Tecnológica de Costa Rica.

Hafkenscheid, R. L. L. J., L. A. Bruijnzeel, R. A. M. deJeu, and N. J. Bink (2002). Water budgets of two upper montane rain forests of contrasting stature in the Blue Mountains, Jamaica. In *Proceedings of the 2 International Colloquium on Hydrology and Water Management in the Humid Tropics*, ed. J. S. Gladwell, pp. 399–424. Paris: IHP-UNESCO, and Panama City: CATHALAC.

Hamilton, L. S., J. O. Juvik, and F. N. Scatena (eds.) (1995). *Tropical Montane Cloud Forests*. New York: Springer-Verlag.

Harden, C. P. (1992). Incorporating roads and footpaths in watershed-scale hydrologic and soil erosion models. *Physical Geography* **13**: 368–385.

Harden, C. P., and P. D. Scruggs (2003). Infiltration on mountain slopes: a comparison of three environments. *Geomorphology* **55**: 5–24.

Hermann, R. (1971). Die zeitlichen Anderung der Wasserbindung im Boden unter verschiedenen Vegetationsformationen der Hohenstufen eines tropischen Hochgebirges (Sierra Nevada de Santa Marta, Kolumbien). *Erdkunde* **25**: 90–102.

Holwerda, F. (2005). Water and energy budgets of rain forests along an elevation gradient under maritime tropical conditions. Ph.D. thesis, VU University Amsterdam, Amsterdam, the Netherlands. Also available at www.falw.vu.nl/nl/onderzoek/earth-sciences/geo-environmental-science-and-hydrology/hydrology-dissertations/index.asp.

Holwerda, F., R. Burkard, W. E. Eugster, *et al.* (2006). Estimating fog deposition at a Puerto Rican elfin cloud forest site: comparison of the water budget and eddy covariance methods. *Hydrological Processes* **20**: 2669–2692.

Huwe, B., B. Zimmermann, J. Zeilinger, M. Quizhpe, and H. Elsenbeer (2008). Gradients and patterns of soil physical parameters at local, field and catchment scales. In *Gradients in a Tropical Mountain Ecosystem of Ecuador*, eds. E. Beck, J. Bendix, I. Kottke, F. Makeschin, and R. Mosandl, pp. 391–402. Berlin: Springer-Verlag.

Ito, T., S. Shoji, and M. Saigusa (1991). Classification of volcanic ash soils from Konsen district, Hokkaido, according to the last *Keys to Soil Taxonomy* (1990). *Japanese Journal of Soil Science and Plant Nutrition* **62**: 237–247 (in Japanese, with English abstract).

Juvik, J. O., and D. Nullet (1995). Relationships between rainfall, cloud water interception and canopy throughfall in a Hawaiian montane forest. In *Tropical Montane Cloud Forests*, eds. L. S. Hamilton, J. O. Juvik, and F. N. Scatena, pp. 165–182. New York: Springer-Verlag.

Kappelle, M., and A. D. Brown (2001). *Bosques nublados del neotropico*. Heredia, Costa Rica: INBIO.

Klute, A., and C. Dirksen (1986). Hydraulic conductivity and diffusivity: laboratory methods. In *Methods of Soil Analysis*, Part 1, *Physical and Mineralogical Methods*, ed. A. Klute, pp. 687–734. Madison, WI: American Society of Agronomy.

Köhler, L., C. Tobón, K. F. A. Frumau, and L. A. Bruijnzeel (2007). Biomass and water storage dynamics of epiphytes in old-growth and secondary montane cloud forest stands in Costa Rica. *Plant Ecology* **193**: 171–184.

Kubota, T. (1976). Surface chemical properties of volcanic ash soils: especially on phenomenon and mechanism of irreversible aggregation of the soil by drying. *Bulletin of National Agricultural Science* **28**B: 1–74 (in Japanese, with English abstract).

Lal, R. (1987). *Tropical Ecology and Physical Edaphology*. New York: John Wiley.

Lazarovitch, N., A. Ben-Gal, J. Simunek, and U. Shani (2007). Uniqueness of soil hydraulic parameters determined by a combined Wooding inverse approach. *Soil Science Society of America Journal* **71**: 860–865.

Luxmoore, R. J. (1981). Micro-, meso-, and macroporosity of soil. *Soil Science Society of America Journal* **45**: 671–672.

Marshall, T. J., J. W. Holmes, and C. W. Rose (1996). *Soil Physics*. Cambridge, UK: Cambridge University Press.

McJannet, D. L., P. G. Fitch, M. G. Disher, and J. Wallace (2007a). Measurements of transpiration in four tropical rainforest types of north Queensland, Australia. *Hydrological Processes* **21**: 3549–3564.

McJannet, D. L., J. S. Wallace, and P. Reddell (2007b). Precipitation interception in Australian tropical rainforests. II. Altitudinal gradient of cloud interception, stemflow, throughfall and interception. *Hydrological Processes* **21**: 1703–1718.

Motzer, T., Munz, N., Küppers, M., Schmitt, D., and Anhuf, D (2005). Stomatal conductance, transpiration and sap flow of tropical montane rain forest trees in the southern Ecuadorian Andes. *Tree Physiology* **25**: 1283–1293.

Parasuraman, K., A. Elshorbagy, and S. Cheng (2006). Estimating saturated hydraulic conductivity in spatially variable fields using neural network ensembles. *Soil Science Society of America Journal* **70**: 1851–1859.

Patel, M. S., and N. T. Singh (1981). Changes in bulk density and water intake rate of a coarse textured soil in relation to different levels of compaction. *Journal of the Indian Society of Soil Science* **29**: 110–112.

Ping, C. T., S. Shoji, and T. Ito (1988). Properties and classification of three volcanic ash-derived pedons from Aleutian Islands and Alaska Peninsula, Alaska. *Soil Science Society of America Journal* **52**: 455–462.

Podwojewski, P., J. L. Janeau, and Y. Leroux (2008). Effects of agricultural practices on the hydrodynamics of a deep tilled hardened volcanic ash–soil (Cangahua) in Ecuador. *Catena* **72**: 179–190.

Regalado, C., and A. Ritter (2006). Characterizing water dependent soil repellency with minimal parameter requirement. *Soil Science Society of America Journal* **69**: 1955–1966.

Reynolds, W. D., and D. E. Elrick (1985). In situ measurements of field-saturated hydraulic conductivity, sorptivity, and the alpha-parameter using the Guelph Permeameter. *Soil Science* **140**: 292–302.

Reynolds, W. D., and D. E. Elrick (1991). Determination of hydraulic conductivity using a tension infiltrometer. *Soil Science Society of America Journal* **55**: 633–639.

Reynolds, W. D., D. E. Elrick, and B. E. Clothier (1985). The constant-head well permeameter: effect of unsaturated flow. *Soil Science* **139**: 172–180.

Reynolds, W. D., B. T. Bowman, R. R. Brunke, C. F. Drury, and C. S. Tan (2000). Comparison of tension infiltrometer, pressure infiltrometer, and soil core estimates of saturated hydraulic conductivity. *Soil Science Society of America Journal* **64**: 478–484.

Riverside Salinity Laboratory (2001). *Riverside Salinity Laboratory: Agricultural Research Service*. Available at www.ars.usda.gov/aboutus/aboutus.htm.

Saghafian, B., P. Y. Julien, and F. L. Ogden (1995). Similarity in catchment response. I. Stationary rainstorms. *Water Resources Research* **31**: 1533–1541.

Saigusa, M., S. Shoji, and H. Nakaminami (1987). Measurement of water retention at 15 bar tension by pressure membrane method and available moisture of Andosols. *Japanese Journal of Soil Science and Plant Nutrition* **58**: 374–377 (in Japanese, with English summary).

Santiago, L. S., G. Goldstein, F. C. Meinzer, J. H. Fownes, and D. Mueller-Dombois (2000). Transpiration and forest structure in relation to soil waterlogging in a Hawaiian montane cloud forest. *Tree Physiology* **20**: 673–681.

Schwartz, R. C., P. W. Unger, and S. R. Evett (2000). Land use effects on soil hydraulic properties. In *Proceedings of the 15th Conference of the International Soil Tillage Research Organization*. Washington, DC: Conservation and Production Research Laboratory, U.S. Department of Agriculture.

Schwärzel, K., and P. Punzel (2007). Hood infiltrometer: a new type of tension infiltrometer. *Soil Science Society of America Journal* **71**: 1438–1447.

Shoji, S., M. Nanzyo, and R. A. Dahlgren (1993). *Volcanic Ash Soils: Genesis, Properties and Utilization*. Amsterdam, the Netherlands: Elsevier.

Soil Survey Staff (1996). Keys to Soil Taxonomy, 7th edn. Washington, DC: Natural Resource Conservation Service of the U.S. Department of Agriculture.

Stakman, W. P., and G. G. van der Harst (1969). *Determination of Soil Moisture Retention Curves, Vol. 2, Pressure Membrane Apparatus, Range* pF 3.0 to 4.2. Wageningen, the Netherlands: Institute for Land and Water Management Research.

Stephens, D. B., S. Tyler, K. Lambert, and S. Yates (1983). Field experiments to determine saturated hydraulic conductivity in the vadose zone. In *Role of the Unsaturated Zone in Radioactive and Hazardous Waste Disposal*, ed. J. W. Mercer, pp. 113–126. Ann Arbor, MI: Ann Arbor Science.

Stolte, J., J. I. Freijer, W. Bouten, *et al.* (1994). Comparison of six methods to determine unsaturated soil hydraulic conductivity. *Soil Science Society of America Journal* **58**: 1596–1603.

Tobón, C. (1999). *Monitoring and Modelling Hydrological Fluxes in Support of Nutrient Cycling Studies in Amazonian Rain Forest Ecosystems*. Wageningen, the Netherlands: Tropenbos Foundation.

Tomasella, J., M. G. Hodnett, and L. Rossato (2000). Pedotransfer functions for the estimation of soil water retention in Brazilian soils. *Soil Science Society of America Journal* **64**: 327–338.

Van Genuchten, M. Th. (1980). A closed-form equation for predicting the hydraulic conductivity of unsaturated soils. *Soil Science Society of America Journal* **44**: 892–898.

Vargas, R. (2001). Geología, hidrogeoquímia y modelo conceptual de reservorio para la prefactibilidad del campo geotérmico poco sol, San Ramón – San Carlos, Costa Rica. M.Sc. thesis, University of Costa Rica, San José, Costa Rica.

Vogeler, I., B. E. Clothier, S. R. Green, D. R. Scotter, and R. W. Tillman (1996). Characterizing water and solute movement by time domain reflectometry and disk permeameter. *Soil Science Society of America Journal* **60**: 5–12.

Wang, D., S. R. Yates, and F. F. Ernst (1998). Determining soil hydraulic properties using tension infiltrometers, Time Domain Reflectometry, and tensiometers. *Soil Science Society of America Journal* **62**: 318–325.

Warrick, A. W., and D. R. Nielsen (1980). Spatial variability of soil physical properties in the field. In *Applications of Soil Physics*, ed. D. Hillel, pp. 319–344. New York: Academic Press.

Wooding, R. A. (1968). Steady infiltration from a shallow circular pond. *Water Resources Research* **4**: 1259–1273.

Woolhiser, D. A., R. E. Smith, and J. V. Giraldez (1996). Effects of spatial variability of saturated hydraulic conductivity on Hortonian overland flow. *Water Resources Research* **32**: 671–678.

Xiang, J. (1994). Improvements in evaluating constant-head permeameter test data. *Journal of Hydrology* **162**: 77–97.

Xiang, J., B. R. Scanlon, W. F. Mullican, L. Chen, and R. S. Goldsmith (1997). A multistep constant-head borehole test to determine field saturated hydraulic conductivity of layered soils. *Advances in Water Resources* **20**: 45–57.

Zadroga, F. (1981). The hydrological importance of a montane cloud forest area of Costa Rica. In *Tropical Agricultural Hydrology*, eds. R. Lal and E. W. Russell, pp. 59–73. New York: John Wiley.

Ziegler, A. D., T. W. Giambelluca, L. T. Tran, *et al.*, (2004). Hydrological consequences of landscape fragmentation in mountainous northern Vietnam: evidence of accelerated overland flow generation. *Journal of Hydrology* **287**: 124–146.

Zimmermann, B., and H. Elsenbeer (2008). Spatial and temporal variability of soil saturated hydraulic conductivity in gradients of disturbance. *Journal of Hydrology* **361**: 78–95.

53 Hydrology and land-cover change in tropical montane environments: the impact of pattern on process

M. Mulligan
King's College London, London, UK

J. Rubiano
Universidad Nacional de Colombia, Valle del Cauca, Colombia

M. Rincón-Romero
Universidad del Valle, Cali, Colombia

ABSTRACT

This chapter examines the details of land-use impacts on the hydrology of tropical montane forest at the watershed scale, with particular reference to the Tambito area in south-western Colombia. In a series of model experiments a watershed-scale, GIS-based, dynamic hydrological model was used to simulate the hydrological impacts of progressive deforestation as a result of expansion of the agricultural frontier. The simulations were driven by 52 separate iterations of a cellular automaton-based land-use change model. Model results were analyzed to understand the changing *sensitivity* of total runoff and erosion to forest loss as deforestation occurred in different parts of the watershed. Sensitivity was calculated for each output variable as the change in the value of the output variable between two successive model iterations divided by the area of land deforested in those iterations. The results indicated an exponential increase in the sensitivity of these processes to deforestation beyond a threshold of 60% forest loss in the watershed. The patterns of sensitivity at a particular iteration of the cellular automaton tended to reflect: (i) the landscape properties of the area remaining under forest, (ii) the soil properties of both forested and deforested areas, and (iii) the location of the deforestation front relative to high-order streams and important lateral hydrological links within the watershed.

INTRODUCTION

The hydrological consequences of land-use change within the tropical lowland forest zone have been much discussed (see Bruijnzeel, 2004 and Bonell and Bruijnzeel, 2005 for recent overviews). Much less hydrological impact research has been carried out in tropical montane cloud forest (TMCF) settings (Bruijnzeel, 2005; Schellekens, 2006; Muñoz-Villers, 2008; cf. Schrump *et al.*, this volume; Tobón *et al.*, this volume #52) and the results obtained in different areas differed greatly. An early "direct" comparison of dry-season flows from areas with TMCF and agricultural cropping in the Sierra de las Minas in Guatemala suggested major *reductions* in streamflow after forest conversion that were attributed to the associated loss of horizontal precipitation inputs (Brown *et al.*, 1996). Conversely, a similar approach in the eastern Sierra Madre in Mexico – where inputs via cloud water interception have been demonstrated to be low (Holwerda *et al.*, 2009) and cloud forest water use is comparatively high (Gomez-Cardenas, 2009) – suggested major *increases* in flows after converting TMCF to pasture (Muñoz-Villers, 2008). Finally, a *near-neutral result* was obtained after the conversion of TMCF to pasture in northern Costa Rica, which was attributed to the compensatory effects of the loss of horizontal precipitation inputs on the one hand, and the lower water use of the pasture compared to that of the cloud forest on the other (Schellekens, 2006).

Although recent FAO assessments of tropical forest loss do not distinguish between lowland and montane tropical forests (FAO, 2001), earlier estimates (FAO, 1993) suggested that montane

Tropical Montane Cloud Forests: Science for Conservation and Management, eds. L. A. Bruijnzeel, F. N. Scatena, and L. S. Hamilton. Published by Cambridge University Press.

forests are being converted at a higher pace (1.1% $year^{-1}$) than lowland forests (0.8% $year^{-1}$). Colombia provides a case in point. Only 10% of the country's original TMCF has been estimated to remain (Castaño, 1991), and although this accounts for only 3% of the national territory, Andean TMCF are rich in both hydrological (Mulligan, 2007) and biological resources (Leo, 1995; Myers *et al.*, 2000; Barthlott *et al.*, 2005; Brehm *et al.*, 2008). As such, these unique ecosystems need to be managed carefully. This chapter describes the development and application of a simple, spatially distributed hydrological model to better understand the soil and water impacts of TMCF conversion to pasture in the Tambito area of south-western Colombia. It is pertinent to note that this chapter is not so much concerned with predicting the future rates of land-use or cover change. Rather, the focus is on providing a spatially plausible scenario for the *pattern* of land-cover change with which to drive the hydrological model to better understand the associated hydrological sensitivity to this change.

MATERIALS AND METHODS

Study watersheds

The Tambito experimental watersheds are located close to the agricultural frontier on the Pacific slopes of the Western (Occidental) Cordillera of the Colombian Andes in the Department of Cauca, Colombia (2° 30′ N, 77° 00′ W; see Letts *et al.*, this volume and Gonzalez, this volume for further site details). The watersheds are situated between 1374 and 2894 m.a.s.l. and together cover an area of 1424 ha. Slopes are extremely steep but normally distributed, with a maximum of 68° measured at a spatial resolution of 25 m (average 31° ± 8.5°). Two sub-watersheds were investigated, viz. the Palo Verde, which is largely under primary montane forest, and the Tambito, which has a mixture of primary forest, secondary forest, and pasture. The two sub-watersheds have similar slopes, geology, soils, and aspects. A floristic survey of the watersheds conducted in 1996 indicated a total area of 861 ha of primary forest, 527 ha of secondary forest, and 36 ha of pasture (A. Cortes, personal communication). The Tambito watersheds have been the subject of an intensive hydrological monitoring, experimentation, and modeling effort since 1997. This has included the installation of automatic climate and streamflow monitoring stations, the development of a GIS database for climate, terrain and soil variables, and various basin-wide studies to quantify the spatial variability of soil and vegetation properties (see www.ambiotek.com/herb).

Land use in the Tambito area

The historic land use of the Tambito watersheds is complex. Tambito means "small market" and the site was almost certainly a stopover on the route from the Pacific lowlands to the markets of El Tambo and Popayan throughout the twentieth century. The pasture and secondary forest areas are a relic of this legacy. There is also evidence of timber removal by Carton de Colombia at the head of the Palo Verde watershed but the very steep terrain and poor communications ensured that this remained a very small-scale exercise. In addition, there is evidence of very light selective felling close to the main streams. For the past 20 years the area has been a Reserve of the University of Cauca and Fundación Proselva. Parts of the secondary forest have developed from pasture during this time. The area currently under pasture remains as pasture due to frequent cutting, as well as grazing by horses used to transport food and materials into and out of the Reserve. Apart from these localized and small-scale impacts it is likely that most of the remaining forest is primary. The absence of large trees is considered to reflect frequent land-sliding and tree-falls triggered by heavy rainfall (Wilcke *et al.*, this volume).

Models for deforestation and hydrological impacts

In order to understand the hydrological significance of the *pattern* of deforestation in the study watersheds, a dynamic land-cover scenario of increasing deforestation was produced. Each iteration of the scenario represented the actual vegetation cover for an integration of the distributed hydrological model. The model for land-cover change used here is a cellular automaton (CA) designed specifically for this study. The CA uses only a Digital Elevation Model (DEM), an initial land-cover map, and maps of the network of rivers and roads. It simulates the conversion of forest to pasture only, and does not model regeneration of forest or more complex land-cover transitions. The automaton uses a neighborhood function which simulates deforestation spreading in an epidemiologic fashion from roads and agricultural frontiers. In this way an agricultural frontier moves through the landscape from all access points. The resistance to the spread of this frontier in a particular direction is assumed proportional to the sine of the slope gradient in that direction. The cellular automaton is not intended to model the *rate* of deforestation, just its *form*. No parameters of the model are time-dependent and the rate of forest loss is solely a function of the slope and total length of the forest–pasture front. It should be noted that the patterns of deforestation produced by this cellular automaton are applicable only in tropical montane environments at the forest–agriculture frontier without complex patterns of land ownership which drive the much more spatially structured patterns of deforestation that one tends to find at lower elevations (Drigo, 2005).

In addition, to better understand how hydrological dynamics vary with progressive deforestation, a dynamic hydrological model was applied to each of the land-cover transitions (iterations) separately, off-line, and in a non-transient manner. The model will not be described in detail here since a full description has been given by Mulligan (1999). Briefly, the model is a distributed GIS-based hydrological process model describing non-riparian surface and shallow subsurface flows and including process equations for erosion and plant growth (Figure 53.1). The model was integrated at a spatial resolution of 25 m and at an hourly time-step across the entire watershed. The response of the model to land-cover change is conditioned by both the model processes and the spatial variability of field-measured soil properties (Rincón-Romero, 2001) and the GIS-measured landscape properties in the forested and deforested parts of the watershed.

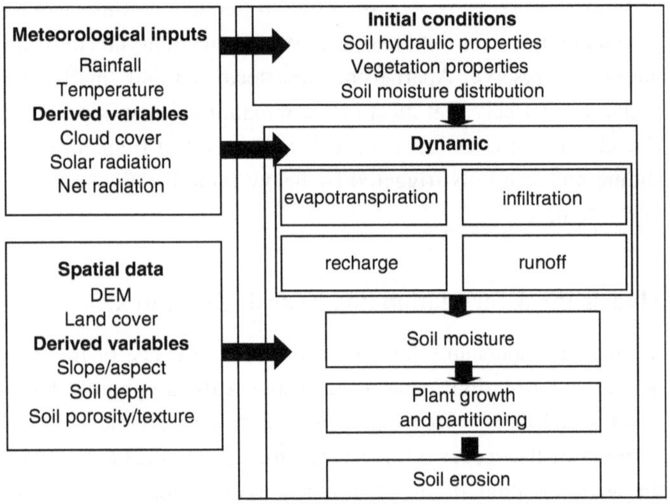

Figure 53.1. Basic outline of the TAMBITO hydrological model used in the present study.

LAND-USE CHANGE IN THE TAMBITO AREA

From an examination of the situation in surrounding areas within the Cauca district it is entirely plausible that the Tambito watersheds could be fully converted from forest to pasture, with the slopes having the lowest gradients and those nearest the roads being the first to be converted. The important property of the cellular automaton in terms of understanding the impact of pattern on process is not so much the *rate* of deforestation but rather the *location* of deforestation with respect to hydrologically important landscape variables such as slope gradient, elevation, topographic wetness index, and the surface flow drainage network. The land-cover situations associated with iterations 1, 4, 8, and 12 of the cellular automaton are shown in Figure 53.2; after 42 iterations the entire watershed was deforested.

Analysis of the pattern of deforestation shows, not surprisingly, that the more moderate slopes were deforested first (in iterations 1–10), with steeper slopes being deforested later (Figure 53.3). The pattern is more complex with respect to elevation: in the initial iterations the distribution of elevations deforested was bimodal with a peak at 1450 m.a.s.l. and another at 2500 m.a.s.l., reflecting the locations of existing roads and pastures. As the cellular automaton continued to iterate,

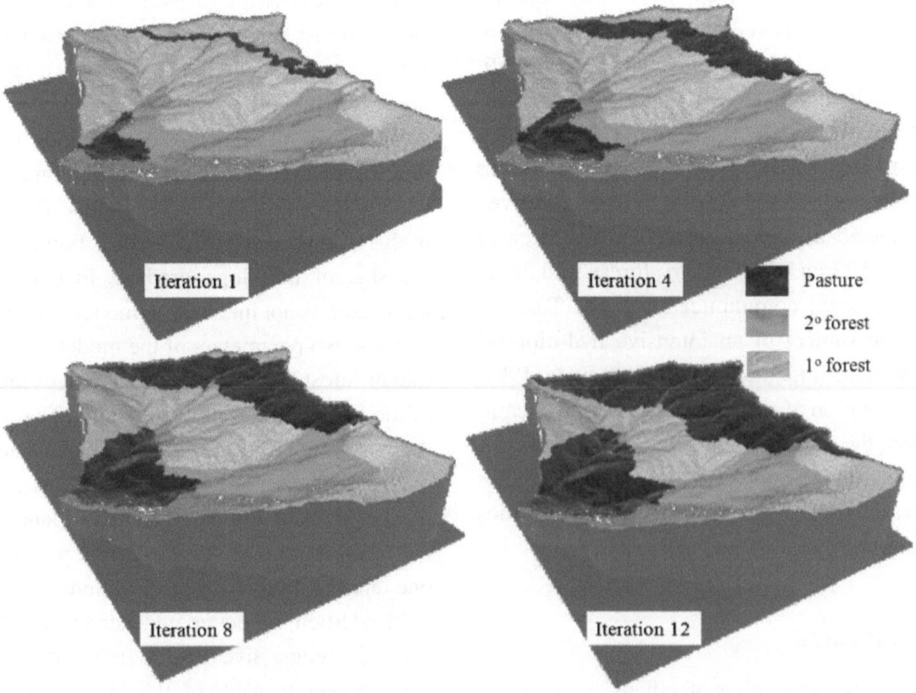

Figure 53.2. Patterns of modeled land-use change for the Tambito watersheds.

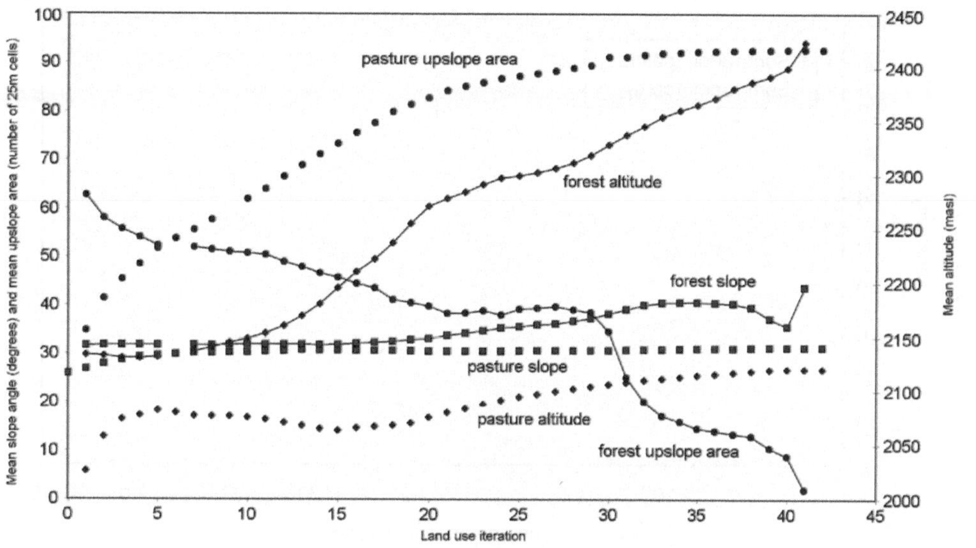

Figure 53.3. Mean landscape properties of the forested and deforested areas for 42 iterations of the land-use change model as applied to the Tambito area.

deforestation occurred more equally across the range of elevations (Figure 53.3). In the initial iterations the lowest slopes were deforested, with the average slope gradient remaining below 30° until iteration 5 before reaching an asymptote at 30° by iteration 12. The average elevation of deforested areas increased sharply from 2010 to 2060 m.a.s.l. between iterations 1 and 5, then decreased a little until iteration 15 because of a greater expansion of pasture around the existing lower-elevation pasture zone compared with the higher-elevation road, before increasing again until iteration 42 (Figure 53.3).

AN ANALYSIS OF THE IMPACT OF PATTERN ON PROCESS

The hydrological model was integrated once (for one month of meteorological data representing average conditions for January 1998), for each iteration of the cellular automaton land-use model. Sensitivity was calculated for each output variable and for each integration of the hydrological model, as the change in the value of the output relative to the previous integration of the model, divided by the area of land deforested in the corresponding cellular automaton iteration. This metric of sensitivity provides a measure of the watershed's hydrological response per unit of land-cover change and thus indicates the importance of the pattern or location of change in terms of its impact on runoff or erosion at the watershed scale. Runoff is considered here to represent all effective rainfall finding its way to the river channel, whether on the slopes or in the channel, since the hydrological model used cumulates rainfall infiltration along a drainage network.

Causes of non-linear response of runoff and erosion to progressive deforestation

The inferred sensitivity of runoff at the watershed outlet to progressive deforestation (Figure 53.4) shows that runoff remained relatively insensitive until iteration 17, by which time 58% of the watershed was deforested. At that point the sensitivity – or change per unit change in land cover – began to rise sharply with further deforestation. Results were the same for model integrations where forest was replaced directly with pasture and where forest was replaced with bare soil, although the rise in sensitivity was steeper for the pasture case (Figure 53.4).

The parameterized differences between forest and non-forest included differences in leaf area and rooting depth (affecting evapotranspiration) and differences in infiltration rates, with a lower infiltration capacity assigned to soils under grazed pasture compared with those under forest, as measured in Tambito (Rincón-Romero, 2001) and elsewhere (Meehan and Platts, 1978; Holl, 1999; Zimmermann and Elsenbeer, 2008; Tobón *et al.*, this volume #52).

Comparing the patterns of change in sensitivity with those of average slope and elevation of the areas deforested (as shown in Figure 53.3), it becomes clear that these factors do not explain the exponential increase in sensitivity of runoff to continued deforestation beyond the 58% threshold. The pattern of change in sensitivity proved very similar for watershed-wide soil erosion (Figure 53.5), partly because wash erosion in the model is largely a function of surface runoff (Mulligan, 1999). Again, the slope gradients and elevations of the areas deforested did not explain the inferred pattern of sensitivity of erosion.

Instead, variables that represent the connectivity of the flow network may explain the obtained patterns of sensitivity better, if

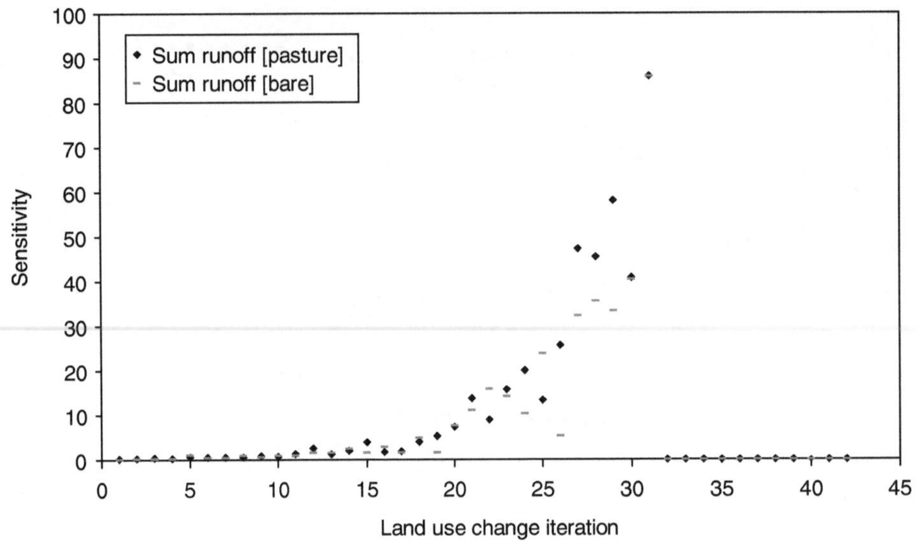

Figure 53.4. Sensitivity of runoff to land-use change in the Tambito area (forest replaced by pasture or bare soil).

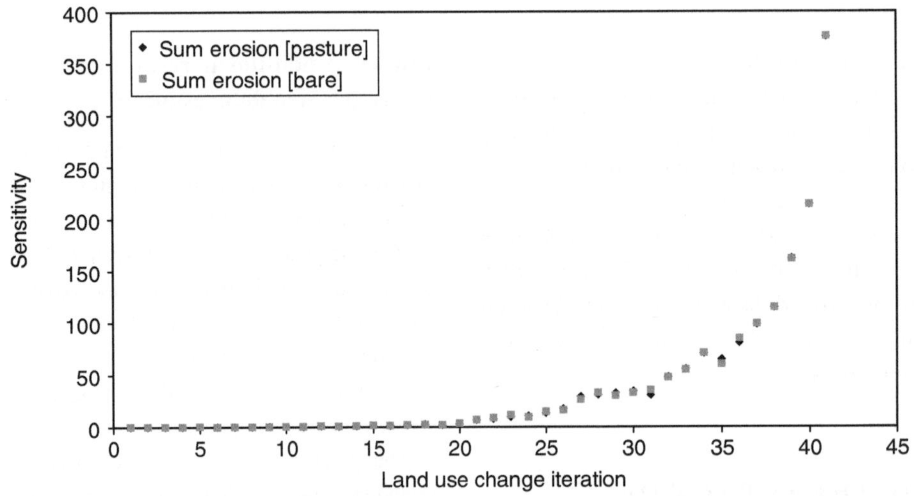

Figure 53.5. Sensitivity of soil erosion to land-use change in the Tambito area (forest replaced by pasture or bare soil).

only because surface runoff and erosion are the only hydrological variables that are directly affected by lateral connectivity or lateral accumulation along the drainage network and show an exponential increase in sensitivity with progressive deforestation (Molina *et al.*, 2007; Vanacker *et al.*, 2007). Soil moisture, evapotranspiration and deep recharge all exhibited more complex patterns of change (data not shown).

The role of flow connectivity

An examination of changes in flow connectivity as deforestation progressed (Table 53.1) indicated that the average stream order of the deforested area (calculated from flow directions following the method of Strahler (1964) as implemented in the PCRASTER GIS) increased perfectly linearly with the number of cells deforested ($r^2 = 1.0$), since the deforested area grows along the stream network. The same was found for the average contributing area of the deforested segment. Whilst neither stream order nor total upslope area of the deforested part show an exponential

Table 53.1 *Response of runoff sensitivity to measures of contributing area and measures of contributing area to deforestation*

Variable (Y,X)	Equation	R^2
Stream order vs. deforestation	$Y = 0.00006X + 0.0359$	1.00
Upslope cells vs. deforestation	$Y = 0.0025X + 38.546$	0.98
Runoff vs. upslope cells	$Y = 0.002e^{0.1039x}$	0.89
Runoff vs. stream order	$Y = 0.0796e^{4.3254x}$	0.93

relationship with the area deforested, the sensitivity of surface runoff and erosion do (Table 53.1).

FORCING BY OTHER HYDROLOGICAL VARIABLES

Since landscape variables such as slope, aspect, stream order and contributing area of the deforested area did not explain the inferred non-linear response of surface runoff to progressive deforestation, it is possible that this response is the result of either:

- The non-linear response of another hydrological variable, such as soil moisture, which then forces the non-linear response of overland flow and erosion;
- The landscape characteristics of the area remaining under forest (as opposed to the deforested area), or
- A response to other as yet undefined properties of the forested or deforested areas.

These three possibilities are examined briefly in the following paragraphs.

Watershed-wide averaged volumetric soil moisture content θ ($m^3\ m^{-3}$) changed positively and linearly with area deforested (A, ha) as $\theta = 5 \times 10^{-6} + 0.268A$ ($r^2 = 0.99$). Although soil moisture sensitivity to land-cover change at Tambito was non-linear (data not shown), and surface runoff was related to soil moisture both functionally within the model and in the results ($Runoff = -313550 + 11617$, $r^2 = 0.85$), soil moisture *sensitivity* to progressive deforestation and runoff *sensitivity* to progressive deforestation were very poorly related to each other.

Contributions by forest and pasture

Some interesting patterns emerge if one looks at the average properties of the remaining forested rather than the deforested areas for each iteration of the land-cover model. Table 53.2 shows the relationships between runoff sensitivity and various landscape properties in the forested and deforested parts of the watershed for the full range of land-cover change iterations. Beyond iteration 30, high runoff sensitivity coincided with a steep increase in the average elevation of the forested area, an increase in the average slope of the forested area, and a sharp decrease in upslope contributing area (which may comprise both forested and non-forested parts) of the forested area. Deforestation by the cellular automaton led to a relatively stable mean elevation for deforested areas after the first few iterations during which the lowest parts of the watershed were deforested (Figure 53.3). However, the mean elevation of forested areas increased steadily as the agricultural frontier continued to break up the forest cover leaving only high-elevation remnants. Similarly, the mean slope gradient for the deforested areas was relatively stable after the first few iterations, whereas that for the remaining forested areas increased continually (Figure 53.3). Finally, as the forested area became reduced to a few small remnants, large fluctuations in average slope gradient occurred. The upslope contributing area of the deforested areas increased most sharply during the early stages of deforestation because runoff on higher-order flow paths (that is, with large upslope areas) were "captured" and contributed runoff to the deforested part. During later stages, further deforestation captured the flowpaths at lower orders and so the overall increase in contributing area per unit of forest loss was less.

Table 53.2 *Relationships between properties of forested and deforested areas* (X, *units given*) *and runoff sensitivity* (Y, *mm year^{-1} ha^{-1}*) *for the full range of land-use change iterations*

Variable	Equation	R^2
Forested area		
Mean elevation (X, m.a.s.l.)	$Y = 8.735X - 279.09$	0.86
Mean slope (X, degrees)	$Y = 8.735X - 279.09$	0.86
Mean contributing area (X, hectares)	$Y = 50924e^{-0.2131X}$	0.86
Mean soil thickness (X, meters)	$Y = -1257.2X + 1694.3$	0.87
Mean Ks_{sfc} (X, mm hr^{-1})	N/A	-----
Mean Ks_{bedrock}	$Y = 215.26X - 152.95$	0.88
Mean soil b-value	$Y = -110.24X + 1381.4$	0.72

The contribution of soil properties

It follows from the previous section that patterns of change in the landscape properties of forest remnants as well as of the deforested areas may contribute to the overall runoff response of the watershed to land-cover change. However, averaged soil properties of the forested and deforested segments within the watersheds may also contribute to its overall runoff response (cf. Molina *et al.*, 2007), as indicated by the significant relationships between runoff sensitivity and various soil parameters used in the model across the range of land-cover change iterations (Table 53.2). The soil parameters that varied spatially in the model included soil thickness, saturated hydraulic conductivity at the soil surface (Ks_{sfc}) and at the soil–bedrock interface (Ks_{bedrock}), and the slope of the soil moisture–matric potential curve (the so-called b-value). The b-values were calculated as a function of soil texture following the method of Campbell (1985) and regionalized on the basis of field measurements (see Mulligan, 1999 for details).

Analyzing the patterns of change in these parameters as deforestation progresses reveals some interesting results (Figure 53.6).

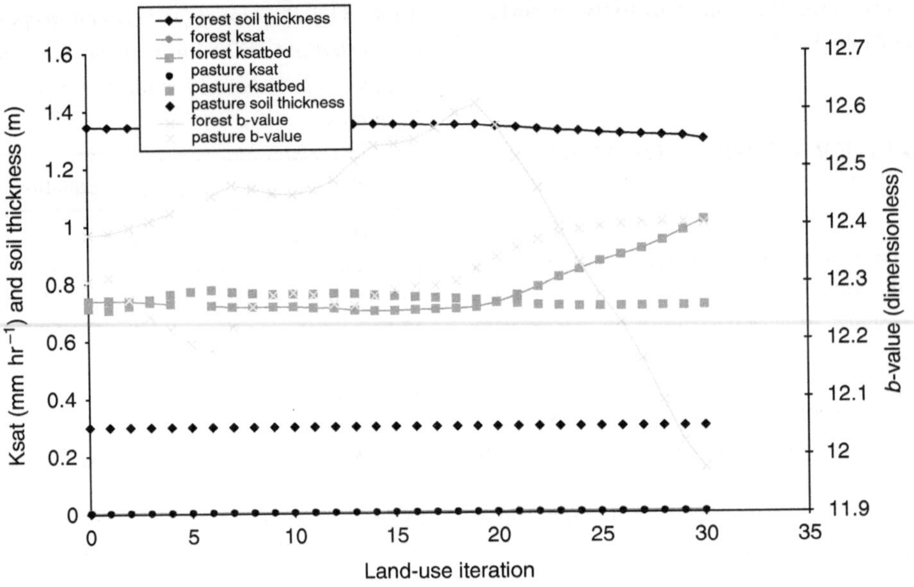

Figure 53.6. Mean soil properties of forested and deforested areas for 42 iterations of the land-use change model as applied to the Tambito area.

For example, predicted average soil thickness in the forested area decreased as deforestation progressed, since the remnant forests occurred mainly on the upper and steeper slopes of the watersheds. Since soil thickness was taken as the minimum of either soil thickness or rooting depth, soil thickness in the deforested part of the watershed converged toward a constant value equal to the rooting depth for pasture (0.3 m). Average values of surface hydraulic conductivity Ks_{sfc} in both forested and deforested areas increased slightly as deforestation progressed (reflecting the concentration of remaining forest areas on the higher slopes and the increasing inclusion of higher slopes also in the deforested average for Ks_{sfc}). However, the difference between forested and non-forested Ks_{sfc} is much more significant and the predicted increases in runoff generation with forest loss result from the increase of the absolute area under the lower infiltration-rate pasture land cover (cf. Molina *et al.*, 2007; Zimmermann and Elsenbeer, 2008; Tobón *et al.*, this volume #52). Values of Ks_{bedrock} remained relatively stable for the deforested area but increased steeply after iteration 20 for the forested segment. Such changes are thought to reflect changes in soil thickness as the remaining slopes under forest are on average steeper, with consequences for the bulk density profile and thus Ks_{bedrock}.

The other major change found beyond iteration 20 was related to the slope of the soil moisture characteristic (*b*-value). This value increased significantly for the deforested section but decreased significantly for the forested section as deforestation progressed. These findings suggest that the major changes in average soil properties with progressive deforestation are those of hydraulic conductivity and water retention. Changes of soil bulk density distributions of the forested and deforested areas – which are, in turn, a function of soil thickness distributions – induce these changes in the model.

Since hydraulic conductivity and *b*-value are major controls on infiltration, the advance of the wetting front through the soil, and the recharging of deeper layers in the model (Mulligan, 1999), it is likely that changes in K_s and *b*-value will dominate the hydrological impact of land-cover change and thus the sensitivity of surface runoff in the Tambito watersheds to deforestation. This latter point is also borne out by Table 53.2 in which Ks_{sfc}, Ks_{bedrock}, and *b*-values of the deforested and forested parts of the watershed exhibit clear but complex effects on the sensitivity of surface runoff to land-cover change.

CONCLUDING REMARKS

Tropical mountains are hydrologically complex and dynamic (Mulligan and Burke, 2005). They are important in the provision of water resources to their often substantial downstream populations (Viviroli *et al.*, 2003; Muñoz-Piña *et al.*, 2008; cf. Calvo-Alvarado *et al.*, this volume; Tognetti *et al.*, this volume), particularly so under seasonally dry climatic conditions. Despite this importance, detailed studies of the hydrological impacts of tropical montane forest conversion in general and tropical montane cloud forests in particular (Bruijnzeel, 1990, 2005), are exceedingly scarce. In addition, results from the literature have been seemingly contradictory with increases in flows in some areas and reductions or no change in others. There is thus a great deal of uncertainty surrounding the interactions between land use and hydrological change in these spatially heterogeneous tropical mountains. Also, lessons learned from impact studies at one site apparently do not necessarily hold for the different temporal and spatial scales, environments, land covers, and land uses encountered at other sites. Arguably, therefore, it is important to facilitate the type of monitoring and modeling studies that will provide site-specific solutions.

A specific example highlighted in the present chapter is that the per-hectare impact of forest conversion on runoff depends critically on the *location* of land-cover change (relative to spatially varying soil and landscape properties) and the *pattern* of land-cover change relative to upstream forested and deforested patches, as well as the *extent* of the land-cover change. Many studies have only considered the latter (Bruijnzeel, 2004). In addition, non-linearities in the sensitivity of key processes such as runoff generation and erosion to land-cover change have been shown to occur as deforestation accumulates along the major flow-lines. Where forest patches remain present along flow-lines, water accumulating along deforested patches is more likely to "sink" and transpire in downslope forested patches. As these patches disappear, flows tend to accumulate exponentially in a downslope direction along the flow-lines (cf. Ziegler *et al.*, 2004).

Arguably, along with variations climate and geology between sites, this non-linearity and threshold behavior of hydrological response to land-cover change may go some way toward explaining the variety in watershed responses observed in studies of montane (cloud) forest conversion (Bruijnzeel, 1990; Brown *et al.*, 1996; Schellekens, 2006; Muñoz-Villers, 2008). It certainly indicates the necessity for policy formulation to be based on detailed local knowledge rather than generic rules of thumb.

ACKNOWLEDGEMENTS

This research was funded by grants from The Royal Society (MM, RSRG 18/19443), The British Council (MM, JR), Colciencias and the University of London Central Research Fund (MR) whose support is gratefully acknowledged. The authors would like to express their gratitude to the late Alvaro Jose Negret, Fundación Proselva and the University of Cauca, Popayan for the use of the Tambito field site and its facilities. They are also grateful to other Colombian organizations that provided logistic support, in particular the International Centre for Tropical Agriculture (CIAT) and the Regional Corporation for Cauca (CRC). Thanks also go to those who provided assistance in the field, particularly Andrew Jarvis, Koulla Pallaris, Quintin, and Olga. This paper was much improved through reviews by Fred Scatena and Sampurno Bruijnzeel.

REFERENCES

Barthlott, W., J. Mutke, M. D. Rafiqpoor, G. Kier, and H. Kreft (2005). Global centres of vascular plant diversity. *Nova Acta Leopoldina NF* **92**, **342**: 61–83.

Bonell, M., and L. A. Bruijnzeel (eds.) (2005). *Forests, Water and People in the Humid Tropics*. Cambridge, UK: Cambridge University Press.

Brehm, G., J. Homeier, K. Fiedler, *et al.* (2008). Mountain rain forests in Southern Ecuador as a hotspot of biodiversity: limited knowledge and diverging patterns. In *Gradients in a Tropical Mountain Ecosystem of Ecuador*, eds. E. Beck, J. Bendix, I. Kottke, F. Makeschin, and R. Mosandl, pp. 15–24. Berlin: Springer-Verlag.

Brown, M. B., I. de la Roca, A. Vallejo, *et al.* (1996). *A Valuation Analysis of the Role of Cloud Forests in Watershed Protection, Sierra de las Minas Biosphere Reserve, Guatemala, and Cusuco N.P., Honduras*. Philadelphia, PA: RARE Center for Tropical Conservation.

Bruijnzeel, L. A. (1990). *Hydrology of Moist Tropical Forests and Effects of Conversion: A State of Knowledge Review*. Paris: UNESCO-IHP, and Amsterdam: VU University.

Bruijnzeel, L. A. (2004). Hydrological functions of tropical forests: not seeing the soil for the trees? *Agriculture Ecosystems and Environment* **104**: 185–228.

Bruijnzeel, L. A. (2005). Tropical montane cloud forest: a unique hydrological case. In *Forests, Water and People in the Humid Tropics*, eds. M. Bonell and L. A. Bruijnzeel, pp. 462–483. Cambridge, UK: Cambridge University Press.

Campbell, G. S. (1985). *Soil Physics with BASIC: Transport Models for Soil–Plant Systems*. Amsterdam, the Netherlands: Elsevier.

Castaño, C. (1991). Bosque de niebla y hombre tradicional. In *Bosques de niebla de Colombia*, pp. 151–170. Bogotá, Colombia: Banco de Occidente.

Drigo, R. (2005). Trends and patterns of tropical land use change. In *Forests, Water and People in the Humid Tropics*, eds. M. Bonell and L. A. Bruijnzeel, pp. 9–39. Cambridge, UK: Cambridge University Press.

FAO (1993). *Forest Resources Assessment 1990: Tropical Countries*, Forestry Paper No. 112. Rome: FAO.

FAO (2001). *Forest Resources Assessment 2000: Tropical Countries*, Forestry Paper No. 140. Rome: FAO.

Gomez-Cardenas, M. (2009). *Transpiration in contrasting vegetation cover types in the tropical montane cloud forest belt in eastern Mexico*. Ph.D. thesis, University of Iowa, Ames, IA, USA.

Holl, K. D. (1999). Factors limiting tropical rain forest regeneration in abandoned pasture: seed rain, seed germination, microclimate and soil. *Biotropica* **31**: 229–242.

Holwerda, F., L. A. Bruijnzeel, L. E. Muñoz, M. Equihua, and H. Asbjornsen (2009). Rainfall and cloud water interception in mature and secondary lower montane cloud forests of central Veracruz, Mexico. *Journal of Hydrology* (in press).

Leo, M. (1995). The importance of tropical montane cloud forests for preserving vertebrate endemism in Peru: the Río Abiseo National Park as a case study. In *Tropical Montane Cloud Forests*, eds. L. S. Hamilton, J. O. Juvik, and F. N. Scatena, pp. 198–211. New York: Springer-Verlag.

Meehan, W. R., and W. S. Platts (1978). Livestock grazing and the aquatic environment. *Journal of Soil and Water Conservation* **33**: 274–278.

Molina, A., G. Govers, V. Vanacker, J. Poesen, and E. Zeelmaekers (2007). Runoff generation in a degraded Andean ecosystem: interaction of vegetation cover and land use. *Catena* **71**: 357–370.

Mulligan, M. (1999). *A Simple Distributed Hydrological Model for Tropical Montane Cloud Forests*, Online Working Paper, King's College London. Available at Permalink: www.ambiotek.com/publications.

Mulligan, M. (2007). *The Hydrological Significance of Mountains in Tropical Continental South America*, Online Working Paper, King's College London. Available at Permalink: www.ambiotek.com/publications.

Mulligan, M., and S. M. Burke (2005). *FIESTA: Fog Interception for the Enhancement of Streamflow in Tropical Areas*, Annex 4a to Final Technical Report DFID-FRP Project No. R7991. Leigh-on-Sea, UK: AMBIOTEK. Also available at www.ambiotek.com/fiesta.

Muñoz-Piña, C., A. Guevara, J. M. Torres, and J. Braña (2008). Paying for the hydrological services of Mexico's forests: analysis, negotiations and results. *Ecological Economics* **65**: 725–736.

Muñoz-Villers, L. E., (2008). Efecto del cambio en el uso de suelo sobre la dinámica hidrológica y calidad de agua en el trópico húmedo del centro de Veracruz, México. Ph.D. thesis, Autonomous University of Mexico, Mexico City.

Myers, N., C. G. Mittermeier, G. A. B. da Fonseca, and J. Kent (2000). Biodiversity hotspots for conservation priorities. *Nature* **403**: 853–858.

Rincón-Romero, M. (2000). The effects of surface hydrological connectivity on hydrological response to land use change. *Advances in Environmental Monitoring and Modelling* **1**: 80–96.

Rincón-Romero, M. (2001). Modelling the sensitivity of Colombian hillsides to land use change. Ph.D. thesis, King's College, University of London, London.

Schellekens, J. (2006). *CQ-FLOW: A Distributed Hydrological Model for the Prediction of Impacts of Land-Cover Change, With Special Reference to the Rio Chiquito Catchment, Northwest Costa Rica*, Annex 3 to Final Technical Report DFID-FRP Project No. R7991. Amsterdam, the Netherlands: Vrije Universiteit. Also available at www.geo.vu.nl/~fiesta/.

Strahler, A. N. (1964). Quantitative geomorphology of drainage basins and channel networks. In *Handbook of Applied Hydrology*, Vol. **4**, ed. V. T. Chow, pp. 4:39–4:76. New York: McGraw-Hill.

Vanacker, V., F. von Blandenburg, G. Govers, *et al.* (2007). Restoring dense vegetation can slow mountain erosion to near natural benchmark levels. *Geology* **35**: 303–306.

Viviroli, D., R. Weingartner, and B. Mersserli (2003). Assessing the hydrological significance of the world's mountains. *Mountain Research and Development* **23**: 32–40.

Ziegler, A. D., T. W. Giambelluca, L. T. Tran, *et al.*, (2004). Hydrological consequences of landscape fragmentation in mountainous northern Vietnam: evidence of accelerated overland flow generation. *Journal of Hydrology* **287**: 124–146.

Zimmermann, B., and H. Elsenbeer (2008). Spatial and temporal variability of soil saturated hydraulic conductivity in gradients of disturbance. *Journal of Hydrology* **361**: 78–95.

Part VI
Effects of climate variability and climate change

L. A. Bruijnzeel and L. S. Hamilton

Part VI
Effects of climate variability and climate change

54 Meso-scale climate change due to lowland deforestation in the maritime tropics

M. K. van der Molen, H. F. Vugts and L. A. Bruijnzeel
VU University, Amsterdam, the Netherlands

F. N. Scatena
University of Pennsylvania, Philadelphia, Pennsylvania , USA

R. A. Pielke Sr.
CIRES, University of Colorado, Boulder, USA

L. J. M. Kroon
Wageningen University and Research Centre, Wageningen, the Netherlands

ABSTRACT

Annual precipitation on the Caribbean island of Puerto Rico has decreased steadily during the twentieth century, on average by 16%. The reduced rainfall manifested itself in the form of regular water rationings for millions of inhabitants during the 1990s. This chapter examines the link between the reduction in precipitation and the land-cover change using a combination of energy balance measurements and meso-scale atmospheric modeling. The explanation of the reduction in precipitation proved to be different than expected. Based on measurements made earlier over rain forest and pasture in Amazonia, a forest-covered island was expected to be cooler because of the higher transpiration of forest compared to grassland. The opposite proved to be the case: transpiration by a coastal wetland forest was less than that of an adjacent, well-watered grassland. In addition, the forest's albedo was 8% lower than that for the grassland. Together, these two factors caused the sensible heat flux over the forest to be twice that over the grassland. The surface energy balance observations over forest and grassland were used in a meso-scale atmospheric circulation model (RAMS) to simulate the meteorological effects of island-wide deforestation. The simulations indicated that the development of a sea breeze during the day dominates the climate on the island. In model runs in which the island was assumed to be completely covered with forest, the sea breeze was considerably stronger than when the vegetation had been transformed to grassland. Along the sea breeze front, convergence caused upward air motions. As sea breeze convergence was stronger over a forested island, more clouds were formed, albeit at a higher elevation, and there was an estimated 10–20% enhancement of precipitation compared to a deforested island. In the deforested scenario, the cloud base was typically lowered by 200 m. Refinement of the model is required to obtain more accurate estimates of the changes in precipitation, although most likely the relevant processes have been included.

INTRODUCTION

Evidence that land cover influences cloud formation and precipitation is increasing (Cutrim *et al.*, 1985; Lawton *et al.*, 2001; Ray *et al.*, 2006; Makareva and Gorshkov, 2007; Pielke *et al.*, 2007; cf. Meesters *et al.*, 2009; Nair *et al.*, this volume). Precipitation decreases exponentially with distance from the ocean over extensive non-forested areas, whereas no such decline is observed where there are extensive tracts of natural forest (Makareva and Gorshkov, 2007). In continental Amazonia, 30–50% of the

Tropical Montane Cloud Forests: Science for Conservation and Management, eds. L. A. Bruijnzeel, F. N. Scatena, and L. S. Hamilton. Published by Cambridge University Press.

precipitation is estimated to originate from locally evaporated moisture, the remainder being supplied by the ocean (Eltahir and Bras, 1994; Costa and Foley, 1999). Hence, an effect of large-scale forest conversion on precipitation in the region is to be expected and many modeling studies (reviewed by Lean *et al.*, 1996; Costa, 2005) have predicted a reduction in the amount of precipitation after complete conversion of Amazonian forest to pasture. Increases in surface albedo, reductions in transpiration as a consequence of the shallower root zone of the replacing grassland vegetation, and reduced interception evaporation are considered to be the major causes for these reductions (Nobre *et al.*, 1991).

It is of interest to explore to what extent such effects also apply under maritime conditions where the climatic influence of the nearby ocean can be expected to be much more pronounced (Malkus, 1955) and land-cover change typically extends over much smaller scales. The Caribbean island of Puerto Rico (180 × 60 km) provides a case in point. After the passage of Hurricane Hugo over the island in September 1989 – which caused widespread defoliation – the cloud base was observed to rise by a few hundred meters to well above the island's highest peaks (>1050 m.a.s.l.) (Scatena and Larsen, 1991). The phenomenon disappeared after several months as the forest canopy regererated. Changes in the Puerto Rican climate are also evident in long-term precipitation records. Bisselink (2003) found summer precipitation totals (May–October) for the majority of 24 stations between 1931 and 1996 to be decreasing significantly (by 0.6–2.3 mm $year^{-1}$), although winter precipitation increased by 0.3–1.7 mm $year^{-1}$. The eight stations with the longest period of record (~100 years) all had negative trends in annual precipitation (−1.6 to −4.9 mm $year^{-1}$). Furthermore, 1997, 1994, and 1991 were the second, third, and sixth driest years in the twentieth century (Larsen, 2000). Mandatory water rationings were imposed six times in the 1990s for periods up to 6 months, with strong societal and ecological impacts and an estimated economic loss of US$165 million in 1994 alone (Lugo and García-Martinó, 1996). The negative precipitation trends roughly coincide with the deforestation of the island from the coastal plains toward the interior mountains, reaching 80% in 1899 and 94% in the 1940s. The cleared land was used primarily for agricultural purposes – pasture and sugar cane in the coastal plain, and coffee and subsistence farms in the uplands. After World War II, many degraded farmlands were abandoned, while government policy caused the island's economy to shift from agriculture to industry. Consequently, secondary forest cover gradually increased, along with urbanization and industrial development in the lowlands (Dietz, 1986; Aide *et al.*, 1995; Franco *et al.*, 1997). By 2000, urban and developed land covered 15.4% of the island while non-wetland forests covered 44.8% (Kennaway and Helmer, 2007). The total pasture area on the island remained constant from 1951 to 2000, covering about one-third of the island.

The objective of this chapter is to determine whether land-cover transformation in the coastal plains may have caused the observed reduction in precipitation totals on Puerto Rico. A micrometeorological field campaign was carried out to quantify the water and energy budgets associated with the original lowland forest and a nearby pasture considered representative of post-forest conditions. Next, the observations were used in a meso-scale atmospheric circulation model to assess the effects of island-wide land-cover change on climate. Given that the frequency, height, density, and spatial distribution of clouds constitute important boundary conditions for the occurrence and functioning of montane cloud forests (cf. Jarvis and Mulligan, this volume), the chapter pays special attention to the effect of lowland land-cover change on cloud occurrence in the adjacent Luquillo Mountains. Here, forests above *c.* 800 m.a.s.l. (Weaver, 1995) are frequently enshrouded in fog and low cloud (cf. Holwerda *et al.*, this volume #28 and 29).

MATERIALS AND METHODS

Field experiment

Two sites were selected near Sabana Seca (18° 27′ 34″ N, 66° 12′ 30″ W), some 15 km west of the capital of San Juan and about 1000 m from the coast line. Here, old-growth *Pterocarpus officinalis* forest (14 m canopy height) and grassland (up to 2 m tall) were found within a kilometer from each other and subject to an unobstructed flow of air from the ocean in the mean direction of the trade winds (NE). Prior to the arrival of the Europeans in Puerto Rico, *P. officinalis* forest constituted the main freshwater wetland ecosystem on the northern coastal plains. *Pterocarpus* was also (one of) the most abundant tree species in mixed forests along rivers on the floodplains (Holdridge, 1940). In addition, *P. officinalis* is an important tree species all along the Atlantic coast from Brazil to Mexico, along the Pacific coast from Costa Rica to Ecuador, and on most Caribbean islands except Cuba (Janzen, 1978; Cintrón, 1983). According to a land survey in 1990, pasture is the most common post-deforestation land-cover type in Puerto Rico (52% of the deforested area). Croplands (sugar cane, coffee and banana) account for another 14%, with the remaining land used for urban and industrial areas, roads, and water (Franco *et al.*, 1997). As such the selected sites are considered representative of the general pre- and post-deforestation land-cover types in the island's lowlands.

Temperature, relative humidity, and wind speeds at the two sites exhibit clear diurnal and seasonal patterns. Temperatures range between 25 and 30 °C in summer and between 22 and 28 °C in winter. Atmospheric humidity is close to saturation at night but drops to 75% and 70% at midday in summer and winter, respectively. Wind speeds are low at night (~1 m s^{-1}) but increase to 6–8 m s^{-1} during the day under the influence of the sea breeze. Mean annual precipitation in the area is about 1700 mm, with monthly totals being somewhat lower in February and March (Van der Molen, 2002).

Micrometeorological instrumentation for the determination of surface energy budgets and the fluxes of sensible heat, latent heat (evaporation), and momentum according to the Euroflux methodology (Aubinet *et al.*, 2000) were installed on a scaffolding tower at the forest site and on masts at the pasture site. Measurements were conducted from May 1997 until May 1998. Closure of the energy balance was very good at 103 ± 11% for the forest site vs. 105 ± 5% for the pasture (Giambelluca *et al.*, 2009). Full instrumental and methodological details can be found in Van der Molen (2002).

Modeling the effect of land-surface type on atmospheric circulation

To analyze the influence of the changes in energy partition and surface fluxes after forest conversion on atmospheric circulation, the meso-scale RAMS (Regional Atmospheric Modeling System) version 4.2 model (Pielke *et al.*, 1992) was used. The effect of the conversion was determined by comparing the atmospheric fields from a model run in which all vegetation up to 300 m elevation was parameterized as lowland forest ("forest run"), and another model run with pasture as the vegetation ("pasture run"), under otherwise identical conditions. The vegetation above 300 m.a.s.l. ("upland forest", comprising 30% of the land surface) was kept constant in the two runs and had its own parameterization based on measurements made over montane palm forest in the Luquillo Mountains. Because the exact parameterization of the upland forest is not essential in the present context, the reader is referred to Van der Molen (2002) for details.

The upper-air model results were verified with rawinsonde data, paying particular attention to the trade winds and atmospheric stability, to ensure a realistic reference case. Simulating the effect of (changes in) surface fluxes on atmospheric circulation at this level of scale requires that the largest of turbulent motions are resolved explicitly – in this case thunderstorms, convective cells, and meso-scale circulations such as sea breezes and mountain-valley winds – which typically have a horizontal extent of kilometers. Therefore, the finest horizontal grid resolution used in the model was 1 km, with a domain stretching 40 km out over the ocean from the north and south coasts. This fine, inner grid was nested within coarser grids (4 and 20 km spacing) with a larger domain, the coarsest of which covered 300 km of ocean on all sides of the island, thereby effectively removing the lateral boundaries from the area of interest. The finest grid had a relatively small extent of 21 km in an east–west direction to reduce computing time. Turbulent mixing was parameterized using the Smagorinsky-type local deformation scheme (Smagorinsky, 1963) and short- and long-wave radiation according to the Chen and Cotton scheme (1983, 1987) in which long-wave emissivities are tuned to match observed long-wave radiation at the surface. Cloud- and rain-water mixing ratios were parameterized using the cloud micro-physics scheme (with other particles disabled) of Walko *et al.* (1995). Initialization and forcing of the lateral boundaries of the coarsest grid were performed using National Center for Environmental Prediction reanalysis data.

The parameterization of the vegetation part of the model was given particular attention. The model frame, LEAF-2 (Walko *et al.*, 2000) has been constructed around the concept that the surface consists of vegetation (represented by a single "big leaf") and soil. All fluxes between vegetation, soil, and atmosphere are transferred through the canopy air, with the fluxes described as a function of gradients and resistances. Three types of fluxes and associated resistances are distinguished: (i) from vegetation to canopy air, (ii) from soil to canopy air, and (iii) from atmosphere to canopy air. The most important resistances are the aerodynamic (r_a) and surface resistance (r_s) because these determine the rate of transfer from the leaf surface to canopy air, and through the plant's stomata, respectively. In the present study, r_a was related to wind speed and r_s to global radiation, rendering the parameterization empirical rather than physical. However, successful tests of the parameterization ensured accurate representation of the respective fluxes and of vegetation/canopy–air temperatures and water-vapor mixing ratios by the model (see Van der Molen, 2002 for details).

The performance of the model in general and the vegetation module in particular was successfully verified for a relatively clear day (5 September 1997), considered typical for Puerto Rican summertime conditions in terms of strength and direction of the trade winds and the partitioning of net radiant energy into sensible and latent heat fluxes. In addition to the "forest" and "pasture" model runs, additional simulations were carried out to test the influence of (i) the presence of the ocean, (ii) the island's topography, and (iii) synoptic-scale weather conditions on atmospheric circulation and cloud formation.

RESULTS

Energy budgets over lowland forest and pasture

The most prominent changes in the surface energy balance as a result of forest conversion to pasture were: (i) an increase in albedo from 13–15% for forest to 18–23% for pasture (Figure 54.1), (ii) a reduction of *c.* 9% in net radiation R_n, and (iii) a reduction in sensible heat flux H and a concomitant increase in latent heat flux LE_v (Figures 54.2 and 54.3). The seasonal variations in albedo shown in Figure 54.1 are related to grazing and cutting of the grass

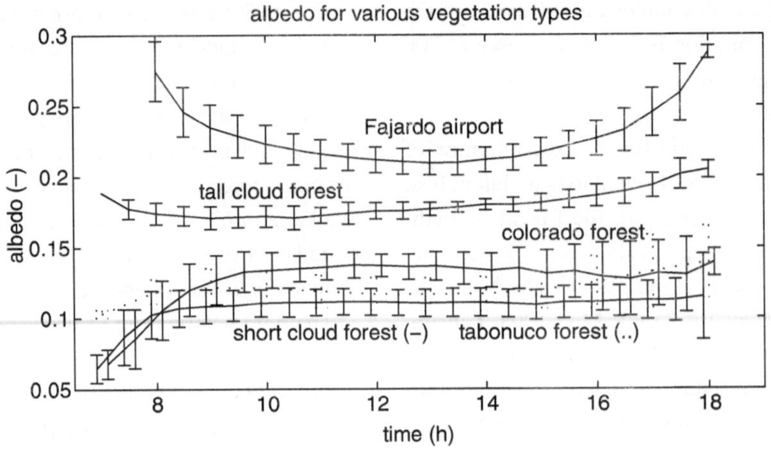

Figure 54.1. Seasonal course of daily albedo for wetland forest and pasture near Sabana Seca, northern Puerto Rico between May 1997 and May 1998.

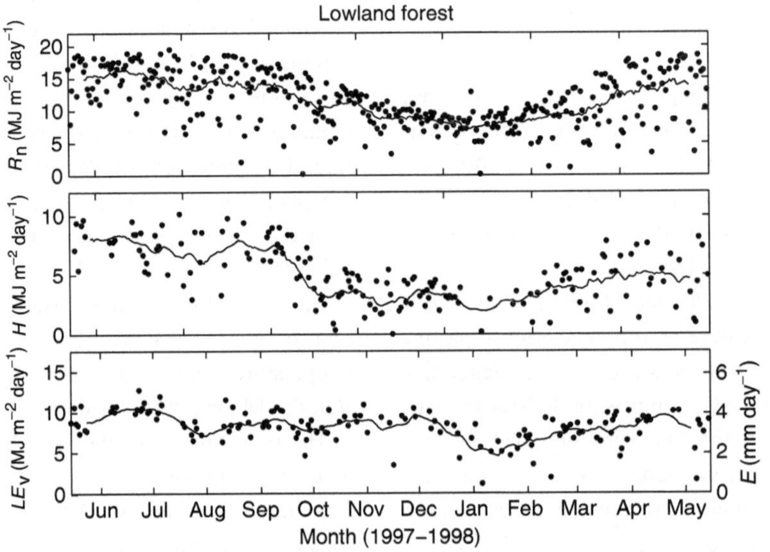

Figure 54.2. Seasonal course of daily totals of net radiation, and sensible and latent heat fluxes above a wetland forest near Sabana Seca, northern Puerto Rico between May 1997 and May 1998.

at the pasture site, and flowering and canopy closure after storm damage at the forest site.

The fluxes of sensible and latent heat constitute two important links between the land surface and the overlying atmosphere. The momentum flux (influenced by the aerodynamic roughness of the surface) is another link but this proved to be less important (Van der Molen, 2002). The seasonal changes in R_n, H, and LE_v for forest and pasture are shown in Figures 54.2 and 54.3, respectively. The 9% difference in R_n is clearly borne out but the most impressive change after forest conversion is the reduction in H from 4–10 MJ m^{-2} day^{-1} over forest to 2–3 MJ m^{-2} day^{-1} over pasture, whereas LE_v shows an opposite pattern (albeit to a lesser extent because of the reduction in R_n). The mean daily evapotranspiration rates (based on LE_v) determined for the forest and pasture were 3.3 and 3.8 mm day^{-1} or 69 and 81% of R_n, respectively.

The changes in energy budget after land-cover transformation were even more distinct on a diurnal basis. As shown in Figures 54.4 (forest) and 54.5 (pasture) the energy storage terms G (net soil heat flux), S (heat storage in vegetation), and J (idem in the air column below the top of the vegetation) were small compared to the radiant (R_n) and turbulent (H, LE_v) fluxes. Apart from the apparent difference in Bowen ratio (i.e. H/LE_v), the rates at which H and LE_v increased in the early morning differed as well: at the forest site they increased simultaneously, whereas in the pasture H started to increase *c.* 1.5 hours later than LE_v.

Model results

The agreement between modeled and observed vegetation and canopy air temperatures and water vapor mixing ratios, wind speeds, and momentum fluxes, as well as net long-wave and

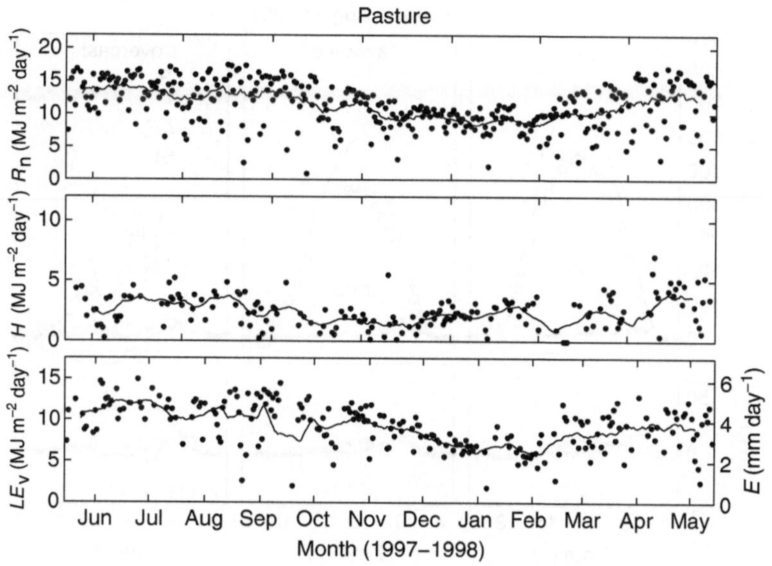

Figure 54.3. Seasonal course of daily totals of net radiation, and sensible and latent heat fluxes above a pasture near Sabana Seca, northern Puerto Rico between May 1997 and May 1998.

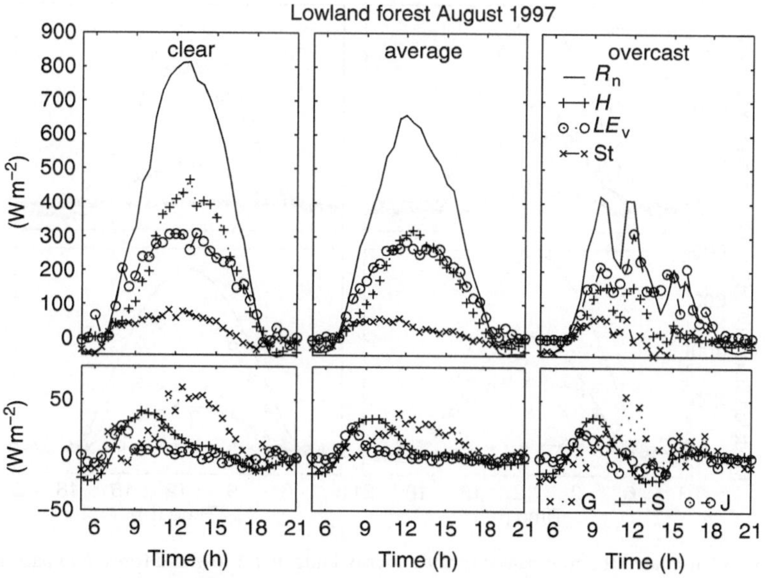

Figure 54.4. Monthly mean diurnal course of net radiation partitioning at the Sebana Seca wetland forest in August 1997 for clear, average, and cloudy days.

short-wave radiation totals was good, with deviations being much smaller than the respective diurnal variations (Van der Molen, 2002). Whilst the latter are not shown here for reasons of space, the strong agreement between the observed and modeled diurnal fluxes presented in Figure 54.6 is the result of the good matches obtained for the respective variables.

The results of the respective model runs are presented below in the form of averaged cross sections of atmospheric fields in a north–south direction over the island near the location of the experimental sites. A number of neighboring cross-sections were averaged to give a general impression of the atmospheric circulation over the island. Considering that both "forest" and "pasture" runs were initialized with identical atmospheric conditions, the model results are best compared for the early afternoon situation. By that time of the day the cumulative differences in atmospheric heat and moisture content between the two land covers have become sufficiently pronounced, whilst turbulence is still active, for instance at 15.00 h local time (Figure 54.7).

In the "forest" run, the potential temperature (top panel in Figure 54.7) near the land surface and above the island is about

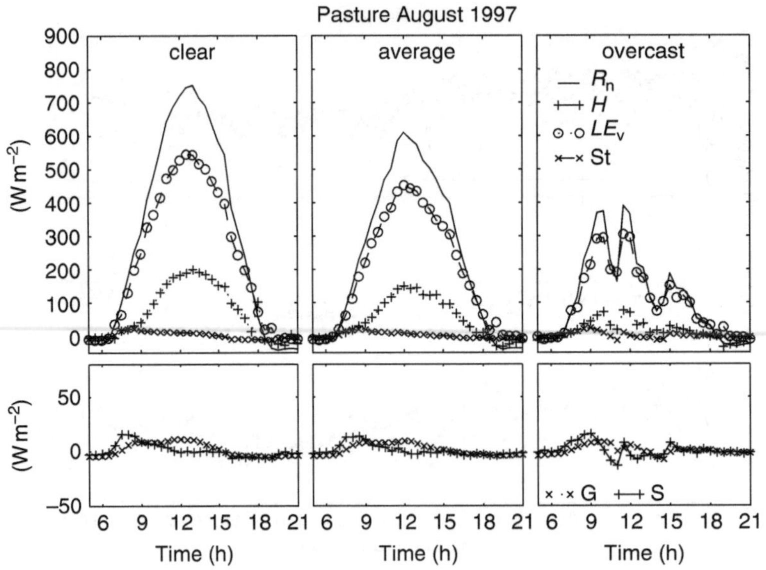

Figure 54.5. Monthly mean diurnal course of net radiation partitioning at the Sebana Seca pasture in August 1997 for clear, average, and cloudy days.

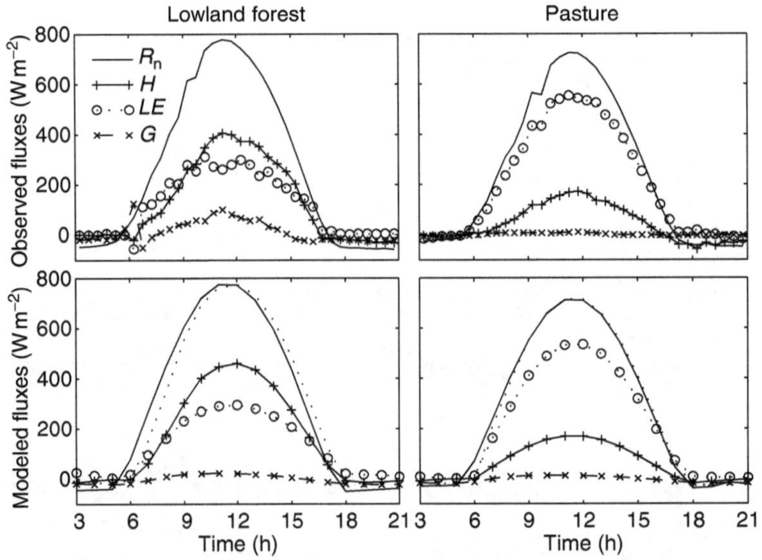

Figure 54.6. Observed (top panel) vs. simulated (bottom panel) surface energy budgets for lowland forest and pasture at Sebana Seca. The dotted line in the lower panel indicates the sum of sensible heat H, latent heat LE_v, and net soil heat flux G.

2 °C higher than in the "pasture" run. This renders the atmosphere above the island more unstable and well mixed to a higher elevation for the forested situation, as is also apparent from the distortion of the temperature contour lines at higher levels. The stronger convection associated with forest is also reflected in the number, size, and strength of convective cells above the island, indicated by the contour lines of positive vertical velocity (Figure 54.7, center of bottom panel). Directly related to this increased upward movement of air above the island, is the convergence of moist air from the ocean to the island from both sides (Figure 54.7, center of top panel) through the formation of a sea breeze. This sea breeze convergence determines the amount of moisture transported upward to levels where condensation (and therefore cloud formation) occurs (cf. Foster, this volume). Indeed, the bottom panel of Figure 54.7 indicates that more moisture is transported to higher levels in the "forest" run, creating denser clouds, albeit at a higher lifting condensation level (LCL) because of the higher temperatures (Figure 54.8).

Amounts of liquid water in the atmosphere as a function of time and height above the surface for the "forest" and "pasture" runs are shown in Figure 54.8. The amount of cloud condensate proves to be several times larger over an island with forested lowlands than with pasture. Moreover, the LCL is several hundred meters higher in the "forest" run than in the "pasture" run.

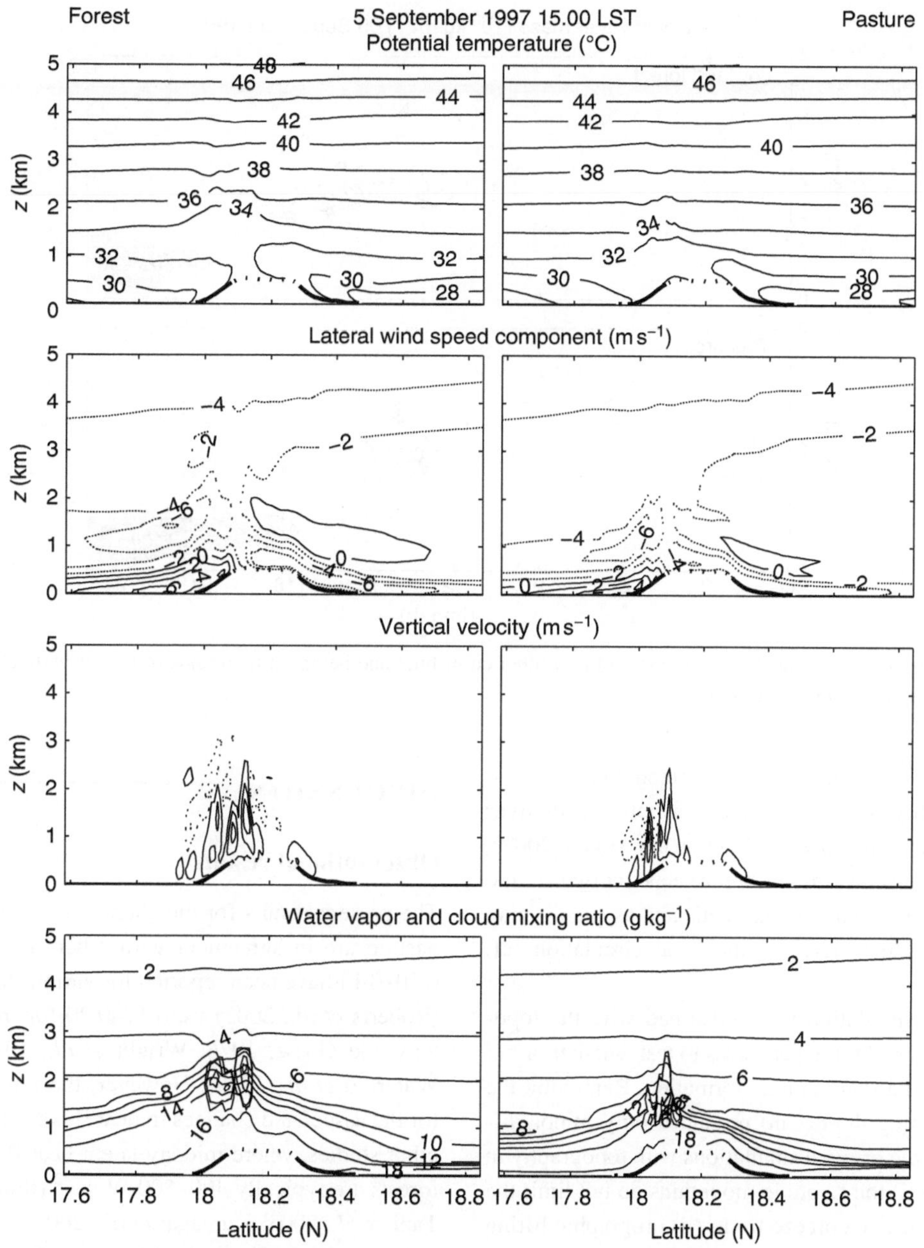

Figure 54.7. Averaged cross-sections of the atmospheric fields of potential temperature (top panel), lateral (S–N) wind speed (center top), vertical velocity (center bottom), and water vapor and cloud mixing ratio (bottom panel) in the model runs with lowland forest vegetation (left) and pasture vegetation (right). Negative contours are dotted, vertical velocity contours drawn every 0.25 m s^{-1} starting from 0.125 m s^{-1}, cloud water contours drawn at 0.001, 0.01, 0.1, 0.2, 0.3, . . . g kg^{-1}. The thick solid line indicates the location of the island with either lowland forest or pasture vegetation; where the line becomes dashed the vegetation is replaced by upland forest. Cloud water mixing ratio contours indicate values averaged over the longitudinal dimensions of the grid.

For all model runs, a fairly linear relationship was found between the liquid water mass in the model domain and the precipitation rate.

Proof of the earlier statement that the sea breeze is the main driver of the atmospheric circulation above Puerto Rico may be obtained by performing a simulation in which the ocean is replaced by extensive and identical lowland forest, thereby essentially removing the cause for the development of a sea breeze. Replacing the surrounding ocean by lowland forest showed the development of numerous small convective cells at regular intervals of about 10 km and reaching a height of 1–2 km. These cells quite effectively mixed water vapor and heat throughout the boundary layer, but failed to reach the LCL. In the case of "pasture" vegetation, the convective cells were less numerous and considerably weaker. However, in one of the cells some condensation occurred, which did not happen in the

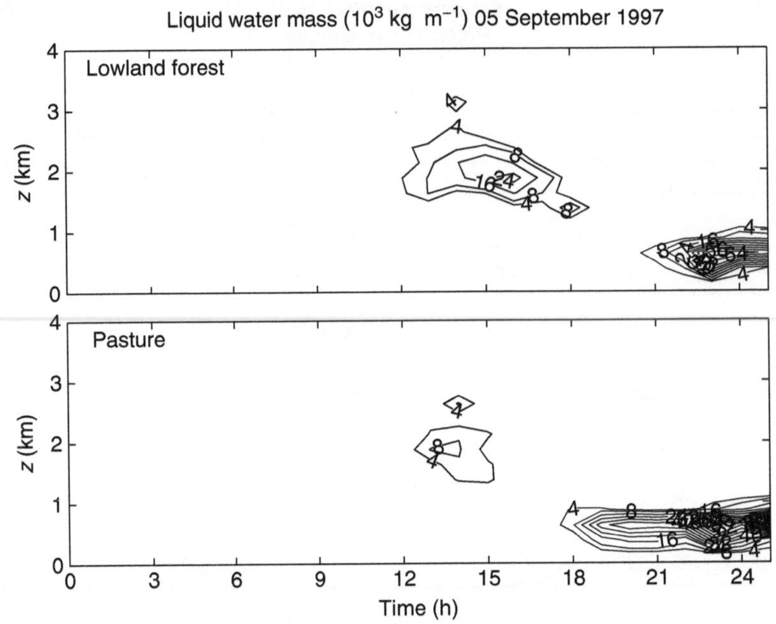

Figure 54.8. Liquid water mass over land in the finest grid as a function of time and height in thousands of kg per vertical meter. The land area over which the cloud mass is summed is 1452 km^2.

corresponding "forest" run. This may be attributed directly to the higher relative humidity of the air in the convective cells over "pasture," which resulted in a lower LCL (Van der Molen, 2002). This experiment indicates that the sea breeze acts to organize the convective cells into fewer but stronger cells and as such it may indeed be regarded as the driver of the local circulation (cf. Malkus, 1955).

Similarly, a model simulation was performed with the topographic height of the island set to zero so as to test whether or not mountain winds played a role in cloud formation. Removing the topography (but retaining the ocean) produced only minor differences compared to the original simulations with topography in place, indicating that: (i) the Luquillo mountains do not limit the inward penetration of the sea breeze front, (ii) orographic lifting does not have much effect on the meso-scale circulation, and (iii) mountain-valley winds play a minor role compared to the sea breeze circulation.

Simulations carried out under varying synoptic conditions indicated that the liquid water content of clouds was higher above an island with forested coastal plains, as long as the sea breeze dominated the circulation. In conditions with strong wind, a highly stable free atmosphere, or heavy cloud cover, the sea breeze did not develop. Under conditions of strong winds or overcast skies, turbulence is driven mainly by wind shear, resulting in smaller differences between "forest" and "pasture" runs. Conversely, when the boundary layer was topped by a highly stable free atmosphere, the convective cells simply were not able to penetrate high enough to reach the LCL and again the difference between "forest" and "pasture" simulations remained small.

DISCUSSION

Observational results

The present results for the change in albedo upon conversion to pasture are in agreement with other studies. Albedo values of 0.10–0.14 have been reported for various lowland tropical forests (Roberts *et al.*, 2005) vs. 0.17–0.20 for tropical pastures (Monteny and Gosse, 1976; Wright *et al.*, 1992; Culf *et al.*, 1995; Waterloo *et al.*, 1999). However, the presently obtained results for the associated changes in heat fluxes differ substantially from other studies, where midday latent heat fluxes over tropical rain forests are typically 400–500 W m^{-2} (Shuttleworth *et al.*, 1984; Tani *et al.*, 2003; Kumagai *et al.*, 2004), with associated Bowen ratios <1. For pastures in Amazonia, Bowen ratios of 0.43 have been reported under moist conditions, increasing to 0.96 after 20 days without precipitation (Wright *et al.*, 1992). Typical evapotranspiration rates (*ET*) for Amazonian forest and pasture are *c.* 4 and 2 mm day^{-1}, respectively (Shuttleworth *et al.*, 1984; Wright *et al.*, 1992; Gash *et al.*, 1996).

At mean *ET* rates of 3.3 and 3.8 mm day^{-1}, respectively, the water use of the lowland forest and pasture in Puerto Rico are at odds with the findings cited above for Amazonian forests and pastures. This has several reasons. First of all, due to frequent precipitation and high groundwater levels, water supply to the Puerto Rican pasture was never limiting, whereas in Amazonia moisture stress occurs during the dry season (Wright *et al.*, 1992). Furthermore, the *Pterocarpus* wetland forest at Sabana Seca grade into mangrove forests and frequently experienced brackish water at the surface, which may have caused physiological water stress. Also, the forest was relatively short-statured (14 m) compared to most other lowland tropical forests (typically 30–50 m high; Roberts *et al.*, 2005), whereas the Sebana Seca pasture had relatively tall grass, reaching maximum heights of 1.5–2 m in the absence of grazing (compared to <0.3 m in Amazonian pasture; Wright *et al.*, 1992). Such factors tend to reduce the normally found contrasts in evaporation between tropical grasslands and forest. Nevertheless, the similarity in average daily *ET* for the *Pterocarpus* forest and pasture is striking, given that the midday latent heat flux LE_v above the forest is almost half that above the pasture (300 vs. 550 W m^{-2} on clear days). This is because the maximum of LE_v at the forest site decreased much less going from clear to cloudy conditions, whereas at the pasture site this was associated with a decrease of almost 50% (Figures 54.4 and 54.5). This is yet another indication that the transpiration rate at the forest site was indeed limited by factors other than available energy.

Meso-scale modeling

The modeled sensitivity of the strength of the sea breeze to land-cover change is the logical consequence of the associated changes in surface fluxes and appears to be rather robust. The predicted changes in strength of the circulation and in the degree of cloud condensation are significant and in line with the observed trends in decreasing precipitation (Bisselink, 2003). However, the present modeling results cannot be translated directly to quantitative changes in amounts of cloud condensation or precipitation, because: (i) these have not been verified with observations, and (ii) the sea breeze circulation does not depend on the magnitude of *H* and LE_v alone, but also on atmospheric stability, strength of the trade winds, and other synoptic-scale weather conditions (Malkus, 1955).

In a similar modeling study of the effects of forest conversion to pasture in northern Costa Rica and adjacent Nicaragua, Lawton *et al.* (2001) derived higher cloud bases over pasture than over forested areas, a finding that at first sight would seem contradictory to the present results. However, in the latter study, *H* was higher for pasture and LE_v lower than for forest. As such, the underlying processes causing the predicted atmospheric changes appear to agree in the two studies. Nair *et al.* (2003) and Nair *et al.* (this volume) present further evidence in the form of satellite observations showing reduced cloud development over deforested areas in Costa Rica.

In the model run where ocean was replaced by lowland forest, the sea breeze did not develop; instead, numerous convective cells were formed that proved less than effective in transporting moisture upward to a level required for condensation. Some cloud condensation occurred in the associated "pasture" run, mainly because relative humidity was high enough for the convective cells to reach the lifting condensation level despite the fact that convection was relatively weak. This result is quite similar to modeling results obtained for Amazonia (Nobre *et al.*, 1991; Lean *et al.*, 1996), where, as in northern lowland Puerto Rico, higher cloud condensation and precipitation are associated with the vegetation having the highest *ET*, i.e. lowland rain forest in Amazonia and tall grassland in Puerto Rico. Thus, comparing continental and maritime tropical conditions, the main driver behind convection and cloud condensation appears to shift from LE_v to *H*. For the maritime case, this stands to reason because the moisture-laden trade winds blowing from the ocean bring in abundant moisture regardless of land-cover type.

The present study has focused mainly on dry canopy conditions. Indeed, including rainfall interception and subsequent recycling of re-evaporated water introduces so many new processes that the model used would not be applicable without a full verification and parameterization of the interception process (cf. Holwerda *et al.*, 2006). However, it is important to note that the sea breeze circulation depends on the presence of a significant sensible heat flux, a condition that may not be met under conditions with wet canopies where advected energy from the ocean seems to be the main driver of wet-canopy evaporation (Schellekens *et al.*, 2000; Holwerda *et al.*, 2006; Wallace and McJannet, 2006). Furthermore, the differences between "forest" and "pasture" runs were smaller under conditions of heavy cloud cover (Figures 54.4 and 54.5).

The model simulations further indicated that the conversion of wetland forest to pasture in Puerto Rico is associated with a lowering of the cloud base by up to several hundred meters, because the air feeding the convective cells is cooler and contains more water vapor. Whilst in reality clouds are frequently observed at altitudes of 600–1000 m.a.s.l., in the model simulations this was 1400–2000 m.a.s.l. Although actual cloud base observations are not available for the specific day for which the modeling was performed, the difference may reflect some combination of the prevailing weather conditions and inaccuracies of the model. Furthermore, it should be realized that the vertical grid spacing used in the model was about 100 m, i.e. a comparable order of magnitude as the differences in predicted LCL between the "forest" and "pasture" runs. Therefore, conclusions on the sensitivity of the cloud base to changes in vegetation cover are mainly qualitative at this stage.

The presence of the ocean appeared to be of crucial importance in this study, because of the dominant role of the sea breeze circulation which caused much stronger upward air movements than there would have been under continental conditions. An extension of the Puerto Rican work has demonstrated that the effects of meso-scale land-cover transformations on atmospheric circulation under maritime conditions across the tropics are much more pronounced than under more continental conditions (Van der Molen *et al.*, 2006).

CONCLUSIONS

The surface fluxes of radiation, sensible heat (*H*) and latent heat (LE_v) were measured above a wetland forest and a pasture considered representative of the pre- and post-deforestation situation in the coastal plain of northern Puerto Rico. Forest conversion to pasture increased the albedo from 13–15% to 18–23%. In combination with changes in net long-wave radiation, net radiation decreased on average by 9%. Average daily evapotranspiration (*ET*) amounted to 3.3 and 3.8 mm day^{-1} for the forest and pasture, respectively, whereas *H* decreased drastically from 4–10 MJ m^{-2} day^{-1} to 2–3 MJ m^{-2} day^{-1}. The change was also manifest in the diurnal course of surface energy fluxes: on a clear summer day in August, *H* and LE_v reached maxima of 420 and 300 W m^{-2}, respectively, above forest, vs. 200 and 550 W m^{-2} above pasture. The occurrence of frequent showers, a high groundwater table, and high foliar biomass provided optimum evaporation conditions in the pasture, in contrast to the wetland forest site which carried relatively short and open vegetation that was regularly inundated with brackish water as well.

The sea breeze proved to play a major role in the atmospheric circulation on the island. In a simulation with the lowland parts of the island forested, more sensible heat was predicted to be transferred from the surface to the atmosphere, resulting in stronger convection and consequently a stronger sea breeze than with lowlands under pasture. Also, the warmer and drier air above the forest (compared to that above pasture) led to a higher lifting condensation level (LCL). Because the first effect dominates, significantly more cloud condensation, and consequently precipitation, was predicted in simulations over a forested island. These results appear typical for clear days without a pronounced temperature inversion or strong winds, i.e. when a sea breeze circulation develops. Under cloudier or windier conditions, the contrasts between "forest" and "pasture" runs were less pronounced and even became reversed when atmospheric stability was too strong for convective cells to reach the LCL in the case of "forest", although not for "pasture."

REFERENCES

Aide, T. M., J. K. Zimmerman, L. Herrera, M. Rosario, and M Serrano (1995). Forest recovery in abandoned tropical pastures in Puerto Rico. *Forest Ecology and Management* **77**: 77–86.

Aubinet, M., A. Grelle, P. Berbigier, *et al.* (2000). Estimation of the annual net CO_2 and H_2O exchanges of European forests: the Euroflux methodology. *Advances in Ecological Research* **30**: 113–175.

Bisselink, B. (2003). Precipitation trends in Puerto Rico: quantification and explanation of complex patterns. M.Sc. thesis, VU University Amsterdam, Amsterdam, the Netherlands.

Chen, C., and W. R. Cotton (1983). A one-dimensional simulation of the stratocumulus capped mixed layer. *Boundary-Layer Meteorology* **25**: 289–321.

Chen, C., and W. R. Cotton (1987). The physics of the marine stratocumulus capped mixed layer. *Journal of Atmospheric Science* **44**: 2951–2977.

Cintrón, B. B. (1983). Coastal freshwater swamp forests: Puerto Rico's most endangered ecosystem. In *Los bosques de Puerto Rico*, ed. A. E. Lugo, pp. 249–275. Rio Piedras, Puerto Rico: U.S. Department of Agriculture Forest Service, Southern Forest Experiment Station, Institute of Tropical Forestry, and Puerto Rico Department of Natural Resources.

Costa, M. H. (2005). Large-scale impacts of tropical forest conversion. In *Forests, Water and People in the Humid Tropics*, eds. M. Bonell and L. A. Bruijnzeel, pp. 590–597. Cambridge, UK: Cambridge University Press.

Costa, M. H., and J. A. Foley (1999). Trends in the hydrological cycle of the Amazon basin. *Journal of Geophysical Research D* **104**: 14 189–14 198.

Culf, A. D., G. Fisch, and M. G. Hodnett (1995). The albedo of Amazonian forest and ranchland. *Journal of Climate* **8**: 1544–1554.

Cutrim, E., D. W. Martin, and R. Rabin (1985). Enhancement of cumulus clouds over deforested lands in Amazonia. *Bulletin of the American Meteorological Society* **76**: 1801–1805.

Dietz, J. L. (1986). *Economic History of Puerto Rico*. Princeton, NJ: Princeton University Press.

Eltahir, E. A. B., and R. L. Bras (1994). Precipitation recycling in the Amazon basin. *Quarterly Journal of the Royal Meteorological Society* **120**: 861–880.

Franco, P. A., P. L. Weaver, and S. Eggen-McIntosh (1997). *Forest Resources of Puerto Rico, 1990*, Resource Bulletin SRS 22. Asheville, NC: U.S. Department of Agriculture Forest Service, Southern Forest Research Station.

Gash, J. H. C., C. A. Nobre, J. M. Roberts, and R. L. Victoria (eds.) (1996). *Amazon Deforestation and Climate*. Chichester, UK: John Wiley.

Giambelluca, T. W., R. E. Martin, G. P. Asner, *et al.* (2009). Evapotranspiration and energy balance of native wet montane cloud forest in Hawai'i. *Agricultural and Forest Meteorology* **149**: 230–243.

Holdridge, L. R. (1940). Some notes on the mangrove swamps in Puerto Rico. *Caribbean Forester* **1**: 19–29.

Holwerda, F., F. N. Scatena, and L. A. Bruijnzeel (2006). Throughfall in a Puerto Rican lower montane rain forest: a comparison of sampling strategies. *Journal of Hydrology* **327**: 592–602.

Janzen, D. H. (1978). Description of a *Pterocarpus officinalis* (Leguminosae) monoculture in Cocovado National Park, Costa Rica. *Brenesia* **14–**15: 305–309.

Kennaway, T., and E. H. Helmer (2007). The forest types and ages cleared for land development in Puerto Rico. *GIScience and Remote Sensing* **4**: 356–382.

Kumagai, T., G. G. Katul, T. M. Saitoh, *et al.* (2004). Water cycling in a Bornean tropical rain forest under current and projected precipitation scenarios. *Water Resources Research* **40**, W01104, doi: 10.1029/2003WR002226.

Larsen, M. C. (2000). Analysis of 20th century rainfall and streamflow to characterize drought and water resources in Puerto Rico. *Physical Geography* **21**: 494–521.

Lawton, R. O., U. S. Nair, R. A. Pielke Sr., and R. M. Welch (2001). Climatic impact of tropical lowland deforestation on nearby cloud forests. *Science* **294**: 584–587.

Lean, J., C. R. Bunton, C. A. Nobre, and P. R. Rowntree (1996). The simulated impact of Amazonian deforestation on climate using ABRACOS vegetation characteristics. In *Amazonian Deforestation and Climate*, eds. J. H. C. Gash, C. A. Nobre, J. M. Roberts, and R. L. Victoria, pp. 549–576. Chichester, UK: John Wiley.

Lugo, A. E., and A. R. García-Martinó (1996). Cartilla del agua para Puerto Rico. *Acta Cientifica* **10**: 15–89.

Makareva, A. M., and V. G. Gorshkov (2007). Biotic pump of atmospheric moisture as driver of the hydrological cycle on land. *Hydrology and Earth System Sciences* **11**: 1013–1033.

Malkus, J. S. (1955). The effects of a large island upon the trade-wind air stream. *Quarterly Journal of the Royal Meteorological Society* **81**: 538–550.

Meesters, A. G. C. A., A. J. Dolman, and L. A. Bruijnzeel (2009). Comment on "Biotic pump of atmospheric moisture as driver of the hydrological cycle on land" by A.M. Makareva and V.G. Gorshkov, *Hydrol. Earth Syst. Sci.*, **11**, 1013–1033. Hydrology and Earth System Sciences 13: 1299–1305.

Monteny, B., and G. Gosse (1976). Analyse et estimation du rayonnement net d'une culture de *Panicum maximum* en zone tropical humide. *Oecologia Plantarum* **11**: 173–191.

Nair, U. S., R. O. Lawton, R. M. Welch, and R. A. Pielke Sr. (2003). Impact of land use on tropical montane cloud forests: sensitivity of cumulus cloud field characteristics to lowland deforestation. *Journal of Geophysical Research* **108**(D7): 4206–4218.

Nobre, C. A., P. J. Sellers, and J. Shukla (1991). Amazonian deforestation and regional climate change. *Journal of Climate* **4**: 957–988.

Pielke, R. A., W. R. Cotton, R. L. Walko, *et al.* (1992). A comprehensive meteorological modeling system: RAMS. *Meteorological and Atmospheric Physics* **49**: 69–91.

Pielke, R. A., J. Adegoke, A. Beltran-Przekurat, *et al.* (2007). An overview of regional land use and land cover impacts on rainfall. *Tellus Series B* **59** : 587–601.

Ray, D. K., U. S. Nair, R. O. Lawton, R. M. Welch, and R. A. Pielke Sr. (2006). Impact of land use on Costa Rican tropical montane cloud forests: sensitivity of orographic cloud formation to deforestation in the plains. *Journal of Geophysical Research* **11**, D02108, doi: 10.1029/2005JD006096.

Roberts, J. M., J. H. C. Gash, M. Tani, and L. A. Bruijnzeel (2005). Controls on evaporation in lowland tropical rain forest. In *Forests, Water and People in the Humid Tropics*, eds. M. Bonell and L. A. Bruijnzeel, pp. 590–597. Cambridge, UK: Cambridge University Press.

Scatena, F. N., and M. C. Larsen (1991). Physical aspects of Hurricane Hugo in Puerto Rico. *Biotropica* **23**: 317–323.

Schellekens, J., L. A. Bruijnzeel, F. N. Scatena, Bink, N. J., and F. Holwerda (2000). Evaporation from a tropical rain forest, Luquillo Experimental Forest, Puerto Rico. *Water Resources Research* **36**: 2183–2196.

Shuttleworth, W. J., J. H. C. Gash, C. R. Lloyd, *et al.* (1984). Eddy correlation measurements of energy partition for Amazonian forest. *Quarterly Journal of the Royal Meteorological Society* **110**: 1143–1162.

Smagorinsky, J. (1963). General circulation experiments with the primitive equations. I. The basic experiment. *Monthly Weather Review* **91**: 99–164.

Tani, M., A. Rahim Nik, Y. Ohtani, *et al.* (2003). Characteristics of energy exchange and surface conductance of a tropical rain forest in Peninsular Malaysia. In *Pasoh: Ecology of a Lowland Rain Forest in Southeast Asia*, eds. T. Okuda, N. Manokaran, Y. Matsumoto, *et al.* pp. 73–88. Berlin: Springer-Verlag.

Van der Molen, M. K. (2002). Meteorological impacts of land use change in the maritime tropics. Ph.D. thesis, VU University Amsterdam, Amsterdam, the Netherlands. Also available at www.falw.vu.nl/ images_upload/ 702F8F36–45DC-488E-B2C550BFDCAC2B9D.pdf.

Van der Molen, M. K., A. J. Dolman, M. J. Waterloo, and L. A. Bruijnzeel (2006). Climate is affected more by maritime than by continental land use change: a multiple-scale analysis. *Global and Planetary Change* **54**: 128–149.

Walko, R. L., W. R. Cotton, M. P. Meyers, and J. Y. Harrington (1995). New RAMS microphysical parameterization. I. The single moment scheme. *Atmospheric Research* **38**: 29–62.

Walko, R. L., L. E. Band, J. Baron, *et al.* (2000). Coupled atmosphere–biophysics–hydrology models for environmental modeling. *Journal of Applied Meteorology* **39**: 931–944.

Wallace, J. S, and D. L. McJannet (2006). On interception modelling of a lowland coastal rainforest in northern Queensland, Australia. *Journal of Hydrology* **329**: 477–488.

Waterloo, M. J., L. A. Bruijnzeel, H. F. Vugts, and T. T. Rawaqa (1999). Evaporation from *Pinus caribaea* plantations on former grassland soil under maritime tropical conditions. *Water Resources Research* **35**: 2133–2144.

Weaver, P. L. (1995). The Colorado and dwarf forests of Puerto Rico's Luquillo Mountains. In *Tropical Forests: Management and Ecology*, eds. A. E. Lugo and C. Lowe, pp. 109–141. New York: Springer-Verlag.

Wright, I. R., J. H. C. Gash, H. R. da Rocha, *et al.* (1992). Dry season micrometeorology of Central Amazonian ranchland. *Quarterly Journal of the Royal Meteorological Society* **118**: 1083–1099.

55 The impact of deforestation on orographic cloud formation in a complex tropical environment

U. S. Nair, D. K. Ray, R. O. Lawton, and R. M. Welch
University of Alabama in Huntsville, Huntsville, Alabama, USA

R. A. Pielke Sr.
Colorado State University, Fort Collins, Colorado, USA

J. Calvo-Alvarado
Instituto Tecnológico de Costa Rica, Cartago, Costa Rica

ABSTRACT

Ecological changes observed in cloud forests in the Monteverde area, northern Costa Rica, including disappearance of anuran populations and expansion of bird and bat ranges to higher elevations, have been linked to an increasing trend in dry-season mist-free days. Prior studies suggest that this trend may be influenced by both large-scale processes of climate change and regional-scale changes in land cover. Preliminary investigations exploring the impact of land use on cloud formation indicated that drying and warming of boundary layer air in response to lowland deforestation leads to increased cloud base heights. In the present study, numerical model experiments utilizing realistic land-use scenarios and atmospheric conditions are used to further explore the impact of land-use change on orographic cloud formation. The Regional Atmospheric Modeling System (RAMS) was used to simulate orographic cloud formation during the time period of 1–14 March 2003 in the Monteverde region for pristine, current, and future land-use scenarios. The simulations were initiated from the same atmospheric conditions and subject to similar lateral boundary conditions. Comparisons against observations showed that RAMS was capable of realistically simulating the nature of orographic cloud formation and boundary-layer thermodynamics. Numerical simulations indicated that deforestation in the lowlands and adjacent pre-montane areas results in an increase in average cloud base height and a consequent decrease in the areal extent of montane forests immersed in clouds. In the current and future land-use scenarios, warmer and drier air is found over the lowlands and pre-montane areas. The simulated differences in cloud formation and air-mass thermodynamics were positively correlated with the degree of deforestation in the lowland and pre-montane regions.

INTRODUCTION

Analysis of meteorological observations at Monteverde, Costa Rica, indicates a pattern of increases in dry-season mist-free days since the early 1970s (Pounds *et al.*, 1999). The dry-season input of moisture from orographic clouds is considered crucial for the hydrological (Guswa *et al.*, 2007) as well as ecological functioning of the area's leeward montane cloud forest (Nadkarni and Wheelwright, 2000). The increase in mist-free days may also be co-responsible for the ecological changes observed in this region, including an expansion of bird and bat ranges to higher elevations (Pounds *et al.*, 1999; LaVal, 2004). At the same time, changes in the diurnal temperature ranges since the 1970s show a decreasing trend, implying increased cloudiness, particularly at night, in the Monteverde region (Pounds *et al.*, 2006). Under these conditions, a decrease in dry-season mist frequency implies an increase in cloud base height.

Recent studies have explored potential processes that might contribute to an elevation of orographic cloud base height at Monteverde and montane sites in other parts of the world that

Tropical Montane Cloud Forests: Science for Conservation and Management, eds. L. A. Bruijnzeel, F. N. Scatena, and L. S. Hamilton. Published by Cambridge University Press.

harbor cloud forest ecosystems (Still *et al.*, 1999; Lawton *et al.*, 2001; Van der Molen, 2002; Nair *et al.*, 2003; Ray *et al.*, 2006; Van der Molen *et al.*, 2006). These studies suggest that sudden warming of tropical oceans in the mid-1970s (Pounds *et al.*, 1999), climate change forced by increasing atmospheric carbon dioxide (Still *et al.*, 1999) and land-cover change in adjacent lowlands (Lawton *et al.*, 2001; Nair *et al.*, 2003; Ray *et al.*, 2006; Van der Molen *et al.*, 2006) could all potentially contribute to increased orographic cloud base heights in areas with montane cloud forest.

Still *et al.* (1999) used Global Circulation Model (GCM) simulations of climate with current and doubled atmospheric CO_2 concentrations to examine the effect of climate change on cloud base heights at various high-altitude locations around the world. Using height of relative humidity surfaces in these simulations as a proxy for cloud base height, Still *et al.* (1999) calculated increases in cloud base heights of up to 300 m in response to doubled atmospheric CO_2. Pounds *et al.* (1999) hypothesized that the effect of increased tropical sea surface temperatures is similar in nature to GCM simulations of surface warming resulting from doubled atmospheric CO_2, namely an increase in cloud base height. Still *et al.* (1999) suggested further research using meso-scale models that better resolve the local topography and land-use change.

Lawton *et al.* (2001) suggested that, through the alteration of surface heat and moisture transfers, deforestation in the lowlands upwind of elevated areas leads to warming and drying of the air masses that are ultimately responsible for orographic cloud formation. Satellite imagery over forested and deforested lowland areas upwind of Monteverde shows distinct differences in formation of fair weather cumulus clouds indicating modification of air masses in this region by land cover (Lawton *et al.*, 2001; Nair *et al.*, 2003; Ray *et al.*, 2006). Numerical simulations conducted using the Regional Atmospheric Modeling System (RAMS; Pielke *et al.*, 1992) showed convective clouds forming over deforested areas to have higher base heights due to a warmer and drier boundary layer (Lawton *et al.*, 2001; Nair *et al.*, 2003; Ray *et al.*, 2006). The preliminary numerical simulations of Lawton *et al.* (2001) also suggested that lowland forest conversion to pasture leads to an elevation of the orographic cloud bank in the Monteverde region.

The preliminary investigations by Lawton *et al.* (2001) and Nair *et al.* (2003) assumed scenarios in which the entire model domain is completely forested (evergreen broadleaf forest) or deforested (pasture) or in which just the lowland areas are deforested. Since deforestation is usually accompanied by changes in hydrological budgets, these studies also assumed different soil water storage values for forested and deforested areas that were consistent with observations from prior studies. Recently, two coordinated research projects have collected a series of ground-based observations with the aim of conducting more realistic numerical model experiments in this area. The present study extends the prior works of Lawton *et al.* (2001) and Nair *et al.* (2003) and uses numerical modeling experiments with realistic specification of atmospheric initial conditions, topography and land-use change scenarios to assess the impact of deforestation in upwind areas on the orographic cloud formation in Monteverde, Costa Rica.

STUDY AREA AND DATA-SETS

The present study focuses on the Monteverde cloud forests, located along the crest of the Cordillera de Tilarán (Figure 55.1) in northern Costa Rica. The north-easterly trade winds flowing over the Cordillera force the formation of frequent and persistent orographic cloud banks along the windward Caribbean slopes. These cloud banks spill over the crest onto the uppermost Pacific slopes and are responsible for the sustenance of one the most diverse cloud forest ecosystems in the world (Nadkarni and Wheelwright, 2000; cf. Häger and Dohrenbusch, this volume).

Even though the forest along the Caribbean slopes of the Cordillera de Tilarán has remained relatively intact, forest cover in the Caribbean plains has steadily declined during the past century. Sader and Joyce (1988) estimated a decrease in primary forest cover from 67% of the land area in 1940 to 17% in 1983, while more recently Carlson and Sanchez-Azofeifa (1999) reported 29% forest cover for 1991, with deforestation proceeding at an average rate of 4.2% year^{-1}. The estimates of Carlson and Sanchez-Azofiefa (1999) pertain to the country as a whole; the majority of the remaining forest is montane, while the lowland and pre-montane regions are largely deforested except in a few National Parks and Wildlife Refuges (cf. Calvo-Alvarado *et al.*, this volume). The areas upwind of Monteverde, such as the San Carlos and Tortuguero plains, have a long history of deforestation, with the exception of Tortuguero National Park located south of the Costa Rica–Nicaragua border.

Various data-sets were used to both initialize and validate the Colorado State University Regional Atmospheric Modeling System (RAMS) in the present study. The National Center for Environmental Prediction (NCEP) reanalysis data (Kalnay *et al.*, 1996) plus rawinsonde observations obtained during the Land Use Cloud Interaction (LUCIE) field experiment were used to derive the initial atmospheric conditions and the temporally varying lateral boundary conditions used in the RAMS simulations. During the LUCIE field experiment (March 2003), atmospheric conditions in northern Costa Rica were sampled over several paired forested and deforested sites at 3-hour intervals starting from 06.00 Local Time (LT) to 18.00 LT using coincident rawinsonde launches (Ray *et al.*, 2006).

The United State Geological Survey (USGS) topography data and the University of Maryland global land use categorization

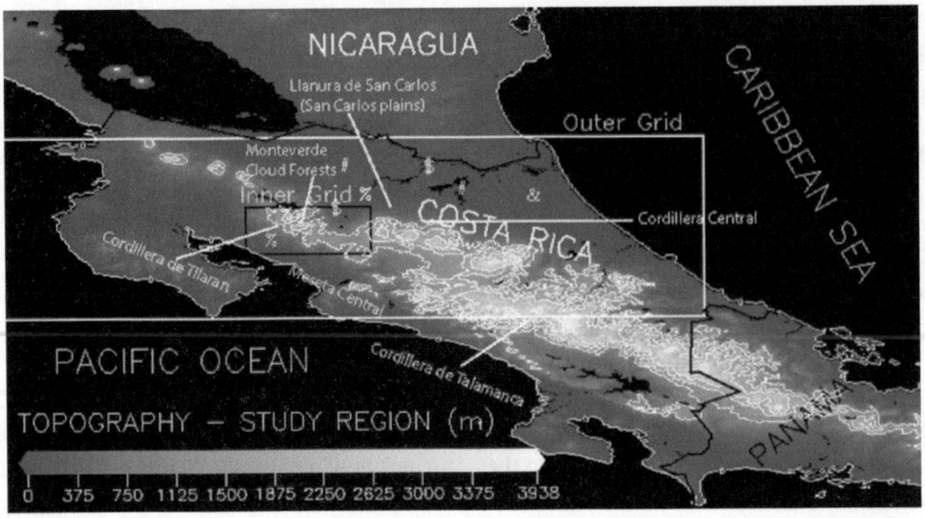

Figure 55.1. Topography, major mountains, water bodies, and model domain referred to in the text. Note the location of outer and inner grids of the model domains with respect to the major mountain ranges and water bodies. The Cordillera de Tilarán is at the center of the inner grid. The outer grid extends from the Caribbean Sea in the east to the Pacific Ocean in the west. Paired symbols for the two LUCIE rawinsonde launch locations are as follows: first site, 1–4 March 2003 indicated by " & " and second site, 1–4 March by " %"; both sites, 5–7 March by " # ", 8–11 March by " $ ", and 12–17 March by " % ". Note that second site has the same symbol " % " for 1–4 March and 12–17 March, because of nearly coincident locations.

(Hansen *et al.*, 2000), both of 1-km spatial resolution, were used to specify terrain and surface vegetation type in the RAMS simulations. Leaf area index (*LAI*), a crucial input to the RAMS surface vegetation parameterization, was specified using 1-km spatial resolution *LAI* composite maps derived from Moderate Resolution Imaging Spectroradiometer (MODIS) data (Myneni *et al.*, 1997; Knyazikhin *et al.*, 1998) which were available at 8-day intervals. The spatial distribution of surface soil types was specified in RAMS using the $1° \times 1°$ resolution Food and Agricultural Organization (FAO) soils database (FAO, 1971–1981). Geostationary Operational Environmental Satellite (GOES-8) imagery was used to assess the ability of RAMS to simulate orographic cloud banks. The automated cloud classification algorithm of Nair *et al.* (1999) was used to identify clouds for comparison to RAMS-simulated cloud fields.

METHODOLOGY

The present study used RAMS to conduct numerical simulations similar in nature to those used in the prior studies of Lawton *et al.* (2001) and Nair *et al.* (2003) with the aim of further investigating the hypothesis that deforestation in the lowland and pre-montane regions upwind of the Cordillera de Tilarán affects the nature of orographic cloud banks, especially base heights and spatial extent (Nair *et al.*, 2008; Welch *et al.*, 2008; Lawton *et al.*, this volume). Three sets of numerical experiments using RAMS, initiated from the same atmospheric conditions and subject to the same temporally varying lateral boundary forcing, were used to simulate orographic cloud formation in the study area during the period of 1–14 March 2003 for three land-use scenarios: (i) pristine conditions, i.e. completely forested, including the lowland and pre-montane regions, (ii) current conditions specified using the MODIS land-cover classification, and (iii) future scenarios where lowland and pre-montane regions are completely deforested, but the mountain tops still remain forested (see Figure 55.2 below).

Numerical modeling experiments

RAMS (Pielke *et al.*, 1992) is a versatile non-hydrostatic numerical modeling system, used for simulating a wide range of atmospheric phenomenon ranging from micro-scale flow around buildings to synoptic-scale features. RAMS uses finite-difference techniques to solve a coupled set of partial differential equations for conservation of mass, momentum, heat, and solid and liquid phases of water on a grid that uses a polar stereographic projection in the horizontal, and a terrain-following sigma coordinate system in the vertical (Mahrer and Pielke, 1975). Sophisticated parameterization schemes are available in RAMS to account for complex micro-physical processes (Walko *et al.*, 2000), land–atmosphere interactions (Tremback and Kessler, 1985; Walko *et al.*, 2000), atmospheric radiative transfer (Harrington *et al.*, 2001), sub-grid scale turbulence, and top and lateral boundary conditions.

The numerical simulations used a nested grid configuration (Figure 55.2) with a fine grid of 1-km spacing and a domain of 62 $\times$ 42 km being nested within a coarser grid of 4-km spacing and

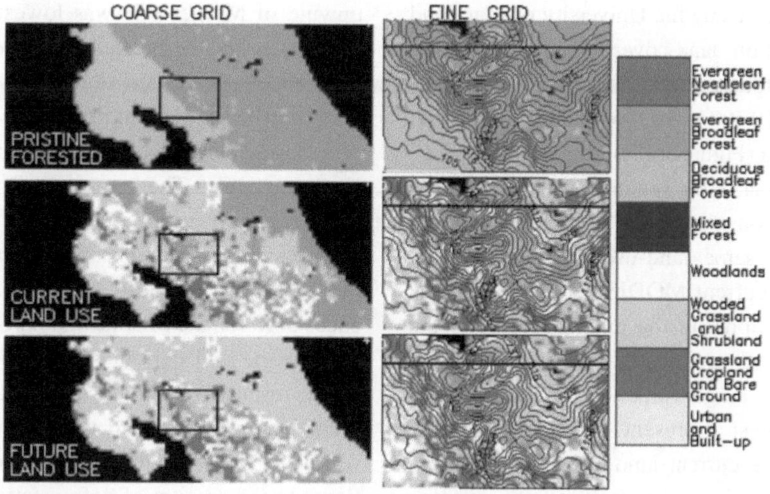

Figure 55.2. Outer coarse grid (left-hand panel) and inner fine grid (right-hand panel) for the three model simulations: (top) completely forested, (center) current conditions, and (bottom) all lowland and pre-montane regions deforested. Forested scenario had evergreen broadleaf forests on the wetter Caribbean, and deciduous broadleaf forests on the drier Pacific side. In the future deforestation scenario, deforestation is assumed to progress up to 1000 m.a.s.l. on the Caribbean side and 1400 m.a.s.l. on the Pacific side. The Pacific side currently is mostly deforested. (See also color plate.)

covering a domain of 400 × 160 km. The grids are centered approximately on the Monteverde region (10.25° N, 84.7° W). The inner grid allows simulation of orographic cloud formation at fine spatial resolution, while the outer coarse grid allows propagation of larger-scale atmospheric flow through the model domain, and insulates the fine grid from lateral boundary forcing effects. A stretched grid was utilized in the vertical for both grids, with a grid stretch ratio of 1.2, and grid spacing that varied from 20 m near the surface to 750 m higher up in the atmosphere. The topography within the two grids is shown in Figure 55.1, and the spatial distribution of land use assumed in the three sets of numerical modeling experiments is shown in Figure 55.2.

The initial atmospheric conditions and a time series of dynamic and thermodynamic fields used for forcing the lateral boundaries in the numerical simulations were drawn from the NCEP reanalysis pressure grids and rawinsonde profiles obtained during the LUCIE field campaign. A Newtonian relaxation scheme was used to nudge the atmospheric conditions along the lateral boundaries of the outer coarse grid toward observations. The variables along the lateral boundaries were adjusted by amounts proportional to the differences between the values predicted by the current model and those at the same point given by one of the LUCIE–NCEP analysis fields (available at 6-hour intervals) at a future time, with the proportionality constant given by the inverse of a user-specified nudging timescale. The nudging scheme was applied to a border region along the lateral boundaries with a width of five grid points, and the strength of the Newtonian forcing term decreased exponentially toward the domain interior. The strength of the forcing term was controlled by the nudging timescale, specified to be 900 s. The Klemp and Wilhelmson (1978) lateral boundary conditions were applied to the coarse grid, allowing disturbances to propagate out of the model domain without strongly reflecting back into the interior.

Based on the values reported in the FAO soil database (FAO–UNESCO, 1971–1981; Webb *et al.*, 1992; Gerakis and Baer, 1999), an average value of 2.5 m was chosen as the depth of the soil layer. *In situ* observations from the study area for March of 2003 showed volumetric moisture content to vary in the ranges 10–15%, 10–20%, and 25–30% at 20, 50, and 100 cm soil depth for both forested and deforested areas. However, field observations suggested pasture grasses in deforested areas were more stressed during the dry season than trees, implying that the trees had access to water stored in deeper soil layers. This is consistent with prior studies that showed enhanced access to soil water by deeply rooted forest vegetation compared to more shallowly rooted vegetation in deforested areas. For example, Nepstad *et al.* (1994) reported that in eastern Amazonia, 75% of all the water extracted during the dry season originated from soil layers below 2 m. In the present numerical model experiments, soil moisture at depths between 0 and 1 m was specified to be consistent with the averaged observed values. In order to represent enhanced soil water access by more deeply rooted vegetation, soil saturation was allowed to increase from 0.3 at 1 m depth to 0.8 at 2.5 m depth.

Land-use scenarios assumed in the three sets of numerical model simulations were as follows: (i) in the pristine scenario, broadly consistent with the original distribution of forests (Holdridge, 1967; Gomez, 1986), evergreen broadleaf forest was assigned to deforested areas above 750 m.a.s.l. on the Pacific slope and the entire Caribbean slope; deciduous broadleaf forest was assigned to deforested areas below 750 m on the Pacific slopes (Figure 55.2a); (ii) the current scenario was prescribed using the University of Maryland (UMD) 1-km spatial resolution land-cover type classification (Figure 55.2b); and (iii) the deforested scenario assumed that deforestation extended to 1000 and 1400 m.a.s.l. on the Caribbean and Pacific slopes of the Cordillera (Figure 55.2c), respectively.

In the current land-use scenario, the spatial distribution of *LAI* was specified using values from the MODIS-derived *LAI* dataset. Grid points having the same land-use type in all three scenarios were assigned the current MODIS-derived *LAI* value at that location in all simulations. In the pristine scenario, the forests that replaced the currently deforested areas were assigned *LAI* values consistent with average values of MODIS-derived *LAI* observed for remnant forest at present. Similarly, the locations that were forested in the current land-cover scenario, but deforested in the future scenario, were assigned the average MODIS-derived *LAI* values presently found over deforested regions.

The explicit micro-physical parameterization (Walko *et al.*, 2000), the two-stream atmospheric radiative transfer scheme of Harrington *et al.* (2001), and the horizontal diffusion scheme based on horizontal deformation and vertical diffusion parameterization of Mellor and Yamada (1982) were used in all the simulations. A time-step of 4 s was used for the coarse grid vs. 1 s for the fine, inner grid. Radiation calculations were updated every 300 s (Ray *et al.*, 2006).

RESULTS

All three simulations (pristine, current, and future land-cover scenarios) were conducted for a 14-day period and were initialized using 1 March 2003, 12:00 UTC observations. The simulations are used to understand the differences in boundary-layer thermodynamic and surface energetics considered responsible for the differences in simulated cloud formation associated with the respective scenarios. Results from the simulation of current land use are first compared against satellite observations and *in situ* atmospheric observations to assess the performance of RAMS in predicting orographic cloud distribution. The three simulations are then inter-compared to determine the differences in orographic cloud characteristics, especially cloud base height and areal extent, in response to varied land use.

Diurnal evolution of surface temperature, dew point, and energy fluxes

Comparison of domain-averaged surface temperature, dew point, and energy fluxes from the simulations, assuming the three different land-use scenarios, showed significant differences in diurnal evolution. The mean averaged surface temperature over the lowlands and pre-montane regions (1000 m.a.s.l. and below) upwind of Monteverde was lowest in the pristine scenario and highest in the deforested scenario (Figure 55.3a). Maximum differences in mean domain-averaged surface temperature between the pristine and deforested scenarios, and between the pristine and current scenarios were approximately -1.1 and -0.9 K, respectively. The mean domain-averaged surface dew-point temperature over the lowlands and pre-montane regions upwind of Monteverde was highest in the pristine scenario and lowest in the deforested scenario (Figure 55.3b). The maximum differences between the pristine and deforested scenarios, and between the pristine and current scenarios were approximately 1.2 and 1.0 K, respectively. Thus, deforestation in the lowlands and pre-montane regions caused the air masses over these areas to be warmer and drier, with the degree of modification being proportional to the amount of deforestation.

These results are due to the impact of deforestation on surface energy budgets. Sensible heat fluxes H were significantly higher in both current and future land-use scenarios compared to the pristine scenario (Figure 55.3c). Maximum differences in mean domain-averaged H of approximately -100 W m^{-2} and -85 W m^{-2} occurred between pristine and future scenarios, and between pristine and current scenario, respectively. As expected, latent heat fluxes LE_v showed a pattern opposite to that of H and were significantly higher in the pristine scenario compared to current and future scenarios (Figure 55.3d). Maximum differences in mean, domain-averaged LE_v of approximately 100 W m^{-2} and 15 W m^{-2} were found between pristine and future, and between pristine and current scenarios, respectively. Differences in diurnal variation of surface energy fluxes indicated enhanced day-time transpiration over forested areas to be responsible for cooler and moister air over these regions.

Comparison of simulated cloud distribution for current land use with satellite-based observations

The ability of RAMS to realistically simulate orographic cloud distributions was evaluated by comparing simulated cloud fields in the 1-km space fine grid to GOES-8 observed cloud distribution at similar spatial resolution. The automated cloud detection algorithm of Nair *et al.* (1999) was used to detect clouds in GOES-8 imagery acquired at 1-hour intervals starting at 12:15 UTC (06:15 LT) and ending at 22:15 UTC. Comparison was restricted to daytime hours, since detection of low-level clouds without visible channel imagery is problematic. Comparison was made between each grid point in the inner fine grid to the corresponding location in the cloud mask generated from the hourly GOES-8 observations. From this comparison the following statistics were derived: (i) the number of grid points for which both observations and the RAMS simulation showed occurrence of cloud (n), (ii) the number of grid points for which the RAMS simulation showed cloud but satellite imagery did not

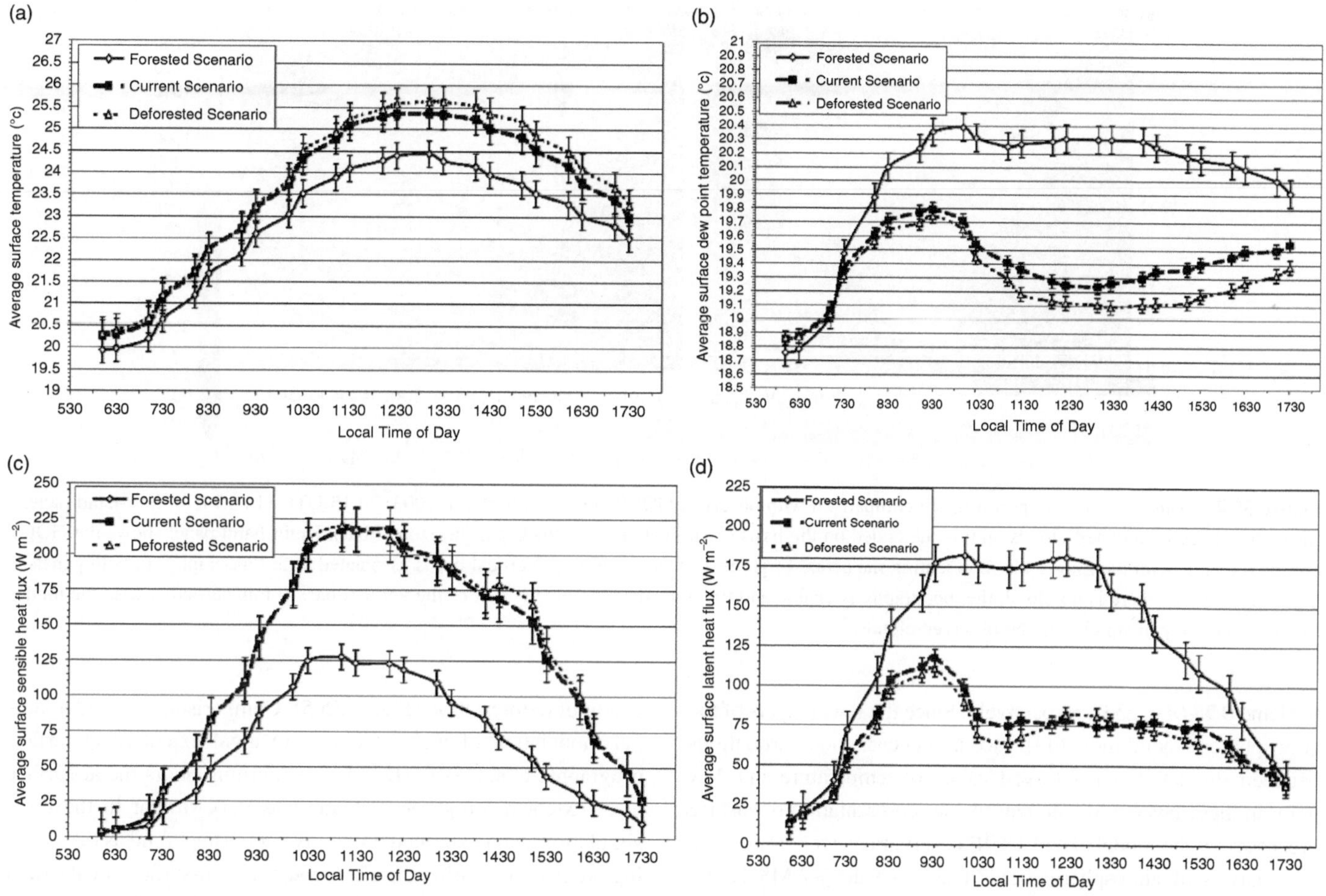

Figure 55.3. (a) Modeled lowest-level air temperature for the three land-cover scenarios of Figure 55.2; (b) idem for dew-point temperatures; (c) idem for sensible heat flux; (d) idem for latent heat flux. Error bars represent the standard error of the mean. Shown are domain-averaged values for lowland and pre-montane regions below 1000 m.a.s.l.

(n^+), and (iii) the number of grid points where cloud occurred in satellite imagery but not in the RAMS simulation (n^-). An accuracy measure (A) was computed from these statistics using the following expression:

$$A = \frac{n}{n + n^+ + n^-} \times 100\%. \tag{55.1}$$

Accuracy measures computed from a set of 142 comparisons showed a wide range of values from 0 to 87%. Visual inspection suggested that the low-accuracy values occurred under observed conditions of low cloud cover in which the model failed to simulate cloud, or in cases in which the model formed cloud at locations different from those observed. Most frequently, model simulations had accuracy values ranging from 20% to 70%, with 102 out of the 142 comparisons (72%) showing values greater than 40%. Visual comparison between numerically simulated cloud field and satellite observations showed very good correspondence, especially in terms of the location of the lee edge of the orographic cloud bank (Figure 55.4). A detailed analysis by Ray (2005) showed that lower values of accuracy resulted mostly from the inability of the model to generated isolated convection at the same locations as observed in the satellite imagery. Based on the accuracy measure, out of the 14 days of simulations, six days (2, 3, 4, 5, 12, and 13 March 2003) exhibited the best performance. Discussion from this point on is restricted to these six days for which the simulations best resembled the observed cloudiness.

Comparison of temperature and dew point in the current scenario with LUCIE observations

Boundary-layer temperature and dew-point observations from LUCIE radiosonde measurements on the six selected days showed good correlation over areas upwind of Monteverde with corresponding values from simulation assuming current land use. Air masses upwind of Monteverde are responsible for the formation of orographic clouds when forced to flow over the terrain, and thus realistic simulation of boundary-layer thermodynamics in these upwind areas is important. Comparisons between observation and model simulations of temperature and dew point in the atmosphere below the 850 hPa pressure level showed correlation coefficients of

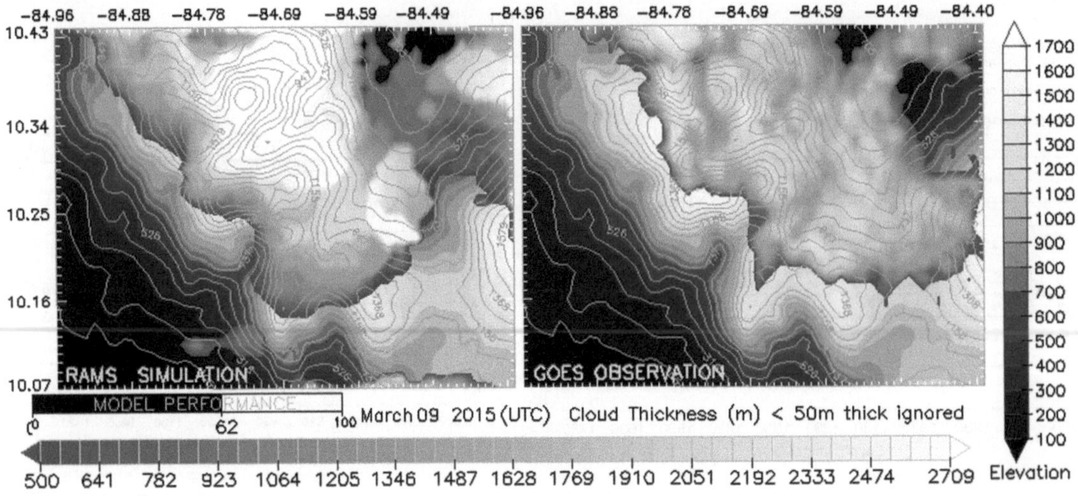

Figure 55.4. Example of model performance comparison with observed GOES clouds on 9 March 2003, 20:15 UTC (14:15 LT). Left-hand panel shows the RAMS-simulated clouds, gray-scale coded on the basis of simulated cloud thickness (horizontal bar). Right-hand panel shows the GOES observed clouds with the gray-scale representing the actual brightness values observed. Vertical bar is associated with topography. In both panels, at locations where there is no cloud, the topography is visible and the gray-scale scheme is that of the vertical bar. In this particular case the model has 62% accuracy in simulating the observed clouds.

0.94 and 0.78 ($n = 253$), respectively. Since the lower layers of the atmosphere respond most to surface influences, good correlations between simulated and observed values of temperature and dew point in these layers indicate reasonable representation of surface processes by the model. The LUCIE observations were utilized in the analysis of atmospheric fields that force the RAMS lateral boundary conditions, so it might appear that good correlation between the simulation and observations is an artifact of assimilating the LUCIE-derived information. However, it should be noted that the Newtonian nudging was applied only to five grid points bordering the lateral boundaries of the outer coarse grid, that were far removed from the inner grid, whereas the strength of the nudging term decreased exponentially toward the domain interior. Thus, within the domain of the inner fine grid, atmospheric conditions were determined mainly by the physical processes simulated by the model rather than by the data assimilation process.

Impact of land-use change on orographic cloud bank characteristics

The RAMS simulations showed consistent differences in the characteristics of orographic cloud banks between the three land-use scenarios. Cloud-cover characteristics were analyzed over the area which would remain forested even in the scenario involving increased future deforestation. These forests are present from 1000 m.a.s.l. to the mountain tops on the Caribbean slope and from 1400 m.a.s.l. to the crests on the Pacific slope. For these areas, simulations showed a decrease in the areal extent of orographic cloud present at the surface in response to lowland and pre-montane deforestation, while the amount of decrease was proportional to the extent of deforestation (Figure 55.5). Comparisons showed a diurnal modulation of the differences in areal extent of at-surface orographic cloud cover. During the morning hours the at-surface areal extent of orographic clouds was very similar in the three simulations, and exhibited a steadily decreasing trend as the morning progressed. During the afternoon hours, at-surface areal extent showed increasing trends in all three simulations (Figure 55.5a), although absolute values of the areal extent differed (Figure 55.5b). At-surface areal extent of orographic clouds in the afternoon was highest for the forested scenario, followed successively by the current and deforested scenarios (Figure 55.5a).

Base heights of orographic cloud banks also differed among the three simulations. Vertical cross-sections of the RAMS-simulated orographic cloud bank showed it to be extending to lower altitudes, both along the Caribbean and Pacific slopes, for the pristine land-use scenario (Figure 55.6a) compared to current (Figure 55.6b) and deforested (Figure 55.6c) scenarios. Averaged cloud base heights exhibited a pattern very similar to that observed for at-the-surface orographic cloud cover. Cloud base heights in all the simulations increased steadily during daytime, reaching a maximum at 13:00 LT and then decreased later in the day (Figure 55.7a). Averaged cloud base heights in the three simulations were similar earlier in the day but started to differ at about 11:30 LT. The largest differences in cloud base heights were found between the deforested and forested simulations, with the average cloud base height being higher by a maximum of 170 m in the deforested simulation (Figure 55.7b). The next largest difference was found between the deforested and current scenarios, with the average cloud base height being higher in the current scenario by maximum values of approximately 120 m.

The smallest differences were found between the current and forested scenarios: cloud base heights were higher in the current scenario by up to 70 m.

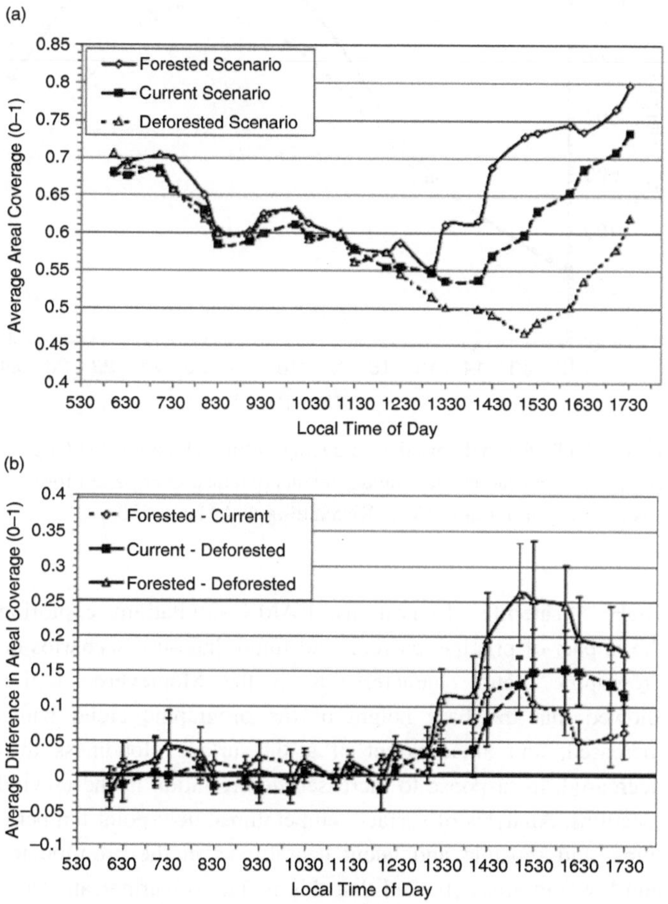

Figure 55.5. (a) Diurnal variation of the relative area covered by clouds with bases touching the ground (6-day averages) for three land-cover scenarios. (b) Difference in relative area covered by clouds with bases touching the ground. The bars equal the standard error in the estimation of the mean. In both cases, the total area analyzed is that over the forested areas shown in the inner fine grid for the lowland deforestation scenario.

The numerical simulations thus showed a consistent pattern in which average orographic cloud base height and daytime areal extent of the higher-elevation, at-the-ground cloud cover both showed declines proportional to the degree of deforestation in the lowlands and pre-montane areas. Diurnal evolution of lifting condensation level (LCL), a surrogate for cloud base height (cf. Foster, this volume), was computed from a series of atmospheric profiles measured at 6:00, 12:00, 15:00, and 18:00 LT over paired forested and deforested sites during LUCIE, and showed patterns very similar to averaged evolution of cloud base heights in the pristine and deforested scenarios. It should be noted that direct comparison between LCL derived from the soundings and average model-simulated cloud base is not possible, since the latter is an average computed over an area with significant variations in terrain. Elevated terrain restricts the minimum possible cloud base height at a location to be greater than or equal to its surface elevation. However, it is valid to compare the trends in LCL and the mean domain-averaged cloud base heights. Like average cloud base heights in the pristine and deforested scenarios, LCL was similar over both forested and deforested sites earlier during the day, but started to differ at 12:00 LT, with the values being higher over deforested areas (Figure 55.8). A maximum in LCL was observed at 15:00 LT, but the radiosonde measurements were made at 3-hour intervals and hence in reality the maximum value may have been achieved at an earlier hour. Later in the day, the LCL started to decrease over both forested and deforested sites, with the LCL still lower over the forest compared to deforested site.

DISCUSSION AND CONCLUSIONS

Recent studies have suggested that changes in distributions of dry-season mist at Monteverde in northern Costa Rica have contributed to anuran population crashes and shift of bat and bird species ranges to higher elevations (Pounds *et al.*, 1999;

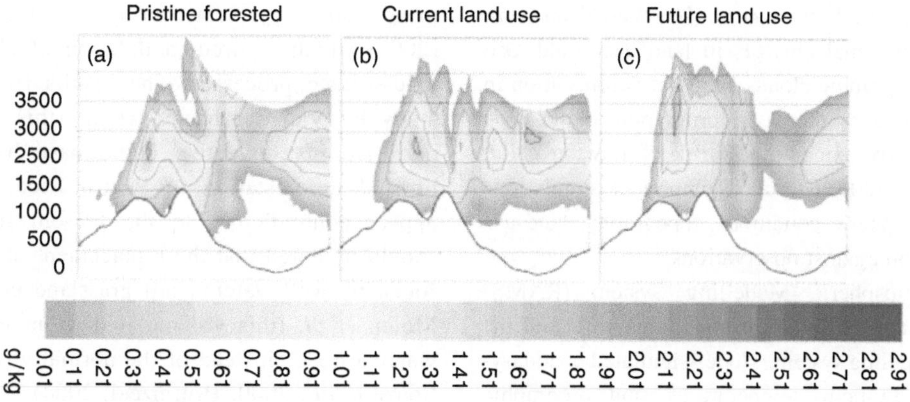

Figure 55.6. *Y–Z* cross-sectional plots (along the line in Figure 55.2's fine grid) for the three model scenarios at 14:30 LT. Clouds reach the lowest elevations for the pristine forested scenario and the highest for the deforested scenario.

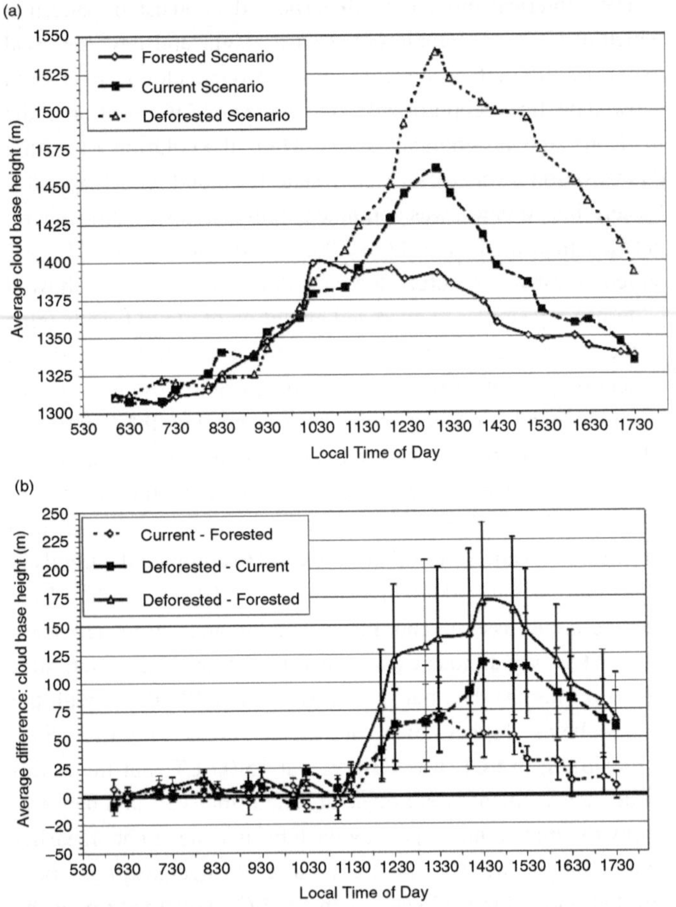

Figure 55.7. (a) Diurnal variation of 6-day averaged cloud base heights over the remnant cloud forest regions shown in Figure 55.2 for the deforestation scenario for the inner fine grid. (b) Difference in cloud base heights over the same region as in (a). The error bars are one standard error of the mean.

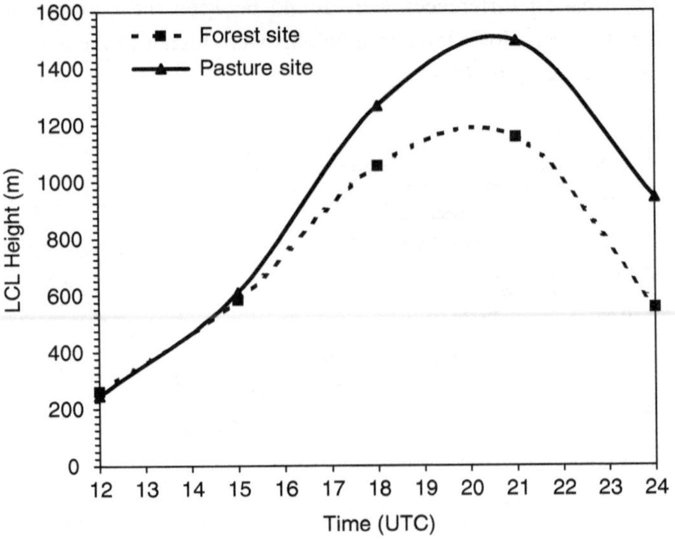

Figure 55.8. Diurnal variation of average lifting cloud level (LCL) computed from radiosonde measurements obtained over paired forest–pasture sites in northern Costa Rica during 6–8 March 2003.

LaVal, 2004). There could be fewer misty dry-season days because of a rise in the elevation of the orographic cloud bank (Pounds *et al.*, 1999) resulting, in turn, from large-scale changes in climate (Still *et al.*, 1999). Lawton *et al.* (2001) and Nair *et al.* (2003) proposed that regional changes in land use could also impact formation of orographic clouds, and that deforestation in areas upwind might lead to increased orographic cloud base height. The present study extends the work of Lawton *et al.* (2001) and Nair *et al.* (2003) through numerical simulations utilizing realistic atmospheric conditions, topography, land-use scenarios, and validation against observations.

The Regional Atmospheric Modelling System (RAMS) simulations assuming current land-use conditions validated the performance of the modeling system. Model-simulated boundary-layer temperatures, dew-point temperatures, and orographic cloud characteristics agreed well with *in situ* radiosonde measurements over both forested and deforested sites, and with GOES-8 satellite observations. RAMS simulations exploring the impact of pristine, current, and future land-use scenarios on orographic cloud characteristics in the Monteverde region showed that the base height of the orographic cloud banks increased, and areal extent of at-the-surface cloudiness thus decreased, in response to increased deforestation in the upwind locations. Analysis of surface temperatures, dew-point temperatures, and sensible and latent heat fluxes in the pre-montane and lowland areas showed that this is due to warmer and drier air masses forming over deforested areas than over forested ones (cf. Ray *et al.*, 2006). These results were consistent with lifting condensation level (LCL) values calculated from rawinsonde data showing that throughout the diurnal course of variation the LCL above deforested areas was higher than over forested ones.

These findings are also consistent with results from prior studies by Schrieber *et al.* (1996) and Golaz *et al.* (2001) which showed modulation of cloud base heights by land-surface processes. The results from a similar tropical study by Van der Molen *et al.* (this volume), reporting a *decrease* in convective cloud base height in response to upwind deforestation in Puerto Rico, may, at first sight, appear contradictory to the present findings. However, it should be noted that the replacement of coastal zone wetland forest by well-watered tall grassland considered by Van der Molen *et al.* (this volume) led to an increase in vegetation water use LE_v instead of the normally observed decrease (cf. Souza *et al.*, 2000; Bruijnzeel, 2004), probably because transpiration by the wetland forest was suppressed by inundation with brackish water (Van der Molen, 2002). Thus, although the conversion of lowland forest to pasture studied by Van der Molen *et al.* (this volume) led to an *observed* decrease in the Bowen ratio (i.e. H/LE_v) instead of the increase *predicted* in the present study, the fact remains that in both studies, the lowest LCL was associated with the land-cover type exhibiting the highest LE_v and the lowest H (cf. discussion in Foster, this volume). Even though direct measurements of energy fluxes are presently lacking over paired forested–deforested sites in Costa Rica's lowlands, there are several other indications that the nature of deforestation in this region does lead to an increase in temperature and a decrease in dew-point temperature. Diurnal observations of the LCL suggest a cooler, moister boundary layer over forested locations. Nair *et al.* (2003) noted that patterns of cloud formation over forested lowland regions in Costa Rica exhibited linear organization opposed to unorganized convection observed over deforested areas. Organized convection over forested lowland areas is indicative of reduced sensible heat fluxes over forests compared to deforested areas (Nair *et al.*, 2003). Measurements of dry-season surface energy fluxes over paired forested–deforested sites in Amazonia (Souza *et al.*, 2000) showed the same pattern as derived for Costa Rica but differences in the magnitudes of sensible heat fluxes, which were significantly larger than those reported in the present study. Since there are no obvious reasons to expect the impacts on surface energy fluxes due to deforestation in Amazonia and Costa Rica to be drastically different, the presently simulated impacts are considered plausible. Also, the RAMS-simulated temperature- and dew-point temperature profiles compared well to several *in situ* radiosonde measurements over paired forest–deforested sites, which is another indication of satisfactory model performance (Ray *et al.*, 2006). However, until direct measurements confirm the validity of the simulated surface energy flux patterns, the present results should be viewed primarily as a sensitivity analysis of orographic cloud formation to hypothesized changes in Bowen ratio resulting from deforestation.

The findings of this study have important implications for conservation of cloud forest preserves since conservation efforts also need to address effects of more remote land-use changes. The type of changes discussed in this study is also relevant to local hydrology in the vicinity of cloud forests (cf. Rhodes *et al.*, 2006; Guswa *et al.*, 2007; Frumau *et al.*, this volume; Rhodes *et al.*, this volume #24; Schmidt *et al.*, this volume). Since, at several cloud forest sites, horizontal precipitation accounts for a significant proportion of the total precipitation, changes in orographic cloud characteristics might have important consequences to local water resources (Bruijnzeel, 2001; cf. Bruijnzeel *et al.*, this volume). Quantification of such impacts could provide further economic incentives to decision makers for conserving cloud forest ecosystems.

ACKNOWLEDGEMENTS

This project was supported by NASA grant NAG5–11941 and NSF grant ATM-0128924. Deepak K. Ray was supported by a NASA Earth System Science Fellowship. The GOES-8 data were acquired by the GOES ground station at the Global Hydrology and Climate Center, Huntsville, Alabama. We are thankful for the helpful comments and suggestions of the reviewers.

REFERENCES

Bruijnzeel, L. A. (2001). Hydrology of tropical montane cloud forests: a reassessment. *Land Use and Water Resources Research* **1**: 1–18.

Bruijnzeel, L. A. (2004). Hydrological functions of tropical forests: not seeing the soil for the trees? *Agriculture, Ecosystems and Environment* **104**: 185–228.

Carlson, T. N., and G. A Sanchez-Azofeifa (1999): Satellite remote sensing of land use changes in and around San José, Costa Rica. *Remote Sensing and Environment* **70**: 247–256.

Food and Agriculture Organization – United Nations Educational, Scientific, and Cultural Organization (FAO-UNESCO) (1971)–1981. *Soil Map of the World, 1:5 000 000*, Vol. II–X. Rome: FAO, and Paris: UNESCO.

Freedman, J. M., D. R. Fitzjarrald, K. E. Moore, and R. K. Sakai (2001). Boundary layer clouds and vegetation–atmosphere feedbacks. *Journal of Climate* **14**: 180–197.

Gerakis, A., and B. Baer (1999). A computer program for soil textural classification. *Soil Science Society of America Journal* **63**: 807–808.

Golaz, J.-C., H. Jiang, and W. R. Cotton (2001). A large-eddy simulation study of cumulus clouds over land and sensitivity to soil moisture. *Atmosperic Research* **59–60**: 373–392.

Gomez, L. D. (1986). *Vegetacion de Costa Rica*. San José, Costa Rica: Editorial Universidad Estatal a Distancia.

Guswa, A. J., A. L. Rhodes, and S. E. Newell (2007). Importance of orographic precipitation to the water resources of Monteverde, Costa Rica. *Advances in Water Resources* **30**: 2098–2112.

Hansen, M. C., R. S. DeFries, J. R. G. Townshend, and R. Sohlberg (2000). Global land cover classification at the 1km spatial resolution using a classification tree approach. *International Journal of Remote Sensing* **21**: 1331–1364.

Harrington, J. Y., and P. Q. Olsson (2001). A method for the parameterization of cloud optical properties in bulk and bin microphysical models: implications for arctic cloudy boundary layers. *Atmospheric Research* **57**: 51–80.

Holdridge, L. R. (1967). *Life Zone Ecology*, revd edn. San José, Costa Rica: Tropical Science Center.

Kalnay, E., M. Kanamitsu, R. Kistler, *et al.* (1996). The NCEP/NCAR 40-year Reanalysis Project. *Bulletin of the American Meteorological Society* **77**: 437–471.

Klemp, J. B., and R. B. Wilhelmson (1978). The simulation of three-dimensional convective storm dynamics. *Journal of Atmospheric Science* **35**: 1070–1096.

Knyazikhin, Y., J. V. Martonchik, R. B. Myneni, D. J. Diner, and S. W. Running (1998). Synergistic algorithm for estimating vegetation canopy leaf area index and fraction of absorbed photosynthetically active radiation from MODIS and MISR data. *Journal of Geophysical Research* **103**: 32 257–32 275.

LaVal, R. (2004). Impact of global warming and locally changing climate on tropical cloud forest bats. *Journal of Mammalogy* 85, doi:10.1644/BWG-016.

Lawton, R. O., U. S. Nair, R. A. Pielke Sr., and R. M. Welch (2001). Climatic impact of tropical lowland deforestation on nearby montane cloud forests. *Science* **294**: 584–587.

Mahrer, Y., and R. A. Pielke (1975). A numerical study of air flow over mountains using the two-dimensional version of the University of Virginia mesoscale model. *Journal of Atmospheric Science* **32**: 2144–2155.

Mellor, G. L., and T. Yamada (1982). Development of a turbulence closure model for geophysical fluid problems, *Review of Geophysics* **20**: 851–875.

Myneni, R. B., R. R. Nemani, and S. W. Running (1997). Estimation of global leaf area index and absorbed PAR using radiative transfer model. *IEEE Transactions of Geosciences and Remote Sensing* **35**: 1380–1393.

Nadkarni, N. M., and N. T. Wheelwright (eds.) (2000). *Monteverde: Ecology and Conservation of a Tropical Cloud Forest*. New York: Oxford University Press.

Nair, U. S., J. A. Rushing, R. Ramachandran, *et al.* (1999). Detection of cumulus cloud fields in satellite imagery. In *Earth Observing Systems IV*, ed. W. L. Barnes, pp. 345–355. Bellingham, WA: SPIE.

Nair, U. S., R. O. Lawton, R. M. Welch, and R. A. Pielke Sr. (2003). Impact of land use on tropical montane cloud forests: sensitivity of cumulus cloud field characteristics to lowland deforestation. *Journal of Geophysical Research* **108** (D7): 4206–4218, doi:10.1029/2001JD001135.

Nair, U. S., S. Asefi, R. M. Welch, *et al.*, (2008). Biogeography of tropical montane cloud forests. II. Mapping of orographic cloud immersion. *Journal of Applied Meteorology and Climatology* **47**: 2183–2197.

Nepstad, D. C., C. R. de Carvalho, E. A. Davidson, *et al.* (1994). The role of deep roots in the hydrological and carbon cycles of Amazonian forests and pastures. *Nature* **372**: 666–669.

Pielke, R. A, W. R. Cotton, R. L. Walko, *et al.* (1992). A comprehensive meteorological modeling system: RAMS. *Meteorological and Atmospheric Physics* **49**: 69–91.

Pounds, J. A., M. P. L. Fogden, and J. H. Campbell (1999). Biological response to climate change on tropical mountain. *Nature* **389**: 611–614.

Pounds, J. A., M. R. Bustamante, L. A. Coloma, *et al.* (2006). Widespread amphibian extinctions from epidemic disease driven by global warming. *Nature* **439**: 161–167.

Ray, D. K. (2005). Climatic impact of land use change on the Mesoamerican Biological Corridor. Ph.D. thesis, University of Alabama in Huntsville, Huntsville, AL, USA.

Ray, D. K., U. S. Nair, R. O. Lawton, R. M. Welch, and R. A. Pielke Sr. (2006). Impact of land use on Costa Rican tropical montane cloud forests: sensitivity of orographic cloud formation to deforestation in the plains. *Journal of Geophysical Research* **111**: D02108, doi:10.1029/2005JD006096.

Rhodes, A. L., A. J. Guswa, and S. E. Newell (2006). Seasonal variation in the stable isotopic composition of precipitation in the tropical montane forests of Monteverde, Costa Rica. *Water Resources Research* **42**, W11402, doi:10.1029/2005WR004535.

Sader, S. A., and A. T. Joyce (1988). Deforestation rates and trends in Costa Rica, 1940 to 1983. *Biotropica* **20**: 11–19.

Schrieber, K., R. Stull, and Q. Zhang (1996). Distributions of surface-layer buoyancy vs. lifting condensation level over a heterogeneous land surface. *Journal of Atmospheric Science* **53**: 1086–1107.

Souza, E. P., N. O. Rennó, and M. A. F. Silva Dias (2000). Convective circulations induced by surface heterogeneities. *Journal of Atmospherc Science* **57**: 2915–2922.

Still, C. J., P. N. Foster, and S. H. Schneider (1999). Simulating the effects of climate change on tropical montane cloud forests. *Nature* **389**: 608–610.

Tremback, C. J., and R. Kessler (1985). A surface temperature and moisture parameterization for use in mesoscale numerical models. *Proceedings of the 7th AMS Conference on Numerical Weather Prediction,* June 17–20, 1985, Montreal, Quebec, Canada, 355–358. Boston, MA: American Meteorological Society.

Van der Molen, M. K. (2002). Meteorological impacts of land use change in the maritime tropics. Ph.D. thesis, VU University Amsterdam, Amsterdam, the Netherlands. Also available at www.falw.vu.nl/ images_upload/ 702F8F36–45DC-488E-B2C550BFDCAC2B9D.pdf.

Van der Molen, M. K., A. J. Dolman, M. J. Waterloo, and L. A. Bruijnzeel (2006). Climate is affected more by maritime than by continental land use change: a multiple-scale analysis. *Global and Planetary Change* **54**: 128–149.

Walko, R. L., L. E. Band, J. Baron, *et al.* (2000). Coupled atmosphere-biophysics-hydrology models for environmental modeling. *Journal of Applied Meteorology* **39**: 931–944.

Webb, R. W., C. E. Rosenzweig, and E. R. Levine (1992). *A Global Data Set of Soil Particle Size Properties,* NASA Techical Memorandum No. **4286**. Greenbett, MD: NASA.

Welch, R. M., S. Asefi, J. Zeng, *et al.* (2008). Biogeography of tropical montane cloud forests. J. Remote sensing of cloud base heights. *Journal of Applied Meterology and Climatology* **47**: 960–975.

56 Meso-scale climate change in the central mountain region of Veracruz State, Mexico

V. L. Barradas
Universidad Nacional Autónoma de México, Mexico City, Mexico

J. Cervantes-Pérez
Universidad Veracruzana, Xalapa, Veracruz, Mexico

R. Ramos-Palacios, C. Puchet-Anyul, P. Vázquez-Rodríguez, and R. Granados-Ramirez
Universidad Nacional Autónoma de México, Mexico City, Mexico

ABSTRACT

Trend analyses of precipitation and fog frequency, and energy balance measurements over typical examples of forest and grassland were made in the central mountain region of the State of Veracruz in eastern Mexico to examine the possibility of meso-scale climate change. It was hypothesized that changes in precipitation and fog frequency with elevation would reflect changes in the position of the lifting condensation level (LCL) and that these could be either positive or negative, depending on site elevation. The energy balance measurements were made to quantify the changes in sensible heat flux – considered the main driving force of cloud lifting – associated with the conversion of forest to pasture. In conclusion, the observed negative tendencies in dry-season precipitation (February) and overall fog frequency at lower elevations, as well as the positive tendencies seen at intermediate and higher elevations, are in line with a postulated shift in the LCL that may have been brought about by the increase in post-forest sensible heat fluxes.

INTRODUCTION

The Grandes Montañas region in the State of Veracruz in eastern Mexico is part of the mountain system where the eastern end of the Eje Neovolcanico and the Sierra Madre Oriental meet. One of the main features of this region is an extremely steep topographic gradient, ranging from sea level to more than 5500 m.a.s.l., over a horizontal distance of less than 100 km. The associated climatic gradients enable the occurrence of highly diverse plant communities, ranging from relatively wet montane coniferous and cloud-affected forests to (semi-)arid vegetation types (Gómez-Pompa, 1978; Barradas, 1983). The prevailing north-easterly winds gather moisture as they pass over the Gulf of Mexico. Upon rising up the slopes of the Sierra Madre Oriental, fog is produced throughout the year, commonly accompanied by drizzle and rain during the summer and early fall months. In winter, the formation of a stratus cloud deck in late morning is often observed over the coastal plain, drifting toward the mountains later in the day and intersecting the topography between *c.* 1000 and 2000 m.a.s.l. (García-García and Montañez, 1991; García-García and Zarraluqui, 2008). Whilst Vogelmann (1973) still reported substantial horizontal precipitation inputs in the montane area (i.e. drizzle and fog), the local people's perception has it that the Xalapa region is now drier than 20 years ago, in particular because of a decline in drizzle events.

Lara-García (2000) and Cervantes *et al.* (2001), in a study trying to detect changes in the precipitation regime for the city of Xalapa (1400 m.a.s.l., with ~450,000 inhabitants and growing steadily), selected the village of Las Vigas (2400 m.a.s.l., with a fairly stable population of ~7900) as a control. Precipitation trend analysis for the period 1923 to 1997 was carried out for the two areas in the expectation that precipitation in Xalapa was increased due to a possible urban heat effect (Jauregui *et al.*, 1992) whereas that in Las Vigas was not expected to show any

Tropical Montane Cloud Forests: Science for Conservation and Management, eds. L. A. Bruijnzeel, F. N. Scatena, and L. S. Hamilton. Published by Cambridge University Press.

major tendency. Instead, negative tendencies were found for both locations, suggesting possible regional climate change. Annual decreases of 3.13 mm for Xalapa and 4.4 mm for Las Vigas were derived, implying potential reductions in precipitation of as much as 30% and 50%, respectively, by the year 2023. Thus, local perceptions of a decline in precipitation were confirmed by measurements (Lara-García 2000; Cervantes *et al.*, 2001; cf. Ponette-González *et al.*, 2009).

The observed decrease in precipitation in the region might be explained by an increase in the height of the lifting condensation level (LCL), which is controlled by two main factors: (i) the sensible heat flux (H) emanated by the land surface – which tends to increase the height of the LCL – and (ii) synoptic-scale subsidence of air which tends to reduce it (Carson, 1973; cf. Foster, this volume). Values of H varys markedly with land-cover type, and are generally (much) higher over deforested areas (pasture, urban) than over primary forest vegetation, whereas the latent heat flux (LE_v) is typically reduced compared to a forested situation (Nobre *et al.*, 1991; Jauregui *et al.*, 1992; Culf *et al.*, 1996; Maitelli and Wright, 1996). The region of the Grandes Montañas has experienced a systematic and extensive change in land cover from primarily forest-dominated land in the 1950s to cattle raising (pasture), urban areas, and coffee plantations in more recent years. Roughly 90% of the original forest cover in the region has been estimated to have disappeared (Williams-Linera *et al.*, 2002), whereas montane cloud forest in particular continues to be depleted (Muñoz-Villers and López-Blanco, 2008). This can be expected to have had a significant effect on the partitioning of surface energy, by decreasing LE_v and consequently raising H and, as a result, also the LCL (Foster, this volume).

Recent studies using regional atmospheric circulation models in nearby Costa Rica and Puerto Rico have confirmed the possibility of a rise in LCL following land-cover change (Nair *et al.* 2003; Ray *et al.*, 2006; Van der Molen *et al.*, 2006; Nair *et al.*, this volume; Van der Molen *et al.*, this volume). With this in mind, trend analyses for precipitation and fog frequency were performed for the Grandes Montañas in order to detect a possible change in the regional climate. Furthermore, the partitioning of net radiation over mixed temperate pine–oak forest and pasture was determined to elucidate the implications for the height of the LCL and thus the occurrence of drizzle and fog.

MATERIALS AND METHODS

Study area

Figure 56.1 shows the locations of the examined climatic stations in relation to the general topography of the eastern end of the Eje Neovolcanico between the States of Puebla and Veracruz. The regional climate is the result of complex interactions between the dominant synoptic systems, the mountainous topography, plant–atmosphere interactions, and the proximity to the Gulf of Mexico (cf. Ponette-González *et al.*, 2009). Precipitation is highly seasonal, with almost 80% of the annual total occurring during the rainy season (June–July to September–October) and only 20% during the dry season (October–November to May–June). This pattern can be explained as follows: in summer, the area is under the control of the trade easterlies along the south-western end of the Bermuda anticyclone (Alemán and García, 1974). As the trades encounter the mountainous wall, forced ascent leads to a strong gradient in precipitation. Conversely, in winter the sub-tropical anticyclone is situated closer to Veracruz and a subsidence inversion persists. The sea and mountain breezes, now unaided by the trade winds, encounter this inversion, producing a stratus deck late in the morning that manifests itself as fog at elevations between *c.* 1000, and 2000, m.a.s.l. (Fitzjarrald, 1986; cf. Vogelmann, 1973; García-García and Zarraluqui, 2008).

Figure 56.2 shows the distribution of vegetation communities and land-cover types in the study area in 1995, illustrating the high degree of forest fragmentation, the dominance of agriculture and grasslands, as well as the significant area occupied by shaded coffee plantations.

Trend analyses

A number of climate stations having at least 50 years of record and less than 5% missing data were selected for a trend analysis of precipitation and fog frequency (see Figure 56.1 for locations). Increasing or decreasing trends were determined using the Mann–Kendall test (Nasrallah *et al.*, 1990). This approach consists of fitting the data series to a distribution function for the derivation of a test parameter (x); next, the value of a theoretical parameter (y) is determined on the basis of the associated function, after which the test and theoretical parameters are compared to establish the trend ($p < 0.05$) which is calculated with a simple linear regression ($y = ax + b$, where a is the mean trend of the analyzed series). Trend analyses were performed for monthly, seasonal, and annual precipitation data. Fog frequency was determined simply from the number of days with fog in each analyzed climatic station (cf. García-García and Zarraluqui, 2008).

Energy balance measurements

The chief components of the energy balance were measured at a grassland and a pine–oak forest near the village of La Joya (19° 36' N, 97° 05' W, 2250 m.a.s.l.; see Figure 56.1 for location). Net radiation (R_n) was measured with a net radiometer (REBS, Campbell Scientific Inc.), and the net soil heat flux (G) with two

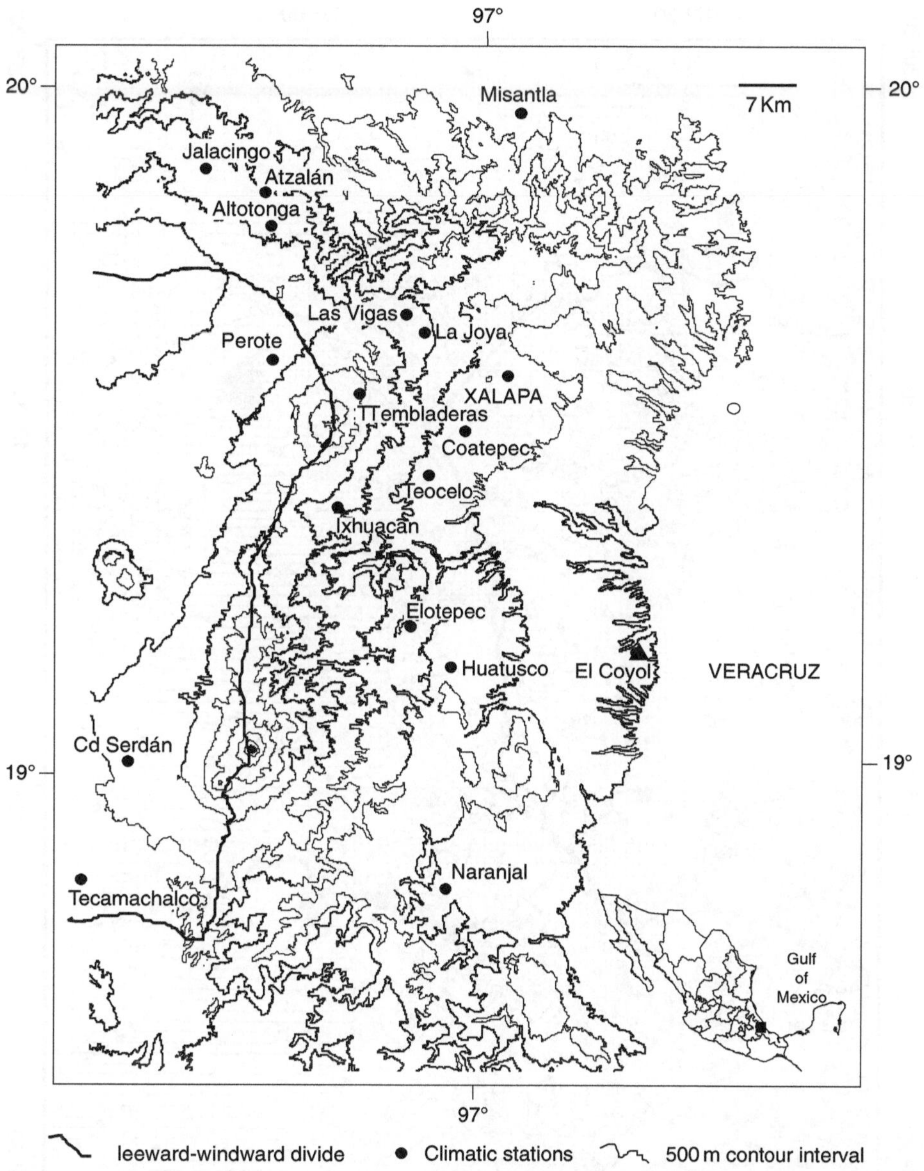

Figure 56.1. Study area in the Grandes Montañas region in the States of Veracruz and Puebla (eastern Mexico) showing general topography and stations for which precipitation and fog frequency trend analyses were performed. Measurements of the chief components of the energy balance over forest and pasture were made at La Joya.

plates (HFP01SC, Campbell Scientific). Air temperature and relative humidity were measured with a thermo-hygrometric probe (HMP35C, Vaisala), and wind direction and wind speed with a Wind Monitor (05103, R.M. Young, USA). All sensors were read by a CR10 data-logger (Campbell Scientific) at intervals of 5 s and averaged every 20 min.

Latent heat flux (LE_v) was estimated from sapflow measurements and the Penman–Monteith equation. Sapflow was measured every 20 min in 10 trees using sapflow meters (ThermalLogic) controlled by a 21X data-logger (Campbell Scientific). Stomatal resistance was measured with a steady-state porometer (LI-1200, LI-COR) whereas aerodynamic resistance was estimated from the wind profile every 20 min. Instruments were installed at 2 m above the ground in the grassland and ~2 m above the forest canopy. Measurements were taken from 06:00 to 18:00 local time during June, August, October, and November 2003, and in March 2004. Measured and Penman–Monteith-based transpiration values (in mm of water) were converted to latent heat flux (LE_v in W m^{-2}). Sapflow (T) and Penman–Monteith-based transpiration (E_{PM}) were similar and related by a simple linear regression ($T = 1.197\ (E_{PM}) - 5.3$; $r^2 = 0.7$; $p < 0.05$; $n = 4335$,). Sensible heat flux H was calculated as the residual of the energy balance equation as $H = R_n - LE_v - G$.

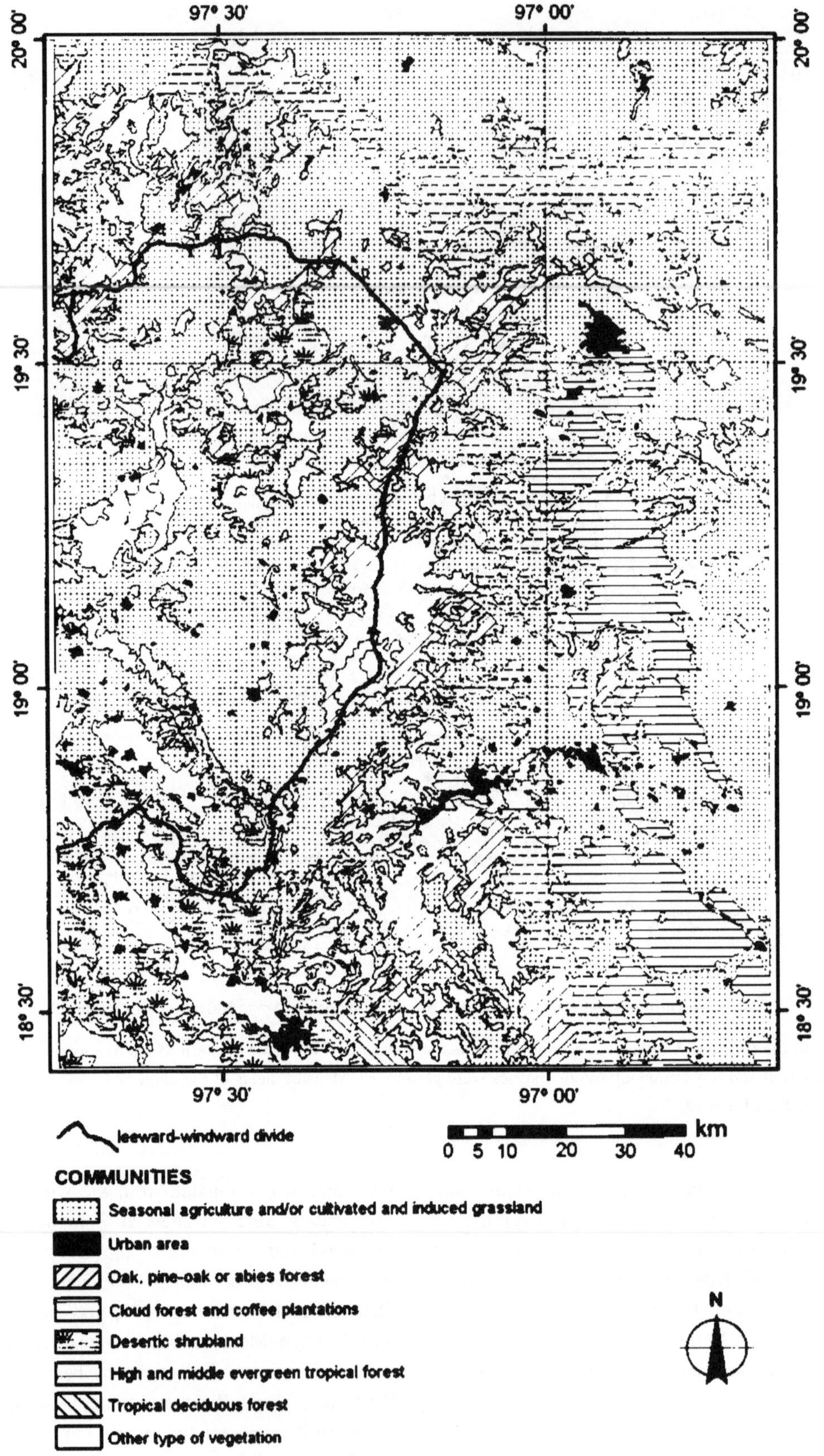

Figure 56.2. Spatial distribution of land-cover types in the study area around 1995.

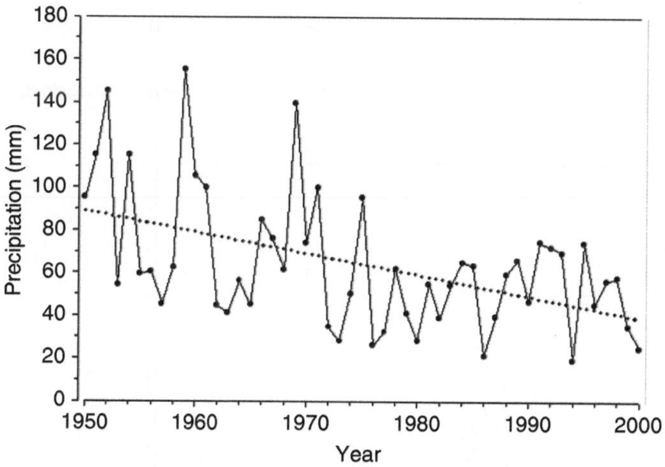

Figure 56.3. Monthly precipitation totals for February between 1950 and 2000 at Naranjal climatic station (continuous line) and the simple linear regression $Pp = -1.0022\ (year) + 2043.5$ ($p < 0.05$) describing the general trend in precipitation (dotted line).

RESULTS

Trends in precipitation and fog frequency

No significant trends in precipitation and fog frequency were detected for the majority of analyzed stations throughout the year. However, significant trends were detected for some sites, mainly for precipitation in February (dry season) (Table 56.1). Figure 56.3 shows a typical example in which the constant decrease in precipitation over time is noteworthy. Interestingly, the trend in precipitation in February was negative at several windward climatic stations, but positive at leeward or high-altitude windward stations, such as Altotonga (1880 m.a.s.l.) and Jalacingo (1944 m.a.s.l., cf. Table 56.1 and Figure 56.1). Trends in fog frequency were concomitant with those in precipitation, even though the analysis for fog frequency used annual data only. Trends in fog frequency were generally positive at high altitudes (Altotonga, 1899 m.a.s.l.; Atzalan, 1842 m.a.s.l.) and negative at low altitudes (Coatepec, 1252 m.a.s.l.; Las Minas, 1365 m.a.s.l.). Nevertheless, in some cases trends were positive at lower altitudes (e.g. Huatusco, 1344 m.a.s.l.) and negative at much higher altitudes (e.g. La Tembladera, 2727 m.a.s.l.).

Energy balance comparison

Comparisons of the energy balance components for the two contrasting vegetations are given in Table 56.2. During the rainy season, net radiation in the forest was largely dissipated by the latent heat flux LE_v (64% and 48% in June and August, respectively), whereas in the grassland the sensible heat flux H could be as high as 90%, even in the rainy season. Bowen ratios ($B = H/LE_v$) during the late rainy season (October and November) were similar between forest and pasture, but became much higher in the pasture (up to 6.5-fold) during the dry season. Increased values of B (3.3-fold) were also observed occasionally during the rainy season (August) in case of a reduction in the number of rainy days (Table 56.2).

DISCUSSION

Overall, no significant trends in annual precipitation were found. However, for February (when the high pressure cell is affecting rainfall in the area) significant trends in rainfall were observed, both positive and negative, depending on location. Generally, trends were negative at windward stations and positive at leeward stations or high-elevation windward stations close to the water divide (e.g. Jalacingo; cf. Figure 56.1). Conversely, in April and May, when the regional weather is in transition from the dry to the wet season reflecting a change from mid-latitude to tropical atmospheric mechanisms (Alemán and García, 1974), precipitation tendencies were positive, as observed also for most of the rainy season. However, for October and November the trends were negative again. These different trends observed for rainfall in different parts of the year probably cause the lack of any detectable trends in annual precipitation.

Bisselink (2003) found summer precipitation totals (May–October) in Puerto Rico for the majority of 24 stations to be decreasing significantly (by 0.6–2.3 mm year^{-1}) between 1931 and 1996, although winter precipitation increased by 0.3–1.7 mm year^{-1}. In addition, the eight stations with the longest period of record (~100 years and all situated in the lowlands) had negative trends in annual precipitation as well (−1.6 to −4.9 mm year^{-1}). Van der Molen *et al.* (this volume) speculated that this long-term decrease in precipitation might be related to conversion of coastal wetland forest to pasture as well as to continued urbanization.

Fog frequency is another variable that presented interesting trends (Table 56.1). Whilst fog frequency in the study area increased at medium and higher altitudes, it generally decreased at lower altitudes. This behavior suggests that it is very likely that the altitude of the lifting condensation level (LCL) has increased by 200–400 m. At Monteverde, Costa Rica, meteorological observations also indicate that the number of mist-free days in the dry season has increased since the early 1970s, partly in response to variations in sea-surface temperatures (Pounds *et al.*, 1999). In addition, the possibility of a maximum rise in LCL by 170 m as a result of complete conversion of lowland forest to pasture in northern Costa Rica and the associated increases in H, has been confirmed by meso-scale atmospheric modeling studies (Nair *et al.*, 2003; Ray *et al.*, 2006; Nair *et al.*, this volume).

Table 56.1 *Trends in annual precipitation and annual fog frequency at climatic stations situated on windward and leeward slopes of the central mountain region of the States of Veracruz and Puebla, Mexico. Statistically significant* ($p < 0.05$) *trends in precipitation (mm year*$^{-1}$*) and fog frequency (day year*$^{-1}$*) indicated by positive or negative signs; figures represent the slope of the linear regression for precipitation*

Climatic stations	Altitude (m.a.s.l.)	Mean annual rainfall (mm)	Trends in precipitation												Fog (day year^{-1})
			F	M	A	M	J	J	S	O	N	D	Rainy season	Dry season	Annual
Leeward side															
Ciudad Serdán	2278	802	**+0.4**												
Huauaxtla	1625	1988													
Perote	2400	494										**+0.6**		**+2.4**	
San Antonio Cañada	1600	512								+2					
Tecamachalco	2055	580	**+0.3**												
Tehuacan	1648	408													
Tlachichuca	2590	547			**−1.5**										
Tlacotepec	1950	486	**+0.3**	**+0.3**											
Tlatlauquitepec	2025	1257													
Zacapoaxtla	2045	1314													
Zautla	1940	750													
Windward side															
Altotonga	1899	1428	**+1.1**							+2					**T+**
Atzalan	1842	1947													**T+**
Coatepec	1252	1703													**T−**
El Coyol	610	1143	**−0.6**												
Elotepec	1860	1978	**−0.6**				**+0.6**	**+0.6**	**+0.6**				**+1.9**		
Huatusco	1344	2031									**−1.6**				**T+**
Jalacingo	1944	1710	**+0.8**												**T+**
Las Animas	1399	1346	**−1.6**												**T+**
Las Minas	1365	1387							+2						**T−**
Los Pescados	2938	779												**+1.4**	**T−**
Naranjal	697	2509	**−1.0**												
Naolinco	1605	1639	**−0.8**		**+2.1**	+2									
Tembladeras	2727	1633													**T−**
Teocelo	1218	2044													**T−**

Table 56.2 *Comparison of the average maxima values of the energy balance components* per month *for a pine–oak forest and a grassland in the Grandes Montañas region in central Veracruz, México*

	Forest	Pasture	Forest	Pasture	Forest	Pasture	Forest	Pasture	Forest	Pasture
	Jun	Jun	Aug	Aug	Oct	Oct	Nov	Nov	Mar	Mar
R_n	679	534	546	950	488	787	338	399	1121	944
LE_v	433	114	263	212	98	160	105	107	245	89
H	246	420	283	738	390	627	234	291	876	856
LE_v/R_n	0.64	0.21	0.48	0.22	0.20	0.20	0.31	0.26	0.22	0.09
H/R_n	0.36	0.79	0.51	0.78	0.78	0.80	0.69	0.73	0.78	0.90
H/LE_v	0.57	3.68	1.07	3.49	3.96	3.91	2.23	2.72	3.58	9.66

Similarly, there has been an inferred increase in H during, especially, the dry season following large-scale forest conversion in the study area (Table 56.2). This is likely to have increased the height of the atmospheric mixing layer and the LCL, thereby transporting moisture to the leeward side of the high mountain wall, as also suggested by the spatial distribution in observed trends in February precipitation (Table 56.1). However, differences in energy partition between forest and pasture were not significant during the rainy season, probably because atmospheric humidity was high regardless of land-cover type, as a result of moisture-laden easterly winds coming from the Gulf of Mexico (Alemán and García, 1974; Báez *et al.*, 1997).

The observed changes in precipitation and fog frequency in the study region could also be the result of large-scale atmospheric mechanisms acting in concert, including the El Niño Southern Oscillation (ENSO) phenomenon, and possibly trends in sea-surface temperature (cf. Pounds *et al.*, 1999). ENSO years in Mexico tend to have higher dry-season rainfall (Cavazos, 1997; Magaña *et al.*, 2003) and fog incidence (Ponette-González *et al.*, 2009). Also, the position and movement of the Bermuda anticyclone exerts a strong influence on the climate in the region (Alemán and García, 1974). If the anticyclone were to be displaced to the south, e.g. due to global atmospheric mechanisms produced by global warming, subsidence over the region would be increased, thereby counteracting the effect of any increases in H associated with forest conversion on precipitation and fog frequency. Conversely, if the anticyclone were to move to the north, subsidence presumably would be reduced and the height of the mixing layer and the LCL would be increased. This last mechanism would provoke greater dryness on the windward side of the mountains and more humid conditions on the leeward side (cf. Sperling *et al.*, 2004). It is possible that any externally induced changes in regional climate might not be so easily detectable due to the extreme topography of the Grandes Montañas which creates a strong rainfall gradient with elevation and which tends to reduce the transport of moisture from the windward to the leeward side of the water divide. Nevertheless, given the observed contrasts in H associated with forest and pasture (Table 56.2), it cannot be excluded that the widespread conversion of forest to pasture in the region (Figure 56.2) has produced, or at least enhanced, a change in rainfall and fog regime. Consequently, if current rates of conversion of cloud-affected forest continue to persist (Muñoz-Villers and López-Blanco, 2008), the observed trends in precipitation and fog frequency might become more pronounced. Shaded coffee plantations, which exhibit similar energy partition as montane forest in the study area (Holwerda *et al.*, 2007) still occupy a considerable area (Figure 56.2) and therefore should exert a moderating influence on overall post-forest values of H. However, for macro-economic reasons (including declining coffee prices and increased demands for biofuels), shaded coffee plantations in the area are increasingly transformed into sugarcane fields (Muñoz-Villers and López-Blanco, 2008). Sugarcane has an even higher albedo – and therefore values of H – than pasture (Chia, 1967).

CONCLUSIONS

Available evidence suggests a possible climatic effect of the widespread conversion of cloud-affected montane forest to agricultural and cattle-raising systems (pasture) in the form of an altitudinal shift in dry-season precipitation (notably in February) and overall fog frequency in central Veracruz, eastern Mexico. The underlying mechanism is hypothesized to be an increase in sensible heat over latent heat following forest conversion. Currently, the effect is probably still moderated by the presence of large areas of shaded coffee plantations, but this might well change if ongoing rates of conversion of coffee plantations to sugarcane are stepped up in response to a world-wide increase in demands for biofuels.

This study has potentially important implications for regional water resources planning. Amongst others, the preservation and restoration of the temperate forests of the area, including montanc cloud forest remnants (cf. Ponette-González *et al.*, 2009; Williams-Linera *et al.*, this volume) and shaded coffee plantations, should be promoted if further warming of the atmosphere is to be mitigated. Future research along the lines of Nair *et al.* (2003), Ray *et al.* (2006), and Van der Molen *et al.* (2006) is needed to understand better the influence of land-cover change (if any) on climatic variability or change in this part of the world.

ACKNOWLEDGEMENTS

The authors thank an anonymous reviewer whose comments helped to improve the manuscript. This project was supported by grant no. 36892-B from the Mexican National Council of Science and Technology (CONACYT).

REFERENCES

Alemán, P. A. M., and E. García (1974). The climate of Mexico. In *World Survey of Climatology*, Vol. 11, ed. H. Landsberg, pp. 345–391. Amsterdam, the Netherlands: Elsevier.

Báez, A. P., H. Padilla, J. Cervantes, D. Pereyra, and R. Belmont (1997). Rainwater chemistry at the eastern flanks of the Sierra Madre Oriental, Veracruz, Mexico. *Journal of Geophysical Research* **102**: 23 329–23 336.

Barradas, V. L. (1983). Capacidad de captación de agua a partir de la niebla en *Pinus montezumae* Lambert, de la región de las Grandes Montañas del Estado de Veracruz. *Biótica* **8**: 427–431.

Bisselink, B. (2003). Precipitation trends in Puerto Rico: Quantification and explanation of complex patterns. M.Sc. thesis, VU University Amsterdam, Amsterdam, the Netherlands.

Carson, D. J. (1973). The development of a dry inversion-capped convectively unstable boundary layer. *Quarterly Journal of the Royal Meteorological Society* **99**: 450–467.

Cavazos, T. (1997). Downscaling large-scale circulation to local winter rainfall in north-eastern Mexico. *International Journal of Climatology* **17**: 1069–1082.

Cervantes-Pérez, J., V. L. Barradas, A. Tejeda-Martínez, and D. Pereyra (2001). Clima urbano, bioclima humano, hidrología superficial y evaluación de riesgos por hidrometeoros en Xalapa. In *Unidades ambientales urbanas: Bases metodológicas para la comprensión integrada del espacio urbano*, eds. C. Capitanachi-Moreno, E. M. Utrera Barillas, and C. B. Smith, pp. 1–57. Xalapa, Veracruz, México: Universidad Veracruzana and Instituto de Ecología, A. C.

Chia, L. S. (1967). Albedos of natural surfaces in Barbados. *Quarterly Journal of the Royal Meteorological Society* **93**: 116–120.

Culf, A. D., J. L. Esteves, A. de O. Marques-Filho, and H. R. da Rocha (1996). Radiation, temperature and humidity over forest and pasture in Amazonia. In *Amazonian Deforestation and Climate*, eds. J. H. C. Gash, C. A. Nobre, J. M. Roberts, and R. L. Victoria, pp. 175–191. Chichester, UK: John Wiley.

Fitzjarrald, D. R. (1986). Slope winds in Veracruz. *Journal of Climate and Applied Meteorology* **25**: 133–144.

García-García, F., and R. A. Montañez (1991). Warm fog in eastern Mexico: a case study. *Atmósfera* **4**: 53–64.

García-García, F., and V. Zarraluqui (2008). A fog climatology for Mexico. *Die Erde* **139**: 45–60.

Gómez-Pompa, A. (1978). *Ecología de la vegetación del estado de Veracruz*. México, DF: Compañía Editorial Continental, Sociedad Anónima.

Holwerda, F., V. L. Barradas, J. Cervantes, and L. A. Bruijnzeel (2007). Balances hídricos y de energía de un cafetal de sombra en el centro de Veracruz, México. In *Reporte técnico final del proyecto INE/A1–064/2007*. Xalapa, Veracruz, México: Instituto de Ecología, and Amsterdam, the Netherlands: VU University Amsterdam.

Jauregui, E., L. Godinez, and F. Cruz (1992). Aspects of heat island development in Guadalajara, Mexico. *Atmospheric Environment* **268**: 391–396.

Lara-García, F. (2000). *Régimen termopluviométrico y aspectos microclimá ticos y bioclimáticos de Xalapa, Veracruz*. B.Sc. thesis, Universidad Veracruzana, Xalapa, Veracruz, México.

Magaña, V. O., J. L. Vázquez, J. L. Pérez, and J. B. Pérez (2003). Impacto of El Niño on precipitation in Mexico. *Geofísica Internacional* **42**: 313–330.

Maitelli, G. T., and I. R. Wright (1996). The climate of a riverside city in the Amazon Basin: urban–rural differences in temperature and humidity. In *Amazonian Deforestation and Climate*, eds. J. H. C. Gash, C. A. Nobre, J. M. Roberts, and R. L. Victoria, pp. 193–206. Chichester, UK: John Wiley.

Muñoz-Villers, L. E., and J. López-Blanco (2008). Land use/cover changes using Landsat TM/ETM images in a tropical and biodiverse mountainous area of central-eastern Mexico. *International Journal of Remote Sensing* **29**: 71–93.

Nair, U. S., R. O. Lawton, R. M. Welch, and R. A. Pielke Sr. (2003). Impact of land use on tropical montane cloud forests: sensitivity of cumulus cloud field characteristics to lowland deforestation. *Journal of Geophysical Research* **108** (D7): 4206–4218, doi:10.1029/2001JD001135.

Nasrallah, H., A. Brazel, and R. C. Balling Jr. (1990). Analysis of the Kuwait City urban heat island. *International Journal of Climatology* **10**: 401–405.

Nobre, C. A., P. J. Sellers, and J. Shukla (1991). Amazonian deforestation and regional climate change. *Journal of Climate* **4**: 957–988.

Ponette-González, A. G., K. C. Weathers, and L. M. Curran (2009). Water inputs across a tropical montane landscape in Veracruz, Mexico: synergistic effects of land cover, rain and fog seasonality, and interannual precipitation variability. *Global Change Biology*, doi: 10.1111/j.1365–2486.2009.01985.x.

Pounds, A. J., M. P. L. Fogden, and J. H. Campbell (1999). Biological responses to climate change on tropical mountain. *Nature* **389**: 611–614.

Ray, D. K., U. S. Nair, R. O. Lawton, R. M. Welch, and R. A. Pielke Sr. (2006). Impact of land use on Costa Rican tropical montane cloud forests: sensitivity of orographic cloud formation to deforestation in the plains. *Journal of Geophysical Research* **111**: D02108, doi:10.1029/2005JD006096.

Sperling, F. N., R. Washington, and R. J. Whittaker (2004). Future climate change of the subtropical north Atlantic: implications for the cloud forests of Tenerife. *Climatic Change* **65**: 103–123.

Van der Molen, M. K., A. J. Dolman, M. J. Waterloo, and L. A. Bruijnzeel (2006). Climate is affected more by maritime than by continental land use change: a multiple-scale analysis. *Global and Planetary Change* **54**: 128–149.

Vogelmann, H. W. (1973). Fog precipitation in the cloud forests of Eastern Mexico. *BioScience* **23**: 96–100.

Williams-Linera, G., R. H. Manson, and E. Isunza Vera (2002). La fragmentación del bosque mesófilo de montaña y patrones del uso del suelo en la región oeste de Xalapa, Veracruz, México. *Madera y Bosques* **8**: 69–85.

57 Potential effects of global climate change on epiphytes in a tropical montane cloud forest: an experimental study from Monteverde, Costa Rica

N. M. Nadkarni
The Evergreen State College, Olympia, Washington, USA

ABSTRACT

Epiphytes are physiologically dependent upon atmospheric sources of nutrients and water delivered in rain and cloud water. In addition, their physical location in tree canopies places epiphytes at the immediate interface between the atmosphere and the forest. Both factors render epiphytes particularly vulnerable to predicted changes in cloud water deposition in montane tropical forests. The sensitivity of epiphytes such as lichens, bryophytes, and poikilohydric vascular plants has been documented for many bioregions. Hence, canopy-dwelling plants constitute excellent candidates for the monitoring of climate change in regions where micro-climatic measurements may be difficult or impossible. The vulnerability of epiphytes to climatic change has broad implications for other ecosystem components due to the many ecological roles that canopy epiphytes perform – including intercepting and retaining nutrients, providing wildlife habitat, and serving as a carbon sink on branch areas not occupied by host-tree foliage. This chapter summarizes the existing literature on this topic, which mainly comprises descriptive studies. In addition, results are presented from experimental work conducted at Monteverde, Costa Rica. Transplant experiments were carried out using epiphyte mats from upper-elevation to mid-elevation and lower-elevation sites along an altitudinal gradient over an 18-month period. Leaves of individual plants of four target taxa were marked and checked at monthly intervals to compare plant longevity, leaf production, and leaf mortality between controls left in the upper cloud forest (intact and disturbed controls) versus transplants that were exposed to drier conditions. Exposure to drier, less misty conditions resulted in significantly higher leaf mortality, lower leaf production, and reduced longevity. Certain epiphyte taxa were more sensitive than others to this exposure. Field and greenhouse epiphyte removal and regrowth experiments revealed that predicted climate change effects could also alter seed-bank dynamics in canopy epiphyte mats. After epiphytes died, seedlings of terrestrial gap-colonizing tree species grew from the seed banks within the residual mats of arboreal soil. Greenhouse experiments confirmed that the death of epiphytes can result in radical compositional changes of canopy communities. These results suggest that cloud forest epiphytes may serve as effective biotic "early warning signals" for conditions of predicted future climate change.

INTRODUCTION

The diverse assemblage of organisms in tropical montane forests face threats from deforestation, fragmentation, and habitat degradation (Bruijnzeel and Hamilton, 2000; Bubb *et al.*, 2004; Mulligan, this volume), and are likely to encounter further challenges from impending changes in climate (Loope and Giambelluca, 1998; Lawton *et al.*, 2001; Karmalkar *et al.*, 2008). Lifting clouds caused by rising sea- and land-surface temperatures (Benisten *et al.*, 1997; Still *et al.*, 1999; Nair *et al.*, 2003; Ray *et al.*, 2006; Nair *et al.*, this volume) have been linked with changes in hydrological inputs – notably a reduction in the amount and frequency of critical dry-season inputs of drizzle and cloud water (cf. Guswa *et al.*, 2007; Ponette-González *et al.*, 2009; Barradas *et al.*, this volume) – and population declines and extinctions of vertebrate species in tropical cloud forests (Connelly and

Tropical Montane Cloud Forests: Science for Conservation and Management, eds. L. A. Bruijnzeel, F. N. Scatena, and L. S. Hamilton. Published by Cambridge University Press.

Crump, 1998; Pounds *et al.*, 1999, 2006). The decline of vagile organisms (reptiles, amphibians, and birds) attributed to these patterns have been based on population studies during unusual dry spells associated with El Niño events (Pounds *et al.*, 1999). Little is known, however, about direct effects of the predicted shifts in cloud-bank levels (Ray *et al.*, 2006; Karmalkar *et al.*, 2008; cf. Nair *et al.*, 2008; Nair *et al.*, this volume) on plants or sessile animals dwelling in cloud forests, many of whom depend upon water and nutrients delivered in cloud water, and who cannot physically escape changes in climate by emigration (Bawa and Markham, 1995; Zotz and Bader, 2009).

Cloud forest epiphytes (canopy-dwelling plants with no vascular connections to the ground or host trees) are expected to be negatively affected by predicted climate changes because their dwelling-places are restricted to the atmospheric–terrestrial interface (Nadkarni and Solano, 2002; Zotz and Bader, 2009). Experimental studies that quantified the interception of host-tree litter (Nadkarni and Matelson, 1991) and other research that partitioned atmospheric vs. terrestrial nitrogen uptake using stable isotopes (Feild and Dawson, 1998; Hietz *et al.*, 1999) have shown that epiphytes absorb and retain atmospheric water and nutrients directly. They are known to contribute substantially to the capture and retention of atmospheric nutrients by the canopy, particularly nitrogen and aerosols (Asbury *et al.*, 1994; Clark *et al.*, 1998a; Hafkenscheid, 2000; cf. Boy *et al.*, 2008; Oesker *et al.*, this volume). With healthy epiphytic communities, cloud forest ecosystems have the benefit of added nutrient and organic matter fluxes originating at the atmospheric–terrestrial interface. Without epiphytes, these inputs might be lost because clouds would pass through with minimized interception and retention, and the forests might become potentially less biologically active and resilient (Clark *et al.*, 1998a,b).

In two seminal papers (Lugo and Scatena, 1992; Benzing, 1998), epiphytes in tropical montane cloud forests were proposed as superlative indicators of climate change. In addition to their tight coupling with atmospheric inputs, their "miniature" size scale, their high diversity of species and life forms, and their independence from the substrate all make them conducive to manipulations that would be impossible with large terrestrial plants. Certain canopy-dwelling plants have long been used as indicators of atmospheric conditions. For example, epiphytes have provided empirical evidence for the connection of global airsheds from their retention of fallout originating from nuclear explosions (Odum *et al.*, 1970). Epiphytes – particularly lichens – have also been used as indicators of heavy-metal air pollution, especially in urban and industrial settings (Rao, 1982). However, very little experimental work has been done to directly test predicted effects of climate change on canopy-dwelling plants in tropical forests (Nadkarni and Solano, 2002; cf. Hietz, this volume).

As part of a long-term study on the ecological roles of canopy biota (Nadkarni *et al.*, 2000), this chapter explores the sensitivity of vascular epiphytes to the conditions predicted for future climatic scenarios having decreased cloud water input to the canopy (Still *et al.*, 1999; Nair *et al.*, 2003; Ray *et al.*, 2006; Karmalkar *et al.*, 2008). Transplant experiments were carried out at three sites along the existing gradient of cloud water input in the tropical montane landscape in and around the Monteverde Cloud Forest Preserve in northern Costa Rica (cf. Häger and Dohrenbusch, this volume), for which important changes in climate have been modeled (Nair *et al.*, 2003; Ray *et al.*, 2006; Karmalkar *et al.*, 2008; cf. Nair *et al.*, this volume). Effects of the transplant experiment on the resulting colonization of the epiphyte mats are included as well.

STUDY AREA

The forest lies within the Monteverde Cloud Forest Preserve, Monteverde, Puntarenas Province, Costa Rica (10° 18′ N, 84° 48′ W). The primary forest is a lower montane cloud forest situated at 1480 m.a.s.l. within the overall Leeward Cloud Forest zone of Lawton and Dryer (1980), but is more windward in appearance because of its exposure to frequent and intense wind and fog events (Häger and Dohrenbusch, this volume). Mean annual rainfall is 2155 mm (33-year average as measured at 1480 m.a.s.l. within 2 km from the main forest site; Clark *et al.*, 2000). The canopy is exposed to frequent and intense wind and mist events throughout much of the year, especially during the windy–misty season (November–March) and the dry season (April–May) (Clark *et al.*, 1998b). Mean diameter of trees (>10 cm diameter at breast height) is 65.5 cm; forest stature 20–33 m, mean tree density 154 individuals ha^{-1}, and forest species diversity ~115 ha^{-1} (Nadkarni *et al.*, 1995; cf. Häger and Dohrenbusch, this volume).

Branch surfaces of the crown interior of nearly all mature trees support epiphytes (bryophytes, herbs, woody shrubs, and hemi-epiphytes) in interwoven root–humus mats up to 25 cm thick. The "epiphyte mats" on the inner branches contain a high diversity of vascular and non-vascular plants (Ingram and Nadkarni, 1993), as well as arboreal soil ("crown humus") (cf. Köhler *et al.*, 2007; Köhler *et al.*, this volume). The dead organic matter is high in recalcitrant humic compounds, extremely acidic, and experiences greater frequencies of "dry-downs," especially during the dry season, than the upper horizons of terrestrial soil (Bohlman *et al.*, 1995; cf. Tobón *et al.*, this volume #26). Canopy soil supports the same major groups of invertebrates, but at reduced densities (Nadkarni and Longino, 1990). Crown humus has equivalent amounts of microbial biomass carbon and nitrogen, fosters similar rates of net mineralization, but has much reduced rates of nitrification relative to forest floor soil (Vance and Nadkarni, 1990). These mats can be peeled back and rolled up like a carpet, leaving the underlying bark substrate intact.

In addition, a diverse seed bank – comprising both terrestrial and epiphytic taxa – is present in these arboreal mats, including terrestrial members of the Melastomataceae, Cecropiaceae, and Moraceae. In 1998, seeds were extracted quantitatively from nine samples of canopy humus mats (10-cm circular cores taken from three *Ficus tuerckheimii* trees) at the 1480 m.a.s.l. site. Samples were air-dried, sieved through a 2-mm mesh screen, and sorted to species. Identifications were made using the seed collection created for Monteverde by R. Solano and G. G. Murray. The five most common plant taxa represented in the canopy-level seed bank included both terrestrial ("ter") families and obligate epiphytic ("epi") groups: *Cecropia* [Cecropicaceae (ter)], *Phytolacca* [Phytolaccaceae (ter)], *Lycianthes* [Solanaceae (epi)], *Ficus* [Moraceae (ter)], and *Oreopanax* [Araliaceae (epi)]. There were large numbers of terrestrial gap-colonizing early-successional species such as *Hampea appendiculata, Conostegia oerstediana*, and *Witheringia* spp. in the canopy mats (N. Nadkarni and R. Solano, unpublished data).

METHODS

The general approach was to move pieces of epiphyte mats from the upper cloud forest (where cloud water is frequent throughout the dry season; Clark *et al.*, 1998a) to trees that grow at a slightly lower elevation and are thus currently naturally exposed to less cloud water during the more protracted dry season (cf. Häger and Dohrenbusch, this volume) – the very conditions predicted for tropical cloud forests by general circulation models (GCMs) under conditions of increasing CO_2 (Still *et al.*, 1999). Three experimental sites were established along a short topographic gradient (<150 m in elevation and 1.5 km in distance). Cloud water inputs are observed to progressively decrease during the dry season as one moves from the upper end of the gradient ("upper site," 1480 m.a.s.l.) to the two transplant sites – one located at the midpoint of the gradient ("mid-elevation," 1410 m.a.s.l.), and the other at the lower end ("lower elevation," 1330 m.a.s.l.) (Figure 57.1; cf. Häger and Dohrenbusch, this volume).

Large, mature *Ficus tuerckheimii* (Moraceae, "strangler fig") trees were climbed using single-rope techniques (Perry, 1978) and the responses of the following four common epiphyte taxa present in the experimental mats were monitored: *Guzmania pachystylis* (Bromeliaceae), a tank bromeliad; *Clusia* (Clusiaceae), a woody shrub; *Peperomia* sp. (Piperaceae), a scandent herb; and *Elaphoglossum* sp. (Lomariopsidaceae), a strap-leafed fern.

A critical question in this type of experimental design is whether epiphytes can be moved from one tree to another without effects on plant productivity or mortality. This was tested with a set of control experiments, comparing plants in undisturbed mats with mats that were transplanted within the upper site. Sets of "intact" controls (75-cm long segments of epiphyte mats that were not cut at all, but flagged around each end) and "disturbed" controls (similar-sized segments of epiphyte mats that were cut, lowered to the ground, driven around in a car, and then replaced in the same tree) were established. Each leaf of each plant was marked with a small dot of paint, all leaves were counted at the beginning of the experiment, and trees were re-climbed and all leaves recounted at monthly intervals over a period of a year. Each month, any new leaves were marked with paint. Leaf production was calculated as the number of new leaves produced during each sampling interval, and leaf mortality as the number of leaves missing from the former sampling interval. Because each plant started out with a different number of leaves, the data were expressed as percent leaf change: [(number of new leaves – number of dead leaves)/number of original leaves] × 100. Plant longevity equaled the number of months the whole plant survived from the onset of the study. For the control experiments, a total of 138 plants were used from four trees (Table 57.1).

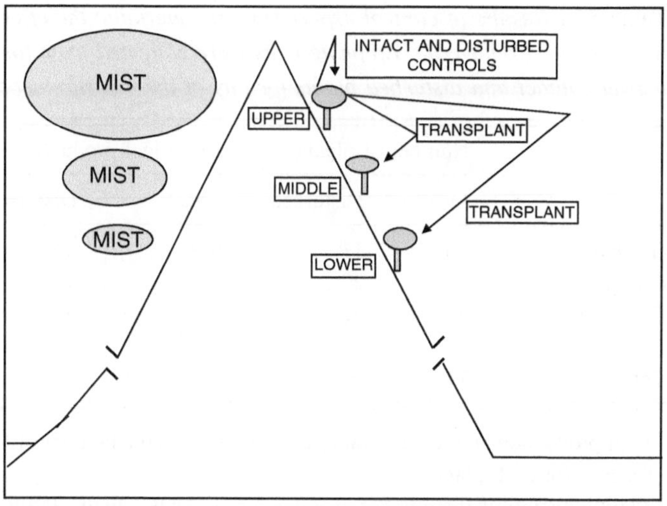

Figure 57.1. Schematic of the field experiment design. The general approach was to move epiphyte mats from the upper cloud forest to mid- and lower-elevation sites where cloud water input is progressively less abundant, symbolized by the diminishing areas of circles. "Intact control" mats were marked with flagging, and were left undisturbed on trees at the upper cloud forest site (1480 m.a.s.l.). "Disturbed control" mats were cut from trees in the upper site, lowered to the ground, moved around, and then replaced in the same trees. Epiphyte mats used for transplants were cut from the upper site and moved to trees that were in the mid-elevation (1410 m.a.s.l.) and lower-elevation sites (1330 m.a.s.l.). Note that the vertical scale at the top of the mountain is expanded; the three sites are within 150 m in elevation and 1.5 km in distance of each other.

There were no significant differences in net rates of leaf production or leaf or plant mortality (*t*-tests) between intact and disturbed controls over the 12-month period (Table 57.1). The individual plants were treated as the independent sample unit (regardless of the mat or tree of origin), because a previous study at Monteverde on epiphyte survivorship had shown that there

Table 57.1 *Results of control experiments to determine the effects of moving epiphyte mats on leaf production, mortality and plant longevity at Monteverde. Epiphyte mats were collected from four* Ficus tuerckheimii *trees. There were no significant differences between intact and disturbed plants for any of the measurements*

	Number of plants		Mean leaf production (%)[a]		Mean leaf mortality (%)		Mean plant longevity (mo)[b]	
	Intact	Disturbed	Intact	Disturbed	Intact	Disturbed	Intact	Disturbed
Guzmania	6	11	9.0	8.8	9.7	9.8	11.7	11.8
Clusia	26	22	4.4	4.5	6.5	3.7	11.1	12.6
Elaphoglossum	24	22	2.5	6.7	3.5	6.2	11.5	12.6
Peperomia	15	12	6.2	4.3	7.6	4.1	10.6	11.2
Total	71	67						

[a] Leaf production and leaf mortality are expressed as the number of leaves that were produced and died as a proportion of the original number of leaves for each plant.
[b] Plant longevity is the number of months that plants remained alive over the 12-month monitoring interval.

Table 57.2 *Number of trees, target plants, and leaves sampled for each plant taxon*

	Number of trees	Number of plants	Original leaves	New leaves produced	Dead leaves	Total leaves
Guzmania	12	46	528	177	465	705
Clusia	11	82	783	273	461	1056
Elaphoglossum	12	81	528	224	437	752
Peperomia	12	63	618	174	517	792
Total		272	2457	848	1880	3305

was no effect of tree or mat on epiphyte longevity (Matelson *et al.*, 1993). This indicated that transplantation alone did not affect the viability or growth of epiphytes, so any differences in plant performance between the upper forest location and the two transplant sites should be due to the environmental differences rather than the impact of moving the mats.

Next, similar-sized experimental epiphyte mats were cut from *Ficus tuerckheimii* trees in the upper site, and tied to branches of three to four individuals of conspecific trees in each of the mid- and lower-elevation sites. Plants were marked as done for the controls and monitored for 18 months after transplantation (Table 57.2). Two sets of transplant experiments were carried out, one initiated during the dry season (January 1998), and another during the wet season (June 1998).

In addition, amount of cloud water input and moisture content of the crown humus substrates of sample mats were measured during the 4-month dry season at each of the three sites. Cloud water (and wind-driven rain and drizzle) was collected with passive collectors (Falconer and Falconer, 1980) placed within the canopy (18–23 m above the forest floor) in four trees at each of the three sites. Volume of input was recorded at 3–4 day intervals ($n = 36$) for the 4-month dry season (13 February–15 June 1998). Gypsum blocks (Delmhurst KS-D1 Soil Moisture Tester) were used to detect relative moisture content of the mats. Two gypsum blocks were buried in the arboreal soil mats of three branches per tree in each of three trees per site. Moisture levels were checked every 3–4 days, using wires that connected each sensor to the ground. Gypsum blocks are sensitive to acid environments and proper readings require optimum contact with the surrounding medium (Campbell and Gee, 1986). Therefore, each time the trees were climbed, the status of the blocks was checked.

At the end of the 18-month period of monitoring, after epiphyte mortality occurred in the lower sites, the patterns of "revegetation" of the transplanted plots were documented. Two-sample t-tests were used between each possible pair of sites, with a Bonferroni correction for multiple comparisons. All statistical analyses were carried out with SYSTAT (SPSS Science, Inc., Chicago, Illinois). The observations on terrestrial plant "invasion" of epiphyte mats were supplemented with greenhouse experiments. In June 1998, 24 canopy mats (0.75 × 0.5 m) were collected from four *F. tuerckheimii* trees in the upper forest. All of the "above-mat" stems of vascular epiphytes were clipped and all of the non-vascular plants from half the mats were pulled out ("pruned" treatment); the other half of the samples served as controls ("intact"). All epiphyte mats were placed in a greenhouse located near the mid-elevation site, at 1420 m.a.s.l. The mats were watered (misted two times per day for 5 min), and temperatures were monitored with a

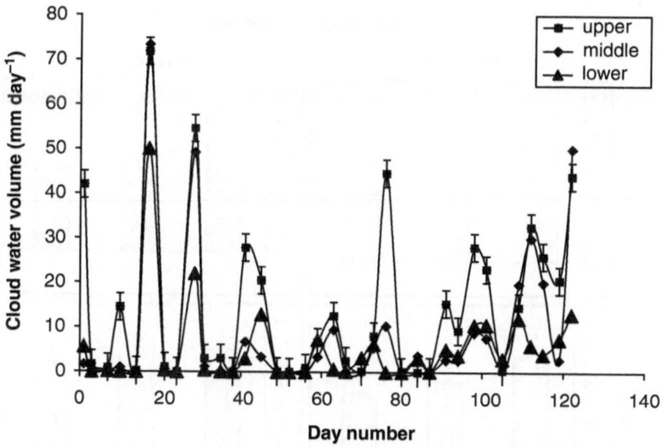

Figure 57.2. Moisture input to the epiphyte mats: volume of cloud water during four dry-season months, measured every 3–4 days with passive cloud water collectors ($n = 29$) in the canopy of four study trees at each of the three sites. Each point is the mean of four collectors; bars are standard errors.

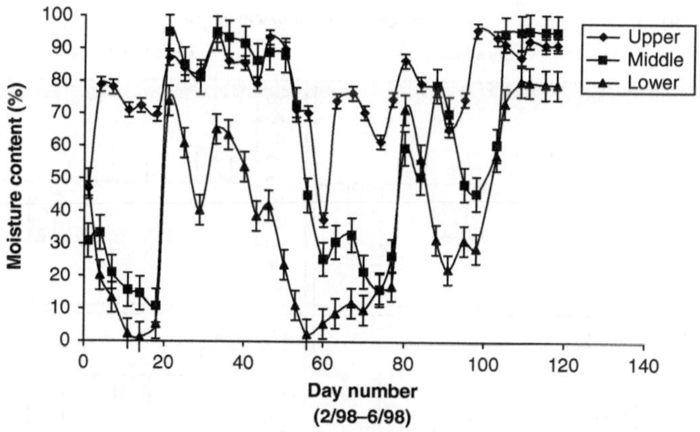

Figure 57.3. Moisture retention by the epiphyte mats: moisture content (%) of crown humus from experimental epiphyte mats at each of the three sites, measured with gypsum blocks every 3–5 days during four dry-season months. Each point is the mean of eight gypsum blocks (two per tree for four trees); bars are standard errors.

maximum–minimum thermometer. Conditions mimicked ambient conditions of the upper site (Nadkarni and Solano, 2002). Over the subsequent 12 months, each emerging individual seedling was identified to genus or species level, its height was measured, and the number of individuals of each taxon recorded. Seedlings as young as 2 months old could be identified using the existing voucher seed collection and photographic matching of seedlings of over 80 species native to the Monteverde Reserve.

RESULTS

The measurements of cloud water inputs and epiphyte-mat moisture contents confirmed that although the dynamics of cloud water followed similar patterns, epiphyte mats received progressively smaller amounts of cloud water inputs (Figure 57.2) and contained progressively smaller amounts of moisture (Figure 57.3) going from the highest to lower-elevation sites. Over the 4-month collection period, cloud water collectors at the lower-elevation site received a total of 660 mm, less than one-third of the cloud water collected at the uppermost site (1840 mm), with intermediate amounts retrieved at the mid-elevation location (1057 mm). Similarly, the relative moisture content of the mats of crown humus differed significantly, with mean values of 90.8%, 61.7%, and 35.3% for the mats of the upper, middle, and lower plots, respectively (one-way ANOVA, $p < 0.05$).

At the upper site, all four taxa distinguished showed positive net leaf production and high longevity during the 18-month study period. However, there were striking and significant differences between plants in the upper forest vs. mid- and lower-elevation sites for net leaf change and plant longevity (Figure 57.4). Effects were consistently more pronounced for transplants that were initiated in the dry season, for which all four taxa showed significantly reduced net leaf production and longevity at the mid- and lower-elevation sites relative to the upper site. Differences in net leaf change ranged between −50% and −110% (Figure 57.4a). Also, different plant taxa showed different degrees of sensitivity. For *Clusia* and *Elaphoglossum*, leaf change was significantly more negative at the lower elevation than at the mid-elevation site, whereas the other two taxa did not show a significant difference between these two sites. Leaf longevity for *Elaphoglossum* and *Peperomia* (but not for the other two taxa) was found to be significantly reduced at the lower elevation relative to the mid-elevation site.

Differences between transplants and controls were much less striking for the wet season transplants. Only one taxon (*Guzmania*) showed a significant reduction in plant longevity, whereas three of the four taxa (all but *Peperomia*) exhibited a significant reduction in net leaf change at the lower-elevation site, compared to the upper elevation site (Figure 57.4b).

At the end of the 18-month experiment, following the death of nearly all of the epiphytes transplanted to the lower- and mid-elevation plots, unexpected changes in composition of the canopy communities were observed. Numerous new seedlings had sprouted in the remaining mats of canopy humus that were still tied to the tree branches. These seedlings were neither resprouting epiphytes nor epiphyte seedlings, as one might expect, but rather, they were terrestrial seedlings that had grown out of the seed bank that existed in the residual arboreal soil. The most numerous were gap-colonizing terrestrial species of the genera *Cecropia* (Moraceae), *Witheringia* (Solanaceae), and *Conostegia* (Melastomataceae). These species are encountered only extremely rarely as seedlings in primary forest and are virtually never seen in the canopy, although they are occasionally

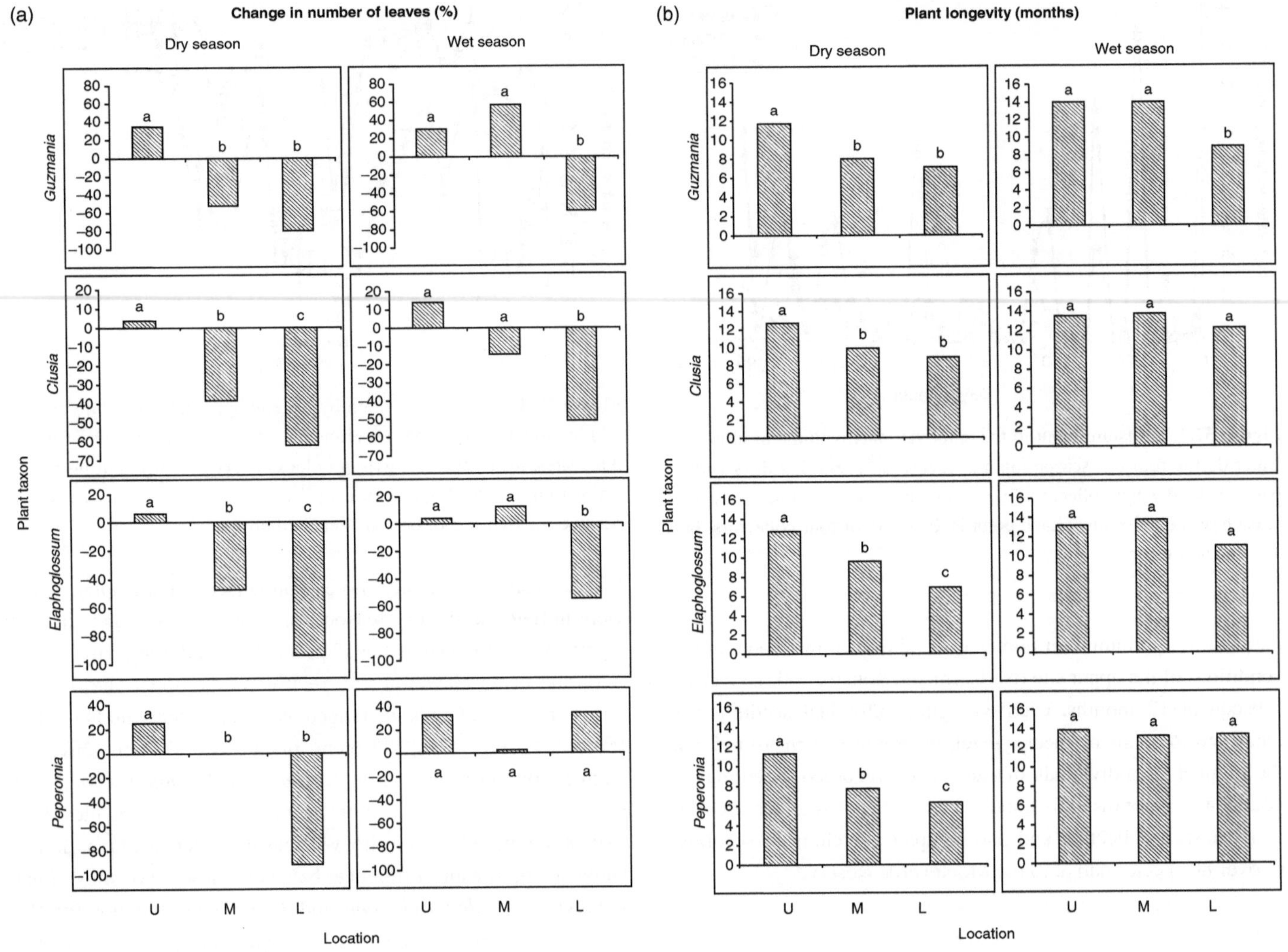

Figure 57.4. Effects of experimental movement of epiphytic taxa from upper to mid- and lower-elevation sites where they were exposed to progressively less cloud water. (a) Percent change of number of leaves was calculated as: [(number of new leaves – number of dead leaves)/number of original leaves]×100. (b) Plant longevity was recorded as the number of months from the outset of the study that the plant remained alive. The different letters indicate a significant difference at the 0.05 level with a Bonferroni correction factor for multiple comparisons. The number of plants and leaves is given in Table 57.2.

present when epiphyte mats fall to the forest floor in canopy gaps that are large enough to receive direct light (N. M. Nadkarni, personal observation).

These observations of terrestrial "plant invasion" of epiphyte mats were confirmed by the results of the greenhouse experiment. The abundance of seedlings was significantly greater in the "pruned" than the "intact" mats (1531 vs. 326 seedlings, or 127.6 per mat vs. 27.1 per mat, respectively; t-test, $p < 0.01$). Also, the number of taxa for all of the mats (taxon density) was significantly higher in the pruned than in the intact mats (37 vs. 25 species, respectively). Individuals reached a mean height of 47 cm (± 16.3 cm) with a maximum of 1.6 m.

The composition of individuals in the pruned mats was overwhelmingly dominated by terrestrial taxa (90.4%), whereas the intact mats had only a negligible number of terrestrial individuals (0.9%). The three most common taxa in the pruned mats (*Cecropia, Conostegia, Solanum*) were terrestrial gap-colonizing species, which appear in light gaps following tree-falls, and exhibit early-successional characteristics such as rapid growth and intolerance to shade. This contrasted with the three most common seedlings that appeared in the control mats (*Oreopanax, Neomirandea, Lysianthes*), which are obligate scandent epiphytic herbs or shrubs. Thus, the removal of an overtopping "epiphyte canopy" – whether by death due to desiccation (as in the field experiment) or by artificial removal from clipping (as in the greenhouse experiment) – appears to change the composition of the epiphyte mats from epiphyte- to terrestrial-plant dominated.

DISCUSSION AND CONCLUSIONS

The present experiments showed that exposure to environments with progressively less cloud water inputs – i.e. conditions associated with future climate scenarios predicted for tropical montane forest environments by GCMs – had a generally negative effect on growth of the epiphyte taxa, although the severity depended on the season of transplant and taxon of plant. The results further predict that effects of climate change will not only result in decreased rates of epiphyte production and increased mortality, but may also radically alter the composition of canopy communities by releasing the latent seed bank of terrestrial plants within arboreal soils. This result is reminiscent of the release of terrestrial seed banks after removal of herbs and bryophytes in other ecosystems, such as high arctic meadows (Sohlberg and Bliss, 1987) and *alvar* grasslands (Zamfir, 2000). Studies in various plant communities and greenhouse experiments have demonstrated that herbs, mosses, and lichens affect the germination and/or seedling emergence of vascular plants. The effects may be positive or negative; the latter has been related to their structure, allelopathy, water retention, and/or shading effects that reduce light intensity or affect the red to far-red ratio (During and Tooren, 1990). Indeed, understanding the mechanisms for the release of the latent seed bank in the crown humus of canopy-held mats that are also dominated by bryophytes offers a rich area for future study of seed-bank dynamics.

Similar experiments that were conducted in Bolivia corroborated these trends. In the high Andes, Jacome *et al.* (2005) carried out a translocation experiment with whole epiphytic communities. Branch segments (1-m long) covered with dense mats of epiphytic bryophytes were cut and moved to a drier climate down slope, from 3000 m.a.s.l. to 2700 and 2500 m.a.s.l., respectively (20 branches at each elevation). At each elevation, the cut branches were hung from trees so that the original growth conditions (light, exposition) were restored. After 2 years, considerable changes in the structure of the communities and the abundances of individual species were noted. Some species had clearly increased in abundance after the 2-year period, at the cost of others, which had become scarcer. The observed changes were gradual and most pronounced at 2500 m.a.s.l., where temperatures were 2.5 °C higher than at the control site. Thus, as in Monteverde, exposure to warmer, drier conditions had a strong effect on many of the epiphyte species, as well as on canopy community composition and structure.

The following scenario may be envisaged if GCM predictions for Monteverde and other cloud forests (Still *et al.*, 1999; Ray *et al.*, 2006; Karmalkar *et al.*, 2008) are borne out: vascular and non-vascular epiphytes will die because of insufficient exposure to cloud water, followed by germination and temporary occupation of the remaining crown humus by invading terrestrial plants from the residual seed bank. Over time, canopy-held soil will fall away and will not be replaced due to the lack of the cloud water that originally formed and maintained this material. Ultimately, the canopy of the upper cloud forest will become bare of crown humus and support only the most desiccation-resistant epiphytes (i.e. those with succulent or poikilohydric leaves, tank bromeliads), such as now exist in lower-elevation forests (Rhoades, 1995; Küper *et al.*, 2004; cf. Hietz, this volume).

Sufficient data now exist to demonstrate the importance of epiphytes to overall ecosystem functioning (Weathers, 1999) and to show that they derive a significant fraction of their nutrition from atmospheric sources (Nadkarni and Matelson, 1991; Clark *et al.*, 1998a; Hietz *et al.*, 1999; Hafkenscheid, 2000; cf. Lakatos *et al.*, 2007; Hietz, this volume). With epiphytes, the canopy can support a rich array of species and contribute to ecosystem biodiversity, productivity, and resiliency. Without epiphytes, environments in which they are prevalent could lose an important mechanism to capture and store nutrients and water. Thus, the present results confirm early concerns that have been articulated about the future health of tropical montane forests (Bawa and Markham, 1995; Benisten *et al.*, 1997; Loope and Giambelluca, 1998; cf. Zotz and Bader, 2009).

This study also confirms the usefulness of epiphytes as indicator species for even subtle changes in climate characteristics (cf. Lugo and Scatena, 1992). Optimal indicator species react quantitatively to relevant stress, respond within appropriate time scales, are manipulable to calibrate responses, and occupy a critical or "keystone" position in the community (Benzing, 1998). Canopy communities provide a unique combination for both single-species approaches and ecosystem approaches to study environmental change. Individual plants or plant parts can be monitored for changes in growth, nutrient status, or physiology. In addition, whole canopy communities – their arboreal soil, seed banks, invertebrates, vascular and non-vascular plants, and vertebrates – can also be monitored for emergent properties (e.g. net primary productivity, water retention capacity, species diversity) to provide insights beyond the reaction of individuals or particular taxa. As the atmosphere continues to warm, tropical montane forest canopy communities may be the world's best first biotic indicators of change.

ACKNOWLEDGEMENTS

The Monteverde Cloud Forest Preserve, the Tropical Science Center, and B. Carlson are acknowledged for protecting the study sites. Rodrigo Solano was a major help with collecting field data throughout the study. W. A. Haber provided critical help with plant identification. J. A. Pounds provided information on climate change in montane forests, J. Campbell recorded long-term weather data in Monteverde, and J. T. Longino made suggestions on the manuscript. This work was supported by Research Grants and Research Experience for Undergraduate Grant Supplements from the National Science Foundation (DEB 9615341, BIR 9975510, DEB 9977035), an OPUS grant from the National Science Foundation (DEB 05–42130), the Committee for Research and Exploration of the National Geographic Society, and a Guggenheim Fellowship. The Helen R. Whiteley Center provided support for reflection and writing.

REFERENCES

Asbury, C. E., W. H. McDowell, R. Trinidad-Pizarro, and S. Berrios (1994). Solute deposition from cloud water to the canopy of a Puerto Rican montane forest. *Atmospheric Environment* **28**: 1773–1780.

Bawa, K. S., and A. Markham (1995). Climate change and tropical forests. *Trends in Ecology and Evolution* **10**: 348–349.

Benisten, H., F. Diaz, and R. S. Bradley (1997). Climatic change at high elevations sites: an overview. *Climate Change* **36**: 233–251.

Benzing, D. H. (1998). Vulnerabilities of tropical forest to climate change: the significance of resident epiphytes. *Climate Change* **39**: 519–540.

Bohlman, S., T. Matelson, and N. Nadkarni (1995). Moisture and temperature patterns of canopy humus and forest floor soils of a montane cloud forest, Costa Rica. *Biotropica* **27**: 13–19.

Boy, J., R. Rollenbeck, C. Valarezo, and W. Wilcke (2008). Amazonian biomass burning-derived acid and nutrient deposition in the north Andean montane forest of Ecuador. *Global Biogeochemical Cycles* **22**, GB4011, doi:10.1029/2007GB003158.

Bruijnzeel, L. A., and L. S. Hamilton (2000). *Decision Time for Cloud Forests*, IHP Humid Tropics Programme Series No. 13. Paris: UNESCO, Amsterdam, the Netherlands: IUCN-NL, and Gland, Switzerland: WWF International.

Bubb, P., I. May, L. Miles, and J. Sayer (2004). *Cloud Forest Agenda*. Cambridge, UK: UNEP-WCMC. Also available at: http://sea.unep-wcmc.org/forest/cloudforest/index.cfm.

Campbell, G. S., and G. W. Gee (1986). Water potential: miscellaneous method. In *Methods of Soil Analysis*, Part 1, *Physical and Mineralogical Methods*, ed. A. Klute, pp. 619–633. Madison, WI: American Society of Agronomy.

Clark, K. L., N. M. Nadkarni, D. Schaefer, and H. L. Gholz (1998a). Cloud water and precipitation chemistry in a tropical montane forest, Monteverde, Costa Rica. *Atmospheric Environment* **32**: 1595–1603.

Clark, K. L., N. M. Nadkarni, D. Schaefer, and H. L. Gholz (1998b). Atmospheric deposition and net retention of ions by the canopy in a tropical montane forest, Monteverde, Costa Rica. *Journal of Tropical Ecology* **14**: 27–45.

Clark, K. L., R. O. Lawton, and P. R. Butler (2000). The physical environment. In *Monteverde: Ecology and Conservation of a Tropical Cloud Forest*, eds. N. Nadkarni and N. Wheelwright, pp. 15–38. New York: Oxford University Press.

Connelly, M. A., and M. L. Crump (1998). Potential effects of climate change on two neotropical amphibian assemblages. *Climatic Change* **39**: 541–561.

During, J., and B. F. van Tooren (1990). Bryophyte interactions with other plants. *Botanical Journal of the Linnean Society* **140**: 79–98.

Falconer, E., and P. D. Falconer (1980). Determination of cloud water acidity at a mountain observatory in the Adirondack Mountains in New York State. *Journal of Geophysical Research* **85**: 7565–7470.

Feild, T. S., and T. E. Dawson (1998). Water sources used by *Didymopanax pittieri* at different life stages in a tropical cloud forest. *Ecology* **79**: 1448–1452.

Guswa, A. J., A. L. Rhodes, and S. E. Newell (2007). Importance of orgraphic precipitation to the water resources of Monteverde, Costa Rica. *Advances in Water Resources*, **30**: 2098–2112.

Hafkenscheid, R. L. L. J. (2000). Hydrology and biogeochemistry of tropical montane rain forests of contrasting stature in the Blue Mountains, Jamaica. Ph.D. thesis, VU University Amsterdam, Amsterdam, the Netherlands. Also available at http://dare.ubvu.vu.nl/bitstream/1871/12734/1/tekst.pdf.

Hietz, P., W. Wanek, and M. Popp (1999). Stable isotopic composition of carbon and nitrogen, and nitrogen content in vascular epiphytes along an altitudinal transect. *Plant, Cell, and Environment* **22**: 1435–1443.

Ingram, S., and N. M. Nadkarni (1993). Composition and distribution of epiphytic organic matter in a Neotropical cloud forest, Costa Rica. *Biotropica* **25**: 370–383.

Jacome Reyes, J. H. (2005). *Factors controlling the lower elevational limits in tropical montane plants in the Andes and their implications under the current climatic change*. Ph.D. thesis, University of Göttingen, Göttingen, Germany.

Karmalkar, A. V., R. S. Bradley, and H. F. Diaz (2008). Climate change scenario for Costa Rican montane forests. *Geophysical Research Letters* **25**, L11702, doi: 10.1029/2008GL033940.

Köhler, L., C. Tobón, K. F. A. Frumau, and L. A. Bruijnzeel (2007). Biomass and water storage of epiphytes in old-growth and secondary montane rain forest stands in Costa Rica. *Plant Ecology* **193**: 171–184.

Küper, W., H. Kreft, J. Nider, N. Köster, and W. Barthlott (2004). Large-scale diversity of vascular epiphytes in Neotropical montane rain forests. *Journal of Biogeography* **31**: 1477–1487.

Lakatos, M., B. Hartard, and C. Máguas (2007). The stable isotopes $\delta^{13}C$ and $\delta^{18}O$ can be used as tracers of micro-environmental carbon and water sources. In *Isotopes as Tracers*, eds. T. E. Dawson and R. T. W. Siegwolf, pp. 73–88. Amsterdam, the Netherlands: Elsevier.

Lawton, R. O., and V. Dryer (1980). The vegetation of the Monteverde Cloud Forest Preserve. *Brenesia* **18**: 101–116.

Lawton, R. O., U. S. Nair, R. A. Pielke Sr., and R. M. Welch (2001). Climatic impact of tropical lowland deforestation on nearby montane cloud forests. *Science* **294**: 584–587.

Loope, L. L., and T. W. Giambelluca (1998). Vulnerability of island tropical montane forest to climate change, with special reference to East Maui, Hawaii. *Climatic Change* **39**: 503–517.

Lugo, A. E., and F. N. Scatena (1992). Epiphytes and climate change research in the Caribbean: a proposal. *Selbyana* **13**: 123–130.

Matelson, T. J., N. M. Nadkarni, and J. T. Longino (1993). Survivorship of fallen epiphytes in a neotropical cloud forest, Monteverde, Costa Rica. *Ecology* **74**: 265–269.

Nadkarni, N. M., and J. T. Longino (1990). Macroinvertebrate communities in canopy and forest floor organic matter in a montane cloud forest, Costa Rica. *Biotropica* **22**: 286–289.

Nadkarni, N. M., and T. J. Matelson (1991). Litter dynamics within the canopy of a neotropical cloud forest, Monteverde, Costa Rica. *Ecology* **72**: 2071–2082.

Nadkarni, N. M., and R. Solano (2002). Potential effects of climate change on canopy communities in a tropical cloud forest: an experimental approach. *Oecologia* **131**: 580–584.

Nadkarni, N. M., T. J. Matelson, and W. A. Haber (1995). Structural characteristics and floristic composition of a neotropical cloud forest, Monteverde, Costa Rica. *Journal of Tropical Ecology* **11**: 481–95.

Nadkarni, N. M., R. O. Lawton, K. L. Clark, T. J. Matelson, and D. A. Schaefer (2000). Ecosystem ecology. In *Monteverde: Ecology and Conservation of a Tropical Cloud Forest*, eds. N. M. Nadkarni and N. T. Wheelwright, pp. 303–350. New York: Oxford University Press.

Nair, U. S., R. O. Lawton, R. M. Welch, and R. A. Pielke Sr. (2003). Impact of land use on tropical montane cloud forests: sensitivity of cumulus cloud field characteristics to lowland deforestation. *Journal of Geophysical Research* **108** (D7): 4206–4218.

Nair, U. S., S. Asefi, R. M. Welch, *et al.* (2008). Biogeography of tropical montane cloud forests. II. Mapping of orographic cloud immersion. *Journal of Applied Meteorology and Climatology* **47**: 2183–2197.

Odum, H. T., G. A. Briscoe, and C. B. Briscoe (1970). Fallout radioactivity and epiphytes. In *A Tropical Rainforest*, eds. H. T. Odum and R. F. Pigeon, Chapter H-23. Springfield, VA: National Technical Information Service.

Perry, D. (1978). A method of access into the crowns of emergent and canopy trees. *Biotropica* **10**: 155–157.

Ponette-González, A. G., K. C. Weathers, and L. M. Curran (2009). Water inputs across a tropical montane landscape in Veracruz, Mexico: synergistic effects of land cover, rain and fog seasonality, and interannual precipitation variability. *Global Change Biology*, doi: 10.1111/j.1365-2486.2009.01985.x.

Pounds, J. A., M. P. L. Fogden, and J. H. Campbell (1999). Biological response to climate change on a tropical mountain. *Nature* **789**: 611–614.

Pounds, J. A., M. R. Bustamante, L. A. Coloma, *et al.* (2006). Widespread amphibian extinctions from epidemic disease driven by global warming. *Nature* **439**: 161–167.

Rao, D. N. (1982). Response of bryophytes to air pollution. In *Bryophyte Ecology*, ed. A. J. E. Smith, pp. 445–471. London: Chapman and Hall.

Ray, D. K., U. S. Nair, R. O. Lawton, R. M. Welch, and R. A. Pielke Sr. (2006). Impact of land use on Costa Rican tropical montane cloud forests: sensitivity of orographic cloud formation to deforestation in the plains. *Journal of Geophysical Research* **111**: D02108, doi: 10.1029/2005JD006096.

Rhoades, F. M. (1995). Non-vascular epiphytes in forest canopies: worldwide distribution, abundance, and ecological roles. In *Forest Canopies*, eds. M. D. Lowman and N. M. Nadkarni, pp. 353–408. San Diego, CA: Academic Press.

Sohlberg, H., and L. C. Bliss (1987). Responses of *Ranunculus sabinei* and *Papaver radicatum* to removal of the moss layer in a high-arctic meadow. *Canadian Journal of Botany* **65**: 1224–1228.

Still, C. J., P. N. Foster, and S. H. Schneider (1999). Simulating the effects of climate change on tropical montane cloud forests. *Nature* **789**: 608–610.

Vance, E., and N. M. Nadkarni (1990). Microbial biomass and activity in canopy organic matter and the forest floor of a tropical cloud forest. *Soil Biology and Biochemistry* **22**: 677–684.

Weathers, K. C. (1999). The importance of cloud and fog in the maintenance of ecosystems. *Trends in Ecology and Evolution* **14**: 214–215.

Zamfir, M. (2000). Effects of bryophytes and lichens on seedling emergence of alvar plants: evidence from greenhouse experiments. *Oikos* **88**: 603–611.

Zotz, G., and M. Y. Bader (2009). Epiphytic plants in a changing world: global change effects on vascular and non-vascular epiphytes. *Progress in Botany* **70**: 147–170.

58 Climatic change impacts on tropical montane cloud forests: fire as a major determinant in the upper zones of Mount Kilimanjaro, Tanzania

A. Hemp
University of Bayreuth, Bayreuth, Germany

ABSTRACT

Cloud forests are of great importance for the hydrological functioning of watersheds in sub-humid East Africa. However, the montane forests of Mt. Kilimanjaro are heavily threatened, at lower elevations by illegal logging and at higher elevations by an increasing fire risk. Based on an evaluation of over 1400 vegetation plots and interpretation of satellite imagery from 1976 and 2000, land-cover changes on Kilimanjaro were evaluated and their impact on the water balance estimated. In recent years, a drier climate has facilitated the spreading of human-caused fires dramatically, and has led to the almost complete destruction of the former closed forest between 3000 m.a.s.l. and the potential tree-line at 4000 m.a.s.l. Therefore, the current upper closed forest limit at 3200 m.a.s.l. has become suppressed by nearly 800 m. Since 1976, about 150 km^2 of cloud forests have been destroyed by fire, possibly representing an estimated annual loss of 20 million m^3 of water potentially captured by "fog stripping" under forested conditions. Such losses are potentially much more harmful hydrologically speaking than those associated with the loss of the showy ice cap on Kilimanjaro and highlight the importance of quantitative hydrological investigations in support of montane forest conservation and management in the region.

INTRODUCTION

During the last 120 years, annual precipitation on Mt. Kilimanjaro (Tanzania) has decreased by 600–1200 mm (Hemp, 2005a), whereas since 1976 temperatures have increased drastically (Altmann *et al.*, 2002). Both these climatic changes do not only cause the glaciers on the mountain to retreat (Kaser *et al.*, 2004), but also cause fires on the higher slopes of the mountain to become more aggressive and devastating, resulting in a potentially much greater impact on the overall Kilimanjaro ecosystem than that of the melting glaciers. Although most of these fires are lit by the carelessness of humans (e.g. honey collectors or poachers; Hemp and Beck, 2001), they would not be so devastating had the climate not become drier. Whilst the lower and middle montane forests on Kilimanjaro are heavily threatened by illegal logging (Lambrechts *et al.*, 2002), the drier upper montane and sub-alpine zones in particular are strongly influenced by fire, as will be shown further below. The aim of this chapter is to assess the fire-caused changes in land cover in the upper forest zone on the mountain during the last three decades, and their respective impacts on the water balance of Kilimanjaro, particularly with respect to possible changes in fog water inputs and hence groundwater recharge and streamflow.

STUDY AREA

Natural forests cover an area of about 1000 km^2 on Mt. Kilimanjaro (~25% of the cross-sectional area of the mountain). The lower parts of the forests on the (wetter) southern slope are characterized by the dominance of the camphor tree (*Ocotea usambarensis*) and higher up by *Podocarpus latifolius*, *Hagenia abyssinica*, and *Erica excelsa* (cf. Table 58.2 below). On the drier northern slope, forests are composed of *Croton–Calodendrum* forests at lower elevations, followed by *Cassipourea* forest at mid-altitudes and *Juniperus* forest at higher altitudes. Above 3200 m.a.s.l. the cloud forests are replaced by *Erica* heathlands

Tropical Montane Cloud Forests: Science for Conservation and Management, eds. L. A. Bruijnzeel, F. N. Scatena, and L. S. Hamilton. Published by Cambridge University Press.

with *E.arborea*, *E. trimera*, *Protea caffra*, and *Euryops dacrydioides* comprising the dominant species. At an altitude of about 3900 m.a.s.l. the *Erica* heathlands grade into a *Helichrysum* cushion vegetation which extends up to 4500 m.a.s.l. The higher altitudes are very poor in vegetation and the top of Kibo, the main peak, is covered with glaciers. For a more detailed description of these vegetation types see Hemp (2001, 2006b) and Hemp (this volume, #12). Basic climatic information for the montane zone can be found in the latter papers as well as in Hemp (2005a).

METHODS

Between 1996 and 2006, over 1400 vegetation plots were examined along 34 elevational transects (see Figure 12.1 in Hemp, this volume #12). Based on these inventories and an interpretation of a Landsat ETM image from 2000 (using IDRISI 3.2 software), a vegetation map was produced (Hemp, 2006b), which was then compared with a Landsat MSS image from 1976 to evaluate changes in vegetation cover between the two years.

Furthermore, since 1997 basic climatic measurements (rainfall, temperature, and humidity) have been taken at 38 stations distributed along six transects inside the forest belt. These data, together with data provided by the Tanzania Meteorological Agency for the lower areas, were used to produce a general annual rainfall map (Hemp, 2006b). Based on the vegetation map, the altitudinal forest zonation, and rainfall data, the Kilimanjaro forest belt was divided into 10 eco-climatic zones for which annual rainfall inputs were compiled. In the absence of direct data on forest water use (cf. Schrumpf *et al.*, this volume), evapotranspiration (*ET*) was estimated from the literature as follows: (i) for altitudes between 1300 and 2500 m.a.s.l., data from various East African mountains were used (Blackie, 1979a, b; Edwards, 1979); (ii) for the upper montane zone between 2500 and 2800 m.a.s.l., the 980 mm derived by Steinhardt (1979) for a cloud forest site at 2300 m.a.s.l. in Venezuela with a similar annual rainfall of 1575 mm was assumed to be appropriate; (iii) for the altitudinal zone between 3200 and 3500 m.a.s.l., the 390 mm estimated for a sub-alpine cloud forest at 3500 m.a.s.l. in Costa Rica (Dohrenwend, 1979) was used, whereas (iv) a value of 680 mm year^{-1} was interpolated for the 2800–3200 m.a.s.l. interval (cf. Table 58.2 below).

No data on annual amounts of cloud water interception appear to exist for East Africa. In other tropical areas, fog incidence (usually measured by a passive fog gage of some kind; Bruijnzeel *et al.*, 2005; cf. Frumau *et al.*, this volume; Giambelluca *et al.*, this volume) has been estimated at 4–35% of ordinary rainfall (i.e. between 100 and over 800 mm year^{-1}; Bruijnzeel and Proctor, 1995; Bruijnzeel, 2001). Since strong winds are rare on Kilimanjaro, fog water interception is likely to be low, and perhaps does not exceed 10% of rainfall. Based on the author's visual observations of fog frequency on Mt. Kilimanjaro and the elevational distribution of certain moisture-loving epiphytes (e.g. filmy ferns, lichens, moss balls; Hemp, 2001, cf. Hemp, this volume #12), tentative fog water inputs to the respective zones were estimated to be as follows (cf. Table 58.2 below): (i) negligible in eco-climatic zones 1, 5, and 6 between 1300 and 2200 m.a.s.l. (cf. Schrumpf *et al.*, this volume); (ii) 5% in zones 2 and 7 between 2200 and 2500 m.a.s.l.; and (iii) 10% ($\leq$200 mm) in the remaining zones above 2500 m.a.s.l. which are commonly enveloped by fog during much of the day. These estimates are considered to be conservative, since preliminary measurements made at 1800 m.a.s.l. within eco-climatic zone 1 from September 2005 to June 2006 using the adapted Juvik-type gage described by Frumau *et al.* (this volume) suggested a horizontal precipitation fraction of about 10% of ordinary rainfall instead of being negligible as assumed in Table 58.2 below (A. Hemp and L. A. Bruijnzeel, unpublished data).

RESULTS

Impact of fire on the upper forest belt

Induced by a reduction in precipitation above the major cloud belt (Hemp, 2005a, 2006a), fire has caused sharp discontinuities in composition and structure of the tall (20–30 m) upper montane *Hagenia–Podocarpus* cloud forests at 2800–3000 m.a.s.l. The giant heather *Erica excelsa* becomes dominant at this altitude, forming dense mono-specific stands of about 10 m height that structurally resemble elfin cloud forest. The occurrence of *Erica* forest is an obvious indication of the prevalence of fire. During longer periods of a dry climate with recurrent fire, *Erica* forest moves downslope, replacing the *Podocarpus* forest, which advances upslope again during wetter periods. The presence of *Erica* enhances the fire risk, since even fresh *Erica* wood burns well, and this, in turn, prevents the *Podocarpus* forest from re-establishing (Hemp and Beck, 2001). At high fire frequencies, the closed *Erica excelsa* forest degrades into open shrub land (height 1.5 m) dominated by *E. trimera* and *E. arborea* between 3200 and 4000 m.a.s.l. (the potential tree-line; Hemp, this volume #12; Figure 58.1). A very high fire frequency may completely destroy this shrubland as well, which is then replaced by *Helichrysum* cushion vegetation (Figure 58.2), the climatic climax vegetation at altitudes above 4000 m.a.s.l. (cf. Figure 58.3 below).

Comparison of the 1976 and 2000 Landsat images reveals enormous changes in the upper vegetation zones of Mt. Kilimanjaro during the intervening 24 years (Hemp, 2005a; Figure 58.3). In 1976, the *Erica trimera* shrubs – today occurring only below 3800 m.a.s.l. – still reached an altitude of 4100 m.a.s.l., in part forming a continuous belt in areas that today are covered by

Figure 58.1. *Erica trimera* and *E. arborea* (flowering on the right-hand side) shrubs at 3200 m.a.s.l. substituting sub-alpine forest on Mt. Kilimanjaro (photograph by A. Hemp). (See also color plate.)

Figure 58.2. *Helichrysum* cushion vegetation with *H. newii*, *H. citrispinum*, and *H. forskahlii* at 4100 m.a.s.l. on Mt. Kilimanjaro. The grasses are *Pentaschistis borussica* and *Festuca abyssinica*. *Helichrysum newii* cushions form lines exposed to the windward direction. This growth form, which cannot be explained by solifluction, apparently enables the plants to capture fog moving upslope (photograph by A. Hemp). (See also color plate.)

Helichrysum cushion vegetation. Closed *Erica* forests in 1976 covered nearly six times the area of today (187 and 32 km^2, respectively), extending in many places up to 3800 m.a.s.l. (Table 58.1). Summarizing, nearly 15% of Kilimanjaro's forest cover has been destroyed by fire since 1976.

Tentative water budget for the forest belt

Maximum amounts of annual rainfall occur in the middle montane zone of the southern slope, where about 3000 mm were recorded at 2200 m.a.s.l., thereby clearly exceeding precipitation totals on other East African high mountains (Hemp, 2001, 2005a; cf. Røhr, 2003; Schrumpf *et al.*, this volume). With increasing elevation, rainfall expressed as a proportion of this maximum declined steadily: 80% at 2400 m.a.s.l., 70% at 2700 m, 50% near the upper forest border at 3000 m, and only 20% at 4000 m. From 2200 m.a.s.l. upward, fog occurs frequently, and may well represent a substantial water input (Table 58.2). Based on their estimated fog and rainfall inputs, the forests below 2200 m.a.s.l. may be classified as "ordinary" montane rain forests (cf. Schrumpf *et al.*, this volume). The cloud forests, comprising different floristically defined middle and upper montane and sub-alpine forest associations above 2200 m.a.s.l. (see Hemp (this volume #12) for details), can be divided into a wet zone between 2200 and 2800 m.a.s.l. with high rainfall and a wealth of vascular epiphytes, and a comparatively dry zone above 2800 m.a.s.l. (Hemp, 2002, 2005a).

Table 58.1 *Summary of land-cover changes between 1976 and 2000 on the upper slopes of Mt. Kilimanjaro*

Vegetation type	1976 (ha)	2000 (ha)	Change (%)
Montane forest	106 624	97 410	−9
Sub-alpine *Erica* cloud forest	18 657	3212	−83
Erica bush	20 166	25 675	27
Helichrysum cushion vegetation	6928	21 755	214
Grassland	9039	4396	−51

Taking the data presented for eco-climatic zones 1–9 in Table 58.2 at face value, the forests of Kilimanjaro above 1300 m.a.s.l. receive nearly 1600 million m^3 water annually, of which 95% is estimated to be supplied by rainfall and *c.* 5% by fog (indicated as "rain" and "fog water," respectively, in Table 58.2). About 500 million m^3 of water (31%) are estimated to percolate toward groundwater or streams ("water outflow" in Table 58.2).

Reduction in fog water input to burnt forest areas

Assuming fog capture to be close to zero once the forest is destroyed, the loss of 150 km^2 of high-altitude forests since 1976 (100 km^2 in eco-climatic zone 4 and 50 km^2 in zone 10, Table 58.1) would correspond to an estimated volume of ~19 million m^3 of non-captured fog water per year (Table 58.2). This loss represents more than a quarter of the estimated annual fog capture by the entire forest belt (73 million m^3) and is roughly equivalent to the annual water demand of the 1 million inhabitants on Kilimanjaro (CES, 2002). However, it is pertinent to note that the previous calculation is based on estimated amounts of fog capture by passive fog gages rather than by a live canopy (Ponette-González *et al.*, 2009; see also discussion below).

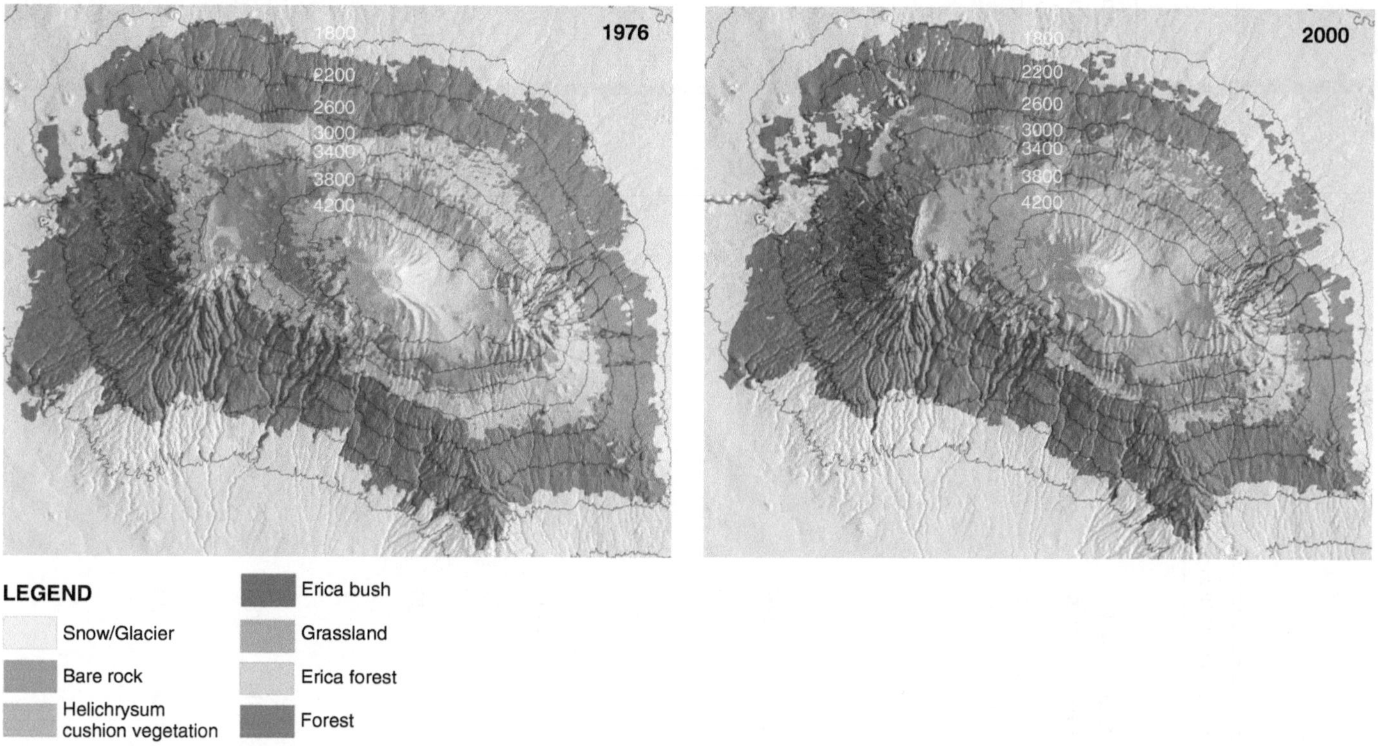

Figure 58.3. Changes in vegetation cover in the montane and alpine zone on Mt. Kilimanjaro. The maps result from supervised classification of Landsat MSS images taken on 24 January 1976 and Landsat ETM images taken on 29 January and 21 February 2000 (source: USGS/UNEP-GRID-Sioux Falls) using the software IDRISI 3.2. The entire sub-alpine ericaceous belt was shifted downslope and became replaced by *Helichrysum* cushion vegetation. Closed *Erica excelsa* forests were reduced to 17% of their land cover in 1976. (See also color plate.)

DISCUSSION

Impact of fire on the composition of the upper vegetation zones

Fire not only transforms land cover, it also maintains certain land-cover types (Eva and Lambin, 2000) and influences species diversity and vegetation structure in the different altitudinal zones on Mt. Kilimanjaro in various ways, particularly in the upper forest and shrub belt (Hemp and Beck, 2001; Hemp, 2005b). From field observations and historic descriptions (Jaeger, 1909; Klute, 1920) it can be assumed that different types of tall forest extended up to 3800 m.a.s.l. in many areas of Kilimanjaro at the beginning of the twentieth century and even until 1996 (Hemp, 2005a). In 2003, isolated remnants of closed 10-m tall *Erica* forest were even found as high as 4000 m.a.s.l. (Hemp, 2006a), i.e. ~800 m higher than the present closed-forest line. The fires of 1996 and 1997 destroyed vast areas of old-growth *Erica* forest, which were replaced by low-stature *Erica* shrub and which moved the upper closed-forest line down-slope by about 600 m of altitude (Figure 58.3).

Whilst alpine vegetation on mountains in the temperate zone has been found to migrate upslope due to rising temperatures and enrichment of atmospheric CO_2 and nitrogen, in the case of pioneer species (Grabherr and Pauli, 1994), or has remained more or less stable in the case of late-successional communities (Körner, 2003), alpine vegetation on Kilimanjaro also migrates downslope, thereby replacing upper montane and sub-alpine cloud forests. This is due to a typical feature of high mountains in Africa: the presence of a vast ericaceous belt (Hedberg, 1951) that is highly inflammable and increasingly susceptible to fire under warmer and drier conditions.

The present findings are in stark contrast to those of Shugart *et al.* (2001) who reported an upslope movement of vegetation zones on Mt. Kilimanjaro, stressing the above-mentioned atmospheric and climatic factors and comparing satellite images from 1984 and 1993. However, it seems that Shugart *et al.* (2001) only studied two very small areas and did not consider the fact that the vegetation in the sub-alpine zone of Mt. Kilimanjaro comprises a mosaic of fire-induced regeneration stages. As can be seen in Figure 58.3, it is always possible to find places within such a mosaic where the vegetation is recovering. Extensive ground surveys are needed to "calibrate" such remotely sensed data. A full coverage of the upper forest belt by satellite- and ground-based data leaves no doubt about the massive decline in forest cover due to fires.

In addition to the loss of 150 km^2 of upper montane and sub-alpine forests by fire since 1976, and of an additional estimated

Table 58.2 *Eco-climatic forest zones on Mt. Kilimanjaro and their approximate water budget*

	Landscape characteristics					Climatic features			Water balance (Mill.m^3)				
									Input			Output	
Eco-climatic zone	**Vegetation**	**Altitudinal zone**[a]	**Altitude (m asl)**[a]	**Location**	**Area (km^2)**[b]	**Rain (mm)**[c]	**ET (mm)**	**Fog (%)**[d]	**Rain**	**Fog-water**	**ET**	**Water outflow**	**Water outflow km^{-2}**
1	*Ocotea–Agauria* forest, *Ocotea–Syzygium* forest, consisting of *Ocotea usambarensis* associated with *Agauria salicifolia, Syzygium guineense, Macaranga kilimandscharica*, and *Polyscias fulva*	Lower montane	1800 – 2200	South/ south-east	153	2400	1280 [e]	0	367	0	196	171	1,1
2	*Ocotea–Podocarpus* forest, consisting of *Ocotea usambarensis* associated with *Podocarpus latifolius* and the tree fern *Cyathea manniana*	Middle montane	2200 – 2500	South/ south-east	108	2500	1280[e]	5	270	14	139	145	1,3
3	*Podocarpus–Ocotea* forest with prevailing *Podocarpus latifolius*	Upper montane	2500 – 2800	South/ south-east	99	2000	980[f]	10	198	20	97	121	1,2
4	*Hagenia–Podocarpus* forest with *Podocarpus latifolius, Hagenia abyssinica*, and *Prunus africana; Erica excelsa* forest	Upper montane	2800 – 3200	South/ south-east	66	1450	680[g]	10	96	10	45	61	0,9
5	*Croton–Calodendrum* forest with *Olea europaea* ssp. *africana, Croton megalocarpus, Calodendrum capense*, and *Diospyros abyssinica*	Sub-montane	1300 – 1600 1600 – 2000	West North	81	1000	1280[e]	0	81	0	104	−23	−0,3
6	*Cassipourea forest* with *Cassipourea malosana, Teclea*	Lower and middle montane	1600 – 2100	West	93	1100	1280[e]	0	102	0	119	−17	−0,2
7	*simplicifolia, Fagaropsis angolensis*, and *Olea capensis*		2000 – 2500	North	192	1200	1280[e]	5	230	12	246	−4	0,0

8	*Juniperus–Podocarpus* forest with *Juniperus procera*,	Upper montane	2500 – 2800	North	117	1100	980[f]	10	129	13	115	27	0,2
9	*Podocarpus latifolius*, and *Hagenia abyssinica*	Upper montane	2800 – 3100	North	36	1000	680[g]	10	36	4	24	16	0,4
Destroyed forest types since 1976													
10	*Erica* and *Hagenia* forest	Upper montane	3200 – 3500	Around the mountain	50	950	390[g]	10	48	5	20	33	0,7
4	*Hagenia–Podocarpus* forest with *Podocarpus latifolius*, *Hagenia abyssinica*, and *Prunus africana; Erica excelsa* forest	Upper montane	2800 – 3200	South/ south-east	100	1450	680[g]	10	145	15	68	92	0,9

[a] According to Hemp (2005c).
[b] Based on the vegetation map (Hemp, 2006b).
[c] Mean annual rainfall, based on own measurements.
[d] Fog water interception in relation to rainfall, estimation based on measured data presented by Bruijnzeel (1990) and Bruijnzeel (pers. com.).
[e] Evapotranspiration according to data for East Africa given by Bruijnzeel (1990).
[f] According to data by Steinhardt (1979).
[g] According to data by Dohrenwend (1979).

150 km^2 lost since 1880, losses due to clear-cutting of lower-elevation forests since 1929 amount to *c*. 450 km^2, bringing the total loss to *c*. 750 km^2 (or about half the mountain's original forest cover; Hemp *et al.*, in press). It is pertinent to note that this total does not include the massive logging going on inside the current forest belt as documented during an aerial survey in 2001, when about 8000 newly cut stumps were identified (Lambrechts *et al.*, 2002).

Impact of fire and subsequent landscape changes on the water balance

Cloud forests are of great importance for the functioning of watersheds in sub-humid East Africa (Pócs, 1976; Røhr, 2003). Fog interception constitutes an important additional source of water at (sub-)tropical montane sites with regular and frequent occurrence of fog but low rainfall (e.g. Cavelier and Goldstein, 1989; Juvik and Nullet, 1995; Juvik, 2009; Giambelluca *et al.*, this volume). Whilst the greatest absolute additions have been observed at wet and windy locations, the largest relative increases tend to be reported for sub-humid locations (Bruijnzeel and Proctor, 1995; Bruijnzeel, 2001; cf. García-Santos, 2007; Marzol-Jaén *et al.*, this volume). Thus, the loss of cloud forests from the drier, upper zones of Mt. Kilimanjaro due to climate-induced fires might cause a reduction in overall water yield due to the loss of the associated cloud water capture.

Whilst the inferred loss of cloud water capture on Mt. Kilimanjaro is substantial (Table 58.2), it is important to remember that the underlying calculations are based on potential amounts of fog capture as would be measured by passive fog gages rather than actually intercepted by a live forest canopy. Although the latter tend to be much less than those measured by a typical fog gage (Bruijnzeel *et al.*, 2005; Ponette-González *et al.*, 2009), the coniferous leaves of *Erica* should render it a highly efficient catcher of fog droplets (Kämmer, 1974; Prada *et al.*, 2009; cf. García-Santos and Bruijnzeel, 2007). Also, the mono-dominant *E. excelsa* forests considered here have very low epiphyte biomass that otherwise might have absorbed and retained a substantial portion of the intercepted cloud water (cf. Köhler *et al.*, 2007; Tobón *et al.*, this volume #26). *Erica excelsa*, with its thin and fragile twigs and branches, is not strong enough to carry heavy epiphytic loads. Furthermore, its branches ascend steeply and do not spread horizontally whereas they have a peeling bark, all features that are not conducive to colonization by epiphytes (Hemp, 2002). Thus, epiphyte-related fog water retention is most likely of minor importance in this forest type and even more so if the old-growth *E. excelsa* forest (typical canopy height ~10 m) is replaced by short *E. trimera* and *E. arborea* bush (~2 m high).

When *Erica* shrubs are replaced by 20–30-cm-high *Helichrysum* cushion vegetation, the fog stripping capacity is lost almost entirely, at least in quantitative terms. However, even for this low-stature vegetation type, horizontal precipitation seems to be of ecological importance. *Helichrysum newii* cushions were found to form lines exposed to the windward direction (Figure 58.2). Circular and crescent-shaped growth can be observed in several afro-alpine taxa on Mt. Kilimanjaro (cf. Furrer and Freund, 1973), when tussocks die in the center and expand outward. In this case the *Helichrysum* cushions form well-spaced crescents and bands parallel to the contour lines. This growth form and orientation, which cannot be explained by soil movement (solifluction), would act to maximize the capturing of fog moving upslope. It is also reminiscent of the self-organized patches of "tiger bush" in arid West Africa where bands of vegetation orientated along the contour capture any overland flow generated on bare intermediate patches of ground (Rietkerk *et al.*, 2004; Van de Koppel and Rietkerk, 2004).

Consideration of the overall effect of vegetation changes on soil and groundwater recharge also needs to take into account any differences in water use by the respective vegetation types. Transpiration in stunted fog-ridden upper montane and sub-alpine cloud forests is generally low (Bruijnzeel and Proctor 1995; Santiago *et al.*, 2000; Holwerda, 2005; McJannet *et al.*, 2007; cf. García-Santos, 2007; Santiago *et al.*, this volume). Walter and Breckle (1984) measured very low transpiration values for four species within ericaceous vegetation on Mt. Kilimanjaro. The higher leaf biomass of *Erica* forests compared to *Erica* shrubs is likely to result in higher total transpiration by the former but presumably values would still be relatively low (Walter and Breckle, 1984). Furthermore, evaporation from the ground layer within *Erica* forest is also expected to be lower due to the denser canopy. Therefore, unlike the suggestion made in Table 58.2 for eco-climatic zones 10 and 4, fog water inputs to *Erica* cloud forest might in fact exceed its water use although this would need to be substantiated by field measurement.[1]

CONCLUSIONS

Drastic land-cover changes have occurred during the last decades in the cloud forest belt on Mt. Kilimanjaro due to increasingly frequent fires induced in turn by changes in climate. Combining newly measured rainfall data, literature data on tropical montane forest evaporation, and expert judgement on potential amounts of cloud water interception on Mt. Kilimanjaro, an attempt was made to assess the hydrological implications of the loss of the

[1] Planning of new research on the climate dynamics (including fog incidence), and the chief controls on water and carbon fluxes along the elevational gradient in the Kilimanjaro region is in an advanced stage (Nauss *et al.*, 2009; Tenhunen *et al.*, 2009).

upper cloud forests. Whilst only approximate at this stage, the available information suggests that the disappearance of large tracts of high-elevation cloud forest may indeed have caused a significant reduction in the amount of cloud water captured, possibly affecting the water supply to over 1 million people living on the mountain. However, to obtain a more accurate assessment, measurements of actual fog water inputs to – and evaporative losses from – the mountain forests of Mt. Kilimanjaro remain most necessary. At any rate, compared with the roughly 500 million m^3 of water emanating annually from the 1000 km^2 of forest remaining to date, the hydrological consequences of the loss of the residual 2.6 km^2 of glaciers on the mountain are almost negligible (Kaser *et al.* 2004; Hemp 2005a). Nevertheless, the melting glaciers are an alarming indicator of the substantial changes in the overall climate of Kilimanjaro. Also, at current rates of incidence, fires will cause the extinction of most of Kilimanjaro's high-altitude forests within the next decades. Along with this, its most effective water source in the fog interception zone will be substantially reduced at the very least. Such findings highlight the importance of proper forest conservation and management measures for the mountain as already proposed in a case study for the Organization for Economic Co-operation and Development (OECD) (Hemp 2003).

ACKNOWLEDGEMENTS

Field work was carried out with financial support from the Deutsche Forschungsgemeinschaft and with permission from the Tanzanian Commission for Science and Technology. For further support I owe gratitude to the Chief Park Warden of Kilimanjaro National Park, Mr. Moirana and Mr. Mafuru, to my research counterpart Mr. Jacob Mushi, and to Mr. Christian Lambrechts (Nairobi). For their valuable comments on the manuscript I am grateful to Dr. Heidi Asbjornsen, and to Professors Lawrence Hamilton and Sampurno Bruijnzeel.

REFERENCES

Altmann, J., S. C. Alberts, S. A. Altmann, and S. B. Roy (2002). Dramatic change in local climate patterns in the Amboseli basin, Kenya. *African Journal of Ecology* **40**: 248–251.

Blackie, J. R. (1979a). The water balance of the Kericho catchments. *East African Agricultural and Forestry Journal* **43**: 55–84.

Blackie, J. R. (1979b). The water balance of the Kimakia catchments. *East African Agricultural and Forestry Journal* **43**: 155–174.

Bruijnzeel, L. A. (1990). *Hydrology of Moist Tropical Forests and Effects of Conversion: A State-of-Knowledge Review.* Paris: UNESCO International Hydrological Programme.

Bruijnzeel, L. A. (2001). Hydrology of tropical montane cloud forests: a reassessment. *Land Use and Water Resources Research* **1**: 1–18.

Bruijnzeel, L. A., and J. Proctor (1995). Hydrology and biogeochemistry of tropical montane cloud forests: what do we really know? In *Tropical Montane Cloud Forests*, eds. L. S. Hamilton, J. O. Juvik, and F. N. Scatena, pp. 38–78. New York: Springer-Verlag.

Bruijnzeel, L. A., W. Eugster, and R. Burkard (2005). Fog as an input to the hydrological cycle. In *Encyclopaedia of Hydrological Sciences*, eds. M. G. Anderson and J. J. McDonnell, pp. 559–582. Chichester, UK: John Wiley.

Cavelier, J., and G. Goldstein (1989). Mist and fog interception in elfin cloud forests in Columbia and Venezuela. *Journal of Tropical Ecology* **5**: 309–322.

CES (Consulting Engineers) (2002). *Hydrometic Report of the Hai District Water Supply Phase IV (2002).* Dar es Salaam, Tanzania: United Republic of Tanzania, Ministry of Water.

Dohrenwend, R. E. (1979). *Hydrologic Behavior at the Top of a Tropical Mountain.* Houghton, MI: Michigan Technological University Forestry Center.

Edwards, K. A. (1979). The water balance of the Mbeya experimental catchments. *East African Agricultural and Forestry Journal* **43**: 231–247.

Eva, H., and E. F. Lambin (2000). Fires and land-cover change in the tropics: a remote sensing analysis at the landscape scale. *Journal of Biogeography* **27**: 765–776.

Furrer, G., and R. Freund (1973). Beobachtungen zum subnivalen Formenschatz am Kilimanjaro. *Zeitschrift für* Geomorphologie N. F. **16**: 180–203.

García-Santos, G. (2007). *An ecohydrological and soils study in a montane cloud forest in the National Park of Garajonay, La Gomera (Canary Islands, Spain).* Ph.D. Thesis, VU University Ámsterdam, Ámsterdam, The Netherlands. [http://www.falw.vu.nl/nl/onderzoek/earth-sciences/geo-environmental-science-and-hydrology/hydrology-dissertations/index.asp].

Grabherr, G., and H. Pauli (1994). Climate effects on mountain plants. *Nature* **369**: 448.

Hedberg, O. (1951). Vegetation belts of the East African mountains. *Svensk Botaniker Tidskrift* **45**: 140–202.

Hemp, A. (2001). Ecology of the pteridophytes on the southern slopes of Mt. Kilimanjaro. II. Habitat selection. *Plant Biology* **3**: 493–523.

Hemp, A. (2002). Ecology of the pteridophytes on the southern slopes of Mt. Kilimanjaro. I. Altitudinal distribution. *Plant Ecology* **159**: 211–239.

Hemp, A. (2003). Climate impacts and responses on Mount Kilimanjaro. In *Development and Climate Change in Tanzania: Focus on Mount Kilimanjaro*, eds. S. Agrawala and A. Moehner. Paris: OECD.

Hemp, A. (2005a). Climate change driven forest fires marginalize the impact of ice cap wasting on Kilimanjaro. *Global Change Biology* **11**: 1013–1023.

Hemp, A. (2005b). The impact of fire on diversity, structure and composition of Mt. Kilimanjaro's vegetation. In *Land Use Changes and Mountain Biodiversity*, eds. E. Spehn, M. Liberman, and C. Körner, pp. 51–68. Boca Raton, FL: CRC Press.

Hemp, A. (2006a). Continuum or zonation? Altitudinal diversity patterns in the forests on Mt. Kilimanjaro. *Plant Ecology* **184**: 27–42.

Hemp, A. (2006b). Vegetation of Kilimanjaro: hidden endemics and missing bamboo. *African Journal of Ecology* **44**: 305–328.

Hemp, A., and E. Beck (2001). *Erica excelsa* as a fire-tolerating component of Mt. Kilimanjaro's forests. *Phytocoenologia* **31**: 449–475.

Hemp, A., C. Lambrechts, and C. Hemp (in press). *Global Trends and Africa: The Case of Mt. Kilimanjaro.* Nairobi, Kenya: UNEP.

Holwerda, F. (2005). Water and energy budgets of rain forests along an elevational gradient under maritime tropical conditions. Ph.D. thesis, VU University Amsterdam, Amsterdam, the Netherlands. Also available at www.falw.vu.nl/images_upload/B321FA03–1279-D040–8B25E94A8EBD B962.pdf.

Jaeger, F. (1909). Forschungen in den Hochregionen des Kilimandscharo. Mitteilungen aus den Deutschen Schutzgebieten **22**: 113–146; 161–196.

Juvik, J. O. (2009). The seminal Hawaiian cloud forest "Lâna'i fog drip study": 50th anniversary reassessment. In *Tropical Hydrology Symposium*, eds. F. L. Ogden, R. S. Harmon, and R. F. Stallard, p. 229. Panama City: Smithsonian Tropical Research Institute.

Juvik, J. O., and D. Nullet (1995). Relationships between rainfall, cloud-water interception, and canopy throughfall in a Hawaian montane forest. In *Tropical Montane Cloud Forests*, eds. L. S. Hamilton, J. O. Juvik, and F. N. Scatena, pp. 165–182. New York: Springer-Verlag.

Kämmer, F. (1974). Klima und Vegetation auf Tenerife, besonders im Hinblick auf den Nebelniederschlag. *Scripta Geobotanica* **7**: 1–78.

Kaser, G., D. R. Hardy, T. Mölg, R. S. Bradley, and T. M. Hyera (2004). Modern glacier retreat on Kilimanjaro as evidence of climate change: observations and facts. *International Journal of Climatology* **24**: 329–339.

Klute, F. (1920). *Ergebnisse der Forschungen am Kilimandscharo 1912.* Berlin: Reimer.

Köhler, L., C. Tobón, K. F. A. Frumau, and L. A. Bruijnzeel (2007). Biomass and water storage dynamics of epiphytes in old-growth and secondary montane cloud forest stands in Costa Rica. *Plant Ecology* **193**: 171–184.
Körner, C. (2003). *Alpine Plant Life*. Berlin: Springer-Verlag.
Lambrechts, C., B. Woodley, A. Hemp, C. Hemp, and P. Nnyiti (2002). *Aerial Survey of the Threats to Mt. Kilimanjaro Forests*. Dar es Salaam, Tanzania: UNDP.
McJannet, D., P. Fitch, M. Disher, and J. Wallace (2007). Measurements of transpiration in four tropical rainforest types of north Queensland, Australia. *Hydrological Processes* **21**: 3549–3564.
Nauss, T., J. Bendix, and L. A. Bruijnzeel (2009). Climate dynamics of the Kilimanjaro region. *Kilimanjaro Ecosystems under Global Change: Linking Biodiversity, Biotic Interactions and Biogeochemical Ecosystem Processes*, ed. I. Steffan-Dewenter. Proposal for the establishment of a Research Unit submitted to the Deutsche Forschungsgemeinschaft.
Pócs, T. (1976). The role of the epiphytic vegetation in the water balance and humus production of the rain forests of the Uluguru Mountains, East Africa. *Boissiera* **24**: 499–503.
Ponette-González, A. G., K. C. Weathers, and L. M. Curran (2009). Water inputs across a tropical montane landscape in Veracruz, Mexico: synergistic effects of land cover, rain and fog seasonality, and interannual precipitation variability. *Global Change Biology*, doi:10.1111/j.1365-2486.2009.01985.x.
Prada, S., M. Menezes de Sequeiro, C. Figueira, and M. Oliveira da Silva (2009). Fog precipitation and rainfall interception in the natural forests of Madeira Island (Portugal). *Agricultural and Forest Meteorology* **149**: 1179–1187.
Rietkerk, M. G., S. C. Dekker, P. C. de ruiter, and J. van de Koppel (2004). Self-organized patches and catastrophic shifts in ecosystems. *Science* **305**: 1826–1829.
Røhr, P. C. (2003). A hydrological study concerning the southern slopes of Mt. Kilimanjaro, Tanzania. Ph.D. thesis, Norwegian University of Science and Technology, Trondheim, Norway.
Santiago, L. S., G. Goldstein, F. C. Meinzer, J. Fownes, and D. Mueller-Dombois (2000). Transpiration and forest structure in relation to soil waterlogging in a Hawaiian montane cloud forest. *Tree Physiology* **20**: 673–681.
Shugart, H. H, N. H. F French, E. S. Kasischke, *et al.* (2001). Detection of vegetation change using reconnaissance imagery. *Global Change Biology* **7**: 247–252.
Steinhardt, U. (1979). Untersuchungen über den Wasser- und Nährstoffhaushalt eines andinen Wolkenwaldes in Venezuela. *Göttinger Bodenkundliche Berichte* **56**: 1–185.
Tenhunen, J., D. Otieno, B. Huwe, and B. Glaser (2009). Patterns in water and carbon flux controls along land use and climate gradients. *Kilimanjaro Ecosystems under Global Change: Linking Biodiversity, Biotic Interactions and Biogeochemical Ecosystem Processes*, ed. I. Steffan-Dewenter. Proposal for the establishment of a Research Unit submitted to the Deutsche Forschungsgemeinschaft.
Van de Koppel, J., and M. G. Rietkerk (2004). Spatial interactions and resilience in arid ecosystems. *American Naturalist* **163**: 113–121.
Walter, H., and S. W. Breckle (1984). *Spezielle Ökologie der tropischen und subtropischen Zonen: Ökologie der Erde*. Berlin: Fischer-Verlag.

59 Historical ^{14}C evidence of fire in tropical montane cloud forests in the Chimalapas region of Oaxaca, southern Mexico

Y. Wård and A. Malmer
Swedish University of Agricultural Sciences, Umeå, Sweden
H. Asbjornsen
Iowa State University, Ames, Iowa, USA

ABSTRACT

Tropical montane cloud forests (TMCF) in the Chimalapas region of Oaxaca, Southern Mexico were subjected to large-scale fires during the El Niño events of 1997/98. This raised the question as to whether fire occurrence in this type of forest is cyclic. Charcoal fragments were collected for ^{14}C dating throughout mineral soil profiles down to 50 cm depth at two sites with contrasting geology. Three distinct clusters of fires were found, the first dating from 2350–1900 yr BP, the second from 960–670 yr BP, and the third from 670–470 yr BP. These new data suggest that there have been fires in these TMCF in the past, but at very long time intervals. The identified clusters of enhanced fire activity in the Chimalapas correlate with other paleo-climatic data from Central America and the Caribbean, suggesting periods when the climate was distinctly drier throughout the region. The clustering of fire occurrence at certain times indicates an enhanced risk of repeated fires in secondary TMCF, as has been found for other humid tropical forests.

INTRODUCTION

Sanford *et al.* (1985) suggested that the fire ecology of tropical rain forests should be considered in both historic and present-day contexts. During the last decades a number of severe large-scale fires have occurred throughout the humid tropical forest domain – usually related to changes in vegetation (fuel and moisture preservation) – and these fires have rendered also shorter droughts critical (Goldammer, 2007). Major El Niño/Southern Oscillation (ENSO) events accompanied by extreme droughts, e.g. in 1982/83 in South-East Asia (Leighton and Wirawan, 1986) and in 1997/98 also in Amazonia and Central America (Malhi and Wright, 2005), have exercerbated fire activity. With respect to the latter event, a report by the UN Food and Agricultural Organization stated that "Ecosystems generally not subjected to fires, such as the Amazon rain forest in Brazil and the cloud forests in Chiapas, Mexico sustained considerable damage" (FAO, 2001). Apart from this note and others on the 1998 fires in the States of Chiapas and Oaxaca in southern Mexico (Anta and Plancarte, 2001; Asbjornsen *et al.*, 2005), there is comparatively little historical evidence of fires in tropical montane cloud forest (TMCF) environments (Hemp, 2005; Asbjornsen and Wickel, 2009; Hemp, this volume #58). Schrumpf *et al.* (2001) reported "frequent occurrence" of charcoal fragments in soil profiles along an elevational transect (1010–3050 m.a.s.l.) in southern Ecuador, particularly in the drier *páramo* zone. The ^{14}C age of most fragments was in the order of 710–980 ($\pm$ 50) yr BP. Whilst taking this as evidence that fires in the upper montane and *páramo* zones are not just a recent phenomenon (cf. Mosandl *et al.*, 2008), Schrumpf *et al.* (2001) were unable to state whether these ancient fires had natural or anthropogenic causes. Horn *et al.* (2000) provided evidence of frequent fire activity over the last 40 000 years in the upper parts of the Cordillera Central in the Dominican Republic, i.e. well before human settlement on the island (estimated at 7000 yr BP). Martin and Fahey (2006) documented fire frequency in the same area across meso-climatic gradients as determined by topographic exposure to the trade winds and height of the trade wind inversion using dendrochronological techniques.

Tropical Montane Cloud Forests: Science for Conservation and Management, eds. L. A. Bruijnzeel, F. N. Scatena, and L. S. Hamilton. Published by Cambridge University Press.

The Chimalapas comprises a 600 000-ha mountainous area in the southern part of the State of Oaxaca which has experienced comparatively minor encroachment by humans (Asbjornsen and Garnica-Sánchez, this volume; cf. Cayuela *et al.*, 2006). The region was severely damaged by fires during the ENSO-related drought of 1998. Approximately 126 000 ha of primary forests were affected. TMCF were among the most severely damaged forest types, with approximately 60% (38 000 ha) of the total TMCF area burned (Anta and Plancarte, 2001) and over 80% of the living above-ground biomass destroyed, while root biomass was reduced by 49–77% some 4 years after the fires (Asbjornsen *et al.*, 2005).

Based on paleo-climatic data derived from the sediments in Lake Miragoane, Haiti, Curtis and Hodell (1993) showed that warmer and drier periods alternated with wetter and cooler periods in the Caribbean during the last 10 500 years. There is also evidence from Laguna Pallcacocha, a lake located at 4200 m.a.s.l. near the western Andean continental divide in Ecuador, for the occurrence of ENSO-induced dry periods during the last 12 000 years (Moy *et al.*, 2002). Increased aridity during the early to mid-Holocene caused a reduction in the extent of Andean cloud forest species, both near their lower (Lago Consuelo, Perú; Bush *et al.*, 2004) and upper elevational limits (Siberia, Bolivia; Mourguiart and Ledru, 2003).

The aim of the present study was to investigate the occurrence and frequency of possible historic fires in the TMCF of the Chimalapas, and to what extent the intervals between clusters of fire related to other evidence of climatic fluctuations and human activity in the region (see Asbjornsen *et al.*, 2005 and Asbjornsen and Wickel, 2009 for more details).

STUDY AREA

The mountainous Chimalapas region has a highly diverse geology and, due to its strongly dissected topography with peaks between 2000 and 3000 m.a.s.l. and its location between two oceans at the Isthmus of Tehuantepec, a highly variable climate as well. Depending on elevation, mean annual temperature is in the range 14–20 °C, whereas annual precipitation exceeds 2000 mm on average for the Chimalapas as a whole, and probably reaches 6000 mm in places (Lorence and Mendoza, 1989).

Two sites with contrasting geology were selected for this study, viz. the Benito Juárez (BJ) site (16° 44′ 02–30″ N, 94° 11′ 30–46″ W) at 1585–1700 m.a.s.l. on crystalline (intrusive) bedrock, and the San Antonio (SA) site (16° 39′ 52–41″ N, 94° 13′ 57–49″ W) at 1640–1740 m.a.s.l. on karstic rock. Both sites are part of the Sierra Madre mountain range but situated more than 10 km from each other. Soils at the BJ sites consisted of humic Acrisols (Ultisols), while the SA sites supported both luvic Calcisols (typic Ustropepts) and rendzic Leptosols (lithic Troporthents) (Wård, 2003). Soils at SA were much more shallow and therefore expected to be more drought- and fire-prone. Forest structure was highly variable at both sites, with tree height reaching 30–40 m at more sheltered and productive foot-slope and valley sites, and low-stature elfin cloud forest (4–6 m in height) at the most exposed ridge-top sites (Asbjornsen *et al.*, 2005).

METHODS

Charcoal samples were collected in May 2002 from soil pits at two locations in BJ and six locations in SA. Maximum distances between pit locations in BJ were 1 km and 1.5 km in SA. The pits were randomly located in the two areas in relation to the main project's permanent plots (Asbjornsen *et al.*, 2005), covering ridge-top and high-elevation slope and local low slope/valley positions. In BJ intact forest situated close to the ridge top was sampled, plus a slope location that had been impacted by the fires of 1998. In SA three pits were in slope positions and three in footslope positions. One each of the topographical positions in SA was in burned forest, whereas the rest were in unburned forest. All pits were 50 × 50 cm and excavated down to 50 cm mineral soil depth. Sampled soil material was sieved and charcoal was collected and labeled by horizon. Charcoal was found in both areas and at all depths down to 50 cm, both in the locations that burned in 1998 and in the ones that escaped the recent fires. A total of 156 pieces of charcoal, with smallest diameters greater than 2 mm, were found of which 48 pieces (20 from BJ, and 28 from SA) were selected for ^{14}C dating (one charcoal fragment did not contain enough organic material to be dated and was discarded). Selected fragments came from all pits, and represented all landscape positions and soil depths. Prior to ^{14}C AMS (accelerator mass spectrometry at the Ångström Laboratory, Uppsala, Sweden) dating, all visible fine roots were removed from the charcoal and charcoal pieces were boiled in a solution of 10% KOH (KOH Proanalysis, Merck Germany) using SuperQ-purified water (SuperQ purification system, AKA-filter, Stockholm). The precipitate, which mainly consists of humus material (SOL), was washed. The remaining insoluble part (INS), which consists largely of original plant material and therefore gives the most accurate and relevant age determination, was acidified to pH 4 to avoid contamination from modern CO_2. Next, the sample was exposed to an ion source and the ^{14}C age determined using the calibrating program OxCal 3.5 (http://rlaha.ox.ac.uk/oxcal) (Bronk Ramsey, 1995, 2001). No correlations between mineral soil depth and charcoal age were found at any of the eight soil pits investigated and therefore samples from each site were treated as individual samples.

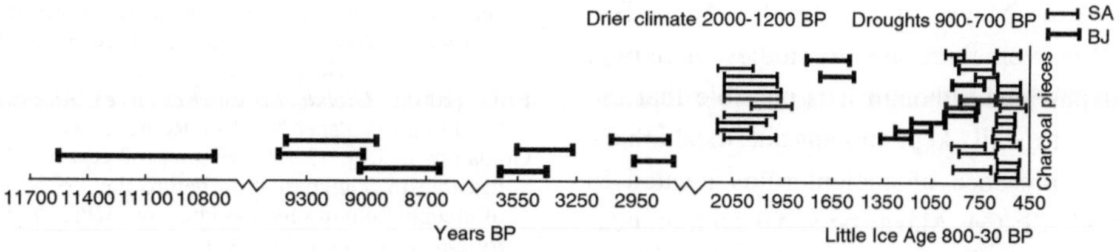

Figure 59.1. Carbon-14 datings and their confidence intervals of charcoal fragments found in mineral soils under tropical montane cloud forest at two sites in the Chimalapas, Oaxaca, Mexico. Each bar represents a single sample. Present day in BP = AD 1950. SA = San Antonio site (shallow soil on limestone), BJ = Benito Juarez site (better-developed soil on crystalline rock).

RESULTS AND DISCUSSION

Past climatic variation in the northern hemisphere is comparatively well known due to the availability of paleo-climatic information contained in lake sediments, as well as in glaciers and ice caps on Greenland and Iceland. The evidence suggests that the climate in the northern hemisphere over the last 2000 years has been quite diverse. The period between AD 1100 and 1920 (850–30 yr BP, taking 1950 as the reference year) is generally called the Little Ice Age (LIA). During this period, glaciers grew on more than one occasion, but the LIA also had warmer subperiods. The first cold period during the LIA occurred in AD 1150–1350 (800–600 yr BP) and the second in AD 1600–1875 (350–75 yr BP) (Sugden and John, 1976). Correlated climate variations have been demonstrated in northern South America, where precipitation minima occurred during the LIA (Haug *et al.*, 2001).

The present results for the ^{14}C age determinations of charcoal fragments from the Chimalapas cloud forests show three distinct clusters of fires: viz. (i) 2350–1900 yr BP, (ii) 960–670 yr BP, and (iii) 670–470 yr BP (Figure 59.1).

The youngest cluster of fires (dated 670–470 yr BP) appeared to be more limited in time and space because all datings had overlapping time intervals, whereas only one charcoal sample was retrieved from the Benito Juarez area (BJ) (soils on crystalline rock) for this period. For the second and generally better-documented cool period within the LIA (Sugden and John, 1976) no evidence of fires was obtained for either the SA or BJ site. The intermediate cluster of fire occurrence derived in Chimalapas (960–670 yr BP), which coincided with the first cold period during the LIA (800–600 yr BP), was well represented at both sites whereas in addition the datings were more spread out in time. This cluster also coincides with other recorded drought data, e.g. from the Cariaco Basin in the southern Caribbean (Haug *et al.*, 2001), southern Ecuador (Schrumpf *et al.*, 2001), and early LIA datings from Iceland (Bergthórsson, 1969). The oldest main cluster of fires indicated by the Chimalapas samples (2350–1900 yr BP), found only at SA (shallow soils on limestone), coincides with the earlier part of dry periods (2200–1200 yr BP) on the Yucatan Peninsula (Hodell *et al.*, 1995). The driest conditions in Yucatan occurred between 1300 and 1200 yr BP, but only a few single charcoal samples retrieved at the BJ site represented this period. There were eight pieces of charcoal that were older than 2300 yr BP, all of them from BJ. Two of these fragments were dated between 3335 and 3150 yr BP and another two between 2875 and 2735 yr BP (Figure 58.1).

It is known that between 7100 and 3000 yr BP the evaporation/precipitation ratios in Mexico were low, representing a wetter climate (Hodell *et al.*, 1995). In accordance, no charcoal at all was found from this period at the Chimalapas sites. The four oldest charcoal findings (again all from BJ) were spread out in time. The oldest piece was dated at 11 550–10 750 yr BP, two others at 9430–8950 yr BP, and the fourth from 9030–8630 yr BP (Figure 59.1).

One explanation for the clustering in time of major fire events in the Chimalapas could be the higher sensitivity of secondary vegetation to drying and fire, which is often given as the main reason for the repeated widespread fires in the humid tropics during the last decades (Goldammer, 2000; Malmer *et al.*, 2005). Hence, the most significant finding of the present study is the clustering of charcoal dates with very long intervals of few fire indications in between (Figure 59.1). This supports the general perception that wet TMCF are amongst the ecosystems on a global basis that have the lowest incidence of fire (Asbjornsen and Wickel, 2009). As indicated, the two sampling sites in the Chimalapas had fundamentally different geologies. The SA soils on limestone were much thinner than the BJ soils on metamorphic rock, rendering them more vulnerable to water shortage and drying out. As a result, a higher frequency of fires is to be expected at SA. However, charcoal fragments from the BJ area indicated a much wider distribution and frequent occurence of forest fires in time than that derived for the SA area (Figure 59.1). A potential explanation for this unexpected pattern may relate to greater human activity within TMCF regions supporting more fertile and productive soils, such as those found in BJ, which in turn might be associated with greater fire ignitions (e.g. Roman-Cuesta *et al.*, 2003).

The issue to what extent human activities interact with, and exacerbate, the effects of droughts and fires in the tropics has only recently become a topic of debate (Hodell *et al.*, 1995; Haug *et al.*, 2003; Malmer *et al.*, 2005; cf. Martin and Fahey, 2006). The present study found only limited indications of fire during the period of the fall of the Maya culture (1300–1150 yr BP; Hodell *et al.*, 1995). However, there are no studies of human artifacts in the Chimalapas, and although it is possible that the region's inaccessibility historically kept humans out, local inhabitants often mention the presence of ancient ruins located in remote regions of the TMCF (M. Maya, personal communication). Today, new villages and pastures are established along the margins of the region and this has been proposed as one reason for the ignition of many of the fires in 1998 (Anta and Plancarte, 2001). Given the current increase in the use of fire in and around the Chimalapas (cf. Asbjornsen and Garnica Sánchez, this volume), and the possibly increasing climatic variability under a global change scenario for the region (Bates *et al.*, 2008; Karmalkar *et al.*, 2008; cf. Asbjornsen and Wickel, 2009; Ponette-González *et al.*, 2009; Barradas *et al.*, this volume), the present results should be taken as an historic indication of the risk of repeated and enhanced fire occurrence in the region. In addition, successional vegetation dries more rapidly and may result in more high-severity fires, imposing a further constraint to the uninterrupted development toward mature forests and restoration of soil fertility in the area (cf. Malmer, 2004; Malmer *et al.*, 2005; Goldammer *et al.*, 2007).

ACKNOWLEDGEMENTS

We thank the Swedish International Development Agency (SIDA) for funding, and the people of the villages of Benito Juarez and San Antonio for their assistance in the field.

REFERENCES

Anta, S. F., and A. B. Plancarte (2001). The forest fires: Chimalapas – the last opportunity. In *The Castor Workshop*, p. 249. Mexico City: World Wildlife Fund and the Mexican Secretariat of the Environment.

Asbjornsen, H., and A. J. Wickel (2009). Changing fire regimes in tropical montane cloud forests: a global synthesis. In *Fire and Tropical Ecosystems*, ed. M. A. Cochrane, pp. 607–622. New York: Springer-Verlag.

Asbjornsen, H., C. Gallardo-Hernández, N. Velázques-Rosas, and R. García-Soriano (2005). Deep ground fires cause massive above- and below-ground biomass losses in tropical montane cloud forests, Oaxaca, Mexico. *Journal of Tropical Ecology* **21**: 427–434.

Bates, B. C., Z. W. Kundzewicz, S. Wu, and J. P. Palutikof (eds.) (2008). *Climate Change and Water*. Geneva, Switzerland: IPCC Secretariat.

Bergthórsson, P. (1969). An estimate of drift ice and temperature in Iceland in 1000 years. *Jökull* **19**: 94–101.

Bronk Ramsey, C. (1995). Radiocarbon calibration and the analysis of stratigraphy. *Radiocarbon* **37**: 425–430.

Bronk Ramsey, C. (2001). Development of the radiocarbon calibration program. *Radiocarbon* **43**: 355–363.

Bush, M. B., M. R. Silman, and D. H. Urrego (2004). 48,000 years of climate and forest change in a biodiversity hotspot. *Science* **303**: 827–829.

Cayuela, L., J. M. R. Benayas, and C. Echevarría (2006). Clearance and fragmentation of tropical montane forests in the Highlands of Chiapas, Mexico (1975–2000). *Forest Ecology and Management* **226**: 208–218.

Curtis, J. H., and D. A. Hodell (1993). An isotopic and trace element study of Ostracods from Lake Miragoane, Haiti: a 10,500 year record of paleosalinity and paleotemperature changes in the Caribbean. In *Climate Change in Continental Isotopic Records*, ed. P. K. Swart, pp. 135–152. Washington, DC: American Geophysical Union.

FAO (2001). *Global Forest Resources Assessment 2000*, Main Report, FAO Forestry Paper No. 140. Rome: FAO.

Goldammer, J. G. (2007). History of equatorial vegetation fires and fire research in Southeast Asia before the 1997–98 episode: a reconstruction of creeping environmental changes. *Mitigation and Adaptation Strategies for Global Change* **12**: 13–32.

Haug, G. H., K. A. Hughen, D. M. Sigman, L. C. Peterson, and U. Rohl (2001). Southward migration of the intertropical convergence zone through the Holocene. *Science* **293**: 1304–1308.

Haug, G. H., D. Gunther, L. C. Peterson, *et al.* (2003). Climate and the collapse of Maya civilization. *Science* **299**: 1731–1735.

Hemp, A. (2005). Climate change driven forest fires marginalize the impact of ice cap wasting on Kilimanjaro. *Global Change Biology* **11**: 1013–1023.

Hodell, D. A., J. H. Curtis, and M. Brenner (1995). Possible role of climate in the collapse of classic Maya civilization. *Nature* **375**: 391–394.

Horn, S. P., K. H. Orvis, L. M. Kennedy, and G. M. Clark (2000). Prehistoric fires in the highlands of the Dominican Republic: evidence from charcoal in soils and and sediments. *Caribbean Journal of Science* **36**: 10–18.

Karmalkar, A. V., R. S. Bradley, and H. F. Diaz (2008). Climate change scenario for Costa Rican montane forests. *Geophysical Research Letters* **25**, L11702, doi:10.1029/2008GL033940.

Leighton, M., and N. Wirawan (1986). Catastrophic drought and fire in Borneo tropical rain forest associated with the 1982–1983 El Nino Southern Oscillation Event. In *Tropical Forests and the World Atmosphere*, ed. G. T. Prance, pp. 75–102. Washington DC: American Association for the Advancement of Science.

Lorence, D. H., and A. G. Mendoza (1989). Oaxaca, Mexico. In *Floristic Inventory of Tropical Countries*, eds. D. G. Campell and D. H. Hammond, pp. 253–309. New York: New York Botanical Gardens.

Malhi, Y., and J. Wright (2005). Late twentieth-century patterns and trends in the climate of tropical forest regions. In *Tropical Forests and Global Atmospheric Change*, eds. Y. Malhi and O. Phillips, pp. 3–16. Oxford, UK: Oxford University Press.

Malmer, A. (2004). Streamwater quality as affected by wild fires in natural and man-made vegetation in Malaysian Borneo. *Hydrological Processes* **18**: 853–864.

Malmer, A., M. van Noorwijk, and L. A. Bruijnzeel (2005). Effects of shifting cultivation and forest fire. In *Forests, Water and People in the Humid Tropics*, eds. M. Bonell and L. A. Bruijnzeel, pp. 533–560. Cambridge, UK: Cambridge University Press.

Martin, P. H., and T. J. Fahey (2006). Fire history along environmental gradients in the subtropical pine forests of the Cordillera Central, Dominican Republic. *Journal of Tropical Ecology* **22**: 289–302.

Mosandl, R., S. Günter, B. Stimm, and M. Weber (2008). Ecuador suffers the highest deforestation rate in South America. In *Gradients in a Tropical Mountain Ecosystem of Ecuador*, eds. E. Beck, J. Bendix, I. Kottke, F. Makeschin, and R. Mosandl, pp. 37–40. Berlin: Springer-Verlag.

Mourguiart, P., and M. P. Ledru (2003). Last glacial maximum in an Andean cloud forest environment (eastern Cordillera, Bolivia). *Geology* **31**: 195–198.

Moy, C. M., G. O. Seltzer, D. T. Rodbell, and D. M. Anderson (2002). Variability of El Nino/Southern Oscillation activity at millennial timescales during the Holocene epoch. *Nature* **420**: 162–165.

Ponette-González, A. G., K. C. Weathers, and L. M. Curran (2009). Water inputs across a tropical montane landscape in Veracruz, Mexico: synergistic effects of land cover, rain and fog seasonality, and interannual precipitation variability. *Global Change Biology*, doi:10.1111/j.1365-2486.2009.01985.x.

Roman-Cuesta, R. M., M. Garcia, and J. Retana (2003). Environmental and human factors influencing fire trends in ENSO and non-ENSO years in tropical Mexico. *Ecological Applications* **13**: 1177–1192.

Sanford, R. L. J., J. Saldarriaga, K. E. Clark, C. Uhl, and R. Herrera (1985). Amazon rain-forest fires. *Science* **227**: 53–55.

Schrumpf, M., G. Guggenberger, C. Valarezo, and W. Zech (2001). Tropical montane rain forest soils: development and nutrient status along an altitudinal gradient in the south Ecuadorian Andes. *Die Erde* **132**: 43–59.

Sugden, D. E., and B. S. John (1976). *Glaciers and Landscape*. London: Edward Arnold.

Wård, Y. (2003). *Soil Descriptions in Burned and Unburned Cloud Forest and Indications of Fire History in the Chimalapas, Oaxaca State, Mexico*. Umeå, Sweden: Swedish University of Agricultural Sciences.

60 Biennial variation in tree diameter growth during eight years in tropical montane cloud forests on Mount Kinabalu, Sabah, Malaysia

S. Aiba
Kagoshima University, Kagoshima, Japan
M. Takyu
Tokyo University of Agriculture, Tokyo, Japan
K. Kitayama
Kyoto University, Shiga, Japan

ABSTRACT

Biennial variation in tree diameter growth was studied during eight years (1995–2003) in nine tropical rain forests on Mt. Kinabalu, Sabah, Malaysian Borneo. These forests are located at four altitudes (700, 1700, 2700, and 3100 m.a.s.l.) on contrasting geological substrates (ultrabasic versus non-ultrabasic). The seven forests at and above 1700 m.a.s.l. may be classified as montane cloud forests, whereas the other two at 700 m.a.s.l. are lowland forests. A severe drought associated with an El Niño–Southern Oscillation event occurred from late 1997 to early 1998. Mean absolute growth rates of stem diameter in each of four 2-year periods increased from smaller (4.8–10 cm diameter) to larger ($\geq$10 cm diameter) stems in each forest, decreased with increasing altitude on the same substrate, and were greater on non-ultrabasic than on ultrabasic substrate at the same altitudes. There were significant, concordant biennial variations in growth rates among the nine forests for both smaller and larger stems. The period of least growth included the drought in all forests for larger stems. Overall, the patterns of variation were similar among forests on different substrates at the same altitude (especially for larger stems), suggesting climatic rather than edaphic control of tree growth. Growth rates exhibited a consistent pattern in that: (i) rates for individual stems changed substantially from pre-drought (1995–1997) to drought periods (1997–1999), and (ii) the altered growth rates persisted during post-drought periods (1999–2001 and 2001–2003), especially in the montane forests. In summary, tree growth in montane cloud forests on Mt. Kinabalu showed substantial spatial and temporal variation. The spatial variation (i.e. between forests) was related to both geology and climate. The drought was one of the major causes of temporal variation in tree growth, but other climatic factors should be responsible for the variations observed during periods with normal rainfall.

INTRODUCTION

Climatically speaking, two types of cloud forests may be distinguished: (i) forests in wet areas receiving relatively persistent rainfall throughout the year, and (ii) forests in drier areas with more seasonal rainfall where fog capture (horizontal precipitation) constitutes a substantial fraction of dry-season water inputs (Stadtmüller, 1987; Dawson, 1998; Bruijnzeel, 2001; Ewing *et al.*, 2009; Marzol-Jaén *et al.*, this volume). In recent years, droughts associated with El Niño–Southern Oscillation (ENSO) events have been shown to significantly affect lowland tropical forests across the globe (Leighton and Wirawan, 1986; Proctor *et al.*, 1989; Laurance *et al.*, 2000; Nakagawa *et al.*, 2000; Williamson *et al.*, 2000; Potts, 2003; Condit *et al.*, 2004; Malhi and Wright, 2005). Similar effects have been demonstrated even

Tropical Montane Cloud Forests: Science for Conservation and Management, eds. L. A. Bruijnzeel, F. N. Scatena, and L. S. Hamilton. Published by Cambridge University Press.

Table 60.1 *Description of the nine study forests at four common altitudes on different geologic substrates on Mount Kinabalu, Sabah, Malaysia*

Common altitude (m)	Geological substrate[a]	Abbreviation	Exact altitude (m)	Forest type[b]	Plot area (ha)	Stem density (ha^{-1})[c]	Fisher's diversity index[c]	Kendall's coefficient of concordance[d]
700	T	07S	650	LRF	1.00[e]	1064	75.0	0.08
700	U	07U	700	LRF	1.00	1175	69.0	0.17***
1700	T	17S	1560	LM(C)F	0.50	1730	38.3	0.39***
1700	Q	17Q	1860	LM(C)F	1.00	2045	24.1	0.36***
1700	U	17U	1860	LM(C)F	0.20	3535	10.1	0.01
2700	T	27S	2590	UMCF	0.25	2116	4.9	0.28*
2700	U	27U	2700	UMCF	0.20	3775	5.2	0.28**
3100	G	31S	3080	SACF	0.20	3665	4.8	0.18*
3100	U	31U	3050	SACF	0.06	4383	1.8	0.23

[a] T, Tertiary sedimentary rock; U, ultrabasic rock; Q, Quaternary sediment; G, granite.

[b] LRF, lowland rain forest; LM(C)F, lower montate (cloud) forest; UMCF, upper montane cloud forest; SACF, sub-alpine cloud forest (after Bruijnzeel, 2001).

[c] For stems ≥4.8 cm diameter.

[d] Kendall's coefficient of concordance among rankings of species' mean growth rate for abundant ($n \geq 5$) species (*, $p < 0.05$; **, $p < 0.01$; ***, $p < 0.001$; df = 3).

[e] Trees in a large canopy gap (0.14 ha area) were excluded from analysis.

for perhumid montane cloud forest (Aiba and Kitayama, 2002), as well as for more seasonal cloud forests (cf. Asbjornsen *et al.*, 2005; Ponette-González *et al.*, 2009; Wård *et al.*, this volume). It has also been shown that tree diameter growth exhibits marked temporal variation during normal (non-drought) periods in tropical lowland forest (Clark and Clark, 1994; Clark *et al.*, 2003). In this chapter, biennial variation in tree diameter growth during eight years (1995–2003) is discussed for nine tropical rain forests on Mt. Kinabalu (4095 m.a.s.l.) in Sabah, Malaysian Borneo. The investigated forests are located at four elevations (700, 1700, 2700, and 3100 m.a.s.l.) on contrasting geological substrates (ultrabasic versus non-ultrabasic) (Aiba and Kitayama, 2002; Kitayama and Aiba, 2002; Kitayama *et al.*, 2004). The seven forests located at and above 1700 m.a.s.l. may be classified as montane (or sub-alpine) cloud forests (Kitayama, 1995; cf. Bruijnzeel, 2001), whereas the other two forests at 700 m.a.s.l. are lowland mixed dipterocarp forests near their upper limit (cf. Kumaran *et al.*, this volume). A major drought associated with an ENSO event occurred from late 1997 to early 1998, during which the climatic departure from normal conditions was greater at higher altitudes (Aiba and Kitayama, 2002).

MATERIALS AND METHODS

Nine primary forests were selected at four common altitudes (700, 1700, 2700, and 3100 m.a.s.l.) on ultrabasic and non-ultrabasic substrates (Table 60.1). Exact altitudes differed slightly between two (or three) forests at the same common altitude. In each forest, a plot of variable size (0.06–1.00 ha) was established. Smaller plots were used for less diverse forests with higher stem density. The first tree census was conducted in 1995. In the seven plots at elevations ≥1700 m.a.s.l., all stems ≥4.8 cm diameter were measured. In two 1-ha plots at 700 m.a.s.l., all stems ≥10 cm diameter were measured, and stems 10 cm > diameter ≥ 4.8 cm were measured in parts (0.2 ha) of the plots. This necessitated the separate analysis of smaller (4.8–10 cm diameter) and larger (≥10 cm diameter) stems for comparison of mean growth rates among plots. Subsequent censuses were conducted biennially (1997, 1999, 2001, and 2003). Detailed information on the study sites (including geology, soils, and climate) and the methods used can be found in Aiba and Kitayama (2002), Kitayama and Aiba (2002), and Aiba *et al.* (2005).

RESULTS AND DISCUSSION

Congruence among forests

In each of the four 2-year periods, mean absolute growth rates of stem diameter increased from smaller (4.8–10 cm diameter) to larger (≥10 cm diameter) stems in each forest, decreased with increasing altitude on the same substrates, and were greater on non-ultrabasic compared to ultrabasic substrates at the same altitudes (Figure 60.1). For example, mean growth rates of larger

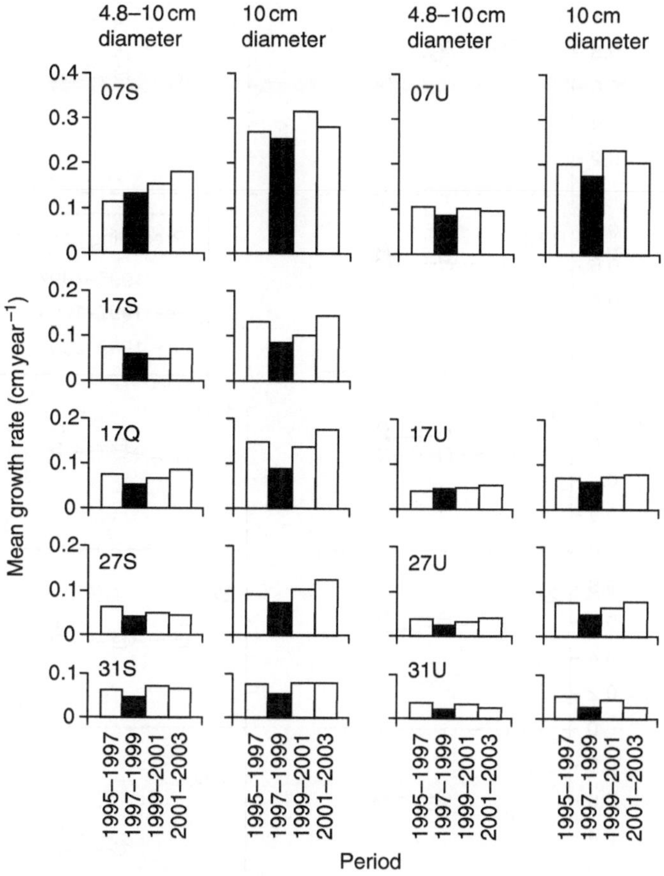

Figure 60.1. Biennial variation in mean diameter growth rates of surviving stems in relation to initial size classes (i.e. in 1995) in nine rain forests on Mount Kinabalu, Sabah, Malaysia. Shaded intervals include a drought. See Table 60.1 for abbreviations of study sites.

stems ranged from 0.18 to 0.32 cm year^{-1} in the two lowland forests, and varied from 0.03 to 0.18 cm year^{-1} in the seven montane forests. Within the 1700 m.a.s.l. group on non-ultrabasic substrates, the forest on Quaternary sediment (17Q; see Table 60.1 for site abbreviations) showed higher growth rates for larger stems than the forest on sedimentary rock (17S), even though the former (1860 m.a.s.l.) was located at a higher elevation than the latter (1560 m.a.s.l.). This might be because the site at 1860 m.a.s.l. is geologically younger and therefore has more nutrient-rich soils than the site at 1560 m.a.s.l. (Kitayama *et al.*, 2004; cf. Roman *et al.*, this volume).

Biennial variations in tree growth rates were significant for both smaller and larger stems in all forests (Friedman ANOVA controlling for individual stems, $p < 0.05$, df $= 3$) except for smaller stems in plot 17U ($p = 0.4$). The patterns of variation were significantly congruent among the nine forests for both size classes (Kendall's coefficient of concordance among rankings of mean biennial growth rate in each forest, $p < 0.05$, df $= 3$). For larger stems, the period of least mean growth rate (1997–1999) included the drought in all forests, but the period of highest growth differed among forests. For smaller stems, the pattern of biennial variations was similar to that for larger stems in each forest, but the period of least growth did not include the drought in three forests (07S, 17S, and 17U). Overall, the patterns in biennial variation in mean growth rates were similar between two (or three) forests on different substrates at the same altitude, especially for larger stems, suggesting the predominantly climatic control of tree growth (Clark and Clark, 1994; Clark *et al.*, 2003).

Magnitude of variation

The ratio of the highest to least mean growth rates among four 2-year periods in each forest (*H*/*L* ratio) was 1.2–1.7 for smaller stems and 1.2–2.0 for larger stems, and tended to be greater with increasing altitude, especially for larger stems on ultrabasic substrate. This may reflect the fact that the climatic departure from normal conditions during the drought was more pronounced at higher elevations (Aiba and Kitayama, 2002). Also, trees in the summit area (>3300 m.a.s.l.) on Mt. Kinabalu have been reported dying following the severe drought from late 1972 to early 1973 (Lowry *et al.*, 1973). However, these findings would seem to be at odds with those reported by Proctor *et al.* (1989) for Mt. Silam, a small coastal mountain mostly formed by ultrabasic rock, elsewhere in Sabah, as will be discussed later.

When different substrates were compared at the same altitude, the *H*/*L* ratio was greater on non-ultrabasic than on ultrabasic substrate at 1700 m.a.s.l. for both size classes, but did not show consistent differences at the other elevations. When the drought period (1997–1999) was excluded, the *H*/*L* ratio declined as expected (1.1–1.6 for smaller stems and 1.1–1.8 for larger stems), being generally greater on non-ultrabasic than on ultrabasic substrates at the same altitudes (except at 3100 m.a.s.l.).

Clark and Clark (1994) found significant, concordant annual variations in growth rates of six canopy species in a lowland forest in Costa Rica. The *H*/*L* ratio of mean annual growth rates was greatest for saplings (3.0–10.0 for <1 cm diameter), intermediate for small trees (1.6–3.6 for 1–10 cm diameter; 1.3–2.5 for 10–30 cm diameter) and least for large trees (1.2–2.1 for ≥30 cm diameter) over a period of eight years. For stems ≥30 cm, the *H*/*L* ratio increased to 1.6–3.8 over 16 years (Clark *et al.*, 2003). In the present case, there were no consistent differences among forests in *H*/*L* ratios for larger and smaller stems when the drought period was included, but the ratio was greater for smaller stems when the drought period was excluded (except at sites 07U and 31U). The *H*/*L* ratios on Mt. Kinabalu appeared to be smaller than those derived for the Costa Rican lowland forest, although the comparison was confounded by differences in methodology (census interval and size range). The ratios for Mt. Kinabalu's forests would be elevated somewhat if annual intervals would be used because biennial intervals tend to mask the variation within the two years.

Temporal correlations

When smaller and larger stems were combined, correlations between individual stem growth during one period and subsequent periods were significant in all forests (Figure 60.2; Spearman's rank correlation, $p < 0.05$). Correlation between growth rates for 1995–1997 and 1997–1999 was less than that between growth rates for 1997–1999 and 1999–2001, and that between results for 1999–2001 and 2001–2003 in all forests, except at plot 07S; growth rates in 1995–1997 and 1997–1999 were also less strongly correlated than growth rates in 1997–1999 and 2001–2003 in the seven montane forests. Similar results were obtained when the results for smaller and larger stems were examined separately. Therefore, growth rates of individual stems showed consistent patterns in that: (i) they changed substantially from pre-drought (1995–1997) to drought periods (1997–1999), and (ii) altered growth rates persisted during post-drought periods (1999–2001 and 2001–2003), especially in the montane forests. This implies that a long-term (3–5 years) impact of the drought could be recognized although mean growth rates in post-drought periods were not necessarily reduced compared to those of the pre-drought period (Figure 60.1). It is possible that the drought caused defoliation of some trees, which may have reduced the growth rate of these trees and increased the growth of neighboring trees by reduced competition. On Mt. Silam (889 m.a.s.l.), Proctor *et al.* (1989) showed amounts of leaf litter on the forest floor in four forests below the usual level of cloud condensation (700 m.a.s.l.; Bruijnzeel *et al.*, 1993) in July 1984 to be distinctly higher than in July 1983, whereas only minor differences were observed for two forests within the cloud belt. Prior to sampling in July 1983, the area had suffered one of the worst droughts in history (Leighton and Wirawan, 1986), which apparently caused extra shedding of litter in forests below, but not within the cloud cap.

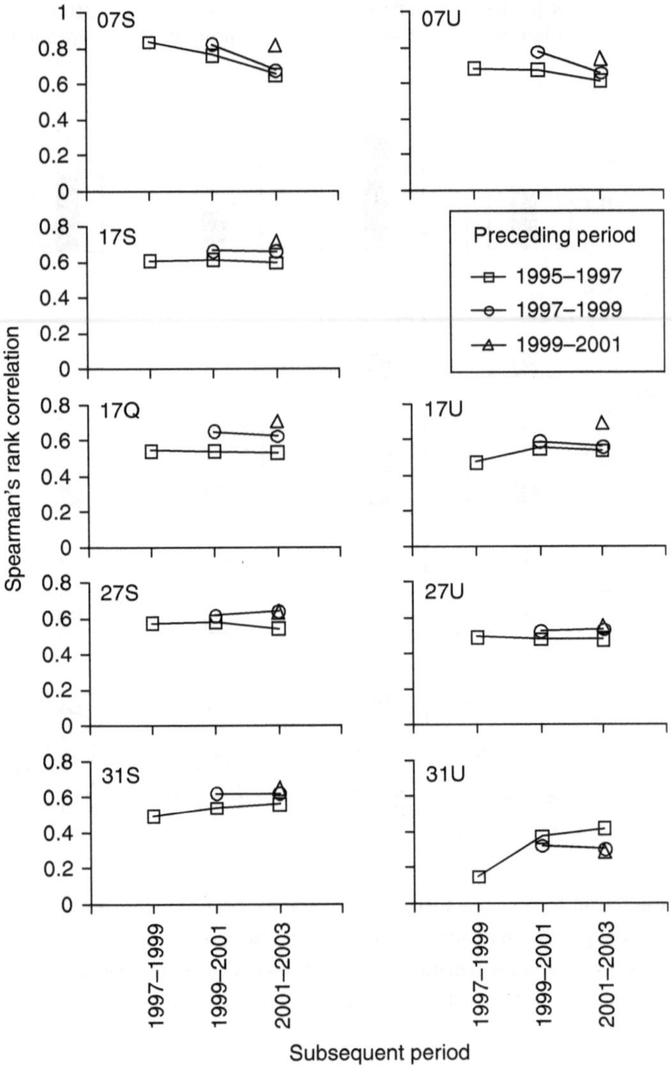

Figure 60.2. Spearman's rank correlations between preceding periods and subsequent periods for surviving stems in each forest. All correlations were significant ($p < 0.05$). See Table 60.1 for abbreviations of study sites.

Congruence among species

To examine the congruence in biennial growth variations among abundant species ($n \geq 5$ per plot) in each forest, mean growth rates were calculated combining smaller and larger stems (due to small sample sizes of individual species). Biennial variations were significantly concordant among species in all forests except three (sites 07S, 17U, and 31U) (Table 60.1; Kendall's coefficient of concordance among rankings of species' mean growth rate, $p < 0.05$, df $= 3$). The exceptions may reflect the fact that the lowland forests did not experience as severe a drought as the montane forests did, and that many species in the montane forests on ultrabasic substrates were more resistant to drought and therefore did not show reduced growth rate during the drought (Aiba and Kitayama, 2002). The insignificant concordance found for site 31U is more likely to be due to small sample size (there were only four species with $n \geq 5$).

CONCLUSION

In summary, tree growth in tropical montane cloud forests on Mt. Kinabalu showed substantial spatial and temporal variation. Both geological substrate and climate (or altitude) appeared to account for the observed spatial variation (i.e. between forests). The similar temporal patterns in variation among forests on different substrates at the same altitude suggested that climate was the primary factor explaining the temporal variation, although other factors related to climate (e.g. mass fruiting) may also be involved. The severe drought occurring in 1997–1998 was one of the major causes of temporal variation. However, other climatic factors such as variation in photosynthetically active radiation in this high-rainfall area should be responsible for the variations observed during periods with normal rainfall (cf. Clark and

Clark, 1994). Based on longer-term records, Clark *et al.* (2003) also demonstrated that annual growth variation in lowland rain forest in Costa Rica was negatively correlated with annual mean daily minimum temperature. A similar analysis is needed to determine what factor(s) explain(s) the growth variations during periods with normal precipitation on Mt. Kinabalu. Finally, it should be noted that only stem growth was studied, and that below-ground production may exhibit different patterns (cf. Röderstein *et al.*, 2005; Hertel and Leuschner, this volume).

REFERENCES

Aiba, S., and K. Kitayama (2002). Effects of the 1997–98 El Niño drought on rain forests of Mount Kinabalu, Borneo. *Journal of Tropical Ecology* **18**: 215–230.

Aiba, S., M. Takyu, and K. Kitayama (2005). Dynamics, productivity and species richness of tropical rainforests along elevational and edaphic gradients on Mount Kinabalu, Borneo. *Ecological Research* **20**: 279–286.

Asbjornsen, H., C. Gallardo-Hernández, N. Velázques-Rosas, and R. García-Soriano (2005). Deep ground fires cause massive above- and below-ground biomass losses in tropical montane cloud forests, Oaxaca, Mexico. *Journal of Tropical Ecology* **21**: 427–434.

Bruijnzeel, L. A. (2001). Hydrology of tropical montane cloud forests: a reassessment. *Land Use and Water Resources Research* **1**: 1–18.

Bruijnzeel, L. A., M. J. Waterloo, J. Proctor, A. T. Kuiters, and B. Kotterink (1993). Hydrological observations in montane rain forests on Gunung Silam, Sabah, Malaysia, with special reference to the "*Massenerhebung*" effect. *Journal of Ecology* **81**: 145–167.

Clark, D. A., and D. B. Clark (1994). Climate-induced annual variation in canopy tree growth in a Costa Rican tropical rain forest. *Journal of Ecology* **82**: 865–872.

Clark, D. A., S. C. Piper, C. D. Keeling, and D. B. Clark (2003). Tropical rain forest tree growth and atmospheric carbon dynamics linked to interannual temperature variation during 1984–2000. *Proceedings of the National Academy of Sciences USA* **100**: 5852–5857.

Condit, R., S. Aguilar, A. Hernandez, *et al.* (2004). Tropical forest dynamics across a rainfall gradient and the impact of an El Niño dry season. *Journal of Tropical Ecology* **20**: 51–72.

Dawson, T. E. (1998). Fog in the California redwood forest: ecosystem inputs and use by plants. *Oecologia* **117**: 476–485.

Ewing, H. A., K. C. Weathers, P. H. Templer, *et al.* (2009). Fog water and ecosystem function: heterogeneity in a California redwood forest. *Ecosystems* **12**: 417–433.

García-Santos,, G. (2007). *An ecohydrological and soils study in a montane cloud forest in the National Park of Garajonay, La Gomera (Canary Islands, Spain).* PhD Thesis, VU University Ámsterdam, Ámsterdam, The Netherlands. [http://www.falw.vu.nl/nl/onderzoek/earth-sciences/geo-environmental-science-and-hydrology/hydrology-dissertations/index.asp].

Kitayama, K. (1995). Biophysical conditions of the montane cloud forests of Mount Kinabalu, Sabah, Malaysia. In *Tropical Montane Cloud Forests*, eds. L. S. Hamilton, J. O. Juvik, and F. N. Scatena, pp. 183–197. New York: Springer-Verlag.

Kitayama, K., and S. Aiba (2002). Ecosystem structure and productivity of tropical rain forests along altitudinal gradients with contrasting soil phosphorus pools on Mount Kinabalu, Borneo. *Journal of Ecology* **90**: 37–51.

Kitayama, K., S. Aiba, M. Takyu, N. Majalap, and R. Wagai (2004). Soil phosphorus fractionation and phosphorus-use efficiency of a Bornean tropical montane rain forest during soil aging with podzolization. *Ecosystems* **7**: 259–274.

Laurance, W. F., P. Delamonica, S. Laurance, H. Vasconcelos, and T. E. Lovejoy (2000). Rainforest fragmentation kills big trees. *Nature* **404**: 836.

Leighton, M., and N. Wirawan (1986). Catastrophic drought and fire in Borneo tropical rain forest associated with the 1982–1983 El Nino Southern Oscillation event. In *Tropical Forests and the World*, ed. G. T. Prance, pp. 75–102. Washington, DC: American Association for the Advancement of Science.

Leuschner, Ch., G. Moser, C. Bertsch, M. Röderstein, and D. Hertel (2007). Large altitudinal increase in tree root/shoot ratio in tropical mountain forests of Ecuador. *Basic and Applied Ecology* **8**: 219–230.

Lowry, J. B., D. W. Lee, and B. C. Stone (1973). Effects of drought on Mount Kinabalu. *Malayan Nature Journal* **26**: 178–179.

Malhi, Y., and J. Wright (2005). Late twentieth-century patterns and trends in the climate of tropical forest regions. In *Tropical Forests and Global Atmospheric Change*, eds. Y. Malhi and O. Phillips, pp. 3–16. Oxford, UK: Oxford University Press.

Nakagawa, M., K. Tanaka, T. Nakashizuka, *et al.* (2000). Impact of severe drought associated with the 1997–1998 El Niño in a tropical forest in Sarawak. *Journal of Tropical Ecology* **16**: 355–367.

Ponette-González, A. G., K. C. Weathers, and L. M. Curran (2009). Water inputs across a tropical montane landscape in Veracruz, Mexico: synergistic effects of land cover, rain and fog seasonality, and interannual precipitation variability. *Global Change Biology*, doi:10.1111/j.1365–2486.2009.01985.x.

Potts, M. D. (2003). Drought in a Bornean everwet rain-forest. *Journal of Ecology* **91**: 464–474.

Proctor, J., C. Philips, G. K. Duff, A. Heaney, and F. M. Robertson (1989). Ecological studies on Gunung Silam, a small ultrabasic mountain in Sabah, Malaysia. II. Some forest processes. *Journal of Ecology* **77**: 317–331.

Röderstein, M., D. Hertel, and Ch. Leuschner (2005). Above- and below-ground litter production in three tropical mountain forests (Ecuador). *Journal of Tropical Ecology* **21**: 483–492.

Stadtmüller, T. (1987). *Cloud Forests in the Humid Tropics*. Tokyo: United Nations University, and Turrialba, Costa Rica: Centro Agronómico Tropical de Investigación y Enseñanza.

Williamson, G. B., W. F. Laurance, A. A. Oliveira, *et al.* (2000). Amazonian tree mortality during the 1997 El Niño drought. *Conservation Biology* **14**: 1538–1542.

61 Modeling the dynamics of tropical montane cloud forest in central Veracruz, Mexico

N. Rüger
UFZ Centre for Environmental Research, Leipzig, Germany

G. Williams-Linera
Instituto de Ecología, A.C., Xalapa, Veracruz, Mexico

A. Huth
Helmholtz Centre for Environmental Research – UFZ, Leipzig, Germany

ABSTRACT

The area covered by tropical montane cloud forest (TMCF) in central Veracruz, eastern Mexico, has decreased rapidly over the last 50 years. Deforestation has been accompanied by fragmentation of the remaining forest. Restoring the TMCF and the important ecological services it provides (e.g. high-quality water, soil protection, biodiversity conservation) requires an understanding of ecosystem dynamics. This study investigates the dynamics of fragments of old-growth TMCF in central Veracruz, with particular reference to regeneration after abandonment of other land uses. A modified version of the process-based forest growth model FORMIND was used. FORMIND is individual-tree-oriented and simulates the spatio-temporal dynamics of an uneven-aged mixed forest stand. Model modifications included: (i) grouping of tree species according to their light demands and maximum height, (ii) defining regeneration, growth, and mortality parameters for each species group, and (iii) developing allometric relations of tree geometry. Model verification was achieved by comparing model outcomes and field data. The model was able to reproduce the structure of old-growth TMCF. Simulations of forest regeneration revealed that aggregated variables (e.g. total stem number and total basal area) reached values similar to those of old-growth forest after approximately 80 years, whereas the proportion of basal area of the different species groups continued to change until ~300 years after the start of succession. These insights can be used to support regional decision-making in forest conservation and restoration planning.

INTRODUCTION

Compared to temperate forests, forest modeling in the tropics faces several difficulties: the richness in tree species, lack of data on tree ages, and the lack of long-term data on forest structure. Nevertheless, in tropical lowland rain forests a variety of phenomena has successfully been studied with the help of forest models. Model applications include the investigation of impacts of alternative logging regimes on forest structure and timber yield in the long term, fragmentation, and edge effects, etc. (e.g. Vanclay, 1995; Chave, 1999; Liu and Ashton, 1999; Huth and Ditzer, 2001; Köhler *et al.*, 2003). Tropical montane cloud forests (TMCF) pose similar problems, but compared to tropical lowland rain forests there are even fewer data available (e.g. Aiba *et al.*, 2005; Leuschner *et al.*, 2007; Moser *et al.*, 2007; Homeier and Breckle, 2008; cf. Aiba *et al.*, this volume; Hertel and Leuschner, this volume). Nevertheless, it can be useful to combine available knowledge on individual tree growth and forest dynamics with a simulation model to address questions that are not easily answered by empirical research. Such questions include long-term phenomena like forest succession after disturbance, or impacts of selective logging on forest structure and composition over time.

Tropical Montane Cloud Forests: Science for Conservation and Management, eds. L. A. Bruijnzeel, F. N. Scatena, and L. S. Hamilton. Published by Cambridge University Press.

As a starting point, this chapter focuses on the TMCF of central Veracruz, eastern Mexico. The area covered by TMCF in central Veracruz has decreased rapidly over the last 50 years, whereas deforestation has been accompanied by fragmentation of the remaining forest (Williams-Linera *et al.*, 2002; Muñoz-Villers and López-Blanco, 2008; cf. Barradas *et al.*, this volume). Restoring the TMCF and the important ecological services it provides (e.g. streamflow of high quality, soil protection, biodiversity conservation) requires an understanding of the dynamics of this ecosystem. Remnant old-growth forest fragments in the area have been studied intensively during the last decade in terms of their floristics, structure, trunk increment rates, leaf area index (*LAI*), and seedling dynamics (e.g. Williams-Linera, 1991, 1993, 1996, 2002; Williams-Linera *et al.*, 2002; Álvarez-Aquino *et al.*, 2004, 2005; Ponette-González *et al.*, 2009) and these data provide an initial basis for the modeling of long-term cloud forest stand dynamics.

This chapter aims to address questions related to forest regeneration and succession using the process-based simulation model FORMIND (Köhler and Huth, 1998, 2007), which is based in part on the more aggregated forest model FORMIX3 (Huth *et al.*, 1998; Ditzer *et al.*, 2000; Huth and Ditzer, 2000; Kammesheidt *et al.*, 2002; Glauner *et al.*, 2003). The model is individual-tree-oriented and simulates the spatio-temporal dynamics of an uneven-aged mixed forest stand. Tree species are aggregated into plant functional types (Köhler *et al.*, 2000). FORMIND has been used successfully to study the dynamics of disturbed tropical lowland rain forests in Malaysia, Venezuela, and French Guiana (Kammesheidt *et al.*, 2001; Köhler *et al.*, 2001, 2003; Köhler and Huth, 2004; Huth *et al.*, 2005) and of temperate rain forests in Chile (Rüger *et al.*, 2007).

This chapter describes the main processes and principal assumptions of FORMIND, as well as their parameterization for TMCF in central Veracruz. Next, the ability of the model to reproduce adequately the structure and composition of TMCF in central Veracruz is tested by comparing model results with inventory data and other field observations. Finally, the model is used to predict the progression of succession after large-scale disturbance (e.g. clear-cutting) or abandonment of previous land use, and the time required for the forest to return to old-growth characteristics.

MATERIALS AND METHODS

Study sites

Five forest fragments in the TMCF region of central Veracruz (19° 30′ N, 96° 54′ W) were selected for this study. The sites are located at altitudes between 1250 and 1875 m.a.s.l. The climate is mild and humid throughout the year with three seasons. A relatively dry–cool season extends from November to March, a dry–warm season from April to May, and a wet–warm season from June to October. Annual precipitation varies between 1350 and 2200 mm; mean annual temperature is between 12 °C and 18 °C (Williams-Linera, 2002; Ponette-González *et al.*, 2009). The soil has been classified as Andisol (Rossignol, 1987). The number of tree species with a diameter at breast height (DBH) ≥ 5 cm varies between 16 and 28 per study sites (58 in total). Dominant tree species include *Carpinus caroliniana*, *Clethra mexicana*, *Fagus grandifolia*, *Liquidambar styraciflua*, *Quercus germana*, *Q. leiophylla*, *Q. xalapensis*, and *Turpinia insignis*.

Model description

MAIN PROCESSES

The individual-oriented forest growth model FORMIND2.3 simulates the spatial and temporal dynamics of uneven-aged mixed forest stands (Köhler and Huth, 1998; Köhler, 2000). The model simulates a forest (in annual time-steps) of several hectares as a mosaic of interacting grid cells with a size of 20 × 20 m, corresponding to the crown size of mature trees. It is assumed that light availability is the main driving force for individual tree growth and forest succession while it is further assumed that the trees do not compete for water or nutrients. Within the grid cells, trees compete for light and space following the gap model approach (Shugart, 1998). For the explicit modeling of the competition for light, each grid cell is divided into small horizontal layers. Leaf area is summed up per height layer and the light climate in the forest interior is calculated via an extinction law (Monsi and Saeki, 1953). The carbon balance of each individual tree is modeled explicitly, including the main physiological processes (photosynthesis, respiration) and litterfall. Growth process equations are partly taken from the FORMIX3-Q model (Ditzer *et al.*, 2000). Allometric functions relate above-ground biomass, stem diameter, tree height, crown diameter, and stem volume. Tree mortality can occur either through self-thinning in dense grid cells, through senescence, or gap formation by large falling trees. Gap formation links neighboring grid cells. Regeneration rates are effective rates concerning the recruitment of small trees at a DBH threshold of 1 cm, with seed loss through predation and other processes already incorporated implicitly.

SPECIES GROUPING

To enable an individual-based simulation of forest dynamics, the 58 tree species that occur in the study sites were grouped into plant functional types (PFTs) based on light demand and maximum attainable height. Three levels of shade tolerance were distinguished: shade-intolerant (i), intermediate (m), and shade-tolerant (t). The following tree height groups were considered: small trees (≤15 m tall, DBH ≤ 35 cm), canopy trees (≤25 m,

DBH ≤ 80 cm), and emergent trees (≤35 m, DBH ≤ 100 cm). This classification resulted in six PFTs, because some of the combinations were rare (Table 61.1).

Model parameters

In the present description of model parameters the focus is on biologically relevant parameters; more technical details can be found in Köhler (2000) and the online appendix of Grimm *et al.* (2006).

ENVIRONMENTAL PARAMETERS

The climate of central Veracruz is mild and humid and it is assumed that trees can grow throughout the year (Gomez-Cardenas, 2009). Mean annual light intensity above the canopy was estimated at 600 μmol m^{-2} s^{-1}. Hafkenscheid (2000) measured average values of 1150 μmol m^{-2} s^{-1} on a sunny day and 260 μmol m^{-2} s^{-1} on a totally overcast day in the Jamaican Blue Mountains at comparable elevation and latitude as the present sites, with an annual average light intensity of 650 μmol m^{-2} s^{-1}. Measurements of the light extinction coefficient k for TMCF are rare. In tall tropical lowland forests a commonly observed value is 0.7 (Kira, 1978). Hafkenscheid (2000) estimated k for a short-statured (12 m) upper montane cloud forest in Jamaica at 0.6, whereas Motzer (2003) derived a k value of 0.45 for a tall (18 m) lower montane rain forest in Ecuador (cf. Motzer *et al.*, this volume). An intermediate value of 0.5 was used in the present simulations.

PARAMETERS RELATED TO TREE GROWTH

Growth characteristics were assumed to be related to shade tolerance and to apply for all PFTs with the same shade tolerance, except in the case of respiration, where parameters were considered to be unique to each PFT.

The rate of photosynthesis P was modeled as a saturating function of the irradiance (I_0) available at the crown of the tree:

$$P(I_0) = \frac{\alpha I_0 A_{max}}{\alpha I_0 + A_{max}} \quad (61.1)$$

with A_{max} being the maximum rate of photosynthesis and α the initial slope of the light-response curve. In a Panamanian lowland rain forest Ellis *et al.* (2000) measured A_{max} for some tree species (or species of the same genus) that also occur in the cloud forest of the study area: *Trema micrantha* (shade-intolerant, 18 μmol (CO_2) m^{-2} s^{-1}), *Palicourea guianensis* (intermediate, 16 μmol m^{-2} s^{-1}), *Zanthoxylum beliziense* (intermediate, 15 μmol m^{-2} s^{-1}), *Beilschmiedia pendula* (shade-tolerant, 12 μmol m^{-2} s^{-1}). Dillenburg *et al.* (1995) reported the maximum photosynthetic rates of sun leaves of *Liquidambar styraciflua*, which is classified as an intermediate species, to be higher than 20 μmol m^{-2} s^{-1} at an irradiance of 2000 μmol (photons) $m^{-2}s^{-1}$. In a Colombian lower montane cloud forest, Letts *et al.* (this volume) measured A_{max} values ranging from 4 to 11 μmol m^{-2} s^{-1}. For the three light groups, A_{max} was estimated to be 20 μmol m^{-2} s^{-1} for shade-intolerant PFTs, 16 μmol m^{-2} s^{-1} for intermediate, and 10 μmol m^{-2} s^{-1} for shade-tolerant PFTs (Figure 61.1). The slope of the light-response curve α was assumed to be 0.15, 0.2, and 0.25 μmol (CO_2) m^{-2} s^{-1} for shade-intolerant, intermediate, and shade-tolerant PFTs, respectively. Self-shading of the canopy was accounted for using the approach of Thornley and Johnson (1990). The proportion of light transmitted by leaves was estimated to be 0.1 (Larcher, 2001).

Respiration processes can be divided into growth respiration during the build-up of new biomass, and maintenance respiration

Table 61.1 *Definition of plant functional types (PFTs) according to shade tolerance (T) and maximum attainable height (H_{max})*

Plant functional type[a]	PFT	T[b]	H_{max}	Examples
Early-successional small trees	1	i	15 m	*Heliocarpus, Myrsine*
Mid-successional small trees	2	m	15 m	*Miconia, Oreopanax*
Late-successional small trees	3	t	15 m	*Cinnamomum, Ilex*
Mid-successional canopy trees	4	m	25 m	Some *Quercus* spp.
Late-successional canopy trees	5	t	25 m	*Magnolia, Beilschmiedia*
Emergents	6	m	35 m	*Liquidambar, Clethra*

[a] The successional status refers to the stage of succession in which a PFT attains maximum basal area values.

[b] Three levels of shade tolerance are distinguished: i, shade-intolerant; m, intermediate; t, shade-tolerant.

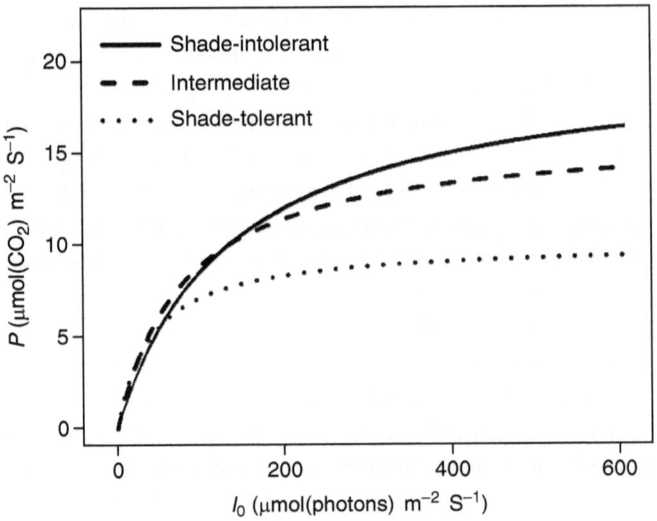

Figure 61.1. Relationship between irradiance (I_0) and photosynthetic production (P) for three levels of shade tolerance.

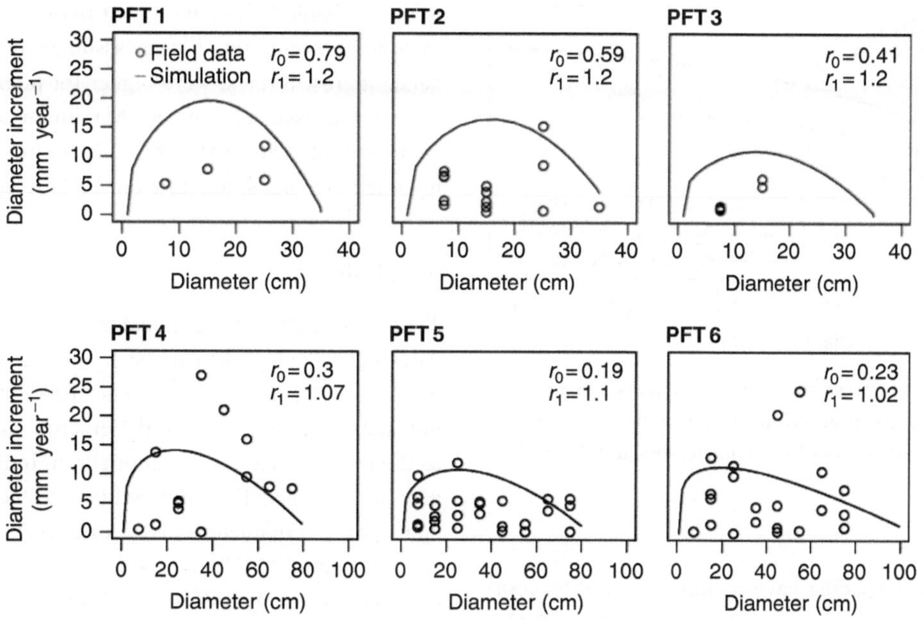

Figure 61.2. Measured and simulated annual diameter increment for the six plant functional types (PFTs) distinguished in the present study. Simulations were carried out under full light conditions (600 μmol m^{-2} s^{-1}) and represent maximum potential growth. Values of r_0 and r_1 are parameters of the formula for maintenance respiration ($r = r_0 \cdot b^{r_1}$). Field measurements were taken in different light environments (Williams-Linera, 1996).

of living biomass. Growth respiration was estimated at 25% of net production of the tree according to Ryan (1991). Maintenance respiration r was assumed to be exponentially dependent on the living biomass (b) of the tree according to $r = r_0 \cdot b^{r_1}$. Parameters of the relationship were fitted such that the simulated diameter increment of each PFT matched the available diameter increment data. The growth of a single tree of each PFT under full sunlight conditions was simulated and the resulting annual diameter increment was compared to measured increment of TMCF species from the Botanical Garden "Francisco Javier Clavijero" in Xalapa, Mexico, and various TMCF fragments in the area (Williams-Linera, 1996). Simulated and observed diameter increment values are shown in Figure 61.2.

For 26 species that occur in the TMCF of central Veracruz, or for species of the same genus, stem wood densities are available (G. Bárcenas, unpublished data; Aguilar-Rodríguez *et al.*, 2001). Wood densities of species with the same shade tolerance were averaged. Resulting arithmetic means were 0.55, 0.65, and 0.7 g cm^{-3} for shade-intolerant, intermediate, and shade-tolerant species, respectively. No marked differences between wood densities of the respective groups were found. This corresponds with the finding of Aguilar-Rodríguez *et al.* (2001) that the majority of cloud forest species exhibit intermediate wood densities.

TREE GEOMETRY

Above-ground biomass (b) was calculated as

$$b(d) = \frac{\pi}{4} d^2 \cdot h(d) \cdot \frac{f \cdot \rho}{s}, \tag{61.2}$$

with d being tree DBH, h the height of the tree, f the form factor that corrects the deviation of the stem from the idealized conical shape, ρ the wood density, and s the fraction of stem wood biomass of total tree biomass. Height (h) was calculated as

$$h(d) = \frac{d}{\frac{1}{h_0} + \frac{d}{h_1}}, \tag{61.3}$$

with h_0 and h_1 being parameters that were fitted to measured diameter and height for single trees (Aguilar-Rodríguez *et al.*, 2001; Figure 61.3). The form factor f was estimated at 0.5 in accordance with previous model parameterizations for tropical lowland forests (Köhler, 2000). Stem wood biomass was assumed to account for 70% of total above-ground biomass, hence $s = 0.7$. The crown of a tree was assumed to have a cylindrical shape and the crown diameter to be proportional to the stem diameter. The ratio between stem diameter and crown diameter was estimated at 1:20 for all PFTs. In inventory data from central Veracruz the average stem-to-crown diameter ratio of small trees was 1 : 23 (G. Williams-Linera, unpublished data), but the ratio seems to be lower for larger trees. The length of the crown was defined as one-tenth of total tree height for all PFTs. When higher values were used (which might be more realistic), too many tree individuals were killed by the space competition mechanism (see below under *Mortality*). *LAI* for individual trees was assumed to be 2.

MORTALITY

A basic mortality was assumed to affect all trees randomly. Simulated mortality rates were estimations based on the

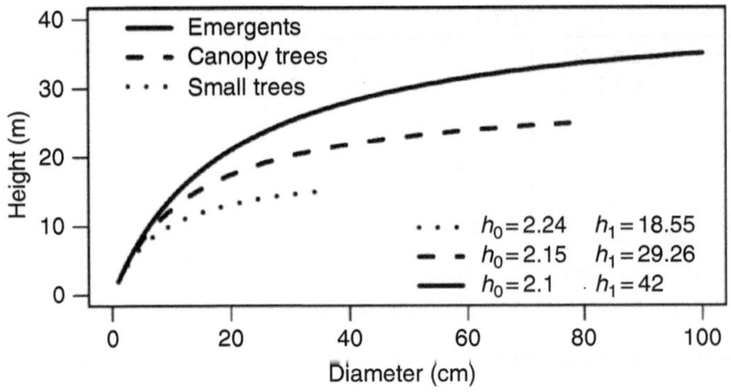

Figure 61.3. Relationship between tree diameter at breast height and tree height for the three height groups distinguished in the present study. Values of h_0 and h_1 are parameters of the diameter–height relationship. See text for details.

assumption that pioneer species have a higher mortality compared to late successional species (e.g. Poorter and Arets, 2003) and that mortality decreases with increasing maximum height. Estimated annual mortality rates were 5%, 1.5%, 1.5%, 1%, 0.8%, and 1% for PFTs 1 to 6, respectively. Small trees up to 10 cm DBH suffered additional mortality, which was assumed to decrease linearly from 2.25% at 1 cm DBH to 0 at 10 cm DBH. Mortality due to crowding was included directly in the model. If in a given height layer the total crown area (i.e. of all trees having their crown in this height layer) exceeded the area of the grid cell, trees were removed randomly from the grid cell until the crown area was smaller than the area of the grid cell.

In Mexican cloud forests trees die mainly through uprooting and less often through snapping or while standing (Bracho and Puig, 1987; Williams-Linera, 1991). In central Veracruz, the fraction of fallen dead trees was 84% (Williams-Linera, 2002), whereas in Tamaulipas 75% of the dead trees had fallen (Arriaga, 1987). Most of the fallen trees belong to upper canopy species that are taller than 15 m (Arriaga, 1987, 2000) or to larger-diameter classes (Williams-Linera, 2002). Based on these data, 80% of dying trees with a DBH > 35 cm (PFTs 4, 5, and 6) were considered in the model to fall and damage the vegetation in the grid cell where their crown hit the ground. Damage was assumed to be proportional to the crown area of the downed tree.

RECRUITMENT

At every time-step new individuals with a diameter of 1 cm appear, if the available light at the forest floor exceeds the minimum light intensity required for germination. The latter light level was estimated to be 10% of full sunlight for shade-intolerant PFTs, 3% for intermediate, and 1% for shade-tolerant PFTs. In shade-house experiments species with intermediate shade tolerance grew well in 5% of full sunlight (Álvarez-Aquino, 2002). The number of new individuals predicted is proportional to the available light at the forest floor, with maximum ingrowth rates at full sunlight. It was assumed that seed numbers, and hence ingrowth rates, were higher for pioneer species compared to late successional species. Maximum annual ingrowth rates were estimated at 1000, 400, and 250 ha^{-1} for shade-intolerant, intermediate, and shade-tolerant PFTs, respectively.

Simulations

To test the ability of the model to reproduce observed characteristics of the old-growth forest in the study area, inventory data were used from five fragments (1000 m^2 each) of relatively undisturbed TMCF in central Veracruz (Williams-Linera, 2002) as the initial situation for a model run. Individuals were grouped according to their PFT and assigned to the corresponding diameter class (20 diameter classes of 5 cm). Ten simulations were run for 1 ha of TMCF over 400 years. In the year 400, diameter distributions for the different PFTs were compared to the inventory data.

To simulate the regeneration of TMCF, model simulations were started from a treeless situation. Again, ten simulations were run for 1 ha of TMCF over 400 years. It was assumed that seed input was not limited by a lack of seed sources and that no further disturbances – other than gap creation by falling trees – occurred during the course of succession.

RESULTS

Comparison of model predictions with field observations

SIMULATION OF OLD-GROWTH FOREST

Figure 61.4 shows the simulated development of the old-growth forest using the inventory data as initial situation. At the beginning of the simulation period stem numbers declined due to the lack of small individuals (<5 cm DBH) in the inventory data, before recovering and fluctuating around the observed value (Figure 61.4a). Total basal area (Figure 61.4b) decreased by 10 m^2 ha^{-1}, but the relative importance of the respective PFTs was maintained. Simulated total stem numbers and basal area are within the range of observations in the forest fragments of the study area (Table 61.2).

DIAMETER DISTRIBUTIONS

A more detailed examination of model results can be achieved through comparison of diameter distributions of the different PFTs. Such a comparison ensures that the size structure of simulated and real forest is similar. Figure 61.5 shows stem number–diameter distributions of inventory data and simulated old-growth forest for all PFTs. For PFTs 5 and 6, which

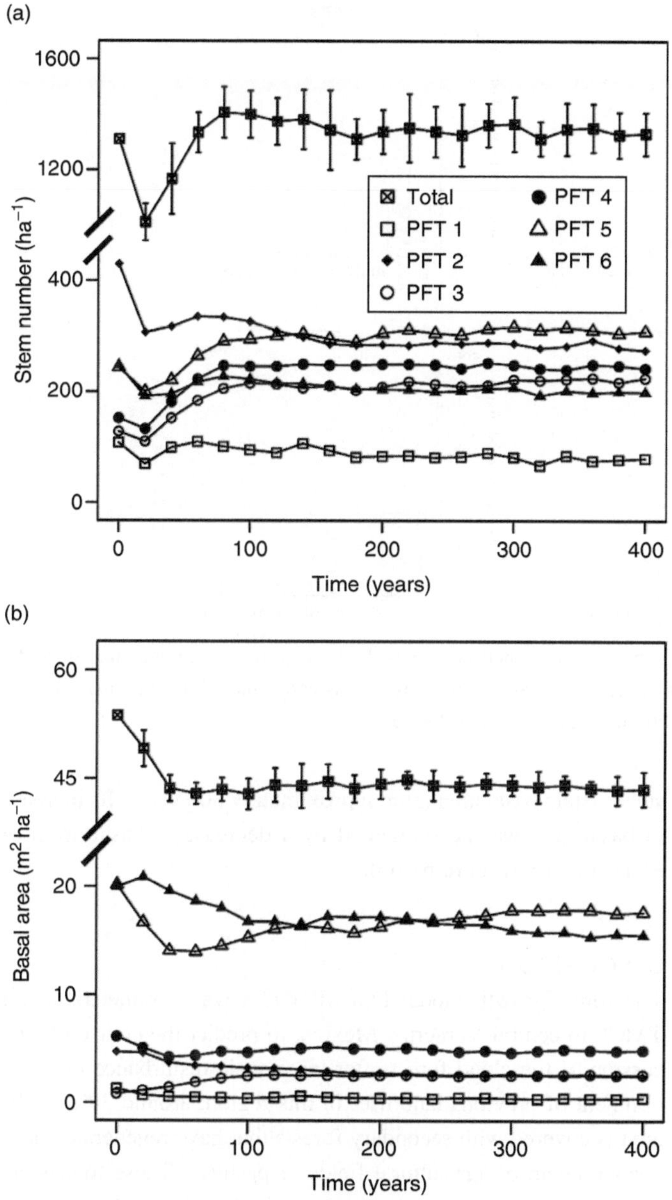

Figure 61.4. Simulation of the dynamics of old-growth montane cloud forest in central Veracruz. Inventory data (individuals ≥ 5 cm DBH) from five forest fragments (0.5 ha) are used as initial situation. Stem numbers (a) and basal area (b) are means of ten simulations for 1 ha and 400 years (individuals ≥ 5 cm DBH). Standard deviation is shown for total stem numbers and basal area.

accounted for the largest trees and represented most of the forest's basal area, the diameter distributions were reproduced well. Similarly, mean and variability of stem numbers of pioneer species (PFT 1) were well captured. In the model, stem numbers for PFT 2 were underestimated, and those for PFT 3 and PFT 4 in the smallest diameter class overestimated. However, such differences would seem acceptable, given the high variability among the study sites.

OTHER OBSERVATIONS

Tree mortality in the model is caused by different processes, and therefore, resulting mortality rates are not known *a priori*. In field studies in the same forest fragments in central Veracruz, observed annual mortality rates ranged between 1% and 12% (Williams-Linera, 2002; individuals ≥5 cm DBH, 2 years observation period, 0.1 ha). Mean annual mortality rate in the model simulation was approximately 5.5%. In several studies of montane forests, available light levels at the forest floor have been measured (Zuill and Lathrop, 1975; Ramírez-Marcial *et al.*, 1998; Hafkenscheid, 2000; Motzer, 2005). Zuill and Lathrop (1975) measured 1% of incoming radiation at the forest floor in an evergreen montane rain forest and 8.4% in pine–oak–*Liquidambar* forest in Chiapas, southern Mexico. Also in Chiapas, Ramírez-Marcial *et al.* (1998) measured a light intensity of 31 μmol m^{-2} s^{-1} in TMCF, which would correspond to 5.2% of 600 μmol m^{-2} s^{-1}. Very similar values have been reported for an upper montane cloud forest of intermediate stature (12 m) in Jamaica (5.0% of the radiation measured at the top of the canopy; Hafkenscheid, 2000) and for a taller (18 m) lower montane rain forest in Ecuador (5.5%; Motzer, 2005). The presently simulated mean percentage of available light at the forest floor was approximately 10%, but also included gaps. In an Ecuadorian lower montane rain forest, Fleischbein *et al.* (2006) derived *LAI* values ranging from 5.2 to 9.3 based on light transmission measurements. Hafkenscheid (2000) obtained *LAI* values of 3.4–5.1 for various upper montane rain forests in Jamaica based on data of specific leaf area, leaf fall, and leaf life span. The mean simulated *LAI* for the study forest was ~5, close to the value of 4.6 measured recently for similar forest fragments in the study area by Ponette-González *et al.* (2009)

Table 61.2 *Comparison of observed and simulated old-growth forest characteristics. Field observations mostly correspond to small areas (e.g. 0.1 ha), whereas simulation results are mean values for ten simulations (1 ha) and ten points in time*

Forest characteristics	Simulation	Observation	Reference
Total stem number[a]	1325 ha^{-1}	810–1700 ha^{-1}	Williams-Linera (2002)
Total basal area[a]	44 m^2 ha^{-1}	35–89 m^2 ha^{-1}	Williams-Linera (2002)
Mortality rate[a]	5.5%	1%–12%	Williams-Linera (2002)
Available light on forest floor	10%	1%–8.4%	Ramírez-Marcial *et al.* (1998), Zuill and Lathrop (1975)
LAI	5	3.4–9.3	Hafkenscheid (2000), Fleischbein *et al.* (2006)

[a] Individuals ≥5 cm DBH.

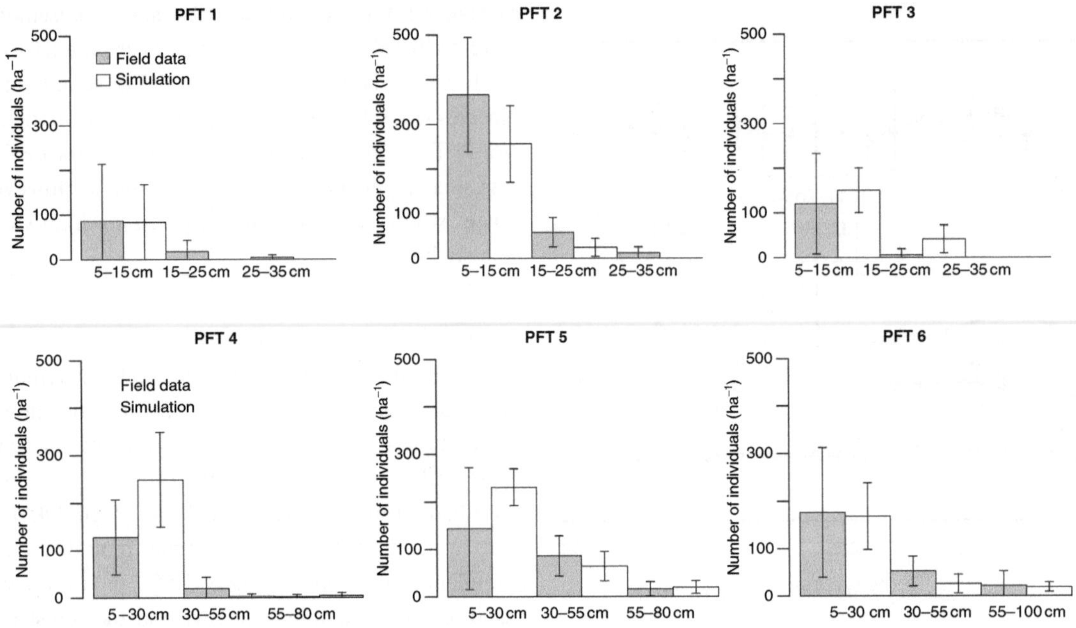

Figure 61.5. Measured and simulated stem number–diameter distributions for the six plant functional types (PFTs) distinguished in the present study. Field data are means and standard deviations for five forest fragments (0.1 ha each). Simulated values are means and standard deviations for ten patches (0.1 ha each) of a simulated old-growth forest, i.e. after 400 years without large-scale disturbance.

and within the range of values observed for tropical montane forests in general (Table 61.2).

Simulation of forest regeneration after disturbance

Figure 61.6 shows the course of succession over 400 years starting from a treeless area. Both total stem numbers ($\geq$5 cm DBH) and total basal area were predicted to reach their steady state after approximately 80–90 years. Simulated stem numbers leveled off at approximately 1350 ha^{-1} whereas total basal area attained a value of ~44 m^2 ha^{-1}.

During the first 20 years all PFTs showed a peak in stem number due to the high light and space availability (Figure 61.6a). Pioneer species (PFT 1) accounted for most of the newly established individuals. Shade-tolerant PFTs (nos. 3 and 5) showed the lowest stem numbers. Self-thinning already started after 10 years, and stem numbers rapidly declined to their steady-state values (Figure 61.6a). In their steady state, pioneer species (PFT 1) were represented by only few individuals, because their establishment was possible in gaps only. All other PFTs maintained similar stem numbers.

Basal area exhibited different dynamics. During the first 20 years pioneer species (PFT 1) accounted for most of the stand's basal area, due to their fast growth. Subsequently, they were rapidly replaced by PFTs with intermediate shade tolerance (PFTs 2, 4, 6), and these reached their maximum basal area after approximately 50 years. PFT 5 (comprising the slow-growing shade-tolerant canopy species) was the last group to attain its steady-state basal area after approximately 300 years. Its increase in basal area was accompanied by a decrease in basal areas of PFTs 4 and 6 (Figure 61.6b).

DISCUSSION

The forest growth model FORMIND2.3 was parameterized for TMCF in central Veracruz, Mexico, to predict the course of succession of the cloud forest after large-scale disturbance or abandonment of previous land use. In this region, around 30% of the area is covered with secondary forests that have regenerated after abandonment of agricultural fields or pastures. These forests are important for conservation of native biodiversity, soil protection, and provision of fuelwood for the local population. It is pertinent, therefore, to know how long the recovery of forest structure and composition after abandonment will take. Before applying the model, it was tested by comparing model outcomes with inventory data of old-growth forest fragments and other field observations (e.g. *LAI*, overall mortality rate, available light on the forest floor). Simulation results indicated that structural variables (such as total stem number and basal area) recovered within a few decades. However, the relative importance of different PFTs continued to change until 300 years after the beginning of the succession.

MODEL PARAMETERS

FORMIND was parameterized on the basis of data on TMCF available in the literature, measurements carried out in the study area, experience gained during former model applications for

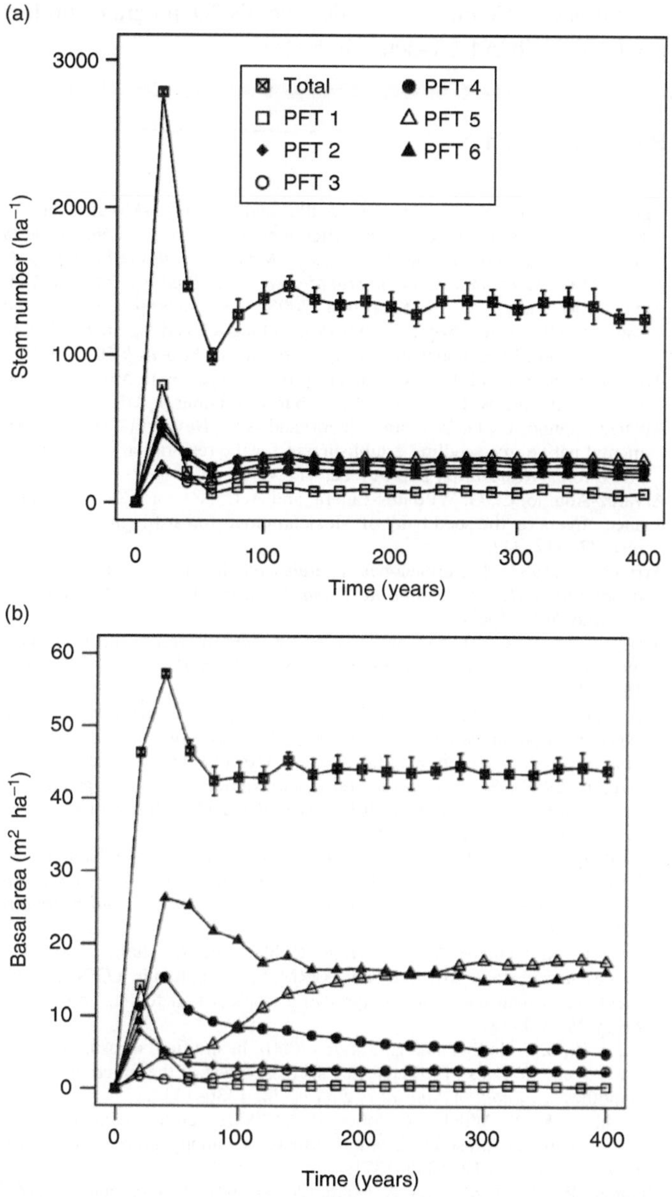

Figure 61.6. Simulation of montane cloud forest regeneration in central Veracruz after large-scale disturbance. Stem numbers (a) and basal area (b) are means of ten simulations for 1 ha and 400 years (individuals ≥5 cm DBH). Standard deviation is shown for total stem numbers and basal area.

tropical lowland rain forest, and parameter estimations by researchers who are well acquainted with central Veracruzan TMCF. However, certain information is lacking and major uncertainty remains about the true value of most model parameters. For some model parameters, measurements became available only after model parameterization. For example, the adopted radiation level of 600 µmol m^{-2} s^{-1} is slightly higher than radiation levels determined subsequently in the area (INECOL-VUA, 2007; F. Holwerda and L. A. Bruijnzeel, personal communication).

Nevertheless, the model includes the state-of-the-art knowledge on this forest type and its application is justified in that it can help to detect gaps in knowledge, and be a useful tool to investigate the importance of certain parameters to the questions addressed (through sensitivity analysis). During model parameterization it already turned out that growth parameters (e.g. initial slope of light response curve) are particularly important for the predicted outcome of the succession. It is suggested, therefore, to focus future field studies on growth characteristics (e.g. photosynthetic rate, respiration, and diameter increment) of important tree species. In particular, additional information on diameter growth would help to adjust model growth parameters.

VERIFICATION OF MODEL RESULTS

Comparison of simulation results with observed patterns confirmed that the simulated forests comply with typical characteristics of old-growth TMCF in central Veracruz. Inventory data of little disturbed old-growth forest constitute an important source of information on forest structure and composition. It is assumed that the inventory data of forest fragments used in this study represent the high variability of TMCF in this region. Some model parameters were adjusted in such a way that the model reproduced important characteristics of the inventory data. If future work is to reveal that some of the fragments were subject to major human disturbance in the past, the current model parameterization will have to be revised.

Field data from a chronosequence study in the study area covering forest ages of 0.5 to 80 years (Muñiz-Castro *et al.*, 2006) allowed for model validation with independent data on forest regeneration after the abandonment of cattle grazing. Forest regeneration was predicted well by the model but a slight overestimation of the speed of forest recovery was observed. Trees in the model seemed to grow slightly faster than in reality, because stem numbers were predicted correctly, whereas basal area was slightly overestimated (Rüger *et al.*, 2008).

SIMULATION OF FOREST REGENERATION AFTER DISTURBANCE

The presently predicted recovery times of less than 100 years for aggregated variables have also been predicted for other tropical ecosystems. Hughes *et al.* (1999) estimated that above-ground biomass of a Mexican humid lowland forest would recover after 30 to 80 years after disturbance. Kammesheidt *et al.* (2001, 2002) simulated the regeneration of a logged-over forest stand in Venezuela and a young secondary forest in Paraguay. In both cases, variables describing forest structure attained a steady state after 50–100 years. On the other hand, forest regeneration on nutrient-poor soils following slash-and-burn agriculture can take much longer. Saldarriaga *et al.* (1988) estimated recovery times of up to 200 years for the basal area and biomass of a mature forest in the Amazon basin. The present study simulated the regeneration of TMCF after a large-scale disturbance assuming unlimited seed dispersal and undisturbed regeneration. The assumption of unlimited seed dispersal may hold true for light wind-dispersed species, because in the TMCF region of central Veracruz diverse land use types with trees (e.g. forest fragments, shaded coffee plantations, pastures with trees) occur side by side (cf. Ponette-González *et al.*, 2009; Barradas *et al.*, this volume). However, for heavy-seeded shade-tolerant species this assumption might be too optimistic. Thus, the simulated scenarios should be regarded as best-case scenarios, especially for shade-tolerant PFTs.

The model has been used to investigate the impact of selective fuelwood harvesting on forest structure and composition (Rüger *et al.*, 2008). Results showed that the overall impact on forest structure and community composition increased linearly with the amount of harvested wood volume. Even at low levels of harvesting, the size structure of the forest became more homogeneous in the long term because large old trees disappeared from the forest. Although recruitment of harvested species benefited from harvesting, the species composition shifted to tree species that are not used for fuelwood.

Conclusions

The process-based forest simulation model FORMIND contains the most important processes to successfully simulate TMCF dynamics after disturbance. With the presented model parameterization it proved possible to reproduce relevant characteristics of TMCF in central Veracruz as shown by a comparison of several observed patterns with simulation results. Simulation of forest regeneration after disturbance suggested that aggregated variables, such as total stem number and total basal area, reached values of old-growth forest after approximately 80 years, whereas the relative importance of different species groups continued to change until 300 years after disturbance. The model can be applied to address questions such as the long-term ecological impacts of selective logging on forest structure and composition and offers the possibility to support the design of management strategies for the natural species-rich forests which achieve a balance between economic needs and ecological goals of the stakeholders.

ACKNOWLEDGEMENTS

We would like to thank Miguel Ángel Muñiz Castro and Guadalupe Bárcenas from the Instituto de Ecología, Xalapa, A.C., Veracruz, Mexico, for providing data on forest regeneration and wood density, respectively. We are grateful to Peter Köhler for support with the source code of FORMIND2.2 and Ludwig Kammesheidt for a literature review. The work was funded by the European Commission under the INCO program, project BIOCORES (ICA4-CT-2001–10095).

REFERENCES

Aguilar-Rodríguez, S., L. Abundiz-Bonilla, and J. Barajas-Morales (2001). Comparación de la gravedad específica y características anatómicas de la madera de dos comunidades vegetales en México. *Anales del Instituto de Biología, Universidad Autónoma de México, Serie Botánica* **72**: 171–185.

Aiba, S., M. Takyu, and K. Kitayama (2005). Dynamics, productivity and species richness of tropical rainforests along elevational and edaphic gradients on Mount Kinabalu, Borneo. *Ecological Research* **20**: 279–286.

Álvarez-Aquino, C. (2002). Regeneration of tree species in Mexican cloud forest. Ph.D. thesis, University of Edinburgh, Edinburgh, UK.

Álvarez-Aquino, C., G. Williams-Linera, and A. C. Newton (2004). Experimental native tree seedling establishment for the restoration of a Mexican cloud forest.*Restoration Ecology* **12**: 412–418.

Álvarez-Aquino, C., G. Williams-Linera, and A. C. Newton (2005). Disturbance effects on the seed bank of Mexican cloud forest fragments. *Biotropica* **37**: 337–342.

Arriaga, L. (1987). Perturbaciones naturales por la caída de árboles. In *El Bosque mesófilo de montaña de Tamaulipas*, pp. 133–152. México, DF: Instituto de Ecología.

Arriaga, L. (2000). Types and causes of tree mortality in a tropical montane cloud forest of Tamaulipas, Mexico. *Journal of Tropical Ecology* **16**: 623–636.

Bracho, R., and H. Puig (1987). Producción de hojarasca y fenología de ocho especies importantes del estrato arbóreo. In *El Bosque mesófilo de montaña de Tamaulipas*, pp. 81–106. México, DF: Instituto de Ecología.

Chave, J. (1999). Study of structural, successional and spatial patterns in tropical rain forests using TROLL, a spatially explicit forest model. *Ecological Modelling* **124**: 233–254.

Dillenburg, L. R., A. H. Teramura, I. N. Forseth, and D. F. Whigham (1995). Photosynthetic and biomass allocation responses of *Liquidambar styraciflua* (Hamamelidaceae) to vine competition. *American Journal of Botany* **82**: 454–461.

Ditzer, T., R. Glauner, M. Förster, P. Köhler, and A. Huth (2000). The process-based stand growth model FORMIX3-Q applied in a GIS environment for growth and yield analysis in a tropical rain forest. *Tree Physiology* **20**: 367–381.

Ellis, A. R., S. P. Hubbel, and C. Potvin (2000). In situ field measurements of photosynthetic rates of tropical tree species: a test of the functional group hypothesis. *Canadian Journal of Botany* **78**: 1336–1347.

Fleischbein, K., W. Wilcke, R. Goller, *et al.* (2006). Rainfall interception in a lower montane forest in Ecuador: effects of canopy properties. *Hydrological Processes* **19**: 1355–1371.

Glauner, R., T. Ditzer, and A. Huth (2003). A new approach for AAC calculation in tropical moist forest: an example from Sabah, Malaysia. *Canadian Journal of Forest Research* **33**: 1–15.

Gomez-Cardenas, M. (2009). Transpiration in contrasting vegetation cover types in the tropical montane cloud forest belt of eastern Mexico. Ph.D. thesis, University of Iowa, Ames, IA, USA.

Grimm, V., U. Berger, F. Bastiansen, *et al.* (2006). A standard protocol for describing individual-based and agent-based models. *Ecological Modelling* **198**: 115–126.

Hafkenscheid, R. (2000). Hydrology and biogeochemistry of tropical montane rain forests of contrasting stature in the Blue Mountains, Jamaica. PhD thesis, VU University Amsterdam, Amsterdam, the Netherlands. Also available at http://dare.ubvu.vu.nl/bitstream/1871/12734/1/tekst.pdf.

Homeier, J., and S. W. Breckle (2008). Gap dynamics in a tropical lower montane forest in South Ecuador. In *Gradients in a Tropical Mountain Ecosystem of Ecuador*, eds. E. Beck, J. Bendix, I. Kottke, F. Makeschin, and R. Mosandl, pp. 311–317. Berlin: Springer-Verlag.

Hughes, R. F., J. B. Kauffman, and V. J. Jaramillo (1999). Biomass, carbon, and nutrient dynamics of secondary forests in a humid tropical region of Mexico. *Ecology* **80**: 1892–1907.

Huth, A., and T. Ditzer (2000). Simulation of the growth of a dipterocarp lowland rain forest with FORMIX3. *Ecological Modelling* **134**: 1–25.

Huth, A., and T. Ditzer (2001). Long-term impacts of logging in a tropical rain forest: a simulation study. *Forest Ecology and Management* **142**: 33–51.

Huth, A., T. Ditzer, and H. Bossel (1998). *The Rain Forest Model FORMIX3*. Göttingen, Germany: Erich Goltze Verlag.

Huth, A., M. Drechsler, and P. Köhler (2005). Using multicriteria decision analysis and a forest growth model to assess impacts of tree harvesting in dipterocarp lowland rain forests. *Forest Ecology and Management* **207**: 215–232.

INECOL-VUA (2007). *Medición y análisis de los efectos de distintos tipos de cubierta forestal sobre los procesos climáticos, hidrológicos y erosivos en Veracruz, México*. Xalapa, Mexico: Instituto de Ecologia, and Amsterdam, the Netherlands: VU University.

Kammesheidt, L., P. Köhler, and A. Huth (2001). Sustainable timber harvesting in Venezuela: a modeling approach. *Journal of Applied Ecology* **38**: 756–770.

Kammesheidt, L., P. Köhler, and A. Huth (2002). Simulating logging scenarios in secondary forest embedded in a fragmented neotropical landscape. *Forest Ecology and Management* **170**: 89–105.

Kira, T. (1978). Community architecture and organic matter dynamics in tropical lowland rainforests of Southeast Asia with special reference to Pasoh Forest, West Malaysia. In *Tropical Trees as Living Systems*, eds. P. B. Tomlinson and M. H. Zimmermann, pp. 26–30. Cambridge, UK: Cambridge University Press.

Köhler, P. (2000). Modelling antropogenic impacts on the growth of tropical rain forests. Ph.D. thesis, University of Kassel, Kassel, Germany.

Köhler, P., and A. Huth (1998). The effect of tree species grouping in tropical rain forest modelling: simulation with the individual based model FORMIND. *Ecological Modelling* **109**: 301–321.

Köhler, P., and A. Huth (2004). Simulating growth dynamics in a South-East Asian rainforest threatened by recruitment shortage and tree harvesting. *Climate Change* **67**: 95–117.

Köhler, P., and A. Huth (2007). Impacts of recruitment limitation and canopy disturbance on tropical tree species richness. *Ecological Modelling* **203**: 511–517.

Köhler, P., A. Huth, and T. Ditzer (2000). Concepts for the aggregation of tropical tree species into functional types and the application on Sabah's dipterocarp lowland rain forests. *Journal of Tropical Ecology* **16**: 591–602.

Köhler, P., T. Ditzer, R. C. Ong, and A. Huth (2001). Comparison of measured and modelled growth on permanent plots in Sabahs rain forest. *Forest Ecology and Management* **144**: 101–111.

Köhler, P., J. Chave, B. Riera, and A. Huth (2003). Simulating long-term response of tropical wet forests to fragmentation. *Ecosystems* **6**: 129–143.

Larcher, W. (2001). *Ökophysiologie der Pflanzen*. Stuttgart, Germany: Verlag Eugen Ullmer.

Leuschner, Ch., G. Moser, C. Bertsch, M. Röderstein, and D. Hertel (2007). Large altitudinal increase in tree root/shoot ratio in tropical mountain forests of Ecuador. *Basic and Applied Ecology* **8**: 219–230.

Liu, J., and P. S. Ashton (1999). Simulating effects of landscape context and timber harvest on tree species diversity. *Ecological Applications* **9**: 186–201.

Monsi, M., and T. Saeki (1953). Über den Lichtfaktor in den Pflanzengesellschaften und seine Bedeutung für die Stoffproduktion. *Japanese Journal of Botany* **14**: 22–52.

Moser, G., D. Hertel, and Ch. Leuschner (2007). Altitudinal change in LAI and stand leaf biomass in tropical montane forests: a transect study in Ecuador and a pan-tropical meta-analysis. *Ecosystems* **10**: 924–935.

Motzer, T. (2003). *Bestandesklima, Energiehaushalt und Evapotranspiration eines neotropischen Bergregenwaldes. Forstmeteorologische und ökophysiologische Untersuchungen in den Anden Süd-Ecuadors*. Mannheim, Germany: Department of Geography, University of Mannheim.

Motzer, T. (2005). Micrometeorological aspects of a tropical mountain forest. *Agricultural and Forest Meteorology* **135**: 230–240.

Muñiz-Castro, M. A., G. Williams-Linera, and J. M. Rey-Benayas (2006). Distance effect from cloud forest fragments on plant community structure in abandoned pastures in Veracruz, Mexico. *Journal of Tropical Ecology* **22**: 431–440.

Muñoz-Villers, L. E., and J. López-Blanco (2008). Land use/cover changes using Landsat TM/ETM images in a tropical and biodiverse mountainous area of central-eastern Mexico. *International Journal of Remote Sensing* **29**: 71–93.

Ponette-González, A. G., K. C. Weathers, and L. M. Curran (2009). Water inputs across a tropical montane landscape in Veracruz, Mexico: synergistic effects of land cover, rain and fog seasonality, and interannual precipitation variability. *Global Change Biology*, doi:10.1111/j.1365-2486.2009.01985.x.

Poorter, L., and E. J. M. M. Arets (2003). Light environment and tree strategies in a Bolivian tropical moist forest: an evaluation of the light partitioning hypothesis. *Plant Ecology* **166**: 295–306.

Ramírez-Marcial, N., S. Ochoa-Gaona, M. González-Espinosa, and P. F. Quintana-Ascencio (1998). Análisis florístico y sucesional en la Estación Biológica Cerro Huitepec, Chiapas, México. *Acta Botánica Mexicana* **44**: 59–85.

Rossignol, J. P. (1987). Los estudios morfoedafológicos en el área Xalapa-Coatepec, Veracruz. In *La morfoedafología en la ordenación de los paisajes rurales*, eds. D. Geissert and J. P. Rossignol, pp. 23–35. Xalapa, México: INIREB, ORSTOM.

Rüger, N., A. G. Gutiérrez, W. D. Kissling, J. J. Armesto, and A. Huth (2007). Ecological impacts of different harvesting scenarios for temperate evergreen rain forest in southern Chile: a simulation experiment. *Forest Ecology and Management* **252**: 52–66.

Rüger, N., G. Williams-Linera, W. D. Kissling, and A. Huth (2008). Long-term impacts of fuelwood extraction on a tropical montane cloud forest. *Ecosystems* **11**: 868–881.

Ryan, M. G. (1991). Effects of climate change on plant respiration. *Ecological Applications* **1**: 157–167.

Saldarriaga, J. G., D. C. West, M. L. Tharp, and C. Uhl (1988). Long-term chronosequence of forest succession in the upper Rio Negro of Colombia and Venezuela. *Journal of Ecology* **76**: 983–958.

Shugart, H. H. (1998). *Terrestrial Ecosystems in Changing Environments*. Cambridge, UK: Cambridge University Press.

Thornley, H. M. J., and I. R. Johnson (1990). *Plant and Crop Modelling: A Mathematical Approach to Plant and Crop Physiology*. Oxford, UK: Clarendon Press.

Vanclay, J. K. (1995). Growth models for tropical forests: a synthesis of models and methods. *Forest Science* **41**: 7–42.

Williams-Linera, G. (1991). Nota sobre la estructura del estrato arbóreo del bosque mesófilo de montaña en los alrededores del campamento "El Triunfo", Chiapas. *Acta Botánica Mexicana* **13**: 1–7.

Williams-Linera, G. (1993). Vegetación de bordes de un bosque nublado en el Parque Ecológico Clavijero, Xalapa, Veracruz, México. *Revista de Biología Tropical* **41**: 443–453.

Williams-Linera, G. (1996). Crecimiento diamétrico de árboles caducifolios y perennifolios del bosque mesófilo de montaña en los alrededores de Xalapa. *Madera y Bosques* **2**: 53–65.

Williams-Linera, G. (2002). Tree species richness complementarity, disturbance and fragmentation in a Mexican tropical montane cloud forest. *Biodiversity and Conservation* **11**: 1825–1843.

Williams-Linera, G., R. H. Manson, and E. Isunza-Vera (2002). La fragmentación del bosque mesófilo de montaña y patrones de uso del suelo en la región oeste de Xalapa, Veracruz, México. *Madera y Bosques* **8**: 73–89.

Zuill, H. A., and E. W. Lathrop (1975). The structure and climate of a tropical montane rain forest and an associated temperate pine–oak–liquidambar forest in the Northern Highlands of Chiapas, Mexico. *Anales del Instituto de Biología, Universidad Nacional Autónoma de México, Serie Botánica* **46**: 73–118.

Part VII
Cloud forest conservation, restoration, and management issues

L. S. Hamilton, L. A. Bruijnzeel, and F. N. Scatena

62 Environmental history and forest regeneration dynamics in a degraded valley of north-west Argentina's cloud forests

H. R. Grau, J. Carilla, and R. Gil-Montero
Universidad Nacional de Tucumán, Tucumán, Argentina

R. Villalba
Instituto Argentino de Nivología, Glaciología y Ciencias Ambientales, Mendoza, Argentina

E. Araoz
Universidad Nacional de Tucumán, Tucumán, Argentina

G. Masse
Instituto Nacional de Estadísticas y Censos, Buenos Aires, Argentina

M. de Membiela
Instituto Argentino de Nivología, Glaciología y Ciencias Ambientales, Mendoza, Argentina

ABSTRACT

Extensive areas of montane cloud forests have been transformed into degraded grasslands due to intensive land use in the past. As a consequence of economic modernization and rural-to-urban migration, land-use intensity is decreasing in many of these areas. This chapter combines analysis of historic land use with dendrochronologic estimates of climate, fire, and tree establishment to explore the interactions between climate, socio-economic changes, and vegetation dynamics in a degraded valley in the cloud forest life zone of NW Argentina. During the twentieth century, population increased and became concentrated in the local capital township. State and services employment increased while density of domestic grazers decreased in the second half of the century. Rainfall increased; the period post-1970 was moister than the previous 250 years. Despite these trends, secondary tree species are not colonizing degraded grasslands in this area. The increase in rainfall and decrease in grazing intensity is negatively associated with tree recruitment, particularly with respect to *Podocarpus parlatorei*, the dominant tree species in secondary forests adjacent to pasture. The interpretation offered here is that decreased grazing and increased rainfall has favored grassland over shrubland. Grasslands are maintained by frequent fire, which eliminates *Podocarpus* seedlings and unpalatable shrubs that, in turn, facilitate *Podocarpus* recruitment by providing perches for seed dispersal and generate a less stressful micro-environment. Only in particular years following periods of intense fire activity, *Alnus acuminata*, a highly light-demanding tree species, recruits. This study suggests that feedbacks between fire, land use, climate, and vegetation may promote resistant degraded grasslands that do not become invaded by forest species even when land-use intensity decreases and climatic conditions become favorable for tree establishment.

INTRODUCTION

In many areas of the tropical Andes, intensive past land use has resulted in the replacement of the original montane forests by grasslands in various stages of degradation (Scott, 1977;

Tropical Montane Cloud Forests: Science for Conservation and Management, eds. L. A. Bruijnzeel, F. N. Scatena, and L. S. Hamilton. Published by Cambridge University Press.

Sarmiento, 1997; Cavelier *et al.*, 1998; Vanacker *et al.*, 2003; Mosandl *et al.*, 2008). Current trends in economic development and globalization are stimulating rural to urban migration (Preston, 1996). One consequence of this pattern is a decrease in land-use intensity in areas that are not suitable for modern agriculture, potentially favoring ecosystem recovery (Aide and Grau, 2004). Cloud forest environments, which are usually characterized by steep slopes and a wet climate (cf. Jarvis and Mulligan, this volume), tend to be unfavorable for modern agriculture. In several such areas out-migration and a reduction in land-use intensity have been observed to occur (e.g. Jokisch and Lair, 2002; Grau *et al.*, 2003a, 2008).

Decreasing land-use intensity should provide opportunities for forest recovery, particularly during climatically favorable years. However, degraded grasslands may persist as a "resilient state" where certain positive feedbacks reinforce their maintenance; for example, where grasslands are frequently burnt, where degraded soils make tree establishment difficult, or where long distances to forest remnants limit seed arrival (Aide and Cavelier, 1994; Hobbs and Norton, 1996; Suding *et al.*, 2004; cf. Hartig and Beck, 2003). Moreover, these factors frequently act to reinforce one another: extensive grasslands favor the spreading of fire; fire reduces soil fertility; poor soils and the dry micro-climates of grasslands favor highly flammable grasses; and fire, a dry micro-climate and extensive distance to seed sources all reduce opportunities for tree recruitment (cf. Aide *et al.*, this volume).

In order to analyze patterns and processes of forest recovery, there is a need for long-term studies. Extra-tropical studies of forest history rely heavily on dendrochronology (Kipfmueller and Swetnam, 2001) but dendrochronologic methods are difficult to apply in tropical environments where the lack of climatic seasonality limits the production of annual tree rings (cf. Martin and Fahey, 2006). In the sub-tropical mountains of NW Argentina, however, the formation of annual tree rings is induced by strong rainfall and moderate seasonality in temperature. During the past decades, the authors have developed dendrochronologic techniques to analyze long-term patterns of climate (Villalba, 1995) and forest disturbance and regeneration (Grau *et al.*, 2003b). In this chapter, dendrochronology is combined with an analysis of historic land use to explore the patterns of forest recovery in the cloud forest Life Zone of NW Argentina, and to relate these to trends in climate and land use. More specifically, the aim is to answer the following questions: (i) how did grazing intensity change in relation to socio-economic trends during the twentieth century?, (ii) what were the local climatic patterns during the last centuries?, and (iii) what were the patterns of tree establishment and how were they related to the land-use and climatic history?

STUDY AREA

The study was conducted in the southern sector of the valley of Los Toldos (22° 24′ S), an area of approximately 4 × 4 km of degraded grasslands and secondary forests. The valley is located in the northern extreme of Argentina (Figure 62.1). Elevation ranges between 1600 and 1900 m.a.s.l. Mean annual temperature is ~13 °C. Frost and snowfalls occur during the winter (May–September). Average annual rainfall is *c.* 1600 mm, and distributed in a monsoonal regime (rainy summers, dry winters). The closest long-term climatic record is that for the city of Oran, located approximately 150 km south of the study area at 600 m.a.s.l. Consistent with regional patterns (Minetti and Vargas, 1997), preliminary dendroclimatologic reconstructions for the Los Toldos area have indicated an increasing trend in rainfall during the second half of the twentieth century (Villalba *et al.*, 1998).

The area is part of the upper montane forest biogeographic unit, and can be considered the southernmost extension of the Neotropical cloud forests along the eastern slopes of the Andes. It is located in the upper Bermejo basin, an area of high conservation priority at the regional and national scale (Brown *et al.*, 2001). The original vegetation, which persists as old-growth forest remnants on steep slopes surrounding the valley, is a montane cloud forest with approximately 25 species of tree per hectare and abundant epiphytes and climbers (Malizia, 2003).

The valley bottom (where this project concentrated) consists of a mosaic of grasslands, shrublands, secondary forests dominated by *Podocarpus parlatorei*, *Alnus acuminata*, and *Cletra scabra*, and areas of bare soil due to severe erosion. Other studies in the region have indicated that secondary forests dominated by *Podocarpus parlatorei* are characteristic of areas where grazing was the main land use; they tend to occur in drier topographic positions (Ramadori *et al.*, 1998; Carilla *et al.*, 2006). *Alnus acuminata*, on the other hand, is a typical pioneer tree that colonizes landslides or recently burnt areas at moister topographic locations (Grau, 1985; Grau and Veblen, 2000).

The dominant land use in the valley is grazing by cattle and sheep. Fire is frequently used to promote resprouting of grasslands of good forage quality and to reduce the density of unpalatable shrubs (Grau, 2003). Fires are ignited during the dry winter, to take advantage of the availability of dry fuel material. At higher elevations, fire is related to above-average rainfall during the previous growing season, suggesting that it is fuel-limited (Grau and Veblen, 2000).

The valley, originally settled by the native Chiriguanos, was under Spanish rule at least since the end of the eighteenth century. At the beginning of the nineteenth century it was a large property that became subdivided into small properties over the century. The main township of the valley is the locality of Los Toldos. The Los Toldos municipality was under Bolivian

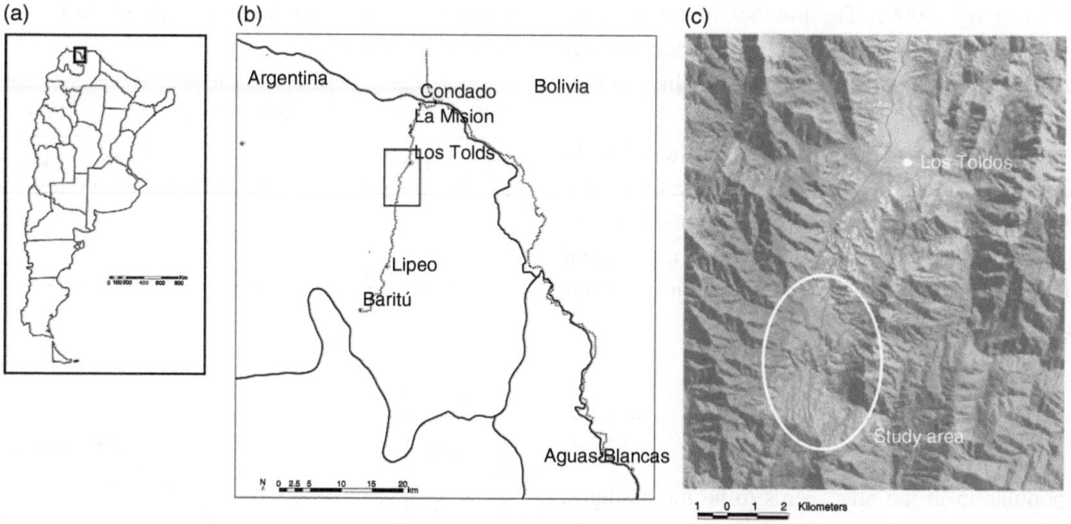

Figure 62.1. (a) Location of the study area in Argentina. (b) Map of the Department of Santa Victoria (including the municipality of Los Toldos), with inset indicating the location of (c). (c) Landsat satellite image of the study area: circle indicates the area of age structure sampling.

rule until 1938, when it was transferred to Argentina as part of a new borders treaty. Throughout its history, Los Toldos has been a marginal area in socio-economic terms (Gil-Montero, 2005). Until the 1930s it was the southern and most isolated municipality of the Tarija Department in southern Bolivia. Since it became part of Argentina, it has remained highly isolated: the only access road comes from Bolivia, it receives very little tourism, and the only exchange with modern economy is through the export of small quantities of high-quality timber, mostly *Cedrela lilloi* (Reboratti, 1998).

METHODS

This chapter presents five complementary data-sets on: (i) human population, (ii) abundance of grazing animals, (iii) rainfall estimates, (iv) tree establishment, and (v) patterns of fire. To analyze trends in human demography official data were used from the Department of Santa Victoria. Los Toldos constitutes one of the three municipalities of Santa Victoria, representing *c.* 20% of the population. In addition, population data were obtained for the township of Los Toldos from the Bolivian census of 1900 and the last three Argentine censuses (1980, 1991, and 2001). Historic data on land use were obtained from the Catastro de Propiedades Rusticas de Tarija, Vicecanton Los Toldos from 1906. Data on domestic animals are based on Argentina's national agricultural censuses for the Department of Santa Victoria in 1952, 1960, 1977, 1988, and 2002.

Rainfall estimates for the past 250 years were based on tree-ring chronologies of *Juglans australis*. Early tree-ring collections in the Los Toldos valley were conducted in the mid-1990s. To increase the consistency of the precipitation estimates around Los Toldos, three additional sites were sampled in upper-elevation, relatively dry valleys. Previous research had indicated that inter-annual variations in tree growth at these sites reflected regional precipitation variations (Villalba *et al.*, 1998). Following standard dendrochronologic procedures (Stokes and Smiley, 1968), tree rings were dated to the exact year of formation and measured with a precision of 0.001 mm. Dating and measuring quality were checked using the program COFECHA (Holmes, 1983). The chronology included 56 samples and covered the past 300 years (1709–2000). However, sample replication was low (< four samples) for the period prior to 1750, hence data will be presented for the post-1750 period only.

Patterns of tree establishment of the two most abundant tree species in secondary forests (*Podocarpus parlatorei* and *Alnus acuminata*) were assessed using dendro-ecological techniques (i.e based on tree rings; Grau *et al.*, 2003b). Twenty transects were distributed in the southern sector of the valley (Figure 62.1c), where a total of 1038 trees were sampled. Trees were sampled at the base, aiming to reach the pith of the tree at its first year of establishment using increment borers. Establishment dates were determined by standard dendrochronologic sample processing and analysis (Stokes and Smiley, 1968). *Alnus acuminata* shows clear annual tree rings, whereas *Podocarpus parlatorei* produces false rings in some years, and partial rings in others. However, these rings can be used for coarse estimates (e.g. 5 to 10 years resolution) of establishment dates (Grau *et al.*, 2003b). The present structure of a tree population is expected to reflect past patterns of both establishment and mortality. To describe patterns of establishment, in addition to the age structure of stands, empirical data on tree mortality as measured in permanent plots were used (Carilla *et al.*, 2006). The number of trees established each year was estimated by adding the expected number of dead trees since the year of establishment to the present, based on constant mortality rates.

Fire dates were also obtained using dendro-ecological techniques (Arno and Sneck, 1977; Grau *et al.*, 2003b). Triangular wedges were extracted using a chainsaw from trees showing fire scars. The wedges were sanded and analyzed under a binocular microscope, and fire dates were determined by counting rings backward from the outermost ring.

RESULTS

Whilst human population in the area seems to be increasing, a major shift in distribution can be observed. Population in the municipality was 766 in 1900 and had grown to 2226 by 2001. But the most marked pattern is a trend toward concentration of population in the capital township. Los Toldos grew from 122 to 833 inhabitants between 1980 and 1990 (Figure 62.2a), which implies a shift of about one-third of the municipal population from rural to a semi-urban location. Santa Victoria, the departmental capital, shows the same trend (Figure 62.2a), suggesting a regional pattern. Age distribution shows a comparative reduction in the most active working age classes (20–29 years), particularly for males (Figure 62.2b), suggesting a pattern of out-migration, likely toward larger urban centers such as Oran, Salta, or Buenos Aires.

At present, although Los Toldos township still does not qualify as "urban" (i.e. having more than 2000 people), a significant proportion of this population has adopted an "urban" lifestyle. For example, in 1900, more than 50% of the workforce was classified as being employed in the agricultural sector and there were no State employees; in 2001, 39% were State employees and a significant proportion of the population worked in the township in the services sector.

The population of domestic animals in Los Toldos in 1900 was dominated by cattle (Table 62.1), a pattern that still persists at present, although in the Department of Santa Victoria as a whole there are more sheep than cattle. The trends in numbers of domestic animals in the Department during the twentieth century also show a clear pattern of a shift toward an urban-based economy. The total number of sheep decreased from more than 60 000 in the 1960s to less than half in 2002 (Figure 62.3a). In the same period, the number of cattle decreased approximately by 30% (Figure 62.3b); numbers of horses, mules, and donkeys also decreased steeply (Figure 62.3c). Although numbers of both cattle and sheep have increased somewhat during the last decade, current densities are much lower than those during the first half of the twentieth century.

Table 62.1 *Census of domestic animals in Los Toldos "vice canton" in 1906*

Cattle	Horses	Donkeys	Mules	Sheep	Pigs	Goats	Poultry
2,531	158	175	36	1346	247	155	598

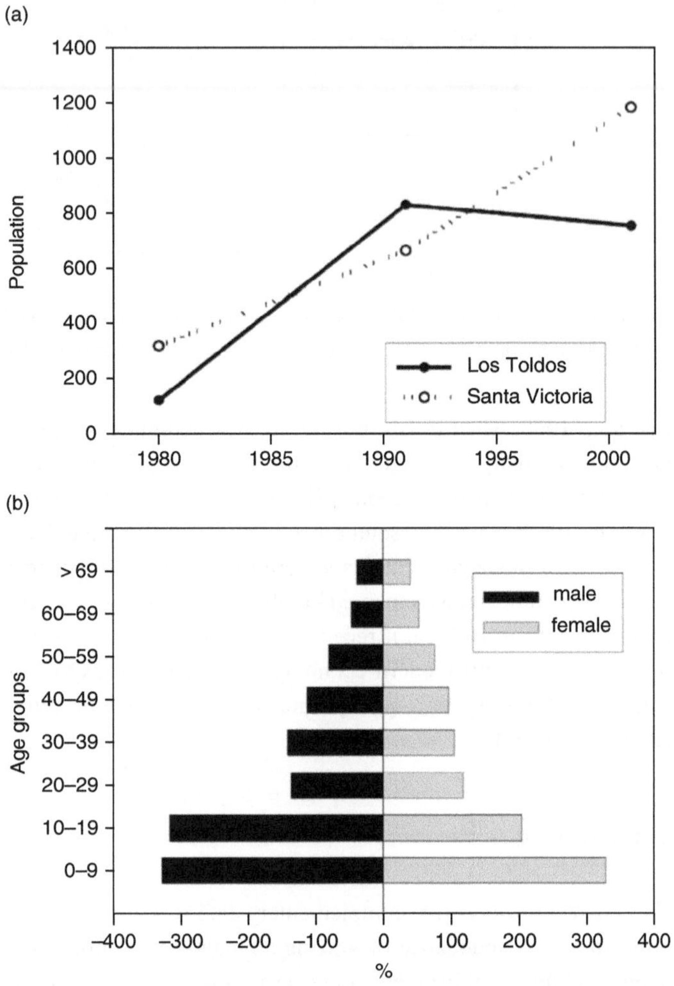

Figure 62.2. (a) Population trend in the townships of Los Toldos and Santa Victoria between 1980 and 2001. (b) Age distribution of the population in Los Toldos in 2001.

Rainfall in Oran showed an increasing trend during the twentieth century. A major increase occurred during the 1950s whereas a further increase was observed in the late 1970s, with high rainfall persisting during the 1980s (Figure 62.4a). Consistent with these observations, the ring-width index for rainfall shows an increasing trend during the twentieth century (Figure 62.4b). The second half of the twentieth century included the moistest 30-year period since 1850, and the peak in rainfall in the 1980s was the highest for the last 250 years. Between 1860 and 1930 there was a comparatively dry period.

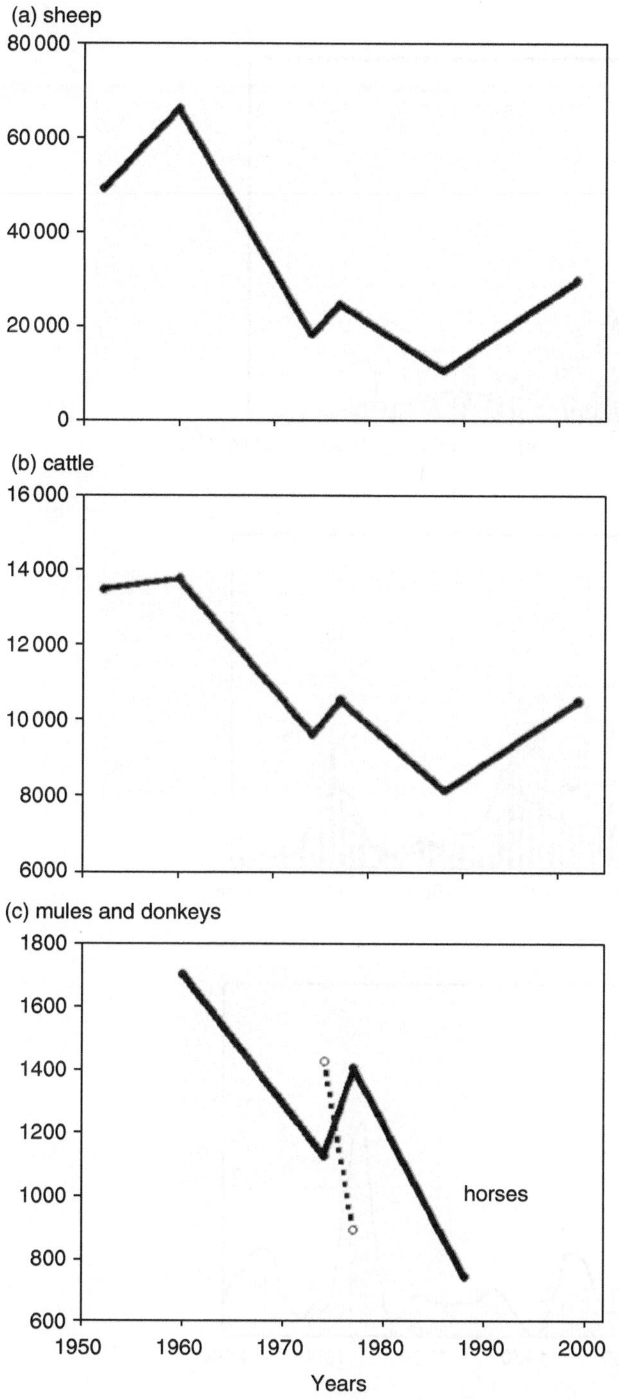

Figure 62.3. Numbers of chief domestic grazing animals in the Department of Santa Victoria during the second half of the twentieth century: (a) sheep, (b) cattle, and (c) horses, and mules and donkeys.

The two tree species analyzed showed different temporal patterns of age structure and estimated tree establishment dates. The age structure of *Podocarpus parlatorei* (Figure 62.5a) shows a relatively even distribution of individuals between 1890 and 1960, followed by a gradual decrease. The curve of estimated tree establishment shows

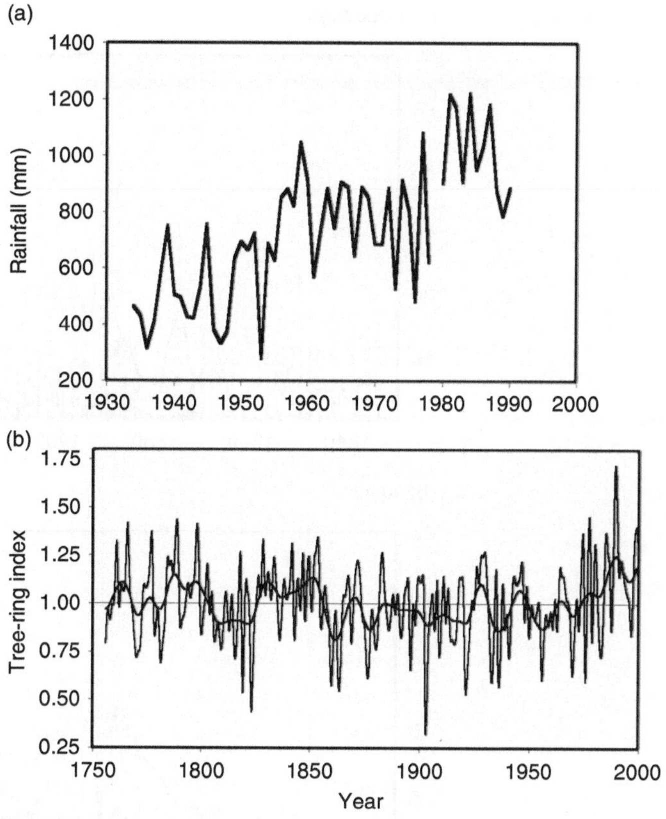

Figure 62.4. (a) Annual rainfall at Oran. (b) Chronology of annual precipitation in the Los Toldos area between the years ~1750 and 2000 based on tree-ring analysis (*Juglans australis*). The smoothed line is a 5-year moving average.

a peak in establishment between 1890 and 1930 (the period characterized by low rainfall) and a clear decrease during the last 70 years, when rainfall increased and grazing intensity decreased. The age structure of *Alnus acuminata* (Figure 62.5b) shows a relatively even distribution of trees established between 1942 and 1974, and a later peak in trees established between 1988 and 1992 coinciding with high rainfall. The curve of estimated tree establishment reflects the same pattern, but in this case the later peak seems to be less important than the previous peaks; for example, the number of estimated trees in 1991 is lower than the peaks in 1942, 1946, 1956, 1972, and 1973 (Figure 62.5b). The fire chronology shows a peak in the late 1960s and a second one in the late 1980s (Figure 62.5c). Both peaks preceded establishment peaks of *A. acuminata*.

DISCUSSION AND CONCLUSIONS

The data on land-use history suggest a clear pattern of decreasing land-use intensity, including a reduction in the density of domestic animals (Figure 62.3). This trend seems to be associated with

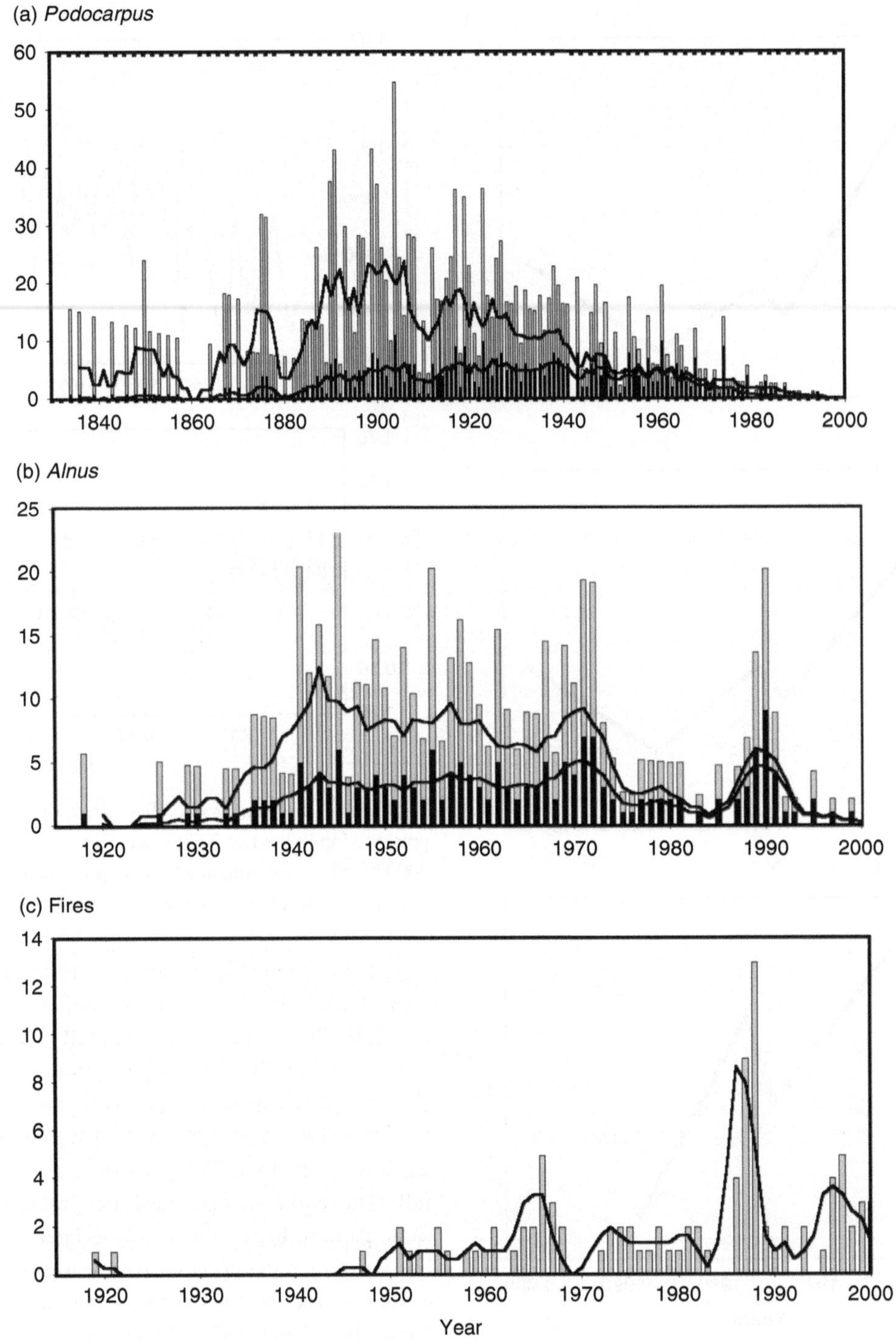

Figure 62.5. Age structure (black bars) and estimated number of trees in each year (gray bars) of (a) *Podocarpus parlatorei*, and (b) *Alnus acuminata*. (c) Fire history. Lines represent 5-year moving averages.

socio-economic changes that promote population out-migration from the region (Figure 62.2b), and a local economy that is based more on services and public-sector employment than on the use of natural resources; this, in turn, promoted population migration from the countryside to the local villages (Figure 62.2a). The trend toward urbanization and reduction in land-use intensity has been observed in other areas of the Neotropical mountains as well (Preston, 1996; Jokisch and Lair, 2002), and represents a significant opportunity for the recovery of natural ecosystems (Aide and Grau, 2004). In line with the regional pattern (Minetti and Vargas, 1997; Villalba *et al.*, 1998), the study area has experienced an increase in rainfall during the twentieth century (Figure 62.4).

Despite these trends in land use and rainfall, forests are not seen to be regenerating in the degraded grassland areas.

Moreover, analysis of the age structure of secondary forest suggests that *P. parlatorei*, the dominant tree species, was established mostly during the drier and more intensively grazed period (i.e. from 1892 to 1920; Figure 62.5a). Although *A. acuminata* showed some more recent peaks in establishment, these include a comparatively small number of trees, and the age structure also suggests an overall reduction in recruitment during the last three decades (Figure 62.5b).

This pattern is likely to be related to the influence of disturbances that maintain the system in the "resilient degraded state" (Hobbs and Norton, 1996; Suding *et al.*, 2004; Aide *et al.*, this volume). *Podocarpus parlatorei* is a species that is partially favored by grazing. It is not palatable to domestic grazers, it is bird-dispersed, and recruits well under toxic shrubs (mostly *Baccharis* spp.) which act as perches for bird dispersal and are characteristic of overgrazed areas (Ramadori, 1998). High establishment of *P. parlatorei* prior to 1960 could therefore be associated with high grazing pressure favoring an abundance of unpalatable shrubs. On the other hand, *P. parlatorei* does not recruit on open grasslands (personal observations of the authors), probably because of slow dispersal to areas without perches for birds, because it is very sensitive to burning, and because of the dry surface micro-climate that characterizes the unshaded grasslands (cf. Aide *et al.*, this volume). Therefore, frequent fires seem to have limited the expansion of *P. parlatorei*, even though the climatic conditions have become more favorable. Reduction in grazing after 1960 may have favored the expansion of grasslands and the accumulation of fuel material (also favored by the increase in rainfall), which has led to a stable fire-maintained system.

Alnus acuminata does not establish under shrubs because it is extremely light-demanding (Grau, 1985; Carilla *et al.*, 2006). Instead, it does establish in recently burned grasslands and it has high growth rates which allow it to achieve fire resistance in approximately 10 years (Grau and Veblen, 2000). The present observations indicate that the two later peaks in *Alnus* establishment (Figure 62.5b) occurred following peaks of fire activity (Figure 62.5c), suggesting that the combined effect of high rainfall with fire may have promoted establishment in the early 1970s and around 1990. One may expect that *Alnus* trees would expand into grasslands at intermediate fire frequencies, i.e. when fires favor recruitment by reducing competition with pre-existing vegetation, but are not too frequent to eliminate juveniles before they achieve fire resistance. Interviews with local people and analysis of multi-scarred trees indicated that fire frequency in the area is very high (every 2–4 years). Some areas, however, may persist without fire for longer periods, either by chance or by reduction of fuel loads during periods of intense fire activity. This may allow the expansion of *A. acuminata*, favored by the increasing moisture availability.

This study suggests that when disturbances interact with vegetation and land use, decreasing land-use intensity is not necessarily followed by forest recovery. Reduction in grazing may not be an efficient way to promote cloud forest recovery on degraded grasslands when these are subject to recurrent fire. Dynamic models should be used to explore the interactions between fire, grazing, climate, and vegetation dynamics (e.g. Higgins and Richardson, 1999; Bai *et al.*, 2004); and to understand the mechanisms leading to feedbacks between disturbances and vegetation. The improved understanding of the complex dynamics of the system may then be used to explore strategies for ecological restoration (e.g. management of fire and grazing patterns) under these favorable scenarios of land use and climate change (cf. Aide *et al.*, this volume).

The present case study provides an analysis of more than a century of environmental history of a system that is comparable to many areas in the tropical Andes. Current efforts to expand this data-set and model the main interactions will contribute to the development of a more general theoretical framework which should help to understand and manage the dynamics of degraded systems in similar tropical cloud forest life zones (cf. Martínez *et al.*, 2008; Asbjornsen and Garnica-Sanchez, this volume; Báez *et al.*, this volume; Hofstede *et al.*, this volume).

ACKNOWLEDGEMENTS

This study was funded with grants from the National Geographic Society (USA), the International Foundation for Science (Sweden), the Consejo de Investigaciones de la Universidad Nacional de Tucumán (Argentina), CONICET, the InterAmerican Institute (IAI), and the Agencia Nacional de Promoción Científica y Tecnológica (FONCYT). Jose Gallo and Javier Foguet helped with the field work. Fundación PROYUNGAS provided logistic support. Larry Hamilton provided helpful comments on the manuscript.

REFERENCES

Aide, T. M., and J. Cavelier (1994). Barriers to lowland tropical forest restoration in the Sierra Nevada de Santa Marta, Colombia. *Restoration Ecology* **2**: 219–229.

Aide, T. M., and H. R. Grau (2004). Globalization, migration, and Latin American ecosystems. *Science* **305**: 1915–1916.

Arno, S. F., and K. M. Sneck (1977). *A Method for Determining Fire History in Coniferous Forests of the Mountain West*, Technical Report No. INT-42. Washington, DC: U.S. Department of Agriculture Forest Service.

Bai, Y., Broersma, K., Thompson, D., and T. J. Ross (2004). Landscape-level dynamics of grassland-forest transitions in British Columbia. *Rangeland Ecology and Management* **57**: 66–75.

Brown, A. D., H. R. Grau, L. R. Malizia, and A. Grau (2001). Argentina. In *Bosques nublados del Neotropico*, eds. M. Kappelle and A. D. Brown, pp. 456–464. Heredia, Costa Rica: INBIO.

Carilla, J., H. R. Grau, and A. Malizia (2006). Successional rates in post-grazing secondary forests in NW Argentina mountains. In *Land Use*

Change and Mountain Biodiversity, eds. E. Spehn, M. Liebermann, and C. Korner, pp. 265–276. Boca Raton, FL: CRC Press.

Cavelier, J., T. M. Aide, C. Santos, A. M. Eusse, and J. M. Dupuy (1998). The savannization of moist forests in Sierra Nevada de Santa Marta, Colombia. *Journal of Biogeography* **25**: 901–912.

Grau, A. (1985). La expansión del Aliso del cerro (*Alnus acuminata* HBK, subsp. *acuminata*) en el noroeste de Argentina. *Lilloa* **36**: 237–247.

Grau, H. R. (2003). Ecología del fuego en los ecosistemas montanos superiores del subtrópico. In *Fuego en los ecosistemas de Argentina*, eds. C. Kuntz, S. Bravo, and J. L. Paniglati, pp. 181–188. Santiago del Estero, Argentina: Instituto Nacional de Tecnología Agropecuaria.

Grau, H. R., and T. T. Veblen (2000). Rainfall variability, fire, and vegetation dynamics in subtropical montane forests in northwestern Argentina. *Journal of Biogeography* **27**: 1107–1121.

Grau, H. R., T. M Aide, J. K. Zimmerman, *et al.* (2003a). The ecological consequences of socioeconomic and land use changes in post agricultural Puerto Rico. *BioScience* **53**: 1159–1168.

Grau, H. R., T. A. Easdale, and L. Paolini (2003b). Subtropical dendroecology: dating disturbances and forest dynamics in subtropical mountains of NW Argentina. *Forest Ecology and Management* **177**: 131–143.

Grau, H. R., M. Perez-Ceballos, S. Martinuzzi, X. Encarnación, and T. M. Aide (2008). Cambios socioeconómicos y regeneración del bosque en República Dominicana. In *Restauración ecológica en América Latina*, eds. M. Gonzalez-Espinosa, J. M. Rey-Benayas, and N. Ramírez-Marcial, pp. 211–227. México, DF: Editorial Mundi Prensa.

Gil Montero, R. (2005). Poblaciones de frontera: Los Toldos en los siglos XIX y XX. *Anuario de Estudios Bolivianos Archivisticos y Bibliográficos* **11**: 125–140.

Hartig, K., and E. Beck (2003). The Bracken fern (*Pteridium arachnoideum* Kaulf.) dilemma in the Andes of Southern Ecuador. *Ecotropica* **9**: 3–13.

Higgins, S. I., and D. M. Richardson (1999). Pine invasions in the southern hemisphere: modelling interactions between organisms, environment and disturbance. *Plant Ecology* **135**: 79–93.

Hobbs, R. J., and D. A. Norton (1996). Towards a conceptual framework for restoration ecology. *Restoration Ecology* **4**: 93–110.

Holmes, R. L. (1983). Computer assisted quality control in tree ring dating and measuring. *Tree Ring Bulletin* **43**: 69–76.

Jokisch, B. D., and B. M. Lair (2002). One last stand: forests and change in Ecuador's eastern Cordillera. *Geographical Review* **92**: 235–256.

Kipfmueller, K. F., and T. W. Swetnam (2001). Using dendrochronology to reconstruct the history of forest and woodland ecosystems. In *The Historical Ecology Handbook: A Restorationist's Guide to Reference Ecosystems*, eds. D. Egan and E. A. Howell, pp. 199–228. Washington, DC: Island Press.

Malizia, A. (2003). Host tree preference of vascular epiphytes and climbers in a subtropical montane cloud forest of Northwest Argentina. *Selbyana* **24**: 196–205.

Martin, P. H., and T. J. Fahey (2006). Fire history along environmental gradients in the subtropical pine forests of the Cordillera Central, Dominican Republic. *Journal of Tropical Ecology* **22**: 289–302.

Martínez, A., M. D. Mahecha, G. Lischeid, and E. Beck (2008). Succession stages of vegetation regeneration: secondary tropical mountain forests. In *Gradients in a Tropical Mountain Ecosystem of Ecuador*, eds. E. Beck, J. Bendix, I. Kottke, F. Makeschin, and R. Mosandl, pp. 425–431. Berlin: Springer-Verlag.

Minetti, J. L., and V. M. Vargas (1997). Trends and jumps in the annual rainfall in South America, South of 15 S. *Atmosfera* **11**: 205–221.

Mosandl, R., S. Günter, B. Stimm, and M. Weber (2008). Ecuador suffers the highest deforestation rate in South America. In *Gradients in a Tropical Mountain Ecosystem of Ecuador*, eds. E. Beck, J. Bendix, I. Kottke, F. Makeschin, and R. Mosandl, pp. 37–40. Berlin: Springer-Verlag.

Preston, D. (1996). People on the move: migrations past and present. In *Latin America Development: Geographical Perspectives*, ed. D. Preston, pp. 165–187. Harlow, UK: Longman.

Ramadori, D. (1998). Sucesión secundaria en bosques montanos del noroeste Argentino. Ph.D. Thesis, Universidad Nacional de La Plata, La Plata, Argentina.

Reboratti, C. (1998). *El Alto Bermejo: Realidades y conflictos*. Buenos Aires, Argentina: La Colmena.

Sarmiento, F. (1997). Arrested succession in pastures hinders regeneration of Tropandean forests and shreds mountain landscapes. *Environmental Conservation* **24**: 14–23.

Scott, G. A. J. (1977). The role of fire in the creation and maintenance of savanna in the Montaña of Peru. *Journal of Biogeography* **4**: 143–208.

Stokes, M. A., and T. L. Smiley (1968). *An Introduction to Tree Ring Dating*. Chicago, IL: University of Chicago Press.

Suding, K. N., Gross, L. G., and G. R. Houseman (2004). Alternative states and positive feedbacks in restoration ecology. *Trends in Ecology and Evolution* **19**: 46–53.

Vanacker, V., S. Barros, J. Poesen, and J. Deckers (2003). The effect of short-term socio-economic and demographic changes on landuse dynamics and its corresponding geomorphic response with relation to water erosion in a tropical mountain catchment. *Landscape Ecology* **18**: 1–15.

Villalba, R. (1995). Estudios dendrocronólogicos en la selva subtropical de montaña, implicaciones para su conservación y desarrollo. In *Investigación, conservación y desarrollo en selvas subtropicales de montaña*, eds. A. D. Brown and H. R. Grau, pp. 59–68. Tucumán, Argentina: Laboratorio de Investigaciones Ecológicas de las Yungas.

Villalba, R., H. R. Grau, J. Boninsegna, G. Jacoby, and A. Ripalta (1998). Tree ring evidence for long term precipitation changes in subtropical South America. *International Journal of Climatology* **18**: 1463–1478.

63 Impact of deforestation and forest regrowth on vascular epiphyte diversity in the Andes of Bolivia

T. Krömer
Universidad Veracruzana, Xalapa, México

S. R. Gradstein
University of Göttingen, Göttingen, Germany

ABSTRACT

Species richness of vascular epiphytes in mature sub-montane and montane forests is compared that in adjacent 15-year old fallow forests at two sites in the Andes of Bolivia. These forests rank among the richest worldwide in terms of epiphyte diversity. Approximately 500 species (25 families, 110 genera) were recorded in total, whereas 0.33 ha of moist montane forest had up to 175 species. Fallows had 60–70% fewer epiphyte species than did mature forest, but species reductions varied considerably among different groups of epiphytes. Species richness of orchids, bromeliads, and grammitid and filmy ferns was much lower in fallows than in mature forest but was not reduced for hemi-epiphytic aroids, nor for polypodioid and asplenioid ferns. The reduced epiphyte diversity in fallows is explained by structural characteristics of the fallow trees, including the lack of dense epiphytic moss mats, and by the drier micro-climate in the fallows.

INTRODUCTION

Vascular epiphytes, including orchids, aroids, bromeliads, and ferns, are important components of moist tropical montane forests, both in terms of species richness (Gentry and Dodson, 1987; Benzing, 1990; Nieder *et al.*, 1999; Krömer *et al.*, 2005; Liede-Schumann and Breckle, 2008; Catchpole and Kirkpatrick, this volume), and their role in rainfall and cloud water interception and release (Hölscher *et al.*, 2004; Köhler *et al.*, 2007; Köhler *et al.*, this volume; Tobón *et al.*, this volume #26) and nutrient cycling (Nadkarni, 1984; Coxson and Nadkarni, 1995; cf. Chang *et al.*, this volume; Oesker *et al.*, this volume). Moist montane forests have largely disappeared from the Andes due to increasing human pressures in the region, with an estimated 10% still remaining (Hamilton *et al.*, 1995; cf. Mulligan, this volume), although forests are on the return in areas where land is abandoned and people move to urban centers (Jokisch and Lair, 2002; Aide and Grau, 2004; Grau *et al.*, this volume). Until recently, comparatively little was known about the impact of deforestation and subsequent regrowth on epiphytic diversity of montane forest land (e.g. Hietz-Seifert *et al.*, 1996; Dunn, 2000; Barthlott *et al.*, 2001), but information is now becoming increasingly available, e.g. for cryptogamic epiphytes in Costa Rica (Holz and Gradstein, 2005; Köhler *et al.*, 2007; Köhler *et al.*, this volume), cryptogamic and vascular epiphytes in Ecuador (Nöske, 2005; Werner *et al.*, 2005; Nöske *et al.*, 2008; Werner and Gradstein, 2009), and vascular epiphytes in Mexico (Wolf, 2005; Pérez-Peña and Krömer, in press; cf. Hietz, this volume; Wolf, this volume). The primary goal of the present study is to compare vascular epiphyte diversity in mature sub-montane and lower montane forests with that of adjacent 15-year-old secondary forests on abandoned old fields or "fallows" in the eastern Andes of Bolivia.

MATERIAL AND METHODS

Study sites

The study was conducted at two sites located on the humid eastern slopes of the Andes in the Department of La Paz, Bolivia. One site is near Sapecho (450 m.a.s.l.; 15° 27′ 32″ S, 67° 18′ 23″ W), located along the Alto Beni river at the north-eastern base of the Serranía Marimonos. Temperature at Sapecho averages

Tropical Montane Cloud Forests: Science for Conservation and Management, eds. L. A. Bruijnzeel, F. N. Scatena, and L. S. Hamilton. Published by Cambridge University Press.

Table 63.1 *Altitude, number of species/families, and percentage of total number of (hemi-)epiphytes contributed by aroids (Arac), bromeliads (Brom), orchids (Orch), piperoids (Pip), ferns (Pter), and other families (other) for different locations in South America.*

Location	Altitude (m.a.s.l.)	Number of (hemi-) epiphytes/families	Arac	Brom	Orch	Pip	Pter	other
Río Suromoni, Venezuela [(1)] (1061 trees, 1.5 ha)	100	53/10	26	4	36	2	21	11
Madidi, Bolivia [(2)] (8 trees, 0.32 ha)	350	89/14	21	8	30	11	24	6
Sapecho PII, Bolivia [(6)] (8 trees, 0.32 ha)	600	127/18	15	6	38	8	27	6
Cotapata PI, Bolivia [(6)] (8 trees, 0.32 ha)	1600	175/23	6	5	38	6	37	8
Río Guajalito, Ecuador [(3)] (22 branches, surroundings)	1900	166/18	15	8	42	5	21	9
Otonga, Ecuador [(3)] (10 trees, surroundings)	2000	196/21	18	5	25	4	22	26
Sehuencas, Bolivia [(4)] (71 trees, 0.08 ha)	2100	156 /17	1	10	48	8	26	7
La Carbonera, Venezuela [(1)] (71 trees, 0.09 ha)	2300	120/18	3	6	49	8	28	6
La Montaña, Venezuela [(5)] (12 trees, 1.5 ha)	2600	117/12	4	11	53	3	26	3

Sources: Data from [(1)]Engwald (1999), [(2)]Acebey and Krömer (2001), [(3)]Nieder *et al.* (1999), [(4)]Ibisch (1996), [(5)]Kelly *et al.* (1994), [(6)]this study.

c. 20–25 °C and annual rainfall varies from 1500 mm in the valley to 2000, mm on the slopes. The rainy season lasts from October to May, alternating with a marked dry period during June–September (Elbers, 1995). The natural forest is an evergreen or semi-evergreen sub-montane forest comprised of 30–40 m tall trees and with a well-developed shrub layer. As a result of increased population pressure, the lower slopes of the Serranía Marimonos are covered by a mosaic of arable fields and fallow land with recovering forests up to 15 years old and with a canopy height of 12–15 m (Krömer, 2003).

The second study site was located in the Parque Nacional Cotapata (16° 11′ 13″ S, 67° 51′ 54″ W), north of the road from La Paz to Coroico and extending between elevations of 1100 and 5600 m.a.s.l. At 1500–2500 m.a.s.l., temperatures average 13–19 °C and annual rainfall is ~2500 mm. The dry period, June–July, is less severe than in Sapecho (cf. Schawe *et al.*, this volume). The very steep slopes at these elevations are covered by evergreen lower montane rain forest in transition to a gradually more mossy lower montane cloud forest made up of 15–25-m tall trees with dense shrub and herb layers underneath. Due to frequent landslides and conversion of small forest patches to agricultural use, the lower slopes are covered by a mosaic of forests in different stages of succession including 15-year-old fallows with a canopy height of 8–12 m (Krömer, 2003).

Sampling

Vascular epiphyte diversity was studied based on representative sampling of 0.32 ha in 1.0-ha stands of natural forest and in neighboring 15-year-old fallows following the protocol developed by Gradstein *et al.* (2003) (see also discussion in Wolf *et al.*, 2009). One fallow and five sub-montane forest stands were sampled in Sapecho (500–1200, m elevation range), and two fallows and three lower montane forest stands in Cotapata (1500–2200 m elevation range). In each forest stand, eight mature canopy trees were sampled from the base to the outer portion of the tree crown, using single-rope techniques (Perry, 1978). Because the epiphyte flora in the forest understory differed from that on the canopy trees, epiphytes on understory shrubs and young trees were sampled within a 20 × 20 m plot around each sampled canopy tree, using collecting poles and binoculars (Gradstein *et al.*, 2003; Krömer *et al.*, 2007). Fallows, generally made up of trees that are too fragile to climb, were sampled in eight randomly placed 400-m^2 plots in each stand.

RESULTS AND DISCUSSION

Vascular epiphyte diversity in mature forest

Almost 500 species of vascular epiphytes (distributed over 25 families and 110 genera) were recorded (Krömer, 2003): 255 in the sub-montane forest of Sapecho (23 families, 87 genera) and 292 in the lower montane forest of Cotapata (24 families, 76 genera). Species assemblages between the two sites were markedly different; only 54 species (11%) occurred in both localities, indicating high β-diversity of vascular epiphytes. A comparison of the present data with inventories carried out in other Andean countries at similar elevations and using similar plot sizes, shows that Cotapata ranks among the richer sites in the world in terms of vascular epiphyte diversity (Table 63.1). With 175 species in 0.32 ha, the lower montane rain forest at 1600 m.a.s.l. in Cotapata is as speciose as lower montane stands inventoried elsewhere in Bolivia (Ibisch, 1996) and Ecuador (Nieder *et al.*, 1999; Werner *et al.*, 2005), and richer than lower and upper montane cloud forests in Venezuela (Kelly *et al.*, 1994; Engwald,

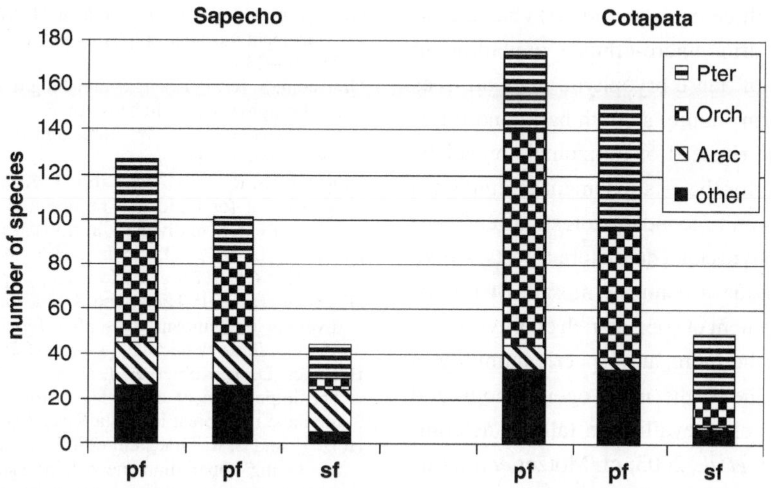

Figure 63.1. Number of epiphytic species of aroids (Arac), orchids (Orch), ferns (Pter), and other families (other) in two natural forests (pf) and one adjacent fallow (sf) stand at Sapecho (sub-montane forest, 600–650 and 550 m.a.s.l.) and at Cotapata (lower montane forest, 1600–1800 and 1650 m.a.s.l.), Bolivia.

1999). In general, epiphyte richness in the Bolivian Andes shows a typical mid-elevation bulge that peaks around 1500 m.a.s.l. (Krömer *et al.*, 2005), similar to other Neotropical sites: in Ecuador (Küper *et al.*, 2004), Perú (Van der Werff and Consiglio, 2004), Costa Rica (Cardelús *et al.*, 2006), and Chiapas, Mexico (Wolf and Flamenco-Sandoval, 2006). Assessment of vascular epiphyte species richness at Sapecho is hampered by a lack of data from other sub-montane forests. It is lower than that found for montane forests in the same area but significantly higher than reported for Amazonian lowland forest or Andean foothill forests (Ter Steege and Cornelissen, 1989; Van Ek *et al.*, 1997; Engwald, 1999; Acebey and Krömer, 2001; Krömer *et al.*, 2005).

Orchids are the most important epiphytic family in terms of species number and may contribute up to 50% of total vascular epiphyte diversity (Kelly *et al.*, 1994; Ibisch, 1996; Engwald, 1999). Much of this diversity is due to the very speciose genera *Epidendrum*, *Maxillaria*, *Pleurothallis*, and *Stelis*. Ferns are the second most speciose group, with diversity at Cotapata almost equaling that of orchids. In Ecuador, total species richness of orchids is more than five times higher than that of ferns (Jørgensen and León-Yánez, 1999), vs. three times higher in Bolivia (Ibisch, 1996). Species richness of epiphytic aroids, bromeliads, and piperoids, as well as that of all other families, is low compared to that of orchids and ferns, and usually contributes less than 10% to total diversity for each group.

Vascular epiphyte diversity in forest and fallows

Forest destruction impacts may be more severe for vascular epiphytes than for other groups of plants (Turner *et al.*, 1994; cf. Werner and Gradstein, 2009; Nadkarni, this volume). In lowland rain forest fragments in Singapore, epiphyte diversity was affected twice as much as that of all vascular plants taken together. The severe reduction of species richness of epiphytes following deforestation was corroborated by the present study. Species richness in 15-year-old fallows was dramatically reduced as compared with mature forest, especially for orchids, bromeliads, and grammitid and filmy ferns. Fallows had 60–70% fewer species than neighboring natural forest (Krömer and Gradstein, 2003) (Figure 63.1). Reduction in species richness varied between groups, however, and was ~90% for orchids and ~45% for ferns. Species richness was not reduced for hemi-epiphytic aroids nor for polypodioid and asplenioid ferns.

The presently found marked differences in species reduction for orchids and ferns agree with earlier findings from secondary forests in Singapore (Turner *et al.*, 1994) but differ from reports for lower montane cloud forest in Venezuela (Barthlott *et al.*, 2001), where the reduction was ~60% for both groups. Moreover, species reductions for bromeliads and for grammitid and filmy ferns were much less at the Venezuelan study site than observed presently. This is probably due to the older age of the secondary forest (25–50 years) and the inclusion of remnant trees in the Venezuelan study. In eastern Mexico and elsewhere in tropical America, remnant forest trees may be very rich in bromeliads and hemi-epiphytic aroids and so these may contribute considerably to the epiphytic diversity of open, secondary habitats (Hietz-Seifert *et al.*, 1996; cf. Dunn, 2000; Werner *et al.*, 2005; Werner and Gradstein, 2009).

CONCLUSIONS

The presently found reduction of vascular epiphyte diversity in fallows compared to that in mature forest in the Bolivian Andes may be explained in terms of three main factors: (i) characteristics of fallow trees, (ii) the drier micro-climate prevailing in fallows, and (iii) the absence of dense bryophytic mats on trees in fallows. Due to their uniform stature, smooth bark, and little-branched crowns with oblique (instead of horizontal) branches, fallow trees are not as suitable for the establishment of epiphytes as mature trees. Also, fallows and secondary forests offer fewer suitable micro-habitats to epiphytes than do mature forests, where old canopy trees tend to provide a complex mosaic of micro-habitats promoting the development of species-rich epiphyte communities (Benzing, 1995). Furthermore, atmospheric humidity in fallows is typically reduced due to the more open canopy and associated higher radiation levels prevailing in fallows as compared to mature forest (Acebey *et al.*, 2003; cf. Motzer *et al.*, this volume). This negatively affects the establishment of many epiphytes, with the exception of xero-tolerant taxa such as hemi-epiphytic aroids and polypodioid ferns (Benzing, 1990). Finally, the absence of dense bryophyte mats on the branches of fallow trees (Acebey *et al.*, 2003; Köhler *et al.*, this volume) also has a negative impact on epiphyte diversity, because these moss layers provide a suitable substrate for germination and establishment of many species of vascular epiphytes (Nadkarni, 1984; Gradstein, 1992; Zotz and Vollrath, 2003) owing to their substantial water retention capacity (Pócs, 1980; Köhler *et al.* 2007; Köhler *et al.*, this volume; cf. Oesker *et al.*, this volume).

REFERENCES

Acebey, A., and T. Krömer (2001). Diversidad y distribución vertical de epífitas en los alrededores del campamento río Eslabón y de la laguna Chalalán, Parque Nacional Madidi, Dpto. La Paz, Bolivia. *Revista de la Sociedad Boliviana de Botánica* **3**: 104–123.

Acebey, A., S. R. Gradstein, and T. Krömer (2003). Species richness and habitat diversification of bryophytes in submontane rain forest and fallows of Bolivia. *Journal of Tropical Ecology* **19**: 9–18.

Aide, T. M., and H. R. Grau (2004). Globalization, migration, and Latin American ecosystems. *Science* **305**: 1915–1916.

Barthlott, W., V. Schmit-Neuerburg, J. Nieder, and S. Engwald (2001). Diversity and abundance of vascular epiphytes: a comparison of secondary vegetation and primary montane rain forest in the Venezuelan Andes. *Plant Ecology* **152**: 145–156.

Benzing, D. H. (1990). *Vascular Epiphytes: General Biology and Related Biota.* Cambridge, UK: Cambridge University Press.

Benzing, D. H. (1995). The physical mosaic and plant variety in forest canopies. *Selbyana* **16**: 159–168.

Cardelús, C. L., R. K. Colwell, and J. E. Watkins (2006). Vascular epiphyte distribution patterns: explaining the mid-elevation richness peak. *Journal of Ecology* **94**: 144–156.

Coxson, D., and N. M. Nadkarni (1995). Ecological roles of epiphytes in nutrient cycles of forest ecosystems. In *Forest Canopies*, eds. M. Lowman and N. M. Nadkarni, pp. 495–546. San Diego, CA: Academic Press.

Dunn, R. R. (2000). Bromeliad communities in isolated trees and three successional stages of an Andean cloud forest in Ecuador. *Selbyana* **21**: 137–143.

Elbers, J. (1995). Estudio de suelos en la zona de colonización Alto Beni, La Paz, Bolivia. *Ecología en Bolivia* **25**: 37–69.

Engwald, S. (1999). *Diversität und Ökologie der vaskulären Epiphyten eines Berg- und eines Tieflandregenwaldes in Venezuela.* Hamburg, Germany: Books on Demand.

Gentry, A. H., and C. H. Dodson (1987). Diversity and biogeography of neotropical vascular epiphytes. *Annals of the Missouri Botanical Garden* **74**: 205–233.

Gradstein, S. R. (1992). The vanishing tropical rain forest as an environment for bryophytes and lichens. In *Bryophytes and Lichens in a Changing Environment*, eds. J. W. Bates and A. W. Farmer, pp. 234–258. Oxford, UK: Clarendon Press.

Gradstein, S. R., N. M. Nadkarni, T. Krömer, I. Holz, and N. Nöske. (2003) A protocol for rapid and representative sampling of vascular and non-vascular epiphyte diversity in tropical rain forests. *Selbyana* **24**: 105–111.

Hamilton, L. S., J. O. Juvik, and F. N. Scatena (eds.) (1995). *Tropical Montane Cloud Forests.* New York: Springer-Verlag.

Hietz-Seifert, U., P. Hietz, and S. Guevara (1996). Epiphyte vegetation and diversity on remnant trees after forest clearance in southern Veracruz, Mexico. *Biology and Conservation* **75**: 103–111.

Hölscher, D., L. Köhler, A. I. J. M. Van Dijk, and L. A. Bruijnzeel (2004). The importance of epiphytes to total rainfall interception by a tropical montane rain forest in Costa Rica. *Journal of Hydrology* **292**: 308–322.

Holz, I., and S. R. Gradstein (2005). Cryptogamic epiphytes in primary and recovering upper montane oak forests of Costa Rica: species richness, community composition and ecology. *Plant Ecology* **178**: 89–109.

Ibisch, P. L. (1996). *Neotropische Epiphytendiversität: Das Beispiel Bolivien.* Wiehl, Germany: Martina Galunder-Verlag.

Jokisch, B. D, and B. M. Lair (2002). One last stand? Montane forests and change on Ecuador's Eastern Cardillera. *Geographical Review* **92**: 235–256.

Jørgensen, P. M., and S. León-Yánez (1999). *Catalogue of the Vascular Plants of Ecuador.* St. Louis, MO: Missouri Botanical Garden Press.

Kelly, D. L., E. V. J. Tanner, E. M. Niclughadha, and V. Kapos (1994). Floristics and biogeography of a rain forest in the Venezuelan Andes. *Journal of Biogeography* **21**: 223–241.

Köhler, L., C. Tobón, K. F. A. Frumau, and L. A. Bruijnzeel (2007). Biomass and water storage dynamics of epiphytes in old-growth and secondary montane cloud forest stands in Costa Rica. *Plant Ecology*, doi:10.1007/511258-006-9256-7.

Krömer, T. (2003). *Diversität und Ökologie der vaskulären Epiphyten in primären und sekundären Bergwäldern Boliviens.* Göttingen, Germany: Cuvillier-Verlag.

Krömer, T., and S. R. Gradstein (2003). Species richness of vascular epiphytes in two primary forests and fallows in the Bolivian Andes. *Selbyana* **24**: 190–195.

Krömer, T., M. Kessler, S. R. Gradstein, and A. Acebey (2005). Diversity patterns of vascular epiphytes along an elevational gradient in the Andes. *Journal of Biogeography* **32**: 1799–1809.

Krömer, T., M. Kessler, and S. R. Gradstein (2007). Vertical stratification of vascular epiphytes in submontane and montane forest of the Bolivian Andes: the importance of the understory. *Plant Ecology* **189**: 261–278.

Küper, W., H. Kreft, J. Nieder, N. Köster, and W. Barthlott (2004). Large-scale diversity patterns of vascular epiphytes in Neotropical montane rain forests. *Journal of Biogeography* **31**: 1477–1487.

Liede-Schumann, S., and S. Breckle (eds.) (2008). *Provisional Checklists of Flora and Fauna of the San Francisco Valley and its Surroundings, Southern Ecuador.* Bonn, Germany: Society of Tropical Ecology.

Nadkarni, N. M. (1984). Epiphyte biomass and nutrient capital of a neotropical elfin forest. *Biotropica* **16**: 249–256.

Nieder, J., S. Engwald, and W. Barthlott (1999). Patterns of neotropical epiphyte diversity. *Selbyana* **20**: 66–75.

Nöske, N. (2005). Effekte anthropogener Störung auf die Diversität kryptogamischer Epiphyten (Flechten, Moose) in einem Bergregenwald in Süd Ecuador. Ph.D. thesis, University of Göttingen, Göttingen, Germany.

Nöske, N., N. Hilt, F. Werner, *et al.* (2008). Disturbance effects on diversity in montane forest of Ecuador: sessile epiphytes versus mobile moths. *Basic and Applied Ecology* **9**: 4–12.

Pérez-Peña, A., and T. Krömer (2009). ¿Qué pueden aportar los acahuales y las plantaciones de cítricos a la conservación de las epífitas vasculares en Los Tuxtlas, Veracruz? In *Avances y perspectivas en la investigación de bosques tropicales y sus alrededores: Los Tuxtlas*, eds. V. H. Reynoso-Rosales and R. Coates. México, DF: UNAM.

Perry, D. R. (1978). A method of access into the crowns of emergent and canopy trees. *Biotropica* **10**: 155–157.

Pócs T. (1980). The epiphytic biomass and its effect on the water balance of two rain forest types in the Uluguru Mountains (Tanzania, East Africa). *Acta Botanica Academiae Scientiarum Hungaricae* **26**: 143–167.

Ter Steege, H., and J. H. C. Cornelissen (1989). Distribution and ecology of vascular epiphytes in lowland rain forest of Guyana. *Biotropica* **21**: 331–339.
Turner, I. M., H. T. W. Tan, Y. C. Wee, *et al.* (1994). A study of plant species extinction in Singapore: lessons for the conservation of tropical biodiversity. *Conservation Biology* **8**: 705–712.
Van der Werff, H., and T. Consiglio (2004). Distribution and conservation significance of endemic species of flowering plants in Peru. *Biodiversity and Conservation* **13**: 1699–1713.
Van Ek, R. C., H. ter Steege, and K. C. Biesmeijer (1997). Vertical distribution and associations of vascular epiphytes in four different forest types in the Guianas. In *Botanical Diversity in the Tropical Rain Forest of Guyana*, ed. R. C. van Ek, pp. 65–89. Wageningen, the Netherlands: Tropenbos Foundation.
Werner, F., and S. R. Gradstein (2009). Diversity of dry forest epiphytes across a gradient of human disturbance in the tropical Andes. *Journal of Vegetation Science* **20**: 59–68.
Werner, F., J. Homeier, and S. R. Gradstein (2005). Diversity of vascular epiphytes on isolated remnant trees in the montane forest belt of southern Ecuador. *Ecotropica* **11**: 21–40.
Wolf, J. H. D. (2005). The response of epiphytes to anthropogenic disturbance of pine–oak forests in the highlands of Chiapas, Mexico. *Forest Ecology and Management* **212**: 376–393.
Wolf, J. H. D., and A. Flamenco-Sandoval (2006). Vascular epiphytes and their potential as a tool in pine–oak forests of Chiapas, Mexico. In *Ecology and Conservation of Neotropical Montane Oak Forests*, ed. M. Kappelle, pp. 375–391. Berlin: Springer-Verlag.
Wolf, J. H. D., S. R. Gradstein, and N. M. Nadkarni (2009). A protocol for sampling vascular epiphyte richness and abundance. *Journal of Tropical Ecology* **25**: 107–121.
Zotz, G., and B. Vollrath (2003). The epiphyte vegetation of the palm *Socratea exorrhiza*: correlations with tree size, tree age, and bryophyte cover. *Journal of Tropical Ecology* **19**: 81–90.

64 Ecology and use of old-growth and recovering montane oak forests in the Cordillera de Talamanca, Costa Rica

D. Hölscher and L. Köhler
University of Göttingen, Göttingen, Germany
M. Kappelle
The Nature Conservancy, San José, Costa Rica
Ch. Leuschner
University of Göttingen, Göttingen, Germany

ABSTRACT

During the last century, substantial parts of old-growth forest (OGF) in the Cordillera de Talamanca, Costa Rica, were clear-cut. However, in many areas, secondary forests (SEC) have become established on abandoned agricultural land. Based on new and published data, this chapter compares upper montane OGF and SEC stands on the Pacific slope of the Cordillera in terms of differences in plant diversity, stand structure, biogeochemical cycles, and forest use. Furthermore, an attempt is made to assess the potential for, and timescale of, recovery of various forest characteristics during regrowth. Oak species (*Quercus* spp.) are present in all successional stages. Alpha-diversity of terrestrial, vascular plant species is significantly higher in early- and mid-successional stands than in OGF, probably due to downslope migration of sub-alpine and alpine terrestrial herb species to cleared and abandoned sites. Beta-diversity is seen to decline during succession. Values of stand leaf area in early-successional stages were similar to those observed in OGF, but biomass of epiphytes was significantly lower in SEC (160–520 kg ha^{-1}) compared to OGF (3400 kg ha^{-1}). Rainfall interception was much higher in OGF (25% of gross precipitation) than in SEC (9% and 15% for early- and mid-successional SEC, respectively), despite similar leaf area for OGF and SEC. A combined experimental and modeling study suggested that the epiphyte layer contributed little (6%) to overall rainfall interception in the OGF and could, therefore, not explain the observed hydrological differences between the various successional stages. Instead, these are thought rather to reflect differences in canopy roughness. Water and nutrient fluxes via stemflow were much higher in SEC than in OGF. For nitrogen, calcium, and magnesium, the sum of nutrient fluxes associated with stemflow and throughfall was exceeded by the corresponding fluxes in leaf litterfall in all successional stages, mainly due to very high leaf fall (up to 12 590 kg ha^{-1} $year^{-1}$ in mid-successional SEC). At present, most montane oak forests in the Cordillera de Talamanca are located within protected areas and have become economically important because of ecotourism. OGF were historically also preferred for charcoal production whereas important non-timber forest products are still gathered by local farmer communities. Presently, SEC only partly fulfill the economic and ecological functions of OGF, although their species composition and the observed rates of change suggest that many characteristics may ultimately recover. Estimated and as yet speculative recovery rates range between a decade for stand leaf area, to more than a century (probably several centuries) for epiphytic biomass.

Tropical Montane Cloud Forests: Science for Conservation and Management, eds. L. A. Bruijnzeel, F. N. Scatena, and L. S. Hamilton. Published by Cambridge University Press.

INTRODUCTION

Next to the well-known, more stunted tropical montane cloud forest formations found around the globe (Stadtmüller, 1987), other montane cloud forest types exist in which huge, 35–60-m tall trees prevail. An example is the tropical montane fagaceous–bamboo forest type found in Central and South America, as well as in certain parts of South-East Asia (Kappelle, 2004). In Costa Rica, mature forest of this type is dominated by different, often gigantic species of black and white oak (*Quercus costaricensis* and *Q. copeyensis*) in the canopy layer, and by *Chusquea* bamboo in the 3–6-m tall understory (Blaser, 1987; Kappelle, 1996).

The old-growth oak forests of the Cordillera de Talamanca, Costa Rica, cover a large area. Today, most of these forests are protected in Biosphere Reserves, National Parks, and surrounding buffer zones. In the past, a substantial part of the oak forest area was converted to pastures or other land-use types, a process that began in the 1940s with the construction of the Pan-American Highway. Where agricultural land was subsequently abandoned, woody regrowth could establish and forest succession followed (cf. Helmer, 2000). In other tropical regions, forest recovery appears to be slower in the montane zone than in the lowlands (Ewel, 1980). This study compares old-growth forest (OGF) and secondary forest (SEC) stands in the upper montane zone of the Cordillera de Talamanca, with the objective to: (i) analyze differences in plant diversity, stand structure, and biogeochemical cycles, (ii) estimate the potential for, and timescale of, recovery of various forest characteristics, and (iii) place these in the context of forest use.

STUDY AREA

Most of the presently discussed work was conducted in the upper part of the Río Savegre watershed on the Pacific slope of the Cordillera. Most sites were located in the Los Santos Forest Reserve which serves as a buffer zone to the La Amistad Biosphere Reserve, a World Heritage Site. Using aerial photographs from 1996, Van Omme (1998) estimated that OGF constituted the dominant land cover (71% of the total) in this region. If one assumes that originally the entire area was forested, then half the area converted to pasture or other land uses is presently covered by SEC.

The studied stands were located at about 2900 m.a.s.l. and belonged to three successional stages. Early-successional stands (ESF) were estimated to be 10–15 years old and mid-successional stands (MSF) ~40 years old. Both were compared to nearby OGF. Vegetation surveys for terrestrial vascular plants were conducted in four replicates per regrowth stage. In contrast, the epiphyte and biogeochemical studies concentrated on one stand per growth stage, each considered "typical" on the basis of vegetation surveys. Basic characteristics of the three focal sites (9° 35′ 40″ N, 83° 44′ 30″ W) are listed in Table 64.1. With respect to tree basal area, they were largely dominated by *Quercus copeyensis*. Besides *Q. copeyensis* some early-successional tree species, mainly from the Asteraceae family, were present in the ESF whereas another oak species, *Q. costaricensis*, co-dominated in the OGF.

Table 64.1 *Characteristics of the three main study sites in the Cordillera de Talamanca, Costa Rica. All sites located at c. 2900 m.a.s.l.*

	Early-successional forest (ESF)	Mid-successional forest (MSF)	Old-growth forest (OGF)
Aspect	S	S	NE
Plot inclination (°)	30	25	31
Maximum tree height (m)	5–9	11–15	30–35
Tree age (year)	10–15	40	–
Stem density (n ha^{-1})[a]	5730	4800	3460
Basal area (m^2 ha^{-1})[a]	4.8	19.3	57.4

[a] Stem density and basal area given for tree stems ≥3 cm diameter at breast height (DBH).

The average annual temperature at a station 15 km east of the study sites (Villa Mills, 3000 m.a.s.l.) is 10.9 °C, and average annual rainfall there is ~2800 mm (Instituto Meteorológico Nacional, 1988). A dry season with precipitation less than 100 mm per month occurs between January and April. The studied stands are considered to be relatively little affected by low cloud or fog incidence. They are not exposed to such strong winds as some cloud forests elsewhere in Costa Rica, such as the ones on either side of the Continental Divide near Monteverde (cf. Clark *et al.*, 1998; Frumau *et al.*, this volume; Häger and Dohrenbusch, this volume). Soils are mainly humic Andisols derived from andesitic volcanic ashes. In the upper horizon, the pH_{KCl} is higher in the ESF (4.9) than in either MSF (3.5) or OGF (3.4). In addition, extractable concentrations of calcium, potassium, and magnesium are significantly higher in the ESF than in the MSF and OGF soils (Hertel *et al.*, 2003). Such differences are probably caused by burning and ash deposition during forest conversion, land preparation, and charcoal production in those areas now covered by SEC. The average thickness of the humus layer increases significantly going from ESF (4.2 ± 1.7 cm) via MSF (9.9 ± 3.8 cm) to OGF (18.8 ± 6.5 cm) (Hertel *et al.*, 2003). If humus accumulation is assumed to increase linearly with stand age, it would recover to OGF levels within a century (77 years based on linear regression analysis).

VEGETATION

Plant species diversity

A total of 176 terrestrial, vascular plant species (122 genera, 75 families) were recorded in a series of 12 0.1-ha forest plots arranged along a successional gradient of recovering forest to OGF (Kappelle *et al.*, 1995a). The Asteraceae were the most speciose plant family. Oak species (*Quercus*) were present in all studied successional stages. Based on presence/absence and abundance data, three successional phases could be distinguished: early-successional forest (ESF, 145 species), mid-successional forest (MSF, 130 species), and old-growth forest (OGF, 96 species). A species ordination using de-trended correspondence analysis (DECORANA; Jongman *et al.*, 1987) illustrated the occurrence of different ecological species groups along the time sequence. Based on the Shannon–Wiener index, α-diversity was significantly higher in ESF and MSF than in OGF, probably due to downslope migration of numerous sub-alpine and alpine terrestrial herb species to cleared and abandoned sites. The invasion of opportunistic sun-loving species was less important than the downward migration of nearby alpine herb species looking for open niches at cleared sites. However, some native invasive species, such as the bracken fern *Pteridium aquilinum* and different species of blackberry (*Rubus* spp.), were occasionally observed to behave very aggressively and cover large areas (so-called *helechales* and *morales*, respectively; Kappelle *et al.*, 1995b; cf. Hartig and Beck, 2003).

Based on Sørensen's similarity coefficients, β-diversity declined during succession. Using linear regression, the theoretical minimum time required for recovery of the terrestrial vascular flora, following clear-cutting and abandonment, was estimated at 65 years (Kappelle *et al.*, 1995a).

Stand structure

Changes in stand structure were recorded in the same 12 0.1-ha forest plots (Kappelle *et al.*, 1996). In the total area of 1.2 ha, 2854 stems $\geq$ 3.0 cm diameter at breast height (DBH) were recorded, distributed over 42 tree species. Tree species richness per 0.1 ha ranged from 15 to 30. Stem density was significantly higher in MSF than in OGF, while basal area was significantly higher in OGF than in ESF or MSF (Table 64.1). Maximum canopy height and basal area increased linearly during the first three decades of stand recovery. Height and DBH showed a significant, logarithmic relationship for all three forest phases. The highest family importance values (FIV) were found for Asteraceae in ESF and for Fagaceae in MSF and OGF. Density of seedlings/saplings <3.0 cm DBH was highest in ESF. A period of 85 years was estimated as the theoretical minimum time needed for recovery of the terrestrial stand structure (including vascular plants only; Kappelle *et al.*, 1996).

Fine root biomass

The fine root biomass in the humus layer increased from ESF (31 g dry weight m^{-2}) via MSF (337 g m^{-2}) to OGF (1128 g m^{-2}) (Hertel *et al.*, 2003). This can be explained partly by the differences in thickness of the humus layer between the three forest stages signaled earlier. However, the organic layer was a preferred substrate for fine roots in the OGF, as indicated by four times higher fine root densities (per unit volume of soil) in the organic layer than in the upper mineral soil. In the two secondary forests, root densities in the organic layer were equal to or lower than those in the mineral soil (see Hertel *et al.*, 2006 for a more detailed analysis and comparison with other montane forests). Using linear regression, the theoretical time required for recovery of fine root biomass in the humus layer would be more than a century (136 years).

Leaf area index

The leaf area index (*LAI*) of the three stands was determined by two different methods. Measurements were conducted with an optical canopy analyzer (LAI-2000, Licor) at 30 replicates per stand in February 2000, whereas leaf area was also estimated from data on leaf lifespan, annual leaf litterfall, and specific leaf area (*SLA*) (cf. Roberts *et al.*, 1996; Moser *et al.*, 2007). With the LAI-2000, no significant difference was found for average *LAI* between the ESF and OGF (3.8 m^2 m^{-2} in both forests). However, a significantly greater leaf area was found for the MSF (4.7 m^2 m^{-2}). The results obtained with the second method gave a similar pattern but values were substantially higher. Based on an average leaf lifespan of 18 months, *LAI* was estimated at 8.8 in the ESF, 13.8 in the MSF, and 7.7 in the OGF. The LAI-2000 has been shown to underestimate *LAI* for complex forest canopies (e.g. Prskawetz and Lexer, 2000; Hölscher, 2004; Moser *et al.*, 2007) but the device may allow a relative comparison between stands. The low values presently obtained with this method may also have been caused in part by the fact that leaf litterfall peaked around the time of the measurements and, as a result, leaf density was temporarily lower. The leaf-mass-related method suffers from the problem that it requires an accurate estimation of leaf lifespan. Most analyzed leaves (n = 332) were alive for more than 15 months but if actual leaf lifespan exceeded the assumed lifespan of 18 months, computed *LAI* values would be even greater as there is little doubt on the accuracy of the determinations of litterfall and *SLA*. For the OGF, the 3.8 m^2 m^{-2} derived with the LAI-2000 seems quite low for a 35-m tall forest compared to literature data, whereas the 7.7 m^2 m^{-2} estimated with the *SLA*-based method is quite high. For various tall Neotropical lowland forests *LAI* values between 4.4 and 8.2 have been reported (Roberts *et al.*, 2005), whereas for tropical montane (cloud) forests of somewhat smaller stature (17–25 m) than the presently studied forest, values of 5.1–7.2 have been derived (Odum, 1970; Steinhardt, 1979; Tanner, 1980; Fleischbein *et al.*, 2005; cf. Moser *et al.*, 2007). A recovery of leaf area within 10 to 15 years (or even less) of forest succession, as indicated by the relatively high values derived for the ESF, does not necessarily disagree with the few observations available for other sites in the tropics (reviewed by Hölscher *et al.*, 2005). A peak in mid-successional *LAI* as observed in the Cordillera de Talamanca also corresponds with modeling results by Rüger *et al.* (this volume) for regenerating cloud forest stands in eastern Mexico.

Epiphyte biomass

In each stand, epiphyte biomass and canopy humus were sampled *in situ* on six oaks (*Q. copeyensis*) reaching the upper canopy level. Epiphytic biomass was largely dominated by non-vascular epiphytes (mosses, liverworts, and lichens) in all stands. They made up 99% of the estimated total epiphyte biomass in the ESF, 89% in the MSF, and 70% in the OGF (Köhler *et al.*, this volume). The estimated total epiphyte biomass (cryptogams and phanerogams) was significantly lower in the secondary forests (ESF 160 kg ha^{-1}; MSF 520 kg ha^{-1}) than in the OGF (3400 kg ha^{-1}). Total epiphytic biomass in the Talamancan OGF falls in the lower range reported for undisturbed tropical montane cloud forests (2130– 44 000 kg ha^{-1}; see review by Köhler *et al.*, 2007). An explanation for the relatively low epiphyte biomass in the Talamancan OGF may be related to the prevailing climatic conditions, notably the pronounced dry season (3 months with less than 100 mm precipitation each) and the low degree of fog incidence. Epiphytic biomass data at the stand level are rare for montane secondary forests. On the windward side of Monteverde, Köhler *et al.* (2007) determined a value of 1.035 t ha^{-1} for *c.* 30-year-old secondary growth vs. 16.2 t ha^{-1} in nearby OGF. Also at Monteverde, but under more sheltered conditions, Nadkarni *et al.* (2004) obtained epiphytic biomass values of 33.1 t ha^{-1} in OGF and only 0.2 t ha^{-1} in 50-year-old regrowth. Similarly, branch-stripping experiments conducted at Monteverde indicated slow recovery of the epiphyte layer even in OGF (Nadkarni, 2000). Although epiphytic biomass increased from early- to mid-successional SEC at Monteverde, it was still very much lower than in OGF. All this leads one to speculate that it may well take more than a century, probably several centuries, for epiphytic biomass to reach OGF values again (cf. Hietz, this volume).

Summarizing, the observed developments in basic stand structural characteristics suggests that the timescale of recovery may be 10–15 years for leaf area, many decades (if not more than a century) for fine root biomass and tree basal area, and probably several centuries for epiphytic biomass.

Table 64.2 *Partitioning of gross precipitation* (P_g) *into throughfall* (TF), *stemflow* (SF), *and interception* (I) *in three upper montane forest plots over a 12-month period in the Cordillera de Talamanca, Costa Rica*

	P_g (mm year^{-1})	*TF* (% of P_g)	*SF* (% of P_g)	*I* (% of P_g)
Early-successional forest (ESF)	2900	69	16	15
Mid-successional forest (MSF)	2900	75	16	9
Old-growth forest (OGF)	2830	73	2	25

ABOVE-GROUND WATER AND NUTRIENT FLUXES

In the three core stands water fluxes via throughfall (*TF*) and stemflow (*SF*) and their associated nutrient fluxes were measured for a year (Hölscher *et al.*, 2003). Gross precipitation at the OGF was 2830 mm, of which *TF* made up 73%, *SF* 2%, and interception 25% (Table 64.2). The secondary forests received a somewhat higher rainfall (2900 mm year^{-1}) and differed from the OGF in having a much higher proportion of *SF* (16% in both stands) and lower rainfall interception. The low *SF* fraction for the OGF agrees with results obtained for most other tall OGFs, both in montane (Bruijnzeel, 2001) and lowland settings (Roberts *et al.*, 2005). The high *SF* values observed in the two SEC fall within the range reported for several relatively small-statured mossy montane OGFs (10–30%; Weaver, 1972; Hafkenscheid, 2000; McJannet *et al.*, 2007; Kumaran, 2008; McJannet *et al.*, this volume) having a relatively high proportion of small or multiple-stemmed trees. However, unlike the Talamancan forests, some of these forests were exposed to considerable amounts of wind-driven rain and correcting for this "occult" input lowered these high apparent stemflow values to only 4–6% (Holwerda *et al.*, 2006; McJannet *et al.*, 2007). Published comparative data for secondary montane forests appear to be almost non-existent. Holwerda *et al.* (2010) observed an *SF* fraction of 4–5% in 19-year-old regrowth dominated by *Alnus jorullensis* (alder) in the cloud forest belt of eastern Mexico under very similar climatic conditions as those prevailing in the Talamancan study area. Epiphyte density in the Mexican SEC was negligible, as was *SF* in nearby tall mossy OGF (Holwerda *et al.*, 2010). As such, it is not possible to conclude that all regenerating montane forests do have high *SF*. On the other hand, high *SF* (23%) has also been reported for a secondary stand in lowland eastern Amazonia (Hölscher *et al.*, 1998) whereas *SF* may be expected to be more frequent in SEC due to high stem densities and the often inclined branches of young trees (Hölscher *et al.*, 2005). The presence or absence of epiphytes does not seem to decisively affect *SF* in montane tropical forests. Steinhardt (1979) did not find a statistically significant difference in *SF* after removing all epiphytes from a plot in a tall lower montane cloud forest in Venezuela having little *SF* under undisturbed conditions. Given the similarity in *SF* between ESF and MSF, it is not possible to estimate the time required for the stemflow fraction to return to similarly low values as observed in the OGF.

At 25%, rainfall interception by the Talamancan OGF falls within the range reported for other montane old-growth forests not subject to appreciable cloud incidence (15–37%; Bruijnzeel and Proctor, 1995; Bruijnzeel, 2001; Bruijnzeel *et al.*, this volume). The only comparative interception study for other tropical secondary forests is that by Holwerda *et al.* (2010) for 19-year-old regrowth in eastern Mexico. However, although the overall interception value of 10% for the Mexican forest is very similar to that obtained presently for the MSF, this must be considered a coincidence as both *SF* (5% vs. 16% in the MSF) and *TF* (85% vs. 75% in the MSF) were rather different in the Mexican forest. The high *TF* fraction observed in the latter forest was considered partly attributable to the deciduous nature of the stand (Holwerda *et al.*, 2010).

High interception values obtained for montane OGF have been attributed repeatedly to high epiphyte loading (Cavelier *et al.*, 1997; Ataroff, 1998; Clark *et al.*, 1998; Fleischbein *et al.*, 2005). The contribution of (mossy) epiphytes to overall rainfall interception in the Talamancan OGF was tested by Hölscher *et al.* (2004). The seasonal variation in moss water contents was measured *in situ* (cf. Köhler *et al.*, this volume) and was reproduced satisfactorily by a running-water balance model. In a second step, the moss water balance model was incorporated in the adapted analytical interception model of Van Dijk and Bruijnzeel (2001). The model predicted observed *TF* values well and enabled the quantification of the contribution by non-vascular epiphytes to total interception as being only 6%. Hence, the hydrological importance of epiphytes in the OGF was rather limited despite their considerable potential water storage capacity (0.81 mm). The main reason for this lies in the fact that during rainy periods (i.e. frequent storms) only part of the total storage capacity was effectively available to absorb additional moisture as the mosses were persistently close to saturation (cf. Köhler *et al.*, this volume; Tobón *et al.*, this volume #26). Therefore, the difference in estimated interception of 10–16% (of total rainfall) between the OGF and the two SEC can be attributed only partially (if at all) to differences in epiphyte biomass between stands. Rather, other factors such as differences in site exposure, the lower trunk storage capacities in the SEC, and the larger aerodynamic roughness of the OGF (allowing a more effective energy exchange with the atmosphere) must be held responsible.

Table 64.3 *Nutrient fluxes in gross precipitation* (P_g) *and in the sum of throughfall and stemflow in early-successional forest (ESF), mid-successional forest (MSF), and old-growth forest (OGF) in the Cordillera de Talamanca, Costa Rica*

	Flux	Water (mm year^{-1})	K	Ca	Mg	Na	N[a]
			(kg ha^{-1} year^{-1})				
Open area (ESF+MSF)[b]	P_g	2900	5.9	5.2	0.7	2.3	4.2
Open area (OGF)[b]	P_g	2830	5.8	4.0	0.7	3.0	3.1
ESF	*Tf*+*Sf*	2469	66.1	14.4	4.4	3.2	3.7
MSF	*Tf*+*Sf*	2651	55.5	18.5	5.1	4.8	7.3
OGF	TF+SF	2121	65.3	15.6	5.6	3.4	4.0

[a] N is the sum of ammonium- and nitrate-nitrogen.
[b] "Open area (OGF)" and "Open area (ESF+MSF)" denote measurements made in clearings close to the OGF, and near the ESF and MSF, respectively.
Source: After Hölscher *et al.* (2003).

Amounts of nutrients carried to the forest floor via *SF* and *TF* (i.e. net precipitation) did not show large differences between the three stands (Table 64.3; Hölscher *et al.*, 2003). The increases in nutrient fluxes during crown passage (enrichment factor between gross and net precipitation; cf. Oesker *et al.*, this volume) of the elements studied were within the range observed for other montane tropical forest stands as reviewed by Hafkenscheid (2000). In the OGF, potassium was enriched by a factor of 9–11, compared to 5–27 in other montane OGFs. Conversely, the ratio of nutrient transport via *SF* to that in net precipitation differed substantially between the OGF and the two SEC. The potassium flux associated with *SF* accounted for 5% of the flux in net precipitation in the OGF vs. 17% in the ESF and 26% in the MSF. Stemflow contributions for the other studied elements were also higher in the two SEC than in the OGF (Table 64.3).

Litterfall was collected over a 12-month period and divided into four fractions (tree leaves, non-vascular epiphytes, vascular epiphytes, twigs and other material). Total litterfall was 9340 kg ha^{-1} year^{-1} in the ESF, 17 200 kg ha^{-1} year^{-1} in the MSF, and 12 880 kg ha^{-1} year^{-1} in the OGF (Table 64.4; Köhler *et al.*, 2008). The contribution of tree leaves was highest in the ESF (80%) and lowest in the OGF (56%). Leaf-fall in all three stands exceeded the reported range (2300–6200 kg ha^{-1} year^{-1}; Hafkenscheid, 2000) for other tropical montane forests, which may be due in part to the marked seasonality in precipitation. Interestingly, whilst similar values were observed in the ESF and OGF (difference of 9% only), leaf-fall peaked in the MSF (75% higher than for the OGF). Not only the amounts of leaf-fall

Table 64.4 *Annual litterfall in early successional forest (ESF), mid-successional-forest (MSF), and old-growth forest (OGF) in the Cordillera de Talamanca, Costa Rica*

	ESF	MSF (kg ha^{-1} year^{-1})	OGF
Tree leaves	7870	12590	7200
Non-vascular epiphytes	30	80	350
Vascular epiphytes	20	40	440
Twigs and others	1420	4490	4890
Total	9340	17200	12880

Source: After Köhler *et al.* (2008).

derived for the Talamancan forests but also the associated nutrient fluxes were high compared to those observed at other montane forest sites. Nutrient fluxes via leaf-fall exceeded those via net precipitation in the case of nitrogen, calcium, and magnesium but were lower in the case of potassium (Hölscher *et al.*, 2003; Köhler *et al.*, 2008).

FOREST USE

Although indigenous peoples have used the OGF on the slopes of the Cordillera de Talamanca for hundreds or even thousands of years, the forests only started to suffer from more intense use during the nineteenth century, when farmers of Spanish descent migrated from Costa Rica's Central Valley into the Talamanca region. They established large plantations of coffee between 1000 and 1800 m.a.s.l. along the seasonally drier Pacific slope, thereby clearing pure stands of oak forests between 1500 (the approximate lower limit of this forest type) and 1800 m.a.s.l. (Kappelle, 1996). However, it was not until the 1950s, after the establishment of the Pan-American Highway, that large-scale cutting began to take place in the core area of the oak forest belt (1800–2600 m.a.s.l.). Since then, uncontrolled exploitation of oak forest for timber and charcoal production, and conversion for the establishment of agricultural cropping (potatoes and cabbage, among other vegetables) and cattle-ranching occurred (Kappelle and Juárez, 1995). Historically, charcoal producers preferred old-growth trees that had to be split into smaller pieces for good combustion. However, as charcoal production from living old-growth trees is prohibited in Costa Rica, charcoal producers now prefer smaller stems as these are not considered "trees" and can be cut without legal permission.

During the mid-1970s, the Costa Rican government became aware of the need for montane oak forest protection and decided to establish two major protected areas: (i) the 62 000-ha Los Santos Forest Reserve, a heavily impacted buffer zone on the Pacific slope which included a mosaic of oak forest patches, coffee plantations, potato fields, grasslands for cattle-ranching, and small villages (Kappelle and Juárez, 1994); and (ii) the 50 000-ha Chirripó National Park, a pristine, fully intact core area, containing montane oak forests, sub-alpine dwarf forest, and alpine *paramo* grasslands at both the Pacific and Atlantic slopes of the Cordillera (Kappelle, 1991; Kappelle *et al.*, 1995b). In 1982, an additional 194 000 ha of adjacent montane forest was set aside as the La Amistad International Park (PILA), straddling the border with Panamá. Recognizing the global importance of the last large tract of tropical montane forest in Central America (i.e. in the high Talamancas), UNESCO declared the Chirripó National Park and the La Amistad International Park, together with neighboring indigenous reserves and the 1964-established Rio Macho Forest Reserve (today a National Park), a Biosphere Reserve and World Heritage Site (Kappelle, 1996). The Los Santos Forest Reserve, a more densely populated area, was kept out of the La Amistad Biosphere Reserve, and since then has served as a true buffer zone, vital for maintaining the integrity of the Biosphere Reserve (Kappelle and Juárez, 1994).

The Talamancan montane forests were not only important to settlers for their precious oak timber, charcoal, or as potential agricultural land. They also housed a wealth of useful plant species. Recent research has revealed that a third of the 590 vascular plant species known from only one small montane valley (San Gerardo de Dota along the Savegre River) have a specific use within the local farmer community (Kappelle *et al.*, 2000). About 24% of the species were used for medicinal purposes, 40% for food, and 25% for construction (timber) or as combustibles (fuelwood, charcoal). Less important uses included dye, ornament (including mosses for Christmas purposes), fodder, gum, oil, and poison. Trunks (53%) and fruits (47%) were the main plant organs used, followed by leaves (33%) and branches (30%) (in some cases more than one organ per plant was used). However, today the use of indigenous species is becoming less common; the trend is to favor synthetic medicines as well as introduced and economically important species. In the medium term, local traditional knowledge of medicinal plants is likely to disappear from rural human communities like that of the San Gerardo de Dota valley as they undergo globalization-driven, modernization processes (cf. Báez *et al.*, this volume; Grau *et al.*, this volume). There is a new focus on rainbow trout hatcheries, apple and peach tree plantations, and, above all, forest-based ecotourism (Kappelle *et al.*, 2000). Ecotourism may additionally contribute to the protection of OGF (cf. Calvo-Alvarado *et al.*, this volume; Tognetti *et al.*, this volume) as the tourists want to see wildlife – especially the resplendent quetzal – and many wildlife species including the quetzal prefer OGF or are even old-growth obligates (Garcia-Rojas, 2006).

CONCLUSIONS

The observed development of stand structure and tree species composition during regeneration of Talamancan upper montane forest, with oak species present in all successional stages, indicated that secondary forests (SEC) will eventually approach old-growth forest (OGF) conditions if left undisturbed for a sufficiently long time. This may be expected to occur within different time periods for different characteristics. The data suggest that stand leaf area may approach values typical of the OGF within roughly 10–15 years, but leaf area may exceed that of OGF during the mid-successional stage as suggested by observations of very high leaf-fall for forest of intermediate age. The minimum time required for recovery of the terrestrial vascular flora was estimated at 65 years vs. at least 85 years for stand basal area and tree height, and 100–150 years for fine root mass. These recovery times appear to be longer than those derived for lower montane or lowland Neotropical forests. Although epiphytic biomass increased from early- to mid-successional SEC, it was still very much lower than in OGF, which leads one to speculate that recovery of epiphyte biomass may take several centuries.

Rainfall partitioning into throughfall, stemflow, and interception differed between SEC and OGF, with much higher stemflow and lower interception values for SEC. The results do not indicate at what timescale rainfall partitioning in SEC may again approach that in OGF. Any changes over time may be related more closely to the development of tree structure than to epiphyte biomass as the present results suggest that epiphytes contributed little to overall rainfall interception in OGF. Nutrient fluxes carried in net precipitation were very similar in the studied stands regardless of age. Leaf-fall and associated nutrient fluxes were high in all stands but peaked in mid-successional forest. Hence, the present data do not allow an estimate of the time-span associated with a return to OGF conditions. In terms of usefulness to the local farmer communities the OGF are more important than SEC. In earlier times the OGF was preferred for charcoal production whereas today people still gather non-timber products such as medicinal plants. OGF and their wildlife are also the major attraction for the increasingly important ecotourism sector in the area. It is concluded that, at present, SEC fulfil only partly the many economic and ecological functions of OGF. However, species composition and the observed course of forest succession suggest that these functions may be largely regained if forest recovery is allowed to continue for a sufficiently long period (cf. Rüger *et al.*, this volume).

REFERENCES

Ataroff, M. A. (1998). Importance of cloud water in Venezuelan Andean cloud forest water dynamics. In *Proceedings of the 1st International Conference on Fog and Fog Collection*, eds. R. S. Schemenauer and H. A. Bridgman, pp. 25–28. Ottawa, Canada: IDRC.

Blaser, J. (1987). *Standörtliche und waldkundliche Analyse eines Eichen-wolkenwaldes* (Quercus *spp.) der Montanstufe in Costa Rica.* Göttinger Beiträge zur Land- und Forstwirtschaft in den Tropen und Subtropen No. 27. Göttingen, Germany: Institute for Silviculture, University of Göttingen.

Bruijnzeel, L. A. (2001). Hydrology of tropical montane cloud forests: a reassessment. *Land Use and Water Resources Research* **1**: 1–18.

Bruijnzeel, L. A., and J. Proctor (1995). Hydrology and biochemistry of tropical montane cloud forests: what do we really know? In *Tropical Montane Cloud Forests*, eds. L. S. Hamilton, J. O. Juvik, and F. N. Scatena, pp. 38–78. New York: Springer-Verlag.

Cavelier, J., M. Jaramillo, D. Solis, and D. DeLeon (1997). Water balance and nutrient inputs in bulk precipitation in tropical montane cloud forest, Panama. *Journal of Hydrology* **193**: 83–96.

Clark, K. L., N. M. Nadkarni, D. Schaeffer, and H. L. Gholz (1998). Atmospheric deposition and net retention of ions by the canopy in a tropical montane forest, Monteverde, Costa Rica. *Journal of Tropical Ecology* **14**: 27–45.

Ewel, J. J. (1980). Tropical succession: manifold routes to maturity. *Biotropica* **12**: 2–7.

Fleischbein, K., W. Wilcke, R. Goller, *et al.* (2005). Rainfall interception in a lower montane forest in Ecuador: effects of canopy properties. *Hydrological Processes* **19**: 1355–1371.

García-Rojas, M. (2006). Diet and habitat preference of the Resplendent Quetzal (*Pharomachrus mocinno costaricensis*) in Costa Rican montane oak forests. In *Ecology and Conservation of Neotropical Montane Oak Forests*, ed. M. Kappelle, pp. 325–336. Berlin: Springer-Verlag.

Hafkenscheid, R. L. L. J. (2000). Hydrology and biogeochemistry of tropical montane rain forests of contrasting stature in the Blue Mountains, Jamaica. Ph.D. thesis, VU University Amsterdam, Amsterdam, the Netherlands. Also available at http://dare.ubvu.vu.nl/bitstream/1871/12734/1/tekst.pdf.

Hartig, K. and E. Beck (2003). The bracken fern (*Pteridium arachnoideum* Kaulf.) dilemma in the Andes of Southern Ecuador. *Ecotropica* **9**: 3–13.

Helmer, E. H. (2000). The landscape ecology of tropical secondary forest in montane Costa Rica. *Ecosystems* **3**: 98–114.

Hertel, D., Ch. Leuschner, and D. Hölscher (2003). Size and structure of fine root systems in old-growth and secondary tropical montane forests (Costa Rica). *Biotropica* **35**: 143–153.

Hertel, D., D. Hölscher, L. Köhler, and Ch. Leuschner (2006). Changes in fine root system size and structure during secondary succession in a Costa Rican montane oak forest. In *Ecology and Conservation of Neotropical Montane Oak Forests*, ed. M. Kappelle, pp. 283–297. Berlin: Springer-Verlag.

Hölscher, D. (2004). Leaf traits and photosynthetic parameters of saplings and adult trees of co-existing species in a temperate broad-leaved forest. *Basic and Applied Ecology* **5**: 163–172.

Hölscher, D., T. D. de A. Sá, R. F. Möller, M. Denich, and H. Fölster (1998). Rainfall partitioning and related hydrochemical fluxes in a diverse and in a mono specific (*Phenakospermum guyannense*) secondary vegetation stand in Eastern Amazonia. *Oecologia* **114**: 251–257.

Hölscher, D., L. Köhler, Ch. Leuschner, and M. Kappelle (2003). Nutrient fluxes in stemflow and throughfall in three successional stages of an upper montane rain forest in Costa Rica. *Journal of Tropical Ecology* **19**: 557–565.

Hölscher, D., L. Köhler, A. I. J. M. van Dijk, and L. A. Bruijnzeel (2004). The importance of epiphytes in rainfall interception by a tropical montane rainforest in Costa Rica. *Journal of Hydrology* **292**: 308–322.

Hölscher, D., J. Mackensen, and J. M. Roberts (2005). Forest recovery in the humid tropics: changes in vegetation structure, nutrient pools and the hydrological cycle. In *Forests, Water and People in the Humid Tropics*, eds. M. Bonell and L. A. Bruijnzeel, pp. 598–621. Cambridge, UK: Cambridge University Press.

Holwerda, F., R. Burkard, W. Eugster, *et al.* (2006). Estimating fog deposition at a Puerto Rican elfin cloud forest site: comparison of the water budget and eddy covariance methods. *Hydrological Processes* **20**: 2669–2692.

Holwerda, F., L. A. Bruijnzeel, L. E. Muñoz, M. Equihua, and H. Asbjornsen (2010). Rainfall and cloud water interception in mature and secondary lower montane cloud forests of central Veracruz, Mexico. *Journal of Hydrology* **384**: 84–96.

Instituto Meteorológico Nacional (1988). *Catastro de las series de precipitaciones medidas en Costa Rica.* San José, Costa Rica: Ministerio de Recursos Naturales, Energía y Minas.

Jongman, R. H. G., C. F. J. ter Braak, and O. F. R. van Tongeren (1987). *Data Analysis in Community and Landscape Ecology.* Wageningen, the Netherlands: Pudoc Publishers.

Kappelle, M. (1991). Distribución altitudinal de la vegetación del Parque Nacional Chirripó, Costa Rica. *Brenesia* **36**: 1–14.

Kappelle, M. (1996). *Los bosques de roble* (Quercus) *de la Cordillera de Talamanca, Costa Rica: Biodiversidad, ecología, conservación y desarrollo.* Santo Domingo de Heredia: Instituto Nacional de Biodiversidad, and Amsterdam, the Netherlands: University of Amsterdam.

Kappelle, M. (2004). Tropical forests: tropical montane forests. In *Encyclopaedia of Forest Sciences*, eds. J. Burley, J. Evans, and J. A. Youngquist, pp. 1782–1793.Oxford, UK: Elsevier.

Kappelle, M., and M. E. Juárez (1994). The Los Santos Forest Reserve: a bufferzone vital for the Costa Rican La Amistad Biosphere Reserve. *Environmental Conservation* **21**: 166–169.

Kappelle, M., and M. E. Juárez (1995). Agroecological zonation along an altitudinal gradient in the montane belt of the Los Santos Forest Reserve in Costa Rica. *Mountain Research and Development* **15**: 19–37.

Kappelle, M., P. A. F. Kennis, and R. A. J. de Vries (1995a). Changes in diversity along a successional gradient in a Costa Rican upper montane *Quercus* forest. *Biodiversity and Conservation* **4**: 10–34.

Kappelle, M., J. G. van Uffelen, and A. M. Cleef (1995b). Altitudinal zonation of montane *Quercus* forests along two transects in the Chirripó National Park, Costa Rica. *Vegetatio* **119**: 119–153.

Kappelle, M., T. Geuze, M. Leal, and A. M. Cleef (1996). Successional age and forest structure in a Costa Rican upper montane *Quercus* forest. *Journal of Tropical Ecology* **12**: 681–698.

Kappelle, M., G. Avertin, M. E. Juárez, and N. Zamora (2000). Useful plants within a campesino community in a Costa Rican montane cloud forest. *Mountain Research and Development* **20**: 162–171.

Köhler, L., C. Tobón, K. F. A. Frumau, and L. A. Bruijnzeel (2007). Biomass and water storage dynamics of epiphytes in old-growth and secondary montane cloud forest stands in Costa Rica. *Plant Ecology* **193**: 171–184.

Köhler, L., D. Hölscher, and Ch. Leuschner (2008). High litterfall in old-growth and secondary upper montane forest of Costa Rica. *Plant Ecology*, doi:10.1007/s11258–008–9421–2.

Kumaran, S., (2008). Hydrometeorology of tropical montane rain forests of Gunung Brinchang, Pahang Darul Makmur, Malaysia. Ph.D. thesis, University of Putra Malaysia, Serdang, Malaysia.

McJannet, D., J. S. Wallace, and P. Reddell (2007). Precipitation interception in Australian tropical rainforests. II. Altitudinal gradient of cloud interception, stemflow, throughfall and interception. *Hydrological Processes* **21**: 1703–1718.

Moser, G., D. Hertel, and Ch. Leuschner (2007). Altitudinal change in LAI and stand leaf biomass in tropical montane forests: a transect study in Ecuador and a pan-tropical meta-analysis. *Ecosystems* **10**: 924–935.

Nadkarni, N. M. (2000). Colonization of stripped branch surfaces by epiphytes in a lower montane cloud forest, Monteverde, Costa Rica. *Biotropica* **32**: 358–363.

Nadkarni, N. M., D. Schaefer, T. J. Matelson, and R. Solano (2004). Biomass and nutrient pools of canopy and terrestrial components in a primary and a secondary montane cloud forest, Costa Rica. *Forest Ecology and Management* **198**: 223–236.

Odum, H. T. (1970). Summary: an emerging view of the ecological system at el Verde. In *A Tropical Rainforest*, eds. H. T. Odum and R. F. Pidgeon, pp. I-199–I-289. Washington, DC: U.S. Atomic Energy Commission.

Prskawetz, M., and M. J. Lexer (2000). Evaluierung des LI-COR LAI-2000 zur Ermittlung des Blattflächenindex in Buchenjungbeständen. *Allgemeine Forst und Jagd-Zeitung* **171**: 185–191.

Roberts, J. M., O. M. R. Cabral, J. P. da Costa, A.-L. C. McWilliam, and T. D. de A. Sá (1996). An overview of the leaf area index and physiological measurements during ABRACOS. In *Amazonian Deforestation and Climate*, eds. J. H. C. Gash, C. A. Nobre, J. M. Roberts, and R. Victoria, pp. 287–305. New York: John Wiley.

Roberts, J. M., J. H. C. Gash, M. Tani, and L. A. Bruijnzeel (2005). Controls on evaporation in lowland tropical rainforest. In *Forests, Water and People in the Humid Tropics*, eds. M. Bonell and L. A. Bruijnzeel, pp. 287–313. Cambridge, UK: Cambridge University Press.

Stadtmüller, T. (1987). *Cloud Forests in the Humid Tropics: A Bibliographic Review.* Turrialba, Costa Rica: CATIE, and Tokyo: The United Nations University.

Steinhardt, U. (1979). Untersuchungen über den Wasser- und Nährstoffhaushalt eines andinen Wolkenwaldes in Venezuela. Göttingen, Germany: *Göttinger Bodenkundliche Berichte* **56**: 1–185.

Tanner, E. V. J. (1980). Studies on the biomass and productivity in a series of montane rain forests in Jamaica. *Journal of Ecology* **68**: 573–588.

Van Dijk, A. I. J. M., and L. A. Bruijnzeel (2001). Modelling rainfall interception by vegetation of variable density using an adapted analytical model. I. Model description. *Journal of Hydrology* **247**: 230–238.

Van Omme, L. (1998). *Present Vegetation Patterns and Deforestation Trends over the Past 55 Years in a High Altitude Cloud Forest Watershed Area, Costa Rica*, Internal Report No. 311. Amsterdam, the Netherlands: Hugo de Vries Laboratory, University of Amsterdam.

Weaver, P. L. (1972). Cloud moisture interception in the Luquillo Mountains of Puerto Rico. *Caribbean Journal of Science* **12**: 129–144.

65 Forest restoration in the tropical montane cloud forest belt of central Veracruz, Mexico

G. Williams-Linera
Instituto de Ecología, Xalapa, Veracruz, Mexico

C. Alvarez-Aquino and R. A. Pedraza
Universidad Veracruzana, Xalapa, Veracruz, Mexico

ABSTRACT

Restoration practices assist in the reversal of ecological degradation of forest remnants and deforested areas. The enormous biodiversity in the cloud forests of central Veracruz, Mexico, poses a challenge to its restoration. The objectives of this study were to determine the ecological restoration potential of native tree species, and to define criteria for matching species to particular site conditions. In 1998, seedlings of *Carpinus caroliniana*, *Fagus grandifolia* var. *mexicana*, *Juglans pyriformis*, *Liquidambar styraciflua*, *Podocarpus matudae*, *Quercus acutifolia*, and *Symplocos coccinea* were planted in three forest fragment interiors, three adjacent agricultural fields, and two old-fields. Survival, height and diameter increment were evaluated using analysis of covariance with initial plant age as covariate. Initial age had a significant effect on survival but not on height or diameter increment across all species and sites. In 2004, overall survival was highest in old-field (70%), followed by forest interior (42%), and adjacent field (36%). *Juglans*, *Podocarpus*, and *Quercus* exhibited the greatest survival (62–80%) but intermediate height increase (26–57 cm year^{-1}). *Carpinus* and *Liquidambar* showed intermediate survival (50–54%) but high growth rates (45–96 cm year^{-1}). *Fagus* and *Symplocos* displayed both low survival (18–20%) and low height increments (13–29 cm year^{-1}). It is concluded that performance of different tree species depends on site disturbance level. *Juglans* and *Quercus* have potential to be used in restoration of disturbed areas, *Podocarpus* in plantation enrichment, and *Liquidambar* and *Carpinus* to expand the cloud forest area. *Fagus* and *Symplocos* can survive and grow in forests other than those in which they are naturally present.

INTRODUCTION

It is widely recognized that restoration practices assist in the reversal of ecological degradation at several scales, from small forest remnants to entire landscapes, and that establishing native tree species, either artificially or through secondary successional processes, contributes significantly to the maintenance of biodiversity (Brown and Lugo, 1994; Lamb, 1998; Holl *et al.*, 2000; Alvarez-Aquino, 2002; Society for Ecological Restoration, 2002; Pedraza, 2003; Quintana-Ascencio *et al.*, 2004; Martinez *et al.*, 2008; Weber *et al.*, 2008). However, insufficient understanding of disturbance ecology and the processes of ecological succession has limited the design of effective restoration programs (Brown and Lugo, 1994; Hartig and Beck, 2003; Martinez *et al.*, 2008; Weber *et al.*, 2008; Günter *et al.*, 2009). Furthermore, the use of most native tree species for restoration has been limited due to a lack of basic information on seed propagation and requirements for successful establishment and growth (Arriaga *et al.*, 1994; Butterfield and Spinoza, 1995; Lamb, 1998; Vázquez-Yañes *et al.*, 1999; Stimm *et al.*, 2008; Günter *et al.*, 2009). Some restoration initiatives have emphasized the importance of matching particular species with specific site conditions (Evans, 1986; Cluzeau *et al.*, 2002). However, this concept has not been applied to tropical montane cloud forest restoration efforts. In addition, there is very little information on the performance of native tree species artificially established under different site conditions (cf. Aguirre *et al.*, 2006; Mosandl and Günter, 2008; Weber *et al.*, 2008; Günter *et al.*, 2009).

Tropical Montane Cloud Forests: Science for Conservation and Management, eds. L. A. Bruijnzeel, F. N. Scatena, and L. S. Hamilton. Published by Cambridge University Press.

Tropical montane cloud forest (TMCF or *bosque mesófilo de montaña* as it is called in Mexico) potentially occurred on up to 20 000 km^2 or between 0.5% and 1% of Mexico's surface area (Rzedowski, 1996; cf. Mulligan, this volume). The Mexican National Forest Inventory of 1993 reported 14 119 km^2 of cloud forest to be remaining, although part of this area carried disturbed forest or traditional coffee plantations. A more realistic estimate would be closer to *c.* 8000 km^2 (Challenger, 1998; cf. Mulligan, this volume). In Mexico, cloud forest is the most diverse type of vegetation per unit area, since it occupies less than 1% of the territory but harbors 10% of the country's plant biodiversity (Rzedowski, 1996; Challenger, 1998; see also Vázquez-García, 1995 for a discussion of regional differences). Mexican MCF contains a mixture of tropical and temperate floristic elements. The dominant trees have temperate affinities, but most plants in the understory – trees, shrubs, herbs – and the epiphytic community have Neotropical affinities (cf. Vázquez-García, 1995). The great diversity in species composition and structure occurring over short geographic distances – related to high heterogeneity in micro-environmental conditions (Williams-Linera, 2002; Williams-Linera *et al.*, 2005) – poses a major challenge to conservation and restoration efforts. However, this great site-to-site variability may also be viewed as an asset rather than as a constraint to ecological restoration at the landscape level (Holl *et al.*, 2003). Because natural succession is a long-term process (Rüger *et al.*, this volume), tree planting has been considered the main tool to catalyze native forest regeneration on degraded tropical lands (Parrotta *et al.*, 1997; Holl *et al.*, 2000) including mountainous areas formerly carrying cloud forest (Stimm *et al.*, 2008; Weber *et al.*, 2008; Günter *et al.*, 2009; cf. Aide *et al.*, this volume).

In the region of Xalapa, in the State of Veracruz, eastern Mexico, until the beginning of the last century, large tracts of continuous cloud forest were reported in a number of chronicles written by travelers and botanical collectors. In recent decades, however, the forest has been reduced to patches (together constituting ~10% of the area). Fragments are surrounded by disturbed forest and shade coffee plantations (17%), secondary vegetation (17%), pastures (37%), and human settlements (18%) (Williams-Linera *et al.*, 2002; cf. Muñoz-Villers and López-Blanco, 2008; Barradas *et al.*, this volume). Given that urban expansion is threatening the remaining cloud forests, conservation and restoration must form part of an active development strategy for the region. Opportunities for restoration do exist, because urban and rural populations, including both public and private land-owners, are increasingly interested in the conservation of the remnant forests. Yet, it is a challenge to offer realistic options and economically feasible alternatives to forest cutting, in order to preserve and restore these forests. The objective of this study was to determine the ecological restoration potential of native tree species at different micro-sites in the cloud forest belt of central Veracruz. Particular objectives were to determine selected tree species survival and growth response to different vegetation cover types, and to define the best criteria for matching tree species to particular site conditions.

MATERIALS AND METHODS

Study area

The study area is located in central Veracruz, Mexico (19° 30′ N, 96° 54′ W). The cloud forest belt occurs between 1250 and 2200 m.a.s.l. The climate is mild and humid with three well-defined seasons: a relatively dry–cool season (November to March), a dry–warm season (April to May), and a wet–warm season (June to October) (Williams-Linera, 2002; cf. Ponette-González *et al.*, 2009). Precipitation, fog incidence, and relative humidity increase with elevation (Vogelmann, 1973) although the climate of the lower parts of the cloud belt seems to have become drier in recent years (Barradas *et al.*, this volume; cf. Ponette-González *et al.*, 2009). Soils are derived from andesitic volcanic ash of recent origin and are classified as Andisols. The prevailing soil texture is a sandy clay loam. The A-horizon is thick and black, with high organic matter content and low bulk density (Rossignol, 1987). Some of the most abundant tree species include *Carpinus caroliniana*, *Clethra mexicana*, *Liquidambar styraciflua*, *Quercus germana*, *Q. leiophylla*, *Q. xalapensis*, and *Turpinia insignis*, yet tree species dominance changes from site to site. A few forest fragments are dominated by a single tree species, such as *Fagus grandifolia* var. *mexicana* or *Oreomunnea mexicana* (Williams-Linera *et al.*, 2005).

Experimental sites and species used

Restoration trials were conducted in three different vegetation cover types: (i) forest fragment interior (sites 1, 3, and 5), (ii) agricultural field adjacent to forest (sites 2, 4, and 6), and (iii) old-field (mixed-species plantations; sites 7 and 8). Also studied were two on-farm plantations established by private land-owners in 1995 and 1991 (sites 9 and 10, respectively); some of the 14 planted tree species were *Cedrela odorata*, *Fraxinus uhdei*, *Juglans pyriformis*, *Liquidambar styraciflua*, and *Persea americana*. All sites were located on private properties with owners interested in conservation, making the sites more secure for long-term experiments. Basic site characteristics and disturbance history are summarized in Table 65.1. Further details can be found in Pedraza and Williams-Linera (2003), and Alvarez-Aquino *et al.* (2004).

Seven native tree species that occurred abundantly in nearby forest fragments were selected and tested for their suitability in restoration activities: *Carpinus caroliniana*, *Fagus grandifolia* var. *mexicana*, *Juglans pyriformis*, *Liquidambar styraciflua*,

Table 65.1 *Characteristics of ten study sites in central Veracruz, Mexico. Trees were established in four treatments represented by the following cover types: forest interior, agricultural field adjacent to forest, old-field, and on-farm plantation. The sites were at locations Las Cañadas, Huatusco (LC, 19° 11′ N, 96° 59′ W), Xolostla, Tlalnelhuayocan (XO, 19° 32′ N, 96° 58′ W), Rancho Viejo, Tlalnelhuayocan (RV, 19° 30′ N, 97° 00′ W), and Mesa de la Yerba, Acajete (MY, 19° 33′ N, 97° 01′ W). Variables are altitude (m.a.s. l.), mean temperature (*T, °C*), annual precipitation (*Pp *mm), and slope (°). Disturbance history is type of human activity at the location, date when the area was deforested and state of the site when the trials were started*

Cover type	Site	Location	Altitude	*T*	*Pp*	Slope	Disturbance history
Forest interior	1	LC	1340	17	2200	27	Forest, private reserve for ecotourism
	3	XO	1450	16	1650	33	Forest, medium fuelwood collection
	5	RV	1500	14	1650	35	Forest, some timber extraction and fuelwood collection
Adjacent field	2	LC	1340	17	2200	16	Young old field in a private reserve
	4	XO	1450	16	1650	12	Between *Macadamia* plantation and forest
	6	RV	1550	14	1650	31	Pasture with a few cattle
Old-field	7	MY	1875	12	1350	37	Clear-cut in 1990 for timber and charcoal then abandoned; arrested succession with *Pteridium aquilinium*
	8	XO	1400	16	1650	24	Deforested *c.* 1980, pasture, then abandoned, 10-year-old succession very diverse, with a tree stratum
On-farm plantation	9	LC	1340	17	2200	6	Pasture since 1970s, plantation started in 1995, left without maintenance with forest recovery. Five tree species planted, 14 tree species in 2004
	10	XO	1325	16	1650	6	Pasture since 1970s, plantation started in 1991, periodically weeded but new trees allowed to grow. 14 tree species planted, 21 tree species in 2004

Podocarpus matudae, *Quercus acutifolia*, and *Symplocos coccinea*. Nomenclature follows that of the *Flora of Veracruz* (Sosa and Gómez-Pompa, 1994). Hereafter, the respective species will be referred to by genus only.

Carpinus and *Liquidambar* are characteristic of the native cloud forest and may also occur in almost pure stands following forest disturbance. Both species have small, wind-dispersed seeds (7 × 1 mm and 5 × 3.6 mm, respectively). *Fagus* is the most shade-tolerant of all the species, and the nut measures 1.5 cm in length. *Quercus* has intermediate shade tolerance, with acorns of 1 cm diameter and 1.5–2 cm length. *Podocarpus* and *Symplocos* have large seeds (2 and 3 cm long, respectively) and have been observed growing in the forest interior. *Juglans* is shade-intolerant and produces large nuts (3.5–4.5 cm diameter). *Fagus*, *Symplocos*, and *Juglans* are threatened taxa and have restricted distributions in Mexico (SEMARNAT, 2002; Alvarez-Aquino *et al.*, 2004).

The plants used for the trials were germinated in the laboratory from seed collected in forests around Xalapa. Seedlings were raised in a greenhouse located in the Botanical Garden at the Instituto de Ecología or were donated by the nursery of the Municipality of Xalapa. The initial plant age was 6–14 months, and initial height was 10.6–85.2 cm at the inception of the field experiments (Table 65.2).

Experimental design

In forest interiors (sites 1, 3, and 5) and on adjacent agricultural fields (sites 2, 4, and 6), plants of four species were transplanted into 10 × 16 m plots (Alvarez-Aquino *et al.*, 2004). Forty seedlings (10 of each species) were transplanted into each plot in a grid with 2 m spacing between adjacent seedlings. In the forest interiors, four plots were positioned randomly at a distance of more than 30 m from the forest edge. In the adjacent fields, four plots were established 5 m from the forest edge. In each old-field, mixed plantations were established (sites 7 and 8) in three plots of 30 × 36 m (Pedraza and Williams-Linera, 2003). One hundred and twenty trees (30 per species) were planted within each plot, 10 plants of each species in three lines with 3 m spacing between them (Table 65.3). In each of the on-farm plantations (sites 9 and 10), all trees ≥2 m tall in five randomly located 10 × 10 m permanent plots were recorded, identified at the species level, and their height and diameter measured.

Between July and September 1998, all plants were permanently numbered with metal tags and their locations marked. In the on-farm plantations, individual trees were not tagged, but permanent plots were marked to facilitate locating them for re-measurements in 2000 and 2004.

Seedling survival and growth

Survival, height and diameter growth were monitored in July–September 1998, February–May 2000, and March–April 2004 for plants established in forest interiors, adjacent fields, old-fields, and on-farm plantations. Survival

Table 65.2 *Tree species used in the trials planted in 1998 in central Veracruz, Mexico. Data are age (months) and mean height in cm (SE in parentheses) at the beginning of seedling establishment trials in forest interior, adjacent field, and old-field, and height attained in 2000 and 2004. On-farm plantations data are mean height in 1998, 2000, and 2004. Last column is mean height (1 SE)*

	Species				
	Carpinus	***Fagus***	***Quercus***	***Symplocos***	Mean
Seedling establishment					
Initial age	8	12	14	6	
Initial height	21.9 (2.3)	10.6 (0.6)	85.2 (1.5)	18.4 (1.3)	34 (3.1)
Forest interior					
Height in 2000	38.5 (3.1)	18.7 (0.7)	102.0 (1.6)	26.0 (1.8)	46.3 (4.9)
Height in 2004	63.7 (6.9)	35.5 (3.3)	105.5 (4.9)	43.1 (4.2)	61.9 (4.6)
Adjacent agricultural field					
Height in 2000	71.3 (12.8)	28.3 (5.0)	123.8 (8.9)	47.8 (11.8)	67.8 (7.1)
Height in 2004	461.9 (35.5)	238.9 (20.6)	534.1 (39.1)	506.0 (54.8)	465.8 (28.1)
	Species				
	Carpinus	***Juglans***	***Liquidambar***	***Podocarpus***	
Old-field					
Initial age	23	17	11	16	
Initial height	40.4 (2.7)	70.8 (2.2)	20.5 (1.1)	72.9 (2.0)	51.1 (1.9)
Height in 2000	158.9 (13.4)	88.3 (7.1)	149.8 (7.0)	57.4 (4.2)	113.6 (9.7)
Height in 2004	444.6 (31.2)	315.3 (69.1)	533.5 (64.6)	160.0 (16.9)	363.3 (37.6)
	Site				
	LC	**XO**			
On-farm plantation					
Establishment	1995	1991			
Height in 1998	253.6 (29.6)	380.6 (30.0)			317.1 (26.8)
Height in 2000	404.2 (36.7)	477.0 (64.7)			440.6 (37.2)
Height in 2004	718.2 (101.1)	640.0 (42.2)			679.2 (48.0)

Table 65.3 *Number of plants per treatment and species planted in 1998. Number of plants alive in 2000 and 2004, and total survival (%) for each species and treatment in 2004. The treatments are three paired forest fragment interior and adjacent field, and two old-fields*

Species	Forest interior			Adjacent field			Old-field			Survival (%)
	1998	2000	2004	1998	2000	2004	1998	2000	2004	
Carpinus	120	94	64	120	41	22	180	157	141	**54.0**
Fagus	120	80	38	120	34	5	–	–	–	**17.9**
Juglans	–	–	–	–	–	–	180	144	128	**71.0**
Liquidambar	–	–	–	–	–	–	180	109	90	**50.0**
Podocarpus	–	–	–	–	–	–	180	153	144	**80.0**
Quercus	120	106	71	120	109	67	–	–	–	**61.7**
Symplocos	120	61	27	120	32	22	–	–	–	**20.4**
Total	**480**	**341**	**200**	**480**	**216**	**116**	**720**	**563**	**503**	
Survival (%)			**41.7**			**36.2**			**69.9**	

percentage, and annual increments in height (cm year^{-1}) or diameter (cm year^{-1}) were calculated per species and site. This chapter compares results for two periods, viz. 1998–2000 and 2000–2004.

Soil

In 1998 and 2004, four soil samples (0–10 cm depth) were collected at each site to measure pH and organic matter content. Soil compaction was measured 10 times at four positions within

Table 65.4 *ANCOVA results for the main effect of site and species on growth rate in height and diameter, and survival of plants. (A) Using the age of the plants when transplanted into the sites as the covariate. (B) Using the height of the plants when transplanted in the sites as the covariate. The model included repeated measurements (Time) since two periods were considered (1998–2000 and 2000–2004)*

		Height		Diameter		Survival	
Source	DF	*F*	*p*	*F*	*p*	*F*	*p*
(A) Initial plant age as covariate							
Species	6,6	0.89	0.55	0.76	0.63	4.91	0.03
Age	1,200	3.01	0.08	3.69	0.06	12.42	0.0005
Species × Time	6,200	6.46	<0.0001	5.31	<0.0001	1.89	0.08
Site	7,7	2.52	0.12	3.25	0.07	0.94	0.53
Age	1,198	0.61	0.44	0.62	0.43	50.85	<0.0001
Site × Time	7,198	19.15	<0.0001	16.64	<0.0001	6.56	<0.0001
(B) Initial height of the plants as covariate							
Species	6,7	1.11	0.45	0.90	0.54	2.21	0.14
Height 1	1,200	3.01	0.08	3.69	0.06	12.42	0.0005
Species × Time	6,200	6.46	<0.0001	5.31	<0.0001	1.89	0.08
Site	7,7	2.44	0.13	3.12	0.08	1.14	0.43
Height 1	1,198	1.62	0.20	4.28	0.04	114.41	<0.0001
Site × Time	7,198	19.32	<0.0001	17.00	<0.0001	7.80	<0.0001

each site with a penetrometer, pH was determined in a soil-water mixture of 1 : 2.5, and organic matter content was estimated using the Walkley–Black method.

Statistical analyses

Percent survival, and annual height and diameter increment per species and treatment (forest interior, adjacent field, and old-field) were averaged within each plot. Survival data were arcsine square root transformed, and height and diameter increment data were Box–Cox transformed. Repeat-measure ANCOVAs were conducted, with initial plant age as the covariate, on data from 2000 and 2004 to analyze survival and height and diameter growth at the different sites and among species. Initial plant height was also used as the covariate with similar results. Since not all species were grown at all sites, species and sites had to be analyzed separately as main factors. Soil variables were compared using a two-way repeat-measure ANOVA. Once relationships were derived, contrasts were expressed as a function of the least square means of the pertinent effect. Statistical analyses were conducted using JMP (SAS, 1997). All reported values are means ±1 standard error.

RESULTS

In 1998, a total of 1680 plants comprising seven species were planted at eight sites (1–8, Table 65.1). By 2000, 67% of these plants had survived, and by 2004, 73% of the individuals recorded in 2000 were still alive (Table 65.3). In 2004, average plant height was greater in the adjacent agricultural fields (4.6 m) and in old-fields (3.6 m) than in the forest fragment interiors (0.62 m). Average height of the on-farm plantations was 7.2 m after 9 years (site 9), and 6.4 m after 13 years (site 10) (Table 65.2).

Species performance

Tree species planted in the forest interior, adjacent field, and old-field treatments were pooled together to analyze species performance. Whilst species survival was significantly affected by the initial age of plants (covariate) when transplanted into the field, neither height nor diameter increment were affected by initial age. Similarly, average initial height of the species used as covariate affected only species survival. The ANCOVA results are shown in Table 65.4. Species used showed distinct survival patterns. Overall survival of *Quercus* (90% and 69% in 2000 and 2004, respectively) was highest and similar to that of *Podocarpus* (83% and 78%) and *Juglans* (74% and 67%), followed by *Liquidambar* (67% and 51%) and *Carpinus* (63% and 50%). *Fagus* (47% and 22%) and *Symplocos* (39% and 24%) had the lowest survival rates (Figure 65.1a).

Increment in height was statistically similar for all species; however, the interaction between species × time was significant

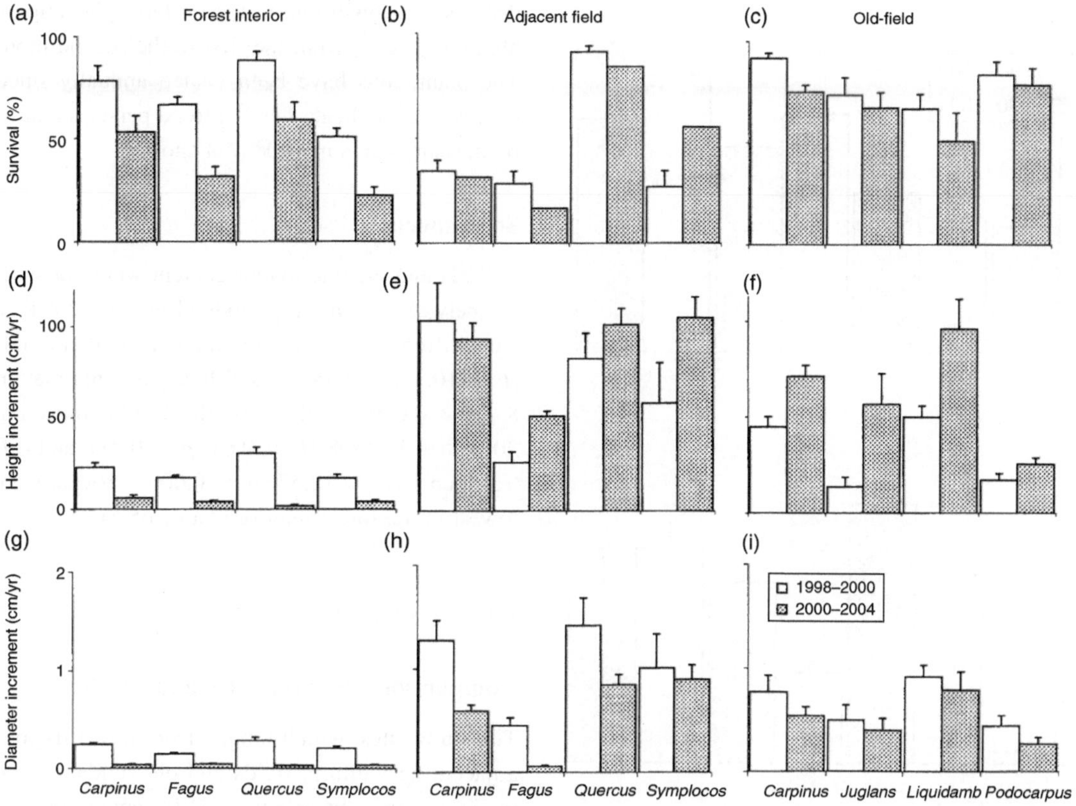

Figure 65.1. Survival percentage (a–c), mean height increment (d–f), and mean diameter increment (g–i) for selected tree species during two periods (1998–2000 and 2000–2004) in forest fragment interior, adjacent agricultural field, and old-field treatments in central Veracruz, Mexico. Error bars represent 1 standard error.

(Table 65.4). During the second sampling period (2000–2004), *Liquidambar* exhibited the greatest height increment (95.9 cm year^{-1}). Whilst the growth rate of *Juglans* increased significantly (from 13.7 cm year^{-1} in 2000 to 56.7 cm year^{-1} in 2004), that for *Quercus* decreased significantly (from 55.5 to 40.3 cm year^{-1}) (Figure 65.1b).

Increment in diameter was also statistically similar among species (Table 65.4). The observed significant interaction of species × time was due to a decrease in diameter growth rate for *Carpinus* (from 0.76 cm year^{-1} in 2000 to 0.32 cm year^{-1} in 2004), *Fagus* (from 0.30 to 0.05 cm year^{-1}), *Juglans* (from 0.50 to 0.40 cm year^{-1}), *Quercus* (from 0.85 to 0.36 cm year^{-1}), and *Symplocos* (from 0.61 to 0.25 cm year^{-1}) during the second sampling period (Figure 65.1c).

Site differences

Plant survival at the different sites can be explained to a certain extent by the age of the plants when transplanted, since this covariate was significant in the model. Using initial height of the plants as covariate also significantly explained plant survival. Percent survival and increments in height and diameter did not differ significantly among sites but the interaction site × time was highly significant (Table 65.4, Figure 65.2). Survival over the two sampling periods was similar for the two old-fields (84% and 73% in 2000 and 2004 at site 7, vs. 92% and 81% at site 8), but was lower in 2004 (42%) than in 2000 (72%) in the forest fragment interiors (sites 1, 3, and 5). Survival in the forest interior was higher than in the adjacent fields (45% in 2000, 36% in 2004; Figure 65.2a).

The maximum height increment rate was recorded in fields adjacent to the forest during the second period, followed by the second sampling period in old-fields. Height increment was lower in forest interior sites (22.1 and 4.2 cm year^{-1} in 2000 and 2004, respectively) than in adjacent field- (65.7 and 86.1 cm year^{-1}) and old-field sites (31.4 and 62.4 cm year^{-1}). Also, height increment was significantly different between the two sampling periods for all treatments (Figure 65.2b). The largest diameter increment rate was recorded during the first period in the adjacent field sites, followed by old-field sites (Figure 65.2c). Annual diameter increment was lower in forest interior sites (0.22 and 0.04 cm year^{-1} in 2000 and 2004, respectively) than in adjacent field (1.04 and 0.64 cm year^{-1}) and old-field sites (0.66 and 0.50 cm year^{-1}). However, diameter increment differed significantly between sampling periods only for forest interior and adjacent field sites (Figure 65.2c).

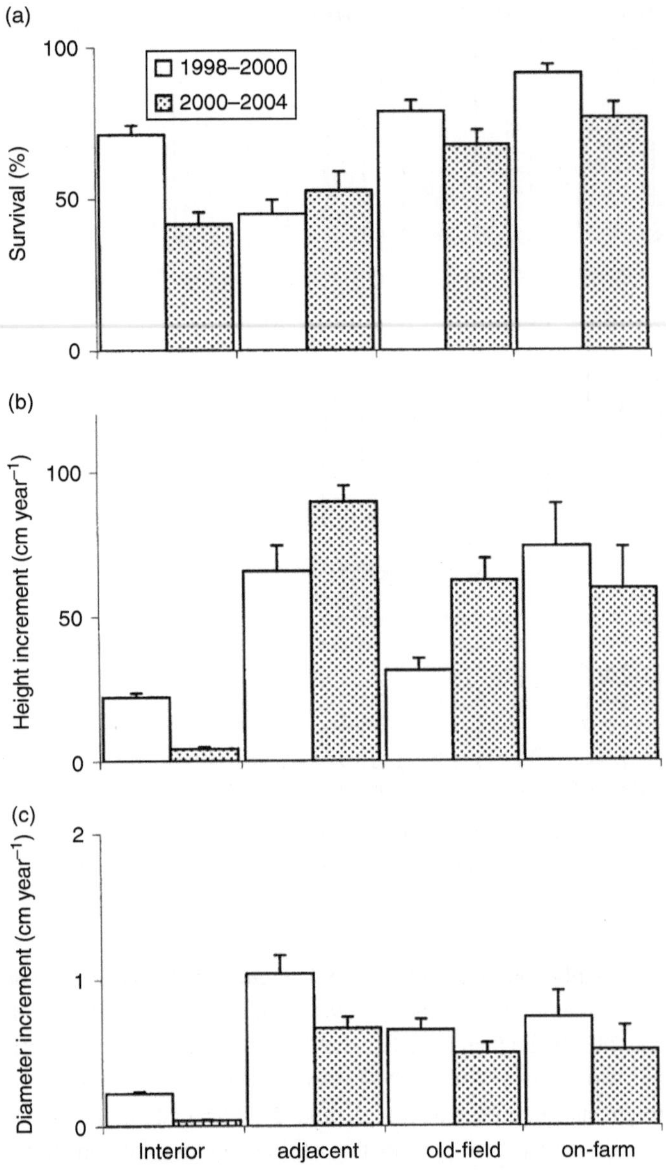

Figure 65.2. (a) Survival percentage, (b) mean height increment, and (c) mean diameter increment during two periods (1998–2000 and 2000–2004) in forest interior, adjacent agricultural field, old-field, and on-farm plantation treatments in central Veracruz, Mexico. Error bars represent 1 standard error.

On-farm plantations

Because individual species were not tagged at the two on-farm sites (9 and 10), tree performance could not be compared between species but only between sites. In 2004, survival was 70% of the individuals recorded in 1998 (Figure 65.2a). When comparing the experimental plantations established in old-fields with the on-farm plantations, increments in height and diameter were not significantly different (Figure 65.2b, c). There was, however, a trend of increasing height growth rate in the experimental plantations (from 31.4 to 62.4 cm year^{-1}) and one of decreasing growth rate in the on-farm plantations (from 74.1 to 59.6 cm year^{-1}) from the first to the second monitoring period. The plantations have been visited annually since 2004; visual observations indicate that the trend persists, and were confirmed by measurements in 2008 (not shown).

Soil aspects

Soil pH and organic matter content were not statistically different between sampling periods (1998 and 2004), although there were differences between treatments, with a trend for higher pH ($F = 10.3$, $p < 0.0001$) and lower organic matter content ($F = 8.4$, $p < 0.0001$) in the old-fields. Soil compaction was lower in 2004 than in 1998 ($F = 9.61$, $p = 0.01$), and differed between treatments ($F = 12.57$, $p = 0.0004$) because compaction was lowest in the forest interior (Table 65.5).

DISCUSSION

Comparisons between sites and species

The study sites, which ranged from forest fragment interiors to adjacent agricultural fields and old-fields, represent a gradient of increasing disturbance. Differences in soil compaction revealed contrasts in past site management history, with the least disturbed forest interiors exhibiting the lowest compaction (cf. Zimmermann and Elsenbeer, 2008). Although plantations established in old-fields had lower organic matter content and less acidic soil than the other treatments, a period of 5.5 years of planted trees proved insufficient for the detection of consistent changes in these soil characteristics as well as in compaction between sites (cf. Ilstedt *et al.*, 2007). Microclimates differed in the respective habitats, and this may have affected seedling establishment differentially. Microclimate is mainly related to canopy openness, hence irradiance levels were lower in the forest interior than in adjacent fields and mixed plantations established on old-fields, whereas soil water content was higher in shaded habitats (cf. Alvarez-Aquino, 2002; Pedraza, 2003).

Plant survival was similar during the two monitoring periods, except in the forest interior. Seedling survival and growth in the forest interior were the lowest, but within the range of values reported for other forest interiors (reviewed in Alvarez-Aquino and Williams-Linera, 2002). Sites adjacent to forest had the greatest height increment, possibly due to more favorable microclimatic conditions compared to the forest interior (e.g. higher levels of photosynthetically active radiation) and the open plantations established in old-fields (e.g. higher humidity and lower temperature) (Alvarez-Aquino, 2002; Pedraza, 2003). Similar trends have been observed in Chiapas, Mexico – where survival and growth differed

Table 65.5 *Soil characteristics of the study sites in central Veracruz, Mexico. Variables were measured in 1998 before the trials were started in sites 1–8, and again in 2004. Data are mean pH (H_2O), organic matter (OM, %), and soil compaction (kg cm^{-2}) (SE in parentheses)*

Site	pH (H_2O)		OM (%)		Compaction (kg cm^{-2})	
	1998	2004	1998[a]	2004	1998	2004
Forest interior						
1	4.6 (0.0)	5.1 (0.1)	14.3	26.1 (0.7)	14.8 (1.2)	15.2 (0.6)
3	4.2 (0.0)	4.7 (0.0)	21.8	22.4 (1.7)	12.3 (1.4)	12.5 (0.4)
5	3.9 (0.0)	4.6 (0.1)	24.7	26.3 (4.2)	14.0 (1.4)	13.2 (0.2)
Adjacent field						
2	4.5 (0.1)	5.0 (0.0)	16.4	24.9 (2.5)	20.9 (0.1)	18.5 (0.2)
4	5.6 (0.3)	5.3 (0.1)	8.9	13.9 (0.6)	17.1 (1.1)	15.9 (0.1)
6	5.0 (0.0)	4.7 (0.0)	10.4	28.2 (1.0)	15.4 (0.0)	16.0 (0.1)
Old-field						
7	4.2 (0.2)	4.9 (0.2)	12.4	20.0 (1.6)	14.2 (1.9)	14.4 (0.7)
8	5.6 (0.1)	5.8 (0.3)	8.9	15.6 (4.6)	17.1 (1.1)	15.7 (0.1)
On-farm plantation						
9	4.5 (0.1)	5.2 (0.1)	16.4	11.1 (0.3)	20.9 (0.0)	17.8 (0.5)
10	5.1 (0.0)	5.3 (0.0)	15.2	18.0 (1.4)	17.6 (1.2)	14.4 (0.3)

[a] SEs were missed.

between successional habitats within human-disturbed forest (Quintana-Ascencio *et al.*, 2004) – and in nearby Xalapa, where young seedlings survived and grew higher at the forest edge and in clearings than in forest interior habitats (Pedraza and Williams-Linera, 2005).

Old-field conditions were different since the sites presented different disturbance level and land-use history. Site 7 was the least disturbed because it was never used as pasture, and its initial soil conditions were similar to those in a contiguous *Fagus*-dominated forest (Williams-Linera, 2002; Pedraza and Williams-Linera, 2003). Here, the planted trees had shaded out *Pteridium aquilinium* plants after 5.5 years. Tree canopy and tree species diversity at site 8 recovered rapidly (Pedraza, 2003) because it had the advantage of the presence of several mature trees remaining from the original forest. The on-farm plantations with their trees being 3–7 years older than the trees planted in the other plots may serve as indicators of the direction that the restoration assays may follow. Tree mortality in the two on-farm plantations was low during the 5–10 and 9–14 years covered, indicating a lower risk of tree mortality with time. Also, height and diameter increments of the on-farm plantations began to level off after 8–13 years from establishment.

Age of the seedlings significantly affected initial mortality. Seedlings are typically most vulnerable during the early (first to second year) seedling stage (Kitajima and Wright, 2002; Hewitt and Kellman, 2004). After surviving 2 years in the field, plants can be considered established. Survival was higher for older plants in old-fields, but height increment was higher for younger plants growing on adjacent fields. Although initial age is important, survival of the species examined in this study may also be related to seed size. Large seeds tend to produce large or reserve-rich seedlings capable of surviving in the forest understory, whereas seedling survival has been demonstrated to be inversely related to seed mass. High seedling survivorship under shade conditions has been shown to be related to large-seeded species, but this comes at the cost of reduced maximum growth rates (Kitajima and Wright, 2002; Hewitt and Kellman, 2004; Moles and Westoby, 2004).

Matching species to site

The performance of different species under forest interior, adjacent field, and old-field conditions can be used to match each species to the range of potential restoration sites within the Veracruz landscape. Matching ecological features and performance of the selected species with site quality suggests some trends.

The species with the greatest seed size (*Juglans*) was able to survive under stressful conditions. It was planted in a degraded old-field that had been over-used as a pasture since 1950, and was characterized by grass dominance and compacted soils. Therefore, *Juglans* may have potential for use in the restoration of degraded sites (Pedraza and Williams-Linera, 2003). *Quercus* seedlings, which obtain carbohydrate and nutrient reserves from their large seeds, also grow faster than saplings. *Quercus* is intermediate in terms of its shade tolerance, and the fact that survival of *Quercus* appeared to be relatively unaffected by the environmental conditions of areas adjacent to forest (since survival was similar in forest interior and adjacent field), highlights the great potential of this species for forest restoration programs (Alvarez-Aquino *et al.*, 2004; Quintana-Ascencio *et al.*, 2004).

Podocarpus had high survival rates but grew relatively slowly, although it performed well in sites of intermediate disturbance. *Podocarpus* can be used to enrich forests and plantations (Stadtmüller, 1987; cf. Mosandl and Günter, 2008) and to protect the soil while other species are developing. Also, most of the 6-year-old individuals growing in the old-fields were already producing abundant fruits with showy red arillus (G. Williams-Linera, C. Alvarez-Aquino, and R. A. Pedraza, personal observation), suggesting that *Podocarpus* can attract native fauna such as birds at an early age.

Fagus and *Symplocos* have restricted distributions in Mexican TMCF (Williams-Linera, 2002; Alvarez-Aquino *et al.*, 2004). They also have large seeds, but their seedlings displayed low survival rates in the trials, and their performance was more successful under closed canopies. Others attempts at reintroducing these species in temperate forests have not been successful but it has been suggested that high-light environments may provide colonization benefits (Hewitt and Kellman, 2004). Based on the present results, it is emphasized that *Fagus* and *Symplocos* are able to survive and grow in forest sites other than the kind of forests in which they are naturally present, and even at disturbed sites such as locations adjacent to forest. *Symplocos* also started reproducing at an early age and copious fruit production was observed in individuals growing in adjacent fields. It is important to include threatened trees in the pool of species to be used in restoration initiatives, in order to couple these initiatives with the preservation of regional biodiversity (Society for Ecological Restoration, 2002).

Finally, *Carpinus* and *Liquidambar*, which have small wind-dispersed seeds that facilitate dispersal into adjacent and open areas, exhibited intermediate survival rates but high growth rates. Their ability to colonize open areas is of particular value in enhancing their potential for restoration of disturbed forest fragments and deforested areas.

CONCLUDING REMARKS

Initial age of the planted trees was observed to affect early survival but maintaining plants in greenhouses and nurseries would add to the price of restoration. Additional greenhouse and field experiments are needed to determine the best age for transplanting and to find a compromise between survival and growth. To complicate matters, the best age has to be defined for each species and overall site condition, including the level of disturbance. Most papers on restoration report results derived from only one or two years of observations but when working with trees, conclusions should preferably be based on at least a decade of data.

Regional tree diversity in central Veracruz is high and several species have the potential to be used in areas that need to be restored. It is especially advisable to work in collaboration with local farmers and land-owners to incorporate their knowledge when selecting a set of trees for restoration activities (McDonald *et al.*, 2003). Most of the species used in the present study had already been selected by the land-owners and were growing in on-farm trials. Much of silviculture is concerned with achieving the best match between species and the planting site (Evans, 1986; Günter *et al.*, 2009). A first approach to match species to site would be to combine performance and site quality. Performance of different tree species depends on the specific level of disturbance exhibited by each site, suggesting the importance of accurate species–site matching to obtain optimum rates of survival and growth in particular scenarios. Plants of the seven species selected for this study can be established successfully and appear to be good candidates for restoration programs focusing on expanding the cloud forest in central Veracruz.

ACKNOWLEDGEMENTS

We thank Marychu Peralta from the Instituto de Ecología in Xalapa for help during the field work. We are grateful to Mitch Aide, Heidi Asbjornsen, and Sampurno Bruijnzeel for helpful comments and suggestions that greatly improved the manuscript. The study was funded by the European Commission under the INCO program; project BIOCORES, contract number ICA4-CT-2001-10095.

REFERENCES

Aguirre, N., S. Günter, M. Weber, and B. Stimm (2006). Enriquecimiento de plantaciones de *Pinus patula* con especies nativas en el sur de Ecuador. *Lyonia* **10**: 17–29.

Alvarez-Aquino, C. (2002). Regeneration of tree species in Mexican cloud forest. Ph.D. thesis, University of Edinburgh, Edinburgh, UK.

Alvarez-Aquino, C., and G. Williams-Linera (2002). Seedling bank dynamics of *Fagus grandifolia* var. *mexicana* before and after a mast year in a Mexican tropical montane cloud forest. *Journal of Vegetation Science* **13**: 179–184.

Alvarez-Aquino, C., G. Williams-Linera, and A. C. Newton (2004). Experimental native tree seedling establishment for restoration of Mexican cloud forest. *Restoration Ecology* **12**: 412–418.

Arriaga, V., V. Cervantes, and A. Vargas-Mena (1994). *Manual de reforestación con especies nativas*. Mexico, DF: Secretaría de Desarrollo Social, Instituto Nacional de Ecología y Universidad Autónoma de México.

Brown, S., and A. E. Lugo (1994). Rehabilitation of tropical lands: a key to sustaining development. *Restoration Ecology* **2**: 97–111.

Butterfield, R. P., and M. Spinoza (1995). Screening trial of 14 tropical hardwoods with an emphasis on species native to Costa Rica: fourth year results. *New Forests* **9**: 35–45.

Challenger, A. (1998). *Utilización y conservación de los ecosistemas terrestres de México pasado, presente y futuro*. Mexico, DF: CONABIO, Instituto de Biología UNAM, Agrupación Sierra Madre, S.C.

Cluzeau, C., J. Drapier, and E. Bruno (2002). Species-site suitability indicators. In *Indicators of Sustainable Management:"Tree Species and Forest Sites" and "Forest Habitats"*, Vol. 2, *France*. eds. J. Drapier and A. Colinot. Available at www.svo.se/life.

Evans, J. (1986). *Plantation Forestry in the Tropics*. Oxford, UK: Oxford University Press.

Günter, S., P. Gonzalez, G. Alvarez, *et al.* (2009). Determinants of successful regeneration of abandoned pastures in the Andes: soil conditions and vegetation cover. *Forest Ecology and Management* **258**: 81–91.

Hartig, K. and E. Beck (2003). The bracken fern (*Pteridium arachnoideum* Kaulf.) dilemma in the Andes of Southern Ecuador. *Ecotropica* **9**: 3–13.

Hewitt, N., and M. Kellman (2004). Factors influencing tree colonization in fragmented forest: an experimental study of introduced seeds and seedlings. *Forest Ecology and Management* **191**: 39–59.

Holl, K. D., M. E. Loik, E. H. V. Lin, and I. A. Samuels (2000). Tropical montane forest restoration in Costa Rica: overcoming barriers to dispersal and establishment. *Restoration Ecology* **8**: 339–349.

Holl, K. D., E. E. Crone, and C. B. Schultz (2003). Landscape restoration: moving from generalities to methodologies. *BioScience* **53**: 491–502.

Ilstedt, U., A. Malmer, E. Verbeeten, and D. Murdiyarso (2007). The effect of afforestation on water infiltration in the tropics: a systematic review and meta-analysis. *Forest Ecology and Management* **251**: 45–51.

Kitajima, K., and S. J. Wright (2002). Growth–survival trade-offs tropical tree seedlings in field and nurseries, *Association of Tropical Biology Annual Meeting Proceedings*, p. 57.

Lamb, D. (1998). Large-scale ecological restoration of degraded tropical forest lands: the potential role of timber plantations. *Restoration Ecology* **6**: 271–279.

Martínez, A., M. D. Mahecha, G. Lischeid, and E. Beck (2008). Succession stages of vegetation regeneration: secondary tropical mountain forests. In *Gradients in a Tropical Mountain Ecosystem of Ecuador*, eds. E. Beck, J. Bendix, I. Kottke, F. Makeschin, and R. Mosandl, pp. 425–431. Berlin: Springer-Verlag.

McDonald, M. A., A. Hofny-Collins, J. R. Healey, and T. C. R. Goodland (2003). Evaluation of trees indigenous to the montane forest of the Blue Mountains, Jamaica for reforestation and agroforestry. *Forest Ecology and Management* **175**: 379–401.

Moles, A. T., and M. Westoby (2004). Seedling survival and seed size: a synthesis of the literature. *Journal of Ecology* **92**: 372–383.

Mosandl, R., and S. Günter (2008). Sustainable management of tropical mountain forests in Ecuador. In *The Tropical Mountain Forest: Patterns and Processes in a Biodiversity Hotspot*, eds. S. R. Gradstein, J. Homeier, and D. Gansert, pp. 177–193. Göttingen, Germany: Göttingen Centre for Biodiversity and Ecology.

Muñoz-Villers, L. E., and J. López-Blanco (2008). Land use/cover changes using Landsat TM/ETM images in a tropical and biodiverse mountainous area of central-eastern Mexico. *International Journal of Remote Sensing* **29**: 71–93.

Parrotta, J. A., J. W. Turnbull, and N. Jones (1997). Catalyzing native forest regeneration on degraded tropical lands. *Forest Ecology and Management* **99**: 1–7.

Pedraza, R. A. (2003). Árboles nativos para plantaciones: una estrategia de restauración en áreas deforestadas. Ph.D. thesis, Instituto de Ecología, A.C., Xalapa, Veracruz, Mexico.

Pedraza, R. A., and G. Williams-Linera (2003). Evaluation of native tree species for the rehabilitation of deforested areas in a Mexican cloud forest. *New Forests* **26**: 83–99.

Pedraza, R. A., and G. Williams-Linera (2005). Microhabitat conditions for germination and establishment of two native temperate tree species in the Mexican montane cloud forest. *Agrociencia* **39**: 457–467.

Ponette-González, A. G., K. C. Weathers, and L. M. Curran (2009). Water inputs across a tropical montane landscape in Veracruz, Mexico: synergistic effects of land cover, rain and fog seasonality, and interannual precipitation variability. *Global Change Biology*, doi: 10.1111/j.1365-2486.2009.01985.x.

Quintana-Ascencio, P. F., N. Ramírez-Marcial, M. González-Espinosa, and M. Martínez-Ico (2004). Sapling survival and growth of coniferous and broad-leaved trees in successional highland habitats in Mexico. *Applied Vegetation Science* **7**: 81–88.

Rossignol, J. P. (1987). Los estudios morfoedafológicos en el área Xalapa–Coatepec, Veracruz. In *La morfoedafología en la ordenación de los paisajes rurales*, eds. D. Geissert and J. P. Rossignol, pp. 23–35. Xalapa, Mexico: INIREB, ORSTOM.

Rzedoswki, J. (1996). Análisis preliminar de la flora vascular de los bosques mesófilos de montaña de México. *Acta Botánica Mexicana* **35**: 25–44.

SAS (1997). *JMP User's Guide*. Cary, NC: SAS Institute.

SEMARNAT (2002). *Norma oficial mexicana NOM-059-ECOL-2001*, Protección ambiental – especies nativas Mexico de flora y fauna silvestres – categorías de riesgo y especificaciones para su inclusión, exclusión o cambio en lista de especies en riesgo. *Mexico, DF: Diario Oficial*.

Society for Ecological Restoration (2002). *The SER Primer on Ecological Restoration*. Society for Ecological Restoration Science and Policy Working Group. Available at www.ser.org/.

Sosa, V., and A. Gómez-Pompa (1994). *Flora de Veracruz: Lista florística. Fascículo 82, Xalapa, Veracruz: Instituto de Ecología, A.C.* Riverside, CA: University of California.

Stadtmüller, T. (1987). *Los bosques nublados en el trópico húmedo*. Turrialba, Costa Rica: CATIE, and Tokyo: United Nations University.

Stimm, B., E. Beck, S. Günter, *et al.* (2008). Reforestation of abandoned pastures: seed ecology of native species and production of indigenous plant material. In *Gradients in a Tropical Mountain Ecosystem of Ecuador*, eds. E. Beck, J. Bendix, I. Kottke, F. Makeschin, and R. Mosandl, pp. 433–445. Berlin: Springer-Verlag.

Vázquez-García, J. A. (1995). Cloud forest archipelagos: preservation of fragmented montane ecosystems in tropical America. In *Tropical Montane Cloud Forests*, eds. L. S. Hamilton, J. O. Juvik, and F. N. Scatena, pp. 314–332. New York: Springer-Verlag.

Vázquez-Yañes, C., A. I. Batis, M. I. Alcocer, M. Gual, and C. Sánchez (1999). *Árboles y arbustos potencialmente valiosos para la restauración ecológica y la reforestación*, Reporte técnico del proyecto J084. Mexico, DF: CONABIO – Instituto de Ecología, UNAM. Also available at www.conabio.gob.mx/arboles/introd-J084.html

Vogelmann, H. W. (1973). Fog precipitation in the cloud forests of Eastern Mexico. *BioScience* **23**: 96–100.

Weber, M., S. Günter, N. Aguirre, B. Stimm, and R. Mosandl (2008). Reforestation of abandoned pastures: silvicultural means to accelerate forest recovery and biodiversity. In *Gradients in a Tropical Mountain Ecosystem of Ecuador*, eds. E. Beck, J. Bendix, I. Kottke, F. Makeschin, and R. Mosandl, pp. 447–457. Berlin: Springer-Verlag.

Williams-Linera, G. (2002). Tree species richness complementarity, disturbance and fragmentation in a Mexican tropical montane cloud forest. *Biodiversity and Conservation* **11**: 1825–1843.

Williams-Linera, G., R. Manson, and E. Izunza (2002). La fragmentación del bosque mesófilo de montaña y patrones de uso del suelo en la región oeste de Xalapa, Veracruz, México. *Madera y Bosques* **8**: 73–89.

Williams-Linera, G., A. M. Lopez, and M. A. Muñiz (2005). Complementariedad y patrones de anidamiento de especies de árboles en el paisaje de bosque de niebla del centro deVeracruz (México). In *Sobre diversidad biológica: el significado de las diversidades alfa, beta y gamma*, eds. G. Halffter, J. Soberón, P. Koleff, and A. Melic, pp. 153–164. Zaragoza, Mexico: m3m-Monografías Tercer Milenio.

Zimmermann, B., and H. Elsenbeer (2008). Spatial and temporal variability of soil saturated hydraulic conductivity in gradients of disturbance. *Journal of Hydrology* **361**: 78–95.

66 Ecological and social bases for the restoration of a High Andean cloud forest: preliminary results and lessons from a case study in northern Ecuador

S. Báez and K. Ambrose
Ecopar, Quito, Ecuador

R. Hofstede
Ecopar and IUCN Sur, Quito, Ecuador

ABSTRACT

The High Andean cloud forests of northern Ecuador have been exploited by humans for centuries. Presently, ecological restoration is urgently needed to enhance biodiversity conservation, and ecosystem processes and services. This chapter presents the basis for a strategy that integrates ecological knowledge and local people's perspectives on the utilization of biodiversity for the restoration and management of a High Andean landscape. The objectives were: (i) to evaluate the restoration needs of the landscape, (ii) to identify groups within society that are interested in forest restoration, (iii) to select the species of plants with potential for forest restoration, and species of mammals for managed production, and (iv) to provide recommendations for integrating groups within society to achieve participative restoration. It was found that: (i) upper montane cloud forests (2900–3400 m.a.s.l.) are most in need of restoration, (ii) more affluent groups within the society are more willing to use biodiversity, with gender and ethno-biological knowledge influencing these trends, and (iii) seven species of plants and five species of mammals have the highest potential for restoration and management. Thus, an initial phase for a restoration project should encourage and guide the more affluent parts of society to restore their lands first. Less affluent people are likely to become interested in restoration once they can explore concrete examples of the uses they can make of local biodiversity, and if support for these activities is available through participatory learning. This case study shows how ecological, social, and ethno-biological knowledge can be combined to design a participative forest restoration strategy for a human-dominated landscape.

INTRODUCTION

The High Andean cloud forest belt, corresponding to the upper montane rain forest formation of Grubb (1977), constitutes one of the most important centers of biodiversity in the world (Dinerstein *et al.*, 1995; Barthlott *et al.*, 2005; Beck *et al.*, 2007). However, the conservation status of these forests is extremely poor due to the intense exploitation that has taken place especially during the last century (Mosandl *et al.*, 2008). Currently, less than 10% of the original forest cover remains in scattered fragments distributed between Venezuela and Bolivia (Laegaard, 1992; Kessler, 1995; Wille *et al.*, 2002; Mulligan, this volume). Over large areas of the High Andes, the expansion of urban settlements and agricultural lands, and inappropriate timber exploitation is reflected in the loss of biodiversity (Pitman *et al.*, 2002; Mosandl *et al.*, 2008; Pohle and Gerique, 2008), and altered hydrological regulation (Buytaert *et al.*, 2006; Molina *et al.*, 2007; cf. Bruijnzeel, 2004). From a socio-economic perspective, the poor land-management practices that lead to the loss of these ecosystems contribute to poverty, malnutrition, and migration (Anand, 2002).

During the last decade, ecological restoration has emerged as an option to increase the cover of terrestrial ecosystems that are highly impacted by human activities (Hobbs and Harris, 2001;

Tropical Montane Cloud Forests: Science for Conservation and Management, eds. L.A. Bruijnzeel, F.N. Scatena, and L.S. Hamilton. Published by Cambridge University Press.

Günter *et al.*, 2009; Williams-Linera *et al.*, this volume). In northern Ecuador, the conversion of degraded lands into natural forest could contribute to the conservation of the area's unique flora and fauna. Increased forest cover could also diminish the risk of extinction of endangered species, and improve habitat conditions for wildlife (cf. Jordan *et al.*, 1987; Cavelier *et al.*, this volume; Hofstede *et al.*, this volume). At the same time, higher forest cover may enhance the provision of ecological services that contribute to the stabilization of agricultural production, including more dependable streamflow during the dry season and lower stream sediment loads; see Bruijnzeel (2004), Scott *et al.* (2005), and Malmer *et al.* (2009) for a fuller discussion of the circumstances under which reforestation may be expected to improve streamflow regulation. In particular, a higher availability of forest resources would provide local communities with a variety of timber and non-timber forest products that could both diminish expenses and raise their income (see also Pohle and Gerique, 2008). Thus, from a holistic perspective, restoring degraded areas in human-dominated landscapes could contribute to the alleviation of local ecological and socio-economic problems.

In landscapes that are highly populated and where most of the land is privately owned, restoration efforts must be tailored on the basis of people's perceptions, resource dependence, and need of biodiversity and ecological services (Cook *et al.*, 2004; Armitage, 2005). Because different groups in society – including men and women, wealthy and poor – tend to have different perceptions and attitudes toward the utilization of forest resources (Phillips and Gentry, 1993; Phillips *et al.*, 1994; Byg and Balslev, 2001), their reasons for supporting forest restoration will vary accordingly. Thus, the connection between people and nature needs to be evaluated carefully in order to develop effective strategies for ecosystem restoration. In addition, it is critical to assess the possibilities of forest restoration in the context set by the prevailing local production systems.

This study aimed at developing a holistic strategy for the restoration of High Andean forests in a human-dominated landscape in northern Ecuador. Livelihoods in the study area are based on the production of potatoes and extensive cattle ranching; both activities promote deforestation and require high human and agrochemical inputs, including excessive use of health-threatening pesticides (Yanggen *et al.*, 2004; Hofstede *et al.*, this volume). The present study integrated the needs of restoration of native forests with local peoples' perspectives on restoration, and made specific suggestions about ways in which each group of society could contribute (Engel, 1997; FAO, 1999; Leeuwis and Rhiannon, 2002). An ethno-biological approach was taken to evaluate people's interest in forest restoration in relation to their socio-economic status, gender, knowledge, and dependence on forest resources. Given the dependence of the local population on agricultural production, the possibilities for using agricultural experimentation to improve the use of local biodiversity and promote forest restoration were explored as well (see also Pohle and Gerique, 2008 for an example from southern Ecuador). Hence, the objectives of the project were: (i) to evaluate the restoration needs of the local forests, (ii) to identify groups of society interested in forest restoration, (iii) to select the species of plants with potential for forest restoration, and of mammals for managed production, and (iv) to provide specific recommendations for integrating groups of society to achieve participative restoration. The investigation was carried out as part of the Biodiversity-Based Livelihoods in High Andean Cloud Forests project investigating ways in which local biodiversity can contribute to sustainable livelihoods, including agricultural diversity and ecological restoration.

STUDY AREA

The study was carried out in a remnant of the humid montane Andean forest known locally as *bosque de Ceja Andina* or *bosque siempreverde montano alto* (Sierra *et al.*, 1999; Balslev and Ølgard, 2002). The study area is located on the western flanks of the volcanic Eastern Andean Cordillera in the Carchi Province of northern Ecuador (Figure 66.1). The investigated forest remnant stretched for approximately 30 km from north to south and covered an altitudinal range between 2800 and 3700 m.a.s.l. (Sierra *et al.*, 1999). Details on the respective forest types found in the area are given in the Results section below. The region has a steep topography, with slopes up to 40°, as well as numerous ravines, rivers, and valleys. The area has a mean annual precipitation of approximately 2000 mm, with two wetter seasons occurring from February to May, and from October to December; the most pronounced dry season is from June to August. The annual mean daily temperature is about 12°C, and daily variations in temperature exceed seasonal variations. Insolation ranges between 1000 and 2000 hours of sunshine per year (Hofstede *et al.*, 1998).

The population is composed mainly of *mestizo* people that during the last 40 years have moved to the area from neighboring Ecuadorian and Colombian highlands. This study was carried out in the localities of Mariscal Sucre and Jesús del Gran Poder, which had 1750 and 400 inhabitants in 2004, respectively.

METHODS

Classification and restoration of forest cover

To define the restoration needs of the natural forest of the area, the various forest types were classified, and their conservation status evaluated. To classify and estimate the extent of the

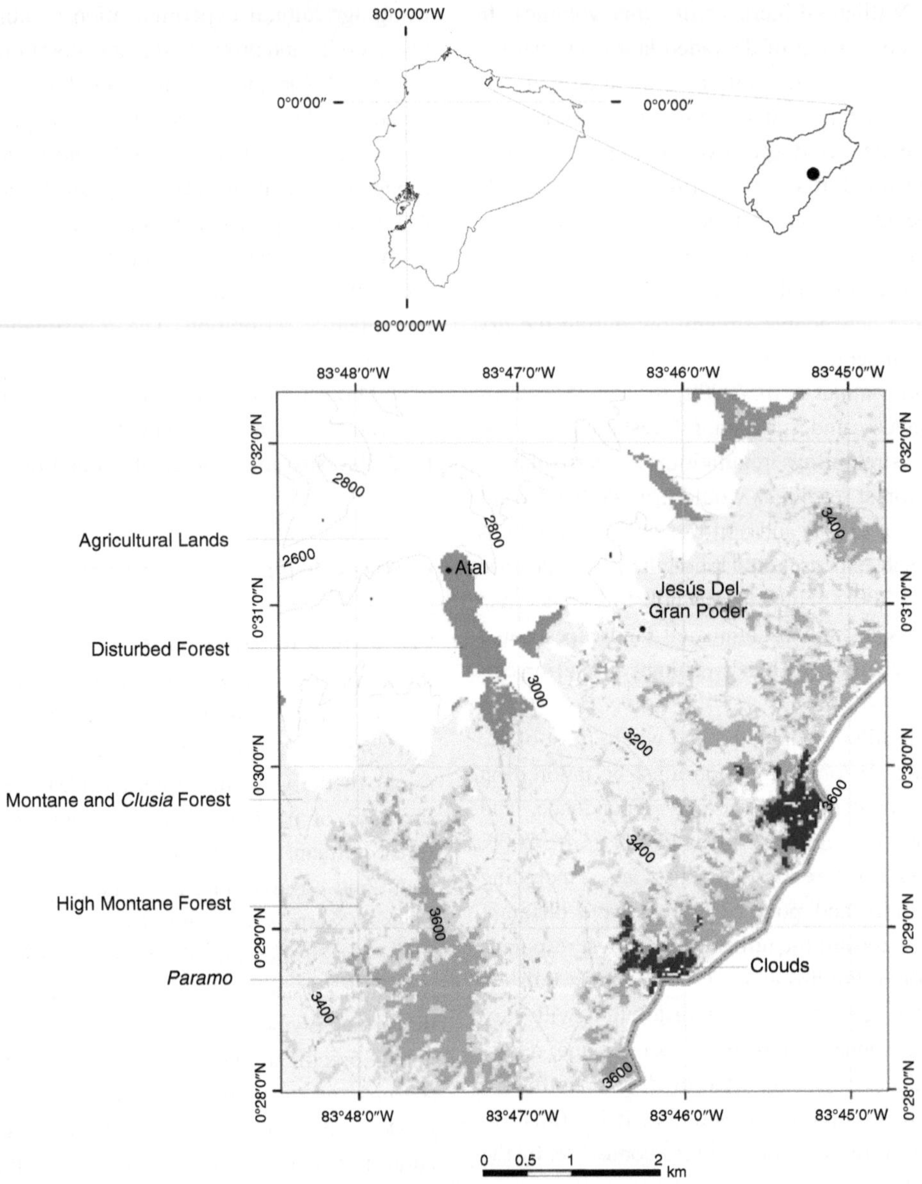

Figure 66.1. Map of a section of the study area in northern Ecuador indicating the distribution of different vegetation types and agricultural lands.

different vegetation types a Landsat TM image was used from November 1999 (scene P10 R60, complete, bands 2, 4, and 5, re-sampled at 30 × 30 m). The image was corrected using coordinate points, so distortion was less than 30 m. The vegetation types present in the image were classified using the method of supervised digital classification Maxlike (IDRISI 2.0, 1997).

To evaluate differences in the floristic composition of the respective forest types, 20 plots of 10 × 20 m were established along an altitudinal gradient between 2900 and 3600 m.a.s.l. at 50–80-m intervals. Plots were set up within each of the major vegetation types distinguished, choosing stands that showed little or no human alteration. For each plot, altitude and coordinates were recorded and each tree >10 cm DBH (diameter at breast height) was measured and identified. For the vegetation classification analysis floristic data were included from previous studies in the study area by Van Geloof (2003) (20 plots of 10 × 10 m), and by Mora (1996) (four plots of 50 × 50 m).

A multivariate method was used to evaluate the similarity of species composition between vegetation plots to distinguish different vegetation types. The Euclidean distance (*ED*) method calculates the absolute distance between points in one or many dimensions and is a measure of the magnitude of difference between samples of interest (Gasgrain and Legendre, 2000). Individuals of unclear taxonomy at the species level were grouped within their corresponding genus (e.g. *Shefflera*, *Oreopanax*). The calculated *ED*-values between plots were used to construct a cluster diagram illustrating similarities of species composition between plots. The conservation status of the

Table 66.1 *Indices of types of ethno-biological information*

Measure	Calculation	Description
Use value (UV_s)	$UV_s = \sum VU_{i,s}/n$; UV is the number of uses that the informant *i* knows for the species *s*; *n*=total number of informants	Indicates the average number of uses that the informants know for a species
Use value per category (UV_c)	$VU_s = \sum UV_{i,s,c}/n$; UV is the number of uses that the informant *i* knows for the species *s* in the use category *c*; *n* = total number of informants	Indicates the average number of uses in one use category that the informants know for a species in a category of use
Index of diversity of use (DU_s)	$DU_s = 1/\sum P_c^2$; P_c=contribution of the use categories *c* to the total utility of a species *s* (= number of times a species *s* was mentioned within a category of use, divided by the total number of reports of use of the species s among all the categories)	Indicates the number of categories of use a species is used, and how homogeneously this contributes to their use; values range from 0 to the total number of categories a plant is used for
Use frequency (UF_i)	$UF_i = \sum UF_{i,s}/n$; UF is the number of times an informant *i* uses a species *s* in a year; *n* = total number of species	Indicates the average number of times an informant uses a species in a 1-year period

respective forest types was mapped using a GIS (IDRISI 2.0). The present distribution of a given forest type was compared with its potential altitudinal distribution, but excluding areas occupied by other types of natural vegetation.

Peoples' interest in forest restoration and their relationship with biodiversity

In total, 155 people (93 men, 62 women) were interviewed to obtain information on their socio-economic status (notably education, land tenure, and income), their knowledge of the use of local plants and mammals, and their interest in restoring and managing wildlife species on their lands. The informants were randomly chosen during visits to their homes, but care was taken to have a representative sample in terms of gender, age, and socio-economic status. The interviews had a fixed format and used a color picture of each tree or mammal species considered in the study (Tirira, 1999). The informants answered questions related to the use of 17 species of timber trees that are naturally abundant in the local forest and considered to have high potential for reforestation projects (Fehse *et al.*, 1998; Loján, 2003). The pictures of the plants and mammals did not include their local names to avoid obtaining information belonging to other species that share a local name. The first question referred to each person's knowledge of the particular plant/animal indicated in the picture. If positively answered, the following questions were related to the species' use, the frequency of use, and its potential to be used for forest restoration. A partly different set of informants provided information about the use of 18 species of mammals native to the High Andean forests of the area. The questions inquired about the use of the animal, the frequency of use, and people's personal interest in managing the species. The interviews also included casual questions related to personal perspectives on restoring and managing species.

The data analysis was designed to identify the socio-economic groups and gender that had an interest in restoring local forests and managing wildlife. It was considered particularly important to know how peoples' interest in restoration related to their ethno-biological knowledge and to what degree they used local biodiversity. Various indices were calculated to standardize this information (Table 66.1). The socio-economic status of the informants was evaluated using an index ranging from 0 to 1, calculated by weighing values of degree of education (ranging from 0=no education, 1=primary school, 2=secondary school, and 3=college education); land tenure (0=no land owned, 1=renting land, 2=land owned by relatives, and 3=land owned by the family); and total monthly income in cash (weighting values relative to the observed minimum (0) and maximum (1) values. The ethno-biological information was used to calculate indices that evaluated different aspects of people's knowledge, and frequency of use of the species. These were log-transformed to achieve normality and homogeneity of variance; those that did not meet these conditions after the transformations were not used in the statistical analyses. One-way ANOVAs were used to identify gender-related differences.

The ethno-botanical and ethno-zoological data were analyzed separately. For the ethno-botanical data, a multiple regression approach with backward selection of variables was used to evaluate the relationship between an informant's socio-economic status (independent variable) and the Use Value, Use Diversity, and Use Frequency, reported for all the plant species (dependent variables). The same analysis was carried out for the ethno-zoological data, but the Use Frequency variable was not included, given the reluctance of informants to answer this question. Finally, a two-way ANOVA was used to search for relationships between socio-economic status of the informants, gender, and interest in restoring one's land with timber species or to manage native species of mammals. The index of Use Value (Table 66.1) was used to evaluate the plant species' potential to be used in restoration activities, and the animal species' potential to be used in projects of wildlife management.

RESULTS

Classification and conservation status of forest cover

The GIS and field classification gave complementary results that indicated the existence of three types of forest that had remained largely unaltered by human activity, plus disturbed forests that were lumped into a single category (Figures 66.1 and 66.2).

The 44 plots used in the vegetation analysis together covered an area of 16 040 m^2 and had 71 species of trees with DBH > 10 cm (Appendix 66.1). The plots were classified into three main vegetation types corresponding to: (i) (upper) montane cloud forest – locally known as *bosque de encinos*, (ii) upper montane *Clusia* forest – locally known as *guandera* forest, and (iii) High Andean forest (sub-alpine cloud forest) – with no local name registered. Figure 66.1 shows the distribution of the respective forest types whereas Table 66.2 lists their areal extent. The upper montane cloud forest tended to occur at somewhat lower altitudes (2900–3400 m.a.s.l.), while *Clusia* forest was typically found at slightly higher elevations (3220–3633 m.a.s.l.) although there was considerable altitudinal overlap between the two forest types (Figures 66.1 and 66.2). The High Andean forest was distributed along the upper boundaries of the forest belt (3540–3675 m.a.s.l.).

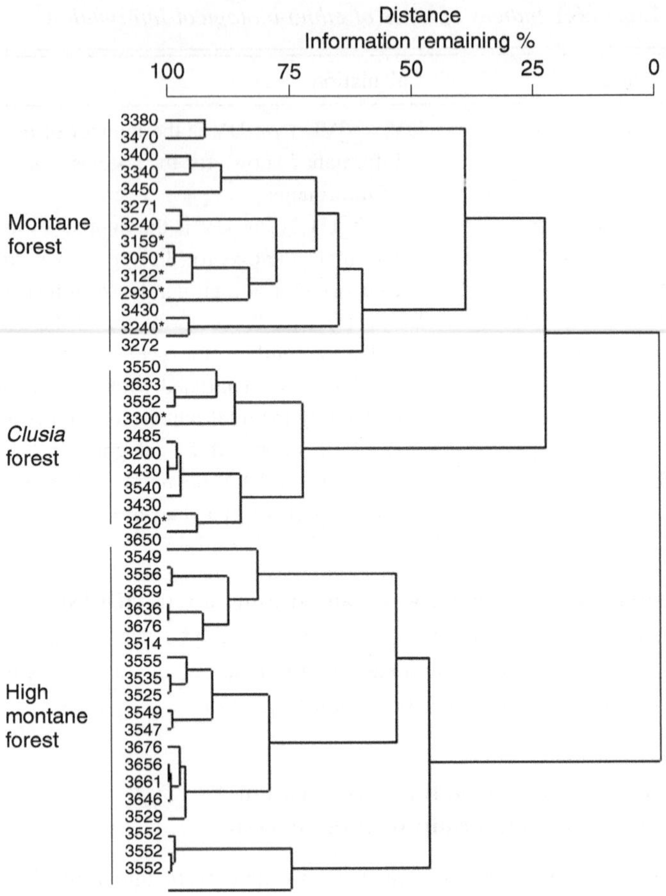

Figure 66.2. Classification analysis based on Euclidean distance of 44 vegetation plots within the upper montane and sub-alpine forest zones in Northern Ecuador. * indicates plots established in the area of Jesús del Gran Poder.

UPPER MONTANE CLOUD FOREST

This forest (2900–3400 m.a.s.l.) is dominated by various species of the genus *Weinmannia* (Cunnoniaceae), locally known as *encinos*. Trees in the *Weinmannia* genus have straight boles that reach 20–25 m height. The understory is dense, and covered with shrubs and herbs. The lowest altitudes with this forest type (e.g. the plot at 2930 m.a.s.l.) included species not found at higher elevations (e.g. *Clethra crispa, Freziera, Hesperomeles, Miconia, Saurauia, Myrciantes rophaloides, Myrsine, Prunus, Symploccos quitensis, Oreopanax* spp., *Schefflera* spp.).

CLUSIA FOREST

This forest (3220–3630 m.a.s.l.) is characterized by the dominance of large *Clusia flaviflora* and *C. multiflora* (Clusiaceae) trees and is perhaps the most impressive and unique forest formation in the area, with trees reaching up to 25 m and crown areas of up to 70 m^2. In addition, abundant aerial roots extend from the large branches to the ground. The understory is open and largely dominated by the clonal rosette *Guzmania* spp. (Bromeliaceae) carpeting the soil surface and tree trunks.

HIGH MONTANE FOREST

This sub-alpine cloud forest (3540–3675 m.a.s.l.) forms the treeline and its border with the *páramo*. Its thick and twisted trunks are covered with mosses and reach a height of 6–10 m, becoming more stunted and elfin-forest-like in appearance at the highest elevations. The most common species are *Gaiadendron punctatum, Ilex* spp., and *Ocotea infrafoveolata*.

It was not possible to evaluate the conservation status of the upper montane and *Clusia* forests separately because they were not distinguishable from each other in the satellite image. Also, their altitude-overlapping distribution prohibited separate evaluation. Taken together, upper montane and *Clusia* forest made up 71.7% of the total forested area in the study area, whereas High Andean forest contributed 21.9%, and disturbed forest the remaining 6.3% (Table 66.2). Old-growth upper montane and *Clusia* forests occupied only 8.7% of their potential distribution (below 3450 m.a.s.l.) and most of their potential distribution is presently covered by agricultural lands and human settlements (89%); disturbed montane forest represented only a very small

Table 66.2 *Extents of different types of forest in the study area*

Altitudinal range (m.a.s.l.)	Montane forest (km^2)	High montane forest (km^2)	Disturbed forest (km^2)	Agricultural land and towns (km^2)	Area total (km^2)
<3450	24.58 (8.7)	2.99 (1.0)	3.61 (1.3)	255.01 (89.1)	286.2
3450–3700	15.96 (62.7)	9.37 (98.5)	0 (0)	0.14 (0.6)	25.4
Total	40.5	12.4	3.6	255.2	311.7

Table 66.3 *Ethno-biological knowledge of men and women of the High Andean forests in northern Ecuador. Differences between men's and women's knowledge were calculated using one-way ANOVA*[a]

	Men (mean (min, max)/standard deviation)	Women (mean (min, max) /standard deviation)	*p*
17 species of plants			
Number of informants/reports	53/898	35/589	
Index of socio-economic status	0.68 (0, 1)/0.220	0.58 (0.2, 93)/0.200	***
Use value (UV)	0.10 (0, 1)/0.229	0.05 (0, 1)/0.169	*
Index of diversity of use (DU)	0.11 (0, 1)/0.229	0.05 (0, 1)/0.169	ns
UV medicinal	0.05 (0, 1.4)/0.095	0.14 (0, 1)/0.234	*
UV edible	0.10 (0, 0.5)/0.154	0.10 (0, 0.5)/0.173	ns
UV timber	0.35 (0, 1)/0.249	0.19 (0, 1)/0.170	ns
UV firewood	0.37 (0, 1)/0.227	0.33 (0, 1)/0.307	ns
UV crafts	0.09 (0, 0.4)/0.118	0.15 (0, 1)/0.231	ns
UV ritual	0.001 (0, 0.02)/0.003	0.01 (0, 0.5)/0.085	nc
12 species of mammals			
Number of informants/reports	46/585	28/377	
Index of socio-economic status	0.57 (0.06, 1)/0.19	0.57 (0.2, 0.9)/0.2	***
Use value (UV)	0.19 (0, 2)/0.45	0.23 (0, 3)/0.5	*
Index of diversity of use (DU)	0.003(0, 1)/0.03	0.006 (0, 0.1)/0.005	ns
UV medicinal	0.09 (0, 1)/0.288	0.009 (0, 1)/0.28	ns
UV edible	0.07 (0, 1)/0.26	0.100 (0, 0.5)/0.31	*
UV pets	0.004 (0, 1)/0.06	0.006 (0, 1)/0.09	nc
UV decoration	0.001 (0, 1)/0.003	0.0 (0,0)/0.0	nc

[a]Statistical significance indicated as *, $p \leq 0.05$; ***, $p \leq 0.001$; ns, not significant; nc, not calculated due to lack of normal distribution or homogeneity of variance.

portion (1.2%). Conversely, the high montane forest still occupied most of its potential distribution (98.5%), with agricultural land and human settlements representing the remaining 1.5% (Table 66.2).

PEOPLE'S RELATIONS WITH BIODIVERSITY AND THEIR INTEREST IN FOREST RESTORATION

The informants provided socio-economic information that resulted in a total of 2854 records of use for 17 species of trees and 18 species of mammals. The data were used to calculate socio-economic and ethno-biological indices (Table 66.3). Four species of mammals were not used in the statistical analyses, as no ethno-biological information was recorded for them: puma (*Puma concolor*), pampas cat (*Lynchailurus pajeros*), pudu deer (*Pudu mephistophiles*), and southern river otter (*Lutra longicaulis*).

Examination of the variation in socio-economic standards and ethno-biological knowledge between men and women indicated that men always had significantly higher socio-economic standards than women (Table 66.3). Also, men reported higher general ethno-botanical knowledge (in terms of use value, UV) than did women, but women had a higher knowledge of the medicinal uses of plants (UV medicinal). The analysis did not detect any differences in the diversity of use (DU), nor within other categories of use of plants. Use frequency (UF) was markedly higher for women although this could not be tested statistically because the data did not meet the statistical requirements.

The ethno-zoological data showed that women had a significantly higher knowledge of the use value, and uses for edible mammals (category UV edible) (ANOVA, $p < 0.05$, $n = 74$) indicating that women know more total uses per species, and more species used for food. The other aspects of ethno-biological knowledge were not significantly different between genders or could not be tested because the data-set did not meet the statistical requirements. In addition, the UF of species was significantly higher for women than for men (Table 66.3).

In the ethno-botanical section, there was a significant relation between people's socio-economic status, and their UV, DU, and UF indices, but the explanatory power of the relationship was very low (multiple regression, $r^2 = 0.09$, $p < 0001$, $n = 81$). However, there was a strong positive relationship between socio-economic status and interest in reforestation and forest restoration, regardless of gender (two-way ANOVA, $p < 0.001$, $n = 88$). In total, 85% of the people interviewed had some interest in reforesting their land, but the inclination was more marked among people with better socio-economic conditions. People's motivations for restoring their land were related to species utility, and need of environmental goods and services such as water, soil conservation, wind protection, and landscape preservation. On the other hand, only 5% of the people that had no interest in restoration expressed a clear reason for their lack of interest, such as inaccessibility to land, slow growth of the species, or little time to dedicate to restoration activities.

In the ethno-zoological section, there was no significant relation between people's socio-economic status and use value or use diversity (multiple regression, $r^2 = 0.06$, $p > 0.05$, $n = 74$). However, when investigating the relationships between socio-economic status, gender, and interest in managing species of mammals, a positive interaction term was found that linked the relationship between gender and interest in management of mammals (two-way ANOVA, $p < 0.001$, $n = 74$). This interest in managing wildlife species was restricted to 24 men with high socio-economic status (30% of the sample). An additional trend indicated that people reporting a higher use value tended to be interested in managing mammals (simple correlation 12%, $p < 0.001$, $n = 74$.)

Species with potential to restore forest cover and for wildlife management

Differences were found in the use values (UV) for species of plants and mammals (Table 66.4). Plant species with a high UV included: *Prunus huantensis, Weinmannia pinnata, W. cochensis, Freziera canescens, Myrica rophaloides*, and to a lesser degree *W. rollonii* and *Vallea stipularis*. The animal species that recorded high UV values were agouti (*Nasuella* sp.), armadillo (*Dasypus novemcinctus*), striped hog-nosed skunk (*Conepatus semistriatus*), Andean tapir (*Tapirus pinchaque*; cf. Cavelier *et al.*, this volume), and white-eared Andean opossum (*Didelphis pernigra*). Most of the plant species recognized as useful were abundant in the upper montane forest. In general, no specific mammals were associated with a given forest type.

DISCUSSION

Restoration of forest cover

Restoration needs for the respective forest types vary greatly. Upper montane and *Clusia* forests in the study area are very poorly preserved because agricultural lands and human settlements already occupy 89% of the potential distribution for these forests. On the other hand, the High Andean montane forests enjoy relatively good conservation status (Table 66.2). Agricultural land in the distribution range of this forest type is still very limited (<1.5%), presumably due to the lack of roads entering these areas, and perhaps due to increased public awareness of the importance of protecting the adjacent *páramo* grasslands and High Andean montane forests from intentional fire (cf. Sarmiento, 2002; Buytaert *et al.*, 2006). Thus, restoration efforts should focus on the upper montane and *Clusia* forests rather than on the high montane forest. Interestingly, the fact that the tree species with the highest UV values all occurred in the upper montane forest may facilitate the use of these species in restoration activities (*P. huantensis, W. pinnata, W. cochensis, F. canescens, M. rophaloides, W. rollonii*, and *V. stipularis*).

People's relations with biodiversity and their interest in forest restoration

The present results suggest that decisions made by men on land use and restoration will be critical in defining whether and how families are to be involved in forest restoration activities. Gender proved to be an important factor in defining socio-economic status and to some extent ethno-biological knowledge, and as a result, decision-making on land use is likely to be influenced mostly by the men's perspectives. Men had more access to secondary education, managed the production and commercialization of crops and cattle produced on their farms, and usually had the family property registered under their names. Female roles were more restricted to the maintenance of homes, small gardens, domestic mammals, and cattle.

The fact that people's interest in forest restoration was positively related to ethno-biological knowledge and socio-economic status indicates that more affluent groups of society could be involved in restoration activities more readily than disadvantaged groups. A positive relationship between people's socio-economic status and their ethno-botanic knowledge (in terms of use value, diversity of use, and use frequency) has also been found in other ethno-botanic studies (Byg and Balslev, 2001) and

Table 66.4 *Differences in the use values for species of plants and animals*

Plant species	Common name	Uses	Use value	UV medical	UV edible	UV wood	UV fuel	UV crafts	UV ritual
Cedrela montana	cedro	F W	0.281	0.000	0.025	0.514	0.088	0.373	0.000
Clethra crispa	palo de león cuguaca	F M W	0.100	0.125	0.208	0.125	0.458	0.0083	0.000
Freziera canescens	pisique guayusa	C E F M W	1.138	0.007	0.366	0.188	0.430	0.009	0.000
Freziera reticulata	palo de rosa, yalte	C F M R W	0.420	0.000	0.171	0.267	0.471	0.092	0.000
Gaiadendron punctatum	popa	C F W	0.115	0.350	0.025	0.150	0.150	0.025	0.000
Myrsine coriacea	yalte, hojarasca	C F	0.113	0.147	0.125	0.125	0.333	0.167	0.000
Myrcianthes rhopaloides	arrayán	C E F M R W	2.039	0.168	0.377	0.115	0.221	0.120	0.000
Nectandra laurel	pandala rojo	C, F, W	0.323	0.000	0.273	0.179	0.547	0.000	0.000
Ocotea heterochroma	yalte	W	0.392	0.063	0.120	0.341	0.400	0.077	0.000
Ocotea infrafoveolata	yalte negro	C M W	0.362	0.050	0.036	0.450	0.428	0.036	0.000
Prunus huantensis	aguacatillo, pandala, canelo	C E F M W	1.141	0.011	0.000	0.434	0.163	0.392	0.000
Ruagea hirsuta	cedrillo	F W	0.051	0.000	0.000	0.375	0.000	0.000	0.125
Valea stipularis	sacha rosa, rosa de monte	E, M, R, W	0.962	0.211	0.000	0.082	0.281	0.421	0.006
Weinmannia cochensis	encino rosado o colorado	F, C, M, W	0.619	0.118	0.000	0.313	0.468	0.102	0.000
Weinmannia fagaroides	encino blanco	C W	0.583	0.104	0.008	0.448	0.088	0.000	0.000
Weinmannia pinnata	encino	F C W	0.950	0.114	0.000	0.336	0.472	0.078	0.000
Weinmannia rollonii	encino común o amarillo	F C M W	0.380	0.167	0.000	0.264	0.569	0.000	0.000

Animal species	Common name in Spanish (English)	Uses	Use value	UV medical	UV edible	UV tools	UV ritual	UV pets	UV decorative
Coendou quichua	irizo (porcupine)	E M	0.380	0.675	0.300	0.000	0.000	0.025	0.000
Conepatus semistriatus	zorro (striped long-nosed skunk)	M	0.511	0.960	0.019	0.021	0.000	0.000	0.000
Cuniculus taczanowskii	sacha cuy (mountain paca)	E	0.090	0.000	0.333	0.000	0.000	0.000	0.000
Dasypus novemcinctus	armadillo cachicambo (armadillo)	E M C O	0.395	0.120	0.742	0.138	0.000	0.000	0.000
Didelphis pernigra	zariguella raposa (white-eared opossum)	E C M	0.752	0.802	0.135	0.063	0.000	0.000	0.000
Mazama rufina	venado soche (red brocket)	E C M O	0.230	0.042	0.596	0.042	0.000	0.071	0.000
Mustela frenata	chucuri (long-tailed weasel)	R	0.010	0.000	0.000	0.000	0.250	0.000	0.000
Nasuella sp.	cusumbe (coati)	E M P	0.326	0.378	0.567	0.056	0.000	0.000	0.000
Odocoileus peruvianus	venado de paramo (deer)	E O	0.227	0.000	0.381	0.119	0.000	0.000	0.000
Pseudalopex culpaeus	lobo de monte (colored fox)	R F	0.266	0.031	0.000	0.156	0.781	0.031	0.000
Sciurus granatensis	ardilla (squirrel)	E O	0.160	0.792	0.042	0.000	0.167	0.000	0.000
Sylvilagus brasiliensis	conejo (tapeti)	E	0.230	0.125	0.583	0.000	0.000	0.042	0.000
Tapirus pinchaque	tapir andino (Andean tapir)	E M O	0.483	0.172	0.772	0.015	0.000	0.000	0.042
Tremarctos ornatus	oso de anteojos (spectacled bear)	E C M O	0.284	0.483	0.258	0.167	0.050	0.000	0.000

contradicts the general belief that poorer people are more dependent on natural products, and therefore know more about them. Rather, the observed trend can be understood in terms of individual attitudes toward exploiting the natural environment, which may result in marked entrepreneurial behavior and higher living standards. In addition, it is also possible that higher exploitation of agro-biodiversity improves the socio-economic standards of families in rural areas (Byg and Balslev, 2001). Higher diversity of agricultural and forest crops is certainly a valid strategy to stabilize the economy of the local families in this area because the market for potatoes, the main agricultural product, suffers from drastic fluctuations in price. This reasoning is supported further by the fact that people of higher socio-economic status showed higher UF of plants.

The influence of gender on ethno-botanic knowledge revealed different patterns in the use of biodiversity by the local population. Such patterns provide an opportunity to exploit a broader variety of resources of the forest. Although men and women had similar knowledge on the diversity of uses per species of plant (DU), women knew more about medicinal uses (UV medicinal), and tended to use them more frequently (UF). Nonetheless, men knew significantly more uses per species (UV). These results show that men and women depended equally on biodiversity, but that gender influenced the specialization of their knowledge about the use of plants (Phillips and Gentry, 1993). The women's greater knowledge and frequency of use of medicinal plants reflect their expertise with respect to meeting the day-to-day health care needs of family members. On the other hand, the men's higher UV may be related to the higher number of ecosystems they are exploiting (e.g. forests, pastures, *páramo*), and the broader spectrum of needs that they have to cover for their families, including wood exploitation for construction of houses, fences, and tools. As a consequence, restoration through the use of biodiversity by men and women could be achieved if men would focus on species that are more broadly distributed in the forest and have a broad variety of uses. Similarly, women may be more interested in species that are useful to their households.

Bias in the present data prevented a characterization of how ethno-zoologic knowledge varied with gender and socio-economic status, or with the opportunities to use wildlife resources to promote the restoration of the Andean forest. The fact that women had significantly higher knowledge (UV) and knew more species of mammals used for food (UV edible) implied that men did not hunt the mammals, or only participated in their consumption and use when prepared. This must be regarded as an anomaly in the data because men are traditionally charged with hunting and fishing activities whilst women cook the hunted animals, prepare medicines extracted from them, and help the men to prepare decorations or tools (e.g. dried skins, leather, dried claws to comb domestic mammals). This bias in the data was likely due to the male informants' reluctance to provide information related to their hunting activities to people involved in local conservation activities, as men and women never responded to questions related to frequency of hunting activities or observation of mammals.

Nonetheless, the present information still reveals general trends in the use of species, the diversity of uses they have, and their potential for production in local farms or to be considered in silvo-pastoral systems. It was found that the interest to use wild mammals was related to gender and socio-economic status. Wealthier men were more eager to try and breed certain species of native mammals on their land. In fact, a few persons in this group mentioned attempts to farm various species of native animals. The mammals with the highest UV (cuchucho, armadillo, striped hog-nosed skunk, Andean tapir, and white-eared opossum) could be farmed locally for the commercialization of meat and possibly contribute to a reduction of hunting pressure on the wild populations (Medellín *et al.*, 2005). Some of these species share their genus with species used by native communities elsewhere in the Neotropics (Orejuela, 1992; Stearman, 1992; Medeiros Costa-Neto, 1999; Cavelier *et al.*, this volume). This consistency in regional trends of use of these species may facilitate the development of management techniques, and of strategies for their conservation in the wild. Nonetheless, it is interesting to note that people were not interested in managing large mammals, such as deer or Andean tapir, which may yield relatively larger benefits, but also require a higher investment of time and energy for their management. On the other hand, people's interest to manage small mammals may be related to their need to fulfill their requirements in the short term.

RECOMMENDATIONS FOR PARTICIPATIVE RESTORATION OF FOREST COVER

A locally designed and implemented restoration plan should contribute to the fulfillment of needs and interests of the men and women in the various socio-economic strata present in this society (Conway *et al.*, 1998). During the course of this study local farmers expressed a favorable view toward forest restoration to improve water regulation, stream protection, soil recovery, and scenic beauty; and as a means to recover timber and non-timber forest resources. However, people were only interested in restoring areas that were not currently used for agricultural production, including riparian areas and fence lines. They often mentioned that converting their agricultural lands into forest was not feasible due to their need to continue with their potato crops and cattle ranching. Therefore, large-scale restoration of the area is unlikely to happen in the short term unless substantial economic benefits (e.g. tax deductions or payments for forest restoration) are provided to compensate the loss of the higher revenues from the agricultural land. Such a system has proved to be effective in a nearby county where a payment scheme for water regulation (ecosystem services) was introduced, and therefore the protection of montane cloud forests was secured (Yaguache and Carrion, 2004; cf. Muñoz-Piña *et al.*, 2008; Calvo-Alvarado *et al.*, this volume; Tognetti *et al.*, this volume).

Clearly, the main cause of forest destruction in the study area is agricultural expansion, which is also the main impediment for the restoration of forest cover. Therefore, promoting the restoration of the original forest cover has to be connected with modifications of agricultural practices. These must help to reduce the need for more agricultural land, and improve opportunities to increase the amount of land available for the recovery of cloud forests (Pohle and Gerique, 2008; Grau *et al.*, this volume; Hofstede *et al.*, this volume). Because both men and women are familiar with and make use of local biodiversity, a viable strategy for forest restoration would be to increase the awareness of their dependence on forest products (plants and mammals) and to integrate useful species into their agricultural practices. It is thought that such a strategy would diversify people's crops, thereby reducing their dependence on the external market and rendering their livelihoods less vulnerable to economic fluctuation.

The inclination of the better-off part of society to become involved in restoration activities, and the fact that this stratum has more knowledge of the use of plant species (UV), provides important insights into understanding people's perceptions regarding which resources have positive probabilities of successful exploitation and management. It is suggested that an initial phase for the successful restoration of this area should encourage and guide the more affluent parts of society to restore their lands using species with high use value that will provide them with services they are familiar with. People in the lower strata of society are then likely to become gradually interested in restoring their lands and managing wildlife species once they witness concrete examples of the direct uses they can obtain from local biodiversity, particularly if support for such activities is made available through participatory research and learning systems. Presently, Farmers, Field Schools (FFS) and Local Agricultural Research Committees (CIALs) are being established in the study area to promote learning processes that enhance the use of local biodiversity in livelihoods through participatory research and experimentation (Hofstede *et al.*, this volume). These local learning centers attract young people of many different socio-economic backgrounds and give them the opportunity to learn alternative ways of agricultural production, including species that could be used in agro-forestry or silvo-pastoral systems. It is expected that these activities may change poorer peoples' perceptions and benefits of forest biodiversity and forest restoration.

ACKNOWLEDGEMENTS

The authors are grateful to Josette Moncayo, Radim Hedl and Luis González, Pool Segarra, Margarita Mancero, Fernanda Encalada, Horacio Narváez, Juan Monge, and Fausto Maldonado for assisting in the development of this project. Thanks are due to Diego Tirira for helping with the ethno-zoologic component; to municipal staff in Montúfar and Huaca, who provided logistic and personnel support; and to the Jatun Sacha Foundation and the staff of the Guandera Biological Station who provided hospitality and support during the field research. Special gratitude goes to the people of Mariscal Sucre and Jesús del Gran Poder for sharing their time and knowledge that made this project possible. Mitch Aide and Larry Hamilton gave very insightful reviews of an earlier version of the manuscript. The Ceja Andina project was financed by the International Development Research Center (IDRC – Canada) and executed by the Ecopar Corporation (Ecuador).

Appendix 66.1

Species	Altitudinal location of the plots (meters above sea level)																																									
	2930	3050	3122	3159	3200	3220	3240	3240	3271	3272	3300	3340	3380	3400	3430	3430	3430	3450	3470	3485	3514	3525	3529	3535	3540	3547	3549	3549	3550	3552	3552	3552	3552	3555	3556	3633	3636	3646	3650	3656	3659	3661	3676
Aequatorium asterotrichum B. Nord.										x		x						x																									
Aristeguietia budleaefolia (Benth) R.M. King & Rob.												x																															
Aequatorium campii B. Nord.												x																															
Ardisia colombiana Lundell.												x																															
Blechnum aureatum						x														x																x			x				
Brunellia pauciflora Cuatrec. & C.I. Orozco					x							x					x																										
Blechnum schomburgkii (Klotzch) C. Chr.																	x																										
Brunellia tomentosa Bompl.				x	x																																						
Clethra crispa C. Gust.	x				x	x											x								x																		
Clusia flaviflora Engl.					x	x					x	x		x	x		x		x	x	x	x	x	x	x	x	x	x	x	x				x	x	x		x			x	x	
Clethra ovaliforia (R. & P.) Dom.					x												x								x	x																	
Cyathea caracasana (Klotzch) Dom.		x	x	x	x		x	x	x	x		x	x																x														
Dicksonia sellowiana Hook.					x																																						
Disterigma alaternoides (Kunth) Nied.					x																																						
Diplostephium floribundum (Benth) Wedd.					x																																						

Escallonia myrtilloides L.f.					x																																	x						
Escallonia paniculata (Ruiz & Pav.) Roem. & Schult.																	x																											
Freziera canescens Bonpl.	x																																											
Freziera microphylla Sandwith								x																																				
Gaiadendron punctatum (R. & P.) G. Don.		x	x	x					x												x	x	x	x			x			x	x	x	x	x			x	x		x	x	x	x	x
Geissanthus spp.	x	x	x	x																																								
Grosvenoria rimbachii (B.L. Rob.) R.M. King. & H. Rob.												x																																
Hedyosmum cuatrecasanum	x	x		x								x																																
Hedyosmum cumbalense Swartz					x												x					x		x	x	x	x	x																
Hedyosmum translucidum														x				x																										
Hesperomeles ferruginea (Pers.) Benth.																		x																										
Ilex colombiana Cuatrec.						x															x	x	x		x	x	x	x		x	x	x	x	x	x	x	x		x	x	x	x	x	x
Indet. 1															x																													
Indet. 2	x																																											
Indet. 3										x																																		
Lauraceae sp. 1													x																															
Lauraceae sp. 2			x	x									x																															
Lauraceae sp. 3													x																															
Macleania rupestris (Kunth) AC. Sm.					x												x																											
Melastomataceae sp. 1													x																															
Melastomataceae sp. 2		x	x																																									

Appendix 66.1 (*cont.*)

Species	2930	3050	3122	3159	3200	3220	3240	3240	3271	3272	3300	3340	3380	3400	3430	3430	3430	3450	3470	3485	3514	3525	3529	3535	3540	3547	3549	3549	3550	3552	3552	3552	3552	3555	3556	3633	3636	3646	3650	3656	3659	3661	3676
Miconia bracteolata (Bompl.) DC.					x				x	x	x	x		x	x	x		x																									
Miconia chlorocarpa Cogn.															x			x																									
Miconia coriacea (Sw.) DC.							x		x																																		
Miconia nodosa Cogn.																		x																									
Miconia ochracea Triana					x							x																															
Miconia ochreotoma Naudin					x							x					x								x																		
Miconia theazans (Bompl.) Cogn.	x		x	x				x				x	x																														
Miconia tinifolia Naudin					x							x					x				x	x	x	x	x	x	x	x		x		x	x	x	x			x		x	x	x	x
Myrica pubescens Humb. & Bonpl. ex Willd.		x																																									
Myrsine andina (Mez.) Pipoly															x																												
Myrsine coriacea (Sw.) R. Br. ex Roem. & Schult.		x																																									
Myrsine dependens (R. & P.) Spreng.		x		x		x		x										x																									
Ocotea infrafoveolata Van der Welf.					x							x		x			x								x																		
Ocotea cf. *serisea*		x																																									
Oreopanax spp.			x		x							x	x		x	x	x					x			x						x											x	x
Palicourea amethystina (R. & P.) DC.					x							x																															

Pentacalia campii (Cuatrec.) Cuatrec.					x																																							
Pernettia prostrata spp.																									x																			
Piper barbatum Kunth												x																																
Prunus huantensis Pilg.		x	x	x								x		x																														
Prunus rugosa Koehne.					x																																							
Saurauia bullosa Wawra					x				x																																			
Shefflera spp.			x	x	x		x		x	x							x																											
Solanum asperolanatum R. & P.					x																																							
Solanum hypalaurotricum Bitter												x																																
Symplocos quitensis Brand				x	x							x		x		x		x																										
Turnefortia fuliginosa Kunt												x																																
Turnefortia cf. *scabrida*															x																													
Weinmannia brachystachia L.																					x	x	x	x		x	x	x		x	x	x	x	x	x		x	x		x	x	x	x	x
Weinmannia cochensis Hieron					x												x			x					x														x					
Weinmannia glabra L.f.	x					x	x	x	x				x		x			x	x																									
Weinmannia pinnata L.					x							x					x			x					x				x							x			x					
Weinmannia rollotti Killip.		x	x	x			x	x	x	x		x		x		x	x	x																										

REFERENCES

Anand, P. (2002). *The Geography of Poverty in Ecuador*. Berkeley, CA: University of California, Department of Agricultural, Environmental and Development Economics.

Armitage, D. R. (2003). Traditional agroecological knowledge, adaptative management and the socio-politics of conservation in Central Sulawesi, Indonesia. *Environmental Conservation* **30**: 79–90.

Armitage, D. R. (2005). Adaptative capacity and community-based natural resource management. *Environmental Management* **35**: 703–715.

Balslev, H., and B. Ølgaard (2002). Mapa de vegetación del sur de Ecuador. In *Botánica Austroecuatoriana: Estudios sobre los recursos vegetales en las Provincias de El Oro, Loja y Zamora-Chinchipe*, eds. M. Z. Aguirre, J. E. Madsen, E. Cotton, and H. Balslev, pp. 51–64. Quito, Ecuador: Abya-Yala.

Barthlott, W., J. Muthke, M. D. Rafiqpoor, G. Kier, and H. Kreft (2005). Global centres of vascular plant diversity. *Nova Acta Leopoldina N.F.* **92**, 342: 61–83.

Beck, E., J. Bendix, I. Kottke, F. Makeschin, and R. Mosandl (eds.) (2007). *Gradients in a Tropical Mountain Ecosystem of Ecuador*. Berlin: Springer-Verlag.

Bruijnzeel, L. A. (2004). Hydrological functions of tropical forests: not seeing the soil for the trees? *Agriculture, Ecosystems and Environment* **104**: 185–228.

Buytaert, W., R. Célleri, B. de Bièvre, *et al.* (2006). Human impact on the hydrology of the Andean páramos. *Earth-Science Reviews* **79**: 53–72.

Byg, A., and H. Balslev (2001). Diversity and use of palms in Zahamena, eastern Madagascar. *Biodiversity and Conservation* **10**: 951–970.

Conway, S., and F. Steward (1998). Mapping innovation networks. *International Journal of Innovation Management* **2**: 223–254.

Cook, W. M., D. G. Casagrande, D. Hope, P. M. Groffman, and S. L. Collins (2004). Learning to roll with the punches: adaptive experimentation in human dominated systems. *Frontiers in Ecology and the Environment* **2**: 467–474.

Crissman, C. C., P. Espinosa, C. E. F. Ducrot, D. C. Cole, and F. Carpio (1998). The case study site: physical, health and potato farming systems in Carchi province – economic, environmental and health tradeoffs. In *Agriculture: Pesticides and the Sustainability of Andean Potato Production*, eds. C. C. Crissman, M. J. Antle and S. M. Capalbo, pp. 85–120. Dordrecht, the Netherlands: Kluwer.

Dinerstein, E., D. M. Olson, D. J. Graham, *et al.* (1995). *A Conservation Assessment of the Terrestrial Ecoregions of Latin America and the Caribbean*. Washington, DC: World Wildlife Fund and World Bank.

Engel, P. (1997). *The Social Organization of Innovation: A Focus on Stakeholder Interactions*. Amsterdam, the Netherlands: Royal Tropical Institute.

FAO (1999). *The Participatory Process for Supporting Collaborative Management of Natural Resources: An Overview*. Rome: FAO.

Fehse, J., N. Aguirre, C. Paladines, *et al.* (1998). *Caracterización de los bosques naturales de la sierra del Ecuador*, Programa FACE de Forestación (PROFAFOR). Quito, Ecuador: Abya-Yala.

Frolich, L. M., and E. Guevara (1999). The role of family-based agricultural innovation in conserving tropical montane cloud forest: The Guandera Project in Northern Ecuador. In *Entendiendo las interfaces ecológicas para la gestión de paisajes culturales en los Andes: Memorias del tercer simposio internacional sobre desarrollo sustentable en los Andes*, eds. F. Sarmiento and J. Hidalgo, pp. 45–49. Quito, Ecuador: Corporación Editora Nacional.

Gasgrain, P., and P. Legendre (2000). *The R Package for Multivariate and Spatial Analysis*. Montréal, Canada: University of Montréal, Dèpartment of Biological Sciences.

Grubb, P. J. (1977). Control of forest growth and distribution on wet tropical mountains: with special reference to mineral nutrition. *Annual Review of Ecology and Systematics* **8**: 83–107.

Günter, S., P. Gonzalez, G. Alvarez, *et al.* (2009). Determinants of successful regeneration of abandoned pastures in the Andes: soil conditions and vegetation cover. *Forest Ecology and Management* **258**: 81–91.

Hobbs, R. J., and J. A. Harris (2001). Restoration ecology: repairing the Earth's ecosystems in the new millennium. *Restoration Ecology* **9**: 239–246.

Hofstede, R., J. Lips, W. Jongsma, and J. Sevink. (1998). *Geografía, ecología y forestación en la Sierra Alta del Ecuador: Revisión de literatura*. Quito, Ecuador: Abya-Yala.

Jordan, W. R. III, M. E. Gilpin, and D. Aber (1987). *Restoration Ecology: A Synthetic Approach to Ecological Research*. Cambridge, UK: Cambridge University Press.

Kessler, M. (1995). Present and potential distribution of *Polylepis* (Rosaceae) forests in Bolivia. In *Biodiversity and Conservation of Neotropical Montane Forests*, eds. S. P. Churchill, H. Balslev, E. Forero, and J. Luteyn, pp. 281–294. New York: New York Botanical Garden.

Laegaard, S. (1992). Influence of fire in the paramo vegetation of Ecuador. In *Paramo: An Andean Ecosystem under Human Influence*, eds. H. Balslev and J. Luteyn, pp. 151–170. London: Academic Press.

Leeuwis, C., and P. Rhiannon (2002). *Wheelbarrows Full of Frogs: Social Learning in Rural Resource Management*. Amsterdam, the Netherlands: Koninklijke Uitgeverij van Gorcum.

Loján, L. (2003). *El verdor de los Andes Ecuatorianos: Realidades y promesas*. Quito, Ecuador: FAO.

Malmer, A., D. Murdiyarso, L. A. Bruijnzeel, and U. Ilstedt (2009). Carbon sequestration in tropical forests and water: a critical look at the basis for commonly used generalisations. *Global Change Biology*, doi:10.1111/j.1365-2486.2009.01984.x.

Medeiros Costa-Neto, E. (1999). Traditional use and sale of animals as medicines in Feira de Santana City, Bahia, Brazil. *Indigenous Knowledge and Development Monitor*. Available at www.luchy.ukc-ac.

Medellín, R., C. Manterola, M. Valdéz, *et al.* (2005). History, ecology, and conservation of the Pronghorn Antelope, Bighorn Sheep, and Black Bear in Mexico. In *Biodiversity, Ecosystems, and Conservation in Northern Mexico*, eds. J. L. E. Cartron, G. Ceballos, and R. Stephen, pp. 387–404. Oxford, UK: Oxford University Press.

Molina, A., G. Govers, V. Vanacker, J. Poesen, and E. Zeelmaekers (2007). Runoff generation in a degraded Andean ecosystem: interaction of vegetation cover and land use. *Catena* **71**: 357–370.

Mora, L. (1996). Vegetacion de la Reserva Guandera. Graduate thesis, Facultad de Filosofia, Universidad Central del Ecuador, Quito, Ecuador.

Mosandl, R., S. Günter, B. Stimm, and M. Weber (2008). Ecuador suffers the highest deforestation rate in South America. In *Gradients in a Tropical Mountain Ecosystem of Ecuador*, eds. E. Beck, J. Bendix, I. Kottke, F. Makeschin, and R. Mosandl, pp. 37–40. Berlin: Springer-Verlag.

Muñoz-Piña, C., A. Guevara, J. M. Torres, and J. Braña (2008). Paying for the hydrological services of Mexico's forests: analysis, negotiations and results. *Ecological Economics* **65**: 725–736.

Orejuela, J. E. (1992). Traditional productive systems of the Awa (Cuaiquer) Indians of Southwestern Colombia and neighboring Ecuador. In *Conservation of Neotropical Forests: Working from Traditional Resource Use*, eds. K. H. Redford and C. Padoch, pp. 379–399. New York: Columbia University Press.

Phillips, O., and A. H. Gentry (1993). The useful plants of Tambopata, Peru: statistical hypothesis test with a new quantitative technique. *Economic Botany* **47**: 15–32.

Phillips, O., A. H. Gentry, P. Wilking, and D. B. Gálvez-Durand (1994). Quantitative ethno botany and Amazonian conservation. *Conservation Biology* **8**: 225–248.

Pitman, N. C. A., P. M. Jørgensen, S. R. Williams, S. León-Yánez, and R. Valencia (2002). Extinction-rate estimates for a modern Neotropical flora. *Conservation Biology* **16**: 1427–1431.

Pohle, P., and A. Gerique (2008). Sustainable and non-sustainable use of natural resources by indigenous and local communities. In *Gradients in a Tropical Mountain Ecosystem of Ecuador*, eds. E. Beck, J. Bendix, I. Kottke, F. Makeschin, and R. Mosandl, pp. 347–361. Berlin: Springer-Verlag.

Sarmiento, F. (2002). Anthropogenic change in the landscapes of highland Ecuador. *Geographical Review* **92**: 213–234.

Scott, D. F., L. A. Bruijnzeel, and J. Mackensen (2005). The hydrological and soil impacts of forestation in the tropics. In *Forests, Water and People in the Humid Tropics*, eds. M. Bonell and L. A. Bruijnzeel, pp. 622–651. Cambridge, UK: Cambridge University Press.

Sierra, R., C. Cerón, W. Palacios, and R. Valencia (1999). *Mapa de vegetación del Ecuador continental*. Quito, Ecuador: EcoCiencia, INEFAN/GEF.

Stearman, A. (1992). Neotropical indigenous hunters and their neighbors: Sirionо, Chimane, and Yuquí hunting on the Bolivian frontier. In *Conservation of Neotropical Forests: Working from Traditional Resource Use*, eds. K. H. Redford and C. Padoch, pp. 108–130. New York: Columbia University Press.

Tirira, D. (1999). *Mamíferos del Ecuador*. Quito, Ecuador: Pontificia Universidad Católica del Ecuador and SIMBIOE.

Van Geloof, I. (2003). Natural regeneration of tropical cloud forest trees around the treeline ecotone in the Guandera Biological Reserve, Ecuador. M.Sc. thesis, Institute of Biodiversity and Ecosystem Dynamics, University of Amsterdam, Amsterdam, the Netherlands.

Wille, M., R. Hofstede, J. Fehse, H. Hooghiemstra, and J. Sevink (2002). Upper forest line reconstruction in a deforested area in northern Ecuador based on pollen and vegetation analysis. *Journal of Tropical Ecology* **18**: 409–440.

Yaguache, R., and R. Carrion (2004). *Construyendo una experiencia de desarrollo: El manejo de recursos naturales en Pimampiro.* Quito, Ecuador: CEDERENA.

Yanggen, D., D. Cole, C. Crissman, and S. Sherwood (2004). Pesticide use in commercial potato production: reflections on research and intervention efforts towards greater ecosystems health in Northern Ecuador. *EcoHealth*, doi:10.1007/s10393–004–0056-z.

67 Biodiversity-based livelihoods in the *ceja andina* forest zone of northern Ecuador: multi-stakeholder learning processes for the sustainable use of cloud forest areas

R. Hofstede
Ecopar and IUCN Sur, Quito, Ecuador

K. Ambrose, S. Báez, and K. Cueva
Ecopar, Quito, Ecuador

ABSTRACT

In the northern part of the Ecuadorian Andes, people are not profiting from the benefits that biological diversity may yield, and highland ecosystems continue to be deforested. At present, this trend is not likely to change due to a regional economy driven by an intensive potato–dairy production system. In highly populated areas, only few remnants of the uppermost forests, the *ceja andina* (a type of sub-alpine montane cloud forest), survive. Due to the implicit spatial relationship between the agricultural zone and the remaining forest area, and the disconnectedness of the rural population to the forest, a change in attitude is required toward more sustainable agricultural production systems and use of biodiversity if forest conservation is to be achieved. This chapter describes and evaluates the change in attitude of a wide range of stakeholders with respect to alternative production systems and forest conservation. Endogenous learning processes were facilitated within farmer communities through the establishment of learning centers in which two participatory learning and research methodologies were applied. The two methods resulted in acceptable participation and adoption by local farmers. Results showed a visible change toward less contaminating and more diverse production systems and the formation of institutionalized commercialization of organic produce and non-timber forest products. Local government-led inter-institutional and social learning processes were also facilitated. The process has been successful in one municipality where decision-makers and municipal staff showed high motivation and increased technical capacity, but not in another municipality where such capacity was lacking. The second municipality suffered from constant staff turnover as well as differing concepts and strategies regarding natural resource management. Outcome Mapping proved to be a useful tool for evaluation of the entire learning process and for the promotion of concerted action by different stakeholder groups.

INTRODUCTION

The northern Andes is one the most diverse eco-regions of the world due to the presence of a broad variety of ecosystems within a relatively small area (Dinerstein *et al.*, 1995; Barthlott *et al.*, 2005). Nevertheless, land-owners are not profiting optimally from the variety of benefits that this high biological diversity may yield (cf. Báez *et al.*, this volume). In the Province of Carchi in northern Ecuador, an economy based largely on mono-cropping has resulted in continued conversion of fragile upper montane and sub-alpine ecosystems – including the stunted forest type near the tree-line called *ceja andina* and the adjacent herbaceous *páramo* vegetation – to make way for cultivation and pastures (Crissman *et al.*, 1998). The *ceja andina* (literally: Andean eyebrow) is a zonal type of tropical montane sub-alpine cloud

Tropical Montane Cloud Forests: Science for Conservation and Management, eds. L. A. Bruijnzeel, F. N. Scatena, and L. S. Hamilton. Published by Cambridge University Press.

forest (*sensu* Bruijnzeel and Hamilton, 2000; cf. Grubb, 1977) found originally between the tall upper montane ("Andean") cloud forest belt below, and the *páramo* above, typically at elevations between 3500 and 4000 m.a.s.l. (Hofstede, 1995; Wille *et al.*, 2002).

The zonal vegetation type to which the *ceja andina* belongs once extended (practically uninterrupted) on all sides of the tropical Andean Cordillera, from Venezuela down to Perú and Bolivia (Acosta-Solis, 1977; Kessler, 1995; Kappelle and Brown, 2001). However, altitudinal expansion of agricultural activities has resulted in large-scale clearance of the upper montane forests (cf. Mosandl *et al.*, 2008; Baez *et al.*, this volume; Mulligan, this volume) whereas the *páramo* is frequently burned, thereby diminishing the upper limit of the *ceja andina* (Laegaard, 1992; Hofstede, 1995) and impairing local hydrological functioning (Buytaert *et al.*, 2006). Due to this pressure from both sides, only 5–10% of the original *ceja andina* now remains as small scattered fragments in Ecuador, Perú, and Bolivia (Kessler, 1995; Sarmiento and Frolich, 2002; cf. Beck *et al.*, 2007).

The remaining small forest remnants do not constitute an important resource for local people. Post-forest land management is typically unsustainable. After a highly productive first year, soils suffer from fertility decline through erosion whereas harvests decline due to gradual resistance of pests, thereby causing increased dependence on agrochemical inputs. Farmers continue to clear the remaining pristine Andean forests and step up their investments in pesticides in order to maintain or increase crop yields so as to allow for the continued payment of labor and agrochemicals. However, low potato prices due to overproduction, lack of alternative income sources, and severe health problems related to the use of highly toxic pesticides have led to a search for alternative practices. Such alternatives should take both the ecological constraints of the land and the need for a decent livelihood into account (Crissman *et al.*, 1998; Yanggen *et al.*, 2003).

Previous studies in the Ecuadorian Andes have shown that supporting biodiversity-based livelihoods by creating alternative, more diverse production systems present a viable opportunity to curb negative social, environmental, and economic impacts on the forest as well as the local population (Kenny-Jordan *et al.*, 1999). Differentiation of crops decreases the dependence on the high investments typically associated with intensive agriculture and unstable market prices. Low-investment crops may improve public health and lessen environmental contamination, whereas less expansive production systems may decrease the pressure on the last remnants of *ceja andina* forests (Yanggen *et al.*, 2003). To plan and promote alternative production systems such as agro-forestry, family orchards, value-added production chains, and organic horticulture, two key factors need consideration: (i) local stakeholders need to learn what are the unsustainable factors of their present production systems, and to recognize the strategic value of the natural vegetation, and (ii) farmers, in collaboration with other key actors (local government, NGOs, researchers, and citizens in general) need to change their attitudes toward such production systems (Kenny-Jordan *et al.*, 1999; Braun *et al.*, 2000; Rist *et al.*, 2003; Yanggen *et al.*, 2003).

The goal of the current study was to examine the behavioral changes of farmers, from the current intensive potato–dairy-based production system toward alternative forms of livelihood, including biodiversity-rich forms of agriculture and sustainable use of forest remnants. To facilitate this behavioral change, the project developed: (i) a knowledge base of people's dependence on biodiversity and on their interest to preserve and recover the High Andean forests (Báez *et al.*, this volume), (ii) endogenous learning processes based on biodiversity-rich production systems including sustainable use of forest remnants, and (iii) an inter-institutional social learning process, led by local governments. These were considered key features for establishing a dialogue among key actors, for achieving engagement in and consensus regarding natural resource management and forest conservation, as well as the creation of policy instruments to support such action.

The present chapter concentrates on the intermediate results of the endogenous learning and inter-institutional processes. It demonstrates concerted action and allows some conclusions regarding behavioral change of the stakeholders after one year of project implementation. The more complete results presented by Ambrose *et al.* (2006) are incorporated to demonstrate the outcomes of the process once the initial behavioral change was consolidated.

STUDY AREA AND METHODS

This study was carried out in the Life Zone originally covered by *ceja andina* at elevations of 2800–3700 m.a.s.l. in the Andean region of Carchi Province in northern Ecuador (Figure 67.1). General background information on the climate, vegetation, and biodiversity of this area is presented in Báez *et al.* (this volume). Several *ceja andina* forest remnants of considerable extent have been preserved in the Carchi area. In addition, there are farmers willing to experiment with natural resource management whereas there is previous experience with community-based natural resource management initiatives (Frolich and Guevara, 1999; Poats, 2001). This provided an apt and attractive context in which to explore and develop biodiversity-based livelihoods that could complement or even replace non-sustainable, high-input production systems.

The project's strategy was two-fold: (i) participatory research initiatives in learning centers which allowed for endogenous learning, and the exploration of various production alternatives through hands-on experience with biodiversity-rich forms of

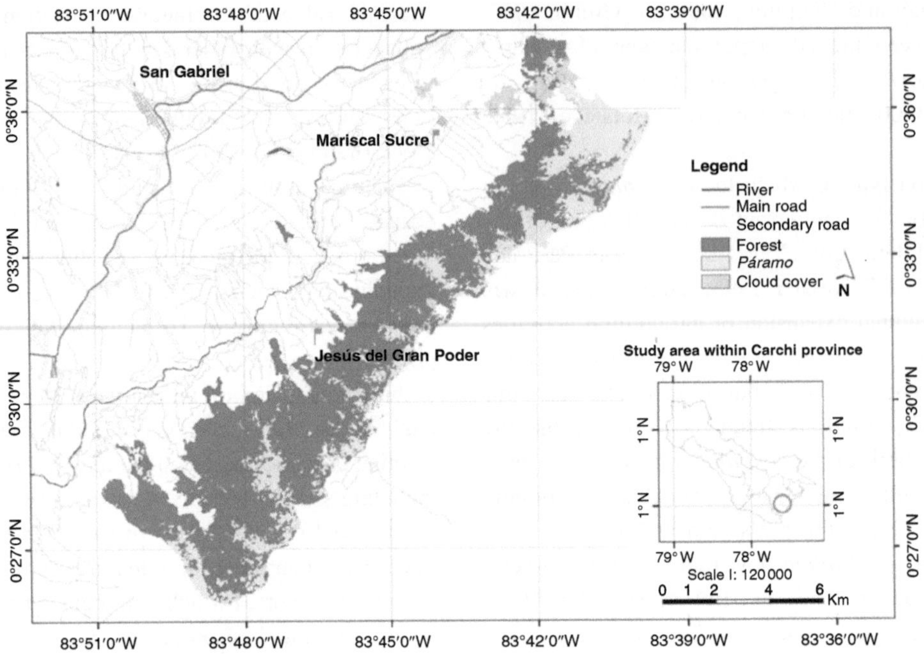

Figure 67.1. Map of the study area in northern Ecuador (Eastern Andean Cordillera, Carchi Province), indicating the present cover of *ceja andina* and *páramo* on the inter-Andean side of the Andes. (See also color plate.)

agriculture; and (ii) facilitation of social learning platforms for inter-institutional dialogue and concerted action, with a focus on local government and the construction of policies focusing on natural resource management. Because the examination of behavioral change toward the sustainable use of the *ceja andina's* agro-biodiversity was the primary goal of the study, Outcome Mapping (OM) methodology and tools were applied to measure such change among various stakeholders. OM is an innovative monitoring and evaluation tool that incorporates learning and reflection into development programs to analyze – through open and closed interviews, surveys, and focal groups – behavioral change in stakeholder groups and their relationships to their natural, social, and cultural surroundings (Earl *et al.*, 2002; Watts *et al.*, 2003).

Since unsustainable agricultural practices form the main threat to the *ceja andina* forest, farmers and farmer groups were considered key stakeholders in the promotion of sustainable resource management. In addition, it was recognized early on that learning processes must extend to a wider audience of stakeholders and decision-makers, and that each group should have their own social learning spaces. A stakeholder analysis (Conway *et al.*, 1998; Biggs *et al.*, 2004) was carried out to identify the interrelations between all local governmental and societal groups and their direct and indirect relationships with the *ceja andina* forest. Next, the project identified the various stakeholder groups that were to participate actively and whose behavioral change was to be studied (the so-called "boundary partners"; Earl *et al.*, 2002). These included two municipalities and the provincial council, as well as the non-governmental municipal assembly. Furthermore, two farmer associations, two rural environmental clubs, one rural farmer youth club, three entire communities (involved in community management plans), six groups of farmers from six watersheds, two Water Boards, two parish boards, four locally-based NGOs (including one regional network), two rural schools and one urban school, one mixed-community group (group of community trainers), and one large land-owner were involved.

Social learning spaces were identified and facilitated for each individual stakeholder group. These spaces could be formal or informal meetings, field events associated with implementation and evaluation of the participatory methodologies for farmer experimentation, gatherings of youth clubs, etc. The main condition for defining an effective learning space for each stakeholder group was that it should allow the development of dialogue, consensus, and strategic action with respect to natural resource management within the *ceja andina* Life Zone (cf. Maarleveld *et al.*, 2002; Röling, 2002; Rist *et al.*, 2003).

Learning centers to facilitate different participatory methodologies for farmer experimentation and research were developed in two villages: Mariscal Sucre, a parish of the municipality of Huaca, and Jesús del Gran Poder, a parish of Montúfar (Figure 67.1). Participatory methodologies included Farmer Field Schools (FFS) and Local Agricultural Research Committees (commonly known by their Spanish acronym, CIAL). The focus of FFS is on ecological literacy: farmers engage in problem-oriented discovery with the support of facilitators, who base learning on a diverse range of activities, as farmers exchange ideas, share experiences, and make their own decisions regarding crop production (Braun *et al.*, 2000; Borja, 2004; Van den Berg, 2004). In the CIAL participatory research methodology, participants decide what they want to test and apply this in an experimental plot, including replica and control (Ashby *et al.*, 2000). During the process, participants at each learning center identified their interests and learning goals, and chose their research methods. Because of the initial focus of local interest in agricultural systems, the process started with learning to develop more biodiverse and less chemical-dependent agricultural systems. This was considered a first step toward more diverse livelihoods options on the one hand, and toward less direct pressure on the forest on the other (through diminished need for expanding intensive agricultural systems). After this initial phase, participants started to experiment with agro-forestry and the use of non-timber forest products as the next step toward better valuation of biodiversity and forest conservation (Ambrose *et al.*, 2006).

RESULTS AND DISCUSSION

Endogenous learning processes

The learning centers proved to be a promising platform that enabled small-scale farmers to lead the development of alternative production systems. Prior to the start of the participatory process, a few young farmers that were still helping out on their parents' farms showed an interest in crops other than potatoes or were open to production systems involving lower pesticide usage. However, adult farmers were reluctant to change. During stakeholder analysis in the town of Mariscal Sucre, only two farmers out of 25 were interested in changing their production system. On their farms, two participatory research plots were established; one FFS and one plot with a CIAL. Once these two farmers had decided to set aside a piece of land for on-farm investigations, other farmers became interested in participating as well. After the FFS was implemented, 21 (adult) farmers participated actively in its implementation, whereas a local previously existing environmental group of 10 (young) farmers participated in the CIAL. In the village of Jesús del Gran Poder, 14 members of a farmer youth group participated in a CIAL that explored the production of native Andean crops and agro-forestry systems. At a later stage, the project developed an FFS in agroforestry with a total of 18 participants from both communities.

Farmers initially decided to investigate integrated pest management (IPM) during potato cultivation, so as to better determine alternatives for the prevailing high-intensity management of certain pests and diseases. Some 21 farmers had an interest in IPM, a number that remained more or less stable during the first cropping cycle of 6 months. A field day organized by the project for the exchange of experiences with other farmers, other communities, and other stakeholders was attended by approximately 50 people. The FFS methodology was useful in that it continuously involved farmers in alternative production systems, sparked regular reflection, evaluation, and discussion, and led to the adoption of alternative practices by other groups. After two production cycles, at least 15 farmers from the village were observed to apply IPM in large parts of their potato fields, and outside the FFS plots. In addition to the people's perception of having a cleaner production system, the observation that IPM required less financial investment and gave a higher economic return (1.69) than did conventional practice (1.46) was decisive for the adoption of the improved practice (Ambrose *et al.*, 2006). A drawback of the FFS approach in the present case was that its initial objectives were not very novel and its goals formed only a minute contribution toward the development of a more biodiversity-based livelihood. Rather, it should be considered a small initial step toward a long process toward attaining "ecological literacy," which would finally lead to awareness of the values of, and concrete action toward, the conservation of the cloud forest remnants.

The CIAL approach was initially used to test various non-traditional fodder and food crops, such as legumes, native Andean tubers, grains, and fruits, household crops (vegetables and fruits), and medicinal and aromatic plants. This methodology proved to be most successful with younger farmers (who, in contrast to their parents, generally had secondary education), mostly because of the relatively quick results. Some of the tested food and fodder crops were immediately adopted in the family orchards of 60% of the participants (e.g. *Vicia* for small household animals, medicinal plants and native potato varieties for own consumption). Compost made of the organic waste of these crops was applied for potato field recuperation during the second year of experimentation (Ambrose *et al.*, 2006). During a second stage, the CIAL plots included native trees and medicinal and aromatic plants extracted as seedlings from the forest. Although the CIAL initially also focused on farming systems – rather than on the actual use and conservation of the forest – the change toward more diverse agricultural systems was more explicit than in the FFS approach whereas progress toward actual forest conservation goals might be more realistic.

After three years of implementation of the learning centers, farmer participation reached a constant level and there was a lively interchange of ideas between the CIAL and FFS participants (Ambrose *et al.*, 2006). Innovative ideas for alternative production systems were implemented at the household scale. Specific attention was paid to post-harvest practices and market access. As a tangible result, a cooperative (APRONOR) was formed by the participating farmers which processed and commercialized organic crops (vegetables, fruits), consumer products (marmalade, handicrafts), and medicinal and aromatic plants (dried mixtures, teas). These products originated from the participants' fields (vegetable and fruit) as well as from the forests (non-timber forest products). The municipality of Montúfar collaborated by providing free space at the weekly market in San Gabriel (the local capital town). Motivated by these positive results, the young farmers group volunteered to involve local schools in the learning centers. Environmental education in relation to organic farming and forest conservation is now a reality in all schools of the two villages and in San Gabriel (Ambrose *et al.*, 2006).

In general, both methods applied in the learning centers proved to influence awareness of local farmers and helped to involve them in alternative production systems. The participation level was average but constant: 21 farmers during the first round of FFS (potato) and 18 in the second round (agro-forestry). A typical FFS usually involves between 15 and 25 farmers vs. 8–10 for a CIAL (Van den Berg, 2004). However, the type of participants differed between FFS and CIAL: adult farmer–landowners were more attracted to the FFS and its study object of potato crop protection, whereas young farmers without a family of their own tended to get more easily involved in alternative cropping systems and the faster results associated with the CIAL. A drawback with both methods was that they required a relatively long period to actually change the daily practice of the farmers. Nevertheless, change was actually observed already during the first year of implementation, something which should be considered as being a critical indicator of the potential of the two methodologies with respect to changing the behavior of local farmers toward more final conservation goals. The final result of the process showed that this change was continued and expanded to other farmers, another municipality and other schools. These participatory research and training approaches proved to be a necessary element in establishing endogenous knowledge creation processes, and therefore a key element and a first step toward more sustainable natural resource and forest management. However, due to their slow effect on tangible change, complementary activities for conservation should be considered, including academic studies on natural resource management, policy debate, and development, forest restoration strategies, direct and indirect financial incentives for forest conservation, and stimulation of market opportunities for diverse produce.

Inter-institutional social learning process led by local governments

Complementary to the local processes with farmer "boundary partners," formal support for conservation of the *ceja andina* forest remnants was obtained by taking advantage of existing platforms as social learning spaces. Two of these were the *Mancomunidad de Carchi* – the joint meetings on environmental issues of the local governments – and the municipal assembly, a platform of non-government entities. In addition, new opportunities were created in collaboration with local actors in the form of an informal monthly multi-stakeholder platform to discuss issues regarding *ceja andina* management (the so-called *Plataforma Ceja Andina*) with the local government, NGOs, and farmers.

One of the key stakeholder groups were local governments, particularly the municipalities to which the villages of the study area pertained. In Huaca (Carchi's youngest municipality), no official Environmental Unit had yet been created and it proved difficult to change attitudes within the municipality toward conservation strategies, also because of constantly changing personnel and differing concepts and strategies regarding the natural resource management. Conversely, in Montúfar – a more established municipality with a well-functioning Environmental Unit – the project was received immediately, and the municipality became its strongest ally throughout the implementation process. Municipality staff actively collaborated in the selection of communities and farmers, field studies in forest and farm areas, and decision-making on project activities. Within one year, they started to support the organization and financed the *Plataforma Ceja Andina*.

Initially, the decision-making bodies like the Mayor and the Municipal Council of Montúfar did not show an active interest in the conservation process. However, when the Municipality became aware that small but significant activities were being successfully executed, it began to assist official community meetings and visit learning centers. In doing so, the Mayor and Council became aware that environmental activities indeed received public support and as a result they began to prioritize conservation projects within their municipality. After one year of contact with the Municipality of Montúfar, a *ceja andina* conservation strategy was applied to all the forests under their jurisdiction and direct investments were made to protect a communal forest reserve (*Bosque de Arrayanes*, one of the only surviving *Myrciantes* stands in Carchi). In general, there has been a positive response of farmers and citizens toward the Municipality's environmental actions and strategies.

Parallel to the activities at the municipal level, the Provincial Government of Carchi proceeded independently with a regional *ceja andina* strategy. They developed a provincial decree aimed at protecting the last remnants of forests and *páramo*, aiming at protecting vital water resources (Gobierno Provincial del Carchi, 2004). This decree, although demonstrating a positive indication of change in an important stakeholder, proved ineffective because farmers felt restricted in their freedom of the use of the land and resentful for not having been included in the process of developing the decree. In certain areas, even a decrease in interest in environmental protection was encountered after publication of the provincial decree.

Besides local governments, several other stakeholders were identified, including their perspectives on and actions toward the forest. These opinions and proposals were brought together in social learning spaces where different actors were able to engage in dialogue and debate, working toward consensus and concerted actions regarding the sustainable use of agro-biodiversity (Poats, 2001). The *Plataforma Ceja Andina* is one such space, which disseminated relevant information to community members, urban citizens, youth groups, women's groups, educational institutions, the municipal assembly, parish boards, and other local government entities. It provided a space for facilitated discussion around issues of interest related to natural resource management in general, and planning of concerted action for restoration and conservation activities. In the initial stages of the project, the project team took a lead role in facilitating this monthly platform. Later on, the Municipality of Montúfar took over the responsibility of the sessions.

After three years of the inter-institutional process, the *Plataforma Ceja Andina* was still functioning with continued municipal support. The Municipality of Montúfar not only supported the APRONOR farmers association, but also implemented a management plan for two watersheds that included most remnants of the *ceja andina* forest within the Municipality's territory. To achieve this, the Municipality established an effective working relationship with the Environmental Department of Carchi Province to apply the regional *ceja andina* conservation decree in collaboration with the farmers. The *Bosque de Arrayanes* was made the emblem of ecotourism promotion and its facilities for visitors were considerably improved. In the other municipality (Huaca) the process never received priority, although other municipalities in Carchi (particularly Bolivar, which shares several watersheds with Montúfar) started several activities aimed toward better conservation of the *ceja andina* (Ambrose *et al.*, 2006).

Outcome Mapping

The project used Outcome Mapping (OM) not only as an evaluation and monitoring tool, but also for initial and continued stakeholder analysis. Outcome Mapping proved useful because it actually measured behavioral change of target group members, and not just tangible products (cf. Earl *et al.*, 2002). Thus, it was possible to monitor the level of collaboration and awareness of farmers, the attitudes of local governments and other key actors, as well as the actual value of field information. Outcome Mapping also helped to identify initial perception, and therefore behavior of farmers toward the *ceja andina* (Báez *et al.*, this volume). It clearly demonstrated the small steps of change: farmers first needed to come to a meeting, then participate in a discussion, get involved with a CIAL, produce, harvest, sell, and then replicate and advocate the case with others. It also allowed continuous adaptation of activities and indicators of progress, which is useful for action research that combines both ecological and social foci. At the end of the three-year project, stakeholders (farmers and local governments) concluded that whilst development strategies could not be fully planned, if the process is guided by OM there is more concentration on monitoring and evaluation rather than on impacts only. All stakeholders agreed that OM promoted concerted action (Ambrose *et al.* 2006).

A challenge in the application of OM in the present case was that the goals of the process and their indicators (outcomes and progress markers; *sensu* Earl *et al.*, 2002) were defined by stakeholders rather than the executors of the project. This forced the project to adapt to current demands (from the local community) without sacrificing previously established objectives and project governance (by the executors of the project). Another complication encountered was that while there are certain advantages to the methodology being open and flexible, it also makes for a complex and time-consuming method of project monitoring. This complexity was overcome by putting emphasis on the measurable added value of OM: in the end, stakeholders considered monitoring and evaluation not as an extra activity but as a critical and useful ingredient of the process of learning, planning, decision-making, and collective action.

CONCLUDING REMARKS

This study in the *ceja andina* Life Zone of northern Ecuador has confirmed that for forest conservation to occur, the attitudes of people, including their activities, relationships, and interactions, toward sustainable agriculture and the use of biodiversity need to change. Involving farmers in participatory research activities increased awareness of certain actual environmental problems (e.g. pests) and encouraged a change in behavior in a pilot group of farmers that became interested in more diverse farming systems, including the use of native crops and agro-forestry. The applied participatory methodologies proved to be promising tools for different target groups (especially established farmer families and younger farmer groups), and concrete changes in attitude and behavior toward environmental protection were attained. In general, the Farmer Field Schools and Local Agricultural Research Committees introduced by the project contributed to a re-established connection of the participating farmer community to environmental issues and to a basic understanding of the relationship between better, more diverse farming techniques and the conservation of the remaining forest. However, the change in behavior through these methodologies was slow and it may take much more time before alternative and more diverse production systems will actually contribute to forest conservation.

Stakeholder analysis and Outcome Mapping helped to reveal the different needs, interests, behavior, and alliances existing between different interest groups in the area. Corresponding platforms for dialogue and concerted action were established. However, halting deforestation completely remains a challenging goal, which depends not only on participatory methodologies, but also on the technical and analytical capacity and motivation of municipal staff. In particular, supporting activities are needed to provide the necessary additional incentives for environmental protection. Such activities include studies on natural resource management (cf. Pohle and Gerique, 2008), policy development, and forest restoration strategies (cf. Weber *et al.*, 2008; Günter *et al.*, 2009; Williams-Linera *et al.*, this volume), as well as financial incentives for conservation and sustainable production systems (cf. Yaguache and Carrion, 2004; Tognetti *et al.*, this volume).

ACKNOWLEDGEMENTS

Staff and collaborators of Ecopar in Quito are acknowledged for their operational collaboration in the field. We are especially indebted to Josette Moncayo, Luis González, Marta Muñoz, and Gerardo Canacuán. Fernando Espíndola drafted the map of the study area. Staff of Guandera Biological Station (Jatun Sacha Foundation), especially Christopher James, provided access to their premises and helped during the entire research project. The municipalities of Montúfar and San Pedro de Huaca collaborated in all stages of the work, especially Emerson Bravo, Hever Racines, and Edison Jiménez of the Environmental Unit, Montúfar. We thank the farmer families of Mariscal Sucre and Jesús del Gran Poder for their endless hospitality and for their collaboration with the field studies, their leading role in the participatory research activities in the learning centers, and their participation in numerous workshops. The Ceja Andina project was financed by the International Development Research Center (IDRC, Canada). The authors acknowledge IDRC Staff, especially Merle Faminow, for guidance and practical suggestions.

REFERENCES

Acosta Solis, M. (1977). *Ecología y fitoecología del Ecuador*. Quito, Ecuador: Casa de la Cultura Ecuatoriana.

Ambrose, K., K. Cueva Rojas, L. Ordóñez Guamán, L. González Villalta, R. M. Borja (2006). *Aprendizaje participativo en el bosque de Ceja Andina*. Quito, Ecuador: Ecopar–IDRC. Also available at www.infoandina.org/apc-aa-files/237543fdce333f3a56026e59e60adf7b/ Aprendizaje_Participativo.pdf.

Ashby, J. A., A. R. Braun, T. Gracia, *et al.* (2000). *Investing in Farmers as Researchers: Experiences with Local Agricultural Research Committees in Latin America*. Cali, Colombia: Centro Internacional de Agricultura Tropical (CIAT).

Barthlott, W., J. Muthke, M. D. Rafiqpoor, G. Kier, and H. Kreft (2005). Global centres of vascular plant diversity. *Nova Acta Leopoldina N.F.* **92**, 342: 61–83.

Beck, E., J. Bendix, I. Kottke, F. Makeschin, and R. Mosandl (eds.) (2007). *Gradients in a Tropical Mountain Ecosystem of Ecuador*. Berlin: Springer-Verlag.

Biggs, S., and H. Matsaert (2004). *Strengthening Poverty Reduction Programmes Using an Actor-Oriented Approach: Examples from Natural Resources Innovation Systems*, Network Paper No. 134. London: ODI/Agricultural Research and Extension Network.

Borja, R. M. (2004). *Documenting Farmer Field Schools in the Ecuadorian Highlands: A Case Study of the Province of Carchi*. Ithaca, NY: Cornell University Press.

Braun, A., G. Thiele, and M. Fernández (2000). *Farmer Field Schools and Local Agricultural Research Committees: Complementary Platforms for Integrated Decision-Making in Sustainable Agriculture*, Network Paper No. 105. London: ODI/Agricultural Research and Extension Network.

Bruijnzeel, L. A., and L. S. Hamilton (2000). *Decision Time for Cloud Forests*, IHP Humid Tropics Programme Series No. 13. Paris: UNESCO–IHP.

Buytaert, W., R. Célleri, B. de Bièvre, *et al.* (2006). Human impact on the hydrology of the Andean páramos. *Earth-Science Reviews* **79**: 53–72.

Conway, S., and F. Steward (1998). Mapping innovation networks. *International Journal of Innovation Management* **2**: 223–254.

Crissman, C. C., P. Espinosa, C. E. H. Ducrot, D. C. Cole, and F. Carpio (1998). The case study site: physical, health and potato farming systems in Carchi province. In *Economic, Environmental and Health Tradeoffs in Agriculture: Pesticides and the Sustainability of Andean Potato Production*, eds. C. C. Crissman, J. M. Antle, and S. M. Capalbo, pp. 85–120. Dordrecht, the Netherlands: Kluwer.

Dinerstein, E., D. M. Olson, D. J. Graham, *et al.* (1995). *A Conservation Assessment of the Terrestrial Ecoregions of Latin America and the Caribbean*. Washington, DC: World Wildlife Fund and World Bank.

Earl, S., F. Carden, and T. Smutylo (2002). *Mapeo de Alcances: Incorporando aprendizaje y reflexión en programas de desarrollo*. Ottawa, Canada: IDRC.

Frolich, L. M., and E. Guevara (1999). The role of family-based agricultural innovation in conserving tropical montane cloud forest: the Guandera Project in Northern Ecuador. In *Entendiendo las interfaces ecológicas para la gestión de paisajes culturales en los Andes, Memorias del Tercer Simposio Internacional sobre Desarrollo Sustentable en los Andes*, eds. F. Sarmiento and J. Hidalgo, pp. 45–49. Quito, Ecuador: Corporación Editora Nacional.

Gobierno Provincial del Carchi (2004). *Ordenanza para la protección, conservación, y regulación del recurso hídrico de la provincia, Reglamento para precautelar el funcionamiento y buen uso de los sistemas de agua potable*. Tulcán, Ecuador: Dirección de Productividad y Desarrollo Sustentable, Area de Medio Ambiente, Gobierno Provincial del Carchi.

Grubb, P. J. (1977). Factors controlling the distribution of forest-types on wet tropical mountains: with special reference to mineral nutrition. *Annual Review of Ecology and Systematics* **8**: 83–107.

Günter, S., P. Gonzalez, G. Alvarez, *et al.* (2009). Determinants of successful regeneration of abandoned pastures in the Andes: soil conditions and vegetation cover. *Forest Ecology and Management* **258**: 81–91.

Hofstede, R. G. M. (1995). The effects of grazing and burning on soil and plant nutrient concentrations in Colombian paramo grasslands. *Plant and Soil* **173**: 111–132.

Kappelle, M., and A. Brown (eds.) (2001). *Bosques nublados del neotrópico*. Santo Domingo de Heredia, Costa Rica: Instituto Nacional de Biodiversidad (INBIO).

Kenny-Jordan, C., C. Herz, M. Añazco, and M. Andrade (1999). *Pioneering Change: Community Forestry in the Andean Highlands*. Rome: FAO.

Kessler, M. (1995). Present and potential distribution of *Polylepis* (Rosaceae) forests in Bolivia. In *Biodiversity and Conservation of Neotropical Montane Forests*, eds. S. P. Churchill, H. Balslev, E. Forero, and J. Luteyn, pp. 281–294. New York: New York Botanical Garden.

Laegaard, S. (1992). Influence of fire in the paramo vegetation of Ecuador. In *Paramo: An Andean Ecosystem under Human Influence*, eds. H. Balslev and J. Luteyn, pp. 151–170. London: Academic Press.

Maarleveld, M., and C. Dangbégon (2002). Social learning: major concepts and issues. In *Wheelbarrows Full of Frogs: Social Learning in Rural Resource Management*, eds. C. Leeuwis and R. Pyburn, pp. 67–84. Assen, the Netherlands: Koninklijke Van Gorcum.

Mosandl, R., S. Günter, B. Stimm, and M. Weber (2008). Ecuador suffers the highest deforestation rate in South America. In *Gradients in a Tropical Mountain Ecosystem of Ecuador*, eds. E. Beck, J. Bendix, I. Kottke, F. Makeschin, and R. Mosandl, pp. 37–40. Berlin: Springer-Verlag.

Poats, S. (2001). El consorcio Carchi: Un ejemplo de coordinación institucional a nivel local. In *Los páramos del Ecuador: Particularidades, problemas y perspectivas*, eds. P. Mena, V. G. Medina, and R. Hofstede, pp. 267–278. Quito, Ecuador: Abya Yala.

Pohle, P., and A. Gerique (2008). Sustainable and non-sustainable use of natural resources by indigenous and local communities. In *Gradients in a Tropical Mountain Ecosystem of Ecuador*, eds. E. Beck, J. Bendix, I. Kottke, F. Makeschin, and R. Mosandl, pp. 347–361. Berlin: Springer-Verlag.

Rist, S., F. Delgado Burgoa, and U. Wiesmann (2003). The role of social learning processes in the emergence and development of Aymara land use systems. *Mountain Research and Development* **23**: 263–270.

Röling, N. (2002). Beyond the aggregation of individual preferences: moving from multiple to distributed cognition in resource dilemmas. In *Wheelbarrows Full of Frogs: Social Learning in Rural Resource Management*, eds. C. Leeuwis and R. Pyburn, pp. 25–48. Assen, the Netherlands: Koninklijke Van Gorcum.

Sarmiento, F. O., and L. M. Frolich (2002). Andean cloud forest tree lines: naturalness, agriculture and the human dimension. *Mountain Research and Development* **22**: 278–287.

Van den Berg, H. (2004). *IPM Farmer Field Schools: A Synthesis of 25 Impact Evaluations*. Wageningen, the Netherlands: Global Integrated Pest Management Facility.

Watts, J., R. Mackay, D. Horton, *et al.* (2003). *Institutional Learning and Change: An Introduction*. The Hague, the Netherlands: International Service for National Agricultural Research (ISNAR).

Weber, M., S. Günter, N. Aguirre, B. Stimm, and R. Mosandl (2008). Reforestation of abandoned pastures: silvicultural means to accelerate forest recovery and biodiversity. In *Gradients in a Tropical Mountain Ecosystem of Ecuador*, eds. E. Beck, J. Bendix, I. Kottke, F. Makeschin, and R. Mosandl, pp. 447–457. Berlin: Springer-Verlag.

Wille, M., R. Hofstede, J. Fehse, H. Hooghiemstra, and J. Sevink (2002). Upper forest line reconstruction in a deforested area in northern Ecuador based on pollen and vegetation analysis. *Journal of Tropical Ecology* **18**: 409–440.

Yaguache, R., and R. Carrion (2004). *Construyendo una experiencia de desarrollo: El manejo de recursos naturales en Pimampiro*. Quito, Ecuador: CEDERENA.

Yanggen, D., C. C. Crissman, and P. Espinosa (2003). *Los plaguicidas: Impactos en producción, salud y medio ambiente en Carchi, Ecuador*. Quito, Ecuador: Abya Yala.

68 Embracing epiphytes in sustainable forest management: a pilot study from the Highlands of Chiapas, Mexico

J. H. D. Wolf
University of Amsterdam, Amsterdam, the Netherlands

ABSTRACT

Vascular epiphyte biomass and species richness were investigated in 16 anthropogenically disturbed pine–oak forests within an area of ~400 km^2 in the Highlands of Chiapas, southern Mexico. Epiphyte biomass on 35 host oak trees in six diameter classes varied from 0.8 to 243 kg dry-weight and comprised 13–34 species. The observed variation in epiphytes could be attributed to type and intensity of past forest disturbance as it affects present-day stand structure, as well as to site position within the landscape. To help preserve the diverse regional epiphyte vegetation it is recommended to abstain from cyclic clear-cutting, to spare a sufficient number of large "rescue" trees, and to consider epiphyte conservation at a large spatial scale. As an alternative to logging, various prerequisites are proposed for the sustainable harvesting of bromeliads from natural populations.

INTRODUCTION

In the footsteps of early explorers, successive epiphyte researchers have concentrated primarily on old-growth forests. From these studies it is evident that in pristine wet mountain forests epiphytes may contribute up to more than half of the vascular plant diversity and green biomass in the forest (Hofstede *et al.*, 1993; Wolf and Flamenco-Sandoval, 2003; cf. Köhler *et al.*, 2007; Gradstein 2008). Thus, epiphytes are possibly key to high-mountain forest ecosystem functioning. At the same time, in many populated mountain areas altered forests prevail over old-growth forests which are increasingly trimmed down to isolated fragments (Sarmiento, 1997; Cavelier *et al.*, 1998; Williams-Linera *et al.*, 2002; Cayuela *et al.*, 2006; Muñoz-Villers and López-Blanco, 2008; cf. Barradas *et al.*, this volume; Baéz *et al.*, this volume). Information on epiphytes in montane cloud forests in succession is still scanty (Barthlott *et al.*, 2001; Köhler *et al.*, this volume) and there is an equally urgent need to obtain more insight into the response of epiphytes to anthropogenic disturbance (Hietz-Seifert *et al.*, 1996; Werner *et al.*, 2005; Nöske *et al.*, 2008) so that this important ecosystem component may be included in forest management plans aiming for ecological sustainability (cf. Hietz, this volume; Nadkarni, this volume).

The felling of trees inevitably reduces epiphytes on a ground-surface area basis. Of particular interest is the question of how epiphytes on remnant or regenerating trees respond to the disturbance (cf. Werner *et al.*, 2005). This chapter reports the results from epiphyte inventories made on a comparable number of trees in forests that vary in the manner and intensity of anthropogenic disturbance in the Highlands of Chiapas, southern Mexico. Attention is also paid to the geographic position of the respective sites within the landscape. The results are used to develop recommendations for improved forest management to help preserve epiphytes. Offering an alternative to wood extraction, the results are also employed to help define criteria for the ecologically sustainable harvesting of ornamental epiphytes (bromeliads). The chapter integrates the results from two previously published papers in which more details can be found (Wolf and Konings, 2001; Wolf, 2005).

MATERIALS AND METHODS

Study area

The Highlands of Chiapas form a well-delimited physiographic area in the south-west of Mexico (Breedlove, 1978). The south-western part of the Highlands around San Cristóbal de Las Casas is known as the Central Plateau (*c.* 16° 42' N, 92° 37' W). This undulating plain has an elevation between 2100 and 2500 m.a.s.l. with a few peaks up to 2900 m.a.s.l. It is composed of marine limestones with extrusions of volcanic rock on the higher peaks.

Tropical Montane Cloud Forests: Science for Conservation and Management, eds. L.A. Bruijnzeel, F.N. Scatena, and L.S. Hamilton. Published by Cambridge University Press.

The vegetation on the peaks has been classified as cloud forest (Breedlove, 1978). On the plain, however, clouds are less frequent. Climatic records (1978–1995) from San Cristóbal de Las Casas at an elevation of 2276 m.a.s.l. show an annual average temperature of 14.8 °C and an average annual precipitation of 1042 mm with less than 30 mm $month^{-1}$ between December and April (Comisión Nacional del Agua, unpublished data).

Pine–oak forests and anthropogenic disturbance

On the Central Plateau, pine–oak forest prevails. In the Highlands, this formation occurs at altitudes between 1300 and 2500 m.a.s.l. wherever the dry season exceeds 3 months (Breedlove, 1978). Pines and oaks dominate the woody vegetation, but other temperate genera such as *Alnus*, *Carpinus*, *Crataegus*, *Ostrya*, *Prunus*, *Sambucus*, and *Viburnum* are also frequently present. Vascular epiphytes, particularly bromeliads, are most conspicuous because of their abundance and variety, especially during the dry season when the oaks shed their leaves. In total, 608 species of vascular epiphytes are known from the pine–oak forest (Wolf and Flamenco-Sandoval, 2003). Lichens and mosses are more abundant than liverworts which are mostly confined to the understory.

The pine–oak forests on the Central Plateau have been logged to create arable lands and pastures at an estimated annual rate of >2% (1984–1990; Ochoa-Gaona and González-Espinosa, 2000). The remaining stands fulfil the demand for wood and humans have altered all residual forests on the Central Plateau to some degree. Pines are the most important source of timber, and trees that exceed commercial sizes are rare nowadays. Oaks, the preferred host tree for epiphytes, are exclusively used for fuelwood. Two main types of harvesting are practiced in the area: (i) periodic (20–30 years) clear-cutting, mostly for the production of charcoal, and (ii) the continuously and apparently haphazard extraction of trees for subsistence purposes. Both types of exploitation take place at varying intensities and typically in small areas of less than a hectare. This has resulted in a fine mosaic of forest stands of variable structure.

Forest regeneration is left to natural processes. Oak tree stumps may produce several stems. In contrast, pines rely on seed dispersal and an open forest canopy favors seedling establishment. Grazing by sheep and surface fires in the understory also affect forest regeneration (cf. Asbjornsen and Garnica-Sánchez, this volume). The net result of frequent anthropogenic disturbance is mostly that the dominance of pines over oaks is increased (Ramírez-Marcial *et al.*, 2001). However, pure (though highly disturbed) oak stands are also present, possibly due to pine seed limitation or selective cutting. In this study, the proportion of sprouted oaks is used to approximate the degree of disturbance, although it is recognized that to obtain a good understanding of the management history of a site, additional structural parameters would also have to be taken into account.

Sampling

Between 1995 and 1999, 16 sites were selected along an extended anthropogenic disturbance gradient, ranging from old-growth forest to heavily disturbed forest. Site selection was based on forest structural parameters and field observations with respect to the occurrence of recent logging, fires, and grazing of the understory. Variation in altitude between sites was kept at a minimum (2160–2490 m.a.s.l.) whereas geographic distance between any two sites varied from about 1 to 25 km. At each site, the structure of the forest was determined in a flat area of 30 × 30 m. All trees with a diameter at breast height (DBH) > 5 cm were identified and measured. Because it was not possible to confidently distinguish *Quercus crispipilis* Trel. from *Q. segoviensis* Liebm., the two species were lumped. Similarly, *Q. crassifolia* H. & B. was grouped together with *Q. rugosa* Nee. Tree heights, tree densities, basal areas, diameter frequency distributions, and the frequency of sprouted oaks were used to parameterize site disturbance history. Observations on the occurrence of logging, grazing, and forest fires were also recorded.

For epiphytes, the tree was used as the sampling unit, because at this scale it is easier to attain a sample of similar size between structurally divergent forest types (cf. Gradstein *et al.*, 2003; Wolf *et al.*, 2009). Only oak trees were considered since epiphytes on pines are rare, possibly due to the lower water-holding capacity of the bark (Castro-Hernández *et al.*, 1999).

Thirty-five oak trees per stand were sampled for epiphytes, 560 in total. To obtain a good distribution of tree sizes, five host trees were chosen in each of five classes of increasing DBH, at 5-cm intervals ranging from 5 to 30 cm. In addition, ten host trees with DBH > 30 cm were sampled. After sampling an initial randomly chosen first tree, the nearest neighbor trees were selected until a particular size class was filled. For each tree, three size parameters were obtained: height, number of bifurcation points with a branch diameter >5 cm, and DBH. Field trials on climbed trees showed that these variables could be estimated with little error.

Epiphyte abundance was expressed as dry-weight. The clonal growth of many species made it impossible to discern individuals. An essentially non-destructive sampling method was applied in which dry-weight was estimated from counts of leaves, fronds, or rosettes, depending on the growth form of the species. Taller trees were climbed using rope-climbing techniques (Mitchell *et al.*, 2002). Vouchers were deposited in the herbarium of El Colegio de La Frontera Sur at San Cristóbal de Las Casas.

Analysis

To compare epiphyte dry-weight and the number of species between sites, an ANCOVA was carried out, controlling for differences in tree size. Richness (alpha diversity) and biomass values were subjected to square-root transformation to homogenize variances. A biomass model ($r^2 = 0.75, p < 0.001$) included

DBH and the number of bifurcation points as covariates, after elimination of tree size parameters that did not contribute significantly ($p < 0.05$), whereas a richness model ($r^2 = 0.69, p < 0.001$) only included DBH (see Wolf, 2005 for details).

The spatial distribution of epiphytes is influenced by spatially structured environmental variables, historic biogeography, and biological processes that are spatially heterogeneous, such as dispersal (e.g. Legendre *et al.*, 2005). Multivariate data analysis was used to discriminate between the contribution of environmental and spatial variables, following Borcard *et al.* (1992). Environmental and spatial variables are partially correlated and pure space is defined as that part of the spatial variation that is not correlated with the environmental variables used in the analysis. The following quantitative environmental variables were used: altitude, forest height, the proportion of sprouted oaks, oak host tree species frequency, and the basal areas and densities of pines, oaks, and other broad-leaf trees. Forest height was taken as the average of the five tallest oak trees in the inventory. Large oaks with trunk DBH > 45 cm were separated from smaller oaks. In addition, some nominal values were considered: grazing, fire, and the presence of tree stumps. For spatial descriptors of the sites, the terms of a cubic trend-surface equation were used that was derived from the geographic coordinates of the sites (Borcard *et al.*, 1992; Legendre and Legendre, 1998). Species biomass values were log-transformed.

Spatial heterogeneity was analyzed further with Mantel's test. Mantel's regression coefficients were determined from a Euclidean distance matrix, computed from the geographic coordinates of the sites, and a species distance (dissimilarity) matrix. For the latter, the probabilistic coefficient of Raup and Crick was used, as suggested by Legendre and Legendre (1998).

The computations were performed using Systat (version 5.03), the R Package for Multivariate and Spatial Analysis (4.0), and CANOCO (4.02) (Ter Braak, 1988; Casgrain and Legendre, 2001).

RESULTS AND DISCUSSION

Forest disturbance, epiphyte richness and epiphyte biomass

Epiphyte biomass on the 35 oak trees varied between 0.8 and 243 kg, and richness between 13 and 34 species (Table 68.1). Forest disturbance, measured as the proportion of sprouted oaks, not only had a negative effect on epiphyte biomass and alpha diversity per ground surface area (Wolf, 2005) but also on biomass (Pearson's $r = -0.55, p = 0.02$) and alpha diversity (Pearson's $r = -0.75, p = 0.001$) of the epiphyte vegetation on the remaining and regrowing trees (Table 68.1). This pattern was sustained after correcting for differences in tree size, as shown by the post-ANCOVA Tukey classification of the sites in cohorts of similar biomass and species alpha diversity; the less disturbed sites near the top of Table 68.1 belong to the epiphyte-high biomass and species-rich cohorts (As). Conversely, the coppiced stands listed at the bottom of the table are poor in epiphytes, particularly sites that have experienced nearly complete clear-cuts on a cyclic basis and comprise more than 80% of sprouted trunks. This form of management is similar to that of tree plantations, which are also poor in epiphytes (Merwin *et al.*, 2003). Colonization from outside the coppice is apparently a slow process, which is in accordance with the rapid decay in seed rain with distance from the source (García-Franco and Gray, 1988).

Dispersal limitation may also explain why the epiphytic vegetation at site Chilil-2 was more affected as compared with the stands at Costik and Mitzitón. The forests of these three sites had a similar proportion of sprouted oaks and comparable oak and pine tree basal areas, suggesting a similar management history. Nevertheless, at Chilil-2 epiphyte alpha diversity and biomass were significantly lower than at the two other sites where several large trees were apparently spared (Figure 68.1). Such heavily loaded "rescue" trees possibly serve as a nearby seed source, thereby facilitating the recovery of the epiphyte vegetation.

Spatial heterogeneity of epiphytes

Twenty-five percent of the species variation between sites was left unexplained by the multivariate analysis (Figure 68.2). This amount is similar to values reported for other ecosystems and may be attributed to unmeasured environmental variables and random fluctuations (Borcard *et al.*, 1992). Interestingly, spatial variables explained most of the distribution of epiphytes in the landscape. Pure space, i.e. that part of the spatial variation not correlated with environmental variables, even explained 32.3%. Space is usually not considered in epiphyte landscape ecology. A Mantel correlogram (Mantel's $r = 0.44, p = 0.002$) visualizes the spatial dependence of epiphytes in the landscape (Figure 68.3). Thus, sites on the Central Plateau at distances less than 10 km proved to be similar in terms of their epiphytes. This finding is of considerable importance for the scale at which epiphyte conservation efforts should be contemplated.

Epiphytes and sustainable harvesting: theory

After the felling of oaks to be used as fuelwood, the attached epiphytes are simply discarded, despite a worldwide growing market for ornamental plants such as bromeliads (Dimmitt, 2000). Could the harvesting of bromeliads from the pine–oak and oak forests of Chiapas offer an alternative source of income to wood extraction and thus be employed as a tool for forest conservation? An answer requires studies of the commercial, social, and

Table 68.1 *Site forest structure and epiphyte biomass and richness (alpha diversity) on 35 oak trees in six trunk diameter classes; the sites are arranged by increasing proportion of sprouted oaks, which approximates a gradient of increased anthropogenic disturbance*

Locality	Altitude (m.a.s.l.)	Sprouted oaks (%)	Forest structure: Height (m)	Basal area (m^2 ha^{-1}): Oaks	Pines	Others	Total	Big oaks	Epiphyte community: *S*	Biomass (kg)	Richness cohort	Biomass cohort
La Florecilla-1	2350	8.6	22.6	31.6	6.5	0.2	38.3	17.5	27	243.9	B	A1
Chilil-1	2300	11.4	23.2	20.3	7.0	2.6	29.9	3.2	35	74.7	A	A2
El Chivero	2360	11.4	22.2	17.7	15.2	3.3	36.2	9.3	34	97.6	A	A2
Basom-1	2490	14.3	29.2	45.3	4.3	10.2	59.8	32.6	23	103.3	A	A2
Costik	2346	14.3	21.8	8.7	26.9	4.2	39.9	1.3	24	72.5	B	B
Chilil-2	2290	20.0	19.4	8.0	21.2	0.1	29.3	0.6	16	0.8	C	C2
San Antonio	2370	31.4	21.8	25.9	8.2	12.5	46.6	9.9	29	98.9	A	A2
Basom-2	2450	31.4	28.6	14.6	56.8	1.1	72.4	4.9	24	113.7	B	A2
Mitzitón	2420	34.3	17.2	7.2	24.6	1.8	33.7	2.4	24	108.1	B	A2
Carrizal	2160	40.0	19.4	27.4	0.0	0.5	27.9	9.5	16	69.7	C	A2
San José	2350	51.4	15.9	21.4	0.2	0.7	22.3	0.0	18	83.8	B	A2
Rancho Nuevo	2400	51.4	23.0	10.4	31.5	0.1	42.0	10.4	23	9.7	C	C2
Las Flores	2375	60.0	17.4	37.9	10.7	4.5	53.0	7.8	18	41.3	B	B
La Florecilla-2	2350	74.3	17.6	21.9	13.5	7.5	42.9	8.0	15	65.5	C	B
La Florecilla-3	2350	82.9	20.4	22.6	20.8	0.0	43.4	0.0	14	16.7	C	C1
Milpoleta	2425	94.3	28.8	41.6	0.0	9.5	51.1	10.9	13	16.0	B	C1

Notes: Forest height is the average of the five tallest oak trees in the inventory. Big oaks are oaks with a trunk DBH > 45 cm. The division of sites in terms of epiphyte richness and biomass cohorts is based on the probabilities of a Tukey pairwise comparison of the means, following an ANCOVA that factored out the differences in tree size between the sites (r^2 =0.69, $p < 0.001$). As to species richness (*S*), cohort A comprises species-rich sites and cohort C species-poor sites. Within each of the two cohorts, site species richness is not significantly different, in contrast to a comparison between cohorts A and C ($p < 0.05$). Cohort B assumes an intermediate position. The biomass at La Florecilla-1 (cohort A1) is significantly higher than at any other site. The sites in cohort A2 showed a significantly higher biomass than the low biomass sites of cohorts C1 and C2. Again, cohort B is intermediate

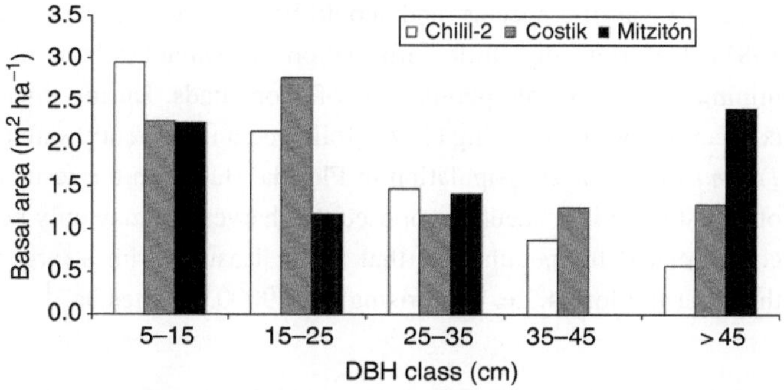

Figure 68.1. Basal area (m^2 ha^{-1}) of oak trees in various trunk diameter classes (DBH) at three disturbed sites, based on 900-m^2 inventories on the Central Plateau, Chiapas Highlands, southern Mexico.

ecological sustainability of such activities. The remainder of this chapter addresses the ecological aspects of epiphyte harvesting.

When defining criteria for sustainable harvesting, two approaches may be followed. By means of demographic and/or genetic studies and modeling one may attempt to establish the minimum viable population size (MVP) and propose

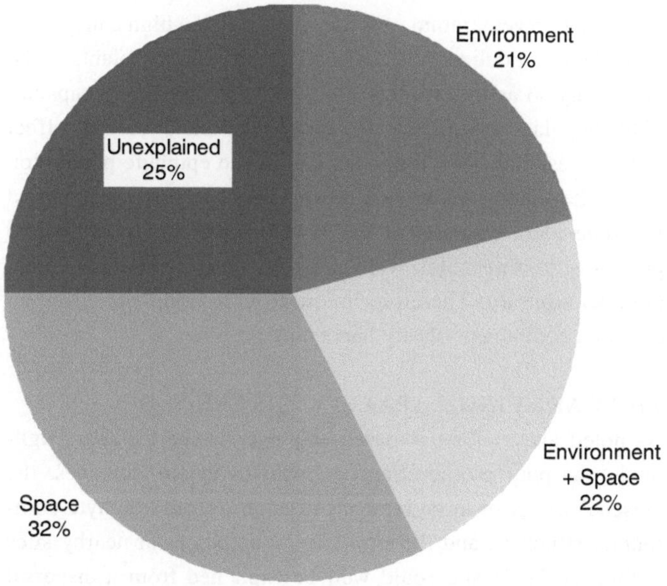

Figure 68.2. Partitioning of the variation in epiphyte communities between sites that may be explained by environmental and spatial variables, following Borcard *et al.* (1992).

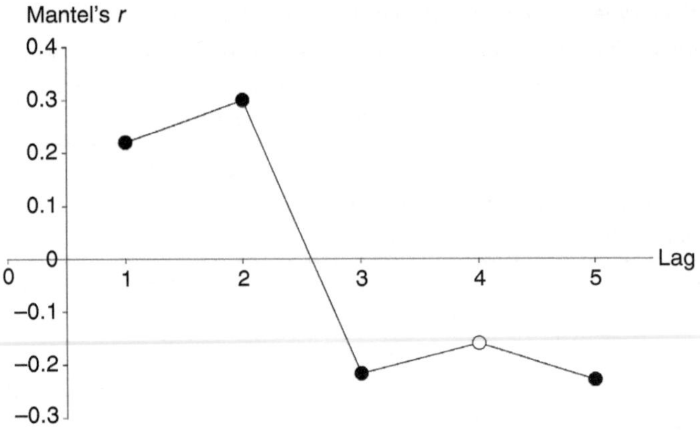

Figure 68.3. Mantel correlogram showing epiphyte spatial dependence between sites. The correlations (i.e. Mantel's r) are calculated from a species similarity matrix, using the Raup and Crick probabilistic coefficient as a measure of species similarity, and a Euclidian distance matrix computed from the geographic coordinates of the 16 sites. Species biomass values were log-transformed. The lags are equidistant at 5 km. Black symbols indicate a significant value of Mantel's r ($p < 0.05$).

management interventions based on population viability analysis (PVA) (Menges, 2000). However, the MVP has no universal value (Soulé, 1987) whereas a reliable PVA also requires observations to be made over a long period of time. For the immediately threatened bromeliads, Wolf and Konings (2001) therefore advocated a more empirical approach to guarantee sustainability of yield. On the basis of generally accepted ecological principles, they proposed that harvesting may only be considered for populations that: (i) are close to carrying capacity (see explanation below), (ii) have high densities (see also below), and (iii) grow both in the lower stratum of the forest and in the high canopy. As to the latter guideline, it is advisable to harvest only plants in the understory so as to limit the effect on the reproductive capacity of the population. Similar to the above-mentioned "rescue" effect of large residual trees, the intact part of the epiphyte population within the canopy serves as a nearby source of seeds that should facilitate recolonization of the lower strata. For the first two prerequisites thresholds were set that, whilst being stringent, are also arbitrary. Therefore, frequent monitoring should be an essential component of any harvesting program.

THE CARRYING CAPACITY THRESHOLD

As noted above, for oak-based epiphytes in the Chiapas Highlands, the poor proliferation of epiphytes in oak coppices, the "rescue" effect of large remnant trees on overall epiphyte community structure, and the similarity in epiphytes at nearby sites within the landscape could well be explained from a dispersal assembly perspective. In this view, the absence of epiphytes on oaks results primarily from a lack of arrivals rather than unfavorable habitat properties (Hubbell, 2001). Accordingly, when a bromeliad species is less abundant in certain parts of the forest, i.e. not distributed homogeneously in space, the population is considered not to be at carrying capacity (yet). For an ecosystem with its epiphytes at carrying capacity this would mean that in any part of a (relatively environmentally homogeneous) forested area the same correlation between tree size and its epiphyte abundance should exist. In other words, the better the size of the tree explains the amount of epiphytic biomass that it supports, the closer is the epiphyte vegetation to its carrying capacity. The corresponding squared correlation coefficient may therefore be used as an index of spatial heterogeneity (ISH) to assess the closeness to carrying capacity of epiphyte populations (Wolf and Konings, 2001). It should be noted that a perfect correlation between tree size and epiphyte abundance is not to be expected, amongst others because the suitable bark area on a tree will not depend linearly on the size of the tree. The level of the ISH threshold is set by its value in an old-growth stand and will depend on species as well as type of habitat.

The anthropogenic disturbance study by Wolf (2005) supported the assertion that the tree-size–abundance correlation may be employed to assess the closeness to carrying capacity of epiphyte populations in the study area. Epiphyte biomass was positively correlated ($p < 0.001$) with the size of the host tree at all sites, except at the heavily disturbed Chilil-2 plot ($r^2 = 0.16$, $p = 0.02$). More significantly, at the nine sites with high epiphytic biomass (i.e. cohorts A1 and A2), the correlation was substantially higher ($r^2 = 0.85$, on average) than at the more disturbed sites which had fewer epiphytes ($r^2 = 0.59$, on average).

THE POPULATION DENSITY THRESHOLD

Harvesting from a small population might negatively affect the local survival of a species, and should be avoided (e.g. Shaffer, 1981). Unfortunately, little information is available about the minimal size of viable populations of bromeliads. During three consecutive years, Benzing (1978) followed an apparently stable *Tillandsia circinnata* population in Florida. Based on the results of that study it is prudently proposed that harvesting may only be contemplated for populations that are at least ten times larger than that in Florida, i.e. comprising over 9000, rosettes ha^{-1}.

Epiphytes and sustainable harvesting in practice: a bromeliad pilot study

In a pilot study at La Florecilla, the above-mentioned guidelines to attain a sustainable yield of epiphytes have been put into practice (Wolf and Konings, 2001). La Florecilla comprises 160 ha of continuous pine–oak forest at various degrees of disturbance by humans. *Tillandsia vicentina* Standley is the dominant epiphyte and the subject of this study. Only species of oak having similar bark properties were considered as host trees. First, the correlation between tree size and square-root

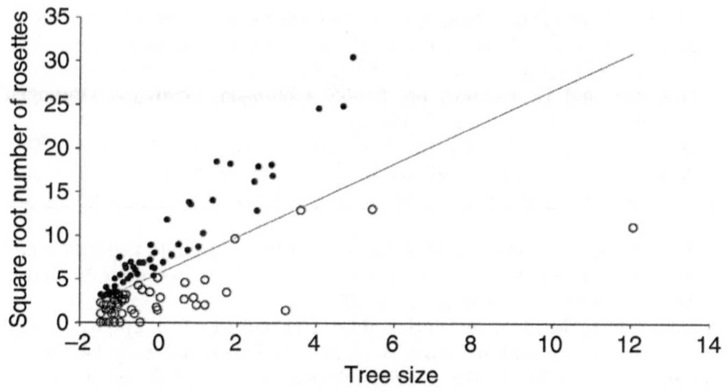

Figure 68.4. The square root of the number of *Tillandsia vicentina* rosettes >20 cm plotted against Tree Size. Tree Size is the sum of diameter at breast height and the number of bifurcation points, after standardization to zero mean and unit variance. Only sample points above the fitted line (solid circles) were used in the calculation of the correlation index ($r^2 = 0.92$).

transformed epiphyte biomass was determined at La Florecilla-1, an old-growth patch of forest in the area: $r^2 = 0.91$. This is the carrying capacity prerequisite threshold value (ISH). Next, the species density and the correlation index for the whole area were estimated from a transect inventory. A random sample of 35 points was selected along parallel transects that were evenly distributed over the forest. At each point, four oak trees were selected applying the plotless point-centered quarter method (Cottam and Curtis, 1956). The DBH of each of the trees was measured and the number of bifurcation points with a branch diameter >5 cm counted. In addition, the number of *T. vicentina* rosettes >20 cm tall were counted per tree, using binoculars. To simplify the analysis, the variables DBH and the number of bifurcation points were added only after standardization to zero mean and unit variance, to obtain a single variable to represent tree size, called Tree Size (Figure 68.4).

At La Florecilla, small patches of forest of a few hectares each – which varied greatly in structure due to differences in degree and type of previous anthropogenic disturbance – formed a small-scale heterogeneous mosaic. The suggestion to restrict harvesting to old-growth forests was rejected because it was problematic for the local population to fairly rank and map stands. Instead, epiphytes in the entire forest were considered for harvesting. Non-stratified sampling of randomly selected transect points is more objective, but because heavily disturbed stands of forest such as that of La Florecilla-3 (Table 68.1) were also sampled, the ISH threshold determined for old-growth stands would certainly not be reached. On the other hand, it would be unrealistic to require that the carrying capacity prerequisite is met everywhere in such a large area as La Florecilla (160 ha). Therefore, it was, again arbitrarily, decided that harvesting might be contemplated if the prerequisite would be met for only approximately half the trees in the forested area. To select the trees to be used for the calculation of the Tree Size–epiphyte abundance correlation, a partitioned regression analysis was performed. First, a line was fit through the entire cloud of data points representing all trees in the transect inventory. Second, the data sub-set above the line was selected (i.e. the positive residuals). Next, the correlation between Tree Size and the number of rosettes was analyzed to yield ISH.

The transect analysis showed that *T. vicentina* met all prerequisites, having both a satisfactory population density of *c.* 24 000 large (>20 cm) rosettes ha^{-1} on oaks, and an ISH of 0.92. The population density value corresponds to the lower limit of the 95% confidence interval of the mean bromeliad density (39 718 rosettes ha^{-1}), as calculated from the average occupation and density per hectare of oak trees.

A separate inventory of *T. vicentina* on all species of host tree in the forest and including terrestrial rooted plants showed that 27% of all rosettes occurred in the understorey up to a height of 6 m. A little less than half of the epiphytic *T. vicentina* plants in the understory (2680, rosettes ha^{-1}) occurred there as solitary rosettes (genets). These are especially attractive because of their symmetric growth. Field observations suggested that *T. vicentina* plants need about 5 years to develop from seed to the commercial size of 20 cm. Because many of the smaller juveniles are already present at the time of harvesting, it is assumed that a period of 4 years will allow such rosettes to attain at least 20 cm. Adhering to a 4-year harvesting rotation scheme, it would be possible in La Florecilla to sustain an annual harvest from the lower stratum of *c.* 700 epiphytic single rosettes of *T. vicentina* ha^{-1}. Thus, from the 160-ha forest at La Florecilla an annual yield of 112 000 particularly attractive rosettes may be contemplated, equivalent to ~3% of the number of large-sized rosettes in the forest.

The large number of annually harvestable plants suggests that such an activity may be commercially viable and that *T. vicentina* may be used as a tool to help conserve the forest. It should be remembered, however, that the thresholds of the prerequisites were, whilst stringent, arbitrarily set. The same is true for the omission of half of the trees in the carrying capacity assessment. Moreover, no data are available on the restoration capacity of the bromeliad population, nor on the effect of harvesting *T. vicentina* on other epiphyte species, amongst others. Therefore, yearly monitoring should be an essential component in any management plan for sustainable harvesting.

IMPLICATIONS FOR FOREST MANAGEMENT

On the Central Plateau of Chiapas, pine–oak forests are a source of timber and fuel for subsistence purposes. This vital function is not likely to change in the near future. This study has shown, however, that tree harvesting practices of some indigenous communities (whereby several large trees are left standing) have preserved the rich epiphyte vegetation better than the nearly complete clear-cutting practiced by other communities. Accordingly, it is recommended to abstain from the cyclic clear-cutting of the oak forest and, in particular, to spare large "rescue" trees during logging that may serve as a source of epiphytes during the recolonization process. Furthermore, based on the observed spatial heterogeneity it is wise to address epiphyte conservation issues at larger spatial scales, e.g. per physiographic unit. Finally, it is proposed to further explore whether and to what extent ornamental epiphytes may be employed as a tool for forest conservation.

ACKNOWLEDGEMENTS

The assistance rendered in the field by Teresa Santiago-Vera, Mariana T. Toledo-Aceves, and Henry E. Castañeda-Ocaña is gratefully acknowledged. Funds and other support were provided by Stichting Het Kronendak, El Colegio de la Frontera Sur (ECOSUR), and the Comisión Nacional para el Conocimiento y Uso de la Biodiversidad (CONABIO), grants B060 and L050.

REFERENCES

Barthlott, W., V. Schmit-Neuerburg, J. Nieder, and S. Engwald (2001). Diversity and abundance of vascular epiphytes: a comparison of secondary vegetation and primary montane rain forest in the Venezuelan Andes. *Plant Ecology* **152**: 145–156.

Benzing, D. H. (1978). The population dynamics of *Tillandsia circinnata* (Bromeliaceae): cypress crown colonies in southern Florida. *Selbyana* **5**: 256–263.

Borcard, D., P. Legendre, and P. Drapeau (1992). Partialling out the spatial component of ecological variation. *Ecology* **73**: 1045–1055.

Breedlove, D. E. (1978). The phytogeography and vegetation of Chiapas (Mexico). In *Vegetation and Vegetational History of Northern Latin America*, ed A. Graham, pp. 149–165. San Francisco, CA: California Academy of Sciences.

Casgrain, P., and P. Legendre (2001). *The R-Package for Multivariate and Spatial Analysis*, version 4.0 d5, *User's Manual*. Montréal, Canada: Département de Sciences Biologiques, University of Montréal. Also available at www.fas.umontreal.ca/BIOL/legendre/.

Castro-Hernández, J. C., J. H. D. Wolf, J. G. García-Franco, and M. González-Espinosa (1999). The influence of humidity, nutrients and light on the establishment of the epiphytic bromeliad *Tillandsia guatemalensis* in the highlands of Chiapas, Mexico. *Revista de Biologia Tropical* **47**: 763–773.

Cavelier, J., T. M. Aide, C. Santos, A. M. Eusse, and J. M. Dupuy (1998). The savannization of moist forests in Sierra Nevada de Santa Marta, Colombia. *Journal of Biogeography* **25**: 901–912.

Cayuela, L., J. M. R. Benayas, and C. Echevarría 2006 Clearance and fragmentation of tropical montane forests in the Highlands of Chiapas, Mexico. *Forest Ecology and Management* **226**: 206–218.

Cottam, G., and J. T. Curtis. (1956). The use of distance measures in phytosociological sampling. *Ecology* **37**: 451–460.

Dimmitt, M. (2000). Endangered Bromeliaceae. In *Bromeliaceae: Profile of an Adaptive Radiation*, ed. D. H. Benzing, pp. 609–620. Cambridge, UK: Cambridge University Press.

García-Franco, J. G., and V. R. Gray (1988). Experiments on seed dispersal and deposition patterns of epiphytes: the case of *Tillandsia deppeana* Steudel (Bromeliaceae). *Phytologia* **65**: 73–79.

Gradstein, S. R. (2008). Epiphytes of tropical montane forests: impacts of deforestation and climate change. In *The Tropical Mountain Forest: Patterns and Processes in a Biodiversity Hotspot*, eds. S. R. Gradstein, J. Homeier, and D. Gansert, pp. 51–65. Göttingen, Germany: Göttingen Centre for Biodiversity and Ecology.

Gradstein, S. R., N. M. Nadkarni, T. Krömer, I. Holz, and N. Nöske (2003). A protocol for rapid and representative sampling of vascular and non-vascular epiphyte diversity in tropical rain forests. *Selbyana* **24**: 105–111.

Hietz-Seifert, U., P. Hietz, and S. Guevara (1996). Epiphyte vegetation and diversity on remnant trees after forest clearance in southern Veracruz, Mexico. *Biology and Conservation* **75**: 103–111.

Hofstede, R. G. M., J. H. D. Wolf, and D. H. Benzing (1993). Epiphytic mass and nutrient status of an Upper Montane Rain Forest. *Selbyana* **14**: 37–45.

Hubbell, S. P. (2001). *The Unified Neutral Theory of Biodiversity and Biogeography*. Princeton, NJ: Princeton University Press.

Köhler, L., C. Tobón, K. F. A. Frumau, and L. A. Bruijnzeel (2007). Biomass and water storage dynamics of epiphytes in old-growth and secondary montane cloud forest stands in Costa Rica. *Plant Ecology* **193**: 171–184.

Legendre, P., and L. Legendre (1998). *Numerical Ecology*, 2nd edn. Amsterdam, the Netherlands: Elsevier.

Legendre, P., D. Borcard, and P. R. Peres-Neto (2005). Analyzing beta diversity: partitioning the spatial variation of community composition data. *Ecological Monographs* **75**: 435–450.

Menges, E. S. (2000). Population viability analyses in plants: challenges and opportunities. *Trends in Ecology and Evolution* **15**: 51–56.

Merwin, M. C., S. A. Rentmeester, and N. M. Nadkarni (2003). The influence of host tree species on the distribution of epiphytic bromeliads in experimental monospecific plantations, La Selva, Costa Rica. *Biotropica* **35**: 37–47.

Mitchell, A. W., K. Secoy, and T. Jackson (eds.) (2002). *The Global Canopy Handbook: Techniques of Access and Study in the Forest Roof.* Oxford, UK: Global Canopy Programme.

Muñoz-Villers, L. E., and J. López-Blanco (2008). Land use/cover changes using Landsat TM/ETM images in a tropical and biodiverse mountainous area of central-eastern Mexico. *International Journal of Remote Sensing* **29**: 71–93.

Nöske, N., N. Hilt, F. Werner, *et al.* (2008). Disturbance effects on diversity in montane forest of Ecuador: sessile epiphytes versus mobile moths. *Basic and Applied Ecology* **9**: 4–12.

Ochoa-Gaona, S., and M. González-Espinosa (2000). Land use and deforestation in the highlands of Chiapas, Mexico. *Applied Geography* **20**: 17–42.

Ramírez-Marcial, N., M. Gonzalez-Espinosa, and G. Williams-Linera (2001). Anthropogenic disturbance and tree diversity in Montane Rain Forests in Chiapas, Mexico. *Forest Ecology and Management* **154**: 311–326.

Sarmiento, F. (1997). Arrested succession in pastures hinders regeneration of Tropandean forests and shreds mountain landscapes. *Environmental Conservation* **24**: 14–23.

Shaffer, M. L. (1981). Minimum population sizes for species conservation. *BioScience* **31**: 131–134.

Soulé, M. E. (1987). Introduction. In *Viable Populations for Conservation*, ed. M. E. Soulé, pp. 1–10. Cambridge, UK: Cambridge University Press.

Ter Braak, C. J. F. (1988). CANOCO: an extension of DECORANA to analyze species–environment relationships. *Vegetatio* **75**: 159–160.

Werner, F., J. Homeier, and S. R. Gradstein (2005). Diversity of vascular epiphytes on isolated remnant trees in the montane forest belt of southern Ecuador. *Ecotropica* **11**: 21–40.

Williams-Linera, G., R. Manson, and E. Izunza (2002). La fragmentación del bosque mesófilo de montaña y patrones de uso del suelo en la región oeste de Xalapa, Veracruz, México. *Madera y Bosques* **8**: 73–89.

Wolf, J. H. D. (2005). The response of epiphytes to anthropogenic disturbance of pine–oak forests in the highlands of Chiapas, Mexico. *Forest Ecology and Management* **212**: 376–393.

Wolf, J. H. D., and A. Flamenco-Sandoval (2003). Patterns in species richness and distribution of vascular epiphytes in Chiapas, Mexico. *Journal of Biogeography* **30**: 1689–1707.

Wolf, J. H. D., and C. J. F. Konings (2001). Toward the sustainable harvesting of epiphytic bromeliads: a pilot study from the highlands of Chiapas, Mexico. *Biological Conservation* **101**: 23–31.

Wolf, J. H. D., S. R. Gradstein, and N. M. Nadkarni (2009). A protocol for sampling vascular epiphyte richness and abundance. *Journal of Tropical Ecology* **25**: 107–121.

69 Fire dynamics and community management of fire in montane cloud forests in south-eastern Mexico

H. Asbjornsen
Iowa State University, Ames, Iowa, USA

Z. Garnica-Sánchez
Rainforest Alliance, Oaxaca City, Oaxaca, Mexico

ABSTRACT

Wildfires are occurring with increasing frequency and extent in montane cloud forests (MCF) worldwide. Fires in MCF pose challenges with respect to both fire behavior and organization of suppression efforts that are often unique from other forest types. This chapter assesses the technical and organizational aspects involved in suppressing severe wildfires occurring in MCF in southern Mexico during the 1998 El Niño Southern Oscillation. In particular, it highlights the valuable lessons learned about fire suppression techniques for the extreme fire behavior observed for these MCF ecosystems (both subterranean and crown fires), and on approaches toward organizing fire suppression efforts and preventing fire occurrence that involve local community participation. Given the importance of human activities in driving fire dynamics in typically ever-wet MCF, it is suggested that integrating local community-based and external government-based fire management approaches is critical to effective fire protection and suppression in MCF regions. The chapter offers various specific recommendations for achieving such integration.

INTRODUCTION

The perception that montane cloud forests (MCFs) are resistant to fire was shattered in 1998 when widespread fires occurred in these ecosystems throughout south-eastern Mexico. Although these fires were proclaimed to be unprecedented events – shocking and surprising the nation, reaching global news, and eliciting international response – they were by far not the first reports of fire in MCFs. On geological timescales, research using ^{14}C-dating techniques has documented the occurrence of fire in MCF regions during periods of climatic warming and drying (Behling, 1997; Colinvaux *et al.*, 1997; Horn *et al.*, 2000; Bush *et al.*, 2004; Urrego *et al.*, 2005; Wård *et al.*, this volume). These fires have left their footprints in the pollen record suggesting the predominance of fire-adapted genera (e.g. *Pinus, Quercus, Alnus*, Ericaceae) immediately following fire activity and their eventual successional replacement (over hundreds to thousands of years) by MCF species (Conserva and Byrne, 2002; Dull, 2004; Weng *et al.*, 2004; Hemp, 2005a). On decadal timescales, fires have occurred in African (Hemp 2005a,b), Central American (Kappelle, 1993; Conserva and Byrne, 2002), Caribbean (Sherman *et al.*, 2005; Martin and Fahey, 2006), and South American (Sarmiento and Frolich, 2002; Grau *et al.*, this volume) MCFs. Thus, large, infrequent, high-severity fires were likely an important determinant of ecosystem development on geological timescales in regions currently supporting MCF. However, under human-induced processes of accelerated climate change within the past century (IPCC, 2001) and the probability of synergistic feedbacks between fire and increasing land-use pressures and drought frequencies (Cochrane *et al.*, 1999; cf. Karmalkar *et al.*, 2008), the risk of fire in these typically non-flammable MCF ecosystems is increasing (Loope and Giambelluca, 1998; Román-Cuesta *et al.*, 2004; Asbjornsen *et al.*, 2005; cf. Foster, this volume).

In response to the trend of escalating fire in moist tropical forests worldwide, preventing and controlling wildfires in these ecosystems has become a critical management and conservation priority (Goldammer, 1999; Meyers, 2006). Whereas international fire initiatives during the 1980s and 1990s emphasized technological solutions and were primarily operated by

Tropical Montane Cloud Forests: Science for Conservation and Management, eds. L. A. Bruijnzeel, F. N. Scatena, and L. S. Hamilton. Published by Cambridge University Press.

government agencies, in recent years the important role of community participation and knowledge in ensuring the success of fire management programs has received increased attention (Goldammer *et al.*, 2002). In south-eastern Mexico, the El Niño Southern Oscillation (ENSO) 1998 fires left their legacy not only in the catastrophic ecological effects on MCFs – resulting in the almost immediate loss of 80–100% of the living biomass in these ecosystems (Asbjornsen *et al.*, 2005) – but also through the new knowledge gained about fire behavior in MCFs and fire management organization. In particular, the MCF fires posed unique challenges to both the technical and organizational aspects of fire suppression efforts, which led to the development of some novel approaches, such as the adaptation of fire suppression techniques to deep ground fires and the integration of community and professional firefighting brigades. Although the tremendous diversity in the structure and composition of MCF ecosystems worldwide (Barthlott *et al.*, 2005; Richter, 2008) precludes broad generalizations regarding fire behavior and management, the lessons learned from the 1998 MCF fires in Mexico can provide important guidance as to how to prepare for this growing threat in other MCF regions around the globe.

This chapter first presents the ecological and historic context of fire in MCFs in south-eastern Mexico, focusing on how fire regimes vary across different ecosystems and are influenced by local communities' use of fire. Next, the technical and organizational aspects of the 1998 fire suppression effort are analyzed, highlighting the challenges created by the unique conditions and behavior of fire in MCF ecosystems, and the participation of diverse local and external firefighting agents. The post-fire consequences in terms of how the 1998 fires affected perceptions of fire as an ecological threat are then assessed, as well as how this experience influenced fire management systems in the region after 1998. Finally, by drawing on the theoretical and practical literature on fire management, combined with the local Mexican experience, recommendations for reducing future fire occurrence in MCFs through co-management approaches are given.

THE ECOLOGICAL AND HISTORIC CONTEXT OF THE 1998 MCF FIRES IN SOUTH-EASTERN MEXICO

Mexico, a mega-diverse and topographically complex country that historically harbored one of the world's largest MCF areas, has already lost approximately 68% of its original MCF cover to deforestation, leaving an estimated 130 000 ha remaining today (Mulligan, this volume). These MCFs generally occur in remote, high mountain areas, extending from the State of Tamaulipas in the north to the State of Chiapas in the south. One of the largest expanses of well-conserved MCF is found in the south-eastern State of Oaxaca in a region known as the Chimalapas, which supports an area of approximately 60 000 ha of relatively continuous MCF cover (Wendt, 1997). Although MCFs throughout Mexico were affected by fire in 1998, the most severe and extensive fires occurred in the Chimalapas (Anta and Plancarte, 2001).

The diverse climate and mountainous topography of the Chimalapas give rise to a complex mosaic of forest formations, ranging from temperate pine–oak–liquidambar forests and deciduous dry forests to tropical evergreen rain forest and MCF (Breedlove, 1981; Salas *et al.*, 2001). Fire regimes vary markedly across these ecosystems. In pine forests, fire is an important and relatively frequent disturbance needed to promote regeneration. For example, in other Mexican pine forest regions, reported mean fire return intervals range between 3 and 6 years (Fulé and Covington, 1997; Mathews, 2003). Fires occur relatively frequently but generally at longer return intervals in oak, pine–oak, and lowland deciduous dry forests. Dominant plant species in these ecosystems typically possess fire-adapted mechanisms (e.g. thick bark, capacity to resprout, fire-resistant seeds) that enable them to persist under frequent fire (e.g. Pyne *et al.*, 1996).

In contrast, there are no modern records of extensive or repeated fires in MCF in the Chimalapas (Anta and Plancarte, 2001; but see Wård *et al.*, this volume), and similar to rain forest species, MCF species do not typically possess fire-adapted traits. A study in the Sierra de Manantlán (Jalisco), where the climate is drier and human population pressure higher compared to the Chimalapas, only 3% of all recorded fires occurred in MCF, whereas over 60% of the fires occurred in pine, oak, and mixed pine–oak systems (Jardel *et al.*, 2001a). In this region, repeated surface fires in pine forests are believed to interrupt the succession of pine forests to MCF, suggesting that pine forests may have derived from former MCF as a result of introduced fire at high elevations (Jardel *et al.*, 2001b). The lower human population densities and more remote location of MCF in the Chimalapas likely protected them from exposure to human ignition sources in the past (Wård *et al.*, this volume). Nevertheless, vegetation in transitional areas between MCF and the pine, oak, and liquidambar forests occurring at lower elevations in the Chimalapas region may reflect successional processes in response to changing fire regimes over longer time intervals.

Two communities, Santa María Chimalapas (460 000 ha) and San Miguel Chimalapas (134 000 ha), comprise the Chimalapas region (Figure 69.1). Their combined population of approximately 15 000 inhabitants is distributed across numerous small villages that fall under the legal jurisdiction of the communities' political authority or "county seat." These villages are located in the periphery of a core area of approximately 300 000 ha that supports a relatively well-conserved, continuous forest cover (including approximately 80% of the region's MCF). As a result of the extremely rugged topography and remote location, this core area contains no permanent human settlements. Each village has a recognized "area of influence," which, although not

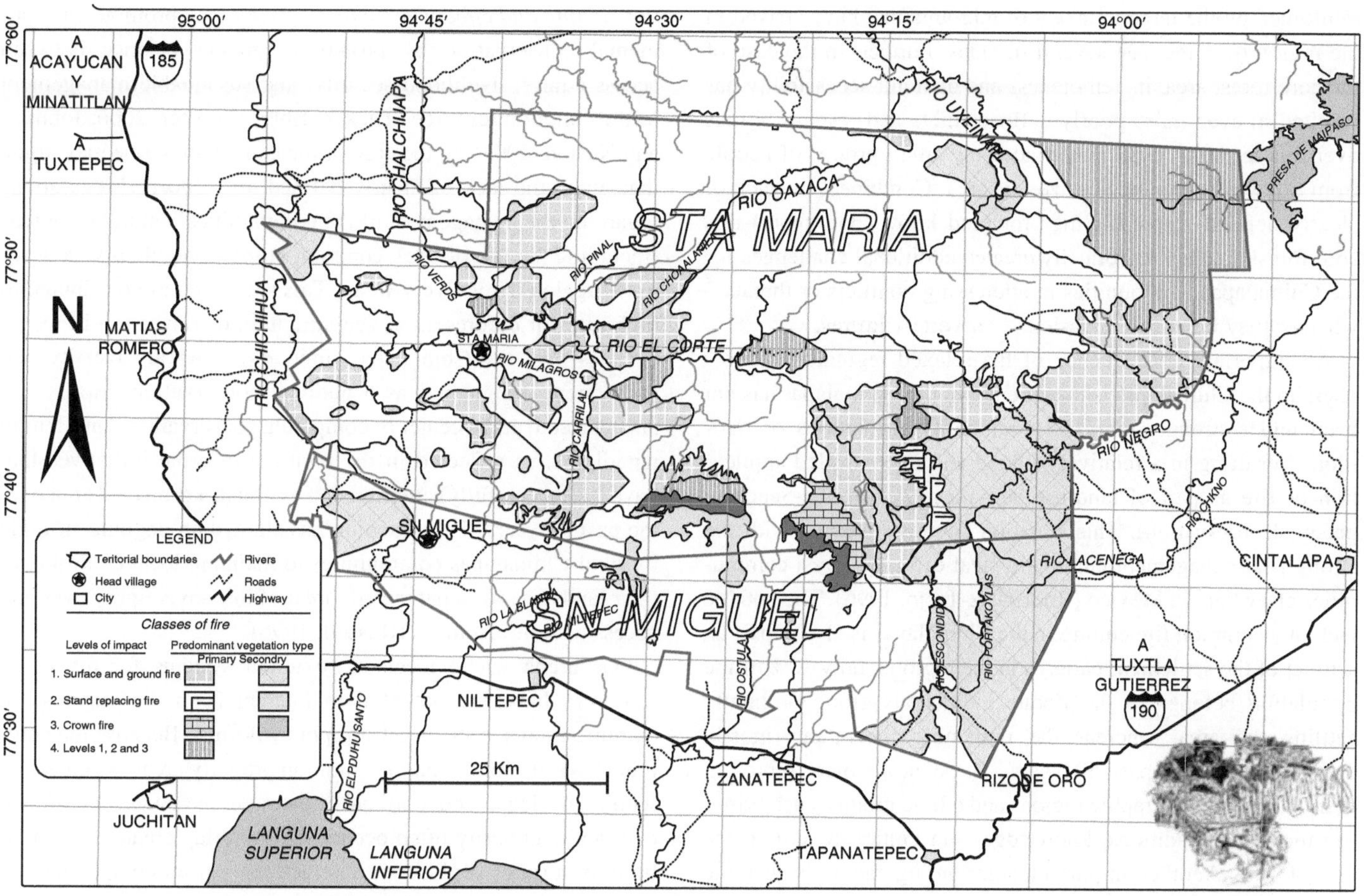

Figure 69.1. Map showing the boundaries of the two Chimalapas communities (San Miguel and Santa María) and the location of their villages along the periphery of the core forest area in the Chimalapas region of Oaxaca, Mexico.

delineated by specific boundaries, is actively managed and regulated by the villagers.

Land tenure in the Chimalapas follows the communal system, an ancient form of collective land-ownership originating during pre-Hispanic times. The other two forms of land-ownership in Mexico, *ejido* (also a form of collective ownership)[1] and private property are not present in the Chimalapas. Under the communal land tenure system, community members have a set of rights and responsibilities that dictate access to and use of common pool resources, and are upheld by strong notions of collective action and social institutions (Johnson and Nelson, 2004). Typically, communal laws designate land as having specific uses or functions, such as agricultural (generally the most fertile and flat lands, allotted for individual use), pasture (usually common access areas), forest (often utilized for extraction of timber and non-timber resources), and in some cases, protected area (minimal usage due to ecological or cultural value).

In addition to gaining benefits associated with having access to and use of communal resources, community members also have many responsibilities related to managing communal lands, such as protecting them against invasions and illegal extraction of resources, and suppressing wildfires when they occur. Many of these responsibilities are accomplished through the *tequio* system, a social institution of collective action whereby community members are required under communal law to participate in specific community work activities – ranging from construction of schools, churches, and health clinics, to maintaining roads, trails, fire breaks, irrigation systems, and communication services, and managing communal lands. However, it is important to note that in the case of the Chimalapas, although community laws regarding management of communal resources exist on paper, to a large extent these have not been widely practiced nor enforced. Although the reasons for the Chimalapas' state of "lawlessness" are manifold and complex (e.g. Russell, 1996), they largely stem from a long history of internal political conflict, lack of clear authority, and a prevailing frontier mentality

[1] The communal land tenure system differs from the *ejido* system, in that communal lands were established by presidential resolutions honoring indigenous populations title to territories they had occupied since pre-Hispanic times, whereas *ejidos* resulted from land redistribution after the Mexican Revolution and often involved designation of marginal lands to groups of people originating from other regions outside of the *ejido* lands.

reinforced by the large numbers of migrants that have arrived in the region over the past several decades. Further, in the case of the core forest area, its remoteness and difficult accessibility has resulted in even more poorly defined rights and responsibilities over this territory, especially since only small groups of people from a few communities utilize this area. Combined, the lack of clearly defined responsibilities for (and knowledge about) this core forest area has historically created additional challenges for the Chimalapas communities in addressing conflicts or threats – whether they be hostile invasions or severe wildfires.

An important consequence of the relaxed regulatory atmosphere in the Chimalapas is that fire use in the Chimalapas has not been under strong control by community regulations or traditions, resulting in a relatively liberal and uncontrolled application of fire as part of land-use activities (Z. Garnica-Sánchez, personal observation). This situation contrasts sharply with the strong fire management cultures typical of many forest communities elsewhere in Mexico (Rodríguez-Trejo, 1996). The notable lack of a common fire culture in the Chimalapas is also in part an artifact of the settlement patterns in the region – only 10% of the population belongs to the Zoque indigenous group originally settling the area, whereas the remaining inhabitants mostly arrived during the past 50 years. These newcomers represent many different geographic regions and ethnic groups, each bringing their own traditions, knowledge, and culture regarding the use of fire. Another important contributing factor is that the typically humid conditions in the region generally resulted in fires quickly self-extinguishing upon reaching the edge between agricultural and forested lands, especially when bordering closed-canopy MCF and rain forest that in most years support ever-moist conditions in the understory vegetation and soils. Thus, when fires occasional escaped into these forests they were usually extinguished quickly on their own or by minimum local effort.

Slash-and-burn practices have been widely applied in the Chimalapas to maintain crop production under a rotational system whereby following 2–3 years of crop production, the land is abandoned and new fallow areas are burned and planted again with crops (Quintana-Ascencio *et al.*, 1996). Annual burning is generally required in pastures and open-canopy pine forests to invigorate growth of grasses and prevent encroachment by woody vegetation. Historically, lowland rain forests and dry deciduous forests were preferred for both crop production and cattle ranching, whereas pine forests were generally used for grazing. Burning typically occurs in April and May just prior to the onset of the rains. Over the past several decades growing populations and land scarcity in the Chimalapas have led to greater encroachment of agricultural activities at the periphery of the core forest area, often including MCF.

Although some systems of common property management can lead to overexploitation and resource degradation (e.g. Cayuela *et al.*, 2006), certain common property regimes – *when properly functioning and effectively implemented* – may promote environmental conservation by promoting greater balance between access equity, livelihood security, and sustainable management of the common resource (Tucker, 1999; Klooster, 2000; Johnson and Nelson, 2004). In the case of the Chimalapas, one may argue that given the lack of clearly defined *and enforced* regulations regarding the management of the communities' natural resources (including the use of fire), common property regulatory capacity in the region is relatively weak. This situation has contributed to high rates of deforestation and agricultural conversion in recent decades. Yet the Chimalapas' unique ecological diversity and subsequent designation as a national conservation priority has also placed it at the center of competing conservation–land-tenure paradigms, as reflected in the multiple external initiatives that have (unsuccessfully) attempted to establish a protected area over the past 30 years. This situation has elicited an ongoing struggle by rural Chimalapas communities to maintain local control over their lands against a perceived threat of lost sovereignty over and access to their resources (Russell, 1996).

One arena where local and external systems for managing resources have the potential to interact is in the response to extreme events, such as catastrophic wildfires. Because the communal system for managing common property areas is embedded within the legal and political system of the State, significant overlap in authority often occurs. This overlap creates the possibility of a range of different interactions – from complementary to antagonistic – especially when situations arise that require participation of both internal and external agents (Young, 2002; Johnson and Nelson, 2004). The different cultures and approaches of the local community and external government to fire management in the Chimalapas region were sharply revealed in 1998 when these groups were forced to cooperate in order to face the common threat of an intensifying wildfire disaster. Whereas the communities were motivated by a traditional system of rights and obligations, communal organization, and collective action, external firefighting brigades operated under a highly structured, top–down, and technologically driven system. Understanding how communal fire management systems function and their relationship to external institutions can provide important insights about their critical but often overlooked role in fire prevention and suppression efforts, and inspire alternative systems that better integrate top–down with bottom–up approaches.

THE ENSO FIRES OF 1998 IN SOUTH-EASTERN MEXICO

The situation in 1998 was very different from the normal burning pattern in the Chimalapas region. A complex set of factors contributed to the extreme severity and extent of the fires that season. An extended dry season coinciding with an ENSO event – in which the rains stopped in January rather than continuing through March as in most years – resulted in drying of the forest understory vegetation and forest floor (Anta and Plancarte, 2001). In response to the drought, evergreen and semi-evergreen trees dropped their leaves, forming an unusually thick, dry litter layer (Asbjornsen *et al.*, 2005). These environmental conditions, combined with a gradual trend over the previous decade of increased expansion of agriculture and pasture along the periphery of the core forest area and thus a greater number of potential ignition sources, significantly increased the susceptibility of the MCF to fire (Román-Cuesta *et al.*, 2003). When burning of agricultural and pasture lands was resumed as usual in April of 1998, fires easily escaped into the surrounding forests. In many cases, local efforts to control these fires were ineffective. The fires not only occurred near the periphery of the forests where human settlements and agricultural activities were concentrated, but also originated within the forest core area in extremely remote locations. The remote fires were presumably caused by dry lightning (Z. Garnica-Sánchez, personal observation), or by escaped fires that had been ignited to establish small openings in the forest for narcotics production. Thus, the 1998 fires initiated and spread from two fronts: along the periphery of the core forest area and from within the core area. Once they were burning freely as wildfires, control of these fires was made especially difficult by the steep terrain, remote locations, thick organic layer, and strong winds.

Fighting fire in the cloud forests: technical aspects

The first wildfires of the 1998 season were initially confronted by local people from the villages located nearest to the fires; however, their efforts quickly proved insufficient and additional firefighter brigades were recruited from external units at both the state and federal levels. Firefighters applied the traditional tools and strategies used to fight fires in pine forests: rakes and machetes to scrape away the surface litter and other organic matter to create 2-m wide firebreaks with a deep trench (40 × 40 cm) in the center, and occasionally chainsaws and axes to remove larger pieces of woody debris. However, the rakes and machetes were largely ineffective at cutting through thick roots and digging into the deep organic soils, and with the additional physical effort required, firefighters quickly became fatigued. Helicopter drops of water and retardants were hampered by low visibility in thick smoke, and even when drops were successfully made, the closed-canopy MCF (very different from open pine stands) prevented the materials from reaching the burning fire below. Strong winds (often exceeding 60 km hour^{-1}) were particularly problematic to fire suppression efforts. Further, many of the 1998 wildfires occurred in MCF growing on karstic substrates where large cavities in the bedrock containing dry organic soil and roots quickly ignited, smoldering for days to weeks and often reigniting to produce a flaming fire on the other side of constructed firebreaks. The lack of a detectible flaming front meant that traditional fire suppression strategies involving the removal of surface fuel or direct attack on flames could not be applied. These ground-fires made it impossible to predict the direction and rate of fire spreading, thereby preventing effective planning of firefighting activities. Another serious obstacle to direct attack in these MCF ecosystems was the large amount of smoke produced by smoldering and glowing fires burning in thick organic soils. Firefighters suffered from respiratory problems and were often unable to continue working on the line. To complicate the situation even more, several areas were highly susceptible to crown fires, particularly large expanses of elfin forest (~7500, ha) growing on an exposed plateau of karst topography (Figure 69.2), as well as pockets of MCF growing on either metamorphic or limestone-derived soils on exposed ridges and hilltops. The problems associated with deep ground-fires, excessive smoke from smoldering fires, and explosive and running crown fires were distinct from typical fire behavior conditions associated with surface fires in pine and oak forests, rendering the firefighting system based on flaming front models of fire behavior developed for pine–oak forests largely ineffective in MCF fires (E. Alvarado, personal communication).

Following the failure of initial direct and indirect attack approaches, the incident commander of the now relatively large and growing fire suppression effort made a crucial decision: since firebreaks were not proving effective at constraining the spread of fire in MCF on karst topography and deep organic soils, firebreak construction efforts were redirected to surrounding MCF areas growing on metamorphic substrates supporting mineral soils where deep ground-fires would be less problematic. Although this approach meant abandoning efforts to try to save large areas of MCF directly threatened by the fire, it offered the possibility of confining the fire and protecting other important and valuable forested areas. Thus, lines were drawn on topographic maps based on local knowledge of the geology and topography of the region. A local community firefighter was placed in charge of field operations to delineate on the ground where the firebreaks would be constructed. The possibility of setting backfires inside the new firebreaks (a common practice in fire-dependent ecosystems) in order to prevent slop-overs (i.e. fire spreading across breaks) and release crews for remobilization elsewhere was considered and then quickly abandoned because backfires would require burning of intact MCF, thereby potentially increasing the total loss of forest cover and wildlife. Despite the greater risk involved, holding down the firebreaks with hand crews allowed the possibility that the fires would either be extinguished by rains or die out on their own, thereby saving more of the MCF ecosystem. Most of these firebreaks proved successful in confining the fires, but in some cases deeply burning ground-fires crossed beneath the breaks, eventually

Figure 69.2. Large expanse of elfin cloud forest on karst topography affected by crown and subterranean fires in 1998 (photograph taken 2 years after the fires) in the Chimalapas region of Oaxaca, Mexico (photograph by H. Asbjornsen). (See also color plate.)

igniting new surface or crown fires on the other side. In other areas the fires never reached the firebreaks and were eventually extinguished by the rains that began on 19 July 1998, leaving unburned areas of intact forest.

A total of 78 fires occurred during the 1998 fire season, of which 68 were actively burning simultaneously. Five of these fires were completely inaccessible to fire suppression efforts due to their remote location, the excessive smoke and ash, and strong winds, all of which complicated access by helicopters and firefighters. These fires were eventually extinguished by the rains. According to evaluations of satellite images, the 1998 fires affected a total of 210 565 ha, equivalent to 35% of the total area of the Chimalapas region, including 126 067 ha of MCF and evergreen rain forest and 84 498 ha of secondary forest (Figure 69.1). At its peak, the fire suppression effort involved over 600 military recruits, 122 federal firefighters, over 1100, local village firefighters, and approximately 20 firefighters from the U.S.D.A. Forest Service. No casualties were incurred among firefighters or other participants, although many firefighters suffered health effects from severe smoke inhalation. Unfortunately, no in-depth analysis of fire burning patterns in relation to fire suppression tactics has been implemented, precluding a more detailed assessment of the success of the 1998 effort.

Fighting fire in the cloud forests: organizational aspects

Once it became clear that the environmental conditions in 1998 were exceptional and that the wildfires in the forests were spreading and unlikely to self-extinguish, communities began actively to mobilize to participate in the fire suppression effort. Initially, only a few villages located in the eastern part of the Chimalapas and in close proximity to the core area joined the effort. Notably, these villages had participated in a capacity building program in fire management during the previous 2-year period organized by the NGO *Maderas del Pueblo Sureste*,[2] and therefore were already trained in basic firefighting techniques and were sensitive to the extreme fire situation. However, as the magnitude of the fire incident escalated, eventually most villages within the Chimalapas joined the effort. Following a village-level decision to implement a *tequio* to support fire suppression activities, its members were obligated to participate under communal law, rendering this a highly powerful system of collective action in response to a perceived immediate threat (see above).

Perhaps one of the most critical issues influencing the effectiveness of the overall mobilization effort was the remarkable success in recruiting participation from the surrounding villages in the region. Many of these participating villages did not consider the burning areas to be part of their jurisdiction and responsibility. Further, several villages had been involved in a complex 40-year socio-political conflict having roots in agrarian disputes over inter-state and inter-village territorial boundaries. For example, one of the villages closest to the fire, Díaz Ordáz, was inhabited by Chiapaneco farmers who claimed to be living on land owned by Chiapas, despite legal records indicating that the land belonged to the State of Oaxaca. Over recent years, this situation had led to frequent escalations of conflict and violence, both between neighboring villages and between the state governments of Oaxaca and Chiapas. Strong socio-political tensions also existed among individual communities and between some communities and the municipality. In particular, communities located in the more remote and marginalized eastern portion of San Miguel Chimalapas (nearest to the 1998 fires) had a long-standing resentment toward the municipality stemming from a perceived lack of financial and political support from the municipality. Historically over the past several decades, these differing opinions and interests resulted in hostility manifested through road blockages, kidnappings, and in some cases, even killings.

Considering the history of conflict and the socio-politically volatile situation at the time of the 1998 fires, it is especially noteworthy that active participation in the fire suppression effort

[2] *Maderas del Pueblo Sureste* had perceived the potential threat of forest fires when it initiated its community development and natural resource management program in the region in 1988. The establishment of a fire management training and education program in the eastern Chimalapas villages was highly fortuitous in that it provided a strong baseline from which the communities were able quickly to perceive and respond to the 1998 situation. Additionally, the Incident Commander on the 1998 fire incident was recruited from the *Maderas del Pueblo Sureste* staff, as he had served as the director of their fire program and had gained valuable experience of the local conditions.

occurred among most of the villages in the region. In fact, the 1998 fire event was by some accounts the only incident in which villages in the Chimalapas region and in the neighboring state of Chiapas (including Díaz Ordáz and several other nearby Chiapaneco villages) united together to confront a common threat. Thus, collective action in the 1998 fire suppression effort functioned as a socially and culturally unifying force against a common threat, reinforced by strong social institutions designed to promote the common good. The galvanizing effect that a common threat can have on a group of people has been widely documented in the natural disaster social science literature (e.g. Carroll *et al.*, 2005). However, the situation with fire events can be particularly complex, because the causes of wildfires are typically more ambiguous than for other natural disasters, often deriving from a mixture of natural and human origins (Carroll *et al.*, 2005). In the case of the Chimalapas, although many of the fires were attributed to land clearing operations, the general perception was that to some extent everyone shared in the responsibility, since no one recognized the extreme vulnerability of the forests to fire during the abnormally dry conditions during the ENSO year – until it was too late. Similarly, there was a shared feeling that the extreme climatic conditions played an overriding effect in promoting the fires – that therefore no single person or group of people could be blamed. However, despite this initial unifying effect of the 1998 fires amongst Chimalapas communities during the actual suppression effort, the longer-term effects of the fires on post-fire community relations was more mixed, as will be shown below.

The emerging perception of fire as a common threat in 1998 and consequently strong incentives to protect communal forests from fire likely related to several factors. Many communities obtain essential resources from the forests, including wood for charcoal production, non-timber products (particularly the *Chamaedorea* palm, which is sold in regional and international markets), wildlife (hunted for local consumption), and high-quality timber (which is either sold directly or processed in community sawmills into furniture that can be sold at higher prices). Additionally, communal forests are culturally important, as they are often associated with a collective sense of community pride, prestige, and status, while MCFs in particular have a special place in traditional myths, stories, poetry, and spiritual and religious practices (J. Castañeda, personal communication). Further, because of the value of forests to community livelihoods and cultural traditions, protecting forests from fire may also have significant symbolic meaning and importance.

A second major organizational challenge faced by the fire suppression effort in 1998 involved how to organize such an immense effort in an area where there had been no previous experience with wildfires, and where knowledge of the local conditions be external agencies was extremely limited. Moreover, only a relatively small group of local hunters and *Chamaedorea* palm collectors were familiar with the remote MCF. This situation led to another major challenge: not only to prevent injuries and fatalities among firefighters, but also to ensure that they did not become lost or disoriented in the extensive area of unknown territory with poor visibility and limited access by trails or roads. To address these challenges, the strategic application of local resources and knowledge proved invaluable. When the 600 Mexican soldiers arrived to assist with the effort, five volunteer firefighters from local villages were assigned to each brigade of 30 soldiers prior to entering the field. The same approach was used when brigades from the U.S. Forest Service were later deployed on the incident. These local villagers served as expert guides in unknown territory. They were also well adapted to the local climatic and biophysical conditions and adept at finding water and food in remote mountain locations. Thus, professional and local firefighters worked effectively side by side during the 1998 fire suppression effort despite the different management structures and cultures of the institutions to which they belonged. The massive effort by local villagers also included critical contributions by women, who prepared large amounts of food and in some cases walked long distances to deliver food, water, and other materials to the firefighters in the field.

In summary, the 1998 ENSO fire suppression effort had a strong galvanizing effect on the communities within the Chimalapas region. Because the fires were perceived as a common threat, they provided a mechanism that enabled people to put aside their differences and mobilize efforts to confront the situation as a unified front. Additionally, the fires served to strengthen awareness of the value of the forests, as this was the first time in recent history that the Chimalapas communities had experienced the existence of their cloud forests directly threatened by a natural disaster. Similarly, at the state and national levels – where a strong sense of pride prevails over the ecologically important and globally recognized Chimalapas MCF – concerns over the potential threats from fire were enhanced.

POST-1998 FIRE SITUATION: PROGRESS AND CHALLENGES

Changing perceptions and attitudes about fire as an ecological threat

The fires of 1998 significantly affected people's perceptions about the risk of fires in MCF. Fire was suddenly viewed as a real danger among local and professional fire managers as well as the general public, and the belief that these moist forest ecosystems were inflammable could no longer be upheld. Fire threatened tangible resources – products and services obtained from the forests, as well as more abstract but nevertheless important social values and beliefs – such as local autonomy over decision-making processes and self-sufficiency in the face of sudden and unexpected change.

The 1998 fires resulted in the immediate loss of large forested areas important for non-timber forest products and hunting of wildlife. Palm collectors, which included up to 80% of the male population in some Chimalapas villages, experienced sharp decreases in the availability of palm plants following the 1998 fires. Many of these people were forced to migrate or find other sources of income. Similarly, almost all members of villages in close proximity to the fires relied to some extent on hunting of wild animals as an important food source, and the diminished supply of this resource affected local diets and economies. Six years after the fires palm-growing regions had still not fully recovered, while areas that experienced high-severity (crown and ground) fires supported predominantly early-successional species and few mature canopy species, suggesting extremely slow recovery rates (Asbjornsen *et al.*, 2005; Gallardo-Hernández *et al.*, 2006). Further, according to numerous testimonies by local people, during the initial post-fire years, water availability (from local streams and rivers) was severely reduced in the dry season while flooding was more frequent and severe during the wet season (Gutierrez-Montes, 2005). These perceived effects on water resources were reported by communities located in both the highlands and in the lower-lying regions near the Pacific Coast. Water scarcity combined with loss of the local labor force due to diversion to firefighting efforts also resulted in lower agricultural yields. Along with these experiences came a heightened recognition within the communities of the relationship between forests and water and thus a growing interest in forest conservation and fire prevention (Gutierrez-Montes, 2005). Although no hydrological studies have been conducted in the region, continued alterations in streamflow are likely, since replacement of mature tall forest vegetation by short-statured but fast-growing secondary growth can lead to changes in rainfall and cloud water interception, rainfall infiltration into the soil, and plant water uptake (Giambelluca 2002; Bruijnzeel, 2004; Gomez-Cárdenas, 2009).

Following the 1998 fires, local communities also felt threatened in their sovereignty over their natural resources as a result of increasing pressures and interventions from external institutions, all seeking to participate in various ways in post-fire forest protection and restoration efforts. These perceived external pressures came from diverse groups – government agencies interested in establishing a biosphere reserve and productive projects, researchers interested in studying the forests, and non-governmental organizations offering a range of rural development projects. Amidst these growing pressures, communities were simultaneously suffering the effects of the fires, through losses in agricultural production, declining water supplies, as well as the psychological and physical stress caused by the fires themselves. Although some programs, projects, and studies were initiated during the immediate post-fire years, local perceptions that these interventions threatened their autonomy, compounded with a long history of unfavorable experiences with external groups, led to growing suspicion, fear, and eventual rejection of most of these interventions (Gutierrez-Montes, 2005). Similar responses of local communities to external groups in Mexico have been documented elsewhere (Russell, 1996; Johnson and Nelson, 2004). In the Chimalapas, the post-fire process of disintegration of both internal social structures and relations with external groups has been described as a "downward spiral" (feedback) response (Gutierrez-Montes, 2005). This response was first initiated following the direct threat (by fire) to the communities' natural resource base, but with time affected many other aspects of community life – including their social, political, human, cultural, financial, and built capitals (Gutierrez-Montes, 2005). The rapid deployment of a government-based fire management program in the region further reinforced the growing separation between local communities and the external world (see details below). Although there was also some evidence for resilience and empowerment through communities' capacities to apply their experience towards creating stronger social institutions (for example, the formation of a local NGO which led a community initiative to negotiate government support for community fire brigades), the overwhelming response during the first five post-fire years was increasing community isolation and livelihood insecurity (Gutierrez-Montes, 2005).

Technical capacity for fire prevention, detection, and suppression

By increasing global and national awareness of the risks of fire in MCF ecosystems, the 1998 fires sparked a greater commitment to preventing future disasters in the Chimalapas. This commitment was manifested in significant changes in the state and federal fire programs and improvements in technical capacities for preventing, detecting, and suppressing wildfires in these ecosystems. In 1999, the state government of Oaxaca created a forest fire program specifically to look after the Chimalapas region, consisting of two fire brigades that contracted local community members for 6 months each year. A similar action was taken by the federal government, which in addition to creating one more brigade (a unionized institution contracting at the national level), also provided a helicopter for use on larger fires. Along with these official brigades came enhanced access to firefighting equipment and clothing, vehicles, and communication networks. Although modest in terms of their capacity to adequately attend to the more than 600 000 ha in the region, the creation of these government-supported brigades significantly strengthened official involvement in fire suppression activities.

Capacity for early fire detection was greatly enhanced through the establishment of an extensive monitoring system that included daily revision of satellite "hot spots" obtained from NOAA, and further supported by a network of fire towers, a small plane available to conduct flyovers to verify fire activity and, if necessary, a helicopter to rapidly deploy firefighter brigades for initial attack strategies. These fire detection systems were maintained by Mexico's National Commission for Forestry (CONAFOR), the agency responsible for operational aspects related to fire management. More extensive training of firefighters in cartography, GPS, use of firefighting equipment, and firefighting tactics and strategies was also initiated. Based on knowledge gained during the 1998 fire experience, the firefighting toolbox was expanded to include more chainsaws, axes, shovels, and the specialized firefighting tool, *pulaski* – which has a flat sharp blade that is particularly effective at creating firebreaks in deep organic soils and large amounts of woody debris. Other advancements in fire management capability included the identification of high-risk areas and establishment of annually maintained permanent firebreaks, the creation of helicopter landing pads in remote areas, and an organizational system for rapidly deploying fire brigades. The success of these efforts was highlighted by the awarding of the "Diploma and Public Recognition for Excellence" to the San Antonio Community Brigade in 2003, honoring it as one of the best firefighting brigades state-wide. It is also noteworthy to mention that the lack of large-scale, high-intensity fire incidents in the years following 1998, despite the occurrence of severe droughts during this period, may at least in part be attributed to improved detection and rapid response systems in the region (Asbjornsen and Wickel, 2009).

Fire management organization and community participation

Prior to 1998, when wildfires occurred (largely in pine forests), communities relied almost exclusively on their communally-based *tequio* system for suppressing wildfires. However, the post-1998 expanded presence and deployment of professional firefighting brigades had a tremendous impact on how local communities perceived firefighting activities and their participation in those efforts. No longer was firefighting viewed as a communal responsibility, but rather, it had become a job for which professional firefighters were contracted under employment arrangements that were often unavailable to local people. This situation weakened the traditional *tequio* system of firefighting, as community members expected monetary compensation beyond the conventional system of providing *in kind* support in the form of food, clothing, tools, and medical assistance (Gutierrez-Montes, 2005). Jealousies and resentments were also fueled, especially between individuals and entire villages invited to participate in the official brigades and those having desires to participate but not being given the opportunity. Similar social and political conflicts were observed by Johnson and Nelson (2004) for communities in Chiapas, when boundaries of the government-purchased and administered protected area overlapped with communal territories, requiring shared – though not necessarily coordinated – participation in monitoring and enforcement of activities within the "common" area between the common property and the external government systems.

Juxtaposed with this conflictive situation, the response of the five communities in eastern Chimalapas (nearest to the core area) was to strongly direct efforts toward strengthening their organizational and technical capacities to actively engage in fire management activities. This region adopted a modern system of fire suppression, while retaining some traditional organization structures, thereby effectively blending modern and traditional approaches. Under this system, response to a wildfire incident comprises three stages. Stage I entails initial detection, in which a small village brigade is sent out to investigate observations of smoke, and if the fire is sufficiently small in size, the brigade engages in direct attack to extinguish it. Stage II occurs when the fire is more distant, larger, or burning with high intensity, in which case the villages located nearest to the fire have primary responsibility to control the fire. Stage III, which applies in cases of very large complex fires that have several active fronts and are often located in remote mountain areas, involves solicitation of all surrounding villages to participate in the fire suppression efforts and additional assistance is often requested from the government. Villages in the eastern Chimalapas also began promoting specific practices and procedures for conducting prescribed burns in order to reduce fire risk, for example, maintaining firebreaks, considering topography and climatic conditions, informing neighbors when planning a burn, assisting neighbors with burning operations, and establishing rules for compensating neighbors if losses are incurred.

As a result of these organizational changes leading to more formalized firefighting activities in the Chimalapas, fire suppression efforts after 1998 reflect a newly emerging division of responsibilities between local and external agents. Whereas wildfires occurring in the vicinity of local populations continue to be handled at the village-level through the traditional *tequio* system (especially in the eastern Chimalapas), fires occurring in more remote and inaccessible locations are generally addressed exclusively by the government-supported professional brigades. The establishment of both external and local organizational structures for responding to fire incidents suggests a potential transition toward a more integrated system of shared responsibilities for fire management in the region. However, the specific guidelines for how responsibilities will be shared and under what terms have not been explicitly formulated and agreed upon by all participants. In particular, some local groups would prefer greater local participation in fire prevention and suppression efforts, based on a system of direct compensation or contractual mechanisms between the government and community authorities. Interestingly, similar proposals to integrate community participation into forest fire management are emerging in other regions of the world (e.g. Goldammer *et al.*, 2002). For example, in Chiapas, community residents claimed that more effective monitoring and enforcement and thus, more successful conservation of forest lands, could be conducted by paying communities to undertake this activity (Johnson and Nelson, 2004). A critical element in creating a joint fire management system in the Chimalapas will be to avoid the potential for weakening of local institutions as a result of increased intervention and oversight by external institutions, while at the same maximizing opportunities for complementarity between external and community authorities. Additionally, important logistical constraints to joint management systems also need to be addressed, such as the usual desire on the part of communities to have a rotating fire crew, which poses unique challenges with respect to training and assignment of responsibilities (E. Alvarado, personal communication).

CONCLUDING REMARKS

The 1998 experience with fire in MCF in the Chimalapas generated valuable information about wildfire behavior and effective firefighting techniques in these unique ecosystems for which little previous knowledge existed. A salient feature of the 1998 fire suppression effort was the critical role played by local communities, which reflected the capacity of the traditional community-based system of collective action to effectively mobilize to confront a common threat, irrespective of underlying differences or conflicts. However, increasing post-1998 control by external institutions over fire management operations could compromise local capacities to respond to future fire threats if traditional social structures dictating common property management and collective action are weakened. Alternatively, this situation also opens up unique opportunities to merge the practical knowledge, experience, and mobilizing capacity of local communities with the financial, technical, and organizational resources available through government fire programs into a joint system with shared responsibilities and mutual agreement.

It is proposed that the explicit integration of traditional community-based management systems and government agency programs into joint fire management institutions offers the most effective solution to achieving the common goal of fire prevention and forest protection in the Chimalapas region. Specific actions that could promote such coordination include training of a broad base of villagers in basic firefighting techniques; equipping villages near vulnerable areas with fire suppression tools, equipment, and communication networks; providing compensation to villages participating in a wide range of fire management activities (e.g. creation and maintenance of firebreaks, manning fire towers, fire suppression, etc.); formulation of agreements and enforcement systems for regulating prescribed burning activities; providing positive incentives to communities with low occurrence of wildfires; promoting natural resource management systems that do not require burning (e.g. timber and non-timber forest products); and preventing establishment of roads within the core forest area.

To date, the Chimalapas region has not been confronted with a large-scale fire incident similar to the 1998 situation (Asbjornsen and Wickel, 2009), and therefore the influence of post-1998 changes in technical capacity and organizational structure on the effectiveness of fire suppression efforts remains to be seen. On the other hand, the absence of major wildfires may also reflect, in part, the success of enhanced fire detection and initial response systems established after 1998. Although more in-depth analysis of the complex social, political, and technical processes underlying fire prevention and control is needed, it is suggested that in the long term, efforts aimed at promoting partnerships and co-management frameworks between local communities and external institutions will likely be the most effective at achieving common goals of MCF conservation.

ACKNOWLEDGEMENTS

The authors wish to acknowledge Javier Castañeda, Isabel Gutierrez-Montes, Vance Russell, and Jan Wolf for their valuable suggestions on earlier versions of this manuscript, and the many farmers in the Chimalapas region and representatives from government agencies in Oaxaca who shared with us their personal experiences during the 1998 cloud forest fires.

REFERENCES

Anta, S. F., and A. B. Plancarte (2001). Forest fires. In *Chimalapas: The Last Opportunity*, eds. S. F. Anta, and A. B. Plancarte, pp. 121–138. Oaxaca City, Mexico: El Castor Publishers.

Asbjornsen, H., and B. Wickel (2009). Changing fire regimes in tropical montane cloud forests: a global synthesis. In *Fire and Tropical Ecosystems*, ed. M. A. Cochrane, pp. 607–622. New York: Springer-Verlag.

Asbjornsen, H., C. Gallardo-Hernández, N. Velázquez-Rosas, and R. García-Soriano (2005). Deep ground fires cause massive above- and below-ground biomass losses in tropical montane cloud forests in Oaxaca, Mexico. *Journal of Tropical Ecology* **21**: 427–434.

Barthlott, W., J. Muthke, M. D. Rafiqpoor, G. Kier, and H. Kreft (2005). Global centres of vascular plant diversity. *Nova Acta Leopoldina N.F.* **92**, 342: 61–83.

Behling, H. (1997). Late Quaternary vegetation, climate and fire history from the tropical mountain region of Morro de Itapeva, SE Brazil. *Palaeogeography, Palaeoclimatology and Palaeoecology* **129**: 407–422.

Breedlove, D. E. (1981). Introduction to the flora of Chiapas, Part I. In *Flora of Chiapas*, ed. D. E. Breedlove, pp. 1–35. San Francisco, CA: California Academy of Sciences.

Bruijnzeel, L. A. (2004). Hydrological functions of tropical forests: not seeing the soil for the trees? *Agriculture, Ecosystems and Environment* **104**: 185–228.

Bush, M. B., M. R. Silman, and D. H. Urrego (2004). 48 000 years of climate and forest change in a biodiversity hot spot. *Science* **303**: 827–829.

Carroll, M. S., P. J. Cohn, D. N. Seesholtz, and L. L. Higgins (2005). Fire as a galvanizing and fragmenting influence on communities: the case of the Rodeo-Chediski fire. *Society and Natural Resources* **18**: 301–320.

Cayuela, L., D. J. Golicher, and J. M. Rey-Benayas (2006). The extent, distribution and fragmentation of vanishing montane cloud forest in the highlands of Chiapas, Mexico. *Biotropica* **38**: 544–554.

Cochrane, M. A., A. Alencar, M. D. Schulze, *et al.* (1999). Positive feedbacks in the fire dynamics of closed canopy tropical forest. *Science* **284**: 1832–1835.

Colinvaux, P. A., M. B. Bush, M. Steinitz-Kannan, and M. C. Miller (1997). Glacial and postglacial records from the Ecuadorian Andes and Amazon. *Quaternary Research* **48**: 69–78.

Conserva, M. E., and R. Byrne (2002). Late Holocene vegetation change in the Sierra Madre Oriental of central Mexico. *Quaternary Research* **58**: 122–129.

Dull, R. A. (2004). An 8000-year record of vegetation, climate, and human disturbance from the Sierra de Apaneca, El Salvador. *Quaternary Research* **61**: 159–167.

Fulé, P., and W. Covington (1997). Fire regimes and forest structure in the Sierra Madre Occidental, Durango, Mexico. *Acta Botanica Mexicana* **41**: 43–79.

Gallardo-Hernández, C., N. Velázquez-Rosas, and H. Asbjornsen (2006). Floristic and structural changes following severe fires in tropical montane cloud forests in the Chimalapas mountains. In *Ecology, Management and Conservation of Mountain Ecosystems in Mexico*, ed. L. R. Sanchez-Velasquez, pp. 167–183. Xalapa, Mexico: University of Veracruz.

Giambelluca, T. W. (2002). The hydrology of altered forests. *Hydrological Processes* **16**: 1665–1669.

Goldammer, J. G. (1999). *Forests on fire. Science* **284**: 1782–1785.

Goldammer, J. G., P. G. H. Frost, M. Jurvélius, *et al.* (2002). Community participation in integrated forest fire management: experiences from Africa, Asia and Europe. In *Communities in Flames, International Conference on Community Involvement in Fire Management*, eds. P. Moore, D. Ganz, L. Cheng Tan, T. Enters, and P. B. Durst, pp. 33–52. Bangkok, Thailand: Food and Agriculture Organization of the United Nations Regional Office for Asia and the Pacific.

Gomez-Cárdenas, M. (2009). Transpiration by contrasting vegetative cover types in the tropical montane cloud forest zone of Eastern Mexico. Ph.D. thesis, Iowa State University, Ames, IA, USA.

Gutierrez-Montes, I. (2005). Healthy communities equal healthy ecosystems? Evolution (and breakdown) of a participatory ecological research project towards a community natural resource management process, San Miguel Chimalapas (Mexico). Ph.D. thesis, Iowa State University, Ames, IA, USA.

Hemp, A. (2005a). The impact of fire on diversity, structure and composition of Mt. Kilimanjaro's vegetation. In *Land use Changes and Mountain Biodiversity*, eds. E. Spehn, M. Liberman and C. Körner, pp. 51–68. Boca Raton, FL: CRC Press.

Hemp, A. (2005b). Climate change driven forest fires marginalize the impact of ice cap wasting on Kilimanjaro. *Global Change Biology* **11**: 1013–1023.

Horn, S. P., K. H. Orvis, L. M. Kennedy, and G. M. Clark (2000). Prehistoric fires in the highlands of the Dominican Republic: evidence from charcoal in soils and and sediments. *Caribbean Journal of Science* **36**: 10–18.

IPCC (2001). *Climate Change 2001: Impacts, Adaptations, and Vulnerability. Contribution of Working Group II to the Third Assessment Report of the Intergovernmental Panel on Climate Change. Cambridge, UK: Cambridge University Press.*

Jardel, E. J. P., A. L. Ezcurra, C. Santiago, C. Cortés, and M. Ramírez (2001a). Succession in pine–oak forests and montane cloud forests in the Manantlán mountain range. In *Conference Proceedings, Vth Mexican congress on forest resources*, Guadalajara, Jalisco, Mexico.

Jardel, E. J. P., R. V. Ramírez, A. A. Saldana, *et al.* (2001b). Fire management and ecological restoration in the biosphere reserve Manantlán. In *Conference Proceedings, Vth Mexican Congress on Forest Resources*, Guadalajara, Jalisco, Mexico.

Johnson, K. A., and K. C. Nelson (2004). Common property and conservation: the potential for effective communal forest management within a national park in Mexico. *Human Ecology* **32**: 703–733.

Kappelle, M. (1993). Recovery following clearing of an upper montane *Quercus* forest in Costa Rica. *Revista Biológica Tropical* **41**: 47–56.

Karmalkar, A. V., R. S. Bradley, and H. F. Diaz (2008). Climate change scenario for Costa Rican montane forests. *Geophysical Research Letters* **25**, L11702, doi:10.1029/2008GL033940.

Klooster, D. (2000). Institutional choice, community, and struggle: a case study of forest co-management in Mexico. *World Development* **28**: 1–20.

Loope, L. L., and T. W. Giambelluca (1998). Vulnerability of island tropical montane cloud forests to climate change, with special reference to East Maui, Hawaii. *Climatic Change* **39**: 503–517.

Martin, P. H., and T. J. Fahey (2006). Fire history along environmental gradients in the subtropical pine forests of the Cordillera Central, Dominican Republic. *Journal of Tropical Ecology* **22**: 289–302.

Mathews, A. S. (2003). Suppressing fire and memory: environmental degradation and political restoration in the Sierra Juárez of Oaxaca 1887–2001. *Environmental History* **8**: 77–108.

Meyers, R. (2006). *Living with Fire: Sustaining Ecosystems and Livelihoods through Integrated Fire Management.* Tallahassee, FL: The Nature Conservancy.

Pyne, S. J., P. L. Andrews, and R. D. Laven (1996). *Introduction to Wildland Fire.* New York: John Wiley.

Quintana-Ascencio, P. F., M. González-Espinosa, N. Ramírez-Marcial, G. Domínguez-Vázquez, and M. Martínez-Icó (1996). Soil seed bank and regeneration of tropical rain forest in milpa fields at Selva Lacandona, Chiapas, Mexico. *Biotropica* **28**: 192–209.

Richter, M. (2008). Tropical mountain forests: distribution and general features. In *The Tropical Mountain Forest: Patterns and Processes in a Biodiversity Hotspot*, eds. S. R. Gradstein, J. Homeier, and D. Gansert, pp. 7–24. Göttingen, Germany: Göttingen Centre for Biodiversity and Ecology.

Rodríguez-Trejo, D. A. (1996). *Incendios forestales.* México, DF: Universidad Autónoma Chapingo/Mundiprensa.

Román-Cuesta, R. M., and J. Martinez-Vilalta (2006). Effectiveness of protected areas in mitigating fire within their boundaries: case study of Chiapas, Mexico. *Conservation Biology* **20**: 1074–1086.

Román-Cuesta, R. M., M. Gracía, and J. Retana (2003). Environmental and human factors influencing fire trends in ENSO and non-ENSO years in tropical Mexico. *Ecological Applications* **13**: 1177–1192.

Román-Cuesta, R. M., J. Retana, and M. García (2004). Fire trends in tropical Mexico: a case study of Chiapas. *Journal of Forestry* **102**: 26–32.

Russell, V. A. (1996). The Chimalapas Ecological Campesino Reserve: the golden gourd of conflict and its role in protected area management. Masters thesis, Cornell University, Ithaca, NY, USA.

Salas-Morales, S. H., L. Schiblim, and E. Torres-Bahena (2001). La importancia ecológica y biológica. In *Chimalapas: La última oportunidad*, eds. S. F. Anta and A. B. Plancarte, pp. 27–47. Oaxaca City, Mexico: World Wildlife Fund Programa México y la Secretaría de Medio Ambiente, Recursos Naturales y Pesca.

Sarmiento, F. O., and L. M. Frolich (2002). Andean cloud forest tree lines: human drivers of landscape change – treeline dynamics in neotropical montology. *Mountain Research and Development* **22**: 278–287.

Sherman, R. E., P. H. Martin, and T. J. Fahey (2005). Vegetation–environment relationships in forest ecosystems of the Cordillera Central, Dominican Republic. *Journal of the Torrey Botanical Society* **132**: 293–310.

Suyanto, S., P. P. Rizki, N. Khususiyah, and L. Joshi (2005). Land tenure, agroforestry adoption, and reduction of fire hazard in a forest zone: a case study from Lampung, Sumatra, Indonesia. *Agroforestry Systems* **65**: 1–11.

Tucker, C. M. (1999). Private versus common property forests: forest conditions and tenure in a Honduran community. *Human Ecology* **27**: 201–230.

Urrego, D. H., M. R. Silman, and M. B. Bush (2005). The Last Glacial Maximum: stability and change in a western Amazonian cloud forest. *Journal of Quaternary Science* **20**: 693–701.

Wendt, T. (1997). Uxpanapa–Chimalapa Region Mexico: Mexico CPD Site MA2. In *Centres of Plant Diversity*, Vol. 3, *The Americas*, eds. D. S. Davis, V. H. Heywood, O. Herrera-MacBryde, J. Villa-Lobos, and A. C. Hamilton, pp. 130–134. Gland, Switzerland: World Wildlife Fund for Nature and The World Conservation Union.

Weng, C., M. B. Bush, and A. J. Chepstow-Lusty (2004). Holocene changes of Andean alder (*Alnus acuminata*) in highland Ecuador and Peru. *Journal of Quaternary Science* **19**: 685–691.

Young, O. R. (2002). Institutional interplay: the environmental consequences of cross-scale interactions. In *The Drama of the Commons*, eds. E. Ostrom, T. Dietz, and N. Dolsaket, pp. 263–291. Washington, DC: National Academy Press.

70 Assessment needs to support the development of arrangements for Payments for Ecosystem Services from tropical montane cloud forests

S. S. Tognetti
Environmental Science and Policy Consultant, Silver Spring, Maryland, USA

B. Aylward
Ecosystem Economics LLC, Bend, Oregon, USA

L. A. Bruijnzeel
VU University, Amsterdam, the Netherlands

ABSTRACT

The willingness of beneficiaries to pay for "ecosystem services" will ultimately depend on confidence in the effectiveness of actions taken to ensure their continued delivery at the watershed scale. Key aspects of effectiveness include: (i) the integrity of ecosystem functioning or processes that support service provision, (ii) whether the impacts or benefits are economically significant at the relevant (watershed) scale, and (iii) the effectiveness of institutional arrangements needed to ensure access to benefits by those who pay for them. The biophysical characteristics of the relevant watershed/ecosystem also have implications for the types of institutional arrangements that are possible or necessary to control access to services. Given the dynamic nature of ecosystems, and the potentially long lag time between multiple causes and effects over large spatial areas, the outcomes of management actions are inherently variable, uncertain, and site-specific. In the absence of adequate site-specific assessment, Payments for Ecosystem Services (PES) initiatives are often based on inappropriate generalizations or even myth regarding the hydrological impact of certain types of land-cover changes. Cloud forests in particular are generally assumed to increase the supply of water because they capture water from clouds, beyond that provided by precipitation. However, there is a need for further site-specific assessments, to better define and quantify the extent to which cloud forests account for additional supplies of water compared to other land-cover types that might replace them (e.g. pasture) under various climatic conditions. Such assessments can also provide a basis for the improved management and monitoring that is necessary to determine whether and to what extent objectives of PES schemes are actually being achieved, as well as for understanding trade-offs and complementarities among multiple services. This chapter presents a brief review of PES initiatives in tropical montane cloud forest (TMCF) areas, highlighting the types of information needed for decision-making, identifying special issues that pertain specifically to TMCF, and discussing recent research findings.

INTRODUCTION

"Ecosystem services" include the direct and indirect benefits that ecosystems supply to humans, provided that these benefits are economically significant, and that the presumed users actually have access to them (Aylward, 2005). In a watershed context, these benefits range from provision of fresh water for various uses, to regulation of the flows of water and sediment that drive

Tropical Montane Cloud Forests: Science for Conservation and Management, eds. L. A. Bruijnzeel, F. N. Scatena, and L. S. Hamilton. Published by Cambridge University Press.

ecosystem functions and support livelihoods. The amount and quality of the water captured by an ecosystem that is potentially available for direct human use depends on the regulatory capacity of the ecosystem (including the soil). To the extent that this regulatory capacity is altered by human use of the land, it will also depend on the willingness of potential beneficiaries to pay for the costs of changing (or avoiding) certain practices so as to achieve optimal services, or on regulatory measures that impose charges on land users. So-called "Payments for Ecosystem or Environmental Services" (PES) are widely considered to be a powerful tool for ecosystem conservation because payments are, in theory, conditional on the provision of services, and because they allow service providers to recover their costs (Pagiola *et al.*, 2002). In practice, however, there are a number of scientific and institutional challenges encountered in their implementation, which also have costs, and most payments are for adoption of practices rather than for the services themselves. Ultimately, the willingness of potential beneficiaries to pay, and therefore the value of ecosystem services at the watershed scale, will depend on whether people have confidence in the effectiveness of certain management strategies. Given the public-good and common property characteristics of ecosystem services at the watershed scale – which make it difficult to control access to what is becoming an increasingly limited supply – the effectiveness of management strategies depends as much on institutions and governance as it does on the biophysical conditions necessary for service provision (e.g. presence of forest).

Proposals for PES schemes have been considered in the context of financing the protection of tropical montane cloud forests amongst others in Costa Rica (Calvo-Alvarado, 2000; Rodriguez-Zuñiga, 2003; Calvo-Alvarado *et al.*, this volume), Ecuador (Yaguache and Carrion, 2004), Mexico (Muñoz-Piña *et al.*, 2008), and Bolivia (Asquith and Vargas, 2007) whereas many other initiatives are in various stages of development (reviewed by Porras *et al.*, 2008). Cloud forests are typically found in wet, remote steepland areas at high elevations, and although not exclusively, can be distinguished by some common environmental characteristics. These include: the occurrence of frequent low cloud (fog), (usually) high rainfall, (sometimes) high winds, and lower temperatures (Jarvis and Mulligan, this volume), as well as wet and often (but not always) water-logged peaty and acid soils (Roman *et al.*, this volume). As do other forest types, cloud forests provide a range of benefits (cf. Hamilton *et al.*, 1995; Bruijnzeel and Hamilton, 2000), but they are generally thought to have greater direct economic value due to their increased water yield compared to other forests not affected by fog. This is mostly because cloud forests intercept water from clouds in addition to that received from normal rainfall (Zadroga, 1981). Whilst it is true that during dry seasons, even a modest additional amount of streamflow can be significant in some places (Brown *et al.*, 1996; Hemp, 2005), reliable quantitative data on the extent to which intercepted "occult precipitation" (cloud water and wind-driven rain) contributes to groundwater reserves and streamflow in cloud forest areas are exceedingly sparse (Bruijnzeel, 2006; García-Santos, 2007; McJannet *et al.*, 2007; McJannet *et al.*, this volume; cf. Bruijnzeel *et al.*, this volume). Hemp (2005) speculated that, on Mt. Kilimanjaro in Tanzania, there is an annual loss of 20 million m^3 of cloud water that would otherwise have been captured by the 15 000 ha of subalpine cloud forests that were destroyed by wildfires associated with climate change (cf. Hemp, this volume #58). If true, this would represent a significant source of water supply in northern Tanzania, but in the absence of hard data it remains to be seen how much of the intercepted cloud water actually becomes streamflow, and how much is trapped by epiphytes and evaporated again (cf. Schrumpf *et al.*, this volume; Tobón *et al.*, this volume #26).

The key challenge then is to demonstrate and quantify the actual costs and benefits that can be expected as a result of changes in specific land-use practices and institutional arrangement in particular places, such that payments for the ecosystem services rendered by the scheme can be linked to demonstrable outcomes (e.g. a reliable supply of high-quality streamflow). However, given the complexity and natural variability of watershed processes, which are dominated by randomly timed and extreme events distributed over large and – in the case of TMCF – mostly remote areas, it is difficult to link unequivocally all upstream causes and downstream effects on watershed functioning. Therefore, making these inherent uncertainties explicit is also critical to managing buyer expectations, and to maintaining their cooperation in the long term (Aylward, 2005). This is harder to sell than the more common practice of developing watershed policies that may be based on myths, or at best on scientific generalizations that are not always appropriate (Kaimowitz, 2004). However, given that it has a cost, uncertainty is a critical consideration, if the goal is to identify and negotiate an equitable distribution of costs and benefits (Tognetti, 2001). Also, demonstrating links between land-management practices (including e.g. cloud forest conservation) and hydrological outcomes will ultimately require continued monitoring over time, in specific places. Equally important is the capacity to adjust PES initiatives as lessons are learned (Tognetti *et al.*, 2005; Asquith and Wunder, 2008).

It is generally easier to establish links between hydrological behavior and (changes in) land-use practices, in watersheds of 100 km^2 or less (FAO, 2002; Bruijnzeel, 2004). This can provide a basis for linking particular costs and benefits with identifiable user groups, and for establishing more direct contractual arrangements between buyers and sellers of ecosystem services, that reflect unique local conditions. However, given the broader social benefits, along with the range of variability and uncertainty of watershed processes at local to intermediate scales, including underground leakages, a case can also be made for the spreading of risk across a larger population of users, at regional or even national levels (Rodriguez-Zuñiga, 2003; Mulligan and Burke, 2005; Muñoz-Piña *et al.*, 2008). Although it is much more difficult to link land-use practices to hydrological behavior at larger scales, a potential advantage of large-scale initiatives is a larger pool of buyers, as well as greater resources to support assessment activities that can demonstrate aggregate benefits, as seen in the U.S. Conservation Reserve Program (Ruhl *et al.*, 2007). However, at this scale, implementation of payments for performance remains a challenge (Heimlich, 2006).

The challenge for science is to translate scientific research findings into protocols and rules of thumb that can be used by practitioners to guide site-specific assessments of the situation. Ideally, science should also provide comparatively user-friendly models that can be used to anticipate the outcomes of alternative management actions (Mulligan and Burke, 2005; Schellekens, 2006; Thamapakpawin *et al.*, 2006). These kinds of tools are needed to support the type of continuous, cooperative, and integrated process of assessment that is needed to identify and evaluate trade-offs between conflicting uses, to engage and inform stakeholders regarding the distribution of costs and benefits associated with proposed management strategies, and to determine the most cost-effective and acceptable means of achieving conservation objectives.

This chapter provides a brief overview of the different types of ecosystem services at the watershed scale, the various kinds of instruments used in PES initiatives, and the services that are being paid for as well as the kinds of scientific information on which they are based. This provides background for discussion of what is known regarding the role of cloud forests with respect to water provision, and the kinds of information needed to better support the development of PES initiatives in cloud forest areas. The following is based in part on Tognetti *et al.* (2004), an assessment guide prepared for the World Bank Environment Department as part of the World Bank–Netherlands Watershed Partnership Program, as well as on inputs from various participants of the 2004 Waimea Tropical Montane Cloud Forest Symposium and more recent literature on the subject (cf. Porras *et al.*, 2008).

ECOSYSTEM SERVICS AT THE WATERSHED SCALE

Within the framework developed by the Millennium Ecosystem Assessment (MEA, 2003), the various types of ecosystem services can be distinguished as provisioning, regulatory, cultural, and supporting, as outlined in the following examples of ecosystem services associated with watershed processes:

- Provision of fresh water for various kinds of human uses.
- Flow regulation and filtration by the control of rates of surface runoff – at local scales, this can provide a buffer against peak or flood flows, help to maintain base or dry-season flows, allow the recharge of soil water reserves, aquifers, wetlands, and floodplains, and control the level of groundwater tables. Control of surface runoff (overland flow) also controls the transport of any pollutants and sediment carried by the water, and therefore water quality (Van Noordwijk *et al.*, 1998; FAO, 2002; Van Noordwijk, 2004).
- Cultural services associated with recreation, tourism, aesthetic values, indigenous ceremonial uses, livelihoods, and ways of life that depend on natural resources.
- Supporting services which include nutrient cycling and primary production. Combined, the regulatory services take on the characteristics of supporting services in that they maintain natural flow and disturbance regimes that drive all ecosystem processes, thereby maintaining diversity of habitats and ecosystem resilience. River channels, wetlands, riparian habitats, mangroves, estuaries, and coastal zone processes all depend on timing of flows of water and sediment and, in turn, support livelihoods in numerous ways.

Ultimately, ecosystem services are the foundation of human well-being. Drawing directly on the work of Sen (1999), human well-being is defined to include: the abilities to earn a livelihood, to maintain health and good social relations, and to be secure. However, underlying all of these aspects of well-being is a more basic one, to have freedom of choice and action as to how these different kinds of needs are met (MEA, 2003). These choices, and ways of life, are both enabled and constrained by ecosystem conditions. Ecosystem conditions are, in turn, influenced by natural and human-induced drivers. One of the key human-induced drivers are the institutions or "rules of the game" that govern the management of soil, water, and other natural resources, and influence land-use decisions (see Figure 70.1).

Whether particular ecosystem services are provided at a specific location to identifiable stakeholders will depend on the complex interactions between land cover (or land-use practices), climatic factors, and other environmental variables (e.g. topography, soil conditions). For example, surface infiltration rates may remain low in reforested areas, depending on the extent of ground cover, and the degree of soil compaction associated with prior land-use practices (Scott *et al.*, 2005; Ilstedt *et al.*, 2007; Zimmermann and Elsenbeer, 2008). Disproportionate impacts are typically associated with land uses covering relatively small areas, such as paths, roads, and settlements (Ziegler and Giambelluca, 1997; Ziegler *et al.*, 2004; Rijsdijk *et al.*, 2007), or with infrastructural developments, such as large-scale conversion of forests to agriculture, urbanization, or mining operations (Bruijnzeel, 2004). Therefore, site-specific assessments are

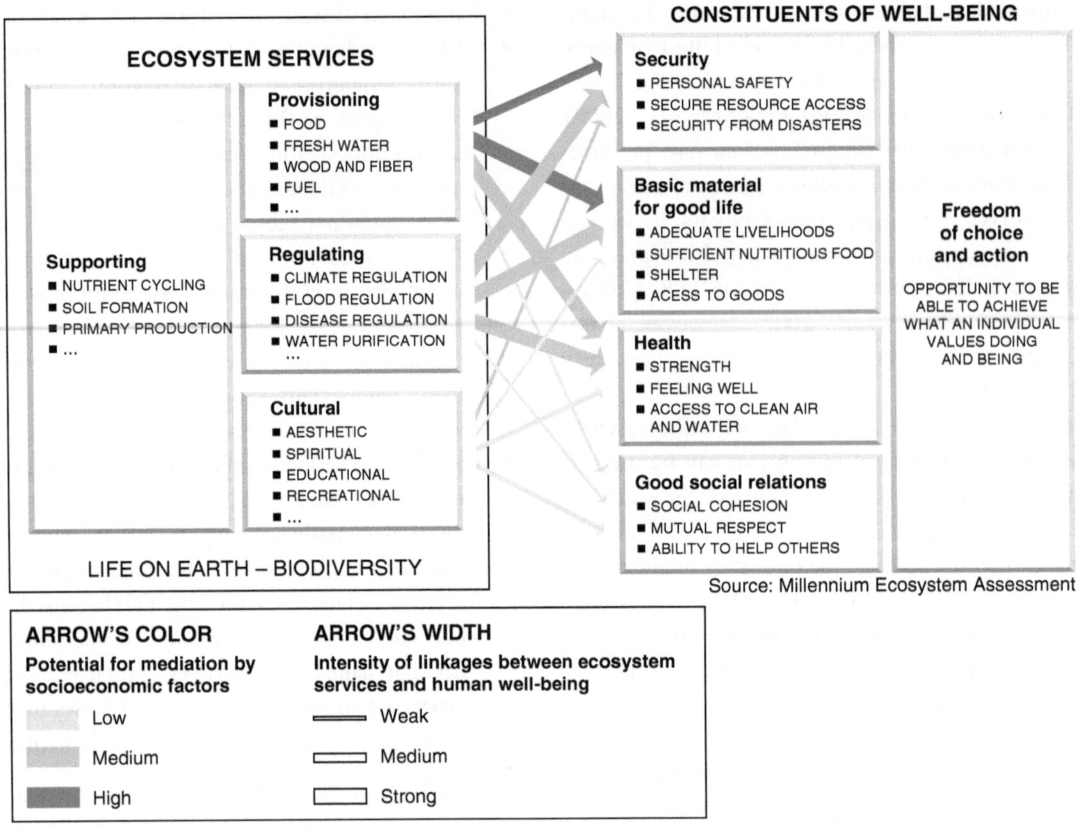

Figure 70.1. Relationship between ecosystem services and human well-being (after MEA, 2003).

necessary to identify the scales at which changes can be detected and whether they have economic significance in a particular social and economic context (Aylward, 2005).

A second key consideration is that these various kinds of services are interdependent, in that there is a trade-off not only between different uses, but also between the different types of services. In other words, the capacity of a watershed to *provide* freshwater for direct uses depends on processes that *regulate* the flow of water, sediment, and pollutants. Therefore, as the supply of any of these services becomes more limiting, human well-being will increasingly depend on achieving an acceptable or optimal balance between these trade-offs (Aylward *et al.*, 2005).

TYPES OF PAYMENT ARRANGEMENTS FOR ECOSYSTEM SERVICES AT THE WATERSHED SCALE

General considerations and definition of terms

Specific arrangements for PES can take various forms, depending on the characteristics of the individual services, the scale of the relevant ecosystem processes by which they are produced, and the socio-economic and institutional context (Tognetti, 2001; Landell-Mills and Porras, 2002; Tognetti *et al.*, 2004; Porras *et al.*, 2008). These forms range from informal, community-based initiatives, to more formal voluntary contractual arrangements (VCAs) between individual parties, and to complex arrangements among multiple parties that are usually facilitated by intermediary organizations. They may also include a mix of market instruments, along with regulatory and policy incentives, that are more likely to become necessary at larger scales, when threats are beyond the response capacities of individual communities.

Compensation to land users is often in the form of transfer payments (TPs) from governments or intermediary organizations, using funds that have been pooled from diverse sources that may include user and licensing fees, taxes, and donations. As a general rule, user and licensing fees are more feasible when it is possible to limit services to those who pay for them, such as domestic and industrial water users. As will be seen in case studies involving cloud forests, compensation is not always in the form of direct payments; co-benefits, such as clarification or establishment of rights of land users in surrounding agricultural frontier areas, may be equally if not more important, and also necessary as a prerequisite to negotiating any form of agreement.

Governments may play different kinds of roles, ranging from the enforcement of contractual agreements, to the creation of regulatory caps, monitoring compliance, contracting with service providers, providing technical assistance, and identification of priority conservation areas as a basis for funding allocation decisions. Some of these roles are often filled by private and non-governmental organizations (NGOs), which may have the advantage of greater flexibility, and the ability to act more expediently. Such NGOs and stakeholder associations may also play a role as advocates on behalf of marginalized stakeholders, so as to create political pressure that may be necessary for governments to recognize rights and respond to the concerns of marginalized groups (cf. Asbjornsen and Garnica-Sánchez, this volume; Hofstede *et al.*, this volume). Fundamental to all of these instruments is the need for a consistent set of criteria and a transparent process for decision-making with respect to the establishment of priorities for the allocation of funds (Tognetti *et al.*, 2005; Asquith and Wunder, 2008; Porras *et al.*, 2008).

As a general rule, benefits tend to be more tangible at smaller scales, where property rights and stakeholders can also be better defined. Smaller scales also make it possible to tailor agreements to unique local conditions. At larger scales, in which it is harder to link causes and effects, and to define rights and responsibilities, there will be a greater need for the involvement of government and/or other intermediaries (cf. Asbjornsen and Garcia-Sánchez, this volume). These groups are needed to facilitate transactions among numerous stakeholders, to support assessment needs, and to establish priorities areas for conservation. At larger scales, given the more diverse range of conditions and actors involved, there will also be a greater need for standardized agreements. An advantage is that they also offer a larger pool of buyers and sellers (Rose, 2002; Porras *et al.*, 2008).

THE USE OF PES TO PROTECT CLOUD FORESTS: CASE STUDIES

Case studies that involve the use of PES in or in the vicinity of cloud forest areas are found primarily in Latin America. In addition to providing examples of the diverse approaches highlighted above, they also illustrate many of the kinds of challenges faced in practice (cf. Hofstede *et al.*, this volume). The cases presented here include: (i) an informal community-based arrangement in Ecuador, (ii) more formal VCAs with varying degrees of complexity – as seen in cases in Costa Rica, Bolivia, Ecuador, Guatemala, and Colombia, and (iii) the national FONAFIFO program in Costa Rica. The latter two categories combine VCAs with TPs. An additional case study from Ecuador illustrates the use of the PES approach as part of a broader integrated watershed management strategy.

At the smaller scale, an example of an informal arrangement is that of the Comuna of Loma Alta, a farming community in western Ecuador. In this case, information about the importance of the forest for water provision, provided by scientists who had been researching the area, led the community to place a greater value on the forest and to establish a community-run forest reserve. Technically speaking, this was not a payment arrangement, because the community already owned the forest under a communal land title. However, there were trade-offs between expected benefits to the community as a whole and benefits to individuals within it who had been using the forested area, and for whom the community was able to provide other opportunities (Becker, 2000; Becker and Ghimire, 2003).

Voluntary contractual arrangements are most straightforward when there is a single buyer and seller, as was the case in an agreement reached between the La Esperanza Hydropower Company and the Monteverde Conservation League (MCL) in northern Costa Rica, the sole owner of the forested area upstream of the company's hydropower plant (Rojas and Aylward, 2002). The payment is essentially meant to assure the hydropower company that the streamflow emanating from the MCL cloud forest will be maintained, or at least will not be subject to changes in land use. These payments are simply based on the value to the company of reducing the uncertainty that would accompany any change in upstream land use (Rojas and Aylward, 2002). In practice, this is unlikely to happen, as the land is permanently under protection as a private reserve. Rather, in this particular case, the hydro-electric company had additional vested interests in accessing land belonging to the MCL to build part of their infrastructure (Rojas and Aylward, 2002).

A small-scale but more formal set of VCAs is represented by an initiative in the Rio Los Negros watershed in south-central Bolivia, having its headwaters in the Amboró National Park, much of which is cloud forest. Key threats are clearing for slash-and-burn agriculture and grazing practices of the inhabitants of the town of Santa Rosa, as occurring on the boundary of the Park, and also by landless and recently arrived colonists. Beginning in 2003, the land-owners in the boundary area who have accepted contractual agreements with Fundación Natura Bolivia received one beehive with a value of US$30 for each 10 ha of forest that they protect per year, or the equivalent in barbed wire to prevent encroachment. In addition, they receive training in apiculture, at an estimated cost of US$35 per participant. The payments and upfront costs are co-financed by international donors and the Municipal Government, on behalf of downstream irrigators (Robertson and Wunder, 2005; see also Asquith and Vargas, 2007; Le Tellier *et al.*, 2009).

As of 2006, 34 land-owners have signed agreements to protect 2073, ha of forest, of which ~1000, ha carry cloud forest. So far, the downstream irrigators themselves have been reluctant to contribute directly to the payments because they do not have confidence that these will result in reduced deforestation. There is also fear that it could lead to charges for irrigation water that they currently receive for free (Robertson and Wunder, 2005). However, irrigators had noted a decline in the water supply, which they attributed both to forest clearance and to the spread of irrigation into the upper parts of the watershed. After being presented with the results of hydrological modeling conducted as part of a series of diagnostic studies supported by the International Institute for Environment and Development (IIED), they formed an association, and are expected to be willing to pay in the future, as they are presented with evidence that the scheme is effective (Asquith and Vargas, 2007; however, see also Le Tellier *et al.*, 2009 who were unable to demonstrate a direct link between forest and streamflow based on (short-term) measurements).[1]

Equally important to the watershed communities are the indirect or co-benefits of increased cooperation and reduction of conflict, evident in the formation of an Association of Beekeepers and four environmental committees, as well as the diversification of livelihoods, and the creation of jobs for the landless, to assist in honey processing (Asquith and Vargas, 2007). It is also clear that a key motivation for the participation of landowners was the perception that contractual agreements for the protection of specific forest areas, documented with GPS mapping, would strengthen their tenure security. Existing property rights in the area are *de facto* based on the occupation and "use" of the land, as demonstrated through clearing of forested land in an area where borderlines with the Park are disputed (Robertson and Wunder, 2005). However, the program also increased conflicts with the landless, and raised the issue of whether the contracts and maps drawn up by Fundación Natura Bolivia were substituting for the lack of an official land titling system (Robertson and Wunder, 2005). On the other hand, initiatives based on local rights reinforce local decentralized authorities, thereby increasing their institutional capacity (Asquith and Vargas, 2007). Decentralization, based on the Bolivian Law of Popular Participation, created the conditions that enabled the Municipality to act because it gave municipalities the responsibility for water supply and sanitation services, as well as management of other natural resources. They also manage their own budgets and have authority for decision-making (Asquith and Vargas, 2007).

Transaction costs in the Los Negros PES initiative were high compared to the amounts provided in actual payments because of the need to build trust in a country where there is a general distrust of "markets" and top–down schemes imposed by outside organizations. However, as noted by Asquith and Vargas (2007), "reciprocal agreements" have a long tradition in the area. In this case, property owners themselves asked for beehives – a form of reciprocal exchange – rather than cash payments because they can potentially have more lasting benefits from the beehives than from cash, which tends to be spent quickly. It should also be noted that the returns from bee-keeping were found to be lower than the typical opportunity costs of forgone land uses (e.g. cropping or cattle ranching), and the approach is therefore unlikely to work in more productive areas. However, the land enrolled in the scheme was located primarily on steep slopes where opportunity costs tend to be low (Robertson and Wunder, 2005). This suggests that the Fundación's strategy could also be applied to help stabilize marginal agricultural frontier zones elsewhere, and so reduce encroachment into more remote cloud forest core areas.

The importance of stabilizing the agricultural frontier zones is seen also in a recently achieved agreement for the Teculután watershed in Guatemala, as part of an initiative of WWF Central America and CARE. This 216-km^2 watershed is one of several that originate in the Sierra de las Minas Biosphere Reserve, which includes important headwater areas with tall pristine cloud forest that receive substantial cloud water input during the dry season (Brown *et al.*, 1996; cf. Holder, 2003). Precipitation exhibits a very steep gradient toward semi-arid conditions in the Motagua valley bordering the Sierra to the south where industrial and agricultural demands for water are high. However, the main problem relates to excess sediment in the stream which increases the costs to the downstream Municipality of Teculután for treating and supplying water. A preliminary survey identified the main source of the sediment to be outside the reserve in the mid-watershed area (Martínez-Tuna and Juarez-Quim, 2008), where land users do not have formal user rights. In this case, formal recognition of user rights to land in regions that border the Reserve was negotiated as part of an agreement between the Municipality and five communities in the mid-watershed area. The communities were granted user rights to a forested area in exchange for their labor to reforest 224 ha with native species, as well as to protect and promote natural regeneration of an additional 2150 ha. In addition, they will receive access to 250 ha of more productive lands in the lower watershed, along with technical assistance in more intensive and sustainable agricultural practices. Over the first four years, half of the upfront costs are to be provided by DANIDA and DGIS – the Danish and Dutch international development agencies (WWF-CARE Guatemala, 2007).

Among the first examples of a formal PES initiative is one in Quito (Ecuador), where agreement was required among a more diverse group of stakeholders that rely on water from two headwater ecological reserves, Cayambe-Coca and Antisana. To facilitate the process, a trust fund, FONAG (*Fondo del Agua*), was established in 2000, which is overseen by a stakeholder board, and allocates pooled funds from water users in Quito, farmers, hydropower companies, industry, and tourism operators as well as international donors (Echavarria, 2002a, 2003). Initial support was provided by The Nature Conservancy, USAID, and

[1] Funding has been acquired in summer 2009 by Fundación Natura Bolivia, UNESCO-IHE (Delft) and VU University Amsterdam for in-depth field research into the hydrological functioning of the (cloud) forests of the area and to evaluate the soil and water impacts of forest conversion to degraded pasture (UNESCO-IHE, 2009).

the Fundación Antisana. As of 2005, the Fund had a total of US $3 million, of which the interest (12% or US$360 000 in 2005), is invested in watershed management programs rather than payments to individuals. Up to 20% may go toward transaction costs (Porras and Neves, 2006).

The Costa Rican nation-wide FONAFIFO program also relies on VCAs, but these are between land-owners and the government, which *transfers* payments to land-owners from different kinds of buyers, to cover the costs of multiple services (Rodriguez-Zuñiga, 2003; Calvo-Alvarado *et al.*, this volume). For example, hydro-electric companies and municipalities may pay for watershed benefits, tourism agencies for landscape beauty, and foreign energy companies may purchase carbon offsets (cf. Pagiola *et al.*, 2002).

In the Mexican government's nationwide PES program (Muñoz-Piña *et al.*, 2008), a design team had initially recommended various actions, including amongst others: (i) setting water use fees at a level that would send a signal to beneficiaries regarding the value of the service, (ii) earmarking a percentage of these fees for payments to service providers, and (iii) targeting forested areas that were most at risk of being converted. Other areas of forest that were eligible for the payment scheme included those in close proximity to a Protected Area, those located in "priority mountain areas," in recharge areas of overexploited aquifers or zones of water scarcity or high flooding risk, on the land of indigenous communities, or simply upstream from potential buyers of services. Additional qualifying factors were forest type, and whether a forest was included in a watershed protection plan. The highest priority was given to the protection of cloud forests, for which higher payments were offered because they were thought to have a greater value for water supply. Payments are made at a fixed rate and for a set period, in support of a transition toward sustainable forest use. The payments are also intended to stimulate the future development of more direct market-based arrangements between providers and consumers of forest ecosystem services at the local watershed level. At the start of the program in 2003, cloud forests constituted 6.6% of the area eligible for payment, and 12% of the areas that actually received payments. In 2003, 11% of the forest for which payments were made were at a high risk of conversion. This figure was more than doubled in the second year (28%) (Muñoz-Piña *et al.*, 2008). Despite initial conflicts with the National Water Commission (because of insufficient funds to cover the costs of water-supply-related infrastructural works) and the political difficulty to increase water use charges to municipalities, the Mexican program was ultimately accepted by all stakeholders on accounts of the adequate political support generated by overlapping interests in environmental protection, poverty reduction, water supply, and forest conservation (Muñoz-Piña *et al.*, 2008).

An arrangement established by irrigation farmers in the Cauca Valley in Colombia, in which the farmers voluntarily pay additional water-use fees to support watershed conservation, also constitutes a system of transfer payments. This is because the farmers themselves do not have the authority to directly fund watershed management activities. Instead, the fees are paid to a governmental entity – the Corporación Autonóma Regional del Valle del Cauca – which uses the funds to implement existing watershed management plans (Echavarría, 2002b).

As in Costa Rica, individual VCAs may also be part of a comprehensive program or plan intended to protect multiple services, which, in addition to those associated with freshwater supply, may include other benefits of forests, such as carbon storage, aesthetic values, and biodiversity maintenance (Calvo-Alvarado, 2000; Rodriguez-Zuñiga, 2003). This approach may also be necessary when watershed benefits alone are not sufficient to offset the opportunity costs of forgone land uses (Landell-Mills and Porras, 2002). The more comprehensive approaches are usually developed over time, as new problems become evident, and as capacity is developed. An example is the city of Cuenca in sub-humid southern Ecuador, where a growing awareness of threats to the water supply due to environmental degradation in the surrounding mountains (cf. Molina *et al.*, 2007) led to the development of a master plan in the early 1980s that included a program of land acquisition beginning in the 1990s. These activities eventually led to a program of Integrated Water Resource Management (IWRM) designed to address multiple threats which included: over-allocation of the available water supply, sedimentation of hydropower plants, water treatment facilities and irrigation canals, and wastewater pollution. This program was carried out through a variety of activities that included scientific assessment needed to identify response options, improvement of monitoring capacity, and training and technical assistance with the development and implementation of management plans (Echavarría *et al.*, 2004). The Machangara Watershed Council was created to provide a legal framework for participation of diverse stakeholder groups. Under a nationwide decentralization initiative, the Municipality was also given a concession to manage the Cajas National Park, which contains 230 lakes and is a key source of water to the city. As part of a plan for expansion of the water supply, there is also interest in developing a payment program to address threats from cattle ranching and agriculture (cf. Molina *et al.*, 2007; Zimmermann and Elsenbeer, 2008), for which the opportunity costs of changing management practices are high (Echavarría *et al.*, 2004). Activities implemented by the Machangara Council include pasture improvement and soil conservation programs. Rather than direct payments, farmers in the mid-watershed are given training, technical assistance, and loans to improve water use efficiency by switching to drip irrigation. As they are paid back, loans are made available to other farmers (Porras and Neves, 2006).

CHALLENGES TO IMPLEMENTATION OF PES

Experience to date with the use of market-based instruments to cover the cost of ecosystem services at the watershed scale suggests that little is known about the effectiveness of these initiatives for actually delivering the services. This is in part because there is a time lag between changes in land-use management practices and their hydrological outcomes downstream, particularly at larger scales. This makes it difficult not only to isolate links between individual causes and effects in terms of changes in water yield (FAO, 2002; Bruijnzeel, 2004; Van Noordwijk, 2004; Le Tellier *et al.*, 2009), but also to design institutional arrangements that can provide assurance of equitable access to any benefits (whether perceived or real). Therefore, one of the key challenges to implementation of PES is to question the underlying assumptions (or even myths) that often make up for a lack of data, and to demonstrate actual benefits. Equally important is to recognize the limits to quantification, and to manage expectations of stakeholders by making the uncertainties explicit, and considering them in the design of PES schemes.

Dispelling myths about forest–water relationships

In the absence of factual local, regional, or national assessments of forest–water relationships that demonstrate the provision of specific services (notably amounts and regulation of streamflow), PES arrangements are often developed based on assumptions or even myths (cf. Hamilton and King, 1982; Hamilton, 1987; Calder, 1999; FAO, 2002; Kaimowitz, 2004; Van Noordwijk, 2004; FAO and CIFOR, 2005; NRI, 2005; Bradshaw *et al.*, 2007; Van Dijk *et al.*, 2008; Garriguata and Balvanera, 2009). Briefly, these fall broadly into three general categories:

- *Inappropriate generalizations* from one site to another, and in particular, application of research results from humid temperate to sub-humid and humid tropical zones (cf. Bruijnzeel *et al.*, 2005; NRI, 2005).
- *Forest and water myths* – notably the expectation that forests significantly reduce or even prevent *extreme* flooding, or that reforestation *always* increases dry-season flows. Whether or not these benefits occur depends on numerous site-specific factors as discussed in more detail by Bruijnzeel (2004), Scott *et al.* (2005), Van Dijk *et al.* (2008), and Malmer *et al.* (2009). In addition, there is a scale aspect involved. For example, the presence of a good forest cover may significantly reduce all but the most extreme flooding in a localized area, but tends to have an insignificant impact or is averaged out at larger scales, where runoff is received at different rates from many different sources in the watershed (Hamilton, 1987).
- *Erosion myths* – notably that stream sediment loads in forested areas are *always* very low. In actual fact, stream sediment yields in wet tropical mountain areas may be very high when sediment production is dominated by natural mass wasting processes (Bruijnzeel, 2004; Sidle *et al.*, 2006; cf. Wilcke *et al.*, this volume). Under conditions of widespread and intensive mass wasting, contributions from surface erosion on agricultural land are relatively unimportant and improved land-use practices will not reduce, let alone prevent sedimentation of reservoirs, although it may have on-site benefits for the farmers (Critchley, 2004). Forestation also reduces the risk of shallow (but not deep) landslides (Bruijnzeel, 2004; Sidle *et al.*, 2006). Nevertheless, as shown by Bruijnzeel (2004), for a given topographic and geological setting, stream sediment loads under pristine forest tend to be (much) lower than after forest conversion. Similarly, Vanacker *et al.* (2007) have demonstrated that sediment yields from formerly heavily degraded but now reforested small to medium-sized watersheds in the Ecuadorian Andes are as low as those associated with natural forest on the same geological substrate, implying that positive changes are indeed possible (cf. Molina *et al.*, 2007).

Scrutiny of specific case studies confirms the general absence of the scientific data needed to confirm that the desired services are actually provided, and to support the evaluation of trade-offs (see also Porras *et al.*, 2008). For example, a detailed review of PES initiatives in upland Costa Rica by Rojas and Aylward (2003) found that most were based instead on conventional wisdom, secondary sources of information, and selective references to literature on forest hydrology supporting statements like "the protection of forests will increase water yields," which is not necessarily the case (see also discussion below). Instead, payments in the Costa Rican initiative are based on opportunity costs of returning cleared land to forest cover, with no attempts to model relationships between land use and hydrology. Alternatively, the value of the resulting change in the amount of streamflow is estimated with respect to specific consumption and production activities (Rojas and Aylward, 2003). A more general review, not limited to Costa Rica, also identified initiatives based on the expected costs of alternative courses of action, willingness to pay or to accept compensation, and consistency with comprehensive management plans (Tognetti *et al.*, 2004).

In the more specific context of cloud forests, recent results of detailed hydrological field and modeling studies in cloud forest in the Arenal area in northern Costa Rica raise questions about the validity of some of the standard assumptions as to why cloud forests produce such high amounts of streamflow (Bruijnzeel, 2006; Schellekens, 2006), and also, whether the benefits of water provision by these cloud forests are economically significant

(Porras, 2008). Cloud forests, especially in Costa Rica, are found in areas with high rainfall and low evaporation; hence, these areas are very water-productive anyway, though largely independently of the presence or absence of cloud forest (Mulligan and Burke, 2005; Bruijnzeel, 2006; cf. McJannet *et al.*, 2007; McJannet *et al.*, this volume). These new findings support earlier conclusions for the same area inferred by Aylward and Echeverria (2001) and Aylward (2005). According to the latter, the relatively small effect of the fog interception component, the similar water use of pasture and cloud forest under the prevailing super-humid climatic conditions, the small relative extent of cloud forest in the overall watershed areas draining into Lake Arenal (turned into an inter-annual storage reservoir), and the large size of the reservoir itself, all resulted in only a small economic contribution to hydro-electricity from cloud forests (Aylward, 2005). However, considering the effects of cloud forest presence on water flows into a (much smaller) daily reservoir in the nearby Peñas Blancas watershed – which also had a much larger proportion of cloud forest cover – led to potentially positive (albeit small) benefits in terms of hydro-electricity production (Porras, 2008).

Thus, conversion of cloud forest to pasture in northern Costa Rica did not lead to the expected decreases in annual or even seasonal water yield; rather the effect on streamflow was more or less neutral. This surprising finding was attributed to the compensatory effects of: (i) low additional inputs by fog interception, and (ii) the lower water use of pasture compared to cloud forest (Bruijnzeel, 2006).

The obvious drawback of basing one's actions on general beliefs or even myths about land–water relationships is that they may instead lead to partial and inappropriate solutions, in which a disproportionate share of the blame for water scarcity or watershed degradation is placed on marginal groups in remote upper watershed areas (Kaimowitz, 2004; FAO and CIFOR, 2005). For example, when dry-season flows in Thailand were observed to be diminishing, the focus of regulatory measures to address the problem was primarily on the financially more vulnerable farmers cultivating rain-fed slopes in areas where significant forest cover remained. Closer scrutiny revealed that a dramatic increase in dry-season cultivation and irrigation of soybeans by more affluent farmers owning paddy fields was to be held responsible for the decline in flows (Forsyth and Walker, 2008). In sum, land use and hydrology interact in complex ways, in which the presence of trees is only one of several factors that should be considered in the context of the entire flow regime. Even in small sub-basins, there may be complex interactions between various biophysical aspects, and between them and human management systems that make it difficult to accurately predict management outcomes. Therefore, whether any of the common presumptions hold depends on the outcome of multiple interactions, and is therefore highly contextual and site-specific. For example, in contrast to the cited lack of streamflow impacts after cloud forest conversion in northern Costa Rica, the same activity in central Veracruz, eastern Mexico, would lead to a major *increase* in flows (Muñoz-Villers, 2008; Gomez-Cárdenas, 2009), amongst other reasons because of low cloud water interception by the forest (Holwerda *et al.*, 2010; cf. Ponette-González *et al.*, 2009) and much higher water consumption by the cloud forest than by pasture (Holwerda *et al.*, 2007; Gomez-Cárdenas, 2009). Conversely, partial clearing of Douglas fir forest in the Pacific Northwest of the USA – an area subject to heavy fog incidence (Harr, 1982) – caused a temporary *decrease* in baseflows during the dry summer months. Pertinently, the effect gradually disappeared after 5–6 years, when the regenerating trees were apparently capturing sufficient amounts of cloud water again (Ingwersen, 1985).

In addition to the assumptions that are often made about the physical land–water relationships, PES initiatives may also be based on socio-economic assumptions that are not always justified, or even downright unrealistic. An additional underlying myth is that there is a panacea or single solution that will solve a complex problem, be it markets, regulations, public or private ownership, or that complex problems such as those associated with land-use change are driven by any single factor. What is important here is to ask the right questions, in a diagnostic approach that draws on diverse areas of knowledge, so as to better understand the conditions, and determine how policy goals can be most effectively achieved (Ostrom *et al.*, 2007). This can, in turn, support a more nuanced and adaptive learning approach to policy.

Whilst science can provide a better approximation of the magnitude and direction of the impacts incurred by a PES initiative, as well as enable more informed decision-making (typically through the use of predictive models, e.g. Uhlenbrook *et al.*, 2004; Mulligan and Burke, 2005; Schellekens, 2006; Thanapakpawin *et al.*, 2006), market-based instruments are critical to finding ways to cover the costs of providing the more valued services. Because of complexity and uncertainty, as well as the amount of time and effort required for gathering scientific information, generalized assumptions are unavoidable to some extent and may even be appropriate, provided that the full range of known uncertainties are made clear to prospective buyers of the services and that monitoring is a built-in component of the initiative. However, these initial assumptions need to be continuously questioned as new knowledge becomes available, and be replaced when they have outlived their usefulness or are not relevant to a particular context.

Demonstrating measurable benefits

Continued downstream or other external support for the costs of upland conservation practices will depend on the ability to demonstrate the magnitude of the hydrological impacts of changed or avoided land-use practices and their associated economic benefits, and the scale at which they can be detected. As discussed further below, it will also depend on the extent to which they are perceived and valued by downstream beneficiaries, as well as on the rights and responsibilities of all stakeholders (Tognetti *et al.*, 2004; Asquith and Wunder, 2008).

Several of the many difficulties associated with demonstrating downstream impacts of changes in upstream land use have been hinted at in previous sections. Of particular importance is the dampening effect of scale, both in terms of watershed size (controlling residence time of water and sediment on the move) and the complementary or compensatory influences brought about by the spatial variability in rainfall, topography, and land use – which tends to be very high in mountainous areas (Hamilton, 1987; Hamilton and Bruijnzeel, 1997; Mulligan and Burke, 2005).

Another key characteristic of watersheds, which is of special relevance for purposes of developing payment arrangements, is that significant changes in conditions are rarely gradual, and are usually linked to extreme and randomly timed events. For example, most sediment and pollutants are transported during a limited number of large storms. Similarly, development of infrastructure (roads, towns) can lead to rapid changes in the way these episodic events manifest themselves through much enhanced stormflow volumes and sedimentation (La Marche and Lettenmaier, 2001; Bruijnzeel, 2004; Rijsdijk *et al.*, 2007). Randomly timed and extreme events are the dominant process in upstream hillslope areas (making up 70–80% of most watershed areas) which, because of their tightly coupled land–water interactions, tend to have greater variability in discharge than downstream areas (Gomi *et al.*, 2002). Therefore, it is arguably equally important to identify known ranges and patterns of variability and uncertainty in water and sediment flows than to identify average values (Tognetti *et al.*, 2004). It is also important to identify expected changes in land use and climate (including frequency of cloud incidence; Nair *et al.*, 2003; Ray *et al.*, 2006; Bates *et al.*, 2008; Karmalkar *et al.*, 2008; cf. Foster, this volume; Mulligan, this volume; Mulligan *et al.*, this volume #53).

Given the difficulties associated with assessing the impact of land-cover change on watershed hydrological behavior it will not come as a great surprise to learn that few, if any, PES initiatives have actually been able to demonstrate a measurable hydrological effect of their activities (Rojas and Aylward, 2003; Porras *et al.*, 2008; Le Tellier *et al.*, 2009; cf. Garriguata and Balvanera, 2009). In addition, monitoring activities are often confined to changes in land use only, whereas hydrological monitoring efforts (if any) are rarely of sufficient rigor and duration to allow solid conclusions to be drawn (Porras *et al.*, 2008; Le Tellier *et al.*, 2009). As such, there is as yet little to no firm evidence that ongoing PES schemes are truly effective in achieving their *hydrological* goals and this may well have consequences for buyers their continued willingness to pay (cf. Aylward, 2005; Muñoz-Piña *et al.*, 2008; Porras *et al.*, 2008; Garriguata and Balvanera, 2009).

IDENTIFYING AND EVALUATING TRADE-OFFS

Identifying the *economic* significance of changes in watershed processes provides a basis for identifying actual or potential trade-offs between meeting various objectives and economic interests. This will require consideration of downstream land uses and stakeholder vulnerability, which depends, in turn, on what options stakeholders have, and on their opportunity costs. Just as with biophysical processes in watersheds, vulnerability, options available to stakeholders, and conflicts all become more evident under extreme conditions. Impacts may be both positive and negative, depending on what is measured and valued. For example, whether flooding increases or decreases welfare will depend on the presence of development in floodplains, and of valued habitats that rely on periodic flooding (FAO and CIFOR, 2005; cf. Bradshaw *et al.*, 2007).

Most payment initiatives have focused on links between upper watershed land-use practices and downstream urban water supplies, or on sedimentation of hydropower dams and irrigation canals. However, it has generally been found difficult to demonstrate the economic significance of impacts at this scale. Even when it is possible to demonstrate and quantify links between land use and hydrology, a literature review by Aylward (2005) raises questions about whether the magnitude of changes in water quantity is likely to be economically significant. Again, this will largely depend on downstream economic interests that rely on water, and on the scale at which the hydrological impacts can be detected. Changes in a watershed may also be significant at a smaller scale than that employed in an investigation, or when considered as one of many complementary goods and services provided by an area, for example, including carbon and biodiversity (Rodriguez-Zuñiga, 2003; Calvo-Alvarado *et al.*, this volume). Others have suggested that, at the smaller scales, where land–water relationships are more detectable, land-use interventions may be justifiable as part of an integrated community resource-management strategy, even when their value for water quantity is modest (Johnson and Baltodano, 2004). Vanacker *et al.* (2007) demonstrated major improvements in stream sediment loads for small to meso-scale watersheds after reforesting at least 30% of heavily degraded watersheds in the Inter-Andean basin area near Cuenca in southern Ecuador. Interestingly, the sediment accumulated behind gully plugs and other imposed barriers acted as a storage reservoir for water running off the hillsides, leading in turn to enhanced baseflows during the dry season despite the presumably much higher water use of the planted trees (G. Govers, personal communication; cf. Scott *et al.*, 2005; Malmer *et al.*, 2009).

Ultimately, the extent of information needed for decision-making will depend on what is required to justify an expenditure in a particular case, or, in other words, to convince the buyers of the ecosystem services. To the extent that uncertainty regarding links between actions and outcomes can be reduced – or at least be made transparent so that it can be factored into the decision-making – stakeholders will have greater economic incentive to participate in these initiatives and in helping to ensure that objectives are actually achieved. However, given the period of time that may be required to detect the results of changes in management practices, and the impossibility of obtaining complete information, it will usually be necessary to begin with rules of thumb, and a working hypothesis that can provide a basis for monitoring and transparency (cf. Le Tellier *et al.*, 2009).

ENSURING ACCESS TO BENEFITS

As discussed in the introduction, willingness to pay is inextricably linked to confidence in the effectiveness of management actions, which include the institutional arrangements needed to ensure access to benefits by those who pay their costs. These are essentially the *rules of the game*, which are necessary to avoid the depletion or degradation of limited resources for which there are competing demands. In their absence, economic value is no more than hypothetic, as there would be no incentive to pay or to take actions needed to ensure provision of the service. Perhaps the most critical amongst these are various forms of property rights and tenure security (cf. Robertson and Wunder, 2005; Asquith and Wunder, 2008). As was illustrated in some of the case studies reviewed, these are critical for PES because they enable land users to enter into contractual agreements that define the rights and responsibilities of all players, and determine how the costs and benefits are distributed (cf. Asquith and Vargas, 2007). Without property rights, land users have little incentive to engage in practices that increase the value of the land, and also risk displacement to more marginal areas, as values are placed on ecosystem services. Unless these issues are addressed, PES initiatives are also likely to overlook areas of highest priority for protection because small and marginalized land-holders tend to occupy the steeper slopes (Aylward and Fernández González, 1998).

In a number of PES case studies reviewed by Rosa *et al.* (2003) it was found that payments for watershed services of forests have disproportionately benefited those who own larger tracts of forests or forest plantations, and who intend to maintain forested land uses (cf. Zbinden and Lee, 2005). A case study conducted in the Virilla watershed in Costa Rica (Miranda *et al.*, 2003) reported that smaller land-owners faced higher opportunity costs because they rely more heavily on small areas of cleared forest, and combine forestry with other activities such as shelter for cattle and shade coffee. The Costa Rican FONAFIFO program, of which this case was a part, has evolved over time to allow greater participation, as lessons have been learned from implementation, and as goals have changed, from ensuring the supply of timber to provision of ecosystem services (Pagiola, 2002). The most recent adjustments were made in 2002 to include agro-forestry and indigenous reserves (Rosa *et al.*, 2003; see also Calvo-Alvarado *et al.*, this volume).

Decentralization of responsibility for management of public funds and natural resources, including provision of water services, can be complementary to property rights, because it enables local governments to establish the rights of stakeholders, as well as to enter into PES agreements, and to play other intermediary roles. Both the Bolivian and Guatemalan cases discussed previously were enabled by recent decentralization laws. Whether governments, be they local or national, are held downwardly accountable for local and livelihood concerns, and for the way funds are allocated, will also depend on whether there is effective participation of stakeholders in decision-making (Ribot, 2004; cf. Hofstede *et al.*, this volume). Effective participation also enables stakeholders to bring local knowledge to the assessment process, which is necessary for a site-specific approach (cf. Báez *et al.*, this volume; Hofstede *et al.*, this volume).

A key question to ask in evaluating PES initiatives is whether stakeholders regard the distribution of costs and benefits as fair and equitable. This is an important aspect of feasibility because, if significant stakeholders are excluded or disadvantaged, and regard existing rights as inequitable, they will have little incentive to pay for services or to cooperate with management plans designed to ensure their provision. Given the political constraints on any form of institutional change, or simply inertia that may need to be overcome, the development of payment arrangements for ecosystem services should generally be regarded as the end result of a long-term process (Tognetti *et al.*, 2005; Asquith and Wunder, 2008; cf. Hofstede *et al.*, this volume). However, the approach is spreading rapidly (Porras *et al.*, 2008), and may be gaining momentum from broader trends of institutional changes in water resource management – from the development of new water supplies toward the reallocation of existing ones, and toward recovery of costs (Saleth and Dinar, 1999; Bauer, 2004).

CONCLUSIONS

The effective use of market-based instruments to provide incentives for actions needed to ensure the provision of ecosystem services at the watershed scale, and to cover their costs, implies the ability to make payments contingent on achieving desired outcomes. This is also critical for building and sustaining stakeholder confidence and willingness to pay for services. Although there is a general scientific understanding of forest–water relationships, given the unique conditions found in each watershed, little is known about whether the services paid for are actually delivered in specific places, and about the overall cost-effectiveness of these initiatives (Porras *et al.*, 2008; Garriguata and Balvanera, 2009). This is in part because of the need for long-term data for large and remote watershed areas where there is usually no or only poor monitoring of basic climatic and hydrological conditions. However, regardless of the amount of data available, uncertainty is inherent in complex watershed processes, which are driven by randomly timed events as well as by changes in land use.

In the absence of site-specific information that links specific land-use practices to specific outcomes, the use of PES instruments is often based on common and generalized beliefs, notably with respect to the regulatory ability of forests vis-à-vis streamflow and water quality. Cloud forests in particular have been thought to increase the supply of water because they capture and store water from clouds in addition to that provided by rainfall. However, the few available site-specific studies of the hydrological impacts of cloud forest conversion to pasture so far have yielded contradictory results. Flows increased in one case but decreased in another, whereas no change was observed under particularly wet conditions. Further field and modeling studies are necessary to determine the extent to which the additional water captured from passing clouds enters into streamflow – which makes it available for human use – or is evaporated by the vegetation (cf. Mulligan and Burke, 2005; Guswa *et al.*, 2007; Bruijnzeel *et al.*, this volume).

However, hydrological issues aside, many cloud forests are located in protected areas where it is equally (if not more) important to stabilize agricultural frontier areas in the buffer zones, so as to prevent further encroachment and forest degradation. In at least some cases, these mid-watershed areas were shown to be the major sources of sediment and other impacts that are of downstream concern (Robertson and Wunder, 2005; Martínez-Tuna and Juarez-Quim, 2008).

This chapter has provided a brief review of PES initiatives in cloud forest areas, highlighting the kinds of information needed to support decision-making. As a point of departure, a site-specific process of assessment is necessary both to demonstrate the economic significance of links between ecosystems and human well-being, to understand what choices are available to land users as well as to beneficiaries of ecosystem services, and to develop initiatives that are appropriate to their context.

An estimate of changes in the water balance provides a framework that can be used to estimate relationships between the extent of changes in land use and the level of benefits provided. However, the specific services provided by ecosystems at the watershed scale will depend on whether benefits or avoided impacts have economic significance. This, in turn, depends on downstream uses of land and water, on the scale at which impacts can be detected, and on what options are available to stakeholders for maintaining their livelihoods.

Demonstrating relationships between ecosystem functioning and services at the watershed scale, remains a major challenge for the establishment (and continued justification) of PES schemes (Muñoz-Piña *et al.*, 2008; Porras *et al.*, 2008), as well as for applied hydrological research (Gariguata and Balvanera, 2009). To the extent that these links will become better understood through further research, it will be possible not only to better quantify the hydrological and economic benefits associated with improved land management and ecosystem conservation, but also to engage local populations more in watershed management. It may also make it possible to better demonstrate links or complementarities between ecosystem functions that support provision of water, and those that support biodiversity, which is inseparable from ecosystem functions. As unique and rare habitats that are highly vulnerable to changes in climate and regional land use (cf. Foster, this volume; Nair *et al.*, this volume), cloud forests are particularly important for the conservation of biodiversity (Bunker *et al.*, 2005; Beck *et al.*, 2008). The global benefits of biodiversity and also carbon storage have generally been the main concern of international donors that are providing support for the upfront costs of developing PES initiatives. Whilst payments for water-related ecosystem services are expected to generate more sustained funding, by themselves, they are unlikely to be sufficient to cover the costs of conservation needs. A more integrated multiple-services approach (Asquith and Wunder, 2008) is required, and more consistent with management of ecosystems to support overall human well-being.

REFERENCES

Asquith, N., and M. T. Vargas (2007). *Fair Deals for Watershed Services in Bolivia*, Natural Resource Issues No. 7. London: International Institute for Environment and Development.

Asquith, N., and S. Wunder (eds.) (2008). *Payments for Watershed Services: The Bellagio Conversations*. Santa Cruz, Bolivia: Fundación Natura Bolivia, Jakarta, Indonesia: CIFOR, and London: International Institute for Environment and Development.

Aylward, B. (2005). Land-use, hydrological function and economic valuation. In *Forests, Water and People in the Humid Tropics*, eds. M. Bonell and L. A. Bruijnzeel, pp. 99–120. Cambridge, UK: Cambridge University Press.

Aylward, B., and J. Echeverria (2001). Synergies between livestock production and hydrological function in Arenal, Costa Rica. *Environment and Development Economics* **6**: 359–382.

Aylward, B., and A. Fernández González (1998). *Institutional Arrangements for Watershed Management: A Case Study of Arenal, Costa Rica*, CREED Working Paper No. 21. London: International Institute for Environment and Development.

Aylward, B., J. Bandyopadhyay, J. C. Belausteguigotia, *et al.* (2005). Freshwater ecosystem services. In *Ecosystems and Human Well-Being: Policy Responses*, Vol. 3, *Findings of the Responses Working Group of the*

Millennium Ecosystem Assessment, eds. K. Chopra, R. Leemans, P. Kumar, and H. Simons, pp. 213–255. Washington, DC: Island Press.

Bates, B. C., Z. W. Kundzewicz, S. Wu, and J. P. Palutikof (eds.) (2008). *Climate Change and Water*, Technical Paper of the Intergovernmental Panel on Climate Change. Geneva, Switzerland: IPCC Secretariat.

Bauer, C. J. (2004). *Siren Song: Chilean Water Law as a Model for International Reform.* Washington, DC: Resources for the Future.

Beck, E., J. Bendix, I. Kottke, F. Makeschin, and R. Mosandl (eds.) (2008). *Gradients in a Tropical Mountain Ecosystem of Ecuador.* Berlin: Springer-Verlag.

Becker, C. D. (2000). IFRI: A springboard to tropical forest conservation and co-management in western Ecuador. In *8th Biennial Conference of the International Association for the Study of Common Property.* Bloomington, IN: Indiana University.

Becker, C. D., and K. Ghimire (2003). Synergy between traditional ecological knowledge and conservation science supports forest preservation in Ecuador. *Conservation Ecology* **8**: 1, www.consecol.org/vol8/iss1/art1/.

Bradshaw, C. J. A., N. S. Sodhi, K. S-H. Peh, and B. W. Brook (2007). Global evidence that deforestation amplifies flood risk and severity in the developing world. *Global Change Biology* **13**: 1–17, doi/abs/10.1111/j.1365–2486.2007.01446.x.

Brown, M. B., I. de la Roca, A. Vallejo, *et al.* (1996). *A Valuation Analysis of the Role of Cloud Forests in Watershed Protection: Sierra de las Minas Biosphere Reserve, Guatemala and Cusuco N.P. Honduras.* Philadelphia, PA: RARE Center for Tropical Conservation.

Bruijnzeel, L. A. (2004). Hydrological functions of tropical forests: not seeing the soil for the trees? *Agriculture Ecosystems and Environment* **104**: 185–228.

Bruijnzeel, L. A. (ed.) (2006). *Hydrological Impacts of Converting Tropical Montane Cloud Forest to Pasture, with Initial Reference to Northern Costa Rica*, Final Technical Report DFID-FRP Project No. R7991. Amsterdam, the Netherlands: VU University Amsterdam, and Aylesford, UK: Forestry Research Programme of the UK Department for International Development.

Bruijnzeel, L. A., and L. S. Hamilton (2000). *Decision Time for Cloud Forests*, IHP Humid Tropics Programme Series No. 13. Paris: UNESCO–IHP.

Bruijnzeel, L. A., M. Bonell, D. A. Gilmour, and D. Lamb (2005). Forests, water and people in the humid tropics: an emerging view. In *Forests, Water and People in the Humid Tropics*, eds. M. Bonell and L. A. Bruijnzeel, pp. 906–925. Cambridge, UK: Cambridge University Press.

Bunker, D. E., F. DeClerck, J. C. Bradford, *et al.* (2005). Species loss and aboveground carbon storage in a tropical forest. *Science* **310**: 1029–1031.

Calder, I. R. (1999). *The Blue Revolution: Land Use and Integrated Water Resources Management.* London: Earthscan.

Calvo-Alvarado, J. (2000). Case study: payments for watershed services in Costa Rica. In *Forests in Sustainable Mountain Development: A State-of-Knowledge Report for 2000*, eds. M. F. Price and N. Butt, pp. 428–429. Wallingford, UK: CAB International.

Critchley, W. R. S. (2004). From nature to nurture: soil and water management for rainfed steeplands in the humid tropics. In *Forests, Water and People in the Humid Tropics*, eds. M. Bonell and L. A. Bruijnzeel, pp. 881–905. Cambridge, UK: Cambridge University Press.

Echavarría, M. (2002a). Financing watershed conservation: the FONAG water fund in Quito, Ecuador. In *Selling Forest Environmental Services: Market-Based Mechanisms for Conservation and Development*, eds. S. Pagiola, J. Bishop, and N. Landell-Mills, pp. 91–102. London: Earthscan.

Echavarría, M. (2002b). *Water User Associations in the Cauca Valley, Colombia: A Voluntary Mechanism to Promote Upstream–Downstream Cooperation in the Protection of Rural Watersheds.* Rome: FAO.

Echavarría, M. (2003). Algunas lecciones sobre la aplicación de pagos por la protección del agua con base en experiencias en Colombia y Ecuador. *3rd Congreso Latinoamericana de Manejo de Cuencas Hidrográficas, Foro Regional sobre Sistemas de Pago por Servicios Ambientales*, Arequipa, Peru. Rome: FAO.

Echavarría, M., J. Vogel, A. Montserrát, and F. Meneses (2004). *The Impacts of Payments for Watershed Services in Ecuador: Emerging Lessons from Pimampira and Cuenca.* London: International Institute for Environment and Development, Environmental Economics Programme.

FAO (2002). *Land–Water Linkages in Rural Watersheds Electronic Workshop*, 18 September – 27 October 2000, Land and Water Bulletin No. 9. Rome: FAO.

FAO and CIFOR (2005). *Forests and Floods: Drowning in Fiction or Thriving on Facts?* RAP Publication 2005/03; Forest Perspectives No. 2. Rome: FAO, and Jakarta, Indonesia: CIFOR.

Forsyth, T., and A. Walker (2008). *Forest Guardians, Forest Destroyers: The Politics of Environmental Knowledge in Northern Thailand.* Seattle, WA: University of Washington Press.

García-Santos, G. (2007). *An ecohydrological and soils study in a montane cloud forest in the National Park of Garajonay, La Gomera (Canary Islands, Spain).* Ph.D. thesis, VU University Amsterdam, Amsterdam, the Netherlands. Also available at www.falw.vu.nl/nl/onderzoek/earth-sciences/geo-environmental-science-and-hydrology/hydrology-dissertations/index.asp.

Garriguata, M. R., and P. Balvanera (2009). Tropical forest service flows. *Forest Ecology and Management*, doi: 10.1016/j.foreco.2009.06.025.

Gomez-Cárdenas, M. (2009). Transpiration in contrasting vegetation cover types in the tropical montane cloud forest belt in eastern Mexico. Ph.D. thesis, University of Iowa, Ames, IA, USA.

Gomi, T., R. C. Sidle, and J. S. Richardson (2002). Understanding processes and downstream linkages of headwater systems. *BioScience* **52**: 905–916.

Guswa, A. J., A. L. Rhodes, and S. E. Newell (2007). Importance of orographic precipitation to the water resources of Monteverde, Costa Rica. *Advances in Water Resources* **30**: 2098–2112.

Hamilton, L. S. (1987). What are the impacts of Himalayan deforestation on the Ganges–Brahmaputra lowlands and delta? Assumptions and facts. *Mountain Research and Development* **7**: 256–263.

Hamilton, L. S., and L. A. Bruijnzeel (1997). Mountain watersheds: integrating water, soils, gravity, vegetation, and people. In *Mountains of the World: A Global Priority*, eds. B. Messerli and J. D. Ives, pp. 337–370. London: Parthenon.

Hamilton, L. S., and P. N. King (1982). *Tropical Forested Watersheds: Hydrological and Soils Response to Major Uses or Conversions.* Boulder, CO: Westview Press.

Hamilton, L. S., J. O. Juvik, and F. N. Scatena (eds.) (1995). *Tropical Montane Cloud Forests.* New York: Springer-Verlag.

Harr, R. D. (1982). Fog drip in the Bull Run municipal watershed, Oregon. *Water Resources Bulletin* **18**: 785–789.

Heimlich, R. (2006). Green payments now – what's missing? In *Building the Scientific Basis for Green Payments: A Report on a Workshop Sponsored by the U.S. Department of Agriculture, CSREES, World Wildlife Fund, and The Elton R. Smith Endowment at Michigan State University*, eds. S. S. Batie and S. Lynch, pp. 12–14. Washington, DC: World Wildlife Fund.

Hemp, A. (2005). Climate change-driven forest fires marginalize the impact of ice cap wasting on Kilimanjaro. *Global Change Biology* **11**: 1013–1023.

Holder, C. D. (2003). Fog precipitation in the Sierra de las Minas Reserve, Guatemala. *Hydrological Processes* **17**: 2001–2010.

Holwerda, F., V. L. Barradas, J. Cervantes, and L. A. Bruijnzeel (2007). Balances hidricos y de energía de un cafetal de sombra en el centro de Veracruz, México. In *Reporte técnico final del proyecto INE/A1-064/2007.* Xalapa, Veracruz, México: Institute de Ecología, and Amsterdam, the Netherlands: VU University Amsterdam.

Holwerda, F., L. A. Bruijnzeel, L. E. Muñoz, M. Equihua, and H. Asbjornsen (2010). Rainfall and cloud water interception in mature and secondary lower montane cloud forests of central Veracruz, Mexico. *Journal of Hydrology* **384**: 84–96.

Ilstedt, U., A. Malmer, E. Verbeeten, and D. Murdiyarso (2007). The effect of afforestation on water infiltration in the tropics: a systematic review and meta-analysis. *Forest Ecology and Management* **251**: 45–51.

Ingwersen, J. B. (1985). Fog drip, water yield, and timber harvesting in the Bull Run municipal watershed, Oregon. *Water Resources Bulletin* **21**: 469–473.

Johnson, N. L., and M. E. Baltodano (2004). The economics of community watershed management: Some evidence from Nicaragua. *Ecological Economics* **49**: 57–71.

Kaimowitz, D. (2004). Useful myths and intractable truths: the politics of the link between forests and water in Central America. In *Forests, Water, and People in the Humid Tropics*, eds. M. Bonell and L. A. Bruijnzeel, pp. 86–98. Cambridge, UK: Cambridge University Press.

Karmalkar, A. V., R. S. Bradley, and H. F. Diaz (2008). Climate change scenario for Costa Rican montane forests. *Geophysical Research Letters* **25**, L11702, doi:10.1029/2008GL033940.

La Marche, J. L., and D. P. Lettenmaier (2001). Effect of forest roads on flood flows in the Deschutes River, Washington. *Earth Surface Processes and Landforms* **26**: 115–134.

Landell-Mills, N., and I. T. Porras (2002). *Silver Bullet or Fools' Gold? A Global Review of Markets for Forest Environmental Services and their Impact on the Poor.* London: International Institute for Environment and Development.

Le Tellier, V., A. Carrasco, and N. Asquith (2009). Attempts to determine effects of forest cover on streamflow by hydrological measurements in Los Negros, Bolivia. *Forest Ecology and Management*, doi:10.1016/j.foreco.2009.04.031.

Malmer, A., D. Murdiyarso, L. A. Bruijnzeel, and U. Ilstedt (2009). Carbon sequestration in tropical forests and water: a critical look at the basis for commonly used generalisations. *Global Change Biology*, doi:10.1111/j.1365–2486.2009.01984.x

Martínez-Tuna, M., and D. Juarez-Quim (2008). *Pagos equitativos por Servicios Hidrológicos (PESH) en Teculután, Zacapa, Guatemala. Resumen estudio hidrológico Teculután.* Guatemala City, Guatemala: WWF-Central America and CARE Guatemala.

McJannet, D. L., J. S. Wallace, P. Fitch, M. Disher, and P. Reddell (2007). Water budgets of tropical rainforests in northern Queensland, Australia. *Hydrological Processes* **21**: 3473–3483.

MEA (2003). *Ecosystems and Human Well-Being: A Framework for Assessment.* Washington, DC: The Millennium Ecosystem Assessment and Island Press.

Miranda, M., I. T. Porras, and M. L. Moreno (2003). *The Social Impacts of Payments for Environmental Services in Costa Rica: A Quantitative Field Survey and Analysis of the Virilla Watershed.* London: International Institute for Environment and Development.

Molina, A., G. Govers, V. Vanacker, J. Poesen, and E. Zeelmaekers (2007). Runoff generation in a degraded Andean ecosystem: interaction of vegetation cover and land use. *Catena* **71**: 357–370.

Mulligan, M., and S. M. Burke (2005). *FIESTA: Fog Interception for the Enhancement of Streamflow in Tropical Areas*, Annex 4a Final Technical Report DFID–FRP Project No. R7991. Leigh-on-Sea, UK: AMBIOTEK. Also available at www.ambiotek.com/fiesta.

Muñoz-Piña, C., A. Guevara, J. M. Torres, and J. Braña (2008). Paying for the hydrological services of Mexico's forests: analysis, negotiations and results. *Ecological Economics* **65**: 725–736.

Muñoz-Villers, L. E. (2008). *Effecto del cambio en el uso de suelo sobre la dinámica hidrológica y la calidad de agua en el trópico húmedo del Centro de Veracruz, Mexico.* Ph.D. thesis, Universidad Nacional Autónoma Metropolitana, Mexico City, Mexico.

Nair, U. S., R. O. Lawton, R. M. Welch, and R. A. Pielke Sr. (2003). Impact of land use on tropical montane cloud forests: sensitivity of cumulus cloud field characteristics to lowland deforestation. *Journal of Geophysical Research* **108** (D7): 4206–4218, doi:10.1029/2001JD001135.

Natural Resources International (NRI) (2005). *From the Mountain to the Tap: How Land Use and Water Management Can Work for the Rural Poor.* Aylesford, UK: NRI.

Ostrom, E., M. A. Janssen, and J. M. Anderies (2007). Going beyond panaceas. *Proceedings of the National Academy of Sciences USA* **104**: 15 176–15 178.

Pagiola, S. (2002). Paying for water services in Central America: learning from Costa Rica. In *Selling Forest Environmental Services: Market-Based Mechanisms for Conservation and Development*, eds. S. Pagiola, J. Bishop, and N. Landell-Mills, pp. 37–61. London: Earthscan.

Pagiola, S., N. Landell-Mills, and J. Bishop (2002). Making market-based mechanisms work for forests and people. In *Selling Forest Environmental Services: Market-Based Mechanisms for Conservation and Development*, eds. S. Pagiola, J. Bishop, and N. Landell-Mills, pp. 1–14. London: Earthscan.

Ponette-González, A. G., K. C. Weathers, and L. M. Curran (2009). Water inputs across a tropical montane landscape in Veracruz, Mexico: synergistic effects of land cover, rain and fog seasonality, and interannual precipitation variability. *Global Change Biology*, doi:10.1111/j.1365–2486.2009.01985.x.

Porras, I. T. (ed.) (2008). *Assessing and Valuing Watershed Environmental Services: Methods and Two Case Studies.* London: International Institute for Environment and Development.

Porras, I. T., and N. Neves (2006). *Markets for watershed services case study summaries.* Retrieved on 30 August 2007, from www.watershedmarkets.org/casestudies.html.

Porras, I. T., M. Grieg-Gran, and N. Neves (2008). *All That Glitters: A Review of Payments for Watershed Services in Developing Countries.* London: International Institute for Environment and Development.

Ray, D. K., Nair, R., O. Lawton, *et al.* (2006). Impact of land use on Costa Rican tropical montane cloud forests: sensitivity of orographic cloud formation to deforestation in the plains. *Journal of Geophysical Research* **11**, D02108. doi:10.1029/2005JD006096.

Ribot, J. C. (2004). *Waiting for Democracy: The Politics of Choice in Natural Resource Decentralization.* Washington, DC: World Resources Institute.

Rijsdijk, A., L. A. Bruijnzeel, and C. Kukuh Sutoto (2007). Runoff and sediment yield from rural roads, trails and settlements in the upper Konto catchment, East Java, Indonesia. *Geomorphology* **87**: 28–37.

Robertson, N., and S. Wunder (2005). *Fresh Tracks in the Forest: Assessing Incipient Payments for Environmental Services Initiatives in Bolivia.* Bogor, Indonesia: Center for International Forestry Research (CIFOR).

Rodriguez-Zuñiga, J. M. (2003). Paying for forest environmental services: the Costa Rican experience. *Unasylva* **212**: 31–33.

Rojas, M., and B. Aylward (2002). *The Case of La Esperanza: A Small Private, Hydropower Producer and a Conservation NGO in Costa Rica*, Land and Water Linkages in Rural Watersheds Case Study Series. Rome: FAO.

Rojas, M., and B. Aylward (2003). *What Are We Learning from Experiences with Markets for Environmental Services in Costa Rica? A Review and Critique of the Literature.* London: International Institute for Environment and Development.

Rosa, H., S. Kandel, and L. Dimas (2003). *Compensation for Environmental Services and Rural Communities: Lessons from the Americas and Key Issues for Strengthening Community Strategies.* San Salvador, El Salvador: PRISMA.

Rose, C. M. (2002). Common property, regulatory property, and environmental protection: comparing community-based management and tradable environmental allowances. In *The Drama of the Commons*, eds. E. Ostrom, T. Dietz, N. Dolšak, *et al.*, pp. 233–257. Washington, DC: National Academy Press.

Ruhl, J. B., S. D. Kraft, and C. L. Lant (2007). *The Law and Policy of Ecosystem Services.* Washington, DC: Island Press.

Saleth, R. M., and A. Dinar (1999). *Water Challlenge and Institutional Response: A Cross-Country Perspective*, Policy Research Working Paper No. (2045). Washington, DC: The World Bank Development Research Group on Rural Development.

Schellekens, J. (2006). *CQ-FLOW: A Distributed Hydrological Model for the Prediction of Impacts of Land-Cover Change, with Special Reference to the Rio Chiquito Catchment, Northwest Costa Rica*, Annex 3 to Final Technical Report DFID–FRP Project No. R7991. Amsterdam, the Netherlands: Vrije Universiteit. Also available at www.geo.vu.nl/~fiesta/.

Scott, D. F., L. A. Bruijnzeel, and J. Mackensen (2005). The hydrological and soil impacts of forestation in the tropics. In *Forests, Water and People in the Humid tropics*, eds. M. Bonell and L. A. Bruijnzeel. pp. 622–651. Cambridge, UK: Cambridge University Press.

Sen, A. K. (1999). *Development as Freedom.* New York: Anchor Books.

Sidle, R. C., A. D. Ziegler, J. N. Negishi, *et al.* (2006). Erosion processes in steep terrain: truths, myths, and uncertainties related to forest management in Southeast Asia. *Forest Ecology and Management* **224**: 199–225.

Thamapakpawin, P., J. Richey, D. Thomas, *et al.* (2006). Effects of landuse change on the hydrological regime of the Mae Chaem river basin, NW Thailand. *Journal of Hydrology* **334**: 215–218.

Tognetti, S. (2001). *Creating Incentives for River Basin Management as a Conservation Strategy: A Survey of the Literature and Existing Initiatives.* Washington, DC: World Wildlife Fund. Also available at www.biodiversityeconomics.org/applications/library_documents/ lib_document.rm?document_id=193.

Tognetti, S. S., G. Mendoza, B. Aylward, D. Southgate, and L. Garcia (2004). *A Knowledge and Assessment Guide to Support the Development of Payment Arrangements for Watershed Ecosystem Services (PWES)*, Working Paper. Washington, DC: World Bank Environment Department. Also available at www.flowsonline.net/data/pes_assmt_guide_en.pdf.

Tognetti, S. S., B. Aylward, and G. M. Mendoza (2005). Markets for watershed services. In *Encyclopedia of Hydrological Sciences*, eds. M. G. Anderson and J. J. McDonnell, pp. 2987–3002. Chichester, UK: John Wiley.

Uhlenbrook, S., S. Roser, and N. Tilch (2004). Development of a distributed, but conceptual catchment model to represent hydrological processes adequately at the meso-scale. *Journal of Hydrology* **291**: 278–296.

UNESCO–IHE (2009). *CHE: Conserving Hydrological and Ecological Functions through Payment for Watershed Services, with Special Reference to South-Central Bolivia*, UPaRF Tender 2009. Paris: UNESCO–IHE.

Vanacker, V., F. von Blandenburg, G. Govers, *et al.* (2007). Restoring dense vegetation can slow mountain erosion to near natural benchmark levels. *Geology* **35**: 303–306.

Van Dijk, A. I. J. M., M. van Noordwijk, I. R. Calder, *et al.* (2008). Forest-flood relation still tenuous: comment on "Global evidence that deforestation amplifies flood risk and severity in the developing world" by C.J.A. Bradshaw, N.S. Sodhi, K. S-H. Peh, and B.W. Brook. *Global Change Biology* **14**, doi:10.1111/j.1365–2486.2008. 01708.x.

Van Noordwijk, M. 2004. Quantifying off-site effects of land use change: filters, flows and fallacies. *Agriculture, Ecosystems and Environment* **104**: 19–34.
Van Noordwijk, M., M. van Roode, E. L. McCallie, and B. Lusiana (1998). Erosion and sedimentation as multiscale, fractal processes: implications for models, experiments and the real world. In *Soil Erosion at Multiple Scales*, eds. F. W. T. Penning de Vries, F. Agus, and J. Kerr, pp. 223–253. Wallingford, UK: CAB International.
WWF–CARE (2007). *Carta de compromisos mutuos para la implementación del mecanismo de compensación Por Servicios Hidrológicos entre la Municipalidad de Teculután y la Asociación de Desarrollo Integral Agropecuario Comunidades de Teculután, Cuenca del Río Teculután, Iniciativa Conjunta de Compensación Equitativa por Servicios Hidrológicos*. Guatemala City, Guatemala: WWF–CARE Guatemala.
Yaguache, R., and R. Carrion (2004). *Construyendo una experiencia de desarrollo: El manejo de recursos naturales en Pimampiro*. Quito, Ecuador: CEDERENA.
Zadroga, F. (1981). The hydrological importance of a montane cloud forest area of Costa Rica. In *Tropical Agricultural Hydrology*, eds. R. Lal and E. W. Russell, pp. 59–73. New York: John Wiley.
Zbinden, S., and D. Lee (2005). Paying for environmental services: an analysis of participation in Costa Rica's PSA Program. *World Development* **33**: 255–272.
Ziegler, A. D., and T. W. Giambelluca (1997). Importance of rural roads as source areas for runoff in mountainous areas of northern Thailand. *Journal of Hydrology* **196**: 204–229.
Ziegler, A. D., T. W. Giambelluca, L. T. Tran, *et al.*, (2004). Hydrological consequences of landscape fragmentation in mountainous northern Vietnam: evidence of accelerated overland flow generation. *Journal of Hydrology* **287**: 124–146.
Zimmermann, B., and H. Elsenbeer (2008). Spatial and temporal variability of soil saturated hydraulic conductivity in gradients of disturbance. *Journal of Hydrology* **361**: 78–95.

71 Conservation strategies for montane cloud forests in Costa Rica: the case of protected areas, payments for environmental services, and ecotourism

J. C. Calvo-Alvarado
Instituto Tecnológico de Costa Rica, Cartago, Costa Rica

G. A. Sánchez-Azofeifa
University of Alberta, Edmonton, Alberta, Canada

A. Méndez
Fondo de Financiamiento Forestal (FONAFIFO), San José, Costa Rica

ABSTRACT

During the 1970s and 1980s, a series of technical reports predicted that, based on the rate of deforestation taking place in Costa Rica at the time, most of the country's forest cover would disappear before the end of the century. The first response to counteract this undesirable situation was the creation of a system of National Parks. These protected areas now constitute the main repositories of the remaining cloud forest of Costa Rica. However, not all cloud forest is protected by the National Park System; other areas are protected by private reserves whose main income is ecotourism. Recently, Costa Rica introduced a novel system for the payment of environmental services (PES) provided by forests as compensation to forest owners for conserving their forests instead of converting them to economically more profitable land uses. The PES system acknowledges the following services of forest ecosystems: (i) greenhouse gas effect mitigation via carbon sequestration, (ii) water resources protection for urban, rural, or hydro-electric uses, (iii) biodiversity protection, and (iv) scenic value. This chapter describes the potential and current distribution of cloud forest in Costa Rica in relation to National Protected Areas, Private Reserves, and the PES system. A brief description of these and other financial mechanisms to support conservation of cloud forest is also presented.

INTRODUCTION

Conservation of areas of montane cloud forest (MCF) is a major concern in Costa Rica due to their high level of biodiversity, their protection of water resources, and their importance as an attraction for ecotourism. According to Hamilton *et al.* (1995) MCF is a type of evergreen mountain forest distributed throughout the tropical belt, typically at elevations between 1500 and 3000 m.a.s.l. and within a wide range of rainfall regimes (500–10 000 mm year^{-1}; cf. Jarvis and Mulligan, this volume). According to the Life Zone System of Holdridge (1967) – widely used in Costa Rica for forest classification – the potential distribution of MCF in Costa Rica may be associated with the Lower Montane (1400–2800 m.a.s.l on the Pacific Slope vs. 1600–2800 m.a.s.l on the Caribbean Slope) and Montane belts (2800–3600 m.a.s.l), which contain the following Life Zones: (i) Tropical Pluvial Lower Montane Forest (TpLM), (ii) Tropical Wet Lower Montane Forest (TwLM), (iii) Tropical Pluvial Montane Forest (TpM), and (iv) Tropical Wet Montane Forest (TwM). The Holdridge system is based on temperature and rainfall regimes and is therefore essentially bioclimatically controlled. Holdridge's lower montane and montane forests (i.e. including the wet and pluvial types within each) correspond with Grubb's (1977) lower montane (LMRF) and upper montane rain forest (UMRF) formations, respectively, but Holdridge's further distinction between wet and pluvial forest types is not maintained in Grubb's approach. The total area of the four Life Zones that potentially carry MCF is 569 700 ha or ~12% of the national

Tropical Montane Cloud Forests: Science for Conservation and Management, eds. L. A. Bruijnzeel, F. N. Scatena, and L. S. Hamilton. Published by Cambridge University Press.

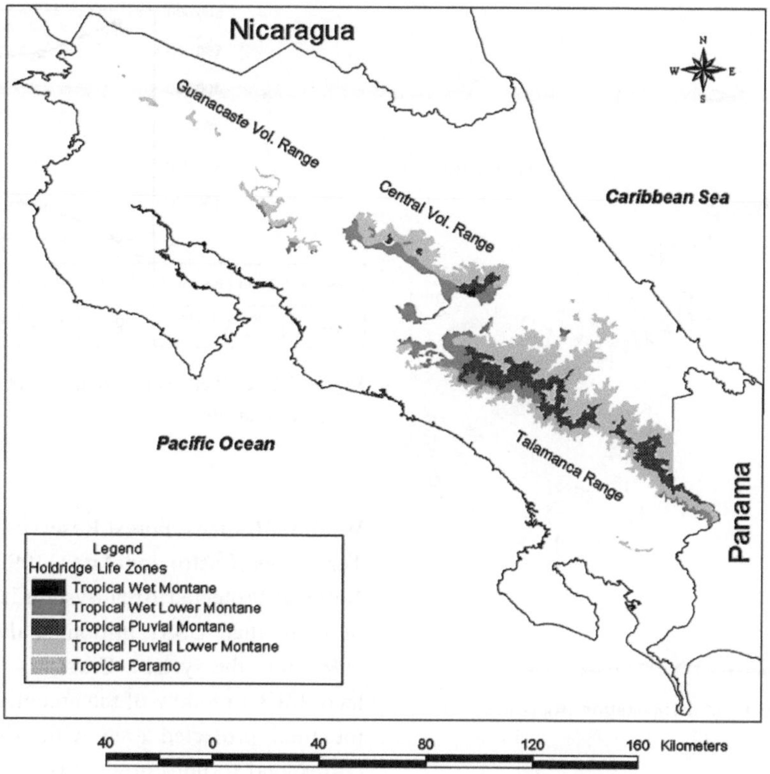

Figure 71.1. Potential distribution of montane cloud forests in Costa Rica according to the Holdridge Life Zone system.

territory (see also the estimate by Mulligan (this volume) based on the frequency of low cloud). These forests are typically located between 1400 and 3600 m.a.s.l. in the chief mountain ranges of Costa Rica, viz. the Talamanca Range (cf. Hölscher *et al.*, this volume), the Central, and the Guanacaste Volcanic Range (Figure 71.1). According to a classification of Landsat 7 ETM+ satellite imagery taken in the year 2000 (Sánchez-Azofeifa *et al.*, 2002), almost 83% of the total original area of MCF estimated in this way had remained under forest cover in 2000 (Figure 71.2). The study did not distinguish secondary growth from old-growth forest. However, because the deforestation process was largely under control around the year 1980 and because the steep terrain and high rainfall of these headwater areas are not conducive to forest clearing, most of the detected forest is probably old-growth or mature forest.

Deforestation in the MCF zone is mostly associated with the expansion of dairy and horticulture farming on gentle to moderate slopes with fertile volcanic soils, mainly within the Wet Lower Montane Life Zone (TwLM) on the Pacific Slope (Figures 71.2 and 71.3). Up to the year 2000, the cumulative conversion of MCF was estimated to be around 100 000 ha, or about 17% of the assumed original MCF area (cf. Figure 71.2). Most of the remaining MCF in the year 2000, estimated to cover ~470 000 ha, is well preserved through the National Protected Areas System, as well as through Private Forest Reserves and conservation efforts promoted by the country's national Payments for

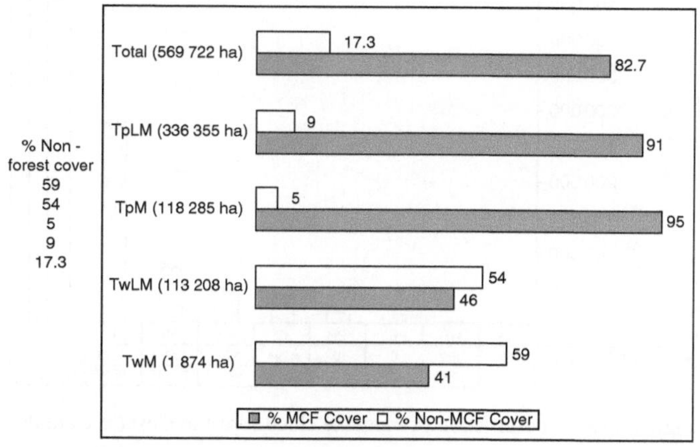

Figure 71.2. Distribution of actual and potential montane cloud forest per Life Zone (ha, %) in Costa Rica.

Environmental Services (PES) scheme (Figure 71.4). Each of these three categories will be discussed briefly below.

CONSERVATION OF MCF THROUGH NATIONAL PARKS AND PROTECTED AREAS

During the 1970s and 1980s, a series of technical reports predicted that, based on the high rates of deforestation taking place

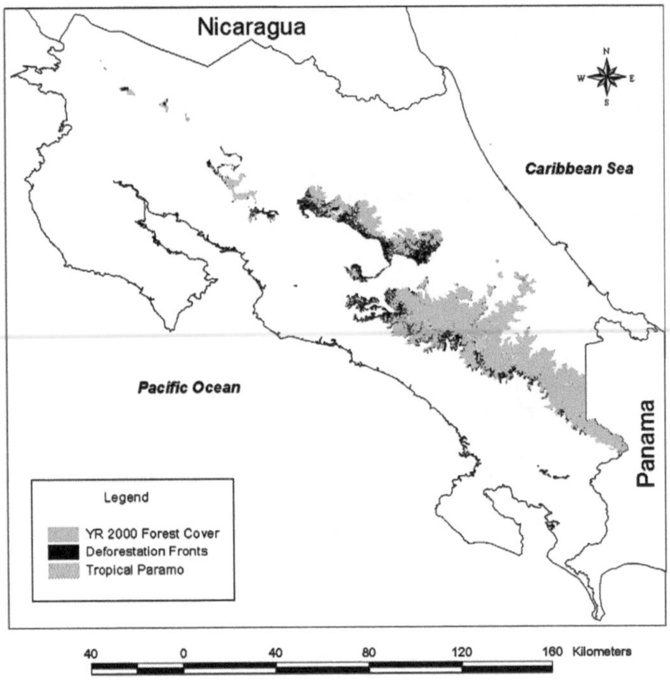

Figure 71.3. Montane cloud forest and deforestation fronts in Costa Rica in 2000.

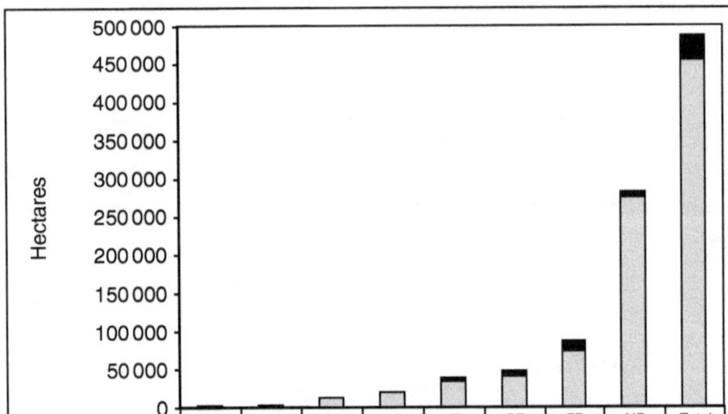

	BR	WR	PES	PR	IT	PZ	FR	NP	Total
■ Non-MCF Cover (ha)	12	152	0	0	4294	7255	13241	7150	32104
□ MCF Cover (ha)	1687	1794	12000	19000	33498	39774	72646	274028	454427

Figure 71.4. Conservation of montane cloud forest in Costa Rica under different protection systems (ha). BR, Biological Reserves; WR, Wildlife Refuges; PES, Payment for Environmental Services; PR, Private Reserves; IT, Indian Territories; PZ, Protected Zones; FR, Forest Reserves; NP, National Parks.

at the time (estimated to be more than 50 000 ha year^{-1}), most of Costa Rica's forest cover would disappear before the end of the year 2000 (Myers, 1981; Sader and Joyce, 1988; Calvo, 1990). The first response to counteract this undesirable situation was the creation of a System of National Parks and other Protected Areas in the year 1972. Currently, some 25% of the national territory (about 1 300 000 ha) is dedicated to conservation in the form of more than 132 different protected areas, grouped into the following categories: National Parks, Biological Reserves,

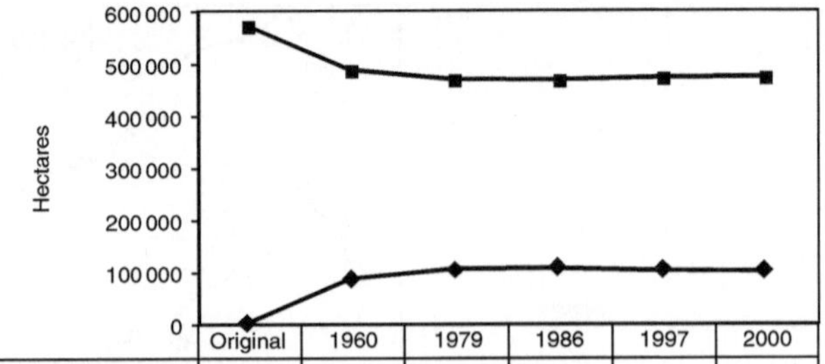

	Original	1960	1979	1986	1997	2000
MCF Cover Area per yr	569722	484798	466958	465925	469975	471431
Non-CMF Cover area per yr	0	84924	102764	103797	99747	98291

Figure 71.5. Changes in montane cloud forest cover (ha) in Costa Rica up to the year 2000.

Wildlife Refuges, Forest Reserves, Protected Zones, and Indian Territories (Castro and Arias, 1999). The decision to create the National Protected Area System has since led to an investment of more than US$1 billion (Calvo, 1990; Castro and Arias, 1999) into the system. Currently, the system protects 455 531 ha of MCF or ~80% of the potential MCF area (Figure 71.4). Of the total protected area with MCF, about 32 000 ha (7%) correspond to non-forest cover (mainly pasture); this is either due to the fact that most of this land is still under restoration or owned by private land-owners whose lands have not yet been expropriated. As shown in Figure 71.5, MCF deforestation processes became largely stabilized after 1979 as a direct consequence of the establishment of the National Protected Area System (Quesada *et al.*, 1998; Calvo *et al.*, 1999; Sánchez-Azofeifa *et al.*, 2002).

CONSERVATION OF MCF THROUGH PRIVATE RESERVES

Not all of Costa Rica's MCF is protected by the National Protected Areas System; some forests are protected by private biological reserves. The network of Costa Rican Private Reserves currently includes more than 75 associates that have established more than 50 000 ha of forest dedicated to conservation and ecotourism (Castro and Arias, 1999). Of these private reserves two institutions have led the effort to conserve a significant portion of MCF in Costa Rica, viz. the Monteverde Cloud Forest Biological Preserve of the Tropical Science Center, and the Children's Eternal Forest of the Monterverde Conservation League (cf. Nadkarni and Wheelwright, 2000). Together these two institutions preserve more than 27 000 ha of forest in the Tilarán Mountain Range in northern Costa Rica (of which 70% or 19 000 ha is situated within the MCF belt). The two Reserves also protect strategic water resources used by the country's major hydro-electricity generation project (Arenal Reservoir) and agricultural irrigation system

(Arenal–Tempisque irrigation project). Both institutions have been able to operate and maintain their Reserves through income generated by ecotourism, research, and payment for environmental services provided to third parties (see also below) as well as private donations (cf. Tognetti *et al.*, this volume)

CONSERVATION OF MCF BY PAYMENTS FOR ENVIRONMENTAL SERVICES

A national system for the Payment for Environmental Services (PES) provided by forest ecosystems was established and developed from 1996 onward as a compensation for forest owners wanting to conserve their forests instead of converting them to economically more profitable land uses (Rodriguez-Zuñiga, 2003; cf. Tognetti *et al.*, this volume). The PES system acknowledges such forest ecosystem services as: (i) greenhouse gas effect mitigation through carbon sequestration, (ii) water resources protection for urban, rural, or hydro-electric uses, (iii) biodiversity protection, and (iv) scenic value. All private forest owners committed to forest protection and sustainable management can benefit from this system (Castro and Arias, 1999).

The PES system is managed by the National Fund of Forest Financing of Costa Rica (FONAFIFO), which is the entity in charge of securing resources and paying the land-owners for environmental services provided by their forests. FONAFIFO's funds come mainly from three different sources: (i) a selective tax on consumption of fuels and other hydrocarbons, (ii) funds collected internationally based on the concept of selling global environmental services such as carbon storage (avoided deforestation) and biodiversity conservation, and (iii) direct negotiation with hydro-electric companies, hotels, and industries that directly benefit from the protection of forest (cf. Rodriguez-Zuñiga, 2003). The pioneering Costa Rican PES system has become a source of lessons for those considering PES as a conservation tool. A great deal of discussion about the generated experience can be found in the literature (e.g Landell-Mills and Porras, 2002; Pagiola, 2002; Rojas and Aylward, 2003; Hope *et al.*, 2007; Porras *et al.*, 2008; cf. Tognetti *et al.*, this volume)

From 1997 to 2003, FONAFIFO has involved more than 5468 small and medium forest holdings of farmers with a total of 378 096 ha, of which 23 1843 ha correspond to tree plantations, 326 845 ha to protected forests, and 28 065 ha to sustainable management of forests. Of this total more than 12 000 ha of MCF have been protected by this innovative program. FONAFIFO pays on average US\$200 ha^{-1} based on a 5-year contract; no distinction is made as yet between MCF and non-MCF (as practiced e.g. in Mexico; Muñoz-Piña *et al.*, 2008). The total amount paid to land-owners participating in the scheme between 1997 and 2003 is US\$75 million (www.fonafifo.com).

FONAFIFO has also been able to negotiate payments for water protection services rendered to several hydro-electric corporations. The first company to negotiate a PES arrangement was the Compañía Energía Global, which owns two hydro-electric projects in the Central Volcanic Range. On average, the company pays US\$10 ha^{-1} $year^{-1}$ whereas the transactions are carried out by FONAFIFO along with the Foundation for the Development of the Central Volcanic Range (FUNDECOR). Other companies to join this effort include the Compañía Nacional de Fuerza y Luz – which agreed to pay up to US\$40 ha^{-1} $year^{-1}$ for a hydro-electric project in the Aranjuez River watershed in the Tilarán Range – and Platanar Hydro-electric Company, which pays US\$15–30 ha^{-1} $year^{-1}$ to forest owners in the Platanar Watershed in the Central Volcanic Range.

There is a lack of quantitative information as to the actual extent to which (cloud) forests protect water resources, and to base valuation efforts upon (Pagiola, 2002; Rojas and Aylward, 2003; Porras *et al.*, 2008; Tognetti *et al.*, this volume). Therefore, the various prices fixed for the payment of water resources protection services in Costa Rica have resulted from negotiations between FONAFIFO and the respective corporations, as a balance between their willingness to pay and the perceived importance of the forests for the protection of these water resources. Involving all companies, the final goal is to protect approximately 18 000 ha of forest, most of which is classified as MCF (Calvo, 2000; ONF/PNUD/MINAE/PROFOR, 2002; Ortiz, 2002). To date, according to FONAFIFO's own records, it has been able to sign PES contracts covering more than 3000 ha of forest land and involving several hydro-electric companies and numerous farmers.

CONCLUSIONS

Montane cloud forests are amongst the best-preserved forest ecosystems in Costa Rica due to the timely establishment of National Parks and to the fact that MCF is geographically located in mountainous zones that are relatively inaccessible or have inappropriate biophysical conditions for the expansion of agriculture. The beauty and unique biodiversity of the MCF have attracted ecotourism and research opportunity, to the point that nowadays there are several success stories of private conservation in the Monteverde area. Forest conservation in Costa Rica has been advanced further by the creation of institutional, legal, and operative capacities to develop a Payments for Environmental Services scheme and a domestic water market, benefiting many farmers with small and medium-sized holdings who are willing to conserve the MCF. The authors believe that Costa Rica's accomplishments can be repeated in other Mesoamerican countries, with appropriate adjustments to fit other ecological, social, and political situations (see also Porras *et al.*, 2008).

REFERENCES

Calvo, J. (1990). The Costa Rican national conservation strategy for sustainable development: exploring the possibilities. *Environmental Conservation* **17**: 355–358.

Calvo, J. (2000). Case study: payments for watershed services in Costa Rica. In *Forests in Sustainable Mountain Development: A State-of-Knowledge Report for 2000*, eds. M. F. Price and N. Butt, pp. 428–429. Wallingford, UK: CAB International.

Calvo, J., C. Quesada, A. Sánchez-Azofeifa, *et al.* (1999). La cobertura forestal en las Áreas Protegidas de Costa Rica entre 1986/7 y 1996/7. *Ciencias Ambientales* **16**: 19–25.

Castro, R., and G. Arias (1999). *Costa Rica: Toward the Sustainability of its Forest Resources.* San José, Costa Rica: Fondo Nacional de Financiamiento Forestal (FONAFIFO).

Grubb, P. J. (1977). Control of forest growth and distribution on wet tropical mountains: with special reference to mineral nutrition. *Annual Review of Ecology and Systematics* **8**: 83–107.

Hamilton, L. S., J. O. Juvik, and F. N. Scatena (1995). The Puerto Rico tropical cloud forest symposium: Introduction and workshop synthesis. In *Tropical Montane Cloud Forests*, eds. L. S. Hamilton, J. O. Juvik, and F. N. Scatena, pp. 1–8. New York: Springer-Verlag.

Holdridge L. (1967). *Life Zone Ecology.* San José, Costa Rica: Tropical Science Center.

Hope, R. A., I. T. Porras, M. Borgoyary, *et al.* (2007). *Negotiating Watershed Services.* London: International Institute for Environment and Development.

Landell-Mills, N., and I. T. Porras (2002). *Silver Bullet or Fools' Gold? A Global Review of Markets for Forest Environmental Services and Their Impact on the Poor.* London: International Institute for Environment and Development.

Muñoz-Piña, C., A. Guevara, J. M. Torres, and J. Braña (2008). Paying for the hydrological services of Mexico's forests: analysis, negotiations and results. *Ecological Economics* **65**: 725–736.

Myers, N. (1981). The hamburger connection: how Central America's forests became North America's hamburgers. *Ambio* **10**: 3–8.

Nadkarni, N. M., and N. T. Wheelwright (eds.) (2000). *Monteverde: Ecology and Conservation of a Tropical Cloud Forest.* New York: Oxford University Press.

ONF/PNUD/MINAE/PROFOR (2002). *El éxito forestal de Costa Rica en cinco casos.* San José, Costa Rica: Oficina Nacional Forestal.

Ortiz, E. (2002). *Sistema de cobro y Pago de Servicios Ambientales*, Serie de Apoyo Académico no. 34. Cartago, Costa Rica: Escuela de Ingeniera Forestal, Instituto Tecnológico de Costa Rica.

Pagiola, S., (2002). Paying for water services in Central America: learning from Costa Rica. In *Selling Forest Environmental Services: Market-Based Mechanisms for Conservation and Development*, eds. S. Pagiola, J. Bishop, and N. Landell-Mills, pp. 36–60. London: Earthscan.

Porras, I. T., M. Grieg-Gran, and N. Neves (2008). *All That Glitters: A Review of Payments for Watershed Services in Developing Countries.* London: International Institute for Environment and Development.

Quesada, C., A. Sánchez-Azofeifa, and J. Calvo-Alvarado (1998). *Estudio de cambios de cobertura forestal de Costa Rica, 1987–1997.* San José, Costa Rica: Centro Científico Tropical, Universidad de Costa Rica, and Conservation International.

Rodriguez-Zuñiga, J. M. (2003). Paying for forest environmental services: the Costa Rican experience. *Unasylva* **212**: 31–33.

Rojas, M., and B. Aylward (2003). *What Are We Learning from Experiences with Markets for Environmental Services in Costa Rica? A Review and Critique of the Literature.* London: International Institute for Environment and Development.

Sader, S. A., and A. T. Joyce (1988). Deforestation rates and trends in Costa Rica, 1940 to 1983. *Biotropica* **20**: 11–19.

Sánchez-Azofeifa, A., J. Calvo-Alvarado, S. Foley, *et al.* (2002). *Estudio de cambios de cobertura forestal de Costa Rica 1997–2000.* Edmonton, Canada: Alberta University, and San José, Costa Rica: Centro Científico Tropical.

72 Tropical montane cloud forests: state of knowledge and sustainability perspectives in a changing world

L. A. Bruijnzeel
VU University, Amsterdam, the Netherlands

M. Kappelle
University of Tennessee, Knoxville, Tennessee, USA and The Nature Conservancy, San José, Costa Rica

M. Mulligan
King's College London, London, UK

F. N. Scatena
University of Pennsylvania, Philadelphia, Pennsylvania, USA

INTRODUCTION

As indicated in the introductory chapter to this book, knowledge of tropical montane cloud forest (TMCF) occurrence, biodiversity, hydrology, and ecological functioning has increased considerably since the ground-breaking publications of Zadroga (1981), Stadtmüller (1987), and the proceedings of the first international symposium on TMCF held in 1993 in Puerto Rico (Hamilton *et al.*, 1995a). Cloud forests continue to be threatened in several ways, notably by their conversion to pasture and various forms of agriculture, as well as by climatic drying – the numerous hydrological and ecological consequences of which are only poorly understood as yet (Bubb *et al.*, 2004; Mulligan and Burke, 2005a; Pounds *et al.*, 2006; Zotz and Bader, 2009).

The collection of chapters in the present volume further advances our knowledge in the three broad and interrelated areas that were defined in the introduction, viz. (i) cloud forest biogeography and biodiversity, (ii) biophysical and ecological processes, and (iii) management and conservation strategies.

As the recognition of the value of TMCFs as treasure houses of biodiversity and as providers of high-quality water continues to increase, an array of initiatives aimed at their conservation has emerged in recent years, often within a Payment for Ecosystem Services (PES) context (Asquith and Wunder, 2008; Muñoz-Piña *et al.*, 2008; Porras *et al.*, 2008; Garriguata and Balvanera, 2009; Tognetti *et al.*, this volume). Such PES schemes, but also land and forest managers and policy-makers in general, need to determine (amongst others) which forests under their jurisdiction are the most diverse and valuable biologically, which ones provide the best water supplies, which forests are the most vulnerable to climate change or most threatened by encroachment, and which degraded TMCFs have the best chances for rehabilitation. In short, there is a great need for site-specific information on TMCFs for incorporation into conservation and management plans at various spatial scales.

This final chapter summarizes the current state of knowledge with respect to the three broad categories distinguished above (biodiversity, biophysical functioning, and conservation and management issues including response to a variety of threats). In doing so, a distinction will be made (where possible and necessary) between the different types of cloud forests defined in the introductory chapter (Table 1.1), viz. tall lower montane cloud forest (LMCF), upper montane cloud forest (UMCF) of intermediate stature, and stunted sub-alpine (SACF) and elfin cloud forests (ECF). The rationale for making such a distinction lies in the wetter and cooler conditions that are generally encountered as one moves up from the lower montane to the upper montane and sub-alpine belts, and which are believed to affect the hydrological and ecological

Tropical Montane Cloud Forests: Science for Conservation and Management, eds. L. A. Bruijnzeel, F. N. Scatena, and L. S. Hamilton. Published by Cambridge University Press.

functioning of the respective forest types (Grubb, 1977; Silver *et al.*, 1999; Hafkenscheid, 2000; Bruijnzeel, 2001; Gerold *et al.*, 2008; Roman *et al.*, this volume). The chapter also makes a deliberate attempt at answering the key questions raised in the introductory chapter for each of the above-mentioned broad categories (Scatena *et al.*, this volume) and concludes with a brief discussion of the main global cloud forest conservation initiatives, key remaining information needs, and future research challenges.

CURRENT STATE OF KNOWLEDGE

Global extent, distribution, and climate of tropical montane cloud forests

Key questions regarding the current extent, distribution, and climates of TMCFs include:

- Where do we find cloud forests and under what climatic conditions?
- How much is left of the original forest?

Producing a global map of actual TMCF distribution has proved a daunting task, as described in more detail in the introductory chapter (Scatena *et al.*, this volume). However, knowledge of where TMCFs occur has increased considerably since the 1993 Puerto Rico symposium when the first map was made by having symposium participants place tacks on maps in areas where they were "reasonably confident" that TMCF existed (Hamilton *et al.*, 1995b). In the pioneering analysis of Aldrich *et al.* (1997b) and the second-generation reiteration by Bubb *et al.* (2004), global forest cover assessments were combined with global elevation data and regionally specific altitudinal limits of cloud forests (see Table 1.2 in Scatena *et al.*, this volume) to derive estimates of the potential (i.e. assuming the entire area between the specified altitudinal limits to be forested) and actual areas of TMCF. Using this approach, the *potential* area of TMCF was estimated to be in the order of 380 000 km^2 (2.5% of the world's tropical forest area), whereas the *actual* area with TMCF (in the year 2000) was estimated at *c.* 215 000 km^2 (1.4% of the world's tropical forests; cf. Tables 1.3 and 1.4 in this volume). The area under cloud forest estimated in this way constitutes 6.6% of the area covered by tropical montane forest in the world as defined by Kapos *et al.* (2000). Taking these estimates of actual and potential cloud forest occurrence at face value implies that roughly 56% of the original forest still remains and that *c.* 165 000 km^2 of cloud forest have been converted. As to their relative distribution, approximately 43% of all remaining cloud forests can be found in Asia (including northern Australia) and Oceania, 41% in the Americas (including the Hawaiian archipelago) and 16% in Africa (cf. Table 1.3).

No distinction was made by Bubb *et al.* (2004) between different types of cloud forest in terms of forest stature, degree of cloud incidence or rainfall regime (see Table 1.1 and Scatena *et al.* (this volume) for general descriptions). Nevertheless, the more than 560 sites with confirmed cloud forest presence (Aldrich *et al.*, 1997b; Figures 72.1a–c) represent a wide range of climatic conditions (rainfall and temperature, wind) and landscape settings (elevation and exposure, mountain size, distance to sea). Jarvis and Mulligan (this volume) employed spatial data-sets derived from the WorldClim database, to describe the climate at 477 "cloud forest" sites as identified by UNEP–WCMC. Further, comparisons were made between the climate of cloud forest sites and that of randomly generated sites covering forested areas throughout the montane tropics, with the aim of identifying the climatic variables most important in distinguishing TMCF from other tropical forests. TMCFs were found to be wetter (by 184 mm $year^{-1}$ on average), cooler (by 4.2 °C on average), and less seasonally variable than other montane forests. The most statistically significant differences in climate between TMCFs and other montane forests were: maximum temperature, mean temperature, rainfall, and rainfall seasonality (in order of significance). Cloud forests also tend to be located closer to the coast (particularly in Asia) and at higher altitudes than montane forests not affected by cloud. Furthermore, cloud forests occupy more topographically exposed areas than other montane forests. Interestingly, cloud forest sites in Africa tend to be drier (average annual rainfall less than 1500 mm) and more seasonal than their counterparts in Asia (*c.* 2150 mm of rain on average), whereas the cloud forests of the Latin American and Caribbean region experience a wide range of rainfall conditions (Jarvis and Mulligan, this volume).

Using a hydro-climatic definition of cloud forest as "forest under frequent and/or persistent exposure to ground-level cloud," Mulligan (this volume) modeled their distribution using new tropics-wide data-sets on satellite-observed cloud frequency, climate and land cover. This approach yielded a much greater global extent of cloud forest cover (2.21 Mkm^2) than the previously discussed altitude-based estimates, suggesting that "significantly cloud-affected" forests (CAF) may represent as much as 14.2% of all tropical forests (cf. Table 2.4). The hydro-climatic approach was tested with a high level of success against the more than 560 sites listed by WCMC-UNEP as having confirmed cloud forest presence (Aldrich *et al.*, 1997b), with the best fit between actual and modeled cloud forest presence for sites experiencing ground-level cloud (i.e. fog) conditions for more than 70% of the time (Mulligan, this volume; Figure 72.2). Comparing the potential and actual distributions of CAF obtained in this way suggests that ~45% of the original CAF still remains and that some 2.72 Mkm^2 of CAF have been lost (Mulligan, this volume).

Differences in results obtained with the two approaches (i.e. elevation- vs. climate-based) are to be expected. The greater

(a)

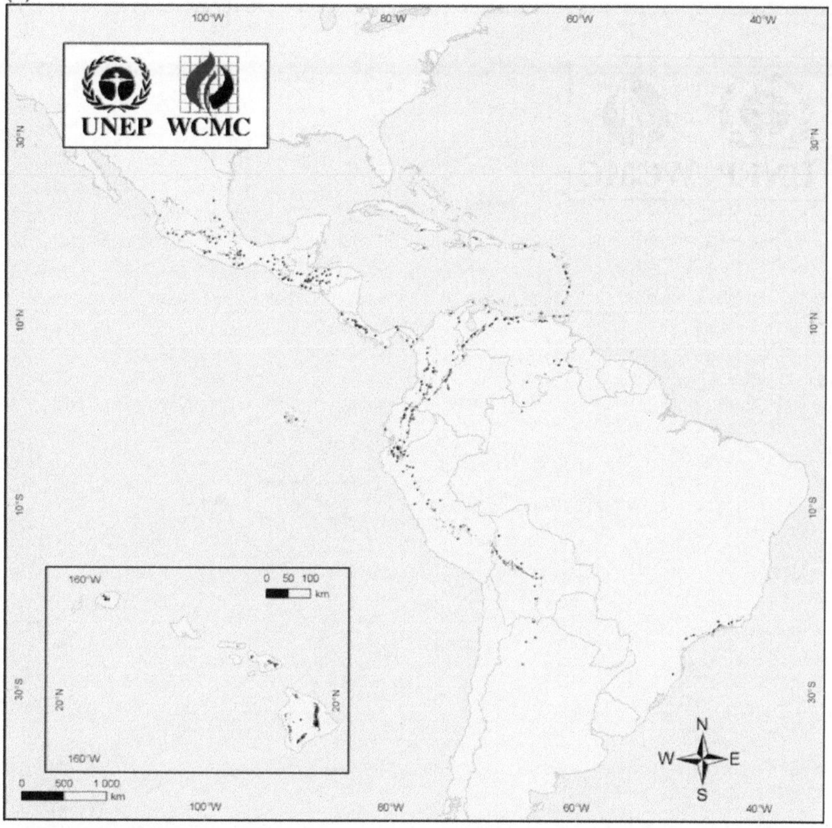

(b)

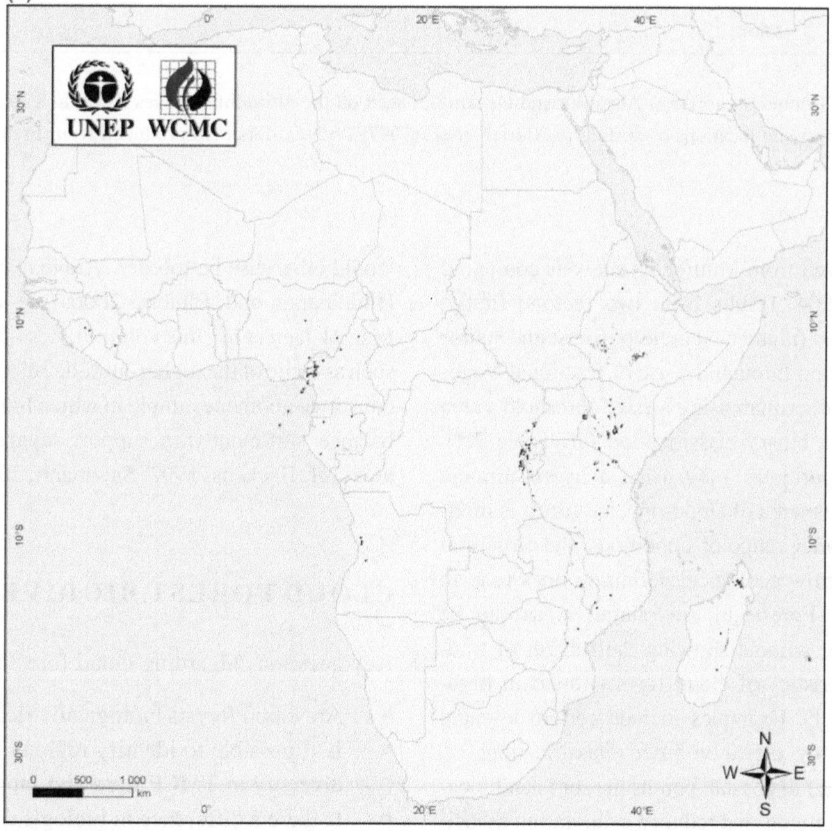

Figure 72.1. Estimated tropical montane cloud forest distribution in (a) the Americas (including the Hawaiian archipelago), (b) Africa and

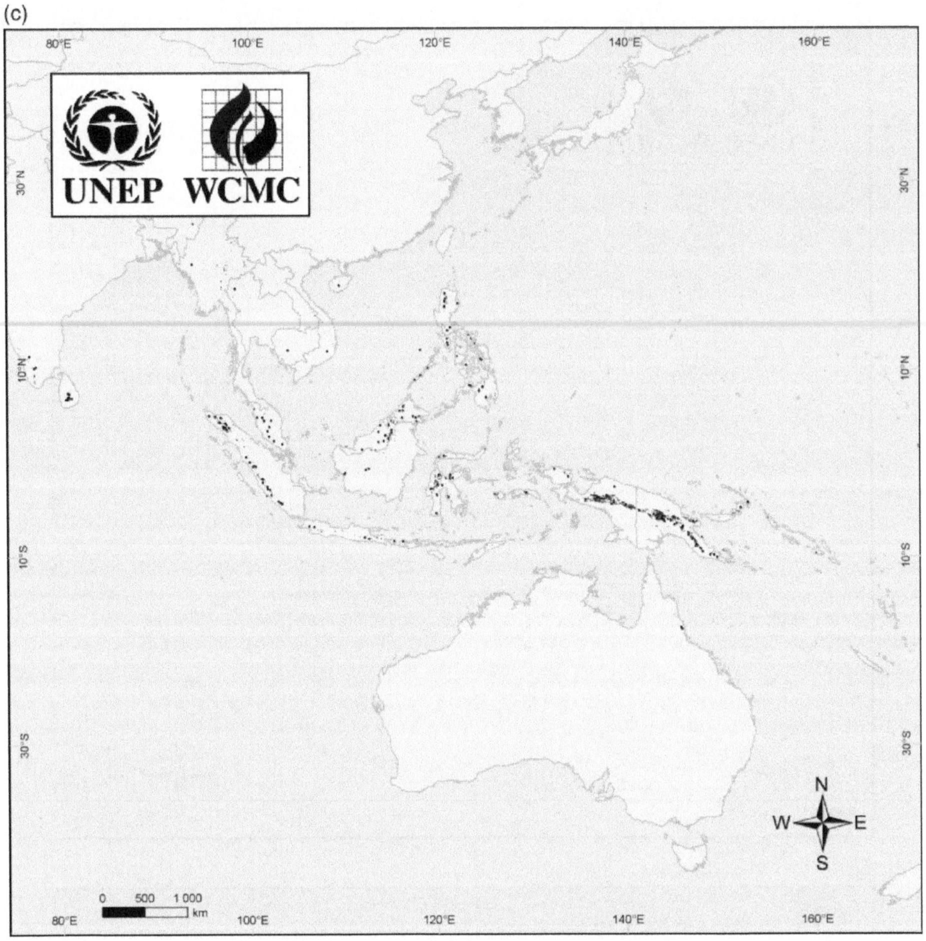

Figure 72.1. (*cont.*) (c) Asia (including northern Australia and Oceania), based on the altitudinal ranges listed in Table 1.2 (areas indicated by red shading) and known cloud forest site locations according to Aldrich *et al.* (1997b) (green dots). (Reproduced from Figures 1.3, 1.4, and 1.5; See also color plate.)

extent of cloud forest derived from Mulligan's analysis compared to that of Bubb *et al.* (2004) results from two factors; firstly, Mulligan's analysis included fragmented areas of forest and sparse woodland at 500-m resolution through the use of fractional Vegetation Continuous Fields rather than using a fixed threshold value to denote forest cover in a binary classification (cf. Table 2.4). Secondly – and more importantly – by using a hydro-climatic definition of cloud forest, the area of cloud-*affected* forest is modeled. This may cover a wider range of conditions and structural features than the more narrow ecological definition does (e.g. in terms of mossiness, etc.). Forests in low-rainfall areas can be significantly cloud-affected without showing the full set of traditional ecological characteristics of cloud forests found in high-rainfall areas (cf. Table 1.1). Examples include certain lowland forests (not included in any elevation-based classification; cf. Gradstein *et al.*, this volume) in coastal, peninsular, or island locations where the extra water inputs and reduced evaporation associated with fog occurrence support forest or woodland in areas that would otherwise be too dry (Aravena *et al.*, 1989; Dawson, 1998; Hildebrandt and Eltahir, 2006; see also García-Santos, 2007; Marzol-Jaén *et al.*, this volume). Less-dense woodland ecosystems such as some of the higher-altitude Miombo woodlands of Tanzania constitute another example in which fog may improve the site water balance sufficiently to support savanna in grassland-dominated areas (cf. Backeus, 1992; Salzmann, 2000).

CLOUD FOREST BIODIVERSITY

Key questions regarding cloud forest biodiversity include:

- Are cloud forests biologically richer than non-cloud forests?
- Is it possible to identify regions and areas with the greatest diversity in TMCF flora and fauna?
- Is there a difference in biological richness between different altitudinal belts on single mountains?

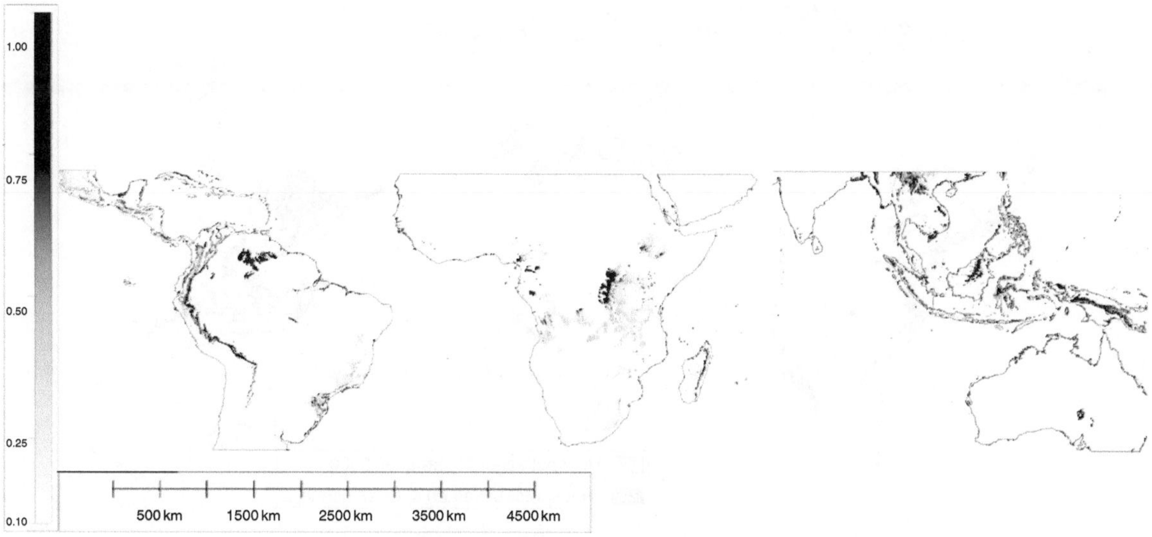

Figure 72.2. Modeled distribution of cloud-affected tropical montane forests (with UNEP–WCMC listed cloud forest sites indicated in red). The darker the shade the greater the fractional cover of forest within the 1-km pixel. (Reproduced from Figure 2.11; See also color plate.)

General considerations

Although the global area of (narrowly defined) TMCF is only about 0.26% of the Earth's land surface (Bubb *et al.*, 2004), it is widely acknowledged that these particular forests are amongst the most species-rich and species-dense ecosystems worldwide, both in terms of flowering plants (Grubb, 1977; Prance, 1982; Stadtmüller, 1987; Groombridge, 1992; Barthlott *et al.*, 1996, 2005; Kappelle, 2004; Mutke and Barthlott, 2005; Richter, 2008; Hemp, this volume #12; Meyer, this volume), mosses (Benzing, 1990; Gradstein *et al.*, 2001), birds (Terborgh, 1977; Long, 1995; Fjeldså and Rahbek, 1997; Watson and Townsend Peterson, 1999), mammals (Leo, 1995; Heaney, 2001; McCain, 2005; cf. Owiunji and Plumptre, this volume), amphibians (e.g. frogs, toads, lizards, salamanders; Scott, 1976; Duellman, 1988; Pineda and Halffter, 2004; Cortés *et al.*, 2008), and reptiles (Scott, 1976; Wilson and McCranie, 2004).

TMCFs also appear to be particularly numerous in at least six of the most important biodiversity "hot spots" around the globe, viz. the tropical Andes and Atlantic Forests in South America, the Caribbean Islands, the Eastern Afro-montane peaks scattered along the eastern edge of Africa, the Asian Sundaland which includes the high mountains of Borneo, and Polynesia–Micronesia which includes the islands of Fiji, Hawai'i, and Tahiti (Myers *et al.*, 2000; Figure 72.3).

In addition, TMCFs occur within a significant portion of the 200 biologically most distinct "Ecoregions" of the planet as identified by WWF-US (Olson and Dinerstein, 2002). Ecoregions are defined as large areas of land or water that contain a geographically distinct assemblage of natural communities that share a large majority of their species and ecological dynamics, share similar environmental conditions, and interact ecologically in ways that are critical for their long-term persistence (Dinerstein *et al.*, 1995; Olson *et al.*, 2001). These "Global 200" Ecoregions were selected on the basis of various features of biodiversity, including species richness, number of endemic species,[1] presence of unusual higher taxa, unusual ecological or evolutionary phenomena, and the global rarity of habitats (Olson and Dinerstein, 2002). The 18 most important "Global 200" Ecoregions (i.e. 9% of the total) that contain tropical cloud forests (and which partly overlap with the above-mentioned six hot spot areas) are, listed per biome: *Afrotropical*: Cameroon Highland forests, Albertine Rift montane forests, Eastern Arc montane forests; *Australasian*: New Guinea montane forests; *Indo-Malayan*: lowland and montane forests in Sumatra and surrounding islands, Peninsular Malaysia, and Borneo, and the montane forests of West Java; *Neotropical*: Greater and LesserAntillean moist forests, Talamancan–Isthmian Pacific forests, northern Andean montane forests, coastal Venezuelan montane forests, Guayana Highland moist forests, Central Andean Yungas, Brazilian Atlantic forests (*Mata Atlantica*); and *Oceania*: South Pacific Island forests, and Hawaiian moist forests (Olson and Dinerstein, 2002; cf. Figure 72.3).

[1] Endemism refers to the phenomenon that the spatial distribution of a taxon (usually a species) is restricted to a narrowly defined local geographic area. Species that occur only in that specific location and nowhere else in their natural state are called endemics.

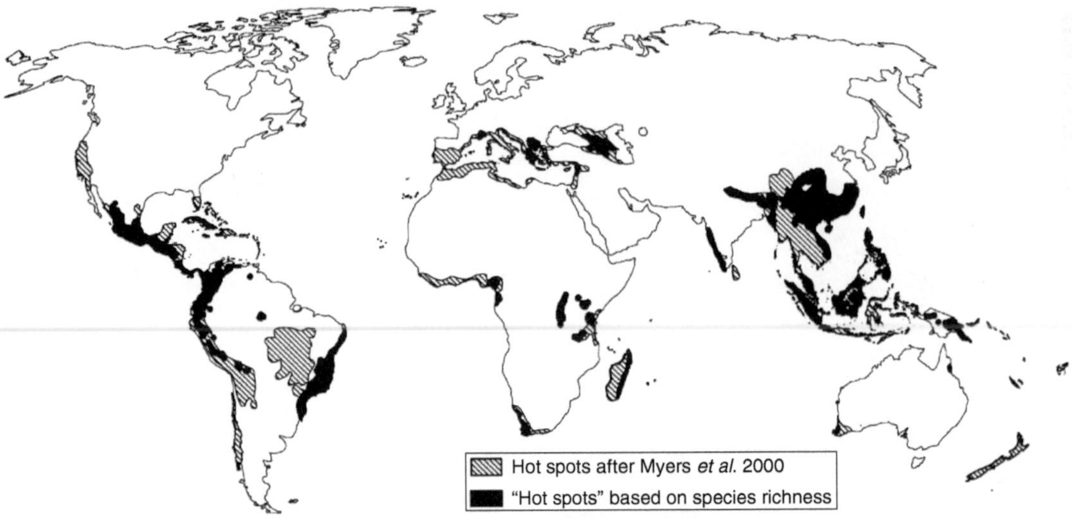

Figure 72.3. Global biodiversity hot spots according to Myers *et al.* (2000) compared with areas with at least 3000 vascular plant species per 10 000 km^2 (Mutke and Barthlott, 2005).

Overall diversity tends to decline away from the equator and – at least for vascular plants – has been explained in terms of temperature, moisture availability, and topographic heterogeneity, with the latter becoming particularly important in humid montane areas (Kreft and Jetz, 2007). The extraordinary diversity of such taxonomic groups as bryophytes, vascular plants, birds, and amphibians in cloud forests is often the result of past explosive evolutionary radiation (e.g. Gentry, 1982, 1988, 1992; Duellman, 1988; Benzing, 1990; Long, 1995; Gradstein *et al.*, 2001). The tropical Andes, for example, appears to have served as an important center of speciation for a variety of taxa including mosses and liverworts (Benzing, 1990; Wolf, 1993, 1995; Nadkarni *et al.*, 1995), orchids and bromeliads (Sugden and Robins, 1979; Wolf, 1994), ericaceous herbs and shrubs (Luteyn, 1989), bamboos (Clark, 1995), ferns (Kessler, 2002; Kessler and Lehnert, 2009), frogs and salamanders (Duellman, 1988), hummingbirds (Bleiweiss, 1998), and other avian groups (Fjeldså and Rahbek, 1997), as well as geometrid moths (Brehm *et al.*, 2007, 2008). Because of the high levels of endemism in TMCFs, new species are still regularly discovered. For example, only two decades ago a whole new tree family (Ticodendraceae, restricted to Neotropical TMCFs) was discovered in Costa Rica (Gómez-Laurito and Gómez, 1991), a country where the study of TMCFs may be considered as being amongst the most advanced in the tropical world. Similarly, even in well-known groups such as birds, several new narrowly endemic species were discovered in the Ecuadorian, Peruvian, and Colombian Andes in the 1990s (Krabbe *et al.*, 1999; O'Neill *et al.*, 2000; Cuervo *et al.*, 2001; Salaman *et al.*, 2003), while Brehm *et al.* (2008) expect to find at least 200 additional species of geometrid moths along an elevational transect (1040–2670 m.a.s.l.) in the montane forests of southern Ecuador, despite previous intensive sampling over 4 years during which some 1223 species of geometrid moths had been identified already. Such extraordinary richness in species and endemics is thought to reflect the interaction of isolation and the creation of numerous new habitat niches during the uplift of the Andes (Gentry, 1982, 1988). Also, past climatic fluctuations (including glacial cycles) caused the montane forest altitudinal belts in Central and South America to move up and down over considerable elevational distances (Van der Hammen, 1974; Van der Hammen and Hooghiemstra, 2001; Islebe and Hooghiemstra, 2006). As a result of this dynamic geoclimatic history, the modern-day flora of Neotropical montane cloud forests is made up of a blend of pan-tropical, Neotropical, Andean-endemic, northern temperate (Holarctic), southern temperate (Austral-Antarctic), and cosmopolitan plant genera (e.g. Kappelle *et al.*, 1992; Van der Hammen and Hooghiemstra, 2001; Holz and Gradstein, 2005a), thereby adding to their overall diversity (cf. Kreft and Jetz, 2007). Other examples of high diversity are given in the subsequent section on regional patterns of species and endemism richness.

Are cloud forests biologically richer than non-cloud forests?

There are as yet no pan-tropical analyses that explicitly indicate the most important cloud forests with respect to overall floristic or faunal biodiversity. However, the hot spots of vascular plants identified by Mutke and Barthlott (2005) and the overall biodiversity hot spots identified by Myers *et al.* (2000) shown in Figure 72.3 exhibit a striking overlap with the distribution of cloud-affected forests derived by Mulligan (this volume) in Figure 72.2.

There are several approaches through which one can go beyond the individual plot- or transect-scale study of species occurrence or richness in particular taxonomic groups in cloud forests so as to obtain a pan-tropical overview. Quantification of the total biodiversity at many TMCF sites is a near-impossibility (Brehm *et al.*, 2008), let alone for all cloud forests worldwide. However, by concentrating on some of the better-known groups (such as flowering plants, mammals, amphibians, and birds) one can make use of the best available data on their known ranges to obtain an indication of biogeographic patterns of richness and endemism. Pan-tropical to global studies of this nature exist for vascular plants (Mutke and Barthlott, 2005; Kreft and Jetz, 2007), and plant diversity in general (Kier *et al.*, 2005), for threatened vertebrates (Grenyer *et al.*, 2006), as well as for various environmental proxies for biodiversity (e.g. Kleidon and Mooney, 2008). Range data are available for threatened amphibians (IUCN *et al.*, 2008a) and mammals (IUCN *et al.*, 2008b), as well as for birds of the western hemisphere (Ridgely *et al.*, 2007) and these were used as a means of understanding the relative biological importance of cloud forests and the potential geographic variation in biological importance across the known cloud forests. The range maps included each of the 6260 species of amphibians known to science (as of the end of 2007), 5488 species (including *Homo sapiens*), 412 sub-species, and 21 sub-populations of mammals, and 4273 species of birds. This information was developed through an extensive process of scientific contribution and review coordinated by the World Conservation Union (IUCN), Conservation International (CI), The Nature Conservancy (TNC), and various universities (see Appendix 72.1 for details). Despite their various limitations (described in Appendix 72.1) and despite the fact that the bird data pertain to the western hemisphere only, these are some of the best available data for a pan-tropical assessment of cloud forest faunal biodiversity. There are no similar data for vascular plants in the public domain though previous analyses confirm the broad patterns of diversity observed for the groups analyzed here (cf. Kier *et al.*, 2005; Mutke and Barthlott, 2005).

Kier and Barthlott (2001) defined endemism richness as the "specific contribution of an area to global biodiversity" and regarded this as a better measure of conservation value than species richness per se as endemism richness accounts for range-size rarity. The contribution of a specific area to the global species inventory is known as its *C*-value. For the 10×10 km raster grids used here the *C*-value of a pixel for a given species was calculated as $1/G$, where G is the global range size (i.e. the number of pixels in which the species occurs). Where species have a large range, the *C*-value is low, and where their range is restricted, the *C*-value is high. Endemism richness of a site is then the sum of *C*-values for all the species for which range data are available. Species richness and endemism richness were calculated for each group and then averaged within areas defined

Table 72.1 *Measures of species richness and endemism richness inside and outside of cloud forests*

Group	Mean species richness: Cloud forest	Mean species richness: Non-cloud forest	Mean endemism richness: Cloud forest	Mean endemism richness: Non-cloud forest
Amphibians	24.69	23.50	0.027	0.0034
Mammals	107.75	74.65	0.014	0.0037

as cloud-affected forest (according to the hydro-climatic definition of Mulligan, this volume) and also for areas outside of these CAFs.

Table 72.1 compares mean species- and endemism-richness values inside and outside of cloud-affected forests (i.e. comparing cloud forests as an ecosystem against all other terrestrial ecosystems) for mammals and amphibians across the tropics and for birds within the Neotropics. For amphibians, CAFs do not appear to be more species-rich on average than the (much more extensive) non-cloud forest ecosystems, but they are an order of magnitude richer in endemics. *Thus, for amphibians, CAFs contribute an order of magnitude more to global biodiversity than do non-cloud forest areas.* As for mammals, species richness in CAF is distinctly higher than outside of cloud forests (by more than 40%), whereas *endemism richness for mammals in CAFs is almost four times higher than in non-cloud-affected forests* (Table 72.1). Although the mean bird species richness in tropical Latin American cloud forests is similar to that outside of cloud forests, CAFs cover only 4.73% of the land surface of tropical Latin America and thus support many more bird species per unit area. Also, endemism richness for birds is 3.4 times higher within cloud-affected forests (Table 72.1).

Can regions and areas with the greatest diversity in TMCF flora and fauna be identified?

FLORA

As indicated earlier, there are no global surveys of plant diversity focusing explicitly on TMCF. However, the general patterns of species richness of vascular plants shown in Figure 72.4 gives some indication and confirms the importance of the various biodiversity hot spots containing areas with cloud forest as identified earlier in Figure 72.3. The following paragraphs briefly discuss the floristic richness of several of these areas in the respective regions.

Costa Rican TMCF is estimated to hold up to at least 5000 vascular plant species (Kappelle, 2001, 2006), while similar forests in Asia (e.g. in Peninsular Malaysia and Borneo) may harbor up to 4000 species (Kumaran *et al.*, this volume). In the Costa Rican cloud forest area of Monteverde alone

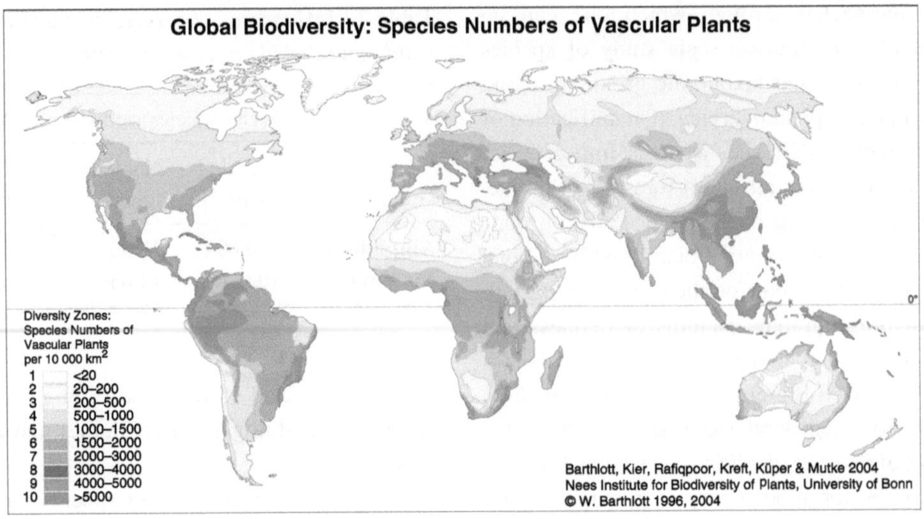

Figure 72.4. World map of species richness of vascular plants (Mutke and Barthlott, 2005). (See also color plate.)

(700–1840 m.a.s.l.) a total of 3020 different plant species have been collected – roughly corresponding with 25–30% of all Costa Rica's plant species – including at least 755 species of tree and 350 fern species (Haber, 1991, 2000; Nadkarni *et al.*, 1995; Bigelow and Kukle, 2000). Further south in Costa Rica, along the Pacific and Atlantic slopes of Mt. Chirripó (3819 m.a.s.l.), almost a hundred fern species, several gymnosperms (Podocarpaceae), nearly 300 terrestrial dicots (trees, shrubs, herbs, lianas, etc.), as well as some 50 monocots (bamboos, palms, cyclanths, orchids, bromeliads, aroids, etc.) were collected from oak-dominated cloud forest stands between 2000 and 3200 m.a.s.l. (Kappelle *et al.*, 1995b; Kappelle and Van Uffelen, 2006).

In Mexico, cloud forests cover less than 1% of the country's territory but they contain about 12% of its 3000 plant species (Rzedowski, 1996). Up to 30% of these are endemic to the country. All in all, about half of the 90 000 known higher plant species in the Neotropics – the world's richest flora – are found in the montane zone (Churchill *et al.*, 1996).

African tropical high mountains (Mt. Kilimanjaro, Mt. Elgon, Mt. Kenya, Mt. Meru, and the Ruwenzori Mountains) and their cloud forests are also characterized by a high biodiversity and a high degree of plant endemism (Hemp, this volume #12). The Eastern Arc Mountains in Tanzania are considered true "hot spots" of biodiversity and endemism due to (more or less) constantly favorable climatic conditions and the presence of a permanent forest cover over very long periods of time (Rodgers and Homewood, 1982; Newmark and Senzota, 2003; C. Hemp, this volume). Over 1200 plant species have been reported for the TMCFs on the slopes of Mt. Kilimanjaro (5895 m.a.s.l.), representing about half of the vascular flora on this mountain (Hemp, this volume #12). On Mt. Kinabalu (4101 m.a.s.l.) – the highest mountain in Asia between the Himalayas and the snow-capped peaks of New Guinea – plant diversity is equally remarkable. Together with a portion of adjacent sub-montane forest, the 745 km^2 Kinabalu Park contains representatives from more than half the families of all flowering plants. The mountain boasts 1000 or more species of orchids. This represents about half the number of orchid species found on the entire island of Borneo, which itself counts about one-tenth of the world total. In addition, over 600 species of fern are found on Mt. Kinabalu, representing one-fifth of the world's pteridophytes. The particular biological richness found on the mountain – despite its comparatively young age (*c.* 1 million years) – is believed to reflect the fact that Mt. Kinabalu acted as a refuge for montane plants from other high-elevation parts of Borneo that were eroded away at a time when Mt. Kinabalu was thrust up (Luping *et al.*, 1978; McKinnon *et al.*, 1996; cf. Kumaran *et al.*, this volume).

In the Pacific Ocean, many Polynesian island groups (e.g. Samoa, Tahiti in the Society Islands, and the Cook, Hawaiian, Austral, and Marquesas Islands) share a common set of flowering plants (Meyer, this volume). Furthermore, TMCFs in French Polynesia (e.g. Tahiti) contain the floristically most diverse of all plant communities, with the highest rate of endemism and the highest number of biological types (Meyer, this volume).

FAUNA

The distribution of species richness and endemism richness (or at least current knowledge of these) within cloud-affected forests for amphibians and mammals is shown in Figures 72.5 and 72.6, respectively. With respect to *amphibians*, the importance of the Atlantic cloud forests of Brazil, the eastern slopes of the Andes, the Guyana shield, eastern Madagascar, the Guinean cloud forests of West Africa, and the cloud forests of northern Borneo is confirmed, both in terms of species richness and endemism richness. Locally, endemism richness for amphibians is high in

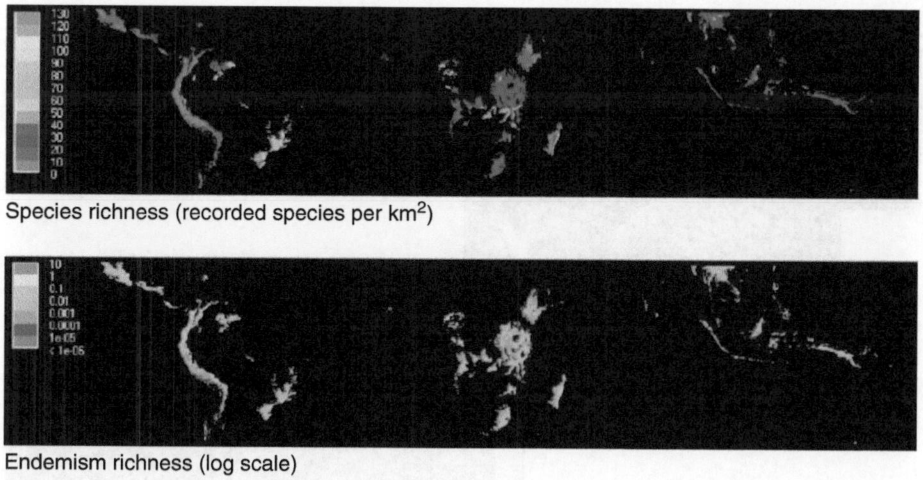

Figure 72.5. The distribution of species richness and endemism richness for amphibians in areas with cloud-affected forest. See text for explanation. (See also color plate.)

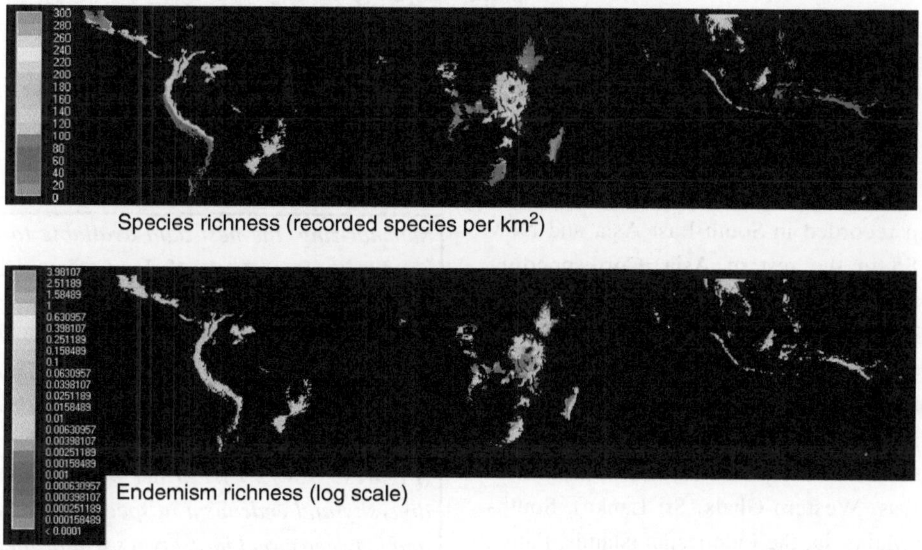

Figure 72.6. The distribution of species richness and endemism richness for mammals in areas with cloud-affected forest. See text for explanation. (See also color plate.)

parts of Costa Rica, the Andean peaks of Colombia, Ecuador, and northern Perú, in the eastern Atlantic forests of Brazil, the Eastern Arc Mountains of Tanzania and Kenya, and in isolated areas in northern Borneo, Papua New Guinea, the Philippines, Malaysia, and Indonesia (Figure 72.5).

The overall patterns derived for *mammals* are broadly similar to those for amphibians, with some notable differences in South-East Asia and Madagascar. Mammal species richness (or at least current knowledge of it) appears to be greatest in South-East Asia and in the north-eastern Andean cloud forests. Endemism richness for mammals is highest in parts of the Andes, in various East African cloud forests (see also Owiunji and Plumptre, this volume), and in parts of Indonesia (Figure 72.6).

The spatial distribution of species richness for *birds* in the Neotropics is greatest in the cloud-affected forests of the Atlantic Coast of Brazil, the Guyana shield, and the eastern Andes. Endemism richness is greatest in the CAFs of Costa Rica, the Sierra Nevada de Santa Marta (Colombia), and the Andean mountains of Venezuela, Colombia, Ecuador, and Perú (Figure 72.7).

The present analysis by and large confirms the results of an early pan-tropical survey of avifaunal diversity of TMCF by Long (1995) who reported over 2500 species of birds confined to TMCF, of which nearly 1000 species occurred in the Neotropics (675 in South America, 180 in Central America, and 140 in the Caribbean). Outside the Neotropics, more than

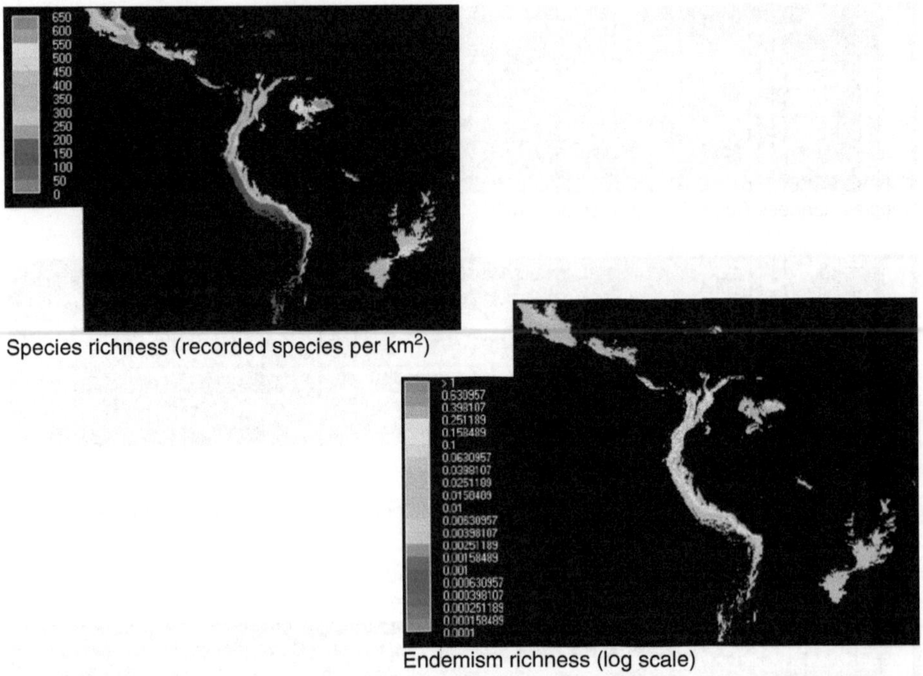

Figure 72.7. The distribution of species richness and endemism richness for birds in Latin American cloud-affected forests. See text for explanation. (See also color plate.)

600 species have been recorded in South-East Asia and New Guinea, and some 185 in the rest of Asia. Corresponding numbers for Africa and the Pacific are 390 in the Pacific, and over 320 in Africa. Areas of primary importance in terms of bird diversity and endemism (known as Endemic Bird Areas, EBAs) are found in the TMCF regions of Africa (e.g. Albertine Rift area, Eastern Arc Mountains, Kenyan Mountains, Cameroon Mountains, Canary Islands), northern/central Asia (Eastern Himalayas, Western Ghats, Sri Lanka), South-East Asia (Peninsular Malaysia, the Indonesian islands, Papua New Guinea, and the Philippines), the Pacific (Micronesian islands, Fiji, Samoa, Solomon, Hawaiian, and Cook Islands, New Caledonia), Latin America (southern Mexico, Central American highlands, northern tropical Andes, southern Brazilian coastal cloud forests), and the insular Caribbean, e.g. Hispaniola (Long, 1995).

Superimposed upon these large-scale diversity patterns are local-scale variations incurred by topography and micro-climatic conditions (e.g. exposure to strong winds). Studies carried out at Monteverde, Costa Rica, have demonstrated differences in bird diversity between windward and leeward slopes. Here, avifaunal richness ranged from 165 species in drier pre-montane forest on the leeward Pacific slope to 315 species in wetter pre-montane forest on the windward Caribbean slope, with 121–187 species occurring in various forest formations at the very wet and windy Continental Divide (1600–1800 m.a.s.l.) (Young and McDonald, 2000; see also Häger and Dohrenbusch, this volume).

Summarizing, the best data available to date for biodiversity confirm the importance of cloud-affected forests (CAF) in terms of species richness per unit area and, particularly, in terms of endemism richness, thereby confirming the very high conservation value of CAFs. In addition, whilst there is considerable geographic variation in species and endemism richness values, a number of well-defined hot spots of diversity can be identified (Figures 72.3–72.7). Further work is needed on mapping plant diversity and endemism in specific tropical cloud forest areas and between cloud forest types within specific areas.

Does biological richness differ between different altitudinal belts on single mountains?

PLANTS

Within individual mountain systems the general trend observed in most TMCFs is a decrease in overall woody plant species diversity and forest stature with increasing elevation (e.g. Grubb, 1977; Gentry, 1988, 1995; Proctor *et al.*, 1988; Kappelle *et al.*, 1995b; Cavelier, 1996; Lieberman *et al.*, 1996; Aiba and Kitayama, 1999; Hemp, 2002, 2005a; Homeier *et al.*, 2008; Paulsch *et al.*, 2008). Gradient analysis has further shown that cloud forest plant species diversity is typically highest at mid-elevations (500–2000 m.a.s.l.) and decreases toward higher, sub-alpine altitudes (Gentry, 1988; Kitayama, 1995; Kappelle *et al.*, 1996; Lieberman *et al.*, 1996; Hemp, this volume #12). Average altitudinal range widths of individual plant species have been reported as being close to 500 m in Bolivian TMCF (Bach *et al.*, 2007) vs. 200–800 m on Mt. Kinabalu, Borneo (Kitayama, 1995) and Mt. Chirripó in Costa Rica (Kappelle *et al.*, 1995b). The causes determining species turnover patterns with elevation are still poorly understood but most probably involve climatic variables such as temperature and humidity, rainfall excess (and therefore soil water status; Kitayama, 1995; Schawe *et al.*, this volume), light and nutrient availability and therefore ecosystem productivity (cf. Grubb, 1977; Raich *et al.*, 1997; Schuur and Matson, 2001; Letts *et al.*, this volume; Benner *et al.*, this volume), as well as historical and evolutionary processes, and dispersal limitations (Kessler and Kluge, 2008). According to Kessler and Kluge (2008), the distribution patterns of endemic species along tropical elevation gradients differ from those of species richness in that richness generally decreases with altitude whilst the number of endemics tends to increase with elevation. In addition, on large mountains endemism tends to peak above 2000 m.a.s.l. This is probably due to a combination of the small size of the area available for speciation (in turn determined by topography and erosion and landslide hazards), long-term climatic stability, and taxon-specific traits like host specificity among epiphytes and very specific, co-evolved plant–pollinator relations in the case of orchids (see also Brehm *et al.*, 2008).

Whilst the number of flowering epiphytes decreases with altitude (lower temperatures), pteridophytic plant groups such as ferns and club mosses (Lycopodiaceae) become more abundant at higher elevations. Non-vascular epiphytes such as bryophytes – which include mosses (Musci) and liverworts (Hepaticae) – become highly diverse, abundant, and dominant in upper montane cloud forests, and cover the trunks, branches, and twigs of numerous trees and shrubs (e.g. Frahm and Gradstein, 1991; Zotz, 1999; Richardson *et al.*, 2000; Gradstein *et al.*, 2001; Nieder *et al.*, 2001; Holz *et al.*, 2002). Epiphylls – bryophytic epiphytes and lichens that grow on the surface of the leaves of vascular plants – may abound in perhumid parts of the forest understory at mid-elevation, e.g. close to shaded rocky outcrops near rivulets and cascades. Of an estimated 4000 species (2700 mosses, 1300 liverworts) occurring in tropical America, about 80% of the liverworts and 50% of the mosses are confined to montane cloud forests (Gradstein and Pócs, 1989).

FAUNA

A recent global analysis of elevational diversity trends for non-volant small mammals revealed a clear pattern of mid-elevational peaks in species richness, while diversity peaked at higher elevations on taller mountains (McCain, 2005). Shukor (2001) assessed diversity patterns of small mammals along an elevational transect on Mt. Kinabalu in Malaysia, utilizing data from previously existing sources and a new field study. Species richness of small mammals (e.g. shrews, squirrels, murid rodents) showed a mid-elevation bulge at about 1200–1400 m.a.s.l. (i.e. below the cloud forest belt that starts at about 1800–2000 m.a.s.l. on Mt. Kinabalu; Kitayama, 1995), with numbers declining rapidly upon entering the wet upper montane cloud forest. Generally speaking, the number of animal species (e.g. amphibians, birds, butterflies – but not moths) on Mt. Kinabalu exhibits a downward trend with increasing elevation; cf. McKinnon *et al.*, 1996; Kumaran *et al.*, this volume). Likewise, on Mt. Mulu (2371 m.a.s.l.) in Sarawak, there is a progressive decline in the number of bird species with increasing altitude, ranging from 171 species in the lowland rain forest to only 12 species at the lower boundary of the UMCF at 1300 m.a.s.l., where a different avifauna takes over. Most species dropped out above 900 m.a.s.l. (McKinnon *et al.*, 1996). A similar discontinuity in bird species found in New Guinean LMRF and UMCF has been reported by Kikkawa and Williams (1971) who also noted that the niches occupied by forest birds changed with elevation. With increasing altitude the proportion of tree-nesting insectivorous species increased, whereas tree-nesting frugivores decreased and omnivores stayed about the same. In addition, the proportion of predatory birds decreased, while the proportion of ground-living birds increased with elevation. These changes were explained in terms of decreasing food resources, especially of fruit and small vertebrates, with increasing altitude (Kikkawa and Williams, 1971).

At mid-elevations in Latin American TMCF at least ten orders of mammals – volant and non-volant – have been recorded, with some 121 species in 25 families found in Costa Rican highland cloud forest (Monteverde): Marsupialia (opossums), Insectivora (shrews), Chiroptera (bats), Primates (monkeys), Xenarthra (sloths, armadillos, anteaters), Lagomorpha (rabbits), Rodentia (porcupines, squirrels, rats, mice, pacas, agoutis), Carnivora (large cats, coyote, gray fox, skunks, raccoons), Perissodactyla (tapir), and Artiodactyla (peccaries, deer) (Timm and LaVal, 2000). Over 40% of the Monteverde cloud forest mammals are bats. Preliminary observations suggest a decline in vertebrate diversity with increasing altitude in Costa Rica (M. Kappelle, personal observations) but more work is needed on the abundance, zonation, and biogeography of most tropical highland mammals (Timm and LaVal, 2000; cf. Williams *et al.*, 2003). Also at Monteverde, some 161 species of reptiles and amphibians are known, with a total of 79 species thriving in the higher and wetter TMCF zones (Pounds, 2000). On Mt. Kinabalu, 40 species of amphibians have been found above 1000 m.a.s.l., as were 27 species of lizards and 36 species of snakes, but no reptiles have been recorded above 2400 m.a.s.l. (i.e. in the UMCF belt). Similar observations were made on Mt. Kerinci (3805 m.a.s.l., Sumatra; McKinnon *et al.*, 1996). Faunal biodiversity associated with tank bromeliads was studied along an elevational gradient in the Luquillo Mountains of Puerto Rico (Richardson *et al.*, 2000). Animal species richness within the bromeliads showed a mid-elevation peak while animal abundance declined with elevation.

Finally, the altitudinal distribution of the main groups of soil macrofauna (ants, termites, etc.) has been studied on a few locations only, e.g. on Mt. Mulu (Sarawak, East Malaysia, 2371 m.a.s.l.; Collins *et al.*, 1984) and Mt. Silam (889 m.a.s.l., Sabah, East Malaysia; Leakey and Proctor, 1987). On Mt. Mulu, strong decreases were observed in total numbers of soil invertebrates with elevation, with earthworms and some beetles replacing termites as the chief detritivores in lower montane forest just below the main cloud condensation level. In the peaty soils of the upper montane forest, beetles became dominant in association with earthworms, fly larvae, and cockroaches. In the most exposed, stunted vegetation zones near the summit, several groups fell away entirely (Collins *et al.*, 1984). On Mt. Silam – a small ultrabasic coastal mountain where a daily cloud cap develops around 700 m.a.s.l. (Bruijnzeel *et al.*, 1993) – no clear trends emerged for individual soil faunal groups with elevation except for a strong decline in the number and biomass of earthworms above the altitude at which the dipterocarps disappeared (610 m.a.s.l.; Leakey and Proctor, 1987). However, no such decline was observed on Mt. Mulu where the disappearance of the dipterocarps between 800 and 1130 m.a.s.l. was not accompanied by a reduction in earthworm densities and biomass (Collins, 1980). Likewise, both earthworm biomass and species richness increased with elevation from LMRF to elfin forest in the Luquillo Mountains of Puerto Rico and was highest in the ECF (Gonzalez *et al.*, 2007).

Summarizing, with the exception of ferns and non-vascular epiphytes (mosses and liverworts), species richness and abundance (numbers of individuals) of both plants and animals (including birds) tend to decline with increasing elevation. A mid-elevation maximum (located within or beneath the cloud belt) tends to be observed for plant, bird, and mammal diversity.

CLOUD FOREST HYDROMETEOROLOGY AND HYDROLOGY

With respect to the hydrological functioning of TMCF, the following questions assume key importance:

- Do lower montane, upper montane, and sub-alpine/elfin cloud forests differ in terms of their rainfall and cloud water inputs, net precipitation, and general hydrometeorology?
- Are there regional contrasts in water inputs via cloud water interception and wind-driven rain (e.g. "maritime" vs. "continental" situations)?
- Does water use (evapotranspiration) differ between different types of TMCF?
- How are annual and seasonal water yields affected by converting different types of TMCF to pasture, annual crops, or coffee plantations?

Hydrometeorology of different types of TMCF

WATER INPUTS AND NET PRECIPITATION

The proper quantification of the total amounts of water inputs received by TMCF and the relative contributions by precipitation, cloud water interception (*CWI*), and horizontal (wind-driven) rain is fundamental to assessing the hydrological importance of intact and converted cloud forest areas. This is illustrated by the results obtained by several early studies from Central America that arguably have contributed to the reputation of TMCFs as suppliers of high amounts of streamflow throughout the year. Zadroga (1981) compared the rainfall and streamflow regimes for two groups of watersheds in northern Costa Rica, one located on the windward Atlantic side of the Continental Divide and the other on the leeward Pacific side. Annual streamflow from the Pacific watersheds amounted to 34% of the rainfall and showed a clear seasonal flow pattern that followed that for precipitation, while annual streamflow from the Atlantic watersheds roughly equaled precipitation (102%) and even exceeded rainfall inputs for seven out of 12 months (Figure 72.8). Whilst acknowledging that the high runoff coefficient derived for the Atlantic watersheds was partly due to underestimation of rainfall inputs in the higher, rainier parts of the watersheds that lacked measurement stations, Zadroga (1981) attributed the very high streamflows primarily to unmeasured inputs of cloud water, emphasizing the fact that months with excess streamflow over precipitation coincided with the dominant occurrence of moisture-laden clouds that were brought in from the

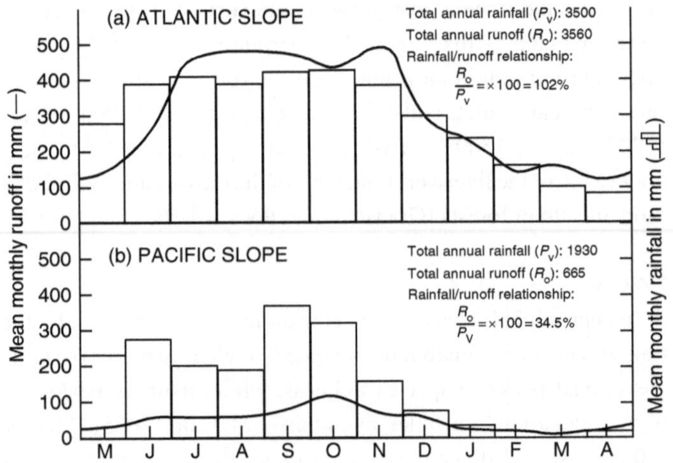

Figure 72.8. Contrasting rainfall and streamflow regimes for watersheds on the Atlantic and Pacific slopes of northern Costa Rica (after Zadroga, 1981).

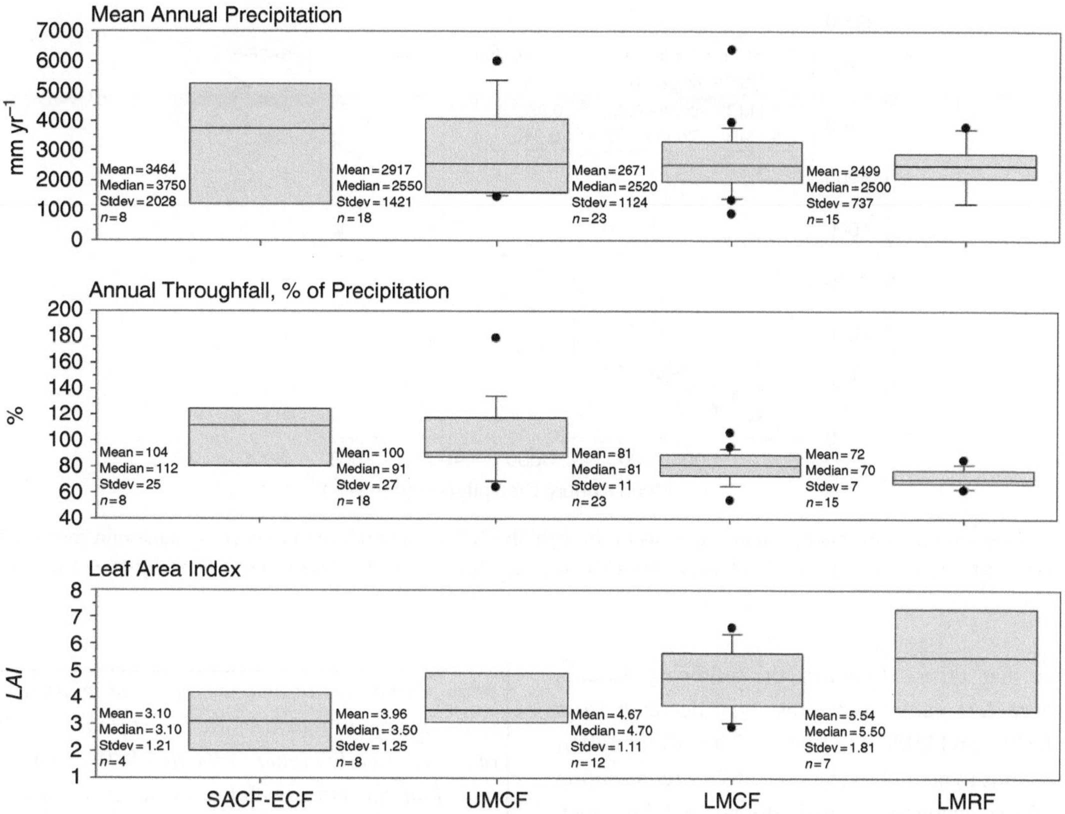

Figure 72.9. Mean and median annual rainfall, relative amounts of throughfall (% of rainfall), and leaf area index (*LAI*) for lower montane rain forest (LMRF), lower montane cloud forest (LMCF), upper montane cloud forest (UMCF), and sub-alpine/elfin cloud forest (SACF–ECF). Error bars denote 1 standard deviation from the mean. Data listed in Appendix 72.2.

Caribbean by the trade winds. In addition, he recognized that evaporative losses from these cloud-ridden slopes would be low.

Zadroga's contentions were confirmed by measurements of rainfall, streamflow, and climatically based estimates of evapotranspiration for another Atlantic watershed located further south in Costa Rica (Calvo, 1986). Whilst both of these early investigations must be considered "black box" studies that did not quantify the underlying hydrological processes, further support came from comparative observations of rainfall and throughfall (*TF*, also called crown drip) in Atlantic cloud forests in Costa Rica (Caceres, 1981) and Honduras (Stadtmüller and Agudelo, 1990). These studies showed that annual *TF* at cloud-exposed locations could attain values of as much as 110–180% of measured rainfall.

To what extent are these observations exemplary for the hydrological behavior of TMCFs in general, and how reliable are such direct comparisons of rainfall and *TF* in view of such potentially disturbing factors as wind losses around the rain gage and ungaged contributions of inclined rainfall (as opposed to horizontally moving clouds or rain)? Earlier reviews of the hydrometeorological literature on TMCF (Bruijnzeel and Proctor, 1995; Bruijnzeel, 2001, 2005) lacked sufficient data for a meta-analysis but the proliferation of local studies of net precipitation (i.e. amounts of water arriving at the forest floor as *TF* plus stemflow, *SF*) – and to a lesser extent forest water use – during the last 5–10 years (many of which are reported upon in this volume) allows an analysis of some of these questions and whether different types of TMCF do exhibit different net precipitation fractions or water use.

Appendix 72.2 lists net precipitation data for lower montane rain forests little or not affected by fog and low cloud (LMRF, $n = 15$), tall lower montane cloud forest subject to moderate cloud incidence (LMCF, $n = 23$), upper montane cloud forest of intermediate stature and subject to frequent cloud incidence (UMCF, $n = 18$), and stunted sub-alpine and elfin cloud forests (SACF–ECF, $n = 8$)[2]. Figure 72.9 shows the average amounts of rainfall, the fraction of rainfall becoming *TF*, and the leaf area index (*LAI*) for the respective forest types, while Figure 72.10 shows scatter-plots of annual *TF* vs. rainfall amounts at individual study sites grouped per forest type.

[2] Note that elfin cloud forest (ECF) represents a small-statured forest irrespective of elevation. As such, ECF is azonal and may refer to the most stunted facies of forest near the timber-line at high elevations, or to stunted wind-exposed forest on ridges and summits. The latter may occur at high or low elevation, with the lowermost occurrences reported at 300–500 m.a.s.l. (cf. Scatena *et al.*, this volume).

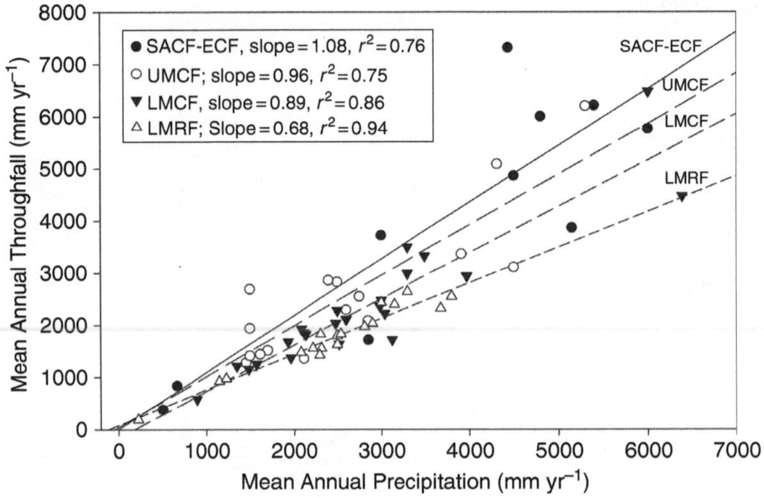

Figure 72.10. Linear regression relationships linking amounts of throughfall (*TF*) and rainfall (*P*) in lower montane rain forest (LMRF), lower montane cloud forest (LMCF), upper montane cloud forest (UMCF), and sub-alpine/elfin cloud forest (SACF–ECF). Data listed in Appendix 72.2.

Based on more than 60 local studies, the following patterns emerge: (i) average *LAI* values per forest type decrease from LMRF through LMCF and UMCF to SACF–ECF; (ii) rainfall at SACF–ECF sites tends to be higher than that at sites representing the other three forest categories which do not exhibit much difference between one another; and (iii) the *TF* fraction of rainfall (*P*) increases steadily from LMRF to SACF, viz. from 72 ± 7% (SD) in LMRF, to 81 ± 11% in LMCF, 100 ± 27% in UMCF, and 104 ± 25% in SACF–ECF (Figure 72.9).

Rigorous comparisons of the statistical differences in *TF*/*P* between the different forest types is limited by the small and uneven sample sizes and differences in sampling methodologies of the various studies. Nevertheless, comparisons of means and medians using *t*-tests, Mann–Whitney rank sum tests, and analysis of variance (ANOVA) where appropriate support the patterns observed in Figure 72.9. Moreover, there are significant differences (at $p = 0.05$) between the means or medians of *TF*/*P* for UMCF and both LMCF and LMRF (but not SACF–ECF). The median *TF*/*P* value for LMCF is also significantly higher than the median value for LMRF.

Comparison of the slopes of rainfall vs. *TF* graphs per forest type also indicates that *TF* exceeds rainfall as measured in the open at SACF–ECF sites, whereas the two are similar at UMCF sites (Figure 72.10). To obtain the total amount of precipitation reaching the forest floor (i.e. net precipitation), the fraction of the rainfall arriving as stemflow (*SF*) should be added to the amount of *TF*. Not all studies of net precipitation measured *SF* but values observed in LMRF and LMCF are typically very low (<1–2% of incident rainfall), although occasionally higher values have been reported (see Appendix 72.2 for details). *SF* fractions obtained for UMCF vary widely (0.1–30.5%) with an overall average approaching 10% of rainfall and a median of 3%; the corresponding range and average for SACF–ECF read 3–18% and 6.7%, respectively (Appendix 72.2).

> *Thus, whilst overall amounts of net precipitation (i.e. throughfall + stemfow) reaching the forest floor in LMRF and LMCF are only slightly higher (73% and 83%, respectively) on average than the more widely measured amounts of throughfall alone, net precipitation in UMCF and SACF–ECF may well exceed throughfall by 5–10% on average and by more than 10% in specific cases.*

The rather low TF fraction observed for most LMRF (Figure 72.9) is often attributed to the high epiphyte loads of these forests (e.g. Cavelier *et al.*, 1997; Ataroff, 1998; cf. Fleischbein *et al.*, this volume). Interestingly, although both LMCF and UMCF typically have a much higher epiphytic biomass than LMRF (Frahm and Gradstein, 1991; Wolf, 1993; Köhler *et al.*, 2007), this does not seem to have a reducing effect on the magnitude of the *TF* fraction in the various types of cloud forests (Figure 72.9 and Appendix 72.2). Based on measurement and modeling of the water dynamics of epiphytes and bryophytes in a montane forest in Costa Rica, Hölscher *et al.* (2004) concluded that – whilst the *potential* water storage capacity of the epiphytes is very high – their *actual* (effective) capacity to absorb additional moisture is modest at best because the epiphytes are usually kept wet by the frequent rain and fog occurring in these forests (cf. Köhler *et al.*, this volume; Tobón *et al.*, this volume #26; Oesker *et al.*, this volume).

Although the high *TF* fractions obtained for UMCF and SACF–ECF must partly reflect "occult" contributions by *CWI* and wind-driven rain captured by the canopy but missed by a standard-type rain gage (cf. Zadroga, 1981), it should be noted that wind speeds at exposed UMCF and SACF–ECF sites tend to be (much) higher than those experienced lower down on the slopes by the more sheltered LMCF and LMRF (Holwerda, 2005; Motzer, 2005; Bendix *et al.*, 2008). Therefore, rainfall amounts measured at exposed, windy sites are in danger of being (grossly) underestimated and need to be corrected for wind losses around the gage (Førland *et al.*, 1996; Yang *et al.*, 1998; Nešpor and Sevruk, 1999) and for the interactive effects of inclined rainfall falling onto steeply sloping terrain (Sharon, 1980; Herwitz and Slye, 1992). To date, only a few studies in TMCF have applied such corrections (Holwerda *et al.*, 2006; McJannet *et al.*, 2007a,b; cf. García-Santos, 2007; Giambelluca *et al.*, this volume). The magnitude of such corrections may be substantial, depending on the local situation. As such, the higher *TF* fractions derived for UMCF and SACF–ECF are almost certainly influenced by an underestimation of incident rainfall by many studies (see e.g. Prada *et al.*, 2009 for a recent example), although the data do not permit a quantification of the effect for different sites.

Local topography and slope aspect also exert a major influence on atmospheric inputs, as demonstrated by the high small-scale variability in relative and absolute amounts of *TF* vis-à-vis topographic position reported by Weaver (1972) for cloud forests in Puerto Rico (cf. Holwerda *et al.*, this volume #29), by Juvik and Ekern (1978) in Hawai'i (cf. DeLay and Giambelluca, this volume; Giambelluca *et al.*, this volume), by Stadtmüller and Agudelo (1990) in Honduras (cf. Häger and Dohrenbusch, this volume) and by García-Santos (2007) in La Gomera (Canary Islands; cf. Marzol-Jaén *et al.*, this volume). Future work will need to explicitly quantify inputs of wind-driven rain and fog in a spatially distributed manner. Although several powerful modeling approaches of various degrees of complexity are available (e.g. Yin and Arp, 1994; Walmsley *et al.*, 1996, 1999; Blocken *et al.*, 2005, 2006; Mulligan and Burke, 2005b), validation of model predictions through field measurements remains a challenge, largely because of the difficulties of adequately measuring rainfall inputs under windy conditions in complex terrain (Blocken *et al.*, 2006). Arguably, the recently developed spherical rain gage of Chang and Flannery (2001) and the "potential rain gage" of Frumau *et al.* (2006a) – which allows quantification of the angle of incidence of the rain – may prove useful in this regard. In addition, use can be made of conventional fog gages for comparative site characterization (cf. Juvik and Nullet, 1995a; Rollenbeck *et al.*, 2007; Frumau *et al.*, this volume; Marzol-Jaén *et al.*, this volume), rainfall radar systems for the quantification of the spatial extent of rain fields (Mulligan, 2006a; Rollenbeck *et al.*, 2007; Rollenbeck *et al.*, this volume), and remotely sensed cloud extent and characteristics (Mulligan, 2006b; Nair *et al.*, 2008; Welch *et al.*, 2008; Lawton *et al.*, this volume).

The estimation of amounts of *CWI* by a forest canopy has evolved considerably in recent years, from a simplistic equating of the catch by some type of fog gage to that by a live canopy (e.g. Ataroff, 1998), or simply assigning all excess *TF* over rainfall to *CWI* (e.g. Holder, 2003), via comparisons of *TF* vs. rainfall relationships for times with and without fog (e.g. Harr, 1982; Hafkenscheid *et al.*, 2002) or times with and without strong winds (Tanaka *et al.*, this volume), to the more recently developed process-based wet-canopy water-budget method (Holwerda *et al.*, 2006; McJannet *et al.*, 2007b,d; Holwerda *et al.*, this volume #29; Giambelluca *et al.*, this volume; Schmid *et al.*, this volume), and direct measurement using sophisticated eddy covariance and spectrometer equipment (Eugster *et al.*, 2006; Holwerda *et al.*, 2006; Beiderwieden *et al.*, 2008; Schmid *et al.*, this volume). Arguably, results obtained with the wet-canopy water budget method may be considered the most realistic but the confidence limits of the estimates are wide due to the accumulation of error associated with the measurement of the respective components (see discussion in Holwerda *et al.*, 2006; McJannet *et al.*, 2007b).

Whilst the currently available measurements of (absolute) amounts of *CWI* do not permit the identification of regional patterns other than higher values being associated with windier and wetter places (Bruijnzeel and Proctor, 1995; Bruijnzeel, 2005; DeLay and Giambelluca, this volume; Häger and Dohrenbusch, this volume), these can be modeled using some of the approaches referred to in the previous paragraphs. An example is given in Figure 72.11 for the whole of Costa Rica, showing the clear influence of topographic exposure and elevation (Mulligan and Burke, 2005b; cf. Bendix *et al.*, 2008; Lawton *et al.*, this volume). At the same time, there remains a clear need to validate such model predictions using field measurements of *CWI* and wind-driven rain. Arguably, this could be done most profitably in data-rich areas such as (amongst others) Puerto Rico, northern Costa Rica, southern Ecuador, several of the Canary Islands, and Hawai'i. Furthermore, when trying to unravel the relative importance of inputs via cloud water capture and wind-driven rain under different climatic and topographic circumstances, use can be made of the contrast in stable isotope signatures of fog and wind-driven rain observed at several (but not all) locations (Scholl *et al.*, this volume; cf. Rhodes *et al.*, 2006; Guswa *et al.*, 2007).

FOREST WATER USE

Comprising only 15 studies in total, the current data-set on cloud forest water use (both for transpiration E_t and total evapotranspiration *ET*, i.e. including wet-canopy evaporation E_i) is much smaller than that for throughfall and stemflow. Nevertheless, considerable progress has been made in the last 5 years with the quantification of E_t and *ET* in montane tropical forests. Whilst Bruijnzeel and Proctor (1995) were unable to report any annual totals for *Et* other than indirect (and therefore highly approximate) estimates derived with the water budget method

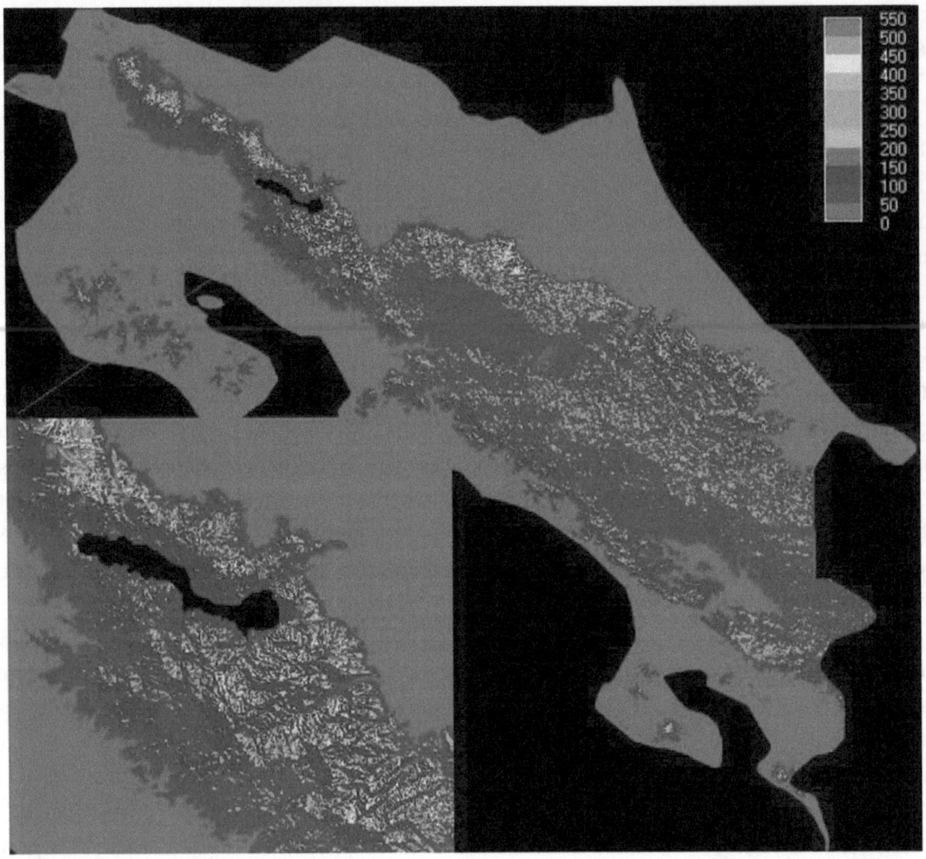

Figure 72.11. Modeled annual fog interception input (mm) at 90-m resolution across Costa Rica (after Mulligan and Burke, 2005b; reproduced with permission). (See also color plate.)

at the watershed scale, the number of studies using sapflow equipment – and to a lesser extent micrometeorological methods – has been increasing steadily in recent years (Appendix 72.3). A preliminary analysis of the available data is offered below, although the conclusions need to be viewed with caution given the small number of replications for UMCF and SACF–ECF in particular.

Figure 72.12 summarizes the average annual totals of *ET* and its two main components (transpiration E_t and wet-canopy evaporation E_i) for the three types of cloud forests distinguished. No comparative data on transpiration for LMRF are available (cf. Bruijnzeel and Proctor, 1995) but average total *ET* for these forests is 1280 ± 72 mm year^{-1} ($n = 7$). At 1188 ± 239 mm year^{-1} ($n = 9$) the average value for LMCF is not significantly different from that for LMRF, but *ET* totals for UMCF (783 ± 112 mm year^{-1}, $n = 3$) and ECF (547 ± 25 mm year^{-1}, $n = 2$) are distinctly lower. And although the small number of observations for the latter forest types precludes statistical testing, the data do confirm the need to distinguish between lower montane (cloud) forests on the one hand, and upper montane (and elfin) forests on the other (cf. Appendix 72.3). Combining averaged data on precipitation inputs (Figure 72.9) with averaged *ET* values per cloud forest type (Figure 72.12) gives some indication of the corresponding precipitation-excess values – and therefore the probability of waterlogged conditions. Although such estimates are obviously approximate, the derived average precipitation-excess values show a clear increase from ~1200 mm and ~1500 mm year^{-1} in LMRF and LMCF, respectively, to ~2100 mm year^{-1} in UMCF and ~2900 mm year^{-1} in ECF (cf. Kitayama, 1995; Schawe *et al.*, this volume). The potential consequences of this increasing wetness with elevation for forest productivity and nutrient availability will be discussed in the corresponding section below.

As was found earlier for *ET*, average annual amounts of *transpiration* (E_t) also decrease gradually with elevation from ~645 mm in LMCF to ~385 mm in UMCF and ~355 mm in ECF (Figure 72.12). Interestingly, both *Et* and *ET* (not shown) are significantly related to forest *LAI*, especially after normalization for net radiation input (cf. Figure 72.13 below). In addition, the fraction of available energy used by the respective forest types for transpiration (E_t/R_n) shows a marked reduction from 0.61 ± 0.08 in lowland evergreen rain forest (LERF), to 0.48 ± 0.14 in LMCF, 0.37 ± 0.06 in UMCF, and 0.26 ± 0.01 in ECF. These important findings will be revisited in the section on forest

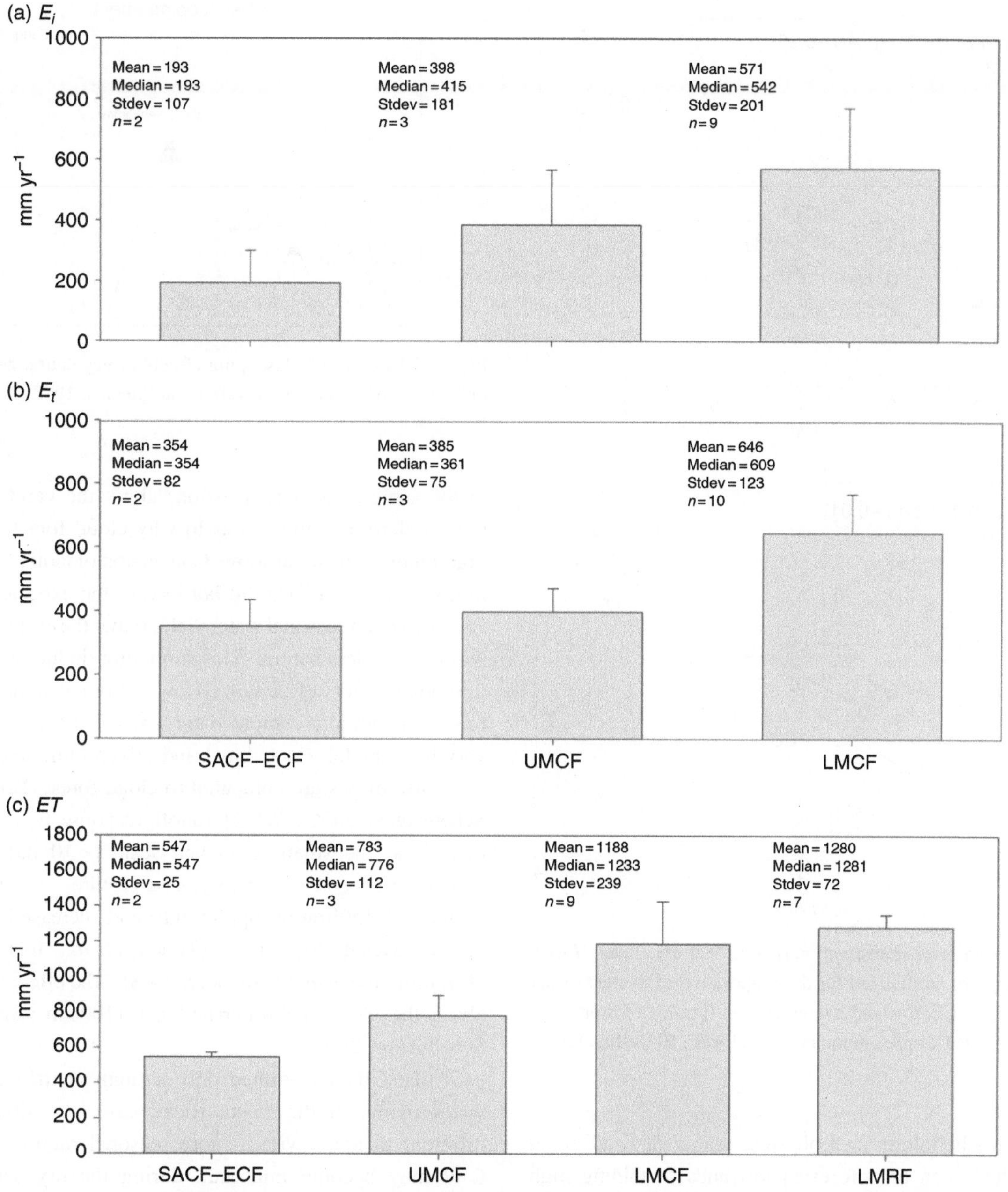

Figure 72.12. Mean and median annual (a) rainfall interception, (b) transpiration, and (c) total evapotranspiration for lower montane rain forest (LMRF, *ET* only), lower montane cloud forest (LMCF), upper montane cloud forest (UMCF), and sub-alpine/elfin cloud forest (SACF–ECF). Error bars denote 1 standard deviation from the mean. Data listed in Appendix 72.3.

productivity below. Across all sites listed in Appendix 72.3, mean annual precipitation (*MAP*) is negatively correlated with net radiant inputs, forest water use (E_t), and *LAI*. Overall, up to 42% of the variance in E_t between sites can be explained by the combination of *MAP*, mean air temperature, and net radiation (which is controlled in turn by the frequency and degree of cloud cover). In addition, the relationship between E_t and *LAI* is particularly strong, also after normalizing E_t for the evaporation equivalent of available net radiant energy (Figure 72.13).

Overall, the available information indicates that relative contributions by transpiration E_t to total evaporative water loss *ET* are roughly as important as those by wet-canopy evaporation E_i (*c.* 50% each) in both LMCF and UMCF. However, there are insufficient data for meaningful statements in this regard for SACF–ECF (see Appendix 72.3 for details). Occasionally, however, very high values have been derived for E_i (e.g. for several near-coastal forests in northern Queensland; McJannet *et al.*, 2007d). It is not clear to what extent this finding reflects

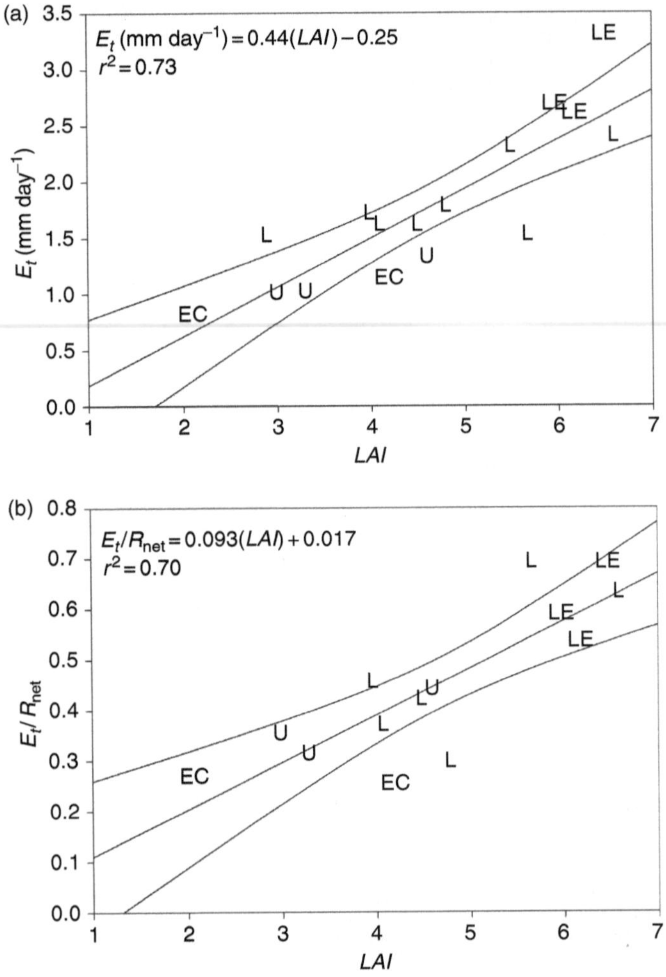

Figure 72.13. Cloud forest transpiration (E_t) vs. leaf area index (LAI): (a) daily E_t values, (b) normalized for the evaporative equivalent of net radiant energy (R_{net}). LE, lowland evergreen rain forest; L, lower montane cloud forest; U, upper montane cloud forest; EC, elfin cloud forest.

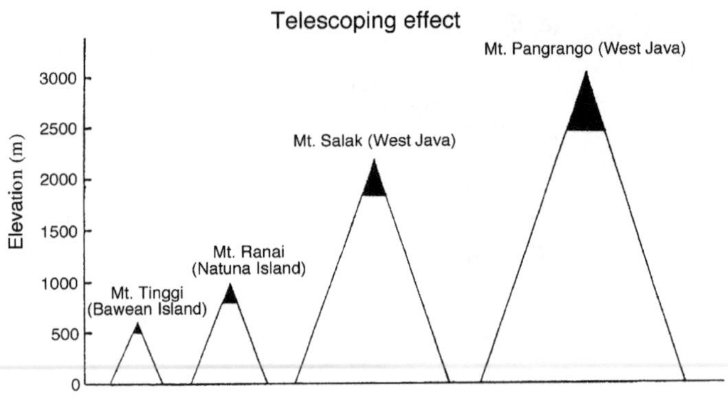

Figure 72.14. The "telescoping effect" of vegetation zonation on differently sized mountains (after Van Steenis, 1972).

differences in methodology or a physical reality (e.g. advected energy from the ocean or deforested lowlands sustaining high evaporation rates during rainfall) given that various other "maritime" sites (Puerto Rico, Hawai'i, Costa Rica) do not exhibit similarly high values (Appendices 72.2 and 72.3). There is a need for additional observations of evaporation in UMCF and SACF–ECF in particular.

Cloud forest conversion and water yield

Based on the very large contrast in streamflow produced by windward and leeward forests in northern Costa Rica (Figure 72.8), Zadroga (1981) postulated that conversion of Atlantic cloud forest might well result in significantly reduced streamflows. However, the recent results of detailed hydrological field and modeling studies in cloud forest and pasture in the same area raise questions about the validity of some of the standard assumptions as to why cloud forests produce such high amounts of streamflow. Conversion of cloud forest to pasture in northern Costa Rica did *not* lead to the expected decreases in annual or even seasonal water yield; rather the effect on streamflow was more or less neutral. This surprising finding was attributed to the compensatory effects of: (i) low additional inputs by *CWI* under forested conditions compared to rough pasture, under conditions of very high rainfall (including wind-driven rain), and (ii) the lower water use of pasture compared to cloud forest (Bruijnzeel, 2006; Schellekens, 2006). Whilst runoff response to rainfall ("quickflows") at the small watershed scale (<10 ha) were roughly doubled after conversion to grazed pasture – presumably due to diminished infiltration opportunity and increased overland flow on compacted cow trails (Tobón *et al.*, this volume #54; cf. Zimmermann and Elsenbeer, 2008) – the effect was not detectable at the operational watershed scale (100 km^2; Bruijnzeel, 2006; Schellekens, 2006).

Whilst *CWI* constituted only a small contribution to overall water inputs in the Costa Rican case, the situation may be different in areas with a more seasonal rainfall regime where *CWI* may become important during the dry season – as for instance in parts of Pacific Central America (Holder, 2003; Mulligan and Burke, 2005b; cf. Figure 72.11). Recent work on *CWI* and forest and pasture water use in central Veracruz, eastern Mexico – an area with strongly seasonal rainfall and reportedly receiving substantial amounts of occult precipitation (Vogelmann, 1973) – demonstrated that cloud forest conversion would lead to a major local *increase* in flows (Muñoz-Villers, 2008; Gomez-Cárdenas, 2009), amongst other reasons because of low *CWI* by the local cloud forest (Holwerda *et al.*, 2010; cf. Ponette-González *et al.*, 2009) and a much higher water consumption by the cloud forest than by pasture (Holwerda *et al.*, 2007; Gomez-Cárdenas, 2009). Finally, partial clearing of Douglas fir forest in the Pacific Northwest of the USA – an area subject to heavy fog incidence (Harr, 1982) – caused a temporary *decrease* in baseflows during the dry summer months. Interestingly, the effect gradually disappeared after 5–6 years, as the regenerating trees were apparently effective at capturing sufficient amounts of cloud water again (Ingwersen, 1985).

Reliable information on the effect of cloud forest conversion to agricultural cropping (with or without associated soil degradation and ensuing negative effects on dry-season flows; cf. Bruijnzeel, 2004, 2005) or conversion to shade-coffee plantations is not available, although the effect on water yield could be modeled from information on plantation water use (Holwerda *et al.*, 2007) and rainfall and cloud water interception (Ponette-González *et al.*, 2009).

Summarizing, the direction and magnitude of changes in water yield (if any) after TMCF conversion depend on the outcome of multiple interactions, and are therefore highly contextual and site-specific. Where cloud water interception is low and forest water use comparatively high, water yields are likely to increase upon cloud forest conversion, and vice versa.

CLOUD FOREST PRODUCTIVITY: SOIL VS. CLIMATIC EFFECTS

The changes in forest structure and physiognomy observed as one ascends wet tropical mountains have, in the words of Whitmore (1989) "puzzled and irritated" scientists for a long time. Of these changes the most notable are a decrease in forest stature and leaf size from tall lower montane forest to increasingly stunted upper montane and sub-alpine cloud forests with increasing altitude (Grubb, 1977; Moser *et al.*, 2007). A related phenomenon is the occurrence of (often stunted) mossy forest at lower altitudes on small outlying mountains (often on islands or in coastal areas) compared to larger mountains away from the coast (Van Steenis, 1972; Richards, 1996; Figure 72.14). This is generally thought to reflect a combination of the higher atmospheric moisture contents prevailing under "maritime" conditions (cf. Foster, this volume; Jarvis and Mulligan, this volume) and the smaller heat-storing capacity of smaller mountains compared to larger ones.

Regardless of the elevation at which these mossy forests occur, the common denominator appears to be the high frequency of fog (cf. Mulligan, this volume; Jarvis and Mulligan, this volume), which is thought to influence an array of ecological processes controlling forest stature (Grubb, 1977; Bruijnzeel *et al.*, 1993; Cavelier, 1996; Silver *et al.*, 1999; Benner *et al.*, this volume). This leads to the following key questions:

- What causes the lower stature of certain types of cloud forests compared to forest not affected appreciably by frequent fog? Are the dominant controls primarily soil- or climate-related or both?
- Do soils under lower and upper montane cloud forests differ and if so, in what respect and what are the consequences for nutrient availability and forest productivity?

Possible reasons that have been advanced to explain the stunting of UMCF and SACF–ECF on wet tropical mountains (summarized and discussed by Grubb, 1977; Flenley, 1995; Cavelier, 1996; Bruijnzeel and Veneklaas, 1998; Tanner *et al.*, 1998; see also Benner *et al.*, this volume; Roman *et al.*, this volume) are manifold and include:

- Periodic water shortage, especially in the case of shallow and stony soils;
- Chronic soil saturation and impeded root respiration due to oxygen deficiency;
- Reduced leaf temperatures and reduced photosynthetic capacity associated with low radiation inputs;
- Limited nutrient uptake due to (i) climatically controlled reduction in soil water uptake rates (transpiration, *Et*), (ii) extreme soil acidity or low fertility or both, often coupled with (iii) reduced decomposition and mineralization of leaf litter, and (iv) elevated levels of toxic substances in the soil water, notably "free" aluminum and polyphenols;
- Exposure to strong winds, notably in the trade-wind belt and at exposed upper slopes and summit locations; and
- Exposure to high intensities of UV-B radiation.

Waide *et al.* (1998) drew attention to the fact that several of these factors are known to co-vary and thus may act in a synergistic manner such that it is difficult to unravel the relative importance of individual factors. However, in the last decade a number of important studies have been made of the changes in forest structural characteristics, and of the corresponding changes in climate, soils, eco-physiology, and nutrient dynamics along elevational or soil fertility gradients. Much of this work has been concentrated at a relatively small number of intensively studied sites, both in the maritime tropics (notably Hawai'i, Jamaica, Sabah (East Malaysia), and Puerto Rico; see Benner *et al.*, this volume for a fuller discussion) and under more continental conditions, e.g. the Yungas of Bolivia (Schawe *et al.*, 2007; Gerold *et al.*, 2008; Schawe *et al.*, this volume), the eastern Andes of southern Ecuador (Beck *et al.*, 2008a; cf. Wilcke *et al.*, this volume), and to a lesser extent the Cauca region of south-western Colombia (Letts *et al.*, this volume), northern Thailand (Tanaka *et al.*, 2003; Tanaka *et al.*, this volume), and Mt. Kilimanjaro, Tanzania (Schrumpf, 2004).

Whilst it remains difficult to identify the relative importance of several co-varying factors (e.g. radiation, temperature, rainfall, humidity vs. elevation), this new work does allow the formulation of a hypothesis that goes a long way toward explaining cloud forest stunting, both at higher elevations on large mountain massifs (UMCF and SACF–ECF) and at much lower elevations on small outlying mountains (ECF) across the humid tropics. First, however, the various hypotheses listed above are discussed briefly in the light of the new evidence.

DROUGHT VS. WATERLOGGING

Both early (as reviewed by Cavelier, 1990, 1996; Bruijnzeel and Proctor, 1995; Bruijnzeel and Veneklaas, 1998) and more recent data on soil water dynamics in different cloud forest types (Silver *et al.*, 1999; Cavelier *et al.*, 2000; Hafkenscheid, 2000; Santiago *et al.*, 2000; García-Santos, 2007; Moser *et al.*, 2008; Schawe *et al.*, this volume) confirm that drought is not an important determinant of tropical montane forest stature, even under conditions of low rainfall, as long as fog inputs are consistent (Cavelier *et al.*, 2000; García-Santos, 2007). Productivity (e.g. trunk increment) has been shown to be reduced during (and after) major drought years (Aiba *et al.*, this volume) but observations of leaf litter standing crop after a severe drought did not suggest major water stress in cloud-affected forest, in contrast to forests below the cloud belt which showed strong increases in leaf litterfall after a drought (Bruijnzeel *et al.*, 1993). Also, whilst many cloud forests have xeromorphic-looking leaves ("sclerophylls"), these leaves are not particularly drought-tolerant (reviewed by Turner, 1994). Reports of severe water stress in cloud-affected forest (as discussed in Bruijnzeel and Proctor, 1995) appear to pertain exclusively to forests located on very shallow soils.

Conversely, the averaged precipitation-excess values cited in the previous section for the respective cloud forest types suggest the soils of UMCF experience roughly twice the amount of water percolating through soils under LMRF. The corresponding volume for SACF–ECF is nearly three times that in LMRF. As such, soils of UMCF tend to be wet to saturated for most of the time and those of SACF and ECF *in high-rainfall areas* tend to be frequently, if not persistently, saturated (see examples in Bruijnzeel and Proctor, 1995; cf. Silver *et al.*, 1999; Hafkenscheid, 2000; Santiago *et al.*, 2000; Moser *et al.*, 2008; Schawe *et al.*, this volume). This contention is confirmed by numerous soil profile descriptions made along elevational transects in South-East Asia (e.g. Tie *et al.*, 1979; Van Reuler, 1987; Bruijnzeel *et al.*, 1993; Mangen, 1993), East Africa (e.g. Schrumpf, 2004), the Caribbean (e.g. Wadsworth and Bonnet, 1951; Lyford, 1969), Hawai'i (Kitayama and Müller-Dombois, 1994), and South America (e.g. Hetsch and Hoheisel, 1976; Schrumpf *et al.*, 2001; Schawe *et al.*, 2007) which all report increased occurrence of hydromorphic characteristics beneath the more stunted forest types. In addition, at a given elevation (and therefore climatic conditions), forest stature is often reported to decline with increasing soil wetness (typically in association with a reduction in slope steepness; e.g. Hetsch and Hoheisel, 1976; Lawton and Dryer, 1980; Mangen, 1993; Silver *et al.*, 1999; Santiago *et al.*, 2000; Vernimmen *et al.*, 2007; cf. Häger and Dohrenbusch, this volume; Roman *et al.*, this volume). However, on young soils derived from volcanic ash – which typically have very high porosity and permeability values and which tend to be very well-drained (cf. Tobón *et al.*, this volume #52) – waterlogging (and associated forest stunting) is observed only under conditions of extreme rainfall (>5000 mm year^{-1}) and low topographic gradients (cf. Häger and Dohrenbusch, this volume). Also, where SACF occurs above the level of most frequent cloud occurrence and maximum rainfall – such as in the Hawaiian archipelago above the trade-wind inversion (cf. Kitayama and Müller-Dombois, 1994; Juvik and Nullet, 1995b) and on Mt. Kilimanjaro above 2800 m.a.s.l. in the *Erica* forest belt (Hemp, this volume #12) – rainfall-excess values will be much smaller due to a combination of comparatively high evaporation and reduced rainfall (Giambelluca and Nullet, 1992; Kitayama and Müller-Dombois, 1994; cf. Røhr, 2003). Nevertheless, even in the sub-humid higher parts of Mt. Kilimanjaro, Schrumpf (2004) observed waterlogged and poorly drained soils in the SACF zone due to the presence of bedrock at shallow depths. Similarly, soil moisture levels in low-elevation ECF have been reported to remain very high (although not reaching full saturation) due to regular nocturnal cloud immersion even under conditions of low rainfall (Cavelier *et al.*, 2000).

Summarizing, high and often excessive soil moisture levels are present in most UMCF and, especially, SACF and ECF settings below the trade-wind inversion. Also, forest stature and leaf area at a given elevation tend to be inversely related to the degree of soil waterlogging.

TEMPERATURE, RADIATION LEVELS, AND PLANT PHOTOSYNTHETIC CAPACITY

The observation that very similar-looking waterlogged ECF can develop at vastly different elevations – and therefore experience highly different temperature and radiation regimes (including UV-B radiation) – already suggests that precipitation excess (read: soil water status) overrides any temperature or radiation effects (Cavelier *et al.*, 2000; Gerold *et al.*, 2008; cf. Schawe *et al.*, this volume). Further support for this contention comes from observations of very tall (30–50 m) montane forests at 2200–3000 m.a.s.l. on well-drained volcanic soils in Papua New Guinea despite reportedly frequent fog (Hyndman and Menzies, 1990; Mangen, 1993). Indeed, although radiation inputs and photosynthetic photon flux densities (*PPFD*) at sites with UMCF or SACF–ECF are typically reduced compared to sites with LMCF (by 30–40%) or compared to lowland forest (by >50%) (Cavelier and Mejia, 1990; Hafkenscheid, 2000; Holwerda, 2005; Gerold *et al.*, 2008), it is pertinent to note that *PPFD* values in low-statured SACF–ECF have been reported to be very similar to those measured in taller UMCF lower down on the same mountain (Schawe *et al.*, this volume). More importantly, at 650–800 μmol m^{-2} s^{-1}, average *PPFD* values observed at a number of UMCF sites are well above the light-saturation level of most plants (~400–500 μmol m^{-2} s^{-1}; Aylett, 1985; Cordell *et al.*, 1998; Hafkenscheid, 2000) and should therefore not be limiting photosynthesis and growth in these places (e.g. Turton, 1990; Hafkenscheid, 2000; cf. Cordell *et al.*, 1998; Schuur and Matson, 2001; Leuschner *et al.*, 2007). In addition, in areas where *average* values of *PPFD* on cloudy days are around the light-saturation level, *maximum* values still occur for 3–4 hours day^{-1} in UMCF and SACF–ECF vs. ~ up to 4–5 hours day^{-1} in LMRF (Cavelier and Mejia, 1990; Gerold *et al.*, 2008; Giambelluca *et al.*, 2009; Schawe *et al.*, this volume) although Letts *et al.* (this volume) report very low *PPFD* values for an LMCF in Colombia during the wet season.

Further support for the contention that light is not limiting intrinsic plant productive capacity in LMCF, UMCF, or ECF comes from the observation that maximum stomatal conductances in various types of TMCF (Aylett, 1985; Cavelier, 1990; Bruijnzeel and Veneklaas, 1998; Cordell *et al.*, 1998; Santiago *et al.*, 2000; Tanaka *et al.*, 2003; Küppers *et al.*, 2008; Letts *et al.*, this volume) are not lower than those found in tall evergreen lowland rain forest (Roberts *et al.*, 2005). Also, values of net photosynthetic capacity *per unit leaf surface area* (A_{max}) of LMCF and of several UMCF do not exhibit any significant trends with elevation or forest stature (Cordell *et al.*, 1998; Hafkenscheid, 2000; cf. Bruijnzeel and Veneklaas, 1998; Letts *et al.*, this volume). However, because forest *LAI* decreases markedly with increasing elevation – and therefore from LMRF to LMCF, UMCF, and SACF–ECF (Moser *et al.*, 2007; Figure 72.9) – rates of photosynthesis (A) and water uptake (Et) at the stand scale also decline with elevation (cf. Figure 72.12). In addition, it is pertinent to note that although Et is strongly related to *LAI* (Figure 72.13a), the *fraction* of available radiation used by the respective forest types for transpiration (Et/R_{net}) shows a gradual reduction with elevation (see previous section on hydrometeorology). In other words, factors other than radiation and (by implication) temperature, must be responsible for the observed decrease in Et with altitude. Further support for this contention comes from the observation that forest *LAI* (and therefore Et and A) at a given elevation (and therefore experiencing similar climatic conditions) decreases sharply as the degree of persistent soil wetness increases (Santiago *et al.*, 2000; Vernimmen *et al.*, 2007).

Summarizing, light levels do not appear to be limiting plant productive capacity in UMCF or SACF–ECF but the declining leaf area with increasing elevation leads to lowered rates of photosynthesis and soil water uptake at the stand level. Also, soil water uptake decreases more rapidly with increasing elevation than do levels of available net radiant energy. Similar trends in leaf area and soil water uptake are found along gradients of soil wetness, with the smallest values being observed under the wettest conditions.

NUTRIENT AVAILABILITY

Foliar nutrient analyses in the more stunted cloud forest types (UMCF, SACF–ECF) and various fertilization experiments involving comparatively high doses (compared to atmospheric inputs) of nutrient additions in various combinations have indicated various nutrient deficiencies (notably of nitrogen; Hafkenscheid, 2000; Soethe *et al.*, 2008a,b; Benner *et al.*, this volume). However, unlike foliar nutrient concentrations, amounts of plant-available nutrients in *mineral* soil horizons do not necessarily decrease with increasing elevation and have sometimes even been reported to increase with altitude (e.g. Soethe *et al.*, 2008b; cf. Roman *et al.*, this volume). Conversely, C : N ratios and C : P ratios in the *organic* layer generally increase with elevation, suggesting mineralization of plant litter to be slowed down, thereby reducing the availability of nitrogen and phosphorus to plant roots (e.g. Soethe *et al.*, 2008b; cf. Benner *et al.*, this volume). Also, the foliar N : P ratios found at different elevations suggest nitrogen supply becomes more critical than phosphorus supply at higher elevations (Soethe *et al.*, 2008a,b; Benner *et al.*, this volume). Whilst such reduced rates of litter decomposition have often been attributed to the lower temperatures prevailing at higher elevations, there is increasing evidence that moisture levels override any temperature effects (Schuur, 2001; cf. Silver *et al.*, this volume). This would explain the slow decomposition rates that have been observed in stunted ECFs on permanently wet soils at low elevations despite favorable ambient temperatures (Cavelier *et al.*, 2000; Benner *et al.*, this volume; Silver *et al.*, this volume). As indicated earlier, excessive soil moisture levels tend to limit soil oxygen availability (Silver *et al.*, 1999; Santiago *et al.*, 2000), which, in turn, may affect the activity of various decomposer communities, particularly fungi (Schuur, 2001). In addition, leaves from wetter sites have higher lignin concentrations (possibly in response to lowered nitrogen availability) rendering them even less amenable to decomposition (Schuur, 2001; Benner *et al.*, this volume). Pertinently, experimental aeration has been shown to greatly increase rates of nitrogen mineralization (Cavelier *et al.*, 2000; cf. Marrs *et al.*, 1988; Schuur and Matson, 2001). Cavelier *et al.* (2000) pointed out that for low mineralization rates to occur, the soil does not necessarily have to be saturated all the time – consistently high moisture levels will produce the same effect.

On the other hand, the few available studies of cloud forest soil nutrient *budgets* indicate that the trees take up only a mere fraction of the amounts of readily available nutrients in net precipitation arriving at the forest floor or contained in leaf litter leachate (Bruijnzeel *et al.*, 1993; Hafkenscheid, 2000). This suggests that the roots have difficulty in making full use of what nutrients are available. Although the precise mechanism is still being debated (Benner *et al.*, this volume; see also below) and although the cited estimates of nutrient uptake concerned the production of above-ground biomass only, *it is clear that the persistently wet to saturated conditions prevailing in UMCF and SACF–ECF soils severely limit opportunities for exploitation of mineral soil horizons by the roots* (Silver *et al.*, 1999; Santiago *et al.*, 2000; Soethe *et al.*, 2006). The occurrence of adventitious roots (i.e. above the soil surface) capturing nutrient-rich stemflow before it reaches the soil – as reported for some excessively wet forests – may be interpreted along similar lines (Herwitz, 1991). Various other adaptations in root architecture are required to improve tree anchorage on such shallow and saturated slopes, particularly under conditions of high wind (Soethe *et al.*, 2006, 2007; see also below).

Alternatively, toxic concentrations of "free" aluminum or hydrogen ions in the soil solution have also been invoked to explain reduced nutrient uptake (Hafkenscheid, 2000). Aluminum concentrations have been reported to increase with site wetness (Schuur and Matson, 2001) and reduced site fertility (Hafkenscheid, 2000; cf. Schrumpf *et al.*, 2001; Schawe *et al.*, 2007). To neutralize potentially toxic levels of aluminum (or hydrogen), the roots are considered to produce phenolic compounds that, in addition, immobilize nitrogen by organic complexation, thereby preventing leaching of precious nitrogen in dissolved form (Northup *et al.*, 1995, 1998; Hafkenscheid, 2000). Interestingly, although concentrations of aluminum in the soil solution in a series of montane forests in Hawai'i increased with site wetness, foliar concentrations of aluminum decreased along the gradient, suggesting the roots were effectively excluding aluminum uptake (Schuur and Matson, 2001). In addition, lignin concentrations in leaf litter at these sites were higher – and nitrogen concentrations lower – as the soils became wetter (Schuur, 2001). Although no measurements were made of phenolic concentrations, these are known to be related positively to lignin concentrations and inversely to nitrogen concentrations (Bruijnzeel *et al.*, 1993; Hafkenscheid, 2000). The mechanisms linking plant phenol production in montane tropical forests – both in leaves and roots – with soil biological and chemical processes, site fertility, and forest productivity require more work before generalizations can be made (Northup *et al.*, 1995, 1998; Hafkenscheid, 2000; Hättenschwiler *et al.*, 2003).

The trees of the more stunted tropical montane forest types appear to respond to these adverse soil conditions in several ways: increased leaf scleromorphism (Turner, 1994), increased leaf longevity (Moser *et al.*, 2007), and increased investment of carbon in below-ground biomass and productivity (Leuschner *et al.*, 2007).

Although the leaves of many of the more stunted types of cloud forest look distinctly xeromorphic, there is ample evidence (reviewed by Turner, 1994) that this is not an adaptation to better withstand drought conditions but rather a response to reduced nutrient availability. Scleromorphic features appear to be closely associated with long-lived leaves (Turner, 1994). Whilst information on leaf lifespan in tropical montane forests is comparatively rare, recent data from an elevational transect in Ecuador suggest that leaves remain significantly longer on the trees in cloud-affected forests than below the cloud belt, although no differences were observed between different types of cloud forest (Moser *et al.*, 2007). However, a meta-analysis of fine-root production data from more than 80 tropical moist forest stands from the lowlands and the montane belt indicates that upper montane forests in the Neotropics have significantly higher average fine-root production (1721 g m^{-2} $year^{-1}$) than do either lowland forests (405 g m^{-2} $year^{-1}$) or lower montane forests not affected by significant cloud (540 g m^{-2} $year^{-1}$; Hertel and Leuschner, this volume). This increased investment of carbon in below-ground biomass under the more adverse (acid, wet, poor in nitrogen) soil conditions prevailing in UMCF and ECF (Roman *et al.*, this volume) is apparently realized at the expense of above-ground productivity and biomass, and represents a marked shift in carbon allocation from shoot to root with increasing elevation or degree of soil wetness (Leuschner *et al.*, 2007; Moser *et al.*, 2007; Hertel and Leuschner, this volume).

WIND

Although wind is rarely a definitive factor determining montane forest stature, the occurrence of persistently high winds on wet, exposed upper slopes and summit locations acts as an added stress factor (Lawton, 1982; Cordero, 1999; cf. Fetcher *et al.*, 2000). The combination of high wind loading and the shallow rooting patterns caused by shallow soil depths or prolonged saturated conditions described above produces various architectural adaptations, including greater occurrence of stilt and aerial roots, a greater proportion of coarse roots, asymmetrical extensions of the soil root plate, and increased sprouting from fallen logs (Gill, 1969; Soethe *et al.*, 2006, 2007). On the other hand, it may well be disadvantageous for trees to be tall and highly productive in such wet and windy environments. Increased height and canopy growth will result in increased wind stress and in enhanced capture of cloud water and wind-driven rain (and thus wetter soils?), which might ultimately be counter-productive. Future research on the causes of cloud forest stunting should also take into account the ecological costs of forest productivity.

CLOUD FORESTS AND CLIMATE CHANGE

During the "anthropocene" (Crutzen and Stoermer, 2000) most of Earth's ecosystems – cloud forests included – have become subject to sustained human impact. This raises a number of

In summary, regardless of elevation, reduced forest stature on wet tropical mountains is typically associated with persistently wet and shallow soils, reduced rooting depth, reduced nitrogen availability, and possibly enhanced soil toxicity. Wind and steep topography constitute additional stress factors at exposed locations. The forests respond to these adverse soil conditions by investing much more carbon in below-ground biomass (i.e. the root system) at the expense of above-ground biomass, and to a lesser extent in the production of harder, thicker, and longer-lasting leaves.

important questions with regard to the impacts of climate change on cloud forest biodiversity and the ecosystem services that these forests provide. These questions include:

- Where and how are different types of TMCF expected to be affected by climatic drying (reduction in precipitation or cloud frequency?
- What ecosystem component or functions are most affected by climatic change?

Observed and projected climate change for cloud forests

Before the above questions can be answered it is helpful first to look at climate change scenarios for cloud forests. Based on preliminary modeling for selected cloud forest sites, Still *et al.* (1999) and Foster (2001) concluded that climate change resulting from anthropogenic greenhouse gas emissions will significantly affect the temperature and humidity of cloud forest environments worldwide. Similarly, both observational (e.g. Pounds *et al.*, 1999; Barradas *et al.*, this volume) and experimental evidence is on the increase that montane areas carrying cloud forest may warm up locally as a result of deforestation in adjacent lowlands (Lawton *et al.*, 2001; Van der Molen, 2002; Ray *et al.*, 2006; Nair *et al.*, this volume). Scatena (1998) interpreted the presence of isolated stands of large and very old (>600 years) Colorado trees (*Cyrilla racemiflora*) in the Luquillo Mountains of eastern Puerto Rico at elevations well below the current cloud base that experience comparatively lower rainfall (<3000 mm year^{-1}) as evidence of a gradual upward shift in vegetation zonation over the past several centuries. *Cyrilla* is currently a dominant tree in areas above the cloud base (>600 m.a.s.l.) and is most common where mean annual rainfall exceeds 4000 mm. Similarly, Brown *et al.* (1996) reported the occurrence of pockets of mossy cloud forest below the current average cloud base in Honduras. According to the temperature lapse rates prevailing on most tropical mountains (0.5–0.65 °C per 100 m rise in elevation), a projected rise in temperature of 3 °C would imply a rise in cloud-base level of 450–600 m (Foster, 2001).

Since cloud forest environments are defined by the presence of ground-level cloud (i.e. fog), the rises in cloud-base levels implied by global or regional warming (Foster, 2001; Foster, this volume) can be expected to have profound impacts upon many aspects of the cloud forest ecosystem, ranging from changes in biological diversity (Pounds *et al.*, 1999; Nadkarni and Solano, 2002; Williams *et al.*, 2003; Colwell *et al.*, 2008) to changes in ecological (Raich *et al.*, 1997; Aiba *et al.*, this volume) and, possibly, hydrological functioning (Hemp, 2005b; Ponette *et al.*, 2009; Hemp, this volume #58;). Generally speaking, because of the fragmented nature of TMCF distributions and their occurrence along narrow elevational bands in heterogeneous mountain landscapes (Vázquez-García, 1995; Scatena *et al.*, this volume), cloud forests must be considered highly vulnerable to both local and regional climate change (Mulligan, this volume; Nair *et al.*, this volume). This is true both in terms of the very real possibility of "mountain-top extinctions" (where summit ecotones are driven out of existence upon lifting of the cloud base), and in terms of spatial discontinuities between current and projected elevational ranges for groups of species (Williams *et al.*, 2003; Colwell *et al.*, 2008). In addition, due to their often exposed locations (Jarvis and Mulligan, this volume), cloud forests in areas affected by hurricane passage (notably the Caribbean, Central America, East Asia, North Queensland, and the southwestern Pacific; cf. Scatena *et al.*, 2005) may become subjected to increasingly frequent or intense disturbance by tropical cyclones due to the global warming of the oceans, although the topic is still a matter of much debate (e.g. Emanuel, 2005; Webster *et al.*, 2005; McBride *et al.*, 2006; cf. Mann *et al.*, 2007).

Because of the lack of precise information on the location of (different types of) cloud forests, there has been little research to date looking at the potential impacts of climate change on cloud forests as a whole and which areas are more likely to be impacted than others (cf. Still *et al.*, 1999; Foster, 2001; Williams *et al.*, 2007). Mulligan and Burke (2005a) used the modeled distributions of cloud-affected forest (CAF) discussed in Chapter 2 to analyze both historically observed and future projected climate change in areas with CAF (cf. Mulligan, this volume) and their hydrological impacts. Mulligan and Burke first used the climatic data-set of New *et al.* (2000) to examine whether a change was detectable between the mean temperatures in areas with CAF for the periods 1901–1946 and 1947–1993. The observed change was a cooling of around 1 °C over all CAF except those in the southern Andes, in contrast to the warming observed for tropical lowlands over the same period. This cooling may be insignificant or could be the result of reduced nocturnal cloud cover leading to lower night-time temperatures. The pattern for historic rainfall change in areas with CAF as derived by Mulligan and Burke (2005a) was less clear, with some areas indicating stationarity, others drying and yet others wetting. In almost all cases the observed changes were modest and within 20% of the 1901–1993 mean rainfall.

Cloud frequency is perhaps the most significant climate variable for cloud-affected forest. To assess changes in cloud cover, Mulligan and Burke (2005a) used the HIRS cloud climatology data-set of the University of Wisconsin (cf. Jin *et al.*, 1996) – which at that time spanned a period of 22 years (1979–2001). They extracted annual cloud frequency data for all areas with known cloud forest according to the UNEP–WCMC database (Aldrich *et al.*, 1997b) and calculated the change in the frequency of clear-sky (i.e. no cloud) observations between the periods 1979–1990 and 1991–2001. The results indicated a decrease in the number of clear days (i.e. an increase in cloudiness) for cloud forests near the equator, and the reverse (decreased cloudiness) for the 10–20° latitudinal belts to the north and south. Observed cloud frequency declined particularly over cloud forest areas in South-East Asia whereas in Latin America there was a decline in Mexico, Central America, and northern Colombia, and an increase in cloud frequency in Perú and Bolivia. The maximum observed changes in cloud-free days were around 4% per decade.

Having found mixed signals of climate change in the limited historic data (a modest temperature decline, small-scale positive or negative changes in rainfall, and small-scale positive or negative changes in cloud frequency), Mulligan and Burke (2005a) examined the results of two different general circulation models (GCMs) – ECHAM5 and HADCM3 – using the SRES A2 CO_2 scenarios prepared for the Intergovernmental Panel on Climate Change Third Assessment Report published in 2001. The monthly gridded GCM data were split into two periods, 1950–2000 and 2000–2050, and mean values for temperature and rainfall were calculated for the two periods and then compared (see also Mulligan, this volume). HADCM3 predicted a mean global temperature increase of 3.3 °C between the two periods, with CAF seeing changes between +2 and +5 °C. The greatest temperature changes were derived for the most continental cloud forests (in the African interior), and the smallest for cloud forests on small islands (cf. Figure 2.14). A similar spatial pattern was predicted by ECHAM5 although the magnitude of the changes was somewhat less than those obtained with HADCM3 (cf. Figure 2.15). Given the typical elevational range of ~500 m observed for many tree species (Kappelle *et al.*, 1995b; Kitayama, 1995; Bach, 2004) and the often narrower ranges for a host of animals (cf. Williams *et al.*, 2003; Colwell *et al.*, 2008; Rovito *et al.*, 2009), such changes in temperature are bound to have major ecological consequences (see corresponding section below).

The GCM results for changes in rainfall were both more extreme and more uncertain than those for temperature, especially at the regional and sub-regional scale. Predicted changes in rainfall over cloud forest areas between 1950–2000 and 2000–2050 indicate a potential loss in rainfall (of 100 to 1000 mm $year^{-1}$ over the period) for some areas (e.g. Colombia, Mexico, Central Africa) and a potential gain in others (e.g. Perú, Madagascar, and the South-East Asian islands (+100 to +1000 mm $year^{-1}$ over the period) according to HADCM3 (cf. Figure 2.16). Conversely, according to ECHAM5 almost all cloud forest areas will experience an increase in rainfall of a few hundred mm $year^{-1}$, with the exception of Madagascar, West Africa, and a few of the South-East Asian islands (cf. Figure 2.17). Thus, differences in predicted changes in rainfall between different GCMs are large but so are the expected magnitudes of change in relation to current mean rainfall (Mulligan and Burke, 2005a; cf. Mulligan, this volume).

IMPACTS OF CLIMATE CHANGE ON TMCF WATER BALANCE

Mulligan and Burke (2005a) developed a simple model to illustrate spatial and temporal patterns in water balance across the tropics (23.5° N – 35° S). The model has a spatial resolution of 1 km and operates at a monthly time-step. Grid-cell water balance is calculated as precipitation (including modeled contributions by wind-driven rain but not cloud water) minus potential evapotranspiration (*PET*). Using grids of current climate data, baseline conditions were simulated (see map in Mulligan and Burke, 2005a, p. 59) to allow preliminary evaluation of the spatial changes in monthly water budget as predicted by the model after applying the changes in temperature, solar radiation and rainfall projected by the ECHAM5 GCM for the period 2000–2050 (Figure 72.15).

The predicted impact of projected climate change on the water balance of cloud-affected forests is variable, with some areas having projected increased precipitation surplus (e.g. the Andes and Central Africa) whilst others show a significant drying (West Africa, Madagascar, South-East Asia, the Brazilian Atlantic cloud forests, and the CAFs of the Guyana shield; Figure 72.15). These results should be viewed as highly preliminary. First of all, because rainfall dominates the water balance signal, the simulated impacts are highly dependent on the rainfall projections of the GCM used. As shown in the previous section, these tend to differ between GCMs, both in terms of direction and magnitude. Moreover, the present simulations do not take into account the effect of seasonal changes in rates of glacial melting at higher elevations (cf. Kaser *et al.*, 2005; Juen *et al.*, 2007), nor the complex relationships between changes in lowland energy budget and upland cloud-base height (cf. Ray *et al.*, 2006; Van der Molen *et al.*, 2006; Nair *et al.*, this volume; Van der Molen *et al.*, this volume). However, Mulligan and Burke (2005a) considered the hydrological impacts of the applied rainfall change scenario to be more important than those simulated using a scenario for all observed forest conversion until 2001.

ECOLOGICAL IMPACTS OF CLIMATE CHANGE

Apart from any changes in water budget, climate change is also likely to affect cloud forest biodiversity (both in terms of abundance and elevational zonation), and ecological functioning.

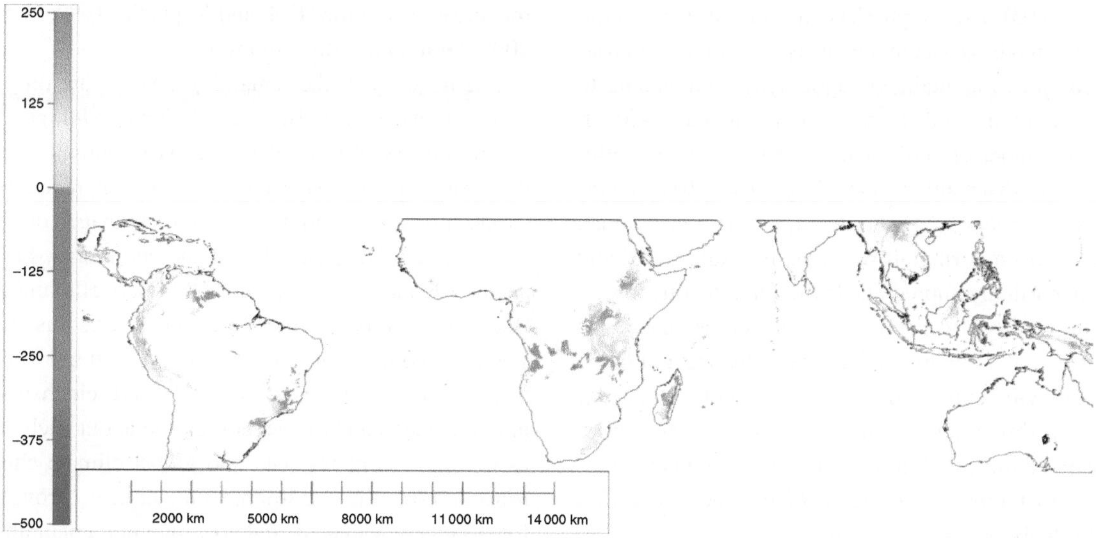

Figure 72.15. Impact of projected climate change (2000–2050) on the water balance (in mm $year^{-1}$) of cloud-affected forest areas across the tropics. (See also color plate.)

Whilst comparatively little is known as yet of such ecological effects, there are various signs on the wall that do not bode well. Nadkarni and Solano (2002) and Nadkarni (this volume) have documented the major changes in epiphyte and bryophyte diversity that may occur upon climatic drying or opening up of cloud forest canopies. Indeed, given their exposed position and dependence on rainfall and fog inputs and the nutrients contained therein (Hietz, this volume; cf. Chang *et al.*, this volume; Liu *et al.*, this volume), epiphytes and bryophytes in the upper parts of the canopy must be regarded particularly vulnerable to climatic drying (Zotz and Bader, 2009; cf. Lugo and Scatena, 1992; Colwell *et al.*, 2008; Köhler *et al.*, this volume) and their demise or disappearance is likely to change the canopy's cloud-water intercepting capacity in ways that are as yet not fully understood (cf. Chang *et al.*, this volume; Mulligan *et al.*, this volume #25; Tobón *et al.*, this volume #26).

Also, amphibian declines and extinctions have become a common feature in many habitats around the world, including protected TMCF reserves, suggesting that the phenomenon cannot be explained solely by localized habitat destruction (Stuart *et al.*, 2004; Rovito *et al.*, 2009; cf. Pineda and Halffter, 2004). At Monteverde (Costa Rica), for instance, a species of harlequin frog (*Atelopus* sp.) and the famous golden toad (*Bufo periglenes*) have vanished due to pathogen outbreaks that have been linked to global warming (Pounds *et al.*, 2006). The agent is a pathogenic chytrid fungus which causes a skin disease amongst the frogs and toads, ultimately leading to mass extinctions (Pounds and Crump, 1994; Pounds *et al.*, 1999, 2006; Lips *et al.*, 2003). Similarly, Rovito *et al.* (2009) recently reported dramatic declines in salamander populations in UMCF in Mexico and Guatemala. The species in decline comprise several major evolutionary lineages of tropical salamanders, underscoring the fact that significant portions of the phylogenetic diversity of Neotropical salamanders are at risk (Rovito *et al.*, 2009). Cloud forest frogs on other continents are also significantly threatened. For instance, the frog *Philautus poppiae*, a native of Sri Lankan closed canopy cloud forest, has been classified as Endangered on the IUCN *Red List of Threatened Species* due to the continued decline in the quality and extent of its habitat. As many as 19 Sri Lankan frog species have gone extinct so far. Apart from habitat loss caused by subsistence farming activities and logging, climatic drying and the use of agrochemicals in cardamom cultivation are considered additional causes. Apparently, no other country in the world has more documented amphibian extinctions, including a number of cloud forest frog species (IUCN, 2009). In short, given their often narrow elevational range of occurrence, cloud forest amphibians must be considered particularly sensitive to climate change (Williams *et al.*, 2003; Rovito *et al.*, 2009).

As for vertebrates, Williams *et al.* (2003) developed bioclimatic models of the current spatial distribution of regionally endemic vertebrates in the rain forests of northern Australia – a World Heritage Site and therefore well protected – to aid in the prediction of distributional changes after various degrees of climatic warming (examined range: 1–7 °C). Even after a 1-°C rise in temperature – a change that is considered inevitable in the next few decades – 63 out of the 65 considered species were predicted to experience significant changes in their distribution already, whereas species losses increased rapidly beyond a temperature increase of 2 °C due to the loss of "core environments." It is pertinent to note that these distribution models only considered the effects of increasing temperature. Apart from the expected deleterious effects of a raised cloud base on amphibians as documented earlier for Monteverde (Pounds *et al.*, 1999, 2006),

Williams *et al.* (2003) also emphasized the increase in habitat fragmentation that tends to accompany a rise in cloud condensation levels in complex mountainous topography (and leading to separation of populations on different peaks). Last but not least, they identified a number of additional adverse effects on vertebrate diversity and distribution, including raised CO_2 levels rendering canopy tree leaves tougher to eat and less nutritional, thereby affecting endemic ringtail possums, tree kangaroos, and many insects; increasingly unpredictable rainfall affecting insectivorous and frugivorous bird populations, as well as changes in litter decomposition and mineralization rates affecting soil fauna and microbial activity and, ultimately, forest nutrient supply (cf. Benner *et al.*, this volume). On a related note, Boy *et al.* (2008) demonstrated increased nutrient loading of montane tropical forests due to biomass burning in (sometimes far away) lowlands (cf. Rollenbeck *et al.*, this volume).

Summarizing, whilst the extent of, and impacts on, cloud forest biodiversity, hydrology, and ecology of impending climate change are only poorly documented as yet, they are likely to be highly significant.

CLOUD FOREST CONSERVATION, MANAGEMENT, AND RESTORATION

Cloud forests face many threats, including exploitation for timber and other forest products, fragmentation, and conversion to other forms of land use – notably cattle ranching, cultivation of temperate vegetables, and coffee plantations (Amelung and Diehl, 1992; Etter and Van Wijngaarden, 2000; Brown and Kappelle, 2001; Cayuela *et al.*, 2006; Beck *et al.*, 2008a; Ponette-González *et al.*, 2009). In addition, there are the more subtle ecological changes associated with regional and global climate change (Scatena, 1998; Pounds *et al.*, 1999; Foster, 2001; Nadkarni and Solano, 2002; Williams *et al.*, 2003; Rovito *et al.*, 2009) and the increased risk of fire due to climatic drying (Asbjornsen and Garnica-Sánchez, this volume). Apart from direct clearing, several studies have shown that the construction of roads and power lines (Olander *et al.*, 1998; Beck *et al.*, 2008c) and the continued harvesting of fuelwood (Rüger *et al.*, 2008), as well as recreational visits (Honey, 2008) can all have long-term impacts on TMCF. In some areas, old-growth forest continues to be cleared to make way for pastures that replace other pastures suffering from decreasing productivity or invasion by bracken (Paulsch *et al.*, 2001; Hartig and Beck, 2003). In other areas, abandonment follows outmigration from rural areas to urban centers as a result of changes in economic conditions, thereby potentially favoring ecosystem recovery (Kappelle *et al.*, 1995a, 1996; Aide and Grau, 2004) and offering opportunities for forest restoration (Holl and Kappelle, 1999; Grau *et al.*, 2003, 2008; Grau *et al.*, this volume).

The majority of the remaining TMCFs are only fragments of their original extent (Bubb *et al.*, 2004; Mulligan, this volume) whereas many of those that still exist continue to be exposed to the various threats listed above. It is widely recognized that, to be successful, any effort to conserve, manage, or restore TMCF needs to be both science-based and community-based (Hamilton *et al.*, 1995a; Aldrich *et al.*, 2000; Kappelle and Brown, 2001; Báez *et al.*, this volume; Hofstede *et al.*, this volume). Thus, science should guide conservation strategies and tactics to achieve sustainable cloud forest use, and scientists should inform land managers and the general public about such diverse themes as the impacts of regional and global climate change on cloud forests, how the environmental services provided by cloud forests can be valuated, and how integrated planning of protected forest areas and surrounding land is to be organized most effectively.

This section discusses various key elements of cloud forest conservation and management for which research has produced relevant scientific knowledge in the past few decades.

Sustainable use and management of cloud forests

For centuries trees in the taller types of TMCF have been felled to be used in local construction, whereas fuelwood has been harvested both for subsistence and for commercial purposes (Kappelle and Brown, 2001). However, over the past few decades the concept of forest management has broadened significantly, evolving from the traditional focus on timber harvesting toward more sustainable practices reflecting changes in the societal perception of nature (Sheil and Lawrence, 2004; Wolf, 2005; Herrera and Chaverri, 2006). Indeed, modern society has come to recognize that tropical highland forests are not just a source of timber, but are also important, multifunctional ecosystems offering an array of non-timber forest products (NTFPs; Kappelle *et al.*, 2000; Wolf, 2005) as well as a variety of ecosystem services including high-quality water, carbon sequestration, biodiversity, and scenic beauty (Scatena, 1995; Calvo, 2000; Pagiola, 2002; Aylward, 2005; Barrantes Moreno, 2006). This new perception is reflected in current definitions of sustainable forest management (Herrera and Chaverri, 2006), which highlight the importance of managing permanent forest land without undue reduction of its inherent values and future productivity, and without undue undesirable effects on the biophysical and social environment (ITTO, 1992). Given the very slow growth rates of most upper montane and elfin cloud forests (Weaver and Murphy, 1990; Kappelle *et al.*, 1996; Raich *et al.*, 1997; Aiba *et al.*, 2005; Aiba *et al.*, this volume) and the generally excessively wet soil conditions prevailing in these forests (Bruijnzeel and Veneklaas, 1998; Roman *et al.*, this volume; Schawe *et al.*, this volume), logging and timber extraction is not an option for this type of cloud forest (Scatena, 1995; Hamilton, 2005).

Reduced-impact logging (RIL) can help enhance juvenile recruitment and growth of economically important tree species in the more productive types of cloud forest and therefore should be considered when developing more sustainable silvicultural systems for TMCFs (Guariguata *et al.*, 2006; Günter *et al.*, 2008). Likewise, measures like RIL, the adoption of longer cutting cycles, and increasing the traditional minimum felling diameter generally leads to higher timber yields, reduced damage to the soil and the remaining stand, and improved conservation of biodiversity compared to standard logging techniques (Brünig, 1996; Cassells and Bruijnzeel, 2005; Günter *et al.*, 2008). However, although various silvicultural principles and approaches (e.g. improvement felling) are universally applicable, the sustainable management of highly diverse tropical montane forest ecosystems requires case-specific silvicultural approaches meeting local socio-economic, biophysical, and ecological conditions. In contrast to the situation for lowland evergreen rain forests rich in high-value timber species (such as the dipterocarp forests of South-East Asia), for which various nominally sustainable timber harvesting schemes have been developed (Brünig, 1996; Thang and Chappell, 2005) there is almost no knowledge on the ecological and economic viability of silvicultural systems in montane cloud forests which tend to have low numbers of large, harvestable trees (cf. Guariguata *et al.*, 2006; Günter *et al.*, 2008). The first results obtained during a recently initiated comprehensive study of the silvicultural and ecological consequences of light selective logging in an Ecuadorian LMCF suggested no significant changes in the overall diversity of various animal groups, although changes were observed at the level of population structure and species composition (Günter *et al.*, 2008). Günter *et al.* (2008) concluded that ecologically sustainable management of Neotropical montane forest is possible as long as the disturbance intensity is kept low (i.e. few gaps of modest size) and provided the disturbed areas are embedded within a matrix of undisturbed forest to allow recruitment and recovery of the affected taxa. Epiphytes on trees located in the vicinity of newly created gaps showed a reduction in the number and size of new leaves during the first year after the disturbance, suggesting a response to gap creation. However, variability within disturbed and non-disturbed experimental groups was very high and results sometimes differed in direction as well (Günter *et al.*, 2008). In view of the slow growth rates of many epiphytes, much longer observation periods are required to demonstrate the eventual effects of selected logging on epiphyte assemblages (Barthlott *et al.*, 2001; cf. Gradstein, 2008).

In a related context, Wolf and colleagues assessed the possibility of harvesting ornamental epiphytes (bromeliads) in the Highlands of Chiapas, Mexico, in an ecologically sustainable manner as an alternative to traditional, destructive wood extraction (Wolf and Konings, 2001; Wolf and Flamenco-Sandoval, 2006; Wolf, this volume). Field observations demonstrated that *Tillandsia vicentina* plants need *c.* 5 years to develop from seed to a commercially viable size of 20 cm, allowing for a 4-year rotational harvesting scheme with an annual harvest from the lower forest stratum of *c.* 700 epiphytic rosettes (Wolf, this volume). These results have since been used to help define criteria for sustainable bromeliad harvesting in the cloud forests of Chiapas, thereby improving overall forest management by indigenous communities in the area.

Sustainable forest management must take into consideration the main threats affecting the forest's viability. Amongst these, fire is typically not perceived as a key threat to TMCFs since these forests' regular immersion in cloud would theoretically render them fire-resistant, although charcoal fragments found in various cloud forest soils indicate that fire may have occurred in the past during times of exceptional drought (Wård *et al.*, this volume), However, recent widespread fires in south-eastern Mexico have shown that the ecological integrity of modern-day TMCF in seasonal climates may very well be at risk as wildfires may become more frequent and severe due to climatic drying (Asbjornsen and Garnica-Sánchez, this volume; cf. Mulligan, this volume). The chapter by Asbjornsen and Garnica-Sánchez assesses the technical and organizational aspects involved in suppressing the severe wildfires that occurred in Mexican cloud forests during the 1998 El Niño Southern Oscillation (ENSO). In doing so, they highlight some valuable lessons learned about fire-suppression techniques during extreme fire behavior (including persistent below-ground fires) and suggest that the integration of local community-based and external government-based fire management approaches is critical to achieve effective forest protection and fire suppression (Asbjornsen and Garnica-Sánchez, this volume).

Where and what are the priorities for cloud forest conservation?

As indicated previously in the section on biodiversity, many cloud forest sites occur in six of the most important biodiversity hot spots around the globe (Myers *et al.*, 2000) – viz. the Tropical Andes and Atlantic Forests in South America, the Caribbean Islands, the Eastern Afro-montane peaks, the Asian Sundaland (including the mountains of Borneo), and the Polynesia–Micronesia hot spot that includes the islands of Fiji, Hawai'i, and Tahiti. Cloud forests are also found in 18 of the 200 biologically most distinct "Ecoregions" of the planet (Olson and Dinerstein, 2002) which partly overlap with, and complement these six hot spot areas. Listed per biome, these include the following (cf. Figure 72.3):

- *Afrotropical*: Cameroon Highland forests, Albertine Rift montane forests, Eastern Arc montane forests;
- *Australasian*: New Guinea montane forests;

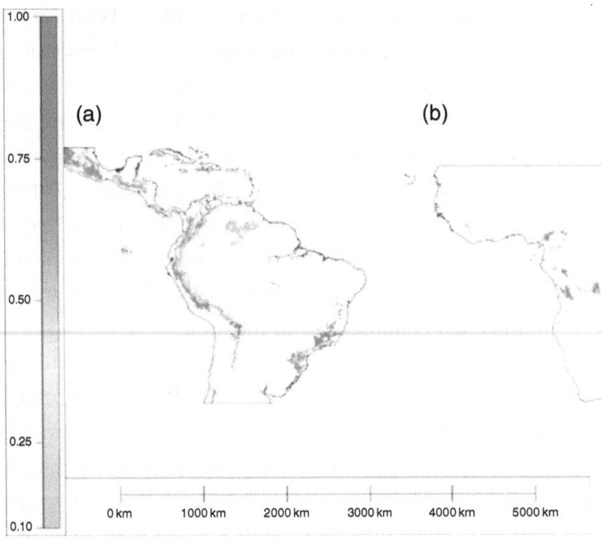

Figure 72.16. Fractional cloud-forest loss according to MODIS_VCF satellite product for (a) Latin America, (b) Africa, and (c) South-East Asia. The darker the shade of red, the greater the forest loss. Red: > 60% forest loss, green < 25% forest loss. (Reproduced from Figure 2.13; See also color plate.)

- *Indo-Malayan*: montane forests in Sumatra and surrounding islands, Peninsular Malaysia, and Borneo, and the montane forests of West Java;
- *Neotropical*: Greater and LesserAntillean moist forests, Talamancan–Isthmian Pacific forests, northern Andean montane forests, coastal Venezuelan montane forests, Guayana Highland moist forests, Central Andean Yungas, Brazilian Atlantic forests (*Mata Atlantica*); and
- *Oceania*: South Pacific Island forests, and Hawaiian moist forests.

Summarizing, more than half of the original cover of CAF has been converted to other land uses and some countries have very little of their original cloud forest remaining. The largest and most intact areas of CAF are found in southern Venezuela, eastern Democratic Republic of Congo, and maritime South-East Asia (Borneo, Celebes, Irian Jaya, and Papua New Guinea). Losses can be expected to continue in all areas with CAF but particularly in those that have shown rapid loss to date.

The hydro-climatically based distribution of cloud-affected forest (CAF) derived by Mulligan (this volume) may be combined with the *World Database on Protected Areas* (WDPA Consortium, 2004) to examine the conservation status and priorities of CAF in more spatial detail than is possible on the basis of the general hot spot areas cited above. Similarly, by comparing hydro-climatically based potential and current distributions of CAF, estimates and spatial patterns of cloud forest loss may be obtained (see Mulligan and Burke (2005a) and Mulligan (this volume) for details).

Mulligan's assessment suggests a total cloud forest loss to date of 2.72 Mkm^2 (or 55% of the originally estimated cover of 4.93 Mkm^2). By comparison, application of the same methodology indicates that 47% of all tropical forests have been lost (Mulligan and Burke, 2005b). Extensive loss is inferred to have occurred throughout Latin America, with intensive losses in much of Mexico, the Colombian western and central Cordilleras, western Ecuador, Perú and Bolivia, and the Brazilian Atlantic forests. The most intact remaining Latin American CAFs are found on the eastern flanks of the Andes and in southern Venezuela (Figures 72.2 and 72.16). Significant inferred losses in Africa are concentrated in Madagascar, Ethiopia and Eritrea, South Africa, Uganda, Kenya and Tanzania. In South-East Asia, the most extensive losses are inferred for Indonesia, southern China, Laos and Vietnam (cf. Table 2.7 and Figure 72.16).

Comparing the current extent and distribution of cloud-affected and non-cloud-affected forests with those of Protected Areas (WDPA Consortium, 2004) suggests that 10% (or 1.57 Mkm^2) of the tropical *tree-covered* area (15.64 Mkm^2) is at least nominally protected, compared with 14.7% (0.38 Mkm^2) of the forest area classified as CAF (2.21 Mkm^2). Thus, protected CAF makes up ~9% of the total tropical tree-covered area and almost one-fourth (24%) of the total protected tropical tree-covered area (Mulligan, this volume).

A number of Protected Areas contain mostly cloud forest, particularly in Venezuela, Costa Rica, the Philippines, Madagascar, Vietnam, Laos, Cameroon, and Indonesia. The largest individual protected cloud forests are found in Venezuela, Indonesia, and Colombia (see Mulligan and Burke, 2005b for details). To indicate which Protected Areas with CAF are particularly threatened, a ranking was made of the absolute and relative losses of CAF.

Protected Areas with the greatest *proportion* of deforested land formerly under CAF occur throughout Central America and the Caribbean, as well as in Thailand and Indonesia. Protected Areas that have lost the greatest *absolute* areas of CAF are largely located in more continental settings, where one also finds some of the largest Protected Areas and the largest areas CAF (cf. Table 2.8; see Mulligan (this volume) for a discussion of the limitations of the analysis with respect to results obtained for individual sites).

Finally, by examining on a national basis, countries that have significant areas of CAF outside of Protected Areas (as listed in WPDA Consortium, 2004) it is possible to identify areas with further potential for (and likely need of) cloud forest protection. Countries with the largest absolute areas of unprotected CAF include the Democratic Republic of Congo, Brazil, Indonesia, Perú, and Colombia. Similarly, those with the greatest proportion of their CAF unprotected include Laos, Burundi, Papua New Guinea, Taiwan, Honduras, and Ecuador (see Table 2.9 in Mulligan (this volume) for details). In these countries, conservation of CAF assumes particular importance. The same holds for countries where very small areas of remaining and unprotected CAF represent the sole refuge for threatened endemic species. Examples include the Comoros, Eritrea, and Cambodia.

Last, but certainly not least, the host of wild varieties of agricultural crop species found in CAF are of particular interest to society and their source areas require adequate protection to ensure the preservation of gene pools needed for continued improvement of agricultural crops (Debouck and Libreros Ferla, 1995; McNeely and Scherr, 2003; Bubb *et al.*, 2004). Examples from Latin American CAFs include avocado, bean, blackberry, papaya, passion fruit, pepper, potato, sweet cucumber, tomato, and tree tomato (Brown and Kappelle, 2001).

Regeneration and restoration of cloud forests

With the exception of some early work on cloud forest regeneration after disturbance by hurricanes (Weaver, 1986, 1999; cf. Bellingham, 1991) or roadside clearing (Ewel, 1980), most research on the regeneration and restoration of TMCFs is comparatively recent and limited to only a few locations (e.g. Cavelier, 1995; Kappelle *et al.*, 1995a, 1996; Olander *et al.*, 1998; González-Espinosa *et al.*, 2008; Günter *et al.*, 2008, 2009; Martínez *et al.*, 2008; Aide *et al.*, this volume). Whilst the leaf area of regenerating cloud forest stands tends to approach values observed for old-growth forests after 10–15 years (Gomez-Cárdenas, 2009; Hölscher *et al.*, this volume), full recovery in terms of forest structure and terrestrial flora may take up to a century for UMCF at altitudes of 2500 m.a.s.l. and higher, even if seed sources are near (Kappelle *et al.*, 1995a, 1996). Recovery of epiphyte biomass and diversity may well take much longer, although data on which to base such estimates are few (Van Dunné and Kappelle, 1998; Dunn, 2000; Holz and Gradstein, 2005b; Gradstein, 2008; Hölscher *et al.*, this volume; Köhler *et al.*, this volume; Krömer and Gradstein, this volume). Similarly, under the adverse climatic and soil conditions prevailing in most upper montane, elfin, and sub-alpine cloud forests, growth – and therefore regeneration – is inherently slow (Weaver and Murphy, 1990; Raich *et al.*, 1997; Aiba *et al.*, 2005). Based on 18 years of observations of forest succession after an airplane crashed into low-elevation ECF in Puerto Rico, Weaver (1990) estimated that full recovery of the original biomass might take at least 200 years. Olander *et al.* (1998) considered a period of 200–300 years to be necessary for disturbed roadsides within the same forest type to return to near-original conditions. Again, the more stunted and wetter types of cloud forest should thus be treaded upon as lightly as possible (cf. Scatena, 1995; Hamilton, 2005).

Forest recovery on some abandoned crop fields and pastures – particularly after repeated use of fire during the period of agricultural usage – may be hampered severely by bracken fern or similarly fire-resistant grasses, particularly on slopes of low gradients where the bracken fern tends to shade out other species. The effect is less pronounced on steep, exposed slopes where wind-dispersed shrub species manage to germinate underneath the ferns, creating a successional mosaic of dense patches of bracken interspersed with low bush (Kappelle *et al.*, 1994; Hartig and Beck, 2003; Beck *et al.*, 2008c). Invasion by such fire-resistant species creates a so-called "resilient degraded system" in which forest recovery is essentially inhibited by positive feedbacks maintaining the degraded state (Hobbs and Norton, 1996; Suding *et al.*, 2004; Aide *et al.*, this volume). For example, the grass *Melinis minutiflora* invades recently burnt land and its resinous and highly flammable leaves tend to promote additional fires, thereby effectively inhibiting forest regeneration (D'Antonio and Vitousek, 1992; cf. Sarmiento, 1997; Cavelier *et al.*, 1998). Grau *et al.* (this volume) suggest similar feedbacks to be important in north-western Argentina where resilient degraded grasslands are not invaded by cloud forest species even after land-use intensity decreases and rainfall conditions become favorable for tree establishment. In addition, seeds and fruits of many cloud forest tree species are predominantly dispersed by birds and bats which tend to avoid such extended tree-less landscapes (Wilms and Kappelle, 2006; Gomes *et al.*, 2008). Finally, as distance to any remnant patches of old-growth forest increases, chances of seed inputs are diminished accordingly (Cubiña and Aide, 2001; Günter *et al.*, 2006; Ten Hoopen and Kappelle, 2006; Weber *et al.*, 2008).

As for cloud forest *restoration*, few restoration projects have been conducted in TMCF as such, and conclusions have to be based primarily on studies conducted in tropical montane forests less affected by fog (Aide *et al.*, this volume). As stated above, competition with invasive grasses and ferns plus poor seed dispersal appear to be the most important factors limiting natural cloud forest regeneration. To overcome these barriers, one of the most cost-effective ways to accelerate vegetation recovery is to promote the establishment of shrubs, which help to shade out invasive grasses and ferns and create more appropriate conditions for seedling growth. Although this strategy can reduce competition, additional tree planting will generally be required to recover a species composition similar to that of intact forest, because most forest species are rarely dispersed far from forest stands (Holl *et al.*, 2000; Günter *et al.*, 2006; Aide *et al.*, this volume). As such, tree planting – including enrichment planting – is generally regarded as the main restoration tool to catalyze native forest regeneration on degraded tropical lands (Parrotta *et al.*, 1997; Zimmerman *et al.*, 2000), including montane cloud forest settings (Holl *et al.*, 2000; Mosandl and Günter, 2008; Stimm *et al.*, 2008; Günter *et al.*, 2009; Williams-Linera *et al.*, this volume). However, current efforts are often based on mono-specific tree plantings and frequently involve fast-growing exotic species (notably pines and eucalypts) that may have various adverse environmental impacts (González-Espinosa *et al.*, 2008; see also discussion in Günter *et al.*, 2009).

Forest restoration studies in the cloud forest belt that have employed native tree species are comparatively rare, mostly because the ecological requirements of such species are still poorly known. However the use of indigenous tree species in restoration efforts is on the increase, because they are not only potentially better adapted to local site conditions but also better at fulfilling the needs of the local people (e.g. Wilms and Kappelle, 2006; González-Espinosa *et al.*, 2008; Stimm *et al.*, 2008; Weber *et al.*, 2008; Günter *et al.*, 2009; Williams-Linera *et al.*, this volume). Clearly, restoration efforts targeting degraded, human-dominated cloud forest landscapes are likely to be more successful if biophysical, ecological, and social criteria are combined with local ethno-biological knowledge and integrated in the project design (cf. Pohle, 2008; Günter *et al.*, 2009; Báez *et al.*, this volume).

In Mexico and elsewhere, a major challenge to cloud forest restoration efforts is the large spatial diversity in micro-environmental conditions, species composition, and structure (Williams-Linera, 2002; Williams-Linera *et al.*, 2005; González-Espinosa *et al.*, 2008; cf. Günter *et al.*, 2009). Tree seedling planting experiments in Mexican cloud-affected forest interiors, adjacent agricultural fields, and old-fields indicated that the performance of different tree species depends on site disturbance level. Some species appear to have the potential to be used effectively in restoration of disturbed areas and others in plantation enrichment efforts; another sub-set of species may be used to expand the cloud forest area, and yet others can survive and grow in forests other than those in which they are naturally present (Williams-Linera *et al.*, this volume). These findings strengthen the point made by Holl *et al.* (2003) that the great site-to-site variability in tropical montane cloud forests may also be viewed as an asset rather than merely as a constraint to ecological restoration at the landscape level. Different tree species respond differently to different site characteristics, thereby enhancing floristic heterogeneity.

Summarizing, it is clear that compared with many other ecosystems, cloud forests take much longer to recover and are much more costly to restore. As such, cloud forests (particularly the more stunted and wetter variants) should be treaded upon as lightly as possible.

Maintaining the ecosystem services provided by cloud forests

Arguably, among the many demonstrated "ecosystem services" that cloud forested headwaters provide to humans, the most important in economic terms are a stable supply of high-quality water that may be used for downstream irrigation, drinking water, industry, or hydropower generation (Scatena, 1995; Aylward, 2005; Barrantes Moreno, 2006; cf. Rhodes *et al.*, this volume #45) – and their unique biodiversity with its considerable ecological (e.g. regional crop pollination; Ricketts *et al.*, 2004), pharmaceutical (e.g. Kappelle *et al.*, 2000; Setzer *et al.*, 2003), and ecotourism potential (Honey, 2008). The taller and more productive types of cloud-affected forest (mostly LMCF) also provide timber (Guariguata *et al.*, 2006; Herrera and Chaverri, 2006; Günter *et al.*, 2008), fuelwood (e.g. Rüger *et al.*, 2008), and various non-timber forest products, such as honey, ornamental plants, game, etc. (Doumenge *et al.*, 1995; Kappelle *et al.*, 2000; Bubb *et al.*, 2004; Shiel and Lawrence, 2004; Wolf, 2005; Báez *et al.*, this volume; Wolf, this volume).

The presence of cloud forest is widely assumed to increase streamflow volumes because of the extra amounts of water captured from passing clouds, beyond that provided by regular precipitation (cf. Zadroga, 1981; Calvo, 1986; Bubb *et al.*, 2004). In addition, the forest helps to reduce the number of shallow landslides and prevents surface erosion, thereby maintaining better water quality (Sidle *et al.*, 2006; cf. Bruijnzeel, 2004). Such considerations lie at the heart of many Payment for Ecosystem Services (PES) schemes in which downstream users pay a certain fee for (mostly hydrological) services rendered by montane cloud forests as compensation to upstream forest owners for conserving their forests instead of converting them to economically more profitable land uses (Pagiola, 2002; Rodrigues-Zuñiga, 2003). Numerous PES schemes have been initiated in recent years in upland settings containing cloud forest, mostly in Central and South America (reviewed by Porras *et al.*, 2008; cf. Tognetti *et al.*, this volume).

Whilst potentially a powerful tool to help conserve threatened upland forests, current PES schemes tend to be based on generalized beliefs with respect to the hydrological role of forests in general, and of cloud forests in particular. In addition, monitoring activities are often confined to changes in land use only, whereas hydrological monitoring efforts (if any) are rarely of sufficient rigor and duration to allow solid conclusions to be drawn with respect to the hydrologic effectiveness of the scheme (Porras *et al.*, 2008; Le Tellier *et al.*, 2009). As discussed earlier in this chapter, the few available site-specific studies of the hydrological impacts of cloud forest conversion to pasture have yielded contradictory results. Flows increased in one case (eastern Mexico) but decreased in another (Pacific Northwest of USA), whereas no change was observed under particularly wet conditions (Costa Rica). Given the difficulties associated with predicting the impact of cloud forest conversion on watershed hydrological behavior it will not come as a great surprise to learn that few, if any, PES initiatives have actually been able to demonstrate a measurable hydrological effect of their activities (Rojas and Aylward, 2003; Porras *et al.*, 2008; Le Tellier *et al.*, 2009). This may well have consequences for buyers and influence their continued willingness to pay (Aylward, 2005; Muñoz-Piña *et al.*, 2008; Porras *et al.*, 2008; Tognetti *et al.*, this volume).

A related problem concerns the difficulty of adequately quantifying – in economic and social terms – the various environmental services provided by cloud forests (Scatena, 1995; Aylward, 2005; Barrantes Moreno, 2006; Porras, 2008); additional challenges relate to the long time priods needed to build sufficient trust between service buyers and providers (Asquith *et al.*, 2008), securing equitable access to (and avoiding conflict about) ecosystem services amongst different groups of stakeholders, and establishing the proper institutional arrangements for the payments (see Tognetti *et al.*, this volume for a fuller discussion). Arguably, each situation is more or less unique and requires tailor-made arrangements (cf. Asquith and Wunder, 2008).

Nevertheless, Calvo-Alvarado *et al.* (this volume) describe the creation of the institutional, legal, and operational capacities that were required for the successful development and implementation of the nationwide PES scheme in Costa Rica, along with the development of a domestic water market benefiting many farmers with small- and medium-sized holdings who are willing to conserve the forest on their land (see also Pagiola, 2008; Porras, 2008). Calvo-Alvarado *et al.* are convinced that the Costa Rican accomplishments can be repeated in other Meso-American countries, with appropriate adjustments to fit other ecological, social, and political situations. However, because the economic value of hydrological ecosystem benefits alone are unlikely to cover all conservation needs, a truly integrated multiple-services approach is needed (Tognetti *et al.*, this volume).

In this regard, ecotourism appears to be a particularly promising additional "service" that cloud forests may offer to local populations (Kappelle and Brown, 2001), as long as the limited carrying capacity of these often fragile sites is taken duly into account (Honey, 2008; cf. Scatena, 1995). The famous Monteverde Cloud Forest Preserve in northern Costa Rica (10 500 ha), for example, received around 3100 ecotourists annually in the early 1980s, a number that increased to ~50 000 in the early 1990s, and exceeded 200 000 by 2005 (Honey, 2008). Whilst such numbers are indeed excessive and not representative of the situation encountered at the majority of cloud forest sites, they do illustrate the need to limit the influx of visitors and the excessive use of local resources. Only in this way will it be possible to achieve a form of sustainable ecotourism that contemplates socially equitable and ecologically sustainable economic development (Lindbergh and Hawkins, 1993; Honey, 2008). It goes without saying that local communities should be involved at all stages of planning and implementing cloud forest ecotourism and conservation efforts (Lindbergh and Hawkins, 1993; Bubb *et al.*, 2004; Shiel and Lawrence, 2004; cf. Báez *et al.*, this volume; Hofstede *et al.*, this volume).

Summarizing, whilst Payment for Ecosystem Services (PES) schemes – in which such key services as a stable freshwater supply from montane cloud forests are valued in monetary terms – constitute a potentially powerful tool to help conserve these forests, demonstrating the hydrological effectiveness and economic sustainability of PES schemes remains a major challenge. The most suitable arrangements are likely to vary from place to place in response to the local biophysical, social, and economic situation.

OUTLOOK

Global cloud forest conservation efforts

Although the hydrological and biological importance of cloud forests – as well as the recognition that these unique forests are seriously at risk of disappearing altogether – has been realized for decades (e.g. Daugherty, 1973; LaBastille and Pool, 1978), global efforts to promote the conservation and sustainable use of montane cloud forests were initiated comparatively recently only (Hamilton, 1995; Aldrich *et al.*, 1997a,b). The Mountain Cloud Forest Initiative was established in 1999 as a collaborative effort by the UNEP–World Conservation Monitoring Centre (WCMC), the World-Wide Fund for Nature (WWF International), the World Conservation Union (IUCN), and UNESCO's International Hydrological Programme (IHP) (Aldrich *et al.*, 2000), extending and consolidating IUCN's earlier *Campaign for Cloud Forests* (Hamilton, 1995). The Initiative envisioned a world in which mountain communities and downstream users value and protect cloud forests for their unique plant and animal species, and for their role in supplying clean water; and in which the global community recognizes the value of cloud forests as sensitive indicators of global climate change (Aldrich *et al.*, 2000). As a first step toward this long-term goal the Initiative paid particular attention to raising awareness about the existence and uniqueness of cloud forests, the important services they provide to humanity, and the various threats that imperil their continued existence in various parts of the world (Aldrich *et al.*, 1997a, 2000; Bruijnzeel and Hamilton, 2000). In doing so, the Initiative directed its efforts primarily to organizations active in tropical forest conservation, management, and research, to policy- and decision-makers in countries with cloud forests, and to the international environment and development community, urging for immediate action. In 2004, the Initiative produced a *Cloud Forest Agenda* which focused on the integration of cloud forest conservation and sustainable livelihoods for the populations living in and around cloud forests (Bubb *et al.*, 2004). Whilst currently no longer active, it is probably fair to say that the Initiative has played a significant role in raising global awareness of the importance of, and threats faced by TMCFs, as manifested by their inclusion in various recent UN Conventions on Biological Diversity (e.g. SCBD, 2003a,b,c, 2004). In addition, the publication of several alarming findings in high-impact scientific journals in recent years was followed by much publicity and debate, adding further general awareness of the threats faced by cloud forests. Examples include the massive decline in numbers of many species of amphibians, and of the extinction of several others (Pounds *et al.*, 1999, 2006), and the fact that lowland deforestation may well impact cloud formation in adjacent uplands carrying cloud forests (Lawton *et al.*, 2001).

On a related note, following the proclamation of the International Year of the Mountains in 2002, a series of actions to promote sustainable mountain development were undertaken by the UN Convention on Biological Diversity (CBD), including the adoption of a Programme of Work on Mountain Biological Diversity (PoW-MBD) and the recommendation to include a host of other international agreements and bodies, institutions, and program initiatives in the PoW-MBD (SCBD, 2004). The latter include (amongst others): the Ramsar Convention on Wetlands of International Importance, the UN Framework Convention on Climate Change (UNFCCC), the UN Food and Agriculture Organization (FAO), UNESCO, the International Centre for Integrated Mountain Development (ICIMOD), the International Partnership for Sustainable Development in Mountain Regions, the International Human Dimensions Programme on Global Environmental Change (IHDP), the Centre for Mountain Studies, the *Consorcio para el Desarrollo de la Ecoregion Andina* (CONDESAN), the Mountain Research Initiative (MRI), the Global Mountain Biodiversity Assessment (GMBA), the International Union of Forest Research Organizations (IUFRO), the Alpine Convention, and UNEP's World Conservation Monitoring Centre (UNEP–WCMC) (SCBD, 2004). To this long list can be added the UN Intergovernmental Forum on Forests (IFF), which stated recently that "*mountain cloud forests are of particular concern*" with respect to soil and watershed protection and the conservation of biological diversity in environmentally critical areas.

Furthermore, within the context of cloud forest research and conservation, the GMBA under the auspices of DIVERSITAS is of particular interest (Körner and Spehn, 2002; Spehn *et al.*, 2006). The GMBA serves as a cross-cutting network that actively seeks to explore and explain the great biological richness of the mountains of the world –with an emphasis on cloud forests, amongst others in the Andes and the south-eastern Himalayas (cf. Wangda and Ohsawa, this volume). It also provides scientific inputs to policy-makers and various stakeholder groups involved in the conservation and sustainable use of biodiversity in mountain regions. Synthesizing often hidden and fragmented results of research on mountain biodiversity is one of the initiative's key activities (Spehn *et al.*, 2006). A similar global initiative to stimulate and link cloud forest hydrological research is lacking.

Research for the advancement of cloud forest conservation and sustainable use

The information contained in the many chapters of this book is clear evidence that knowledge of cloud forest environments, biodiversity, soils, productivity, and hydrological functioning has greatly increased since the 1993 Puerto Rico cloud forest symposium. We not only have a much better idea now of where cloud-affected forests (CAFs) may be found (based on an analysis of satellite imagery of cloud frequency across the tropics; Figure 72.2) but also where the largest contingent areas of CAF still exist, what the proportional losses are in different regions (Figure 72.16), and where significant areas of CAF are not included as yet in any protected area system (Mulligan, this volume). In addition, global analyses of vascular plant diversity (Figure 72.4), and pan-tropical comparisons of species richness and level of endemism in CAF for amphibians, mammals, and (Neotropical) birds (Figures 72.5–72.7) all point to the existence of a number of well-defined biodiversity hot spots that include significant amounts of CAF (Figure 72.3).

Knowledge of the hydro-meteorology of different types of cloud forest has equally increased in recent years. Whilst total evaporative losses (*ET*) from tall lower montane cloud forest (LMCF) do not appear to be significantly lower than those from lower montane rain forest (LMRF) not affected by fog and low cloud, *ET* from upper montane cloud forest (UMCF) of intermediate stature, and from stunted elfin cloud forest (ECF) in particular is significantly reduced (Figure 72.12). This, together with the higher rainfall observed at many ECF locations, produces significantly greater precipitation excess values – and therefore greater degree of soil wetness and streamflow volumes – in areas with UMCF and ECF compared with LMRF and, to a lesser extent

LMCF. *Average* estimated increases in total annual water production (i.e. including possible underground flows not detected by stream gaging stations) for different types of cloud forest compared to LMRF range from ~300 mm in the case of LMCF, to ~900 mm for UMCF, and as much as ~2900 mm for ECF (cf. Appendix 72.3). Furthermore, amounts of net precipitation arriving at the forest floor (i.e. throughfall plus stemflow) increase steadily from LMRF to LMCF, UMCF, and ECF, with values for the latter two categories equaling or exceeding conventionally measured rainfall amounts (Figure 72.9 and Appendix 72.2) whereas soil water uptake (transpiration, E_t) decreases steadily for the respective forest types (Figure 72.12 and Appendix 72.3).

Our capacity to measure on-site fog deposition, interception of wind-driven fog and rain by the vegetation, and evaporative losses from cloud forests has improved enormously in recent years (see chapters in Parts III–V in this book; cf. Bruijnzeel, 2005; Holwerda, 2005; Frumau *et al.*, 2006a; McJannet *et al.*, 2007a,c; Beiderwieden *et al.*, 2008). In addition, the increased availability of remotely sensed information on topography, vegetation type, and surface characteristics (including reflection coefficient, leaf area index, surface temperature, etc.), as well as rainfall and cloud characteristics has greatly expanded the possibilities for modeling the spatial and temporal variations in hydrological processes (e.g. Walmsley *et al.*, 1996, 1999; Mulligan and Burke, 2005a,b; Mulligan, 2006a,b; Wu *et al.*, 2006; Welch *et al.*, 2008; Ray *et al.*, 2009; Lawton *et al.*, this volume). Arguably, what is needed now is the integration of on-site measurements of climate, hydrological processes, and streamflow volumes emanating from cloud forest headwater areas, with observations made by national networks in countries with cloud forest (e.g. García-García and Zarraluqui, 2008) and with spatially explicit model predictions (cf. Mulligan and Burke, 2005b; Bruijnzeel, 2006; Wu *et al.*, 2006).

The reduced stature and productivity of certain types of cloud forest – as well as their occurrence at lower elevations on small, isolated mountains compared to larger mountain chains (Figure 72.14) – has puzzled scientists for decades and numerous theories have been advanced over the years to explain the phenomenon. However, much has been learned in this regard too in recent years. Amongst others, there is a major shift in carbon allocation by the plants from above-ground biomass to below-ground biomass (i.e. roots) with elevation, i.e. going from LMRF to UMCF and ECF (Hertel and Leuschner, this volume). This implies, amongst others, that overall ecosystem productivity decreases much less with elevation than previously thought. Whilst foliar nutrient analysis in the more stunted cloud forest types and various fertilization experiments suggest certain nutrient deficiencies (notably for nitrogen; Benner *et al.*, this volume), the few available studies of cloud forest nutrient *budgets* all show that the trees take up only a mere fraction of the amounts of nutrients dissolved (and therefore readily available) in net precipitation and leaf litter leachate (Bruijnzeel *et al.*, 1993; Hafkenscheid, 2000). This suggests that the roots have difficulty to make full use of available nutrients. Although the precise mechanism is still being debated (Benner *et al.*, this volume), it is clear that the persistently wet to saturated conditions prevailing in most if not all UMCF and ECF, cause strong reductions in soil aeration levels (Silver *et al.*, 1999) – which in turn have various ecophysiological consequences (Santiago *et al.*, 2000; Santiago *et al.*, this volume) and affect rates of decomposition (Schuur, 2001). In addition, opportunities for exploitation of mineral soil horizons by the roots are strongly reduced in waterlogged soils (Soethe *et al.*, 2006). Finally, nitrogen availability has been shown to be much improved upon aeration (Cavelier *et al.*, 2000; cf. Marrs *et al.*, 1988; Schuur and Matson, 2001). An added stressor at many exposed ridge-top and summit locations is the persistence of high winds (Lawton, 1982).

On the downside, there is still considerable uncertainty about the hydrological consequences of converting (different types of) cloud forest to pasture or vegetable cropping. In some cases an increase in flows has been inferred, in others a decrease, and in yet another case no change was found. There are no rules of thumb in this regard, although the greatest effect would be expected where inputs from wind-driven fog and rain during the dry season are highest and where contrasts in evaporative losses from forest and pasture are largest. More site-specific work is needed. Likewise, there is great uncertainty about the magnitude and even the direction of changes in rainfall associated with climatic warming (Mulligan, this volume), and therefore about the potential hydrological (cf. Figure 72.15) and ecological consequences. Finally, although information on the carbon dynamics and productivity of various types of cloud forest is on the increase (Weaver and Murphy, 1990; Raich, 1998; Röderstein *et al.*, 2005; Moser *et al.*, 2007; Hertel and Leuschner, this volume), overall carbon budgets in different types of cloud forests have not been quantified as yet (cf. Schuur *et al.*, 2001).

Sutherland *et al.* (2009) presented a list of 100 questions of importance to the conservation of global biological diversity, following a strict protocol involving over 750 individuals that generated almost 2300 questions which were shortlisted and refined by workshop participants representing 24 international organizations. On the basis of Sutherland *et al.*'s global list a sub-set of practical research questions was prepared,[3] that, if answered, may be considered to contribute the most toward conserving and sustainably using the biological diversity of cloud forests in practice. These core questions relate to the following strategic themes: ecosystem function and services

[3] Questions prepared by M. Kappelle.

(6 questions), climate change (9), protected areas (4), ecosystem management and restoration (6), and species management (6).

Ecosystem function and services

(1) Do critical thresholds exist at which the loss of cloud forest species diversity, or the loss of particular species, disrupts ecosystem functions and services, and how can these thresholds be predicted? (2) How can cloud forest biodiversity considerations be integrated into economic policies to reflect the monetary and non-monetary value of cloud forest biodiversity, ecosystem processes, goods, and services? (3) To what extent can (and need) cloud forest ecosystems be managed to increase protection of humans and biodiversity from extreme events (e.g. floods and droughts, hurricanes)? (4) How, where, and when has cloud forest biodiversity loss affected human welfare? (5) What strategies for distributing the material benefits derived from cloud forests most effectively foster environmental stewardship and the conservation of biological and hydrological values? (6) How can cloud forest protected area networks be designed so as to increase carbon storage benefits and help to mitigate climate impacts, with these benefits acting as possible incentives to further support conservation actions?

Climate change

(1) Which elements of cloud forest biodiversity, and in which locations, are most vulnerable to climate change, including extreme events (hurricanes, droughts, massive rainfall, and landsliding)? (2) How is the resilience of cloud forest ecosystems to climate change affected by local and regional human activities and interventions? (3) How will climate change, together with other environmental stressors, alter the distribution and prevalence of diseases of wild species (e.g. as in the case of the pathogenic chytrid fungus that affects frogs' health and survival)? (4) How will climate change affect the hydrological behavior of different types of cloud forest? (5) How might cloud forest biodiversity policies and management practices be modified and implemented to accommodate climate change? (6) How might emerging carbon markets affect cloud forest conservation through their impacts on the protection, management, and creation of cloud forest habitats? (7) What are the potential effects of feedbacks between climate change and cloud forest ecosystem dynamics (e.g., drought, forest dieback) on the effectiveness of policy measures to sequester carbon and protect biodiversity? (8) How much carbon is sequestered by cloud forest ecosystems, including their soils, and how can these lands be managed – and if necessary restored – to contribute most effectively to the mitigation of climate change? (9) How will climate change affect the distribution and impacts of climate-dependent disturbance regimes in cloud forests, such as forest fires and hurricanes?

Protected areas

(1) How effective are different types of protected areas (e.g. strict nature reserves, hunting reserves, and national parks) at conserving cloud forest biodiversity and providing ecosystem services? (2) What are the costs per hectare of cloud forest required to manage cloud forests effectively, and how do these vary with management category, geography, and levels of threats? (3) What are the costs and benefits of cloud forest protected areas in terms of human well-being, how are these distributed geographically, and how do they vary with governance, resource tenure arrangements, and site characteristics? (4) How does the management of cloud forest protected areas affect conservation beyond the boundaries of the protected area, such as through the displacement of human populations, hunting, etc.?

Ecosystem management and restoration

(1) What are the trade-offs for cloud forest biodiversity, streamflow regime and water quality under different forest management systems, such as timber harvesting from natural forest or plantation forests? (2) What was the condition of cloud forest ecosystems before significant human disruption in different geographic areas, and how can this knowledge be used to improve current and future management? (3) What, and where, are the most significant opportunities for large-scale cloud forest ecosystem restoration that benefits both biodiversity and human well-being (including improved hydrological functioning)? (4) How can cloud forest ecosystem management systems be designed to better emulate natural processes, notably natural disturbance regimes (including fire), and to what extent does this improve cloud forest conservation effectiveness? (5) What is the contribution of cloud forest areas that are intensively managed for production of commodities (such as food, timber, etc.) to conservation of cloud forest biodiversity at the landscape scale? (6) Under what circumstances can afforestation, reforestation, and reduced emissions from deforestation and degradation (REDD) benefit cloud forest biodiversity conservation, reduce emissions, and help to achieve more sustainable livelihoods?

Species management

(1) Under what conditions is trade in captive or wild-harvested cloud forest species beneficial for wild populations of the traded species? (2) What is the relative effectiveness of different methods for facilitating movement of a cloud forest species between disjunct patches of its habitat? (3) What is the cost-effectiveness of different contributions to cloud forest species conservation programs such as education, captive breeding, and habitat management? (4) What are the ecosystem benefits of efforts to conserve charismatic, flagship, or umbrella cloud forest

species? (5) What are the likely risks, costs, and benefits of reintroducing or translocating cloud forest species as a response to climate change? (6) What are the most effective approaches for reversing range and population collapse in top predators (e.g. large cats), large herbivores (e.g. tapirs), and other species that exert disproportionate effects on cloud forest ecosystem structure and function?

CONCLUDING REMARKS

With populations and pressure on the land on the increase in many tropical montane areas, the wet, windy, and steep environments characterizing mossy forests often no longer provide the *de facto* protection they did historically (Mosandl *et al.*, 2008). Climate change aside, cloud forests continue to be seriously at risk and may even disappear completely during this century if no major conservation action is taken in the coming decades (Bubb *et al.*, 2004; Mulligan, this volume). The many underlying issues such as poverty, insecurity of land ownership, failed forest policies, and the lack of effective law enforcement continue to challenge our globalized society. As a result, montane forest clearing, accelerated soil erosion and landslips, as well as disruptions of the hydrological regime, and, ultimately, species extinction will continue to take place.

Fortunately, awareness of the importance and value of cloud forests has increased considerably in recent decades, both amongst people living in and near cloud forests and at national and global levels. There is now a better understanding of the importance of cloud forests in terms of the various environmental goods and services they provide, such as biodiversity, recreation and tourism, protection against soil erosion, and perhaps most importantly, the stable hydrological flow regimes that are indispensable for agricultural crop irrigation, hydro-electric power generation, and drinking water for downstream populations.

In addition, during the last 50 years an increasing number of tropical montane cloud forest sites have received some kind of protected area status, ranging from multiple-use protected landscapes and forest reserves, to restricted-use national parks and absolute reserves. However, if one is to preserve a large part of cloud forest variety of life in the long term, a conservation strategy will be needed that not only goes beyond the current networks of protected core areas, buffer zones, and corridors, but also considers the implications of climate change for the future elevational range of cloud forest occurrence. This will require a concerted, global effort with focused local action, in which government agencies, local authorities, scientists, NGOs, private enterprises, and civil society all work together, to develop and implement strategies that halt the threats of invasive species, deforestation, and fragmentation, and so ensure that cloud forests can continue to serve both nature and humankind in the long run. An essential element of such an effort is the institutionalized participation of all relevant stakeholder groups in the decision-making process at all levels of scale and, where necessary, across borders, so as to establish a broad-based, consensus-oriented conservation framework. Such consensus is a prerequisite for long-term conservation and sustainable forest use, for it is the recognition and valuation of the overall set of environmental goods and services offered by cloud forests to local, regional, and even global communities which will make their preservation and wise management an economic necessity, socio-politically feasible, culturally rewarding, and ecologically balanced. It is hoped that the contents of this book will inspire readers to contribute to this goal.

APPENDIX 72.1 BIOGEOGRAPHIC PATTERNS OF BIODIVERSITY IN CLOUD-AFFECTED FORESTS: BACKGROUND INFORMATION

Range data available for threatened amphibians (IUCN *et al.*, 2008a), birds of the western hemisphere (Ridgely *et al.*, 2007) and mammals (IUCN *et al.*, 2008b) were used as a means of understanding the relative biological importance of cloud forests for biological conservation and the potential geographic variation in biological importance across the known cloud forests. The data-sets were developed through an extensive process of scientific contribution and review coordinated by the World Conservation Union (IUCN), Conservation International, and NatureServe (for amphibians); by IUCN, Conservation International, Arizona State University, Texas A&M University, University of Rome, University of Virginia, and the Zoological Society London (for mammals); and by NatureServe in collaboration with Robert Ridgely, James Zook, The Nature Conservancy's Migratory Bird Program, Conservation International–CABS, World Wildlife Fund–US, and Environment Canada (for birds).

Ranges were determined by interpolation of occurrence between known locations if the ecological conditions seemed appropriate, but not allowing extrapolation beyond known locations. Some species may therefore occur more widely than has been mapped so that some regions are recorded as having lower diversity (and higher endemism) than may actually be the case. Also, for parts of the world that remain poorly known the current data-sets miss any species that are as yet undiscovered. Nevertheless, these are some of the best available data for pan-tropical assessment of cloud forest biodiversity.

Appendix 72.2 Throughfall (*TF*), stemflow (*SF*) and *apparent* rainfall interception (*Ei*) fractions as measured in different types of tropical montane rain forest

Location and forest type	Elevation (m.a.s.l.)	Mean annual precipitation (mm)	*LAI*/H –/(m)	Epiphyte biomass (t ha^{-1})	*TF*	*SF* (% of *P*)	E_i	Remarks
Montane rain forests little affected by fog								
Bolivia, Yungas[1]	1850	2310	–/20	–	79	–	<21	10 roving gages, 3 mo, weekly
Colombia, Cordillera Central[2]	1700	3150	–/35	–	76	–	<24	6 roving gages, 15 mo, weekly
	1950	2200	–/35	–	85	–	<15	Idem
Ecuador[3]	1950	2320	7.3/25	–	67	1.1	32	Transect 2 only; 15–36 fixed gages, 36 mo, weekly
Idem[4]	1960	2080	–/27	13.4	71	–	<29	31 fixed gages, 12 mo, weekly
Costa Rica, Monteverde[5]	1200	2500	3.5/29	26.0	72	–	<28	5 fixed gages, 12 mo, biweekly
Costa Rica, Talamanca[6]	2900	2810	7.7/35	2.96	70	2.0	28	UMRF, 30 fixed gages, 46 weeks, weekly
Guatemala[7]	2100	2500	–	–	65	–	<35	36 fixed gages, 44 weeks, weekly; *P* likely overestimated since measured at 2550 m; forest grading into LMCF
Indonesia (Sulawesi)[8]	1042	2900	6.4/23	–	70	0.3	30	Site NF4, 30 fixed gages, 15–17 events
Malaysia, Peninsular[9]	1600	2300	–/14	–	62	2.2	36	20 roving gages, 68 weeks, 2–3 days
Panama[10]	1200	3680	–/30	–	63	0.4	37	50 troughs (0.05 m^2), 12 mo, daily
Papua New Guinea[11]	2500	3800	5.5/30	2.0	67	<1	32	32 fixed gages, 8 mo, bi-weekly
Perú, Central Cordillera[12]	2470	2220	2.9/20	–	70	0.1	30	20 troughs (0.1 m^2), 54 weeks, weekly
Tanzania, Usambara Mts.[13]	1500	1230	–/35	2.1[14]	79	1	20	8 fixed gages, 30 mo, daily
Lower montane cloud forests								
Australia, SE Queensland[15]	1000	1350	5.0/35	–	90	2	8	25 troughs, 31 mo, 2–3 weekly
Australia, N Qld, UB site[16,17]	1050	2985	4.1/25	–	79	8.8	12	4–6 troughs (2.4–3.6 m^2), 25 mo, continuous recording
Idem, Mt. Lewis site[16,17]	1100	3040	4.5/32	–	73	3	24	Idem, 23 mo
Bolivia, Yungas[1]	2600	3970	–/18	–	74	–	<26	10 roving gages, 3 mo, weekly, forest grading into UMCF
China, Yunnan, Ailao Mt.[18]	2500	1930	5.8/25	1.4	87	2	11	12 fixed gages, 24 mo, 1–3×/day
Idem, Xishuang banna[19]	750	1485	5.0/36	–	78	4.8	17	6 fixed collectors (0.6 m^2), 36 mo, event-based
Colombia, Central Cordillera[20,21]	2550	2115	–/23	–	87.5	–	12.5	20 fixed gages, 12 mo, weekly

Appendix 72.2 (*cont.*)

Location and forest type	Elevation (m.a.s.l.)	Mean annual precipitation (mm)	*LAI*/H –/(m)	Epiphyte biomass (t ha^{-1})	*TF*	*SF* (% of *P*)	E_i	Remarks
Costa Rica, Monteverde								
Windward forest[5]	1200	6390	3.4/28	–	70	–	<30	5 fixed gages, 12 mo, biweekly
Leeward forest[22,5]	1500	2520	3.6/25	33	65	–	<35	20 fixed gages, 12 mo, 1–3 days
Costa Rica, Monte de los Olivos[23]								
Windward forest	1500	3300	–40	–	90.5	–	<9.5	20 troughs (0.1 m^2), 12 mo, daily
Idem	1500	3300	–/25	–	106	–	–7	Idem
Ecuador[4]	2030	2140	5.7/20	5.3	85	1	14	31 fixed gages, 12 mo, weekly
Idem[4]	2175	2500	4.9/15	11.6	91	1	8	Idem
Guatemala[24] (windward)	2200	2600	–/–	–	81	<1	18	3–6 fixed collectors (0.22 m^2), 12 mo, 2×/week
Hawaii, Maui (leeward)[25]	1220	890	–/2.5	–	65	–	<35	4 fixed collectors (0.24 m^2), 6 mo, continuous recording
Honduras[24]	900–1400	>3500	–/–	–	95	–	<5	3–6 fixed collectors (0.22 m^2), 12 mo, 2×/week?
Indonesia (W Java)[26]	1750	3300	–/–	–	80	–	<20	4 collectors of 1 m^2, 12 mo, weekly
Mexico, Veracruz[27]	2100	3000	6.6/27	–	82.5	–	17	4 roving collectors (1.2 m^2 each), 12 mo, continuous recording
Tanzania, Mt. Kilimanjaro[28]	2250	2480	–/40	–	82	<1	17	10 fixed gages, 12 mo, weekly
Idem, dry year		1960	–/40	–	70	<1	29	Idem
Thailand, Kog Ma[29]	1265	2085	4.0/25	–	89	1.5	9.5	30 fixed gages, 48 mo, daily
Venezuela, San Eusebio[30]	2300	1575	–/27	2.9	79.5	1	19.5	30 fixed gages, 12 mo, 1–2×/week
Venezuela, La Mucuy[31]	2350	3125	2.9/25	–	54	0.2	46	6 fixed collectors (0.6 m^2), 12 mo, weekly
Upper montane cloud forests								
Australia, Queensland, BK[16,17]	1560	5300	3.3/8	–	117	14	–31	4–6 troughs (totaling 2.4–3.6 m^2), 21 mo, continuous recording
Colombia, Central Cordillera[3]	3000	1700	–/–	–	89	–	<11	6 roving gages, 15 mo, weekly
Idem, Central Cordillera[20,21]	3370	1455	–/22	12.0	87.5	0.1	12.5	20 fixed gages, 12 mo, weekly
Idem, Zipaquira[32]	3250	1615	3.6/15	–	89	2	9	20 roving gages, 24 mo, weekly
Costa Rica, Monteverde[5,33]	1500	4310	3.4/10	4.7	118	–	–18	5 fixed gages, 12 mo, biweekly
Idem[34]	1450	6000	3.0/22	16.2[35]	108	1	–9	30 roving gages, 12 mo, daily[a]

Appendix 72.2 (*cont.*)

Location and forest type	Elevation (m.a.s.l.)	Mean annual precipitation (mm)	*LAI*/H –/(m)	Epiphyte biomass (t ha^{-1})	*TF*	*SF* (% of *P*)	E_i	Remarks
Guatemala[24] (windward)	2400	2500	–/–	–	113	1	−14	3–6 fixed collectors (0.22 m^2), 12 mo, 2×/week
Guatemala[7] (windward)	2550	2600	–/–	–	88	1	11	36 fixed gages, 44 weeks, weekly
Hawaii[25] (windward)	1950	2400	–/6	–	119	–	−19	4 fixed collectors (0.24 m^2), 9 mo, continuous recording
Honduras[36]	1795	1500	–/–	–				4 fixed collectors (0.25 m^2), 12 mo, daily
Concave slope					94	–	6	
Convex slope					129	–	−29	
Ridge top					179	–	−79	
Jamaica, PMull forest[37]	1810	2850	5.0/12	–	73	12	14	12 roving gages, 12 mo, 3–4 days; + 1 recording trough
Malaysia, Peninsular[9]	2030	2115	–/7.5	–	64	30.5	5.5	20 roving gages, 68 weeks, weekly
Perú, Central Cordillera[12]	2815	2750	2.5/14	–	92.5	<0.1	7.5	20 fixed collectors (0.115 m^2), 12 mo, weekly
Philippines, Luzon[38]	2200	3910	–/9	–	86	12	2	16 fixed collectors (0.041 m^2), 60 mo, daily
Puerto Rico Palm forest[39]	900	4500	4.6/11	–	69	26	5	20 roving collectors (0.0254 m^2), 6 months, daily
Elfin and sub-alpine cloud forests								
Bolivia, Yungas[1]	3050	5150	–/10	–	75	–	<25	10 roving gages, 3 mo, weekly
Hawaii[40]	2600	500	–/8	–	75	–	<25	2 fixed collectors (0.125 m^2), 12 mo, continuous recording
Jamaica, MMor Forest[37, 41]	1825	2855	4.1/8	2.8	60^b	18	22	12 roving gages, 12 mo, 3–4 days; + 1 recording trough
Puerto Rico, East Peak								
Elfin, windward[42,43]	1000	5400	2.0/ 2.5	7.4	115	5	−20	60 fixed collectors, 8 mo, daily
Idem, ridge	1015	4800	–/–	4.4	125	9.5	−35	Idem
Idem, leeward	930	6000	–/–	4.8	96	3	1	Idem
Idem, sheltered[44]	970	4500	2.1/3	(4.4)	108	(5)	−13	45 roving and 22 fixed gages, 97 d, twice/week; + 3 troughs (1.0 m^2)
Réunion, France[45]	2150	3000	–/3	–	124	–	−24	8 fixed gages, 8 mo, continuous
Spain, La Gomera[46]	1290	660	4.2/9	–	126	2.8	−29	2 fixed collectors (0.20 m^2), 19 mo, continuous

[1] Gerold *et al.* (2008).
[2] Vis (1986).
[3] Fleischbein *et al.* (2006).
[4] Oesker *et al.* (this volume).
[5] Haeger and Dohrenbusch (this volume).

Appendix 72.2 (*cont.*)

[6] Hoelscher *et al.* (2004).
[7] Holder (2004).
[8] Dietz *et al.* (2006).
[9] Kumaran (2008).
[10] Cavelier *et al.* (1997).
[11] Edwards (1982).
[12] Gomez-Peralta *et al.* (2008).
[13] Lundgren and Lundgren (1979).
[14] Pócs (1980).
[15] Hutley *et al.* (1997).
[16] McJannet *et al.* (2007a).
[17] McJannet *et al.* (2007b).
[18] Liu *et al.* (2002).
[19] Liu *et al.* (2007).
[20] Veneklaas and van Ek (1990).
[21] Veneklaas *et al.* (1990).
[22] Clark *et al.* (1998).
[23] Fallas (2002).
[24] Brown *et al.* (1996).
[25] Giambelluca *et al.* (this volume).
[26] Gonggrijp (1941).
[27] Holwerda *et al.* (2009).
[28] Schrumpf *et al.* (this volume).
[29] Tanaka *et al.* (2005).
[30] Steinhardt (1979).
[31] Ataroff and Rada (2000).
[32] Tobón (2009).
[33] Nadkarni (1984).
[34] K. F. A. Frumau (personal communication).
[35] Köhler *et al.* (2007).
[36] Stadtmueller and Agudelo (1990).
[37] Hafkenscheid *et al.* (2002).
[38] Mamanteo and Veracion (1985).
[39] Holwerda (2005).
[40] Juvik and Nullet (1995b).
[41] Tanner (1980).
[42] Weaver (1972).
[43] Weaver *et al.* (1986).
[44] Holwerda *et al.* (this volume #29).
[45] Gabriel and Jauze (2008).
[46] García-Santos (2007).

[a] Although classified as LMCF on the basis of forest stature alone, this forest is considered to behave as an UMCF due to very wet and windy conditions; relatively tall stature may be explained by well-drained volcanic soils despite excessive rainfall.

[b] Probably underestimated.

Appendix 72.3 Transpiration (E_t), wet canopy (interception) evaporation (E_i), and total evapotranspiration (ET) as measured in different types of tropical rain forest

Location and forest type	Elevation (m.a.s.l.)	Mean annual precipitation (mm)	*LAI* (m^2 m^{-2})	E_t (mm $year^{-1}$)	E_i (mm $year^{-1}$)[a]	ET(mm $year^{-1}$)	E_i/E_t (%)	$E_t/R_n(-)$	Setting/method for E_t or ET
Lowland evergreen rain forest									
Central Amazonia, Ducke[1]	80	2500	6.0	980	330	1310	25	0.59	Continental (C), micrometeo
Pasoh, Peninsular Malaysia[2]	100	1800	6.5	1206	242	1448	17	0.69	Maritime (M), micrometeo
Lambir Hills, Sarawak[3]	50	2740	6.2	946	357	1303	27	0.54	M, micrometeo
Lower montane rain forest little affected by fog									
Madagascar, Périnet[4]	1010	2080			1295				M, watershed budget for *ET*
Colombia, Sierra Nevada[5]	1150	1985			1265				Idem
Ecuador[6]	1950	2050	7.3	626	655	1281			C, watershed budget for *ET* and E_t
Kenya, Kericho[7]	2200	2130			1337				C, watershed budget for *ET*
Idem, sub-basin[7]	2350	2015			1240				Idem
Kenya, Kimakia[8]	2440	2305			1156				Idem
Tanzania, Mbeya[9]	2500	1925			1381				Idem
Lower montane cloud forest									
Queensland, Gambubal[10]	1000	1350	5.5	845	414	1259	33	–	C/M, sapflow, soil water budget
Idem, Upper Barron[11]	1050	2985	4.1	591	854	1445	59	0.37	M, sapflow
Idem, Mt. Lewis[11]	1100	3040	4.5	579	880	1459	60	0.42	M, sapflow
Hawai'i, Volcano NP[12]	1219	2500	4.8	645	588[13]	1233	48	0.30	M, micrometeo
Indonesia, Java[14]	1750	3300	–	510	660	1170	56	–	M, watershed budget for *ET* and E_t
Ecuador[15,16]	2030	2140	5.7	555	364	919	40	0.69	C, sapflow
Mexico, Veracruz[17,18]	2100	3100	6.6	876	542	1418	38	0.63	C/M, sapflow
Thailand, Kog-Ma[19,20]	1268	1768	4.0	626	530	812	46	0.46	C, sapflow and micrometeo
Venezuela, San Eusebio[21]	2300	1465	–	675	305	980	31	0.44	C/M, energy and water budgets
Venezuela, La Mucuy[22]	2350	3125	2.9	558					C, porometry for E_t
Upper montane cloud forest									
Costa Rica, Monteverde[23,24]	1450	6000	3.0	361	415[b]	776	53	0.36	M, micrometeo
Queensland, Bellenden Ker[11]	1560	7470	3.3	349	549	898	61	0.31	M, sapflow

Appendix 72.3 (*cont.*)

Location and forest type	Elevation (m.a.s.l.)	Mean annual precipitation (mm)	*LAI* (m^2 m^{-2})	E_t (mm $year^{-1}$)	E_i (mm $year^{-1}$)[a]	*ET*(mm $year^{-1}$)	E_i/E_t (%)	$E_t/R_n(-)$	Setting/method for E_t or *ET*
Puerto Rico, Luquillo Mts.[25]	900	4450	4.6	484	190	674	28	0.44	M, micrometeo
Elfin cloud forest									
Puerto Rico, Luquillo Mts.[25]	1010	4500	2.1	296	268[b]	564	47	0.27	M, micrometeo
La Gomera, Canary Islands[26]	1270	900	4.2	412	117	529	22	0.26	M, sapflow

[1] Shuttleworth (1988).
[2] Tani *et al.* (2003).
[3] Kumagai *et al.* (2005).
[4] Bailly *et al.* (1974).
[5] Hermann (1971).
[6] Fleischbein *et al.* (2006).
[7] Blackie (1979a).
[8] Blackie (1979b).
[9] Edwards (1979).
[10] Hutley *et al.* (1997).
[11] McJannet *et al.* (2007d).
[12] Giambelluca *et al.* (2009).
[13] T. W. Giambelluca (personal communication).
[14] Gonggrijp (1941).
[15] Motzer *et al.* (this volume).
[16] Oesker *et al.* (this volume).
[17] Gomez-Cardenas (2009).
[18] Holwerda *et al.* (2009).
[19] Tanaka *et al.* (2003).
[20] Tanaka *et al.* (this volume).
[21] Steinhardt (1979).
[22] Ataroff and Rada (2000).
[23] Bruijnzeel (2006).
[24] K.F.A. Frumau (personal communication).
[25] Holwerda (2005).
[26] García-Santos (2007).
[a] Based on wet-canopy water budget method.
[b] Wet variant of Penman–Monteith equation.

REFERENCES

Aiba, S., and K. Kitayama (1999). Structure, composition and species diversity in an altitude–substrate matrix of rain forest tree communities on Mount Kinabalu, Borneo. *Plant Ecology* **140**: 139–157.

Aiba, S., M. Takyu, and K. Kitayama (2005). Dynamics, productivity and species richness of tropical rainforests along elevational and edaphic gradients on Mount Kinabalu, Borneo. *Ecological Research* **20**: 279–286.

Aide, T. M., and H. R. Grau (2004). Globalization, migration, and Latin American ecosystems. *Science* **305**: 1915–1916.

Aldrich, M., C. Billington, M. Edwards, and R. Laidlaw (1997a). Tropical Montane Cloud Forests: An Urgent Priority for Conservation, WCMC Biodiversity Bulletin No. 2. Cambridge, UK: UNEP–World Conservation Monitoring Centre.

Aldrich, M., C. Billington, M. Edwards, and R. Laidlaw (1997b). *A Global Directory of Tropical Montane Cloud Forests.* Cambridge, UK: UNEP–World Conservation Monitoring Centre.

Aldrich, M., P. Bubb, S. Hostettler, and H. van der Wiel (2000). *Tropical Montane Cloud Forests: Time for Action.* Gland, Switzerland: WWF International and IUCN.

Amelung, T., and M. Diehl (1992). *Deforestation of Tropical Rain Forests: Economic Causes and Impacts on Development*, Kieler Studien No. 241. Tübingen, Germany: JCB Mohr Verlag.

Aravena, R., O. Suzuki, and A. Pollastri (1989). Coastal fog and its relation to ground-water in the IV region of northern Chile. *Chemical Geology* **79**: 83–91.

Asquith, N., and S. Wunder (eds.) (2008). *Payments for Watershed Services: The Bellagio Conversations.* Santa Cruz, Bolivia: Fundación Natura Bolivia, Jakarta, Indonesia: CIFOR, and London: International Institute for Environment and Development.

Asquith, N. M., M. T. Vargas, and S. Wunder (2008). Selling two environmental services: in-kind payments for bird habitat and watershed protection in Los Negros, Bolivia. *Ecological Economics* **65**: 675–684.

Ataroff, M. (1998). Importance of cloud water in Venezuelan Andean cloud forest water dynamics. In *Proceedings of the 1st International Conference on Fog and Fog Collection*, eds. R. S. Schemenauer and H. A. Bridgman, pp. 25–28. Ottawa, Canada: IDRC.

Ataroff, M., and F. Rada (2000). Deforestation impact on water dynamics in a Venezuelan Andean cloud forest. *Ambio* **29**: 440–444.

Aylett, G. P. (1985). Irradiance interception, leaf conductance and photosynthesis in Jamaican upper montane rain forest trees. *Photosynthetica* **19**: 323–337.

Aylward, B. (2005). Land-use, hydrological function and economic valuation. In *Forests, Water and People in the Humid Tropics*, eds. M. Bonell and L. A. Bruijnzeel, pp. 99–120. Cambridge, UK: Cambridge University Press.

Bach, K. (2004). Vegetationskundliche Untersuchungen zur Höhenzonierung tropischer Bergregenwälder in den Anden Boliviens. Ph.D. thesis, University of Göttingen, Göttingen, Germany.

Bach, K., M. Kessler, and S. R. Gradstein (2007). A simulation approach to determine statistical significance of species turnover peaks in a species-rich tropical cloud forest. *Diversity and Distributions* **13**: 863–870.

Backeus, I. (1992). Distribution and vegetation dynamics of humid savannas in Africa and Asia. *Journal of Vegetation Science* **3**: 345–356.

Bailly, C., G. Benoit de Cognac, C. Malvos, J. M. Ningre, and J. M. Sarrailh (1974). Etude de l'influence du couvert naturel et de ses modifications à Madagascar: expérimentations en bassins versants élémentaires. *Cahiers Scientifiques du Centre Technique Forestier Tropical* **4**: 1–114.

Barrantes Moreno, G. (2006). Economic valuation of water supply as a key environmental service provided by montane oak forest watershed areas in Costa Rica. In *Ecology and Conservation of Neotropical Montane Oak Forests*, ed. M. Kappelle, pp. 435–448. Berlin: Springer-Verlag.

Barthlott, W., W. Lauer, and A. Placke (1996). Global distribution of species diversity in vascular plants: towards a world map of phytodiversity. *Erdkunde* **50**: 317–327.

Barthlott, W., V. Schmit-Neuerburg, J. Nieder, and S. Engwald (2001). Diversity and abundance of vascular epiphytes: a comparison of secondary vegetation and primary montane rain forest in the Venezuelan Andes. *Plant Ecology* **152**: 145–156.

Barthlott, W., J. Mutke, M. D. Rafiqpoor, G. Kier, and H. Kreft (2005). Global centers of vascular plant diversity. *Nova Acta Leopoldina NF* **92** (342): 61–83.

Beck, E., K. Hartig, and K. Roos (2008a). Forest clearing by slash and burn. In *Gradients in a Tropical Mountain Ecosystem of Ecuador*, eds. E. Beck, J. Bendix, I. Kottke, F. Makeschin, and R. Mosandl, pp. 387–390. Berlin: Springer-Verlag.

Beck, E., J. Bendix, I. Kottke, F. Makeschin, and R. Mosandl (eds.) (2008b). *Gradients in a Tropical Mountain Ecosystem of Ecuador*. Berlin: Springer-Verlag.

Beck, E., K. Hartig, K. Roos, M. Preuszing, and M. Nebel (2008c). Permanent removal of the forest: construction of roads and power supply lines. In *Gradients in a Tropical Mountain Ecosystem of Ecuador*, eds. E. Beck, J. Bendix, I. Kottke, F. Makeschin, and R. Mosandl, pp. 377–386. Berlin: Springer-Verlag.

Beiderwieden, E., V. Wolff, Y. J. Hsia, and O. Klemm (2008). It goes both ways: measurements of simultaneous evapotranspiration and fog droplet deposition at a montane cloud forest. *Hydrological Processes* **22**: 4181–4189.

Bellingham, P. J. (1991). Landforms influence patterns of hurricane damage: evidence from Jamaican montane forests. *Biotropica* **23**: 427–433.

Bendix, J., R. Rollenbeck, M. Richter, P. Fabian, and P. Emck (2008). Climate. In *Gradients in a Tropical Mountain Ecosystem of Ecuador*, eds. E. Beck, J. Bendix, I. Kottke, F. Makeschin, and R. Mosandl, pp. 63–74. Berlin: Springer-Verlag.

Benzing, D. H. (1990). *Vascular Epiphytes*. Cambridge, UK: Cambridge University Press.

Bigelow, S., and P. Kukle (2000). Ferns. In *Monteverde: Ecology and Conservation of a Tropical Cloud Forest*, eds. N. M. Nadkarni and N. T. Wheelwright, 89. Oxford, UK: Oxford University Press.

Blackie, J. R. (1979a). The water balance of the Kericho catchments. *East African Agricultural and Forestry Journal* **43**: 55–84.

Blackie, J. R. (1979b). The water balance of the Kimakia catchments. *East African Agricultural and Forestry Journal* **43**: 155–174.

Bleiweiss, R. (1998). Origin of hummingbird faunas. *Biological Journal of the Linnean Society* **65**: 77–97.

Blocken, B., J. Carmeliet, and J. Poesen (2005). Numerical simulation of the wind-driven rainfall distribution over small-scale topography in space and time. *Journal of Hydrology* **315**: 252–273.

Blocken, B., J. Poesen, and J. Carmeliet (2006). Impact of wind on the spatial distribution of rain over micro-scale topography: numerical modeling and experimental verification. *Hydrological Processes* **20**: 345–368.

Boy, J., R. Rollenbeck, C. Valarezo, and W. Wilcke (2008). Amazonian biomass burning-derived acid and nutrient deposition in the north Andean montane forest of Ecuador. *Global Biogeochemical Cycles* **22**, GB4011, doi:10.1029/2007GB003158.

Brehm, G., R. K. Colwell, and J. Kluge (2007). The role of environment and mid-domain effect on moth species richness along a tropical elevational gradient. *Global Ecology and Biogeography* **16**: 205–219.

Brehm, G., J. Homeier, K. Fiedler, *et al.* (2008). Mountain rain forests in Southern Ecuador as a hotspot of biodiversity: limited knowledge and diverging patterns. In *Gradients in a Tropical Mountain Ecosystem of Ecuador*, eds. E. Beck, J. Bendix, I. Kottke, F. Makeschin, and R. Mosandl, pp. 15–24. Berlin: Springer-Verlag.

Brown, A. D., and M. Kappelle. (2001). Introducción a los bosques nublados del Neotrópico: una síntesis regional. In *Bosques nublados del Neotrópico*, eds. M. Kappelle and A. D. Brown, pp. 25–40. Santo Domingo de Heredia, Costa Rica: Editorial INBio and IUCN.

Brown, M. B., I. de la Roca, A. Vallejo, *et al.* (1996). *A Valuation Analysis of the Role of Cloud Forests in Watershed Protection: Sierra de las Minas Biosphere Reserve, Guatemala and Cusuco N.P. Honduras*. Philadelphia, PA: RARE Center for Tropical Conservation.

Bruijnzeel, L. A. (2001). Hydrology of tropical montane cloud forests: a reassessment. *Land Use and Water Resources Research* **1**: 1–18.

Bruijnzeel, L. A. (2004). Hydrological functions of tropical forests: not seeing the soil for the trees? *Agriculture, Ecosystems and Environment* **104**: 185–228.

Bruijnzeel, L. A. (2005). Tropical montane cloud forest: a unique hydrological case. In *Forests, Water and People in the Humid Tropics*, eds. M. Bonell and L. A. Bruijnzeel, pp. 462–483. Cambridge, UK: Cambridge University Press.

Bruijnzeel, L. A. (ed.) (2006). *Hydrological Impacts of Converting Tropical Montane Cloud Forest to Pasture, with Initial Reference to Northern Costa Rica*, Final Technical Report for Project R7991, DFID Forestry Research Programme. Amsterdam, the Netherlands: VU University Amsterdam. Also available at www.ambiotek.com/fiesta.

Bruijnzeel, L. A., and L. S. Hamilton (2000). *Decision Time for Cloud Forests*, IHP Humid Tropics Programme Series No. 13. Paris: UNESCO Division of Water Sciences; Gland, Switzerland: WWF and IUCN. Also available at http://sea.unep-wcmc.org/forest/cloudforest/index.cfm.

Bruijnzeel, L. A. and J. Proctor (1995). Hydrology and biochemistry of tropical montane cloud forests: what do we really know? In *Tropical Montane Cloud Forests*, eds. L. S. Hamilton, J. O. Juvik, and F. N. Scatena, pp. 38–78. New York: Springer-Verlag.

Bruijnzeel, L. A., and E. J. Veneklaas (1998). Climatic conditions and tropical montane forest productivity: the fog has not lifted yet. *Ecology* **79**: 3–9.

Bruijnzeel, L. A., M. J. Waterloo, J. Proctor, A. T. Kuiters, and B. Kotterink (1993). Hydrological observations in montane rain-forests on Gunung-Silam, Sabah, Malaysia, with special reference to the "*Massenerhebung*" effect. *Journal of Ecology* **81**: 145–167.

Brünig, E. F. (1996). *Conservation and Management of Tropical Rainforests: An Integrated Approach to Sustainability*. Wallingford, UK: CAB International.

Bubb, P., I. May, L. Miles, and J. Sayer (2004). *Cloud Forest Agenda*. Cambridge, UK: UNEP–World Conservation Monitoring Centre. Also available at http://sea.unep-wcmc.org/forest/cloudforest/index.cfm.

Caceres, G. (1981). Importancia hidrologica de la intercepción horizontal en un bosque muy humedo premontano en Balalaica, Turrialba, Costa Rica. M.Sc. thesis, University of Costa Rica, Turrialba, Costa Rica.

Calvo, J. C. (1986). An evaluation of Thornthwaite's water balance technique in predicting stream runoff in Costa Rica. *Hydrological Sciences Journal* **31**: 51–60.

Calvo, J. (2000). Case study: payments for watershed services in Costa Rica. In *Forests in Sustainable Mountain Development: A State-of-Knowledge Report for 2000*, eds. M. F. Price and N. Butt, pp. 428–429. Wallingford, UK: CAB International.

Cassells, D. S., and L. A. Bruijnzeel (2005). Guidelines for controlling vegetation, soil and water impacts of timber harvesting in the humid tropics. In *Forests, Water and People in the Humid Tropics*, eds. M. Bonell and L. A. Bruijnzeel, pp. 840–851. Cambridge, UK: Cambridge University Press.

Cavelier, J. (1990). Tissue water relations in elfin cloud forest tree species of Serranía de Macuira, Guajira, Colombia. *Trees: Structure and Function* **4**: 155–163.

Cavelier, J. (1995). Reforestation with the native tree *Alnus acuminata* : effects on phytodiversity and species richness in an upper montane rain forest area of Columbia. In *Tropical Montane Cloud* Forests, eds. L. S. Hamilton, J.O. Juvik, and F.N. Scatena, pp. 125–137. New York: Springer-Verlag.

Cavelier, J. (1996). Environmental factors and ecophysiological processes along altitudinal gradients in wet tropical mountains. In *Tropical Forest Plant Ecophysiology*, eds. S. S. Mulkey, R. L. Chazdon, and A. P. Smith, pp. 399–439. New York: Chapman and Hall.

Cavelier, J., and C. A. Mejia (1990). Climatic factors and tree stature in the elfin cloud forest of Serranía de Macuira, Colombia. *Agricultural and Forest Meteorology* **53**: 105–123.

Cavelier, J., M. Jaramillo, D. Solis, and D. De Leon (1997). Water balance and nutrient inputs in bulk precipitation in tropical montane cloud forest in Panama. *Journal of Hydrology* **193**: 83–96.

Cavelier, J., T. M. Aide, C. Santos, A. M. Eusse, and J. M. Dupuy (1998). The savannization of moist forests in Sierra Nevada de Santa Marta, Colombia. *Journal of Biogeography* **25**: 901–912.

Cavelier, J., E. Tanner, and J. Santamaria (2000). Effect of water, temperature and fertilizers on soil nitrogen net transformations and tree growth in an elfin cloud forest of Colombia. *Journal of Tropical Ecology* **16**: 83–99.

Cayuela, L., D. J. Golicher, and J. M. Rey-Benayas (2006). The extent, distribution, and fragmentation of vanishing montane cloud forest in the highlands of Chiapas, Mexico. *Biotropica* **38**: 544–554.

Chang, M., and L. A. Flannery (2001). Spherical gages for improving the accuracy of precipitation measurements. *Hydrological Processes* **15**: 643–654.

Churchill, S. P., H. Balslev, E. Forero, and J. L. Luteyn (eds.) (1996). *Biodiversity and Conservation of Neotropical Montane Forests.* New York: New York Botanical Garden.

Clark, K. L., N. M. Nadkarni, D. Schaefer, and H. L. Gholz (1998). Atmospheric deposition and net retention of ions by the canopy in a tropical montane forest, Monteverde, Costa Rica. *Journal of Tropical Ecology* **14**: 27–45.

Clark, L. G. (1995). Diversity and distribution of the Andean woody bamboos (Poaceae: Bambuseae). In *Biodiversity and Conservation of Neotropical Montane Forests*, eds. S. P. Churchill, H. Balslevee, E. Forero, and J. L. Luteyn, pp. 501–512. New York: New York Botanical Garden.

Collins, N. M. (1980). The distribution of soil macrofauna on the west ridge of Gunung (Mount) Mulu, Sarawak. *Oecologia* **44**: 263–275.

Collins, N. M., J. M. Anderson, and H. W. Vallack (1984). Studies on the soil invertebrates of lowland and montane rain forest in the Gunung Mulu National Park. *Sarawak Museum Journal* **30**(51): 19–33.

Colwell, R. K., G. Brehm, C. L. Cardelús, A. C. Gilman, and J. T. Longino (2008). Global warming, elevational range shifts, and lowland biotic attrition in the wet Tropics. *Science* **322**: 258–261.

Cordell, S., G. Goldstein, D. Müeller-Dombois, D. Webb, and P. M. Vitousek (1998). Physiological and morphological variation in *Metrosideros polymorpha*, a dominant Hawaiian tree species, along an altitudinal gradient: the role of phenotypic plasticity. *Oecologia* **113**: 188–196.

Cordero, R. A. (1999). Ecophysiology of *Cecropia schreberiana* saplings in two wind regimes in an elfin cloud forest: growth, gas exchange, architecture, and stem biomechanics. *Tree Physiology* **19**: 153–165.

Cortés, A. M., M. P. Ramírez-Pinilla, H. A. Suárez, and E. Tovar (2008). Edge effects on richness, abundance and diversity of frogs in Andean cloud forest fragments. *South American Journal of Herpetology* **3**: 213–222.

Crutzen, P. J., and E. F. Stoermer (2000). "The Anthropocene." *IGBP Newsletter* **41**: 17–18.

Cubiñá, A., and T. M. Aide (2001). The effect of distance from forest edge on seed rain and soil seed bank in a tropical pasture. *Biotropica* **33**: 260–267.

Cuervo, A. M., P. Salaman, T. Donegan, and J. M. Ochoa (2001). A new species of *Piha* (Cotingidae: *Lipaugus*) from the Cordillera Central of Colombia. *Ibis* **143**: 353–368.

D'Antonio, C. M. and P. M. Vitousek. (1992). Biological invasions by exotic grasses, the grass/fire cycle, and global change. *Annual Review of Ecology and Systematics* **23**: 63–87.

Daugherty, H. E. (1973). The Montecristo cloud-forest of El Salvador: a chance for protection. *Biological Conservation* **5**: 227–230.

Dawson, T. E. (1998). Fog in the California redwood forest: ecosystem inputs and use by plants. *Oecologia* **117**: 476–485.

Debouck, D. G., and D. Libreros Ferla. (1995). Neotropical montane forests: a fragile home of genetic resources of wild relatives of new world crops. In *Biodiversity and Conservation of Neotropical Montane Forests*, eds. S. P. Churchill, H. Balslev, E. Forero, and J. L. Luteyn, pp. 561–577. New York: New York Botanical Garden.

Dietz, J., D. Hölscher, Ch. Leuschner, and Hendrayanto. (2006). Rainfall partitioning in relation to forest structure in differently managed montane forest stands in Central Sulawesi, Indonesia. *Forest Ecology and Management* **237**: 170–178.

Dinerstein, E., D. M. Olson, D. J. Graham, *et al.* (1995). *A Conservation Assessment of the Terrestrial Ecoregions of Latin America and the Caribbean.* Washington, DC: The World Bank.

Doumenge, C., D. Gilmour, M. Ruiz-Perez, and J. Blockhus (1995). Tropical montane cloud forests: conservation status and management issues. In *Tropical Montane Cloud Forests*, eds. L. S. Hamilton, J. O. Juvik, and F. N. Scatena, pp. 24–37. New York: Springer-Verlag.

Duellman, W. E. (1988). Patterns of species diversity in anuran amphibians in the American tropics. *Annals of the Missouri Botanical Garden* **75**: 79–104.

Dunn, R. R. (2000). Bromeliad communities in isolated trees and three successional stages of an Andean cloud forest in Ecuador. *Selbyana* **21**: 137–143.

Edwards, K. A. (1979). The water balance of the Mbeya experimental catchments. *East African Agricultural and Forestry Journal* **43**: 231–247.

Edwards, P. J. (1982). Studies of mineral cycling in a montane rain forest in New Guinea. V. Rates of cycling in throughfall and litter fall. *Journal of Ecology* **70**: 807–827.

Emanuel, K. A. (2005). Increasing destructiveness of tropical cyclones over the past 30 years. *Nature* **436**: 686–688.

Etter, A., and W. van Wijngaarden (2000). Patterns of landscape transformation in Colombia,with emphasis on the Andean region. *Ambio* **29**: 432–439.

Eugster, W., R. Burkard, F. Holwerda, F. N. Scatena, and L. A. Bruijnzeel (2006). Characteristics of fog and fog-water fluxes in a Puerto Rican elfin cloud forest. *Agricultural and Forest Meteorology* **139**: 288–306.

Ewel, J. J. (1980). Tropical succession: manifold routes to maturity. *Biotropica* **12**: 2–7.

Fallas, J. (2002). Net precipitation patterns in undisturd and fragmented Costa Rican cloud forests. In *Proceedings of the 2nd International Colloquium on Hydrology and Water Management*, ed. J. S. Gladwell, pp. 389–398. Panamá City, Panamá: CATHALAC, and Paris: IHP–UNESCO.

Fetcher, N., R. A. Cordero, and J. Voltzow (2000). Lack of ecotypic differentiation: plant response to elevation, population origin, and wind in the Luquillo Mountains, Puerto Rico. *Biotropica* **32**: 225–234.

Fjeldså, J., and C. Rahbek (1997). Species richness and endemism in South American birds: implications for the design of networks of nature reserves. In *Tropical Forest Remnants*, eds. W. F. Laurance and R. D. B. Bierregaard, pp. 466–482. Chicago, IL: University of Chicago Press.

Fleischbein, K., W. Wilcke, C. Valarezo, W. Zech, and K. Knoblich (2006). Water budgets of three small catchments under montane forest in Ecuador: experimental and modelling approach. *Hydrological Processes* **20**: 2491–2507.

Flenley, J. R. (1995). Cloud forest, the *Massenerhebung* effect, and ultraviolet insolation. In *Tropical Montane Cloud Forests*, eds. L. S. Hamilton, J. O. Juvik, and F. N. Scatena, pp. 150–155. New York: Springer-Verlag.

Førland E. J., P. Allerup, B. Dahlström, *et al.* (1996). *Manual for Operational Correction of Nordic Precipitation Data.* Oslo: Norwegian Meteorological Institute.

Foster, P. (2001). The potential negative impacts of global climate change on tropical montane cloud forests. *Earth-Science Reviews* **55**: 73–106.

Frahm, J. P., and S. R. Gradstein (1991). An altitudinal zonation of tropical rain forests using bryophytes. *Journal of Biogeography* **18**: 669–676.

Frumau, K. F. A., L. A. Bruijnzeel, and C. Tobón (2006a). *Hydrological Measurement Protocol for Montane Cloud Forests*, Appendix 2 to Final Technical Report for Project R7991, DFID Forestry Research Programme. Amsterdam, the Netherlands: VU University Amsterdam. Also available at www.ambiotek.com/fiesta.

Frumau, K. F. A., L. A. Bruijnzeel, and C. Tobón (2006b). Measurement of precipitation in montane tropical catchments: comparative performance of conventional, spherical and "potential" rain gages. In *Forest and Water in a Changing Environment*, eds. S.R. Liu, G. Sun, and P.S. Sun, pp. 104–108. Vienna: IUFRO, and Beijing: Chinese Academy of Forestry.

Gabriel, G., and L. Jauze (2008). Fog water interception by *Sophora denudata* trees in a Réunion upper montane forest, Indian Ocean. *Atmospheric Research* **87**: 338–351.

García-García, F., and V. Zarraluqui (2008). A fog climatology for Mexico. *Die Erde* **139**: 45–60.

García-Santos, G. (2007). An ecohydrological and soils study in a montane cloud forest in the National Park of Garajonay, La Gomera (Canary Islands, Spain). Ph.D. thesis, VU University Amsterdam, Amsterdam, the Netherlands. Also available at www.falw.vu.nl/nl/onderzoek/earth-sciences/geo-environmental-science-and-hydrology/hydrology-dissertations/index.asp.

Garriguata, M.R., and P. Balvanera. (2009). Tropical forest service flows. *Forest Ecology and Management*, doi:10.1016/j.foreco.2009.06.025.

Gentry, A.H. (1982). Neotropical floristic diversity: phytogeographical connections between Central and South America, Pleistocene climatic fluctuations, or an accident of the Andean orogeny? *Annals of the Missouri Botanical Garden* **69**: 557–593.

Gentry, A.H. (1988). Changes in plant community diversity and floristic composition on environmental and geographical gradients. *Annals of the Missouri Botanical Garden* **75**: 1–34.

Gentry, A.H. (1992). Diversity and floristic composition of Andean forests of Peru and adjacent countries: implications for their conservation. In *Biogeografía, ecología y conservación del bosque montano en el Perú*, Memorias Museo de Historia Natural No. 21, eds. K.R. Young and N. Valencia, pp. 11–29. Lima, Perú: UNMSM.

Gentry, A.H. (1995). Patterns of diversity and floristic composition in neotropical montane forests. In *Biodiversity and Conservation of Neotropical Montane Forests*, eds. S.P. Churchill, H. Balslev, E. Forero, and J.L. Luteyn, pp. 103–126. New York: New York Botanical Garden.

Gerold, G., M. Schawe, and K. Bach (2008). Hydrometeorologic, pedologic and vegetation patterns along an elevational transect in the montane forest of the Bolivian Yungas. *Die Erde* **139**: 141–168.

Giambelluca, T.W., and D.A. Nullet (1992). Evaporation at high elevations in Hawaii. *Journal of Hydrology* **136**: 219–235.

Giambelluca, T.W., R.E. Martin, G.P. Asner, *et al.* (2009). Evapotranspiration and energy balance of native wet montane cloud forest in Hawai'i. *Agricultural and Forest Meteorology* **149**: 230–243.

Gill, A.M. (1969). The ecology of an elfin forest in Puerto Rico. VI. Aerial roots. *Journal of the Arnold Arboretum* **50**: 197–209.

Gomes, L.G.L., V. Oostra, V. Nijman, A.M. Cleef, and M. Kappelle (2008). Tolerance of frugivorous birds to habitat disturbance in a tropical cloud forest. *Biological Conservation* **141**: 860–871.

Gomez-Cardenas, M. (2009). Transpiration by contrasting vegetation cover types in the montane cloud forest belt of eastern Mexico. Ph.D. thesis, Iowa State University, Ames, IA, USA.

Gómez-Laurito, J., and L.D. Gómez-P. (1991). Ticodendraceae: a new family of flowering plants. *Annals of the Missouri Botanical Garden* **78** 87–88.

Gomez-Peralta, D., S.F. Oberbauer, M.E. McClain, and T.E. Philippi (2008). Rainfall and cloud-water interception in tropical montane forests in the eastern Andes of Central Peru. *Forest Ecology and Management* **255**: 1315–1325.

Gonggrijp, L. (1941). Evaporation from montane forest in West Java at an altitude of 1750–2000 m a.s.l. *Tectona* **34**: 437–447 (in Dutch, with a summary in English).

Gonzalez, G., E. Garcia, V. Cruz, *et al.* (2007). Earthworm communities along an elevation gradient in Northeastern Puerto Rico. *European Journal of Soil Biology* **43**: S24–S32.

González-Espinosa M., J.M. Rey-Benayas, and N. Ramírez-Marcial (eds.) (2008). *Restauración de bosques en América Latina*. México City, Mexico: Mundi-Prensa.

Gradstein, S.R. (2008). Epiphytes of tropical montane forests: impact of deforestation and climate change. In *The Tropical Mountain Forest: Patterns and Processes in a Biodiversity Hotspot*, eds. S.R. Gradstein, J. Homeier, and D. Gansert, pp. 51–65. Göttingen, Germany: Göttingen Centre for Biodiversity and Ecology.

Gradstein, S.R., and T. Pócs. (1989). Bryophytes. In *Tropical Rain Forest Ecosystems*, eds. H. Lieth and M.J.A. Werger, pp. 311–325. Amsterdam, the Netherlands: Elsevier.

Gradstein, S.R., S.P. Churchill, and N. Salazar Allen (2001). *Guide to the Bryophytes of Tropical America*, Memoirs of the New York Botanical Garden No. 86. New York: New York Botanical Garden.

Grau, H.R., T.M. Aide, J.K. Zimmerman, *et al.* (2003). The ecological consequences of socioeconomic and land-use changes in post-agriculture Puerto Rico. *BioScience* **53**: 1159–1168.

Grau, H.R., M. Perez-Ceballos, S. Martinuzzi, X. Encarnación, and T.M. Aide (2008). Cambios socioeconómicos y regeneración del bosque en República Dominicana. In *Restauración ecológica en América Latina*, eds. M. Gonzalez-Espinosa, J.M. Rey-Benayas, and N. Ramírez-Marcial, pp. 211–227. México City, Mexico: Mundi-Prensa.

Grenyer, R., C.D.L. Orme, S.F. Jackson, *et al.* (2006). Global distribution and conservation of rare and threatened vertebrates. *Nature* **444**: 93–96.

Groombridge, B. (ed.) (1992). *Global Biodiversity: Status of the Earth's Living Resources*. London: Chapman and Hall.

Grubb, P.J. (1977). Control of forest growth and distribution on wet tropical mountains: with special reference to mineral nutrition. *Annual Review of Ecology and Systematics* **8**: 83–107.

Guariguata, M.R., G.P. Sáenz, and L. Pedroni (2006). Regeneration dynamics in a Costa Rican montane oak forest after reduced-impact logging. In *Ecology and Conservation of Neotropical Montane Oak Forests*, ed. M. Kappelle, pp. 235–244. Berlin: Springer-Verlag.

Günter, S., M. Weber, R. Erreis, and N. Aguirre (2006). Influence of distance to forest edges on natural regeneration of abandoned pastures: a case study in the tropical montane rain forest of Southern Ecuador. *European Journal of Forest Research*, doi:10.1007/s10342–006–0156–0.

Günter, S., O. Cabrera, M. Weber, *et al.* (2008). Natural forest management in Neotropical mountain rain forests: an ecological experiment. In *Gradients in a Tropical Mountain Ecosystem of Ecuador*, eds. E. Beck, E.J. Bendix, I. Kottke, F. Makeschin, and R. Mosandl, pp. 363–375. Berlin: Springer-Verlag.

Günter, S., P. Gonzalez, G. Alvarez, *et al.* (2009). Determinants of successful regeneration of abandoned pastures in the Andes: soil conditions and vegetation cover. *Forest Ecology and Management* **258**: 81–91.

Guswa, A.J., A.L. Rhodes, and S.E. Newell (2007). Importance of orographic precipitation to the water resources of Monteverde, Costa Rica. *Advances in Water Resources* **30**: 2098–2112.

Haber, W. (1991). Lista provisional de las plantas de Monteverde, Costa Rica. *Brenesia* **34**: 63–120.

Haber, W. (2000). Plants and vegetation. In *Monteverde: Ecology and Conservation of a Tropical Cloud Forest*, eds. N.M. Nadkarni and N.T. Wheelwright, pp. 39–70. Oxford, UK: Oxford University Press.

Hafkenscheid, R.L.L.J. (2000). Hydrology and biogeochemistry of tropical montane rain forests of contrasting stature in the Blue Mountains, Jamaica. Ph.D. thesis, VU University Amsterdam, Amsterdam, the Netherlands. Also available at http://dare.ubvu.vu.nl/bitstream/1871/12734/1/tekst.pdf.

Hafkenscheid, R.L.L.J., L.A. Bruijnzeel, R.A.M. de Jeu, and N.J. Bink (2002). Water budgets of two upper montane rain forests of contrasting stature in the Blue Mountains, Jamaica. In *Proceedings of the 2 International Colloquium on Hydrology and Water Management in the Humid Tropics*, ed. J.S. Gladwell, pp. 399–424. Panamá City, Panamá: CATHALAC, and Paris: IHP–UNESCO.

Hamilton, L.S. (1995). *A Campaign for Cloud Forests*. Gland, Switzerland: IUCN.

Hamilton, L.S. (2005). Red flags of warning in land clearing. In *Forests, Water and People in the Humid Tropics*, eds. M. Bonell and L.A. Bruijnzeel, pp. 866–880. Cambridge, UK: Cambridge University Press.

Hamilton, L.S., J.O. Juvik, and F.N. Scatena (eds.) (1995a). *Tropical Montane Cloud Forests*. New York: Springer-Verlag.

Hamilton, L.S., J.O. Juvik, and F.N. Scatena (1995b). The Puerto Rico tropical montane cloud forest symposium: introduction and workshop synthesis. In *Tropical Montane Cloud Forests*, eds. L.S. Hamilton, J.O. Juvik, and F.N. Scatena, pp. 1–16. New York: Springer-Verlag.

Harr, R.D. (1982). Fog-drip in the Bull Run municipal watershed, Oregon. *Water Resources Bulletin* **18**: 785–789.

Hartig, K., and E. Beck (2003). The bracken fern (*Pteridium arachnoideum* Kaulf.) dilemma in the Andes of Southern Ecuador. *Ecotropica* **9**: 3–13.

Hättenschwiler, S., A.E. Hagerman, and P.M. Vitousek (2003). Polyphenols in litter from tropical montane forests across a wide range in soil fertility. *Biogeochemistry* **64**: 129–148.

Heaney, L.R. (2001). Small mammal diversity along elevational gradients in the Philippines: an assessment of patterns and hypotheses. *Global Ecology and Biogeography* **10**: 15–39.

Hemp, A. (2002). Ecology of the pteridophytes on the southern slopes of Mt. Kilimanjaro. I. Altitudinal distribution. *Plant Ecology* **159**: 211–239.

Hemp, A. (2005a). Continuum or zonation? Altitudinal gradients in the forests on Mt. Kilimanjaro. *Plant Ecology* **184**: 27–42.

Hemp, A. (2005b). Climate change driven forest fires marginalize the impact of ice cap wasting on Kilimanjaro. *Global Change Biology* **11**: 1013–1023.
Herrera, B., and A. Chaverri (2006). Criteria and indicators for sustainable management of Central American montane oak forests. In *Ecology and Conservation of Neotropical Montane Oak Forests*, ed. M. Kappelle, pp. 421–434. Berlin: Springer-Verlag.
Herrmann, R. (1971). Die zeitliche Anderung der Wasserbindung im Boden unter verschiedenen Vegetationsformationen der Höhenstufen eines tropischen Hochgebirges (Sierra Nevada de Sta. Marta, Kolumbien). *Erdkunde* **25**: 90–102.
Herwitz, S. (1991). Aboveground adventitious roots and stemflow chemistry of *Ceratopetalum virchowii* in an Australian montane tropical rain forest. *Biotropica* **23**: 210–218.
Herwitz, S., and R. E. Slye (1992). Spatial variation in the interception of inclined rainfall by a tropical rainforest canopy. *Selbyana* **13**: 62–71.
Hetsch, W., and H. Hoheisel (1976). Standorts- und Vegetationsgliederung in einem tropischen Nebelwald. *Allgemeine Forst- und Jagdzeitung* **147**: 200–207.
Hildebrandt, A., and E. A. B. Eltahir (2006). Forest on the edge: seasonal cloud forest in Oman creates its own niche. *Geophysical Research Letters* **33**, L11401, doi:10.1029/2006GL026022.
Hobbs, R. J., and D. A. Norton (1996). Towards a conceptual framework for restoration ecology. *Restoration Ecology* **4**: 93–110.
Holder, C. D. (2003). Fog precipitation in the Sierra de las Minas Biosphere Reserve, Guatemala. *Hydrological Processes* **17**: 2001–2010.
Holder, C. D. (2004). Rainfall interception and fog precipitation in a tropical montane cloud forest of Guatemala. *Forest Ecology and Management* **190**: 373–384.
Holl, K. D., and M. Kappelle (1999). Tropical forest recovery and restoration. *Trends in Ecology and Evolution* **14**: 378–379.
Holl, K. D., M. E. Loik, E. H. V. Lin, and I. A. Samuels (2000). Tropical montane forest restoration in Costa Rica: overcoming barriers to dispersal and establishment. *Restoration Ecology* **8**: 339–349.
Holl, K. D., E. E. Crone, and C. B. Schultz (2003). Landscape restoration: moving from generalities to methodologies. *BioScience* **53**: 491–502.
Hölscher, D., L. Köhler, A. I. J. M. Van Dijk, and L. A. Bruijnzeel (2004). The importance of epiphytes to total rainfall interception by a tropical montane rain forest in Costa Rica. *Journal of Hydrology* **292**: 308–322.
Holwerda, F. (2005). Water and energy budgets of rain forests along an elevation gradient under maritime tropical conditions. Ph.D. thesis, VU University Amsterdam, Amsterdam, the Netherlands. Also available at www.falw.vu.nl/nl/onderzoek/earth-sciences/geo-environmental-science-and-hydrology/hydrology-dissertations/index.asp.
Holwerda, F., R. Burkard, W. E. Eugster, *et al.* (2006). Estimating fog deposition at a Puerto Rican elfin cloud forest site: comparison of the water budget and eddy covariance methods. *Hydrological Processes* **20**: 2669–2692.
Holwerda, F., Barradas, V. L., Cervantes, J., and Bruijnzeel, L. A. (2007). Balances hídricos y de energía de un cafetal de sombra en el centro de Veracruz, México. In *Reporte técnico final del proyecto INE/A1–064/2007*. Xalapa, Veracruz, México: Instituto de Ecología, and Amsterdam, the Netherlands: VU University Amsterdam.
Holwerda, F., L. A. Bruijnzeel, L. E. Muñoz, M. Equihua, and H. Asbjornsen (2010). Rainfall and cloud water interception in mature and secondary lower montane cloud forests of central Veracruz, Mexico. *Journal of Hydrology* **384**: 84–96.
Holz, I., and S. R. Gradstein (2005a). Phytogeography of the bryophyte floras of oak forests and paramo of the Cordillera de Talamanca, Costa Rica. *Journal of Biogeography* **32**: 1591–1609.
Holz, I., and S. R. Gradstein (2005b). Cryptogamic epiphytes in primary and recovering upper montane oak forests of Costa Rica: species richness, community composition and ecology. *Plant Ecology* **178**: 89–109.
Holz, I., S. R. Gradstein, J. Heinrichs, and M. Kappelle (2002). Bryophyte diversity, microhabitat differentiation and distribution of life forms in Costa Rican upper montane *Quercus* forest. *The Bryologist* **105**: 334–348.
Homeier, J., F. A. Werner, S. R. Gradstein, S.-W. Breckle, and M. Richter (2008). Potential vegetation and floristic composition of Andean forests in South Ecuador, with a focus on the RBSF. In *Gradients in a Tropical Mountain Ecosystem of Ecuador*, eds. E. Beck, J. Bendix, I. Kottke, F. Makeschin, and R. Mosandl, pp. 425–431. Berlin: Springer-Verlag.
Honey, M. (2008). *Ecotourism and Sustainable Development: Who Owns Paradise?* Washington, DC: Island Press.
Hutley, L. B., D. Doley, D. Yates, and A. Boonsaner (1997). Water balance of an Australian subtropical rainforest at altitude: the ecological and physiological significance of intercepted cloud and fog. *Australian Journal of Botany* **45**: 311–329.
Hyndman, D. C., and J. I. Menzies (1990). Rain forests of the Ok Tedi headwaters, New Guinea: an ecological analysis. *Journal of Biogeography* **17**: 241–273.
Ingwersen, J. B. (1985). Fog drip, water yield and timber harvesting in the Bull Run Municipal Watershed, Oregon. *Water Resources Bulletin* **21**: 469–473.
Islebe, G. A., and H. Hooghiemstra (2006). Effects of the Younger Dryas cooling event on late Quaternary montane oak forest in Costa Rica. In *Ecology and Conservation of Neotropical Montane Oak Forests*, ed. M. Kappelle, pp. 29–37. Berlin: Springer-Verlag.
ITTO (1992). *Criteria for the Measurement of Sustainable Tropical Forest Management*, ITTO Policy Development Series No. 2. Yokohama, Japan: International Timber Trade Organization.
IUCN, Conservation International, and NatureServe (2008a). *An Analysis of Amphibians on the 2008 IUCN Red List*. Gland, Switzerland: IUCN. Also available at www.iucnredlist.org/amphibians.
IUCN, Conservation International, Arizona State University, Texas A&M University, University of Rome, University of Virginia, and Zoological Society London. (2008b). *An Analysis of Mammals on the 2008 IUCN Red List*. Available at www.iucnredlist.org/mammals.
IUCN (2009). *Fighting for Forest Frogs*, IUCN Species Survival Commission, Amphibian Specialist Group. Gland, Switzerland: IUCN. Also available at www.iucn.org/about/work/programmes/species/red_list/?3061/Fighting-for-forest-frogs.
Jin, Y., W. B. Rossow, and D. P. Wylie (1996). Comparison of the climatologies of high-level clouds from HIRS and ISCCP. *Journal of Climate* **9**: 2850–2879.
Juen, I., C. Georges, and G. Kaser (2007). Modelling observed and future runoff from a glacierized tropical catchment (Cordillera Blanca, Perú). *Global and Planetary Change* **59**: 37–48.
Juvik, J. O., and P. C. Ekern (1978). *A Climatology of Mountain Fog on Mauna Loa, Hawaii Island*, Technical Report No. 118. Honolulu, HI: Water Resources Research Center, University of Hawai'i.
Juvik, J. O., and D. Nullet (1995a). Comments on "a proposed standard fog collector for use in high elevation regions." *Journal of Applied Meteorology* **34**: 2108–2110.
Juvik, J. O., and D. Nullet (1995b). Relationships between rainfall, cloud-water interception and canopy throughfall in a Hawaiian montane forest. In *Tropical Montane Cloud Forests*, eds. L. S. Hamilton, J. O. Juvik, and F. N. Scatena, pp. 165–182. New York: Springer-Verlag.
Kapos, V., J. Rhind, M. Edwards, and M. F. Price (2000). Developing a map of the world's mountain forests. In *Forests in Sustainable Mountain Development: A State-of-Knowledge Report for 2000*, eds. M. F. Price and N. Butt, pp. 4–9. Wallingford, UK: CAB International. Also available at www.unep-wcmc.org/habitats/mountains/homepage. htm.
Kappelle, M. (2001). Bosques nublados de Costa Rica. In *Bosques nublados del Neotrópico*, eds. M. Kappelle and A. D. Brown, pp. 301–370. Santo Domingo de Heredia, Costa Rica: Editorial INBio and IUCN.
Kappelle, M. (2004). Tropical montane forests. In *Encyclopedia of Forest Sciences*, Vol. **4**, eds. J. Burley, J. Evans, and J. A. Youngquist, pp. 1782–1793. Oxford, UK: Elsevier.
Kappelle, M. (2006). Neotropical montane oak forests: overview and outlook. In *Ecology and Conservation of Neotropical Montane Oak Forests*, ed. M. Kappelle, pp. 449–467. Berlin: Springer-Verlag.
Kappelle, M., and A. D. Brown (eds.) (2001). *Bosques nublados del Neotrópico*. Santo Domingo de Heredia, Costa Rica: Editorial INBio and IUCN.
Kappelle, M., and G. J. van Uffelen (2006). Altitudinal zonation of montane oak forests along climate and soil gradients in Costa Rica. In *Ecology and Conservation of Neotropical Montane Oak Forests*, ed. M. Kappelle, pp. 39–54. Berlin: Springer-Verlag.
Kappelle, M., A. M. Cleef, and A. Chaverri (1992). Phytogeography of Talamanca montane *Quercus* forests, Costa Rica. *Journal of Biogeography* **19**: 299–315.
Kappelle, M., H. P. van Velzen, and W. H. Wijtzes (1994). Plant communities of montane secondary vegetation in the Cordillera de Talamanca, Costa Rica. *Phytocoenologia* **22**: 449–484.
Kappelle, M., P. A. F. Kennis, and R. A. J. de Vries (1995a). Changes in diversity along a successional gradient in a Costa Rican upper montane *Quercus* forest. *Biodiversity and Conservation* **4**: 10–34.

Kappelle, M., J. G. van Uffelen, and A. M. Cleef (1995b). Altitudinal zonation of montane *Quercus* forests along two transects in the Chirripó National Park, Costa Rica. *Vegetatio (Plant Ecology)* **119**: 119–153.

Kappelle, M., T. Geuze, M. Leal, and A. M. Cleef (1996). Successional age and forest structure in a Costa Rican upper montane *Quercus* forest. *Journal of Tropical Ecology* **12**: 681–698.

Kappelle, M., G. Avertin, M. E. Juárez, and N. Zamora. (2000). Useful plants within a campesino community in a Costa Rican montane cloud forest. *Mountain Research and Development* **20**: 162–171.

Kaser, G., C. Georges, I. Juen, and T. Moelg (2005). Low latitude glaciers: unique global climate indicators and essential contributors to regional fresh water supply – a conceptual approach. In *Global Change and Mountain Regions: A State of Knowledge Overview*, eds. U. Huber, H. K. M. Bugmann, and M. A. Reasoner, pp. 185–196. New York: Kluwer.

Kessler, M. (2002). The elevational gradient of Andean plant endemism: varying influences of taxon-specific traits and topography at different taxonomic levels. *Journal of Biogeography* **29**: 1159–1166.

Kessler, M., and J. Kluge (2008). Diversity and endemism in tropical montane forests: from patterns to processes. In *The Tropical Mountain Forest: Patterns and Processes in a Biodiversity Hotspot*, eds. S.R. Gradstein, J. Homeier, and D. Gansert, pp. 35–50. Göttingen, Germany: Göttingen Centre for Biodiversity and Ecology.

Kessler, M., and M. Lehnert (2009). Do ridge habitats contribute to pteridophyte diversity in tropical montane forests? A case study from southeastern Ecuador. *Journal of Plant Research* **122**: 421–428.

Kier, G., and W. Barthlott (2001). Measuring and mapping endemism and species richness: a new methodological approach and its application on the flora of Africa. *Biodiversity and Conservation* **10**: 1513–1529.

Kier, G., J. Mutke, E. Dinerstein, *et al.* (2005). Global patterns of plant diversity and floristic knowledge. *Journal of Biogeography* **32**: 1–10.

Kikkawa, J., and W. T. Williams (1971). Altitudinal distribution of land birds in New Guinea. *Search* **2**: 64–65.

Kitayama, K. (1995). Biophysical conditions of the montane cloud forests of Mount Kinabalu, Sabah, Malaysia. In *Tropical Montane Cloud Forests*, eds. L. S. Hamilton, J. O. Juvik, and F. N. Scatena, pp. 183–197. New York: Springer-Verlag.

Kitayama, K., and D. Müller-Dombois (1994). An altitudinal transect analysis of the windward vegetation on Haleakala, a Hawaiian island mountain. I. Climate and soils. *Phytocoenologia* **24**: 111–133.

Kleidon, A., and H. A. Mooney (2008). A global distribution of biodiversity inferred from climatic constraints: results from a process-based modelling study. *Global Change Biology* **6**: 507–523.

Köhler, L., C. Tobón, K. F. A. Frumau, and L. A. Bruijnzeel (2007). Biomass and water storage dynamics of epiphytes in old-growth and secondary montane cloud forest stands in Costa Rica. *Plant Ecology* **193**: 171–184.

Körner, C., and E. M. Spehn (eds.) (2002). *Mountain Biodiversity: A Global Assessment*. Boca Raton, FL: Parthenon.

Krabbe, N., D. J. Agro, N. H. Rice, *et al.* (1999). A new species of *Antpitta* (Formicariidae: Grallaria) from the southern Ecuadorian Andes. *The Auk* **116**: 882–890.

Kreft, H., and W. Jetz (2007). Global patterns and determinants of vascular plant diversity. *Proceedings of the National Academy of Sciences USA* **104**: 5925–5930.

Kumagai, T., T. M. Saitoh, Y. Sato, *et al.* (2005). Annual water balance and seasonality of evapotranspiration in a Bornean tropical rainforest. *Agricultural and Forest Meteorology* **128**: 81–92.

Kumaran, S. (2008). Hydrometeorology of Tropical montane rain forests of Gunung Brinchang, Pahang Darul Makmur, Malaysia. Ph.D. thesis, University of Putra Malaysia, Serdang, Malaysia.

Küppers, M., T. Motzer, D. Schmitt, *et al.* (2008). Stand structure, transpiration responses in trees and vines, and stand transpiration of different forest types within the mountain rainforest. In *Gradients in a Tropical Mountain Ecosystem of Ecuador*, eds. E. Beck, J. Bendix, I. Kottke, F. Makeschin, and R. Mosandl, pp. 243–258. Berlin: Springer-Verlag.

LaBastille, A., and D. J. Pool (1978). On the need for a sytem of cloud-forest parks in Middle America and the Caribbean. *Environmental Conservation* **5**: 183–190.

Lawton, R. O. (1982). Wind stress and elfin stature in a montane rain forest tree: an adaptive explanation. *American Journal of Botany* **69**: 1224–1230.

Lawton, R. O., and V. Dryer (1980). The vegetation of the Monteverde Cloud Forest Preserve. *Brenesia* **18**: 101–116.

Lawton, R. O., U. S. Nair, R. A. Pielke Sr., and R. M. Welch (2001). Climatic impact of tropical lowland deforestation on nearby montane cloud forests. *Science* **294**: 584–587.

Leakey, R. J. G., and J. Proctor (1987). Invertebrates in the litter and soil at a rnge of altitudes on Gunng Silam, a small ultrabasic mountain in Sabah. *Journal of Tropical Ecology* **3**: 119–129.

Leo, M. (1995). The importance of tropical montane cloud forests for preserving vertebrate endemism in Peru: the Río Abiseo National Park as a case study. In *Tropical Montane Cloud Forests*, eds. L. S. Hamilton, J. O. Juvik, and F. N. Scatena, pp. 198–211. New York: Springer-Verlag.

Le Tellier, V., A. Carrasco, and N. Asquith (2009). Attempts to determine effects of forest cover on streamflow by hydrological measurements in Los Negros, Bolivia. *Forest Ecology and Management*, doi:10.1016/j.foreco.2009.04.031.

Leuschner, C h., G. Moser, C. Bertsch, M. Röderstein, and D. Hertel (2007). Large altitudinal increase in tree root/shoot ratio in tropical mountain forests of Ecuador. *Basic and Applied Ecology* **8**: 219–230.

Lieberman, D., M. Lieberman, R. Peralta, and G. S. Hartshorn (1996). Tropical forest structure and composition on a large-scale altitudinal gradient in Costa Rica. *Journal of Ecology* **84**: 137–152.

Lindberg, K., and D. E. Hawkins (eds.) (1993). *Ecotourism: A Guide for Planners and Managers*. North Bennington, VT: The Ecotourism Society.

Lips, K. R., D. E. Green, and R. Pappendick (2003). Chytridiomycosis in wild frogs from southern Costa Rica. *Journal of Herpetology* **37**: 215–218.

Liu, W., J. E. D. Fox, and Z. Xu (2002). Nutrient fluxes in bulk precipitation, throughfall and stemflow in montane subtropical moist forest on Ailao Mountain in Yunnan, south-west China. *Journal of Tropical Ecology* **18**: 527–541.

Liu, W. J., W. Y. Liu, P. J. Li, *et al.* (2007). Using stable isotopes to determine sources of fog drip in a tropical seasonal rain forest of Xishuangbanna, SW China. *Agricultural and Forest Meteorology* **143**: 80–91.

Long, A. (1995). The importance of tropical montane cloud forests for endemic and threatened birds. In *Tropical Montane Cloud Forests*, eds. L. S. Hamilton, J. O. Juvik, and F. N. Scatena, pp. 79–106. New York: Springer-Verlag.

Lugo, A. E., and F. N. Scatena (1992). Epiphytes and climate change in the Caribbean: a proposal. *Selbyana* **13**: 123–130.

Lundgren, L., and B. Lundgren (1979). Rainfall, interception and evaporation in the Mazumbai forest reserve, West Usambara Mts., Tanzania and their importance in the assessment of land potential. *Geografiska Annaler* **61**: 157–178.

Luping, M., N. Chin, and E. R. Dingley (eds.) (1978). *Kinabalu, Summit of Borneo*. Kota Kinabalu, Sabah, Malaysia: The Sabah Society.

Luteyn, J. L. (1989). Speciation and diversity of Ericaceae in neotropical montane vegetation. In *Tropical Forests: Botanical Dynamics, Speciation and Diversity*, eds. L. B. Holm-Nielsen, I. Nielsen, and H. Balslev, pp. 297–310. London: Academic Press.

Lyford, W. H. (1969). The ecology of an elfin forest in Puerto Rico. VII. Soil, root, and earthworm relationships. *Journal of the Arnold Arboretum* **50**: 210–224.

Mamanteo, B. P., and V. P. Veracion (1985). Measurements of fog drip, throughfall and stemflow in the mossy and Benguet pine (*Pinus kesiya* Royle ex Gordon) forests in the upper Agno river basin. *Sylvatrop (Philippines Forestry Research Journal)* **10**: 271–282.

Mangen, J. M. (1993). Ecology and vegetation of Mt Trikora, New Guinea (Irian Jaya/Indonesia). *Travaux Scientifiques du Musée National d'Histoire Naturelle de Luxembourg* **21**: 1–216.

Mann, M. E., K. A. Emanuel, G. L. Holland, and P. J. Webster (2007). Atlantic tropical cyclones revisited. *Eos, Transactions of the American Geophysical Union* **88**(36): 349–350.

Marrs, R., J. Proctor, A. Heaney, and M. Mountford (1988). Changes in soil nitrogen mineralization and nitrification along an altitudinal transect in tropical rain forest in Costa Rica. *Journal of Ecology* **76**: 466–482.

Martínez, A., M. D. Mahecha, G. Lischeid, and E. Beck (2008). Succession stages of vegetation regeneration: secondary tropical mountain forests. In Gradients in a Tropical Mountain Ecosystem of Ecuador, eds. E. Beck, J. Bendix, I. Kottke, F. Makeschin, and R. Mosandl, pp. 425–431. Berlin: Springer-Verlag.

McBride, J., J. Kepert, J. Chan, *et al.* (2006). *Statement on Tropical Cyclones and Climate Change*, WMO/CAS Tropical Meteorological Research Program, Steering Committee for Project TC-2. Available at www.bom.gov.au/info/CAS-statement.pdf.

McCain, C. M. (2005). Elevational gradients in diversity of small mammals. *Ecology* **86**: 366–372.

McJannet, D., J. S. Wallace, and P. Reddell (2007a). Precipitation interception in Australian tropical rainforests. I. Measurement of stemflow, throughfall and cloud interception. *Hydrological Processes* **21**: 1692–1702.

McJannet, D., J. Wallace, and P. Reddell (2007b). Precipitation interception in Australian tropical rainforests. II. Altitudinal gradients of cloud interception, stemflow, throughfall and interception. *Hydrological Processes* **21**: 1703–1718.

McJannet, D. L., P. G. Fitch, M. G. Disher, and J. S. Wallace (2007c). Measurements of transpiration in four tropical rainforest types of north Queensland, Australia. *Hydrological Processes* **21**: 3549–3564.

McJannet, D. L., J. S. Wallace, P. Fitch, M. Disher, and P. Reddell (2007d). Water budgets of tropical rainforests in northern Queensland, Australia. *Hydrological Processes* **21**: 3473–3483.

McKinnon, K., G. Hatta, H. Halim, and A. Mangalik (1996). *The Ecology of Kalimantan: Indonesian Borneo, The Ecology of Indonesia Series, Vol. III*. Hong Kong, China: Periplus Editions.

McNeely, J. A., and S. J. Scherr (2003). *Ecoagriculture: Strategies to Feed the World and Save Wild Biodiversity*. Washington, DC: Island Press.

Mosandl, R., and S. Günter (2008). Sustainable management of tropical mountain forests in Ecuador. In *The Tropical Mountain Forest: Patterns and Processes in a Biodiversity Hotspot*, eds. S. R. Gradstein, J. Homeier, and D. Gansert, pp. 177–193. Göttingen, Germany: Göttingen Centre for Biodiversity and Ecology.

Mosandl, R., S. Günter, B. Stimm, and M. Weber (2008). Ecuador suffers the highest deforestation rate in South America. In *Gradients in a Tropical Mountain Ecosystem of Ecuador*, eds. E. Beck, J. Bendix, I. Kottke, F. Makeschin, and R. Mosandl, pp. 37–40. Berlin: Springer-Verlag.

Moser, G., D. Hertel, and Ch. Leuschner (2007). Altitudinal change in LAI and stand leaf biomass in tropical montane forests: a transect study in Ecuador and a pan-tropical meta-analysis. *Ecosystems* **10**: 924–935.

Moser, G., M. Röderstein, N. Soethe, D. Hertel, and Ch. Leuschner (2008). Altitudinal changes in stand structure and biomass allocation of tropical mountain forest in relation to microclimate and soil chemistry. In *Gradients in a Tropical Mountain Ecosystem of Ecuador*, eds. E. Beck, J. Bendix, I. Kottke, F. Makeschin, and R. Mosandl, pp. 229–242. Berlin: Springer-Verlag.

Motzer, T. (2005). Micrometeorological aspects of a tropical mountain forest. *Agricultural and Forest Meteorology* **135**: 230–240.

Mulligan, M. (2006a). *Global Gridded 1km TRMM Rainfall Climatology and Derivatives*, Version 1.0. Database available at www.ambiotek.com/1kmrainfall.

Mulligan, M. (2006b). *MODIS MOD35 Pan-Tropical Cloud Climatology*, Version 1, September 2006. Available at www.ambiotek.com/clouds.

Mulligan, M., and S. M. Burke (2005a). *Global Cloud Forests and Environmental Change in a Hydrological Context*, Final Report of DFID FRP Project ZF0216. Available at www.ambiotek.com/cloud forests.

Mulligan, M., and S. M. Burke (2005b). *FIESTA: Fog Interception for the Enhancement of Streamflow in Tropical Areas*, Appendix 4a to Final Technical Report of DFID-FRP Project no. R7991. Available at www.ambiotek.com/fiesta).

Muñoz-Piña, C., A. Guevara, J. M. Torres, and J. Braña (2008). Paying for the hydrological services of Mexico's forests: analysis, negotiations and results. *Ecological Economics* **65**: 725–736.

Muñoz-Villers, L. E., (2008). Efecto del cambio en el uso de suelo sobre la dinámica hidrológica y calidad de agua en el trópico húmedo del centro de Veracruz, México. Ph.D. thesis, Autonomous University of Mexico, Mexico City.

Mutke, J., and W. Barthlott (2005). Patterns of vascular plant diversity at continental to global scales. *Biologiske Skrifter* **55**: 521–537.

Myers, N., R. Mittermeier, C. G. Mittermeier, G. A. B. Da Fonseca, and J. Kent. (2000). Biodiversity hotspots for conservation priorities. *Nature* **403**: 853–858.

Nadkarni, N. M. (1984). Epiphyte biomass and nutrient capital of a neotropical elfin forest. *Biotropica* **16**: 249–256.

Nadkarni, N. M., and R. Solano (2002). Potential effects of climate change on canopy communities in a tropical cloud forest: an experimental approach. *Oecologia* **131**: 580–584.

Nadkarni, N., T. J. Matelson, and W. A. Haber (1995). Structural characteristics and floristic composition of a neotropical cloud forest, Monteverde, Costa Rica. *Journal of Tropical Ecology* **11**: 481–494.

Nair, U. S., S. Asefi, R. M. Welch, *et al.* (2008). Biogeography of tropical montane cloud forests. II. Mapping of orographic cloud immersion. *Journal of Applied Meteorology and Climatology* **47**: 2183–2197.

Nespor, V., and B. Sevruk (1999). Estimation of wind-induced error of rainfall gauge measurements using a numerical simulation. *Journal of Atmospheric and Oceanic Technology* **16**: 450–464.

New, M., M. Hulme, and P. D. Jones (2000). *Global Monthly Climatology for the Twentieth Century*. Oak Ridge, TN: Oak Ridge National Laboratory Distributed Active Archive Center. Data-set available at www.daac.ornl.gov.

Newmark, W. D., and R. B. M. Senzota (2003). Power to detect trends in ecological indicators in the East Usambara Mountains, Tanzania. *African Journal of Ecology* **41**: 294–298.

Nieder J., J. Prosperí, and G. Michaloud (2001). Epiphytes and their contribution to canopy diversity. *Plant Ecology* **153**: 51–63.

Northup, R., Z. Yu, R. A. Dahlgren, and K. A. Vogt (1995). Polyphenol control of nitrogen release from pine litter. *Nature* **377**: 227–229.

Northup, R., R. A. Dahlgren, and J. G. McColl (1998). Polyphenols as regulators of plant-litter-soil interactions: a positive feedback. *Biogeochemistry* **42**: 189–220.

Olander, L. P., F. N. Scatena, and W. L. Silver (1998). Impacts of disturbance initiated by road construction in a subtropical cloud forest in the Luquillo Experimental Forest, Puerto Rico. *Forest Ecology and Management* **109**: 33–49.

Olson, D. M., and E. Dinerstein (2002). The Global 200: Ecoregions for global conservation. *Annals of the Missouri Botanical Garden* **89**: 199–224.

Olson, D. M, E. Dinerstein, E. D. Wikramanayake, *et al.* (2001). Terrestrial ecoregions of the world: a new map of life on Earth. *BioScience* **51**:933–938. Also available at www.worldwildlife.org/science/data/terreco.cfm.

O'Neill, J. P., D. F. Lane, A. W. Kratter, A. P. Capparella, and C. F. Joo (2000). A striking new species of barbet (Capitoninae: *Capito*) from the eastern Andes of Peru. *The Auk* **117**: 569–577.

Pagiola, S. (2002). Paying for water services in Central America: learning from Costa Rica. In *Selling Forest Environmental Services: Market-Based Mechanisms for Conservation and Development*, eds. S. Pagiola, J. Bishop, and N. Landell-Mills, pp. 36–60. London: Earthscan.

Pagiola, S. (2008). Payments for environmental services in Costa Rica. *Ecological Economics* **65**: 712–724.

Parrotta, J. A., J. W. Turnbull, and N. Jones (1997). Catalyzing native forest regeneration on degraded tropical lands. *Forest Ecology and Management* **99**: 1–7.

Paulsch, A., R. Schneider, and K. Hartig (2001). Land-use induced vegetation structure in a montane region of southern Ecuador. *Die Erde* **132**: 93–102.

Paulsch, A., D. Piechowski, and K. Müller-Hohenstein (2008). Forest structure along an altitudinal gradient in Southern Ecuador. In *Gradients in a Tropical Mountain Ecosystem of Ecuador*, eds. E. Beck, J. Bendix, I. Kottke, F. Makeschin, and R. Mosandl, pp. 113–122. Berlin: Springer-Verlag.

Pineda, E., and G. Halffter (2004). Species diversity and habitat fragmentation: frogs in a tropical montane landscape in Mexico. *Biological Conservation* **117**: 499–508.

Pócs, T. (1980). The epiphytic biomass and its effect on the water balance of two rain forest types in the Uluguru Mountains (Tanzania, East Africa). *Acta Botanica Academiae Scientiarum Hungariae* **26**: 143–167.

Pohle, P. (2008). Indigenous land use practices and biodiversity conservation in southern Ecuador. In *The Tropical Mountain Forest: Patterns and Processes in a Biodiversity Hotspot*, ed. S. R. Gradstein, J. Homeier, and D. Gansert, pp. 163–176. Göttingen, Germany: Göttingen Centre for Biodiversity and Ecology.

Ponette-González, A. G., K. C. Weathers, and L. M. Curran (2009). Water inputs across a tropical montane landscape in Veracruz, Mexico: synergistic effects of land cover, rain and fog seasonality, and interannual precipitation variability. *Global Change Biology*, doi:10.1111/j.1365-2486.2009.01985.x.

Porras, I. (2008). Forests, flows and markets for watershed environmental services: evidence from Costa Rica and Panama. Ph.D. thesis, University of Newcastle-upon-Tyne, Newcastle, UK.

Porras, I. T., M. Grieg-Gran, and N. Neves (2008). *All That Glitters: A Review of Payments for Watershed Services in Developing Countries*. London: International Institute for Environment and Development.

Pounds, J. A. (2000). Amphibians and reptiles. In *Monteverde: Ecology and Conservation of a Tropical Cloud Forest*, eds. N. M. Nadkarni and N. T. Wheelwright, pp. 149–177. Oxford, UK: Oxford University Press.

Pounds, J. A., and M. L. Crump (1994). Amphibian declines and climate disturbance: the case of the golden toad and the harlequin frog. *Conservation Biology* **8**: 72–85.

Pounds, J. A., M. P. L. Fogden, and J. H. Campbell (1999). Biological response to climate change on a tropical mountain. *Nature* **398**: 611–615.

Pounds, J. A., M. R. Bustamante, L. A. Coloma, *et al.* (2006). Widespread amphibian extinctions from epidemic disease driven by global warming. *Nature* **439**: 161–167.

Prada, S., M. Menezes de Sequeiro, C. Figueira, and M. Oliveira da Silva (2009). Fog precipitation and rainfall interception in the natural forests of Madeira Island (Portugal). *Agricultural and Forest Meteorology* **149**: 1179–1187.

Prance, G. T. (ed.) (1982). *Biological Diversification in the Tropics*. New York: Columbia University Press.

Proctor, J., Y. F. Lee, A. M. Langley, W. R. C. Munro, and T. Nelson (1988). Ecological studies on Gunung Silam, a small ultrabasic mountain in Sabah, Malaysia. I. Environment, forest structure, and floristics. *Journal of Ecology* **76**: 320–340.

Raich, J. W. (1998). Aboveground productivity and soil respiration in three Hawaiian rainforests. *Forest Ecology and Management* **107**: 309–318.

Raich, J. W., A. E. Russell, and P. M. Vitousek (1997). Primary productivity and ecosystem development along an elevational gradient on Mauna Loa, Hawai'i. *Ecology* **78**: 707–721.

Ray, D. K., U. S. Nair, R. O. Lawton, R. M. Welch, and R. A. Pielke Sr. (2006). Impact of land use on Costa Rican tropical montane cloud forests: sensitivity of orographic cloud formation to deforestation in the plains. *Journal of Geophysical Research* **111**: D02108, doi:10.1029/2005JD006096.

Ray, D. K., R. A. Pielke Sr., U. S. Nair, R. M. Welch, and R. O. Lawton (2009). Importance of land use versus atmospheric information verified from cloud simulations from a frontier region in Costa Rican. *Journal of Geophysical Research* **114**: D08113, doi:10.1029/2007JD009565.

Rhodes, A. L., A. J. Guswa, and S. E. Newell (2006). Seasonal variation in the stable isotopic composition of precipitation in the tropical montane forests of Monteverde, Costa Rica. *Water Resources Research* **42**, W11402, doi:10.1029/2005WR004535.

Richards, P. W. (1996). *The Tropical Rain Forest*, 2nd edn. Cambridge, UK: Cambridge University Press.

Richardson, B. A., M. J. Richardson, F. N. Scatena, and W. H. McDowell (2000). Effects of nutrient availability and other elevational changes on bromeliad populations and their invertebrate communities in a humid tropical forest in Puerto Rico. *Journal of Tropical Ecology* **16**: 167–188.

Richter, M. (2008). Tropical mountain forests: distribution and general features. In *The Tropical Mountain Forest: Patterns and Processes in a Biodiversity hotspot*, eds. S. R. Gradstein, J. Homeier, and D. Gansert, pp. 7–24. Göttingen, Germany: Göttingen Centre for Biodiversity and Ecology.

Rickets, T. H., G. C. Daily, P. R. Ehrlich, and C. D. Michener (2004). Economic value of tropical forest to coffee production. *Proceedings of the National Academy of Sciences USA* **101**: 12 579–12 582.

Ridgely, R. S., T. F. Allnutt, T. Brooks, *et al.* (2007). *Digital Distribution Maps of the Birds of the Western Hemisphere*, Version 3.0. Arlington, VA: NatureServe. Also available at www.natureserve.org/getData/birdMaps.jsp.

Roberts, J. M., J. H. C. Gash, M. Tani, and L. A. Bruijnzeel (2005). Controls on evaporation in lowland tropical rainforest. In *Forests, Water and People in the Humid Tropics*, eds. M. Bonell and L. A. Bruijnzeel, pp. 287–313. Cambridge, UK: Cambridge University Press.

Röderstein, M., D. Hertel, and Ch. Leuschner (2005). Above- and below-ground litter production in three tropical montane forests in southern Ecuador. *Journal of Tropical Ecology* **21**: 483–492.

Rodgers, W. A., and K. M. Homewood (1982). Species richness and endemism in the Usambara mountain forests, Tanzania. *Biological Journal of the Linnean Society* **18**: 197–242.

Rodriguez-Zuñiga, J. M. (2003). Paying for forest environmental services: the Costa Rican experience. *Unasylva* **212**(54): 31–33.

Røhr, P. C. (2003). A hydrological study concerning the southern slopes of Mt. Kilimanjaro, Tanzania. Ph.D. thesis, Norwegian University of Science and Technology, Trondheim, Norway.

Rojas, M., and B. Aylward (2003). *What Are We Learning from Experiences with Markets for Environmental Services in Costa Rica? A Review and Critique of the Literature*. London: International Institute for Environment and Development.

Rollenbeck, R., J. Bendix, P. Fabian, *et al.* (2007). Comparison of different techniques for the measurement of precipitation in tropical montane rain forest regions. *Journal of Atmospheric and Oceanic Technology* **24**: 156–168.

Rovito, S. M., G. Parra-Olea, C. R. Vásquez-Alamazán, T. J. Papenfuss, and D. B. Wake (2009). Dramatic declines in neotropical salamander populations are an important part of the global amphibian crisis. *Proceedings of the National Academy of Sciences USA* **106**: 3231–3236.

Rüger, N., G. Williams-Linera, W. D. Kissling, and A. Huth (2008). Long-term impacts of fuelwood extraction on a tropical montane cloud forest. *Ecosystems* **11**: 868–881.

Rzedowski, J. (1996). Análisis preliminar de la flora vascular de los bosques mesófilos de montaña de México. *Acta Botanica Mexicana* **35**: 25–44.

Salaman, P., T. M. Donegan, M. Mulligan, *et al.* (2003). A new species of wood-wren (Troglodytidae: Henicorhina) from the western Andes of Colombia. *Ornitología Colombiana* **1**: 4–21.

Salzmann, U. (2000). Are modern savannas degraded forests? A Holocene pollen record from the Sudanian vegetation zone of NE Nigeria. *Vegetation History and Archaeobotany* **9**: 1–15.

Santiago, L. S., G. Goldstein, F. C. Meinzer, J. Fownes, and D. Müeller-Dombois. (2000). Transpiration and forest structure in relation to soil waterlogging in a Hawaiian montane cloud forest. *Tree Physiology* **20**: 673–681.

Sarmiento, F. (1997). Arrested succession in pastures hinders regeneration of Tropandean forests and shreds mountain landscapes. *Environmental Conservation* **24**: 14–23.

Scatena, F. N. (1995). The management of Luquillo cloud forest ecosystems: irreversible decisions in a non-substitutable ecosystem. In *Tropical Montane Cloud Forests*, eds. L. S. Hamilton, J. O. Juvik, and F. N. Scatena, pp. 296–308. New York: Springer-Verlag.

Scatena, F. N. (1998). An assessment of climatic change in the Luquillo Mountains of Puerto Rico. *In American Water Resources Association* ***TPS-98-2***: 193–198.

Scatena, F. N., E. O. Planos-Gutierrez, and J. Schellekens (2005). Natural disturbances and the hydrology of humid tropical forests. In *Forests, Water and People in the Humid Tropics*, eds. M. Bonell and L. A. Bruijnzeel, pp. 489–512. Cambridge, UK: Cambridge University Press.

SCBD (Secretariat of the Convention on Biological Diversity) (2003a). Status and trends of, and threats to, mountain biodiversity and marine, coastal and inland water ecosystems. Abstracts of Poster Presentations at the *8th Meeting of the Subsidiary Body on Scientific, Technical and Technological Advice (SBSTTA-8) of the Convention on Biological Diversity* , CBD Technical Series No. 8. Montréal, Canada: SCBD. Also available at www.cbd.int/doc/publications/cbd-ts-08.pdf.

SCBD (Secretariat of the Convention on Biological Diversity). (2003b). *Status and Trends of, and Threats to, Mountain Biological Diversity* Document No. UNEP/CBD/SBSTTA/8/5. Montréal, Canada: SCBD. Also available at www.cbd.int/doc/meetings/sbstta/sbstta-08/official/sbstta-08–05-en.pdf.

SCBD (Secretariat of the Convention on Biological Diversity) (2003c). *Measures Taken for the Conservation and Sustainable Use of Mountain Biological Diversity*, Document No. UNEP/CBD/SBSTTA/8/6. Montréal, Canada: SCBD. Also available at www.cbd.int/doc/meetings/sbstta/sbstta-08/official/sbstta-08–06-en.pdf.

SCBD (Secretariat of the Convention on Biological Diversity) (2004). *Programme of Work on Mountain Biological Diversity*. Montréal, Canada: SCBD. Also available at www.cbd.int/mountain/pow.shtml.

Schawe, M., S. Glatzel, and G. Gerold (2007). Soil development along an altitudinal transect in a Bolivian tropical montane rainforest: podzolization vs. hydromorphy. *Catena* **69**: 83–90.

Schellekens, J. (2006). *CQ-FLOW: A Distributed Hydrological Model for the Prediction of Impacts of Land-Cover Change, with Special Reference to the Rio Chiquito Catchment, Northwest Costa Rica*, Annex 3 to Final Technical Report DFID-FRP Project No. R7991. Amsterdam, the Netherlands: Vrije Universiteit. Also available at (www.geo.vu.nl/~fiesta/.

Schrumpf, M. (2004). Biogeochemical investigations in old growth and disturbed forest sites at Mount Kilimanjaro. Ph.D. thesis, University of Bayreuth, Bayreuth, Germany.

Schrumpf, M., G. Guggenberger, C. Valarezo, and W. Zech (2001). Tropical montane rain forest soils: development and nutrient status along an altitudinal gradient in the south Ecuadorian Andes. *Die Erde* **132**: 43–59.

Schuur, E. (2001). The effect of water on decomposition dynamics in mesic to wet Hawaiian montane forests. *Ecosystems* **4**: 259–273.

Schuur, E., and P. Matson (2001). Net primary productivity and nutrient cycling across a mesic to wet precipitation gradient in Hawaiian montane forest. *Oecologia* **128**: 431–442.

Schuur, E. A. G., O. A. Chadwick, and P. A. Matson (2001). Carbon cycling and soil carbon storage in mesic to wet Hawaiian montane forests. *Ecology* **82**: 3182–3196.

Scott, N. J., Jr. (1976). The abundance and diversity of the herpetofaunas of tropical forest litter. *Biotropica* **8**: 41–58.

Setzer, M. C., D. M. Moriarity, R. O. Lawton, *et al.* (2003). Phytomedicinal potential of tropical cloudforest plants. *Revista de Biología Tropical* **51**: 647–674.

Sharon, D. (1980). Distribution of hydrologically effective rainfall incident on sloping ground. *Journal of Hydrology* **46**: 165–188.

Sheil, D., and A. Lawrence (2004). Tropical biologists, local people and conservation: new opportunities for collaboration. *Trends in Ecology and Evolution* **19**: 634–638.

Shukor, M. N. (2001). Elevational diversity patterns of small mammals on Mount Kinabalu, Sabah, Malaysia. *Global Ecology and Biogeography* **10**: 41–62.

Shuttleworth, W. J. (1988). Evaporation from Amazonian rain forest. *Philosophical Transactions of the Royal Society of London Series B* **233**: 321–346.

Sidle, R. C., A. D. Ziegler, J. N. Negishi, *et al.* (2006). Erosion processes in steep terrain: truths, myths, and uncertainties related to forest management in Southeast Asia. *Forest Ecology and Management* **224**: 199–225.

Silver, W. L., A. E. Lugo, and M. Keller (1999). Soil oxygen availability and biogeochemistry along rainfall and topographic gradients in upland wet tropical forest soils. *Biogeochemistry* **44**: 301–328.

Soethe, N., J. Lehmann, and C. Engels (2006). The vertical pattern of rooting and nutrient uptake at different altitudes of a south Ecuadorian montane forest. *Plant and Soil* **286**: 287–299.

Soethe, N., J. Lehmann, and C. Engels (2007). Carbon and nutrient stocks in roots of forests at different altitudes in the Ecuadorian Andes. *Journal of Tropical Ecology* **23**: 319–328.

Soethe, N., W. Wilcke, J. Homeier, J. Lehmann, and C. Engels (2008a). Plant growth along the altitudinal gradient: role of plant nutritional status, fine root activity, and soil properties. In *Gradients in a Tropical Mountain Ecosystem of Ecuador*, eds. E. Beck, J. Bendix, I. Kottke, F. Makeschin, and R. Mosandl, pp. 259–266. Berlin: Springer-Verlag.

Soethe, N., J. Lehmann, and C. Engels (2008b). Nutrient availability at different altitudes in a tropical montane forest in Ecuador. *Journal of Tropical Ecology* **24**: 397–406.

Spehn, E. M., M. Liberman, and Ch. Körner (eds.) (2006). *Land Use Change and Mountain Biodiversity*. Boca Raton, FL: CRC Press.

Stadtmüller, T. (1987). *Cloud Forests in the Humid Tropics: A Bibliographic Review*. Turrialba, Costa Rica: CATIE, and Tokyo: United Nations University. Also available at www.unu.edu/unupress/unupbooks/80670e/80670E00.htm.

Stadtmüller, T., and N. Agudelo. (1990). Amount and variability of cloud moisture input in a tropical cloud forest. *International Association of Hydrological Sciences Publication* **193**: 25–32.

Steinhardt, U. (1979). Untersuchungen über den Wasser- und Nährstoffhaushalt eines andinen Wolkenwaldes in Venezuela. *Göttinger Bodenkundliche Berichte* **56**: 1–185.

Still, C. J., P. N. Foster, and H. Schneider (1999). Simulating the effects of climate change on tropical montane cloud forests. *Nature* **398**: 608–610.

Stimm, B., E. Beck, S. Günter, *et al.* (2008). Reforestation of abandoned pastures: seed ecology of native species and production of indigenous plant material. In *Gradients in a Tropical Mountain Ecosystem of Ecuador*, eds. E. Beck, J. Bendix, I. Kottke, F. Makeschin, and R. Mosandl, pp. 433–445. Berlin: Springer-Verlag.

Stuart, S., J. S. Chanson, N. A. Cox, *et al.* (2004). Status and trends of amphibian declines and extinctions worldwide. *Science* **306**: 1783–1786.

Suding, K. N., L. G. Gross, and G. R. Houseman (2004). Alternative states and positive feedbacks in restoration ecology. *Trends in Ecology and Evolution* **19**: 46–53.

Sugden, A. M., and R. J. Robins (1979). Aspects of the ecology of vascular epiphytes in Colombian cloud forests. I. The distribution of the epiphytic flora. *Biotropica* **11**: 173–188.

Sutherland, W. J., W. M. Adams, R. B. Aronson, *et al.* (2009). One hundred questions of importance to the conservation of global biological diversity. *Conservation Biology* **23**: 557–567.

Tanaka, K., H. Takizawa, N. Tanaka, *et al.* (2003). Transpiration peak over a hill evergreen forest in northern Thailand in the late dry season: assessing the seasonal changes in evapotranspiration using a multilayer model. Journal of Geophysical Research *108, **D17***, 4533, doi:10.1029/2002JD003028.

Tanaka, N., C. Tantasirin, K. Kuraji, M. Suzuki, and N. Tangtham (2005). Inter-annual variation in rainfall interception at a hill evergreen forest in northern Thailand. *Bulletin of the Tokyo University Forest* **113**: 11–44.

Tani, M., A. Rahim Nik, Y. Yasuda, *et al.* (2003). Long-term estimation of evapotranspiration from a tropical rain forest in Peninsular Malaysia. *International Association of Hydrological Sciences Publication* **280**: 267–274.

Tanner, E. V. J. (1980). Studies on the biomass and productivity in a series of montane rain forests in Jamaica. *Journal of Ecology* **68**: 573–588.

Tanner, E. V. J., P. Vitousek, and E. Cuevas (1998). Experimental investigation of nutrient limitation of forest growth on wet tropical mountains. *Ecology* **79**: 10–22.

Ten Hoopen, G. M., and M. Kappelle (2006). Soil seed bank size and composition in disturbed and old growth montane oak forests in Costa Rica. In: *Ecology and Conservation of Neotropical Montane Oak Forests*, ed. M. Kappelle, pp. 299–308. Berlin: Springer-Verlag.

Terborgh, J. (1977). Bird species diversity on an Andean elevational gradient. *Ecology* **58**: 1007–1019.

Thang, H. C., and N. Chappell (2005). Minimizing the hydrological impact of forest harvesting in Malaysia's rainforests. In *Forests, Water and People in the Humid Tropics*, eds. M. Bonell and L. A. Bruijnzeel, pp. 852–865. Cambridge, UK: Cambridge University Press.

Tie, Y. L., I. C. Baille, C. M. S. Phang, and C. P. Lim (1979). *Soils of Gunung Mulu National Park*. Kuching, Sarawak, Malaysia: Department of Agriculture.

Timm, R. M., and R. K. LaVal (2000). Mammals. In *Monteverde: Ecology and Conservation of a Tropical Cloud Forest*, eds. N. M. Nadkarni and N. T. Wheelwright, pp. 223–244 and 553–557. Oxford, UK: Oxford University Press.

Tobón, C. (2009). *Los bosques andinos y el agua*, Serie Investigación y Sistematización No. 4. Quito, Ecuador: Programa Regional ECOBONA–INTERCOOPERATION–CONDESAN.

Turner, I. M. (1994). Sclerophylly: primarily protective? *Functional Ecology* **8**: 669–675.

Turton, S. M. (1990). Light environments within tropical rainforest, Mt. Bellenden Ker, North Queensland. *Austral Ecology* **15**: 35–42.

Van der Hammen, T. (1974). The Pleistocene changes of vegetation and climate in tropical South America. *Journal of Biogeography* **1**: 3–26.

Van der Hammen, T., and H. Hooghiemstra (2001). Historia y paleoecología de los bosques montanos andinos neotropicales. In *Bosques nublados del Neotrópico*, eds. M. Kappelle and A. D. Brown, pp. 63–84. Santo Domingo de Heredia, Costa Rica: Editorial INBio and IUCN.

Van der Molen, M. K. (2002). Meteorological impacts of land use change in the maritime tropics. Ph.D. thesis, VU University Amsterdam, Amsterdam, the Netherlands. Also available at (www.falw.vu.nl/images_upload/702F8F36-45DC-488E-B2C550BFDCAC2B9D.pdf.

Van der Molen, M. K., A. J. Dolman, M. J. Waterloo, and L. A. Bruijnzeel (2006). Climate is affected more by maritime than by continental land use change: a multiple-scale analysis. *Global and Planetary Change* **54**: 128–149.

Van Dunné, H. J. F., and M. Kappelle (1998). Biomass–diversity relations of epiphytic bryophytes on small *Quercus copeyensis* stems in a Costa Rican montane cloud forest. *Revista de Biología Tropical* **46**: 35–42.

Van Reuler, H. (1987). Soil studies in the Bukit Raya nature reserve. In *Report of the 1982–1983 Bukit Raya Expedition*, ed. H. P. Nooteboom, pp. 7–23. Leiden, the Netherlands: Rijksherbarium.

Van Steenis, C. G. G. J. (1972). *The Mountain Flora of Java*. Leiden, the Netherlands: E.J. Brill.

Vázquez-García, J. A. (1995). Cloud forest archipelagos: preservation of fragmented montane ecosystems in tropical America. In *Tropical Montane Cloud Forests*, eds. L. S. Hamilton, J. O. Juvik, and F. N. Scatena, pp. 314–332. New York: Springer-Verlag.

Veneklaas, E. J., and R. van Ek (1990). Rainfall interception in two tropical montane rain forests, Colombia. *Hydrological Processes* **4**: 311–326.

Veneklaas, E. J., R. J. Zagt, A. van Leerdam, *et al.* (1990). Hydrological properties of the epiphyte mass of a montane tropical rain forest, Colombia. *Vegetatio* **89**: 183–192.

Vernimmen, R. R. E., L. A. Bruijnzeel, D. Romdoni, and J. Proctor (2007). Rainfall interception in three contrasting rain forest types in Central Kalimantan, Indonesia. *Journal of Hydrology* **340**: 217–232.

Vis, M. (1986). Interception, drop size distributions and rainfall kinetic energy in four Colombian forest ecosystems. *Earth Surface Processes and Landforms* **11**: 591–603.

Vogelmann, H. W. (1973). Fog precipitation in the cloud forests of Eastern Mexico. *BioScience* **23**: 96–100.

Wadsworth, F. H., and J. A. Bonnet (1951). Soil as a factor in the occurrence of two types of montane forest in Puerto Rico. *Caribbean Forester* **12**: 67–70.

Waide, R. B., J. K. Zimmerman, and F. N. Scatena (1998). Controls of primary productivity: lessons from the Luquillo Mountains in Puerto Rico. *Ecology* **79**: 31–37.

Walmsley, J. L., R. S. Schemenauer, and H. A. Bridgman (1996). A method for estimating the hydrologic input from fog in mountainous terrain. *Journal of Applied Meteorology* **35**: 2237–2249.

Walmsley, J.L., W.R. Burrows, and R.S. Schemenauer (1999). The use of routine observations to calculate liquid water content in summertime high-elevation fog. *Journal of Applied Meteorology* **38** : 369–384.

Watson, D. M, and A. Townsend Peterson (1999). Determinants of diversity in a naturally fragmented landscape: humid montane forest avifaunas of Mesoamerica. *Ecography* **22**: 582–589.

WDPA Consortium (2004). *World database on Protected Areas 2004*. Gland, Switzerland: World Conservation Union (IUCN), and Cambridge, UK: UNEP–World Conservation Monitoring Centre. (Source for this data-set was the Global Land Cover Facility.)

Weaver, P. L. (1972). Cloud moisture interception in the Luquillo mountains of Puerto Rico. *Caribbean Journal of Science* **12**: 129–144.

Weaver, P. L. (1986). Hurricane damage and recovery in the montane forests of Puerto Rico. *Caribbean Journal of Science* **15**: 21–30.

Weaver, P. L. (1990). Succession in the elfin woodland of the Luquillo Mountains of Puerto Rico. *Biotropica* **22**: 83–89.

Weaver, P. L. (1999). Impacts of Hurricane Hugo on the dwarf cloud forest of Puerto Rico's Luquillo Mountains. *Caribbean Journal of Science* **35**: 101–111.

Weaver, P. L., and P. G Murphy (1990). Forest structure and productivity in Puerto Rico's Luquillo Mountains. *Biotropica* **22**: 69–82.

Weber, M., S. Günter, N. Aguirre, B. Stimm, and R. Mosandl (2008). Reforestation of abandoned pastures: silvicultural means to accelerate forest recovery and biodiversity. In *Gradients in a Tropical Mountain Ecosystem of Ecuador*, eds. E. Beck, J. Bendix, I. Kottke, F. Makeschin, and R. Mosandl, pp. 447–457. Berlin: Springer-Verlag.

Webster, P. J., G. J. Holland, J. A. Curry, and H. R. Chang (2005). Changes in tropical cyclone number, duration and intensity in a warming environment. *Science* **309**: 1844–1846.

Welch, R. M., S. Asefi, J. Zeng, *et al.* (2008). Biogeography of tropical montane cloud forests. I. Remote sensing of cloud base heights. *Journal of Applied Meterology and Climatology* **47**: 960–975.

Whitmore, T. C. (1989). Tropicasl forest nutrients, where do we stand? A tour de horizon. In *Mineral Nutrients in Tropical Forest and Savanna Ecosystems*, ed. J. Proctor, pp. 1–13. Oxford, UK: Blackwell Scientific.

Williams, J. W., S. T. Jackson, and J. E. Kutzbach (2007). Projected distributions of novel and disappearing climates by 2100 AD. *Proceedings of the National Academy of Sciences USA* **104**: 5738–5742.

Williams, S., E. Bolitho, and S. Fox (2003). Climate change in Australian tropical rainforests: an impending environmental catastrophe. *Proceedings of the Royal Society of London Series B* **270**: 1887–1893.

Williams-Linera, G. (2002). Tree species richness complementarity, disturbance and fragmentation in a Mexican tropical montane cloud forest. *Biodiversity and Conservation* **11**: 1825–1843.

Williams-Linera, G., A. M. Lopez, and M. A. Muñiz (2005). Complementariedad y patrones de anidamiento de especies de árboles en el paisaje de bosque de niebla del centro deVeracruz (México). In *Sobre diversidad biológica: el significado de las diversidades alfa, beta y gamma*, eds. G. Halffter, J. Soberón, P. Koleff, and A. Melic, pp. 153–164. Zaragoza, Mexico: m3m-Monografías Tercer Milenio.

Wilms, J. J. A. M., and M. Kappelle (2006). Frugivorous birds and seed dispersal in disturbed and old growth montane oak forests in Costa Rica. In *Ecology and Conservation of Neotropical Montane Oak Forests*, ed. M. Kappelle, pp. 309–324. Berlin: Springer-Verlag.

Wilson, L. D., and J. R. McCranie (2004). The conservation status of the herpetofauna of Honduras. *Amphibian and Reptile Conservation* **3**: 6–33.

Wolf, J. H. D. (1993). Diversity patterns and biomass of epiphytic bryophytes and lichens along an altitudinal gradient in the northern Andes. *Annals of the Missouri Botanical Garden* **80**: 928–960.

Wolf, J. H. D. (1994). Factors controlling the distribution of vascular and nonvascular epiphytes in the northern Andes. *Vegetatio* **112**: 15–28.

Wolf, J. H. D. (1995). Non vascular epiphyte diversity patterns in the canopy of an Upper Montane Rain Forest, Central Cordillera, Colombia. *Selbyana* **16**: 185–195.

Wolf, J. H. D. (2005). The response of epiphytes to anthropogenic disturbance of pine–oak forests in the highlands of Chiapas, Mexico. *Forest Ecology and Management* **212**: 376–393.

Wolf, J. H. D., and A. Flamenco-Sandoval (2006). Vascular epiphytes and their potential as a conservation tool in pine-oak forests of Chiapas, Mexico. In *Ecology and Conservation of Neotropical Montane Oak Forests*, ed. M. Kappelle, pp. 375–391. Berlin: Springer-Verlag.

Wolf, J. H. D., and C. J. F. Konings (2001). Toward the sustainable harvesting of epiphytic bromeliads: a pilot study from the highlands of Chiapas, Mexico. *Biological Conservation* **101**: 23–31.

Wu, W., C. A. Hall, F. N. Scatena, and L. J. Quackenbush (2006). Spatial modeling of evapotranspiration in the Luquillo Experimental Forest of Puerto Rico using remotely-sensed data. *Journal of Hydrology* **328**: 733–752.

Yang, D., B. E. Goodison, S. Ishida, and C. S. Benson (1998). Adjustment of daily precipitation data at 10 climate stations in Alaska: application of World Meteorological Organization intercomparison results. *Water Resources Research* **34**: 241–256.

Yin, X. W., and P. A. Arp (1994). Fog contributions to the water budget of forested watersheds in the Canadian Maritime Provinces: a generalized algorithm for low elevations. *Atmosphere–Ocean* **32**: 553–566.

Young, B. E., and D. B McDonald (2000). Birds. In *Monteverde: Ecology and Conservation of a Tropical Cloud Forest*, eds. N. M. Nadkarni and N. T. Wheelwright, pp. 179–222. Oxford, UK: Oxford University Press.

Zadroga, F. (1981). The hydrological importance of a montane cloud forest area of Costa Rica. In *Tropical Agricultural Hydrology*, eds. R. Lal and E. W. Russell, pp. 59–73. New York: John Wiley.

Zimmermann, B., and H. Elsenbeer (2008). Spatial and temporal variability of soil saturated hydraulic conductivity in gradients of disturbance. *Journal of Hydrology* **361**: 78–95.

Zimmerman, J. K., J. B. Pascarella, and T. M. Aide (2000). Barriers to forest regeneration in an abandoned pasture in Puerto Rico. *Restoration Ecology* **8**: 350–360.

Zotz, G. (1999). Altitudinal changes in diversity and abundance of non-vascular epiphytes in the tropics: an ecophysiological explanation. *Selbyana* **20**: 256–260.

Zotz, G., and M. Y. Bader (2009). Epiphytic plants in a changing world: global change effects on vascular and non-vascular epiphytes. *Progress in Botany* **70**: 147–170.